St He

IB

Water

This bo

PHYLOGENY OF THE LIVING WORLD—OVERVIEW

UNIVERSAL PHYLOGENETIC TREE. This tree is derived from comparative sequencing of 16S or 18S ribosomal RNA. Note the three major domains of living organisms: the *Bacteria*, the *Archaea*, and the *Eukarya*. The evolutionary distance between two groups of organisms is proportional to the cumulative distance between the end of the branch and the node that joins the two groups. See Sections 11.5–11.9 for further information on ribosomal RNA-based phylogenies. The phylogenetic relationships depicted in this tree have been supported by several other genotypic and phenotypic relationships. *Data for the tree obtained from the Ribosomal Database project* **http://rdp.cme.msu.edu**

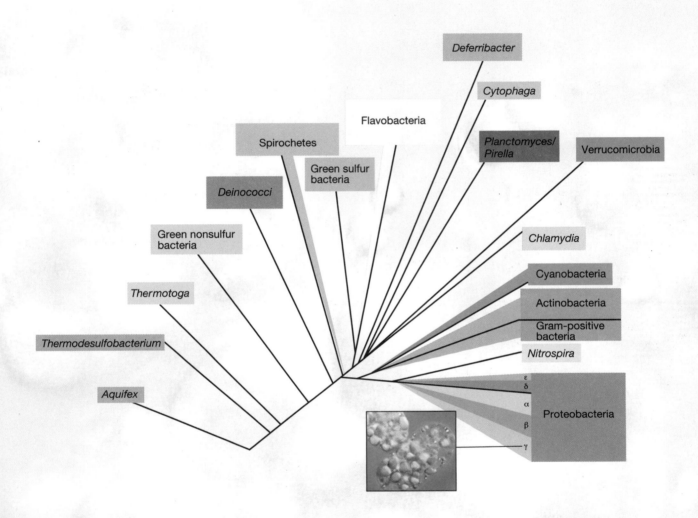

PHYLOGENETIC TREE OF *BACTERIA*. This tree is derived from 16S ribosomal RNA sequences. At least 17 major groups of *Bacteria* can be defined as indicated. See Sections 11.5–11.9 for further information on ribosomal RNA-based phylogenies. *Data for the tree obtained from the Ribosomal Database project* **http://rdp.cme.msu.edu**

BROCK

BIOLOGY OF MICROORGANISMS

MICHAEL T. MADIGAN dedicates this book to the two old friends pictured with him below: Willie (left) and Plum (right). For the past 11 years these wonderful animals have given me the comfort and companionship that only a dog lover could understand. Willie is, and always has been, the paragon of a sweet dog. Plum (deceased April 16, 2004), by contrast, was your basic junkyard dog. But in reality, she had a heart of gold and loved people, and our paths never crossed that she wasn't thrilled to see me. Rest in peace amigo.

JOHN M. MARTINKO dedicates this book to his students, past and present. The best students continually present problems from new perspectives, inspiring me to not only teach, but to expand my knowledge and enhance my understanding. To all who I have had the pleasure to teach, thank you for teaching me!

MICHAEL T. MADIGAN received a bachelor's degree in biology and education from Wisconsin State University at Stevens Point in 1971 and M.S. and Ph.D. degrees in 1974 and 1976, respectively, from the University of Wisconsin, Madison, Department of Bacteriology. His graduate work involved the study of hot spring phototrophic bacteria under the direction of Thomas D. Brock. Following three years of postdoctoral training in the Department of Microbiology, Indiana University, where he worked on phototrophic bacteria with Howard Gest, he moved to Southern Illinois University Carbondale, where he is a Professor of Microbiology. He has been a coauthor of *Biology of Microorganisms* since the fourth edition (1984) and teaches courses in introductory microbiology, bacterial diversity, and diagnostic and applied microbiology. In 1988 he was selected as the outstanding teacher in the College of Science, and in 1993 its outstanding researcher. In 2001 he received the university's Outstanding Scholar Award. In 2003 he received the Carski Award for Distinguished Teaching of Undergraduates from the American Society of Microbiology. His research has dealt almost exclusively with anoxygenic phototrophic bacteria, especially those species that inhabit extreme environments. He has published over 100 research papers, has coedited a major treatise on phototrophic bacteria, and has served as editor and chief editor of the journal *Archives of Microbiology*. His nonscientific interests include reading, hiking, tree planting, and caring for his dogs and horses. He lives beside a quiet lake about five miles from the SIU campus with his wife, Nancy, two dogs, Willie and Pupagano, and Springer and Feivel (horses).

JOHN M. MARTINKO received his B.S. in Biology from The Cleveland State University. As an undergraduate student he participated in a cooperative education program, gaining experience in several microbiology and immunology laboratories. He then worked for two years at Case Western Reserve University as a laboratory manager, conducting research on the structure, serology and epidemiology of Streptococcus pyogenes. He did his graduate work at the State University of New York at Buffalo, investigating antibody specificity and antibody idiotypes for his M.A. and Ph.D. in Microbiology. As a postdoctoral fellow, he worked at Albert Einstein College of Medicine in New York on the structure of major histocompatibility complex proteins. Since 1981, he has been in the Department of Microbiology at Southern Illinois University Carbondale where he is an Associate Professor and Chair. His research interests include the effects of growth hormone in the immune response, the development of immunodiagnostic tests for soybean brown stem rot disease, and the investigation of structural mutations that alter function in peptide-major histocompatibility protein complexes. His teaching interests include undergraduate and graduate courses in immunology. He also teaches immunology, host defense, and infectious disease topics in a general microbiology course. In 2004, he was selected the outstanding teacher in the College of Science. He is also an avid golfer and cyclist. He lives in Carbondale with his wife, Judy, a high school science teacher.

Eleventh Edition

BROCK
BIOLOGY OF
MICROORGANISMS

Michael T. Madigan

John M. Martinko

Southern Illinois University Carbondale

PEARSON

Benjamin Cummings

San Francisco   Boston   New York
Cape Town   Hong Kong   London   Madrid   Mexico City
Montreal   Munich   Paris   Singapore   Sydney   Tokyo   Toronto

Executive Editor: Gary Carlson
Editor-in-Chief: John Challice
Project Manager: Crissy Dudonis
Production Editor: Debra A. Wechsler
Executive Managing Editor: Kathleen Schiaparelli
Assistant Managing Editor: Beth Sweeten
Managing Editor, Media: Nicole M. Jackson
Editor-in-Chief, Development: Carol Trueheart
Development Editor: Jonathan Haber
Senior Media Editor: Patrick Shriner
Marketing Manager: Andrew Gilfillan
Assistant Manufacturing Manager: Michael Bell
Manufacturing Buyer: Alan Fischer
Director of Creative Services: Paul Belfanti
Art Director: Kenny Beck
Interior Design: Wanda España
Cover Design: Geoffrey Cassar
Managing Editor, Audio and Visual Assets: Patricia Burns

AV Production Manager: Ronda Whitson
AV Production Editor: Denise Keller/Sean Hogan
Art Studio: J/B Woolsey Associates
Manager, Composition: Allyson Graesser
Desktop Administration: Clara Bartunek
Electronic Page Makeup: Clara Bartunek, Julie Nazario, Jude Wilkens
Director, Image Resource Center: Melinda Reo
Manager, Rights and Permissions: Zina Arabia
Interior Image Specialist: Beth Brenzel
Cover Image Specialist: Karen Sanatar
Image Permission Coordinator: Debbie Latronica
Editorial Assistant: Jennifer Hart
Cover Image: Cells of a filamentous cyanobacterium (yellow) from a soda lake biofilm (see back cover text). Photo courtesy of Gernot Arp and Christian Boeker, Carl Zeiss, Jena, Germany

First through seventh editions titled *Biology of Microorganisms*

Editions of Biology of Microorganisms
First Edition, 1970, Thomas D. Brock
Second Edition, 1974, Thomas D. Brock
Third Edition, 1979, Thomas D. Brock
Fourth Edition, 1984, Thomas D. Brock, David W. Smith, and Michael T. Madigan
Fifth Edition, 1988, Thomas D. Brock and Michael T. Madigan
Sixth Edition, 1991, Thomas D. Brock and Michael T. Madigan
Seventh Edition, 1994, Thomas D. Brock, Michael T. Madigan, John M. Martinko, and Jack Parker

Eighth Edition, 1997, Michael T. Madigan, John M. Martinko, and Jack Parker
Ninth Edition, 2000, Michael T. Madigan, John M. Martinko, and Jack Parker
Tenth Edition, 2003, Michael T. Madigan, John M. Martinko, and Jack Parker
Eleventh Edition, 2006, Michael T. Madigan and John M. Martinko

Pearson Prentice Hall™ is a trademark of Pearson Education, Inc.

Printed in the United States of America

10 9 8 7 6

ISBN 0-13-196893-9

Pearson Education LTD., *London*
Pearson Education Australia PTY, Limited, *Sydney*
Pearson Education Singapore, Pte. Ltd.
Pearson Education North Asia Ltd., *Hong Kong*
Pearson Education Canada, Ltd., *Toronto*
Pearson Educación de Mexico, S.A. de C.V.
Pearson Education—Japan, *Tokyo*
Pearson Education Malaysia, Pte. Ltd.
Pearson Education, *Upper Saddle River, New Jersey*

BRIEF CONTENTS

ELEVENTH EDITION OVERVIEW

Each edition creates an opportunity to clarify concepts through the graphics, and *BBOM* has the most exceptional **Illustration Program** in microbiology. Students rely on graphics more each year, and *BBOM* remains dedicated to providing figures, photos and tables that are not only visually engaging, but focused on helping students to better understand the material. ▼

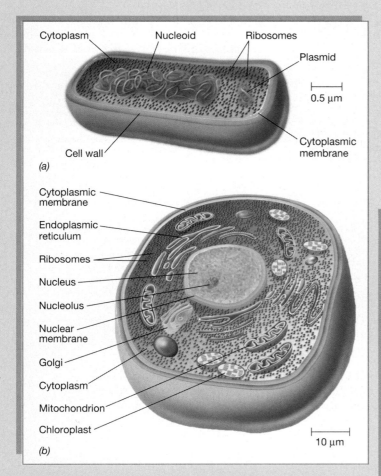

Cytoplasm Nucleoid Ribosomes

Plasmid

0.5 μm

Cytoplasmic membrane

Cell wall

(a)

Cytoplasmic membrane

Endoplasmic reticulum

Ribosomes

Nucleus

Nucleolus

Nuclear membrane

Golgi

Cytoplasm

Mitochondrion

Chloroplast

10 μm

(b)

▲ Geared toward today's visual learners, hundreds of graphics in the 11th edition have been completely redesigned and improved upon to provide greater depth and realism.

(a) *(b)*

(c) *(d)*

Attractant

Control

Repellent

Cells per tube

Time

(e)

Nicholas Blackburn

(f)

Wherever possible, techniques and processes include photomicrographs that link abstract representations to biological reality. ▶

COMPREHENSION

The **Working Glossary** is the students' guide to the language of microbiology. By beginning each chapter with the glossary, students can more easily master terminology and increase their understanding of key concepts. ▶

WORKING GLOSSARY

Bacteriophage a virus that infects prokaryotic cells

Early protein a protein synthesized soon after virus infection

Late protein a protein synthesized toward the end of virus infection

Lysogen a bacterium containing a prophage

Lysogenic pathway a series of steps that, after virus infection, lead to a state (lysogeny) where the viral genome is replicated as a prophage along with that of the host

Lytic pathway a series of steps after virus infection that leads to virus replication and the destruction (lysis) of the host cell

Minus (negative)-strand virus a virus with an RNA genome in which the RNA strand has the opposite sense of (is complementary to) the mRNA of the virus

Nucleocapsid the complex of nucleic acid and proteins of a virus

Oncogene a gene whose expression causes formation of a tumor

Plaque a zone of lysis or cell inhibition caused by virus infection of a lawn of sensitive cells

Plus (positive)-strand virus a virus with an RNA or DNA genome in which the genome has the same complementarity as the mRNA of the virus

Prion an infectious protein whose extracellular form contains no nucleic acid

Provirus (prophage) the genome of a temperate virus when it is replicating with, and usually integrated into, the host chromosome

Retrovirus a virus whose RNA genome has a DNA intermediate as part of its replication cycle

Reverse transcription the process of copying information found in RNA into DNA by the enzyme reverse transcriptase

Temperate virus a virus whose genome is able to replicate along with that of its host and not cause cell death in a state called lysogeny

Transformation in eukaryotes, a process by which a normal cell becomes a cancer cell (but see alternative usage in Chapter 10)

Virion the complete virus particle; the nucleic acid surrounded by a protein coat and in some cases other material

Virulent virus a virus that lyses or kills the host cell after infection; a nontemperate virus

Virus a genetic element containing either RNA or DNA that replicates in cells but is characterized by having an extracellular state

Viroid small, circular, single-stranded RNA that causes various plant diseases

---

1.7 Concept Check

Beijerinck and Winogradsky studied bacteria in soil and water and developed the enrichment culture technique for the isolation of representatives of various physiological groups. Major new concepts in microbiology emerged during this period, including enrichment cultures, chemolithotrophy, chemoautotrophy, and nitrogen fixation.

◆ What is the enrichment culture technique?

◆ In examining Figure 1.16, describe why sulfur oxidation and nitrification are considered chemolithotrophic processes while nitrogen fixation is not. (*Hint:* Look at the reactions involving ATP in each case.)

◀ **Concept Checks** encourage students to stop and assess their understanding of key concepts before moving on. A new "stop sign" icon indicates when a concept check appears in the narrative.

Microbial Sidebars replace the boxed inserts. These illustrated vignettes were designed and written to be interesting, timely and related to each chapter's central theme. ▶

Microbial Sidebar ◆ RNA Editing

In Chapters 7 and 14 we saw that certain genes have coding regions that are split by noncoding regions called *introns*. Typically, introns are removed after transcription to form a mature mRNA, a process called *splicing* (⚬ Section 14.8). Interestingly, there is a phenomenon found in the genomes of organelles that is almost the opposite of splicing: *RNA editing*.

RNA editing involves either the insertion or deletion of nucleotides *into the final mRNA* that were not present in the DNA transcribed. Editing can also involve the chemical *modification* of a base in the mRNA that changes it from one base to another. In either case, RNA editing can alter codons in such a way that one or more different amino acids are inserted in a polypeptide than those encoded by its gene.

In the mitochondria of trypanosomes and related protozoa (⚬ Section 14.10) some mitochondrial transcripts are edited such that large numbers (hundreds in some cases) of uridylates are added or, more rarely, deleted. An example of this type of RNA editing is shown in Figure 1●. RNA editing is precisely controlled by short sequences present in the mRNA that "guide" the enzymes involved in their specific edits. Obviously this process must be very precisely controlled. Inserting too many or too few bases would yield a frameshift product that would likely be nonfunctional.

The other type of RNA editing, the changing of one base into another, is common in the mitochondria and chloroplasts of higher plants. At specific sites in some mRNAs, a C will be converted to a U by oxidative deamination (the opposite modification is more rare). There are at least 25 sites of C to U conversion in the maize chloroplast. Although mostly found in organellar genomes, an example of the programmed conversion of a C to a U is also known for a mammalian nuclear gene. Depending on the location of the edit, a new codon may be formed, leading to formation of a protein sequence not predictable from the gene that encodes it.

RNA editing, although a curious phenomenon, was not a significant obstacle in analyzing *organellar* genomes. This is because the number of proteins they encode is small and the proteins highly conserved. By contrast, had RNA editing been a widespread phenomenon in *cells*, genomic annotations and the identification of orthologous genes in different organisms could have been an even more formidable challenge than it has been to date.

The function and origin of RNA editing is unknown. But some scientists have pointed out that this process may be yet another remnant, along with ribozymes (⚬ Section 14.8) and other catalytic RNAs, of the RNA World (⚬ Section 11.2). ∎

Protein	...Leu Cys Phe Trp Phe Arg Phe Phe Cys...
mRNA	...uuG uGu UUU UGG uuu AGG uuu uuu uGu...
DNA	... G G TTT TCC AGG G ...
	... C C AAA AGG TCC C ...

Figure 1 RNA editing. The upper part of the figure shows a portion of the amino acid sequence of subunit III of the enzyme cytochrome oxidase from the protozoan Trypanosoma brucei (⚬ Section 14.10). This protein is encoded by mitochondria. Beneath the amino acid sequence, the sequence of the messenger RNA (mRNA) for this region is shown. The bases in uppercase letters are those transcribed from the gene, which is shown below. The bases in the mRNA in lowercase have been inserted into the transcript by RNA editing. Although the DNA has many informational gaps, there are no actual gaps in the molecule itself. The spaces between the base pairs are simply to aid in visualization.

NAVIGATION

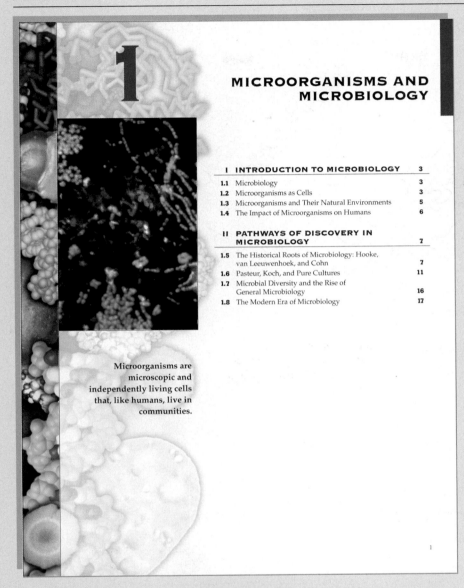

1

MICROORGANISMS AND MICROBIOLOGY

Microorganisms are microscopic and independently living cells that, like humans, live in communities.

◀ **Section Numbers** keyed to page references provide an easy way to assign readings as well as an organized outline for students to review and take notes.

---

rier that separates the inside from the outside called the **cytoplasmic membrane** (Figure 2.1●). It is through the cytoplasmic membrane that nutrients and other substances needed by the cell enter, and waste materials and other cell products exit. Within a cell, and bounded by the cytoplasmic membrane, is a complex mixture of sub-

◀ **Figure Reference Locators** occur next to key figure references in the narrative. Requested by students, these red dots help them study more effectively by allowing them to quickly return to the narrative after viewing a figure.

Concept Links, signaled with a blue chain link (∞), alert students that a concept is related to material from other areas of the text. Each link refers students to a section number for quickly reviewing the related material, helping them to continually make connections between concepts. ▶

In a typical Winogradsky column a mixture of organisms develops. Algae and cyanobacteria appear quickly in the upper portions of the water column; by producing O_2 these organisms help to keep this zone oxic. Decomposition processes in the mud lead to the production of organic acids, alcohols, and H_2, suitable substrates for sulfate-reducing bacteria (∞ Sections 12.18, 13.7, 17.15, and 19.13). Sulfide from the sulfate reducers triggers development of purple and green sulfur bacteria (anoxygenic phototrophs, ∞ Sections 12.2 and 12.32) that use sulfide as a photosynthetic electron donor. These organisms typically appear in

INSTRUCTOR RESOURCE CENTER CD/DVD

Instructor Resource Center on CD/DVD (0-13-144340-2)

The multi-disc IRC on CD/DVD package offers everything an instructor needs to prepare and present lectures. Easy-to-use menus and exporting functions enable you to locate resources and build presentations efficiently. IRC on CD/DVD resources include:

- JPEG files of all drawings, photos, and tables from the text optimized for projection purposes

- JPEG files of all text figures and tables in comprehensive PowerPoint® presentations for each chapter

- A second set of PowerPoint® presentations consisting of lecture notes and key figures for each chapter

- ALL NEW Instructor Animations adding depth and clarity to key topics, dynamic processes, and techniques described in the 11th Edition

- A comprehensive set of CRS/In-Class questions

- Word files of the Instrutor's Manual and the assessment materials including the Test Bank

- Printable PDF files of a Media Integration Guide describing each chapters media offerings

COMPANION WEBSITE

Home > Macromolecules > Online Study Guide > Review Question

Chapter 3: Online Study Guide
Review Question 2

Hydrophobic interactions are important in maintaining [Hint]

- the structure of the cytoplasmic membrane.
- interactions between proteins in multisubunit enzymes.
- protein structure.
- all of the above.

Submit Answer for Grading

Companion Website

www.prenhall.com/madigan

Valuable study tools to aid student comprehension of the 11th Edition are offered on the Companion Website. Features include:

- Online Study Guide provides a focused section-by-section review of topic coverage featuring summaries, key illustrations, and review questions

- "Track Your Progress" tool helps students get a better sense of where to focus study time shows them how successful their efforts have been

- **ALL NEW** Web Tutorials help students visualize key topics, processes, and techniques

ANIMATION RESOURCES

Instructor Animation and Student Web Tutorial topics:

Pasteur's Experiment
Koch's Postulates
The Gram Stain
The Prokaryotic Flagellum
Aseptic Transfer and the Streak Plate Method
Direct Microscopic Counting Procedure (Petroff-Hausser Chamber)
DNA Replication
The Polymerase Chain Reaction (PCR)
Transcription
Translation
Enzyme Regulation
Negative Control of Transcription and the *lac* Operon
Attenuation and the Tryptophan Operon
A Temperate Bacteriophage
The Molecular Basis for Mutations
Replica Plating
The Molecular Basis for Mutations
Generating Phylogenetic Trees from RNA Sequences
Cell Division in Conventional, Budding, and Stalked Bacteria
Bacteriorhodopsin and Light-Mediated ATP Synthesis
Life Cycle and Mating Type Switching in a Typical Yeast
DNA Chips
Replication of Poliovirus
Electron Transport Processes: Aerobic and Anaerobic Conditions
Enrichment Cultures
Serial Dilutions and a Most-Probable Number Analysis

Root Nodule Bacteria and Symbiosis with Legumes
Antibiotic Modes of Action
Diphtheria and Cholera Toxins
Antigen Presentation
Producing Monoclonal Antibodies
The ELISA Test
HIV Replication
Life Cycle of the Malaria Parasite
Isolation and Screening of Antibiotic Producers
Production of Recombinant Vaccinia Virus

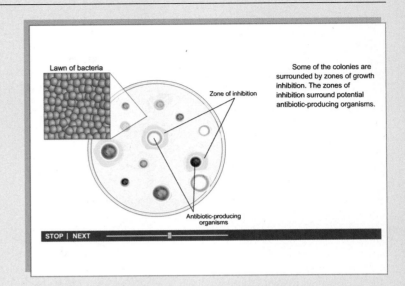

Lawn of bacteria

Zone of inhibition

Some of the colonies are surrounded by zones of growth inhibition. The zones of inhibition surround potential antibiotic-producing organisms.

Antibiotic-producing organisms

STOP | NEXT

OTHER RESOURCES

Instructor's Resource Manual with Tests (0-13-144342-9)
This manual and test bank contains over 2000 questions an instructor can use to prepare exams. This resource also provides chapter summaries as well as the answers to the end of chapter review and application questions.

Test Gen EQ Computerized Testing Software (0-13-144337-2)
In addition to the printed volume, the test questions are available as part of the Test Gen EQ Testing Software, a text specific testing program that is networkable for administering tests. It also allows instructors to view and edit questions, export the questions as tests, and print them out in a variety of formats.

Transparencies (0-13-144341-0)
400 figures from the text are included in this transparency package. The font sizes of the labels have been increased for easy viewing from the back of the classroom.

Research Navigator www.researchnavigator.com
Prentice Hall's Research Navigator™ gives your students access to the most current information available for a wide array of subjects via EBSCO's Content Select™ Academic Journal Database, The New York Times Search by Subject Archive,

"Best of the Web" Link Library, and information on the latest news and current events. This valuable tool helps students find the most useful articles and journals, cite sources, and write effective papers for research assignments.

SafariX Textbooks Online is an exciting new choice for students looking to save money. As an alternative to purchasing the print textbook, students can subscribe to the same content online and

save up to 50% off the suggested list price of the print text. With a SafariX WebBook, students can search the text, make notes online, print out reading assignments that incorporate lecture notes, and bookmark important passages for later review. For more information, or to subscribe to the SafariX WebBook, visit www.safarix.com.

Course Management
Prentice Hall offers all of the instructor and student media for the 11th Edition through your preferred course management platform. For details visit http://cms.prenhall.com.

CONTENTS

UNIT IV
IMMUNOLOGY, PATHOGENICITY, AND HOST RESPONSES

PREFACE

We are living in the age of microbiology. Almost daily come reports of new discoveries in this exciting science— new emerging infections, new organisms, and new tools to facilitate discovery. Such is the pace of the science of microbiology today. And here to bring you the up-to-the-minute picture of microbiology today is the eleventh edition of *Brock Biology of Microorganisms (BBOM)*

This textbook has roots going back over 35 years. Since the publication of the first edition of *Biology of Microorganisms* by Thomas D. Brock (Prentice Hall, 1970), this book has had a single mission: to present the principles of microbiology within the framework of modern science. *BBOM 11/e* maintains this tradition, and speaks with the same accuracy and authority of the previous ten editions.

Microbiology today places unusual demands on students and instructors alike. The amount of information is enormous, the background required in supporting sciences is significant, and introductory classes in microbiology are bursting at the seams. The authors of *BBOM 11/e* are keenly aware of these problems and have worked hard to craft a textbook of microbiology where the principles stand up and shout, the details are complementary, and the supporting concepts are well integrated. We hope you will agree.

What's New in the 11ᵗʰ Edition?

Those who have taught from *BBOM* in the past will find the new edition the same old friend they knew before. However, instructors will find *BBOM 11/e* more teachable, and students will find it a more invaluable learning resource, than ever before.

BBOM 11/e is pedagogically geared towards today's visual learner. The design of *BBOM 11/e* starts where the tenth edition left off, but charts new ground in terms of presentation, use of color, and art. The chapters are organized around the logical and helpful numbered outline system, present in this book from the first edition. But we have now organized each chapter into several blocks of related information, integrating the concepts into more digestible bites. As usual, a working glossary—the student's dictionary of essential terms—opens each chapter, to bring the language of microbiology where it needs to be, up front and center. Concept checks remain and are now signaled by a bright red "stop sign!" Concept checks signal the student to stop, review, and assess, before proceeding to the next concept. As usual, challenging review and application study questions are present at the end of each chapter. A comprehensive glossary and index in the back of the book, wrap up the package.

Traditional "boxed" material is presented in *BBOM 11/e* in our new *Microbial Sidebars*. These richly illustrated vignettes were designed and written to be "fun reads" of enrichment material related to a chapter's central theme. Tables have been completely redesigned in *BBOM 11/e* to make the information in them easier to follow and better organized. Since a science like microbiology relies heavily on tabular resources, the new table design should be a winner with both students and instructors alike. Many other pedagogically useful features will make themselves obvious to the reader as s/he proceeds through this book. These include more distinctive heads, an eye-catching red dot icon that leads the reader's eye from text to figures and back again, and review questions keyed to section number. The latter will make it easier for students to refresh their memories before answering each question.

Pervading the entire book is a spectacular art program, with every piece of art redone by a new art studio. The result is brighter, more distinctive art, which is also more colorful, appealing, impeccably consistent, and instructive than ever before. Moreover, the use for the first time of a high quality, glossy paper in the eleventh edition, has brought out the best in the outstanding photomicrographs and other photos that have been a tradition in this book since the first edition. Users will quickly recognize new pedagogical aids, such as our "energy arrows," built right into the art. Cellular reactions that produce or consume ATP are often key ones. Energy arrows—bright red wavy arrows—signal these reactions and bring them to a student's attention.

Although *BBOM 11/e* is actually shorter than the previous edition, it contains substantial new content. Indeed, new material can be found in every chapter and we give only a taste of what's in store here: Toxic Forms of Oxygen (Chapter 6); Diversity of Sigma Factors, Consensus Sequences, and Other RNA Polymerases (Chapter 7); The Stringent Response (Chapter 8); RNA Regulation and Riboswitches (Chapter 8); Sub-Viral Particles (Chapter 9); The Carbon and Energy Metabolism of Primitive Life Forms (Chapter 11); The Biology of *Nanoarchaeum* (Chapter 13); RNA Processing and Ribozymes (Chapter 14); Replication of Linear DNA (Chapter 14); Annotating the Genome (Chapter 15); Bioinformatic Analyses and Gene Distribution in Prokaryotes (Chapter 15); Microarrays and the Transcriptome (Chapter 15); Viruses of *Archaea* (Chapter 16); Environmental Genomics (Chapter 18); Host Risk Factors for Infection (Chapter 21); Inflammation, Fever, and Septic Shock (Chapter 22); Natural Immunity (Chapter 22); Receptors and Immunity (Chapter 23); West Nile Virus (Chapter 27); Microbial Sampling and Food Poisoning (Chapter 29); Severe Acute Respiratory Syndrome (SARS) (Chapter 25); Anthrax as a Biological Weapon (Chapter 25); and Fermented Foods (Chapter 29).

Several supplements accompany *BBOM 11/e*. These include a website (***www.prenhall.com/madigan***) containing online media resources (flagged by an icon in the text), practice exam questions, and additional content resources. For instructors, a CD is available that contains *all* of the tables and figures in the book arranged in PowerPoint format for ease in organizing classroom learning activities. Overhead transparencies are also available for those who use this format in the classroom. Indeed, the *BBOM 11/e* instructor package offers every necessary tool for developing clear, compelling, and stimulating presentations.

Acknowledgments

The final product you see before you is the collective effort of many people. These include a number of folks at Prentice Hall/Pearson Publishing, but especially Executive Editor Gary Carlson and his assistants Susan Zeigler and Jennifer Hart, and our outstanding production editor, Debra Wechsler. Gary was the guiding light for this edition, while Susan/Jennifer and Debra maintained the pace of the project and were the "glue" in editorial and production, respectively. Ed Thomas (production) is also acknowledged for production assistance early in the project. The authors give a hearty thanks to the input of our design and art editors, Kenny Beck and Jay McElroy, and our superb media editors, Patrick Shriner and Crissy Dudonis. Excellent copy-editing of the entire manuscript was done by Jane Loftus (Clackamas, OR).

The authors especially wish to thank the thorough and very helpful input early on in the project of the development editor, Jon Haber (New York, NY), and the expert assistance with electronic photographs provided by Deborah O. Jung (Carbondale, IL). The excellent input of Elizabeth McPherson (University of Tennessee) and David Crowley (University of California-Riverside) as accuracy checkers during the production of *BBOM 11/e* is heartily acknowledged. Special thanks are also extended to Gernot Arp and Christian Boeker, Universität Göttingen, Germany and Carl Zeiss, Inc., Jena, Germany for the gorgeous confocal micrograph that graces the covers of this book.

Both authors also wish to thank their graduate students, colleagues, and Department of Microbiology staff for their assistance and patience during the busy time of preparing a book of this scope. The seemingly endless patience of our wives, Nancy and Judy, for the countless hours spent away from them during the development and production of *BBOM 11/e*, is also acknowledged. Their love, understanding, and support allowed the authors to devote the time necessary for such a massive project.

Finally, we are extremely grateful for the kind help of so many individuals that provided reviewer input into the draft manuscript or who supplied new photos for the 11th edition. They are listed below.

Laurie Achenbach, *Southern Illinois University*
Richard Adler, *University of Michigan-Dearborn*
Karen Aguirre, *Clarkson University*

Stephen Aley, *University of Texas-El Paso*
Mary Allen, *Hartwick College*
Ricardo Amils, *University Autonoma, Madrid, Spain*
Robert Andrews, *Iowa State University*
Michael Benedik, *Texas A&M University*
David Boone, *Portland State University*
Matt Boulton, *University of Michigan*
Cheryl Broadie, *Southern Illinois University*
Jean Cardinale, *Alfred University*
Jannice Carr, *Centers for Disease Control and Prevention-Atlanta*
David Clark, *Southern Illinois University*
Rhonda Clark, *University of Calgary*
Morris Cooper, *Southern Illinois University School of Medicine*
David Crowley, *University of California-Riverside*
Mark Davis, *University of Evansville*
Michael Davis, *Central Connecticut State University*
Dennis Dean, *Virginia Tech University*
Arvind Dhople, *Florida Institute of Technology*
Biao Ding, *Ohio State University*
Rodney Donlan, *Centers for Disease Control and Prevention-Atlanta*
Paul Dunlap, *University of Michigan*
Paul Edmonds, *Georgia Institute of Technology*
Elizabeth Ehrenfeld, *Southern Maine Community College*
Bruce Farnham, *Metropolitan State University-Denver*
Rebecca Ferrell, *Metropolitan State University-Denver*
Doug Fix, *Southern Illinois University*
Niels-Ulrik Frigaard, *Pennsylvania State University*
George Garrity, *Michigan State University*
Claire Geslin, *Université de Bretagne Occidentale, France*
Eric Grafman, *Centers for Disease Control and Prevention-Atlanta*
Bonita Glatz, *Iowa State University*
Ricardo Guerrero, *University of Barcelona, Spain*
John Haddock, *Southern Illinois University*
Martin Hanczyc, *Harvard University*
Ernest Hanning, *University of Texas-Dallas*
Pamela Hathorn, *Midwestern University*
John Hayes, *Woods Hole Oceanographic Institution*
Lee Hughs, *University of North Texas*
Michael Ibba, *Ohio State University*
Johannes Imhoff, *Universität Kiel, Germany*
Mary Johnson, *Indiana University*
Deborah Jung, *Southern Illinois University*
Judy Kandel, *California State University-Fullerton*
Patrick Keeling, *University of British Columbia*
Joan Kiely, *SUNY-Stonybrook*
Arthur Koch, *Indiana University*
Allan Konopka, *Purdue University*
Vikki Kourkouliotis, *National Renewable Energy Laboratory*
Susan F. Koval, *University of Western Ontario*
Harry Kurtz, *Clemson University*
Sharon Long, *University of Massachusetts*

Bonnie Lustigman, *Montclair State University*
Mark Martin, *Occidental College*
William McCleary, *Brigham Young University*
Elizabeth McPherson, *University of Tennessee*
Ohad Medalia, *Max Planck Institut für Biochemie, Germany*
Jianghang Meng, *University of Maryland*
Eric Miller, *North Carolina State University*
Abraham Minisky, *Weizmann Institute of Science, Israel*
Ivan Oresnik, *University of Manitoba*
Jörg Overmann, *University of Munich, Germany*
Norm Pace, *University of Colorado*
Jack Parker, *Southern Illinois University*
Laurence Pelletier, *Max Planck Institut für Cell Biologie und Genetiks, Germany*
Michael Pfaller, *University of Iowa*
Reinhard Rachel, *Universität Regensburg, Germany*
Michael Rappe, *Oregon State University*
Chris Rensing, *University of Arizona*
Frank Roberto, *Idaho Environmental and Engineering Labs*
Craig Rouskey, *Southern Illinois University*
Jill Zeilstra Ryalls, *Oakland University*
Herb Schelhorn, *McMaster University*
Bernhard Schink, *Universität Konstanz, Germany*
Heide Schulz, *University of California-Davis*
Kate Scow, *University of California-Davis*
James Shapleigh, *Cornell University*
Jolynn Smith, *Southern Illinois University*

Jerry Sipe, *Anderson University*
Nancy Spear, *Murphysboro, Illinois*
Julia Thompson, *American Society for Microbiology*
Sonia Tiquia, *University of Michigan*
David Tison, *Multicare Health Systems*
Paul Tomasek, *California State University-Northridge*
Amy Treonis, *Creighton University*
Michael Wagner, *University of Vienna, Austria*
David Ward, *Montana State University*
Joy Watts, *University of Maryland*
Mary Watwood, *Northern Arizona University*
Susan Wells, *Affymetrix, Santa Clara, California*
Carl Woese, *University of Illinois*
Gordon Wolfe, *California State University-Chico*
Alexander Worden, *University of Miami*
Mark Young, *Montana State University*
Stephen Zinder, *Cornell University*

Any errors in this book, either ones of commission or omission, are solely the responsibility of the authors. In past editions, users have been so kind as to contact us when they came across an error. Although we hope that *BBOM 11/e* is error-free, no book ever is. So users should feel free to contact either author about errors or concerns. We hope you enjoy *BBOM 11/e*.

Michael T. Madigan (madigan@micro.siu.edu)
John M. Martinko (martinko@micro.siu.edu)

BROCK

BIOLOGY OF
MICROORGANISMS

1

MICROORGANISMS AND MICROBIOLOGY

Microorganisms are microscopic and independently living cells that, like humans, live in communities.

WORKING GLOSSARY

Cell the fundamental unit of living matter
Cytoplasm the fluid portion of a cell, bounded by the cell membrane but excluding the nucleus (if present)
DNA deoxyribonucleic acid, the hereditary material of cells and some viruses
Ecology the study of organisms in their natural environments
Ecosystem organisms plus their nonliving environment
Enrichment culture a method for isolating microorganisms from nature using specific culture media and incubation conditions

Enzyme a protein catalyst that functions to speed up chemical reactions
Habitat the location in an environment where a microbial population resides
Koch's Postulates a set of criteria for proving that a given microorganism causes a given disease
Metabolism all biochemical reactions in a cell
Microorganism a microscopic organism consisting of a single cell or cell cluster, including the viruses

Pathogen a disease-causing microorganism
Pure culture a culture containing a single kind of microorganism
RNA ribonucleic acid, involved in protein synthesis as messenger RNA, transfer RNA, and ribosomal RNA
Spontaneous generation the hypothesis that living organisms can originate from nonliving matter
Sterile absence of all living organisms and viruses

Welcome to microbiology—the study of microorganisms. Microorganisms include a large and diverse group of microscopic organisms that exist as single cells or cell clusters, and the viruses, which are microscopic but not cellular.

What is microbiology all about? Microbiology is about cells and how they work, especially the bacteria, a large group of cells of enormous basic and practical significance (Figure 1.1●). Microbiology is about microbial diversity and evolution, about how different kinds of microorganisms arose and why. It is about what microorganisms do in the world at large, in human society, in the human body, and in animals and plants. One way or the other, microorganisms affect all other life forms on Earth (Figure 1.1*b*), and thus the science of microbiology is of enormous importance.

Microorganisms are distinct from the cells of animals and plants—macroorganisms. Cells of plants and animals are unable to live alone in nature and exist only as parts of multicellular structures, such as the organ systems of animals or the structural components of plants. By contrast, most microorganisms can carry out their life processes of growth, energy generation, and reproduction independently of other cells, either of the same kind or of a different kind.

(a) (b)

(c)

● **Figure 1.1 Microorganisms.** (a, b) A single microbial cell can have an independent existence. Shown are photomicrographs of phototrophic (photosynthetic) microorganisms called (a) *purple bacteria*, and (b) *cyanobacteria*. Purple bacteria were among the first phototrophs on Earth, while cyanobacteria were the first oxygen-evolving phototrophs on Earth. Cyanobacteria oxygenated the atmosphere, allowing for the evolution of higher life forms. (c, d) In nature or in the laboratory, bacterial cells can grow to form extremely large populations. Shown are (c) a bloom of purple bacteria (see Figure 1.15) in a small lake in Spain, Lake Cisó, and (d) bioluminescent (light-emitting) cells of the bacterium *Photobacterium leiognathi* grown in laboratory culture. One milliliter of lakewater (c) or one colony from the plate (d) contains more than one billion (10^9) individual cells.

(d)

This chapter begins our journey to the microbial world. We discover what microorganisms are and their impact on life. We set the stage for a consideration of the structure and evolution of microorganisms that will unfold in the next chapter. We also place microbiology in historical perspective, as a process of scientific discovery. Thanks to the landmark contributions of both early microbiologists and scientists practicing today, we can see the ramifications of microbiology for medicine, agriculture, and the environment.

I INTRODUCTION TO MICROBIOLOGY

In the first four sections of this chapter we introduce the subject of microbiology, look at microorganisms as cells, examine where and how microorganisms live in nature, and review the impact that microorganisms have had and continue to have on human affairs.

1.1 Microbiology

Microbiology revolves around two basic themes: (1) the basic science of understanding life, and (2) the applications of science to human needs.

As a *basic biological science*, microbiology provides tools for probing the processes of life. Our most sophisticated understanding of the chemical and physical basis of life has arisen from studies of microorganisms. Microbial cells share many biochemical properties with cells of multicellular organisms; indeed, *all* cells have much in common. Moreover, microbial cells can grow to extremely high densities in laboratory culture (Figure 1.1*d*) and are readily amenable to biochemical and genetic study. All of these features make microorganisms excellent models for understanding cell function in higher organisms, including humans.

As an *applied biological science*, microbiology deals with many important practical problems in medicine, agriculture, and industry. Some of the most important diseases of humans, other animals, and plants are caused by microorganisms. Microorganisms also play major roles in soil fertility and domestic animal production. In addition, many large-scale industrial processes, such as the production of antibiotics or human proteins, are microbially based.

The Importance of Microorganisms

As this book unfolds, we will see the central role that microorganisms play in both human activities and the web of life on Earth. In the absence of microorganisms, for instance, higher life forms would never have arisen and could not now be sustained. Indeed, the very oxygen we breathe is the result of past microbial activity (Figure 1.1*b*). Moreover, we will see how humans, plants, and animals are intimately tied to microbial activities for the recycling of key nutrients and for degrading organic matter. No

other life forms are as important as microorganisms for the support and maintenance of life on Earth.

Microorganisms existed on Earth for billions of years before plants and animals appeared. Thus, their evolutionary diversity has far outpaced that of higher organisms. This huge diversity accounts for some of the spectacular properties of microorganisms. For example, we will see how microorganisms can live in places unsuitable for higher organisms and how their diverse physiological capacities rank them as Earth's greatest chemists. We will also see how microorganisms have established relationships with higher organisms that can be either beneficial or harmful. In humans, for instance, the very health we enjoy and many of the diseases we suffer are due to microorganisms.

Microorganisms are clearly central to the very functioning of the biosphere. Hence, the science of microbiology is the foundation of all the biological sciences. We begin our study of this important science with a consideration of microorganisms as cellular entities.

1.2 Microorganisms as Cells

The **cell** is the fundamental unit of life. A single cell is an entity, isolated from other cells by a membrane (and perhaps a cell wall) and contains within it a variety of chemicals and subcellular structures (Figure 1.2●). Compartmentalization is a prerequisite for life. This is because the chemical components of life must remain at concentrations sufficient to allow necessary chemical reactions to occur. But the cell is not a closed system. Instead, the cell is an open, dynamic entity. Cells communicate and exchange materials with their environments, and they are constantly undergoing change. We will study cell structure and function in detail in Chapters 2, 4, and 14.

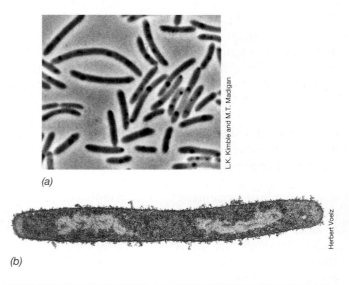

(a)

(b)

● **Figure 1.2   Cells.** (a) Photomicrograph of rod-shaped bacterial cells as seen in the light microscope; a single cell is about $1\,\mu$m in diameter. (b) Longitudinal section through a bacterial cell as viewed with an electron microscope. The two lighter areas are the nucleoid, regions in the cell containing aggregated DNA.

Cell Chemistry and Key Structures

Cells are highly organized structures consisting primarily of four chemical components: **proteins, nucleic acids, lipids**, and **polysaccharides**. Collectively, these large molecules are called **macromolecules**. It is the chemistry and arrangement of macromolecules in different cells that make organisms distinct from one another. Thus, chemically speaking, all cells share much in common, whether they are parts of a plant or animal or whether they are microorganisms.

There are a number of key structures in a cell. The **cytoplasmic membrane** is the barrier that separates the inside of the cell from the outside. Inside the cell membrane are various structures and chemicals suspended or dissolved in the **cytoplasm**. The latter is where the *machinery* for cell growth and function is present. Key structures here include the **nucleus** or **nucleoid**, where the cells' *genetic information*—deoxyribonucleic acid (**DNA**)—is stored, and **ribosomes**, the structures upon which new proteins are made in the cell.

Characteristics of Living Systems

What are the essential characteristics of life? What differentiates cells from inanimate objects? Our concept of what is alive is constrained by what we observe on Earth today or can deduce from the fossil record. But from our knowledge of biology thus far, we can identify several characteristics shared by most living systems. These are summarized in Figure 1.3●.

All cellular organisms show some form of **metabolism**. That is, cells take up nutrients from their environment and transform them—conserving some of the energy present in these substances in a form the cell can use—and then eliminate waste products. All cells show **reproduction**. That is, a cell can direct a series of biochemical events that results in growth and division to form two cells. Many cells undergo **differentiation**, a process by which new substances or structures are formed. Cell differentiation is often part of a cellular life cycle in which cells form special structures, such as spores, involved in reproduction, dispersal, or survival.

Cells respond to chemical signals in their environment, including those produced by other cells. Cells can thus undergo **communication**. Cells can even assess their own numbers by way of small diffusible molecules passed between neighboring cells, a process call *quorum sensing*. Although not universal, living organisms are often capable of **movement** by self-propulsion; in the microbial world we will see several different mechanisms of motility. Finally, unlike nonliving structures, cells can *evolve*. Through the process of **evolution**, cells can change their characteristics and transmit these new properties to their offspring.

Cells as Machines and as Coding Devices

Cells can be viewed in two ways. On one hand, cells can be considered *machines* that carry out chemical

1. Metabolism

Uptake of nutrients from the environment, their transformation within the cell, and elimination of wastes into the environment. The cell is thus an *open* system.

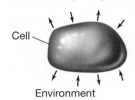

2. Reproduction (growth)

Chemicals from the environment are turned into new cells under the direction of preexisting cells.

3. Differentiation

Formation of a new cell structure such as a spore, usually as part of a cellular *life cycle*.

4. Communication

Cells *communicate* or *interact* primarily by means of chemicals that are released or taken up.

5. Movement

Living organisms are often capable of self-propulsion.

6. Evolution

Cells contain genes and *evolve* to display new biological properties. Phylogenetic trees show the evolutionary relationships between cells.

● **Figure 1.3** **The hallmarks of cellular life.** Differentiation and motility are not properties of all microbial cells.

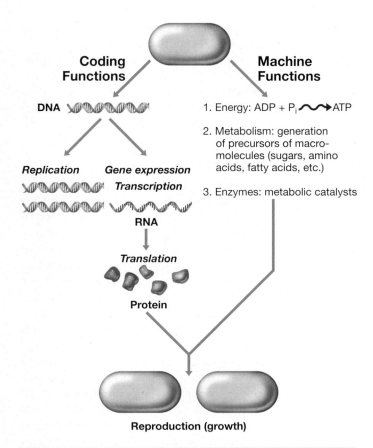

● **Figure 1.4   The machine and coding functions of the cell.** In order for a cell to reproduce itself there must be (1) an adequate supply of energy and precursors for the synthesis of new macromolecules, (2) the genetic instructions replicated such that upon division each cell receives a copy, and (3) gene expression (the processes of transcription and translation) to form the proper amounts of necessary proteins and other macromolecules that will make up the new cell.

transformations. The catalysts of this chemical machine are **enzymes**, proteins capable of greatly accelerating the rate of chemical reactions (Figure 1.4●). On the other hand, cells can be considered *coding devices*, analogous to computers, which store and process genetic information (DNA) that is eventually passed on to offspring during reproduction (Figure 1.4). Replication and processing of the stored genetic information will be considered in Chapter 7, where we cover the core cellular functions of *DNA replication, transcription*—the production of **RNA**—and *translation*—the production of **protein**.

In reality, cells are both chemical machines *and* coding devices. The link between these two cellular attributes is *growth*. Under proper conditions, a cell will grow larger and then divide to form two cells (Figure 1.4). In the orderly process that results, the amount of all the constituents of the cell must *double*. This requires the chemical machinery of the cell to supply energy and precursors for biosynthesis of macromolecules. Also, when a cell divides, each of the two resulting cells must contain all of the genetic information necessary for the formation of more cells. Thus, during the growth process

there must also be a replication of DNA (Figure 1.4). The machine and coding functions of the cell must therefore be highly coordinated in order for the cell to reproduce itself faithfully. Indeed, we will see later that in addition to *coordination*, the various machine and coding functions of the cell are subject to *regulation*. This ensures that the cell stays optimally attuned to its environment.

The First Cells

Where did the first cells come from? Because all cells are constructed in similar ways, it is hypothesized that all cells have descended from a common ancestor, the *universal ancestor* of all life (⌀⌀Chapter 11). In some way the first cell must have come from a noncell, something before the cell, a *procellular* structure. Evolution of the first cell on Earth—over 3.8 billion years ago—may have taken several hundred million years to occur. Yet once the first cells arose, their growth and division formed populations. Evolution could then select for improvements and diversification of these early life forms. Through billions of years of evolution, a tremendous diversity of cell types appeared. We will get a taste of this microbial diversity in Chapter 2, and then consider the topic in detail in Chapters 12–16.

⬢ 1.2   Concept Check

The cell has a barrier, the cytoplasmic membrane, that separates the cytoplasm from the environment. Other cell features include the nucleus or nucleoid, and cytoplasm. Metabolism and reproduction are associated with the living state, and cells can be considered both chemical machines and coding devices. Life has been present on Earth for nearly 4 billion years.

◆   List the four classes of cellular macromolecules.

◆   List six features associated with living organisms. Why might each feature be important to the survival of a cell?

◆   Compare the machine and coding functions of a microbial cell. Why is neither of value to a cell without the other?

1.3   Microorganisms and Their Natural Environments

In nature, cells live in association with other cells in assemblages called **populations**. Populations are composed of groups of cells derived from successive cell divisions from a single parent cell. The location in an environment where a microbial population lives is called the **habitat**. In microbial habitats, a population of cells rarely lives alone. Rather, a population of cells lives and interacts with other populations in **microbial communities** (Figure 1.5●). The components and cell numbers in a microbial community are governed by the resources and conditions that exist in the habitat. The study of microorganisms in their natural habitats is called **microbial ecology**.

(a)   (b)

D. E. Caldwell

Jiri Snaidr

● **Figure 1.5 Examples of microbial communities.** (a) Photomicrograph of a bacterial community that developed in the depths of a small lake (Wintergreen Lake, Michigan), showing cells of various sizes. (b) A bacterial community in a sewage sludge sample. The sample was stained with a series of dyes, each of which stained a different bacterial group (∞Section 18.4 and Figure 18.11*b* for details of the staining process). From R. Amann, J. Snaidr, M. Wagner, W. Ludwig, and K.-H. Schleifer, 1996. *Journal of Bacteriology* 178: 3496–3500, Fig. 2b. © 1996 American Society for Microbiology.

The Effect of Organisms on Each Other and on Their Habitats

Populations in microbial communities interact in various ways, both harmful and beneficial. In many cases microbial populations interact and cooperate. For example, the waste products of the metabolic activities of some cells can be nutrients for others. Organisms in a habitat also interact with their physical and chemical environment. Habitats differ markedly in their characteristics, and a habitat that is favorable for the growth of one organism may actually be harmful for another organism. Collectively, we speak of the living organisms together with the physical and chemical constituents of their environment as an **ecosystem**. Major microbial ecosystems include aquatic (oceans, ponds, lakes, streams, ice, hot springs), terrestrial (soil, deep subsurface), and higher organisms, both plant and animal.

Ecosystems are controlled to a significant extent by microbial activities. Organisms carrying out metabolic processes remove nutrients from the environment and use them to build new cells. At the same time, organisms excrete waste products of their metabolism into the environment. Thus, over time, a microbial ecosystem will gradually change, both chemically and physically, through the cycling of nutrients by microorganisms. The environmental changes may allow for other microorganisms to grow. For example, molecular oxygen (O_2) is a vital nutrient for some microorganisms, but a poison to others. However, the oxygen-consuming activities of one group of organisms (aerobes) can make an oxic habitat

anoxic (O_2-free) and suitable for growth of anaerobic organisms that were previously unable to grow.

In later chapters, after we have learned some of the basic features of microbial structure and function, genetics, evolution, and diversity, we will return to a discussion of the ways in which microorganisms affect animals, plants, and the whole global ecosystem.

The Extent of Microbial Life

Microorganisms are small, but their biomass on Earth is huge, even when compared with the biomass of higher organisms. Examination of natural material such as soil or water always reveals microbial cells. Although such tiny cells may seem inconsequential, single cells are capable of multiplying rapidly and producing large populations that may have a major effect on the habitat (see Figure 1.1*c*). Microorganisms are thus extremely important and quantitatively significant parts of virtually every ecosystem.

Estimates of total microbial cell numbers on Earth and specifically, the numbers of **prokaryotes (bacteria,** ∞Chapter 2) show this number to be on the order of 5×10^{30} cells. The total amount of carbon present in this very large number of very small cells equals that of all plants on Earth (and plant carbon far surpasses animal carbon). In addition, the collective contents of nitrogen and phosphorous in prokaryotic cells is over 10 times that in all plant biomass.

Thus prokaryotic cells, small as they are, constitute the *major portion of biomass on Earth*, and are key reservoirs of essential nutrients for life. An equally startling revelation is the realization that most prokaryotic cells do not reside on Earth's surface, but instead *lie underground* in the oceanic and terrestrial subsurfaces. Because these habitats are relatively unexplored, there is much left for microbiologists to discover and understand about the life forms that dominate Earth.

 1.3 Concept Check

Microorganisms exist in nature in populations that interact with other populations in microbial communities. The activities of microbial communities can greatly affect the chemical and physical properties of their habitats. Most of the biomass on Earth is microbial.

◆ What is a *microbial habitat?*

◆ How do microorganisms change the chemical and physical properties of their habitats?

◆ Where are most prokaryotic cells located on Earth?

1.4 The Impact of Microorganisms on Humans

One goal of the microbiologist is to understand how microorganisms work and devise ways in which the benefits of microorganisms can be increased and their harmful

● **Figure 1.6   The impact of microorganisms on human affairs.** Although many people think of microorganisms in the context of infectious diseases, few microorganisms actually cause disease. Microorganisms affect many aspects of our lives in addition to playing a role as disease agents.

effects curtailed. Microbiologists have been highly successful in achieving these goals, and microbiology has played a major role in the advancement of human health and welfare. An overview of the impact of microorganisms on human affairs is shown in Figure 1.6●.

Microorganisms as Disease Agents

One measure of the microbiologist's success in controlling microorganisms is shown by the statistics in Figure 1.7●. These data compare the present causes of death in the United States to those of 100 years ago. At the beginning of the twentieth century, the major causes of death were infectious diseases (Figure 1.7). Infectious diseases are caused by **pathogens**. Children and the aged in particular succumbed in large numbers to microbial diseases. Today, infectious diseases

are of much less importance in terms of mortality rates in developed countries. Control of infectious disease has come as a result of our comprehensive understanding of disease processes, as well as improved sanitary practices and the discovery and use of antimicrobial agents. As we will see later in this chapter, microbiology as a science had its roots in the study of infectious disease.

Although now many infectious diseases can be controlled, microorganisms can still be a major threat to survival. Consider here the individual dying slowly of a microbial infection as a consequence of acquired immune deficiency syndrome (AIDS) or the individual infected with a multiple-drug-resistant pathogen. Furthermore, microbial diseases are still the major causes of death in many developing countries of the world. While eradication of smallpox from the world was a stunning triumph for medical science, millions still die yearly from such pervasive microbial diseases as malaria, tuberculosis, cholera, African sleeping sickness, and severe diarrheal syndromes.

Clearly, microorganisms are still serious threats to human health. However, microbiology has shown that most microorganisms are *not* harmful to humans. In fact, most microorganisms cause no harm to higher organisms and instead are beneficial, carrying out processes that are of immense value to human society. We consider some of these processes now.

Microorganisms and Agriculture

Our whole system of *agriculture* depends in many important ways on microbial activities (Figure 1.6). For example, a number of major crops are members of a plant group called the **legumes**, which live in close association with bacteria that form structures called **nodules** on their roots. In the root nodules, these bacteria convert atmospheric nitrogen (N_2) into fixed nitrogen (NH_3) that the plants use for growth. In this way, the activities of the bacteria reduce the need for costly and polluting plant fertilizer.

Also of major agricultural importance are the microorganisms that are essential for the digestive process in ruminant animals such as cattle and sheep. These important farm animals have a special digestive vessel called the **rumen** in which microorganisms carry out the digestion of cellulose, the major component of plants. Without these microorganisms, cattle and sheep could not thrive on cellulose-rich but otherwise nutrient-poor substances like grass and hay.

Microorganisms also play key roles in the cycling of important nutrients in plant nutrition, particularly those of *carbon, nitrogen,* and *sulfur.* Microbial activities in soil and water convert these elements to forms that are readily accessible to plants. In addition to benefitting agriculture, microorganisms can also have harmful effects. Microbial diseases of plants and animals have major economic impacts on the agricultural industry. For example,

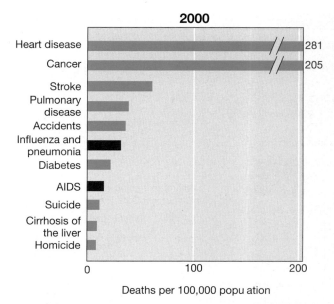

● **Figure 1.7** **Death rates for the 10 leading causes of death in the United States: 1900 and 2000.** Infectious diseases were the leading causes of death in 1900, whereas today they are much less significant. Microbial diseases are shown in red, nonmicrobial causes of death in green. Data from the United States National Center for Health Statistics.

the mad cow disease (∞Section 29.10) incident in the United States in 2003 shut down exports of beef to several major foreign markets, causing a temporary yet major impact to the U.S. beef industry.

Microorganisms and Food

Once food crops and animals are produced, they must be delivered in wholesome form to consumers. Microorganisms play important roles in the food industry (Figure 1.6). *Food spoilage* alone results in huge economic losses each year. Indeed, the canning, frozen-food, and dried-food industries exist to prepare foods in such ways that they will not undergo microbial spoilage. *Foodborne disease* is also a consideration. Because food fit for human consumption can support the growth of many microorganisms, foods must be properly prepared and monitored to avoid transmission of disease.

However, not all microorganisms in foods have harmful effects on food products or those who eat them. For example, dairy products that benefit from microbial activity include cheese, yogurt, and buttermilk, all products of major economic value. Similarly, sauerkraut, pickles, and some sausages also owe their existence to microbial fermentations. Moreover, baked goods and alcoholic beverages are based on the fermentative activities of yeast. Many of these topics are covered in Chapter 30 of this book.

Microorganisms, Energy, and the Environment

Microorganisms play major roles in energy production (Figure 1.6). Natural gas (**methane**) is a product of bacterial activity, arising from the metabolism of *methanogenic* mi-

croorganisms. Phototrophic microorganisms can harvest light energy for the production of **biomass**, energy stored in living organisms. Microbial biomass and existing waste materials such as domestic refuse, surplus grain, and animal wastes, can be converted to "biofuels," such as methane and ethanol, by the activities of microorganisms.

Microorganisms can also be used to help clean up pollution created by human activities, a process called **bioremediation** (Figure 1.6). Various microorganisms can be used to consume spilled oil, solvents, pesticides, and other environmentally toxic pollutants. The great diversity of microorganisms on Earth contains vast genetic resources to mediate solutions for cleaning up the environment, and much research in this area is taking place at present.

Microorganisms and the Future

Biotechnology entails the use of microorganisms in industrial biosyntheses, typically by microorganisms that have been genetically modified to synthesize products of high commercial value (∞Chapter 31). Biotechnology is highly dependent on **genetic engineering**—the artificial manipulation of genes and their products (Figure 1.6). Genes from any source can be manipulated and modified using microorganisms and their enzymes as molecular tools. For instance, human insulin, a hormone found in abnormally low amounts in people with the disease diabetes, can be produced microbiologically from a human insulin gene engineered into a microorganism. And now, using genomic techniques (∞Chapter 15), one can search hundreds of genetic blueprints for genes encoding proteins of interest, clone the genes into a suitable host, and then produce the proteins commercially.

The overwhelming influence of microorganisms in human society is clear. Indeed, we have many reasons to be aware of microorganisms and their activities (Figure 1.6). As the eminent French scientist Louis Pasteur, one of the founders of microbiology, expressed it: "The role of the infinitely small in nature is infinitely large." Before we begin our study of microbiology in earnest, let us briefly consider the contributions that Pasteur and other early microbiologists made to the development of the science of microbiology.

 1.4 Concept Check

Microorganisms can be both beneficial and harmful to humans. Although we tend to emphasize harmful microorganisms (infectious disease agents), many more microorganisms in nature are beneficial than harmful.

◆ In what ways are microorganisms important in the food and agricultural industries?

◆ What fuels can be made by microorganisms?

◆ What is *biotechnology* and how might it improve the lives of humans?

 II PATHWAYS OF DISCOVERY IN MICROBIOLOGY

Like any science, modern microbiology owes much to its past. Although claiming very early roots, the science of microbiology didn't really develop until the nineteenth century. Since that time, the field has exploded and spawned several new but related fields. We retrace these pathways of discovery now.

1.5 | The Historical Roots of Microbiology: Hooke, van Leeuwenhoek, and Cohn

Although the existence of creatures too small to be seen with the naked eye had long been suspected, their discovery was linked to the invention of the microscope. Robert Hooke described the fruiting structures of molds in 1665 (Figure 1.8●), and by doing so was the first person to describe microorganisms. The first person to see bacteria was the Dutch draper and amateur microscope builder Antoni van Leeuwenhoek. In 1684, van Leeuwenhoek, who was aware of the work of Hooke, used extremely simple microscopes of his own construction to examine a variety of natural substances for their microbial content (Figure 1.9●).

Van Leeuwenhoek's microscopes were crude by today's standards, but by careful manipulation and focusing he was able to see organisms as small as bacteria. His discovery of bacteria occurred in 1676 while

(a)

(b)

● **Figure 1.8 Early microscopy.** (a) A drawing of the microscope used by Robert Hooke in 1664. The objective lens was fitted at the end of an adjustable bellows (G), with illumination focused on the specimen by a single lens (1). (b) A drawing by Robert Hooke. This drawing, published in *Micrographia* in 1655, is the first description of a microorganism. The organism is a bluish-colored mold growing on the surface of leather. The round structures (sporangia) contain spores of the mold.

(a)

T. D. Brock

(b)

(c)

Brian J. Ford

● **Figure 1.9 The van Leeuwenhoek microscope.** (a) Photograph of a replica of van Leeuwenhoek's microscope. The lens is mounted in the brass plate adjacent to the tip of the adjustable focusing screw. (b) van Leeuwenhoek's drawings of bacteria, published in 1684. Even from these relatively crude drawings we can recognize several morphological types of common bacteria. A, C, F, and G, rod–shaped; E, spherical or coccus–shaped; H, cocci packets (⌀⌀Figure 4.11). (c) Photomicrograph of a human blood smear taken through a van Leeuwenhoek microscope. Red blood cells are clearly apparent.

studying pepper-water infusions. van Leeuwenhoek reported his observations in a series of letters to the Royal Society of London, which published them in 1684 in English translation. Drawings of some of van Leeuwenhoek's "wee animalcules," as he referred to them, are shown in Figure 1.9b.

As years went by, van Leeuwenhoek's observations were confirmed by others, but progress in understanding the nature and importance of these tiny organisms remained slow for nearly the next 150 years. Only in the nineteenth century did improved microscopes become widely distributed, and about this time the extent and nature of microbial life forms became more apparent.

In the mid- to late nineteenth century major advances in the new science of microbiology occurred, primarily because of a focus on two major questions that pervaded biology and medicine at the time: *spontaneous generation* and *the nature of infectious disease*. Answers to these seminal questions were provided by two giants in the fledgling field of microbiology: the French chemist *Louis Pasteur* and the German physician *Robert Koch*. But before we explore their work, let us briefly consider the work of a German botanist, *Ferdinand Cohn*, a contemporary of Pasteur and Koch, and the founder of the field we now call *bacteriology*.

Ferdinand Cohn and the Science of Bacteriology

Ferdinand Cohn (1828–1898) was trained as a botanist, and his interests in microscopy naturally led him toward the study of unicellular plants—the algae—and later to photosynthetic bacteria. Cohn believed that all bacteria, even those lacking photosynthetic pigments, were members of the plant kingdom, and his microscopic studies gradually drifted away from plants and algae to a variety of different bacteria, including the large sulfur bacterium *Beggiatoa* (Figure 1.10●).

Cohn became particularly interested in heat-resistant forms of bacteria, which led him to discover the genus *Bacillus* and the process of endospore formation. We now know that bacterial endospores are extremely heat-resistant. Cohn described the entire life cycle of *Bacillus* (vegetative cell → endospore → vegetative cell; ⌀⌀Section 4.15) and discovered that vegetative cells but not endospores are killed by boiling. Indeed, Cohn's findings helped explain why earlier scientists, such as John Tyndall, had found boiling to be an unreliable means of sterilization.

Cohn continued to work with bacteria until his retirement. During this time he laid the groundwork for a scheme of bacterial classification and founded a major scientific journal. Cohn was also a strong advocate of the techniques and research of the first medical microbiologist, Robert Koch. Cohn also is credited with helping devise simple but very effective methods for preventing the contamination of sterile culture media, such as the use of cotton for closing flasks and tubes. These methods

Tabula I. der Hedwigia

Fig. 1.

Beggiatoa mirabilis Cohn.

● **Figure 1.10 Drawing by Ferdinand Cohn made in 1866 of the filamentous sulfur-oxidizing bacterium *Beggiatoa mirabilis.*** The small granules inside the cell consist of elemental sulfur, produced from the oxidation of hydrogen sulfide (H_2S). Cohn was the first to identify the granules as sulfur.

were later used by Koch and allowed him to make rapid progress in the isolation and characterization of several disease-causing bacteria (see Section 1.6).

⬢ 1.5 Concept Check

Robert Hooke was the first to describe microorganisms and Antoni van Leeuwenhoek was the first to describe bacteria. Ferdinand Cohn founded the field of bacteriology and discovered bacterial endospores.

◆ About what period was van Leeuwenhoek discovering bacteria? When was Cohn working on endospores?

◆ What prevented the science of microbiology from developing before the era of van Leeuwenhoek?

1.6 Pasteur, Koch, and Pure Cultures

The mid- to late nineteenth century saw the science of microbiology blossom. The idea of spontaneous generation was crushed, and pure culture microbiology came to the forefront.

Pasteur and the Downfall of Spontaneous Generation

The concept of **spontaneous generation** existed since biblical times. The basic idea of spontaneous generation can easily be understood. For example, if food is allowed to stand for some time, it putrefies. When the putrefied material is examined microscopically, it is found to be teeming with bacteria, and perhaps even higher organisms like maggots and worms. Where do

these organisms come from, since they are not apparent in the fresh food? Some people said they developed from seeds or germs that had entered the food from the air, whereas others said they arose spontaneously from nonliving materials. Keen insight was needed to solve this problem.

Louis Pasteur (1822–1895) was a powerful opponent of spontaneous generation. Pasteur first showed that microorganisms were present in air that closely resembled those seen in putrefying materials. Pasteur concluded that the organisms found in putrefying materials originated from microorganisms present in the air. He postulated that these cells are constantly being deposited on all objects and grow when conditions are favorable. Pasteur reasoned that if food were treated in such a way as to destroy all living organisms contaminating it, that is, rendered **sterile**, and then protected from further contamination, it should not putrefy.

Since it had already been established that heat effectively kills microorganisms, Pasteur used heat to eliminate contaminants. In fact, other workers had shown that when a nutrient solution was sealed in a glass flask and heated to boiling, it did not support microbial growth. Killing all the bacteria or other microorganisms in or on objects is a process we now call **sterilization**.

Proponents of spontaneous generation criticized such experiments by declaring that fresh air was necessary for spontaneous generation. Boiling, so they claimed, in some way affected the air in the sealed flask so that it could no longer support spontaneous generation. In 1864 Pasteur side-stepped this objection simply and brilliantly by constructing a swan-necked flask, now called a *Pasteur flask* (Figure 1.11●). In such a flask nutrient solutions could be heated to boiling. However, after the flask was cooled, air could reenter, but bends in the neck (the "swan neck" design) prevented particulate matter (containing microbial contaminants) from entering the main body of the flask and growing.

Broth sterilized in a Pasteur flask did not putrefy, and microorganisms never appeared in the flask as long as the neck did not contact the sterile liquid. If, however, the flask was tipped to allow the sterile liquid to contact the contaminated neck of the flask (Figure 1.11c), putrefaction occurred and the liquid soon teemed with microorganisms. This simple experiment effectively settled the controversy surrounding the theory of spontaneous generation, and the science of microbiology was able to move ahead on firm footing. Not incidentally, Pasteur's work also led to the development of effective sterilization procedures, which were eventually refined and carried over into both basic and applied research. Food science also owes a debt to Pasteur, as his principles are applied today in the canning and preservation of many foods (*pasteurization*).

Pasteur went on to many other triumphs in microbiology and medicine. Chief among these was his development of vaccines for the diseases anthrax, fowl

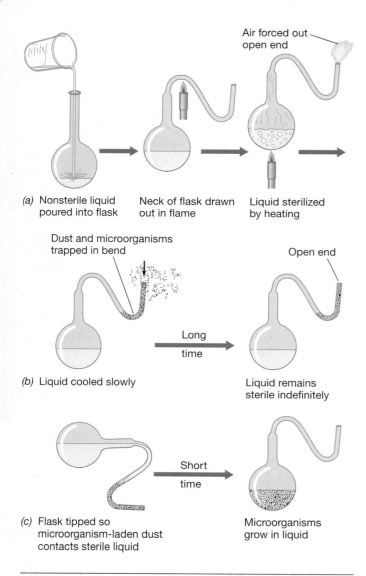

(a) Nonsterile liquid poured into flask

Neck of flask drawn out in flame

Air forced out open end

Liquid sterilized by heating

Dust and microorganisms trapped in bend

Open end

Long time

(b) Liquid cooled slowly

Liquid remains sterile indefinitely

Short time

(c) Flask tipped so microorganism-laden dust contacts sterile liquid

Microorganisms grow in liquid

● **Figure 1.11** **Pasteur's experiment with the swan-necked flask.** (a) Sterilizing the contents of the flask. (b) If the flask remained upright, no microbial growth occurred. (c) If microorganisms trapped in the neck reached the sterile liquid, microbial growth ensued.

cholera, and rabies during a very scientifically productive period in his life from 1880 to 1890. These medical and veterinary breakthroughs were not only highly significant in their own right, but helped solidify the concept of the *germ theory of disease* whose principles were being developed at this time by a contemporary of Pasteur, Robert Koch.

Koch and the Germ Theory of Disease: The Development of Koch's Postulates

Proof that microorganisms could cause disease provided the greatest impetus for the development of the science of microbiology. Even in the sixteenth century it was thought that something that induced a disease could be transmitted from a diseased person to a healthy person. After the discovery of microorgan-

isms, it was widely believed that these organisms were responsible, but definitive proof was lacking. Discoveries in sanitation by Ignaz Semmelweis and Joseph Lister provided indirect evidence for the importance of microorganisms in causing human diseases, but it was not until the work of Robert Koch (1843–1910), a physician by training, that the *germ theory of disease* was clearly conceptualized and given experimental support.

In his early work Koch studied *anthrax*, a disease of cattle that occasionally occurs in humans. Anthrax is caused by an endospore-forming bacterium called *Bacillus anthracis*, and the blood of an infected animal teems with cells of this large bacterium (∞Section 25.13 and Figure 25.13). By careful microscopy, Koch established that the bacteria were always present in the blood of an animal that was succumbing to the disease. However, mere association of the bacterium with the disease did not prove that it actually *caused* the disease. Instead, the bacterium might be a *result* of the disease.

Using proper experimental controls, Koch demonstrated that a small amount of blood from a diseased mouse, which he used as an experimental animal, injected into a healthy mouse quickly transferred the disease anthrax. He took blood from this second animal, injected it into another, and again obtained the characteristic disease symptoms. However, Koch carried this experiment a critically important step further. He found that the bacteria could be grown in nutrient fluids *outside the animal body* and that even after many transfers in culture, the bacteria still caused the disease when reinoculated into an animal.

On the basis of these and other experiments Koch formulated the following criteria, now called **Koch's Postulates**, for proving that a specific microorganism causes a specific disease:

1. The organism must always be present in animals suffering from the disease and should not be present in healthy individuals.

2. The organism must be cultivated in a *pure culture* away from the animal body.

3. Such a pure culture, when inoculated into susceptible animals, must initiate the characteristic disease symptoms.

4. The organism must be reisolated from these experimental animals and cultured again in the laboratory, after which it should still be the same as the original organism.

Koch's postulates are summarized in Figure 1.12●. Koch's postulates were a monumental step forward in the study of infectious diseases. The postulates not only offered a means of linking cause and effect in infectious disease, but also served to stress the importance of laboratory culture. With these postulates as a guide, Koch and subsequent microbiologists discovered the causes

● **Figure 1.12**   **Koch's postulates for proving that a specific microorganism causes a specific disease.**
Note how it is essential that following isolation of a pure culture of the suspected pathogen, a laboratory culture
of the organism should both initiate the disease *and* be recovered from the diseased animal. Establishing the
correct conditions for growing the pathogen is essential, otherwise it will be missed.

of many important diseases of humans and other animals. These discoveries led to the development of successful treatments for the prevention and cure of many infectious diseases, thereby greatly improving the scientific basis of clinical medicine and general human health (Figure 1.7).

Koch and Pure Cultures

To link a specific microorganism to a specific process such as a disease, the organism must first be isolated in a culture; that is, the culture must be *pure*. This concept was not lost on Robert Koch in formulating his famous postulates (Figure 1.12), and he developed several simple but ingenious methods of obtaining bacteria in pure culture (see the Microbial Sidebar, Solid Media, the Petri Plate, and Pure Cultures).

Koch started in a crude way by using solid nutrients such as a potato slice to culture bacteria. But he quickly developed more reliable methods, many of which are still in use today. Koch observed that when a solid surface such as a potato slice was incubated in air, bacterial colonies developed, each having a characteristic shape and color. He inferred that each colony had arisen from a single bacterial cell that had fallen on the surface, found suitable nutrients, and multiplied. In other words, each colony represented a **pure culture**. Koch realized that this discovery provided a simple way of obtaining pure cultures. Since not all organisms grow on potato slices, however, Koch devised more uniform and reproducible nutrient solutions solidified with gelatin and later, agar (see the Microbial Sidebar).

A Test of Koch's Postulates: Tuberculosis

Koch's greatest accomplishment in medical bacteriology was his discovery of the causative agent of tuberculosis. At the time Koch began this work (1881), one-seventh of all reported human deaths were caused by tuberculosis (Figure 1.7). There was strong evidence that tuberculosis was a contagious disease, but the suspected causal organism had never been seen, either in diseased tissues or in culture. Koch was determined to demonstrate the causal agent of tuberculosis, and to this end he brought together all of the methods he had so carefully developed in his previous studies: microscopy, staining of tissues, pure culture isolation, and an animal model system.

As is now well known, the bacterium *Mycobacterium tuberculosis* is very difficult to stain because of large amounts of waxy lipid present in its cell wall. But Koch devised a staining procedure for *M. tuberculosis* in tissue samples using alkaline methylene blue in conjunction with a second stain (bismarck brown) that stained only the tissue (Koch's method was the forerunner of the acid-fast stain used today for staining bacteria such as *M. tuberculosis*; ⚭Section 12.23). Using this method, Koch observed bright-blue, rod-shaped cells of *M. tuberculosis* in tuberculous tissues, the tissue itself staining a light brown (Figure 1.13●). However, from his previous work on anthrax, Koch realized that simply *identifying* an organism associated with tuberculosis was not enough. He knew he must *culture* the organism in order to prove that it was the specific cause of tuberculosis.

Producing cultures of *M. tuberculosis* was not easy, but eventually Koch was successful in growing colonies of this organism on a medium containing coagulated blood serum. Later he used agar, which had just been introduced as a solidifying agent (see the Microbial Sidebar). Under the best of conditions, *M. tuberculosis* grows slowly in culture, but Koch's persistence and patience eventually led to pure cultures of this organism from a variety of human and animal sources.

From here it was relatively easy to use his postulates to obtain definitive proof that the organism was the cause of the disease tuberculosis. Guinea pigs can be readily infected with *M. tuberculosis* and eventually succumb to systemic tuberculosis. Koch showed that diseased guinea pigs contained masses of *M. tuberculosis* cells in their tissues and that pure cultures obtained from such animals transmitted the disease to uninfected animals. Thus, Koch successfully satisfied

(a)　　　　　　　　　　(b)

(c)　　　　　　　　　　(d)

● **Figure 1.13 Robert Koch's drawings of cells of *Mycobacterium tuberculosis* in tissues and in laboratory culture.** Robert Koch was the first to isolate *M. tuberculosis* and show it to cause tuberculosis. (a) Section through a tubercle from lung tissue. Cells of *M. tuberculosis* stain blue, whereas the lung tissue stains brown. (b) Cells of *M. tuberculosis* in a sputum sample of a tuberculous patient. (c, d) Growth of *M. tuberculosis* in pure culture. (c) Growth of *M. tuberculosis* on a glass plate of coagulated blood serum inside a glass box (with lid open). (d) A colony of *M. tuberculosis* cells taken from the plate in (c) and observed microscopically at 700×; cells appear as long "cordlike" forms (compare with Figure 12.70b). Original drawings appeared in Koch, R. 1884. "Die Aetiologie der Tuberkulose." *Mittheilungen aus dem Kaiserlichen Gesundheitsamte* 2:1–88.

all four of his postulates (Figure 1.12), and the cause of tuberculosis was understood. For his contributions on tuberculosis, Robert Koch was awarded the 1905 Nobel Prize for physiology or medicine.

 1.6 Concept Check

Louis Pasteur's work on spontaneous generation led to the development of methods for controlling the growth of microorganisms. Robert Koch developed criteria for the study of infectious microorganisms and developed the first methods for growth of pure cultures of microorganisms.

◆ How did Pasteur's famous experiment defeat the theory of spontaneous generation?

◆ How can Koch's postulates prove cause and effect in a disease?

◆ What advantages do solid media offer for the culture of microorganisms?

Microbial Sidebar ◆ Solid Media, the Petri Plate, and Pure Cultures

Robert Koch was the first to grow bacteria on solid culture media. Koch's early use of potato slices as solid media was fraught with problems. Besides being rather selective in terms of which bacteria would grow on the slices, the slices would frequently get overgrown with fungi. Koch thus needed a more reliable and reproducible means of growing bacteria on solid media, and he found the answer in agar.

Koch initially employed gelatin as a solidifying agent for the various nutrient fluids he used to culture pathogenic bacteria and developed a method for preparing horizontal slabs of solid media that were kept free of contamination by covering them with a bell jar or glass box (see Figure 1.13c). Nutrient gelatin was a good culture medium for the isolation and study of various bacteria, but it had several drawbacks, the most important being that it did not remain solid at 37°C, the optimum temperature for growth of most human pathogens. Thus, a more versatile solidifying agent was needed, and this turned out to be agar.

Agar is a polysaccharide derived from red algae. It was used widely in the nineteenth century as a gelling agent. Walter Hesse first used agar as a solidifying agent for bacteriological culture media. The actual suggestion that agar be used instead of gelatin was made by Hesse's wife, Fannie. Fannie Hesse had used agar in the preparation of fruit jellies, and when it was tried as a solidifying agent in nutrient media, its superior qualities were immediately evident. Hesse wrote to Koch about this discovery, and Koch quickly adapted agar to his own studies, including his classic studies on the isolation of the bacterium *Mycobacterium tuberculosis,* the cause

of the disease tuberculosis (see text and Figure 1.13).

Agar had many other properties that made it desirable as a gelling agent for microbial culture media. In particular, agar remained solid at body temperature, and after melting during the sterilization process, remained liquid to about 45°C, at which time it could be poured into sterile vessels. In addition, unlike gelatin, which many bacteria can hydrolyze causing the medium to liquify, agar is not degraded by the vast majority of bacteria. Hence, agar found its place early in the annals of microbiology and is still used today for obtaining and maintaining pure cultures of bacteria.

In 1887 Richard Petri published a brief paper describing a modification of Koch's flat plate technique. Petri's enhancement, which turned out to be amazingly useful, was the development of the double-sided dishes that bear his name. The advantages of Petri dishes were immediately apparent. They could easily be stacked and sterilized separately from the medium, and, following the addition of molten medium to the smaller of the two double dishes, the larger dish could be used as a cover to prevent contamination. Colonies that formed on the surface of the agar in the Petri dish remained fully exposed to air and could easily be manipulated for further study. The original idea of Petri has not been improved on to this day, as the Petri dish, made either of reusable glass and sterilized by dry heat or of disposable plastic and sterilized by ethylene oxide (a gaseous sterilant), is a mainstay of the microbiology laboratory.

Finally, it should also be noted that Koch was keenly aware of the implications his pure culture methods had for the study of microbial

systematics. Koch observed that different colonies (differing in color, morphology, size, and the like, see Fig. 1) developed on solid media exposed to a contaminated object, and that these colonial forms bred true and could be distinguished from one another by their colony characteristics. Cells from different colonies also differed microscopically and often in their temperature or nutrient requirements as well. Koch realized that these differences among microorganisms met all the requirements that taxonomists had established for the classification of larger organisms, such as plant and animal species. In Koch's own words (translated from the German): "*All bacteria which maintain the characteristics which differentiate one from another when they are cultured on the same medium and under the same conditions, should be designated as species, varieties, forms, or other suitable designation.*" Koch also realized from the study of pure cultures that one could show that specific organisms have specific effects, not only in causing disease, but in other capacities as well. Such insightful thinking was significant in the relatively rapid acceptance of microbiology as an independent biological science.

Koch's discovery of solid culture media and his emphasis on pure culture microbiology reached far beyond the realm of medical bacteriology. His discoveries supplied critically needed tools for development of the fields of bacterial taxonomy, genetics, and several other subdisciplines. Indeed, the entire field of microbiology owes much to Robert Koch and his associates for the intuition they displayed in grasping the significance of pure cultures and developing some of the most basic methods in microbiology. ∎

Figure 1 *A hand-colored photograph of colonies formed on agar taken by Walter Hesse, an associate of Robert Koch. The colonies include those of fungi (molds) and bacteria and were obtained during studies Hesse initiated on the microbiological content of air in Berlin, Germany, in 1882.* From Hesse, W. 1884. "Ueber quantitative Bestimmung der in der Luft enthaltenen Mikroorganismen," in Struck (ed.), *Mittheilungen aus dem Kaiserlichen Gesundheitsamte.* August Hirschwald.

1.7 Microbial Diversity and the Rise of General Microbiology

As microbiology advanced from the nineteenth to the twentieth century, our understanding of microbial diversity improved significantly. During this period several subdisciplines arose in microbiology, leading to the present era of "molecular microbiology."

Two giants in the field led this transition, the Dutchman Martinus Beijerinck and the Russian Sergei Winogradsky. Both of these early microbiologists were interested in bacteria that inhabit soil and water, and both are remembered primarily for their contributions to bacterial diversity. This period and the years thereafter until the era of molecular biology was ushered in were the heyday of *general microbiology*. The latter deals primarily with the nonmedical aspects of microbiology, with an emphasis on the diversity and physiology of microorganisms.

Martinus Beijerinck and the Enrichment Culture Technique

Martinus Beijerinck (1851–1931) was a professor at the Delft Polytechnic School in Holland but was originally trained in botany. He thus began his career in microbiology studying the microbiology of plants. Perhaps Beijerinck's greatest contribution to the field of microbiology was his clear formulation of the **enrichment culture**. In the enrichment culture, microorganisms are isolated from natural samples in a *selective* fashion, paying careful attention to nutrient and incubation requirements (ꝏSection 18.1).

Using his enrichment culture technique, Beijerinck isolated the first pure cultures of many soil and aquatic microorganisms, including aerobic nitrogen-fixing bacteria (Figure 1.14●), sulfate-reducing and sulfur-oxidizing bacteria, nitrogen-fixing root nodule bacteria, *Lactobacillus* species, green algae, and many other microorganisms. Using selective filter techniques in his studies of tobacco mosaic disease, Beijerinck showed that the infectious agent (a virus) was not bacterial, but somehow became incorporated into the cells of the host plant and required the living plant to reproduce. In this insightful work, Beijerinck described not only the first virus, but also the basic tenets of virology.

Sergei Winogradsky and the Concept of Chemolithotrophy

Sergei Winogradsky (1856–1953) had scientific interests similar to Beijerinck's and was also successful in isolating several key bacteria for the first time. Winogradsky was particularly interested in bacteria involved in the cycling

of nitrogen and sulfur compounds (Figures 1.15 and 1.16●). He showed in this work that bacteria can be important biogeochemical agents. Moreover, Winogradsky's keen insight revealed the *metabolic significance* of biogeochemical processes. For example, from his studies of the sulfur-oxidizing bacteria, Winogradsky proposed the concept of **chemolithotrophy**, the oxidation of *inorganic* compounds linked to energy conservation (Figure 1.16*a*). And from his studies of the chemolithotrophic process of *nitrification* (the oxidation of ammonia to nitrate), Winogradsky concluded that the organisms responsible—the nitrifying bacteria—

(a)

● **Figure 1.14   Martinus Beijerinck and *Azotobacter.*** (a) Portion of a page from the laboratory notebook of M. Beijerinck dated December 31, 1900, that describes his observations on the aerobic nitrogen-fixing bacterium *Azotobacter chroococcum* (name circled in red). It is on this page that Beijerinck uses this name for the first time. Compare Beijerinck's drawings of pairs of *A. chroococcum* cells with a photomicrograph of cells of *Azotobacter* shown in Figure 12.19*a*. (b) A painting by M. Beijerinck's sister, Henriëtte Beijerinck, showing cells of *Azotobacter chroococcum*. Beijerinck used such paintings to illustrate his lectures, because this was long before the days of the overhead, slide, and computer projectors used in lectures today.

● **Figure 1.15 Hand-colored drawings of cells of purple sulfur phototrophic bacteria.** The original drawings were made by Sergei Winogradsky about 1887 and then copied and hand-colored by his wife Hélène. These drawings included cells of the genus *Chromatium*, such as *C. okenii* (Figs. 3 and 4). This species is still recognized today. Compare with a photomicrograph of cells of *C. okenii* shown in Figure 12.4a.

(a)

(b)

● **Figure 1.16 Major concepts developed by Sergei Winogradsky.** (a) Chemolithotrophy and chemoautotrophy. Oxidation of the sulfur or nitrogen compounds yield energy (ATP), while cell carbon is obtained from CO_2. Photos, left, *Achromatium*; right, *Nitrobacter*. (b) Nitrogen fixation. This process consumes ATP but allows the cell to use gaseous nitrogen for all of its nitrogen needs. Photo, *Azotobacter*, an aerobic nitrogen-fixer (see also Figure 1.14).

obtained their carbon from CO_2. Winogradsky thus proposed that these organisms were autotrophs, now called *chemoautotrophs*, to distinguish them from autotrophic organisms that are phototrophic (Figure 1.16a).

Using an enrichment method, Winogradsky also isolated the first nitrogen-fixing bacterium (the anaerobe *Clostridium pasteurianum*) and formulated the concept of bacterial N_2 fixation (Figure 1.16b). Winogradsky lived to be almost 100, publishing many scientific papers, along with a major monograph, *Microbiologie du Sol* (*Soil Microbiology*). The latter work, a milestone in microbiology, contained original drawings of many of the organisms he had isolated or otherwise studied in enrichment culture or natural material during his career (Figure 1.15).

Table 1.1 summarizes some of the important discoveries in the field of microbiology, from van Leeuwenhoek to the present.

 1.7 Concept Check

Beijerinck and Winogradsky studied bacteria in soil and water and developed the enrichment culture technique for the isolation of representatives of various physiological groups. Major new concepts in microbiology emerged during this period, including enrichment cultures, chemolithotrophy, chemoautotrophy, and nitrogen fixation.

◆ What is the enrichment culture technique?

◆ In examining Figure 1.16, describe why sulfur oxidation and nitrification are considered chemolithotrophic processes while nitrogen fixation is not. (*Hint:* Look at the reactions involving ATP in each case.)

1.8 The Modern Era of Microbiology

In the twentieth century, the field of microbiology developed rapidly in two distinct directions—applied microbiology and basic microbiology.

Development of the Major Subdisciplines of Applied Microbiology

The practical advances of Koch led to extensive developments in applied microbiology. For example, the fields of *medical microbiology* and *immunology* emerged. Along with these disciplines came the discovery of many new bacterial pathogens and elucidation of the mechanisms by which these pathogens infect the body and are resisted by the body's defenses. Other early practical advances, bolstered by the discoveries of Beijerinck and Winogradsky, were in the field of *agricultural microbiology*, which led to an understanding of microbial processes in the soil, such as nitrogen fixation, that benefit plant growth. Later in the

Table 1.1	Three hundred years of microbiology: Some key papers in microbiology, 1684–2000[a]	
Year	**Investigator(s)**	**Discovery**
1684	Antoni van Leeuwenhoek	Discovery of bacteria
1798	Edward Jenner	Smallpox vaccination
1857	Louis Pasteur	Microbiology of lactic acid fermentation
1860	Louis Pasteur	Role of yeast in alcoholic fermentation
1864	Louis Pasteur	Settled spontaneous generation controversy
1867	Robert Lister	Antiseptic principles in surgery
1876	Ferdinand Cohn	Discovery of endospores
1881	Robert Koch	Methods for study of bacteria in pure culture
1882	Robert Koch	Discovery of cause of tuberculosis
1882	Élie Metchnikoff	Phagocytosis
1884	Robert Koch	Koch's postulates
1884	Christian Gram	Gram-staining method
1885	Louis Pasteur	Rabies vaccine
1889	Sergei Winogradsky	Concept of chemolithotrophy
1889	Martinus Beijerinck	Concept of a virus
1890	Emil von Behring and Shibasaburo Kitasato	Diphtheria antitoxin
1890	Sergei Winogradsky	Autotrophic growth of chemolithotrophs
1901	Martinus Beijerinck	Enrichment culture method
1901	Karl Landsteiner	Human blood groups
1908	Paul Ehrlich	Chemotherapeutic agents
1911	Francis Rous	First cancer virus
1915/1917	Frederick Twort/Felix d'Hérelle	Discovery of bacterial viruses (bacteriophage)
1928	Frederick Griffith	Discovery of pneumococcus transformation
1929	Alexander Fleming	Discovery of penicillin
1931	Cornelius van Niel	H_2S (sulfide) as electron donor in anoxygenic photosynthesis
1935	Gerhard Domagk	Sulfa drugs
1935	Wendall Stanley	Crystallization of tobacco mosaic virus
1941	George Beadle and Edward Tatum	One gene–one enzyme hypothesis
1943	Max Delbruck and Salvador Luria	Inheritance of genetic characters in bacteria
1944	Oswald Avery, Colin Macleod, Maclyn McCarty	Explanation of Griffith's work—DNA is genetic material
1944	Selman Waksman and Albert Schatz	Discovery of streptomycin
1946	Edward Tatum and Joshua Lederberg	Bacterial conjugation
1951	Barbara McClintock	Discovery of transposable elements
1952	Joshua Lederberg and Norton Zinder	Bacterial transduction
1953	James Watson, Francis Crick, Rosalind Franklin	Structure of DNA
1959	Arthur Pardee, François Jacob, Jacques Monod	Gene regulation by a repressor protein
1959	Rodney Porter	Immunoglobulin structure
1959	F. Macfarlane Burnet	Clonal selection theory
1960	François Jacob, David Perrin, Carmon Sanchez, Jacques Monod	Concept of an operon
1960	Rosalyn Yalow and Solomon Bernson	Development of radioimmunoassay (RIA)
1961	Sydney Brenner, François Jacob, and Matthew Meselson	Messenger RNA and ribosomes as the site of protein synthesis
1966	Marshall Nirenberg and H. Gobind Khorana	Discovery of the genetic code
1967	Thomas Brock	Discovery of bacteria growing in boiling hot springs
1969	Howard Temin, David Baltimore, Renato Dulbecco	Discovery of retroviruses/reverse transcriptase
1969	Thomas Brock and Hudson Freeze	Isolation of *Thermus aquaticus*, source of *Taq* DNA polymerase
1970	Hamilton Smith	Specificity of action of restriction enzymes
1973	Stanley Cohen, Annie Chang, Robert Helling, and Herbert Boyer	Recombinant DNA
1975	Georges Kohler, Cesar Milstein	Monoclonal antibodies
1976	Susumu Tonegawa	Rearrangement of immunoglobulin genes
1977	Carl Woese and George Fox	Discovery of the *Archaea*
1977	Fred Sanger, Steven Niklen, Alan Coulson	Methods for sequencing DNA
1981	Stanley Prusiner	Characterization of prions
1982	Karl Stetter	Isolation of first prokaryote with temperature optimum $>100°C$
1983	Luc Montagnier	Discovery of HIV, the cause of AIDS
1985	Kary Mullis	Invention of the polymerase chain reaction (PCR)
1986	Norman Pace	Molecular microbial ecology
1992	Jed Fuhrman and Edward DeLong	Discovery of marine *Archaea*
1995	Craig Venter and Hamilton Smith	Complete sequence of a bacterial genome
1999	The Institute for Genomic Research (TIGR), and others	Over 100 microbial genomes sequenced or in progress
2000	Edward DeLong	Discovery of proteorhodopsin
2004	Craig Venter and others	First large scale environmental genome: the Sargasso Sea

[a] Major reference sources here include Brock, T. D. (1961), *Milestones in Microbiology*, Prentice Hall, Englewood Cliffs, NJ; Brock, T. D. (1990). *The Emergence of Bacterial Genetics*, Cold Spring Harbor Press, Cold Spring Harbor, NY. *Year* refers to the year in which the discovery was published.

twentieth century, studies of soil microorganisms led to the discovery of antibiotics and other important chemicals. This led to the field of *industrial microbiology*, the large-scale growth of microorganisms for the production of commercial products.

Advances in soil microbiology also provided the foundation for studies on microorganisms in lakes, rivers, and the oceans—the fields of *aquatic microbiology* and *marine microbiology*. One branch of aquatic microbiology deals with the important processes of treating sewage and providing safe water for humans. As interest in the biodiversity and activities of microorganisms in their natural environments grew, the field of *microbial ecology* emerged in the 1960s and 1970s. Microbial ecology is now enjoying a second "golden era" with the influx of molecular biology into the discipline (Figure 1.17●).

Basic Science Subdisciplines in Microbiology

In addition to advances in *applied* areas of microbiology, the twentieth century saw many new *basic science* areas develop, particularly those employing the concepts or tools of molecular biology. Some of the landmarks in the past 65 years are summarized in Figure 1.17.

Since World War II many new microorganisms have been discovered and classified, resulting in considerable refinement of *microbial systematics* and construction of the phylogenetic tree of life (Figure 1.17). Study of the nutrients that microorganisms require and the products that they make has advanced the field of *microbial physiology*. Enhanced understanding of the physical and chemical structure of microorganisms (*cytology*) and the discovery of microbial enzymes and the chemical reactions they carry out (*microbial biochemistry*), have also influenced how microbiology is practiced today.

A key area of basic research that moved forward rapidly in the mid-twentieth century was the study of heredity and variation in bacteria, the discipline of *bacterial genetics*. Although some aspects of bacterial genetics were known early in the twentieth century, it was not until the discovery of genetic exchange in bacteria around 1950 that bacterial genetics became a major field of study. Bacterial genetics, biochemistry, and physiology developed mainly during the 1950s. By the early 1960s, these fields had provided an advanced understanding of DNA, RNA, and protein synthesis. The field of *molecular biology* arose to a great extent from these bacterial studies (Figure 1.17).

The study of viruses also blossomed in the twentieth century. Although Beijerinck discovered the first virus over 100 years ago, it was not until the middle of the twentieth century that the true nature of viruses was understood. Much of this work involved the study of viruses that infect bacteria, called *bacteriophages*. Scientists realized that virus infection was analogous to genetic transfer, and the relationship between viruses and other genetic elements was worked out primarily from research on bacteriophages.

The Era of Molecular Microbiology

By the 1970s, our knowledge of bacterial physiology, biochemistry, and genetics had advanced to such an extent that it was possible to experimentally manipulate the genetic material of cells. With the discovery of restriction enzymes, it also became possible to introduce DNA from foreign sources into bacteria and control its replication. This led to development of the field of *biotechnology*. At about this same time, nucleic acid sequencing was developed, and its ramifications were felt in all areas of biology. In microbiology, sequencing technology helped to reveal phylogenetic (evolutionary) relationships among prokaryotes, which led to revolutionary new concepts in the field of biological classification. Sequencing also gave birth to the field of **genomics**—the comparative analysis of the genes of different organisms.

●**Figure 1.17 Some landmarks in microbiology in the past 65 years.** The icons or photos above the dates are symbolic of the discoveries; each has been covered here or will be revisited in later chapters.

The huge amounts of genomic information now in hand are fueling major advances in medicine, microbial ecology, industrial microbiology, and many other areas of biology. Indeed, the genomics era is upon us in full force (Figure 1.17) and has already given birth to a new subdiscipline, **proteomics**, the study of protein expression in cells. Both genomics and proteomics are considered in more detail in Chapter 15.

New Frontiers

Not only has nearly 350 years of microbiology brought us startling insight into the biology of microorganisms, it has brought new challenges, both good and bad. On the one hand, new emerging diseases, such as SARS, and before that AIDS, seem to appear without warning and have challenged even our most sophisticated understanding of microbial diseases. On the other hand, new research discoveries have put us on the doorstep of understanding how a cell works at the most fundamental

of levels, and newly discovered bacteria stretch our already overwhelming picture of microbial diversity. With genomics poised to reveal the minimalist genome—the minimum complement of genes necessary for life—we may soon be able to precisely define the prerequisites for life. Can the laboratory creation of a cell be that far off?

As the late evolutionary biologist Stephen Jay Gould put it, this is the "age of bacteria." What an exciting time to be learning the science! Welcome to microbiology!

 1.8 Concept Check

In the middle to latter part of the twentieth century basic and applied microbiology worked hand-in-hand to usher in the current era of molecular microbiology.

◆ List the subdisciplines of microbiology whose focus is the following: metabolism; enzymology; nucleic acid and protein synthesis; microorganisms and their natural environments.

REVIEW QUESTIONS

1. List six key properties associated with the living state. Which of these are properties of *all* cells? Which are properties of only *some types* of cells (꩜ Sections 1.1 and 1.2)?

2. Cells can be thought of as both machines and coding devices. Explain how these two attributes of a cell differ (꩜ Section 1.2).

3. What is needed for translation to occur in a cell? What is the product of the translational process (꩜ Section 1.2)?

4. What is an ecosystem? Do microorganisms live in pure cultures in an ecosystem? What effects can microorganisms have on their ecosystems (꩜ Section 1.3)?

5. How would you convince a friend that microorganisms are much more than just agents of disease (꩜ Section 1.4)?

6. For what contributions are Hooke and van Leeuwenhoek remembered in microbiology? How did Ferdinand Cohn contribute to bacteriology (꩜ Section 1.5)?

7. What is a pure culture and how can one be obtained? Why was knowledge of how to obtain a pure culture important for development of the science of microbiology (꩜ Section 1.6)?

8. Explain the principle behind the use of the Pasteur flask in studies on spontaneous generation (꩜ Section 1.6).

9. Explain why the invention of solid culture media was of great importance to the development of microbiology as a science (꩜ Section 1.6).

10. What are Koch's postulates and how did they influence the development of microbiology? Why are they still relevant today (꩜ Section 1.6)?

11. Describe a major contribution to microbiology of the early microbiologist Martinus Beijerinck (꩜ Section 1.7).

12. What major concepts do we owe to Sergei Winogradsky (꩜ Section 1.7)?

13. What major advances in microbiology have occurred since World War II (꩜ Section 1.8)?

APPLICATION QUESTIONS

1. Pasteur's experiments on spontaneous generation were of enormous importance for the advance of microbiology, having an impact on the methodology of microbiology, ideas on the origin of life, and the preservation of food, to name just a few. Explain briefly how the impact of his experiments was felt on each of these topics.

2. Describe the various lines of proof Robert Koch used to definitively associate the bacterium *Mycobacterium tuberculosis* with the disease tuberculosis How would his proof have been flawed if any of the tools he developed for studying bacterial diseases had not been available for his study of tuberculosis?

2

AN OVERVIEW OF MICROBIAL LIFE

The scope of microbial diversity is enormous, and microorganisms have exploited every means of making a living consistent with the laws of chemistry and physics.

WORKING GLOSSARY

Archaea phylogenetically related prokaryotes distinct from *Bacteria*

Bacteria phylogenetically related prokaryotes distinct from *Archaea*

Chemolithotroph an organism obtaining its energy from the oxidation of inorganic compounds

Chemoorganotroph an organism obtaining its energy from the oxidation of organic compounds

Chromosome a genetic element containing genes essential to cell function

Cytoplasm cellular contents inside the cytoplasmic membrane, excluding the nucleus (if present)

Domain the highest level of biological classification

Endosymbiosis the process by which mitochondria and chloroplasts originated from descendants of *Bacteria*

Eukarya the domain of life that includes all eukaryotic cells

Eukaryote a cell having a membrane-bound nucleus and usually other organelles

Evolution change in a line of descent over time leading to new species or varieties within a species

Extremophile an organism that grows optimally under one or more environmental extremes

Genome the complement of genes in an organism

Morphology cell shape

Nucleoid the aggregated mass of DNA that constitutes the chromosome of cells of *Bacteria* and *Archaea*

Nucleus a membrane-enclosed structure that contains the chromosomes in eukaryotic cells

Organelle a unit membrane-enclosed structure such as a mitochondrion or chloroplast present in the cytoplasm of eukaryotic cells

Phototroph an organism that obtains its energy from light

Phylogeny the evolutionary relationships between organisms

Plasmid an extrachromosomal genetic element nonessential for growth

Prokaryote a cell that lacks a membrane-enclosed nucleus and other organelles

Proteobacteria a large phylum of *Bacteria* that includes many of the common gram-negative bacteria, such as *Escherichia coli*

Ribosome a cytoplasmic particle that functions in protein synthesis

CELL STRUCTURE AND EVOLUTIONARY HISTORY

This chapter introduces key concepts of cell structure and function and microbial diversity that will carry through the remainder of this book. Here we compare the internal architecture of microbial cells, differentiate cells from viruses, explore the evolutionary tree of life, and consider some of the major groups of microorganisms that affect our lives and our planet.

Much of what we know about cell structure has come from microscopy, including studies with both light and electron microscopes. So in this chapter we present several light and electron *micrographs* (photographs taken under the microscope). However, we reserve formal discussion of the microscopes themselves until Chapter 4, using it as a prelude to a more detailed consideration of cell structure.

2.1 Elements of Cell and Viral Structure

All cells have much in common and contain many of the same or functionally same elements. All cells have a barrier that separates the inside from the outside called the **cytoplasmic membrane** (Figure 2.1●). It is through the cytoplasmic membrane that nutrients and other substances needed by the cell enter, and waste materials and other cell products exit. Within a cell, and bounded by the cytoplasmic membrane, is a complex mixture of substances and structures called the **cytoplasm**. These materials and structures, either dissolved or suspended in water, carry out the functions of the cell.

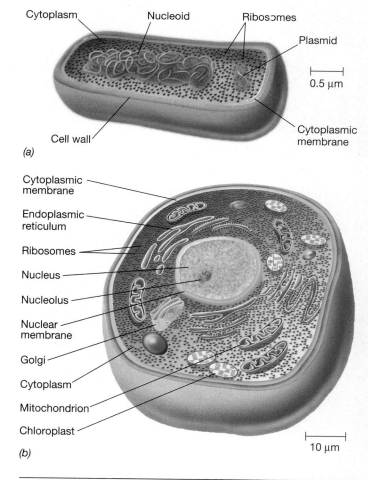

● **Figure 2.1   Internal structure of microbial cells.** (a) Diagram of a prokaryotic cell. (b) Diagram of a eukaryotic cell. Note differences in scale and internal structure.

The major components dissolved in the cytoplasm include **macromolecules** (two especially important classes of which are the *proteins* and *nucleic acids*), small organic molecules (mainly precursors of macromolecules), and various inorganic ions. **Ribosomes**—the cell's protein-synthesizing factories—are particulate structures composed of ribonucleic acid (RNA) and various proteins and are suspended in the cytoplasm. Ribosomes interact with several cytoplasmic proteins and messenger and transfer RNAs in the key process of *protein synthesis* (translation) (⚭ Figure 1.4).

The **cell wall** gives structural strength to a cell. The cell wall is relatively permeable and located outside the membrane (Figure 2.1*a*); it is a much stronger layer than the membrane itself. Plant cells and most microorganisms have cell walls, while animal cells for the most part do not. Animal cells are instead reinforced by scaffolding within the cytoplasm called the *cytoskeleton*.

Eukaryotic Cells

Examination of the internal structure of cells reveals two structural types: the **prokaryote** and the **eukaryote** (Figure 2.1). Eukaryotic cells are generally larger and structurally more complex than prokaryotic cells. Eukaryotic microorganisms include **algae**, **fungi**, and **protozoa** (see Figures 2.23 and 2.24). All multicellular plants and animals are constructed of eukaryotic cells. We consider eukaryotic cells in more detail in Chapter 14.

A major feature of eukaryotic cells, absent from prokaryotic cells, is the presence of membrane-enclosed structures called **organelles.** These include, first and foremost, the *nucleus*, but also *mitochondria*, and *chloroplasts* (the latter in photosynthetic cells only) (Figures 2.1*b*, 2.2*c*●). The nucleus is the repository of the cells' genetic information (DNA, "the genome"), and is also the site of transcription in eukaryotic cells. Mitochondria and chloroplasts play specific roles in energy generation by carrying out respiration and photosynthesis, respectively.

Prokaryotic Cells

In contrast to eukaryotic cells, prokaryotic cells have a simpler internal structure, lacking membrane-enclosed organelles (Figures 2.1*a* and 2.2*a, b*). Prokaryotes consist of the **Bacteria** and the **Archaea**. Although species of *Bacteria* and *Archaea* share a prokaryotic cell structure, they differ dramatically in their evolutionary history. In this book, the term *bacteria*, written with a lower case "b," is synonymous with the term *prokaryote*. By contrast, the term *Bacteria* (written with a capital "B" and set in italics) refers to the phylogenetically related group of prokaryotes distinct from the *Archaea* (see Section 2.3).

In general, microbial cells are very small, particularly prokaryotic cells. For example, a rod-shaped prokaryote is typically about 1–5 micrometers (μm) long and about 1 μm wide (a micrometer is 10^{-6} of a meter) and thus is invisible to the naked eye. To conceive of how small a bacterium is, consider that 500 bacteria each 1 μm long could be placed end-to-end across the period at the end of this sentence. Eukaryotic cells are typically much larger than prokaryotic cells, but the range of sizes in eukaryotic cells can vary dramatically, from as small as three to several hundred micrometers in diameter. We revisit the subject of cell size in more detail in Chapter 4.

(a) John Bozzola and M.T. Madigan *(b)* R. Rachel and K.O. Stetter

Cytoplasmic membrane · Nucleus · Cell wall · Internal membrane · Mitochondrion

S.F. Conti and T.D. Brock *(c)*

● **Figure 2.2 Electron micrographs of sectioned cells from each of the domains of living organisms.** (a) *Heliobacterium modesticaldum* (*Bacteria*); the cell measures 1 × 3 μm. (b) *Methanopyrus kandleri* (*Archaea*); the cell measures 0.5 × 4 μm. [Reinhard Rachel and Karl O. Stetter, 1981. *Archives of Microbiology* 128:288–293. © Springer-Verlag Gmb H & Co. KG]. (c) *Saccharomyces cerevisiae* (*Eukarya*); the cell measures 8 μm in diameter.

(a)

Erskine Caldwell

(b)

D. Kaiser

Eukaryotic cell

Virus

Prokaryotic cell

Nucleus

1000 nm (1 μm)

(c)

● **Figure 2.3** **Virus structure and size comparisons of viruses and cells.** (a) Rhabdovirus (a virus of eukaryotes) particles. A single virus particle is about 65 nm (0.065 μm) in diameter. (b) Bacterial virus (bacteriophage) lambda. The head of each particle is about 65 nm in diameter. (c) The size of the viruses shown in (a) and (b) in comparison to a bacterial and eukaryotic cell.

Viruses

Viruses are a major class of microorganisms, but they are not cells (Figure 2.3●). Viruses lack many of the attributes of cells, the most important of which is that they are not dynamic open systems, taking in nutrients and expelling wastes. Instead, a virus particle is a static structure, quite stable and unable to change or replace its parts. Only when it infects a cell does a virus acquire the key attribute of a living system—reproduction. Unlike cells, viruses have no metabolic abilities of their own. And, although they contain their own genomes, viruses lack ribosomes and therefore depend totally on the cell's biosynthetic machinery for protein synthesis.

Viruses are known to infect all cells, including microbial cells. Many viruses cause disease in the organisms they infect. However, viral infection can have many profound effects on cells, including genetic alterations that can actually improve the capabilities of the cell. Viruses are also much smaller than cells, even much smaller than prokaryotic cells (Figure 2.3), with the smallest known viruses only some 10 nanometers (0.010 μm) in diameter.

 2.1 *Concept Check*

All microbial cells share certain basic structures in common such as a cytoplasmic membrane, ribosomes, and (usually) a cell wall. Two structural types of cells are recognized: the prokaryote and the eukaryote. Viruses are not cells but depend on cells for their replication.

◆ By looking inside a cell how could you tell if it was *prokaryotic* or *eukaryotic*?

◆ What important function do *ribosomes* play in cells?

◆ How long is a typical rod-shaped bacterial cell? How much larger are you than this single cell?

2.2 **Arrangement of DNA in Microbial Cells**

The living processes of all cells are governed by their complement of genes (the **genome**). In cells, a gene can be defined as a segment of DNA that encodes a protein (via messenger RNA) or another RNA molecule, such as a ribosomal RNA or transfer RNA. In Chapter 15 we will consider the rapid advances that have been made in sequencing and analyzing the genomes of living organisms, from bacteria to humans. These advances have yielded detailed genetic blueprints of hundreds of different organisms, and have allowed for extensive and rather revealing comparisons to be made. Here we just consider how genomes are organized in prokaryotic and eukaryotic cells.

Nucleus vs. Nucleoid

The genomes of prokaryotic and eukaryotic cells are organized differently. In prokaryotic cells, DNA is present in a large double-stranded molecule called the *bacterial chromosome*. The chromosome aggregates to form a visible

(a)

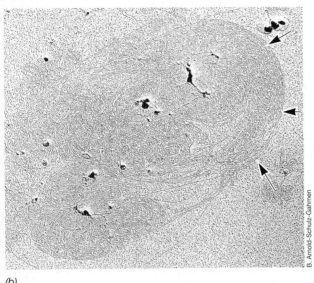

(b)

● **Figure 2.4 The nucleoid.** (a) Photomicrograph of cells of *Escherichia coli* treated in such a way as to make the nucleoid visible. A single cell is about 3 μm in length. (b) Electron micrograph of an isolated nucleoid released from a cell of *E. coli*. The cell was gently lysed to allow the highly compacted nucleoid to emerge intact. Arrows point to the edge of DNA strands.

mass called the **nucleoid** (Figure 2.4●). We will see in Chapter 7 that DNA is *circular* in most prokaryotes. Most prokaryotes have only a *single* chromosome. Because of this, they typically contain only a *single copy* of each gene and are therefore genetically *haploid*. Many prokaryotes also contain small amounts of circular extrachromosomal DNA called **plasmids**. Plasmids typically contain genes that confer special properties (such as unique metabolic properties) on a cell. This is in contrast to essential ("housekeeping") genes, which are needed for basic survival and are located on the chromosome.

In eukaryotes, DNA is present in linear molecules within the nucleus, packaged and organized in **chromosomes**. Chromosome number varies with the organism. For example, the baker's yeast *Saccharomyces cerevisiae* contains 16 chromosomes arranged in 8 pairs, while human cells contain 46 (23 pairs). Chromosomes in eukaryotes contain more than just DNA, however. They

also include proteins that assist in folding and packing the DNA and other proteins that are required for gene expression. A key genetic difference between prokaryotes and eukaryotes is that eukaryotes typically contain *two copies* of each gene and are thus genetically *diploid*. During cell division in eukaryotic cells the nucleus divides (following a doubling of chromosome number) in the process called **mitosis** (Figure 2.5●). Two identical daughter cells result, and each daughter cell receives a nucleus with a full complement of genes.

The diploid genome of eukaryotic cells is halved in the process of **meiosis** to form haploid gametes for sexual reproduction. Fusion of two gametes during zygote formation restores the cell to the diploid state. We discuss these processes in more detail in Chapter 14.

Genes, Genomes, and Proteins

How many genes and proteins does a cell have? The genome of *Escherichia coli*, a typical bacterium, is a single circular chromosome of 4.68 million base pairs of DNA. Because the *E. coli* genome has been completely sequenced, we also know that it contains about 4,300 genes. The genomes of some bacterial cells have nearly three times this number of genes while the genomes of others have fewer than one-eighth this number. Eukaryotic cells typically have much larger genomes than prokaryotes. A human cell, for example, contains over 1,000 times as much DNA as a cell of *E. coli* and about 7 times as many genes (we will see later that much of the DNA in eukaryotic cells is noncoding DNA).

A single cell of *E. coli* contains about 1,900 *different kinds* of proteins and about 2.4 million *total* protein molecules (⌀ Table 3.2). Some proteins in *E. coli* are very abundant, others are only moderately abundant, and some are present in only one or a very few copies. Thus,

● **Figure 2.5 Mitosis in stained kangaroo rat cells.** The cell was photographed while in the *metaphase* stage of mitotic division. The green color stains a protein called *tubulin*, important in pulling chromosomes apart (⌀ Section 14.5). The blue color is from a DNA-binding dye and shows the chromosomes. Although an integral part of the cell cycle of eukaryotic cells, mitosis does not occur in prokaryotic cells.

E. coli has mechanisms for *controlling the expression* of its genes so that not all genes are expressed (transcribed and translated, ∞ Figure 1.4) to the same extent or at the same time. This is a common observation in all cells—prokaryotic and eukaryotic—and we focus on the major mechanisms of gene expression in Chapter 8.

 2.2 Concept Check

Genes govern the properties of cells, and a cell's complement of genes is called its genome. DNA is arranged in cells to form chromosomes. In prokaryotes there is usually a single circular chromosome, while in eukaryotes, several linear chromosomes exist.

◆ Differentiate between the *nucleus* and the *nucleoid*.

◆ How do *plasmids* differ from *chromosomes*?

◆ Why does it make sense that a human cell would have more genes than a bacterial cell?

2.3 The Tree of Life

Evolution is the change in a line of descent over time leading to new species or varieties. Is cell structure an evolutionary determinant? The answer to this question is both "yes" and "no." The evolutionary relationships between life forms are the subject of the science of **phylogeny**. On the one hand, we can say that all known prokaryotic cells are phylogenetically distinct from eukaryotic cells. But on the other hand, not all prokaryotic cells are closely related in an evolutionary sense; *Bacteria* and *Archaea* are themselves phylogenetically distinct.

Phylogenetic relationships can be deduced by comparing sequences of certain macromolecules. For reasons to be discussed in Chapter 11, macromolecules that form the ribosome, in particular *ribosomal RNAs*, are excellent tools for determining evolutionary relationships. And because all cells contain ribosomes (and thus ribosomal RNA), this molecule can and has been used to construct a phylogenetic tree of all life forms, prokaryotic and eukaryotic (see Figure 2.7). Recognition of ribosomal RNA as a tool for constructing phylogenetic relationships was first made by Carl Woese, an American microbiologist.

The steps involved in generating an RNA-based phylogenetic tree are outlined in Figure 2.6●. In brief, genes encoding ribosomal RNA from two or more organisms are sequenced and the sequences aligned and inspected, base by base, in a computer. The greater the difference in ribosomal RNA gene sequence between two or more organisms, the greater their evolutionary distance. These distances are then expressed in the form of a phylogenetic tree (Figure 2.6).

The Three Domains of Life

From comparative ribosomal RNA sequencing, three phylogenetically distinct lineages of cells have been identified. Two of these lineages contain only prokaryotic cells, while the third is composed of eukaryotes. The lineages, called **domains**, are the *Bacteria*, the *Archaea*, and the *Eukarya* (eukaryotes) (Figure 2.7●). The domains are thought to have diverged from a common ancestral organism or community of organisms early in the history of life on Earth (∞ Section 11.7).

Besides clearly showing that all prokaryotes are *not* phylogenetically closely related, the tree of life reveals

● **Figure 2.6   Ribosomal RNA (rRNA) gene sequencing and phylogeny.** (a) Cells, either from a pure culture or from an environmental sample, are broken open; (b) the gene-encoding rRNA is isolated, and many identical copies are made by the technique called the polymerase chain reaction (PCR); ∞ Section 7.9. (c) The gene is sequenced (∞ Section 10.13), and (d) the sequences obtained are aligned by computer. An algorithm makes pairwise comparisons and generates a tree (e) that depicts the differences in rRNA sequence between the organisms analyzed. In the example shown, the sequence differences are as follows: organism 1 versus organism 2, three changes; 1 versus 3, two changes; 2 versus 3, four changes. Thus organisms 1 and 3 are closer relatives than are 2 and 3 or 1 and 2. If an environmental sample is used, the isolated rRNA genes from the different microorganisms in the sample must first be sorted out (cloned) before copies are made and sequencing is performed. For further discussion of these methods, see Sections 11.5 and 18.5.

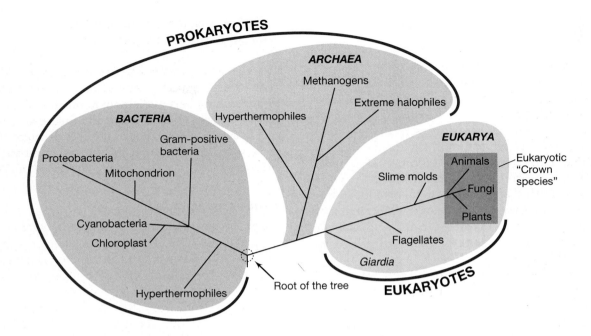

● **Figure 2.7** **The phylogenetic tree of life as defined by comparative rRNA gene sequencing.** The tree consists of three domains of organisms: the *Bacteria* and the *Archaea*, cells of which are prokaryotic, and the *Eukarya* (eukaryotes). Only a few of the groups of organisms within each domain are shown. See more detailed trees of each domain in Figures 2.9, 2.18, and 2.22, and the phylogenetic trees in Chapters 11–14. Hyperthermophiles are prokaryotes that grow best at temperatures of 80°C or higher. The groups shaded in red are *macro*organisms. All other organisms on the tree of life are *micro*organisms.

another important evolutionary fact: *Archaea* are more closely related to eukaryotes than are species of *Bacteria* (Figure 2.7). Thus, evolutionary diversification from the universal ancestor initially went in two directions: *Bacteria* and a second main branch. The latter eventually diverged to yield the *Archaea* and the *Eukarya*.

Eukarya

Because the cells of higher animals and plants are all eukaryotic, it follows that eukaryotic microorganisms were the ancestors of multicellular organisms. The tree of life clearly bears this out. As expected, microbial eukaryotes branch off early on the eukaryotic lineage while plants and animals branch near the crown (Figure 2.7). However, molecular sequencing, as well as several other lines of evidence, have shown that eukaryotic cells contain genes from cells of two domains. In addition to the genome in the chromosomes of the nucleus, mitochondria and chloroplasts of eukaryotes contain their own genomes (DNA arranged in circular fashion, as in *Bacteria*) and ribosomes. Using ribosomal RNA sequencing technology (Figure 2.6), these organelles have been shown to be highly derived ancestors of specific lineages of *Bacteria* (Figure 2.7). Mitochondria and chloroplasts were thus once free-living cells that, for protection or other reasons, established stable residency in cells of *Eukarya* eons ago. The process by which this stable arrangement developed is known as *endosymbiosis* and is discussed in later chapters (∞ Sections 11.3 and 14.4).

Contributions of Molecular Sequencing to Microbiology

Molecular phylogenies have revealed the evolutionary connections between all cells. Equally important, however, the technology has created an evolutionary framework for the prokaryotes, something that the science of microbiology had been without since its founding. In addition, RNA-based phylogenies have spawned new tools that have impacted many subdisciplines of microbiology. These include, in particular, microbial classification, microbial ecology, and clinical diagnostics. Within these areas molecular phylogeny has helped shape our concept of a bacterial species and has allowed microbial ecologists and clinical microbiologists to identify organisms without the need to culture them. We will consider these developments in later chapters (∞ Chapters 11, 18, and 24).

 2.3 Concept Check

Comparative ribosomal RNA sequencing has defined the three domains of life: *Bacteria*, *Archaea*, and *Eukarya*. Molecular sequencing has also shown that the major organelles of *Eukarya* have evolutionary roots in the *Bacteria* and has yielded new tools for microbial ecology and clinical microbiology.

◆ How can *Bacteria* and *Archaea* be differentiated? In what ways are they similar?

◆ What molecular evidence supports the theory of endosymbiosis?

II MICROBIAL DIVERSITY

Evolution has molded all life on Earth. The diversity we see in microbial cells today is the result of nearly 4 billion years of evolutionary change. Microbial diversity can be seen in many ways. For example, variations occur in cell size and **cell morphology (shape)**, metabolic strategies (physiology), motility, mechanisms of cell division, pathogenicity, developmental biology, adaptation to environmental extremes, and many other aspects of cell biology. Indeed, taken collectively, the diversity of microorganisms is overwhelming. And each year new research reveals ever more spectacular examples of microbial ingenuity, many of which we will cover in this book.

In the following sections we paint a picture of microbial diversity with a broad brush. We will return to the theme of microbial diversity in more detail in Chapters 12–15. We preface our discussion of *microbial* diversity in this chapter with a brief consideration of *metabolic* diversity. The two are closely linked. Microorganisms have exploited every conceivable means of "making a living" consistent with the laws of chemistry and physics. This great metabolic capacity made available new habitats for microbial colonization, and this in turn helped fuel evolutionary diversification. Metabolic diversity will be covered in more detail in Chapters 5, 6, and 17.

● **Figure 2.8 Metabolic options for obtaining energy.** The organic and inorganic chemicals listed here are just a few of the many different chemicals used by various chemotrophic organisms. Oxidation of the organic or inorganic chemicals yields ATP in chemotrophic organisms while conversion of solar energy to chemical energy (again, in the form of ATP) occurs in phototrophic organisms.

2.4 Physiological Diversity of Microorganisms

All cells require energy and a means to trap it. All cells also require genetic mechanisms to allow for replication and to adapt to variations in their environments. These processes are highly energy demanding, and thus energy sources are of prime importance for all cells. Energy can be obtained in three ways in nature, and Figure 2.8● summarizes the options: *organic* chemicals, *inorganic* chemicals, or *light*.

Chemoorganotrophs

Many of the thousands of different organic chemicals present on Earth can be used by one microorganism or another to obtain energy. All natural and even most synthetic organic compounds can be broken down by one or more microorganisms. Energy is obtained by *oxidizing* (removing electrons from) the compound and is conserved in the cell as the energy-rich compound, **adenosine triphosphate** (**ATP**) (Figure 2.8). Some microorganisms can extract energy from the compound only in the presence of oxygen; these organisms are called **aerobes**. Others can extract energy only in the absence of oxygen (**anaerobes**). Still others can break down organic compounds in either the presence or absence of oxygen. Organisms that obtain energy from *organic* compounds are called **chemoorganotrophs** (Figure 2.8). Most microorganisms that have been cultured are chemoorganotrophs.

Chemolithotrophs

A number of prokaryotes can tap the energy available in *inorganic* compounds. This is a form of metabolism called *chemolithotrophy* (discovered by Winogradsky, ∞ Section 1.7) and is carried out by organisms called **chemolithotrophs** (Figure 2.8). This form of energy-yielding metabolism is found only in prokaryotes and is widely distributed among *Bacteria* and *Archaea*. The spectrum of different inorganic compounds used is quite broad, but as a rule, a particular prokaryote specializes in the use of one or a related group of inorganic compounds.

It should be obvious why the capacity to extract energy from inorganic chemicals might be advantageous: Competition with chemoorganotrophs is not an issue. In addition, many of the inorganic compounds oxidized, for example H_2 (Figure 2.8) and H_2S, are actually the *waste products* of chemoorganotrophs. Thus, chemolithotrophs have evolved strategies for exploiting resources that most organisms are unable to use.

Phototrophs

Phototrophic microorganisms contain pigments that allow them to use light as an energy source, and thus their cells are colored (see Figure 2.10*a*). Unlike chemotrophic organisms, **phototrophs** do not require chemicals as a source of energy; ATP is made from the energy of sunlight. This is obviously a significant advantage, since competition for energy with chemotrophs is not a problem and light is available in a wide variety of microbial habitats.

Two major forms of phototrophy are known in prokaryotes. In one form, called *oxygenic photosynthesis*, O_2 is evolved. Oxygenic photosynthesis is characteristic of cyanobacteria and their phylogenetic relatives. The other form, *anoxygenic photosynthesis*, occurs in the purple and green bacteria, and does not result in O_2 evolution. Both groups of phototrophs use light to make ATP, however, and the similarities in their mechanisms of ATP synthesis are quite striking. Indeed, the evolutionary roots of oxygenic photosynthesis lie in the much simpler anoxygenic process. We consider photosynthesis in more detail in Chapter 17.

Heterotrophs and Autotrophs

All cells require *carbon* as a major nutrient. Microbial cells are either **heterotrophs**, requiring one or more organic compounds as their carbon source, or **autotrophs**, where CO_2 is the carbon source. Chemoorganotrophs are also clearly heterotrophs. By contrast, many chemolithotrophs and virtually all phototrophs are autotrophs. Autotrophs are sometimes called *primary producers* because they synthesize organic matter from CO_2 for both their own benefit and that of chemoorganotrophs. The latter either feed directly on the primary producers or live off products they excrete. All organic matter on Earth has been synthesized by primary producers, in particular by phototrophic organisms.

Habitats and Extreme Environments

Microorganisms are present everywhere on Earth that will support life. These include common habitats we are all familiar with—soil, water, animals, and plants—as well as on and in virtually any structures made by humans. Indeed, sterility (the absence of life forms) in a natural sample is a very rare occurrence.

Some microbial environments are ones that we as humans would find too extreme for life. Although these environments pose particular challenges to microorganisms, in many cases extreme environments are awash in microbial life. Organisms inhabiting extreme environments are called **extremophiles**, a remarkable group of primarily prokaryotes that, collectively, define the physiochemical limits of life.

Extremophilic prokaryotes abound in such harsh environments as boiling hot springs, on or within the ice covering lakes, glaciers, or the polar seas, in extremely salty bodies of water, and in soils and waters having a pH lower than 0 or as high as 12. Interestingly, these prokaryotes are not just *tolerant* of these extremes, but actually *require* the environmental extreme in order to grow. That is why they are called *extremophiles* (the suffix "phile" means "loving"). Table 2.1 summarizes the current "record holders" for extremophilic prokaryotes and lists the types of habitats in which they reside. We will revisit many of these species again in later chapters (∞ Chapters 6, 12, and 13).

2.4 Concept Check

Carbon and energy sources are needed by all cells. The terms chemoorganotroph, chemolithotroph, and phototroph refer to cells that use organic chemicals, inorganic chemicals, or light, respectively, as their source of energy. Autotrophic organisms use CO_2 as their carbon source, while heterotrophs use organic carbon. Extremophiles thrive under environmental conditions that higher organisms cannot.

Table 2.1	Classes and examples of extremophiles[a]						
Extreme	**Descriptive term**	**Genus/species**	**Domain**	**Habitat**	**Minimum**	**Optimum**	**Maximum**
Temperature							
High	Hyperthermophile	*Pyrolobus fumarii*	*Archaea*	Hot, undersea hydrothermal vents	90°C	**106°C**	113°C[b]
Low	Psychrophile	*Polaromonas vacuolata*	*Bacteria*	Sea ice	0°C	**4°C**	12°C
pH							
Low	Acidophile	*Picrophilus oshimae*	*Archaea*	Acidic hot springs	−0.06	**0.7**[c]	4
High	Alkaliphile	*Natronobacterium gregoryi*	*Archaea*	Soda lakes	8.5	**10**[d]	12
Pressure	Barophile	*Moritella yayanosii*[e]	*Bacteria*	Deep ocean sediments	500 atm	**700 atm**	>1000 atm
Salt (NaCl)	Halophile	*Halobacterium salinarum*	*Archaea*	Salterns	15%	**25%**	32% (saturation)

[a] In each category the organism listed is the current "record holder" for requiring a particular extreme condition for growth.
[b] A newly isolated archaeon can apparently grow up to 121°C.
[c] *P. oshimae* is also a thermophile, growing optimally at 60°C.
[d] *N. gregoryi* is also an extreme halophile, growing optimally at 20% NaCl.
[e] *Moritella yayanosii* is also a psychrophile, growing optimally at about 4°C.

◆ How might you distinguish a *phototrophic* microorganism from a *chemotrophic* one by simply looking at it under a microscope?

◆ What are *extremophiles*?

2.5 Prokaryotic Diversity

As we have seen, prokaryotes cluster into two phylogenetically distinct domains, the *Archaea* and the *Bacteria* (Figure 2.7). In this section we will explore the phylogenetic tree and briefly consider some major organisms found there. Most of the prokaryotes that are familiar to the beginning student of microbiology are found in the domain *Bacteria*, and so we begin here.

Bacteria

The domain *Bacteria* contains an enormous variety of prokaryotes. All known disease-causing (pathogenic) prokaryotes are *Bacteria*, as well as thousands of nonpathogenic species. Moreover, a large variety of morphologies and physiologies occur in this domain. The **Proteobacteria** is the largest division (called a *phylum*) of *Bacteria* (Figure 2.9●). Within the Proteobacteria are found many chemoorganotrophic bacteria, such as *Escherichia coli*, the model organism of microbial physiology, biochemistry, and molecular biology. Several phototrophic (Figure 2.10*a*●) and chemolithotrophic (Figure 2.10*b*) species are Proteobacteria, as well. Many chemolithotrophs use hydrogen sulfide (H_2S, the smell from rotten eggs) in their metabolism, pro-

(a)

D. E. Caldwell

(b)

Hans-Dietrich Babenzien

● **Figure 2.10 Phototrophic and chemolithotrophic Proteobacteria.** (a) The phototrophic purple sulfur bacterium, *Chromatium* (large, red, rod-shaped cells in this photomicrograph of a natural microbial community). A cell is about 10 μm in diameter. (b) The large chemolithotrophic sulfur-oxidizing bacterium, *Achromatium*. A cell is about 20 μm in diameter. Globules of elemental sulfur can be seen in the cells (arrows). Both of these organisms oxidize hydrogen sulfide (H_2S) produced by sulfate-reducing bacteria. Sulfate-reducing bacteria are chemoorganotrophs that oxidize organic compounds or H_2 and couple this to the reduction of sulfate (SO_4^{2-}) tc H_2S, thus completing the sulfur cycle (⟳ Section 19.13).

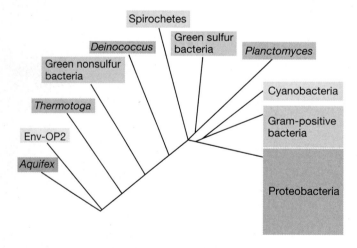

● **Figure 2.9 Detailed phylogenetic tree of *Bacteria*.** Not all known groups of *Bacteria* are depicted on this tree. The relative sizes of the colored boxes are an indication of the number of known genera and species in each of the groups. The Proteobacteria are currently the largest group of *Bacteria* known. The lineage on the tree labeled "Env" (for Environmental) does not represent a cultured organism but instead is a sequence of rRNA genes isolated from an organism in a natural sample (see text). In this example, the closest known relative of Env-OP2 would be *Aquifex*. Although not shown, there are many other such Env's known, located all over the tree.

ducing elemental sulfur that is stored within or outside the cell (Figure 2.10*b*). The sulfur is an oxidation product of H_2S and can be further oxidized to sulfate (SO_4^{2-}). The sulfide and sulfur are oxidized to fuel important metabolic functions such as CO_2 fixation (autotrophy) or energy generation (Figure 2.8).

Several other common prokaryotes of soil and water, and species that live in or on plants and animals in both casual and disease-causing ways, are members of the Proteobacteria. These include species of *Pseudomonas*, many of which can degrade complex and otherwise toxic natural and synthetic organic compounds, and *Azotobacter*, a nitrogen-fixing bacterium. A number of key pathogens are Protoeobacteria, including *Salmonella, Rickettsia, Neisseria*, and many others.

Some bacteria can be distinguished by use of the *Gram-staining* procedure, a technique that stains cells either *gram-positive* or *gram-negative* (staining properties of bacteria will be discussed in Chapter 4). The **Gram-positive lineage** of *Bacteria* contains a number of species united by a common phylogeny and cell wall structure. Here we find the endospore-forming *Bacillus* (discovered by Ferdinand Cohn, ⌖ Section 1.5) (Figure 2.11a●) and *Clostridium* and related spore-forming prokaryotes such as the antibiotic-producing *Streptomyces*. Also included here are the lactic acid bacteria, common inhabitants of decaying plant material and dairy products, that include such organisms as *Streptococcus* (Figure 2.11b) and *Lactobacillus*. Other interesting gram-positive relatives are the mycoplasmas. These prokaryotes lack a cell wall, have very small genomes, and are often pathogenic. *Mycoplasma* is a major genus of organisms in this medically important group (⌖ Section 12.21 and Figure 12.62).

The **Cyanobacteria** (Figure 2.12●) are phylogenetic relatives of gram-positive bacteria (Figure 2.9) and are oxygenic phototrophs. Cyanobacteria were critical in the evolution of life, as they were the first oxygenic phototrophs to have evolved on Earth (⌖ Figure 1.1b). The production of O_2 on an originally anoxic Earth paved the way for the evolution of prokaryotes that could respire using oxygen. The development of "higher organisms," such as the plants and animals, followed billions of years later when Earth had a more oxygen-rich environment.

(a)

(b)

● **Figure 2.12 Filamentous cyanobacteria.** (a) *Oscillatoria*, (b) *Spirulina*. Eons ago, cyanobacteria produced the oxygen now present on Earth. Many other morphological forms of cyanobacteria are known, including unicellular, colonial, and heterocystous. The latter contain special structures called *heterocysts* that carry out nitrogen fixation (⌖ Sections 12.25 and 17.28).

Several lineages of *Bacteria* contain species with unique morphologies. These include the aquatic **Planctomyces** group, characterized by cells with a distinct stalk that allows the organisms to attach to a solid substratum (Figure 2.13●), and the helically shaped **Spirochetes**

(a)

(b)

● **Figure 2.11 Gram-positive bacteria.** (a) The rod-shaped endospore-forming bacterium *Bacillus*, here shown as cells in a chain. Note the presence of endospores (bright refractile structures) inside the cells. Endospores are extremely resistant to heat, chemicals, and radiation. (b) *Streptococcus*, a spherical cell that exists in chains. Streptococci are widespread in dairy products and some are potent pathogens.

● **Figure 2.13 The morphologically unusual stalked bacterium** *Planctomyces*. Shown are several cells attached by their stalks to form a rosette.

John Breznak

● **Figure 2.14** **Spirochetes.** A cell of *Spirochaeta zuelzerae*. These morphologically distinct prokaryotes are also phylogenetically distinct (see Figure 2.9). Spirochetes are widespread in nature and some cause diseases such as syphilis and Lyme disease.

(Figure 2.14●). Several diseases, most notably syphilis and Lyme disease (Sections 26.12 and 27.4), are caused by spirochetes.

Two other major lineages of *Bacteria* are phototrophic: the **green sulfur bacteria** and the **green nonsulfur bacteria** (*Chloroflexus* group) (Figure 2.15●). Species in both lineages contain similar photosynthetic pigments and can grow as autotrophs. *Chloroflexus* is a filamentous prokaryote that inhabits hot springs and shallow marine bays and is often the dominant organism in stratified microbial mats containing a community of microorgan-

isms. *Chloroflexus* is also noteworthy because it is believed to be an important link in the evolution of photosynthesis (Sections 12.35 and 17.7).

Other major lineages of *Bacteria* include the **Chlamydia** and **Deinococcus** groups (Figure 2.9). The genus *Chlamydia*, most species of which are pathogens, harbors a variety of respiratory and venereal pathogens of humans (Sections 12.27 and 26.13). Chlamydia are *obligate intracellular parasites*, meaning that they live inside the cells of higher organisms, in this case, human cells. Several other pathogenic prokaryotes (for example, species of *Rickettsia*, a member of the Proteobacteria that causes diseases such as typhus and Rocky Mountain spotted fever, or *Mycobacterium tuberculosis*, a gram-positive bacterium that causes tuberculosis) are also intracellular pathogens. The intracellular location of these pathogens provides a means by which they can avoid destruction by the host's immune response.

The *Deinococcus* lineage contains species with unusual cell walls and an innate resistance to high levels of radiation; *Deinococcus radiodurans* (Figure 2.16●) is a major species in this group. This organism can survive doses of radiation many-fold greater than that sufficient to kill animals and can even reassemble its chromosome after it has been shattered by radiation treatment. We learn more about this amazing organism in Section 12.34.

Finally, several lineages of *Bacteria* branch off very early on the phylogenetic tree, very near the root (Figure 2.9). Although phylogenetically distinct from one another, these groups are unified by the common property of growth at very high temperature (*hyperthermophily*). Organisms such as *Aquifex* (Figure 2.17●) and *Thermotoga* grow in environments that are near the boiling point of water. As you might imagine, their habitats are hot springs. The early branching nature of

Norbert Pfennig

(a)

M. T. Madigan

(b)

● **Figure 2.15** **Phototrophic green bacteria.** (a) *Chlorobium* (green sulfur bacteria); a single cell is about 0.8 μm wide. (b) *Chloroflexus* (green nonsulfur bacteria); a filament is about 1.3 μm wide. Despite sharing many features like pigments and membrane structures in common (Section 17.2), these organisms are phylogenetically quite distinct (Figure 2.9).

● **Figure 2.17** *Aquifex.* This "early-branching" species of *Bacteria* (see Figure 2.9) is a hyperthermophile, growing optimally above 80°C.

● **Figure 2.16** **The highly radiation-resistant bacterium *Deino-coccus radiodurans.*** This organism can withstand radiation levels far above that sufficient to kill a human. See Section 12.34 for further discussion of this amazing radiation resistance.

these lineages (Figures 2.7 and 2.9) is consistent with the idea that the early Earth was much hotter than it is today (∞ Section 11.1). If life first evolved on a very warm planet, it makes good sense that the closest extant relatives of early organisms would be hyperthermophiles themselves. Interestingly, the phylogenetic trees of both the *Bacteria* and the *Archaea* show just this (Figures 2.9 and 2.18●). Organisms such as *Aquifex*, *Methanopyrus*, and *Pyrolobus* can therefore be considered modern descendants of very ancient cell lineages.

Archaea

Inspection of the domain *Archaea* (Figure 2.18) shows that two subdivisions exist, the *Euryarchaeota* and the *Crenarchaeota*. Each subdivision forms a major branch on the archaeal tree (Figure 2.8). Many *Archaea* are extremophiles, with species capable of growth at the highest temperatures and extremes of pH of all known organisms (Table 2.1). The organism *Pyrolobus* (Figures 2.18 and 2.19●), for example, is the most heat-loving of all known prokaryotes (Table 2.1).

All *Archaea* are chemotrophic, although *Halobacterium* (to be discussed shortly) can use light to make ATP (but not in the way that phototrophic organisms do). Some *Archaea* use organic compounds in their energy metabolism. Many others are chemolithotrophs, with hydrogen gas (H_2) being a widely used energy source (Figure 2.8). Chemolithotrophy is particularly widespread among hyperthermophilic *Archaea*.

The Euryarchaeota branch on the tree of *Archaea* (Figure 2.18) contains three groups of organisms with dramatically different physiologies. Some species require O_2 while others are killed by it, and some grow at the upper or lower extremes of pH (Table 2.1). *Methanogens* like *Methanobacterium* are strict anaer-

obes. Their metabolism is unique in that energy is obtained by the *production* of an energy-rich substance, methane (natural gas). Methanogens are ecologically important organisms in the biodegradation of organic matter in nature (∞ Sections 13.4, 17.17, and 19.10), and most if not all of the natural gas found on Earth has accumulated from their metabolism.

The extreme halophiles are relatives of the methanogens (Figure 2.18), but are physiologically distinct from them. Unlike methanogens, which are killed by oxygen, extreme halophiles *require* oxygen and are unified by their requirement for very large amounts of salt (NaCl) for metabolism and reproduction. These organisms are therefore called *halophiles* (salt lovers). In

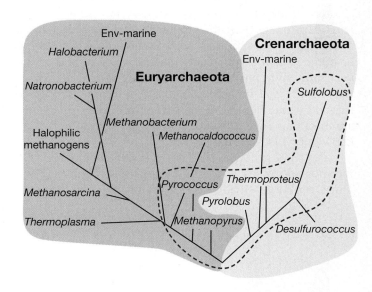

● **Figure 2.18** **Detailed phylogenetic tree of *Archaea.*** Not all known groups of *Archaea* are depicted on this tree. There are two major subgroups of *Archaea*, Euryarchaeota and Crenarchaeota. The organisms circled in red are hyperthermophiles, growing at very high temperatures. In pink are shown methanogens and the extreme halophiles and acidophiles. Each major group has its own Env lineages (see legend to Figure 2.9 and text) as well, most of which are marine species. There are roughly the same number of species in both subgroups of *Archaea*, but the total number of cultured *Archaea* is far lower than for *Bacteria*.

R. Rachel and K. O. Stetter

● **Figure 2.19** *Pyrolobus.* A hyperthermophilic archaeon that grows optimally above the boiling point of water!

T. D. Brock

● **Figure 2.21** **Extremely acidophilic prokaryotes.** The cell wall-less archaeon, *Thermoplasma*, shown here. and its close relative *Picrophilus* (see Table 2.1) grow at moderate.y high temperature and extremely low pH. Although members of the *Bacteria*, the genus *Mycoplasma* also contains species lacking a cell wall. Cell wall-less prokaryotes are discussed in Sections 12.21 and 13.5.

fact, organisms like *Halobacterium* are so salt loving that they can actually grow on and within salt crystals (Figure 2.20●).

As previously mentioned (see Section 2.4), many prokaryotes are phototrophic and can generate ATP from light. Although *Halobacterium* species do not produce chlorophyll like true phototrophs, they nevertheless contain a class of light-sensitive pigments that can absorb light and trigger ATP synthesis (⌀ Section 13.3). Extremely halophilic *Archaea* inhabit salt lakes, salterns, and other very salty environments. Some extreme halophiles, such as *Natronobacterium*, inhabit soda lakes, environments characterized by both high levels of salt *and* pH. Such organisms are therefore *alkaliphilic* and grow at the highest pH of all known organisms (Table 2.1).

The final group of *Archaea* we will consider is the thermoacidophiles, such as *Thermoplasma* (Figure 2.21●). These prokaryotes have a cell membrane but lack cell

William D. Grant

● **Figure 2.20** **Extremely halophilic *Archaea*.** A vial of brine at the point of NaCl precipitation and containing cells of the extreme halophile, *Halobacterium*. The organism contains pigments that absorb light and lead to ATP production. Cells of *Halobacterium* can also live *within* salt crystals themselves (⌀ Microbial Sidebar, Chapter 4, How Long Can an Endospore Survive?).

walls (resembling *Mycoplasma* in this respect) and grow best at moderately high temperatures and extremely low pH. This group includes *Picrophilus*, the most *acidophilic* of all known prokaryotes (Table 2.1).

Phylogenetic Analyses of Natural Microbial Communities

Not all *Archaea* are extremophiles. Many can be found in lakes, soils, and oceans. Unfortunate.y, however, most of these *Archaea* have thus far defied laboratory culture, and so we know very little about what they are doing in their habitats. Much the same can be said about the legion of uncultured species of *Bacteric*. If these organisms have not been cultured, how do we know that they are there? We know this because it is possible to isolate *ribosomal RNA genes* from cells in a natural sample, such as soil. If a natural sample contains ribosomal RNA, it is because the organisms that made this ribosomal RNA were present in the sample. If we process such a sample for ribosomal RNA genes, isolate and sequence them, we can place them on the phylogenetic tree just as if we had started from cultures of the different organisms (see Figures 2.9 and 2.18).

Using these methods of *molecular microbial ecology* initially devised by Norman Pace, an American microbiologist, it is clear that the extent of prokaryotic diversity is far greater than originally thought. A sampling of virtually any microbial habitat reveals that the vast majority of prokaryotes in that habitat have never been cultured. The challenge now is to try to discover enough about the biology of these uncultured prokaryotes that creative culturing techniques can be devised to grow them. Genomic analyses of uncultured *Archaea* and

Bacteria (environmental genomics, ⌘ Section 18.6), will help in this regard. This is because identification of the complement of genes present in these uncultured organisms will reveal their metabolic capacities. Knowledge of the metabolic blueprint of these organisms should then help in the design of specific isolation methods for their culture.

 2.5 Concept Check

Several lineages are present in the domains *Bacteria* and *Archaea*, and an enormous diversity of cell morphologies and physiologies are represented there. Retrieval and analysis of ribosomal RNA genes from cells in natural samples have shown that many phylogenetically distinct but as yet uncultured prokaryotes exist in nature.

◆ What important bacterial species that resides in your gut is a member of the Proteobacteria?

◆ Why can it be said that the cyanobacteria prepared Earth for the evolution of higher life forms?

◆ What is unusual about the genus *Halobacterium*?

◆ How do we know a particular lineage of prokaryote exists in a natural habitat without first isolating and growing it in laboratory culture?

2.6 Eukaryotic Microorganisms

Eukaryotic microorganisms are related by their distinct cell structure (Figure 2.1) and phylogenetic history (Figure 2.7). Inspection of the domain *Eukarya* (Figure 2.22●) shows a long branch in the tree of life culminating in the most derived **eukaryotes**, the plants and animals. Interestingly, however, and consistent with their phylogenetic location on the tree, some of the "early-branching" *Eukarya* turn out to be structurally simple eukaryotes, lacking mitochondria and some other key organelles. These cells, such as the diplomonad *Giardia* (Figure 2.22), appear to be modern descendents of primitive eukaryotic cells that did not engage in endosymbiosis or perhaps had, but then lost their endosymbionts (⌘ Sections 2.3, 11.3, and 14.4). Most of these early eukaryotes are parasites of humans and other animals, and are unable to live a free and independent existence.

Eukaryotic Microbial Diversity

Like prokaryotes, a diverse array of eukaryotic microorganisms is known. Collectively, microbial eukaryotes are known as the **Protista**, or *protists* for short. Some protists, such as the algae (Figure 2.23*a*●), are phototrophic. Algae contain chlorophyll-rich organelles called *chloroplasts* and can live in environments containing only a few minerals (for example, K, P, Mg, N, S), H_2O, CO_2, and light.

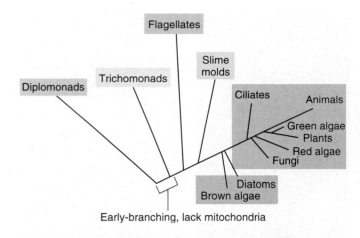

● **Figure 2.22 Detailed phylogenetic tree of *Eukarya*.** Not all known lineages of *Eukarya* are depicted. Some early-branching species of *Eukarya* lack organelles other than the nucleus. Note how "higher organisms" (plants and animals) branch near the apex of the tree.

Algae inhabit both soil and aquatic habitats and are major primary producers in nature. Fungi (Figure 2.23*b*) lack photosynthetic pigments and are either unicellular (yeasts) or filamentous (molds). Fungi are major agents of biodegradation in nature and recycle much of the organic matter in soils and other ecosystems.

Cells of algae and fungi have cell walls, whereas the protozoa (Figure 2.23*c*) do not. Most protozoans are motile, and different species are widespread in nature in aquatic habitats and as pathogens of humans and other animals. Different protozoa are also spread about the phylogenetic tree of *Eukarya*. Some, like the flagellates, are fairly early branching species, while others, like the ciliated species *Paramecium* (Figure 2.23), are phylogenetically more derived (Figure 2.22). The *slime molds* resemble protozoa in that they are motile and lack cell walls. Slime molds differ from protozoa, however, not only in phylogeny but by the fact that their cells undergo a life cycle. During the slime mold life cycle, motile cells aggregate to form a multicellular structure called a *fruiting body* from which spores are produced that yield new motile cells (⌘ Section 14.11). Slime molds are the earliest known protists to show such cellular cooperation to form microscopic structures.

Lichens are leaf-like structures often found growing on rocks, trees, and other surfaces (Figure 2.24●). Lichens are an example of microbial *mutualism*, a situation where two organisms live together for mutual benefit. Lichens consist of a fungus and a phototrophic partner, either an alga (eukaryote) or a cyanobacterium (a prokaryote). The phototrophic component is the primary producer while the fungus provides the phototroph with an anchor and protection from the elements. The lichen is thus a dynamic organism that has evolved a successful strategy of mutualistic interaction between two quite different microorganisms.

(a)

(b)

(c)

(a)

(b)

●**Figure 2.23** **Microbial *Eukarya*.** (a) Algae; the colonial green alga, *Volvox* (⚫ Section 14.13). Each spherical-shaped cell contains several chloroplasts, the photosynthetic organelle of phototrophic eukaryotes. (b) Fungi; the spore-bearing structures of a typical mold. Each spore can give rise to a new filamentous fungus (⚫ Section 14.12). (c) Protozoa; the ciliated protozoan *Paramecium* (⚫ Section 14.10). Cilia function like oars in a boat, conferring motility on the cell.

●**Figure 2.24** **Lichens.** (a) An orange-pigmented lichen growing on a rock, and (b) a yellow-pigmented lichen growing on a dead tree stump, Yellowstone National Park, USA. The color of the lichen comes from the pigmented (algal) component of the lichen structure. Besides chlorophylls, the algal components of lichens contain carotenoid pigments (⚫ Section 17.3), which can be yellow, orange, brown, red, green, or purple.

Final Remarks

Our tour of microbial diversity here has only served up an overview of the subject. The story is much larger than this and will continue in Chapters 12–15. Note also that the viruses were intentionally left out of our coverage here. Recall that viruses are not cells, yet require cells for their replication (see Section 2.1). Cells in all domains of life have viral parasites, and we will deal with viral diversity in later chapters (⚫ Chapters 9 and 16).

Before we can proceed to a more detailed consideration of microbial diversity, however, we must become familiar with the molecular features of cells, especially prokaryotic cells. Along the way we will learn of the great chemical diversity of living organisms, a direct result of

the evolutionary breadth of the tree of life. Then we will be well prepared to revisit the subject of microbial diversity and expand on our coverage here.

2.6 *Concept Check*

Microbial eukaryotes are a diverse group that includes algae, protozoa, fungi, and slime molds. Some algae and fungi have developed mutualistic associations called *lichens*.

◆ List at least two ways algae differ from cyanobacteria.

◆ List at least two ways algae differ from protozoa.

◆ How do the components of a lichen benefit each other?

REVIEW QUESTIONS

1. Why does a cell need a cytoplasmic membrane (∞ Section 2.1)?

2. Which domains of life have a prokaryotic cell structure? Can prokaryotic cell structure predict evolutionary relationships (∞ Section 2.1)?

3. How do viruses resemble cells? How do they differ from cells (∞ Section 2.1)?

4. What is meant by the word *genome*? How does the arrangement of the genome of prokaryotes differ from that of eukaryotes (∞ Section 2.2)?

5. For what reason do the processes of **mitosis** and **meiosis** occur in eukaryotic cells (∞ Section 2.2)?

6. How many genes does an organism such as *Escherichia coli* have? How does this compare with the number of genes in one of your cells (∞ Section 2.2)?

7. What is the theory of endosymbiosis (∞ Section 2.3)?

8. Molecular studies have shown many macromolecules in species of *Archaea* resemble their counterparts in various eukaryotes more closely than in species of *Bacteria*. Explain (∞ Section 2.3).

9. How do *chemoorganotrophs* differ from *chemolithotrophs* from the standpoint of energy metabolism? What carbon sources do members of each group use? Are they therefore *heterotrophs* or *autotrophs* (∞ Section 2.4)?

10. What is unusual about the organism *Pyrolobus* (∞ Section 2.5)?

11. What similarities and differences exist between the following three organisms: *Pyrolobus*, *Halobacterium*, and *Thermoplasma* (∞ Section 2.5)?

12. Examine Figure 2.18. What does the lineage "Env-marine" mean (∞ Section 2.5)?

13. How does *Giardia* differ from a human cell, both structurally and phylogenetically (∞ Section 2.6)?

APPLICATION QUESTIONS

1. Prokaryotic cells containing plasmids can often be "cured" of their plasmids (that is, the plasmids can be permanently removed) with no ill effects, while removal of the chromosome would be lethal. Explain.

2. It has been said that knowledge of the evolution of *macroorganisms* greatly preceded that of *microorganisms*. Why do you think that reconstruction of the evolutionary lineage of horses, for example, might have been an easier task than doing the same for any group of prokaryotes?

3. Examine the phylogenetic tree shown in Figure 2.6. Using the sequence data shown, describe why the tree would be incorrect if its branches remained the same but the positions of organisms 2 and 3 on the tree were switched?

4. Microbiologists have cultured a great diversity of microorganisms but know that an even greater diversity exists, despite the fact that they have never seen these organisms or grown them in the laboratory. Explain.

5. Which data from this chapter could you use to convince your friend that extremophiles are not just organisms that were "hanging on" in their respective habitats?

6. Defend this statement: If cyanobacteria had never evolved, life on Earth would have remained strictly microbial.

MACROMOLECULES

The structure of proteins, a key class of macromolecules, dictates their function.

WORKING GLOSSARY

Covalent bond a chemical bond in which electrons are shared between two atoms

Denaturation destruction of the folding properties of a protein leading (usually) to loss of biological activity

Enantiomer one form of a molecule that is the mirror image of another form of the same molecule

Glycosidic bond a type of covalent bond that links sugar units together in a polysaccharide

Hydrogen bond a weak chemical bond between a hydrogen atom and a second, more electronegative element, usually an oxygen or nitrogen atom

Isomers two molecules with the same molecular formula but which differ structurally

Lipid glycerol bonded to fatty acids or other hydrophobic molecules by ester or ether linkage. Often contain other groups, such as phosphate as well

Macromolecule polymer of covalently linked monomeric units

Molecule two or more atoms chemically bonded to one another

Nonpolar possessing hydrophobic (water-repelling) characteristics and not easily dissolved in water

Nucleic acid DNA or RNA

Nucleoside a nucleotide without its phosphate group

Nucleotide a monomer of a nucleic acid containing a nitrogen base (adenine, guanine, cytosine, thymine, or uracil), a molecule of phosphate, and a sugar, either ribose (in RNA) or deoxyribose (in DNA)

Peptide bond a type of covalent bond linking amino acids in a polypeptide

Phosphodiester bond a type of covalent bond linking nucleotides together in a polynucleotide

Polar possessing hydrophilic characteristics and generally water-soluble

Polymer a chemical compound formed by polymerization and consisting of repeating units called monomers

Polynucleotide a polymer of nucleotides bonded to one another by phosphodiester bonds

Polypeptide a polymer of amino acids bonded to one another by peptide bonds

Polysaccharide a polymer of sugar units bonded to one another by glycosidic bonds

Primary structure in an informational macromolecule, such as a polypeptide, the precise sequence of monomeric units

Protein a polypeptide or group of polypeptides that form a molecule of specific biological function

Quaternary structure in proteins, the number and types of individual polypeptides in the final protein molecule

Secondary structure the initial pattern of folding of a polypeptide or a polynucleotide, usually dictated by opportunities for hydrogen bonding

Tertiary structure the final folded structure of a polypeptide that has previously attained secondary structure

CHEMICAL BONDING AND WATER IN LIVING SYSTEMS

To understand how a cell works, an understanding of the molecules present and chemical processes that take place within cells is essential. Molecules, especially **macromolecules**, are the "guts" of the cell and are the subject of this chapter. It is assumed that the reader has some background in elementary chemistry, especially regarding the nature of atoms and atomic bonding. Here we will expand on this background with a primer on relevant biochemical bonds, followed by a discussion of the structure and function of the four classes of macromolecules: **polysaccharides, lipids, nucleic acids**, and **proteins**.

3.1 Strong and Weak Chemical Bonds

The major chemical elements of life include *hydrogen, oxygen, carbon, nitrogen, phosphorus,* and *sulfur;* these elements can bond in various ways to form the molecules of life. A **molecule** is two or more atoms chemically bonded to one another. Thus, two oxygen (O) *atoms* can combine to form a *molecule* of oxygen (O_2). Or, carbon (C), hydrogen (H), and O atoms can combine to form glucose, $C_6H_{12}O_6$, a hexose sugar.

The chemical elements of life are able to form strong bonds in which electrons are shared more or less equally between atoms. These are called **covalent bonds**. To en-vision a covalent bond, consider the formation of a molecule of water from its constituent elements, O and H:

$$\ddot{O} + 2H\cdot \longrightarrow H\!:\!\ddot{O}\!:\!H$$

Oxygen contains six electrons in its outermost shell, while hydrogen has but a single electron. When they combine to form H_2O, covalent bonds maintain the three atoms in tight association. Depending on the molecule, double and even triple covalent bonds can form, and the strength of these bonds increases dramatically with their number (Figure 3.1●).

The chemical elements of life bond in various combinations to form the individual constituents of macromolecules, called **monomers**. Macromolecules are thus **polymers**, chemical compounds formed from repeating monomeric units. Thousands of distinct monomers are known, but only a relatively small number play important roles in the four classes of macromolecules. To a large extent it is the chemical properties of monomers that, once combined to form macromolecules, give the latter their distinctive structure and function.

Hydrogen Bonding

In addition to covalent bonds, a variety of much weaker chemical bonds also play an important role in biological molecules. Foremost among these are *hydrogen bonds*. **Hydrogen bonds** (Figure 3.2●) form between

Ethylene, a double-bonded organic compound

Acetylene, a triple-bonded organic compound

Some inorganic compounds with double or triple bonds

Peptide bond of proteins

Cytosine (nitrogen base of DNA and RNA)

Phenylalanine (amino acid in proteins)

Organic compounds with double bonds

● **Figure 3.1   Covalent bonding of some molecules containing double or triple bonds.** For acetylene and ethylene, the electronic configuration of the molecules is shown.

hydrogen atoms and more *electronegative* (electron attracting) elements, such as oxygen or nitrogen. For example, an oxygen atom is rather electronegative (electron withdrawing), while a hydrogen atom is not. In the covalent bond between oxygen and hydrogen, the shared electrons therefore orbit slightly nearer to the oxygen nucleus than to the hydrogen nucleus. Because electrons carry a negative charge, this creates a slight charge separation, oxygen negative and hydrogen positive (Figure 3.2a). An individual hydrogen bond by itself is very weak. However, when many hydrogen bonds form within and between molecules, stability of the molecules can increase greatly.

Water molecules readily undergo hydrogen bonding (Figure 3.2a), and this contributes to the distinct *polarity* of water. Because it is **polar**, water molecules readily associate with one another and apart from **nonpolar** (hydrophobic) molecules. As water molecules orient themselves in solution, the slight positive charge on a hydrogen atom can bridge negative charges on two oxygen atoms. This bridge is the hydrogen bond.

Hydrogen bonds also form between atoms in macromolecules (Figure 3.2b and c). When these weak electrical forces accumulate in a large molecule such as a protein, they increase the stability of the molecule and can also affect its overall structure. We will see later that hydrogen bonds play major roles in the biological properties of proteins (Figure 3.2b) and nucleic acids (Figure 3.2c) (see Sections 3.5, 3.7, and 3.8).

Other Weak Bonds

Biomolecules form other types of weak interactions. For instance, **van der Waals forces** are weak attractive forces that occur between atoms when they become closer than about 3–4 angstroms (Å). Van der Waals forces can play significant roles in the binding of substrates to enzymes (⚭ Section 5.5) and in protein-nucleic acid interactions. **Ionic bonds**, such as that between Na^+ and Cl^- in NaCl, are weak electrostatic interactions that allow for ionization to occur in aqueous solution. Many important biomolecules, such as carboxylic acids and phosphates (see Table 3.1), are ionized at cytoplasmic pH levels (generally about pH 6–8) and can thus be dissolved to high levels in the cytoplasm.

Hydrophobic interactions are also important in biomolecules. Hydrophobic interactions occur because nonpolar molecules or nonpolar regions of molecules tend to associate tightly in a polar environment. Hydrophobic interactions can play a major role in the folding of proteins. Hydrophobic interactions also play important roles in the binding of substrates to enzymes (⚭ Section 5.5). In addition, hydrophobic interactions often control how differ-

(a) Water

(b) Amino acids in a protein chain

(c) Nitrogen bases in DNA

● **Figure 3.2   Hydrogen bonding.** In nucleic acids, hydrogen bonds are often depicted as lines rather than dots, with two lines between adenine/thymine pairs and three lines between guanine/cytosine pairs (see Figure 3.11). In (b), the R represents the side chain of each amino acid (see Figure 3.12).

Table 3.1 Some functional groups of biochemical importance

Chemical species	Structure[a]	Biological relevance
Carboxylic acid	$$\begin{array}{c} O \\ \parallel \\ -C-OH \end{array}$$	Organic, amino, and fatty acids; lipids; proteins
Aldehyde	$$\begin{array}{c} O \\ \parallel \\ -C-H \end{array}$$	Functional group of reducing sugars such as glucose; polysaccharides
Alcohol	$$\begin{array}{c} H \\ \mid \\ -C-OH \\ \mid \\ H \end{array}$$	Lipids; carbohydrates
Keto	$$\begin{array}{c} O \\ \parallel \\ -C- \end{array}$$	Pyruvate, citric acid cycle intermediates
Ester	$$\begin{array}{c} H \qquad O \\ \mid \qquad \parallel \\ -C-O-C- \\ \mid \\ H \end{array}$$	Lipids of *Bacteria* and *Eukarya*; amino acid attachment to tRNAs
Phosphate ester	$$\begin{array}{c} O^- \qquad \\ \mid \qquad \mid \\ {}^-O-P-O-C- \\ \parallel \qquad \mid \\ O \end{array}$$	Nucleic acids, DNA and RNA
Thioester	$$\begin{array}{c} O \\ \parallel \\ -C{\sim}S- \end{array}$$	Energy metabolism; biosynthesis of fatty acids
Ether	$$\begin{array}{c} H \qquad H \\ \mid \qquad \mid \\ -C-O-C- \\ \mid \qquad \mid \\ H \qquad H \end{array}$$	Lipids of *Archaea*; sphingolipids
Acid anhydride	$$\begin{array}{c} O \qquad O^- \\ \parallel \qquad \mid \\ -C{\sim}O-P{=}O \\ \mid \\ O_- \end{array}$$	Energy metabolism, for example, acetyl phosphate
Phosphoanhydride	$$\begin{array}{c} O^- \qquad O^- \\ \parallel \qquad \mid \\ {}^-O-P{\sim}O-P-O^- \\ \parallel \qquad \parallel \\ O \qquad O \end{array}$$	Energy metabolism, for example, ATP

[a] A squiggle-type bond depiction ($\sim$) indicates an "energy-rich" bond (Section 5.8).

ent subunits in a multisubunit protein associate with one another to form the biologically active molecule and are also important in stabilizing RNA.

Bonding Patterns in Biomolecules

The element **carbon** is a major component of all macromolecules. Carbon can bond not only with itself, but with many other elements as well, to yield large structures of considerable diversity and complexity. In different *organic* (carbon containing) compounds, a variety of bonding patterns are possible. Each of these *functional groups*, as they are called, has unique chemical properties that are important in determining their biological role within the cell. An awareness of these functional groups will make our later discussion of macromolecular structure, cell physiology, and biosynthesis easier to follow. Table 3.1 lists several functional groups of biochemical importance and examples of molecules and macromolecules that contain them.

3.1 Concept Check

Covalent bonds are strong bonds that bind elements in macromolecules. Weak bonds, such as hydrogen bonds, van der Waals forces, and hydrophobic interactions, also affect macromolecular structure, but through more subtle atomic interactions. A variety of functional groups containing carbon atoms are common in biomolecules.

◆ Why are *covalent* bonds stronger than *hydrogen* bonds?

◆ How can a hydrogen bond play a role in macromolecular structure?

3.2 An Overview of Macromolecules and Water as the Solvent of Life

If you were to chemically analyze a prokaryotic cell like the common intestinal bacterium *Escherichia coli*, what would you find? You would find water as the major constituent, but after removing the water you would find large amounts of macromolecules, much smaller amounts of monomers and a variety of inorganic ions (Table 3.2). About 95% of the dry weight of a cell consists of macromolecules, and of these, *proteins* are by far the most abundant class by weight (Table 3.2).

Proteins are polymers of monomers called *amino acids*. Proteins are found throughout the cell, in both structural as well as catalytic (enzymatic) roles (Figure 3.3*a*●), and an average cell will have many thousands of different types of proteins (Table 3.2).

Nucleic acids are polymers of *nucleotides* and are found in the cell in two forms, **RNA** and **DNA**. After proteins, *ribonucleic acids* (RNAs) are the next most abundant macromolecule in an actively growing cell (Table 3.2 and Figure 3.3*b*). This is because there are thousands of ribosomes (the "machines" that make new proteins) in each cell, and ribosomes are composed of RNA and protein. In addition, smaller amounts of RNA are present in the form of messenger and transfer RNAs, other key players in protein synthesis. In contrast to RNA, *DNA* makes up a relatively insignificant (by weight) fraction of the bacterial cell (Table 3.2). However, although quantitatively minor, DNA is central to cell function as the repository of genetic information.

Lipids have both hydrophobic and hydrophilic properties and play crucial roles in membrane structure and as storage depots for excess carbon (Figure 3.3*d*). **Polysaccharides** are polymers of *sugars* and are present primarily in the cell wall. As with lipids, however, polysaccharides such as glycogen (see later in this chapter) can be major forms of carbon and energy storage in the cell (Figure 3.3*c*).

Water as a Biological Solvent

Macromolecules and all other molecules in cells are bathed in water. Water has several important chemical features that make it an ideal biological solvent. Indeed, water is a necessary prerequisite for life as we know it.

(a) **Proteins**

(b) **Nucleic Acids:** DNA RNA

(c) **Polysaccharides**

(d) **Lipids**

● **Figure 3.3 The locations of macromolecules in the cell.** (a) *Proteins* (brown) are found throughout the cell both as parts of cell structures and as enzymes. The flagellum is a structure involved in swimming motility. (b) *Nucleic acids.* DNA (green) is found in the nucleoid of prokaryotic cells and in the nucleus of eukaryotic cells. RNA (orange) is found in the cytoplasm (mRNA, tRNA) and in ribosomes (rRNA). (c) *Polysaccharides* (yellow) are located in the cell wall and occasionally in internal storage granules. (d) *Lipids* (blue) are found in the cytoplasmic membrane, the cell wall, and in storage granules.

Table 3.2	Chemical composition of a prokaryotic cell[a]	
Molecule	**Percent of dry weight**[b]	**Molecules per cell (different kinds)**
Total macromolecules	96	24,610,000 (~2500)
Protein	55	2,350,000 (~1850)
Polysaccharide	5	4,300 (2)[c]
Lipid	9.1	22,000,000 (4)[d]
Lipopolysaccharide	3.4	1,430,000 (1)
DNA	3.1	2.1 (1)
RNA	20.5	255,500 (~660)
Total monomers	3.0	—[e](~350)
Amino acids and precursors	0.5	—(~100)
Sugars and precursors	2	—(~50)
Nucleotides and precursors	0.5	—(~200)
Inorganic ions	1	—(18)
Total	100%	—

[a] Data from Neidhardt, F.C., et al. (eds.), 1996. *Escherichia coli* and *Salmonella typhimurium—Cellular and Molecular Biology*, 2nd edition. American Society for Microbiology, Washington, DC.

[b] Dry weight of an actively growing cell of *E. coli* ≅ 2.8×10^{-13} g; total weight (70% water) = 9.5×10^{-13} g.

[c] Assuming peptidoglycan and glycogen to be the major polysaccharides present.

[d] There are several classes of phospholipids, each of which exists in many kinds because of variability in fatty acid composition between species and because of different growth conditions.

[e] Reliable estimates of monomer and inorganic ion composition are lacking.

Two key properties of water that make it such a good solvent are its *polarity* and *cohesiveness*.

The polar properties of water are important because many biologically important molecules (Table 3.2) are themselves polar and thus readily dissolve in water. As we will see in Chapter 4, dissolved substances are continually passing into and out of the cell through transport activities of the cytoplasmic membrane (⚬⚬ Sections 4.6 and 4.7). These substances include nutrients needed to build new cell material as well as waste products of metabolic processes.

The polar properties of water also promote the aggregation of large molecules, thanks to the increased opportunities for hydrogen bonding. Water forms three-dimensional networks, both with itself (Figure 3.2a) and within macromolecules. By so doing, water molecules help to position atoms within biomolecules for potential interactions. The high polarity of water is also beneficial to the cell because it forces *nonpolar* substances to aggregate and remain together. Membranes, for example, contain large amounts of lipids, which have major nonpolar (hydrophobic) components, and these aggregate in such a way as to prevent the unrestricted flow of polar molecules into and out of the cell.

In addition to hydrogen bonding, the polar nature of water makes it highly *cohesive*. This means that water molecules have a high affinity for one another and form chemically ordered arrangements in which hydrogen bonds (Figure 3.2) are constantly forming, breaking, and re-forming. The cohesive nature of water is responsible for some of its biologically important properties, such as *high surface tension* and *high specific heat* (heat required to raise the temperature 1°C). Also, the fact that water expands on freezing to yield a less dense solid form (ice) has a profound effect on life in temperate and polar aquatic environments. In a lake, for example, ice on the surface insulates the water beneath the ice and prevents it from freezing, thus allowing aquatic organisms to survive under the overlying ice.

Life originated in an aqueous milieu nearly 4 billion years ago, and virtually anywhere on Earth where liquid water exists, microorganisms are likely to be found. With these important properties of water in mind, we now consider the structure of the major macromolecules of life (Table 3.2 and Figure 3.3).

⬣ 3.2 Concept Check

Proteins are the most abundant class of macromolecule in the cell. Other macromolecules include the nucleic acids (DNA and RNA), lipids, polysaccharides, and lipopolysaccharides. Water is a particularly good solvent for living organisms because of its polarity and cohesiveness.

◆ Why do protein and RNA make up such a large proportion of an actively growing cell?

◆ Why does the high polarity of water make it useful as a biological solvent?

▓▓ II NONINFORMATIONAL MACROMOLECULES

In this unit we examine the structure and function of *noninformational* macromolecules—polysaccharides and lipids. Although the sequences of monomers in these macromolecules do not carry genetic information, the macromolecules themselves still play important roles in the cell, primarily as structural or reserve materials.

3.3 Polysaccharides

Carbohydrates (sugars) are organic compounds containing carbon, hydrogen, and oxygen in a ratio of 1:2:1. The structural formula for glucose, the most abundant of all sugars, is $C_6H_{12}O_6$ (Figure 3.4●). The most biologically relevant carbohydrates are those containing 4, 5, 6, and 7 carbon atoms (designated as C_4, C_5, C_6, and C_7). C_5 sugars (*pentoses*) are of special significance because of their role as structural backbones of nucleic acids. Likewise, C_6 sugars (*hexoses*) are the monomeric constituents of cell wall

● **Figure 3.4   Structural formulas of a few common sugars.** The formulas can be depicted in two alternate ways, open chain and ring. The open chain is easier to visualize, but the ring form is the commonly used structure. Note the numbering system on the ring.

polymers and energy reserves. Figure 3.4 shows the structural formulas of a few common sugars.

Derivatives of simple carbohydrates can be formed by replacing one or more of the hydroxyl groups by other chemical species. For example, the important bacterial cell wall polymer **peptidoglycan** (⚭ Section 4.8) contains the glucose derivative *N*-acetylglucosamine (Figure 3.5●). Besides sugar derivatives, sugars having the same *structural* formula can differ in their *stereoisomeric properties* (see Section 3.6). Hence, a large number of different sugars are potentially available to the cell for the construction of polysaccharides.

The Glycosidic Bond

Polysaccharides are carbohydrates containing many (sometimes hundreds or even thousands) of monomeric units called *monosaccharides*. The latter are connected by covalent bonds called **glycosidic bonds** (Figure 3.6●). If two monosaccharides are joined by a glycosidic linkage, the resulting molecule is called a *disaccharide*. The addition of one more monosaccharide unit yields a *trisaccharide*, and several more an *oligosaccharide*. An extremely long chain is then a *polysaccharide*.

The glycosidic bond can exist in two different geometric orientations, referred to as alpha (α) and beta (β) (Figure 3.6a). Polysaccharides with a repeating structure composed of glucose units bonded between carbons 1 and 4 in the *alpha* orientation (for example, glycogen and starch, Figure 3.6b) function as important carbon and energy reserves in bacteria, plants, and animals. Alternatively, glucose units joined by β-1,4 linkages are present in cellulose (Figure 3.6b), a stiff plant and algal cell wall component. Thus, even though starch and cellulose are both composed solely of glucose units, their functional properties differ because of the different configurations, α or β, of their glycosidic bonds.

Polysaccharides can also combine with other classes of macromolecules, such as proteins and lipids, to form *complex polysaccharides*—**glycoproteins** and **glycolipids**. These compounds play important roles in cells, in particular as cell surface receptor molecules in cytoplasmic membranes. The compounds typically reside on the external surfaces of the membrane where they are in con-

α-1,6-Glycosidic bond   β-1,4-Glycosidic bond

(a)

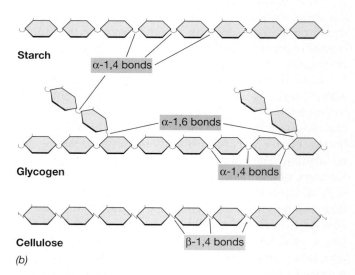

(b)

● **Figure 3.6 Polysaccharides.** (a) Structure of different glycosidic bonds. Note that both the *linkage* (position on the ring of the carbon atoms bonded) and the *geometry* (α or β) of the linkage can vary about the glycosidic bond. (b) Structures of some common polysaccharides. Compare color coding to (a).

tact with the environment. Glycolipids also constitute a major portion of the cell wall of gram-negative bacteria, and as such, impart a number of unique surface properties to these organisms (⚭ Section 4.9).

3.3 *Concept Check*

Sugars combine into long polymers called polysaccharides. The two different orientations of the glycosidic bonds that link sugar residues impart different properties to the resultant molecules. Polysaccharides can also contain other molecules such as protein or lipid, forming complex polysaccharides.

◆ How can glycogen and cellulose differ so much in their physical properties when they both consist of 100% glucose?

● **Figure 3.5 *N*-acetylglucosamine, a derivative of glucose.** *N*-acetylglucosamine is an important component of the cell wall polysaccharide *peptidoglycan* in species of *Bacteria* (⚭ Section 4.8).

3.4 Lipids

Lipids are essential components of all cells and are *amphipathic* macromolecules, meaning they contain both hydrophilic and hydrophobic properties. Lipid structure varies between the domains of life, and, within a domain, many different lipids are known. *Fatty acids* are major constituents of the lipids of *Bacteria* and *Eukarya*. By contrast, the lipids of *Archaea* are constructed from the hydrophobic molecule *phytane* (∞ Section 4.5).

Fatty acids contain both hydrophobic and hydrophilic components. Palmitate (the ionized form of palmitic acid) (Figure 3.7●), for example, is a 16-carbon fatty acid composed of a chain of 15 saturated (fully hydrogenated and highly hydrophobic) carbon atoms and a single carboxylic acid group (the hydrophilic portion). Other common fatty acids in the lipids of *Bacteria* include saturated or monounsaturated forms, from C_{12} to C_{20} (Figure 3.7).

Triglycerides and Complex Lipids

Simple lipids (fats) consist of fatty acids (or phytanyl units in *Archaea*) bonded to the C_3 alcohol *glycerol* (Figure 3.7). Simple lipids are also called **triglycerides** because three fatty acids are linked to the glycerol molecule.

Complex lipids are simple lipids that contain additional elements such as phosphorus, nitrogen, or sulfur, or small hydrophilic organic compounds such as sugars, ethanolamine (Figure 3.7), serine, or choline. Lipids containing a phosphate group, called **phospholipids**, are an important class of complex lipids because they play a major structural role in the cytoplasmic membrane (∞ Section 4.5).

The amphipathic properties of lipids make them ideal as structural components of membranes. Lipids aggregate to form membranes; the *hydrophilic* (glycerol) portion remains in contact with the cytoplasm and the external environment while the *hydrophobic* portion remains buried away inside the membrane (∞ Section 4.5 and Figure 4.16). Because of this property, membranes are ideal permeability barriers. The inability of polar substances to flow through the hydrophobic region of the lipids renders the membrane impermeable and prevents leakage of cytoplasmic constituents. However, this also means that polar substances necessary for cell function do not leak *in*, either, but we reserve this story for the next chapter (transport, ∞ Section 4.5)

Common fatty acids:

C_{16} saturated (palmitic)

C_{16} monounsaturated (palmitoleic)

Simple lipids (triglycerides):
Fatty acids linked to glycerol by ester linkage

Glycerol

Fatty acids

Ester linkage

Complex lipid:
Phosphatidyl ethanolamine (a phospholipid)

Fatty acids

Phosphate

Ethanolamine

Complex lipid:
Monogalactosyl diglyceride (a glycolipid)

Galactose

Fatty acids

● **Figure 3.7 Fatty acids, simple lipids (fats), and complex lipids.** Simple lipids are formed by a dehydration reaction between fatty acids and glycerol to yield the ester linkage. The fatty acid composition of a cell varies with growth temperature.

3.4 Concept Check

Lipids contain both hydrophobic and hydrophilic components; their chemical properties make them ideal structural components for cytoplasmic membranes.

◆ What part of a fatty acid molecule is hydrophobic? Hydrophilic?

◆ How does a *phospholipid* differ from a *triglyceride*?

◆ Draw the chemical structure of butyrate, a C_4 fully saturated fatty acid.

III INFORMATIONAL MACROMOLECULES

Unlike polysaccharides and lipids, the sequence of monomers in nucleic acids and proteins carry genetic information. Nucleic acids and proteins are thus the *informational* macromolecules.

3.5 Nucleic Acids

The nucleic acids deoxyribonucleic acid (DNA) and ribonucleic acid (RNA) are macromolecules composed of monomers called *nucleotides*. Therefore, DNA and RNA are **polynucleotides**. As we already know, DNA carries the genetic blueprint for the cell while RNA is the intermediary molecule that converts the blueprint into defined amino acid sequences in proteins (∞ Figure 1.4).

A nucleotide is composed of three components: a five-carbon sugar, either ribose (in RNA) or deoxyribose (in DNA), a nitrogen base, and a molecule of phosphate, PO_4^{3-}. The general structure of nucleotides of DNA and RNA is very similar (Figure 3.8●).

Nucleotides

The nitrogen bases of nucleic acids belong to either of two chemical classes. *Purine* bases—**adenine** and **guanine**—contain two fused heterocyclic rings (rings containing more than one kind of atom). *Pyrimidine* bases—**thymine, cytosine,** and **uracil**—contain a single six-membered heterocyclic ring (Figure 3.9●). Guanine, adenine, and cytosine occur in both DNA and RNA. Thymine is present (with minor exceptions) only in DNA, and uracil is present only in RNA.

Nucleotides consist of a nitrogen base attached to a pentose sugar by a glycosidic linkage between carbon atom 1 of the sugar and a nitrogen atom of the base, either the nitrogen atom labeled 1 (pyrimidine base) or 9 (purine base). Without a phosphate, a base bonded to its

● Figure 3.9   Structure of the bases of DNA and RNA. Note the numbering system of the rings. In attaching the base to the 1′ carbon of the sugar phosphate shown in Figure 3.8, pyrimidine bases are bonded through N–1 of the ring and purine bases through N–9 of the ring.

sugar is referred to as a **nucleoside.** Nucleotides are thus *nucleosides* containing one or more phosphates (Figure 3.10●).

Nucleotides play other roles in the cell besides their major role as constituents of nucleic acids. Nucleotides, especially adenosine triphosphate (ATP) (Figure 3.10), are key sources of chemical energy, releasing sufficient energy during the cleavage of a phosphate bond to drive energy-requiring reactions in the cell (∞ Section 5.8). Other nucleotides or nucleotide derivatives function in oxidation-reduction reactions in the cell (∞ Section 5.7) as carriers of sugars in the biosynthesis of polysaccharides (∞ Section 5.15) and as regulatory molecules inhibiting or stimulating the activities of certain enzymes or metabolic events. However, we discuss here only the role of nucleotides as building blocks of nucleic acids, the major *informational* function of nucleotides.

Nucleic Acids

The nucleic acid backbone is a polymer of alternating sugar and phosphate molecules (Figure 3.11●). Polynu-

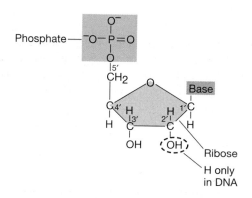

● Figure 3.8   Nucleotides. The numbers on the sugar contain a prime (′) after them because the ring structure in the nitrogen base is also numbered (1, 2, 3, etc.) (see Figure 3.9).

● Figure 3.10 Components of the important nucleotide, adenosine triphosphate. The energy of hydrolysis of a phosphoanhydride bond (shown as squiggles) is greater than that of a phosphate ester and will have significance for bioenergetics in Chapter 5 (∞ Section 5.8). Adenosine (a nucleoside) consists of a sugar bonded to a nitrogenous base but lacks the phosphate group.

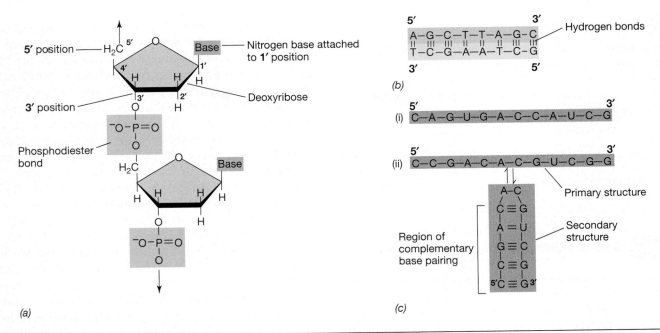

● **Figure 3.11 DNA and RNA.** (a) Structure of part of a DNA chain. The nitrogen bases can be adenine, guanine, cytosine, or thymine. In RNA, an OH group is present on the 2′ carbon of the pentose sugar (see Figure 3.8), and uracil replaces thymine. (b) Simplified structure of DNA in which only the nitrogen bases are shown. Note how the two strands are complementary in base sequence (A = T; G≡C) and bonded by hydrogen bonds. (c) RNA: (i) A sequence showing only primary structure; (ii) A sequence that allows for secondary structure. In RNA, secondary structures form when opportunities for *intra*strand base pairing arise, as shown here. In certain very large RNA molecules, such as ribosomal RNA (∞ Sections 7.15 and 11.5), some parts of the molecule contain only primary structure while others contain both primary and secondary structure. This can lead to highly folded and twisted molecules (∞Figure 11.11c) whose biological function is dependent on their final three–dimensional shape.

cleotides consist of nucleotides covalently bonded via phosphate from carbon 3 [referred to as the 3′ (3 prime) carbon] of one sugar to carbon 5 (5′) of the adjacent sugar (Figure 3.11a). Chemically, the phosphate linkage is a **phosphodiester,** since a single phosphate is connected by ester linkage to two separate sugars (Figure 3.11a).

The *sequence* of nucleotides in a DNA or RNA molecule is referred to as its **primary structure**. As we have discussed, the sequence of bases in a DNA or RNA molecule is informational, encoding the sequence of amino acids in proteins or encoding specific ribosomal or transfer RNAs. The replication of DNA and the synthesis of RNA are key events in the life of a cell (∞ Section 1.2 and Figure 1.4). We will see later that a virtually error-free mechanism is employed to ensure the faithful transfer of genetic traits from one generation to another (∞ Chapter 7).

DNA

In cells, DNA is present in *double-stranded* form. Each chromosome contains two strands of DNA, with each strand containing hundreds of thousands to several million nucleotides linked by phosphodiester bonds. The strands themselves associate with one another by hydrogen bonds that form between the nucleotides of one strand and the complementary nucleotides of the other. When positioned adjacent to one another, purine and pyrimidine bases can undergo hydrogen bonding (see Figure 3.2c).

The most stable hydrogen bonding occurs when guanine (G) forms hydrogen bonds with cytosine (C), and adenine (A) forms hydrogen bonds with thymine (T) (see Figure 3.2c). Specific base pairing, A with T and G with C, means that the two strands of DNA are *complementary* in base sequence. That is, wherever a G is found in one strand, a C is found in the other, and wherever a T is present in one strand, its complementary strand has an A (Figure 3.11b).

RNA

With a few exceptions, all ribonucleic acids are *single-stranded* molecules. However, RNA molecules can fold back upon themselves in regions where complementary base pairing is possible to form folded structures. This pattern of folding in RNA is referred to as its **secondary structure** (Figure 3.11c).

RNA plays three crucial roles in the cell. **Messenger RNA** (mRNA) contains the genetic information of DNA in a single-stranded molecule *complementary* in base sequence to that of DNA. **Transfer RNAs** (tRNAs) are the "adaptor" molecules in protein synthesis. Transfer RNAs convert the genetic information from the language of nucleotides to the language of amino acids, the building blocks of proteins. **Ribosomal RNAs** (rRNAs), of which there are several types, are important structural and catalytic components of the *ribosome*, the protein-synthesizing system of the cell. These various RNA molecules are discussed in detail in Chapters 7 and 11.

3.5 *Concept Check*

The informational content of a nucleic acid is determined by the sequence of nitrogen bases along the polynucleotide chain. Both RNA and DNA are informational macromolecules. RNA can fold into various configurations to obtain secondary structure.

◆ What components are found in a *nucleotide*?

◆ How does a nucleo*side* differ from a nucleo*tide*?

◆ Distinguish between the *primary* and *secondary* structure of RNA.

3.6 Amino Acids and the Peptide Bond

Amino acids are the monomers of **proteins**. Most amino acids consist of carbon, hydrogen, oxygen, and nitrogen only, but 2 of the 21 common amino acids found in cells also contain sulfur, and 1 contains selenium. All amino acids contain two important functional groups, a *carboxylic acid* group ($-COOH$) and an *amino* group ($-NH_2$) (Table 3.1 and Figure 3.12a●). These groups are important because covalent bonds can form between the carboxyl carbon of one amino acid and the amino nitrogen of a second amino acid (with elimination of a molecule of water) to form the **peptide bond** (Figure 3.13●).

All amino acids have the general structure shown in Figure 3.12a. But in addition, each amino acid is unique due to a side group (abbreviated R in Figure 3.12a) attached to the α-carbon. The α-carbon is the carbon atom *immediately adjacent* to the carboxylic acid group. The side chains vary considerably in structure, from as simple as a hydrogen atom in the amino acid glycine to aromatic ringed structures in amino acids such as phenylalanine (Figure 3.12b).

The chemical properties of an amino acid are to a major degree governed by the nature of the side chain, and thus amino acids that show similar chemical properties can be grouped into amino acid "families" as shown in Figure 3.12b. For example, the side chain may itself contain a carboxylic acid group, such as in aspartic acid or glutamic acid, rendering the amino acid *acidic*. Others contain additional amino groups, rendering them *basic*.

(a) **General structure of an amino acid**

(b) **Structure of the amino acid "R" groups**

● **Figure 3.12   Structure of the 21 common amino acids.** (a) General structure. (b) R group structure. The three-letter codes for the amino acids are to the left of the names, and the one-letter codes are in parentheses to the right of the names. See the Microbial Sidebar in Chapter 7 for description of a possible 22nd amino acid.

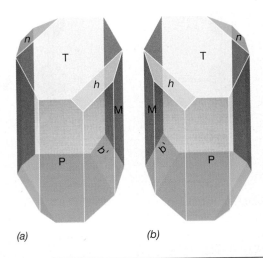

(a)   (b)

● **Figure 3.13 Peptide bond formation.** R_1 and R_2 refer to the variable portion (side chain) of the amino acid (see Figure 3.12). Note how, following peptide bond formation, a free OH group is present for formation of the next peptide bond (◯◯Figure 7.38).

● **Figure 3.14 Louis Pasteur's drawings of tartaric acid ($C_4H_6O_6$) crystals used to illustrate his famous paper on optical activity.** (a) Left-handed crystal (L form). (b) Right-handed crystal (D form). Note that the two crystals are mirror images (that is, they are *enantiomers*). The letters on the faces of the crystals were Pasteur's way of labeling the mirror image faces of the two crystals. Color has been added here to show the mirror image faces more clearly. Pasteur went on to show that the mold *Aspergillus* would metabolize only D-tartrate, emboldening his belief in a link between chemical asymmetry and life.

Alternatively, several amino acids contain hydrophobic side chains and are grouped together as *nonpolar* amino acids. The amino acid cysteine contains a sulfhydryl group ($—SH$), which can connect one chain of amino acids to another by *disulfide linkage* ($R—S—S—R$).

The diversity of chemically distinct amino acids (Figure 3.12*b*) makes it possible for cells to produce an enormous number of unique proteins with widely different biochemical properties. For instance, a cell of *Escherichia coli* contains almost 2000 different kinds of proteins (Table 3.2). These include soluble and membrane-integrated enzymes, structural proteins, transport proteins, sensor proteins, and many others. The *function* of a protein is in large part dictated by its *structure*. The converse is also true. Proteins of a particular functional class often show structural relatedness.

Isomers

Two molecules may have the same molecular formula but exist in different structural forms. These related but not identical molecules are called **isomers**. Louis Pasteur, the famous early microbiologist who crushed the theory of spontaneous generation (◯◯ Section 1.5), began his scientific career as a chemist studying a class of isomers called *optical isomers*. The asymmetry that Pasteur first discovered in tartaric acid crystals (Figure 3.14●) was the foundation for his later work that showed that living organisms could produce optically active molecules (such as amino acids and sugars) as well. (A molecule is optically active if a pure solution or crystal of the molecule defracts light in only one direction.)

Isomers are important in cell structure. For example, many isomers of common sugars are found as constituents of the cell walls of *Bacteria* and *Archaea* (◯◯Section 4.8). Isomers that contain the same molecular and structural formulas, except that one is a "mirror image" of the other, just as the left hand is a mirror

image of the right, are called **enantiomers,** and are given the designations D and L (Figure 3.15*b*●). Sugars of the D isomer predominate in biological systems.

Amino acids also exist as D or L enantiomers. However, in proteins, cells employ the L-amino acid rather than the D form (Figure 3.15*c*). Nevertheless, D-amino acids are occasionally found in cells, most notably in the cell wall polymer *peptidoglycan* (◯◯ Section 4.8) and in certain peptide antibiotics (◯◯ Section 20.9). Cells can interconvert many enantiomers by way of enzymes, called *racemases*, that specifically catalyze the necessary transformation. For instance, some prokaryotes can grow on L-sugars or D-amino acids because they can convert these forms into the opposite enantiomer.

3.6 Concept Check

Twenty-one common amino acids are found in cells and can bond to each other via the *peptide bond*. Mirror image (enantiomeric) forms of sugars and amino acids exist, but only one optical isomer of each is found in most cell polysaccharides and proteins, respectively.

◆ Why can it be said that all amino acids are structurally similar yet different at the same time?

◆ Draw the complete structure of a dipeptide containing the amino acids alanine and tyrosine. Outline the peptide bond.

◆ What enantiomeric form of sugars and amino acids are commonly found in living organisms? Why doesn't the amino acid *glycine* have different enantiomers?

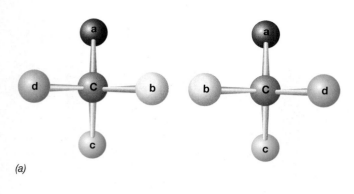

(a)

D-Glucose   **L-Glucose**   **L-Alanine**   **D-Alanine**

(b)

Planar projection

Three-dimensional projection

(c)

● **Figure 3.15   Isomers.** (a) Ball-and-stick model showing mirror images. (b) Enantiomers of glucose. (c) Enantiomers of the amino acid alanine. Note that no matter how the three-dimensional views are rotated, the L and D forms can never be superimposed. In the three-dimensional projection the arrow should be understood as coming *toward* the viewer whereas the dashed line indicates a plane *away* from the viewer.

3.7 Proteins: Primary and Secondary Structure

Proteins play key roles in cell function. In essence, a cell is what it is and does what it does because of the kinds and amounts of proteins it contains. That is, every different type of cell will have a different complement of proteins. An understanding of protein structure is therefore essential for understanding cells of all types.

Two major classes of proteins are *catalytic* proteins (*enzymes*) and *structural* proteins. **Enzymes** are the catalysts for the wide variety of chemical reactions that occur in cells (∞Chapters 5 and 17). By contrast, structural proteins become integral parts of the cell structures that make up membranes, walls, and cytoplasmic components. However, all proteins show certain structural features, and we discuss these now.

Primary Structure

Proteins are polymers of amino acids covalently bonded by peptide bonds (Figure 3.13). Two amino acids

bonded by peptide linkage constitute a *dipeptide*, three amino acids a *tripeptide*, and so on. When *many* amino acids are covalently linked via peptide bonds, they form a **polypeptide**.

Proteins consist of one or more polypeptides. The number of amino acids in a protein varies from one protein to another. Proteins with as few as 15 and as many as 10,000 amino acids are known. Since proteins may vary in their composition, sequence, and number of amino acids, it is easy to see that enormous variation in protein structure (and thus function) is possible.

The linear array of amino acids in a polypeptide is called its **primary structure**. The primary structure of a polypeptide is very important, because a given primary structure is consistent with only certain types of folding patterns. And it is only the final, folded polypeptide that assumes biological activity.

Secondary Structure

Interactions of the R groups on the amino acids in a polypeptide force the molecule to twist and fold in a specific way. This leads to formation of **secondary structure** (Figure 3.16●). Hydrogen bonds, the weak noncovalent linkages discussed earlier (see Section 3.1), play important roles in polypeptide secondary structure. One common type of secondary structure is the *α-helix*. To envision an *α*-helix, imagine a linear polypeptide wound around a cylinder (Figure 3.16a). In this twisted structure, oxygen and nitrogen atoms from different amino acids become positioned close enough to allow for hydrogen bonding. This opportunity for hydrogen bonding gives the *α*-helix its inherent stability (Figure 3.16a).

The primary structure of some polypeptides allows for a different type of secondary structure, called a *β-sheet*. In the *β*-sheet, the chain of amino acids in the polypeptide folds back and forth upon itself instead of forming a helix. However, as in the *α*-helix, the folding in a *β*-sheet exposes hydrogen atoms that can undergo hydrogen bonding (Figure 3.16b). Typically, a *β*-sheet secondary structure yields a rather rigid polypeptide while *α*-helical secondary structures are more flexible. Thus, an enzyme, for example, whose activity may depend on it being rather flexible, may contain a high degree of *α*-helix secondary structure. By contrast, a structural protein that functions in cellular scaffolding may contain large regions of *β*-sheet secondary structure.

Many polypeptides contain regions of *α*-helix and regions of *β*-sheet secondary structure, the type of folding and its location in the molecule being determined by the available opportunities for hydrogen bonding and hydrophobic interactions (see Figure 3.17). These structural regions, called *domains*, are typically segments of the polypeptide that have specific functions in the final protein molecule.

(a) Insulin   (b) Ribonuclease

● **Figure 3.17 Tertiary structure of polypeptides showing where regions of α-helix or β-sheet secondary structure are located.** (a) *Insulin*, a protein containing two polypeptide chains (∞ Section 31.6); note how the B chain contains both α-helix and β-sheet secondary structure and how disulfide linkages (shown in blue) help in dictating folding patterns (tertiary structure). (b) *Ribonuclease*, a large protein with several regions of α-helix and β-sheet secondary structure.

Hydrogen bonds between nearby amino acids

(a)

Hydrogen bonds between distant amino acids

(b)

● **Figure 3.16 Secondary structure of polypeptides.** (a) *α*-Helix secondary structure. Note that hydrogen bonding does not involve the R groups but instead occurs between atoms in the peptide bonds. (b) *β*-Sheet secondary structure.

<table>
<tr><td>3.8</td><td>**Proteins: Higher Order Structure and Denaturation**</td></tr>
</table>

Once a polypeptide has achieved secondary structure it will continue to fold to form an even more stable molecule. This folding leads to formation of a unique three-dimensional shape, called the *tertiary structure* of the protein.

Like secondary structure, **tertiary structure** is ultimately determined by primary structure. However, tertiary structure is also governed to some extent by the secondary structure of the molecule, because the side chain of each amino acid in the polypeptide is positioned in a specific way (Figure 3.16). If additional hydrogen bonds, covalent bonds, hydrophobic interactions, or other atomic interactions are able to form, the polypeptide will fold to accommodate them (Figure 3.17●). The tertiary folding of the polypeptide ultimately forms exposed regions or grooves in the molecule (Figures 3.17 and 3.18●) that are important for binding other molecules (for example, in the binding of a substrate to an enzyme or the binding of DNA to a specific regulatory protein) (∞ Sections 5.5 and 8.4).

Frequently a polypeptide folds in such a way that adjacent sulfhydryl groups of cysteine residues are exposed. These free —SH groups can join covalently to form a *disulfide bond* between the two amino acids. If the two cysteine residues are located in different polypeptides in a protein, the disulfide bond physically links the two molecules (Figure 3.17a). In addition, a single polypeptide can spontaneously fold and bond to itself if a disulfide bond can form within the molecule.

α Chains

β Chains

(a)   (b)

● **Figure 3.18   Quaternary structure of human hemoglobin.** (a) There are two *kinds* of polypeptide in hemoglobin, α chains (shown in blue and red) and β chains (shown in orange and yellow), but a total of four polypeptides in the final protein molecule (α and β refer here to chain names, not to polypeptide secondary structures). Separate colors are used to distinguish the four distinct chains. (b) Molecular structure of human hemoglobin as determined by X-ray crystallography. In this view each chain (α and β) is in its own color.

If a protein consists of two or more polypeptides, and many proteins do, the number and type of polypeptides that form the final protein molecule are referred to as its **quaternary structure** (Figure 3.18). In proteins showing quaternary structure, each polypeptide, called a *subunit*, contains primary, secondary, and tertiary structure. Some proteins contain multiple copies of a single subunit. A protein containing two identical subunits, for example, would be called a *homodimer*. Other proteins may contain nonidentical subunits, each present in one or more copies (a *heterodimer*, for example, contains one copy each of two different polypeptides). The subunits in multisubunit proteins are held together by noncovalent interactions (hydrogen bonding, van der Waals forces, and hydrophobic interactions) or by covalent linkages, typically intersubunit disulfide bonds.

Gentle denaturation; urea   **Active protein**   Harsh denaturation; 100°C

Inactive   Inactive

Remove urea; reactivation   Cool

Active protein   Inactive

● **Figure 3.19   Denaturation of the protein ribonuclease.** The structure of ribonuclease was discussed in Figure 3.17b. Note how harsh denaturation yields a permanently destroyed molecule (from the standpoint of biological function) because of improper folding.

Denaturation

When proteins are exposed to extremes of heat or pH or to certain chemicals or metals that affect their folding, they may undergo **denaturation** (Figure 3.19●). Denaturation causes the polypeptide chain to unfold, destroying the higher order (secondary, tertiary, and quaternary, if relevant) structure of the molecule. Depending on the severity of the denaturing conditions, refolding of the polypeptide may occur after removal of the denaturant (Figure 3.19).

Typically, the biological properties of a protein are lost when it is denatured. Peptide bonds (Figure 3.13) are unaffected, however, and so a denatured molecule *retains its primary structure*. This shows that biological activity is not inherent in the primary structure of a protein, but instead is a function of the uniquely folded form of the molecule as ultimately directed by primary structure. In other words, folding of a polypeptide confers upon it a unique shape that is compatible with a *specific* biological function.

Denaturation of proteins is of more than just academic interest, since this process is a major means of destroying microorganisms. For example, alcohols such as phenol and ethanol are effective disinfectants because they readily penetrate cells and irreversibly denature their proteins. Such chemical agents are thus useful for disinfecting inanimate objects, such as surfaces, and have enormous practical value in household, hospital, and industrial disinfectant applications (�Chapter 20).

The Road Ahead

Now that we have reviewed the chemistry of life, we can better understand the structural relationships of cells. In the next chapter we will see how the different macromolecules come together to form major structures of the cell, including the membrane, wall, and flagellum. From there we will consider the metabolic properties of cells in Chapter 5. Metabolism, the *machine* function of a cell, drives the biosynthesis and assembly of macromolecules, the *coding* function of a cell (∞ Section 1.2 and Figure 1.4). The combination of these processes results in cell growth (Chapter 6). Cell growth results in *cell populations*, and groups of populations form *microbial communities*. It is at the community level that microorganisms typically have major ecological effects in their habitats.

As we travel through this book, it may be useful to return to Chapter 3 periodically to reinforce the basic principles surrounding the chemistry of life. Although collectively, microorganisms have evolved a nearly limitless chemical diversity of macromolecules, this diversity simply boils down to variations on the four classes of macromolecules discussed in this chapter. And as any successful microbiologist knows, an understanding of the biochemistry of proteins, lipids, nucleic acids, and polysaccharides is essential and pays big dividends in grasping both the basic and more advanced principles of microbiology.

 3.7 and 3.8   Concept Check

The primary structure of a protein is determined by its amino acid sequence, but it is the folding (higher order structure) of the polypeptide that determines how the protein functions in the cell.

◆ Define the terms *primary*, *secondary*, and *tertiary* with respect to protein structure.

◆ How does a *polypeptide* differ from a *protein*?

◆ What secondary structural features tend to make β-sheet proteins more rigid than α-helices?

◆ Describe the *number* and *kinds* of polypeptides present in a homotetrameric protein.

◆ Describe the structural and biological effects of the denaturation of a protein. Of what *practical* value is knowledge of protein denaturation?

REVIEW QUESTIONS

1. Which are the major elements found in living organisms? Why are oxygen and hydrogen particularly abundant in living organisms (∞ Section 3.1)?

2. Define the word *molecule*. How many atoms are in a molecule of hydrogen gas? In a molecule of glucose (∞ Sections 3.1 and 3.3)?

3. Refer to the structure of the nitrogen base *cytosine* shown in Figure 3.1. Draw this structure and then label the positions of all single bonds and double bonds in the cytosine molecule (∞ Section 3.1).

4. Compare and contrast the words *monomer* and *polymer*. Give three examples of biologically important polymers and list the monomers of which they are composed. Which classes of macromolecules are most abundant (by weight) in a cell (∞ Sections 3.1 and 3.2)?

5. List the components that would make up a simple lipid. How does a triglyceride differ from a complex lipid (∞ Section 3.4)?

6. Examine the structures of the triglyceride and of phosphatidyl ethanolamine shown in Figure 3.7. How might the substitution of phosphate and ethanolamine for a fatty acid alter the chemical properties of the lipid (∞ Section 3.4)?

7. RNA and DNA are similar types of macromolecules but show distinct differences as well. List three ways in which RNA differs chemically or physically from DNA. What is the cellular function of DNA and RNA (∞ Section 3.5)?

8. Why are *amino acids* so named? Write a general structure for an amino acid. What is the importance of the R group to final protein structure? Why does the amino acid *cysteine* have special significance for protein structure (∞ Section 3.6)?

9. Chemically, what type of reaction between two amino acids leads to formation of the peptide bond (∞ Section 3.6)?

10. Define the terms *primary*, *secondary*, *tertiary*, and *quaternary* as they apply to proteins. Which of these is (are) affected by the denaturation process (∞ Sections 3.7 and 3.8)?

11. Fill in the blanks. A *glycosidic* bond is to a _____ as a _____ bond is to a polypeptide, and a _____ is to a nucleic acid. All of these bonds are examples of _____ bonds, which are chemically much stronger than weak bonds, such as _____, _____, and _____ (∞ Chapter 3).

APPLICATION QUESTIONS

1. Observe the following nucleotide sequences of RNA: (a) GUCAAAGAC, (b) ACGAUAACC. Can either of these RNA molecules have secondary structure? If so, draw the potential secondary structure(s).

2. A few soluble (cytoplasmic) proteins contain a high content of hydrophobic amino acids. How would you predict these proteins would fold as to their tertiary structure and why?

3. Cells of the genus *Halobacterium*, an archaeon that lives in very salty environments, contain over 5 molar (M) potassium (K^+). Because of this high K^+ content, many cytoplasmic proteins of *Halobacterium* cells are enriched in two specific amino acids that are present in much higher proportions in *Halobacterium* proteins than in functionally similar proteins from *Escherichia coli* (which has only very low levels of K^+ in its cytoplasm). Which amino acids are enriched in *Halobacterium* proteins and why? (*Hint:* Which amino acids could best neutralize the *positive* charges due to K^+?)

4. When a culture of the bacterium *Escherichia coli*, an inhabitant of the human gut, is placed in a beaker of boiling water, significant changes in the cells occur almost immediately. However, when a culture of *Pyrodictium*, a hyperthermophilic bacterium that grows optimally in boiling hot springs is put in the same beaker, similar changes do not occur. Explain.

5. Review Figure 3.6b and then describe the differences that make each of these polymers unique. If all of the glycosidic bonds in these polymers were hydrolyzed, what single molecule would remain?

6. Review Figure 3.12b. Of all the amino acids shown in blue, what is it about their chemistry that unites them as a "family"?

4

CELL STRUCTURE/ FUNCTION

Prokaryotic cells are typically very small, but are occasionally very large. Shown is a photomicrograph of a cell of *Epulopiscium*, a prokaryote, aside four cells of the eukaryote *Paramecium*.

WORKING GLOSSARY

ABC (ATP-Binding Cassette) transporter a membrane transport system consisting of three proteins, one of which hydrolyzes ATP, to transport specific nutrients into the cell

Capsule a polysaccharide or protein outermost layer, usually rather slimy, present on some bacteria

Chemotaxis directed movement of an organism toward (*positive*) or away from (*negative*) a chemical gradient

Cytoplasmic membrane the permeability barrier of the cell, separating the cytoplasm from the environment

Endospore a highly heat-resistant, thick-walled, differentiated structure produced by certain gram-positive *Bacteria*

Flagellum a long, thin cellular appendage capable of rotation in prokaryotic cells and responsible for swimming motility

Gas vesicles gas-filled cytoplasmic structures bounded by protein and conferring buoyancy on cells

Gram-negative a prokaryotic cell whose cell wall contains small amounts of peptidoglycan, and an outer membrane, containing lipopolysaccharide, lipoprotein, and other complex macromolecules

Gram-positive a prokaryotic cell whose cell wall consists chiefly of peptidoglycan and lacks the outer membrane of gram-negative cells

Group translocation an energy-dependent transport process in which the substance transported is chemically modified during the transport process

Lipopolysaccharide (LPS) lipid in combination with polysaccharide and protein, which forms the major portion of the outer membrane in gram-negative *Bacteria*

Magnetosomes particles of magnetite (Fe_3O_4) organized into nonunit membrane-enclosed structures in the cytoplasm of magnetotactic *Bacteria*

Morphology the *shape* of a cell—rod, cocus, spirillum, and the like

Outer membrane a phospholipid and polysaccharide containing unit membrane that lies external to the peptidoglycan layer in cells of gram-negative *Bacteria*

Peptidoglycan a polysaccharide composed of alternating repeats of acetylglucosamine and acetylmuramic acid arranged in adjacent layers and cross-linked by short peptides

Periplasm a gel-like region between the outer surface of the cytoplasmic membrane and the inner surface of the lipopolysaccharide layer of gram-negative *Bacteria*

Peritrichous a pattern of flagellation where flagella are located in many places around the surface of the cell

Phototaxis movement of an organism toward light

Polar in reference to flagellation, having flagella emanating from one or both poles of the cell

Poly-β-hydroxybutyrate (PHB) a common storage material of prokaryotic cells consisting of a polymer of β-hydroxybutyrate or another β-alkanoic acid or mixtures of β-alkanoic acids

Protoplast an osmotically protected cell whose cell wall has been removed

Resolution the ability to distinguish two objects as distinct and separate

S-layer an outermost cell surface layer composed of protein or glycoprotein present on some *Bacteria* and *Archaea*

Sterols hydrophobic heterocyclic ringed molecules that strengthen the cytoplasmic membrane of eukaryotic cells and a few prokaryotes

MICROSCOPY AND CELL MORPHOLOGY

In Chapter 2 we introduced the concept of prokaryotic and eukaryotic cells. In this chapter we examine the structure of the prokaryotic cell, reserving discussion of eukaryotic cell structure for Chapter 14.

The basic design of all cells, *Bacteria*, *Archaea*, and *Eukarya*, share much in common. Like houses, cells are built by connecting building blocks to create more complex structures. Monomers are used to construct macromolecules, and these form into complexes to yield structures with defined functions—ribosomes, membranes, cell walls, and the like. We will see that despite great diversity in the *chemistry* of these macromolecules, cells face similar architectual problems and solve them in common ways.

We begin this chapter with a consideration of microscopy and then look at cell membranes and walls, cell surface structures and inclusions, and locomotion, respectively. We begin with microscopy because, historically, it was the microscope that first revealed the secrets of cell structure, and even today it remains a powerful tool in microbiology.

4.1 Light Microscopy

Visualization of microorganisms requires use of either the *light microscope* or the *electron microscope*. In general, light microscopes are used to look at intact cells at low magnification, while electron microscopes are used to look at internal structure or the details of cell surfaces.

All microscopes employ lenses to magnify the image of a cell such that details of its structure are more apparent. In addition to magnification, however, is the concept of **resolution**, the ability to distinguish two adjacent objects as distinct and separate. Although magnification can be increased virtually without limit, resolution cannot; resolution is dictated by the physical properties of light. It is thus *resolution* and not *magnification* that ultimately dictates what we are able to see with a microscope.

We begin our discussion with the light microscope, for which the limits of resolution are about $0.2\ \mu m$ [0.2 micrometer or 200 nanometers (nm)]. We then proceed to describe the electron microscope, for which resolution is improved over that of the light microscope by about 1000-fold, allowing the examination of individual molecules.

The Compound Light Microscope

The **light microscope** uses visible light to illuminate cell structures. Several types of light microscopes are commonly used in microbiology: *bright-field, phase-contrast, dark-field,* and *fluorescence.*

With the **bright-field microscope** specimens are visualized because of the differences in contrast (density) that exist between them and the surrounding medium. Contrast differences arise because cells absorb or scatter light to varying degrees. The bright-field microscope is most commonly used in laboratory courses in biology and microbiology and consists of two series of lenses (objective lens and ocular lens), which function together to reveal the image (Figure 4.1●). Many bacterial cells are difficult to see well with the bright-field microscope because of their lack of contrast with the surrounding medium. Pigmented organisms are an exception, because the color of the organism adds contrast, thus improving visualization of the cells (Figure 4.2●).

Magnification and Resolution

The total magnification of a compound microscope is the *product* of the magnification of its objective and ocular lenses (Figure 4.1*b*). Magnifications of about 1500× are the upper limit with a compound light microscope. Above this limit, resolution does not improve. Resolution is a function of the wavelength of light used and a char-acteristic of the objective lens known as its *numerical aperture* (a measure of light-gathering ability). In general, there is a correspondence between the magnification of a lens and its numerical aperture: Lenses with higher magnification typically have higher numerical apertures (the numerical aperture of a lens is stamped on the lens alongside the magnification). The diameter of the smallest object resolvable by any lens is equal to 0.5λ/numerical aperture, where λ is the wavelength of light used. Based on this formula, resolution is greatest when *blue light* is used to illuminate a specimen and the objective that is used has a *very high* numerical aperture.

As mentioned, the highest resolution possible in a compound light microscope is about 0.2 μm. This means that two objects closer together than 0.2 μm are not resolvable as distinct and separate. Most microscopes used in microbiology have oculars that magnify 10–15× and objectives of 10–100× (Figure 4.1*b*). At 1000×, objects 0.2 μm in diameter can then just be resolved. With the 100× objective, and with certain other objectives of very high numerical aperture, a high-grade optical oil is placed between the specimen and the objective. Lenses on which oil is used are called *oil-immersion lenses.* Immersion oil increases the light-gathering ability of a lens by allowing rays emerging from the specimen at angles that would otherwise be lost to the objective lens to be collected and viewed.

Ocular —
Objective —
Stage —
Condenser —
Focusing knobs —
Light —

ZEISS

Carl Zeiss, Inc.

(a)

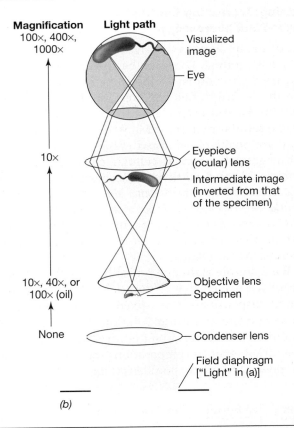

Magnification **Light path**

100×, 400×, 1000× — Visualized image
— Eye

10× — Eyepiece (ocular) lens
— Intermediate image (inverted from that of the specimen)

10×, 40×, or 100× (oil) — Objective lens
— Specimen

None — Condenser lens
— Field diaphragm ["Light" in (a)]

(b)

● **Figure 4.1 Microscopy.** (a) A compound light microscope. Various key parts of the microscope are labeled. (b) Path of light through a compound light microscope. Besides 10×, eyepieces (oculars) are available in 15–30×.

● **Figure 4.2** **Photomicrographs of pigmented microorganisms by bright-field microscopy.** (a) A green alga (eukaryote). (b) A purple phototrophic bacterium (prokaryote). The algal cells are about 15 μm in diameter, and the bacterial cells are about 5 μm in diameter.

Staining: Increasing Contrast for Bright-Field Microscopy

One of the limitations of bright-field microscopy is insufficient contrast. Dyes can be used to stain cells and increase their contrast so that they can be more easily seen in the bright-field microscope. Dyes are organic compounds, and each class of dye has an affinity for specific cellular materials. Many dyes used in microbiology are positively charged (*basic dyes*) and combine with negatively charged cellular constituents such as nucleic acids and acidic polysaccharides. Examples of basic dyes include *methylene blue, crystal violet*, and *safranin*. Because cell surfaces also tend to be negatively charged, these dyes combine with high affinity to structures on the surfaces of cells and hence are excellent general-purpose stains.

For a **simple stain** one typically uses dried preparations of cell suspensions (Figure 4.3●). A slide containing a dried suspension of heat-fixed cells is flooded for a minute or two with a dilute solution of a dye, rinsed several times in water, and blotted dry. It is typical to observe dried stained preparations of bacteria with a high-power (oil-immersion) lens (Figure 4.3).

Differential Stains: The Gram Stain

Differential stains are so named because they do not stain all kinds of cells the same color. An important differential staining procedure widely used in microbiology is the **Gram stain** (Figure 4.4*a*●). On the basis of their reac-

I. Preparing a smear
Spread culture in thin film over slide → Dry in air

II. Heat fixing and staining
Pass slide through flame to fix → Flood slide with stain; rinse and dry

III. Microscopy
Place drop of oil on slide; examine with 100× objective — 100×, Slide, Oil

● **Figure 4.3** **Staining cells for microscopic observation.** Stains improve the contrast between cells and their background.

tion to the Gram stain, bacteria can be divided into two major groups: *gram-positive* and *gram-negative*. After Gram staining, **gram-positive** bacteria appear purple and **gram-negative** bacteria appear red (Figure 4.4*b*). This difference in reaction to the Gram stain arises because of differences in the cell wall structure of gram-positive and gram-negative cells (as discussed later in this chapter). This leads to ethanol decolorizing gram-negative, but not gram-positive, cells (Figure 4.4).

The Gram stain is one of the most useful staining procedures in microbiology. Typically, one begins the characterization of a new bacterium by determining whether it is gram-positive or gram-negative. If a fluorescent microscope is available (see later discussion of fluorescence microscopy), the Gram stain can be reduced to a one-step procedure where gram-positive and gram-negative cells fluoresce different colors (Figure 4.4*c*).

Phase-Contrast, Dark-Field, and Fluorescence Microscopy

The **phase-contrast microscope** was developed to improve contrast differences between cells and the surrounding medium, making it possible to see cells without staining them (Figure 4.5●). The phase-contrast microscope is widely employed in research because it can be used to observe wet-mount (living) preparations.

The Gram Stain

Step 1

Flood the heat-fixed smear with crystal violet for 1 min

Result:
All cells purple

Step 2

Add iodine solution for 1 min

Result:
All cells remain purple

Step 3

Decolorize with alcohol briefly — about 20 sec

Result:
Gram-positive cells are purple; gram-negative cells are colorless

Step 4

G⁻

Counterstain with safranin for 1–2 min

Result:
Gram-positive (G⁺) cells are purple; gram-negative (G⁻) cells are pink to red

G⁺

(a)

(b)

Leon J. Lebeau

Molecular Probes, Inc., Eugene, Oregon

(c)

Staining, on the other hand, although a widely used procedure in light microscopy, kills cells and can distort their features.

Phase-contrast microscopy was discovered by the Dutch mathematical physicist Frits Zernike in 1936. It is based on the principle that cells slow the speed of light passing through them and thus differ in *refractive index* from their surroundings. This results in a difference in "phase" between the cell itself and its surroundings, and this subtle difference is amplified by a special ring in the objective lens of a phase-contrast microscope. This leads to the formation of a dark image on a light background (Figure 4.5*b*). The ring consists of a "phase plate," the key discovery of Zernike, that amplifies this minute variation in phase. Zernike's discovery of differences in contrast between cells and background stimulated later innovations in microscopy, such as fluorescence and confocal microscopy (see later), where, like phase-contrast, differences in contrast form the heart of the technique. For his discovery of phase-contrast microscopy, Zernike was awarded the 1953 Nobel Prize in Physics.

The **dark-field microscope** is a light microscope in which the lighting system has been modified to reach the specimen from the sides only. The only light reaching the lens is scattered by the specimen, and thus the specimen appears light on a dark background (Figure 4.5*c*). Resolution by dark-field microscopy is somewhat better than by light microscopy, and thus objects can often be resolved by dark-field that are not resolvable in bright-field or even phase-contrast microscopes. Dark-field microscopy is also an excellent way to observe the motility of microorganisms, as bundles of flagella are often resolvable with this technique (see Figure 4.55*a*).

The **fluorescence microscope** is used to visualize specimens that *fluoresce*, that is, emit light of one color when light of another color shines upon them (Figure 4.6●). Fluorescence occurs either because of the presence within cells of naturally fluorescent substances such as chlorophyll or other fluorescing components (*autofluorescence*) (see Figure 4.6*a, b*), or because the cells have been treated with a fluorescent dye (Figures 4.4*c* and 4.6*c*). DAPI (diamidino-2-phenylindole) is a widely used fluorescent dye, staining cells a bright blue (∞Figure 18.6). Using DAPI, cells can be identified in a complex milieu, such as soil, water, food, or a clinical specimen (∞Section 18.3). Fluorescence microscopy is widely used in clinical diagnostic microbiology and also in microbial ecology (∞Chapters 18–20 and 24).

● **Figure 4.4 The Gram stain.** (a) Steps in the Gram stain procedure. (b) Photomicrograph of Gram-stained *Bacteria* that are gram-positive (blue-purple) and gram-negative (pink-red). The species are *Staphylococcus aureus* and *Escherichia coli*, respectively. (c) Photomicrograph of cells of *Pseudomonas aeruginosa* (gram-negative, green) and *Bacillus cereus* (gram-positive, orange) stained with a one-step fluorescent staining method. This method allows for differentiating gram-positive from gram-negative cells in a single staining step.

(a) M.T. Madigan *(b)* M.T. Madigan *(c)* M.T. Madigan

● **Figure 4.5** **Photomicrographs of the same field of cells of the baker's yeast *Saccharomyces cerevisiae*, taken by different types of light microscopy.** (a) Bright-field. (b) Phase contrast. (c) Dark-field. Cells average 8–10 μm in diameter.

 4.1 *Concept Check*

Microscopes are essential for microbiological studies. Various types of light microscopes exist, including bright-field, dark-field, phase contrast, and fluorescence microscopes. In bright-field microscopy, stains are necessary to increase contrast.

◆ Define the term *resolution*.

◆ What is the upper limit of magnification for a light microscope?

◆ What light microscopic techniques can sometimes improve resolution?

◆ What color would a gram-negative bacterium be after Gram staining by the conventional method?

4.2 Three-Dimensional Imaging: Interference Contrast, Atomic Force, and Confocal Scanning Laser Microscopy

In the types of light microscopy just considered, the images obtained are essentially two dimensional. How can this limitation be overcome? We will see in the next section that the scanning electron microscope offers one solution to this problem, but so can certain forms of light microscopy.

Differential Interference Contrast Microscopy

Differential interference contrast (DIC) is a type of light microscopy that employs a polarizer to produce polarized light. The polarized light then passes through a prism that generates two distinct beams. These beams traverse the specimen and enter the objective lens where they are recombined into one. Because the two beams pass through different substances with slightly different

(a) R.W. Castenholz *(b)* R.W. Castenholz *(c)* T.D. Brock

● **Figure 4.6** **Photomicrographs of various microorganisms as visualized by fluorescence microscopy.** (a, b) Cyanobacteria. (a) Cells observed by bright-field microscopy. (b) Same cells observed by fluorescence (cells exposed to light of 546 nm). The red color is due to autofluorescence of chlorophyll *a* and other pigments. (c) Cells of the filamentous bacterium *Leucothrix mucor* stained with the fluorescent dye, acridine orange, which fluoresces green. Cells are 3 μm in diameter and may reach lengths of greater than 100 μm.

refractive indices, the combined beams are not totally in phase but instead create an interference effect. This effect intensifies subtle differences in cell structure. Thus, by DIC microscopy, structures such as the nucleus of eukaryotic cells (Figure 4.7a●), and endospores, vacuoles, and granules of prokaryotic cells, attain a three-dimensional appearance. DIC microscopy is also particularly useful for observing *unstained* cells because it can reveal internal cell structures that are less apparent (or even invisible) by bright-field techniques (compare Figure 4.5a with Figure 4.7a).

Atomic Force Microscopy

Another type of microscopy useful for three-dimensional imaging of biological structures is the **atomic force microscope (AFM)**. In atomic force microscopy, a tiny stylus is positioned extremely close to the specimen such that weak repulsive atomic forces are established between the probe and the specimen. As the specimen is scanned in both the horizontal and vertical directions, the stylus rides up and down the hills and valleys, constantly recording its interactions with the surface. This pattern is processed by a series of detectors that feed the digital information into a computer that generates an image (Figure 4.7b).

Although the images obtained from an atomic force microscope appear similar to those from the scanning electron microscope (compare Figure 4.7b with Figure 4.10b), the AFM has one advantage—no fixatives or coatings are required. The AFM thus allows living and hydrated specimens to be viewed, something that is generally not possible with electron microscopes.

Confocal Scanning Laser Microscopy

Confocal scanning laser microscopy (CSLM) is a computerized microscope that couples a laser source to a light microscope. This technique allows for the generation of three-dimensional digital images of microorganisms and other biological specimens (Figure 4.8●).

In CSLM, a laser beam is bounced off a mirror that directs the beam through a scanning device. Then the laser beam is directed through a pinhole that precisely adjusts the plane of focus of the beam to a given vertical layer within a specimen. By precisely illuminating only a single plane of the specimen, illumination intensity drops off rapidly above and below the plane of focus. Because of this, stray light from other planes of focus is minimized. Thus, in a relatively thick specimen such as a microbial biofilm (Figure 4.8a), not only are cells on the surface of the biofilm apparent, as would be the case with conventional light microscopy, but cells in the various layers can also be observed by adjusting the plane of focus of the laser beam.

Cells in CSLM preparations are frequently stained with fluorescent dyes to make them more distinct

(a)

Linda Barnett and James Barnett

(b)

Suzanne Kelly

● **Figure 4.7 Three-dimensional imaging of cells.** (a) Interference contrast microscopy and (b) atomic force microscopy. The yeast cells in (a) are about 8 μm in diameter. Note how the nucleus is clearly visible here (compare Figure 4.7a with Figure 4.5a). The bacterial cells in (b) are about 2.2 μm in length, and the micrograph was taken from a natural biofilm that developed on the surface of a glass slide immersed for 24 h in a dog's water bowl. The slide was air dried before viewing with an atomic force microscope.

(Figure 4.8a). Alternatively, false color images can be generated by adjusting the microscope in such a way as to make different layers take on different colors. The laser confocal microscope is equipped with computer software to assemble digital images for subsequent image processing. Thus, images obtained from different layers can be stored and then digitally overlaid to reconstruct a three-dimensional image of the entire specimen (Figure 4.8a).

CSLM has found widespread use in microbial ecology, especially for identifying phylogenetically distinct populations of cells present in a microbial habitat (see for example, Figure 18.11b), or for resolving the different layered components as in a biofilm (Figure 4.8a; ∞Figure 19.4b). However, CSLM is useful anywhere thick specimens need to be examined to see how their microbial content varies with depth.

(a)

(b)

● **Figure 4.8   Confocal scanning laser microscopy.**   (a) Confocal image of a mixed microbial biofilm community cultivated in the laboratory. The green, rod-shaped cells are *Pseudomonas aeruginosa* experimentally introduced into the biofilm. Other cells that are different colors are present at different depths in the biofilm. (b) Confocal micrograph of a filamentous cyanobacterium growing in a soda lake.

 4.2   Concept Check

Interference contrast (DIC) and confocal scanning (CSLM) are forms of light microscopy that allow for greater three-dimensional imaging than other forms of light microscopy, and confocal microscopy allows imaging through thick specimens. The atomic force microscope yields a detailed three-dimensional image of live preparations.

◆   What structure in eukaryotic cells is more easily seen using DIC than bright-field microscopy? (Hint: compare Figures 4.5*a* and 4.7*a*).

◆   How is CSLM able to view different layers in a thick preparation?

4.3   Electron Microscopy

Electron microscopes use electrons instead of photons to image cells or cell structures. In the **transmission electron microscope (TEM)** electromagnets function as lenses, and

the whole system operates in a vacuum (Figure 4.9●). Electron microscopes are fitted with cameras to allow a photograph, called an *electron micrograph*, to be taken.

The transmission electron microscope is typically used to examine internal cell structure. The resolving power of the electron microscope is much greater than that of the light microscope, enabling one to view structures at the molecular level (∞Figure 2.4*b*). For example, whereas the resolving power of a good light microscope is about 0.2 *micrometers*, the resolving power of a good TEM is about 0.2 *nanometers*. Thus, individual molecules, such as proteins and nucleic acids, can be visualized in the electron microscope.

Unlike visible light, electron beams do not penetrate very well; even a single cell is too thick to be viewed directly. Consequently, special techniques of *thin sectioning* are needed to prepare specimens for the electron microscope. A single bacterial cell, for instance, is cut into many very thin (20–60 nm) slices, which are then examined individually with the electron microscope (Figure 4.10*a*●). To obtain sufficient contrast, the preparations are treated with stains such as osmic acid, or permanganate, uranium, lanthanum, or lead salts. Because these substances are composed of atoms of high atomic weight, they scatter electrons well and thus improve contrast (Figure 4.10*a*).

Scanning Electron Microscopy

If only the *external* features of an organism need to be observed, thin sections are unnecessary. Intact cells or cell components can be observed directly by TEM with a technique called *negative staining* (see, for example, Figure 4.54). Alternatively, one can use the **scanning electron microscope (SEM)** (Figures 4.9 and 4.10*b*).

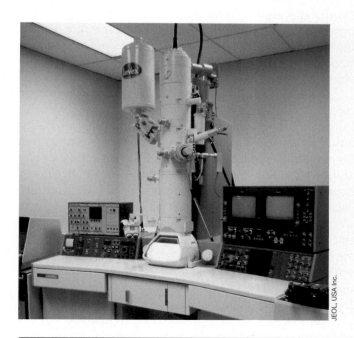

● **Figure 4.9   The electron microscope.**   This instrument encompasses both transmission and scanning electron microscope functions.

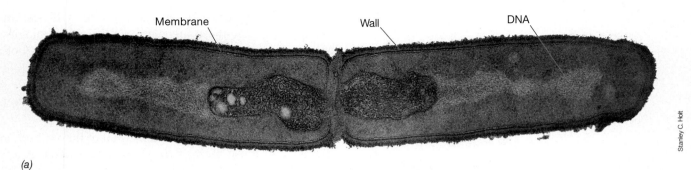

Membrane   Wall   DNA

Stanley C. Holt

(a)

● **Figure 4.10 Electron micrographs of bacterial cells.** (a) Transmission electron micrograph, and (b) scanning electron micrograph. (a) Thin section of a typical gram-positive bacterium, *Bacillus subtilis*. The cell has just divided, and two membrane-containing structures are attached to the cross-wall. Note the light region in the middle (DNA or the *nucleoid*). The cell is about 0.8 μm in diameter. (b) Cells of the phototrophic bacterium *Rhodo-vibrio sodomensis*. A single cell is about 0.75 μm wide. Note how the scanning micrograph allows for great depth of field and thus excellent three-dimensional imaging.

(b)

F. R. Turner

In scanning electron microscopy, the specimen is coated with a thin film of a heavy metal such as gold. An electron beam from the SEM is then directed onto the specimen and scans back and forth across it. Electrons scattered by the metal are collected, and they activate a viewing screen to produce an image (Figure 4.10*b*). In the SEM, even fairly large specimens can be observed, and the depth of field is extremely good. A wide range of magnifications can be obtained with the SEM, from as low as 15× up to about 100,000×, but only the *surface* of an object can be visualized.

⬡ *4.3 Concept Check*

Electron microscopes have far greater resolving power than do light microscopes, the limits of resolution being about 0.2 nm. Two major types of electron microscopy are performed: transmission electron microscopy, for observing internal cell structure down to the molecular level, and scanning electron microscopy, useful for three-dimensional imaging and for examining surfaces.

◆ What is an *electron micrograph?* How does an electron micrograph differ from a photomicrograph in terms of how the image is obtained?

◆ Keeping in mind that chemical fixatives are necessary and that electron microscopes must operate in a vacuum, what major *disadvantage* do electron microscopes have compared with light microscopes?

◆ What type of electron microscope would you use to observe the bacterial nucleoid?

<table>
<tr><td>**4.4**</td><td>**Cell Morphology and the Significance of Being Small**</td></tr>
</table>

In biology, the term **morphology** refers to cell *shape*. Several morphologies are known among prokaryotes and most have recognized terms to describe them. We explore morphology here and then look at some of the biological benefits of small cells.

Major Cell Morphologies

Schematic examples of typical bacterial shapes along with phase photomicrographs of example organisms are shown in Figure 4.11●. A bacterium that is spherical or ovoid in morphology is called a **coccus** (plural, **cocci**). A bacterium with a cylindrical shape is called a **rod**. Some rods are curved, frequently forming spiral-shaped patterns, and are then called **spirilla**. In many prokaryotes, cells remain together in groups or clusters after division, and the arrangements in these groups are often characteristic of different organisms. For instance, cocci or rods may occur in long chains. Some cocci form sheets of cells, whereas others occur in three-dimensional cubes or irregular cubelike clusters.

Several groups of bacteria are immediately recognizable by their unusual shapes. Examples include **spirochetes**, which are tightly coiled bacteria, **appendaged bacteria**, which possess extensions of their cells as long tubes or stalks, and **filamentous bacteria**, which form long, thin cells or chains of cells (Figure 4.11). The

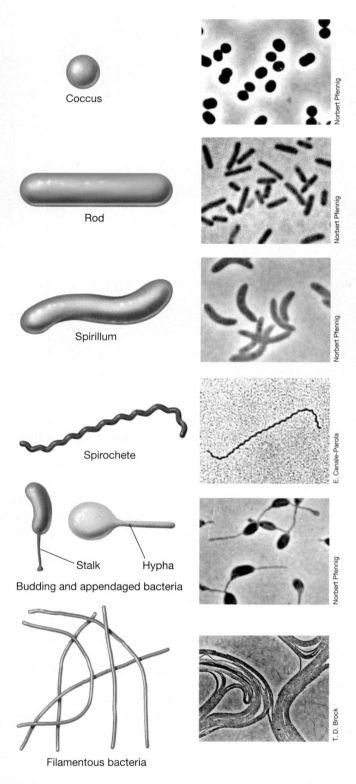

● **Figure 4.11 Representative cell shapes (morphology) in prokaryotes.** Next to each drawing is a phase photomicrograph showing an example of that morphology. Organisms are coccus, *Thiocapsa roseopersicina* (diameter of a single cell = 1.5 μm); rod, *Desulfuromonas acetoxidans* (diameter = 1 μm); spirillum, *Rhodospirillum rubrum* (diameter = 1 μm); spirochete, *Spirochaeta stenostrepta* (diameter = 0.25 μm); budding and appendaged, *Rhodomicrobium vannielii* (diameter = 1.2 μm); filamentous, *Chloroflexus aurantiacus* (diameter = 0.8 μm).

cell shapes in Figure 4.11 should be studied with the understanding that these are *representative* morphologies. Many variations of these basic morphological types, both subtle and distinct, are known in the microbial world.

The Size of Microbial Cells and the Significance of Being Small

Prokaryotes vary in size from cells smaller than 0.2 μm in diameter to those more than 50 μm in diameter. A few very large prokaryotes, such as the surgeonfish symbiont *Epulopiscium fishelsoni* (Figure 4.12a●), are up to 80 μm in

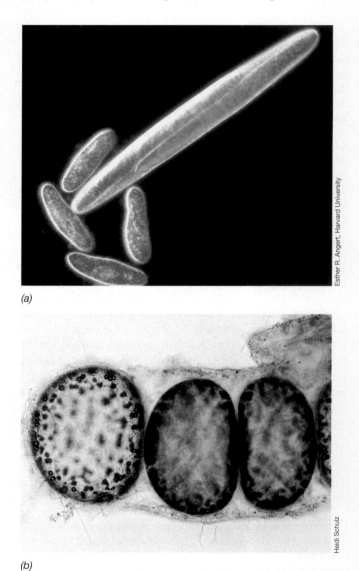

● **Figure 4.12 Some very large prokaryotes.** (a) Dark-field photomicrograph of a giant prokaryote, the surgeonfish symbiont *Epulopiscium fishelsoni*. The rod-shaped *E. fishelsoni* cell in this field is about 600 μm (0.6 mm) long and 75 μm wide and is shown with four cells of the protozoan (eukaryote) *Paramecium*, each of which measures about 150 μm in length. *E. fishelsoni* is a member of the *Bacteria* and phylogenetically related to *Clostridium* species. (b) *Thiomargarita namibiensis*, a large sulfur chemolithotroph (phylum Proteobacteria of the *Bacteria*) and currently the largest known prokaryote. A single coccus-shaped cell is about 400 μm wide. See also Table 4.1 for some large and small prokaryotes.

diameter and can be more than 0.6 millimeters (mm) in length (Table 4.1). The largest known prokaryote in terms of total cell volume, the sulfur chemolithotroph *Thiomargarita* (Figure 4.12b) can be 0.75 mm (750 μm) in diameter, nearly visible with the naked eye.

Prokaryotic cells vary dramatically in both size and volume (Table 4.1). Most very large prokaryotes are either sulfur chemolithotrophs or cyanobacteria (∞ Chapter 2 and Table 4.1). Why these cells are so large is not well understood, although for sulfur bacteria large cell size may be a mechanism for storing substrates to high levels. It is thought that limitations in diffusion and metabolism ultimately dictate the upper limit for the size of a prokaryotic cell. The metabolic rate of a cell varies inversely with the square of its size. Thus, for very large cells, diffusion processes may eventually limit metabolism such that the cell is no longer competitive.

Very large cells are not the norm in the prokaryotic world. By contrast to *Thiomargarita* or *Epulopiscium* (Figure 4.12), the dimensions of an average rod-shaped prokaryote, the bacterium *Escherichia coli*, for example, are about 1×3 μm and are typical for the vast majori-

ty of prokaryotes. For comparison, eukaryotic cells may be 2 μm to more than 200 μm in diameter. In general, prokaryotes are thus very small cells compared with eukaryotes.

The likely reason most prokaryotes are very small is that there are significant advantages to being small. For example, nutrients and waste products pass more readily into and out of a small cell than a large cell, thus accelerating cellular metabolism and growth. This is because relative to cell volume, small cells contain more *surface area* than do large cells. Consider the simple case of a sphere, in which the *volume* is a function of the cube of the radius ($V = 4/3$ πr^3), while the *surface area* is a function of the square of the radius ($S = 4\pi r^2$). The surface-to-volume (S/V) ratio of a sphere can thus be expressed as 3/r (Figure 4.13●). A cell with a smaller r value therefore has a *higher* S/V ratio than a cell with a larger r value.

Besides the exchange of nutrients, the S/V ratio affects other aspects of a cell's biology. For instance, because growth rate depends to some extent on the rate of nutrient exchange, the higher S/V of small cells typically supports more rapid growth than for larger cells.

Table 4.1	Cell size and volume of prokaryotic cells, from the largest to the smallest		
Organism	**Characteristics**	**Size^a (μm)**	**Cell volume (μm^3)**
Thiomargarita namibiensis	Spherical sulfur chemolithotroph	750	200,000,000
Epulopiscium fishelsoni	Chemoorganotrophic bacterium	80×600	3,000,000
Beggiatoa sp.	Filamentous sulfur chemolithotroph	50×160	1,000,000
Achromatium oxaliferum	Ellipsoid sulfur bacterium	35×95	80,000
Lyngbya majuscula	Filamentous cyanobacterium	8×80	40,000
Prochloron sp.	Prochlorophyte	30	14,000
Thiovulum majus	Spherical sulfur chemolithotroph	18	3,000
Staphylothermus marinus	Hyperthermophile	15	1,800
Titanospirillum velox	Rod-shaped sulfur chemolithotroph	5×30	600
Magnetobacterium bavaricum	Magnetotactic bacterium	2×10	30
Escherichia coli	Chemoorganotrophic bacterium	1×2	2
Mycoplasma pneumoniae	Pathogenic bacterium	0.2	0.005

^{a}Where only one number is given, this is the diameter of spherical cells. The values given are for the largest cell size observed in each species. For example, for *T. namibiensis*, an average cell is only about 200 μm in diameter. But on occasion, giant cells of 750 μm are observed. Likewise, an average cell of *S. marinus* is about 1 μm in diameter.

Source: Data obtained from Schulz, H.N., and B.B. Jørgensen. 2001. *Ann. Rev. Microbiol.* 55: 105–137.

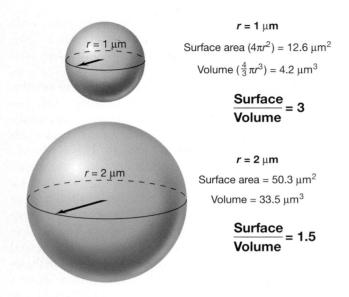

$r = 1 \, \mu m$

Surface area $(4\pi r^2) = 12.6 \, \mu m^2$

Volume $(\frac{4}{3}\pi r^3) = 4.2 \, \mu m^3$

$$\frac{\text{Surface}}{\text{Volume}} = 3$$

$r = 2 \, \mu m$

Surface area $= 50.3 \, \mu m^2$

Volume $= 33.5 \, \mu m^3$

$$\frac{\text{Surface}}{\text{Volume}} = 1.5$$

● **Figure 4.13   Surface area and volume relationships in cells.**
As a cell *increases* in size, its surface area-to-volume ratio *decreases*.

Moreover, per unit of available resources, small cells will typically develop larger populations than will large cells. And since mutation rates are the same for all organisms, larger cell populations mean more cell divisions and this means more mutations accumulate from spontaneous errors in DNA replication. Mutations are the "raw material" that drives evolutionary change. Coupling this with the fact that prokaryotes are genetically haploid (∞Section 2.2), a condition that allows beneficial mutations to be immediately expressed, the small size of prokaryotes may ultimately explain why these organisms tend to adapt rapidly to changing environmental conditions and easily exploit new habitats.

Lower Limits of Cell Size

From the foregoing it may seem that smaller and smaller bacteria will have greater and greater selective advantages in nature. Obviously there are lower limits to cell size, but what are these? Some microbiologists have proposed that *very* small bacteria exist in nature, cells referred to as *nanobacteria*. The size of putative nanobacteria are on the order of 0.1 μm in diameter for coccus-shaped structures, or even smaller. This is extremely small, even by prokaryotic standards (Table 4.1). Are nanobacteria really cells?

Most reports of nanobacteria have been associated with their supposed formation of precipitates and biofilms (∞Section 19.3) in environments as diverse as mineral surfaces and human tissues. Although some microbiologists firmly believe in the nanobacteria concept, others have claimed that nanobacteria are simply artifacts of chemical or geochemical reactions of nonliving materials and that even the smallest known bacterial cells are significantly larger than reported nanobacteria (Table 4.1). Moreover, if one considers the space needed to house all of the essential biomolecules of life, it is

highly unlikely that these could exist within the volume available to a structure of 0.1 μm or less. Thus, whether nanobacteria are living organisms or not is an unsettled question. If such very small cells actually exist, they would be the smallest known living structures.

Regardless of the status of nanobacteria, many microbial habitats clearly contain very small cells. The open oceans, for example, contain 10^4–10^5 prokaryotic cells per milliliter, and these tend to be very small cells, 0.2–0.4 μm in diameter. We will see that many pathogenic bacteria are very small as well. Thus one can conclude that very small cells do exist in nature, but that structures smaller than about 0.2 μm may not be viable (capable of reproducing) cells.

 4.4   Concept Check

Prokaryotes are typically smaller in size than eukaryotes, and prokaryotic cells can have a wide variety of morphologies. The small size of prokaryotic cells affects their physiology, growth rate, and ecology. Cell-like structures smaller than about 0.2 μm may or may not be living organisms.

◆   List three morphological types of prokaryotes.

◆   What physical property of cells *increases* as cells become smaller?

II CELL MEMBRANES AND CELL WALLS

We now consider two extremely important cell structures in prokaryotes: the cytoplasmic membrane and the cell wall. Each carries out well-defined and critical functions for the cell, including the transport of nutrients (membrane) and the prevention of osmotic lysis (wall).

4.5   Cytoplasmic Membrane: Structure

The **cytoplasmic membrane** is a thin structure that surrounds the cell. Only about 8 nm thick, this vital structure is the barrier separating the inside of the cell (the cytoplasm) from its environment. If the membrane is broken, the integrity of the cell is destroyed, the cytoplasm leaks into the environment, and the cell dies. The cytoplasmic membrane is also a *highly selective permeability barrier*, enabling a cell to concentrate specific metabolites and excrete waste materials.

Chemical Composition of Membranes

The general structure of biological membranes is a **phospholipid bilayer** (Figure 4.14●). As discussed in Section 3.4, phospholipids contain both hydrophobic (fatty acid) and hydrophilic (glycerol-phosphate) components and can exist in many different chemical forms as a result of variation in the groups attached to the glycerol backbone. As phospholipids aggregate in an aqueous solution, they naturally form bilayer structures spontaneously. In a phos-

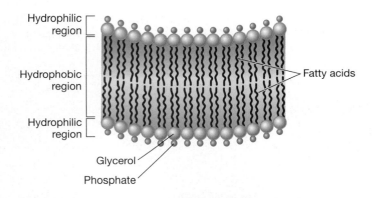

Hydrophilic region

Hydrophobic region

Hydrophilic region

Fatty acids

Glycerol

Phosphate

● **Figure 4.14   Structure of a phospholipid bilayer.** The cytoplasmic membrane is about 8 nm (80 Å) wide. Review Figure 3.7 for the structure of a phospholipid.

pholipid, the fatty acids point inward toward each other to form a hydrophobic environment, while the hydrophilic portions remain exposed to the aqueous external environment (Figure 4.14). The bilayer character of membranes represents the most stable arrangement of lipid molecules in an aqueous environment.

Thin sections of the cytoplasmic membrane can be seen with the electron microscope (Figure 4.15*a*●). The cytoplasmic membrane appears as two light-colored lines separated by a darker area (Figure 4.15*a*). This **unit membrane**, as it is called (because each phospholipid leaf forms half of the "unit"), consists of a phospholipid (⌒⌒ Figure 3.7) bilayer with proteins embedded in it (Figure 4.16●). The major proteins of the cytoplasmic membrane typically have very hydrophobic external surfaces in the regions of the protein that span the membrane, and hydrophilic surfaces that make contact with the environment and the cytoplasm (Figure 4.16). The overall structure of the cytoplasmic membrane is stabilized by hydrogen bonds and hydrophobic interactions (⌒⌒ Section 3.1). In addition, cations such as Mg^{2+} and Ca^{2+} help stabilize the membrane by combining ionically with negative charges of the phospholipids.

Membrane Proteins

The outer surface of the cytoplasmic membrane faces the environment and in certain bacteria makes contact with a variety of proteins that bind substrates or process large molecules for transport into the cell (periplasmic proteins, see discussion in Sections 4.7 and 4.9). The inner side of the cytoplasmic membrane faces the cytoplasm and interacts with proteins involved in energy-yielding reactions and other important cellular functions.

Many membrane proteins are firmly embedded in the membrane and are called *integral membrane proteins*. Other proteins are not embedded in the membrane but associate quite firmly with one of the membrane surfaces and actually function as if they were membrane-bound proteins. These include proteins in the periplasm (a region between the cytoplasmic membrane and the outer membrane of gram-negative bacteria, see Section 4.9) and some cytoplasmic proteins. Some of these *peripheral membrane proteins*, as they are called, are lipoproteins, which contain a lipid tail on the amino terminus of the protein that anchors the protein in the membrane. These proteins interact directly with integral membrane proteins in important cellular processes such as energy metabolism.

In a diagram, the cytoplasmic membrane may appear somewhat rigid. In reality, however (Figures 4.15 and 4.16), the cytoplasmic membrane is quite fluid—phospholipid and protein molecules have significant freedom to move about within the membrane. Membranes have a viscosity approximating that of a light-grade oil. Thus, integral membrane proteins likely span a highly mobile, yet highly ordered, phospholipid bilayer. We will see in the next section how this *fluid mosaic* structure guides the function of the membrane as well.

Membrane Strengthening Agents: Sterols and Hopanoids

One major difference in chemical composition of membranes between eukaryotic and prokaryotic cells is that eu-

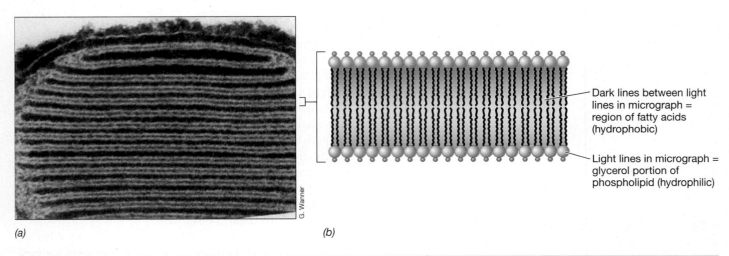

Dark lines between light lines in micrograph = region of fatty acids (hydrophobic)

Light lines in micrograph = glycerol portion of phospholipid (hydrophilic)

G. Wanner

(a)   (b)

● **Figure 4.15   The cytoplasmic membrane.**   (a) Electron micrograph of photosynthetic membrane stacks derived from the cytoplasmic membrane in the phototrophic bacterium *Halorhodospira halochloris*. Note the multiple lipid bilayers. Each bilayer is about 8 nm thick. (b) Enlarged schematic view of a single unit membrane shown in (a).

● **Figure 4.16** **Structure of the cytoplasmic membrane.** The inner surface (**In**) faces the cytoplasm and the outer surface (**Out**) faces the environment. The matrix of the cytoplasmic membrane is composed of phospholipids, with the hydrophobic groups directed inward and the hydrophilic groups toward the outside, where they associate with water. Embedded in the matrix are proteins that have considerable hydrophobic character in the region that traverses the fatty acid bilayer. Hydrophilic proteins and other charged substances, such as metal ions, may be attached to the hydrophilic surfaces. Although there are some chemical differences, the overall structure of the cytoplasmic membrane shown is similar in both prokaryotes and eukaryotes (but see an exception to the bilayer design in Figure 4.19*d*).

karyotes have **sterols** in their membranes (Figure 4.17*a*, *b*●). Sterols are absent from the membranes of virtually all prokaryotes (methanotrophic bacteria and the mycoplasmas are exceptions, ⧜Sections 12.6 and 12.21). Depend-

(a)

(b)

● **Figure 4.17** **Sterols and hopanoids.** (a) The structure of cholesterol, a typical sterol. (b) The structure of the hopanoid diploptene. Sterols are found in the membranes of eukaryotes and hopanoids in the membranes of some prokaryotes. The intra-ring labels (1, 2, and 3) are meant to highlight similarities in the parent structure of sterols and hopanoids.

ing on the cell type, sterols can make up from 5 to 25% of the total lipids of eukaryotic membranes.

Sterols and related molecules are rigid, planar molecules, whereas fatty acids are flexible. The presence of sterols in a membrane thus stabilizes it and makes it less flexible. Molecules similar to sterols, called *hopanoids*, are present in the membranes of many *Bacteria*, and likely play a role similar to that of sterols in eukaryotic cells. One widely distributed hopanoid is the C_{30} hopanoid *diploptene* (Figure 4.17*b*). As far as is known, hopanoids are not present in species of *Archaea*.

Archaeal Membranes

The lipids of *Archaea* differ from those of other organisms. In contrast to the lipids of *Bacteria* and *Eukarya* in which *ester* linkages bond the fatty acids to glycerol (Figure 4.18*a*●; ⧜Section 3.4), the lipids of *Archaea* have *ether* linkages between glycerol and their hydrophobic side chains. In addition, archaeal lipids lack fatty acids. Instead, their side chains are composed of repeating units of the five-carbon hydrocarbon *isoprene* (Figure 4.18*c*). Nevertheless, the overall architecture of the cytoplasmic membrane of *Archaea*, forming inner and outer hydrophilic surfaces with a hydrophobic interior, is the same as in *Bacteria* and *Eukarya*.

Glycerol *diethers* and glycerol *tetraethers* (Figure 4.19*a*, *b*●) are the major lipids present in *Archaea*. In the *tetraether* molecule, the phytanyl (composed of four linked isoprenes)

● Figure 4.18 Chemical bonds in lipids. (a) The *ester* linkage as found in the lipids of *Bacteria* and *Eukarya*. (b) The *ether* linkage of lipids from *Archaea*. (c) Isoprene, the parent structure of the hydrophobic side chains of archaeal lipids. By contrast, in lipids of *Bacteria* and *Eukarya*, the side chains are composed of fatty acids.

4.5 Concept Check

The cytoplasmic membrane is a highly selective permeability barrier constructed of lipids and proteins that forms a bilayer with hydrophilic exteriors and a hydrophobic interior. Other molecules, such as sterols and hopanoids, may strengthen the membrane, and integral proteins involved in transport and other functions traverse it. Unlike *Bacteria* and *Eukarya*, *Archaea* contain ether-linked lipids, and some species have membranes of monolayer instead of bilayer construction.

◆ Draw the basic structure of a lipid bilayer.

◆ Why are compounds like sterols and hopanoids good at stabilizing the cytoplasmic membrane?

◆ Contrast the linkage between glycerol and the hydrophobic portion of lipids in *Bacteria* and *Archaea*.

side chains from each glycerol molecule are *covalently bonded* together (Figure 4.19*b*). This structure therefore yields a lipid *monolayer* instead of a lipid *bilayer* cytoplasmic membrane (Figure 4.19*d*). Lipid monolayers are quite resistant to peeling apart. This membrane structure is therefore widespread among hyperthermophilic *Archaea*, prokaryotes that grow at very high temperatures (∞Sections 6.10 and 13.5–13.10).

We will encounter many other features that set the prokaryotic organisms *Bacteria* and *Archaea* apart, but the chemistry of their membrane lipids is a major defining feature of each phylogenetic group.

4.6 Cytoplasmic Membrane: Function

The cytoplasmic membrane is more than just a barrier separating the inside from the outside of the cell. The membrane plays several critical roles in cell function. First and foremost, the membrane functions as a *permeability barrier*, preventing the passive leakage of cytoplasmic constituents into or out of the cell (Figure 4.20●). In addition, the membrane is the location of many proteins. Some of these are enzymes involved in bioenergetic functions and others are involved in the transport of substances into and out of the cell.

● Figure 4.19 Major lipids of *Archaea* and the structure of archaeal membranes. (a) Glycerol diethers. (b) Diglycerol tetraethers. Note that in both cases, the hydrocarbon is attached to the glycerol by *ether* linkages. Hydrocarbon in (a) phytanyl (C_{20}) and (b) dibiphytanyl (C_{40}). (c, d) Membrane structure in *Archaea*. (c) Lipid bilayer. (d) Lipid monolayer.

(a) Glycerol diether

(b) Diglycerol tetraether

(c) Lipid bilayer

(d) Lipid monolayer

1. **Permeability Barrier**—Prevents leakage and functions as a gateway for transport of nutrients into and out of the cell

2. **Protein Anchor**—Site of many proteins involved in transport, bioenergetics, and chemotaxis

3. **Energy Conservation**—Site of generation and use of the proton motive force

● **Figure 4.20** **The major functions of the cytoplasmic membrane.** Although structurally weak, the cytoplasmic membrane has many important cellular functions.

We will learn in Chapter 5 that the cytoplasmic membrane is a major site of *energy conservation* in the cell. The membrane can exist in an energetically "charged" form in which a separation of protons (H^+) from hydroxyl ions (OH^-) occurs across its surface (Figure 4.20). This charge separation is a form of energy, analogous to the potential energy present in a charged battery. This energized state of the membrane, called the *proton motive force* (PMF), is responsible for driving many energy-requiring functions in the cell, including some forms of transport, motility, and the biosynthesis of the cell's energy currency, ATP.

The Cytoplasmic Membrane as a Permeability Barrier

The interior of the cell (the cytoplasm) consists of an aqueous solution of salts, sugars, amino acids, nucleotides, vitamins, coenzymes, and a variety of other soluble materials. The hydrophobic nature of the internal portion of the cytoplasmic membrane (Figure 4.16) makes it a tight diffusion barrier. Although some small hydrophobic molecules may pass through the membrane by diffusion, hydrophilic and charged molecules do not pass through but instead must be specifically transported. Because it is charged, even a substance as

small as a hydrogen ion (H^+) cannot diffuse across the cytoplasmic membrane.

One molecule that does freely penetrate the membrane is water, which is sufficiently small to pass between phospholipid molecules in the lipid bilayer (Table 4.2). However, water transport through the membrane can be greatly accelerated by transport proteins called *aquaporins*. These proteins form membrane-spanning channels that specifically transport water into or out of the cytoplasm. Aquaporin AqpZ of *Escherichia coli*, for example, is a well-studied water channel. The synthesis of AqpZ protein is greatly increased under low osmotic conditions. Under these conditions, AqpZ functions as a water *exporter*, preventing the cell from experiencing hypoosmotic shock. Under high osmotic conditions, lower amounts of AqpZ are present, but here the protein functions to transport water *into* the cell to counteract the tendency for water to flow to higher solute conditions.

The relative permeability of a few biologically important substances is shown in Table 4.2. As can be seen, most substances do not passively enter the cell, and thus *transport* is required. The data of Table 4.2 should also be viewed with the understanding that water flow in prokaryotic cells is assisted by aquaporins and is not entirely due to diffusion through the membrane.

The Necessity for Transport Proteins

Transport proteins do more than just ferry things across the membrane—they accumulate solutes *against* the concentration gradient. The necessity for carrier-mediated transport in microorganisms is easy to understand. If diffusion was the only way that solutes could enter the cell, cells would never achieve the intracellular concentrations necessary to carry out biochemical reactions. This is because both the rate of uptake and the intracellular level of diffusible solutes are proportional to their external concentration (Figure 4.21●). The concentration of nutrients in nature, however, is often very low, much lower than typical microbial culture media (the solutions used to grow microorganisms in the laboratory). Hence, cells

Table 4.2	**Comparative permeability of membranes to various molecules**
Substance	**Rate of permeability** [a]
Water	100
Glycerol	0.1
Tryptophan	0.001
Glucose	0.001
Chloride ion (Cl^-)	0.000001
Potassium ion (K^+)	0.0000001
Sodium ion (Na^+)	0.00000001

[a] Relative scale—permeability with respect to permeability of water given as 100. Permeability of the membrane to water may be affected by aquaporins (see text).

Rate of solute entry — External concentration of solute

Carrier saturated with substrate

Carrier-mediated transport

Simple diffusion

● **Figure 4.21   Relationship between uptake rate and external concentration in diffusion and transport.** Note that in the carrier-mediated process, the uptake rate shows saturation at relatively low external concentrations.

must have mechanisms for accumulating nutrients to levels higher than those in nature, and this is the function of transport systems.

Unlike simple diffusion, carrier-mediated transport systems also show a *saturation effect*. If the concentration of substrate is high enough to saturate the carrier, which typically occurs at even very low substrate concentration, the rate of uptake becomes maximal and the addition of more substrate does not increase the rate (Figure 4.21). This characteristic greatly assists cells in concentrating nutrients in the cytoplasm from an often very dilute nutrient environment.

Another characteristic of carrier-mediated transport processes is the *highly specific* nature of the transport event. Many carrier proteins react only with a single molecule while others show affinities for a related class of molecules. For instance, there are carriers that transport a variety of related sugars or amino acids. This economy in uptake reduces the need for separate transport proteins for every single amino acid or every single sugar.

The synthesis of transport proteins is also *regulated* by the cell such that the specific complement of transporters present in the membrane is a function of both the nutrients present in the environment and their concentration. The latter is an important factor because transport of a particular nutrient often occurs via one type of transporter when the nutrient is present at high concentration and by a different, higher affinity transporter, when present at low concentration.

 4.6   *Concept Check*

The major function of the cytoplasmic membrane is as a permeability barrier, preventing leakage of cytoplasmic metabolites into the environment. Selective permeability also prevents the diffusion of most solutes. To accumulate nutrients against the concentration gradient, specific transport mechanisms are employed.

◆ Besides permeability, what other functions does the cytoplasmic membrane have?

◆ List two reasons why a cell cannot depend on diffusion as a means of getting nutrients into the cell.

◆ Why is physical damage to the cytoplasmic membrane a more critical problem for the cell than damage to some other cell component?

4.7   Membrane Transport Systems

As just discussed, transport of nutrients and expulsion of wastes are key cellular events. Different mechanisms for transport have evolved in prokaryotes, each with its own unique features. We explore this subject here.

Structure and Function of Membrane Transport Proteins

There are at least three systems for transporting substances in prokaryotes: **simple transport, group translocation**, and the **ABC system**. Simple transporters require only a membrane-spanning protein. Group translocation involves a series of proteins in the transport event. The ABC system involves a *substrate-binding protein*, a *membrane transporter*, and an *ATP-hydrolyzing protein* (Figure 4.22●). All of these transport systems require energy, either in the form of the proton motive force, ATP, or some other energy-rich compound.

Simple transport: Driven by the energy in the proton motive force

Transported substance

Group translocation: Chemical modification of the transported substance driven by phosphoenolpyruvate

The ABC system: Periplasmic binding proteins are involved and energy comes from ATP

● **Figure 4.22   The three classes of membrane-transporting systems.** Note how simple transporters and the ABC system transport substances *without* chemically modifying them, while group translocation results in the chemical modification (phosphorylation) of the transported substance. The three proteins of the ABC system are labeled 1, 2, and 3.

The membrane-spanning proteins of virtually all bacterial transport systems show significant homologies in both their primary and secondary structure, a testament to their common evolutionary roots. Structurally, these transporters form 12 α-helices (⬭Section 3.7) that wind back and forth through the membrane to form a channel through which the transported substance is carried into the cell (Figure 4.23●). The actual transport event involves a conformational change in the protein following binding of its substrate, and this event shuttles the compound across the membrane.

At least three *classes* of transporters are known (Figure 4.24). *Uniporters* transport a molecule in a unidirectional fashion across the membrane. *Symporters* transport a substance *along with* another substance, typically a proton (H⁺). *Antiporters*, as their name implies, transport one substance across the membrane in one direction while simultaneously transporting a second substance in the *opposite* direction (Figure 4.23).

Lactose Uptake in *Escherichia coli*: The Lac Permease

The bacterium *Escherichia coli* metabolizes the disaccharide sugar lactose. Lactose is transported by cells of *E. coli* through the activity of a symporter called the *Lac permease*. Lac permease is a simple transporter. This is shown in Figure 4.24●, where the activity of the lac permease is compared with other simple transporters, including ones that function as uniporters or antiporters.

The lac permease requires energy. Note that as each lactose molecule is transported, the energy in the proton motive force is slowly diminished by the co-transport of protons into the cell. However, the proton motive force is reestablished in the cell through energy-yielding reactions that we will describe in later chapters (⬭

● **Figure 4.24 Function of the Lac permease (a symporter) of** *Escherichia coli* **and several other well-characterized simple transporters.** Although for simplicity the membrane-spanning proteins are drawn here in globular form, note that their structure is actually as depicted in Figure 4.23. Also note that symporters and antiporters are driven by the energy of the proton motive force. The latter is regenerated by energy-yielding (catabolic) reactions in the cells.

● **Figure 4.23 Structure of membrane-spanning transporters and types of transport events.** In prokaryotes, membrane-spanning transporters typically contain 12 α-helices that align with each other in a circle to form a channel through the membrane. Shown here are three individual transporters, each showing a different type of transport event. For antiporters and symporters, the cotransported molecule is shown in yellow.

Chapters 5 and 17). The net result of the activity of the lac permease is the accumulation of lactose to a sufficiently high concentration such that its metabolism can yield useful energy for the cell.

Group Translocation: The Phosphotransferase System

Group translocation is a transport mechanism in which the transported substance is *chemically altered* during passage across the membrane. The best-studied group translocation system involves transport of the sugars glucose, mannose, and fructose in *E. coli*. These compounds are *phosphorylated* during transport by the **phosphotransferase system**.

The phosphotransferase system consists of a family of proteins, five of which are necessary to transport any given sugar. Before the sugar is transported, the proteins in the phosphotransferase system are themselves alternately phosphorylated and dephosphorylated in a cascading fashion until the membrane-spanning protein, called *Enzyme II_c*, receives the phosphate group and phosphorylates the sugar in the actual transport event (Figure 4.25●). A small protein called HPr, the enzyme

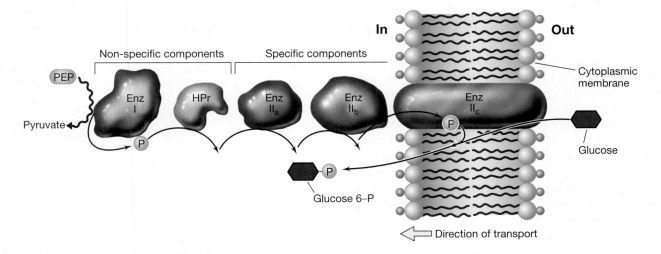

● **Figure 4.25 Mechanism of the phosphotransferase system of *Escherichia coli.*** For glucose uptake, the system consists of five proteins: Enzyme (Enz) I; Enzymes II$_a$, II$_b$, and II$_c$, and HPr. Sequential phosphate transfer occurs from phosphoenolpyruvate (PEP) through the proteins shown to Enzyme II$_c$. The latter actually transports (and phosphorylates) the sugar. Proteins HPr and Enz I are nonspecific and involved in the transport of any sugar. The Enz II components are specific for a particular sugar.

that phosphorylates it (Enzyme I), and Enzyme II$_a$ are cytoplasmic proteins. By contrast, Enzyme II$_b$ lies on the inner membrane surface, and Enzyme II$_c$ is an integral membrane protein (Figure 4.25). HPr and Enzyme I are nonspecific components of the phosphotransferase system and participate in the uptake of various sugars, while specific Enzymes II exist for each individual sugar transported (Figure 4.25).

Energy for the phosphotransferase system comes from the energy-rich compound *phosphoenolpyruvate*. However, it should be noted that although energy in the form of one energy-rich phosphate bond is consumed in the process of transporting the glucose molecule (Figure 4.25), the phosphorylation of glucose to glucose-6-P is the first step in its intracellular metabolism anyway (glycolysis, ∞∞Section 5.10). Thus, the phosphotransferase system prepares glucose for immediate entry into a central metabolic pathway.

Periplasmic Binding Proteins and the ABC System

We will learn a bit later in this chapter (see Section 4.9) that gram-negative bacteria contain a space called the *periplasm* between the cytoplasmic membrane and a lipid-rich outer membrane layer (see Figure 4.35). The periplasm contains various proteins, many of which function in transport. These are called *periplasmic-binding proteins*. Transport systems of this type actually employ three components: (1) periplasmic-binding proteins; (2) membrane-spanning proteins; and (3) ATP-hydrolyzing proteins (kinases). The latter supply the necessary energy.

Transporters of this type have been called *ABC transport systems*, the *ABC* standing for *ATP-binding cassette* (Figure 4.26●). More than 200 different ABC transport systems have been identified in prokaryotes, and structural studies have shown that they are clearly a *family* of related proteins. ABC transporters exist for organic nu-

trients such as sugars and amino acids, for a variety of inorganic nutrients, such as sulfate and phosphate, and for trace metals.

One of the interesting properties of ABC-type transport systems is the extremely high substrate affinity of the periplasmic-binding proteins. These proteins can

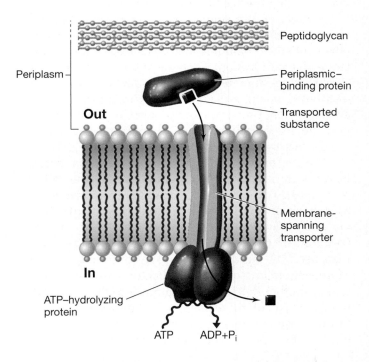

● **Figure 4.26 Mechanism of an ATP-Binding Cassette (ABC-type) transporter.** The periplasmic binding protein has high affinity for substrate, the membrane-spanning protein is the transport channel, and the cytoplasmic ATP-hydrolyzing protein supplies the energy for the transport event. In *Escherichia coli*, the maltose (a disaccharide sugar) transport system is an example of an ABC system. We discuss the structure of the gram-negative cell wall, including the periplasm, in Section 4.9.

move within the periplasm and bind substrates even when they are present at extremely low concentration. For example, substrate concentrations of 1 micromolar (10^{-6} M) or less can easily be bound and transported by periplasmic-binding proteins. Once trapped, the complex interacts with its respective membrane-spanning component, and the actual transport event occurs driven by the energy of ATP (Figure 4.26).

Interestingly, even though gram-positive bacteria lack a periplasm, binding protein-dependent transport systems have also been found in many of these organisms. However, in gram-positive bacteria, specific binding proteins are not mobile but instead are anchored to the cytoplasmic membrane. Nevertheless, as in gram-negative bacteria, once these binding proteins bind their substrate, they interact with a membrane-spanning component where, at the expense of ATP, transport across the membrane occurs.

Protein Export

Thus far our discussion of transport has focused on small molecules. What about the transport of large molecules, such as proteins? To function properly, many proteins need to be transported *outside* the cytoplasmic membrane or inserted into the membrane in specific ways. Protein translocation occurs in prokaryotic cells through the activities of proteins called *translocases*, a key one being the Sec (for secretory) system.

SecYEG, for example, is a membrane-associated translocase that exports certain proteins while inserting others into the membrane in a specific orientation consistent with their function. Some translocases are very specific in the types of proteins exported, but SecYEG is widely distributed among prokaryotes and can translocate a variety of different proteins. How proteins destined for transport are recognized as such is another story, and we discuss this issue in a later chapter (∞Section 7.17).

Protein export is important to bacteria because many bacterial enzymes function outside the cell (exoenzymes). For example, hydrolytic exoenzymes such as amylase or cellulase are excreted directly into the environment where they cleave starch or cellulose (∞Figure 3.6*b*), respectively, into glucose. The latter is then used by the cell as a carbon and energy source. Moreover, many pathogenic bacteria excrete protein toxins or other deleterious proteins into the host during infection. All of these large molecules need to move through the cytoplasmic membrane, and translocases like SecYEG assist in these transport events.

⬢ 4.7 Concept Check

At least three types of transporters are known: simple transporters, phosphotransferase-type transporters, and ABC systems, the latter of which contains three interacting components. Transport requires energy from either the proton motive force, ATP, or some other energy-rich substance.

◆ Contrast the energy requirements of simple transporters, the phosphotransferase system, and the ABC transport system.

◆ Contrast the three classes of transport systems in terms of any chemical alterations that occur in the transported molecule.

◆ Which transport system is best suited for the transport of nutrients present in the environment in extremely low amounts, and why?

◆ How are proteins exported from the cell?

4.8 The Cell Wall of Prokaryotes: Peptidoglycan and Related Molecules

Bacterial cells contain a high concentration of dissolved solutes. This causes a considerable turgor pressure to develop—about 2 atmospheres in a bacterium like *Escherichia coli*. This is roughly the same as the pressure in an automobile tire. To withstand these pressures, bacteria possess **cell walls**, which also function to some extent to give shape and rigidity to the cell.

We have seen that species of *Bacteria* can be divided into two major groups, called **gram-positive** and **gram-negative**. The original distinction between gram-positive and gram-negative was based on the *Gram stain* (see Section 4.1). But differences in cell wall structure are at the heart of the Gram-staining reaction. The appearance of the cell walls of gram-positive and gram-negative cells in the electron microscope differs markedly, as is shown in Figure 4.27●. The gram-negative cell wall is a multilayered structure and quite complex, whereas the gram-positive cell wall primarily consists of a single type of molecule, and is often much thicker.

The focus of this section is on the polysaccharide component of the cell walls of prokaryotes, both *Bacteria* and *Archaea*. These include, in particular, peptidoglycan, but also a variety of related and unrelated polysaccharides found in *Archaea*. In Section 4.9 we describe the special wall components found in gram-negative *Bacteria*.

Peptidoglycan

The cell walls of *Bacteria* have a rigid layer that is primarily responsible for the strength of the wall. In gram-negative *Bacteria*, additional layers are present outside this rigid layer. The rigid layer of both gram-negative and gram-positive *Bacteria* is very similar in chemical composition. Called **peptidoglycan**, this polysaccharide is composed of two sugar derivatives—*N-acetylglucosamine* and *N-acetylmuramic acid*—and a small number of special amino acids including L-alanine, D-alanine, D-glutamic acid, and either lysine or diaminopimelic acid (DAP) (Figure 4.28●). These con-

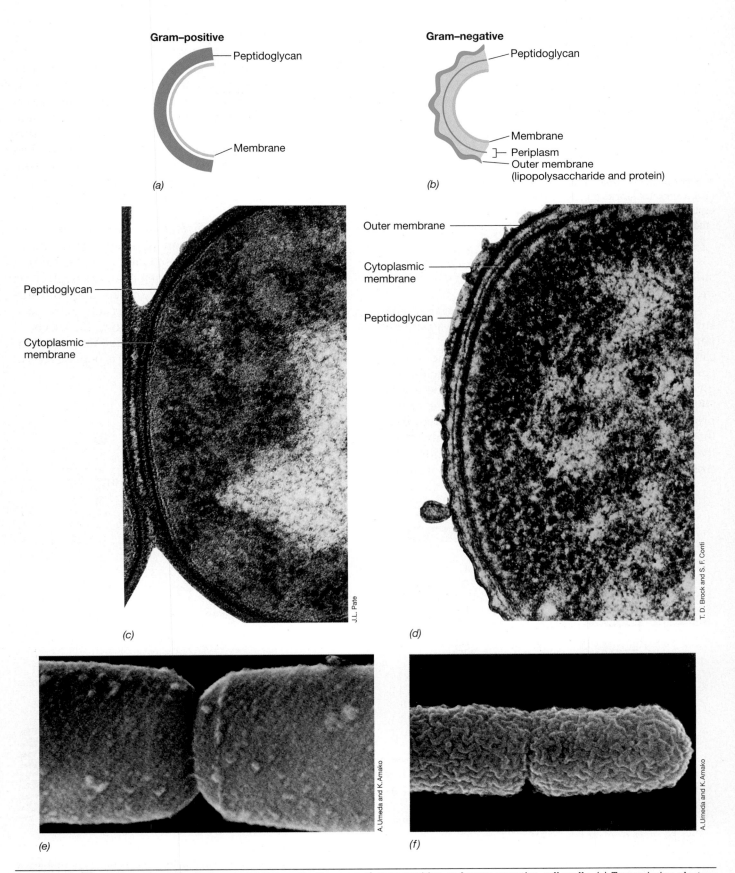

● **Figure 4.27 Cell walls of *Bacteria*.**   (a, b) Schematic diagrams of gram-positive and gram-negative cell walls. (c) Transmission electron micrographs showing the cell wall of a gram-positive bacterium, *Arthrobacter crystallopoietes*, and (d) Gram-negative bacterium, *Leucothrix mucor*. (e, f) Scanning electron micrographs of gram-positive (*Bacillus subtilis*) and gram-negative (*Escherichia coli*) *Bacteria*. Note the surface texture in the cells shown in (e) and (f). A single cell of *B. subtilis* or *E. coli* is about 1 μm in diameter.

COOH
|
H₂N—CH
|
CH₂
|
CH₂
|
CH₂
|
H₂N—CH
|
COOH

COOH
|
H₂N—CH
|
CH₂
|
CH₂
|
CH₂
|
H₂N—CH
|
H

(a)   (b)

● **Figure 4.28 Cross-linking amino acids in peptidoglycan.**
(a) Diaminopimelic acid. (b) Lysine. The only difference in the two molecules is highlighted in color. Besides these two amino acids, several other amino acids are found in peptidoglycan cross-links.

stituents are connected to form a repeating structure, the *glycan tetrapeptide* (Figure 4.29●).

The basic structure of peptidoglycan is a sheet that surrounds the cell formed from individual strands of peptidoglycan lying adjacent to one another. The glycan chains that form the sheet are connected by *tetrapeptide cross-links* formed by the amino acids. The glycosidic bonds connecting the sugars in the glycan strands are very strong, but these chains alone cannot provide rigidity in all directions. The full strength of the peptidoglycan structure is realized only when these chains are cross-linked by amino acids. This cross-linking occurs to different extents in different

Bacteria, with greater rigidity coming from more complete cross-linking.

In *gram-negative Bacteria*, cross-linkage occurs by peptide linkage of the amino group of diaminopimelic acid to the carboxyl group of the terminal D-alanine (Figure 4.30a●). In *gram-positive Bacteria*, cross-linkage occurs by way of a *peptide interbridge*, the kinds and numbers of amino acids varying from organism to organism. For example, in *Staphylococcus aureus*, a well-studied gram-positive bacterium, the interbridge peptide consists of five molecules of glycine (Figure 4.30b). The overall structure of a peptidoglycan molecule is shown in Figure 4.30c.

In gram-positive *Bacteria*, as much as 90% of the cell wall consists of peptidoglycan, although another molecule called *teichoic acid* (discussed later in this section), is usually present in small amounts. And, although some bacteria have only a single layer of peptidoglycan surrounding the cell, many *Bacteria*, especially gram-positive *Bacteria*, have several (up to about 25) sheets of peptidoglycan stacked upon one another. In gram-negative *Bacteria* only about 10% of the wall is peptidoglycan,

● **Figure 4.29 Structure of one of the repeating units of the peptidoglycan cell wall structure, the glycan tetrapeptide.** The structure given is that found in *Escherichia coli* and most other gram-negative *Bacteria*. In some *Bacteria*, other amino acids are found.

(a) *Escherichia coli*
(gram-negative)

(b) *Staphylococcus aureus*
(gram-positive)

(c)

● **Figure 4.30 How the peptide and glycan units are connected to form the peptidoglycan sheet in *Escherichia coli* and *Staphylococcus aureus*.** (a) No interbridge is present in *E. coli* and other gram-negative *Bacteria*. (b) The glycine interbridge in *S. aureus* (gram-positive). (c) Overall structure of peptidoglycan. G, *N*-acetylglucosamine; M, *N*-acetylmuramic acid.

the majority of the wall consisting of the outer membrane as discussed in Section 4.9. However, the shape of both gram-positive and gram-negative cells is thought to be determined to some extent by the lengths of the peptidoglycan chains and by the manner and extent of cross-linking of the chains.

Diversity in Peptidoglycan

Peptidoglycan is present only in species of *Bacteria*—the sugar *N*-acetylmuramic acid and the amino acid diaminopimelic acid have never been found in the cell walls of *Archaea* or *Eukarya*. However, not all *Bacteria* examined have DAP in their peptidoglycan. This amino acid is present in peptidoglycan in all gram-negative *Bacteria* and in some gram-positive species, but most gram-positive cocci have lysine instead of DAP (Figure 4.30b), and a few other gram-positive *Bacteria* have other amino acids. Another unusual feature of peptidoglycan is the presence of two amino acids that have the D configuration, D-alanine and D-glutamic acid. As we saw in Chapter 3, in cellular proteins amino acids are always of the L enantiomeric form (∞ Section 3.6).

More than 100 different peptidoglycan types are known, with the variations occurring in the interbridge. In each different peptidoglycan structure the glycan portion is uniform, with only the sugars *N*-acetylglucosamine and *N*-acetylmuramic acid being present. These sugars are always connected in β-1,4 linkage (Figure 4.29). The tetrapeptide of the repeating unit shows major variation in only one amino acid, the lysine–diaminopimelic acid alternation. However, the D-glutamic acid at position 2 is hydroxylated in some or-

ganisms, whereas substitutions occur in amino acids at positions 1 and 3 in some others. Any of the amino acids present in the tetrapeptide can also occur in the interbridge. But in addition, a number of other amino acids such as glycine, threonine, serine, and aspartic acid can be found in the interbridge. However, branched-chain amino acids, aromatic amino acids, sulfur-containing amino acids, and histidine, arginine, and proline (∞ Figure 3.12) are never found in the interbridge.

Thus, although the peptide chemistry of peptidoglycan can vary, the backbone of peptidoglycan—alternating repeats of glucosamine and muramic acid—is the same in all species of *Bacteria*.

Teichoic Acids and a Summary of the Gram-Positive Wall

Many gram-positive *Bacteria* have acidic substances called **teichoic acids** embedded in their cell wall. The term *teichoic acid* includes all wall, membrane, or capsular polymers containing glycerophosphate or ribitol phosphate residues. These polyalcohols are connected by phosphate esters and usually have other sugars and D-alanine attached (Figure 4.31a●). Because they are negatively charged, teichoic acids are partially responsible for the negative charge of the cell surface as a whole. Teichoic acids also function to bind divalent cations such as Ca^{2+} and Mg^{2+}, some of which are transported into the cell. Certain teichoic acids are covalently bound to membrane lipids, and because of this, have been called **lipoteichoic acids.**

Figure 4.31b summarizes the structure of the cell wall of gram-positive *Bacteria* and shows how teichoic

● **Figure 4.31 Teichoic acids and the overall structure of the gram-positive cell wall.** (a) Structure of the ribitol teichoic acid of *Bacillus subtilis*. The teichoic acid is a polymer of the repeating ribitol units shown here. (b) Summary diagram of the gram-positive cell wall.

acids and lipoteichoic acids are arranged in the overall wall structure.

Cells with No Walls

Peptidoglycan—the signature molecule of species of *Bacteria*—can be destroyed by certain agents. One such agent is the enzyme **lysozyme**, a protein that breaks the β-1,4-glycosidic bonds between *N*-acetylglucosamine and *N*-acetylmuramic acid in peptidoglycan (Figure 4.29), thereby weakening the wall. Water then enters the cell, and the cell swells and eventually bursts, a process called **lysis** (Figure 4.32*a*●). Lysozyme is found in animal secretions including tears, saliva, and other body fluids, and presumably functions as a major line of defense against infection by *Bacteria*.

If a solute that does not penetrate the cell, such as sucrose, is added to a cell suspension, the solute concentration outside the cell balances that inside (conditions called *isotonic*). Under these conditions, lysozyme still digests peptidoglycan, but water does not enter the cell and lysis does not occur. Instead, a **protoplast** (a bacterium that has lost its cell wall) is formed (Figure 4.32*b*). If such sucrose-stabilized protoplasts are placed in water, lysis occurs immediately. The word *spheroplast* is often used as a synonym for protoplast, although the two words have slightly different meanings. Protoplasts are cells that are free of residual cell wall material, whereas

(a)

(b)

● **Figure 4.32 Protoplasts and their formation.** (a) In dilute solution breakdown of the cell wall releases the protoplast, but it immediately lyses because the cytoplasmic membrane is structurally very weak. (b) In a solution containing an isotonic concentration of a solute such as sucrose, water does not enter the protoplast and it remains stable. Lysozyme breaks the β-1,4 glycosidic bonds in peptidoglycan (see Figure 4.29).

spheroplasts contain pieces of wall material attached to the otherwise membrane-enclosed structure.

Although most prokaryotes cannot survive in nature without their cell walls, some are able to do so. These include the mycoplasmas, a group that causes certain infectious diseases (⌦ Section 12.21), and the *Thermoplasma* group, *Archaea* that naturally lack cell walls (⌦ Section 13.5). These prokaryotes are free-living protoplasts and are able to survive without cell walls either because they have unusually tough membranes or because they live in osmotically protected habitats, such as the animal body. Certain mycoplasmas have sterols (see Section 4.5) in their cell membranes, which lends strength and rigidity to this structure.

Pseudopeptidoglycan, S-Layers, and Other Cell Walls of *Archaea*

Some species of *Archaea* contain cell walls constructed of a polysaccharide very similar to that of peptidoglycan. This material is called *pseudopeptidoglycan* (Figure 4.33*a*●). The backbone of pseudopeptidoglycan is composed of alternating repeats of *N*-acetylglucosamine and *N*-acetyltalosaminuronic acid (the latter replaces the *N*-acetylmuramic acid of peptidoglycan) (compare Figures 4.29 and 4.33*a*). The backbone of pseudopeptidoglycan also varies from peptidoglycan in that the glycosidic bonds are β-1,3 instead of the β-1,4 bonds found in true peptidoglycan (compare Figures 4.29 and 4.33*a*).

Cell walls of other *Archaea* lack both peptidoglycan and pseudopeptidoglycan and consist of polysaccharide, glycoprotein, or protein. For example, *Methanosarcina* species contain thick polysaccharide walls composed of glucose, glucuronic acid, galactosamine, and acetate. Extremely halophilic (salt-loving) *Archaea* such as *Halococcus* contain cell walls similar to that of *Methanosarcina* but that contain in addition, sulfate (SO_4^{2-}) residues.

The most common cell wall type among *Archaea* is the paracrystalline surface layer (S-layer) (see Section 4.10 for further discussion). The S-layer consists of protein or glycoprotein and generally has a hexagonal symmetry. S-layers have been found among species of all groups of *Archaea*, the extreme halophiles, the methanogens, and the hyperthermophiles (⌦ Section 2.5). Several species of *Bacteria* also have S-layers on their outer surfaces (Figure 4.33*b*).

In species of *Archaea* we thus see a variety of cell wall chemistries, varying from molecules that closely resemble peptidoglycan to walls totally lacking a polysaccharide component. But with rare exception, all *Archaea* contain a cell wall of some sort, and as in *Bacteria*, the archaeal cell wall functions to prevent osmotic lysis and to define cell shape. In addition, because they lack peptidoglycan in their cell walls, all *Archaea* are naturally resistant to the action of lysozyme (see earlier) and penicillin, agents that destroy this molecule or prevent its proper synthesis (⌦ Section 6.2), respectively.

(a) *(b)*

● **Figure 4.33 Pseudopeptidoglycan and S-layers.** (a) Structure of pseudopeptidoglycan, the cell wall polymer of *Methanobacterium* species. Note the resemblance to the structure of peptidoglycan shown in Figure 4.29, especially the peptide cross-links, in this case between *N*-acetyltalosaminuronic acid (NAT) residues instead of muramic acid residues. NAG, *N*–Acetylglucosamine. (b) Transmission electron micrograph of a portion of an S-layer showing the paracrystalline nature of this cell wall layer. Shown is the S-layer from the prokaryote *Aquaspirillum serpens* (a member of the *Bacteria*); this S-layer displays hexagonal symmetry as do many of the S-layers found in *Archaea*.

4.8 Concept Check

The cell walls of *Bacteria* contain a polysaccharide called peptidoglycan. This material consists of strands of alternating repeats of *N*-acetylglucosamine and *N*-acetylmuramic acid, with the latter cross-linked between strands by short peptides. Many sheets of peptidoglycan can be present, depending on the organism. *Archaea* lack peptidoglycan but contain walls made of other polysaccharides or of protein. The enzyme lysozyme destroys peptidoglycan, leading to cell lysis.

◆ List the monomeric components of peptidoglycan.

◆ Why is peptidoglycan such a strong molecule?

◆ How do some cells live without cell walls?

◆ How does pseudopeptidoglycan resemble peptidoglycan? How do the two molecules differ?

4.9 The Outer Membrane of Gram-Negative *Bacteria*

In addition to peptidoglycan, gram-negative *Bacteria* contain an additional wall layer, the **outer membrane**. This layer is effectively a second lipid bilayer, but it is not constructed solely of phospholipid and protein as is the cytoplasmic membrane (Figure 4.16). The outer membrane also contains polysaccharide. The lipid and polysaccharide are linked in the outer membrane to form a *lipopolysaccharide complex*. Because of this, the outer membrane is often called the **lipopolysaccharide layer**, or simply **LPS**.

Chemistry of LPS

Although complex, the chemistry of LPS from several bacteria is now understood. As seen in Figure 4.34●, the polysaccharide portion of LPS consists of two components, the *core polysaccharide* and the *O-polysaccharide*. In *Salmonella* species, where LPS has been best studied, the **core polysaccharide** consists of ketodeoxyoctonate (KDO), seven-carbon sugars (heptoses), glucose, galactose, and *N*-acetylglucosamine. Connected to the core is the *O-specific polysaccharide*, which typically contains galactose, glucose, rhamnose, and mannose (all hexoses), as well as one or more unusual dideoxy sugars such as abequose, colitose, paratose, or tyvelose. These sugars are connected in four- or five-membered sequences, which often are branched. When the sequences repeat, the long *O*-polysaccharide is formed.

The relationship of the *O*-polysaccharide to the rest of the LPS is shown in Figure 4.35●. The lipid portion of the

● **Figure 4.34** **Structure of the lipopolysaccharide of gram-negative *Bacteria*.** The precise chemistry of lipid A and the polysaccharide components varies among species of gram-negative *Bacteria*, but the sequence of major components (lipid A–KDO–core–O-specific) is generally uniform. The O-specific polysaccharide varies among species. KDO, ketodeoxyoctonate; Hep, heptose; Glu, glucose; Gal, galactose; GluNac, *N*-acetylglucosamine; GlcN, glucosamine; P, phosphate. Glucosamine and the lipid A fatty acids are linked by an amine ester bond. The lipid A portion of LPS can be toxic to animals and comprises the *endotoxin complex* (∞ Section 21.12). Compare this figure with Figures 4.35 and 4.36, and note the color coding of different portions of the LPS in Figures 4.34 and 4.35.

lipopolysaccharide, referred to as **lipid A** (Figure 4.34), is not a glycerol lipid, but instead the fatty acids are connected by amine ester linkage to a disaccharide composed of *N*-acetylglucosamine phosphate (Figure 4.34). The disaccharide is attached to the core polysaccharide through KDO (Figure 4.34). Fatty acids commonly found in lipid A include caproic, lauric, myristic, palmitic, and stearic acids.

In the outer membrane, LPS associates with several proteins to form the *outer* leaflet of the membrane. A **lipoprotein** complex is found on the *inner* leaflet of the LPS of a number of gram-negative *Bacteria* (Figure 4.35*a*). Lipoprotein functions as an anchor between the outer membrane and peptidoglycan. Finally, in the *outer* leaflet of the outer membrane, LPS replaces phospholipids. The latter are found only in the inner leaf (Figure 4.35*a*). Thus, although the outer membrane can be considered a lipid bilayer, its structure is quite distinct from that of the cytoplasmic membrane (compare Figures 4.16 and 4.35*a*).

Endotoxin

Although the major function of the outer membrane is structural, one of its important biological properties is its *toxicity* to animals. Gram-negative *Bacteria* that are pathogenic for humans and other mammals include members of the genera *Salmonella*, *Shigella*, and *Escherichia*, among others, and some of the symptoms these pathogens elicit in their hosts are due to their toxic outer membrane.

The toxic properties are associated with part of the lipopolysaccharide layer, in particular, *lipid A*. The term **endotoxin** refers to this toxic component of LPS, as we will discuss in Section 21.12. Some endotoxins cause violent symptoms in humans, including severe gastrointestinal distress (gas, diarrhea, vomiting). Endotoxins are responsible for a number of bacterial illnesses, including in particular, *Salmonella* food infection (∞ Section 29.7). Interestingly, LPS from several nonpathogenic bacteria have also been shown to have endotoxin activity. Thus, the organism itself need not be pathogenic to contain toxic outer membrane components.

Porins and the Periplasm

Unlike the cytoplasmic membrane, the outer membrane of gram-negative *Bacteria* is relatively permeable to small molecules even though it is basically a lipid bilayer. This is because proteins called **porins** are present in the outer membrane that function as channels for the entrance and exit of hydrophilic low-molecular-weight substances (Figure 4.35). Several different porins are known, including both specific and nonspecific classes. *Nonspecific porins* form water-filled channels through which any small substance can pass. By contrast, some porins are highly specific because they contain a specific binding site for one or a group of structurally related substances. Structural studies have shown that most porins are proteins that contain *three* identical subunits. Porins are transmembrane proteins (Figure 4.35*a*) and associate to form small holes about 1 nm in diameter in the outer membrane (Figure 4.35*b*).

Although permeable to small molecules, the outer membrane is *not* permeable to enzymes or other large molecules. In fact, one of the major functions of the outer membrane is to keep proteins that are present outside the cytoplasmic membrane from diffusing away from the cell. These enzymes are present in a region called the **periplasm** (see Figures 4.35 and 4.36). This space, located between the outer surface of the cytoplasmic membrane and the inner surface of the outer membrane, is about 12–15 nm wide. The periplasm contents are gel-like in consistency because of the abundance of periplasmic proteins found there (Figure 4.36●).

Depending on the organism, the periplasm can contain several proteins. These include *hydrolytic enzymes*, which function in the initial degradation of food molecules, *binding proteins*, which begin the process of transporting substrates (see Section 4.7), and *chemoreceptors*, which are proteins involved in the chemotaxis response (see Sections 4.16 and 8.12). As previously discussed, most of these proteins reach the periplasm via transport by the SecYEG system (see Section 4.7).

(a)

● **Figure 4.35** **The gram-negative cell wall.** Note that although the outer membrane is often called the "second lipid bilayer," the chemistry and architecture of this layer differ in many ways from that of the cytoplasmic membrane. (a) Arrangement of lipopolysaccharide, lipid A, phospholipid, porins, and lipoprotein in the outer membrane. See Figure 4.34 for details of the structure of LPS. Lipid A can be toxic in humans, and if so, is referred to as endotoxin (⌒⌒⌒ Section 21.12). (b) Molecular model of porin proteins. Note the four pores present, one formed from each of the proteins forming a porin molecule and, a central pore between the porin proteins. The view is perpendicular to the plane of the membrane. Model based on X-ray diffraction studies of *Rhodobacter blasticus* porin.

(b)

Relationship of Cell Wall Structure to the Gram Stain

The structural differences between the cell walls of gram-positive and gram-negative *Bacteria* are thought to be responsible for differences in the Gram stain reaction. In the Gram stain (see Section 4.1), an insoluble crystal violet-iodine complex is formed inside the cell. This complex is extracted by alcohol from gram-*negative* but not from gram-*positive Bacteria* (see Figure 4.4). As we have seen, gram-positive *Bacteria* have very thick cell walls consisting of several layers of peptidoglycan. These become dehydrated by the alcohol, causing the pores in the walls to close and preventing the insoluble crystal violet-iodine complex from escaping. By contrast, in gram-negative *Bacteria*, alcohol readily penetrates the lipid-rich outer membrane and extracts the crystal violet-iodine complex from the cell.

● **Figure 4.36** **The cell wall of *Escherichia coli*.** High-magnification thin section of the cell envelope of *E. coli* showing the periplasmic gel bounded by the outer membrane and the cytoplasmic membrane. The large, dark particles in the cytoplasm are ribosomes.

4.9 Concept Check

In addition to peptidoglycan, gram-negative *Bacteria* contain an outer membrane consisting of lipopolysaccharide, protein, and lipoprotein. Proteins called porins allow for permeability across the outer membrane. The space between the membranes is the periplasm, which contains various proteins involved in important cellular functions.

◆ What components constitute the LPS layer of gram-negative *Bacteria*?

◆ What is the function of porins and where are they located in a gram-negative cell wall?

◆ What component of the cell has endotoxin properties?

◆ Why does alcohol readily decolorize gram-negative but not gram-positive *Bacteria*?

III SURFACE STRUCTURES AND INCLUSIONS OF PROKARYOTES

In addition to the cell wall and related surface layers just discussed, some prokaryotic cells have other outer layers or structures in contact with the environment. Moreover, most cells can form one or more types of cellular inclusions of one sort or another. The inclusions are then later metabolized as nutrient sources. We examine some of these surface and internal inclusions here.

4.10 Bacterial Cell Surface Structures

Prokaryotes can produce a variety of structures that are attached to or in some way protrude from the cell surface. These include fimbriae, pili, S-layers, capsules, and other slime layers. We examine these structures now.

Fimbriae and Pili

Fimbriae and **pili** are short filamentous structures composed of protein that extend from the surface of a cell. *Fimbriae* (Figure 4.37●) enable organisms to stick to surfaces, including animal tissues in the case of some pathogenic bacteria, or to form pellicles or biofilms (◌◍ Section 19.3) on surfaces. Notorious among these pathogens include *Salmonella typhimurium* (salmonellosis), *Neisseria gonorrhoeae* (gonorrhea), and *Bordetella pertussis* (whooping cough).

Pili are structurally similar to fimbriae but are typically longer, and only one or a few pili are present on the surface. Because they serve as receptors for certain types of viruses, pili can be seen under the electron microscope when they become coated with virus particles (Figure 4.38●). Although possibly involved in attachment as for fimbriae, pili are clearly involved in the process of conjugation (a form of genetic exchange) in prokaryotes, as will be discussed in Section 10.9.

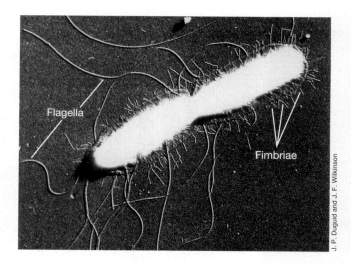

● **Figure 4.37 Fimbriae.** Electron micrograph of a dividing cell of *Salmonella typhi*, showing flagella and fimbriae. A single cell is about 0.9 μm in diameter.

Many classes of fimbriae/pili are known, distinguished by their structure and function. One class, called *type IV fimbriae/pili*, is involved in an unusual form of motility in certain bacteria, called *twitching motility*. Twitching motility is a type of movement on solid surfaces, where it is thought that rapid and reversible extension and retraction of the fimbriae allow the cell to crawl along the surface. Unlike other fimbriae, type IV fimbriae are found only at the poles of the cells, and besides motility, have been implicated as key host colonization factors in a variety of pathogens including *Vibrio cholerae* (cholera), and *Neisseria menigitidis* (bacterial meningitis). Type IV fimbriae are also thought to mediate genetic transfer by the process of transformation (◌◍ Section 10.7) in a wide variety of bacteria.

Paracrystalline Surface Layers

Many prokaryotes contain a cell surface layer composed of a two-dimensional array of protein. These layers are called **S-layers**. S-layers have been detected in representatives of virtually every phylogenetic grouping of *Bacteria* and are widespread among *Archaea*. In some

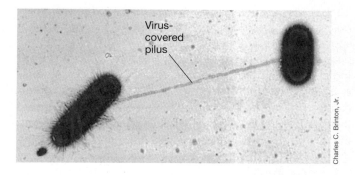

● **Figure 4.38 Pili.** The presence of pili on an *Escherichia coli* cell is revealed by the use of viruses that specifically adhere to the pilus. The cell is about 0.8 μm in diameter.

species of *Archaea* the S-layer is also the cell wall (see Section 4.8). S-layers have a crystalline appearance and show various symmetries, such as hexagonal, tetragonal, or trimeric, depending upon the number and structure of the protein or glycoprotein subunits of which they are composed (see Figure 4.33*b* to view an electron micrograph of an S-layer).

The major function of S-layers is unknown. However, as the interface between the cell and its environment, it is likely that the S-layer functions as a selective sieve, allowing the passage of low-molecular-weight substances while excluding large molecules and structures (such as viruses). The S-layer would then function to retain proteins near the cell, much like the outer membrane does in gram-negative bacteria. Evidence also exists that in pathogenic bacteria that contain S-layers, this structure may provide protection against certain host defense mechanisms.

Capsules and Slime Layers

Many prokaryotic organisms secrete slimy or gummy materials on their surfaces (Figure 4.39●). A variety of these structures consist of polysaccharide, and a few consist of protein. The terms **capsule** and **slime layer** are frequently used to describe these polysaccharide layers. The composition of these layers varies in different organisms but may be thick or thin, rigid or flexible, depending on their chemical nature. The rigid layers are organized in a tight matrix that excludes small particles, such as india ink; this form is called a *capsule* (Figure 4.39*a*). If the layer is more easily deformed, it will not exclude particles and is more difficult to see; this form is called a *slime layer*.

Polysaccharide layers have several functions in bacteria. Surface polysaccharides assist in the *attachment* of microorganisms to solid surfaces. As we will see (∞ Section 21.6), pathogenic microorganisms that enter the animal body by specific routes usually do so by first binding specifically to surface components of host tissues. This binding is often mediated by surface polysaccharides on the bacterial cell. Many non-pathogenic bacteria also bind to solid surfaces in nature, sometimes forming a thick layer of cells called a **biofilm**. Polysaccharides play a key role in the development of biofilms (∞ Section 19.3).

Slime layers play other roles as well. For example, encapsulated pathogenic bacteria are typically more difficult for phagocytic cells of the immune system (∞ Sections 22.2) to recognize and subsequently destroy. In addition, because outer polysaccharide layers bind a significant amount of water, it is likely that these layers play some role in resistance to desiccation.

⬣ 4.10 Concept Check

Prokaryotic cells often contain various surface structures. These include fimbriae and pili, S-layers, capsules, and slime layers. These structures have several functions, but a key one is in attaching cells to a solid surface.

◆ How do fimbriae differ from pili, both structurally and functionally?

◆ Although they lack lipid, how could S-layers play a role similar to that of the outer membrane?

4.11 Cell Inclusions

Granules or other inclusions are often seen within cells. Their nature differs in different organisms, but they commonly function as energy reserves or as a reservoir of structural building blocks. Inclusions can often be seen directly with the light microscope (see Figure 4.41). Most cellular inclusions are enclosed by a thin *nonunit* membrane consisting of lipid separating the inclusion from the cytoplasm proper.

● **Figure 4.39 Bacterial capsules.** (a) A capsule in *Acinetobacter* species observed by negative staining with india ink and phase-contrast microscopy. The india ink does not penetrate the capsule, and so it is revealed in outline as a light structure on a dark background. (b) Electron micrograph of a thin section of a *Rhizobium trifolii* cell stained with ruthenium red to reveal the capsule. The diameter of the cell proper (not including the capsule) is about 0.7 μm. Although most capsules consist of polysaccharide, some bacteria contain protein capsules. Cells of *Bacillus anthracis*, for example, an organism associated with both animal disease and bioterrorism (∞ Sections 25.12 and 25.13), contain a poly-D-glutamic acid capsule that is effective in preventing cell destruction by host defenses.

(a)

(b)

Elliot Juni

Frank Dazzo and Richard Heinzen

Carbon Storage Polymers

In prokaryotic organisms, one of the most common inclusion bodies consists of **poly-β-hydroxbutyric acid (PHB)**, a lipid that is formed from β-hydroxbutyric acid units (Figure 4.40a●). The monomers of this acid are connected by ester linkages, forming the long PHB polymer, and these polymers aggregate into granules (Figure 4.40b). The length of the monomer in the polymer can vary considerably, from as short as C_4 to as long as C_{18} in certain organisms. Thus, the more generic term *poly-β-hydroxyalkanoate* (PHA) is used to describe this whole class of carbon/energy storage polymers. PHAs are synthesized when carbon is in excess and are broken down for use as carbon skeletons for biosynthesis or to make ATP when conditions warrant. A wide variety of prokaryotes, including species of both *Bacteria* and *Archaea*, produce PHAs.

Another storage product formed by prokaryotes is **glycogen**, which is a polymer of glucose (∞ Section 3.3 and Figure 3.6). Like PHAs, glycogen is a storage depot for carbon and energy. Glycogen is produced when carbon is in excess in the environment and is consumed when carbon is limiting. Glycogen resembles starch, the major storage reserve of plants, but differs from starch in the way the glucose units are linked together (see Figure 3.6b).

Other Storage Materials and Inclusions

Many microorganisms accumulate inorganic phosphate in the form of granules of **polyphosphate**. These granules can be degraded and used as sources of phosphate for nucleic acid and phospholipid. In addition, many prokaryotes are capable of oxidizing reduced sulfur compounds such as hydrogen sulfide (H_2S). These oxidations are linked to either reactions of energy metabolism (∞ Sections 17.8 and 17.10) or biosynthesis (∞ Section 17.6). In either case, **elemental sulfur** may accumulate inside the cell in readily visible globules (Figure 4.41●). These globules of elemental sulfur remain as long as the source of reduced sulfur is still present. However, as the reduced sulfur source becomes limiting, the sulfur in the granules is oxidized to sulfate (SO_4^{2-}), and the granules slowly disappear as this reaction proceeds.

Interestingly, the sulfur globules themselves actually reside in the *periplasm* rather than the *cytoplasm* (sulfur-oxidizing phototrophic or chemolithotrophic prokaryotes are gram-negative organisms). The periplasm expands outwards to accommodate the globules as H_2S is oxidized ($H_2S \rightarrow S^0$). The periplasm then contracts inward as sulfur is oxidized ($S^0 \rightarrow SO_4^{2-}$). It is considered likely that certain other "intracytoplasmic" inclusions may be periplasmic in gram-negative bacteria, as well. Thus far, it appears that at least poly-β-hydroxyalkanoates (see Figure 4.40) fall into this category.

Magnetosomes are intracellular particles of the iron mineral magnetite—Fe_3O_4 (Figure 4.42●). Magnetosomes impart a magnetic dipole to a cell, allowing it to respond to a magnetic field. Bacteria that produce magnetosomes (Figure 4.42a) exhibit *magnetotaxis*, the

(a)

β-carbon

(b)

Poly-β-hydroxybutyrate

F. R. Turner and M. T. Madigan

● **Figure 4.40 Poly-β-hydroxybutyrate (PHB).** (a) Chemical structure of PHB, a common poly-β-hydroxyalkanoate. A monomeric unit is shown in color. Other alkanoate polymers are made by substituting longer-chain hydrocarbons for the $-CH_3$ group on the β carbon. (b) Electron micrograph of a thin section of cells of the phototrophic bacterium *Rhodovibrio sodomensis* (see also Figure 4.10b) containing granules of PHB.

Norbert Pfennig

● **Figure 4.41 Sulfur globules.** Bright-field photomicrograph of cells of the purple sulfur bacterium *Isochromatium buderi*. Note the sulfur globules inside the cell, obtained from the oxidation of hydrogen sulfide (H_2S). A single cell measures about $4 \times 7 \ \mu m$.

Stefan Spring

R. Blakemore and W. O'Brien

(a)   (b)

Dennis Bazylinski

(c)

● **Figure 4.42   Magnetotactic bacteria and magnetosomes.** (a) Interference contrast micrograph of coccoid magnetotactic bacteria. Note magnetosomes. A single cell is about 2.2 μm in diameter. (b) Magnetosomes isolated from the magnetotactic bacterium *Magnetospirillum magnetotacticum*. Each particle is about 50 nm in length (⚬⚬ Figure 12.32). (c) Transmission electron micrograph of magnetosomes from a magnetic coccus. The arrow points to the membrane surrounding each magnetosome. A single magnetosome is about 90 nm wide. Although containing lipids and proteins, the magnetosome membrane, like the PHB membrane (see Figure 4.40), is a monolayer and not a true "unit" membrane.

process of orienting and migrating along Earth's magnetic field lines (⚬⚬ Section 12.14). Although the suffix -*taxis* is used in the word *magnetotaxis*, there is no evidence that magnetotactic bacteria employ the sensory systems of chemotactic or phototactic bacteria (see Sections 4.16 and 8.12). Instead, the alignment of magnetosomes in the cell simply imparts magnetic properties to it, which then orient the cell in a particular direction in its environment.

The major function of magnetosomes is unknown. However, magnetosomes have been found in a variety of aquatic *Bacteria* (Figure 4.42*a*) that grow best at low O_2 concentrations. It has thus been hypothesized that a function of magnetosomes is to guide these aquatic cells toward the sediments (since the major magnetic field line points downwards) where O_2 levels are lower.

Magnetosomes are surrounded by a membrane containing phospholipids, proteins, and glycoproteins (Figure 4.42*b, c*). This membrane is not a unit membrane, as is the cytoplasmic membrane (Figure 4.16), but instead is a nonunit membrane, like that surrounding

granules of poly-β-hydroxybutyrate (PHB, Figure 4.40). Magnetosome membrane proteins probably play a role in precipitating Fe^{3+} (brought into the cell in soluble form by chelating agents) as Fe_3O_4 in the developing magnetosome. The morphology of magnetosomes appears to be species-specific, varying in shape from square to rectangular to spike-shaped in certain bacteria, forming into chains inside the cell (Figure 4.42).

🛑 **4.11   Concept Check**

Prokaryotic cells often contain internal granules such as sulfur, polyphosphate, PHAs, and magnetosomes. These substances function as storage materials or in magnetotaxis.

◆ Under what growth conditions would you expect PHAs or glycogen to be produced?

◆ Why would it be impossible for *gram-positive* bacteria to store sulfur like sulfur-oxidizing chemolithotrophs can?

◆ What form of iron is present in magnetosomes?

4.12   Gas Vesicles

A number of prokaryotic organisms are *planktonic*, living a floating existence within the water column of lakes and the oceans. Many planktonic prokaryotes produce **gas vesicles**. These structures confer buoyancy on cells by decreasing their density. Gas vesicles are actually a means of motility, allowing cells to float up and down in a water column in response to environmental factors.

The most dramatic instances of flotation due to gas vesicles are seen in cyanobacteria that form massive accumulations (blooms) in lakes (Figure 4.43●). Gas-vesiculate cells rise to the surface of the lake and are blown by winds into dense masses. Gas vesicles are also present in certain purple and green phototrophic bacteria (⚬⚬ Sections 12.2 and 12.32) and in some nonphototrophic bacteria that live in lakes and ponds. Some species of *Archaea* also contain gas vesicles.

T. D. Brock

● **Figure 4.43   Gas vesicles in nature.** Flotation of gas vesiculate cyanobacteria from a bloom on a nutrient-rich lake, Lake Mendota, Madison, Wisconsin.

A. E. Konopka and J.T. Staley

● **Figure 4.44 Isolated gas vesicles.** Transmission electron micrographs of gas vesicles purified from the bacterium *Ancyclobacter aquaticus* and examined in negatively stained preparations. A single gas vesicle is about 100 nm in diameter. [Reproduced with permission from *Archives of Microbiology* 112:133–140 (1977).]

Structure of Gas Vesicles

Gas vesicles are spindle-shaped gas-filled structures made of protein. They are hollow yet rigid and of variable length and diameter (Figure 4.44●). Gas vesicles in different organisms vary in length from about 300 to over 1000 nm and in width from 45 to 120 nm, but the vesicles of any given organism are more or less of constant size.

Gas vesicles may number from a few to hundreds per cell. The gas vesicle membrane is composed of protein, is about 2 nm thick, and is impermeable to water and solutes but permeable to gases. The presence of gas vesicles in cells can be determined either by light microscopy (where clusters of vesicles, called *gas vacuoles*, show up as irregular bright inclusions) or by electron microscopy (Figure 4.45●).

Molecular Structure of Gas Vesicles

Gas vesicles are composed of two different proteins (Figure 4.46●). The major gas vesicle protein, called *GvpA*, is a small, highly hydrophobic and very rigid protein. The rigidity of the gas vesicle membrane is essential for the structure to resist the pressures exerted on it from outside. GvpA makes up the gas vesicle shell and composes 97% of total gas vesicle protein. The minor protein, called *GvpC*, functions to strengthen the shell of the gas vesicle (Figure 4.46).

Gas vesicles are constructed of copies of the GvpA protein aligned as parallel "ribs" forming a watertight shell. The GvpA ribs are strengthened by GvpC, which cross-links GvpA at an angle, binding several GvpA molecules together like a clamp (Figure 4.46). The final shape of the gas vesicle, which can vary in different organisms from long and thin to short and fat (compare Figures 4.44 and 4.45*b*), is a function of how the GvpA and GvpC proteins are arranged to form the intact vesicle.

Because the gas vesicle membrane is freely permeable to gases, the composition and pressure of the gas inside a gas vesicle is the same as that of the gas in which the organism is suspended. And, because the gas

(a)

A. E. Walsby

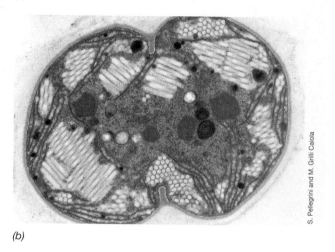

(b)

S. Pellegrini and M. Grilli Caiola

● **Figure 4.45 Gas vesicles of the cyanobacteria *Anabaena* and *Microcystis.*** (a) *Anabaena flos-aquae.* The dark cell in the center (a heterocyst) lacks gas vesicles. In the other cells, the vesicles group together as phase-bright gas vacuoles (arrows). (b) Transmission electron micrograph of the cyanobacterium *Microcystis.* Gas vesicles are arranged in bundles, here observable in both longitudinal and cross section.

vesicle attains a density of about 5–20% of that of the cell proper, gas vesicles decrease the density of the cell, thereby increasing its buoyancy. Aquatic phototrophic organisms in particular benefit from this because it allows them to adjust their position vertically in a water column to regions where the light intensity for photosynthesis is optimal.

GvpA

4.6 nm

GvpC

● **Figure 4.46 Gas vesicle proteins.** Model of how the two proteins that make up the gas vesicle, GvpA and GvpC, interact to form a watertight but gas-permeable structure. GvpA makes up the rib and is a rigid β-sheet. GvpC is the cross-linker and is of an α-helix structure (⬭⬭ Section 3.7 and Figure 3.16).

(a)　　　　　　　　　　　(b)　　　　　　　　　　　(c)

● **Figure 4.47** **The bacterial endospore.** Phase contrast photomicrographs illustrating several types of endospore morphologies and intracellular locations. (a) Terminal. (b) Subterminal. (c) Central.

 4.12 **Concept Check**

Gas vesicles are small gas-filled structures made of protein that function to confer buoyancy on cells. Gas vesicles contain two different proteins arranged to form a gas permeable, but watertight structure.

◆ What might be the benefit of gas vesicles to phototrophic cells? (*Hint*: What do phototrophs need to grow?).

◆ How are the two proteins that make up the gas vesicle, GvpA and GvpC, arranged to form such a water-impermeable structure?

4.13　Endospores

Certain *Bacteria* produce structures called **endospores** (Figure 4.47●) during a process called *sporulation* (see Figure 4.50). Endospores (the prefix "endo" means "within") are differentiated cells that are highly resistant to heat and are difficult to destroy, even by harsh chemicals or radiation. The biological function of endospores is undoubtedly to enable the organism to endure difficult times, including extremes of temperature, drying, and nutrient depletion. Endospores are also ideal structures for dispersal by wind, water, or through the animal gut. Endospore-forming bacteria are found most commonly in the soil, and the genera *Bacillus* and *Clostridium* are the best studied of endospore-forming bacteria.

The discovery of bacterial endospores was of immense importance to microbiology because the knowledge of such remarkably heat-resistant forms was essential for the development of adequate methods of sterilization, not only of culture media but also of foods and other perishable products. Although many microorganisms form spores, the bacterial endospore is unique in its degree of heat resistance. Highly resistant bacterial endospores survive heating to temperatures as high as 150°C, although an autoclave operating at 121°C should kill the endospores of most species. Endospores are also resistant to other harmful agents such as drying, ultraviolet radiation, strong acids or bases, and chemical disinfectants, and can remain dormant for extremely long periods (see the Microbial Sidebar, How Long Can an Endospore Survive?).

Endospore Structure

Endospores stand out under the light microscope as strongly refractile structures (see Figure 4.47). Endospores are impermeable to most dyes, so occasionally they are seen as unstained regions within cells that have been stained with basic dyes such as methylene blue. To stain endospores, special staining procedures must be used.

The structure of the endospore as seen with the electron microscope differs distinctly from that of the vegetative cell (Figure 4.48●). In particular, the endospore is

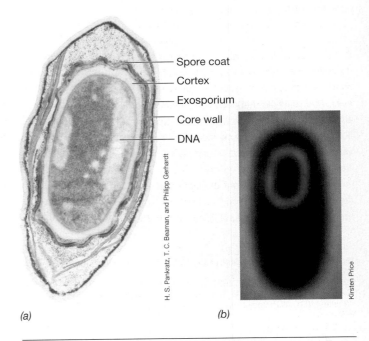

Spore coat
Cortex
Exosporium
Core wall
DNA

(a)　　　　　　　　　　　(b)

● **Figure 4.48** **Dissecting the bacterial endospore.** (a) Transmission electron micrograph of a mature endospore from *Bacillus megaterium*. (b) Fluorescent photomicrograph of a cell of *Bacillus subtilis* undergoing sporulation. The green area is due to a dye that specifically stains a sporulation protein in the spore coat.

Microbial Sidebar ◆ How Long Can an Endospore Survive?

In this chapter we have discussed the dormancy and resistance properties of bacterial endospores and have pointed out that endospores can survive for long periods in a dormant state. But how long is *long*?

Evidence for endospore longevity suggests that these structures can remain viable (that is, capable of germination into vegetative cells) for at least several decades and probably for much longer than that. A suspension of endospores of the bacterium *Clostridium aceticum* (Figure 1*a*) prepared in 1947 was placed in sterile growth medium in 1981, 34 years later, and in less than 12 h growth commenced, leading to a robust culture. *Clostridium aceticum* was originally isolated by the Dutchman K. T. Wieringa in 1940 but was thought to have been lost until this vial of *C. aceticum* endospores was found in a storage room at the University of California at Berkeley and revived.[1] *C. aceticum* is a homoacetogen, capable of making acetate from $CO_2 + H_2$ or from glucose (⚬⚬⚬ Section 17.16).

Other more extreme examples of endospore longevity have been documented. Bacteria of the genus *Thermoactinomyces* are thermophilic endospore formers that are widespread in nature in soil, plant litter, and fermenting plant material. Microbiological examination of a Roman archaeological site in the United Kingdom that was dated to over *2000* years ago yielded significant numbers of viable *Thermoactinomyces* endospores in various pieces of debris. Additionally, *Thermoactinomyces* endospores were recovered in fractions of sediment cores from a Minnesota lake known to be over *7000* years old. Although contamination is always a possibility in such studies, samples in both of these cases were processed in such a way as to virtually rule out contamination with "recent" endospores.[2]

What factors could limit the age of an endospore? Cosmic radiation has been considered a major factor because it can introduce

Gerhard Gottschalk

William D. Grant

(a)   *(b)*

Figure 1 Longevity of endospores. *(a) Photograph of a tube containing endospores from the bacterium* Clostridium aceticum *prepared on May 7, 1947. After remaining dormant for over 30 years, the endospores were suspended in a culture medium after which growth occurred within 12 h.[1] (b) Halophilic bacteria trapped within salt crystals. These crystals (about 1 cm in diameter) were grown in the laboratory in the presence of* Halobacterium *cells (orange) that remain viable in the crystals. Crystals similar to these but of Permian age (≈250 million years old) were reported to contain viable halophilic endosporulating bacteria.*

mutations in DNA.[2] It has been hypothesized that over periods of thousands of years, the cumulative effects of cosmic radiation could introduce so many mutations into the genome of an organism that even highly radiation-resistant structures such as endospores would succumb to the genetic damage. However, extrapolations from actual experimental assessments of the effect of natural radiation on endospores suggest that if suspensions of endospores were partially shielded from cosmic radiation, for example, by being embedded in layers of organic mat-

ter, they could retain viability over periods as great as *several hundred thousand years* and perhaps even longer. Amazing, but is this the upper limit?

In 1995 a group of scientists reported the revival of bacterial endospores they claimed were 25–40 million years old.[3] The endospores were allegedly preserved in the gut of an extinct bee trapped in amber of known geological age. The presence of endospore-forming bacteria in these bees was previously suspected from electron microscopic studies of the insect gut that showed endospore-like structures and because *Bacillus*-like DNA was recovered from the insect. Incredibly, samples of bee tissue incubated in a sterile culture medium quickly yielded endospore-forming bacteria. Rigorous precautions were taken to demonstrate that the endospore-forming bacterium revived from the amber-encased bee was not a modern-day contaminant. An even more spectacular claim was made that halophilic (salt-loving) endospore-forming bacteria had been isolated from fluid inclusions in salt crystals of Permian age, over 250 million years old.[4] Presumably these cells were trapped in brines within the crystal (Figure 1*b*) as it formed eons ago, and remained viable for over a quarter billion years!

If these claims of almost unbelievable endospore longevity are supported by repetition of the results in independent laboratories (and such confirmation is crucial for verifying such highly controversial findings), then endospores stored under the proper conditions can likely remain viable indefinitely. This is a remarkable testimony to the endospore, a structure that undoubtedly evolved to maintain viability for relatively short periods but that turned out to be such a well-designed structure that dormancy for hundreds of thousands, if not millions, of years may be possible. ◾

[1]Braun, M., F. Mayer, and G. Gottschalk. 1981. *Clostridium aceticum* (Wieringa), a microorganism producing acetic acid from molecular hydrogen and carbon dioxide. *Arch. Microbiol.* 128: 288–293.
[2]Gest, H., and J. Mandelstam. 1987. Longevity of microorganisms in natural environments. *Microbiol. Sci.* 4: 69–71.
[3]Cano, R. J., and M. K. Borucki. 1995. Revival and identification of bacterial spores in 25- to 40-million-year-old Dominican amber. *Science* 268: 1060–1064.
[4]Vreeland, R.H., W.D. Rosenzweig, and D.W. Powers. 2000. Isolation of a 250 million-year-old halotolerant bacterium from a primary salt crystal. *Nature* 407: 897–900.

structurally more complex in that it has many layers that are absent from the vegetative cell. The outermost layer is the **exosporium**, a thin protein covering. Within this are the **spore coats**, composed of layers of spore-specific proteins (Figure 4.48*b*). Below the spore coat is the **cortex**, which consists of loosely cross-linked peptidoglycan, and inside the cortex is the **core** or **spore protoplast**, which contains the core wall, cytoplasmic membrane, cytoplasm, nucleoid, ribosomes, and other cellular essentials (∞ Section 2.1 and Figure 2.1*a*). Thus, the endospore differs structurally from the vegetative cell primarily in the kinds of structures found *outside* the core wall.

One chemical substance that is characteristic of endospores but absent from vegetative cells is **dipicolinic acid** (Figure 4.49●). This substance has been found in endospores from all endospore-forming bacteria examined and is located in the core. Endospores are also enriched in calcium ions, most of which are combined with dipicolinic acid. The calcium-dipicolinic acid complex of the core represents about 10% of the dry weight of the endospore. The complex functions to reduce water availability within the endospore, thus helping to dehydrate it. In addition, the complex intercalates (inserts between bases) in DNA, and in so doing, stabilizes it to heat denaturation.

Properties of the Endospore Core

The core of a mature endospore differs greatly from the vegetative cell from which it was formed. Besides the high levels of calcium dipicolinate (Figure 4.49), which reduces the water content of the core, the core becomes further dehydrated during the sporulation process. The core of a mature endospore contains only 10–25% of the water content of the vegetative cell, and thus the consistency of the core cytoplasm is that of a gel. This dehydration of the core greatly increases the heat resistance of the endospore. However, dehydration has also been shown to confer resistance in the endospore to chemicals, such as hydrogen peroxide (H_2O_2), and causes enzymes remaining in the core to become inactive.

In addition to the low water content of the endospore, the pH of the core is about one unit lower than that of the vegetative cell cytoplasm. Moreover, the core contains high levels of proteins called *small acid-soluble proteins* (SASPs). These are made during the sporulation process and have at least two functions. SASPs bind tightly to DNA in the core and protect it from potential damage from ultraviolet radiation, desiccation, and dry heat. Ultraviolet resistance is conferred when SASPs change the molecular structure of DNA from the normal "B" form to the more compact "A" form. A-form DNA is more resistant to the formation of pyrimidine dimers by UV radiation, a means of mutation (∞ Section 10.4), and to the denaturing effects of dry heat. In addition, SASPs function as a carbon and energy source for the outgrowth of a new vegetative cell from the endospore, a process called *germination* (discussed later in this section).

Endospore Formation

During endospore formation, a vegetative cell is converted to a nongrowing, heat-resistant structure (Figure 4.50●). As previously described and as summarized in Table 4.3, the differences between the endospore and the vegetative cell are profound. Sporulation involves a complex series of events in *cellular differentiation*. Bacterial sporulation does not occur when cells are dividing exponentially, but only when growth ceases owing to the exhaustion of an essential nutrient. Thus, cells of *Bacillus*, a typical endospore-forming bacterium, cease vegetative growth and begin sporulation when a key nutrient such as the carbon or nitrogen source becomes limiting.

Many genetically directed changes in the cell underlie the conversion from vegetative growth to sporulation. The structural changes occurring in sporulating cells of *Bacillus* are shown in Figure 4.51●. The process of sporulation can be divided into several stages. In *Bacillus subtilis*, where detailed studies have been done, the entire sporulation process takes about 8 hours. Genetic studies of mutants of *Bacillus*, each blocked at one

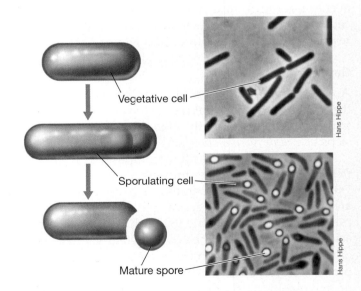

● Figure 4.49 Dipicolinic acid (DPA). (a) Structure of DPA. (b) How Ca^{2+} cross-links DPA molecules to form a complex.

(a)

(b) Carboxylic acid groups

Vegetative cell

Sporulating cell

Mature spore

Hans Hippe

Hans Hippe

● Figure 4.50 Sporulation. Phase contrast photomicrographs are of cells of *Clostridium pascui*.

Table 4.3	Differences between endospores and vegetative cells	
Characteristic	**Vegetative cell**	**Endospore**
Structure	Typical gram-positive cell; a few gram-negative cells	Thick spore cortex Spore coat Exosporium
Microscopic appearance	Nonrefractile	Refractile
Calcium content	Low	High
Dipicolinic acid	Absent	Present
Enzymatic activity	High	Low
Metabolism (O_2 uptake)	High	Low or absent
Macromolecular synthesis	Present	Absent
mRNA	Present	Low or absent
Heat resistance	Low	High
Radiation resistance	Low	High
Resistance to chemicals (for example, H_2O_2) and acids	Low	High
Stainability by dyes	Stainable	Stainable only with special methods
Action of lysozyme	Sensitive	Resistant
Water content	High, 80–90%	Low, 10–25% in core
Small acid-soluble proteins (product of *ssp* genes)	Absent	Present
Cytoplasmic pH	About pH 7	About pH 5.5–6.0 (in core)

of the stages of sporulation shown in Figure 4.51, indicate that more than 200 genes are involved in the sporulation process. Sporulation requires that the synthesis of many proteins involved in vegetative cell functions cease and that specific endospore proteins be made (see Figure 4.48b). This is accomplished by activation of a variety of endospore-specific genes including *spo*, *ssp* (which encodes SASPs), and many other genes in response to an environmental trigger to sporulate. The proteins encoded by these genes catalyze the series of events leading from a moist, metabolizing vegetative cell to a relatively dry, metabolically inert but extremely resistant, endospore (Table 4.3 and Figure 4.51).

Germination

An endospore can remain dormant for many years (see the Microbial Sidebar), but it can convert back to a vegetative cell relatively rapidly. This process involves three steps: *activation, germination,* and *outgrowth* (Figure 4.52●).

Activation is most easily accomplished by heating freshly formed endospores for several minutes at an elevated but sublethal temperature. Activated endospores are then conditioned to germinate when placed in the presence of specific nutrients, such as certain amino acids (for example, alanine is a particularly good trigger of endospore germination). *Germination*, usually a rapid process (on the order of several minutes), involves loss of microscopic refractility of the endospore, increased ability to be stained by dyes, and loss of resistance to

heat and chemicals. Loss of calcium dipicolinate and cortex components from the endospores occurs during this stage, and the SASPs are degraded. The final stage, *outgrowth*, involves visible swelling due to water uptake and synthesis of new RNA, proteins, and DNA. The cell emerges from the broken spore coat and eventually begins to divide (Figure 4.52). The cell then remains in vegetative growth until environmental signals once again trigger sporulation .

Diversity and Phylogenetic Aspects of Endospore Formation

How widespread is the process of endospore formation? Nearly 20 genera of *Bacteria* have been shown to form endospores (∞ Table 12.25), although the process has only been studied in detail in a few species of *Bacillus* and *Clostridium*. Nevertheless, many of the secrets to endospore survival, like the formation of calcium dipicolinate complexes (Figure 4.49) and the possession of endospore-specific genes, seem to be universal. It is thus likely that although some of the details may vary, the general principles of endospore construction are the same in all endospore-forming bacteria.

From a phylogenetic perspective, the capacity to produce endospores is uniquely tied to a particular sublineage of the gram-positive bacteria (∞ Sections 2.5 and 12.20). Despite this, the physiologies of endospore-forming bacteria are highly diverse and include anaerobes, aerobes, phototrophs, and chemolithotrophs.

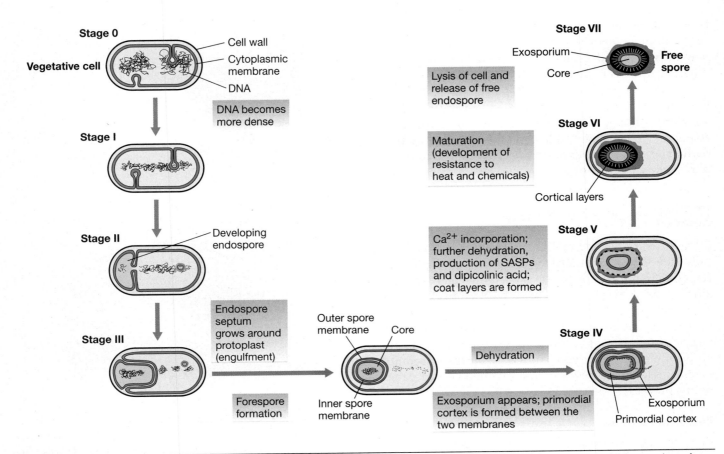

● **Figure 4.51 Stages in endospore formation.** The stages listed (0 through VII) are defined from both genetic studies and microscopic analyses.

Considering this physiological diversity, the actual triggers for endospore formation may vary with different species and could include signals other than simple nutrient star-vation, the major trigger for endospore formation in *Bacillus* and *Clostridium* species. Interestingly, no species of *Archaea* have been shown to sporulate, suggesting that the capacity to produce endospores evolved sometime *after* the major prokaryotic lineages diverged billions of years ago (⬚ Section 2.3 and Figure 2.7).

 4.13 Concept Check

The endospore is a highly resistant differentiated bacterial cell produced by certain gram-positive *Bacteria*. Endospore formation leads to a highly dehydrated structure that con-tains essential macromolecules and a variety of substances such as calcium dipicolinate and small acid-soluble proteins, absent from vegetative cells. Endospores can remain dor-mant indefinitely but germinate quickly when the appropri-ate trigger is applied.

◆ What is *dipicolinic acid* and where is it found?

◆ What are *SASPs* and what is their function?

◆ What happens when an endospore germinates?

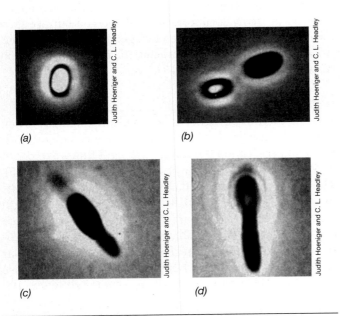

● **Figure 4.52 Endospore germination in *Bacillus*.** Conversion of the mature endospore (a) to a vegetative cell (d); photomicrographs showing the sequence of events starting from a highly refractile mature endospore. In (b) (activation) refractility is being lost, while in (c) and (d) the new vegetative cell is emerging (outgrowth).

IV MICROBIAL LOCOMOTION

We finish our survey of microbial structure and function by examining cell locomotion. Many cells can move under their own power. Motility allows cells to reach different regions of their environment. In the struggle for survival, movement to a new location may offer a cell new resources and growth opportunities and spell the difference between life and death.

In the last few sections of this chapter we examine the different types of cell movement, including swimming and gliding. We then consider how motile cells are able to move in a directed fashion toward or away from particular stimuli (phenomena called *taxes*), and examples of simple behavioral responses.

4.14 Flagella and Motility

Many prokaryotes are motile, and this function is typically due to a special structure, the **flagellum** (plural, **flagella**) (Figure 4.53●). Certain bacterial cells can move along solid surfaces by *gliding* (see Section 4.14), and planktonic microorganisms can regulate their position in a water column by gas-filled structures called gas vesicles (see Sections 4.12 and 12.25). However, the majority of motile prokaryotes move by means of rotating flagella, and so we begin with a detailed consideration of this process.

Bacterial Flagella

Bacterial flagella are long thin appendages free at one end and attached to the cell at the other end. Flagella are so thin (about 20 nm) that a single flagellum can only be seen with the light microscope after staining with special flagella stains that increase their diameter (Figure 4.53). Flagella are easily seen with the electron microscope (Figure 4.54●).

Flagella are arranged differently on different bacteria. In **polar** flagellation, the flagella are attached at one or both ends of the cell (Figures 4.53b and 4.54a). Occasionally a tuft (group) of flagella may arise at one end of the cell, an arrangement called **lophotrichous** (*lopho* means "tuft"; *trichous* means "hair") (Figure 4.53c). Tufts of flagella of this type can be seen in living cells by dark-field microscopy (see Section 4.1 and Figure 4.55a●), where the flagella appear light and are attached to light-colored cells against a dark background. In extremely large prokaryotes, tufts of flagella can also be observed

(a)

Carl E. Bauer

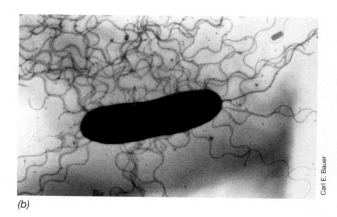

(b)

Carl E. Bauer

● **Figure 4.54 Fine structure of bacterial flagella.** Bacterial flagella as observed by negative staining in the transmission electron microscope. (a) A single polar flagellum. (b) Peritrichous flagella. Both micrographs are of cells of the phototrophic bacterium *Rhodospirillum centenum*. Cells of *R. centenum* are normally polarly flagellated but under certain growth conditions form peritrichously flagellated "swarmer" cells. See also Figure 4.63b.

by phase contrast microscopy (Figure 4.55b). In **peritrichous** flagellation (Figures 4.53a and 4.54b), the flagella are inserted at many locations around the cell surface (*peri* means "around"). The type of flagellation, polar or peritrichous, is used as one characteristic in the classification of bacteria.

Flagellar Structure

Flagella are not straight but are helically shaped. When flattened, flagella show a constant distance between adjacent curves, called the *wavelength*, and this wavelength is characteristic for any given species (Figures 4.53–4.55). The filament of bacterial flagella is composed of subunits of a protein called **flagellin**. The shape and wavelength of the flagellum are in part determined by the structure of the flagellin protein and also to some extent by the direction of rotation of the filament. The basic flagellar structure to be described here varies little among species of *Bacteria*. However, in *Archaea*, several different flagellins are known, and the flagellar structure seems to be quite different from that of *Bacteria*, although flagellar function may be quite similar. In species of *Bacteria*, flagellin is highly conserved, suggesting that flagellar motility has deep evolutionary roots within this evolutionary domain.

(a)   (b)   (c)

E. Leifson

● **Figure 4.53 Bacterial flagella.** Light photomicrographs of prokaryotes containing different flagellar arrangements. Cells are stained with Leifson flagella stain. (a) Peritrichous. (b) Polar. (c) Lophotrichous.

(a)

(b)

R. Jarosch

Norbert Pfennig

● **Figure 4.55 Bacterial flagella as observed in living cells.** (a) Dark-field photomicrograph of a group of large rod-shaped bacteria with flagellar tufts at each pole. A single cell is about 2 μm wide. Dark-field microscopy uses horizontal illumination to yield reflected light (see Section 4.1
and Figure 4.5c). (b) Phase contrast photomicrograph of the large phototrophic purple bacterium *Rhodospirillum photometricum*. A single cell measures about 3 × 30 μm. Note the lophotrichous flagella that emanate from one of the poles.

A flagellum consists of several components and
functions by *rotation*, much like a propeller in a motor
boat. The base of the flagellum is different in structure
from that of the filament (Figure 4.56a●). There is a
wider region at the base of the flagellum called the
hook. The hook consists of a single type of protein and
functions to connect the filament to the motor portion
of the flagellum.

The *motor* is anchored in the cytoplasmic membrane
and cell wall. The motor consists of a small central rod
that passes through a series of rings. In gram-negative
Bacteria, an outer ring, called the *L ring*, is anchored in
the lipopolysaccharide layer. A second ring, called the *P
ring*, is anchored in the peptidoglycan layer of the cell
wall. A third set of rings, called the *MS and C rings*, are
located within the cytoplasmic membrane and the cytoplasm, respectively (Figure 4.56a). In gram-positive
Bacteria, which lack an outer membrane, only the inner
pair of rings is present. Surrounding the inner ring and
anchored in the cytoplasmic membrane are a series of
proteins called *Mot proteins* (Figure 4.56a). A final set of
proteins, called the *Fli proteins* (Figure 4.56a) function as
the motor switch, reversing the direction of rotation of
the flagella in response to intracellular signals.

Flagellar Movement

The flagellum is a tiny rotary motor. How does this
motor work? Rotary moters contain two components:
the rotor and the stator. In the flagellar motor, the *rotor*
consists of the C, MS, and P rings. Collectively, these
structures make up the **basal body**. The *stator* consists of
Mot proteins that surround the MS and C rings and
functions to generate torque.

The rotary motion of the flagellum is imparted from
the basal body. The energy required for rotation of the flagellum comes from the proton motive force (see Sections
4.6 and 5.12). Proton movement across the cytoplasmic
membrane through the Mot complex (Figure 4.56a)
drives rotation of the flagellum, and calculations have
shown that about 1000 protons must be translocated per
single rotation of the flagellum. How this actually occurs
is not yet known. However, a "proton turbine" model has
been proposed to explain the available experimental data
(Figure 4.56b). In this model, protons flowing through
channels in the stator exert electrostatic forces on helically arranged charges on the rotor proteins. Attractions between positive and negative charges would cause the
basal body to rotate as protons flow though the stator
(Figure 4.56b).

Flagellar Synthesis

Several genes are required for flagellar synthesis and
subsequent motility. In *Escherichia coli* and *Salmonella typhimurium*, where studies have been most extensive,
over 40 genes are necessary for motility. These genes
have several functions, including encoding structural
proteins of the flagellar apparatus, export of flagellar
components through the cytoplasmic membrane to the
outside of the cell, and regulation of the many biochemical events surrounding the synthesis of new flagella.

An individual flagellum grows not from its base, as
does an animal hair, but from the tip. The MS ring is synthesized first and inserted into the cytoplasmic membrane.
Then other anchoring proteins are synthesized along with
the hook before filament formation occurs (Figure 4.57●).
Flagellin molecules synthesized in the cytoplasm pass up

Web Tutorial 4.2

The Prokaryotic Flagellum

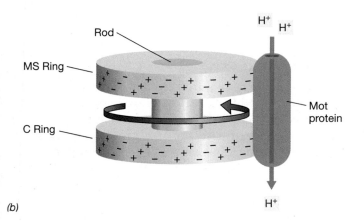

(a)

(b)

through a 3-nm channel inside of the filament and add on at the terminus to form the mature flagellum. At the end of the growing flagellum a protein "cap" exists. Cap proteins assist flagellin molecules that have diffused through the channel to organize at the flagellum termini to form new filament (Figure 4.57). Growth of the flagellum occurs more-or-less continuously until the flagellum reaches its final length. Broken flagella still rotate and can be repaired with new flagellin units passed through the filament channel to replace the lost ones.

Cell Speed

Flagella do not rotate at a constant speed but instead can increase or decrease their speed in relation to the strength of the proton motive force. Flagella rotation can move bacteria through liquid media at speeds of up to 60 cell lengths/second (sec). Although this is only about 0.00017 kilometer/hour (km/h), when comparing this speed with that of higher organisms in terms of the *number of lengths* moved per second, it is extremely fast. The fastest animal, the cheetah, moves at a maximum rate of about 110 km/h, but this represents only about 25 body lengths/sec. Thus, when size is accounted for, prokaryotic cells swimming at 50–60 lengths/sec are actually moving much faster than larger organisms!

The motions of polarly and lophotrichously flagellated organisms are different from those of peritrichously flagellated organisms, and these can be differentiated under the microscope. Peritrichously flagellated organisms typically move in a straight line in a slow, stately fashion. Polarly flagellated organisms, on the other hand, move more rapidly, spinning around and dashing from place to place. The different behavior of flagella on polar and peritrichous organisms, including differences in reversibility of the flagellum, is illustrated in Figure 4.58●.

4.14 *Concept Check*

Motility in most microorganisms is due to flagella. In prokaryotes the flagellum is a complex structure made of several proteins, most of which are anchored in the cell wall and cytoplasmic membrane. The flagellum filament, which is made of a single kind of protein, rotates at the expense of the proton motive force, which drives the flagellar motor.

◆ What is *flagellin* and where is it found?

◆ How does a bacterial flagellum move a cell forward?

◆ How does *polar flagellation* differ from *peritrichous flagellation*?

● **Figure 4.56** **Structure and function of the prokaryotic flagellum in gram-negative *Bacteria*.** (a) Structure. The L ring is embedded in the LPS, and the P ring in peptidoglycan. The MS ring is embedded in the cytoplasmic membrane and the C ring in the cytoplasm. A narrow channel exists in the rod and filament through which flagellin molecules diffuse to reach the site of flagellar synthesis. The Mot proteins function as the flagellar motor, whereas the Fli proteins function as the motor switch. The flagellar motor rotates the filament to propel the cell through the medium. (b) Function. A "proton turbine" model has been proposed to explain rotation of the flagellum. Protons, flowing through the Mot proteins, may exert forces on charges present on the C and MS rings, thereby spinning the rotor.

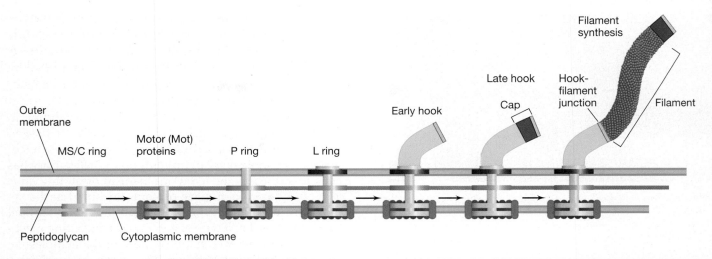

● **Figure 4.57** **Summary of steps in flagella biosynthesis.** Synthesis begins with MS/C ring assembly in the cytoplasmic membrane. This is followed by formation of the other rings, the hook and the cap. At this point, flagellin protein (approximately 20,000 copies are needed to make one filament) flows through the hook to form the filament. Flagellin molecules are guided into position by cap proteins to ensure that the growing filament develops evenly.

4.15 Gliding Motility

Some prokaryotes are motile but lack flagella. These nonswimming yet motile bacteria move across solid surfaces in a process called **gliding**. Unlike flagellar motility, where cells often stop and then start off in a different direction, gliding motility is a smoother form of movement and typically occurs along the long axis of the cell. Gliding motility is widely distributed among *Bacteria* but has been well studied in only a few groups. The gliding movement itself—up to 10 μm/sec in some gliding bacteria—is considerably slower than propulsion by flagella, but still offers the cell a means of moving about its habitat.

Gliding prokaryotes are filamentous or rod-shaped cells (Figure 4.59●), and the gliding process requires that the cells be in contact with a solid surface. The morphology of colonies of a typical gliding bacterium (colonies are masses of bacterial cells that form from successive cell divisions of a single cell, ∞ Section 5.3) are distinctive, since cells glide out and move away from the center of the colony (Figure 4.59*c*). Perhaps the most well-known gliding bacteria are the filamentous cyanobacteria (Figure 4.59*a, b* and ∞ Section 12.25), certain gram-negative *Bacteria*, such as *Myxococcus* and other myxobacteria (∞ Section 12.17), and species of *Cytophaga* and *Flavobacterium* (Figure 4.59*c, d* and ∞ Section 12.31).

Mechanisms of Gliding Motility

Although no gliding mechanism is well understood, it is likely that more than one mechanism occurs. In cyanobacteria (Figure 4.59*a, b*) it is known that a polysaccharide slime is secreted on the outer surface of the cell

(a) **Peritrichous**

(b) **Polar**

● **Figure 4.58** **Manner of movement in polarly and peritrichously flagellated prokaryotes.** (a) Peritrichous: Forward motion is imparted by all flagella rotating counterclockwise (CCW) in a bundle. Clockwise (CW) rotation causes the cell to tumble, and then a return to counterclockwise rotation leads the cell off in a new direction. (b) Polar: Cells change direction by reversing flagellar rotation (thus pulling instead of pushing the cell), or in unidirectional flagella, by stopping periodically to reorient, and then moving forward by clockwise rotation of its flagella. The yellow arrows show the direction the cell is traveling.

(a)

Richard W. Castenholz

(b)

Richard W. Castenholz

(c)   *(d)*

Mark J. McBride

● **Figure 4.59 Gliding bacteria.** (a, b) The large filamentous cyanobacterium *Oscillatoria princeps*. (a) Photomicrograph. A cell is about 35 μm wide. (b) Photograph of filaments gliding on an agar surface. Cells can move by gliding against a solid surface or one filament can glide using a second filament as the solid surface. (c, d) The gram-negative gliding bacterium *Flavobacterium johnsoniae*. (c) Masses of cells gliding away from the center of the colony (the colony is about 2.7 mm wide). (d) Nongliding mutant strain showing typical colony morphology of nongliding bacteria (the colonies are 0.7–1 mm in diameter). For the proposed mechanism of gliding in *F. johnsoniae* cells, see Figure 4.60.

as it glides. The slime appears to contact both the cell surface and the solid surface against which the gliding cell moves. As the excreted slime adheres to the surface, the cell is pulled along. This hypothesis is supported by the observation of slime-excreting pores on the cell surface of several filamentous cyanobacteria.

Slime extrusion is not the mechanism of gliding in nonphototrophic gliding bacteria. In *Flavobacterium johnsoniae* (Figure 4.59*c*), for example, no slime is excreted. Instead, the movement of proteins on the cell surface has been identified as the likely mechanism of gliding. In *F. johnsoniae*, specific motility proteins anchored in the cytoplasmic and outer membranes are thought to propel the cell forward in a type of continuous ratcheting mechanism (Figure 4.60●). The movement of the cytoplasmic membrane proteins is driven by energy released from the proton motive force (∞ Sections 4.6 and 5.12), and they somehow transmit this energy to outer membrane proteins located along the cell surface. It is hypothesized that movement of the proteins against the solid surface literally pulls the cell forward (Figure 4.60).

Like other forms of motility, gliding motility has significant ecological relevance. Gliding allows a cell to exploit new resources or to interact in some beneficial way with other cells. In the latter regard, it is of interest that the myxobacteria, classic examples of gliding bacteria, have a very social and cooperative lifestyle, where gliding motility may play an important role in cell-to-cell interactions (∞ Section 12.17).

⬡ **4.15 Concept Check**

Prokaryotes that move by gliding motility do not employ rotating flagella, but instead creep along a solid surface by any of several possible mechanisms.

◆ How does *gliding* motility differ from *swimming* motility in both mechanism and requirements?

● **Figure 4.60 Model for how gliding motility may occur in *Flavobacterium johnsoniae* and some other gliding bacteria.** Tracks (yellow) are thought to exist in the peptidoglycan that connect cytoplasmic proteins (brown) to outer membrane proteins (orange) and propel them along the solid surface. Note the difference in direction of movement of outer membrane proteins and of the cell proper. Protein propulsion is somehow linked to dissipation of the proton motive force.

♦ Contrast the mechanism of gliding motility in a filamentous cyanobacterium and in *Flavobacterium*.

Cell Motion as a Behavioral Response: Chemotaxis and Phototaxis

Prokaryotes often encounter *gradients* of physical or chemical agents in nature, and cells have evolved means to respond to these gradients by moving either toward or away from the signal molecule, depending on whether it is a useful or harmful substance. Such directed movements are called *taxes*. **Chemotaxis**, a response to chemicals, and **phototaxis**, a response to light, are two well-known taxes. Here we cover taxes in a general way. We will return in Section 8.13 to cover the details of chemotaxis and its regulation as a model for all prokaryotic taxes.

Chemotaxis as a phenomenon has been well studied in swimming bacteria, and much is known at the genetic level concerning how the chemical status of the environment is communicated to the flagellar assembly. Our discussion here will thus deal solely with swimming bacteria. However, some gliding bacteria have also been shown to be chemotactic, and phototactic movements in filamentous cyanobacteria also occur. Thus, although the mechanisms are unknown, like swimming bacteria, gliding bacteria must in some way be able to communicate with their motility apparatus.

Chemotaxis

To understand chemotaxis let us focus on the behavior of a single bacterial cell faced with a chemical gradient of an attractant (Figure 4.61●). Unlike larger organisms, prokaryotes are too small to sense a gradient along the length of a single cell. Instead, while moving, prokaryotes compare the chemical or physical state of their environment with that sensed a few seconds before. In other words, bacteria respond to the *temporal* (rather than *spatial*) *gradient* of signal molecules as they swim along.

Much research on chemotaxis has been done with the peritrichously flagellated bacterium, *Escherichia coli*. In the absence of a gradient, cells of *E. coli* move in a random fashion that includes **runs**, where the cell is swimming forward in a smooth fashion, and **tumbles**, when the cell stops and jiggles about (Figure 4.61a). Following a tumble, the direction of the next run is random (Figure 4.61a). Thus, by means of runs and tumbles, the cell moves about randomly in its environment but does not really go anywhere.

However, if a gradient of a chemical attractant is present, these random movements become biased. As the organism senses that it is moving toward higher concentrations of the attractant (through periodic sampling of the concentration of the chemical in its environment), runs become longer and tumbles less frequent. The net result of this behavioral response is that the organism moves up the concentration gradient of the attractant (Figure 4.61b).

If the organism is sensing a repellent, the same general mechanism applies, although in this case it is the *decrease* in concentration of the repellent (rather than the *increase* in concentration of an attractant) that promotes runs. Forward movement in a run occurs when the flagellar motor is rotating *counterclockwise*. When the flagella rotate *clockwise*, the bundle pushes apart, forward motion ceases, and the cells tumble (Figures 4.58 and 4.61).

The situation with polarly flagellated cells is somewhat different. Many polarly flagellated bacteria, such as *Pseudomonas* species, can reverse the direction of rotation of their flagella like peritrichously flagellated cells and reverse direction in this fashion (Figure 4.58b). However,

(a) **No attractant**   (b) **Attractant present**

● **Figure 4.61   Chemotaxis in a peritrichously flagellated bacterium such as *Escherichia coli*.**   (a) In the absence of a chemical attractant the cell swims randomly in runs, changing direction during tumbles. (b) In the presence of an attractant runs become biased, and the cell moves up the gradient of the attractant.

some polarly flagellated bacteria, such as the phototrophic bacterium *Rhodobacter sphaeroides*, have unidirectional flagella that rotate only in a clockwise direction. How do such cells change direction and are they chemotactic?

In *R. sphaeroides*, which has only a single flagellum inserted subpolarly, rotation of the flagellum stops periodically. During this time the cell becomes randomly reoriented by Brownian motion. As the flagellum begins to rotate once again, the cell moves off in a new direction. Cells of *R. sphaeroides* are strongly chemotactic to a variety of carbon compounds and also show tactic responses to oxygen and light (see later discussion of these taxes). And although it cannot reverse its flagellar motor, chemotaxis in *R. sphaeroides* works in a similar way to that of *E. coli:* An increased gradient of attractants keeps the *R. sphaeroides* flagellum rotating while a decreased gradient of an attractant or an increased gradient of a repellant tends to make the cells stop and reorient.

Measuring Chemotaxis

How do bacteria use temporal changes in chemical concentrations to control flagellar rotation? This is a complex story and involves several events at the genetic and biochemical levels (⚭ Section 8.13). For now, suffice it to say that the molecular mechanism of chemotaxis involves sensory proteins in the cytoplasmic membrane. These proteins, called *chemoreceptors*, sense the chemical gradient over time and interact with cytoplasmic proteins to affect flagellar motor direction. Thus, because the direction of flagellar rotation governs whether the cell runs or tumbles, chemotaxis can be thought of as a chemically driven *sensory response system* affecting flagellar function.

Bacterial chemotaxis can be demonstrated by immersing a small glass capillary tube containing an attractant in a suspension of motile bacteria that does not contain the attractant. From the tip of the capillary, a gradient is set up into the surrounding medium, with the concentration of chemical gradually decreasing with distance from the tip (Figure 4.62*a*●). When an attractant is present, the bacteria will move toward it, forming a swarm around the open tip (Figure 4.62*b*). Subsequently, many of the motile bacteria will move into the capillary. Of course some bacteria will move into the capillary even if it contains a solution of the same composition as the medium because of random movements (Figure 4.62*c*). However, when an attractant is present, the concentration of bacteria within the capillary can be many times higher than the external concentration.

On the other hand, if the capillary contains a repellent, the concentration of bacteria within the capillary will be considerably less than the concentration outside (Figure 4.62*d*). In this case, the cell senses an increasing gradient of repellent, and the appropriate chemoreceptors affect flagellar rotation to move the cell away from the repellent (Figure 4.62*d*). Using the capillary method, it is possible to screen chemicals for their ability to act as attractants or repellents for a given bacterium.

(a) (b)

(c) (d)

(e)

(f)

● **Figure 4.62 Chemotaxis.** (a–e) Techniques for measuring chemotaxis in bacteria. (a) Insertion of capillary into a bacterial suspension. As the capillary is inserted, gradient formation begins. (b) Accumulation of bacteria in a capillary containing an attractant. (c) Control capillary contains a salt solution that is neither an attractant nor a repellent. Cell concentration inside the capillary becomes the same as that outside. (d) Repulsion of bacteria by a repellent. (e) Time course showing cell numbers in capillaries containing various chemicals. (f) Tracks of motile bacteria in seawater swarming around an algal cell (large white spot, center) taken using a tracking video camera system attached to a microscope. Note how the bacterial cells are showing positive aerotaxis by moving toward the oxygen-producing algal cell. The average velocity of the cells was about 25 μm/sec. The alga is about 60 μm in diameter.

Chemotaxis can also be observed microscopically. Using a video camera that captures the position of bacterial cells with time and shows the tracks of each cell, it is possible to see the chemotactic movements of cells (Figure 4.62*f*). This method has been adapted to studies of chemotaxis of bacteria in natural environments. In nature it is thought that the major chemotactic agents for bacteria are nutrients excreted from larger microbial cells or from live or dead macroorganisms. In algae, for example, both organic compounds and oxygen (O_2, from photosynthesis) are produced and can stimulate chemotactic movements toward an algal cell (Figure 4.62*f*).

Phototaxis

Many phototrophic microorganisms move toward light, a process called *phototaxis*. The advantage of phototaxis is that it allows a phototrophic organism to orient itself most efficiently to receive light for photosynthesis. This can be shown if a light spectrum is spread across a microscope slide on which there are motile phototrophic bacteria. On such a slide the bacteria accumulate at wavelengths at which their photosynthetic pigments absorb (Figure 4.63*a*●; ⌒⌒ Sections 17.2 and 17.3 for a discussion of photosynthetic pigments).

Two different taxes are observed in phototrophic prokaryotes. One, called *scotophobotaxis*, can be observed only microscopically and occurs when a phototrophic bacterium by chance happens to swim outside the illuminated field of view of the microscope into darkness. Entering darkness negatively affects the energy state of the cell and signals the cell to tumble, reverse direction, and once again swim in a run, thus reentering the light.

In addition to scotophobotaxis, phototrophic microorganisms can carry out true phototaxis, a directed movement up a light gradient toward an increasing intensity of light. This is analogous to positive chemotaxis except that the attractant is *light* instead of a chemical. In some species, such as the highly motile phototrophic organism *Rhodospirillum centenum* (see Figure 4.54), *entire colonies* of cells show phototaxis and move in unison toward the light (Figure 4.63*b*).

There is good reason to believe that several parts of the regulatory system that govern chemotaxis are also involved in phototaxis. These include in particular cytoplasmic proteins (Che proteins) that control the direction of rotation of the flagella (⌒⌒ Section 8.13). This conclusion has emerged from the study of mutants of phototrophic bacteria defective in phototaxis; such mutants typically have defective chemotaxis systems as well. A *photoreceptor*, analogous to a chemoreceptor but able to sense a gradient of *light* instead of chemicals, orchestrates the phototaxis response. The photoreceptor interacts with cytoplasmic proteins that affect flagella rotation to maintain the cell in a run if it is swimming toward an increasing intensity of light. Thus, although the stimulus in chemotaxis and phototaxis is different, the *same cytoplasmic machinery* is employed to process both types of signals.

Other Taxes

Other bacterial taxes, such as movement toward or away from oxygen (*aerotaxis*, see Figure 4.62*f*) or toward or away from conditions of high ionic strength (*osmotaxis*), are also beginning to be understood in molecular terms. These are all simple forms of behavior, and in all cases it appears that a common mechanism applies—cells periodically sample their environment and process this information through a network of sensory and response proteins that controls the direction of flagellar rotation. From a behavioral point of view, therefore, it is obvious that motile prokaryotes are well attuned to the chemical and physical state of their environment. By moving toward or away from various stimuli, the cell can compete successfully with other species in its microbial community.

(a)

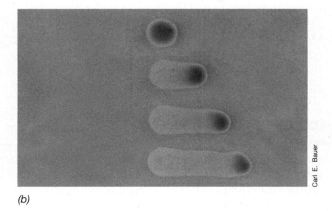

(b)

● **Figure 4.63 Phototaxis.** (a) Scotophobic accumulation of the phototrophic bacterium *Thiospirillum jenense* at light wavelengths at which its pigments absorb. A light spectrum was displayed on a microscope slide containing a dense suspension of the bacteria; after a period of time, the bacteria had accumulated selectively and the photomicrograph was taken. The wavelengths at which accumulations occur are those at which bacteriochlorophyll *a* absorbs (compare with Figure 17.3*b*). (b) Phototaxis of an entire colony of the purple phototrophic bacterium *Rhodospirillum centenum* over a 2-h time course (time 0 at top). These strongly phototactic cells move in unison toward the light source on the right. See Figure 4.54 for electron micrographs of *R. centenum* cells.

4.16 Concept Check

Motile bacteria can respond to chemical and physical gradients in their environment. In the processes of chemotaxis and phototaxis, random movement of a prokaryotic cell can be biased either toward or away from a stimulus by controlling the degree to which runs or tumbles occur. The latter are controlled by the direction of rotation of the flagellum, which in turn is controlled by a network of sensory and response proteins.

◆ Define the word *chemotaxis*.

◆ What causes a *run* versus a *tumble*?

◆ How does *scotophobotaxis* differ from *phototaxis*?

REVIEW QUESTIONS

1. What is the function of staining in light microscopy? Why are cationic dyes used for general staining purposes (∞ Section 4.1)?

2. What is the advantage of a *differential interference contrast* microscope over a bright-field microscope? A *phase-contrast* microscope over a bright-field microscope (∞ Section 4.2)?

3. What is the major advantage of electron microscopes over light microscopes? What type of electron microscope would be used to view the three-dimensional features of a cell (∞ Section 4.3)?

4. What are the major morphologies of prokaryotes? Draw cells for each morphology you list. How large can a prokaryote be? How small? Why is it that we likely know the *lower* limit more accurately than the *upper* limit? What are the dimensions of the rod-shaped bacterium *Escherichia coli* (∞ Section 4.4)?

5. Describe in a single sentence the structure of a unit membrane (∞ Section 4.5).

6. Explain in a single sentence why ionized molecules do not readily pass through the cytoplasmic membrane barrier of a cell. How do such molecules get through the cytoplasmic membrane (∞ Sections 4.5 and 4.6)?

7. Describe a major chemical difference between membranes of *Bacteria* and of *Archaea* (∞ Section 4.5).

8. Cells of *Escherichia coli* take up lactose via the Lac permease system, glucose via the phosphotransferase system, and maltose via an ABC-type transporter. For each of these sugars describe: (1) the components of their transport system, and (2) the source of energy that drives the transport event (∞ Section 4.7).

9. Why is the bacterial cell wall rigid layer called *peptidoglycan?* What are the chemical reasons for the rigidity that is conferred on the cell wall by the peptidoglycan structure (∞ Section 4.8)?

10. Why is sucrose able to stabilize bacterial cells from lysis by lysozyme (∞ Section 4.8)?

11. List several functions for the outer membrane in gram-negative *Bacteria*. What is the chemical composition of the outer membrane (∞ Section 4.9)?

12. What function(s) does polysaccharide layers have in prokaryotes (∞ Section 4.10)?

13. What types of cytoplasmic inclusions are formed by prokaryotes? How does an inclusion of poly-β-hydroxybutyric acid (PHB) differ from a magnetosome in composition and metabolic role (∞ Section 4.11)?

14. What is the function of gas vesicles? How are these structures made such that they can remain gas tight (∞ Section 4.12)?

15. In a few sentences, indicate how the bacterial endospore differs from the vegetative cell in structure, chemical composition, and ability to resist extreme environmental conditions (∞ Section 4.13).

16. Define the following terms: mature endospore, vegetative cell, and germination (∞ Section 4.13).

17. How long may endospores remain viable in a state of "suspended animation," so to speak, and what is the evidence for this (∞ Section 4.13)?

18. Describe the structure and function of a bacterial flagellum. What is the energy source for the flagellum (∞ Section 4.14)?

19. How does the mechanism for gliding motility in *Flavobacterium* differ from the motility in *Escherichia coli* (∞ Section 4.15)?

20. In a few sentences, write an explanation for how a motile bacterium is able to sense the direction of an attractant and move toward it (∞ Section 4.16).

APPLICATION QUESTIONS

1. Calculate the size of the smallest resolvable object if 600-nm light is used to observe a specimen with a 100× oil-immersion lens having a numerical aperture of 1.32. How could resolution be improved using this same lens?

2. Calculate the surface-to-volume ratio of a spherical cell 15 μm in diameter and a cell 2 μm in diameter. What are the consequences of these differences in surface-to-volume ratio for cell function?

3. Assume you are given two cultures, one of a species of gram-negative *Bacteria* and one of a species of *Archaea*. Other than by sequencing ribosomal RNA (∞ Section 2.3), discuss at least four different ways you could tell which culture was which.

4. Calculate the amount of time it would take a cell of *Escherichia coli* ($1 \times 3\mu$m) swimming at maximum speed to travel up a 3-cm-long capilliary tube containing a chemical attractant.

5

NUTRITION, LABORATORY CULTURE, AND METABOLISM OF MICROORGANISMS

An understanding of nutrition and bioenergetics is necessary to grow microorganisms in the laboratory and to understand how microbial cells "make a living" in their natural habitats.

WORKING GLOSSARY

Activation energy the energy required to bring an enzyme's substrate(s) to the reactive state

Anabolism the sum total of all biosynthetic reactions in the cell

Aseptic technique the series of manipulations used to prevent contamination of sterile objects or microbial cultures during handling

ATP synthase (ATPase) a multiprotein enzyme complex embedded in the cytoplasmic membrane that catalyzes the synthesis of ATP coupled to dissipation of the proton motive force

Autotroph an organism capable of biosynthesizing all cell material from CO_2 as the sole carbon source

Catabolism biochemical reactions leading to the production of usable energy (usually ATP) by the cell

Catalyst a substance that accelerates a chemical reaction but is not consumed in the reaction

Chemiosmosis the process by which ATP synthesis is linked to dissipation of a proton motive force

Citric acid cycle a cyclical series of reactions resulting in the conversion of acetate to two CO_2

Coenzyme a small nonprotein molecule that participates in a catalytic reaction as part of an enzyme

Complex medium a culture medium composed of digests of chemically undefined substances such as yeast and meat extracts

Culture medium an aqueous solution of various nutrients suitable for the growth of microorganisms

Defined medium a culture medium whose precise chemical composition is known

Electron acceptor a substance that can accept electrons from an electron donor, becoming reduced in the process

Electron donor a substance that can donate electrons to an electron acceptor, becoming oxidized in the process

Endergonic energy requiring

Enzyme a protein that can speed up (catalyze) a specific chemical reaction

Exergonic energy releasing

Fermentation anaerobic catabolism in which an organic compound serves as both an electron donor and an electron acceptor and ATP is produced by substrate-level phosphorylation

Free energy (G) *energy* available to do work; $G^{0'}$ is free energy under standard conditions

Glycolysis a biochemical pathway in which glucose is fermented yielding ATP

and various fermentation products; also called the Embden-Meyerhof pathway

Oxidative phosphorylation the production of ATP at the expense of a proton motive force formed by electron transport

Photophosphorylation the production of ATP from a proton motive force formed from light-driven electron transport

Proton motive force an energized state of a membrane resulting from the separation of charge and the elements of water (H^+ versus OH^-) across the membrane

Pure culture a culture that contains a single kind of microorganism

Reduction potential (E_0') the inherent tendency, measured in volts, of a compound to donate electrons; E_0' is the reduction potential under standard conditions

Respiration the process in which a compound is oxidized with O_2 (or an O_2 substitute) as the terminal electron acceptor, usually accompanied by ATP production by oxidative phosphorylation

Siderophore an iron chelator that can bind iron present at very low concentrations

Substrate-level phosphorylation production of ATP by the direct transfer of an energy-rich phosphate molecule from a phosphorylated organic compound to ADP

NUTRITION AND CULTURE OF MICROORGANISMS

Before it can replicate, a cell must coordinate many different chemical reactions and organize molecules into specific structures. Collectively, these reactions are referred to as **metabolism**. Metabolic reactions are either *energy releasing,* called **catabolic reactions (catabolism)**, or *energy requiring,* called **anabolic reactions (anabolism)**. Several classes of catabolic and anabolic reactions occur in cells, and we will examine some of the key ones in this and future chapters.

However, before we do this, we will discuss how microorganisms are grown in the laboratory. Most of what we know about the metabolic activities of microorganisms has emerged from the study of laboratory cultures. Our focus here will be on *chemoorganotrophs,* organisms that rely on organic compounds for carbon and energy (∞Section 2.4). Later in this chapter we will consider energy-generating mechanisms other than chemoorganotrophy. Organisms that employ these bioenergetic alternatives can also be grown in the laboratory, and many of the basic principles of nutrition and cell culture discussed now will apply to them as well.

5.1 Microbial Nutrition

Recall from Chapter 3 that cells consist mainly of macromolecules and water and that macromolecules are polymers of smaller units called *monomers* (∞Section 3.2). Microbial nutrition is that aspect of microbial physiology that deals with the supply of monomers (or the precursors of monomers) that cells need for growth. Collectively, these required substances are called **nutrients**.

Different organisms need different sets of nutrients, often in one or another specific form. And not all nutrients are required in the same amounts. Some nutrients, called *macronutrients* (Table 5.1), are required in large amounts, while others, called *micronutrients,* are required in lesser, sometimes even trace amounts. We begin with a consideration of the major macronutrients *carbon* and *nitrogen*.

Carbon and Nitrogen

All cells require **carbon,** and most prokaryotes require an *organic compound* of some sort as their source of carbon. On a dry weight basis, a typical cell is about 50% carbon, and carbon is the major element in all classes of macro-

Table 5.1	Macronutrients in nature and in culture media	
Element	**Usual form of nutrient found in the environment**	**Chemical form supplied in culture media**
Carbon (C)	CO_2, organic compounds	Glucose, malate, acetate, pyruvate, amino acids, hundreds of other compounds, or complex mixtures (yeast extract, peptone, and so on)
Hydrogen (H)	H_2O, organic compounds	H_2O, organic compounds
Oxygen (O)	H_2O, O_2, organic compounds	H_2O, O_2, organic compounds
Nitrogen (N)	NH_3, NO_3^-, N_2, organic nitrogen compounds	*Inorganic:* NH_4Cl, $(NH_4)_2SO_4$, KNO_3, N_2 *Organic:* Amino acids, nitrogen bases of nucleotides, many other N-containing organic compounds
Phosphorus (P)	PO_4^{3-}	KH_2PO_4, Na_2HPO_4
Sulfur (S)	H_2S, SO_4^{2-}, organic S compounds, metal sulfides (FeS, CuS, ZnS, NiS, and so on)	Na_2SO_4, $Na_2S_2O_3$, Na_2S, cysteine, or other organic sulfur compounds
Potassium (K)	K^+ in solution or as various K salts	KCl, KH_2PO_4
Magnesium (Mg)	Mg^{2+} in solution or as various Mg salts	$MgCl_2$, $MgSO_4$
Sodium (Na)	Na^+ in solution or as NaCl or other Na salts	NaCl
Calcium (Ca)	Ca^{2+} in solution or as $CaSO_4$ or other Ca salts	$CaCl_2$
Iron (Fe)	Fe^{2+} or Fe^{3+} in solution or as FeS, $Fe(OH)_3$, or many other Fe salts	$FeCl_3$, $FeSO_4$, various chelated iron solutions (Fe^{3+} EDTA, Fe^{3+} citrate, and so on)

molecules. Bacteria can assimilate different organic carbon compounds and use them to make new cell material. Amino acids, fatty acids, organic acids, sugars, nitrogen bases, aromatic compounds, and countless other organic compounds have been shown to be used by one or another bacterium. By contrast, some prokaryotes are *autotrophs,* able to build all of their cellular structures from *carbon dioxide* (CO_2). The energy needed for this process is obtained from either light or inorganic chemicals.

After carbon, the next most abundant element in the cell is **nitrogen**. A typical bacterial cell is about 12% nitrogen (by dry weight), and nitrogen is important in proteins, nucleic acids, and several other cell constituents. In nature, nitrogen is available in both organic and inorganic forms (Table 5.1). However, the bulk of available nitrogen is in *inorganic* form, either as ammonia (NH_3), nitrate (NO_3^-), or N_2. Most bacteria are capable of using ammonia as the sole nitrogen source, and many can also use nitrate. However, nitrogen gas (N_2) can only be a nitrogen source for certain bacteria, the *nitrogen-fixing bacteria,* and we discuss the properties of these organisms in detail later (∞Sections 12.9, 17.28, and 19.12).

Other Macronutrients: P, S, K, Mg, Ca, Na

Phosphorus occurs in nature in the form of organic and inorganic phosphates and is required by the cell primarily for synthesis of nucleic acids and phospholipids. **Sulfur** is required because of its structural role in the amino acids cysteine and methionine (∞Section 3.6) and because it is present in a number of vitamins, such as thiamine, biotin, and lipoic acid, as well as in coenzyme A. Sulfur is available to organisms in a variety of forms and undergoes a number of chemical transformations in nature, many of which are carried out exclusively by microorganisms (∞Section 19.13). Most cell sulfur originates from inorganic sources, either sulfate (SO_4^{2-}) or sulfide (HS^-) (Table 5.1).

Potassium is required by all organisms. A variety of enzymes, including some of those involved in protein synthesis, specifically require potassium for activity. **Magnesium** functions to stabilize ribosomes, membranes, and nucleic acids, and is also required for the activity of many enzymes. **Calcium**, which is not an essential nutrient for the growth of many microorganisms, helps stabilize cell walls and plays a key role in the heat stability of endospores (∞Section 4.13). **Sodium** is required by some but not all organisms, and its need typically reflects the habitat of the organism. For example, seawater has a high sodium con-tent and marine microorganisms usually require sodium for growth. By contrast, closely related freshwater species are typically able to grow in the absence of sodium.

Iron

Iron plays a major role in cellular respiration, being a key component of cytochromes and iron-sulfur proteins involved in electron transport (see Section 5.11 and Table 5.2). Under anoxic conditions, iron is generally in the +2 oxidation state (ferrous, Fe^{2+}) and soluble. However, under oxic conditions, iron is generally in the +3 oxidation state (ferric, Fe^{3+}) and forms various insoluble minerals. To obtain iron from such minerals, cells produce iron-binding agents called **siderophores** that bind iron and transport it into the cell. One major group of siderophores consists of derivatives of *hydroxamic acid,* which chelate ferric iron very strongly (Figure 5.1a●). Once the iron-hydroxamate complex has passed into the cell, the iron is released and the hydroxamate can be excreted and used again for iron transport.

Bacteria such as *Escherichia coli* and *Salmonella typhimurium* produce structurally complex phenolic siderophores called **enterobactins** (Figure 5.1b). These siderophores are derivatives of the aromatic compound catechol and have an extremely high binding affinity for

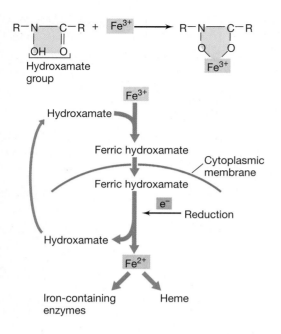

(a)

(b)

(c)

Peptide ferric-binding head group

Hydrophobic tail

● **Figure 5.1 Iron–chelating agents produced by microorganisms.** (a) Hydroxamate. Iron is bound as Fe^{3+} and released inside the cell as Fe^{2+}. The hydroxamate then exits the cell and repeats the cycle. (b) Ferric enterobactin of *Escherichia coli*. The oxygen atoms of each catechol molecule are shown in yellow. (c) The peptidic and tailed siderophore aquachelin, showing Fe^{3+} binding. The hydrophobic tail assists in threading aquachelin through the membrane into the cell.

iron. Without such tenacious iron-binding agents, many pathogens would be unable to initiate an infection. Iron is often in short supply in animal tissues because iron scavenging systems tie up iron and make it unavailable to microorganisms (⚙Sections 21.7 and 21.8).

In marine waters, iron can be virtually undetectable. For example, surface ocean waters typically contain only a few *picograms* (one picogram is 10^{-12} g) of iron per milliliter. To counter this, many marine bacteria produce structurally complex siderophores that can sequester iron present in these vanishingly small amounts. These siderophores contain a peptide head group that complexes Fe^{3+} and a lipid tail that can associate with the cytoplasmic membrane. Once compounds such as *aquachelin* (Figure 5.1c) bind iron, they aggregate into lipidlike micelles and transport the bound iron into the cell.

Some prokaryotes can grow in the total *absence* of iron. For example, cells of the bacteria *Lactobacillus plantarum* and *Borrelia burgdorferii* (the latter is the causative agent of Lyme disease, ⚙Section 27.4) do not contain iron. In these bacteria, Mn^{2+} substitutes for iron as the metal component of enzymes that normally contain Fe^{2+}.

Micronutrients (Trace Elements)

Although required in just tiny amounts, *micronutrients* are nevertheless also critical to cell function. Like iron, micronutrients are metals, and are thus often called *trace elements*. Micronutrients typically play a role as components of various enzymes, the cells' catalysts. Table 5.2 summarizes the major micronutrients of cells and gives examples of enzymes in which each plays a role.

For the routine laboratory culture of microorganisms it is often unnecessary to add trace elements to the culture medium because the requirement for trace elements is so small. However, if a culture medium contains highly purified chemicals dissolved in high-purity distilled water, a trace element deficiency can occur. In such cases a dilute solution of trace metals (Table 5.2) is added to the medium to make available the necessary metals.

Growth Factors

Growth factors are *organic* compounds that, like micronutrients, are required in very small amounts and then only by some cells. Growth factors include vitamins, amino acids, purines, and pyrimidines. Although most microorganisms are able to synthesize all of these compounds, some microorganisms require one or more of them preformed from the environment and thus must be supplied with these compounds in laboratory culture.

Vitamins are the most commonly required growth factors. Most vitamins function as components of coenzymes (see, for instance, Figures 5.10, 5.12, and 5.15), and these are summarized in Table 5.3. Vitamin requirements are highly variable among microorganisms, ranging from none to several. Lactic acid bacteria, which include the genera *Streptococcus*, *Lactobacillus*, *Leuconostoc*, among others (⚙Section 12.19), are renowned for their multiple vitamin requirements, which are even more extensive than those of humans (see Table 5.4). The vitamins most commonly required by microorganisms are thiamine (vitamin B_1), biotin, pyridoxine (vitamin B_6), and cobalamin (vitamin B_{12}).

5.1 Concept Check

The hundreds of chemical compounds present inside a living cell are formed from starting materials called nutrients. Elements required in fairly large amounts are called macronutrients, while metals and organic compounds needed in very small amounts are called micronutrients and growth factors, respectively.

◆ What two classes of macromolecules contain the bulk of the *nitrogen* in a cell?

◆ Why is an element like Co^{2+} considered a *micro*nutrient whereas an element like C is considered a *macro*nutrient?

◆ What roles does iron play in cellular metabolism? How do cells sequester iron?

5.2 Culture Media

Culture media are the nutrient solutions used to grow microorganisms in the laboratory. Because laboratory culture is required for the detailed study of a microorganism, careful attention must be paid to both the selection and preparation of media for successful culture to take place.

Classes of Culture Media

Two broad classes of culture media are used in microbiology: *chemically defined* and *undefined (complex)*. **Chemically defined media (defined media,** for short**)** are prepared by adding precise amounts of highly purified inorganic or organic chemicals to distilled water. Therefore, the *exact chemical composition* of a defined medium is known. Of paramount importance in any culture medium is the *carbon source*, since all cells need large amounts of carbon to make new cell material. In a *simple defined medium* (Table 5.4), a *single* carbon source is present. The nature of the carbon source and its concentration depends on the organism to be cultured.

For growing many organisms, knowledge of the exact composition of a medium is not essential. In these

Table 5.2	Micronutrients (trace elements) needed by living organisms[a]
Element	**Cellular function**
Boron (B)	Present in an autoinducer for quorum sensing in bacteria; also found in some polyketide antibiotics
Chromium (Cr)	Required by mammals for glucose metabolism; no known microbial requirement
Cobalt (Co)	Vitamin B_{12}; transcarboxylase (propionic acid bacteria)
Copper (Cu)	Respiration, cytochrome *c* oxidase; photosynthesis, plastocyanin, some superoxide dismutases
Iron (Fe)[b]	Cytochromes; catalases; peroxidases; iron-sulfur proteins; oxygenases; all nitrogenases
Manganese (Mn)	Activator of many enzymes; present in certain superoxide dismutases and in the water-splitting enzyme in oxygenic phototrophs (Photosystem II)
Molybdenum (Mo)	Certain flavin-containing enzymes; some nitrogenases, nitrate reductases, sulfite oxidases, DMSO-TMAO reductases; some formate dehydrogenases
Nickel (Ni)	Most hydrogenases; coenzyme F_{430} of methanogens; carbon monoxide dehydrogenase; urease
Selenium (Se)	Formate dehydrogenase; some hydrogenases; the amino acid selenocysteine
Tungsten (W)	Some formate dehydrogenases; oxotransferases of hyperthermophiles
Vanadium (V)	Vanadium nitrogenase; bromoperoxidase
Zinc (Zn)	Carbonic anhydrase; alcohol dehydrogenase; RNA and DNA polymerases; and many DNA-binding proteins

[a] Not every micronutrient listed is required by all cells; some metals listed are found in enzymes present in only specific microorganisms.
[b] Needed in greater amounts than other trace metals.

Table 5.3	Growth factors: Vitamins and their functions
Vitamin	**Function**
p-Aminobenzoic acid	Precursor of folic acid
Folic acid	One-carbon metabolism; methyl group transfer
Biotin	Fatty acid biosynthesis; β-decarboxylations; some CO_2 fixation reactions
Cobalamin (B_{12})	Reduction of and transfer of single carbon fragments; synthesis of deoxyribose
Lipoic acid	Transfer of acyl groups in decarboxylation of pyruvate and α-ketoglutarate
Nicotinic acid (niacin)	Precursor of NAD^+ (see Figure 5.10); electron transfer in oxidation-reduction reactions
Pantothenic acid	Precursor of coenzyme A; activation of acetyl and other acyl derivatives
Riboflavin	Precursor of FMN (see Figure 5.15), FAD in flavoproteins involved in electron transport
Thiamine (B_1)	α-Decarboxylations; transketolase
Vitamins B_6 (pyridoxal-pyridoxamine group)	Amino acid and keto acid transformations
Vitamin K group; quinones	Electron transport; synthesis of sphingolipids
Hydroxamates	Iron-binding compounds; solubilization of iron and transport into cell

Table 5.4	Examples of culture media for microorganisms with simple and demanding nutritional requirements[a]		

Defined culture medium for *Escherichia coli*	Defined culture medium for *Leuconostoc mesenteroides*	Complex culture medium for either *E. coli* or *L. mesenteroides*
K_2HPO_4 7 g KH_2PO_4 2 g $(NH_4)_2SO_4$ 1 g $MgSO_4$ 0.1 g $CaCl_2$ 0.02 g Glucose 4–10 g Trace elements (Fe, Co, Mn, Zn, Cu, Ni, Mo) 2–10 μg each Distilled water 1000 ml pH 7	K_2HPO_4 0.6 g KH_2PO_4 0.6 g NH_4Cl 3 g $MgSO_4$ 0.1 g Glucose 25 g Sodium acetate 20 g Amino acids (alanine, arginine, asparagine, aspartate, cysteine, glutamate, glutamine, glycine, histidine, isoleucine, leucine, lysine, methionine, phenylalanine, proline, serine, threonine, tryptophan, tyrosine, valine) 100–200 μg of each Purines and pyrimidines (adenine, guanine, uracil, xanthine) 10 mg of each Vitamins (biotin, folate, nicotinic acid, pyridoxal, pyridoxamine, pyridoxine, riboflavin, thiamine, pantothenate, *p*-aminobenzoic acid) 0.01–1 mg of each Trace elements (see first column) 2–10 μg each Distilled water 1000 ml pH 7	Glucose 15 g Yeast extract 5 g Peptone 5 g KH_2PO_4 2 g Distilled water 1000 ml pH 7

(a)

(b)

[a] The photos are tubes of (a) the defined medium described, and (b) the complex medium described. Note how the complex medium is colored from the various organic extracts and digests that it contains. Photos courtesy of Cheryl L. Broadie and John Vercillo, Southern Illinois University at Carbondale.

instances complex media may suffice and may even be advantageous. **Complex media** employ digests of animal or plant products, such as casein (milk protein), beef, soybeans, yeast cells, or any of a number of other highly nutritious (yet chemically undefined) substances. These digests are commercially available in powdered form and can be quickly weighed out and dissolved in distilled water to yield a medium. However, a major concession in using a complex medium is loss of control of its precise nutrient composition.

In particular situations, especially in clinical microbiology, culture media are often made to be *selective* or *differential* (or both). A **selective medium** contains compounds that selectively inhibit the growth of some microorganisms but not others. By contrast, a **differential medium** is one in which some sort of indicator, typically a dye, is added, that allows for the differentiation of particular chemical reactions occurring during growth. Differential media are quite useful for distinguishing between species of bacteria, some of which may carry out the particular reaction while others do not. Differential and selective media are further discussed in Chapter 24.

Nutritional Requirements and Biosynthetic Capacity

Table 5.4 shows three recipes for culture media, two defined and one complex. The complex medium is easiest to prepare and supports good growth of either organism shown in the table, the enteric bacterium *Escherichia coli* or the lactic acid bacterium *Leuconostoc mesenteroides*, an extremely *fastidious* (nutritionally demanding) bacterium.

The simple defined medium shown in Table 5.4 supports excellent growth of *E. coli* but not of *L. mesenteroides*. Growth of the latter organism in defined medium requires the addition of several organic nutrients and growth factors not needed by *E. coli* (Table 5.4). Which organism, *E. coli* or *L. mesenteroides*, has the greater *biosynthetic* capacity? Obviously, it is *E. coli*, since its ability to grow on a simple defined culture medium means that it can synthesize *all* of its organic constituents from a single carbon compound, in this case glucose (Table 5.4).

In contrast to *E. coli*, *L. mesenteroides* has multiple growth factor and major organic nutrient (for example, amino acid) requirements, indicative of its limited biosynthetic capacity. The complex nutritional needs of *L. mesenteroides* can be satisfied by either preparing a defined medium as shown in Table 5.4 (although such a medium could take hours to prepare) or by using a complex medium (Table 5.4), which can usually be prepared quickly.

Some pathogenic microorganisms have even more stringent nutritional requirements than those shown for *L. mesenteroides*. For culturing these organisms, an *enriched medium* may be needed. An **enriched medium** starts as a complex medium to which additional nutrients, such as serum or whole blood, are added. These added nutrients better mimic conditions in the host and are required for the successful laboratory culture of organisms such as *Streptococcus pyogenes* (strep throat) and *Neisseria gonorrhoeae* (gonorrhea).

A major conclusion to be drawn from Table 5.4 is that *different microorganisms can have vastly different nutri-*

tional requirements. Thus, for successful culture of a given microorganism it is necessary to understand its nutritional requirements and then supply the nutrients in the proper form and proportions in a culture medium. If care is taken in preparing culture media, it is fairly easy to culture many different types of microorganisms in the laboratory. We discuss the procedures involved in culturing microorganisms now.

 5.2 Concept Check

Culture media supply the nutritional needs of microorganisms and can be either chemically defined or undefined (complex). Selective, differential, and enriched are terms that describe media used for the isolation of particular species or for comparative studies of microorganisms.

◆ Why is the routine culture of *Leuconostoc mesenteroides* easier in a complex medium than in a chemically defined medium?

◆ In which medium, defined or complex (shown in Table 5.4), do you think *Escherichia coli* would grow the fastest? Why?

5.3 Laboratory Culture of Microorganisms

Once a culture medium has been prepared and sterilized to render it free of all microorganisms, it can be *inoculated* (that is, organisms added) and incubated under conditions that will support microbial growth. In a laboratory situation, inoculation will typically be of a **pure culture**, a culture containing only a *single kind* of microorganism.

It is essential that other organisms are prevented from entering a pure culture. Such unwanted organisms, called *contaminants,* are ubiquitous (as Pasteur discovered over 125 years ago, ⌖Section 1.6), and microbiological techniques are designed to avoid contaminants. A major method for obtaining pure cultures and for assessing the purity of a culture is the use of solid media, specifically, solid media prepared in the Petri plate, and we consider this now.

Solid versus Liquid Culture Media

Culture media are sometimes prepared in a semisolid form by the addition of a gelling agent to liquid media. Such solid culture media immobilize cells, allowing them to grow and form visible, isolated masses called *colonies* (Figure 5.2●). Bacterial colonies can be of various

● **Figure 5.2 Examples of bacterial colonies.** Colonies are visible masses of cells formed from the subsequent division of one or a few cells. The size, shape, texture, and color of a bacterial colony is a function of the organism that made them. Depending on its size and the arrangement of cells within the colony, a colony can have highly variable cell numbers; colonies containing over one *billion* individual cells are not uncommon. (a) *Serratia marcescens,* grown on MacConkey agar. (b) Close-up of colonies outlined in (a). (c) *Pseudomonas aeruginosa,* grown on Trypticase-Soy agar. (d) *Shigella flexneri,* grown on MacConkey agar.

(a)

(b)

(c)

(d)

James A. Shapiro, University of Chicago

shapes and sizes depending on the organism, the culture conditions, the nutrient supply (including the amount of oxygen present), and several other physiological parameters. Some bacteria produce pigments that cause the colony to be colored (Figure 5.2). Colonies permit the microbiologist to visualize the purity of the culture. Plates that contain more than one colony type are indicative of a contaminated culture. In this way, the Petri plate has been used as one criterion of culture purity for over 100 years (∞Section 1.6).

Solid media are prepared the same way as for liquid media except that before sterilization, *agar* (Microbial Sidebar, Solid Media, the Petri Plate, and Pure Cultures ∞Chapter 1) is added as a gelling agent, usually at a concentration of 1.5%. The agar melts during the sterilization process and the molten medium is then poured into sterile glass or plastic plates and allowed to solidify before use (Figure 5.2).

Aseptic Technique

Because microorganisms are everywhere, culture media must be sterilized before use. For most culture media this is done by heating, typically by moist heat in a large pressure cooker called an *autoclave*. We discuss the operation and principles of the autoclave later, along with other methods of sterilization (∞Section 20.1).

Once a sterile culture medium has been prepared, it is ready to receive an *inoculum* from a previously grown pure culture to start the growth process once again. This manipulation requires the practice of **aseptic technique**, the series of steps used to prevent contamination during manipulations of cultures and sterile culture media (Figures 5.3● and 5.4●). A mastery of aseptic technique is required for success in the microbiology laboratory, and it is one of the first methods learned by the novice microbiologist. Airborne contaminants are the most common problem because laboratory air contains minute dust particles that will have a community of microorganisms on them. When containers are opened, they must be handled in such a way that contaminant-laden air does not enter (Figures 5.3 and 5.4).

Aseptic transfer of a culture from one tube of medium to another is usually accomplished with an inoculating loop or needle that has been sterilized in a flame (Figure 5.3). Cells from liquid cultures can also be transferred to the surface of agar plates (Figure 5.4), where colonies develop from the growth and division of single cells. Picking and restreaking from an isolated colony is a major method of obtaining pure cultures from microbial communities containing many different organisms.

 5.3 Concept Check

Microorganisms can be grown in the laboratory in culture media containing the nutrients they require. Successful cultivation and maintenance of pure cultures of microorganisms can only be done if aseptic technique is practiced.

◆ What is meant by the word *sterile?* What would happen if freshly prepared culture media were not sterilized and left at room temperature?

◆ Why is aseptic technique necessary for successful cultivation of pure cultures in the laboratory?

II ENERGETICS AND ENZYMES

We learned in Chapter 2 of the different energy classes of microorganisms—*chemoorganotrophs, chemolithotrophs,* and *phototrophs* (∞Section 2.4). Regardless of how an organism makes a living, however, it must be able to conserve some of the energy it obtains as ATP. Here we discuss the principles of energy conservation, using some simple laws of chemistry and physics to guide our understanding. We then consider enzymes, the cell's biocatalysts.

5.4 Bioenergetics

Energy is defined as the ability to do work. In microbiology, energy is calculated in units of *kilojoules* (kJ), a measure of heat energy. Chemical reactions are accompanied by *changes* in energy. Although in any chemical reaction

(a)   (b)   (c)   (d)   (e)   (f)

● **Figure 5.3 Aseptic transfer.** (a) Loop is heated until red-hot and cooled in air briefly. (b) Tube is uncapped. (c) Tip of tube is run through the flame. (d) Sample is removed on sterile loop. (e) After removing sample on loop, the tube is reflamed and then the sample is transferred to a sterile medium. (f) The tube is recapped. Loop is reheated before being taken out of service.

● **Figure 5.4** **Method of making a streak plate to obtain pure cultures.** (a) Loop is sterilized, and then a loopful of inoculum is removed from tube. (b) Streak is made over a sterile agar plate, spreading out the organisms. Following the initial streak, subsequent streaks are made at angles to it, the loop being resterilized between streaks. (c) Appearance of a well-streaked plate after incubation. Colonies of the bacterium *Micrococcus luteus* grown on a blood agar plate. It is from such well-isolated colonies that pure cultures can usually be obtained.

some energy is lost as heat, in microbiology we are interested in **free energy** (abbreviated G), which is defined as the energy released *that is available to do useful work*. The change in free energy during a reaction is expressed as $\Delta G^{0\prime}$, where the symbol Δ should be read "change in." The "0" and "prime" superscripts mean that the free-energy value was obtained under standard conditions: pH 7, 25°C, one atmosphere of pressure, and all reactants and products at 1 M concentration.

Consider the reaction:

$$A + B \rightarrow C + D$$

If $\Delta G^{0\prime}$ for this reaction is *negative,* then the reaction will proceed with the *release* of free energy, energy that the cell may be able to conserve in the form of ATP. Such energy-yielding reactions are called **exergonic**. However, if $\Delta G^{0\prime}$ is *positive,* the reaction *requires* energy in order to proceed. Such reactions are called **endergonic**. Thus, from the standpoint of the microbial cell, exergonic reactions *yield* energy while endergonic reactions *require* energy.

Free Energy of Formation and Calculating $\Delta G^{0\prime}$

To calculate the free energy yield of a reaction, one needs to know the free energy of its reactants and products. This is the *free energy of formation* (G_f^0), the energy released or required during the *formation* of a given molecule from the elements. Table 5.5 gives a few examples of G_f^0. By convention, the free energy of formation of the elements in their elemental form (for instance, C, H_2, N_2) is zero. The G_f^0 of compounds, however, is not zero. If

the formation of a compound from its elements proceeds exergonically, then the G_f^0 of the compound is *negative* (energy is released). If the reaction is endergonic (energy is required), then the G_f^0 of the compound is *positive*.

For most compounds G_f^0 is *negative*. This reflects the fact that compounds tend to form spontaneously (that is, with energy being released) from their elements. However, the positive G_f^0 for nitrous oxide (+104.2 kJ/mol, Table 5.5) tells us that this molecule will *not* form spontaneously. Instead, it will decompose spontaneously to nitrogen and oxygen. The free energies of formation of a variety of compounds of microbiological interest are given in Appendix 1.

Table 5.5	Free energy of formation for a few compounds of biological interest
Compound	**Free energy of formation[a]**
Water (H_2O)	−237.2
Carbon dioxide (CO_2)	−394.4
Hydrogen gas (H_2)	0
Oxygen gas (O_2)	0
Ammonium (NH_4^+)	−79.4
Nitrous oxide (N_2O)	+104.2
Acetate ($C_2H_3O_2^-$)	−369.4
Glucose ($C_6H_{12}O_6$)	−917.3
Methane (CH_4)	−50.8
Methanol (CH_3OH)	−175.4

[a] The free energy of formation values (G_f^0) are in *kJ/mol*. See Table A1.1 for a more complete list of free energies of formation.

Using free energies of formation, it is possible to calculate the *change in standard free energy* ($\Delta G^{0\prime}$) that occurs in a given reaction. For the reaction $A + B \rightarrow C + D$, $\Delta G^{0\prime}$ is calculated by subtracting the *sum* of the free energies of formation of the reactants (A and B) from that of the products (C and D). Thus:

$$\Delta G^{0\prime} = G_f^{0}[C + D] - G_f^{0}[A + B]$$

The phrase "products minus reactants" is a simple way to recall how to calculate changes in free energy during chemical reactions. However, it is first necessary to balance the reaction chemically before free-energy calculations can be made. Appendix 1 details the steps involved in calculating free energies for any hypothetical reaction.

$\Delta G^{0\prime}$ versus ΔG

Free energy calculations using *standard* conditions are only approximations of the free energy changes that actually occur when a reaction takes place in nature. Although calculations of $\Delta G^{0\prime}$ are usually good estimates of actual free energy changes, under some circumstances they are not. We will see later in this book that the actual concentrations of products and reactants in nature, which are rarely at 1 M levels, can alter the bioenergetics of reactions, and sometimes in significant ways (Sections 17.21, 19.10, and Appendix 1). Thus, what is most relevant to a bioenergetic calculation is not $\Delta G^{0\prime}$, but ΔG, the free energy change that occurs *under the actual conditions* in which the organism is growing. ΔG is a form of the free energy equation that takes into account the actual concentrations of reactants and products in the reaction, and is calculated as:

$$\Delta G = \Delta G^{0\prime} + RT \ln K$$

where R and T are physical constants and K is the equilibrium constant for the reaction in question (Appendix 1). We will pick up the distinction between $\Delta G^{0\prime}$ and ΔG in Chapter 17 where we consider metabolic diversity in more detail. But for now, we will focus on what the expression $\Delta G^{0\prime}$ can tell us about chemical reactions catalyzed by microorganisms.

 5.4 Concept Check

The chemical reactions of the cell are accompanied by changes in energy, expressed in kilojoules. A chemical reaction can occur with the release of free energy (exergonic) or with the consumption of free energy (endergonic).

◆ What is free energy?

◆ In general, are *catabolic* reactions exergonic or endergonic?

◆ Using the data in Table 5.5, calculate $\Delta G^{0\prime}$ for the reaction $CH_4 + \frac{1}{2}O_2 \rightarrow CH_3OH$. How does $\Delta G^{0\prime}$ differ from ΔG?

5.5 Catalysis and Enzymes

A free energy calculation reveals only whether energy is released or required in a given reaction. The number obtained says nothing about the *rate* of the reaction. Consider the formation of water from gaseous oxygen and hydrogen. The energetics of this reaction are quite favorable: $H_2 + \frac{1}{2}O_2 \rightarrow H_2O$, $\Delta G^{0\prime} = -237$ kJ. However, if we were to mix O_2 and H_2 together in a bottle, no measurable formation of water would occur for years. This is because the rearrangement of oxygen and hydrogen atoms to form water first requires that the chemical bonds of the reactants be broken. The breaking of bonds requires energy, and this energy is called **activation energy**.

Activation energy is that required to bring all molecules in a chemical reaction into the reactive state. For a reaction that proceeds with a net release of free energy (that is, an exergonic reaction), the situation is as diagrammed in Figure 5.5●. Although the activation energy barrier is virtually insurmountable in the absence of catalysis, in the presence of the proper catalyst it is much less formidable (Figure 5.5).

Enzymes

The idea of activation energy leads us to the concept of catalysis and enzymes. In a biochemical sense, a **catalyst** is a substance that *lowers* the activation energy of a reaction, thereby *increasing* the reaction rate. Catalysts facilitate reactions but are themselves not consumed or transformed by the reactions. Moreover, catalysts do not affect the energetics or the equilibrium of a reaction; catalysts affect only the *rate* at which reactions proceed.

Most reactions in living organisms would not occur at appreciable rates without catalysis. Biological catalysts

● **Figure 5.5 Progress of a hypothetical exergonic reaction: $A + B \rightarrow C + D$, and the concept of activation energy.** Chemical reactions may not proceed spontaneously even though energy would be released, because the reactants must first be activated. Once activation has occurred, the reaction then proceeds spontaneously. Catalysts such as enzymes lower the required activation energy.

are called **enzymes**. Enzymes are proteins (or in a few cases, RNAs) that are highly specific in the reactions they catalyze. That is, each enzyme catalyzes only a *single type* of chemical reaction, or in the case of some enzymes, a *class* of closely related reactions. This specificity is a function of the precise three-dimensional structure of the enzyme molecule. In an enzyme-catalyzed reaction, the enzyme temporarily combines with the reactant, called a **substrate** (S), forming an **enzyme-substrate complex**. Then, as the reaction proceeds, the **product** (P) is released and the enzyme (E) is returned to its original state:

$$E + S \leftrightarrow E—S \leftrightarrow E + P$$

The enzyme is generally much larger than the substrate(s), and the combination of enzyme and substrate(s) typically depends on weak bonds, such as hydrogen bonds, van der Waals forces, and hydrophobic interactions (∞Section 3.1) to join the enzyme to the substrate. The portion of the enzyme to which substrates bind is called the **active site** of the enzyme.

Enzyme Catalysis

The catalytic power of enzymes is impressive. Enzymes typically increase the rate of chemical reactions from 10^8 to 10^{20} times the rate that would occur spontaneously. To catalyze a specific reaction, an enzyme must do two things: (1) bind the correct substrate, and (2) position the substrate relative to the catalytically active amino acids in the enzyme's active site. The enzyme-substrate complex (Figure 5.6●) aligns reactive groups and places strain on specific bonds in the substrate(s). The net result of enzyme-substrate complex formation is a reduction in

the activation energy required to make the reaction proceed from substrate(s) to product(s) (Figure 5.5). These steps are summarized diagrammatically in Figure 5.6 for the key glycolytic enzyme *fructose bisphosphate aldolase* (see Section 5.10).

Note that the reaction depicted in Figure 5.5 is exergonic because the free energy of formation of the substrates is *greater* than that of the products. That is, product formation proceeds with the *release* of free energy. Enzymes can also catalyze reactions that require energy, converting energy-poor substrates to energy-rich products. In these cases, however, not only must an activation energy barrier be overcome, but sufficient free energy must also be put *into* the reaction to raise the energy level of the substrates to that of the products. This is done by *coupling* the energy-requiring reaction to an energy-yielding one, such as the hydrolysis of ATP (see Figure 5.12).

Theoretically all enzymes will catalyze their reaction in either direction. In practice, however, enzymes that catalyze highly exergonic or highly endergonic reactions are essentially unidirectional. If a particularly exergonic or endergonic reaction needs to be reversed during cellular metabolism, a different enzyme is usually involved in the reaction.

Structure and Nomenclature of Enzymes

Most enzymes are proteins, polymers of amino acids (∞Sections 3.6–3.8), and every protein has a specific three-dimensional shape. A given protein thus assumes specific binding and physical properties. The structure of an enzyme may be visualized in a computer-generated space-filling model (Figure 5.7●). In this example of the

● **Figure 5.6   The catalytic cycle of an enzyme as depicted for the enzyme fructose bisphosphate aldolase.** This enzyme catalyzes the reaction: fructose 1,6-bisphosphate → glyceraldehyde 3-phosphate + dihydroxyacetone phosphate in glycolysis (see Figure 5.14). Following binding of fructose 1,6-bisphosphate in the formation of the enzyme–substrate complex, the conformation of the enzyme is altered, placing strain on certain bonds of the substrate, which break and yield the two products.

Richard Feldmann

● **Figure 5.7 Computer-generated space-filling model of the enzyme lysozyme.** The substrate (peptidoglycan) binding site (active site) is in the large cleft (arrow) on the left side of the model (∞ Section 4.8).

peptidoglycan-cleaving enzyme *lysozyme* (∞ Section 4.8), the large cleft is the site where the substrate binds (the active site).

Many enzymes contain small nonprotein molecules that participate in catalysis but are not themselves considered substrates. These small enzyme-associated molecules are divided into two classes based on the way they associate with the enzyme: *prosthetic groups* and *coenzymes*.

Prosthetic groups are bound very tightly to their enzymes, usually covalently and permanently. The heme group present in cytochromes (see Section 5.11) is an example of a prosthetic group. **Coenzymes**, by contrast, are loosely bound to enzymes, and a single coenzyme molecule may associate with a number of different enzymes. Most coenzymes are derivatives of vitamins, and $NAD^+/NADH$ is a good example, being a derivative of the vitamin niacin (Table 5.3).

Enzymes are named either for the substrate they bind or for the chemical reaction they catalyze, by addition of the suffix *-ase.* Thus, cellul*ase* is an enzyme that attacks cellulose, glucose oxid*ase* is an enzyme that catalyzes the oxidation of glucose, and ribonucle*ase* is an enzyme that decomposes ribonucleic acid. A more formal enzyme nomenclature system is employed by biochemists, in which a specific numbering system is used to name enzymes according to the class of chemical reaction that they catalyze. But for the purposes of this book, trivial names ending in "ase" are sufficient and will be used exclusively.

 5.5 Concept Check

The reactants in a chemical reaction must first be activated before the reaction can take place, and this requires a catalyst. Enzymes are catalytic proteins that speed up the rate of biochemical reactions. Enzymes are highly specific in the reactions they catalyze, and this specificity resides in the three-dimensional structure of the polypeptide(s) in the protein.

◆ What is the function of a *catalyst?*

◆ What *class* of macromolecules are enzymes?

◆ Where on an enzyme does the substrate bind?

◆ What is *activation energy?*

III OXIDATION–REDUCTION AND ENERGY–RICH COMPOUNDS

Energy conservation in cells involves **oxidation–reduction (redox) reactions**. The energy released in these reactions is conserved in the production of energy-rich storage compounds, such as ATP. Here we consider oxidation–reduction reactions and the major electron carriers present in both the cytoplasm and the cytoplasmic membrane of a cell. We will then examine the nature of the compounds that actually conserve the energy released in oxidation–reduction reactions.

5.6 Oxidation–Reduction

Chemically, an oxidation is defined as the *removal* of an electron (or electrons) from a substance. A reduction is defined as the *addition* of an electron (or electrons) to a substance. In biochemistry, oxidations and reductions frequently involve the transfer of not just electrons, but both an electron (e^-) plus a proton (H^+). We will, on occasion, need to distinguish between oxidation–reduction reactions involving electrons only or electrons plus protons, but reserve this distinction for the appropriate time (see Sections 5.11 and 5.12).

Electron Donors and Acceptors

Redox reactions occur when electrons removed from an electron donor are taken up by an electron acceptor. For example, the electron donor hydrogen gas (H_2) can release electrons and protons and become oxidized:

$$H_2 \rightarrow 2\,e^- + 2\,H^+$$

However, electrons cannot exist alone in solution; they must be part of atoms or molecules. Thus, the equation as drawn provides chemical information but does not itself represent an independent reaction. The reaction is only a *half reaction,* a term that implies the need for a second half reaction. This is because for any *oxidation* to occur, a subsequent *reduction* must also occur. For example, the oxidation of H_2 could be coupled to the reduction of many different substances, including O_2, in a second reaction:

$$\tfrac{1}{2}O_2 + 2\,e^- + 2\,H^+ \rightarrow H_2O$$

This half reaction, which is a reduction, when coupled to the oxidation of H_2 above, yields the following overall balanced reaction:

$$H_2 + \tfrac{1}{2}O_2 \rightarrow H_2O$$

In reactions of this type, we refer to the substance *oxidized,* in this case H_2, as the **electron donor**, and the substance *reduced,* in this case O_2, as the **electron acceptor** (Figure 5.8●). The key to understanding biological oxidations and reductions is to keep the proper half reactions

1. $H_2 \rightarrow 2\,e^- + \boxed{2\,H^+}$

Electron-donating
half reaction

2. $\frac{1}{2}O_2 + 2\,e^- \rightarrow \boxed{O^{2-}}$

Electron-accepting
half reaction

3. $\boxed{2\,H^+} + \boxed{O^{2-}} \rightarrow H_2O$

Formation of water

4.

Electron   Electron
donor   acceptor

$H_2 + \frac{1}{2}O_2 \rightarrow H_2O$

Net reaction

● **Figure 5.8 Example of an oxidation–reduction reaction.** The formation of H_2O from the electron donor H_2 and the electron acceptor O_2.

straight. In a redox reaction there must always be one reaction involving an electron *donor* and another reaction involving an electron *acceptor*.

Reduction Potentials

Substances vary in their tendency to become oxidized or reduced. This tendency is expressed as the **reduction potential** (E_0', standard conditions) of the half reaction. This potential is measured electrically in volts (V) in reference to a standard substance, H_2. By convention, reduction potentials are expressed for half reactions written as *reductions*. If protons are involved in the reaction, as is often the case, then the reduction potential is to some extent influenced by the hydrogen ion concentration (pH). By convention in biology, however, reduction potentials are given for pH 7 because the cytoplasm of most cells is neutral or nearly so. Using these conventions, at pH 7 the E_0' of

$$\frac{1}{2}O_2 + 2\,H^+ + 2\,e^- \rightarrow H_2O$$

is +0.816 volts (V), and the E_0' of

$$2\,H^+ + 2\,e^- \rightarrow H_2$$

is −0.421 V. We will see in the following text that these values of E_0' mean that O_2 is an excellent electron acceptor, while H_2 is an excellent electron donor (Figure 5.8).

Oxidation–Reduction Couples and Complete Redox Reactions

Many molecules can be either electron donors or electron acceptors under different circumstances, depending on the substances with which they react. The chemical constituents on each side of the arrow in half reactions can be thought of as representing a *redox couple*, such as $2\,H^+/H_2$ or $\frac{1}{2}O_2/H_2O$. By convention, when writing a redox couple, the *oxidized* form of the couple is always placed on the left.

In constructing complete oxidation–reduction reactions from their constituent half reactions, the reduced substance of a redox couple whose E_0' is more negative *donates* electrons to the oxidized substance of a redox couple whose E_0' is more positive. Thus, in the couple $2\,H^+/H_2$ (E_0' −0.42 V), H_2 has a greater tendency to

donate electrons than do protons to accept them. On the other hand, in the couple $\frac{1}{2}O_2/H_2O$ (E_0' +0.82 V), H_2O has a very weak tendency to donate electrons, while O_2 has a great tendency to accept them. It follows then that in a reaction of H_2 and O_2, H_2 will be the electron *donor* and become oxidized, and O_2 will be the electron *acceptor* and become reduced (Figure 5.8).

By convention, all half reactions are written as *reductions*. However, in an actual redox reaction, one of the two half reactions will proceed as an oxidation and therefore will be written in the *opposite* direction. Thus, in the reaction shown in Figure 5.8, the oxidation of H_2 to $2\,H^+ + 2\,e^-$ is reversed from the formal half reaction, written as a reduction.

The Electron Tower

A convenient way of viewing electron transfer reactions in biological systems and their importance to bioenergetics is to imagine a vertical tower (Figure 5.9●). The tower represents the range of reduction potentials possible for redox couples in nature, from those with the most negative E_0's on the top to those with the most positive E_0's at the bottom. The *reduced* substance in the redox pair at the top of the tower has the greatest tendency to donate electrons, whereas the *oxidized* substance in the couple at the bottom of the tower has the greatest tendency to accept electrons.

As electrons from the electron donor at the top of the tower fall, they can be "caught" by acceptors at various levels. The difference in reduction potential between two substances on the tower is expressed as $\Delta E_0'$. The farther the electrons drop from a donor before they are caught by an acceptor, the greater the amount of energy released. That is, $\Delta E_0'$ *is proportional to* $\Delta G^{0'}$ (Figure 5.9). Oxygen (O_2), at the bottom of the tower, is the most favorable electron acceptor commonly found in nature. In the middle of the tower, redox couples can be either electron donors or acceptors, depending on which redox couples they react with. For instance, the $2\,H^+/H_2$ couple (−0.42 V) can react with the fumarate/succinate couple (+0.02 V) as follows:

$$H_2 + \text{fumarate}^{2-} \rightarrow \text{succinate}^{2-}$$

By contrast, the oxidation of succinate to fumarate can be coupled to the reduction of NO_3^- or $\frac{1}{2}O_2$:

$$\text{Succinate}^{2-} + NO_3^- \rightarrow \text{fumarate}^{2-} + NO_2^- + H_2O$$

$$\text{Succinate}^{2-} + \frac{1}{2}O_2 \rightarrow \text{fumarate}^{2-} + H_2O$$

Hence, in the presence of H_2 and under anoxic conditions, fumarate can be an electron acceptor (producing succinate). Under other conditions (for example, anoxic in the presence of NO_3^-, or under oxic conditions) succinate can be an electron donor (producing fumarate). Indeed, all the transformations involving fumarate and succinate described here are actually carried out by various microorganisms (including *Escherichia coli*) under certain nutritional and environmental conditions.

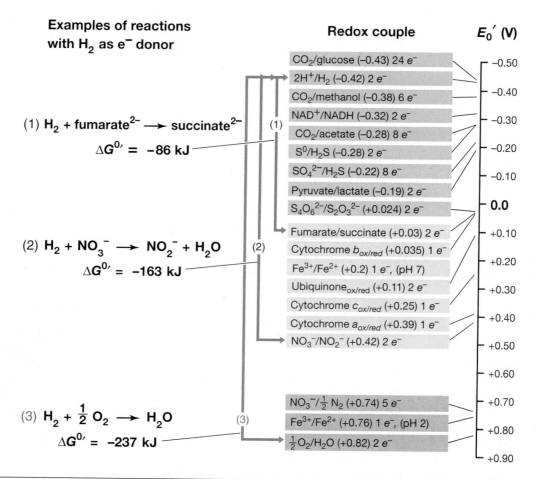

Examples of reactions with H_2 as e^- donor

(1) $H_2 + fumarate^{2-} \longrightarrow succinate^{2-}$

$\Delta G^{0'} = -86$ kJ

(2) $H_2 + NO_3^- \longrightarrow NO_2^- + H_2O$

$\Delta G^{0'} = -163$ kJ

(3) $H_2 + \frac{1}{2} O_2 \longrightarrow H_2O$

$\Delta G^{0'} = -237$ kJ

Redox couple E_0' **(V)**

CO_2/glucose (−0.43) 24 e^- — −0.50

$2H^+/H_2$ (−0.42) 2 e^-

CO_2/methanol (−0.38) 6 e^- — −0.40

NAD^+/NADH (−0.32) 2 e^- — −0.30

CO_2/acetate (−0.28) 8 e^-

S^0/H_2S (−0.28) 2 e^- — −0.20

SO_4^{2-}/H_2S (−0.22) 8 e^- — −0.10

Pyruvate/lactate (−0.19) 2 e^-

$S_4O_6^{2-}/S_2O_3^{2-}$ (+0.024) 2 e^- — **0.0**

Fumarate/succinate (+0.03) 2 e^- — +0.10

Cytochrome $b_{ox/red}$ (+0.035) 1 e^-

Fe^{3+}/Fe^{2+} (+0.2) 1 e^-, (pH 7) — +0.20

Ubiquinone$_{ox/red}$ (+0.11) 2 e^- — +0.30

Cytochrome $c_{ox/red}$ (+0.25) 1 e^-

Cytochrome $a_{ox/red}$ (+0.39) 1 e^- — +0.40

NO_3^-/NO_2^- (+0.42) 2 e^- — +0.50

— +0.60

$NO_3^-/\frac{1}{2} N_2$ (+0.74) 5 e^- — +0.70

Fe^{3+}/Fe^{2+} (+0.76) 1 e^-, (pH 2)

$\frac{1}{2}O_2/H_2O$ (+0.82) 2 e^- — +0.80

— +0.90

● **Figure 5.9** **The electron tower.** Redox couples are arranged from the strongest reductants (negative reduction potential) at the top to the strongest oxidants (positive reduction potentials) at the bottom. As electrons are donated from the top of the tower, they can be "caught" by acceptors at various levels. The farther the electrons fall before they are caught, the greater the difference in reduction potential between electron donor and electron acceptor and the more energy released. As an example of this, on the left is shown the differences in energy released when a single electron donor, H_2, reacts with any of three different electron acceptors, fumarate, nitrate, and oxygen.

Electron Donor ↔ Energy Source

Electron donors are often called **energy sources** because energy is released when they are oxidized. Many potential electron donors exist in nature, including a wide variety of organic and inorganic compounds (∞Chapters 17 and 19). However, it is not the electron donor *per se* that contains energy, but the *chemical reaction* in which the electron donor gets oxidized that actually releases energy. The presence of a suitable electron acceptor is thus just as important as an electron donor. Moreover, the amount of energy released in a redox reaction depends on the nature of *both* the electron donor and the electron acceptor. The greater the difference between reduction potentials of the two half reactions, the more energy will be released when they react (Figure 5.9) (see also Appendix 1).

 5.6 Concept Check

Oxidation–reduction reactions involve the transfer of electrons from electron donor to electron acceptor. The tendency of a compound to accept or release electrons is expressed quantitatively by its reduction potential, E_0'.

◆ In the reaction $H_2 + \frac{1}{2}O_2 \rightarrow H_2O$, what is the electron *donor* and what is the electron *acceptor*?

◆ What is the E_0' of the $2 H^+/H_2$ couple? Are protons good electron acceptors? Why or why not?

◆ Why is nitrate (NO_3^-) a better electron acceptor than fumarate?

5.7 NAD as a Redox Electron Carrier

The transfer of electrons in redox reactions typically involves one or more intermediates called *carriers*. When such carriers are used, we refer to the initial donor as the **primary electron donor** and to the final acceptor as the **terminal electron acceptor**. The net energy change of the complete reaction sequence is determined by the *difference* in reduction potentials between the primary donor and the terminal acceptor.

Electron carriers can be divided into two classes: those that are *freely diffusible* (coenzymes) and those that are firmly attached to enzymes in the cytoplasmic membrane (prosthetic groups). The fixed carriers function in membrane-associated electron transport reactions and are discussed in Section 5.11. Common diffusible carriers include the coenzymes nicotinamide-adenine dinucleotide (NAD^+) and NAD-phosphate ($NADP^+$) (Figure 5.10●). NAD^+ and $NADP^+$ are *electron plus H^+ carriers*, transporting $2e^-$ and $2H^+$ at a time.

The reduction potential of the NAD^+/NADH (or $NADP^+$/NADPH) couple is -0.32 V, which places it fairly high on the electron tower. That is, NADH (or NADPH) is a good electron *donor*. However, although the NAD^+ and $NADP^+$ couples have the same reduction potentials, they generally function in different capacities in the cell. NAD^+/NADH is directly involved in energy-generating (catabolic) reactions, whereas $NADP^+$/NADPH is involved primarily in biosynthetic (anabolic) reactions.

NAD/NADH Cycling

Coenzymes increase the diversity of redox reactions possible in a cell by allowing chemically dissimilar molecules to interact as primary electron donor and terminal electron acceptor, with the coenzyme acting as an intermediary. In the case of NAD^+/NADH, for example, electrons removed from one molecule can reduce NAD^+ to NADH, while the latter can be converted back to NAD^+ by donating electrons to a second molecule. Figure 5.11● is a diagram showing the functioning of NAD^+/NADH

● **Figure 5.10 Structure of the oxidation–reduction coenzyme nicotinamide adenine dinucleotide (NAD^+).** In $NADP^+$, a phosphate group is present, as indicated. Both NAD^+ and $NADP^+$ undergo oxidation–reduction as shown, are freely diffusible, and are $2e^- + 2H^+$ carriers. "R" in the top portion of the art is the adenine dinucleotide portion of NAD^+, shown in full in the bottom portion of the art.

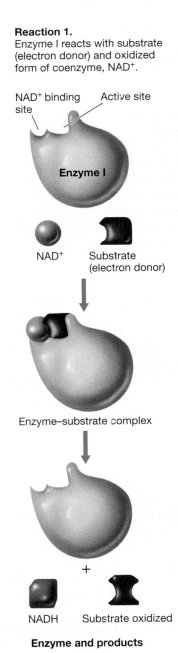

Reaction 1.
Enzyme I reacts with substrate (electron donor) and oxidized form of coenzyme, NAD^+.

NAD^+ binding site   Active site

Enzyme I

NAD^+ Substrate (electron donor)

Enzyme–substrate complex

NADH   Substrate oxidized

Enzyme and products

Reaction 2.
Enzyme II reacts with substrate (electron acceptor) and reduced form of coenzyme, NADH.

NADH binding site   Active site

Enzyme II

NADH   Substrate (electron acceptor)

Enzyme–substrate complex

NAD^+ Substrate reduced

Enzyme and products

● **Figure 5.11 Schematic example of an oxidation–reduction reaction.** This reaction involves the oxidized and reduced forms of the coenzyme nicotinamide-adenine dinucleotide, NAD^+ and NADH.

in such a two-part reaction. NAD$^+$ and NADH are simply intermediaries in the process, facilitating the reaction but not being consumed in the process. Thus, unlike a primary electron donor or a terminal electron acceptor, where relatively large amounts of each are required, the cell needs only a tiny amount of the coenzymes NAD$^+$/NADH. All that is needed is an amount sufficient to service the redox enzymes in the cell that use these coenzymes in their reaction mechanisms.

5.7 *Concept Check*

The transfer of electrons from donor to acceptor in a cell typically involves one or more electron carriers. Some electron carriers are membrane-bound, whereas others, such as NAD$^+$/NADH, are freely diffusible, transferring electrons from one place to another in the cell.

◆ Is NADH a better electron donor than H$_2$? Using the data of Figure 5.9, how can you tell this?

5.8 Energy-Rich Compounds and Energy Storage

Once energy is released from redox reactions it must be conserved by the cell if it is to be used to drive various energy-requiring functions. In living organisms, chemical energy released in redox reactions is conserved primarily in the form of phosphorylated compounds, in particular,

ATP. Such compounds are said to be **energy-rich**. This is because the potential energy released upon hydrolysis of the phosphate bonds is significantly greater than that of the average covalent bond in the cell.

In phosphorylated compounds, phosphate groups are attached via oxygen atoms by *ester* or *anhydride* bonds, as illustrated in Figure 5.12●. However, not all phosphate bonds are energy-rich. The energy of phosphate bonds is expressed in terms of the free energy released when the phosphate is hydrolyzed. As seen in Figure 5.12, the $\Delta G^{0'}$ of hydrolysis of the phosphate *ester* bond in glucose 6-phosphate is only -13.8 kJ/mol. By contrast, the $\Delta G^{0'}$ of hydrolysis of the phosphate *anhydride* bond in phosphoenolpyruvate is -51.6 kJ/mol, almost four times that of glucose 6-phosphate. Thus, phosphoenolpyruvate, a phosphoanhydride (∞Table 3.1), is an energy-rich compound, while glucose 6-phosphate, a phosphate ester, is not. Although theoretically either compound could be hydrolyzed to yield energy, cells typically use compounds whose $\Delta G^{0'}$ is *greater than* $-30\,kJ/mol$ as energy "currencies" in the cell (Figure 5.12 and see Table 17.6).

Adenosine Triphosphate (ATP)

The most important energy-rich phosphate compound in cells is **adenosine triphosphate (ATP)**. ATP consists of the ribonucleoside adenosine to which three phosphate molecules are bonded in series (Figure 5.12). ATP is the

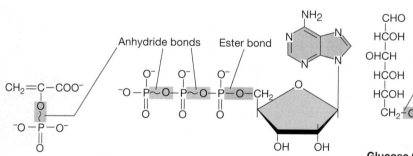

Compound	$G^{0'}$ kJ/mol
$\Delta G^{0'}$ > 30kJ	
Phosphoenolpyruvate	−51.6
1,3-Bisphosphoglycerate	−52.0
Acetyl phosphate	−44.8
ATP	−31.8
ADP	−31.8
Acetyl CoA	−31
$\Delta G^{0'}$ < 30kJ	
AMP	−14.2
Glucose 6-phosphate	−13.8

● **Figure 5.12** **Some compounds important in energy transformations in cells.** The table shows the free energy of hydrolysis of some of the key phosphate esters and anhydrides, indicating that some of the phosphate ester bonds are of higher energy than others. Structures of four of the compounds are given to indicate the position of bonds. ATP contains three phosphates, but only two of them have free energies of hydrolysis >30 kJ. Also shown is the structure of the coenzyme acetyl-CoA. The thioester bond between C and S has a free energy of hydrolysis of >30 kJ. The "R" group of acetyl-CoA is a 3' phospho ADP group.

prime energy currency in cells, being generated during exergonic reactions and consumed in certain endergonic reactions. From the structure of ATP (Figure 5.12) it can be seen that two of the phosphate bonds are phosphoanhydrides and thus have free energies of hydrolysis greater than 30 kJ. Thus the reactions ATP → ADP + P_i, and ADP → AMP + P_i, each release roughly 32 kJ/mol of energy (Figure 5.12).

Although the energy released in ATP hydrolysis is about −32 kJ, a caveat must be introduced here to more precisely define the energy requirements for the *synthesis* of ATP. In an actively growing cell, the ratio of ATP:ADP is kept high, roughly 1000. This significantly affects the energy requirements for ATP synthesis. In an actively growing cell, the actual energy expenditure (that is, the ΔG, see Section 5.4) for the synthesis of one mole of ATP is on the order of −55 to −60 kJ. Nevertheless, for the purposes of learning the basic principles of bioenergetics, we will remain under "standard conditions" ($\Delta G^{0\prime}$) and take the energy required for synthesis or hydrolysis of ATP to be the same, 32 kJ/mol.

Coenzyme A

In addition to energy-rich *phosphate* compounds, the cell produces certain other compounds that can conserve the energy released in exergonic reactions. These include derivatives of coenzyme A (for example, acetyl-CoA; see structure in Figure 5.12). **Coenzyme A** derivatives contain *thioester* bonds (Figure 5.12, and ◯Table 3.1). These yield sufficient free energy upon hydrolysis to drive the synthesis of an energy-rich phosphate bond. For example, in the reaction:

acetyl-S-CoA + H_2O + ADP + P_i →

acetate⁻ + HS-CoA + ATP + H⁺,

the energy released in the hydrolysis of coenzyme A is conserved in the synthesis of ATP. Coenzyme A derivatives (acetyl-CoA is just one of many) are especially important to the energetics of *anaerobic* microorganisms, in particular those whose energy metabolism involves fermentation (◯Sections 5.10, 17.19, and 17.20). We will return to the importance of these compounds in Chapter 17.

Long-Term Energy Storage

ATP is a dynamic entity in the cell; it is continuously being broken down to drive biosynthetic reactions and resynthesized at the expense of catabolic reactions. However, ATP is present in relatively low concentration in an actively growing cell, about 2 millimolar (mM). For longer-term energy *storage,* microorganisms typically produce insoluble polymers that can be oxidized later for the production of ATP.

Examples of long-term energy storage in prokaryotes include the polyglucose polymer glycogen (◯Figure 3.6), the lipid polymer poly-β-hydroxybutyrate and other polyhydroxyalkanoates (◯Figure 4.52), and elemental sulfur, stored by many sulfur chemolithotrophs (◯Figure 2.10*b*). These polymers are deposited within the cell as large granules that can be seen with the light or electron microscope (◯Section 4.11). In eukaryotic microorganisms, polyglucose in the form of starch (◯Figure 3.6), or lipid in the form of simple fats (◯Figure 3.7), are the major reserve materials. In the absence of an external energy source, cells oxidize these polymers in order to make new cell material or to supply the energy needed to maintain cell integrity, called *maintenance energy,* when in a nongrowing state.

⬣ **5.8 Concept Check**

The energy released in redox reactions is conserved in the formation of certain compounds that contain energy-rich phosphate or sulfur bonds. The most common of these compounds is ATP, the prime energy carrier in the cell. Long-term storage of energy is linked to the formation of polymers, which can be consumed to yield ATP.

◆ How much energy is released per mole of ATP converted to ADP + P_i? Per mole of AMP converted to adenosine and P_i?

◆ Following periods of nutrient abundance, how can cells prepare for periods of nutrient starvation?

IV MAJOR CATABOLIC PATHWAYS, ELECTRON TRANSPORT, AND THE PROTON MOTIVE FORCE

We next look at some common catabolic reactions in microorganisms. First we will compare the processes of fermentation and respiration, followed by a more detailed look at each process. We then move to a discussion of the proton motive force, the key product of respiration and other catabolic processes, including anaerobic respiration, photosynthesis, and chemolithotrophy. Besides yielding ATP, the proton motive force also drives a number of other functions in the cell, including many different transport reactions and both flagellar and gliding motility (◯Sections 4.7 and 4.14–4.16).

5.9 Energy Conservation: Options

Two mechanisms for energy conservation are known in chemoorganotrophs: **fermentation** and **respiration**. In each mechanism, the synthesis of ATP is driven by the energy released in oxidation–reduction reactions. However, the redox reactions in fermentation and respiration differ significantly. In *fermentation* the redox process occurs in the *absence* of exogenous electron acceptors. By contrast, in *respiration,* molecular oxygen or some other electron acceptor is present as a terminal electron acceptor.

Because every oxidation has to be coupled to a reduction, the oxidation reaction in a fermentation is coupled to the reduction of a compound *derived from the electron donor*. In respiration, on the other hand, an *exogenous* electron acceptor is reduced. This reality of fermentation places constraints on the amount of energy that can be obtained, as we will see when we compare energy yields from fermentation and respiration in a later section.

Substrate-Level Phosphorylation and Oxidative Phosphorylation

In addition to differences in the nature of electron acceptors, fermentation and respiration differ in the *mechanism* by which ATP is synthesized. In fermentation, ATP is produced by **substrate-level phosphorylation**. In this process, ATP is synthesized during steps in the catabolism of an organic compound (Figure 5.13a●). This is in contrast to **oxidative phosphorylation,** in which ATP is produced at the expense of the proton motive force

(a) **Substrate-level phosphorylation**

(b) **Oxidative phosphorylation**

● **Figure 5.13 Energy conservation in fermentation and respiration.** (a) In fermentation, ATP synthesis occurs as a result of *substrate-level phosphorylation;* a phosphate group gets added to some intermediate in the biochemical pathway and eventually gets transferred to ADP to form ATP. (b) In respiration, the cytoplasmic membrane, energized by the proton motive force, dissipates some of that energy in the formation of ATP from ADP and inorganic phosphate (P_i) in the process called *oxidative phosphorylation.* The coupling of the proton motive force to ATP synthesis occurs by way of a membrane protein enzyme complex called ATP synthase (ATPase) (see Section 5.12 and Figure 5.21).

(Figure 5.13b). A third form of ATP synthesis, **photophosphorylation**, occurs in phototrophic organisms. However, the basic mechanism of photophosphorylation is similar to that of oxidative phosphorylation except that *light* rather than a chemical compound drives the redox reactions that generate the proton motive force (∞Section 17.2).

 5.9 Concept Check

Fermentation and respiration are the two means by which chemoorganotrophs can conserve energy from the oxidation of organic compounds. During these catabolic reactions, ATP synthesis occurs by way of either substrate-level phosphorylation (fermentation) or oxidative phosphorylation (respiration).

◆ Which form of ATP synthesis requires cytoplasmic membrane participation? Why?

◆ How does substrate-level phosphorylation differ from oxidative phosphorylation?

5.10 Glycolysis as an Example of Fermentation

As we mentioned, a fermentation is an *internally balanced* oxidation–reduction reaction. In a fermentation, some atoms of the electron donor become more reduced whereas others become more oxidized, and energy is conserved by substrate-level phosphorylation. A common biochemical pathway for the fermentation of glucose is **glycolysis**, also named the **Embden-Meyerhof pathway** for its major discoverers. Several other fermentations are also known (∞Chapter 17).

Glycolysis is an anoxic process and can be divided into three major stages, each involving a series of individually catalyzed enzymatic reactions (Figure 5.14●). *Stage I* of glycolysis is a series of preparatory reactions that do not involve oxidation–reduction and do not release energy, but that lead to the production of two molecules of the key intermediate, *glyceraldehyde 3-phosphate* from glucose. In *Stage II*, oxidation–reduction reactions occur, energy is conserved in the form of ATP, and two molecules of pyruvate are formed. In *Stage III*, oxidation–reduction reactions occur once again and *fermentation products* are formed (Figure 5.14).

Stages I and II: Preparatory and Redox Reactions

In Stage I, glucose is phosphorylated by ATP, yielding glucose 6-phosphate. The latter is then converted to an isomeric form, fructose 6-phosphate. A second phosphorylation leads to the production of *fructose 1,6-bisphosphate*, a key intermediate of glycolysis. The enzyme **aldolase** splits fructose 1,6-bisphosphate into two three-carbon molecules, *glyceraldehyde 3-phosphate* and its isomer, *dihydroxyacetone phosphate*, which is converted into glyceraldehyde 3-phosphate. Note that thus far,

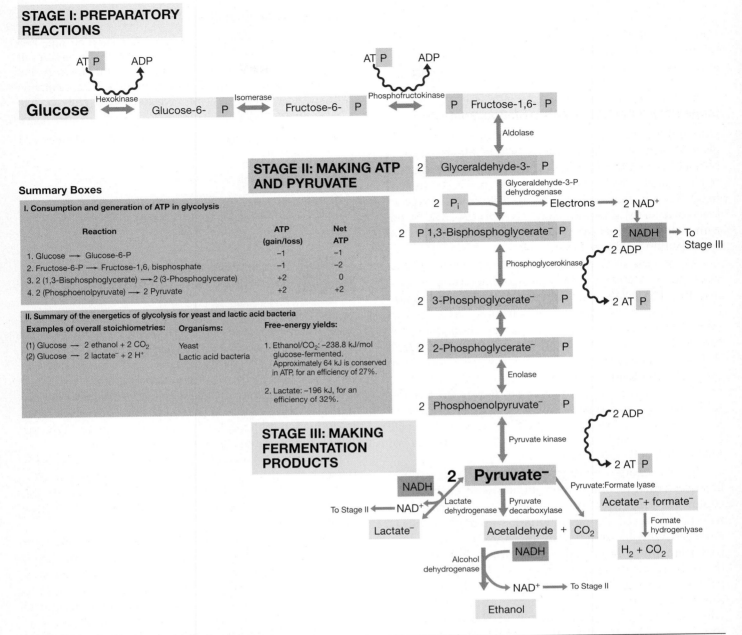

● Figure 5.14 Embden—Meyerhof pathway (glycolysis). The sequence of enzymatic reactions in the conversion of glucose to pyruvate and then to fermentation products (enzymes are shown in small type). The product of aldolase is actually glyceraldehyde 3-P and dihydroxyacetone P, but the latter is converted to glyceraldehyde 3-P. Note how pyruvate is the central "hub" of glycolysis—all fermentation products are made from pyruvate, and just a few common examples are given.

all of the reactions, including the consumption of ATP, have proceeded without redox reactions.

The first redox reaction of glycolysis occurs in Stage II during the conversion of glyceraldehyde 3-phosphate to 1,3-bisphosphoglyceric acid. In this reaction (which occurs twice, once for each molecule of glyceraldehyde 3-phosphate), an enzyme containing NAD^+ as a coenzyme accepts $2\,e^- + 2\,H^+$ to form NADH. The enzyme catalyzing this reaction is **glyceraldehyde-3-phosphate dehydrogenase**. Simultaneously, each glyceraldehyde-3-P molecule is phosphorylated by the addition of a molecule of inorganic phosphate. This reaction, in which

inorganic phosphate is converted to organic form, sets the stage for later energy conservation by substrate-level phosphorylation. ATP formation is possible because each of the phosphates on a molecule of 1,3-bisphosphoglyceric acid has a free energy of hydrolysis of >30 kJ (see Figure 5.12). The synthesis of ATP occurs when (1) each molecule of 1,3-bisphosphoglyceric acid is converted to 3-phosphoglyceric acid, and (2) when each molecule of phosphoenolpyruvate is converted to pyruvate (Figure 5.14).

In glycolysis, *two* ATP molecules are consumed in the two phosphorylations of glucose in Stage I, and *four*

ATP molecules are synthesized (two from each 1,3-bis-phosphoglyceric acid converted to pyruvate) in the reactions of Stage II (Figure 5.14). Thus, the net gain to the organism in glycolysis is *two molecules of ATP per molecule of glucose fermented.*

Stage III: Production of Fermentation Products

During the formation of two molecules of 1,3-bisphospho-glyceric acid, two molecules of NAD^+ are reduced to NADH (see Figure 5.14). However, the continued oxidation of glyceraldehyde 3-phosphate can only proceed if there is free NAD^+ available to accept released electrons. This problem is overcome by the oxidation of NADH to NAD^+ by enzymes that reduce pyruvate to any of a variety of **fermentation products** (Figure 5.14).

In the case of yeast, pyruvate is reduced to *ethanol* with the release of CO_2. In lactic acid bacteria, pyruvate is reduced to *lactate.* Many routes of pyruvate reduction in various fermentative prokaryotes are known (∞Chapters 12 and 17), but the net result is the same: NADH is reoxidized to NAD^+ during the process. As a diffusible coenzyme, NADH can diffuse away from glyceraldehyde-3-phosphate dehydrogenase, attach to an enzyme that reduces pyruvate to lactate (lactate dehydrogenase), and diffuse away once again following the oxidation of NADH to NAD^+ to repeat the cycle (Figure 5.14, and see Figure 5.11 for details of NAD^+/NADH shuttling).

In any energy-yielding process, oxidations must balance reductions, and there must be an electron acceptor for each electron removed. In this case, the *reduction* of NAD^+ at one enzymatic step in glycolysis is balanced with its *oxidation* at another. The final product(s) must also be in redox and atomic balance with the starting substrate, glucose. Hence, the fermentation products discussed here, ethanol plus CO_2, or lactate plus protons, are in both electrical and atomic balance with the starting substrate, glucose, as shown here: glucose ($C_6H_{12}O_6$) = 2 ethanol (C_2H_5OH) + 2 CO_2; glucose ($C_6H_{12}O_6$) = 2 lactate$^-$ ($C_3H_{10}O_3$) + 2 H^+.

Glucose Fermentation: Net and Practical Results

The ultimate result of glycolysis is the consumption of glucose, the net synthesis of two ATPs, and the production of fermentation products. For the organism the crucial product is ATP, which is used in a wide variety of energy-requiring reactions, and fermentation products are merely waste products. However, the latter substances are hardly considered waste products by the distiller, the brewer, the cheese-maker, or the baker (see the Microbial Sidebar, The Products of Yeast Fermentation). Thus, fermentation is more than just an energy-yielding process. It is a means of producing natural products useful to humans. We discuss industrial production of fermentation products in more detail in Chapter 30.

5.10 *Concept Check*

Glycolysis is a major pathway of fermentation and is a widespread means of anaerobic metabolism. The end result of glycolysis is the release of a small amount of energy that is conserved as ATP and the production of fermentation products. For each glucose consumed in glycolysis, two ATPs are produced.

◆ Which reaction(s) in glycolysis involve oxidations and reductions?

◆ What is the role of NAD^+/NADH in glycolysis?

◆ Why are fermentation products made during glycolysis?

5.11 Respiration and Membrane-Associated Electron Carriers

Fermentation occurs in the absence of external electron acceptors, and it releases only a relatively small amount of energy. Only a few ATP molecules are synthesized as a result. This small energy release can be understood in terms of the principles of oxidation–reduction reactions: (1) the carbon atoms in the starting compound are only partially oxidized (see Figure 5.14), and (2) the difference in reduction potentials between the primary electron donor and terminal electron acceptor is small.

By contrast, if O_2 or some other terminal acceptor is present, the glucose can be oxidized completely to CO_2 and a far higher yield of ATP is possible. Oxidation using O_2 as the terminal electron acceptor is called **aerobic respiration**.

Our discussion of aerobic respiration will deal with both the carbon transformations and redox reactions that occur. Thus, our focus will be on two issues: (1) the way electrons are transferred from the organic compound to the terminal electron acceptor, and (2) the biochemical pathways involved in the transformation of organic carbon to CO_2. During the former, ATP synthesis occurs at the expense of the proton motive force (Figure 5.13b). We begin with a discussion of electron flow.

Electron Transport Carriers

Electron transport systems are *membrane-associated* electron carriers. These systems have two basic functions. First, electron transport systems mediate the transfer of electrons from primary donor to terminal acceptor. And second, these systems conserve some of the energy released during electron transfer for synthesis of ATP.

Several types of oxidation–reduction enzymes are involved in electron transport. These include *NADH dehydrogenases, flavoproteins, iron-sulfur proteins,* and *cytochromes.* In addition, a class of *nonprotein* electron carriers is known, the *quinones.* We now consider each of these classes of electron transport components in slightly more detail.

Microbial Sidebar   ◆   The Products of Yeast Fermentation

Every home wine maker or brewer is an amateur microbiologist, perhaps without even realizing it. Indeed, the anaerobic processes of energy generation are at the heart of some of the most striking discoveries of the human race: fermented foods and beverages (Figure 1●).

In the production of breads and most alcoholic beverages, it is the yeast *Saccharomyces cerevisiae* that is exploited to produce ethanol plus CO_2. Found in various sugar-rich environments such as fruit juices and nectar, yeasts can carry out the two opposing modes of chemoorganotrophic metabolism discussed in this chapter, *fermentation* and *respiration*. When oxygen is present, yeasts grow efficiently on various sugars, making yeast cells and CO_2 (the latter from the citric acid cycle, see Section 5.13) in the process. However, when conditions are anoxic, yeasts switch to fermentative metabolism. This results in a reduced cell yield but significant amounts of alcohol plus CO_2.

When grapes are squeezed to make juice, small numbers of yeast cells present on the grapes in the vineyard are transferred to the must. During the first several days of the wine-making process, these yeast cells grow by respiration and consume O_2, making the juice anoxic. As soon as the oxygen is depleted, fermentation can begin, and the process of alcohol formation takes over. This switch from aerobic to anaerobic metabolism is crucial, and special care must be taken to make sure air is kept out of the fermenting vessel.

Wine is only one of many products made with yeast. Others include beer (Microbial

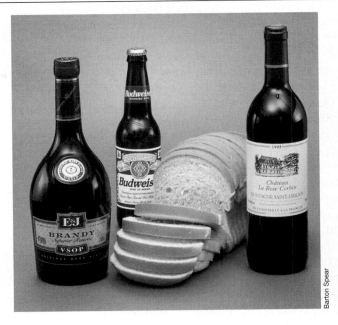

Figure 1 *Major products in which fermentation by the yeast* Saccharomyces cerevisiae *is critical.*

Sidebar, Home Brew, ∽Chapter 30) and distilled spirits such as brandy, whisky, vodka, and gin (see Figure 1). In distilled spirits, the ethanol, produced in relatively low amounts (10–15% by volume) by the yeast, is concentrated by distilling to make a beverage containing 40–60% alcohol. Even alcohol for motor fuel is made with yeast in parts of the world where sugar is plentiful but petroleum is in short supply (such as Brazil). In the United States, major production of ethyl alcohol for use as an industrial solvent occurs using corn starch as the fermentable substrate. Yeast also serves as the leavening agent in bread, al-

though here it is not the *alcohol* that is important, as it volatilizes in the baking process, but CO_2, the *other* product of the alcohol fermentation (see Figure 5.14). We discuss yeast and yeast products in some detail in Chapter 30.

We can thus appreciate how the yeast cell, forced to carry out a fermentative lifestyle because the oxygen it needs for respiration is absent, has impacted the lives of humans. Besides being "waste products" of the glycolytic pathway in yeast, ethanol and CO_2 are the key ingredients in the products of the alcoholic beverage and baking industries, respectively. ∎

NADH dehydrogenases are proteins bound to the inside surface of the cytoplasmic membrane. They have an active site that binds NADH and accepts $2 e^- + 2 H^+$ when NADH is converted to NAD^+ (Figure 5.10). The $2 e^- + 2 H^+$ are passed on to flavoproteins.

Flavoproteins are proteins containing a derivative of riboflavin (Figure 5.15●). The flavin portion, which is bound to a protein, is a prosthetic group that is alternately reduced as it accepts hydrogen atoms and oxidized when electrons are passed on. Note that flavoproteins *accept* $2 e^- + 2 H^+$ but *donate* electrons only. We will consider what happens to the two protons later. Two flavins are commonly observed in cells, *flavin mononucleotide*

(FMN) and *flavin-adenine dinucleotide (FAD)*. In the latter, FMN is bonded to ribose and adenine through a second phosphate. Riboflavin, also called vitamin B_2, is a source of the parent flavin molecule in flavoproteins and is a required growth factor for some organisms (see Section 5.2 and Table 5.3).

The **cytochromes** are proteins with iron-containing porphyrin ring (heme) prosthetic groups attached to them (Figure 5.16●). Cytochromes undergo oxidation and reduction through loss or gain of a *single electron* by the iron atom in the heme of the cytochrome:

$$\text{Cytochrome} - Fe^{2+} \longleftrightarrow \text{Cytochrome} - Fe^{3+} + e^-$$

Several classes of cytochromes are known, differing widely in their reduction potentials (see Figure 5.9). Different classes of cytochromes are designated by letters, such as cytochrome *a*, cytochrome *b*, cytochrome *c*, depending upon the type of heme they contain. The cytochromes of one organism may differ slightly from those of another, and so there are designations such as cytochromes a_1, a_2, a_3, and so on among cytochromes of the same class. Occasionally, cytochromes form tight complexes with other cytochromes or with iron-sulfur proteins. An example of such a complex is the cytochrome bc_1 complex, which contains two different *b*-type cytochromes and one *c*-type cytochrome. The cytochrome bc_1 complex plays an important role in energy metabolism, as we will see later (see Section 5.12 and Figure 5.20).

In addition to the cytochromes, in which iron is bound to heme (Figure 5.16), one or more **nonheme iron-proteins** are typically present in electron transport chains. These proteins contain clusters of iron and sulfur atoms, and Fe_2S_2 and Fe_4S_4 clusters are the most common (Figure 5.17●). The iron atoms are bonded to free sulfur and to the protein via sulfur atoms from cysteine residues (Figure 5.17). *Ferredoxin*, a common iron-sulfur protein in biological systems, has an Fe_2S_2 configuration.

The reduction potentials of iron-sulfur proteins vary over a wide range depending on the number of iron and sulfur atoms present and how the iron centers are embedded in the protein. Thus, different iron-sulfur proteins can function at different locations in the electron

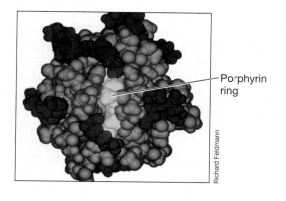

● **Figure 5.16 Cytochrome and its structure.** (a) Structure of the pyrrole ring. (b) Four pyrrole rings are condensed, leading to formation of the porphyrin ring. Various metals can be incorporated into the porphyrin ring system. For example, in chlorophyll pigments, the metal is Mg^{2+} (⚬⚬Section 17.2 and Figure 17.3); in vitamin B_{12}, it is Co^{2+} (⚬⚬Section 30.7 and Figure 30.12); and in certain unique porphyrin coenzymes, Ni^{2+} may be the central metal atom (⚬⚬ Section 17.17 and Figure 17.42). (c) In some cytochromes, like cytochrome *c*, the porphyrin ring is covalently linked via disulfide bridges to cysteine molecules in the protein. Note the presence of iron in the center of the ring. (d) Computer-generated model of cytochrome *c*. The protein completely surrounds the porphyrin ring (light color) in the center. Cytochromes carry electrons only; the redox site is the iron atom, which can alternate between the Fe^{2+} and Fe^{3+} oxidation state. The reduction potential of different cytochromes varies widely (see Figure 5.9) depending on the cytochrome type and binding to the parent protein.

Isoalloxazine ring

Oxidized

Ribitol

2 H (2 e⁻ + 2 H⁺)

Reduced

● **Figure 5.15 Flavin mononucleotide (FMN) (riboflavin phosphate, a hydrogen atom carrier).** The site of oxidation–reduction is the same in FMN and flavin–adenine dinucleotide (FAD).

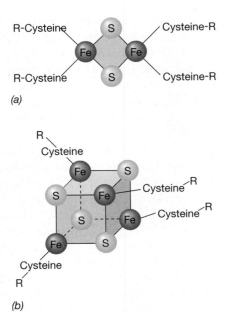

(a)

(b)

● **Figure 5.17 Arrangement of the iron-sulfur centers of nonheme iron-sulfur proteins.** (a) Fe_2S_2 center. (b) Fe_4S_4 center. The cysteine linkages are from the protein portion of the molecule. Iron-sulfur proteins typically carry electrons only.

● **Figure 5.18 Structure of oxidized and reduced forms of coenzyme Q, a quinone.** The five-carbon unit in the side chain (an isoprenoid) occurs in a number of multiples. In prokaryotes, the most common number is $n = 6$; in eukaryotes, $n = 10$. Note that oxidized quinone requires $2\,e^-$ and $2\,H^+$ to become fully reduced. An intermediate form, the semiquinone (one H more reduced than oxidized quinone), is formed during the reduction of a quinone.

transport process. Like cytochromes, iron-sulfur proteins carry *electrons only*.

The **quinones** (Figure 5.18●) are highly hydrophobic nonprotein-containing molecules involved in electron transport. Some quinones found in bacteria are related to vitamin K, a growth factor for higher animals. Like flavoproteins, quinones accept $2\,e^- + 2\,H^+$, but pass on only electrons to the next carrier in the chain.

 5.11 Concept Check

Electron transport systems consist of a series of membrane-associated electron carriers that function in an integrated fashion to carry electrons from the primary electron donor to oxygen as terminal electron acceptor.

◆ In what major way do quinones differ from other electron carriers in the membrane?

◆ Which electron carriers described in this section accept $2\,e^- + 2\,H^+$? Which accept e^- only?

5.12 Energy Conservation from the Proton Motive Force

The major components of electron transport chains are shown in Figure 5.19● against a backdrop of E_0' values. Note that each component has its own unique reduction potential. During electron transport, ATP is produced by the process of *oxidative phosphorylation*. The production of ATP is linked to the establishment of a **proton motive force (pmf)** (◯◯Section 4.6) across the membrane, and electron transport reactions establish this energized state. We now consider the details of this key cellular process.

The Proton Motive Force: Chemiosmosis

To understand how electron transport is linked to ATP synthesis, we must first understand how the electron transport system is oriented in the cytoplasmic membrane. The overall structure of the membrane was outlined in Section 4.5 (◯◯Figure 4.16). Recall that proteins are embedded in the lipid bilayer of membranes such that most have access to both the outside and the inside of the cell. That is, integral membrane proteins are typically *transmembrane* proteins.

Electron transport carriers are oriented so that a *separation* of protons from electrons occurs across the membrane during the transport process. Two electrons plus two protons, removed from substances such as NADH, are transported through the chain by specific carriers. During this process, protons are released into the environment (in gram-negative prokaryotes, protons are released into the periplasm). This extrusion of protons results in a slight acidification of the external surface of the membrane. At the end of the electron transport chain, the electrons are passed to the terminal electron acceptor. In the case of aerobic respiration, O_2 is the terminal acceptor and it is reduced to water.

To reduce O_2 to H_2O, protons from the cytoplasm are needed to complete the reaction. These protons originate from the dissociation of water into H^+ and OH^-. The use of H^+ in the reduction of O_2 to H_2O and the extrusion of H^+ to the environment causes OH^- to accumulate on the *inside* of the membrane. Despite their small size, neither H^+ nor OH^- can diffuse through the membrane because they are charged. Thus, equilibrium cannot be spontaneously restored.

The net result of electron transport is thus the generation of a *pH gradient* and an *electrochemical potential* across the membrane. The *inside* surface of the membrane becomes electrically negative and alkaline and the *outside* surface of the membrane electrically positive and acidic. This pH gradient and electrochemical potential (collectively called the *proton motive force*) cause the membrane to be energized much like a battery. Some of this electrical energy can be conserved by the cell in the formation of ATP.

The energized state of the membrane formed during electron transport can be used directly to do work, such as ion transport (∞Section 4.7) or flagellar rotation (∞Section 4.14), or it can be used to drive the formation of ATP, as will be described later. The concept of a proton gradient driving ATP synthesis was first proposed as the *chemiosmotic theory* in 1961 by the British scientist Peter Mitchell. Mitchell later received the Nobel Prize for this important contribution, a concept that has far reaching implications for every form of energy metabolism except for fermentation. Moreover, the process of **chemiosmosis** occurs in *all* cells, whether they are *Bacteria, Archaea,* or *Eukarya.*

Generation of the Proton Motive Force: Complexes I and II

The proton motive force develops from the activities of flavin enzymes, quinones, the cytochrome bc_1 complex, and the terminal oxidase. Here we examine the carriers sequentially.

Following the donation of $2 e^- + 2 H^+$ from NADH to FAD to form FADH, $2 H^+$ are extruded to the outer surface of the membrane when FADH donates $2 e^-$ to a nonheme iron protein, part of membrane protein *Complex I*, shown in Figure 5.20●. These carriers are referred to as "complexes" because each contains several proteins. For example, Complex I in *Escherichia coli* actually contains 14 distinct proteins. Complex I has the formal name of *NADH:quinone oxidoreductase*, because the net reaction is one in which NADH becomes oxidized and quinone becomes reduced. Two protons are taken up from the dissociation of water in the cytoplasm when the nonheme iron protein of Complex I reduces coenzyme Q (Figure 5.20).

Complex II simply bypasses the steps in Complex I and feeds electrons and protons (through FAD/FADH) directly into the quinone pool. Complex II is also called the *succinate dehydrogenase complex* because of the specific substrate, *succinate* (a product of the citric acid cycle, see Figure 5.22), that it oxidizes. However, because Complex II bypasses Complex I, *fewer protons* are pumped per two electrons that enter the electron transport chain here than for electrons that enter from NADH (Figure 5.20).

Complexes III and IV: bc_1 and a-type Cytochromes

Reduced coenzyme Q passes electrons one at a time to the *cytochrome bc_1* complex, shown as membrane protein Complex III in Figure 5.20. The cytochrome bc_1 complex is an association of several proteins containing hemes or

● **Figure 5.19 Electron transport chains and their relation to E_0'.** Shown here is one example of an electron transport system, leading to the transfer of electrons from the primary electron donor (labeled as substrates) to O_2 (the terminal electron acceptor). This particular sequence is typical of the electron transport chain of the mitochondrion of eukaryotic cells and that of some *Bacteria* (for example, *Paracoccus denitrificans*), and also the mitochondrion of eukaryotic cells. The electron transport chain in *Escherichia coli* lacks cytochromes *c* and aa_3, and instead electrons go directly from cytochrome *b* to cytochrome *o* or *d* (the latter has a similar E_0' to cytochrome aa_3), which is the terminal oxidase (∞Figure 17.37a). By breaking up the complete oxidation into a series of discrete steps, energy conservation is possible through proton motive force formation leading to ATP synthesis. Compare color-coding here with those in Figure 5.9.

metal centers. These include two *b* type hemes (b_L and b_H), one *c* type heme (c_1), and one Fe/S protein (called the *Rieske protein*). The bc_1 complex is present in the electron transport chain of most organisms that can respire. It also plays a role in photosynthetic electron flow (∞Sections 17.4 and 17.5).

The major function of the cytochrome bc_1 complex is to transfer electrons from quinones to cytochrome *c* (Figure 5.20). Electrons travel from the bc_1 complex to a molecule of cytochrome *c* in the periplasm, and from there to the high potential cytochromes *a* and a_3 (Complex IV, Figure 5.20). The latter is the *terminal oxidase* and reduces O_2 to H_2O in the final step of electron transport. Complex IV also pumps two H^+ per each 2 e^- consumed in the reaction (Figure 5.20).

The cytochrome bc_1 complex can also interact with the quinone pool to pump two additional protons at the Q-bc_1 site (Figure 5.20). This is due to a series of electron exchanges called the *Q cycle*. Because quinone and bc_1 have roughly the same E_0' (near 0 V, Figure 5.19). different molecules of quinone can alternately oxidize and reduce one another using electrons fed back from the bc_1 complex. This mechanism, the details of which needn't concern us here, allows for a total of *4 H^+* to be pumped to the outer surface of the membrane at the Q-bc_1 site for every *2 e^-* that enter the chain in Complex I (Figure 5.20).

The electron transport scheme shown in Figure 5.20 is just one of many different sequences known from different organisms. However, several features are characteristic of *all* electron transport chains: (1) the presence of a series of membrane-associated electron carriers arranged in order of increasingly more positive E_0'; (2) an alternation in the chain of electron only and electron plus proton carriers; and (3) the generation of a proton motive force as a result of charge separation across the membrane, acidic outside and alkaline inside.

As we will see now, it is the proton motive force that drives ATP synthesis.

ATPases, the Proton Motive Force, and ATP Formation

How does the proton motive force generated by electron transport (Figure 5.20) actually drive ATP synthesis? Interestingly, there is a strong parallel here between the mechanism of ATP synthesis and the mechanism of the motor that drives rotation of the bacterial flagellum (∞Section 4.14 and Figure 4.56). The catalyst for conversion of the proton motive force into ATP is a large membrane enzyme complex called **ATP synthase**, or **ATPase** for short. ATPases are also known as *Complex V* of the overall electron transport system.

ATPases consists of two major parts, a multi-subunit headpiece called F_1, located on the *cytoplasmic* side of the membrane, and a proton-conducting channel, called F_o, that *spans the membrane* (Figure 5.21●). The F_1/F_o complex catalyzes a reversible reaction between ATP and ADP + P_i as shown in Figure 5.21. The structure of ATPase proteins is highly conserved throughout the

● **Figure 5.20 Generation of the proton motive force during aerobic respiration.** The figure shows the orientation of electron carriers in the membrane of an organism like *Paracoccus denitrificans*, a model prokaryote for studies of respiration. The + and − charges across the membrane represent H^+ and OH^-, respectively. Abbreviations are as follows: FMN, flavoprotein; FAD, flavin adenine dinucleotide; Q, quinone; Fe/S, iron sulfur protein; cyt *a, b, c,* cytochromes (b_L and b_H, low and high potential *b*-type cytochromes, respectively). At the quinone site a recycling of electrons occurs during the "Q cycle." This is because electrons from QH_2 can be split in the bc_1 complex (Complex III) between the Fe/S protein and the *b*-type cytochromes. Electrons that travel through the latter reduce Q (in two, one-electron steps) back to QH_2, thus increasing the number of protons pumped at the Q-bc_1 site. Electrons that travel to Fe/S proceed to reduce cytochrome c_1 and then cytochrome *c*, and then *a*-type cytochromes in Complex IV, eventually reducing O_2 to H_2O (two electrons and two protons are required to reduce $\frac{1}{2}O_2$ to H_2O, and these come from electrons through cyt *c* and cytoplasmic protons, respectively). Complex II, the succinate dehydrogenase complex, bypasses Complex I and feeds electrons directly into the quinone pool. The numbers of the complexes shown are those used by convention by scientists in the field of membrane bioenergetics. Compare this electron transport chain with that of *Escherichia coli*, shown in Figure 17.37. Differences in growth conditions can significantly affect the components present in any electron transport chain.

● **Figure 5.21** **Structure and function of ATP synthase (ATPase, Complex V).** F_1 consists of five different polypeptides present as an $\alpha_3\beta_3\gamma\varepsilon\delta$ complex. F_1 is the catalytic complex responsible for the interconversion of ADP + P_i and ATP. F_o is integrated in the membrane and consists of three polypeptides in an ab_2c_{12} complex. Subunit a is responsible for channeling protons across the membrane, while subunit b protrudes outside the membrane and forms, along with the b_2 and δ subunits, the stator. As protons enter, the dissipation of the proton motive force drives ATP synthesis. The ATPase is reversible in its action; that is, ATP hydrolysis can drive formation of a proton motive force.

domains of life, suggesting that this mechanism of energy conservation was a very early evolutionary invention (∞Section 11.2).

The F_1/F_o ATPase is the smallest known biological motor. Proton movement through subunit a of F_o drives rotation of the c proteins, generating a torque that is transmitted to F_1 by the $\gamma\varepsilon$ subunits (Figure 5.21). Energy is transferred to F_1 through the coupled rotation of the $\gamma\varepsilon$ subunits. The latter causes conformational changes in the β subunits, and this is a form of potential energy that can be tapped to make ATP. This is possible because the conformational changes in the β subunits allow for the sequential binding of ADP + P_i to each subunit.

Conversion to ATP occurs when the β subunits return to their original conformation. In analogy to the flagellar motor (∞Figure 4.56), the primary function of the $b_2\delta$ subunits of F_1 is to serve as a fixture (stator), preventing the α and β subunits from rotating with $\gamma\varepsilon$ so that conformational changes in β can be affected. But unlike the flagellum, rotation of the ATPase is not used for cell propulsion, but instead to synthesize ATP (Figure 5.21). ATPase-catalyzed ATP synthesis is referred to as **oxidative phosphorylation** in respiratory systems or **photophosphorylation** in phototrophic organisms. Measurements of the stoichiometry between the number of protons consumed by the ATPase per ATP produced yield a number between 3 and 4.

Reversibility of the ATPase

The tiny F_1/F_o molecular motor is reversible. The hydrolysis of ATP supplies torque for γ to rotate in the opposite direction, causing protons to be pumped from inside to outside the cell through the a subunit. These events *generate* instead of *dissipate* the proton motive force. Reversibility of the ATPase explains why strictly fermentative organisms that lack electron transport chains and are unable to carry out oxidative phosphorylation still make these enzyme complexes. Many important reactions in the cell, such as motility and transport, require energy from the proton motive force rather than from ATP (∞Section 4.6). Thus, ATPase in nonrespiratory organisms such as the lactic acid bacteria, for example, functions unidirectionally to *generate* a proton motive force to drive these necessary cell functions.

Inhibitors and Uncouplers

Electron transport reactions can be studied with certain chemicals that affect electron flow or oxidative phosphorylation. Two such classes of chemicals are known: *inhibitors* and *uncouplers*. Inhibitors block electron flow and, thus, establishment of the proton motive force. Examples include *carbon monoxide* (CO) and *cyanide* (CN^-), both of which bind tightly to a-type cytochromes (Figure 5.20) and prevent their functioning. By contrast, uncouplers prevent ATP synthesis *without* affecting electron transport. Uncouplers are lipid soluble substances, such as *dinitrophenol* and *dicumarol*. These substances make membranes leaky, thereby destroying the proton motive force and its ability to drive ATP synthesis. Loss of the proton motive force thus makes ATP synthesis impossible (Figure 5.21).

5.12 Concept Check

When electrons are transported through an electron transport chain, protons are extruded to the outside of the membrane forming the proton motive force. Key electron carriers include flavins, quinones, the cytochrome bc_1 complex, and other cytochromes, depending on the organism. The cell uses the proton motive force to make ATP through the activity of ATPase.

◆ How do electron transport reactions generate the proton motive force?

◆ What is the ratio of protons extruded per two electrons run through the electron transport chain of *P. denitrificans*? At which sites in the chain is the proton motive force being established?

◆ What structure in the cell converts the proton motive force to ATP? How does it operate?

 V CARBON FLOW IN RESPIRATION AND CATABOLIC ALTERNATIVES

Now that we have a grasp of how ATP is made in respiration, we need to deal with aspects of carbon metabolism associated with it. We will also take a brief look at alternative energy-generating schemes, including anaerobic respirations, photosynthesis, and chemolithotrophy.

5.13 Carbon Flow in Respiration: The Citric Acid Cycle

The early steps in the respiration of glucose involve the same biochemical steps as those of glycolysis (see Figure 5.14). As we have noted, a key intermediate in glycolysis is *pyruvate*. However, whereas in fermentation, pyruvate is reduced and converted into fermentation products (Figure 5.14), in respiration pyruvate is oxidized fully to CO_2. A major and widely distributed pathway by which pyruvate is completely oxidized to CO_2 is called the **citric acid cycle** (CAC), summarized in Figure 5.22●.

Pyruvate is first decarboxylated, leading to the production of one molecule of NADH and an acetyl molecule coupled to coenzyme A (*acetyl-CoA*, see Figure 5.12). The acetyl group of acetyl-CoA combines with the four-carbon compound oxalacetate, leading to the formation of citric acid, the energy of the acetyl-CoA thioester bond (Figure 5.12) being used to drive this synthesis. Hydration, decarboxylation, and oxidation reactions follow, and two additional CO_2 molecules are released (Figure 5.22). Ultimately, oxalacetate is regenerated and can function again as an acetyl acceptor, thus completing the cycle.

CO_2 Release and Fuel for Electron Transport

For each pyruvate molecule oxidized through the cycle, three CO_2 molecules are released, one during the formation of acetyl-CoA, one by the decarboxylation of isocitrate, and one by the decarboxylation of α-ketoglutarate (Figure 5.22). As in fermentation, the electrons released during the oxidation of intermediates in the CAC are transferred to enzymes containing the coenzyme NAD^+ or FAD. However, respiration differs from fermentation in the manner in which NADH and FADH are oxidized. Instead of being used to reduce pyruvate as in fermentation (Figure 5.14), in respiration electrons from NADH are transferred to oxygen or other terminal electron acceptors through the activity of the electron transport chain (see Section 5.12). Thus, unlike fermentation, the presence of an *electron acceptor* in respiration allows for the complete oxidation of glucose to CO_2 with a much greater yield of energy (Figure 5.22b). Indeed, while only 2 ATP are produced per glucose in the alcoholic or lactic acid fermentations (Figure 5.14), a total of 38 ATP can be made by respiring the same glucose molecule (Figure 5.22).

Biosynthesis and the Citric Acid Cycle

Besides playing a key role in catabolism, the citric acid cycle is important for biosynthetic reasons as well. This is because the cycle contains a number of key intermediates that can be drawn off for biosynthetic purposes

when needed. Particularly important in this regard are α-ketoglutarate and oxalacetate, which are the precursors of a number of amino acids (see Section 5.16), and succinyl-CoA, needed to form the porphyrin ring of the cytochromes, chlorophyll, and several other tetrapyrrole compounds (see Figure 5.16). Oxalacetate is also important because it can be converted to phosphoenolpyruvate, a precursor of glucose (see Figure 5.25). In addition to these, acetyl-CoA provides the starting material for fatty acid biosynthesis (see Section 5.17). Thus, the citric acid cycle plays two major roles in the cell: *bioenergetic* and *biosynthetic*. Much the same can be said about the glycolytic pathway, as certain intermediates from this pathway are drawn off for various biosynthetic needs as well (see Figure 5.26).

 5.13 Concept Check

Respiration involves the complete oxidation of an organic compound with much greater energy release than during fermentation. The citric acid cycle plays a major role in the respiration of organic compounds.

◆ How many molecules of CO_2 and pairs of electrons are released *per acetate consumed* in the citric acid cycle?

◆ What two major roles do the citric acid cycle and glycolysis have in common?

5.14 Catabolic Alternatives

As emphasized in Chapter 2, the heart of microbial diversity is *metabolic diversity*, in particular, the diverse strategies microorganisms have evolved for making ATP. Thus far in this chapter, we have dealt only with the reactions of chemoorganotrophs. We now briefly consider some of the alternatives to the use of organic compounds as energy sources, with an emphasis on electron flow and carbon metabolism.

Figure 5.23● summarizes the mechanisms by which cells may generate energy other than via fermentation and aerobic respiration. These include *anaerobic respiration, chemolithotrophy*, and *phototrophy*.

Anaerobic Respiration

One alternate mode of respiratory energy generation is one in which electron acceptors *other than* oxygen are used. In contrast to aerobic respiration, these processes are called **anaerobic respiration**. Some of the electron acceptors used in anaerobic respiration include nitrate (NO_3^-), ferric iron (Fe^{3+}), sulfate (SO_4^{2-}), carbonate (CO_3^{2-}), and even certain organic compounds. Because of their positions on the electron tower (none of these acceptors has as positive an E_0' as the O_2/H_2O couple) (see Figure 5.9), less energy is released when these electron acceptors are used instead of oxygen.

Alternative electron acceptors permit microorganisms to respire in environments where oxygen is absent.

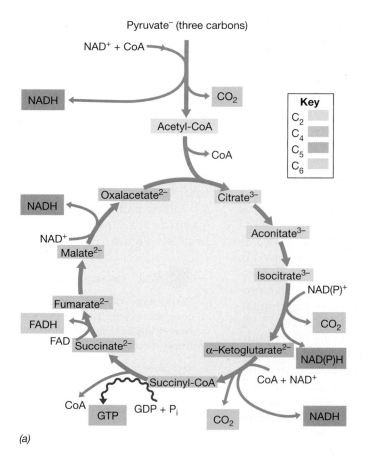

(a)

Energetics Balance Sheet for Aerobic Respiration

(1) **Glycolysis:** Glucose + 2NAD$^+$ + 2 ATP ➤ 2 Pyruvate$^-$ + 4 ATP + 2 NADH
+ 4 ADP ↓   ↓
to CAC   to Complex I

(a) Substrate-level phosphorylation
2 ADP + Pi ➤ 2 ATP (× 2)
(b) Oxidative phosphorylation   ⎱ **8 ATP**
2 NADH ➤ 6 ATP

(2) **CAC:** Pyruvate$^-$ + 4NAD$^+$ + GDP + FAD ➤ 3 CO$_2$ + 4 NADH + FADH + GTP
↓   ↓
to Complex I   to Complex II

(a) Substrate-level phosphorylation
1 GDP + Pi ➤ 1 GTP
1 GTP + 1 ADP ➤ 1 ATP + 1 GDP   ⎱ **15 ATP (×2)**
(b) Oxidative phosphorylation
4 NADH ➤ 12 ATP
1 FADH ➤ 2 ATP

(3) **Sum: Glycolysis plus CAC** ➤ **38 ATP per glucose**

● **Figure 5.22** **The citric acid cycle (CAC).** (a) The CAC begins when the two-carbon compound acetyl-CoA (formed from pyruvate) condenses with the four-carbon compound oxalacetate to form the six-carbon compound citrate. Through a series of oxidations and transformations, this six-carbon compound is ultimately converted back to the four-carbon compound oxalacetate, which then begins another cycle with addition of the next molecule of acetyl-CoA. (b) The overall balance sheet of fuel (NADH/FADH) for the electron transport chain and CO$_2$ generated in the CAC. NADH and FADH feed into the electron transport chain complexes (Figure 5.20) shown.

(a) **Chemoorganotrophic metabolism**

(b) **Chemolithotrophic metabolism**

(c) **Phototrophic metabolism**

● **Figure 5.23** **Energetics and carbon flow.** (a) Chemoorganotrophic respiratory metabolism, (b) chemolithotrophic metabolism, and (c) phototrophic metabolism. Note how in phototrophic metabolism carbon for biosynthesis can come from CO_2 (photoautotrophy) or organic compounds (photoheterotrophy). Note also the importance of electron transport leading to proton motive force formation in each case.

Because the solubility of O_2 in water is rather low, and because O_2 is in such high demand as an electron acceptor, anaerobic respirations are thus ecologically extremely important. We discuss anaerobic respiration in more detail in Section 17.13.

Chemolithotrophy

A second mode of energy generation involves the use of *inorganic* rather than *organic* chemicals. Organisms able to use inorganic chemicals as electron donors are called **chemolithotrophs**. Examples of relevant inorganic electron donors include hydrogen sulfide (H_2S), hydrogen gas (H_2), ferrous iron (Fe^{2+}), and ammonia (NH_3).

Chemolithotrophic metabolism is typically aerobic, but begins with the oxidation of an inorganic rather than an organic electron donor (Figure 5.23). Chemolithotrophs have electron transport complexes similar to those of chemoorganotrophs and form a proton motive force from electron flow in exactly the same way. The latter ultimately drives ATP synthesis. However, one other important distinction between chemolithotrophs and chemoorganotrophs is their sources of *carbon* for biosynthesis. Chemoorganotrophs use organic compounds (glucose, acetate, and the like) as carbon sources. By contrast, chemolithotrophs use *carbon dioxide* (CO_2) as a carbon source and are, hence, **autotrophs**. We discuss many forms of chemolithotrophy in Chapter 17.

Phototrophy

A large number of microorganisms are *phototrophic,* using light as an energy source in the process of photosynthesis. The mechanisms by which light is used as an energy source are unique and complex, but the underlying result is the generation of a proton motive force that can be used in the synthesis of ATP, a process called **photophosphorylation**. Most phototrophs use energy conserved in ATP for the assimilation of carbon dioxide as the carbon source for biosynthesis; they are called **photoautotrophs**. However, some phototrophs use organic compounds as carbon sources with light as the energy source; these are the **photoheterotrophs** (Figure 5.23).

As we mentioned in Chapter 2, photosynthesis in microorganisms is of two different but related types: oxygenic and anoxygenic. *Oxygenic photosynthesis*, carried out by cyanobacteria and their relatives, is similar to that of higher plants and results in O_2 evolution. *Anoxygenic photosynthesis* is a much simpler form found in purple and green bacteria in which O_2 is not evolved. The bioenergetics of both forms of photosynthesis have strong parallels, as we will see when we consider the processes in more detail in Chapter 17.

Importance of the Proton Motive Force to Alternate Bioenergetic Strategies

It should now be clear that in terms of energy metabolism, microorganisms show an amazing diversity of bioenergetic strategies. Thousands of organic compounds, many inorganic compounds, and light can be used by one or another microorganism as an energy source. However, with the exception of fermentations, where substrate-level phosphorylation occurs, metabolic diversity in respiration and photosynthesis revolves around a common theme: *generation of a proton motive force.*

Whether electrons come from the oxidation of organic or inorganic chemicals or from phototrophic processes, in membrane-mediated bioenergetic processes electrons all traverse a membrane-associated electron transport chain. In so doing, they generate a proton motive force (Figure 5.20). Energy conservation in all cases occurs through activity of the ATPase (Figure 5.21) by either oxidative phosphorylation (chemoorganotrophs and chemolithotrophs) or photophosphorylation (phototrophs) (Figure 5.23).

5.14 Concept Check

Electron acceptors other than O_2 can function as terminal electron acceptors for energy generation. Because O_2 is absent under these conditions, the process is called anaerobic respiration. Chemolithotrophs use inorganic compounds as electron donors, while phototrophs use light to form a proton motive force. The proton motive force is involved in all forms of respiration and photosynthesis.

◆ In terms of their electron donor(s), how do *chemoorganotrophs* differ from *chemolithotrophs*?

◆ What is the carbon source for autotrophic organisms?

◆ How do *photoautotrophs* differ from *photoheterotrophs*?

VI BIOSYNTHESIS

We close this chapter with a summary of monomer biosynthesis. Monomers are the building blocks of polymers, and for organisms growing in culture media or in nature that are not provided with these building blocks they must be biosynthesized from simpler components, a process called **anabolism**. We provide an overview of anabolism here, bearing in mind that the details of these processes are beyond the scope of a microbiology textbook and can be found in textbooks of biochemistry.

Energy for anabolism is provided by ATP or the proton motive force (Figure 5.24●). Energy conserved during catabolic reactions as ATP or the proton motive force may be consumed in the formation of monomers and is readily consumed during the polymerization of monomers to form macromolecules. The bulk of a cell's energy expenditure goes toward the synthesis of proteins, although nucleic acid synthesis is energy intensive as well. Comparatively little energy is required to biosynthesize lipids and polysaccharides. Although most of these polymeric biosyntheses are driven by ATP or some related phosphorylated compound, it should always be kept in mind that, provided a cell has adequate resources, it can quickly regenerate ATP from fermentations or respiratory reactions that generate a proton motive force. Thus, ATP and the proton motive force are interchangeable forms of cellular energy (Figure 5.24).

5.15 Biosynthesis of Sugars and Polysaccharides

Polysaccharides are key constituents of the cell walls of many organisms, and in *Bacteria*, the peptidoglycan cell wall (∞Section 4.8) has a polysaccharide backbone. In addition, cells often store carbon and energy reserves in the form of the polysaccharides *glycogen* or *starch* (∞Sections 3.3 and 4.13). The monomeric units of these polysaccharides are six-carbon sugars called *hexoses*, in particular, glucose or glucose derivatives. In addition to hexoses, five-carbon sugars called *pentoses* are common in the cell. Most notably,

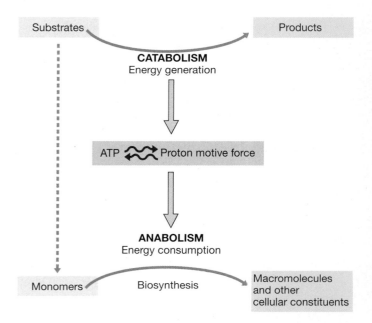

● **Figure 5.24 Scheme of anabolism and catabolism showing the key role of ATP and the proton motive force in integrating the processes.** Monomers can come preformed as nutrients from the environment or from catabolic pathways like glycolysis and the citric acid cycle. In many cases the monomers need to be biosynthesized from catabolic intermediates or nutrients available from nature, and these biosyntheses are anabolic reactions as well.

these include ribose and deoxyribose, present in the backbone of RNA and DNA, respectively.

In prokaryotes, polysaccharides are synthesized from either *uridine diphosphoglucose* (UDPG, Figure 5.25a●) or *adenosine diphosphoglucose* (ADPG), both of which are *activated* forms of glucose. ADPG is the precursor for the biosynthesis of glycogen. UDPG is the precursor of various glucose derivatives needed for the biosynthesis of other polysaccharides in the cell, such as N-acetylglucosamine and N-acetylmuramic acid in peptidoglycan, or the lipopolysaccharide component of the gram-negative outer membrane (∞ Sections 4.8 and 4.9).

When a cell is growing on a hexose such as glucose, obtaining glucose for polysaccharide synthesis is obviously not a problem. But when the cell is growing on other carbon compounds, glucose must be synthesized. This process, called **gluconeogenesis**, uses *phosphoenolpyruvate*, one of the key intermediates of glycolysis, as starting material (see Figure 5.14). Phosphoenolpyruvate can be synthesized from oxalacetate, a citric acid cycle intermediate. An overview of gluconeogenesis is shown in Figure 5.25c.

Biosynthesis of Pentoses

Pentoses are formed by the removal of one carbon atom from a hexose, typically as CO_2 (Figure 5.25d). The pentoses needed for nucleic acid synthesis, ribose and deoxyribose, are formed as shown in Figure 5.25d. The important enzyme *ribonucleotide reductase* converts ribose into deoxyribose by reduction of the 2' carbon on the ring. Interestingly, this reaction occurs *after*, not before, synthe-

(a)

Uridine diphosphoglucose (UDPG)

(b)

ADPG + Glycogen ⟶ ADP + Glycogen-Glucose

(c)

C_2, C_3, C_4, C_5, Compounds

Citric acid cycle

Oxalacetate

Phosphoenolpyruvate + CO_2

Reversal of glycolysis

Glucose-6-P

(d)

Glucose-6-P

Ribulose-5-P + CO_2

Ribose-5-P

Ribonucleotides → RNA

Ribonucleotides — NADPH / Ribonucleotide reductase → Deoxyribonucleotides ⟶ DNA

● **Figure 5.25 Sugar metabolism.** (a) Polysaccharides are synthesized from activated forms of hexoses such as UDPG. Glucose is shown here in blue. UDPG is primarily involved in the biosynthesis of glucose derivatives, such as *N*-acetylglucosamine. (b) Glycogen is biosynthesized from adenosine-phosphoglucose (ADPG) by the sequential addition of glucose. (c) Gluconeogenesis. When glucose is needed it can be biosynthesized from other carbon compounds, generally by the reversal of steps in glycolysis. (d) Pentoses for nucleic acid synthesis are formed by decarboxylation of hexoses like glucose-6-phosphate. Note how the precursors of *DNA* are *produced* from the precursors of *RNA* by the enzyme *ribonucleotide reductase.*

sis of nucleotides. Thus, *ribo*nucleotides are biosynthesized, and from them some of each type later are reduced to *deoxyribo*nucleotides for use as precursors of DNA.

5.15 Concept Check

Polysaccharides are important structural components of cells and are biosynthesized from activated forms of their monomers. Gluconeogenesis is the production of glucose from nonsugar precursors.

◆ In what form is glucose used to form glycogen?

5.16 Biosynthesis of Amino Acids and Nucleotides

The monomeric constituents of proteins and nucleic acids are amino acids and nucleotides, respectively. Their biosyntheses often involve long, multistep pathways requiring many individual enzymes to complete. We approach their biosyntheses here by identifying the carbon skeletons needed to begin the biosynthetic pathways.

Monomers of Proteins: Amino Acids

Organisms that cannot obtain some or all of their amino acids preformed from the environment must synthesize them from other sources. Amino acids can be grouped into structurally related *families* that share biosynthetic steps. The carbon skeletons for amino acids come almost exclusively from intermediates of glycolysis or the citric acid cycle (Figure 5.26●).

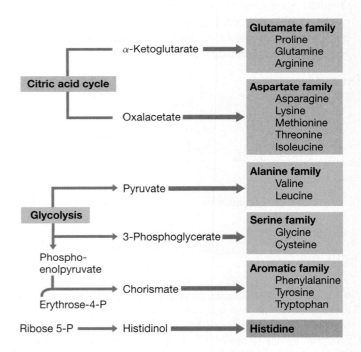

● **Figure 5.26 Amino acid families.** Note how the carbon skeletons for most amino acids are derived from either the citric acid cycle or from glycolysis. Synthesis of the various amino acids in a family frequently requires many separate enzymatically catalyzed steps starting from the parent amino acid (shown in bold).

The *amino group* of amino acids is typically derived from some inorganic nitrogen source in the environment, such as ammonia (NH_3). Ammonia is most often incorporated through formation of the amino acids *glutamate* or *glutamine* by the enzymes *glutamate dehydrogenase* and *glutamine synthetase,* respectively (Figure 5.27a,b●).

Once ammonia is incorporated, it can be used to form other nitrogenous compounds. For example, glutamate can donate its amino group to oxalacetate in a *transaminase* reaction, producing α-ketoglutarate and aspartate (Figure 5.27c). Alternatively, glutamine can react with α-ketoglutarate to form two molecules of glutamate in an *aminotransferase* reaction (Figure 5.27d). The end result of these types of reactions is the shuttling of ammonia into various carbon skeletons from which further biosynthetic reactions can occur to form all the 21 amino acids (∞Figure 3.12) needed to make proteins.

Monomers of Nucleic Acids: Nucleotides

The biosynthesis of the purines and pyrimidines involves quite complex biochemistry. Purines are constructed literally atom by atom from several distinct carbon and nitrogen sources, including CO_2 (Figure 5.28a●). The first key purine, *inosinic acid* (Figure 5.28b), is the precursor of the purine nucleotides *adenine* and *guanine* (∞Figure 3.9). Once these are synthesized (in their triphosphate forms) and have been attached to their correct pentose sugar, they are ready to be incorporated into DNA or RNA (∞Sections 7.5–7.11).

Like the purine ring, the pyrimidine ring is also constructed from several sources (Figure 5.28c). The first key pyrimidine is the compound *uridylate,* and from this the pyrimidines *thymine, cytosine,* and *uracil* (∞ Figure 3.9) are derived (Figure 5.28d).

(a)

(b)

(c)

(d)

● **Figure 5.28** **Biosynthesis of purines and pyrimidines.** (a) The precursors of the purine skeleton. (b) Inosinic acid, the precursor of all purine nucleotides. (c) The precursors of the pyrimidine skeleton, orotic acid. (d) Uridylate, the precursor of all pyrimidine nucleotides. Uridylate is formed from orotate following a decarboxylation and the addition of ribose–5–phosphate.

(a)   α-Ketoglutarate + NH_3 →(NADH)(Glutamate dehydrogenase) Glutamate–NH_2

(b)   Glutamate–NH_2 + NH_3 →(ATP ADP + P_i)(Glutamine synthetase) Glutamine–NH_2–NH_2

(c)   Glutamate–NH_2 + Oxalacetate →(Transaminase) α-Ketoglutarate + Aspartate–NH_2

(d)   Glutamine–NH_2–NH_2 + α-Ketoglutarate →(NADH)(Glutamate synthase) 2 Glutamate–NH_2

● **Figure 5.27 Ammonia incorporation in bacteria.** Free ammonia as well as the amino groups of all amino acids are shown in blue. Two major pathways for NH_3 assimilation in bacteria are those catalyzed by the enzymes (a) glutamate dehydrogenase and (b) glutamine synthetase. (c) Transaminase reactions simply transfer an amino group from an amino acid to an organic acid. (d) In the reaction catalyzed by the enzyme glutamate synthase, two glutamates are formed from one glutamine and one α-ketoglutarate.

5.16 Concept Check

Amino acids are formed from carbon skeletons generated during catabolism while nucleotides are biosynthesized using carbon from several sources.

◆ What form of nitrogen is commonly used to form the amino group of amino acids?

◆ Which bases are *purines* and which are *pyrimidines*?

5.17 Biosynthesis of Fatty Acids and Lipids

Although the lipids of *Archaea* contain hydrophobic constituents *other* than fatty acids, the lipids of species of *Bacteria* and *Eukarya* contain fatty acids (∞Section 4.5). Lipids are important constituents of cells, as they are major structural components of membranes. Lipids are also carbon and energy reserves. Other lipids function in and around the cell surface, including, in particular, the lipopolysaccharide layer of gram-negative *Bacteria* (∞ Section 4.9). A cell can produce many different types of lipids, some of which are produced only under certain conditions, or play special roles in only certain cell structures. The biosynthesis of fatty acids is thus a major series of reactions in most cells.

Fatty Acid Biosynthesis

Fatty acids are biosynthesized two carbon atoms at a time with the help of a small protein called the *acyl carrier protein* (ACP). ACP holds the growing fatty acid as it is being synthesized and releases it once it has reached its final length (Figure 5.29●; ∞ Figure 3.7). Interestingly, although the fatty acid chain is constructed *two* carbons at a time, the two carbons are donated from a *three*-carbon compound called *malonate* attached to ACP to form malonyl-ACP. As each malonyl residue is donated, one molecule of CO_2 is released (Figure 5.29).

The fatty acid composition of cells varies from species to species and can also vary somewhat within a species due to temperature (growth at low temperatures favors shorter-length fatty acids, while growth at higher temperatures favors longer-length fatty acids). The most common fatty acids in lipids of *Bacteria* contain C_{12}–C_{20} fatty acids.

In addition to saturated, even-carbon-number fatty acids, fatty acids can also be unsaturated, contain branches, or have an odd number of carbon atoms. *Unsaturated* fatty acids contain one or more double bonds in the long hydrophobic portion of the molecule. The number and position of these double bonds is often species or group specific, and the double bonds are typically added by desaturating a saturated fatty acid. *Branched chain* and *odd-carbon-number* fatty acids are made using an initiating molecule that contains a branched chain fatty acid or a propionyl (C_3) group, respectively.

Lipids

The final assembly of lipids in *Bacteria* and *Eukarya* involves the addition of fatty acids to a glycerol molecule. For simple triglycerides, all three glycerol carbons are esterified with fatty acids. In complex lipids, one of the carbons of glycerol contains a molecule of phosphate, ethanolamine, a sugar, or some other polar substance (∞ Figure 3.7). In *Archaea*, lipids contain phytanyl side chains instead of fatty acid side chains (∞ Section 4.5). However, as in *Bacteria* or *Eukarya*, the third carbon of the glycerol backbone in archaeal lipids typically contains a polar group of some sort.

Lipids have been used as biomarkers in microbial ecology studies. For example, ether-linked lipids are typical of species of *Archaea* but not *Bacteria* or *Eukarya*. Other lipids are highly specific for particular groups of organisms, and this extensive lipid diversity has been used to detect these groups in natural samples (∞ Section 11.10). Lipids from dead organisms are often more refractory to degradation than other cell macromolecules. This fact, along with their specificity, make lipid analyses an ideal method for assessing the bacterial composition of ancient materials. For example, lipid analyses done on samples from lake or marine sediments can trace the kinds and amounts of different microorganisms in the individual layers of a sediment core. Since each layer has a different age, and the age can be readily determined, the lipid profile gives a record of the types of organisms present in the habitat through time.

● **Figure 5.29 The biosynthesis of fatty acids.** Shown is the biosynthesis of the C_{16} fatty acid, *palmitate* (∞Figure 3.7). The condensation of acetyl-ACP and malonyl-ACP forms acetoacetyl-CoA. Each successive addition of an acetyl unit comes from malonyl-ACP.

 5.17 Concept Check

Fatty acids are synthesized two carbons at a time and then attached to glycerol to form lipids.

◆ Explain why in fatty acid synthesis fatty acids get added *two* carbon atoms at a time, while the immediate donor for these carbons contains *three* carbon atoms.

REVIEW QUESTIONS

1. Why are *carbon* and *nitrogen* macronutrients while *cobalt* is a micronutrient (∞Section 5.1)?

2. What are *siderophores* and why are they necessary (∞Section 5.1)?

3. Why would the following medium *not* be considered a chemically defined medium: glucose, 5 grams (g); NH_4Cl, 1 g; KH_2PO_4, 1 g; $MgSO_4$, 0.3 g; yeast extract, 5 g; distilled water, 1 liter (∞Section 5.2)?

4. What is aseptic technique and why is it necessary (∞Section 5.3)?

5. Describe how you would calculate $\Delta G^{0\prime}$ for the reaction: glucose $+ 6 O_2 \rightarrow 6 CO_2 + 6 H_2O$. If you were told that this reaction is highly *exergonic*, what would be the sign (negative or positive) of the $\Delta G^{0\prime}$ you would expect for this reaction (∞Section 5.4)?

6. Distinguish between $\Delta G^{0\prime}$, ΔG, and G_f^0 (∞Section 5.4).

7. Why are enzymes needed by the cell (∞Section 5.5)?

8. Describe the difference between a *coenzyme* and a *prosthetic group* (∞Section 5.5).

9. The following is a series of coupled electron donors and electron acceptors (written as donor/acceptor). Using just the data given in Figure 5.9, order this series from most energy-yielding to least energy-yielding. H_2/Fe^{3+}, H_2S/O_2, methanol/NO_3^- (producing NO_2^-), H_2/O_2, Fe^{2+}/O_2, NO_2^-/Fe^{3+}, H_2S/NO_3^- (∞Section 5.6).

10. What is the reduction potential of the $NAD^+/NADH$ couple (∞Section 5.7)?

11. Why is acetyl phosphate considered an energy-rich compound while glucose 6-phosphate is not (∞Section 5.8)?

12. How is ATP made in fermentation and respiration (∞Section 5.9)?

13. Where in glycolysis is NADH *produced*? Where is NADH *consumed* (∞Section 5.10)?

14. What is meant by the term *proton motive force* and why is this concept so important in biology (∞Sections 5.11 and 5.12)?

15. How is rotational energy in the ATPase used to produce ATP (∞Section 5.12)?

16. The chemicals dinitrophenol and cyanide are both cellular poisons but act in quite different ways. Compare and contrast the modes of action of these two chemicals (∞Section 5.12).

17. Work through the energy balance sheets for fermentation and respiration, and account for all sites of ATP synthesis. Organisms can obtain nearly 20 times more ATP when growing aerobically on glucose than by fermenting it. Write one sentence that accounts for this difference (∞Section 5.13).

18. Why can it be said that the citric acid cycle plays *two* major roles in the cell (∞Section 5.13)?

19. What are the differences in electron donor and carbon source used by *Escherichia coli* and *Acidithiobacillus thioparus* (a sulfur chemolithotroph) (∞Section 5.14)?

20. What two catabolic pathways supply carbon skeletons for sugar and amino acid biosyntheses (∞Sections 5.15 and 5.16)?

21. Describe the process by which a fatty acid such as palmitate (a C_{16} straight chain saturated fatty acid) is synthesized in a cell (∞Section 5.17).

APPLICATION QUESTIONS

1. Design a defined culture medium for an organism that can grow aerobically on acetate as a carbon and energy source. Make sure all the nutrient needs of the organism are accounted for and in the correct relative proportions.

2. *Desulfovibrio* can grow anaerobically with H_2 as electron donor and SO_4^{2-} as electron acceptor (which is reduced to H_2S). Based on this information and the data in Table A1.2 (Appendix 1), indicate which of the following components *could not* exist in the electron transport chain of this organism and why: cytochrome c, ubiquinone, cytochrome c_3, cytochrome aa_3, ferredoxin.

3. Again, using the data in Table A1.2, predict the sequence of electron carriers in the membrane of an organism growing aerobically and producing the following electron carriers: ubiquinone, cytochrome aa_3, cytochrome b, NADH, cytochrome c, FAD.

4. Explain the following observation: cells of *Escherichia coli* fermenting glucose grow faster when NO_3^- is supplied to the culture (NO_2^- is produced), and then grow even faster (and stop producing NO_2^-) when the culture is highly aerated.

6

MICROBIAL GROWTH

Cell division is the series of events that generates two cells from one.

WORKING GLOSSARY

Acidophile an organism that grows best at low pH

Aerobe an organism that can use oxygen (O_2) in respiration; some require oxygen for growth

Aerotolerant anaerobe a microorganism unable to respire oxygen (O_2) but whose growth is unaffected by the presence of oxygen

Alkaliphile an organism that grows best at high pH

Anaerobe an organism that cannot use oxygen in respiration and whose growth may be inhibited by oxygen

Autolysis spontaneous cell lysis, usually due to the activity of lytic proteins called autolysins

Batch culture a closed-system microbial culture of fixed volume

Binary fission cell division following enlargement of a cell to twice its minimum size

Cardinal temperatures the minimum, maximum, and optimum growth temperatures for a given organism

Chemostat a device that allows for the continuous culture of microorganisms in which both growth rate and cell number can be controlled independently

Compatible solute a molecule that is accumulated in the cytoplasm for adjustment of water activity but that does not inhibit biochemical processes

Divisome a complex of proteins involved in cell division processes in prokaryotes

Exponential growth growth of a microorganism where the cell number doubles within a fixed time period

Extreme halophile a microorganism that requires very large amounts of salt (NaCl), usually greater than 10% and in some cases near to saturation, for growth

Extremophile an organism that grows optimally under one or more chemical or physical extremes, such as high or low temperature or pH

Facultative with respect to oxygen, an organism that can grow in either its presence or absence

FtsZ a key cell division protein that forms a ring along the division plane to initiate cell elongation

Generation time the time required for a population of microbial cells to double

Growth an increase in cell number

Halophile a microorganism that requires NaCl for growth

Halotolerant an organism that does not require NaCl for growth but that can grow in the presence of salt, in some cases, substantial levels of salt

Hyperthermophile a microorganism that has a growth temperature optimum of 80°C or greater

Lag phase a period preceding the exponential growth phase when cells may be metabolizing but are not yet growing

Mesophile an organism that grows best at temperatures between 20 and 45°C

Microaerophile an aerobic organism that can grow only when oxygen tensions are reduced from that in air

pH the negative logarithm of the hydrogen ion (H^+) concentration of a solution

Psychrophile an organism with a growth temperature optimum of 15°C or lower and a maximum growth temperature below 20°C

Psychrotolerant an organism capable of growth at low temperatures but whose growth temperature optimum is above 20°C

Stationary phase the period immediately following exponential growth when the growth rate of the population falls to zero

Thermophile an organism whose growth temperature optimum lies between 45 and 80°C

Transpeptidation formation of peptide cross-links between muramic acid residues in peptidoglycan synthesis

Viable capable of reproducing

Xerophile an organism that is able to live, or that lives best, in very dry environments

BACTERIAL CELL DIVISION

In the past three chapters we have discussed the structure of cellular macromolecules (∞Chapter 3), cell structure and function (∞Chapter 4), and the general principles of microbial nutrition and metabolism (∞Chapter 5). Before we begin our study of the biosynthesis of macromolecules and the molecular genetics of microorganisms (∞Chapter 7), we consider here some aspects of microbial growth. Growth is the ultimate process in the life of a cell—one cell becoming two.

6.1 Cell Growth and Binary Fission

In microbiology, **growth** is defined as *an increase in the number of cells.* Growth is an essential component of the circle of life. Any given cell has a finite life span, and the species is maintained only as a result of continued growth of the population. In addition to understanding the basic science of microbial growth, many practical situations call for the *control* of microbial growth. Knowledge of how microbial populations can rapidly expand is useful for designing methods to control microbial growth. We will study these control methods in Chapter 20.

The bacterial cell is a living machine capable of duplicating itself. The process of bacterial cell growth involves as many as 2000 chemical reactions of a wide variety of types. Some of these reactions involve energy transformations. Other reactions involve the biosynthesis of small molecules—the building blocks of macromolecules. Still others provide the various cofactors and coenzymes needed for enzymatic reactions. However, the main reactions of cell synthesis are *polymerization* reactions, the processes by which macromolecules are made from monomers. As macromolecules accumulate in the cytoplasm of a cell, they are assembled into new structures, such as the cell wall, cytoplasmic membrane, flagella, ribosomes, inclusion bodies, enzyme complexes, and so on, eventually leading to cell division.

Binary Fission

In most prokaryotes, growth of an individual cell continues until the cell divides into two new cells. This is a process called **binary fission** (*binary* to express the fact

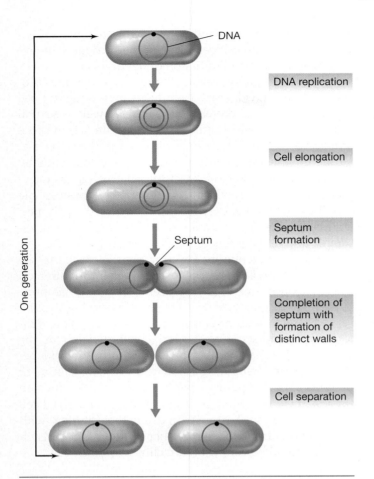

DNA

DNA replication

Cell elongation

Septum formation

Septum

Completion of septum with formation of distinct walls

Cell separation

One generation

● **Figure 6.1 The general process of binary fission in a rod-shaped prokaryote.** For simplicity, the nucleoid is depicted as a single circle in green.

conditions the bacterium *E. coli* can complete the cycle in about 20 min. A few bacteria can grow even faster than this, but many grow much slower. As we will see, cell division is intimately tied to chromosomal replication events.

 6.1 Concept Check

Microbial growth involves an increase in the *number* of cells. Growth of most microorganisms occurs by the process of binary fission.

◆ Define the term *generation*.

6.2 Fts Proteins, the Cell Division Plane, and Cell Morphology

Several proteins called **Fts proteins** are essential for cell division. The acronym *Fts* stands for "*f*ilamentous *t*emperature *s*ensitive", which describes the properties of cells harboring mutations in the genes that encode Fts proteins. Such cells have great difficulty in dividing. **FtsZ**, a key protein in the group, has been well studied in *Escherichia coli* and several other bacteria.

Fts proteins are universally distributed among prokaryotes, including the *Archaea,* and FtsZ-type proteins have even been found in mitochondria and chloroplasts, further emphasizing the evolutionary ties that these organelles have to the *Bacteria* (∞Sections 2.3, 14.4, and 14.5). Interestingly, FtsZ shows structural similarities to *tubulin*, the important cell division protein in eukaryotes (∞Section 14.4).

Fts Proteins and Cell Division

Fts proteins interact to form a *division apparatus* in the cell called the **divisome**. In rod-shaped cells, formation of the divisome begins with the attachment of molecules of FtsZ in a ring around the cell cylinder in the precise center of the cell (Figure 6.2●). This spot will eventually become the cell division plane. In a cell of *Escherichia coli* about 10,000 FtsZ molecules polymerize to form the ring, and then the ring attracts other cell division proteins, including *FtsA* and *ZipA* (Figure 6.2). FtsA is an ATP-hydrolyzing enzyme that provides energy for assembly of the many proteins into the divisome. ZipA is an anchor that attaches the FtsZ ring to the cytoplasmic membrane (Figure 6.2).

The divisome also contains Fts proteins involved in peptidoglycan synthesis, such as FtsI (Figure 6.2). FtsI is also called a *penicillin-binding protein*, because its activity is blocked by the antibiotic penicillin (see Section 6.3). The divisome is thought to orchestrate synthesis of new cytoplasmic membrane and cell wall material in both directions until a cell reaches a length twice its original length. Following this, constriction occurs to form the septum yielding two daughter cells as was shown in Figure 6.1.

that *two* cells have arisen from one). In a growing culture of a rod-shaped bacterium such as *Escherichia coli,* cells elongate to approximately twice their original length and then form a partition that eventually separates the cell into two daughter cells (Figure 6.1●). This partition is called a *septum* and is a result of the inward growth of the cytoplasmic membrane and cell wall from opposing directions until the two daughter cells are pinched off (Figure 6.1). By definition, when one cell divides to form two, one **generation** has occurred (Figure 6.1).

During the growth cycle all cellular constituents increase proportionally. Each daughter cell receives a chromosome and sufficient copies of ribosomes and all other macromolecular complexes, monomers, and inorganic ions to exist as an independent cell. Partitioning of the replicated DNA molecule between the two daughter cells depends on the DNA remaining attached to membranes during division, with septum formation leading to separation of the chromosomes, one to each daughter cell (Figure 6.1).

The time required for a generation to occur in bacteria is highly variable and is dependent on a number of factors, both nutritional and genetic. Under the best nutritional

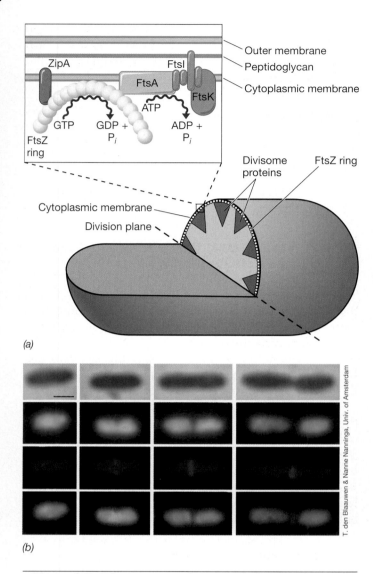

(a)

(b)

T. den Blaauwen & Nanne Nanninga, Univ. of Amsterdam

● **Figure 6.2   The FtsZ ring and cell division.** (a) Cut-away view of a rod-shaped cell showing the ring of FtsZ molecules around the division plane. Blowup shows the arrangement of individual divisome proteins. ZipA is an FtsZ anchor, FtsI is a peptidoglycan biosynthesis protein, FtsK assists in chromosome separation, and FtsA is an ATPase. (b) Appearance and breakdown of the FtsZ ring during the cell cycle of *Escherichia coli.* Microscopy: upper row, phase contrast; second row, nucleoid stain; third row, cells stained with a specific reagent against FtsZ; fourth row, combination nucleoid and FtsZ staining. Cell division events: first column, FtsZ ring not yet formed; second column, FtsZ ring appears as nucleoids start to segregate; third column, full FtsZ ring forms as cell elongates; fourth column, breakdown of the FtsZ ring and cell division. Marker bar in upper left photo, 1 μm.

DNA Replication and Cell Division

DNA replication occurs prior to the formation of the FtsZ ring. Indeed, cessation of DNA synthesis appears to be the signal for FtsZ ring formation. The ring forms in the space between the duplicated nucleoids. Location of the actual cell midpoint by FtsZ is assisted by a series of proteins called *Min* proteins, especially *MinC* and *MinE*. MinC is an inhibitor of cell division that prevents the FtsZ ring from forming until the precise cell center has

been found. MinE is an inhibitor of MinC activity and attaches itself to the cell center. This event triggers the recruitment of FtsZ and initiation of the divisome.

As cell elongation occurs, the two copies of the chromosome are pulled apart, each to its own daughter cell (Figure 6.1). An Fts protein called *FtsK,* as well as several other proteins, assist in this process (Figure 6.2). As constriction occurs, the FtsZ ring begins to depolymerize, triggering the inward growth of wall materials to eventually seal off one daughter cell from the other. FtsZ has enzymatic activity and can hydrolyze guanosine triphosphate (GTP) to yield the energy necessary to fuel the polymerization and depolymerization of FtsZ and subsequent assembly and disassembly of the FtsZ ring (Figure 6.2).

Properly functioning Fts proteins are crucial to the bacterial cell division process. Much new information on cell division has emerged in recent years, and genomic studies have confirmed that Fts proteins are highly conserved across broad phylogenetic lines. There is great interest in understanding bacterial cell division at the molecular level, not only for basic reasons, but also because such understanding could lead to the development of new drugs targeting specific steps in the division process. Like penicillin (a drug that targets bacterial cell wall synthesis, see Section 6.3), drugs that interfere with the function of specific Fts or other bacterial cell division proteins could be very useful in clinical medicine.

Cell Shape and Actin-Like Proteins in Prokaryotes

Although FtsZ plays a major role in the cell *division* process, what factor(s) affect the *morphology* of a prokaryotic cell? For many years it was thought that peptidoglycan was synthesized in such a way as to define the morphology of a cell. Now it is clear that specific proteins exist in prokaryotes that define cell shape and that peptidoglycan plays only a minor role. Interestingly, these shape-determining proteins show significant homology to the eukaryotic cell protein *actin,* a key component of the cytoskeleton of eukaryotic cells (∞Section 14.4).

The major shape-determining protein in prokaryotes is called *MreB.* This protein forms an actin-like cytoskeleton in species of *Bacteria* and probably also in *Archaea.* MreB forms into filamentous spiral-shaped bands around the inside of the cell, just underneath the cytoplasmic membrane. Presumably, the MreB cytoskeleton defines cell shape by in some way generating a force against the cytoplasmic membrane. Interestingly, coccus-shaped bacteria lack MreB and the gene that encodes it. This indicates that the "default" shape for a bacterium is a sphere. Variations in the arrangement of MreB filaments in cells of nonspherical bacteria are probably responsible for the rod and other cell morphologies found among various prokaryotes (∞Figure 4.11).

Between FtsZ and MreB, prokaryotic cells produce proteins structurally similar to tubulin and actin, respectively, the key eukaryotic proteins involved in cell division and internal cell scaffolding. Because of the importance of cell division to the biology of the cell, it is not surprising that biological solutions to cell division and shape in eukaryotic cells have their evolutionary roots in prokaryotic cells. Like other core functions of cells—DNA replication, transcription, and translation—cell division mechanisms were likely rather early inventions in microbial evolution.

 6.2 *Concept Check*

Cell division and chromosome replication are coordinately regulated, and the Fts proteins are the keys to these processes. The protein FtsZ defines the division plane in prokaryotes, while Mre proteins help define cell shape.

◆ When does the bacterial chromosome replicate in the binary fission process?

◆ How does FtsZ find the cell midpoint?

◆ What eukaryotic proteins are homologues of FtsZ and MreB?

6.3 Peptidoglycan Synthesis and Cell Division

Before cell division can occur, new cell wall synthesis must take place. And most importantly, new wall material must be added to the preexisting cell wall without loss of structural integrity. This critical cellular process occurs as shown in Figure 6.3●.

(a)

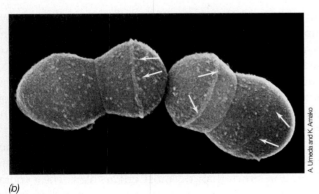

(b)

● **Figure 6.3 Cell wall synthesis in gram-positive *Bacteria.*** (a) Localization of new cell wall synthesis during cell division. In cocci, new cell wall synthesis (shown in green) is localized at only one point. The FtsZ ring (see Figure 6.2) defines the cell division plane. (b) Scanning electron micrograph of cells of *Streptococcus hemolyticus* showing wall bands (arrows). A single cell is about 1 μm in diameter.

Beginning at the FtsZ ring (Figure 6.2*a* and Figure 6.3), small openings in the wall are created by enzymes called *autolysins*, enzymes similar in function to lysozyme (∞Section 4.8). Autolysins are present within the divisome protein complex. New cell wall material is then added across the openings (Figure 6.3*a*). The junction between new and old peptidoglycan forms a ridge on the cell surface of gram-positive *Bacteria* (Figure 6.3*b*), analogous to a scar. It is essential in peptidoglycan synthesis that new cell wall precursors (muramic acid/glucosamine/tetrapeptide units, see Figure 6.5) be spliced into existing peptidoglycan in a coordinated manner to prevent a breach in peptidoglycan integrity at the splice point. If this does not take place, a process of spontaneous cell lysis called **autolysis** can occur.

Biosynthesis of Peptidoglycan

We discussed the general structure of peptidoglycan in Chapter 4 (∞Section 4.8). The peptidoglycan layer can be thought of as a stress-bearing fabric, much like a thin sheet of rubber. Synthesis of new peptidoglycan during growth requires controlled cutting of preexisting peptidoglycan by autolysins along with the simultaneous insertion of peptidoglycan precursors. A lipid carrier molecule called **bactoprenol** (Figure 6.4●) plays a major role in this process. Bactoprenol is a very hydrophobic C_{55} alcohol that bonds to the *N*-acetyl glucosamine/*N*-acetylmuramic acid/pentapeptide peptidoglycan precursor (Figure 6.5*a*●). Bactoprenol transports peptidoglycan precursors across the cytoplasmic membrane by rendering them sufficiently hydrophobic to pass through the interior of the cytoplasmic membrane. Once in the periplasm, bactoprenol interacts with enzymes that insert cell wall precursors into the growing point of the cell wall and catalyze glycosidic bond formation (Figure 6.5*b*).

Transpeptidation: The Penicillin Target

The final step in cell wall synthesis is called **transpeptidation**. This process involves formation of the peptide cross-links between muramic acid residues in adjacent glycan chains (∞Section 4.8 and Figures 4.29 and 4.30). Transpeptidation is medically noteworthy because it is

$$H_3C-\underset{CH_3}{\underset{|}{C}}=CHCH_2(CH_2\underset{CH_3}{\underset{|}{C}}=CHCH_2)_9CH_2\underset{CH_3}{\underset{|}{C}}=CHCH_2$$

$$\begin{array}{c} O \\ | \\ O=P-O^- \\ | \\ O \\ | \\ O=P-O^- \\ | \\ \boxed{\textit{N-Acetylmuramic acid}} - O \end{array}$$

● **Figure 6.4 Bactoprenol (undecaprenol diphosphate).** This highly hydrophobic molecule carries cell wall peptidoglycan precursors through the cytoplasmic membrane.

Figure 6.5 Peptidoglycan synthesis. (a) Transport of peptidoglycan precursors across the cytoplasmic membrane to the growing point of the cell wall. (b) The transpeptidation reaction that leads to the final cross-linking of two peptidoglycan chains. Penicillin inhibits this reaction. The terms **Out** and **In** represent the environment and the cytoplasm, respectively.

the reaction inhibited by the antibiotic *penicillin*. Several penicillin-binding proteins have been identified in the periplasm of gram-negative *Bacteria*, including the previously mentioned FtsI (see Figure 6.2a). When penicillin binds to these proteins, they are no longer catalytically active. In the absence of new wall synthesis, the continued activity of autolysins eventually weakens the cell wall and leads to cell lysis. Penicillin has been such a widely used drug in clinical medicine for two reasons. First, humans are *Eukarya* and therefore lack peptidoglycan; the drug can thus be administered in high doses. And second, virtually all bacterial pathogens contain peptidoglycan and are thus potential targets of the drug.

Transpeptidation involves peptide bond formation with one of several different amino acids, depending on the cell wall structure of the organism involved. In gram-negative *Bacteria* such as *Escherichia coli*, cross-linking typically occurs between diaminopimelic acid on one peptide and D-alanine on the adjacent peptide (Figure 6.5b). Ini-

tially, there are *two* D-alanine residues at the end of the peptidoglycan precursor, but one D-alanine molecule is removed during the transpeptidation reaction (Figure 6.5b). This supplies the energy necessary to drive the reaction forward (transpeptidation occurs outside the cytoplasmic membrane where ATP is unavailable). In *E. coli*, the protein FtsI (Figure 6.2a) is thought to be the key protein involved in transpeptidation. In gram-positive *Bacteria*, where a glycine interbridge is commonly present (∞ Section 4.8 and Figure 4.31), cross-links occur across the interbridge, typically from an L-lysine of one peptide to a D-alanine on the other.

6.3 Concept Check

New cell wall is synthesized during bacterial growth by inserting new glycan units into preexisting wall material. A hydrophobic alcohol called bactoprenol facilitates transport of new glycan units through the cytoplasmic membrane to become part of the growing cell wall. Transpeptidation bonds the precursors into the peptidoglycan fabric.

◆ What are *autolysins* and why are they necessary?

◆ What is the function of bactoprenol?

◆ What is *transpeptidation* and why is it important?

II GROWTH OF BACTERIAL POPULATIONS

As we have mentioned, *microbial growth* is defined as an increase in the *number* of cells in a population. We now move on from growth and division of an individual cell to consider the dynamics of growth in bacterial populations.

6.4 Growth Terminology and the Concept of Exponential Growth

During the cell division cycle (Figure 6.1), all structural components of the cell double. While the interval for the formation of two cells from one is called a **generation**, the *time required* for this to occur is called the **generation time** (see Figure 6.1). The generation time is the time required for the cell population to double (the cell *mass* doubles during this period as well). Because of this, the generation time is also called the *doubling time.*

Generation times vary widely among microorganisms. In general, most bacteria have shorter minimum generation times than do most eukaryotes. The generation time of a given organism is quite dependent on the growth medium and the incubation conditions employed. Many bacteria have minimal generation times of 1–3 h under the best of growth conditions, but a few very rapidly growing organisms are known whose doubling times are less than 10 min and a few slow-growing organisms whose doubling times are as long as several days.

In nature, microbial doubling times may be much longer than those obtained in laboratory culture. This is because in nature, ideal growth conditions for a given organism may exist only intermittently. Thus, depending on resource availability, physiochemical conditions (temperature, pH, and the like), moisture availability, and seasonal changes, bacterial populations in nature may double only once every few weeks, or even longer.

Exponential Growth

A growth experiment beginning with a single cell having a doubling time of 30 min is presented in Figure 6.6●. This pattern of population increase, where the number of cells *doubles* during a regular time interval, is called **exponential growth**. When the cell number from such an experiment is graphed on arithmetic (linear) coordinates as a function of time, one obtains a curve with a continuously increasing slope (Figure 6.6*b*).

By contrast, when the number of cells is plotted on a logarithmic ($\log_{10}$) scale and time is plotted arithmetically (a *semilogarithmic* graph), as shown in Figure 6.6*b*, the points follow a straight line. This straight-line function indicates that the cells are growing exponentially— the cell population is doubling in a constant period of time. Semilogarithmic graphs are also convenient and simple to use for estimating *generation times* from a set of

(a)

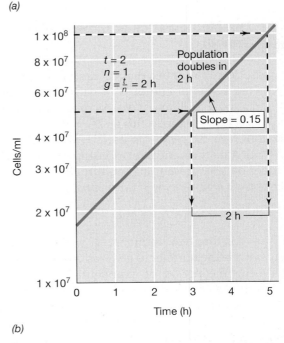

(b)

● **Figure 6.7 Calculating growth parameters.** Method of estimating the generation times (g) of exponentially growing populations with generation times of (a) 6 h and (b) 2 h from data plotted on semilogarithmic graphs. The slope of each line is equal to $0.301/g$ and n equals the number of generations that have occurred in the time, t. All numbers are expressed in scientific notation; that is, 10,000,000 is 1×10^7, 60,000,000 is 6×10^7, and so on.

Time (h)	Total number of cells	Time (h)	Total number of cells
0	1	4	256 (2^8)
0.5	2	4.5	512 (2^9)
1	4	5	1,024 (2^{10})
1.5	8	5.5	2,048 (2^{11})
2	16	6	4,096 (2^{12})
2.5	32	.	.
3	64	.	.
3.5	128	10	1,048,576 (2^{19})

(a)

(b)

● **Figure 6.6 The rate of growth of a microbial culture.** (a) Data for a population that doubles every 30 min. (b) Data plotted on an arithmetic (left ordinate) and a logarithmic (right ordinate) scale.

growth data. The generation time may be read directly from the graph (Figure 6.7●; see also Figure 6.12*b*).

In exponential growth, the rate of increase in cell number is slow initially but increases at an ever faster rate. In the later stages of growth, this results in an explosive increase in cell numbers. For example, in the experiment in Figure 6.6, the rate of cell production in the first 30 min of growth is one cell per 30 min. However, between 4 and 4.5 h of growth, the rate of cell production is considerably higher, 256 cells per 30 min (Figure 6.6). A practical implication of exponential growth is that when a nonsterile and nutrient rich food product such as milk is allowed to stand under ideal bacterial growth conditions, a few hours during the early stages of exponential growth are not detrimental, whereas standing *for the same length of time* during the later stages is disastrous.

6.4 *Concept Check*

Microbial populations show a characteristic type of growth pattern called exponential growth, which is best seen by plotting the number of cells over time on a semilogarithmic graph.

◆ Why does exponential growth lead to large cell populations in so short a period of time?

◆ What is a *semilogarithmic* plot?

6.5 The Mathematics of Exponential Growth

The increase in cell number that occurs in an exponentially growing bacterial culture is a geometric progression of the number 2 (Figure 6.6*a*). As one cell divides to become two cells, we express this as $2^0 \rightarrow 2^1$. As two cells become four, we express this as $2^1 \rightarrow 2^2$, and so on (Figure 6.6*a*). A mathematical relationship exists between the number of cells present in a culture *initially* and the number present *after a period of exponential growth*:

$$N = N_0 2^n$$

where N is the final cell number, N_0 is the initial cell number, and n is the number of generations that have occurred during the period of exponential growth. The generation time g of the exponentially growing population is t/n, where t is the duration of exponential growth expressed in days, hours, or minutes, depending on the organism and the growth conditions. From a knowledge of the initial and final cell numbers in an exponentially growing cell population, it is possible to calculate n, and from n and knowledge of t, the generation time g.

The Relationship of N and N_0 to n

The equation $N = N_0 2^n$ can be expressed in terms of n as follows:

$$N = N_0 2^n$$

$$\log N = \log N_0 + n \log 2$$

$$\log N - \log N_0 = n \log 2$$

$$n = \frac{\log N - \log N_0}{\log 2} = \frac{\log N - \log N_0}{0.301}$$

$$= 3.3 \, (\log N - \log N_0)$$

With this formula, we can calculate generation times in terms of readily measurable quantities, N and N_0. As an example, consider the actual data from the lower graph in Figure 6.7, in which $N = 10^8$, $N_0 = 5 \times 10^7$, and $t = 2$:

$$n = 3.3 \, [\log(10^8) - \log(5 \times 10^7)] = 3.3(8 - 7.69)$$

$$= 3.3(0.301) = 1$$

Thus, in this example, the generation time $t/n = 2/1 = 2$ h. If exponential growth occurred for another 2 h, the cell number would be 2×10^8.

Related Growth Parameters

The generation time g of an exponentially growing culture can also be calculated from the *slope of the line* obtained in the semilogarithmic plot of exponential growth. The slope is equal to $0.301n/t$ (log $2n/t$) and in the above example would be $0.301(1)/2$, or 0.15. Since g is equal to $0.301/\text{slope}$, we arrive at the same value of 2 for g. The term $0.301n/t$ is called the *specific growth rate*, abbreviated k.

Another index of growth is the reciprocal of the generation time, called the *division rate*, abbreviated ν. The division rate is equal to $1/g$ and has units of reciprocal hours (h^{-1}). While the term g is a measure of the *time* it takes for a population to double in cell number, ν is a measure of the *number of generations* that occur per unit time in an exponentially growing culture. The slope of the line relating log cell number to time (Figure 6.7) is equal to $\nu/3.3$.

Armed with knowledge of n and t, one can calculate g, k, and ν for different microorganisms growing under different culture conditions. This is often useful for optimizing culture conditions for a particular organism and also for testing the positive or negative effect of some treatment on the bacterial culture.

6.5 *Concept Check*

From knowledge of the initial and final cell numbers and the time of exponential growth, the generation time and growth rate constant of a cell population can be calculated directly. Key parameters here are n, g, ν, k, and t.

◆ Distinguish between the terms *specific growth rate* and *generation time*.

◆ If in 8 h an exponentially growing cell population increases from 5×10^6 cells/ml to 5×10^8 cells/ml, calculate g, n, ν, and k.

6.6 The Growth Cycle

The data presented in Figure 6.6 reflect only part of the growth cycle of a microbial population, the part called *exponential growth*. In an enclosed vessel, such as a tube or a flask, a condition called a *batch culture*, exponential growth cannot occur indefinitely. Instead, a typical *growth curve* for a population of cells is obtained, as illustrated in Figure 6.8●. The growth curve describes an entire growth cycle, including the **lag phase**, **exponential phase**, **stationary phase**, and **death phase**.

Lag Phase

When a microbial population is inoculated into a fresh medium, growth usually begins only after a period of

● **Figure 6.8   Typical growth curve for a bacterial population.** A viable count measures the cells in the culture that are capable of reproducing. See Sections 6.5 and 6.6 for further descriptions of the counting methods employed. Turbidity, or optical density, is a quantitative measure of light-scattering by a liquid culture (see Figure 6.12).

time called the *lag phase*. This interval may be brief or extended, depending on the history of the culture and the growth conditions. If an exponentially growing culture is transferred into the same medium under the same conditions of growth, a lag does not occur and exponential growth begins immediately. However, if the inoculum is taken from an old (stationary phase) culture and transferred into the same medium, a lag usually occurs even if all the cells in the inoculum are *viable*, that is, able to reproduce. This is because the cells are depleted of various essential constituents and time is required for their resynthesis. A lag also ensues when the inoculum consists of cells that have been damaged (but not killed) by treatment with heat, radiation, or toxic chemicals, because of the time required for the cells to repair the damage.

A lag is also observed when a population is transferred from a rich culture medium to a poorer one. For growth to occur in a particular culture medium the cells must have a complete complement of enzymes for synthesis of the essential metabolites not present in that medium. Hence, upon transfer to a medium where biosyntheses will be required, time is needed for synthesis of the new enzymes that will carry out these reactions.

Exponential Phase

As we have seen, during the *exponential phase* of growth each cell divides to form two cells, each of which also divides to form two more cells, and so on, for a brief or extended period, depending on the available resources and other factors. Cells in exponential growth are usually in their healthiest state. Hence, cells in "mid-exponential" phase are often desirable for studies of enzymes or other cell components.

Most unicellular microorganisms grow exponentially, but rates of exponential growth vary greatly. The rate of exponential growth is influenced by environmental conditions (temperature, composition of the culture medium), as well as by genetic characteristics of the organism itself. In general, prokaryotes grow faster than eukaryotic microorganisms, and small eukaryotes grow faster than large ones. This harkens back to the surface-to-volume ratio discussion in Chapter 4 (Section 4.4). There we saw how small cells have an increased capacity for nutrient and waste exchange compared with larger cells, and this metabolic advantage can dramatically affect their growth rate.

Stationary Phase

In a *batch culture*, such as in a tube or a flask, exponential growth is limited. Consider the fact that a single bacterium with a 20-min generation time could produce, if allowed to grow exponentially for 48 h, a population of cells that weighed 4000 times the weight of Earth! This is particularly impressive because a single bacterial cell weighs only about one-trillionth (10^{-12}) of a gram (Table 3.2).

Obviously, this scenario is impossible. Something must happen to limit growth of the population long before this occurs. Typically, one or both of two things occurs to limit growth: (1) an essential nutrient of the culture medium is used up, or (2) some waste product(s) of the organism accumulates in the medium to inhibit growth. Either way, exponential growth ceases, and the population reaches the **stationary phase**.

In the stationary phase, there is no net increase or decrease in cell number. Although growth usually does not occur in the stationary phase, many cell functions may continue, including energy metabolism and some biosynthetic processes. In some organisms, slow growth may actually occur during the stationary phase, but no net increase in cell number occurs. This is because some cells in the population grow, whereas others die, the two processes balancing each other out. This is a phenomenon called *cryptic growth*.

Death Phase

If incubation continues after a population reaches the stationary phase, the cells may remain alive and continue to metabolize, but they will eventually die. When this occurs, the population enters the **death phase** of the growth cycle. In some cases death is accompanied by actual cell **lysis**. Figure 6.8 indicates that the death phase of the growth cycle is also exponential. Typically, however, the rate of cell death is much slower than that of exponential growth.

The phases of bacterial growth shown in Figure 6.8 are reflections of the events in a *population* of cells, not in individual cells. The terms *lag phase, exponential phase, stationary phase,* and *death phase* have no meaning with respect to individual cells but only to *cell populations.*

 6.6 Concept Check

Microorganisms show a characteristic growth pattern when inoculated into a fresh culture medium. There is usually a lag phase, and then growth commences in an exponential fashion. As essential nutrients are depleted or toxic products build up, growth ceases, and the population enters the stationary phase. If incubation continues, cells may begin to die.

◆ In what phase of the growth curve are cells dividing in a regular and orderly process?

◆ When does a lag phase usually *not* occur?

◆ Why do cells enter the stationary phase?

III MEASURING MICROBIAL GROWTH

Population growth is measured by following changes in the *number* of cells or in the *weight* of some component of cell mass. The latter could be protein, nucleic acids, or the dry weight of cells themselves. Different methods for counting cells or estimating cell mass exist that are suited to different organisms or different problems, and we consider these methods here.

6.7 Direct Measurements of Microbial Growth: Total and Viable Counts

Total and viable counting methods are two widely practiced procedures for enumerating single-celled microorganisms. The procedures each have their strengths and weaknesses and can often yield quite different results for counts of a single bacterial culture.

Total Cell Count

The number of cells in a population can be measured by counting a sample under the microscope, a method called the **direct microscopic count**. Two kinds of direct microscopic counts are done, either on samples dried on slides or on samples in liquid. With liquid samples, special *counting chambers* must be used. In such a counting chamber, a grid is marked on the surface of the glass slide, with squares of known area (Figure 6.9●). Over each square on the grid is a volume of known amount, very small but precisely measured. The number of cells per unit area of grid can be counted under the microscope, giving a measure of the number of cells per small chamber volume. Converting this value to the number of cells per milliliter of suspension is easily done by multiplying by a conversion factor based on the volume of the chamber sample (Figure 6.9).

Direct microscopic counting is a quick way of estimating microbial cell number. However, it has several limitations: (1) Dead cells are not distinguished from living cells; (2) small cells are difficult to see under the microscope, and some cells are probably missed; (3) precision is difficult to achieve; (4) a phase-contrast microscope is required when the sample is not stained; (5) the method is not usually suitable for cell suspensions of low density. (With bacteria, if a cell suspension has less than 10^6 cells/milliliter (ml), few if any bacteria will be

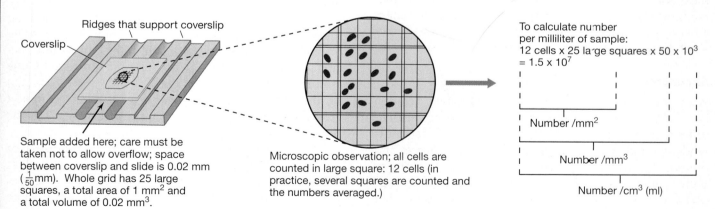

Ridges that support coverslip

Coverslip

Sample added here; care must be taken not to allow overflow; space between coverslip and slide is 0.02 mm ($\frac{1}{50}$mm). Whole grid has 25 large squares, a total area of 1 mm^2 and a total volume of 0.02 mm^3.

Microscopic observation; all cells are counted in large square: 12 cells (in practice, several squares are counted and the numbers averaged.)

To calculate number per milliliter of sample:
12 cells x 25 large squares x 50 x 10^3
= 1.5 x 10^7

Number /mm^2

Number /mm^3

Number /cm^3 (ml)

● **Figure 6.9 Direct microscopic counting procedure using the Petroff–Hausser counting chamber.** A phase-contrast microscope is typically used to count the cells to avoid the necessity for staining.

seen in the microscope field; however, dilute suspensions may be counted if a sample is first concentrated and resuspended in a small volume.), (6) motile cells must be immobilized before counting.

In microbial ecology, total cell counts are often performed on natural samples using general or specific stains to visualize the cells. The DNA-specific stain called *DAPI* stains all cells in a sample (∞ Figure 18.6). By contrast, fluorescent stains can be made highly specific for certain organisms or groups of organisms by attaching them to nucleic acid probes (∞ Figure 18.11). If cells are present at low densities, for example in an open ocean water sample, they can first be concentrated on a filter and then counted after staining.

Viable Count

In direct microscopic counting, both living and dead cells are counted. However, in many cases we are interested in counting only live cells, and for this purpose, **viable count** methods are appropriate. A **viable** cell is one that is able to divide and form offspring. The usual way to perform a viable count is to determine the number of cells in the sample capable of forming *colonies* on a suitable agar medium. For this reason, the viable count is often called the **plate count**, or **colony count**. The assumption made in this type of counting procedure is that *each viable cell can grow and divide to yield one colony*. Some staining procedures can also indicate viability (∞ Figure 18.7), but the colony method is the usual way viable counts are done.

There are two ways of performing a plate count: the *spread plate method* and the *pour plate method* (Figure 6.10●).

In the **spread plate method**, a volume (usually 0.1 ml or less) of an appropriately diluted culture is spread *over* the surface of an agar plate using a sterile glass spreader. The plate is then incubated until the colonies appear, and the number of colonies is counted. It is important that the surface of the plate be fairly dry so that the spread liquid soaks in. Volumes greater than 0.1 ml are rarely used in this method because the excess liquid does not soak in and may cause the colonies to coalesce as they form, making them difficult to count.

In the **pour plate method** (Figure 6.10), a known volume (usually 0.1–1.0 ml) of culture is pipetted *into* a sterile Petri plate. Melted agar medium is then added and mixed well by gently swirling the plate on the benchtop. Because the sample is mixed with the molten agar medium, a larger volume can be used than with the spread plate. However, with this method the organism to be counted must be able to briefly withstand the temperature of molten agar (~45–50°C). In this method, colonies will form throughout the plate, and not just on the agar surface as in the spread plate method. The plate must therefore be examined closely to make sure all colonies are counted (Figure 6.10).

Diluting Cell Suspensions Before Plating

With both the spread plate and pour plate methods, it is important that the *number* of colonies developing on the plates not be too large. On crowded plates some cells may not form colonies, and some colonies may fuse, leading to erroneous measurements. It is also essential that the number of colonies not be too small, or the statistical significance of the calculated count will be low.

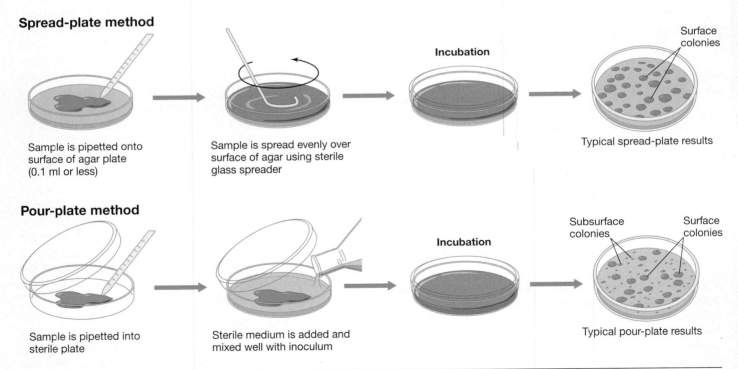

Spread-plate method

Sample is pipetted onto surface of agar plate (0.1 ml or less)

Sample is spread evenly over surface of agar using sterile glass spreader

Incubation

Surface colonies

Typical spread-plate results

Pour-plate method

Sample is pipetted into sterile plate

Sterile medium is added and mixed well with inoculum

Incubation

Subsurface colonies

Surface colonies

Typical pour-plate results

● **Figure 6.10   The viable count.** Two methods of performing a viable count (plate count). In either case the sample must usually be diluted before plating. Note how in the pour plate method, colonies form *within* the agar as well as on the agar surface.

The usual practice, which is the most valid statistically, is to count colonies only on plates that have between 30 and 300 colonies. Moreover, it is usual to determine the incubation conditions (medium, temperature, time) that will give the maximum number of colonies of a given organism and then use these conditions throughout.

To obtain the appropriate colony number, the sample to be counted must almost always be *diluted*. Since one rarely knows the approximate viable count ahead of time, it is usually necessary to make more than one dilution. Several 10-fold dilutions of the sample are commonly used (Figure 6.11●). To make a 10-fold (10^{-1}) dilution, one can mix 0.5 ml of sample with 4.5 ml of diluent, or 1.0 ml sample with 9.0 ml diluent. If a 100-fold (10^{-2}) dilution is needed, 0.05 ml can be mixed with 4.95 ml diluent, or 0.1 ml with 9.9 ml diluent. Alternatively, a 10^{-2} dilution can be made by making two successive 10-fold dilutions. In most cases, such *serial dilutions* are needed to reach the final dilution desired. Thus,

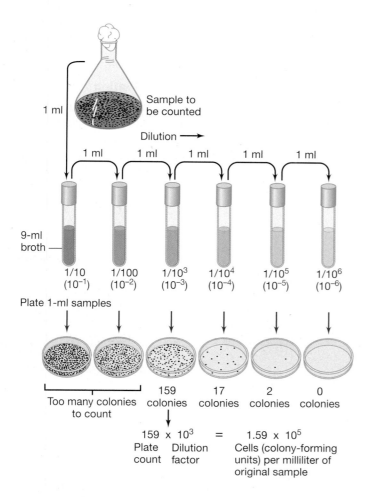

1 ml — Sample to be counted

Dilution ⟶

1 ml | 1 ml | 1 ml | 1 ml | 1 ml

9-ml broth

| 1/10 (10^{-1}) | 1/100 (10^{-2}) | 1/10³ (10^{-3}) | 1/10⁴ (10^{-4}) | 1/10⁵ (10^{-5}) | 1/10⁶ (10^{-6}) |

Plate 1-ml samples

| Too many colonies to count | 159 colonies | 17 colonies | 2 colonies | 0 colonies |

$$159 \times 10^3 = 1.59 \times 10^5$$

Plate count — Dilution factor — Cells (colony-forming units) per milliliter of original sample

● **Figure 6.11 Procedure for viable counting using serial dilutions of the sample and the pour plate method.** The sterile liquid used for making dilutions can simply be water, but a balanced salt solution or growth medium may yield a higher recovery. The dilution factor is the reciprocal of the dilution. For spread plating (Figure 6.10), further dilutions may be necessary in order to spread samples of 0.1 ml.

if a $10^{-6}(1/10^6)$ dilution is needed, it can be achieved by making three successive $10^{-2}(1/10^2)$ dilutions or six successive 10^{-1} dilutions (Figure 6.11).

Sources of Error in Plate Counting

The number of colonies obtained in a viable count experiment depends not only on the inoculum size and its viability, but also on the suitability of the culture medium and the incubation conditions. The colony number can also change with the length of incubation. For example, if a mixed culture is used, the cells deposited on the plate will not all develop into colonies at the same rate; if a short incubation time is used, fewer than the maximum number of colonies will be obtained. Furthermore, the size of colonies often varies. If some tiny colonies develop, they may be missed during the counting. With pure cultures, colony development is a more synchronous process.

Viable counts can be subject to rather large errors for several reasons. These include pipetting inconsistencies, inhomogeneity of the sample, insufficient mixing, and several other factors. Hence, if accurate counts are to be obtained, great care and consistency must be taken in sample preparation and pipetting, and replicate plates of key dilutions must be prepared. Note also that two or more cells in a clump will form only a single colony. So if many clumps are present in a sample, a viable count of that sample may be erroneously low. To more clearly state the results of a viable count, the data are often expressed as the number of *colony-forming units* (*cfu*) obtained rather than as the number of *viable cells* (since a colony-forming unit may contain one or more cells).

Despite the difficulties associated with viable counting, the procedure gives the best information on the number of viable cells in a sample and so is widely used in many areas of microbiology. For example, in food, dairy, medical, and aquatic microbiology, viable counts are employed routinely. The method has the virtue of *high sensitivity*, as a sample containing only a single viable cell can in theory be counted. This feature allows for the sensitive detection of microbial contamination of food products or other materials. Moreover, the use of highly selective culture media and growth conditions (∞Sections 5.2 and 24.2) in viable counting procedures allows for the counting of only particular species in a mixed population of microorganisms present in the sample. For example, in practical applications such as in the food industry, viable counting on selective media allows for both quantitative and qualitative assessments of the microbial content of a food product.

The Great Plate Count Anomaly

Although a very sensitive technique, plate counts can be highly unreliable when used to assess total cell numbers of natural samples such as soil and water. Indeed, direct microscopic counts of natural samples typically reveal far more organisms than are recoverable on plates of *any* given culture medium (⚬⚬Sections 18.3 and 18.4). Some microbiologists have referred to this as "the great plate count anomaly."

Why do plate counts reveal lower numbers of cells than direct microscopic counts? One factor is that microscopic methods count dead cells, whereas viable methods do not. Furthermore, different organisms in even a very small sample may have vastly different requirements for resources and conditions in laboratory culture (⚬⚬Chapters 5, 6, 17–19). Thus, one medium and set of growth conditions will allow for growth of only a *subset* of the total population. If this subset makes up, for example, 10^6 cells/g in a total viable population of 10^9 cells/g, the plate count will reveal at best only 0.1% of the total population, a vast underestimation of the actual number of viable cells.

The conclusion here, should be clear. Targeted plate counts using highly selective media, as in, for example, the microbial analysis of sewage or food, can yield rather reliable data (⚬⚬Sections 28.1, 29.1, and 29.4). In contrast, "total cell counts" of the same samples using a single medium and set of growth conditions may be, and usually are, underestimates by one to several orders of magnitude.

 6.7 Concept Check

Growth is measured by the change in number of cells with time. Cell counts done microscopically measure the total number of cells in a population, whereas viable cell counts (plate counts) measure only the living population.

◆ Why is a *viable count* more sensitive than a *microscopic count*?

◆ What is the major assumption made in relating plate count results to cell number?

◆ Describe how you would dilute a bacterial culture by 10^{-7}.

◆ What is the "great plate count anomaly"?

6.8 Indirect Measurements of Microbial Growth: Turbidity

A rapid and quite useful method of estimating cell numbers is by *turbidity measurements*. A cell suspension looks cloudy (turbid) to the eye because cells scatter light passing through the suspension. The more cells that are present, the more light is scattered, and hence the more turbid the suspension.

Turbidity can be measured with a *photometer* or a *spectrophotometer*, devices that pass light through a cell suspension and detect the amount of unscattered light that emerges (Figure 6.12●). The major difference between these two instruments is that a photometer em-

ploys a single broad bandpass filter to generate incident light, whereas a spectrophotometer employs a more precise prism or diffraction grating to generate incident light (Figure 6.12*a*). Commonly used wavelengths for bacterial turbidity measurements include 540 nm (green), 600 nm (orange), or 660 nm (red). As mentioned, both spectrophotometers and photometers share the fact that they measure only *unscattered* light. The *decrease* in unscattered light that occurs as cells *increase* in number is measured in photometer units (for example, "Klett units" for the Klett-Summerson photometer) or optical density (OD) units for a spectrophotometer (Figure 6.12*b*).

Generating a Standard Curve

For unicellular organisms, photometer units or OD are proportional (within certain limits) to cell number. Turbidity readings can therefore be used as a substitute for direct counting methods. However, before doing so, a *standard curve* must first be prepared, relating some direct measurement of cell number (microscopic or viable count) or mass (dry weight) to the *indirect* measurement obtained by turbidity (Figure 6.12*c*). For example, the standard curve can contain data for both cell number and cell mass, allowing for an estimate of both parameters from a single turbidity reading (Figure 6.12*c*).

At high cell concentrations, light scattered away from the photocell by one cell can be rescattered back by another. To the photocell this makes it appear as if light had never been scattered in the first place. At such high cell densities, the correspondence between cell number and turbidity therefore drifts from linearity (Figure 6.12*c*). Nevertheless, within limits, turbidity measurements can be reasonably accurate, and they have the virtue of being quick and easy to perform.

Turbidity measurements can usually be made without destroying or significantly disturbing the sample. For these reasons, turbidity measurements are widely employed to follow the growth rate of microbial cultures. The same sample can be checked repeatedly and the measurements plotted on a semilogarithmic plot versus time (Figure 6.12*b*). From these, it is easy to calculate the generation time and other growth parameters of the growing culture.

 6.8 Concept Check

Turbidity measurements are an indirect but very rapid and useful method of measuring microbial growth. However, in order to relate a direct cell count to a turbidity value, a standard curve must first be established.

◆ List two advantages of using turbidity as a measure of cell growth.

◆ Describe how you could use a turbidity measurement to tell how many colonies you would expect from plating a culture of a given OD.

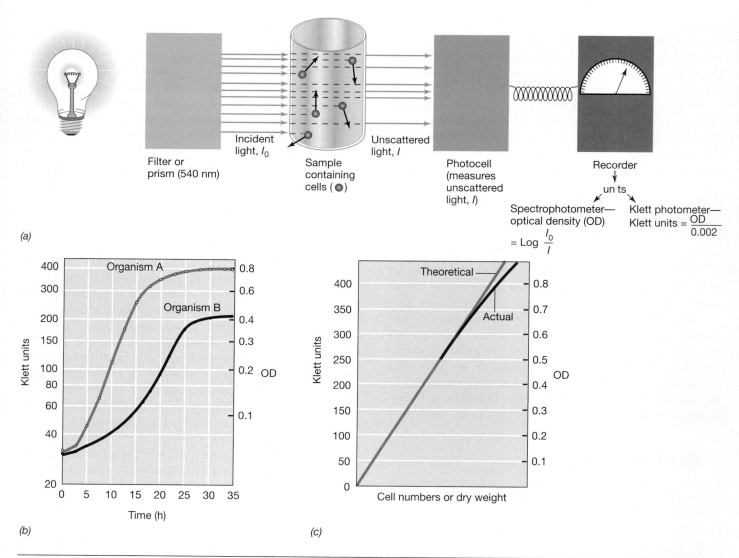

(a)

(b)

(c)

● **Figure 6.12 Turbidity measurements of microbial growth.** (a) Measurements of turbidity are made in a spectrophotometer or photometer. The photocell measures incident light unscattered by cells in suspension and gives readings in optical density or photometer units. (b) Typical growth curve data obtained in Klett units or optical density (OD) for two organisms growing at different growth rates. For practice, calculate the generation time (*g*) of the two cultures using the formula $n = 3.3 (\log N - \log N_0)$ where N and N_0 are two different Klett values or OD readings taken between a time interval *t*. Which organism is growing faster, A or B? (c) Relationship between cell number or dry weight and turbidity readings. Note that the one-to-one correspondence between these relationships breaks down at high turbidities.

6.9 Continuous Culture: The Chemostat

Our discussion of population growth thus far has been confined to **batch cultures**. A batch culture is a fixed volume of culture medium that is continually being altered by the metabolic activities of growing organisms and is therefore a *closed system*. In the early stages of exponential growth in batch cultures, conditions may remain relatively constant, but in later stages when cell numbers become quite large, drastic changes in the chemical composition of the culture medium occur.

For many studies it is desirable to keep cultures in constant environments for long periods. For example, if one is studying a physiological process such as the synthesis of a particular enzyme, the ready availability of exponentially growing cells may be very useful. Such systems are only possible with a *continuous culture*. A con-

tinuous culture is an *open system* of constant volume to which fresh medium is added continuously and spent culture medium removed continuously, both at a constant rate. Once such a system is in equilibrium, the chemostat volume, cell number, and nutrient status remain *constant*, and the system is said to be in **steady state**.

The Chemostat

The most common type of continuous culture device is the **chemostat** (Figure 6.13●). The chemostat controls both the *growth rate* and the *population density* of the culture simultaneously. Two factors are important in such control: the *dilution rate* and the *concentration of a limiting nutrient*, such as a carbon or nitrogen source.

In a batch culture, nutrient concentration can affect both the growth rate and the growth yield of a culture

● **Figure 6.13** **Schematic for a continuous culture device (chemostat).** In such a device, the population density is controlled by the concentration of limiting nutrient in the reservoir, and the growth rate is controlled by the flow rate (see Figure 6.15). Both parameters can be set by the experimenter.

● **Figure 6.15** **Steady-state relationships in the chemostat.** The dilution rate is determined from the flow rate and the volume of the culture vessel. Thus, with a vessel of 1000 ml and a flow rate through the vessel of 500 ml/h, the dilution rate would be 0.5 h⁻¹. Note that at high dilution rates, growth cannot balance dilution, and the population washes out. Note also that although the population density remains constant during steady state, the growth rate (doubling time) can vary over a wide range. Thus, the experimenter can obtain populations with widely varying growth rates without affecting population density.

(Figure 6.14●). At very low concentrations of a given nutrient, the growth rate is reduced, probably because the nutrient cannot be transported into the cell fast enough to satisfy metabolic demand. At moderate or higher nutrient levels, the growth *rate* may not be affected while the cell *yield* continues to increase (Figure 6.14). In contrast to a batch culture, in a chemostat, growth rate and growth yield can be controlled *independently of each other*, the former by adjusting the dilution rate and the latter

by varying the concentration of a nutrient present in a limiting amount.

The effects of varying dilution rate and concentration of the growth-limiting nutrient are shown in Figure 6.15●. As seen, there are rather wide limits over which the dilution rate controls growth rate, although at both very low and very high dilution rates the steady state breaks down. At *high* dilution rates, the organism cannot grow fast enough to keep up with its dilution, and the culture is washed out of the chemostat. By contrast, at very *low* dilution rates, a large fraction of the cells may die from starvation because the limiting nutrient is not being added fast enough to permit maintenance of cell metabolism.

The *cell density* (cells/milliliter) in the chemostat is controlled by the level of the limiting nutrient, just as cell yield was controlled in a batch culture (Figure 6.14). If the concentration of this nutrient in the incoming medium is raised, with the dilution rate remaining constant, the cell density will increase. Thus, by adjusting dilution rate and nutrient level, the experimenter can obtain dilute (for example, 10^5 cells/ml), moderate (for example, 10^7 cells/ml), or dense (for example, 10^9 cells/ml) populations growing at slow, moderate, or rapid growth rates.

Experimental Uses of the Chemostat

As previously mentioned, the chemostat allows the experimenter to control growth rate and population density *independently* of each other. As was shown in Figure 6.15, even over rather wide ranges, different growth rates can be achieved by simply varying the dilution rate. Similarly,

● **Figure 6.14** **The effect of nutrients on growth.** Relationship between nutrient concentration, growth rate (green curve), and growth yield (red curve) in a batch culture (closed system). At low nutrient concentrations both growth rate and growth yield are affected.

the population density may be set by varying the concentration of a single nutrient (the growth-limiting nutrient) in the medium reservoir. Independent control of these two crucial growth parameters is impossible with batch cultures because the batch culture is a closed system where growth conditions are constantly changing with time.

A practical advantage to the chemostat is that a population may be maintained in the exponential growth phase for long periods, for days and even weeks. Because exponential phase cells are usually most desirable for physiological experiments, the experimenter using the chemostat can have such cells available at any time. Thus, experiments can be planned in detail and then performed whenever most convenient. Moreover, repetition of experiments can be done with the knowledge that the cell population will be as close to being the same each time as possible. For some applications, such as the study of a particular enzyme, enzyme activities may be quite lower in stationary phase cells than in exponential phase cells and thus chemostat-grown cultures are ideal.

The chemostat can also be used in microbial ecology. For example, because the chemostat can easily mimic the low substrate concentrations that often prevail in nature, it is possible to study mixed bacterial populations in a chemostat and ask questions about the competitiveness of different organisms at particular nutrient concentrations. Using cultural methods as well as the powerful tools of molecular ecology such as phylogenetic stains and gene tracking (∞∞Chapter 18), changes in the microbial community can be monitored as a function of chemostat conditions. Such experiments often reveal interactions among individual components of the population that are not obvious from growth studies in batch culture.

Chemostats have also been used for enrichment and isolation of bacteria (∞∞Sections 1.7, 18.1, and 18.2 for discussion of enrichment and isolation of bacteria). From a mixed inoculum, one can select a stable population under the nutrient and dilution rate conditions chosen and then slowly increase the dilution rate until a single organism remains. In this way, microbiologists recently isolated a bacterium with a 6-minute doubling time—the fastest growing bacterium known.

 6.9 *Concept Check*

Continuous culture devices (chemostats) are a means of maintaining cell populations in exponential growth for long periods. In a chemostat, the rate at which the culture is diluted governs the growth rate, and the population size is governed by the concentration of the growth-limiting nutrient entering the vessel.

◆ How do microorganisms in a *chemostat* differ from microorganisms in a *batch culture*?

◆ What happens in a chemostat if the dilution rate exceeds the maximal growth rate of the organism?

◆ Do pure cultures have to be used in a chemostat?

IV ENVIRONMENTAL EFFECTS ON MICROBIAL GROWTH: TEMPERATURE

Up to now we have described the growth of microorganisms under ideal laboratory conditions. However, the activities of microorganisms are greatly affected by the chemical and physical conditions of their environments. Understanding environmental influences helps explain the distribution of microorganisms in nature and makes it possible to devise methods for controlling or enhancing microbial activities.

Many environmental factors can be considered. However, four key factors play major roles in controlling the growth of all microorganisms: *temperature, pH, water availability*, and *oxygen*. Some other factors can potentially affect the growth of microorganisms, such as pressure and radiation. These more specialized environmental factors will be considered later in this book when we encounter microbial habitats in which they play major roles.

6.10 Effect of Temperature on Growth

Temperature is one of the most, if not *the* most, important environmental factor affecting growth and survival of microorganisms. At either too cold or too hot a temperature, microorganisms will not be able to grow. But the minimum and maximum temperatures vary greatly among different microorganisms, usually reflecting the temperature range and average temperature of their habitats.

Cardinal Temperatures

Temperature can affect living organisms in either of two opposing ways. As the temperature rises, chemical and enzymatic reactions in the cell proceed at more rapid rates, and growth becomes faster. However, *above* a certain temperature, particular proteins may be irreversibly denatured. Thus, as the temperature is increased within a given range, growth and metabolic function increase up to a point where denaturation reactions set in. Above this point, cell functions fall sharply to zero.

For every organism there is thus a *minimum* temperature below which growth no longer occurs, an *optimum* temperature at which growth is most rapid, and a *maximum* temperature above which growth is not possible (Figure 6.16●). The optimum temperature is always nearer the maximum than the minimum. These three temperatures, called the **cardinal temperatures**, are characteristic of each organism but are not completely fixed entities, as they can be modified slightly by other factors of the environment, in particular, by the composition of the growth medium.

The cardinal temperatures of different microorganisms differ widely—some organisms have temperature optima as low as 4°C and some higher than 100°C. The temperature range throughout which growth occurs by

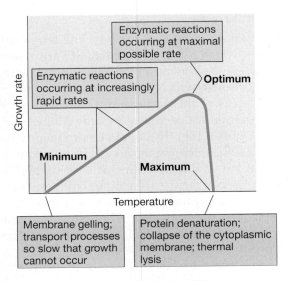

● **Figure 6.16 Effect of temperature on growth rate and the molecular consequences for the cell.** The three cardinal temperatures vary by organism.

various organisms is even wider than this, from below freezing to greater than boiling. However, no single organism can grow over this whole temperature range, and the typical range for any given organism is 30–40 degrees.

The *maximum* growth temperature of an organism most likely reflects the denaturation of one or more essential proteins in the cell. However, the factors controlling an organism's *minimum* growth temperature are not as clear. As mentioned earlier (∞Section 4.5), the cytoplasmic membrane must be in a fluid state for proper functioning. An organism's minimum temperature may be the result of stiffening of its cytoplasmic membrane such that it no longer functions properly in

nutrient transport or in developing a proton motive force. This explanation is supported by experiments in which the minimum temperature for an organism can be altered to some extent by adjustments in cytoplasmic membrane lipid composition (see Section 6.11). In this regard, the maximum and minimum temperatures supporting growth of an organism are higher and lower, respectively, when tested in complex rather than defined media.

Temperature Classes of Organisms

Although there is a continuum of organisms, from those with very low temperature optima to those with high temperature optima, it is possible to broadly distinguish *four groups* of microorganisms in relation to their temperature optima: **psychrophiles**, with low temperature optima, **mesophiles**, with midrange temperature optima, **thermophiles**, with high temperature optima, and **hyperthermophiles**, with very high temperature optima (Figure 6.17●).

Mesophiles are found in warm-blooded animals and in terrestrial and aquatic environments in temperate and tropical latitudes. Psychrophiles and thermophiles are found in unusually cold and unusually hot environments, respectively. Hyperthermophiles are found in extremely hot habitats such as hot springs, geysers, and deep-sea hydrothermal vents (see Sections 6.12, 13.8, and 19.8).

In *Escherichia coli*, a typical mesophile, a detailed study of growth as a function of temperature has precisely defined its cardinal temperatures. The optimum temperature of *E. coli* in a complex medium (∞Section 5.2) is 39°C, the maximum is 48°C, and the minimum is 8°C. Thus, the temperature *range* for *E. coli* is 40 degrees, near the high end for the average prokaryote.

● **Figure 6.17 Relation of temperature to growth rates of a typical psychrophile, a typical mesophile, a typical thermophile, and two different hyperthermophiles.** The temperature optima of the example organisms are shown on the graph. Hyperthermophiles show growth temperature optima above 80°C.

6.10 *Concept Check*

Temperature is a major environmental factor controlling microbial growth. The cardinal temperatures describe the minimum, optimum, and maximum temperatures at which each organism grows. Microorganisms can be grouped by the temperature ranges they require.

◆ What are the cardinal temperatures for *Esherichia coli*? To what temperature class does it belong?

◆ How does a *hyperthermophile* differ from a *psychrophile*?

◆ *Escherichia coli* can grow at a higher temperature in a complex medium than in a defined medium. Why?

6.11 Microbial Growth at Cold Temperatures

Because humans live and work on the surface of Earth where temperatures are generally moderate, it is natural to consider very hot and very cold environments as being "extreme." And they are extreme for human habitation because humans would die quickly if immersed in boiling or freezing water. However, the natural habitats of many microorganisms can be either extremely hot or extremely cold, and the organisms that live there, referred to as **extremophiles** (∞Section 2.4 and Table 2.1), have evolved to grow *optimally* under these conditions. We consider first organisms that grow at cold temperatures.

Cold Environments

Much of Earth's surface experiences low temperatures. The oceans, which make up over half of Earth's surface, have an average temperature of 5°C, and the depths of the open oceans have constant temperatures of 1–3°C. Vast land areas of the Arctic and Antarctic are permanently frozen or are unfrozen for only a few weeks in summer (Figure 6.18●). These cold environments are not sterile, and some microorganisms can be found alive and growing at any low temperature at which liquid water still exists. Even in many frozen materials there are small pockets of liquid water present where microorganisms can metabolize and grow. Within glaciers, for example, there exists a network of small liquid water channels in which prokaryotes thrive and reproduce.

In considering cold environments, it is important to distinguish between environments that are *constantly* cold and those that are cold *only* in winter. The latter, characteristic of continental temperate climates, may have summer temperatures as high as 40°C and winter temperatures of −20°C or colder. A temperate lake, for example, may have a period of ice cover in the winter, but the time that the water is at 0°C is relatively brief. Such highly variable environments are less favorable for cold-adapted organisms than are the *constantly cold* environments found in polar regions, at high altitudes, and in the depths of the oceans. For example, freshwater lakes in the Antarctic Dry Valleys contain a *permanent* ice

(a)

(b)

John Gosink and James T. Staley

(c)

Deborah Jung and Michael T. Madigan

● **Figure 6.18 Antarctic microbial habitats.** (a) A core of permanently frozen seawater from McMurdo Sound, Antarctica. The core is about 8 cm wide. Note the dense coloration due to pigmented microorganisms. (b) Phase contrast micrograph of phototrophic microorganisms from the core shown in (a). Most organisms are either diatoms or green algae (both eukaryotic microorganisms, ∞Section 14.13). (c) Photo of the surface of Lake Bonney, McMurdo Dry Valleys, Antarctica. Like many other Antarctic lakes, Lake Bonney, which is about 40 meters deep, remains permanently frozen and has an ice cover of about 5 meters. The water column of Lake Bonney remains near 0°C and contains both oxic and anoxic zones (∞Section 19.5 and Figure 19.9); thus both aerobic and anaerobic microorganisms are present. However, no higher eukaryotic organisms inhabit Dry Valley lakes, making them unique microbial ecosystems.

cover of several meters in thickness (Figure 6.18c). The water column below the ice in these lakes remains at 0°C or colder year round and is thus an ideal habitat for cold-adapted organisms.

Psychrophilic and Psychrotolerant Microorganisms

As noted earlier, organisms with low temperature optima are called **psychrophiles**. A psychrophile can be defined as an organism with an *optimal* growth temperature of *15°C or lower*, a maximum growth temperature below 20°C, and a minimal growth temperature at 0°C or lower. Organisms that grow at 0°C but have optima of 20–40°C are called **psychrotolerant**.

Psychrophiles are found in environments that are constantly cold, such as in polar regions or in marine sediments from polar regions, and they may be rapidly killed by warming to room temperature. For this reason, their laboratory study requires that great care be taken to ensure that they never warm up during sampling, transport to the laboratory, isolation, or other manipulations.

Some of the best-studied psychrophiles have been algae that grow in dense masses within and under the ice in polar regions or other permanent ice fields (Figure 6.18). Psychrophilic algae are also often seen on the surfaces of snowfields and glaciers at such densities that they impart a distinctive red or green coloration to the surface (Figure 6.19a●). The most common snow alga is *Chlamydomonas nivalis*; its brilliant red spores are responsible for the red color (Figure 6.19b). This green alga grows within the snow as a green-pigmented vegetative cell and then sporulates. As the snow dissipates by melting, erosion, and vaporization, the spores become concentrated on the surface. Snow algae are most commonly seen on melting permanent snowfields in midsummer to late summer and are especially common in sunny, dry areas. In addition to snow algae, several psychrophilic chemoorganotrophic bacteria are known, many from the Antarctic (Figure 6.18). Some of these, particularly isolates from sea ice (frozen seawater that forms seasonally in polar regions), show the lowest maximum growth temperature of all known microorganisms.

Psychrotolerant microorganisms are much more widely distributed than psychrophiles and can be isolated from soils and water in temperate climates as well as from meat, milk and other dairy products, cider, vegetables, and fruit stored at refrigeration temperatures (~4°C). As noted, psychrotolerant microorganisms grow best at a temperature between 20° and 40°C. Because temperate environments warm up in summer, they cannot support the heat-sensitive psychrophiles because the warming provides selection against them. It should be emphasized that although psychrotolerant microorganisms do grow at 0°C, most do not grow very well at this temperature, and one must often wait several weeks before visible growth is seen in culture media. Various genera of *Bacteria*, fungi, algae, and protozoa have members that are psychrotolerant.

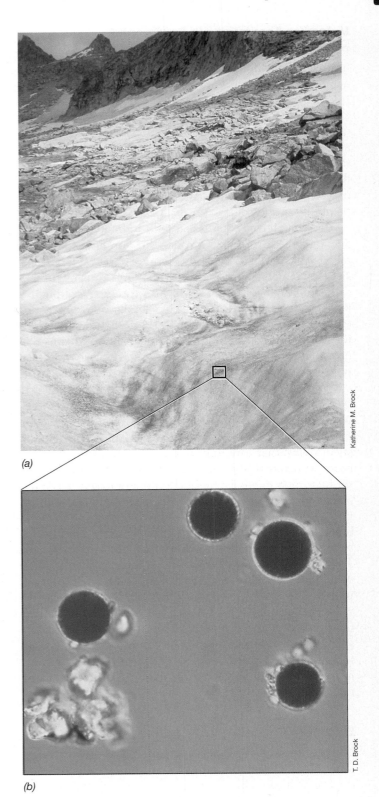

(a)

(b)

● **Figure 6.19 Snow algae.** (a) Snow bank in the Sierra Nevada, California, with red coloration caused by the presence of snow algae (eukaryotic cells, ⌀ Section 14.13). Pink snow such as this is common on summer snow banks at high altitudes throughout the world. (b) Photomicrograph of red-pigmented spores of the snow alga *Chlamydomonas nivalis*. The spores germinate to yield motile green algal cells. Related species of snow algae contain different carotenoid pigments (⌀ Section 17.3), and thus, fields of snow algae can also be green, orange, brown, or purple in color.

Molecular Adaptations to Psychrophily

Psychrophiles produce enzymes that function optimally in the cold and that are often denatured or otherwise inactivated at even very moderate temperatures. The molecular basis for this is not entirely understood, but it is known that cold-active enzymes have greater amounts of α-helix and lesser amounts of β-sheet secondary structure (∞Section 3.7 and Figure 3.16) than enzymes that are inactive in the cold. Because the β-sheet tends to form a more rigid structure, the greater α-helix content of cold-active enzymes allows these proteins greater flexibility in the cold.

Cold-active enzymes also tend to have greater polar and lesser hydrophobic amino acid contents than their mesophilic and thermophilic counterparts. This also assists in keeping the protein flexible (and thus enzymatically active) at cold temperatures. Moreover, cold-active proteins tend to have decreased levels of weak bonds (∞Section 3.1) and decreased interactions between their domains compared to proteins from organisms that grow best at higher temperatures. Again, these modifications probably favor protein flexibility.

Another feature of psychrophiles is that compared to mesophiles, transport processes (∞Section 4.7) occur optimally at low temperature, an indication that the cytoplasmic membranes of psychrophiles are constructed in such a way that low temperatures do not inhibit membrane functions. Cytoplasmic membranes from psychrophiles tend to have a higher content of *unsaturated* fatty acids (∞Section 5.17). This helps to maintain a semifluid state of the membrane at low temperatures (membranes composed of predominantly saturated fatty acids become waxy and nonfunctional at low temperatures). The lipids of some psychrophilic bacteria also contain *polyunsaturated* fatty acids and long-chain hydrocarbons with multiple double bonds. For example, a hydrocarbon with nine double bonds ($C_{31:9}$) has been identified from the lipids of some Antarctic bacteria, and the bacterium *Psychroflexus* has been shown to contain fatty acids with four and five double bonds. These fatty acids remain more fluid at low temperatures than do saturated or monounsaturated fatty acids.

Freezing

Despite the ability of some organisms to grow at low temperatures, there is a lower limit below which reproduction is impossible. Pure water freezes at 0°C and seawater at −2.5°C, but freezing is not a continuous process. Microscopic pockets of water continue to exist at these and even much lower temperatures. As long as liquid water is available, microbial growth is possible.

Although freezing prevents microbial growth, it does not necessarily cause microbial death. In addition, the medium in which the cells are suspended considerably affects sensitivity to freezing. Water-miscible liquids such as glycerol and dimethylsulfoxide (DMSO), when added at about 10% (final concentration) to the suspending medium, penetrate the cells and protect them by reducing the severity of dehydration effects and preventing ice crystal formation. In fact, the addition of such agents, called *cryoprotectants*, is a common way of *preserving* microbial cultures at very low temperatures (usually at −70 to −196°C). Properly prepared frozen cells can remain viable for long periods (decades or longer).

6.11 Concept Check

Organisms with cold temperature optima are called *psychrophiles*, and the most extreme representatives inhabit permanently cold environments. Psychrophiles have evolved biomolecules that function best at cold temperatures but that can be unusually sensitive to warm temperatures.

◆ How do *psychrotolerant* organisms differ from *psychrophilic* organisms?

◆ What molecular adaptations to the cytoplasmic membrane are seen in psychrophiles, and why are they necessary?

6.12 Microbial Growth at High Temperatures

Microbial life flourishes in high-temperature environments, up to and including boiling water. Above about 65°C, only *prokaryotic* life forms are found, but even here, a huge diversity of both *Bacteria* and *Archaea* exist. We consider some of these hot environments and their microbial life here.

Thermal Environments

Recall that organisms whose growth temperature optimum is *above* 45°C are called **thermophiles**, and those whose optimum is *above* 80° C are called **hyperthermophiles** (Figure 6.17). Temperatures as high as these are found in nature only in certain areas. For example, soils subject to full sunlight are often heated to temperatures above 50°C at midday, and some soils may become warmed to even 70°C, although a few centimeters under the surface the temperature is much lower. Fermenting materials such as compost piles and silage can reach temperatures of 70°C. However, the most extensive and extreme high-temperature environments found in nature are in association with volcanic phenomena. These include, in particular, hot springs.

Many hot springs have temperatures at or near boiling, and steam vents (fumaroles) may reach 150–500°C. Hydrothermal vents in the bottom of the ocean have temperatures of 350°C or greater (∞Section 19.8). Hot springs occur throughout the world, but are especially concentrated in the western United States, New Zealand, Iceland, Japan, Italy, Indonesia, Central America, and central Africa. The world's largest single concentration of hot springs is in Yellowstone National Park, Wyoming (USA).

Although some hot springs vary in temperature, others are very constant, not varying more than 1–2°C over many years. In addition, different springs have different chemical compositions and pH values, but generally contain sufficient levels of nutrients to support populations, often very large populations, of hyperthermophilic chemoorganotrophs and chemolithotrophs.

Hyperthermophiles in Hot Springs

Many hot springs are at the boiling point for the altitude (92–93°C at Yellowstone, 99–100°C at locations where the springs are close to sea level). In boiling hot springs (Figure 6.20●), a variety of hyperthermophiles are typically present. The growth of such organisms can be studied by immersing microscope slides into the spring and retrieving them after a few days. Microscopic examination of the slides reveals colonies of prokaryotes (Figure 6.20b) that have developed from single bacterial cells that attached to and grew on the glass surface.

Ecological studies of organisms living in boiling springs have shown that growth rates are fairly rapid, and doubling times of as short as 1 h have been recorded. Cultures of many of these prokaryotes have been obtained, and a variety of morphological and physiological

(a)

(b)

● **Figure 6.20   Growth of hyperthermophiles in boiling water.** (a) Boulder Spring, a small boiling spring in Yellowstone National Park. This spring is superheated, having a temperature 1–2°C above the boiling point. The mineral deposits around the spring consist mainly of silica and sulfur. (b) Photomicrograph of a microcolony of prokaryotes that developed on a microscope slide immersed in a boiling spring such as that shown in (a).

types exist (∞Chapter 13). Phylogenetic studies using ribosomal RNA sequencing (∞Sections 2.3, 11.5, and 11.6) have shown great evolutionary diversity among these hyperthermophiles. Species of both *Bacteria* and *Archaea* are present. Some hyperthermophilic *Archaea* show growth-temperature optima *greater than 100°C* and must be grown in pressurized vessels in the laboratory in order to reach temperatures above the boiling point.

Thermophiles

Many thermophiles (optima 45–80°C) are also present in hot springs as well as other thermal environments. In hot springs, as boiling water overflows the edges of the spring and flows away from the source, it gradually cools, setting up a *thermal gradient*. Along this gradient, various microorganisms grow (Figure 6.21●), with different species growing in the different temperature ranges. By studying the species distribution along such thermal gradients and by examining hot springs and other thermal habitats at different temperatures around the world, it has been possible to determine the upper temperature limits for each type of organism (Table 6.1). From this information we can conclude that (1) prokaryotic organisms in general are able to grow at temperatures higher than those at which eukaryotes can grow; (2) the most thermophilic of all

● **Figure 6.21   Growth of thermophilic cyanobacteria in a hot spring in Yellowstone National Park.** Characteristic V-shaped pattern formed by cyanobacteria at the upper temperature for phototrophic life, 70–74°C, in the thermal gradient formed from a boiling hot spring. The pattern develops because the water cools more rapidly at the edges than in the center of the channel. The spring flows from the back of the picture toward the foreground. The light-green color is from a high temperature strain of the cyanobacterium *Synechococcus*. As water flows down the gradient, the density of cells increases, less thermophilic strains enter, and the color becomes more intensely green.

Table 6.1	Presently known upper temperature limits for growth of living organisms	
Group		**Upper temperature limits (°C)**
Animals		
Fish and other aquatic vertebrates		38
Insects		45–50
Ostracods (crustaceans)		49–50
Plants		
Vascular plants		45
Mosses		50
Eukaryotic microorganisms		
Protozoa		56
Algae		55–60
Fungi		60–62
Prokaryotes		
Bacteria		
Cyanobacteria		70–74
Anoxygenic phototrophs		70–73
Chemoorganotrophic/chemolithotrophic *Bacteria*		95
Archaea		
Chemoorganotrophic/chemolithotrophic *Archaea*		113[a]

[a] The upper temperature limit for growth of the organism *Pyrolobus fumarii*. Related species of *Pyrodictium* may be able to grow up to as high as 121°C.

prokaryotes are certain species of *Archaea*; and (3) non-phototrophic organisms are able to grow at higher temperatures than can phototrophic forms.

Thermophilic prokaryotes have also been found in artificial thermal environments. The hot water heater, domestic or industrial, has a temperature of 55–80°C and is therefore a favorable habitat for the growth of thermophilic prokaryotes. Organisms resembling *Thermus aquaticus*, a common hot spring thermophile (⌘Section 12.34), have been isolated from hot water heaters. Electric power plants, industrial hot water discharges, and other artificial thermal sources also provide sites where thermophiles can grow. Many of these organisms can be isolated easily using the appropriate enrichment techniques.

Molecular Adaptations to Thermophily

How can thermophiles and hyperthermophiles thrive at high temperatures? First, their enzymes and other proteins are much more stable to heat than are those of mesophiles and actually function *optimally* at high temperatures. How is heat stability achieved? Amazingly, studies of several thermostable enzymes have shown that they often differ very little in amino acid sequence from heat-labile forms of the enzymes that catalyze the same reaction in mesophiles. It appears that a critical amino acid substitution in only a few locations in the enzyme allows it to fold in a way that is consistent with heat stability.

Heat stability of proteins in hyperthermophiles is also improved as a result of an increased number of ionic bonds between the positive and negative charges of various amino acids and because of the densely packed highly hydrophobic interiors of the proteins, which naturally resist unfolding in the aqueous cytoplasm. Finally, the solutes di-inositol phosphate, diglycerol phosphate, and mannosylglycerate that are produced in high levels in the cytoplasm of certain hyperthermophiles may help to stabilize their proteins against thermal degradation.

In addition to enzymes and other components of the cell, the cytoplasmic membranes of thermophiles and hyperthermophiles must be heat stable. We mentioned earlier that psychrophiles have membrane lipids rich in *unsaturated* fatty acids, thus making the membranes fluid and functional at low temperatures. Conversely, thermophiles typically have lipids rich in *saturated* fatty acids, thus allowing the membranes to remain stable and functional at high temperatures. Saturated fatty acids form a stronger hydrophobic environment than do unsaturated fatty acids, which helps account for the membrane stability.

Hyperthermophiles, most of which are *Archaea*, do not contain fatty acids at all in their membranes, but instead have C_{40} hydrocarbons composed of repeating units of the five-carbon compound isoprene (⌘Figure 4.19c) bonded by ether linkage to glycerol phosphate. In addition, the overall structure of these cytoplasmic membranes forms a *lipid monolayer* (⌘ Figure 4.19d), and this structure is undoubtedly much more heat stable than the lipid bilayer of species of *Bacteria* and *Eukarya*. We discussed the details of the unique membrane architecture of hyperthermophilic *Archaea* in Section 4.5 and will consider other aspects of heat stability, including that of DNA, in hyperthermophiles in Section 13.12.

Thermophily and Biotechnology

Thermophilic and hyperthermophilic microorganisms are interesting for more than just basic biological reasons. These organisms offer some major advantages for industrial and biotechnological processes, many of which run more rapidly and efficiently at high temperatures. Enzymes from thermophiles and hyperthermophiles are capable of catalyzing biochemical reactions at high temperatures (⌘Section 30.9 and Figure 30.15b) and are typically more stable than enzymes from mesophiles, thus prolonging the shelf life of enzyme preparations.

A practical example of a heat-stable enzyme of great applied importance is the DNA polymerase isolated from the thermophile *Thermus aquaticus*. *Taq polymerase*, as this enzyme is known, has been used to automate the repetitive steps in the *polymerase chain reaction (PCR)* technique, an extremely important method for amplifying specific DNA sequences (⌘Section 7.9). Several other uses of heat-stable enzymes and other thermostable cell products are also known or are being developed for industrial applications.

6.12 Concept Check

Organisms with growth temperature optima between 45 and 80°C are called *thermophiles*, while those with optima greater than 80°C are called *hyperthermophiles*. These organisms inhabit hot environments up to and including boiling hot springs, as well as undersea hydrothermal vents that can have temperatures in excess of 100°C. Thermophiles and hyperthermophiles produce heat-stable macromolecules.

◆ What is the current upper temperature limit for growth of a prokaryote? Is the organism that grows at this temperature a member of the *Bacteria* or the *Archaea*?

◆ What is the structure of membranes of hyperthermophilic *Archaea*, and why might this structure be useful for growth at high temperature?

◆ What is *Taq polymerase* and why has it been of use to biotechnology?

V ENVIRONMENTAL EFFECTS ON MICROBIAL GROWTH: pH, OSMOLARITY, AND OXYGEN

Temperature has a major effect on the growth of microorganisms. But many other factors do as well; chief among them are pH, osmolarity, and oxygen. We will see that just as organisms showed distinct optima for growth temperature, optima for these other parameters exist as well, running from one extreme to the other. We begin with pH, the expression of acidity in an organism's environment.

6.13 Microbial Growth at Low or High pH

Acidity or alkalinity of a solution is expressed by its **pH** on a scale on which neutrality is pH 7 (Figure 6.22●). pH values less than 7 are said to be *acidic*, and those greater than 7 are *alkaline* (or *basic*). It is important to remember that pH is a *logarithmic* function—a change of 1 pH unit represents a *10-fold* change in hydrogen ion concentration. Thus, vinegar (pH near 2) and household ammonia (pH near 11) differ in hydrogen ion concentration by a billion-fold.

pH and Microbial Growth

Each organism has a pH range within which growth is possible and usually has a well-defined pH optimum. Most organisms show a growth pH range of 2–3 units. Most natural environments have pH values between 5 and 9, and organisms with optima in this range are most common. Only a few species can grow at pH values of less than 2 or greater than 9. Organisms that grow best at low pH are called **acidophiles**. Fungi as a group tend to be more acid-tolerant than bacteria. Many fungi grow op-

● **Figure 6.22 The pH scale.** Note that although some microorganisms can live at very low or very high pH, the cell's internal pH remains near neutrality.

timally at pH 5 or below, and a few grow well at pH values as low as 2. Several bacteria are also acidophilic. In fact, some of these are *obligate* acidophiles, unable to grow at neutral pH. Obligately acidophilic bacteria include several species of *Acidithiobacillus* (∞Section 12.4) and several genera of *Archaea*, including *Sulfolobus*, *Thermoplasma*, and *Ferroplasma* (∞Sections 13.5 and 13.9).

A critical factor governing acidophily is the stability of the cytoplasmic membrane. When the pH is raised to neutrality, the cytoplasmic membranes of strongly acidophilic bacteria are destroyed and the cells lyse. This indicates that high concentrations of hydrogen ions are required for membrane stability. For example, the most acidophilic prokaryote known, *Picrophilus oshimae*, grows optimally at pH 0.7. Above pH 4, cells of *P. oshimae* spontaneously lyse. *P. oshimae* inhabits extremely acidic thermal soils associated with volcanic activity (∞Section 13.5).

A few extremophiles have very high pH optima for growth, sometimes as high as pH 10; these are known as **alkaliphiles**. Alkaliphilic microorganisms are typically found in highly basic habitats such as soda lakes and high carbonate soils. The most well-studied alkaliphilic prokaryotes have been *Bacillus* species, such as *B. firmus*. This organism is alkaliphilic but has an unusually broad pH range for growth, from 7.5 to 11. Some extremely alkaliphilic bacteria are also halophilic (salt-loving), and most of these are *Archaea*

(∞Section 13.3). Some alkaliphiles have found industrial uses because they produce hydrolytic enzymes, such as proteases and lipases, which function well at alkaline pH and are used as supplements for household detergents (∞Section 30.9).

Alkaliphiles are also of interest because of the bioenergetic problems they face living at such high pH. For example, can a proton motive force (∞Section 5.12) be established when the external surface of the cytoplasmic membrane is so alkaline? From studies of *B. firmus* it has been shown that a Na^+ gradient (instead of the usual proton motive force) supplies the energy for transport and motility, but that a proton motive force is also established and is responsible for driving respiratory ATP synthesis. Exactly how this occurs is an interesting problem in alkaliphile research today.

Internal Cell pH

The optimal pH for growth represents the pH of the *extracellular* environment only. The *intracellular* pH must remain relatively close to neutral in order to prevent destruction of acid- or alkali-labile macromolecules in the cell. For the majority of microorganisms whose pH optimum for growth is between pH 6 and 8 (called **neutrophiles**), the cytoplasm remains neutral or very nearly so (Figure 6.22). However, in acidophiles and alkaliphiles the internal pH can vary from this. For example, in the previously mentioned acidophile *P. oshimae*, the internal pH has been measured at pH 4.6, and in extreme alkaliphiles an intracellular pH of as high as 9.5 has been measured. If these are not the lower and upper limits of cytoplasmic pH, respectively, they must be extremely close to the limits, since macromolecular stability would almost certainly be compromised above or below these pH values.

Buffers

In a batch culture, the pH can change during growth as the result of metabolic reactions that consume or produce acidic or basic substances. Thus, *buffers* are frequently added to microbial culture media to keep the pH relatively constant. However, a given buffer works over only a narrow pH range. Hence, different buffers must be used to buffer at different pH values.

For near-neutral pH ranges (pH 6–7.5), potassium phosphate (KH_2PO_4) is an excellent buffer. Many other buffers for use in microbial growth media or for the assay of enzymes extracted from microbial cells are available, and the best buffering system for one organism or enzyme may be considerably different from that of another. Thus, the optimal buffer for use in a particular situation must usually be determined empirically. For assaying enzymes *in vitro*, though, a buffer that works well in an assay for the enzyme from one organism will usually work well for assaying the same enzyme from other organisms.

6.13 Concept Check

The acidity or alkalinity of an environment can greatly affect microbial growth. Some organisms have evolved to grow best at low or high pH, but most organisms grow best between pH 6 and 8. The internal pH of a cell must stay relatively close to neutral even though the external pH is highly acidic or basic.

◆ What is the increase in concentration of protons when going from pH 7 to pH 4?

◆ What are *buffers* and why are they needed?

6.14 Osmotic Effects on Microbial Growth

Water is the solvent of life. As far as is known, all organisms require water, and water availability is an important factor affecting the growth of microorganisms in nature. Water availability not only depends on the water content of an environment, that is, how moist or dry a solid microbial habitat may be, but is also a function of the concentration of solutes such as salts, sugars, or other substances that are dissolved in water. This is because dissolved substances have an affinity for water, which makes the water associated with solutes unavailable to organisms.

Water Activity and Osmosis

Water availability is generally expressed in physical terms such as **water activity**. Water activity, abbreviated a_w, is the ratio of the vapor pressure of the air in equilibrium with a substance or solution to the vapor pressure of pure water. Thus, values of a_w vary between 0 and 1, and some representative values are given in Table 6.2. Water activities in agricultural soils generally range between 0.90 and 1.

Table 6.2	Water activity of several substances	
Water activity (a_w)	**Material**	**Example organisms**[a]
1.000	Pure water	*Caulobacter, Spirillum*
0.995	Human blood	*Streptococcus, Escherichia*
0.980	Seawater	*Pseudomonas, Vibrio*
0.950	Bread	Most gram-positive rods
0.900	Maple syrup, ham	Gram-positive cocci such as *Staphylococcus*
0.850	Salami	*Saccharomyces rouxii* (yeast)
0.800	Fruit cake, jams	*Saccharomyces bailii*, *Penicillium* (fungus)
0.750	Salt lakes, salted fish	*Halobacterium, Halococcus*
0.700	Cereals, candy, dried fruit	*Xeromyces bisporus* and other xerophilic fungi

[a] Selected examples of prokaryotes or fungi capable of growth in culture media adjusted to the stated water activity.

Water diffuses from a region of high water concentration (low solute concentration) to a region of lower water concentration (higher solute concentration) in the process of *osmosis* (∞Section 4.8 and Figure 4.32). In most cases, the cytoplasm of a cell has a higher solute concentration than the environment, so water tends to diffuse *into* the cell. Under such conditions, the cell is said to be in *positive water balance*. However, when a cell is in an environment of low water activity, there is a tendency for water to flow out of the cell. This can cause serious problems if a cell has no way to deal with it.

Halophiles and Related Organisms

In nature, osmotic effects are of interest mainly in habitats with high concentrations of salts. Seawater contains about 3% NaCl plus small amounts of many other minerals and elements. Microorganisms found in the sea usually have a specific requirement for the sodium ion in addition to growing optimally at the water activity of seawater (Figure 6.23●). Such organisms are called **halophiles**. The growth of halophiles requires at least some NaCl, but the optimum varies with the organism. For example, the terms *mild halophile* and *moderate halophile* are used to describe halophiles with low (1–6%) and moderate (7–15%) NaCl requirements, respectively (Figure 6.23).

Most microorganisms are unable to cope with environments of very low water activity and either die or become dehydrated and dormant under such conditions. **Halotolerant** organisms can tolerate some reduction in the a_w of their environment but generally grow best in the absence of the added solute (Figure 6.23). By contrast, some organisms thrive at very low water activity. These organisms are of interest not only from the standpoint of their adaptation to life under these conditions, but also from an applied standpoint, for example, in the food industry, where solutes such as salt and sucrose are commonly used as preservatives to inhibit microbial growth.

Organisms capable of growth in very salty environments are called **extreme halophiles** (Figure 6.23). These

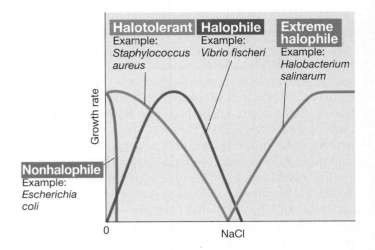

● **Figure 6.23 Effect of sodium chloride concentration on growth of microorganisms of different salt tolerances or requirements.** The optimum NaCl concentration for marine microorganisms such as *V. fischeri* is about 3%; for extreme halophiles, it is between 15 and 30%, depending on the organism.

organisms require 15–30% NaCl, depending on the species, for optimum growth (∞Section 13.3). Organisms able to live in environments high in sugar are called **osmophiles**, and those able to grow in very dry environments (made dry by lack of water) are called **xerophiles**. Examples of these various organisms are given in Table 6.2.

Compatible Solutes

How do organisms grow under conditions of low water activity? When an organism grows in a medium with a low water activity, it can obtain water from its environment only by increasing its *internal* solute concentration. An increase in internal solute concentration can be accomplished by either (1) pumping inorganic ions into the cell from the environment, or (2) synthesizing or concentrating an organic solute. Many organisms are known that employ one or the other of these mechanisms, and several examples are given in Table 6.3.

Table 6.3	Compatible solutes of microorganisms	
Organism	**Major solute(s) accumulated**	**Minimum a_w for growth**
Bacteria, nonphototrophic	Glycine betaine, proline (mainly gram-positive), glutamate (mainly gram-negative)	0.97–0.90
Freshwater cyanobacteria	Sucrose, trehalose	0.98
Marine cyanobacteria	α-Glucosylglycerol	0.92
Marine algae	Mannitol, various glycosides, proline, dimethylsulfoniopropionate	0.92
Salt lake cyanobacteria	Glycine betaine	0.90–0.75
Halophilic anoxygenic phototrophic *Bacteria* (*Ectothiorhodospira/Halorhodospira* and *Rhodovibrio* species)	Glycine betaine, ectoine, trehalose	0.90–0.75
Extremely halophilic *Archaea* (for example, *Halobacterium*) and some *Bacteria* (for example, *Haloanaerobium*)	KCl	0.75
Dunaliella (halophilic green alga)	Glycerol	0.75
Xerophilic yeasts	Glycerol	0.83–0.62
Xerophilic filamentous fungi	Glycerol	0.72–0.61

The solute used inside the cell for adjustment of cytoplasmic water activity must be noninhibitory to biochemical processes within the cell. Such compounds are called **compatible solutes**. Several different compatible solutes are known in microorganisms (Table 6.3 and Figure 6.24●). These substances are all highly water-soluble sugars or sugar alcohols, other alcohols, or amino acids or their derivatives (Figure 6.24). In the case of extremely halophilic *Archaea* and a very few extremely halophilic *Bacteria,* the compatible solute is K^+ (as KCl).

Compatible solutes are either synthesized by the microorganisms directly or in some cases (such as glycine betaine or KCl) accumulated from the environment. The concentration of compatible solutes in the cell is a function of the level of external solutes, and in each organism the maximal amount of compatible solute(s) made or that can be accumulated is a genetically directed characteristic. Thus, different organisms can tolerate different ranges of water potential (Tables 6.2 and 6.3). Nonhalotolerant, halotolerant, halophilic, and extremely halophilic microorganisms are to a major extent defined by their genetic capacity to produce or accumulate compatible solutes.

Gram-positive cocci of the genus *Staphylococcus* are notoriously halotolerant (in fact, a common isolation procedure for them is to use media containing 7.5% NaCl), and these organisms use the amino acid *proline* as a compatible solute. *Glycine betaine* is a derivative of the amino acid glycine in which the hydrogens on the amino group are replaced by methyl groups. This leaves a permanent positive charge on the N atom (Figure 6.24), which increases its solubility. Glycine betaine is widely distributed as a compatible solute, especially among halophilic *Bacteria* and cyanobacteria (Table 6.3).

Some extremely halophilic bacteria produce the compatible solute *ectoine* (Figure 6.24), which is a cyclic derivative of the amino acid aspartate. A variety of glycosides and the compound dimethylsulfoniopropionate (Figure 6.24) are produced by marine algae, but with rare exception they accumulate in only low amounts because the cells themselves are not very halophilic. Xerophilic yeasts and halophilic green algae mainly produce *glycerol* as a compatible solute. Other examples of compatible solutes are listed in Table 6.3, and structures are shown in Figure 6.24.

1. Amino acid–type and related solutes:

Glycine betaine   Ectoine

Dimethylsulfoniopropionate

2. Carbohydrate-type solutes:

Sucrose

Trehalose

3. Alcohol-type solutes:

Glycerol   Mannitol

● **Figure 6.24** **Structures of some common and highly soluble compatible solutes in microorganisms.** The structures of glutamate and proline, other common solutes, were shown in Figure 3.12. The formal name of ectoine is 1,4,5,6-tetrahydro-2-methyl-4-pyrimidine carboxylate. Glycine betaine and dimethylsulfoniopropionate are common solutes in marine algae.

 6.14 **Concept Check**

Water activity becomes limiting to an organism when the dissolved solute concentration in its environment increases. To counteract this situation organisms produce or accumulate intracellular compatible solutes that function to maintain the cell in positive water balance. Some microorganisms have evolved to grow best at reduced water potential, and some even require high levels of salts in order to grow.

◆ What is the a_w of pure water?

◆ What is a *compatible solute* and why is it needed?

◆ What is the compatible solute for *Halobacterium* species?

6.15 **Oxygen and Microbial Growth**

Since humans absolutely require molecular oxygen (O_2) for life, it is easy to assume that all life forms require oxygen. However, this is not true, because many microorganisms can and some must live in the total absence of oxygen.

Oxygen is weakly soluble in water, and because of the respiratory activities of microorganisms in aquatic or other moist habitats (especially those containing an abundance of organic materials), O_2 can quickly become exhausted. Thus, *anoxic* microbial habitats are common in nature, including muds and other sediments, bogs and marshes, water-logged soils, intestinal tracts of animals, sewage sludge, the deep subsurface of Earth, and many other environments. In these anoxic habitats, microorganisms, particularly various prokaryotes, thrive.

Oxygen Classes of Microorganisms

Microorganisms vary in their need for, or tolerance of, oxygen. In fact, microorganisms can be divided into several groups depending on how oxygen affects them, as outlined in Table 6.4. **Aerobes** are species capable of growth at full oxygen tensions (air is 21% O_2) and respire oxygen in their metabolism. Many aerobes can even tolerate elevated concentrations of oxygen (hyperbaric oxygen). **Microaerophiles**, by contrast, are aerobes that can use oxygen only when it is present at levels reduced from that in air (*microoxic* conditions). This is because of their limited capacity to respire or because they contain some oxygen-sensitive molecule such as an oxygen-labile enzyme. Many aerobes are **facultative**, meaning that, under the appropriate nutrient and culture conditions, they can grow under *either* oxic or anoxic conditions.

Some organisms cannot respire oxygen; such organisms are called **anaerobes**. There are two kinds of anaerobes: **aerotolerant anaerobes**, which can tolerate oxygen and grow in its presence even though they cannot use it, and **obligate** (or **strict**) **anaerobes**, which are inhibited or even killed by O_2 (Table 6.4). The reason obligate anaerobes are killed by oxygen is unknown, but it may be because they are unable to detoxify some of the products of oxygen metabolism (see Section 6.16).

So far as is known, *obligate anaerobiosis* occurs in only three groups of microorganisms: a wide variety of prokaryotes, a few fungi, and a few protozoa. The best-known group of obligately anaerobic *Bacteria* belongs to the genus *Clostridium*, a group of gram-positive endospore-forming rods. Clostridia are widespread in soil, lake sediments, and intestinal tracts, and are often responsible for spoilage of canned foods (∞Sections 12.20

and 29.2). Other obligately anaerobic bacteria are found among the methanogens and many other species of *Archaea* (∞Chapter 13), the sulfate-reducing and homoacetogenic bacteria (∞Chapters 12 and 13), and many of the bacteria that inhabit the animal gut (∞Section 21.4). Among obligate anaerobes, however, the sensitivity to oxygen varies greatly. Some species are able to tolerate traces of oxygen or even full exposure to oxygen, while others are not.

Culture Techniques for Aerobes and Anaerobes

For the growth of many aerobes, it is necessary to provide extensive aeration. This is because the oxygen that is consumed by the organisms during growth is not replaced fast enough by diffusion from the air. Forced aeration of cultures is therefore frequently needed and can be achieved either by vigorously shaking the flask or tube on a shaker or by bubbling sterilized air into the medium through a fine glass tube or porous glass disc. Aerobes usually grow much better with forced aeration than when oxygen is provided by simple diffusion.

For the culture of anaerobes, the problem is to *exclude*, not provide, oxygen. And because oxygen is present in the air, special methods are needed to culture anaerobic microorganisms. Obligate anaerobes vary in their sensitivity to oxygen, and a number of procedures are available for reducing the oxygen content of cultures. Some of these techniques are simple and suitable mainly for less-sensitive organisms, while others are more complex but necessary for growth of strict anaerobes.

Bottles or tubes filled completely to the top with culture medium and provided with tightly fitting stoppers provide suitably anoxic conditions for organisms not too sensitive to small amounts of oxygen. It is also possible to add to culture media a chemical called a *reducing agent* that reacts with oxygen and reduces it to H_2O. A typical example is *thioglycolate*, which is added to a medium

Table 6.4	Oxygen relationships of microorganisms			
Group	**Relationship to O_2**	**Type of metabolism**	**Example**[a]	**Habitat**[b]
Aerobes				
Obligate	Required	Aerobic respiration	*Micrococcus luteus* (B)	Skin, dust
Facultative	Not required, but growth better with O_2	Aerobic respiration, anaerobic respiration, fermentation	*Escherichia coli* (B)	Mammalian large intestine
Microaerophilic	Required but at levels lower than atmospheric	Aerobic respiration	*Spirillum volutans* (B)	Lake water
Anaerobes				
Aerotolerant	Not required, and growth no better when O_2 present	Fermentation	*Streptococcus pyogenes* (B)	Upper respiratory tract
Obligate	Harmful or lethal	Fermentation or anaerobic respiration	*Methanobacterium* (A) *formicicum*	Sewage sludge digestors, anoxic lake sediments

[a] Letters in parentheses indicate phylogenetic status (B, *Bacteria*; A, *Archaea*). Representatives of either domain of prokaryotes are known in each category. Most eukaryotes are obligate aerobes, but facultative aerobes (for example, yeast) and obligate anaerobes (for example, certain protozoa and fungi) are known.
[b] Listed are typical habitats of the example organism.

(a) (b) (c) (d) (e)

Oxic zone

Anoxic zone

● **Figure 6.25 Growth versus oxygen concentration.** Aerobic, anaerobic, facultative, microaerophilic, and aerotolerant anaerobe growth, as revealed by the position of microbial colonies (depicted here as black dots) within tubes of thioglycolate broth culture medium. A small amount of agar has been added to keep the liquid from becoming disturbed and the redox dye, resazurin, which is pink when oxidized and colorless when reduced, is added as a redox indicator. (a) Oxygen penetrates only a short distance into the tube, so obligate aerobes grow only at the surface. (b) Anaerobes, being sensitive to oxygen, grow only away from the surface. (c) Facultative aerobes are able to grow in either the presence or the absence of oxygen and thus grow throughout the tube. However, better growth occurs near the surface because these organisms can respire. (d) Microaerophiles grow away from the most oxic zone. (e) Aerotolerant anaerobes grow throughout the tube. However, growth is no better near the surface because these organisms can only ferment.

called *thioglycolate broth*, commonly used to test an organism's requirements for oxygen (Figure 6.25●).

After thioglycolate reacts with oxygen throughout the tube, oxygen can penetrate only near the top of the tube where the medium contacts air. Obligate aerobes grow only at the top of such tubes. Facultative organisms grow throughout the tube but best near the top. Microaerophiles grow near the top but not right at the top. Anaerobes grow only near the bottom of the tube, where oxygen cannot penetrate (Figure 6.25). A redox indicator dye called *resazurin* is added to the medium because the dye changes color in the presence of oxygen (pink when oxygenated, colorless when reduced) and thereby indicates the degree of penetration of oxygen into the medium (Figure 6.25).

To remove all traces of O_2 for the culture of anaerobes, it is possible to place an oxygen-consuming system in a jar holding the tubes or plates. One of the simplest devices for this is an *anoxic jar*, a heavy-walled jar with a gastight seal within which tubes, plates, or other containers to be incubated are placed (Figure 6.26a●). The air in the jar is replaced with a mixture of H_2 and CO_2, and in the presence of a chemical catalyst the traces of O_2 left in the jar and culture medium are consumed by the H_2 ($H_2 + O_2 \rightarrow H_2O$), eventually leading to anoxic conditions.

For strict anaerobes, such as the methanogens (∞ Sections 13.4 and 17.17), it is necessary not only to carefully remove all traces of O_2 but also to carry out all manipulations of cultures in an anoxic atmosphere. Strict anaerobes can be killed by even a brief exposure to oxygen. In these cases, a culture medium is first boiled to render it oxygen-free, and then a reducing agent such as H_2S is added, and the mixture is sealed under an oxygen-free gas. All manipulations are carried out under a jet of oxygen-free hydrogen or nitrogen gas that is direct-

(a)

(b)

Deborah O. Jung and M. T. Madigan

Coy Laboratory Products

● **Figure 6.26 Incubation under anoxic conditions.** (a) Anoxic jar. A chemical reaction in the envelope in the jar generates $H_2 + CO_2$. The H_2 reacts with O_2 in the jar on the surface of a palladium catalyst to yield H_2O; the final atmosphere contains N_2, H_2, and CO_2. (b) Anoxic glove bag for manipulating and incubating cultures under anoxic conditions. The airlock on the right, which can be evacuated and filled with O_2-free gas, serves as a port for adding and removing materials to and from the glove bag.

ed into the culture vessel when it is open, thus driving out any oxygen that might enter. For extensive research on anaerobes, special boxes fitted with gloves, called *anoxic glove boxes*, permit work with open cultures in completely anoxic atmospheres (Figure 6.26*b*).

6.15 Concept Check

Aerobes require oxygen to live, whereas anaerobes do not and may even be killed by oxygen. Facultative organisms can live with or without oxygen. Special techniques are needed to grow aerobic and anaerobic microorganisms.

◆ What is a *facultative aerobe*?

◆ How does a reducing agent work?

6.16 Toxic Forms of Oxygen

Oxygen is a powerful oxidant and the best electron acceptor for respiration (Section 5.11). But at the same time, oxygen can be a poison to some microorganisms. Why? It turns out that oxygen, *per se*, is not the poison, but instead it is certain *oxygen derivatives* that are toxic to microorganisms. We consider this topic here.

Oxygen Chemistry

Oxygen in its ground state is referred to as **triplet oxygen** ($^3O_2$). However, other electronic configurations of oxygen are possible. One major form of toxic oxygen is called **singlet oxygen** ($^1O_2$), a higher energy form of oxygen in which outer shell electrons surrounding the nucleus become highly reactive and are able to carry out a variety of spontaneous and undesirable oxidations within the cell. Singlet oxygen is produced both photochemically and biochemically, the latter through the action of various peroxidase enzymes. Organisms that frequently encounter singlet oxygen, such as airborne bacteria and phototrophic microorganisms, often contain pigments called **carotenoids**, which function to convert singlet oxygen to nontoxic forms (Section 17.3).

Other highly toxic forms of oxygen include **superoxide anion** (O_2^-), **hydrogen peroxide** (H_2O_2), and **hydroxyl radical** ($OH \cdot$). All of these are produced as by-products of the reduction of O_2 to H_2O in respiration (Figure 6.27●). Flavoproteins, quinones, thiols, and iron-sulfur proteins (Section 5.11), found in virtually all cells, can also carry out the reduction of O_2 to O_2^-. Thus, whether or not they can use oxygen (Table 6.3), virtually all cells are faced with exposure to toxic oxygen species from time to time.

Superoxide is a strong oxidizing agent and can oxidize virtually any organic compound in the cell, including macromolecules. Peroxides such as H_2O_2 can damage cell components but are generally not as toxic as superoxide or hydroxyl radical. The latter is the most reactive of all toxic oxygen species and can, like superoxide, quickly oxidize any organic substance in the cell.

$$O_2 + e^- \longrightarrow O_2^- \quad \text{Superoxide}$$
$$O_2^- + e^- + 2\,H^+ \longrightarrow H_2O_2 \quad \text{Hydrogen peroxide}$$
$$H_2O_2 + e^- + H^+ \longrightarrow H_2O + OH \cdot \quad \text{Hydroxyl radical}$$
$$OH \cdot + e^- + H^+ \longrightarrow H_2O \quad \text{Water}$$

$$\text{Overall: } O_2 + 4\,e^- + 4\,H^+ \longrightarrow 2\,H_2O$$

● **Figure 6.27 Four-electron reduction of O_2 to H_2O by stepwise addition of electrons.** All the intermediates formed are reactive and toxic to cells except for water, of course.

However, the hydroxyl radical is only a transient species in most cells because its major source is ionizing radiation, a substance to which most cells are rarely exposed. Small amounts of hydroxyl radical can also be produced from H_2O_2 (Figure 6.27). But if peroxides are removed from the cell (through the activity of the enzyme catalase, which is discussed later), this source of hydroxyl radical is virtually eliminated. In later chapters we will see that some toxic oxygen species can be produced by certain immune cells in the animal body and used to kill microbial invaders (Section 22.2).

Enzymes That Destroy Toxic Oxygen

With such an array of toxic oxygen derivatives, it is not surprising that organisms have evolved enzymes that destroy these compounds (Figure 6.28●). The most common enzyme in this regard is **catalase**, which attacks hydrogen peroxide. The activity of catalase is illustrated in Figures 6.28*a* and 6.29. Another enzyme that destroys hydrogen peroxide is **peroxidase** (Figure 6.28*b*), which differs from catalase in requiring a reductant for activity, usually NADH, producing H_2O as a product. Superoxide is destroyed by the enzyme **superoxide dismutase** (Figure 6.28*c*), which combines two molecules of superoxide to form

(a) **Catalase:**
$$H_2O_2 + H_2O_2 \longrightarrow 2\,H_2O + O_2$$

(b) **Peroxidase:**
$$H_2O_2 + \text{NADH} + H^+ \longrightarrow 2\,H_2O + NAD^+$$

(c) **Superoxide dismutase:**
$$O_2^- + O_2^- + 2\,H^+ \longrightarrow H_2O_2 + O_2$$

(d) **Superoxide dismutase/catalase in combination:**
$$4\,O_2^- + 4\,H^+ \longrightarrow 2\,H_2O + 3\,O_2$$

(e) **Superoxide reductase:**
$$O_2^- + 2\,H^+ + \text{cyt } c_{\text{reduced}} \longrightarrow H_2O_2 + \text{cyt } c_{\text{oxidized}}$$

● **Figure 6.28 Enzymes that destroy toxic oxygen species.** (a) Catalases and (b) peroxidases are porphyrin-containing proteins, although some flavoproteins (Section 5.11) may consume toxic oxygen species as well. (c) Superoxide dismutases are metal-containing proteins, the metals being copper and zinc, manganese, or iron. (d) Combined reaction of superoxide dismutase and catalase. (e) Superoxide reductase catalyzes the one electron reduction of O_2^- to H_2O_2 using reduced cytochrome *c* as the electron donor.

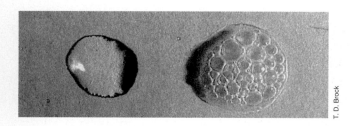

● **Figure 6.29 Method for testing a microbial culture for the presence of catalase.** A heavy loopful of cells from an agar culture was mixed on a slide (right) with a drop of 30% hydrogen peroxide. The immediate appearance of bubbles is indicative of the presence of catalase. The bubbles are O_2 produced by the reaction $H_2O_2 + H_2O_2 \rightarrow 2\,H_2O + O_2$.

one molecule of hydrogen peroxide and one molecule of oxygen. Superoxide dismutase and catalase work in tandem to bring about the conversion of superoxide back to oxygen (Figure 6.28d).

Aerobes and facultative aerobes typically contain both superoxide dismutase and catalase, although a few obligate aerobes lack catalase. Superoxide dismutase is indispensable to aerobic cells, and the absence of this enzyme in obligate anaerobes was originally thought to be the reason why oxygen is toxic to them (but see next paragraph). Some aerotolerant anaerobes, such as lactic acid bacteria, also lack superoxide dismutase, but they use protein-free manganese (Mn^{2+}) complexes to carry out the dismutation of O_2^- to H_2O_2 and O_2. Such a reaction may have functioned as a primitive form of superoxide dismutase in ancient organisms. This is supported by the fact that all known superoxide dismutases contain a metal cofactor, usually Mn^{2+}, but also Fe^{2+}, or Cu^{2+} plus Zn^{2+}, at the enzyme's active site.

Another means of superoxide disposal is present in certain obligately anaerobic prokaryotes. In *Pyrococcus furiosus* (a member of the *Archaea*), for example, superoxide dismutase is absent, but a unique enzyme,

superoxide reductase, is present and is responsible for superoxide removal. Unlike superoxide dismutase, superoxide reductase reduces superoxide to H_2O_2 *without* the production of O_2 (Figure 6.28e), thus avoiding exposure of the organism to O_2. *P. furiosus* also lacks catalase, an enzyme that like superoxide dismutase, also generates O_2 (Figure 6.28a). In *P. furiosus*, the H_2O_2 produced by superoxide reductase is removed by the activity of peroxidase-like enzymes that yield H_2O as a final product (Figure 6.28b).

Superoxide reductase may be widely distributed among obligate anaerobes, since genomic studies have revealed superoxide reductase-like genes in the genomes of several obligate anaerobes. This means that these organisms, previously thought to be strict anaerobes because of a lack of superoxide dismutase, do indeed have a mechanism to deal with the highly toxic superoxide anion. The sensitivity of these organisms to oxygen in culture media may therefore be for entirely other and as yet unknown reasons.

Interestingly, many obligately anaerobic hyperthermophiles are quite tolerant of cold, oxic conditions. Although they do not grow under such conditions, superoxide reductase presumably prevents their killing. It is thought that this oxygen tolerance is important in the transfer of these organisms from one deep sea hydrothermal system to another (Section 19.3).

6.16 Concept Check

Several toxic forms of oxygen can be formed in the cell, but enzymes are present that can neutralize most of them. Superoxide in particular seems to be a common toxic oxygen species.

◆ How does superoxide dismutase protect a cell from toxic oxygen?

◆ How does the activity of superoxide *dismutase* differ from that of superoxide *reductase*?

REVIEW QUESTIONS

1. Describe the key molecular processes that occur when a cell grows and divides. What proteins assist in this cell division process (◯◯Section 6.1)?

2. Describe the role that Fts proteins play in the cell division process (◯◯Section 6.2).

3. In what way do derivatives of the rod-shaped bacterium *Escherichia coli* that carry mutations that inactivate the protein MreB look different microscopically from wild-type (unmutated) cells? What is the reason for this (◯◯Section 6.2)?

4. How does the antibiotic penicillin kill bacterial cells (◯◯Section 6.3)?

5. What is the difference between the specific growth rate (k) of an organism and its generation time (g) (◯◯Sections 6.4 and 6.5)?

6. Describe the growth cycle of a population of bacterial cells from the time this population is first inoculated into fresh medium (◯◯Section 6.6).

7. Describe one direct and one indirect method by which microbial growth can be measured. Make sure that the methods you choose agree with your definition (◯◯Sections 6.7 and 6.8).

8. Briefly describe the process by which a single cell develops into a visible colony on an agar plate. With this explanation as a background, describe the principle behind the viable count method (◯◯Section 6.7).

9. How can a chemostat regulate growth rate and cell numbers independently (◯◯Section 6.9)?

10. Examine the graph describing the relationship between growth rate and temperature (Figure 6.17). Give an explanation, in biochemical terms, of why the optimum temperature for an organism is usually closer to its maximum than its minimum (◯◯Section 6.10).

11. Describe a habitat where you would find a psychrophile. A hyperthermophile (◯◯Sections 6.11 and 6.12).

12. Concerning the pH of the environment and of the cell, in what ways are acidophiles and alkaliphiles different? In what ways are they similar (◯◯Section 6.13)?

13. Write an explanation in molecular terms for how a cell of a halophile is able to make water molecules flow *into* itself (◯◯Section 6.14).

14. List three *chemical classes* of compatible solutes produced by various microorganisms. List at least two things they all have in common (◯◯Section 6.14).

15. Contrast an *aerotolerant* and an *obligate anaerobe* in terms of sensitivity and ability to grow in the presence of oxygen (O_2). How does an aerotolerant anaerobe differ from a microaerophile (◯◯Section 6.15)?

16. Compare and contrast the enzymes *catalase, superoxide dismutase*, and *superoxide reductase* from the following points of view: substrates, oxygen products, organisms containing them, role in oxygen tolerance of the cell (◯◯Section 6.16).

APPLICATION QUESTIONS

1. Starting with four bacterial cells per milliliter in a rich nutrient medium, with a 1-h lag phase and a 20-min generation time, how many cells will there be in 1 liter of this culture after 1 h? After 2 h? After 2 h if one of the initial four cells was dead?

2. Calculate g and k in a growth experiment in which a medium was inoculated with 5×10^6 cells/ml of *Escherichia coli* cells and, following a 1-h lag, grew exponentially for 5 h, after which the population was 5.4×10^9 cells/ml.

3. Return to Chapter 3 and locate a figure that best describes what happens to enzymes from a cell of a mesophile like *Escherichia coli* when placed in a culture medium at 80°C. Contrast this with a figure from Chapter 6 that best describes what would happen if cells of *Pyrolobus fumarii* were placed under the same conditions. Describe why neither organism would grow.

4. Would you expect to find a psychrophilic microorganism alive in a hot spring? Why? It is frequently possible to isolate hyperthermophilic microorganisms from cold-water environments. Supply an explanation for this.

7

ESSENTIALS OF MOLECULAR BIOLOGY

Although many forms of nucleic acid are known in nature, all cells contain double-stranded DNA genomes. The genome directs all of the molecular events in the cell.

WORKING GLOSSARY

Aminoacyl-tRNA synthetase an enzyme that catalyzes attachment of an amino acid to its cognate tRNA

Anticodon a sequence of three bases in a tRNA molecule that base-pairs with a codon during protein synthesis

Antiparallel in reference to double-stranded DNA, one strand runs $5' \rightarrow 3'$ and the complementary strand $3' \rightarrow 5'$

Chromosome a genetic element, usually circular in prokaryotes and linear in eukaryotes, carrying genes essential to cellular function

Codon a sequence of three bases in mRNA that encodes an amino acid

Complementary nucleic acid sequences that can base-pair with each other

DNA gyrase an enzyme found in most prokaryotes that introduces negative supercoils in DNA

DNA polymerase an enzyme that synthesizes a new strand of DNA in the $5' \rightarrow 3'$ direction using an antiparallel DNA strand as a template

Exon the coding DNA sequences in a split gene (contrast with *intron*)

Gene a segment of DNA specifying a protein (via mRNA), a tRNA, or an rRNA

Genome the total complement of genes contained in a cell or virus

Hybridization formation of a duplex nucleic acid with strands derived from different sources by complementary base pairing

Intron the intervening noncoding DNA sequences in a split gene (contrast with *exon*)

Messenger RNA (mRNA) an RNA molecule that contains the genetic information to encode one or more polypeptides

Molecular chaperone a protein that helps other proteins fold or refold from a partially denatured state

Operon a cluster of genes whose expression is controlled by a single operator

Polymerase chain reaction (PCR) a method for the amplification of a specific DNA sequence *in vitro* by repeated cycles of synthesis using specific primers and DNA polymerase

Primary transcript an unprocessed RNA molecule that is the direct product of transcription

Primer an oligonucleotide to which DNA polymerase can attach the first deoxyribonucleotide during DNA replication

Promoter a site on DNA to which RNA polymerase binds to commence transcription

Replication synthesis of DNA using DNA as a template

Restriction enzyme an enzyme that recognizes and breaks DNA at specific sequences

Ribosomal RNA (rRNA) types of RNA found in the ribosome; some participate actively in the process of protein synthesis

Ribosome a cytoplasmic particle composed of ribosomal RNA and protein whose function is to synthesize proteins

RNA polymerase an enzyme that synthesizes RNA in the $5' \rightarrow 3'$ direction using a complementary and antiparallel DNA strand as a template

Semiconservative replication DNA synthesis yielding new double helices, each consisting of one parental and one progeny strand

Transcription the synthesis of RNA using a DNA template

Transfer RNA (tRNA) an adaptor molecule used in translation that has specificity for both a particular amino acid and for one or more codons

Translation the synthesis of protein using the genetic information in messenger RNA as a template

---

In Chapter 1 we characterized cells as *chemical machines* and *coding devices*. As chemical machines, cells accumulate and transform their vast array of macromolecules into new cells. As coding devices, they store, process, and use genes, the genetic information of the cell. Genes and gene processing are the subject of *molecular biology*, the focus of this chapter. Here we will study genes, DNA/RNA structure and function, and DNA replication. We then consider the biosynthesis of proteins, the macromolecules that play important roles in both cell structure and the machine function of the cell.

GENES AND GENE EXPRESSION

7.1 Macromolecules and Genetic Information

The functional unit of genetic information is the **gene**. All microorganisms, indeed all life forms, contain genes, and thus a fundamental understanding of what a gene is and what it does is important for understanding the structure, function, and behavior of cells and viruses. Moreover, because biology is rapidly moving toward

defining cells in terms of their gene complement (the "genomic era"), we must understand the process of biological information flow if we are to understand the biology of microorganisms.

Genes and the Steps in Information Flow

In all cells genes are composed of *deoxyribonucleic acid (DNA)*. The information in the gene is present as the sequence of bases—purine (adenine and guanine) and pyrimidines (thymine and cytosine)—in the DNA. The information stored in DNA is transferred to *ribonucleic acid (RNA)*. RNA can be either an informational intermediate (a messenger), or, in some cases, a more active part of the cell's machinery. Because all three of these molecules, DNA, RNA, and protein, contain genetic information in their sequences, they are called **informational macromolecules** (∞Section 3.2).

The molecular processes underlying genetic information flow can be divided into three stages (Figure 7.1●).

1. *Replication*. The DNA molecule is a **double helix** of two long chains (∞Section 3.5). During replication, DNA is duplicated, producing two double helices (Figure 7.1).

2. *Transcription*. DNA participates in protein synthesis through an RNA intermediate. Transfer of the informa-

tion to RNA is called **transcription**, and the RNA molecule that encodes one or more polypeptides is called **messenger RNA** (mRNA). Some genes contain information for other types of RNA, such as **transfer RNA** (tRNA) and **ribosomal RNA** (rRNA). These play roles in protein synthesis but do not themselves encode the genetic information for making proteins.

3. *Translation.* The sequence of amino acids in a polypeptide is determined by a specific sequence of bases in the mRNA. There is a linear correspondence between the base sequence of a gene and the amino acid sequence of a polypeptide (Figure 7.1). *Three* bases on a mRNA molecule encode a single amino acid, and each such triplet of bases is called a **codon**. This **genetic code** is translated into protein by means of the protein-synthesizing system. This system consists of **ribosomes** (which are themselves made up of proteins and rRNA), tRNA, and a number of enzymes.

The three transfer steps shown in Figure 7.1 are those used in all cells. In Chapter 9 we will learn that some viruses violate this **central dogma of molecular biology** (DNA → RNA → protein). This includes instances where RNA is the viral genetic material and either functions as mRNA directly or encodes mRNA, and also a situation in retroviruses such as HIV—the causative agent of AIDS—where its RNA genome encodes the production of DNA (reverse transcription). Note that in both of these cases information transfer remains nucleic acid to nucleic acid, but not in the way defined by the central dogma. RNA viruses are just one of many molecular surprises that await us in the viral world!

● **Figure 7.1 Synthesis of the three types of informational macromolecules.** Note that in any particular region, only one of the two strands of the DNA double helix is transcribed.

Prokaryotic and Eukaryotic Genetics

Replication, transcription, and translation occur in all organisms. However, there are some differences in the mechanisms of these processes in prokaryotes and in eukaryotes. In part this is due to differences in the organization of DNA in these two classes of organisms and to the fact that eukaryotes have a nucleus

We emphasized in Chapter 2 the basic differences in the organization of DNA in prokaryotes and eukaryotes. To review, the typical prokaryotic genome consists of a single, covalently closed *circular* molecule of DNA present in the cytoplasm of the cell. In contrast, the eukaryotic genome consists of several *linear* pieces of DNA present in individual chromosomes in the cell nucleus. In virtually all prokaryotes there is no membrane separating the chromosome from the cytoplasm (∞ Section 2.2). In eukaryotes, however, the chromosomes are located inside the nucleus while the ribosomes are in the cytoplasm, so transcription and translation are spatially separated processes.

We also emphasized in Chapter 2 how the two domains of prokaryotes—*Bacteria* and *Archaea*—are phylogenetically distinct. This distinction is also obvious in many aspects of the molecular biology of these organisms. In many cases we will see closer parallels between molecular processes in *Archaea* and *Eukarya* than in *Bacteria* and *Archaea*.

In all cells the definition of a gene is the same: a segment of DNA specifying a protein (via mRNA), a tRNA, or an rRNA. However, in eukaryotes, protein-encoding genes are typically split into two or more coding regions, with noncoding regions separating coding regions. The *coding sequences* are called **exons**, and the *intervening noncoding regions*, **introns**. Both introns and exons are transcribed into the **primary transcript**. From here, the mature (functional) mRNA is formed by excision of noncoding regions and transport to the cytoplasm for translation (∞ Section 14.8). A few genes in prokaryotes also contain introns, but the vast majority do not. A summary contrasting molecular events in prokaryotes and eukaryotes is given in Figure 7.2●.

🛑 *7.1 Concept Check*

The three key processes of macromolecular synthesis are (1) DNA replication; (2) transcription (the synthesis of RNA from a DNA template); and (3) translation (the synthesis of proteins using messenger RNA as template). Although the basic processes are the same in both prokaryotes and eukaryotes, the organization of genetic information is more complex in eukaryotes. Most eukaryotic genes have both coding regions (exons) and noncoding regions (introns).

◆ What three informational macromolecules are involved in genetic information flow?

◆ In all cells there are *three* processes involved in genetic information flow. What are they?

(a) **Prokaryote**

(b) **Eukaryote**

● **Figure 7.2 Contrast of information transfer in prokaryotes and eukaryotes.** (a) Prokaryote. A single mRNA often contains more than one coding region. (b) Eukaryote. Noncoding regions (*introns*) are removed from the primary RNA transcript before translation, leaving *exons* to be joined to form the mature mRNA. For details of RNA processing in eukaryotes, see Section 14.8.

▌▌DNA STRUCTURE

We dealt with the structure of nucleic acids in a general way in Chapter 3. In the next few sections of this chapter we discuss the details of DNA structure including the types of genetic elements containing DNA that are found in cells. With this information as a foundation, we can then discuss how DNA is replicated, transcribed into RNA, and translated into protein.

7.2 DNA Structure: The Double Helix

Four nucleic acid bases are found in DNA: *adenine* (A), *guanine* (G), *cytosine* (C), and *thymine* (T). As already shown in Figure 3.11, the *backbone* of the DNA chain consists of alternating repeats of phosphate and the pentose sugar *deoxyribose*; connected to each sugar is one of the nucleic acid bases. Recall especially the numbering system for the positions of sugar and base. The phosphate connecting two sugars spans from the 3′-carbon of one sugar to the 5′-carbon of the adjacent sugar (see Figure 7.4). At one end of the DNA molecule the sugar has a phosphate on the 5′-hydroxyl, whereas at the other end the sugar has a free hydroxyl at the 3′-position.

DNA as a Double Helix

As we will discuss in Chapter 9, the chromosomes of some viruses are single-stranded. However, in all *cells*, DNA exists as two polynucleotide strands whose base sequences are **complementary**. The complementarity of DNA arises because of the specific pairing of the purine and pyrimidine bases: Adenine always pairs with thymine, and guanine always pairs with cytosine (Figure 7.3●). The two strands in the resulting *double-stranded* molecule are arranged in an **antiparallel** fashion (distinguished in Figures 7.3 and 7.4● as two shades of green). For example, in Figure 7.4 the strand on the left is arranged 5′ to 3′ (top to bottom), whereas the other strand is 5′ to 3′ (bottom to top).

The two strands are wrapped around each other forming a **double helix** (Figure 7.5●). In the double helix, DNA forms two distinct grooves, the *major groove* and the *minor groove*. Of the many proteins that interact

● **Figure 7.3 Specific pairing between adenine (A) and thymine (T) and between guanine (G) and cytosine (C) via hydrogen bonds.** These are the typical base pairs found in double-stranded DNA. Atoms that are found in the major groove of the double helix and that interact with proteins are highlighted in pink. The deoxyribose phosphate backbones of the two strands of DNA are also indicated. Note the different shades of green for the two strands of DNA, a convention used throughout this book.

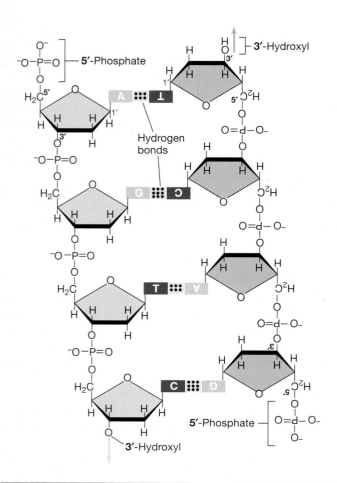

● **Figure 7.4 DNA structure.** Complementary and antiparallel nature of DNA. Note that one chain ends in a 5′-phosphate group, whereas the other ends in a 3′-hydroxyl. The red bases represent the pyrimidines cytosine (C) and thymine (T), and the yellow bases represent the purines adenine (A) and guanine (G).

● **Figure 7.5 A computer model of a short segment of DNA showing the overall arrangement of the double helix.** One of the sugar–phosphate backbones is shown in blue and the other in green. The pyrimidine bases are shown in red and the purines in yellow. Note the locations of the major and minor grooves. One helical turn contains 10 base pairs.

specifically with DNA (as we shall see in Chapter 8), most engage predominantly with the *major* groove, where there is a considerable amount of space. Because of the regularity of the double helix, some atoms of the bases are always exposed in the major groove (and some in the minor groove). Key regions of nucleotides that are known to be important in interactions with proteins are shown in Figure 7.3.

Size of a DNA Molecule

The size of a DNA molecule can be expressed in terms of the *number of thousands* of nucleotide bases or base pairs per molecule. Thus, a DNA molecule with 1000 bases contains 1 *kilobase (kb)* of DNA. If the DNA is a double helix, then one speaks of *kilobase pairs (kbp)*. Thus, a double helix 5000 base pairs in size would be 5 kbp. The bacterium *Escherichia coli* has about 4640 kb pairs of DNA in its chromosome. Given the huge amount of genomic sequence information now becoming available, it is often useful to speak of *millions* of base pairs, or *megabase pairs (Mbp)*. The genome of *E. coli* is 4.64 Mbp.

In terms of actual length, each base pair takes up 0.34 nanometers (nm) in length along the helix, and each

turn of the helix contains approximately 10 base pairs. Therefore, 1 kilobase of DNA has 100 turns of the helix and is 0.34 μm long. Calculations such as this can be very interesting, as we shall soon see (see Section 7.3).

Inverted Repeats, Secondary Structure, and Stem-Loops

Long DNA molecules are quite flexible, but stretches of DNA less than 100 base pairs are more rigid. Some short segments of DNA can be bent by proteins that interact with them. However, certain sequences themselves result in bends in the DNA. The sequences of this *bent DNA* often involve several runs of five or six adenines (in the same strand), each separated by four or five other bases. In addition, some sequences contribute to the ability of DNA to bend when certain proteins interact with the DNA. DNA bending is one mechanism for the *regulation* of gene expression (the turning on and off of genes), as we shall discuss in Chapter 8.

Short, repeated sequences are often found in DNA molecules. Many proteins interact with regions of DNA containing repeated sequences (� Chapter 8) that are repeated in inverse orientation. This type of repeat is called an **inverted repeat**. Inverted repeats give the DNA sequence a twofold symmetry. As shown in Figure 7.6●, nearby inverted repeats can lead to the formation of **stem-**

ATCGTCAGCAGTTCGCCGCTGCTGACAGC
TAGCAGTCGTCAAGCGGCGACGACTGTCG 5′

Inverted repeats

Loop

Stem-loop structure

Stem

● **Figure 7.6** **Inverted repeats and the formation of a stem-loop structure.** (a) Nearby inverted repeats in DNA. The arrows indicate the symmetry around the imaginary axis (dashed line). (b) Formation of stem-loop structures by pairing of complementary bases on the *same* strand. Note how if RNA was formed from either strand of DNA, the RNA could form a single stem-loop.

loop structures. These are short double-helical regions with normal base pairing and antiparallel strands.

The production of stem-loop structures in DNA may not be a widespread process in cells. However, the production of stem-loop structures in *RNA produced from DNA* following transcription is common. Such *secondary structure* (⚭Section 3.5) formed by base pairing within a single strand of nucleic acid is critical to the functioning of transfer RNA (see Section 7.15) and ribosomal RNA (⚭Section 11.5). Moreover, even if a stem-loop does not form in DNA, inverted repeats in DNA are often binding sites for specific DNA-binding proteins that function as transcriptional regulators (⚭Chapter 8).

The Effect of Temperature on DNA Structure

Although the hydrogen bonds between the base pairs are individually very weak (⚭Section 3.1), many such bonds in a long DNA molecule effectively hold the two strands together. There may be millions or even hundreds of millions of hydrogen bonds in a DNA molecule, depending on the number of base pairs present. Recall from Figure 7.3 that each adenine-thymine base pair has two such bonds, while each guanine-cytosine base pair has three. This makes GC pairs stronger than AT pairs.

When isolated from cells and kept near room temperature and at physiological salt concentrations, DNA remains in a double-stranded form. However, if the temperature is raised, the hydrogen bonds will break but the covalent bonds holding a chain together will not, and so the two DNA strands will separate. This process is called *denaturation* (*melting*) and can be measured experimentally because single-stranded and double-stranded nucleic acids differ in their ability to absorb ultraviolet radiation at 260 nm (Figure 7.7●).

DNA with a high percentage of GC pairs melts at a *higher* temperature than a similar-sized molecule with more AT pairs. If the heated DNA is allowed to cool slowly, a process called *annealing*, the double-stranded native DNA can reform. Such a process can be used not only to reform native DNA but also to form *hybrid* molecules whose two strands come from different sources. **Hybridization**, the artificial construction of a double-stranded nucleic acid by complementary base pairing of two single-stranded nucleic acids, can be a powerful and useful technique in molecular biology (see Section 7.7).

⯃ 7.2 Concept Check

DNA is a double-stranded molecule that forms a helical configuration and is measured in terms of numbers of base pairs. The two strands in the double helix are antiparallel, but inverted repeats allow for the formation of secondary structure. The strands of a double-helical DNA molecule can be denatured by heat and allowed to reassociate following cooling.

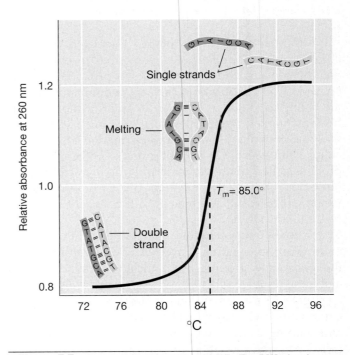

Relative absorbance at 260 nm

Single strands

Melting

$T_m = 85.0°$

Double strand

72   76   80   84   88   92   96
°C

● **Figure 7.7** **Thermal denaturation of DNA.** The DNA absorbs more ultraviolet radiation at 260 nm as the double helix is denatured. The transition is quite abrupt, and the temperature of the midpoint, T_m, is proportional to the GC content of the DNA. Although the denatured DNA can be renatured by a slow cooling, the process does not follow a similar curve. Renaturation becomes progressively more complete at temperatures well below the T_m, and then only after a considerable incubation time.

◆ Explain what the word *antiparallel* means as regards the structure of double-stranded DNA.

◆ Define the term *complementary* as it is used to refer to two strands of DNA.

◆ Define the terms, *denaturation*, *reannealing*, and *hybridization* as they apply to nucleic acids.

7.3 DNA Structure: Supercoiling

A *relaxed* DNA molecule is one with exactly the number of turns of the helix one would predict by knowing the number of base pairs. An example is shown in Figure 7.8●. However, if linearized, the *Escherichia coli* chromosome would be more than 1 mm in length, about 400 times longer than the *E. coli* cell itself! How is it possible to pack so much DNA into such a little space? The solution is the imposition of a kind of "higher order" structure on the DNA, in which the double-stranded DNA is further twisted in a process called *supercoiling*.

Figure 7.8 diagrams how supercoiling occurs in a circular DNA duplex. Supercoiling puts the DNA molecule under torsion, much like the added tension to a rubber band that occurs when it is twisted. DNA can be supercoiled in either a *positive* or *negative* direction. **Negative supercoiling** occurs when the DNA is twisted about its axis in the *opposite* direction from that of the right-handed double helix. It is in this form that supercoiled DNA is predominantly found in nature, with some interesting exceptions among the *Archaea* (see below)

Although there are proteins associated with the chromosomes of prokaryotes, compared to eukaryotes one can consider the chromosomes of prokaryotes to be "naked" DNA. In eukaryotes, large amounts of protein are bound to the DNA in a very regular fashion. As has been mentioned (see Section 7.1), each eukaryotic chromosome contains a *linear* double-stranded DNA molecule. This linear DNA molecule is wound around proteins called **histones**, to form structures called **nucleosomes** (Figure 7.9●). In eukaryotic chromosomes, the formation of the nucleosome introduces negative supercoils into the DNA.

Histones are positively charged proteins that can neutralize the negative charge of DNA (resulting from the phosphate groups). The nucleosomes are spaced along the double helix at very regular intervals, but can aggregate to form a fibrous material called *chromatin*. Chromatin itself can be further compacted by folding and looping to eventually form very dense structures called *chromosomes*. It is these chromosomes that are

(a) Relaxed, covalently closed circular DNA

Break one strand ⇅ Seal

Nick

(b) Relaxed, nicked circular DNA

Break one strand ⇅ Rotate one end of broken strand around helix and seal

(c) Supercoiled circular DNA

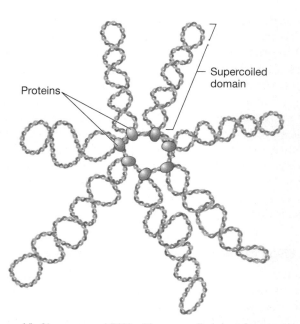

Proteins

Supercoiled domain

(d) Chromosomal DNA with supercoiled domains

● **Figure 7.8 Supercoiled DNA.** Parts (a), (b), and (c) show supercoiled circular DNA and relaxed, nicked circular DNA interconversions. A nick is a break in a phosphodiester bond of one strand. (d) In actuality, the double-stranded DNA in the bacterial chromosome is arranged not in one supercoil but in several *supercoiled domains*, as shown here. In *Escherichia coli* over 50 supercoiled domains are thought to exist, each of which is stabilized by binding to specific proteins.

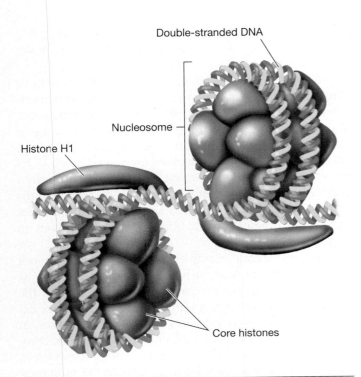

Double-stranded DNA

Nucleosome

Histone H1

Core histones

● **Figure 7.9 Packaging of DNA around a core of histone proteins to form a nucleosome.** Nucleosomes are arranged along the DNA strand somewhat like beads on a string. This arrangement is typical of DNA in eukaryotic cells. In eukaryotes, "core" histones are composed of eight proteins, two copies each of histones H2A, H2B, H3, and H4, and one copy of histone H1. A single nucleosome contains about 200 base pairs of DNA.

most easily visible during cell division in eukaryotic cells (∞Section 14.7).

Topoisomerases: DNA Gyrase

In *Bacteria* and most *Archaea*, there is an enzyme called **DNA gyrase**, which introduces negative supercoils into DNA. The process can be thought to occur in several stages. First, the circular DNA molecule is twisted, then a break occurs where the two chains come together, and then the broken double helix is resealed on the opposite side of the intact strand (Figure 7.10●). DNA gyrase is a type of enzyme called a *topoisomerase*, specifically a topoisomerase II. It is worth noting that some of the antibiotics that affect *Bacteria*, such as the quinolones (e.g., *nalidixic acid*), fluoroquinolones (e.g., *ciprofloxacin*), and *novobiocin*, inhibit the activity of DNA gyrase. Novobiocin is also effective against several species of *Archaea*, where it also seems to inhibit DNA gyrase.

Another enzyme is able to *remove* supercoiling in DNA. This enzyme, called *topoisomerase I*, introduces a single-strand break in the DNA and causes the rotation of one single strand of the double helix around the other. As was shown in Figure 7.8, a break in the backbone (a *nick*) of either strand allows the DNA to return to the relaxed state. However, to prevent the entire bacterial chromosome from becoming relaxed every time a nick is made, the chromosome contains *supercoiled domains* as shown in Figure 7.8*d*. A nick in the DNA in one of these domains does not relax DNA in the others. It is unclear what holds the DNA in these domains, but it is likely to involve specific proteins.

Through the activity of topoisomerases, the DNA molecule can be alternately supercoiled and relaxed. Because supercoiling is necessary for packing the DNA into the cell and relaxing is necessary for DNA to be replicated, these two complementary processes play an important role in the biology of the cell. In most prokaryotes, the extent of negative supercoiling is the result of a balance between the activity of DNA gyrase and topoisomerase I. In addition to replication, however, supercoiling is known to affect *gene expression*. Certain genes are more actively transcribed when DNA is supercoiled, whereas transcription of other genes is inhibited by excessive supercoiling.

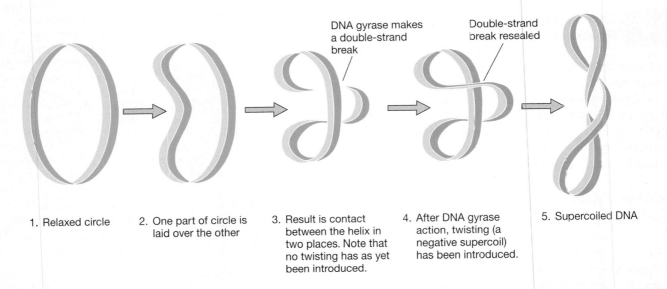

DNA gyrase makes a double-strand break

Double-strand break resealed

1. Relaxed circle

2. One part of circle is laid over the other

3. Result is contact between the helix in two places. Note that no twisting has as yet been introduced.

4. After DNA gyrase action, twisting (a negative supercoil) has been introduced.

5. Supercoiled DNA

● **Figure 7.10 DNA gyrase.** Introduction of supercoiling into a circular DNA by activity of DNA gyrase (topoisomerase II), which makes double-strand breaks. DNA gyrase introduces a *negative* supercoil into DNA.

Archaea and Reverse Gyrase

A few prokaryotes contain an enzyme called *reverse gyrase*. This topoisomerase introduces *positive* supercoils in DNA. The organisms that contain this enzyme are species of *Archaea* that grow at extremely high temperatures (hyperthermophiles, ⮌Section 6.12 and Chapter 13). Interestingly, one example, the hyperthermophile *Methanothermus fervidus*, a member of the *Archaea*, also contains histone-like proteins, and forms nucleosome-like structures in which the DNA is *positively* supercoiled (⮌Section 13.12).

Positive supercoiling is probably why DNA does not heat denature in hyperthermophiles. Supporting this hypothesis is the fact that with rare exception, reverse gyrase is only present in these very heat-loving organisms. However, as in all cells, in order to be of use, the genetic information in DNA must be accessible to the cell's replication and transcriptional machinery. Therefore, whether or not supercoiling is controlled by the activity of DNA gyrase or reverse gyrase, the structure of DNA within the cell likely remains quite dynamic.

 7.3 Concept Check

The very long DNA molecule can be packaged into the cell because it is supercoiled. In prokaryotes this supercoiling is brought about by enzymes called topoisomerases. In eukaryotic chromosomes, DNA is wound around proteins called histones, forming structures called nucleosomes. DNA gyrase is a key enzyme in prokaryotes, introducing negative supercoils to the DNA. Reverse gyrase introduces positive supercoiling.

◆ Why is supercoiling important?

◆ What function do topoisomerases serve inside cells?

◆ How do the activities of DNA gyrase and reverse gyrase differ?

7.4 Chromosomes and Other Genetic Elements

Structures containing genetic material (DNA in most organisms but RNA in some viruses) are called *genetic elements*. The **genome** is the total complement of genes in a cell or virus. Although the main genetic element in prokaryotes is the **chromosome**, other genetic elements are known and play important roles in gene function in both prokaryotes and eukaryotes (Table 7.1).

The Chromosome

In Section 2.2 we discussed the fact that a typical prokaryote has a single chromosome containing all (or most) of the genes found inside the cell. Although a single chromosome is the rule among prokaryotes, there are exceptions. A few prokaryotes, both species of *Bacteria* and *Archaea*, contain two chromosomes (Table 7.2). Eukaryotes have multiple chromosomes as a part of their genome. Also, the typical prokaryotic chromosome is a circular DNA molecule, whereas the DNA in all known eukaryotic chromosomes is linear.

In Table 7.2 the number, size, and configuration of chromosomes in a few microorganisms, both prokaryotic and eukaryotic, are given. Note that the chromosome of the bacterium *Borrelia burgdorferi*, the causative agent of Lyme disease (⮌Section 27.4), is linear. Although uncommon, linear chromosomes are now known to be present in several other *Bacteria*, including the important antibiotic-producing genus *Streptomyces* (⮌Sections 12.24 and 30.5). More detailed information on the sizes and characteristics of microbial genomes is given in Chapter 15.

The prokaryotic examples listed in Table 7.2 are not random. They include species of *Archaea* as well as *Bacteria* and examples of the smallest and some of the largest known prokaryotic chromosomes. Only a few examples of eukaryotic microorganisms are also given in Table 7.2. The linear DNA molecule in a eukaryotic chromosome has a special DNA sequence called a *telomere* at each end and a *centromere* located between the telomeres. Centromeres are important for partitioning the chromosomes during cell division. As we shall see in Chapter 14 (Eukaryotic Microorganisms), telomeres play an important role in the replication of linear DNA molecules (⮌Section 14.6 and Figure 14.12).

Eukaryotes typically contain much more DNA than is needed to encode all the proteins required for cell function. For instance, in the human genome only about 3% of the total DNA encodes protein, whereas in *Bacteria* this fraction often exceeds 90%. The "extra" DNA in eu-

Table 7.1	**Kinds of genetic elements**	
Organism	**Element**	**Description**
Prokaryote	Chromosome	Extremely long, usually circular, double-stranded DNA molecule
	Plasmid	Typically a relatively short, usually circular, double-stranded DNA molecule, which is extrachromosomal
Eukaryote	Chromosome	Extremely long, linear, double-stranded DNA molecule
	Plasmid[a]	Typically a relatively short circular or linear double-stranded DNA molecule, which is extrachromosomal
All Organisms	Transposable elements	Double-stranded DNA molecule always found within another DNA molecule
Mitochondrion or chloroplast	Chromosome	Intermediate-length DNA molecules, usually circular
Virus	Genome	Single- or double-stranded DNA or RNA molecule

[a]Plasmids are uncommon in eukaryotes.

Table 7.2 Sizes, shapes, and numbers of chromosomes in selected microorganisms from each domain of life

Organism	Comments	Chromosome Size (Mbp)[a]	Chromosome Number	Chromosome Geometry
Bacteria				
Mycoplasma genitalium	Smallest known cellular genome in Bacteria	0.58	1	
Borrelia burgdorferi	Causes Lyme disease (Chapter 27)	0.91[b]	1	
Haemophilus influenzae	Gram-negative, can cause disease (Chapter 26)	1.83[c]	1	
Rhodobacter sphaeroides	Gram-negative, phototrophic	4.00[d]	2	
Bacillus subtilis	Gram-positive, genetic model	4.21	1	
Escherichia coli K-12	Gram-negative, genetic model	4.64[e]	1	
Streptomyces coelicolor	Actinomycete, produces antibiotics (Chapter 12)	8.66	1	
Archaea				
Nanoarchaeum equitans	Parasite of Ignicoccus (Chapters 13 and 15)	0.49	1	
Methanococcus jannaschii	Methanogen, which grows at high temperature (Chapters 6 and 13)	1.66	1	
Pyrococcus abyssi	Grows at high temperature (Chapters 6 and 13)	1.77	1	
Halobacterium sp. NRC1	Grows in high salt (Chapters 6 and 13)	2.57[f]	3	
Sulfolobus solfatarius	Grows at high temperature and high acidity (Chapters 6 and 13)	2.99	1	
Eukarya[g]				
Giardia lamblia	Flagellated protozoan that causes acute gastroenteritis (Chapter 14)	12.00	4	
Saccharomyces cerevisiae	Yeast, widely used in science and industry (Chapters 14 and 30)	12.06[h]	16	
Dictyostelium discoideum	Cellular slime mold, developmental model (Chapter 14)	34.0	6	
Tetrahymena thermophila	Ciliated protozoan (Chapter 14)	210.0	5	

[a]Mbp is megabase pairs. In the case of the Eukarya the genome sizes and chromosome number are for the haploid form. (All the prokaryotic genomes listed have been completely sequenced.)

[b]This is for the linear chromosome. The genome of this organism also contains at least 17 circular and linear plasmids, which themselves have a combined size of more than 0.5 Mbp.

[c] Haemophilus influenzae strain Rd was the first cellular organism to have its genome entirely sequenced.

[d]Chromosome I is 3.1 Mbp and chromosome II is 0.90 Mbp. The strain sequenced also contains five plasmids.

[e]The reported sequence does not contain that of the F-plasmid (Section 10.9) nor that of bacteriophage lambda (Section 9.11), both of which would be present in a typical K-12 strain.

[f]There is one large chromosome of 2.01 Mbp and two minichromosomes of 0.19 Mbp and 0.37 Mbp.

[g]All the organisms listed are single-celled.

[h] Saccharomyces cerevisiae was the first eukaryote to have its genome completely sequenced. The number given here does not include the mitochondrial genome and all the copies of some repeated sequences.

karyotes is present as introns (see Figure 7.2) or repetitive sequences, some repeated hundreds of thousands of times. Eukaryotic microorganisms have fewer introns than do "higher" eukaryotes. For example, about 70% of DNA in the yeast Saccharomyces cerevisiae encodes protein. Eukaryotes also often have multiple copies of certain genes, such as those that encode transfer RNAs and ribosomal RNAs. These latter may also be found repeated in prokaryotes but usually only a few copies are present at most.

Nonchromosomal Genetic Elements

Besides the chromosome, a number of other genetic elements are known, referred to collectively as *nonchromosomal genetic elements*. These include *viruses, plasmids, organellar genomes*, and *transposable elements*.

Viruses contain genomes, either DNA or RNA, that control their own replication and transfer from cell to cell. The viral genome is also called a *chromosome*. However, it contains genes essential to the virus, not the host cell, and is therefore clearly functionally distinct from cellular chromosomes. Both linear and circular viral chromosomes are known. Viruses are of special interest because they often cause disease. We discuss viruses in Chapters 9 and 16, and viral diseases in Chapters 26 and 27.

Plasmids are genetic elements that exist and replicate separately from the chromosome. The great majority of plasmids are double-stranded DNA, and although most plasmids are circular, some are linear. Plasmids differ from viruses in two ways: (1) they do not cause cellular damage (generally they are beneficial), and (2) plasmids do not have extracellular forms, whereas viruses do. Although plasmids have been recognized in only a few eukaryotes, they have been found in most prokaryotic species and can be of profound importance to the biology of the organism. We discuss plasmids in Chapter 10.

Many prokaryotes contain one or more plasmids in addition to their chromosome. Some plasmids contain genes whose protein products confer important properties on the host cell, such as resistance to antibiotics. What is the difference then, between a large plasmid and a chromosome? A chromosome is a genetic element that contains genes whose products are involved in *essential* cellular functions. Such essential genes are referred to as *housekeeping genes*. Some housekeeping genes encode essential proteins, such as DNA and RNA polymerases, while others encode essential RNAs, such as ribosomal and transfer RNAs. By contrast to the chromosome, plasmids are often expendable and do not contain genes necessary for growth under all conditions. There are many genes on a chromosome that are unessential, as well, but the presence of essential genes is necessary for a genetic element to be called a *chromosome*.

The major eukaryotic cell organelles are the **mitochondrion** and the **chloroplast**. Each of these contains a small chromosome. However, these organelles contain not only DNA, but also a complete machinery for protein synthesis, including ribosomes, transfer RNAs, and all the other components necessary for translation and formation of functional proteins. Indeed, these organelles share many traits in common with prokaryotic cells, to which they are phylogenetically related (∞Sections 2.3 and 14.4)

Transposable elements are molecules of DNA that can move from one site on a chromosome to another. Transposable elements are found in prokaryotes and eukaryotes and play important roles in genetic variation. In prokaryotes there are three types of transposable elements: insertion sequences, transposons, and some special viruses.

Insertion sequences are the simplest type of transposable element and carry no genetic information other than that required for them to move about the chromosome. *Transposons* are larger and contain other genes. We discuss both of these in more detail in Chapter 10. In Chapter 16 we discuss a bacterial virus, Mu, that is itself a transposable element. The unique feature common to all transposable elements is that *they all replicate as part of some other molecule of DNA.*

7.4 Concept Check

In addition to the chromosome, a number of other genetic elements exist in cells. Plasmids are DNA molecules that exist separately from the chromosome of the cell. Mitochondria and chloroplasts contain their own DNA chromosomes. Viruses contain a genome, either DNA or RNA, that controls their own replication. Transposable elements exist as a part of other genetic elements.

◆ What is a *genome*?

◆ What genetic material is found in all cellular chromosomes?

◆ What defines a chromosome in prokaryotes?

 III DNA REPLICATION

The flow of biological information begins with **DNA replication**. DNA replication is necessary for cells to divide, whether to reproduce new organisms, as in unicellular microorganisms, or to produce new cells as part of a multicellular organism. The process of DNA replication must occur at high fidelity such that a daughter cell is genetically identical to the mother cell. This involves a host of special cell enzymes and processes.

7.5 DNA Replication: Templates and Enzymes

DNA exists in cells as a double helix with complementary base pairing (see Figures 7.3 and 7.4). If the DNA double helix is opened up, a new strand can be synthesized as the *complement* of each of the parental strands. As shown in Figure 7.11●, replication is thus **semiconservative**, meaning that the two resulting double helices consist of one newly synthesized strand and one parental strand. The DNA molecule that is copied is called the *template*, and in DNA replication, each parental strand is a template for one newly synthesized (and complementary) strand (Figure 7.11).

The chemistry of DNA, the nature of its precursors, and the activities of the enzymes involved in replication place some important restrictions on the manner in which new strands are synthesized. The precursor of each new nucleotide in the chain is a nucleoside 5'-*tri*phosphate, from which the two terminal phosphates are removed and the internal phosphate is attached covalently to deoxyribose of the growing chain (Figure 7.12●). The addition of the nucleotide to the growing chain requires the presence of a free hydroxyl group, and such is available only at the 3'-end of the

● **Figure 7.11 Overview of DNA replication.** DNA replication is a semiconservative process in both prokaryotes and eukaryotes. Note that the new double helices each contain one new and one old strand.

molecule. This restriction leads to an important principle that is at the basis of DNA replication: *DNA replication always proceeds from the 5' end to the 3'-hydroxyl end, the 5'-phosphate of the incoming nucleotide being attached to the 3'-hydroxyl of the previously added nucleotide.*

DNA Polymerase and Primase

The enzymes that catalyze addition of the nucleotides are called **DNA polymerases**. Several such enzymes exist and each has a specific function. All known DNA polymerases synthesize DNA in the 5' → 3' direction. However, no known DNA polymerase can synthesize DNA *de novo*; all of these enzymes can only add a nucleotide onto a preexisting 3'-OH group. Therefore, for a *new* chain to be started, there must be a **primer**, a nucleic acid molecule to which DNA polymerase can attach the first nucleotide. In most cases this primer is a short stretch of *RNA*.

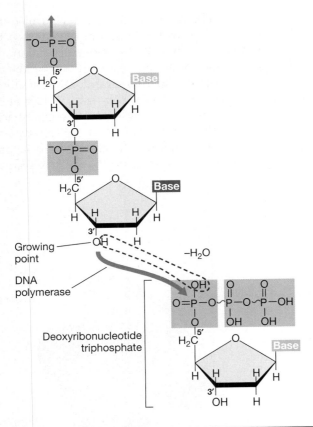

● **Figure 7.12 Structure of the DNA chain and mechanism of growth by addition of a deoxyribonucleoside triphosphate at the 3'-end of the chain.** Growth proceeds from the 5'-phosphate to the 3'-hydroxyl end. The enzyme DNA polymerase catalyzes the addition reaction. The four deoxyribonucleotides that serve as precursors are deoxythymidine triphosphate (dTTP), deoxyadenosine triphosphate (dATP), deoxyguanosine triphosphate (dGTP), and deoxycytidine triphosphate (dCTP). Upon nucleotide insertion, the two terminal phosphates of the triphosphate are split off as pyrophosphate (PP_i). Thus, two energy-rich phosphate bonds are consumed in the addition of a single nucleotide.

● **Figure 7.13 The RNA primer.** Structure of the RNA–DNA combination that results at the initiation of DNA synthesis.

When the double helix is opened up at the beginning of replication, an RNA-polymerizing enzyme catalyzes formation of the RNA primer. This enzyme, called *primase*, participates in primer synthesis by synthesizing a short (<15 nucleotides) stretch of RNA complementary in base pairing to the template DNA. At the growing end of this RNA primer is a 3'-OH group to which DNA polymerase can add the first deoxyribonucleotide. Therefore, continued extension of the molecule occurs as *DNA* rather than RNA. Thus, the newly synthesized molecule has a structure like that shown in Figure 7.13●. The primer must eventually be removed and replaced with DNA, and we consider this later.

7.5 Concept Check

Both strands of the DNA helix serve as templates for the synthesis of two new strands (semiconservative replication). The two progeny double helices each contain one parental strand and one new strand. The new strands are elongated by addition to the 3'-end. DNA polymerases require a primer, which is composed of RNA.

◆ To which end (5' or 3') of a newly synthesized strand of DNA does polymerase add a base?

◆ Why is a primer required for DNA replication?

7.6 DNA Replication: The Replication Fork

Much of the information on the mechanism of DNA replication was obtained from the bacterium *Escherichia coli*, and the following discussion deals primarily with this organism. However, the process of DNA replication is probably quite similar in all *Bacteria*. By contrast, in *Archaea*, although most species have circular chromosomes, the general series of events in DNA replication resembles the process in eukaryotic cells more than it does that in *Bacteria*. This again is a reflection of the phylogenetic affiliation between *Archaea* and *Eukarya* (∞Section 2.3).

Initiation of DNA Synthesis and Leading/Lagging Strands

In prokaryotes, there is a single location on the chromosome where DNA synthesis is initiated, the **origin of replication**. The origin of replication consists of a specific sequence of about 300 bases that is recognized by specific initiation proteins, including a key initiation protein (see Table 7.3) that binds to this region and opens up the helix to expose the strands to the replication machinery. Initiation of DNA replication then begins on the two single strands. As replication proceeds, the site of replication, called the **replication fork**, appears to move down the DNA (Figure 7.14●).

There are five different DNA polymerases in *Escherichia coli*, called *DNA polymerases* I, II, III, IV, and V. It is *DNA polymerase III* (Pol III) that is the primary enzyme of replication at the replicating forks. However, several other non-polymerase enzymes are also involved (Table 7.3). For example, at the replication fork, the DNA double helix is unwound and a small single-stranded region is exposed by the activity of specific proteins called *helicases*. Helicases are ATP-dependent enzymes that move down the helix and separate the strands in advance of the replicating fork (Figures 7.14 and 7.15●). The single-stranded region generated is complexed with a special protein, the *single-strand binding protein*, which stabilizes the single-stranded DNA, preventing the formation of intrastrand hydrogen bonds and reversion to the helix.

Note in Figure 7.14 an important distinction in replication of the two strands, which arises from the fact that DNA replication always proceeds from 5'-phosphate to 3'-hydroxyl (always adding a *new* nucleotide to the 3'-OH of the growing chain). On the strand growing

Table 7.3	Major enzymes involved in DNA replication in *Bacteria*	
Enzyme	**Encoding genes**	**Function**
DNA polymerase III	*polC; dnaE,Q,N,X; holA–E; mutD*	Main polymerizing enzyme
DNA polymerase I	*polA*	Excises RNA primer and fills in gaps
Helicase	*dnaB*	Unwinds helix at the replication fork
Primase	*dnaG*	Primes new strands of DNA
Origin-binding protein	*dnaA*	Binds to origin of replication site; facilitates melting to open the complex
Single-strand binding protein	*ssb*	Prevents opened helix from annealing
DNA ligase	*ligA, lig B*	Seals nicks in DNA

from the 5'-phosphate to the 3'-hydroxyl, called the **leading strand**, DNA synthesis occurs *continuously* because there is always a free 3'-OH at the replication fork to which a new nucleotide can be added. But on the opposite strand, called the **lagging strand**, DNA synthesis occurs *discontinuously* (because there is no 3'-OH at the replication fork to which a new nucleotide can attach). Where is the 3'-OH on this strand? At the *opposite* end, *away* from the replication fork. Therefore, on the lagging strand, an RNA primer must be synthesized by primase

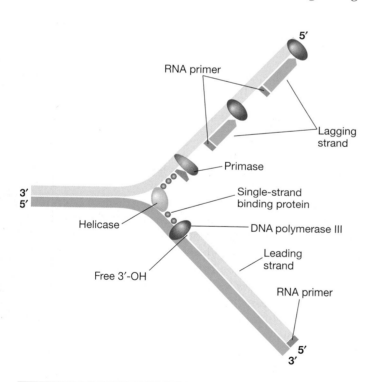

● **Figure 7.14 Events at the DNA replication fork.** Note the polarity and antiparallel nature of the DNA strands. The substrates for primase are ribonucleotide triphosphates, while for DNA polymerase, they are *deoxy*ribonucleotide triphosphates. Single–strand binding protein keeps the template strands from reforming a helix following their separation by helicase. Compare with Figure 7.19, which shows the replisome protein complex.

● **Figure 7.15 DNA helicase unwinding a double helix.** In this figure, the protein and DNA molecules are drawn to scale. Simple diagrams are often used to give an idea of molecular events in the cell, but they may give the incorrect impression that most proteins are relatively small compared to DNA. This scale model shows otherwise.

multiple times to provide free 3′-OH groups. By contrast, the leading strand is primed only once, at the origin.

After synthesizing the primer, primase is replaced by Pol III. This enzyme is a complex of ten proteins, including the polymerase core complex itself. Other proteins include those that function as a sliding clamp to maintain the polymerase on each template strand of DNA as well as proteins that assist in replisome (see later) assembly. Pol III adds deoxyribonucleotide triphosphates until it reaches previously synthesized DNA (Figure 7.16●). At this point, Pol III stops. The next enzyme that is involved, *DNA polymerase I* (Pol I), has more than one enzymatic activity. Besides synthesizing DNA, Pol I has a 5′ → 3′ *exonuclease* activity that removes the RNA primer preceding it (Figure 7.16). When the primer has been removed and replaced with DNA, Pol I is released. The last phosphodiester bond is made by an enzyme called *DNA ligase*.

This enzyme can seal nicks in DNAs that have a 5′-phosphate and 3′-OH, and along with Pol I, is also involved in DNA repair.

Bidirectional Chromosome Replication and the Replisome

In *Escherichia coli*, and probably in all prokaryotes that contain a circular chromosome, replication is *bidirectional* from the origin of replication, as shown in Figures 7.17● and 7.18●. There are thus *two* replication forks on each chromosome, replicating in opposite directions. In circular DNA, bidirectional replication leads to the formation of characteristic structures called *theta structures* (Figure 7.17). Most large DNA molecules, whether from prokaryotes or eukaryotes, have bidirectional replication from fixed origins. In fact, a single eukaryotic chromosome has many origins. Close inspection of the replicating DNA shows that synthesis is occurring in both a leading and lagging manner *on each template strand* (Figure 7.18).

Bidirectional DNA synthesis along a circular chromosome allows DNA replication to occur as rapidly as

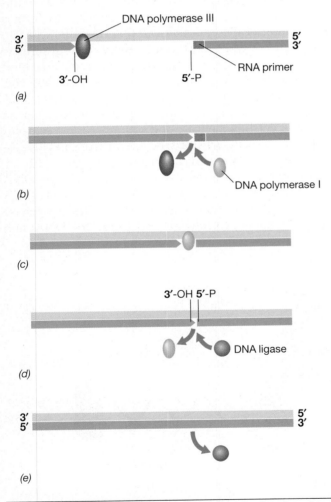

(a)

(b)

(c)

(d)

(e)

● **Figure 7.16 Sealing two fragments on the lagging strand.** (a) DNA polymerase III is synthesizing DNA in the 5′ → 3′ direction toward the RNA primer of a previously synthesized fragment on the lagging strand. (b) On reaching the fragment, DNA polymerase I replaces III. (c) DNA polymerase I continues synthesizing DNA while removing the RNA primer from the previous fragment. (d) DNA ligase replaces DNA polymerase I after the primer has been removed. (e) DNA ligase seals the two fragments together.

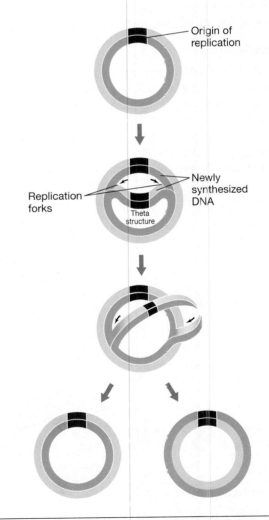

● **Figure 7.17 Replication of circular DNA: the theta structure.** In circular DNA, bidirectional replication from an origin leads to the formation of replication intermediates resembling the Greek letter theta (Θ).

Figure 7.18 Dual replication forks in the circular chromosome. At an origin of replication that directs bidirectional replication, two replication forks must start. Therefore, two leading strands must be primed, one in each direction. In *Escherichia coli*, the origin of replication is identified by the binding of a specific protein, Dna A (see Table 7.3).

possible. Even taking this into account and considering that Pol III can add nucleotides to a growing DNA strand at the rate of about 1000 per second, chromosome replication in *E. coli* still takes about 40 minutes. Under the best growth conditions, *E. coli* can grow with a doubling time of about 20 minutes. However, under these conditions, chromosome replication still takes 40 minutes. The solution to this conundrum is that cells of *E. coli* growing at doubling times shorter than 40 minutes contain *multiple DNA replication forks.* That is, a new round of DNA replication begins before the last round has been completed. Only in this way can a generation time shorter than the chromosome replication time be maintained.

The Replisome

While DNA synthesis is continuing at the replication fork, changes in the coiling of the DNA are occurring, modified by the activities of topoisomerases (see Section 7.3). Unwinding is an essential feature of DNA replication, and because supercoiled DNA is under tension, it unwinds more easily than DNA that is not supercoiled. Thus, by regulating the degree of supercoiling, topoisomerases

regulate the process of replication (and also transcription, as discussed later).

Figure 7.14 shows the differences in replication of the leading and the lagging strands and the various enzymes involved. From such a simplified drawing it would appear that each replication fork contains a host of different proteins all working independently. Actually, this is not the case. The proteins involved in this process are highly dynamic and aggregate to form a large complex called the *replisome* (Figure 7.19●). The lagging strand of DNA loops out to allow the replisome to move smoothly on both strands, and the replisome literally pulls the DNA template through it as replication occurs (Figure 7.19). Therefore, it is the *DNA*, and not DNA polymerase, that moves during replication. Note also in the replisome how helicase and primase form a complex (called the *primosome*) since they must work in close association during the replication process (Figure 7.19).

In summary, the replisome contains, besides DNA polymerase, several key replication proteins: (1) DNA gyrase, to remove supercoils; (2) DNA helicase/primase (the primosome), to unwind and prime the DNA; and

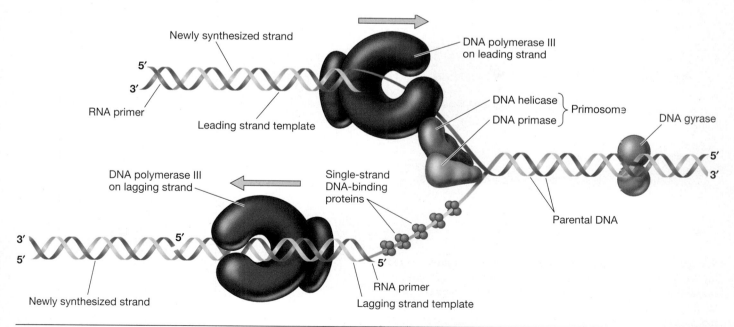

● **Figure 7.19 The replisome.** The replisome consists of two copies of DNA polymerase III, helicase and primase (together forming the primosome), and many copies of single-strand DNA-binding protein. Just upstream of the replisome DNA gyrase is removing supercoils in the DNA to be replicated. Note how the lagging-strand template loops out to keep the components of the replisome away from each other and moving forward smoothly.

(3) Single-strand binding protein, to prevent the separated template strands from re-forming a double helix (Figure 7.19). Table 7.3 summarizes the properties of the essential proteins involved in DNA replication.

Fidelity of DNA Replication: Proofreading

Errors in DNA replication introduce *mutations*, changes in DNA sequence. Mutation rates in cells are remarkably low, between 10^{-8} and 10^{-11} errors per base pair inserted. This accuracy is possible partly because DNA polymerases get *two* chances to incorporate the correct base at a given site. The first chance occurs when complementary bases are inserted by base-pairing rules, A with T and G with C, using the template strand as the pattern. The second chance occurs because of a second enzymatic activity of both Pol I and III, called **proofreading** (Figure 7.20●).

How does proofreading work? In addition to inserting nucleotides in the replicating strand, Pol I and III also contain a $3' \rightarrow 5'$ *exonuclease* (*exo* means "end") activity that can remove a misinserted nucleotide and replace it with the correct nucleotide. Proofreading activity occurs if an incorrect base has been inserted because misinsertion creates a mismatch in base pairing. The polymerase is signaled to this problem by virtue of the fact that a misinserted nucleotide is unable to form the correct (most stable) hydrogen-bonding pattern with its mismatched complement. This proofreading activity gives the polymerase activity a second chance to insert the correct base (Figure 7.20).

Proofreading exonuclease activity is distinct from the $5' \rightarrow 3'$ exonuclease activity of Pol I that is used to remove the RNA primer from both the leading and lagging strands (Figure 7.16). Only Pol I has this latter activity. Exonuclease proofreading occurs in prokaryotes, eukaryotes, and viral DNA replication systems. However, many organisms have other mechanisms for reducing errors made during DNA replication, and we will discuss some of these in Chapter 10.

Termination of Replication

Details about the process of termination of the replicating forks are not completely known. However, it is clear that certain DNA sequences and several proteins are involved in this process. When the replication of the circular molecule is complete (Figure 7.17), the two circular molecules are linked together, much like the links of a chain. These can be unlinked by a topoisomerase. In Chapter 6 we outlined how cell wall synthesis and DNA replication is coupled with cell division (∞Section 6.2). Obviously, it is critical that, after DNA replication, the DNA is partitioned so that each daughter cell has a copy of the chromosome. This process may be assisted by the important cell division protein *FtsZ*, a critical protein that helps orchestrate several key events in the cell division process (∞Section 6.2).

 7.6 Concept Check

DNA synthesis begins at a unique location called the origin of replication. The double helix is unwound by helicase and is stabilized by single-strand binding protein. Extension of the DNA occurs continuously on the leading strand but discontinuously on the lagging strand. Most errors in base pairing are corrected by proofreading functions associated with the activities of DNA polymerases.

◆ Why are there *leading* and *lagging* strands?

◆ What is the *replisome* and what are its components?

◆ Why is proofreading so important to the cell?

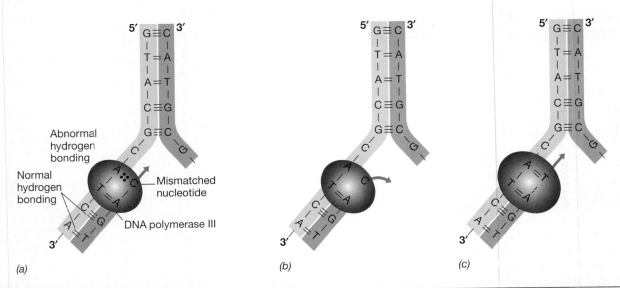

● **Figure 7.20 Proofreading by the $3' \rightarrow 5'$ exonuclease activity of DNA polymerase III.** (a) A mismatch in base pairing at the terminal base pair causes the polymerase to pause briefly. This is a signal for the proofreading activity (b) to excise the mismatched nucleotide, after which the correct base is incorporated (c) by polymerase activity.

IV TOOLS FOR MANIPULATING DNA

7.7 Restriction Enzymes and Hybridization

Prokaryotic cells typically contain one or more enzymes that can chemically modify DNA. A major class of such enzymes are the *restriction endonucleases*, or **restriction enzymes** for short. Although such systems are widespread among prokaryotes (both *Bacteria* and *Archaea*), they are very rare in eukaryotes. Restriction enzymes recognize certain sequences of DNA and cut the DNA. Besides playing important roles in the cells, restriction enzymes are essential for *in vitro* DNA manipulation, and their discovery gave birth to the field of *genetic engineering* (∞Chapter 31).

Mechanism of Restriction Enzymes

A major class of restriction endonucleases are the *type II restriction enzymes*. Most of the DNA sequences recognized by type II restriction enzymes exhibit twofold symmetry around a given point. Figure 7.21● shows the sequence that is recognized and cleaved by the restriction endonuclease from *Escherichia coli* called **Eco**RI (this acronym stands for *Escherichia coli*, **r**estriction enzyme I). The cleavage sites are indicated by arrows and the axis of symmetry by a dashed line. Note that the two strands have the same sequence if one is read from the left and the other from the right (or, in terms of polynucleotide strands, if both are read $5' \rightarrow 3'$ or both read $3' \rightarrow 5'$). Such a sequence pattern is called a **palindrome**. (The term *palindrome* is derived from the Greek meaning "to run back again.")

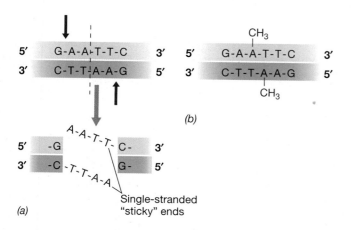

(b)

(a)

● **Figure 7.21 Restriction and modification of DNA.** (a) (Top panel) The sequence of DNA recognized by the restriction endonuclease *Eco*RI. The red arrows indicate the bonds cleaved by the enzyme. The dashed line indicates the axis of symmetry of the sequence. (Bottom panel) Appearance of DNA after cutting with restriction enzyme *Eco*R1. Note the single-stranded "sticky ends" (∞Section 10.15). (b) The same sequence after modification by the *Eco*RI methylase. The methyl groups added by this enzyme are shown and these protect the restriction site from cutting by *Eco*R1.

Most restriction enzymes are homodimeric proteins composed of two identical polypeptide subunits, and each subunit recognizes and cuts the DNA on one of the strands. Since the sequences recognized by restriction enzymes are typically short and palindromic, such enzymes typically make *double-stranded* breaks, leaving short, single-stranded regions of DNA free instead of severing the DNA and yielding blunt ends (Figure 7.21). The recognition sequences and cut sites for a few restriction enzymes are given in Table 7.4.

Well over 2000 restriction enzymes with over 200 different specificities are known. Others are being sought because these enzymes are of great importance in genetic research. Consider that the enzyme *Eco*RI can cut *any* double-stranded DNA that has its recognition sequence and cuts only at that sequence. For example, a specific 6 base-pair sequence, such as that cut by the restriction enzyme *Eco*RI (Figure 7.21), should appear in a genome about once every 4096 nucleotides (four bases taken six at a time, 4^6) assuming that the DNA has a "random" sequence. As we will explain in Chapters 10 and 31, such fragments with defined termini have many uses, especially in gene-cloning technologies. Restriction enzymes are such important tools in modern molecular genetic research that they have become widely available commercially.

Table 7.4	Recognition sequences of a few restriction endonucleases	
Organism	**Enzyme designation**[a]	**Recognition sequence**[b]
Bacillus globigii	*Bgl*II	A↓GATCT
Bacillus subtilis	*Bsu*RI	GG↓CC
Brevibacterium albidum	*Bal*I	TGG↓CCA
Escherichia coli	*Eco*RI	G↓AATTC[c]
Haemophilus haemolyticus	*Hha*I	GCG↓C
Haemophilus influenzae	*Hind*II	GTPy↓PuAC
Haemophilus influenzae	*Hind*III	A↓AGCTT
Klebsiella pneumoniae	*Kpn*I	GGTAC↓C
Nocardia otitidiscaviarum	*Not*I	GC↓GGCCGC
Proteus vulgaris	*Pvu*I	CGAT↓CG
Serratia marcescens	*Sma*I	CCC↓GGG
Thermus aquaticus	*Taq*I	T↓CGA

[a] Nomenclature: The first letter of the three letter abbreviation of a restriction endonuclease designates the genus from which the enzyme originates, the second two letters, the species. The roman numeral designates the order of discovery of enzymes in that particular organism, and any additional letters are strain designations.

[b] Arrows indicate the sites of enzymatic attack. Asterisks indicate the site of methylation (modification). G, guanine; C, cytosine; A, adenine; T, thymine; Pu, any purine; Py, any pyrimidine. Only the $5' \rightarrow 3'$ sequence is shown.

[c] See Figure 7.21a

Modification: Protection from Restriction

The major function of restriction enzymes in prokaryotes is probably to protect the cell from invasion by foreign DNA, for example, viral DNA. If such DNA enters the cell, the cell's restriction enzyme(s) will destroy it. However, a cell must protect its own DNA from inadvertent destruction by its own restriction enzyme(s), and such protection is conferred by a **modification system**. This consists of a modifying enzyme, which chemically **modifies** the specific sequences on its *own* DNA so these sequences are not attacked by the cell's own restriction enzymes. Such modification typically involves *methylation* of specific bases within the recognition sequence so that the restriction nuclease can no longer bind. For example, the sequence recognized by the *Eco*RI restriction enzyme (Figure 7.21*a*) can be modified by methylation of the two most interior adenines (Figure 7.21*b*). The enzyme that performs this modification is called *Eco*RI *methylase*. If even a single strand is modified, the sequence is no longer a substrate for the restriction enzyme *Eco*RI.

Electrophoresis and Restriction Analysis of DNA

Because the base sequences recognized by many restriction enzymes are four to six nucleotides long (Table 7.4), there will generally be only a limited number of such sequences in a single DNA molecule (for example, see Figure 7.23). After cleaving the DNA, the fragments generated can be separated from each other by **gel electrophoresis** and analyzed.

Electrophoresis is a procedure by which charged molecules migrate in an electrical field, the rate of migration being determined by the charge on the molecule and by its size and shape. In gel electrophoresis (Figure 7.22*a*●) the molecules are separated in an agarose gel. When an electrical current is applied, nucleic acids (which are negatively charged) will migrate through the gel, and small or compact molecules migrate more rapidly than large molecules. After a certain period of time the gel can be "stained" with a compound that binds to DNA, such as *ethidium bromide* (∞Section 10.3), and the DNA will then fluoresce under ultraviolet light (Figure 7.22*b*).

A typical agarose gel is shown in Figure 7.22*b*. In lane A a restriction enzyme digestion of a standard DNA sample has been applied, and the size of each band is thus known. The other lanes contain a purified source of DNA (in this case a plasmid) cut by one or more restriction enzymes. Because a given restriction enzyme always cuts at the same site, the banding pattern of a given DNA is reproducible. By comparison with the standard, the size of the fragments can therefore be determined. Such a technique can be used to generate what is called a **restriction map** of the DNA. Combining restriction enzyme digestion of a single DNA molecule with fluorescent microscopy allows a restriction map to be made directly

(a)

(b)

● **Figure 7.22 Agarose gel electrophoresis of DNA.** (a) DNA samples are loaded into wells in a submerged agarose gel. (b) A photograph of a stained agarose gel. The DNA has been loaded into wells toward the top of the gel (negative pole) as shown, and the positive pole of the electrical field is at the bottom. The sample in lane A is used as a standard where the size of the fragments was known. Using the standards, one can determine the sizes of the fragments in the other lanes. Bands stain less intensely at the bottom of the gel because the fragments are smaller and chemically there is less DNA to stain.

from the pattern of cuts along the DNA (Figure 7.23●). This has been called *optical mapping*.

Restriction analyses have many uses in microbiology, primarily in comparative studies of two or more DNAs. This includes studies in the classification of microorganisms. For example, the banding patterns generated from restriction analyses of either whole chromosomes or specific genes from a series of organisms can yield an indication of their genetic relationships (∞Section 11.11). Multiple restriction digests, either using different enzymes one at a time or in combination, can generate patterns that allow one to readi-

● **Figure 7.23 Optical mapping of restriction fragments.** The figure shows a digital fluorescence micrograph of a portion of the *Escherichia coli* chromosome digested with the restriction enzyme *Xho*I (isolated from *Xanthomonas holica*). Sites of cutting by the restriction enzyme are indicated by arrows. Although the DNA strand itself is too small to be seen by light microscopy, *fluorescence* of the DNA is visible. The length of the DNA shown here is about 260 kb pairs. Optical mapping has been used to create restriction maps of the complete chromosomes of *E. coli* and other *Bacteria* such as *Deinococcus radiodurans*.

ly interpret the relative relationships between different DNA molecules.

Nucleic Acid Hybridization and the Southern Blot

DNA fragments can be purified from gels and used for many purposes including *nucleic acid hybridization*, or **hybridization** for short. Recall from Section 7.2 that when DNA is denatured (i.e., made single-stranded), it can be used to form hybrid molecules with other single-stranded DNA (or RNA) molecules having complementary (or almost complementary) base sequences. These latter molecules are called **nucleic acid probes** or, simply, **probes**.

Hybridization can be very useful for finding similar sequences from different genetic elements, or to find the location of a specific gene. Techniques are available for hybridizing probes to DNA fragments that have been separated by gel electrophoresis, and a very common one is the **Southern blot**, named for its inventor, E.M. Southern.

In a Southern blot the DNA fragments in the gel are transferred to a membrane and then denatured to yield single strands. The membrane is then exposed to a labeled probe to allow hybrids to form if the probe is complementary to any of the fragments (the fragments are fixed on the membrane and are thus prevented from reannealing with themselves). Hybridization can be detected by assaying for the label that has become bound to the membrane. Probes can be made radioactive or labeled with agents that yield colored products or that fluoresce. The hybridization procedure when *DNA* is in the gel and RNA or DNA is the probe is called a *Southern blot*. By contrast, when *RNA* is in the gel and DNA or RNA is the probe, the procedure is called a *Northern blot*. Figure 7.24● shows how a Southern blot can be used to identify fragments of DNA containing sequences that hybridize to the probe.

● **Figure 7.24 Southern blotting.** (Left panel) Agarose gel electrophoresis of DNA molecules. Purified molecules of DNA from several different plasmids were treated with restriction enzymes and then subjected to electrophoresis. (Right panel) Southern blot of the DNA gel shown to the left. After blotting, hybridization with a radioactively labeled probe was carried out. The positions of the bands have been detected by X-ray autoradiography. Note that only some of the DNA fragments (circled in yellow) have sequences complementary to the labeled probe. Lane 6 contained DNA used as a size marker and none of the bands hybridized to the probe.

 7.7 Concept Check

Restriction enzymes recognize specific short sequences in DNA and make breaks in the DNA. The products of restriction enzyme digestion can be separated using gel electrophoresis and complementary sequences detected by hybridization.

◆ Why are *restriction enzymes* useful to the geneticist?

◆ What is a *Southern blot* and what does it tell you?

7.8 Sequencing and Synthesizing DNA

In addition to restriction enzyme and hybridization analyses, another common analysis of DNA is **sequencing**, determining the precise order of nucleotides in a DNA molecule. To assist in sequencing and also for hybridization studies, procedures are available to synthesize DNA molecules with defined sequences for use as *primers* and *probes*. Such short fragments of DNA (typically 10–20 nucleotides) are called *oligonucleotides*. Along with restriction enzymes (see Section 7.7), nucleic acid sequencing is a key tool for molecular biology, and we describe the essentials of the sequencing process here.

DNA Sequencing

Most DNA sequencing today is done by the *Sanger dideoxy* method. This method generates DNA fragments that terminate at each of the four bases and that are labeled in some way, for example, with radioactivity. These fragments are then subjected to gel electrophoresis so that molecules with one-nucleotide difference in length are separated on the gel. The procedure requires four separate reactions (and four separate gel lanes) for each sequence determination, one for fragments ending at each of the four bases in DNA: *adenine, guanine, cytosine,* and *thymine.* The positions of the fragments are located by autoradiography (or by fluorescent probes, see later), and the sequences can then be read off the gel.

In the Sanger procedure the sequence is actually determined by making a *copy* of the single-stranded DNA, using the enzyme *DNA polymerase.* As we have previously seen, this enzyme uses deoxyribonucleoside triphosphates as substrates and adds them to a *primer* (∞ Section 7.6). In each of the incubation mixtures (four separate test tubes) are small amounts of a different *dideoxy* analog of the deoxyribonucleoside triphosphates (Figure 7.25●). Because the dideoxy sugar lacks the 3'-hydroxyl, when it is inserted, continuation of the chain cannot occur. The dideoxy analog thus acts as a specific *chain-termination reagent.* Oligonucleotide fragments of variable length are obtained, depending on the incubation conditions. Electrophoresis of these fragments is then carried out, and the positions of the bands are determined by exposing X-ray film or by fluores-

cence. By aligning the four dideoxynucleotide lanes and noting the vertical position of each fragment relative to its neighbor, the sequence of the DNA copy can be read directly from the gel (Figure 7.26●).

RNA and Automated Sequencing

The Sanger method can be used to sequence RNA as well as DNA. To sequence RNA, a single-stranded *DNA copy* is first made (using the RNA as the template) by an enzyme found in certain viruses called *reverse transcriptase* (∞ Section 9.13). By making the single-stranded DNA in the presence of dideoxynucleotides, various-sized DNA fragments are generated suitable for Sanger-type sequencing. From the sequence of the DNA, the RNA sequence is deduced by base-pairing rules.

The demands of large-scale sequencing projects (∞ Chapters 15 and 31) have led to the development of *automated* DNA-sequencing systems. With such systems, the sequencing reactions are still based on the dideoxy method, but fluorescent dye-labeled primers (or nucleotides) are used instead of radioactivity. The products are separated by automated electrophoresis and the bands detected by fluorescence spectroscopy. Each of the four different reactions uses a different fluorescent label, and the lanes are scanned by a fluorescence-detecting laser. In this way, all four reactions can be run on a single lane. The results are analyzed by computer and a sequence generated, with each of the four bases being distinguished by a separate color (Figure 7.26c).

Synthetic DNA

Procedures for making **synthetic DNA** are completely automated so that an oligonucleotide of 30–35 bases can easily be made in a few hours, and oligonucleotides of well over 100 bases in length can be made if necessary. For the synthesis of longer polynucleotides, the oligonucleotide fragments can be joined enzymatically using DNA ligase (see Section 7.6).

DNA is synthesized *in vitro* in a *solid-phase procedure* in which the first nucleotide in the chain is fastened to an insoluble porous support (such as silica gel with particles about 50 μm in size). The overall procedure is shown in Figure 7.27●. Several steps are needed for the addition of each nucleotide, but the chemistry need not concern us here. After each step is completed, the reaction mixture is flushed out of the solid support and the series of reactions repeated for the addition of the next nucleotide. Once the desired length is achieved, the oligonucleotide is cleaved from the solid-phase support by a specific reagent and purified to eliminate by-products and contaminants.

Synthetic DNA molecules are widely used for various purposes. In addition to their use as *primers* for DNA sequencing, they are used as primers for the polymerase chain reaction (see Section 7.9). They are also used as *probes* to detect, via nucleic acid hybridization, specific

Normal deoxynucleotide

Dideoxy analog

(a)

(b)

● **Figure 7.25 Dideoxynucleotides and Sanger sequencing.** (a) A normal deoxynucleotide has a hydroxyl group on the 3' carbon and a dideoxynucleotide does not. (b) Chain termination will occur after incorporation of a dideoxynucleotide. Compare this figure with Figure 7.12.

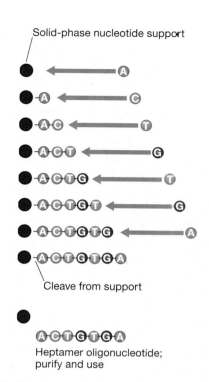

DNA strand to be sequenced

To each tube add one dideoxynucleotide triphosphate (ddGTP, ddATP, ddTTP, or ddCTP).

Reaction products are separated by electrophoresis on polyacrylamide gel and identified by autoradiography.

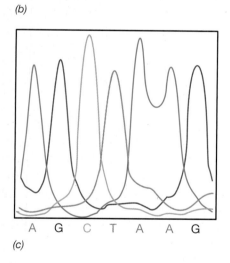

Sequence reads from bottom of gel as A G C T A A G. Sequence of unknown is **3′** T C G A T T C **5′**.

(b)

(c)

● **Figure 7.26 DNA sequencing using the Sanger method.** (a) The template DNA strand is the DNA whose sequence is to be determined. Note that four different reactions must be run, one with each dideoxynucleotide. Also note that since these reactions are run *in vitro*, the primer for DNA synthesis need not be RNA, and for convenience is DNA. (b) A portion of a gel containing the reaction products from (a). (c) Results of sequencing the same DNA as shown in (a) and (b), but using an automated sequencer and fluorescent labels.

Solid-phase nucleotide support

Cleave from support

Heptamer oligonucleotide; purify and use

● **Figure 7.27 Solid-phase procedure for synthesis of a DNA fragment of defined sequence.** Chemical synthesis proceeds by adding one nucleotide at a time to the growing chain.

DNA or RNA sequences in Southern (Figure 7.24) or Northern blots, respectively. We will describe later (see Section 10.18) how synthetic DNA is also used in a procedure called **site-directed mutagenesis** to create mutations of interest at specific locations within a gene.

7.8 Concept Check

DNA can be sequenced by the Sanger method, which involves copying the DNA to be sequenced in the presence of chain-terminating dideoxynucleotides The final products are separated by electrophoresis and the sequence read. The short DNA primers required in this method can be synthesized chemically.

◆ Why do the methods involved in DNA sequencing involve the use of *four* different reactions?

◆ What is the term given to a nucleic acid molecule containing several nucleotides?

7.9 Amplifying DNA: The Polymerase Chain Reaction

The development of synthetic DNA and DNA sequencing spawned a new method for the rapid amplification of DNA *in vitro*, the **polymerase chain reaction (PCR)**. The polymerase chain reaction can multiply DNA molecules by up to a billionfold in the test tube, yielding large amounts of specific genes for a host of applications in molecular biology. PCR makes use of the enzyme *DNA polymerase*, which copies DNA molecules (⟳Section 7.6).

The steps in PCR amplification of DNA can be summarized as follows:

1. Two DNA oligonucleotide primers flanking the target DNA (Figure 7.28a●) are made on an oligonucleotide synthesizer and added in great excess to heat-denatured target DNA.

2. As the mixture cools, the excess of primers relative to the target DNA ensures that most target strands anneal to a primer and not to each other (Figure 7.28a).

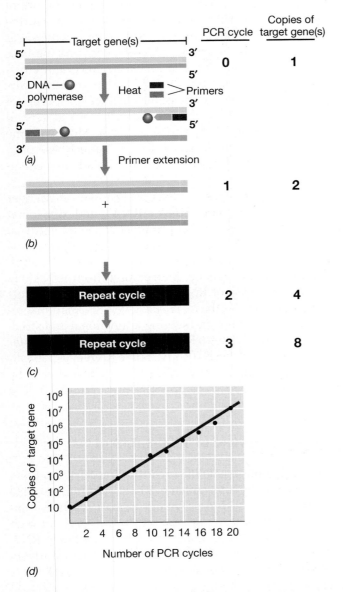

3. DNA polymerase then extends the primers using the target strands as templates (Figure 7.28b).

4. After an appropriate incubation period, the mixture is heated again to separate the strands. The mixture is then cooled to allow the primers to hybridize with complementary regions of newly synthesized DNA, and the whole process is repeated (Figure 7.28c).

The secret to PCR is that the *products* of one primer extension serve as the *template* in the next cycle (Figure 7.28). Indeed, the beauty of the PCR technique is that each cycle literally *doubles* the content of the original target DNA. In practice, 20–30 cycles are usually run, yielding a 10^6- to 10^9-fold increase in the target sequence (Figure 7.28d).

PCR at High Temperature

The original PCR technique employed *Escherichia coli* Pol III, but because of the high temperatures needed to denature the double-stranded copies of DNA being made, the enzyme was also denatured and had to be replenished every cycle. This problem was solved by employing a thermal stable DNA polymerase isolated from the thermophilic hot spring bacterium *Thermus aquaticus*. DNA polymerase from *T. aquaticus*, known as **Taq polymerase**, is stable to 95°C and thus is unaffected by the denaturation step employed in the PCR reaction. The use of *Taq* DNA polymerase also increased the *specificity* of the PCR reaction because the DNA is copied at 72°C rather than 37°C. At high temperatures, nonspecific hybridization of primers to nontarget DNA rarely occurs, thus making the product of *Taq* PCR more homogeneous than that obtained using the *E. coli* enzyme.

DNA polymerase from the hyperthermophile *Pyrococcus furiosus* (growth temperature optimum, 100°C) (∞Sections 6.12 and 13.6), called *Pfu polymerase* (or "vent polymerase" because *P. furiosus* was isolated from a hydrothermal vent, ∞Section 19.8), is also widely used and is even more thermally stable than *Taq* polymerase. Moreover, unlike *Taq* polymerase, *Pfu* polymerase has proofreading activity (see Section 7.6), making it a particularly good enzyme when high accuracy is crucial.

Because a number of highly repetitive steps are involved in the PCR technique, PCR machines, called *thermocyclers*, have been developed that run through the heating and cooling cycles automatically. Because each cycle requires only about 5 minutes, the automated procedure allows for large amplifications in only a few hours. To supply the demand for thermostable DNA polymerases in the PCR and DNA sequencing markets, the genes for these enzymes have been cloned into *E. coli* and produced commercially in large quantities. As a result, the cost of the PCR method is now just a fraction of what it was when the technique was first introduced.

● **Figure 7.28 The polymerase chain reaction (PCR) for amplifying specific DNA sequences.** (a) Target DNA is heated to separate the strands, and a large excess of two oligonucleotide primers, one complementary to the target strand and one to the complementary strand, is added along with DNA polymerase. (b) Following primer annealing, primer extension yields a copy of the original double-stranded DNA. (c) Two additional PCR cycles yield 4 and 8 copies, respectively, of the original DNA sequence. (d) Effect of running 20 PCR cycles on a DNA preparation originally containing 10 copies of a target gene. Note that the graph is semilogarithmic (∞Figure 6.6).

Web Tutorial 7.2

Polymerase Chain Reaction (PCR)

Applications and Sensitivity of PCR

PCR is a powerful tool. It is easy to perform, extremely sensitive and specific, and highly efficient. During each round of amplification the amount of product *doubles*, leading to an exponential increase in the desired DNA (Figure 7.28) This means not only that a large amount of amplified DNA can be produced in just a few hours, but that only a few molecules of target DNA need be present in the sample to start the reaction. The reaction is so specific that, with primers of 15 or so nucleotides and high annealing temperatures, almost no "false priming" occurs, and therefore the PCR product is virtually homogeneous.

PCR is extremely valuable for obtaining DNA for cloning genes or for sequencing purposes because the gene or genes of interest can easily be amplified if flanking sequences are known. PCR is also used routinely in comparative or phylogenetic studies to amplify genes from various sources. In these cases the primers are made to regions of the gene thought to be conserved throughout a wide variety of organisms. For example, because 16S rRNA, a molecule used for phylogenetic analyses (∞Section 2.3), has both highly conserved and highly variable regions, primers specific for the 16S rRNA gene from *Bacteria* or *Archaea* can be synthesized and used to survey the organisms in a habitat for the presence of species of each group. Moreover, if more specific primers are used, only certain subgroups within each domain can be targeted. This technique is in widespread use in microbial ecology and has revealed the enormous diversity of the microbial world, much of it still uncultured (∞Chapters 2, 11–14, 18).

Because it is such a sensitive technique, PCR can be used to amplify very small quantities of DNA. For example, PCR has been used to amplify and clone DNA from sources as varied as mummified human remains and fossilized plants and animals. The ability of PCR to amplify and analyze DNA from cell mixtures has also made it a common tool of diagnostic microbiology. For example, if a clinical sample shows evidence of a gene specific to a particular pathogen, then it can be assumed that the pathogen existed in the sample. With this information, treatment of the patient can begin (∞Section 24.12). In this way, PCR alleviates the need to culture the organism, a time-consuming and often fruitless process. PCR has also been used in DNA *fingerprinting*, a powerful forensic tool that permits identification of individuals, or relationships between individuals, from very small samples of their DNA (Microbial Sidebar DNA Fingerprinting, ∞Chapter 31). PCR can also be coupled with a reverse transcription step in a process called *RT-PCR*. This can be very useful when one wants to make large amounts of DNA using RNA as a template.

7.9 Concept Check

The polymerase chain reaction is a procedure for amplifying DNA *in vitro* and employs a heat-stable DNA polymerase from thermophilic prokaryotes. Heat is used to denature the DNA into two single-stranded molecules, each of which is copied by the polymerase. After each cycle, the newly formed double strands are separated by heat again, and a new round of copying proceeds. At each cycle, the amount of target DNA doubles.

◆ Why does PCR require primers?

◆ Why is a primer needed at each "end" of the DNA to be amplified?

V RNA SYNTHESIS: TRANSCRIPTION

Ribonucleic acid (RNA) plays a number of important roles in the cell. There are three key differences in the chemistry of RNA and DNA: (1) RNA contains the sugar *ribose* instead of *deoxyribose*; (2) RNA contains the base *uracil* instead of *thymine*; and (3) except in certain viruses, RNA is not double-stranded. A change from *deoxyribose* to *ribose* affects some of the chemical properties of a nucleic acid, and enzymes that affect DNA in general have no effect on RNA, and vice versa. Moreover, the change from *thymine* to *uracil* does not affect base pairing, as the two nucleotide bases pair with adenine equally well.

Three major types of RNA are known: **messenger RNA (mRNA)**, **transfer RNA (tRNA)**, and **ribosomal RNA (rRNA)**. These are all products of the *transcription* of DNA. It should be emphasized that RNA plays a role at two levels, *genetic* and *functional*. At the *genetic* level, RNA carries the genetic information from DNA via mRNA (in the case of RNA viruses, RNA plays a direct genetic function). At the *functional* level, RNA is a macromolecule in its own right, serving a functional and structural role in ribosomes (rRNA) or an amino acid transfer role in protein synthesis (tRNA). Some RNA even has catalytic (enzymatic) activity (ribozymes, ∞Section 14.8). In this section we focus on how RNA is synthesized.

7.10 Overview of Transcription

Transcription of genetic information from DNA to RNA is carried out by the enzyme **RNA polymerase**. Like DNA polymerase, RNA polymerase catalyzes the formation of phosphodiester bonds, but in this case between *ribo*nucleotides rather than *deoxyribo*nucleotides. RNA polymerase requires DNA as a template. The precursors of RNA are the ribonucleoside triphosphates ATP, GTP, UTP, and CTP.

The chemistry of RNA synthesis is much like the chemistry of DNA synthesis (see Figure 7.12). That is, during elongation of an RNA chain, ribonucleotide triphosphates are added to the 3'–OH of the ribose of the preceding nucleotide and polymerized with the release of the two energy-rich phosphate bonds. Thus, as in DNA synthesis, in RNA synthesis the overall direction of chain growth is from the 5'-end to the 3'-end, and the *template* strand is antiparallel to the newly synthesized strand. Unlike DNA polymerase, however, RNA polymerase can initiate synthesis *de novo*; that is, no *primer* is necessary.

RNA Polymerases

The DNA template for RNA polymerase is a double-stranded DNA molecule, but only *one* of the two strands is transcribed for any given gene. Nevertheless, genes are present on either strand of DNA and thus regions of both strands are transcribed at various times during the transcription process. While these principles are true for RNA polymerases from all organisms, RNA polymerase differs markedly among *Bacteria, Archaea,* and *Eukarya. Bacteria* and *Archaea* each have a single RNA polymerase while in eukaryotes, the nucleus contains three such enzymes: RNA polymerases I, II, and III, each involved in transcribing different types of genes. RNA polymerases from all sources contain some subunits that are evolutionarily conserved (∞Figure 11.16). However, true to their phylogenetic roots, archaeal and eukaryal RNA polymerases are similar and structurally more complex than those of *Bacteria.* The following discussion deals only with RNA polymerase from *Bacteria,* which has the simplest structure and about which the most is known.

All RNA polymerases from species of *Bacteria* are closely related proteins. The enzyme from *Escherichia coli* has four different subunits, designated β, β', α, and σ (sigma), with α present in two copies. The subunits interact to form the active enzyme, called the *RNA polymerase holoenzyme,* but the sigma factor is not as tightly bound as the others and easily dissociates, leading to the formation of what is called the *RNA polymerase core enzyme* ($\alpha_2\beta\beta'$). The core enzyme alone can catalyze the formation of RNA, and the role of sigma is in *recognition* of the appropriate site on the DNA for the initiation of RNA synthesis. The process of RNA synthesis involving RNA polymerase and sigma is illustrated in Figure 7.29●.

Promoters

RNA polymerase is a large protein and makes contact with the DNA over many bases simultaneously. Proteins like RNA polymerase can interact specifically with DNA because portions of the base pairs are exposed in the major groove (see Figure 7.5). However, in order to *start* an RNA chain correctly, RNA polymerase must first recognize the proper regions on the DNA. These important sites are called **promoters**.

Once the RNA polymerase has bound to the promoter, transcription can proceed. In this process, the DNA double helix at the promoter is *opened up* by the RNA polymerase (Figure 7.29). As the polymerase moves, it causes the DNA to unwind in short segments. As a result of this transient unwinding, the template strand is exposed and is copied into the RNA complement. Thus, the promoter can be thought of as *pointing* the RNA polymerase in one direction or the other along the DNA. If a region of DNA has two nearby promoters pointing in opposite directions, then it follows that transcription

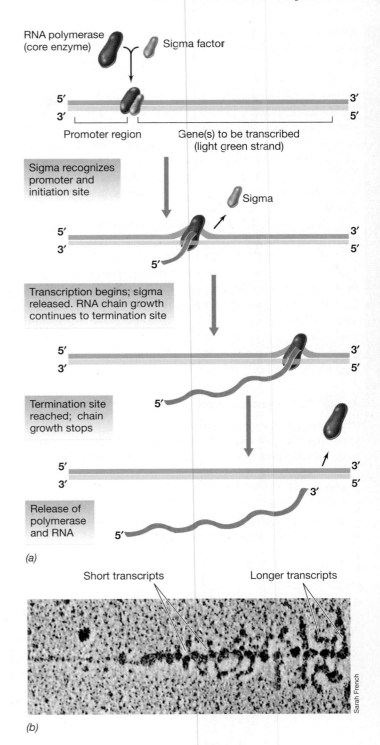

● **Figure 7.29 Transcription.** (a) Steps in RNA synthesis. The initiation and termination sites are specific nucleotide sequences on the DNA. The sigma factor allows RNA polymerase to recognize the initiation site (the promoter). The sigma factor is released during elongation. RNA polymerase moves down the DNA chain, causing temporary opening of the double helix and transcription of one of the DNA strands. When a termination site is reached, chain growth stops, and the mRNA and polymerase are released. (b) Electron micrograph of transcription occurring along a gene on the *Escherichia coli* chromosome. The region of active transcription represents about 2 kb pairs of DNA. Transcription proceeds from left to right, with the shorter transcripts on the left becoming longer transcripts as transcription proceeds.

from one of the promoters will occur in one direction (on one of the strands) while transcription from the other will occur in the opposite direction (on the other strand).

Once a short stretch of RNA has been formed, the sigma factor dissociates. Elongation is then carried out by the core enzyme alone (Figure 7.29). Thus, sigma is only involved in the formation of the initial RNA polymerase-DNA complex at the promoter. As the newly synthesized RNA dissociates from the DNA, the opened DNA closes back into the original double helix. Transcription stops at specific regions called *transcription terminators* (see Section 7.12).

Thus, unlike DNA replication, which involves copying an entire genome, transcription involves much smaller units of DNA, often as little as a single gene. This system allows the cell to transcribe different genes at different frequencies depending on the need within the cell for different proteins. As we shall see in Chapter 8, regulation of transcription in prokaryotes is an important and elaborate process involving many different mechanisms and is very efficient at controlling gene expression and conserving cell resources.

7.10 Concept Check

The three major types of RNA are messenger RNA (mRNA), transfer RNA (tRNA), and ribosomal RNA (rRNA). Transcription of RNA from DNA involves the enzyme RNA polymerase, which adds bases onto 3'-ends of growing chains. Unlike DNA polymerase, RNA polymerase needs no primer and recognizes a specific start site on the DNA called the promoter.

◆ Which direction (5' → 3' or 3' → 5') *along the template strand* does transcription occur?

◆ What is a *promoter*?

7.11 Diversity of Sigma Factors, Consensus Sequences, and Other RNA Polymerases

As we have noted, the promoter plays a central role in transcription. Promoters are specific DNA sequences where RNA polymerase binds. The sequences of a large number of promoters from a variety of organisms have been determined, and Figure 7.30● shows the sequence of a few promoters from *Escherichia coli*. It is the *sigma factor*, as part of the RNA polymerase, that is primarily involved in recognition of these promoters.

Sigma Factor Diversity

A single organism can synthesize several different sigma factors. For example, the *Escherichia coli* genome encodes 7 distinct sigma factors while the *Bacillus subtilis* genome encodes 14. These *alternative sigma factors* allow the RNA polymerase to recognize several different promoter sequences. Although in any given organism a single sigma factor is typically involved in the transcription of the bulk of the organism's genes, genes for special or rarely needed functions may require a different sigma factor. This is, of course, one way of controlling expression of these genes. If a needed sigma factor is not present, genes requiring this factor are effectively "turned off."

All the DNA sequences in Figure 7.30 are recognized by the same sigma factor, the major sigma factor in *E. coli*, called σ^{70}. If you carefully examine the sequences, you will see that they are not identical. However, two sequences within the promoter region are *highly conserved* between promoters, and it is these that are recognized by sigma. Both sequences precede (are *upstream* of) the site where

● **Figure 7.30 The interaction of RNA polymerase with the promoter.** Shown below the top of the diagram are six different promoter sequences identified in *Escherichia coli*, a species of *Bacteria*. The contacts of the RNA polymerase with the −35 sequence and the Pribnow box (−10 sequence) are shown. Transcription begins at a unique base just downstream from the Pribnow box. Below the actual sequences at the −35 and Pribnow box regions are consensus sequences derived from comparing many promoters. Note in this example that although the promoter sequences shown are from the 5' → 3' (dark green) strand of DNA, it is the light green strand running 3' → 5' that would actually be transcribed because RNA polymerase works only in a 5' → 3' direction.

transcription starts. One is a region 10 bases before the start of transcription, the −10 region (called the *Pribnow box*). Notice that although each promoter is slightly different, most bases are the same. When comparing the −10 regions of all the promoters recognized by this sigma factor to determine which base occurs most often at each position, one arrives at the *consensus sequence*: TATAAT. In our example, each promoter has from three to five matches for these bases. The second region of conserved sequence is about 35 bases from the start of transcription. The consensus sequence in the −35 region is TTGACA (Figure 7.30). Once again, most of the sequences are not *exactly* the same as the consensus sequence, but are very close.

Note that in Figure 7.30 the promoter sequence on the DNA of only one strand is given. By convention, the strand shown is the one oriented with its 5′-end upstream (therefore, it is *not* the strand used as the template by RNA polymerase, see Figure 7.30). Showing only the sequence of one strand is simply "shorthand" for describing promoters. In reality, promoters are actually double-stranded entities. That is, RNA polymerase recognizes and binds to double-stranded DNA even though transcription itself occurs using only one or the other strand of DNA as template.

Sigma factors in other organisms can sometimes be much more specific as regards binding sequences than the σ^{70} example shown in Figure 7.30. In such cases, very little leeway is allowed in the critical bases that are recognized. In *E. coli*, promoters that are most like the consensus are usually more effective in binding RNA polymerase. These more effective promoters are called *strong promoters* and are very useful in genetic engineering, as will be discussed in Chapter 31.

RNA Polymerases in *Eukarya* and *Archaea*

Recall that the eukaryotic nucleus contains *three* different RNA polymerases. Each of these enzymes recognizes a promoter that is associated with a particular class of gene. **RNA polymerase I** synthesizes most types of rRNA. **RNA polymerase II** synthesizes all the mRNA, and **RNA polymerase III** synthesizes tRNA and one type of rRNA. The basis for this specificity is that each type of RNA polymerase recognizes only those promoters that occur with the particular class of gene. By contrast, in *Bacteria* the promoter for a gene encoding a protein could well be identical to a promoter for a gene encoding a tRNA. Eukaryotic RNA polymerases also require accessory proteins in order to recognize specific promoters but, unlike in *Bacteria*, these eukaryotic initiation factors (and those of the *Archaea*) recognize the promoter elements independently, not as part of a polymerase holoenzyme (Figure 7.31●).

A depiction of RNA polymerase II binding to a promoter on the DNA is shown in Figure 7.31. This promoter contains a *TATA box*, a conserved sequence resembling in some respects the Pribnow box in the promoters of *Bacteria* (see Figure 7.30), as well as an *initiator element* very near the transcription start site. These two regions

are key elements of eukaryotic promoters, although a variety of other sequences may also be involved in any given promoter.

In species of *Archaea*, which contain only a single RNA polymerase most closely resembling that of eukaryotic RNA polymerase II, promoter structure resembles that of eukaryotes. Archaeal promoters that have been characterized contain a 6–8 base pair TATA box 18–27 nucleotides upstream of the transcriptional start site. Archaeal transcription also requires some accessory factors resembling those in eukaryotic transcription (Figure 7.31). Thus, the phylogenetic relationships between *Archaea* and *Eukarya* (◠Section 2.3) are also reflected in several details of a central molecular process, transcription. We will see many other such molecular relationships between *Archaea* and *Eukarya* as we proceed through this book.

　7.11 Concept Check

In *Bacteria*, promoters are recognized by the sigma subunit of RNA polymerase. Promoters recognized by a specific sigma factor have very similar sequences. In the *Eukarya* the major classes of RNA are transcribed by different RNA polymerases, with RNA polymerase II producing most mRNA. The single RNA polymerase of *Archaea* resembles in both structure and function this RNA polymerase II.

◆ What is a *consensus sequence*?

◆ In a eukaryote, which type of RNA polymerase transcribes genes that encode proteins?

7.12　Transcription Terminators

As important as initiation of transcription is, *termination* of transcription is also important to the fidelity of protein synthesis. **Termination** of RNA synthesis occurs at

● **Figure 7.31 The interaction of eukaryotic RNA polymerase II with a promoter.** The polymerase itself (brown) is positioned at the initiator element (INR) of the promoter. A TATA box-binding protein (yellow) is shown bound at the TATA box. The polymerase has a repetitive amino acid sequence at one end (shown as a tail-like structure) that can be phosphorylated, affecting activity of the polymerase. The other proteins shown in blue are a few of the large number of accessory factors required for initiation of transcription in eukaryotes.

specific base sequences on the DNA. In *Bacteria* a common termination sequence on the DNA is one containing an inverted repeat with a central nonrepeating segment (see Section 7.2 and Figure 7.6 for an explanation of inverted repeats). When such a DNA sequence is transcribed, the *RNA* can form a stem-loop structure by intrastrand base pairing (Figure 7.32●). When such stem-loop structures in the RNA are followed by runs of uridines, they are effective transcription terminators. Other sequence-specific termination sites are regions where a GC-rich sequence is followed by an AT-rich sequence. Such sequence patterns lead to termination without addition of any extra factors and are referred to as *intrinsic terminators*.

Rho-Dependent Termination

Other types of terminators require specific protein factors. In *Escherichia coli* one type of transcription terminator requires a protein called *Rho*. Rho does not bind to RNA polymerase or to DNA, but binds tightly to RNA and moves down the chain toward the RNA polymerase–DNA complex. Once RNA polymerase has paused at a *Rho-dependent termination site*, Rho causes both the RNA and the polymerase to be released from the DNA, thus terminating transcription. Other proteins involved in transcription termination are, like Rho, RNA-binding proteins. In all cases, the sequences involved in termination function at the level of RNA. However, remember that RNA is transcribed from DNA, and so transcription termination is ultimately determined by *specific nucleotide sequences on the DNA*.

Less is known about the transcription termination signals in the *Archaea*, but in some genes it seems clear that inverted repeats followed by an AT-rich sequence

are involved, sequences very similar to those found in many bacterial transcription terminators. However, such sequences are not found in other archaeal genes. One other type of suspected transcription terminator lacks inverted repeats but contains repeated runs of T's. In some way, this signals the archaeal termination machinery to terminate the transcription process. No Rho-like proteins have been found in *Archaea*.

7.12 Concept Check

RNA polymerase stops transcription at specific sites called transcription terminators. Although encoded by DNA, these signals function at the level of RNA. Some are intrinsic terminators and require no accessory proteins beyond the polymerase. In *Bacteria* these sequences are often stem-loops followed by a run of U's. Other terminators require proteins, such as Rho.

◆ What is an *intrinsic terminator*?

◆ What is a *stem-loop structure*?

7.13 The Unit of Transcription

Chromosomes are organized into units that are bounded by sites where transcription of DNA into RNA is initiated and terminated: *units of transcription*. One might assume that each unit of transcription includes only a single gene. While such can occur, it is not always the case. Some transcription units contain two or more genes. These genes are then said to be *cotranscribed*, yielding a single RNA molecule.

Ribosomal and Transfer RNAs and RNA Longevity

As we saw in Section 7.1, most genes encode proteins, but others encode RNAs that are not translated, such as ribosomal RNA (rRNA) and transfer RNA (tRNA). There are several different types of rRNA in an organism (with a ribosome having one copy of each type; see Section 7.16). Prokaryotes have three types: 16S rRNA, 23S rRNA, and 5S rRNA (we discussed the importance of 16S rRNA in evolutionary studies of prokaryotes in Chapter 2). As shown in Figure 7.33●, there are clusters containing one gene for each of these rRNAs, and the genes in such a cluster are cotranscribed. A similar situation occurs in eukaryotes. Therefore, in all organisms the unit of transcription for most rRNA is longer than a single gene. In prokaryotes tRNA genes are often cotranscribed with each other or even, as shown in Figure 7.33, with genes for rRNA. However, all such transcripts must be processed (cleaved into individual units) to yield mature (functional) rRNAs or tRNAs (Figure 7.33).

In prokaryotes, most messenger RNAs have a short half-life (on the order of a few minutes), after which they are degraded by cellular ribonucleases. This is in contrast to rRNAs and tRNAs, which are *stable RNAs*. The stability occurs because tRNAs and rRNAs attain highly

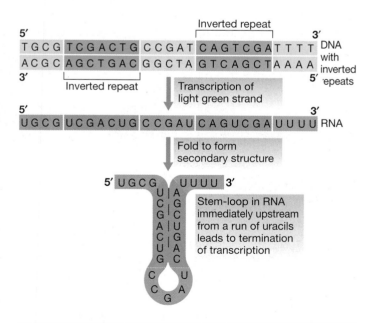

● **Figure 7.32** **Inverted repeats and transcription termination.** Inverted repeats in transcribed DNA lead to formation of a stem-loop structure in the RNA, which can result in termination of transcription.

folded structures that prevent them from being degraded by ribonucleases. By contrast, a normal mRNA does not form such structures and is susceptible to ribonuclease attack. The rapid turnover of prokaryotic mRNAs is probably a mechanism for the cell to quickly adapt to new environmental conditions and not continue translating messages whose products are no longer needed.

Polycistronic mRNA and the Operon

In *prokaryotic* genetic elements, genes encoding related enzymes are often clustered together (∞Figure 10.42). In these situations the RNA polymerase proceeds down the chain and transcribes the whole series of genes into a single, long mRNA molecule. An mRNA encoding such a group of cotranscribed genes is called a **polycistronic mRNA** (∞Section 10.13). Subsequently, when this polycistronic mRNA is translated, several polypeptides encoded by a single mRNA can be synthesized at one time.

We will discuss regulation of mRNA synthesis in Chapter 8, but introduce here the concept of the operon. An **operon** is a complete unit of gene expression, often involving genes encoding several polypeptides (or genes encoding rRNA, see Figure 7.33) on a polycistronic mRNA. In some cases, the transcription of the mRNA for an operon is under the control of a specific region of the DNA, the **operator**, which is adjacent to the coding region of the first gene in the operon. As we shall see in Chapter 8, the operator functions by binding specific proteins that regulate the transcriptional process.

● **Figure 7.33   A ribosomal rRNA transcription unit (rRNA operon) from *Bacteria* and its subsequent processing.** In *Bacteria* all such operons have the genes for the rRNAs in the order 16S rRNA, 23S rRNA, and 5S rRNA (shown approximately to scale). Note that in this particular operon the "spacer" between 16S and 23S rRNA genes contains a tRNA gene. In other operons this region may contain more than one tRNA gene and often one or more tRNA genes follow the 5S rRNA gene, which are cotranscribed. *Escherichia coli* contains seven such operons.

For the most part, polycistronic mRNA does not occur in eukaryotes. Because of split genes in eukaryotes (see Section 7.1 and Figure 7.2), processing of virtually *all* mRNA is required in eukaryotes. We cover these topics in Chapter 14, where we take a detailed look at the eukaryotic cell.

 7.13   Concept Check

The unit of transcription often contains more than a single gene. Transcription of several genes into a single mRNA molecule may occur in prokaryotes, and so the mRNA may contain the information for more than one polypeptide. Such genes that are transcribed together from a single promoter constitute an operon. In all organisms, genes encoding rRNA are cotranscribed but then processed to form the final rRNA species.

◆   What is *messenger RNA (mRNA)*?

◆   What is a *polycistronic mRNA*?

VI PROTEIN SYNTHESIS

The first two steps in biological information transfer, *replication* and *transcription*, involve synthesis of nucleic acids using nucleic acid templates. The last step, **translation**, involves a nucleic acid template but in this case the final product is a *protein* rather than a *nucleic acid*.

7.14   The Genetic Code

Before we discuss the mechanism of translation, we need to discuss the correspondence between the nucleic acid template and the amino acid sequence of the polypeptide product: the **genetic code**. As we mentioned in Section 7.1, a triplet of three bases called a **codon** encodes a specific amino acid. The genetic code is written as mRNA rather than as DNA because it is with mRNA that the translation process occurs. The 64 possible codons (4 bases taken 3 at a time, 4^3) of mRNA are shown in Table 7.5. Note that in addition to the codons specifying the various amino acids, there are also specific codons for *starting* and for *stopping* translation.

Perhaps the most interesting feature of the genetic code is that many amino acids are encoded by *several* different but related codons. This means that in many cases there is no one-to-one correspondence between the amino acid and the codon. That is, knowing the amino acid at a given location does not mean that the codon at that location is automatically known. A code such as this in which there is no one-to-one correspondence between word and code is called a **degenerate code**. The converse is true, however. Knowing the DNA sequence and the correct reading frame, one can specify the amino acid in the protein. This permits the determination of amino

Table 7.5	The genetic code as expressed by triplet base sequences of mRNA[a]						
Codon	**Amino acid**	**Codon**	**Amino acid**	**Codon**	**Amino acid**	**Codon**	**Amino acid**
UUU	Phenylalanine	UCU	Serine	UAU	Tyrosine	UGU	Cysteine
UUC	Phenylalanine	UCC	Serine	UAC	Tyrosine	UGC	Cysteine
UUA	Leucine	UCA	Serine	UAA	None (stop signal)	UGA	None (stop signal)
UUG	Leucine	UCG	Serine	UAG	None (stop signal)	UGG	Tryptophan
CUU	Leucine	CCU	Proline	CAU	Histidine	CGU	Arginine
CUC	Leucine	CCC	Proline	CAC	Histidine	CGC	Arginine
CUA	Leucine	CCA	Proline	CAA	Glutamine	CGA	Arginine
CUG	Leucine	CCG	Proline	CAG	Glutamine	CGG	Arginine
AUU	Isoleucine	ACU	Threonine	AAU	Asparagine	AGU	Serine
AUC	Isoleucine	ACC	Threonine	AAC	Asparagine	AGC	Serine
AUA	Isoleucine	ACA	Threonine	AAA	Lysine	AGA	Arginine
AUG (start)[b]	Methionine	ACG	Threonine	AAG	Lysine	AGG	Arginine
GUU	Valine	GCU	Alanine	GAU	Aspartic acid	GGU	Glycine
GUC	Valine	GCC	Alanine	GAC	Aspartic acid	GGC	Glycine
GUA	Valine	GCA	Alanine	GAA	Glutamic acid	GGA	Glycine
GUG	Valine	GCG	Alanine	GAG	Glutamic acid	GGG	Glycine

[a]The boxes of codons are colored according to the scheme: ▢ ionizable: acidic, ■ ionizable: basic, ▢ nonionizable polar, and ▢ nonpolar (∞ Figure 3.12). The nucleotide on the left is at the 5′-end of the triplet. Note that certain stop (nonsense) codons do not always function as such in certain organisms (see text and the Microbial Sidebar, Unconventional Amino Acids).

[b]AUG encodes N-formylmethionine at the beginning of mRNAs of *Bacteria*.

acid sequences from DNA base sequences and is at the heart of the genomics revolution (∞Chapter 15).

As we shall see (Section 7.16), a codon is recognized following specific base-pairing with a sequence of three bases on a tRNA called the **anticodon**. If this base-pairing was always the standard pairing of A with U and G with C, then one would expect at least one specific tRNA to exist for each codon. In some cases, this is true. For instance, there are six different tRNAs in *Escherichia coli* that carry the amino acid leucine, one for each codon (Table 7.5). By contrast, some tRNAs can recognize more than one codon. For instance, in *E. coli*, although there are two lysine codons, there is only one lysyl tRNA, and its anticodon can base pair with either AAA or AAG (see Table 7.5). This is possible because in these cases tRNA molecules form standard base pairs at only the *first two* positions of the codon, while tolerating irregular base pairing at the third position. This apparent mismatch phenomenon, called **wobble**, is illustrated in Figure 7.34● where it can be seen that the pairing between G and U (rather than G with C) is allowed at the wobble position.

Stop and Start Codons

As seen in Table 7.5, a few codons do not encode an amino acid. These triplets (UAA, UAG, and UGA) are the **stop codons**, and they signal the termination of translation of the mRNA encoding a specific protein (see Section 7.16). Stop codons are also called **nonsense codons**, because they interrupt (by terminating translation) the "sense" of the growing polypeptide.

The message is read by reading the **start codon, AUG**, which, at the beginning of the message, codes for the amino acid N-formylmethionine (or methionine in *Eukarya* and *Archaea*). The importance of having a precise starting point is apparent if we consider that with a triplet code it is critical that translation begin at the correct location. If it does not, the whole **reading frame** will be shifted and an entirely different protein (or no protein at all if the shift introduces a stop codon in the correct reading frame) will be formed. By convention the "correct" reading frame, that is, the one that can be translated to give the protein encoded by the gene, is called the *0 frame*. As can be seen in Figure 7.35●, the other two pos-

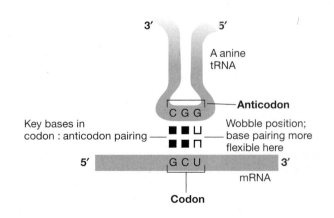

● **Figure 7.34 The wobble concept.** Base pairing is more flexible for the third base of the codon than for the first two. Only a portion of the tRNA is shown (see Figure 7.36).

● Figure 7.35 Possible reading frames in an mRNA. An interior sequence of an mRNA is shown. (a) The amino acids that would be encoded if the ribosome were in the correct reading frame (designated the "0" frame). (b) The amino acids that would be encoded by this region of the mRNA if the ribosome were in the −1 reading frame. (c) The amino acids that would be encoded if the ribosome were in the +1 reading frame.

sible reading frames (−1 and +1) do not encode the same amino acid sequence. Therefore it is essential that the ribosome finds the *correct start codon* to begin translation, and once it has, that it moves down the mRNA *exactly* three bases at a time.

As we will discuss in Section 7.16, ribosomes from *Bacteria* recognize a specific AUG on the mRNA as a start codon with the aid of an upstream sequence called the *Shine–Dalgarno sequence*. This additional alignment requirement at initiation explains why a few messages from *Bacteria* can use other codons, such as GUG, for a start codon. However, even these unusual start codons specify N-formylmethionine as the initiator amino acid.

Open Reading Frames

Today the genomes of many organisms have been sequenced (∞Chapter 15). But these mountains of sequence data would be useless unless scientists could use it to determine the location of protein-encoding genes. One common method of identifying protein encoding genes is to examine each strand of the DNA sequence for *open reading frames*. Remember that mRNA is transcribed from DNA, so that if one knows the sequence of DNA, one also knows the sequence of RNA that could be transcribed from it. If an RNA can be translated, it must contain an **open reading frame (ORF)**: a start codon (typically AUG) followed by some number of codons and then a stop codon in the same reading frame as the start codon.

A computer can be programmed to scan long base sequences in DNA databases to look for open reading frames. In addition to looking for start/stop codons, the search may include promoters and Shine–Dalgarno

ribosome binding sequences as well. The search for ORFs is very important in genomics (∞Chapter 15) and genetic engineering (∞Chapter 31). If an unknown piece of DNA has been isolated and sequenced, the presence of an open reading frame indicates that it can encode protein.

Other Genetic Codes

As far as is known, virtually all cells use the same genetic code. Therefore, the genetic code is nearly a **universal code**. However, this view has been tempered a bit from the discovery that some organelles and a few cells use genetic codes that are slight variations of the "universal" genetic code (see the Microbial Sidebar, Unconventional Amino Acids).

Alternative genetic codes were first discovered in the genomes of animal mitochondria. These modified codes typically involve using nonsense codons as sense codons. For example, animal (but not plant) mitochondria use the codon UGA as a tryptophan codon instead of as a stop codon (Table 7.5). Several cells are known that also use slightly different genetic codes. For example, in the genus *Mycoplasma* (Bacteria) and the genus *Paramecium* (Eukarya), certain nonsense codons encode actual amino acids. These organisms simply have fewer actual nonsense codons because one or two of them are read as sense codons. In a few cases, nonsense codons encode unusual amino acids rather than one of the common core of 20 amino acids (see the Microbial Sidebar).

Codon Bias

Codon usage was originally thought to be a random process in cells. That is, after the genetic code had been cracked and before any genes had actually been sequenced, it was assumed that the degenerate codons for an amino acid would be used at equal frequencies. However, this assumption has proven to be incorrect. Genomics has shown us that codon usage is highly biased and that this bias varies from one organism to the next. In *Escherichia coli,* for instance, only about 1 out of 20 isoleucine residues in proteins is encoded by the isoleucine codon AUA, the other 19 being encoded by the other isoleucine codons, AUU and AUC.

The origin of codon biases is unclear, but these genomic peculiarities are now easily recognized and taken into account in practical uses of gene sequence information. For example, in biotechnology, a gene from one organism whose codon usage differs dramatically from that of another may not be translated well if cloned into the latter. Using genetic engineering tools, however, this problem can be corrected or at least compensated for such that translation of even very unusual mRNAs can occur efficiently. We will return to the issue of codon bias in our discussion of biotechnology in Chapter 31.

Microbial Sidebar ◆ Unconventional Amino Acids

The genetic code has codons for 20 amino acids (Table 7.5). However, many proteins contain "other" amino acids. In fact, there are well over 100 different amino acids that have been found in different proteins. It was originally thought that all of these unconventional amino acids were made by modifying one of the standard amino acids *after* it was incorporated into protein, a process called *post-translational modification*. Although this can occur, at least two of these "other" amino acids are inserted into proteins by the translational machinery itself. These exceptions are *seleno-cysteine* and *pyrrolysine*.

Selenocysteine has the same structure as cysteine except that it contains a *selenium* rather than a *sulfur* atom (Figure 1●). It was known for some time that several proteins, including some from humans, contained this unusual amino acid. For example, *Escherichia coli* makes two different formate dehydrogenase enzymes and both contain a selenocysteine residue. When the gene encoding one of these enzymes was sequenced, it was found that the codon encoding the selenocysteine was UGA. Normally, UGA is a nonsense codon in *E. coli* (Table 7.5); but UGA is translated directly as selenocysteine in certain mRNA molecules in *E. coli* and many other prokaryotes and eukaryotes, including humans. Therefore, selenocysteine became the twenty-first genetically encoded amino acid.

Pyrrolysine is a lysine analog that contains an aromatic ring that lysine itself lacks (Figure 1). This amino acid has been found in certain *Archaea* and *Bacteria* but was first discovered in species of *Archaea* called *methanogens*, organisms whose metabolism generates natural gas (methane, ⌒⊃Section 13.4). In certain methanogens the enzyme *methylamine methyltransferase* contains a pyrrolysine residue. When analyzing the gene for this enzyme, it was found that the way in which pyrrolysine is encoded bears a striking resemblance to the selenocysteine story. A nonsense codon, in this case UAG instead of UGA, encodes pyrrolysine.

How can a codon sometimes be a nonsense codon and sometimes a sense codon? The answer in the selenocysteine case is well understood and lies in the *context* of the codon, that is, the sequence of the bases surrounding the UGA codon. This differs from the mitochondrial "alternative code," where UGA encodes tryptophan regardless of sequence context. In certain contexts, the translational machinery decodes UGA as "selenocysteine," while in all other contexts UGA stops translation. Selenocysteine has its own tRNA (as do all the standard amino acids) containing the predicted anticodon (UCA) and also has a special protein factor that brings only this tRNA to the ribosome when called for during the translational process. Pyrrolysine also has its own tRNA, but differs from the selenocysteine situation in a key way. Selenocysteine is formed by modifying a serine that has been attached to the selenocysteine tRNA by the serine aminoacyl-tRNA synthetase. By contrast, in the case of pyrrolysine. a unique aminoacyl-tRNA synthetase exists that charges the pyrrolysyl tRNA directly with pyrrolysine.

Evolution has obviously allowed for some "play" in the genetic code. The incorporation of selenocysteine and pyrrolysine are excellent examples of this, and one can only wonder what other unusual amino acids may be genetically encoded from either reassigned nonsense codons or perhaps even from reassigned sense codons. The genomic era has simplified the search for examples of the former, as computer inspection of nucleic acid sequences can quickly pinpoint nonsense codons that appear out of place in an otherwise normal open reading frame. Detection of reassigned sense codons, if they occur, is another matter, although computer analysis of correlations between unusual amino acids, their cognate codons, and sequence context, may yield more interesting surprises. ■

Figure 1 *Comparisons of the structures of cysteine/selenocysteine and lysine/pyrrolysine*

7.14 Concept Check

The genetic code is expressed in terms of RNA, and a single amino acid may be encoded by several different but related codons. In addition to the nonsense codons, there is also a specific start codon that signals where the translation process should begin.

◆ Why is it important for the ribosome to read "in frame"?

◆ Describe an *open reading frame*. If you were given a nucleotide sequence, how would you find the open reading frame?

7.15 Transfer RNA

Recall from Section 7.14 that it is the *anticodon* portion of the tRNA that base pairs with the codon. However, a tRNA is much more than simply an anticodon (Figure 7.36●). A tRNA has a specificity for both a codon *and* its cognate amino acid. The tRNA and its specific amino acid are brought together by specific enzymes that ensure that a particular tRNA receives its correct amino acid. These enzymes, called **aminoacyl-tRNA synthetases**, have the important function of recognizing *both* the cognate amino acid *and* the specific tRNA for that amino acid.

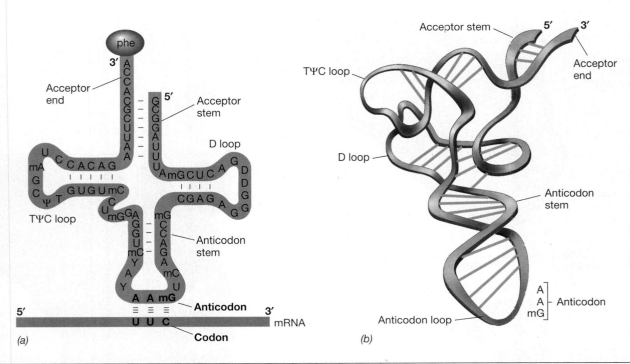

● **Figure 7.36   Structure of a transfer RNA.** (a) The conventional cloverleaf structure of yeast phenylalanine tRNA. The amino acid is attached to the ribose of the terminal A at the acceptor end. A, adenine; C, cytosine; U, uracil; G, guanine; ψ, pseudouracil; D, dihydrouracil; m, methyl; Y, a modified purine. (b) In actuality, the tRNA molecule folds so that the D loop and TψC loops are close together and associate by hydrophobic interactions. Note the unusual bases present in the tRNA and also the presence of an occasional thymine, absent from mRNA.

Structure of tRNAs

There are about 60 different tRNAs in bacterial cells and 100–110 in mammalian cells. Transfer RNA molecules are short, single-stranded molecules that contain extensive secondary structure and have lengths of 73–93 nucleotides. When compared, it has been found that certain bases and secondary structures are constant for all tRNAs, whereas other parts are variable. Transfer RNA molecules also contain some purine and pyrimidine bases differing slightly from the normal bases found in RNA in that they are chemically modified. These modifications are made to the bases after transcription. Some of these unusual bases are *pseudouridine, inosine, dihydrouridine, ribothymidine, methyl guanosine, dimethyl guanosine,* and *methyl inosine.* The final mature and active tRNA not only contains unusual bases, but also extensive double-stranded regions within the molecule as a result of internal base pairing when the molecule folds back on itself (Figure 7.36).

The structure of a tRNA can be drawn in a cloverleaf fashion, as in Figure 7.36*a*. Some regions of tRNA secondary structure are given names having to do either with the bases most often found there (the TψC and D loops, for example) or with specific functions (anticodon loop and acceptor end). The three-dimensional structure of a tRNA is shown in Figure 7.36*b*. Note that bases that appear widely separated in the cloverleaf model are actually closer together when viewed in three dimensions. This means that some of the bases in the "loops" are actually paired with bases in other loops.

One of the key variable parts of the tRNA molecule contains the **anticodon**, the site that recognizes the codon on the mRNA. The anticodon is found in the *anticodon loop* (Figure 7.36). There are *three* nucleotides in the anticodon loop that are specifically involved in the recognition process and that base-pair with the codon (see Section 7.14 and Figure 7.34). Other portions of the tRNA interact with the ribosome (both rRNA and protein components), nonribosomal translation proteins, and the activating synthetase enzyme. At the 3′-end, or **acceptor end**, of all tRNAs, are three unpaired nucleotides. The sequence of these three nucleotides is always cytosine-cytosine-adenine (CCA), and it is to the ribose sugar of the terminal adenine that the amino acid is covalently attached via an ester linkage. From this location on the tRNA, the amino acid is incorporated into the growing polypeptide chain on the ribosome by a mechanism that will be described in the next section.

Recognition, Activation, and Charging of tRNAs

Recognition of the correct tRNA by an aminoacyl-tRNA synthetase involves specific contacts between key regions of the nucleic acid and particular amino acids of its respective synthetase (Figure 7.37●). As might be expected because of the unique sequence in this region, the *anticodon* of the tRNA is important in recognition by the synthetase. However, other contact sites between the tRNA and the synthetase are also important. Studies of tRNA binding to aminoacyl-tRNA synthetases in which specific bases in the tRNA have been changed

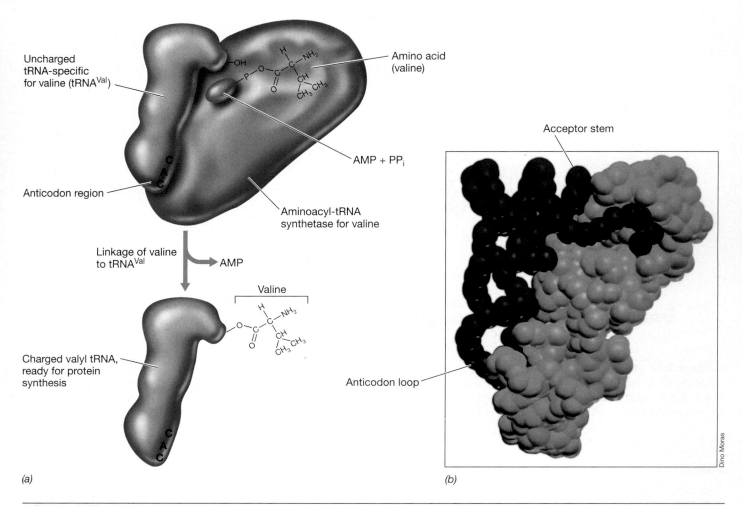

● Figure 7.37 Aminoacyl–tRNA synthetases. (a) Mode of action of an aminoacyl–tRNA synthetase. Recognition of the correct tRNA by a particular synthetase involves contacts between specific nucleic acid sequences in the D-loop and acceptor stem of the tRNA and specific amino acids of the synthetase. In this diagram, valyl-tRNA synthetase is shown catalyzing the final step of the reaction, where the valine in valyl–AMP is transferred to tRNA. (b) A computer model showing the interaction of glutaminyl–tRNA synthetase (blue) with its tRNA (red). Reprinted with permission from M. Ruff et al. 1991. *Science* 252: 1682–1689. © 1991, AAAS.

by mutation have shown that only a small number of key nucleotides in a tRNA besides the anticodon region are involved in recognition. These other key recognition nucleotides are often part of the acceptor stem or D-loop of the tRNA molecule (see Figure 7.36). It should be emphasized that the fidelity of this recognition process is crucial, for if the wrong amino acid becomes attached to the tRNA, it will be inserted into the polypeptide, likely leading to the synthesis of a faulty protein.

The specific reaction between amino acid and tRNA catalyzed by the aminoacyl-tRNA synthetase first involves *activation* of the amino acid by reaction with ATP:

Amino acid + ATP $\longleftrightarrow$ aminoacyl—AMP + P—P

The aminoacyl-AMP intermediate formed normally remains bound to the enzyme until collision with the appropriate tRNA molecule, and, as shown in Figure 7.37a, the activated amino acid is then transferred to the tRNA to form a *charged* tRNA:

Aminoacyl—AMP + tRNA $\longleftrightarrow$

aminoacyl—tRNA + AMP

The pyrophosphate (PP$_i$) formed in the first reaction is split by a pyrophosphatase, forming two molecules of inorganic phosphate. Hence, since *ATP* is used and *AMP* is formed, a total of *two* energy-rich phosphate bonds are required for the activation of an amino acid and its cognate tRNA. Once activation and charging have occurred, the aminoacyl-tRNA leaves the synthetase and travels to the ribosome where polypeptide synthesis occurs. The mechanism of protein synthesis is discussed in the next section.

 7.15 Concept Check

One or more transfer RNAs exist for each amino acid found in protein. Enzymes called aminoacyl-tRNA synthetases function to attach an amino acid to a tRNA. Once the correct amino acid is attached to its tRNA, further specificity resides primarily in the codon-anticodon interaction.

◆ What is the function of the *anticodon* of a tRNA?

◆ What is the function of the *acceptor end* of a tRNA?

7.16 Translation: The Process of Protein Synthesis

We learned in Chapter 3 that it is the amino acid *sequence* that determines the structure (and ultimately the function) of a protein. Thus, it is critical to the fidelity of translation that the proper amino acid be inserted at the proper place in the polypeptide chain. This is the role of the protein-synthesizing machinery of the cell, including, in particular, the ribosome.

Ribosomes

Ribosomes are the site of protein synthesis. A cell may have many thousands of ribosomes, the number being positively correlated with growth rate. Each ribosome is constructed of two subunits. In prokaryotes, the ribosome subunits are of 30S and 50S, yielding intact 70S ribosomes. The numbers 30S, 50S, and 70S refer to *Svedberg units*, which are units of sedimentation coefficients of ribosome subunits (30S and 50S) or intact ribosomes (70S) when subjected to centrifugal force in an ultracentrifuge.

Each ribosomal subunit is a ribonucleoprotein complex made up of specific ribosomal RNAs and ribosomal proteins. The 30S subunit contains 16S rRNA and about 21 proteins, while the 50S subunit contains 5S and 23S rRNA and about 34 proteins (Table 7.6 and Figure 7.38a•). In *Escherichia coli*, there are at least 53 distinct ribosomal proteins, most present at one copy per ribosome.

Since the genome of *E. coli* has been completely sequenced and the ribosome has been studied for some time, one might think that we should know exactly how many ribosomal proteins there are overall and in each subunit. However, some proteins are tightly associated with the ribosome, some less so. In addition, some ribosomal proteins are associated only with one subunit, some with both. There are also proteins that are absolutely essential for ribosome function and that interact with the ribosome at various stages in the trans-

lational process but are not considered "ribosomal proteins," per se. Moreover, it should be kept in mind that the ribosome is a very dynamic structure whose parts alternately associate and dissociate and that interact with many other proteins in the cell. Thus, determining the total number of "ribosomal proteins," even in *E. coli*, is not easy.

Steps in Protein Synthesis

The synthesis of a protein involves a complex cycle in which the various ribosomal components play specific roles. Although a continuous process, protein synthesis can be broken down into a number of discrete steps: **initiation**, **elongation**, and **termination–release**. In addition to mRNA, tRNA, and ribosomes, the process involves a number of proteins designated initiation, elongation, and termination factors, while the energy-rich compound guanosine triphosphate provides energy for the process. The key steps in protein synthesis are shown in Figure 7.38.

Initiation

In prokaryotes initiation of protein synthesis always begins with a free 30S ribosome subunit. From this an **initiation complex** forms consisting of a 30S ribosome subunit, mRNA, formylmethionine tRNA, and several initiation proteins called *IF1*, *IF2*, and *IF3*. Guanosine triphosphate is also required for this step. To this initiation complex a 50S ribosomal subunit is added to make the active 70S ribosome. At the end of the translation process, the released ribosome separates again into 30S and 50S subunits.

Just preceding the initiation codon on the mRNA is a sequence of from three to nine nucleotides called the **Shine–Dalgarno sequence** that is involved in the binding of the mRNA to the ribosome. This *ribosome binding site* at the 5′-end of the mRNA is complementary to regions in the 3′-end of the 16S RNA of the ribosome. Base pairing between these two molecules ensures effective formation of the ribosome–mRNA complex *in the correct reading frame*. The presence of the Shine–Dalgarno site on the mRNA and its specific interaction with 16S rRNA allow prokaryotic ribosomes to translate *polycistronic* mRNA because the ribosome can find each initiation site within a message after binding to the Shine-Dalgarno site (see Section 7.13).

Initiation always begins with a special initiator aminoacyl-tRNA binding to the **start codon**, AUG. In *Bacteria* this is **formylmethionine** tRNA. Subsequently, the formyl group at the N-terminal end of the polypeptide is removed; the terminal amino acid of the completed protein is hence methionine. Because the Shine–Dalgarno sites (and other possible interactions between the rRNA and the mRNA) direct the ribosome to the proper start site, prokaryotic messages can use a start codon other than AUG. The most common alternative start codon is GUG. When used in this context, however,

Table 7.6 Ribosome structure[a]		
Property	**Prokaryote**	**Eukaryote**
Overall size	70S	80S
Small subunit	30S	40S
Number of proteins	~21	~30
RNA size (number of bases)	16S (1500)	18S (2300)
Large subunit	50S	60S
Number of proteins	~34	~50
RNA size (number of bases)	23S (2900)	28S (4200)
	5S (120)	5.8S (160)
		5S (120)

[a] Ribosomes of mitochondria and chloroplasts of eukaryotes are similar to prokaryotic ribosomes (☞ Section 14.4).

(a)

(i)

INITIATION
(requires several initiation proteins, IF-1, 2, and 3)

(ii)

ELONGATION
(requires proteins EF-Tu and EF-Ts)

(iii)

(iv)

TRANSLOCATION
(requires protein EF-G)

(v)

(b)

GUG calls for formylmethionine initiator tRNA (and not valine, see Table 7.5). In *Eukarya* and *Archaea*, initiation begins with *methionine* instead of *formylmethionine*. Although all proteins in cells are *initiated* with a methionine or formylmethionine, this amino acid is not always found as the first amino acid in a cell's proteins. Methionine (or formylmethionine in *Bacteria*) is removed from many mature proteins by a specific protease.

Elongation, Translocation, and Termination

The mRNA is threaded through the ribosome primarily bound to the 30S subunit. The ribosome contains other sites where the tRNAs interact. Two of these sites are located primarily on the 50S subunit, and they are termed the *P-site* and the *A-site* (see Figure 7.38b). The A-site, the **acceptor site**, is the site where the new charged tRNA first attaches. Travel and attachment of a tRNA to the A-site is assisted by one of a series of elongation factors (EF) proteins called *EF-Ta*.

The P-site, the **peptide site**, is the site where the growing peptide is held by a tRNA. During peptide bond formation, the peptide moves to the tRNA at the A-site as a new peptide bond is formed. Several nonribosomal proteins are required for **elongation**, especially other proteins of the EF family, such as EF-Tu and EF-Ts, as well as additional molecules of GTP (to simplify Figure 7.38b, the elongation factors are omitted and only a portion of the ribosome is shown). Following elongation, the tRNA that holds the peptide must now be moved (translocated) from the A-site to the P-site, thus opening up the A-site for another charged tRNA (Figure 7.38b).

Translocation requires a specific EF protein called EF-G and one molecule of GTP per each translocation event. At each translocation step the ribosome advances three nucleotides, exposing a new codon at the ribosome A-site. Translocation pushes the now empty tRNA to a third site, called the *E-site*. It is from this **exit site** that the tRNA is actually released from the ribosome (Figure 7.38b). The precision of the translocation step is critical to the accuracy of protein synthesis. The ribosome must move exactly one codon at each step. Although during this process mRNA appears to be moving through the ribosome complex, in reality, the *ribosome* is moving along the mRNA. Thus, the

● **Figure 7.38 The ribosome and protein synthesis.** (a) Structure of the ribosome, showing the position of the A (acceptor) site, the P (peptide) site, and the E (exit) site. (b) Translation. Initiation and elongation. (i, ii) Interaction between the codon and anticodon brings into the position the correct charged tRNA—in this case the initiator tRNA and the second charged tRNA. (iii) The formation of a peptide bond between amino acids on adjacent tRNA molecules completes elongation. (iv) Translocation of the ribosome from one codon to the next occurs with release of the tRNA from the E-site. (v) The next charged tRNA binds to the A-site.

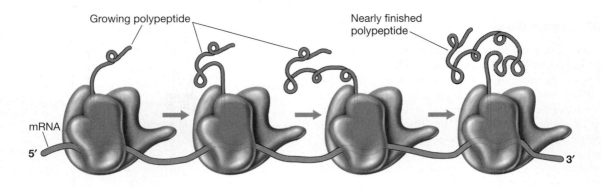

● **Figure 7.39** **Polysomes.** Translation by several ribosomes on a single messenger RNA forms the polysome. Note how the ribosomes nearest the 5'-end of the message are at an earlier stage in the translation process and thus only a portion of the final polypeptide has been made.

three sites on the ribosome that we have identified in Figure 7.38 are not static locations but instead are *moving parts* of a complex biomolecular machine.

Several ribosomes can simultaneously translate a single mRNA molecule forming a complex called a **polysome** (Figure 7.39●). Polysomes increase the speed and efficiency of translation, and because the activity of each ribosome is independent of that of its neighbors, each ribosome in a polysome complex makes a complete polypeptide. Note in Figure 7.39 how ribosomes closest to the 5'-end (the beginning) of the mRNA molecule have short polypeptides attached to them because only a few codons have been read, while ribosomes closest to the 3'-end of the message have nearly finished polypeptides.

The **termination** of protein synthesis occurs when a nonsense codon is reached. No tRNA binds to a nonsense codon. Instead, specific proteins called *release factors* (RF factors) recognize this chain-terminating signal and cleave the attached polypeptide from the terminal tRNA, releasing the finished product. Following this, the ribosome subunits dissociate, and the 30S/50S subunits are then free to form new initiation complexes and repeat the process.

Role of Ribosomal RNA in Protein Synthesis

Ribosomal RNA plays a critical functional role in all stages of protein synthesis, from initiation to termination. The role of the many proteins present in the ribosome, although less clear, may be more as facilitators of RNA function by stabilizing or positioning key sequences in the various ribosomal RNAs.

As already discussed, in prokaryotes it is clear that 16S rRNA is involved in initiation through base pairing with the ribosome binding sequence (the Shine–Dalgarno sequence) on the mRNA. Besides this, other mRNA/rRNA interactions occur during elongation. Ribosomal RNA also plays a role in ribosome subunit association, as well as in tRNA positioning in the A and P sites (see Figure 7.38b) on the ribosome. Although charged tRNAs that enter the ribosome recognize the correct codon by codon:anticodon base pairing, they are also physically attached to the ribosome by interactions of the anticodon stem-loop of the tRNA with specific sequences within 16S rRNA. Moreover, the acceptor end of the tRNA (see Figure 7.36) base pairs with sequences in the 23S rRNA.

In addition to all of this, the *actual formation of peptide bonds* is catalyzed by rRNA. This process that occurs on the 50S subunit of the ribosome and is called the **peptidyl transferase reaction** is not catalyzed by *any* of the multitude of ribosomal or ribosomal-associated proteins, but instead by the activity of 23S rRNA itself. 23S rRNA also plays a role in the translocation process, and in this regard, the EF proteins are known to interact specifically with 23S rRNA. Finally, 16S rRNA is also involved in a catalytic way in termination reactions, probably by interacting with the mRNA or through interactions with release proteins. Thus, besides its role as a structural backbone of the ribosome, ribosomal RNA plays a major catalytic role in the translation process as well.

Effect of Antibiotics on Protein Synthesis

A large number of antibiotics inhibit protein synthesis by interacting with the ribosome. These interactions are quite specific, and many have been shown to involve rRNA. Several of these antibiotics are clinically useful, and several are also effective research tools because they are specific for different steps in protein synthesis. For instance, *streptomycin* inhibits initiation, whereas *puromycin, chloramphenicol, cycloheximide*, and *tetracycline* inhibit elongation.

Adding to their clinical usefulness is the fact that many antibiotics specifically inhibit ribosomes of organisms from only one or two of the phylogenetic domains. Of the antibiotics just listed, for example, chloramphenicol and streptomycin are specific for the ribosomes of *Bacteria* and cycloheximide for ribosomes of *Eukarya*. The mode of action of these and other antibiotics will be discussed in Chapter 20.

7.16 Concept Check

The ribosome plays a key role in the translation process, bringing together mRNA and aminoacyl tRNAs. There are three sites on the ribosome: the acceptor site, where the charged tRNA first combines; the peptide site, where the growing polypeptide chain is held; and an exit site. During each step of amino acid addition, the ribosome advances three nucleotides (one codon) along the mRNA and the tRNA moves from the acceptor to the peptide site. Termination of protein synthesis occurs when a nonsense codon, which does not encode an amino acid, is reached.

◆ What are the components of a ribosome?

◆ What functional roles does rRNA play in protein synthesis?

7.17 Folding and Secreting Proteins

We have now discussed how the genetic information present in the sequence of bases in DNA is *replicated* into an identical copy, *transcribed* into a sequence of bases in RNA, and *translated* into a sequence of amino acids in protein. For a protein to function, however, it must be *folded* correctly (∞Sections 3.6–3.8), and it must also end up in the correct location in the cell. Here we briefly deal with these two processes.

Protein Folding

Many polypeptides fold spontaneously into their active form while they are being synthesized (Figure 7.39). However, many others do not and require assistance from other proteins called **molecular chaperones** for proper folding or for assembly into larger complexes. The chaperones themselves do not become part of the assembled proteins, but only assist in the folding process. Indeed, one important function of chaperones is to prevent improper aggregation of proteins.

There are many different kinds of these chaperones and some are even associated with ribosomes. Some are also extremely abundant in the cell, especially under certain growth conditions when protein stability is at risk (for example, at high temperatures). Molecular chaperones seem to be both extremely widespread and their sequences highly conserved.

Four key chaperones in *Escherichia coli* are the proteins *DnaK, DnaJ, GroEL,* and *GroES*. DnaK and DnaJ are ATP-dependent enzymes that bind to newly formed polypeptides and keep them from folding too abruptly, a process that increases the risk of improper folding (Figure 7.40●). Slower folding thus improves the chances of correct folding. If the DnaKJ complex is unable to fold the protein properly, it may transfer the partially folded protein to the two multi-subunit proteins GroEL and GroES. The protein first enters GroEL, a large barrel-shaped protein that, using the energy of ATP hydrolysis, properly folds the protein. GroES assists in this process (Figure 7.40).

● **Figure 7.40 The activity of molecular chaperones.** An improperly folded protein can be refolded by either the DnaKJ complex or by the GroEL/ES complex. In both cases, energy for refolding comes from ATP.

In addition to folding newly synthesized proteins, chaperones can also refold proteins that have partially denatured in the cell. Such protein denaturation can occur because the organism has temporarily experienced high temperatures in its environment. Chaperones are thus a type of **heat shock protein**, and their synthesis is greatly accelerated when a cell is stressed by excessive heat (∞Section 8.9). The heat shock response is an attempt by the cell to refold its partially denatured proteins for re-use before proteases recognize them as improperly folded and destroy them. Refolding is not always successful, and cells contain a variety of proteases whose function is to specifically target and destroy misfolded proteins in order to generate a pool of amino acids for use in making new proteins.

Protein Secretion and the Signal Recognition Particle

Many proteins carry out their function in the cytoplasmic membrane, in the periplasm of gram-negative cells (∞Section 4.9) or even *outside* the cell proper. Such proteins must get from their site of synthesis on ribosomes into or through the cytoplasmic membrane. How is it possible for a cell to selectively transfer some proteins across a membrane while leaving most proteins in the cytoplasm?

Most proteins that must be transported into or through membranes are synthesized with an extra peptide sequence (about 15–20 amino acids long) at the beginning of the molecule, called the **signal sequence**. Signal sequences are quite variable but typically have a few positively charged residues at the beginning, a central region of hydrophobic residues, and then a more polar region. The signal sequence "signals" the cell's secretory system that this particular protein is to be exported and also helps prevent the protein from completely folding, a process that could interfere with its secretion. Since the signal sequence is the first part of the protein to be synthesized, the early steps in export may begin before the protein is completely synthesized (Figure 7.41●).

A principal role in the identification of proteins to be secreted is played by the **signal recognition particle (SRP)** (Figure 7.41). SRPs are found in all cells. In *Bacteria*, they contain a single protein and a small RNA molecule called 4.5S RNA. This *small RNA* (∞Section 8.14) is not a transfer RNA, a ribosomal RNA, or a messenger RNA. The SRP recognizes the signal sequence-containing protein and delivers it to a special membrane protein complex (for example, to the SecYEG complex, ∞Section 4.7) where the protein is secreted through a pore and into the periplasm or the environment (Figure 7.41). During the transport process the signal sequence is usually removed by a protease, an example of the process of **post-translational modification**.

Less is known about protein translocation through the membrane in species of *Archaea*. However, genomic analyses indicate that *Archaea* produce a SRP and share several aspects of the translocation process with *Bacteria*. Interestingly, and in line with their phylogenetic relationships (∞Section 2.3), *Archaea* share several aspects of the protein translocation process with *Eukarya* as well. In particular, the small RNA in the archael SRP is 7S in size, as it is in eukaryotic cells.

Secretion of Folded Proteins: The TAT System

In the Sec system of protein transport, mediated by the signal recognition particle, the transported proteins are threaded through the cytoplasmic membrane in an unfolded state and are only folded upon reaching the periplasm (in gram-negative cells) (Figure 7.41). However, there are a number of proteins that must be transported outside the cell that contain small cofactors. The latter usually have to be inserted into the protein as it folds into its final form. This occurs in the cytoplasm, and then the folded proteins are transported by a transport system distinct from Sec, called the *Tat protein export system*.

The acronym *Tat* stands for "*t*win *a*rginine *t*ranslocase" because the transported proteins contain a short signal sequence containing a pair of arginine residues. This signal sequence on a folded protein is recognized by the TatBC proteins, and they carry the protein to TatA, the membrane transporter. Energy is required for the actual transport event, and this is sup-

plied by the proton motive force. A wide variety of proteins are transported by the Tat system, especially proteins involved in energy metabolism that function in the periplasm. This includes a variety of iron-sulfur proteins and several other "redox" type proteins. In addition, the Tat pathway transports proteins involved in outer membrane (Section 4.9) biosynthesis and a variety of proteins that do not contain cofactors but that only fold properly within the cytoplasm.

In this chapter we have covered the essentials of the key molecular processes that occur in all cells. We next turn our attention to how cells *regulate* the expression of a particular gene or set of genes, a process as we will see, that is rather diverse and highly elaborate.

🛑 *7.17* **Concept Check**

Proteins must be properly folded in order to function correctly. Folding may occur spontaneously but may also involve other proteins called molecular chaperones. Many proteins also need to be transported into or through cell membranes. Such proteins are synthesized with a signal sequence that is recognized by the cellular export apparatus and is removed either during or after export.

♦ What is a *molecular chaperone*?

♦ Why do some proteins have a *signal sequence*?

♦ What is a signal recognition particle?

● **Figure 7.41 Secretory proteins and the signal recognition particle (SRP).** The signal sequence is recognized by the SRP and the protein transported to the membrane secretion system. In gram-negative bacteria, the protein is secreted into the periplasm (∞Section 4.9). In species of *Bacteria*, the SRP is 4.5S in size, while in *Archaea* and *Eukarya*, it is larger, about 7S.

REVIEW QUESTIONS

1. Describe the *central dogma* of molecular biology (∞Section 7.1).

2. Genes were discovered before their chemical nature was known. Define a gene without mentioning its chemical nature. Of what is a gene composed (∞Section 7.1)?

3. Inverted repeats can give rise to stem-loops. Show this by giving the sequence of a double-stranded DNA containing an inverted repeat and show how the transcript from this region can form a stem-loop (∞Section 7.2).

4. Is the sequence 5′-GCACGGCACG-3′ an inverted repeat? Explain your answer (∞Section 7.2).

5. DNA molecules that are AT-rich separate into two strands more easily when the temperature is raised than do DNA molecules that are GC-rich. Write an explanation for this observation based on the properties of AT and GC base pairing (∞Section 7.2).

6. Describe how DNA, which when linearized, is many times the length of a cell, fits into the cell (∞Section 7.3).

7. List the major genetic elements known in microorganisms (∞Section 7.4).

8. A structure commonly seen in circular DNA during replication is the *theta structure*. Draw a diagram of the replication process and show how a theta structure could arise (∞Sections 7.5 and 7.6).

9. Why are errors in DNA replication so rare? What additional enzyme activity (other than polymerization) is associated with DNA polymerase III and how does it serve to reduce errors (∞Section 7.6)?

10. What are restriction enzymes? What is the probable function of a restriction enzyme in the cell that produces it? How is it that the restriction enzyme in a cell does not cause the degradation of that cell's DNA (∞Section 7.7)?

11. Why do dideoxynucleotides function as chain terminators? In Figure 7.26b, why do the different bands in the gel each migrate to a unique position (∞Section 7.8)?

12. Describe the basic principles of gene amplification using the polymerase chain reaction (PCR). How have thermophilic and hyperthermophilic prokaryotes simplified the use of PCR (∞Section 7.9)?

13. Do genes for tRNAs have promoters? Do they have start codons? Explain (∞Sections 7.10 and 7.11).

14. The start and stop sites for mRNA synthesis (on the DNA) are different from the start and stop sites for protein synthesis (on the mRNA). Explain (∞Sections 7.10–7.13).

15. What is "wobble," and what makes it necessary in protein synthesis (∞Section 7.14)?

16. What are aminoacyl-tRNA synthetases and what types of reactions do they carry out? Approximately how many different types of these enzymes are present in the cell? How does a synthetase recognize its correct substrates (∞Section 7.15)?

17. The activity that forms peptide bonds on the ribosome is called *peptidyl transferase*. What catalyzes this reaction (∞Section 7.16)?

18. Sometimes misfolded proteins can be correctly refolded, but sometimes they cannot and are destroyed. What kinds of proteins are involved in refolding misfolded proteins? What kinds of enzymes are involved in destroying misfolded proteins (∞Section 7.17)?

19. How does a cell know which of its proteins are designed to function *outside* of the cell (∞Section 7.17)?

APPLICATION QUESTIONS

1. The genome of the bacterium *Neisseria gonorrhoeae* consists of a single double-stranded DNA molecule that contains 2220 kilobase pairs. Calculate the length of this DNA molecule in centimeters. If 85% of this DNA molecule is made up of the open reading frames of genes encoding proteins and the average protein is 300 amino acids long, how many protein-encoding genes does *Neisseria* have? What kind of information do you think might be present in the other 15% of the DNA?

2. Compare and contrast the activity of DNA and RNA polymerases. What is the function of each? What are the substrates of each? Which is required for PCR? What else is required for PCR?

3. What would be the result (in terms of protein synthesis) if RNA polymerase initiated *transcription* one base upstream of its normal starting point? Why? What would be the result (in terms of protein synthesis) if *translation* began one base downstream of its normal starting point? Why?

4. In Chapter 10 we will learn about *mutations*, inheritable changes in the sequence of nucleotides in the genome. In inspecting Table 7.5, discuss how the genetic code has evolved to help minimize the impact of mutations.

8

METABOLIC REGULATION

Cells regulate the expression of their genes so that proteins and other molecules are made in the correct amounts and at the correct times in the cell cycle. Regulation of gene expression is greatly influenced by the cell's environment.

WORKING GLOSSARY

Activator protein a regulatory protein that binds to specific sites on DNA and stimulates transcription; involved in positive control

Allosteric enzyme an enzyme that contains two binding sites, the active site (where the substrate binds) and the allosteric site (where an effector molecule binds)

Attenuation a mechanism for controlling gene expression; typically transcription is terminated after initiation but before a full-length mRNA is produced

Feedback inhibition a decrease in the activity of the first enzyme of a pathway caused by the final product of the pathway

Gene expression transcription and/or translation of genes

Heat shock proteins a series of proteins induced by a sudden upshift in temperature or certain other stress factors that function to refold partially denatured proteins

Induction production of an enzyme only when its substrate is present.

Kinase an enzyme that adds a phosphoryl group to a compound

Negative control a mechanism for regulating gene expression in which a *repressor protein* functions to prevent transcription of a gene or genes

Operon one or more genes transcribed into a single RNA and under the control of a single regulatory site

Positive control a mechanism for regulating gene expression in which an *activator protein* functions to promote transcription of a gene or genes

Quorum sensing a regulatory system in an organism that requires a certain density of cells of the same species be present before the regulatory events occur

Repression prevention of an enzyme's synthesis when the product of its reaction is present in excess

Repressor protein a regulatory protein that binds to specific sites on DNA and blocks transcription; involved in negative control

Response regulator protein one of the members of a two-component system; a regulatory protein that is phosphorylated by a sensor kinase protein (see sensor kinase protein)

Riboswitch a messenger RNA that can bind a specific small molecule near its 5' end that alters its secondary structure and makes it unavailable for translation

Sensor kinase protein one of the members of a two-component system; a membrane-integrated protein that phosphorylates itself in response to an external signal and then transfers the phosphoryl group to a response regulator protein (see response regulator protein)

Stringent response a global regulatory control that is activated by amino acid starvation

Two-component regulatory system a regulatory system containing two proteins: a sensor kinase and a response regulator (see sensor kinase protein and response regulator protein)

In the previous chapter we saw how the genetic information stored as a sequence of nucleotides in a gene can be *transcribed* into RNA. This information is then *translated* to yield a specific polypeptide. Collectively, these processes are called **gene expression**.

Most proteins are enzymes (∞Section 5.5), which carry out the hundreds of different enzymatic reactions that occur during cell growth. However, to be successful in nature, microorganisms must respond rapidly to changes in their environment. To efficiently orchestrate the numerous chemical reactions in a cell and make maximal use of available resources, cells need to *regulate* the kinds and amounts of macromolecules they make. Such *metabolic regulation* is the focus of this chapter.

OVERVIEW OF REGULATION

Some enzymes are needed in the cell at about the same level under all growth conditions. Such proteins are said to be *constitutive*. However, it is more common for an enzyme to be needed under some conditions but not others. For instance, enzymes required for the catabolism of the sugar lactose are useful to the cell only if lactose is present in its environment. Microbial genomes encode many more proteins than are actually present in the cell under any particular condition. Thus, regulation is a major process in all cells, and a mechanism for conserving energy and resources.

8.1 Major Modes of Regulation

There are two major levels of regulation in the cell. One controls the *activity* of preexisting enzymes and one controls the *amount* (or even the complete presence or absence) of an enzyme (Figure 8.1●). Regulation of the *activity* of an enzyme can happen only *after* the protein has been synthesized (that is, *posttranslationally*). By contrast, regulation of the *amount* of enzyme synthesized can occur at either the level of transcription (varying the amount of messenger RNA [mRNA] made) or at the level of translation (translating or not translating the mRNA).

In prokaryotes, regulation of enzyme *activity* is typically a very rapid process (occurring in seconds or less), while the regulation of enzyme *synthesis* is a relatively slow process (taking a few minutes). If a new enzyme needs to be synthesized, it will take some time before it is present in the cell in amounts sufficient to affect metabolism. Alternatively, if synthesis of an enzyme is stopped, a considerable amount of time may elapse before the existing enzyme is diluted sufficiently to no longer affect metabolism. However, working together, controls on activity and controls on synthesis efficiently regulate cell metabolism.

Systems that control the level of expression of particular genes are varied, and several genes can be regulated by more than one system. We begin by briefly discussing the processes involved in regulating the *activity* of preformed enzymes before considering how the synthesis of enzymes is controlled.

● **Figure 8.1 An overview of the mechanisms that can be used in regulation.** The product of gene A is enzyme A, which in this case is synthesized constitutively and carries out its reaction. Enzyme B is also synthesized constitutively but its *activity* can be inhibited. The *synthesis* of the product of gene C can be prevented by control at the level of translation. The *synthesis* of the product of gene D can be prevented by control at the level of transcription.

 8.1 Concept Check

Most genes encode proteins and most proteins are enzymes. The expression of such a gene can be regulated by controlling the activity of the enzyme or controlling the amount of enzyme produced.

◆ What steps in the synthesis of protein might be subject to regulation?

◆ Which is likely to be more rapid, the regulation of activity or the regulation of synthesis? Why?

II REGULATION OF ENZYME ACTIVITY

There are many ways to regulate a preformed enzyme (posttranslational regulation). In the next two sections we discuss reversible forms of regulation as well as more drastic changes to the enzyme molecule caused by protein processing.

8.2 Noncovalent Enzyme Inhibition

Some proteins are subject to a reduction in their activity, or even complete inhibition, by specific compounds in the cell. These compounds are typically components of the metabolic pathway in which the enzyme functions. Inhibition can be the result of either covalent or noncovalent changes in the enzyme. We begin with feedback inhibition and isoenzymes, examples of *noncovalent* interactions.

Feedback Inhibition

A major mechanism for the control of enzymatic activity involves the phenomenon of **feedback inhibition**. Feedback inhibition occurs primarily in the regulation of entire biosynthetic pathways, such as the pathway for the biosynthesis of an amino acid or nucleotide. As we have seen, such pathways involve many enzymatic steps, and the final product, the amino acid or nucleotide, is many steps removed from the starting substrate (∞Section 5.16). Yet in pathways subject to feedback inhibition, the final product can communicate with an earlier step in the pathway, and in so doing, regulate its own biosynthesis.

The regulation in feedback inhibition occurs because the amino acid or other end product of the biosynthetic pathway inhibits the activity of the *first* enzyme in this pathway. Inhibiting the first step effectively shuts down the entire pathway, since no intermediates are generated as substrates for other enzymes in the pathway (Figure 8.2●). Thus, as the end product of the pathway accumulates in the cell, its further synthesis is inhibited. Feedback inhibition is reversible, however, because once the inhibitor becomes limiting, synthesis resumes (Figure 8.2).

● **Figure 8.2 Feedback inhibition of enzyme activity.** The activity of the first enzyme of the pathway is inhibited by the end product, thus controlling production of end product.

How can an end product of a pathway inhibit the activity of an enzyme whose substrate is quite unrelated to it? This is possible because of a property of the inhibited enzyme known as **allostery.** An **allosteric enzyme** has *two* binding sites, the *active site,* where the substrate binds (∞Section 5.5), and the *allosteric site,* where the inhibitor (called an *effector*) binds reversibly. When the effector binds—generally noncovalently—at the allosteric site, the conformation of the enzyme changes such that the substrate no longer binds at the active site (Figure 8.3●). When the concentration of the effector falls in the cell, equilibrium favors its dissociation from the allosteric site, returning the active site to its catalytic shape. When this occurs, the enzyme is once again active.

Allosteric enzymes are common in both anabolic and catabolic pathways and are especially important in branched pathways. For example, the amino acids proline and arginine are both synthesized from glutamic acid. These two amino acids can control the first enzyme unique to their own synthesis without affecting the other. Thus, a surplus of proline, for example, does not cause starvation for arginine (Figure 8.4●).

The mechanism of feedback inhibition is of interest to more than just basic science. We will see in Chapter 30 how an understanding of the biochemistry of feedback inhibition has allowed industrial microbiologists to isolate mutant strains of amino acid-producing organisms that have lost the ability to feedback inhibit the production of specific amino acids. These mutants "overproduce" amino acids

● **Figure 8.4 Feedback inhibition in a branched biosynthetic pathway.** A key intermediate in each pathway is shown in pink. The enzymes being inhibited (shown by dashed red arrows) are *N*-acetyl glutamate synthase (AGS) and γ-glutamyl kinase (GK).

and are used for the large-scale production of amino acids as a food supplement (∞Section 30.7 and Figure 30.13).

Isoenzymes

Some biosynthetic pathways under feedback inhibition employ **isoenzymes** (*iso* means "same"). Isoenzymes are different proteins that catalyze the same reaction, but which are subject to different regulatory controls. An example is the synthesis of aromatic amino acids in *Escherichia coli* (Figure 8.5●; ∞Figure 5.26).

The enzyme DAHP synthase plays an important role in aromatic amino acid biosynthesis. In *E. coli,* three different DAHP synthase isoenzymes catalyze the first reaction in this pathway, and each is regulated independently by one of the three different end-product amino acids. However, unlike the earlier examples of feedback inhibition where inhibitors completely stopped an enzyme activity, in this case the enzyme activity is diminished in a stepwise fashion; activity falls to zero only when *all three* products are present in excess. Some organisms use only a *single enzyme* to accomplish the same thing. In these organisms **concerted feedback inhibition** occurs. Each product feedback inhibits the enzyme's activity partially; complete inhibition occurs only when all three products are present in excess.

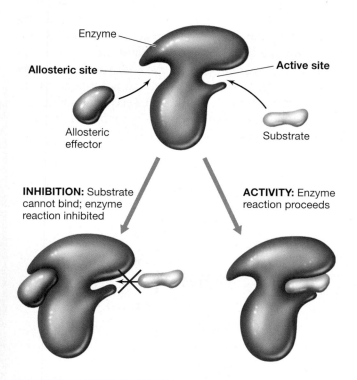

● **Figure 8.3 Allostery.** The mechanism of enzyme inhibition by an allosteric effector. When the effector combines with the allosteric site, the conformation of the enzyme is altered so that the substrate can no longer bind to the active site.

🛑 **8.2 Concept Check**

Many metabolic reactions can be regulated through control of the activities of the enzymes that catalyze them. An important type of regulation of enzyme activity is feedback inhibition, in which the final product of a biosynthetic pathway inhibits the first enzyme unique to that pathway.

◆ What is *feedback inhibition?*

◆ What is an *allosteric enzyme?*

● Figure 8.5 Isoenzymes and feedback inhibition. The common pathway leading to the synthesis of the aromatic amino acids contains three isoenzymes of DAHP synthase (DAHP is 3-deoxy-D-*arabino*-heptulosonate 7-phosphate). Each of these enzymes is specifically feedback-inhibited by one of the aromatic amino acids. Note how an excess of all three amino acids is required to completely shut off the synthesis of DAHP.

8.3 Covalent Modification of Enzymes

Several examples are known in bacteria in which an enzyme is regulated by **covalent modification,** usually by attaching or removing some small molecule to the protein. As in the case of allosteric proteins, the binding of the modifying group changes the conformation of the protein, affecting its catalytic activity. Removal of the modifying group then returns the enzyme to an active state. Common mechanisms of covalent modification include attachment of the nucleotides *adenosine monophosphate (AMP)* or *adenosine diphosphate (ADP)*, *inorganic phosphate* (PO_4^{2-}), or *methyl* (CH_3) *groups.* We consider here the well-studied case of glutamine synthetase, whose activity is modulated by AMP.

Glutamine Synthetase and Adenylylation

The enzyme **glutamine synthetase** (GS) can be covalently modified by *adenylylation,* the addition of AMP. GS plays a key role in the assimilation of ammonia in the cell (∞Section 5.16). The activity of GS is controlled at two levels. First, GS activity is controlled by feedback inhibition by nine different compounds, including several amino acids and compounds involved in nucleotide metabolism. Inhibition is concerted in the sense that each additional effector reduces GS activity proportionally; the presence of all nine inhibitors completely inhibits GS activity.

However, the activity of GS can be regulated independently of feedback inhibition by covalent modification.

This form of regulation is controlled by the concentrations of glutamine and a glutamine precursor, α-ketoglutarate, a citric acid cycle intermediate (∞Figure 5.22) that contains no amino groups.

Each GS molecule contains 12 identical subunits, and each can be adenylylated at a particular site. When the enzyme is completely adenylylated (that is, the enzyme contains 12 AMP groups), it is catalytically inactive. When it is partially adenylylated, it is partially active. Interestingly, the enzyme that adds and removes AMP to GS, an enzyme called P_{II}, is itself regulated by covalent modification. As glutamine levels decrease in the cell, P_{II} becomes covalently modified, and this promotes the deadenylylation of GS and an increase in its activity. As the glutamine pool in the cell increases, GS becomes more highly adenylylated and its activity further diminishes (Figure 8.6●). The activity level of GS can thus be thought of as a nitrogen barometer for the cell. When nitrogen levels are low, GS shows high activity, and when nitrogen levels are high, GS activity decreases. Because the catalytic reaction of GS consumes ATP (∞Figure 5.27b), its regulation is important to the energetic state of the cell.

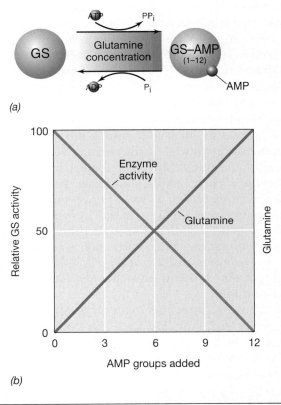

● Figure 8.6 Regulation of glutamine synthetase by covalent modification. (a) When cells are grown in a medium rich in fixed nitrogen, glutamine synthetase (GS) is covalently modified by becoming progressively adenylylated. As many as 12 adenyl (AMP) groups can be added. When the medium becomes nitrogen poor, the groups are removed, forming ADP. (b) Adenylylated GS subunits are catalytically inactive, so the overall GS activity *decreases* as more subunits are adenylylated.

Protein Processing and Inteins

Covalent modification of enzymes for regulatory purposes, such as the adenylylation of GS (Figure 8.6), is readily *reversible*. However, not all posttranslational modification of proteins is reversible. In some cases, newly translated polypeptides must be *processed* before they become active. We look at some mechanisms for this here.

As we learned in Chapter 7, all proteins are synthesized beginning with a methionine (a formylated methionine in cells of *Bacteria*). Therefore, one might expect that all proteins would have a methionine or a formylmethionine on their N-terminal ends. However, this is not true, as most proteins have the initiating methionine removed by a specific enzyme (in *Bacteria* the formyl group of the methionine is removed first, and it is removed from all proteins). We also discussed in Chapter 7 that secretory proteins are synthesized with a leader sequence that is removed as part of the export process from the cell (∞Section 7.17). Both of these are examples of posttranslational modification.

In some posttranslational processing events, a protein may be the result of extensive processing of the original product of translation. For example, some genes in viruses are transcribed and then translated to yield a "polyprotein," a long polypeptide that is subsequently cleaved to yield several different active products. We will see examples of this in our discussion of poliovirus (∞Section 16.8) and retroviruses (∞Section 9.13).

A rather unusual type of posttranslational processing involves removing and discarding *portions* of a protein and then reconnecting the active protein domains. Recall when we briefly discussed the intervening sequences in eukaryotic split genes, called *introns*. These noncoding nucleic acids are removed during processing of the mRNA to yield the final mature RNA (∞Sections 7.1 and 14.8). A few instances of introns are known in prokaryotes, as well. However, a number of instances are known where "extra" information is present in the gene encoding a specific protein, but unlike the case with introns, this information is removed at the level of the *protein* instead of the *RNA*. This process is called **protein splicing** and the peptide removed is called an **intein**.

Protein splicing occurs in a variety of proteins from *Archaea*, *Bacteria*, and *Eukarya*. Figure 8.7 shows how this works in the production of DNA gyrase (topoisomerase II, ∞Section 7.3) in the gram-positive bacterium *Mycobacterium leprae*, the causative agent of the disease leprosy (∞Section 26.5). The gene that encodes the A subunit of the *M. leprae* DNA gyrase is called *gyrA*; it encodes the protein GyrA (Figure 8.7). Note that the flanking sequences in the GyrA polypeptide, referred to as *exteins*, are ligated together to form the final active protein, while the inteins in the pre-processed GyrA are discarded (Figure 8.7). Interestingly, inteins are *self-splicing* entities and thus have the enzymatic activity of a highly specific protease.

● **Figure 8.7 Protein splicing.** The protein synthesized from the *gyrA* mRNA in *Mycobacterium leprae* is 1273 amino acid residues in length. The N-extein is the amino terminal extein and the C-extein is the carboxyl terminal extein. Residues 131 to 550 make up an intein, which removes itself in a self-splicing reaction that generates the free intein and the DNA gyrase A subunit.

Why GyrA requires this unusual processing to form the active protein is unknown. Good evidence exists (at least in eukaryotes) that introns play important regulatory roles in gene expression. It is thus possible that similar regulatory events surround the inteins involved in protein splicing as well.

8.3 *Concept Check*

Covalent modification is a regulatory mechanism for changing the activity of an enzyme. Enzymes regulated in this way can be reversibly modified. One type of modification is adenylylation (the addition of AMP). Protein splicing is a form of posttranslational modification.

◆ What does covalent modification do to enzyme activity?

◆ How can covalent modification be a mechanism that "temporarily" affects an enzyme?

◆ What is the difference between an *intein* and an *intron*?

 III DNA-BINDING PROTEINS AND REGULATION OF TRANSCRIPTION BY NEGATIVE AND POSITIVE CONTROL

In most of the remainder of this chapter we explore mechanisms by which cells control the *amount* of a protein synthesized. Our discussion will be focused on control at the level of *transcription*, where several particularly elegant mechanisms are known in prokaryotes.

Recall that the half-life of a typical mRNA in prokaryotes is short, only a few minutes at best. This is a key strategy that prokaryotes employ to respond quickly to changing environmental parameters. Although there are energy costs in resynthesizing mRNAs that have been translated only a few times and then degraded, this strategy removes unneeded messages rapidly from the cell and prevents production of unneeded proteins. In the growing cell, transcription and mRNA degradation are thus coexisting processes.

For transcription to occur, RNA polymerase must recognize a specific promoter on the DNA and begin functioning. Regulation of transcription, both the turning on and turning off, typically requires proteins that can bind to DNA. Thus, before discussing specific regulatory mechanisms, we must consider the proteins that actually bind to DNA.

8.4 DNA-Binding Proteins

Small molecules are often involved in the regulation of enzyme activity. For instance, in the example given in Figure 8.4, the amino acids proline and arginine bind directly to enzymes and inhibit them. The situation with regard to regulating enzyme *synthesis* is quite different. Although small molecules are often involved in regulating transcription, they rarely do so directly. Instead, they typically influence the binding of certain proteins, called *regulatory proteins,* to specific sites on the DNA. It is these proteins that actually regulate transcription.

Interaction of Proteins with Nucleic Acids

Protein–nucleic acid interactions are central to replication, transcription, and translation, as well as to the regulation of these processes. Protein–nucleic acid interactions may be nonspecific or specific, depending on whether the protein attaches *anywhere* along the nucleic acid or whether the interaction is *sequence specific.* Histones (∞Section 7.3) are good examples of the former. Because they are positively charged, histones combine strongly and relatively nonspecifically with negatively charged DNA. If the DNA is covered with histones, RNA polymerase will be unable to bind and transcription cannot occur. However, removal of histones does not automatically lead to transcription, but simply leaves genes in a position to be activated by other factors. Even after binding and beginning transcription, eukaryotic RNA polymerases need several protein factors to help them elongate through the histone-coated DNA (∞Section 7.11). Histones are universally present in *Eukarya* and are also present in several *Archaea.*

Most DNA-binding proteins interact with DNA in a *sequence-specific* manner. These interactions occur by association of the amino acid side chains of the proteins with the bases and the sugar/phosphate backbone of the DNA. The major groove in DNA, because of its size, is an important site of protein binding. Figure 7.3 identified several of the atoms of the base pairs found in the major groove that are known to interact with proteins. To achieve *specificity* in such interactions, the binding protein must interact simultaneously with more than

one nucleotide, frequently several. This means that a specific binding protein will bind only to DNA containing a *specific base sequence.*

We have already described a structure in DNA called an *inverted repeat* (∞Figure 7.6). Such inverted repeats are frequently the locations at which protein molecules combine specifically with DNA (Figure 8.8●). Note that this interaction does not involve the formation of stem-loop structures in the DNA, as was shown in Figure 7.6. DNA-binding proteins are typically *homodimeric,* meaning that they are composed of two identical polypeptides. On each polypeptide chain is a domain that interacts specifically with a region of DNA in the major groove. When protein dimers interact with inverted repeats on DNA, *each* of the polypeptides in the dimer combines with *each* of the DNA strands (Figure 8.8). The protein does not recognize base sequences, per se, but instead recognizes specific molecular contacts *associated* with specific base sequences.

Structure of DNA-Binding Proteins

Studies of the structure of several DNA-binding proteins from both prokaryotes and eukaryotes have revealed several classes of protein domains that are critical for proper binding of these proteins to DNA. One of these is called the *helix-turn-helix motif* (Figure 8.9●). The helix-turn-helix consists of a polypeptide that attains an α-helix secondary structure; this is the *recognition* helix that interacts specifically with DNA. The recognition helix is linked to three amino acids—the first is typically

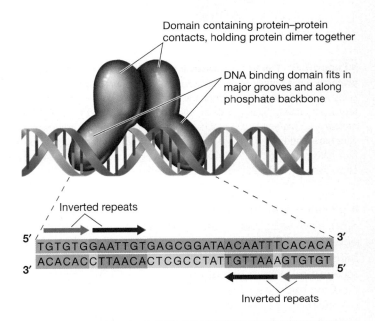

Domain containing protein–protein contacts, holding protein dimer together

DNA binding domain fits in major grooves and along phosphate backbone

Inverted repeats

5′ TGTGTGGAATTGTGAGCGGATAACAATTTCACACA 3′
3′ ACACACCTTAACACTCGCCCTATTGTTAAAGTGTGT 5′

Inverted repeats

●**Figure 8.8 DNA-binding proteins.** Many such proteins are dimers that combine specifically with *two sites* on the DNA. The specific DNA sequences that interact with the protein are *inverted repeats.* The nucleotide sequence of the operator gene of the lactose operon is shown, and the inverted repeats, which are sites at which the *lac* repressor makes contact with the DNA, are shown in shaded boxes.

Helix-turn-helix

(a)

(b)

Stephen Edmondson

● **Figure 8.9** **The helix-turn-helix structure of some DNA-binding proteins.** (a) A simple model of the helix-turn-helix components. (b) A computer model of the bacteriophage lambda repressor, a typical helix-turn-helix protein, bound to its operator gene. DNA is red/blue and the protein is brown/yellow. One subunit of the dimeric repressor is shown in dark brown and the other in dark yellow. Each subunit contains a helix-turn-helix structure. The coordinates used to generate this image were downloaded from the Protein Data Base, Brookhaven, NY.

a glycine—that functions to "turn" the protein (Figure 8.9*a*). The other end of the "turn" is connected to a second helix, the *stabilizing* helix, which stabilizes the first helix by interacting hydrophobically with it. Sequence recognition occurs through noncovalent interactions including hydrogen bonds and van der Waals contacts (∞ Section 3.1) between the protein and specific contacts in the sequence of base pairs on the DNA.

Many different DNA-binding proteins from *Bacteria* contain the helix-turn-helix structure, including many repressor proteins, such as the *lac* and *trp* repressors of *Escherichia coli* (see Section 8.5), and some bacteriophage proteins, such as the phage lambda repressor (Figure 8.9*b*). Indeed, there are over 250 different proteins

known with this motif that bind to DNA to regulate transcription in *E. coli.*

Two other types of protein domains are commonly found in DNA-binding proteins. One of these, the *zinc finger,* is frequently found in regulatory proteins in eukaryotes. The zinc finger is a protein that, as its name implies, binds a zinc ion (Figure 8.10*a* ●). Part of the "finger" of amino acids that is created forms an α-helix, and this recognition helix interacts with DNA in the major groove. There are typically at least two such fingers on the protein involved in binding.

The other protein domain commonly found in DNA-binding proteins is the *leucine zipper.* These proteins contain regions in which leucine residues are spaced every seven amino acids, somewhat resembling a zipper. Unlike the helix-turn-helix and the zinc finger, the leucine zipper does not interact with DNA itself but functions to hold two recognition helices in the correct orientation to bind DNA (Figure 8.10*b*).

Once a protein combines at a specific site on the DNA, a number of outcomes are possible. In some cases, the DNA-binding protein is an enzyme that catalyzes a specific reaction on the DNA, such as transcription by RNA polymerase. In other cases, however, the binding event can *block* transcription (negative regulation, see Section 8.5) or *activate* it (positive regulation, see Section 8.6).

 8.4 Concept Check

Certain proteins can bind to DNA because of interactions between specific domains of the proteins and specific regions of the DNA molecule. In most cases the interactions are sequence-specific. Proteins that bind to nucleic acid may be enzymes that use nucleic acid as substrates, or they may be regulatory proteins that affect gene expression.

◆ What is a *protein domain?*

◆ Why are some interactions specific to certain DNA sequences?

8.5 **Negative Control of Transcription: Repression and Induction**

Transcription is the first step in biological information flow (Figure 8.1); because of this, it is relatively easy to affect gene expression at this point. If one gene is transcribed more frequently than another, there will be a greater abundance of its mRNA in the cell available for translation and therefore a greater amount of its protein product. Several different mechanisms for controlling enzyme synthesis are known in bacteria, and all of them are greatly influenced by the *environment* in which the organism is growing, in particular by the presence or absence of specific small molecules. These molecules signal the physiological status of the cells' environment and can interact with specific proteins like the DNA-binding

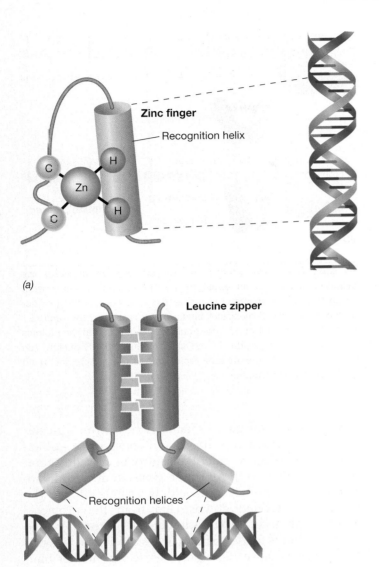

(a)

(b)

● **Figure 8.10 Simple models of protein substructures found in eukaryotic DNA-binding proteins.** α-Helices are represented by cylinders. Recognition helices are the domains involved in DNA binding. (a) The zinc finger structure. The amino acids holding the Zn^{2+} ion always include at least two cysteine residues (C), with the other residues being histidine (H). (b) The leucine zipper structure. The leucine residues (shown in yellow) are spaced exactly every seven amino acids. The interaction of the leucine side chains helps hold the two helices together.

proteins just described. The result is the control of transcription or, more rarely, translation.

We begin our discussion by describing *repression* and *induction*, simple forms of regulation that govern gene expression at the level of *transcription*. In this section we will deal only with **negative control** of transcription, a regulatory mechanism that stops transcription (hence, the term *negative*).

Enzyme Repression and Induction

Often the enzymes that catalyze the synthesis of a specific product are not synthesized if the product is present in the medium in sufficient amounts. For example, the en-

zymes involved in formation of the amino acid arginine are synthesized only when arginine is *absent* from the culture medium; an excess of arginine *represses* the synthesis of these enzymes. This event is called enzyme **repression**.

As can be seen in Figure 8.11●, if arginine is added to a culture growing exponentially in a medium devoid of arginine, growth continues at the previous rate, but the formation of the enzymes involved in arginine synthesis stops. Note—and this is quite important—that this is a *specific* effect, as the synthesis of all other enzymes in the cell continues at the same rate as previously. This is because the enzymes affected by a particular repression event make up only a tiny fraction of the entire complement of proteins in the cell at that time.

Enzyme repression is widespread in bacteria as a means of controlling the synthesis of a variety of enzymes involved in the biosynthesis of amino acids and the nucleotide precursors purines and pyrimidines. In almost all cases, it is the *final product* of a particular biosynthetic pathway that represses the enzymes of the pathway. In these cases repression is quite specific, and the process usually has no effect on the synthesis of enzymes other than those involved in the specific biosynthetic pathway (Figure 8.11). The benefit to the organism of enzyme repression should be obvious. It effectively ensures that the organism does not waste energy and nutrients synthesizing unneeded enzymes.

Enzyme **induction** is conceptually the opposite of enzyme repression. In enzyme induction, an enzyme is made only when its substrate is *present*. Enzyme repression typically involves *biosynthetic* (*anabolic*) enzymes. In contrast, enzyme induction usually involves *catabolic* enzymes.

Consider, for example, the utilization of the sugar lactose as a carbon and energy source by *Escherichia coli*. Figure 8.12● shows this process as regards β-galactosidase, the enzyme that cleaves lactose into glucose and galactose. This enzyme is required for *E. coli* to grow on

● **Figure 8.11 Enzyme repression.** Repression of enzymes involved in arginine synthesis by addition of arginine to the medium. Note that the rate of total protein synthesis remains unchanged.

● **Figure 8.12 Enzyme induction.** Induction of the enzyme β-galactosidase on the addition of lactose to the medium. Note that the rate of total protein synthesis remains unchanged.

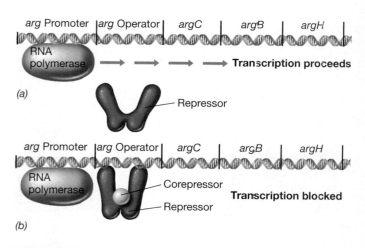

● **Figure 8.13 The process of enzyme repression using the arginine operon as an example.** (a) Transcription of the operon occurs because the repressor is unable to bind to the operator. (b) After a corepressor (small molecule) binds to the repressor, the repressor binds to the operator and blocks transcription; mRNA and the proteins it encodes are not made. In the case of the *argCBH* operon, the repressor would be the arginine repressor and the corepressor would be the amino acid arginine.

lactose. If lactose is *absent* from the medium, the enzyme is not synthesized, but synthesis begins almost immediately after lactose is added. One can immediately see the value to the organism of such a control mechanism, as it provides a means to synthesize specific enzymes *only when they are needed.*

The substance that initiates enzyme induction is called an **inducer,** and a substance that represses enzyme synthesis is called a **corepressor.** These substances, which are always small molecules, are collectively called **effectors.** Not all inducers and corepressors are actual substrates or end products of the enzymes involved. For example, structural *analogs* may induce or repress even though they are not substrates of the enzyme. Isopropyl-thiogalactoside (IPTG), for instance, is an inducer of β-galactosidase even though IPTG cannot be hydrolyzed by the enzyme. In nature, however, inducers and corepressors are probably normal cell metabolites. Interestingly, though, the actual inducer of β-galactosidase, which is one of three proteins encoded by genes in the *lac operon,* is not lactose, but instead the structurally similar compound *allolactose,* a derivative made by the cell from lactose.

Mechanism of Repression and Induction

How can inducers and corepressors affect transcription in such a specific manner? They do this indirectly by combining with specific DNA-binding proteins, which, in turn, affect transcription. In the case of a repressible enzyme (Figure 8.11), the corepressor (in this case, arginine) combines with a specific **repressor protein,** the *arginine repressor,* that is present in the cell (Figure 8.13●). The repressor protein is itself an allosteric protein (see Sections 8.2 and 8.4); that is, its conformation can be altered when the corepressor combines with it.

By binding its effector, the repressor protein becomes *active* and can then combine with a specific region of the DNA near the promoter of the gene, the **operator**

region. This region gave its name to the **operon**, a cluster of genes arranged in a linear and consecutive fashion whose expression is under the control of a single operator (∞Section 7.13). All of the genes in an operon are transcribed as a single unit yielding a single mRNA. The operator is located downstream from the promoter where synthesis of mRNA is initiated (Figure 8.13). If the repressor binds to the operator, transcription is physically blocked because RNA polymerase can neither bind nor proceed. Hence, the polypeptide(s) encoded by the genes in the operon cannot be synthesized. If the mRNA is polycistronic (∞Section 7.13), *all* the polypeptides encoded by this mRNA will be repressed.

Enzyme induction may also be controlled by a repressor. The specific repressor protein is *active* in the *absence* of the inducer, completely blocking transcription. When the inducer is added, it combines with the repressor protein and inactivates it. When this occurs, inhibition is overcome and transcription can proceed (Figure 8.14●).

All regulatory systems involving repressors have the same underlying mechanism: *inhibition* of mRNA synthesis by the activity of specific repressor proteins that are themselves under the control of specific small-molecule inducers and corepressors. And, as previously noted, because the repressor's role is inhibitory, regulation involving repressors is called *negative control.*

⬡ **8.5 Concept Check**

The amount of an enzyme in the cell can be controlled by increasing (induction) or decreasing (repression) the amount of mRNA that encodes the enzyme. This transcriptional regulation involves allosteric regulatory proteins that bind to DNA. For negative control of transcription, the regulatory protein is called a repressor and it functions by inhibiting mRNA synthesis.

● **Figure 8.14 The process of enzyme induction using the lactose operon as an example.** (a) A repressor protein binds to the operator region and blocks the action of RNA polymerase. (b) An inducer molecule binds to the repressor and inactivates it so that it no longer can bind to the operator. Transcription by RNA polymerase occurs, and an mRNA for that operon is formed. In the case of the *lac* operon, the repressor would be the *lac* repressor, and the inducer would be allolactose.

◆ Why is "negative control" so named?

◆ How does a repressor inhibit the synthesis of a specific mRNA?

8.6 Positive Control of Transcription

In negative control, the controlling element—the repressor protein—brings about *repression* of mRNA synthesis. By contrast, in **positive control** of transcription, a regulator protein *activates* the binding of RNA polymerase, hence the term *positive*. An excellent example of positive regulation is the catabolism of the disaccharide sugar maltose in *Escherichia coli*.

Maltose Catabolism in *Escherichia coli*

The enzymes for maltose catabolism in *Escherichia coli* are synthesized only after the addition of maltose to the medium. The expression of these enzymes thus follows the pattern shown for β-galactosidase in Figure 8.12 except that maltose rather than lactose is required to affect gene expression. However, control of the synthesis of maltose-degrading enzymes is not under negative control as in the *lac* operon, but under positive control; transcription requires the activity of an **activator protein**.

The *maltose activator protein* cannot bind to the DNA unless this protein first binds maltose, the effector. When the maltose activator protein binds to DNA, it allows RNA polymerase to begin transcription (Figure 8.15●). Like repressors, activators bind specifically to only certain sequences on the DNA. However, the region on the

DNA that is the site of the activator is not called an operator (Figures 8.13 and 8.14), but instead an *activator-binding site* (Figure 8.15). Nevertheless, the genes controlled by this activator-binding site are still called an *operon*.

Binding of Activator Proteins

In negative control, the repressor binds to the operator and blocks transcription. How does an activator protein work? The promoters of positively controlled operons have nucleotide sequences that are not close matches to the consensus sequence (∞Figure 7.27). Thus, even with the correct sigma factor, the RNA polymerase has difficulty recognizing these promoters.

The activator protein, when bound to DNA, helps the RNA polymerase recognize the promoter and begin transcription. For example, the activator protein may cause a change in the structure of the DNA by bending it (Figure 8.16●), allowing the RNA polymerase to make the correct contacts with the promoter to begin transcription. Alternatively, the activator protein may interact directly with the RNA polymerase. This can happen either when the activator binding site is close to the promoter (Figure 8.17a●) or when it is several hundred base pairs away from the promoter, a situation where DNA looping is required to make the necessary contacts (Figure 8.17b).

Many genes in *Escherichia coli* have promoters under positive control, and many have promoters under negative control. In addition, many operons have a promoter with multiple types of control, while some even have more than one promoter, each with its own control system! Thus, the rather simple picture outlined in Figures 8.13–8.17 is not

● **Figure 8.15 Positive control of enzyme induction using the maltose operon as an example.** (a) In the absence of an inducer, neither the activator protein nor the RNA polymerase can bind to the DNA. (b) An inducer molecule binds to the activator protein, which in turn binds to the activator-binding site. This allows RNA polymerase to bind to the promoter and begin transcription. In the case of the *malEFG* operon, the activator protein would be the maltose activator protein, and the inducer would be the sugar maltose.

DNA

Protein

Thomas A. Steitz and Steve Schultz

●**Figure 8.16 Computer model of the interaction of a positive regulatory protein with DNA.** This figure shows the cyclic AMP-binding protein (CAP protein; see Section 8.7), a regulatory protein involved in the control of several operons. The α-carbon backbone of this protein is shown in blue and purple. The protein is shown binding to a DNA double helix, which is shown in yellow and light blue. Note that binding of this protein to DNA has caused the DNA to be significantly bent.

Operons versus Regulons

In *Escherichia coli*, the genes required for maltose utilization are spread out over the chromosome in several operons, each of which has an activator-binding site to which a copy of the maltose-activator protein can bind. Therefore, the maltose-activator protein actually controls more than one operon. When more than one operon is under the control of a single regulatory protein, these operons are collectively known as a **regulon.** Therefore, the enzymes for maltose utilization are encoded by the *maltose regulon*.

Regulons are also known for operons under negative control. For example, the arginine biosynthetic enzymes (see Section 8.5) are encoded by the *arginine regulon,* whose operons are all under the control of the arginine repressor protein (only one of the arginine operons was shown in Figure 8.13). However, whether functioning as an activator or a repressor, in regulon control, a specific DNA-binding protein binds *only* at the operons it controls; other operons are not affected.

8.6 *Concept Check*

Positive regulators of transcription are called activator proteins. They bind to activator-binding sites on the DNA and stimulate transcription. As in repressors, activator protein activity is modified by effectors. For positive control of enzyme induction, the effector promotes the binding of the activator protein and thus stimulates transcription.

Activator binding site   Promoter

RNA polymerase

Transcription proceeds

Activator protein

(a)

Promoter

RNA polymerase

Transcription proceeds

Activator protein

Activator binding site

(b)

●**Figure 8.17 Activator protein interactions with RNA polymerase.** (a) The activator-binding site is near the promoter. (b) The activator-binding site is several hundred base pairs from the promoter. In this case, the DNA must be looped to allow the activator and the RNA polymerase to contact.

◆ Compare and contrast the activities of an *activator* protein and a *repressor* protein.

◆ Distinguish between an *operon* and a *regulon*.

IV GLOBAL REGULATORY MECHANISMS

An organism often needs to regulate many unrelated genes simultaneously in response to a change in its environment. Regulatory mechanisms that respond to environmental signals by regulating expression of many different genes are called *global control systems*. Both the lactose operon and the maltose regulon respond to global controls. Thus, we begin a consideration of this regulatory system by revisiting the *lac* operon.

8.7 Global Control and the *lac* Operon

Our discussions of negative and positive control did not consider the quite realistic possibility that the environment in which the cells are growing might contain several different carbon sources that the bacteria could use. For example, *Escherichia coli* can use a wide array of carbon sources. When faced with several sugars, including glucose, do cells of *E. coli* use them simultaneously or one at a time? The answer is that *glucose is always used first*. Indeed it would be wasteful to induce enzymes for the metabolism of other sugars if a better carbon source—glucose—was available. One mechanism of global control, *catabolite repression*, ensures that this occurs.

typical of all operons. Multiple control features are common in the operons of virtually all prokaryotes and thus their regulation can be very complex.

Catabolite Repression

In **catabolite repression** the syntheses of a variety of unrelated, primarily catabolic, enzymes are repressed when cells are grown in a medium that contains *glucose*. Catabolite repression is also called the *glucose effect* because glucose was the first substance shown to initiate it. In some organisms, however, carbon sources other than glucose cause catabolite repression. The key point is that the substrate that catabolite represses the utilization of other substrates, simply needs to be the *better* energy source. In this way, catabolite repression ensures that the organism uses the *best* available carbon and energy source first.

One consequence of catabolite repression is that it leads to two exponential growth phases, a situation called *diauxic growth*. If two utilizable energy sources are available to the cell, in diauxic growth the organism grows first on the best energy source. Then, following a lag period, growth resumes on the other energy source. Diauxic growth is illustrated in Figure 8.18● for growth of *Escherichia coli* on a mixture of glucose and lactose.

As we have seen, the enzyme β-galactosidase, which is required for utilization of lactose, is inducible (Figures 8.12 and 8.14). But, in addition, its synthesis is also subject to catabolite repression. Thus, as long as glucose is present, β-galactosidase is not synthesized and lactose is not used. However, when glucose is exhausted, catabolite repression is abolished, and after a brief lag period, β-galactosidase is synthesized and growth on lactose occurs. Notice that Figure 8.18 shows that the cells grow more rapidly on glucose than on lactose. Even though glucose and lactose are both excellent energy sources for *E. coli*, glucose is the *best* of all possible carbon sources for this organism, and thus growth is fastest on this substrate.

Cyclic AMP and CAP

How does catabolite repression work? Catabolite repression involves control of transcription by an activator protein and is therefore a form of *positive* control (see Section 8.6). In the case of catabolite-repressible enzymes, binding of RNA polymerase to the DNA that encodes them occurs only if another protein, called the **catabolite activator protein (CAP),** has bound first. An allosteric protein itself, CAP binds to DNA only if it has first bound a small molecule called *cyclic adenosine monophosphate* or **cyclic AMP** (see Figure 8.16).

Cyclic AMP (Figure 8.19●) is a key molecule in a variety of metabolic control systems, not only in prokaryotes but in eukaryotes as well. Cyclic AMP is synthesized from ATP by an enzyme called *adenylate cyclase*. However, glucose *inhibits* the synthesis of cyclic AMP and *stimulates* cyclic AMP transport out of the cell. When glucose enters the cell, the cyclic AMP level is lowered, and binding of RNA polymerase to the promoters of CAP-specific operons does not occur. Thus, catabolite repression is really an indirect result of the presence of a better carbon source (glucose). The direct cause of catabolite repression is a *cellular deficiency of cyclic AMP.*

Global Aspects of Catabolite Repression

Why is catabolite repression considered a mechanism of *global* control? In the case of *Escherichia coli* and other organisms in which glucose is the prime energy source, as long as glucose is present, catabolite repression prevents expression of *all other* catabolic operons affected by this control mechanism. This may total dozens of catabolic operons. So, not only is lactose catabolism affected, but also affected are the catabolism of maltose, a host of other sugars, and most of the commonly used carbon and energy sources of *E. coli.*

Let us return to the *lac* operon to put aspects of catabolite repression in context with the entire regulatory picture. To do this, the entire regulatory region of the *lac* operon is shown in Figure 8.20●. For transcription of *lac* genes to occur, two requirements must be met: (1) the level of cyclic AMP must be high enough so that the CAP protein binds to the CAP-binding site (*positive control*), and (2) lactose must be present so that the lactose repressor does not block transcription by binding to the operator (*negative control*). Once these two conditions have been met, the cell is signaled that glucose is absent and lactose is present; then and only then does transcription of the *lac* operon begin.

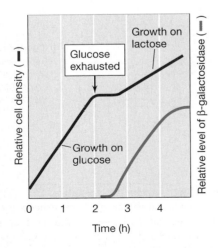

● **Figure 8.18   Diauxic growth on a mixture of glucose and lactose.** Glucose represses the synthesis of β-galactosidase. After glucose is exhausted, a lag occurs until β-galactosidase is synthesized, and then growth can resume on lactose but at a slower rate.

● **Figure 8.19   Cyclic AMP.** Cyclic adenosine monophosphate (cyclic AMP, cAMP) is produced from ATP by the enzyme adenylate cyclase.

(a)

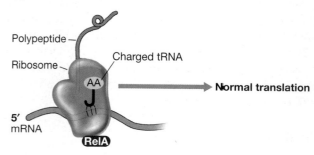

● Figure 8.20 Overall regulation of the lactose operon. The first structural gene in this operon, *lacZ*, encodes the enzyme β-galactosidase, which breaks down lactose (see Figure 8.14). The operon contains two other genes that are also involved in lactose metabolism. The two halves of the operator (where the repressor would bind) are almost perfect inverted repeats. There are also inverted repeats in the CAP binding site, although these are less perfect. The transcriptional start site would be located on the DNA exactly at the 5′-end of the mRNA. The location of the −35 sequence and the Pribnow box, which are part of the promoter (oooFigure 7.30), are also shown. In addition, the location of the base pairs encoding the Shine–Dalgarno sequence and the start codon are also given. These two sequences are critical sequences on the mRNA (oooSection 7.16).

(b)

⬣ **8.7 Concept Check**

Global control systems regulate the expression of many genes simultaneously. Catabolite repression is a global control system, and it helps cells make the most efficient use of carbon sources. The *lac* operon is under the control of catabolite repression as well as its own specific negative regulatory system.

◆ Explain how catabolite repression can involve an activator protein.

◆ Explain how the *lac* operon is both positively and negatively controlled.

(c)

8.8 The Stringent Response

Often times in nature bacterial cells will experience a transient but significant change in nutrient levels. Such conditions can easily be simulated in the laboratory, of course, and much work has been done with *Escherichia coli* and other prokaryotes on the regulation of gene expression following a "shift down" or "shift up" in nutrient status. These include, in particular, the regulatory events triggered by starvation for amino acids.

As a result of a shift down from amino acid excess to limitation—as would occur when a culture is transferred from a rich complex medium to a defined medium with a single carbon source—the syntheses of ribosomal RNA and tRNA cease almost immediately. As a result, no new ribosomes will be produced. These events cause protein and DNA syntheses to be curtailed, while the biosynthesis of new amino acids is activated (Figure 8.21a●). Following such a shift, a variety of new proteins need to be made in order to synthesize the amino acids no longer available in the environment; these are

● Figure 8.21 The stringent response. (a) Upon nutrient downshift, rRNA, tRNA, and protein syntheses temporarily cease. Some time later growth eventually occurs at a new (decreased) rate from that before the response. (b) Structure of guanosine tetraphosphate (ppGpp), a trigger of the stringent response. (c) Synthesis of ppGpp. Normal translation (upper panel) requires charged tRNAs. When amino acid starvation occurs, an uncharged tRNA can bind to the ribosome (lower panel), causing a cessation in ribosome activity. This event triggers the RelA protein to synthesize ppGpp.

made on existing ribosomes. After a time period, rRNA synthesis (and hence, new ribosomes) commences again, but at a new rate commensurate with the cell's reduced growth rate (Figure 8.21*a*). This course of events is called the **stringent response** (or *stringent control*), and is another example of global control.

ppGpp and the Mechanism of the Stringent Response

The triggers of the stringent response are two modified nucleotides, *guanosine tetraphosphate* (ppGpp) and *guanosine pentaphosphate* (pppGpp) (Figure 8.21*b*). In *Escherichia coli*, these nucleotides, which are called **alarmones,** rapidly accumulate during a shift down from amino acid excess to amino acid starvation conditions. Alarmones are synthesized by a specific protein, called *RelA,* using ATP as a phosphate donor (Figure 8.21*b, c*). RelA is associated with the 50S subunit of the ribosome (Table 7.6) and is activated by a signal from the ribosome during nutrient limitation. This signal arises from cessation of ribosome function due to amino acid limitation. When starved for amino acids, *uncharged* tRNAs increase in abundance in the cell pool. Eventually one of these uncharged tRNAs binds to the ribosome. When it does, the ribosome stalls, and this leads to ppGpp/pppGpp synthesis by RelA (Figure 8.21*c*).

The alarmones ppGpp and pppGpp have global control effects. They strongly inhibit rRNA and tRNA synthesis by interfering with RNA polymerase initiation of transcription of genes for these RNAs. On the other hand, alarmones positively activate certain amino acid biosynthetic operons and a variety of catabolic operons that yield macromolecular precursors. By contrast, operons that encode biosynthetic proteins whose amino acid products are present in sufficient amounts remain shut down. Many secondary effects are also observed during the stringent response, including inhibition of the initiation of new rounds of DNA synthesis, membrane lipid synthesis, and cell division processes.

The stringent response can be thought of as a mechanism for adjusting the cell's biosynthetic machinery to conditions of amino acid limitation. By so doing, the cell achieves a new balance between anabolism and catabolism. For most prokaryotes, resources in nature likely appear on an intermittent basis. Thus a global mechanism such as the stringent response that balances a cell's metabolic state with amino acid availability likely improves its competitive state in nature.

● 8.8 Concept Check

The stringent response is a global control mechanism triggered by amino acid starvation. The alarmones ppGpp and pppGpp are produced by RelA, a protein that monitors ribosome activity. The stringent response achieves balance within the cell between protein production and protein requirements.

◆ Which genes are activated during the stringent response, and why?

◆ How are alarmones synthesized?

8.9 Other Global Control Networks

In Section 8.7 we discussed catabolite repression in *Escherichia coli,* and in Section 8.8 we discussed the stringent response. Both are examples of global control. However, there are several other global control systems in *E. coli* (and probably in all prokaryotes), and a few of these are shown in Table 8.1.

Global control systems may include regulation of more than one regulon (see Section 8.7). The term *modulon* is used to describe a group of genes that are regulated by the same regulatory protein even though they may be members of different regulons (and therefore have at least one other type of control). The term *stimulon* is used to describe a group of genes that all respond to the same environmental signal, even though the genes may encode proteins whose functions are

Table 8.1	Examples of global control systems known in *Escherichia coli*[a]		
System	**Signal**	**Primary activity of regulatory protein**	**Number of genes regulated**
Aerobic respiration	Presence of O_2	Repressor (ArcA)	50+
Anaerobic respiration	Lack of O_2	Activator (FNR)	70+
Catabolite repression	Cyclic AMP concentration	Activator (CAP)	300+
Heat shock	Temperature	Alternative sigma (σ^{32})	36
Nitrogen utilization	NH_3 limitation	Activator (NR_I)/alternative sigma (σ^{54})	12+
Oxidative stress	Oxidizing agent	Activator (OxyR)	30+
SOS response	Damaged DNA	Repressor (LexA)	20+

[a] For many of the global control systems, regulation is complex. A single regulatory protein can play more than one role. For instance, the regulatory protein for aerobic respiration is a repressor for many promoters but an activator for others, whereas the regulatory protein for anaerobic respiration is an activator protein for many promoters but a repressor for others. Regulation can also be indirect or require more than one regulatory protein. Some of the regulatory proteins involved are members of two-component systems (see Section 8.12). Many genes are regulated by more than one global system. (For a discussion of the SOS response, Section 10.4.)

unrelated. Examples of these broad global control systems are shown in Table 8.1.

Alternative Sigma Factors

Genes belonging to global control systems do not all use a simple combination of repressors or activators to achieve regulation. Several are controlled by *alternative sigma factors* (Table 8.2). In these cases, regulation is brought about by changing the amount or activity of these sigma factors, since each alternative sigma factor recognizes only a certain subset of genes (for example, particular modulons or stimulons) in the genome.

Recall that sigma is the subunit of RNA polymerase responsible for promoter recognition (Section 7.10). Most genes in *Escherichia coli* require the sigma factor σ^{70} (the superscript 70 indicates the size of this protein, 70 kilodaltons) for transcription and have promoters like those that were shown in Figure 7.30. In total there are seven different sigma factors in *E. coli,* and each recognizes different promoter consensus sequences (Table 8.2). Most of these have counterparts in other *Bacteria*. The endospore-forming bacterium *Bacillus subtilis* has 14 sigma factors, with 4 different sigma factors alone dedicated to the transcription of endospore-specific genes (Section 4.13) alone.

The key to controlling transcription with alternative sigma factors is to control the synthesis and activity of each sigma factor. The *concentration* of each sigma factor in the cell can be modulated by the transcriptional controls we have previously discussed, or by the rate of their degradation within the cell by specific proteases. By contrast, the *activity* of preformed alternative sigma factors can be controlled by other proteins called *anti-sigma factors*, which can temporarily inactivate one sigma factor or another in response to changes in environmental signals (Table 8.1).

Heat Shock Response

Most proteins are relatively stable. Once made, they continue to perform their functions and are passed along at cell division. However, some proteins are recognized by protease enzymes in the cell as being improperly folded and for this or other reasons are rapidly degraded. For example, in *Escherichia coli*, the sigma factor σ^{32} is normally degraded within a minute or two of its synthesis. However, when cells experience a significant and sudden increase in temperature (a *heat shock*), degradation of σ^{32} is inhibited. This means there will be more σ^{32} in the cell and more transcription of operons that contain its specific promoter (Table 8.2). A major set of genes controlled by σ^{32} in *E. coli* are the *heat shock genes*, which encode the **heat shock proteins** expressed during the **heat shock response.**

Heat shock proteins assist the cell in recovering from stress. They are not only induced by heat, but by several other stress factors that the cell can encounter. These include exposure to high levels of certain chemicals—such as ethanol—or from exposure to high doses of ultraviolet radiation.

In *E. coli* and in most prokaryotes examined, there are three major classes of heat shock protein, *Hsp70, Hsp60,* and *Hsp10.* Although not by these names, we have encountered these proteins before. The Hsp70 protein of *E. coli* is *DnaK*, a protein that functions to prevent aggregation of newly synthesized proteins and to stabilize unfolded proteins (Section 7.17). DnaK also plays a role in the degradation of σ^{32}, thus serving as a feedback loop for keeping the amount of this sigma factor in check in unstressed cells. During a stress situation, the role of DnaK is primarily in protein salvaging/repair, so σ^{32} becomes more abundant and transcription of heat shock genes occurs at a high rate.

Major representatives of the Hsp60 and Hsp10 families in *E. coli* are the proteins *GroEL* and *GroES*, respectively. These are *molecular chaperones* that catalyze the correct folding of misfolded proteins (Section 7.17 and Figure 7.40). Another class of heat shock proteins includes various proteases that function in the cell to remove denatured or irreversibly aggregated proteins. Heat shock proteins are apparently quite ancient and are present in all cells. Molecular sequencing of heat shock proteins, especially Hsp70, has been used to help unravel the phylogeny of eukaryotes (Section 14.9).

Table 8.2	**Sigma factors in *Escherichia coli***	
Name[a]	**Upstream (-35) Consensus Recognition Sequence**[b]	**Function**
σ^{70}	TTGACA	For most genes, major sigma factor during normal growth
σ^{54}	TTGGCACA	Nitrogen assimilation
σ^{38}	CCGGCG	Major sigma factor during stationary phase, also for genes involved in oxidative and osmotic responses
σ^{32}	TNTCNCCTTGAA[c]	Heat shock response
σ^{28}	TAAA	For genes involved in flagella synthesis
σ^{24}	GAACTT	Response to misfolded proteins in periplasm
σ^{19}	AAGGAAAAT	For certain genes in iron transport

[a] Superscript number in name indicates size of protein in kilodaltons. Most factors also have other names, i.e., σ^{70} is also sometimes called σ^{D}.

[b] For a discussion of consensus sequences, Sections 7.10 and 7.11 and Figure 7.30.

[c] N, any nucleotide.

When the stress situation has passed, for example, upon a temperature downshift, σ^{32} is rapidly inactivated by DnaK, and the synthesis of heat shock proteins is greatly reduced. Since heat shock proteins perform vital functions in the cell, there is always a low level of these proteins present, even when cells are growing under optimal conditions. However, the rapid synthesis of heat shock proteins in stressed cells emphasizes their importance for survival from excessive heat, chemicals, or physical agents. Heat or other stresses can generate large amounts of inactive proteins that need to be refolded (and in the process, reactivated) or degraded to release free amino acids for the biosynthesis of new proteins.

Cold Shock

Cellular stress induced by cold temperatures also induces a number of genes (the *cold shock response*), but these are not Hsp proteins. Instead, the cold shock response is triggered by the effect that cold has on retarding ribosome function. Cold shock proteins include helicases, nucleases, and ribosome-associated proteins that directly or indirectly interact with DNA, RNA, and ribosomes to decrease macromolecular syntheses. Also induced by cold shock are proteins that increase the production of compatible solutes (oo Section 6.14) in the cell to help defray ice formation in the cytoplasm. Ice nucleation proteins are synthesized if conditions approach freezing; these catalyze the formation of an ice lattice that is less destructive to cell components (such as membranes) than is amorphous ice.

 8.9 Concept Check

Cells can control several regulons by employing alternative sigma factors. These recognize only certain promoters and thus allow transcription of a select category of genes. Other global signals include heat and cold shock proteins that function to help the cell overcome temperature stress.

◆ Why do cells have more than one type of sigma factor?

◆ Why might the proteins induced during a heat shock not be needed during a cold shock?

8.10 Quorum Sensing

As previously mentioned, global control systems allow an organism to regulate many different genes at one time in response to signals in its environment. One potential "signal" that prokaryotes can respond to is the presence in their surroundings of other cells *of the same species*. Some prokaryotes have regulatory pathways that are controlled by the *density* of cells of their own kind. This type of control is called **quorum sensing** (the word *quorum* in this sense means "sufficient numbers").

Mechanism of Quorum Sensing

Quorum sensing is a mechanism to ensure that sufficient cell numbers of a given species are present before eliciting a particular biological response. For example, a pathogenic bacterium that secretes a toxin can have no effect as a single cell; production of the toxin by that cell would be a waste of resources. However, if there is a sufficiently high population of cells present, the coordinated expression of the toxin may successfully initiate disease.

Quorum sensing is widespread among gram-negative *Bacteria*. Each species that employs this type of regulation synthesizes a specific acylated *homoserine lactone* (*AHL*) (Figure 8.22*a*). This molecule is freely diffusible to the outside of the cell. Because of this, the AHL reaches high concentrations in the cell only if there are many cells nearby, each making the same AHL. The specific AHL functions as an inducer that combines with a specific activator protein, thus triggering transcription of specific genes (Figure 8.22*b*).

Quorum sensing was first discovered as the mechanism of regulating bioluminescence in bacteria. Several prokaryotes can emit light in this way, including the marine bacterium *Vibrio fischeri* (oo Section 12.12 and Figure 12.28). Figure 8.23● shows bioluminescent colonies of *V. fischeri*. The light is a by-product of the activity of a

Acyl homoserine lactone (AHL)

(a)

(b)

● **Figure 8.22 Quorum sensing.** (a) General structure of an acyl homoserine lactone (AHL). Different AHLs are structural variants of this parent structure. R, keto group or hydroxyl group. (b) A cell capable of quorum sensing transcribes and translates genes for acyl homoserine lactone synthase at basal levels. This enzyme makes the cells' specific AHL. When numbers of cells of the same species reach a certain level, the concentration of AHL rises sufficiently that it can bind to the activator protein. The latter activates transcription of quorum-specific proteins.

Timothy C. Johnston

● **Figure 8.23 Bioluminescent bacteria producing the enzyme luciferase** (∞Section 12.12). Cells of the bacterium *Vibrio fischeri* were streaked on nutrient agar in a Petri dish and allowed to grow overnight. The photograph was taken in a darkened room using only the light generated by the bacteria.

specific enzyme called *luciferase*. The *lux* operons, which encode the proteins involved in bioluminescence, are under control of an activator protein called *LuxR* and are induced when the concentration of the specific *V. fischeri* AHL, *N*-3-oxohexanoyl homoserine lactone, becomes high enough. This AHL is synthesized by the protein encoded by the *luxI* gene.

Examples of Quorum Sensing

A variety of genes are controlled by a quorum-sensing system, including some in disease-causing (pathogenic) bacteria. In *Pseudomonas aeruginosa*, for instance, quorum-sensing triggers the expression of a large number of unrelated genes when the population density becomes sufficiently high. These genes assist cells of *P. aeruginosa* in the transition from growing freely suspended in liquid to growing in a semisolid matrix called a *biofilm* (∞Section 19.3). The biofilm, formed from specific polysaccharides produced by *P. aeruginosa*, increases the pathogenicity of this organism and prevents the penetration of antibiotics.

The pathogenesis of *Staphylococcus aureus* (∞Sections 25.7 and 26.9) involves, among many other things, the production and secretion of small cell surface and extracellular peptides that damage host cells or that interfere with the immune system. The genes encoding these virulence factors are under the control of a quorum sensing system that responds to a peptide (rather than an AHL) produced by the organism. The regulation of these genes is quite complex and, in part, involves a *regulatory RNA molecule*. Although we have discussed only regulatory proteins in this chapter, reg-

ulatory RNA molecules also exist (see Section 8.14). The *S. aureus* quorum-sensing system also involves regulatory proteins that are part of a two-component, signal-transducing regulatory system. We discuss this type of regulation in Section 8.12.

Quorum-sensing has been discovered in at least one species of *Archaea*. In the haloalkaliphile *Natronococcus occultus* (an organism that requires both high salt concentrations and high pH, ∞Section 13.3), synthesis of an extracellular protease is under control of a specific AHL. This control ensures that the protease is not produced and excreted until enough cells are present for the level of protease to be sufficient to degrade proteins at a reasonable rate. Although this is the only instance of quorum sensing currently known among the *Archaea*, it would be surprising if other *Archaea* do not employ quorum-sensing control for reactions that can benefit them only when cell densities are high enough to have an impact.

 8.10 Concept Check

Quorum sensing allows cells to survey their environment for cells of their own kind. Quorum sensing involves the sharing of specific small molecules. Once a sufficient concentration of the signaling molecule is present, specific gene expression is triggered.

◆ What is the major class of quorum sensing signaling molecules?

◆ In terms of signaling molecules, how does quorum sensing differ in *Pseudomonas* from that in *Staphylococcus*?

V OTHER MECHANISMS OF REGULATION

We have discussed several mechanisms cells use to regulate the transcription of genes: repressors, activators, alternative sigma factors, quorum-sensing molecules. Could there be anything left? A few other important regulatory mechanisms fall outside the scope of our discussion thus far, and we describe them here.

8.11 Attenuation

Some control systems do not employ regulatory proteins binding to DNA. **Attenuation** is such a system. Moreover, in attenuation, the control occurs *after* initiation of transcription but *before* its completion. That is, the number of *completed* transcripts from an operon is reduced, even though the number of *initiated* transcripts is not. The best examples of attenuation involve regulation of genes controlling the biosynthesis of certain amino acids in gram-negative *Bacteria*. The first such system to be described was in the *tryptophan operon* in *Escherichia coli*, and we focus on it here.

(a)

Threonine	Met-Lys-Arg-Ile-Ser-Thr-Thr-Ile-Thr-Thr-Thr-Ile-Thr-Ile-Thr-Thr-Gly-Asn-Gly-Ala-Gly
Histidine	Met-Thr-Arg-Val-Gln-Phe-Lys-His-His-His-His-His-His-His-Pro-Asp
Phenylalanine	Met-Lys-His-Ile-Pro-Phe-Phe-Phe-Ala-Phe-Phe-Phe-Thr-Phe-Pro

(b)

● **Figure 8.24 Attenuation and the leader peptide.** Structure of the tryptophan operon and of tryptophan and other leader peptides in *Escherichia coli.* (a) Arrangement of the tryptophan operon. Note that the leader (L) encodes a short peptide containing two tryptophan residues near its terminus (there is a stop codon following the Ser codon). The promoter is labeled P, and the operator is labeled O. The genes labeled *trpE* through *trpA* encode the enzymes involved in tryptophan biosynthesis. (b) Amino acid sequence of leader peptides synthesized in some other amino acid biosynthetic operons. Because isoleucine is made from threonine, it is an important constituent of the threonine leader peptide.

Attenuation and the Tryptophan Operon

The tryptophan operon contains structural genes for five proteins of the tryptophan biosynthetic pathway plus the usual promoter and regulatory sequences at the beginning of the operon (Figure 8.24●). Like many operons, the tryptophan operon has more than one type of regulation. The first enzyme in the pathway, *anthranilate synthase* (a multisubunit enzyme encoded by *trpD* and *trpE*) is subject to feedback inhibition by tryptophan (see Section 8.2). Transcription of the entire tryptophan operon is also under negative control (see Section 8.5).

However, in addition to the promoter (P) and operator (O) regions needed for negative control, there is a sequence in the operon called the **leader sequence.** The leader encodes a polypeptide that contains tandem tryptophan codons near its terminus and functions as an **attenuator** (Figure 8.24).

The basis of control is as follows. If tryptophan is plentiful in the cell, there will be a sufficient pool of charged tryptophan tRNAs and the leader peptide will be synthesized. On the other hand, if tryptophan is in short supply, the tryptophan-rich leader peptide will *not* be synthesized. Synthesis of the leader peptide results in *termination* of transcription of the remainder of the *trp* operon, which includes the structural genes for the biosynthetic enzymes. By contrast, if synthesis of the leader peptide is blocked by tryptophan deficiency, transcription of the rest of the operon occurs.

Mechanism of Attenuation

How does *translation* of the leader peptide regulate *transcription* of the tryptophan genes downstream? Consider that in prokaryotic cells transcription and translation are simultaneous processes; as mRNA is released from the DNA the ribosome binds to it and translation begins (∞Figure 7.39). That is, while *transcription* of downstream DNA sequences is still proceeding, *translation* of sequences transcribed has already begun (Figure 8.25●).

Attenuation occurs (transcription stops) because a portion of the newly formed mRNA folds into a unique stem-loop that causes cessation of RNA polymerase activity. The

(a)

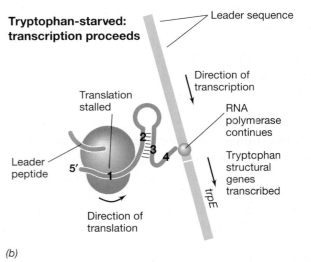

(b)

● **Figure 8.25 Mechanism of attenuation.** Control of transcription of tryptophan operon structural genes by attenuation in *Escherichia coli.* The leader peptide is encoded by regions 1 and 2 of the mRNA. Two regions of the growing mRNA chain are able to form double-stranded loops, shown as 2:3 and 3:4. (a) Under conditions of excess tryptophan, the ribosome translates the complete leader peptide, and so region 2 cannot pair with region 3. Regions 3 and 4 then pair to form a loop that terminates RNA polymerase. (b) If translation is stalled because of tryptophan starvation, loop formation via 2:3 pairing occurs, loop 3:4 does not form, and transcription proceeds past the leader sequence.

stem-loop structure formed by mRNA occurs because two stretches of nucleotides near each other are complementary and can thus base pair. If tryptophan is plentiful, the ribosome will translate the leader sequence until it comes to the leader stop codon. The remainder of the leader RNA then assumes the stem-loop, a *transcription pause site*, which is followed by a uracil-rich sequence that actually causes termination (Figure 8.25).

If tryptophan is in short supply, transcription of genes encoding tryptophan biosynthetic enzymes is obviously desirable. During transcription of the leader, the ribosome pauses at a tryptophan codon because of a shortage of charged tryptophan tRNAs. The presence of the stalled ribosome at this position allows a stem-loop to form that differs from the terminator stem-loop (sites 2 and 3 in Figure 8.25); this alternative stem-loop is *not* a transcription termination signal. Instead, it prevents the terminator stem-loop (sites 3 and 4 in Figure 8.25) from forming. This allows RNA polymerase to move past the termination site and begin transcription of tryptophan structural genes. Thus, in attenuation control, transcription and translation communicate with one another, with the rate of transcription being influenced by the rate of translation.

Attenuation has also been shown to occur in *Escherichia coli* in the biosynthetic pathways for histidine, threonine-isoleucine, phenylalanine, and several other amino acids and essential metabolites as well. As shown in Figure 8.24*b*, the leader peptide for each of these amino acid biosynthetic operons is rich in that particular amino acid. The *his* operon is dramatic in this regard because its leader contains *seven* histidines in a row near the end of the peptide (Figure 8.23*b*). This longer stretch of regulatory codons gives attenuation a major effect in regulation, which may compensate for the fact that unlike the *trp* operon, the *his* operon in *E. coli* is not under negative control.

Translation-Independent Attenuation Mechanisms

Gram-positive *Bacteria*, such as *Bacillus*, also use attenuation to regulate certain amino acid biosynthetic operons. As in gram-negative *Bacteria*, the mechanism also involves attenuation of transcription and alternative mRNA secondary structures, which in one configuration lead to termination. However, the mechanism is *translation independent*, and involves an RNA-binding protein.

In the *Bacillus subtilis* tryptophan operon the binding protein is called the *trp attenuation protein*. In the presence of sufficient amounts of the amino acid tryptophan, this regulatory protein binds to the leader mRNA and causes transcription termination. By contrast, if tryptophan is limiting, the protein does not bind to the mRNA. This allows the favorable secondary structure to form and transcription proceeds.

Attenuation also occurs with genes unrelated to amino acid biosynthesis. These mechanisms obviously do not involve the cell assessing its amino acid pools. Some of the operons involved in pyrimidine biosynthesis (the *pyr* operons) in *Escherichia coli* are regulated by attenuation, and the same is true for *Bacillus*. The mechanisms in the two organisms are, however, quite different, although each employs a system to assess the level of pyrimidines in the cell. In *E. coli* the mechanism involves monitoring the rate of *transcription*, not *translation*. If pyrimidines are not limiting, RNA polymerase can move along and transcribe the leader DNA at a normal rate; this allows for a terminator type stem-loop to form in the mRNA. By contrast, if pyrimidines are in short supply, the polymerase pauses at pyrimidine-rich sequences, and this leads to formation of a nonterminator stem-loop that allows for further transcription.

In *Bacillus*, a different mechanism is employed. For *pyr* attenuation, an RNA-binding protein controls the alternative stem-loop structures of the *pyr* mRNA, terminating transcription when pyrimidines are in excess in the cell. In this way the cell can maintain levels of pyrimidines—compounds that require significant cell resources to biosynthesize (Section 5.16)—at levels needed to balance biosynthetic needs.

8.11 Concept Check

Attenuation is a mechanism whereby gene expression (typically at the level of transcription) is controlled *after* initiation of RNA synthesis. Attenuation mechanisms involve a coupling of transcription and translation and can therefore occur only in prokaryotes.

◆ Why can attenuation not occur in *eukaryotes*?

◆ Explain how the formation of one stem-loop in the RNA can block the formation of another.

◆ How does attenuation of the tryptophan operon differ between *Escherichia coli* and *Bacillus subtilis*?

8.12 Signal Transduction and Two-Component Regulatory Systems

Bacteria regulate cell metabolism in response to a wide variety of environmental fluctuations, including temperature changes, changes in pH and oxygen availability, changes in the availability of nutrients, and even changes in the number of cells present. Therefore, there must be mechanisms by which bacteria receive *signals from the environment* and transmit them to the specific target to be regulated.

We have seen in preceding sections that some signals can be small molecules that enter the cell (often by specific uptake mechanisms) and function as *effectors*. However, in many cases the external signal is not transmitted directly to the regulatory protein, but instead to a *sensor* that transmits the signal to the rest of the regulatory machinery, a process called **signal transduction**.

Sensor Kinases and Response Regulators

Most signal transduction systems contain two parts and are thus also called **two-component regulatory systems**. Characteristically, such systems include two different proteins: (1) a specific **sensor kinase protein** located in the cell membrane, and (2) a partner **response regulator protein**.

The sensor kinase is an enzyme that phosphorylates compounds, typically using phosphate from ATP. Sensor kinases detect a signal from the environment and phosphorylate *themselves* (a process called *autophosphorylation*) at a specific histidine residue on their cytoplasmic surface (Figure 8.26●). Sensor kinases are thus also called *histidine kinases*. The phosphoryl group is then transferred to another protein inside the cell, the *response regulator*. The response regulator is typically a DNA-binding protein that regulates transcription, either in a positive or negative fashion, depending on the system (Figure 8.26).

In order to have a balanced regulatory system, there must be a feedback loop; that is, a way to complete the regulatory circuit and terminate the initial signal. This resets the system for another cycle. This feedback loop involves a *phosphatase*, an enzyme that removes the phosphoryl group from the response regulator protein at a constant rate. In some cases this reaction is carried out by the sensor kinase itself, while in other systems there is a separate protein that carries out this reaction (Figure 8.26). Phosphatase activity is typically a slower process than response regulator phosphorylation. However, if phosphorylation ceases due to reduced sensor kinase activity, phosphatase activity eventually returns the response regulator to the fully unphosphorylated state.

Examples of Two-Component Regulatory Systems

Two-component systems regulate a large number of genes in many different bacteria. Interestingly, two-component systems seem to be less widely distributed among species of *Archaea* than *Bacteria*, and are all but absent in parasitic *Bacteria*. A few key examples of two-component systems include those for phosphate assimilation in *Escherichia coli*, nitrogen fixation in *Klebsiella* and *Rhizobium*, and sporulation in *Bacillus*. In *E. coli* it is estimated that almost 50 different two-component systems operate, and a few of these are listed in Table 8.3.

Many two-component regulatory systems also involve global regulatory phenomena, as well (see Table 8.1). For example, the osmolarity of the environment controls which of the two proteins, OmpC or OmpF, is synthesized as part of the *E. coli* outer membrane. OmpC and OmpF are *porins*, proteins that allow metabolites to cross the outer membrane of gram-negative bacteria (⊂⊃Section 4.9 and Figure 4.29). If osmotic pressure is low, OmpF, a porin with a large pore, is synthesized, while if osmotic pressure is higher, OmpC, a porin with a smaller pore, is synthesized. The response regulator of this system is OmpR. When OmpR is phosphorylated, it *activates* transcription of the *ompC* gene and *represses* transcription of the *ompF* gene. The *ompF* gene in *E. coli* is also controlled by antisense RNA (see Section 8.14).

In some cases multiple regulatory elements are involved. For instance, in the *Ntr regulatory system*, which regulates nitrogen assimilation in many *Bacteria* including *E. coli*, the response regulator is an activator protein, NR_I (nitrogen regulator I). NR_I activates transcription from promoters recognized by RNA polymerase using σ^{54}, an alternative sigma factor (see Section 8.9 and Table 8.2). The sensor kinase in the Ntr system, NR_{II} (nitrogen regulator II), fills a dual role, as both kinase and phosphatase. The activity of NR_{II} is in turn regulated by the state of phosphorylation of another protein, P_{II}.

The *Nar regulatory system* (Table 8.3) controls a set of anaerobically induced genes and contains two different sensor kinases and two different response regulators. In addition, all the genes regulated by this system are also under global control of the anaerobically activated transcriptional regulatory protein FNR (see Table 8.1). It is easy to see from all of this that regulation of gene expression in prokaryotes can quickly get very complicated!

Genomic analyses allow for the ready detection of genes encoding two component regulatory systems, since histidine kinases show significant amino acid sequence conservation. Two-component systems closely related to those in a wide variety of *Bacteria* are also present in microbial

● **Figure 8.26 The control of gene expression by a two-component regulatory system.** The main components of the system include a *sensor kinase* in the cell membrane that phosphorylates itself in response to an environmental signal. The phosphoryl group is then transferred to the other main component, a *response regulator*. In the system diagrammed in this figure the phosphorylated response regulator is a repressor protein. The phosphatase slowly removes phosphate from the response regulator and closes the circuit.

Table 8.3	Examples of two-component regulatory systems that regulate transcription in *Escherichia coli*			
System	**Environmental signal**	**Sensor kinase**	**Response regulator**	**Activity of response regulator**[a]
Arc system	O_2	ArcB	ArcA	Repressor/Activator
Nitrate and nitrite anaerobic regulation (*Nar*)	Nitrate and nitrite	NarX and NarQ	NarL	Activator/Repressor
			NarP	Activator/Repressor
Nitrogen utilization (*Ntr*)	NH_4^+	NR_{II}, the product of *glnL*	NR_I, the product of *glnG*	Activates RNA polymerase at promoters requiring σ^{54}
Pho regulon	Inorganic phosphate	PhoR	PhoB	Activator
Porin regulation	Osmotic pressure	EnvZ	OmpR	Activator/Repressor

[a] Note that several of the response regulator proteins act as both activators and repressors depending on the genes being regulated. Although ArcA can function as either an activator or a repressor, it functions as a repressor on most operons that it regulates.

eukaryotes, such as the yeast *Saccharomyces cerevisiae*. Higher eukaryotes also use phosphorylation as a mechanism of signal transduction in order to respond to environmental changes. Some species of *Archaea,* including in particular the extreme halophiles (*Halobacterium* species, ⊙⊙Section 13.3) and methanogens (⊙⊙Section 13.4), employ two-component systems to regulate certain genes, including those encoding chemotaxis proteins (see Section 8.13).

 8.12 Concept Check

Signal transduction systems transmit environmental signals to the cell. In prokaryotes signal transduction typically involves two-component regulatory systems, which include a membrane-integrated sensor kinase and a cytoplasmic response regulator. The activity of the response regulator depends on its state of phosphorylation.

◆ What are *kinases* and what is their role in two-component regulatory systems?

◆ After depletion of an environmental signal, how are the two components reset for a second stimulus response?

8.13 Regulation of Chemotaxis

Not all two-component systems regulate transcription. We have previously seen how prokaryotes can move toward attractants or away from repellants, a process called *chemotaxis* (⊙⊙Section 4.12). We noted that prokaryotes are too small to actually sense *spatial* gradients of a chemical, but that they can respond to *temporal* gradients. That is, they can sense the *change in concentration* of a chemical over time rather than the *absolute* concentration of the chemical stimulus. Prokaryotes use a two-component system to sense temporal changes in attractants or repellants and process this information to regulate flagellar rotation.

Step One: Response to Signal

The mechanism of chemotaxis is complex and involves a variety of different proteins. A number of *sensory proteins*

reside in the cell membrane, and these sense the presence of attractants and repellants. These sensor proteins are not themselves sensor kinases but are proteins that interact with cytoplasmic sensor kinases. These sensory proteins allow the cell to monitor the concentration of the substance over time.

The sensory proteins are called **methyl-accepting chemotaxis proteins (MCPs).** In *Escherichia coli*, five different MCPs have been identified, and each is a transmembrane protein (Figure 8.27●). Each MCP can sense a variety of compounds. For example, the *Tar* transducer of *E. coli* can sense the attractants aspartate and maltose as well as repellents, such as the heavy metals cobalt and nickel.

MCPs bind attractants or repellents directly, or in some cases indirectly through interactions with periplasmic binding proteins. Binding of an attractant or repellent initiates a series of interactions with cytoplasmic proteins that eventually affects flagellar rotation. Recall that if rotation of the flagellum is *counterclockwise*, the cell will continue to move in a run, whereas if the flagellum rotates *clockwise*, the cell will tumble, reorient itself, and begin to move in a random direction (⊙⊙Section 4.14).

MCPs are in contact with the cytoplasmic proteins CheW and CheA (Figure 8.27). CheA is the *sensor kinase* in chemotaxis. When an MCP has bound a chemical, it changes conformation and (with help from CheW) causes CheA to autophosphorylate, forming CheA-P. Attractants *decrease* the rate of autophosphorylation, whereas repellents *increase* this rate. CheA-P then phosphorylates CheY (forming CheY-P); this is the *response regulator*. CheA-P can also phosphorylate CheB, another response regulator, but this is a much slower reaction than the phosphorylation of CheY. We will discuss the activity of CheB-P later.

Step Two: Controlling Flagellar Rotation

CheY is a key protein in the system because it governs the direction of rotation of the flagellum. CheY-P interacts with the flagellar motor to induce clockwise flagellar rotation and tumbling (the motor switch itself consists of proteins encoded by *fla* genes; ⊙⊙Section 4.14). If unphosphorylated, CheY cannot bind, the flagellar motor

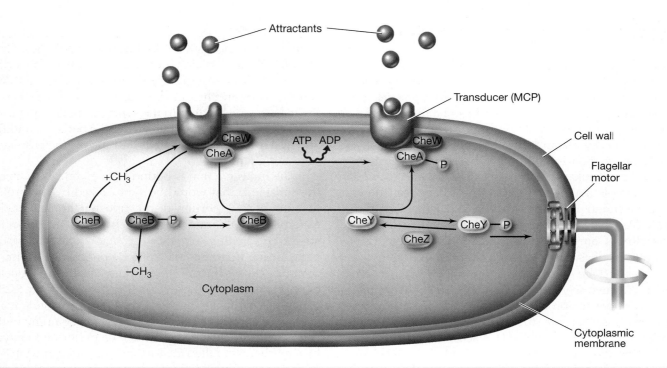

● **Figure 8.27** **Interactions of MCPs, chemotaxis (Che) proteins, and the flagellar motor in bacterial chemotaxis.** The MCP forms a complex with the *sensor kinase* CheA and the coupling protein CheW. This combination results in a signal-regulated autophosphorylation of CheA to CheA-P. CheA-P can then phosphorylate the *response regulators* CheB and CheY. Phosphorylated CheY (CheY-P) interacts directly with the flagellar motor switch. CheZ dephosphorylates CheY-P. CheR continually adds methyl groups to the transducer. CheB-P (but not CheB) removes them. The degree of methylation of the MCPs controls their ability to respond to attractants and repellants and leads to adaptation. The structure of the flagellar motor was shown in Figure 4.56.

continues counterclockwise rotation, and the cell continues its run. Another protein, CheZ, dephosphorylates CheY, returning it to a form that allows runs instead of tumbles. Because repellents increase the level of CheY-P, they lead to tumbling, whereas attractants lead to a lower level of CheY-P and smooth swimming.

Step Three: Adaptation

In addition to processing a signal and regulating flagellar rotation, the system must be able to detect a *change* in concentration of the chemotactic agent with time. This problem is solved by yet another aspect of chemotaxis, the process of **adaptation.** Adaptation is the feedback loop necessary to reset the system. This involves covalent modification of the MCPs by CheB, mentioned earlier.

As their name implies, MCPs can be methylated. There is a cytoplasmic protein, CheR (Figure 8.27), that continually adds methyl groups to the MCPs at a slow rate using S-adenosylmethionine as a methyl donor. The phosphorylated form of the response regulator CheB is a *demethylase* that removes methyl groups from the MCPs. The level of methylation of the MCPs affects their conformation and controls adaptation to a sensory signal. This occurs as follows.

If the level of an attractant remains high, the level of phosphorylation of CheA (and, therefore, of CheY and CheB) will remain low, and the cell will swim smoothly. The level of methylation of the MCPs during this period will increase (because CheB-P is not present to demethylate). However, MCPs no longer respond to the attractant when they are fully methylated. Therefore, even though the level of attractant might remain high, the level of CheA-P (and CheB-P) increases and the cell begins to tumble. However, now the MCPs can be demethylated by CheB-P, and when this happens, the receptors are "reset," and can once again respond to attractants.

The course of events is just the opposite with regard to repellants. Fully methylated MCPs respond best to an increasing gradient of repellants and send a signal for cell tumbling to begin. The cell then moves off in a random direction while MCPs are slowly demethylated. With this mechanism for adaptation, chemotaxis successfully achieves the ability to monitor changing concentrations of both attractants and repellants over time.

 8.13 *Concept Check*

Chemotaxis is under complex regulation involving signal transduction in which regulation occurs in the activity of the proteins involved rather than their synthesis. Adaptation by methylation allows the system to reset itself to the continued presence of a signal.

◆ What is the primary *response regulator* and the primary *sensor kinase* involved in regulating chemotaxis?

◆ Why is adaptation important?

8.14 RNA Regulation and Riboswitches

A major theme in this chapter has been *transcriptional control* of protein synthesis. Less often, genes are controlled at the level of *translation*. Either way, *proteins* are the regulatory elements. However, in some forms of regulation, it is a regulatory *RNA*, not a regulatory *protein*, that plays the key role in controlling gene expression.

RNA regulation is a rapidly expanding area in both prokaryotic and eukaryotic molecular biology. In *Escherichia coli*, for example, a number of **small RNAs** have been found to regulate various aspects of cell physiology by binding to other RNAs or even to small molecules in some cases. Small RNAs (sRNAs) are 40–400 nucleotides long and do not encode proteins, ribosomal RNA, or transfer RNA.

At least three mechanisms for sRNA activity are known. We have seen one sRNA in the *signal recognition particle,* a ribonucleoprotein that identifies newly synthesized proteins that are to be excreted from the cell (∞Section 7.17 and Figure 7.41). A second mechanism of sRNA activity is the binding of the sRNA to a mRNA by complementary base pairing. This ties up the mRNA and prevents its translation (Figure 8.28*a●*). Small RNAs that show this activity are called *antisense RNA*, so-named because the sRNA has a sequence complementary to the coding "sense" of the mRNA. Transcription of antisense RNA is enhanced by conditions that no longer favor expression of the genes they regulate, and the antisense RNAs are encoded by genes complementary to those that produce the mRNA they bind to. Antisense sRNAs survey the cytoplasm for their mRNA complements and form double-stranded RNAs from them. Double-stranded RNAs are not translatable, and these RNAs are soon degraded by specific ribonucleases (Figure 8.28*a*).

Riboswitches

A unique form of small RNAs are the **riboswitches**. These are *mRNAs* that contain sequences upstream of the coding sequences that can bind small molecules (Figure 8.28*b*). Riboswitches are thus far known in a variety of biosynthetic pathways leading to the synthesis of enzymatic cofactors, such as the vitamins thiamine, riboflavin, and B_{12}, and also for a few amino acids and the purine bases adenine and guanine.

The molecule that is bound by a given riboswitch is the substance that would be synthesized by the protein that the mRNA portion of the riboswitch encodes. For example, the thiamine riboswitch binds thiamine upstream of a coding sequence for an enzyme that participates in the thiamine biosynthetic pathway. Binding of thiamine causes the mRNA to attain a stem and loop secondary structure that prevents translation (Figure 8.28*b*).

In riboswitches we see a control mechanism analogous to one we have seen before. At the beginning of this chapter we discussed the control of preformed enzyme by a mechanism called *feedback inhibition* (see Section 8.1). In that case, conformational changes in an allosteric protein following the binding of a specific metabolite shut down enzymatic activity of the entire pathway (see Figure 8.5). A riboswitch can assume either of two structures as well. In the absence of the metabolite, the Shine–Dalgarno ribosome-binding site on the mRNA can bind to the ribosome and translation can begin. However, when the metabolite binds to the mRNA—a signal that a sufficient amount of the metabolite is present—the secondary structure of the riboswitch prevents mRNA binding to the ribosome, and thus translation is prevented.

How widespread are riboswitches and how did they evolve? Thus far riboswitches have only been found in

(a)   (b)

● **Figure 8.28 Regulatory RNAs.** (a) Antisense RNA. Gene A is transcribed from its promoter (∞Section 7.10) to yield an mRNA that can be translated to form protein A. Gene X is a small gene whose sequence is identical to that of a portion of Gene A but has its promoter at the opposite end. Therefore, if it is transcribed, the resulting RNA will be complementary to the mRNA of gene A. If these two RNAs base pair, translation will be blocked. (b) Riboswitches. Binding of a specific metabolite affects the secondary structure of the mRNA portion of the riboswitch, preventing translation.

some bacteria and a few plants and fungi. Some scientists believe that riboswitches are remnants of the *RNA world,* a period eons ago before cells, DNA, and protein, when it is hypothesized that catalytic RNAs were the only self-replicating life forms (∞Section 11.2). In such an environment, riboswitches may have been a primitive mechanism of metabolic control—a simple means by which RNA life forms could have controlled the synthesis of other RNAs. As proteins evolved, riboswitches might have been the first control mechanisms for their synthesis, as well. If true, the riboswitches that remain today are the last vestiges of this simple form of control, since, as we have seen in this chapter, most have been replaced by regulatory proteins.

REVIEW QUESTIONS

1. If an enzyme can be effectively inhibited by feedback inhibition, why would cells also have mechanisms to regulate its synthesis (∞Section 8.1)?

2. Contrast DAHP synthase regulation in *Escherichia coli* with that of an organism employing a concerted feedback inhibition. Which employs more proteins? How can the final result be the same in each case (∞Section 8.2)?

3. In what ways are the events leading to the production of the protein GyrA in *Mycobacterium leprae* similar and dissimilar to gene splicing in eukaryotes (∞Section 8.3)?

4. Describe why a protein that binds to a specific sequence of double-stranded DNA is unlikely to bind to the same sequence if the DNA is single-stranded (∞Section 8.4).

5. Most biosynthetic operons need only be under negative control for effective regulation, while most catabolic operons need to be under *both* negative and positive control. Why (∞Sections 8.5 and 8.6)?

6. Describe the mechanism by which catabolite activator protein (CAP), the regulatory protein for catabolite repression, functions using the lactose operon as an example. For this operon the CAP protein is not a repressor. Describe the regulatory region of a gene for which the CAP protein *is* a repressor (∞Section 8.7).

7. What events trigger the stringent response? Why are the events that occur in the stringent response a logical consequence of the trigger of the response (∞Section 8.8)?

8. Describe the proteins produced when cells of *Escherichia coli* experience a heat shock. Of what value are they to the cell (∞Section 8.9)?

9. How can quorum sensing be considered a regulatory mechanism for conserving cell resources (∞Section 8.10)?

10. Describe how transcriptional attenuation works. What is actually being "attenuated"? Why hasn't the type of attenuation that controls several different amino acid biosynthetic pathways in *Escherichia coli* also been found in eukaryotes (∞Section 8.11)?

11. What are the two components that give the name to signal transduction regulation in prokaryotes? What is the function of each of the components (∞Section 8.12)?

12. Adaptation allows the regulatory mechanism controlling flagellar rotation to be reset. How is this accomplished (∞Section 8.13)?

13. How does regulation by antisense RNA differ from that of riboswitches? Which form of regulation is better suited to control of synthesis of enzymes in a biosynthetic pathway? Why (∞Section 8.14)?

14. Although completely different forms of regulations, a form discussed in Section 8.2 shows similarities to a form discussed in Section 8.14. Explain.

APPLICATION QUESTIONS

1. What would happen to regulation from a promoter under negative control if the region where the regulatory protein binds were deleted? What if the promoter were under positive control?

2. Promoters from *Escherichia coli* under positive control are not close matches to the DNA consensus sequence for *E. coli* (∞Section 7.9). Why?

3. The attenuation control of some of the pyrimidine biosynthetic pathway genes in *Escherichia coli* actually involves coupled transcription and translation. Can you describe a mechanism whereby the cell could somehow make use of translation to help it measure the level of pyrimidine nucleotides?

4. Most of the regulatory systems described in this chapter involve regulatory proteins. However, regulatory RNA is known to exist. Describe how one could achieve negative control of the *lac* operon using either of two different types of regulatory RNA.

5. Many amino acid biosynthetic operons under attenuation control are also under negative control. Considering that the environment of a bacterium can be highly dynamic, what advantage could be conferred by having attenuation as a second layer of control?

ESSENTIALS OF VIROLOGY

Viruses are infectious agents that require cells for their reproduction. Although the genome of many viruses is double-stranded deoxyribonucleic acid (DNA), for others it is single-stranded DNA and for still others, RNA. The infectious form of a virus is called a *virion*, as shown here for bacteriophage T4.

WORKING GLOSSARY

Bacteriophage a virus that infects prokaryotic cells

Early protein a protein synthesized soon after virus infection

Late protein a protein synthesized toward the end of virus infection

Lysogen a bacterium containing a prophage

Lysogenic pathway a series of steps that, after virus infection, lead to a state (lysogeny) where the viral genome is replicated as a prophage along with that of the host

Lytic pathway a series of steps after virus infection that leads to virus replication and the destruction (lysis) of the host cell

Minus (negative)-strand virus a virus with an RNA genome in which the RNA strand has the opposite sense of (is complementary to) the mRNA of the virus

Nucleocapsid the complex of nucleic acid and proteins of a virus

Oncogene a gene whose expression causes formation of a tumor

Plaque a zone of lysis or cell inhibition caused by virus infection of a lawn of sensitive cells

Plus (positive)-strand virus a virus with an RNA or DNA genome in which the genome has the same complementarity as the mRNA of the virus

Prion an infectious protein whose extracellular form contains no nucleic acid

Provirus (prophage) the genome of a temperate virus when it is replicating with, and usually integrated into, the host chromosome

Retrovirus a virus whose RNA genome has a DNA intermediate as part of its replication cycle

Reverse transcription the process of copying information found in RNA into DNA

by the enzyme reverse transcriptase

Temperate virus a virus whose genome is able to replicate along with that of its host and not cause cell death in a state called lysogeny

Transformation in eukaryotes, a process by which a normal cell becomes a cancer cell (but see alternative usage in Chapter 10)

Virion the complete virus particle; the nucleic acid surrounded by a protein coat and in some cases other material

Virulent virus a virus that lyses or kills the host cell after infection; a nontemperate virus

Virus a genetic element containing either RNA or DNA that replicates in cells but is characterized by having an extracellular state

Viroid small, circular, single-stranded RNA that causes various plant diseases

---

Viruses are genetic elements that replicate independently of a cell's chromosome(s) but not independently of cells themselves (∞Section 7.4). However, unlike genetic elements such as plasmids (∞Sections 7.4 and 10.9), viruses have an *extracellular form* that enables them to exist outside the host for long periods and that facilitates transmission from one host to another. In order to multiply, viruses must enter a cell in which they can replicate, a process called **infection**. The infected cell is called the **host**. The study of viruses is a key subdiscipline of microbiology called *virology*, and we introduce the essentials of the field in this chapter.

Viruses, like plasmids and some other genetic elements, exploit the metabolic machinery of the cell. Like these other elements, viruses can confer important new properties on their host cell. These properties will be inherited when the host cell divides if each new cell also inherits the viral genome. These changes are often not harmful and may even be beneficial. Viruses can replicate in a way that is destructive to the host cell, and this accounts for the fact that some viruses are agents of disease. We cover a number of human diseases caused by viruses in later chapters (∞Chapters 26–29).

This chapter is divided into four parts. The first part introduces basic concepts of virus structure. The second part deals with the host cell and how viruses can be quantitated. The third part deals with the basic molecular biology of virus multiplication. The final part provides an overview of a few of the viruses that infect bacteria and animals. Further coverage of viral diversity can be found in Chapter 16. The topics in this chapter also expand on the concepts of macromolecular synthesis and gene regulation covered in Chapters 7 and 8.

Viruses are among the most numerous microorganisms on our planet and infect all types of cellular organisms. Therefore, they are interesting to study in their own right. However, scientists also study viruses for what they can tell us about the genetics and biochemistry of cellular processes and, in the case of many viruses, the development of disease. Furthermore, as we shall see in Chapters 10, 15, and 31, viruses are also important tools for the microbial geneticist and the genetic engineer.

 VIRUS AND VIRION

9.1 General Properties of Viruses

Viruses can exist in either extracellular or intracellular forms. In the *extracellular* form, a virus is a minute particle containing nucleic acid surrounded by protein and, occasionally, depending on the specific virus, other macromolecules. In the extracellular form, the virus particle, also called the **virion**, is metabolically inert and does not carry out respiratory or biosynthetic functions. The virion is the structure by which the virus genome moves from the cell in which it has been produced to another cell where the viral nucleic acid can be introduced. Once in the new cell, the *intracellular state* is initiated. In the intracellular state, virus replication occurs: New copies of the virus genome are produced, and the components that make up the virus coat are synthesized.

Viral genomes are very small, and they encode primarily those functions that they cannot adapt from their hosts. Therefore, during replication inside a cell, viruses depend heavily on host cell structural and metabolic components. The virus redirects host metabolic functions to support virus replication and the assembly of new virions. Eventually, new viral particles are released, and the process can repeat itself.

Viral Genomes

As we have seen (∞Section 7.1), all cells contain double-stranded DNA genomes. By contrast, viruses contain *either* DNA *or* RNA genomes. One group of viruses uses both DNA *and* RNA as their genetic material but at different stages of their reproductive cycle. Viruses can be classified according to whether the nucleic acid in the virion is DNA or RNA and further subdivided according to whether the nucleic acid is single- or double-stranded, linear or circular (Figure 9.1●).

Despite their diverse genome structures, viruses follow the *central dogma* of molecular biology (∞Section 7.1): *Genetic information flows from nucleic acid to protein.* In addition, all viruses use the cell's translational machinery, and so no matter what the genome structure of the virus, messenger RNA (mRNA) must be generated that can be translated on the host's ribosomes.

Viral Hosts and Taxonomy

In addition to their genomes, viruses can also be classified on the basis of the hosts they infect. Thus, we have animal viruses, plant viruses, and bacterial viruses. Bacterial viruses, sometimes called **bacteriophages** (or *phage* for short; from the Greek *phagein*, meaning "to eat"), have been intensively studied as model systems for the molecular biology and genetics of virus reproduction. Species of both *Bacteria* and of *Archaea* are infected by specific bacteriophages. Indeed, many of the basic concepts of virology were first worked out with bacterial viruses and subsequently applied to viruses of higher organisms. Because of their frequent medical importance, *animal viruses* also have been extensively studied, while *plant viruses*, although of enormous importance to modern agriculture, have been the least studied of all viruses.

There exists a formal system of viral classification that groups viruses within various taxa, such as orders, families, and even genus/species, much as has been done with prokaryotes (∞Section 11.10). The *family* taxon seems particularly useful. Members of a family of viruses all have a similar virion morphology, genome structure, and/or strategy of replication. Virus families have names, which include the suffix *-viridae* (as in *Poxviridae*). We will discuss a few of these in Chapter 16.

 9.1 Concept Check

A virion is the extracellular form of a virus and contains either an RNA or a DNA genome. The virus genome is introduced into a new host cell by infection. The virus redirects the host metabolism in order to support virus replication. Viruses are classified by replication strategy as well as by type of host.

◆ How does a *virus* differ from a *plasmid*?

◆ How does a *virion* differ from a *cell*?

◆ What is a *bacteriophage*?

9.2 Nature of the Virion

Viral virions come in many sizes and shapes. Viruses are smaller than prokaryotic cells, ranging in size from 0.02 to 0.3 μm (20–300 nm). A common unit of measure for viruses is the *nanometer*, which is one thousandth of a micrometer. Smallpox virus, one of the largest viruses, is about 200 nm in diameter (about the size of the smallest cells of *Bacteria*). Poliovirus, one of the smallest viruses, is only 28 nm in diameter (about the size of a ribosome).

Viral *genomes* are smaller than those of most cells. Most bacterial genomes are between 1000 and 5000 kbp of DNA, with the smallest known being about 500 kbp. (Interestingly, *Bacteria* with the smallest genomes are, like viruses, parasites that replicate in other cells; ∞Sections 12.13, 12.27, and 13.11). The largest known viral genome, that of bacteriophage G, is 670 kbp. This virus, which infects *Bacillus megaterium*, is, however, one of only a few viruses currently known whose genome is larger than some cellular genomes. More typical virus genome sizes are listed in Table 9.1. Some viruses have genomes so small they contain fewer than five genes. Also, as can be seen in the table, the genome of some viruses, such as reovirus, is *segmented* into more than one molecule.

● **Figure 9.1 Viral genomes.** The genomes of viruses can be composed of either DNA or RNA, and some use both as their genomic material at different stages in their life cycle. However, only one type of nucleic acid is found in the virion of any particular type of virus. This can be single-stranded (ss), double-stranded (ds), or in the case of the hepadnaviruses, partially double-stranded. Some viral genomes are circular, but most are linear.

Table 9.1   **Some types of viral genomes**[a]

| Virus | Host | Viral genome | | | |
		Type of nucleic acid in virion	Structure	Number of molecules	Size[a]
H-1 parvovirus	Animals	Single-stranded DNA	Linear	1	5,176 bases
φX174	*Bacteria*	Single-stranded DNA	Circular	1	5,386 bases
Simian virus 40 (SV40)	Animals	Double-stranded DNA	Circular	1	5,243 base pairs
Poliovirus	Animals	Single-stranded RNA	Linear	1	7,433 bases
Cauliflower mosaic virus	Plants	Double-stranded DNA	Circular	1	8,025 base pairs
Cowpea mosaic virus	Plants	Single-stranded RNA	Linear	2 different	9,370 bases (total)
Reovirus type 3	Animals	Double-stranded RNA	Linear	10 different	23,549 base pairs (total)
Bacteriophage lambda	*Bacteria*	Double-stranded DNA	Linear	1	48,514 base pairs[b]
Herpes simplex virus type I	Animals	Double-stranded DNA	Linear	1	152,260 base pairs
Bacteriophage T4	*Bacteria*	Double-stranded DNA	Linear	1	168,903 base pairs
Human cytomegalovirus	Animals	Double-stranded DNA	Linear	1	229,351 base pairs

[a] The sizes of the viral genomes chosen for this table are known accurately because they have been sequenced. However, this accuracy can be misleading because only a particular strain or isolate of a virus was sequenced. Therefore, the sequence and exact number of bases for other isolates may be slightly different. No attempt has been made to choose the largest and smallest viruses known, but rather to give a fairly representative sampling of the sizes and structures of the genomes of viruses containing both single- and double-stranded RNA and DNA.

[b] This total includes single-stranded extensions of 12 nucleotides at either end of the linear form of the DNA (see Section 9.11).

Viral Structure

The structures of virions are quite diverse, varying widely in size, shape, and chemical composition. The nucleic acid of the virion is always located within the particle, surrounded by a protein shell called the **capsid**. The protein coat is composed of a number of individual protein molecules, called *structural subunits*, which are arranged in a precise and highly repetitive pattern around the nucleic acid (Figure 9.2●).

The small genome size of most viruses restricts the number of different viral proteins. A few viruses have only a single kind of protein in their capsid, but most viruses have several chemically distinct structural subunits that are themselves associated in specific ways to form larger assemblies called **capsomers** (Figures 9.2 and 9.3●). The capsomer is the smallest morphological unit that can be seen with the electron microscope (Figures 9.2 and 9.4●). The information for proper folding and aggregation of the proteins into capsomers is contained within the structure of the proteins themselves, and the overall process of virion assembly is one called **self-assembly**.

A single virion can have a large number of capsomers. The complete complex of nucleic acid and protein, packaged in the virion, is called the virus **nucleocapsid**. Inside the virion are often one or more virus-specific *enzymes*. Such enzymes play a role during the infection and replication process, as we will discuss later in this chapter. Some viruses are *naked* while others contain a membrane around the nucleocapsid (Figure 9.3). The *enveloped viruses* contain a lipid bilayer membrane (●Section 4.5) surrounding the nucleocapsid. Associated with these membranes are typically *virus-specific proteins* that may be critical for attachment of the virion to or release of the virion from the host cell.

18 nm

Structural subunits (capsomers)

Virus RNA

J.T. Finch

(a)   *(b)*

● **Figure 9.2   An example of the arrangement of virus nucleic acid and protein coat in a simple virus, tobacco mosaic virus.** (a) Electron micrograph at high resolution of a portion of the virus particle. (b) Assembly of the tobacco mosaic virion. The RNA assumes a helical configuration surrounded by the protein capsid. The center of the particle is hollow. The Dutchman Martinus Beijerinck, father of the enrichment culture technique (●Section 1.7), was the discoverer of tobacco mosaic virus. Beijerinck studied the disease itself and clearly concluded that the infectious agent was something smaller than a bacterium.

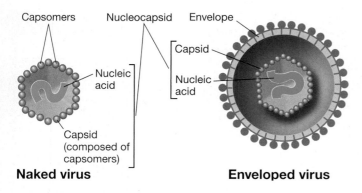

Naked virus

Enveloped virus

● **Figure 9.3** **Comparison of naked and enveloped virus, two basic types of virus particles.** Many enveloped viruses are so because they become covered with host cytoplasmic membrane as they are released from the cell.

Virus Symmetry

The nucleocapsids of viruses are constructed in highly symmetric ways. *Symmetry* refers to the way in which the capsomeres are arranged in the virus capsid. When a symmetric structure is rotated around an axis, the same form is seen again after a certain number of degrees of rotation. Two kinds of symmetry are recognized in viruses, which correspond to the two primary shapes, rod and spherical. Rod-shaped viruses have helical symmetry, and spherical viruses have icosahedral symmetry. In all cases, the characteristic structure of the virus is determined by the structure of the protein subunits of which it is constructed.

A typical virus with **helical symmetry** is the *tobacco mosaic virus* (TMV) illustrated in Figure 9.2. It is an RNA virus in which the 2130 identical capsomers are arranged in a helix. The overall dimensions of the TMV virion are 18 × 300 nm. The lengths of helical viruses are determined by the length of the nucleic acid, but the width of the helical virion is determined by the size and packaging of the protein subunits.

An **icosahedron** is a symmetric structure containing 20 faces and is roughly spherical in shape. Icosahedral symmetry is the most efficient arrangement of subunits in a closed shell because it uses the smallest number of units to build the shell. The simplest arrangement of capsomers is three per face, for a total of 60 units per virion. Most viruses have more nucleic acid than can be packed into a shell made of just 60 morphological units. The next possible structure that permits close packing contains 180 units, and many viruses have shells with this configuration. Other common configurations contain 240 units and 420 units. Figure 9.4a shows a model of an icosahedron. Figure 9.4b shows an electron micrograph of a typical icosahedral virus, human papilloma virus, that contains 360 units. Figure 9.4c shows a computer model of the same virus.

Enveloped Viruses

Many viruses have membranous structures surrounding the nucleocapsid (Figure 9.5a●). Most enveloped viruses infect animals (for example, influenza virus), but a few enveloped bacterial and plant viruses are also known. The virus envelope consists of a lipid bilayer with proteins, usually glycoproteins, embedded in it. The lipids of the membrane are derived from the membranes of the host cell, but the proteins are encoded by the virus. The symmetry of enveloped viruses is expressed not in terms of the virion as a whole but in terms of the nucleocapsid present inside the virus membrane.

What is the function of the membrane in a virus particle? We will discuss this in detail later, but note that the membrane is the structural component of the virus particle that makes initial contact with the cell. The specificity of virus infection, and some aspects of virus penetration, are controlled in part by characteristics of virus membranes.

(a)

W. F. Noyes

(b)

Tim Baker and Norm Olson

(c)

● **Figure 9.4** **Icosahedral symmetry.** (a) A model of an icosahedron. (b) Electron micrograph of human papilloma virus, a virus with icosahedral symmetry. The individual particles are about 55 nm in diameter. (c) Three-dimensional reconstruction of human papilloma virus calculated from images of frozen hydrated virions.

(a)

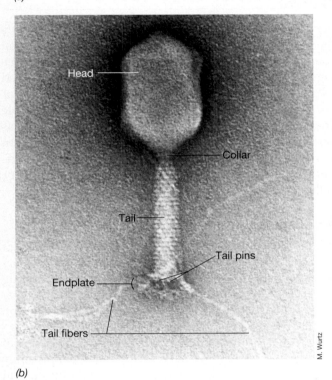

Head

Collar

Tail

Tail pins

Endplate

Tail fibers

(b)

● **Figure 9.5 Electron micrographs of animal and bacterial viruses.** (a) Influenza virus, an enveloped virus. The individual particles are about 80 nm in diameter, but have no defined shape (◯◯Section 16.8 and Figure 16.15). Influenza virus contains a single-stranded RNA genome. (b) Bacteriophage T4 of *Escherichia coli*. The tail components are involved in attachment of the virion to the host and injection of the nucleic acid (see Figure 9.10). The head is about 85 nm in diameter. Phage T4 has a double-stranded DNA genome.

Complex Viruses

Some virions are even more complex than anything discussed so far, being composed of several parts, each with separate shapes and symmetries. The most complicated viruses in terms of structure are some of the bacterial viruses, which possess not only icosahedral heads but also helical tails. In some bacterial viruses, such as the T4 virus of *Escherichia coli* (Figure 9.5*b*), the tail itself has a complex structure. The complete T4 tail

has almost 20 different proteins, and the T4 head has several more proteins. In such complex viruses, assembly is also complex. For instance, in T4 the complete tail is formed as a subassembly, and then the tail is added to the DNA-containing head. Finally, tail fibers formed from another protein are added to make the mature, infectious virion.

Enzymes in Virions

Virions do not carry out metabolic processes and thus outside a host cell, a virion is metabolically inert. However, some virions do contain enzymes that play important roles in the infection process. Some of these enzymes are required for very early events in the infection process. For example, some bacteriophages contain *lysozyme* (◯◯Section 4.8), an enzyme that makes a small hole in the bacterial cell wall. This allows the viral nucleic acid to enter. Lysozyme is again produced in large amounts in the later stages of infection, causing lysis of the host cell and release of the virions.

Some viruses contain their own nucleic acid polymerases. For example, retroviruses are RNA viruses that replicate as DNA intermediates. These viruses possess an RNA-dependent DNA polymerase, called *reverse transcriptase*, that transcribes off of the incoming RNA to form a DNA intermediate. Other viruses contain RNA genomes and require their own RNA polymerase. These virion enzymes are necessary because cells do not make DNA or RNA off of an RNA template (◯◯Sections 7.5 and 7.10).

Some viruses contain enzymes that aid in their release from the host. For example, certain animal viruses contain surface proteins called *neuraminadases*, enzymes that cleave glycosidic bonds in glycoproteins and glycolipids of animal cell connective tissue, liberating the virions. Thus, although most viruses lack their own enzymes, those that contain them do so for good reason: The cell would not be able to produce virions in the absence of these added enzymes.

🛑 **9.2 Concept Check**

In the virion of the naked virus, only nucleic acid and protein are present, with the nucleic acid on the inside; the whole unit is called the nucleocapsid. Enveloped viruses have one or more lipoprotein layers surrounding the nucleocapsid. The nucleocapsid is arranged in a symmetric fashion, with a precise number and arrangement of structural subunits surrounding the virus nucleic acid. Although viruses are metabolically inert, in some viruses, one or more key enzymes are present within the virion.

◆ What is the difference between a *naked* virus and an *enveloped* virus?

◆ What kinds of enzymes can be found within the virions of specific viruses?

GROWTH AND QUANTIFICATION

9.3   The Virus Host

Because viruses replicate only inside *living* cells, the growth of viruses requires use of appropriate hosts. Of the three types of hosts—prokaryotes, plants, and animals—viruses infecting prokaryotes are typically the easiest to grow in the laboratory. For the study of bacterial viruses, pure cultures are used either in liquid or on semisolid (agar) media. Plant viruses can be more difficult to work with, since their study sometimes requires use of the whole plant. This is a problem because plants grow much slower than bacteria, and plant viruses also often require a break in the thick plant cell wall in order to infect. Most animal viruses and many plant viruses can be cultivated in *tissue* or *cell cultures,* and the use of such cultures has enormously facilitated research on these viruses.

Cell Cultures

A cell culture is obtained by promoting growth of cells taken from an organ of an experimental animal. Cell cultures are obtained by aseptically removing pieces of tissue, dissociating the cells by treatment with an enzyme that breaks apart the intercellular cement, and spreading the resulting suspension out on the bottom of a flat surface, such as a bottle or a Petri dish. The thin layer of cells adhering to the glass or plastic dish, called a *monolayer,* is overlaid with a suitable culture medium and incubated at a suitable temperature (see Figure 9.7). The culture media used for cell cultures are generally quite complex, employing a number of amino acids and vitamins, salts, glucose, and a bicarbonate buffer system. To obtain best growth, addition of a small amount of blood serum is usually necessary, and several antibiotics are added to prevent bacterial contamination.

Some cell cultures prepared in this way can be subcultured and grown indefinitely as *permanent cell lines.* Cell lines are convenient for virus research because cell material is continuously available. In other cases, indefinite growth does not occur, but the culture may remain alive for a number of days. Such cultures, called *primary cell cultures,* may still be useful for growing virus, although new cultures need to be prepared from fresh sources from time to time, an expensive and time consuming process. In some cases, primary or permanent cell lines cannot be obtained, but whole organs or pieces of organs can successfully replicate the virus. Such **organ cultures** may still be useful in virus research because they permit growth of viruses under more-or-less controlled laboratory conditions.

9.3   *Concept Check*

Viruses can only replicate in certain types of cells or in whole organisms. Bacterial viruses have proved useful as model systems because the host cells are easy to grow and manipulate in culture. Many animal and plant viruses can be grown in cultured cells.

◆   In virology, what is a *host*?

◆   Why is it helpful to use *cell culture* for viral research?

9.4   Quantification of Viruses

For many purposes in virology it is necessary to *quantify* the number of virions in a suspension. Although one can count virion numbers in an electron microscope (see Figure 9.4), in general, viruses are more easily quantified by measuring effects on their hosts. In this context a *virus infectious unit* is the smallest unit that causes a detectable effect when added to a susceptible host. This can be as few as one virion, although more often a larger inoculum is required. By determining the number of infectious units per volume of fluid, a measure of virus quantity, called **titer**, can be obtained (Figure 9.6●).

Plaque Assay

When a virion initiates an infection on a layer or lawn of host cells growing on a flat surface, a zone of *lysis* may occur that results in a clear area in the lawn of growing host cells. This clearing is called a **plaque**, and it is assumed that each plaque has originated from replication events that began with one virion.

Plaques are essentially "windows" in the lawn of confluent cell growth. With bacteriophages, plaques may be obtained when virus particles are mixed into a thin layer of host bacteria that is spread out as an agar overlay on the surface of an agar medium (Figure 9.6a). During incubation of the culture, the bacteria grow and form a turbid layer that is visible to the naked eye. However, wherever a successful viral infection has been initiated, lysis of the cells occurs, resulting in the formation of a plaque (Figure 9.6b). By counting the number of **plaque-forming units**, one can calculate the number of virus infectious units present in the original sample.

The plaque procedure also permits the isolation of pure virus strains. This is because if a plaque has arisen from a single virion, all the virions in this plaque should be genetically identical. Some of the virions from this plaque can be picked and inoculated into a fresh bacterial culture to establish a pure virus line. The development of the plaque assay technique was as important for the advance of virology as Koch's development of solid media (∞Section 1.6) was for pure culture microbiology.

Plaques may be obtained for animal viruses by using animal cell culture systems as hosts. A monolayer of cul-

(a)   *(b)*

● **Figure 9.6 Quantification of bacterial virus by plaque assay using the agar overlay technique.** (a) A dilution of a suspension containing the virus material is mixed in a small amount of melted agar with the sensitive host bacteria and the mixture poured on the surface of an agar plate of the appropriate medium. The host bacteria, which have been spread uniformly throughout the top agar layer, begin to grow, and after overnight incubation form a *lawn* of confluent growth. Each virion that attaches to a cell and reproduces may cause cell lysis, and the virions released can spread to adjacent cells in the agar, infect them, multiply, and again lead to lysis and release. The size of the plaque formed depends on the virus, the host, and conditions of culture. (b) Photograph of a plate showing plaques formed by a bacteriophage on a lawn of sensitive bacteria. The plaques shown are about 1–2 mm in diameter.

tured animal cells is prepared on a plate or flat bottle, and the virus suspension overlaid. Plaques are revealed by zones of destruction of the animal cells, and from the number of plaques produced, an estimation of the virus titer can be made (Figure 9.7●).

Plating Efficiency

An important concept in quantitative virology involves the concept of *plating efficiency*. In any given viral system, the plaque-forming units are always lower than counts made of the viral suspension with an electron microscope. The efficiency with which virions infect host cells is thus rarely 100% and may often be considerably less. This does not mean that virions that have not caused infection are inactive, although this is sometimes the case. It may merely mean that under the conditions used, successful infection with these particles has not occurred. Although with bacterial viruses, efficiency of plating is often higher than 50%, with many animal viruses it may be much lower, 0.1 or 1%. Knowledge of plating efficiency is useful in growing viruses because it allows one to gauge how dense a viral suspension needs to be (that is, its *titer*) to yield a certain number of plaques.

Whole Animal Methods

Some viruses do not cause recognizable effects in cell cultures yet cause death in the whole animal. In such

cases, quantification can be done only by titration in infected animals. The general procedure is to carry out a serial dilution of the unknown sample (⚭Section 6.5), generally at 10-fold dilutions, and to inject samples of each dilution into several sensitive animals. After a suitable incubation period, the fraction of dead and live animals at each dilution is tabulated and an *end point dilution* is calculated. This is the dilution at which, for example, *half* of the injected animals die (the LD_{50}, ⚭Section 21.8). Although such serial dilution methods are much more

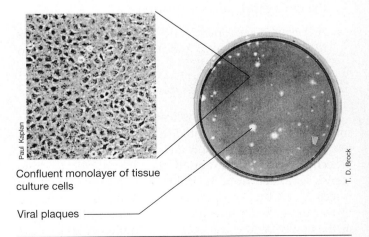

Confluent monolayer of tissue culture cells

Viral plaques ————

● **Figure 9.7 Cell cultures in monolayers grown on a Petri plate.** Note the presence of plaques where virus-induced cell lysis has occurred. Also shown is a photomicrograph of a cell culture.

cumbersome and much less accurate than cell culture methods, they may be essential for the study of certain types of viruses.

 9.4 Concept Check

Although it requires only a single virion to initiate an infectious cycle, not all virions are equally infectious. One of the most accurate ways of measuring virus infectivity is by the plaque assay. Plaques are clear zones that develop on lawns of host cells. Theoretically, each plaque is due to infection by a single virus particle. The virus plaque is analogous to the bacterial colony.

◆ Give a definition of *plating efficiency*.

◆ What is a *plaque-forming unit*?

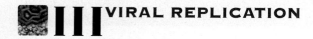 **VIRAL REPLICATION**

9.5 General Features of Virus Replication

For a virus to replicate it must induce a living host cell to synthesize all the essential components needed to make more virions. These components must then be assembled into new virions and escape from the cell in order to infect other cells. The various phases of this replication process in a bacteriophage can be categorized in five steps (Figure 9.8●).

1. *Attachment* (adsorption) of the virion to a susceptible host cell.

2. *Penetration* (injection) of the virion or its nucleic acid into the cell.

3. *Synthesis of nucleic acid and protein* by cell metabolism as redirected by the virus early in the infection. Late in infection, structural proteins that are subunits of the virus capsid are synthesized.

4. *Assembly* of capsomers (and membrane components in enveloped viruses) and *packaging* of nucleic acid into new virions.

5. *Release* of mature virions from the cell.

These stages in virus replication begin when virions infect cells and are illustrated in Figure 9.9●. In the first few minutes after infection the virus is said to undergo an **eclipse**. During this period, the virus nucleic acid becomes separated from its protein coat, and so even if the infected cell breaks open at this point, the virion no longer exists as an infectious entity. **Maturation** begins as the newly synthesized nucleic acid molecules become packaged inside protein coats. During the maturation phase, the titer of active virions *inside* the cell rises dramatically. However, because newly synthesized virions have not yet appeared *outside* the cell, the eclipse and maturation periods are also called the **latent period**.

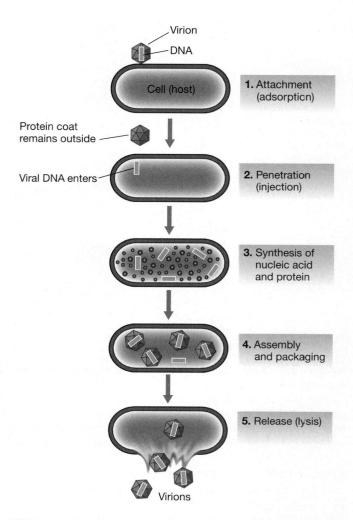

● **Figure 9.8 The replication cycle of a bacterial virus.** The general stages of virus replication are indicated.

At the end of maturation, *release* of mature virions occurs, either as a result of cell lysis or by budding or excretion, depending on the virus. The number of virions released, called the *burst size*, varies with the particular virus and the particular host cell and can range from a few to a few thousand. The timing of this overall virus replication cycle varies from 20–60 min in many bacterial viruses to 8–40 h in most animal viruses. Because the release of virions is a more or less simultaneous process, the virions are said to undergo a **one-step growth curve** (Figure 9.9). In the next two sections we consider a few key steps of the virus multiplication cycle in more detail.

 9.5 Concept Check

The virus life cycle can be divided into five stages: attachment (adsorption), penetration (injection), protein and nucleic acid synthesis, assembly and packaging, and virion release.

◆ What is *packaged* into the virions?

◆ What does the term *eclipse* refer to?

◆ What events occur during the *latent period* of viral replication?

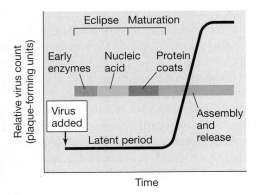

● **Figure 9.9** **The one-step growth curve of virus replication.** This graph displays the results of a single round of viral multiplication in a population of cells. Following adsorption, the infectivity of the virus particles disappears, a phenomenon called *eclipse*. This is due to the uncoating of the virus particles. During the *latent period*, replication of viral nucleic acid and protein occurs. The *maturation period* follows, when virus nucleic acid and protein are assembled into mature virus particles. At this time, if the cells are broken open, active virus can be detected. Finally, *release* occurs, either with or without cell lysis. The timing of the one-step growth cycle varies with the virus and host. Compare this general picture and color scheme with specific replication events shown for bacteriophage T4 in Figure 9.15.

| 9.6 | **Virus Multiplication: Attachment and Penetration** |

In this section we focus on virus attachment and penetration, the first steps in multiplication. In addition, we consider the mechanism by which some bacteria react to penetration by bacteriophage DNA.

Attachment

The most common basis for the host specificity of a virus involves attachment. The virion itself (whether it is naked or enveloped) has one or more proteins on the outside that interact with specific cell surface components called *receptors*. The receptors on the cell surface are normal surface components of the host, such as proteins, carbohydrates, glycoproteins, lipids, lipoproteins, or complexes of these, to which the virion attaches. The receptors carry out normal functions for the cell. For example, the receptor for the bacteriophage T1 is an iron-uptake protein and that for bacteriophage lambda normally participates in maltose uptake. Animal virus receptors may include macromolecules involved in cell–cell contact or the immune system.

In the absence of its specific receptor, the virus cannot adsorb and hence cannot infect. Moreover, if the receptor site is altered, for example, by mutation, the host may become resistant to virus infection. However, mutants of the virus can also arise that are able to adsorb to resistant hosts. In addition, some animal viruses may be able to use more than one receptor, so the loss of one may not necessarily prevent attachment.

Penetration

The attachment of a virus to the cell results in changes to the virus and/or the cell that in turn result in **penetration**. Viruses must replicate within cells. Therefore, at a minimum, the viral genome must enter the cell (Figure 9.8). However, we mentioned (see Section 9.2) that for some viruses to replicate, certain proteins must also enter the host cell. It is important to note that attachment to and even penetration of a susceptible cell will not lead to virus multiplication if the information in the viral genome cannot be read. A cell that allows multiplication of a virus to take place is said to be *permissive* for that virus.

Different viruses have different strategies for penetration. For some animal viruses the enveloped virus is uncoated at the cell membrane. In the case of many animal viruses the entire virion enters the cell. In such cases the virus must subsequently be *uncoated* (partially or completely) so that the genome is exposed and replication can proceed. In some viruses this uncoating takes place in the cytoplasm, while in others, uncoating occurs at the nuclear membrane.

Tailed Bacteriophage Attachment and Penetration

Cells that contain cell walls, such as most bacteria, are infected in a different manner from animal cells, which lack a cell wall. The most complex penetration mechanisms have been found in viruses that infect bacteria. The bacteriophage T4, which infects *Escherichia coli*, is a good example.

The structure of the bacterial virus T4 was shown in Figure 9.5b. The virion has a **head**, within which the viral linear double-stranded DNA is folded, and a long, fairly complex **tail**, at the end of which is a series of tail fibers and tail pins. During attachment, the virions first attach to cells by means of the tail fibers (Figure 9.10●). The ends of the fibers interact specifically with core polysaccharides that are part of the outer layer of the gram-negative cell wall (∞Section 4.8). These tail fibers then retract, and the core of the tail makes contact with the cell wall of the bacterium through a series of fine tail pins at the end of the tail. The activity of a lysozyme-like enzyme results in the formation of a small pore in the peptidoglycan. The tail sheath then contracts, and the viral DNA passes into the cytoplasm through a hole in the tip of the phage tail, with the majority of the coat protein remaining outside (Figure 9.10).

Virus Restriction and Modification by the Host

Animals can often eliminate invading viruses by a variety of immune defense mechanisms before the infection becomes widespread or sometimes even before the virus has penetrated target cells. We will discuss such mechanisms in Chapters 21 and 22. Prokaryotes lack these defenses.

However, prokaryotes do have specific DNA destruction systems that can destroy double-stranded viral

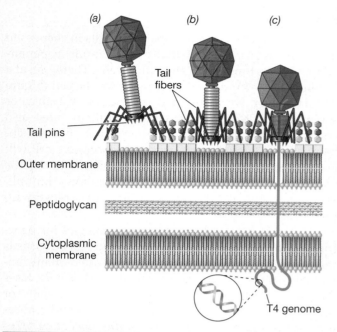

(a) (b) (c)

Tail fibers

Tail pins

Outer membrane

Peptidoglycan

Cytoplasmic membrane

T4 genome

● **Figure 9.10** **Attachment of T4 bacteriophage virion to the cell wall of *Escherichia coli* and injection of DNA.** (a) Attachment of a virion to the cell wall by the long tail fibers interacting with core polysaccharide. (b) Contact of cell wall by the tail pins. (c) Contraction of the tail sheath and injection of the T4 genome. For a detailed description of the gram-negative cell wall, see Section 4.9.

DNA after it has been injected. This destruction is brought about by host restriction endonucleases (Section 7.7), enzymes that cleave the viral DNA at one or several places, thus preventing its replication. This phenomenon is called **restriction** and is part of a general host mechanism to prevent the invasion of foreign nucleic acid. For such a system to be effective, however, the host must have a mechanism for protecting its *own* DNA. This is accomplished by specific **modification** of its DNA at the sites where the restriction enzymes cut (Section 7.7).

Restriction enzymes are specific for double-stranded DNA, and thus single-stranded DNA viruses and all RNA viruses are unaffected by restriction systems. Although the host restriction system confers significant protection, some DNA viruses have overcome host restriction mechanisms by modifying their own nucleic acids so that they are no longer subject to restriction enzyme attack.

Two patterns of chemical modifications of viral DNA have been recognized: *glucosylation* and *methylation*. For instance, the T-even bacteriophages (T2, T4, and T6) have their DNA glucosylated to varying degrees, and the glucosylation prevents endonuclease attack. Many other viral nucleic acids have been found to be modified by methylation. Modification of viral nucleic acid occurs following genomic replication by modification proteins encoded by the virus. Other viruses, such as the bacteriophages T3 and T7, avoid restriction by encoding proteins that inhibit the host restriction systems. To counter this, some hosts have multiple restriction and

methylation systems that help in preventing infection by viruses that can circumvent only one of them.

Hosts also contain other DNA methylases in addition to those involved in protecting themselves from their own restriction enzymes. Some of these methylases may be involved in DNA repair or in gene regulation, but others confer protection on the host from *foreign* endonucleases. This is because some viruses encode restriction systems designed to destroy host DNA. Evolution has richly fueled this cat and mouse game between prokaryotic hosts and their viruses!

 9.6 Concept Check

The attachment of a virion to a host cell is a highly specific process involving complementary receptors on the surface of a susceptible host cell and its infecting virus. Resistance of the host to infection by the virus can involve restriction-modification systems that recognize and destroy double-stranded foreign DNA.

◆ How does the attachment process contribute to virus–host specificity?

◆ Why do some viruses need to be uncoated after penetration while others do not?

9.7 **Virus Multiplication: Production of Viral Nucleic Acid and Protein**

Once infection of a host has occurred, new copies of the viral genome must be replicated and virus-specific proteins must be synthesized in order for virus multiplication to occur. Typically, the production of at least some viral proteins begins very early after the viral genome has been taken up by the cell. The production of these proteins requires virus-specific messenger RNA. For certain types of RNA viruses the genome *itself* is the mRNA. For most viruses, however, the mRNA must first be synthesized from the DNA or RNA genome and then the genome must be replicated. We consider these important events here.

The Baltimore Classification Scheme and DNA Viruses

The virologist David Baltimore, a Nobel Laureate for the discovery of retroviruses and reverse transcriptase, developed a classification scheme for viruses that is based on the relationship of the viral genome to its mRNA. In the Baltimore Classification scheme (Table 9.2), seven classes of viruses are recognized. Double-stranded (ds) DNA viruses are in Class I. The production of mRNA and genome replication in such viruses occurs as it would from the host genome, although different viruses use different strategies to ensure that viral mRNA is expressed in preference to host mRNA.

Table 9.2 **The Baltimore classification system of viruses**

Class	Description of genome and replication strategy	Examples	
		Bacterial viruses	Animal viruses
I	Double-stranded DNA genome	Lambda, T4	Herpesvirus, pox virus
II	Single-stranded DNA genome	φX174	Chicken anemia virus
III	Double-stranded RNA genome	φ6	Reoviruses (◯◯Section 16.9)
IV	Single-stranded RNA genome of plus sense	MS2	Poliovirus
V	Single-stranded RNA genome of minus sense		Influenza virus, rabies virus
VI	Single-stranded RNA genome that replicates with DNA intermediate		Retroviruses
VII	Double-stranded DNA genome that replicates with RNA intermediate		Hepatitis B virus

Class II viruses are single-stranded (ss) DNA viruses. Before mRNA can be produced from such viruses, a *complementary* DNA strand must be synthesized, since RNA polymerase requires double-stranded DNA as a template (◯◯Sections 7.8 and 7.9). These viruses form a double-stranded DNA intermediate during replication, and it is this intermediate that is used for transcription (Figure 9.11●). The synthesis of the ds DNA intermediate and subsequent transcription can be carried out by cellular enzymes (although viral proteins may also be involved). The double-stranded intermediate is also used to replicate the genome; one strand becomes the genome while the other is discarded (Figure 9.11).

Positive and Negative Strand RNA Viruses

The production of mRNA and genome replication is much different with RNA viruses (Classes III–VI). How-ever, before discussing these viruses, some nomenclature related to nucleic acid strand orientation is needed. Recall that mRNA is *complementary* in base sequence to the template strand of DNA. By convention in virology, mRNA is said to be in the *plus* (+) configuration. Its complement is said to be in the *minus* (−) configuration. This nomenclature is also used to describe the configuration of the genome of a single-stranded virus, whether its genome contains RNA or DNA (Figure 9.11). For example, a virus that has a single-stranded RNA genome with the same orientation as its mRNA is said to be a **positive-strand RNA virus**. A virus whose single-stranded RNA genome is complementary to its mRNA is said to be a **negative-strand RNA virus**. How do these viruses replicate?

Cellular RNA polymerases do not catalyze the formation of RNA from an RNA template, but instead require a DNA template. Therefore, RNA viruses, whether plus, negative, or double stranded, require a specific

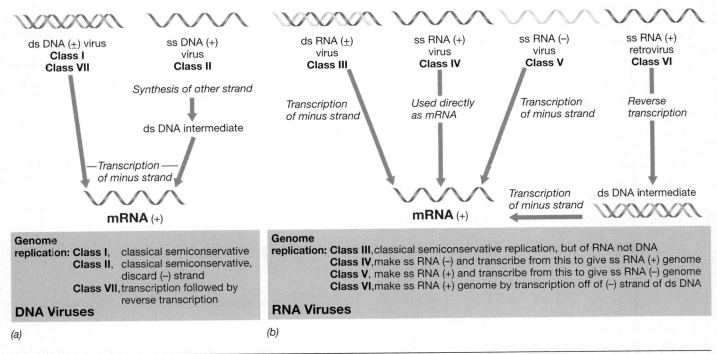

● **Figure 9.11** **Formation of mRNA and new genomes in (a) DNA viruses, and (b) RNA viruses.** By convention, the mRNA is always considered to be of the plus (+) complementarity. Examples of each class of virus are given in Table 9.2.

RNA-dependent RNA polymerase. The simplest case is the positive-strand RNA viruses (Class IV) in which the viral genome is of the plus configuration and hence serves directly as mRNA (Figure 9.11). In addition to other required proteins, this mRNA encodes a *virus-specific and RNA-dependent RNA polymerase* (also called *RNA replicase*). Once synthesized, this polymerase first makes complementary minus strands of RNA and then uses them as templates to make more plus strands. The plus strands produced can either be translated as mRNA or packaged as the genome in newly synthesized virions (Figure 9.11).

For negative-strand RNA viruses (Class V) or double-stranded RNA viruses (Class III), the situation is more complex. In neither case can the incoming RNA serve as mRNA, and therefore mRNA must be synthesized first. However, as mentioned earlier, cells do not have an RNA polymerase capable of this. To circumvent this problem, these viruses contain some of this enzyme in their virions, and it enters into the cell along with the genomic RNA. Therefore, in these cases, the complementary plus strand of RNA is synthesized by this RNA-dependent RNA polymerase and used as mRNA. This plus-strand mRNA is also used as a template to make more negative-strand genomes (Figure 9.11).

Retroviruses

Retroviruses (causal agents of certain kinds of cancers and acquired immunodeficiency syndrome, AIDS) are RNA animal viruses that replicate through a DNA intermediate (Class VI). The process of copying the information found in RNA into DNA is called **reverse transcription**, and thus these viruses require an enzyme called **reverse transcriptase**. Although the incoming RNA of retroviruses is the plus strand, it is not used as message, and therefore these viruses must carry reverse transcriptase in their virions. After infection, the virion ss RNA is copied to a double-stranded DNA through an ss DNA intermediate. The ds DNA is then the template for mRNA synthesis (thus, ss RNA → ss DNA → ds DNA → mRNA).

Finally, Class VII viruses are those that have double-stranded DNA in their virions but replicate through an RNA intermediate. These unusual viruses also use reverse transcriptase (∞Section 16.15). The strategy these viruses use to produce mRNA is the same as that of Class I viruses (Figure 9.11), although DNA replication is very unusual (∞Section 16.15).

While the Baltimore scheme may seem to cover all possibilities, there are exceptions. For example, *ambiviruses* contain a single-stranded RNA genome, half of which is in the plus orientation (and can thus be used as mRNA) and half in the minus configuration (which cannot). A complementary strand must be synthesized from the latter half before the genes there can be translated. Evolution has obviously pushed viral diversity to the limits of possibilities!

Viral Proteins

Once viral mRNA is made (Figure 9.11), viral proteins can be synthesized. The proteins synthesized as a result of virus infection can be grouped into two broad categories:

1. Proteins synthesized soon after infection, called the **early proteins**, which are necessary for the replication of virus nucleic acid, and

2. Proteins synthesized later, called the **late proteins**, which include the proteins of the virus coat.

Generally, both the *timing* and *amount* of virus proteins are highly regulated. The *early* proteins are enzymes that, because they act catalytically, are synthesized in smaller amounts. By contrast, the *late* proteins, usually structural components of the virion, are made in much larger amounts.

Virus infection upsets the regulatory mechanisms of the host because there is a marked overproduction of viral nucleic acid and protein in the infected cell. In some cases, virus infection causes a complete shutdown of host macromolecular synthesis, whereas in other cases, host synthesis proceeds concurrently with virus synthesis. In either case, regulation of virus synthesis is under the control of the *virus* rather than the host. There are several elements of this control that are similar to the regulatory mechanisms discussed in Chapter 8, but there are also some uniquely viral regulatory mechanisms. We discuss these regulatory mechanisms when we next consider some typical viruses.

 9.7 Concept Check

Before replication of viral nucleic acid can occur, new virus proteins are needed. These are encoded by messenger RNA molecules transcribed from the virus genome. In some RNA viruses, the viral RNA itself is the mRNA. In others, the virus genome is a template for the formation of viral mRNA, and in certain cases, essential transcriptional enzymes are contained in the virion.

◆ Why must some types of virus contain enzymes in the virion in order for mRNA to be produced?

◆ Distinguish between a *positive-strand* RNA virus and a *negative-strand* RNA virus.

◆ Both plus strand viruses and retroviruses contain plus configuration RNA genomes. Contrast mRNA production in these two classes of viruses.

 IV VIRAL DIVERSITY

9.8 Overview of Bacterial Viruses

Bacteriophages are quite diverse, and examples of the various classes of bacterial viruses are illustrated in Figure 9.12●. Most of the bacterial viruses that have been studied in detail infect species of *Bacteria* of the enteric

● **Figure 9.12 Schematic representations of the main types of bacterial viruses.** Those discussed in detail either here or in Chapter 16 are M13, φX174, MS2, T4, lambda, T7, and Mu. Sizes are to approximate scale. The nucleocapsid of φ6 is surrounded by a membrane.

group, such as *Escherichia coli* and *Salmonella typhimurium*. However, viruses are known that infect a variety of prokaryotes, both *Bacteria* and *Archaea*.

Most well-studied bacteriophages contain double-stranded DNA genomes, and this type of bacteriophage is thought to be the most common type in nature. However, there are many other kinds known, including those with single-stranded RNA genomes, double-stranded RNA genomes, and single-stranded DNA genomes (Figure 9.11).

A few bacterial viruses have lipid envelopes, but most are naked. However, many bacterial viruses are structurally complex. All examples of bacteriophages with double-stranded DNA genomes shown in Figure 9.12 have tails. The tails of bacteriophages T2, T4, and Mu are contractile and are involved in nucleic acid penetration of the host (Figure 9.10). By contrast, the tail of phage lambda is flexible.

Although tailed bacterial viruses were first studied as *model systems* for understanding general features of virus multiplication, some of them are now used as convenient tools for *genetic engineering*. Thus, the information on bacterial viruses is not only valuable as background for the discussion of animal viruses but also is essential for the material presented in the chapters on microbial genetics (∞Chapter 10) and genetic engineering (∞Chapter 31).

In the next two sections of this chapter we will briefly examine two different types of viral life cycles: **virulent** and **temperate**. Virulent viruses lyse or kill their hosts after infection, while temperate viruses can achieve a state where their genome replicates along with the host genome without killing their hosts.

 9.8 Concept Check

Bacterial viruses, or bacteriophages, are very diverse. The best-studied bacteriophages infect enteric bacteria such as *Escherichia coli* and are structurally quite complex, containing heads, tails, and other components.

◆ What is thought to be the most common type of bacteriophage genome?

9.9 Virulent Bacteriophages and T4

The first viruses to be studied in any detail were a number of bacteriophages with linear, double-stranded DNA genomes that infect *Escherichia coli* and a number of related *Bacteria*. Virologists began isolating and studying these viruses as model systems for virus replication and used them to establish many of the fundamental principles of molecular biology and genetics. These phages were given designations of T1, T2, and so on, up to T7. Already in this chapter we have briefly mentioned how one of these viruses, T4, attaches to its host and how its DNA penetrates the host (see Section 9.6, Figure 9.10). In this section we consider this virus in more detail to illustrate the replication cycle of virulent viruses.

The Genome of T-Even Bacteriophages

Bacteriophages T2, T4, and T6 are closely related viruses, but T4 is the most extensively studied. The virion of phage T4 is structurally complex (see Figure 9.5b). It consists of an elongated icosahedral head whose overall dimensions are 85 × 110 nm. To this head is attached a complex tail consisting of a helical tube (25 × 110 nm) to which are connected a sheath, a connecting "neck" with "collar," and a complex end plate, to which are attached long, jointed tail fibers (see Figure 9.5b). Altogether, the virus contains over 25 distinct types of structural proteins.

The genome of T4 is a double-stranded *linear* DNA molecule of 168,903 base pairs. The T4 genome encodes over 250 different proteins, and although no known virus encodes its own translational apparatus, T4 does encode several of its own tRNAs. While the T4 genome has a unique linear sequence, the genome in one virion may not be exactly the same as that of another. This is because the DNA of phage T4 is *circularly permuted*. Molecules that are circularly permuted appear to have been linearized by opening a circle, but at different locations. In addition to circular permutation, the DNA in each T4 virion has repeated sequences at each end called *terminal repeats* of about 3–6 kbp. Both of these factors affect genome packaging.

The terminally redundant T4 DNA infecting a single host cell is first replicated as a unit, and then several genomic units are recombined end-to-end to form a long DNA molecule called a *concatemer* (Figure 9.13●). The packaging mechanism of T4 DNA involves cutting a segment of DNA from the concatemer sufficient to fill a

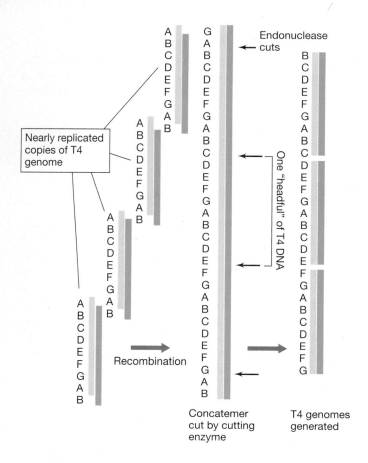

A B C D E F G

Endonuclease cuts

Nearly replicated copies of T4 genome

One "headful" of T4 DNA

Recombination

Concatemer cut by cutting enzyme

T4 genomes generated

● **Figure 9.13 Circular permutation.** Generation of virus length T4 DNA molecules with permuted sequences by an endonuclease that cuts off constant lengths of DNA irrespective of the sequence. Left: nearly replicated copies of infecting T4 genome are recombined to form a concatemer. Middle: arrows, sites of endonuclease cuts. Right: molecules generated. Note how each of the T4 genomes formed on the right contains genes A–G, but that the termini are unique in each molecule.

phage head (at least one genomic equivalent) rather than cutting the DNA at a specific sequence. Since the T4 head holds slightly more than a genome length, this "headful mechanism" leads to circular permutation and terminal redundancy. T4 DNA contains the modified base 5-hydroxymethylcytosine instead of cytosine (Figure 9.14●). It is these residues that are glucosylated (see Section 9.6), and DNA with this modification is resistant to virtually all known restriction enzymes. Consequently, the incoming T4 DNA is well protected from host defenses.

Events During T4 Infection

Things happen rapidly in a T4 infection. Early in infection T4 directs the synthesis of its own RNA and also begins to replicate its unique DNA (Figure 9.14). About 1 minute after attachment and penetration of the host by T4 DNA, the synthesis of host DNA and RNA ceases, while transcription of specific phage genes begins. Translation of viral mRNA begins soon after, and within 4 minutes of infection, phage DNA replication has begun.

Site of glucosylation ———— HOH₂C

NH₂

N

O

N

H

● **Figure 9.14 The unique base in the DNA of the T-even bacteriophages, 5-hydroxymethylcytosine.** The site of glucosylation is shown.

The T4 genome can be divided into three parts, encoding early proteins, middle proteins, and late proteins, respectively (Figure 9.15●). The **early** and **middle proteins** are primarily enzymes involved in DNA replication and transcription, while the **late proteins** are the head and tail proteins and the enzymes involved in liberating the mature phage particles from the cell. The time course of events during T4 infection is shown in Figure 9.15.

Although T4 has a very large genome for a virus, it does not encode its own RNA polymerase. The control of T4 mRNA synthesis involves the production of proteins that sequentially modify the specificity of the host RNA polymerase so that it recognizes phage promoters. The *early* promoters are read directly by the host RNA polymerase and involve the function of host *sigma* factor. Host transcription is shut down shortly thereafter by a phage encoded *anti-sigma factor* that binds to host σ^{70} (∞Section 8.9) and interferes with its recognition of host promoters.

Phage-specific proteins synthesized from the early genes also carry out covalent modifications on the host RNA polymerase α subunits (∞Section 7.10), and a few phage-encoded proteins also bind to the polymerase. These modifications change the specificity of the polymerase so that it now recognizes T4 *middle* promoters. One of the T4 early proteins, called *MotA*, recognizes a particular DNA sequence in middle promoters and guides RNA polymerase to these sites. Transcription from the *late* promoters requires a new T4-encoded sigma factor. Sequential modification of host cell RNA polymerase as described here for phage T4 is used to regulate gene expression by many other bacteriophages as well.

T4 encodes over 20 new proteins that are synthesized early after infection. These include enzymes for the synthesis of the unusual base 5-hydroxymethylcytosine (Figure 9.14) and for its glucosylation, as well as an enzyme that degrades the normal DNA precursor deoxycytidine triphosphate. In addition, T4 encodes a number of enzymes that have functions similar to those of host enzymes in DNA replication but that are formed in larger amounts, thus permitting faster synthesis of T4-specific DNA. Additional early proteins include those involved in the processing of newly replicated phage DNA (see Figure 9.13).

● **Figure 9.15** **Time course of events in phage T4 infection.** Following injection of DNA, early and middle mRNA is produced that codes for nucleases, DNA polymerase, new phage-specific sigma factors, and various other proteins involved in DNA replication. Late mRNA codes for structural proteins of the phage virion and for T4 lysozyme, needed to lyse the cell and release new phage particles.

Most of the *late* genes encode structural proteins for the virion, including those for the head and tail. Assembly of heads and tails occurs independently; DNA is packaged into the assembled head (Figure 9.15) and the tail and tail fibers are added later. Exit of the virus from the cell occurs as a result of cell lysis. The phage codes for a lytic enzyme, *T4 lysozyme*, which attacks the peptidoglycan of the host cell. After a lytic cycle, which takes only about 25 min (Figure 9.15), over 100 new virions will be released from each host cell, which itself has now been almost completely destroyed.

 9.9 **Concept Check**

After a virion of T4 attaches to a host cell and the DNA penetrates into the cytoplasm, the expression of viral genes is regulated so as to redirect the host synthetic machinery to the reproduction of viral nucleic acid and protein. New virions are then assembled and released from the cell by cell lysis. T4 has a double-stranded DNA genome that is circularly permuted and terminally redundant.

◆ What does it mean that the bacteriophage T4 genome is both *circularly permuted* and *terminally redundant*?

◆ Give an example of a mechanism used by T4 to ensure that its genes rather than those of the host are transcribed.

9.10 Temperate Bacteriophages

Bacteriophage T4 is **virulent**. However, some other viruses, although also able to kill cells through a lytic cycle, have the option of undergoing a different life cycle resulting in a stable genetic relationship with the host. Such viruses are called **temperate viruses**. These viruses can enter into a state called **lysogeny**, where most virus genes are not expressed and the virus genome, called a **prophage**, is replicated in synchrony with the host chromosome.

The temperate phage genome can be replicated along with that of the host and during cell division be passed from one generation to the next. Under certain conditions cells that harbor a temperate virus, called **lysogens**, can spontaneously produce and release virions.

Lysogeny is probably of ecological importance because most bacteria isolated from nature are lysogens for one or more bacteriophages. Lysogeny can also confer new genetic properties on the bacterial cell, and we will see several examples in later chapters of pathogenic bacteria whose virulence is due, at least in part, to the lysogenic bacteriophage they harbor. Also, we will see that lysogeny is not limited to bacteriophages. Many animal viruses establish similar relationships with their hosts.

The Life Cycle of a Temperate Phage

In a typical lytic virus it is not the presence (or even the replication) of viral DNA that leads to the production of new virions and host cell death. Rather it is *expression* of the viral genome that is deleterious. Host cells can harbor viral genomes without harm if the expression of the viral genes can be controlled. This is the situation found in lysogens. However, if this control is lost, the virus enters the **lytic pathway** and produces new virions, eventually lysing the host cell. Lysogeny can thus be considered a genetic trait of a bacterial strain.

An overall view of the life cycle of a temperate bacteriophage is shown in Figure 9.16●. The temperate virus does not exist in its extracellular form inside of the cell.

● **Figure 9.16 The consequences of infection by a temperate bacteriophage.** The alternatives upon infection are *replication* and release of mature virus (lysis) or *integration* of the virus DNA into the host DNA (lysogenization). The lysogenic cell can also be induced to produce mature virus and lyse.

Instead, the prophage is integrated into the bacterial chromosome and replicates along with the host cell as long as the genes controlling its lytic pathway are not expressed. Typically this control is maintained by a phage-encoded repressor protein (the gene encoding the repressor protein *is* expressed). The virus repressor protein not only controls the lytic genes on the prophage but also prevents the expression of any incoming genes of the same virus. This results in the lysogens having **immunity** to infection by the same type of virus.

If the phage repressor is inactivated or if its synthesis is prevented, the prophage is induced (Figure 9.16). Induction results in the production of new virions and the lysis of the host cell. In some cases (as we shall see later), induction can be brought about by environmental conditions. If the virus loses the ability to leave the host genome (because of mutation), it becomes a cryptic virus. Genomic studies (⌒Chapter 15) have shown that many bacterial chromosomes contain DNA se-

quences that were clearly once part of a viral genome. Thus the establishment and breakdown of the lysogenic state is likely a dynamic process in prokaryotes.

We now focus on a bacteriophage whose lysogenic properties are known in great detail, bacteriophage lambda.

■ **9.10 Concept Check**

Lysogeny is a state in which viral genes become integrated into the host chromosome and lytic events are repressed. Viruses capable of inducing the lysogenic state are called temperate viruses. In lysogeny, the virus genome becomes a prophage; however, lytic events can be induced by certain environmental stimuli.

◆ What are the two pathways available to a temperate virus?

◆ What might the advantage be to hosts of temperate viruses?

9.11 Bacteriophage Lambda

One of the best-studied temperate phages is **lambda**, which infects *Escherichia coli*. As with other temperate viruses, both the *lytic* and the *lysogenic* pathways are possible (Figure 9.16). Morphologically, lambda virions look like those of many other tailed bacteriophages, although unlike phage T4, no tail fibers are present (Figure 9.17●; see also Figure 9.12). The genome of lambda consists of a linear *double-stranded* DNA molecule, but at the 5′-terminus of each of the strands is a *single-stranded* tail 12 nucleotides long. These single-stranded ends are complementary (the ends of the DNA are said to be *cohesive*). Thus, when the two ends of the DNA are free in the host cell they associate (the *cos* site), and the DNA is ligated to form a double-stranded circle of 48,502 base pairs. Figure 9.18● shows this process along with the genetic map of lambda.

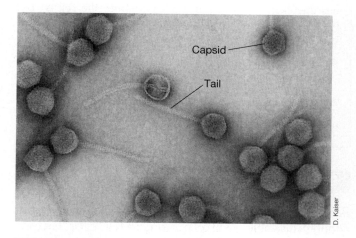

● **Figure 9.17 Bacteriophage lambda.** Electron micrograph by negative staining of phage lambda virions. The head of each virion is about 65 nm in diameter.

(a)

(b)

● **Figure 9.18   The lambda genome.** (a) Conversion of lambda DNA from a linear to a circular molecule upon infection. (b) Genetic and molecular map of lambda. The genes are designated by letters; *att*, attachment site for phage to host chromosome. Regulatory genes of special interest: *cI*, repressor protein; O_R, operator right; P_R, promoter right; O_L, operator left; P_L, promoter left; *cro*, gene for second repressor; N, antiterminator protein. The regulatory region of lambda (shown in yellow) is positioned at the top of this circular map. The site created when the cohesive ends of the lambda genome join is called *cos* (shown in red). Early transcription in lambda is primarily leftward (counterclockwise, L1) from P_L and rightward (clockwise, R1) from P_R. The three transcription terminators affected by the N-protein are shown as blue boxes on the DNA. The late rightward transcript, which encodes head and tail proteins and proteins for lytic function, is labeled R2 and begins at a promoter near the Q gene. The transcript labeled L2 is the positively regulated transcript from P_E that encodes the repressor protein.

Lambda Infection and the Lytic Pathway

The lambda virion attaches to a specific receptor protein in the cell wall of *Escherichia coli* (see Section 9.6) and injects its DNA. The DNA circularizes almost immediately (Figure 9.18*a*), and expression of the phage genome begins. The first steps in gene expression are the same whether the final result is lysis or lysogeny.

Production of RNA using host RNA polymerase begins at two key promoters called P_L (*promoter left*) and P_R (*promoter right*) (Figure 9.18*b*). These yield short transcripts that are translated to give the products of the *N* and the *cro* genes, both proteins being involved in regulatory events. The *Cro* protein regulates whether the lytic or the lysogenic pathway is followed, and we discuss its function later. The N protein is an *antiterminator* that allows the RNA polymerase to transcribe past specific terminators (Figure 9.18*b*), extending the transcripts from P_L and P_R. These longer transcripts can be translated to yield more proteins, including the products of the *cII* and *cIII* genes. The N protein is not completely effective at the terminator before the Q gene, and so only a small amount of the Q protein is made.

The Q protein is also an antiterminator protein. If its concentration becomes high enough, it will allow the transcript from a nearby promoter to synthesize the transcript labeled R2 in Figure 9.18. This transcript is translated to yield the late proteins, all the necessary structural proteins to construct a virion, and the proteins necessary for cell lysis. However, at the same time the Q protein is accumulating, the Cro protein also reaches levels where it can block transcription from both P_L and P_R by binding to both O_L (operator left) and O_R (operator right). Therefore, Cro functions as a repressor protein (∞Section 8.5).

The mechanism of Cro protein binding at O_R is detailed in Figure 9.19●. Note that there are three similar but nonidentical sites at this operator where the Cro protein can bind. It does so first at site 3, and then site 2, and only when those two sites are filled, at site 1. Note that only when bound to site 1 does it block P_R. Note also that once P_L and P_R are blocked, no more cII or cIII proteins can be synthesized. These proteins are needed to enter the lysogenic pathway (see later). Thus, *when Cro is made in high amounts, lambda is irreversibly committed to the lytic pathway.*

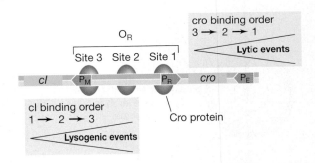

● **Figure 9.19   Lambda operator right (O_R).** Both regulatory proteins Cro and the lambda repressor bind to O_R to carry out regulatory functions. The Cro protein binds to the three sites in the order site 3, then site 2, and then site 1, turning off cI synthesis first and its own synthesis last. The lambda repressor (cI) binds to these sites in the opposite order, turning off Cro synthesis first and its own synthesis last. Compare this figure with the summary of lambda lytic/lysogenic steps in Figure 9.21.

Rolling Circle DNA Replication

The shutoff of P_L and P_R results in a change of lambda DNA replication. Early DNA synthesis is bidirectional from a single origin and gives rise to typical theta-like intermediates (∞Section 7.6). However, by the time late proteins are being made, long, linear concatamers of DNA are synthesized by **rolling circle replication**. In this mechanism and unlike semiconservative replication, replication proceeds *in only one direction* and can result in very long chains of replicated DNA (Figure 9.20●). Rolling circle replication permits rapid DNA replication and thus is a useful mechanism for rapidly accumulating copies of the lambda genome to package into mature virions. The long concatemers are cut into virus-sized lengths by a DNA-cutting enzyme. In the case of lambda (unlike that in T4; see Section 9.9), the cutting enzyme cuts at *specific* sites on the two strands (the *cos* sites), 12 nucleotides apart, providing the cohesive ends involved in the cyclization process (Figure 9.18*a*). The linear DNA is then packaged into phage heads and the tails and other proteins added. The cell is then lysed by the activity of phage-encoded enzymes, and the lambda lytic cycle is complete.

Lysis or Lysogenization: The Genetic Switch

What controls whether bacteriophage lambda will take the lytic or the lysogenic route? Lambda and other temperate viruses have a *genetic switch* that controls which pathway will occur. In order to establish lysogeny, two events must happen: (1) the production of all late proteins must be prevented; and (2) a copy of the lambda genome must be integrated into the host chromosome.

In order to prevent synthesis of lambda late proteins, the product of the *cI* gene must be produced. This protein is the **lambda repressor** (∞Figure 8.9*b*). If cI is synthesized, *it will repress the synthesis of all other lambda-encoded proteins, and lysogeny will occur*. The *cI* gene is located between P_L and P_R (see Figure 9.18), but these promoters are oriented in such a way that neither transcribes the *cI* gene. The promoter that can produce

mRNA from the *cI* gene during infection is called P_E (promoter establishment) and is located on the map slightly to the right of the *cro* gene but in the opposite direction from P_R (Figures 9.18*b* and 9.19). However, unlike the other promoters we have described, P_E must first be *activated*. Once this occurs, lambda repressor protein is synthesized and the **lysogenic pathway** is followed.

The product of the *cII* gene is the P_E *activator protein* (∞Section 8.6) that activates this promoter (and another promoter required for the production of integrase) (see later). Although the cII protein is made early after infection, it is typically unstable in *Escherichia coli* because it is degraded by a host protease. If the cII protein is degraded, then the lysogenic pathway is impossible. However, cII can be stabilized by the phage-encoded cIII protein if there is no excess of host protease (or if there is an excess of cIII protein). If the cII protein is stabilized, then it will activate P_E and lambda repressor protein (cI) will be made. Lambda repressor binds to O_L and O_R, as does the Cro protein, but it binds to the sites within these operators in the order opposite that of Cro (see Figure 9.19); that is, it first binds to site 1, turning off P_R (and P_L by a similar mechanism). When this happens, the synthesis of all other lambda proteins is stopped, and lambda cannot enter the lytic pathway.

Without the cII protein P_E no longer functions and lambda repressor is no longer made. Therefore, if the lysogenic state is to be maintained, there must be another way to transcribe the *cI* gene. Note that the lambda genome reveals yet another promoter, P_M (promoter maintenance, Figure 9.19). This promoter is facing toward the *cI* gene (in the same direction as P_E). It is *activated* when lambda repressor binds to site 1 and is repressed only when lambda repressor is bound to all three sites. Therefore, the lambda repressor is both a *repressor* and an *activator* when it binds to site 1, *repressing* P_R and *activating* P_M. When P_M is active, more lambda repressor is made, and this maintains the lysogenic state.

A summary of the steps in lambda infection leading to the lytic versus the lysogenic state is shown in Figure 9.21●. In a nutshell, the direction taken—lysis or lysogeny—is determined by whether or not the Cro protein or the lambda repressor protein (cI) takes the upper hand in regulatory events. If Cro dominates regulatory events, the outcome is lysis, while if the lambda repressor dominates, lysogeny will occur.

Integration

Integration of lambda DNA occurs at a unique site on the *Escherichia coli* chromosome and is required to complete the lysogenic state. Integration occurs by insertion of the virus genome into the host genome (thus effectively lengthening the host genome by the length of the virus DNA). As illustrated in Figure 9.22●, upon injection, the cohesive ends of the linear lambda molecule find each other and form a circle, and it is this circular

● **Figure 9.20 Rolling circle replication of the lambda genome.** As the dark green strand rolls out, it is being replicated at its opposite end. Note that this synthesis is *asymmetric* because one of the parental strands continues to serve as a template and the other is used only once.

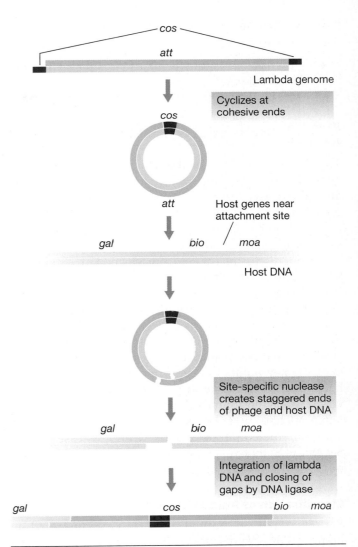

● **Figure 9.21 Summary of the steps in lambda infection.** Lysis versus lysogeny is governed by whether or not the lambda repressor (cI) is made. High Cro activity prevents sufficient cII/cIII protein from being made, and the result is lysis. If sufficient cII is present, the promoter for cI is activated and cI is made, resulting in lysogeny. See text for gene designations.

DNA that becomes integrated into the host genome (the site created when these ends join is called *cos*). To establish lysogeny, genes *cI* (lambda repressor) and *int* (encoding *integrase*) must be expressed. The integration process requires lambda integrase, which is a site-specific nuclease/integrase-catalyzing recombination of the phage and bacterial attachment sites (labeled *att* in Figures 9.18 and 9.22). The *int* gene has a promoter that, like P_E, is activated by the cII protein.

During cell growth, the lambda repression system prevents expression of the integrated lambda genome except for gene *cI* encoding the lambda repressor. During host DNA replication, the integrated lambda DNA is replicated along with the rest of the host genome and transmitted to progeny cells. When release from repression occurs, the lambda lytic cycle occurs. In order to be excised from the chromosome, *excisionase* (the product of the *xis* gene) and the *int* gene product are required.

Lytic Growth of Lambda After Induction

Agents that induce lambda lysogens to produce phage are agents that damage DNA. These include ultraviolet irradiation, X-rays, and DNA-damaging chemicals such as the nitrogen mustards. When DNA damage occurs, a host defense mechanism called the SOS response (∞Section 10.3) is activated. An array of bacterial genes is turned on by this response, some of which help the bacterium survive radiation. However, one result of the SOS response is that a bacterial protein called *RecA*, which is normally involved in genetic recombination, is converted into a protease (∞Figure 10.7) that destroys, among other things, the lambda repressor. Once this repressor has been inactivated, its control is abolished and

● **Figure 9.22 Integration of lambda DNA into the host.** See the genetic map, Figure 9.18b, for details of the gene order. Integration always occurs at a specific site on the host DNA, involving a specific attachment site (*att*) on the phage. Some of the host genes near the attachment site are given: *gal* operon, galactose utilization; *bio* operon, biotin biosynthesis; *moa* operon, molybdenum cofactor biosyntheses. A site-specific enzyme (integrase) is involved, and specific pairing of the complementary ends results in integration of phage DNA.

phage-specific transcription can begin. This inevitably leads to the completion of lytic events and cell lysis because even if the lambda repressor is made, it is quickly inactivated. Induction of bacteriophage lambda is thus an indirect consequence of DNA damage to the host's chromosome.

9.11 *Concept Check*

Lambda is a double-stranded DNA temperate phage. Regulation of lambda lytic versus lysogenic events is under the control of several promoters and regulatory proteins. Repression of lambda lytic events is caused by the cI protein (the lambda repressor), while activation of lytic events is under control of the Cro protein. Although the genome of lambda is linear, the genome circularizes inside the cell, where DNA synthesis occurs by a rolling circle mechanism.

◆ What events need to happen for lambda to become a prophage?

◆ What regulatory events occur at the O_R site on the lambda chromosome?

9.12 Overview of Animal Viruses

The first few sections of this chapter were devoted to general properties of all viruses, and little was said about animal viruses. Here we will consider some of the attributes of animal viruses, setting the stage for a discussion of the retroviruses, one of which causes *acquired immunodeficiency syndrome* (AIDS).

In our discussions of viral reproduction in Sections 9.5–9.7, we also dealt with the virus "host" in a very general way. However, it is important to remember that the bacteriophage host is a prokaryotic cell while the host of an animal virus is a eukaryotic cell. We will emphasize these important differences in Chapter 16, where we discuss several types of animal viruses in more detail. However, the key points to keep in mind for our discussion here are that (1) unlike in prokaryotes, the entire virus virion enters the animal cell, and (2) eukaryotic cells contain a nucleus, where most viruses have to replicate. In addition, there are details of RNA processing (∞ Section 14.8) that affect the replication of animal viruses, but again, these are details that need not bother us here.

Classification of Animal Viruses

Various types of animal viruses are illustrated in Figure 9.23●. We discussed the principles by which viruses are classified in Sections 9.1, 9.2, and 9.7. Note that there are many more kinds of enveloped animal viruses than enveloped bacterial viruses (see Section 9.7). This likely relates to the differences in host cell exteriors. Unlike prokaryotic cells, animal cells lack a cell wall, and thus viruses are more easily secreted from the cell. As they are, they remove part of the cell's lipid bilayer as they pass through the membrane.

Most of the animal viruses that have been studied in any detail are those that have been amenable to growth in cell cultures. Animal viruses are classified according to the same schemes as bacterial viruses, including the Baltimore Classification System (see Table 9.2), which classifies viruses on the basis of genome type and reproductive strategy. Animal viruses are known in all replication categories and many of these will be discussed in Chapter 16.

Consequences of Virus Infection in Animal Cells

Viruses can have several different effects on animal cells. **Lytic infection** results in the destruction of the host cell (Figure 9.24●). In the case of enveloped viruses, however, release of virions, which occurs by a kind of budding process, may be slow, and the host cell may not be lysed.

(a) **DNA viruses**

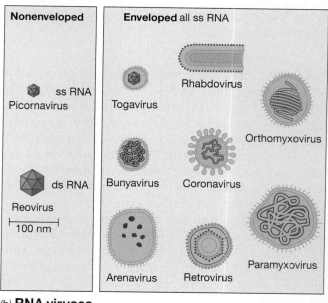

(b) **RNA viruses**

● **Figure 9.23 Diversity of animal viruses.** The shapes and relative sizes of vertebrate viruses of the major taxonomic groups. The hepadnavirus genome has one complete DNA strand and part of its complement (∞ Section 16.15).

The infected cell may therefore remain alive and continue to produce virus indefinitely. Such infections are called **persistent infections** (Figure 9.24).

Viruses may also cause **latent infection** of a host. In a latent infection, there is a delay between infection by the virus and lytic events. Fever blisters (cold sores), caused by the herpes simplex virus (∞ Section 16.12), are a typical example of a latent viral infection; the symptoms (the result of lysed cells) reappear sporadically as the virus emerges from latency. The latent stage in viral infection of an animal cell is generally not due to in-

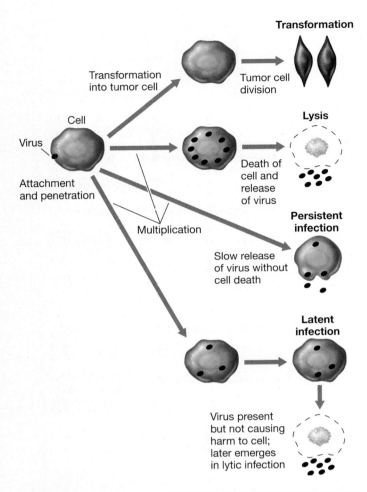

Transformation

Transformation into tumor cell — Tumor cell division

Cell

Virus

Attachment and penetration

Multiplication

Lysis — Death of cell and release of virus

Persistent infection — Slow release of virus without cell death

Latent infection — Virus present but not causing harm to cell; later emerges in lytic infection

● **Figure 9.24 Possible effects that animal viruses may have on cells they infect.** Note that most viruses are lytic and that only a very few are known to cause cancer. Many tumorigenic viruses belong to the Herpesvirus group (⚙Section 16.12).

tegration of the viral genome into the genome of the animal cell, as is the case with latent infections by temperate bacteriophages.

Finally, certain animal viruses can catalyze the conversion of a normal cell into a tumor cell, a process called **transformation** (Figure 9.24). We discuss cancer-causing viruses in Section 16.12.

Many different viruses are known. But of all the viruses listed in Figure 9.23, one stands out as having an absolutely unique mode of replication. These are the retroviruses. We explore them here as an example of a complex and highly unusual animal virus with significant medical implications.

⬡ **9.12 Concept Check**

Animal viruses are known that cover all known modes of viral genome replication. Many animal viruses are enveloped, picking up portions of the host cytoplasmic membrane as they leave the cell. Not all infections of animal host cells result in cell lysis or death; latent or persistent infections are common, and some animal viruses can cause cancer.

◆ Differentiate between a *persistent* and a *latent* viral infection.

◆ Contrast the mechanisms by which animal viruses enter cells with those used by bacterial viruses.

9.13 Retroviruses

Retroviruses contain an RNA genome that is replicated through a DNA intermediate (see Section 9.1 and Figure 9.11). The term *retro* means "backward," and the name of this class of virus is derived from the fact that these viruses transfer information backward from RNA to DNA (⚙Section 7.1). Retroviruses employ the enzyme **reverse transcriptase** to carry out this interesting process. The use of reverse transcriptase is not restricted to the retroviruses. Hepatitis B virus (a human virus) and cauliflower mosaic virus (a plant virus) also use **reverse transcription** in their replication processes (⚙Section 16.15). But in contrast to the retroviruses, these latter viruses encapsidate *DNA* as their genome rather than *RNA*.

The retroviruses are interesting for several other reasons. For example, they were the first viruses shown to cause *cancer* and have been studied most extensively for their carcinogenic characteristics. Also, one retrovirus, *human immunodeficiency virus* (HIV) causes *acquired immunodeficiency syndrome (AIDS)*. This virus infects a specific kind of T lymphocyte (T-helper cell) in humans that is vital for proper functioning of the immune system. In later chapters we discuss the medical and immunological aspects of AIDS (⚙Sections 25.6 and 26.14).

Retroviruses are enveloped viruses (Figure 9.25a●). There are several proteins in the virus coat and typically seven internal proteins, four of which are structural and three enzymatic. The enzymes found in the virus particle are *reverse transcriptase, DNA endonuclease (integrase),* and a *protease*. The virion also contains specific cellular tRNA molecules used in *replication* (see later in this section).

Features of Retroviral Genomes and Replication

The genome of the retrovirus is unique. It consists of *two* identical single-stranded RNA molecules of plus complementarity. A genetic map of a typical retrovirus genome is shown in Figure 9.25b. Although there are differences between the genetic maps of different retroviruses, all contain the following genes and are arranged in the same order: *gag*, encoding structural proteins; *pol*, encoding reverse transcriptase and integrase; and *env*, encoding envelope proteins. Some retroviruses, such as Rous sarcoma virus, carry a fourth gene downstream from *env* that is involved in cellular transformation and cancer. The terminal repeats shown on the map play an essential role in the replication process (see later).

The overall process of replication of a retrovirus can be summarized in the following steps (Figure 9.26●):

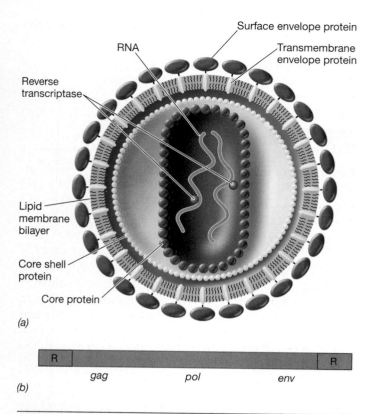

(a)

(b)

● **Figure 9.25   Retrovirus structure and function.** (a) Structure of a retrovirus. (b) Genetic map of a typical retrovirus genome. Each end of the genomic RNA contains direct repeats (R). See text for more details.

1. **Entrance** into the cell via fusion with cell membrane at sites of specific receptors.

2. **Uncoating of the virion** at the membrane, but such that the genome and enzymes remain in the core.

3. **Reverse transcription** of *one* of the two RNA genomes into a single-stranded DNA that is subsequently converted to a linear double-stranded DNA by reverse transcriptase, which then enters the nucleus.

4. **Integration** of retroviral DNA into the host genome.

5. **Transcription** of retroviral DNA, leading to the formation of viral mRNAs and viral genomic RNA.

6. **Assembly and encapsidation** of the genomic RNA into nucleocapsids in the cytoplasm.

7. **Budding** of enveloped virions at the cytoplasmic membrane and release from the cell.

Activity of Reverse Transcriptase

A very early step after the entry of the RNA genome into the cell is reverse transcription: conversion of RNA into a DNA copy using the enzyme reverse transcriptase present in the virion. The DNA formed is a linear double-stranded molecule and is synthesized in the cytoplasm within an uncoated viral core particle. Details on this process can be found in Section 16.15. Here we note only that reverse transcriptase is a DNA polymerase and, like all DNA polymerases, must have a primer (⚬⚬Section 7.5). The

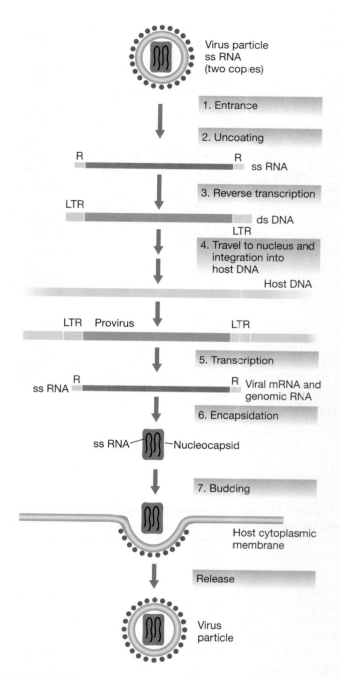

● **Figure 9.26   Replication process of a retrovirus.** R, direct repeats; LTR, long terminal repeats. For more detail on conversion of RNA to DNA (step 3) refer to Figure 16.25.

primer for retrovirus reverse transcription is a specific *cellular transfer RNA (tRNA)*. The type of tRNA used as primer depends on the virus and is brought into the virion from the previous host cell.

The overall process of reverse transcription results in a product that has long terminal repeats (LTRs, Figure 9.26) that are longer than the terminal repeats on the genome itself (Figure 9.25). This entire double-stranded DNA molecule enters the nucleus along with the integrase protein; here the viral DNA is integrated into the host DNA. The LTRs contain strong promoters of transcription and are also involved in the integration process. The *integration* of the viral DNA into the host

9.14 ◆ **Viroids and Prions** ◆ 253

genome is analogous to the integration of phage DNA into a bacterial genome to form the lysogenic state (see Section 9.10). Integration of viral DNA can occur anywhere in the cellular DNA, and once integrated, the element, now called a **provirus**, is a stable genetic element and can remain this way indefinitely.

If the promoters in the right LTR are activated, the integrated proviral DNA is transcribed by a cellular RNA polymerase into transcripts. These RNA transcripts either may be encapsidated into virus particles (as the genome) or may be translated into virus proteins. Some virus proteins are made initially as a large primary *gag* protein, which is split by proteolysis into the capsid proteins (➾Figure 16.26). Occasionally, read-through past the *gag* region (involving either inserting an amino acid in response to a stop codon or a shift in reading frame by the ribosome) leads to the translation of *pol*, the reverse transcriptase gene. This allows for reverse transcriptase to be produced for insertion into virions.

When virus proteins have accumulated in sufficient amounts, assembly of nucleocapsids occurs. Encapsidation of the RNA genome leads to the formation of mature nucleocapsids, which move to the cytoplasmic membrane for final assembly into the enveloped virus particles. As nucleocapsids bud through the membrane they are sealed and then released to infect neighboring cells (Figure 9.26).

9.13 Concept Check

Retroviruses are RNA viruses that replicate through a DNA intermediate. The retrovirus called human immunodeficiency virus causes AIDS. The retrovirus virion contains an enzyme, reverse transcriptase, that copies the information from its RNA genome into DNA. The DNA becomes integrated into the host chromosome in the manner of a temperate virus. The retrovirus DNA can be transcribed to yield mRNA (and new genomic RNA) or may remain in a latent state.

◆ Why are these viruses called *retro*viruses?

◆ How does the life cycle of a temperate bacteriophage differ from that of a retrovirus?

V SUB-VIRAL PARTICLES

If our brief tour of viral diversity, which we expand upon in Chapter 16, didn't exhaust all possibilities for replication schemes and viral idiosyncracies, we end this chapter on viral essentials by considering two viral-like infectious agents that are not viruses: viroids and prions. Both are best described as sub–viral particles.

9.14 Viroids and Prions

Recall that our definition of a virus is of a genetic element that subverts normal cellular processes for its own replication and that has a mature infectious extracellular form. There are a few infectious agents that resemble viruses but whose properties are at odds with this definition and are thus not viruses. Two of the most important are *viroids* and *prions*, and we consider these here.

Viroids

Viroids are small, circular, *single-stranded* RNA molecules that are the smallest known pathogens. Viroids range in size from 246 to 399 nucleotides and show a considerable degree of sequence homology, suggesting that they have common evolutionary roots. Viroids cause a number of important plant diseases and can have a severe agricultural impact. A few well-studied viroids include coconut cadang-cadang viroid (246 nucleotides), citrus exocortis viroid (375 nucleotides), and potato spindle tuber viroid (359 nucleotides; see Figure 9.28).

The extracellular form of the viroid is *naked RNA*—there is no protein capsid of any kind. Even more interestingly, *the RNA molecule contains no protein-encoding genes*, and therefore the viroid is totally dependent on host function for its replication. Although the viroid RNA is a single-stranded circle, there is such considerable secondary structure possible that it resembles a short double-stranded molecule with closed ends (Figure 9.27●). The viroid molecule enters the plant through a wound, as from insect or other mechanical damage. The viroid is replicated in the host cell nucleus by one of the plant RNA polymerases.

Viroid-infected plants can be symptomless or develop symptoms that range from mild to lethal, depending on the viroid (Figure 9.28●). Most severe symptoms are growth related, suggesting that viroids are a type of "regulatory RNA" (➾Section 8.14), examples of which are widely known in both plants and animals. Thus, viroids could be regulatory RNAs that have evolved away from constructive roles in the cell to now catalyzing destructive events in the cell. No viroid diseases of animals are known, and the precise mechanisms by which viroids cause plant diseases remain unclear.

Prions

Prions represent the other extreme from viroids. They have a distinct extracellular form, but the extracellular form is *entirely protein*. The prion particle does not contain any nucleic acid. However, it is infectious, and prions are known to cause a variety of diseases in animals, such as scrapie in sheep, *bovine spongiform encephalopathy* in cattle (BSE or "mad cow disease"), chronic wasting disease in

● **Figure 9.27   Viroids.** Structure of viroids, showing how single-stranded circular RNA can form a seemingly double-stranded structure by intrastrand base-pairing.

● **Figure 9.28 Viroids.** Photograph of healthy (left) and potato spindle tuber viroid (PSTV)–infected (right) tomato plants. The host range of most viroids is quite restricted. However, PSTV will infect tomato as well as potato, causing growth stunting, a flat top, and premature plant death (right).

● **Figure 9.29 Mechanism of action of a prion.** Neuronal cells produce prion protein (PrP^c), which carries out some normal function in the cell. Abnormally folded prion protein (PrP^{sc}) can catalyze refolding of PrP^c into PrP^{sc}. The PrP^{sc} form is protease resistant, insoluble, and forms aggregates in neural cells. This eventually leads to destruction of neural tissues and neurological symptoms.

deer and elk, and kuru and a form of Creutzfeldt-Jakob Disease (CJD) in humans. No prion diseases of plants are known. Collectively, prion diseases have come to be known as **transmissible spongiform encephalopathies**, or TSEs. In 1997 the American scientist Stanley B. Pruisner won the Nobel Prize for his pioneering work with these diseases and with the prion proteins.

In 1996 it became clear from disease tracking in England that the prion that causes BSE in cattle can also infect humans, resulting in a variant of CJD, called *new variant CJD (nvCJD)*. Because transmission was from contaminated beef products, in just a few years nvCJD became a worldwide health concern with a major impact on the animal husbandry industry (∞Section 29.11).

Most such instances of BSE occurred in the United Kingdom or other European Union (EU) countries and were linked to improper feeding practices in which protein supplements containing rendered cattle and sheep were used to feed uninfected animals. Since 1994 this practice has been banned in all EU countries, and cases of BSE have dropped dramatically. Thus far, TSE transmission via other domesticated animals, such as swine, chicken, or fish, has not been found.

Although a consensus has not been reached on how prions cause disease, the most likely scenario is as follows (Figure 9.29●). Prion infection results in production of more copies of the prion protein, and this leads to disease symptoms. Symptoms are invariably neurological, and in most cases are due to destruction of brain or related nervous tissue by the accumulation of aggregated prion protein (∞Section 29.10 and Figure 29.10). But if prions lack nucleic acid, how is the protein they consist of encoded?

The host cell contains a gene that encodes a glycoprotein called *PrP^C*, which is nearly identical to the prion protein. PrP^C is produced in the healthy animal and is found mostly in neurons. The incoming prion, called *PrP^{Sc}*, is a structurally modified form of the PrP^C protein and can itself modify PrP^C, converting it into PrP^{Sc}. The modification involves an alternative pattern of folding. This causes the protein to lose its normal function, to become partially resistant to proteases, and to become insoluble in the neural cell, leading to aggregation (Figure 9.29). In this state, prion protein accumulates (∞Figure 29.10) and neurological symptoms commence. Therefore, prions do not simply subvert host enzymes in the cell, but convert a normal cell protein into a *self-propagating conformational state.*

Viroids and prions do more than stretch our definition of a virus. They also demonstrate both the unexpected ways that genetic elements can replicate and the unexpected ways they can subvert the host cells. However, prions stand out among all of the genetic elements we have considered in this chapter because the infectious transmissible agent lacks nucleic acid.

 9.14 Concept Check

Viroids are circular single-stranded RNA molecules that do not encode proteins and are completely dependent on host-encoded enzymes. By contrast, prions consist of protein but have no nucleic acid. Collectively, prions and viroids are the smallest known pathogens.

◆ If viroids are circular molecules, why are they drawn as shown in Figure 9.27?

◆ How does a prion induce disease?

REVIEW QUESTIONS

1. In what ways do viral genomes differ from those of cells (∞Section 9.1)?

2. Define *virus*. What are the minimal features needed to fit your definition (∞Section 9.2)?

3. Define the term *host* as it relates to viruses (∞Section 9.3).

4. Describe the events that occur on an agar plate containing a bacterial lawn when a single bacteriophage particle causes the formation of a *bacteriophage plaque* (∞Section 9.4).

5. Under some conditions, it is possible to obtain nucleic acid-free protein coats (*capsids*) of certain viruses. Under the electron microscope these capsids look very similar to complete virions. What does this fact tell you about the role of the virus nucleic acid in the virus assembly process? Would you expect such virions to be infectious? Why (∞Section 9.5)?

6. Describe how a *restriction endonuclease* might play a role in resistance to bacteriophage infection. Why could a restriction endonuclease play such a role whereas a generalized DNase could not (∞Section 9.5)?

7. One can divide the replication process of a virus into five steps. Describe the events associated with each of these steps (∞Sections 9.6 and 9.7).

8. Specifically, why are both the life cycle and the virion of a positive-strand RNA virus likely to be simpler than those of a negative-strand RNA virus (∞Section 9.8)?

9. In terms of structure, how does the genome of bacteriophage T4 resemble and differ from that of *Escherichia coli* (∞Section 9.9)?

10. Many of the viruses we have considered have *early* genes and *late* genes. What is meant by these two classifications? What types of proteins tend to be encoded by early genes? What type of proteins by late genes? For bacteriophages T4 and lambda describe how expression of the late genes is controlled (∞Section 9.9).

11. Define the following: virulent, lysogeny, prophage (∞Section 9.10).

12. A strain of *Escherichia coli* that is missing the outer membrane protein responsible for maltose uptake is *resistant* to lambda infection. A lambda lysogen is *immune* to lambda infection. Describe the functional difference between resistance and immunity. Explain how these conditions are brought about in the examples given (∞Sections 9.10 and 9.11).

13. Describe and differentiate the effects animal virus infection can have on an animal (∞Section 9.12).

14. Typically, transfer RNA is used in translation. However, it also plays a role in the replication of retroviral nucleic acid. Explain this role (∞Section 9.13).

15. What are the similarities and differences between prions, viruses, and viroids (∞Section 9.14)?

APPLICATION QUESTIONS

1. What causes the viral plaques that appear on a bacterial lawn to stop growing larger?

2. The promoters for mRNA encoding *early* proteins in viruses like T4 have a different sequence than the promoters for mRNA encoding *late* proteins in the same virus. Explain why this might be true.

3. One characteristic of *temperate bacteriophages* is that they cause turbid rather than clear plaques on bacterial lawns. Can you think why this might be? (Remember the process by which a bacterial plaque develops.)

4. There are three lambda genes that, when rendered nonfunctional, turn lambda from a temperate to a virulent virus. What are these three genes and how do they normally function?

5. Figure 9.19 shows two promoters, P_R and P_E, on either side of the *cro* gene. Both transcribe through the *cro* gene, but only the transcript from P_R can be translated to yield a functional Cro protein. Explain.

6. Contrast the enzyme(s) present in the virions of a retrovirus and a positive-strand RNA bacteriophage. Why do they differ if each has plus complementarity single-stranded RNA as their genome?

7. Explain why as viral infection leads to more viral particles being formed, that the "growth curve" for viruses shows the pattern in Figure 9.9 rather than that in Figure 6.8.

BACTERIAL GENETICS

To understand an organism's growth, development, and reproduction, it is necessary to understand its genetics. Genetic analyses contribute to advances in both basic science and applied science. Genetic exchange, as shown here in two conjugating bacteria, is at the heart of bacterial genetics.

WORKING GLOSSARY

Auxotroph an organism that has developed a nutritional requirement through mutation

Cloning vector genetic element into which genes can be recombined and replicated

Conjugation transfer of genes from one prokaryotic cell to another by a mechanism involving cell-to-cell contact and a plasmid

Gene disruption use of genetic techniques to inactivate a gene by inserting within it a DNA fragment containing a selectable marker. The inserted fragment is called a *cassette*, and the process of insertion, *cassette mutagenesis*. Also called *gene knockout*

Genetic map the arrangement of genes on a chromosome

Genotype the precise genetic makeup of an organism; the complete sequence of a cell's chromosome(s) and plasmids if present

Molecular cloning isolation and incorporation of a fragment of DNA into a vector where it can be replicated

Mutagen an agent that causes mutation

Mutant an organism whose genome carries a mutation

Mutation an inheritable change in the base sequence of the genome of an organism

Phenotype the observable characteristics of an organism

Plasmid an extrachromosomal genetic element that has no extracellular form

Point mutation a mutation that involves a single base pair

Recombination the process by which parts or all of the DNA molecules from two separate sources are exchanged or brought together into a single DNA molecule

Screening any of a number of procedures that permit the identification of organisms by phenotype or genotype, but do not inhibit or enhance the growth of particular phenotypes or genotypes

Selection placing organisms under conditions where the growth of those with a particular genotype will be favored

Shotgun cloning making a gene library by random cloning of DNA fragments

Site-directed mutagenesis a technique whereby a gene with a specific mutation can be constructed *in vitro*

Transduction transfer of host genes from one cell to another by a virus

Transformation transfer of bacterial genes involving free DNA (but see alternative usage in Chapter 9)

Transposable element a genetic element that has the ability to move (transpose) from one site on a chromosome to another

Transposon a type of transposable element that carries genes in addition to those involved in transposition

Vector (as in cloning vector) a DNA molecule that, on being replicated in a cell, brings about the replication of other genes inserted into the DNA molecule

Wild type a bacterial strain isolated from nature

In this chapter we present the basic principles and techniques of bacterial genetics. First we will discuss mutation and genetic recombination, and then we consider how genes can be transferred from one organism to another. Next, we will explore some of the basic techniques of gene cloning that have revolutionized the study of genetics of all organisms, from bacteria to humans. We close with a detailed look at the *Escherichia coli* chromosome.

MUTATION AND RECOMBINATION

All organisms contain a specific sequence of nucleotides in their **genome**, their genetic blueprint. **Mutation** is an *inherited change* in the nucleotide base sequence of that genome. Mutations can lead to changes—some good, some bad—in an organism. Change can also be brought about by **genetic recombination**. This is the process by which genes contained in two separate genomes are brought together in one molecule. Through this mechanism, new combinations of genes can arise even in the absence of mutation. Whereas mutation usually brings about only a very *small* amount of genetic change in a cell, genetic recombination typically involves much *larger* changes. Entire genes, sets of genes, or even whole chromosomes, can be transferred between organisms. Taken together, mutation and recombination fuel the evolutionary process.

Unlike most eukaryotes, prokaryotes do not reproduce sexually. However, there are mechanisms of genetic exchange in prokaryotes that allow for both gene transfer and recombination. These mechanisms are based on *lateral gene flow* (also called *horizontal gene flow*), because genes are transferred from donor to recipient rather than vertically, from mother cell to daughter cell. To detect genetic exchange between two organisms, it is necessary to employ *genetic markers* whose transfer can be detected. Genetically altered strains are used for this purpose, the alteration(s) being due to one or more mutations in the DNA of the organism. These mutations may involve changes in only one or a few base pairs or even the insertion or deletion of entire genes.

We begin this chapter on microbial genetics with a consideration of the molecular mechanism of mutation and the properties of mutant microorganisms as a prelude to our discussion of genetic exchange.

10.1 Mutations and Mutants

As previously mentioned, a *mutation* is a heritable change in the base sequence of the nucleic acid in the genome of an organism. In all cells this nucleic acid is double-stranded DNA (∞ Section 7.1). A strain carrying such a change is called a **mutant**. A mutant by definition differs from its parental strain in **genotype**, the nucleotide sequence of the genome. But in addition, the *observable properties* of the mutant—its **phenotype**—may also be altered relative to the parental strain. One refers to this altered phenotype as a *mutant phenotype*. It is common to refer to a strain isolated from nature as a **wild-type strain**. Mutant derivatives can be obtained either from wild-type strains or from a strain derived from the wild type, for example, another mutant.

Genotype Versus Phenotype

Depending on the mutation, a mutant strain may or may not differ in phenotype from its parent. By convention in bacterial genetics, the *genotype* of an organism is designated by three lowercase letters followed by a capital letter (all in italics) indicating the particular gene involved. For example, the *hisC* gene of *Escherichia coli* encodes a protein called the HisC protein, involved in biosynthesis of the amino acid histidine. Mutations in the *hisC* gene would be designated as *hisC1*, *hisC2*, and so on, the numbers referring to the order of isolation of the mutant strains. Each *hisC* mutation would be different and would likely affect the HisC protein in different ways.

The *phenotype* of an organism is designated by a capital letter followed by two lowercase letters, with either a plus or minus superscript to indicate the presence or absence of that property. For example, a His$^+$ strain of *E. coli* is capable of making its own histidine whereas a His$^-$ strain is not. The His$^-$ strain would require a histidine supplement for growth. Returning to our previous example, mutations in the *hisC* gene may lead to a His$^-$ phenotype if they eliminate the function of the HisC protein.

Isolation of Mutants: Screening Versus Selection

Virtually any characteristic of an organism can be changed by mutation. However, some mutations are *selectable*, conferring some type of advantage on organisms possessing them, while others are *nonselectable*, even though they may lead to a very clear change in the phenotype of an organism. A *selectable* mutation confers a clear advantage on the mutant strain under certain environmental conditions, so the progeny of the mutant cell are able to outgrow and replace the parent. A good example of a selectable mutation is drug resistance: An antibiotic-resistant mutant can grow in the presence of antibiotic concentrations that inhibit or kill the parent (Figure 10.1*a*●) and is thus *selected* for under these conditions. It is relatively easy to detect and isolate selectable mutants by choosing the appropriate environmental conditions. **Selection** is therefore an extremely powerful genetic tool, allowing the isolation of a single mutant from a population containing millions or even billions of parental organisms.

An example of a nonselectable mutation is that of loss of color in a pigmented organism (Figure 10.1*b*). Nonpigmented cells usually have neither an advantage nor a disadvantage over the pigmented parent cells when grown on agar plates, although there may be a selective advantage for pigmented organisms in nature. We can detect such mutations only by examining large numbers of colonies and looking for the "different" ones, a process called **screening**.

(a)

(b)

(c)

● **Figure 10.1 Selectable and nonselectable mutations.** (a) Development of antibiotic-resistant mutants, a type of easily selectable mutation, within the inhibition zone of an antibiotic assay disc. (b) Nonselectable mutations. Spontaneous pigmented and nonpigmented mutants of the fungus *Aspergillus nidulans*. The wild type has a green pigment. The white or colorless mutants make no pigment, whereas the yellow mutants cannot convert the pigment they do make to the normal (green) color. (c) Colonies of mutants of a species of *Halobacterium*, a member of the *Archaea*. The colonies that appear white are the wild type. The orange/brown colonies are mutants that are missing gas vesicles (Section 4.12). The gas vesicles scatter light and mask the color of the colony.

Isolation of Nutritional Auxotrophs and Penicillin Selection

Although screening is always more tedious than selection, for certain types of mutations methods are available for screening large numbers of colonies. For instance, nutritionally defective mutants can be detected by the technique of **replica plating** (Figure 10.2●). With the use of sterile velveteen cloth or filter paper, an imprint of colonies from a master plate is made onto an agar plate lacking the nutrient. Colonies of the parental type will grow normally, whereas those of the mutant will not. Thus, the inability of a colony to grow on the replica plate containing minimal medium (Figure 10.2) is a signal that it is a mutant. The colony on the master plate corresponding to the vacant spot on the replica plate (Figure 10.2*b*) can then be picked, purified, and characterized.

A mutant that has a nutritional requirement for growth is called an **auxotroph**, and the parent from which the auxotroph was derived is called a **prototroph**. (A prototroph may or may not be the wild type. An auxotroph may be derived from the wild type or from a mutant derivative of the wild type.) For instance, mutants of *Escherichia coli* with a His⁻ phenotype are said to be *histidine auxotrophs*. Although of great utility, replica plating is nevertheless a *screening* process, and it can be laborious to isolate mutants by screening.

An ingenious method widely used to isolate auxotrophs is the **penicillin-selection method**. Ordinarily, mutants that require specific nutrients are at a disadvantage in competition with the parent cells, and so there is no direct way of isolating them. For instance, histidine mutants are rare in a mutagenized culture, and it could take a great deal of time to obtain them by replica plating alone. However, penicillin selection can be used to *enrich* a population of mutagenized cells in His⁻ mutants, after which replica plating can be much more effective. How does penicillin selection work?

Penicillin is an antibiotic that kills only *growing* cells. If penicillin is added to a population growing in a medium lacking the nutrient required by the desired mutant, the parent cells will be killed, whereas the (nongrowing) mutant cells will be unaffected. Thus, after preliminary incubation in the absence of the nutrient in a penicillin-containing medium, the population is washed free of the penicillin and transferred to plates containing the nutrient. The colonies that appear include some wild-type cells that escaped penicillin killing, but also include some of the desired mutants. Penicillin selection is thus a kind of *negative selection*; the selection is not *for* the mutant but *against* the parental type. Penicillin selection is often used as a prelude to replica plating to increase the chances of obtaining auxotrophic mutants.

Examples of common classes of mutants and the means by which they are detected are listed in Table 10.1.

 10.1 Concept Check

Mutation is a heritable change in DNA sequence that can lead to a change in phenotype. Selectable mutations are those that give the mutant a growth advantage under certain environmental conditions and are especially useful in genetic research. If selection is not possible, mutants must be identified by screening.

◆ Distinguish between the words *mutation* and *mutant*.

◆ Distinguish between the words *screening* and *selection*.

10.2 Molecular Basis of Mutation

Mutations can be either *spontaneous* or *induced*. **Spontaneous mutations** can occur as a result of exposure to natural radiation (cosmic rays, and so on) that alters the structure of bases in the DNA. Also, oxygen radicals (∞ Section 6.16) can affect DNA structure by chemically modifying DNA. For example, oxygen radicals can convert guanine into 8-hydroxyguanine, and this causes spontaneous mutations. However, the bulk of spontaneous mutations occur during DNA replication as a result of errors in the pairing of bases.

● **Figure 10.2 Screening for nutritional auxotrophs.** The replica plating method can be used for the detection of nutritional mutants. Photos: The photograph on the left shows the master plate. The colonies not appearing on the replica plate are marked with an X. The replica plate (right) lacked one nutrient (leucine) present in the master plate. Therefore, the colonies marked with an X on the master plate are leucine auxotrophs.

Table 10.1	Kinds of mutants	
Phenotype	**Nature of change**	**Detection of mutant**
Auxotroph	Loss of enzyme in biosynthetic pathway	Inability to grow on medium lacking the nutrient
Cold-sensitive	Alteration of an essential protein so it is inactivated at low temperature	Inability to grow at a low temperature (for example, 20°C) that normally supports growth
Drug-resistant	Alteration of permeability to drug or drug target or detoxification of drug	Growth on medium containing a growth-inhibitory concentration of the drug
Nonencapsulated	Loss or modification of surface capsule	Small, rough colonies instead of larger, smooth colonies
Nonmotile	Loss of flagella; nonfunctional flagella	Compact colonies instead of flat, spreading colonies
Pigmentless	Loss of enzyme in biosynthetic pathway leading to loss of one or more pigments	Presence of different color or lack of color
Rough colony	Loss or change in lipopolysaccharide outer layer	Granular, irregular colonies instead of smooth, glistening colonies
Sugar fermentation	Loss of enzyme in degradative pathway	Lack of color change on agar containing sugar and a pH indicator
Temperature-sensitive	Alteration of an essential protein so it is more heat-sensitive	Inability to grow at a temperature normally supporting growth (for example, 40°C) but still growing at a lower temperature (for example, 30°C)
Virus-resistant	Loss of virus receptor	Growth in presence of large amounts of virus

Mutations involving a change in one base pair are referred to as **point mutations**. Point mutations occur from *base-pair substitutions* in the DNA. As is the case with all mutations, the phenotypic change that results from a point mutation depends on exactly where the mutation occurs in the gene, what the nucleotide change is, and what product the gene encodes.

Base-Pair Substitutions

If a point mutation occurs within the coding region of a gene that encodes a polypeptide, any change in the phenotype of the cell is almost certainly the result of a change in the amino acid *sequence* of the polypeptide. The error in the DNA is transcribed into mRNA, and the erroneous mRNA in turn is translated to yield the polypeptide. The triplet code that directs the insertion of an amino acid via a transfer RNA (⌘ Sections 7.15 and 7.16) will thus be incorrect. Figure 10.3● shows the consequences of various base-pair substitutions.

In interpreting the results of a mutation, we must first recall that the genetic code is degenerate (⌘ Section 7.14). Because of this degeneracy, not all mutations in polypeptide-encoding genes result in changes in the polypeptide. This is illustrated in Figure 10.3, which shows several possible results when the DNA that encodes a single tyrosine codon in a polypeptide undergoes mutation. First, a change in the RNA from UAC to UAU would have no apparent effect because UAU is also a tyrosine codon. Mutations that give rise to such changes are indeed still mutations, but because they do not affect the primary sequence of the encoded polypeptide, they are called **silent mutations**. Note that silent mutations in coding regions almost always occur in the *third base* of the codon (arginine and leucine can also have silent mutations in the first position).

Second, changes in the first or second base of the triplet more often lead to significant changes in the polypeptide. For instance, a single base change from UAC

to AAC (Figure 10.3) would result in an amino acid change within the polypeptide from tyrosine to asparagine at a specific site. This is referred to as a **missense mutation** because the informational "sense" (sequence of amino acids) in the ensuing polypeptide has changed. If the change occurs at a critical point in the polypeptide chain, the protein could be inactive or have reduced activity. However, not all mutations that cause amino acid substitution necessarily lead to nonfunctional proteins. The outcome depends on where in the polypeptide chain the substitution has occurred and on how it affects its folding and activity. For example, mutations in the active site of

● **Figure 10.3 Possible effects of base-pair substitution in a gene encoding a protein:** Three different protein products are possible from changes in the DNA for a single codon.

an enzyme are more likely to destroy activity than mutations that affect other parts of the protein.

A third possible result of a base pair substitution is the formation of a *nonsense (stop) codon*. This results in premature termination of translation, leading to an incomplete polypeptide that would almost certainly not be functional (Figure 10.3). Mutations of this type are called **nonsense mutations** because the change is from a codon for an amino acid (sense codon) to a nonsense codon (◠◡ Table 7.5). Unless the nonsense mutation occurs very near the end of the gene, the incomplete product will be completely inactive.

The terms *transition* and *transversion* are used to describe the nature of the base substitution that occurs in a point mutation. **Transitions** involve mutations in which one purine base (A or G) is substituted for another purine, or one pyrimidine base (C or T) is substituted for another pyrimidine. **Transversions** are point mutations where a purine base is substituted for a pyrimidine base, or vice versa.

Frameshifts and Other Insertions or Deletions

Because the genetic code is read from one end of the nucleic acid in consecutive blocks of three bases (that is, codons), any deletion or insertion of a base pair results in a *reading frameshift*. These **frameshift mutations** can have serious consequences. Single base insertions or deletions change the primary sequence of the encoded polypeptide, typically in a major way (Figure 10.4●). Such microinsertions or microdeletions result from replication errors.

Insertions or deletions can also involve the gain or loss of hundreds or even thousands of base pairs. Such changes inevitably result in complete loss of gene function. Some deletions are so large that they may involve several genes. If any of the deleted genes are essential, the mutation will be lethal. Such deletions cannot be restored through further mutations but only through genetic recombination. Indeed, one way in which large deletions are distinguished from point mutations is that the latter are reversible through further mutations, whereas the former are not. Larger insertions typically arise as a result of errors during genetic recombination. Many insertion mutations are due to the insertion of specific identifiable DNA sequences 700–1400 bp in length called **insertion sequences**, a type of transposable element (◠◡ Section 7.4). The behavior of such insertion sequences is discussed in detail in Section 10.14.

Other types of large-scale mutations seem to involve rearrangements brought about by mistakes in recombination. These include **translocations**, in which a large section of chromosomal DNA is moved to a new location (and in eukaryotes often to a different chromosome), and **inversions**, in which the orientation of a particular segment of DNA is reversed with respect to the surrounding DNA.

Transposon and Site-Directed Mutagenesis

Mutations can be introduced through the process of *transposon mutagenesis*. We discuss the details of transpo-

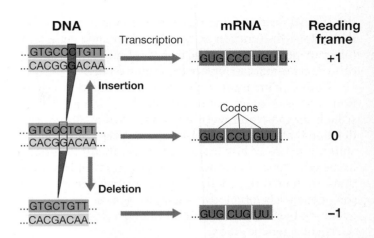

● **Figure 10.4** **Shifts in the reading frame of messenger RNA caused by insertion or deletion mutations in DNA.** The reading frame in mRNA is established by the ribosome that begins at the 5′ end (toward the left in the figure) and proceeds by units of three bases (codons, ◠◡ Section 7.14). The normal reading frame is referred to as the 0 frame, that missing a base the −1 frame, and that with an extra base the +1 frame. To determine the effects of a frameshift, translate the codons using Table 7.5.

son mutagenesis later in this chapter (see Section 10.14) but only note here that if insertion of a transposable element occurs *within* a gene, loss of gene function generally results. Because transposable elements can enter the chromosome at various locations, transposons are widely used by microbial geneticists as mutagenic agents.

The mutations that we have been discussing thus far have been random, that is, not directed at any particular gene. However, recombinant DNA technology (see Sections 10.15–10.17) and the use of synthetic DNA (◠◡ Section 7.8) make it possible to induce *specific* mutations in *specific* genes. The procedures for carrying out mutagenesis of specific sites in the genome are called *site-directed mutagenesis* and will be discussed later in this chapter (see Section 10.18).

Back Mutations or Reversions

Point mutations are typically reversible by the process of **reversion**. A *revertant* is a strain in which the wild-type (or prototrophic, depending on the strain) phenotype that was lost in the mutant is restored. Revertants can be of two types. In *same-site revertants*, the mutation that restores activity occurs at the same site at which the original mutation occurred. (If the back mutation is not only at the same site but also leads to the wild-type sequence, it is called a *true revertant*.)

In *second-site revertants*, the mutation occurs at a *different* site in the DNA. Second-site mutations cause restoration of a wild-type phenotype because they function as **suppressor** mutations—mutations that compensate for the effect of the original mutation and restore the original phenotype. Several classes of suppressor mutations are known. These include (1) a mutation somewhere else in the same gene that restores enzyme function, such as a second frameshift mutation that occurs near the first and restores the original reading frame; (2) a mutation in another gene that restores the function of the original mutated gene; and (3) a mutation in another gene that results in the production of an enzyme that can replace the mutated one.

Unlike point mutations, large-scale deletion mutations are essentially nonrevertable. By contrast, large-scale insertions can revert as the result of a subsequent deletion that removes the insertion. But typically, frameshift mutations of any magnitude are difficult to restore to the wild type, and mutants carrying frameshift mutations are therefore genetically quite stable. For this reason, geneticists often use them in genetic crosses to avoid inadvertent reversion of mutant strains during the course of a genetic study.

 10.2 Concept Check

Mutations, which can be either spontaneous or induced, arise because of changes in the base sequence of the nucleic acid of an organism's genome. A point mutation, which is due to a change in a single base pair, can lead to a single amino acid change in a polypeptide or to no change at all, depending on the particular codon involved. In a nonsense mutation, the codon becomes a stop codon and an incomplete polypeptide is made. Deletions and insertions cause more dramatic changes in the DNA, including frameshift mutations and often result in complete loss of gene function.

◆ What does it mean to say that point mutations can spontaneously revert?

◆ Do *missense* mutations occur in genes encoding transfer RNAs (tRNAs)?

10.3 Mutation Rates

The *rates* at which various kinds of mutations actually occur vary widely. Some types of mutations occur so rarely that they are almost impossible to detect, whereas others occur so frequently that they present difficulties for an experimenter trying to maintain a genetically stable stock culture.

Errors in DNA replication occur at a frequency of 10^{-7}–10^{-11} per base pair during a single round of replication. A typical gene has about 1000 base pairs. Therefore, the frequency of a mutation occurring in a given gene is in the range of 10^{-4} to 10^{-8} per generation. For instance, in a bacterial culture having 10^8 cells/ml there are likely to be a number of different mutants for a given gene in each milliliter of culture.

When any single-base error occurs during DNA replication, it will more likely lead to *missense* mutations than to *nonsense* mutations because most single-base substitutions yield codons that encode other amino acids (∞ Table 7.5). The next most frequent type of codon change caused by a single-base change would lead to a *silent* mutation. This is because for the most part, alternate codons for a given amino acid differ from each other by a single base change in the "silent" third position (∞ Table 7.5). A given

codon can be changed to any of 27 other codons by a single-base substitution, and on average, about two of these will be silent mutations, about one a nonsense mutation, and the rest will be missense mutations. There are also DNA sequences, usually involving short repeats, which are *hot spots* for mutations because the error frequency of DNA polymerase can be quite high at such sequences. Hot spots are defined by the context of nucleotides in and around the spot that affect normal polymerase function.

Unless a mutation can be selected for, its experimental detection is difficult, and much of the skill of the microbial geneticist involves increasing the efficiency of mutation detection. As we will see in the next section, it is possible to significantly increase the *rate* of mutation by the use of mutagenic treatments. In addition, we will also see that the mutation rate of a gene may change in certain situations. For example, mutations called *adaptive mutations* are strongly selected for under high-stress conditions. Adaptive mutations are generated by mechanisms activated by the organism itself in order to survive a particular stress condition (see Section 10.4).

Mutations in RNA Genomes

Whereas all cells have *DNA* as their genetic material, some viruses have *RNA* genomes (∞ Sections 9.1, 16.1, 16.7–16.10, and 16.15). These genomes can also undergo mutation. Interestingly, however, the mutation rate in RNA genomes is about *1000-fold higher* than in DNA genomes. Why should this be so?

Some RNA polymerases have *proofreading* activities like those of DNA polymerases (∞ Section 7.6), thus limiting the total number of polymerase errors. However, while there are several repair systems for DNA that can correct changes before they become fixed in the genome as mutations (see Section 10.4), comparable RNA repair mechanisms do not exist. This leads to heightened mutation rates for RNA genomes. This high mutation rate in RNA viruses has serious consequences. For example, the RNA genomes of viruses that cause disease can mutate very rapidly, presenting a constantly changing and evolving population of viruses. Such changes are one of many challenges to human medicine posed by the AIDS virus, HIV, an RNA virus with a notorious ability to undergo genetic changes that affect its virulence (∞ Sections 16.15 and 26.14).

 10.3 Concept Check

Different types of mutations can occur at different frequencies. For a typical bacterium, mutation rates of 10^{-7}–10^{-11} per base pair are generally seen. Although RNA and DNA polymerases make errors at about the same rate, RNA genomes typically accumulate mutations at much higher frequencies than DNA genomes.

◆ Which class of mutation, *missense* or *nonsense*, is more common, and why?

◆ Why are RNA viruses genetically unstable?

10.4 Mutagenesis

While the spontaneous rate of mutation is very low, there are a variety of chemical, physical, and biological agents that can increase the mutation rate and are therefore said to *induce* mutations. These agents are called **mutagens**. We discuss some of the major categories of mutagens and their activities here.

Chemical Mutagens

An overview of some of the major chemical mutagens and their modes of action is given in Table 10.2. Several classes of chemical mutagens exist. One class is the **nucleotide base analogs**, molecules that resemble DNA purine and pyrimidine bases in structure yet display faulty pairing properties (Figure 10.5●). When one of these base analogs is incorporated at a site in DNA in place of the natural nucleotide, replication may occur normally most of the time. However, DNA replication errors occur at higher frequencies at these sites, resulting in the more frequent incorporation of the wrong base into the copied strand of DNA and thus introduction of a mutation. During subsequent segregation of this strand in cell division, the mutation is revealed.

Other chemical mutagens induce chemical modifications in one base or another, resulting in faulty base pairing or related changes (Table 10.2). For example, *alkylating agents* (chemicals that react with amino, carboxyl, and hydroxyl groups in proteins and nucleic acids, substituting them with alkyl groups) such as *nitrosoguanidine*, are powerful mutagens and generally induce mutations at higher frequency than base analogs. Alkylating agents differ from the base analogs in that the chemicals are able to introduce changes even in nonreplicating DNA (base analogs have an effect only when incorporated during DNA replication). Both base analogs and alkylating agents tend to induce base-pair substitutions (see Section 10.2).

Analog	Substitutes for
5-Bromouracil	Thymine
2-Aminopurine	Adenine

(a)
(b)

● **Figure 10.5 Nucleotide base analogs.** Structure of two common nucleotide base analogs used to induce mutations and the normal nucleic acid bases they substitute for. (a) 5-Bromouracil can base pair with guanine causing AT to GC substitutions. (b) 2-Aminopurine can base pair with cytosine, causing AT to GC substitutions.

Another group of chemicals, the acridines, are planar molecules that function as *intercalating agents*. These mutagens become inserted between two DNA base pairs and in the process push them apart. During replication, this abnormal conformation can lead to insertions or deletions in acridine-containing DNA. Thus, acridines typically induce *frameshift* mutations (see Section 10.2). Ethidium bromide, which is often used to detect DNA in electrophoresis (see Section 10.12), is also an intercalating agent that functions as a mutagen.

Table 10.2	Chemical and physical mutagens and their modes of action	
Agent	**Action**	**Result**
Base analogs		
5-Bromouracil	Incorporated like T; occasional faulty pairing with G	AT pair → GC pair, occasionally GC → AT
2-Aminopurine	Incorporated like A; faulty pairing with C	AT → GC; occasionally GC → AT
Chemicals reacting with DNA		
Nitrous acid (HNO_2)	Deaminates A and C	AT → GC and GC → AT
Hydroxylamine (NH_2OH)	Reacts with C	GC → AT
Alkylating agents		
Monofunctional (for example, ethyl methane sulfonate)	Puts methyl on G; faulty pairing with T	GC → AT
Bifunctional (for example, nitrogen mustards, mitomycin, nitrosoguanidine)	Cross-links DNA strands; faulty region excised by DNase	Both point mutations and deletions
Intercalative dyes		
Acridines, ethidium bromide	Inserts between two base pairs	Microinsertions and microdeletions
Radiation		
Ultraviolet	Pyrimidine dimer formation	Repair may lead to error or deletion
Ionizing radiation (for example, X-rays)	Free-radical attack on DNA, breaking chain	Repair may lead to error or deletion

Radiation

Several forms of radiation are highly mutagenic. We can divide mutagenic radiation into two main categories, *nonionizing* and *ionizing* (electromagnetic) (Figure 10.6●). Although both kinds of radiation are used in microbial genetics to generate mutations, *nonionizing* radiation has the widest use and will be discussed first.

The purine and pyrimidine bases of the nucleic acids absorb ultraviolet (UV) radiation strongly, and the absorption maximum for DNA and RNA is at 260 nm (∞ Figure 7.9). Killing of cells by UV radiation is due primarily to its effect on DNA. Although several effects are known, one well-established effect is the production in DNA of **pyrimidine dimers**. This is a state in which two adjacent pyrimidine bases (cytosine or thymine) on the same strand of DNA become covalently bonded to one another in such a way that during replication, the probability of DNA polymerase misreading the sequence at this point is greatly increased.

The type of UV radiation source most commonly used for mutagenesis is the *germicidal lamp*, which emits large amounts of UV radiation in the 260-nm region. A dose of UV radiation is used that brings about 50–90% killing of the cell population (∞ Section 20.2), and mutants are then selected or screened for among the survivors. If much higher doses of radiation are used, the number of viable cells will be too low. If lower doses are used, insufficient damage to DNA in the cell population will be induced to make recovery of the desired mutant(s) possible. When used at the correct dose, UV radiation is a very useful and convenient tool for isolating mutants and avoids the necessity of handling toxic chemicals.

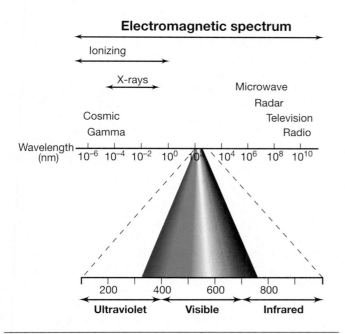

● **Figure 10.6 Wavelengths of radiation.** Note that ultraviolet radiation consists of wavelengths just shorter than visible light. For any electromagnetic radiation, the shorter the wavelength, the higher the energy. DNA absorbs strongly at 260 nm.

Ionizing Radiation

Ionizing radiation is a more powerful form of radiation than UV and includes short-wavelength rays such as X-rays, cosmic rays, and gamma rays (Figure 10.6). These radiations cause water and other substances to ionize, and mutagenic effects are brought about indirectly through this ionization. Among the potent chemical species formed by ionizing radiation are chemical free radicals, of which the most important is the hydroxyl radical, OH· (∞ Section 6.16).

Free radicals react with and inactivate macromolecules in the cell, of which the most important is DNA. At low doses of ionizing radiation only a few "hits" on DNA occur, but at higher doses, multiple hits occur, leading to the death of the cell. In contrast to UV radiation, ionizing radiation penetrates readily through glass and other materials. Because of this, ionizing radiation is used frequently to induce mutations in animals and plants (where its penetrating power makes it possible to reach the gamete-producing cells of these organisms readily). However, because ionizing radiation is more dangerous to use and is less readily available than UV, it finds less use in microbial genetics.

Mutations That Arise from DNA Repair: The SOS System

Recall that a mutation is an *inheritable* change in the genetic material. Therefore, if an error in DNA synthesis can be corrected before the cell divides, no mutation has occurred. Furthermore, some DNA damage cannot be easily replicated. For instance, if a DNA molecule contains pyrimidine dimers, it will not be replicated to give two DNA molecules, each containing pyrimidine dimers. If such DNA damage cannot be repaired, replication can stall and the cell will die. Most cells have a variety of different DNA repair processes to correct mistakes or repair damage. Many of these DNA repair systems are virtually error-free. However, other processes seem to be *error-prone*, and it is the repair process itself that introduces the mutation.

A complex cellular mechanism called the **SOS regulatory system** is activated as a result of some types of DNA damage and initiates a number of DNA repair processes. However, in the SOS system, some DNA repair occurs *in the absence of template instruction*, that is, without base pairing. As you might expect, this results in many errors, hence many mutations.

In the SOS regulatory system, DNA damage serves as a distress signal to the cell, resulting in the coordinate expression of a number of cellular functions involved in DNA repair. The SOS system is normally repressed by a protein called *LexA*. However, this repressor protein is inactivated by a second protein called *RecA*, a protease that is activated as a result of DNA damage (Figure 10.7●). Since one of the DNA repair mechanisms of the SOS system is inherently error-prone, many mutations arise. Thus, through the SOS regulatory system, DNA damage

● **Figure 10.7 Mechanism of the SOS response.** DNA damage results in conversion of RecA protein into a protease that cleaves LexA protein. LexA protein normally represses the activities of the *recA* gene and the DNA repair genes *uvrA* and *umuD* (the UmuD protein is part of DNA polymerase V). However, repression is not complete. Some RecA protein is produced even in the presence of LexA protein. With LexA inactivated, these genes become active.

when cells are exposed to various agents such as chemicals and radiation, leads to mutagenesis. In *Escherichia coli*, where mutagenesis has been studied in great detail, much of this is brought about by *DNA polymerase V*, an enzyme encoded by the *umuCD* genes (see Figure 10.7). The highly error-prone DNA polymerase IV is also induced as part of the SOS response. Collectively, these enzymes are referred to as **DNA mutases**.

Once the DNA damage has been repaired, the SOS system is repressed and further mutagenesis ceases. In addition to its effect on the cellular mutation rate, the SOS regulatory system plays a central role in the regulation of temperate virus replication (⟳ Section 9.10).

It should be emphasized that not all DNA repair occurs in the absence of template instruction. Cells generally have many DNA repair systems that require template instruction and lead to proper DNA repair. These systems apparently work most of the time but are not sufficient to repair the large scale damage done by some very mutagenic chemicals or by large doses of radiation.

Changes in Mutation Rate

High fidelity (low error frequency) in DNA replication is essential if organisms are to enjoy genetic stability. On the other hand, perfect fidelity is impractical, because it would prevent evolution. Therefore, a mutation rate has evolved in cells that is very low yet detectable. This allows organisms to balance the need for genetic stability with that for evolutionary improvement.

The fact that organisms as phylogenetically disparate as hyperthermophilic *Archaea* (⟳ Chapter 13) and *Escherichia coli* have about the same mutation rate might make one believe that evolutionary pressure has selected for organisms with the *lowest possible* mutation rates. However, this is not so. The mutation rate in an organism is subject to change. For example, mutants of some organisms have been selected in the laboratory that are hyperaccurate in DNA replication and repair. However, in these strains, the improved proofreading/repair mechanisms has a significant metabolic cost; thus, a *hyperaccurate* mutant might actually be at a disadvantage in its natural environment. On the other hand, some organisms seem to benefit from a hyperaccurate phenotype that enables them to occupy particular niches in nature. A good example is the bacterium *Deinococcus radiodurans* (⟳ Section 12.34). This organism is 20 times more resistant to UV radiation and 200 times more resistant to ionizing radiation than *E. coli*. This resistance, dependent in part upon redundant DNA repair systems and on a mechanism for exporting damaged nucleotides, allows the organism to survive in environments where other organisms cannot, such as near concentrated sources of radiation or on the surfaces of dust particles exposed to intense sunlight.

In contrast to the hyperaccurate phenotype, some organisms actually benefit from an *increased* mutation rate. DNA repair systems are themselves encoded by genes and thus subject to mutation. For example, the protein subunit of DNA polymerase III involved in proofreading (⟳ Section 7.6) is encoded by the gene *dnaQ*. Certain mutations in this gene lead to mutant strains that are still viable but have an increased rate of mutation. Such a strain is said to be hypermutable or a **mutator**. Mutations leading to a mutator phenotype are known in several other DNA repair systems as well. The mutator phenotype is apparently selected for in complex and changing environments, since strains of bacteria with mutator phenotypes appear to be more abundant under these conditions. Presumably, whatever disadvantage an increased mutation rate may have in such environments is offset by the ability to generate greater numbers of useful mutations. These mutations ultimately increase evolutionary fitness of the population and make the organism more successful in its ecological niche.

As was indicated earlier, a mutator phenotype may be induced in wild-type strains by stress situations. For instance, the SOS response induces error-prone repair. Therefore, when the SOS response is activated, the mutation rate increases. While in some cases this may be a necessary by-product of DNA repair, in other cases the increased mutation rate may itself be of value. That is, the organism may generate *adaptive mutations*. **Adaptive mutations** lead to a phenotype in the mutant that allows it to survive a particular stress. Selection is obviously strong under these conditions because, depending on the stress condition, the only option to potential survival from an increased mutation rate may be death.

10.4 *Concept Check*

Mutagens are chemical, physical, or biological agents that increase the mutation rate. Mutagens can alter DNA in many different ways. However, alterations in DNA are not mutations unless they can be inherited. Some DNA damage can lead to cell death if not repaired, and both error-prone as well as high-fidelity DNA repair systems exist.

◆ How do mutagens work?

◆ Why might a mutator phenotype be successful in an environment experiencing rapid changes?

10.5 Mutagenesis and Carcinogenesis: The Ames Test

A practical use of bacterial mutants has been developed to identify potentially hazardous chemicals in the environment. Because the sensitivity with which selectable mutants can be detected in large populations of bacteria is very high, bacteria can be used as screening agents for the potential mutagenicity of chemicals. This is relevant because many mutagenic chemicals are also *carcinogenic*, capable of causing cancer in humans or other animals.

The variety of chemicals, both natural and artificial, that humans encounter through agricultural and industrial exposure is enormous. There is good evidence that a large proportion of human cancers have environmental causes, most likely from various chemicals, making the detection of chemical carcinogens an urgent and very important matter. It does not necessarily follow that because a compound is mutagenic it is also carcinogenic. The correlation, however, is quite high, and the knowledge that a compound is mutagenic in a bacterial system serves as a warning of possible danger. The development of bacterial tests for carcinogenic screening was carried out primarily by a group at the University of California in Berkeley under the direction of Bruce Ames, and the mutagenicity test for carcinogens is called the **Ames test** (Figure 10.8●).

Protocol for an Ames Test

The standard way to test chemicals for mutagenesis is to look for an increase in the rate of *back* mutation (reversion) in auxotrophic strains of bacteria in the presence of the suspected mutagen. It is important that the mutation to be examined be a point mutation because reversion will occur in such a strain at the same rate as the forward mutation did. When cells of such an auxotrophic strain are spread on a medium lacking the required nutrient (for example, an amino acid), no growth occurs, and even very large populations of cells can be spread on the plate without formation of visible colonies. However, if back mutants (revertants) are present, those cells will form colonies. Thus, if 10^8 cells are spread on the surface of a single plate, even as few as 10–20 revertants can be detected by the 10–20 colonies they form (Figure 10.8, left photo). If the reversion

rate has been *increased* by addition of a chemical mutagen, the number of revertant colonies will increase. Histidine auxotrophs of *Salmonella enterica* (Figure 10.8) and tryptophan auxotrophs of *Escherichia coli* have been the major tools of the Ames test.

Two additional elements have been introduced in the Ames test to make it much more powerful. The first of these is to use test strains that almost exclusively use error-prone pathways to repair DNA damage; normal repair mechanisms are thus thwarted (see Section 10.3). The second important element in the Ames test is the addition of liver enzyme preparations to convert the chemicals to be tested into their active mutagenic (and potentially carcinogenic) forms. It has been well established that many carcinogens are not directly carcinogenic or mutagenic themselves but undergo modifications in the human body that convert them into active substances. These changes take place primarily in the liver, where enzymes called *mixed-function oxygenases* normally involved in detoxification cause the formation of epoxides or other activated forms of the compounds; these are then highly reactive (and thus mutagenic) with DNA.

In the Ames test, a preparation of enzymes from rat liver is first used to activate the test compound. The activated complex is then taken up on a filter-paper disk, which is placed in the center of a plate on which the proper bacterial strain has been overlaid. After overnight incubation, the mutagenicity of the compound can be detected by looking for a halo of back mutations in the area around the paper disk (Figure 10.8). It is always necessary to carry out this test with several different concentrations of the compound and with appropriate positive (known mutagens) and negative (no mutagen) controls, because compounds vary in their mutagenic activity and may be lethal at higher levels. A wide variety of chemicals have been subjected to the Ames test, and it has become one of the most useful screens for determining the potential carcinogenicity of a compound.

● **Figure 10.8 The Ames test is used to evaluate the mutagenicity of a chemical.** Two plates were inoculated with a culture of a histidine-requiring (auxotrophic) mutant of *Salmonella enterica*. The medium does not contain histidine, so only cells that revert back to wild type can grow. Spontaneous revertants appear on both plates, but the chemical on the filter paper disc in the test plate (right) has caused an increase in the mutation rate, as shown by the large number of colonies surrounding the disc. The plate on the left was the negative control, as the disc contained only water. Revertants are not seen very close to the test disc because the concentration of the mutagen is so high there that it is lethal.

10.6 Genetic Recombination

Recombination is the physical exchange of genes between genetic elements. In this section, we focus on **homologous recombination**, which results in genetic exchange between *homologous* DNA sequences from two different sources. Homologous DNA sequences are those that have nearly the same sequence; therefore, base pairing can occur over an extended length of the two DNA molecules. This type of recombination is involved in the process referred to as "crossing over" in classical genetics.

Molecular Events in Homologous Recombination

Recombination has been intensively studied in prokaryotes, viruses, and some fungi, especially yeast. In *Bacteria*, homologous recombination involves the participation of the RecA protein, previously mentioned in regard to the error-prone SOS repair system (see Section 10.4). The RecA protein has been shown to be essential in nearly every homologous recombination pathway. RecA-like proteins have been identified in all prokaryotes examined, including the *Archaea*, as well as in yeast and in the higher *Eukarya*.

A molecular mechanism of homologous recombination is shown in Figure 10.9●. The process begins with a *nick* (generated by an endonuclease) in one of the DNA molecules. This nicked strand must be displaced from the other strand by proteins having helicase activity (∞ Section 7.6). In some pathways specialized enzymes, such as the RecBCD enzyme of *Escherichia coli*, have both endonuclease and helicase activities. Single-stranded binding protein (∞ Section 7.6) then binds to the resulting single-stranded segment. Next, the RecA protein binds to the single-stranded fragment, forming a complex that facilitates annealing with a complementary sequence in the adjacent DNA duplex, simultaneously displacing the resident strand (Figure 10.9). This process is referred to as *strand invasion* and leads to the *pairing* of DNA molecules over long stretches. Following pairing, *exchange* of homologous DNA molecules occurs, leading to the formation of recombination intermediates containing extensive **heteroduplex** regions, where each strand has originated from a different chromosome. This process also requires DNA polymerase and ligase activities. Finally, the linked molecules are *resolved* by nucleases and DNA ligase to form two recombinant DNA molecules.

Effect of Homologous Recombination on Genotype

For new genotypes to arise from homologous recombination, it is essential that the two homologous sequences be genetically distinct. This is obviously the case in a diploid eukaryotic cell (∞ Section 14.7), which has two sets of chromosomes, one from each parent. However, in prokaryotes, genetically distinct but homologous DNA molecules are brought together in different ways, but the process of genetic recombination is no less important. Genetic recombination in prokaryotes occurs because *fragments* of homologous DNA from a donor chromosome are transferred to a recipient cell by one of three processes: *transformation*

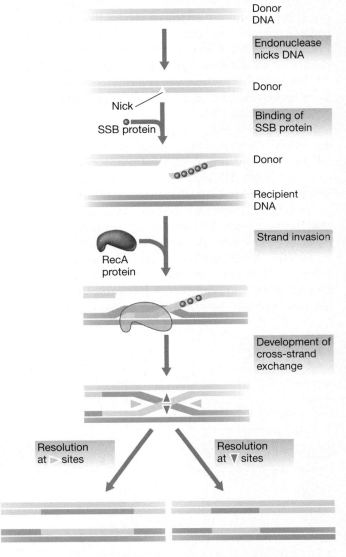

● **Figure 10.9 A simplified version of one molecular mechanism of genetic recombination: homologous recombination.** Homologous DNA molecules pair and exchange DNA segments. The mechanism involves breakage and reunion of paired segments. Two of the proteins involved, a single-stranded binding (SSB) protein and the RecA protein, are shown. The other proteins involved are not shown. The diagram is not to scale: Pairing can occur over hundreds or thousands of bases. Resolution occurs by cutting and ligating the cross-linked DNA molecules. Note that there are two possible outcomes, depending on where strands are cut during the resolution process.

(see Section 10.7), *transduction* (see Section 10.8), or *conjugation* (see Section 10.11). It is only *after* the transfer, when the DNA fragment from the host is in the recipient cell, that homologous recombination occurs. Because in prokaryotes only a chromosomal fragment is transferred, if recombination does not occur, the fragment will be lost because it cannot replicate independently. Therefore, it is important to remember that in prokaryotes, transfer is just the first step in obtaining recombinant organisms.

Detection of Recombination

In order to detect physical exchange of DNA segments, the cells resulting from recombination must be phenotypically different from either parent. In genetic crosses, one must usually use recipient strains that lack some selectable characteristic that the recombinants will possess. For instance, the recipient may not be able to grow on a particular medium, and genetic recombinants are selected that can. Various kinds of selectable markers, such as drug resistance and nutritional requirements, were discussed in Section 10.1.

The exceedingly great sensitivity of the selection process allows for even a few recombinant cells to be detected in a large population of nonrecombinant cells (Figure 10.10●). The only requirement for effective detection of recombination is that the *reverse* mutation rate for the selected characteristic be low, because revertants will also form colonies. This problem can often be overcome by using *double mutants*—strains that carry two different mutations—because it is very unlikely that two back mutations will occur in the same cell. Alternatively, frameshift mutants can be used, because their reversion rates are typically very low. Much of the skill of the bacte-

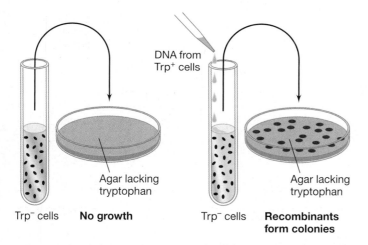

● **Figure 10.10 How a selective medium can be used to detect rare genetic recombinants among a large population of nonrecombinants.** On the selective medium only the rare recombinants form colonies. Procedures such as this, which offer high resolution for genetic analyses, can ordinarily be used only with microorganisms. The type of genetic exchange being illustrated is transformation, which is discussed in Section 10.7.

rial geneticist is exhibited in the choice of proper mutants and selective media for efficient detection of genetic recombination. Because selection is so powerful and because crosses can be made using billions of individual cells, recombinational analysis is a very important tool for the microbial geneticist (see Sections 10.7–10.12).

 10.6   Concept Check

Homologous recombination arises when closely related DNA sequences from two distinct genetic elements are combined together in a single element. Recombination is an important evolutionary process, and cells have specific mechanisms for ensuring that recombination takes place. Mechanisms of recombination that occur in prokaryotes involve DNA transfer during the processes of transformation, transduction, and conjugation.

◆   Which protein, found in all prokaryotes, facilitates the pairing required for homologous recombination?

◆   In eukaryotes, recombination involves entire chromosomes, but this is not true in prokaryotes. Explain.

▐▐ GENETIC EXCHANGE IN PROKARYOTES

For genetic analyses, the microbial geneticist must cross strains of an organism that have different genotypes (and phenotypes) and look for recombinants (Figure 10.10) Three mechanisms of genetic exchange are known in prokaryotes: (1) **transformation**, in which DNA released from one cell is taken up by another (see Section 10.7), (2) **transduction**, in which donor DNA transfer is mediated by a virus (see Section 10.8); and (3) **conjugation**, in which the transfer involves cell-to-cell contact and a *conjugative plasmid* in the donor cell (see Sections 10.9–10.12). These processes are contrasted in Figure 10.11● .

10.7   Transformation

Transformation is a process by which free DNA is incorporated into a recipient cell and brings about genetic change. A number of prokaryotes are naturally transformable, including certain species of both gram-negative and gram-positive *Bacteria* and some species of *Archaea*. Since the DNA of prokaryotes is present in the cell as a large single molecule, when the cell is gently lysed, the DNA pours out (Figure 10.12●). Because of their extreme length (1700 μm in *Bacillus subtilis*, for example), bacterial chromosomes break easily. Even after gentle extraction, the *B. subtilis* chromosome of 4.2 Mbp is converted to fragments of about 10 kbp. Because the DNA that corresponds to an average gene is about 1000 nucleotides, each of the fragments of *B. subtilis* DNA therefore contains about 10 genes. This is a typical transformable size. A single cell usually incorporates only one or a few DNA fragments, so only a small proportion of

● **Figure 10.12 The prokaryotic chromosome, as shown in the electron microscope.** The circular chromosome is from the hyperthermophile *Sulfolobus,* a member of the *Archaea* (◌◌◌ Section 13.9). See also Figure 10.17.

M. Shioda and S. Takayanago

● **Figure 10.11 Processes by which DNA is transferred from donor to recipient bacterial cell.** Just the initial steps in transfer are shown. For details of how the DNA is integrated into the recipient, see text.

the genes of one cell can be transferred to another by a single transformation event.

Transformation in the History of Molecular Biology

Although genetic recombination in eukaryotes had been known for a long time, the discovery of genetic recombination in bacteria has been a more recent event. However, the discovery of transformation was one of the seminal events in biology, as it led to experiments demonstrating that DNA was the genetic material (Figure 10.13●). This discovery became the cornerstone of molecular biology and modern genetics, and so we review it here.

The first evidence of bacterial transformation was obtained by the British scientist Frederick Griffith in the late 1920s. Griffith was working with *Streptococcus pneumoniae* (pneumococcus), a bacterium that owes its ability to in-vade the body in part to the presence of a polysaccharide capsule (◌◌◌ Section 4.10). Mutants can be isolated that lack this capsule and are thus unable to cause infection. Such mutants are called *R strains* because their colonies appear rough on agar, in contrast to the smooth appearance of encapsulated strains, called *S strains*. A mouse infected with only a few cells of an S strain succumbs in a day or two to a massive pneumococcus infection. By contrast, even large numbers of R cells do not cause death when injected. Griffith showed that if heat-killed S cells were injected along with living R cells, the mouse developed a fatal infection and the bacteria isolated from the dead mouse were of the S type (Figure 10.13). Since the S cells isolated in such an experiment always had the capsule type of the heat-killed S cells, Griffith concluded that the R cells had been *transformed* into a new type. The process had all the properties of a genetic event and so set the stage for the discovery of DNA.

The molecular explanation for the transformation of pneumococcus types was provided by Oswald T. Avery and his associates at the Rockefeller Institute in New York in a series of studies carried out during the 1930s, culminating in the classic paper by Avery, C. M. MacLeod, and M. McCarty in 1944. Avery and his coworkers showed that the transformation process could be carried out in the test tube rather than the mouse and that a cell-free extract of heat-killed cells could induce transformation. In a series of painstaking biochemical experiments, the active fraction of cell-free extracts was purified and was shown to be DNA. The transforming activity of purified DNA preparations was very high, and only a very small amount of material was necessary. Subsequently, others at Rockefeller showed that transformation could occur in pneumococcus not only for capsular characteristics but for other genetic characteristics of the organism as well, such as antibiotic resistance and sugar fermentation.

In 1953, James Watson and Francis Crick announced their model of the structure of DNA, providing a theoretical framework for how DNA could serve as the genetic

| Heat-killed S cells | Live S cells | Live R cells | Live R cells +
heat-killed S cells |

● **Figure 10.13** **Griffith's experiments with pneumococcus.** Live smooth (S) cells contain a capsule and kill mice because immune cells cannot kill the encapsulated bacteria; the cells proliferate in the lung and cause a fatal pneumonia. Rough (R) cells have no capsule and are not pathogenic. But a combination of *live* R and *dead* S cells kill mice, and live S type cells can be isolated from the animals. DNA-encoding capsule production is released from dead S cells and taken up by R cells, thus *transforming* them into S-type cells.

material. Thus, three types of studies, the bacteriological ones of Griffith, the biochemical ones of Avery, and the physical-chemical ones of Watson and Crick, solidified the concept of DNA as the genetic material. In subsequent years, this work led to the whole field of molecular biology and molecular genetics.

Competence

Even within transformable genera, only certain strains or species are transformable. A cell that is able to take up DNA and be transformed is said to be **competent**, and the capacity is genetically determined. Competence in most naturally transformable bacteria is regulated, and special proteins play a role in the uptake and processing of DNA. These competence-specific proteins include a membrane-associated DNA-binding protein, a cell wall autolysin, and various nucleases. One pathway to natural competence in *Bacillus subtilis*—an easily transformed species—is part of a quorum-sensing system (a regulatory system that responds to cell density, ∞ Section 8.10) controlled by a two-component regulatory system (∞ Section 8.12). Cells produce and excrete a small peptide during growth, and the accumulation of this peptide to high concentrations induces the cells to become competent. In *Bacillus*, about 20% of the cells become competent and stay that way for several hours. However, in *Streptococcus*, 100% of the cells can become competent, but only for a brief period during the growth cycle.

High-efficiency natural transformation is known in only a few *Bacteria*. For example, *Acinetobacter, Azotobacter, Bacillus, Streptococcus, Haemophilus, Neisseria,* and *Thermus* are naturally competent and easily transformable. By contrast, many prokaryotes are poorly transformed, if at all, under natural conditions. Interestingly, *Escherichia coli* and many other gram-negative bacteria fall into this category. However, when cells of *E. coli* are treated with high concentrations of calcium ions and then chilled for several minutes, they become transformable. Cells of *E. coli* treated in this manner take up double-stranded DNA, and therefore transformation of this organism by plasmid DNA is relatively efficient. As we will see later, this discovery was important for getting DNA into *E. coli*—the workhorse of genetic engineering—for applied reasons in the field of biotechnology (∞ Chapter 31).

Uptake of DNA in Transformation

During transformation, competent bacteria reversibly bind DNA. Soon, however, the binding becomes irreversible. Competent cells bind much more DNA than do noncompetent cells—as much as 1000 times more. As we noted earlier, the sizes of the transforming fragments are much smaller than that of the whole genome, and the fragments are further degraded during the uptake process. In *Streptococcus pneumoniae* each cell can bind only about 10 molecules of double-stranded DNA of 10–15 kbp each. However, as these fragments are taken up, they are converted into single-stranded pieces of about 8 kb, with the complementary strand being degraded. The DNA fragments in the mixture compete with each other for uptake, and if excess DNA that does not contain the genetic marker is added, a decrease in the number of transformants occurs.

In preparations of transforming DNA, typically only about 1 out of 100–300 DNA fragments contains the genetic marker being studied. Thus, at high concentrations of DNA, the competition between DNA molecules results in saturation of the system, so even under the best of conditions it is impossible to transform all the cells in a population for a given marker. The maximum frequency of transformation that has so far been obtained is about 20% of the population; the values usually obtained are between 0.1 and 1.0%. But with the recipient population sizes being very high, even this low frequency is easy to detect. The minimum concentration of DNA yielding detectable transformants is about 0.01 ng/ml, which is so low that it is chemically undetectable.

Interestingly, in transformation in *Haemophilus influenzae* there is a requirement that the DNA fragment have a particular 11-bp sequence for irreversible binding and uptake to occur. This sequence is found at an unexpectedly high frequency in the *Haemophilus* genome, which has been completely sequenced (∞ Section 15.4). Evidence such as this, and the fact that certain bacteria have been shown to become competent in their natural environment, suggests that transformation is not a laboratory artifact but plays an important role in lateral gene transfer in nature. By promoting new combinations of genes, naturally transformable bacteria increase diversity and fitness of the microbial community as a whole.

Integration of Transforming DNA

Transforming DNA is bound at the cell surface by a DNA-binding protein. Following this, either the entire double-stranded fragment is taken up, or a nuclease degrades one strand and the remaining strand only is taken up, depending on the organism (Figure 10.14●). After uptake, the DNA becomes attached to a competence-specific protein. This prevents the DNA from nuclease attack until it reaches the chromosome, where the RecA protein takes over. The DNA is integrated into the genome of the recipient by recombination (Figure 10.14; see also Figure 10.9). During replication of this heteroduplex DNA, one parental and one recombinant DNA molecule are formed.

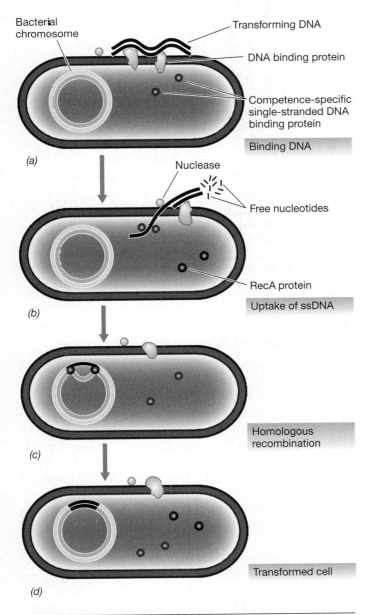

●Figure 10.14   Mechanism of transformation in a gram-positive bacterium. (a) Binding of double-stranded DNA by a membrane-bound DNA binding protein. (b) Passage of one of the two strands into the cell while nuclease activity degrades the other strand. (c) The single strand in the cell is bound by specific proteins, and recombination with homologous regions of the bacterial chromosome is mediated by RecA protein. (d) Transformed cell.

On segregation at cell division, the latter is present in the transformed cell, which is now genetically altered as compared to the parental type. The preceding pertains to only small pieces of *linear* DNA. Many naturally transformable *Bacteria* are transformed only poorly by plasmid DNA because the plasmid must remain double-stranded and circular in order to replicate.

Transfection

Bacteria can be transformed with DNA extracted from a bacterial virus rather than from another bacterium. This process is known as **transfection**. If the DNA is from a lytic bacteriophage, transfection leads to virus production and can be measured by the standard phage plaque assay (∞ Section 9.4). Transfection is useful for studying the mechanism of transformation and recombination because the small size of phage genomes allows for the isolation of a nearly homogeneous population of DNA molecules. By contrast, in conventional transformation, the transforming DNA is typically a random assortment of chromosomal DNA of various lengths, and this tends to complicate experiments designed to study the mechanism of transformation.

 10.7   Concept Check

Certain prokaryotes exhibit competence, a state in which cells are able to take up free DNA released by other bacteria. Incorporation of donor DNA into a recipient cell requires the activity of single-stranded binding protein, RecA protein, and several other enzymes. Only competent cells are transformable.

◆ The donor bacterial cell in a transformation is probably dead. Explain.

◆ Even in naturally transformable cells, competency is usually inducible. What does this mean?

10.8   Transduction

In transduction, DNA is transferred from cell to cell by a bacterial virus (bacteriophage). Genetic transfer of host genes by viruses can occur in two ways. In the first, called **generalized transduction**, host DNA derived from virtually any portion of the host genome becomes a part of the DNA of the mature virion in place of the virus genome. The second, called **specialized transduction**, occurs only in some temperate viruses. In specialized transduction, DNA from a specific region of the host chromosome is integrated directly into the virus genome—usually replacing some of the virus genes. The transducing virion in both generalized and specialized transduction is usually noninfective as a virus because bacterial genes have replaced some or all necessary viral genes.

In generalized transduction, if the donor genes do not undergo homologous recombination with the recipient bacterial chromosome, they will be lost. They cannot replicate independently and are not part of a viral genome.

In specialized transduction, homologous recombination may also occur. However, since the donor bacterial DNA is now actually a part of a temperate phage genome, there are two other possibilities: (1) the DNA may be integrated into the host chromosome during lysogenization (🔗 Section 9.10), and (2) the DNA may be replicated in the recipient as part of a lytic infection.

Transduction occurs in a variety of *Bacteria*, including genera of *Desulfovibrio*, *Escherichia*, *Pseudomonas*, *Rhodococcus*, *Rhodobacter*, *Salmonella*, *Staphylococcus*, and *Xanthobacter*, as well as *Methanothermobacter thermoautotrophicus*, a species of *Archaea*. Not all phages can transduce, and not all bacteria are transducible, but the phenomenon is sufficiently widespread that it likely plays an important role in genetic transfer in nature.

Generalized Transduction

In generalized transduction, virtually any gene on the donor chromosome can be transferred to the recipient. Generalized transduction was first discovered and extensively studied in the bacterium *Salmonella enterica* with phage P22 and has also been studied with phage P1 in *Escherichia coli*. An example of how *transducing particles* may be formed is given in Figure 10.15●. When a bacterial cell is infected with a phage, the events of the phage lytic cycle may be initiated. However, during the lytic infection, the enzymes responsible for packaging viral DNA into the bacteriophage sometimes package host DNA accidentally. The resulting virion is called a *transducing particle*. On lysis of the cell, these particles are released along with normal (that is, potentially lytic) virions, and so the lysate contains a mixture of normal virions and transducing particles.

Because transducing particles cannot lead to a normal viral infection (they contain no viral DNA), they are said to be *defective*. When this lysate is used to infect a population of recipient cells, most of the cells become infected with normal virus. However, a small proportion of the population receives transducing particles that inject the DNA they packaged from the previous host bacterium. While this DNA cannot replicate, it can undergo genetic recombination with the DNA of the new host. Because only a small proportion of the particles in the lysate are defective, and each of these contains only a small fragment of donor DNA, the probability of a given transducing particle containing a particular gene is quite low. Typically, only about 1 cell in 10^6 to 10^8 is transduced for a given marker.

Phages that form transducing particles can be either temperate or virulent, the main requirements being that they have a DNA-packaging mechanism that accepts host DNA and that DNA packaging occurs before the host genome is completely degraded. The detection of transduction is most certain when the multiplicity of phage to host is low, so a host cell is infected with only a single phage particle; with multiple infection, the cell will likely be killed by the normal virions in the lysate.

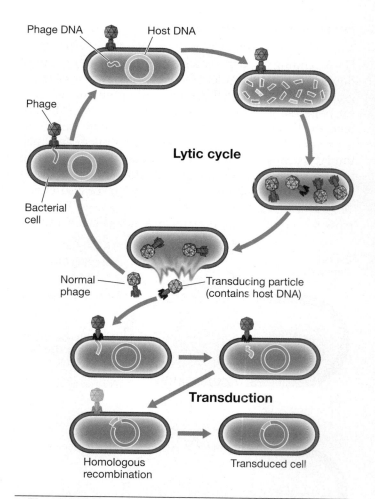

● **Figure 10.15 Generalized transduction.** Note that "normal" virions contain phage genes while a transducing particle contains host genes.

Phage Lambda and Specialized Transduction

Generalized transduction allows the transfer of DNA from one bacterium to another at a low frequency. However, specialized transduction can allow extremely efficient transfer while also allowing a small region of a bacterial chromosome to be selectively transduced. The example we use to discuss specialized transduction was the first to be discovered and involves transduction of the galactose genes by the temperate phage **lambda** of *Escherichia coli*.

As we discussed (🔗 Section 9.10), when a cell is lysogenized by lambda, the phage genome becomes integrated into the host DNA at a *specific site*. The region in which lambda integrates in the *E. coli* chromosome is immediately adjacent to the cluster of genes that encode the enzymes involved in galactose utilization (🔗 Figure 9.21). After insertion, viral DNA replication is under control of the bacterial host chromosome. Upon induction, the viral DNA separates from the host DNA by a process that is the reverse of integration (Figure 10.16●). Ordinarily when the lysogenic cell is induced, the lambda DNA is excised as a unit. Under rare conditions, however, the phage genome is excised incorrectly. Some of the adja-

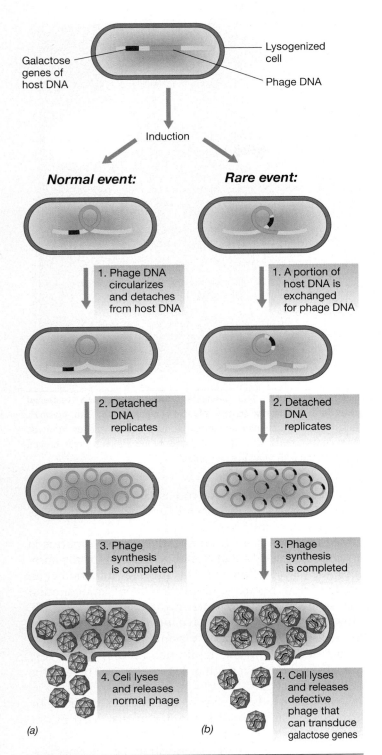

● **Figure 10.16 Specialized transduction.** (a) Normal lytic events, and (b) the production of particles transducing the galactose genes in an *Escherichia coli* cell containing a lambda prophage.

cent bacterial genes (for example, the galactose operon) are excised along with phage DNA. At the same time, some phage genes are left behind (Figure 10.16).

One type of altered phage particle, called **lambda dgal** (λ*dgal*; *dgal* means "defective, galactose"), is defective because of the phage genes lost. It will not make mature phage in a subsequent infection. However, a viable lambda virion, in this case called a **helper phage**, can

provide those functions missing in the defective particle. When cells are *coinfected* with λ*dgal* and the helper phage, the culture lysate contains a few λ*dgal* particles mixed in with a large number of normal lambda virions. When a galactose-negative (Gal⁻) bacterial culture is infected at high multiplicity with such a lysate and Gal⁺ transductants selected, many are double lysogens, carrying both lambda and λ*dgal*. When such a double lysogen is induced, the lysate can contain large numbers of λ*dgal* virions and can transduce at high efficiency, although only for the restricted group of *gal* genes.

If a lambda virion is to be viable, there is a limit to the amount of phage DNA that can be replaced with host DNA. Sufficient phage DNA must be retained to provide for production of the phage protein coat and for other phage proteins needed for lysis and lysogenization. However, if a helper phage is used together with the defective phage in a mixed infection, then even fewer phage-specific genes are needed in the defective phage for transduction. Only the *att* (attachment) region, the *cos* site (cohesive ends, for packaging), and the replication origin of the lambda genome are absolutely needed for production of a transducing particle, provided that a helper phage is used (the genetic map of lambda, Figure 9.18 and Sections 9.10 and 9.11). In this way, specialized transducing phages covering many specific regions of the *E. coli* genome have been isolated. In addition, lambda transducing phages can be constructed by the techniques of genetic engineering to contain genes from any organism (see Section 10.17).

Phage Conversion

When a normal temperate phage (that is, a nondefective phage) lysogenizes a cell and its DNA is converted to the prophage state, the cell is immune to further infection by the same type of phage. This acquisition of immunity can be considered a change in phenotype. However, other phenotypic alterations can often be detected in the lysogenized cell that are unrelated to phage immunity. Such a change, which is brought about through lysogenization by a normal temperate phage, is called **phage conversion**.

Two cases of phage conversion have been especially well studied. One involves a change in structure of a polysaccharide on the cell surface of *Salmonella anatum* on lysogenization with bacteriophage ε¹⁵. The second involves the conversion of nontoxin-producing strains of *Corynebacterium diphtheriae* (the bacterium that causes the disease diphtheria) to toxin-producing (pathogenic) strains upon lysogenization with phage β (Section 26.3). In both of these situations, the genes encoding the necessary molecules are an integral part of the phage genome and hence are automatically (and exclusively) transferred upon infection by the phage and lysogenization.

Lysogeny probably carries a strong selective value for the host cell because it confers resistance to infection by viruses of the same type. Phage conversion may also be of considerable evolutionary significance because it

results in efficient genetic alteration of host cells. Many bacteria isolated from nature are natural lysogens. It seems reasonable to conclude, therefore, that lysogeny is a common condition and may often be essential for survival of the host in nature.

10.8 Concept Check

Transduction involves transfer of host genes from one bacterium to another by bacterial viruses. In generalized transduction, defective virus particles randomly incorporate fragments of the cell's chromosomal DNA, but the efficiency is low. In specialized transduction, the DNA of a temperate virus excises incorrectly and takes adjacent host genes along with it; transducing efficiency here may be very high.

◆ What is the major difference between generalized transduction and transformation?

◆ In specialized transduction, the donor DNA can replicate inside the recipient cell without homologous recombination taking place, but this is not true in generalized transduction. Explain.

10.9 Plasmids: General Principles

Before we discuss the third method of genetic transfer, **conjugation**, we must first discuss *plasmids*. **Plasmids** are genetic elements that replicate independently of the host chromosome (∞ Section 7.4). Unlike viruses, plasmids do not have an extracellular form and exist inside cells simply as free, and typically circular, DNA. Plasmids and chromosomes can be differentiated on the basis that plasmids carry only *unessential* (but often very helpful) genes. Essential genes reside on chromosomes. Literally thousands of different plasmids are known. Indeed, over 300 different naturally occurring plasmids have been isolated from strains of *Escherichia coli* alone. In this section we discuss the properties of a few of them.

Physical Nature and Replication of Plasmids

Almost all known plasmids consist of double-stranded DNA. Most plasmids are circular, but many linear plasmids are also known. Naturally occurring plasmids vary in size from approximately 1 kilobase to more than 1 megabase. The typical plasmid is a circular double-stranded DNA molecule less than 5% the size of the chromosome (Figure 10.17●). Most of the plasmid DNA isolated from cells is in the supercoiled configuration, which is the most compact form for DNA to exist within the cell (∞ Figure 7.10).

The enzymes involved in actual plasmid replication are normal cell enzymes. Therefore, the genes carried by the plasmid itself are concerned primarily with *control* of the replication initiation process and with apportionment of the replicated plasmids between daughter cells. Also, different plasmids are present in cells in different numbers; this is called the *copy number*. Some plasmids are present in the cell in only 1–3 copies, whereas others

● Figure 10.17 The bacterial chromosome and bacterial plasmids, as shown in the electron microscope. The plasmids (arrows) are the circular structures, and are much smaller than the main chromosomal DNA. The cell (large, white structure) was broken gently so the DNA would remain intact.

may be present in over 100 copies. Copy number is controlled by genes on the plasmid and by interactions between the host and the plasmid.

Most plasmids in gram-negative *Bacteria* replicate in a manner similar to that already described for the chromosome (∞ Section 7.6). This involves initiation at an origin of replication and bidirectional replication around the circle, giving a *theta* intermediate (∞ Figure 7.17). However, some plasmids have *unidirectional* replication. Because of the small size of plasmid DNA relative to the chromosome, the whole replication process occurs very quickly, perhaps in a tenth or less of the total time of the cell division cycle.

Most plasmids of gram-positive *Bacteria* replicate by a rolling circle mechanism similar to that used by the phage φX174 (∞ Section 16.2 and Figure 16.4). This mechanism gives rise to a single-stranded intermediate, and thus these plasmids are sometimes referred to as *single-stranded DNA plasmids*. Most linear plasmids replicate using a mechanism involving a protein bound to the 5'-end of each strand that is used in priming DNA synthesis (∞Section 14.6).

Plasmid Incompatibility and Plasmid Curing

Some bacteria may contain several different types of plasmids. For example, *Borrelia burgdorferi* (the Lyme disease pathogen, ∞ Section 27.4) contains 17 different circular and linear plasmids! The ability of two different plasmids to each replicate in the same cell is controlled by plasmid genes involved in regulating DNA replication. When a plasmid is transferred into a cell that already carries another plasmid, a common observation is that the second plasmid may not be maintained and is lost during subsequent cell replication. If this occurs, the two plasmids are said to be **incompatible**. A number of incompatibility (Inc) groups have been recognized. The plasmids of one incompatibility group exclude each other from replicating in the cell but can coexist with plasmids from other groups. Plasmids of an incompatibility group share a common mechanism of regulating their replication and are thus *related* to one another. Therefore, although a bacterial cell may contain different kinds of plasmids, each is genetically distinct.

Some plasmids, called *episomes*, can integrate into the chromosome, and under such conditions their replication comes under control of the chromosome. This situation is remarkably like that of several viruses whose genomes can become incorporated into the host genome (prophages, ⟁ Sections 9.10, 16.5, 16.11, and 16.15). Plasmids can sometimes be eliminated from host cells by various treatments. This removal, called **curing**, results from inhibition of plasmid replication without parallel inhibition of chromosome replication. As a result of cell division, the plasmid is diluted out. Curing may occur spontaneously, but it is greatly increased by treatments with certain chemicals such as acridine dyes, which become inserted into DNA, or other treatments that seem to interfere more with plasmid replication than with chromosome replication.

Many of these characteristics are exemplified in a very well-characterized plasmid of *Escherichia coli*, called the F *plasmid*. The F plasmid ("F" stands for "fertility") is a circular DNA molecule of 99,159 bp. Cells containing it can easily be cured with acridine orange. Figure 10.18● shows a genetic map of the F plasmid. One region of the plasmid contains genes involved in regulating DNA replication. It also contains a number of transposable elements (see Section 10.14) involved in its ability to function as an episome. Last, it has a large region of DNA, the *tra* region, containing genes that permit it to be transferred from one cell to another (see later).

Cell-to-Cell Transfer of Plasmids

Because one of the defining characteristics of a plasmid is the lack of a distinct extracellular form, one might imagine that plasmids are transferred only during cell division. Since some prokaryotic cells can take up free DNA from the environment (see Section 10.6), it is possible that lysis of the host, however this may happen, may bring the plasmid in contact with a new host. However, this process occurs naturally in only a few bacterial species and is unlikely to account for much cell-to-cell plasmid transfer. The main mechanism of cell-to-cell transfer is **conjugation**, a function encoded by some plasmids themselves. Conjugation is a replicative process, and both cells end up with copies of the plasmid (Figure 10.19●).

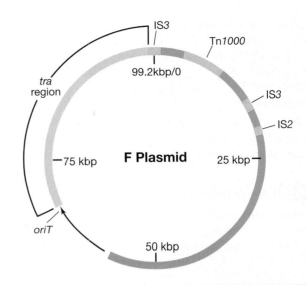

● **Figure 10.18   Genetic map of the F (fertility) plasmid of *Escherichia coli*.** The numbers on the interior show the size of the plasmid in kilobase pairs (the exact size is 99,159 bp). The region shown in dark green at the bottom of the map contains genes primarily responsible for the replication and segregation of the F plasmid in normally growing cells. The light green region, the *tra* region, contains the genes involved in conjugative transfer. The *oriT* sequence is the origin of transfer during conjugation. The arrow indicates the direction of transfer (the *tra* region would be transferred last). The regions shown in yellow on F are transposable elements where integration into identical elements on the bacterial chromosome can occur and lead to the formation of different Hfr strains (see Section 10.12).

Plasmids that govern their own transfer by cell-to-cell contact are called *conjugative*. Not all plasmids are conjugative. Transmissibility by conjugation is controlled by a set of genes within the plasmid called the *tra* (for *transfer*) region. The *tra* region contains genes encoding proteins that function in DNA transfer and replication and others that function in mating pair formation. The presence of a *tra* region in a plasmid can have another important consequence if the plasmid becomes integrated into the chromosome. In that case, the plasmid can *mobilize* the transfer of chromosomal DNA from one cell to another. The use of conjugation to transfer host genes is discussed further in the next section.

Some conjugative plasmids from *Pseudomonas* have a *broad host range*. This means that they are transferable to a wide variety of other gram-negative *Bacteria*. Such

● **Figure 10.19   Plasmid transfer from cell to cell during conjugation.** Note that as in all gene transfer processes in prokaryotes, DNA transfer is *unidirectional*.

plasmids can transfer genetic information between distantly related organisms. Conjugative plasmids have been shown to transfer between gram-negative and gram-positive *Bacteria*, between *Bacteria* and plant cells, and between *Bacteria* and fungi. Even if the plasmid cannot replicate independently in the new host, transfer of the plasmid itself could have important evolutionary consequences if genes from the plasmid recombine with the genome of the new host.

10.9 Concept Check

Plasmids are small circular or linear DNA molecules that carry any of a variety of unessential genes. Although a cell can contain more than one plasmid, they cannot be genetically closely related. Plasmids can be transferred by lateral transfer in the process of conjugation.

◆ How can a *large* plasmid be differentiated from a *small* chromosome?

◆ What function do the *tra* genes of the F plasmid carry out?

<table>
<tr><td>**10.10**</td><td>**Types of Plasmids and Their Biological Significance**</td></tr>
</table>

Clearly all plasmids must carry genes that ensure their own replication. As we have seen, some plasmids also carry genes necessary for conjugation, and they can sometimes be detected biologically, either by the transfer functions themselves or by sensitivities to certain viruses (∞ Section 16.1).

Although plasmids do not carry genes that are essential to the host, the plasmids can carry genes that have a profound influence on the cell's phenotype. For example, in some cases plasmids encode properties considered fundamental to the ecology of the bacterium. For example, the ability of *Rhizobium* to interact and form nitrogen-fixing root nodules with plants requires certain plasmid functions (∞ Section 19.22). Other plasmids have been shown to confer special metabolic properties on a cell, such as for the degradation of toxic pollutants. Indeed, plasmids seem to be a major mechanism for conferring special properties on bacteria, and in many instances for also exporting these properties by horizontal gene flow. The only limitation to the types of genes present on a plasmid is that they do not interfere with their own replication or with the survival of the host. A few of the phenotypes conferred on prokaryotes by plasmids are summarized in Table 10.3.

Resistance Plasmids

Among the most widespread and well-studied groups of plasmids are the *resistance plasmids* (R plasmids), which confer resistance to antibiotics and various other growth inhibitors. R plasmids were first discovered in Japan in strains of enteric bacteria that had acquired resistance to a number of antibiotics (multiple resistance) and have since been found throughout the world. The emergence of bacteria resistant to several antibiotics is of considerable medical significance and was correlated with the increasing use of antibiotics for the treatment of infectious diseases. Soon after these resistant strains were isolated it

Table 10.3	Some phenotypes conferred by plasmids in prokaryotes
Phenotype class[a]	**Organisms**[b]
Antibiotic production	*Streptomyces*
Conjugation	*Escherichia, Pseudomonas, Rhizobium, Staphylococcus, Streptococcus, Sulfolobus, Vibrio*
Physiological functions	
Degradation of octane, camphor, naphthalene	*Pseudomonas*
Degradation of herbicides	*Alcaligenes*
Formation of acetone and butanol (∞ Section 12.20)	*Clostridium*
Lactose, sucrose or urea utilization and nitrogen fixation	Enteric bacteria
Nodulation and symbiotic nitrogen fixation (∞ Section 19.22)	*Rhizobium*
Pigment production	*Erwinia, Staphylococcus*
Resistance	
Antibiotic resistance (∞ Section 20.12)	*Campylobacter*, Enteric bacteria, *Neisseria, Staphylococcus*
Resistance to cadmium, cobalt, mercury, nickel, and/or zinc (∞ Section 19.16)	*Acidocella, Alcaligenes, Listeria, Pseudomonas, Staphylococcus*
Bacteriocin resistance (and production)	*Bacillus*, Enteric bacteria, *Lactococcus, Propionibacterium*
Virulence	
Host cell invasion	*Salmonella, Shigella, Yersinia*
Coagulase, hemolysin, enterotoxin (∞ Sections 21.9 and 21.11)	*Staphylococcus*
Enterotoxin, K antigen (∞ Sections 12.11 and 21.11)	*Escherichia*
Tumorigenicity in plants (∞ Section 19.21)	*Agrobacterium*

[a] Only a few of the many phenotypes known to be associated with plasmids are given.
[b] Only a few well-characterized examples are given. All of the organisms given in the list are *Bacteria* except for *Sulfolobus*, which is a member of the *Archaea*.

was shown that they could transfer resistance to sensitive strains via cell-to-cell contact. The infectious nature of the conjugative R plasmids permitted their rapid spread through cell populations. Resistance plasmids are now a major problem in clinical medicine (∞ Section 20.12).

Several antibiotic resistance genes can be carried by an R plasmid. In general, these genes encode proteins that either inactivate the antibiotic or prevent its uptake into the cell. Plasmid R100, for example, is a 94.3-kbp plasmid (Figure 10.20●) that carries genes encoding resistance to sulfonamides, streptomycin and spectinomycin, fusidic acid, chloramphenicol, and tetracycline. Plasmid R100 also carries several genes conferring resistance to mercury (∞ Section 19.16). Plasmid R100 can be transferred between enteric bacteria of the genera *Escherichia*, *Klebsiella*, *Proteus*, *Salmonella*, and *Shigella*, but does not transfer to the nonenteric gram-negative bacterium *Pseudomonas*. Different R plasmids with genes for resistance to most antibiotics are known. Many drug-resistant elements on R plasmids, such as those on R100, are also transposable elements (see Section 10.14) and this, plus the fact that many of these plasmids are conjugative, have made them a serious threat to traditional antibiotic therapies.

Plasmids Encoding Toxins and Other Virulence Characteristics

We will discuss in Chapter 21 the characteristics of pathogenic microorganisms that enable them to colonize hosts and establish infections. In the present context, we merely note the two major characteristics involved in the virulence (disease-causing ability) of pathogens: (1) the ability of the pathogen to attach to and colonize specific host tissues; and (2) the formation of substances (toxins, enzymes, and other molecules) that cause damage to the host.

In several pathogenic bacteria each of these virulence characteristics is carried on plasmids. For example, enteropathogenic strains of *Escherichia coli* are characterized by an ability to colonize the small intestine and to produce a toxin that causes symptoms of diarrhea (∞ Section 21.11). Colonization requires the presence of a cell surface protein called the *colonization factor antigen*, encoded by a plasmid. This protein confers on cells the ability to attach to epithelial cells of the intestine. At least two toxins in enteropathogenic *E. coli* are known to be encoded by a plasmid: the *hemolysin*, which lyses red blood cells, and the *enterotoxin*, which induces extensive secretion of water and salts into the bowel. It is the enterotoxin that is responsible for diarrhea, as will be discussed in Chapter 21.

Some virulence factors are encoded on plasmids, while others are encoded by other types of *mobile genetic elements*, such as transposons and bacteriophages; some virulence factors, of course, are chromosomal. Several examples are known where virulence genes are present on different genetic elements within the same cell. For instance, the genes encoding the virulence determinants of shigatoxin-producing strains of *E. coli* are distributed between the chromosome, a bacteriophage, and a plasmid.

Bacteriocins

Many bacteria produce proteins that inhibit or kill closely related species or even different strains of the same species. These agents, called **bacteriocins** to distinguish them from the antibiotics, have a more narrow spectrum of activity than antibiotics. The genes encoding bacteriocins and the proteins involved with processing and transporting them (and for conferring immunity on the producing organism) are often carried on a plasmid or a transposon. Bacteriocins are named after the species of organism that produces them. Thus, in *Escherichia coli* we have *colicins*, encoded by Col plasmids; *Bacillus subtilis* produces *subtilisin*, and so on.

The Col plasmids of *Escherichia coli* encode various colicins. Colicins released from a cell bind to specific receptors on the surface of susceptible cells. The receptors for colicins are generally entities whose normal function is to transport some substance, frequently a growth factor or micronutrient, through the outer membrane (the lipopolysaccharide layer) of the cell. Colicins kill cells by disrupting some critical cell function. For example, many colicins form channels in the cell membrane that allow potassium ions and protons to leak out, leading to a loss of the cell's energy-forming ability. However, colicin E2 is a DNA endonuclease that can cleave DNA, and colicin E3 is a nuclease that cuts at a specific site in 16S rRNA and inactivates ribosomes. Col plasmids can be either conjugative or nonconjugative.

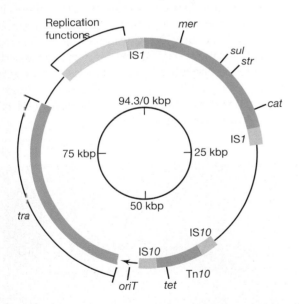

● **Figure 10.20 Genetic map of the resistance plasmid R100.** The inner circle shows the size of the plasmid in kilobase pairs. The outer circle shows the location of major antibiotic resistance genes and other key functions: *cat*, chloramphenicol resistance; *oriT*, origin of conjugative transfer; *mer*, mercuric ion resistance; *sul*, sulfonamide resistance; *str*, streptomycin resistance; *tet*, tetracycline resistance; *tra*, transfer functions. The locations of insertion sequences (IS) and the transposon Tn10 are also shown. Several genes related to plasmid replication are found in the region from 88–92 kbp.

The bacteriocins or bacteriocin-like agents of *gram-positive* bacteria are quite different from the colicins but are also often encoded by plasmids; some even have commercial value. For instance, lactic acid bacteria produce the bacteriocin Nisin A, which strongly inhibits the growth of a wide range of gram-positive bacteria and is used as a preservative in the food industry.

Engineered Plasmids

The basic tools of genetic engineering, discussed later in this chapter, have made possible the construction in the laboratory of countless new, artificial plasmids. Incorporation into such plasmids of genes from a wide variety of sources allows for the transfer of genetic material across any species barrier. Moreover, genes can be synthesized and introduced into plasmids. The only requirements for artificial plasmids are that (1) they contain genes controlling their own replication; (2) if conjugative, they contain transfer (*tra*) functions that facilitate this process; and (3) they are stably maintained in the host of choice.

We now turn our attention to the details of conjugation and show how certain plasmids can mobilize the bacterial chromosome, allowing transfer of chromosomal genes from donor to recipient.

 10.10 Concept Check

The genetic information that plasmids carry is not essential for cell function under all conditions but may confer a selective growth advantage under certain conditions. Examples include antibiotic resistance, enzymes for degradation of unusual organic compounds, and special metabolic pathways. Virulence factors of many pathogenic bacteria are often plasmid encoded.

◆ How does an *R plasmid* differ from the *F plasmid* discussed previously? How are they similar?

◆ How do *bacteriocins* differ from *antibiotics*?

10.11 Conjugation: Essential Features

Bacterial **conjugation** (mating) is a process of genetic transfer that involves cell-to-cell contact. As we discussed (see Section 10.9), conjugation is a plasmid-encoded mechanism. A conjugative plasmid uses this mechanism to transfer a copy of itself to a new host. However, sometimes other genetic elements can be *mobilized* (meaning *transferred*) during conjugation. These other genetic elements can be other plasmids, or the host chromosome itself. Indeed, conjugation was discovered because the F plasmid of *Escherichia coli* (see Figure 10.17) can mobilize the host chromosome. Mechanisms of conjugative transfer may differ depending on the plasmid involved, but most plasmids in gram-negative *Bacteria* seem to employ a mechanism similar to that used by the F plasmid.

The process of conjugation involves a *donor* cell, which contains a particular type of conjugative plasmid, and a *recipient* cell, which does not. The genes that con-

trol conjugation are contained in the *tra* region of the plasmid (see Section 10.9). Many genes in the *tra* region are involved in mating pair formation, and most of these have to do with the synthesis of a surface structure, the **sex pilus** (Figure 10.21●). Only *donor* cells produce these pili. Different conjugative plasmids may have slightly different *tra* regions, and the pili may also be different. The F plasmid and its relatives encode *F pili*.

Pili allow specific pairing to take place between the donor cell and the recipient cell. All conjugation in gram-negative *Bacteria* is thought to depend on cell pairing brought about by pili. The pili make specific contact with a receptor on the recipient and then retract, pulling the two cells together. The contacts between the donor and recipient cells then become stabilized, probably from fusion of the outer membranes, and DNA is then transferred from one cell to another.

Mechanism of DNA Transfer During Conjugation

DNA synthesis is necessary for DNA transfer to occur during conjugation. A mechanism of DNA synthesis in certain bacteriophages, called **rolling circle replication** (⌾ Section 9.11), best explains DNA transfer during conjugation, and this process is described in Figure 10.22●. Conjugation is triggered by cell-to-cell contact, at which time *one strand* of the plasmid DNA circle is nicked and is transferred to the recipient. The nicking enzyme required to initiate the process, TraI, is encoded by the *tra* operon of the F plasmid. This protein also has helicase activity and is thus also involved in unwinding the strand to be transferred. As this transfer occurs, DNA synthesis by the rolling circle mechanism replaces the transferred strand in the donor, while the complementary DNA strand is made in the recipient. Therefore, at the end of the process, both donor and recipient possess completely formed plasmids. For transfer of the F plasmid then, an F-containing cell, which is designated F$^+$, can mate with a cell lacking the plasmid, designated F$^-$, to yield two F$^+$ cells (Figure 10.22).

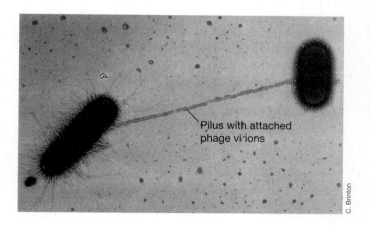

Pilus with attached phage virions

C. Brinton

● **Figure 10.21 Direct contact between two conjugating bacteria is first made via a pilus.** The cells are then drawn together to form a mating pair for the actual transfer of DNA. This occurs by retraction (depolymerization) of the pilus within the donor cell. Note the F-specific bacteriophages on the pilus (⌾ Section 16.1).

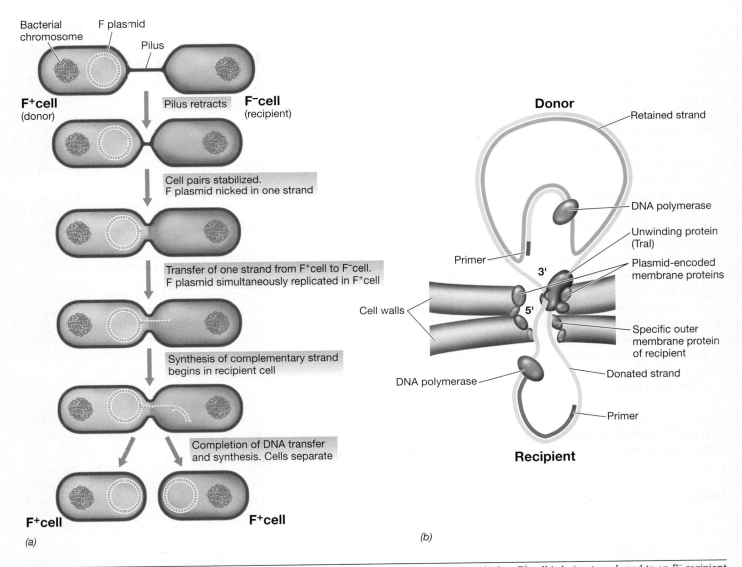

● **Figure 10.22   Transfer of plasmid DNA by conjugation.** (a) In this example, the F plasmid of an F⁺ cell is being transferred to an F⁻ recipient cell. Note the mechanism of rolling circle replication (Figure 9.20 and Figure 16.4). (b) Details of the replication and transfer process.

The plasmid DNA transfer process is highly efficient; under appropriate conditions virtually every recipient cell that pairs acquires a plasmid. When the plasmid genes can be expressed in the recipient, the recipient itself becomes a donor and can transfer the plasmid to other recipients. In this fashion, conjugative plasmids can spread rapidly between populations, behaving much like infectious agents. The infectious nature of plasmid transfer is of major ecological significance because a few plasmid-containing cells introduced into an appropriate population of recipients can, if they contain genes that confer a selective advantage, convert the entire recipient population into a plasmid-bearing (and thus donating) population in a short period of time. Plasmids can also be lost from a cell by a process called *curing*. This could happen spontaneously in natural populations when there is no selection pressure to maintain the plasmid. For example, plasmids conferring antibiotic resistance can be lost without affecting the cell's viability if there are no antibiotics in the cell's environment.

 10.11   Concept Check

Conjugation is a mechanism of DNA transfer in prokaryotes that requires cell-to-cell contact. Conjugation is controlled by genes carried by certain plasmids (such as the F plasmid) and involves transfer of the plasmid from a donor cell to a recipient cell. Plasmid DNA transfer involves replication via the rolling circle mechanism.

◆ In conjugation, how are donor and recipient cells brought into contact with each other?

◆ How does rolling circle DNA replication differ from the process by which chromosomal DNA replication occurs, described in Sections 7.5 and 7.6?

10.12 The Formation of Hfr Strains and Chromosome Mobilization

In conjugation discussed thus far, no mention was made of the transfer of *chromosomal* genes. However, the F plasmid of *Escherichia coli* (see Section 10.9) can, under certain circumstances, mobilize the chromosome for transfer during cell-to-cell contact. The F plasmid is an episome, a plasmid that can integrate into the host chromosome (see Section 10.9). When the F plasmid is integrated into the chromosome, the chromosome becomes mobilized and can lead to transfer of chromosomal genes. Following genetic recombination between donor and recipient, lateral gene transfer by this mechanism can be very extensive.

Cells possessing an *unintegrated* F plasmid are called F⁺. Those that have a *chromosome-integrated* F plasmid are called **Hfr** (for *h*igh *f*requency of *r*ecombination). Both F⁺ and Hfr cells are donors and are unable to take up a second copy of the F plasmid or genetically related plasmids. But unlike conjugation between an F⁺ and an F⁻, conjugation between an F⁻ and an Hfr donor leads to transfer of genes from the host chromosome because the plasmid *is* part of the chromosome. Following recombination, the F⁻ recipient cell may express a new phenotype. Plasmid integration is thus *a mechanism for mobilizing the cell's main genetic resources.* The term "high frequency of recombination" refers to the high rates of genetic recombination between genes on the donor chromosome and that of the recipient.

The presence of the F plasmid therefore results in three distinct alterations in the properties of a cell: (1) the ability to synthesize the F pilus, (2) the mobilization of DNA for transfer to another cell, and (3) the alteration of surface receptors so the cell is no longer able to behave as a recipient in conjugation.

Integration of F and Chromosome Mobilization

Integration of the F plasmid into the host chromosome can occur at several specific sites, called *IS* (for *insertion sequences*) sites. These sites are regions of DNA sequence homology between chromosomal and F plasmid DNA (see Section 10.13 for a discussion of insertion sequences). Figure 10.23● shows the integration of an F plasmid at an IS site. Once integrated, the plasmid no longer controls its own replication, but the *tra* operon still functions normally and the strain synthesizes pili. When a recipient is encountered, conjugation is triggered just as in an F⁺ cell, and DNA transfer is initiated at the *oriT* (*origin of transfer*) site. However, since the plasmid is now part of the chromosome, after part of the plasmid DNA is transferred, *chromosomal genes* begin to be transferred (Figure 10.24●). As is the case of conjugation with just the F plasmid itself (Figure 10.22), chromosomal DNA transfer also involves replication. Thus, after transfer, the Hfr strain remains genetically Hfr because it retains a copy of the transferred genes.

Because a number of distinct insertion sites are present on the chromosome, a number of distinct Hfr strains are possible. A given Hfr strain always donates genes in the same order, beginning with the same position. However, Hfr strains that differ in the position of integration of the F plasmid in the chromsmome transfer genes in different orders (Figure 10.25●). Because breakage of the DNA strand typically occurs during transfer, only *part* of

● **Figure 10.23   Integration of an F plasmid into the chromosome with the formation of an Hfr.** The insertion of the F plasmid occurs at a variety of specific sites where IS elements are located, the one here being an IS*3* located between the chromosomal genes *pro* and *lac*. Some of the genes on the F plasmid are shown. The arrow indicates the origin of transfer, *oriT*, with the arrow as the leading end. Thus, in this Hfr *pro* would be the first chromosomal gene to be transferred and *lac* would be among the last (see Figure 10.24).

the donor chromosome is transferred. And, since only part of the chromosome is transferred, it cannot replicate in the recipient cell. Therefore, donor genes normally cannot be detected in the recipient cells unless recombination between the incoming fragment and the recipient chromosome takes place. Finally, although Hfr strains transmit chromosomal genes at high frequency, they generally do not convert F⁻ cells to F⁺ or Hfr because the entire F plasmid is only rarely transferred. Instead, an Hfr × F⁻ cross yields the original Hfr and an F⁻ cell that

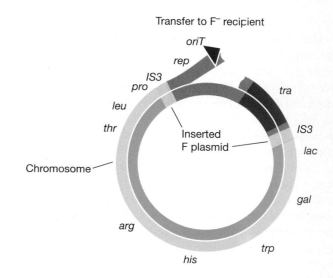

● **Figure 10.24   Breakage of the Hfr chromosome at the origin of transfer and the beginning of DNA transfer to the recipient.** Replication occurs during transfer (see Figure 10.22). The figure is not drawn to scale; the inserted F plasmid is actually less than 3% of the size of the *Escherichia coli* chromosome.

(a)

Hfr 1 — CDE ⟵ XYZAB : *Gene C donated first; clockwise order*

Hfr 2 — LKJ BAZYX ONM : *Gene L donated first; counterclockwise order*

Hfr 3 — XYZAB ⟵ UVW : *Gene X donated first; clockwise order*

Hfr 4 — GFE BAZYX JIH : *Gene G donated first; counterclockwise order*

(b)

● **Figure 10.25   Manner of formation of different Hfr strains, which donate genes in different orders and from different origins.** The bacterial chromosome (a) opens at various insertion sequences, at which F plasmids can become inserted. The gene orders are shown in part (b).

now has a new genotype and possibly phenotype. As in transformation and transduction, genetic recombination between Hfr genes and F⁻ genes involves *homologous recombination* in the recipient cell.

At some insertion sites, the F plasmid is integrated with the origin pointing in one direction, whereas at other sites the origin points in the opposite direction. The direction in which the F plasmid is inserted determines which of the chromosomal genes will be transferred into the recipient first (Figure 10.25). By use of various Hfr strains in mating experiments, it was possible to determine the arrangement and orientation of virtually all chromosomal genes in *Escherichia coli* (see Figure 10.42) long before the chromosome of this bacterium was sequenced.

Use of Hfr Strains in Genetic Crosses

As is the case for any system of bacterial gene transfer, one *selects* recombinants from conjugation. However, unlike the situation in transformation and transduction, during conjugation both the donor and recipient cells are viable. It is thus necessary to chose selection conditions where the desired recombinants can grow but where neither of the parental strains can form colonies. Typically a recipient is used that is resistant to an antibiotic but is auxotrophic for some substance, and a donor is used that is sensitive to the antibiotic but is prototrophic for the

same substance. Thus, on minimal medium containing the antibiotic, only recipient cells will grow following the mating event.

For instance, in the experiment shown in Figure 10.26●, an Hfr donor that is sensitive to streptomycin (Str^s) and is wild-type for synthesis of the amino acids threonine and leucine (Thr$^+$ and Leu$^+$) and for utilization of lactose (Lac$^+$), is mated with a recipient cell that cannot make these amino acids, but that is resistant to streptomycin (Str^r). The selective medium is a minimal medium containing streptomycin so that only recombinant cells can grow. The composition of each selective medium is varied depending on which genotypic characteristics are desired in the recombinant, as shown in Figure 10.26. The frequency of the process is measured by counting the colonies grown on the selective medium.

The order in which genes are present on the donor chromosome can also be determined by the kinetics of transfer of individual markers. For example, in the process called **interrupted mating**, conjugating cells can be separated by agitation in a mixer or blender. If mixtures of Hfr and F⁻ cells are agitated at various times after mixing and the genetic recombinants scored, it is found that the longer the time between pairing and agitation, the greater the number of genes of the Hfr that will appear in the F⁻ recombinant. As shown in Figure 10.27●, genes present closer to the origin of transfer enter the F⁻ first and are present in a higher percentage of the recombinants than genes that are transferred later. In addition to showing that gene transfer from donor to recipient is a sequential process, experiments of this kind provide a

Hfr donor: **Thr⁺ Leu⁺ Lac⁺ Str^s**

X

F⁻ recipient: **Thr⁻ Leu⁻ Lac⁻ Str^r**

Mating, to allow conjugation, followed by plating onto agar media

Thr⁺ Leu⁺ recombinants

Lac⁺ recombinants

Selection for recombinants

Agar mimimal medium with streptomycin and glucose; selects for markers Thr⁺ Leu⁺; does not select for Lac

Agar mimimal medium with streptomycin, lactose, threonine, leucine; selects for marker Lac⁺; does not select for Thr or Leu

●**Figure 10.26   Example experiment for the detection of conjugation.** Thr, threonine; Leu, leucine; Lac, lactose; Str, streptomycin. Note that each medium selects for specific classes of recombinants. The controls for the experiment are to plate samples of the donor and the recipient *before* they are mixed. Neither should be able to grow on the selective media used.

● **Figure 10.27 Rate of formation of recombinants containing different genes after mixing Hfr and F⁻ bacteria by interrupted mating.** The location of the genes along the Hfr chromosome is shown at the upper left. Note that the genes closest to the origin (0 min) are the first ones whose transfer is detected. The experiment is done by mixing Hfr and F⁻ cells under conditions in which essentially all Hfr cells find recipients. At various times, samples of the mixture are shaken violently to separate the mating pairs and plated on a selective medium in which only the recombinants can form colonies.

method of *determining the order* of the genes on the bacterial chromosome (genetic mapping). The arrangement of gene loci on the chromosome is called a **genetic map** (see Section 10.19 and Figure 10.42).

Transfer of Chromosomal Genes to the F Plasmid

Occasionally integrated F plasmids may be excised from the chromosome, and the possibility exists for the incorporation at that time of *chromosomal* genes into the liberated F plasmid. This can happen because both the integrated F plasmid and the chromosome contain a number of identical IS at which recombination can occur (Figure 10.23). Such F plasmids containing chromosomal genes are called F′ (*F prime*) *plasmids*. F′ plasmids differ from normal F plasmids in that they contain identifiable chromosomal genes, and they transfer these genes at high frequency to recipients. F′-mediated transfer resembles specialized transduction (see Section 10.8) in that only a restricted group of chromosomal genes can be transferred. Transferring a known F′ into a recipient allows one to establish diploids (two copies of each gene) for a limited region of the chromosome. Such partial diploids are called *merodiploids*. This procedure is important for doing complementation tests, as we will see in the next section.

Other Conjugation Systems

Although we have discussed conjugation almost exclusively as it occurs in *Escherichia coli*, conjugative plasmids have been found in many other gram-negative *Bacteria*. Conjugative plasmids of the incompatibility group (see Section 10.8) IncP can be maintained in virtually all gram-negative *Bacteria* and even transferred between different genera. Conjugative plasmids are also

known in gram-positive *Bacteria* (for example, in *Streptococcus*, *Enterococcus*, and *Staphylococcus*).

Several conjugative plasmids have also been found in *Sulfolobus*, a genus of *Archaea*. Little is known about conjugation in *Sulfolobus*, although it is known that cell-pairing occurs before plasmid transfer and that transfer is unidirectional. However, with one exception, the genes involved seem to have little similarity to those in gram-negative *Bacteria*. The exception is a gene similar to *traG*, whose protein product in F plasmid-mediated conjugation seems to be involved in stabilizing mating pairs. It thus seems likely that the mechanism of conjugation in *Archaea* is quite different from that in *Bacteria*.

 10.12 Concept Check

The donor cell chromosome can be mobilized for transfer to a recipient cell. This requires that the F plasmid integrate into the chromosome to form the Hfr phenotype. Transfer of the host chromosome is rarely complete but can be used to map the order of the genes on the chromosome. F′ plasmids are previously integrated F plasmids that have deintegrated and excised some chromosomal genes.

◆ In conjugation involving the F plasmid of *Escherichia coli*, how is the host chromosome mobilized?

◆ Why does an Hfr × F⁻ mating not yield two Hfr cells?

◆ At which sites in the chromosome can the F plasmid integrate?

10.13 Complementation

All methods of bacterial gene transfer involve only a *portion* of the donor chromosome. Therefore, unless recombination takes place with the recipient chromosome, the donor DNA will be lost in the recipient because it cannot replicate independently. In only two instances have we seen that a state of partial diploidy can be stably maintained. One was in the case of a specialized transducing phage where the donor genes are maintained as part of the viral genome (see Section 10.8). The other was the use of F′ plasmids where donor genes have become a part of the F plasmid genome on an episome (see Section 10.12). Since it is possible to create specialized transducing phage or specific plasmids using recombinant DNA techniques (see Sections 10.15–10.17), it is possible to put essentially *any* portion of the bacterial chromosome on a phage or plasmid. This can be quite useful in bacterial genetic analyses, and we consider this now.

Complementation Test and the Cistron

When two mutant strains are genetically crossed (mated), homologous recombination can yield a wild-type recombinant *unless* both of the mutations occur in exactly the same base pairs. For example, if two different Trp⁻ *Escherichia coli* auxotrophs (strains that require the amino acid tryptophan in the medium) are crossed and Trp⁺ recombinants are obtained, it is obvious that the mutations in the two strains were not in the same base pairs. However, this kind of experiment cannot detect whether two mutations are in different regions of *the same gene*. This can be determined by a **complementation test**.

Continuing with our tryptophan auxotroph example, if one of the tryptophan genes has been inactivated in a particular strain by a mutation leading to the Trp⁻ phenotype, then inserting a copy of the wild-type gene on a plasmid or viral genome into the same cell should restore the wild-type phenotype of the cell; that is, the resulting partially diploid cell should be Trp⁺. One would then say that the wild-type gene *complements* the mutation. How can a complementation test be done to look for mutations *in the same gene*?

Even though only a chromosomal fragment is transferred in conjugation, this fragment may contain many genes. Again, from our tryptophan example, recall that the genes for the biosynthesis of tryptophan form an operon (☞ Figure 8.24), and therefore one can easily transfer the entire operon from the donor cell to a recipient cell. Note then that even if the operon in the donor also has a mutation, it will complement the mutated gene in the recipient *if the two mutations are in different genes*. The two mutations are then said to complement one another. This is shown diagrammatically in Figure 10.28●. If each homologous DNA molecule contributes a different required gene, then the cell will have all the enzymes it requires to synthesize tryptophan.

It is important to note that complementation *does not* involve recombination. To do a complementation test, the mutations must be located in two different DNA molecules: the chromosome and the gene carrier (plasmid or viral DNA). Such mutations are referred to as being in *trans* with respect to one another. If one molecule has both mutations, a condition called *cis*, the second molecule can complement *only* if it is wild-type for the given gene. Therefore, having the mutations in *cis* serves as a positive control in such a complementation experiment. This type of complementation test, called a *cis-trans test*, is used to determine whether two mutations are in the same genetic unit. The genetic unit defined by the cis-trans test is called a **cistron** (a term essentially equivalent to a gene). If two mutations occur in genes encoding different enzymes, or even different subunits of the same enzyme, complementation of the two mutations is possible, and the mutations are therefore not in the same cistron (Figure 10.28).

Although genetic crosses involving complementation analyses are still done in bacterial genetics, in many cases it is simply easier to sequence the gene in question and to analyze the sequence to identify the nature and location of any mutations. This is especially true if the sequence of the wild-type gene is known, since highly specific primers can be used and the DNA quickly sequenced. The term *cistron* is now rarely used in microbial genetics except when describing whether an mRNA has

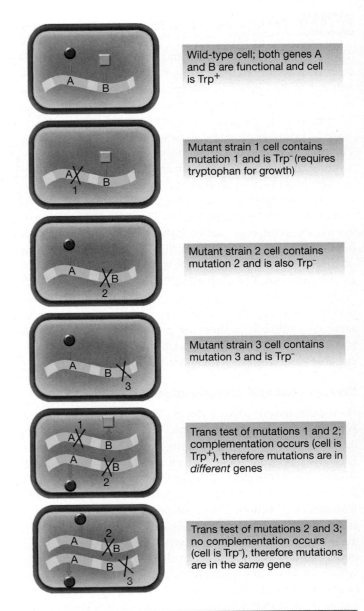

Wild-type cell; both genes A and B are functional and cell is Trp⁺

Mutant strain 1 cell contains mutation 1 and is Trp⁻ (requires tryptophan for growth)

Mutant strain 2 cell contains mutation 2 and is also Trp⁻

Mutant strain 3 cell contains mutation 3 and is Trp⁻

Trans test of mutations 1 and 2; complementation occurs (cell is Trp⁺), therefore mutations are in *different* genes

Trans test of mutations 2 and 3; no complementation occurs (cell is Trp⁻), therefore mutations are in the *same* gene

● **Figure 10.28 Complementation analysis.** In this example, the protein products of both genes A and B are required to synthesize tryptophan. Mutations 1, 2, and 3 each lead to the same phenotype, a requirement for tryptophan (Trp⁻). Complementation analysis indicates that mutations 2 and 3 are in the same gene while mutation 1 is in a separate gene.

the genetic information from one gene (*monocistronic* mRNA) or from more than one gene (*polycistronic* mRNA) (☞ Section 7.13).

 10.13 Concept Check

If a cell is treated so as to to contain two copies for a region of its genome, one can do complementation tests to determine if two mutations are in the same or different genes. This is often necessary because mutations in different genes in the same pathway may give the same phenotype. Complementation tests do not involve recombination.

◆ What is a *merodiploid*?

◆ Complementation tests have been referred to as *cis-trans* tests. Explain.

10.14 Transposons and Insertion Sequences

The order of genes on a bacterial chromosome can be determined by the methods of gene transfer we have considered, just as genes in eukaryotic chromosomes can be mapped by mating experiments. However, the exact arrangement of the genes along a chromosome is not necessarily permanent; some genes are capable of moving. The process by which a gene moves from one place to another in the genome is called **transposition** and is an important process in evolution and in genetic analysis.

Transposition is a *rare* event, occurring at frequencies of 10^{-5}–10^{-7} per generation. Thus, for the most part, genes are relatively stable entities. Moreover, not all genes are capable of transposition. Rather, transposition is linked to the presence of special genetic elements called **transposable elements**. Transposition was originally discovered in corn (maize) and then later in *Bacteria* owing to the extremely sensitive types of genetic analysis available in these organisms. It has now been shown that DNA sequences with the properties of transposable elements are widespread in nature.

Transposons and Transposable Elements

As we discussed earlier (Section 7.4), there are three types of transposable elements in *Bacteria: insertion sequences, transposons,* and some *special viruses* (such as *Mu*) (Section 16.5). In this section we confine our discussion primarily to insertion sequences (IS) and transposons. Both of these elements have two important features in common: They both carry genes encoding a **transposase**, the enzyme necessary for transposition, and they both have short *inverted terminal repeats* at the ends of their DNA (these "ends" are continuous with whatever DNA the element has inserted into). The repeats range in length from fewer than 20 bp in simple IS elements to more than 1 kbp in some transposons, and each different IS has a specific number of base pairs in its terminal repeats. Such inverted terminal repeats are involved in the transposition process. Figure 10.29● shows a genetic map of a common insertion element called IS2 and of the transposon Tn5.

Insertion sequences are the simplest type of transposable element and carry no genes other than those required for them to move to new locations. Insertion sequences are short segments of DNA, about 1000 nucleotides long, that can become integrated at specific sites on the genome. Insertion sequences are found in both chromosomal and plasmid DNA, as well as in certain bacteriophages. Several hundred distinct IS elements have been characterized, and most are designated by a number identifying its type: IS1, IS2, IS3, and so on.

IS elements are scattered about the chromosome, and strains vary in the number and frequency of these elements. For instance, one strain of *Escherichia coli* has five copies of IS2 and five copies of IS3. Many plasmids also carry these insertion sequences (see Figures 10.18 and 10.20), and it is homologous recombination between

● **Figure 10.29 Maps of the transposable elements IS2 and Tn5.** Inverted repeats are shown in red. The arrows above the maps show the direction of transcription of any genes on the elements. The gene encoding the transposase is *tnp*. (a) IS2 is an insertion sequence of 1327 bp with inverted repeats of 41 bp at its ends. (b) Tn5 is a composite transposon of 5.7 kbp containing the insertion sequences IS50L and IS50R at its left and right ends, respectively. IS50L is not capable of independent transposition because there is a *nonsense mutation* (see Section 10.2) marked by a blue cross in its transposase gene. Otherwise, the two IS50 elements are very nearly identical. The genes *kan, str,* and *bleo* confer resistance to the antibiotics kanamycin (and neomycin), streptomycin, and bleomycin. Tn5 is commonly used to generate mutants in *Escherichia coli* and other gram-negative bacteria.

identical insertion sequences on the F plasmid and the chromosome and not transposition that allows the F plasmid to integrate into the bacterial chromosome and later mobilize it (see Section 10.12 and Figure 10.23). Some species of *Archaea* also have large numbers of IS elements in their chromosomes, suggesting that transposition is characteristic of all prokaryotes.

Transposons are larger than insertion sequences and carry other genes, some of them conferring important properties on the organism carrying them. These often include drug resistance and other easily selectable genes. In addition, there are *conjugative transposons,* transposons that can move between bacterial species by conjugation. These transposons have genes allowing them not only to move from one location on a genome to another, but also *tra* genes necessary to transfer themselves from one bacterium to another (see Section 10.11). Some transposons are actually composite genetic elements, containing a gene or group of genes lying between two identical insertion sequences. The existence of such *composite transposons* indicates that novel transposons likely arise periodically in cells that contain insertion sequences located close to one another.

Transposition and Microbial Evolution

As mentioned previously, the inverted repeats found at the ends of transposable elements and the transposase enzyme are essential for transposition. The transposase recognizes, cuts, and eventually ligates the DNA during transposition (Figure 10.30●). When a transposable ele-

● **Figure 10.30 Transposition.** Insertion of a transposable element generates a duplication of the target sequence. Note the presence of inverted repeats (IRs) at the ends of the transposable element. Figure 10.31 shows more detailed models of the mechanism of transposition.

ment becomes inserted into a target DNA, a short sequence in the target at the site of integration is duplicated during the insertion process. The duplication arises because single-stranded DNA breaks are generated by the transposase. The transposon is then attached to the single-stranded ends that have been generated, and repair of the single-strand portions results in the duplication (see Figure 10.31).

If the duplicated DNA is sufficient to have duplicated an entire gene or group of genes, the organism will contain multiple copies of these particular genes. Such **gene-duplication** events are thought to fuel microbial evolution. This is because mutations occurring in one copy of the gene(s) do not affect the other copy; the function of the faulty protein is still "covered" by the product of the unmutated duplicate gene. However, beneficial mutations in one copy can lead to production of a protein that has superior properties to the normal protein and thereby increase fitness. In this way, evolution can "experiment" with one copy of the gene while the identical copy provides the necessary backup function. Genomic analyses have revealed numerous examples of protein-encoding genes that were clearly derived from gene duplication.

Mechanisms of Transposition

Two mechanisms of transposition are known, called *conservative* and *replicative* (Figure 10.31●). In **conservative transposition**, such as occurs in the transposon Tn5, the transposable element is excised from one location in the chromosome and becomes reinserted at a second location. The copy number of a conservative transposon therefore remains at one. By contrast, in **replicative transposition**, such as occurs in bacteriophage Mu (⚭ Section 16.5), a new copy is produced during transposition and is inserted at another location on the chromosome. Thus, after a replicative transposition event, one copy of the transposing element remains at the original site and another copy is found at the new site.

● **Figure 10.31 Mechanisms of transposition.** For simplicity, donor DNA (carrying the transposon) is shown in light green and DNA carrying the target DNA is shown in dark green. (a) In both conservative and replicative transposition the transposase makes cuts (marked with arrows) in the DNA strands at the end of the transposable element (orange) and at the target site (red). The number and location of the cuts may vary depending on the mechanism. (b) The target site becomes ligated to the transposable element. The black balls indicate free 3′ ends of DNA strands at which replication can occur (⚭ Sections 7.5 and 7.6). (c) In conservative transposition, further cuts are made before DNA replication/repair occurs, and the transposable element is lost from the donor DNA. (d) Repair leads to duplication of the target site and completion of transposition to the new site. (e) In replicative transposition, replication occurs without the cutting of the transposable element from the donor site leading to two copies of the transposable element as part of a cointegrate. Note, however, this has led to the joining of the donor and target DNA molecules together. (f) These molecules are separated (resolved) in a further reaction. Resolution of cointegrates is shown in more detail in Figure 10.32.

Models for both conservative and replicative transposition are illustrated in Figure 10.31. As seen, single-strand cuts are made at the ends of the transposon (at the sites of the inverted repeats), and staggered single-strand cuts are made at the target site. The transposon is now joined to the target site via the single-stranded ends. In conservative transposition, the donor site is now cut, and replication repair then fills in the single-strand gaps in the target site. This process results in the formation of *direct repeats* in the target site at the ends of the transposon. In replicative transposition, the replication repair takes place while the transposable element is still attached to both the original and the target sites. This leads to the formation of a composite structure called a *cointegrate*. The final event in this pathway is *resolution* of the cointegrate structure, leading to release of the original transposon and the presence of a new copy of the transposon at the target site (Figure 10.32●).

Transposition is essentially a *recombination event*, but one that does not occur between homologous sequences or use the general recombination system of the cell. It involves *transposase* rather than the RecA protein that is involved in general recombination (Figure 10.9). Because this recombination involves a *specific* base sequence, it is called *site-specific recombination* (in contrast to *homologous recombination* discussed in Section 10.6).

Mutagenesis with Transposable Elements

If the insertion site for a transposable element is *within* a gene, insertion of the transposon will result in mutation (Figure 10.33●). Transposons thus provide a facile means of creating mutants throughout the chromosome. The most convenient element for **transposon mutagenesis** is one containing an antibiotic resistance gene. Clones containing the transposon can then be selected by the isolation of antibiotic-resistant colonies. If the antibiotic-resistant clones are selected on rich medium on which all auxotrophs can grow, they can subsequently be screened on minimal medium supplemented with various growth factors to determine if a growth factor is required.

Transposons are also useful for incorporating an auxotrophic gene marker into a wild-type organism. Normally, auxotrophic recombinants cannot be isolated by positive selection (see Figure 10.2), but if the auxotrophic marker to be introduced contains a transposable element with an antibiotic resistance marker, then one can select for antibiotic-resistant clones, a positive selection procedure, and automatically obtain clones that have incorporated the auxotrophic marker. Two transposons widely used in microbiology for mutagenesis are Tn5 (see Figure 10.30), which confers neomycin and kanamycin resistance, and Tn10, which contains a marker for tetracycline resistance (see Figure 10.20).

Integrons

Integrons are transposons that can capture and express genes from other sources. However, unlike other transposons, integrons are not inserted at random but are highly selective in their insertion site, frequently becoming inserted into plasmids.

Integrons contain a gene that encodes a protein called *integrase*, needed for *site-specific recombination*. Recall that the lambda genome becomes integrated into the

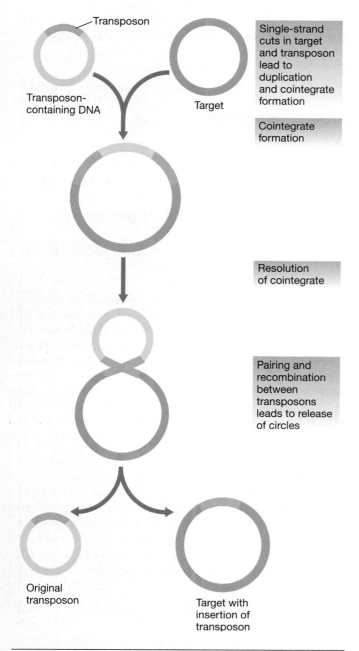

● **Figure 10.32 Replicative transposition.** After the formation of single-strand cuts, a cointegrate structure arises by association of the two molecules (see Figure 10.31). After recombination, resolution of the cointegrate structure leads to the release of the original transposon and duplication of the transposon in the target molecule.

● **Figure 10.33 Transposon mutagenesis.** The transposon moves into the middle of gene 2. Gene 2 is now disrupted by the transposon and is inactivated. Gene A in the transposon will be expressed in both locations.

Escherichia coli genome at a specific site by the activity of the lambda integrase (◯◯ Section 9.11). Integrons also contain a specific DNA sequence that allows the integrase to insert groups of genes called *cassettes*, along with a promoter that allows for expression of the newly integrated gene cassette.

Integrons can become part of transposons, plasmids, or even the bacterial chromosome. Some integrons contain as many as five different gene cassettes. Over 40 different antibiotic resistance genes have been identified on such cassettes, as have some genes associated with virulence in certain pathogenic bacteria. Figure 10.34● shows the structure of two integrons from *Pseudomonas aeruginosa*, a potentially serious pathogen. Integrons have been found in various species of *Bacteria*, often in clinical isolates, and thus their selection by horizontal gene transfer in such antibiotic-rich environments as hospitals and clinics is obvious. What is less obvious is the origin of the gene cassettes themselves. These are not simply random genes that can be captured, but genes that are either (1) bounded by specific DNA sequences that are recognized by the integrase, or (2) apparently not expressed until they become part of an integron and can be transcribed from the promoter on the integron.

10.14 Concept Check

Transposons and insertion sequences are genetic elements that can move from one location on a chromosome to another by a process called transposition, a type of site-specific recombination. Transposition can be either replicative or conservative. Transposons often carry genes encoding antibiotic resistance. Transposons can be used as biological mutagens.

◆ What features do *insertion sequences* and *transposons* have in common?

◆ What are integrons and how do they differ from transposons?

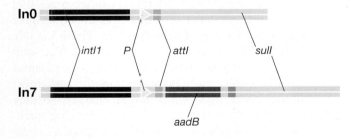

● **Figure 10.34 Structure of two naturally occurring integrons from *Pseudomonas*.** The integron In0 has the basic set of genes: *intI1*, encodes integrase; *attI*, the site where site–specific integration can occur; P, a promoter, and *sulI*, a gene conferring sulfonamide resistance that contains its own promoter. The integron In7 contains all of these genes, but in addition, a gene cassette has been integrated. All cassettes contain a site (blue square) for site-specific recombination. This cassette contains *aadB*, which confers resistance to certain aminoglycoside antibiotics.

III BACTERIAL GENETICS AND GENE CLONING

The three mechanisms of genetic exchange we have discussed, transformation (see Section 10.7), transduction (see Section 10.8), and conjugation (see Sections 10.9–10.11), are used to do bacterial crosses and select recombinants for many reasons. These include making a genetic map of an organism, creating new strains of an organism, or specifically studying the genetics and biochemistry of an organism's metabolism. However, some severe limitations are associated with doing all the crosses *in vivo* using these techniques. Many of these limitations can be overcome by manipulating DNA *in vitro* (in a test tube). The next several sections of this chapter discuss some of these *in vitro* techniques.

The techniques to be described here rely heavily on the molecular biologist's toolkit: (1) restriction enzymes; (2) gel electrophoresis and nucleic acid hybridization; and (3) DNA sequencing, the synthesis of DNA probes, and PCR. These topics were covered in Sections 7.7–7.9, respectively. It might be useful to review these sections before beginning here. We begin with molecular cloning, the capture of specific genes on small molecules of DNA that can be easily manipulated by the experimenter.

10.15 Essentials of Molecular Cloning

Molecular (gene) cloning provides the foundation for genetic-engineering and most molecular genetics procedures and has greatly facilitated the detailed analysis of genomes. The purpose of gene cloning is to isolate multiple copies of specific genes in pure form. Consider the nature of this problem. For a genetically "simple" organism like *Escherichia coli*, an average gene represents 1–2 kbp out of a genome of over 4600 kbp. An average *E. coli* gene is thus less than 0.05% of the total DNA in the cell. In human DNA the problem is even more complicated because the coding regions of average genes are not much larger than in *E. coli*, genes are typically split into pieces, and the genome is almost 1000 times larger! How then can a specific gene be obtained in multiple copies?

If a gene of interest can be isolated onto a fragment of DNA whose replication is under the experimenter's control, making copies of the entire fragment (and thus the gene of interest) is possible. The basic strategy of molecular cloning is thus to move the desired gene(s) from a large, complex genome to a small, simple one (Figure 10.35●). Fortunately, our knowledge of DNA chemistry and enzymology allows us to break and join DNA molecules *in vitro*. This process is known as *in vitro recombination*. Restriction enzymes, DNA ligase, PCR, and synthetic DNA are important tools used for *in vitro* recombination.

Steps in Gene Cloning

Gene cloning can be divided into several steps as summarized here:

Foreign DNA

Cut with restriction enzyme

Sticky ends

Add vector cut with same restriction enzyme

Vector

Add DNA ligase to form recombinant molecules

Cloned DNA

Introduction of recombinant vector into a host

● **Figure 10.35** **Major steps in gene cloning.** The vector can be a plasmid or a viral genome. By cutting the foreign DNA and the vector DNA with the same restriction enzyme, complementary sticky ends are generated that allow foreign DNA to become part of the vector.

1. **Isolation and fragmentation of the source DNA.** This can be total genomic DNA from an organism of interest, DNA synthesized from an RNA template by reverse transcriptase (◐➡ Section 9.13), a gene or genes amplified by the polymerase chain reaction (◐➡ Section 7.9), or even totally synthetic DNA made *in vitro* (◐➡ Section 7.8). If genomic DNA is the source, it is cut with restriction enzymes first to give a mixture of manageable-sized fragments (Figure 10.35).

2. **Joining the DNA fragments to a cloning vector with DNA ligase.** Cloning vectors are small, independently replicating genetic elements used to replicate genes, and most are derived from plasmids or viruses. Cloning vectors are typically designed to allow *in vitro* insertion of foreign DNA at a restriction site that cuts the vector in a way that does not affect its replication (Figure 10.35). If the source DNA and the vector are cut with the same restriction enzyme, joining can be mediated by annealing of the single-stranded regions called "sticky ends" (Figure 10.35 and see Figure 7.21). "Blunt ends" generated by some restriction enzymes can also be joined using synthetic DNA **linkers** or **adapters**. DNA ligase is required to seal the final phosphodiester bond. The properties of cloning vectors are discussed in Sections 10.15 and 10.16, as well as in Chapter 31.

3. **Introduction and maintenance of the cloned DNA in a host organism.** The recombinant DNA molecule made in a test tube is introduced into a host organism, for example, by the process of transformation (see Section 10.7), where it can then replicate (Figure 10.35). Transfer of the DNA into the host usually yields a mixture of clones. Some cells contain the desired cloned gene, whereas other cells contain other clones generated by joining the source DNA to the vector. Such a mixture is known as a **DNA library** or a **gene library** because many different clones can be purified from the mixture, each containing different cloned DNA segments from the source organism. Making a gene library by cloning *random fragments* of a genome is called **shotgun cloning**, and is a widely practiced technique in gene cloning and genomic analyses (◐➡ Sections 15.1 and 15.2).

We now consider in more detail the properties of plasmids and viruses, respectively, used as cloning vectors.

 10.15 **Concept Check**

The isolation of a specific gene or region of a chromosome by molecular cloning is done using a plasmid or virus as the cloning vector. Restriction enzymes and DNA ligase are used in an *in vitro* recombination procedure to produce the hybrid DNA molecule. Once introduced into a suitable host, the target DNA can be produced in large amounts under the control of the cloning vector.

◆ What is the purpose of molecular cloning?

◆ What are the roles of a cloning vector, restriction enzymes, and DNA ligase in molecular cloning?

10.16 Plasmids as Cloning Vectors

Plasmids replicate independently of the host chromosome (see Section 10.9). In addition to carrying genes required for their own replication, most plasmids are natural **vectors** because they often carry other genes that confer important properties on their hosts. Plasmids have very useful properties as **cloning vectors**. These include (1) *small size*, which makes the DNA easy to isolate and manipulate; (2) *independent origin of replication*, so plasmid replication in the cell proceeds independently from direct chromosomal control; (3) *multiple copy number*, so they can be present in the cell in several copies, making *amplification* of the DNA possible; and (4) *presence of selectable markers* such as antibiotic resistance genes, making detection and selection of plasmid-containing clones easier.

Although in the natural environment conjugative plasmids are transferred by cell-to-cell contact, plasmid cloning vectors have been genetically modified to prevent their transfer conjugatively in order to achieve biological containment. However, vector transfer in the laboratory can be brought about by transformation or an artificial type of transformation called *electroporation* (see

Sections 10.7 and 31.2). Depending on the host-plasmid system, replication of the plasmid may be under tight cellular control, in which case only a few copies are made, or under relaxed cellular control, in which case a large number of copies is made. Achievement of high copy number is often important in gene cloning, and by proper selection of the host-plasmid system and manipulation of cellular macromolecule synthesis, plasmid copy numbers of several thousand per cell can be obtained.

The Plasmid pBR322

In Section 10.9, we discussed how the F plasmid could obtain genes from the *Escherichia coli* chromosome *in vivo*, becoming an F′ plasmid. Therefore, one might imagine using F as an *in vitro* cloning vector. However, the F plasmid is far too large (almost 100 kb, see Figure 10.18) to be a useful cloning vector, and it has no easily selectable markers. Even so, the very first plasmid cloning vectors used were natural isolates. These were soon replaced, however, by plasmids that were themselves the result of *in vitro* manipulations. An example of one of these modified plasmid cloning vectors is pBR322, which replicates in *Escherichia coli* (Figure 10.36●). Plasmid pBR322 has a number of characteristics that make it suitable as a cloning vehicle:

1. It is relatively small, only 4361 bp.

2. It is stably maintained in its host (*Escherichia coli*) in relatively high copy number, 20–30 copies per cell.

3. It can be amplified to a very high number (1000–3000 copies per cell, about 40% of the genome!) by inhibiting protein synthesis with the antibiotic chloramphenicol.

4. It is easy to isolate in the supercoiled form using a variety of routine techniques.

5. A reasonable amount of foreign DNA can be inserted into it, although inserts of more than 10 kbp lead to plasmid instability.

6. The complete base sequence of this plasmid is known, making it possible to identify all restriction enzyme cut sites.

7. There are *single* cleavage sites for various restriction enzymes such as *Pst*I, *Sal*I, *Eco*RI, *Hin*dIII, and *Bam*HI. It is crucial that only a single recognition site for a given restriction enzyme exist on the cloning vector so that treatment with that enzyme linearizes the vector but does not cut it into pieces. Individual sites for each of several restriction enzymes increase the versatility and usefulness of the vector.

8. It has genes conferring ampicillin resistance and tetracycline resistance on its host. These permit ready selection of hosts containing the plasmid because such hosts are resistant to both antibiotics. The sites recognized by some of the restriction enzymes are within one or the other of these resistance genes, facilitating the identification of plasmids carrying cloned DNA (see later).

9. It can be inserted into cells easily by transformation or artificial transformation (see Section 10.7).

Cloning Genes into pBR322

The use of plasmid pBR322 in gene cloning is shown in Figure 10.37●. As seen, the *Bam*HI restriction site is within the gene for tetracycline resistance, and the *Pst*I site is within the gene for ampicillin resistance. If foreign DNA is inserted into one of these sites, the antibiotic resistance conferred by the gene containing this site is lost, a phenomenon called **insertional inactivation**. Insertional inactivation is used to detect the presence of foreign DNA within the plasmid. Thus, when pBR322 is digested with *Bam*HI, linked with foreign DNA, and transformed bacterial clones isolated, those clones that are *both* ampicillin resistant and tetracycline resistant *lack* the foreign DNA (the plasmid incorporated into these cells represents vector DNA that had recyclized without picking up foreign DNA). On the other hand, those cells *resistant* to ampicillin but *sensitive* to tetracycline *contain* the plasmid with inserted foreign DNA. Since ampicillin resistance and tetracycline resistance can be determined independently on agar plates using replica plating (Figure 10.2), isolation of bacteria containing the desired clones and elimination of cells not containing the plasmid can readily be accomplished.

Second Generation Plasmid Vectors

Plasmid pBR322 represents an early generation of cloning vectors which were themselves partially constructed by genetic engineering. There are now newer generations of plasmid vectors that have been engineered to have even more useful cloning features and are even simpler to

● **Figure 10.36 The structure of plasmid pBR322, a widely used cloning vector.** Shown are the essential features including antibiotic resistance markers and restriction enzyme cut sites. The arrow indicates the direction of DNA replication from the origin.

● **Figure 10.37 Cloning with plasmid pBR322.** The use of plasmid pBR322 as a cloning vector, showing how insertion of foreign DNA causes inactivation of the tetracycline resistance gene, permitting easy identification of transformants containing the cloned DNA fragment.

use. These new features, as exemplified by the plasmid pUC19, typically include a **multiple cloning site** (also called a **polylinker**), which is a short segment of DNA containing many different restriction enzyme cut sites but ones that are present only in the polylinker portion of the vector. This polylinker is contained within the coding region of a gene where insertional inactivation is very easy to monitor, such as the gene encoding β-galactosidase, an enzyme involved in the catabolism of lactose (∞ Sections 8.5 and 8.7). When a special reagent is added to the medium, activity of this enzyme leads to a blue-colored product. Thus, cells containing pUC19, which *does not* contain cloned DNA (β-galactosidase is therefore active), form blue colonies, while cells containing the plasmid *with* cloned DNA (β-galactosidase is therefore inactive) are colorless. This allows for screening colonies for cloned DNA by inspection. Such color screening features are also found in bacteriophage cloning vectors, and a specific example is discussed in Section 15.1 and shown in Figure 15.1.

Sometimes insertional inactivation can be detected by *selection*, rather than by screening. For example, in some

vectors, the gene carrying the polylinker normally produces a protein that is lethal to the host cell. Therefore, only cells containing a plasmid in which this gene has been *inactivated* can grow, a very rigorous selection tool indeed!

Cloning using plasmid vectors is a versatile and fairly general procedure widely used in genetic engineering, particularly when the fragment to be cloned is fairly small. Also, plasmids are often used as cloning vectors if *expression* of the cloned gene is desired, since regulatory genes can be engineered into the plasmid to obtain expression of the cloned genes under specific conditions (∞ Section 31.4).

 10.16 Concept Check

Plasmids are useful cloning vectors because they are easy to isolate and purify and are able to multiply to high copy numbers in bacterial cells. Antibiotic resistance genes of the plasmid are used to identify bacterial cells containing the plasmid.

◆ Explain why in cloning it is necessary to use a restriction enzyme that cuts the plasmid vector in only one location.

◆ What is *insertional inactivation*?

10.17 Bacteriophage Lambda as a Cloning Vector

Recall that during specialized transduction (see Section 10.8) some host genes become incorporated into a bacteriophage genome. One phage that is used as a specialized transducing phage is bacteriophage lambda (∞ Section 9.11). During specialized transduction lambda acts as a vector, but the recombination occurs in the cell, not in a test tube. Lambda can also be used as a cloning vector for *in vitro* recombination.

Lambda is a particularly useful cloning vector because its biology is well understood, it can hold larger amounts of DNA than most plasmids, and DNA can be efficiently packaged into phage particles *in vitro*. These can be used to infect suitable host cells, and infection is much more efficient than transformation (transfection). Phage lambda has a complex genetic map (∞ Figure 9.18) and a large number of genes. Of particular significance for its use as a cloning vector, however, is the fact that the central third of the lambda genome, the region between genes *J* and *N*, is *unessential* for infectivity and can be replaced with foreign DNA (Figure 10.38●). This allows relatively large DNA fragments, up to about 20 kb, to be cloned into lambda. This is twice the cloning capacity of a small plasmid vector such as pBR322.

Modified Lambda Phages

Wild-type lambda is not suitable as a cloning vector because its genome has too many restriction enzyme sites. To avoid this difficulty, modified lambda phages have been constructed especially for cloning. In one set of modified lambda phages, called *Charon phages,* unwanted restriction

● **Figure 10.38 Molecular cloning with lambda.** Abbreviated genetic map of bacteriophage lambda showing the cohesive ends as circles (🞂 Figure 9.18a). Charon 4A and 16 are both derivatives of lambda, which have various substitutions and deletions in the nonessential region. One of the substitutions in each case is a gene (*lacZ*) that encodes the enzyme β-galactosidase, which permits detection of clones containing this phage. Whereas the wild-type lambda genome is 48.5 kb pairs, that for Charon 4A is 45.4 and that for Charon 16 is 41.7 kb pairs. The arrowheads shown above the maps of each phage indicate the sites recognized by the restriction enzyme *Eco*RI.

enzyme sites have been removed by mutation. In variants that have only a single restriction site, such as Charon 16, a foreign piece of DNA can be *inserted*. By contrast, in variants with two sites, such as Charon 4A, foreign DNA can *replace* a specific segment of the lambda DNA (Figure 10.38). The latter variants, called **replacement vectors**, are especially useful in cloning large DNA fragments. Charon 4A is used as a replacement vector; the two small interior fragments are cut out and discarded during cloning.

Both vectors are also engineered to contain *reporter genes*, such as the gene for β-galactosidase previously discussed. When the vectors replicate in a lactose-negative (Lac⁻) strain of *Escherichia coli*, β-galactosidase is synthesized from the phage gene and the presence of lactose-positive (Lac⁺) plaques can be detected by using a color indicator agar (🞂 Section 15.1). However, if a foreign gene is inserted *into* the β-galactosidase gene, the Lac⁺ character is lost. Such Lac⁻ plaques can be readily detected as colorless plaques among a background of colored plaques (🞂 Figure 15.1).

Steps in Cloning with Lambda

Cloning with lambda replacement vectors involves the following steps (Figure 10.39●):

1. Isolating the vector DNA from phage particles and digestion with the appropriate restriction enzyme.

2. Connecting the two lambda fragments to fragments of foreign DNA using DNA ligase. Conditions are

● **Figure 10.39 The use of bacteriophage lambda as a cloning vector.** The maximum size of inserted DNA is about 20 kb pairs.

chosen so molecules are formed of a length suitable for packaging into phage particles.

3. Packaging of the DNA by adding cell extracts containing the head and tail proteins and allowing the formation of viable phage particles to occur spontaneously.

4. Infecting *E. coli* and isolating phage clones by picking plaques on a host strain.

5. Checking recombinant phage for the presence of the desired foreign DNA sequence using nucleic acid hybridization procedures, DNA sequencing, or observation of genetic properties.

Selection of recombinants is less of a problem with lambda replacement vectors (such as Charon 4A) than with plasmids because (1) the efficiency of transfer of recombinant DNA into the cell by lambda is very high, and (2) lambda fragments that have not received new DNA are too small to be incorporated into phage particles; thus, every viable phage virion will contain cloned DNA.

Cosmids

Like replacement vectors, a **cosmid** employs specific lambda genes. Cosmids are plasmid vectors containing foreign DNA plus the *cos* (cohesive end) site from the lambda genome (👓 Section 9.11). These *cos* sites are required for packaging DNA into lambda virions. Cosmids are constructed from plasmids containing cloned DNA by ligating the lambda *cos* region to the plasmid DNA. The modified plasmid can then be packaged into lambda virions *in vitro* as described previously and the phage particles used to transduce *Escherichia coli*. Cosmid construction avoids the necessity of having to transform *E. coli*, which at best is an inefficient process (see Section 10.7).

One additional advantage of cosmids is that they can be used to clone large fragments of DNA, with inserts as large as 50 kbp accepted by the system. Therefore, with bigger inserts, fewer clones are needed to obtain representation of the whole genetic element. Cosmids also permit storage of the DNA in phage particles instead of in plasmids. Phage particles are much more stable than plasmids, so the recombinant DNA can be kept for long periods of time.

10.17   Concept Check

Bacteriophages such as lambda have been modified to make useful cloning vectors. Larger amounts of foreign DNA can be cloned with lambda than with many plasmids. In addition, the recombinant DNA can be packaged *in vitro* for efficient transfer to a host cell. Plasmid vectors containing the lambda *cos* sites are called cosmids, and they can carry a large fragment of foreign DNA.

◆ Why is the ability to package recombinant DNA in a test tube useful?

◆ What is a *replacement vector*?

10.18   *In Vitro* and Site-Directed Mutagenesis

Methods for *in vitro* DNA manipulation opened up a whole new field of mutagenesis, *in vitro* mutagenesis, better known as **site-directed mutagenesis**. Whereas conventional mutagens (see Section 10.3) introduce mutations *at random* in the intact organism, site-directed mutagenesis makes use of synthetic DNA and DNA cloning techniques to introduce mutations at *precisely determined sites* in genes in a test tube.

Site-Directed Mutagenesis

Site-directed mutagenesis is a powerful tool, as it allows the experimenter to change any base pair in a specific gene. The basic procedure is to synthesize a short DNA oligonucleotide primer (👓 Section 7.8) containing the desired base change and to allow this to base pair with a single-stranded DNA containing the gene of interest. Pairing is complete except for the short region of mismatch. Then the short single-stranded fragment of the synthetic oligonucleotide is extended using DNA polymerase, thus copying the rest of the gene. The double-stranded molecule obtained is then inserted into a host cell by transformation. Mutants are selected by using some sort of positive selection, such as antibiotic resistance (in this case, the modified DNA would also contain an antibiotic resistance marker). The mutant obtained is then used in production of the modified (mutant) protein.

One procedure for site-directed mutagenesis is illustrated in Figure 10.40●. It is convenient to begin the process with the gene of interest cloned into a *single-stranded* DNA vector. The mutagenized DNA can then bind by complementary pairing with the gene of interest (Figure 10.40). A widely used vector for this purpose is *bacteriophage M13*, a single-stranded DNA phage whose genome is easy to manipulate and that replicates in *Escherichia coli* (👓 Section 16.3). Several basic cloning methods are available based on M13, and we discuss it as a vector for cloning DNA for genomics analysis in Chapter 15.

Site-directed mutagenesis can be used to test the efficacy of proteins containing known amino acid substitutions. For example, if one was studying the active site of an enzyme to assess the importance of particular amino acids in this region, site-directed mutagenesis could be used to change a specific amino acid and the modified enzyme assayed and compared to the wild-type enzyme. In such an experiment, the vector containing the mutated DNA would be inserted into a mutant strain unable to make the enzyme in question. In this way, the enzyme activity measured would only be that of the one encoded by the site-directed mutation.

Using site-directed mutagenesis, enzymologists can link virtually any aspect of an enzyme's activity, such as catalysis, resistance or susceptibility to chemical or physical agents, or interactions with other proteins, to *specific amino acids* in the protein. In particular, site-directed mutagenesis has given enzymologists detailed information on which amino acids are critical for an enzyme's function and significant insight into how enzymes catalyze chemical reactions.

Cassette Mutagenesis and Gene Disruption

Because of the large number of restriction enzymes commercially available and therefore the large number of different DNA sequences that can be cut, it is usually possible to find several different restriction sites within a gene of interest. If sites for the appropri-

● **Figure 10.40 Site-directed mutagenesis using short synthetic oligodeoxyribonucleotide fragments.** Note that cloning into bacteriophage M13 yields the single-stranded DNA needed for site-
directed mutagenesis to work.

● **Figure 10.41 Gene disruption using cassette mutagenesis.**
(a) A plasmid containing a cloned wild-type copy of gene X is cut with
the restriction enzyme *Eco*RI and mixed with a DNA fragment (the
kanamycin cassette) that contains a gene capable of conferring
kanamycin resistance on a cell and that has been obtained using the
same restriction enzyme. The cut plasmid and the cassette are ligated.
(b) The product of the ligation is a plasmid that now contains the
kanamycin cassette as an insertion mutation within gene X. This new
plasmid is now cut with a further restriction enzyme, *Bam*HI, and
transformed into a cell containing a wild-type gene X. (c) The transformed cell contains the linearized plasmid with a disrupted gene X
and its own chromosome with a wild-type copy of the gene. In some
cells, homologous recombination (see Figure 10.9) occurs between the
wild-type and mutant forms of gene X. (d) Cells that can grow in the
presence of kanamycin must have the kanamycin cassette recombined
into their chromosome because the linearized plasmid cannot replicate.
These cells now have only a single disrupted copy of gene X. This
disruption typically abolishes all gene X function.

ate enzyme are not found in the gene, or at the precise
location required, they can also be inserted by site-directed mutagenesis (see Figure 10.40). If restriction
sites are close together, the intervening DNA fragment
can be excised and replaced by a synthetic DNA fragment in which one or more of the nucleotides have
been changed. These synthetic fragments are called
cassettes (or cartridges), and the process is known as
cassette mutagenesis.

Insertion mutations can also be generated by simply
inserting a cassette at a single site. When using cassettes
to replace sections of genes, the cassettes are typically
the same size as wild-type DNA fragments. However,
the cassette used for making insertion mutations can be
almost any size and can even be an entire gene. In fact, to
ease the selection process, cassettes that encode proteins
that confer a particular antibiotic resistance on the host
are commonly used. This type of cassette mutagenesis is
used in a process called **gene disruption**.

The process of gene disruption is illustrated in
Figure 10.41●. In this case, a fragment carrying a gene
conferring kanamycin resistance, the *kan* cassette, is inserted at a restriction site in a cloned gene. The vector
carrying this mutant gene is then linearized by being cut
with a different restriction enzyme, and the linear DNA
is transformed into the host with kanamycin resistance
selected. The linearized plasmid cannot replicate, and so
resistant cells likely arise by homologous recombination

(see Section 10.6) between the mutated gene on the plasmid and the wild-type gene on the chromosome.

Note that when a cassette is inserted, the cells have not only gained kanamycin resistance but have also *lost the function* of the gene in which the *kan* cassette was inserted. These mutations are called **knockout mutations**. This process is similar to searching for insertion mutations made by transposons (see Section 10.14), but in this case the experimenter chooses exactly which gene will receive the mutation. Knockout mutations in haploid organisms (such as prokaryotes), yield viable cells only if the disrupted gene is not essential. In fact, gene knockouts are a convenient means of determining whether a given gene is an essential one.

 10.18 Concept Check

Synthetic DNA molecules of desired sequence can be made *in vitro* and used to construct a mutated gene directly or to change specific base pairs within a gene via site-directed mutagenesis. Genes can also be disrupted by inserting DNA fragments, called cassettes, into them. The inserted cassette eliminates the function of the wild-type gene while conferring a new, and usually selectable, phenotype on the cell.

◆ How can site-directed mutagenesis be useful to enzymologists?

◆ What are *knockout mutations*?

IV THE BACTERIAL CHROMOSOME

The three mechanisms of genetic exchange described in this chapter—transformation, transduction, and conjugation—together with the powerful tools of gene cloning, can be used to map the locations of genes on a bacterial chromosome. Genes in *Escherichia coli* were initially mapped using conjugation. By using Hfr strains with origins at different sites (Figure 10.25), it was possible to map the entire *E. coli* genome. However, conjugation does not permit ordering genes that are closely linked. Therefore, generalized transduction was used for fine-structure mapping of the *E. coli* chromosome. Bacteriophage P1 can carry small fragments of DNA and proved very useful for detailed mapping of genes in *E. coli*.

The genetic techniques used to map the chromosomes of other prokaryotes were dictated by the efficiency with which genetic transfer occurs in the particular organism. For example, transformation is a very inefficient process in *E. coli* but is a very efficient process in gram-positive bacteria (see Section 10.7). Thus, transformation proved an effective tool for mapping genes in *Bacillus*. In addition to conjugation, transduction was also used to map genes in the *E. coli* chromosome. However, because transduction only involves the transfer of a few genes—many fewer than is possible using conjugation—transduction was only useful for fine structure mapping purposes.

Although all of these "classical" techniques are still useful for the characterization and construction of strains, they have largely been supplanted in constructing genetic maps by molecular cloning and DNA sequencing. Cloning and sequencing has revolutionized the study of genome organization in prokaryotes and eukaryotes, and we consider the field of genomics in Chapter 15.

10.19 Genetic Map of the *Escherichia coli* Chromosome

The genetic map of *Escherichia coli* strain K-12 is shown in Figure 10.42●. The map distances are given in "minutes" of transfer, with the entire chromosome containing 100 minutes and with "zero time" arbitrarily set as that at which the first genetic transfer (the *threonine* operon) can be detected using the first Hfr strain of *E. coli* that was discovered.

The map shown in Figure 10.42 contains only a few of the many thousands of genes present in the *E. coli* chromosome and shows the location of a few restriction enzyme recognition sites. The size of the chromosome is given in both minutes (the original units determined by conjugation studies) and in kilobase pairs. The *E. coli* chromosome, like that of many other prokaryotes, has been completely sequenced. Sequencing of relatively small bacterial chromosomes of 3 or 4 megabases is now done routinely by automated sequencing of random, or *shotgun*, clones (∞ Section 15.2). Because of the enormous amount of genetic information genomic analyses makes available, this information can only be managed through computer databases. This reality has spawned a whole new field of molecular biology called *bioinformatics* (∞ Chapter 15). Database analyses using powerful gene search software has placed the field of bioinformatics in a central position in genomic studies.

Escherichia coli as a Model Prokaryote

Many factors have favored the use of *Escherichia coli* as the workhorse for studies of biochemistry, genetics, and bacterial physiology. As emphasized in Chapters 9 and 16, even *E. coli* viruses have served as model systems of study. Therefore, although the chromosome of *E. coli* was not among the first prokaryotic chromosomes sequenced, this organism remains the best characterized microorganism. In addition to its important role as a model organism, *E. coli* continues to be the organism of choice for both research and applications of genetic engineering (∞ Chapter 31).

The strain of *E. coli* whose chromosome was originally sequenced, *strain MG1655*, is a derivative of *E. coli* K-12, the traditional strain used for genetic studies. The wild-type *E. coli* K-12 is a lysogen of bacteriophage lambda (∞ Section 9.11) and also contains the F plasmid (see Section 10.9). However, strain MG1655 had been "cured" of lambda (by radiation) and of the F plasmid (by acridine treatment, see Section 10.9) before sequencing began. The chromosome of strain MG1655

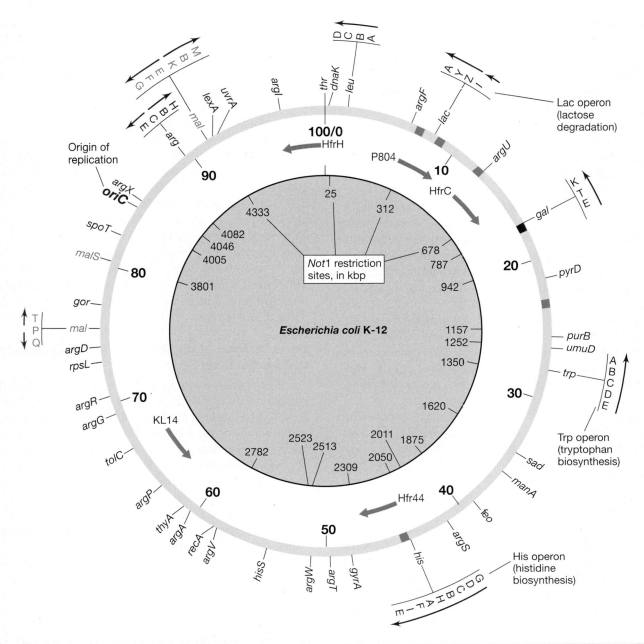

● **Figure 10.42** **Circular linkage map of the chromosome of *Escherichia coli* strain K-12.** On the outer edge of the map, the locations of a few of the mapped genes are indicated. The actual number of ORFs in the *E. coli* chromosome is 4288 and the chromosome contains 4,639,221 bp. A few operons are also shown, along with the direction in which they are transcribed. Along the inner edge of the map, the numbers from 0 to 100 refer to map position in minutes. The origin of DNA replication is marked *oriC* (84.3 min), and replication proceeds bidirectionally from this point (◯◯ Section 7.6 and Figures 7.16–7.20). The inner circle shows the locations, in kilobase pairs, of the sites recognized by the restriction enzyme *Not*I. Note that 0 minutes and 0 kbp are both, by convention, at the *thr* locus. The origins and directions of transfer of a few Hfr strains are also shown (arrows). The positions where five copies of the transposable element IS*3* have been located in a particular strain are shown in blue. This element is also found in two copies on the F plasmid and is involved in Hfr formation (see Section 10.12). The position of the site where the bacteriophage lambda prophage integrates is shown in red (◯◯ Section 9.11). If the prophage was present, it would add an extra 48.5 kbp (slightly over 1 minute) to the map. The genes of the maltose regulon (◯◯ Section 8.6), which includes several operons, are shown in green. Although most genes in this regulon have an abbreviation beginning with *mal*, note that one of the genes is *lamB*. This gene encodes a membrane protein involved in maltose uptake by the cell, but the protein is also the receptor for bacteriophage lambda. The gene *rpsL* (73 minutes) encodes a ribosomal protein. The gene was once called *str* because mutations in this gene lead to streptomycin resistance.

contains 4,639,221 bp. Analysis of the genomic sequence showed there to be 4288 possibly functional *open reading frames* (ORFs, ◯◯ Sections 7.13 and 15.2), which account for about 88% of the genome. Approximately 1% of the genome is composed of genes encoding tRNAs and rRNAs, and about 0.5% is noncoding, repetitive sequences. The remaining 10% of the genome contains all of the regulatory sequences: promoters, operators, origin and terminus of DNA replication, and so forth.

Arrangement and Expression of Genes on the *Escherichia coli* Chromosome

Early mapping experiments and studies on the regulation of the genes that control the enzymes of a single biochemical pathway in *E. coli* had shown that these genes were often clustered. On the genetic map in Figure 10.42 a few such clusters are shown. Notice, for instance, the *gal* gene cluster at 18 min, the *trp* gene cluster at about 28 min, and the *his* cluster at 44 min. Each of these gene clusters comprise an *operon* that is transcribed as a single *polycistronic* mRNA (∞ Section 7.13 and 10.13).

Genes for some biochemical pathways in *E. coli* are not clustered. For example, genes involved in arginine biosynthesis (*arg* genes) are scattered around the chromosome, as part of the *arg* regulon (∞ Section 8.6). Because of the early discovery of multigene operons and their usefulness as models for the study of gene regulation, for example, the *lac* operon (∞ Sections 8.5 and 8.7), one often gets the impression that such operons are the rule in prokaryotes. However, sequence analysis of the *E. coli* chromosome has shown that of the 2584 predicted or known transcriptional units, over 70% have only a *single* gene. Only about 6% of the operons have polycistronic mRNAs encoding four or more genes.

Within a genome, all operons are not transcribed in the same direction. In *E. coli* the transcription of some operons proceeds in one direction along the chromosome, whereas others proceed in the opposite direction. The direction of transcription of a few multigene operons is shown by arrows in Figure 10.42. Genomic analyses have shown that there are about equal numbers of operons on both strands. Many genes that are *highly expressed* in *E. coli* are oriented on the chromosome such that they are transcribed *in the same direction* that the DNA replication fork moves through them. The two replication forks start at the origin, *oriC* at about 84 min on the map shown in Figure 10.42 and move bidirectionally (∞ Section 7.6) toward the terminus, which is located at approximately 34 min on the map. Genomic sequencing has confirmed that all seven of the rRNA operons and 53 of the 86 tRNA genes are transcribed in the direction of replication. Presumably this arrangement for the transcription of highly expressed genes allows RNA polymerase to avoid collision with the replication fork, which would be moving in the same direction as the RNA polymerase.

Almost 2000 *E. coli* proteins, or genes encoding proteins, had been identified before the chromosome had been completely sequenced. From sequence analyses of the *E. coli* chromosome we now know that as many as 4288 different proteins can theoretically be encoded by this organism. Approximately 38% of these proteins are of unknown function and/or are simply hypothetical protein-encoding genes. As was expected, the "average" *E. coli* protein contains slightly more than 300 amino acid residues, but many proteins are smaller and many are much larger. The largest gene in *E. coli* should encode a protein of 2383 amino acids, but this protein has not yet been characterized. However, from comparative genomic analyses, the giant protein shows similarities with proteins found in pathogenic enteric bacteria highly related to *E. coli*. This large protein may thus play some role in the infection process.

Although *E. coli* has very few duplicated genes, computer analyses of its genome have shown that many of the genes that encode proteins have arisen by gene duplication during evolutionary history (see Section 10.14). There are also some large *gene families* in the *E. coli* genome, groups of genes with related sequences encoding products that have related functions (∞ Section 15.8). For example, there is a family of 70 genes whose products are all membrane-transport proteins. Gene families are common, not only within a species, but across even broad taxonomic lines. This emphasizes the fact that gene duplication and lateral gene transfer (see next paragraph) are powerful mechanisms for controlling the genetic complement of prokaryotes.

Other Features of the *Escherichia coli* Chromosome and Lateral Gene Flow

The *E. coli* genome was found to contain many other genetic elements that replicate as part of it. For example, there are copies of 10 different IS elements, including 7 copies of IS2 and 5 copies of IS3 (see Section 10.14). Both of these elements are also found on the F plasmid, and both are involved in the formation of Hfr strains (see Section 10.12 and Figure 10.23). There are also several defective prophages (three of which are related to lambda) present in the *E. coli* genome and partial genomes from several more. It is thus likely that lysogenic bacteriophages have been a part of the biology of *E. coli* for some time.

Genomic analyses have shown that *E. coli* obtained at least a portion of its genome by *horizontal (lateral) gene transfer* from other organisms. In fact, it has been estimated that nearly 20% of the *E. coli* genome originated from such transfers. Laterally transferred genes can be recognized by either their significantly different GC ratios from native *E. coli* genes, or by the fact that the genes encode proteins using a significantly different assortment of codons (codon bias, ∞ Section 7.14) from the pattern used by the host organism to encode its native proteins.

Large-scale changes in an organism's genome can apparently occur by lateral gene flow. For example, strains of *E. coli* are known that contain virulence genes located on large, unstable regions of the chromosome called *pathogenicity islands*, which can be acquired by horizontal transfer. Interestingly, horizontal gene transfer does not necessarily result in an ever-larger genome size. Many of the genes acquired in this way provide no selective advantage and so are lost by deletion. This keeps the chromosome of a given species at roughly the same size over time. For example, comparisons of genome sizes of several strains of *E. coli* have shown them all to be about 4.6 Mb pairs, despite the fact that prokaryotic genomes can vary from under 0.5 to over 10 Mb pairs. Genome size is therefore a species-specific trait.

Deeper Analysis of Prokaryotic Genomes

Analysis of the complete sequence of a genome can yield an incredible amount of information (∞ Chapter 15). Since the molecular genetics of *E. coli* have been seriously explored for several decades, and the genome has now been sequenced, does this mean that further analysis will be done mostly by computer and that traditional genetic and biochemical studies of *E. coli* are over? Absolutely not! Although computer analysis of a sequence can yield much information, in order to understand the *function* of genes, and particularly of regulatory sequences, it is still necessary to isolate mutants, map their mutations, and use biochemical and physiological analyses to determine their effects on the organism. This is especially true of the 20–40% (depending on the species) of genes that show up in all genomic analyses as genes encoding "proteins of unknown function" (∞ Chapter 15).

This huge repository of hypothetical proteins almost certainly holds new biochemical secrets that, collectively, will expand the extensive metabolic capabilities of prokaryotes already known (∞ Chapters 5 and 17). But in addition, because many prokaryotic genes have homologues in eukaryotic organisms including humans, understanding gene function in prokaryotes will have a practical pay off in terms of better understanding human genetics. Many human diseases have a genetic link. Because of this, the more we understand about what proteins do in bacteria, the more we will understand their function in humans. If history is any indication, this will lead, among other things, to the development of new drugs and other treatment regimens to control diseases.

 10.19 Concept Check

The *Escherichia coli* chromosome has been mapped using conjugation, transduction, molecular cloning, and sequencing. *E. coli* has been a useful model organism, and a considerable amount of information has been obtained from it, not only about gene structure but also gene function and regulation.

◆ Genetic maps of prokaryotic chromosomes are now typically made using only molecular cloning and DNA sequencing. Why were other methods also used in the case of *E. coli*?

REVIEW QUESTIONS

1. Write a one-sentence definition of the term *genotype*. Do the same for the term *phenotype*. Does the phenotype of an organism automatically change when a change in genotype occurs? Why or why not? Can phenotype change without a change in genotype? In both cases, give some examples to support your answer (∞ Section 10.1).

2. Explain why an *Escherichia coli* strain that is His⁻ is an auxotroph and one that is Lac⁻ is not. (*Hint:* Think about what *E. coli* does with histidine and lactose.) (∞ Section 10.1).

3. What are *silent mutations*? From your knowledge of the genetic code, why do you think most silent mutations affect the *third* position in a codon (∞ Section 10.2)?

4. Microinsertions occur in promoters but are not frameshift mutations. Define the terms *microinsertion, frameshift, mutation,* and *promoter* (∞ Section 7.10). Explain how the statement can be true (∞ Section 10.2).

5. Explain how it is possible for a frameshift mutation early in a gene to be corrected by another frameshift mutation farther along the gene (∞ Section 10.2).

6. What is the average rate of mutation in a cell? Can this rate change (∞ Section 10.3)?

7. Give an example of one biological, one chemical, and one physical mutagen and describe the mechanism by which each causes a mutation (∞ Section 10.4).

8. How can the Ames test, an assay using bacteria, have any relevance to human cancer (∞ Section 10.5)?

9. How does homologous recombination differ from site-specific recombination (∞ Section 10.6)?

10. Why is it difficult in a single experiment using transformation or transduction to transfer a large number of genes to a cell (∞ Section 10.7 and 10.8)?

11. Explain why in generalized transduction one always refers to a transducing *particle* but in specialized transduction one refers to a transducing *virus* (or transducing phage) (∞ Section 10.8).

12. How do plasmids replicate and how does this differ from chromosomal replication (∞ Section 10.9)?

13. What are R factors and why are they of medical concern (∞ Section 10.10)?

14. What is a sex pilus and which cell type, F⁻ or F⁺, would produce this structure (∞ Section 10.11)?

15. What does an F⁺ cell need to do before it can transfer chromosomal genes (∞ Section 10.12)?

16. What does it mean to complement a mutation "in trans" (∞ Section 10.13)?

17. The most useful transposons for isolating a variety of bacterial mutants are transposons containing antibiotic resistance genes. Why are such transposons so useful for this purpose (∞ Section 10.14)?

18. What are the essential characteristics of a cloning vector? What characteristics of plasmids make them especially useful as vectors for molecular cloning? What characteristic(s) of the F plasmid makes it less useful for use *in vitro* (∞ Sections 10.15 and 10.16)?

19. If insertional inactivation of an antibiotic resistance gene is used to detect the presence of a plasmid containing foreign DNA, why is it desirable to have two antibiotic resistance markers on the plasmid vector (∞ Section 10.16)?

20. What advantages can there be to using a viral cloning vector rather than a plasmid vector (∞ Section 10.17)?

21. What does site-directed mutagenesis allow you to do that normal mutagenesis does not (∞ Section 10.18)?

22. What is the size of the *Escherichia coli* chromosome? About how many proteins can it encode? How much noncoding DNA is present in the *E. coli* chromosome (∞ Section 10.19)?

APPLICATION QUESTIONS

1. A constitutive mutant is a strain that continuously makes a protein that in the wild type is inducible. Describe two ways in which a change in a DNA molecule could lead to the development of a constitutive mutant. How could these two types of constitutive mutants be distinguished genetically?

2. In Chapter 7, we saw that it was critical for the ribosome to translocate with great accuracy in order to maintain the proper reading frame. However, sometimes ribosomes make frameshift errors. Compare the impact on the cell of a ribosome periodically making a frameshift error in the mRNA from a particular gene with the impact of a frameshift mutation in the same gene.

3. Although a large number of mutagenic chemicals are known, no chemical is known to induce mutations in a single gene (gene-specific mutagenesis). From what you know about mutagens, explain why it is unlikely that a gene-specific chemical mutagen will be found. How then is site-specific mutagenesis accomplished?

4. Transposable elements cause mutations when the element is inserted within the gene. These elements disrupt the continuity of a gene. One could also say that introns (⬥ Section 7.13) disrupt the continuity of a gene, yet the gene is still functional. Explain why the presence of an intron in a gene does not inactivate that gene but insertion of a transposable element does.

5. When employing the plasmid vector pUC19, DNA fragments made using the restriction enzyme *Bam*HI inactivate the gene encoding β-galactosidase. When employing pBR322, cloning the same fragments inactivates the gene conferring tetracycline resistance. Both vectors contain a gene for ampicillin resistance that is used to select for bacterial transformants after making the recombinant molecules. Explain why it is much more efficient to use pUC19 rather than pBR322 as a cloning vector. (*Hint*: The increased efficiency has to do with finding the cells that contain vectors with cloned DNA. Figure 15.1 will also help you here.)

11

MICROBIAL EVOLUTION AND SYSTEMATICS

The evolutionary history of prokaryotes mirrors the evolutionary history of Earth itself. Of the 4.5 *billion* years that Earth has existed, prokaryotes have thrived for over 4/5 of this time. Evidence for the antiquity of prokaryotes exists in many forms, including oxidized iron formations widespread on Earth.

WORKING GLOSSARY

Allele one form of a given gene

Archaea a group of phylogenetically related prokaryotes distinct from *Bacteria*

Bacteria a group of phylogenetically related prokaryotes distinct from *Archaea*

Binomial system the system devised by Linnaeus for naming living organisms by which an organism is given a genus name and a species epithet

Domain in a taxonomic sense, the highest level of biological classification

Ecotype a population of genetically identical cells sharing a particular resource within an ecological niche

Endosymbiosis the process by which the mitochondrion and chloroplast, which were originally free-living *Bacteria*, established stable residence in primitive eukaryotic cells, eventually yielding the modern eukaryotic cell

Eukarya all eukaryotic cells: algae, protozoa, fungi, slime molds, plant and animal cells

Evolutionary chronometer a molecule, such as ribosomal RNA, whose molecular sequence can be used as a comparative measure of evolutionary divergence

Evolutionary distance in phylogenetic trees, the sum of the physical distance on a tree separating organisms; this distance is inversely proportional to evolutionary relatedness

FAME fatty acid methyl ester

Family in biological classification, an intermediate level of taxonomic hierarchy. Contains one or more genera, each of which consists of one or more species

FISH fluorescent *in-situ* hybridization

GC ratio in DNA (or RNA) from any organism, the percentage of the total nucleic acid that consists of guanine and cytosine bases

Genomic hybridization the experimental determination of genome similarities by measuring the extent of hybridization of DNA from the genome of one organism with that of another

Genus a collection of different species, each sharing one or more (usually several) major properties

Lateral (horizontal) gene transfer the exchange of genes between and among cells in a microbial community

Metazoa multicellular organisms

Multilocus sequence typing a taxonomic tool for classifying organisms on the basis of gene sequence comparisons of 6–7 housekeeping genes

Organelle unit membrane-enclosed structures of bacterial size and specialized in metabolic function found inside eukaryotic cells

Phylogenetic probe an oligonucleotide, sometimes made fluorescent by attachment of a dye, complementary in sequence to some ribosomal RNA signature sequence

Phylogeny the evolutionary history of organisms

Phylum a major lineage of cells in one of the three domains of life

Proteobacteria a large group of phylogenetically related gram-negative *Bacteria*

Ribosomal Database Project (RDP) a large database of small subunit ribosomal RNA sequences that can be retrieved electronically and used in comparative ribosomal RNA sequencing studies

Ribotyping a means of identifying microorganisms from analysis of DNA fragments generated from restriction enzyme digestion of genes encoding their 16S rRNA

RNA life a life form lacking DNA and protein that may have existed on early Earth and in which RNA served both a genetic coding and a catalytic function

Signature sequence short oligonucleotides of defined sequence in 16S or 18S ribosomal RNA characteristic of specific organisms or a group of phylogenetically related organisms and useful for constructing probes

16S ribosomal RNA a large polynucleotide (~1500 bases) that functions as part of the small subunit of the ribosome of prokaryotes and from whose sequence evolutionary information can be obtained; eukaryotic counterpart, 18S rRNA

Small subunit (SSU) RNA ribosomal RNA from the 30S ribosomal subunit of prokaryotes or the 40S ribosomal subunit of eukaryotes

Species in microbiology, a collection of strains that all share the same major properties but differ in one or more significant properties from other collections of strains

Stromatolite a laminated microbial mat, typically built from layers of filamentous prokaryotes and other microorganisms, which can become fossilized

Taxonomy the science of identification, classification, and nomenclature

Universal phylogenetic tree a tree that shows the position of representatives of all domains of living organisms

EARLY EARTH, THE ORIGIN OF LIFE, AND MICROBIAL DIVERSIFICATION

An integrating theme in all of biology is **evolution**. We begin here a unit on microbial diversity as we know it today, the result of nearly 4 billion years of evolution. This chapter focuses on microbial evolution and how the phylogenetic picture of prokaryotes corresponds to their taxonomy and classification. We will see how our understanding of the evolutionary relationships of microorganisms has emerged from molecular biology and how the concept of a "bacterial species" is now beginning to crystallize. This chapter can be considered the background from which we present the organisms themselves and their genomes in the next four chapters.

11.1 Evolution of Earth and Earliest Life Forms

In the first few sections we travel from a time before cells appeared on Earth through to the modern eukaryotic cell. We will learn that prebiotic chemistry could have supplied the necessary prerequisites for the origin of life, that early self-replicating "life forms" may not have been cellular, that early energy sources may have been inorganic, and that the eukaryotic cell is a genetic chimera consisting of at least two distinct organisms.

Origin of Earth

Earth itself is about 4.6 billion years old. Our solar system formed when a large, very hot star exploded, generating a new star (our sun) and the other components of our solar system. Although no rocks dating to the origin of our planet have yet been discovered, presumably because they have been weathered, rocks dating back to nearly 4 *billion* years old (4 gigayears, GY) have been found in several locations on Earth. The oldest rocks discovered thus far are in the Itsaq gneiss complex, in southwestern Greenland, which date to about 3.86 billion years ago. Other rocks of ancient origin include the Warrawoona series, Towers Formation, and Pilbara supergroups in Western Australia, and the Swaziland and Barberton supergroups in southern Africa; all of these rock formations are about 3.5 billion years old (see Figure 11.1).

Ancient rocks are of three types: *sedimentary*, *volcanic*, and *carbonaceous*. The earliest sedimentary rocks are of particular evolutionary interest because we know that the formation of sedimentary rocks requires liquid water. Thus, since sedimentary rocks nearly 4 GY old exist on Earth, it follows that liquid water was also present at that time, probably in the form of oceans or large lakes. The presence of liquid water in turn implies that conditions were compatible with life as we know it.

Evidence for Microbial Life on Early Earth

Evidence for microbial life in ancient rocks rests on the fossilized remains of cells and the isotopically "light" carbon abundant in these rocks (we discuss the use of isotopic analyses of carbon and sulfur as an indication of living processes in Section 18.8). Some ancient rocks contain microfossils that appear very much like bacteria, typically as simple rods or cocci (Figure 11.1●).

In rocks of 3.5 GY or younger, microbial formations called *stromatolites* are abundant. **Stromatolites** are fossilized microbial mats consisting of layers of filamentous prokaryotes and trapped sediment (Figure 11.2●); we discuss some characteristics of microbial mats in Section 18.7. What kind of organisms were these ancient

(a) (b) (c) (d) (e)

● **Figure 11.2   Ancient and modern stromatolites.** (a) The oldest known stromatolite, found in a rock about 3.5 billion years old, from the Warrawoona Group in Western Australia. Shown is a vertical section through the laminated structure preserved in the rock. Arrows point to the laminated layers. (b) Stromatolites of conical shape from 1.6-billion-year-old dolomite rock of the McArthur basin of the Northern Territory of Australia. (c) Modern stromatolites in a warm marine bay, Shark Bay, Western Australia. (d) Modern stromatolites composed of thermophilic cyanobacteria growing in a thermal pool in Yellowstone National Park. Each structure is about 2 cm high. (e) Another view of modern and very large stromatolites from Shark Bay. Individual structures are 0.5–1 m in diameter.

stromatolitic bacteria? By comparing ancient stromatolites with modern stromatolites growing in shallow marine basins (Figure 11.2c and e) or in hot springs (Figure 11.2d; ∞Figure 18.18a), it has been concluded that ancient stromatolites were formed by filamentous phototrophic bacteria, perhaps relatives of the green nonsulfur bacterium *Chloroflexus* (∞Section 12.35).

Figure 11.3● shows photomicrographs of thin sections of more recent rocks containing cell-like structures remarkably similar to modern filamentous bacteria and green algae (∞Section 14.13). In the oldest stromatolites these organisms were likely anoxygenic (nonoxygen-evolving) phototrophic bacteria (∞Sections 12.2, 12.21, 12.32, and 12.35) rather than the O_2-evolving cyanobacteria (∞Sections 12.25 and 12.26) that dominate modern stromatolites (Figure 11.2c, e). Nevertheless, the conclusion is inescapable that prokaryotic microorganisms had evolved an impressive morphological diversity very early in Earth's history.

● **Figure 11.1   Ancient microbial life.** Scanning electron micrograph of microfossil prokaryotes from 3.45 billion year old rocks of the Barberton Greenstone Belt, South Africa. Note the rod-shaped bacteria (arrow) attached to particles of mineral matter. The cells are about 0.7 μm in diameter.

(a)

(b)

J.W. Schopf

● **Figure 11.3 Fossil prokaryotes and eukaryotes from more recent rocks than those shown in Figure 11.1.** The two photographs in (a) show fossil prokaryotic microorganisms from the Bitter Springs Formation, a rock formation in central Australia about 1 billion years old. These forms bear a striking resemblance to modern filamentous cyanobacteria, anoxygenic phototrophs, or filamentous sulfur chemolithotrophs (∞Chapter 12). Cell diameters, 5–7 μm. (b) Microfossils of eukaryotic cells from the same rock formation. The cellular structure is remarkably similar to that of certain modern green algae, such as *Chlorella* species (∞Section 14.13). Cell diameter, about 15 μm.

Conditions on Early Earth: Hot and Anoxic

The atmosphere of early Earth was devoid of oxygen. Besides water, a variety of gases were present, the most abundant being methane, carbon dioxide, nitrogen, and ammonia. In addition, trace amounts of carbon monoxide and hydrogen likely existed, as well as large reservoirs of sulfide, as a mixture of H_2S and FeS. It is also likely that a considerable amount of hydrogen cyanide, HCN, was produced on early Earth when NH_3 and CH_4 reacted to yield HCN.

Most evidence suggests that the early Earth was hotter than it is today. For its first two hundred million years or so, the surface of Earth may have exceeded 100°C. In addition, the planet was frequently bombarded by meteorites. Under these conditions, free water could not exist. Water only accumulated as Earth cooled. How fast Earth cooled is unknown, but it has been hypothesized that self-replicating entities first appeared at a time when Earth was still fairly hot. Thus, early life forms were likely quite heat tolerant, and may have resembled hyperthermophilic prokaryotes (∞Section 6.12 and Chapter 13) in this regard. We will consider the evidence for this hypothesis in Sections 11.8, 11.9, and 13.13.

Origin of Life

The synthesis of biomolecules can occur spontaneously if reducing atmospheres containing the aforementioned gases are subjected to intense energy sources. Of the energy sources available on primitive Earth, the most important was probably ultraviolet (UV) radiation from the sun, as no UV-absorbing ozone layer was present at this time. However, lightning discharges, radioactivity, heat from meteoritic impacts, and thermal energy from volcanic activity were also likely to have been available.

If gaseous mixtures resembling those thought to be present on primitive Earth are irradiated with UV or subjected to electric discharges in the laboratory today, a variety of biomolecules form. These include sugars, amino acids, purines, pyrimidines, various nucleotides, thioesters, and fatty acids. Moreover, under simulated prebiological conditions in the laboratory some of these biochemical building blocks have been shown to *polymerize*, leading to the formation of polypeptides, polynucleotides, and other important macromolecules. We can therefore imagine that by whatever means, a mixture of organic compounds eventually accumulated on the primitive Earth, and in so doing, set the stage for the origin of life. The sterility of Earth at this time meant that these organic compounds were stable and not subject to biodegradation as they would be today.

Catalysis and the Importance of Montmorillonite Clay

To obtain spontaneous synthesis of macromolecules on early Earth, a catalyst would have had to be available. Good possibilities include the surfaces of clays or pyrite (FeS_2). Such surfaces would have provided a stable, relatively dry environment for the synthesis and accumulation of macromolecules into organic films. From these films and over long periods of time, primitive, self-replicating structures could then have emerged.

Experiments with Montmorillonite clay have shown it to be an excellent catalytic surface. For example, in the presence of fatty acids, particles of Montmorillonite clay trigger the formation of small lipid vesicles that appear very much like cells (Figure 11.4●). These vesicles have been shown to self-enlarge by incorporating more fatty acids and to "divide" if passed through a small-pored filter. Such clay particles also bind RNAs and encapsu-

● **Figure 11.4   Lipid vesicles made in the laboratory from the fatty acid myristic acid and RNA.** The vesicle itself stains green and the RNA complexed inside the vesicle stains red. Vesicle synthesis is catalyzed by the surfaces of Montmorillonite clay particles.

late them within the lipid vesicles (Figure 11.4). In addition, if ribonucleotides are added to particles of Montmorillonite clay, the particles catalyze the spontaneous formation of RNA oligonucleotides.

Although clay-catalyzed vesicles are not modern-day cells and RNA oligonucleotides are not modern day genes (although recall that many viruses do have RNA genomes, ∞Chapters 9 and 16), RNA-containing vesicles (Figure 11.4) show many of the properties of cells. If such self-replicating vesicles were formed on early Earth, improved versions containing other macromolecules and a more complex design would eventually have arisen, given the geological time periods available. Such structures could have been the forerunners of the first cellular organisms.

 11.1   Concept Check

Planet Earth is approximately 4.6 billion years old. The first evidence for microbial life can be found in rocks about 3.86 billion years old. Early Earth was anoxic and much hotter than the present. The first biochemical compounds were made by abiotic syntheses that set the stage for the origin of life.

◆   How did the atmosphere of early Earth compare with that of Earth today?

◆   How were early biochemicals formed?

◆   What role could Montmorillonite clay have played in the origin of life?

11.2   Primitive Life: The RNA World and Molecular Coding

What was the first self-replicating entity like? Based on what we know about life today, we can predict that early life forms were simpler than modern cells and that even the simplest self-replicating entities showed at least two properties: (1) a means of obtaining energy, and (2) some form of heredity. But would these properties have required a *cellular* structure?

It is tempting to extrapolate backward from the present and postulate that early organisms resembled modern cells, differing primarily by having only a few genes and possessing limited transcriptional and translational abilities. However, even a structure like this would have been relatively complex—probably much more complex than the first self-replicating entities. What would the latter have been like then? Following the discovery that certain types of ribonucleic acid (RNA) are catalytic (∞Section 14.8), many scientists now believe that the *earliest* life forms may not have been cells at all, but instead naked RNAs. If true, this would have been the age of "**RNA life**," where RNA was the agent of both catalysis and genetic coding.

RNA Life

In an **RNA world**, if it ever existed, various self-replicating RNAs would have carried out the catalytic reactions necessary for their self-replication (Figure 11.5●). This idea is not far-fetched, because RNAs that do just that are known even today. Various *catalytic RNAs* (∞**ribozymes**, Section 14.8) have been shown to catalyze several different reactions, including self-replication. Moreover, certain ribozymes can synthesize nucleotides from a sugar and a nitrogenous base and polymerize amino acids into polypeptides.

Self-replicating RNA life forms may have evolved into the first *cellular* life forms when they became enclosed within lipoprotein vesicles, perhaps similar to Montmorillonite clay vesicles (Figure 11.4). This step may have occurred to no avail countless times on early Earth, but eventually, the proper constituents and set of circumstances came together and a primitive enclosed structure arose, capable of self-replication (Figures 11.4 and 11.5). Although still lacking DNA and proteins, this protocellular life form would otherwise have resembled a modern cell.

As these structures replicated, natural selection would have pushed for their evolutionary refinement. This would likely have included a selection for more efficient catalysts and a more stable genetic coding process. It is well known that proteins show higher catalytic specificity than do RNAs. Thus, as primitive organisms became more complex and catalytic reactions more biochemically demanding, *proteins* would have replaced *RNAs* as the cells' primary biocatalysts. But this was not an overnight occurrence. It is likely that proteins appeared gradually in cells, perhaps at first complexed with RNA (as ribonucleoprotein complexes, well known in today's cells). However, as evolution selected for more and more precise biochemical catalysts, RNA was eventually replaced almost entirely by proteins as cellular enzymes (Figure 11.5). By this point, the modern cell was clearly on the horizon.

Sterile Earth

Prebiotic syntheses (proteins and RNA made abiotically)

RNA

The RNA World

Self-replicating RNAs

Lipid or lipoprotein vesicle

Early cellular life (RNA as coding and catalytic molecule)

Protein

Cellular Life

Proteins assume catalytic functions (RNA only as coding molecule)

Evolution of DNA from RNA

DNA

Modern cellular life (DNA replaces RNA as coding molecule leading to DNA → RNA → Protein)

● **Figure 11.5 Possible scenario for the evolution of cellular life forms from RNA life forms.** Self-replicating RNAs could have become cellular entities by becoming stably integrated into lipoprotein vesicles. With time, proteins replaced the catalytic functions of RNA, and DNA replaced the coding functions of RNA.

The Modern Cell: DNA → RNA → Protein

Why did the DNA-based cell arise? This may have come about for several reasons. The establishment of DNA as the genome of the cell may have resulted from the need to store genetic information in one place in the cell and keep it there in a stable form. From here, genes could be *expressed* to yield proteins. With further evolutionary input, *selective expression* would have arisen, the latter effecting an energy savings on the cell. Continued evolutionary change in this area would have led to the

extensive and highly sophisticated metabolic controls we see in prokaryotes today (∞Chapter 8).

The instability of RNA as genetic material might also have led to DNA-based systems. For cells to evolve, there has to be a high degree of genetic stability in the system, otherwise favorable genetic traits will not be faithfully transmitted from generation to generation. This requirement for genetic stability in primitive cells was probably the evolutionary push for the conversion from RNA to DNA as the master genetic molecule in early cells. For reasons probably dealing with the chemistry of their substrates, RNA polymerases are inherently less accurate than DNA polymerases. Thus, to ensure the faithful transfer of genetic information from generation to generation, the evolution of a DNA genome would eventually have been selected. RNAs were retained in protein synthesis, however, where many of their roles (even in modern cells) are catalytic (∞Section 7.16).

However it happened, somewhere in the early stages of microbial evolution the three-part system—DNA, RNA, and protein—became fixed in cells as the best solution to biological information processing. That this system was an evolutionary success story is obvious, since without exception, modern cells contain all three types of these macromolecules.

◉ *11.2 Concept Check*

The first life forms may have been self-replicating RNAs. These were both catalytic and informational. Eventually, DNA became the genetic repository of cells and the three part system, DNA, RNA, and protein, became universal among cells.

◆ What evidence supports the concept of a period of RNA life, and why would RNA life forms likely not have survived to the present?

11.3 Primitive Life: Energy and Carbon Metabolism

Life is a highly ordered process. *Energy* is required to assemble an assortment of molecules into a complex biological machine—the cell. How were the energy demands of primitive self-replicating entities met? For catalytic RNAs, the energy required may have come from the exothermic nature of many of the reactions they carried out. But for cells, energy conservation became an issue.

Until the evolution of cyanobacteria, molecular oxygen was unavailable in any significant quantity on Earth (see Figure 11.7). Thus, only energy-generating mechanisms that could occur under *anoxic* conditions could be exploited by primitive cells to fuel their energy needs. However, as we will see in Chapter 17, anoxic conditions place few restrictions on microbial metabolic diversity. A variety of chemoorganotrophic and chemolithotrophic energy-generating mechanisms as well as several variations of photosynthesis occur in the absence of oxygen. All of these mechanisms are rather complex, however, and simpler mechanisms could, and probably did, prevail.

Reactions involving ferrous iron (which was known to have been abundant on early Earth), have often been proposed as energy-yielding reactions for primitive organisms. For example, the reaction:

$$FeS + H_2S \longrightarrow FeS_2 + H_2 \quad \Delta G^{0'} = -42 \text{ kJ/reaction}$$

proceeds exergonically with the release of energy. This reaction yields H_2, and primitive cells could have used this powerful electron donor to generate a proton motive force. The latter could have fueled a primitive ATPase to yield ATP (Figure 11.6●). Other sources of H_2 may also have been available from ferrous iron on the early Earth. For example, Fe^{2+} can reduce protons to hydrogen in the presence of ultraviolet radiation as an energy source (Figure 11.6). Early oceans were thought to be loaded with ferrous iron, and their surface waters were subject to intense UV radiation.

With H_2 as electron *donor*, an electron *acceptor* would also have been required to form a redox pair (∞ Section 5.6); this could have been elemental sulfur (S^0). As shown in Figure 11.6, the oxidation of H_2 with the production of H_2S would have required few enzymes. Moreover, because of the abundance of H_2 and sulfur compounds on the early Earth, cells would have had a nearly limitless supply of energy. Interestingly, many hyperthermophilic *Archaea* (which we will see later are, along with hyperthermophilic

Bacteria, the closest extant relatives of Earth's earliest organisms) can oxidize H_2 with S^0 as electron acceptor.

Primitive organisms may have derived carbon from various sources, including organic carbon from abiotic syntheses or even from CO_2, a gas that was abundant on early Earth. Use of carbon dioxide would have had to follow the evolution of autotrophy, the process where CO_2 is converted into all organic components in the cell (∞ Sections 17.6 and 17.7). Although originally thought to be unlikely, the hypothesis that autotrophy was an early physiological process has gained momentum from the discovery of autotrophic prokaryotes such as *Aquifex*. This member of the *Bacteria* is a hyperthermophile, contains a very small genome, and branches near to the root of the evolutionary tree of life (see Figure 11.16). All of these are properties one would associate with a "primitive" state. Therefore, it is distinctly possible that early prokaryotes used CO_2 as a major or sole carbon source for growth.

Molecular Oxygen: Banded Iron Formations

A milestone in Earth's history occurred with the evolution of *oxygenic photosynthesis* in the cyanobacteria (Figure 11.7●). Cyanobacteria first appeared on Earth about 2.8 billion years ago, but the O_2 that they produced did not accumulate in the atmosphere because of all of the reducing substances (such as FeS) that were still present in the oceans; these materials reacted sponta-

● **Figure 11.6 A possible energy-generating scheme for primitive cells.** Formation of pyrite leads to H_2 production and S^0 reduction, which fuels a primitive ATPase. Note how H_2S plays only a catalytic role; the net substrates would be FeS and S^0. Also note how few different proteins would be required. The $\Delta G^{0'}$ of the reaction FeS + $H_2S \rightarrow FeS_2 + H_2 = -42$ kJ. An alternative source of H_2 could have been the UV-catalyzed reduction of H^+ by Fe^{2+} as shown.

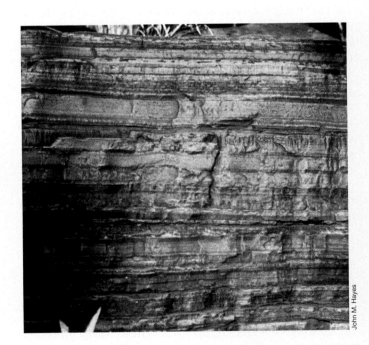

● **Figure 11.7 Banded iron formations.** An exposed cliff about 10 m in height containing layers of iron oxides interspersed with layers containing iron silicates and other silica materials. Brockman Iron Formation, Hammersley Basin, Western Australia. The iron oxides contain iron in the ferric (Fe^{3+}) form produced from ferrous iron (Fe^{2+}) primarily by the oxygen released by cyanobacterial photosynthesis.

neously with O_2 to produce H_2O. In the case of pyrite oxidation, the Fe^{3+} produced became a prominent marker in the geological record. Cyanobacteria almost certainly evolved from anoxygenic phototrophs through the development of a photosystem that could use H_2O as an electron donor for photosynthetic CO_2 reduction; this releases O_2 as a by-product. We discuss the biochemistry of photosynthesis in Chapter 17.

Much of the iron found in rocks of Precambrian origin (>0.5 billion years ago, see Figure 11.8) exists in *banded iron formations* (BIF) (Figure 11.7). These are laminated sedimentary rocks with alternations of iron-rich and silica-rich layers. BIF range in age from 2.7 to 1.9 GY. The evolution of oxygenic photosynthesis (cyanobacteria) is thought to have occurred some 2.75–3 GY ago. The metabolism of cyanobacteria yielded O_2 that oxidized iron present as Fe^{2+} (for example, in FeS_2) to Fe^{3+}. The ferric iron formed various iron oxides that accumulated as BIF (Figure 11.7). Once the abundant Fe^{2+} was consumed, the stage was set for O_2 accumulation in the atmosphere, a major trigger of evolutionary events in both the prokaryotic and eukaryotic worlds (Figure 11.8●).

Oxygenation of the Atmosphere: New Metabolisms and the Ozone Shield

The appearance of oxygenic photosynthesis had enormous consequences for the Earth because as O_2 accumulated, the atmosphere gradually changed from an anoxic to an oxic one (Figure 11.8). With O_2 now available as an electron acceptor, aerobic organisms could evolve. These organisms were able to obtain more energy from the oxidation of organic compounds than anaerobes could (∞Chapter 17), allowing larger populations to develop and increasing the chances for the evolution of new types of organisms and metabolic schemes. There is good evidence from the fossil record that at about the time that Earth's atmosphere became oxidizing there was an enormous burst in the rate of evolution. This triggered the appearance of organelle-containing eukaryotic microorganisms and from them the rapid diversification of multicellular organisms. As evolution proceeded, animals and plants eventually appeared (see Figure 11.16 and Section 11.9).

A major consequence of the appearance of O_2 was the formation of *ozone*, a substance that provides a barrier preventing the intense ultraviolet radiation of the sun from reaching the Earth. When O_2 is subject to short wavelength UV radiation, it is converted to O_3, which strongly absorbs wavelengths up to 300 nm. Until an ozone shield developed, evolution could have continued only in environments protected from direct radiation from the sun, such as under rocks or in the oceans, because the intense UV radiation would have caused lethal DNA damage. However, after the photosynthetic production of O_2 and subsequent development of an ozone shield, organisms could have ranged generally over the

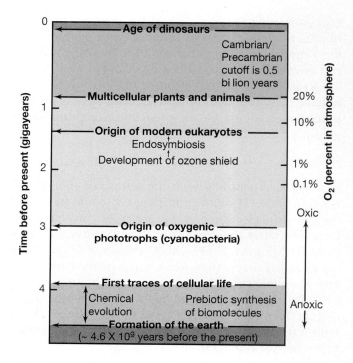

● **Figure 11.8 Major landmarks in biological evolution and Earth's changing geochemistry.** One gigayear (GY) is one billion years. Note how the oxygenation of the atmosphere due to cyanobacterial metabolism was a gradual process, occurring over a period of about 2 billion years. Although full (20%) oxygen levels are required for animals and most other higher organisms, this is not true of prokaryotes, as many are facultative aerobes or microaerophiles (∞ Section 6.15 and Table 6.4). Thus, prokaryotes respiring at reduced O_2 levels may have dominated Earth for the period of a billion years or so before Earth's atmosphere reached current levels of oxygen.

surface of Earth, permitting evolution of a greater diversity of living organisms. Figure 11.8 summarizes some major events in biological evolution and changes in the Earth's geochemistry from a highly reducing to a highly oxidizing planet.

 11.3 Concept Check

Primitive metabolism was anaerobic and likely chemolithotrophic, exploiting the abundant sources of FeS and H_2S present. Carbon metabolism may have included autotrophy. Oxygenic photosynthesis led to development of banded iron formations, an oxic environment, and great bursts of biological evolution.

◆ How could energy have been obtained by early cells from FeS + H_2S?

◆ Why is the advent of cyanobacteria considered a critical step in evolution?

◆ In what form is iron present in BIF?

◆ What role did ozone play in biological evolution, and how did cyanobacteria make the production of ozone possible?

11.4 Eukaryotes and Organelles: Endosymbiosis

Chapter 2 offered a brief tour of eukaryotic microbial diversity (∞Section 2.6). Chapter 14 will examine eukaryotic microorganisms in more detail. Here, we consider how the modern eukaryotic cell evolved its characteristic internal structure: the membrane-enclosed nucleus and **organelles**.

Origin of the Nucleus

Primitive eukaryotic cells likely consisted of structurally simple cells. Like modern-day prokaryotic cells, they lacked mitochondria, chloroplasts, and a membrane-enclosed nucleus. How then did a nucleus first appear?

As cells in the eukaryotic line of descent became larger, the nucleus and mitotic apparatus evolved, along with the partitioning of DNA into distinct units (chromosomes). Chromosomes may have arisen to ensure the orderly replication and separation of DNA. This is because as primitive genomes enlarged, a point was reached where replication of DNA as a single molecule (as in prokaryotes) was no longer feasible. Development of the nucleus also facilitated handling of the larger genomes needed by the more complex eukaryotic cell and also made possible the recombination of genomes through sexual reproduction (∞Sections 2.2 and 14.7).

There is no obvious reason why primitive eukaryotes would have needed organelles *other* than the nucleus, especially if the Earth was still anoxic at the time. Hence, the other major organelles likely evolved much later. Their origin, however, is a very interesting story, and we look at that process now.

Endosymbiosis

The modern (organelle-containing) eukaryotic cell did not come about from the gradual partitioning of cellular functions into enclosed structures, the organelles. Instead, organelles originated from the *stable incorporation* of chemoorganotrophic and phototrophic symbionts from the domain *Bacteria* (Figure 11.9●). This process, called **endosymbiosis** (*endo* means "inside"), probably began when an aerobic bacterium established stable residency within the cytoplasm of a primitive eukaryote and supplied the cell with energy in exchange for a protected environment and a ready supply of nutrients (Figure 11.9). This symbiont was the forerunner of the modern mitochondrion. Similarly, the endosymbiotic uptake of an oxygenic phototroph conferred photosynthetic properties on a primitive eukaryote; such a cell was freed from a reliance on organic compounds for energy production and became phototrophic. The phototrophic endosymbiont was the forerunner of the modern day chloroplast (Figure 11.9).

A few eukaryotic cells either never incorporated endosymbionts or, what appears more likely, had them at

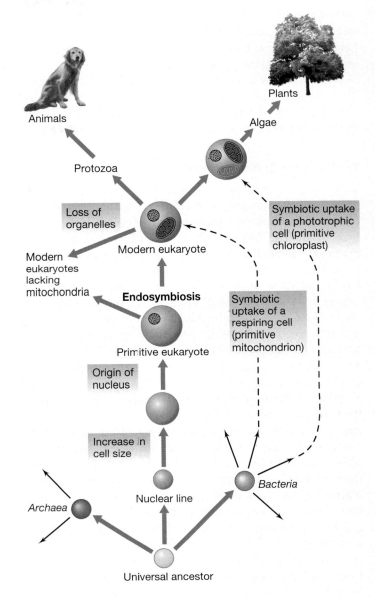

● **Figure 11.9 Origin of the modern eukaryotic cell by endosymbiotic events.** Note how organelles originated from *Bacteria* rather than *Archaea*. Some primitive eukaryotes either never underwent endosymbiotic events or permanently lost their symbionts, but otherwise maintained the basic properties of eukaryotic cells. Compare this figure with Figure 11.16 that shows lineages of *Bacteria* from which organelles originated (∞Section 14.4).

one time and then disposed of them, possibly because their niche was permanently anoxic. Such organelle-less eukaryotes are known today (∞Sections 14.2 and 14.9). Most endosymbiotic associations probably increased the fitness of the host cell, and thus today, organelle-containing eukaryotic cells predominate. Microfossil records (see Figure 11.3) suggest that endosymbiotic events began sometime after about 2 GY ago (see Figure 11.10).

Endosymbiosis was probably not a one-time event, but rather a multiple and perhaps even common occurrence between prokaryotic cells and early eukaryotic cells. Unstable or harmful associations were not maintained while beneficial associations were. Selection helped refine the beneficial associations. This included modifications to the structure of the endosymbiont (loss of a cell wall, for example), genomic modifications, and the transfer of genes from the endosymbiont to the nucleus of the host cell (∞Sections 14.9 and 15.7). From these associations, the modern eukaryotic cell was born.

Evidence for Endosymbiosis

Many lines of evidence support the endosymbiosis hypothesis. For example, both chloroplasts and mitochondria contain ribosomes that are clearly of the prokaryotic type (∞Table 7.4) and show 16S ribosomal RNA sequences (see Section 11.6) characteristic of specific *Bacteria*. Moreover, organellar ribosome function is inhibited by the same antibiotics that affect ribosome function in free-living *Bacteria* (∞Section 14.5). Mitochondria and chloroplasts also contain small amounts of DNA arranged in a covalently closed, circular form, typical of prokaryotes (∞Section 2.2). Indeed, many telltale signs of a prokaryotic background are present in organelles from modern cells today (∞ Section 14.4).

Endosymbiotic events occur even today. For example, there are a number of protozoa, especially anaerobic protozoa, that contain symbiotic prokaryotes (∞Figure 19.25). Although these symbionts appear to be fully functional cells rather than the highly modified cells that we find in the mitochondria and chloroplasts, they show that eukaryotic cells can and still do take up prokaryotic cells, presumably to the advantage of both partners.

Endosymbiosis was a major boost for the eukaryotic lineage of cells. The presence of respiratory and photosynthetic energy factories conferred such important new properties on eukaryotic cells that the stage was set for an explosion in biological diversity. The period from about 1.5 billion years ago to the present saw the rise and diversification of unicellular eukaryotic microorganisms and the metazoa, culminating in the appearance of the structurally complex higher plants and animals (see Figures 11.8, 11.9, and 11.16).

Biological Evolution and Geological Time Scales

The period from the origin of the first metazoan to the present represents about one-sixth of the total time that life has existed on Earth. Put another way, five-sixths of Earth's biological history was restricted to microbial life, the bulk of this period *solely to prokaryotes*. Prior to this period would have been the age of acellular replicating life forms—the RNA world—if such structures actually existed. Prior to this, planet Earth was sterile, in the truest sense of the word. These eras in the evolution of life are summarized using the geologist's time scale in Figure 11.10.

Metazoa have left a considerable and highly diverse fossil record. Because of this, our understanding of biological evolution from the time of the metazoa to the pres-

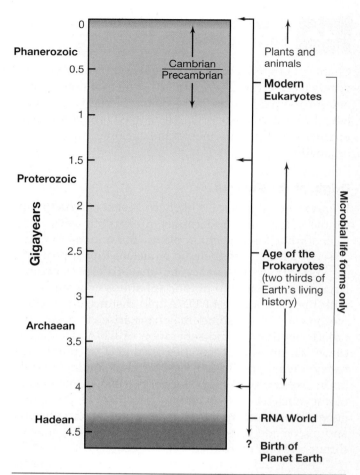

● **Figure 11.10** **The eras of self-replicating entities on Earth.** The left scale relates life forms to the major geological time periods.

ent is quite good. The story for microorganisms, however, is quite incomplete, especially with regards to prokaryotes. In fact, the evolutionary history of prokaryotes was a mystery until just a little more than a quarter century ago. The advent of molecular methods for determining evolutionary relationships has opened many new avenues in evolutionary science, especially in the area of microbial evolution. We examine this story now, and see how these methods allow us to place all cells onto a phylogenetic tree of life.

⬢ 11.4 *Concept Check*

The eukaryotic nucleus and mitotic apparatus probably arose as a necessity for ensuring the orderly partitioning of DNA in large-genome organisms. Mitochondria and chloroplasts, the principal energy-producing organelles of eukaryotes, arose from symbiotic association of prokaryotes of the domain *Bacteria* within eukaryotic cells, a process called endosymbiosis. Assuming that an RNA world existed, self-replicating entities have populated Earth for over 4 billion years.

◆ When in geological time did eukaryotes first appear? Metazoa?

◆ What evidence supports the idea that organelles were once free-living species of the domain *Bacteria*?

II METHODS FOR DETERMINING EVOLUTIONARY RELATIONSHIPS

In the next few sections we consider the **phylogeny** of microorganisms—their evolutionary relationships. The major focus here will be on the comparative sequencing of structural RNAs from ribosomes. From such studies, the first revealing picture of microbial phylogeny has emerged. In addition, the techniques employed have yielded new research tools for microbial ecology and clinical diagnostics.

11.5 Evolutionary Chronometers

Certain genes and proteins are **evolutionary chronometers**—measures of evolutionary change. In other words, differences in nucleotide or amino acid sequence of functionally similar (homologous) macromolecules are a function of their *evolutionary distance*. However, in order to measure evolutionary distances by molecular sequencing, it is essential that the correct molecules be chosen for sequencing studies.

Criteria for Molecular Chronometers

A number of criteria define the perfect molecular chronometer. A molecular chronometer should be *universally distributed* across the group chosen for study. This facilitates the comparison of all organisms. Second, the molecule must be *functionally homologous* in each organism; functionally distinct molecules would not be expected to show sequence similarities. Third, it is crucial that the molecule have regions of sequence conservation for *aligning* the sequences for analysis. And finally, the sequence of the molecule chosen should reflect evolutionary change in the organism as a whole. In fact, the greater the phylogenetic distance to be measured, the *slower* must be the rate at which the sequence changes, as too much change over long time periods will scramble the evolutionary signal.

Many genes and proteins have been proposed as molecular chronometers. However, genes encoding *ribosomal RNAs*, central components in the translational system (∞Sections 7.14–7.16), *ATPase proteins*, the enzyme complex that synthesizes or hydrolyzes ATP (∞Section 5.12), *RecA*, the enzyme that facilitates genetic recombination (∞Section 10.6), and certain translational proteins, have provided the most insightful phylogenetic information about microorganisms. All of these molecules were probably essential for even rather primitive cells, and thus their gene sequence variations allow us to look deeply into the evolutionary past. We focus our discussion here on the most widely used of these chronometers, **ribosomal RNA** (Figure 11.11●).

Ribosomal RNAs as Evolutionary Chronometers

Ribosomal RNAs have several features that make them excellent evolutionary chronometers. Ribosomal RNAs are relatively large, functionally constant, universally distributed, and contain several regions in which the nucleotide sequence is conserved in all cells. There are three ribosomal RNA molecules, which in prokaryotes have sizes of 5S, 16S, and 23S (the "S" stands for "Svedberg," a measure of mass based on the sedimentation of a particle in a centrifuge). The focus of sequencing efforts has been on 16S rRNA, or in eukaryotes, the functionally similar, albeit slightly larger, 18S molecule. Because 16S and 18S rRNAs are part of the *small* (30S or 40S) *subunit* of the ribosome (Figure 11.11*b*), the acronym **SSU** (for small subunit) sequencing is synonymous with 16S or 18S sequencing.

A huge database of rRNA sequences exists. For example, the **Ribosomal Database Project (RDP)** contains a large collection of such sequences, now numbering over 100,000. The RDP can be accessed electronically (http://rdp.cme.msu.edu/html/) and, besides sequences, contains phylogenetic tutorials, reference citations, previews of new releases of sequences, and a host of other features.

Use of SSU rRNA as a phylogenetic tool was pioneered in the early 1970s by Carl Woese at the University of Illinois, and the method is now widely used throughout biology. For his accomplishments, Woese received the 2003 Crafoord Prize, the highest recognition for scientific achievement in biology, from the Royal Swedish Academy of Sciences.

 11.5 Concept Check

Comparisons of sequences of ribosomal RNA can be used to determine the evolutionary relationships between organisms. Phylogenetic trees based on ribosomal RNA have now been prepared for all the major prokaryotic and eukaryotic groups.

◆ Why is ribosomal RNA a good evolutionary chronometer?

◆ What is the *RDP*?

◆ Who is Carl Woese?

11.6 Ribosomal RNA Sequencing as a Tool of Molecular Evolution

The methods for obtaining ribosomal RNA sequences and generating phylogenetic trees are now quite routine, thanks to a combination of molecular biology and computer analyses. Newly generated sequences can be compared with sequences in the RDP and other genetic databases, such as GenBank (USA), DDBS (Japan), or EMBL (Germany). Then, using a treeing algorithm, a phylogenetic tree is produced that best describes the evolutionary information inherent in the sequences.

● **Figure 11.11   Ribosomal RNA.** (a) Electron micrograph of 70S ribosomes from *Escherichia coli*. (b) Parts of the ribosome; 5S, 16S, and 23S refer to different forms of RNA in the small subunit of the ribosome. (c) Primary and secondary structure of 16S ribosomal RNA (rRNA). This is the 16S rRNA from *Escherichia coli* (*Bacteria*); 16S rRNA from *Archaea* is similar in overall secondary structure (folding) but has numerous differences in primary structure (sequence). The counterpart to 16S rRNA in eukaryotes is 18S rRNA present in cytoplasmic ribosomes.

Sequencing Methodology

We begin with the assumption that we are working with a pure culture of a microorganism, although, as we will see, pure cultures are not necessary for comparative SSU rRNA sequencing. To begin the process, the polymerase chain reaction (PCR, ⌾ Section 7.9) is used to amplify the gene encoding 16S ribosomal RNA from genomic DNA. Following this, the PCR product is sequenced by the dideoxy DNA sequencing method (Sanger sequencing; ⌾ Section 7.8) (Figure 11.12●). This procedure is both rapid and specific. Using PCR primers complementary to conserved sequences in SSU ribosomal RNAs, only a tiny amount of cell material can yield a huge amount of DNA product for sequencing purposes (Figure 11.12). Once sequencing is done, the sequence data are ready for computer analyses.

Generating Phylogenetic Trees from RNA Sequences

Several different algorithms for sequence analysis and phylogenetic tree formation are available for comparative ribosomal RNA sequencing. However, regardless of which program is used, raw sequence data must first be *aligned* with previously aligned sequences using a sequence editor. Not all rRNAs are exactly the same length. Thus, during alignment, gaps must be inserted

16S rRNA gene

1. Isolate DNA

2. Heat to separate strands; add specific primers

3. Primer extension with DNA polymerase

4. Repeat above steps to obtain many copies of 16S rRNA gene

5. Run agarose gel and check for correct sized product

6. Purify and sequence PCR product

● **Figure 11.12 Ribosomal RNA sequencing of a pure culture of a microorganism using the polymerase chain reaction (PCR).** The 16S rRNA gene is amplified and then sequenced by the Sanger method (∞Section 7.8). The primers added are complementary to conserved sequences of 16S rRNA (see Figure 11.11*c*).

Organism	Sequence	Analysis
A	CGUAGACCUGAC	For A ⟶ B, three differences occur out of a total of twelve; thus $\frac{3}{12} = 0.25$
B	CCUAGAGCUGGC	
C	CCAAGACGUGGC	
D	GCUAGAUGUGCC	

(a) **Sequence alignment and analysis**

Evolutionary distance · Corrected evolutionary distance

E_D	A ⟶ B	0.25	0.30
E_D	A ⟶ C	0.33	0.44
E_D	A ⟶ D	0.42	0.61
E_D	B ⟶ C	0.25	0.30
E_D	B ⟶ D	0.33	0.44
E_D	C ⟶ D	0.33	0.44

(b) **Calculation of evolutionary distance**

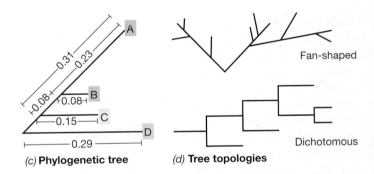

A
0.31
0.23
0.08
0.08 — B
0.15 — C
D
0.29

(c) **Phylogenetic tree**

Fan-shaped

Dichotomous

(d) **Tree topologies**

● **Figure 11.13 Preparing a phylogenetic distance tree from 16S ribosomal RNA sequences.** For illustrative purposes, only short sequences are shown. The evolutionary distance E_D in (b) is calculated as the percentage of *nonidentical* sequences between the RNAs of any two organisms. The corrected E_D is a statistical correction necessary to account for either back mutations to the original genotype or additional forward mutations at the same site that could have occurred. The tree (c) is ultimately generated by computer analysis of the data to give the best fit. (d) Different forms of phylogenetic trees can be generated, including fan-shaped and dichotomous. In either format, the total length of the branches separating any two organisms is proportional to their evolutionary distance.

where necessary in regions where one sequence is shorter than another. The aligned sequences are then imported into a treeing program and comparative analyses done.

Two widely used treeing algorithms are *distance* and *parsimony*. Using distance methods, sequences are aligned and then an **evolutionary distance** (E_D) calculated by having the computer record every position in the dataset in which there is a *difference* in sequence (Figure 11.13●). From these data, a distance matrix can be constructed that shows the E_D between any two sequences in the dataset. Following this, a statistical correction is factored into the E_D that considers the possibility that more than one change may have occurred at any given site (Figure 11.13). Once this is accounted for, a phylogenetic tree is generated in which the lengths of the lines in the tree are proportional to evolutionary distances (Figure 11.13*c, d*).

Parsimony, another popular method of phylogenetic analysis, generates evolutionary trees based on the assumption that the *minimal amount* of sequence change necessary to diverge two lineages from a common ancestor actually occurred during their evolutionary divergence. Like distance programs, this method requires summing the number of sequence differences in a particular data set, but the algorithm handles the analysis somewhat differently. Nevertheless, phylogenetic trees based on parsimony appear similar to distance trees, although the branching order within a parsimonious tree can and sometimes does differ from that in a distance tree generated from the exact same data set. Thus, no single phylogenetic tree provides the final word on the evolutionary relationships of the organisms in question. Any tree can only be considered a close approximation to the true phylogeny of a group.

11.6 *Concept Check*

Comparative ribosomal RNA sequencing is now a routine procedure involving the amplification of the gene encoding 16S rRNA, sequencing it, and analyzing the sequence in reference to other sequences. Two major treeing algorithms are distance and parsimony.

◆ What is an *evolutionary distance?*

◆ How is the *polymerase chain reaction (PCR)* of use in comparative ribosomal RNA sequencing?

11.7 Signature Sequences, Phylogenetic Probes, and Microbial Community Analyses

SSU rRNA sequencing has spawned many research tools that employ this technology. These include signature sequences and ribosomal RNA probes for use in microbial ecology and diagnostic medicine.

Signature Sequences

Computer analyses of ribosomal RNA sequences have revealed so-called **signature sequences**, short oligonucleotides unique to certain groups of organisms. For example, signature sequences specific for each of the domains of cellular life are known (Table 11.1). Moreover, signatures defining a specific group within a domain or, in some cases, a particular genus or even a single species, are also known or can be determined by computer inspection of aligned sequences. Because of their exclusivity, signature sequences are very useful. For example, signature sequences can help place newly isolated or previously misclassified organisms into their correct phylogenetic group. But the most common use of signature sequences is for the design of specific nucleic acid probes called **phylogenetic probes**.

Phylogenetic Probes and FISH

Recall that a *probe* is a strand of nucleic acid that can be labeled and used to hybridize to a complementary nucleic acid from a mixture (⚮ Section 7.7). Probes can be general or specific. For example, universal ribosomal RNA probes can be synthesized that will bind to conserved sequences in the ribosomal RNA of all organisms, regardless of domain. By contrast, specific probes can be designed that will react only with cells of species of the domain *Bacteria* because of unique signatures found in their RNA (Table 11.1). Likewise, archaeal-specific and eukaryl-specific probes will react only with species of *Archaea* or *Eukarya*, respectively. Even major groups within a domain such as genera or families (see Section 11.10) can be targeted with specific probes.

The binding of probes to cellular ribosomes can be seen microscopically when a fluorescent dye is attached to the probe (Figure 11.14●). By treating cells with the appropriate reagents, membranes become permeable and allow penetration of the probe/dye mixture. After hybridization of the probe directly to ribosomal RNA in ribosomes, the cells become uniformly fluorescent and can be observed under a fluorescent microscope (Figure 11.14, see also Figures 11.15● and 11.17). This technique has been nicknamed *FISH*, for *fluorescent in-situ hybridization*, and can be applied directly to cells in culture or in a natural environment (the term *in situ* means "in the environment"). In essence, FISH is a *phylogenetic stain*.

FISH technology is widely used in microbial ecology and clinical diagnostics. In ecology, FISH can be

Table 11.1	Signature sequences from 16S or 18S rRNA defining the three domains of life				
			Occurrence among[c]		
Oligonucleotide signatures[a]	Approximate position[b]	Archaea	Bacteria	Eukarya
CACYYG	315	0	>95	0
AAACUCAAA	910	3	100	0
AAACUUAAAG	910	100	0	100
YUYAAUUG	960	100	<1	100
CAACCYYCR	1110	0	>95	0
UCCCUG	1380	>95	0	100
UACACACCG	1400	0	>99	100
CACACACCG	1400	100	0	0

[a] Y, any pyrimidine; R, any purine.
[b] Refer to Figure 11.11c for numbering scheme of 16S rRNA.
[c] Occurrence refers to percentage of organisms examined in any domain that contain that sequence.

(a)

(b)

Norman Pace

(c)

● **Figure 11.14   Fluorescently labeled ribosomal RNA probes: Phylogenetic stains.** (a) Phase contrast photomicrograph (no probes present) of cells of *Bacillus megaterium* (rod, member of the *Bacteria*) and the yeast *Saccharomyces cerevisiae* (oval-shaped cells, *Eukarya*). (b) Same field, cells stained with a yellow-green universal rRNA probe (this probe reacts with species from any domain). (c) Same field, cells stained with a eukaryal probe (only cells of *S. cerevisiae* react). Cells of *B. megaterium* are about 1.5 μm in diameter and cells of *S. cerevisiae* are about 6 μm in diameter. Section 18.4 discusses genetic stains in more detail, including stains not based on rRNA sequences.

used for the microscopic identification and tracking of organisms directly in the environment. FISH also offers a method for assessing the composition of microbial communities directly by microscopy (Figure 11.15 and Section 18.4). In clinical diagnostics, FISH has been used for the rapid identification of specific pathogens from patient specimens. The technique circumvents the need to grow an organism in culture. Instead, microscopic examination of a specimen can confirm the presence of a specific pathogen. In clinical diagnostics this allows for more rapid diagnosis and treatment of a patient suffering from a microbial infection. By contrast, isolation and identification techniques typically take a minimum of 24 hours to complete and can take much longer.

Microbial Community Analysis

PCR-amplified ribosomal RNA genes do not need to originate from a pure culture grown in the laboratory. Using methods to be described in detail in Chapter 18, a

David A. Stahl

● **Figure 11.15   Use of phylogenetic stains to make nitrifying bacteria visible in a granule of activated sewage sludge.** (Left) Phase photomicrograph. (Right) Color photomicrograph of the same field after applying phylogenetic stains. The probe that fluoresces red is specific for a signature sequence (see Table 11.1) in the 16S rRNA of ammonia-oxidizing bacteria, while the probe that fluoresces green is specific for a sequence present only in nitrite-oxidizing bacteria. Both ammonia-oxidizing and nitrite-oxidizing bacteria are phylogenetically closely related members of the *Bacteria* (Sections 12.3 and 17.12) and carry out an interdependent series of chemolithotrophic reactions (Section 19.12).

phylogenetic snapshot of a *natural microbial community* can be taken using PCR to amplify the genes encoding SSU ribosomal RNA from all members of that community. Such genes can easily be sorted out, sequenced, and aligned. From the data, a phylogenetic tree can be generated of "environmental" sequences that show the different ribosomal RNAs present in the community. From this tree, specific organisms can be inferred even though none of them were actually cultivated or otherwise identified. Such **microbial community analyses**, as they have come to be called, are a major thrust of microbial ecology research today and have revealed many key features of microbial community structure and microbial interactions (Chapter 18 for detailed discussion of this topic).

11.7   Concept Check

Signature sequences, short oligonucleotides found within a ribosomal RNA molecule, can be highly diagnostic of a particular organism or group of related organisms. Signature sequences can be used to generate specific phylogenetic probes, useful for FISH or microbial community analyses.

◆ Using the sequence data in Table 11.1, choose a specific phylogenetic probe that would allow you to distinguish a cell of *Archaea* from a cell of *Bacteria*.

◆ How can oligonucleotide probes be made visible under the microscope? What is this technology called?

III MICROBIAL EVOLUTION

11.8 | Microbial Phylogeny Derived from Ribosomal RNA Sequences

Biologists previously grouped the living world into five kingdoms: *plants, animals, fungi, protists,* and the *monera* (prokaryotes). Molecular phylogeny, on the other hand, has revealed that the five kingdoms do not represent five major evolutionary lines. Instead, as previously outlined in Chapter 2, cellular life on Earth has evolved along *three* major lineages, called *domains,* two of which are exclusively microbial and are composed only of prokaryotic cells. The third lineage contains the eukaryotes (Figure 11.16●) and includes all of the original five kingdoms except for the monera. The two prokaryotic groups are the *Bacteria* and the *Archaea;* the eukaryotic domain is called the *Eukarya* (Figure 11.16). These terms define the three **domains** of life, the domain being the highest of biological taxons. Thus plants, animals, fungi, and protists are all *kingdoms* within the domain *Eukarya.*

The Universal Tree of Life

The **universal phylogenetic tree** (Figure 11.16) is the road map of life. It depicts the evolutionary history of the cells of all organisms and clearly reveals the three domains. The *root* of the universal tree represents a point in evolutionary history when all extant life on Earth shared a common ancestor, the **universal ancestor**.

Complete genomic sequences have confirmed the concept of the *Archaea*—the latter contain a large complement of genes that have no counterparts in *Bacteria* or *Eukarya*—as well as the three-domain concept in general. But equally important has been genome data that show that many genes are *shared* among species of *Bacteria, Archaea,* and *Eukarya.* How can these seemingly conflicting findings be reconciled in light of the universal ancestor concept?

One hypothesis that accounts for these findings is that early in the history of life, before the primary domains had diverged, *lateral (horizontal) gene transfer* (oChapter 10) was extensive. During this time, genes encoding proteins that conferred exceptional fitness, for example, genes for the core cellular functions of transcription and translation, were promiscuously transferred among a population of primitive organisms derived from a common ancestral cell. If true, this would explain why, as genome analyses have shown, *all cells,* regardless of domain, share many core functional genes in common, as would be expected if all cells shared a common ancestor (Figure 11.16).

But what about the genetic *differences* observed from complete genome sequences? It is further hypothesized that with time, barriers to unrestricted lateral transfer evolved, perhaps from the selective colonization of habitats (thereby generating reproductive isolation) or as the result of structural or enzymatic (for example, restriction

● **Figure 11.16 Universal phylogenetic tree as determined from comparative ribosomal RNA sequencing.** Only a few key organisms or lineages are shown in each domain. For detailed domain trees refer to Figures 12.1, 13.1, and 14.20. Of the three domains, two (*Bacteria* and *Archaea*) contain only prokaryotic representatives.

endonuclease) barriers that in some way prevented free genetic exchange. As a result, the previously genetically promiscuous population slowly began to sort out into the primary lines of evolutionary descent (Figure 11.16). As each lineage continued to evolve, certain unique biological traits became fixed within each group. Today, after some 3.8 GY of microbial evolution (Figure 11.10), we see the grand result of cellular evolution: three domains of cellular life that on the one hand share many common features while on the other, display distinctive evolutionary histories of their own.

Primitive Versus Modern Organisms

None of the organisms living today and depicted on the universal tree (Figure 11.16) are *primitive* organisms. All extant life forms are *modern* organisms, well adapted to and successful in their ecological niches. Certain of these organisms may indeed be *phenotypically* similar to primitive organisms, and hyperthermophilic prokaryotes (growth temperature optima >80°C) are a good case in point. For example, the organisms *Aquifex* and *Methanopyrus* branch relatively close to the root of the tree within the domains *Bacteria* and *Archaea*, respectively (Figure 11.16). Both of these organisms are capable of growth at very high temperatures (∞Sections 12.38 and 13.6), conditions that likely faced their relatives growing on a much hotter, early Earth. Nonetheless, these early branching organisms are not themselves primitive. They are modern organisms that are simply phylogenetically less derived than organisms that branch further up the tree of life, such as the Proteobacteria (*Bacteria*) or the extreme halophiles (*Archaea*) (Figure 11.16).

Bacteria

Among *Bacteria*, at least 40 divisions (called **phyla**, singular **phylum**) have been discovered thus far, and several key ones are shown in the universal tree in Figure 11.16. Many phyla have been defined from environmental sequences only (see Section 11.7 and ∞Figure 12.1). Some of the lineages in the domain *Bacteria* are those previously distinguished by some phenotypic property like morphology or physiology; the spirochetes and the cyanobacteria, respectively, are good examples of this. But most of the phyla consist of species that, although specifically related from a phylogenetic standpoint, lack strong phenotypic cohesiveness. The Proteobacteria are a good example here; the mixture of physiologies present within this group runs the gamut of known microbial physiology (∞Chapter 17). This clearly indicates that physiology and phylogeny are not necessarily linked.

Eukaryotic organelles clearly originated from within the domain *Bacteria*. As shown on the universal tree, mitochondria arose from the **Proteobacteria**, a major group of *Bacteria* (Figure 11.16), specifically from relatives like *Rhizobium* and the rickettsias. Interestingly, like mitochondria, these organisms also live an intracellular existence, either within the cells of plants (∞Sections 19.21

and 19.22) or animals (∞Sections 12.13 and 27.3). The chloroplast arose from within the cyanobacterial phylum (Figure 11.16), as could be predicted since both cyanobacteria and the chloroplast carry out oxygenic photosynthesis (∞Section 17.5).

Archaea

From a phylogenetic perspective, the domain *Archaea* consists of four phyla, the Crenarchaeota, Euryarchaeota, Korarchaeota, and Nanoarchaeota (Figure 11.16). Branching close to the root of the universal tree are hyperthermophilic Crenarchaeota, such as *Thermoproteus, Pyrolobus,* and *Pyrodictium* (Figure 11.16). They are followed by the Euryarchaeota, the methane-producing (methanogenic) prokaryotes and the extreme halophiles; *Thermoplasma*, an acidophilic, thermophilic, cell wall-less prokaryote is loosely related to this latter group (Figure 11.16).

There are some branches in the Crenarchaeota lineage (Figure 11.16 and ∞Figure 13.1) that are known only from community sampling of ribosomal genes from the environment (see Sections 11.7, 18.5, and 18.6). Interestingly, however, these sequences have originated from organisms that inhabit the open oceans, including Antarctic marine waters where temperatures are far colder than in the hot-spring or deep sea hydrothermal-vent habitats of known Crenarchaeota (∞Section 13.8). Other Crenarchaeota sequences have emerged from community sampling of soil and lake water. We discuss these cold-adapted Crenarchaeota in more detail in Section 13.8.

Like cold-adapted Crenarchaeota, an entire phylum of *Archaea*, provisionally called the *Korarchaeota*, has been identified from microbial community analysis of an unusual hot spring (Obsidian Pool) in Yellowstone National Park (Wyoming, USA). Stable mixed cultures containing Korarchaeota (identifiable through phylogenetic staining, Figure 11.17●) can be grown in the laboratory. From laboratory culture, it is clear that Korarchaeota are hyperthermophiles. Hence they may have metabolic properties

(a)

(b)

Nicole Eis

● **Figure 11.17 Identifying cells of Korarchaeota with a phylogenetic stain.** (a) Phase-contrast micrograph of an enrichment culture containing korarchaeotan cells (arrows). (b) Fluorescence photomicrograph of the same field as in (a) but treated with a phylogenetic stain designed from a signature sequence (see Section 11.7) from the 16S rRNA of the Korarchaeota. The two red long rod cells in this field are thus korarchaeotans.

similar to those of hyperthermophilic Crenarchaeota (∞Sections 13.8–13.10).

Lying basal to the Korarchaeota are the Nanoarchaeota, a group thus far represented by only a single organism, *Nanoarchaeum*. This tiny bacterium is a parasite of another archaeon, called *Ignicoccus*, and contains the smallest genome of all known prokaryotes (∞Sections 13.11 and 15.4). We consider the Nanoarchaeota and all phyla of *Archaea* in more detail in Chapter 13.

Eukarya

Phylogenetic trees of species in the domain *Eukarya* are generated from comparative sequencing of 18S rRNA, the functional equivalent of 16S rRNA, in eukaryotic cytoplasmic ribosomes. Early eukaryotes may have been similar to present-day microsporidia and diplomonads. These organisms are all obligate parasites that live in association with representatives of various groups of eukaryotes, from microorganisms to humans [for example, the pathogen *Giardia* (∞Section 28.6) is a member of the diplomonad group]. Interestingly, although these organisms contain a membrane-enclosed nucleus, microsporidia and diplomonads lack mitochondria. In this connection they resemble the type of cell that might have first accepted stable endosymbionts (see Figure 11.9 and ∞Sections 14.5 and 14.9). (Several problems have emerged in pinpointing the true phylogeny of the microsporidia and these will be discussed in Section 14.9.)

More phylogenetically derived *Eukarya* include the multicellular organisms, culminating in the largest and most structurally complex of the *Eukarya*, the plants and animals (Figure 11.16). When the fossil record is compared with the phylogenetic tree of *Eukarya*, the beginnings of rapid evolutionary radiation can be traced to about 1.5 billion years ago. Geochemical evidence suggests that this is the period in Earth's history in which significant oxygen levels had accumulated in the atmosphere (Figure 11.8). It is thus a good possibility that the onset of oxic conditions and subsequent development of an ozone shield (which would have greatly expanded the number of surface habitats available for colonization) was a major trigger for the rapid diversification of *Eukarya*. We consider the major groups of microbial *Eukarya* and their biology in Chapter 14.

 11.8 Concept Check

Life on Earth evolved along three major lines, called domains, all derived from a common ancestor. Each domain contains several phyla. Two of the domains, *Bacteria* and *Archaea*, remained prokaryotic, whereas the third line, *Eukarya*, evolved into the modern eukaryotic cell.

◆ How does the universal tree support the idea that early Earth was very hot?

◆ How does the universal tree support the hypothesis of endosymbiosis (see Figure 11.9)?

11.9 Characteristics of the Domains of Life

Although the primary domains (*Bacteria, Archaea,* and *Eukarya*) were founded on the basis of comparative ribosomal RNA sequencing—genetic criteria—each domain can also be characterized by certain *phenotypic* properties. Some of these characteristics are unique to one domain, whereas others are found in two of three domains. We discuss here an overview of major phenotypic traits of phylogenetic value.

Cell Walls

Virtually all *Bacteria* have cell walls containing *peptidoglycan* (∞Section 4.8). The only known exceptions are members of the *Planctomyces–Pirella* group (∞Section 12.28), whose cell walls are composed of protein, and the *Mycoplasma–Chlamydia* groups, which lack cell walls altogether (∞Sections 12.21 and 12.27). Peptidoglycan can thus be considered a "signature" molecule for species of *Bacteria* (when assaying for peptidoglycan, it is *muramic acid* that is actually detected because this is what is unique to peptidoglycan; see Table 11.3).

Cells of *Eukarya* and *Archaea* lack peptidoglycan. In eukaryotes, if cell walls are present, they are typically made of cellulose or chitin (∞Chapter 14). In *Archaea* various cell wall chemistries are known, from the peptidoglycan analog *pseudopeptidoglycan* to walls made of polysaccharide, protein, or glycoprotein (∞Section 4.8). Thus, great diversity exists in the chemistry of microbial cell walls, but it is the presence or absence of *peptidoglycan* that is most useful in distinguishing *Bacteria* from *Archaea*.

Lipids

The chemical nature of membrane lipids is a key feature for differentiating *Archaea* from *Bacteria*. *Bacteria* and *Eukarya* synthesize membrane lipids with a backbone consisting of fatty acids bonded in *ester* linkage to a molecule of glycerol (Figure 11.18●). Although the nature of the fatty acid can be highly variable, the *ester link* to glycerol is the key defining feature. By contrast, archaeal lipids consist of *ether-linked* molecules (Figure 11.18). In ester-linked lipids, the fatty acids are usually straight-chain (linear) molecules, whereas in *Archaea*, long-chain, branched hydrocarbons, either of the phytanyl or biphytanyl type, are present in place of fatty acids and are bonded by ether linkage to glycerol.

A few exceptions to the phylogenetic pattern of lipids are known. For example, the thermophilic sulfate-reducing bacterium *Thermodesulfobacterium* (∞Section 12.36) and a few other sulfate-reducing bacteria (∞Section 12.18), as well as *Propionibacterium* species, contain ether-linked lipids. However, the converse is not true. No species of *Archaea* has been shown to contain ester-linked lipids.

In terms of overall membrane structure, the membranes of some *Archaea*, especially hyperthermophilic

Ester

CH$_2$OH O ┌Fatty acid side chain┐
HC─O─C─CH$_2$─(CH$_2$)$_{13}$─CH$_3$
CH$_2$OH

Bacteria, Eukarya

Ether

Phytanyl side chain

CH$_2$OH   H   H   H   H
HC─O─C─CH$_2$─C─(CH$_2$)$_3$─C─(CH$_2$)$_3$─C─(CH$_2$)$_3$─C─CH$_3$
CH$_2$OH   CH$_3$ CH$_3$ CH$_3$ CH$_3$

Archaea

● **Figure 11.18 Lipids in *Bacteria*, *Eukarya*, and *Archaea*.** In *Bacteria* and *Eukarya*, lipids contain fatty acids (palmitic acid is shown) bonded by *ester* linkages to glycerol. In *Archaea*, the side chains are branched hydrocarbons (phytanyl, C$_{20}$, is shown) bonded by *ether* linkages to glycerol. Phytanyl is synthesized from isoprene (∞Figure 4.18c).

species, form into a lipid *monolayer* instead of a lipid *bilayer* (∞Section 4.5 and Figure 4.20). A lipid monolayer is less likely than a lipid bilayer to denature at the high temperatures at which these organisms grow. The ether linkage between glycerol and the hydrophobic side chains (Figure 11.18) combined with monolayer membrane construction protect the membranes of these extremophilic *Archaea* from the harsh conditions of their habitats.

RNA Polymerase

Transcription is carried out by DNA-dependent RNA polymerases in all organisms; DNA is the template, and RNA is the product. Cells of *Bacteria* contain a single type of RNA polymerase of rather simple quaternary structure. This is the classic RNA polymerase containing *four* polypeptides, α, β, β′, and σ, combined in a ratio of 2:1:1:1, respectively, in the active polymerase (Figure 11.19●) (∞also Section 7.10).

Archaeal RNA polymerases are structurally more complex than those of *Bacteria*. The RNA polymerases of species of *Archaea* contain *eight* or more polypeptides, more closely resembling the pattern in *Eukarya* (Figure 11.19). The major RNA polymerase of eukaryotes (there are three) contains 10–12 polypeptides, and the relative sizes of the peptides coincide most closely with those from species of hyperthermophilic *Archaea* (Figure 11.19). Thus, in terms of phylogenetic signatures, the α$_2$ββ′σ polymerase is diagnostic of *Bacteria*; RNA polymerases with more than four subunits are not phylogenetically definitive, except to say that they do not belong to *Bacteria*.

Features of Protein Synthesis

Because of differences in ribosomal RNA sequences and several protein synthesis factors, it is not surprising that certain aspects of the protein-synthesizing machinery differ in representatives of the three domains. Although ribosomes of *Archaea* and *Bacteria* are the same size (70S,

● **Figure 11.19 RNA polymerase and phylogeny.** RNA polymerases from representatives of the three domains: Ec, *Escherichia coli* (*Bacteria*), Hs, *Halobacterium salinarum* (Euryarchaeota, *Archaea*), Sa, *Sulfolobus acidocaldarius* (Crenarchaeota, *Archaea*), and Sc, *Saccharomyces cerevisiae* (*Eukarya*). The purified components of the RNA polymerases have been separated by electrophoresis on a polyacrylamide gel. The largest polypeptide subunits are on the top, and the smallest subunits are on the bottom. Only members of the *Bacteria* contain the simple (four-polypeptide) RNA polymerase.

as compared with the 80S ribosomes in the cytoplasm of eukaryotes), several steps in archaeal protein synthesis more strongly resemble those in eukaryotes than in *Bacteria*. Recall that translation always begins at a unique codon, the so-called *start codon*. In *Bacteria* this start codon (AUG) calls for the incorporation of an initiator tRNA containing a modified methionine residue, *formyl*-methionine (∞Section 7.16). By contrast, in eukaryotes and in *Archaea*, the initiator tRNA carries an *unmodified* methionine.

The exotoxin produced by *Corynebacterium diphtheriae* is a potent inhibitor of eukaryotic protein synthesis because it ADP-ribosylates (adds ADP-ribose to) an elongation factor required to translocate the ribosome along the mRNA; the modified elongation factor is inactive (∞Section 21.10). Diphtheria toxin also inhibits protein synthesis in species of *Archaea* but not in species of *Bacteria*.

Most antibiotics that specifically affect protein synthesis in *Bacteria* do not affect archaeal or eukaryal protein synthesis. The sensitivity of representatives of the three domains to various protein synthesis inhibitors is shown in Table 11.2 (∞also Section 7.16), where various antibiotics are grouped according to their modes of action in blocking protein synthesis in various domains of organisms.

Table 11.2 **Sensitivity of representatives of the three domains to various protein synthesis inhibitors**[a]

| | | Archaea | | Bacteria | Eukarya |
| | | Euryarchaeota | Crenarchaeota | | |
Antibiotics	**Mode of action**	*Methanobacterium thermoautotrophicum*	*Sulfolobus acidocaldarius*	*Escherichia coli*	*Saccharomyces cerevisiae*
Fusidic acid, sparsomycin	Inhibits elongation steps	+	−	+	+
Anisomycin, narciclasine	Inhibits peptidyl transfer	+	−	−	+
Cycloheximide	Blocks initiation	−	−	−	+
Erythromycin, streptomycin, chloramphenicol	Increases error frequencies and other effects	−	−	+	−
Virginiamycin, pulvomycin	Inhibits elongation steps	+	−	+	−
Neomycin, puromycin	Causes premature termination	+	+	+	+
Rifamycin	Inhibits β subunit of RNA polymerase	−	−	+	−

[a] A + indicates that protein synthesis (and growth) is inhibited.

Collectively, these results suggest that ribosomal proteins from cells of *Archaea* and *Eukarya* are functionally more similar to each other than they are to ribosomal proteins from *Bacteria*. This further supports the position of the domains relative to one another in the universal tree of life (Figure 11.16).

Other Features Defining the Domains

A number of other phenotypic features, physiological and otherwise, can be used to differentiate organisms at the domain level; Table 11.3 summarizes these. When examining this table it should be understood that not all features are universally present in a given domain. For example, chlorophyll-based photosynthesis is characteristic of only some representatives of the *Bacteria* and the *Eukarya* (and in the latter only because of photosynthetic endosymbiotic *Bacteria*, see Figure 11.9). By contrast, other distinguishing features, such as the presence of peptidoglycan in the cell walls of *Bacteria*, may be universal, or nearly so.

 11.9 *Concept Check*

Although the three domains of living organisms were originally defined by ribosomal RNA sequencing, subsequent studies have shown that they differ in many other ways. In particular, the *Bacteria* and *Archaea* differ extensively in cell wall and lipid chemistry and in features of transcription and protein synthesis.

◆ How do the lipids from *Bacteria* and *Eukarya* differ from those of the *Archaea*?

◆ Organisms from which domain of prokaryotes synthesize RNA polymerases that most closely resemble those of the *Eukarya*? What phylogenetic evidence supports this observation?

 IV **MICROBIAL TAXONOMY AND ITS RELATIONSHIP TO PHYLOGENY**

A major discipline in microbiology is microbial classification, or *taxonomy*. Classification allows microbiologists to contrast relationships among different microorganisms and develop systematic procedures for their nomenclature. We consider the basic concepts of bacterial taxonomy in the next four sections.

11.10 Classical Taxonomy

Taxonomy, the science of classification, consists of two major subdisciplines, *identification* and *nomenclature*. It is important to distinguish between *taxonomy* and the main topic of this chapter up to this point, *phylogeny*, for the terms really mean different things.

Bacterial taxonomy traditionally relies on *phenotypic* analyses. These include what an organism looks like, its energy metabolism, its enzymes, and other properties. Although the phylogeny of prokaryotes has emerged from *genotypic* analyses as discussed in the previous sections, phenotypic analyses continue to play an important role in bacterial identification and classification. This is especially true of applied situations where identification

Table 11.3 Summary of major differential features among *Bacteria, Archaea,* and *Eukarya*[a]

Characteristic	*Bacteria*	*Archaea*	*Eukarya*
Morphological and Genetic			
Prokaryotic cell structure	Yes	Yes	No
DNA present in covalently closed and circular form	Yes	Yes	No
Histone proteins present	No	Yes	Yes
Membrane-enclosed nucleus	Absent	Absent	Present
Cell wall	Muramic acid present	Muramic acid absent	Muramic acid absent
Membrane lipids	Ester-linked	Ether-linked	Ester-linked
Ribosomes (mass)	70S	70S	80S
Initiator tRNA	Formylmethionine	Methionine	Methionine
Introns in most genes	No	No	Yes
Operons	Yes	Yes	No
Capping and poly-A tailing of mRNA	No	No	Yes
Plasmids	Yes	Yes	Rare
Ribosome sensitivity to diphtheria toxin	No	Yes	Yes
RNA polymerases (see Figure 11.19)	One (4 subunits)	Several (8–12 subunits each)	Three (12–14 subunits each)
Transcription factors required (∞Section 7.11)	No	Yes	Yes
Promoter structure (∞Sections 7.10 and 7.11)	−10 and −35 sequences (Pribnow box)	TATA box	TATA box
Sensitivity to chloramphenicol, streptomycin, and kanamycin	Yes	No	No
Physiological/Special Structures			
Methanogenesis	No	Yes	No
Dissimilative reduction of S^0 or SO_4^{2-} to H_2S, or Fe^{3+} to Fe^{2+}	Yes	Yes	No
Nitrification	Yes	No[b]	No
Denitrification	Yes	Yes	No
Nitrogen fixation	Yes	Yes	No
Chlorophyll-based photosynthesis	Yes	No	Yes (in chloroplasts)
Rhodopsin-based energy metabolism	Yes	Yes	No
Chemolithotrophy (Fe, S, H_2)	Yes	Yes	No
Gas vesicles	Yes	Yes	No
Synthesis of carbon storage granules composed of poly-β-hydroxyalkanoates	Yes	Yes	No
Growth above 80° C	Yes	Yes	No
Growth above 100°C	No	Yes	No

[a] Note that for many features only particular representatives within a domain show the property.
[b] Environmental genomics studies of prokaryotes in marine waters strongly suggest that nitrifying *Archaea* exist (∞Section 18.6).

may be an end in itself; for example, in clinical diagnostic microbiology. In this section, we discuss classical bacterial taxonomy and in the next section summarize some molecular methods that have been found useful in the classification of bacteria.

In classical bacterial taxonomy, several phenotypic characteristics are assessed, and the data are used to group the organisms up the taxonomic ladder from *species* to *domain* (Tables 11.5 and 11.6). Characteristics of taxonomic value that are widely used in this regard include morphology, nutrition and physiology, and habitat. Table 11.4 breaks these major categories down into specific topics of taxonomic value.

To identify an organism, one must assess several of its phenotypic properties, from general to specific. An example of this is shown using morphological and phys-

iological tests in Figure 11.20● (see also Table 11.5). By collecting data for several individual characteristics, the list of possible organisms is narrowed until a positive identification can be made (Figure 11.20).

GC Ratios

One property that is informative in drawing taxonomic conclusions is an organism's genomic DNA **GC ratio**. The GC ratio is defined as the percentage of guanine plus cytosine in an organism's DNA. This ratio can be determined in several ways, including by measuring the melting temperature of the DNA (∞Section 7.2) or by chromatographic methods.

GC ratios vary over a wide range, with values as low as 20% and as high as nearly 80% known among prokary-

Table 11.4	Some phenotypic characteristics of taxonomic value
Major category	**Components**
I. Morphology	Shape; size; Gram reaction; arrangement of flagella, if present
II. Motility	Motile by flagella; motile by gliding; motile by gas vesicles; nonmotile
III. Nutrition and Physiology	Mechanism of energy conservation (phototroph, chemoorganotroph, chemolithotroph); relationship to oxygen; temperature, pH, and salt requirements/tolerances; ability to use various carbon, nitrogen, and sulfur sources; growth factor requirements
IV. Other factors	Pigments; cell inclusions, or surface layers; pathogenicity; antibiotic sensitivity

otes; this is a somewhat broader range than for eukaryotes (Figure 11.21●). Base compositions of DNA have been determined for a wide variety of organisms, and knowledge of an organism's GC ratio can be useful, depending on the situation. For example, two organisms can have identical GC ratios and yet turn out to be quite unrelated (both taxonomically and phylogenetically), because a variety of base *sequences* is possible with DNA of a single base *composition*. In this case, the identical GC ratios are taxonomically meaningless. By contrast, if two organisms' GC ratio differs by greater than about 5%, they will share few DNA sequences in common and are therefore unlikely to be closely related.

 11.10 Concept Check

Conventional bacterial taxonomy places heavy emphasis on analyses of phenotypic properties of the organism. Determining the guanine plus cytosine (GC) base ratio of the DNA of the organism can be part of this process.

◆ List three phenotypic properties that can be used to distinguish bacteria.

I. Isolation and microscopy

Isolation ⟹ Pure culture ⟹ Gram reaction/ morphology

II. General physiology

Gram-negative rod ⟹ Facultative ⟹ Ferments lactose to acid/gas

III. Detailed physiology

Facultative lactose fermenter ⟹ Perform series of biochemical tests ⟹ Positive: indole, methyl red, mucate; Negative: citrate, Voges-Proskauer, H₂S

IV. Conclusion ⟹ *Escherichia coli*

● **Figure 11.20 Example of methods that would be used for identification of a newly isolated enteric bacterium.** This scheme uses classical microbiological methods (the example given shows the procedures that would be used for identifying *Escherichia coli*). Note that most of the analyses here require that the organisms be grown in pure culture and that solely phenotypic criteria be used in the identification. A description of biochemical tests is presented in Chapter 24 (◯○ Section 24.2, Table 24.3, and Figure 24.7).

◆ How is it possible for two organisms to have very similar DNA GC ratios yet share few genes in common?

11.11 Chemotaxonomy

Molecular taxonomy, or **chemotaxonomy** as it is also called, involves molecular analyses of one or more constituents in the cell. Chief among chemotaxonomic methods that have been used routinely are *genomic DNA:DNA hybridization, ribotyping, multilocus sequence typing,* and *lipid profiling.*

DNA:DNA Hybridization

A GC base ratio describes the percentage of each nucleotide present in genomic DNA of a given species but gives absolutely no information on the *sequence* of those nucleotides. Sequences are critical, of course, because if two organisms have many of the same nucleotide sequences in their DNAs, they likely contain many highly similar (if not identical) genes. Two DNAs would thus be expected to *hybridize* to one another in proportion to the *similarities* in their gene sequences. **Genomic hybridization** measures the degree of sequence similarity in two DNAs and is useful for differentiating very closely related organisms where rRNA sequencing may fail to be definitive.

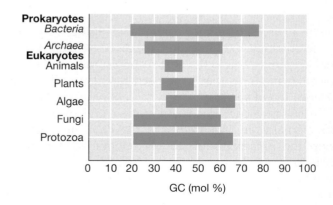

● **Figure 11.21 Ranges of genomic DNA base composition of various organisms.** Note that the greatest range of GC ratios exists with *Bacteria.*

We discussed the theory and methodology of nucleic acid hybridization in Section 7.7. In an actual hybridization experiment, DNA isolated from one organism is made radioactive with ^{32}P or ^{3}H, sheared to a relatively small size, heated to denature, and mixed with an excess of unlabeled DNA prepared in the same way from a second organism (Figure 11.22●). The DNA mixture is then cooled to allow it to reanneal and double-stranded DNA is separated from any remaining un-

Organisms to be compared:

Organism 1 **Organism 2**

DNA preparation

DNA

Shear and label (-(P))

DNA

Shear DNA

Heat to denature

(a)

Hybridization experiment:

Mix DNA from two organisms—unlabeled DNA is added in excess:

1 x 1 ⟶

Hybridized DNA —

1 x 2 ⟶

Hybridized DNA —

Unhybridized DNA

(b)

Results and interpretation:

Same species | Same genus, but different species | Different genera

100 75 50 25 0
Percent hybridization

(c)

1 x 1
100%
Same strain (control)

1 x 2
25%
1 and 2 are likely different genera

● **Figure 11.22 Genomic hybridization as a taxonomic tool.** (a) DNA is isolated from test organisms. One of the DNAs is labeled (shown here as radioactive phosphate in the DNA of Organism 1). (b) Actual hybridization experiment. Excess unlabeled DNA is added to prevent labeled DNA from reannealing with itself. Following hybridization, hybridized DNA is separated from unhybridized DNA before measuring radioactivity in the hybridized DNA only. (c) Results. Radioactivity in the control (Organism 1 DNA hybridizing to itself) is taken as the 100% hybridization value.

hybridized DNA. Following this, the amount of radioactivity in the hybridized DNA is determined and compared with the control, which is taken as 100% (Figure 11.22). In addition to radioactivity, a variety of nonradioactive DNA labels are available, and these have the advantage that the hybridization experiment generates no radioactive wastes.

There is no fixed convention as to how much hybridization between two DNAs is necessary to assign two organisms to the same taxonomic rank. However, hybridization values of 70% or greater are recommended for considering two isolates to be of the same *species*. By contrast, values of at least 25% are required to argue that two organisms should reside in the same *genus* (see Section 11.12 for working definitions of a bacterial genus and species). Hybridization of DNAs from more distantly related organisms, for example, *Clostridium* (gram-positive) and *Salmonella* (gram-negative), will hybridize at only background levels, 10% or less (Figure 11.22).

DNA:DNA hybridization is a sensitive method for revealing subtle differences in the genes of two organisms and is therefore useful for differentiating *closely related* organisms. Indeed, a common application of genomic hybridization in taxonomic studies is to test two organisms that are suspected to be different species even though SSU ribosomal RNA sequencing and phenotypic analyses may have failed to reveal significant differences between them.

Where available, complete genome sequences make hybridization analyses unnecessary. When the genome of a given bacterium is sequenced, its complete assortment of genes can be compared with those of any other sequenced bacterium (∞ Chapter 15). At present, however, with only several hundred prokaryotic genomes completely sequenced, such comparisons are not possible in most cases. In future years, microbiologists will have the ultimate tool for molecular taxonomy in complete genome sequences. However, until taxonomy is based on complete genome sequence comparisons, hybridization will remain useful for differentiating close taxonomic relationships.

Ribotyping

Ribotyping is a technique for bacterial identification that employs some of the methods previously discussed for ribosomal RNA-based phylogenetic characterizations (see Section 11.6). However, unlike comparative *sequencing* methods, ribotyping does not involve sequencing. Instead, it measures the unique *pattern* that is generated when DNA from an organism is digested by a restriction enzyme and the fragments are separated and probed with a ribosomal RNA probe (Figure 11.23a●).

Because differences in ribosomal RNA sequences between two organisms translate into the presence or absence of specific restriction enzyme cut sites (see

(a)

(b)

● **Figure 11.23 Ribotyping and multilocus sequence typing (MLST).** (a) Ribotype results from four different lactic acid bacteria. The pattern of DNA fragments generated from restriction enzyme digestion of DNA taken from a colony of each bacterium and then probed with 16S rRNA genes is unique to a species or even to strains within a species. The patterns generated on a gel with known organisms are digitized and stored in a database for comparisons in identifying environmental or clinical isolates. Variations in both *position* and *intensity* of the bands are important in identification. (b) Steps in multilocus sequence typing (MLST).

Section 7.7 for a discussion of restriction enzymes), the restriction pattern of a particular bacterial species is unique (Figure 11.23*a*). In fact, ribotyping is so specific it has been given the nickname "molecular fingerprinting," because a unique series of bands appears for virtually any organism.

In practice, ribotyping starts with DNA from a colony or liquid culture. Using PCR, genes for 16S rRNA and related molecules are amplified, treated with one or more restriction enzymes, separated by electrophoresis, and then probed (∞Chapter 7 for discussion of these methods). The pattern generated from the fragments of DNA on the gel is then digitized and a computer used to make comparisons of this pattern with patterns from reference organisms available from a database (Figure 11.23*a*). Ribotyping is both a *rapid* (since it bypasses the actual sequencing, sequence alignment, and analysis requirements of ribosomal RNA sequencing methods) and *specific* method of bacterial identification. For these reasons, ribotyping has found

many applications in clinical diagnostics and for the microbial analyses of food, water, and beverages.

Multilocus Sequence Typing

One of the limitations of both ribosomal RNA sequencing and ribotyping is that analyses focus on only a single gene. **Multilocus sequence typing** (**MLST**) circumvents this problem and is a powerful technique for characterizing strains of organisms within a species.

MLST involves sequencing fragments of six to seven "housekeeping genes" from an organism and comparing these with the same gene set from *different strains* of the same organism. Recall that housekeeping genes encode essential functions in cells and are located on the chromosome rather than on a plasmid (∞Section 7.4). For each gene, an approximately 450-bp sequence is amplified using PCR and is then sequenced. The comparative sequencing data is then expressed in a dendrogram (Figure 11.23*b*).

In MLST, strains with *identical sequences* for a given gene are said to have the same **allele** at that gene and are assigned the same number for that gene. A given gene can have many different alleles, and typically, 10–30 alleles exist for a given gene within a group of strains of the same species. Each strain eventually ends up with a series of numbers—its *multilocus sequence type*—characteristic of that strain. The relatedness between each sequence type is then expressed in a dendrogram of linkage distances that vary from 0 (strains are identical) to 1 (strains are only distantly related) (Figure 11.23*b*).

MLST has sufficient resolving power to distinguish even very closely related strains. It is thus a better instrument than rRNA sequencing for differentiating organisms below the species level. With seven genes in the analysis and 20 alleles per gene it has been calculated that *several billion* distinct genotypes can be resolved with MLST. By contrast, MLST is not useful for comparing organisms above the species level because its resolution is too sensitive to yield a meaningful phylogeny of higher order taxa.

Thus far, MLST has made an impact primarily in clinical microbiology for differentiating strains of a particular pathogen. This is serious business when it is considered that within a species—*Escherichia coli*, for example—some strains, such as strain K-12, may be harmless, whereas others, such as strain O157:H7, can cause serious and even fatal infections (∞ Section 29.8). MLST is also quite useful for epidemiological studies. For example, MLST can track a virulent strain of a bacterial pathogen as it moves through a population and can unequivocally identify the strain even if many closely related strains are also present in the population. In addition, since the technique is PCR based, MLST can be used to test clinical samples without first isolating the organism in laboratory culture. That is, if DNA from a strain of a specific pathogen can be obtained from a clinical sample, MLST analyses can be done.

MLST analyses have revealed some interesting genetic patterns in bacteria. For example, some prokaryotes are essentially clonal, with very little MLST variability apparent. This is typical of strains of *Staphylococcus aureus*, for example. Other organisms, such as *Neisseria meningitis*, are only weakly clonal, and show great variability by MLST analyses. These results indicate that for whatever reason, organisms like *N. meningitidis* have undergone much greater lateral gene flow in the past than has *S. aureus*. These differences undoubtedly reflect differences in the ecology of these organisms and the factors that make them competitively successful. Such information may be useful for developing drug strategies and vaccines based on the genetic stability or instability of a particular pathogen.

Fatty Acid Analyses: FAME

Another popular method of bacterial identification is through characterization of the types and proportions of *fatty acids* present in cytoplasmic membrane and outer membrane (gram-negative bacteria, ⟳ Section 4.9) lipids of cells. This technique has been nicknamed *FAME*, for *fatty acid methyl ester* and is in widespread use in clinical, public health, and food and water inspection laboratories where the identification of pathogens or other bacterial hazards needs to be done on a routine basis.

The fatty acid composition of prokaryotes can be highly variable, including differences in the fatty acids in terms of their chain length, the presence or absence of double bonds, rings, branched chains, or hydroxy groups (Figure 11.24*a*●). Hence, a fatty acid profile can often identify a particular bacterial species. For actual analyses, fatty acids, extracted from cell hydrolysates of a culture grown under standardized conditions, are chemically derivatized to form their corresponding methyl esters. These now volatile derivatives are then identified by gas chromatography. A chromatogram showing the types and amounts of fatty acids from the unknown bacterium is then compared with a database containing the fatty acid profiles of thousands of reference bacteria grown under the same conditions. The best matches to that of the unknown are then selected (Figure 11.24*b*).

As a chemotaxonomic tool, FAME does have some drawbacks. In particular, FAME analyses require rigid standardization, since fatty acid profiles of an organism can vary as a function of temperature, growth phase (exponential versus stationary), and to a lesser extent, growth medium. Thus, for consistent results, it is necessary to grow the unknown organism on a specific medium and at a specific temperature in order to compare its fatty acid profile with those of organisms from the database that have been grown in the same way. For many organisms, this is impossible, of course, and thus FAME analyses are limited to those organisms that can be grown under the specified conditions.

Classes of Fatty Acids in *Bacteria*

Class/Example — Structure of example

I. *Saturated:* tetradecanoic acid

II. *Unsaturated:* omega-7-cis hexadecanoic acid

III. *Cyclopropane:* cis 7, 8 methylene hexadecanoic acid

IV. *Branched:* 13-methyltetradecanoic acid

V. *Hydroxy:* 3-hydroxytetradecanoic acid

(a)

(b)

● **Figure 11.24 Fatty acid methyl ester (FAME) analysis in bacterial identification.** (a) Classes of fatty acids in *Bacteria*. Only a single example is given of each class, but in actuality, more than 200 different fatty acids have been discovered from bacterial sources. A methyl ester contains a methyl group (CH_3) in place of the proton on the carboxylic acid group (COOH) of the fatty acid. (b) Procedure. Each peak from the gas chromatograph is due to one particular fatty acid methyl ester and the peak height is proportional to the amount.

 11.11 Concept Check

Molecular taxonomy includes molecular analyses of specific cell components. These include, among others, DNA:DNA hybridization, ribotyping, multilocus sequence typing, and fatty acid analyses.

◆ Hybridization of less than 10% between two organisms' DNA indicates that they are _____.

◆ How does ribotyping differ from 16S ribosomal RNA gene sequencing? From MLST?

◆ What is FAME analysis?

11.12 The Species Concept in Microbiology

In the world of plants and animals, a **species** is defined as a population (1) that can naturally interbreed and produce fertile offspring, and (2) that is reproductively isolated from other species. However, this definition does not hold for prokaryotes. Prokaryotes are haploid and reproduce asexually. Concepts like "the production of fertile offspring" thus have no meaning. Nevertheless, microbiologists traditionally refer to "species" of bacteria and regularly give new isolates of bacteria genus names and species names, the latter called *epithets*.

So what *is* a bacterial species? Although the concept of a prokaryotic species is still evolving, microbiologists today are using SSU ribosomal RNA sequencing, genomic hybridization, and the phenotypic tools discussed in Section 11.11 for discerning prokaryotic species.

Bacterial Species and Higher Taxa

It has been proposed that a prokaryote whose 16S ribosomal RNA sequence differs by more than 3% from that of all other organisms (that is, the sequence is less than 97% identical to any other sequence in the databases), should be considered a new species. This is not an arbitrary number. Support for this proposal includes the observation that genomic DNA from two prokaryotes whose 16S rRNA sequences are less than 97% identical typically hybridize to less than 70%, a minimal value considered evidence for two organisms being of the same species (see Section 11.11 and Figure 11.22). This is depicted in Figure 11.25●.

The data of Figure 11.25 also shows how in some cases 16S sequencing lacks the resolving power of genomic DNA hybridization for differentiating organisms at the species level. Although an organism whose 16S sequence differs from that of all other organisms by more than 3% will likely turn out to be a new species by DNA hybridization criterion as well, some organisms that have very similar (or even identical) ribosomal RNA sequences have genomes that are quite unrelated (Figure 11.25). Thus, in cases where SSU sequencing shows greater than 97% sequence identity, genomic hybridization is an important taxonomic tool for identifying new species.

A new species is usually defined from the characterization of several strains. The species concept is important in microbiology because it gives the collected strains formal taxonomic identity. Groups of species are then collected into genera (singular, **genus**). What constitutes a genus by molecular criteria is more a matter of judgment than for species, but 16S sequence *differences* of more than 5% from all other organisms (this corresponds to less than 95% sequence *identity*) have been taken as support for an organism constituting a new genus. Groups of genera are collected into **families**, families into **orders**, orders into **classes**, and so on up to the high-

● **Figure 11.25 Relationship between 16S ribosomal RNA sequence similarity and genomic DNA:DNA hybridization between different pairs of organisms.** These data are the results from several independent experiments with various species of the domain *Bacteria*. Points in the orange box represent combinations where 16S sequence similarity and genomic hybridization were both very high; thus, in each case, the two organisms tested were clearly the same species. Points in the green box represent combinations where different species were detected, and both methods show this. The blue box shows examples of organisms that were different species as measured by genomic DNA hybridization but not by 16S sequencing. Note that above 70% DNA hybridization, no 16S similarities were found of less than 97%. Data redrawn from Rosselló-Mora, R., and R. Amann. 2001. *FEMS Microbiol. Revs.* 25:39–67.

est level taxon, the **domain** (Tables 11.4 and 11.5). Almost 6500 species of *Bacteria* and *Archaea* were formally recognized as of 2004 (Table 11.5).

In the identification and naming of a newly isolated organism, it is essential that the organism satisfy all the taxonomic criteria of ranks *above* its species designation. Thus, in the example given in Table 11.4 where the taxonomic hierarchy for the phototrophic bacterium *Allochromatium warmingii* is shown, all species of the genus *Allochromatium* must be rod-shaped purple sulfur gram-negative *Bacteria*; if one of these criteria is not satisfied, the organism is not considered a species of *Allochromatium*. Moreover, the genus *Allochromatium* must meet all of the criteria defining the family Chromatiaceae, and so on up the taxonomic ladder. In other words, as one *descends* the taxonomic hierarchy from the level of domain to that of species, the criteria used to distinguish two organisms become *less general* and *more specific* (Table 11.4).

Bacterial Speciation

How do new bacterial species arise? This is currently a hotly debated point among microbiologists. Nevertheless, models of bacteria speciation have been proposed, and we consider a popular one here. Imagine a population of cells originating from the growth of a single cell and that occupies a particular ecological niche. If cells in this population share a particular resource (for example, a key nutrient), this population of cells can be called an

Table 11.5	Taxonomic hierarchy for the purple sulfur bacterium *Allochromatium warmingii*		
Taxonomic division	**Name**	**Properties**	**Confirmed by**
Domain	*Bacteria*	Prokaryotic cells; ribosomal RNA sequences typical of *Bacteria*	Microscopy; 16S ribosomal RNA sequencing; presence of unique biomarkers, for example, peptidoglycan
Phylum	Proteobacteria	Ribosomal RNA sequence typical of Proteobacteria	16S rRNA sequencing
Class	Gammaproteobacteria	Gram-negative bacteria; rRNA sequence typical of Gammaproteobacteria	Gram-staining, microscopy
Order	Chromatiales	Phototrophic purple bacteria	Characterizing pigments (⌀Figure 17.3)
Family	Chromatiaceae	Purple sulfur bacteria	Ability to oxidize H_2S and store S^0 within cells; observe culture microscopically for presence of S^0 (see photo)
Genus	*Allochromatium*	Rod-shaped purple sulfur bacteria	Microscopy (see photo)
Species	*warmingii*	Cells 3.5–4.0 μm × 5–11 μm; store sulfur mainly in poles of cell (see photo)	Measure cells in microscope using a micrometer; look for position of S^0 globules in cells (see photo)

Sulfur (S^0) globules

Photograph of cells of *Allochromatium warmingii*

Norbert Pfennig

ecotype. Different ecotypes can coexist in a habitat, but each is only successful within its niche in the habitat.

Errors in DNA replication (mutations) occur at a low but regular frequency (⌀Section 10.3). Within a given ecotype, the occurrence of a mutation that confers increased fitness on the ecotype (an adaptive mutation), triggers periodic selection. When this occurs, the old ecotype becomes purged by a population of the new ecotype (Figure 11.26●). In this way, populations of cells eventually "move away" from each other in a genetic sense. Repeated rounds of mutation and selection eventually lead to an ecotype that is sufficiently distinct genetically from the original ecotype to be recognized as a new species (Figure 11.26). Note that this series of events *within* an ecotype has no effect on *other* ecotypes, since different ecotypes do not compete for the same resources (Figure 11.26).

The model for speciation shown in Figure 11.26 is based solely on the assumption of *vertical* (mother to daughter) gene flow. However, we know that bacterial speciation is also affected to some degree by **lateral (horizontal) gene transfer**. Lateral flow is the transfer of genes *between* species by conjugation, transduction, and transformation (⌀Chapter 10). Prokaryotes are sexually promiscuous and can exchange genes across broad phylogenetic lines. Thus, a new genetic capability in an ecotype may arise from genes obtained from cells of another ecotype rather than from mutation and selection.

The extent of horizontal gene flow among prokaryotes is variable. Genome sequencing has revealed examples where lateral flow has apparently been rampant and others where it has been scarce (⌀Section 15.8). In addition, multilocus sequence typing (see Section 11.11) has revealed that genetic exchange *within* some species is widespread, while virtually nonexistent within others. Despite the impact of lateral gene flow, *speciation* is thought to be driven primarily by mutation and periodic selection (Figure 11.26) rather than by lateral transfer. This is because the genes obtained by lateral transfers are typically few in number, confer only temporary benefits, and can often be lost if the selective pressure to retain them decreases.

How Many Prokaryotic Species Are There?

The result of nearly 4 billion years of bacterial evolution (see Figures 11.8 and 11.10) is the prokaryotic world we see today. Microbial taxonomists agree that no firm estimate of the number of prokaryotic species can be given at present. However, they also agree that in the final analysis, this number will be very large. Several thou-

One microbial habitat

Ecotype II

Ecotype III

Ecotype I

Cell containing an adaptive mutation

Periodic selection

Adaptive mutant survives. Original Ecotype I wild-type cells out competed

Population of mutant Ecotype I

Repeat process many times

New species of Ecotype I

● **Figure 11.26 A model for bacterial speciation.** Several ecotypes can coexist in a single microbial habitat, each occupying their own prime ecological niche. When a beneficial mutation occurs within an ecotype, the cell containing that mutation will eventually form a population that will replace the original ecotype. As this occurs repeatedly within a given ecotype, a genetically distinct population of cells arises that represents a new species. Because other ecotypes do not compete for the same resources, they are unaffected by genetic and selection events that occur outside of their ecotype or habitat.

Table 11.6	**Taxonomic ranks and numbers of known prokaryotic species**[a]		
Rank	*Bacteria*	*Archaea*	**Total**
Domains	1	1	2
Phyla	25	4[a]	29
Classes	34	9	43
Orders	78	13	91
Families	230	23	243
Genera	1227	79	1306
Species	6740	289	7029

[a] Numbers represent validly named genera and species of *Bacteria* and *Archaea* as of 2005. The phyla category for *Archaea* includes the Korarchaeota and the Nanoarchaeota, not yet officially recognized phyla.

Source: Garrity, G.M., Libum, T.G., and Bell, J.A. 2005. *Bergey's Manual of Systematic Bacteriology*, 2d ed., Vol. 2, part A, pp159–220. Springer-Verlag, New York.

🛑 *11.12 Concept Check*

The species concept applies to prokaryotes as well as eukaryotes, and a similar taxonomic hierarchy exists, with the domain as the highest level taxon. Bacterial speciation may occur from a combination of repeated periodic selection for a favorable trait within an ecotype and lateral gene flow.

◆ If a given genus includes several _____, a _____ includes several genera.

◆ What is an ecotype?

◆ How many species of prokaryotes are already known? How many likely exist?

11.13 Nomenclature and *Bergey's Manual*

Following the **binomial system** of nomenclature used throughout biology, prokaryotes are given genus names and species epithets. The terms used are Latin or Latinized Greek derivations of some descriptive property appropriate for the organism and are set in print in *italics*. For example, over 100 species of the genus *Bacillus* have been described, including *Bacillus (B.) subtilis, B. cereus*, and *B. megaterium*. These species epithets mean "slender," "waxen," and "big beast," respectively, and refer to key morphological, physiological, or ecological traits characteristic of each organism.

The nomenclature of prokaryotes, *Bacteria* as well as *Archaea*, is regulated by the rules of the Bacteriological Code—*The International Code of Nomenclature of Bacteria*. The Code presents the formal framework by which prokaryotes are to be officially named, and the procedures by which existing names can be changed, for example, when new data warrants taxonomic rearrangements. There are even rules for rejecting names if errors were made in the original naming process or a name has otherwise become invalid. The *Code* covers the rules for the naming of all species, genera, families, and orders of prokaryotes.

sand prokaryotic species are already known (Table 11.6), and several thousands more, perhaps as many as 100,000–1,000,000 in total (or 10 times this by some estimates) are suspected to exist. By anyone's count, the final number of prokaryotic species will likely be enormous.

Microbial community analyses (see Sections 11.7 and 18.5) indicate that we have only scratched the surface in our ability to *culture* the diversity of prokaryotes in nature. With more exacting tools, both molecular and cultural, for revealing prokaryotic diversity, it is possible that the large numbers of species already predicted will be an underestimate, perhaps by several fold. The reality today is that an accurate estimate of prokaryotic speciation is simply out of reach of current technology. But as with other things in microbiology, this will likely change. Obviously, much exciting work is left for those whose interests include prokaryotic diversity!

Culture Collections and Publication of New Taxa

When a new organism is isolated and thought to be unique, a decision must be made as to whether it is sufficiently different from other species to be described as novel, or perhaps even sufficiently different from all described genera to warrant description as a new genus (in which case a species is automatically created). To achieve formal taxonomic standing as a new genus or species, a detailed description of the isolate and the proposed name is published, and a viable culture of the organism is deposited in two international culture collections. Examples of the latter include the American Type Culture Collection (ATCC, Manassas, Virginia, USA) or the Deutsche Sammlung von Mikroorganismen und Zellkulturen (DSMZ, German Collection for Microorganisms, Braunschweig, Germany) (⌔Section 30.1). The deposited strain becomes the *type* strain of the new species and the standard by which other strains thought to be the same can be compared.

Culture collections preserve the deposited culture, typically by freezing at very low temperatures (-80 to $-196°C$), or by freeze-drying. This practice differs from the botanical or zoological approach to taxonomy. These disciplines employ preserved (dead) specimens (either dried herbarium material or chemically fixed animal specimens) as the basis for comparison with proposed new species. By contrast, microbiologists have always relied on a *living type strain* that can be distributed to the scientific community, grown in different laboratories, and studied and compared. This approach allows for more detailed and reproducible comparisons of the properties of organisms, especially at the molecular level.

If the description of a new organism is published in a journal other than the *International Journal of Systematic and Evolutionary Microbiology (IJSEM)*, the official publication of record for the taxonomy and classification of prokaryotes, a copy of the published paper must be submitted to this journal and the name validated before it is formally accepted as a new taxon of prokaryotes. In each issue the *IJSEM* publishes an approved list of newly created names and serves as the publication of record for research in prokaryotic taxonomy. A website dedicated to tracking new and official names of prokaryotes can be found at (www.bacterio.cict.fr).

Bergey's Manual and *The Prokaryotes*

By validating newly proposed names, *IJSEM* paves the way for their inclusion in *Bergey's Manual of Systematic Bacteriology*, a major taxonomic treatment of prokaryotes (Figure 11.27●). Widely used, *Bergey's Manual* has served the community of microbiologists since 1923 and is a compendium of information on all recognized species of prokaryotes. Each chapter, written by an expert, contains tables, figures, and other systematic information useful for identification purposes. Volume I of the second edition of *Bergey's Manual* appeared in 2001, Volume 2 in 2005, and three additional volumes will appear by 2007.

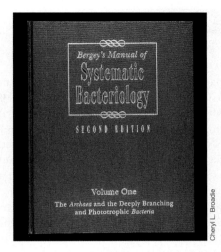

● **Figure 11.27** *Bergey's Manual of Systematic Bacteriology,* **second edition.** In five volumes this source describes the major properties of all known prokaryotes, both *Bacteria* and *Archaea.*

The second edition of *Bergey's Manual* has incorporated many of the concepts that have emerged from ribosomal RNA sequencing and genomic studies and blends this with a wealth of phenotypic information as well.

A second major reference in prokaryotic diversity is *The Prokaryotes*. This work, with more than 4100 pages (four volumes) in its second edition (1992), is now available in a third edition online (http://141.150.157.117:8080/prokPUB/index.htm). The electronic edition incorporates frequent updates to reflect the rapid pace at which new data appear on prokaryotic taxonomy and phylogeny.

Collectively, *Bergey's Manual* and *The Prokaryotes* offer microbiologists the foundations as well as the details of prokaryotic taxonomy and phylogeny as we know it today and are typically the "go-to" sources for microbiologists characterizing newly isolated prokaryotes.

As we will see in Chapter 18, there are many prokaryotes in nature that have thus far eluded laboratory culture. These are probably not "unculturable" prokaryotes, but simply ones that we know too little about to culture at present. However, when these organisms are finally cultured, characterized, and named, *Bergey's Manual* and *The Prokaryotes* will have to expand dramatically to accomodate what will likely be a very large number of prokaryotic species.

 11.13 *Concept Check*

Prokaryotes are given descriptive genus names and species epithets. Formal recognition of a new prokaryotic species requires depositing a sample of the organism in a culture collection and official publication of the new species name and description. *Bergey's Manual of Systematic Bacteriology* is a major taxonomic compilation of *Bacteria* and *Archaea.*

◆ What is the *IJSEM* and what taxonomic function does it fulfill?

◆ Why might living cell material be of more use in taxonomy than preserved specimens?

REVIEW QUESTIONS

1. What is the age of planet Earth? What is the age of the earliest known microfossils (⚬⚬Section 11.1)?

2. What major features would primitive organisms have had to have in order to replicate themselves, and why (⚬⚬Section 11.2)?

3. What properties of RNA could have made possible an era of RNA life? If RNA life forms ever existed, what remnants of them may exist today (⚬⚬Section 12.2)?

4. Discuss the role that ferrous iron, FeS, and H_2 may have played in early life processes (⚬⚬Section 11.3).

5. Why was the evolution of cyanobacteria of such importance to the further evolution of life on Earth? What component of the geological record is used to date the evolution of cyanobacteria (⚬⚬Section 11.3)?

6. What is the evidence that endosymbiosis is responsible for the modern, organelle-containing eukaryotic cell (⚬⚬Section 11.4)? (*Hint*: You may want to review ⚬⚬Sections 14.2–14.5.)

7. Why are ribosomal RNAs better molecules for phylogenetic studies than proteins such as cytochromes (⚬⚬ Section 11.5)? (*Hint*: Think about the distribution of these macromolecules among prokaryotes.)

8. Describe the methods involved in obtaining 16S rRNA sequences. What role does the polymerase chain reaction (PCR) play in molecular phylogeny (⚬⚬Section 11.6)?

9. What are signature sequences and of what phylogenetic value are they? How are signature sequences discerned (⚬⚬Section 11.7)?

10. What is FISH technology? Give an example of how it would be used (⚬⚬Section 11.7).

11. What major evolutionary finding has emerged from the study of ribosomal RNA sequences? How did this modify the classic view of evolution? How has this discovery supported previous beliefs on the origin of eukaryotic organisms (⚬⚬Section 11.8)?

12. What major physiological and biochemical properties do *Archaea* share with *Eukarya*? With *Bacteria* (⚬⚬Section 11.9)?

13. What major phenotypic properties are used to group organisms in classical bacterial taxonomy? Which, if any, of these properties have phylogenetic predictive value (⚬⚬Section 11.10)?

14. Why aren't GC base ratios useful for making phylogenetic determinations? In what situations are GC base ratios of use in taxonomic studies (⚬⚬Section 11.10)?

15. How does ribotyping differ from 16S sequencing as an identification tool? How does ribotyping differ from multilocus sequence typing (⚬⚬Section 11.11)?

16. What is measured in FAME analyses (⚬⚬Section 11.11)?

17. How is it thought that new bacterial species arise? How many bacterial species are there? Why don't we know this number more precisely (⚬⚬Section 11.12)?

18. Examine the following bacterial name: *Pseudomonas aeruginosa*. What part of this is the *species epithet*? What level taxa is the other name? In reference to taxonomic hierarchy, which of the two names might have several other names listed *under* it (⚬⚬Section 11.13)?

APPLICATION QUESTIONS

1. Compare and contrast the physical and chemical conditions on Earth at the time life first arose with conditions today. From a physiological standpoint, discuss at least two reasons why *animals* could not have existed on early Earth.

2. Why is it highly unlikely that life could originate today as it did billions of years ago?

3. Imagine that you are debating someone who is arguing against the theory of endosymbiosis. List five forms of evidence you would use to convince your opponent that endosymbiosis did occur. (You may wish to review Section 14.4 before writing your answer.)

4. On the basis of the following sequences, calculate an evolutionary distance between these three organisms and predict which two of the three are most closely related.

   Organism 1: AGGUACGUUA

   Organism 2: UGCCACGGUU

   Organism 3: AGGUACGGUA

   Sketch a phylogenetic tree that shows the approximate evolutionary relationships of these three organisms.

5. Imagine that you are doing lipid analyses of two prokaryotes along the lines shown in Figure 11.24. Your results on culture A show an abundance of short-chain unsaturated fatty acids. Analyses of cells in culture B show ether-linked phytanyl lipids to be present. Based on this information, to which phylogenetic domain do organisms A and B belong? Also, if you were told that these organisms were both extremophiles (⚬⚬Section 2.4) and that one originated from a boiling hot spring and one from polar sea ice, which would be which? Finally, based on your lipid analyses and knowledge of where most lipid is located in a cell (⚬⚬Section 3.4 and Figure 3.3), describe how the substances you detected might benefit each organism to thrive in its extreme environment. (You will likely need to review material in Sections 6.10–6.12 before answering.)

6. Determine the GC ratio of the following stretch of DNA:

   TAAGCCTGCAAGCTTAGCTA
   ATTCGGACGTTCGAATCGAT

7. What reference resource would you check for information on the taxonomy and phylogeny of prokaryotes? On enrichment, isolation, and culture? If your library has both sources, compare their tables of contents. Which has the greater emphasis on classification and nomenclature?

12

PROKARYOTIC DIVERSITY: THE *BACTERIA*

Species of *Bacteria* show enormous diversity in terms of their morphology, physiology, and phylogeny. The fruiting body of *Chondromyces crocatus*, a fruiting myxobacterium shown here, contains myxospores, produced during an elaborate life cycle. Myxospores can germinate and form a new population of cells.

WORKING GLOSSARY

Acid-fastness a property of *Mycobacterium* species in which cells stained with the dye basic fuchsin resist decolorization with acidic alcohol

Carboxysome a polyhedral cellular inclusion of crystalline ribulose bisphosphate carboxylase (RubisCO), the key enzyme of the Calvin cycle

Chemolithotroph an organism able to oxidize inorganic compounds (such as H_2, Fe^{2+}, S^0, or NH_4^+) as energy sources (electron donors)

Chlorosome a cigar-shaped structure bounded by a nonunit membrane and containing the light-harvesting bacteriochlorophyll (*c*, *d*, or *e*) in green bacteria and *Chloroflexus*

Consortium two- or more-membered association of prokaryotes, usually living in an intimate symbiotic fashion

Cyanobacteria prokaryotic oxygenic phototrophs that contain chlorophyll *a* and phycobilins but not chlorophyll *b*

Enteric bacteria a large group of gram-negative rod-shaped *Bacteria* characterized by a facultatively aerobic metabolism

Green sulfur bacteria anoxygenic phototrophs containing chlorosomes and bacteriochlorophyll *c*, c_s, *d*, or *e* as light-harvesting chlorophyll

Heliobacteria anoxygenic phototrophs containing bacteriochlorophyll *g*

Heterocyst a differentiated cyanobacterial cell that carries out nitrogen fixation but not oxygenic photosynthesis

Heterofermentative in reference to lactic acid bacteria, capable of making more than one fermentation product

Homofermentative in reference to lactic acid bacteria, producing only lactic acid as a fermentation product

Hyperthermophile an organism with a growth temperature optimum of greater than 80°C

Methanotroph an organism capable of oxidizing methane (CH_4) as an electron donor in energy metabolism

Methylotroph an organism capable of oxidizing organic compounds that do not contain carbon-carbon bonds; if able to oxidize CH_4, also a methanotroph

Mixotroph an organism that can conserve energy from the oxidation of inorganic compounds but requires organic compounds as a carbon source

Nitrifying bacteria chemolithotrophs capable of carrying out the transformation $NH_3 \rightarrow NO_2^-$, or $NO_2^- \rightarrow NO_3^-$

Prochlorophyte a prokaryotic oxygenic phototroph that contains chlorophylls *a* and *b* but lacks phycobilins

Prosthecae an extrusion of cytoplasm, often forming a distinct appendage, bounded by the cell wall

Proteobacteria a major lineage of *Bacteria* that contains a large number of gram-negative rods and cocci

Purple nonsulfur bacteria a group of phototrophic prokaryotes containing bacteriochlorophylls *a* or *b* that grows best as photoheterotrophs and has a relatively low tolerance for H_2S

Purple sulfur bacteria a group of phototrophic prokaryotes containing bacteriochlorophylls *a* or *b* and characterized by the ability to oxidize H_2S and store elemental sulfur inside the cells (or in the genera *Ectothiorhodospira* and *Halorhodospira*, outside the cell)

Spirochete a slender, tightly coiled gram-negative prokaryote characterized by possession of endoflagella used for motility

Stickland reaction fermentation of an amino acid pair in which one amino acid serves as an electron donor and a second serves as an electron acceptor

Sulfate-reducing bacteria a large group of anaerobic *Bacteria* that respire anaerobically with SO_4^{2-} as electron acceptor, producing H_2S

THE PHYLOGENY OF BACTERIA

In Chapter 11 we stressed the evolutionary relationships among microorganisms. In this and the next two chapters we expand on these concepts with a discussion of properties of the major microbial groups themselves. In this chapter we focus on species of *Bacteria*, while in the next two chapters we focus on *Archaea* and microbial *Eukarya*, respectively.

With nearly 7000 species of prokaryotes known, obviously we will not be able to consider them all. So, using a phylogenetic tree as the focus of our discussion, we will explore some of the best-known cultured species, particularly ones in which much phenotypic information is available. For more detailed information on prokaryotic diversity the student should refer to *Bergey's Manual of Systematic Bacteriology* and *The Prokaryotes* (∞Section 11.13).

12.1 Phylogenetic Overview of *Bacteria*

At least 18 major lineages (phyla) of *Bacteria* are known from the study of laboratory cultures, and many others have been identified from retrieval and sequencing of ribosomal RNA genes in natural habitats. Figure 12.1●

gives a phylogenetic overview of *Bacteria*. The most phylogenetically ancient (least derived) phylum contains the genus *Aquifex* and relatives, all of which are hyperthermophilic H₂-oxidizing chemolithotrophs. Other "early" phyla such as *Thermodesulfobacterium*, *Thermotoga*, and the green nonsulfur bacteria (the *Chloroflexus* group), also contain thermophilic species.

Continuing past the green nonsulfur bacteria, we see the deinococci and relatives, the morphologically unique spirochetes, the phototrophic green sulfur bacteria, the chemoorganotrophic *Flavobacterium* and *Cytophaga* groups, the budding *Planctomyces-Pirella* and the *Verrucomicrobium* groups, the *Chlamydia*, and the genera *Nitrospira* and *Deferribacter* (Figure 12.1). Each of these groups is discussed in this chapter.

The remaining phyla of cultured *Bacteria* constitute the major thrust of this chapter. They include the gram-positive bacteria, the cyanobacteria, and the Proteobacteria. Each of these is a large group containing many genera and are *Bacteria* about which much phenotypic information is known. The gram-positive bacteria can be separated into two subgroups, called *low* GC and *high* GC (Actinobacteria), the terms referring to the fact that the species tend to have DNA GC base ratios (see Section 11.10) either well below or well above 50%, respectively. The gram-positive bacteria are a large group of primarily chemoorganotrophic *Bacteria* and are discussed in detail

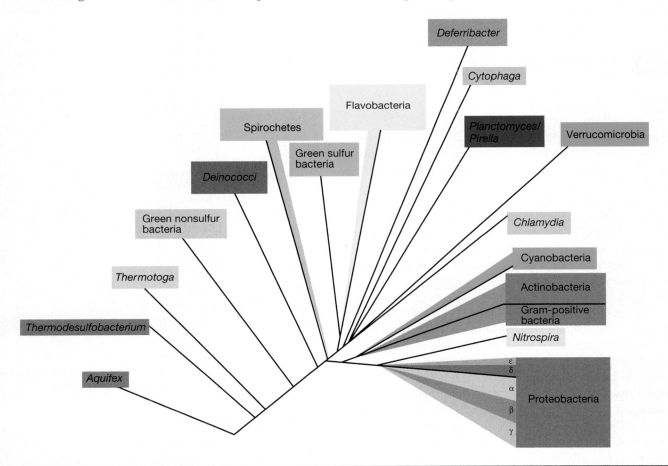

● **Figure 12.1   Detailed phylogenetic tree of the major lineages (phyla) of *Bacteria* based on 16S ribosomal RNA sequence comparisons.**
Including many phyla known only from community sampling (∞Section 11.7), over 40 phyla of *Bacteria* actually exist.

in Sections 12.19–12.24. The cyanobacteria are oxygenic phototrophic prokaryotes with evolutionary roots near those of the gram-positive *Bacteria*; these organisms are covered in Sections 12.25 and 12.26.

The final phylum on the tree of *Bacteria* is the **Proteobacteria** (Figure 12.1). This group is by far the largest of all *Bacteria* (see Sections 12.2–12.18). The Proteobacteria consists of five clusters containing several genera. Each cluster is designated by a Greek letter: *alpha, beta, gamma, delta,* or *epsilon* (see Table 12.1). Physiologically, Proteobacteria can be either phototrophic, chemolithotrophic, or chemoorganotrophic. Indeed, we will see in Chapter 17 the great diversity of energy-generating mechanisms characteristic of representatives of this group.

We proceed now to a description of each phylum. We begin with the largest and most metabolically diverse group of known *Bacteria*: the Proteobacteria.

PHYLUM 1: PROTEOBACTERIA

Table 12.1 lists some key genera of Proteobacteria. As a group these organisms are all gram-negative, show extreme metabolic diversity, and represent the majority of known gram-negative bacteria of medical, industrial, and agricultural significance. We begin our discussion with *phototrophic* Proteobacteria—the purple bacteria.

12.2 Purple Phototrophic *Bacteria*

Key Genera: *Chromatium, Ectothiorhodospira, Rhodobacter, Rhodospirillum*

The purple phototrophic bacteria carry out *anoxygenic* photosynthesis. Thus, unlike the cyanobacteria (see Section 12.25), no O_2 is evolved. The purple bacteria are a morphologically diverse group, and the taxonomy of these organisms has been established along phylogenetic, morphological, and physiological lines. Different genera fall within the alpha, beta, or gamma Proteobacteria.

Purple bacteria contain chlorophyll pigments called *bacteriochlorophylls* and any of a variety of *carotenoid* pigments (∞Figures 17.4 and 17.9). Together, these pigments give purple bacteria their spectacular colors, usually purple, red, or brown (Figure 12.2●). We will examine the structure of these pigments and learn how they actually function in light-mediated energy generation (a process called *photophosphorylation*) in Chapter 17.

Purple bacteria synthesize intracytoplasmic photosynthetic membrane systems into which their pigments are inserted. These membranes can be of various morphologies (Figure 12.3●) but in all cases originate from invaginations of the cytoplasmic membrane. These internal membranes allow purple bacteria to increase their specific pigment content and to thus better utilize the available light. When cells are grown at *high* light inten-

Table 12.1	Major genera of Proteobacteria[a]	
Subdivision	**Genera**	
Alpha	*Acetobacter*	*Nitrobacter*
	Agrobacterium	*Paracoccus*
	Alcaligenes	*Rhodospirillum*
	Azospirillum	*Rhodopseudomonas*
	Beijerinckia	*Rhodobacter*
	Bradyrhizobium	*Rhodomicrobium*
	Brucella	*Rhodovulum*
	Caulobacter	*Rhodopila*
	Ehrlichia	*Rhizobium*
	Gluconobacter	*Rickettsia*
	Hyphomicrobium	*Sphingomonas*
	Methylocystis	*Zymomonas*
Beta	*Aquaspirillum*	*Oxalobacter*
	Bordetella	*Polaromonas*
	Burkholderia	*Ralstonia*
	Chromobacterium	*Rhodocyclus*
	Dechloromonas	*Rhodoferax*
	Gallionella	*Sphaerotilus*
	Leptothrix	*Spirillum*
	Methylophilus	*Thiobacillus*
	Neisseria	*Zoogloea*
	Nitrosomonas	
Gamma	*Acetobacter*	*Photobacterium*
	Acinetobacter	*Pseudomonas*
	Azotobacter	*Methylococcus*
	Chromatium	*Methylobacter*
	Escherichia	*Nitrosococcus*
	Ectothiorhodospira	*Nitrococcus*
	Erwinia	*Thermochromatium*
	Francisella	*Thiomicrospira*
	Halomonas	*Thiospirillum* and
	Halorhodospira	other purple
	Halothiobacillus	sulfur bacteria
	Legionella	*Salmonella* and other
	Leucothrix	enteric bacteria
	Methylomonas	*Vibrio*
	Oceanospirillum	*Xanthomonas*
Delta	*Acinetobacter*	*Geobacter*
	Aeromonas	*Halomonas*
	Bdellovibrio	*Moraxella*
	Desulfuromonas	*Myxococcus* and other
	Desulfovibrio and most	myxobacteria
	other sulfate-	*Pelobacter*
	reducing bacteria	*Syntrophobacter*
	Francisella	
Epsilon	*Campylobacter*	*Thiovulum*
	Helicobacter	*Wolinella*

[a] This table is not meant to be inclusive but only lists some well-described genera of Proteobacteria. For a complete list of genera of Proteobacteria and genera of other lineages of *Bacteria*, see Appendix 2.

sities, internal membranes are few and pigment contents low. By contrast, at low light intensities, the cells are packed with membranes and photopigments.

Purple Sulfur Bacteria

Purple bacteria that utilize hydrogen sulfide (H_2S) as an electron donor for CO_2 reduction in photosynthesis are known as **purple sulfur bacteria** (Table 12.2). The sulfide is oxidized to elemental sulfur (S^0) that is stored in glob-

● **Figure 12.2 Photograph of liquid cultures of phototrophic purple bacteria showing the color of species with various carotenoid pigments.** The blue culture is a carotenoidless mutant derivative of *Rhodospirillum rubrum* showing how bacteriochlorophyll *a* is actually *blue* in color. The bottle on the far right (*Rhodobacter sphaeroides* strain G) lacks one of the carotenoids of the wild type and thus is more green in color.

(a)

(b)

● **Figure 12.3 Membrane systems of phototrophic purple bacteria as revealed by the electron microscope.** (a) Purple phototrophic bacterium, *Ectothiorhodospira mobilis*, showing the photosynthetic membranes in flat sheets (lamellae). (b) *Allochromatium vinosum*, another purple phototrophic bacterium, showing the membranes as individual, spherical-shaped vesicles.

Table 12.2	Genera and characteristics of purple sulfur bacteria[a]
Characteristics	**Genus/DNA (mol % GC)**
Sulfur deposited externally:	
Spirilla, polar flagella	*Ectothiorhodospira* (62–67)
Spirilla, extreme alkaliphiles	*Thiorhodospira* (57)
Spirilla, extreme halophiles	*Halorhodospira* (50–69)
Sulfur deposited internally:	
Do not contain gas vesicles	
Ovals or rods, polar flagella	*Chromatium* (48–70)
	Allochromatium;
	Halochromatium;
	Rhabdochromatium;
	Thermochromatium;
	Isochromatium;
	Marichromatium
Spheres, alkaliphilic	*Thioalkalicoccus* (64)
Spheres, contain bacteriochlorophyll *b*	*Thioflavicoccus* (66)
Spheres	*Thiorhodococcus* (67)
Spheres, diplococci, tetrads, nonmotile; cells 1.2–3 μm in diameter	*Thiocapsa* (63–70)
Spheres or ovals, polar flagella; cells 2.5–3 μm in diameter	*Thiocystis* (61–68)
Spheres, 1.5–2.5 μm in diameter	*Thiohalocapsa* (66)
Spheres, 1–2 μm in diameter	*Thiorhodococcus* (67)
Spheres, 1.2–1.5 μm in diameter	*Thiococcus* (69)
Large spirilla, polar flagella	*Thiospirillum* (45)
Small spirilla	*Thiorhodovibrio* (61–62)
Contain gas vesicles	
Irregular spheres forming platelets of 4–16 cells	*Thiolamprovum*
Rods	*Lamprobacter* (64)
Spheres, ovals, polar flagella	*Lamprocystis* (64)
Rods, nonmotile; forming irregular network	*Thiodictyon* (65–66)
Spheres, nonmotile; forming flat sheets of tetrads	*Thiopedia* (62–64)

[a] From a phylogenetic standpoint, all are members of the gamma subdivision of the Proteobacteria (Figure 12.1).

ules inside the cells (Figure 12.4●); the sulfur later disappears as it is oxidized to sulfate (SO_4^{2-}). Many purple sulfur bacteria can also use other reduced sulfur compounds as photosynthetic electron donors, thiosulfate ($S_2O_3^{2-}$) being a key one commonly used to grow laboratory cultures. All purple sulfur bacteria discovered thus far group with the gamma Proteobacteria.

Purple sulfur bacteria are generally found in illuminated anoxic zones of lakes and other aquatic habitats where H_2S accumulates and also in "sulfur springs," where geochemically or biologically produced H_2S can trigger the formation of blooms of purple sulfur bacteria (Figure 12.5●). The most favorable lakes for development of purple sulfur bacteria are *meromictic* (permanently stratified) lakes. Meromictic lakes stratify because they have denser (usually saline) water in the bottom and less dense (usually freshwater) nearer the surface. If sufficient sulfate is present to support sulfate reduction, the sulfide, produced in the sediments, diffuses upward into the

(a)

(b)

(c)

(d)

● **Figure 12.4 Bright-field photomicrographs of purple sulfur bacteria (see also Table 12.2).** (a) *Chromatium okenii*; cells are about 5 μm wide. Note the globules of elemental sulfur inside the cells. (b) *Thiospirillum jenense*, a very large, polarly flagellated spiral; cells are about 30 μm long. Note the sulfur globules. (c) *Thiopedia rosea*; cells are about 1.5 μm wide. (d) Phase micrograph of cells of *Ectothiorhodospira mobilis*. Cells are about 0.8 μm wide. Note external sulfur globules (arrow). Compare the photo of *Chromatium okenii* with the drawings of purple sulfur bacteria made by the great Russian microbiologist, Sergei Winogradsky, over 115 years ago (⬠Section 1.7 and Figure 1.15).

anoxic bottom waters, and here purple sulfur bacteria can form dense cell masses, called *blooms*, usually in association with green phototrophic bacteria (Figure 12.5*c*).

The genera *Ectothiorhodospira* and *Halorhodospira* are of special interest. Unlike other purple sulfur bacteria, these organisms oxidize H_2S and produce S^0 *outside* of the cell (Figure 12.4*d*). These genera are also interesting because many species are extremely halophilic (salt-loving) and/or alkaliphilic and are among the most extreme in these regards of all known prokaryotes. These organisms are typically found in saline lakes, soda lakes, and salterns.

Purple Nonsulfur Bacteria

Some purple bacteria are called **purple nonsulfur bacteria** because it was originally thought that they were unable to use sulfide as an electron donor for the reduction of CO_2 to cell material. In fact, sulfide *can* be used by most species in this group, although the levels of sulfide ideal for purple *sulfur* bacteria (1–3 mM) are often toxic to most purple *nonsulfur* bacteria. Some purple nonsulfur bacteria can also grow anaerobically in the dark using fermentative or anaerobic respiratory metabolism, and most can grow aerobically in darkness by respiration. Under the latter conditions, synthesis of the photosynthetic machinery is repressed, and the electron donor can be an organic compound or in some species even an inorganic compound, such as H_2.

However, it is the capacity of this group for *photoheterotrophy* (light as the energy source and an organic compound as the carbon source, ⬠Figure 17.1) that likely accounts for their competitive success in nature. Purple nonsulfur bacteria are typically nutritionally diverse using fatty, organic, or amino acids; sugars; alcohols; and even aromatic compounds like benzoate as carbon sources. Most species can also grow photoautotrophically with $CO_2 + H_2$ or $CO_2 +$ low levels of H_2S.

(a)

(b)

(c)

● **Figure 12.5 Blooms of purple sulfur bacteria.** (a) *Thiopedia roseopersicina*, in a sulfide spring in Madison, Wisconsin. The bacteria grow near the bottom of the spring pool and float to the top (by virtue of their gas vesicles) when disturbed (⬠ Section 4.12 for further discussion of gas vesicles). The green color is from cells of the eukaryotic alga *Spirogyra* (⬠Figure 14.35*d*). (b) Sample of water from 7 m in Lake Mahoney, British Columbia. The major organism is *Amoebobacter purpureus*. (c) Phase-contrast photomicrograph of layers of purple sulfur bacteria from a small stratified lake in Michigan. The purple sulfur bacteria include *Chromatium* species (large rods) and *Thiocystis* (small cocci).

Table 12.3	Genera and characteristics of purple nonsulfur bacteria[a]
Characteristics	**Genus/DNA (mol % GC)**
Alpha Proteobacteria Spirilla, polarly flagellated	*Rhodospirillum* (62–68) *Phaeospirillum; Rhodovibrio; Rhodothalassium; Roseospira; Rhodospira;* (66) *Roseospirillum* (71)
Rods, polarly flagellated; divide by budding	*Rhodopseudomonas* (64–72) *Rhodoplanes* (66–69) *Rhodobium* (61–65)
Rods; divide by binary fission	*Rhodobacter* (62–71)
Ovoid to rod-shaped cells	*Rhodovulum* (64–68)
Ovals, peritrichously flagellated; growth by budding and hypha formation	*Rhodomicrobium* (61–63)
Large spheres, acidophilic (pH 5 optimum)	*Rhodopila* (66)
Small spheres, alkaliphilic (pH 9 optimum)	*Rhodobaca* (59)
Beta Proteobacteria Ring-shaped or spirilla	*Rhodocyclus* (64–66)
Curved rods	*Rubrivivax* (70–72)
Curved rods	*Rhodoferax* (59–60)

[a] All are members of the Proteobacteria (see Figure 12.1 and Table 12.1).

● **Figure 12.6 Representatives of several genera of purple nonsulfur bacteria (see also Table 12.3).** (a) *Phaeospirillum fulvum;* cells are about 3 μm long. (b) *Rhodopseudomonas acidophila;* cells are about 4 μm long. (c) *Rhodobacter sphaeroides;* cells are about 1.5 μm wide. (d) *Rhodopila globiformis;* cells are about 1.6 μm wide. (e) *Rhodocyclus purpureus;* cells are about 0.7 μm in diameter. (f) *Rhodomicrobium vannielii;* cells are about 1.2 μm wide.

Enrichment and isolation of purple nonsulfur bacteria is easy using a mineral salts medium supplemented with an organic acid as carbon source. Such media, inoculated with a mud, lake water, or sewage sample and incubated anoxically in the light, invariably select for purple nonsulfur bacteria. Enrichment cultures can be made even more selective by omitting fixed nitrogen sources (for example, NH_4^+) or organic nitrogen sources (for example, yeast extract or peptone) from the medium and supplying a gaseous headspace of N_2; virtually all purple nonsulfur bacteria can fix N_2 (◁◯ Section 17.28) and will thrive under such conditions, usually outcompeting other organisms.

The morphological diversity of purple nonsulfur bacteria is typical of that of purple sulfur bacteria (Table 12.3 and Figure 12.6●), and it is clearly a heterogeneous group in this regard. All purple nonsulfur bacteria isolated thus far are either alpha or beta Proteobacteria (Figure 12.1).

◆ Give a major reason why photosynthesis in purple nonsulfur bacteria does not occur under aerobic conditions.

◆ Can purple bacteria grow in the absence of light?

12.3 The Nitrifying *Bacteria*

Key Genera: *Nitrosomonas, Nitrobacter*

Many species of *Bacteria* can grow as **chemolithotrophs**. Chemolithotrophic bacteria are physiologically united by their ability to utilize *inorganic* electron donors as energy sources (we discuss the conceptual basis of chemolithotrophy in Chapter 17). Most chemolithotrophs are also capable of autotrophic growth and in this way share a major physiological trait with phototrophic bacteria and cyanobacteria. We focus here on the best-studied chemolithotrophs: those capable of oxidizing reduced sulfur or nitrogen compounds, or H_2.

12.2 Concept Check

Purple bacteria are anoxygenic phototrophs that grow phototrophically, obtaining carbon from $CO_2 + H_2S$ (purple sulfur bacteria) or organic compounds (purple nonsulfur bacteria). Purple nonsulfur bacteria are physiologically diverse and most can grow as chemoorganotrophs in darkness. The purple bacteria reside in the alpha, beta, and gamma subdivisions of the Proteobacteria.

◆ What is meant by the term *anoxygenic*?

Nitrosifyers and Nitrifyers

Bacteria able to grow chemolithotrophically at the expense of reduced inorganic nitrogen compounds are called **nitrifying bacteria**. Several genera are recognized on the basis of morphology and phylogeny and the particular steps in the oxidation sequences that they carry out (Table 12.4). Phylogenetically, the nitrifying bacteria scatter among four of the five subdivisions of the Proteobacteria: alpha, beta, gamma, and delta. The genus *Nitrospira* forms its own phylum of *Bacteria* (Figure 12.1 and see Section 12.38), and is related to other nitrifying bacteria in a metabolic sense only.

No single chemolithotroph is known that will carry out the complete oxidation of ammonia to nitrate. Thus, **nitrification** in nature results from the sequential action of two separate groups of organisms, the *ammonia-oxidizing bacteria*, the **nitrosifyers** (Figure 12.7●), and the *nitrite-oxidizing bacteria*, the true **nitrifying** (nitrate-producing) **bacteria** (Figure 12.8●). Nitrosifying bacteria typically have genus names beginning in "Nitroso," while true nitrifyers usually begin with "Nitro"; *Nitrosomonas* and *Nitrobacter* are major genera of nitrifying bacteria (Table 12.4). Historically, the nitrifying bacteria were the first organisms to be shown to grow chemolithotrophically. Winogradsky showed that they were able to produce organic matter and cell mass when provided with CO_2 as sole carbon source (∞Microbial Sidebar, Winogradsky's Legacy, Chapter 17).

Many species of nitrifying bacteria have complex internal membrane systems. In many respects, these are similar to the internal membranes found in their close phylogenetic relatives, the purple phototrophs (see Section 12.2) and the methane-oxidizing (methanotrophic) bacteria (see Section 12.6). The membranes are the location of key

● **Figure 12.7** **Phase-contrast photomicrograph (left) and electron micrograph (right) of the nitrosifying bacterium *Nitrosococcus oceani*.** A single cell is about 2 μm in diameter.

● **Figure 12.8** **Phase-contrast photomicrograph (left) and electron micrograph (right) of the nitrifying bacterium *Nitrobacter winogradskyi*.** A cell is about 0.7 μm in diameter.

enzymes in nitrification: *ammonia monooxygenase*, which oxidizes NH_3 (ammonia) to NH_2OH (hydroxylamine), and *nitrite oxidase*, which oxidizes NO_2^- to NO_3^- (∞ Section 17.12) Hydroxylamine is further oxidized to NO_2^- (nitrite) by the nitrosifying bacteria, leaving a substrate for the true nitrifying bacteria (Figure 12.9●).

Table 12.4	Characteristics of the nitrifying bacteria			
Characteristics	**Genus**	**Phylogenetic group**[a]	**DNA (mol % GC)**	**Habitats**
Oxidize ammonia:				
Gram-negative short to long rods, motile (polar flagella) or nonmotile; peripheral membrane systems	*Nitrosomonas*	Beta	45–53	Soil, sewage, freshwater, marine
Large cocci, motile; vesicular or peripheral membranes	*Nitrosococcus*	Gamma	49–50	Freshwater, marine
Spirals, motile (peritrichous flagella); no obvious membrane system	*Nitrosospira*	Beta	54	Soil
Pleomorphic, lobular, compartmented cells; motile (peritrichous flagella)	*Nitrosolobus*	Beta	54	Soil
Oxidize nitrite:				
Short rods, reproduce by budding, occasionally motile (single subterminal flagellum); membrane system arrange as a polar cap	*Nitrobacter*	Alpha	59–62	Soil, freshwater, marine
Long, slender rods, nonmotile; no obvious membrane system	*Nitrospina*	Delta	58	Marine
Large cocci, motile (one or two subterminal flagella); membrane system randomly arranged in tubes	*Nitrococcus*	Gamma	61	Marine
Helical to vibrioid-shaped cells, nonmotile; no internal membranes	*Nitrospira*	*Nitrospira* group	50	Marine, soil

[a] Phylogenetically, all nitrifying bacteria thus far examined are Proteobacteria, except for *Nitrospira,* which constitutes its own phylogenetic lineage (Figure 12.1).

Nitrosifying bacteria

1. $NH_3 + O_2 + 2 e^- + 2 H^+ \longrightarrow NH_2OH + H_2O$

2. $NH_2OH + H_2O + \frac{1}{2} O_2 \longrightarrow NO_2^- + 2 H_2O + H^+$

Sum: $NH_3 + 1\frac{1}{2} O_2 \longrightarrow NO_2^- + H_2O$

$\Delta G^{0\prime} = -288$ kJ/reaction

Nitrifying bacteria

$NO_2^- + \frac{1}{2} O_2 \longrightarrow NO_3^-$

$\Delta G^{0\prime} = -74.1$ kJ/reaction

● **Figure 12.9 Nitrification.** Reactions involved in the oxidation of inorganic nitrogen compounds by chemolithotrophic nitrifying bacteria (∞also Figures 17.32 and 17.33).

Ecology, Isolation, and Culture

The nitrifying bacteria are widespread in soil and water. They are present in highest numbers in habitats where considerable amounts of ammonia are present, such as sites where extensive protein decomposition occurs (ammonification), and also in sewage treatment facilities (∞Section 28.2). Nitrifying bacteria develop especially well in lakes and streams that receive inputs of sewage or other wastewaters because these are frequently high in ammonia (∞Figure 19.10a).

Enrichment cultures of nitrifying bacteria can be readily established by using mineral salts media containing ammonia or nitrite as electron donor and bicarbonate (HCO_3^-) as sole carbon source. Because of the inefficien-

cy of growth of these organisms (∞Section 17.12), visible turbidity may not develop even after extensive nitrification has occurred. An easy means of monitoring growth is thus to assay for the production of nitrite (with ammonia as electron donor) or the disappearance of nitrite or production of nitrate (with nitrite as electron donor). Most of the nitrifying bacteria are obligate chemolithotrophs and obligate aerobes. Species of *Nitrobacter* are an exception and are able to grow chemoorganotrophically on acetate or pyruvate as sole carbon and energy source. One group, the *anammox* organisms (∞ Section 17.12), are phylogenetically distinct and oxidize ammonia anaerobically.

12.4 Sulfur- and Iron-Oxidizing *Bacteria*

Key Genera: *Thiobacillus, Acidithiobacillus, Achromatium, Beggiatoa*

The ability to grow chemolithotrophically on reduced sulfur compounds is a property of a diverse group of Proteobacteria (Table 12.5). Two broad ecological classes of sulfur-oxidizing bacteria exist, those living at neutral pH and those living at acid pH. Some of the acidophiles also have the ability to grow chemolithotrophically using ferrous iron (Fe^{2+}) as an electron donor. We discuss the biogeochemistry of acidophilic sulfur- and iron-oxidizing bacteria in Sections 19.13–19.16 and the biochemistry of these processes in Sections 17.10 and 17.11.

Table 12.5 Physiological characteristics of sulfur-oxidizing chemolithotrophic prokaryotes

Genus and species	Inorganic electron donor	Range of pH for growth	Phylogenetic group[a]	DNA (mol % GC)
Species growing poorly if at all in organic media:				
Thiobacillus thioparus	H_2S, sulfides, S^0, $S_2O_3^{2-}$	6–8	Beta	61–66
Thiobacillus denitrificans[b]	H_2S, S^0, $S_2O_3^{2-}$	6–8	Beta	63–68
Halothiobacillus neapolitanus	S^0, $S_2O_3^{2-}$	6–8	Gamma	52–56
Acidothiobacillus thiooxidans	S^0	2–4	Gamma	51–53
Acidothiobacillus ferrooxidans	S^0, metal sulfides, Fe^{2+}	2–4	Gamma	55–65
Species growing well in organic media:				
Starkeya novella	$S_2O_3^{2-}$	6–8	Beta	66–68
Thiomonas intermedia	$S_2O_3^{2-}$	3–7	Beta	64
Filamentous sulfur chemolithotrophs:				
Beggiatoa	H_2S, $S_2O_3^{2-}$	6–8	Gamma	37–51
Thiothrix	H_2S	6–8	Gamma	52
Thioploca[c]	H_2S, S^0	—	Gamma	—
Other genera:				
Achromatium	H_2S	—	Gamma	—
Thiomicrospira	$S_2O_3^{2-}$, H_2S	6–8	Gamma	36–44
Thiosphaera[d]	H_2S, $S_2O_3^{2-}$, H_2	6–8	Alpha	66
Thermothrix	H_2S, $S_2O_3^{2-}$, SO_3^-	6.5–7.5	Beta	—
Thiovulum	H_2S, S^0	6–8	Epsilon	—

[a] All are Proteobacteria.

[b] Facultative aerobes; use NO_3^- as electron acceptor anaerobically.

[c] Pure cultures not yet available.

[d] *Thiosphaera pantotropha* has the exact same 16S rRNA sequence as *Paracoccus denitrificans*.

Thiobacillus and *Achromatium*

The genus *Thiobacillus* and close relatives contains several gram-negative, rod-shaped bacteria, indistinguishable morphologically from most other gram-negative rods (Figure 12.10*a*●); they are the best studied of the sulfur chemolithotrophs. Phylogenetically, species of *Thiobacillus* scatter among the Proteobacteria, with different species residing in the alpha, beta, and gamma subdivisions (Table 12.5). The sulfur compounds most commonly used as electron donors in chemolithotrophic metabolism of *Thiobacillus* species are H_2S, S^0, and $S_2O_3^{2-}$. The energy-yielding reactions involved are as follows:

$$H_2S + 2\,O_2 \longrightarrow SO_4^{2-} + 2\,H^+$$

$$\Delta G^{0'} = -798 \text{ kJ/reaction}$$

$$S^0 + H_2O + 1\tfrac{1}{2}O_2 \longrightarrow SO_4^{2-} + 2\,H^+$$

$$\Delta G^{0'} = -587 \text{ kJ/reaction}$$

$$S_2O_3^{2-} + H_2O + 2\,O_2 \longrightarrow 2\,SO_4^{2-} + 2\,H^+$$

$$\Delta G^{0'} = -818 \text{ kJ/reaction}$$

It is obvious that large amounts of energy are available from these reactions, and we will see in Section 17.10 that some of this energy can be trapped as ATP from electron transport reactions that lead to a proton motive force. Moreover, these reactions generate large amounts of sulfuric acid and thus several thiobacilli are acidophilic. One acidophilic species, *Acidithiobacillus ferrooxidans*, can also grow chemolithotrophically by the oxidation of ferrous iron and is a major biological agent for the oxidation of this metal (∞Section 19.14). Iron pyrite (FeS_2) is a major source of Fe^{2+} as well as sulfide. The oxidation of FeS_2, especially in mining operations, can be both beneficial (because leaching of the ore releases the iron from the sulfide mineral) and ecologically disastrous (acidification of the environment occurs along with the release of other heavy metals associated with the pyrite) (∞Sections 19.14 and 19.15).

Achromatium is a spherical sulfur-oxidizing chemolithotroph that is common in freshwater sediments containing sulfide. Cells of *Achromatium* are large cocci that can have diameters of 10–100 μm (Figure 12.10*b*). Phylogenetic analyses of natural populations (∞Sections 11.6 and 18.5) of *Achromatium* have shown that several species likely exist (probably each of distinct size), although pure cultures of this organism have not yet been achieved. Phylogenetically, *Achromatium* belongs to the gamma Proteobacteria and is specifically related to phototrophic purple bacteria, such as its phototrophic counterpart *Chromatium* (see Section 12.2). Like *Chromatium*, cells of *Achromatium* store elemental sulfur internally (Figure 12.10*b*); the granules later disappear as sulfur is oxidized to sulfate. Cells of *Achromatium* also store large granules of calcite ($CaCO_3$) (Figure 12.10*b*), possibly as a carbon source for autotrophic growth.

(a)

Jessup M. Shively

(b)

Hans-Dietrich Babenzien

● **Figure 12.10 Nonfilamentous sulfur chemolithotrophs.** (a) Transmission electron micrograph of cells of the chemolithotrophic sulfur oxidizer *Halothiobacillus neapolitanus*. A single cell is about 0.5 μm in diameter. Note the polyhedral bodies (carboxysomes) distributed throughout the cell (arrows). (b) *Achromatium*. Cells isolated from a small lake in Germany and photographed by Nomarski light microscopy. The small globular structures near the periphery of the cells (arrow) are elemental sulfur, while the large granules consist of calcium carbonate. A single *Achromatium* cell is about 25 μm in diameter.

Culture

Some sulfur chemolithotrophs are *obligate* chemolithotrophs, locked into a lifestyle of using inorganic instead of organic compounds as electron donors. When growing in this fashion they are also autotrophs, converting CO_2 into cell material by reactions of the Calvin cycle (∞Section 17.6). **Carboxysomes** are often present inside the cells of obligate chemolithotrophs (Figure 12.10*a*). These structures contain high levels of Calvin cycle enzymes and probably increase the rate at which these organisms can fix CO_2.

Other sulfur chemolithotrophs are *facultative* chemolithotrophs, facultative in the sense that they can grow either *chemolithotrophically* (and thus, also as autotrophs) or *chemoorganotrophically* (Table 12.5). Most species of *Beggiatoa*, however, can obtain energy from the oxidation of inorganic sulfur compounds but lack enzymes of the Calvin cycle. They thus require organic compounds as carbon sources. Such a nutritional lifestyle is called **mixotrophy**.

Beggiatoa

Organisms of this genus are filamentous, gliding, sulfur-oxidizing bacteria. The cells of *Beggiatoa* are usually quite large in diameter and long, consisting of many short cells attached end to end (Figure 12.11●). Filaments then flex and twist so that many filaments may become intertwined to form a complex tuft.

Beggiatoa is found in nature primarily in habitats rich in H_2S, such as sulfur springs (Figure 12.11*b*), decaying seaweed beds, mud layers of lakes, and waters polluted with sewage. In such environments, filaments of *Beggiatoa* are typically filled with sulfur granules (Figure 12.11*a*). *Beggiatoa* are also common inhabitants of hydrothermal vents (∞Section 19.8). It was with *Beggiatoa* that Winogradsky first demonstrated that a living organism could oxidize H_2S to S^0 and then to SO_4^{2-}, leading him to formulate the concept of *chemolithotrophy* (∞Microbial Sidebar, Winogradsky's Legacy, Chapter 17). Although a few strains of *Beggiatoa* are truly chemolithotrophic autotrophs, most grow best mixotrophically with reduced

sulfur compounds as electron donors and organic compounds as carbon sources.

An interesting habitat of *Beggiatoa* is the rhizosphere of plants (rice, cattails, and other swamp plants) living in flooded, and hence anoxic, soils. Such plants pump oxygen down into their roots so a sharply defined oxic/anoxic boundary develops between the root and the soil. *Beggiatoa* (and probably other sulfur bacteria) develops at this boundary and plays a beneficial role for the plant by oxidizing (and thus detoxifying) the H_2S.

Beggiatoa and other filamentous bacteria like *Sphaerotilus* (see Section 12.15) can cause major settling problems in sewage treatment facilities and in industrial waste lagoons from such waste as canning, paper pulping, brewing, and milling. These problems are referred to as *bulking* and occur when filamentous bacteria such as *Beggiatoa* overgrow the normal flora of the waste system, producing a loose detrital floc instead of the normal and more easily settling tight floc containing organisms such as *Zoogloea* and related sewage bacteria (∞Section 28.2). If bulking occurs, the wastewater remains improperly treated because the effluent discharged is still high in organic matter (∞Section 28.2).

Thioploca and Thiothrix

Other filamentous sulfur-oxidizing bacteria include *Thioploca* and *Thiothrix*. *Thioploca* is a large, filamentous sulfur-oxidizing chemolithotroph that forms cell bundles surrounded by a common sheath (Figure 12.12●). Thick mats of a marine *Thioploca* species have been found on the ocean floor off the coast of Chile and Peru. Studies on the ecology of these organisms have shown that they carry out the *anoxic* oxidation of H_2S coupled to the reduction of nitrate (NO_3^-) to ammonium (NH_4^+) (∞Sections 17.14 and 19.12). Interestingly, it has been shown that cells of *Thioploca* can accumulate huge amounts of nitrate intracellularly and that this nitrate can then support extended periods of anaerobic respiration with H_2S as electron donor. It is postulated that these marine *Thioploca* mats fix substantial amounts of CO_2 and also play a major role in sulfur and nitrogen

(a)

(b)

● **Figure 12.11 Filamentous sulfur-oxidizing bacteria.** (a) Phase-contrast photomicrograph of a *Beggiatoa* species isolated from a sewage treatment plant. Note the abundant elemental sulfur granules in some of the cells. (b) Sulfur-oxidizing bacteria in the outflow of a small sulfide spring. The filamentous cells twist together to form thick streamers, and the white color is due to the abundant elemental sulfur content of the cells.

● **Figure 12.12 Cells of a large marine *Thioploca* species.** Cells contain sulfur granules (yellow) and are about 40–50 μm wide.

cycling. Similar mats consisting primarily of *Beggiatoa* are found near hydrothermal vents (∞Section 19.8), but the connection with nitrate respiration in these cases is not as well established.

Thiothrix is a filamentous sulfur-oxidizing organism in which the filaments group together at their ends by way of a holdfast to form cell arrangements called *rosettes* (Figure 12.13●). Physiologically, *Thiothrix* is an obligately aerobic mixotroph, and in this and most other respects resembles *Beggiatoa*.

12.5 Hydrogen-Oxidizing *Bacteria*

Key Genera: *Ralstonia, Paracoccus*

A wide variety of bacteria can grow with H_2 as sole electron donor and O_2 as electron acceptor using the "knallgas" reaction, the reduction of O_2 with H_2, as their energy metabolism:

(a)

(b)

● **Figure 12.13** **Thiothrix.** (a) A sulfide-containing artesian spring in Florida (USA). The outside of the spring is coated with a mat of *Thiothrix*. The mat is about 1.5 m in diameter. (b) Phase-contrast photomicrograph of a rosette of cells of *Thiothrix* isolated from the spring and grown in pure culture. Note the internal sulfur globules produced from the oxidation of sulfide. Each filament is about 4 μm in diameter.

$$H_2 + \tfrac{1}{2}O_2 \longrightarrow H_2O \qquad \Delta G^{0'} = -237 \text{ kJ}$$

Many, but not all, of these organisms can also grow autotrophically (using reactions of the Calvin cycle to incorporate CO_2) and are grouped together here as the chemolithotrophic *hydrogen-oxidizing* bacteria. Both gram-positive and gram-negative hydrogen bacteria are known, with the best-studied representatives classified in the genera *Ralstonia* (Figure 12.14●), *Pseudomonas*, and *Paracoccus* (Table 12.6). All hydrogen-oxidizing bacteria contain one or more *hydrogenase* enzymes that function to bind H_2 and use it to either produce ATP or for reducing power for autotrophic growth (Table 12.6).

Almost all hydrogen bacteria are *facultative* chemolithotrophs, meaning that they can also grow chemoorganotrophically with organic compounds as energy sources. This is a major distinction between hydrogen chemolithotrophs and many sulfur chemolithotrophs or nitrifying bacteria. Most representatives of these groups are *obligate* chemolithotrophs—growth cannot occur in the absence of the inorganic energy source. By contrast, hydrogen chemolithotrophs can switch between chemolithotrophic and chemoorganotrophic modes of metabolism as nutritional conditions in their habitats warrant.

Physiology and Ecology of Hydrogen Bacteria

When growing chemolithotrophically on H_2, most hydrogen bacteria grow best under microoxic conditions because hydrogenases are oxygen-sensitive enzymes. Typically, oxygen levels of about 5%–10% support best growth. The element *nickel* must be present in the medium for chemolithotrophic growth of hydrogen bacteria because virtually all hydrogenases contain Ni^{2+} as a

● **Figure 12.14** **Hydrogen bacteria**. Transmission electron micrograph of negatively stained cells of the hydrogen-oxidizing chemolithotroph *Ralstonia eutropha*. A cell is about 0.6 μm in diameter and contains several flagella.

Table 12.6 Differential characteristics of a few common species of hydrogen-oxidizing bacteria

Genus and species	Denitri-fication	Growth on fructose	Motility	Phylogenetic group[a]	DNA (mol % GC)	Other characteristics
Gram-negative						
Acidovorax facilis	−	+	+	Beta	64	Membrane-bound hydrogenase
Ralstonia eutropha	+	+	+	Beta	66	Membrane-bound and cytoplasmic hydrogenases
Achromobacter xylosoxidans	−	+	+	Beta	—	Membrane-bound and cytoplasmic hydrogenases
Aquaspirillum autotrophicum	−	−	+	Beta	61	Membrane-bound hydrogenase
Pseudomonas carboxydovorans	−	−	+	Gamma	60	Membrane-bound hydrogenase; also oxidizes CO
Hydrogenophaga flava	−	+	+	Beta	67	Colonies are bright yellow
Paracoccus denitrificans	+	+	−	Alpha	66	Membrane-bound hydrogenase; strong denitrifier
Aquifex pyrophilus	+	−	+	*Aquifex* group[b]	65	Hyperthermophile, grows microaerophilically or anaerobically (with NO_3^-), obligate chemolithotroph; also uses S^0 or $S_2O_3^{2-}$ as electron donor
Hydrogenobacter thermophilus	−	−	−	*Aquifex* group[b]	37–46	As for *Aquifex*, but obligate aerobe (microaerophile)
Gram-positive						
Bacillus schlegelii	−	−	+	Low GC gram-positive[c]	66	Produces endospores; thermophile; also uses CO or $S_2O_3^{2-}$ as electron donor
Arthrobacter sp.	−	+	−	Actinobacteria[d]	70	Membrane-bound hydrogenase
Mycobacterium gordonae	−	?	−	Actinobacteria[d, e]	—	Acid-fast; colonies yellow to orange

[a] Aerobic hydrogen bacteria are Proteobacteria except as indicated.
[b] See Section 12.37.
[c] See Section 12.20.
[d] See Section 12.22.
[e] See Section 12.23.

metal cofactor. A few hydrogen bacteria also fix molecular N_2, and when growing on N_2, the organisms are quite oxygen-sensitive because the enzyme nitrogenase needed for the reduction of molecular nitrogen (∞ Section 17.28) is an oxygen-sensitive enzyme.

Hydrogen-oxidizing bacteria can be enriched if a small amount of mineral salts medium containing trace metals (especially Ni^{2+} and Fe^{2+}) is inoculated with soil or water and incubated in a large, sealed flask containing a head space of 5% O_2, 10% CO_2, and 85% H_2. When the liquid becomes turbid, plates of the same medium are streaked and incubated in a glass jar containing the same gas mixture (one must exercise care here, as mixtures of O_2 and H_2 are potentially explosive).

CO Oxidation

Some hydrogen bacteria can grow on carbon monoxide, CO, as electron donor. Electrons from the oxidation of CO to CO_2 enter the electron transport chain to drive

ATP synthesis. CO-oxidizing bacteria, called *carboxydotrophic* bacteria, grow autotrophically via the Calvin cycle reactions (∞ Section 17.6) to fix the CO_2 generated from the oxidation of CO. Carbon monoxide is oxidized to CO_2 by the enzyme *carbon monoxide dehydrogenase*, which is a molybdenum-containing enzyme. The molybdenum in CO dehydrogenase is bound to a small cofactor consisting of a multiringed structure called a *pterin*, similar to the situation in the enzyme nitrate reductase (∞ Section 17.14).

CO consumption by carboxydotrophic bacteria in nature is a significant ecological process. Although much CO is generated from various human and other sources, CO levels in air have not risen significantly over many years. Because the most significant releases of CO (primarily from automobile exhaust, incomplete combustion of fossil fuels, and the catabolism of lignin) occur in oxic environments, carboxydotrophic bacteria in the upper layers of soil probably represent the most significant sink for CO in nature.

Some of the best-studied carboxydobacteria include species of *Pseudomonas*, such as *P. carboxydovorans* (Table 12.6). At least one carboxydobacterium can grow on CO anaerobically with nitrate as electron acceptor, but this does not seem to be a widespread property of the group. Like the hydrogen bacteria, virtually all isolates of carboxydotrophic bacteria also grow chemoorganotrophically on organic substrates as well as on CO.

 12.3–12.5 Concept Check

Chemolithotrophs are prokaryotes that can oxidize inorganic electron donors and in many cases use CO_2 as sole carbon source.

◆ Compare and contrast the nitrifying bacteria with the sulfur, iron, and hydrogen bacteria in terms of inorganic electron donors used, carbon sources, E_0' of electron donors (◯◯Chapter 17), and habitats.

◆ What major pathway is present for assimilation of CO_2 in many chemolithotrophs?

Table 12.7	**Substrates used by methylotrophic bacteria**[a]

I. Substrates used for growth

Methane, CH_4[b]	Formate, $HCOO^-$
Methanol, CH_3OH	Formamide, $HCONH_2$
Methylamine, CH_3NH_2	Carbon monoxide, CO
Dimethylamine, $(CH_3)_2NH$	Dimethyl ether, $(CH_3)_2O$
Trimethylamine, $(CH_3)_3N$	Dimethyl carbonate,
Tetramethylammonium,	$CH_3OCOOCH_3$
$(CH_3)_4N^+$	Dimethyl sulfoxide,
Trimethylamine N-oxide,	$(CH_3)_2SO$
$(CH_3)_3NO$	Dimethylsulfide,
Trimethylsulfonium, $(CH_3)_3S^+$	$(CH_3)_2S$

II. Substrates oxidized but not used for growth

Ammonium, NH_4^+	Bromomethane, CH_3Br
Ethylene, $H_2C=CH_2$	Higher hydrocarbons
Chloromethane, CH_3Cl	(ethane, propane)

[a] A single isolate does not use all of the above, but at least one methylotrophic bacterium has been reported to oxidize each of the listed compounds.
[b] Methylotrophs able to oxidize methane are called *methanotrophs*.

12.6 Methanotrophs and Methylotrophs

Key Genera: *Methylomonas, Methylobacter*

Methane, CH_4, is found extensively in nature. It is produced in anoxic environments by methanogenic *Archaea* (◯◯Sections 13.4 and 17.17) and is a major gas of anoxic muds, marshes (◯◯Figure 13.5), anoxic zones of lakes, the rumen, and the mammalian intestinal tract. Methane is the major constituent of natural gas and is also present in many coal formations.

Methane is a chemically stable molecule, but a variety of bacteria, the **methanotrophs**, oxidize it readily. These bacteria use methane and a few other one-carbon compounds as electron donors for energy generation and as sole sources of carbon. These bacteria are all aerobes and are widespread in nature in soil and water. They exhibit diverse morphologies and are related both phylogenetically and in their ability to oxidize methane. Methane is scarce in marine sediments, where the major form of anaerobic respiration is sulfate reduction, not methanogenesis (◯◯Section 19.10).

C_1 Metabolism

In addition to methane, a number of other one-carbon compounds can be utilized by microorganisms. A list of these compounds is given in Table 12.7. From a biochemical viewpoint, these compounds share a key characteristic: *they contain no carbon-carbon bonds*. Thus, all carbon-carbon bonds of the cell must be synthesized de novo. Organisms that can grow using only one-carbon organic compounds are called **methylotrophs**. Many, but not all, methylotrophs are also methanotrophs. However, methanotrophs are unique in that they can grow not only on some of the more oxidized one-carbon compounds (Table 12.7), but also on methane.

Methanotrophs possess a specific enzyme, *methane monooxygenase*, for the introduction of an oxygen atom into the methane molecule, leading to the formation of methanol (◯◯Section 17.24). The requirement for O_2 as a reactant in the initial oxygenation of methane thus explains why these methanotrophs are obligate aerobes (we will see in Section 17.24 that methane can be oxidized under anoxic conditions by a consortium of special organisms). All methanotrophs also appear to be *obligate* C_1 utilizers, unable to use compounds containing carbon-carbon bonds. By contrast, many nonmethanotrophic methylotrophs are able to use organic acids, ethanol, and sugars.

Methane-oxidizing bacteria are unique among prokaryotes in possessing relatively large amounts of *sterols*. As we noted in Section 4.5, sterols are found in eukaryotes as a functional part of the membrane system but are absent from most prokaryotes. We will see that in methanotrophs, sterols may be an essential part of the complex internal membrane system involved in methane oxidation. The only other group of prokaryotes where sterols are extensively distributed is in the cell wall-less mycoplasmas (see Section 12.21).

Classification of Methanotrophs

Table 12.8 gives a taxonomic overview of the methanotrophs. These bacteria were initially distinguished on the basis of morphology and formation of resting stages. However, now they are formed into two major groups, based on their internal cell structure, phylogeny, and carbon assimilation pathway. **Type I methanotrophs** assimilate one-carbon compounds via the **ribulose monophosphate cycle**, and are gamma Proteobacteria. By con-

Table 12.8 Some characteristics of methanotrophic bacteria

Organism	Morphology	16S rRNA group[a]	Resting stage	Internal membranes[b]	Citric acid cycle[c]	Carbon assimilation pathway[d]	N₂ fixation	DNA (mol % GC)
Methylomonas	Rod	Gamma	Cystlike body	I	Incomplete	Ribulose monophosphate	No	50–54
Methylomicrobium	Rod	Gamma	None	I	Incomplete	Ribulose monophosphate	No	49–60
Methylobacter	Coccus to ellipsoid	Gamma	Cystlike body	I	Incomplete	Ribulose monophosphate	No	50–54
Methylococcus	Coccus	Gamma	Cystlike body	I	Incomplete	Ribulose monophosphate	Yes	62–64
Methylosinus	Rod or vibrioid	Alpha	Exospore	II	Complete	Serine	Yes	63
Methylocystis	Rod	Alpha	Exospore	II	Complete	Serine	Yes	63
Methylocella[e]	Rod	Alpha	Exospore	II	Complete	Serine	Yes	61

[a] All are Proteobacteria.
[b] Internal membranes: Type I, bundles of disc-shaped vesicles distributed throughout the organism; Type II, paired membranes running along the periphery of the cell. See Figure 12.15.
[c] Organisms with an incomplete citric acid cycle lack the enzyme α-ketoglutarate dehydrogenase and thus cannot oxidize acetate to CO_2.
[d] See Figures 17.59 and 17.60. Unlike other methylotrophs, *Methylococcus* species contain Calvin cycle enzymes.
[e] Acidophilic, growth optimal at pH 5.

trast, **Type II methanotrophs** assimilate C_1 intermediates via the **serine pathway** and are alpha Proteobacteria (Table 12.8). We discuss the biochemical details of these pathways in Section 17.24.

Both groups of methanotrophs contain extensive internal membrane systems involved in methane oxidation. Membranes in Type I methanotrophs are arranged as bundles of disc-shaped vesicles distributed throughout the cell (Figure 12.15b●). By contrast, Type II species possess paired membranes running along the periphery of the cell (Figure 12.15a). The key enzyme *methane monooxygenase* is located in these membranes.

Type I methanotrophs are also characterized by a lack of a complete citric acid cycle (the enzyme *α-ketoglutarate dehydrogenase* is absent), whereas Type II organisms pos-

sess a complete cycle. Absence of a complete citric acid cycle (∞Figure 5.22) greatly diminishes the ability of an organism to grow chemoorganotrophically. If these reactions cannot be run as a cycle, NADH cannot be generated from reactions of the cycle. This prevents growth at the expense of organic compounds metabolized through the citric acid cycle. Type I methanotrophs are therefore obligate methylotrophs.

Ecology and Isolation

Methanotrophs are widespread in aquatic and terrestrial environments, being found wherever stable sources of methane are present. Methane produced in the anoxic regions of lakes rises through the water column, and

(a)

D.W. Ribbons

(b)

D. W. Ribbons

● **Figure 12.15 Electron micrographs of methanotrophs.** (a) A *Methylosinus* species, illustrating a Type II membrane system. Cells are about 0.6 μm in diameter. (b) *Methylococcus capsulatus*, illustrating a Type I membrane system. Cells are about 1 μm in diameter.

methanotrophs are often concentrated in a narrow band at the oxycline, where methane from the anoxic zone meets oxygen from the oxic zone. Methane-oxidizing bacteria therefore play an important role in the carbon cycle, converting methane derived from anoxic decomposition back into cell material (and CO_2).

For the enrichment of methanotrophs all that is needed is a mineral salts medium with an atmosphere of 80% methane and 20% air. Once good growth is obtained, purification is carried out by repeated streaking on mineral salts agar plates, which are incubated in a jar with the methane-air mixture. Colonies appearing on the plates are typically of two types: nonmethanotrophic chemoorganotrophs growing on traces of organic matter in the medium, which appear in 1–2 days, and methanotrophs, which appear after about a week. The colonies of many methanotrophs are pink in color from the production of various carotenoid pigments, and this feature can help in their isolation.

Methanotrophs and Nitrosifying Bacteria

Methanotrophs are able to oxidize ammonia, although they cannot grow chemolithotrophically using ammonia as sole electron donor. In addition to methane oxidation, methane monooxygenase also functions to oxidize ammonia, and a competitive interaction between the two substrates exists. For this reason, ammonia is generally toxic to methanotrophs, and the preferred nitrogen source is nitrate.

It has been speculated that methanotrophic bacteria evolved from the nitrosifying bacteria via mutations that converted an *ammonia* monooxygenase into a *methane* monooxygenase. The fact that both groups of bacteria have elaborate internal membrane systems (see Section 12.3) and are phylogenetically closely related, supports this theory. In addition, however, methano*trophic* bacteria contain some of the same genes and make some of the same proteins as methano*genic* (methane-producing) prokaryotes, all of which phylogenetically are *Archaea* (∞Section 13.4). Such results suggest that lateral (horizontal) gene transfer (∞Section 15.8) has occurred between these groups as the biochemistry of methane metabolism evolved eons ago. We will see how the contrasting processes of methanogenesis and methanotrophy are related in this regard in Sections 17.17 and 17.24.

Methanotrophic Symbionts of Animals

Methanotrophic bacteria and certain marine mussels and sponges develop symbiotic relationships. Some marine mussels live in the vicinity of hydrocarbon seeps on the seafloor, places where methane is released in substantial amounts. Isolated mussel gill tissues consume methane at high rates in the presence of O_2. In these tissues, coccoid-shaped prokaryotes are present in high numbers (Figure 12.16*a*●). The bacterial symbionts contain stacks of intracytoplasmic membranes (Figure 12.16*b*) typical of

(a)

(b)

● **Figure 12.16** **Methanotrophic symbionts of marine mussels.** (a) Electron micrograph of a thin section at low magnification of gill tissue of a marine mussel living near hydrocarbon seeps in the Gulf of Mexico. Note the symbiotic methanotrophs (arrows) in the tissues. (b) High magnification view of gill tissue showing Type I methanotrophs. Note membrane bundles (arrows). The methanotrophs are about 1 μm in diameter. Compare with Figure 12.15*b*.

Type I methanotrophs. The symbionts are found in vacuoles within animal cells near the gill surface, which ensures an effective gaseous exchange with seawater. That these symbionts are indeed methanotrophs has been demonstrated by ribosomal RNA sequencing.

Methane assimilated by the methanotrophs in the mussels is distributed throughout the animal by the excretion of organic compounds by the methanotrophs. Methanotrophic symbioses are therefore conceptually similar to those that develop between sulfide-oxidizing chemolithotrophs and hydrothermal vent tube worms and giant clams (∞Section 19.8). Animal-bacteria symbioses can therefore make prokaryotic cells the basis of a one-step food chain.

 12.6 *Concept Check*

Methylotrophs are prokaryotes able to grow on carbon compounds that lack carbon-carbon bonds. Some methylotrophs are also methanotrophs, able to grow on CH_4. Two classes of methanotrophs are known, each having a number of structural and biochemical properties in common. Methanotrophs reside in water and soil and can also exist as symbionts of marine shellfish.

◆ What is the difference between a *methanotroph* and a *methylotroph*?

◆ What features differentiate Type I from Type II methanotrophs?

◆ What types of animals harbor methanotrophic symbionts and where does the methane the symbionts need come from?

12.7 *Pseudomonas* and the Pseudomonads

Key Genera: *Pseudomonas, Burkholderia, Zymomonas, Xanthomonas*

All the genera in this group are straight or slightly curved chemoorganotrophic aerobic rods with *polar* flagella (Figure 12.17*b*●). The key genera are *Pseudomonas, Comamonas, Ralstonia,* and *Burkholderia,* discussed in some detail here. Other genera include *Xanthomonas, Zoogloea,* and *Gluconobacter. Xanthomonas* is primarily a plant pathogen that is responsible for a number of necrotic plant lesions and that is characterized by its yellow-colored pigments. *Zoogloea* is characterized by its formation of an extracellular fibrillar polymer that causes the cells to aggregate into distinctive flocs (this organism is a major component of activated sewage sludge) (∞Section 28.2). *Gluconobacter* is characterized by its incomplete oxidation of sugars or alcohols to acids, such as the oxidation of glucose to gluconic acid or ethanol to acetic acid (this organism is discussed briefly with the other acetic acid bacteria in Section 12.8). Phylogenetical-

ly, the pseudomonads scatter within the Proteobacteria (see Table 12.1).

Characteristics of Pseudomonads

The distinguishing characteristics of the pseudomonad group are given in Table 12.9. Also listed in this table are the minimal characteristics needed to identify an organism as a pseudomonad. Key identifying characteristics are the absence of gas formation from glucose and the positive oxidase test, both of which help to distinguish pseudomonads from enteric bacteria (see Section 12.11).

Species of the genus *Pseudomonas* and related genera are defined on the basis of phylogeny and various physiological characteristics, as outlined in Tables 12.10 and 12.11. Pseudomonads have very simple nutritional requirements and grow chemoorganotrophically at neutral pH and at temperatures in the mesophilic range.

One of the striking properties of many species of pseudomonads is the wide variety of organic compounds used as carbon and energy sources. Some species utilize over *100* different compounds, and only a few species utilize fewer than 20. As an example of this versatility, a single strain of *Burkholderia cepacia* can use many different sugars, fatty acids, dicarboxylic acids, tricarboxylic acids, alcohols, polyalcohols, glycols, aromatic compounds, amino acids and amines, plus miscellaneous organic compounds not

●**Figure 12.17   Typical pseudomonad colonies and cell morphology and a biochemical pathway common in pseudomonads.** (a) Photograph of colonies of *Burkholderia cepacia* on an agar plate. (b) Shadow-cast TEM preparation of a *Pseudomonas* species. The cell measures about 1 μm in diameter. (c) The Entner–Doudoroff pathway, the major means of glucose catabolism in pseudomonads.

James Shapiro

(a)

Arthur Kelman

(b)

(c)

Table 12.9   Characteristics of pseudomonads

General characteristics:
Straight or curved rods but not vibrioid; size 0.5–1.0 μm by 1.5–4.0 μm; no spores; gram-negative; polar flagella: single or multiple; no sheaths, appendages, or buds; respiratory metabolism, never fermentative, although may produce small amounts of acid from glucose aerobically; use low-molecular-weight organic compounds, not polymers; some are chemolithotrophic, using H_2 or CO as sole electron donor; some can use nitrate as electron acceptor anaerobically; some can use arginine as energy source anaerobically

Minimal characteristics for identification:
Gram-negative, straight or slightly curved; no spores; motile (always); polar flagella (flagellar stain); oxidative-fermentative medium with glucose: tube open, acid produced; tube sealed, acid not produced; gas not produced from glucose (distinguishes them easily from enteric bacteria and *Aeromonas*); oxidase, almost always positive (enterics are oxidase-negative); catalase always positive; photosynthetic pigments absent (distinguishes them from purple nonsulfur bacteria); indole-negative; methyl red-negative; Voges-Proskauer-negative (for discussion of many of these biochemical tests, see Section 24.2)

fitting into any of these categories. On the other hand, pseudomonads generally lack the hydrolytic enzymes necessary to break down polymers into their component monomers. These nutritionally versatile pseudomonads contain numerous inducible operons (∞ Section 8.5) because the catabolism of unusual organic substrates often requires the activity of several different enzymes.

The pseudomonads are ecologically important organisms in soil and water and are probably responsible for the degradation of many soluble compounds derived from the breakdown of plant and animal materials in oxic habitats. They are also capable of breaking down many xenobiotic (not naturally occurring) compounds, such as pesticides and other toxic chemicals, and are thus important agents of bioremediation in the environment.

Many pseudomonads, as well as a variety of other gram-negative *Bacteria*, metabolize glucose via the **Entner–Doudoroff pathway** (Figure 12.17c). Two key enzymes of the Entner–Doudoroff pathway are *6-phosphogluconate de-*

Table 12.10   Subgroups and characteristics of pseudomonads

Group	Phylogenetic group[a]	Characteristics	DNA (mol % GC)
Fluorescent subgroup	Gamma	**Most produce water-soluble, yellow-green fluorescent pigments; do not form poly-β-hydroxybutyrate; single DNA homology group**	
Pseudomonas aeruginosa		Pyocyanin production; growth at up to 43°C; single polar flagellum; capable of denitrification	67
Pseudomonas fluorescens		Does not produce pyocyanin or grow at 43°C; tuft of polar flagella	59–61
Pseudomonas putida		Similar to *P. fluorescens* but does not liquefy gelatin and does grow on benzylamine	60–63
Pseudomonas syringae		Lacks arginine dihydrolase; oxidase-negative; pathogenic to plants	58–60
Pseudomonas stutzeri		Soil saprophyte; strong denitrifier and nonfluorescent	62
Acidovorans subgroup	Beta	**Nonpigmented; form poly-β-hydroxybutyrate; tuft of polar flagella; do not use carbohydrates; single DNA homology group**	
Commamonas acidovorans		Uses muconic acid as sole carbon source and electron donor	67
Commamonas testosteroni		Uses testosterone as sole carbon source	62
Pseudomallei-cepacia subgroup	Beta	**No fluorescent pigments; tuft of polar flagella; forms poly-β-hydroxybutyrate; single DNA homology group**	62
Burkholderia cepacia		Extreme nutritional versatility; some strains pathogenic to plants	67
Burkholderia pseudomallei		Causes melioidosis in animals; nutritionally versatile	69
Burkholderia mallei		Causes glanders in animals; nonmotile; nutritionally restricted	69
Diminuta-vesicularis subgroup	Alpha	**Single flagellum of very short wavelength; require vitamins (pantothenate, biotin, B_{12})**	
Brevundimonas diminuta		Nonpigmented; does not use sugars	66–67
Brevundimonas vesicularis		Carotenoid pigment; uses sugars	66
Ralstonia subgroup	Beta		
Ralstonia solanacearum		Plant pathogen	66–68
Ralstonia saccharophila		Grows chemolithotrophically with H_2; digests starch	69
Stenotrophomonas maltophilia		Requires methionine; does not use NO_3^- as N source; oxidase-negative	67

[a] All pseudomonads are members of the Proteobacteria (see Table 12.1).

hydrase and 2-keto-3-deoxyglucosephosphate aldolase (Figure 12.17c). A survey for the presence of these enzymes in a variety of bacteria has shown that they are absent from gram-positive *Bacteria* but are generally present in bacteria of the genera *Pseudomonas, Rhizobium, Agrobacterium, Zymomonas*, and several other gram-negative *Bacteria*.

Pathogenic Pseudomonads

A number of pseudomonads are pathogenic (Table 12.11). Among the fluorescent pseudomonads, the species *Pseudomonas aeruginosa* is frequently associated with infections of the urinary and respiratory tracts in humans. *Pseudomonas aeruginosa* infections are also common in patients receiving treatment for severe burns or other traumatic skin damage and in people suffering from cystic fibrosis (∞Section 31.7). *Pseudomonas aeruginosa* is not an obligate parasite. Instead, the organism is an *opportunist*, initiating infections in individuals whose resistance is low. In addition to urinary infections, it can also cause systemic infections, usually in individuals who have experienced extensive skin damage.

P. aeruginosa is naturally resistant to many of the widely used antibiotics, so chemotherapy is often difficult. Resistance is due to a *resistance transfer plasmid (R plasmid)* (∞Sections 10.9 and 20.12), which is a plasmid carrying genes encoding proteins that detoxify various antibiotics. *Pseudomonas aeruginosa* is commonly found in the hospital environment and can easily infect patients receiving treatment for other illnesses (∞Section 25.7). Polymyxin, an antibiotic not ordinarily used in human therapy because of its toxicity, is effective against *P. aeruginosa* and is used in certain medical situations.

Certain species of *Pseudomonas, Ralstonia*, and *Burkholderia* and the genus *Xanthomonas* are well-known plant pathogens (phytopathogens) (see Table 12.11). In many cases these organisms are so highly adapted to the plant environment that they are rarely isolated from other habitats, including soil. Phytopathogens frequently inhabit nonhost plants (where disease symptoms are not apparent) and from there become transmitted to host plants and initiate infection. Disease symptoms vary considerably depending on the particular phytopathogen and host plant. The pathogen releases plant toxins, lytic enzymes, plant growth factors, and other substances that destroy or distort plant tissue. In many cases the disease symptoms help identify the phytopathogen. Thus, *Pseudomonas syringae* is typically isolated from leaves showing chlorotic (yellowing) lesions, whereas *P. marginalis* is a typical "soft-rot" pathogen, infecting stems and shoots but rarely leaves.

Zymomonas

The genus *Zymomonas* consists of large, gram-negative rods that carry out a vigorous fermentation of sugars to ethanol. Although strictly fermentative, *Zymomonas* shows phylogenetic affiliation with the pseudomonads and employs the Entner–Doudoroff pathway (Figure 12.17c).

Zymomonas is a common organism involved in alcoholic fermentation of various plant saps, and in many tropical areas of South and Central America, Africa, and Asia it occupies a position in the fermented beverage industry similar to that of *Saccharomyces cerevisiae* (yeast) in North America and Europe. For example, *Zymomonas* is involved in the alcoholic fermentation of agave in Mexico to form the drink *pulque* and the fermentation of palm sap in many other tropical areas. It also carries out an alcoholic fermentation of sugarcane juice and honey. Although *Zymomonas* is rarely the sole organism involved in these alcoholic fermentations, it is often the dominant organism and is probably responsible for the production

Table 12.11	Pathogenic pseudomonads
Species	**Relationship to disease**
Animal pathogens	
Pseudomonas aeruginosa	Opportunistic pathogen, especially in hospitals; in patients with metabolic, hematologic, and malignant diseases; hospital-acquired (nosocomial) infections from catheterizations, tracheostomies, lumbar punctures, and intravenous infusions; in patients given prolonged treatment with immunosuppressive agents, corticosteroids, antibiotics and radiation; may contaminate surgical wounds, abscesses, burns, ear infections, lungs of patients treated with antibiotics; cystic fibrosis; primarily a soil organism
Pseudomonas fluorescens	Rarely pathogenic, as does not grow well at 37°C; may grow in and contaminate blood and blood products under refrigeration
Stenotrophomonas maltophilia	A ubiquitous, free-living organism that is a common nosocomial pathogen
Burkholderia cepacia	Causes onion bulb rot; has also been isolated from humans and from environmental sources of medical importance
Burkholderia pseudomallei	Causes melioidosis, a disease endemic in animals and humans in Southeast Asia
Burkholderia mallei	Causes glanders, a disease of horses that is occasionally transmitted to humans
Pseudomonas stutzeri	Often isolated from humans and environmental sources; may live saprophytically in the body
Plant pathogens	
Ralstonia solanacearum	Causes wilts of many cultivated plants (for example, potato, tomato, tobacco, peanut)
Pseudomonas syringae	Attacks foliage, causing chlorosis and necrotic lesions on leaves; rarely found free in soil
Pseudomonas marginalis	Causes soft rot of various plants; active pectinolytic species
Xanthomonas campestris	Causes necrotic lesions on foliage, stems, fruits; also causes wilts and tissue rots; rarely found free in soil

of most of the ethanol in these beverages. *Zymomonas* is also responsible for spoilage of fruit juices, such as apple and pear ciders, and is also a constituent of the bacterial flora of spoiled beer.

Zymomonas is distinguished from *Pseudomonas* by its fermentative metabolism, microaerophilic to anaerobic nature, oxidase negativity, and other molecular taxonomic characteristics. It also resembles the acetic acid bacteria (see Section 12.8), and it is often found in nature associated with these organisms. This is of interest because, like yeast, *Zymomonas* ferments glucose to ethanol, whereas the acetic acid bacteria oxidize ethanol to acetic acid. Thus, the acetic acid bacteria depend on the activity of yeast and *Zymomonas* for the production of their growth substrate, ethanol.

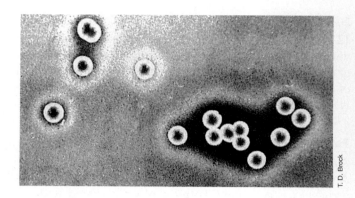

● **Figure 12.18** **Photograph of colonies of *Acetobacter aceti* on calcium carbonate agar containing ethanol as energy source.** Note the clearing around the colonies due to the dissolution of calcium carbonate ($CaCO_3$) by the acetic acid produced by the bacteria.

12.8 Acetic Acid *Bacteria*

Key Genera: *Acetobacter, Gluconobacter*

As originally defined, the *acetic acid bacteria* comprised a group of gram-negative, aerobic, motile rods that carried out the *incomplete* oxidation of alcohols and sugars, leading to the accumulation of organic acids as end products. With *ethanol* as a substrate, *acetic acid* is produced, which give acetic acid bacteria their name.

Acetic acid bacteria have a relatively high tolerance to acidic conditions; most strains can grow well at pH values lower than 5. This *acid tolerance* should of course be essential for an organism producing large amounts of acid. The acetic acid bacteria are a heterogeneous assemblage, comprising both peritrichously and polarly flagellated organisms. The *polarly* flagellated organisms are classified in the genus *Gluconobacter*, while the peritrichously flagellated species are grouped into the genus *Acetobacter*. All known acetic acid bacteria group phylogenetically with the alpha Proteobacteria (see Table 12.1).

In addition to flagellation, *Acetobacter* differs from *Gluconobacter* in being able to further oxidize the acetic acid it forms to CO_2. This difference in ability to oxidize acetic acid is related to the presence of a complete citric acid cycle. *Gluconobacter*, which *lacks* a complete citric acid cycle, is unable to oxidize acetic acid, whereas *Acetobacter*, which has all enzymes of the cycle, can oxidize it.

Ecology and Industrial Uses

The acetic acid bacteria are commonly found in alcoholic juices. Acetic acid bacteria can often be isolated from an alcoholic fruit juice such as hard cider or wine, or from beer. Colonies of acetic acid bacteria can be recognized on $CaCO_3$ agar plates containing ethanol, since the acetic acid produced causes a dissolution and clearing of the otherwise insoluble $CaCO_3$ (Figure 12.18●). Cultures of acetic acid bacteria are used in the commercial production of vinegar (⚭Section 30.11).

In addition to ethanol, these organisms carry out an incomplete oxidation of such organic compounds as higher alcohols and sugars. For instance, glucose is oxidized only to gluconic acid, galactose to galactonic acid, arabinose to arabonic acid, and so on. This property of "underoxidation" is exploited in the manufacture of ascorbic acid (vitamin C). Ascorbic acid can be formed from sorbose, but sorbose is difficult to synthesize chemically. It is, however, conveniently obtainable from acetic acid bacteria, which oxidize sorbitol (a readily available sugar alcohol) only to sorbose, a process called *biotransformation* (⚭Section 30.8).

Another interesting property of some acetic acid bacteria is their ability to synthesize *cellulose*. The cellulose formed does not differ significantly from that of plant cellulose, with the exception that it is pure and not mixed in with other polymers like the hemicelluloses, pectin, or lignins of plants. Cellulose from acetic acid bacteria is formed as a matrix outside the cell wall and causes cells to become embedded in a tangled mass of cellulose microfibrils. When these species of acetic acid bacteria grow in an unshaken vessel, they form a surface pellicle of cellulose in which the bacteria develop. Since these bacteria are obligate aerobes, the ability to form such a pellicle may be a means by which the organisms remain at the surface of the liquid where oxygen is readily available.

12.9 Free-Living Aerobic Nitrogen-Fixing *Bacteria*

Key Genera: *Azotobacter, Azomonas*

A variety of organisms inhabit soil and are capable of fixing N_2 *aerobically* (Table 12.12). The genus *Azotobacter* comprises large, gram-negative, obligately aerobic rods capable of fixing N_2 nonsymbiotically (Figure 12.19●). The first species f this genus was discovered by the Dutch microbiologist Martinus Beijerinck early in the twentieth century (⚭Section 1.7). Beijerinck used an aerobic enrichment culture technique with a medium containing N_2 (air) but devoid of a combined nitrogen source (⚭Section 18.1 and Figure 18.1). Phylogenetical-

Table 12.12	Genera of free-living aerobic nitrogen-fixing bacteria
Characteristics	**Genus/DNA (mol % GC)**
Gamma Proteobacteria	
Large rod; produces cysts; primarily found in neutral to alkaline soils	*Azotobacter* (63–67)
Large rod; no cysts; primarily aquatic	*Azomonas* (52–59)
Alpha Proteobacteria	
Microaerophilic rod; associates with plants	*Azospirillum* (69–71)
Pear-shaped rod with large lipid bodies at each end; produces extensive slime; inhabits acidic soils	*Beijerinckia* (54–59)
Rods; form coarse, wrinkled colonies	*Derxia* (69–73)
Beta Proteobacteria	
Small curved cells; no cysts	*Azoarcus* (65)
Very thin curved cells; no cysts	*Azovibrio* (64–65)
Very thin rods to vibrios; no cysts	*Azospira* (65–66)
Cells form coils up to 50 μm long; no cysts	*Azonexus* (65–66)

ly, most free-living nitrogen-fixing bacteria are alpha, beta, or gamma Proteobacteria (Table 12.12).

Taxonomy

The major free-living nitrogen-fixing bacteria that have been studied include *Azotobacter*, *Azospirillum*, and *Beijerinckia*. *Azotobacter* cells are large, many isolates being almost the size of yeasts, with diameters of 2–4 μm or more. Pleomorphism is common, and a variety of cell shapes and sizes have been described. Some strains are motile by peritrichous flagella. On

carbohydrate-containing media, extensive capsules or slime layers are produced by free-living N_2-fixing bacteria (Figure 12.20●; ∞Figure 17.71).

Despite the fact that *Azotobacter* is an obligate aerobe, its nitrogenase, the enzyme that catalyzes biological N_2 fixation (∞Section 17.28), is O_2-sensitive. It is thought that the high respiratory rate characteristic of *Azotobacter* and the abundant capsular slime help protect nitrogenase from O_2. *Azotobacter* is able to grow on a wide variety of carbohydrates, alcohols, and organic acids. The metabolism of carbon compounds is strictly oxidative, and acids or other fermentation products are rarely produced. All members fix nitrogen, but growth also occurs on simple forms of combined nitrogen: ammonia, urea, and nitrate.

Azotobacter can form resting structures called *cysts*. Like bacterial endospores, *Azotobacter* cysts (Figure 12.19*b*) show negligible endogenous respiration and are resistant to desiccation, mechanical disintegration, and ultraviolet and ionizing radiation. In contrast to endospores, however, cysts are *not* especially heat-resistant, and they are not completely dormant because they rapidly oxidize exogenous energy sources.

The remaining key genera of free-living N_2 fixers include *Azomonas*, a genus of large, rod-shaped bacteria that resemble *Azotobacter* except that they do not produce

(a)

(b)

● **Figure 12.20** **Examples of slime production by free-living N_2-fixing bacteria.** (a) Cells of *Derxia gummosa* encased in slime. Cells are about 1–1.2 μm wide. (b) Colonies of *Beijerinckia* species growing on a carbohydrate-containing medium. Note the raised, glistening appearance of the colonies due to abundant capsular slime.

(a)

(b)

● **Figure 12.19** *Azotobacter vinelandii.* (a) vegetative cells and (b) cysts visualized by phase contrast microscopy. A cell measures about 2 μm in diameter, and a cyst about 3 μm. Compare with Figure 1.14*b*.

cysts and are primarily aquatic, and *Beijerinckia* (Figure 12.20*b*) and *Derxia* (Figure 12.21●), two genera that grow well in acidic soils. *Azospirillum*, a rod to spirillum-shaped nitrogen-fixing bacterium that forms nonspecific symbiotic associations with various plants, in particular, corn (Table 12.12), rounds out this group.

Azotobacter and Alternative Nitrogenases

We study the important process of biological N_2 fixation in Section 17.28. There we will learn of the central importance of the metals molybdenum (Mo) and iron (Fe) to the enzyme nitrogenase. The species *Azotobacter chroococcum* was the first nitrogen-fixing bacterium shown capable of growth on N_2 in the *absence* of molybdenum. In *A. chroococcum*, it was first shown that in place of Mo nitrogenase, two "alternative nitrogenases" containing either vanadium (V) plus Fe or Fe only can be made under certain growth conditions (when Mo is absent). Subsequent investigations of other nitrogen-fixing bacteria have shown that these "backup" nitrogenase systems are widely distributed among nitrogen-fixing bacteria, including *Archaea*, in which a few species fix N_2.

⬡ *12.7–12.9 Concept Check*

Pseudomonads include many gram-negative chemoorganotrophic aerobic rods; many N_2-fixing species are phylogenetically closely related. The acetic acid bacteria are also phylogenetically related to pseudomonads and are characterized by an ability to oxidize ethanol to acetate aerobically.

◆ Compare and contrast the pseudomonads, *Azotobacter*, and the acetic acid bacteria in terms of O_2 and nitrogen requirements, electron donors, pathogenicity, and habitats.

◆ Compare and contrast the organisms *Acetobacter* and *Gluconobacter* in as many ways as you can think of.

(a) Michael K. Ochman   *(b)* J. H. Becking

● **Figure 12.21 Phase-contrast photomicrographs of two genera of acid-tolerant, free-living N_2-fixing bacteria.** (a) *Beijerinckia indica.* The cells are roughly pear-shaped, about 0.8 μm in diameter, and contain a large globule of poly-β-hydroxybutyrate (∞Section 4.11) at each end. (b) *Derxia gummosa.* Cells are about 1 μm in diameter.

12.10 *Neisseria, Chromobacterium, and Relatives*

Key Genera: *Neisseria, Chromobacterium*

This group of beta and gamma Proteobacteria comprises a diverse collection of organisms, related phylogenetically as well as by Gram stain, morphology, lack of motility, and aerobic metabolism. The genera *Neisseria, Moraxella, Branhamella, Kingella,* and *Acinetobacter* are distinguished as outlined in Table 12.13.

In the genus *Neisseria*, the cells are always cocci (Figure 12.22*a*●; ∞Figure 26.28), whereas cells of the other genera are rod-shaped, becoming coccoid only in the stationary phase of growth. This has led to designation of these organisms as **coccobacilli.** Organisms of the genera *Neisseria, Kingella,* and *Moraxella* are commonly isolated from animals, and some of them are pathogenic. We discuss the clinical microbiology of *Neisseria gonorrhoeae,* the causative agent of the disease *gonorrhea,* in Section 24.1 and the pathogenesis of gonorrhea itself in Section 26.12.

Species of *Acinetobacter* are common soil and water organisms, although they are occasionally found as parasites of animals and have been implicated in some nosocomial (hospital-associated) infections. Some strains of *Moraxella* and *Acinetobacter* possess the interesting property of *twitching motility,* exhibited as brief translocative movements or "jumps" covering distances of about 1–5 μm. Twitching bacteria contain special pili (∞ Section 4.10) that facilitate the movement process.

Table 12.13	**Characteristics of the genera of gram-negative cocci[a]**
Characteristics	**Genus/DNA (mol % GC)**
I. Oxidase-positive, penicillin-sensitive:	
Cocci; complex nutrition, utilize carbohydrates, obligate aerobes; Beta Proteobacteria or Gamma Proteobacteria	*Neisseria* (49–55) *Moraxella* (40–46)
Rods or cocci; generally no growth factor requirements, generally do not utilize carbohydrates; do not contain flagella, but some species exhibit twitching motility; many are commensals or pathogens of animals; Beta Proteobacteria	*Branhamella* (40–46) *Kingella* (47–55)
II. Oxidase-negative, penicillin-resistant:	
Some strains can utilize a restricted range of sugars, and some exhibit twitching motility; saprophytes in soil, water, and sewage; Gamma Proteobacteria	*Acinetobacter* (38–47)

[a] All are Proteobacteria.

● **Figure 12.22** *Neisseria* and *Chromobacterium.* (a) Transmission electron micrograph of cells of *Neisseria gonorrhoeae* showing the typical diplococcus cell arrangements. (b) A large colony of *Chromobacterium violaceum.* The purple pigment is an aromatic compound called *violacein.*

(a)

(b)

Chromobacterium is a close phylogenetic relative of *Neisseria* but is rod-shaped in morphology, resembling the pseudomonads or enteric bacteria. The best-known *Chromobacterium* species is *C. violaceum*, a purple-pigmented organism (Figure 12.22b) found in soil and water and occasionally in pus-forming infections of humans and other animals. *Chromobacterium violaceum* and a few other chromobacteria produce the purple pigment *violacein* (Figure 12.22b), a water-insoluble pigment that has antibiotic-like properties and is produced only in media containing the amino acid tryptophan. Like enteric bacteria, *Chromobacterium* is a facultative aerobe, growing fermentatively on sugars and aerobically on a variety of carbon sources.

12.11 Enteric *Bacteria*

Key Genera: *Escherichia, Salmonella, Proteus, Enterobacter*

The **enteric bacteria** comprise a relatively homogeneous phylogenetic group within the gamma Proteobacteria and are characterized phenotypically as gram-negative, nonsporulating rods, nonmotile or motile by *peritrichous* flagella (Figure 12.23a●), facultative aerobes. Enteric bacteria are also oxidase-*negative*, have relatively simple nutritional requirements, and ferment sugars to a variety of end products. The phenotypic characteristics used to separate the enteric bacteria from other bacteria of similar morphology and physiology are given in Table 12.14.

Among the enteric bacteria are many species pathogenic to humans, animals, or plants as well as other strains of industrial importance. Undoubtedly more is known about *Escherichia coli* than about any other bacterial species (∞Section 10.19). Because of the medical importance of the enteric bacteria, an extremely large number of isolates have been studied and characterized, and a fair number of distinct genera have been defined. Despite the fact that there is marked genetic relatedness among many of the enteric bacteria, as shown by high genomic DNA homology (∞Section 11.11) and widespread genetic recombination, separate genera are maintained, largely for practical reasons, such as identification in clinical microbiology. Since these organisms are

(a)

(b)

Stoichiometry:

2 pyruvate + NADH ⟶ 2 CO_2 + butanediol

(c)

● **Figure 12.23** **Butanediol production.** (a) Electron micrograph of a shadow-cast preparation of cells of the butanediol-producing enteric bacterium *Erwinia carotovora.* Cells are about 0.8 μm wide. Note the peritrichously arranged flagella (∞Section 4.14). (b) Biochemical pathway for formation of butanediol from two molecules of pyruvate by butanediol fermenters. (c) Overall stoichiometry. Note how only *one* NADH but *two* pyruvates are required to make butanediol.

Table 12.14	Defining characteristics of the enteric bacteria

General characteristics:
Gram-negative straight rods; motile by peritrichous flagella, or nonmotile; nonsporulating; facultative aerobes, producing acid from glucose; sodium neither required nor stimulatory; catalase-positive; oxidase-negative; usually reduce nitrate to nitrite (not to N_2); 16S rRNA of gamma Proteobacteria (see Table 12.1)

Key tests to distinguish enteric bacteria from other bacteria of similar morphology[a]:
Oxidase test, enterics always negative—separates enterics from oxidase-positive bacteria of genera *Pseudomonas, Aeromonas, Vibrio, Alcaligenes, Achromobacter, Flavobacterium, Cardiobacterium*, which may have similar morphology; nitrate reduced only to nitrite, (assay for nitrite after growth)— distinguishes enteric bacteria from bacteria that reduce nitrate to N_2 (gas formation detected), such as *Pseudomonas* and many other oxidase-positive bacteria; ability to ferment glucose— distinguishes enterics from obligately aerobic bacteria

[a] See Section 24.2 and Figure 24.7.

frequently cultured from diseased states, some means of identifying them is necessary to facilitate rapid treatment, and thus phenotypic characteristics have traditionally been very important in distinguishing between genera of enteric *Bacteria*.

Fermentation Patterns in Enteric Bacteria

One important taxonomic characteristic separating the various genera of enteric bacteria is the type and propor-

tion of *fermentation products* produced by anaerobic fermentation of glucose. Two broad patterns are recognized, the *mixed-acid fermentation* and the *2,3-butanediol fermentation* (Figure 12.24●).

In the **mixed-acid fermentation**, *three* acids are formed in significant amounts—acetic, lactic, and succinic; ethanol, CO_2, and H_2 are also formed, but *not* butanediol. In the **butanediol fermentation**, *smaller* amounts of acids are formed, and butanediol, ethanol, CO_2, and H_2 are the main products. As a result of a mixed-acid fermentation, *equal* amounts of CO_2 and H_2 are produced, whereas with a butanediol fermentation, considerably *more* CO_2 than H_2 is produced. This is because mixed-acid fermenters produce CO_2 only from formic acid by means of the enzyme system *formate hydrogen lyase*:

$$HCOOH \longrightarrow H_2 + CO_2$$

and this reaction results in equal amounts of CO_2 and H_2. The butanediol fermenters also produce CO_2 and H_2 from formic acid, but they produce two additional molecules of CO_2 during the formation of each molecule of butanediol (Figure 12.23*b*).

Identification of Enteric Bacteria

A variety of diagnostic tests and differential media are used to separate the various genera in the two broad groups of enteric bacteria (∞Tables 24.2 and 24.3). On the basis of these and other tests, genera can be defined as outlined in Tables 12.15 and 12.16. Because enteric

● **Figure 12.24 Enteric fermentations.** Distinction between (a) mixed acid and (b) butanediol fermentation in enteric bacteria. The bold arrows indicate reactions leading to major products. Dashed arrows indicate minor products. The upper photo shows the production of acid (yellow color) and gas (in the inverted Durham tube) in a culture of *E. coli* (purple tube was uninoculated). The bottom photo shows the pink-red color in the Voges-Proskauer (VP) test, which indicates butanediol production, following growth of *Enterobacter aerogenes*. The left (yellow) tube was uninoculated. Note the major difference in CO_2 production in the two pathways, butanediol production leading to substantially greater CO_2 yields. Because the production of one molecule of butanediol from two pyruvates consumes only one NADH (see pathway in Figure 12.23 *b, c*), 0.5 molecules of ethanol must be made for each butanediol produced to consume the second NADH generated in glycolysis.

(a) **Mixed acid fermentation** (for example, *Escherichia coli*)

Typical products (molar amounts)

Acidic : neutral
4 : 1
CO_2 : H_2
1 : 1

(b) **Butanediol fermentation** (for example, *Enterobacter*)

Typical products (molar amounts)

Acidic : neutral
1 : 6
CO_2 : H_2
5 : 1

Cheryl L. Broadie and John Vercillo

Table 12.15 Key diagnostic reactions used to separate some key genera of enteric bacteria[a]

Genus	H$_2$S(TSI)	Urease	VP[b]	Indole	Motility	Gas from glucose[b]	β-Galactosidase
Escherichia	−	−	−	+	+ or −	+	+
Enterobacter	−	−	+	−	+	+	+
Shigella	−	−	−	+ or −	−	−	+ or −
Edwardsiella	+	−	−	+	+	+	−
Salmonella	+	−	−	−	+	+	+ or −
Klebsiella	−	+	+ or −	−	−	+	+
Citrobacter	+ or −	−	−	−	+	+	+
Proteus	+ or −	+	−	+ or −	+	+ or −	−
Providencia	−	−	−	+	+	−	−
Yersinia	−	+	−	−	+[c]	−	+
Hafnia	−	−	+	−	+	+	+ or −

Genus	KCN	Citrate	Mucate utilization	Phenyl-methyl red	Tartrate utilization	Alanine deaminase	DNA (mol % GC)
Escherichia	−	−	+	+	+	−	48–52
Enterobacter	+	+	+	−	−	−	52–60
Shigella	−	−	−	+	−	−	50
Edwardsiella	−	−	−	+ or −	−	−	53–59
Salmonella	−	+ or −	+ or −	+	+ or −	−	50–53
Klebsiella	+	+	+	−	+ or −	−	53–58
Citrobacter	+ or −	+	+	+	+	−	50–52
Proteus	+	+ or −	−	+	+	+	38–41
Providencia	+	+	−	+	+	+	39–42
Yersinia	−	−	−	+	−	−	46–50
Hafnia	+	+	−	+	−	−	48–49

[a] See Table 24.1 for the procedures for these diagnostic reactions.
[b] See Figure 12.24 for a photo of this reaction.
[c] Motile when grown at room temperature; nonmotile at 37°C.

bacteria are genetically very closely related, their positive identification often presents considerable difficulty. In clinical laboratories, identification is often based on computer analysis of a large number of diagnostic tests carried out using miniaturized rapid diagnostic media kits and immunological and nucleic acid probes (Figure 24.1), with consideration being given for variable reactions of exceptional strains. Alternatively, molecular methods can be used (Section 24.5–24.12). The separation of genera given in Tables 12.15 and 12.16 must only be considered as approximate, for it is always possible to isolate a strain that does not possess one or another characteristic normally considered positive for the genus as a whole. With these limitations in mind, an even more simplified separation of the key genera is found in Figure 12.25●. This key permits a quick decision on the likely *genus* in which to place a new enteric isolate. We now consider some of the properties of key genera.

The Genus *Escherichia*

Members of the genus *Escherichia* are almost universal inhabitants of the intestinal tract of humans and warm-blooded animals, although they are by no means the dominant organisms in these habitats. *Escherichia* may play a nutritional role in the intestinal tract by synthesizing vitamins, particularly vitamin K. As a facultative aer-

obe, this organism probably also helps consume oxygen, thus rendering the large intestine anoxic. Wild-type *Escherichia* strains rarely show any growth-factor requirements and are able to grow on a wide variety of carbon and energy sources such as sugars, amino acids, organic acids, and so on.

Some strains of *Escherichia* are pathogenic. These have been implicated in diarrhea in infants, occasionally occurring in epidemic proportions in children's nurseries or obstetric wards. *Escherichia* may also cause urinary tract infections in older persons or in those whose resistance has been lowered by surgical treatment or by exposure to ionizing radiation. Enteropathogenic strains of *E. coli* are becoming more frequently implicated in dysentery-like infections and generalized fevers (Sections 21.6, 21.11, and 29.8). As noted there, these strains form a structure called *K antigen*, permitting attachment and colonization of the small intestine, and *enterotoxin*, responsible for the symptoms of diarrhea.

Salmonella and *Shigella*

Salmonella and *Escherichia* are quite closely related, the two genera showing about 50% genomic homology by DNA:DNA hybridization (Section 11.11). However, in contrast to most *Escherichia*, members of the genus *Salmonella* are usually pathogenic, either to humans or to

Table 12.16 Key diagnostic reactions used to separate the various genera of 2,3-butanediol producers[a]

Genus	Ornithine decarboxyl-ase	Gelatin hydrolysis	Temperature optimum (° C)	Pigmentation	Motility	Lactose	DNase	Sorbitol	DNA (mol % GC)
Klebsiella	–	–	37–40	None	–	+	–	+	53–58
Enterobacter	+	Slow	37–40	Yellow (or none)	+	+	–	+	52–60
Serratia	+	+	37–40	Red (or none)	+	–	+	–	52–60
Erwinia[b]	–	+ or –	27–30	Yellow (or none)	+	+ or –	–	+	50–58
Hafnia	+	–	35	None	+	–	–	–	48–49

[a] See Table 24.3 for a description of these diagnostic tests.
[b] See Figure 12.23a.

other warm-blooded animals. In humans the most common diseases caused by salmonellas are *typhoid fever* and *gastroenteritis* (∞Sections 28.8 and 29.7). The salmonellas are characterized immunologically on the basis of three cell surface antigens, the O, or cell wall (somatic) antigen; the H, or flagellar, antigen; and the Vi (outer polysaccharide layer) antigen, found primarily in strains of *Salmonella* causing typhoid fever.

The O antigens are part of the lipopolysaccharides that comprise the outermost layer of the outer membrane of these organisms (∞Sections 4.9 and 21.12). We discussed the chemical structure of lipopolysaccharides in Section 4.9 (∞Figures 4.35 and 4.36). The genus *Salmonella* contains over 1000 distinct serotypes having different antigenic specificities in their O antigens. Additional antigenic subdivisions are based on the antigenic specificities of the flagellar (H) antigens. There is little or no correlation between the antigenic type of a *Salmonella* and the disease symptoms elicited, but immunological typing permits tracking a single strain involved in an epidemic.

The shigellas are also genetically very closely related to *Escherichia*. Tests for DNA homology show that strains of *Shigella* have 70% or even higher genomic homology with *Escherichia coli*. In contrast to *Escherichia*, however, *Shigella* is commonly pathogenic to humans, causing a rather severe gastroenteritis called *bacillary dysentery*. *Shigella dysenteriae* is transmitted by food- and waterborne routes and is capable of invading intestinal epithelial cells (∞Section 29.11). Once established, it produces both an endotoxin and a neurotoxin that exhibits enterotoxic (gastrointestinal) effects.

Proteus

The genus *Proteus* is characterized by rapid motility (Figure 12.26●) and by production of the enzyme *urease*. By genomic DNA homology it shows only a distant relationship to *Escherichia coli*. *Proteus* is a frequent cause of urinary tract infections in humans and probably benefits in this regard from its ready ability to degrade urea. Because of the rapid motility of *Proteus* cells, colonies growing on agar plates often exhibit a characteristic **swarming** phenomenon (Figure 12.26b). Cells at the edge of the growing colony are more rapidly motile than those in the center of the colony. The former move a short distance away from the colony in a mass and then undergo a reduction in motility, settle down, and divide, forming a new population of motile cells that again swarm. As a result, the mature colony appears as a series of concentric rings, with higher concentrations of cells alternating with lower concentrations (Figure 12.26b).

Diagnostic test	Go to number
1 MR+; VP – (mixed-acid fermenters)	2
MR –; VP + (butanediol producers)	7
2 Urease +	*Proteus*
Urease –	3
3 H₂S (TSI) +	4
H₂S (TSI) –	6
4 KCN +	*Citrobacter*
KCN –	5
5 Indole +; citrate –	*Edwardsiella* *Salmonella*
Indole –; citrate +	
6 Gas from glucose	*Escherichia* *Shigella*
No gas from glucose	
7 Nonmotile; ornithine –	*Klebsiella*
Motile; ornithine +	8
8 Gelatin+; DNAse +	*Serratia* (red pigment)
Gelatin slow; DNAse –	*Enterobacter*

Key
Mixed-acid fermenters
Butanediol producers

● **Figure 12.25 A simple key to the main genera of enteric bacteria.** Only the most common genera are given. See text for precautions in the use of this key. Diagnostic tests for use with this figure are given in Table 24.3. Other characteristics of the genera are given in Tables 12.14–12.16. Color coding is as in Figure 12.24.

Butanediol Fermenters: *Enterobacter*, *Klebsiella*, and *Serratia*

The butanediol fermenters are genetically more closely related to each other than to the mixed-acid fermenters, a finding that is in agreement with the observed physiological differences. Their DNA base composition is higher,

(a)

John Vercillo and Cheryl Broadie

● **Figure 12.27 Colonies of *Serratia marcescens*.** The orange-red pigmentation is due to the pyrrole-containing pigment *prodigiosin*.

James Shapiro

(b)

● **Figure 12.26 Swarming in *Proteus*.** (a) Cells of *Proteus mirabilis* stained with a flagella stain: the peritrichous flagella of each cell group into a bundle. (b) Photo of a swarming colony of *Proteus vulgaris*. Note the concentric rings.

53–58% GC, and a classification of this group is outlined in Table 12.16.

Enterobacter aerogenes is a common species in water and sewage as well as the intestinal tract of warm-blooded animals and is an occasional pathogen in urinary tract infections. One species of *Klebsiella*, *K. pneumoniae*, occasionally causes pneumonia in humans, but klebsiellas are most commonly found in soil and water. Most *Klebsiella* strains also fix N_2 (∞Section 17.28), a property not found among other enteric bacteria.

The genus *Serratia* forms a series of red pyrrole-containing pigments called **prodigiosins** (Figure 12.27●). Prodigiosin is produced in stationary phase as a secondary metabolite (∞Section 30.2) and is of interest because it contains the pyrrole ring also found in the pigments involved in energy transfer: porphyrins, chlorophylls, and phycobilins (∞Sections 17.2 and 17.3). However, there is no evidence that prodigiosin plays any role in energy transfer, and its exact function is unknown. Species of *Serratia* can be isolated from water and soil as well as from the gut of various insects and vertebrates and occasionally from the intestines of humans.

12.12 *Vibrio* and *Photobacterium*

Key Genera: *Vibrio, Photobacterium*

The *Vibrio* group contains gram-negative, facultatively aerobic *rods* and *curved rods* that possess a fermentative metabolism. Most of the members of the *Vibrio* group are polarly flagellated, although some are peritrichously flagellated. One key difference between the *Vibrio* group and enteric bacteria is that members of the former are oxidase-*positive* (∞Table 24.3), whereas members of the latter are oxidase-*negative*. Although *Pseudomonas* is also polarly flagellated and oxidase-positive, it is *not* fermentative. Hence, it can be separated from the vibrios by simple sugar fermentation tests. The best known genera of the *Vibrio* group include *Vibrio* and *Photobacterium*.

Most vibrios and related bacteria are aquatic, found either in freshwater or marine habitats. *Vibrio cholerae* is the specific cause of the disease *cholera* in humans (∞Sections 21.11 and 28.5); the organism does not normally infect other hosts. Cholera is one of the most common infectious human diseases in underdeveloped countries and one that has had a long history (the famous early medical microbiologist Robert Koch first isolated *Vibrio cholerae* in 1884). The organism is transmitted almost exclusively via water, and studies on its distribution in the nineteenth century played a major role in demonstrating the importance of water purification in urban areas. We discuss the pathogenesis of *V. cholerae* in Section 21.11.

Vibrio parahaemolyticus is a marine organism. It is a major cause of gastroenteritis in Japan (where raw fish is widely consumed) and has also been implicated in outbreaks of gastroenteritis in other parts of the world, including the United States. The organism can be frequently isolated from seawater or from shellfish and crustaceans, and its primary habitat is probably marine animals, with human infection being a secondary development (∞Section 28.1).

Photobacterium and Bioluminescence

A number of gram-negative, polarly flagellated rods possess the interesting property of emitting light, a process called *bioluminescence*. Most of these bacteria have been classified in the genus *Photobacterium*, but a few *Vibrio* isolates are also bioluminescent (Figure 12.28●). Most **bioluminescent bacteria** are marine, usually found associated with fish. Some fish possess a special organ in which bioluminescent bacteria grow (Figure 12.28*c–f*). Other bioluminescent marine bacteria live saprophytically on dead fish and occasionally form visible colonies on the fish surface. (To see bioluminescence readily, one should observe the material in a completely dark room after the eyes have become adapted to the dark, Figure 12.28*a, b*.)

Although *Photobacterium* isolates are facultative aerobes, they are bioluminescent only when O_2 is present. Several components are needed for bioluminescence. These include the enzyme **luciferase** and a long-chain aliphatic aldehyde (for example, *dodecanal*). Reduced flavin mononucleotide ($FMNH_2$) and O_2 are also required for bioluminescence. The primary electron donor is NADH, and the electrons pass through luciferase as follows:

$$FMNH_2 + O_2 + RCHO \xrightarrow{\text{Luciferase}}$$

$$FMN + RCOOH + H_2O + Light$$

The light-generating system constitutes a bypass route for shunting electrons from $FMNH_2$ to O_2, without involving other electron carriers such as quinones and cytochromes.

Regulation of Bioluminescence

The enzyme luciferase shows a unique kind of regulatory synthesis called **autoinduction**. The luminous bacteria produce a specific organic molecule, the *autoinducer*, which accumulates in the culture medium during growth. When the amount of this substance reaches a critical level, induction of luciferase occurs. The auto inducer in *Vibrio fischeri* has been identified as N-β-ketocaproyl homoserine lactone. Thus, cultures of luminous bacteria at *low* cell density are not luminous but only become luminous when growth reaches a sufficiently *high* density that the autoinducer can accumulate and function. This is a mechanism called **quorum sensing**, because of the density-dependent nature of the phenomenon (∞Section 8.10).

Because the autoinducer cannot accumulate, free-living luminescent bacteria in seawater are not luminous, and luminescence develops only when conditions are favorable for the development of high population densities (Figure 12.28). Although it is not clear why luminescence is density-dependent in free-living bacteria, in symbiotic strains of luminescent bacteria (see

(a)   (b)   (c)

(d)   (e)   (f)

● **Figure 12.28 Bioluminescent bacteria and their role as light organs in the flashlight fish.** (a) Two Petri plates of luminous bacteria photographed by their own light. Note the different colors. Left, *Vibrio fischeri* strain MJ-1, blue light, and right, *V. fischeri* strain Y-1, green light. (b) Colonies of *Photobacterium phosphoreum* photographed by their own light. (c) The flashlight fish *Photoblepharon palpebratus*; the bright area is the light organ containing bioluminescent bacteria. (d) Same fish photographed by its own light. (e) Underwater photograph taken at night of *P. palpebratus* in coral reefs in the Gulf of Eilat. (f) Electron micrograph of a thin section through the light-emitting organ of *P. palpebratus*, showing the dense array of bioluminescent bacteria (arrows).

Figure 12.28c–f) the rationale for density-dependent luminescence is clear: luminescence develops only when sufficiently high population densities are reached in the light organ of the fish to allow a visible flash of light. The genetics of quorum sensing using bioluminescence as a model experimental system has been actively explored, and it is now clear that this form of regulation is not limited to bioluminescent bacteria, but instead is a general regulatory feature of a number of different bacteria for phenomena in which a minimum cell density (a "quorum") is required (∞Section 8.10).

(a)

(b)

● **Figure 12.29 Rickettsias growing within host cells.** (a) *Rickettsia rickettsii* in tunica vaginalis cells of the vole, *Microtus pennsylvanicus*. Cells are about 0.3 μm in diameter. (b) Electron micrograph of cells of *Rickettsiella popilliae* within a blood cell of its host, the beetle *Melolontha melolontha*. Notice that the bacteria are growing in a vacuole within the host cell.

 12.11–12.12 Concept Check

The enteric bacteria are a large group of facultative aerobic rods of medical and molecular biological significance. *Vibrio* and *Photobacterium* species are marine organisms; some species are pathogenic while others are bioluminescent.

◆ How is *Escherichia coli* distinguished from *Enterobacter aerogenes* based on physiology?

◆ Describe two major properties of *Proteus* species that distinguish them from other enteric bacteria.

◆ What is necessary for an organism like *Photobacterium* to give off visible light?

12.13 Rickettsias

Key Genera: *Rickettsia, Wolbachia*

The rickettsias are small, gram-negative, coccoid or rod-shaped Proteobacteria in the size range of 0.3–0.7 μm wide and 1–2 μm long. They are, with one exception, *obligate intracellular parasites* (Figure 12.29a●) and have not yet been cultivated in the absence of host cells. Rickettsias are the causative agents of several human diseases including typhus fever, Rocky Mountain spotted fever, and Q fever (∞Section 27.3).

Electron micrographs of thin sections of rickettsias show cells with a normal bacterial morphology (Figure 12.29b); both cell wall and cell membrane are clearly present. The cell wall contains muramic acid and diaminopimelic acid. Both RNA and DNA are present, and the rickettsias divide by normal binary fission, with doubling times of about 8 h. The penetration of a host cell by a rickettsial cell is an active process, requiring both host and parasite to be alive and metabolically active. Once inside the host phagocytic cell, the bacteria multiply primarily in the cytoplasm and continue replicating until the host cell is loaded with parasites (see Figures 12.29 and Figure 27.3). The host cell then bursts and liberates the bacteria into the surrounding fluid. Several genera of rickettsias are known, and the properties of four key genera are shown in Table 12.17.

Metabolism and Pathogenesis

Much research has been done on the metabolic activities and biochemical pathways of rickettsias in an attempt to explain why they are obligate intracellular parasites. Many rickettsias possess a highly specific energy metabolism: they can oxidize only glutamate or glutamine and cannot oxidize glucose or organic acids. However, *Coxiella burnetii*, the causative agent of the disease Q fever, is able to utilize both glucose and pyruvate as electron donors. Rickettsias possess a respiratory chain complete with cytochromes and are able to carry out electron transport phosphorylation, using NADH as electron donor. They are also able to synthesize at least some of the small molecules needed for macromolecular synthesis and growth, and they obtain the rest of their nutrients from the host cell. Thus, although parasites, rickettsias maintain a number of independent metabolic functions.

Rickettsias do not survive long outside their hosts, and this may explain why they must be transmitted from animal to animal by arthropod vectors. When the arthropod obtains a blood meal from an infected vertebrate, rickettsias present in the blood are inoculated directly into the arthropod, where they penetrate to the

Table 12.17 **Characteristics of rickettsias**

Genus and Species	Rickettsial group	Alternate host	Cellular location	DNA (mol % GC)	Phylogenetic group[a]	DNA hybridization to *R. rickettsii* DNA (%)[b]
Rickettsia						
R. rickettsii	Spotted fever	Tick	Cytoplasm and nucleus	32–33	Alpha	100
R. prowazekii[c]	Typhus	Louse	Cytoplasm	29–30		53
R. typhi	Typhus	Flea	Cytoplasm	29–30		36
Rochalimaea						
R. quintana	Trench fever	Louse	Epicellular	39	Alpha	30
R. vinsonii	—	Vole	Epicellular	39		30
Coxiella						
C. burnetii	Q fever	Tick	Vacuoles	43	Gamma	—
Ehrlichia						
E. chaffensis	Ehrlichiosis (humans)	Tick or	Mononuclear	—	Alpha	—
E. equi	Potomac fever (horses)	domestic animals	leukocytes	—	Alpha	—
Wolbachia[d]						
W. pipientis	—	Arthropods	Cytoplasm	30	Alpha	—

[a] All are Proteobacteria.
[b] For discussion of DNA:DNA hybridization, see Section 11.11.
[c] The genome of this organism has been sequenced and shows several similarities to the mitochondrial genome.
[d] Not a pathogen of humans or other animals.

epithelial cells of the gastrointestinal tract, multiply, and appear later in the feces. When the arthropod feeds on an uninfected individual, it then transmits the rickettsias either directly with its mouthparts or by contaminating the bite with its feces. However, *C. burnetii* (⟲Section 27.3) can also be transmitted to the respiratory system by aerosols. *C. burnetii* is the most resistant of the rickettsias to physical damage, probably because it produces a resistant, sporelike form. This probably explains the ability of *C. burnetii* to survive in air.

Rochalimaea is an atypical rickettsia because it can be grown in culture and is thus not an obligate intracellular parasite. In addition, when growing in tissue culture, cells of *Rochalimaea* grow on the *outside surface* of the eukaryotic host cells rather than within the cytoplasm or the nucleus. *Rochalimaea quintana* is the causative agent of *trench fever*, a disease that decimated troops in World War I. Species of the genus *Ehrlichia* cause disease in humans and other animals, two of which, *ehrlichiosis* in humans and *Potomac fever* in horses, can be quite debilitating (⟲Section 27.3).

Wolbachia

The genus *Wolbachia* contains species of rod-shaped alpha Proteobacteria that are intracellular parasites of arthropod insects (Figure 12.30●). *Wolbachia* are phylogenetically related to the rickettsias and can have any of several effects on their insect hosts. These include inducing parthenogenesis (development of unfertilized eggs), the killing of males, and feminization (the conversion of male insects into females).

Wolbachia pipientis is the best-studied species in the genus. Cells colonize the insect egg (Figure 12.30), where they multiply in vacuoles of host cells surrounded by a

membrane of host origin. *W. pipientis* cells are passed from infected females to her offspring through this egg infection. *Wolbachia*-induced parthenogenesis occurs in a number of species of wasps. In these insects, males normally arise from unfertilized eggs (which contain only one set of chromosomes), while females arise from fertilized eggs (which contain two sets of chromosomes). However, in unfertilized eggs infected with *Wolbachia*, the organism somehow triggers a doubling of the chromosome number, thus yielding only females. Predictably, if female insects are fed antibiotics that kill *Wolbachia*, parthenogenesis ceases.

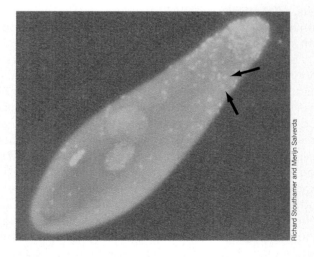

● **Figure 12.30** *Wolbachia.* Photomicrograph of a 4',6-diamidine-2' phenylindole dihydrochloride–stained (DAPI, ⟲Section 18.3) egg of the parasitoid wasp, *Trichogramma kaykai* infected with *Wolbachia pipientis*, which induces parthenogenesis. The *W. pipientis* cells are primarily located in the narrow end of the egg (arrows).

In certain woodlice (pillbugs), a species of *Wolbachia* causes males to develop as females. Here the *Wolbachia* infection damages the male hormone-producing glands. In other insects, such as certain lady beetles and butterflies, *Wolbachia* infection somehow results in the death of male but not female offspring. In addition to these reproductive anomalies, in some insects *Wolbachia* infection may actually be *essential* for survival. For example, in the nematode worms that cause the diseases *elephantiasis* and *river blindness*, antibiotic treatment kills the worms, apparently by killing their *Wolbachia* symbionts.

Like parasitic prokaryotes in general, *W. pipientis* has a small genome (about 1.5 Mbp), which has now been sequenced. Although this parasite does not cause disease in either vertebrates or its invertebrate hosts, its various reproductive effects raise the interesting question of the extent it has affected evolution and speciation in a major class of insects (arthropods make up 70% of all known insects). However, because the association between *Wolbachia* and insects is so common, it could turn out that the reproductive effects are secondary to a true *Wolbachia*–insect symbiosis. That is, it is likely that by infecting insects, the *Wolbachia* cells benefit from a ready source of nutrients and a protected environment. Moreover, the insects may also benefit from the association in some as yet unknown ways.

12.13 Concept Check

The rickettsias are obligate intracellular parasites, many of which cause disease. Rickettsias are deficient in many metabolic functions and obtain key metabolites from their hosts.

♦ Name a disease caused by a *Rickettsia* species.

♦ What is meant by the phrase "obligate intracellular parasite"?

♦ What effects can *Wolbachia* have on its insect hosts?

12.14 Spirilla

Key Genera: *Spirillum, Bdellovibrio, Campylobacter*

The **spirilla** are gram-negative, motile, spiral-shaped Proteobacteria that show a wide variety of physiological attributes. Some of the key taxonomic criteria used are cell shape, size, kind of polar flagellation (single or multiple), relation to oxygen (obligately aerobic, microaerophilic, facultative), relationship to plants (as symbionts or plant pathogens) or animals (as pathogens), fermentative ability, and certain other physiological characteristics (nitrogen-fixing ability, halophilic nature, thermophilic nature). The genera to be covered here are given in Table 12.18, where it can be seen that genera of spirilla can be found in each of the five subdivisions of Proteobacteria.

Spirillum, Aquaspirillum, Oceanospirillum, and Azospirillum

These helically curved rods are motile by means of polar flagella, usually tufts at both poles (see Figure 12.31b●). The number of turns in the helix may vary from less than one complete turn (in which case the organism looks like a vibrio) to many turns. Spirilla with many turns can superficially resemble spirochetes (see Section 12.33) but differ distinctly from the latter phylogenetically. In addition, spirilla do not have the outer sheath and endoflagella of spirochetes, but instead contain typical bacterial flagella.

Some spirilla are very large bacteria and were seen by early microscopists. It is likely that Antoni van Leeuwenhoek first observed a *Spirillum* species in the 1670s (∞Section 1.5), and the genus was first created by the

---

Table 12.18 Characteristics of the genera of spiral-shaped bacteria[a]

Genus	Phylogenetic group[b]	Characteristics	DNA (mol % GC)
Spirillum	Beta	Cell diameter 1.7 μm; microaerophilic; freshwater	36–38
Aquaspirillum	Alpha or beta	Cell diameter 0.2–1.5 μm; aerobic; freshwater	49–66
Magnetospirillum	Alpha	Vibrio to spirillum-shaped; cell diameter about 0.3 μm; contains magnetosomes; microaerophilic	65
Oceanospirillum	Gamma	Cell diameter 0.3–1.2 μm; aerobic; marine (require 3% NaCl)	42–51
Azospirillum	Alpha	Cell diameter 1 μm; microaerophilic; soil and rhizosphere; fixes N_2	68–70
Herbaspirillum	Beta	Cell diameter 0.6–0.7 μm; microaerophilic; soil and rhizosphere; fixes N_2	66–67
Campylobacter	Epsilon	Cell diameter 0.2–0.8 μm; microaerophilic to anaerobic; pathogenic or commensal in humans and animals; single polar flagellum	30–38
Helicobacter	Epsilon	Cell diameter 0.5–1 μm; tuft of polar flagella; associated with pyloric ulcers in humans	36–38
Bdellovibrio	Delta	Cell diameter 0.25–0.4 μm; aerobic; predatory on other bacteria; single polar sheathed flagellum	33–52
Ancyclobacter	Alpha	Cell diameter 0.5 μm; curved rods forming rings; nonmotile, aerobic; sometimes gas-vesiculate	66–69

[a] All are gram-negative and respiratory but never fermentative.
[b] All are Proteobacteria.

(a)

Noel Krieg

(b)

Stanley Erlandsen

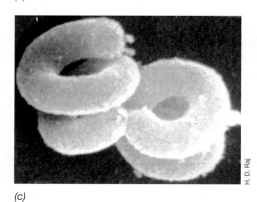

(c)

H. D. Raj

● **Figure 12.31 Spirilla.** (a) *Spirillum volutans*, visualized by dark-field microscopy, showing flagellar bundles and volutin (polyphosphate) granules. Cells are about 1.5 × 25 μm. (b) Scanning electron micrograph of an intestinal spirillum. Note the polar flagellar tufts and the spiral structure of the cell surface. (c) Scanning electron micrograph of cells of *Ancyclobacter aquaticus*. Cells are about 0.5 μm in diameter.

German protozoologist Ehrenberg in 1832. The organism seen by these workers is now called *Spirillum volutans* and is a rather large bacterium (Figure 12.31*a*). A phototrophic organism resembling *S. volutans* is *Thiospirillum* (see Figure 12.4*b*). *S. volutans* is microaerophilic, requiring O_2, but is inhibited by O_2 at normal levels. Another prominent characteristic of cells of *S. volutans* is the formation of granules (volutin granules) consisting of polyphosphate (see Figure 12.31*a* and Section 4.11).

Azospirillum lipoferum is a nitrogen-fixing organism, which was originally described and named *Spirillum lipoferum* by Beijerinck in 1922. *A. lipoferum* is of considerable interest because it enters into a symbiotic relationship with tropical grasses and grain crops. Although not as intimate an association as that between root nodule bacteria and leguminous plants (Section 19.22), the *A. lipoferum*/corn association clearly benefits the corn plant from nitrogen fixation.

The small-diameter spirilla (which are not microaerophilic) have been separated into two genera, *Aquaspirillum* and *Oceanospirillum*. The former includes freshwater species while the latter includes species that inhabit seawater and require NaCl for growth (Table 12.18). Numerous species of *Aquaspirillum* and *Oceanospirillum* have been described, the various species being separated on physiological and phylogenetic grounds. These organisms undoubtedly play an important role in the recycling of organic matter in aquatic environments.

Magnetotactic Spirilla

Highly motile microaerophilic magnetic spirilla have been isolated from freshwater habitats. These organisms demonstrate a dramatic directed movement in a magnetic field referred to as **magnetotaxis**. In an artificial magnetic field magnetic spirilla quickly orient their long axis along the north-south magnetic moment of the field. Within the cells are chains of 5–40 magnetic particles called **magnetosomes,** consisting of magnetite (Fe_3O_4) and greigite (Fe_3S_4) (Section 4.11 and Figure 4.42). Magnetosomes function as internal magnets that orient the cells along a specific magnetic field.

Magnetic bacteria can have one of two magnetic polarities depending on the orientation of magnetosomes within the cell. Cells in the Northern Hemisphere have the north-seeking pole of their magnetosomes forward with respect to their flagella and thus move in a northward direction. Cells in the Southern Hemisphere have the opposite polarity and move southward. Although the ecological role of bacterial magnets is unclear, the ability to orient in a magnetic field may be of selective advantage in maintaining these microaerophilic organisms in zones of low O_2 concentration near the oxic/anoxic interface. The spirillum *Magnetospirillum magnetotacticum* (Figure 12.32● and Table 12.18) is a major organism in this group.

Bdellovibrio

These small vibroid organisms have the unusual property of preying on other bacteria, using as nutrients the cytoplasmic constituents of their hosts. These bacterial predators are small, highly motile cells that stick to the surfaces of their prey cells. Because of this, they have been given the name *Bdellovibrio* (*bdello* is a prefix meaning "leech"). Other predatory bacteria have been isolated and given such suggestive genus names as *Vampirococcus*. However, *Bdellovibrio* has a unique mode of attack and develops intraperiplasmically.

After attachment of a *Bdellovibrio* cell to its prey, the predator penetrates through the prey wall and replicates in the periplasmic space, eventually forming a spherical structure called a **bdelloplast.** The stages of attachment and penetration are shown in electron micrographs in Figure 12.33● and diagrammatically in Figure 12.34●. A wide variety of gram-negative *Bacteria* can be attacked

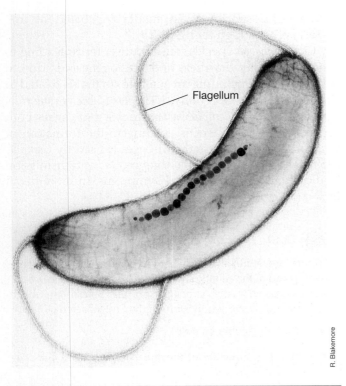

derivatives of predatory strains of *Bdellovibrio* that are prey-independent can be isolated, showing that predation is not obligatory. Prey-independent strains can be grown on complex media containing yeast extract and peptone.

Phylogenetically, bdellovibrios fall into the delta subdivision of Proteobacteria (Figure 12.1) and contain a genome about half the size of that of *E. coli*. Taxonomically, two species of *Bdellovibrio* and one species of the genus *Bacteriovorax* are recognized and supported by the results of genomic DNA:DNA hybridization (◆◆Section 11.11).

Members of the genus *Bdellovibrio* are widespread in soil and water, including the marine environment. Their detection and isolation require methods reminiscent of those used in the study of bacterial viruses (◆◆Section 9.4). Prey bacteria are spread on the surface of an agar plate to form a lawn, and the surface is inoculated with a small amount of soil suspension that has been filtered through a membrane filter; the latter retains most bacteria but allows the small *Bdellovibrio* cells to pass. On incubation of the agar plate, *plaques* analogous to bacteriophage plaques are formed at locations where *Bdellovibrio* cells are growing. However, unlike phage plaques, which continue to enlarge only as long as the bacterial host is growing, *Bdellovibrio* plaques continue to enlarge even after the prey has stopped growing. This results in ever enlarging plaques on the agar surface. Pure cultures of *Bdellovibrio* can then be isolated from these plaques. *Bdellovibrio* cultures have been obtained from a wide variety of soils and are thus common members of the soil population.

● **Figure 12.32 Negatively stained electron micrograph of a magnetotactic spirillum, *Magnetospirillum magnetotacticum*.** A cell measures 0.3 × 2 μm. This bacterium contains particles of Fe₃O₄ (magnetite), called *magnetosomes* (◆◆Figure 4.42), arranged in a chain; the particles align the cell along geomagnetic lines. The organism was isolated from a water treatment plant in Durham, New Hampshire.

by a single *Bdellovibrio* species; gram-positive cells are not attacked.

Bdellovibrio is an obligate aerobe, obtaining its energy from the oxidation of amino acids and acetate. In addition, *Bdellovibrio* assimilates nucleotides, fatty acids, peptides, and even intact proteins, directly from its host without first breaking them down. Thus it is clear that the predatory mode of existence has involved the development in *Bdellovibrio* of interesting and unusual biochemical processes. In addition to a predatory lifestyle, however,

Ancyclobacter

Members of the genus *Ancyclobacter* are ring-shaped, nonmotile, extremely nutritionally diverse chemoorganotrophic bacteria (Figure 12.31c). They resemble very tightly coiled vibrios and are widely distributed in aquatic environments. A phototrophic counterpart to *Ancyclobacter* exists in the purple nonsulfur bacterium *Rhodocyclus purpureus* (see Figure 12.6e).

(a)   (b)

● **Figure 12.33 Stages of attachment and penetration of a prey cell by *Bdellovibrio*.** A *Bdellovibrio* cell measures about 0.3 μm in diameter. (a, b) Electron micrographs of thin sections of *Bdellovibrio* attacking *Escherichia coli*; (a) early penetration; (b) complete penetration. The *Bdellovibrio* cell is enclosed in a membranous infolding of the prey cell (the bdelloplast) and replicates in the periplasmic space between the wall and the membrane (see Figure 12.34).

(a)

(b)

● **Figure 12.34 Developmental cycle of the bacterial predator *Bdellovibrio bacteriovorus*.** (a) Electron micrograph of a cell of *Bdellovibrio bacteriovorus*. Note the very thick flagellum present on this organism (∞Figure 4.54). (b) Events in predation. Following primary contact between a highly motile *Bdellovibrio* cell and a gram-negative bacterium, attachment and penetration into the prey periplasmic space occurs. Once inside, *Bdellovibrio* elongates and within 4 hours progeny cells are released. The number of progeny cells released varies with the size of the prey bacterium. For example, 5–6 bdellovibrios are released from each infected *Escherichia coli* cell, and 20–30 for a larger cell, such as an *Aquaspirillum* sp.

Campylobacter and *Helicobacter*

These two genera are key representatives of the *epsilon* subdivision of the Proteobacteria. They are gram-negative, motile spirilla, most species of which are pathogenic to humans or other animals. *Campylobacter* and *Helicobacter* are both microaerophilic and are cultured from clinical specimens in media incubated at low (3–15%) O_2 and high (3–10%) CO_2.

Campylobacter species cause acute enteritis leading to (usually) bloody diarrhea, and pathogenesis is due to several factors, including an enterotoxin that is related to cholera toxin (∞Section 21.11). *Helicobacter pylori* is closely related to *Campylobacter* species and causes both acute and chronic gastritis, leading to the formation of *peptic ulcers*. We discuss the diseases caused by *Campylobacter* and *Helicobacter*, including modes of transmission of the organisms and clinical symptoms, in more detail in Section 29.9.

⬣ *12.14* **Concept Check**

Spirilla are spiral-shaped, chemoorganotrophic prokaryotes widespread in the aquatic environment. The genera *Helicobacter* and *Campylobacter* are pathogenic spirilla. Spirilla are distributed among all five subdivisions of the Proteobacteria.

◆ What is a *volutin granule*?

◆ What is unique about the spirilla *Bdellovibrio* and *Magnetospirillum*?

12.15 ## Sheathed Proteobacteria: *Sphaerotilus* and *Leptothrix*

Key Genera: *Sphaerotilus*, *Leptothrix*

Sheathed bacteria are filamentous beta Proteobacteria (Table 12.1) with a unique life cycle involving formation of flagellated swarmer cells within a long tube or sheath. Under certain (generally unfavorable) conditions, the swarmer cells move out and become dispersed to new environments, leaving behind the empty sheath. Under favorable conditions, vegetative growth occurs within the filament, leading to the formation of long, cell-packed sheaths. Sheathed bacteria are common in freshwater habitats that are rich in organic matter, such as polluted streams. They are also abundant in trickling filters and activated sludge digestors in sewage treatment plants (∞Section 28.2), generally being found in flowing waters. In habitats where reduced iron or manganese compounds are present, the sheaths may become coated with ferric hydroxide or manganese oxide (∞Figure 17.29). Iron precipitation is probably due to chemical reactions, but some sheathed bacteria have the biochemical ability to oxidize manganous ions to manganese oxide. Two genera are the major organisms here: *Sphaerotilus*, in which manganese oxidation does not occur, and *Leptothrix*, whose members do oxidize Mn^{2+}.

Sphaerotilus

The *Sphaerotilus* filament is composed of a chain of rod-shaped cells with rounded ends enclosed in a closely fitting sheath. This thin, transparent sheath is difficult to see when it is filled with cells, but when it is partially empty, the sheath can easily be seen by phase contrast microscopy (Figure 12.35a●) or by staining. Individual cells are 1–2 μm wide and 3–8 μm long and stain gram negatively.

The cells within the sheath divide by binary fission (Figure 12.35b), and the new cells pushed out at the end synthesize new sheath material. Thus, the sheath is always formed at the *tips* of the filaments. Eventually cells are liberated from the sheaths, probably when the nutrient supply is low. These free cells are actively motile, the flagella being arranged lophotrichously (in a bundle at one pole) (Figure 12.35c). Probably the flagella are synthesized before the cells leave the sheath and, if so, may even aid in their liberation. It is thought that the swarmer cells then migrate, attach to some solid surface, and begin to

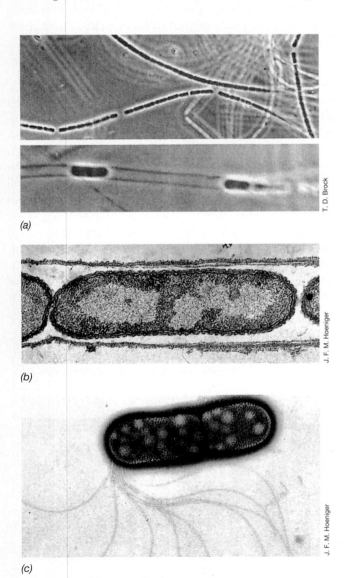

(a)

T. D. Brock

(b)

J. F. M. Hoeniger

(c)

J. F. M. Hoeniger

● **Figure 12.35** *Sphaerotilus natans.* A single cell is about 2 μm wide. (a) Phase contrast photomicrographs of material collected from a polluted stream. Active growth stage (above) and swarmer cells leaving the sheath. (b) Electron micrograph of a thin section through a filament. (c) Electron micrograph of a negatively stained swarmer cell. Notice the polar flagellar tuft.

grow, each swarmer being the forerunner of a new filament. The sheath, which is devoid of muramic acid or other components of the peptidoglycan cell wall, is a protein-polysaccharide-lipid complex. However, it differs from the capsules formed by many gram-negative *Bacteria* (∞Section 4.10) in that it forms a linear structure.

Sphaerotilus cultures are nutritionally versatile, able to use a wide variety of simple organic compounds as carbon and energy sources, along with inorganic nitrogen sources. Befitting its habitat in flowing waters, *Sphaerotilus* is an obligate aerobe. *Sphaerotilus* blooms often occur in the fall of the year in streams and brooks when leaf litter causes a temporary increase in the organic content of the water. Its filaments are the main component of a microbial complex that sanitary engineers call "sewage fungus," which is a filamentous slime found on the rocks in streams receiving sewage pollution. In activated sludge works of sewage treatment plants (∞Section 28.2), *Sphaerotilus* growth, like that of *Beggiatoa* (see Section 12.4), is often responsible for a detrimental condition called "bulking." The tangled masses of *Sphaerotilus* filaments so increase the bulk of the sludge that it does not settle properly, thus presenting difficulties in sludge clarification.

Leptothrix

The ability of *Sphaerotilus* and *Leptothrix* to precipitate iron oxides on their sheaths is well established, and such iron-encrusted sheaths are frequently seen in iron-rich waters (Figure 12.36●). Iron precipitation occurs when iron, chelated to organic materials such as humic or tannic acids, is metabolized. The iron gets precipitated on the sheath while the organic constituents get taken up and used as a carbon or energy source.

Besides Fe^{2+} oxidation, *Leptothrix* is capable of oxidizing Mn^{2+} to Mn^{4+}. The reaction is exergonic:

$$Mn^{2+} + \tfrac{1}{2}O_2 + H_2O \longrightarrow MnO_2 + 2\,H^+$$

$$\Delta G^{0'} = -68 \text{ kJ}$$

There is evidence that Mn^{2+} oxidation is coupled to energy-yielding reactions in the cell as well, probably involving electron transport and formation of a proton motive force.

The gene encoding the manganese-oxidizing protein of *Leptothrix* has been isolated, and biochemical studies have shown that the protein resides in the sheath, not inside the cells. Thus the sheath of *Leptothrix* is critical for energy metabolism in this bacterium.

12.16 Budding and Prosthecate/Stalked *Bacteria*

Key Genera: *Hyphomicrobium, Caulobacter*

This large and heterogeneous group of primarily alpha Proteobacteria contains organisms that form various

W. C. Ghiorse

● **Figure 12.36** ***Leptothrix* and iron precipitation.** Transmission electron micrograph of a thin section of *Leptothrix* sp. in a sample from a ferromanganese film in a swamp in Ithaca, New York. A single cell measures about 0.9 μm in diameter. Note the protuberances of the cell envelope that contact the sheath (arrows).

kinds of cytoplasmic extrusions: *stalks, hyphae,* or *appendages* (Table 12.19). Extrusions of these kinds, which are smaller in diameter than the mature cell, contain cytoplasm, and are bounded by the cell wall, are collectively called **prosthecae** (Figure 12.37●).

Cell division in budding bacteria occurs as a result of *unequal cell growth*. In contrast to cell division in a typical bacterium, which occurs by *binary fission* and results in the formation of two equivalent cells, cell division in the stalked and budding bacteria involves the formation

of a new daughter cell with the mother cell retaining its identity (Figure 12.38●). The major genera of budding bacteria and a summary of their cell division processes are shown in Figure 12.38.

A fundamental difference between these bacteria and conventional bacteria is not the formation of buds or stalks but the formation of new cell wall material from a *single point* (polar growth) rather than throughout the whole cell (intercalary growth). Several genera not normally considered to be budding bacteria show polar growth without differentiation of cell size (Figure 12.38). An important consequence of **polar growth** is that internal structures, such as membrane complexes, are not involved in the cell division process, thus permitting the formation of more complex internal structures than in cells undergoing intercalary growth. And many budding bacteria, particularly phototrophic species, contain extensive internal membrane systems.

Budding Bacteria: *Hyphomicrobium*

Two well-studied budding bacteria are closely related phylogenetically: *Hyphomicrobium*, which is chemoorganotrophic, and *Rhodomicrobium*, which is phototrophic. These organisms release buds from the ends of long, thin hyphae. The **hypha** is a direct cellular extension of the mother cell (Figures 12.39● and 12.40*b*●) and contains cell wall, cytoplasmic membrane, ribosomes, and, occasionally, DNA.

Figure 12.39 shows the process of reproduction in *Hyphomicrobium*. The mother cell, which is often attached by its base to a solid substrate, forms a thin outgrowth that lengthens to become a hypha. At the end of the hypha, a bud forms. This bud enlarges, forms a flagellum, breaks loose from the mother cell, and swims

Table 12.19	Characteristics of stalked, appendaged (prosthecate), and budding bacteria		
Characteristics	**Genus**	**Phylogenetic group**[a]	**DNA (mol% GC)**
Stalked bacteria:			
Stalk an extension of the cytoplasm and involved in cell division	*Caulobacter*	Alpha	62–67
Stalked, fusiform-shaped cells	*Prosthecobacter*	Alpha	54–60
Stalked, but stalk an excretory product not containing cytoplasm:			
Stalk depositing iron, cell vibrioid	*Gallionella*	Beta	55
Laterally excreted gelatinous stalk not depositing iron	*Nevskia*	Gamma	60
Appendaged (prosthecate) bacteria:			
Single or double prosthecae	*Asticcacaulis*	Alpha	55–61
Multiple prosthecae			
Short prosthecae, multiply by fission, some with gas vesicles	*Prosthecomicrobium*	Alpha	64–70
Flat, star-shaped cells, some with gas vesicles	*Stella*	Alpha	69–74
Long prosthecae, multiply by budding, some with gas vesicles	*Ancalomicrobium*	Alpha	70–71
Budding bacteria:			
Phototrophic, produce hyphae	*Rhodomicrobium*	Alpha	61–63
Phototrophic, budding without hyphae	*Rhodopseudomonas*	Alpha	64–72
Chemoorganotrophic, rod-shaped cells	*Blastobacter*	Alpha	59–66
Chemoorganotrophic, buds on tips of slender hyphae			
Single hyphae from parent cell	*Hyphomicrobium*	Alpha	59–65
Multiple hyphae from parent cell	*Pedomicrobium*	Alpha	62–67

[a] All are Proteobacteria.

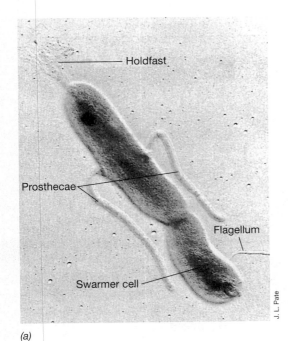

Holdfast

Prosthecae

Flagellum

Swarmer cell

J. L. Pate

(a)

J. T. Staley

(b)

H. Schlesner

(c)

● **Figure 12.37   Prosthecate bacteria.** (a) Electron micrograph of a shadow-cast preparation of *Asticcacaulis biprosthecum*, illustrating the location and arrangement of the prosthecae. Cells are about 0.6 μm wide. Note also the holdfast material and the swarmer cell in the process of differentiation. (b) Electron micrograph of a negatively stained preparation of a cell of the prosthecate bacterium *Ancalomicrobium adetum*. The appendages are cellular (prosthecae) because they are bounded by the cell wall and contain cytoplasm and are about 0.2 μm in diameter. (c) Electron micrograph of the star-shaped bacterium *Stella*. Cells are about 0.8 μm in diameter.

I. Equal products of cell division:

Binary fission: conventional bacteria

II. Unequal products of cell division:

1. Simple budding: *Pirella, Blastobacter*

2. Budding from hyphae: *Hyphomicrobium, Rhodomicrobium, Pedomicrobium*

3. Cell division of stalked organism: *Caulobacter*

4. Polar growth without differentiation of cell size: *Rhodopseudomonas, Nitrobacter, Methylosinus*

● **Figure 12.38   Cell division.** Contrast between cell division in conventional bacteria and in budding and stalked bacteria.

away. Later, the daughter cell loses its flagellum and after a period of maturation forms a hypha and buds. Further buds can also form at the hyphal tip of the mother cell leading to arrays of cells connected by hyphae (Figure 12.40*a*). In some cases, a bud begins to form directly from the mother cell without the intervening formation of a hypha, whereas in other cases a single cell forms hyphae from each end (Figure 12.40*a*).

Nucleoid replication events during the budding cycle are of interest (Figure 12.39). First, the DNA located in the mother cell replicates. Then, once the bud has formed, a copy of the circular chromosome is moved down the length of the hypha and into the bud. A cross-septum then forms, separating the still-developing bud from the hypha and mother cell.

Hyphomicrobium is a methylotrophic bacterium (see Section 12.6) and is widespread in freshwater, marine, and terrestrial habitats. Preferred carbon sources are *one-carbon* compounds such as methanol, methylamine, formaldehyde, and formate. Enrichment cultures of *Hyphomicrobium* can be prepared taking advantage of the ability of this organism to grow in very dilute conditions. All that is needed is a mineral salts medium lacking organic carbon and nitrogen, to which an inoculum of soil, mud, or water is added. After several weeks of incubation, a surface film develops and is streaked out on agar medium containing methylamine or methanol as a sole carbon source. Colonies are then checked microscopically for the characteristic *Hyphomicrobium* cellular

● **Figure 12.39** **Stages in the *Hyphomicrobium* cell cycle.** The single chromosome of *Hyphomicrobium* is circular.

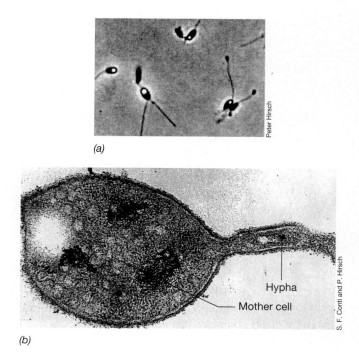

(a)

(b)

● **Figure 12.40** **Morphology of *Hyphomicrobium*.** (a) Phase micrograph of cells of *Hyphomicrobium*. Cells are about 0.7 μm wide. (b) Electron micrograph of a thin section of a single *Hyphomicrobium* cell. The hypha is about 0.2 μm wide.

morphology (Figure 12.40). A fairly specific enrichment procedure for *Hyphomicrobium* uses methanol as electron donor with nitrate as electron acceptor under anoxic conditions. Virtually the only denitrifying (∞Section 17.14) organism using methanol is *Hyphomicrobium*, and so this procedure selects this organism out of a wide variety of environments.

Prosthecate and Stalked Bacteria

Prosthecate and stalked bacteria (Figures 12.37 and 12.41●) are appendaged chemoorganotrophic aerobes that attach to particulate matter, plant material, or other microorganisms in aquatic habitats. Although a major function of these appendages is undoubtedly attachment, they also significantly increase the surface-to-volume ratio of the cells (prosthecae

have large surface areas but almost no volume). Recall from Chapter 4 that the high surface-to-volume ratio of small cells like bacteria confers an increased ability on them to take up nutrients and expel wastes. The unusual morphology of appendaged bacteria (Figure 12.37) carries this theme to an extreme and may be an evolutionary adaptation to life in the oligotrophic (nutrient-poor) waters where these organisms are most commonly found.

Another function of prosthecae may be to decrease sedimentation rates in aquatic environments. It has been shown that centrifugation of prosthecate cells in the laboratory requires greater centrifugal force than for nonprosthecate cells. This inherent tendency to resist sinking is likely due to the prosthecae and might be an advantage for prosthecate cells in their natural habitats. Because these organisms are in general strict aerobes, prosthecae may keep cells from sinking into sediments or other anoxic zones where they would be unable to respire.

Selective isolation of prosthecate/stalked bacteria can be achieved by mixing an inoculum with a very dilute nutrient solution, such as 0.01% peptone, and leaving it to sit undisturbed. Within a few days a surface film often develops containing prosthecae and stalked bacteria. Isolation can then be achieved by streaking on agar plates of the same, very dilute, media.

Caulobacter and *Gallionella*

Two common stalked bacteria are *Caulobacter* (Figure 12.41) and *Gallionella* (see Figure 12.43). The former is a chemoorganotroph that produces a cytoplasm-filled stalk, that is, a prostheca, while the latter is a chemo-lithotrophic iron-oxidizing bacterium whose stalk is composed of ferric hydroxide.

Caulobacter cells are often seen on surfaces in aquatic environments with the stalks of several cells attached to form *rosettes* (Figure 12.41*a*). At the end of the stalk is a structure called a *holdfast* by which the stalk is attached to a surface. The *Caulobacter* cell division cycle (Figure 12.42●) is of special interest because it involves a process of *unequal binary fission*. Much molecular research has been done on this organism as a model system for cell division and developmental events. Cell division of a stalked cell of *Caulobacter* occurs by elongation of the cell followed by fission, a single flagellum forming at the pole opposite the stalk. The flagellated cell so formed, called a *swarmer*, separates from the nonflagellated mother cell, swims around, and attaches to a new surface, forming a new stalk at the flagellated pole; the flagellum is then lost (Figure 12.42). Stalk formation is a necessary precursor of cell division and is coordinated with DNA synthesis (Figure 12.42). The cell division cycle in *Caulobacter* is thus more complex than simple binary fission because the stalked and swarmer cells are structurally different and the growth cycle must include both forms.

● **Figure 12.42 Growth of *Caulobacter*.** Stages in the *Caulobacter* cell cycle beginning with a swarmer cell.

Gallionella forms a twisted stalklike structure containing ferric hydroxide [Fe(OH)$_3$] (Figure 12.43●). However, the stalk of *Gallionella* is not an integral part of the cell but is simply *excreted* from the cell surface. It contains an organic matrix on which the ferric hydroxide accumulates. *Gallionella* is common in the waters draining bogs, iron springs, and other habitats where ferrous iron (Fe^{2+}) is present, usually in association with sheathed bacteria such as *Sphaerotilus*. *Gallionella* is an autotrophic chemolithotroph containing enzymes of the Calvin cycle (∞ Section 17.6) by which CO$_2$ is incorporated into cell material with Fe^{2+} as electron donor.

● **Figure 12.41 Stalked bacteria.** (a) A *Caulobacter* rosette. A single cell is about 0.5 μm wide. The five cells are attached by their stalks (prosthecae). Two of the cells have divided, and the daughter cells have formed flagella. (b, c) Electron micrographs of *Caulobacter* cells. (b) Negatively stained preparation of a cell in division. (c) A thin section. Notice that cytoplasmic constituents are present in the stalk region.

⬡ 12.15–12.16 Concept Check

Sheathed bacteria are filamentous Proteobacteria in which individual cells form chains within an outer layer called the sheath. Budding and prosthecate bacteria are appendaged cells that form stalks or prosthecae used for attachment or nutrient absorption and are primarily aquatic.

◆ Physiologically, what is unique about the sheathed bacterium *Leptothrix*?

◆ How does *budding* division differ from *binary* fission? How does binary fission differ from the division process in *Caulobacter*?

◆ What advantage might a prosthecate organism have in a very nutrient-poor environment?

12.17 Gliding Myxobacteria

Key Genera: *Myxococcus, Stigmatella*

A variety of prokaryotes exhibit a form of motility called *gliding* (∞ Section 4.15). Gliding microorganisms, usually long rods or filaments, lack flagella but are able to move when in contact with surfaces. One group of gliding bacteria, the *fruiting myxobacteria*, form multicellular

(a)

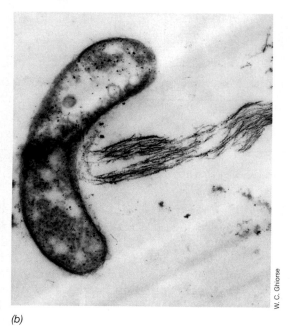

(b)

● **Figure 12.43 The neutrophilic ferrous iron-oxidizer, *Gallionella ferruginea*.** (a) Photomicrograph of cells from an iron seep near Ithaca, New York. Notice the twisted stalk leading to the two bean-shaped cells (arrow). (b) Transmission electron micrograph of a thin section of a cell. Cells are about 0.6 μm wide. Note the twisted stalk of ferric hydroxide emanating from the center of the cell in both photos.

structures called **fruiting bodies** and show complex developmental life cycles involving intercellular communication (see Figure 12.47). Phylogenetically, the gliding myxobacteria are members of the *delta* subdivision of Proteobacteria (Table 12.20).

The fruiting myxobacteria exhibit the most complex behavioral patterns and life cycles of all known prokaryotic organisms. In keeping with this complexity, the chromosome of some myxobacteria is very large. *Myxococcus xanthus*, for example, has a single circular chromosome of 9.2 Mbp, *twice* as large as that of *Escherichia coli* (∞ Sections 15.4 and 10.19). Indeed, this is two-thirds the size of the entire yeast genome, which is contained on 16 chromosomes (∞ Section 15.6). The vegetative cells of the fruiting myxobacteria are simple, nonflagellated, gram-negative rods (Figure 12.44*a*●) that glide across surfaces and obtain their nutrients primarily by causing the lysis of other bacteria. Under appropriate conditions, a swarm of vegetative cells aggregate and construct *fruiting bodies*, within which some of the cells become converted to resting structures called **myxospores** (Figure 12.44*b*).

Fruiting Bodies

The fruiting bodies of the myxobacteria vary from simple globular masses of myxospores in loose slime to complex forms with a fruiting body wall and a stalk (Figure 12.45●). The fruiting bodies are often strikingly colored and morphologically elaborate (Figures 12.45 and 12.46●). They can usually be seen with a hand lens or dissecting microscope on moist pieces of decaying wood or plant material. Fruiting bodies of myxobacteria often develop on dung pellets (for example, rabbit pellets) after they have been incubated for a few days in a moist chamber.

Another method for isolating fruiting myxobacteria is to prepare Petri plates of water agar (1.5% agar in distilled water with no added nutrients) onto which is spread a heavy suspension of virtually any bacterium. In the center of the plate a small amount of soil, decaying bark, or other natural material is placed. Myxobacteria in the inoculum lyse the bacterial cells and use their liberated products as nutrients; as they grow, they swarm out across the plate from the inoculum site. After several days to a week, the plates are examined under a dissecting microscope for myxobacterial swarms or fruiting bodies, and pure cultures are obtained by transfer of cells from the fruiting bodies or from the edge of the

Table 12.20	Classification of the fruiting myxobacteria[a]
Characteristics	**Genus/DNA (mol% GC)**
Vegetative cells tapered:	
Spherical or oval myxospores, fruiting bodies usually soft and slimy without well-defined sporangia or stalks	*Myxococcus* (68–71)
Rod-shaped myxospores: Myxospores not contained in sporangia, fruiting bodies without stalks	*Archangium* (67–68)
Myxospores embedded in slime envelope:	
Fruiting bodies without stalks	*Cystobacter* (68)
Stalked fruiting bodies, single sporangia	*Melittangium* (—)
Stalked fruiting bodies, multiple sporangia	*Stigmatella* (68–69)
Fruiting bodies are dark-brown clusters consisting of tiny spherical or disclike sporangia with an outer wall	*Angiococcus* (—)
Vegetative cells not tapered (blunt, rounded ends); myxospores resemble vegetative cells; sporangia always produced:	
Fruiting bodies without stalks; myxospores rod-shaped	*Polyangium* (69)
Fruiting bodies without stalks; myxospores oval; highly cellulolytic	*Sorangium* (—)
Fruiting bodies without stalks; myxospores coccoid	*Nannocystis* (70–72)
Stalked fruiting bodies	*Chondromyces* (69–70)

[a] Phylogenetically, those species examined fall into the delta subdivision of the Proteobacteria (see Table 12.1).

(a) Herbert Voelz   *(b)* Herbert Voelz

● **Figure 12.44** *Myxococcus.* (a) Electron micrograph of a thin section of a vegetative cell of *Myxococcus xanthus*. A cell measures about 0.75 μm wide. (b) Myxospore of *M. xanthus*, showing the multilayered outer wall. Myxospores measure about 2 μm in diameter.

swarm to organic media. Many myxobacteria can be grown in the laboratory on media containing peptone or casein hydrolysate, which provides organic nutrients in the form of amino acids or small peptides. The organisms are typical aerobes with a complete citric acid cycle (◯◯ Figure 5.22) and respiratory chain.

Life Cycle of a Fruiting Myxobacterium

The life cycle of a typical fruiting myxobacterium is shown in Figure 12.47●. A vegetative cell excretes slime, and as it moves across a solid surface it leaves a slime trail behind (Figure 12.48●). This trail is preferentially used by other cells in the swarm so that a characteristic radiating pattern is soon created, with cells migrating along slime trails (Figure 12.48). The fruiting body ultimately formed (Figures 12.45 and 12.46) is a complex structure produced by the differentiation of cells in the stalk region and in the myxospore-bearing head.

Fruiting body formation does not occur so long as adequate nutrients for vegetative growth are present, but upon nutrient exhaustion, the vegetative swarms begin to fruit. Cells aggregate, likely through a chemotactic response (◯◯ Sections 4.16 and 8.13), with the cells migrating toward each other and forming mounds or heaps (Figure 12.49●); a single fruiting body may have 10^9 or more cells. As the cell mounds become higher, the differentiation of the fruiting body into stalk and head begins (Figure 12.49*b* and *c*). Figure 12.49*d* clearly illustrates the differentiation of the fruiting body into stalk and head. The stalk is composed of slime, within which a few cells may be trapped. The majority of the cells accumulate in the fruiting-body head and undergo differentiation into *myxospores* (Figures 12.44–12.47). In some genera, the myxospores are enclosed in large walled structures called **cysts**.

Compared to the vegetative cell, the myxospore is more resistant to drying, sonic vibration, UV radiation, and heat, but the degree of heat resistance is much less than that of the bacterial endospore (◯◯ Section 4.13). The main function of encysted myxospores is therefore

(a) Hans Reichenbach

(b) David White

● **Figure 12.45** *Stigmatella aurantiaca.* (a) Color photo of a single fruiting body. The structure is about 150 μm high. (b) Scanning electron micrograph of a fruiting body growing on a piece of wood. Note the individual cells visible in each fruiting structure. The color of the fruiting body shown in (a) is due to the production of carotenoid pigments (◯◯ Section 17.3).

(a)

(b)

(c)

● **Figure 12.46** **Fruiting bodies of three species of fruiting myxobacteria.** (a) *Myxococcus fulvus* (125 μm high). (b) *Myxococcus stipitatus* (170 μm high). (c) *Chondromyces crocatus* (560 μm high).

to enable the organism to survive desiccation during dispersal or during drying of the habitat. Upon dissemination to a suitable habitat or restoration of adequate growth conditions, the myxospore eventually germinates by a localized rupture of the capsule, with the growth and emergence of a new vegetative cell.

Myxobacteria are usually colored by carotenoid pigments (see Figures 12.45*a* and 12.46), and the main pigments are carotenoid glycosides. Pigment formation is promoted by light, and at least one function of the pigment is photoprotection. Since in nature the myxobacteria usually form fruiting bodies in the light, the presence of these photoprotective pigments is understandable. In the genus *Stigmatella* (Figure 12.45), light greatly stimulates fruiting body formation and catalyzes production of the lipid pheromone *2,5,8-trimethyl-8-hydroxy-nonane-4-one*. This substance initiates the aggregation step. The fruiting myxo-bacteria are classified on morphological grounds using characteristics of the vegetative cells, the myxo-spores, and fruiting-body structure (Table 12.20), and on phylogenetic grounds, using rRNA sequencing.

● **Figure 12.47** **Life cycle of *Myxococcus xanthus*.** Aggregation serves to assemble vegetative cells for fruiting–body formation. Vegetative cells undergo morphogenesis to resting cells called *myxospores*. The latter germinate under favorable nutritional and physical conditions to yield vegetative cells. Vegetative cells can be converted directly to myxospores without fruiting-body formation by certain chemical inducers, notably high concentrations of glycerol. See the photograph of *Myxococcus* fruiting bodies in Figures 12.45 and 12.46.

(a)　　　*(b)*

● **Figure 12.48** **Swarming in *Myxococcus*.** (a) Photomicrograph of a swarming colony (5-mm radius) of *Myxococcus xanthus* on agar. (b) Single cells of *Myxococcus fulvus* from an actively gliding culture, showing the characteristic slime trails on the agar. A cell of *M. fulvus* is about 0.8 μm in diameter.

(a)   (b)   (c)   (d)

●**Figure 12.49 Scanning electron micrographs of fruiting-body formation in *Chondromyces crocatus*.** (a) Early stage, showing aggregation and mound formation. (b) Initial stage of stalk formation. Slime formation in the head has not yet begun, and so the cells of which the head is composed are still visible. (c) Three stages in head formation. Note that the diameter of the stalk also increases. (d) Mature fruiting bodies. The entire fruiting structure is about 600 μm in height (see Figure 12.46c).

12.17 Concept Check

The fruiting myxobacteria are rod-shaped, gliding bacteria that aggregate to form complex masses of cells called *fruiting bodies*. Myxobacteria are chemoorganotrophic soil bacteria that live by consuming dead organic matter or other bacterial cells.

◆ What environmental conditions trigger fruiting body formation in myxobacteria?

◆ What is a *myxospore* and how does it compare with an *endospore*?

◆ To what specific phylogenetic group do the myxobacteria belong?

12.18 Sulfate- and Sulfur-Reducing Proteobacteria

Key Genera: *Desulfovibrio, Desulfobacter, Desulfuromonas*

Sulfate (SO_4^{2-}) and sulfur (S^0) can be electron acceptors under anoxic conditions for a large group of delta Proteobacteria that utilize organic compounds or H_2 as electron donors. Hydrogen sulfide (H_2S) is the product of both SO_4^{2-} and S^0 reduction. Over 40 genera of these organisms, collectively known as the dissimilative **sulfate-reducing bacteria** and **sulfur-reducing bacteria**, are known, and some of the key ones are shown in Table 12.21.

The genera of sulfate-reducing bacteria in group I, such as *Desulfovibrio* (Figure 12.50a●), *Desulfomonas, Desulfotomaculum*, and *Desulfobulbus* (Figure 12.50c), utilize lactate, pyruvate, ethanol, or certain fatty acids as electron donors, reducing sulfate to hydrogen sulfide. The genera in group II, such as *Desulfobacter* (Figure 12.50d), *Desulfococcus, Desulfosarcina* (Figure 12.50e), and *Desulfonema* (Figure 12.50b), specialize in the oxidation of fatty acids, particularly *acetate*, reducing sulfate to sulfide. The sulfate-reducing bacteria are for the most part obligate anaerobes, and strict anoxic techniques must be used in their cultivation (Figure 12.50g).

Sulfate-reducing bacteria are widespread in aquatic and terrestrial environments that become anoxic as a result of microbial decomposition processes. The best-studied genus is *Desulfovibrio* (Figure 12.50a), which is common in aquatic habitats or waterlogged soils containing abundant organic material and sufficient levels of sulfate. *Desulfotomaculum*, phylogenetically a member of the gram-positive *Bacteria*, consists of endospore-forming rods found primarily in soil. Growth and reduction of sulfate by *Desulfotomaculum* in certain canned foods leads to a type of spoilage called *sulfide stinker*. The remaining genera of sulfate reducers are indigenous to anoxic freshwater (group I) or marine environments (groups I and II), and can occasionally be isolated from the mammalian intestine.

Sulfur Reduction

The dissimilative *sulfur-reducing bacteria* can reduce elemental sulfur to sulfide but are unable to reduce sulfate to sulfide. Members of the genus *Desulfuromonas* (Figure 12.50f) can grow anaerobically by coupling the *oxidation* of substrates such as acetate or ethanol to the *reduction* of elemental sulfur to hydrogen sulfide. However, the ability to reduce elemental sulfur, as well as other sulfur compounds such as thiosulfate, sulfite, or dimethyl sulfoxide (DMSO), is a widespread property of a variety of chemoorganotrophic, generally facultatively aerobic bacteria (for example, *Proteus, Campylobacter, Pseudomonas,* and *Salmonella*). *Desulfuromonas* differs from the latter in that it is an obligate anaerobe and utilizes *only* sulfur as an electron acceptor (see Table 12.21). Dissimilative sulfur-reducing bacteria like *Desulfuromonas* reside in many of the same habitats as sulfate-reducing bacteria and often form associations with bacteria that oxidize H_2S to S^0, such as green sulfur bacteria (see Section 12.32). The sulfur produced is then reduced back to H_2S during metabolism of the sulfur-reducer, completing an abbreviated anoxic sulfur cycle (∞Section 19.13).

Physiology of Sulfate-Reducing Bacteria

The range of electron donors used by sulfate-reducing bacteria is fairly broad. Hydrogen, lactate, and pyruvate are almost universally used, and many group I species utilize malate, sulfonates, and certain primary alcohols

Table 12.21 Characteristics of some key genera of sulfate-and sulfur-reducing bacteria[a]

Genus	Characteristics	DNA (mol% GC)
Group I sulfate reducers: Nonacetate oxidizers		
Desulfovibrio	Polarly flagellated, curved rods, no spores; gram-negative; contain desulfoviridin; one thermophilic	46–61
Desulfomicrobium	Motile rods, no spores; gram-negative; desulfoviridin absent	52–57
Desulfobotulus	Vibrios; gram-negative; motile; desulfoviridin absent	53
Desulfofustis	Motile rods, specializes in the degradation of glycolate and glyoxalate	56
Desulfotomaculum	Straight or curved rods; motile by peritrichous or polar flagellation; gram-negative; desulfoviridin absent; produce endospores; capable of utilizing acetate as energy source	37–46
Desulfomonile	Rod; capable of reductive dechlorination of 3-chlorobenzoate to benzoate (⌖Section 17.18)	49
Desulfobacula	Oval to coccoid cells, marine; can oxidize various aromatic compounds including the aromatic hydrocarbon toluene, to CO_2	42
Archaeoglobus	Archaeon; hyperthermophile, temperature optimum, 83°C; contains some unique coenzymes of methanogenic bacteria, makes small amount of methane during growth; H_2, formate, glucose, lactate, and pyruvate are electron donors, SO_4^{2-}, $S_2O_3^{2-}$, or SO_3^{2-}, electron acceptors (⌖Section 13.7)	41–46
Desulfobulbus	Ovoid or lemon-shaped cells; no spores; gram-negative; desulfoviridin absent; if motile, by single polar flagellum; utilizes propionate as electron donor with acetate + CO_2 as products	59–60
Desulforhopalus	Curved rods, gas vacuolate, psychrophile; uses propionate, lactate, or alcohols as electron donor	48
Thermodesulfobacterium	Small, gram-negative rods; desulfoviridin present; thermophilic, optimum growth at 70°C; a member of the *Bacteria* but contains ether-linked lipids (see Section 12.36)	34
Group II sulfate reducers: Acetate oxidizers		
Desulfobacter	Rods; no spores, gram-negative; desulfoviridin absent; if motile, by single polar flagellum; utilizes only acetate as electron donor and oxidizes it to CO_2 via the citric acid cycle	45–46
Desulfobacterium	Rods, some with gas vesicles, marine; capable of autotrophic growth via the acetyl-CoA pathway	41–59
Desulfococcus	Spherical cells; nonmotile; gram-negative; desulfoviridin present, no spores; utilizes C_1 to C_{14} fatty acids as electron donor with complete oxidation to CO_2; capable of autotrophic growth via the acetyl-CoA pathway	57
Desulfonema	Large, filamentous gliding bacteria; gram-positive, no spores; desulfoviridin present or absent; utilizes C_2 to C_{12} fatty acids as electron donor with complete oxidation to CO_2; capable of autotrophic growth via the acetyl-CoA pathway (H_2 as electron donor)	35–42
Desulfosarcina	Cells in packets (sarcina arrangement); gram-negative; no spores; desulfoviridin absent; utilizes C_2 to C_{14} fatty acids as electron donor with complete oxidation to CO_2; capable of autotrophic growth via the acetyl-CoA pathway (H_2 as electron donor)	51
Desulfoarculus	Vibrios; gram-negative; motile; desulfoviridin absent; utilizes only C_1 to C_{18} fatty acids as electron donor	66
Desulfacinum	Cocci to oval-shaped cells; gram-negative; utilizes C_1 to C_{18} fatty acids, very nutritionally diverse, capable of autotrophic growth; thermophilic	64
Desulforhabdus	Rods; no spores; gram-negative; nonmotile; utilizes fatty acids with complete oxidation to CO_2	52
Thermodesulforhabdus	Gram-negative motile rods; thermophilic; uses fatty acids up to C_{18}	51
Dissimilative sulfur reducers		
Desulfuromonas	Straight rods, single lateral flagellum; no spores; gram-negative; does not reduce sulfate; acetate, succinate, ethanol, or propanol used as electron donor; obligate anaerobe; one species is capable of the reductive dechlorination of trichloroethylene (⌖Section 17.18)	50–63
Desulfurella	Motile short rods; gram-negative; requires acetate; thermophilic	31
Sulfurospirillum	Small vibrios, reduces S^0 with H_2 or formate as electron donors	—
Campylobacter	Curved, vibrio-shaped rods; polar flagella; gram-negative; no spores; unable to reduce sulfate but can reduce sulfur, sulfite, thiosulfate, nitrate, or fumarate anaerobically with acetate or a variety of other carbon or electron donor sources; facultative aerobe	40–42

[a] Phylogenetically, most sulfate- and sulfur-reducing bacteria are delta Proteobacteria.

(for example, ethanol, propanol, and butanol). Some strains of *Desulfotomaculum* utilize glucose, but this is rather rare among sulfate reducers in general. Group I sulfate-reducers oxidize their electron donor to the level of acetate and excrete this fatty acid as an end product.

Group II organisms differ from those in group I by their ability to oxidize fatty acids (including acetate), lactate, succinate, and even benzoate in some species, completely to CO_2. *Desulfosarcina, Desulfonema, Desulfococcus, Desulfobacterium,* *Desulfotomaculum,* and certain species of *Desulfovibrio,* are unique among sulfate-reducers in their ability to grow chemolithotrophically and autotrophically with H_2 as electron donor, sulfate as electron acceptor, and CO_2 as sole carbon source. A few specialized sulfate-reducers can use hydrocarbons, even crude oil itself, as electron donors. This process is noteworthy because until such organisms were recognized, it was thought that hydrocarbons could only be oxidized under oxic conditions (⌖Section 17.23).

In addition to using sulfate as an electron acceptor, many sulfate-reducing bacteria can grow using *nitrate* as an electron acceptor, reducing NO_3^- to NH_3, or reducing sulfonates, such as isethionate ($HO—CH_2—CH_2—SO_3^-$),

(a)

(b)

(c)

(d)

(e)

(f)

(g)

● **Figure 12.50 Phase-contrast photomicrographs of (a–e) representative sulfate-reducing and (f) sulfur-reducing bacteria.** (a) *Desulfovibrio desulfuricans*; cell diameter about 0.7 μm. (b) *Desulfonema limicola*; cell diameter 3 μm. (c) *Desulfobulbus propionicus*; cell diameter about 1.2 μm. (d) *Desulfobacter postgatei*; cell diameter about 1.5 μm. (e) *Desulfosarcina variabilis* (interference contrast microscopy); cell diameter about 1.25 μm. (f) *Desulfuromonas acetoxidans*; cell diameter about 0.6 μm. (g) Enrichment culture of sulfate-reducing bacteria. Left, sterile medium; center, a positive enrichment showing black ferrous sulfide (FeS); right, colonies of sulfate-reducing bacteria in a "shake tube" (🔗 Section 18.2).

to sulfide. Elemental sulfur can also be reduced to sulfide by most sulfate-reducing bacteria.

Certain organic compounds can be fermented by sulfate-reducing bacteria. The most common fermentable compound is *pyruvate*, which is converted via the phosphoroclastic reaction to acetate, CO_2, and H_2 (🔗 Figure 17.51). Moreover, although thought to be *obligate* anaerobes, certain isolates of sulfate-reducing bacteria, primarily ones isolated from microbial mats where they coexist with O_2-producing cyanobacteria, are quite oxygen tolerant and can respire with O_2 as electron acceptor. At least one species of *Desulfovibrio*, *D. oxyclinae*, can actually *grow* with O_2 as electron acceptor under microoxic conditions.

Isolation

The enrichment of *Desulfovibrio* is relatively easy on an anoxic lactate-sulfate medium to which ferrous iron is added. A reducing agent, such as thioglycolate or ascorbate, is also added to achieve a lower E_0'. When sulfate-reducing bacteria grow, the sulfide formed from sulfate reduction combines with the ferrous iron to form black, insoluble ferrous sulfide (Figure 12.50g). This blackening not only indicates sulfate reduction, but the iron also ties up and detoxifies the sulfide, making possible growth to higher cell yields. After some growth has occurred as evidenced by blackening of the medium, purification is accomplished by streaking onto a tube coated on the inside surface with a thin layer of agar (called *roll tubes*) or on Petri plates in an anoxic glove box (🔗 Figure 6.26).

Alternatively, agar shake tubes can be used for purification purposes. In the *shake tube method* a small amount of liquid from the original enrichment is added to a tube of molten agar growth medium, mixed thoroughly, and sequentially diluted through a series of molten agar tubes (🔗 Section 18.2 and Figure 18.3b). On solidification, individual cells of sulfate reducers distributed throughout the agar form black colonies (Figure 12.50g) that can be removed aseptically, and the whole process is repeated until pure cultures are obtained.

⬢ 12.18 Concept Check

Sulfate- and sulfur-reducing bacteria are a large group of delta Proteobacteria unified by their physiological process of reducing either SO_4^{2-} or S^0 to H_2S under anoxic conditions. Two physiological subgroups of sulfate-reducing bacteria are known: group I, which is incapable of oxidizing acetate to CO_2, and group II, which is capable of doing so.

◆ What organic substrate would you use to enrich and isolate a *group II* sulfate reducer from nature?

◆ For sulfate-reducing bacteria capable of chemolithotrophic and autotrophic growth: (1) What is the electron donor? (2) What is the electron acceptor? (3) What is the source of cell carbon?

◆ Physiologically, how does *Desulfuromonas* differ from *Desulfovibrio*?

III | PHYLUM 2 AND 3: GRAM-POSITIVE BACTERIA AND ACTINOBACTERIA

12.19 | Nonsporulating, Low GC, Gram-Positive *Bacteria*: Lactic Acid *Bacteria* and Relatives

Key Genera: *Staphylococcus, Micrococcus, Streptococcus, Lactobacillus, Sarcina*

In the introduction to this chapter, we saw that gram-positive *Bacteria* fall into two major phylogenetic subdivisions, low GC and high GC (Actinobacteria). We consider in this section the genera of the former division. We also consider *Micrococcus* because it is morphologically quite similar to *Staphylococcus* although it is actually a member of the Actinobacteria. We also include here the lactic acid *Bacteria*, classical nonsporulating gram-positive rods and cocci (Table 12.22).

Staphylococcus and *Micrococcus*

Staphylococcus (Figure 12.51●) and *Micrococcus* are both aerobic organisms with a typical respiratory metabolism. They are catalase-positive, and this test permits their distinction from *Streptococcus* and some other genera of gram-positive cocci. Gram-positive cocci are relatively resistant to reduced water potential and tolerate drying and high salt fairly well. Their ability to grow in media with high salt provides a selective means for isolation. For example, if an appropriate inoculum is spread on an agar plate with a rich medium containing 7.5% NaCl and the plate incubated aerobically, gram-positive cocci often form the predominant colonies. Many species are pigmented, and this provides an additional aid in selecting gram-positive cocci.

The genera *Micrococcus* and *Staphylococcus* can easily be separated based on the oxidation–fermentation (O/F) (∞Table 24.3) test. *Micrococcus* is an obligate aerobe and produces acid from glucose only aerobically, whereas *Staphylococcus* is a facultative aerobe and produces acid from glucose both aerobically and anaerobically. *Staphylococcus* also typically forms cell clusters (Figure 12.51a), while *Micrococcus* does not.

Staphylococci are common commensals and parasites of humans and animals, and they occasionally cause serious infections. In humans, two major species are recognized, *Staphylococcus epidermidis*, a nonpigmented, nonpathogenic organism usually found on the skin or mucous membranes, and *Staphylococcus aureus* (Figure 12.51), a yellow pigmented species that is most commonly associated with pathological conditions, including boils, pimples, pneumonia, osteomyelitis, meningitis, and arthritis. We discuss the pathogenesis of *S. aureus* in Chapter 21 and staphylococcal diseases in Chapters 26 and 29. *Micrococcus* species can also be isolated from skin but are much more common on inanimate objects, dust particles, and in soil.

Sarcina

The genus *Sarcina* contains species of bacteria that divide in three perpendicular planes to yield packets of eight cells or more (Figure 12.52a●). *Sarcina* are obligate anaerobes and are extremely acid-tolerant, being able to ferment sugars and grow down to pH 2. Cells of one species, *Sarcina ventriculi*, contain a thick fibrous layer of cellulose surrounding the cell wall (Figure 12.52b). The cellulose layers of adjacent cells become attached, and this functions as a cementing material to hold together packets of *S. ventriculi* cells.

(a)

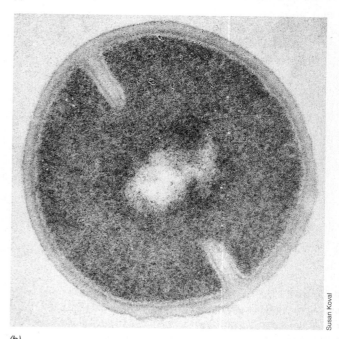

(b)

● **Figure 12.51** *Staphylococcus.* (a) Scanning electron micrograph of typical *Staphylococcus aureus* cells, showing the irregular arrangement of the cell clusters. Individual cells are about 0.8 μm in diameter. (b) Transmission electron micrograph of a dividing cell of *S. aureus*. Note the thick gram-positive cell wall (∞Section 4.8).

Table 12.22 **Distinguishing features of major gram-positive cocci**

Genus	Motility	Arrangement of cells	Growth by fermentation	DNA (mol % GC)	Phylogenetic group[a]	Other characteristics
Micrococcus	−	Clusters, tetrads	−	66–73	Actinobacteria	Strict aerobe
Staphylococcus	−	Clusters, pairs	+	30–39	Low GC	Only genus of this group to contain teichoic acid in cell wall
Stomatococcus	−	Clusters, pairs	+	56–60	Actinobacteria	Only genus of this group containing a capsule
Planococcus	+	Pairs, tetrads	−	39–52	Low GC	Primarily marine
Sarcina	−	Cuboidal packets of eight or more cells	+	28–31	Low GC	Extremely acid-tolerant; cellulose in cell wall
Ruminococcus	+	Pairs, chains	+	39–46	Low GC	Obligate anaerobe; inhabits rumen, cecum, and large intestine of many animals
Peptococcus	−	Clusters, pairs	+	50–51	Low GC	Obligate anaerobe; ferments peptone but not sugars
Peptostreptococcus	−	Clumps, short chains	+	28–37	Low GC	Obligate anaerobe; ferments peptone; common member of human normal flora, skin, intestine, vagina; also isolated from vaginal and purulent discharges

[a] All are members of the gram-positive *Bacteria* (see Figure 12.1).

(a)

(b)

Sarcina species can be isolated from soil, mud, feces, and stomach contents. Because of its extreme acid tolerance, *S. ventriculi* is one of the only bacteria that can actually grow in the stomach of humans and other monogastric animals. Rapid growth of *S. ventriculi* is observed in the stomach of humans suffering from certain gastrointestinal disorders, such as pyloric ulcerations. These pathological conditions retard the flow of food to the intestine and often require surgery to correct.

Lactic Acid Bacteria and Lactic Acid Fermentations

The lactic acid bacteria are gram-positive rods and cocci that produce lactic acid as a major or sole fermentation product. Members of this group lack porphyrins and cytochromes, do not carry out electron transport phosphorylation, and hence obtain energy only by *substrate-level phosphorylation*. All lactic acid bacteria grow anaerobically. Unlike many anaerobes, however, most lactic acid bacteria are not sensitive to O_2 and can grow in its presence as well as in its absence; thus they are **aerotolerant anaerobes**. Most lactic acid bacteria obtain energy only from the metabolism of sugars and hence are usually restricted

● **Figure 12.52** *Sarcina.* (a) Phase-contrast photomicrograph of cells of a typical gram-positive coccus *Sarcina* sp. A single cell is about 2 μm in diameter. (b) Electron micrograph of a thin section. The outermost layer of the cell consists of cellulose.

to habitats in which sugars are present. They typically have only limited biosynthetic ability, and their complex nutritional requirements include needs for amino acids, vitamins, purines, and pyrimidines (∞Table 5.4).

One important difference between subgroups of the lactic acid bacteria lies in the pattern of products formed from the fermentation of sugars. One group, called **homofermentative**, produces a single fermentation product, *lactic acid*. The other group, called **heterofermentative**, produces other products, mainly *ethanol* plus CO_2, as well as

lactate (Table 12.23). Figure 12.53● summarizes pathways for the fermentation of glucose by a homo- and a heterofermentative organism. The differences observed in the fermentation patterns are determined by the presence or absence of the enzyme **aldolase**, the key enzyme in *glycolysis* (∞Figure 5.14). The heterofermenters, lacking aldolase, cannot break down fructose bisphosphate to triose phosphate. Instead, they oxidize glucose 6-phosphate to 6-phosphogluconate and then decarboxylate this to pentose phosphate. The latter is broken down to triose phos-

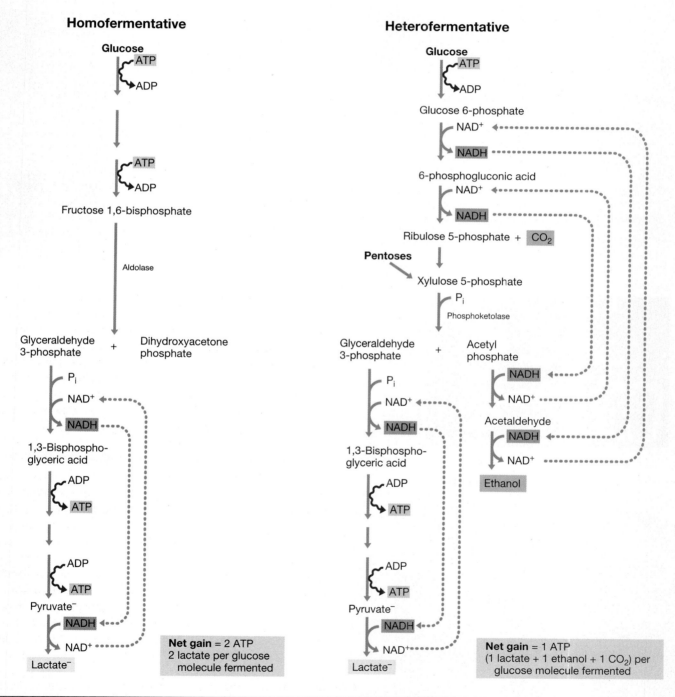

● **Figure 12.53   The fermentation of glucose in homofermentative and heterofermentative lactic acid bacteria.** Note that no ATP is made in reactions leading to ethanol formation. If oxygen is present, many heterofermentative lactic acid bacteria can reduce oxygen with NADH (through flavin enzymes as intermediates), forming water; acetate is then made instead of ethanol, and this allows for the production of one additional ATP.

Table 12.23	Differentiation of the principal genera of lactic acid bacteria[a]
Cell form and arrangement	**Genus/DNA (mol % GC)**
Cocci in chains or tetrads	
Homofermentative	*Streptococcus* (34–46)
	Enterococcus (38–40)
	Lactococcus (38–41)
	Pediococcus (34–42)
Heterofermentative	*Leuconostoc* (38–41)
Rods, typically in chains	
Homofermentative	*Lactobacillus* (32–53)
Heterofermentative	*Lactobacillus* (34–53)

[a] Phylogenetically, all organisms are members of the low GC subdivision of the gram-positive *Bacteria*.

(a)   (b)

● **Figure 12.54 Phase-contrast (a) and scanning electron (b) micrographs of gram-positive cocci.** (a) *Lactococcus lactis*. (b) *Streptococcus* sp. Cells in both cases are 0.5–1 μm in diameter.

phate and acetylphosphate by the key enzyme **phosphoketolase** (Figure 12.53).

In heterofermenters, triose phosphate is converted ultimately to lactic acid with the production of 1 mol of ATP. However, to achieve redox balance, the acetylphosphate produced accepts electrons from the NADH generated during the production of pentose phosphate and is converted to ethanol. This occurs *without* yielding ATP. Because of this, heterofermenters produce only *1 mol* of ATP from glucose instead of the *2 mol* produced by homofermenters. Because the heterofermenters decarboxylate 6-phosphogluconate, they produce CO_2 as a fermentation product, whereas the homofermenters produce little or no CO_2. One simple way of detecting a heterofermenter, therefore, is to observe for production of CO_2 in laboratory cultures.

The various genera of lactic acid bacteria have been defined on the basis of cell morphology, DNA base composition, phylogeny, and type of fermentative metabolism, as is shown in Table 12.23. Members of the genera *Streptococcus, Enterococcus, Lactococcus, Leuconostoc,* and *Pediococcus* have fairly similar DNA base ratio compositions; in addition, there is very little variation from strain to strain. The genus *Lactobacillus*, on the other hand, has members with widely diverse DNA compositions and hence does not constitute a homogeneous group.

Streptococcus and Other Cocci

The genus *Streptococcus* (Figure 12.54●) contains a wide variety of homofermentative species with quite distinct habitats, whose activities are of considerable practical importance to humans. Some members are pathogenic to humans and animals (∞Section 26.2). As producers of lactic acid, other streptococci play important roles in the production of buttermilk, silage, and other fermented products (∞Section 29.2), and certain species play a major role in dental caries formation (∞Section 21.3). To distinguish nonpathogenic streptococci from human pathogenic species, three genera are recognized. The

genus *Lactococcus* contains those streptococci of dairy significance, whereas the genus *Enterococcus* includes streptococci that are primarily of fecal origin.

Organisms in the genus *Streptococcus* have been divided into two groups of related species on the basis of characteristics enumerated in Table 12.24. Hemolysis on blood agar is of considerable importance in the subdivision of the genus. Colonies of those strains producing streptolysin O or S are surrounded by a large zone of complete red blood cell hemolysis, a condition called **β hemolysis** (∞Figure 21.17a). On the other hand, many streptococci, lactococci, and enterococci do not produce hemolysins, but instead cause the formation of a greenish or brownish zone around colonies on blood agar. This is not due to true hemolysis, but to discoloration and loss of potassium from the red cells. This type of reaction is called **α hemolysis**.

Streptococci and related cocci are also divided into *immunological* groups based on the presence of specific carbohydrate antigens. These antigenic groups (or **Lancefield groups** as they are commonly known, named for Rebecca Lancefield, a pioneer in *Streptococcus* taxonomy), are designated by letters; A through O are currently recognized. Those β-hemolytic streptococci found in humans usually contain the group A antigen, while enterococci contain the group D antigen. Group B streptococci, usually found in association with animals, are a cause of *mastitis* (inflammation of the udder) in cows and have also been implicated in certain human infections. Lactococci are of antigen group N and are not pathogenic.

Placed in the genus *Leuconostoc* are heterofermentative cocci. Strains of *Leuconostoc* also produce the flavoring ingredients diacetyl and acetoin from the catabolism of citrate and have been used as starter cultures in dairy fermentations. Some strains of *Leuconostoc* produce large amounts of dextran polysaccharides (α-1,6-glucan) when cultured on sucrose (∞Figure 17.64), and some of these have found medical use as plasma extenders in blood transfusions. Other strains of *Leuconostoc* produce other polysaccharide polymers such as fructose polymers called *levans*.

Table 12.24 Differential characteristics of streptococci, lactococci, and enterococci

Group	Antigenic (Lancefield) groups	Representative species	Type of hemolysis on blood agar	Good growth at 10° C	Good growth at 45° C	Survive 60° C for 30 min	Growth in Milk with 0.1% methylene blue	Growth in Broth with 40% bile	Habitat
Streptococci									
Pyogenes subgroup	A,B,C,F,G	*Streptococcus pyogenes*	Lysis (β)	–	–	–	–	–	Respiratory tract, systemic
Viridans subgroup	–	*Streptococcus mutans*	Greening (α)	–	+	–	–	–	Mouth, intestine
Enterococci	D	*Enterococcus faecalis*	Lysis (β), greening (α), or none	+	+	+	+	+	Intestine, vagina, plants
Lactococci	N	*Lactococcus lactis* (see Figure 12.54a)	None	+	–	+	+	+	Plants, dairy products

Lactobacillus

Lactobacilli are typically rod-shaped, varying from long and slender to short, bent rods (Figure 12.55●). Most species are homofermentative, but some are heterofermentative (Table 12.23). Lactobacilli are common in dairy products, and some strains are used in the preparation of fermented milk products. For instance, *Lactobacillus delbrueckii* (Figure 12.55c) is used in the preparation of yogurt, *L. acidophilus* (Figure 12.55a) in the production of acidophilus milk, and other species in the production of sauerkraut, silage, and pickles (◆Section 29.2). The lactobacilli are usually more resistant to acidic conditions than are the other lactic acid bacteria, being able to grow well at pH values as low as 4. Because of this, they can be selectively isolated from natural materials by use of an acidic rich carbohydrate-containing medium such as tomato juice-peptone agar.

The acid resistance of the lactobacilli enables them to continue growing during natural lactic fermentations, even when the pH value has dropped too low for other lactic acid bacteria to grow. The lactobacilli are therefore responsible for the final stages of most lactic acid fermentations. They are rarely, if ever, pathogenic.

Listeria

Listeria are gram-positive coccobacilli that tend to form chains of three to five cells (◆Figure 29.9). *Listeria* is phylogenetically related to *Lactobacillus* species, and like homofermentative lactic acid bacteria, they produce acid but not gas from glucose. True lactic acid bacteria, however, are capable of growth under strictly anoxic conditions and lack the enzyme catalase. *Listeria*, in contrast, requires microoxic or fully oxic growth conditions and produces catalase.

Although several species of *Listeria* are known, the species *L. monocytogenes* is most noteworthy because it causes a major foodborne illness, *listeriosis* (◆Section 29.10 and Figure 29.9). The organism is transmit-

(a)

(b)

(c)

● **Figure 12.55 Phase-contrast and electron micrographs of *Lactobacillus* species.** (a) *Lactobacillus acidophilus.* Cells are about 0.75 μm wide. (b) *Lactobacillus brevis*, transmission electron micrograph. Cells measure about 0.8 × 2 μm. (c) *Lactobacillus delbrueckii*, scanning electron micrograph. Cells are about 0.7 μm in diameter. Both heterofermentative and homofermentative species of *Lactobacillus* are known (see Figure 12.53 and Table 12.23).

Otto Kandler

Otto Kandler

V. Bottazzi

ted in contaminated, usually ready-to-eat, foods (cheese is a common vehicle) and can cause anything from a mild illness to a fatal form of meningitis.

12.20 Endospore-Forming, Low GC, Gram-Positive *Bacteria*: *Bacillus*, *Clostridium*, and Relatives

Key Genera: *Bacillus, Clostridium, Sporosarcina, Heliobacterium*

Several genera of endospore-forming bacteria have been recognized (Table 12.25), distinguished on the basis of cell morphology, shape, and cellular position of the endospore (Figure 12.56●), relationship to O_2, and energy metabolism. The structure and heat resistance of the bacterial endospore along with the process of endospore formation itself was discussed in Section 4.13. All endospore formers show phylogenetic affiliation to the "low GC" gram-positive *Bacteria*. The two genera most frequently studied are *Bacillus*, species of which are aerobic or facultatively aerobic, and *Clostridium*, which contains the strictly anaerobic, fermentative species. One group of endospore-formers, the *heliobacteria*, are phototrophic (the word *helio* comes from the Greek word for *sun*).

Considerable genetic heterogeneity exists among endospore-formers. For example, the GC ratios of *Bacillus* species alone vary over a range of nearly 40%. Yet all endospore-forming bacteria are *ecologically* related, because they are found in nature primarily in soil. Even those species that are pathogenic to humans or other animals are primarily saprophytic soil organisms and infect hosts only incidentally. Indeed, the ability to produce en-

dospores should be advantageous for a soil microorganism because soil can be a highly variable environment in terms of nutrient levels, temperature, and water activity. Thus, a heat- and desiccation-resistant structure capable of remaining dormant for long periods (perhaps even millions of years, ⊂⊃Section 4.13) should offer considerable survival value in nature.

Endospore formers can be selectively isolated from soil, food, dust, and other materials by treating the sample to 80°C for 10 min (pasteurization), a treatment that effectively kills vegetative cells while the endospores present remain viable. Streaking heat-treated samples on plates of the appropriate media and incubating either aerobically or anaerobically readily yield species of *Bacillus* or *Clostridium*, respectively.

Bacillus and *Paenibacillus*

A list of representatives in the *Bacillus* group is shown in Table 12.26. Species of *Bacillus* and *Paenibacillus* grow well on defined media containing any of a number of carbon sources. Many bacilli produce extracellular hydrolytic enzymes that break down complex polymers such as polysaccharides (⊂⊃Figure 17.63), nucleic acids, and lipids, permitting the organisms to use these products as carbon sources and electron donors. Many bacilli produce antibiotics, including bacitracin, polymyxin, tyrocidin, gramicidin, and circulin. In most cases the antibiotics are released during sporulation, when the culture enters the stationary phase of growth and after it is committed to sporulation.

Several bacilli, most notably *P. popilliae* and *B. thuringiensis*, produce insect larvicides. *Paenibacillus popilliae* causes a fatal condition called *milky disease* in Japanese beetle larvae and larvae of closely related beetles of the family

Table 12.25	**Major genera of endospore-forming bacteria**[a]	
Characteristics	**Genus**	**DNA (mol % GC)**
Rods		
Aerobic or facultative, catalase produced	*Bacillus*	32–69
	Paenibacillus	40–54
Microaerophilic, no catalase; homofermentative lactic acid producer	*Sporolactobacillus*	46–47
Anaerobic:		
Sulfate-reducing	*Desulfotomaculum*	38–50
Does not reduce sulfate, fermentative	*Clostridium* (see Figure 12.56)	21–54
Thermophilic, temperature optimum 65–70° C, fermentative	*Thermoanaerobacter*	31–39
Gram-negative; can grow as homoacetogen on $H_2 + CO_2$	*Sporomusa*	41–49
Halophile, isolated from the Dead Sea	*Sporohalobacter*	31
Produces up to five spores per cell; fixes N_2	*Anaerobacter*	29
Acidophile, pH optimum 3	*Alicyclobacillus*	52–60
Alkaliphile, pH optimum 9	*Amphibacillus*	36–38
Phototrophic	*Heliobacterium, Heliophilum, Heliorestis*	50–58
Syntrophic, degrades fatty acids but only in coculture with a H_2-utilizing bacterium (⊂⊃Section 17.21)	*Syntrophospora*	37
Reductively dechlorinates chlorophenols (⊂⊃Section 17.18)	*Desulfitobacterium*	46
Cocci (usually arranged in tetrads or packets), aerobic		
	Sporosarcina (see Figure 12.60)	40–41

[a] Phylogenetically, all organisms are members of the low GC subdivision of the gram-positive *Bacteria*.

(a)

(b)

(c)

● **Figure 12.56** **Phase-contrast photomicrographs of various *Clostridium* species, showing the different locations of the endospore.** (a) *Clostridium cadaveris,* terminal spores. Cells are about 0.9 μm wide. (b) *Clostridium sporogenes,* subterminal spores. Cells are about 1 μm wide. (c) *Clostridium bifermentans.* central spores. Cells are about 1.2 μm wide.

Scarabaeidae. *Bacillus thuringiensis* causes a fatal disease of larvae of many different groups of insects, although individual strains are specific as to host affected. Some strains are specific for lepidopterans, such as the silkworm, the cabbage worm, the tent caterpillar, and the gypsy moth. Other strains kill dipterans such as mosquitoes and black flies. Still others kill coleopterans such as Colorado potato beetles. Strains of *B. thuringiensis* have also been discovered that are toxic to Japanese beetles. Spore preparations derived from *B. thuringiensis* and *P. popilliae* are commercially available as biological insecticides.

The disease caused by *Paenibacillus popilliae* is a septicemia, whereas the disease caused by *B. thuringiensis* is essentially an intoxication. Both of these insect pathogens form a crystalline protein during sporulation called the *parasporal body*, which is deposited within the sporangium but outside the endospore proper (Figure 12.57●). In the case of *B. thuringiensis*, the crystal (parasporal body) protein is a protoxin that is converted to a toxin by proteolytic cleavage in the larval gut. The toxin binds to intestinal epithelial cells and induces pore formation that causes leakage of the host cells followed by lysis.

Genes encoding crystal proteins from several *B. thuringiensis* strains have been isolated. The genes for the *B. thuringiensis* crystal protein (known commercially as "Bt-toxin") have been introduced into plants to render the plants "naturally" resistant to insects. A variety of genetically altered Bt-toxins have also been developed by genetic engineering to help increase toxicity and reduce resistance (⚙ Section 31.10).

Clostridium

The clostridia lack a respiratory chain and hence, unlike *Bacillus* species, they obtain ATP *only* by substrate-level phosphorylation. A wide variety of anaerobic energy-yielding mechanisms are known in the clostridia (fermentative diversity will be discussed in Section 17.20). Indeed, the separation of the genus *Clostridium* into subgroups is based primarily on these properties and on the nature of the fermentable substrate used (Table 12.27).

A number of clostridia ferment sugars, producing as a major end product *butyric acid*. Some of these also produce *acetone* and *butanol*, and at one time acetone-butanol fermentation by clostridia was industrially important as the main commercial source of these products. Some clostridia of the acetone-butanol type fix nitrogen. The most vigorous N_2 fixer is *Clostridium pasteurianum,*

Table 12.26	**Characteristics of representative species of bacilli**		
Characteristics	**Genus/Species**	**Endospore position**	**DNA (mol % GC)**
I. Endospores oval or cylindrical, facultative aerobes, casein and starch hydrolyzed; sporangia not swollen, endospore wall thin			
Thermophiles and acidophiles	*Bacillus coagulans*	Central or terminal	47
	Alicyclobacillus acidocaldarius	Terminal	60
Mesophiles	*Bacillus licheniformis*	Central	46
	Bacillus cereus	Central	35
	Bacillus anthracis	Central	33
	Bacillus megaterium	Central	37
	Bacillus subtilis	Central	43
Insect pathogen	*Bacillus thuringiensis*	Central	34
Sporangia distinctly swollen, spore wall thick			
Thermophile	*Geobacillus stearothermophilus*	Terminal	52
Mesophiles	*Paenibacillus polymyxa*	Terminal	44
	Bacillus macerans	Terminal	52
	Bacillus circulans	Central or terminal	35
Insect pathogens	*Paenibacillus larvae*	Central or terminal	—
	Paenibacillus popilliae	Central	41
II. Endospores spherical, obligate aerobes, casein and starch not hydrolyzed			
Sporangia swollen	*Bacillus sphaericus*	Terminal	37
Sporangia not swollen	*Sporosarcina pasteurii*	Terminal	38

Endospore   Crystal

● **Figure 12.57 The toxic parasporal crystal in the insect pathogen *Bacillus thuringiensis*.** Electron micrograph of a thin section of a sporulating cell. The crystalline protein (Bt-toxin) is toxic to certain insects by causing lysis of intestinal cells.

which probably is responsible for most anaerobic nitrogen fixation in the soil. One group of clostridia ferments cellulose with the formation of acids and alcohols, and these are likely the major organisms decomposing cellulose anaerobically in soil.

The biochemical steps in the formation of butyric acid and butanol from sugars are well understood (Figure 12.58●). Glucose is converted to pyruvate via the Embden–Meyerhof pathway, and pyruvate is split to acetyl-CoA, CO_2, and hydrogen (through reduced ferredoxin) by the phosphoroclastic reaction (∞Section 17.19 and Figure 17.51). Acetyl-CoA is then reduced to fermentation products using the NADH derived from glycolytic reactions. The proportions of the various products are influenced by the duration and the conditions of the fermentation.

During the early stages of the butyric fermentation, butyric and acetic acids are the predominant products, but as the pH of the medium drops, synthesis of acids ceases and the neutral products acetone and butanol begin to accumulate. However, if the pH of the medium is kept neutral with $CaCO_3$, very little of the neutral products are formed and the fermentation products consist of about three parts butyric and one part acetic acid. This makes good physiological sense because, unlike

Table 12.27 Characteristics of some groups of clostridia

Key characteristics	Other characteristics	Species	DNA (mol % GC)
I. Ferment carbohydrates			
Ferment cellulose	Fermentation products: acetate, lactate, succinate, ethanol, CO_2, H_2	C. cellobioparum[a]	28
		C. thermocellum	38–39
Ferment sugars, starch, and pectin	Fermentation products: acetone, butanol, ethanol, isopropanol, butyrate, acetate, propionate, succinate, CO_2, H_2; some fix N_2	C. butyricum	27–28
		C. acetobutylicum	28–29
		C. pasteurianum	26–28
		C. perfringens	24–27
Ferment sugars primarily to acetic acid	Total synthesis of acetate from CO_2; cytochromes present in some species	C. aceticum	33
		Moorella thermoacetica	54
		C. formicaceticum	34
Ferments only pentoses or methylpentoses	Ring-shaped cells form left-handed, helical chains; fermentation products: acetate, propionate, n-propanol, CO_2, H_2	C. methylpentosum	46
II. Ferment amino acids	Fermentation products: acetate, other fatty acids, NH_3, CO_2, sometimes H_2; some also ferment sugars to butyrate and acetate; may produce exotoxins (∞Sections 21.10, 27.8, and 29.5)	C. sporogenes	26
		C. tetani	25–26
		C. botulinum	26–28
		C. tetanomorphum	25–28
	Ferments three-carbon amino acids (for example, alanine) to propionate, acetate, and CO_2	C. propionicum	35
III. Ferments carbohydrates or amino acids	Fermentation products from glucose: acetate, formate, small amounts of isobutyrate and isovalerate	C. bifermentans	27
IV. Purine fermenters	Ferments uric acid and other purines, forming acetate, CO_2, NH_3	C. acidurici	27–30
V. Ethanol fermentation to fatty acids	Produces butyrate, caproate, and H_2; requires acetate as electron acceptor; does not use sugars, amino acids, or purines	C. kluyveri	30

[a] All listings beginning with a "C." are species of the genus *Clostridium*.

neutral product formation, acid production allows for additional ATP synthesis (Figure 12.58).

Another group of clostridia obtains their energy by fermenting *amino acids*. Some species ferment individual amino acids while others ferment only amino acid *pairs*. In this situation one functions as the electron *donor* and is *oxidized*, whereas the other acts as the electron *acceptor* and is *reduced*. This type of coupled amino acid decomposition is known as the **Stickland reaction**. For instance, *Clostridium sporogenes* catabolizes a mixture of glycine and alanine, as outlined in Figure 12.59●. Various amino acids that can function as either electron donors or acceptors in Stickland reactions are also listed in Figure 12.59. The products of Stickland oxidation are always NH_3, CO_2, and a carboxylic acid with one *less* carbon atom than the amino acid that was oxidized (Figure 12.59).

The amino acids that can be fermented singly are alanine, cysteine, glutamate, glycine, histidine, serine, or

threonine. The products are generally acetate, butyrate, CO_2, and H_2. Many of the products of amino acid fermentation by clostridia are foul-smelling substances, and the odor that results from putrefaction is a result mainly of clostridial action. In addition to butyric acid, other odoriferous compounds produced are isobutyric acid, isovaleric acid, caproic acid, hydrogen sulfide, methylmercaptan (from sulfur amino acids), cadaverine (from lysine), putrescine (from ornithine), and ammonia.

The main habitat of clostridia is the soil, where they live primarily in anoxic "pockets," made anoxic by facultative organisms metabolizing organic compounds. In addition, a number of clostridia inhabit the anoxic environment of the mammalian intestinal tract. Also, as is discussed in Section 21.8, several clostridia are capable of causing disease in humans under specialized conditions. Botulism is caused by *Clostridium botulinum*, tetanus by *C. tetani*, and gas gangrene by *C. perfringens* and a number of other clostridia, both sugar and amino acid fermenters. These pathogenic clostridia seem in no way unusual metabolically but are distinct in that they produce specific toxins or, in the case of those causing gas gangrene, a group of toxins (∞Section 21.10 and Table 21.4). *C. perfringens* and related species can also cause gastroenteritis in humans and domestic animals (∞Section 29.6), and botulism occurs in sheep and ducks and a variety of other animals. An unsolved ecological problem is what role these toxins play in the natural habitat of the organism.

Sporosarcina

The genus *Sporosarcina* is unique among endospore formers because cells are *cocci* instead of rods. *Sporosarcina* consists of strictly aerobic spherical to oval cells that divide in two or three perpendicular planes to form tetrads or packets of eight or more cells (Figure 12.60●). The major species of *Sporosarcina* is *S. ureae*. This organism (Figure 12.60) can be enriched from soil by plating dilutions of a pasteurized soil sample on alkaline nutrient agar supplemented with 8% urea and incubating in air. Most soil bacteria are strongly inhibited by as little as 2% urea. However, *S. ureae* actively decomposes even high levels of urea to CO_2 and NH_3, dramatically raising the pH. *S. ureae* is remarkably alkaline-tolerant and can be grown in media up to pH 10. This feature can be used to advantage in its enrichment from soil.

Sporosarcina ureae is common in soils, and studies on its distribution suggest that numbers of *S. ureae* are greatest in soils that receive inputs of urine (a source of urea), such as soils in which animals periodically urinate. Since many soil organisms are quite urea-sensitive, these results suggest that *S. ureae* is ecologically important as a major urea degrader in nature.

Heliobacteria

Heliobacteria are *phototrophic*, low GC, gram-positive *Bacteria*. **Heliobacteria** are anoxygenic phototrophs and

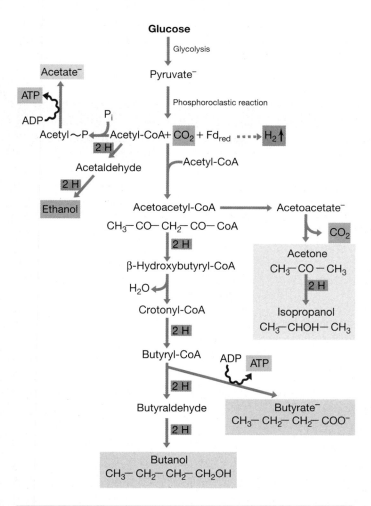

● **Figure 12.58 Pathway of formation of fermentation products from the butyric acid group of clostridia.** The designation "2H" represents two electrons from one molecule of NADH. Note how the production of acetate and butyrate lead to additional ATP by substrate-level phosphorylation (∞Section 17.19). By contrast, formation of butanol and acetone (these products are typically found in a ratio of 1:1 because butanol formation requires two NADH, while acetone formation requires none) reduces the ATP yield because the butyryl-CoA to butyrate step is bypassed.

Amino acids participating in coupled fermentations (Stickland reaction)	
Amino acids oxidized:	**Amino acids reduced:**
Alanine	Glycine
Leucine	Proline
Isoleucine	Hydroxyproline
Valine	Tryptophan
Histidine	Arginine

Overall: Alanine + 2 Glycine + 2 H$_2$O + 3 ADP + 3 P$_i$ ⟶ 3 Acetate$^-$ + CO$_2$ + 3 NH$_4^+$ + 3 ATP

● **Figure 12.59 Coupled oxidation-reduction reaction (Stickland reaction) between alanine and glycine in *Clostridium sporogenes*.** The structures of the key substrates, intermediates, and products are shown (in brackets) to allow the chemistry of the reaction to be followed. Note how in the reaction shown, alanine is the electron *donor*, while glycine is the electron *acceptor*.

produce a unique structural form of bacteriochlorophyll, called *bacteriochlorophyll g* (⚬Section 17.2). The group contains four genera: *Heliobacterium, Heliophilum, Heliorestis,* and *Heliobacillus*. All known heliobacteria are rod-shaped, either short or long rods, frequently with pointed ends (Figure 12.61●). *Heliophilum* is morphologically interesting because its rod-shaped cells form into bundles (Figure 12.61*b*) that are motile as a unit.

Heliobacteria are strict anaerobes, but in addition to anaerobic phototrophic growth, they can grow chemotrophically in darkness by pyruvate fermentation (as can many clostridia, close relatives of the heliobacteria). Like the endospores of *Bacillus* or *Clostridium* species, the endospores of heliobacteria (Figure 12.61*c*) contain elevated Ca^{2+} levels and the signature molecule of the endospore, *dipicolinic acid* (⚬Section 4.13). He-

liobacteria reside in soil, especially paddy field soils, where their strong N$_2$-fixation activities may benefit rice productivity. A large diversity of heliobacteria have also been found in highly alkaline environments, such as soda lakes and their surrounding alkaline soils.

12.19–12.20 Concept Check

The "low GC," gram-positive *Bacteria* are a large phylogenetic group that contains rods and cocci, sporulating and nonsporulating species. Production of endospores is a hallmark of the key genera *Bacillus* and *Clostridium*. Gram-positive *Bacteria* are major agents for the degradation of organic matter in soil, and a few species are pathogenic.

◆ What are the major features that differentiate *Staphylococcus* from *Bacillus*?

◆ How could you distinguish between a *heterofermentative* and a *homofermentative* lactic acid bacterium?

◆ What is the major physiological distinction between *Bacillus* and *Clostridium* species?

◆ Among endospore-producing genera, what is unique about the heliobacteria?

● **Figure 12.60 Phase-contrast photomicrograph of cells of *Sporosarcina ureae*.** A single cell is about 2 μm wide. Note bright refractile endospores. Most cell packets contain eight cells.

Dieter Claus

12.21 Cell Wall-Less, Low GC, Gram-Positive *Bacteria*: The Mycoplasmas

Key Genera: *Mycoplasma, Spiroplasma*

The mycoplasmas are organisms that lack cell walls and that do not revert to walled organisms. They are probably the smallest organisms capable of autonomous growth and are of special evolutionary interest because of their

(a) (b) (c)

● **Figure 12.61 Cells and endospores of heliobacteria.** (a) Electron micrograph of *Heliobacillus mobilis*, a peritrichously flagellated species. (b) *Heliophilum fasciatum*, bundles as observed by electron microscopy. (c) Phase micrograph of endospores from *Heliobacterium gestii*.

extremely simple cell structure and small genomes. Also, although not staining gram-positively (since they lack cell walls), the mycoplasmas are phylogenetically related to low GC, gram-positive *Bacteria*. They are organisms that once likely had cell walls but lost them in colonizing their special habitats. Mycoplasmas are parasitic, inhabiting a variety of animal and plant hosts.

Properties of Mycoplasmas

The absence of cell walls in the mycoplasmas is readily observed by electron microscopy. It has been confirmed by chemical analysis as well, which shows that the key components of peptidoglycan, muramic acid, and di-aminopimelic acid (∞ Section 4.8), are missing. In Chapter 4 we discussed how *protoplasts* can form when cells are treated in an osmotically protected medium. When the osmotic stabilizer is removed, protoplasts take up water, swell, and burst (Figure 4.33). Mycoplasmas resemble protoplasts in their lack of a cell wall, but they are more resistant to osmotic lysis and are able to survive conditions under which protoplasts lyse. This ability to resist osmotic lysis is at least partially determined by the presence of *sterols* (∞ Section 4.5), which make the my-coplasma cytoplasmic membrane more stable than that of other prokaryotes. Indeed, some mycoplasmas re-quire sterols in their growth media, and this sterol re-quirement is a basis for separating the mycoplasmas into two groups (Table 12.28).

In addition to sterols, certain mycoplasmas con-tain compounds called **lipoglycans** (see Table 12.28). Lipoglycans are long-chain heteropolysaccharides co-valently linked to membrane lipids and embedded in the cytoplasmic membrane of many mycoplasmas. Li-poglycans resemble the lipopolysaccharides (LPS) of gram-negative *Bacteria* (∞ Section 4.9) except that they lack the lipid A backbone and the phosphate typ-ical of bacterial LPS. Lipoglycans also function to help stabilize the membrane and have also been identified as facilitating attachment of mycoplasmas to cell sur-face receptors of animal cells. Like LPS, lipoglycans stimulate antibody production when injected into ex-perimental animals.

Growth of Mycoplasmas

Mycoplasma cells are small and highly pleomorphic, a consequence of their lack of a cell wall and hence rigidi-ty. A single culture may exhibit small coccoid elements, larger, swollen forms, and filamentous forms of variable lengths, often highly branched (Figure 12.62●).

The small coccoid elements (0.2–0.3 μm in size) are the smallest mycoplasma units capable of independent growth. Derivatives of these cells with diameters close to 0.1 μm are occasionally seen in mycoplasma cultures, but these are not viable. Even so, the minimum repro-ductive unit of 0.2–0.3 μm probably represents the smallest *free-living* cell (∞ Section 4.4). Additionally, the genome size of mycoplasmas is also smaller than that of most prokaryotes, between 500 and 1100 kbp of DNA in most cases (Table 12.28). This is comparable to that of the obligately parasitic chlamydia and rickettsia (see Sections 12.27 and 12.13) and about one-fifth to one-fourth that of *Escherichia coli*. The genome of one *Mycoplasma* species, *M. genitalium* contains 580 kbp and has been completely sequenced (∞ Section 15.3). This very small genome is thought to be near to the lower limit for the amount of DNA any species of *Bacteria* must have to carry out life processes (we will see a smaller genome in a species of *Archaea*, *Nanoarchaeum*, in Chapter 13).

The mode of growth of mycoplasmas differs in liq-uid and agar cultures. On agar, there is a tendency for the organisms to grow so that they become embedded in the medium. These colonies show a characteristic "fried-egg" appearance; a dense central core, which penetrates downward into the agar, is surrounded by a circular spreading area that is lighter in color (Figure 12.63●). As would be expected, growth of mycoplasmas is not inhib-ited by penicillin, vancomycin, or other antibiotics that inhibit cell wall synthesis. Still, the organisms are as sen-sitive as other *Bacteria* to antibiotics whose targets are other than the cell wall.

Media for the culture of mycoplasmas are typically quite complex. For many species, growth is poor or absent even in complex yeast extract-peptone-beef heart infusion media. Fresh serum or ascitic fluid is needed as well to pro-vide unsaturated fatty acids and sterols. Some mycoplas-

Table 12.28 Major characteristics of mycoplasmas[a]

Genus	Properties	DNA (mol % GC)	Genome size (kilobase pairs)	Presence of lipoglycans
Require sterols				
Mycoplasma	Many pathogenic; facultative aerobes (see Figure 12.62)	23–41	600–1350	+
Anaeroplasma	May or may not require sterols; obligate anaerobes; degrade starch, producing acetic, lactic, and formic acids plus ethanol and CO_2; inhibited by thallium acetate; found in the bovine and ovine rumen	29–33	1500–1600	+
Spiroplasma	Spiral to corkscrew-shaped cells; associated with various phytopathogenic (plant disease) conditions (see Figure 12.64)	25–30	940–2200	−
Ureaplasma	Coccoid cells; occasional clusters and short chains; growth optimal at pH 6; strong urease reaction; associated with certain urinary tract infections in humans; inhibited by thallium acetate	27–30	750	−
Entomoplasma	Facultative aerobe; associated with insects and plants	27–29	790–1140	?
Do not require sterols				
Acholeplasma	Facultative aerobes	27–36	1500	+
Asteroleplasma	Obligate anaerobe; isolated from the bovine or ovine rumen	40	1500	+
Mesoplasma	Phylogenetically and ecologically related to *Entomoplasma*	27–30	870–1100	?

[a] Phylogenetically, all known mycoplasmas are members of the "Low GC" gram-positive *Bacteria*.

mas can be cultivated on relatively simple culture media, however, and defined media have been developed for some species. Most mycoplasmas use carbohydrates as carbon and energy sources and require a range of vitamins, amino acids, purines, and pyrimidines as growth factors. The energy metabolism of mycoplasmas is variable but not unique, as some species are strictly respiratory while others are facultative or even obligate anaerobes (Table 12.28).

● **Figure 12.62 Electron micrograph of a metal-shadowed preparation of *Mycoplasma mycoides*.** Note the coccoid and hyphalike elements. The average diameter of cells in chains is about 0.5 μm.

Alan Rodwell

Spiroplasma

The genus *Spiroplasma* consists of helical or spiral-shaped cell wall-less cells (Figure 12.64). Although they lack a cell wall and flagella, they are motile by means of a rotary (screw) motion or a slow undulation. Intracellular fibrils that are thought to play a role in motility have been demonstrated. The organism has been isolated from ticks, the hemolymph (Figure 12.64) and gut of insects, vascular plant fluids and insects that feed on fluids, and the surfaces of flowers and other plant parts.

Spiroplasma citri has been isolated from the leaves of citrus plants, where it causes a disease called *citrus stubborn disease*, and from corn plants suffering from *corn stunt disease*. A number of other mycoplasma-like bodies have been detected in diseased plants by electron microscopy, which indicates that a large group of plant-associated mycoplasmas may exist. Some species of *Spiroplasma* are known that cause a variety of insect diseases, such as *honeybee spiroplasmosis* and *lethargy disease* of the beetle *Melolontha*.

12.21 Concept Check

The mycoplasma group contains organisms that lack cell walls and contain a very small genome. Many species require sterols to strengthen their membranes, and several are pathogenic for humans, other animals, and plants.

● **Figure 12.63 Typical "fried egg" appearance of mycoplasma colonies on agar.** The colonies are about 0.5 mm in diameter.

◆ Why do mycoplasmas need to have stronger cell membranes than other prokaryotes?

◆ Where do the mycoplasmas group phylogenetically?

◆ Why couldn't motile spiroplasmas contain a normal bacterial flagellum?

12.22 High GC, Gram-Positive *Bacteria* (Actinobacteria): Coryneform and Propionic Acid *Bacteria*

Key Genera: *Corynebacterium, Arthrobacter, Propionibacterium*

High GC, gram-positive *Bacteria* group into their own phylum, the **Actinobacteria**. This group is very large, consisting of over 30 taxonomic *families* of organisms (∞Section 11.10). Species of Actinobacteria are typically rod-shaped to filamentous, primarily aerobic prokaryotes that are common inhabitants of the soil and of plant materials. For the most part they are harmless commensals, species of *Mycobacterium* (for example *Mycobacterium tuberculosis*) being notable exceptions. Some are of great economic value in either the production of antibiotics or certain fermented dairy products. We begin with the rod-shaped representatives.

Corynebacteria

The coryneform bacteria are gram-positive, aerobic, nonmotile, rod-shaped organisms with the characteristic of forming irregular-shaped, club-shaped, or V-shaped cell arrangements during normal growth. V-shaped cell groups arise as a result of a snapping movement that occurs just after cell division (called postfission snapping movement or, simply, *snapping division*) (Figure 12.65●). Snapping division occurs because the cell wall consists of two layers; only the inner layer participates in cross-wall formation, and so after the cross-wall is formed, the two daughter cells remain attached by the outer layer of the cell wall. Localized rupture of this outer layer on one side results in a bending of the two cells away from the ruptured side (Figure 12.66●) and thus development of V-shaped forms.

The main genera of coryneform bacteria are *Corynebacterium* and *Arthrobacter*. The genus *Corynebacterium*

● **Figure 12.64 Dark-field micrograph of the "sex ratio" spiroplasma removed from the hemolymph of the fly *Drosophila pseudoobscura*.** Female flies infected with the sex ratio spiroplasma bear only female progeny. Individual spiroplasma cells are about 0.15 μm in diameter. Dark-field microscopy makes these extremely thin cells visible (∞Section 4.1).

consists of an extremely diverse group of bacteria, including animal and plant pathogens as well as saprophytes. Some species, such as *C. diphtheriae*, are pathogenic (diphtheria, ∞Section 26.3). The genus *Arthrobacter*, consisting primarily of soil organisms, is distinguished from *Corynebacterium* on the basis of a developmental cycle involving conversion from rod to sphere and back to rod again (Figure 12.67●). However, some corynebacteria are pleomorphic and form coccoid cells during growth, and so the distinction between the two genera on the basis of life cycle is not absolute. The *Corynebacterium* cell frequently has a swollen end, so it has a club-shaped appearance (hence the name of the genus: *koryne* is the Greek word for "club"), whereas *Arthrobacter* species are less commonly club-shaped.

Organisms of the genus *Arthrobacter* are among the most common of all soil bacteria. They are remarkably resistant to desiccation and starvation, despite the fact that they do not form spores or other resting cells. Arthrobacters are a heterogeneous group that have considerable nutritional versatility, and strains have been isolated that decompose herbicides, caffeine, nicotine, phenols, and other unusual organic compounds.

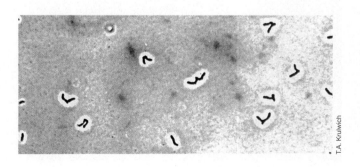

● **Figure 12.65 Snapping division in *Arthrobacter*.** Photomicrograph of characteristic V-shaped cell groups in *Arthrobacter crystallopoietes*, resulting from snapping division. Cells are about 0.9 μm in diameter.

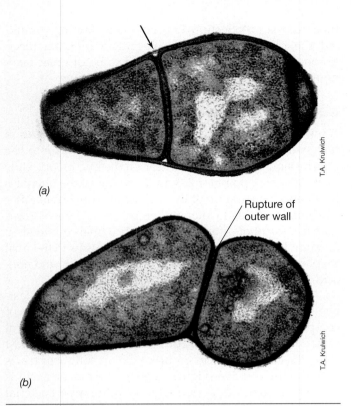

(a)

Rupture of
outer wall

(b)

● **Figure 12.66 Cell division in *Arthrobacter*.** Electron micrograph of cell division in *Arthrobacter crystallopoietes*, illustrating how snapping division and V-shaped cell groups arise. (a) Before rupture of the outer cell wall layer (arrow). (b) After rupture of the outer layer on one side. Cells are 0.9–1 μm in diameter.

Propionic Acid Bacteria

The propionic acid bacteria (genus *Propionibacterium*) were first discovered in Swiss (Emmentaler) cheese, where their fermentative production of CO_2 produces the characteristic holes. Moreover, the propionic acid they produce is at least partly responsible for the unique flavor of the cheese. Although this acid is produced by some other bacteria, its production in large amounts by the propionic acid bacteria is a distinguishing characteristic of the genus. The bacteria in this group are grampositive anaerobes that ferment lactic acid, carbohydrates, and polyhydroxy alcohols, producing primarily propionic acid, acetic acid, and CO_2. Their nutritional requirements are complex, and they usually grow rather slowly.

Figure 12.68● shows the enzymatic reactions leading from glucose to propionic acid. The initial catabolism of glucose to pyruvate follows the glycolytic pathway, as in the lactic acid bacteria, but the NADH formed is oxidized as one part of a cycle in which *propionic acid* is formed. Pyruvate accepts a carboxyl group from methylmalonyl-CoA by a transcarboxylase reaction, leading to the formation of oxalacetate and propionyl-CoA. The latter substance reacts with succinate in a step catalyzed by a CoA transferase, producing succinyl-CoA and propionate. The succinyl-CoA is then isomerized to methylmalonyl-CoA, and the cycle is complete (Figure 12.68). Oxidation of NADH occurs in the steps between oxalacetate and succinate, and the oxidation-reduction balance is restored.

Propionibacterium also ferments *lactate* with the production of propionate, acetate, and CO_2. The anaerobic fermentation of lactic acid to propionate is of interest because lactic acid itself is an end product of fermentation for many bacteria (see Section 12.19). The propionic acid bacteria are thus able to obtain energy anaerobically from a fermentation product that other bacteria have produced. This metabolic strategy is called a *secondary fermentation*.

It is the fermentation of lactate to propionate that is important in Swiss cheese manufacture. The starter culture consists of a mixture of homofermentative streptococci and lactobacilli, plus propionic acid bacteria. The initial fermentation of lactose to lactic acid during formation of the curd is carried out by the homofermentative organisms. After the curd (protein and fat) has been drained, the propionic acid bacteria develop rapidly. The eyes (or holes) characteristic of Swiss cheese are formed by the accumulation of CO_2, the gas diffusing through the curd and gathering at weak points. In the fermentation, lactate is oxidized to pyruvate, from which it is converted to propionate as shown in Figure 12.68.

Propionate is also formed in the fermentation of succinate by the bacterium *Propionigenium*. This organism is phylogenetically and ecologically unrelated to *Propionibacterium*, but energetic aspects of its fermentation are of considerable interest. We discuss the mechanism of the *Propionigenium* fermentation in Section 17.20.

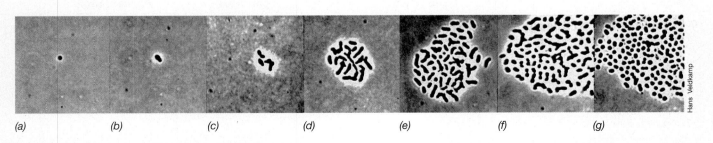

(a)   (b)   (c)   (d)   (e)   (f)   (g)

● **Figure 12.67 Stages in the life cycle of *Arthrobacter globiformis* as observed in slide culture.** (a) Single coccoid element; (b–e) conversion to rod and growth of a microcolony consisting predominantly of rods; (f–g) conversion of rods to coccoid forms. Cells are about 0.9 μm in diameter.

Stoichiometry:

$$3 \text{ Lactate} \longrightarrow 2 \text{ Propionate}^- + \text{Acetate}^- + CO_2 + 3\text{-}5\text{ATP}$$

● **Figure 12.68   The formation of propionic acid from lactate by** *Propionibacterium.* Note that the four NADH made from the oxidation of 3 lactate to 2 pyruvate/acetate + CO_2 are reoxidized in the reduction of oxalacetate and fumarate. Also, note that the CoA group from propionyl~CoA is transferred to succinate during the formation of propionate in an energy neutral fashion.

12.23   Actinobacteria: *Mycobacterium*

Key Genus: *Mycobacterium*

The genus *Mycobacterium* consists of rod-shaped organisms that at some stage of their growth cycle possess the distinctive staining property called **acid-fastness**. This property is due to the presence on the surface of the mycobacterial cell of unique lipids called **mycolic acids**, found only in the genus *Mycobacterium*. First discovered by Robert Koch during his pioneering investigations on

tuberculosis (Section 1.6), acid-fast staining permitted the identification of the organism in tuberculous lesions. It has subsequently proved to be of great taxonomic use in defining the genus *Mycobacterium*.

Acid-Fastness (Ziehl–Neelsen Stain)

A mixture of the dye basic fuchsin and phenol is used in this staining procedure, the stain being driven into the cells by slow heating of the smear on the microscope slide to the steaming point for 2–3 min. The role of the phenol is to enhance penetration of the fuchsin into the lipids. After washing in distilled water, the preparation is decolorized with acid-alcohol. After another wash, a final counterstain of methylene blue is used. Acid-fast organisms in the final preparation appear *red*, whereas the background and non-acid-fast organisms appear *blue* (Figure 26.7).

As noted, the key component necessary for acid-fastness is *mycolic acid*. Mycolic acid is actually a group of complex branched-chain hydroxy lipids (the general structure is shown in Figure 12.69*a*●). In the acid-fast stain, the carboxylic acid group of the mycolic acid reacts with the fuchsin dye (Figure 12.69*b*). The mycolic acid is covalently bound to peptidoglycan in the mycobacterial wall, and this complex leaves the cell surface with a waxy, hydrophobic consistency. Mycobacteria are not readily stained by the Gram stain method because of the high surface-lipid content. If the lipoidal portion of the cell is removed with alkaline ethanol, however, the intact cell remaining is non-acid-fast but instead is gram-positive.

Characteristics of Mycobacteria

Mycobacteria are rather pleomorphic and may undergo branching or filamentous growth. However, in contrast to those of the actinomycetes (see Section 12.24), filaments of the mycobacteria become fragmented into rods or coccoid elements upon slight disturbance; a true mycelium is not formed. In general, mycobacteria can be separated into two major groups, *slow growers* and *fast growers* (Table 12.29). *Mycobacterium tuberculosis* is a typical slow grower, and visible colonies are produced from

Table 12.29	Some characteristics of representative mycobacteria				
Species	**Growth in 5% NaCl**	**Nitrate reduction**	**Growth at 45°C**	**Human pathogen**	**Pigmentation**
Slow-growing species					
Mycobacterium tuberculosis	–	+	–	+	None
Mycobacterium avium	–	–	–	+	Old colonies pigmented (see Figure 12.70*c*)
Mycobacterium bovis	–	–	+	+	None
Mycobacterium kansasii	–	+	–	+	Photochromogenic
Fast-growing species					
Mycobacterium smegmatis	+	+	+	–	None
Mycobacterium phlei	+	+	+	–	Pigmented
Mycobacterium chelonae	+	–	–	+	None
Mycobacterium parafortuitum	+	+	–	–	Photochromogenic

$$R_1 - \overset{\overset{\displaystyle H}{|}}{C} - \overset{\overset{\displaystyle H}{|}}{C} - COO^-$$
$$\underset{OH}{} \quad \underset{R_2}{}$$

$$H_2N - \text{⬡} - \overset{}{C} - \text{⬡} = NH_2^+ \ Cl^-$$

$$NH_2$$

(a) Mycolic acid; R_1 and
R_2 are long-chain
aliphatic hydrocarbons

(b) Basic fuchsin

● **Figure 12.69   Acid-fast staining.** Structure of (a) mycolic acid and (b) basic fuchsin, the dye used in the acid-fast stain. The fuchsin dye combines with the mycolic acid via ionic bonds between COO^- and NH_2^+.

dilute inoculum only after days to weeks of incubation. (Koch was successful in first isolating *M. tuberculosis* because he waited long enough after inoculating media; ∞Section 1.6). When growing on solid media, mycobacteria form tight, compact, often wrinkled colonies (Figure 12.70*a*●). This colony morphology is probably due to the high lipid content and hydrophobic nature of the cell surface that make cells stick together.

For the most part, mycobacteria have relatively simple nutritional requirements. Growth often occurs in simple mineral salts medium with ammonium as nitrogen source and glycerol or acetate as sole carbon source and electron donor incubated in air. Growth of *Mycobacterium tuberculosis* is stimulated by lipids and fatty acids, and egg yolk (a good source of lipids) is often added to culture media to achieve more luxuriant growth. A glycerol-whole egg medium (Lowenstein–Jensen medium) is often used in primary isolation of *M. tuberculosis* from pathological materials.

Perhaps because of the high lipid content of its cell walls, *M. tuberculosis* is able to resist such chemical agents as alkali and phenol for considerable periods of time, and this property is used in the selective isolation of the organism from patient sputum and other materials that are grossly contaminated. The sputum is first treated with 1 N NaOH for 30 min and then neutralized and streaked onto an isolation medium.

A characteristic of many mycobacteria is their ability to form yellow carotenoid pigments (Figure 12.70*c*). Based on pigmentation, the mycobacteria can be classified into three groups: nonpigmented (including *Mycobacterium tuberculosis*, *M. bovis*); forming pigment only when cultured in the light, a property called *photochromogenesis* (including *M. kansasii*, *M. marinum*); and forming pigment even when cultured in the dark, a property called *scotochromogenesis* (for example, *M. gordonae*; Table 12.29). Photoinduction of carotenoid formation requires short-wavelength (blue) light and occurs only in the presence of O_2. The evidence indicates that the critical event in photoinduction is a light-catalyzed oxidation event, and it appears that one of the early enzymes in carotenoid biosynthesis is photoinduced. As with other carotenoid-containing bacteria, it has been suggested that carotenoids protect mycobacteria against oxidative damage involving singlet oxygen (∞Section 6.16).

The virulence of *Mycobacterium tuberculosis* cultures has been correlated with the formation of long, cordlike structures (Figure 12.70*b*) on agar or in liquid medium, due to side-to-side aggregation and intertwining of long chains of bacteria. Growth in cords reflects the presence on the cell surface of a characteristic glycolipid, the **cord factor** (Figure 12.71●). The pathogenesis of the disease tuberculosis is discussed in detail in Section 26.5.

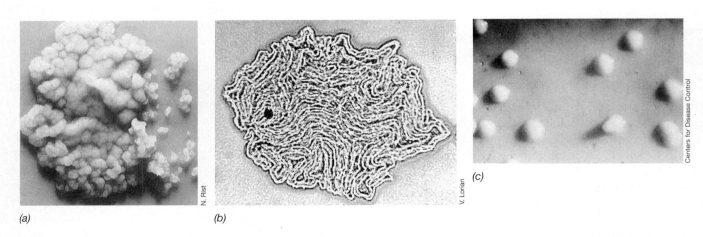

(a) N. Rist *(b)* V. Lorian *(c)* Centers for Disease Control

● **Figure 12.70   Characteristic colony morphology of mycobacteria.** (a) *Mycobacterium tuberculosis*, showing the compact, wrinkled appearance of the colony. The colony is about 7 mm in diameter. (b) A colony of virulent *M. tuberculosis* at an early stage, showing the characteristic cordlike growth. Individual cells are about 0.5 μm in diameter. (See also the historic drawings of *M. tuberculosis* cells made by Robert Koch, ∞Figure 1.13). (c) Colonies of *Mycobacterium avium* from a strain of this organism isolated as an opportunistic pathogen from an AIDS patient.

● **Figure 12.72** ***Nocardia.*** A young colony of an actinomycete of the genus *Nocardia*, showing typical filamentous cellular structure (mycelium). Each filament is about 0.8–1 μm in diameter.

● **Figure 12.71** **Structure of cord factor, a mycobacterial glycolipid: 6,6′-dimycolyltrehalose.** The two identical long-chain dialcohol groups are shown in purple.

 12.22–12.23 Concept Check

High GC, gram-positive *Bacteria* include such organisms as *Corynebacterium*, *Arthrobacter*, *Propionibacterium*, and *Mycobacterium*. They are mainly harmless soil saprophytes but *M. tuberculosis* is the causative agent of the disease tuberculosis. *M. tuberculosis* cells have a lipid-rich, waxy outer surface layer that requires special staining procedures (the acid-fast stain) in order to observe the cells microscopically.

◆ What is snapping division and what organism practices it?

◆ What organism is involved in the ripening of Swiss cheese and what chemical compounds does it make that helps to flavor the cheese and make the holes?

◆ What is mycolic acid, what organism produces it, and what properties does this substance confer on cells that make it?

12.24 Filamentous Actinobacteria: *Streptomyces* and Other Actinomycetes

Key Genera: *Streptomyces, Actinomyces*

The actinomycetes are a large group of filamentous, gram-positive *Bacteria* that form branching filaments. As a result of successful growth and branching, a ramifying network of filaments is formed, called a *mycelium* (Figure 12.72●). Although it is of bacterial dimensions, the mycelium is analogous to the mycelium formed by the filamentous fungi (∞Figure 14.25).

Most actinomycetes form spores; the manner of spore formation varies and is used in separating subgroups, as outlined in Table 12.30. The DNA base compositions of most members of the actinomycetes fall within the range of 63–78% GC, and organisms at the upper end of this range have the highest GC percent-

age of any bacteria known. Phylogenetically, the filamentous actinomycetes form a coherent group. Thus, the mycelial spore-forming habit is of both phylogenetic as well as taxonomic importance. We focus here on the genus *Streptomyces*.

Streptomyces

The genus *Streptomyces* contains a large number of species and varieties. Over 500 species of *Streptomyces* are recognized, although their GC base ratios cluster tightly between 69 and 73 mol %. *Streptomyces* filaments are typically 0.5–1.0 μm in diameter, are of indefinite length, and often lack cross-walls in the vegetative phase.

Growth of *Streptomyces* occurs at the tips of the filaments, often accompanied by branching. Thus, the vegetative phase consists of a complex, tightly woven matrix, resulting in a compact, convoluted mycelium and subsequent colony. As the colony ages, characteristic aerial filaments called *sporophores* are formed, which project above the surface of the colony and give rise to spores (Figure 12.73●). *Streptomyces* spores, called **conidia**, are quite distinct from the endospores of *Bacillus* and *Clostridium*. Unlike the elaborate cellular differentiation that leads to the formation of an endospore, streptomycete spores are produced by the formation of cross-walls in the multinucleate sporophores followed by separation of the individual cells directly into spores (Figure 12.74●).

Differences in the shape and arrangement of aerial filaments and spore-bearing structures of various species are among the fundamental features used in classifying the *Streptomyces* species (Figure 12.75● page 393). The conidia and sporophores are often pigmented and contribute a characteristic color to the mature colony (Figure 12.76a●, page 393). The dusty appearance of the mature colony, its compact nature, and its color make detection of *Streptomyces* colonies on agar plates relatively easy (Figure 12.76b).

Ecology and Isolation of *Streptomyces*

Although a few streptomycetes can be found in aquatic habitats, they are primarily *soil* organisms. In fact, the

Table 12.30 Representative Actinomycetes and related genera of Actinobacteria (all gram-positive)[a]

Major groups	DNA (mol % GC)
Coryneform group of bacteria: rods, often club-shaped, morphologically variable; not acid-fast or filamentous; snapping cell division	
Corynebacterium: irregularly staining segments, sometimes granules; club-shaped swelling frequent; animal and plant pathogens, also soil saprophytes	51–65
Arthrobacter: coccus-rod morphogenesis; soil organisms	59–70
Cellulomonas: coryneform morphology; cellulose digested; facultative aerobe	71–73
Kurthia: rods with rounded ends occurring in chains; coccoid later	36–38
Brevibacterium: coccus-rod morphogenesis; cheese, skin	60–67
Propionic acid bacteria: anaerobic to aerotolerant; rods or filaments, branching	
Propionibacterium: nonmotile; anaerobic to aerotolerant; produce propionic acid and acetic acid; dairy products (Swiss cheese); skin, may be pathogenic	53–68
Eubacterium: obligate anaerobes; produce mixture of organic acids, including butyric, acetic, formic, and lactic; intestine, infections of soft tissue, soil; may be pathogenic; probably the predominant member of the intestinal flora	26–48
Obligate anaerobes	
Bifidobacterium: smooth microcolony, no filaments; coryneform cells common; found in intestinal tract of breast-fed infants	55–67
Acetobacterium: homoacetogen; sediments and sewage	39–43
Butyrivibrio: curved rods; rumen	36–42
Thermoanaerobacter: rods, thermophilic, found in hot springs	37–39
Actinomycetes: filamentous, often branching; highly diverse	
Group I. Actinomycetes: not acid-fast; facultatively aerobic; mycelium not formed; branching filaments may be produced; rod, coccoid, or coryneform cells	
Actinomyces: anaerobic to facultatively aerobic; filamentous microcolony, but filaments transitory and fragment into coryneform cells; may be pathogenic for humans or animals; found in oral cavity	57–69
Other genera: *Arachnia, Bacterionema, Rothia, Agromyces*	
Group II. Mycobacteria: acid-fast, filaments transitory	
Mycobacterium: pathogens, saprophytes; obligate aerobes; lipid content of cells and cell walls high; waxes, mycolic acids; simple nutrition; growth slow; tuberculosis, leprosy, granulomas, avian tuberculosis; also soil organisms; hydrocarbon oxidizers	62–70
Group III. Nitrogen-fixing actinomycetes: nitrogen-fixing symbionts of plants; true mycelium produced	
Frankia: forms nodules of two types on various plant roots; probably microaerophilic; grows slowly; fixes N_2	67–72
Group IV. Actinoplanes: true mycelium produced; spores formed, borne inside sporangia	
Actinoplanes, Streptosporangium	69–71
Group V. Dermatophilus group: mycelial filaments divide transversely, and in at least two longitudinal planes, to form masses of motile, coccoid elements; aerial mycelium absent; occasionally responsible for epidermal infections	
Dermatophilus, Geodermatophilus	56–75
Group VI. Nocardias: mycelial filaments commonly fragment to form coccoid or elongate elements; aerial spores occasionally produced; sometimes acid-fast; lipid content of cells and cell wall very high	
Nocardia: common soil organisms; obligate aerobes; many hydrocarbon utilizers	61–72
Rhodococcus: soil saprophytes, also common in gut of various insects; utilize hydrocarbons	59–69
Group VII. Streptomycetes: mycelium remains intact, abundant aerial mycelium and long spore chains	
Streptomyces: Nearly 500 recognized species, many produce antibiotics	69–75
Other genera (differentiated morphologically): *Streptoverticillium, Sporichthya, Kitasatoa, Chainia*	67–73
Group VIII. Micromonosporas group: mycelium remains intact; spores formed singly, in pairs, or short chains; several thermophilic; saprophytes found in soil, rotting plant debris; one species produces endospores	
Micromonospora, Microbispora, Themobispora, Thermoactinomyces, Thermomonospora	54–79

[a] Phylogenetically, all species (except for *Acetobacterium, Butyrivibrio,* and *Thermoanaerobacter*) fall into the Actinobacteria.

characteristic earthy odor of soil is caused by the production of a series of streptomycete metabolites called *geosmins*. These substances are sesquiterpenoid compounds—unsaturated ring compounds of carbon, oxygen, and hydrogen. A common geosmin is trans-1, 10-dimethyl-trans-9-decalol. Geosmins are also produced by some cyanobacteria (see Section 12.25).

Alkaline and neutral soils are more favorable for the development of *Streptomyces* than are acid soils. Higher numbers of *Streptomyces* are usually found in well-drained soils (such as sandy loams, or soils covering limestone), and there is some evidence to suggest that *Streptomyces* require a lower water potential for growth than many other soil bacteria. Isolation of *Streptomyces* from soil is relatively easy: A suspension of soil in sterile water is diluted and spread on selective agar medium, and the plates are incubated at 25°C (∞Figure 30.7a). Media often selective for *Streptomyces* contain the usual

● **Figure 12.73 Photomicrographs of several spore-bearing structures of actinomycetes.** (a) *Streptomyces*, a monoverticillate type. (b) *Streptomyces*, a closed spiral type. Filaments are about 0.8 μm wide in both cases. Compare these photos with the art in Figure 12.75.

assortment of inorganic salts to which starch, asparagine, or calcium malate is added as a carbon source and undigested casein or potassium nitrate as a nitrogen source. After incubation for 5–7 days in air, the plates are examined for the presence of the characteristic *Streptomyces* colonies (Figures 12.76 and 12.77●), and spores of interesting colonies can be streaked to isolate pure cultures.

Nutritionally, the streptomycetes are quite versatile. Growth-factor requirements are rare, and a wide variety of carbon sources, such as sugars, alcohols, organic acids, amino acids, and some aromatic compounds, can be utilized. Most isolates produce extracellular hydrolytic enzymes that permit utilization of polysaccharides (starch, cellulose, and hemicellulose), proteins, and fats, and some strains can use hydrocarbons, lignin, tannin, or even rubber. *Streptomyces* can often be obtained by spreading a soil dilution on an alkaline agar medium containing polymers such as casein and starch (Figure 12.76a). A single isolate may be able to break down over 50 distinct carbon sources.

Streptomycetes are strict aerobes whose growth in liquid culture is usually markedly stimulated by forced aeration. Sporulation usually does not take place in liquid culture but only when the organism is growing on the surface of agar or another solid substrate; it can occur, however, when organisms form a pellicle on the surface of an unshaken liquid culture.

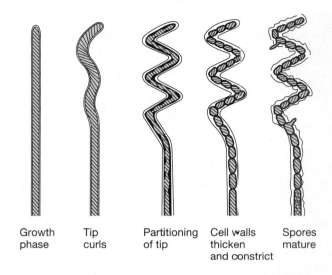

| Growth phase | Tip curls | Partitioning of tip | Cell walls thicken and constrict | Spores mature |

● **Figure 12.74 Spore formation in *Streptomyces*.** Diagram of stages in the conversion of a streptomycete's aerial hypha (sporophores) into spores (conidia).

Antibiotics of *Streptomyces*

Perhaps the most striking property of the streptomycetes is the extent to which they produce *antibiotics* (Table 12.31). Evidence for antibiotic production is often seen on the agar plates used in the initial isolation of *Streptomyces*: Adjacent colonies of other bacteria show zones of inhibition (Figures 12.76a and 12.77a; ∞Figure 30.7a). About 50% of all *Streptomyces* isolated have proved to be antibiotic producers.

Because of the great economic and medical importance of many streptomycete antibiotics, an enormous amount of work has been done on these producers. Over 500 distinct antibiotic substances have been shown to be produced by streptomycetes, and many more are suspected (∞Sections 30.5 and 30.6); most of these have been identified chemically (Figure 12.77b). Some organisms produce more than one antibiotic, and often the several kinds produced by one organism are not chemically related molecules. The same antibiotic may be formed by different species found in widely scattered parts of the world. And, although an antibiotic-producing organism is resistant to its own antibiotics, it usually remains sensitive to antibiotics produced by other streptomycetes. Many genes are often required to encode the enzymes involved in antibiotic synthesis, and because of this, the genomes of *Streptomyces* species are typically quite large (8 Mbp and larger, ∞Table 15.1).

More than 60 streptomycete antibiotics have found practical application in human and veterinary medicine, agriculture, and industry. Some of the more common antibiotics of *Streptomyces* origin are listed in Table 12.31. They are grouped into classes based on the chemical structure of the parent molecule. The search for new streptomycete antibiotics continues because many infectious diseases are still not adequately controlled by existing antibiotics. Also, the development of antibiotic-resistant pathogens requires the continual discovery of new agents.

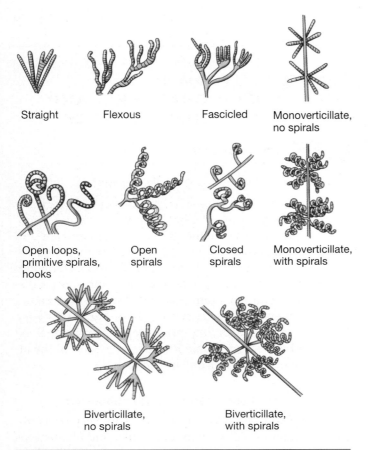

Straight · Flexous · Fascicled · Monoverticillate, no spirals

Open loops, primitive spirals, hooks · Open spirals · Closed spirals · Monoverticillate, with spirals

Biverticillate, no spirals · Biverticillate, with spirals

● **Figure 12.75** **Various types of sporebearing structures in the streptomycetes.** A given species of *Streptomyces* produces only one morphological type of sporebearing structure. Compare with Figure 12.73.

(a)

(b)

● **Figure 12.76** *Streptomyces.* (a) Colonies of *Streptomyces* and other soil bacteria derived from spreading a soil dilution on a casein-starch agar plate. The *Streptomyces* colonies are of various colors (several black *Streptomyces* colonies are in the foreground) but can easily be identified by their opaque, rough, nonspreading morphology. (b) Close-up photo of colonies of *Streptomyces coelicolor.*

Ironically, despite the extensive practical studies on antibiotic-producing streptomycetes by the antibiotic industry and the fact that antibiotics from *Streptomyces* species are a multibillion dollar a year industry, the ecology of *Streptomyces* remains poorly understood. The interactions of these organisms with other prokaryotes and the ecological rationale for why antibiotics are pro-duced in the first place, is still an area where we know very little. One hypothesis for why *Streptomyces* species produce antibiotics is that antibiotic production, which is linked to sporulation (a process itself triggered by nutrient depletion), might be a mechanism to inhibit the growth of other organisms competing with *Streptomyces* cells for limiting nutrients. This would allow the

Table 12.31	Some common antibiotics synthesized by species of *Streptomyces*		
Chemical class	**Common name**	**Produced by**	**Active against**[a]
Aminoglycosides	Streptomycin	*S. griseus*[b]	Most gram-negative *Bacteria*
	Spectinomycin	*Streptomyces* spp.	*M. tuberculosis*, penicillinase-producing *N. gonorrhoeae*
	Neomycin	*S. fradiae*	Broad spectrum, usually used in topical applications because of toxicity
Tetracyclines	Tetracycline	*S. aureofaciens*	Broad spectrum, gram-positive and gram-negative *Bacteria*, rickettsias and chlamydias, *Mycoplasma*
	Chlortetracycline	*S. aureofaciens*	As for tetracycline
Macrolides	Erythromycin	*Saccharopolyspora erythraea*	Most gram-positive *Bacteria*, frequently used in place of penicillin, *Legionella*
	Clindamycin	*S. lincolnensis*	Effective against obligate anaerobes, especially *Bacteroides fragilis*
Polyenes	Nystatin	*S. noursei*	Fungi, especially *Candida* infections
	Amphocetin B	*S. nodosus*	Fungi
None	Chloramphenicol	*S. venezuelae*	Broad spectrum; drug of choice for typhoid fever

[a] Most antibiotics are effective against several different *Bacteria*. The entries in this column refer to the common clinical application of a given antibiotic. The structures and mode of action of many of these antibiotics are discussed in Sections 20.7–20.9.

[b] All listings beginning with an "S." are species of the genes *Streptomyces*.

(a)

(b)

● **Figure 12.77 Antibiotics from *Streptomyces*.** (a) Antibiotic action of soil microorganisms on a crowded plate. The smaller colonies surrounded by inhibition zones are streptomycetes; the larger, spreading colonies are *Bacillus* species. (b) The red-colored antibiotic undecylprodigiosin is being excreted by colonies of *Streptomyces coelicolor*.

Streptomyces to complete the sporulation process and form a dormant structure that would have increased chances of survival.

 12.24 Concept Check

The streptomycetes are a large group of filamentous, gram-positive *Bacteria* that form spores at the end of aerial filaments. Many clinically useful antibiotics like tetracycline and neomycin have come from *streptomyces* species.

◆ How do spores and the process of sporulation in a *Streptomyces* species differ from that in a *Bacillus* species?

◆ What energy class of organism is a *Streptomyces* and from what types of compounds do these organisms obtain their energy?

◆ Why might antibiotic production be of advantage to Streptomycetes?

IV PHYLUM 4: CYANOBACTERIA AND PROCHLOROPHYTES

12.25 Cyanobacteria

Key Genera: *Synechococcus, Oscillatoria, Nostoc*

Cyanobacteria comprise a large and morphologically heterogeneous group of phototrophic *Bacteria*. Cyanobacteria differ in fundamental ways from purple and green bacteria (see Sections 12.2 and 12.32), most notably in that they are *oxygenic* phototrophs. Cyanobacteria represent one of the major phyla of *Bacteria* and show a distant relationship to gram-positive *Bacteria* (see Figure 12.1). As we saw in Section 11.1, these organisms were the first oxygen-evolving phototrophic organisms on Earth and were responsible for the conversion of the atmosphere of the Earth from anoxic to oxic.

Structure and Classification of Cyanobacteria

The morphological diversity of the cyanobacteria is impressive (Figure 12.78●). Both unicellular and filamentous forms are known, and considerable variation within these morphological types occurs. Still, cyanobacteria can be divided into just five morphological groups: (1) unicellular dividing by binary fission (Figure 12.78*a*); (2) unicellular dividing by multiple fission (colonial) (Figure 12.78*b*); (3) filamentous containing differentiated cells called *heterocysts* that function in nitrogen fixation (Figure 12.78*d* and see Figure 12.80); (4) filamentous nonheterocystous forms (Figure 12.78*c*); and (5) branching filamentous species (Figure 12.78*e*). Table 12.32 lists the major genera currently recognized in each group. Cyanobacterial cells range in size from those of typical bacteria (0.5–1 μm in diameter) to cells as large as 40 μm in diameter (in the species *Oscillatoria princeps*, ∞Figure 4.44*a*).

The structure of the cell wall of cyanobacteria is similar to that of gram-negative *Bacteria*, and peptidoglycan is present in the walls (Figure 12.79●). Many cyanobacteria produce extensive mucilaginous envelopes, or sheaths, that bind groups of cells or filaments together (see, for example, Figure 12.78*a*). The photosynthetic membrane system is often complex and multilayered (∞Figure 17.10*b*), although in some of the simpler cyanobacteria the lamellae are regularly arranged in concentric circles around the periphery of the cytoplasm (Figure 12.79).

Cyanobacteria have only one form of chlorophyll, *chlorophyll a*, and all of them also have characteristic biliprotein pigments, **phycobilins** (∞Figure 17.10*a*), which function as accessory pigments in photosynthesis. One class of phycobilins, *phycocyanins*, are blue, and together with the green chlorophyll *a*, are responsible for the blue-green color of the bacteria. However,

© Eli Lilly & Co. Used with permission.

David A. Hopwood

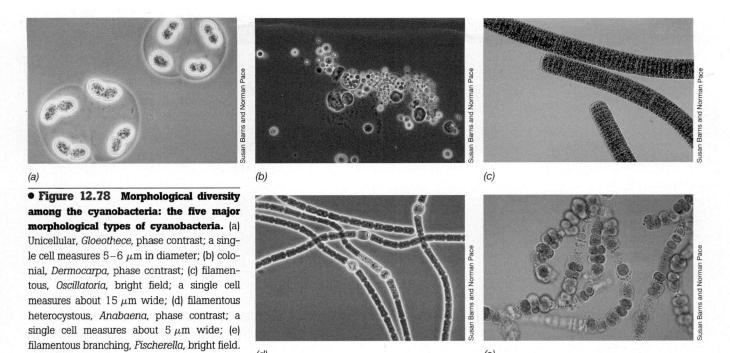

(a) (b) (c)

● **Figure 12.78 Morphological diversity among the cyanobacteria: the five major morphological types of cyanobacteria.** (a) Unicellular, *Gloeothece*, phase contrast; a single cell measures 5–6 μm in diameter; (b) colonial, *Dermocarpa*, phase contrast; (c) filamentous, *Oscillatoria*, bright field; a single cell measures about 15 μm wide; (d) filamentous heterocystous, *Anabaena*, phase contrast; a single cell measures about 5 μm wide; (e) filamentous branching, *Fischerella*, bright field.

(d) (e)

some cyanobacteria produce *phycoerythrin*, a red phycobilin, and species possessing this pigment are red or brown in color.

Structural Variations: Gas Vesicles and Heterocysts

Among the cytoplasmic structures seen in many cyanobacteria are **gas vesicles** (⬿Section 4.12), which are especially common in species that live in open waters (planktonic species). Their function is to regulate cell buoyancy such that cells can remain in a position in the water column where light intensity is optimal for photosynthesis. Some filamentous cyanobacteria form *heterocysts*, which are rounded, seemingly empty cells, usually distributed regularly along a filament or at one end of a filament (Figure 12.80a●). Heterocysts arise from differentiation of vegetative cells and are the sole sites of *nitrogen fixation* (the reduction of N_2 to NH_3, ⬿Section 17.28) in heterocystous cyanobacteria. In *Anabaena*, a well-studied heterocystous cyanobacterium, complex gene rearrangements occur within the heterocyst to yield a contiguous cluster of *nif* genes that can be expressed as a unit (*nif* genes encode nitrogenase, ⬿Section 17.28).

Heterocyst

(a)

O_2 O_2 O_2 O_2 O_2 O_2

CO_2 $N_2 \rightarrow NH_3$ CO_2

CH_2O Glutamine CH_2O

Heterocyst

Vegetative cells Vegetative cells

(b)

● **Figure 12.80 Heterocysts.** (a) Heterocysts in the cyanobacterium *Anabaena*. Heterocysts are the sole site of nitrogen fixation in heterocystous cyanobacteria. (b) Model for the operation of a heterocyst. The heterocyst lacks oxygen–producing ability (Photosystem II, ⬿Section 17.5) and obtains the needed reductant for nitrogen fixation from organic matter produced by adjacent vegetative cells. Glutamine is the form of fixed nitrogen transported from heterocysts to vegetative cells.

● **Figure 12.79 Thylakoids in cyanobacteria.** Electron micrograph of a thin section of the cyanobacterium *Synechococcus lividus*. A cell is about 5 μm in diameter. Note thylakoid membranes running parallel to the cell wall.

Table 12.32 Genera and grouping of cyanobacteria

Group	Genera	DNA (mol % GC)
Group I. Unicellular: single cells or cell aggregates	*Gloeothece* (Figure 12.78a), *Gloeobacter, Synechococcus, Cyanothece, Gloeocapsa, Synechocystis, Chamaesiphon, Merismopedia*	35–71
Group II. Pleurocapsalean: reproduce by formation of small spherical cells called baeocytes produced through multiple fission	*Dermocarpa* (Figure 12.78b), *Xenococcus, Dermocarpella, Pleurocapsa, Myxosarcina, Chroococcidiopsis*	40–46
Group III. Oscillatorian: filamentous cells that divide by binary fission in a single plane	*Oscillatoria* (Figure 12.78c), *Spirulina, Arthrospira, Lyngbya, Microcoleus, Pseudanabaena*	40–67
Group IV. Nostocalean: filamentous cells that produce heterocysts	*Anabaena* (Figure 12.78d), *Nostoc, Calothrix, Nodularia, Cylindrosperum, Scytonema*	38–46
Group V. Branching: cells divide to form branches	*Fischerella* (Figure 12.78e), *Stigonema, Chlorogloeopsis, Hapalosiphon*	42–46

Heterocysts have intercellular connections with adjacent vegetative cells, and there is mutual exchange of materials between these cells. The products of photosynthesis move from vegetative cells to heterocysts, and products of nitrogen fixation move from heterocysts to vegetative cells (Figure 12.80b). Heterocysts are low in phycobilin pigments and *lack* photosystem II, the oxygen-evolving photosystem that generates reducing power from H_2O (∞Section 17.5). Without photosystem II heterocysts are unable to fix CO_2 and thus lack the necessary electron donor to reduce N_2 to NH_3; fixed carbon imported to the heterocyst from an adjacent vegetative cell solves this problem (Figure 12.80b).

Heterocysts are surrounded by a thickened cell wall containing large amounts of glycolipid, which serves to slow the diffusion of O_2 into the cell. Because of the oxygen lability of the enzyme nitrogenase (∞Section 17.29), the heterocyst maintains an anoxic environment, and by doing so stabilizes the nitrogen-fixing system in organisms that are not only aerobic but also oxygen producing. Indeed, some nonheterocystous filamentous cyanobacteria produce nitrogenase and fix nitrogen in normal vegetative cells if they are grown anaerobically by vigorous bubbling with N_2 to remove O_2.

Cyanophycin and Other Structures

A structure called **cyanophycin** can be seen in electron micrographs of many cyanobacteria. This structure is a copolymer of aspartic acid and arginine:

```
Asp — Asp — Asp — Asp —Asp —
 |     |     |     |     |
Arg   Arg   Arg   Arg   Arg
```

and can constitute up to 10% of the cell mass. Cyanophycin is a nitrogen storage product, and when nitrogen in the environment becomes deficient, this polymer is broken down and used as a cellular nitrogen source. Cyanophycin is also an energy reserve in cyanobacteria. Arginine, derived from cyanophycin, can be hydrolyzed to yield ornithine, with the production of ATP through the activity of the enzyme *arginine dihydrolase*, with carbamyl phosphate (∞Section 17.19) occurring as an intermediate:

$$\text{Arginine} + \text{ADP} + P_i + H_2O \rightarrow$$
$$\text{Ornithine} + 2\,NH_3 + CO_2 + \text{ATP}$$

Arginine dihydrolase is present in many cyanobacteria and may function as a source of ATP for maintenance purposes during dark periods.

Many cyanobacteria exhibit gliding motility; true rotating flagella have never been found. Gliding occurs only when the cell or filament is in contact with a solid surface or with another cell or filament. In some cyanobacteria gliding is not a simple translational movement but is accompanied by rotations, reversals, and flexings of filaments. Most gliding species exhibit directional movement toward light (phototaxis), and chemotaxis (∞Section 4.16) may occur as well.

Among the filamentous cyanobacteria, fragmentation of the filaments often occurs by formation of **hormogonia** (Figure 12.81a,b●), which break away from the filaments and glide off. In some species, resting spores or **akinetes** (Figure 12.81c) are formed, which protect the organism during periods of darkness, drying, or freezing. These are cells with thickened outer walls; they germinate through the breakdown of the outer wall and outgrowth of a new vegetative filament. However, even the vegetative cells of many cyanobacteria are relatively resistant to drying or low temperatures.

Physiology of Cyanobacteria

The nutrition of cyanobacteria is simple. Vitamins are not required, and nitrate or ammonia is used as nitrogen source. Nitrogen-fixing species are common. Most species tested are obligate phototrophs, being unable to grow in the dark on organic compounds. However, some cyanobacteria can assimilate simple organic compounds such as glucose and acetate if light is present (photoas-

(a)

(b)

(c)

Labels on figure: Separation of hormogonium; Hormogonium; Akinete; T. D. Brock

● **Figure 12.81** **Structural differentiation in filamentous cyano-bacteria.** (a) Initial stage of hormogonium formation in *Oscillatoria*. Notice the empty spaces where the hormogonium is separating from the filament. (b) Hormogonium of a smaller *Oscillatoria* species. Notice that the cells at both ends are rounded. Nomarski interference contrast microscopy. (c) Akinete (resting spore) of *Anabaena* by phase contrast.

similation). A few cyanobacteria, mainly filamentous species, can actually grow in the dark on glucose or su-crose, using the sugar as both carbon and energy source.

Several metabolic products of cyanobacteria are of considerable practical importance. Many cyanobacteria produce potent neurotoxins, and during water blooms when massive accumulations of cyanobacteria may de-velop, animals ingesting such water may be killed. Many cyanobacteria are also responsible for the production of earthy odors and flavors in fresh waters, and if such wa-ters are used as drinking water sources, aesthetic prob-lems may arise. The major compound produced is *geosmin* (trans-1,10-dimethyl-trans-9-decalol). This sub-stance is also produced by many actinomycetes (see the discussion in Section 12.24) and is responsible for the dis-tinctive "earthy" odor of moist, freshly turned soil.

Ecology and Phylogeny of Cyanobacteria

Cyanobacteria are widely distributed in nature in terres-trial, freshwater, and marine habitats. In general, they are more tolerant of environmental extremes than are algae and are often the dominant or sole oxygenic phototrophic organisms in hot springs (🔗Table 6.1), saline lakes, and other extreme environments. Many species are found on the surfaces of rocks or soil and occasionally even within rocks themselves (🔗Figure 14.38). In desert soils subject to intense sunlight, cyanobacteria often form extensive crusts over the surface, remaining dormant during most of the year and growing during the brief winter and spring rains. In shallow marine bays, where relatively warm seawater temperatures exist, cyanobacterial mats of considerable thickness may form. Freshwater lakes, espe-cially those that are rich in nutrients, may develop blooms of cyanobacteria (🔗Figure 19.10b). A few cyanobacteria are symbionts of liverworts, ferns, and cycads; a number are found as the phototrophic component of lichens. In the case of the water fern *Azolla* (🔗Sections 17.28 and 19.22), it has been shown that the cyanobacterial endo-phyte (a species of *Anabaena*) fixes nitrogen that becomes available to the plant.

Base compositions of genomic DNA of a variety of cyanobacteria have been determined. Those of the uni-cellular forms vary from 35 to 71% GC, a range so wide as to suggest that this group contains many members with little genetic relationship to each other. On the other hand, the values for the heterocyst formers vary much less, from 38 to 46% GC. Phylogenetically, cyanobacteria group along morphological lines in most cases. Filamen-tous heterocystous and nonheterocystous species form distinct groups, as do the branching forms. However, unicellular cyanobacteria are phylogenetically highly di-verse, with different representatives showing phyloge-netic relationships to different morphological groups.

12.26 Prochlorophytes and Chloroplasts

Key Genera: *Prochlorococcus, Prochloron, Prochlorothrix*

Prochlorophytes are oxygenic phototrophs that contain chlorophyll *a* and *b* but do *not* contain phycobilins. Prochlorophytes therefore resemble both cyanobacteria (because they are prokaryotic and have chlorophyll *a*) and the green plant/green alga chloroplast (because they contain chlorophyll *b* instead of phycobilins). Phy-logenetically, prochlorophytes show specific relation-ships to cyanobacteria.

Prochloron

Prochloron was the first prochlorophyte discovered. It is found in nature as a symbiont of marine inverte-brates (didemnid ascidians), but it has not been cul-tured in the laboratory. Cells of *Prochloron* expressed from the cavities of didemnid tissue are roughly spherical (Figure 12.82●), and 8–10 μm in diameter.

● **Figure 12.82 Electron micrograph of the prochlorophyte *Prochloron*.** Note the extensive intracytoplasmic membranes (thylakoids). Cells are about 10 μm in diameter. Unlike cyanobacteria, prochlorophytes lack phycobiliproteins.

Electron micrographs of thin sections of *Prochloron* (Figure 12.82) show an extensive thylakoid membrane system similar to that observed in the chloroplast (∞Figure 14.6). Further evidence that *Prochloron* is phylogenetically a member of the *Bacteria* is the presence of muramic acid in the cell walls, indicating that peptidoglycan is present (∞Section 4.8). The carotenoids of *Prochloron* are similar to those of cyanobacteria, predominantly β-carotene and zeaxanthin. The GC ratio of genomic DNA from samples of *Prochloron* obtained from different ascidians varies from 31 to 41%, indicating genetic heterogeneity. Thus, different species of *Prochloron* probably exist, but confirmation must await laboratory culture and study of pure cultures.

Prochlorothrix and *Prochlorococcus*

Prochlorothrix is a filamentous prochlorophyte (Figure 12.83●) that can be grown in pure culture. Like *Prochloron*, *Prochlorothrix* contains chlorophylls *a* and *b* and lacks phycobilins, although the thylakoid membranes are less well developed than in *Prochloron* (compare Figures 12.82 and 12.83*b*).

A novel prochlorophyte, *Prochlorococcus*, inhabits the euphotic zone of the open oceans. Cells of these phototrophs are small cocci, measuring less than 1 μm in diameter (∞Figure 19.11*a*). Like other prochlorophytes, cells of *Prochlorococcus* contain chlorophyll *b*. However, *Prochlorococcus* lacks true chlorophyll *a* and produces instead a modified form of chlorophyll *a* called *divinyl chlorophyll a*. Cells of *Prochlorococcus* also contain α-(instead of β-) carotene, a pigment previously unknown in prokaryotes. Because their numbers in the oceans are relatively large ($10^4 - 10^5$ cells/ml), prochlorophytes like *Prochlorococcus* probably have considerable ecological significance as primary producers in open ocean waters. A variety of other prochlorophytes have been isolated including *Acaryochloris* (Figure 12.83*c*), which contains chlorophyll *d* as its major pigment. Chlorophyll is also present in a variety of algae (eukaryotic cells; ∞Section 14.13), some species of which are extremely small (∞Figure 19.11*c*).

Prochlorophytes, Chloroplasts, and Evolution

Based on our discussion of endosymbiosis (∞Sections 2.6, 11.4, and 14.4), the evolutionary significance of prochlorophytes should be apparent. Until the discovery of prochlorophytes, it was always assumed that the chloroplast originated from endosymbiotic association of *cyanobacteria* with a primitive eukaryotic cell. However,

(a)   *(b)*   *(c)*

● **Figure 12.83 Phase and electron micrographs of the filamentous prochlorophytes *Prochlorothrix* and *Acaryochloris*.** (a) Phase contrast. (b) Electron micrograph of thin section showing arrangement of membranes. The diameter of cells is about 2 μm. (c) *Acaryochloris*. This prochlorophyte contains chlorophyll *d* as its main chlorophyll pigment. A cell is about 1.5 μm in diameter.

this hypothesis has been scientifically unsatisfying for at least one major reason: How did the green plant chloroplast evolve the pigment complement it has today if it originated from a cyanobacterial endosymbiont that contained *phycobilins* instead of chlorophyll *b*? The hypothesis that *prochlorophytes* instead of cyanobacteria were the ancestors of the green plant chloroplast eliminates this major point of contention. However, phylogenetic analyses do not show *Prochloron*, *Prochlorococcus*, or *Prochlorothrix* to be the *immediate* ancestors of the green plant chloroplast. Instead, prochlorophytes, cyanobacteria, and the plant chloroplast all *shared* a common ancestor, and remain distinct lineages in the cyanobacterial phylum.

Another hypothesis does account for the pigmentation patterns observed in prokaryotic oxygenic phototrophs. Cyanobacteria and prochlorophytes may have evolved from ancestors that contained phycobilins and chlorophylls other than just chlorophyll *a*, either chlorophyll *b* or *d*. From here, cyanobacterial lineages evolved to lose accessory chlorophylls while prochlorophyte lineages dispensed with phycobilins. The complement of pigments that we find in each group today, then, likely represents the best combination of photosynthetic pigments for fitness in their particular habitat.

 12.25–12.26 Concept Check

Cyanobacteria and prochlorophytes are oxygenic phototrophic prokaryotes. Prochlorophytes differ most clearly from cyanobacteria in that prochlorophytes contain chlorophyll *b* or *d* and lack phycobilins. Oxygen in Earth's atmosphere is thought to have originated from cyanobacterial photosynthesis.

◆ Describe at least three ways in which cyanobacteria differ from purple bacteria.

◆ What is a *heterocyst* and what is its function?

◆ How are cyanobacteria, prochlorophytes, and the chloroplasts of corn plants similar and how do they differ?

◆ Of what ecological significance is *Prochlorococcus*?

 V PHYLUM 5: CHLAMYDIA

12.27 The Chlamydia

Key Genera: *Chlamydia*, *Chlamydophila*

Organisms of the genera *Chlamydia* and *Chlamydophila* are obligately parasitic bacteria with poor metabolic capacities that form a distinct phylum of *Bacteria* (see Figure 12.1). Several key species are recognized (Table 12.33): *Chlamydophila psittaci*, the causative agent of the disease *psittacosis*; *Chlamydia trachomatis*, the causative agent of *trachoma* and a variety of other human diseases; and *Chlamydophila pneumoniae*, the cause of a variety of respiratory syndromes (Table 12.33).

Psittacosis is an epidemic disease of birds that is occasionally transmitted to humans and causes pneumonia-like symptoms. **Trachoma** is a debilitating disease of the eye characterized by vascularization and scarring of the cornea. Trachoma is the leading cause of blindness in humans. Other strains of *Chlamydia trachomatis* infect

Table 12.33 Differential characteristics of species of the genera *Chlamydia* and *Chlamydophila*

Characteristic	*Chlamydia trachomatis*	*Chlamydophila psittaci*	*Chlamydophila pneumoniae*
Hosts	Humans	Birds, mammals, occasionally humans	Humans
Usual site of infection	Mucous membrane	Multiple sites	Respiratory mucosa
Human-to-human transmission	Common	Rare	Probable
Mol % GC of DNA	42–45	39–43	40
Percent homology to *C. trachomatis* DNA by DNA:DNA hybridization[a]	100	10	10
DNA, kilobase pairs/genome (*Escherichia coli* = 4600)	1000	550	~1000
Human diseases	Trachoma, otitis media, nongonococcal urethritis (males), urethral inflammation (females), lymphogranuloma venereum, cervicitis	Psittacosis	Respiratory syndromes
Domesticated animal diseases	—	Avian chlamydiosis (parrots, parakeets), pneumonia, synovial tissue arthritis, or conjunctivitis (kittens, lambs, calves, piglets, foals)	—

[a] For discussion of DNA:DNA hybridization, see Section 11.11.

the genitourinary tract, and chlamydial infections are currently one of the leading sexually transmitted diseases (◌◌Section 26.13). A comparison of the properties of *Chlamydophila psittaci, Chlamydia trachomatis*, and *Chlamydophila pneumoniae* and the diseases they cause is shown in Table 12.33.

Molecular and Metabolic Properties

Besides being disease entities, the chlamydias are intriguing because of the biological, evolutionary, and metabolic problems they pose. Biochemical studies show that the chlamydias have gram-negative-type cell walls, and they have both DNA and RNA, that is, they are clearly cellular. Electron microscopy of thin sections of infected cells shows cells dividing by binary fission (Figure 12.84◉).

The biosynthetic capacities of the chlamydias are much more limited than even the rickettsias, the other group of obligate intracellular parasites known among the *Bacteria* (see Section 12.13). Indeed, for some time it was thought that chlamydias were "energy parasites," obtaining not only biosynthetic intermediates from their hosts, as do the rickettsias, but also ATP. However, this hypothesis has been questioned following the sequencing of the genome of *C. trachomatis* (◌◌Section 15.3). The approximately 1 Mbp chromosome of *C. trachomatis* contains easily recognizable genes for ATP synthesis and even contains a complement of genes encoding peptidoglycan biosynthetic functions. This suggests that this organism may well contain peptidoglycan even though chemical analyses for this cell-wall polymer have been negative. Nevertheless, the chlamydias still probably have the simplest biochemical capacities of all known *Bacteria*, and a summary of this is shown in Table 12.34.

Interestingly, the *C. trachomatis* genome lacks a gene encoding the protein FtsZ, a key protein involved in septum formation during cell division (◌◌Section 6.2). This

● **Figure 12.84** *Chlamydia.* Electron micrograph of a thin section of a dividing cell (reticulate body, see Figure 12.85) of *Chlamydia psittaci*, a member of the psittacosis group, within a mouse tissue culture cell. A single chlamydial cell is about 1 μm in diameter.

protein was previously thought to be indispensable for growth of all prokaryotes, both *Archaea* and *Bacteria*. Moreover, genes are present in *C. trachomatis* that have a distinct "eukaryotic look" to them, suggesting that *C. trachomatis* has picked up some host genes that may encode functions that assist it in its pathogenic lifestyle (Table 12.33; ◌◌Sections 21.7 and 26.13).

Life Cycle of *Chlamydia*

The life cycle of a typical chlamydial species is shown in Figure 12.85◉. Two cellular types are seen in the life cycle: (1) a small, dense cell, called an **elementary body**, which is relatively resistant to drying and is the means of *dispersal* of the agent, and (2) a larger, less dense cell, called a **reticulate body**, which divides by binary fission and is the *vegetative* form.

Elementary bodies are nonmultiplying cells specialized for infectious transmission. By contrast, reticulate bodies are noninfectious forms that function only to

Table 12.34	**Comparison of obligate intracellular parasites: rickettsias, chlamydias, and viruses**		
Property	**Rickettsias**	**Chlamydias**	**Viruses**
Structural			
Nucleic acid	RNA and DNA	RNA and DNA	Either RNA or DNA (single- or double-stranded), never both
Ribosomes	Present	Present	Absent
Cell wall	Peptidoglycan present	Peptidoglycan present[a]	No wall
Structural integrity during multiplication	Maintained	Maintained	Lost
Metabolic capacities			
Macromolecular synthesis	Carried out	Carried out	Only with use of host machinery
ATP-generating system	Present	Present [a]	Absent
Capable of oxidizing glutamate	Yes	No	No
Sensitivity to antibacterial antibiotics	Sensitive	Sensitive (except for penicillin)	Resistant
Phylogeny	Alpha Proteobacteria	Chlamydial phylum	Not cells

[a] The genome of one chlamydial species, *Chlamydia trachomatis*, has been entirely sequenced (◌◌ Section 15.4), and genes for peptidoglycan synthesis and ATP synthesis are present. However, the lack of penicillin sensitivity of the chlamydia raises doubt whether peptidoglycan is synthesized.

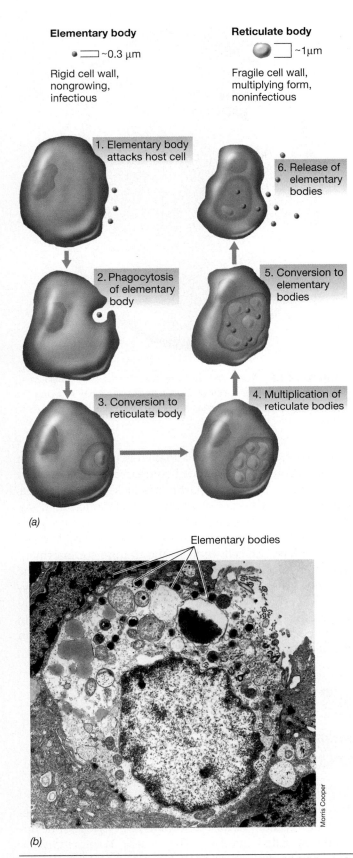

Elementary body

●⌁▭ ~0.3 µm

Rigid cell wall,
nongrowing,
infectious

Reticulate body

◯ ▭ ~1µm

Fragile cell wall,
multiplying form,
noninfectious

1. Elementary body attacks host cell

2. Phagocytosis of elementary body

3. Conversion to reticulate body

4. Multiplication of reticulate bodies

5. Conversion to elementary bodies

6. Release of elementary bodies

(a)

Elementary bodies

Morris Cooper

(b)

● **Figure 12.85 The infection cycle of chlamydia.** (a) Schematic diagram of the cycle: the whole cycle takes about 48 h. (b) Human chlamydial infection. An infected fallopian tube cell is bursting, releasing mature elementary bodies.

multiply inside host cells to form a large inoculum for transmission. Unlike the rickettsias (see Section 12.13),

the chlamydias are not transmitted by arthropods but are primarily *airborne* invaders of the respiratory system—hence the significance of resistance to drying of the elementary bodies. A dividing reticulate body can be seen in Figure 12.84. After a number of cell divisions, these vegetative cells are converted into elementary bodies that are released when the host cell disintegrates and can then infect other cells. Generation times of 2–3 h have been measured for reticulate bodies, which are considerably faster than those found for the rickettsias.

In sum, the chlamydias appear to have evolved an efficient and effective survival strategy including parasitizing the resources of the host (Table 12.34) and the production of resistant cell forms for transmission. It is thus not surprising that chlamydias have been associated with so many different disease syndromes (Table 12.33; ∞Section 26.13).

🛑 *12.27 Concept Check*

Chlamydia are extremely small parasitic bacteria that cause a variety of human diseases. Chlamydia contain a very small genome and are apparently deficient in many metabolic functions.

♦ Using the data of Table 12.34 as a guide, how can chlamydias be differentiated from rickettsias? From viruses?

♦ What is the difference between an *elementary* body and a *reticulate* body?

♦ What surprises have emerged from sequencing of the chlamydial genome?

VI PHYLUM 6: PLANCTOMYCES/ PIRELLULA

12.28 *Planctomyces: A Phylogenetically Unique Stalked Bacterium*

Key Genera: *Planctomyces, Pirellula, Gemmata*

This phylum contains a number of morphologically unique bacteria including the genera *Planctomyces, Pirellula, Gemmata,* and *Isosphaera.* The best studied of these has been *Planctomyces* (Figure 12.86●). In Section 12.16 we considered stalked bacteria such as *Caulobacter. Planctomyces* is also a stalked bacterium. However, unlike *Caulobacter,* the stalk of *Planctomyces* is made of protein and does not contain a cell wall or cytoplasm (compare Figure 12.86 with Figure 12.41). The *Planctomyces* stalk presumably functions in attachment but it is a much narrower and finer structure than the prosthecal stalk of *Caulobacter.*

Other Features of the *Planctomyces* Group

Planctomyces and relatives are also of interest because they lack peptidoglycan and their cell walls are of an S-layer type (∞Sections 4.8 and 4.10), consisting of protein containing

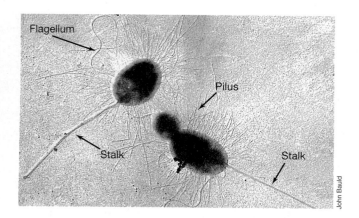

● **Figure 12.86** **An electron micrograph of a metal-shadowed preparation of *Planctomyces maris*.** A single cell is about $1-1.5 \mu m$ long. Note the fibrillar nature of the stalk. Pili are also abundant. Note also the flagella (curly appendages) on each cell and the bud that is developing from the nonstalked pole of one cell.

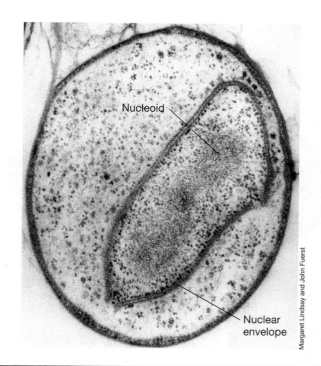

● **Figure 12.87** **Nucleated prokaryotes.** Thin-section transmission electron micrograph of a cell of *Gemmata obscuriglobus* showing the nucleoid surrounded by a nuclear envelope. The cell is about $1.5 \mu m$ in diameter.

large amounts of cysteine (as cystine) and proline. As would be expected of organisms lacking peptidoglycan, these organisms are resistant to antibiotics that disrupt peptidoglycan synthesis, such as penicillin and cephalosporin.

Like in *Caulobacter* (Figure 12.41), *Planctomyces* is also a budding bacterium and shows a type of "life cycle," wherein motile swarmer cells attach to a surface, grow a stalk from the attachment point, and generate a new cell from the opposite pole by budding. This daughter cell grows a flagellum, breaks away from the attached mother cell, and begins the cycle anew. Physiologically, *Planctomyces* species are typical facultatively aerobic chemoorganotrophs, growing either by fermentation or respiration of sugars.

The habitat of *Planctomyces* is primarily aquatic, both freshwater and marine, and the genus *Isosphaera* is a filamentous, gliding hot spring bacterium. Like for *Caulobacter* (see Section 12.16), isolation of *Planctomyces* and relatives requires dilute media, and, since all known members of this group lack peptidoglycan, enrichments can be made even more selective by the addition of penicillin.

We learned in Chapter 2 of the major structural differences between prokaryotic and eukaryotic cells. In particular, eukaryotes have a membrane-bound nucleus (∞Section 2.2). However, the Planctomycetes are unique among all known prokaryotes in that they show extensive cell compartmentalization, including a membrane-bound nuclear structure. For example, in the bacterium *Gemmata* (Figure 12.87●), the nucleoid is surrounded by a nuclear envelope. DNA in *Gemmata* remains in a covalently closed, circular, and supercoiled form, typical of prokaryotes (∞Section 7.3), but it is highly condensed and remains partitioned from the remaining cytoplasm by a true unit membrane (Figure 12.87). Another interesting compartment exists in the anammoxosome of *Brocadia anammoxidans*, a relative of *Planctomyces*. This bacterium carries out the anaerobic oxidation of ammonia to N_2 within the enclosed anammoxosome structure (∞Section 17.12).

Thus far, all species of Planctomycetes have been found to contain internal cell compartments of one sort or the other. Some compartments lack DNA and thus have other functions (for example, metabolic functions, ∞Section 17.12). In no other group of known prokaryotes do internal compartments so closely resemble those of the eukaryotic cell. Indeed, the unique compartmentalization of cells of Planctomycetes tends to blur the structural distinction between prokaryotes and eukaryotes. Nevertheless, Planctomycetes are distinct in many ways other than cell structure, including, in particular, the specific phylogenetic position they occupy within the heart of the domain *Bacteria* (Figure 12.1).

VII PHYLUM 7: THE VERRUCOMICROBIA

12.29 *Verrucomicrobium* and *Prosthecobacter*

Key Genera: *Verrucomicrobium*

This phylum of bacteria shares with prosthecate Proteobacteria (see Section 12.16) the formation of cytoplasmic appendages called *prostheca*. The genera *Verrucomicrobium* and *Prosthecobacter* produce two to several *prosthecae* per cell (Figure 12.88●). Also, unlike cells of *Caulobacter* (Figure 12.41), which contain a single prostheca and produce flagellated and nonprosthecate swarmer cells (see Section 12.16), *Verrucomicrobium* and *Prosthecobacter* divide symmetrically, and both mother and daughter cells contain

● **Figure 12.88 Negatively stained transmission electron micrograph of a dividing cell of *Verrucomicrobium spinosum*.** Note the warty-like prosthecae. A single cell is about 1 μm in diameter.

prostheca at the time of cell division. The genus name *Verrucomicrobium* derives from Greek roots meaning "warty," and the appearance of cells of *V. spinosum* with its multiple projecting prosthecae (Figure 12.88) is very appropriate.

Members of the Verrucomicrobia share with other prosthecate bacteria the presence of peptidoglycan in their cell walls and are aerobic to facultatively aerobic bacteria, capable of fermenting various sugars. Verrucomicrobia are widespread in nature, inhabiting freshwater and marine environments as well as forest and agricultural soils. From a phylogenetic standpoint, Verrucomicrobia are distinct from all known *Bacteria* (Figure 12.1). The group shows some phylogenetic affiliation with the *Planctomyces* and *Chlamydia* phyla (Figure 12.1), but is clearly sufficiently distinct from both of these groups to form its own separate lineage.

Species of the genus *Prosthecobacter* have been found to contain two genes that show significant homology to the tubulin genes of eukaryotic cells. Tubulin is the key protein that makes up the cytoskeleton of eukaryotic cells (∞ Section 14.5). Although the important cell division protein FtsZ (∞ Section 6.2) is also a tubulin homologue, the *Prosthecobacter* proteins are structurally more similar to eukaryotic tubulin than is FtsZ. The role of the tubulin proteins in *Prosthecobacter* is unknown, because a eukaryotic-like cytoskeleton has not been observed in these organisms. However, these genes likely signal that a specific relationship exists between Verrucomicrobia and eukaryotic cells, either in terms of shared ancestry or from lateral gene transfer between cells of the two domains.

VIII PHYLUM 8: THE FLAVOBACTERIA

Flavobacteria range from obligate aerobes to obligate anaerobes, unified by a common phylogenetic thread.

The organisms inhabit many different types of environments, and we focus here on the two main genera in the group.

12.30 *Bacteroides* and *Flavobacterium*

Key Genera: *Bacteroides, Flavobacterium*

The genus *Bacteroides* contains obligately anaerobic, nonsporulating species that are saccharolytic, fermenting sugars or proteins, depending on the species, to primarily acetate and succinate as fermentation products. *Bacteroides* are normally commensals, found in the intestinal tract of humans and other animals (∞ Sections 19.11 and 21.4). In fact, *Bacteroides* species are the numerically dominant bacteria in the human large intestine, where measurements have shown that over 10^{10} cells are present per gram of human feces. However, species of *Bacteroides* can also be pathogens and are the most important anaerobic bacteria associated with human infections.

Species of *Bacteroides* are unusual among *Bacteria* in that they are one of the few groups of organisms to synthesize *sphingolipids*, a heterogeneous collection of lipids characterized by the long-chain amino alcohol *sphingosine* in place of glycerol (Figure 12.89●). Sphingolipids such as sphingomyelin, cerebrosides, and gangliosides are common in mammalian tissues, especially in the brain and other nervous tissues.

In contrast to *Bacteroides*, *Flavobacterium* species are primarily found in aquatic habitats, both freshwater and marine, as well as in foods and food-processing plants. Colonies of *Flavobacterium* are frequently yellow-pigmented, and physiologically these organisms are aerobes and rather nutritionally restricted, using glucose as carbon and energy source but very few other carbon compounds. Flavobacteria are rarely pathogenic; however, one species, *F. meningosepticum*, has been associated with cases of infant meningitis.

Other important genera in this group are psychrophilic or psychrotolerant (∞ Section 6.11). These include, in particular, the genera *Polaribacter* and *Psychroflexus*. Many other genera in the group are also capable of good growth below 20°C.

(a)

$$H_2C-OH$$
$$HC-OH$$
$$H_2C-OH$$

(b)

$$H_3C-(CH_2)_{12}-C=C-C-C-CH_2OH$$
$$OH\ NH_3^+$$

● **Figure 12.89 Sphingolipids.** Comparison of (a) glycerol with (b) sphingosine. In sphingolipids, characteristic of *Bacteroides* species, sphingosine is the esterifying alcohol; a fatty acid is bonded by peptide linkage through the N atom (shown in red), and the terminal —OH group (shown in green) can contain any of a number of compounds including phosphatidyl choline (sphingomyelin) or various sugars (cerebrosides and gangliosides).

Heinz Schlesner

IX PHYLUM 9: THE CYTOPHAGA GROUP

Key Genera: *Cytophaga, Flexibacter, Rhodothermus, Salinibacter*

12.31 *Cytophaga* and Relatives

Organisms of the *Cytophaga* group are long, slender, gram-negative rods, often with pointed ends, that move by gliding (Figure 12.90*a,b*●). The related genus, *Sporocytophaga*, is similar to *Cytophaga* in morphology and physiology, but the cells form resting spherical structures called *microcysts* (Figure 12.90*d*), similar to those produced by some fruiting myxobacteria (see Section 12.17). They are widespread in soil and water, often in great abundance.

Many cytophagas digest polysaccharides like cellulose (Figure 12.90*c*), agar (Figure 12.90*a*), or chitin. The cellulose decomposers can be easily isolated by placing small crumbs of soil on pieces of cellulose filter paper laid on the surface of mineral salts agar. The bacteria attach to and digest the cellulose fibers, forming spreading colonies (Figure 12.90*c*). The cytophagas do not produce soluble, extracellular, cellulose-digesting enzymes (cellulases). Instead, their cellulases remain attached to the cell envelope, which probably explains why the cells must adhere to cellulose fibrils in order to digest them.

In pure culture, *Cytophaga* can be cultured on agar containing embedded cellulose fibers, the presence of the organism being indicated by the clearing that occurs as the cellulose is digested (Figure 12.90*c*; see also Figure 17.62).

Species of *Cytophaga* and *Sporocytophaga* are obligately aerobic and probably account for much of the cellulose digestion that occurs by prokaryotes in oxic environments in nature. A number of *Cytophaga* species are also fish pathogens and can cause serious problems in the cultivated fish business. Two of the most important diseases are *columnaris disease*, caused by *C. columnaris*, and *cold-water disease*, caused by *C. psychrophila*. Both diseases preferentially affect stressed fish, such as those living in waters receiving pollutant discharges or living in high-density confinement situations such as fish hatcheries and aquaculture operations. Infected fish show tissue destruction, frequently around the gills, and this may stem from the fact that *Cytophaga* species isolated from infected fish are typically strongly proteolytic.

The genus *Flexibacter* differs from the cytophagas in that the species usually require complex media for good growth and are not cellulolytic. Cells of some *Flexibacter* species also undergo changes in cell morphology from long, gliding, threadlike filaments lacking cross-walls, to short, nonmotile rods. Many species are pigmented due to carotenoids located in the cytoplasmic membrane, or related pigments called *flexirubins*, located in the gram-negative outer membrane. *Flexibacter* species are com-

(a)

(b)

(c)   (d)

●**Figure 12.90** *Cytophaga* and *Sporocytophaga*. (a) Streak of an agarolytic marine *Cytophaga* species hydrolyzing agar in the Petri dish. (b) Colonies of *Sporocytophaga* growing on cellulose. Note the clearing zones (arrows) where the cellulose has been degraded. (c) Phase-contrast photomicrograph of cells of *C. hutchinsonii* grown on cellulose filter paper (cells are about 1.5 μm in diameter). (d) Phase-contrast photomicrograph of the rod-shaped cells and spherical microcysts of *Sporocytophaga myxococcoides* (cells are about 0.5 μm and microcysts about 1.5 μm in diameter).

mon soil and freshwater saprophytes, and none have been identified as pathogens.

Rhodothermus/Salinibacter

The genera *Rhodothermus* and *Salinibacter* fall into the *Cytophaga* phylum, but are only distantly relatives. *Rhodothermus* and *Salinibacter* consist of gram-negative, red- or yellow-pigmented, obligatory aerobic chemoorganotrophic bacteria. *Rhodothermus* is thermophilic, with a temperature optimum around 60–65°C. *Rhodothermus* grows best on sugars or simple and complex polysaccharides. The organism inhabits shallow-water submarine hot springs and has also been detected in terrestrial hot springs. *Rhodothermus* produces thermal stable hydrolytic enzymes of biotechnological interest, including an amylase (that degrades starch), a cellulase (cellulose), and a xylanase (that degrades hemicelluloses, abundant in plant cell walls), among many others.

Salinibacter is a genus of extremely halophilic red bacteria that is perhaps the most salt-tolerant and salt-requiring of all *Bacteria*. In fact, its salt requirements rival those of extremely halophilic *Archaea*, such as *Halobacterium* (∞ Section 13.3). *Salinibacter ruber*, the only known species, lives in saltern cystallization ponds and related highly saline environments. *Salinibacter* shares with *Halobacterium* the use of K^+ as a compatible solute, a property that is rarely found among halophiles of the domain *Bacteria* (most halophilic *Bacteria* either synthesize or accumulate organic solutes to maintain water balance in their salty environments, ∞ Section 6.14). In contrast to the highly hydrolytic *Rhodothermus*, *Salinibacter* grows best with amino acids as electron donors, similar to *Halobacterium*. Thus, although phylogenetically unique, we see in *Halobacterium* and *Salinibacter* a likely case of convergent evolution, where physiological strategies have converged along the same pattern, probably because of the unique demands of the extreme environment that these organisms share.

12.28–12.31 Concept Check

The *Planctomyces* group contains stalked, budding bacteria while the flavobacteria contain a variety of gram-negative *Bacteria* motile by either flagella or by gliding associated with animals, food, and the soil. Members of the Verrucomicrobia are distinguished by their multiple prosthecate cells. The *Cytophaga* group includes a variety of obligately aerobic chemoorganotrophic bacteria in soil, water, hot springs, and thermal environments.

◆ What is unique about the cell wall and arrangement of DNA in *Planctomyces*?

◆ How does the stalk of *Planctomyces* differ from the stalk of *Caulobacter*?

◆ Where might you find large numbers of *Bacteroides* cells in nature?

◆ Describe a method for isolating *Cytophaga* species from nature.

◆ Contrast the habitat and physiology of *Rhodothermus* and *Salinibacter*.

X PHYLUM 10: GREEN SULFUR BACTERIA

12.32 *Chlorobium* and Other Green Sulfur *Bacteria*

Key Genera: *Chlorobium, Chlorobaculum, Prosthecochloris, "Chlorochromatium"*

Green sulfur bacteria are a phylogenetically distinct group of nonmotile anoxygenic phototrophic bacteria that contain only obligately anaerobic and phototrophic species among cultured isolates. The group is morphologically restricted and includes short to long rods (Table 12.35 and Figure 12.91●). Like purple sulfur bacteria they utilize H_2S as an electron donor, oxidizing it first to S^0 and then to SO_4^{2-}. But unlike purple sulfur bacteria, the sulfur produced by green sulfur bacteria resides *outside* the cell (Figure 12.91a; ∞Figure 17.17b). Most species can also assimilate a few organic compounds in the light (that is, *photoheterotrophy*, ∞ Section 17.4). Strict autotrophy, however, is supported not by the reactions of the Calvin cycle as in purple bacteria, but instead by a reversal of steps in the citric acid cycle (∞Section 17.7, reverse citric acid cycle), a unique means of autotrophy.

(a)

(b)

● **Figure 12.91 Phototrophic green sulfur bacteria.** (a) *Chlorobium limicola*; cells are about 0.8 μm wide. Note the sulfur granules deposited extracellularly. (b) *Chlorobium clathratiforme*, a bacterium forming a three-dimensional network; cells are about 0.8 μm wide.

Table 12.35	Genera and characteristics of phototrophic green sulfur bacteria	
Characteristics	**Genus (mol % GC)**	
Straight or curved rods, some branching; nonmotile;	*Chlorobium* (49–58)	
color green or brown; some contain gas vesicles	*Chlorobaculum* (56–58)	
Spheres and ovals, nonmotile; forming prosthecae; green or brown	*Prosthecochloris* (50–56)	
Rods, motile by gliding; green	*Chloroherpeton* (47)	

Pigments and Ecology

The bacteriochlorophylls found in green sulfur bacteria include bacteriochlorophyll *a* and either bacteriochlorophylls *c*, *d*, or *e*. The latter pigments function only in light-harvesting reactions (∞Section 17.2) and are located in unique structures called **chlorosomes** (Figure 12.92●). Chlorosomes are oblong bacteriochlorophyll-rich bodies bounded by a thin, nonunit membrane and lie attached to the cytoplasmic membrane in the periphery of the cell (Figure 12.92; see also Figure 17.7).

Studies of energy transfer in green sulfur bacteria (∞Section 17.2) have shown that light energy absorbed by bacteriochlorophylls *c*, *d*, or *e* in the chlorosome gets funneled to bacteriochlorophyll *a*, which resides in the cytoplasmic membrane. It is here where photosynthetic energy conversion and ATP synthesis actually occur (∞Figure 17.7). Both green- and brown-colored species of green sulfur bacteria are known, the brown-colored species containing bacteriochlorophyll *e* and carotenoids that render the cells brown in color (Figure 12.93●; ∞Figure 17.9).

Like purple sulfur bacteria (see Section 12.2), green sulfur bacteria live in anoxic aquatic environments, especially those where H_2S is abundant (in general, green sulfur bacteria are more tolerant of sul-

● **Figure 12.93   Green and brown chlorobia.** Tube cultures of (a) *Chlorobium tepidum* and (b) *Chlorobium phaeobacteroides*. Cells of *C. tepidum* contain bacteriochlorophyll *c* and a series of green-colored carotenoids, while cells of *C. phaeobacteroides* contain bacteriochlorophyll *e* and isorenieratene, a brown-colored carotenoid. For structures of the specific green and brown carotenoids, see Figure 17.9.

fide than are purple bacteria). Because the chlorosome is such an efficient light-harvesting structure, little light is required to support the photosynthetic activities of green sulfur bacteria, and they are thus typically found at the greatest depths in lakes of any phototrophic organisms.

One species, *Chlorobaculum tepidum* (Figure 12.92), is thermophilic and forms dense microbial mats in high-sulfide hot springs. *C. tepidum* is also notable because its genome (2.1 Mbp) has been completely sequenced—the first genome of any anoxygenic phototroph. *C. tepidum* also grows rapidly and is amenable to genetic manipulation by both conjugation and transformation. It has become the model organism for studying the molecular biology of green sulfur bacteria.

Green Sulfur Bacteria Consortia

Certain green sulfur bacteria can form an intimate two-membered association with a chemoorganotrophic bacterium called a **consortium**. In the consortium, each organism benefits the other, and thus a variety of such consortia containing different phototrophic and chemotrophic components probably exist in nature. The phototrophic component, called an *epibiont*, appears physically attached to the nonphototrophic cell (Figure 12.94●), although the mechanism of attachment is not clear.

The name "*Chlorochromatium aggregatum*" has been used to describe a commonly observed consortium, although the name has no basis in formal taxonomy because it refers to two separate organisms rather than one. The "*C. aggregatum*" consortium is green in color because the epibionts are green sulfur bacteria that contain bacteriochlorophyll *c* or *d* and green-colored carotenoids and surround a central nonphototrophic cell (Figure 12.94*e*) A structurally similar organism called "*Pelochro-matium roseum*" is brown in color. In yet other consortia, the epibiont cells are vibroid in morphology (Figure 12.94*b,c*).

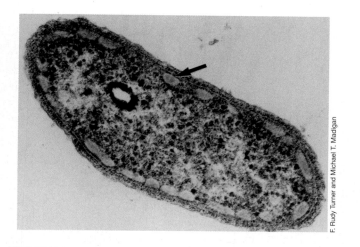

● **Figure 12.92   Thin-section electron micrograph of a cell of the green sulfur bacterium** *Chlorobium tepidum.* Note chlorosomes (arrow) in the cell periphery. A cell is about 0.7 μm wide.

(a)   (b)   (c)

Douglas Caldwell

Charles Abella

Charles Abella

(d)

Douglas Caldwell

(e)   (f)   (g)

Jörg Overmann

Jörg Overmann

Jörg Overmann

● **Figure 12.94 Green sulfur bacteria consortia.** (a–c) Phase contrast micrographs and (d) transmission electron micrograph of the green or brown bacterial consortia "*Chlorochromatium aggregatum*," or "*Pelochromatium roseum*." In (a–c) the nonphototrophic central organism is much lighter in color than the pigmented phototrophic bacteria. Note the chlorosomes (arrows) in (d). The entire consortium is about 3 μm in diameter. (e) Differential contrast micrograph of cells of "*Chlorochromatium aggregatum*." (f, g) Fluorescence micrographs. (f) DAPI-stained cells (stains DNA). (g) Same preparation as in (f) stained with a phylogenetic probe for green sulfur bacteria (yellow).

Some green bacterial consortia have been grown in laboratory culture. On average, the "*C. aggregatum*" consortium (Figure 12.94) contains about 12 epibionts per central cell while the "*P. roseum*" consortium contains about 20. Solid evidence that the epibionts are indeed green sulfur bacteria comes from the fact that chlorosomes are visible in thin sections of the epibionts (Figure 12.94*d*) and from molecular evidence. Treatment of the consortium with a fluorescent oligonucleotide probe specific for green sulfur

bacterial 16S rRNA (FISH technology, ∞Sections 11.7 and 18.4) causes the epibionts, but not the central cell, to fluoresce (Figure 12.94*f,g*). Studies of laboratory cultures have also shown that the central cell and the epibiont divide in synchrony, suggesting that the two components have some means of intercommunicating.

It is unclear why and how these associations form. However, laboratory and field studies suggest that the epibionts in these consortia are adapted to a very narrow regimen of light intensities and sulfide concentrations. The central cell may thus allow the otherwise nonmotile phototrophs to position themselves in a water column for optimal photosynthesis. The central cell in the consortium likely benefits from the arrangement by the uptake of nutrients formed by photosynthetic CO_2 formation and then excreted by the green sulfur bacteria.

⬢ **12.32   Concept Check**

Green sulfur bacteria are obligately anaerobic anoxygenic phototrophs that produce unique structures called *chlorosomes*. These organisms can grow at very low light intensities and oxidize H_2S to S^0 and SO_4^{2-}. Consortia containing phototrophic green bacteria and a nonphoto-trophic central cell are common in sulfidic aquatic environments.

◆   What pigments are found in the chlorosome?

◆   What is unique about autotrophy in green sulfur bacteria (∞Section 17.7)?

◆   What evidence supports the idea that the epibionts of green bacterial consortia are truly green sulfur bacteria?

XI PHYLUM 11: THE SPIROCHETES

12.33 Spirochetes

Key Genera: *Spirochaeta, Treponema, Cristispira, Leptospira, Borrelia*

Spirochetes are gram-negative, motile, tightly coiled *Bacteria*, typically slender and flexuous in shape (Figure 12.95●). These morphologically unique prokaryotes form a major phylogenetic lineage of *Bacteria* (see Figure 12.1). Spirochetes are widespread in aquatic environments and in animals. Some cause diseases, including syphilis, an important human sexually transmitted disease (∞Section 26.12).

The spirochete cell is made up of a *protoplasmic cylinder*, consisting of the regions enclosed by the cell wall and membrane (Figure 12.96●). Motility is conferred by a single to many flagella that emerge from each pole (Figure 12.96). However, unlike typical bacteria flagella (∞Section 4.14), spirochete flagella fold back from each pole upon the protoplasmic cylinder

(a)   (b)

● **Figure 12.95 Morphology of spirochetes.** Two spirochetes at the same magnification, showing the wide size range in the group. (a) *Spirochaeta stenostrepta*, by phase-contrast microscopy. A single cell is 0.25 μm in diameter. (b) *Spirochaeta plicatilis*. A single cell is 0.75 μm in diameter and can be up to 250 μm (0.25 mm) in length.

and remain located in the periplasm of the cell; thus, they have been called **endoflagella**. Both the endoflagella and the protoplasmic cylinder are surrounded by a multilayered but flexible membrane called the *outer sheath* (Figure 12.96).

Motility of Spirochetes

Spirochetes have an unusual mode of motility. Each endoflagellum is anchored at one end and extends about two-thirds of the length of the cell. Endoflagella rotate, as do typical bacterial flagella (◌◌Section 4.14). However, because the protoplasmic cylinder is rigid whereas the outer sheath is flexible, when both endoflagella rotate in the same direction, the protoplasmic cylinder rotates in the opposite direction, placing torsion on the cell as illustrated in Figure 12.96b. This causes the spirochete cell to move by flexing or lashing motions due to torque exerted at the ends of the protoplasmic cylinder by the rotating endoflagella (Figure 12.96b). Thus, despite the fact that endoflagella do not extend away from the cell, their rotation provides

motility, albeit of a more irregular and jerky form than do the flagella of other prokaryotes.

Classification

Spirochetes are classified into eight genera primarily on the basis of habitat, pathogenicity, phylogeny, and morphological and physiological characteristics. Table 12.36 lists the major genera and their characteristics.

Spirochaeta and *Cristispira*

The genus *Spirochaeta* includes free-living, anaerobic, and facultatively aerobic spirochetes. These organisms are common in aquatic environments, such as the water and mud of rivers, ponds, lakes, and oceans. One species of the genus *Spirochaeta* is *S. plicatilis* (Figure 12.95b), a fairly large spirochete found in freshwater and marine H_2S-containing habitats. The endoflagella of *S. plicatilis* are arranged in a bundle that winds around the coiled protoplasmic cylinder. From 18 to 20 endoflagella are inserted at each pole of this spirochete. Another species, *Spirochaeta stenostrepta*, has been cultured and is shown

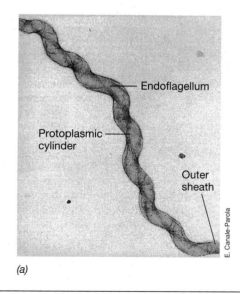

Endoflagellum

Protoplasmic cylinder

Outer sheath

(a)

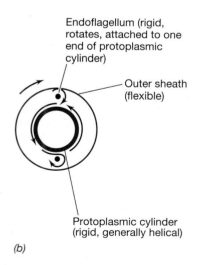

Endoflagellum (rigid, rotates, attached to one end of protoplasmic cylinder)

Outer sheath (flexible)

Protoplasmic cylinder (rigid, generally helical)

(b)

● **Figure 12.96 Motility in spirochetes.** (a) Electron micrograph of a negatively stained preparation of *Spirochaeta zuelzerae*, showing the position of the endoflagellum. A single cell is about 0.3 μm in diameter. (b) Cross section of a spirochete cell, showing the arrangement of the protoplasmic cylinder, endoflagella, and external sheath and the manner in which the rotation of the rigid endoflagellum can generate rotation of the protoplasmic cylinder and (in the opposite direction) rotation of the external sheath. If the sheath is free, the cell will rotate about its longitudinal axis and move along it. If the sheath is in contact with a solid surface, the cell will creep forward.

Table 12.36 Genera of spirochetes and their characteristics

Genus	Dimensions (μm)	General characteristics	Number of endoflagella	DNA (mol % GC)	Habitat	Diseases
Cristispira	30–150 × 0.5–3.0	3–10 complete coils; bundle of endoflagella visible by phase contrast microscopy	>100	—	Digestive tract of molluscs; has not been cultured	None known
Spirochaeta	5–250 × 0.2–0.75	Anaerobic or facultatively aerobic; tightly or loosely coiled	2–40	50–65	Aquatic, free-living, freshwater and marine	None known
Treponema	5–15 × 0.1–0.4	Microaerophilic or anaerobic; helical or flattened coil amplitude up to 0.5 μm	2–32	25–53	Commensal or parasitic in humans, other animals	Syphilis, yaws, swine dysentery, pinta
Borrelia	8–30 × 0.2–0.5	Microaerophilic; 5–7 coils of approx. 1 μm amplitude	7–20	46	Humans and other mammals, arthropods	Relapsing fever, Lyme disease, ovine and bovine borreliosis
Leptospira	6–20 × 0.1	Aerobic, tightly coiled, with bent or hooked ends; requires long-chain fatty acids	2	33–43	Free-living or parasitic in humans, other mammals	Leptospirosis
Leptonema	6–20 × 0.1	Aerobic; does not require long-chain fatty acids	2	54	Free-living	None known
Brachyspira	7–10 × 0.35–0.45	Anaerobe	8–28	25–27	Intestine of warm-blooded animals	Causes diarrhea in chickens and swine
Brevinema	4–5 × 0.2–0.3	Microaerophile, by 16S rRNA sequencing, forms deep branch in spirochete lineage (see Figure 12.1)	2	34–36	Blood and tissue of mice and shrews	Infectious for laboratory mice

in Figure 12.95*a*. It is an obligate anaerobe commonly found in H_2S-rich, black muds. It ferments sugars via the glycolytic pathway to ethanol, acetate, lactate, CO_2, and H_2. The species *Spirochaeta aurantia* is an orange-pigmented facultative aerobe, fermenting sugars via the glycolytic pathway under anaerobic conditions and oxidizing sugars aerobically mainly to CO_2 and acetate.

The genus *Cristispira* (Figure 12.97●) contains organisms with a unique distribution, being found in nature primarily in the *crystalline style* of certain molluscs, such as clams and oysters. The crystalline style is a flexible, semisolid rod seated in a sac and rotated against a hard surface of the digestive tract, thereby mixing with and grinding the small particles of food. Being large spirochetes, the cristispiras can readily be seen microscopically within the mollusc style as they rapidly rotate forward and backward in corkscrew fashion. *Cristispira* may occur in both freshwater and marine molluscs, but not all species of molluscs

possess them. Unfortunately, *Cristispira* has not been cultured, and so the physiological rationale for its restriction to this unique habitat is unknown.

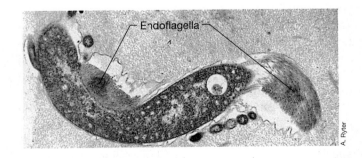

● **Figure 12.97** *Cristispira.* Electron micrograph of a thin section of *Cristispira*, a large spirochete. A cell measures about 2 μm in diameter. Notice the numerous endoflagella.

Treponema

Anaerobic, host-associated spirochetes that are commensals or parasites of humans and animals reside in the genus *Treponema*. *Treponema pallidum*, the causal agent of syphilis (∞Section 26.12), is the best-known species of *Treponema*. It differs in morphology from other spirochetes; the cell is not helical but has a flat wave form. The *T. pallidum* cell is remarkably thin, measuring only 0.2 μm in diameter. Because of this, dark-field microscopy has long been used to examine exudates from suspected syphilitic lesions (∞Figure 26.28).

In nature, *T. pallidum* is restricted to humans, although artificial infections have been established in rabbits and monkeys. Although never grown in laboratory culture, it has been established from animal studies that virulent *T. pallidum* cells (purified from infected rabbits) contain a cytochrome system and are in fact microaerophiles. Such cells have also yielded sufficient DNA for the genome of *T. pallidum* (1.14 Mbp) to be sequenced (∞Section 15.3).

Other species of the genus *Treponema* are common commensal organisms in the oral cavity of humans and can generally be seen in material scraped from between the teeth and from the narrow space between the gums and the teeth. *Treponema denticola* is a major oral treponeme and ferments amino acids such as cysteine and serine, forming acetate as the major fermentation acid, as well as CO_2, NH_3 and H_2S.

Metabolically unusual species of the genus *Treponema* have been isolated from the hindgut of the termite. This environment is highly cellulolytic, as termites live mostly on wood and wood products. H_2 and CO_2 is produced from the fermentation of glucose released from the cellulose. *Treponema primitia* converts this H_2 plus CO_2 to acetate, that is, it is a *homoacetogen* (∞ Section 17.16 for a discussion of homoacetogenesis). This is the first instance of this form of energy metabolism in a phylogenetic group of *Bacteria* outside of the clostridia and their relatives (see Section 12.20). The hindgut spirochete *Treponema azotonutricium* fixes molecular nitrogen (N_2 fixation, ∞Section 17.28), a property not previously found in spirochetes, as well.

Spirochetes are also found in the rumen, the digestive organ of ruminant animals (∞Section 19.11). *Treponema saccharophilum* (Figure 12.98●) is a large, *pectinolytic* spirochete found in the bovine rumen. *Treponema saccharophilum* is an obligate anaerobe that ferments pectin, starch, inulin, and other plant polysaccharides. This and other spirochetes may play an important role in the conversion of plant polysaccharides to volatile fatty acids, usable as energy sources by the ruminant (∞Section 19.11). Although the genus *Treponema* is phylogenetically a unit, the true relationship between *T. pallidum* and other *Treponema* species may be a distant one. This is because the GC base ratio of *T. pallidum* genomic DNA is about 53%, whereas other species of this genus cluster between 38 and 40%, or near 25%.

(a)

(b)

● **Figure 12.98** ***Treponema* and *Borrelia*.** (a) Phase-contrast photomicrographs of *Treponema saccharophilum*, a large pectinolytic spirochete from the bovine rumen. A cell measures about 0.4 μm in diameter. Left, regularly coiled cells; right, irregularly coiled cells. (b) A scanning electron micrograph of *Borrelia burgdorferii*, the causative agent of Lyme disease.

Borrelia

The majority of species in the genus *Borrelia* are animal or human pathogens. *Borrelia recurrentis* is the causative agent of **relapsing fever** in humans and is transmitted via an insect vector, usually by the human body louse. Relapsing fever is characterized by a high fever and generalized muscular pain, which lasts for 3–7 days followed by a recovery period of 7–9 days. Left untreated, the fever returns in two to three more cycles (hence the name, relapsing fever) and causes death in up to 40% of those infected. Fortunately, the organism is quite sensitive to tetracycline, and if the disease is correctly diagnosed, treatment is straightforward. Other borrelia are of veterinary importance, causing diseases in cattle, sheep, horses, and birds. In most of these diseases the organism is transmitted by ticks.

Borrelia burgdorferi (Figure 12.98b) is the causative agent of the tick-borne disease called *Lyme disease*, which infects humans and animals. Lyme disease is discussed in Section 27.4. *Borrelia burgdorferi* is also of interest because it is as yet one of the few known prokaryotes that has a *linear* (as opposed to a *circular*) chromosome. The rather small genome of *B. burgdorferi* (1.44 Mbp) has been completely sequenced (∞Section 15.3).

Leptospira and Leptonema

The genera *Leptospira* and *Leptonema* contain strictly *aerobic* spirochetes that use long-chain fatty acids (for example, oleic acid) as electron donor and carbon sources. With few exceptions, these are the only substrates utilized by leptospiras for growth. The leptospira cell is thin, finely coiled, and usually bent at each end into a semicircular hook. At present, several species are recognized in this group, some free-living and many parasitic. Two major species are *Leptospira interrogans* (parasitic) and *L. biflexa* (free-living). Strains of *L. interrogans* are parasitic for humans and animals. Rodents are the natural hosts of most leptospiras, although dogs and pigs are also important carriers of certain strains.

In humans the most common leptospiral syndrome is **leptospirosis**, a disorder in which the organism localizes in the kidney and can cause renal failure and death. Leptospiras ordinarily enter the body through the mucous membranes or through breaks in the skin. After a transient multiplication in various parts of the body, the organism localizes in the kidney and liver, causing nephritis and jaundice. The organism then passes out of the body in the urine, and infection of another individual is most commonly by contact with infected urine.

Therapy with penicillin, streptomycin, or the tetracyclines is possible, but it may take extended courses to eliminate the organism from the kidney. Domestic animals such as dogs are vaccinated against leptospirosis with a killed virulent strain in the combined distemper-leptospira-hepatitis vaccine. In humans, prevention is effected primarily by elimination of the disease from animals.

12.33 Concept Check

Spirochetes are tightly coiled, motile, helical prokaryotes that contain both free-living as well as pathogenic species.

◆ How do the endoflagella of spirochetes compare with the flagella of *Escherichia coli* in both structure and function?

◆ Name two diseases of humans caused by spirochetes.

◆ What is the habitat of the spirochete *Cristispira*?

◆ List two human diseases caused by spirochetes.

XII PHYLUM 12: DEINOCOCCI

12.34 Deinococcus/Thermus

Key Genera: *Deinococcus, Thermus*

This phylum of *Bacteria* contains only a few genera, the best studied being *Deinococcus* and *Thermus*. The latter genus contains thermophilic chemoorganotrophic bacteria including *Thermus aquaticus*, the organism from which *Taq DNA polymerase* is obtained. Because it is so heat stable, this enzyme is the major one used in the polymerase chain reaction (PCR) technique for amplifying DNA, as was discussed in Section 7.9.

Members of this phylum stain gram-negatively and contain a rare form of peptidoglycan in which ornithine is present in place of diaminopimelic acid in the muramic acid cross-bridges (∞Section 4.8). A number of species of *Thermus* and *Deinococcus* have been described, and all grow aerobically by catabolism of sugars, amino and organic acids, or various complex mixtures. We focus the rest of the discussion here on *Deinococcus*.

The genus *Deinococcus* contains four species of gram-positive cocci; *D. radiodurans* is the best-studied species. The *D. radiodurans* cell wall is structurally complex and consists of several layers including an outer membrane (Figure 12.99●), normally present only in gram-negative bacteria (∞Section 4.9). However, unlike the outer membrane of gram-negative bacteria such as *E. coli*, the outer membrane of *D. radiodurans* lacks lipid A. The organism is found in soil and can also be isolated from dust particles.

Radiation Resistance of *Deinococcus radiodurans*

Most deinococci are red or pink in color due to carotenoids, and many strains are highly resistant to UV radiation and to desiccation. Resistance to radiation can be used to advantage in isolating deinococci. These remarkable organisms can be isolated from soil, ground meat, dust, and filtered air following exposure of the sample to intense UV (or even gamma) radiation and plating on a rich medium containing tryptone and yeast extract. Because many strains of *Deinococcus radiodurans* are even more resistant to radiation than bacterial endospores, treatment of a sample with strong doses of radiation effectively sterilizes the sample of organisms other than *D. radiodurans*, making isolation of deinococci relatively straightforward. For example, *D. radiodurans* cells can survive exposure to up to 15,000 grays (Gy) of ionizing radiation (1 Gy = 100 rad). This is sufficient to shatter the organism's chromosome into hundreds of fragments (by contrast, a human can be killed by exposure to less than 10 Gy).

In addition to impressive radiation resistance, *D. radiodurans* is resistant to the mutagenic effects of many other mutagenic agents. The only chemical mutagens that seem to work on *D. radiodurans* are agents like nitrosoguanidine, which tend to induce *deletions* in DNA. Deletions are apparently not repaired as efficiently as point mutations in this organism, and mutants of *D. radiodurans* can be isolated in this way.

Studies of *D. radiodurans* have shown it to be highly efficient in repairing damaged DNA. Several different DNA repair enzymes exist in *D. radiodurans*. In addition

(a)

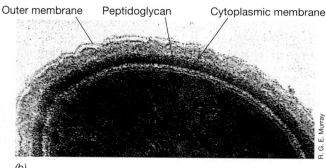

Outer membrane   Peptidoglycan   Cytoplasmic membrane

(b)

(c)

● **Figure 12.99   The radiation-resistant coccus *Deinococcus radiodurans.*** An individual cell is about 2.5 μm in diameter. (a) Transmission electron micrograph of *D. radiodurans.* Note the outer membrane layer. (b) High-magnification micrograph of wall layer. (c) Transmission electron micrograph of cells of *D. radiodurans* colored to show the toroidal morphology of the nucleoid (green).

to the DNA repair enzyme RecA (○○ Section 10.6), several RecA-independent DNA systems exist in *D. radiodurans* that can repair breaks in single- or double-stranded DNA and excise and repair misincorporated bases. In fact, repair processes are so effective, that the chromosome can even be reassembled from a fragmented state.

It is also thought that the unique arrangement of DNA in *D. radiodurans* cells plays a role in radiation resistance. Cells of *D. radiodurans* always exist as pairs or tetrads (Figure 12.99*a*). Instead of scattering DNA within the cell as in a typical nucleoid, DNA in *D. radiodurans* is ordered into a toroidal ringlike structure (Figure 12.99*c*). Repair is then facilitated by the fusion of nucleoids from adjacent compartments, as this provides a platform for homologous recombination. From this extensive recombination, a single repaired chromosome emerges, and the cell containing this chromosome can then grow and divide.

XIII PHYLUM 13: THE GREEN NONSULFUR BACTERIA

This phylum of *Bacteria* is phylogenetically distinct and contains just a few genera, the best known being the anoxygenic phototroph *Chloroflexus;* all cultured representatives are thermophilic. *Thermomicrobium* is a chemotrophic member of this group and is a strictly aerobic, gram-negative rod, growing optimally in complex media at 75°C. Besides its phylogenetic novelty, *Thermomicrobium* is also of interest because of its membrane lipids. Recall that the lipids of *Bacteria* and *Eukarya* contain fatty acids esterified to glycerol (○○ Sections 3.4 and 4.5). By contrast, the lipids of *Thermomicrobium* contain 1,2-dialcohols instead of glycerol, and have neither ester *nor* ether linkages (Figure 12.100●). In addition, cells of *Thermomicrobium* are unusual for species of *Bacteria* in that they lack peptidoglycan, the signature cell-wall polymer of this domain of prokaryotes (○○Section 11.9).

12.35   *Chloroflexus* and Relatives

Key Genera: *Chloroflexus, Heliothrix, Roseiflexus*

Chloroflexus and most other *green nonsulfur bacteria* (so named because of the physiology of *Chloroflexus,* see later) are filamentous prokaryotes that, along with cyanobacteria, form thick microbial mats in neutral to alkaline hot springs (Figure 12.101●; see also Figure 18.18*a*). *Chloroflexus*-like organisms have also been found in marine microbial mats. Although an anoxygenic phototroph, *Chloroflexus* is a "hybrid" phototroph in the sense that its photosynthetic mechanism shows

(a)

(b)

● **Figure 12.100   The unusual lipids of *Thermomicrobium.*** (a) Membrane lipids from *T. roseum* contain long chain diols like the one shown here (13-methyl-1,2 nonadecanediol). Note that unlike the lipids of other *Bacteria* or of *Archaea* (◯◯Section 4.5), neither ester- nor ether-linked side chains are present. (b) To form a bilayer membrane, dialcohol molecules presumably oppose each other at the methyl groups, with the OH groups being the inner and outer hydrophilic surfaces. Small amounts of the diols have fatty acids esterified to the secondary —OH group (shown in red) while the primary —OH group (shown in green) can bond a hydrophilic molecule like phosphate.

(a)   (b)

(c)   (d)

● **Figure 12.101   Green nonsulfur bacteria.** (a) Phase photo- micrograph of the thermophilic phototroph, *Chloroflexus aurantiacus.* Cells are about 1 μm in diameter. (b) Phase micrograph of the large phototroph *Oscillochloris.* Cells are about 5 μm wide. The brightly contrasting material is a holdfast, used for attachment. (c) Color photo- micrograph of filaments of *Chloronema species* growing in a stratified Michigan lake. These cells of *Chloronema* are wavy filaments and about 2.5 μm in diameter. (d) Tube cultures of *Chloroflexus aurantiacus* (right) and *Roseiflexus* (left). *Roseiflexus* lacks bacteriochlorophyll *c* and chlorosomes (◯◯Section 17.2 and Figure 17.7) and so lacks the green color of *Chloroflexus.*

features characteristic of both purple bacteria and green sulfur bacteria. Like the latter, *Chloroflexus* con- tains bacteriochlorophyll *c* and chlorosomes (see Figure 12.92 for an electron micrograph of chlorosomes). How- ever, the bacteriochlorophyll *a* located in the cytoplas- mic membrane of cells of *Chloroflexus* is arranged to form a photosynthetic reaction center structurally simi- lar to those of purple bacteria (by contrast, the reaction center of green sulfur bacteria is structurally quite dif- ferent, ◯◯Figure 17.18).

Because of its phylogeny and unique mechanism of autotrophy (see below), it is possible that *Chloroflexus* may be a vestige of an early form of phototroph that perhaps first evolved a photosynthetic reaction center and then received chlorosome-specific genes later by lateral transfer. From a phylogenetic standpoint, *Chlor- oflexus* is clearly the earliest known phototrophic bac- terium (Figure 12.1).

Physiology of *Chloroflexus*

Physiologically, *Chloroflexus* resembles purple nonsulfur bacteria in that photoautotrophy can occur; however, phototrophic growth occurs best when organic com- pounds are added as carbon sources (*photoheterotrophy*). *Chloroflexus* also grows well in the dark as a chemo- organotroph by aerobic respiration. Interestingly, and this should be considered in light of the evolutionary po- sition of *Chloroflexus* as the most phylogenetically an- cient of anoxygenic phototrophs (see Figure 12.1), autotrophy in *Chloroflexus* is based on a CO$_2$ incorpora-

tion pathway, the *hydroxypropionate pathway*, unique to this and a few other phylogenetically "ancient" organ- isms. We consider the biochemistry of this novel au- totrophic pathway in Section 17.7.

Other Green Nonsulfur Bacteria

In addition to *Chloroflexus*, other phototrophic green nonsulfur bacteria include the thermophile *Heliothrix* and the large-celled mesophiles *Oscillochloris* (Figure 12.101*b*) and *Chloronema* (Figure 12.101*c*). *Oscillochloris* and *Chloronema* are unusual because they are rather large cells, 2–5 μm wide and can be up to several hundred micrometers long (Figure 12.101c). Both organisms inhabit freshwater lakes containing low levels of sulfide, together with other species of green sulfur and purple sulfur bacteria.

Two genera in the green nonsulfur phylum, *Roseiflexus* and *Heliothrix*, differ quite dramatically from *Chloroflexus*. These phototrophs lack bacteriochlorophyll *c* and chlorosomes, and so more closely resemble phototrophic purple bacteria (∞ Section 12.2) than *Chloroflexus*. Despite this clear difference in pigment complement, *Chloroflexus*, *Roseiflexus*, and *Heliothrix* share many other things in common. These include a filamentous morphology and thermophilic lifestyle. And if *Chloroflexus* obtained chlorosomes and bacteriochlorophyll *c* biosynthesis genes by lateral transfer from green sulfur bacteria, as is suspected, it is possible that *Chloroflexus* and *Roseiflexus* were at one point phenotypically indistinguishable. *Heliothrix* is not available in pure culture, but *Roseiflexus* is. Cells of *Roseiflexus* are filamentous, and cultures are yellow/orange in color from their extensive carotenoid pigments (Figure 12.101d).

 12.34–12.35 Concept Check

Deinococcus and *Chloroflexus* are each key genera in separate major lineages of *Bacteria*. *Deinococcus radiodurans* is the most radiation resistant of all known organisms, and *Chloroflexus* is an anoxygenic phototroph that shows photosynthetic properties characteristic of both purple bacteria and green bacteria.

◆ How does *Deinococcus radiodurans* prevent being killed by high levels of radiation?

◆ In what ways do *Chloroflexus* and *Roseiflexus* resemble an organism like *Chlorobium*? An organism like *Rhodobacter*?

◆ What is unique about the organism *Thermomicrobium*?

XIV PHYLUM 14–16: DEEPLY BRANCHING HYPERTHERMOPHILIC BACTERIA

These three phyla of *Bacteria* cluster deep in the phylogenetic tree of *Bacteria*, near the hypothetical "root" (∞ Figures 11.16 and 12.1). Each phylum consists of one or two major genera, and a key physiological feature of most of them is *hyperthermophily*, that is, optimal growth at temperatures *above* 80°C (∞ Sections 6.10 and 6.12).

12.36 *Thermotoga* and *Thermodesulfobacterium*

Key Genera: *Thermotoga, Thermodesulfobacterium*

Thermotoga is a rod-shaped hyperthermophile capable of growth to 90°C (optimum, 80°C). Cells of *Thermotoga* contain a sheath-like envelope (the "toga," see Figure 12.102a●), stain gram-negatively, and are nonsporulating. *Thermotoga* is an anaerobic, fermentative chemoorganotroph, catabolizing sugars and polymers such as starch, and producing lactate, acetate, CO_2, and H_2 as fermentation products. The organism can also grow by anaerobic respiration using H_2 as electron donor and ferric iron (Fe^{3+}) as electron acceptor. Species of *Thermotoga* have been isolated from terrestrial hot springs as well as marine hydrothermal vents.

The genome of *Thermotoga* has been completely sequenced, and interestingly, this organism contains many archaeal genes. In fact, over 20% of the genes of *Thermotoga* probably originated from hyperthermophilic species of *Archaea* by horizontal transfers (∞ Section 15.8). In its very hot habitat where many *Archaea* coexist with it, it is hypothesized that *Thermotoga* underwent extensive lateral gene exchange with species of *Archaea*, perhaps picking up several genes useful for survival under such extreme conditions. Although a few archaeal-like genes have been identified in the genomes of

(a)

(b)

● **Figure 12.102 Hyperthermophilic *Bacteria*.** (a) *Thermotoga maritima*—temperature optimum, 80°C. Note the outer covering on the cell (the "toga"). (b) *Aquifex pyrophilus*—temperature optimum, 85°C. Cells of *Thermotoga* (thin section) measure 0.6 × 3.5 μm; cells of *Aquifex* (freeze-fracture micrograph) measure 0.5 × 2.5 μm. Both *Thermotoga* and *Aquifex* form their own phylogenetic lineages on the tree of *Bacteria* (see Figure 12.1).

other *Bacteria* and vice versa, thus far, only in *Thermotoga* do we see such large-scale lateral transfer of genes between domains.

Thermodesulfobacterium

Thermodesulfobacterium (Figure 12.103*a*●) is a thermophilic, sulfate-reducing bacterium, positioned on the phylogenetic tree as a separate phylum between *Thermotoga* and *Aquifex* (Figure 12.1). Although not a true hyperthermophile because its growth temperature optimum is only 70°C, *Thermodesulfobacterium* is the most thermophilic of all known sulfate-reducing *Bacteria* (the archaeon sulfate-reducer *Archaeoglobus* is a true hyperthermophile, ∞ Section 13.7). Like other "group I" sulfate reducers (see Section 12.18), *Thermodesulfobacterium*, a strict anaerobe, cannot utilize acetate as an electron donor in its energy metabolism and instead uses compounds like lactate, pyruvate, and ethanol, reducing SO_4^{2-} to H_2S.

An unusual biochemical feature of species of *Thermodesulfobacterium* is the presence of *ether-linked lipids*. Recall that the latter are a hallmark of the *Archaea* and that a poly-isoprenoid C_{20} hydrocarbon (phytanyl) replaces fatty acids as the side chains in archaeal lipids (∞Sections 4.5 and 11.9). The ether-linked lipids in *Thermodesulfobacterium* are unusual because the glycerol side chains are not phytanyl groups, as they are in *Archaea* (∞ Section 4.5), but instead are composed of a unique C_{17} hydrocarbon along with some fatty acids (Figure 12.103*b*).

(a)

(b)

● **Figure 12.103 *Thermodesulfobacterium*.** (a) Phase photomicrograph of cells of *T. mobile*. (b) Structure of one of the lipids of *T. mobile*. Note that although these are ether-linked, the two hydrophobic side chains are *not* phytanyl units as in *Archaea* (∞Section 4.5). The designation "R" is for a hydrophilic residue, such as a phosphate group.

Thus we see in *Thermodesulfobacterium* both a deep phylogenetic lineage (Figure 12.1) and a lipid profile that combines features of both the *Archaea* and the *Bacteria* (Figure 12.103*b*). Like in *Thermotoga*, perhaps this is another case of lateral gene flow from *Archaea* to *Bacteria*. However, the bacterium *Ammonifex*, a thermophilic member of the low GC, gram-positive *Bacteria* that grows anaerobically by H_2 oxidation coupled to the reduction of NO_3^- to NH_3, also contains lipids like those of *Thermodesulfobacterium*. Thus, ether-linked lipids may be more common among *Bacteria* than previously thought.

12.37 *Aquifex, Thermocrinis,* and Relatives

Key Genera: *Aquifex, Thermocrinis*

The genus *Aquifex* (Figure 12.102b) is an obligately chemolithotrophic and autotrophic hyperthermophile and is the most thermophilic of all known *Bacteria*. Various *Aquifex* species utilize H_2, S^0, or $S_2O_3^{2-}$ as electron donors and O_2 or NO_3^- as electron acceptors; growth occurs up to 95°C (optimum, 85°C). Only very low O_2 concentrations are tolerated by *Aquifex*, but it remains (along with a few other species of *Archaea*, ∞Sections 13.5 and 13.9) one of the few aerobic (or more exactly, microaerophilic) hyperthermophiles known. Nutritional studies of *Aquifex* species have shown them totally unable to grow chemoorganotrophically on organic compounds, including complex mixtures like yeast or meat extract. *Hydrogenobacter*, a relative of *Aquifex*, shows most of the same properties as *Aquifex* but is an obligate aerobe.

Autotrophy in *Aquifex* is supported by enzymes of the reverse citric acid cycle, a series of reactions previously found only in green sulfur bacteria (see Sections 12.32 and 17.7) within the domain *Bacteria*. The complete genome sequence of *Aquifex aeolicus* has been determined (∞Section 15.3), and its entirely chemolithotrophic/autotrophic lifestyle is supported by an amazingly small genome of only 1.55 Mbp (one-third the size of the *E. coli* genome). This fact was previously discussed in Section 11.2 with regards to the physiological properties of early life forms.

The discovery that so many hyperthermophiles, both species of *Archaea* and *Bacteria* like *Aquifex*, are H_2 chemolithotrophs, coupled with the finding that they branch as very early lineages on their respective phylogenetic trees (∞Figures 11.13, 12.1, and 13.1), suggests that H_2 was a key electron donor for primitive organisms under early Earth conditions (∞Section 13.13). A simple model for the utilization of H_2 to generate a proton motive force and hence, usable energy as ATP, was shown in Figure 11.6.

Thermocrinis

Thermocrinis (Figure 12.104●) is an interesting relative of *Aquifex* and *Hydrogenobacter*. This organism is a hyperthermophilic (temperature optimum, 80°C) chemolithotroph capable of oxidizing either H_2, thiosulfate, or sulfur as electron donors, with O_2 as electron acceptor.

Thermocrinis ruber, the only known species, grows in the outflow of certain hot springs in Yellowstone National Park, where it forms pink "streamers" consisting of a filamentous form of the cells attached to siliceous sinter (Figure 12.104*a*). In static culture, cells of *T. ruber* grow as individual rod-shaped cells (Figure 12.104*b*). However, when cultured in a flowing system in which growth medium is trickled over a solid glass surface to which cells can attach, *Thermocrinis* assumes the streamer morphology, just as it exists in its constantly flowing habitat in nature (Figure 12.104*a*).

Thermocrinis ruber is of historical significance in microbiology because it was one of the organisms studied in the 1960s by Thomas Brock, a pioneer in the field of thermal biology. The discovery by Brock that the pink streamers (Figure 12.104*a*) contained protein and nucleic acids clearly indicated that they were living organisms and not just mineral debris. Moreover, the presence of streamers in the outflow of hot springs at 80–90°C but not at lower temperatures supported Brock's hypothesis that hot spring microorganisms actually *required* high temperatures for growth and were likely to be present in even boiling water. Both of these conclusions were subsequently supported by the discovery by Brock and others of literally dozens of genera of hyperthermophilic prokaryotes inhabiting hot springs, hydrothermal vents, and other thermal environments. For more coverage of hyperthermophiles ∞Sections 6.10, 6.12, 13.4–13.13, and 19.8.

(a)

(b)

● **Figure 12.104   *Thermocrinis.*** (a) Cells of *Thermocrinis ruber* growing as filamentous streamers (arrow) attached to siliceous sinter in the outflow (85°C) of Octopus Spring, Lower Geyser Basin, Yellowstone National Park. The pink color is due to a carotenoid pigment. (b) Scanning electron micrograph of rod-shaped cells of *T. ruber* grown on a silicon-coated cover glass. The hairlike structures are silicon. A single cell of *T. ruber* is about 0.4 μm in diameter and from 1 to 3 μm long.

XV PHYLUM 17 AND 18: NITROSPIRA AND DEFERRIBACTER

12.38 *Nitrospira, Deferribacter,* and Relatives

Key Genera: *Nitrospira, Deferribacter*

Several other phyla of *Bacteria* have been identified by ribosomal RNA sequencing of cultures, although relatively little is known about them. Two such phyla contain the organisms *Nitrospira* and *Deferribacter* (Figure 12.1). Physiologically, these organisms are either chemolithotrophs or chemoorganotrophs and are mesophiles to thermophiles.

Like nitrifying Proteobacteria (see Sections 12.3 and 17.12), *Nitrospira* oxidizes NO_2^- to NO_3^- and grows autotrophically. Despite this close physiological relationship to the classical nitrifying bacteria, *Nitrospira* is phylogenetically quite distinct from them. Also, *Nitrospira* lacks the extensive internal membranes found in species of nitrifying Proteobacteria (see Figures 12.7 and 12.8). Nevertheless, *Nitrospira* inhabits many of the same environments as nitrifying Proteobacteria, so it has been suggested that its physiological capacities may have been transferred to it by horizontal gene flow from nitrifying Proteobacteria (or vice versa). As we know, this mechanism for acquiring physiological traits has been widely exploited in the prokaryotic world (∞Section 10.6–10.12).

Other genera in the *Nitrospira* group include *Leptospirillum*, an iron-oxidizing chemolithotroph responsible for much of the acid mine drainage associated with the mining of coal and iron (∞Section 19.15), and *Thermodesulfovibrio*, a thermophilic sulfate-reducing bacterium that inhabits hot spring microbial mats (∞Section 18.7). Note that although similar in name and physiology, *Thermodesulfovibrio* and *Thermodesulfobacterium* (see Section 12.36) are distinct organisms.

Deferribacter

The genus *Deferribacter* forms its own distinct lineage (Figure 12.1) and is composed of species that specialize in *anaerobic* energy metabolism. Other genera in this group include *Geovibrio* and *Flexistipes*; the latter genus is an obligately anaerobic and fermentative bacterium.

Deferribacter and *Geovibrio* are extremely versatile at anaerobic respiration using a variety of electron acceptors, including the metals Fe^{3+} and Mn^{4+}. We discuss anaerobic respiration in Chapter 17 and show that the process can be linked to many different terminal electron acceptors. Members of the *Deferribacter* group appear to be unusual in the vast number of alternative electron acceptors they can use and in the fact that they are obligate anaerobes. By contrast, most organisms capable of growing by anaerobic respiration with nitrate or metals as electron acceptors are facultative aerobes, able to grow fully aerobically as well as by anaerobic respiration (∞Section 17.13).

12.36–12.38 Concept Check

Thermotoga, Thermodesulfobacterium, and *Aquifex* grow at high temperature, and each spearhead a major lineage of *Bacteria. Aquifex* and *Thermocrinis* are H_2-oxidizing chemolithotrophs, while *Thermotoga* and *Thermodesulfobacterium* are both anaerobic chemoorganotrophs. *Nitrospira* and *Deferribacter* each form their own phylum.

◆ Compare the catabolic metabolism of *Thermotoga* and *Thermodesulfobacterium*.

◆ What is unusual about the lipids of *Thermodesulfobacterium*?

◆ From a genomic perspective, why does the fact that *Aquifex* is able to grow on the gases $H_2 + CO_2 + O_2$ seem surprising?

◆ Contrast the metabolic features of *Nitrospira* and *Deferribacter*.

REVIEW QUESTIONS

1. Of all of the phyla of *Bacteria* studied in this chapter, which has groups with the most diverse physiologies? Give examples of this in terms of O_2 requirements, carbon sources used, and energy metabolism.

2. List examples of how the Gram reaction has phylogenetic predictive value.

3. What do cyanobacteria have in common with prochlorophytes? With chloroplasts? How are all of these thought to be phylogenetically related?

4. What do species in the *Planctomyces* phylum share in common with *Archaea*? With eukaryotic cells?

5. In what ways are *Chlorobium* and *Chloroflexus* similar? In what ways do they differ?

6. Describe a key *physiological* feature of the following *Bacteria* that would differentiate each from the others: *Acetobacter, Methylococcus, Azotobacter, Desulfovibrio, Lactobacillus, Nitrobacter, Oscillatoria.*

7. Describe a key *morphological* feature that would differentiate each of the following *Bacteria: Streptococcus, Spirillum, Streptomyces, Verrucomicrobium,* and *Spirochaeta.*

8. What key features could be used to differentiate the following genera of gram-positive *Bacteria: Bacillus, Mycoplasma,* and *Mycobacterium?*

9. What traits do the chlamydia and the rickettsias share in common? In what ways do they differ?

10. What major physiological property unites species of *Thermotoga, Aquifex,* and *Thermocrinis?*

11. Compare and contrast the metabolism, morphology, and phylogeny of purple nonsulfur and green nonsulfur bacteria.

12. List an electron donor for energy metabolism for each of the following *Bacteria* and state whether the organism is an aerobe or an anaerobe: *Thiobacillus, Nitrosomonas, Ralstonia eutropha, Methylomonas, Acetobacter, Gallionella,* and *Propionibacterium.*

APPLICATION QUESTIONS

1. Defend the following statement using phylogenetic, structural, and physiological arguments: "*Escherichia coli* is a much more evolved bacterium than *Thermodesulfobacterium*."

2. Defend or refute the following statements: using examples from this chapter: (a) "Cell morphology has absolutely no phylogenetic predictive value." (b) "Physiology has absolutely no phylogenetic value."

13

PROKARYOTIC DIVERSITY: THE *ARCHAEA*

Species of *Archaea* include even hyperthermophilic parasites such as *Nanoarchaeum* (red cells). *Nanoarchaeum* contains the smallest genome of all known prokaryotes.

WORKING GLOSSARY

Acetotrophic acetate consuming

Acetyl-CoA pathway a pathway of autotrophic CO_2 fixation widespread in obligate anaerobes including methanogens, homoacetogens, and sulfate-reducing bacteria

Bacteriorhodopsin a membrane protein containing retinal produced by certain extreme halophiles and capable of light-mediated proton motive force formation

Compatible solutes an organic or inorganic substance accumulated in the cytoplasm of a halophilic organism in order to maintain osmotic pressure

Crenarchaeota a phylum of *Archaea* that contains both hyperthermophilic and cold-dwelling prokaryotes

Euryarchaeota a phylum of *Archaea* that contains primarily methanogens, the extreme halophiles, *Thermoplasma*, and some marine species

Extreme halophile an organism whose growth is dependent on large concentrations (generally 10% or more) of NaCl

Halorhodopsin a light-driven chloride pump that accumulates Cl^- within the cytoplasm

Hydrothermal vent a deep sea hot spring emitting warm (~20°C) to superheated (>300°C) water

Hyperthermophile a prokaryote with a growth temperature optimum of 80°C or greater

Korarchaeota a phylum of hyperthermophilic *Archaea* that branches closest to the archaeal root

Methanogen a methane-producing prokaryote; CH_4 is produced by either reduction of CO_2 with H_2 or from certain organic compounds

Nanoarchaeota a phylum of very small, parasitic *Archaea* that branches closest to the archaeal root

Phytanyl a branched-chain hydrocarbon containing 20 carbon atoms and commonly found in the lipids of *Archaea*

Reverse DNA gyrase a protein universally present in hyperthermophiles that introduces positive supercoils into circular DNA

Solfatara a hot, sulfur-rich, generally acidic environment commonly inhabited by hyperthermophilic *Archaea*

Thermosome a heat-shock (chaperonin) protein complex that functions to refold partially heat-denatured proteins in hyperthermophiles

W e now consider the domain *Archaea*. In Chapter 11 we emphasized the profound phylogenetic as well as phenotypic differences between *Bacteria* and *Archaea*. Here we examine the organisms themselves. Some major characteristics of the *Archaea* (described in Table 11.3) include the absence of peptidoglycan in cell walls, and the presence of ether-linked lipids and complex RNA polymerases. But beyond this, *Archaea* show enormous phenotypic diversity, and we consider this here. As in Chapter 12 on *Bacteria*, we begin this chapter on *Archaea* with a phylogenetic overview to show the evolutionary relationships within the domain.

16S ribosomal RNA-based tree of life (∞ Figures 11.13 and 13.1). The latter suggests that these organisms are more slowly evolving than other lineages in the domain. Such organisms should thus be good models for "early" *Archaea* and perhaps early life forms in general; we return to this theme at the end of this chapter (see Section 13.13). By contrast, cold-dwelling relatives of hyperthermophilic crenarchaeotes have been identified by community sampling of ribosomal RNA (rRNA) genes (∞ Sections 18.5 and 19.6) from *Archaea* in the oceans. From a phylogenetic perspective these are more rapidly evolving species and thus occupy longer branches on the tree (Figure 13.1). We consider the Crenarchaeota in more detail in Sections 13.8–13.10.

Euryarchaeota comprise a physiologically diverse group of *Archaea*. Like crenarchaeotes, many inhabit extreme environments of one sort or the other. This phylum includes methanogenic *Archaea*—organisms whose metabolism is linked to the production of methane (CH_4)—and several genera of extremely halophilic *Archaea*, the "halobacteria" (Figure 13.1). As a study in physiological contrasts, these two groups are remarkable. Methanogens are the strictest of anaerobes, while extreme halophiles are for the most part obligate aerobes.

Other groups of euryarchaeotes include the hyperthermophiles *Thermococcus* and *Pyrococcus* and the methanogen *Methanopyrus*, all of which branch near the root (Figure 13.1) and the cell wall-less prokaryote *Thermoplasma*, an organism phenotypically similar to the mycoplasmas (∞ Section 12.21). Finally, in parallel to the Crenarchaeota, a large group of as yet uncultured marine euryarchaeotes forms long branches near the top of the archaeal tree (Figure 13.1). Many euryarchaeotes also inhabit freshwater and terrestrial habitats. We consider Euryarchaeota in more detail in Sections 13.3–13.7.

PHYLOGENY AND GENERAL METABOLISM

13.1 Phylogenetic Overview of the *Archaea*

A phylogenetic tree of *Archaea* is shown in Figure 13.1●. The tree splits into two major phyla called the *Crenarchaeota* and the *Euryarchaeota*. Two other phyla, the *Korarchaeota* and the *Nanoarchaeota*, branch off close to the root (Figure 13.1). Among cultured *Archaea*, the Crenarchaeota contain mostly hyperthermophilic species—organisms whose growth temperature optimum is greater than 80°C—including those able to grow at the highest temperatures of all known organisms. Many hyperthermophiles are chemolithotrophic autotrophs, and because their habitats are devoid of phototrophic life, these organisms are the only primary producers in these harsh environments.

Hyperthermophilic species of **Crenarchaeota** tend to cluster closely together and occupy short branches on the

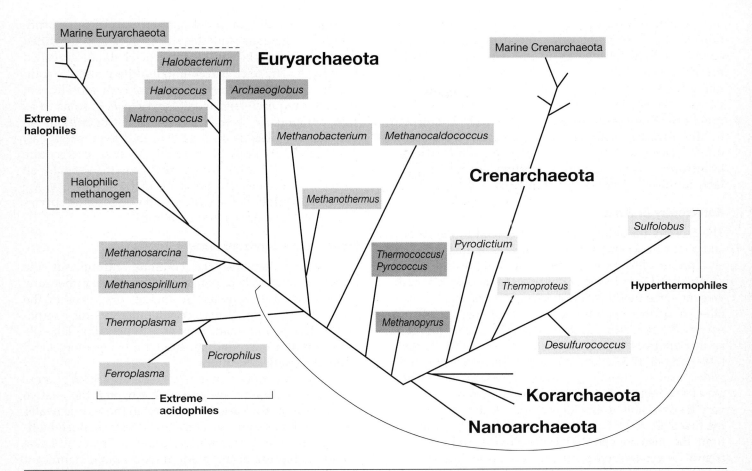

● Figure 13.1 **Detailed phylogenetic tree of the *Archaea* based on 16S ribosomal RNA gene sequence comparisons.** The marine Euryarchaeota and marine Crenarchaeota are known thus far only from community sampling (⚬➣⚬ Section 11.7, 18.5, and 18.6). Not all *Methanobacterium* species are hyperthermophiles.

The **Korarchaeota** (Figure 13.1) were originally discovered by community sampling of rRNA genes from organisms inhabiting an unusual Yellowstone hot spring; no pure cultures of these organisms exist. Korarchaeota are not yet officially recognized in taxonomy, but the organisms clearly branch on the archaeal tree close to the root. For this reason their biological properties may mimic those of ancient *Archaea*.

The **Nanoarchaeota** are the latest addition to the domain *Archaea*. *Nanoarchaeum*, the only genus in this phylum, is a very small parasitic *prokaryote* that lives attached to cells of *Ignicoccus*, a crenarchaeote. The very small cell and genome size (⚬➣⚬ Table 15.1) of this organism coupled with its ancient phylogeny make it a very interesting organism indeed. We discuss the implications of this and the Nanoarchaeota in general in Section 13.11.

With this overview to the phylogeny of the *Archaea*, we proceed to a brief description of the metabolic features of *Archaea*. From there we will describe the major cultured representatives.

<table>
<tr><td>**13.2**</td><td>**Energy Conservation and Autotrophy in *Archaea***</td></tr>
</table>

Energy metabolism in methanogens, a major group of *Archaea*, is unlike that of any other group of prokaryotes,

Bacteria or *Archaea*. We reserve our discussion of the many novel features of methanogenesis for Chapter 17 and focus here on the metabolism of nonmethanogenic *Archaea*, which resembles more that of the *Bacteria*.

Chemoorganotrophy and Chemolithotrophy in *Archaea*

Several *Archaea* are chemoorganotrophic and thus use organic compounds as energy sources for growth. Catabolism of glucose in *Archaea* proceeds via slight modifications of the Entner–Doudoroff (⚬➣⚬ Section 12.7) or glycolytic (⚬➣⚬ Section 5.10) pathways. Oxidation of acetate to CO_2 in *Archaea* proceeds through the citric acid cycle (⚬➣⚬ Section 5.13) or some slight variation of this reaction series, or by the acetyl-CoA pathway (⚬➣⚬ Section 17.16). Little is known concerning biosynthesis of amino acids and other macromolecular precursors in *Archaea*, but presumably key monomers are produced from the central biosynthetic intermediates discussed previously for *Bacteria* (⚬➣⚬ Sections 5.15–5.17).

Electron transport chains including cytochromes of the *a*, *b*, and *c* types exist in some *Archaea*. Employing these and other electron carriers, chemoorganotrophic metabolism in most *Archaea* proceeds by introduction of electrons from organic electron donors into an electron transport chain, leading to the reduction of O_2, S^0, or some other electron acceptor. Concurrent with these events is the establishment of a proton motive force that drives adenosine triphosphate (ATP) synthesis through membrane-bound ATPases (∞ Section 5.12 for a description of ATPase function). Chemolithotrophy is also well established in the *Archaea*, with H_2 being a common electron donor (see Section 13.13). We examine chemolithotrophic metabolism of hyperthermophilic *Archaea* later in this chapter (see Section 13.8).

Autotrophy in *Archaea*

The capacity for autotrophy is widespread in the *Archaea* and occurs by several different pathways. In methanogens, and presumably in most chemolithotrophic hyperthermophiles, CO_2 is incorporated via the **acetyl-CoA pathway**, or some modification thereof (∞ Section 17.16). In other hyperthermophiles, CO_2 fixation occurs via the *reverse citric acid cycle*, a reaction series that also functions as the autotrophic pathway in the green sulfur bacteria (∞ Sections 12.32 and 17.7), or the *Calvin cycle*, the most widespread autotrophic pathway in *Bacteria* and eukaryotes (∞ Section 17.6). Genes encoding functional and very thermostable RubisCO enzymes (RubisCO catalyzes the first step in the Calvin cycle) have been characterized from the methanogen *Methanocaldococcus jannaschii* and from a *Pyrococcus* species, both hyperthermophiles.

We thus see that many of the catabolic and anabolic sequences in the *Archaea* are familiar ones from our study of these processes in various *Bacteria*, a good reminder that metabolism has a long evolutionary history. With this as background we now begin our study of the organismal diversity of this fascinating domain of life.

 13.1–13.2 Concept Check

Archaea form four major phyla, the Euryarchaeota, the Crenarchaeota, the Korarchaeota, and the Nanoarchaeota. With the exception of methanogenesis, bioenergetics and intermediary metabolism in species of *Archaea* are much the same as that of various species of *Bacteria*.

◆ In which phylum of *Archaea* do the majority of cultured hyperthermophiles reside?

◆ What autotrophic pathways are found in *Archaea*?

 PHYLUM EURYARCHAEOTA

13.3 Extremely Halophilic *Archaea*

Key Genera: *Halobacterium, Haloferax, Natronobacterium*

Extremely halophilic *Archaea* (Figure 13.1) are a diverse group that inhabits highly saline environments, such as

solar salt evaporation ponds, natural salt lakes, or artificial saline habitats and the surfaces of heavily salted foods like certain fish and meats. Such habitats are called *hypersaline*. The term **extreme halophile** is used to indicate that these organisms are not only halophilic, but that their requirement for salt is *very high*; in some cases this amounts to near saturation. By definition, an extreme halophile is an organism that requires at least 1.5 M (about 9%) NaCl for growth. Most species require 2–4 M NaCl (12–23%) for optimal growth. Virtually all extreme halophiles can grow at 5.5 M NaCl (32%, the limit of saturation for NaCl), although some species grow only very slowly at this salinity.

Hypersaline Environments

Hypersaline habitats are common throughout the world, but *extremely* hypersaline habitats are rather rare. Most such environments are in hot, dry areas of the world. Salt lakes can vary considerably in ionic composition. The predominant ions in a hypersaline lake depend on the surrounding topography, geology, and general climatic conditions.

Great Salt Lake in Utah (USA) (Figure 13.2*a*●), for example, is essentially concentrated seawater; the relative proportions of the various ions remain those of seawater although the overall concentration of ions is much higher. Sodium is the predominant cation in Great Salt Lake, whereas chloride is the predominant anion; significant levels of sulfate are also present at a slightly alkaline pH (Table 13.1). By contrast, another hypersaline basin, the Dead Sea, is relatively low in sodium but contains high levels of magnesium (Table 13.1).

The water chemistry of soda lakes resembles that of hypersaline lakes such as Great Salt Lake, but because high levels of *carbonate* minerals are present in the surrounding rocks, the pH of soda lakes is quite high. Waters of pH 10–12 are not uncommon in these environments (Table 13.1 and Figure 13.2*c*). In addition, Ca^{2+} and Mg^{2+} are virtually absent from soda lakes because they precipitate out at high pH and carbonate concentrations (Table 13.1).

Table 13.1	Ionic composition of some highly saline environments[a]		
	Concentration (g/l)		
Ion	**Great Salt Lake[b]**	**Dead Sea**	**Lake Zugm[c]**
Na^+	105	40.1	142
K^+	6.7	7.7	2.3
Mg^{2+}	11	44	<0.1
Ca^{2+}	0.3	17.2	<0.1
Cl^-	181	225	155
Br^-	0.2	5.3	—
SO_4^{2-}	27	0.5	23
HCO_3^-	0.7	0.2	67
pH	7.7	6.1	11

[a] For comparison, seawater contains (grams per liter): Na^+, 10.6; K^+, 0.38; Mg^{2+}, 1.27; Ca^{2+}, 0.4; Cl^-, 19; Br^-, 0.065; SO_4^{2-}, 2.65; HCO_3^-, 0.14; pH 7.8.
[b] See Figure 13.2*a*.
[c] Wadi El Natroun, Egypt (see Figure 13.2*c*).

● **Figure 13.2 Hypersaline habitats for halophilic *Archaea*.** (a) Great Salt Lake, Utah, a hypersaline lake in which the ratio of ions is similar to that in seawater but in which absolute concentrations of ions are about 10 times that of seawater. The green color is primarily from cells of the halophilic green alga, *Dunaliella salina*. (b) Aerial view near San Francisco Bay, California, of a series of seawater evaporating ponds where solar salt is prepared. The red–purple color is predominantly due to bacterioruberins and bacteriorhodopsin in cells of *Halobacterium*. (c) Lake Hamara, Wadi El Natroun, Egypt. A bloom of pigmented haloalkaliphiles is growing in this pH 10 soda lake. Note the deposits of trona ($NaHCO_3 \cdot Na_2CO_3 \cdot 2\ H_2O$) around the edge of the lake. (d) Scanning electron micrograph of halophilic prokaryotes including square bacteria present in a Spanish saltern.

The diverse chemistries of hypersaline habitats has selected for a large diversity of halophilic microorganisms. Some organisms are only known from one environment while others are widespread in several habitats. Moreover, despite what may seem like rather harsh conditions, salt lakes can be highly productive ecosystems (the term *productive* here means high levels of CO_2 fixation, autotrophy). *Archaea* are not the only microorganisms present. The eukaryotic alga *Dunaliella* is the major, if not sole, oxygenic phototroph in most salt lakes. In highly alkaline soda lakes where *Dunaliella* is absent, anoxygenic phototrophic purple bacteria of the genera *Ectothiorhodospira* and *Halorhodospira* (∞ Section 12.2) predominate. Organic matter originating from primary production by oxygenic or anoxygenic phototrophs then sets the stage for development of the extremely halophilic *Archaea*, all of which are chemoorganotrophic. In addition, a few extremely halophilic anaerobic chemoorganotrophic *Bacteria*, such as *Haloanaerobium* and *Halobacteroides*, thrive in such environments.

Marine salterns are also habitats for extremely halophilic prokaryotes. Marine salterns are small enclosed basins filled with seawater that are left to evaporate, yielding solar sea salt (Figure 13.2b, d). As salterns approach the minimum salinity limits for extreme halophiles, the waters turn a reddish purple color due to the massive growth—called a *bloom*—of halophilic *Archaea* (the red coloration apparent in Figures 13.2b and c comes from carotenoids and other pigments discussed later). Morphologically unusual *Archaea* are often present in salterns, including species with a square morphology (Figure 13.2d). Extreme halophiles have also been found in high-salt foods such as certain sausages, marine fish, and salted pork.

Taxonomy and Physiology of Extremely Halophilic *Archaea*

Table 13.2 lists the currently recognized species of extremely halophilic *Archaea*. 16S ribosomal RNA gene

Table 13.2	Some genera of extremely halophilic *Archaea*		
Genus	**Morphology**	**DNA (mol % GC)**	**Habitat**
Extreme Halophiles			
Halobacterium	Rods	66–71	Salted fish; hides; hypersaline lakes; salterns
Halorubrum	Rods	62–71	Dead Sea; salterns
Halobaculum	Rods	70	Dead Sea
Haloferax	Flattened discs	60–66	Dead Sea; salterns
Haloarcula	Irregular discs	63–65	Salt pools, Death Valley, CA; marine salterns
Halococcus	Cocci	59–66	Salted fish; salterns
Halogeometricum	Rods	59–60	Solar salterns
Haloterrigena	Rods, ovals	59–60	Saline soil
Haloalkaliphiles			
Natronobacterium	Rods	65	Highly saline soda lakes
Natrinema	Rods	70	Salted fish; hides
Natrialba	Rods	60–63	Soda lakes; beach sand
Natronomonas	Rods	61–64	Soda lakes
Natronococcus	Cocci	63–64	Soda lakes
Natronorubrum	Flattened cells	59–60	Soda lakes

sequencing and other criteria have been used to define 14 genera of extreme halophiles (Table 13.2). The extremely halophilic *Archaea* are frequently referred to collectively as "halobacteria," because the genus *Halobacterium* (Figure 13.3●) was the first in this group to be described and is still the best-studied representative of the group. *Natronobacterium*, *Natronomonas*, and their relatives dif-fer from other extreme halophiles in being extremely *alkaliphilic* as well as halophilic. As befits their soda lake habitat (see Table 13.1 and Figure 13.2*c*), growth of natronobacteria is optimal at very low Mg^{2+} concentrations and high pH (9–11).

Extremely halophilic *Archaea* stain gram negatively, reproduce by binary fission, and do not form resting

● **Figure 13.3 Electron micrographs of thin sections of the extreme halophile *Halobacterium salinarum*.** A cell is about 0.8 μm in diameter. (a) Longitudinal section of a dividing cell showing the nucleoids. (b) High-magnification electron micrograph showing the glycoprotein subunit structure of the cell wall.

stages or spores. Most halobacteria are nonmotile, but a few strains are weakly motile by flagellar means or are able to float from their gas vesicles. The genomic organization of *Halobacterium* and *Halococcus* is unusual in that large plasmids containing up to 25–30% of the total cellular DNA are present and the GC base ratio of these plasmids (57–60% GC) is significantly different from that of chromosomal DNA (66–68% GC). Plasmids from extreme halophiles are among the largest naturally occurring plasmids known (∞ Section 10.9).

Most species of extremely halophilic *Archaea* are obligate *aerobes*. Most halobacteria use amino acids or organic acids as energy sources and require a number of growth factors (mainly vitamins) for optimal growth. A few *Halobacterium* species oxidize carbohydrates, but this ability is relatively rare. Electron transport chains containing cytochromes of the *a*, *b*, and *c* types are present in *Halobacterium*, and energy is conserved during aerobic growth via a proton motive force arising from membrane-mediated chemiosmotic events. Some halophilic *Archaea* have been shown to grow anaerobically. Anoxic growth at the expense of sugar fermentation and by anaerobic respiration (∞ Section 17.13) linked to the reduction of nitrate or fumarate has been demonstrated in certain species.

Water Balance in Extreme Halophiles

Extremely halophilic *Archaea* require large amounts of sodium for growth. In the case of *Halobacterium*, where detailed salinity studies have been performed, the requirement for Na^+ cannot be satisfied by any other ion, even the chemically related ion K^+. However, cells of *Halobacterium* need *both* Na^+ and K^+ for growth, since each plays an important role in maintaining osmotic balance.

As we learned in Section 6.14, microorganisms must withstand the osmotic forces that accompany life. To do so in a high-solute environment, such as the NaCl-enriched habitats of *Halobacterium*, organisms must either accumulate or synthesize solutes *intracellularly*. These solutes are called **compatible solutes**. These compounds counteract the tendency of the cell to become dehydrated under conditions of high osmotic strength by placing the cell in positive water balance with its surroundings (⚙ Section 6.14). As a compatible solute, however, cells of *Halobacterium* do not employ organic compounds but instead pump large amounts of K⁺ from the environment into the cytoplasm. This ensures that the concentration of K⁺ *inside* the cell is even greater than the concentration of Na⁺ *outside* the cell (Table 13.3). This ionic condition maintains positive water balance.

The *Halobacterium* cell wall (Figure 13.3*b*) is composed of glycoprotein and stabilized by sodium ions. Sodium ions bind to the outer surface of the *Halobacterium* wall and are absolutely essential for maintaining cellular integrity. When insufficient Na⁺ is present, the cell wall breaks apart and the cell lyses. This is a consequence of the exceptionally high content of the *acidic* (negatively charged) amino acids aspartate and glutamate (⚙ Figure 3.12) in the glycoprotein of the *Halobacterium* cell wall. The negative charges contributed by the carboxyl groups of these amino acids are shielded by Na⁺. When Na⁺ is diluted away, the negatively charged parts of the proteins actively repel each other, leading to cell lysis.

Halophilic Cytoplasmic Components

Like cell wall proteins, cytoplasmic proteins of *Halobacterium* are also highly acidic, but it is K⁺, not Na⁺, that is required for activity. This, of course, makes sense, since K⁺ is the predominate internal cation in cells of *Halobacterium* (Table 13.3). Besides a high acidic amino acid composition, halobacterial cytoplasmic proteins typically contain very low levels of *hydrophobic* amino acids and lysine (a *basic* amino acid) compared with those of nonhalophiles. This also makes good sense since in a highly ionic cytoplasm, polar proteins would tend to remain in solution whereas nonpolar proteins would tend to cluster and perhaps lose activity. The ribosomes of *Halobacterium* also require high K⁺ levels for stability, while ribosomes of nonhalophiles have no K⁺ requirement.

Extremely halophilic *Archaea* are thus well adapted, both internally and externally, to life in a highly ionic environment. Cellular components exposed to the external environment require high Na⁺ for stability, whereas internal components require high K⁺. With the exception of a few extremely halophilic members of the *Bacteria* that also use K⁺ as a compatible solute, in no other group of prokaryotes do we find this unique requirement for such high amounts of specific cations.

Bacteriorhodopsin and Light–Mediated ATP Synthesis in Halobacteria

Certain species of extremely halophilic *Archaea* can carry out a *light-driven* synthesis of ATP. This occurs without chlorophyll pigments, so it is not photosynthesis. However, other light-sensitive pigments are present, including red and orange carotenoids—primarily C₅₀ pigments called *bacterioruberins*—and inducible pigments involved in energy conservation as discussed now.

Under conditions of low aeration, *Halobacterium salinarum* and certain other extreme halophiles synthesize and insert a protein called **bacteriorhodopsin** into their membranes. Bacteriorhodopsin was so named because of its structural and functional similarity to the visual pigment of the eye, *rhodopsin*. Conjugated to bacteriorhodopsin is a molecule of *retinal*, a carotenoid-like molecule that can absorb light energy and catalyze formation of a proton motive force (Figure 13.4●). The retinal gives bacteriorhodopsin a purple hue. Cells of *Halobacterium* switched from growth under conditions of high aeration to oxygen-limiting conditions (a trigger of bacteriorhodopsin synthesis) gradually change from an orange or red color to a more reddish-purple as they synthesize bacteriorhodopsin and insert it into the cytoplasmic membrane.

● **Figure 13.4 Model for the mechanism of bacteriorhodopsin activity.** Light near 570 nm (h𝜈₅₇₀) converts the protonated retinal of bacteriorhodopsin from the *trans* form (Ret_T) to the *cis* form (Ret_C), along with translocation of a proton to the outer surface of the membrane, thus establishing a proton motive force. ATPase activity (⚙ Section 5.12) is driven by the proton motive force.

Table 13.3	Concentration of ions in cells of *Halobacterium salinarum*	
Ion	**Concentration in medium (*M*)**	**Concentration in cells (*M*)**
Na⁺	3.3	0.8
K⁺	0.05	5.3
Mg²⁺	0.13	0.12
Cl⁻	3.3	3.3

Bacteriorhodopsin absorbs visible light in the green region of the spectrum (about 570 nm). Following absorption, the retinal of bacteriorhodopsin, which normally exists in an *all-trans* configuration, becomes excited and converted to the *cis* form (Figure 13.4). This transformation results in the translocation of a proton to the *outside* surface of the membrane. The retinal molecule then returns to its more stable all-trans isomer in the dark along with the uptake of a proton from the cytoplasm; this completes the cycle (Figure 13.4). As protons accumulate on the outer surface of the membrane, the proton motive force (∞ Section 5.12) increases until the membrane is sufficiently "charged" to drive ATP synthesis through the activity of a proton translocating ATPase (Figure 13.4).

Light-mediated ATP production in *Halobacterium salinarum* supports slow growth of this organism under anoxic conditions. The light-stimulated proton pump of *H. salinarum* also functions to pump Na^+ out of the cell by activity of a Na^+/H^+ antiport system (∞ Section 4.6) and drives the uptake of a variety of nutrients, including the K^+ needed for osmotic balance. The uptake of amino acids by *H. salinarum* has been shown to be indirectly driven by light because the transport of amino acids occurs with Na^+ uptake by an amino acid–Na^+ symporter (∞ Section 4.6). Continued uptake of amino acids depends on the removal of Na^+ via the (light-driven) Na^+/H^+ antiporter.

Other Rhodopsins

Besides bacteriorhodopsin, at least three other rhodopsins are present in membranes of *Halobacterium salinarum*. **Halorhodopsin** is a light-driven pump that pumps chloride (Cl^-) into the cell as a counterion for K^+. The retinal of halorhodopsin binds Cl^- and transports it into the cell. Two light sensors, called **sensory rhodopsins,** are present in *H. salinarum*. These light sensors control phototaxis (the movement towards light, ∞ Section 4.16) by the organism. Through the interaction of a cascade of proteins similar to those involved in chemotaxis (∞ Sections 4.16 and 8.13), sensory rhodopsins affect flagellar rotation, moving cells of *H. salinarum* toward light where bacteriorhodopsin can then function to make ATP (Figure 13.4).

We will learn in our discussion of marine microbiology (∞ Section 19.6) that species of *Bacteria* exist in open ocean waters that contain bacteriorhodopsin-like proteins called *proteorhodopsins*. As far as is known, proteorhodopsin functions like bacteriorhodopsin, except that several different spectral forms exist, each tuned to the absorption of different wavelengths of light. Proteorhodopsin as a mechanism for energy conservation in marine prokaryotes makes good ecological sense, since concentrations of dissolved organic matter in the oceans are typically very low (∞ Section 19.6), and thus a strictly chemoorganotrophic lifestyle would be difficult.

 13.3 Concept Check

Extremely halophilic *Archaea* require large amounts of NaCl for growth. These organisms accumulate large levels of KCl in their cytoplasm as a compatible solute. These salts affect cell wall stability and enzyme activity. The light-mediated proton pump bacteriorhodopsin helps extreme halophiles make ATP.

◆ What is the major physiological difference between *Halobacterium* and *Natronobacterium*?

◆ If cells of *Halobacterium* require high levels of Na^+, why is this not true of *cytoplasmic enzymes* from *Halobacterium* (see Table 13.3).

◆ What benefit does bacteriorhodopsin confer on a cell of *H. salinarum*? Halorhodopsin?

13.4 Methane-Producing *Archaea*: Methanogens

Key Genera: *Methanobacterium, Methanocaldococcus, Methanosarcina*

A large number of Euryarchaeota produce methane (CH_4) as an integral part of their energy metabolism. Such organisms are called **methanogens** and the process of methane formation **methanogenesis**. Methane was discovered as "combustible air" by the Italian physicist Alessandro Volta, who collected gas from marsh sediments and showed that it was flammable. The *Volta experiment*, as it has come to be known, can be easily reproduced if methane trapped in freshwater sediments is collected and *carefully* ignited (Figure 13.5●).

In later chapters we will study the biochemically unique and amazingly complex process of methane formation (∞ Section 17.17). Moreover, we will learn how methanogenesis is the terminal step in the biodegradation of organic matter in many anoxic habitats in nature, such as the swamp shown in Figure 13.5 (∞ Section 19.10). Table 13.4 lists the major sources of biogenic methane in nature.

Diversity and Physiology of Methanogens

Methanogens show a variety of morphologies (Figure 13.6● and Table 13.5). Their taxonomy is based on both phenotypic as well as phylogenetic analyses (Table 13.5), with several taxonomic *orders* being recognized (in taxonomy, an *order* contains groups of related *families*, each of which contain one or more *genera*; ∞ Table 11.6).

Methanogens show a diversity of cell wall chemistries. These include the pseudopeptidoglycan walls of *Methanobacterium* species and relatives (Figure 13.7a●), the methanochondroitin (so named because of its structural resemblance to chondroitin, the connective tissue polymer of vertebrate animals) walls of *Methanosarcina* and relatives (Figure 13.7b), the protein or glycoprotein walls of

● **Figure 13.5 The Volta Experiment as demonstrated in the Microbial Diversity Summer Course in Woods Hole, Massachusetts.** A large inverted funnel was placed over freshwater sediments in which anaerobic decomposition was occurring in Cedar Swamp, Woods Hole. After water displaced the air in the funnel, the funnel was capped and the sediments were mixed with a stick, allowing trapped bubbles of methane to collect in the inverted funnel. Immediately after uncapping the funnel, a flame was held near the funnel port, thus igniting the methane. This experiment was performed over 200 years ago by the Italian physicist Alessandro Volta, which led him to describe methane as "combustible air."

Methanocaldococcus (Figure 13.8*a*●), and *Methanoplanus* species, respectively, and the S-layer walls of *Methanospirillum* (Figure 13.6; ∞ Section 4.8).

Physiologically, methanogens are obligate anaerobes, and strict anoxic techniques are necessary to culture them. Depending on the species, cultures of methanogens can be established in a mineral salts medium under an atmosphere of H_2 plus CO_2 (in the ratio of 4:1, see the reaction shown later), or in complex media. Most known methanogens are mesophilic

Table 13.4 Habitats of methanogens

I. Anoxic sediments: marsh, swamp (see Figure 13.5), and lake sediments, paddy fields, moist landfills (∞ Section 19.10)
II. Animal digestive tracts:
  (a) Rumen of ruminant animals such as cattle, sheep, elk, deer, and camels (∞ Section 19.11)
  (b) Cecum of cecal animals such as horses and rabbits (∞ Section 19.11)
  (c) Large intestine of monogastric animals such as humans, swine, and dogs (∞ Section 21.4)
  (d) Hindgut of cellulolytic insects (for example, termites)
III. Geothermal sources of H_2 + CO_2: hydrothermal vents (∞ Section 19.8)
IV. Artificial biodegradation facilities: sewage sludge digestors (∞ Section 28.2)
V. Endosymbionts of various anaerobic protozoa (∞ Figure 19.26)

and nonhalophilic, although "extremophilic" species growing optimally at very high (Figure 13.8 and see Figure 13.13) or very low temperatures or at very high salt concentration have also been described.

Substrates for Methanogenesis

At least 11 substrates have been shown to be converted to methane by pure cultures of methanogens (Table 13.6). Interestingly, these substrates do *not* include such common compounds as glucose and organic or fatty acids (other than acetate and pyruvate). Compounds such as glucose can be converted to methane, but only in cooperative reactions involving methanogens and other anaerobic bacteria. With the right mixed culture, virtually any organic compound, even including hydrocarbons, can be converted to methane plus CO_2 (∞ Section 19.10).

Three *classes* of compounds make up the list of methanogenic substrates shown in Table 13.6. These include *CO_2-type substrates*, *methyl substrates*, and *acetotrophic substrates*. *CO_2-type substrates* naturally include CO_2 itself, which is reduced to methane using H_2 as electron donor:

$$CO_2 + 4\ H_2 \rightarrow CH_4 + 2\ H_2O \quad \Delta G^{0\prime} = -131\ \text{kJ}$$

(a)   *(b)*   *(c)*   *(d)*

● **Figure 13.6 Scanning electron micrographs of cells of methanogenic *Archaea*, showing the considerable morphological diversity.**
(a) *Methanobrevibacter ruminantium*. A cell is about 0.7 μm in diameter. (b) *Methanobrevibacter arboriphilus*. A cell is about 1 μm in diameter. (c) *Methanospirillum hungatii*. A cell is about 0.4 μm in diameter. (d) *Methanosarcina barkeri*. A cell is about 1.7 μm wide.

| Table 13.5 | Characteristics of some methanogenic *Archaea*[a] | | |

Genus	Morphology	Substrates for methanogenesis	DNA (mol % GC)
Methanobacteriales			
Methanobacterium	Long rods	$H_2 + CO_2$, formate	30–55
Methanobrevibacter	Short rods	$H_2 + CO_2$, formate	27–31
Methanosphaera	Cocci	Methanol + H_2 (both needed)	23–26
Methanothermus	Rods	$H_2 + CO_2$; can also reduce S^0; hyperthermophile	33
Methanothermobacter	Rods	$H_2 + CO_2$, formate, thermophiles	32–61
Methanococcales			
Methanococcus	Irregular cocci	$H_2 + CO_2$, pyruvate + CO_2, formate	29–35
Methanothermococcus	Cocci	$H_2 + CO_2$, formate	31–34
Methanocaldococcus	Cocci	$H_2 + CO_2$	31–33
Methanotorris	Cocci	$H_2 + CO_2$	31
Methanomicrobiales			
Methanomicrobium	Short rods	$H_2 + CO_2$, formate	49
Methanogenium	Irregular cocci	$H_2 + CO_2$, formate	47–52
Methanospirillum	Spirilla	$H_2 + CO_2$, formate	45–50
Methanoplanus	Plate-shaped cells—occurring as thin plates with sharp edges	$H_2 + CO_2$, formate	39–50
Methanocorpusculum	Irregular cocci	$H_2 + CO_2$, formate, alcohols	48–52
Methanoculleus	Irregular cocci	$H_2 + CO_2$, alcohols, formate	49–61
Methanofollis	Irregular cocci	$H_2 + CO_2$, formate	54–60
Methanolacinia	Irregular rods	$H_2 + CO_2$, alcohols	38–45
Methanosarcinales			
Methanosarcina	Large irregular cocci in packets	$H_2 + CO_2$, methanol, methylamines, acetate	36–43
Methanolobus	Irregular cocci in aggregates	Methanol, methylamines	39–46
Methanohalobium	Irregular cocci	Methanol, methylamines; halophile	37
Methanococcoides	Irregular cocci	Methanol, methylamines	42
Methanohalophilus	Irregular cocci	Methanol, methylamines, methyl sulfides; halophile	39–41
Methanosaeta	Long rods to filaments	Acetate	52–61
Methanosalsum	Irregular cocci	Methanol, methylamines, dimethylsulfide	38–40
Methanopyrales			
Methanopyrus	Rods in chains	$H_2 + CO_2$; hyperthermophile, growth at 110°C	60

[a] Taxonomic orders are listed in bold.

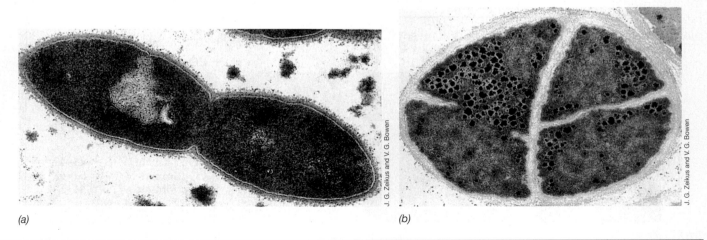

(a) *(b)*

J. G. Zeikus and V. G. Bowen

● **Figure 13.7** **Transmission electron micrographs of thin sections of methanogenic *Archaea*.** (a) *Methanobrevibacter ruminantium*. A cell is 0.7 μm in diameter. (b) *Methanosarcina barkeri*, showing the thick cell wall and the manner of cell segmentation and cross-wall formation. A cell is 1.7 μm in diameter. The cell wall of *M. ruminantium* contains pseudopeptidoglycan (⌀ Figure 4.33*a*), while the *M. barkeri* wall consists of protein and polysaccharides.

(a) Helmut König and K. O. Stetter

(b) R. Rachel and K. O. Stetter

(c) König/Stetter

(d) Stephen Zinder

● **Figure 13.8 Hyperthermophilic and thermophilic methanogens.** (a) *Methanocaldococcus jannaschii* (temperature optimum, 85°C), shadowed preparation electron micrograph. A cell is about 1 μm in diameter. (b) *Methanotorris igneus* (temperature optimum, 88°C), thin section. A cell is about 1 μm in diameter. (c) *Methanothermus fervidus* (temperature optimum, 88°C), thin-sectioned electron micrograph. A cell is about 0.4 μm in diameter. (d) *Methanosaeta thermophila* (temperature optimum, 60°C), phase contrast micrograph. A cell is about 1 μm in diameter. The refractile bodies inside the cells are gas vesicles (⊙⊙ Section 4.12).

Other substrates here include formate (which is simply $CO_2 + H_2$ in combined form) and CO, carbon monoxide.

The second class of methanogenic substrates are *methylated substances* (Table 13.6). Using methanol (CH_3OH) as a model methyl substrate here, the formation of CH_4 can occur in either of two ways. First,

CH_3OH can be reduced using an external electron donor such as H_2:

$$CH_3OH + H_2 \rightarrow CH_4 + H_2O \quad \Delta G^{0'} = -113\text{kJ}$$

Alternatively, in the absence of H_2, some CH_3OH can be oxidized to CO_2 to generate the electrons needed to reduce other molecules of CH_3OH to CH_4:

$$4\,CH_3OH \rightarrow 3\,CH_4 + CO_2 + 2\,H_2O \quad \Delta G^{0'} = -319 \text{ kJ}$$

The final methanogenic process is the cleavage of acetate to CO_2 plus CH_4, called the **acetotrophic** reaction:

$$CH_3COO^- + H_2O \rightarrow CH_4 + HCO_3^- \quad \Delta G^{0'} = -31 \text{ kJ}$$

Only a very few methanogens are acetotrophic (Table 13.5). However, measurements of methane formation in highly methanogenic habitats such as sewage sludge (⊙⊙ Section 28.2) have shown that about two-thirds of the methane formed there originates from acetate and one-third from $H_2 + CO_2$. Thus, although apparently lacking in terms of known diversity, acetotrophic methanogens are very ecologically important.

As can be seen by inspection of each of the above reactions, they are all energy-releasing (exergonic) and can thus be used to synthesize ATP. The biochemical details

Table 13.6	**Substrates converted to methane by various methanogenic *Archaea***

I. CO₂-type substrates
Carbon dioxide, CO_2 (with electrons derived from H_2, certain alcohols, or pyruvate)
Formate, $HCOO^-$
Carbon monoxide, CO

II. Methyl substrates
Methanol, CH_3OH
Methylamine, $CH_3NH_3^+$
Dimethylamine, $(CH_3)_2NH_2^+$
Trimethylamine, $(CH_3)_3NH^+$
Methylmercaptan, CH_3SH
Dimethylsulfide, $(CH_3)_2S$

III. Acetotrophic substrates
Acetate, CH_3COO^-
Pyruvate, CH_3COCOO^-

of methanogenesis will be discussed in Chapter 17. We only note here that the process is coupled to proton motive force formation and ATP synthesis by chemiosmotic mechanisms (◌◌ Section 5.12). Substrate-level phosphorylation, typical of fermentative bacteria, apparently does not occur in methanogens.

Methanocaldococcus jannaschii as a Model Methanogen/Archaeon

As will be discussed in Section 15.3, the genome of the hyperthermophilic methanogen *Methanocaldococcus jannaschii* (Figure 13.8a) and that of several other methanogens have been sequenced. The 1.66-Mbp circular genome of *M. jannaschii* contains about 1700 genes, and from sequence analyses genes encoding enzymes of methanogenesis and several other key cell functions have been identified. Interestingly, the majority of *M. jannaschii* genes encoding functions such as central metabolic pathways and cell division are similar to those in *Bacteria*. By contrast, most of the *M. jannaschii* genes encoding core molecular processes such as transcription and translation more closely resemble those of eukaryotes. These findings can be rationalized from inspection of the phylogenetic tree of life that shows the domain *Archaea* to be positioned *between* the domains *Bacteria* and *Eukarya* (◌◌ Section 11.8 and Figure 11.13). However, analyses of the *M. jannaschii* genome also show that nearly 50% of its genes have no counterparts in known genes from *any group of organisms*. This suggests that there are either many new cellular functions encoded in archaeal DNA that have yet to be discovered or that many existing functions are encoded by redundant (and quite distinct) sets of genes.

 13.4 **Concept Check**

Methanogenic *Archaea* are strictly anaerobic prokaryotes whose metabolism is tied to the production of methane (CH_4).

◆ What did the Volta experiment demonstrate?

◆ What are the major substrates for methanogenesis?

13.5 Thermoplasmatales: *Thermoplasma, Ferroplasma,* and *Picrophilus*

Key Genera: *Thermoplasma, Picrophilus, Ferroplasma*

A phylogenetically distinct line of *Archaea* contains three thermophilic and extremely acidophilic prokaryotes: *Thermoplasma, Ferroplasma,* and *Picrophilus* (Figure 13.1). These microorganisms are among the most acidophilic of all known microorganisms and, in the case of *Picrophilus,* even capable of growth below pH 0. They also form their own *order* of prokaryotes within the Euryarchaeota, the Thermoplasmatales. We begin with a

description of the mycoplasma-like organisms *Thermoplasma* and *Ferroplasma*.

Archaea Lacking Cell Walls

Thermoplasma and *Ferroplasma* are *Archaea* that lack cell walls, resembling the mycoplasmas (◌◌ Section 12.21) in this regard. *Thermoplasma* (Figure 13.9a●) is a chemoorganotroph that grows optimally at 55°C and pH 2 in complex media. Two species of *Thermoplasma* have been described, *T. acidophilum* and *T. volcanium.* Species of *Thermoplasma* are facultative aerobes, growing either aerobically or anaerobically by sulfur respiration. Most strains of *Thermoplasma* have been obtained from self-heating coal refuse piles (Figure 13.10●). Coal refuse contains coal fragments, pyrite (FeS_2), and other organic

(a)

(b)

● **Figure 13.9** *Thermoplasma* **species.** (a) *Thermoplasma acidophilum,* an acidophilic, thermophilic mycoplasma-like archaeon. Electron micrograph of a thin section. The diameter of cells is highly variable from 0.2 to 5 μm. The cell shown is about 1 μm in diameter. (b) Shadowed preparation of cells of *Thermoplasma volcanium* isolated from hot springs. Cells are 1–2 μm in diameter. Notice abundant flagella.

materials extracted from coal. When dumped into piles in surface-mining operations, coal refuse heats by spontaneous combustion (Figure 13.10). This sets the stage for growth of *Thermoplasma,* which apparently metabolizes organic compounds leached from the hot coal refuse. A second species of *Thermoplasma, T. volcanium,* has been isolated in hot acidic soils throughout the world and is highly motile by multiple flagella (Figure 13.9*b*).

To survive the osmotic stresses of life without a cell wall and to withstand the dual environmental extremes of low pH and high temperature, *Thermoplasma* has evolved a unique cell membrane structure. The membrane contains a lipopolysaccharide-like material called *lipoglycan* (∞ Section 12.21). This substance consists of a *tetraether* lipid monolayer membrane with mannose and glucose (Figure 13.11●). This molecule constitutes a major fraction of the total lipid composition of *Thermoplasma.* The membrane also contains glycoproteins but not sterols. These molecules render the *Thermoplasma* membrane stable to hot acid conditions.

The genome of *Thermoplasma* is of interest. Like other mycoplasmas (∞ Sections 12.21 and 15.3), *Thermoplasma* contains a small genome (1.5 Mbp). In addition, however, *Thermoplasma* DNA is complexed with a highly basic DNA-binding protein that organizes the DNA into globular particles resembling the nucleosomes of eukaryotic cells (∞ Section 7.3 discusses the arrangement of DNA in eukaryotes). This protein is homologous to the basic histone proteins of eukaryotic cells. Similar histonelike proteins have been found in several other Euryarchaeota (see Section 13.12).

Ferroplasma

Ferroplasma is a chemolithotrophic relative of *Thermoplasma. Ferroplasma* is a strong acidophile; however, it is not a thermophile, growing optimally at 35°C. *Ferroplasma* (∞ Figure 19.35) oxidizes Fe^{2+} to Fe^{3+} to obtain energy (this reaction generates acid; ∞ Section 19.14 and see Figure 13.15*d*) and uses CO_2 as its carbon source (autotrophy).

T. D. Brock

● **Figure 13.10 A typical self-heating coal refuse pile, habitat of Thermoplasma.** The pile containing coal debris, pyrite, and other microbial substrates, self heats from microbial metabolism.

Ether linkage

$[R]_8$ Glu (α1→1) — O

R = Man (α1 → 2) Man (α1 → 4) Man (α1 → 3)

● **Figure 13.11 Structure of the tetraether lipoglycan of Thermoplasma acidophilum.** Glu, Glucose; Man, mannose. Note the ether linkages (shown in green) and the fact that this lipid would form a monolayer rather than a bilayer membrane (∞ compare with Figure 4.19*b*).

Ferroplasma grows in mine tailings containing pyrite (FeS), which is its energy source. The extreme acidophily of *Ferroplasma* allows it to drive down the pH of its habitat to extremely acidic values. After moderate acidity is generated from Fe^{2+} oxidation by acidophilic organisms such as *Acidithiobacillus ferrooxidans* and *Leptospirillum ferrooxidans, Ferroplasma* becomes active and subsequently generates the very low pH values typical of acid mine drainage. Acidic waters at pH 0 can be generated by the activities of *Ferroplasma* (∞ Figure 19.35 and Sections 19.14 and 19.15).

Picrophilus

A phylogenetic relative of *Thermoplasma* and *Ferroplasma* is *Picrophilus.* Although *Thermoplasma* and *Ferroplasma* are extreme acidophiles, *Picrophilus* is even more so, growing optimally at pH 0.7 and capable of growth to as low as pH −0.06! *Picrophilus* also has a cell wall (an S-layer, ∞ Section 4.10) and a much lower DNA GC base ratio than does *Thermoplasma* or *Ferroplasma.* Thus, although phylogenetically related, *Thermoplasma, Ferroplasma,* and *Picrophilus* have quite distinct genomes. Two species of *Picrophilus* have been isolated from acidic Japanese solfataras, and like *Thermoplasma,* both grow heterotrophically on complex media.

The physiology of *Picrophilus* is of interest as a model for extreme acid tolerance. Studies of its cytoplasmic membrane suggest an unusual arrangement of lipids that forms a highly acid impermeable membrane at optimal pH values. By contrast, at only moderately acidic pH values such as pH 4, the membranes of cells of *Picrophilus* quickly become leaky and disintegrate. Obviously, this organism has evolved to survive only in highly acidic habitats.

⬟ 13.5 Concept Check

Thermoplasma, Ferroplasma, and *Picrophilus* are extremely acidophilic thermophiles that form their own phylogenetic family of *Archaea* inhabiting coal refuse piles and highly acidic solfataras. Cells of *Thermoplasma* and *Ferroplasma* lack cell walls and thus resemble the mycoplasmas in this regard.

◆ In what ways are *Thermoplasma* and *Picrophilus* similar? In what ways do they differ?

◆ How does *Thermoplasma* strengthen its cell membrane to survive life in the absence of a cell wall?

◆ How does *Ferroplasma* obtain energy for growth?

13.6 Hyperthermophilic Euryarchaeota: Thermococcales and *Methanopyrus*

Key Genera: *Thermococcus, Pyrococcus, Methanopyrus*

A few euryarchaeotes thrive in thermal environments and some are **hyperthermophiles**. We consider here three hyperthermophilic euryarchaeotes that branch on the archaeal tree (Figure 13.1) very near the root. Two of these organisms, *Thermococcus* and *Pyrococcus*, show phenotypic properties very similar to those of hyperthermophilic crenarchaeotes to be discussed in Sections 13.8–13.10 and form a distinct order: the Thermococcales (see Table 13.9). The other, *Methanopyrus*, is a methanogen that closely resembles other methanogens (see Section 13.4 and Table 13.5) in its basic physiology but is unusual in its hyperthermophily and phylogenetic position (Figure 13.1).

Thermococcus and *Pyrococcus*

Thermococcus is a spherical hyperthermophilic euryarchaeote indigenous to anoxic thermal waters in various locations throughout the world. The spherical cells contain a tuft of polar flagella and are thus highly motile (Figure 13.12*a*●). *Thermococcus* is an obligately anaerobic chemoorganotroph that grows on proteins and other complex organic mixtures (including some sugars) with S^0 as electron acceptor at temperatures from $70-95°C$.

Pyrococcus is morphologically similar to *Thermococcus* (Figure 13.12*b*). *Pyrococcus* (the Latin derivation

literally means "fireball") differs from *Thermococcus* primarily by its higher temperature requirements; *Pyrococcus* grows between 70 and 106°C with an optimum of 100°C. *Thermococcus* and *Pyrococcus* are also metabolically quite similar. Proteins, starch, or maltose are oxidized as electron donors and S^0 is the terminal acceptor and is reduced to H_2S. More properties of *Thermococcus* and *Pyrococcus* are described later (see Tables 13.8 and 13.9).

Methanopyrus

Methanopyrus is a rod-shaped hyperthermophilic methanogen (Figure 13.13●). *Methanopyrus* was isolated from sediments near submarine hydrothermal vents and from the walls of "black smoker" hydrothermal vent chimneys (👁👁 Sections 13.8 and 19.8 for a discussion of hydrothermal vents). *Methanopyrus* occupies a unique phylogenetic position on the tree of *Archaea* (Figure 13.1). It lies near the base of the archaeal tree and shares phenotypic properties with both the hyperthermophiles (the growth temperature maximum of *Methanopyrus* is 110°C) and the methanogens.

Methanopyrus only produces methane from $H_2 + CO_2$ and grows rapidly for an autotrophic organism (generation time less than 1 h at its temperature optimum of 100°C). Unlike mesophilic methanogens, however, cells of *Methanopyrus* contain large amounts of cyclic 2,3-diphosphoglycerate, a derivative of glycolysis, dissolved in the cytoplasm. This compound, present at more than 1 *molar* concentration in *Methanopyrus*, is

(a)

(b)

● **Figure 13.13** **Methanopyrus.** *Methanopyrus* grows optimally at 100°C and can make CH_4 only from $CO_2 + H_2$. (a) Electron micrograph of a cell of *Methanopyrus kandleri*, the most thermophilic of all known methanogens (upper temperature limit, 110°C). This cell measures $0.5 \times 8\ \mu m$. (b) Structure of the novel lipid of *M. kandleri*. This is the normal ether-linked lipid of the *Archaea* (👁👁 Section 4.5) with the exception that the side chains are an *unsaturated* form of phytanyl called *geranylgeraniol*. It is thought that this unusual lipid predated the appearance of saturated phytanyl lipids and that *Methanopyrus* is therefore a descendant of a very ancient lineage of *Archaea*; the data of Figure 13.1 support this hypothesis.

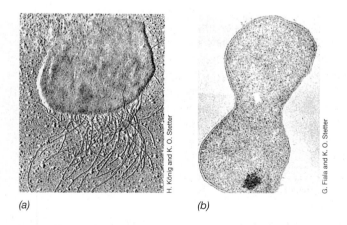
(a)   (b)

● **Figure 13.12** **Spherical hyperthermophilic** *Archaea* **from submarine volcanic areas.** (a) *Thermococcus celer*. Electron micrograph of shadowed cells (note tuft of flagella). (b) Dividing cell of *Pyrococcus furiosus*. Electron micrograph of thin section. Cells of both organisms are about 0.8 μm in diameter.

thought to function as a thermostabilizing agent to prevent denaturation of enzymes and DNA inside the cell (see Section 13.12).

Methanopyrus is also unusual because it contains membrane lipids found in no other known organism. Recall that in the lipids of *Archaea*, the glycerol side chains contain **phytanyl** rather than fatty acids bonded in ether linkage to the glycerol (⬭ Section 4.5). In *Methanopyrus*, this ether-linked lipid is an unsaturated form of the otherwise saturated dibiphytanyl tetraethers found in other hyperthermophilic *Archaea* (Figure 13.13*b*; ⬭ Section 4.5 and Figure 4.19). Because of the biochemistry of the pathway for how these lipids are biosynthesized, this lipid is thought to be a primitive characteristic. Along with its phylogenic position, hyperthermophily and anaerobic metabolism, the primitive lipids of *Methanopyrus* make this organism an intriguing model for the earliest life forms (⬭ Sections 11.1–11.3, and 13.13).

The discovery of *Methanopyrus* offers an explanation for the source of methane in hot oceanic sediments previously thought to be too hot to support biogenic methanogenesis. At the depth at which *Methanopyrus* was found—approximately 2000 m—water remains liquid at temperatures up to 350°C. This leaves open the possibility that other hyperthermophilic methanogens also exist capable of growth at 110°C or at even higher temperatures.

13.7 Hyperthermophilic Euryarchaeota: The Archaeoglobales

Key Genera: *Archaeoglobus, Ferroglobus*

We will see later that a number of hyperthermophilic Crenarchaeota catalyze anaerobic respirations in which elemental sulfur (S^0) is used as an electron acceptor, being reduced to H_2S (see Table 13.8). Curiously, none of these crenarchaeotes are *sulfate*-reducers, that is, capable of reducing SO_4^{2-} to H_2S. However, one hyperthermophilic euryarchaeote, *Archaeoglobus*, is a true sulfate-reducer and forms a phylogenetically distinct lineage within the Euryarchaeota (Figure 13.1).

Archaeoglobus and Its Genome

Archaeoglobus was isolated from hot marine sediments near hydrothermal vents. In its metabolism, *Archaeoglobus* couples the oxidation of H_2, lactate, pyruvate, glucose, or complex organic compounds to the reduction of sulfate to sulfide. Cells of *Archaeoglobus* are irregular cocci (Figure 13.14*a*●) and cultures grow optimally at 83°C.

Archaeoglobus and methanogens have features in common. We will learn in Chapter 17 about the unique biochemistry of methanogenesis. Briefly, this process requires a series of novel coenzymes. With rare exceptions, these coenzymes have only been found in cells of methanogens (⬭ Section 17.17). Surprisingly, however, *Archaeoglobus* also contains these coenzymes, and cultures of this organism actually produce small amounts of

(a)

(b)

● **Figure 13.14 Archaeoglobales.** (a) Transmission electron micrograph of the sulfate-reducing hyperthermophile *Archaeoglobus fulgidus*. The cell measures 0.7 μm in diameter. (b) Freeze-etched electron micrograph of *Ferroglobus placidus*, a ferrous iron-oxidizing, nitrate-reducing hyperthermophile. The cell measures about 0.8 μm in diameter.

methane during growth. Thus, *Archaeoglobus*, which also shows a rather close phylogenetic relationship to methanogens (Figure 13.1), may represent a transitional type of organism in a physiological sense, one that bridged the energy-conserving processes of sulfur reduction and methanogenesis.

The genome sequence of *Archaeoglobus* has revealed a conundrum concerning methanogenesis. Although *Archaeoglobus* can make methane, it lacks genes for a key enzyme of methanogenesis, *methyl-CoM reductase* (⬭ Section 17.17). Thus, the small amounts of methane that are made by *Archaeoglobus* do not come from the activity of this key enzyme and thus their exact origin is unclear. The genome of *Archaeoglobus* contains about 2400 genes and shares a number of genes with methanogens (see Section 13.4). However, the *Archaeoglobus* genome also contains a number of genes unique to this organism, an example of the broad genetic diversity in *Archaea*. As in species of *Bacteria*, roughly one-third of the genes in every archaeal genome are unique, not present in any other organism (⬭ Section 15.3).

Ferroglobus

Ferroglobus (Figure 13.14*b*) is related to *Archaeoglobus* but is not a sulfate-reducing bacterium. Instead, *Ferroglobus* is an iron-oxidizing chemolithotrophic autotroph, conserving energy from the oxidation of Fe^{2+} to Fe^{3+} coupled to the reduction of NO_3^- to NO_2^- plus NO (see Table 13.8). *Ferroglobus* can also use H_2 or H_2S as electron donors in its energy metabolism. *Ferroglobus* was isolated from a shallow marine hydrothermal vent and grows optimally at 85°C.

Ferroglobus is interesting for several reasons, but especially so for its ability to oxidize Fe^{2+} to Fe^{3+} under anoxic conditions. The process might help explain the abundance of Fe^{3+} in ancient rocks, such as the banded iron formations (∞ Section 11.3) previously thought to have originated from the oxidation of Fe^{2+} by O_2 produced by cyanobacteria (oxygenic photosynthesis). The metabolism of *Ferroglobus* thus has implications for dating the origin of cyanobacteria and the subsequent oxygenation of Earth (∞ Sections 11.1 and 11.2). Certain anoxygenic phototrophic bacteria can also oxidize Fe^{2+} under anoxic conditions (∞ Section 17.11). So, several anoxic routes to ancient Fe^{3+} obviously exist, making it a challenge to accurately estimate when cyanobacteria first evolved (∞Figure 11.8).

 13.6–13.7 Concept Check

Hyperthermophilic euryarchaeota include the Thermococcales, Archaeoglobales, and *Methanopyrus*. All organisms in this group have growth temperature optima above 80°C.

◆ How does the metabolism of *Thermococcus* differ from that of *Methanopyrus*?

◆ Which of the hyperthermophilic Euryarchaeota is a true sulfate reducer?

◆ How does the metabolism of *Ferroglobus* differ from that of *Archaeoglobus*?

III PHYLUM CRENARCHAEOTA

Crenarchaeotes are phylogenetically distinct from Euryarchaeotes (Figure 13.1) and contain representatives that live at both ends of nature's temperature extremes: boiling water and freezing water. However, all cultured crenarchaeotes are *hyperthermophiles* (growth temperature optima above 80°C) and some actually have optima *above* the boiling point of water. We start with an overview of the habitats and energy metabolism of crenarchaeotes and then describe the properties of some key genera.

13.8 Habitats and Energy Metabolism of Crenarchaeotes

A summary of the habitats of Crenarchaeota is shown in Table 13.7. They include very hot and very cold environments. Most hyperthermophilic *Archaea* have been isolated from geothermally heated soils or waters containing elemental sulfur and sulfides (Figure 13.15●) and most species metabolize sulfur in one way or another. In terrestrial environments, sulfur-rich springs, boiling mud, and soils may have temperatures up to 100°C and are mildly to extremely acidic owing to production of sulfuric acid (H_2SO_4) from the biological oxidation of H_2S and S^0 (∞ Sections 17.10 and 19.13). Such hot, sulfur-rich environments, called **solfataras**, are found throughout the world (Figure 13.15), including Italy, Iceland, New Zealand, and Yellowstone National Park in Wyoming (USA).

Depending on the surrounding geology, solfataras can be mildly acidic to slightly alkaline, pH 5–8, or extremely acidic, with pH values below 1. Hyperthermophilic crenarchaeotes have been obtained from both types of environments, but the majority of these organisms inhabit neutral or mildly acidic hot habitats. In addition to these natural habitats, hyperther-

---

Table 13.7 Habitats of Crenarchaeota

Characteristic	Thermal area[a]		Nonthermal area[b]
	Terrestrial	Marine	
Locations	Solfataras (hot springs, fumaroles, mudpots, steam-heated soils); geothermal power plants; deep in Earth's crust	Submarine hot springs, hot sediments and vents ("black smokers"); deep oil reservoirs	Planktonic in oceans worldwide; near-shore and deep Antarctic waters; sea ice; symbionts of marine sponges
Temperature	Surface to 100°C; subsurface, above 100°C	Up to 400°C (smokers)	−2 to +4°C
Salinity/pH	Usually less than 1% NaCl; pH 0.5–9	Moderate, about 3% NaCl; pH 5–9	3–8% NaCl; pH 7–9
Gases and other nutrients	CO_2, CO, CH_4, H_2, H_2S, S^0, $S_2O_3^{2-}$, SO_4^{2-}, NH_4^+, N_2	Same as for terrestrial	CO_2, N_2, O_2; for chemolithotrophs, inorganic substrates such as NH_4^+

[a] See Figures 13.10 and 13.15 (∞ Figures 19.19–19.21).
[b] See Figure 13.16. Some species of Euryarchaeota are also present in marine waters.

● **Figure 13.15 Habitats of hyperthermophilic _Archaea_.** (a) A typical solfatara in Yellowstone National Park. Steam rich in hydrogen sulfide rises to the surface of the earth. Because of the heat and acidity, only prokaryotes are present. (b) Sulfur-rich hot spring, a habitat containing dense populations of _Sulfolobus_. The acidity in solfataras and sulfur springs comes from the oxidation of H_2S and S^0 to H_2SO_4 (sulfuric acid) by _Sulfolobus_ and related prokaryotes. (c) A typical boiling spring of neutral pH in Yellowstone Park; Imperial Geyser. Many different species of hyperthermophilic _Archaea_ may reside in such a habitat. (d) An acidic iron-rich geothermal spring, another _Sulfolobus_ habitat. Here the oxidation of Fe^{2+} to Fe^{3+} (rather than S^0 to SO_4^{2-}) causes acidic conditions $(Fe^{3+} + 3\ H_2O \rightarrow Fe(OH)_3 + 3\ H^+)$.

mophilic _Archaea_ also thrive within artificial thermal habitats, in particular the boiling outflows of geothermal power plants.

Hyperthermophilic Crenarchaeota also inhabit undersea hot springs called **hydrothermal vents**. We discuss the geology and microbiology of these habitats in Section 19.8. Here it is only necessary to note that submarine waters can be much hotter than surface waters because the water is under pressure. Indeed, all hyperthermophiles with growth temperature optima above 100°C have come from submarine sources. The latter include both shallow (2–10 meter depth) vents such as those off the coast of Vulcano, Italy, to deep (2000–4000 meter depth) vents near ocean-spreading centers (∞ Section 19.8). Deep vents are the hottest habitats so far known to yield microorganisms.

Cold-Dwelling Crenarchaeotes

In contrast to hyperthermophiles, cold-dwelling crenarchaeotes (Table 13.7) have been identified from community sampling of ribosomal RNA genes from many nonthermal environments. For example, using fluorescent phylogenetic probes (∞ Sections 11.7 and 18.4), crenarchaeotes have been found in marine waters worldwide (Figure 13.16b●). In stark contrast to the hyperthermophiles, marine crenarchaeotes thrive even in frigid waters and sea ice, such as those near the Antarctic (Figure 13.16a). These organisms are _planktonic_ (suspended freely or attached to suspended particles in the water column) and occur in significant numbers ($\sim 10^4$/ml) for waters that are both nutrient poor and very cold (2–4°C in seawater and less than 0°C in sea-ice). Although their physiology is still a mystery, lipid analyses of marine crenarchaeotes filtered from seawater have shown that they contain ether-linked lipids, the hallmark of the _Archaea_ (∞ Sections 4.5 and 11.9). The ecology of marine crenarchaeotes is discussed in Section 19.6.

Cold-dwelling species of Euryarchaeota are also present in marine environments. From studies thus far it appears that these organisms are more abundant in temperate surface waters while crenarchaeotes seem to be more abundant in cold (Figure 13.16a) waters and in waters of great depth (∞ Section 19.6 and Figure 19.13).

(a)   *(b)*

● **Figure 13.16 Cold-dwelling Crenarchaeota.** (a) Photo of the Antarctic peninsula taken from shipboard. The frigid waters that lie under the surface ice shown here are habitats for cold-dwelling crenarchaeotes. (b) Fluorescence photomicrograph of seawater treated with a phylogenetic stain composed of a green fluorescing dye attached to an oligonucleotide complementary to a signature sequence in the 16S rRNA of species of Crenarchaeota (◯◯ Section 11.7 for a description of the methods used here). Blue cells are stained with DAPI, a stain that stains all cells (◯◯ Section 18.3 and Figure 18.6). Cells stained green are thus cold-dwelling Crenarchaeota. See Section 19.6 and Figure 19.13 for a description of *Bacteria* and *Archaea* in open ocean waters.

Energy Metabolism

With a few exceptions, hyperthermophilic Crenarchaeota are obligate anaerobes. Their energy-yielding metabolism is either chemoorganotrophic or chemolithotrophic (or both, for example in *Sulfolobus*) and involves a wide diversity of different electron donors and acceptors (Table 13.8). Fermentation is rare and most bioenergetic strategies are anaerobic respirations (Table 13.8). Energy conservation during these respiratory processes occurs by the same general mechanism widespread in *Bacteria*:

electron transfer within the cytoplasmic membrane leading to the formation of a proton motive force from which ATP is made by way of proton-translocating ATPases (◯◯ Section 5.12).

Many hyperthermophilic crenarchaeotes can grow chemolithotrophically under anoxic conditions with H_2 as electron donor and S^0 or NO_3^- as electron acceptor; a few can also oxidize H_2 aerobically (Table 13.8). H_2 respiration with ferric iron (Fe^{3+}) as electron acceptor also occurs in several hyperthermophiles (Table 13.8). Other chemolitho-trophic lifestyles include the oxidation of S^0 and Fe^{2+} aerobically or Fe^{2+} anaerobically with NO_3^- as acceptor (Table 13.8). Only one sulfate-reducing hyperthermophile is known (the euryarchaeote *Archaeoglobus*, see Section 13.7). The only bioenergetic option apparently ruled out is photosynthesis. The most thermophilic phototrophic organism known can grow up to 73°C (◯◯ Table 6.1), far too cold for most hyperthermophiles (see Table 13.9).

Armed with a basic understanding of the habitats and energy metabolism of hyperthermophilic crenarchaeotes, let us now look at some representative organisms.

13.9 Hyperthermophiles from Terrestrial Volcanic Habitats: Sulfolobales and Thermoproteales

Key Genera: *Sulfolobus, Acidianus, Thermoproteus, Pyrobaculum*

Volcanic habitats can have temperatures as high as 100°C and are thus suitable for hyperthermophilic *Archaea*. Two phylogenetically related organisms isolat-

---

Table 13.8 Energy-yielding reactions of hyperthermophilic *Archaea*

Nutritional class	Energy-yielding reaction	Metabolic[a] type	Example
Chemoorganotrophic	Organic compound + $S^0 \rightarrow H_2S + CO_2$	AnR	*Thermoproteus, Thermococcus, Desulfurococcus, Thermofilum, Pyrococcus*
	Organic compound + $SO_4^{2-} \rightarrow H_2S + CO_2$	AnR	*Archaeoglobus*
	Organic compound + $O_2 \rightarrow H_2O + CO_2$	AeR	*Sulfolobus*
	Organic compound $\rightarrow CO_2 + H_2$ + fatty acids	F	*Staphylothermus, Pyrodictium*
	Organic compound + $Fe^{3+} \rightarrow CO_2 + Fe^{2+}$	AnR	*Pyrodictium*
	Pyruvate $\rightarrow CO_2 + H_2$ + acetate	F	*Pyrococcus*
Chemolithotrophic	$H_2 + S^0 \rightarrow H_2S$	AnR	*Acidianus, Pyrodictium, Thermoproteus, Stygiolobus, Ignicoccus*
	$H_2 + NO_3^- \rightarrow NO_2^- + H_2O$ (NO_2^- is reduced to N_2 by some species)	AnR	*Pyrobaculum*
	$4 H_2 + NO_3^- + H^+ \rightarrow NH_4^+ + 2 H_2O + OH^-$	AnR	*Pyrolobus*
	$H_2 + 2 Fe^{3+} \rightarrow 2 Fe^{2+} + 2 H^+$	AnR	*Pyrobaculum, Pyrodictium, Archaeoglobus*
	$2 H_2 + O_2 \rightarrow 2 H_2O$	AeR	*Acidianus, Sulfolobus, Pyrobaculum*
	$2 S^0 + 3 O_2 + 2 H_2O \rightarrow 2 H_2SO_4$	AeR	*Sulfolobus, Acidianus*
	$2 FeS_2 + 7 O_2 + 2 H_2O \rightarrow 2 FeSO_4 + 2 H_2SO_4$	AeR	*Sulfolobus, Acidianus, Metallosphaera*
	$2 FeCO_3 + NO_3^- + 6 H_2O \rightarrow 2 Fe(OH)_3 + NO_2^- + 2 HCO_3^- + 2 H^+ + H_2O$	AnR	*Ferroglobus*
	$4 H_2 + SO_4^{2-} + 2 H^+ \rightarrow 4 H_2O + H_2S$	AnR	*Archaeoglobus*
	$4 H_2 + CO_2 \rightarrow CH_4 + 2 H_2O$	AnR	*Methanopyrus, Methanocaldococcus, Methanothermus*

[a] AnR, anaerobic respiration; AeR, aerobic respiration; F, fermentation

Table 13.9 Properties of some hyperthermophilic Crenarchaeota

Group/Genus[a]	Morphology	Relationship to O_2[b]	DNA (mol % GC)	Temperature Minimum	Temperature Optimum	Temperature Maximum	Optimum pH
Sulfolobales							
Sulfolobus	Lobed coccus	Ae	37	55	75	87	2–3
Acidianus	Coccus	Fac	31	60	88	95	2
Metallosphaera	Coccus	Ae	45	50	75	80	2
Stygiolobus	Lobed coccus	An	38	57	80	89	3
Stetteria	Coccus	An	65	68	95	102	6
Sulfophobococcus	Disc-shaped	An	54–56	70	85	95	7.5
Thermosphaera	Coccus	An	46	67	85	90	7
Thermoproteales							
Thermoproteus	Rod	An	56	60	88	96	6
Thermofilum	Rod	An	57	70	88	95	5.5
Pyrobaculum	Rod	Fac	46	74	100	102	6
Caldivirga	Rod	An	43	60	85	92	4
Thermocladium	Rod	An	52	60	75	80	4.2
Desulfurococcales							
Desulfurococcus	Coccus	An	51	70	85	95	6
Aeropyrum	Coccus	Ae	56	70	95	100	7
Staphylothermus	Cocci in clusters	An	35	65	92	98	6–7
Pyrodictium	Disc-shaped with filaments	An	62	82	105	110[c]	6
Pyrolobus	Lobed coccus	Fac	53	90	106	113	5.5
Thermodiscus	Disc-shaped	An	49	75	90	98	5.5
Ignicoccus	Irregular coccus	An	35	65	90	103	5
Hyperthermus	Irregular coccus	An	56	75	102	108	7
Sulfurisphaera	Coccus	Fac	33	63	84	92	2
Sulfurococcus	Coccus	Ae	43–46	40	75	85	2.5
Archaeoglobales[c]							
Archaeoglobus	Coccus	An	46	64	83	95	7
Ferroglobus	Irregular coccus	An	43	65	85	95	7
Thermococcales[d]							
Thermococcus	Coccus	An	38–57	70	88	98	6–7
Pyrococcus	Coccus	An	38	70	100	106	6–8

[a] The group names ending in "ales" are order names (◌◌ Section 11.10).

[b] Ae, aerobe; An, anaerobe; Fac, facultative.

[c] One species may grow up to 121°C.

[d] Phylogenetically, genera in this order of hyperthermophiles are members of the Euryarchaeota (see Sections 13.6 and 13.7).

ed from these environments include *Sulfolobus* and *Acidianus*. These genera form the heart of an order called the Sulfolobales (Table 13.9).

Sulfolobales

Sulfolobus grows in sulfur-rich acidic hot springs (Figure 13.15) at temperatures up to 90°C and at pH values of 1–5.* *Sulfolobus* (Figure 13.17a●) is an aerobic chemolithotroph that oxidizes H_2S or S^0 to H_2SO_4 and fixes CO_2 as carbon source. *Sulfolobus* can also grow chemoorganotrophically. Cells of *Sulfolobus* are more or less spherical but contain distinct lobes (Figure 13.17a). Cells adhere tightly to sulfur crystals where they can be seen microscopically using fluorescent dyes (Figure 17.26b). Besides the aerobic respiration of sulfur or organic compounds, *Sulfolobus* can also oxidize Fe^{2+} to Fe^{3+}, and this has been applied in the high-temperature leaching of iron and copper ores (◌◌ Section 19.15).

A facultative aerobe resembling *Sulfolobus* also lives in acidic solfataric springs. This organism, named *Acidianus* (Figure 13.17b), differs from *Sulfolobus* most clearly by its ability to grow anaerobically. Remarkably, *Acidianus* is able to use S^0 *both* aerobically and anaerobically. Under aerobic conditions the organism uses S^0 as an electron *donor*, oxidizing S^0 to H_2SO_4, with O_2 as electron acceptor. Anaerobically, *Acidianus* uses S^0 as an electron *acceptor* (with H_2 as electron donor) forming H_2S as

---

*Historical note: *Sulfolobus* was first discovered by Thomas Brock and colleagues in 1970, and formally described in 1972. The discovery of *Sulfolobus*, along with the previously isolated *Thermus aquaticus* (source of the extremely thermostable *Taq* DNA polymerase, ◌◌ Section 7.9), is generally considered to have launched the field of hyperthermophilic microbiology. *Sulfolobus*, in particular, was the first prokaryote shown to be able to grow above 80°C. Thomas Brock was the senior author of the first seven editions of this book and is now an emeritus professor at the University of Wisconsin in Madison. Since 1980, Karl Stetter and colleagues at the University of Regensburg, Germany, have greatly expanded the field of hyperthermophilic microbiology with the discovery of many new genera and species.

(a)

T. D. Brock

(b)

H. König and K. O. Stetter

● **Figure 13.17   Acidophilic hyperthermophilic *Archaea*, the Sulfolobales.** (a) *Sulfolobus acidocaldarius.* Electron micrograph of a thin section. (b) *Acidianus infernus.* Electron micrograph of a thin section. Cells of both organisms vary from 0.8 to 2 μm in diameter.

the reduced product. Thus, the metabolic fate of S^0 in cultures of *Acidianus* depends on the presence or absence of O_2. Like *Sulfolobus, Acidianus* is roughly spherical in shape but is not as lobed (Figure 13.17*b*). It grows at temperatures from 65°C up to a maximum of 95°C, with an optimum of about 90°C.

Thermoproteales

Key genera within the Thermoproteales are *Thermoproteus, Thermofilum,* and *Pyrobaculum.* The genera *Thermoproteus* and *Thermofilum* consist of rod-shaped cells that inhabit neutral or slightly acidic hot springs. Cells of *Thermoproteus* are rigid rods about 0.5 μm in diameter and highly variable in length, ranging from short cells of 1–2 μm (Figure 13.18*a●*) up to filaments 70–80 μm long. Filaments of *Thermofilum* are thinner, some 0.17–0.35 μm wide with filament lengths ranging up to 100 μm (Figure 13.18*b*).

Both *Thermoproteus* and *Thermofilum* are strict anaerobes that carry out a S^0-based anaerobic respiration (Table 13.8). Unlike most hyperthermophiles, the oxygen sensitivity of *Thermoproteus* and *Thermofilum* is extreme, comparable to that of the methanogens (see Section 13.4). Thus, strict precautions must be taken in their culture. Most *Thermoproteus* isolates can grow chemolithotrophically on H_2 or chemoorganotrophically on complex carbon substrates such as yeast extract, small peptides, starch, glucose, ethanol, malate, fumarate, or formate (see Table 13.8). *Thermoproteus* and *Thermofilum* have similar GC base ratios (56–58% GC), but their genomes are distinct by nucleic acid hybridization analyses (∞ Section 11.11).

(a)

H. König and K. O. Stetter

(b)

H. König and K. O. Stetter

(c)

R. Rachel and K. O. Stetter

● **Figure 13.18   Rod-shaped hyperthermophilic *Archaea*, the Thermoproteales.** (a) *Thermoproteus neutrophilus.* Electron micrograph of a thin section. A cell is about 0.5 μm in diameter. (b) *Thermofilum librum.* A cell is about 0.25 μm in diameter. (c) *Pyrobaculum aerophilum.* Transmission electron micrograph of a thin section; a cell measures 0.5 × 3.5 μm.

Pyrobaculum (Figure 13.18c) is a rod-shaped hyper-thermophile but is physiologically unique from other Thermoproteales in at least one sense; some species of *Pyrobaculum* are capable of aerobic respiration. In addition, cultures of *Pyrobaculum* can carry out anaerobic respirations with NO_3^-, Fe^{3+}, or S^0 as electron acceptors and H_2 as electron donor (that is, chemolithotrophic and autotrophic growth). Other species of *Pyrobaculum* can grow anaerobically on organic electron donors, reducing S^0 to H_2S. The growth temperature optimum of *Pyrobaculum* is 100°C, and species of this organism have been isolated from terrestrial hot springs as well as from hydrothermal vents.

(a)

13.10 Hyperthermophiles from Submarine Volcanic Habitats: Desulfurococcales

Key Genera: *Pyrodictium, Pyrolobus, Ignicoccus, Staphylothermus*

We now turn to the microbiology of *submarine* volcanic habitats, homes to the most thermophilic of all known *Archaea*. These habitats include both shallow water thermal springs and deep-sea hydrothermal vents. We discuss the geology of these fascinating microbial habitats in Section 19.8. The organisms to be described here make up an order-level group of *Archaea* called the Desulfurococcales (Table 13.9).

Pyrodictium and *Pyrolobus*

Pyrodictium and *Pyrolobus* are examples of prokaryotes whose growth temperature optimum lies above 100°C. The optimum for *Pyrodictium* is 105°C and for *Pyrolobus* is 106°C. Cells of *Pyrodictium* are irregularly disc-shaped and grow in culture in a mycelium-like layer attached to crystals of elemental sulfur. The cell mass consists of a network of fibers to which individual cells are attached (Figure 13.19a, b●). The fibers are hollow and consist of protein arranged in a fashion similar to that of bacterial flagella (⌘ Section 4.14). However, the filaments do not function in motility but instead as organs of attachment (see Figure 13.22). The cell walls of *Pyrodictium* are composed of glycoprotein. Physiologically, *Pyrodictium* is a strict anaerobe that grows chemolithotrophically on H_2 with S^0 as electron acceptor or chemoorganotrophically on complex mixtures of organic compounds (see Table 13.8).

Pyrolobus fumarii (Figure 13.19c) currently holds the record for the most thermophilic of all well-characterized

(b)

(c)

● **Figure 13.19 Desulfurococcales with growth temperature optima** >100°C. (a) *Pyrodictium occultum* (growth temperature optimum, 105°C), dark-field micrograph. (b) Thin section electron micrograph of *P. occultum*. Cells are highly variable in diameter from 0.3 to 2.5 μm. (c) Thin section of a cell of *Pyrolobus fumarii*, the most thermophilic of all known prokaryotes (growth temperature optimum, 106° C); a single cell is about 1.4 μm in diameter.

prokaryotes. Its growth temperature maximum is 113°C (Table 13.9). *Pyrolobus* lives in the walls of "black smoker" hydrothermal vent chimneys (👁 Section 19.8 and Figure 19.20), where its autotrophic abilities contribute organic carbon to this otherwise inorganic environment. Other species of *Pyrolobus* have been isolated that can apparently grow even above 113°C, so this genus clearly contains the most hyperthermophilic of all known microorganisms.

Pyrolobus cells are coccoid-shaped (Figure 13.19*c*), and the cell wall is composed of protein. The organism is an obligate H_2 chemolithotroph, growing by the oxidation of H_2 coupled to the reduction of NO_3^- (to NH_4^+), $S_2O_3^{2-}$ (to H_2S), or very low concentrations of O_2 (to H_2O). Besides its extremely thermophilic nature, *Pyrolobus* is also resistant to temperatures substantially above its growth temperature maximum. For example, cultures of *P. fumarii* survive autoclaving (121°C) for 1 h, a condition that even bacterial endospores (👁 Section 4.13) cannot withstand.

Desulfurococcus and *Ignicoccus*

Other notable members of the Desulfurococcales include *Desulfurococcus*, the genus for which the order is named (Figure 13.20*a*●), and *Ignicoccus*. *Desulfurococcus* is a strictly anaerobic S^0-reducing bacterium like *Pyrodictium* but differs from this organism in that it is much less thermophilic, growing optimally at about 85°C.

Ignicoccus grows optimally at 90°C, and its metabolism is H_2/S^0 based, as is that of so many hyperthermophilic *Archaea* (Table 13.8). *Ignicoccus* (Figure 13.20*b*) is a novel hyperthermophile because it contains an *outer membrane*, similar to that of *Bacteria* (👁 Section 4.9). The outer membrane of *Ignicoccus* is unusual, however, in that it is present at some distance from the cytoplasm of the cell. This arrangement allows for an unusually large periplasm (Figure 13.20*b*). Indeed, the volume of the *periplasm* of *Ignicoccus* is some 2–3 times that of its *cytoplasm*, in contrast to that of gram-negative *Bacteria*, where periplasmic volume is about 25% that of the cytoplasm. The periplasm of *Ignicoccus* also contains membrane-bound vesicles (Figure 13.20*b*) that may function in exporting substances outside the cell. In addition, however, some *Ignicoccus* species are hosts to a small, parasitic prokaryote, *Nanoarchaeum*, as discussed in the next section.

Staphylothermus

A morphologically unusual member of the order Desulfurococcales is the genus *Staphylothermus* (Figure 13.21●). Cells of *Staphylothermus* are spherical, about 1 μm in diameter, and form aggregates of up to 100 cells, resembling those of *Staphylococcus* (compare Figure 13.21 with Figure 4.4*b*). *Staphylothermus* is not a chemolithotroph like so many of its hyperthermophilic relatives, but in-

(a)

R. Rachel and K. O. Stetter

(b)

Reinhard Rachel and Karl O. Stetter

● **Figure 13.20 Examples of Desulfurococcales with growth temperature optima below the boiling point.** (a) Thin section of a cell of *Desulfurococcus saccharovorans*; a cell is 0.7 μm in diameter. (b) Thin section of a cell of *Ignicoccus islandicus*. The cell proper is surrounded by an extremely large periplasm (👁 Section 4.9). The cell itself measures about 1 μm in diameter but cell plus periplasm measures 1.4 μm.

stead is a chemoorganotroph that grows optimally at 92°C. Energy is obtained from the fermentation of peptides, producing the fatty acids acetate and isovalerate as fermentation products (Table 13.8).

Isolates of *Staphylothermus* have been obtained from both shallow marine hydrothermal vents, as well as the very hot black smokers (👁 Section 19.8). This organism is thus widely distributed in submarine thermal areas where it is likely a significant consumer of proteins released from dead organisms.

● **Figure 13.21** **The hyperthermophile *Staphylothermus marinus* (growth temperature optimum, 92°C).** Electron micrograph of shadowed cells. A single cell is about 1 μm in diameter.

 13.8–13.10 Concept Check

Hyperthermophilic Crenarchaeota inhabit the hottest habitats currently known to support life. Cold-dwelling phylogenetic relatives of these organisms are also known. A variety of different morphological types of Crenarchaeota are known and several different metabolic strategies used to support growth.

◆ What are the major differences between the organisms *Sulfolobus* and *Pyrolobus*? *Staphylothermus* and *Ignicoccus*?

◆ What is unusual about the metabolic properties of *Acidianus* regarding elemental sulfur (S^0)?

◆ What energy class of organisms do we call those organisms that use H_2 as electron donor? List the genera of Crenarchaeota that can use H_2 and what they use as electron acceptors.

IV PHYLUM NANOARCHAEOTA

A unique phylum of *Archaea* contains tiny parasitic cells, the **Nanoarchaeota**. These fascinating prokaryotes harbor many surprises, including the smallest genomes of all known prokaryotes.

13.11 *Nanoarchaeum*

Nanoarchaeum is a genus of very small coccoid cells that live as parasites, or possibly as symbionts, of the crenarchaeote *Ignicoccus* (see Section 13.10). Cells of *Nanoarchaeum* are about 0.4 μm in diameter and contain only about 1% of the volume of an *Escherichia coli* cell. Cells of *Nanoarchaeum* replicate only when attached to the surface of *Ignicoccus* cells. *Nanoarchaeum* cells can occur

singly, in pairs, or up to 10 or more cells per *Ignicoccus* cell (Figure 13.22*a*●). The cell wall of *Nanoarchaeum* consists of an S-layer (⚬⚬ Section 4.10) that overlays what appears to be a periplasmic space (Figure 13.22*b*).

Nanoarchaeum cannot grow in pure culture but only with *Ignicoccus* as a host. Whether *Ignicoccus* receives any benefit from the association is unknown. *Nanoarchaeum* is hyperthermophilic, with optimal growth at about 90°C. The metabolism of *Nanoarchaeum* is unknown, but its host is an autotroph, growing with H_2 as electron donor and elemental sulfur as electron acceptor. Isolates of *Nanoarchaeum* have been obtained from submarine hydrothermal vents as well as terrestrial hot springs, and the organism appears to be distributed worldwide in suitable hot habitats. Only one species, *N. equitans*, is known.

Phylogeny and Genomics of *Nanoarchaeum*

The Nanoarchaeota represent a deep lineage within the *Archaea*, distinct from all other phyla (Figure 13.1). Some difficulty in pinpointing the exact phylogenetic position of the Nanoarchaeota was originally encountered because the 16S ribosomal RNA of the group is so unique. *Nanoarchaeum* 16S rRNA contains many base substitutions, even in regions of the molecule that are highly conserved among other phyla of *Archaea*. In fact, oligonucleotide primers that amplify (using PCR) 16S rRNA genes from all known species of *Archaea*, fail to amplify the gene from *Nanoarchaeum*.

The genome of *Nanoarchaeum* is only 0.49 Mbp. This is the smallest genome of any cell known. Astonishingly, the *Nanoarchaeum* genome lacks identifiable genes for most metabolic functions, including the biosynthesis of monomers such as amino acids, nucleotides, and coenzymes. Also missing in the *Nanoarchaeum* genome are genes encoding proteins for catabolic pathways, such as glycolysis.

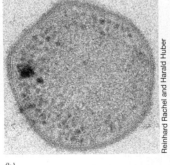

(a)　　　　*(b)*

● **Figure 13.22** *Nanoarchaeum.* (a) Fluorescence micrograph of cells of *Nanoarchaeum* (red) attached to cells of *Ignicoccus* (green). Cells were stained by FISH (⚬⚬ Section 11.7) using specific nucleic acid probes targeted to each organism. (b) Transmission electron micrograph of a thin section of a cell of *Nanoarchaeum*. Note the distinct cell wall. Cells of *Nanoarchaeum* are about 0.4 μm in diameter.

Presumably, all of these functions are carried out for *Nanoarchaeum* by its *Ignicoccus* host. *Nanoarchaeum* contains some but not all of the genes for an ATPase (∞ Section 5.12), and so whether or not the organism has a functional ATPase or not is unknown. If no ATPase is present and substrate-level phosphorylation is impossible, then it is likely that *Nanoarchaeum* is both an energy parasite and a carbon parasite of its *Ignicoccus* host. *Nanoarchaeum* likely parasitizes *Ignicoccus* for most of its needed metabolites, much as *Rickettsia* and *Chlamydia* do by infecting human hosts (∞ Sections 12.13, 12.27, 26.13, and 27.3).

With so many genes missing, what genes does *Nanoarchaeum* have left? *Nanoarchaeum* contains recognizable genes for all of the key enzymes needed for "core" molecular processes—DNA replication, transcription, and translation. Although its genome has been greatly pared down, *Nanoarchaeum* has not (and indeed, could not, and still be a cell) discarded genes necessary to fulfill the central dogma of molecular biology: DNA → RNA → Protein. In addition to the small size of its genome, the genome of *Nanoarchaeum* is the most compact (gene dense) of any organism known. Over 95% of the *Nanoarchaeum* single circular chromosome encodes proteins.

Evolution and *Nanoarchaeum*

Where does *Nanoarchaeum* fit in the evolution of prokaryotes? The organism roots deeply in the *Archaea* (Figure 13.1), which suggests it is the least derived of all known *Archaea*. However, was *Nanoarchaeum* always a parasite? Presumably not, since this would imply that an even less derived free-living host once existed for it to parasitize. Regardless of whether the ancestor of *Nanoarchaeum* was a free-living prokaryote, we can marvel today at its extremely small cell size and genome. Clearly, *Nanoarchaeum* is living at or close to the absolute limits of life in terms of both cell volume and genetic capacity. In fact, the only known self-replicating structures containing fewer genes than *Nanoarchaeum* are viruses, which are, of course, not cells. At present, the best scenario for the evolution of *Nanoarchaeum* is that it is a hyperthermophile that diverged early in the archaeal lineage to exploit a parasitic mode of existence. No free-living Nanoarchaeota are known, and thus until we learn otherwise, parasitism may be considered a hallmark of this phylum as we understand it today.

 13.11 Concept Check

Nanoarchaeum is a small, parasitic, early-branching member of the *Archaea*. Its genome is the smallest of all known organisms. *Nanoarchaeum* lacks genes for all but core molecular processes and thus depends on its host, *Ignicoccus*, for most of its cellular needs.

◆ What aspects of *Nanoarchaeum* make it especially interesting from an evolutionary perspective?

◆ How do we know that *Nanoarchaeum* is a hyperthermophile?

V EVOLUTION AND LIFE AT HIGH TEMPERATURES

The hyperthermophilic *Archaea* grow at temperatures far higher than those supporting the growth of other prokaryotes. What is the nature of this extreme heat tolerance? Are there temperature limits beyond which life is impossible? We will see that the answers offer clues about Earth's early history.

13.12 Heat Stability of Biomolecules

Protein and DNA stability in hyperthermophiles is critical to surviving high temperature. Most proteins denature at high temperatures, so much research has been done to identify the properties of thermostable proteins. The solution to protein thermostability turns on the *folding* of the molecule. As for nucleic acids, temperature is also a factor, and unique solutions have evolved in hyperthermophiles to keep their DNA intact.

Protein Folding and Thermostability

The amino acid composition of thermostable proteins from hyperthermophiles is not particularly unusual. In fact, enzymes from hyperthermophiles often contain the same major structural features, in terms of both their primary and higher-order structure (∞ Sections 3.6–3.8), as their heat labile counterparts from mesophilic bacteria. However, thermostable proteins tend to have additional features that improve thermostability. These include highly hydrophobic cores, which decrease the tendency of the protein to unfold, and more ionic interactions on the protein surfaces, which also help hold the protein together and work against unfolding Ultimately, however, it is the *folding* of the protein itself that most affects its heat resistance. Interestingly, subtle changes in amino acid sequence are often sufficient to affect a significant change in folding in a portion of a molecule that renders heat stable an otherwise heat-labile protein.

Chaperonins: Assisting Proteins to Remain in Their Native State

In Chapter 7 we discussed a class of proteins called *chaperonins* (heat-shock proteins, ∞ Section 7.17) that function to refold partially denatured proteins. Hyperthermophilic prokaryotes typically produce special classes of chaperonins that function only at the highest growth temperatures. In cells of *Pyrodictium* (Figure 13.23●), for example, the major chaperonin is a protein complex called the **thermosome**. This complex functions to keep the cell's other proteins properly folded and functional at high temperature and can help cells survive even above their maximal growth temperature. For example, cells of *Pyrodictium* grown near its maximum temperature (110°C) produce large amounts of thermosome protein. Because of this, the cells can remain viable following a

G. Rieger, R. Hermann, R. Rachel, K.O. Stetter

● **Figure 13.23** *Pyrodictium abyssi,* **scanning electron micrograph.** *Pyrodictium* has been studied as a model of macromolecular stability at high temperatures. Cells are enmeshed in a sticky glycoprotein matrix that binds them together.

heat shock, even a one-hour treatment in an autoclave (121°C). In cells experiencing such a treatment and then returned to the optimum temperature, the thermosome, which is itself quite heat resistant, refolds sufficient copies of key denatured proteins that *Pyrodictium* can once again begin to grow and divide. Thus, due to chaperonin activity, the upper temperature limit for *growth* of many hyperthermophiles is lower than the upper temperature at which they can *survive*. This "safety net" probably insures that cells that experience brief heat treatment above their growth temperature maximum are not killed by the exposure.

DNA Stability at High Temperatures: Solutes and Reverse Gyrase

How does DNA stay intact at high temperatures and keep from melting? A variety of mechanisms may be involved. One such mechanism involves increasing cellular solute levels. For example, as previously mentioned, the cytoplasm of hyperthermophilic methanogens like *Methanopyrus* (see Section 13.6) contains molar amounts of potassium cyclic 2,3-diphosphoglycerate. This solute prevents chemical damage to DNA, such as depurination or depyrimidization (loss of a nucleotide base through hydrolysis of the glycosidic bond) that can occur at high temperatures. These chemical changes can lead to mutation (👁 Section 10.2). Not all hyperthermophiles produce this solute, however, so obviously other DNA-stabilizing mechanisms exist.

It is thought that a unique protein found *only* in hyperthermophiles is the main reason that DNA does not melt in prokaryotes. All hyperthermophiles produce a DNA topoisomerase called **reverse DNA gyrase**. Reverse gyrase introduces *positive* supercoils into DNA (in contrast to the *negative* supercoils introduced by *DNA gyrase,* found in all nonhyperthermophilic prokaryotes 👁 Section 7.3). Positive supercoiling greatly stabilizes

DNA to heat and effectively prevents the DNA helix from undergoing denaturation. Since reverse gyrase is apparently absent from any organism whose growth temperature optimum is below about 80°C, this is a good indication of its importance to DNA stability at high temperatures.

DNA Stability: DNA-Binding Proteins

In addition to salts and reverse DNA gyrase, other proteins in hyperthermophiles may also function to maintain the integrity of duplex DNA. For example, a small heat-stable DNA-binding protein present in cells of *Sulfolobus* binds to the minor groove of DNA (👁 Figure 7.5) in a nonspecific manner and increases its melting temperature by some 40°C. This protein, called *Sac7d,* sharply kinks the DNA and thus may also be involved in gene regulation (👁 bent DNA, Section 8.4).

Species of Euryarchaeota also contain DNA-binding proteins. But unlike Sac7d, these are highly basic proteins that show strong amino acid sequence homology and folding properties with the core histones of the *Eukarya* (👁 Section 7.3). These *archaeal histones* have been particularly well studied from the hyperthermophilic methanogen *Methanothermus fervidus.* Histones from this organism wind and compact DNA into nucleosome-like structures (Figure 13.24●) (👁 Section 7.3). These structures have been shown to maintain the DNA in a double-stranded form at very high temperatures. Archaeal histones are found in most Euryarchaeota, including extremely halophilic *Archaea,* such as *Halobacterium.* However, since the extreme halophiles are not thermophiles, archaeal histones may have other functions besides assisting in DNA stability, such as in regulating gene expression.

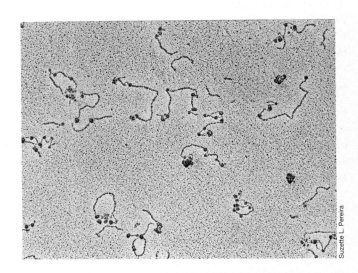

Suzette L. Pereira

● **Figure 13.24 Archaeal histones and nucleosomes.** Electron micrograph of linearized plasmid DNA wrapped around copies of archaeal histone Hmf (from the hyperthermophilic methanogen *Methanothermus fervidus,* see Figure 13.8*c*) to form the roughly spherical, darkly stained nucleosome structures (arrows). Compare this micrograph with an artist's depiction of the histones and nucleosomes of *Eukarya* shown in Figure 7.9.

Lipid Stability

What about cellular lipids? How do hyperthermophiles prevent their membranes from separating at high temperatures? Virtually all hyperthermophiles produce lipids constructed upon the dibiphytanyl tetraether model (∞ Section 4.5 for the structure of such lipids). Dibiphytanyl tetraether lipids are naturally heat resistant because the covalent bond between phytanyl units forms a lipid *monolayer* membrane structure instead of the normal lipid *bilayer* (∞ Figure 4.19). This structure, supported by covalent bonds, resists the tendency of heat to pull apart a lipid bilayer constructed of fatty acids.

Stability of Monomers

Besides the stability of *macromolecules*, the thermal lability of *monomers* is also an important factor in governing the upper temperature limits for growth of hyperthermophiles. No matter how stable macromolecules are, life is impossible if the basic building blocks themselves are unstable. Interestingly, it is the thermal lability of *small* molecules rather than macromolecules that may actually dictate the upper temperature for life.

Even at temperatures as "low" as 120°C, some important small molecules are destroyed at significant rates. For example, two key molecules in energy metabolism, ATP and NAD^+ (∞ Chapter 5), hydrolyze rapidly at these temperatures. For example, the half-life of ATP or NAD^+ *in vitro* is less than 30 minutes at 120°C and shortens dramatically at temperatures above this. (Note, however, that these are measurements done in water. The stability of these molecules within the cytoplasm could be significantly greater than this due to as yet undiscovered heat protective agents). So, is 120°C the upper temperature for life? Despite the problems cells face with the stability of small molecules, the upper temperature limit for life is likely to lie above 120°C.

The Temperature Limits to Microbial Existence

Because life as we know it depends on liquid water, microbial habitats hotter than 100°C are only found in environments under *pressure*, such as the sea floor. Indeed, all hyperthermophilic *Archaea* capable of growth over 100°C seem to be restricted to these superheated environments (see Table 13.9). How hot can these environments be and still support life? Some evidence already exists for a new *Pyrodictium* species capable of growth at 121°C. Thus, even the remarkable upper temperature limit for growth of *Pyrolobus fumarii* (113°C), is not the upper temperature limit for life.

Deep-sea hydrothermal vent black smokers offer models for superheated natural environments and their microflora. Black smokers emit hydrothermal fluid at 250–350°C and form upright metallic structures called *chimneys* from metal sulfides in the fluid (∞ Figure 19.19). Several hyperthermophiles have been isolated from smoker chimney walls, including *Pyrodictium*, *Methanopyrus*, and *Pyrolobus*. The walls show a gradient in temperature from 250 to 350°C inside to 2°C outside, and the organisms probably remain in regions of the wall nearest their temperature optima. Studies of the superheated water itself show it to be sterile, consistent with measurements that have shown that both small molecules and macromolecules would be instantly destroyed at such high temperatures.

Laboratory experiments on the stability of biomolecules point to 140–150°C as the most likely upper temperature limit for life as we know it. Above this, organisms would not be able to overcome the heat lability of the essential biomolecules of life. For example, ATP is instantly degraded at 150°C. Thus, if organisms exist that are capable of growth at temperatures even as "low" as 150°C, their energy economy would almost certainly have to be based on something other than ATP. Is this territory that evolution cannot overcome? It is too early to tell for sure. However, if "super" hyperthermophiles capable of growth at or above 150°C are someday discovered, they will likely reveal fundamental new principles in biology and biochemistry.

13.13 Hyperthermophilic *Archaea*, H₂, and Microbial Evolution

Why do so many *Archaea* seem to inhabit extreme environments? Although molecular probing of nonextreme environments (such as soil and water) indicate that *Archaea* are all around us (and even *in* us, as methanogens in our large intestine), a common theme of cultured *Archaea* is *adaptation to environmental extremes*. Until we have cultured the thus far elusive marine and soil *Archaea*, we cannot say for sure that extremophily is a hallmark of *Archaea*. But let's say for the sake of argument that it is. What would this say about Earth and evolution in general?

Extreme environments of various types existed on early Earth just as they do today, and it is within such environments that life may first have flourished. At the time that cellular life evolved nearly 4 billion years ago, it is almost certain that Earth was far hotter than it is today (∞ Section 11.1), and probably suitable only for hyperthermophiles. So, are extant hyperthermophiles relics of Earth's earliest life forms? Let's see what the phylogenetic tree of *Archaea* tells us about hyperthermophiles.

Hyperthermophiles and Their Slow Evolutionary Clocks

Molecular sequencing of rRNA genes suggests that hyperthermophilic *Archaea* evolved at slower rates than other organisms (Figure 13.1). Much the same can be said about hyperthermophilic *Bacteria* such as *Thermotoga* and *Aquifex*

(∞ Figure 12.1). This conclusion emerges from inspection of the evolutionary trees—lines to the hyperthermophiles are typically rather short and branch near the root (∞ Figures 12.1 and 13.1).

Thermal habitats themselves may explain why hyperthermophiles have such slow evolutionary clocks. Organisms living at very high temperatures may be under unusually strong constraints to maintain the characteristics critical to life there. Any change in amino acid sequence of a protein resulting from a spontaneous mutation must be a change that does not decrease its heat stability. If it does, the mutation could well be lethal. For organisms living in nonextreme environments, such pressures would not be a significant controlling factor in the diversification of their proteins. In other words, beyond a certain point, survival of hyperthermophiles may not be amenable to additional evolutionary change, since attempts to increase fitness must be balanced by the need for all cellular molecules to remain heat stable. If life originated on a hot planet Earth, as most evolutionary scenarios predict, then hyperthermophilic *Archaea* and *Bacteria* are likely the closest living relatives to early life forms that remain today. Therefore the biology of these hyperthermophiles is not only interesting, but may offer us a window into the past.

Hydrogen (H₂) as a Primitive Energy Source

Before we leave this chapter we must point out a final clue to the biology of early life forms. Note how often *H₂ metabolism* enters into the metabolic picture of hyperthermophilic prokaryotes, especially those growing at the highest temperatures (Figure 13.25●). Many hyperthermophiles grow anaerobically with H_2 as an electron donor and one or more of the electron acceptors S^0, NO_3^-, Fe^{3+}, or O_2 (see Table 13.8). H_2 metabolism is likely a physiological relic of ancient metabolic schemes.

These schemes would have evolved in primitive organisms because of the ready availability of H_2 and suitable inorganic electron acceptors in their primordial environments, and because the catabolism of H_2 can be a relatively simple process biochemically (∞ Figure 11.6).

Figure 13.25 shows the known upper temperature limits for growth of prokaryotes showing each of the three energy-conserving processes known in biology. Photosynthesis is obviously the most susceptible to heat, and one can only speculate that this is due to the instability of the large multi-protein complexes that must interact in the process of photophosphorylation (∞ Figure 17.15). Chemoorganotrophy occurs up to at least 110°C, as this is the upper temperature limit for growth of *Pyrodictium occultum*, an organism that can ferment certain organic compounds as well as grow chemolithotrophically on H_2 with S^0 as electron acceptor (Table 13.9). Above 110°C, only H_2-oxidizing *Archaea* are known, including the genera *Pyrolobus* and a *Pyrodictium* species that couples H_2 oxidation to the reduction of Fe^{3+} (Table 13.9 and Figure 13.25). Although there are indications from analyses of hydrothermal vent samples that *Archaea* exist in waters above 120°C, cultures of such organisms have not yet been obtained.

The diversity of H_2-oxidizing hyperthermophilic *Archaea* (and *Bacteria*, ∞ Sections 12.5 and 12.37) known today, attests to the evolution of H_2 oxidation as a metabolic success story. Indeed, even though we know little about the metabolism of the parasitic *Nanoarchaeum* (see Section 13.11), it is known that H_2 stimulates its growth in laboratory culture. Could *Nanoarchaeum* be the genetically simplest of all H_2-oxidizing prokaryotes? The use of H_2 by so many different physiological groups of prokaryotes, including even those living deep in the Earth (∞ Section 19.4), points to the antiquity of this source of energy.

Both hydrothermal vents, the habitat of the most extreme hyperthermophiles, and Earth's hot, deep subsurface have often been suggested as environments where life could have first arisen. These environments contain H_2 and

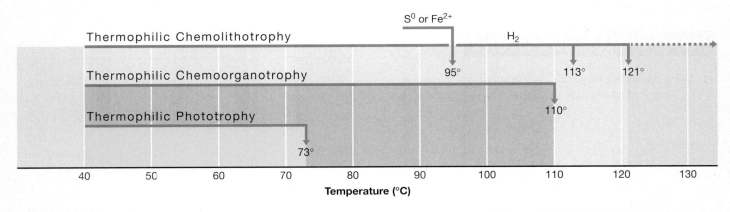

● **Figure 13.25 Upper temperature limits for energy metabolism.** Phototrophy, *Synechococcus lividus* (*Bacteria*, cyanobacteria); chemoorganotrophy, *Pyrodictium occultum* (*Archaea*); chemolithotrophy-S^0 as electron donor, *Acidianus infernus* (*Archaea*); chemolithotrophy-Fe^{2+} as electron donor, *Ferroglobus placidus* (*Archaea*); chemolithotrophy-H_2 as electron donor, *Pyrolobus fumarii* (*Archaea*, 113°C); *Pyrodictium* sp., strain 121 (*Archaea*, 121°C). Strain 121 also grows on formate but this compound may be metabolized through $H_2 + CO_2$.

an abundance of other inorganic electron donors and acceptors. They are also free from exposure to UV radiation. If this hypothesis is true, then the hydrogen-oxidizing hyperthermophiles discussed here and in Chapter 12 are probably relics of Earth's earliest biology.

 13.12–13.13 Concept Check

Although hyperthermophiles live at very high temperatures, in some cases above the boiling point of water, there are temperature limits beyond which no living organism can survive. This limit is likely 140–150°C. Hydrogen (H_2) catabolism may have been the first energy-yielding metabolism of cells.

◆ How do hyperthermophiles keep important macromolecules such as proteins and DNA from being destroyed by high heat?

◆ List at least two reasons why an upper temperature limit to life undoubtedly exists.

◆ What phylogenetic and physiological evidence suggests that extant hyperthermophiles most closely resemble ancient organisms?

REVIEW QUESTIONS

1. What are some features that all *Archaea* have in common (∞ Section 13.1)?

2. Which organism, *Pyrodictium, Thermoplasma*, or *Methanosarcina*, is the closest relative of the extreme halophile *Halobacterium*. (*Hint*: The answer lies in Figure 13.1.) What form of energy metabolism does each of these organisms have (∞ Sections 13.1 and 13.2)?

3. How can organisms such as *Halobacterium* survive in a high-salt environment whereas an organism such as *Escherichia coli* cannot (∞ Section 13.3)?

4. Contrast the roles of bacteriorhodopsin, halorhodopsin, and sensory rhodopsin in *Halobacterium salinarum* (∞ Section 13.3).

5. What is the electron *donor* for methanogenesis when CO_2 is reduced to CH_4 (∞ Section 13.4)?

6. What one major physiological feature unifies all members of the Thermoplasmatales? Why does this allow some of them to successfully colonize mine tailings (∞ Section 13.5 and Sections 19.14 and 19.15)?

7. What is physiologically unique about *Methanopyrus* compared with another methanogen such as *Methanobacterium* (∞ Section 13.6)? What is physiologically unique about *Archaeoglobus* (∞ Section 13.7)?

8. What forms of energy metabolism occur in Crenarchaeota? What form is absent (∞ Section 13.8)?

9. Why is the organism *Sulfolobus* of historical interest in the field of microbiology (∞ Section 13.9)?

10. What is unusual about the organism *Pyrolobus fumarii* (∞ Section 13.10)?

11. How is *Nanoarchaeum* similar to other *Archaea*? How does it differ (∞ Section 13.11)?

12. What is *reverse gyrase* and why is it important to hyperthermophiles (∞ Section 13.12)?

APPLICATION QUESTIONS

1. Using the data of Figure 13.1 as a guide, discuss why bacteriorhodopsin was likely a late evolutionary invention.

2. Defend or refute the following statement: The upper temperature limit to life is unrelated to the stability of macromolecules such as proteins or nucleic acids.

14

EUKARYOTIC CELL BIOLOGY AND EUKARYOTIC MICROORGANISMS

Eukaryotic cells include both microorganisms as well as the cells of plants and animals. The green alga *Dunaliella*, shown here, contains the typical organelles of a phototrophic eukaryotic cell, including a membrane-enclosed nucleus, mitochondria, and chloroplasts.

WORKING GLOSSARY

Algae phototrophic eukaryotic microorganisms

Amoeboid movement a type of motility in which cytoplasmic streaming moves the organism forward

Chitin a polymer of N-acetylglucosamine commonly found in the cell walls of algae and fungi

Chloroplast the photosynthetic organelle of eukaryotic phototrophs

Ciliate a protozoan characterized by rapid motility driven by numerous short appendages called cilia

Conidia asexual spores of fungi

Cristae the internal membranes of a mitochondrion

Cytoskeleton cellular scaffolding typical of eukaryotic cells, in which an array of microfilaments defines the cell's shape

Endosymbiosis the uptake of free-living *Bacteria* and their conversion into energy-generating organelles

Eukarya all eukaryotic organisms

Eukaryote a member of the *Eukarya*

Flagellate a protozoan characterized by motility driven by the whiplike action of one or more flagella

Fungi nonphototrophic eukaryotic microorganisms that contain rigid cell walls

Hydrogenosome an organelle of endosymbiotic origin present in certain anaerobic eukaryotic microorganisms that functions to oxidize pyruvate to H_2, CO_2, and acetate along with the production of one ATP

Meiosis the process of nuclear division during gamete formation when the diploid number of chromosomes is halved to the haploid number, present in the gametes

Microfilaments filamentous polymers of the protein *actin* that help maintain the shape of a cell (see *cytoskeleton*)

Mitochondrion the respiratory organelle of eukaryotic organisms

Mitosis during cell division in eukaryotic cells, the process by which chromosomes are replicated and partitioned to each daughter cell

Mold a filamentous fungus

Mushroom a filamentous fungus that produces large, often edible structures, called fruiting bodies

Nucleus the organelle that contains the genome

Phagocytosis a mechanism for ingesting particulate food in which a portion of the cell membrane surrounds the particle and brings it into the cell

Protozoa nonphototrophic unicellular eukaryotic microorganisms that usually lack cell walls

Ribozyme catalytic RNA

RNA processing the conversion of a precursor RNA to its mature form

Slime mold a nonphototrophic eukaryote related to protozoa that lacks cell walls and that aggregates to form fruiting structures (cellular slime molds) or masses of protoplasm (acellular slime molds)

Spliceosome a complex of ribonucleoproteins that catalyze the removal of introns from RNA primary transcripts

Sporozoa nonmotile parasitic protozoa

Telomerase an enzyme complex that replicates DNA at the end of eukaryotic chromosomes

Thylakoid a membrane layer containing the photosynthetic pigments in chloroplasts

Yeast a unicellular fungus

In this chapter we consider the structure, molecular biology, phylogeny, and diversity of eukaryotic microorganisms. In other words, the basic principles of eukaryotic cell biology will unfold in this chapter. Besides its basic science interest, the material will be useful for understanding several of the medical and applied aspects of microbiology in the remaining chapters of this book.

EUKARYOTIC CELL STRUCTURE/FUNCTION

The following five sections explore the structure of the eukaryotic cell and the ancestral link between organelles and *Bacteria* in the relationship called *endosymbiosis* (∞ Section 11.4).

14.1 Eukaryotic Cell Structure and the Nucleus

The typical eukaryotic cell is shown in Figure 14.1●. In contrast to prokaryotes, eukaryotes contain a *membrane-enclosed nucleus* and several other organelles, the complement of which depends on the organism. For example, *mitochondria* are nearly universal among eukaryotic cells, while the pigmented *chloroplasts* are found only in phototrophic cells. Eukaryotic cells may have a cell wall (as in plant cells, algae, or fungi) or may

not (as in animal cells or most protozoa). Other internal structures typically include the *Golgi apparatus, peroxisomes, lysosomes, endoplasmic reticula,* and *microtubules* and *microfilaments* (Figure 14.1). *Flagella* and *cilia*—organelles of motility—are present in some eukaryotic cells but not others.

Nucleus

The **nucleus** contains the genome of the eukaryotic cell (Figure 14.2●). In eukaryotes, DNA within the nucleus is wound around histones to form *nucleosomes* (∞ Figure 7.9) and from them, *chromosomes* (see Figure 14.13). In many eukaryotic cells the nucleus has a diameter of many micrometers and is easily visible with the light microscope even without staining (∞ Figure 4.7a). In smaller eukaryotes, however, special staining procedures are often required to see the nucleus.

The nucleus is enclosed by a *pair* of membranes, each with its own function, separated by a space. The inner membrane is a simple sac, while the outer membrane is in many places continuous with the endoplasmic reticulum. The inner and outer nuclear membranes specialize in interactions with the *nucleoplasm* and the *cytoplasm*, respectively.

The nuclear membrane also contains **pores** (Figure 14.2), formed from holes where the inner and outer membranes are joined. The pores allow a complex of proteins to import and export other proteins and nucleic acids into

Smooth endoplasmic reticulum

Rough endoplasmic reticulum

Cytoplasmic membrane

Mitochondrion

Microfilaments

Lysosome

Chloroplast

Nuclear envelope

Nuclear pores

Microtubules

Mitochondrion

Flagellum

Ribosomes

Peroxisome

Golgi complex

Nucleus

Nucleolus

● **Figure 14.1** **Schematic, cut-away view of a eukaryotic cell.** Although all eukaryotic cells contain a nucleus, not all organelles or other structures shown are present in all eukaryotic cells.

and out of the nucleus, a process called *nuclear transport*. Like transport across the cytoplasmic membrane (⟨∞⟩ Section 4.7), nuclear transport requires energy, and this comes from hydrolysis of the energy-rich compound, guanosine triphosphate (GTP).

Within the nucleus, the *nucleolus* (Figure 14.1) is the site of ribosomal RNA synthesis. The nucleolus is rich in RNA and often visible under the light microscope. Ribosomal proteins synthesized in the cytoplasm are transported into the nucleolus and combined with ribosomal RNA to form the small and large subunits of the eukaryotic ribosome. These are then exported to the cytoplasm, where they associate to form the intact ribosome and function in protein synthesis.

<table>
<tr><td>**14.2**</td><td>**Respiratory and Fermentative Organelles: The Mitochondrion and the Hydrogenosome**</td></tr>
</table>

The mitochondrion and the hydrogenosome specialize in chemotrophic energy metabolism. Both organelles are enclosed by membranes but have quite distinct functions.

Mitochondria

In aerobic eukaryotic cells, respiration and oxidative phosphorylation (a mechanism of ATP formation) (⟨∞⟩ Sections 5.11 and 5.12) are localized in **mitochondria** (singular, **mitochondrion**). Mitochondria are of prokaryotic dimensions and can be rod-shaped or nearly spherical (Figure 14.3●). A typical animal cell can contain over 1000 mitochondria, but the number per cell depends somewhat on the cell type and size. A yeast cell may have many fewer mitochondria per cell (see Figure 14.2). Mitochondria are surrounded by two membranes. The *outer membrane*, composed of an equal mixture of protein and lipid, is relatively permeable and contains numerous minute channels that allow passage of ions and small organic molecules. The *inner membrane* is more protein rich than the outer membrane and is also less permeable. Mitochondrial membranes lack sterols (⟨∞⟩ Section 4.5), so are much less rigid than the eukaryotic cytoplasmic membrane, which does contain sterols. Mitochondria can thus take on highly varied shapes (Figure 14.3*b, c*).

Mitochondria also possess a series of folded internal membranes called **cristae**. These membranes, formed by invagination of the inner membrane, are the

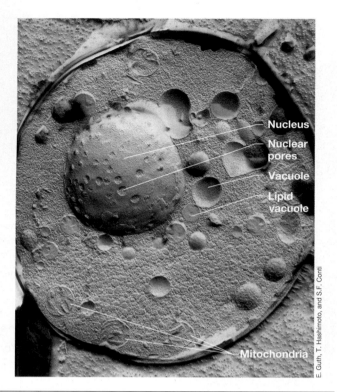

E. Guth, T. Hashimoto, and S.F. Conti

● **Figure 14.2** **The nucleus.** Electron micrograph of a yeast cell by freeze–etching, showing a surface view of the nucleus. Freeze–etching is a form of transmission electron microscopy (∞ Section 4.3), in which the specimen is frozen, fractured, and a thin replica made that is observed under the microscope. The cell is about 8 μm wide and the nucleus about 1.5 μm wide.

(a)

(b)   (c)

D. W. Fawcett

● **Figure 14.3** **Structure of the mitochondrion.** (a) Diagram showing the overall structure of the mitochondrion. Note inner and outermost membranes. (b, c) Transmission electron micrographs of mitochondria from rat tissue, showing the variability in morphology. Note the cristae.

sites of enzymes involved in respiration and ATP production. Cristae also contain specific transport proteins that regulate the passage of metabolites, in particular ATP, into and out of the *matrix* of the mitochondrion (Figure 14.3a). The matrix contains a number of enzymes involved in the oxidation of organic compounds—in particular, enzymes of the *citric acid cycle* (∞ Section 5.13).

The Hydrogenosome

Some *anaerobic* eukaryotic microorganisms lack mitochondria and contain instead a **hydrogenosome** (Figure 14.4a●). Although the size of a mitochondrion, the hydrogenosome lacks cristae and citric acid cycle enzymes. A variety of microbial eukaryotes contain hydrogenosomes and all are either obligate or aerotolerant anaerobes whose metabolism is strictly fermentative. Examples include such parasites as the flagellate *Trichomonas* (see Section 14.10) and various ciliated protozoa that inhabit the rumen of ruminant animals (∞ Section 19.11) or anoxic muds and sediments.

The major biochemical reactions in the hydrogenosome are the oxidation of pyruvate to H_2, CO_2, and acetate (Figure 14.4b). Pyruvate is oxidized to the energy-rich compound acetyl-CoA (from which additional ATP is made during the formation of acetate, ∞ Section 17.19), along with H_2 plus CO_2 (Figure 14.4b). In some anaerobic eukaryotes, H_2-consuming symbiotic pro-

karyotes such as methanogens reside in their cytoplasm (∞Figure 19.25b, c). The symbionts consume the H_2 and CO_2 produced by the hydrogenosome (Figure 14.4b), yielding methane. Because hydrogenosomes lack an electron transport chain and citric acid cycle, they cannot oxidize the acetate produced from pyruvate catabolism as mitochondria do. Acetate is therefore excreted from the hydrogenosome into the cytoplasm of the host cell (Figure 14.4b).

 14.2 *Concept Check*

The mitochondrion and the hydrogenosome are energy-generating organelles of eukaryotic cells. Mitochondria are involved in aerobic respiration, whereas the hydrogenosome, found only in certain obligately anaerobic eukaryotes, ferments pyruvate to yield H_2 plus CO_2, acetate, and ATP.

◆ What key reactions occur in the mitochondrion? What key product does the mitochondrion make for the cell proper?

◆ Compare and contrast the metabolic fate of pyruvate in mitochondria and the hydrogenosome.

(a)

(b)

● **Figure 14.4 The hydrogenosome.** (a) Electron micrograph of a thin section through a cell of the anaerobic flagellate, *Trichomonas vaginalis*, showing five hydrogenosomes in cross-section. Compare their internal structure with that of the mitochondrion in Figure 14.3. (b) Biochemistry of the hydrogenosome. Pyruvate is taken up by the hydrogenosome and H_2, CO_2, acetate, and ATP (all shown in red) are the products. The key enzymes of the hydrogenosome are *pyruvate: ferredoxin oxidoreductase* and *hydrogenase*. Endosymbiotic methanogens (⚬⚬⚬ Section 13.4) are often present in the cytoplasm of hydrogenosome-containing eukaryotes, producing methane from the $H_2 + CO_2$ (⚬⚬⚬ Figure 19.26*c*).

| **14.3** | **Photosynthetic Organelle: The Chloroplast** |

Chloroplasts are chlorophyll-containing organelles found in phototrophic eukaryotes—algae. Chloroplasts of many algae are relatively large and readily visible with the light microscope (Figure 14.5●). The size, shape, and number of chloroplasts per cell vary markedly, and in contrast to mitochondria, chloroplasts are typically much larger than prokaryotic cells.

Like mitochondria, chloroplasts have a permeable outermost membrane, a much less permeable inner membrane, and an intermembrane space. The inner membrane surrounds the lumen of the chloroplast, called the **stroma**,

(a)   (b)

● **Figure 14.5 Photomicrographs of algal cells showing chloroplasts.** (a) Fluorescence photomicrograph of the diatom *Stephanodiscus*. The chlorophyll in the chloroplasts (arrows) absorbs light and fluoresces red. (b) Phase contrast photomicrograph of *Spirogyra* showing the characteristic spiral-shaped chloroplasts (arrows) of this phototroph.

but it is not folded into cristae like the inner membrane of the mitochondrion (Figure 14.3*a*). Instead, chlorophyll and all other components needed for photosynthesis are located in a series of flattened membrane discs called **thylakoids** (Figure 14.6●). The thylakoid membrane is highly impermeable to ions and other metabolites because its function is to establish the proton motive force necessary for ATP synthesis (⚬⚬⚬ Section 5.12). In green algae and green plants, thylakoids are typically stacked into discrete structural units called *grana* (⚬⚬⚬ Figure 17.5).

The chloroplast stroma contains large amounts of the enzyme *ribulose bisphosphate carboxylase* (RubisCO). RubisCO is a key catalyst of the **Calvin cycle**, the series of biosynthetic reactions by which most photosynthetic organisms convert CO_2 to organic compounds (⚬⚬⚬ Section 17.6). RubisCO makes up over 50% of the total chloroplast protein and catalyzes the formation of *phosphoglyceric acid*, a key compound in the biosynthesis of glucose (⚬⚬⚬ Sections 5.10 and 5.15). The permeability of the outermost chloroplast membrane allows glucose and ATP produced during photosynthesis to diffuse into the cytoplasm where they can be used to build new cell material.

● **Figure 14.6 The chloroplast.** Transmission electron micrograph showing a chloroplast of the golden brown alga *Ochromonas danica*. Note the thylakoids.

14.3 *Concept Check*

Chloroplasts are the site of photosynthetic energy production and CO_2 fixation in eukaryotic phototrophs.

◆ Differentiate the *stroma* from *thylakoids*.

◆ What is the function of RubisCO and where is it found?

● **Figure 14.7 Cells of the yeast *Saccharomyces cerevisiae*.** The cells have been stained with 4'6-diamidine-2'-phenylindole dihydrochloride (DAPI) (⚭ Section 18.3) to show mitochondrial DNA. Each mitochondrion has two to four circular chromosomes that stain blue with the fluorescent dye used.

14.4 Endosymbiosis: Relationships of Mitochondria and Chloroplasts to *Bacteria*

On the basis of their relative autonomy, size, and morphological resemblance to prokaryotes, it was suggested long ago that mitochondria and chloroplasts were descendants of ancient bacteria. Establishment of these organelles was postulated to have been by the process of **endosymbiosis** (*endo* means "within"), the engulfment of a prokaryotic cell by a larger cell ("the host") and conversion of the engulfed cell into an organelle (⚭ Section 11.4 and Figure 11.9). Several lines of molecular evidence support the endosymbiotic theory, and we summarize these here:

1. **Mitochondria and chloroplasts contain DNA.**

   Although most of their functions are encoded by nuclear DNA, a few organellar components are encoded by a small genome present within the organelle itself. These include ribosomal RNA, transfer RNAs, and certain proteins of the respiratory chain (mitochondria) and photosynthetic apparatus (chloroplast). Thus, nonphototrophic eukaryotic cells are genetic chimeras containing DNA from *two* different sources, the *endosymbiont* and the *host cell nucleus*. Phototrophic eukaryotes—algae and higher plants—contain DNA from *three* different sources, the mitochondrial and chloroplast endosymbionts, and the nucleus. Most mitochondrial DNA and all chloroplast DNA exists in a covalently closed *circular* form, just as it does in most prokaryotes (⚭ Sections 2.2, 7.2, and 7.3). Mitochondrial DNA can be seen in cells by using special staining methods (Figure 14.7●). We discuss many other very interesting features of organellar genomes in Section 15.7.

2. **The eukaryotic nucleus contains bacterially derived genes.**

   Genomic sequencing (⚭ Chapter 15) and other genetic studies have clearly shown that several nuclear genes encode properties unique to organelles. Because the sequences of these genes more closely resemble those of *Bacteria* than those of *Archaea* or eukaryotes, it is concluded that these genes were transferred to the nucleus from bacterial symbionts during the transition from an initially engulfed cell to a modern day organelle.

3. **Mitochondria and chloroplasts contain their own ribosomes.**

   Ribosomes, cell structures involved in protein synthesis (⚭ Section 7.16), exist in either a large form [80 Svedberg (S) units], typical of the cytoplasm of eukaryotic cells, or in a smaller form (70S), present in prokaryotes. Mitochondria and chloroplasts also contain ribosomes, and they are 70S in size, the same as those of prokaryotes.

4. **Antibiotic specificity.**

   Several antibiotics (streptomycin is one example) kill or inhibit *Bacteria* by specifically interfering with 70S-ribosome function. These same antibiotics also inhibit protein synthesis in mitochondria and chloroplasts.

5. **Molecular phylogeny.**

   Phylogenetic studies using comparative ribosomal RNA sequencing methods (⚭ Sections 11.4–11.8) and organellar genome studies (⚭ Section 15.7) have shown convincingly that the chloroplast and mitochondrion originated from the *Bacteria*. The modern eukaryotic cell thus arose from an association of at least two quite distinct organisms by the process of endosymbiosis (⚭ Section 11.3 and Figure 11.7).

   The evidence that *hydrogenosomes* are endosymbionts is also strong. For example, in the obligately anaerobic ciliated protozoan *Nyctotherus ovalis* that lives in the hindgut of termites (⚭ Section 19.10), mitochondria are absent and hydrogenosomes have been identified that contain DNA and ribosomes. Moreover, the nucleus of hydrogenosome-containing eukaryotes contains genes encoding proteins of bacterial origin. Thus, the hydrogenosome originated by endosymbiosis, just as the mitochondrion did. In fact, hydrogenosomes are thought to be metabolically degenerate mitochondria that dispensed with respiration to exploit pyruvate fermentation as a means of energy conservation in a host cell with an anaerobic lifestyle. Although not rivaling the energy available from respiration in the mitochondrion, acetate production in the hydrogenosome supplies cells with more energy than if the fermentation products were lactate or ethanol (Figure 14.4*b*; ∞ Sections 5.10 and 17.19). The mitochondrion and the hydrogenosome can thus be viewed as functionally related organelles that specialize in different metabolic strategies for making ATP. Other structures called *mitosomes* are present in some eukaryotic cells and are probably even more degenerate mitochondria, having lost virtually all energy-related functions altogether (see Section 14.9).

As we can see, many forms of evidence point to organelles as having arisen from the endosymbiotic uptake of free-living *Bacteria* by eukaryotic host cells. In this cozy arrangement, host cells obtained permanent partners specializing in energy generation, while the symbionts received a stable and supportive growth environment. That endosymbiosis was an evolutionary success story can be attested to by the presence of mitochondria, hydrogenosomes, or chloroplasts in virtually all eukaryotic cells today.

 14.4 Concept Check

Key metabolic organelles of eukaryotes are the chloroplast, involved in photosynthesis, and the mitochondrion or hydrogenosome, involved in respiration or fermentation. These organelles were originally *Bacteria* that established permanent residence inside other cells (endosymbiosis).

◆ Summarize the molecular evidence that supports the relationship of organelles to *Bacteria*.

◆ Why might *Nyctotherus* be better off with hydrogenosomes than with mitochondria?

14.5 Other Organelles and Eukaryotic Cell Structures

A variety of other cytoplasmic structures are typically present in eukaryotic cells. These include the *endoplasmic reticulum, Golgi apparatus, lysosomes, peroxisomes*, and the organelles of motility—*flagella* and *cilia*. In contrast to mitochondria and chloroplasts, however, these structures lack DNA and ribosomes, and are not of endosymbiotic origin.

Endoplasmic Reticulum and Golgi Complex

The **endoplasmic reticulum** (ER) is a network of membranes continuous with the nuclear membrane. Two types of endoplasmic reticulum are recognized: *rough*, which contains attached ribosomes, and *smooth*, which does not (Figure 14.1). Smooth ER participates in the synthesis of lipids and in some aspects of carbohydrate metabolism. Rough ER, through the activity of its ribosomes, is a major producer of glycoproteins and also

produces new membrane material that is transported throughout the cell to enlarge the various cell membrane systems (see Figure 14.1) before cell division.

The **Golgi complex** consists of a stack of membranes distinct from the ER (Figures 14.1 and 14.8●), but which functions in concert with the ER. In the Golgi complex products of the ER are chemically modified and sorted into those destined to be secreted, for example, hormones or digestive enzymes, and those that function in other membranous structures in the cell. Golgi arise from the division of preexisting Golgi and contain various enzymes that modify secretory and membrane proteins differently, depending on their final destination in the cell. Many of the modifications involve *glycosylation* (adding sugar residues) to convert the proteins into specific glycoproteins.

Lysosomes and Peroxisomes

Lysosomes (Figure 14.1) are membrane-enclosed structures that contain various *digestive enzymes* that the cell uses to digest macromolecules such as proteins, fats, and polysaccharides. In doing so, the cell recycles these materials for new biosyntheses. The internal pH of the lysosome is about 5, two units lower than that of the cytoplasm, and the hydrolytic enzymes within the lysosome function optimally at this pH. These hydrolytic enzymes are nonspecific in their activity and could potentially destroy key cellular macromolecules if not contained. Thus, the lysosome allows for lytic activities to be partitioned away from the cytoplasm proper. Following hydrolysis of macromolecules in the lysosome, the resulting monomers pass from the lysosome into the cytoplasm as nutrients for the cell.

The **peroxisome** is a membrane-enclosed structure (Figure 14.1) whose function is to produce hydrogen peroxide (H_2O_2) from the reduction of O_2 by various hydrogen

● **Figure 14.8 The Golgi complex.** Transmission electron micrograph of a portion of a cell of the protozoan *Toxoplasma gondii*. The Golgi is colored in red. Note the multiple folded membranes that make up the Golgi complex (the membrane stacks are 0.5–1.0 μm in diameter). Other structures such as cytoplasmic granules are shown in other colors. *T. gondii* is a model system for the study of the Golgi complex because each cell contains only one such structure.

donors, including alcohols and long chain fatty acids. The H_2O_2 produced in the peroxisome is degraded to H_2O and O_2 by the enzyme *catalase* (◯◯ Section 6.16). Peroxisomes play other roles as well, such as synthesizing bile salts that aid in the absorption and digestion of fats. Peroxisomes originate in the cell by incorporating their proteins and lipids from the cytoplasm, eventually becoming a membrane-enclosed entity that can enlarge and divide in synchrony with the cell.

Microfilaments and Microtubules

Just as houses are supported by structural reinforcement, the large size of eukaryotic cells and their ability to move requires structural reinforcement. This internal structural network comes from proteins that form filamentous structures called **microfilaments** and **microtubules**. Together, these structures form the **cell cytoskeleton**.

Microfilaments are about 8 nm in diameter and are polymers of the protein *actin*. These fibers form scaffolds throughout the cell, defining and maintaining the shape of the cell (Figures 14.1 and 14.9●). *Microtubules* are larger filaments, about 25 nm in diameter, and are composed of the protein *tubulin*. Microtubules assist microfilaments in maintaining cell structure. They also play an important role in cell motility, both in the movement of internal cell structures (for example, in the separation of chromosomes during cell division) and in movement of the organism itself (for example, in movement of the flagellum in flagellated eukaryotic cells, see Figures 14.1 and 14.10).

We saw in an earlier chapter that prokaryotic cells produce structural and functional homologs of actin and tubulin in the form of the proteins MreB and FtsZ, respectively (◯◯ Section 6.1). This indicates that although prokaryotes are typically much smaller than eukaryotes, they still require some minimal level of internal scaffolding. The existence of the proteins MreB and FtsZ in prokaryotes also shows that the eukaryotic cytoskeleton has deep evolutionary roots.

Flagella and Cilia

Flagella and **cilia** are present on species of many groups of eukaryotic microorganisms. Flagella and cilia are organelles of motility, allowing cells to move by swimming motility. Cilia are essentially short flagella that beat in synchrony to propel the cell—usually quite rapidly—through the medium. Flagella are long appendages present singly or in groups that push the cell along—typically more slowly than by cilia—through a whip-like motion (Figure 14.10a●).

The flagella of eukaryotic cells are structurally quite distinct from bacterial flagella (◯◯ Section 4.14) and do not rotate. In cross-section, cilia and flagella are very similar. Each contains a bundle of 9 pairs of microtubules surrounding a central pair of microtubules called the *axoneme* (Figure 14.10b). Microtubules are composed of the protein *tubulin*. A second protein, called *dynein* is attached to the tubulin and functions as an ATPase, hydrolyzing ATP to yield the energy necessary to drive motility. Movement of flagella and cilia is similar. In both cases, movement involves the coordinated sliding of axonemal microtubules (Figure 14.10b) against one another in a direction toward or away from the base of the cell. This movement confers the whip-like action on the flagellum or cilium that results in cell propulsion.

● **Figure 14.9** **Microfilaments and eukaryotic cell architecture.** An electron tomographic image of a cell of the cellular slime mold *Dictyostelium discoideum* (see also Figure 14.28) showing the network of actin microfilaments that functions along with microtubules as the cell *cytoskeleton*. Microfilaments are about 8 nm in diameter. Electron tomography is a method for three-dimensional reconstruction of cells from a series of images taken with a transmission electron microscope.

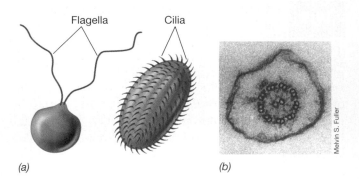

● **Figure 14.10** **Flagella and cilia, motility organelles in eukaryotic cells.** (a) Flagella can be present as single or multiple filaments. Cilia are structurally very similar to flagella but much shorter. Eukaryotic flagella move in a whip-like motion, unlike the flagella of prokaryotes, which propel the cell by rotating, much like a propeller on a motor boat (◯◯Section 4.14 and Figures 4.53–4.58). (b) Cross section through a flagellum of the fungus *Blastocladiella emersonii* showing the outer sheath, the outer nine pairs of microtubules, and the central pair of single microtubules.

A clear distinction can thus be made between the prokaryotic and eukaryotic flagellum. The filament of the prokaryotic flagellum is made from a helical array of a single protein, flagellin, and the structure itself is firmly anchored in a rotary motor complex embedded in the cell wall and membrane (∞ Figure 4.56). In addition, the bacterial flagellum functions as a propeller, being driven by the energy of the proton motive force (∞ Sections 4.14 and 5.12). The eukaryotic flagellum, by contrast, propels the cell by the whip-like motion of sliding microtubules driven by the energy of ATP. Nevertheless, in all motile cells, motility likely has survival value; the ability to move allows motile organisms to explore new habitats and exploit their resources.

 14.5 Concept Check

Besides the major organelles of eukaryotes, several other structures with defined functions are present in the cytoplasm. These include the endoplasmic reticulum, the site of ribosomes and cellular lipid syntheses; the Golgi apparatus involved in protein modification and secretion; lysosomes, which play a role in macromolecular digestion; and the peroxisome, an organelle involved in H_2O_2 production. In addition, proteinaceous tubes called microfilaments and microtubules are present, forming the cell's cytoskeleton. Flagella and cilia are organelles of motility that have extensive microtubular structure.

◆ How does *smooth* ER differ from *rough* ER?

◆ Why are the activities that occur in the lysosome best partitioned away from the cytoplasm proper?

◆ Besides scaffolding, what other functions do microtubules have?

II ESSENTIALS OF EUKARYOTIC GENETICS AND MOLECULAR BIOLOGY

Several aspects of the genetics and molecular biology of eukaryotes lack counterparts in most prokaryotic cells. These include (1) the replication of linear (as opposed to circular) genetic elements; (2) mitosis and meiosis; and (3) processing of messenger RNAs. We briefly explore each of these topics here.

14.6 Replication of Linear DNA

Most prokaryotic chromosomes are circular, as are most plasmids and some viruses, and the genomes of most organelles (∞ Section 15.7). In contrast, eukaryotic cells contain *linear DNA*. Almost all the steps in DNA replication are identical whether the chromosome is linear or circular. However, there is a key problem with the replication of linear genetic elements that is not an issue with circular ones: *replication of DNA at the extreme 5'–end of each strand*.

To understand the nature of the problem, first review Figure 7.13. Imagine that the left end of the DNA in this diagram is actually one end of a linear chromosome. Even if the RNA primer is very short and there is a special enzyme to remove it, no DNA polymerase can replace it with DNA since *all* known DNA polymerases require a primer. Therefore, if nothing is done, the DNA molecule will become shorter each time it is replicated. The replication of linear DNA thus requires special attention and there are at least two solutions to the problem.

Replication of Linear DNA Using a Protein Primer

Viruses that contain linear DNA genomes (this includes most viruses that infect eukaryotes) and many linear plasmids solve the problem of replicating linear DNA by using a *protein* primer instead of an *RNA* primer. Although all DNA polymerases must add each nucleotide to a free hydroxyl (OH) group, some DNA polymerases can add the first base onto an hydroxyl group present on specific proteins that bind to the ends of linear chromosomes (Figure 14.11●). These proteins are encoded by the plasmid or virus and they function to recognize the ends of the chromosomes. These protein primers are not removed, so these plasmids and viruses have proteins permanently attached to the 5'-ends of their DNA. Protein primers are also the means by which some linear chromosomes of *Bacteria* (∞ Section 15.4) are replicated.

Telomeres and Telomerase

A special method is used to replicate the ends of eukaryotic chromosomes, which are called *telomeres*. Telomeres

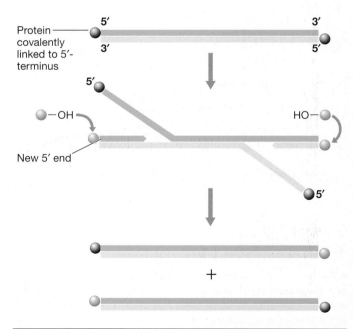

● **Figure 14.11 Replication of linear DNA using protein primers.** New strands of DNA are primed by proteins covalently attached to their 5' ends. Note the free OH group on the protein. DNA polymerase III can add a nucleotide to this OH group.

contain repetitive DNA—a short sequence (often 6 bp) tandemly repeated from 20 to several hundred times (Figure 14.12a●). The sequences from the telomeres of different eukaryotes are closely related, and one strand always has several guanines. In terms of replication, this guanine-rich sequence is present on the *leading* strand of the DNA duplex and can base pair with the 3′ end of a complementary RNA present in an enzyme called **telomerase** (Figure 14.12a).

Telomerase becomes active once bound to the 3′ end of the G-rich overhang of telomeric DNA. Telomerase does not need a template to begin DNA synthesis because the enzyme itself contains a small RNA template as a cofactor. In essence, telomerase is a type of reverse transcriptase (◌◌ Section 9.13), making DNA off of an RNA template. Telomerase works repetitively to make a long extension of the leading strand. Once this extension is made, the other strand (the lagging strand) can be primed with an RNA primer in the normal fashion and

the DNA replicated (Figure 14.12b). Telomeres do not need to be a precise number of repeats long, just long enough to ensure that no portion of a gene becomes lost during DNA replication.

14.6 Concept Check

The ends of linear genetic elements present a problem to the replication machinery that circular genetic elements do not. Some prokaryotic and viral linear elements solve this problem using a protein primer. Eukaryotes solve the problem using a special enzyme called telomerase to extend one strand of the DNA.

◆ How can a protein prime DNA for replication?

◆ What is *telomerase* and what is its function?

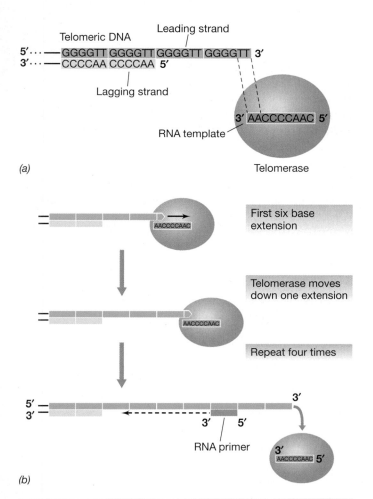

(a)

(b)

● **Figure 14.12 Model for the activity of telomerase at one end of a eukaryotic chromosome.** (a) A diagram of the sequence of the end of the DNA in a telomere, with four of the guanine-rich repeats, and the enzyme telomerase, which contains a short RNA template. (b) Steps in elongation of the guanine-rich strand catalyzed by telomerase. After telomerase finishes, the lagging strand can be primed with an RNA primer by primase followed by completion of the lagging strand by DNA polymerase and ligase.

14.7 Overview of Eukaryotic Genetics

Here we briefly consider some of the principles of eukaryotic genetics. Eukaryotic cells divide by mitosis and then undergo sexual reproduction. During the latter, male and female gametes are formed by meiosis and then fuse to form the zygote. This is fundamentally different from the unidirectional mating processes in prokaryotes (◌◌ Chapter 10), and so we focus on these events in our discussion, using the baker's yeast, *Saccharomyces cerevisiae*, as our model eukaryote.

Mitosis and Meiosis

From a genetic standpoint, eukaryotic cells can exist in two forms: *haploid* or *diploid*. Diploid cells have two copies of each chromosome whereas haploid cells have only one. Cells of *S. cerevisiae* can exist indefinitely in the haploid (vegetative) stage (containing 16 chromosomes). But occasionally, two haploid yeast cells can fuse (mate) to yield a diploid cell (32 chromosomes) (see Figure 14.14). This process should not be confused with the state of affairs in cells of multicellular plants and animals. In these organisms the diploid phase is present in the organism proper, and the haploid phase occurs only in the gametes.

Recall that **mitosis** is the process following DNA replication in a cell in which chromosomes condense, divide, and are separated into two sets, one for each daughter cell (Figure 14.13●). In haploid *S. cerevisiae* cells (Figure 14.14●), mitosis occurs before each cell division to maintain the 16 chromosomes per cell. **Meiosis** is the process by which the conversion from the diploid to the haploid stage occurs. Meiosis involves *two* divisions. In the first meiotic division, homologous chromosomes are segregated into separate cells, changing the genetic state from diploid to haploid. The second meiotic division is essentially the same as mitosis, as the two haploid cells divide to form a total of four haploid gametes (ascospores, Figure 14.14).

(a)

(b)

Carolina Biological Supply Co.

● **Figure 14.13** **Mitosis, as seen in the light microscope.** These are onion root tip cells that have been stained to reveal nucleic acid and chromosomes. (a) Metaphase. Chromosomes are paired in the center of the cell. (b) Anaphase. Chromosomes are separating.

The Mating Process in Yeast: Mating Types

Yeasts have two different forms of haploid cells called *mating types*; these can be considered analogous to male and female gametes. On mating of opposite types, a diploid cell is formed. From a single diploid cell, a structure containing four gametes is formed, two of each mating type. The cell in which the gametes are formed is called an *ascus*, and the cells within the ascus are called *ascospores* (Figure 14.14).

The two mating types of *Saccharomyces cerevisiae* are designated α and a. Cells of type a mate only with cells of type α, and whether a cell is a or α is genetically determined. Some haploid strains of *S. cerevisiae* remain a or α, while other strains are able to *switch* their mating type from one to the other. This switch in mating type is caused by an inserted gene and the process is illustrated in Figure 14.15●.

There is a single location on one of the *S. cerevisiae* chromosomes called the MAT (for *mating type*) locus, at which either gene a or gene α can be inserted. At this locus, the MAT promoter controls transcription of whichever gene is present. If gene a is at that locus, then the cell is mating type a, whereas if gene α is at that locus, the cell is mating type α. Elsewhere in the yeast genome are copies of genes a and α that are not expressed. These *silent copies* are the source of the inserted gene. When the switch occurs (Figure 14.15), the appropriate gene, a or α, is copied from its silent site and inserted into the MAT location, *replacing* the gene already present. The old mating type gene is excised and discarded, and the new gene is inserted. This process has been called the *cassette mechanism* because the a and α genes can be considered analogous to magnetic tape cassettes. Whichever gene is inserted in the MAT locus is the one that will be transcribed.

The α and a genes of *S. cerevisiae* are *regulatory* genes. Among other things, they regulate the production of the peptide hormones α *factor* or a *factor*, which are excreted by yeast cells undergoing mating. These hormones bind to cells of the opposite mating type and bring about changes in their cell surfaces so that fusion can occur (Figure 14.16●). Once cells of opposite mating type have associated, fusion of both the cells and their nuclei occurs, resulting in the formation of a diploid zygote (Figure 14.16). From the latter, meiosis regenerates the haploid vegetative form and the life cycle is complete (Figure 14.14).

14.7 Concept Check

Eukaryotic microorganisms can mate and exchange DNA during sexual reproduction. Mitosis ensures appropriate segregation of the chromosomes during asexual cell division. Haploid cells formed by meiosis can fuse to form a diploid zygote. There are two mating types in yeast, and yeast cells can convert from one type to the other.

Web Tutorial 14.1 Life Cycle and Mating Type in a Typical Yeast

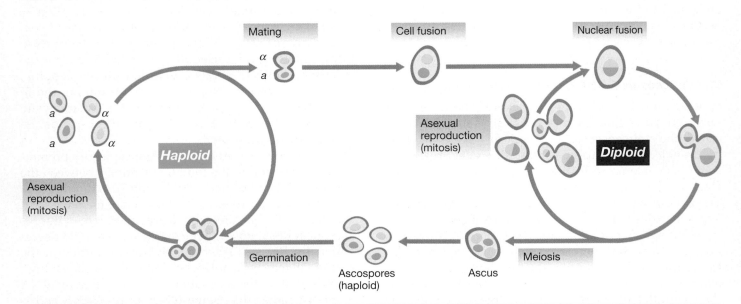

● **Figure 14.14** **Life cycle of a typical yeast, *Saccharomyces cerevisiae*.** A haploid cell of *S. cerevisiae* contains 16 chromosomes.

(a)

(b)

● **Figure 14.15** **The cassette mechanism involved in the switch in yeast from mating type α to a.** Whichever "cassette" is inserted at the active locus (reading head) determines the mating type. The process shown is reversible, so type *a* can also revert to type *α*.

◆ If the diploid number of chromosomes in human cells is 46, how many chromosomes are present in a human sperm cell?

◆ Explain how a *single* haploid cell of *Saccharomyces* can eventually yield a diploid cell.

14.8 RNA Processing and Ribozymes

In Chapter 7 we learned that the process of transcription yields several types of RNA: *messenger RNA, transfer RNA*, and *ribosomal RNA* (∞ Section 7.10). In eukaryotes, and much less commonly in prokaryotes, mRNA is *cleaved* to remove **introns**, the intervening noncoding RNA. This is called **RNA processing**. Processing of mRNA leaves **exons**, the protein encoding RNAs, which are joined to form the mature (functional) mRNA (∞ Section 7.1). The processing by which introns are removed and exons are joined is called **splicing**.

The Spliceosome, RNA Capping, and the Poly-A Tail

Splicing is done by a complex of several ribonucleoproteins (enzymes that contain both RNA and protein), called the **spliceosome**. The spliceosome is a large macromolecular complex—about the size of a ribosome—and along with other nonspliceosome protein factors, over 100 proteins are involved in its activity. The spliceosome functions to remove introns and join adjacent exons to form a mature mRNA (Figure 14.17●).

RNA splicing occurs while the mRNA is still in the nucleus. Note that there are some conserved bases at the splice junctions and that the intron is removed as a *lariat structure* (Figure 14.17). These excised introns are then degraded by the cell, releasing free nucleotides. In many mRNAs there may be more than one intron in a

● **Figure 14.16** **Electron micrographs of the mating process in a yeast, *Hansenula wingei*.** (a) Two cells have fused at the point of contact and have sent out protuberances toward each other. (b) Late stage of mating. The nuclei of the two cells have fused, and a diploid bud has formed at a right angle to the mating cells. This bud eventually separates and becomes the progenitor of a diploid cell line. A single cell of *Hansenula* is about 10 μm in diameter.

single gene, and so it is clearly important that they all be recognized and removed by the spliceosome to generate the final mature mRNA.

Two other unique steps occur in the processing of eukaryotic mRNA. Both steps take place in the nucleus before splicing and transport of the mature mRNA into the cytoplasm for translation. The first step is called **capping** and actually occurs before transcription is complete. Capping is the adding of a *methylated guanine nucleotide* at the 5′-phosphate end of the mRNA. Occasionally other nucleotides near the 5′ end of the eukaryotic mRNA are also modified during capping. The final processing step consists of trimming the 3′ end of the pre-mRNA and adding about 200 adenylate residues as the **poly-A tail**. Neither the cap nor the poly-A tail are translated.

The guanosine cap facilitates translation by promoting the formation of the initiation complex between the mRNA and the ribosome (∞ Section 7.16) through specific cap-binding proteins. The function of the poly-A tail is less clear. It may affect the degradation rate of the eukaryotic mRNA, possibly protecting it from rapid degradation by RNases. Alternatively (or in addition), the poly-A tail may be a molecular signal of "translatability," an indication to the translation machinery that the RNA is mRNA rather than some other form of RNA and is ready for translation.

(a) Exon 1   Intron   Exon 2

Conserved bases

Assembly of spliceosome

(b) Spliceosome

Cutting of **5′** splice site, formation of lariat

(c)

Cutting of **3′** splice site, joining of exons

(d)

(e) Intron (lariat)

Degraded

5′ Exon 1 Exon 2 3′ Mature mRNA

Exported to cytoplasm for translation

● **Figure 14.17 Activity of the spliceosome.** Removal of an intron from the transcript of a protein-encoding gene in a eukaryote. (a) The pre-mRNA with a single intron. The sequence GU is conserved at the 5′ splice site and AG at the 3′ splice site. There is also an interior A that serves as a branch point. (b) Several small ribonucleoprotein particles (shown in brown) assemble on the RNA to form a spliceosome. Each of these particles contains distinct small RNA molecules that are involved in the splicing mechanism. (c) The 5′ splice site has been cut with the simultaneous formation of a branch point. (d) The 3′ splice site has been cut, while the two exons were joined. Note that overall, two phosphodiester bonds were broken but two others were formed. (e) The final products are the joined exons (the mRNA), and the released intron.

The three steps leading to the formation of eukaryotic mRNA are summarized in Figure 14.18●. The synthesis of a mature, functional mRNA in eukaryotes is a complex and dynamic process that involves several steps not seen in the comparatively direct formation of functional mRNAs during transcription of most genes in prokaryotes (⟳ Section 7.13).

Ribozymes

Certain types of RNA function in the cell as *enzymes* as well as RNAs. These *catalytic RNAs*, which are called **ribozymes**, are involved in a number of important cellular

reactions. Although a few ribozymes have been found in prokaryotes, they are more common in eukaryotes. Ribozymes work like protein enzymes in that they have an "active site" that binds the substrate and catalyzes formation of a product (⟳ Section 5.5).

Most eukaryotic ribozymes are **self-splicing introns**, enzymes that excise themselves from an RNA molecule while joining adjacent exons together. One very well-studied ribozyme is present in the protozoan *Tetrahymena*. The ribozyme is a 413-nucleotide intron, that, in the presence of guanosine, splices itself out of a longer precursor rRNA. During this event, the ribozyme joins two adjacent exons to form the mature rRNA (Figure 14.19●). This *Tetrahymena* ribozyme is thus a *sequence-specific endoribonuclease*, and carries out a reaction analogous to that of the spliceosome (Figure 14.17). Once removed from the precursor RNA, the ribozyme circularizes following the removal of a short oligonucleotide fragment from itself (Figure 14.19).

Although catalyzing a specific reaction just like protein enzymes do, self-splicing ribozymes differ from protein enzymes in a key way. Unlike protein enzymes, a self-splicing ribozyme catalyzes its reaction only once. The circular fragment of the *Tetrahymena* ribozyme, for example, is eventually degraded by the cell (Figure 14.19). Other ribozymes that carry out this same self-splicing process (referred to collectively as *group I introns* because of the requirement for an exogenous guanosine in the reaction) have been found in a variety of other genes of eukaryotes and in certain genes of mitochondria and chloroplasts.

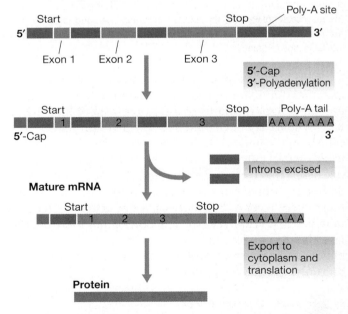

Pre-mRNA (primary transcript)

Start   Stop   Poly-A site

5′   3′

Exon 1   Exon 2   Exon 3

5′-Cap
3′-Polyadenylation

Start   Stop   Poly-A tail

5′-Cap   1   2   3   AAAAAAA   3′

Introns excised

Mature mRNA

Start   Stop

1   2   3   AAAAAAA

Export to cytoplasm and translation

Protein

● **Figure 14.18 An overview of the processing of pre-mRNA into mature mRNA in eukaryotes.** The processing steps include adding a cap at the 5′ end, removing the introns, and clipping of the 3′ end of the transcript while adding a poly-A tail. All these steps are carried out in the nucleus. The location of the start and stop codons to be used during translation are also indicated.

● **Figure 14.19 Self-splicing ribozymal intron of the protozoan _Tetrahymena._** There is considerable secondary structure in such molecules, which is critical for the splicing reaction. (a) A ribosomal RNA precursor contains a 413-nucleotide intron. (b) Following the addition of the nucleoside guanosine, the intron splices itself out and joins the two exons. (c) The intron is spliced out and then circularizes with the loss of a 15-nucleotide fragment.

Why do ribozymes exist in a world where protein enzymes dominate? It has been proposed that ribozymes are the vestigial remains of a simpler form of life, "RNA life," which may have predated the era of proteins as the cell's major catalysts. Indeed it is possible that an _RNA world_ existed long before a cellular structure evolved. We discussed this concept in Chapter 11 (⟳ Section 11.2). Although proteins have replaced RNA enzymes in almost every facet of biocatalysis, a few reactions remain in which RNA catalyzes the reaction. Either these are the last "holdouts" from the RNA world, or alternatively, may be reactions that protein enzymes are incapable of or catalyze only very poorly.

 14.8 Concept Check

The processing of eukaryotic pre-mRNAs is unique and involves three distinct processing steps: splicing, capping, and tailing. Introns in some other transcripts are self-splicing and the RNA itself catalyzes the reaction. RNA molecules with catalytic activity are called ribozymes and play an important role in the cell.

◆ What is _splicing_, where does it occur, and what is required for it to occur?

◆ What is a _ribozyme?_ What substrate do all group I introns require for enzymatic activity?

III EUKARYOTIC MICROBIAL DIVERSITY

We now explore the diversity of microbial _Eukarya._ These include the protozoa, the slime molds, the fungi, and the algae. We preface our tour with a phylogenetic consideration of these organisms. Here we will learn that microbial eukaryotes vary in many aspects of their biology and that some are apparently phylogenetically highly derived while others may be much less so, retaining properties of ancient eukaryotes.

14.9 Phylogeny of the _Eukarya_

From the universal phylogenetic tree of life (⟳ Figure 11.13) we learned that _Eukarya_ form their own phylogenetic domain more closely related to _Archaea_ than to _Bacteria._ This phylogeny of _Eukarya_ was deduced from comparative sequencing of 18S ribosomal RNA, obtained from the small subunit (SSU) of cytoplasmic ribosomes (⟳ Section 11.5).

Most microbiologists agree that phylogenetic trees based on comparative sequencing—particularly of SSU rRNA—are the best molecular trees available and generally yield a valid picture of microbial phylogeny. However, agreement is considerably stronger in this regard for the phylogeny of the prokaryotic domains—_Bacteria_ and _Archaea_—than for the eukaryotes. Comparative sequencing of certain eukaryotic genes and proteins has yielded a rather different tree topology for eukaryotes than that based on 18S rRNA. At this point no consensus has been achieved as to which tree gives the best picture of the phylogeny of _Eukarya._ Thus, we present here two phylogenetic trees that paint a contrasting picture of eukaryote phylogeny (Figure 14.20●). We begin with the SSU tree that clearly distinguishes early from more derived eukaryotes. We then consider an alternative tree that concludes that a major phylogenetic radiation took place during the evolution of the _Eukarya._ Both SSU trees and alternative trees confirm the three domains of life. It is mainly with the details of the eukaryotic lineage that the trees disagree.

The Ribosomal RNA Tree of _Eukarya_

Before the era of molecular phylogeny eukaryotic organisms were grouped into four **kingdoms**: _plants, animals, fungi,_ and a variety of microbial eukaryotes called _protists._ These four kingdoms, along with the _prokaryotes_ (which were simply lumped into one phylogenetic kingdom), formed the so-called _five kingdoms of living organisms._ However, SSU rRNA sequencing has convincingly shown that the five-kingdom system greatly magnified

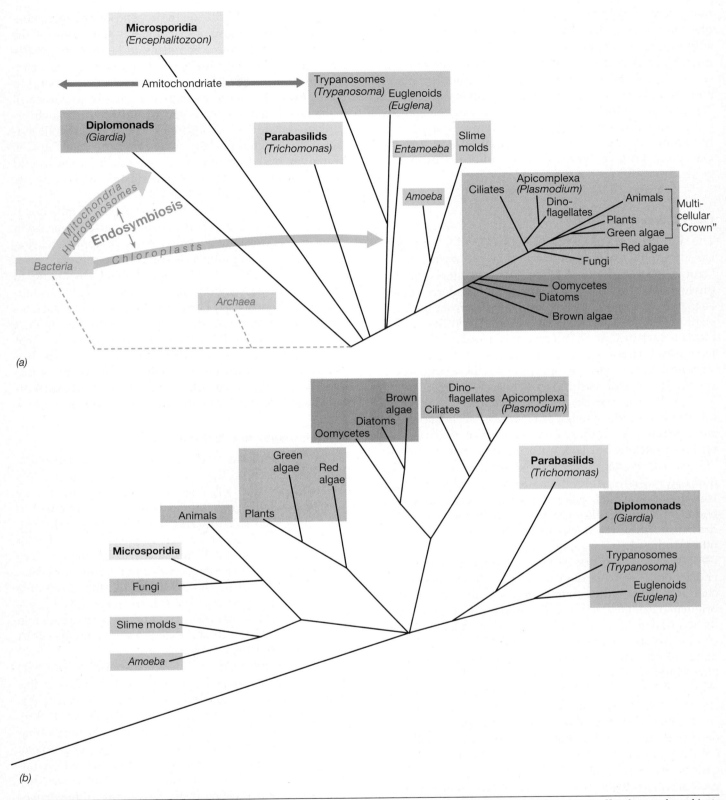

● **Figure 14.20 Phylogenetic trees of *Eukarya*.** (a) Phylogenetic tree based on comparative 18S ribosomal RNA sequences. Note how endosymbionts (mitochondria and chloroplasts) are shown to originate in the domain *Bacteria*. (b) A phylogenetic tree based on other eukaryotic genes and proteins. Phylogenetically "early" lineages in (a) are highlighted in bold in both (a) and (b).

the evolutionary importance of multicellular plants and animals and completely overlooked the broad phylogenetic diversity of the prokaryotes. The SSU tree rectified this situation and, in the process, revealed a very intriguing phylogeny of eukaryotes.

Figure 14.20*a* shows some details of the SSU phylogeny of *Eukarya*. Putting the eukaryotic tree in perspective with the *Bacteria* and *Archaea* lineages reminds us that the major organelles—the mitochondrion and chloroplast—originated within the *Bacteria* and that in many respects, *Archaea* more closely resemble eukaryotes than do the *Bacteria*. An abundance of evidence supports these conclusions. We can also see that several microbial eukaryotes are phylogenetically deep lineages in the SSU tree (Figure 14.20a). By contrast, plants and animals are highly derived organisms, forming a "crown" at the top of the eukaryotic lineage. Algae are scattered within the eukaryotic tree (mainly in relatively recent lineages) while the fungi (except for the Oomycetes) form a very tight, and rather recent, phylogenetic unit (Figure 14.20a).

Early Eukaryotes

The rRNA tree of eukaryotes shows a "ladder" of organisms, from the least derived microbial eukaryotes to the "crown species" (Figure 14.20a). Of great interest are the phylogenetically "early" eukaryotes—modern organisms that are apparently the least derived from their ancient ancestors. The rRNA tree clearly identifies diplomonads such as *Giardia*, microsporidia such as *Encephalitozoon*, and parabasilids such as *Trichomonas* as early eukaryotes. What other properties do these organisms possess that warrants such a conclusion?

Giardia, Trichomonas, and *Encephalitozoon* differ from other eukaryotes in a major way: *they are all amitochondriate eukaryotes*. All of these organisms have a membrane-enclosed nucleus but lack mitochondria. Some, such as *Trichomonas*, contain hydrogenosomes (see Section 14.2 and Figure 14.4), while others do not. Are these organisms relics of ancient eukaryotes that never underwent endosymbiosis? Although this conclusion could be drawn from the topology of the SSU tree, genetic and other studies do not support it.

Using sensitive nucleic acid identification methods and genomic sequence data, it has been shown that amitochondriate eukaryotes contain genes in their nuclei that originated from *Bacteria*. These would be analogous to mitochondrial genes that remain today in the nucleus of aerobic eukaryotes (see Sections 14.4 and 15.7). This suggests that amitochondriate eukaryotes once harbored endosymbionts but for some reason discarded them, leaving only a trace of their previous existence in the form of bacterial-specific genes in the nuclei of these organisms.

Further evidence that amitochondriate eukaryotes once had mitochondria has been obtained visually. In *Giardia*, for example, scientists have discovered that this organism makes mitochondrial-like proteins and that these proteins cluster together within the cell. Using high resolution electron microscopy, these proteins can be shown to be surrounded by tiny double membrane sacs. These relic mitochondria have been called **mitosomes** and have subsequently been found in several other amitochondriate eukaryotes, including *Entamoeba*, an amitochondriate organism considerably more derived than *Giardia* based on SSU rRNA sequencing (Figure 14.20a). It is hypothesized that mitosomes are highly degenerate mitochondria that contain only a few proteins useful for energy generation.

Given the existence of hydrogenosomes, mitochondria, and now, mitosomes, it appears that none of the extant eukaryotic microorganisms thought to be amitochondriate were ever really so. Eukaryotic microorganisms that never established an endosymbiotic relationship with a respiratory or fermentative organelle are either all extinct, or scientists have yet to find them. If such organisms exist, they should lack organelles *and* all evidence for bacterially derived genes in their nuclei. Moreover, based on 18S rRNA sequencing, they should branch earlier on the phylogenetic tree of eukaryotes than even *Giardia* (Figure 14.20a).

Finding cells that never harbored endosymbionts, if they exist, is an exciting challenge for scientists interested in the microbial diversity of eukaryotes. One approach being used to study this problem is community sampling (∞ Section 11.7) of rRNA from various habitats, including, in particular, anoxic, mildly hot habitats (conditions approaching those of early Earth, ∞ Section 11.1). If such organisms are alive today, they may be detected using PCR primers specific for 18S rRNA. Detection of rRNAs that branch earlier on the SSU tree than *Giardia* would then greatly stimulate attempts to obtain cultures of the organisms that produce them. One would predict that such organisms would display many unusual properties.

The Microsporidia Problem

The microsporidia are an evolutionary enigma. These early eukaryotes based on the SSU tree (Figure 14.20a) are tiny (2–5 μm) parasitic cells that are even more structurally stripped down than other amitochondriate eukaryotes. Microsporidia such as *Encephalitozoon* lack *all* organelles, including the Golgi complex and hydrogenosomes (whether mitosomes are present is unknown), and contain very small genomes. The genome of *Encephalitozoon*, for example, is only 2.9 Mbp; this is 1.5 Mbp *smaller* than that of the bacterium *Escherichia coli*! Indeed, microsporidia have a combination of features one would predict to be present in a "primitive" eukaryote, and their placement on the rRNA tree confirms this (Figure 14.20a).

However, the position of the microsporidia on the SSU tree has been questioned by some because rRNA in this group has apparently undergone very rapid evolution (this is indicated by the long branch in the tree in Figure 14.20a). This fact is thought to introduce artifacts in positioning such organisms on a tree. To try and resolve this problem, molecular sequencing of other genes and proteins from microsporidia and other amitochondriate eukaryotes has been carried out and collectively, tells a quite different evolutionary story, as we discuss now.

An Alternative View of Eukaryotic Evolution

Molecular sequencing of certain eukaryotic genes and proteins yield trees that differ dramatically from the rRNA-based tree (Figure 14.20b). The sequences of several molecules have been used to build these trees, including tubulin proteins, RNA polymerase, and ATPase subunits, among many others. But some of the most revealing information has emerged from protein sequences of heat shock proteins and related chaperonins (∞ Sections 7.17 and 8.9), proteins that are thought—like rRNA—to be excellent phylogenetic markers. How do these alternative eukaryotic trees compare to the SSU rRNA tree?

First, the evolutionary ladder from least derived organisms such as *Giardia* to the multicellular organism crown that is a prominent feature of the SSU rRNA tree (Figure 14.20a) disappears. Instead, alternative trees reveal that a major radiation of eukaryotes evolved almost simultaneously; this radiation included the vast majority of known eukaryotic organisms (Figure 14.20b). Although this is a scientifically less satisfying picture than a "line of descent" type of phylogenetic tree (Figure 14.20a), it is what the molecular sequence data say as currently understood. Second, animals and fungi are closely related in alternative trees. Third, the microsporidia, which branch very early in the SSU tree, are shown to be specifically related to the rather highly derived fungi (Figure 14.20). Interestingly, microsporidia are transmitted by small infectious sporelike structures similar in some respects to the spores of fungi. The study of heat shock genes has solidified the fungal/microsporidium connection, as bacterial heat shock genes (which resemble those from the mitochondrion) have been detected in *Encephalitozoon*. This strongly implies that this organism once had mitochondria.

What can we conclude from these two quite different pictures of eukaryotic phylogeny? Well, for one, we can conclude that the true phylogeny of eukaryotes is a work in progress. In other words, neither tree shown in Figure 14.20 can be considered the "final word" on eukaryotic phylogeny. That the two trees disagree is a reflection of the many things we don't yet understand about the evolutionary history of eukaryotes and the phylogenetic information present in different molecules. As more comparative sequencing results come to hand and other studies (such as the previously mentioned discovery of mitosomes) reveal new aspects of eukaryotic biology, the true phylogeny of the eukaryotes will eventually emerge.

The consensus among microbiologists is that both the SSU rRNA tree and alternative trees will provide some of the answers and that a final phylogeny of eukaryotes will draw concepts from both of these trees.

Now that we have a feeling for the phylogeny of microbial eukaryotes and the many unanswered questions in this regard, let us proceed to a consideration of the organisms themselves. We begin with the protozoa.

 14.9 Concept Check

By ribosomal RNA sequencing, eukaryotic cells form their own major line of evolutionary descent (the *Eukarya*). Some microbial eukaryotes, such as *Giardia* and *Trichomonas*, are early branching species, while the eukaryotic "crown" of the tree contains the multicellular plants and animals. Trees based on the comparative sequencing of other genes and proteins yield give a quite different evolutionary picture.

◆ From a *cellular* perspective, what is wrong with the five-kingdom hypothesis for grouping the eukaryotes?

◆ Summarize the evidence that the organism *Giardia* is a closer relative of primitive eukaryotes than is a human cell.

◆ How do alternative phylogenetic trees of *Eukarya* fundamentally differ from the SSU rRNA tree?

14.10 Protozoa

Key Genera: *Amoeba, Paramecium, Trypanosoma*

Protozoa are unicellular eukaryotic microorganisms that lack cell walls (Figure 14.21●). They are generally colorless and motile. **Protozoa** are distinguished from prokaryotes by their eukaryotic nature and typically much greater size, from algae by their lack of chlorophyll, from yeasts and other fungi by their motility and absence of a cell wall in most cases, and from the slime molds by their inability to form fruiting bodies. Also, as previously mentioned,

(a)   (b)   (c)   (d)

● **Figure 14.21 Typical protozoa.** (a) *Amoeba*. (b) A typical ciliate, *Paramecium* (see also Figure 14.25). (c) A flagellate, *Dunaliella* (this flagellate contains chloroplasts and thus can also be considered an alga). (d) *Plasmodium vivax*, an apicomplexan sporozoan, growing in a human red blood cell.

protozoa are phylogenetically diverse, appearing in several lineages on the *Eukarya* tree (Figure 14.20).

Protozoa are found in a variety of freshwater and marine habitats; a large number are parasitic in other animals, including humans, and some are found growing in soil or in aerial habitats, such as on the surface of trees. Most protozoa feed by ingesting particulate matter, usually other cells, by **phagocytosis**, a process of surrounding a food particle with a portion of their flexible cell membrane to engulf the particle and bring it into the cell. Some protozoa can literally swallow bacterial cells or small eukaryotic cells by operation of a special structure called the *gullet* (see Figure 14.25).

As is appropriate for organisms that "catch" their own food, most protozoa are motile. Indeed, their mechanisms of motility are key characteristics used to divide them into taxonomic subgroups (Table 14.1). Protozoa that move by ameboid motion are called *Sarcodina*; those using flagella, the *Mastigophora*; and those using cilia, the *Ciliophora*. A fourth group, the *Apicomplexa*, are generally nonmotile and are all parasitic for higher animals.

Mastigophora: The Flagellates

Members of this protozoal group are all motile by the activity of flagella and are thus referred to collectively as **flagellates** (Figure 14.21*c* and Figure 14.22●). Flagellated protozoa are found throughout the rRNA phylogenetic tree, from early organisms such as *Giardia*, through the euglenoids, up to the dinoflagellates (Figure 14.20*a*). Hence a large diversity of eukaryotic microorganisms employ this means of moving about their habitats.

Although many flagellated protozoa are free-living organisms, a number are parasitic in or pathogenic for animals, including humans. The most important pathogenic Mastigophora are the *trypanosomes*. These organisms cause a number of serious diseases in humans and vertebrate animals, including the disease *African sleeping sickness*. In *Trypanosoma*, a genus infecting humans, the protozoa are rather small, about 20 μm in length, and are thin, crescent-shaped organisms. They have a single flagellum that originates in a basal body and folds back laterally across the cell

Membrane flap   Red blood cell

Trypanosome cell

● **Figure 14.22 Trypanosomes.** Photomicrograph of the flagellated protozoan *Trypanosoma brucei*, the causative agent of African sleeping sickness, from a blood smear.

where it is enclosed by a flap of surface membrane (Figure 14.22). Both the flagellum and the membrane participate in propelling the organism, making effective movement possible even in a rather viscous liquid, such as blood.

Trypanosoma brucei is the species that causes the chronic and usually fatal African sleeping sickness. In humans, the parasite lives and grows primarily in the bloodstream, but in the later stages of the disease, invasion of the central nervous system occurs causing an inflammation of the brain and spinal cord that is responsible for the characteristic neurological symptoms of the disease. The parasite is transmitted from host to host by the tsetse fly, *Glossina* sp., a bloodsucking fly found only in certain parts of Africa. The parasite proliferates in the intestinal tract of the fly and invades the insect's salivary glands and mouthparts, from which it can be transferred to a new human host following a fly bite.

Other parasitic flagellates include *Trichomonas*, a human sexually transmitted pathogen (∞ Section 26.13). This parabasilid protozoan (so-called because it contains a *parabasal body* that, among other functions, gives structural support to the Golgi apparatus) lives in the intestinal

Table 14.1	**Characteristics of the major groups of protozoa**			
Group	**Common name**	**Typical representatives**	**Habitats**	**Common diseases**
Mastigophora	Flagellates	*Trypanosoma, Giardia, Leishmania, Trichomonas*	Freshwater; parasites of animals	African sleeping sickness, giardiasis, leishmaniasis
Euglenoids[a]	Phototrophic flagellates	*Euglena*	Freshwater; some marine	None known
Sarcodina	Amebas	*Amoeba, Entamoeba*	Freshwater and marine; animal parasites	Amebic dysentery (amebiasis)
Ciliophora	Ciliates	*Balantidium, Paramecium*	Freshwater and marine; animal parasites; rumen	Dysentery
Apicomplexa	Sporozoans	*Plasmodium, Toxoplasma*	Primarily animal parasites; insects (vectors for parasitic diseases)	Malaria, toxoplasmosis

[a] This group is also considered with the algae (see Section 14.13 and Table 14.3).

and urogenital tract of vertebrates and invertebrates. Many species of *Trichomonas* inhabit the hindgut of termites and other insects (👓 Section 19.10). Some flagellates are even *phototrophic*. These are the *euglenoids*, flagellates that contain chloroplasts, which allow for photosynthetic growth. However, in darkness, cells of *Euglena* (see Figure 14.33*a*), a typical euglenoid, can grow as chemoorganotrophs. As such, these organisms become phenotypically indistinguishable from flagellated protozoa. Many euglenoids are known and they are exclusively aquatic, inhabiting primarily freshwaters. However, unlike other flagellated protozoa, the euglenoids are nonpathogenic (see Section 14.13).

Sarcodina: The Amoebae

Among the sarcodines are organisms such as *Amoeba*, which are naked in the vegetative phase (see Figure 14.21*a*), and the foraminifera, amoebae that secrete a shell during vegetative growth (Figure 14.23●). *Amoeba* is a freshwater species whose cells vary in size from 15 µm —clearly microsocopic—to over 750 µm—almost visible with the naked eye.

Shelled sarcodines present a variety of interesting morphological forms. The best-known of the shelled forms are the *foraminifera*. Foraminifera are exclusively marine organisms, living primarily in coastal waters. Their shells, called *tests*, show distinctive characteristics and are often quite ornate (Figure 14.23). Tests are usually made of calcium carbonate. The cell is not firmly attached to the test, and the amoeba cell may extend partway out of the shell during feeding. However, because of the weight of the test, the cell usually sinks to the bottom, and it is thought that the organisms feed on particulate deposits in the sediments, primarily bacteria and detritus. The shells of foraminifera are relatively resistant to decay and hence readily become fossilized (the White Cliffs of Dover, England, for example, are composed of foraminiferal shells laid down in an ancient sea). Because these organisms have left an excellent fossil record, we have a better idea of their distribution through geological time than for virtually any other protozoa.

Pathogenic Amoebae

A variety of naked amoebae are parasites of humans and other vertebrates, and their usual habitat is the oral cavity or the intestinal tract. They move in these habitats by **amoeboid movement** (Figure 14.24●), a mechanism also employed by the acellular slime molds (see Section 14.11). *Entamoeba histolytica* (👓Figure 28.13) is a good example of a parasitic amoeba. In many cases infection causes no obvious symptoms. But in some individuals, *E. histolytica* causes ulceration of the intestinal tract, which results in a bloody diarrheal condition called *amebiasis*. *E. histolytica* is transmitted from person to person in the cyst form by fecal contamination of water and food. We discuss the etiology and pathogenesis of amebiasis in Section 28.8.

Ciliophora: The Ciliates

Ciliates are those protozoa that, in some stage of their life cycle, possess *cilia* (Figure 14.25●; see also Figure 14.21*b*), structures that function in motility. **Ciliates** are also unique among protozoa in having two kinds of nuclei: the *micronucleus*, which is concerned only with inheritance and sexual reproduction, and the *macronucleus*, which is involved only in the production of RNA (transcription) or various aspects of cell growth and function (see Figure 14.26).

Probably the best-known and most widely distributed of the ciliates are those of the genus *Paramecium* (Figure 14.25), which will be used here as an example of the group. Most ciliates obtain their food by ingesting particulate materials through a distinct oral region or mouth connected to an underlying gullet (Figure 14.25*b*). Once inside, the food particles are carried down the gul-

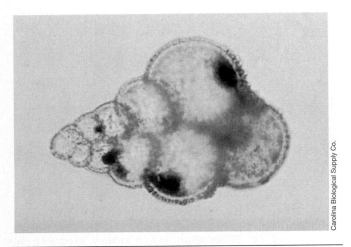

Carolina Biological Supply Co.

● **Figure 14.23 Shelled amoebae: foraminifera.** Note the ornate and multilobed test.

M. Haberey

● **Figure 14.24 Side view of a moving amoeba, *Amoeba proteus*, taken from a film.** The time interval from top to bottom is about 6 sec. The arrows point to a fixed spot on the surface. A single cell is about 80 µm in diameter.

(a)

Yeast cell
(for scale)　　　　Cilia　　　　Mouth (gullet)

(b)

● **Figure 14.25 *Paramecium*, a ciliated protozoan.** (a) Phase photomicrograph. (b) Scanning electron micrograph. Note the cilia in both micrographs. A single *Paramecium* cell is about 60 μm in diameter.

let and into the cytoplasm where they are enclosed in a food vacuole, a structure into which digestive enzymes are secreted.

In addition to cilia, which function in motility, many ciliates have **trichocysts**. These are long, thin filaments of a contractile nature, anchored beneath the surface of the outer cell layer. These enable the protozoa to attach to a surface and can aid in defense by signaling the cell that it is being attacked by a predator, thus stimulating cellular defenses. In the case of predatory ciliates, such as *Didinium*, trichocysts hold onto and paralyze the prey as a prelude to ingestion.

Many *Paramecium* species (as well as many other protozoa) are hosts for endosymbiotic bacteria that reside in the cytoplasm or the macronucleus. In some cases, evidence exists that these endosymbionts play a nutritional role for the host, synthesizing vitamins or other growth factors that would otherwise have to be obtained from the environment. In the case of ciliated protozoa inhabiting the termite gut, endosymbiotic methanogens remove H_2 produced from pyruvate oxidation in the hydrogenosome (Figure 14.4), yielding methane, which is released to the atmosphere (∞ Section 19.10 and Figure 19.26*b, c*).

Although a few ciliates are parasitic for animals, this mode of existence is less extensively developed in the ciliates than it is in other groups of protozoa. The species *Balantidium coli* (Figure 14.26●) is primarily a parasite of domestic animals but occasionally infects the intestinal tract of humans, producing intestinal dysentery symp-

toms similar to those caused by *Entamoeba histolytica*. In addition, there is usually a characteristic fauna of *obligately anaerobic* ciliates in the rumen, the forestomach of ruminant animals (∞ Section 19.11). These protozoa play a beneficial role in the digestive and fermentative processes that occur there.

Apicomplexa (Sporozoans)

The Apicomplexa, or **Sporozoans** as they are also called (Figure 14.21*d*), comprise a large group of obligately parasitic protozoa. Apicomplexan parasites cause severe disease such as malaria (*Plasmodium* species), toxoplasmosis (*Toxoplasma*), and coccidiosis (*Eimeria*). The apicomplexans are characterized by a lack of motile adult stages, and their food is absorbed in soluble form through the membrane, as occurs in prokaryotes and in fungi. Although the name *sporozoa* implies the formation of spores, these organisms do not form true resting spores, like those of bacteria, algae, and fungi, but instead produce analogous structures called *sporozoites*, which function in transmission to a new host.

Numerous vertebrates and invertebrates are hosts for Sporozoa, and in some cases an alternation of hosts takes place, with some stages of the life cycle occurring in one host and some in another. The most important members of the sporozoa are the coccidia, usually parasites of birds, and the genus *Plasmodium* (malaria parasites) (Figure 14.21*d*), which infect birds and mammals, including humans. Because malaria is a major disease of humans, especially in developing countries, we devote a considerable discussion to this disease and the properties of malarial parasites in Section 27.5.

Interestingly, apicomplexan parasites contain a plastid called the *apicoplast*. Plastids are degenerate chloroplasts that lack the pigments and photosynthetic capacity of a chloroplast. Like the chloroplast, however, the apicoplast contains its own genome and expresses a few genes. The majority of apicoplast functions are encoded in the nucleus, similar in this regard to the mitochondrion.

● **Figure 14.26 *Balantidium coli*, a ciliated protozoan that causes a dysentery-like disease in humans.** The dark blue stained structure is the macronucleus.

Apicoplasts carry out fatty acid, isoprenoid, and heme biosyntheses, and export their products to the cytoplasm. It is thought that apicoplasts are derived from endosymbiotic engulfment by an apicomplexan of a red algal cell whose chloroplast eventually degenerated to play a very specialized nonphotosynthetic role in the apicomplexan cell.

14.10 Concept Check

Protozoa are unicellular microbial *Eukarya* that typically lack cell walls and are usually motile by various means. Many protozoa are pathogenic to humans and other animals.

◆ List at least two major ways in which the protozoan *Paramecium* differs from the protozoan *Trypanosoma*.

◆ How do the Sporozoa differ from all other protozoa?

◆ Why is the alga *Euglena* also considered a protozoan?

14.11 Slime Molds

Key Genera: *Dictyostelium, Physarum*

Slime molds are microbial eukaryotes that have phenotypic similarity to both fungi and protozoa. Like fungi (see Section 14.12), slime molds undergo a life cycle and can produce spores. However, like protozoa (see Section 14.10), slime molds are motile and can move across a solid surface rather rapidly (see Figures 14.27–14.29). From a phylogenetic perspective, slime molds are related to amoeboid protozoa (Figure 14.20a).

The slime molds are divided into two groups, the **cellular slime molds**, whose vegetative forms are composed of single amoebae, and the **acellular slime molds**, whose vegetative forms are masses of protoplasm of indefinite size and shape called *plasmodia* (Figure 14.27●). Slime molds live primarily on decaying plant matter, such as leaf litter, logs, and soil. Their food consists mainly of other microorganisms, especially bacteria,

which they ingest by phagocytosis. Slime molds can maintain themselves in a vegetative state for long periods but eventually form differentiated sporelike structures that can remain dormant and then germinate later to once again generate the active amoeboid state.

Acellular Slime Molds

In the vegetative phase, acellular slime molds such as *Physarum* exist as a mass of expanding protoplasm of indefinite size (Figure 14.27). This structure is actively motile by amoeboid motion, the plasmodium flowing over the surface of the substratum, engulfing food particles as it moves. Ameboid motion is the result of cytoplasmic streaming. Cytoplasm flows forward because the tip of the plasmodium is less contracted and viscous, and thus cytoplasm takes the path of least resistance.

Cytoplasmic streaming is facilitated by microfilaments, which exist in a thin layer just beneath the cytoplasmic membrane in eukaryotic cells (see Section 14.4 and Figures 14.1 and 14.9). In acellular slime molds, cytoplasmic streaming occurs in definite strands, each surrounded by the thin cytoplasmic membrane (Figure 14.27). The streaming itself is a mechanism for distributing cellular metabolites.

The acellular slime mold plasmodium (Figure 14.27) is genetically *diploid*. From this mass of protoplasm a sporangium and haploid spores can be produced. Under favorable conditions, spores germinate to yield haploid swarm cells. The fusion of two swarm cells then regenerates the diploid plasmodium.

Cellular Slime Molds: *Dictyostelium*

Dictyostelium discoideum, a cellular slime mold, has been used as a model for the study of intercellular communication and cooperation among microorganisms. This communal organism undergoes a remarkable life cycle in which vegetative cells aggregate, migrate as a cell mass, and eventually produce fruiting bodies in which cells differentiate and form spores (Figures 14.28● and 14.29●). As cells of *Dictyostelium* become starved, they aggregate and form a *pseudoplasmodium*, a structure in which the cells lose their individuality but do not fuse. This aggregation is triggered by the production of two compounds, *cyclic adenosine monophosphate* (cAMP) and a specific glycoprotein, both of which function as chemotactic agents (we discussed the involvement of cAMP in various regulatory systems in prokaryotes in Section 8.7). Those cells that are the first to produce these compounds become centers for the attraction of other vegetative cells, leading to aggregating masses of cells that come together to form a slimy migrating mass called a *slug*.

Fruiting-body formation begins when the slug ceases to migrate and becomes vertically oriented (Figures 14.28 and 14.29). The fruiting body then becomes differentiated into a stalk and a head; cells in the anterior end of the slug become stalk cells while those in the posterior end become spores. Cells that form stalk cells begin to secrete

● **Figure 14.27 Slime molds.** Plasmodia of the acellular slime mold *Physarum* growing on an agar surface.

● **Figure 14.28** **Photomicrographs of various stages in the life cycle of the cellular slime mold *Dictyostelium discoideum.*** (a) Amoebae in preaggregation stage. Note irregular shape and lack of orientation. (b) Aggregating amoebae. Notice the regular shape and orientation. The cells are moving in streams in one direction. (c) Low-power view of aggregating amoebae. (d) Migrating pseudoplasmodia (slugs) moving on an agar surface and leaving trails of slime in their wake. (e, f) Early stage of fruiting body. (g) Mature fruiting bodies. See Figure 14.29 for sizes of these structures.

cellulose, which provides the rigidity of the stalk. Cells from the rear of the slug swarm up the stalk to the tip and form the head. Most of these posterior cells differentiate into spores. On maturation of the head, the spores are released and dispersed. Each spore then germinates and becomes a vegetative amoeba.

The cycle of fruiting-body and spore formation in *Dictyostelium* is an *asexual* process. However, *sexual* spores

called *macrocysts* can also be produced. Macrocysts are formed from aggregates of amoebae that become enclosed in a cellulose wall. Following the conjugation of two amoebae, a single large amoeba develops and proceeds to phagocytize the remaining amoebae. At this point a thick cellulose wall forms around this giant amoeba forming the mature macrocyst; this can remain dormant for long periods. Eventually, the diploid nucleus undergoes meiosis to

● **Figure 14.29** **Stages in fruiting-body formation in the cellular slime mold *Dictyostelium discoideum.*** (A–C) Aggregation of amoebae; aggregation is triggered by cAMP. (D–G) Migration of the slug formed from aggregated amoebae; (H–L) culmination and formation of the fruiting body; (M) mature fruiting body composed of stalk and head. Cells from the rear of the slug form the head and become spores. The fruiting body contains spores that can regenerate vegetative cells and begin the life cycle anew.

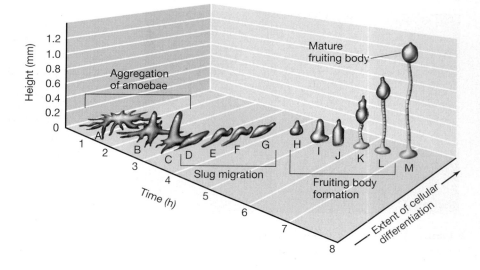

form haploid nuclei that become integrated into new amoebae that can once again initiate vegetative growth (Figures 14.28 and 14.29).

 14.11 Concept Check

Acellular slime molds are masses of motile protoplasm while cellular slime molds are masses of individual cells that aggregate to form fruiting bodies from which spores are released.

◆ Describe the major steps in the life cycle of *Dictyostelium discoideum*.

◆ What is a *macrocyst*?

14.12 Fungi

Key Genera: *Penicillium, Aspergillus, Saccharomyces, Candida*

In contrast to the protozoa, **fungi** contain cell walls and produce spores, among many other differences. Fungi form a tight phylogenetic cluster whether in SSU rRNA trees or trees based on other molecules (Figure 14.20). Three major groups of fungi are recognized: the *molds*, the *yeasts*, and the *mushrooms*.

The habitats of fungi are quite diverse. Some fungi are aquatic, living primarily in freshwater, and a few marine fungi are also known. Most fungi, however, are terrestrial. They inhabit soil or dead plant matter and play crucial roles in the mineralization of organic carbon. A large number of fungi are parasites of terrestrial plants. Indeed, fungi cause the majority of economically significant diseases of crop plants (Table 14.2). A few fungi are parasitic on animals, including humans, although in general fungi are less significant as animal pathogens than are other microorganisms (∞ Section 27.8 for a discussion of pathogenic fungi).

Cell Walls and Metabolism

Fungal cell walls resemble plant cell walls architecturally, but not chemically. Although the plant cell wall polysaccharide cellulose is present in the walls of certain fungi, most fungi contain **chitin**, a polymer of the glucose derivative, *N*-acetylglucosamine (∞ Figure 3.5), in their cell walls. The chitin is laid down in microfibrillar bundles like cellulose. Other polysaccharides such as mannans, galactosans, and chitosans replace chitin in some fungal cell walls.

Fungal cell walls are typically 80–90% polysaccharide, with proteins, lipids, polyphosphates, and inorganic ions making up the wall-cementing matrix. An understanding of fungal cell wall chemistry is important because of the extensive biotechnological uses of fungi (∞ Chapters 30 and 31). The chemical nature of

Table 14.2	**Classification and major properties of fungi**[a]					
Group	**Common name**	**Hyphae**	**Typical representatives**	**Type of sexual spore**	**Habitats**	**Common diseases**
Ascomycetes	Sac fungi	Septate	*Neurospora, Saccharomyces, Morchella* (morels)	Ascospore	Soil, decaying plant material	Dutch elm, chestnut blight, ergot, rots
Basidiomycetes	Club fungi, mushrooms	Septate	*Amanita* (poisonous mushroom), *Agaricus* (edible mushroom)	Basidiospore	Soil, decaying plant material	Black stem, wheat rust, corn smut
Zygomycetes	Bread molds	Coenocytic	*Mucor, Rhizopus* (common bread mold)	Zygospore	Soil, decaying plant material	Food spoilage; rarely involved in parasitic disease
Oomycetes	Water molds	Coenocytic	*Allomyces*	Oospore	Aquatic	Potato blight, certain fish diseases
Deuteromycetes	Fungi imperfecti	Septate	*Penicillium, Aspergillus, Candida*	None known	Soil, decaying plant material, surfaces of animal bodies	Plant wilt, infections of animals such as ringworm, athlete's foot, surface or systemic infections (*Candida*)

[a] With the exception of the Oomycetes, which are phylogenetically distinct, the other groups of fungi are closely related (see Figure 14.20). See photos of many of the organisms listed here in Figures 14.30–14.34.

the fungal cell wall has been used in classifying fungi for research and industrial purposes.

Fungi are chemoorganotrophs and typically have simple nutritional requirements. Many fungi can grow at environmental extremes of low pH or high temperature (up to 62° C) and this, coupled with the ubiquitous nature of fungal spores, makes these organisms common contaminants of food products, microbial culture media, and surfaces. Molds and yeasts are not, however, classified on physiological grounds, but instead by their diverse array of life cycle patterns, including the formation of a variety of different sexual spores (see Table 14.2).

Filamentous Fungi: Molds

The **molds** are *filamentous fungi*. They are widespread in nature and are commonly seen on stale bread, cheese, or fruit. Most molds are obligate aerobes. Each filament grows mainly at the tip, by extension of the terminal cell (Figure 14.30●). A single filament is called a *hypha* (plural, *hyphae*). Hyphae usually grow together across a surface and form compact tufts, collectively called a *mycelium* (Figure 14.31●), which can be seen easily without a microscope. The mycelium arises because the individual hyphae form branches as they grow, and these branches intertwine, resulting in a compact mat. In most cases, the vegetative cell of a fungal hypha contains more than one nucleus—often hundreds of nuclei are present. Thus, a typical hypha is a nucleated tube containing cytoplasm (referred to as *coenocytic*).

From the fungal mycelium, other hyphal branches may reach up into the air above the surface, and on these aerial branches spores called **conidia** are formed (Figure 14.30). Conidia are *asexual* spores (their formation does *not* involve the fusion of gametes), and are often pig-

mented (Figure14.31; ⚭ Figure 10.1*a*) and resistant to drying. Conidia function in the dispersal of the fungus to new habitats. When conidia form, the white color of the mycelium changes, taking on the color of the conidia, which may be black, blue-green, red, yellow, or brown. The presence of these spores gives the mycelial mat a rather dusty appearance (Figure 14.31*c*).

Some molds also produce *sexual* spores, formed as a result of sexual reproduction (Table 14.2). The spores occur from the fusion either of unicellular gametes or of specialized hyphae called *gametangia*. Alternatively, sexual spores can originate from the fusion of two haploid cells to yield a diploid cell, which then undergoes meiosis and mitosis to yield individual spores. Depending on the group (see Table 14.2), different types of sexual spores are produced. Spores formed within an enclosed sac (*ascus*) are called *ascospores* (see Figure 14.14), and those produced on the ends of a club-shaped structure (*basidium*) are *basidiospores* (Table 14.2 and see Figure 14.32*c*). *Zygospores*, produced by zygomycetous fungi like the common bread mold *Rhizopus*, are macroscopically visible structures that result from the fusion of hyphae and genetic exchange. Eventually the zygospore matures and produces asexual spores that are dispersed by air and germinate to form new fungal mycelia.

Sexual spores of fungi are typically resistant to drying, heating, freezing, and some chemical agents. However, fungal sexual spores are not as resistant to heat as are bacterial endospores (⚭ Section 4.13). Either an asexual or a sexual spore of a fungus can germinate and develop into a new hypha and mycelium.

A major ecological activity of many fungi, especially members of the Basidiomycetes (see Table 14.2), is the decomposition of wood, paper, cloth, and other products derived from natural sources. Basidiomy-

(a)

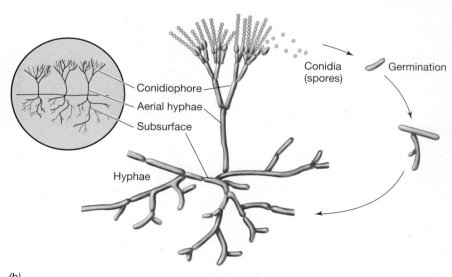

Conidia (spores) — Germination

Conidiophore

Aerial hyphae

Subsurface

Hyphae

(b)

● **Figure 14.30 Mold structure and growth.** (a) Photomicrograph of a typical mold. Conidia are seen as the spherical structures at the ends of aerial hyphae. (b) Diagram of a mold life cycle. Conidia can be either wind-blown or be transported by animals.

(a)

Cheryl L. Broadie

(b)

CDC Public Health Image Library, PHIL

● **Figure 14.31 Fungi.** (a) Colonies of an *Aspergillus* species growing on an agar plate. Note the appearance of the masses of filamentous cells (the mycelium) and asexual spores (see Figure 14.30*b*) that give the colonies a custy, matted appearance. (b) Conidiophore and conidia of *Aspergillus fumigatus*.

cetes that attack these products can use cellulose or lignin from the product as carbon and energy sources. **Lignin** is a complex polymer in which the building blocks are phenolic compounds. It is an important constituent of woody plants, and in association with cellulose, it confers rigidity on them. The decomposition of lignin in nature occurs almost exclusively through the activities of certain Basidiomycetes called *wood-rotting fungi*. Two types of wood rot are known: *brown rot*, in which the cellulose is attacked preferentially and the lignin left unmetabolized, and *white rot*, in which both cellulose and lignin are decomposed. The white rot fungi are of considerable ecological importance because they play such a key role in decomposing woody material in forests.

Macroscopic Fungi: Mushrooms

Mushrooms are filamentous basidiomycetes that form large **fruiting bodies**, the edible part of the mushroom (Figure 14.32*a, b●*). We will discuss in Section 30.14 the commercial growth of mushrooms as a food source.

During most of its existence, the mushroom fungus lives as a simple mycelium, growing in soil, leaf litter, or decaying logs. However, when environmental conditions are favorable, usually following periods of wet and

(a) W. Ormerod

(b) USDA

(c) S. L. Flegler

(d)

● **Figure 14.32 Mushrooms.** (a) *Amanita*, a highly poisonous mushroom. (b) Gills on the underside of the mushroom fruiting body contain the spore-bearing basidia. (c) Scanning electron micrograph of basidiospores released from mushroom basidia. (d) Life cycle of a typical mushroom. The production of mushrooms as a food source will be covered in Section 30.14.

cool weather, the fruiting body develops, beginning first as a small button-shaped structure underground and then expanding into the full-grown fruiting body that we see above ground (Figure 14.32*a*). Sexual spores, called **basidiospores** (Figure 14.32*c*) are formed, borne on the underside of the fruiting body on flat plates called *gills*, which are attached to the cap of the mushroom (Figure 14.32*b, d*). The mushroom basidiospores are dispersed by wind and eventually light on a favorable, usually moist and organic rich soil, and begin the cycle again (Figure 14.32*d*).

Unicellular Fungi: Yeasts

The **yeasts** are *unicellular fungi*, and most of them belong to the Ascomycetes (Table 14.2). Yeast cells are typically spherical, oval, or cylindrical, and cell division typically takes place by budding (Figure 14.33●). In the budding process, a new cell forms as a small outgrowth of the old cell; the bud gradually enlarges and then separates (Figures 14.33 and 14.34●). Although most yeasts reproduce only as single cells, some yeasts can form filaments as well. In these species, certain characteristics may be expressed only by the filamentous form. For example, the filamentous phase is essential for pathogenicity in *Candida albicans*, a yeast that can cause vaginal, oral, or lung infections and, in acquired immunodeficiency (AIDS) patients, systemic tissue damage (⚭ Section 26.14).

Yeast cells are typically much larger than bacterial cells and can be distinguished microscopically from bacteria by their larger size and by the obvious presence of internal cell structures, such as the nucleus (Figure 14.34). Some yeasts exhibit sexual reproduction by a process called *mating*, in which two yeast cells fuse. Within the fused cell, called a *zygote*, ascospores are eventually formed. We discussed the sexual cycle of a typical yeast, *Saccharomyces*, including the important property of *mating types*, in Section 14.7 (see Figures 14.14–14.16).

Yeasts flourish in habitats where sugars are present, such as fruits, flowers, and the bark of trees. Most yeasts are facultative aerobes, capable of fully aerobic as well as

● **Figure 14.34   Growth by budding division in *Saccharomyces cerevisiae*.** Shown is a time-lapse series of photos showing budding division starting from a single cell. Note the pronounced nucleus. Phase-contrast micrograph. A single cell of *S. cerevisiae* is about 8 μm in diameter.

fermentative metabolism. A number of yeast species live symbiotically with animals, especially insects, and a few species are pathogenic for animals and humans (⚭ Section 27.7). The most important commercial yeasts are the baker's and brewer's yeasts, which are members of the genus *Saccharomyces* (⚭ Microbial Sidebar, The Products of Yeast Fermentation, Chapter 5). The original habitats of these yeasts were undoubtedly fruits and fruit juices. But the commercial yeasts of today are probably quite different from wild strains because they have been greatly improved through the years by careful selection and genetic manipulation by industrial microbiologists. The yeast *S. cerevisiae* has been studied as a model eukaryote for many years (see Section 14.7) and was the first eukaryote to have its genome completely sequenced (⚭ Section 15.6).

⬡ 14.12   *Concept Check*

Fungi include the molds and yeasts. Fungi differ from protozoa by virtue of their rigid cell wall, production of spores, lack of motility, and phylogenetic position. Mushrooms are large, often edible fungi that produce fruiting bodies containing basidiospores.

◆ How do *molds* differ from *yeasts*?

◆ What is *chitin* and what is its function in fungi?

◆ In molds, how do *conidia* differ from *ascospores*?

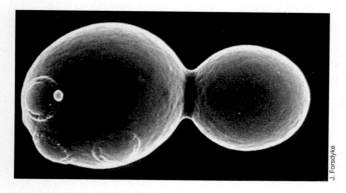

● **Figure 14.33   Scanning electron micrograph of the common baker's and brewer's yeast *Saccharomyces cerevisiae*.** Note the budding division and previous bud scars. A single large cell is about 8 μm in diameter.

14.13   Algae

Key Genera: *Chlamydomonas, Euglena, Gonyaulax*

Algae are a large and diverse group of eukaryotic organisms (Figures 14.35● and 14.36●) that contain chlorophyll and carry out oxygenic photosynthesis (⚭ Section 17.5). *Algae* should be distinguished from *cyanobacteria* and *prochlorophytes* (⚭ Sections 12.25 and 12.26), which are also oxygenic phototrophs but are *Bacteria* and thus evolutionarily quite distinct from algae (*Eukarya*). Although most algae are of microscopic size and hence are clearly microorganisms, a number of forms are multicellular and

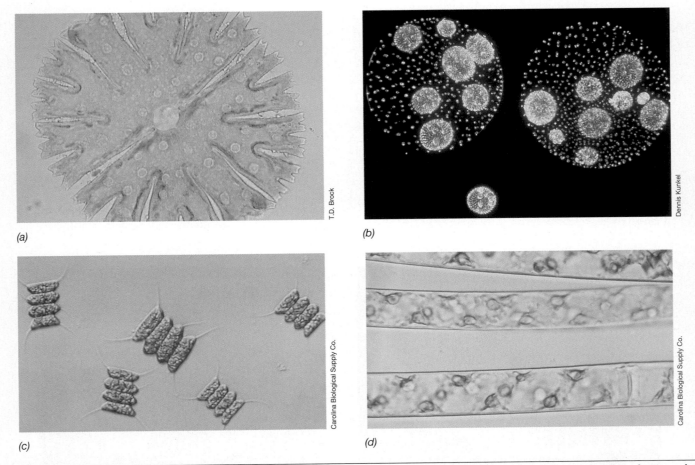

(a)

T.D. Brock

(b)

Dennis Kunkel

(c)

Carolina Biological Supply Co.

(d)

Carolina Biological Supply Co.

● **Figure 14.35** **Light micrographs of representative green algae.** (a) *Micrasterias*. A single cell. (b) *Volvox* colony, containing a large number of cells. (c) *Scenedesmus*. A packet of four cells. (d) *Spirogyra*. A filamentous alga. Note the green spiral-shaped chloroplasts.

macroscopic; kelp, for example, phylogenetic relatives of the diatoms, can grow to over 30 m in length.

Algal Diversity

Most algae are either **unicellular** (Figure 14.35*a* and *c*) or **colonial**, the later existing as aggregates of cells (Figure 14.35*b*). When the cells are arranged end-to-end, the alga is said to be **filamentous** (Figure 14.35*d*). Among the filamentous forms, both species that produce unbranched filaments and those that produce more intricate branched filaments are known.

Algae contain chlorophyll and are thus green in color. However, a few kinds of common algae are not green but appear brown or red because in addition to chlorophyll, other pigments such as xanthophylls and carotenoids are present that mask the green color. Algal cells contain one or more **chloroplasts**, membranous structures that house the photosynthetic pigments (see Section 14.3). Chloroplasts can often be recognized microscopically within algal cells by their distinct green color (Figure 14.35). We discussed the general structure and properties of chloroplasts in Section 14.3, and the phylogeny of the chloroplast in Sections 12.27 and 14.4.

Examination of Figure 14.20 shows the algae to constitute a phylogenetically heterogeneous group. Although green algae (Chlorophyta, Table 14.3) and to a lesser extent red algae (Rhodophyta, Table 14.3) are quite closely related to green plants, other algal groups, such as the brown algae and diatoms, are less derived organisms by 18S rRNA criteria (Figure 14.20*a*). Even less derived than these are the euglenoids, such as the alga *Euglena* (Figure 14.36*a*). Euglenoids show a phylogenetic relationship to flagellated protozoa such as *Trypanosoma* in all phylogenetic trees (Figure 14.20). Not surprisingly then, cells of *Euglena* can spontaneously lose their chloroplasts and exist as completely heterotrophic organisms (see Section 14.10). Red algae (Figure 14.36*b*) are noteworthy in that their chloroplasts lack chlorophyll *b* and contain *phycobiliproteins*, the major light-harvesting pigments of the cyanobacteria (∞ Section 17.3). By contrast, the green algal chloroplast lacks phycobilins and contains chlorophylls *a* and *b*.

Dinoflagellates are flagellated, primarily marine algae that are phylogenetic relatives of several groups of protozoa (Figure 14.20). Some dinoflagellates are free-living (Figure 14.36*f*) while others live a symbiotic existence with

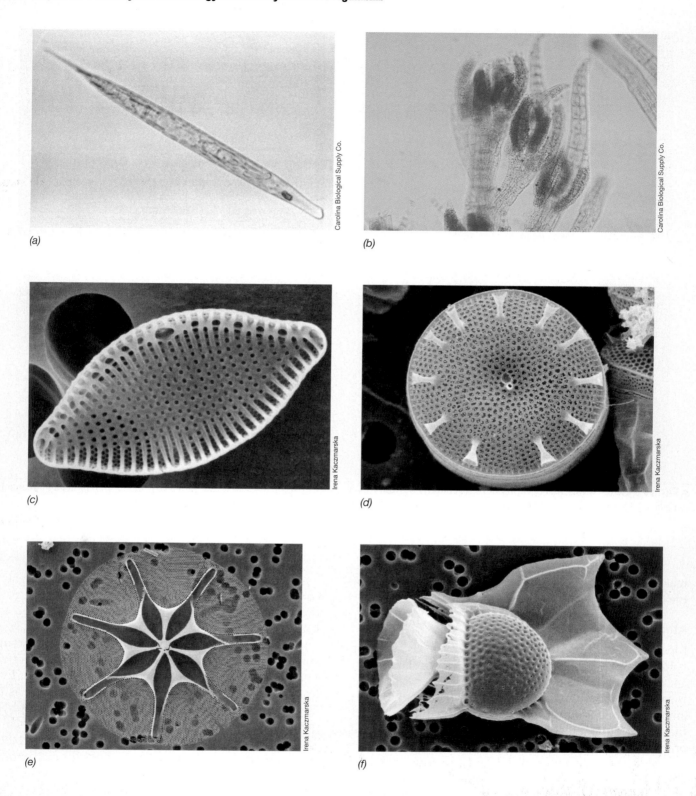

● **Figure 14.36** **Algae other than green algae (Chlorophyta).** (a–b) Light micrographs. (a) *Euglena*, a member of the Euglenophyta (the cell is about 15 μm wide). This organism is phylogenetically much less derived than green algae (see Figure 14.20), and shares a number of other properties in common with flagellated protozoa (see text). (b) *Polysiphonia*, a marine red alga (Rhodophyta) that grows attached to the surfaces of marine plants (cells are about 150 μm wide). (c–f) Scanning electron micrographs. (c–e) Diatom frustules. A frustule is the silica enclosure or wall surrounding the diatom cell proper. (c) Frustule of the marine diatom *Nitzschia*, showing pennate symmetry (the cell is about 10 μm wide). (d) Frustule of the marine diatom *Thalassiosira*, showing radial symmetry (a cell is about 5 μm in diameter). (e) Frustule of the marine diatom *Asteriolampra*, showing radial symmetry (the cell is about 5 μm wide). (f) Cell of the marine dinoflagellate, *Ornithocercus magnificus*. The cell proper is the globular central structure; the ornate structures attached are called *lists* (the cell is about 30 μm wide).

Table 14.3 Properties of major groups of algae

Algal group	Common name	Morphology	Pigments	Typical representative	Carbon reserve materials	Cell wall	Major habitats
Chlorophyta	Green algae	Unicellular to leafy	Chlorophylls *a* and *b*	*Chlamydomonas*	Starch (α-1,4-glucan), sucrose	Cellulose	Freshwater, soils, a few marine
Euglenophyta[a]	Euglenoids	Unicellular, flagellated	Chlorophylls *a* and *b*	*Euglena* (see Figure 14.36a)	Paramylon (β-1,2-glucan)	No wall present	Freshwater, a few marine
Dinoflagellata	Dinoflagellates	Unicellular, flagellated	Chlorophylls *a* and *c*, xanthophylls	*Gonyaulax*; *Pfiesteria*	Starch (α-1,4-glucan)	Cellulose	Mainly marine
Chrysophyta	Golden-brown algae, diatoms	Unicellular	Chlorophylls *a* and *c*	*Nitzschia* (see Figure 14.36c)	Lipids	Two overlapping components made of silica	Freshwater, marine, soil
Phaeophyta	Brown algae	Filamentous to leafy, occasionally massive and plantlike	Chlorophylls *a* and *c*, xanthophylls	*Laminaria*	Lammarin (β-1,3-glucan), mannitol	Cellulose	Marine
Rhodophyta	Red algae	Unicellular, filamentous to leafy	Chlorophylls *a* and *d*, phycocyanin, phycoerythrin	*Polysiphonia* (see Figure 14.36b)	Floridean starch (α-1,4- and α-1,6-glucan)	Cellulose	Marine

[a] This group is also considered with the protozoa (see Section 14.10).

animals that make up coral reefs, obtaining a sheltered and protected habitat in exchange for supplying photosynthetically fixed carbon as a food source for the reef.

Toxic Dinoflagellates

Dinoflagellates such as *Gonyaulax* occasionally grow in dense suspensions called *blooms* that are bright red in color from the xanthophylls present as accessory pigments in this organism. These blooms are the classic "red tides" (Figure 14.37a●) that occasionally develop in warm, usually polluted, coastal waters. Such blooms are often associated with fish kills and common source poisoning (∞ Section 28.4) in humans following consumption of contaminated water or shellfish. Toxicity occurs because of a potent neurotoxin produced by *Gonyaulax* that is sufficiently toxic to kill fish at nanogram levels.

Pfiesteria is another genus of toxic dinoflagellates. Toxic spores of *P. piscicida* (Figure 14.37b) infect fish and eventually kill them from neurotoxins that affect movement and destroy skin. Lesions form on areas of the fish where the skin has been removed, and this allows the growth of opportunistic bacterial pathogens (Figure 14.37c). *P. piscicida* can cause massive fish kills (Figure 14.37c). For example, over a billion fish were killed by a *Pfiesteria* outbreak in the Nuese

Estuary, North Carolina (USA), in 1991, and other outbreaks along the eastern coast of the United States have been suspected. Human toxemia due to *Pfiesteria* poisoning is also thought to occur, but the connection here is not as well established as in fish.

Pigments, Energy Metabolism, and Reserve Polymers

Several characteristics are used to classify algae, including the nature of the chlorophyll(s) present, the carbon reserve polymers produced (for example, starch and several starch derivatives), the cell wall structure, and the type of motility. All algae contain chlorophyll *a*. Some, however, also contain other chlorophylls that differ in minor ways from chlorophyll *a*. The presence of these additional chlorophylls is characteristic of particular algal groups. The distribution of chlorophylls and other photosynthetic pigments in algae is summarized in Table 14.3.

All algae carry out oxygen-evolving (oxygenic) photosynthesis, using H_2O as an electron donor (∞ Section 17.5). In addition, some algae carry out photosynthesis *without* yielding oxygen, using H_2 as an electron donor. Many algae are obligate phototrophs and are thus unable to grow in darkness on organic carbon com-

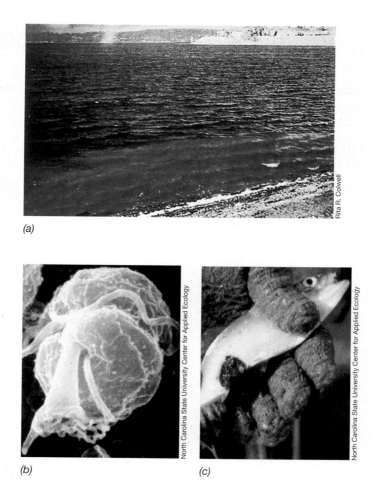

(a)

(b)   *(c)*

● **Figure 14.37** **Toxic dinoflagellates.** (a) Photograph of a "red tide" caused by massive growth of toxin-producing dinoflagellates, such as *Gonyaulax*. The toxin is excreted into the water and also accumulates in shellfish that feed on the dinoflagellates. The toxin is harmless to the shellfish but can poison fish and humans that eat infected shellfish. (b–c) Toxic *Pfiesteria*. (b) Scanning electron micrograph of a toxic spore of *P. piscicida*. (c) Fish killed by *P. piscicida*. Note the flesh-decaying lesions.

pounds. However, some species can grow chemoorgan-otrophically and catabolize simple sugars or organic acids in the dark. One of the organic compounds most widely used by algae is acetate, which can be used as a sole carbon and energy source by many flagellates and chlorophytes. In addition, some algae can assimilate simple organic compounds in the light (photoheterotro-phy; ∞ Figure 17.1 and Section 17.4) but cannot grow on them in darkness.

One of the key characteristics used in the classification of algal groups is the nature of the **reserve polymer** synthe-sized as a result of photosynthesis. Algae of the division Chlorophyta produce starch (α-1,4-linked glucose) in a form very similar to that of higher plants. By contrast, algae of other groups produce a variety of reserve substances, some polymeric and some as free monomers; the major substances are listed in Table 14.3.

Cell Walls of Algae

Algae show considerable diversity in the structure and chemistry of their cell walls. In many cases the cell wall is composed of a network of cellulose fibrils, but it is usually supplemented with other polysaccharides such as pectin, xylans, mannans, alginic acid, or fucinic acid. In some algae, the wall is additionally strengthened by the deposition of calcium carbonate; these forms are called *calcareous* or *coralline* (coral-like) algae. Some-times chitin (a polymer of *N*-acetylglucosamine, ∞ Figure 3.5) is also present in the cell wall. In euglenoids a cell wall is absent (Figure 14.36a).

In diatoms (Figure 14.36c–e) the cell wall is com-posed of *silica*, to which protein and polysaccharide are added. Even after the cell dies and the organic materi-als have disappeared, the external structure, called the *frustule*, (Figure 14.36c–e) often remains, showing that the siliceous component is indeed responsible for the rigidity of the cell. Because of the resistance of these di-atom frustules to decay, they remain intact for long pe-riods of time and constitute some of the best algal fossils known. From this excellent fossil record, it is known that diatoms first appeared on Earth about 200 million years ago. Diatoms typically show some form of symmetry in the morphology of their frustules. Examples of symmetry include *pennate* (having similar parts arranged on opposite sides of an axis) (Figure 14.36c), and *radial* (Figure 14.36d, e).

Algal cell walls contain pores some 3–5 nm wide that are large enough to pass only low-molecular-weight substances such as water, inorganic ions, gases, and other small nutrient molecules needed for metabolism and growth. Phagocytic activities are thus not possible, and in this regard, algae differ from the actively phagocytic protozoa and slime molds.

Algal Motility

A number of algae are motile, usually because of flagella (Figure 14.10); as far as is known, cilia do not occur in algae. Flagellated algae, such as *Euglena* (Figure 14.36a), typically have a single polar flagellum, whereas flagellated species of Chlorophyta have either two or four polar flagella. Dinoflagellates (Figure 14.36f; 14.37) have two flagella of different lengths and with different points of insertion into the cell, transverse and lateral. The *transverse* flagellum is attached laterally, whereas the *longitudinal* flagellum originates from the lateral groove of the cell and extends lengthwise (Figure 14.37b). In many cases, algae are nonmotile in the vegetative state and only form motile gametes. A few algae, some diatoms for example, can also move by *gliding motility*. Gliding is common among filamentous cyanobacteria and a variety of other prokaryotes (◌◌ Section 4.15).

Algal Ecology and Endolithic Algal Communities

Algae abound in aquatic habitats, both freshwater and marine. Algae are also found in artificial aquatic habitats like fish tanks and swimming pools and in temporary pools of water formed from rainwater runoff and exposed to sunlight. Algae are also common in most soils, while a few species thrive in dry and even extremely dry soils where the water potential (◌◌ Section 6.14) can be very low. Algae are often the dominant or sole phototrophic microorganisms in extremely acidic habitats as well; in habitats below pH 4, cyanobacteria are absent and so oxygenic photosynthesis occurs exclusively by the activities of algae and a few acid-tolerant plants. The red alga *Cyanidium*, for example, can grow to below pH 2.

Some algae grow inside of rocks. These *endolithic* (*endo* means "inside") phototrophs inhabit porous rocks, for example, those containing quartz, and are typically found in layers near the surface. Endolithic phototrophic communities are most common in dry environments such as deserts or cold dry environments like the Antarctic (Figure 14.38●). For example, in the Antarctic Dry Valleys, where temperatures and humidity are very low, life within a rock has its advantages. Rocks are heated by the sun and water from snow melt can be absorbed and retained for relatively long periods, supplying needed moisture for growth.

Moreover, water absorbed by a porous rock makes the rock more transparent, thus funneling more light to the algal layers.

A wide variety of phototrophs can form endolithic communities, including cyanobacteria (Figure 14.38b) and various green algae. In addition to free-living phototrophs, green algae and cyanobacteria coexist with fungi in endolithic lichen communities (◌◌ Section 19.20 for discussion of the lichen symbiosis). Metabolism and growth of these internal rock communities slowly weathers the rock, allowing gaps to develop where water can enter, freeze and thaw, and eventually crack the rock, producing new habitats for microbial colonization.

 14.13 Concept Check

Algae are phototrophic *Eukarya* that contain chlorophyll and carotenoid pigments within a structure called the chloroplast. The chloroplast itself has its roots in the domain *Bacteria*.

◆ How do *algae* differ from *cyanobacteria*?

◆ What is unusual about red algae?

◆ What is the primary habitat of algae?

(a)

(b)

● **Figure 14.38 Endolithic cyanobacteria.** (a) Photograph of a limestone rock from Makhtesh Gadol, Negev Desert, Israel, showing a layer of cells of the cyanobacterium *Chroococcidiopsis*. (b) Photomicrograph of cells of *Chroococcidiopsis* isolated from a sandstone rock in the Negev Desert.

REVIEW QUESTIONS

1. List at least three features of eukaryotic cells that clearly differentiate them from prokaryotic cells (∞ Sections 4.1–4.5).

2. How are the mitochondrion and the hydrogenosome similar structurally? How do they differ? How do they differ metabolically (∞ Section 14.2)?

3. What major physiological processes (energy related and carbon related) occur in a chloroplast (∞ Section 14.3)?

4. What does knowing that an antibiotic like streptomycin blocks protein synthesis in organelles tell us about their relationship to *Bacteria* (∞ Section 14.4)?

5. What are the chemical differences between *microfilaments* and *microtubules*? Are such structures really absent from prokaryotic cells? (You will want to review Section 6.2 before answering this question.) (∞ Section 14.5).

6. What are the functions of the following eukaryotic cell structures: endoplasmic reticulum, Golgi complex, lysosome, peroxisome (∞ Section 14.5).

7. Contrast the action of flagella in prokaryotes and eukaryotes (∞ Sections 14.4 and 14.5).

8. Discuss how in the absence of telomerase activity, genes would eventually be lost from eukaryotic DNA (∞ Section 14.6).

9. Compare and contrast the processes of *mitosis* and *meiosis*. Which process is absolutely necessary for growth of *Saccharomyces cerevisiae* and why? How is the mating type of a yeast cell determined (∞ Section 14.7)?

10. Why do eukaryotic mRNAs have to be "processed," while most prokaryotic RNAs do not (∞ Section 14.8)?

11. The phylogeny of microbial eukaryotes is unsettled. Discuss the major disagreements between the two phylogenetic trees shown in Figure 14.20 (∞ Section 14.9).

12. What are the common names of some major groups of protozoa (∞ Section 14.10)?

13 What major differences exist between the organisms *Physarum* and *Dictyostelium* (∞ Section 14.11)?

14. Compare and contrast protozoa with fungi and algae, listing at least two ways you could differentiate a member of one group from the other. Which of these groups most closely resemble slime molds and why (∞ Sections 14.12 and 14.13)?

15. In what ways can different groups of algae be distinguished (∞ Section 14.13)?

16. What is a *frustule* and which group of algae produce them (∞ Section 14.13)?

APPLICATION QUESTIONS

1. If organelles such as the mitochondrion and chloroplast had originally been free-living *eukaryotic* cells, how would the molecular properties of organelles listed in Section 14.4 differ?

2. In the next chapter you will learn that eukaryotic cells have, in general, much larger genomes than prokaryotic cells. List three reasons why this is not surprising.

15

MICROBIAL GENOMICS

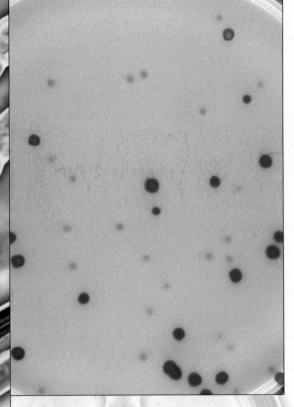

Genome sequencing reveals the genetic blueprint of an organism. Any genomics project begins with cloning a genome into managable-sized pieces, such as within a bacteriophage, as shown here. White plaques of bacteriophage M13 that have replicated in cells of *Escherichia coli* contain cloned DNA.

WORKING GLOSSARY

Annotation in reference to genomics, the conversion of raw sequence data into a list of genes present in the organism (the "assignment" of genes)

Artificial chromosome a cloning vector that can carry very large inserts of foreign DNA and exist in the cell very much like a cellular chromosome. The most widely used are bacterial artificial chromosomes (BACs) and yeast artificial chromosomes (YACs)

Bioinformatics the use of computational tools to acquire, analyze, store, and access DNA and protein sequences

Gene family genes that are related in sequence to other genes within the organism

Genome the total complement of genes of a cell or a virus

Genomics the discipline involving mapping, sequencing, analyzing, and comparing genomes

Homologue a gene that shares evolutionary ancestry with one or more other genes

Horizontal gene transfer the presence of a gene in an organism that originated in another organism and was transferred by genetic exchange

Open reading frame (ORF) a sequence of DNA that contains no in-frame stop codons and, if transcribed, could be translated to yield a protein of known length and composition. A *functional* ORF is one that actually encodes a protein in the cell.

Ortholog a gene found in one organism that is similar to that in another organism but differs because of speciation (see also *Paralog*)

Paralog a gene within an organism whose similarity to one or more other genes in the same organism is the result of gene duplication (see also *Ortholog*)

Proteome the total complement of proteins present in a cell, tissue, or organism at any one time

Proteomics the large-scale or genome-wide study of the structure, function, and regulation of the proteins of an organism

RNA editing modification of the RNA transcript of a protein-encoding gene by a process other than splicing, to achieve the required coding properties

Shotgun sequencing sequencing of previously cloned small fragments of a genome in a random fashion followed by computational methods to reconstruct the entire genome

Transcriptome the complement of mRNAs produced in an organism under a specific set of conditions

An organism's **genome** is its entire complement of genes. Knowledge of the genome sequence of an organism reveals not only its genes, but also yields important clues to how the organism functions and its evolutionary history. Genomic sequences also facilitate the study of *gene expression*, the transcription and translation of the genetic information in the genome. The traditional approach for studying gene expression was to focus on a single gene or group of related genes. Today, expression of the *entire complement* of an organism's genes can be examined in a single experiment. The word **genomics** refers to the discipline of mapping, sequencing, analyzing, and comparing genomes. This chapter is focused on microbial genomics.

The first genome to be sequenced was the 3569-nucleotide RNA genome of the *virus* MS2 (Section 16.1) in 1976. The first DNA genome to be sequenced was the 5386-nucleotide sequence of the small, single-stranded DNA virus, ϕX174 (Section 16.2) in 1977. This feat was accomplished by a group led by Frederick Sanger and debuted the Sanger dideoxynucleotide technique for DNA sequencing (Section 7.8). The first *cellular* genome to be sequenced was the 1,830,137-bp chromosome of *Haemophilus influenzae* described in 1995 by Hamilton O. Smith, J. Craig Venter, and their colleagues at The Institute for Genomic Research (TIGR), Rockville, MD (USA). By spring of the year 2000 a "rough draft" had been established of the sequence of the haploid human genome, which is about 3 billion bp. At present, several hundred genomes have been sequenced and many others are in progress. We are clearly in the "Age of Genomics," and the fallout from this new era in biology promises new insight on topics as diverse as clinical medicine and microbial evolution.

It would have been impossible to sequence and analyze large and complex genomes without a corresponding improvement in technology. A major part of this improvement has been the automation of DNA sequencing (Section 7.8). Another has been the development of powerful computational methods to analyze, store, and access DNA and protein sequences, a field called **bioinformatics**. Yet another major advance has been in cloning large fragments of DNA.

In this chapter we discuss some model microbial genomes, some of the techniques used to analyze these genomes, and what microbial genomics has shown us thus far. We begin with cloning, since success at this stage is necessary before a genomics project can really begin.

GENOMIC CLONING TECHNIQUES

Most of the techniques we will discuss in this section are modifications or expansions of the *in vitro* techniques discussed in Chapter 10. The principles for these *in vitro* technologies remain the same, but now the emphasis is on studying the *entire genome* of an organism, rather than a single gene or set of genes.

15.1 Vectors for Genomic Cloning and Sequencing

We have previously seen some plasmid and viral vectors used for molecular cloning (Sections 10.15–10.17). Phage pBR322 and the Charon derivatives of bacteriophage lambda, for example, have been extensively used for cloning and sequencing, including genomic analysis. Other, more specialized vectors include bacteriophage M13 and bacterial and yeast artificial chromosomes, and we focus on these here.

Vectors Derived from Bacteriophage M13

M13 is a filamentous bacteriophage that replicates without killing its host (⌒ Section 16.3). Mature particles of M13 are released from host cells without lysing by a budding process, and infected cultures can provide continuous sources of phage DNA. Most of the genome of phage M13 contains genes essential for virus replication. However, a small region called the *intergenic sequence* can be used as a cloning site. Variable lengths of foreign DNA, up to about 5 kbp, can be cloned without affecting phage viability. As the genome gets larger, the virion simply grows longer.

Phage M13mp18 is a derivative of M13 in which the intergenic region has been modified to facilitate cloning (Figure 15.1*a* ●). One useful modification is the insertion of a functional fragment of *lacZ*, the *Escherichia coli* gene that encodes the enzyme β-galactosidase. Cells infected with M13mp18 can be detected easily by their color on indicator plates (besides cleaving lactose into glucose and galactose, β-galactosidase cleaves a compound that yields a blue color, Figure 15.1*b*). This *lacZ* gene has itself been modified to contain a 54-bp DNA fragment called a *poly-linker*. The polylinker contains several restriction enzyme cut sites absent from the M13 genome and can therefore be used for cloning. The polylinker is inserted into the beginning of the coding portion of the *lacZ* gene and does not affect enzyme activity. However, insertion of additional DNA into the polylinker during cloning inactivates the gene. Phages that contain additional DNA inserts give rise

(a)

(b)   (c)

● **Figure 15.1 Cloning using bacteriophage M13.** (a) A partial map of M13mp18, a derivative of M13 constructed for use as a cloning vector. The vector contains the *lac* promoter *(lacP)* and a gene, *lacZ'*, which encodes a functional part of β-galactosidase. At the beginning of this gene is a polylinker that contains several restriction sites but maintains the proper reading frame. The amino acids encoded by the polylinker are shown. Most DNA fragments cloned into the polylinker disrupt the *lacZ'* gene and abolish β-galactosidase activity. (b) A plate with plaques formed by M13mp18 and by clones made using this vector on a lawn of sensitive bacteria plated on a medium containing the chemical 5-bromo-4-chloro-3-indolyl-β-D-galactopyranoside, called X-gal. When β-galactosidase hydrolyzes X-gal, it releases a relatively insoluble blue dye. Many plaques on this plate are blue, indicating the presence of vector *without* cloned DNA. Many other plaques are colorless, indicating that foreign DNA has been inserted into the vector and the *lacZ'* gene has been disrupted. In terms of genetic techniques, cloning using bacteriophage M13 combines aspects of both *selection* and *screening*. Using blue/white color differences, phage containing cloned DNA can be *selected*. However, unless PCR product is cloned, one must still *screen* the white plaques to find the ones containing the cloned gene of interest. (c) An enlargement of a portion of this plate.

to colorless plaques (no β-galactosidase activity), and it is therefore very simple to identify clones (Figure 15.1*b, c*). Similar constructs have been used in lambda cloning vectors and plasmid cloning vectors to allow identification of plaques or colonies containing cloned DNA.

Use of M13 in Molecular Cloning

To clone DNA in M13 vectors, replicative double-stranded DNA (∞ Section 16.3) is isolated from the infected host and treated with a restriction enzyme. The foreign DNA is then treated with the same restriction enzyme. On ligation, double-stranded M13 molecules are obtained that contain the foreign DNA. When these molecules are introduced into the cell by transformation (∞ Section 10.7), they are replicated and produce single-stranded DNA bacteriophage particles containing the cloned DNA.

The single-stranded M13 DNA produced can then be used directly in DNA sequencing. Since the base sequence where the foreign DNA is inserted is known (based on the specificity of the restriction enzyme used), it is possible to construct an oligonucleotide primer complementary to this region and with this determine the sequence of the whole DNA downstream from this point using Sanger sequencing. In this way, M13 derivatives have proven extremely useful in sequencing foreign DNA, even rather long molecules, and have featured prominently in the sequencing of several genomes.

Vectors like M13, or plasmid vectors that hold about 2 kb of cloned DNA, are adequate for making *gene libraries* (∞ Section 10.15) for sequencing of prokaryotic genomes. Bacteriophage lambda vectors, which hold 20 kb or more (∞ Section 10.17), are also widely used in genomics projects. However, as the size of the genome to be sequenced increases, so will the number of clones needed to obtain a complete sequence. Therefore, for making libraries of DNA from eukaryotic microorganisms or from higher eukaryotes, such as from humans, it is useful to have vectors that can carry very large segments of DNA. This allows the size of the initial library to be manageable. Such vectors have been developed and are called *artificial chromosomes*.

Bacterial Artificial Chromosomes: BACs

Many bacteria contain large plasmids that are stably replicated within the cell. For example, the *F plasmid* of *Escherichia coli* is such a plasmid (∞ Sections 7.4 and 10.10). F is very stably replicated in *E. coli*, and naturally occurring derivatives, called F′ plasmids, are known that can carry large amounts of chromosomal DNA (∞ Section 10.12). Because of these desirable properties, the F plasmid has been used to construct cloning vectors called **bacterial artificial chromosomes**, or **BACs**.

Figure 15.2● shows the structure of a BAC based on the F plasmid. The vector is only 6.7 kb compared to the 99.2 kb of F itself. The BAC contains only a few genes from F, including *oriS* and *repE*, which are necessary for replication, and *sopA* and *sopB*, which keep the copy num-

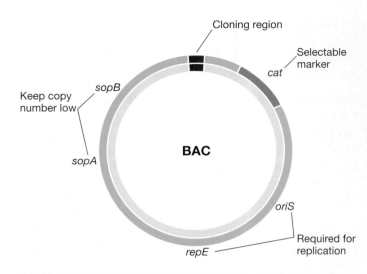

● **Figure 15.2   Genetic map of a bacterial artificial chromosome.** BACs are derivatives of the F plasmid of *Escherichia coli*. The BAC diagrammed is 6.7 kb. At the top of the map is a cloning region that contains several unique restriction enzyme sites. The *cat* gene confers resistance to the antibiotic chloramphenicol. The other genes shown are involved in plasmid replication. In the 99.2-kb F plasmid itself (∞Figure 10.18) all these genes are located in a relatively compact replication region. Therefore, BACs contain only a small fraction of the entire F plasmid.

ber very low (∞ Section 10.9). Inserted into the plasmid is the gene *cat*, which confers chloramphenicol resistance on the host, and a cloning region that includes several restriction sites for cloning DNA. Foreign DNA of *over 300 kb* can be inserted and stably maintained in a BAC vector such as this. The host for a BAC is typically a mutant strain of *E. coli* that is missing the normal restriction and modification systems of the wild type (∞ Sections 7.7 and 9.6). This prevents the BAC from being destroyed. Typically, the host strain will be defective in some recombination pathways (∞ Section 10.6), as well. This prevents recombination and rearrangements of the cloned DNA from the BAC into the host's chromosomes.

Yeast Artificial Chromosomes: YACs

Historically, the term **artificial chromosome** originated not with BACs but with **yeast artificial chromosomes**, or **YACs**. These vectors replicate in yeast like normal chromosomes, but they have sites where very large fragments of DNA can be inserted. To function like normal eukaryotic chromosomes, YACs must have (1) an *origin of DNA replication*, (2) *telomeres* for replicating DNA at the ends of the chromosome (∞ Section 14.6), and (3) a *centromere* (the section of the chromosome required for segregation during mitosis). They must also contain a cloning site and a gene that can be used for selection after transformation into the host, typically the yeast *Saccharomyces cerevisiae*. Figure 15.3● shows a diagram of a YAC vector into which foreign DNA has been cloned. YAC vectors are themselves only about 10 kbp, but they can have 200–800 kbp of cloned DNA inserted.

Figure 15.3 **Diagram of a yeast artificial chromosome (YAC) containing foreign DNA.** The foreign DNA was cloned into the vector at a *Not*I restriction site. The telomeres are labeled TEL and the centromere CEN. The origin of replication is labeled ARS (for autonomous replication sequence). The gene used for selection is called URA3. The host into which the clone is transformed has a mutation in that gene so that it normally requires uracil for growth (¯Ura¯). Host cells containing this YAC become Ura⁺. The diagram is not drawn to scale; the inserted DNA would normally be 200–800 kbp long and the vector only about 10 kbp.

After confirming that a particular fragment of DNA has been cloned on a BAC or a YAC, this region can be *subcloned* into a plasmid or bacteriophage vector for more detailed analysis or sequencing. Although YACs can hold larger DNA inserts than BACs, there is a greater problem with recombination and rearrangement of the cloned DNA within yeast than within *E. coli*. For this reason BACs are now more widely used in genomic cloning than are YACs.

15.1 Concept Check

Specialized cloning vectors have been constructed that are useful for the sequence and assembly of genomes. Some, like the M13 derivatives, are useful for both cloning and for direct DNA sequencing. Others, like artificial chromosomes, are useful for cloning very large fragments of DNA, fragments approaching a megabase in size.

◆ Compare the capacity for cloning foreign DNA in M13, lambda, BACs, and YACs.

◆ The yeast artificial chromosome behaves like a chromosome in a yeast cell. What makes this possible?

15.2 Sequencing the Genome

The analysis of a genome begins with the formation of a genomic library—the molecular cloning of all of the DNA fragments of the genome generated from restriction enzyme cuts. Once this is done, actual sequencing begins.

Shotgun Sequencing

Virtually all genomic sequencing projects today employ **shotgun sequencing**. This technique is made possible by high-throughput sequencing capacity, robotics, and powerful computational capacities. Shotgun sequencing has been used on chromosomes from prokaryotes as well as eukaryotes, including for the privately funded version of the human genome project.

Whole genome shotgun techniques involve cloning the entire genome in a random fashion and then sequencing the resultant clones. That is, the clones are sequenced without knowing the order or orientation of any of the cloned DNA. The sequences are then analyzed by a computer that searches for overlapping sequences.

The overlaps allow the computer to assemble the sequenced fragments in the correct order.

Much of the sequencing in the shotgun method is redundant. To ensure that all sequences are obtained it is necessary to sequence a very large number of clones, many of which will be identical or nearly identical. Typically there will be 7–10 sequences over any given part of the genome. This seven- to tenfold *coverage*, as it is referred to, greatly reduces errors in the sequence because the redundancy in sequencing allows for a consensus nucleotide to be selected at any point in the sequence where ambiguity may exist.

For shotgun sequencing to work effectively, it is essential that the cloning itself be efficient (one needs a large number of clones) and, in so far as possible, the DNA cloned should be randomly generated. Restriction sites are not random, but by cutting the genomic DNA with an enzyme that recognizes a short sequence that occurs commonly in the DNA, one can approach random digestion. To obtain more truly random fragments, DNA can be mechanically sheared by forcing DNA through a *nebulizer*. This is a device with a small opening similar to a nozzle that reduces the DNA solution to a spray; in the process, the DNA is sheared. The DNA fragments can be purified by size using gel electrophoresis (∞ Section 7.7) and then cloned into a vector and transformed into a host.

Genome Assembly

Once shotgun sequencing is completed, all of the fragments must be ordered, a process called **assembly**. Assembly of a circular genome from a prokaryote, for example, involves putting all of the fragments in the correct order and eliminating overlaps and then generating a genome suitable for *annotation*, the process of identifying genes and other functional regions (see Section 15.3).

Sometimes shotgun sequencing and assembly will not yield a complete genome sequence; that is, sometimes there will be *gaps* left in the sequence. In such situations, clones can be sought that are predicted to cover the gap. One method of doing this is to perform PCR reactions using specific primers complementary to sequences that are already known and that flank a given gap. Note that these additional clones are targeted, not random, as in the shotgun method. However, the sequences in these clones are likely to contain overlapping sequences sufficient to close the gap.

Some genome projects have the goal of obtaining a *closed genome*, meaning that the entire genome sequence is determined. Other projects stop at the *draft* stage, dispensing with the sequencing of the small gaps. Since shotgun sequencing and assembly are heavily automated procedures while gap closure is not, a closed genome is considerably more expensive to generate than a draft genome sequence.

⬢ 15.2 Concept Check

Shotgun techniques employ random cloning and sequencing of relatively small genome fragments followed by computer-generated assembly of the genome using overlaps as a guide to the final sequence.

◆ Why is shotgun sequencing considered to give a very *accurate* genome sequence?

◆ What is done during genome *assembly*?

15.3 Annotating the Genome

It is not the sequence of the genome, *per se*, that is the final goal in genome sequencing projects, but instead, determining the *genes* that the sequence contains. Once sequencing and assembly is completed, the next step in genomic analysis is **annotation**, the conversion of raw sequence data into a list of the genes present in the genome.

Locating Putative Genes: Identifying ORFs

The great majority of genes of any organism encode proteins, and in most microbial genomes, the great majority of the genome consists of coding sequences. Because the genomes of microbial eukaryotes typically have fewer introns than plant and animal genomes, and prokaryotes have almost none, microbial genomes essentially consist of hundreds to thousands of **open reading frames (ORFs)** (∞ Section 7.13) separated by regulatory regions and transcriptional terminators. A *functional ORF* is one that actually encodes a protein in the cell. Thus, the simplest way to locate protein-encoding genes, or potential protein-encoding genes, is to have a computer search the sequence of the genome for ORFs.

How Does the Computer Find an ORF?

In the cell, ribosomes establish a reading frame by initiating translation at a *start codon*, usually an AUG. The ribosome then proceeds until it reaches an in-frame *stop codon* (∞ Section 7.13). The first step in finding an ORF is to look for these signals in the sequence. However, the identification of possible *functional* ORFs is typically more complex than just searching for in-frame start and stop codons, since these will appear randomly with reasonable frequency. Thus, other clues are sought.

One hint that an ORF is functional will be its *size*. The majority of cellular proteins contain 100 or more

amino acids, so most functional ORFs are longer than 100 codons (300 nucleotides). However, simply programming the computer to ignore ORFs shorter than 100 codons will miss some functional genes. So other factors must be considered. Because most organisms show preferences among synonymous codons (∞ Section 7.14), *codon bias* can also give a clue as to whether an ORF is functional. If the codon usage in a given ORF is considerably different from the consensus codon usage, the ORF may not be functional or may be functional but obtained by lateral gene transfer (see Section 15.8). It should also be remembered that prokaryotic ribosomes start translation not at the first (most 5′) possible start codon, but at one immediately downstream of a *Shine–Dalgarno sequence* on the mRNA (∞ Section 7.16). Therefore, searching the DNA sequence of a prokaryotic genome for potential Shine–Dalgarno sequences can help establish both whether an ORF is functional and which start codon is actually used.

Figure 15.4● shows a region of DNA with the characteristics of an ORF that a computer would recognize as a "hit." Although any given gene is always transcribed from a single strand, in all but the smallest plasmid or viral genomes, both strands are transcribed in some part of the genome. Thus, computer inspection of *both strands* of DNA is required. Figure 15.4 shows a diagram of a region of a genome in which one gene is transcribed from one strand and one from the other. Of course, some genes encode transfer RNAs and ribosomal RNAs, and these are not recognized by programs that search only for ORFs. However,

● **Figure 15.4   Locating possible functional open reading frames.** The sequence shown for Gene 1 from one strand includes bases for a potential Shine–Dalgarno sequence (AGGA) separated from a start codon (ATG) by about 8 bases. The bases encoding the start codon are followed by about 100 sense codons and then a stop (nonsense) codon (UAA is the most commonly used stop codon). In the DNA molecule with the ORF described as gene 1, note that upstream of this ORF is a promoter and downstream a transcription terminator. Also shown is Gene 2, which has the same components, but that would be transcribed in the opposite direction. Shine–Dalgarno sequences, start codons, sense codons, and stop codons function only at the level of RNA. Here we are showing DNA bases (base pairs) that encode these functional sequences.

genes for tRNAs and rRNAs can usually be located rather easily because the sequences of these RNAs are very highly conserved (∞ Sections 7.15 and 11.5–11.7).

Locating Putative Genes: Genomic Comparisons

An ORF is also likely to be functional if its sequence is similar to sequences of ORFs obtained from the genomes of other organisms (regardless of whether they encode known proteins) or if some part of the ORF has a sequence known to encode a protein functional domain. This is because proteins that carry out the same function in different cells tend to be *homologous*; that is, they are proteins that are related in an evolutionary sense and typically share sequence and structural features. The computer can search for such sequence similarities in the annotation process.

It would be almost impossible to assemble even a small prokaryotic genome and to locate genes and other important functional DNA sequences without the availability of sophisticated computational tools to handle large databases. The use of computers to do sophisticated analyses of this type is called working *in silico*, a term analogous to the terms *in vivo* and *in vitro* (*in silico* refers to the silicon processor chips present in the computer). Figure 15.5● shows a genetic map constructed by computer from shotgun sequencing of the 4.4 Mbp genome of *Mycobacterium tuberculosis*, the causative agent of tuberculosis (∞ Section 26.5). Although a highly detailed figure, all ORFs are indicated, as well as genes encoding tRNAs, rRNAs, and a few other sequences.

15.3 Concept Check

After major sequencing is through, computers search for ORFs and genes encoding protein homologues as part of the annotation process.

◆ What is an *open reading frame (ORF)*?

◆ How can protein homology assist in the annotation process?

II MICROBIAL GENOMES

Over 250 microbial genomes, mostly prokaryotic, have now been sequenced, and hundreds more projects are currently ongoing. Here we will discuss only a few of them and explore what analysis of these genomes tells us.

15.4 **Prokaryotic Genomes: Sizes and ORF Contents**

Several hundred prokaryotic genomes are now available in public databases (for an up-to-date list of genome sequencing projects search the URL *http://www.genomesonline.org/*). These include many species of *Bacteria* and *Archaea* and representatives containing circular as well as linear ge-

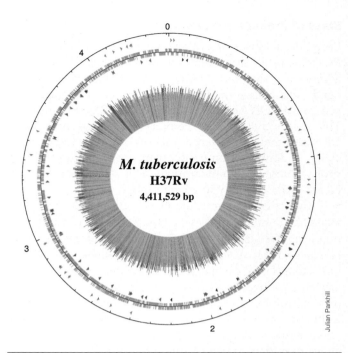

● **Figure 15.5 Chromosome map of *Mycobacterium tuberculosis*.** The outer circle shows the size of the chromosome in Mbp with 0 placed at the point of origin of DNA replication. Inside this is a ring showing the locations of tRNA genes (blue) and the single rRNA operon (orange). The next ring shows ORFs transcribed clockwise (dark green) and counterclockwise (light green). Beneath this is a ring showing the location of some repetitive DNA elements, including insertion sequences (orange). The histogram in the center shows regional GC content (∞ Section 11.10) with yellow rays showing regions below 65% GC and orange rays showing regions above 65% GC. For additional discussion of the organism *M. tuberculosis*, see Section 12.23, and for the disease tuberculosis, Section 26.5. The figure was generated with software from DNASTAR and is reprinted by permission from *Nature 393*: 537–544 (1998). © Macmillan Magazines Ltd.

nomes. Table 15.1 lists a few representative examples. Although the sequencing of prokaryotic genomes is becoming more rapid and routine, sequencing and analysis of a complete genome is still a major undertaking, requiring both considerable financial as well as human efforts. Thus, one still needs important scientific and/or societal reasons and considerable financial resources before initiating a genome sequence.

Note that the list of prokaryotic genomes in Table 15.1 contains several pathogens. Naturally, such organisms would have high priority for sequencing resources. The hyperthermophiles on the list may have important uses in biotechnology since the enzymes in these organisms will be heat stable (∞ Sections 6.10, 6.12, 13.12 and 30.9). Indeed, the needs of the biomedical and biotechnology industries are important considerations for generating the interest and the funding to pursue genomic sequencing. However, the list in Table 15.1 also includes organisms such as *Bacillus subtilis*, *Escherichia coli*, and *Pseudomonas aeruginosa*, all of which remain widely studied genetic model systems, and *Caulobacter crescentus*, a model system for cell differentiation (∞ Section 12.16).

Sizes of Prokaryotic Genomes

Based on analyses of prokaryotic genomes of species of *Bacteria* and *Archaea*, there is a strong correlation between a genome's size and its ORF content (Figure 15.6●). Indeed, regardless of the organism, each *megabase* of prokaryotic DNA encodes about *1000* ORFs (Figure 15.6). This shows that unlike in eukaryotes, where noncoding DNA can make up a large fraction of the genome (especially in large genome organisms, see Table 15.3), in prokaryotes, as genomes increase in size, they also increase proportionally in gene number.

Analyzing genomic sequences allows us to answer some fundamental biological questions. For example, genomic analysis of the small genomes of *Mycoplasma genitalium* and *Mycoplasma pneumoniae* shows that all 470 ORFs found in *M. genitalium* are also present in *M. pneumoniae*. By seeking homologous genes in other organisms and by determining which genes are unessential using such techniques as transposon mutagenesis (∞ Section 10.14), scientists have deduced that about 300 protein-encoding genes may be sufficient to maintain a cellular existence. No known organism has these few genes. However, some remarkable organisms do have relatively small genomes. For example, *Methanocaldococcus jannaschii* (*Archaea*) is an *autotroph* (∞ Sections 13.4 and 17.17) and contains only 1738 ORFs. This is enough, however, to enable it to be not only free living, but also to synthesize all of its organic cellular components from CO_2 (∞ Section 13.4). *Aquifex aeolicus* (*Bacteria*) is also an autotroph and contains the smallest genome of any known autotroph at just 1.5 Mbp (Table 15.1). Both *Methanocaldococcus* and *Aquifex* are also hyperthermophiles. One can thus conclude, rather surprisingly, that an autotrophic lifestyle in near boiling water does not require a huge genome to support it.

It is no coincidence that Table 15.1 lists several relatively small genomes. Sequencing strategies are simpler for small genomes, so there was some emphasis on sequencing small genomes in early sequencing efforts. Note that the genome of *Mycoplasma genitalium* is smaller than that of some viruses, such as the *Chlorella* viruses (∞ Section 16.6) and bacteriophage G (∞ Section 9.2). The smallest known prokaryotic genome, one even smaller than that of *M. genitalium*, is that of *Nanoarchaeum equitans* (*Archaea*). This genome is some 90 kbp smaller that that of *M. genitalium* (Table 15.1). *N. equitans* is a hyperthermophile and a parasite of a second hyperthermophile, *Ignicoccus* (∞ Section 13.11). Analyses of the gene content of *N. equitans* shows it to be free of virtually all genes that encode proteins involved in anabolism or catabolism. Interestingly, however, the genome of *N. equitans* actually contains more ORFs than the larger genome of *M. genitalium* (Table 15.1). This is because the *N. equitans* genome is extremely compact and contains virtually no noncoding DNA.

Some prokaryotes have very large genomes. *Bradyrhizobium japonicum*, the organism that forms nitrogen-fixing root nodules on soybeans (∞ Section 19.22), for example, has 2800 more ORFs than the yeast *Saccharomyces cerevisiae*, a eukaryote (see Section 15.6 and

●Figure 15.6 Correlation between genome size and ORF content in prokaryotes. These data are from analyses of 115 completed prokaryotic genomes and include species of both *Bacteria* and *Archaea*. Data from *Proc. Natl. Acad. Sci. (USA)* 101: 3160–3165 (2004).

Table 15.1). The bacterium *Streptomyces coelicolor*, whose linear chromosome of over 8 Mbp is typical of *Streptomyces* species, has 7846 ORFs, almost 1800 more than *S. cerevisiae*. Some even larger prokaryotic genomes are known, but their sequence is not yet complete. These include the 9.2-Mbp chromosome of *Myxococcus xanthus* and the 12.3-Mbp chromosome of *Sorangium cellulosum*, both gliding bacteria (∞ Section 12.17). These are the largest prokaryotic genomes known. Genomes of comparable size have not yet been found in *Archaea* (Table 15.1). Prokaryotic genomes thus range in size from those of large viruses to those of small eukaryotes.

⬢ 15.4 *Concept Check*

Sequenced prokaryotic genomes range in size from 0.49 Mbp to 9.1 Mbp. The smallest prokaryotic genomes are the size of the largest viruses, and the largest prokaryotic genomes have more genes than some eukaryotes. In prokaryotes, ORF content is proportional to genome size.

◆ How many more genes does *Bradyrhizobium* have than *Nanoarcheaum*? How does this affect the lifestyles of these two organisms?

◆ Approximately how many ORFs will a genome of 4 Mbp contain?

15.5   Prokaryotic Genomes: Bioinformatic Analyses and Gene Distributions

Following assembly and annotation, a typical genomic analysis proceeds to the *gene comparison stage*. The basic question here is "how does the complement of genes in this organism compare with that of other organisms?" Such analyses put the just-sequenced genome in perspective with the genomes of other organisms, similar as well as more distant. These activities are a major part of the field of **bioinformatics**.

Table 15.1 Select prokaryotic chromosomes[a]

Organism	Size (base pairs)	ORFs[b]	Comments
Bacteria			
Mycoplasma genitalium	580,070	470	Smallest known bacterial genome (⚭ Section 12.21)
Mycoplasma pneumoniae	816,394	677	Causes pneumonia (⚭ Section 12.21)
Borrelia burgdorferi	910,725	853	Spirochete, has linear chromosome, causes Lyme disease (⚭ Sections 12.33 and 27.4)
Chlamydia trachomatis	1,042,519	894	Obligate intracellular parasite, common human pathogen (⚭ Sections 12.27 and 26.13)
Rickettsia prowazekii	1,111,523	834	Obligate intracellular parasite, causes epidemic typhus (⚭ Sections 12.13 and 27.3)
Treponema pallidum	1,138,006	1041	Spirochete, causes syphilis (⚭ Sections 12.33 and 26.12)
Aquifex aeolicus	1,551,335	1544	Hyperthermophile, autotroph (⚭ Section 12.37)
Prochlorococcus marinus	1,657,990	1716	Most abundant phototroph in oceans (⚭ Section 12.26 and 19. 6)
Helicobacter pylori	1,667,867	1590	Causes peptic ulcers (⚭ Section 26.10)
Streptococcus pyogenes	1,852,442	1752	Pathogen, causes step throat and scarlet fever (⚭ Section 26.2)
Thermotoga maritima	1,860,725	1877	Hyperthermophile (⚭ Section 12.36). See also Figure 15.7
Chlorobium tepidum	2,154,946	2288	Model phototrophic bacterium (⚭ Section 12.32)
Staphylococcus aureus	2,814,816	2593	Major cause of nosocomial infections (⚭ Sections 12.19 and 25.7)
Deinococcus radiodurans	3,284,156	2185	Radiation resistant, multiple chromosomes (⚭ Section 12.34)
Synechocystis sp.	3,573,470	3168	Cyanobacterium (⚭ Section 12.25)
Bdellovibrio bacteriovorus	3,782,950	3584	Predator of other prokaryotes (⚭ Section 12.14)
Geobacter sulfurreducens	3,814,139	3467	Model for bioremediation (⚭ Section 17.18)
Caulobacter crescentus	4,016,942	3767	Complex life cycle (⚭ Section 12.16)
Bacillus subtilis	4,214,810	4100	Gram-positive genetic model (⚭ Section 12.20)
Mycobacterium tuberculosis	4,411,529	3924	Causes tuberculosis (⚭ Sections 12.23 and 26.5)
Escherichia coli	4,639,221	4288	Gram-negative genetic model (⚭ Section 12.11)
Bacillus anthracis	5,227,293	5738	Pathogen, biowarfare agent (⚭ Section 12.20)
Rhodopseudomonas palustris	5,459,213	4836	Metabolically versatile anoxygenic phototroph (⚭ Section 12.2)
Pseudomonas aeruginosa	6,264,403	5570	Metabolically versatile opportunistic pathogen (⚭ Section 12.7)
Streptomyces coelicolor	8,667,507	7825	Has linear chromosome, produces antibiotics (⚭ Section 12.24)
Bradyrhizobium japonicum	9,105,828	8317	N_2 fixation, nodulates soybeans (⚭ Section 19.22)
Archaea			
Nanoarchaeum equitans	490,885	552	Smallest known genome (⚭ Section 13.11)
Thermoplasma acidophilum	1,564,905	1509	Thermophilic, acidophilic (⚭ Section 13.5)
Methanocaldococcus jannaschii	1,664,976	1738	Methanogen (⚭ Section 13.4)
Aeropyrum pernix	1,669,695	1841	Hyperthermophile (⚭ Section 13.9)
Pyrococcus horikoshii	1,738,505	2061	Hyperthermophile (⚭ Section 13.6)
Methanothermobacter thermoautotrophicus	1,751,377	1855	Methanogen (⚭ Section 13.4)
Archaeoglobus fulgidus	2,178,400	2436	Hyperthermophile (⚭ Section 13.7)
Halobacterium salinarum	2,571,010	2630	Extreme halophile, bacteriorhodopsin (⚭ Section 13.3)
Sulfolobus solfataricus	2,992,245	2977	Hyperthermophile, sulfur chemolithotroph (⚭ Section 13.9)

[a] Information on these and hundreds of other prokaryotic genomes can be found in the TIGR Database (www.tigr.org/tdb), a web site maintained by The Institute for Genomic Research (TIGR), Rockville, MD, a not-for-profit research institute, and at *http://www.genomesonline.org*. Links are listed there to other relevant web sites.

[b] Open reading frames. The purpose of reporting ORFs is to predict the total number of proteins that an organism might encode. Of course, genes encoding known proteins are included, as are all ORF's that could encode a protein greater than 100 amino acid residues. Smaller ORFs are typically not included unless they show similarity to a gene from another organism or unless the codon usage is typical of the organism being studied.

Gene Content of Prokaryotic Genomes

In many ways, the complement of genes in a particular organism defines its biology. In other words, *genomes are molded by an organism's lifestyle*. One might imagine, for instance, that obligately parasitic organisms such as *Treponema pallidum* (⚭ Section 26.12) might require relatively few genes for amino acid biosynthesis, since the amino acids they need can be supplied by their hosts. In fact, this is the case, as computer inspection of the *T. pallidum*

genome showed it to lack recognizable genes for amino acid biosynthesis, although it encodes several proteases that can convert peptides taken up from the host into free amino acids. In contrast, the nonparasitic *Escherichia coli* has 131 genes involved in amino acid biosynthesis and metabolism, and *Bacillus subtilis* has over 200.

Comparative analyses such as these are useful in the search for genes that encode enzymes that almost certainly must exist because of the known properties of an organism. *Thermotoga maritima* (*Bacteria*) for example, is a hyperthermophile found in heated marine sediments and is known to catabolize a large number of sugars. Figure 15.7● summarizes some of the metabolic pathways and transport systems of *T. maritima* that have been derived from analysis of its genome. As might be expected, the genome is particularly rich in genes involved in transport, particularly for carbohydrates and amino acids. A large portion of its genome (7%) is also involved in metabolism of simple and complex sugars. All this suggests that *T. maritima* exists in an environment rich in organic material.

Some analysis on the division of genes and their activities in prokaryotes is given in Table 15.2. Similar data are assembled when each new finished and annotated genome is published. Thus far it is clear that a distinct pattern exists in gene distribution in prokaryotes. Metabolic genes, for example, are typically the most abundant class of gene in prokaryotic genomes, although genes involved in translation overtake metabolic genes on a percentage basis as genome size decreases (Table 15.2, and see Figure 15.8). Surprisingly, as important as these genes are to the biology of a cell, genes involved in DNA replication and transcription make up

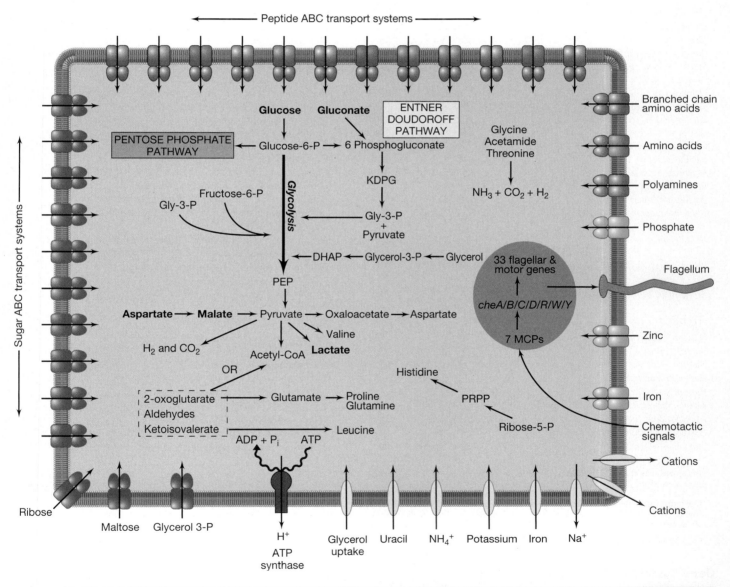

● **Figure 15.7   Overview of some metabolic pathways and transport systems of *Thermotoga maritima*, a member of the domain *Bacteria*.** The figure shows a blueprint for the metabolic capabilities of this organism. These include some of the pathways for energy production and the metabolism of certain organic compounds, as well as proteins involved in transport, that were identified from analysis of the genomic sequence. Gene names are not shown. The genome contains several ABC transport systems (∞ Section 4.7). There are 12 for carbohydrates, 14 for peptides and amino acids, and still others for ions. These are shown as multisubunit structures in the figure. Other types of transport proteins have also been identified and are shown as simple ovals. The flagellum is shown, and this organism has seven transducers (MCPs) and several chemotaxis (*che*) genes (∞ Section 8.13) as well as several genes involved in flagellar assembly (∞ Section 4.14). A few aspects of sugar metabolism are also shown. This figure is adapted from work published by The Institute for Genomic Research (TIGR).

Table 15.2	Gene function in bacterial genomes		
	Percentage of genes on chromosome in that category		
Functional categories	**Escherichia coli (4.64 Mbp)[a]**	**Haemophilus influenzae (1.83 Mbp)[a]**	**Mycoplasma genitalium (0.58 Mbp)[a]**
Metabolism	21.0	19.0	14.6
Structural	5.5	4.7	3.6
Transport	10.0	7.0	7.3
Regulation	8.5	6.6	6.0
Translation	4.5	8.0	21.6
Transcription	1.3	1.5	2.6
Replication	2.7	4.9	6.8
Other, known	8.5	5.2	5.8
Unknown	38.1	43.0	32.0

[a] Chromosome size. Each organism listed contains only a single circular chromosome.

only a minor fraction of the typical prokaryotic genome (Table 15.2).

Unidentified ORFs

Although there are differences from organism to organism, in most cases the number of genes that can actually be identified in a given genome is 70% or less of the total number of ORFs detected. Unidentified ORFs are said to encode *hypothetical proteins*, proteins that likely exist although we do not yet know their function. An unidentified ORF contains all of the signs of a protein-encoding gene—a ribosome binding site (or a location in an operon downstream from a Shine–Dalgarno site), a start codon and a stop codon—but lacks sufficient amino acid sequence homology with any known protein to be readily identified as such.

Unidentified ORFs reflect the fact that there is still much we don't know about prokaryotic genes. On the other hand, as gene functions become identified in one organism, homologous unidentified ORFs in other organisms will likely be found to encode the same protein. It is interesting to note that even in the world's best-understood organism, *Escherichia coli*, functions have been assigned to only about 2700 of its almost 4300 genes. However, most of the genes involved in macromolecular syntheses and central metabolism essential for growth of *E. coli* have been identified. Therefore, as the functions of more unidentified ORFs are identified, it is likely that most of them will encode nonessential or redundant proteins. As progress is made in identifying hypothetical proteins, the percentage of *E. coli* genes involved in macromolecular synthesis or central metabolism will therefore likely decrease (Table 15.2). It has been predicted that many of the unidentified genes in *Escherichia coli* encode proteins that carry out regulatory functions, are those needed only under special conditions or for the catabolism of unusual substrates, or are those that encode redundant proteins as "backup" systems for key metabolic reactions.

Gene Categories as a Function of Genome Size

As the data of Table 15.2 show, the *percentage* of an organism's genes devoted to one or another cell function is to some degree a function of genome size. This is summarized for a large number of genomes from species of *Bacteria* in Figure 15.8●. In this figure it can be seen that the *percentage* of genes devoted to protein synthesis, for example, rises dramatically in *small* genome organisms. By contrast, genes devoted to transcription and to regulation involving two-component regulatory systems (signal transduction, ⬢ Section 8.12), increase significantly in *large* genome organisms. Other core cellular processes, such as DNA replication and energy production, show only minor variations in gene number with genome size (Figure 15.8).

The genomic analyses summarized in Figure 15.8 suggest that although many genes can be dispensed with, genes that encode the protein-synthesizing apparatus cannot. The smaller the genome, the greater the percentage of genes devoted to translational processes (Table 15.2). Large genome organisms, on the other hand, contain more genes involved in transcriptional and other types of regulation than small genome organisms. These regulatory systems likely allow the cell to better respond to its available resources with specific gene expression. In small genome organisms these regulatory processes are dispensable, probably because small genome prokaryotes are often parasitic and can obtain much of what they need from their hosts. Large genome organisms, by contrast, have the coding capacity to include many such regulatory and special metabolic genes in their genome. This likely makes these organisms more competitive in their habitats, which for very large genome prokaryotes is often soil. Soil is a habitat where carbon and energy sources are often scarce or available only intermittently and are of a wide variety of different types (⬢ Sections 19.2 and 19.4). A

● **Figure 15.8 Functional category of genes as a relative percent of the genome versus total ORFs in the genome.** Note how genes encoding translational or DNA replication increase in percentage in *small* genome organisms, while transcriptional regulatory genes increase in percentage in *large* genome organisms. Data from *Proc. Natl. Acad. Sci. (USA)* 101: 3160–3165 (2004).

large genome encoding several metabolic options would thus be strongly selected for in such a habitat. Interestingly, all of the prokaryotes listed in Table 15.1 whose genomes are in excess of 6 Mbp inhabit soil.

Gene Distribution in *Bacteria* and *Archaea*

Analyses of gene categories have been done on several prokaryotes beyond the three species of *Bacteria* shown in Table 15.2, and the results are compared in Figure 15.9●. Note that in this figure the results are the *averages* of the gene content of several individual genomes and include both *Bacteria* and *Archaea*. On average, species of *Archaea* seem to devote a higher percentage of their genomes to energy and coenzyme production than do *Bacteria* (these results are undoubtedly skewed a bit due to the large number of novel coenzymes produced by methanogenic *Archaea*, ∞ Sections 13.4 and 17.17). On the other hand, *Archaea* tend to contain fewer genes devoted to carbohydrate metabolism or to cell membrane functions, such as transport and membrane biosynthesis, than do *Bacteria*.

Both groups of prokaryotes have relatively large numbers of genes whose function is either unknown or that encode only hypothetical proteins, although in both categories, more uncertainty exists among species of *Archaea* than species of *Bacteria* (Figure 15.9). Thus, in addition to controls on gene complement that are related to genome size (Figure 15.8), additional controls appear to operate at the level of cellular domain, as well (Figure 15.9). Data of these types are a likely reflection of lifestyle differences between *Bacteria* and *Archaea* that remain to be elucidated.

 15.5   *Concept Check*

Many genes can be identified by their sequence similarity to genes found in other organisms. However, a significant percentage of sequenced genes are of unknown function. On average, the gene complement of *Bacteria* and *Archaea* are related but distinct. Bioinformatics plays an important role in genomic analyses.

◆ Does every functional ORF encode a protein?

◆ What is a hypothetical protein?

◆ What category of genes do prokaryctes contain the most of on a percentage basis?

15.6   Eukaryotic Microbial Genomes

A large number of microbial eukaryotes are known, and to date, a variety of microbial and higher eukaryotic genomes have been sequenced (Table 15.3). The yeast *Saccharomyces cerevisiae* is an extremely important eukaryote because of its widespread use in industry (∞ Chapter 30) and its use as a model organism for cell biology, and so we focus on it here.

The Yeast Genome

The haploid yeast genome contairs 16 chromosomes ranging in size from 220 kbp to about 2352 kbp. The total yeast nuclear genome (excluding the mitochondria and some plasmid and viruslike genetic elements) is approximately 13,392 kbp. You may wonder why the words *about* and *approximately* are used when this genome has been completely sequenced. Yeast, like many other eukaryotes, has a large amount of repetitive DNA (∞ Section 7.4). When the yeast genome was published in 1997, not all of the "identical" repeats had been sequenced. It is difficult to sequence a very long run of identical or nearly identical sequences and then assemble the data into a coherent framework. For example, yeast chromosome XII contains a stretch of approximately 1260 kb containing 100–200 repeats of yeast rRNA genes. Another repeated sequence follows this long series of rRNA gene repeats. Because this entire region has not actually been completely sequenced, the precise size of the yeast genome is not known.

In addition to having 100–200 identical copies of the rRNA operons, the yeast nuclear genome has 275 genes

● **Figure 15.9   Variations in gene category in *Bacteria* and *Archaea*.** Data are averages from 34 species of *Bacteria* and 12 species of *Archaea*. "Unknown function" represents genes known to encode proteins but whose functions are unknown. "General prediction" encode hypothetical proteins that may or may not exist. Data from *Proc. Natl. Acad. Sci. (USA)* 101: 3160–3165 (2004).

for tRNAs (only a few are identical) and 80 genes for other types of noncoding RNAs (◌◌ Section 7.16). There were 6340 possible ORFs originally predicted in yeast, but more detailed analysis has reduced this number to 5570, less than that of some prokaryotes (Table 15.1). Of these, almost 3400 encode proteins whose functions are known. The wide variety of genetic and biochemical techniques available for studying this organism have resulted in significant advances in the understanding of the function of the remaining proteins as well (see Section 15.11).

Minimal Gene Complement of Yeast and the Presence of Introns

Of all the known genes in yeast, which are absolutely essential? A mechanism for exploring this question in yeast or in any microorganism exists and involves *knockout mutations* (◌◌ Section 10.14). Knockouts are mutations in which a gene has been rendered nonfunctional. Ordinarily such mutations cannot be obtained in a gene essential for cell viability in a haploid organism. However, yeast can also be grown in a diploid state (◌◌ Section 14.7). By obtaining such mutations in a diploid and then investigating whether they can also exist in a haploid, it is possible to determine whether a particular gene is essential for a cell's viability. Using such techniques, it has been shown, surprisingly, that at least 877 of the yeast ORFs are essential, while 3121 clearly are not. Note that this number of essential genes is considerably greater than the approximately 300 that are predicted to be the minimal number required for cellular existence in prokaryotes (see Section 15.3). However, since eukaryotes are more complex organisms than prokaryotes, a larger minimal gene complement would be expected.

Since yeast is a eukaryote, it contains introns (◌◌ Section 7.1). However, in the protein-encoding genes of yeast there are only a total of 225 introns. Most yeast genes that have introns have a single small intron near the 5′ end of the gene. This situation is much different than in the genomes of higher organisms shown in Table 15.3. In the worm *Caenorhabditis elegans*, for example, the average gene has 5 introns, and in the fruitfly *Drosophila*, the average gene has 4 introns. Introns are also very common in the mustard plant *Arabidopsis*, which also averages 5 introns per gene, and over 75% of *Arabidopsis* genes have introns. In humans almost all protein-encoding genes have introns, and it is not uncommon for a single gene to have 10 or more. In these organisms, the introns are typically much larger than the exons, as well.

Other Eukaryotic Microorganisms

The genomes of several other notable eukaryotic microorganisms have been sequenced. *Plasmodium falciparum* is the parasite that causes malaria (◌◌ Section 27.5); the sequencing of its genome was an international effort. The 27-Mbp genome of *P. falciparum* consists of 14 chromosomes ranging in size from 0.7 to 3.4 Mbp. *Encephalitozoon cuniculi* is an intracellular pathogen of humans and other animals that causes lung infections. *E. cuniculi* lacks mitochondria (◌◌ Section 14.9), and although its haploid genome contains 11 chromosomes, the genome size is only 2.9 Mbp (Table 15.3); this is smaller than that of many prokaryotic genomes (see Table 15.1). *Ustilago maydis* is a plant pathogenic fungus with a genome of approximately 20 Mbp. It causes smut disease in corn (maize), a disease with a large economic impact.

Table 15.3 Some eukaryotic nuclear genomes[a]

Organism	Comments	Genome size	Chromosome number	Protein-encoding genes[b]
Encephalitozoon cuniculi	Very small genome; human pathogen	2.9 Mbp	11	1997
Saccharomyces cerevisiae	This yeast is an industrially important organism (◌◌ Sections 30.10 and 30.13) that is also a model for biochemical and genetic studies	13 Mbp	16	5,570
Caenorhabditis elegans	This roundworm is an important model for studying animal development	97 Mbp	6	19,099
Drosophila melanogaster	The fruit fly is an intensively studied model organism	180 Mbp	4	13,601
Arabidopsis thaliana	This plant serves as a model organism for genetic studies	125 Mbp	5	25,498
Mus musculus	Mouse, a model mammalian system	2500 Mbp	23	30,000
Homo sapiens	The human genome is available only in draft form	3000 Mbp	23	25,000–35,000[c]

[a] All data are for the haploid nuclear genomes of these organisms.

[b] The number of protein-encoding genes is in all cases an estimate based on the number of known genes and sequences that seem likely to encode functional proteins.

[c] There is still debate over the number of genes in the human genome, despite the release of "draft" sequences by two different groups. The number of genes could be as high as 60,000.

Considerable progress has been made in sequencing the genomes of other eukaryotic microorganisms, primarily pathogens, including *Leishmania major* (leishmaniasis), *Candida albicans* (various yeast infections; Section 27.8), *Entamoeba histolytica* (amebic dysentery; Sections 14.9 and 28.8), *Giardia lamblia* (giardiasis; Section 28.6), and *Pneumocystis carinii* (AIDS-associated pneumonia; Section 26.14). In addition, the complete sequence of the mouse and rat genomes are known. The human genome has been sequenced of course, but not yet fully annotated (Table 15.3).

⬣ *15.6* *Concept Check*

The complete genomic sequence of the yeast *Saccharomyces cerevisiae* and that of many other microbial eukaryotes has been determined. Yeast may encode up to 5570 proteins of which only 877 appear essential for viability. Relatively few of the protein-encoding genes of yeast contain introns.

◆ How might you show that a gene is essential?

◆ What is unusual about the genome of the eukaryote *Encephalitozoon*?

▨▮▮ OTHER GENOMES AND THE EVOLUTION OF GENOMES

Besides cells, other structures have genomes, including organelles and viruses. We consider here organellar genomes and then some issues of genome evolution. We end with a consideration of how genomes can be "mined" for genes of interest.

15.7 Genomes of Organelles

The major organelles, mitochondria and chloroplasts, contain their own DNA (Section 14.4). Both mitochondria and chloroplasts originated from *Bacteria* by endosymbiosis (Sections 11.4 and 14.4). Therefore, it is not surprising that the proteins that organellar DNA encode are more closely related to those of *Bacteria* than to those of *Eukarya* or *Archaea*. We will see that this is for the most part true; however, mitochondrial genomes are somewhat enigmatic. We begin with the photosynthetic organelle, the chloroplast.

Chloroplast Genomes

Known chloroplast genomes are all *circular* DNA molecules. Moreover, although there are several copies of the genome in each chloroplast, each is identical. The typical chloroplast genome is about 120–160 kbp and contains two inverted repeats of 6–76 kb (Figure 15.10●). Several chloroplast genomes have been completely sequenced and a few of these are summarized in Table 15.4. The flagellated protozoan *Mesostigma viride* belongs to the

earliest diverging green plant lineage. Its chloroplast contains more protein-encoding genes and tRNA genes than any other so far known and has the typical genome structure illustrated in Figure 15.10.

Many of the genes in the chloroplast genome encode proteins involved in photosynthesis and autotrophy. However, the chloroplast genome also encodes rRNAs used in chloroplast ribosomes, tRNAs used by the translational apparatus, and a few of the proteins used in transcription and translation, as well as some other proteins. Some proteins that function in the chloroplast are encoded by genes in the nucleus, presumably genes that migrated there as the chloroplast evolved from a free-living photosynthetic endosymbiont. Unlike free-living prokaryotes, introns are common in chloroplast genes, and they are primarily of the self-splicing type (Section 14.8).

Analyses of chloroplast genomes have firmly supported the endosymbiotic hypothesis (Section 11.4). For example, chloroplast genomes have been shown to contain genes that are homologues of those in *Escherichia coli*, cyanobacteria, and other *Bacteria*. These include, among others, genes encoding proteins involved in cell division (Section 6.2), suggesting that the mechanism of chloroplast division is similar to that of *Bacteria*. In addition, chloroplast genes involved in protein transport through membranes are highly related to those of *Bacteria*.

Mitochondrial Genomes

Mitochondria are involved in energy production and are found in most eukaryotic organisms. Mitochondrial genomes primarily encode proteins involved in oxidative phosphorylation and, as is the case with chloroplast genomes, also encode rRNAs, tRNAs, and translational proteins needed for protein synthesis. However, most mitochondrial genomes encode many fewer proteins than do those of chloroplasts.

● **Figure 15.10 The map of a typical chloroplast genome.** The genomes of chloroplasts are circular double-stranded DNA molecules. Most contain two inverted repeat regions (IR_A and IR_B), which form the borders of a small single copy region (SSC) and a large single copy region (LSC).

Table 15.4 **Some chloroplast genomes**[a]

Organism		Size (bp)	Genes encoding			Inverted repeats[d]
			Proteins[b]	tRNA	rRNA[c]	
Chlorella vulgaris	Green alga	150,613	77	31	1	Absent
Euglena gracilis	Protozoan	143,170	67	27	3	Absent
Mesostigma viride	Protozoan	118,360	92	37	2	Present
Pinus thunbergii	Black pine	119,707	72	32	1	Present[e]
Oryza sativa	Rice	134,525	70	30	2	Present
Zea mays	Corn	140,387	70	30	2	Present

[a] All chloroplast genomes are circular, double-stranded DNA (◯◯◯ Section 14.4).
[b] These include genes encoding proteins of known function and ORFs that might be functional.
[c] Each unit is an rRNA operon, containing genes for each of the rRNAs (◯◯◯ Section 7.13).
[d] See Figure 15.10.
[e] Although the inverted repeats are present, they are greatly truncated.

Well over 200 mitochondrial genomes have been sequenced. The largest mitochondrial genome has 62 protein-encoding genes, while others encode as few as 3 proteins. The mitochondria of almost all animals including humans encode only 13 proteins (plus 22 tRNAs and 2 rRNAs) and that of the yeast *Saccharomyces cerevisiae* only 8 proteins.

Whereas chloroplasts use the "universal" genetic code, mitochondria use slightly different, simplified codes (◯◯◯ Section 7.14). These seem to have arisen from the selection pressures for smaller genomes. The standard set of 22 mitochondrial tRNAs is insufficient to read the entire genetic code, even with the "standard" wobble pairing taken into consideration. Therefore, base-pairing between the anticodon and the codon (◯◯◯ Figure 7.34) is even more flexible in mitochondria than it is in cells.

Unlike chloroplast genomes, which are all single, circular DNA molecules, the genomes of mitochondria are quite diverse. For example, some mitochondrial genomes are linear, including the mitochondria of some species of algae, protozoans, and fungi. In other cases, such as in the yeast *Saccharomyces cerevisiae*, although genetic and restriction mapping have shown the mitochondrial genome to be circular, it seems that the major *in vivo* form is linear. (Recall that bacteriophage T4 has a genetically circular genome structure but is physically linear, ◯◯◯ Section 9.9.) Figure 15.11● shows a map of the 16,569-bp (13 gene) human mitochondrial genome. The yeast mitochondrial genome is larger (85,779 bp) but has only 8 protein-encoding genes. Outside of the genes encoding the RNAs and proteins, the genome of yeast mitochondria contains large stretches of extremely AT–rich DNA that serve no apparent function.

Finally, it should be noted that in the mitochondria of several organisms, small plasmids exist, making mitochondrial genome analysis even more complicated. An additional complication in analyzing some mitochondrial and chloroplast genomes is that it is sometimes difficult to find the gene for a particular protein even when the sequence of both the protein and the organellar DNA

are both known. This is because of **RNA editing** (see the Microbial Sidebar).

●**Figure 15.11** **Map of the human mitochondrial genome.** The circular genome of the human mitochondrion contains 16,569 bp. The genome encodes the 16S and 12S rRNA (corresponding to the prokaryotic 23S and 16S rRNA) and 22 tRNAs. These genes are indicated in two shades of orange, a darker orange for genes that are transcribed counterclockwise from the map as drawn and those in lighter orange that are transcribed clockwise. (The amino acid designations for the tRNA are also on the outside for counterclockwise-transcribed genes and on the inside of the map for clockwise-transcribed genes.) The 13 protein-encoding genes are shown in green (once again with darker green indicating those transcribed counterclockwise and lighter green for those transcribed clockwise). The genes encode: Cyt*b*, cytochrome *b*; ND1-6, components of the NADH dehydrogenase complex; COI-III, subunits of the cytochrome oxidase complex; ATPase 6 and 8, polypeptides of the mitochondrial ATPase complex. The two promoters are in the region called the D-loop, a region also involved in DNA replication.

Microbial Sidebar ◆ RNA Editing

In Chapters 7 and 14 we saw that certain genes have coding regions that are split by noncoding regions called *introns*. Typically, introns are removed after transcription to form a mature mRNA, a process called *splicing* (◯◯ Section 14.8). Interestingly, there is a phenomenon found in the genomes of organelles that is almost the opposite of splicing: *RNA editing*.

RNA editing involves either the insertion or deletion of nucleotides *into the final mRNA* that were not present in the DNA transcribed. Editing can also involve the chemical *modification* of a base in the mRNA that changes it from one base to another. In either case, RNA editing can alter codons in such a way that one or more different amino acids are inserted in a polypeptide than those encoded by its gene.

In the mitochondria of trypanosomes and related protozoa (◯◯ Section 14.10) some mitochondrial transcripts are edited such that large numbers (hundreds in some cases) of uridylates are added or, more rarely, deleted. An example of this type of RNA editing is shown in Figure 1●. RNA editing is precisely controlled by short sequences present in the mRNA that "guide" the enzymes involved in their specific edits. Obviously this process must be very precisely controlled. Inserting too many or too few bases would yield a frameshift product that would likely be nonfunctional.

The other type of RNA editing, the changing of one base into another, is common in the mitochondria and chloroplasts of higher plants. At specific sites in some mRNAs, a C will be converted to a U by oxidative deamination (the opposite modification is more rare). There are at least 25 sites of C to U conversion in the maize chloroplast. Although mostly found in organellar genomes, an example of the programmed conversion of a C to a U is also known for a mammalian nuclear gene. Depending on the location of the edit, a new codon may be formed, leading to formation of a protein sequence not predictable from the gene that encodes it.

RNA editing, although a curious phenomenon, was not a significant obstacle in analyzing *organellar* genomes. This is because the number of proteins they encode is small and the proteins highly conserved. By contrast, had RNA editing been a widespread phenomenon in *cells*, genomic annotations and the identification of orthologous genes in different organisms could have been an even more formidable challenge than it has been to date.

The function and origin of RNA editing is unknown. But some scientists have pointed out that this process may be yet another remnant, along with ribozymes (◯◯ Section 14.8) and other catalytic RNAs, of the RNA World (◯◯ Section 11.2).

◼

Protein	...Leu Cys Phe Trp Phe Arg Phe Phe Cys...
mRNA	...uuG uGu UUU UGG uuu AGG uuu uuu uGu...
DNA	... G G TTT TCC AGG G ...
	... C C AAA AGG TCC C ...

Figure 1 RNA editing. *The upper part of the figure shows a portion of the amino acid sequence of subunit III of the enzyme cytochrome oxidase from the protozoan Trypanosoma brucei (◯◯ Section 14.10). This protein is encoded by mitochondria. Beneath the amino acid sequence, the sequence of the messenger RNA (mRNA) for this region is shown. The bases in uppercase letters are those transcribed from the gene, which is shown below. The bases in the mRNA in lowercase have been inserted into the transcript by RNA editing. Although the DNA has many informational gaps, there are no actual gaps in the molecule itself. The spaces between the base pairs are simply to aid in visualization.*

Organelles and the Nuclear Genome

Chloroplasts and mitochondria require more proteins than they encode. For example, far more proteins are involved in organellar translation alone than are encoded by them. Thus, many organellar functions are orchestrated by nuclear genes.

It is estimated that the yeast mitochondrion contains over 400 different proteins, and as we have seen, only 8 of them are encoded by the yeast mitochondrion. Hence, almost all proteins required by the yeast mitochondrion are encoded by genes *in the nucleus*. Although it might seem reasonable that the genes (and proteins) that function in specific processes in the eukaryotic nucleus/cytoplasm could be put to the same use in organelles, this is not the case. Although the genes for many organellar proteins are present in the nucleus, transcribed there and translated in the cell's cytoplasm, the gene products are used specifically by the organelles and must be transported there.

Among the first mitochondrial proteins to be studied were those involved in translation and it was clear that these nuclear-encoded mitochondrial proteins were closely related to counterparts in *Bacteria*, not to those in *Eukarya*. Thus, it appeared at first that most genes encoding mitochondrial proteins had been transferred from the mitochondrion to the nucleus. However, in the absence of genomics approaches, it was difficult to gain further insight into this problem.

What is required to analyze this problem is a genome sequence of a eukaryote, that of a species of *Bacteria* phylogenetically closely related to the mitochondrial genome, and genomic sequences of other *Bacteria* for comparative purposes. Knowledge of the mitochondrial genome of the eukaryote was actually not required (since it encodes so few proteins), but a knowledge of the proteins that function in the mitochondria was. All these requirements were met in the case of the yeast *Saccharomyces cerevisiae*, and the analysis has been revealing.

Surprisingly, of the 400 nuclear genes encoding mitochondrial proteins, only about 50 were closely related to the phylogenetic lineage in *Bacteria* that led to mitochondria (α Proteobacteria, Section 12.1). Another 150 were clearly related to proteins of *Bacteria*, but not necessarily α Proteobacteria. However, the remaining 200 proteins were encoded by genes that have no identifiable homologues among known genes of *Bacteria*. The *Bacteria*-like proteins were mostly involved in energy conversions, translation, and biosynthesis, whereas the other proteins were mostly involved in membranes, regulation, and transport. Thus, although the mitochondrion has many of the telltale signs of having originated from an ancient endosymbiotic event, genomic analyses have shown that its genetic history is more complicated than previously thought.

15.7 Concept Check

Chloroplasts and mitochondria have small genomes independent of nuclear genomes. These genomes encode rRNAs, tRNAs, and a few proteins involved in energy metabolism. Although the genomes of the organelles are independent of the nuclear genome, the organelles themselves are not. Many genes in the nucleus encode proteins required for organellar function. These genes have various phylogenetic histories.

◆ How is genome size and gene content correlated in yeast and human mitochondria?

◆ What is unusual about the genes that encode mitochondrial functions in yeast?

◆ What is RNA *editing*? How does it differ from RNA *processing*?

15.8 Evolution and Gene Families

The first priority of genomics is, of course, to determine the number, sequence, and function of genes in an organism. However, there is more to genomics than just sequencing and annotating genes and then interpreting the results in terms of how an organism might interact with its environment. Comparative genomics also helps us to understand the *evolutionary relationships* between organisms. Reconstructing evolutionary relationships from genome sequences helps to distinguish primitive from derived characteristics and can resolve ambiguities in phylogenetic trees based on analyses of a single gene (for example, 16S rRNA, Sections 11.5 and 14.9). Such knowledge helps us better understand early life forms and, in time, may answer the most basic question in biology: How did life first arise?

Gene Duplication and Gene Families: Paralogs and Orthologs

Genomes from both prokaryotic and eukaryotic sources contain **gene families**—genes that are related to other genes *within* the organism. Although large gene families are not the rule, comparative genomics has shown that many genes have arisen by *duplication* of other genes. Such genes are called **paralogs**, genes whose similarity is the result of gene duplication at some time in the evolution of an organism. Genes found in one organism that are similar to genes *in another organism* but which differ because of speciation are called **orthologs**. An example of paralogous genes would be genes encoding lactate dehydrogenase (LDH) isoenzymes in humans. These enzymes are structurally distinct yet all highly related and carry out the same enzymatic reaction. By contrast, orthologs of LDH exist between the LDH from *Escherichia coli*, for example, and an LDH isoenzyme from humans.

The study of genes and gene families is a major task in comparative genomics. Because chromosomes from many different microorganisms have already been sequenced, such comparisons can be easily done and the results are often surprising. For instance, genes in *Archaea* involved in DNA replication, transcription, and translation, are more similar to those in *Eukarya* than to those in *Bacteria*. Unexpectedly, however, many other genes in *Archaea*, for example, those encoding metabolic functions other than information processing, are more similar to those in *Bacteria* than those in *Eukarya*. The powerful analytical tools of bioinformatics allow us to deduce such genetic relationships between domains of life very quickly, at either the single gene, groups of genes, or entire genome level. The results obtained thus far lend further support to the phylogenetic picture of life deduced originally by comparative ribosomal RNA sequence analysis (Section 11.5) and suggest that many genes in all organisms have common evolutionary roots. However, such analyses have also revealed instances of horizontal gene flow, an important issue to which we now turn.

Horizontal Gene Transfer

Evolution is premised on the transfer of genetic traits from one generation to the next. However, *horizontal (lateral) gene transfer* also occurs, and it can complicate evolutionary studies, especially of entire genomes. **Horizontal gene transfer** occurs whenever genes are transferred from one cell to another *other than* by the usual inheritance process, from mother cell to daughter cell. In prokaryotes, at least three methods for horizontal gene transfer are known: *transformation, transduction,* and *conjugation* (Chapter 10).

Horizontal gene flow may be extensive in nature and is a process that can even cross domains. However, to be detectable by comparative genomics, the difference between the organisms must be rather large. For example, several genes with eukaryotic origins have been found in *Chlamydia* (∞ Sections 15.3 and 26.13) and *Rickettsia* (∞ Sections 12.13 and 27.3), both human pathogens. Moreover, genomic analyses have shown that *Thermotoga maritima* (a species of *Bacteria*) contains over 400 genes (greater than 20% of its genome) that are clearly of archaeal origin. Of these genes, 81 are found in discrete clusters. This strongly suggests that they were obtained by horizontal gene transfer—presumably from thermophilic *Archaea* that share the hot environments inhabited by *Thermotoga*.

Horizontal gene transfers can be spotted in genomes once the genes have been annotated. The presence of genes that encode proteins typically found only in distantly related species is one signal that the genes originated from horizontal transfer. However, another clue to horizontally transferred genes is the presence of an ORF or ORFs whose GC content (∞ Section 11.10) or codon bias differs significantly from that of the remaining genome. Using these clues many likely examples of horizontal transfer have been documented in the genomes of various prokaryotes. Horizontally transferred genes typically encode metabolic functions *other than* the core molecular processes of DNA replication, transcription, and translation and may account for the previously mentioned similarities of metabolic genes in *Archaea* and *Bacteria*. In addition, there have been several examples of virulence genes of pathogens (∞ Sections 21.8 and 21.9) having been transferred by horizontal means. Obviously, prokaryotes are exchanging genes in nature, and the process likely functions to "fine-tune" an organism's genome to a particular situation or habitat.

The *Chlamydia trachomatis* genome revealed potential horizontal transfer from a eukaryotic source, possibly even its human host. This is the opposite of the situation where genes in the eukaryotic nucleus have been transferred there from the ancestor of the mitochondrion (see Section 15.7). If these *C. trachomatis* genes were actually obtained from humans, it shows that horizontal gene transfer is not limited by the prokaryotic/eukaryotic boundary.

Evolutionary analyses of genomes also have a practical component. The ability to identify genes that are clearly prokaryotic or, in particular, genes characteristic of pathogenic bacteria, has helped develop powerful and extremely specific diagnostic tools and therapeutic agents that target the products of these genes for use in clinical medicine.

15.8 Concept Check

Genomics can be used to study the evolutionary history of an organism. Organisms contain gene families, genes with related sequences. If these arose because of gene duplication, the genes are said to be paralogs; if they arose by speciation, they are called orthologs. Organisms may acquire genes from other organisms in their environment by a process called horizontal gene transfer.

◆ Contrast gene *paralogs* with gene *orthologs*.

◆ Describe a mechanism by which horizontal gene transfer might occur in a prokaryote.

15.9 Genomic Mining

As we have discussed, genomic analysis provides an organism's genetic blueprint. From genomic analyses we can gain valuable insight into the metabolism and ecology of an organism. Often, however, genomic analyses indicate that one or more of the genes usually associated with a given pathway or process seem to be missing. This could mean that the organism uses a novel pathway or that at least one novel reaction is involved. In such cases it would be of interest to search for a gene encoding a protein that can carry out the missing function. This can easily be done by computer. Searching through a database containing the complete sequence of an organism's genome to find such a new gene is referred to as **genomic mining**. The search for the DNA polymerase of the cyanobacterium *Synechocystis* is a good example (Figure 15.12●).

DNA Polymerase of *Synechocystis*: Inteins and the Fruits of Genomic Mining

All species of *Bacteria* contain a DNA polymerase similar to DNA polymerase III of *Escherichia coli*, which is used as the primary polymerase in DNA replication (∞ Section 7.6). This is a complex enzyme containing many different

● **Figure 15.12   Inteins: The split *dnaE* gene of *Synechocystis*.** The two ORFs are transcribed in opposite directions, as shown by the orange arrows. The protein products of these ORFs each contain part of an intein (the protein product of ORF2 has been drawn in the standard amino terminal to carboxyl terminal orientation). Several DNA replication enzymes from *Archaea* also show inteins, so the phenomenon is not limited to *Bacteria*.

subunits, and the subunit that catalyzes the actual polymerase reaction is DnaE, a product of the *dnaE* gene. Because DNA polymerases are highly conserved proteins and of prime importance to all cells, inspection of the genome of *any* member of the *Bacteria* will quickly yield a gene with sequence similarity (an *orthologous* gene) to the *dnaE* gene of *E. coli*. Interestingly, however, the original search of the genome of *Synechocystis* (see Table 15.1) failed to identify such a gene. Instead, what was found were two ORFs of unknown function, which, if combined, formed a gene with high similarity to *dnaE*. But these two ORFs were over 700 kbp apart on the *Synechocystis* chromosome and located on opposite DNA strands, indicating that they were transcribed in opposite directions. Could these two ORFs really be part of *dnaE*?

Careful sequence analysis of the two ORFs revealed that this was indeed the case. The genes encoded complementary halves of an **intein**, a self-splicing protein (∞ Section 8.3). The two *Synechocystis* ORFs were then cloned. When transcribed and translated, this intein catalyzed a splicing reaction between the two halves of DnaE to form a complete and enzymatically active DnaE protein (Figure 15.12).

Why has *Synechocystis* arranged its *dnaE* gene in this split fashion? This is unknown, but evolution would tell us that whatever the reason, it is the best solution for expressing this gene in this particular organism. But *Synechocystis* is not unique in this regard. Several archaeal DNA replication genes also contain inteins, including those of *Nanoarchaeum equitans*, the parasitic archaeon with the tiny genome (Table 15.1). It is difficult to imagine how these highly unusual mechanisms would have been discovered without the ability to scan and compare entire genome sequences, as can be done routinely today. Mining genomes is thus likely to yield lots of surprises and may also lead to discoveries that have practical applications as well.

 15.9 *Concept Check*

Often it is necessary to search carefully through a genomic database to find a particular gene, a process called genomic mining. This can be done to find novel genes or to find genes that one predicts must be present.

◆ How does an *intein* differ from a *self-splicing intron*?

 IV **GENE FUNCTION AND REGULATION**

15.10 Proteomics

An additional aim of genomic studies is to determine which genes are *expressed* to yield protein products and to determine the function of these products. The term **proteome** refers to the collection of proteins present in a cell *at any one time*. Note that the proteome is not simply "all the proteins encoded by an organism's genome." Although the genes of an organism are constantly present, gene expression is regulated (∞ Chapter 8). The number and types of proteins present in a cell constantly change in response to an organism's environment or other factors, such as developmental cycles. The genomewide study of the structure, function, and regulation of an organism's proteins is called **proteomics**, or **functional genomics**.

Two-Dimensional Gel Electrophoresis

One approach to proteomics has been the use of *two-dimensional polyacrylamide gel electrophoresis*. This technique can separate, identify, and measure all the proteins present in a sample of cells. A two-dimensional (2-D) gel separating proteins from *Escherichia coli* is shown in Figure 15.13●. In the first dimension (the horizontal dimension in the figure) the proteins are separated by their *isoelectric points*, the pH at which the net charge on each protein reaches 0. In the second dimension, the proteins are denatured in a way that gives each amino

● **Figure 15.13 Proteomics: Two-dimensional polyacrylamide gel electrophoresis of proteins.** The figure shows an autoradiogram of the proteins of a culture of *Escherichia coli*. Each spot on the gel is a particular protein. The proteins were labeled with radioactive methionine during growth to allow for visualization and quantitation. The proteins were first separated by isoelectric focusing under denaturing conditions, concentrating on the pH range 5–7 where most *E. coli* proteins are found (the basic ribosomal proteins are missing from this separation). The second dimension separates denatured proteins by their mass (M_r; given here in kilodaltons), with the largest proteins being toward the top of the gel.

acid residue a fixed charge. The proteins are then separated by *size* (in much the same way DNA molecules are separated by size; ∞ Figure 7.22).

In the case of *E. coli* and a few other organisms, hundreds of proteins resolved in 2-D gels have been identified by biochemical or genetic means and their regulation studied under a variety of conditions. Using 2-D gels, the appearance or disappearance of a particular protein under different growth conditions can be used to assess the proteome as a function of environmental conditions. One method of connecting an unknown protein with a particular gene using the 2-D gel system is to elute the protein from the gel and sequence the N-terminus region of the protein. This sequence information may be sufficient to design oligonucleotide probes that amplify the gene encoding the protein from genomic DNA by the polymerase chain reaction (∞ Section 7.9). Then, following sequencing and comparative genomics, it is possible that the gene's function can be determined.

Functional and Structural Genomics

Although proteomics often requires intensive experimentation (Figure 15.13), *in silico* techniques can also be quite useful. After obtaining the sequence of an organism's genome, initial analysis can compare its sequence to that of other organisms to locate and identify genes that are similar to ones already known. The "sequence" here that is most important is the *amino acid* sequence of the protein. Because of degeneracy of the genetic code (∞ Section 7.14), differences in the DNA sequence may not necessarily lead to differences in the amino acid sequence (Figure 15.14●).

Proteins with greater than 50% sequence identity frequently have similar functions. Proteins with identities above 70% are all but certain to have similar functions. One important clue to the possible function of a protein is the identification of regions in the coding sequence that are present in other proteins and that encode domains of the protein having known functions. The latter would include regions such as metal-binding domains, nucleotide-binding domains, or binding domains for some other cofactor. For example, a protein with an NAD^+-binding domain, for example, is almost certainly a protein involved in a redox reaction (∞ Section 5.7) of some type.

Specific protein domains have specific sequences, but what makes them *functional* is their three-dimensional structure (∞ Section 3.8). **Structural genomics** is the determination of the three-dimensional structures of proteins representative of the range of protein structure and function found in an organism. The ultimate aim is to build a body of structural information that will allow *in silico* techniques to predict the probable structure and potential function for any protein from knowledge of its primary structure. Although this goal is not yet a reality, as more and more sequence data are generated and analyzed, it is one whose time will come. The link between genomics and proteomics will then be even stronger.

Coupling proteomics with genomics is yielding important clues to how gene expression in different organisms correlates with environmental stimuli. Not only does such information have important basic science benefits, but it also has potential applications. These include advances in medicine, the environment, and agriculture. In all of these areas, understanding the link between the genome and the proteome and how it is regulated could give humans unprecedented control over major aspects of disease, pollution, and agricultural productivity.

 15.10 Concept Check

The proteome encompasses all the proteins present in an organism at any one time. The aim of proteomics is to study these proteins to learn their structure, function, and regulation.

◆ Can an organism have more than one proteome?

◆ What is *functional* genomics?

15.11 Microarrays and the Transcriptome

One major aim of proteomics is to study gene expression. But there are other ways to study this process. Recall that gene expression is often regulated at the level of transcription (∞ Chapter 8). For a gene to be expressed, it must first be transcribed. Knowing the conditions under which a gene is transcribed may also give information about a gene's function. In analogy to the proteome, the entire complement of *mRNAs* produced under a given set of conditions is called a **transcriptome**. And a powerful technique exists for transcriptome analysis: **microarrays**.

●**Figure 15.14 Comparison of nucleic acid and amino acid sequence similarities.** Three different nucleotide sequences are shown (for convenience RNA is shown). Both sequence 2 and sequence 3 differ from sequence 1 in only three positions. However, the amino acid sequence encoded by 1 and 2 are identical, whereas that encoded by sequence 3 is unrelated to the other two.

Microarrays and the DNA Silica Chip

In Chapter 7 we discussed how nucleic acid hybridization techniques help to locate genes on specific fragments of DNA (◔◯ Section 7.7). Hybridization techniques can also be used in conjunction with genomic sequence data to measure the expression of genes by hybridizing mRNA to specific DNA fragments. This technique has been radically enhanced with the development of **microarrays**, or **gene chips** as they are also called.

Microarrays are small solid-state supports to which genes or portions of genes representing the entire genome of an organism are affixed and spatially arrayed in a known pattern (see Figure15.16a). The genes are synthesized by PCR, or, alternatively, oligonucleotides are designed for each gene based on the genomic sequence. Once attached to the solid support, these genes (or gene fragments) can then be hybridized with mRNA from cells grown under a specific condition and scanned and analyzed by a computer. Hybridization between a specific mRNA and the DNA on the chip confirms expression of that gene.

A method for making and using microarrays is shown in Figure 15.15●. The same process used to produce computer chips (photolithography) has been adapted to produce 1 to 2-cm silica microarray chips, each of which can hold thousands of different DNA fragments (Figures 15.15 and 15.16●). For example, one company markets a single array human genome chip that contains the entire human genome on it (Figure 15.16a). The chip can analyze over 47,000 transcripts, plus has room for 6500 additional oligonucleotides for use in clinical medicine! Figure 15.16b shows a part of a chip used to assay expression of the *Saccharomyces cerevisiae* genome.

The gene chip shown in Figure 15.16b easily holds the 5600 protein-encoding genes of *S. cerevisiae*, so that global gene expression in this organism can be measured in a single experiment. To do an experiment, the chip is hybridized with mRNA obtained from yeast cells grown under specific conditions and then tagged with a fluorescent dye. A particular tagged mRNA binds only to the DNA on the chip that is *complementary* to its sequence. To assay hybridization, the chip is scanned with a laser fluorescence detector and the signals analyzed by computer. A distinct pattern of hybridization is typically observed, depending upon which complementary DNA sequences on the chip show hybridization with mRNAs (Figures 15.15 and 15.16b). Although hybridization indicates *expression* of the gene, a qualitative result, the *intensity* of the binding is a quantitative measure of gene expression (Figure 15.16b). Following hybridization and laser scanning, the computer makes a list of which genes were expressed and to what extent. Thus, using gene chips, one knows in an instant the transcriptome of the organism of interest grown under specified conditions.

● **Figure 15.15 Measuring the transcriptome: Making and using DNA chips.** Short single-stranded (ss) oligonucleotides corresponding to all the genes of an organism are synthesized individually (◔◯ Section 7.8) and affixed at known locations to make a DNA chip (microarray). The DNA chip is assayed by hybridizing labeled mRNA obtained from cells grown under a specific condition to the DNA probes on the chip and then scanning the chip with a laser.

Applications of DNA Chips: Gene Expression

Using gene chips one can ask general or very specific types of scientific questions, depending on the genes affixed to the chip. For instance, one can assay an organism's global gene expression by using the entire population of mRNA as a probe (Figure 15.16b). By contrast, one could compare expression of different genes, singly or in sets, under different conditions. The ability to relatively quickly analyze the simultaneous expression of thousands of genes both qualitatively and quantitatively has tremendous potential for sorting through the complexity of metabolism and regulation. This is true for both organisms as "simple" as prokaryotes, with their 400–8000 or more genes (Table 15.1), or as complex as humans, with our approximately 30,000 different genes.

For example, the *S. cerevisiae* gene chip (Figure 15.16b) was used in a detailed study of metabolic con-

(a)

(b)

● **Figure 15.16   Measuring the transcriptome: DNA chips and their use to assay gene expression.** (a) The human genome chip. This chip has over 40,000 gene fragments on it. (b) A hybridized chip. The photo shows fragments from one-fourth of the entire genome of the yeast, *Saccharomyces cerevisiae* affixed to a portion of a silica gene chip. Each gene is present in several copies and has been probed with fluorescently labeled mRNA obtained from yeast cells grown under a specific condition. The background of the chip is blue. Locations where the RNA has hybridized to the DNA are indicated by a gradation of colors up to maximum hybridization, which shows as white. Because the location of different genes on the chip is known, once the chip is scanned it will reveal which specific genes were expressed.

trol in this organism. Yeast can grow by fermentation and by respiration. Using transcriptome analysis it was possible to see which genes were shut down and which were turned on when yeast cells were switched from fermentative (anoxic) to respiratory metabolism. Transcriptome analyses of gene expression in such an experiment showed that yeast undergo a major metabolic "reprogramming" during the switch to aerobic

growth. A whole suite of genes that control ethanol (a key fermentation product) production were strongly repressed while citric acid cycle functions (needed for aerobic growth) were strongly activated by the switch. Overall, over 700 genes were "turned on" and over 1000 "turned off" during this metabolic transition. Moreover, by using a microarray, the expression pattern of several yeast genes of unknown function could be monitored as well during the fermentative to respiratory switch, yielding clues to their possible function. At present, there is no other method available that can come close to giving as much information about gene expression as can microarrays. Clearly, microarrays offer a powerful approach for exploring gene expression patterns on a genomic scale.

Applications in Identification

Besides their use in probing gene expression, microarrays can be used to *identify* microorganisms. For example, one can use fragmented genomic DNA from a particular organism as a probe and differentiate between closely related strains by differences in their hybridization patterns. This allows for very rapid identification of pathogenic viruses or bacteria from clinical samples or detection of specific strains of these organisms in various substances, such as food. Such chips have been used in the food industry to detect particular pathogens, for example *Escherichia coli* O157:H7 (∞ Section 29.8).

DNA chips are also available that will identify *macroorganisms*. A commercially available chip called the *FoodExpert-ID* contains 88,000 gene fragments from a variety of vertebrate animals and is used in the food industry to ensure food purity. For example, the chip can confirm the presence of the meat listed on a food label and can also detect foreign animal meats that may have been added as supplements to or substitutes for the food product listed on the label. The eventual goal is to have each meat product receive an "identity card" listing all the animal species whose tissues were detected in it. This is intended to give consumers more confidence in the wholesomeness of their food products. The FoodExpert-ID can also be used to detect vertebrate byproducts in animal feed, a growing concern with the advent of prion-mediated diseases such as "mad cow disease" (∞ Sections 9.14 and 29.11).

Applications in Environmental Genomics

DNA chip technology is also useful for identifying and assessing the activities of microbial communities in the environment, an important goal of the microbial ecologist (∞ Chapter 19). Indeed a whole new field in microbial ecology is opening up called **environmental genomics** (∞ Section 18.6). Because of the capacity and power of whole genome shotgun sequencing and gene expression microarrays, it is now possible to sequence the collective genomes of the microorganisms present in a habitat, such as a soil or lake water sample, and identify their patterns of gene expression. The collective genomes, called the *metagenome*, define the genetic potential of an ecosystem. Then, using microarrays, one can explore these natural microbial communities for their patterns of gene expression. These techniques have empowered microbial ecologists to ask new and very complex questions about how microbial ecosystems function and to measure the contributions of specific members of a microbial community to the overall activities in an ecosystem. We discuss environmental genomics along with specific examples of the technology in Section 18.6.

 15.11 Concept Check

Microarrays are genes or gene fragments attached to a solid support in a known pattern. These arrays can be used to hybridize to mRNA and analyzed to determine patterns of gene expression. The arrays are large enough and dense enough that the transcription pattern of the entire genome (the transcriptome) can be analyzed.

◆ What do microarrays tell you that studying gene expression by assaying a particular enzyme, for example, cannot?

◆ Why might it be useful to know how gene expression of the entire genome responds to a particular condition?

REVIEW QUESTIONS

1. What is meant when it is said that cloned genes in phage M13 can be detected "visually" (◯ Section 15.1)?

2. Compare and contrast BACs with YACs with respect to: size, features necessary for replication, and amount of cloned DNA held (◯ Section 15.1).

3. Describe how "shotgun sequencing" methods can lead to the sequence of a complete microbial genome when the sequencing is essentially random (◯ Section 15.2)?

4. What is meant by "annotating" the genome? How is this done? How does *annotation* differ from *assembly*? Which process has to occur first (◯ Section 15.3)?

5. The organisms in Table 15.1 are listed in ascending order of chromosome size. What is the relationship between genome size and ORF content (◯ Section 15.4)?

6. How much larger is your genome than that of *Nanoarchaeum*? How many more genes do you have than *Nanoarchaeum* (◯ Sections 15.4 and 15.6)?

7. As a proportion of the total genome, which class of genes predominates in small genome organisms? In large genome organisms (◯ Section 15.5)?

8. In *Bacteria* and *Archaea* the acronym ORF is almost a synonym for the word "gene." However, in eukaryotes this is not, strictly speaking, true. Explain (◯ Section 15.6).

9. Whose genomes are larger, those of chloroplasts or those of mitochondria? Describe one unusual feature about a chloroplast and a mitochondrial genome (◯ Section 15.7).

10. The gene encoding the beta (β) subunit of RNA polymerase from *Escherichia coli* is said to be *orthologous* to the *rpoB* gene of *Bacillus subtilis*. What does that mean about the relationship between the two genes? What protein do you suppose the *rpoB* gene of *Bacillus subtilis* encodes? The genes for the different sigma factors (◯ see Table 8.2) of *Escherichia coli* are *paralogous*. What does that say about the relationship between these genes (◯ Section 15.8)?

11. Explain how horizontally transferred genes can be detected in a genome (◯ Section 15.8).

12. What is an intein? How can inteins be discovered (◯ Section 15.9)?

13. What does a 2-D protein gel show? How can the results of such a gel be tied to protein function (◯ Section 15.10)?

14. Distinguish between the terms *genome, genomics, proteome, proteomics*, and *transcriptome* (◯ Section 15.11).

APPLICATION QUESTIONS

1. When using shotgun sequencing, it is possible that there may be "gaps" left in the sequence. Describe how it is possible to close these gaps and complete the sequence.

2. The sequence of the yeast nuclear genome was published, but the entire sequence was never actually completely determined. The yeast mitochondrial genome proved very difficult to sequence accurately. Describe in both cases the practical difficulties that were encountered in the sequencing.

3. Describe how one might determine which *proteins* in *Escherichia coli* are repressed (◯ Section 8.5) when a culture is shifted from a *minimal* medium (which contains only a single carbon source) to a *rich* medium, containing a large number of amino acids, bases, and vitamins? Describe how one might study which *genes* are expressed during each growth condition.

VIRAL DIVERSITY

Viruses infect all types of organisms, from prokaryotes to humans. Many different replication strategies are known in the viral world, and viral classification is based on these fundamental replication schemes.

WORKING GLOSSARY

Bacteriophage a virus that infects cells of *Bacteria*

Concatemer two or more identical linear nucleic acid molecules in tandem

Enveloped in virology, refers to a virus containing a lipoprotein membrane surrounding the virion

Hepadnavirus a virus whose DNA genome replicates by way of an RNA intermediate

Minus (negative)-strand nucleic acid an RNA or DNA strand that has the opposite sense of (is complementary to) the mRNA of a virus

Nucleocapsid the complete complex of nucleic acid and protein packaged in a virus particle

Overlapping genes two or more genes in which part or all of one gene is embedded in the other

Plus (positive)-strand nucleic acid an RNA or DNA strand that has the same sense as the mRNA of a virus

Polyprotein in poliovirus, a large protein that is produced and subsequently cleaved to form several individual proteins

Replicative form a double-stranded DNA molecule that is an intermediate in the replication of single-stranded DNA viruses

Retrovirus a virus whose RNA genome has a DNA intermediate as part of its replication cycle

Reverse transcription the process of copying genetic information found in RNA into DNA

RNA replicase an enzyme that can produce RNA off of an RNA template

Rolling circle replication in bacteriophage *φ*X174, DNA replication that proceeds by elongation of the 3' end with progressive displacement of the 5' end, the unbroken circular strand acting as a template

Transposase an enzyme that catalyzes the insertion of DNA within other DNA

Virion the complete extracellular virus particle; the nucleic acid surrounded by a protein coat and, in some cases, other materials

Virus a genetic element containing either RNA or DNA that replicates in cells but is characterized by having an extracellular state

---

The previous chapters in this unit have explored the enormous diversity of microbial cells: *Bacteria*, *Archaea*, and *Eukarya*. Here we will focus on the diversity of *viruses* and their genomes. This chapter on **viruses** completes our coverage of viral diversity. It extends Chapter 9, which introduced the principles of virology.

All cells contain *double-stranded DNA* as their genetic material. By contrast, viruses are known that have *single-stranded RNA*, *double-stranded RNA*, *single-stranded DNA*, or *double-stranded DNA* as their genetic material (∞ Section 9.1). This makes for interesting schemes of replication and gene expression, as we will see. In this chapter we will illustrate viral diversity by exploring the replication processes of viruses in each of these groups. However, we will separate our discussion into viruses that infect prokaryotes and those that infect eukaryotes. This is because the differences between these two types of cells place certain constraints on the life cycles of the viruses that infect them.

▌VIRUSES OF PROKARYOTES

Many viruses are known that infect *Bacteria* and increasing numbers are known that infect *Archaea*. We mentioned in Chapter 9 that most known bacterial viruses have double-stranded DNA genomes (∞ Section 9.8). Even so, there are many **bacteriophages** with other types of genomes. The simplest are those with RNA genomes, and we begin our coverage with these.

16.1 RNA Bacteriophages

Many bacteriophages contain genomes of the plus configuration. In such viruses the viral *genome*—**plus strand RNA**—and the *mRNA* are of the same complementarity (∞ Section 9.5). Interestingly, RNA viruses of the enteric bacteria group infect only bacterial cells that contain a type of plasmid, called a *conjugative plasmid*, which allows the bacterial cell to function as a *donor* in the process of conjugation (∞ Sections 10.9 and 10.11). This restriction to only donor cells arises because these viruses infect bacteria by first attaching to *pili* (Figure 16.1●), which are encoded by the plasmid and present only on the donor cell.

Phage MS2

The bacterial RNA viruses are all quite small, about 25 nm in size, and they are all icosahedral (∞ Section 9.2) with 180 copies of coat protein per virus particle. The complete nucleotide sequences of several RNA phage genomes are known. For example, the genome of the RNA phage MS2, which infects *Escherichia coli*, is 3569 nucleotides long.

● **Figure 16.1 Small RNA Bacteriophages.** Electron micrograph of the pilus of a donor bacterial cell of *Escherichia coli* showing virions of a small RNA phage attached to the pilus.

The genetic map of MS2 is shown in Figure 16.2a●, and the flow of events of MS2 multiplication is shown in Figure 16.2b. The small genome encodes only four proteins. These are the **maturation protein** (present in the mature virus particle as a single copy), **coat protein, lysis protein** (involved in the lysis process that results in release of mature virus particles), and a subunit of **RNA replicase**, the enzyme that brings about replication of the viral RNA. Interestingly, RNA replicase is a composite protein, composed partly of a virus-encoded polypeptide and partly of host polypeptides. The maturation protein functions as a protease to process some viral proteins necessary for producing infective virions.

The genome of phage MS2 is of the *plus sense* (∞ Section 9.7) and can thus be translated directly upon entry into the cell. After RNA replicase is synthesized, it in turn can synthesize RNA of the minus sense using the genomic RNA as a template (see Figure 9.11). After minus RNA has been synthesized, more plus RNA is made using this minus RNA as a template. The newly made plus RNA strands are translated for continued virus protein synthesis. The gene for the maturation protein is at the 5′ end of the RNA. Because of extensive folding of the RNA (see below), maturation protein synthesis is limited.

The virus RNA is folded into a complex form with extensive secondary structure. Of the four AUG translational start sites on the mRNA, the most accessible to the translation machinery is that for the coat protein, and translation begins there very early. The replicase mRNA is also translated early. As coat protein molecules increase in number in the cell, they combine with the RNA around the AUG start site for the replicase protein, effectively turning off synthesis of replicase. Access to the maturation protein start site is extremely limited, and thus only a few copies are synthesized. In this way, the major virus protein synthesized is coat protein, which is needed in the highest amounts.

Overlapping Genes and Assembly of MS2

An interesting feature of bacteriophage MS2 is that the fourth virus protein, the *lysis* protein, is encoded by a gene that *overlaps* with both the coat protein gene and the replicase gene (see genetic map in Figure 16.2a). This phenomenon of **overlapping genes** is quite common in very small viral genomes (see Section 16.2) and makes possible more efficient use of small genomes. The start codon of the MS2 lysis gene is not easily accessible to ribosomes because of the secondary structure found in the RNA. However, when the ribosome terminates synthesis of the coat protein gene, the secondary structure in this region of the RNA is disrupted, and sometimes this disruption allows a ribosome to begin reading the lysis gene. By restricting the efficiency of translation in this way, premature lysis of the cell is avoided. Only after sufficient copies of coat protein are available for the assembly of mature virions does lysis commence.

Ultimately, self-assembly of MS2 virions takes place, and release of virions from the cell occurs as a result of cell lysis. The features of replication of small RNA viruses like MS2 are thus fairly simple. The viral RNA itself functions as an mRNA and regulation occurs primarily by way of controlling access of ribosomes to the appropriate start sites on the viral RNA.

Although several positive-strand RNA bacteriophages are known that are similar in their biology to that of MS2, no bacteriophage has yet been discovered that contains a *negative-strand* RNA genome, a strategy that is fairly common among eukaryotic viruses (see Section 16.9). However, there are a few bacteriophages known that have segmented, *double-stranded* RNA as their genetic material. The best studied of these is φ6, whose host is *Pseudomonas syringae*. This virus, which is **enveloped** (enclosed by a lipid membrane, ∞ Figure 9.12), seems very closely related to the reoviruses (see Section 16.10), which infect eukaryotes.

(a) Genetic map of MS2

(b) Flow of events during viral multiplication

●**Figure 16.2 Bacteriophage MS2** (a) Genetic map of the RNA bacteriophage MS2. Note how the lysis protein gene overlaps with both the coat and replicase genes. The numbers in (a) refer to the nucleotide positions on the RNA. (b) Flow of events during multiplication.

16.1 Concept Check

A variety of RNA viruses that infect bacteria are known. The small RNA genome of these bacterial viruses is translated directly and encodes only a few proteins.

◆ What is the difference between the genomes of *positive-strand* RNA viruses and *negative-strand* RNA viruses?

◆ Describe what is meant by *overlapping genes*.

16.2 Icosahedral Single-Stranded DNA Bacteriophages

A number of bacteriophages contain a single-stranded DNA genome of the *plus* complementarity. Before such a genome can be transcribed, a *complementary strand* of DNA must be synthesized, forming a double-stranded molecule (➤ Section 9.7 and Figure 9.11). These bacteriophages then replicate this double-stranded DNA but only package the positive-strand of DNA in progeny virions (there are no known negative-strand DNA bacteriophages). In this section we discuss one of the best known of these phages, **φX174**. Following this we will consider the single-strand DNA bacteriophage M13, which is an important tool in genetic engineering (➤ Section 15.1) and whose life cycle is distinct from that of φX174.

Bacteriophage φX174 is one of a number of viruses that contain a *circular* single-stranded DNA genome and an icosahedral virion. These viruses are very small, about 25 nm in diameter, and the principal building block of the protein coat is a single protein present in 60 copies, the minimum number of protein subunits possible in an icosahedral virus. Attached at the vertices of the icosahedron are several other proteins that make up spikelike structures (➤ Figure 9.12). These small DNA viruses possess only a few genes, and the host cell DNA replication machinery is used exclusively in the replication of virus DNA.

The Genome of Phage φX174

Phage φX174 infects *Escherichia coli*. Its genome consists of a *circular single-stranded* DNA molecule of 5386 nucleotides. The DNA of φX174 was the first DNA molecule to be completely sequenced, a remarkable achievement when it was accomplished by Frederick Sanger and colleagues in 1977. Today, DNA sequencing is a routine and highly automated procedure (➤ Section 7.8). Phage φX174 is also of special interest because it was the first genetic element shown to have *overlapping genes*, a concept just discussed with reference to the RNA phage MS2 (see Section 16.1). For example, in the φX174 genome gene B resides within gene A, and gene K resides within both genes A and C (Figure 16.3●). In very small viruses such as φX174 there is simply insufficient DNA to encode all viral-specific proteins unless parts of the genome

A	Replicative form DNA synthesis	E	Host cell lysis
A*	Shut off of host DNA synthesis	F	Major capsid protein
B	Formation of capsid precursors	G	Major spike protein
C	DNA maturation	H	Minor spike protein
D	Capsid assembly	J	DNA packaging protein
		K	Function unknown

(a) Genetic map of φX174

(b) Flow of events during φX174 replication

● **Figure 16.3 Bacteriophage φX174, a single-stranded DNA phage.** (a) Genetic map. Note the regions of gene overlap (A/B, K/B, K/C, K/A, and D/E). Intergenic regions are not colored. Protein A* is formed using only part of the coding sequence of gene A by reinitiation of translation (see text). (b) Flow of events in φX174 multiplication. The production of progeny ss DNA from replicative form ds DNA involves rolling circle replication and is shown in more detail in Figure 16.4.

are read more than once in *different reading frames* (Figure 16.3).

As seen in the genetic map of φX174, genes D and E also overlap, gene E being contained completely *within* gene D. However, in addition, the termination codon of gene D overlaps the initiation codon of gene J (Figure 16.3*a*). In addition to overlapping genes, a small protein in φX174 called A*-protein is synthesized by reinitiation of *translation* (not *transcription*) within the mRNA for gene A. The A*-protein is read and terminated from the same mRNA reading frame as A-protein but is initiated at a different in-frame start codon and is thus a smaller protein.

DNA Replication by the Rolling Circle Mechanism

Because cellular DNA always replicates in the double-stranded configuration (∞ Section 7.5), the replication process of the single-stranded genome of φX174 is of interest. On infection, the plus sense viral DNA becomes separated from the protein coat. Entrance into the cell is accompanied by the conversion of this single-stranded DNA into a double-stranded molecule called the **replicative form** (RF) (Figure 16.3b). Cell-encoded proteins involved in the conversion of viral DNA to RF DNA include the enzymes *primase, DNA polymerase, ligase,* and *gyrase* (∞ Section 7.5).

In cells, replication of the lagging strand involves the formation of short *RNA primers* by activity of primase (∞ Section 7.5). These RNA primers are made at intervals on the lagging strand and are then removed and replaced with DNA by DNA polymerase I (∞ Figure 7.16). The situation with φX174 DNA is similar, but of course the DNA in this case is a single-stranded closed circle. To begin replication of this DNA, primase brings about the synthesis of a short RNA primer at one or more specific initiation sites. DNA is then synthesized by DNA polymerase III, and the primer is removed and replaced by DNA using DNA polymerase I, exactly as in the case of a lagging strand. This results in the formation of the complete, circular, double-stranded replicative form.

Once the replicative form is completed, copies are made by conventional semiconservative replication involving theta-form intermediates (∞ Figure 7.16). From the replicative form, φX174 mRNA is made, and from the latter, all of the viral proteins. However, the formation of single-stranded viral genomes involves a unique form of DNA replication called **rolling circle replication** (Figure 16.4●). The rolling circle arises because one strand is cut and the 3′ end of the exposed DNA is used to prime synthesis of a new strand. Continued rotation of the circle leads to the synthesis of a linear, single-stranded structure, the φX174 genome. Also, note that synthesis is asymmetric because only one of the strands (the negative strand) is serving as template. Contrast this with DNA replication in phage lambda, which also replicates its DNA by a rolling circle mechanism, but is a double-stranded DNA phage (∞ Figure 9.20).

In φX174, genomic synthesis begins when the protein encoded by gene A, called *gene A protein*, cleaves the plus strand of the RF. When the growing viral strand reaches unit length (5386 residues for φX174), gene A protein cleaves and then ligates the two ends of the newly synthesized single strand to give a circular single-stranded DNA.

Transcription and Translation in φX174

Viral mRNA synthesis in φX174 occurs off of the replicative form DNA. Synthesis of mRNA begins at several major promoters and terminates at a number of sites (see map, Figure 16.3a). The polycistronic mRNA molecules (∞ Section 7.13) are then translated into the various phage proteins. As we have noted, several proteins are

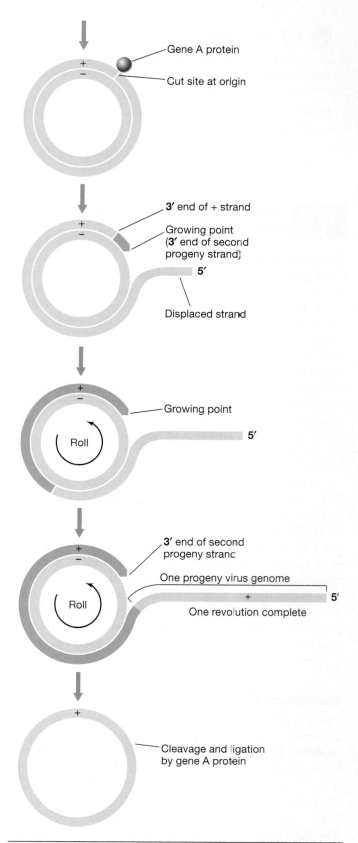

● **Figure 16.4 Rolling circle replication in phage φX174.** Replication begins at the origin of the double-stranded replicative form by cutting the plus strand of DNA by gene A protein (both strands of DNA are shown in light green here to simplify the diagram.). After one new progeny strand has been synthesized (one revolution of the circle), the gene A protein cleaves the new strand and ligates its two ends.

made from mRNA transcripts formed from different reading frames from the same DNA sequences (overlapping genes, Figure 16.3*a*). One can truly be impressed by the efficiency with which such a small genome as that of ϕX174 can encode so many different proteins.

Ultimately, assembly of mature ϕX174 virions occurs. Release of virions from the cell takes place as a result of cell lysis, which involves the participation of gene E protein. Interestingly, protein E catalyzes lysis by inhibiting the activity of one of the enzymes involved in peptidoglycan synthesis in the cell wall (⟳ Section 6.3). Because of the resulting weakness in newly synthesized cell wall material, the cell eventually lyses, releasing the viral particles.

16.2 Concept Check

The single-stranded DNA genome of the virus ϕX174 is so small that overlapping genes are required to encode all of its essential proteins. This virus provided the first example in biology of overlapping genes. The production of progeny viral DNA involves a rolling circle mechanism.

◆ If the nucleic acid genome of ϕX174 is single-stranded and in the plus configuration, why can't it be used directly as mRNA as it is in phage MS2?

◆ How does the replicative form of ϕX174 nucleic acid differ from the form found in the virion?

16.3 Filamentous Single-Stranded DNA Bacteriophages

Quite distinct from bacteriophage ϕX174 are the filamentous DNA phages, which have *helical* rather than *icosahedral* symmetry. The most studied member of this group is phage **M13**, which infects *Escherichia coli*, but related phages include f1 and fd. As with the small RNA bacteriophages, these filamentous DNA phages infect only conjugational donor cells, entering after attachment to the pilus (⟳ Section 10.11 and Figure 16.1). Even though these phages are linear (filamentous) in shape, like ϕX174 they possess *circular* single-stranded DNA.

Phage M13

Phage M13 is the model filamentous bacteriophage. The phage has found extensive use as a cloning vector and DNA-sequencing vehicle in genetic engineering (⟳ Section 15.1). The virion of phage M13 is only 6 nm in diameter but is 860 nm long. These filamentous DNA phages have the additional interesting property of being released from the cell *without* lysing the host cell. Thus, a cell infected with phage M13 can continue to grow, all the while releasing **virions**. Virus infection causes a slowing of cell growth, but otherwise a cell is able to coexist with its virus. Typical plaques (⟳ Figure 9.6*b*) are thus not observed; instead, only areas of reduced turbidity occur within a bacterial lawn.

Many aspects of DNA replication in filamentous phages are similar to those of ϕX174 (see Section 16.2 and Figure 16.4). The property of release without cell killing occurs by *budding*. In this mechanism, the end of the virion containing several copies of a protein known as the *A-protein* is released first, with the remainder of the virion following (Figure 16.5●). With phage M13 there is no accumulation of intracellular virions as with typical bacteriophages. Instead, the assembly of mature M13 virions occurs on the inner surface of the cytoplasmic membrane and virus assembly is coupled with the budding process.

Phage M13 and Genetic Engineering

Several features of phage M13 make it useful as a cloning and DNA sequencing vehicle. First, it has single-stranded DNA, which means that sequencing can easily be carried out by the Sanger dideoxynucleotide method (⟳ Section 7.8). Second, a double-stranded form of genomic DNA essential for cloning purposes is produced naturally when the phage produces the replicative form. Third, as long as infected cells are kept in the growing state, they can be maintained indefinitely with cloned DNA, so a continuous source of the cloned DNA is available. And finally, like phage lambda (⟳ Figures 9.17 and 9.18 and Section 10.17), there is an intergenic space in the genome of phage M13 that does not encode proteins and can be replaced by variable amounts of foreign DNA (⟳ Section 15.1). For these and other reasons, phage M13 is an important part of the biotechnologist's toolbox.

● **Figure 16.5 Release of filamentous phages.** Illustration of the manner in which the virion of a fila-mentous single-stranded phage (such as M13 or fd) leaves an infected cell without lysis. The A protein passes first through the membrane at a site on the membrane where coat protein molecules have first become embedded. The intracellular circular DNA is coated with dimers of outer phage protein, which is displaced by coat protein as the DNA passes through the intact cytoplasmic membrane.

16.3 *Concept Check*

Some single-stranded DNA viruses, such as M13, have filamentous virions. These viruses are very useful tools for DNA sequencing and genetic engineering. They also are released without actually killing the host.

◆ Describe the genome of phage M13. How does the genome relate to the mRNA produced by this phage?

◆ How can M13 virions be released without killing the infected host cell?

16.4 Double-Stranded DNA Bacteriophages: T7

The double-stranded DNA bacteriophages are among the best-studied of all viruses, and we have already discussed two of them, T4 and lambda, in Chapter 9 (⚭ Sections 9.9 and 9.10). Because of their importance as models for understanding molecular biology and gene

regulation, we will consider two more such viruses, T7 in this section and Mu in the next section.

Replication of Bacteriophage T7: Early Events

Bacteriophage T7 and its close relative T3 are relatively small DNA viruses that infect *Escherichia coli* and a few other enteric bacteria, notably *Shigella*. The virion has an icosahedral head and a very short tail (Figure 16.6●).

The T7 genome is composed of a linear double-stranded DNA molecule of 39,936 bp. Gene overlap (see Sections 16.1 and 16.2) also occurs in the T7 genome, as do other translational strategies such as internal translational reinitiation and internal frame shifts within certain genes, all apparently geared toward maximizing genetic economy. We discussed these strategies with respect to φX174 in Section 16.2.

The genetic map of T7 is shown in Figure 16.6. The order of the genes on the T7 chromosome influences the regulation of virus multiplication. When the virion at-

Bacteriophage T7

1. Replication cycle requires 25 minutes

2. Genome is linear double-stranded DNA of 39,737 bp

3. T7 encodes all of its own proteins for DNA replication and transcription

4. Time to complete 100 T7 genome copies from a single copy: 5 minutes

5. Burst size (*Escherichia coli* host): about 100 virions/cell

●**Figure 16.6 Genetic map of bacteriophage T7, showing gene numbers, approximate sizes, and functions of the gene products.** Transcription from the early promoters involves host RNA polymerase. Transcription from all other promoters involves T7 RNA polymerase. The genes are designated by numbers.

taches to the bacterial cell, the DNA is injected with the genes at the "left end" of the genetic map, entering the cell first. Several genes at this end are transcribed immediately by the cellular RNA polymerase, using three closely spaced promoters. One of these early proteins inhibits the host restriction system, a mechanism for protecting the cell from foreign DNA (⊙⊙ Section 7.7). This occurs very rapidly, as the antirestriction protein is synthesized before the entire T7 genome even enters the cell.

Another one of these early proteins is a viral RNA polymerase, called *T7 RNA polymerase*. Two other early mRNA molecules encode proteins that inhibit host RNA polymerase, thus turning off the transcription of the early genes as well as the transcription of host genes. Host RNA polymerase is thus used just to transcribe the first few genes. The T7-specific RNA polymerase then takes over and carries out the major transcription processes of the phage.

Phage T7 RNA polymerase recognizes only phage-specific promoters that are distributed along the genome (Figure 16.6). The T7 RNA polymerase is specific for these promoters (whose sequence is unrelated to typical *E. coli* promoters, ⊙⊙ Sections 7.10 and 7.11) and is also an extremely efficient RNA polymerase. (Genetic engineers have taken advantage of this to fashion genes that can be highly transcribed using the T7 RNA polymerase; ⊙⊙ Section 31.4.). Note that this situation is unlike that in phage T4, which uses the *host* RNA polymerase along with T4-specific sigma factor throughout infection (⊙⊙ Section 9.9).

Genome Replication in T7

DNA replication in T7 begins at a single origin of replication (shown in Figure 16.6) and proceeds *bi-directionally* from this origin (Figure 16.7●). Replicating molecules of T7 DNA can be recognized under the electron microscope by their characteristic structures. Because the origin of replication is near the left end, Y-shaped molecules are typically seen in electron micrographs of replicating T7 DNA. Earlier in replication, bubble-shaped molecules can appear (Figure 16.7). Several virus-encoded proteins including a T7-specific DNA polymerase are involved in T7 DNA replication, again, unlike the situation we found in phage T4 (⊙⊙ Section 9.9) or φX174 (see Section 16.2).

●**Figure 16.7   Replication of the linear, double-stranded DNA genome of bacteriophage T7.** (a) Bidirectional replication of DNA giving rise to intermediate "eye" and "Y" forms (for simplicity, both template strands are shown in light green and both newly synthesized strands in dark green). (b) Formation of concatemers by joining DNA molecules at the unreplicated terminal ends. The designation of the genes is arbitrary. (c) Production of mature viral DNA molecules from T7 concatemers by activity of the cutting enzyme, an endonuclease. Left: The enzyme makes single-stranded cuts of specific sequences (arrows); center: DNA polymerase completes the single-stranded ends; right: the mature T7 molecule with terminal repeats. Bacteriophage T4 also uses recombination to help replicate its linear genome; compare Figure 16.7*c* with Figure 9.13.

(a)

DNA polymerase and ligase activity

G A B C D E F G
                        G′ A′ B′ C′ D′ E′ F′ G′

Pairing of unreplicated terminal repeats

G A B C D E F G
                        G′ A′ B′ C′ D′ E′ F′ G′

Joining of new and old molecules by DNA polymerase and ligase, forming a concatemer

(b)

Cutting enzyme (arrows) makes single-stranded cuts

DNA polymerase completes the single-stranded cuts

Mature T7 molecule, with terminal repeats

(c)

A structural feature of T7 DNA that is important for its replication is a *direct terminal repeat* of 160 bp at both ends of the molecule. To replicate DNA near the 5'-terminus, RNA primer molecules have to be removed before replication is complete. There is thus an unreplicated portion of the T7 DNA at the 5'-terminus of each strand (see lower part of Figure 16.7a). As discussed in Section 14.6, genetic elements composed of linear DNA employ various strategies for solving this problem in DNA replication. The strategy employed by T7 is similar to that used by T4 and involves the repeated sequence at its ends (◯◯ Section 9.9). The opposite single 3'-strands on two separate DNA molecules, being complementary, can pair with these 5'-strands, forming a DNA molecule twice as long as the original T7 DNA (Figure 16.7b). The unreplicated portions of this structure are then completed through the activity of T7 DNA polymerase and ligase, resulting in a linear bimolecule called a **concatemer**.

Continued replication and recombination can lead to concatemers of considerable length, but ultimately a phage-encoded endonuclease cuts each concatemer at a *specific site*, resulting in the formation of virus-sized linear DNA molecules with terminal repeats (Figure 16.7c). Because T7, like lambda but unlike T4, cuts the concatemer at specific sequences, the DNA sequence in each T7 virion is identical. Recall that this is *not* the case in phage T4. This phage processes DNA using a "headful mechanism," which means that its DNA is not only terminally redundant but is also circularly permuted (◯◯ Section 9.9).

  16.4 Concept Check

The bacteriophage T7 double-stranded DNA genome always enters the host cell in the same orientation. The late genes in T7 are transcribed by a virus-encoded RNA polymerase. The replication strategy for the T7 genome employs T7 DNA polymerase and involves terminal repeats and the formation of concatemers.

◆ Of what significance is it that the T7 genome enters the cell in only one orientation?

◆ What is meant by *terminal repeats*?

◆ In what ways are DNA replication similar and in what ways are they different in phages T4 and T7?

16.5 Mu: A Double-Stranded Transposable DNA Bacteriophage

One of the more interesting bacteriophages is phage **Mu**. This virus is temperate, like lambda (◯◯ Section 9.11), but has the unusual property of replicating as a *transposable element* (◯◯ Section 10.14). This phage is called *Mu* because it is a *mutator* phage, inducing mutations in a host genome into which it becomes integrated. The mutagenic property of Mu arises because the genome of the virus can be inserted within host genes, thus inactivating them. Hence, a host cell that has become infected with Mu can assume a mutant phenotype. Mu is a useful phage in bacterial genetics because it can be used to easily generate a wide variety of bacterial mutants.

Transposable elements are sequences of DNA that can move from one location on their host genome to another as discrete genetic units. They are found in both prokaryotes and eukaryotes and play important roles in genetic variation (◯◯ Section 10.14). Although a bacteriophage, Mu is in reality a very large transposable element that replicates its DNA by transposition.

Basic Properties of Phage Mu

Bacteriophage Mu is a large virus with an icosahedral head, a helical tail, and six tail fibers (Figure 16.8●). The genome of Mu contains linear double-stranded DNA and its genetic map is shown in Figure 16.9a●. It can be seen that the bulk of the genes are involved in the synthesis of head and tail proteins and that important genes at each end of the genome are involved in replication and host range.

The DNA in Mu is approximately 39 kbp, but only 37.2 kbp constitute the actual Mu genome. This is because both ends of the Mu genome contain *host* DNA. At the left end of the Mu DNA are 50–150 bp of host DNA, and at the right end are 1–2 kbp of host DNA. These host DNA sequences are not unique but simply represent DNA adjacent to the location where Mu was inserted into the genome of its previous host. How does this happen?

When a Mu phage virion is formed, a length of DNA containing the Mu genome just large enough to fill the phage head is excised from the host. The DNA is packaged until the head is full, but the place at the right end where the DNA is cut varies from one virion to another. For that reason, as shown on the genetic map, there is a variable sequence of host DNA at the right-hand end of the phage (right of the *attR* site). Thus, each virion arising from a single infected cell will be genetically unique, since it will have different host DNA.

Mu and the Invertible G Region

As shown on the genetic map (Figure 16.9a), a specific segment of the Mu genome called *G* is *invertible*, being present in the genome either in the orientation designated

● **Figure 16.8 Bacteriophage Mu.** Electron micrograph of virions of the double-stranded DNA phage Mu, the mutator phage.

(a)

(b)

● **Figure 16.9 Replication of bacteriophage Mu.** (a) Genetic map of the Mu genome. See text for details. Note that there is a lowercase *c* gene, which encodes a repressor, and an uppercase *C* gene, which encodes an activator protein. The region encoding the head and tail genes is not drawn to scale. (b) Integration of Mu into the host DNA, showing the generation of a 5-bp duplication of host DNA.

G$^+$ or in the inverted orientation G$^-$. The orientation of this segment determines the kind of tail fibers that are present on the phage. Since adsorption to the host cell is controlled by molecular interactions between the tail fibers and the cell surface, the *host range* of Mu is determined by which orientation of this invertible segment is present in the phage. For example, if the G segment is in the orientation designated G$^+$, then the phage will make tail fibers that allow it to infect *Escherichia coli* strain K12. By contrast, if the G segment is in the G$^-$ orientation, then the phage will infect *E. coli* strain C or several other species of enteric bacteria. The two tail fiber proteins are encoded on opposite strands within this small G segment.

Left of the G segment is a promoter that directs transcription into the G segment. In the orientation G$^+$, the

promoter directing transcription of genes S and U is active, whereas in the orientation G$^-$, a different promoter directs transcription of genes S' and U' on the opposite strand. Inversion of the G region is a rare event and is under the control of a gene adjacent to the G region. We thus see in the inversion phenomenon a simple mechanism for attacking a variety of different host cells.

Replication of Mu

Note that the bacteriophages we have discussed that have double-stranded DNA genomes, *T4* (∞ Section 9.9), *T7* (see Section 16.4), *lambda* (∞ Section 9.10), and *Mu*, all have *linear* genomes. Despite this, the viruses use three quite distinct strategies to replicate the ends of their genomes (∞ Section 14.6). Both T4 and T7 have terminal repeats and use recombination to form concatemers. By contrast, lambda circularizes its genome after infection. However, Mu has an unusual property not found in any of these other phages, as its genome is replicated as part of a larger DNA molecule.

On infection of a host cell by Mu, the DNA is injected and is protected from host restriction by a modification system in which about 15% of the adenine residues are modified by acetylation. In contrast with lambda, integration of Mu DNA into the host genome is essential for both lytic and lysogenic growth. Integration requires the activity of the gene A product, which is a **transposase** enzyme. At the site where the Mu DNA becomes integrated, a 5-bp duplication of host DNA arises at the target site. As shown in Figure 16.9*b*, this host DNA duplication arises because staggered cuts are made at the point in the host genome where Mu is inserted. The resulting single-stranded segments are converted to the double-stranded form as part of the Mu integration process. Duplication of short stretches of host DNA is typical of transposable element insertion (∞ Section 10.14).

Lytic growth of Mu can occur either upon initial infection, if the Mu repressor (the product of the *c* gene) is not formed, or by induction of a lysogen. In either case, replication of Mu DNA involves repeated transposition of Mu to multiple sites on the host genome. Initially, transcription of only the early genes of Mu occurs, but after the C protein is expressed (C is a positive activator of late transcription), synthesis of the Mu head and tail proteins occurs. Eventually, expression of the lytic function occurs and mature phage particles are released. The lysogenic state in Mu requires the sufficient accumulation of repressor protein to prevent transcription of integrated Mu DNA.

⬡ **16.5 Concept Check**

Bacteriophage Mu is a temperate virus that is also a transposable element. In either the lytic or lysogenic pathway, its genome is integrated into the host chromosome by the activity of a transposase. Even in the lytic pathway, its genome is replicated as part of a larger DNA molecule. The genome is packaged into the virion in such a way that there are short sequences of host DNA at either end.

◆ What is a transposable element?

◆ What mechanism does Mu use to ensure that the ends of its linear genome are completely replicated?

16.6 Viruses of *Archaea*

Several viruses have been discovered whose hosts are species of *Archaea*, including methanogenic (∞ Section 13.4), halophilic and alkaliphilic (∞ Section 13.3), and hyperthermophilic (∞ Sections 13.6–13.10) *Archaea*. Most archaeal viruses that infect methanogenic and halophilic *Archaea* are of the head and tail type, characteristic of phages that infect enteric bacteria, such as phage T4 (∞ Section 9.9). In fact, certain tailed viruses of halophilic and haloalkaliphilic *Archaea*, such as phage ΦH of *Halobacterium salinarum* and phage ΦCh1 of *Natrialba magadii*, have linear double-stranded DNA ge-nomes that are circularly permuted and terminally redundant, as in phage T4. These viruses likely package DNA using the same "head full" mechanism of phage T4 (∞ Section 9.9 and Figure 9.13).

Viruses of *Sulfolobus* and *Pyrococcus*

The most diverse and morphologically unusual archaeal viruses infect hyperthermophiles (species of the phylum Crenarchaeota, ∞ Chapter 13), and they will be our focus here. The hyperthermophilic sulfur-oxidizing organism *Sulfolobus* is host to several morphologically unusual viruses. Considering the habitat of *Sulfolobus*—hot, acidic soils and hot springs (∞ Section 13.8), these viruses must be remarkably resistant to heat and acid denaturation.

A common virus that infects *Sulfolobus* species, nicknamed *SSV*, forms spindle-shaped virions that often cluster in rosettes (Figure 16.10a●). Such viruses are apparently widespread in thermal environments, as they have been isolated from hot springs in Iceland and Japan, as well as Yellowstone National Park. Virions of SSV contain *circular* double-stranded DNA of about 15 kbp in length, considerably smaller than the linear genome of a tailed bacteriophage like T4 (~168 kbp) (no circular double-stranded DNA viruses of *Bacteria* are known).

A second morphological type of *Sulfolobus* bacteriophage is a rigid, helical rod (Figure 16.10b). Viruses in this class, nicknamed *SIF*, contain *linear* double-stranded DNA genomes of about 35 kbp in length. Many variations on the spindle- and rod-shaped patterns have been seen in viral isolation studies. These include very long rods similar in morphology to filamentous bacteriophages (see Section 16.3) and spindle-shaped viruses that contain a large spindle-shaped center with appendages at each end (Figure 16.10). Thus far, no RNA viruses infecting *Sulfolobus* (or any other member of the *Archaea*) have been found.

(a)   (b)

(c)

●**Figure 16.10 Archaeal viruses.** Electron micrographs of (a, b) viruses of Crenarchaeota, and (c) a virus of a Euryarchaeote. (a) Spindle-shaped virus SSV1 that infects *Sulfolobus solfataricus* (temperature optimum, 80°C). (b) Filamentous virus SIFV that infects *S. solfataricus*. (c) Spindle-shaped virus PAV1 that infects *Pyroccccus abyssi* (temperature optimum, 96°C). The dimensions of the viruses are: SSV1, 40 × 80 nm; SIFV, 50 × 900–1500 nm; PAV1, 80 × 120 nm. Besides these morphologies, tailed viruses resembling tailed bacteriophages (∞ Section 9.9) are known to infect some *Archaea*.

A spindle-shaped virus also infects *Pyrococcus*, a species that falls within the archaeal phylum Euryarchaeota (∞ Section 13.6). This virus, named *PAV1*, resembles SSV, but is larger and contains a very short tail (Figure 16.10c). Phage PAV1 contains a *circular* double-stranded DNA genome of about 18 kbp and, interestingly, is released from host cells without cell lysis, probably by a budding mechanism similar to that of the filamentous *Escherichia coli* phage M13 (see Section 16.3).

Pyrococcus is a hyperthermophile with a growth temperature optimum of about 100°C, meaning that PAV1 virions must be extremely heat stable. From genomic comparisons, PAV1 and SSV-type viruses, despite their similar morphologies, show little sequence similarity, indicating that the two types of virions do not have common evolutionary roots.

Replication and Evolution of Archaeal Viruses

Viral replication studies of archaeal viruses still need to be done in order to define the major events in genome replication and virion assembly. However, considering the genomes of these viruses, all double-stranded DNA, it is unlikely that any major new replication strategies will emerge from such studies. However, many molecular details, such as the extent to which viral rather than host polymerases and related enzymes are used in replication events, await further work on these relatively newly discovered viruses. Considering the very small genomes of archaeal viruses, one would predict that they are very host dependent and may well employ DNA economizing features, such as overlapping genes.

Spindle-shaped viruses are unknown among species of *Bacteria*. This, along with limited genomic similarity to viruses of *Bacteria*, suggests that these archaeal viruses have not shared a common ancestor with known bacteriophages. By contrast, structural and genomic studies have shown that *tailed* viruses of *Archaea* may well have shared common ancestry with similar phages from the *Bacteria*. This could be because they share common evolutionary roots or be the result of lateral gene transfer (∞ Chapter 10) of phage genes from *Bacteria* to *Archaea* (or vice versa).

It is clear that species of *Archaea* can replicate viruses. And, although it is not yet known what effects archaeal viruses have on their hosts (other than the obvious in the case of lytic viruses), it would be surprising if at least some archaeal viruses do not affect the genetics of *Archaea* in ways similar to that in *Bacteria* (∞ Chapter 9). Thus, one would predict that both temperate viruses (∞ Section 9.10) and transducing viruses (∞ Section 10.8) exist within the *Archaea*, although at this point, neither category of virus has been found.

 16.6 *Concept Check*

Several viruses infect *Archaea*. Many of these have double-stranded circular DNA genomes not known in bacteriophages of *Bacteria*. Although head/tail-type viruses are known, most archaeal viruses have an unusual spindle-shaped morphology.

◆ In which group(s) of *Archaea* have head/tail-type phages been discovered? Are they related to tailed phages of *Bacteria*?

◆ Compare the morphologies of viruses that infect methanogenic and halophilic *Archaea* with those that grow at high temperatures.

 VIRUSES OF EUKARYOTES

The differences between prokaryotic cells and eukaryotic cells place some constraints on the viruses that infect them. For example, in prokaryotes transcription and translation can be coupled processes. In eukaryotes, by contrast, *transcription* and *DNA replication* occur in the *nucleus* while *translation* occurs in the *cytoplasm*. In addition, mRNAs in eukaryotes are capped and contain poly-A-tails (∞ Sections 14.7 and 14.8). We will see that these details are relevant to viruses that replicate in eukaryotes, as well.

We discussed some animal viruses in Chapter 9 (∞ Section 9.12). In addition to their general biology, our interest in animal viruses springs in part from their role in human disease, and we will focus our attention on animal viruses in the remainder of this chapter. However, not all eukaryotes are animals, and we begin here with a consideration of two well-characterized *plant* viruses.

16.7 Plant Viruses

Plant cell walls are extremely thick and strong. Yet there are a number of plant viruses that infect plants and, in multicellular plants, spread from the infected cell to neighboring cells. The great majority of plant viruses known are *positive-strand RNA viruses*, and it is thought that this is so because these small genomes facilitate transfer from cell to cell within the plant.

Tobacco Mosaic Virus: General Properties

In 1892, the Russian scientist Dmitri Ivanovsky showed that the causative agent of tobacco mosaic disease could pass through filters that retain bacteria. In 1898, the Dutch microbiologist Martinus Beijerinck (∞ Section 1.7) showed that this agent was not only filterable, but that it had many of the properties of a living organism. This agent, the first virus of any type to be recognized, was *tobacco mosaic virus* (TMV).

TMV has a virion with helical symmetry (∞ Section 9.2) and contains 2130 copies of a coat protein plus a single copy of the positive-strand RNA genome (∞ Figure 9.2). It was with TMV that it was first shown that RNA could be the genetic material, just as DNA is in other viruses and in cells. TMV remains a serious agricultural problem because it infects tomato plants as well as tobacco. TMV infection of a plant requires damage to plant cell walls through which the virion enters. Uncoating takes place in the cell.

Genome and Replication of TMV

The RNA genome of TMV contains 6395 nucleotides, and a map of the genome is shown in Figure 16.11●. Like the bacteriophage MS2 (see Section 16.1), TMV encodes only four proteins. The genome has a 5′ cap (∞ Section 14.8) so that it can be translated in the plant cell, and the 3′ end folds into a transfer RNA-like structure (Figure 16.11). In small RNA bacteriophages we saw that the viral genome was also the viral mRNA (see Section 9.7). However, eukaryotes cannot translate polycistronic mRNA (∞

● **Figure 16.11   Genetic map of tobacco mosaic virus.** The genome is a positive-strand RNA that is capped at its 5′ end and has a tRNA-like structure (drawn larger than scale) at its 3′ end. The MTH gene encodes a protein with both methyltransferase and RNA helicase activity. The RNP gene encodes an RNA-dependent RNA polymerase that is only translated as a polyprotein by ribosomes reading through the stop codon following the MTH gene. The MP gene encodes a movement protein, and the CP gene encodes the coat protein. These two proteins are translated from subgenomic mRNAs that are synthesized using the negative-strand RNA after it has been made during infection. Besides tobacco, tobacco mosaic virus can attack a variety of other plants including tomato, pepper, and spinach.

Section 14.7), so the expression of TMV genes differs from that of a small bacteriophage like MS2.

The first gene in TMV encodes an enzyme called *MTH* that has two enzymatic activities, a *methyltransferase* that functions in capping RNA, and an *RNA helicase*. The RNP gene encodes the *RNA-dependent RNA polymerase* (RNA replicase) that the virus must synthesize to replicate a negative-strand copy from which it can then make more copies of the genomic RNA. This protein is synthesized as part of a *polyprotein* (see Section 16.8) including the MTH domains and is only synthesized when a ribosome accidentally reads through the stop codon at the end of the MTH gene. Since this happens rather infrequently, this longer protein is only made at low levels. The remaining two genes encode the *movement protein* (MP) and the *coat protein* (CP). Two small monocistronic mRNAs are made from the negative-strand RNA, and each is translated to yield one of these proteins.

Like coat proteins of other viruses, that of TMV is essential to the formation of the virion. The virion is essential for infecting new plants. However, it is the movement protein that enables TMV to infect neighboring cells in an already infected plant. Plant cells have structures called *plasmodesmata*, which are intercellular connections that span cell walls and connect the plant cells. Plasmodesmata have very narrow channels, so narrow in fact that neither the TMV virion nor free RNA can easily traverse these openings. Movement protein binds to the new genomic plus-strand RNA and forms a complex that is extremely thin (about 2.5 nm), and it is these nucleoprotein complexes that move through the plasmodesmata and infect neighboring cells.

As was mentioned before, there are many plant positive-strand RNA viruses known, and there are also positive-strand RNA viruses of *Bacteria* (see Section 16.1) and animals (see Section 16.8). Genomic analyses indicate that many of these viruses are closely related despite their quite different hosts and that all of their RNA

replicases are related. Nonetheless, as shown with TMV, they have slightly different replication strategies, usually related to differences in the host cells.

DNA Plant Viruses: *Chlorella*

Green algae are microbial green plants, and *Chlorella* is a widely distributed genus of green alga. Most species of *Chlorella* are free living, but some *Chlorella*-like algae are endosymbionts (∞ Section 14.4) of freshwater or marine animals, including protozoa such as *Paramecium*. Many of these phototrophic endosymbionts can be grown in the laboratory independently of the organism in which they reside, and some of these are hosts to viruses; the best studied of these is *Paramecium bursaria* chlorella virus 1 (PBCV-1).

Virus PBCV-1 has large icosahedral virions (Figure 16.12●) and a double-stranded DNA genome. The virions have a lipid component that is essential for infectivity but it is inside the capsid, therefore, the virion is not enveloped. The genomes of the *Chlorella* viruses are extremely large: All are over 300 kb (for a comparison with other viruses see Table 9.1). In many cases the DNA is also extensively modified by methylation. The genome of PBCV-1 has been completely sequenced. This 330,742-bp genome encodes over 370 different proteins and 10 tRNAs. The ends of the genome are incompletely base-paired hairpin loops (∞ Figure 7.8) very similar to those found in poxviruses (see Section 16.13).

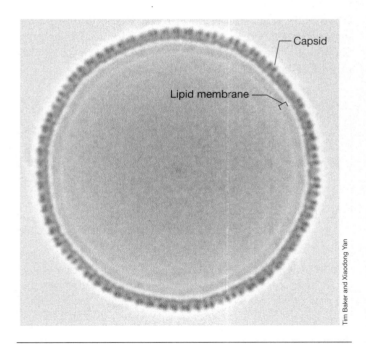

● **Figure 16.12   A cross-section of a virion of the *Chlorella* virus PBCV-1.** A lipid bilayer membrane is visible beneath the capsid shell. The virion has a diameter of approximately 170 nm. The image is reconstructed from several transmission electron micrographs. *Chlorella* is a member of the green algae (∞ Section 14.13). Besides being very large viruses, *Chlorella* viruses also contain very large genomes of greater than 0.3 Mbp.

Replication of *Chlorella* Viruses

Virus PBCV-1 enters cells somewhat like bacteriophages do. The virion binds specifically to the cell wall of the host, then at least five different enzymes carried by the virion digest away the cell wall at the point of contact. Viral DNA is then released into the cell leaving the empty virion behind. As is the case for most double-stranded DNA viruses of eukaryotes, the DNA of PBCV-1 is replicated in the nucleus, and RNA is also synthesized there. PBCV-1 encodes several enzymes involved in DNA replication, including a DNA polymerase. However, although PBCV-1 encodes some transcription factors, it does not encode its own RNA polymerase. A few of the genes of PBCV-1 contain introns that must be removed (Section 7.1). The virus mRNA is capped, and some of the enzymes involved are virus-encoded. Early mRNA, but not late mRNA, has poly-A tails. Like bacteriophage T4, PBCV-1 encodes several of its own tRNAs.

Although the known hosts for viruses such as PBCV-1 are only *Chlorella* strains obtained as endosymbionts from other cells, the viruses themselves are quite widespread in nature. Therefore, there must be abundant natural hosts. In addition, the *Chlorella* viruses are related to other viruses that infect algae. The latter have large double-stranded DNA genomes, but some of the genomes are circular.

There is one other extremely interesting fact to note about the *Chlorella* viruses: Their genomes encode several restriction and modification enzyme systems (Sections 7.7 and 9.9). Indeed, they are the *only* source of restriction enzymes outside of the prokaryotes and a few bacteriophages. Thus we see features in *Chlorella* viruses typical of both prokaryotic and eukaryotic genetic elements.

🛑 16.7 Concept Check

Most plant viruses have positive-strand RNA genomes, and one example is tobacco mosaic virus (TMV), the first virus discovered. The genomes of these viruses can move within the plant through intercellular connections that span the cell walls. Other types of plant viruses are also known, including the *Chlorella* viruses that have very large double-stranded DNA genomes.

◆ Most plant RNA viruses encode a *movement protein*. What is its role in infection?

◆ Although the TMV genome is used as a messenger RNA, not all the proteins encoded by the virus can be translated from it. Explain.

◆ Which features of *Chlorella* viruses are prokaryotic-like?

16.8 Positive-Strand RNA Viruses of Animals: Poliovirus and Coronaviruses

Several positive-strand RNA animal viruses cause disease in humans and other animals, including the *polioviruses*, the *rhinoviruses* that cause the common cold, *coronaviruses* that cause respiratory syndromes including SARS, and the *hepatitis A virus*. The first animal virus discovered, foot-and-mouth disease virus, the causative agent of a debilitating and eventually fatal disease of cloven-hoofed (ruminant) animals, is also in this group. With the exception of the coronaviruses, which are much larger, these viruses have been called *picornaviruses* because they are typically very small (about 30 nm in diameter) (*pico* means "small") and contain single-stranded RNA. We focus in this section on poliovirus and the coronaviruses.

An understanding of the biology of poliovirus is very important. At one time, polio was a major infectious disease of humans, but the development of an effective vaccine (Section 22.13) has brought the disease almost completely under control. The World Health Organization (WHO) has a vaccination program intended to eradicate the disease worldwide, and at present there are reports of the virus in only a few countries in Africa and in the Indian subcontinent.

Replication of Poliovirus: General Features

The virion of poliovirus has an icosahedral structure with 60 morphological units per virion, each unit consisting of four distinct proteins (Figure 16.13a●). The genome of poliovirus is a linear single-stranded RNA molecule of 7440 bases in length (Table 9.1). At the 5' terminus of the viral RNA is a protein, called the *VPg protein*, that is attached covalently to the RNA. At the 3' terminus of the RNA is a poly-A tail.

An overview of the replication of poliovirus is illustrated in Figure 16.13b. The RNA genome of the virus is also the messenger RNA. Interestingly, this occurs even though the RNA is not capped, normally a prerequisite for translation in eukaryotes (Section 14.8). The 5' end of poliovirus RNA has a long sequence that can fold into several stem-loops. The VPg protein and the stem-loops mimic the cap-binding complex, and these permit binding of the poliovirus mRNA to the ribosome. The virus RNA is monocistronic but codes for all the proteins of the virus in a single protein called a **polyprotein**. This is later cleaved into the individual proteins. The whole poliovirus replication process occurs in the cell cytoplasm.

To initiate infection, the poliovirus virion attaches to a specific receptor on the surface of a sensitive cell and enters the cell. Once inside the cell, the virus particle is uncoated, and the free RNA associates with ribosomes. The viral RNA is then translated from a single start codon into the large polyprotein. This giant protein (about 2200 amino acid residues) then undergoes self-cleavage into about 20 smaller proteins (including cleavage intermediates), among which are the four different *structural proteins* of the virion. These include the RNA-linked *VPg protein*, an *RNA polymerase* (RNA replicase) responsible for synthesis of both minus-strand and plus-strand RNA, and at least one *virus-encoded protease*, which carries out the polyprotein cleavage. This cleavage process, called *posttranslational*

(a)

(b)

● **Figure 16.13** **Poliovirus.** (a) A computer model based on electron diffraction analysis of poliovirus virions. The various structural proteins are shown in distinct colors. (b) The replication of poliovirus. The single-stranded RNA of the virus is translated directly as a messenger RNA, with the production of one large protein molecule. This protein is cleaved, leading to production of the active viral proteins, including the structural coat proteins and the RNA polymerase that brings about replication of the poliovirus RNA. The assembly of intact poliovirus from coat protein molecules and RNA then follows. Translation of an mRNA to yield a polyprotein that is subsequently cleaved is a strategy also used by other viruses (see Section 16.15).

cleavage, occurs in a wide variety of animal viruses as well as in normal cell metabolism in animal cells.

Replication of Poliovirus RNA

Replication of viral RNA begins within a short time after infection and is catalyzed by the RNA replicase made in the process just described. This replicase transcribes the plus-sense viral RNA into a complementary RNA molecule of minus complementarity. This minus strand is then the template for repeated transcription of progeny plus strands, again catalyzed by the viral-specific RNA replicase. Some of the progeny plus strands may again be transcribed into minus strands, and as many as 1000 minus strands may subsequently be present in the cell. From these minus strands, as many as a million plus strands may ultimately be formed. Both the plus and the minus strands become covalently linked to the tiny VPg protein (only 22 amino acids long), which functions as a primer for transcription.

Once poliovirus multiplication begins, host RNA and protein syntheses are inhibited. Host protein synthesis is inhibited as a result of the destruction of an important host protein, the cap-binding protein, required for translation of capped mRNAs (∞ Section 14.8). Poliovirus mRNA itself circumvents this limitation as previously mentioned through the extensive secondary structure of the 5′ end of its genome that facilitates binding to the ribosome.

Coronaviruses and SARS

Coronaviruses are single-stranded plus-strand RNA viruses that, like poliovirus, replicate in the cytoplasm, but differ from poliovirus in their larger size and details of replication. Coronaviruses cause a variety of respiratory infections in humans and other animals, including about 15% of common colds (∞ Section 26.8). In 2003, a coronavirus caused several outbreaks of *severe acute respiratory syndrome (SARS)*, an occasionally fatal pneumonia of the lower respiratory tract in humans (∞ Section 25.8).

Coronavirus virions are enveloped and more or less spherical, 60–220 nm in diameter, and contain club-shaped glycoprotein spikes on their surfaces. These give the virus the appearance of having a "crown" (the term *corona* is Latin for crown) (Figure 16.14●) (∞ also Figure 25.9). Coronavirus genomes are noteworthy because they are the largest of any known RNA viruses (27–31 kb; strains of SARS virus average 29,700 nucleotides in length).

The coronavirus genome has a 5′ methylated cap and a 3′ poly-A tail and so can function directly in the animal as mRNA. However, most viral proteins are not

(a)

(b)

● **Figure 16.14 Coronaviruses.** (a) Electron micrograph of a coronavirus. (b) Steps in coronavirus replication. Messenger RNA encoding viral proteins occurs exclusively by transcription off of the (−) strand made by the replicase enzyme from transcription of the viral genome.

made from the translation of genomic RNA. Instead, upon infection, only a portion of the genome is translated, yielding a viral RNA replicase. The replicase uses the genomic RNA as template to produce a full-length negative complementarity RNA from which transcription of several monocistronic mRNAs occur. The latter are then translated to produce viral proteins. Progeny genomes are also made by transcription off of the negative-strand RNA. Viral assembly occurs within the Golgi apparatus, a major secretory organelle in eukaryotic cells (Section 14.5), with fully assembled virions being released at the cell surface.

Coronavirus thus differs from poliovirus in terms of virion size, genome size and capping, lack of the VPg protein, and the absence of polyprotein formation and cleavage. Nevertheless, like poliovirus, coronaviruses can cause severe disease, including a fatal acute infection that is rarely seen in polio. Fatality rates from SARS range from 13% in those under age 60 to nearly 45% in those above 60 (Section 25.8).

🛑 16.8 Concept Check

In small RNA viruses such as poliovirus, the viral RNA is translated directly, causing the production of a long polyprotein that is broken down by enzymes into the numerous small proteins necessary for nucleic acid multiplication and virus assembly. Coronavirus is a large single-stranded RNA virus that resembles poliovirus in some but not all of its replication features.

◆ How can poliovirus RNA be synthesized in the cytoplasm whereas host RNA must be synthesized in the nucleus?

◆ In what ways are protein synthesis and genomic replication processes similar/different in poliovirus and the SARS virus?

16.9 Negative-Strand RNA Viruses of Animals: Rabies, Influenza, and Related Viruses

Poliovirus and coronavirus replication requires conversion of the plus-stranded genome into a negative-stranded intermediate from which new plus strands are synthesized. However, in a number of RNA animal viruses the RNA genome *itself* is **minus-strand RNA**. These are thus called *negative-strand viruses*. We discuss here two important examples: **rhabdoviruses**, including rabies virus, and **orthomyxoviruses**, including influenza virus. The Ebola virus, a human pathogen responsible for an emerging infectious disease (Section 25.10), is also a negative-strand RNA virus. There are no known negative-strand RNA bacteriophages or archaeal viruses.

Rhabdoviruses: General Features

One of the most important negative-strand RNA viral pathogens is the rabies virus, which causes the disease *rabies* in animals and humans. Worldwide, over 30,000 (mostly fatal) cases of rabies occur each year in humans, and countless more in both domesticated and wild animals (Section 27.1). Rabies virus is called a *rhabdovirus*, from *rhabdo* meaning "rod," which refers to the shape of the virus particle. Another rhabdovirus that has been extensively studied is vesicular stomatitis virus (VSV) (Figure 16.15●), a virus that causes the disease *vesicular stomatitis* in cattle, pigs, horses, and sometimes humans. Many rhabdoviruses, such as *potato yellow dwarf virus*, infect both insects and plants and can cause important agricultural problems.

Erskine Caldwell

● **Figure 16.15 Electron micrograph of a rhabdovirus (vesicular stomatitis virus).** A particle is about 65 nm in diameter. Rhabdoviruses also include the rabies virus. The disease rabies is discussed in Section 27.1.

The rhabdoviruses are enveloped viruses, with an extensive and rather complex lipid envelope surrounding the nucleocapsid. In rhabdoviruses the virion is bullet-shaped, about 70 nm in diameter and 175 nm long (Figure 16.15).

The nucleocapsid is helically symmetric and makes up only a small part of the virus particle weight (about 2–3% of the virion is RNA).

Replication of Rhabdoviruses

The rhabdovirus virion contains several enzymes that are essential for the infection process. One of these is an *RNA-dependent RNA polymerase* (a type of RNA replicase). As discussed in Section 9.7, the presence of such an enzyme is essential because the genome of negative-stranded viruses cannot be translated directly but must first be converted into the plus complement; host enzymes capable of transcribing RNA from an RNA template do not exist.

The RNA of rhabdoviruses is transcribed in the cytoplasm into two distinct classes of RNAs (Figure 16.16●). The first is a series of mRNAs encoding the structural genes of the virus (for example, VSV has five genes). Each mRNA is monocistronic, encoding a single protein. The second type of RNA is a plus complementarity RNA that is a copy of the complete viral genome (the VSV genome is 11,162 nucleotides long). These full-length plus-strand RNAs are templates for the synthesis of *negative-strand* genomic RNA molecules for progeny virions. Once mRNA encoding the virus RNA polymerase is made in the primary transcription process, synthesis of more copies of the virus RNA polymerase occurs, leading to the formation of many plus-strand RNA molecules, both mRNAs and genomic RNA templates (Figure 16.16).

Assembly of Rhabdoviruses

Translation of viral mRNAs leads to the synthesis of viral coat proteins. Assembly of an enveloped virus is considerably more complex than assembly of a naked virion. Two kinds of coat proteins are formed, *nucleocapsid proteins* and *envelope proteins*. The **nucleocapsid** is formed first by association of the nucleocapsid protein molecules around the viral RNA.

The envelope proteins possess hydrophobic amino acid leader sequences at their amino-terminal ends (∞ Section 7.16). As these proteins are synthesized, sugar residues are added, leading to the formation of *glycoproteins*. Such glycoproteins, characteristic of membrane-associated proteins, migrate to the cytoplasmic membrane where the leader sequences are removed; here they replace host membrane proteins. Nucleocapsids then migrate to the areas on the cytoplasmic membrane where these virus-specific glycoproteins exist, recognizing the virus glycoproteins with great specificity. The nucleocapsids then become aligned with the glycoproteins and bud through them, becoming coated by the glycoproteins in the process.

The final result is an enveloped virion with a nucleocapsid center and a surrounding membrane whose lipid is derived from the host cell but whose membrane proteins are encoded by the virus. The budding process itself does not cause detectable damage to the cell, which may continue to release virions in this way for a considerable period of time. Host damage does eventually

● **Figure 16.16 Flow of events during multiplication of a negative-strand RNA virus.** Note the importance of the viral RNA polymerase, carried in the virion. This is critical, because animal cells cannot make RNA off of an RNA template.

occur, of course, but is brought about by factors other than just the production of virus.

Influenza and Other Orthomyxoviruses

Another group of negative-strand viruses is the *orthomyxoviruses*, which contain the important human pathogen, **influenza virus**. The term *myxo* refers to the fact that these viruses interact with the *mucus* or *slime* of cell surfaces. In the case of influenza virus, this mucus is at the mucous membrane of the respiratory tract, because these viruses are transmitted primarily by the respiratory route (∞ Section 26.8). The term *ortho* distinguishes this group from another group of negative-strand viruses, the *paramyxovirus* group. The paramyxoviruses, which include such important human pathogens as mumps and measles viruses (Section 26.7), are quite similar in their molecular biology to rhabdoviruses.

The orthomyxoviruses have been extensively studied over many years, beginning with early work during and after the 1918 pandemic of influenza that caused the deaths of millions of people worldwide (∞ Section 26.8). The orthomyxoviruses are enveloped viruses in which the viral RNA is present in the virion in a number of separate pieces. The genome of the orthomyxoviruses is thus said to be **segmented**. In the case of influenza A virus, the genome is segmented into *eight* linear single-stranded molecules ranging in size from 890 to 2341 nucleotides. The influenza virus nucleocapsid is of helical symmetry, about 6–9 nm in diameter and about 60 nm long. This nucleocapsid is embedded in an envelope that has a number of virus-specific proteins as well as lipid derived from the host (Figure 16.17*a*●).

Because of the way influenza virus buds as it leaves the cell, the virus has no defined shape and is said to be *polymorphic* (Figure 16.17*a*). There are proteins on the outside of the envelope that interact with the host cell surface. One of these is called *hemagglutinin*, so named because it causes agglutination (clumping) of red blood cells. The red blood cell is not the type of host cell the virus normally infects, but contains on its surface the same type of membrane component, *sialic acid*, that the mucous membrane cells of the respiratory tract contain. Thus, the red blood cell is merely a convenient cell type for assaying agglutination activity. An important feature of the influenza virus hemagglutinin is that antibodies directed against this hemagglutinin *prevent* the virus from infecting a cell. Thus, antibody directed against the hemagglutinin *neutralizes* the virus, and this is the mechanism by which immunity to influenza is brought about during the immunization process (∞ Sections 22.13 and 26.8).

A second type of protein on the influenza virus surface is an enzyme called *neuraminidase* (∞ Figure 26.19). Neuraminidase breaks down the sialic acid component of the cytoplasmic membrane, which is a derivative of neuraminic acid. Neuraminidase appears to function primarily in the virus assembly process, destroying host

(a)

(b)

● **Figure 16.17   Influenza virus.** (a) Electron micrograph of human influenza virions. (b) Diagram, showing some of the components including the segmented genome. The disease influenza is discussed in Section 26.8.

membrane sialic acid that would otherwise block assembly or become incorporated into the mature virus particle.

Replication of Influenza Virus

In addition to neuraminidase, influenza virions possess two other key enzymes, an *RNA-dependent RNA polymerase* (RNA replicase), which is involved in the conversion of the negative-stranded genome into a positive strand (as already discussed for the rhabdoviruses), and an *RNA endonuclease*, which cuts a primer from the host's capped mRNA precursors. The virion enters the cell, after which the nucleocapsid becomes separated from the envelope and migrates to the nucleus. Replication of the viral nucleic acid then occurs in the nucleus.

Uncoating results in activation of the virus RNA polymerase. The mRNA molecules are then transcribed in the nucleus from the virus RNA, using oligonucleotide primers cut from the 5′ ends of newly synthesized capped cellular mRNAs by the viral endonuclease. Thus, the viral mRNAs have 5′ caps. The poly-A tails of the viral mRNAs are added, and the virus mRNA molecules move to the cytoplasm for translation. Thus, although influenza virus RNA *replicates* in the nucleus, influenza virus proteins, like all viral proteins, are *synthesized* in the cytoplasm.

Ten virus proteins are encoded by the *eight* segments of the virus genome. The mRNAs transcribed from six segments each encode a single protein, whereas the other two segments encode two proteins each. The latter is not done by using true polycistronic mRNA as in prokaryotes because eukaryotic ribosomes typically recognize only the AUG codon closest to the 5′ end of the mRNA as a start codon (∞ Section 7.16). Therefore, they can make only one protein from a given RNA. The original full-length mRNAs transcribed from these two segments are each translated to give one protein. But in each case, an *additional protein* is translated from these messages following processing of the message by the host's RNA splicing machinery. Thus, like overlapping genes (see Section 16.2), this is yet another example of how RNA viruses can maximize the coding potential of a small genome.

Some of the proteins made are involved in influenza virus RNA replication, while others are structural proteins of the virion. The overall strategy of genomic RNA synthesis resembles that of the rhabdoviruses, with primary transcription resulting in the formation of *plus*-strand templates for the formation of progeny *minus*-strand molecules. The formation of the complete enveloped virion occurs by a budding-out process, as was described for the rhabdoviruses.

Influenza and Antigenic Shift/Drift

The segmented genome of the influenza virus has important practical consequences. Influenza virus and other viruses of this family exhibit a phenomenon called **antigenic shift** in which portions of the RNA genome from two genetically distinct strains of virus infecting the same cell are reassorted. This generates virions that express a significantly altered set of surface proteins from that of either original virus. Surface proteins are the major target of antibodies that form as a result of immunization by artificial means (∞ Section 22.13) or through natural infection. However, when antigenic shift occurs, previous immunity to one strain or the other of influenza virus is insufficient to ward off infection by the genetically "new" viruses. Antigenic shift is thought to bring about major pandemics and epidemics of influenza because immunity to the new forms of the virus is essentially absent from the population (∞ Section 26.8).

Antigenic shift can be contrasted with **antigenic drift**. In the latter, the structure of the neuraminidase and hemagglutinin proteins on the surface of the influenza virus virion is altered, usually in a subtle way, by mutation in the genes encoding them. The alterations change the surface properties of influenza virus suffi-

ciently that antibodies that recognized the virus previously no longer do so, or do so less effectively. Thus, for effective immunity to be achieved, new antibodies must be produced. This is a major reason why influenza vaccines rarely confer protection for more than one year (∞ Section 26.8). Genetic drift during the previous year yields slightly modified forms of influenza virus for which new vaccines must be made.

16.9 Concept Check

In negative-strand viruses the virus RNA is not the mRNA but is copied into mRNA by an enzyme present in the virion. Important negative-strand viruses include rabies virus and influenza virus.

◆ Why is it essential that negative-strand viruses carry an enzyme in their virions?

◆ What is a *segmented genome*?

◆ What is the difference between antigenic *shift* and antigenic *drift*?

16.10 Double-Stranded RNA Viruses: Reoviruses

Reoviruses are an important family of animal viruses that have *double-stranded RNA* genomes. *Rotavirus* is a typical reovirus and is the most common cause of diarrhea in infants from 6 to 24 months of age. There are also reoviruses known that cause respiratory infections, and others that infect plants, and we previously mentioned that the bacteriophage φ6 seems closely related to the reoviruses (see Section 16.1). Reovirus virions consist of a nonenveloped nucleocapsid 60–80 nm in diameter, with a *double* shell of icosahedral symmetry (Figure 16.18●). Predictably, the virions of these double-stranded RNA viruses contain virus-encoded enzymes necessary to synthesize mRNA and new RNA genomes.

Replication of Reoviruses

The genome of reoviruses is segmented into 10–12 molecules of *linear* double-stranded RNA. Replication occurs exclusively in the cytoplasm of the host. The double-stranded RNA is inactive as mRNA, and the first step in reovirus replication is transcription by a viral-encoded RNA-dependent RNA polymerase using the minus strand as a template to make plus-sense mRNA. The mRNA is then capped and methylated by viral enzymes and then translated.

Generally, each molecule of RNA in the genome encodes a single protein, although in a few cases the protein formed is cleaved to yield the final product. However, one of the mRNAs produced actually encodes two proteins, but the RNA does *not* have to be processed in order to do so. Instead, a ribosome sometimes "misses" the start codon for the first gene in this message and travels

• **Figure 16.18 Double-stranded RNA viruses: The reoviruses.** (a) An electron micrograph showing several reovirus virions (each with a diameter of about 70 nm). (b) Three-dimensional reconstruction of a reovirus virion calculated from electron micrographs of frozen-hydrated virions.

on to the start codon of the second gene. Therefore, there are exceptions to the generalization that eukaryotic ribosomes initiate at the first AUG codon in an mRNA.

In the initial infection process, the reovirus virion binds to a cellular receptor protein. Once attachment has occurred, the virus enters the cell and is transported into lysosomes (👁 Section 14.5), where normally it would be destroyed. However, within the lysosome the outer shell of the virus particle is modified by removal of two proteins and cleavage of another by lysosomal enzymes. This uncoating process activates the viral RNA polymerase and hence initiates the virus replication process.

Replication of the reovirus occurs within an intracellular equivalent of the viral core, called the *subviral particle*, which remains intact in the cell. Each of the 10 capped, single-stranded plus RNAs is assembled into this double-stranded RNA-synthesizing body. The single-stranded plus RNAs serve as templates for the synthesis of complementary minus-strand RNA, eventually yielding progeny double-stranded viral genomic RNA. The latter are encapsidated, and when enough viral capsid proteins are present, mature virions are assembled and released by cell lysis.

 16.10 Concept Check

Reoviruses contain segmented double-stranded RNA genomes. Like negative-strand RNA viruses, reoviruses contain an RNA-dependent RNA polymerase within the virion.

◆ How does reovirus genome replication resemble that of influenza virus, and how does it differ?

16.11 Replication of Double-Stranded DNA Viruses of Animals

A large number of double-stranded DNA viruses of animals are known, among them several viruses that are human pathogens. These include the polyoma and papilloma viruses, the herpesviruses, the pox viruses, and the adenoviruses. The genomes of all of these replicate in the nucleus, except for the pox viruses, which replicate in the cytoplasm. In this and the following sections, we briefly discuss the replication of each of these groups of viruses.

Polyomaviruses: SV40

Some viruses of the polyomavirus family induce tumors in animals; indeed, the suffix *oma* means tumor. One of these DNA tumor viruses was first isolated from monkeys, and it was thus called simian virus 40 or SV40. Virus SV40 was one of the first genetic elements to be studied by genetic engineering techniques and has been extensively used as a *vector* for moving genes into eukaryotic cells (👁 Section 31.4).

The SV40 virion is a nonenveloped particle 45 nm in diameter with an icosahedral head containing 72 protein subunits. Unlike RNA viruses, there are no enzymes in the virion. The genome of SV40 consists of one molecule of double-stranded DNA of 5243 bp. The DNA is circular (Figure 16.19•) and exists in a supercoiled configuration within the virion. The complete base sequence of SV40 has been determined, and a genetic map is shown in Figure 16.20•.

SV40 nucleic acid is replicated in the nucleus, while the proteins are synthesized in the cytoplasm. Final assembly of the virion occurs in the nucleus. The replication of these viruses can be divided into two distinct stages, *early* and *late*. During the early stage, the *early region* of the viral DNA is transcribed (Figure 16.20). A single RNA molecule, the primary transcript, is made by cellular RNA polymerase, but it is processed into *two* mRNAs, a large one and a small one. Introns are present in the SV40 genome, so they are excised out of the primary RNA transcript. In the cytoplasm, the mRNAs are capped and translated to yield two proteins. One of these proteins, the *T-antigen*, binds to the site on the parental DNA that is the origin of replication; this initiates viral genome synthesis.

The genome of SV40 is too small to encode its own DNA polymerase, so host DNA polymerases are used.

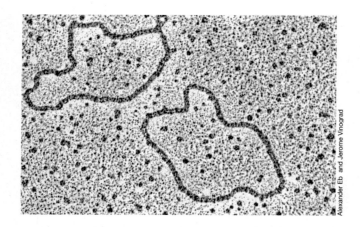

Alexander Eb and Jerome Vinograd

● **Figure 16.19 Polyomaviruses.** Electron micrograph of relaxed (nonsupercoiled) circular DNA from a tumor virus. The contour length of each circle is about 1.5 μm. Polyomavirus DNA is replicated in animal cells using host DNA polymerase.

Replication occurs in a bidirectional fashion (⚬⚬ Figure 7.16) from a single origin of replication. The process involves the same events that have already been described for host cell DNA replication (⚬⚬ Sections 7.5 and 7.6).

Late SV40 mRNA molecules are synthesized using the strand complementary to that used for early mRNA synthesis (see Figure 16.20). Transcription begins at a promoter near the origin of replication. This *late RNA* is then processed by splicing, capping, and polyadenylation to yield mRNA corresponding to the three coat proteins, *VP1*, *VP2*, and *VP3*. Interestingly, the genes for these proteins overlap (Figure 16.20), a phenomenon we have seen many times in this chapter (see Sections 16.1, 16.2, and 16.4). SV40 coat protein mRNAs are then trans-ported to the cytoplasm and translated into the viral coat proteins. The latter are then transported back into the nucleus where virion assembly takes place. Release of new SV40 virions occurs by cell lysis.

Some polyomaviruses cause cancer. When a virus of the polyomavirus group infects a host cell, one of two modes of replication can occur, depending on the type of host cell. In some types of host cells, known as *permissive* cells, virus infection results in the usual formation of new virions and the lysis of the host cell. In other types of host cells, known as *nonpermissive*, efficient multiplication does not occur. Instead, the virus DNA becomes integrated into host DNA, analogous to a prophage (⚬⚬ Section 9.10), in the process genetically altering the cells (Figure 16.21●). Such cells can show loss of growth inhibition and become tumor cells, a process called **transformation** (⚬⚬ Figure 9.24). As in certain tumorigenic retroviruses (⚬⚬ Sections 9.13 and 16.15), expression of specific polyomavirus genes leads cells to the transformed state (Figure 16.21).

16.11 Concept Check

Most double-stranded DNA animal viruses, such as SV40, replicate in the nucleus. SV40 has a tiny genome and employs the strategy of overlapping genes to boost its genetic-coding potential. Some of these viruses cause cancer.

● **Figure 16.21 General scheme of molecular events involved in cell transformation by a polyomavirus such as SV40.** Either all or portions of the viral DNA are incorporated into host cell DNA. The viral genes that encode cell transformation are transcribed and processed to viral mRNA molecules, which are transported to the cytoplasm. Here they are translated to form transforming proteins that encode functions that can convert host cells into tumor cells (⚬⚬ Figure 9.24).

● **Figure 16.20 Genetic map of the polyomavirus SV40.** VP1, VP2, and VP3 are the genes encoding the three proteins that make up the coat of SV40. The arrows show the direction of transcription. Note how genes encoding VP1, VP2, and VP3 overlap.

♦ Why doesn't a virus like SV40 need to carry enzymes in the virion as was discussed with influenza virus?

♦ How can one transcript yield more than one mRNA in SV40?

16.12 Double-Stranded DNA Viruses: Herpesviruses

The herpesviruses are a large group of double-stranded DNA viruses that cause a variety of diseases in humans and animals, including fever blisters (cold sores), venereal herpes, chickenpox, shingles, and infectious mononucleosis; a number of these diseases are discussed in Chapter 26. One of the interesting features of herpesviruses is their ability to remain *latent* in the body for long periods of time, becoming active only under conditions of stress. Both **herpes simplex**, the virus that causes *fever blisters* and *genital herpes*, and **varicella-zoster virus**, the cause of *chickenpox* and *shingles*, are able to remain latent in the neurons of the sensory ganglia, from which they emerge periodically to cause infections of the skin.

An important group of herpesviruses are tumorigenic, causing clinical forms of cancer. For example, **Epstein-Barr virus** causes *Burkitt's lymphoma*, a common tumor among children in Central Africa and New Guinea. Burkitt's lymphoma was among the first human cancers to be linked to virus infection. Epstein-Barr virus can also cause a nonmalignant general malaise in humans called *infectious mononucleosis*.

Herpesviruses: General Features

The *herpesvirus particle* is structurally complex, consisting of four distinct morphologic units. Herpes simplex type I is an enveloped virus about 150 nm in diameter (Figure 16.22*a*●); the center of the virus is called the *core*, consisting of *linear* double-stranded DNA. The herpesvirus nucleocapsid is of icosahedral symmetry and consists of 162 capsomeres, each of which is composed of a number of distinct proteins. Outside the nucleocapsid is an amorphous layer that is called the *tegument*, a fibrous structure unique to the herpesviruses. Surrounding the tegument is an *envelope* whose outer surface contains many small *spikes*.

A large number of separate proteins are present within the virion, but not all of them have been characterized. The genome of herpes simplex type I virus consists of one large linear double-stranded DNA molecule of 152,260 bp (about 30 times larger than the SV40 genome) that encodes at least 84 distinct polypeptides.

Herpesvirus Infection and Replication

Herpesvirus infection occurs following attachment of virions to specific cell receptors. Following fusion of the cytoplasmic membrane with the virus envelope, the nucleocapsid is released into the cell. The nucleocapsids are

● **Figure 16.22 Herpesvirus.** Flow of events in multiplication of herpes simplex virus starting from an electron micrograph of a herpes virion (diameter about 150 nm). DNA replication in herpesviruses is carried out in the cell nucleus by a viral DNA polymerase.

transported to the nucleus, where viral DNA is uncoated. Proteins in the virus particle inhibit macromolecular synthesis by the host.

Following infection, three classes of mRNA are produced: *immediate early*, which encodes five regulatory proteins; *delayed early*, which encodes DNA replication proteins including DNA polymerase; and *late*, which encodes structural proteins of the virus particle (Figure 16.22*b*). During the immediate early stage about one-third of the viral genome is transcribed by a host cell RNA polymerase. Early mRNA encodes certain regulatory proteins that stimulate the synthesis of the delayed early proteins. The delayed early proteins appear only after the immediate early proteins have been made. During this stage, about 40% of the viral genome is transcribed. Among the many key proteins synthesized during the delayed early stage are a viral-specific DNA polymerase, enzymes involved in synthesis of deoxyribonucleotides, and a DNA-binding protein. These enzymes are all involved in viral DNA replication.

Herpesvirus DNA synthesis itself takes place in the nucleus. After infection, the herpesvirus genome circularizes (remarkably, like bacteriophage lambda) (∞ Section 9.11) and replicates by a rolling circle mechanism (∞ Figure 9.20). However, three origins of replication seem to be involved. Long concatemers are formed that become processed into virus-length genomic DNA during the assembly process in a manner similar to that described for DNA bacteriophages (∞ Sections 9.9 and 16.4). Viral nucleocapsids are assembled in the nucleus, and acquisition of the virus envelope occurs via a budding process through the *inner membrane of the nucleus*. Mature virions are subsequently released through the endoplasmic reticulum to the outside of the cell. Thus, the assembly of herpesvirus differs from that of the enveloped RNA viruses, which were assembled on the *cytoplasmic membrane* instead of the nuclear membrane.

Cytomegalovirus

A major and very widespread herpesvirus is **cytomegalovirus** (CMV). CMV replicates as a typical herpesvirus and is present in 50–85% of all adults in the United States by 40 years of age. For healthy individuals, infection with CMV comes with no apparent symptoms or long-term health consequences. However, CMV infection can cause serious disease in immune-compromised individuals, such as those receiving immunosuppressant drugs (for example, organ transplant cases and some cancer and dialysis patients), or those infected with the AIDS virus, HIV. In such individuals, CMV can cause pneumonia, retinitis (an eye condition), and gastrointestinal disease. These conditions can be serious and occasionally cause death of the infected individual.

In the healthy host the immune system keeps CMV in check. However, CMV remains dormant within cells of an infected individual and can become reactivated whenever the immune system is compromised, leading to anywhere from mild to severe symptoms, depending on the degree to which the immune system is kept in check. Like the Epstein-Barr virus, CMV also causes some cases of infectious mononucleosis.

16.12 *Concept Check*

Herpesviruses are large double-stranded DNA viruses. The viral DNA circularizes and is replicated by a rolling circle mechanism. Herpesviruses cause a variety of disease syndromes and can maintain themselves in a latent state in the host indefinitely, initiating viral replication periodically.

◆ Of what use is it to the herpesviruses to circularize their DNA prior to replication?

◆ Where is the nucleocapsid of a herpesvirus assembled and how does this affect the protein content of its envelope?

16.13 Double-Stranded DNA Viruses: Pox Viruses

Pox viruses are among the most complex and largest animal viruses known (Figure 16.23●). These viruses are also unique in that they are DNA viruses that *replicate in the cytoplasm*. Thus, a host cell infected with a pox virus exhibits *DNA synthesis outside the nucleus*, something that otherwise occurs in eukaryotic cells only in organelles.

General Properties of Pox Viruses

Pox viruses have been important medically as well as historically. *Smallpox* was the first virus to be studied in any detail and was the first virus for which a vaccine was developed (by Edward Jenner in 1798). By diligent application of this vaccine on a worldwide basis, the disease smallpox has been *eradicated* in the wild, the first infectious disease to be eliminated in this fashion. Other pox viruses of importance are *cowpox* and *rabbit myxomatosis virus*, an important infectious agent of rabbits and one that was intentionally used in an attempt to control the Australian rabbit population (∞ Section 25.5). Some pox viruses also cause tumors.

The pox viruses are very large, so large in fact, that they can actually be seen under the light microscope. Most research on pox viruses has been done with *vaccinia virus*, a close relative of smallpox virus and the virus used as a smallpox vaccine. The vaccinia virion is a brick-shaped structure about 400 × 240 × 200 nm in dimensions (Figure 16.23). The virion lacks an envelope but is covered on its outer surface with protein tubules arranged in a membrane-like pattern (Figure 16.23). Within the virion there are two lateral bodies of unknown composition and a *nucleocapsid*, which contains DNA bounded by a layer of protein subunits.

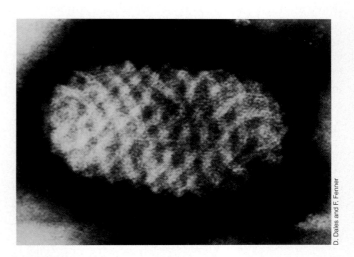

D. Dales and F. Fenner

● **Figure 16.23 Pox viruses.** Electron micrograph of a negatively stained vaccinia virus virion. The virion is approximately 400 nm (0.4 μm) long. Vaccinia virus has been used as both a smallpox vaccine and in genetically engineered vaccines (∞ Section 31.8).

The pox virus genome consists of *linear double-stranded DNA*. The vaccinia virus genome has about 185 kbp and about 180 genes. Pox virus DNA is unique because the two strands of the double helix are cross-linked at their termini as a result of phosphodiester bonds between adjacent strands. The ends of pox virus DNA are therefore very similar to those of the *Chlorella* viruses discussed previously (see Section 16.7).

Replication of Pox Viruses

Vaccinia virions are taken up into cells and the nucleocapsids liberated in the cytoplasm. Interestingly, uncoating of the viral genome requires the activity of a viral protein that is synthesized after infection. This protein is encoded by viral DNA, and the gene encoding it is transcribed by a viral-encoded RNA polymerase *contained within* the virus particle. In addition to this uncoating gene, a number of other viral genes are transcribed. The primary transcripts are turned into mRNAs by capping and polyadenylation while they are still inside the nucleocapsid. Copies of the genome are made by a viral encoded DNA polymerase.

Once the vaccinia DNA is fully uncoated, the formation of *inclusion bodies* within the cytoplasm begins. Within these inclusion bodies, transcription, replication, and encapsidation into progeny virions occur. Each infecting virion initiates its own inclusion body, so the number of inclusions depends on the multiplicity of infection. Progeny DNA molecules form a pool from which individual molecules are incorporated into virions. Mature virions accumulate in the cytoplasm. There seems to be no specific release mechanism, and most virions are released only when the infected cell disintegrates.

Pox Viruses and Recombinant Vaccines

Vaccinia virus has been used as a host for genetically altered proteins of other viruses, permitting the construction of genetically engineered vaccines (⌘ Section 31.8). As we will see in Chapter 22, a vaccine is a substance capable of eliciting an immune response in an animal and serves to protect the animal from future infection with the same agent. Vaccinia virus causes no serious health effects in humans but is highly immunogenic. Therefore, as a carrier of proteins from pathogenic viruses, vaccinia virus is a safe and effective tool for stimulating the immune response.

Molecular cloning methods (⌘ Chapters 10 and 31) have been used to express key viral proteins of various viral pathogens including influenza virus, rabies virus, herpes simplex type I virus, and hepatitis B virus, in vaccinia virions, and then the latter used to develop a vaccine against that pathogen (⌘ Section 31.8). A similar vaccine delivery system using adenovirus (see next section) as a vehicle has been developed because, like vaccinia virus, adenoviruses are also of minor health consequence to humans.

16.14 Double-Stranded DNA Viruses: Adenoviruses

The adenoviruses are a major group of icosahedral *linear double-stranded DNA* viruses. The term *adeno* is derived from the Latin for "gland" and refers to the fact that these viruses were first isolated from the tonsils and adenoid glands of humans. Adenoviruses cause mild respiratory infections in humans, and a number of such viruses can be isolated, even from healthy individuals. The genomes of the adenoviruses consist of linear double-stranded DNA of about 36 kbp. Attached in covalent linkage to the 5' end of the DNA is a protein called the *terminal protein*, essential for replication of the DNA. The DNA also has inverted terminal repeats of 100–1800 bp (the number varies with the virus strain) that are important in the replication process.

Replication of Adenoviruses: Early Events

Unlike poxviruses, replication of adenoviral DNA occurs in the *nucleus* (Figure 16.24●). After the virus particle has been transported to the nucleus, the nucleocapsid is released and converted to a viral DNA-histone complex. Early transcription is carried out by a host RNA polymerase, and a number of primary transcripts are made. The transcripts contain introns and so must first be spliced to yield mature transcripts; the latter are then capped and polyadenylated before translation.

As is typical for many viruses, early adenoviral proteins are involved in regulation of DNA replication while later proteins are structural. Viral DNA replication uses the terminal protein as a primer and another virus-encoded protein as DNA polymerase. The terminal protein contains a covalently bound cytosine residue from which DNA replication proceeds.

Replication of Adenoviruses: Genome Replication

Replication of the adenoviral genome begins at either end, the two strands being replicated asynchronously (Figure 16.24). The products of a round of replication are a double-stranded and a single-stranded molecule (Figure 16.24). However, then a unique replication mechanism proceeds. The single strand cyclizes by means of

Terminal protein

5′
3′
+
−

Plus strand copied

5′
+
−
3′
5′

5′
+
−
−
5′
5′

Direction of cyclization

−
5′

Minus strand cyclizes via inverted terminal repeats

5′
−

Minus strand copied

5′
5′
3′
+
−

Completed linear double strand

5′
+
−
5′

● **Figure 16.24 Replication of adenovirus DNA.** Note how because of loop formation (cyclization), replication occurs *without a lagging strand* being involved; synthesis is leading on both strands.

its inverted terminal repeats, and a new complementary strand is synthesized beginning from the 5′ end (Figure 16.24). This mechanism of replication is noteworthy because it does not involve the formation of a lagging strand as occurs in conventional DNA replication (∞ Section 7.6). Instead, DNA synthesis occurs in leading fashion on both newly synthesized DNA strands (Figure 16.24).

 16.14 Concept Check

Different double-stranded DNA animal viruses have different genome replication strategies. That of the adenoviruses involves protein primers and a mode of replication that avoids the synthesis of a lagging strand and that occurs within the nucleus.

◆ Describe how adenovirus replicates its double-stranded DNA genome without the synthesis of a lagging strand.

16.15 Viruses Using Reverse Transcriptase: Retroviruses and Hepadnaviruses

In Section 9.12 we discussed the **retroviruses**, a group of viruses that replicates through **reverse transcription** using the enzyme *reverse transcriptase*. There are two different types of viruses that use reverse transcriptase, and they are differentiated by the type of nucleic acid in their genomes. The *retroviruses* have *RNA genomes*, while the *hepadnaviruses* have *DNA genomes*. In this section we will deal with examples of both replication patterns.

Retroviruses: General Principles

Recall that retroviruses have enveloped virions that contain *two* copies of the RNA genome (∞ Figure 9.24). The virion also contains several enzymes, including **reverse transcriptase**, and also a specific tRNA. The reason that enzymes are required for retrovirus replication is because although the retroviral genome is of the plus sense and is capped and tailed, it is *not* used directly as messenger RNA. Instead, one of the copies of the genome is converted to DNA by reverse transcriptase and is integrated into the host genome. The DNA that is eventually formed is a linear double-stranded molecule and is synthesized in the cytoplasm within an uncoated viral core particle. An outline of the steps in reverse transcription is given in Figure 16.25●.

Activity of Reverse Transcriptase

Reverse transcriptase is essentially a DNA polymerase, but it actually possesses *three* enzymatic activities: (1) synthesis of DNA from an RNA template (**reverse transcription**), (2) synthesis of DNA from a DNA template, and (3) ribonuclease H activity (an enzymatic activity that degrades the RNA strand of an RNA:DNA hybrid). Like all DNA polymerases, reverse transcriptase needs a primer for DNA synthesis. The primer for retrovirus reverse transcription is a specific *cellular transfer RNA (tRNA)* (Figure 16.25). This is packaged in the virion during its assembly in the previous host cell.

Using the tRNA primer, the 100 or so nucleotides at the 5′ terminus of the RNA are reverse-transcribed into DNA. Once reverse transcription reaches the 5′ end of the RNA, the process stops. To copy the remaining RNA, which is the bulk of the RNA of the virus, a different mechanism comes into play. First, terminally redundant RNA sequences at the 5′ end of the molecule are removed by the ribonuclease H activity of reverse transcriptase. This leads to the formation of a small, single-stranded DNA that is complementary to the RNA segment at the *other end* of the viral

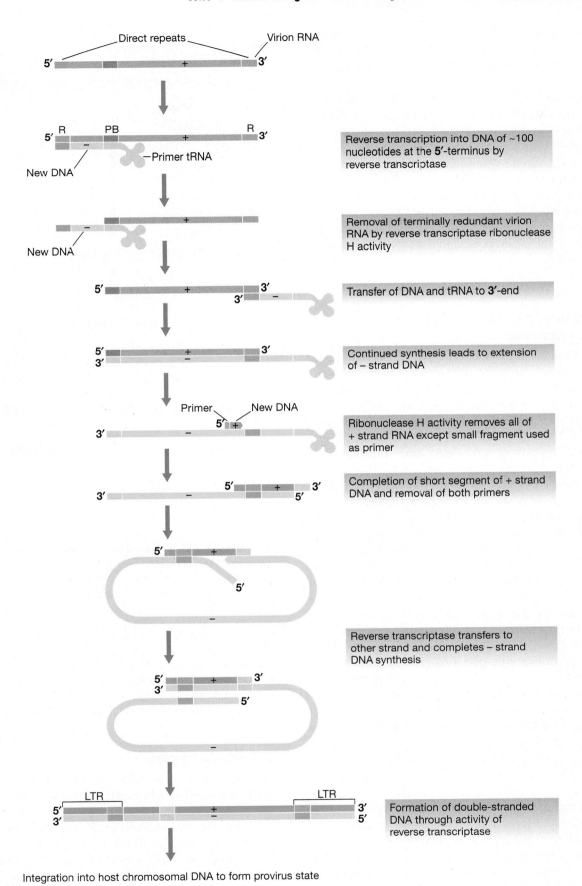

● **Figure 16.25   Overall steps in the formation of double-stranded DNA from retrovirus single-stranded RNA.** The sequences labeled R on the RNA are *direct repeats* found at either end. The sequence labeled PB is where the primer (tRNA) binds. Note that the process of DNA synthesis has yielded longer direct repeats on the DNA than were originally on the RNA. These are called *long terminal repeats (LTRs)*.

RNA. This short, single-stranded piece of DNA then hybridizes with the other end of the viral RNA molecule, where copying of the viral RNA sequences continues.

As summarized in Figure 16.25, continued reverse transcription and ribonuclease H activities leads to the formation of a double-stranded DNA molecule with *long terminal repeats* (*LTRs*) at each end. These LTRs contain strong transcriptional promoters and are also involved in the integration process. The *integration* of the viral DNA into the host genome is analogous to the integration of virus Mu DNA (see Section 16.5). Integration can occur anywhere in the cellular DNA, and once integrated, the retroviral DNA, now called a *provirus*, is a stable genetic element (Figure 16.25).

Retroviral Gene Expression, Processing, and Virion Assembly

As a provirus, the retroviral genome may be expressed, or it may remain in a latent state and not be expressed. If the promoters in the right LTR are activated, the integrated proviral DNA is transcribed by a cellular RNA polymerase into transcripts that are capped and polyadenylated. These RNA transcripts may be either encapsidated into virions or processed and translated into virus proteins.

The translation and processing of such an mRNA from a retrovirus is shown in Figure 16.26●. All retroviruses have at least the three genes *gag*, *pol*, and *env*, arranged in that order in the genome. The *gag* "gene" at the 5' end of the mRNA actually encodes several small viral structural proteins. These are first synthesized as a *polyprotein* that is subsequently processed by a protease (which itself is a part of the polyprotein). The structural proteins make up the capsid, and the protease becomes packaged in the virion.

The *pol* gene is translated next, into a large polyprotein containing the *gag* proteins (Figure 16.26). Compared to structural proteins, the *pol* products are required in only small amounts, and conveniently, this occurs because their synthesis requires that the ribosome make an error. To produce *pol* gene products the ribosome must either read through a stop codon at the end of the *gag* gene or make a precise switch to a different reading frame in this region. These are rather rare events.

Once produced, the *pol* gene product is processed in two ways. First, it is removed from the *gag* proteins, and second, reverse transcriptase is cleaved from integrase, a protein required for DNA integration which is packaged in the virion along with reverse transcriptase (Figure 16.26). For the *env* gene to be translated, the full-length mRNA is first processed to remove the *gag* and *pol* regions. The *env* product is made and then processed into two distinct envelope proteins (Figure 16.26).

(a)

(b)

● **Figure 16.26 Translation of retrovirus mRNA and processing of the proteins.** (a) The full-length mRNA with the three genes *gag, pol,* and *env* is shown at the top. The asterisk shows the site where a ribosome must read through a stop codon or do a precise shift of reading frame to synthesize the GAG-POL polyprotein. The thick blue arrows indicate translation, while the thin blue arrows indicate protein-processing events. One of the *gag* gene products is a protease. The POL product is processed to give reverse transcriptase (RT) and integrase (IN). (b) The mRNA has been processed to remove most of the *gag-pol* region. This shortened message is translated to give the ENV polyprotein, which is cleaved into two envelope proteins (EP), EP1 and EP2.

HIV as a Retrovirus

Although the replication pattern of retroviruses described above is admittedly complex, this is actually the pattern typical of a rather "simple" retrovirus. By contrast, the genome of the retrovirus human immunodeficiency virus 1 (HIV-1), the causative agent of AIDS, is even more involved than this and includes several additional small genes. Its expression not only requires extensive protein processing by proteases but also complex patterns of alternative splicing of introns. However, a hallmark of retroviruses is the *many protease steps* involved in forming mature proteins. In this regard, specific protease inhibitors have been developed for use in treating retroviral infections, including HIV-1 infections (⚯ Sections 20.10 and 26.14).

DNA Reverse-Transcribing Viruses: Hepadnaviruses

We have seen that the life cycles of viruses show a variety of unexpected genome structures and replication schemes. But none could be more unusual than the **hepadnaviruses**, such as *human hepatitis B virus*, a serious blood-borne pathogen (⚯ Section 26.11). The term "hepadnavirus" comes from the fact that the virus infects the liver (thus "hepa") and the genome consists of DNA (thus "dna"). The virions of hepatitis B virus are small, irregular, rod-shaped particles (Figure 16.27*a*●).

The genomes of hepadnaviruses are among the smallest known of any viruses, 3–4 kb, while the virus life cycle is unique and very complex. Like the retroviruses, hepadnaviruses use reverse transcriptase in their replication cycle. However, unlike retroviruses, the *DNA* genome of hepadnaviruses is replicated through an *RNA* intermediate, the opposite of what occurs in retroviruses.

The genomes of hepadnaviruses are only *partially* double-stranded DNA (Figure 16.27*b*). One strand is incomplete, and both strands have gaps. Nevertheless, the two strands are held together in a circular form by hydrogen bonding between complementary base pairs. On entering the cytoplasm a viral DNA polymerase carried in the virion completes the replication of this molecule. This polymerase is an extremely versatile protein. It contains DNA polymerase *plus* reverse transcriptase activities and also functions as a protein primer for synthesis of one of the DNA strands! Despite the small size of the genome, it encodes several proteins, as the strategy of overlapping genes used extensively by small viruses (see Sections 16.1, 16.2, and 16.4) appears once again in the biology of hepadnaviruses (Figure 16.27*b*).

Replication of the hepadnavirus genome involves transcription by host RNA polymerase (in the nucleus), yielding a transcript with terminal repeats. (Figure 16.27*b*). The repeats occur because the polymerase proceeds slightly more than once around the circular molecule. The viral reverse transcriptase then copies this into DNA, very much like the replication of retroviruses, but in this case the *DNA* (rather than the *RNA* of retroviruses) becomes packaged into newly formed virions (Figure 16.27*b*).

Hepadnaviruses are thus incredible examples of making the most of a small genome. The hepatitis B virus, for example, at 3.4 kb has a smaller genome than either the single-stranded RNA or DNA bacteriophages we explored at the beginning of this chapter (see Sections 16.1–16.3). Nevertheless, this small virus can cause a serious human disease, and we examine the hepatitis syndromes in a later chapter (⚯ Section 26.11).

(b)

● Figure 16.27   Hepadnaviruses. (a) Electron micrograph of hepatitis B virions. (b) Hepatitis B genome. The partially double-stranded genome is shown in green. Note that the positive strand is incomplete. The sizes of the transcripts are also shown. All of the genes in the hepatitis B virus overlap and collectively, they cover every base in the genome. Reverse transcriptase produces the DNA genome from a single genome-length mRNA made by host RNA polymerase.

🛑 16.15   Concept Check

The retroviruses contain RNA genomes and use reverse transcriptase to make a DNA copy during their life cycle. The hepadnaviruses contain DNA genomes and use reverse transcriptase to make genomic DNA from an RNA copy. Both types of viruses have complex patterns of gene expression.

◆ Why might protease inhibitors be an effective treatment for human AIDS?

◆ Describe the different role played by reverse transcriptase in the replication cycle of retroviruses and hepadnaviruses.

REVIEW QUESTIONS

1. Describe the types of genomes found in viruses. Give an example of at least one virus (or group of viruses) with each type of genome (entire chapter).

2. What are *overlapping* genes? Give examples of viruses that have overlapping genes (Sections 16.1, 16.2, 16.4, 16.11, and 16.15).

3. Many bacteriophages have single-stranded DNA genomes. Describe how the genes carried by these viruses can be transcribed and translated (Sections 16.2 and 16.3).

4. Positive-strand RNA viruses are known for *Bacteria*, for plants, and for animals. What are the important distinctions between gene expression in these viruses, particularly between those that infect *Bacteria* and those that infect eukaryotes (Sections 16.1 and 16.8)?

5. One end of the T7 genome always enters the cell first during infection. This is necessary for the infection process to proceed. Explain (Section 16.4).

6. Why is bacteriophage Mu mutagenic? What features are necessary for Mu to insert into DNA (Section 16.5)?

7. What is unusual about the genome of archaeal viruses like PAV-1 compared to the genomes of double-stranded DNA viruses of bacteriophages (Section 16.6)?

8. What type of genome does tobacco mosaic virus contain? How does it compare in terms of structure and size to that of *Chlorella*-like viruses (Section 16.7)?

9. What is the function of the VPg protein of poliovirus, and how, since its genome is also plus-sense RNA, can coronaviruses replicate without a VPg protein (Section 16.8)?

10. The rhabdoviruses have an unusual genome structure. What is the nature of their genome and why *must* the virion contain an enzyme involved in its replication (Section 16.9)?

11. The reovirus genome is unique in all of biology. Explain (Section 16.10).

12. Although animal viruses are usually not referred to as *temperate*, as are some bacteriophages (ᗧᐤ Section 9.10), several can set up latent infections. Describe any two such types of animal viruses (Section 16.11 and 16.15).

13. Although the genomes of herpesviruses and adenoviruses both contain double-stranded DNA, their replication features are quite distinct. Elaborate (Sections 16.12 and 16.14).

14. Of all the double-stranded DNA animal viruses, pox viruses stand out concerning one unique aspect of their DNA replication process. What is this unique aspect and how can this be accomplished without special enzymes being packaged in the virion (Section 16.13)?

15. What is there about expression of retroviral genes that makes virus replication sensitive to protease inhibitors (Section 16.15)?

APPLICATION QUESTIONS

1. Not all proteins are made from the RNA genome of bacteriophage MS2 in the same amounts. Can you explain why? One of the proteins functions very much like a repressor (ᗧᐤ Section 8.5), but it functions at the translational level. Which protein is it and how does it function?

2. Give a mechanistic explanation of why the genomes of double-stranded RNA viruses might be segmented.

3. The mechanism of replication of both strands of DNA in some viruses, such as adenoviruses, is continuous (leading). Show how this can be without violating the "rule" learned in Chapter 7 that all DNA synthesis occurs in the overall direction of $5' \rightarrow 3'$.

4. Most genetic elements that express reverse transcriptase make it in small amounts and/or as part of a polyprotein. Can you think of any reason(s) why this might be so?

17

METABOLIC DIVERSITY

The enormous genetic diversity of prokaryotes is mirrored by an equally enormous metabolic diversity. Green sulfur bacteria are phototrophs that contain chlorosomes, shown here as the cigar-shaped structures in the periphery of a cell of *Chlorobium tepidum*. Chlorosomes allow green bacteria to grow at the lowest light intensities of any phototrophic organisms.

WORKING GLOSSARY

Anaerobic respiration respiration in which some substance, such as SO_4^{2-} or NO_3^-, is used as a terminal electron acceptor instead of O_2

Anammox anoxic ammonia oxidation

Anoxic oxygen-free

Anoxygenic photosynthesis photosynthesis in which O_2 is not produced

Antenna in reference to photocomplexes, light harvesting pigment molecules that funnel energy to the reaction center

Autotroph an organism that can use CO_2 as sole carbon source

Bacteriochlorophyll the chlorophyll pigment of anoxygenic phototrophs

Calvin cycle the biochemical route of CO_2 fixation in many autotrophic organisms

Carboxysomes crystalline inclusions of RubisCo

Carotenoids a hydrophobic accessory pigment present along with chlorophyll in photosynthetic membranes

Chemolithotroph a microorganism capable of oxidizing inorganic compounds as energy sources

Chlorophyll the light-sensitive, Mg-containing porphyrin of photosynthetic organisms that initiates the process of photophosphorylation

Chlorosome cigar-shaped structures present in the periphery of cells of green sulfur and green nonsulfur bacteria and that contain the antenna bacteriochlorophylls (c, d, or e)

Denitrification anaerobic respiration in which NO_3^- is reduced to gaseous nitrogen compounds, primarily N_2

Disproportionation splitting of a compound into two new compounds, one more oxidized and one more reduced than the original compound

Fermentation anaerobic catabolism of an organic compound in which the compound serves as both an electron donor and an electron acceptor and in which ATP is produced by substrate-level phosphorylation

Homoacetogenesis (acetogenesis) energy metabolism involving the production of acetate from either H_2 plus CO_2 or from organic compounds

Hydrogenase an enzyme, widely distributed in anaerobic microorganisms, capable of taking up or evolving H_2

Hydroxypropionate pathway an autotrophic pathway found in *Chloroflexus* and a few *Archaea*

Methanogenesis the biological production of methane (CH_4)

Methanotroph an organism that can oxidize methane

Methylotroph an organism capable of growth on compounds containing no C—C bonds

Mixotrophic a nutritional state in which an inorganic compound serves as energy source (electron donor) and organic compounds serve as carbon source

Monooxygenase an enzyme that catalyzes the incorporation of one atom of O_2 into a substrate while the other atom is reduced to H_2O

Nitrification the microbial conversion of NH_3 to NO_3^-

Nitrogenase an enzyme capable of reducing N_2 to NH_3 in the process of nitrogen fixation

Nitrogen fixation the biological reduction of N_2 to NH_3 by nitrogenase

Oxygenic photosynthesis photosynthesis carried out by cyanobacteria and green plants in which O_2 is evolved

Photophosphorylation the production of ATP in photosynthesis

Photosynthesis the series of reactions in which ATP is synthesized by light-driven reactions and CO_2 is fixed into cell material

Phototroph an organism that can use light as an energy source

Phycobiliprotein the accessory pigment complex in cyanobacteria that contains a molecule of phycocyanin or phycoerythrin coupled to proteins

Reaction center a photosynthetic complex containing chlorophyll (or bacteriochlorophyll) and several other components within which occurs the initial electron transfer reactions of photosynthetic electron flow

Reductive dechlorination an anaerobic respiration where a chlorinated organic compound is used as an electron acceptor, usually with release of Cl^-

Reverse electron transport the energy-dependent movement of electrons against the thermodynamic gradient to form a strong reductant from a weaker electron donor

RubisCO the acronym for ribulose bisphosphate carboxylase, a key enzyme of the Calvin cycle

Syntrophy a process whereby two or more microorganisms cooperate to degrade a substance neither can degrade alone

The preceding chapters considered the great *phylogenetic* diversity of microbial life on Earth. Now we focus on their *metabolic* diversity, with special emphasis on the biochemical processes behind this diversity. We can then study microbial ecology—the interactions of microorganisms with each other and their environments. Metabolic diversity and microbial ecology go hand in hand; the metabolic diversity to be described in this chapter fuels the nutrient cycles and other microbial activities that we will discuss in Chapters 18 and 19. Indeed, as we will see, microorganisms and their metabolic reactions play key roles in maintaining the web of life on Earth and are of crucial importance in agriculture and other aspects of human affairs (∞ Section 1.4).

I THE PHOTOTROPHIC WAY OF LIFE

Phototrophy, the use of *light* as an energy source, is widespread in the microbial world. In the first five sections we consider the major forms of phototrophy, including that which forms the very oxygen we breathe. We then proceed to see how most phototrophs satisfy all of their carbon needs from CO_2.

17.1 Photosynthesis

The most important biological process on Earth is **photosynthesis**, the conversion of light energy to chemical energy. Organisms that can carry out photosynthesis are called **phototrophs** (Figure 17.1●). Most phototrophic organisms are also **autotrophs**, capable of growing with CO_2 as sole carbon source. Energy from light is used in the reduction of CO_2 to organic compounds (*photoautotrophy*). However, some phototrophs can use organic carbon as their carbon source; this lifestyle is called *photoheterotrophy* (Figure 17.1).

The ability to photosynthesize requires light-sensitive pigments, the *chlorophylls*, found in plants, algae, and several groups of prokaryotes. Light reaches pho-

```
                    PHOTOTROPHS
                    (use light
                    as energy source)
            ┌──────────────────┴──────────────────┐
     Photoautotrophs                      Photoheterotrophs
     (Use CO₂)                            (Use Organic Carbon)
```

● **Figure 17.1 Classification of phototrophic organisms in terms of energy and carbon sources.** In nature, phototrophic organisms switch their mode of metabolism depending on the resources available in their habitat. Photoheterotrophy typically represses photoautotrophy.

totrophic organisms in distinct units of energy called *quanta*. Absorption of light quanta by chlorophylls begins the process of photosynthetic energy conversion, and the net result is ATP.

Energy Production and CO_2 Assimilation

The growth of a photoautotroph can be characterized by two distinct sets of reactions: (1) ATP production, and (2) CO_2 reduction to organic compounds. For autotrophic growth, energy is supplied from ATP, while electrons for the reduction of CO_2 come from NADH (or NADPH). These are produced by the reduction of NAD^+ (or $NADP^+$) by electrons originating from various electron donors to be discussed later.

To drive autotrophic reactions, some phototrophic bacteria obtain reducing power from electron donors in their environment, typically reduced sulfur sources (H_2S, S^0, $S_2O_3^{2-}$) or H_2. By contrast, green plants, algae, and cyanobacteria use H_2O, an inherently poor electron donor (∞ Figure 5.9), as a source of reducing power to reduce $NADP^+$ to NADPH. The oxidation of H_2O produces molecular oxygen (O_2) as a by-product (Figure 17.2●). Because oxygen is produced, photosyn-

(a)

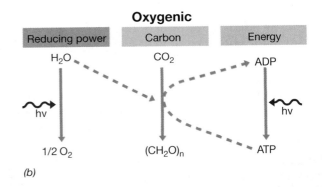

(b)

● **Figure 17.2 Patterns of photosynthesis.** Energy and reducing power synthesis in (a) anoxygenic versus (b) oxygenic phototrophs. Although both types of phototrophs obtain their energy from light (hv), in oxygenic phototrophs light also drives the oxidation of water to oxygen.

thesis in these organisms is called **oxygenic**. In many phototrophic bacteria oxygen is not produced, and thus the process is called **anoxygenic photosynthesis**. The production of NADH from substances like H_2S by anoxygenic phototrophs may or may not be directly driven by light, depending on the organism. By contrast, the oxidation of H_2O to O_2 by oxygenic phototrophs is always driven by light (Figure 17.2). Oxygenic phototrophs thus require light for *both* reducing power and energy conservation.

 17.1 *Concept Check*

Phototrophs obtain energy from light. Photosynthesis involves reactions in which ATP is generated and reactions in which ATP is consumed in the reduction of CO_2 by NADH.

◆ How are *photoautotrophs* and *photoheterotrophs* similar and how do they differ?

◆ What is the fundamental difference between an *oxygenic* and an *anoxygenic* phototroph?

17.2 Photosynthetic Pigments and Their Location Within the Cell

Photosynthesis occurs only in organisms that possess some type of **chlorophyll** (oxygenic phototrophs) or **bacteriochlorophyll** (anoxygenic phototrophs). Chlorophyll is a porphyrin, as are the cytochromes (⊃ Section 5.11). But unlike the cytochromes, chlorophyll contains a *magnesium* atom instead of an *iron* atom at the center of the porphyrin ring. Chlorophyll also contains specific substituents bonded to the porphyrin ring, as well as a hydrophobic alcohol side chain that enables chlorophyll to associate with lipid and hydrophobic proteins of photosynthetic membranes.

The structure of chlorophyll *a*, the principal chlorophyll of higher plants, most algae, and the cyanobacteria, is shown in Figure 17.3*a*. Chlorophyll *a* is green because it *absorbs* red and blue light preferentially and *transmits* green light. We mentioned the electromagnetic spectrum in Section 10.4. The spectral properties of any pigment can best be expressed by its **absorption spectrum,** a measure of the degree to which the pigment absorbs light of different wavelengths. The absorption spectrum of cells containing chlorophyll *a* shows strong absorption of red light (maximum absorption at a wavelength of 680 nm) and blue light (maximum at 430 nm) (Figure 17.3*b*).

Chlorophyll Diversity

There are a number of chemically different chlorophylls and bacteriochlorophylls (bacteriochlorophyll is the form of chlorophyll present in anoxygenic phototrophs) that are

(a)

(b)

• **Figure 17.3** **Structures and spectra of chlorophyll *a* and bacteriochlorophyll *a*.** (a) The two molecules are identical except for those portions contrasted in yellow and green. (b) Absorption spectra of (green curve) cells of the green alga *Chlamydomonas* (⊃ Section 14.13). The peaks at 680 and 430 nm are due to chlorophyll *a*; the peak at 480 nm is due to carotenoids; (red curve) cells of the phototrophic purple bacterium *Rhodopseudomonas palustris* (⊃ Section 12.2). Peaks at 870, 805, 590, and 360 nm are due to bacteriochlorophyll *a* while peaks at 525 and 475 nm are due to carotenoids.

distinguished by their different absorption spectra. Chlorophyll *b*, for instance, absorbs maximally at 660 nm rather than at 680 nm. Many plants have more than one chlorophyll, but the most common are chlorophylls *a* and *b*.

Among prokaryotes, cyanobacteria have chlorophyll *a*, but anoxygenic phototrophs, such as the purple and green bacteria, can have any of a number of bacteriochlorophylls (Figures 17.3 and 17.4●). Bacteriochlorophyll *a* (Figure 17.3), present in most purple bacteria (∞ Section 12.2), absorbs maximally between 800 and 925 nm, depending on the species of phototroph. Different species have different pigment-binding proteins, and the absorption maxima of bacteriochlorophyll *a* depends to some degree on the nature of these proteins and how they are arranged in photocomplexes. Other bacteriochlorophylls, whose distribution runs along phylogenetic lines, absorb in other regions of the visible and infrared spectrum (Figure 17.4).

Why do different organisms have different kinds of chlorophylls absorbing light at different wavelengths? This is likely a strategy to make better use of the energy of the electromagnetic spectrum. Only light energy that is *absorbed* is useful for energy conservation. By having different pigments, two organisms can coexist in a habitat, each using wavelengths of light that the other is not using. Thus, pigment diversity has ecological significance.

Photosynthetic Membranes and Chloroplasts

The chlorophyll pigments and all the other components of the light-gathering apparatus are found inside the cell within special membrane systems, the **photosynthetic membranes**. The location of the photosynthetic membranes differs between prokaryotic and eukaryotic microorganisms. In eukaryotes, photosynthesis is associated with special intracellular organelles, the **chloroplasts** (Figure 17.5*a*●;

Pigment/Absorption maxima (*in vivo*)	R₁	R₂	R₃	R₄	R₅	R₆	R₇
Bchl *a* (purple bacteria)/ 805, 830–890	—C(=O)—CH₃	—CH₃[b]	—CH₂—CH₃	—CH₃	—C(=O)—O—CH₃	P/Gg[a]—H	
Bchl *b* (purple bacteria)/ 835–850, 1020–1040	—C(=O)—CH₃	—CH₃[c]	=C(—CH₃)(H)	—CH₃	—C(=O)—O—CH₃	P	—H
Bchl *c* (green sulfur bacteria)/745–755	—C(H)(OH)—CH₃	—CH₃	—C₂H₅ / —C₃H₇[d] / —C₄H₉	—C₂H₅ / —CH₃	—H	F	—CH₃
Bchl *cₛ* (green nonsulfur bacteria)/740	—C(H)(OH)—CH₃	—CH₃	—C₂H₅	—CH₃	—H	S	—CH₃
Bchl *d* (green sulfur bacteria)/705–740	—C(H)(OH)—CH₃	—CH₃	—C₂H₅ / —C₃H₇ / —C₄H₉	—C₂H₅ / —CH₃	—H	F	—H
Bchl *e* (green sulfur bacteria)/719–726	—C(H)(OH)—CH₃	—C(H)(=O)	—C₂H₅ / —C₃H₇ / —C₄H₉	—C₂H₅	—H	F	—CH₃
Bchl *g* (heliobacteria)/ 670, 788	—C(H)=CH₂	—CH₃[b]	—C₂H₅	—CH₃	—C(=O)—O—CH₃	F	—H

[a]P, Phytyl ester (C₂₀H₃₉O—); F, farnesyl ester (C₁₅H₂₅O—); Gg, geranylgeraniol ester (C₁₀H₁₇O—); S, stearyl alcohol (C₁₈H₃₇O—).

[b]No double bond between C₃ and C₄; additional H atoms are in positions C₃ and C₄.

[c]No double bond between C₃ and C₄; an additional H atom is in position C₃.

[d]Bacteriochlorophylls *c*, *d*, and *e* consist of isomeric mixtures with the different substituents on R₃ as shown.

● **Figure 17.4 Structure of all known bacteriochlorophylls (Bchl).** The different substituents present in the positions R₁ to R₇ in the accompanying structure are listed. Absorption properties can be determined by suspending intact cells in a viscous liquid such as 60% sucrose (this reduces light scattering and smoothes out spectra) and running absorption spectra as shown in Figure 17.3. *In vivo* absorption maxima are the *physiologically relevant* absorption peaks. The spectrum of bacteriochlorophylls dissolved in organic solvents is often quite different.

Outer membrane

Inner membrane

Stroma

Thylakoid membrane

Stacked thylakoids forming grana

Hilda Canter-Lund

(a)

(b)

● **Figure 17.5 The chloroplast.** (a) Photomicrograph of an algal cell showing chloroplasts. (b) Details of chloroplast structure, showing how the convolutions of the thylakoid membranes define an inner space called the stroma and form membrane stacks called grana.

Section 14.3). The chlorophyll pigments are attached to sheetlike (lamellar) membrane structures of the chloroplast (Figure 17.5*b*). These photosynthetic membrane systems are called **thylakoids**; stacks of thylakoids are called *grana* (Figure 17.5*b*). The thylakoids are so arranged that the chloroplast is divided into two regions, the *matrix space* that surrounds the thylakoids, and the *inner space* within the thylakoid array (Figure 17.5*b*). This arrangement makes possible the development of a light-driven proton motive force that can be used to synthesize ATP, as will be described in Section 17.5.

In prokaryotes, chloroplasts are absent. Here photosynthetic pigments are integrated into internal membrane systems. These systems arise (1) from invagination of the cytoplasmic membrane (purple bacteria) (for example, Figure 12.3; see also Figure 17.12), (2) from the cytoplasmic membrane itself (heliobacteria; Section 12.20), (3) in both the cytoplasmic membrane and specialized non-unit membrane-enclosed structures called *chlorosomes* (green bacteria; see Figure 17.7 and Section 12.32), or (4) in thylakoid membranes (cyanobacteria, Section 12.25).

Reaction Centers and Antenna Pigments

Within a photosynthetic membrane chlorophyll or bacteriochlorophyll molecules are associated with proteins to form *complexes* consisting of anywhere from 50 to 300 molecules (Figure 17.6●). Only a very small number of these pigment molecules, called **reaction centers**, participate directly in the conversion of light energy to ATP (Figure 17.6). Reaction center chlorophylls or bacteriochlorophylls are surrounded by the more numerous **light-harvesting** or **antenna** chlorophylls or bacteriochlorophylls. The antenna pigments function to harvest light and funnel the energy of light to the reaction center (Figure 17.6). At the low light intensities that often prevail in nature, this arrangement of pigment molecules allows for the capture and utilization of photons that would otherwise be insufficient to drive reaction center photochemistry by themselves.

The ultimate in low-light efficiency is found in the **chlorosome** of green sulfur bacteria and *Chloroflexus* (Figure 17.7●). This structure is a giant antenna system. But unlike the antenna of other phototrophs, bacteriochlorophyll molecules in the chlorosome are not bound to proteins and instead function much like a solid state circuit. Chlorosome bacteriochlorophylls (bacteriochlorophylls *c, d,* and *e,* see Figure 17.4) absorb light and transfer the energy to bacteriochlorophyll *a* located in the reaction center in the cytoplasmic membrane (Figure 17.7). This arrangement is highly efficient for absorbing light at low

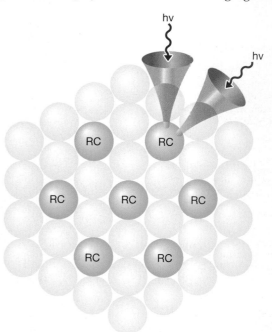

● **Figure 17.6 Model for the arrangement of light-harvesting chlorophylls/bacteriochlorophylls versus reaction centers within a photosynthetic membrane.** Light energy, absorbed by light-harvesting molecules (light green), is transferred to the reaction centers (dark green, RC) where photosynthetic electron transport reactions begin. All pigment molecules are held in place within the membrane by specific pigment-binding proteins. Compare this figure to Figures 17.13 and 17.15.

(a)

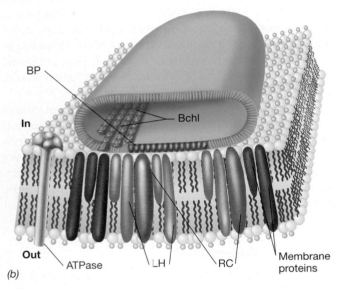

(b)

● **Figure 17.7** **The chlorosome of green sulfur and green non-sulfur bacteria.** (a) Electron micrograph of a cell of the green sulfur bacterium *Chlorobium tepidum*. Note the chlorosomes (arrows). (b) Model of chlorosome structure. The chlorosome (green) lies appressed to the inside surface of the cytoplasmic membrane. Antenna bacteriochlorophyll (Bchl) molecules (Bchls *c*, *d*, or *e*) are arranged in tubelike arrays inside the chlorosome, and energy is transferred from these bacteriochlorophylls through light-harvesting Bchl *a* molecules (LH) to reaction center (RC) Bchl *a* in the cytoplasmic membrane (blue). Base plate (BP) proteins function as connectors between the chlorosome and the cytoplasmic membrane.

intensities. In this regard, it has been shown that green sulfur bacteria can grow at the lowest light intensities of all known phototrophs.

17.2 Concept Check

The central pigment of photosynthesis is chlorophyll (or bacteriochlorophyll). Chlorophylls are located in photosynthetic membranes where the light reactions of photosynthesis are carried out. Antenna chlorophyll molecules harvest light energy and transfer it on to reaction center chlorophylls.

◆ Considering their function, why is it necessary for chlorophyll pigments to be located in membranes?

◆ What is the difference between the numbers of *antenna* and *reaction center* chlorophyll molecules in a photosynthetic complex, and why?

◆ What pigments are found within the *chlorosome*?

17.3 Carotenoids and Phycobilins

Although chlorophyll or bacteriochlorophyll is obligatory for photosynthesis, phototrophic organisms contain various accessory pigments involved in the capture and processing of light energy. These include the *carotenoids* and the *phycobilins*. These pigments primarily play a photoprotective role (carotenoids) or function in light-harvesting (phycobilins).

Carotenoids

The most widespread accessory pigments are the **carotenoids**, which are always found in phototrophic organisms. Carotenoids are hydrophobic pigments that are firmly embedded in the membrane. Figure 17.8● shows the structure of a typical carotenoid, *β-carotene*. Carotenoids have long hydrocarbon chains with alternating C—C and C=C bonds, an arrangement called a *conjugated* double-bond system. Carotenoids are typically yellow, red, brown, or green in color (∞ Figure 12.2) and absorb light in the blue region of the spectrum (see Figure 17.3). The types and structures of carotenoids of various phototrophs have been well studied, and the major carotenoids of anoxygenic phototrophs are shown in Figure 17.9●. These pigments are responsible for the brilliant colors of red, purple, pink, green, yellow, or brown that are observed in different species of anoxygenic phototrophs (∞ Figures 12.2 and 12.5).

Carotenoids are closely associated with chlorophyll or bacteriochlorophyll in photosynthetic pigment complexes but do not function directly in ATP synthesis. They can, however, *transfer* energy to the reaction center, and this transferred energy may be used to make ATP in the same way as light energy captured directly by chlorophyll. Carotenoids also function as photoprotective agents. Bright light can be harmful to cells in that it catalyzes photooxidation reactions that can lead to the production of toxic forms of oxygen, such as singlet oxygen ($^1O_2$) (∞ Section 6.16). The latter can oxidize and thereby damage components of the photosynthetic apparatus itself. Carotenoids quench toxic oxygen species and absorb much of this harmful light. Because

● **Figure 17.8** **Structure of *β*-carotene, a typical carotenoid.** The conjugated double-bond system is highlighted in orange.

I. Aliphatic carotenoids

II. Aryl carotenoids

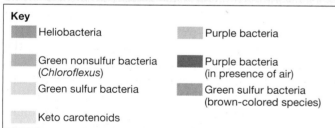

Key

Heliobacteria	Purple bacteria
Green nonsulfur bacteria (*Chloroflexus*)	Purple bacteria (in presence of air)
Green sulfur bacteria	Green sulfur bacteria (brown-colored species)
Keto carotenoids	

● **Figure 17.9 Structures of some common carotenoids found in anoxygenic phototrophs.** Compare the structure of β-carotene shown in Figure 17.8 with how it is drawn here. For simplicity, in the structures shown here, methyl (CH_3) groups are designated by their bonds only. Aryl carotenoids are distinguished from aliphatic carotenoids in that aryl carotenoids contain an aromatic ring on one end. Purple bacteria are discussed in Section 12.2, heliobacteria in Section 12.20, green sulfur bacteria in Section 12.32, and green nonsulfur bacteria in Section 12.35.

phototrophic organisms must by their very nature live in the light, the photoprotective role of carotenoids is thus an obvious advantage.

Phycobiliproteins and Phycobilisomes

Cyanobacteria and red algal chloroplasts contain **phycobiliproteins**. These are the main light-harvesting (antenna)

pigments of these organisms. Phycobiliproteins are red or blue pigments that consist of open-chain tetrapyrroles coupled to proteins (Figure 17.10*a*●). The red pigment, called *phycoerythrin*, absorbs light most strongly at wavelengths around 550 nm, whereas the blue pigment, *phycocyanin* (Figure 17.10*a*), absorbs most strongly at 620 nm (Figure 17.11●). A third pigment, called *allophycocyanin* absorbs at about 650 nm.

Phycobiliproteins occur as high-molecular-weight aggregates, called **phycobilisomes,** attached to the photosynthetic membranes (Figure 17.10*b, c*). Phycobilisomes are constructed such that the allophycocyanin molecules make physical contact with the photosynthetic membrane. Allophycocyanin is surrounded by molecules of phycocyanin or phycoerythrin (or both, depending on the organism). The latter pigments absorb shorter (higher energy) wavelengths of light and transfer the energy to allophycocyanin, which is closely linked to the reaction center chlorophyll and transfers energy to this site. Phycobilisomes thus show very efficient energy transfer from the biliprotein complex to chlorophyll *a*, which allows for growth of cyanobacteria at fairly low light intensities. Indeed, phycobilisome content *increases* in cells of cyanobacteria as light intensity *decreases.*

The light-gathering function of accessory pigments like carotenoids and phycobiliproteins is an obvious advantage for the organism. Light from the sun is distributed over the whole visible range (⚬⚬ Figure 10.6), yet chlorophylls absorb well in only part of this spectrum. Accessory pigments allow the organism to capture more of the available light (Figures 17.3 and 17.11).

🛑 *17.3 Concept Check*

Accessory pigments such as carotenoids and phycobilins absorb light and transfer the energy to reaction center chlorophyll, thus broadening the wavelengths of light usable in photosynthesis. Carotenoids also play an important photoprotective role in preventing photooxidative damage to cells.

◆ In what organisms are phycobiliproteins found?

◆ How does the structure of a phycobilin compare with that of a chlorophyll? With a carotenoid?

◆ Phycocyanin is blue-green in color; what wavelengths of light is it absorbing (⚬⚬ Figure 10.6)?

17.4 Anoxygenic Photosynthesis

The process of light-mediated ATP synthesis in phototrophic organisms involves electron transport through a series of electron carriers. These carriers are arranged in the photosynthetic complex in series from those with electronegative to those with more electropositive reduction potentials (E_0', ⚬⚬ Section 5.6). As we will see, this allows for ATP synthesis from a proton motive force.

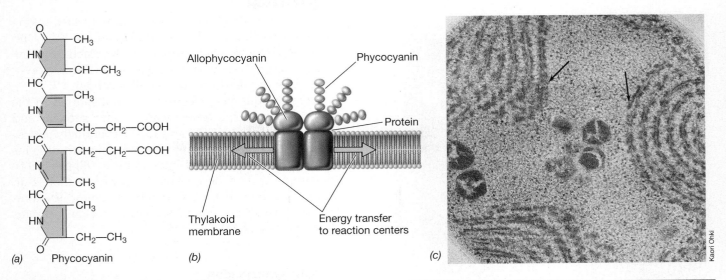

● **Figure 17.10 Phycobiliproteins and phycobilisomes.** (a) A typical phycobilin. This compound is an open–chain tetrapyrrole derived biosynthetically from a closed porphyrin ring by loss of one carbon atom as carbon monoxide. The structure shown is the prosthetic group of phycocyanin, found in cyanobacteria (◯◯ Section 12.25) and red algae (◯◯ Section 14.13). (b) Structure of a phycobilisome. Phycocyanin absorbs at higher energies (shorter wavelengths) than allophycocyanin. Chlorophyll *a* absorbs at longer wavelengths than allophycocyanin. Energy flow is thus: phycocyanin → allophycocyanin → chlorophyll *a*. (c) Electron micrograph of a thin section of the cyanobacterium *Synechocystis*. Note the darkly staining ball-like phycobilisomes (arrows) attached to the lamellar membranes.

Photosynthetic Reaction Centers

The photosynthetic apparatus of purple phototrophic bacteria is contained in intracytoplasmic membrane systems of various morphologies. Membrane vesicles (chromatophores) or lamellae are commonly observed membrane types (Figure 17.12●). Reaction centers of purple bacteria consist of three polypeptides, designated the L, M, and H subunits. These proteins are firmly embedded in the photosynthetic membrane and traverse the membrane several times (Figure 17.13*b*●). The L, M, and H polypeptides bind the reaction center photocomplex. The latter consists of two molecules of bacteriochloro-

● **Figure 17.11   The absorption spectrum of a cyanobacterium that has a phycobiliprotein (phycocyanin) as an accessory pigment.** Note how the presence of phycocyanin broadens the wavelengths of usable light energy (between 600 and 700 nm). Compare with Figure 17.3*b*.

phyll *a*, called the *special pair*, two additional bacteriochlorophyll *a* molecules whose function is unknown, two molecules of *bacteriopheophytin* (bacteriochlorophyll *a* minus its magnesium atom), two molecules of quinone, and two molecules of a carotenoid pigment (Figure 17.13*a*). All components of the reaction center are integrated in such a way that they can interact in very fast electron transfer reactions that ultimately result in ATP production.

Photosynthetic Electron Flow in Purple Bacteria

Recall that the photosynthetic reaction center is surrounded by light-harvesting antenna bacteriochlorophyll *a* complexes that function to funnel light energy to the reaction center (see Figure 17.6). Light energy from photons is transferred from the antenna to the reaction center in packets called *excitons*, mobile forms of energy that migrate through the antenna pigments to the reaction center at high efficiency.

Photosynthesis begins when exciton energy strikes the special pair of bacteriochlorophyll *a* molecules (Figure 17.13*a*). The absorption of energy excites the special pair, converting it to a strong electron donor with a very low reduction potential. Once this strong donor has been produced, the remaining steps in photosynthetic electron flow function to conserve the energy released when electrons are transported through a membrane from carriers of low E_0' to those of high E_0' (Figure 17.14●).

Before excitation, the bacterial reaction center, which is called *P870*, has an E_0' of about +0.5 V; after excitation it has a potential of about −1.0 V (Figure 17.14). The excited electron within P870 proceeds to reduce a molecule

(a)

M.T. Madigan

(b)

Steven J. Schmitt and M.T. Madigan

● **Figure 17.12 Membranes in anoxygenic phototrophs.** (a) Chromatophores. Section through a cell of the phototrophic purple bacterium *Rhodobacter capsulatus* containing an abundance of vesicular photosynthetic membranes. The vesicles arise by invagination of the cytoplasmic membrane. The clear areas are regions in the cell in which the reserve polymer, poly-β-hydroxybutyrate (∞ Section 4.13), was stored. A cell is about 1 μm wide. (b) Lamellar membranes in a halophilic purple bacterium. A cell is about 1.5 μm wide. These membranes also arise from invagination of the cytoplasmic membrane, but instead of forming vesicles, they become arranged as membrane stacks, similar to the thylakoids of cyanobacteria (Figure 17.5).

of bacteriopheophytin *a* within the reaction center (Figures 17.13*a* and 17.14). This transition takes place incredibly fast, taking only about three-trillionths of a second (3×10^{-12}). Once reduced, bacteriopheophytin *a* reduces several intermediate quinone molecules within

the membrane. This transition is also very fast, taking less than one-billionth of a second (Figures 17.14 and 17.15●). Relative to what happens in the reaction center, further electron transport reactions occur rather slowly, on the order of microseconds to milliseconds.

From the quinone, electrons are transported in the membrane through a series of iron-sulfur proteins and cytochromes (Figures 17.14 and 17.15), eventually returning to the reaction center. Key electron transport proteins include cytochrome bc_1 and cytochrome c_2 (Figure 17.14). Cytochrome c_2 is a periplasmic cytochrome that is an electron shuttle between the membrane-bound bc_1 complex and the reaction center (∞ Section 5.11 and Figures 5.20, 17.14, and 17.15).

Photophosphorylation

Synthesis of ATP during photosynthetic electron flow occurs as a result of the formation of a *proton motive force* generated during electron transport, and the activity of ATPases in coupling the dissipation of the proton motive force to ATP formation (∞ Section 5.12). The reaction series is completed when cytochrome c_2 donates an electron to the special pair bacteriochlorophylls (Figure 17.13); this returns these molecules to their original ground state potential ($E_0' = +0.5$ V). The reaction center is then capable of absorbing new energy and repeating the process.

This method of making ATP is called **photophosphorylation**, specifically *cyclic photophosphorylation*, because electrons move within a closed loop. Cyclic photophosphorylation resembles respiration in that electron flow through the membrane establishes a proton motive force. However, unlike respiration, in cyclic photophosphorylation *there is no net input or consumption of electrons*; electrons simply travel a circuitous route.

The spatial relationship of the electron transport components in the purple bacterial photosynthetic membrane is illustrated in Figure 17.15. Note that as in respiratory electron flow (∞ Section 5.11), the cytochrome bc_1 complex interacts with the quinone pool during photosynthetic electron flow (∞ Figure 5.20) as a major means of establishing the proton motive force used to drive ATP synthesis (Figure 17.15).

Genetics of Bacterial Photosynthesis

Purple phototrophic bacteria are gram-negative prokaryotes (∞ Section 12.2), and certain species are readily amenable to genetic manipulation. Species of the genus *Rhodobacter*, especially *R. capsulatus* and *R. sphaeroides*, have been the main subjects of genetic research in bacterial photosynthesis. The genomes of both of these purple nonsulfur bacteria (∞ Section 12.2) have been sequenced. In *Rhodobacter*, most of the genes involved in photosynthesis are clustered in several operons that span a 50-kbp region of the chromosome called the **photosynthetic gene cluster** (Figure 17.16●). Genes in the photosynthetic gene cluster encode proteins involved in (1) bacteriochlorophyll biosyn-

● **Figure 17.13 Structure of the reaction center of purple phototrophic bacteria.** (a) Arrangement of pigment molecules in the reaction center. The "special pair" of bacteriochlorophyll molecules are overlapping and shown in red, and molecules of quinone are in dark yellow and point downward in the figure. The accessory bacteriochlorophylls are in lighter yellow near the special pair, and the bacteriopheophytin molecules are shown in blue. (b) Molecular model of the protein structure of the reaction center. The pigments discussed in (a) are bound to membranes by three reaction center proteins called protein H (blue), protein M (red), and protein L (green). The reaction center pigment–protein complex is integrated into the lipid bilayer.

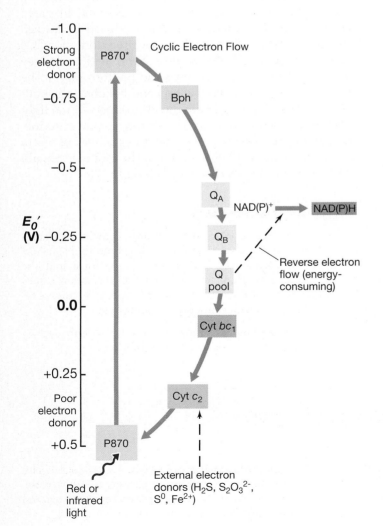

thesis (*bch* genes), (2) carotenoid biosynthesis (*crt* genes), and (3) polypeptides that bind pigment molecules in the reaction center and light-harvesting complexes (*puf* and *puh* genes) (Figure 17.16).

The gene cluster facilitates the coordinated synthesis of bacteriochlorophyll, carotenoids, and pigment-binding proteins, to ensure that the correct proportions of each are synthesized for final assembly into photocomplexes. The master regulatory signal governing transcription of the photosynthetic gene cluster in anoxygenic phototrophs is O_2. Molecular oxygen represses pigment synthesis in these organisms such that photosynthesis occurs only under *anoxic* conditions.

Genetic analysis of photosynthesis in purple bacteria has been greatly assisted by the bioenergetic diversity of *Rhodobacter* species. In addition to their capacity for photosynthesis, these organisms can grow in darkness by respiration in the presence or absence of oxygen. Thus, mutants unable to photosynthesize are easily obtained and have been used in genetic experiments (∞ Chapter 10) to characterize the number, organization, relatedness, and expression of photosynthesis genes (Figure 17.16).

● **Figure 17.14 General scheme of electron flow in anoxygenic photosynthesis in a purple bacterium.** Only a single light reaction occurs. Note how light energy converts a weak electron donor, P870, into a very strong electron donor, P870*, and that following this event, the remaining steps in photosynthetic electron flow are much the same as that of respiratory electron flow (∞ Figure 5.20). Bph, bacteriopheophytin; Q_A, Q_B, intermediate quinones; Q pool, quinone pool in membrane; Cyt, cytochrome.

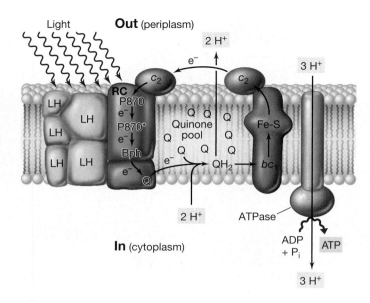

● **Figure 17.15   Arrangement of protein complexes in the photosynthetic membrane of a purple phototrophic bacterium.** The light-generated proton gradient is used in the synthesis of ATP by the ATP synthase (ATPase). LH, Light-harvesting bacteriochlorophyll complexes; RC, reaction center; Bph, bacteriopheophytin; Q, quinone; FeS, iron-sulfur protein; bc_1, cytochrome bc_1 complex; c_2, cytochrome c_2. For description of the functioning of the cytochrome bc_1 complex and the Q cycle, the details of which are not shown here, see Figure 5.20. For a description of ATPase function, see Figure 5.21.

Autotrophy in Purple Bacteria: Electron Donors and Reverse Electron Flow

For a purple bacterium to grow autotrophically, the formation of ATP is not enough. Reducing power (NADH) must also be made so that CO_2 can be reduced to the level of cell material. As previously mentioned, for purple sulfur bacteria this usually comes from H_2S, although S^0, $S_2O_3^{2-}$ and even Fe^{2+} can be used by various species. When H_2S is the electron donor in purple sulfur bacteria, globules of S^0 are stored inside the cells (Figure 17.17a●).

Reduced substances such as H_2S are oxidized by cytochromes of the c type and electrons from them eventually end up in the "quinone pool" of the photosynthetic membrane (see Figure 17.14). However, the E_0' (about 0 volts) of quinone is insufficiently negative to reduce NAD^+ (−0.32 V) directly. Instead, electrons from the quinone pool must be forced *backward*, against the thermodynamic gradient, to reduce NAD^+ to NADH (Figure

17.14). This energy-requiring process is called **reverse electron flow** and is driven by the energy of the proton motive force. Reverse electron flow is also the mechanism by which chemolithotrophs make reducing power, in many cases from electron donors of extremely positive E_0' (see Sections 17.8–17.12).

Photosynthesis in Other Anoxygenic Phototrophs

Our discussion of photosynthetic electron flow has thus far focused on purple bacteria. Although similar membrane components drive photophosphorylation in other anoxygenic phototrophs, there are differences in certain photochemical reactions that impact on reducing power biosynthesis.

Figure 17.18● contrasts photosynthetic electron flow of purple and green bacteria and the heliobacteria. Note that in the latter two groups the excited state of the reaction center bacteriochlorophylls resides at a significantly more negative E_0' than in purple bacteria and that actual *chlorophyll a* (green bacteria) or a structurally modified form of chlorophyll *a* called *hydroxychlorophyll a* (heliobacteria) is present in the reaction center. Thus, unlike in purple bacteria, where the first stable acceptor molecule (quinone) has an E_0' of about 0 V, the acceptors of green bacteria and heliobacteria (FeS proteins) have a much more electronegative E_0' than NADH. In green bacteria, a protein called *ferredoxin* (reduced by an FeS protein, Figure 17.18) is the direct electron donor for CO_2 fixation reactions in the reverse citric acid cycle (see Figure 17.24a). Thus, like oxygenic phototrophs (to be discussed next), in green bacteria and heliobacteria both ATP *and* reducing power are direct products of the light reactions. When sulfide is the electron donor for reducing power synthesis in green bacteria, globules of S^0 are produced as in purple bacteria, but the globules remain *outside*, rather than *inside* the cells (Figure 17.17b).

 17.4   Concept Check

A series of electron transport reactions occur in the photosynthetic reaction center of anoxygenic phototrophs, resulting in the formation of a proton motive force and the synthesis of ATP. Reducing power for CO_2 fixation comes from reductants present in the environment and requires reverse electron transport in purple phototrophs.

◆   How does photophosphorylation compare with electron transport phosphorylation in respiration?

● **Figure 17.16   Map of the photosynthetic gene cluster of the purple phototrophic bacterium, *Rhodobacter capsulatus*.** The *bch* genes, which encode bacteriochlorophyll synthesis proteins, are shown in green, while *crt* genes, which encode proteins that synthesize carotenoids, are shown in red. Genes encoding reaction center polypeptides (*puh* and *puf* genes) are shown in blue, and genes encoding light-harvesting I polypeptides (B870 complex) (*puf* genes) are shown in yellow. Genes shown by diagonal lines are of unknown function. Arrows indicate direction of transcription.

(a)

(b)

Norbert Pfennig

● **Figure 17.17 Photomicrographs of phototrophic sulfur bacteria taken by bright-field microscopy.** (a) Purple bacterium: *Chromatium okenii.* Notice the sulfur granules deposited *inside* the cell (arrows). (b) Green bacterium: *Chlorobium limicola.* The refractile bodies are sulfur granules deposited *outside* the cell (arrows). In both cases the sulfur granules arise from the oxidation of H_2S to obtain reducing power. Cells of *C okenii* are about 5 μm in diameter, and cells of *C. limicola* are about 0.9 μm in diameter.

◆ What is *reverse electron flow* and why is it necessary? Which phototrophs need to use reverse electron flow?

17.5 Oxygenic Photosynthesis

In contrast to anoxygenic phototrophs, electron flow in oxygenic phototrophs involves two distinct, but interconnected, photochemical reactions. Oxygenic phototrophs use light to generate both ATP *and* NADPH, the electrons for the latter arising from the splitting of *water* into oxygen and electrons (see Figure 17.2). The two systems of light reactions are called *photosystem I* and *photosystem II*, each photosystem having a spectrally distinct form of reaction center chlorophyll *a*. **Photosystem I** chlorophyll, called *P700*, absorbs light at long wavelengths (far red light), whereas **photosystem II** chlorophyll, called *P680*, absorbs at shorter wavelengths (near red light).

Like anoxygenic photosynthesis, oxygenic photochemical reactions occur in membranes. In eukaryotic cells, these membranes are found in the *chloroplast* (Figure 17.5), whereas in cyanobacteria, photosynthetic membranes are arranged in stacks within the cytoplasm (Figure 17.6). In both groups of phototrophs the membranes are arranged in a similar way and the two forms of chlorophyll *a* are attached to specific proteins in the membrane and interact as shown in Figure 17.19●.

Electron Flow in Oxygenic Photosynthesis

The path of electron flow in oxygenic phototrophs resembles the letter Z turned on its side, and Figure 17.19 outlines this so-called "Z" scheme. The reduction potential of the P680 chlorophyll *a* molecule in photosystem II is very electropositive, even more positive than that of the O_2/H_2O couple. This facilitates the first step in oxygenic electron flow, the *splitting* of water into oxygen and hydrogen atoms (Figure 17.19), a thermodynamically unfavorable reaction. An electron from water is donated to the oxidized P680 molecule following the absorption of a quantum of light near 680 nm. Light energy converts

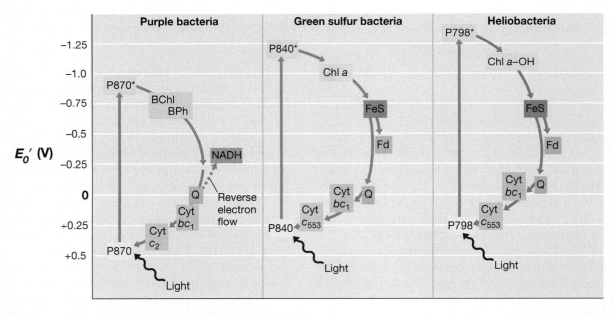

● **Figure 17.18 A comparison of electron flow in purple bacteria, green sulfur bacteria, and heliobacteria.** Note how reverse electron flow in purple bacteria is necessitated by the fact that the primary acceptor (quinone, Q) is more positive in potential than the $NAD^+/NADH$ couple. In green and heliobacteria ferredoxin (Fd), whose E_0' is actually more negative than that of NADH, is produced by light-driven reactions. In green bacteria ferredoxin drives the carboxylation reactions of autotrophy (see Figure 17.24*a*). Bchl, Bacteriochlorophyll; BPh, bacteriopheophytin. P870 and P840 are reaction centers of purple and green bacteria, respectively, and consist of Bchl *a*. The reaction center of heliobacteria (P798) contains Bchl *g*. The reaction center of *Chloroflexus* is similar to that of purple bacteria. Note the presence of forms of chlorophyll *a* in the reaction centers of green bacteria and heliobacteria.

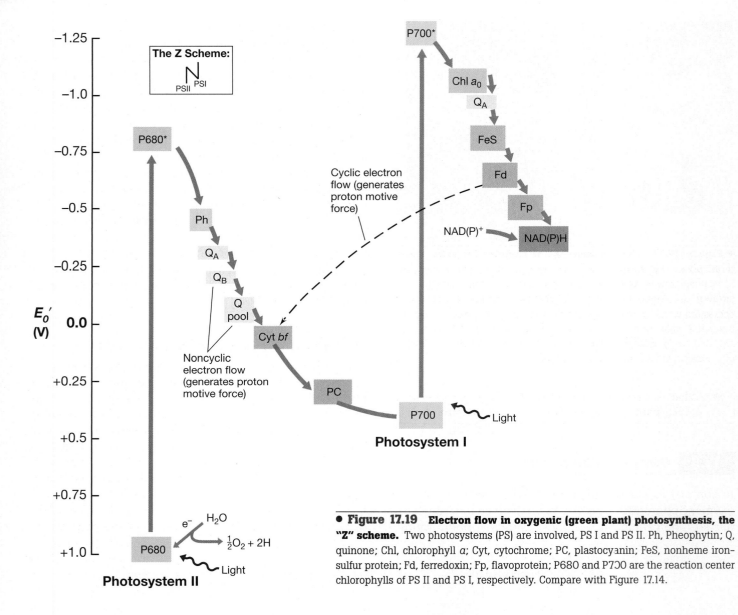

● **Figure 17.19** **Electron flow in oxygenic (green plant) photosynthesis, the "Z" scheme.** Two photosystems (PS) are involved, PS I and PS II. Ph, Pheophytin; Q, quinone; Chl, chlorophyll a; Cyt, cytochrome; PC, plastocyanin; FeS, nonheme iron-sulfur protein; Fd, ferredoxin; Fp, flavoprotein; P680 and P700 are the reaction center chlorophylls of PS II and PS I, respectively. Compare with Figure 17.14.

P680 into a moderately strong reductant, capable of reducing a molecule of E_0' about -0.5 V. The nature of this molecule is uncertain, but it is likely pheophytin a (chlorophyll a minus its magnesium atom).

From here the electron travels through several membrane carriers including quinones, cytochromes, and a copper-containing protein called *plastocyanin*. The latter donates electrons to photosystem I. The electron is accepted by the reaction center chlorophyll of photosystem I, P700, which has previously absorbed light quanta and begun the steps that lead to the reduction of NADP$^+$ (Figure 17.18). These involve electron transfer through several carriers of increasingly positive E_0', terminating with the reduction of NADP$^+$ (Figure 17.19).

ATP Synthesis in Oxygenic Photosynthesis

Besides the net synthesis of reducing power (that is, NADPH), other important events take place while electrons flow in the membrane from one photosystem to another. During transfer of an electron from the acceptor

in photosystem II to the reaction center chlorophyll molecule in photosystem I, electron transport occurs in a thermodynamically favorable (negative-to-positive) direction. This generates a proton motive force from which ATP can be produced. This type of ATP generation is called *noncyclic* photophosphorylation. This is because electrons here do not cycle back to reduce the oxidized P680, but instead are used in the reduction of NADP$^+$.

When sufficient reducing power is present, ATP can also be produced in oxygenic phototrophs by *cyclic* photophosphorylation involving only photosystem I (Figure 17.18). This occurs when electrons travel from ferredoxin to the cytochrome *bf* complex from which electron transport returns the electron to P700. This flow creates a membrane potential and synthesis of additional ATP (see dashed line in Figure 17.19).

Anoxygenic Photosynthesis in Oxygenic Phototrophs

Photosystems I and II normally function together in the oxygenic process. However, under certain conditions some algae and cyanobacteria can carry out cyclic photophosphorylation using *only* photosystem I, obtaining reducing power for CO$_2$ reduction from sources other than water. This is, in effect, *anoxygenic photosynthesis*.

A number of cyanobacteria can use H$_2$S as an electron donor for anoxygenic photosynthesis, while green algae use H$_2$. When H$_2$S is used, it is oxidized to elemental sulfur (S^0), and sulfur granules similar to those produced by green sulfur bacteria (see Figure 17.17*b*) are deposited outside the cells. Figure 17.20● shows an example of this with the cyanobacterium *Oscillatoria limnetica*. This filamentous cyanobacterium lives in sulfide-rich saline ponds where it carries out anoxygenic photosynthesis along with photosynthetic green and purple bacteria and produces sulfur as an oxidation product of sulfide (Figure 17.20). In cultures of *O. limnetica*, electron flow from photosystem II is strongly inhibited by H$_2$S, thus necessitating anoxygenic photosynthesis if the organism is to survive.

From an evolutionary point of view, the existence of cyclic photophosphorylation in both oxygenic and anoxygenic phototrophs is one of many indications of their close relationship. Oxygenic phototrophs such as *Oscillatoria limnetica* have acquired photosystem II, and hence the ability to split water. However, they still retain the ability under certain conditions to use photosystem I alone, just as anoxygenic phototrophs do during phototrophic growth.

17.5 Concept Check

In oxygenic photosynthesis, water donates electrons to drive autotrophy, and oxygen is produced as a by-product. Two separate light reactions are involved, photosystems I and II. Photosystem I resembles the system in anoxygenic photosynthesis. Photosystem II is responsible for splitting H$_2$O to yield O$_2$.

◆ Why is the term *noncyclic* electron flow used in reference to oxygenic photosynthesis?

◆ What are the major differences between the two reaction center chlorophyll molecules in photosystems I and II regarding absorption properties and E_0'?

● **Figure 17.20 Cells of the filamentous cyanobacterium *Oscillatoria limnetica* grown anaerobically on sulfide as photosynthetic electron donor.** Note the globules of elemental sulfur (arrows), the oxidation product of sulfide, formed *outside* the cells.

17.6 Autotrophic CO$_2$ Fixation: The Calvin Cycle

Several biochemical mechanisms are known for the fixation of CO$_2$ into cell material supporting autotrophic growth. The most widespread of these is the **Calvin cycle**, named for its discoverer, Melvin Calvin. The Calvin cycle requires NAD(P)H and ATP and two key enzymes, *ribulose bisphosphate carboxylase* and *phosphoribulokinase*. The remainder of the cycle is driven by a series of enzymes that are present in many organisms, both autotrophs and heterotrophs.

RubisCO and the Formation of PGA

The first step in CO$_2$ reduction in the Calvin cycle is catalyzed by the enzyme **ribulose bisphosphate carboxylase**, or **RubisCO** for short. RubisCO is widely distributed in nature, being present in purple bacteria, cyanobacteria, algae, green plants, most chemolithotrophic *Bacteria*, and even in some halophilic and hyperthermophilic *Archaea*. RubisCO catalyzes the formation of two molecules of *3-phosphoglyceric acid* (PGA) from ribulose bisphosphate and CO$_2$ as shown in Figure 17.21●. The PGA is then phosphorylated and reduced to a key intermediate of glycolysis, *glyceraldehyde 3-phosphate* (⊂⊃ Section 5.10). From here, glucose can be formed by reversal of the early steps in glycolysis (⊂⊃ Figure 5.14).

Stoichiometry of the Calvin Cycle

It is convenient to consider reactions of the Calvin cycle based on the incorporation of 6 molecules of CO$_2$, as this is what is required to make one molecule of glucose (C$_6$H$_{12}$O$_6$). For RubisCO to incorporate 6 molecules of CO$_2$, 6 ribulose bisphosphate molecules are required as acceptor molecules (Figure 17.22●). This yields 12 molecules of 3-phosphoglyceric acid (a total of 36 carbon atoms). These 12 molecules form the carbon skeletons for 6 *new* molecules of ribulose bisphosphate (a total of 30 carbon atoms) and a hexose molecule (6 carbons) for cell biosynthesis. A complex series of rearrangements involving C$_3$, C$_4$, C$_5$, C$_6$, and C$_7$ intermediates finally yields the 6 molecules of ribulose 5-phosphate from which the 6 ribulose bisphosphates are generated. The final step in this regeneration is the phosphorylation of ribulose 5-phosphate with ATP by the enzyme **phosphoribulokinase** (Figure 17.21*c*). Like RubisCO, this enzyme is also unique to the Calvin cycle.

In the *overall* stoichiometry of the Calvin cycle (Figure 17.22), 12 molecules each of ATP and NADPH are required for the reduction of 12 molecules of phosphoglyceric acid (PGA) to glyceraldehyde 3-phosphate. Six more ATP molecules are required for conversion of ribulose phosphate to ribulose bisphosphate. Thus, *12 NADPH and 18 ATP are required to synthesize 1 hexose from 6 CO$_2$ by the Calvin cycle*. The hexose molecules can be converted to *storage polymers* such as glycogen, starch, or

● **Figure 17.21 Key reactions of the Calvin cycle.** (a) Reaction of the enzyme *ribulose bisphosphate carboxylase*. (b) Steps in the conversion of 3-phosphoglyceric acid (PGA) to glyceraldehyde 3-phosphate. Note that both ATP and NADPH are required. (c) Conversion of ribulose 5-phosphate to the CO_2 acceptor molecule ribulose bisphosphate by the enzyme *phosphoribulokinase*.

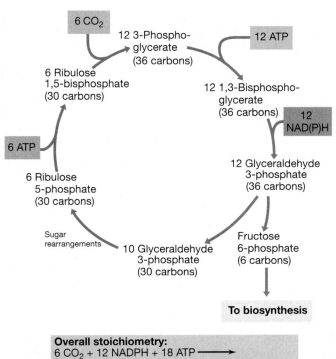

Overall stoichiometry:
$6\ CO_2 + 12\ NADPH + 18\ ATP \longrightarrow$
$C_6H_{12}O_6(PO_3H_2) + 12\ NADP^+ + 18\ ADP + 17\ P_i$

poly-β-hydroxyalkanoates (∞ Section 4.11) and used later to build new cell material.

Carboxysomes

Several autotrophic prokaryotes that use the Calvin cycle for CO_2 fixation produce polyhedral cell inclusions called **carboxysomes**. The inclusions, about 100 nm in diameter, are surrounded by a thin, nonunit membrane and consist of a tightly packed crystalline array of RubisCO molecules (Figure 17.23●; ∞ Figure 12.10*a*). It is thought that carboxysomes are a mechanism to increase the amount of RubisCO in the cell to allow for more rapid CO_2 fixation without affecting the osmolarity of the cytoplasm (osmotic pressure is not affected because the carboxysome is insoluble).

● **Figure 17.22 The Calvin cycle.** Shown is the production of one hexose molecule from CO_2. For each *six* molecules of CO_2 incorporated, *one* fructose 6-phosphate is produced. In phototrophs, ATP comes from photophosphorylation and NAD(P)H from light or reverse electron flow. Use the color coding here to follow the biochemical reactions occurring in Figure 17.21.

● Figure 17.23 Crystalline Calvin cycle enzymes: Carboxysomes. Electron micrograph of carboxysomes purified from the chemolithotrophic sulfur oxidizer *Thiobacillus neapolitanus*. The structures are about 100 nm in diameter. Carboxysomes are present in a wide variety of obligately autotrophic prokaryotes.

Carboxysomes have been found in obligately chemolithotrophic sulfur-oxidizing bacteria, the nitrifying bacteria, and cyanobacteria and prochlorophytes (∞ Sections 12.3, 12.4, 12.26, and 12.27, respectively). They are not present in *facultative* autotrophs (organisms that can grow either as autotrophs or as heterotrophs) such as purple anoxygenic phototrophs. Thus, the carboxysome appears to be an evolutionary adaptation to life under *strictly autotrophic conditions*.

 17.6 Concept Check

The fixation of CO_2 by most phototrophic and other autotrophic organisms occurs via the Calvin cycle, in which the enzyme ribulose bisphosphate carboxylase (RubisCO) plays a key role. The Calvin cycle is an energy-demanding process in which CO_2 is converted into sugar.

◆ What reaction does the enzyme *ribulose bisphosphate carboxylase* carry out?

◆ Why is reducing power needed for autotrophic growth?

◆ What is a *carboxysome*?

17.7 Autotrophic CO_2 Fixation: Reverse Citric Acid Cycle and the Hydroxypropionate Cycle

Alternative mechanisms of autotrophic CO_2 fixation are present in green sulfur bacteria and green nonsulfur bacteria. In the green sulfur bacterium *Chlorobium* (see Figures 17.7a and 17.17b), CO_2 fixation occurs by a reversal of steps in the citric acid cycle (∞ Figure 5.22), a pathway called the *reverse citric acid cycle* (Figure 17.24a●). *Chlorobium* contains two ferredoxin-linked enzymes that catalyze the reductive fixation of CO_2 into intermediates of the citric acid cycle. The two ferredoxin-linked reactions involve the carboxylation of succinyl-CoA to α-ketoglutarate and the carboxylation of acetyl-CoA to pyruvate (Figure 17.24a). Most of the other reactions of the reverse citric acid cycle are cat-

alyzed by enzymes working in reverse of the normal oxidative direction of the cycle. One exception is *citrate lyase*, an ATP-dependent enzyme that cleaves citrate into acetyl-CoA and oxaloacetate in green sulfur bacteria. In the oxidative direction of the cycle, citrate is produced from these same precursors by the enzyme *citrate synthase* (∞ Figure 5.22).

The reverse citric acid cycle as a mechanism of autotrophy has also been found in certain nonphototrophic hyperthermophiles, including *Archaea* such as *Thermoproteus* (∞ Section 13.9), and in *Aquifex*, a very early branching autotroph on the phylogenetic tree of *Bacteria* (∞ Figure 12.1 and Section 12.37). These findings, of course, leave open the possibility that this pathway is more widespread than previously thought and suggest that it may have been an "early" evolutionary form of autotrophy.

Autotrophy in *Chloroflexus*

The green nonsulfur phototroph *Chloroflexus* (∞ Section 12.35) grows autotrophically with either H_2 or H_2S as electron donors. However, neither the Calvin cycle nor the reverse citric acid cycle operates in this organism. Instead, two molecules of CO_2 are reduced to glyoxylate by a unique cyclic pathway, the **hydroxypropionate pathway** (Figure 17.24b). This pathway leads to the synthesis of hydroxypropionate as a key intermediate.

In phototrophic bacteria, the hydroxypropionate pathway has been confirmed only in *Chloroflexus*, the earliest branching anoxygenic phototroph on the tree of *Bacteria* (∞ Figure 12.1). This suggests that the hydroxypropionate pathway may have been the first attempt at autotrophy in anoxygenic phototrophs. If *Chloroflexus* evolved long before other phototrophic organisms, as the tree of *Bacteria* suggests (Figure 12.1), the hydroxypropionate pathway was perhaps the first autotrophic pathway in *any* phototrophic organism. But in addition to *Chloroflexus*, the hydroxypropionate pathway functions in several hyperthermophilic *Archaea*, including *Metallosphaera*, *Acidianus*, and *Sulfolobus* (∞ Section 13.9). These are all nonphototrophic prokaryotes and lie near the base of the archaeal domain. The roots of the hydroxypropionate pathway may thus be very deep; it was perhaps nature's first attempt at autotrophy.

 17.7 Concept Check

The reverse citric acid cycle and the hydroxypropionate cycle are pathways of CO_2 fixation found in green sulfur and green nonsulfur bacteria, respectively.

◆ Including the route of CO_2 fixation, discuss at least three ways that you could distinguish a purple sulfur bacterium from a green sulfur bacterium.

◆ Including the route of CO_2 fixation, what similarities and differences exist between green *sulfur* and green *nonsulfur* bacteria?

(a)

Net reaction:
3 CO₂ + 12 H + 5 ATP ⟶ triose-P

(b)

Net reaction:
2 CO₂ + 6 H + 3 ATP ⟶ glyoxylate

● **Figure 17.24 Unique autotrophic pathways in phototrophic green bacteria.** (a) The reverse citric acid cycle is the mechanism of CO₂ fixation in the green sulfur bacterium *Chlorobium* (see also Figure 17.17*b*; ⚬⚬ Section 12.32). Ferredoxin$_{red}$ indicates carboxylation reactions requiring reduced ferredoxin (2 H each). Reduced ferredoxin is generated in *Chlorobium* by light-driven reactions (see Figure 17.18). Starting from oxalacetate, each turn of the cycle results in three molecules of CO₂ being incorporated and pyruvate as the product. The cleavage of citrate by the ATP-dependent enzyme citrate lyase regenerates the C₄ acceptor oxalacetate and produces acetyl-CoA for biosynthesis. The conversion of pyruvate to phospho-enolpyruvate consumes two high-energy phosphate bond equivalents. (b) The hydroxypropionate pathway is the means of autotrophy in the green nonsulfur bacterium *Chloroflexus* (⚬⚬ Section 12.35). Acetyl-CoA is carboxylated twice to yield methylmalonyl-CoA. This intermediate is rearranged to yield acetyl-CoA and glyoxylate. The latter is converted to cell material probably through a serine or glycine intermediate. The source of reducing power in the hydroxypropionate pathway is NADPH.

II CHEMOLITHOTROPHY: ENERGY FROM THE OXIDATION OF INORGANIC ELECTRON DONORS

Chapter 5 focused on microorganisms whose energy metabolism was based on *chemoorganotrophy*, the use of organic compounds as energy sources. The next four sections focus on the *chemolithotrophs*, highlighting the strategies, problems, and advantages of a lifestyle geared toward the use of *inorganic* chemicals as energy sources.

17.8 Inorganic Electron Donors and Energetics

Organisms that obtain energy from the oxidation of *inorganic* compounds are called **chemolithotrophs**. Most chemolithotrophic bacteria can also obtain their carbon from CO₂, so they are also autotrophs. As we have noted, for growth on CO₂ as sole carbon source an or-

ganism needs (1) energy (ATP), and (2) reducing power. Some chemolithotrophs are **mixotrophic**, meaning that although they are able to obtain energy from the oxidation of an inorganic compound, they require an *organic* compound as carbon source.

In chemolithotrophs, ATP generation is in principle similar to that in chemoorganotrophs, except that the electron donor is *inorganic* rather than organic. ATP synthesis is coupled to *oxidation* of the electron donor. Reducing power in chemolithotrophs is obtained either directly from the inorganic compound, if it has a sufficiently low reduction potential, or by *reverse electron transport* reactions, as discussed in Section 17.5 for phototrophic purple bacteria.

Sources of Inorganic Electron Donors

Chemolithotrophs have many sources of inorganic electron donors, which may be geological, biological, or anthropogenic in nature. Volcanic activity is a major source of reduced sulfur compounds, primarily H₂S. Agricultural and mining operations add inorganic electron donors to the environment, especially nitrogen and iron compounds, as does the burning of fossil fuels and the input of industrial wastes. Biological sources are also quite extensive, especially the production of H_2S, H_2, and NH_3. The ecological success (∞ Chapter 19) and metabolic diversity (this chapter) of chemolithotrophs underscores the abundance of sources and supplies of inorganic electron donors in nature.

Energetics of Chemolithotrophy

A review of reduction potentials listed in Table A1.2 reveals that a number of inorganic compounds can provide sufficient energy for ATP synthesis when oxidized with O_2 as electron acceptor. Recall from Chapter 5 that the further apart two half reactions are in terms of E_0', the greater the amount of energy released. For instance, the difference in reduction potential between the H^+/H_2 couple and the $\frac{1}{2}O_2/H_2O$ couple is -1.23 V, which is equivalent to a free-energy yield of -237 kJ/mol (see Appendix 1 for calculations). On the other hand, the potential difference between the H^+/H_2 couple and the NO_3^-/NO_2^- couple is less, -0.84 V, equivalent to a free-energy yield of -163 kJ/mol. This is still quite sufficient for the production of ATP (the energy-rich phosphate bond of ATP has a free energy of -31.8 kJ/mol, see Table 17.6). However, a similar calculation will show that there is insufficient energy available from, for example, the oxidation of H_2S using CO_2 as electron acceptor.

Such energy calculations make it possible to predict the kinds of chemolithotrophs that should exist in nature. Since organisms must obey the laws of thermodynamics, only reactions that are thermodynamically favorable are potential energy-yielding reactions (see also Section 17.21). Table 17.1 summarizes energy yields for some reactions known to be carried out by chemolithotrophic microorganisms. The organisms themselves were considered in Chapters 12 and 13. We also examine ecological aspects of chemolithotrophy in Chapter 19,

where we will see that chemolithotrophic reactions are at the heart of most nutrient cycles.

 17.8 Concept Check

Chemolithotrophs are able to oxidize inorganic chemicals as their sole sources of energy and reducing power. Most chemolithotrophs are also able to grow autotrophically.

◆ For what *two* purposes are inorganic compounds used by chemolithotrophs?

◆ Why does the oxidation of H_2 yield more energy with O_2 as electron acceptor than with SO_4^{2-} as electron acceptor?

17.9 Hydrogen Oxidation

Hydrogen is a common product of microbial metabolism, and a number of chemolithotrophs are able to use it as an electron donor in energy metabolism. A wide variety of anaerobic H_2-oxidizing *Bacteria* and *Archaea* are known, differing in the electron acceptor they use (for example, nitrate, sulfate, ferric iron, and others), and these organisms are discussed later in this chapter (see Sections 17.14–17.18). Here we consider only the *aerobic* H_2-oxidizing bacteria.

Energetics of H₂ Oxidation

Generation of ATP during H_2 oxidation results from the oxidation of H_2 by O_2 leading to formation of a proton motive force. The overall reaction:

$$H_2 + \tfrac{1}{2}O_2 \longrightarrow H_2O \qquad \Delta G^{0\prime} = -237 \text{ kJ}$$

is highly exergonic and can support the synthesis of at least one ATP. The reaction is catalyzed by the enzyme **hydrogenase**, the electrons from H_2 initially being transferred to a quinone acceptor. From here electrons pass through a series of cytochromes to eventually reduce O_2

Table 17.1	Energy yields from the oxidation of various inorganic electron donors[a]					
Electron donor	**Reaction**	**Type of chemolithotroph**	**E_0' of couple (V)**	**$\Delta G^{0\prime}$ (kJ/reaction)**	**Number of electrons**	**$\Delta G^{0\prime}$ (kJ/2e⁻)**
Phosphite[b]	$4 HPO_3^{2-} + SO_4^{2-} + H^+ \rightarrow 4 HPO_4^{2-} + HS^-$	Phosphite bacteria	-0.69	-91	2	-91
Hydrogen	$H_2 + \frac{1}{2}O_2 \rightarrow H_2O$	Hydrogen bacteria	-0.42	-237.2	2	-237.2
Sulfide	$HS^- + H^+ + \frac{1}{2}O_2 \rightarrow S^0 + H_2O$	Sulfur bacteria	-0.27	-209.4	2	-209.4
Sulfur	$S^0 + 1\frac{1}{2}O_2 + H_2O \rightarrow SO_4^{2-} + 2 H^+$	Sulfur bacteria	-0.20	-587.1	6	-195.7
Ammonium[c]	$NH_4^+ + 1\frac{1}{2}O_2 \rightarrow NO_2^- + 2 H^+ + H_2O$	Nitrifying bacteria	$+0.34$	-274.7	6	-91.6
Nitrite	$NO_2^- + \frac{1}{2}O_2 \rightarrow NO_3^-$	Nitrifying bacteria	$+0.43$	-74.1	2	-74.1
Ferrous iron	$Fe^{2+} + H^+ + \frac{1}{4}O_2 \rightarrow Fe^{3+} + \frac{1}{2}H_2O$	Iron bacteria	$+0.77$	-32.9	1	-65.8

[a] Data calculated from values in Appendix 1; values for Fe^{2+} are for pH 2, and others are for pH 7. At pH 7 the Fe^{3+}/Fe^{2+} couple is about $+0.2$ V.

[b] Except for phosphite, all reactions are shown coupled to O_2 as electron acceptor. The only known phosphite oxidizer couples to SO_4^{2-} as electron acceptor.

[c] Ammonium can also be oxidized with NO_2^- as electron acceptor by anammox organisms (see Section 17.12).

to water (Figure 17.25●). Some hydrogen bacteria have *two* hydrogenases, one soluble and one membrane-bound (∞ Table 12.6). In this case, the membrane-bound enzyme is involved in energetics while the soluble hydrogenase takes up H_2 and reduces NAD^+ to NADH directly (the reduction potential of H_2, -0.42 V, is so low that reverse electron flow as a means of making reducing power is unnecessary) (Figure 17.25). The organism *Ralstonia eutropha* has been a model for studying aerobic H_2 oxidation, and we discussed some of the properties of this organism in Section 12.5.

Autotrophy in H_2 Bacteria

Although most hydrogen bacteria can also grow as chemoorganotrophs, when growing chemolithotrophically they fix CO_2 by the Calvin cycle (see Section 17.6). The stoichiometry observed here is:

$$6 H_2 + 2 O_2 + CO_2 \longrightarrow (CH_2O) + 5 H_2O$$

where (CH_2O) represents cell material. However, when readily useable organic compounds such as glucose are present, synthesis of Calvin cycle and hydrogenase enzymes in typical aerobic H_2-oxidizing bacteria is repressed. In nature, H_2 levels in oxic environments are transient and low at best. It is thus likely that aerobic hydrogen bacteria closely regulate their catabolic enzymes. These organisms probably shift between chemoorganotrophy and chemolithotrophy often, depending on levels of useable organic compounds and H_2 in their habitats. Moreover, many aerobic H_2 bacteria grow best microaerobically. It is therefore likely that these organisms are most successful as H_2 chemolithotrophs in *oxic/anoxic interfaces*, where H_2 from fermentative metabolism would be in greater and more continuous supply than in highly oxic habitats.

17.10 Oxidation of Reduced Sulfur Compounds

Many reduced sulfur compounds can be used as electron donors by a variety of sulfur bacteria, called *colorless* to distinguish them from the bacteriochlorophyll-containing (pigmented) green and purple sulfur bacteria discussed earlier in this chapter (see Figure 17.17). Indeed, the very concept of chemolithotrophy emerged from studies of the sulfur bacteria in the work of the great Russian microbiologist Sergei Winogradsky (see the Microbial Sidebar, Winogradsky and Chemolithoautotrophy, and ∞ Section 1.7).

Energetics of Sulfur Oxidation

The most common sulfur compounds used as electron donors are hydrogen sulfide (H_2S), elemental sulfur (S^0), and thiosulfate ($S_2O_3^{2-}$). The final product of sulfur oxidation in most cases is sulfate (SO_4^{2-}), and the total number of electrons involved between H_2S (oxidation state, -2) and sulfate (oxidation state, $+6$) is eight (see Table 17.3 for a summary of sulfur oxidation states). Less energy is available when one of the intermediate sulfur oxidation states is used:

$$H_2S + 2 O_2 \longrightarrow SO_4^{2-} + 2 H^+$$
$$\text{(final product, sulfate)} \quad \Delta G^{0\prime} = -798.2 \text{ kJ/reaction}$$
$$HS^- + \tfrac{1}{2}O_2 + H^+ \longrightarrow S^0 + H_2O$$
$$\text{(final product, sulfur)} \quad -209.4 \text{ kJ/reaction}$$
$$S^0 + H_2O + 1\tfrac{1}{2}O_2 \longrightarrow SO_4^{2-} + 2 H^+$$
$$\text{(final product, sulfate)} \quad -587.1 \text{ kJ/reaction}$$
$$S_2O_3^{2-} + H_2O + 2 O_2 \longrightarrow 2 SO_4^{2-} + 2 H^+$$
$$\text{(final product, sulfate)} \quad -818.3 \text{ kJ/reaction}$$
$$(-409.1 \text{ kJ/S atom oxidized})$$

The oxidation of the most reduced sulfur compound, H_2S, occurs in stages, and the first oxidation step results in the formation of elemental sulfur, S^0. Some H_2S-oxidizing bacteria deposit this elemental sulfur inside the cell (Figure 17.26a●). The sulfur deposited as a result of the initial oxidation is an energy reserve. When the supply of H_2S has been depleted, additional energy can be obtained from the oxidation of sulfur to sulfate.

When elemental sulfur is provided externally as an electron donor, the organism must grow attached to the sulfur particle because of the extreme insolubility of elemental sulfur (Figure 17.26b). By adhering to the particle, the organism can remove sulfur atoms for oxidation to sulfate. This is thought to occur through the action of membrane or periplasmic proteins that solubilize the

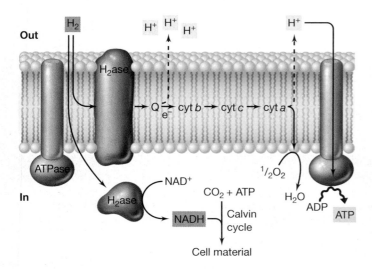

● **Figure 17.25 Bioenergetics and function of the two hydrogenases of aerobic H_2 bacteria.** In *Ralstonia eutropha*, where two hydrogenases are present, the membrane–bound hydrogenase is involved in energetics while the cytoplasmic hydrogenase makes NADH for the Calvin cycle. Note how the membrane–bound hydrogenase begins the flow of electrons leading to formation of a proton motive force. Some H_2 bacteria have only the membrane–bound hydrogenase, and in these organisms reducing power synthesis occurs from reverse electron flow. H_2ase, hydrogenase; cyt, cytochrome; Q, quinone.

Microbial Sidebar ◆ Winogradsky and Chemolithoautotrophy

The concept of chemolithotrophic autotrophy was first conceived by the great Russian microbiologist Sergei Winogradsky (Section 1.7). Winogradsky studied sulfur bacteria because certain colorless sulfur bacteria (*Beggiatoa, Thiothrix*) are very large (Figure 1) and easy to investigate, even in the absence of pure cultures. Springs with waters rich in H_2S are fairly common around the world. In the Bernese Oberland district of Switzerland, vast populations of *Beggiatoa* and *Thiothrix* develop in the outflow channels of sulfur springs. It was here that Winogradsky found suitable material for microscopic and physiological studies by merely lifting up the white filamentous masses of cells (Section 12.4 and Figures 12.11 and 12.13) and performing experiments directly in the field.

Winogradsky first showed that colorless sulfur bacteria are present only in water containing H_2S. As the water flows away from the source, the H_2S gradually dissipates, and sulfur bacteria were no longer present. This suggested to him that their development depends on the presence of H_2S. Winogradsky then showed that when *Beggiatoa* filaments are starved for sulfide, they lost their sulfur granules. He found, however, that the granules are rapidly restored if a small amount of H_2S was added (Figure 1). He concluded that H_2S was being oxidized to elemental sulfur.

But what happens to the sulfur granules when the filaments are starved of H_2S? Winogradsky showed by some clever microchemical tests that when the sulfur granules disappeared, sulfate appeared in the medium. He concluded that *Beggiatoa* (and by inference other colorless sulfur bacteria) oxidize

H_2S to elemental sulfur and subsequently to sulfate ($H_2S \rightarrow S^0 \rightarrow SO_4^{2-}$). Because this organism seemed to require H_2S for development in the springs, he further postulated that this oxidation was the principal source of *energy* for these organisms.

Winogradsky's studies on *Beggiatoa* thus provided the first evidence that an organism could oxidize an *inorganic* substance as an energy source. This was the origin of the concept of chemolithotrophy. From these beginnings, Winogradsky turned to a study of the nitrifying bacteria, and it was with this group that he first showed that autotrophic fixation of CO_2 is coupled to the oxidation of an inorganic compound. The process of nitrification had been known before Winogradsky's work from studies on the fate of sewage when added to soil. For example, ammonia-rich sewage passed through a soil column showed conversion to nitrate. Winogradsky proceeded from here to isolate nitrifying bacteria using completely mineral media in which CO_2 was the sole carbon source and ammonia was the sole electron donor. Because ammonia is chemically stable, it was easy to show that the oxidation of ammonia to nitrite, and subsequently to nitrate, is a strictly bacterial process.

In fact, Winogradsky further showed that nitrification is a *two-step* process, with one group of organisms converting NH_4^+ to NO_2^- and a second, NO_2^- to NO_3^- (see Section 17.12). Because no organic materials were present in the medium, it was also possible to show that organic matter (the bacterial cell material) was formed from CO_2. When the ammonia or nitrite was left out of the medium, no growth occurred. Careful chemical

analyses showed that the amount of organic matter formed by the bacteria was proportional to the amount of ammonia or nitrite they oxidize. Winogradsky concluded, "This [process] is contradictory to that fundamental doctrine of physiology which states that a complete synthesis of organic matter cannot take place in nature except through chlorophyll-containing plants by the action of light." The concept of chemolithoautotrophy was thus born.

We now know that at least in one way, autotrophy in most chemolithotrophs and phototrophs is similar. In both groups, the pathway of CO_2 fixation follows the same biochemical steps (the Calvin cycle) involving the enzyme ribulose bisphosphate carboxylase (see Section 17.6).■

Figure 1 Beggiatoa and chemolithotrophy. *Drawings made by Winogradsky of Beggiatoa and translation (from the French) of the legend accompanying these figures. "Fig. 1. The tip of a filament of Beggiatoa alba: (a) in sulfurous [sulfide-containing] water, (b) after 24 h in water nearly depleted in H_2S, (c) after 48 h in water without H_2S [note depletion of sulfur globules with time]. Fig. 2. The tip of a filament of Beggiatoa media. Fig. 3. The tip of a filament of Beggiatoa minima."* From Winogradsky, S. 1949. *Microbiologie du Sol.* Masson, Paris.

sulfur, probably by reduction of S^0 to HS^-, from which it is transported into the cell and enters chemolithotrophic metabolism (Figure 17.27●).

One of the products of the sulfur oxidation reactions is H^+. Production of protons lowers the pH. Consequently, one result of the oxidation of reduced sulfur compounds is the acidification of the medium. The acid formed by the sulfur bacteria is a stronger acid, *sulfuric acid*—H_2SO_4—and sulfur bacteria are often able to bring about a marked reduction in the pH of the medium.

Biochemistry and Energetics of Sulfur Oxidation

The biochemical steps in the oxidation of various sulfur compounds are summarized in Figure 17.27. Starting with sulfide, sulfite (SO_3^{2-}) is produced, a six-electron oxidation. If S^0 is the starting substrate, sulfite is also produced, although the S^0 must first be reduced to sulfide (Figure 17.27a). There are two ways in which SO_3^{2-} can be oxidized to SO_4^{2-}. The most widespread system is that employing the enzyme *sulfite oxidase*. Sulfite oxidase transfers electrons from SO_3^{2-} directly to cytochrome *c*,

(a)

(b)

T.D. Brock

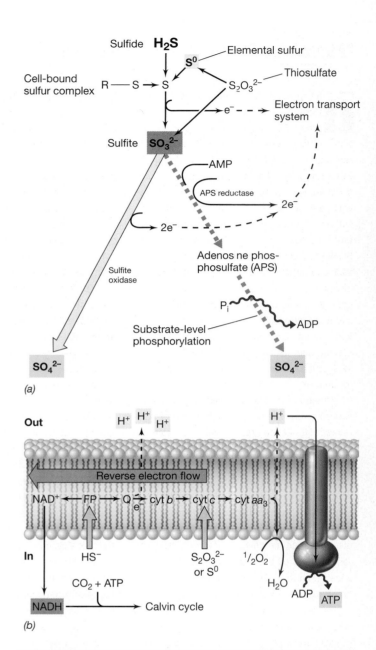

(a)

(b)

● **Figure 17.26 Sulfur bacteria.** (a) Deposition of internal sulfur granules by *Beggiatoa*. (b) Attachment of the sulfur-oxidizing archaeon *Sulfolobus acidocaldarius* to a crystal of elemental sulfur. Cells are visualized by fluorescence microscopy after staining them with the dye acridine orange. The sulfur crystal does not fluoresce. See Figure 17.27 for how sulfide and sulfur are oxidized to yield ATP.

● **Figure 17.27 Oxidation of reduced sulfur compounds by sulfur chemolithotrophs.** (a) Steps in the oxidation of different compounds. The sulfite oxidase pathway accounts for the majority of sulfite oxidized. (b) Electrons from sulfur compounds feed into the electron transport chain to drive a proton motive force; electrons from thiosulfate and elemental sulfur enter at the level of cytochrome *c*. NADH must be made by energy-consuming reactions of reverse electron flow since the electron donors have a more electropositive E_0' than does $NAD^+/NADH$. Cyt, cytochrome; FP, flavoprotein; Q, quinone. For the structure of APS, see Figure 17.38.

and ATP is made from this during electron transport and proton motive force formation (Figure 17.27*b*).

In addition to sulfite oxidase, a few sulfur chemo-lithotrophs oxidize SO_3^{2-} to SO_4^{2-} via a reversal of the activity of *adenosine phosphosulfate (APS) reductase*, an enzyme essential for the metabolism of sulfate-reducing bacteria (compare Figures 17.27*a* and 17.38). This reaction, run in the direction of SO_4^{2-} *production* by sulfur chemolithotrophs, yields one high-energy phosphate bond when AMP is converted to ADP (Figure 17.27*a*). When thiosulfate is the electron donor for sulfur chemolithotrophs, it is split into S^0 and SO_3^{2-}, both of which are eventually oxidized to SO_4^{2-}.

All the electrons from reduced sulfur compounds eventually reach the electron transport system as shown in Figure 17.27*b*. Depending on the E_0' of the couple, electrons enter at either the flavoprotein ($E_0' = \sim-0.2$) or cytochrome *c* ($E_0' = +0.3$) level and are shuttled to O_2, generating a proton motive force that leads to ATP synthesis by ATPase. Electrons for autotrophic CO_2 fixation come from reverse electron flow (see Section 17.5), eventually yielding NADH, and CO_2 is fixed via the Calvin cycle (Figure 17.27*b*). Although the sulfur chemolithotrophs are primarily an aerobic group (⚌ Section 12.4), some species can grow anaerobically using nitrate as an electron acceptor; *Thiobacillus denitrificans* is a classic example of this lifestyle.

⬡ *17.9–17.10 Concept Checks*

Hydrogen (H_2) and reduced sulfur compounds such as H_2S and S^0 are excellent electron donors for chemolithotrophs. These compounds can be oxidized by the hydrogen bacteria or the sulfur bacteria, respectively, thereby generating a proton motive force and ATP synthesis. These chemolithotrophs are also autotrophs and fix CO_2 by the Calvin cycle.

◆ What special enzyme is needed for growth on H_2?

◆ How many electrons are available from the oxidation of H_2S if S^0 is the final product? If SO_4^{2-} is the final product?

17.11 Iron Oxidation

The aerobic oxidation of iron from the ferrous (Fe^{2+}) to the ferric (Fe^{3+}) state is an energy-yielding reaction for some prokaryotes. Only a small amount of energy is available from this oxidation (see Table 17.1), and for this reason the iron bacteria must oxidize large amounts of iron in order to grow. The ferric iron produced forms insoluble ferric hydroxide [$Fe(OH)_3$] precipitates in water (Figure 17.28●). This is in part because at neutral pH ferrous iron rapidly oxidizes nonbiologically to the ferric state. It is thus stable for long periods only under anoxic conditions. At acid pH, however, ferrous iron is stable under oxic conditions. This explains why most iron-oxidizing bacteria are obligately acidophilic.

(a)

(b)

● **Figure 17.28 Iron-oxidizing bacteria.** (a) Acid mine drainage, showing the confluence of a normal river and a creek draining a coal-mining area. The acidic creek is very high in ferrous iron (Fe^{2+}). At low pH values, ferrous iron does not oxidize spontaneously in air, but *Acidithiobacillus ferrooxidans* carries out the oxidation. Insoluble ferric hydroxide and complex ferric salts precipitate, forming the precipitate called "yellow boy" by coal miners. (b) Cultures of *A. ferrooxidans*. Shown is a dilution series, with no growth in the tube on the left and increasing amounts of growth from left to right. Growth is evident from the production of Fe^{3+} from Fe^{2+}, which readily forms $Fe(OH)_3$ and acidity, leading to the yellow–orange color ($Fe^{3+} + 3H_2O \rightarrow Fe(OH)_3 + 3H^+$).

● **Figure 17.29 Iron bacteria at neutral pH: *Sphaerotilus*.** Phase contrast photomicrograph of empty iron-encrusted sheaths of *Sphaerotilus* collected from seepage at the edge of a small swamp. *Sphaerotilus* is a typical interface organism, living in habitats where anoxic ferrous-containing water meets oxic zones.

The best-known iron-oxidizing bacteria, *Acidithiobacillus ferrooxidans* and *Leptospirillum ferrooxidans*, can both grow autotrophically using ferrous iron (Figure 17.28b) as electron donor. Both organisms can grow at pH values below pH 1. These organisms are very common in acid-polluted environments such as coal-mining dumps (Figure 17.28a). *Ferroplasma*, a member of the *Archaea*, is an extremely acidophilic iron oxidizer and is even capable of growing at pH values below 0 (∞ Section 13.5). We will discuss the role of all of these organisms in acid-mine pollution and mineral oxidation in Sections 19.14 and 19.15.

Despite the instability of Fe^{2+} at neutral pH, there are a number of iron-oxidizing bacteria that thrive in such environments, but these are situations where ferrous iron is moving from anoxic to oxic conditions. At interfaces between these zones iron bacteria can oxidize Fe^{2+} as it comes from an anoxic source before the Fe^{2+} oxidizes spontaneously. *Gallionella ferruginea* and *Sphaerotilus natans* are examples of organisms that live at these interfaces. They are typically seen mixed in with the characteristic deposits they form (Figure 17.29●; also ∞ Figure 12.43a).

Energy from Ferrous Iron Oxidation

The bioenergetics of iron oxidation by *Acidithiobacillus ferrooxidans* is of interest because of the very electropositive reduction potential of the Fe^{3+}/Fe^{2+} couple (+0.77 V at pH 2). The respiratory chain of *A. ferrooxidans* contains cytochromes of the *c* and *a* types and a periplasmic copper-containing protein called *rusticyanin* (Figure 17.30●). Because the reduction potential of the Fe^{3+}/Fe^{2+} couple is so high, the route of electron transport to oxygen $\left(\frac{1}{2}O_2/H_2O, E_0' = +0.82 \text{ V}\right)$ can only be very short.

Ferrous iron oxidation begins in the periplasm, where rusticyanin oxidizes Fe^{2+} to Fe^{3+}, a one-electron

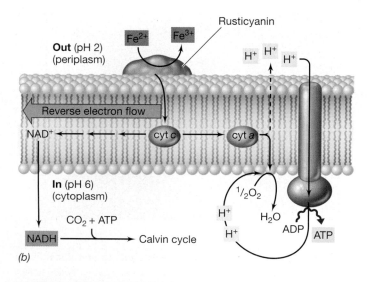

(b)

● **Figure 17.30 Electron flow during Fe^{2+} oxidation by the acidophile _Acidithiobacillus ferrooxidans._** The periplasmic copper-containing protein rusticyanin is the immediate acceptor of electrons from Fe^{2+}. From here, electrons travel a short electron transport chain resulting in the reduction of O_2 to H_2O. Reducing power to drive the Calvin cycle comes from reactions of reverse electron flow. Note the steep pH gradient (~4 units) across the membrane.

transition. This protein then reduces cytochrome _c_, and this subsequently reduces cytochrome _a_. The latter interacts directly with O_2 to form H_2O (Figure 17.30). ATP is then synthesized from proton-translocating ATPases in the membrane; ATP yields are typically low because of the high potential of the electron donor.

Because of the large natural gradient of protons across the _A. ferrooxidans_ membrane (the periplasm is pH 1–2 while the cytoplasm is pH 5.5–6), protons entering the cytoplasm via the ATPase must be consumed in order to maintain the internal pH within acceptable limits (Figure 17.30). The protons are consumed during the production of H_2O, but this reaction also requires electrons; these come from Fe^{2+} as follows:

$$2\,Fe^{2+} + \tfrac{1}{2}O_2 + 2\,H^+ \longrightarrow 2\,Fe^{3+} + H_2O$$

Thus, as long as _A. ferrooxidans_ has Fe^{2+} available, ATP synthesis can occur at the expense of the natural proton motive force that exists across the cytoplasmic membrane (Figure 17.30).

Autotrophy in _A. ferrooxidans_ is driven by the Calvin cycle, and because of the high potential of the electron donor, Fe^{2+}, much energy is consumed in reverse electron flow reactions to obtain the reducing power (NADH) necessary to drive CO_2 fixation. Thus, a relatively poor energetic yield coupled with large energetic demands in biosynthesis means that _A. ferrooxidans_ must oxidize large amounts of Fe^{2+} in order to produce even a very small amount of cell material. Because of this, in environments where acidophilic Fe^{2+}-oxidizing bacteria thrive, their presence is signaled not by the formation of much cell material but by the presence of large amounts of ferric iron

precipitates (Figures 17.28; ∞ Figure 19.37). We consider the important ecological processes connected with the iron-oxidizing bacteria in Sections 19.14 and 19.15.

Ferrous Iron Oxidation by Anoxygenic Phototrophs

Ferrous iron can be oxidized under _anoxic_ conditions by certain anoxygenic phototrophic bacteria (Figure 17.31●). The ferrous iron is used in this case not as an electron donor in energy metabolism, but as an electron donor for CO_2 reduction (autotrophy). At neutral pH where these organisms thrive, the Fe^{3+}/Fe^{2+} couple (about 0.2 V) is much less electropositive than at pH 2. Thus, electrons from Fe^{2+} can reduce cytochrome _c_ in the photosystem of purple bacteria (see Section 17.5 for a discussion of anoxygenic photosynthesis). The organisms involved, which are species of phototrophic purple bacteria (Figure 17.31_b_), can also use FeS as electron donor; under these conditions both Fe^{2+} and S^{2-} are oxidized as electron donors.

Certain phototrophic green sulfur bacteria (genus _Chlorobium_; ∞ Section 12.32) can also use Fe^{2+} as a photosynthetic electron donor. Moreover, various chemotrophic denitrifying bacteria have been isolated that can couple the oxidation of Fe^{2+} to the reduction of NO_3^- to N_2 and grow under anoxic conditions. However, like the aerobic iron bacteria, in these organisms iron is an electron donor for _both_ energy and reducing power needs.

(a) (b)

● **Figure 17.31 Ferrous iron oxidation by anoxygenic phototrophic bacteria.** (a) Fe^{2+} oxidation in anoxic tube cultures. Left to right: Sterile medium, inoculated medium, a growing culture. The brown–red color is mainly due to $Fe(OH)_3$ precipitate. (b) Phase contrast photomicrograph of an iron-oxidizing purple bacterium. The bright refractile areas within cells are gas vesicles (∞ Section 4.12). The granules outside the cells are iron precipitates. This organism is phylogenetically related to the purple sulfur bacterium _Chromatium_ (∞ Section 12.2).

The discovery of Fe^{2+}-oxidizing phototrophs has important implications for both understanding the evolution of photosynthesis and explaining the large deposits of ferric iron found in ancient sediments. Such ferric iron was previously thought to have been formed from the oxidation of Fe^{2+} by O_2 produced by oxygenic phototrophs (∞ Section 11.1). However, because of the age of these sediments, it is more likely that the ferric iron was formed by anoxygenic phototrophs oxidizing Fe^{2+} in anoxic environments.

 17.11 Concept Check

The iron bacteria are chemolithotrophs able to use ferrous iron (Fe^{2+}) as sole energy source. Most iron bacteria grow only at acid pH and are often associated with acid pollution from mineral and coal mining. Some phototrophic purple bacteria can oxidize Fe^{2+} to Fe^{3+} anaerobically.

◆ Why is only a very small amount of energy available from the oxidation of Fe^{2+} to Fe^{3+} at acidic pH?

◆ What is the function of *rusticyanin* and where is it found in the cell?

◆ How can Fe^{2+} be oxidized anoxically?

17.12 Nitrification and Anammox

The most common *inorganic nitrogen compounds* used as electron donors are ammonia (NH_3) and nitrite (NO_2^-). These compounds are oxidized aerobically by the chemolithotrophic **nitrifying bacteria** (∞ Section 12.3) in the process of **nitrification**. The nitrifying bacteria are widely distributed in soil and water. One group, the *nitrosifyers* (*Nitrosomonas* is one genus), oxidizes ammonia to nitrite, and another group (*Nitrobacter*) oxidizes nitrite to nitrate. The complete oxidation of ammonia to nitrate, an eight-electron transfer, is thus carried out by two groups of organisms acting in concert (see the Microbial Sidebar).

Bioenergetics and Enzymology of Nitrification

The electrons from nitrogen compounds enter an electron transport chain, and electron flow establishes a proton motive force linked to ATP synthesis. However, because of the reduction potential of their electron donors, nitrifying bacteria are faced with bioenergetic problems similar to those of the sulfur chemolithotrophs. The E_0' of the NO_2^-/NH_3 couple (the first step in the oxidation of NH_3) is $+0.34$ V. The E_0' of the NO_3^-/NO_2^- couple is even higher, about $+0.43$ V. These relatively high reduction potentials mean that nitrifying bacteria must donate electrons to their electron transport chains at rather late steps in the overall process. This effectively limits the energy released and the amount of ATP that can be produced from each pair of electrons.

Several key enzymes are involved in oxidizing reduced nitrogen compounds. In ammonia-oxidizing bacteria, NH_3 is oxidized by **ammonia monooxygenase** (see Section 17.22 for a discussion of monooxygenase enzymes) that produces NH_2OH and H_2O (Figure 17.32). *Hydroxylamine oxidoreductase* then oxidizes NH_2OH to NO_2^-, removing *four* electrons in the process. Ammonia monooxygenase is an integral membrane protein, whereas hydroxylamine oxidoreductase is periplasmic (Figure 17.32●). In the reaction carried out by ammonia monooxygenase,

$$NH_3 + O_2 + 2 H^+ + 2 e^- \longrightarrow NH_2OH + H_2O$$

there is a need for two exogenously supplied electrons plus two protons to reduce one atom of oxygen to water. These electrons originate from the oxidation of hydroxylamine and are supplied to ammonia monooxygenase from hydroxylamine oxidoreductase via cytochrome c and ubiquinone (Figure 17.32). Thus, for every *four* electrons generated from the oxidation of NH_3 to NO_2^-, only *two* actually reach the terminal oxidase (cytochrome aa_3, Figure 17.32).

Nitrite-oxidizing bacteria employ the enzyme *nitrite oxidoreductase* to oxidize nitrite to nitrate, with electrons traveling a very short electron transport chain (because of the high potential of the NO_3^-/NO_2^- couple) to the terminal oxidase (Figure 17.33●). Cytochromes of the a and c types are present in the electron transport chain of

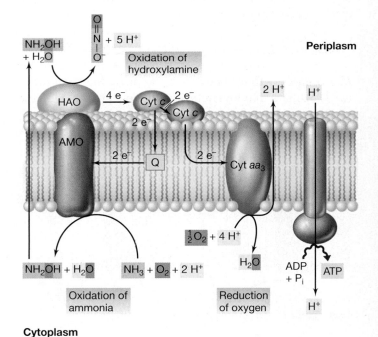

● **Figure 17.32 Oxidation of ammonia and electron flow in ammonia-oxidizing bacteria.** The reactants and the products of this reaction series are highlighted. The cytochrome c (cyt c) in the periplasm is a different form of cyt c than that in the membrane. AMO, ammonia monooxygenase; HAO, hydroxylamine oxidoreductase; Q, ubiquinone.

nitrite oxidizers, and generation of a proton motive force (which ultimately drives ATP synthesis) occurs through the activity of cytochromes aa_3 (Figure 17.33). Like the situation with iron oxidation (see Section 17.11), only small amounts of energy are available in this reaction. Thus, growth yields of nitrifying bacteria are low.

Carbon Metabolism in Nitrifying Bacteria

Like sulfur- and iron-oxidizing chemolithotrophs, aerobic nitrifying bacteria employ the Calvin cycle for CO_2 fixation, and the ATP and reducing power requirements of this process place additional burdens on an already relatively low-yielding energy-generating system (NADH to drive the Calvin cycle is formed by reverse electron flow). The energetic constraints are particularly severe for nitrite oxidizers, and it is perhaps for this reason that most nitrite oxidizers can also grow chemo-organotrophically on glucose and certain other organic substrates (∞ Section 12.3).

Anoxic Ammonia Oxidation: Anammox

Although classical nitrifying bacteria are *strict aerobes*, at least when growing on their reduced nitrogen substrates, ammonia can also be oxidized under **anoxic** conditions. This process, known as **anammox** (for *an*oxic *amm*onia *ox*idation), is highly exergonic and linked to energy metabolism of the organisms involved. Anammox involves the oxidation of ammonia with nitrite as the electron acceptor to yield gaseous nitrogen:

$$NH_4^+ + NO_2^- \longrightarrow N_2 + 2\,H_2O \quad \Delta G^{0'} = -357 \text{ kJ}$$

The organism that catalyzes anammox, *Brocadia anammoxidans*, is a phylogenetically distinct member of the Planctomycetes phylum of *Bacteria* (∞ Section 12.28) (Figure 17.34●). Planctomycetes are unusual members of the *Bacteria*, lacking peptidoglycan and containing membrane-enclosed compartments inside the cell, including a structure analogous to the nucleus of eukaryotic cells (∞ Section 12.28). In *B. anammoxidans* a major compartment is present called the *anammoxosome*; this is the location of the anammox reaction (Figure 17.34b).

The anammoxosome is a unit membrane-enclosed structure; in other words, an *organelle* (Figure 17.34b). Yet, *Brocadia* is clearly a prokaryote. The anammoxosome is another type of membrane-enclosed compartment beside the aforementioned "nucleus." Anammoxosome membrane lipids are unique as well. The lipids consist of fatty acids containing multiple cyclobutane rings (∞ Figure 11.24) that are connected to glycerol by both ester and ether bonds. The lipids aggregate to form an unusually dense membrane structure, highly resistant to diffusion. The strong anammoxosome membrane likely

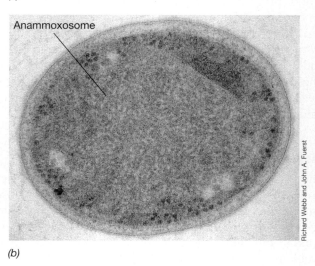

(a)

(b)

● **Figure 17.34 The anammox organism, *Brocadia anammoxidans.*** (a) Phase-contrast photomicrograph. A single cell is about 1 μm in diameter. (b) Transmission electron micrograph of a cell. Notice the membrane enclosed compartments including the large fibrillar anammoxosome in the center portion of the cell (arrow). *Brocadia* is phylogenetically related to the *Planctomyces*, organisms that contain several different types of membrane-enclosed compartments within their cytoplasm (∞ Section 12.28 and Figure 12.87).

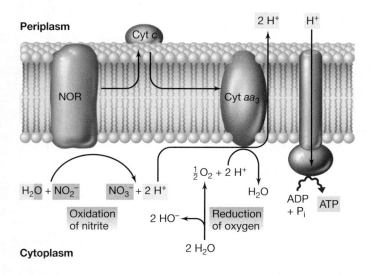

● **Figure 17.33 Oxidation of nitrite to nitrate by nitrifying bacteria.** The reactants and products of this reaction series are highlighted. NOR, nitrite oxidoreductase.

protects the cell from the toxic intermediates produced during the anammox reaction.

The source of NO_2^- in the anammox reaction is the product of ammonia oxidation by aerobic nitrifying bacteria (Figure 17.32). The two groups of nitrifyers, aerobic (for example, *Nitrosomonas*) and anaerobic (*Brocadia*), live together in ammonia-rich habitats such as sewage and other wastewaters. In these environments suspended particles are present that contain both oxic and anoxic zones where the two groups of ammonia oxidizers can coexist. In mixed laboratory cultures, high levels of oxygen inhibit anammox and favor classic nitrification, and thus it is likely that in nature the extent of anammox-mediated ammonia oxidation is governed by the concentration of O_2 in the system.

Like classical nitrifying bacteria, *Brocadia anammoxidans* is also an autotroph. The anammox organism can grow with CO_2 as sole carbon source and uses nitrite as electron donor to produce cell material:

$$CO_2 + 2\,NO_2^- + H_2O \longrightarrow CH_2O + 2\,NO_3^-$$

Although an autotrophic nitrifyer, it appears that *B. anammoxidans* lacks Calvin cycle enzymes, and the mechanism of CO_2 fixation is unclear. However, the use of nitrite as electron donor for CO_2 fixation along with the production of nitrate is exactly the same reaction carried out by aerobic nitrifyers like *Nitrobacter* (see Figure 17.33). Thus, although phylogenetically distinct, aerobic nitrifyers and the anammox organism share common substrates and ecology.

Before anammox was understood, it was thought that ammonia was stable in anoxic environments. Now, however, it is clear that ammonia can be oxidized in the absence of O_2. From an environmental standpoint, anammox is a very beneficial process in the treatment of anoxic wastewaters. The removal of ammonia and related amines helps reduce the fixed nitrogen pollution of rivers and streams, thereby maintaining high water quality. Ecological studies of anammox have shown that this process also occurs in marine sediments by organisms similar to *Brocadia*. This discovery has helped account for a significant fraction of ammonia known to be lost from marine environments but that has been unaccounted for as to mechanism.

 17.12 Concept Check

Ammonia and nitrite can be used as electron donors by the nitrifying bacteria. The ammonia-oxidizing bacteria produce nitrite, which is then oxidized by the nitrite-oxidizing bacteria to nitrate. Anoxic NH_3 oxidation is coupled to both N_2 and NO_3^- production in the anammoxosome.

◆ What is the inorganic electron donor for *Nitrosomonas*? For *Nitrobacter*?

◆ What are the substrates for the enzyme *ammonia monooxygenase*?

◆ What do nitrifying bacteria use as a carbon source?

◆ What is the *anammox* reaction and how does it differ from aerobic nitrification?

III THE ANAEROBIC WAY OF LIFE: ANAEROBIC RESPIRATIONS

In eukaryotes, anaerobic growth is rare. In prokaryotes, by contrast, anaerobic growth is common, and the mechanisms of anaerobic metabolism are highly diverse. In the next several sections we survey anaerobic respiration and see the many ways by which prokaryotes can conserve energy using electron acceptors *other* than O_2.

17.13 Anaerobic Respiration

We examined the process of *aerobic* respiration in some detail in Chapter 5. As we noted there, molecular oxygen functions as a terminal electron acceptor, accepting electrons from electron carriers by way of an electron transport chain. However, we also noted that a variety of other electron acceptors can be used instead of O_2, in which case the process is called **anaerobic respiration**. Here we consider some of these processes.

The bacteria carrying out anaerobic respiration possess electron transport systems containing cytochromes, quinones, iron-sulfur proteins, and other typical electron transport proteins. Their respiratory systems are thus analogous to those of conventional aerobes. In some cases, such as with the denitrifying bacteria, the anaerobic respiration process competes in the same organism with an aerobic one. In such cases, if O_2 is present, aerobic respiration is favored, and when O_2 is depleted from the environment, the alternate electron acceptor is reduced. Other organisms carrying out anaerobic respiration are *obligate* anaerobes and are unable to use O_2.

Alternative Electron Acceptors and the Electron Tower

The energy released from the oxidation of an electron donor using O_2 as electron acceptor is higher than if the same compound is oxidized with an alternate electron acceptor (∞ Figure 5.9). These energy differences are apparent if the reduction potentials of each acceptor are examined (Figure 17.35●). Because the O_2/H_2O couple is the most electropositive, more energy is available when O_2 is used than when another electron acceptor is used. Other electron acceptors that are near the O_2/H_2O couple are Fe^{3+}, NO_3^-, and NO_2^-. More electronegative acceptors are SO_4^{2-}, S^0, and CO_2. A summary of the most common types of anaerobic respiration is given in Figure 17.35.

● **Figure 17.35 Examples of anaerobic respirations.** The couples are arranged in order from most electronegative E_0' (top) to most electropositive E_0' (bottom). Compare with Figure 5.9 to see how the energy yields of these anaerobic respirations vary.

Assimilative and Dissimilative Metabolism

Inorganic compounds such as NO_3^-, SO_4^{2-}, and CO_2 are reduced by many organisms as sources of cellular nitrogen, sulfur, and carbon, respectively. The end products of such reductions are primarily amino groups ($-NH_2$), sulfhydryl groups ($-SH$), and organic carbon compounds, respectively. When an inorganic compound such as NO_3^-, SO_4^{2-}, or CO_2 is reduced for use in biosynthesis, it is said to be *assimilated*, and the reduction process is called **assimilative metabolism**. Assimilative metabolism of NO_3^-, SO_4^{2-}, and CO_2 is conceptually quite different from their use as electron acceptors for *energy* metabolism in anaerobic respiration. To distinguish these two kinds of reduction processes, the use of

these compounds as electron acceptors in energy metabolism is called **dissimilative metabolism**.

Assimilative and dissimilative metabolism differ markedly. In *assimilative* metabolism, only enough of the compound (NO_3^-, SO_4^{2-}, or CO_2) is reduced to satisfy the needs for cell growth. The products are eventually converted to cell material in the form of macromolecules. In *dissimilative* metabolism, a large amount of the electron acceptor is reduced, and the reduced product is *excreted* into the environment. Many organisms carry out assimilative metabolism of compounds such as NO_3^-, SO_4^{2-}, and CO_2 (for example, many *Bacteria*, *Archaea*, fungi, algae, and higher plants), whereas only a restricted variety of organisms, primarily prokaryotes, carry out dissimilative metabolism.

 17.13 Concept Check

Although oxygen is the most widely used electron acceptor in energy-yielding metabolism, a number of other compounds can be used as electron acceptors. This process of anaerobic respiration is less energy efficient but makes it possible for respiration to occur in environments where oxygen is absent.

◆ What is anaerobic respiration?

◆ With H_2 as electron donor, why is the reduction of NO_3^- a more favorable reaction than the reduction of S^0?

17.14 Nitrate Reduction and Denitrification

Inorganic nitrogen compounds are some of the most common electron acceptors in anaerobic respiration. Table 17.2 summarizes the various inorganic nitrogen species with their oxidation states. The most widespread inorganic nitrogen compounds in nature are *ammonia* and *nitrate*, both of which are formed in the atmosphere by inorganic chemical processes, and also nitrogen gas—N_2—the most stable form of nitrogen in nature. We discuss *nitrogen fixation*, the use of N_2 as a biosynthetic nitrogen source, later in this chapter (see Section 17.28).

One of the most common alternative electron acceptors is nitrate, NO_3^-, which is reduced to N_2O, NO, and N_2. Because these products of nitrate reduction are all gaseous, they can easily be lost from the environment, a process called **denitrification** (Figure 17.36●). The pro-

Table 17.2	Oxidation states of key nitrogen compounds	
Compound	**Oxidation state**
Organic N ($R-NH_2$)	-3
Ammonia (NH_3)	-3
Nitrogen gas (N_2)	0
Nitrous oxide (N_2O)	$+1$ (average per N)
Nitrogen oxide (NO)	$+2$
Nitrite (NO_2^-)	$+3$
Nitrogen dioxide (NO_2)	$+4$
Nitrate (NO_3^-)	$+5$

Nitrate (NO_3^-)

    ↓ Nitrate reductase

Nitrite (NO_2^-)

    ↓ Nitrite reductase

Nitric oxide (NO)

    ↓ Nitric oxide reductase

Nitrous oxide (N_2O)

    ↓ Nitrous oxide reductase

Dinitrogen (N_2)

Gases, released to the atmosphere (denitrification)

● **Figure 17.36 Steps in the dissimilative reduction of nitrate.** Some organisms, for example *Escherichia coli*, can carry out only the first step. All enzymes involved are derepressed by anoxic conditions. Also, some prokaryotes are known that can reduce NO_3^- to NH_4^+ in dissimilative metabolism.

cess is the main means by which gaseous N_2 is formed biologically. N_2 is much less available to organisms than nitrate as a source of nitrogen, so for agricultural purposes, at least, denitrification is a detrimental process. For sewage treatment (∞ Section 28.2), however, denitrification is beneficial because it converts NO_3^- to N_2. This effectively decreases the load of fixed nitrogen in the sewage treatment effluent that can stimulate algal growth (∞ Section 19.5).

Biochemistry of Dissimilative Nitrate Reduction

The enzyme involved in the first step of dissimilative nitrate reduction, *nitrate reductase*, is a molybdenum-containing membrane-integrated enzyme whose synthesis is repressed by molecular oxygen. All subsequent enzymes of the pathway (Figure 17.37●) are coordinately regulated and thus also repressed by O_2, but in addition to anoxic conditions, nitrate must also be present before these enzymes are fully expressed.

The first product of nitrate reduction is nitrite (NO_2^-), and the enzyme *nitrite reductase* reduces it to

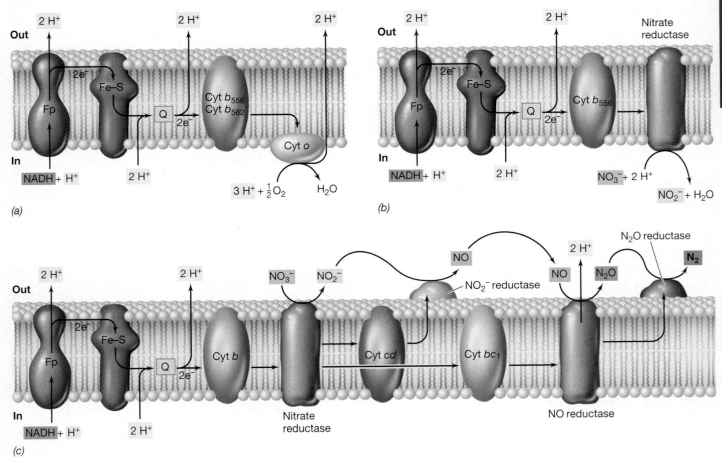

● **Figure 17.37 Respiration.** Electron transport processes in the membrane of *Escherichia coli* when (a) O_2 or (b) NO_3^- is used as an electron acceptor and NADH is the electron donor. Fp, flavoprotein; Q, ubiquinone. Under high oxygen conditions, the sequence of carriers is cyt b_{562} → cyt o → O_2. However, under low oxygen conditions (not shown), the sequence is cyt b_{568} → cyt d → O_2. Note how more protons are translocated per two electrons oxidized aerobically during electron transport reactions than anaerobically with nitrate as electron acceptor, because the aerobic terminal oxidase (cyt o) can pump one proton. (c) Possible scheme for electron transport in membranes of *Pseudomonas stutzeri* during denitrification. Nitrate and nitric oxide (NO) reductases are located in the cytoplasmic membrane whereas nitrite (NO_2^-) and nitrous oxide (N_2O) reductases are periplasmic. The immediate electron donors to the various reductases, with the exception of nitrate reductase, have not been definitively identified.

nitric oxide (NO) (Figure 17.37c). Some organisms can reduce NO_2^- to ammonia (NH_3) in a dissimilative process, but the production of gaseous products—*denitrification*—is of greatest global significance. This is because it consumes a fixed form of nitrogen (NO_3^-) readily available to plants and produces gaseous nitrogen compounds, some of which are of environmental significance. For example, N_2O can be converted to NO by sunlight, and NO reacts with ozone in the upper atmosphere to form nitrite (NO_2^-), which returns to Earth as acid rain (nitrous acid, HNO_2). The remaining biochemical steps in denitrification are shown in Figure 17.37c.

The biochemistry of dissimilative nitrate reduction has been studied in detail in several organisms including *Escherichia coli*, where NO_3^- is reduced only to NO_2^- (Figure 17.37b), and *Paracoccus denitrificans* and *Pseudomonas stutzeri*, where true denitrification occurs (Figure 17.37c). The *E. coli* nitrate reductase accepts electrons from a *b*-type cytochrome, and a comparison of the electron transport chains in aerobic versus anaerobic respiration in *E. coli* is shown in Figure 17.37a, b. As seen here, because of the reduction potential of the NO_3^-/NO_2^- couple (+0.43 V), only two proton translocating steps occur during nitrate reduction while three can occur in aerobic respiration ($\frac{1}{2}O_2/H_2O$, +0.82 V). In *P. denitrificans* and *P. stutzeri* nitrogen oxides are formed from nitrite by *nitrite reductase, nitric oxide reductase,* and *nitrous oxide reductase,* as summarized in Figure 17.37c. During electron transport, a proton motive force is established, and ATP is produced by ATPase in the usual fashion. Additional ATP is available when NO_3^- is reduced to N_2 because the NO reductase is linked to proton extrusion (Figure 17.37c).

Other Properties of Denitrifying Prokaryotes

Most denitrifying prokaryotes are phylogenetically members of the Proteobacteria (◌ Sections 12.2–12.18) and are facultative aerobes. Aerobic respiration occurs when air is present, even if nitrate is also present in the medium. Many denitrifying bacteria will also reduce other electron acceptors anaerobically, such as ferric iron (Fe^{3+}) and certain organic electron acceptors (see Section 17.18). In addition, many denitrifying bacteria can grow by fermentation. Thus, the denitrifying bacteria are quite metabolically diverse in terms of alternative energy-generating mechanisms.

 17.14 Concept Check

Nitrate is a commonly used electron acceptor in anaerobic respiration. Its use requires the enzyme nitrate reductase that reduces nitrate to nitrite. Many bacteria that use nitrate in anaerobic respiration eventually produce N_2, a process called *denitrification*.

◆ For *Escherichia coli*, why is more energy released in aerobic respiration than during NO_3^- reduction?

◆ Where is the dissimilative nitrate reductase found in the cell? What unusual metal does it contain?

◆ Why does an organism like *Pseudomonas stutzeri* derive more energy from NO_3^- respiration than does *Escherichia coli*?

17.15 Sulfate Reduction

Several inorganic *sulfur* compounds are important electron acceptors in anaerobic respiration. A summary of the oxidation states of the key sulfur compounds is given in Table 17.3. Sulfate, the most oxidized form of sulfur, is one of the major anions in seawater and is used by the *sulfate-reducing bacteria*, a group that is widely distributed in nature. The end product of sulfate reduction is H_2S, an important natural product that participates in many biogeochemical processes (◌ Section 19.13).

Again, as with nitrogen, it is important to distinguish between *assimilative* and *dissimilative* metabolism. Many organisms, including higher plants, algae, fungi, and most prokaryotes, use sulfate as a sulfur source for biosynthesis. The ability to use sulfate as an *electron acceptor* for energy-generating processes, however, involves the large-scale reduction of SO_4^{2-} and is restricted to the sulfate-reducing bacteria. In assimilative sulfate reduction, the H_2S formed is immediately converted into organic sulfur in the form of amino acids, and so on, but in dissimilative sulfate reduction, the H_2S is excreted.

As the reduction potentials in Table A1.2 and Figure 17.35 show, sulfate is a much less favorable electron acceptor than either O_2 or NO_3^-. However, sufficient energy to make ATP is available when an electron donor that yields NADH or FADH is used. Because of the less favorable energetics, growth yields are lower for an organism growing on SO_4^{2-} than for one growing on O_2 or NO_3^-. Table 17.3 lists some of the electron donors used by sulfate-reducing bacteria. H_2, lactate, and pyruvate are used by a wide vari-

Table 17.3	**Sulfur compounds and electron donors for sulfate reduction**
Compound	**Oxidation state**
Oxidation states of key sulfur compounds	
Organic S (R—SH)	−2
Sulfide (H_2S)	−2
Elemental sulfur (S^0)	0
Thiosulfate ($S_2O_3^{2-}$)	+2 (average per S)
Sulfur dioxide (SO_2)	+4
Sulfite (SO_3^{2-})	+4
Sulfate (SO_4^{2-})	+6
Some electron donors used for sulfate reduction	
H_2	Acetate
Lactate	Propionate
Pyruvate	Butyrate
Ethanol and other alcohols	Long-chain fatty acids
Fumarate	Benzoate
Malate	Indole
Choline	Hexadecane

ety of sulfate-reducing bacteria, while the others have more restricted use. However, a large variety of morphological and physiological types of sulfate-reducing bacteria are known (⌦ Sections 12.18 and 13.17). With the exception of *Archaeoglobus*, a member of the *Archaea*, all known sulfate-reducing prokaryotes are *Bacteria*.

Figure 17.38 Biochemistry of sulfate reduction. (a) Two forms of *active sulfate* can be made, adenosine 5′-phosphosulfate (APS) and phosphoadenosine 5′-phosphosulfate (PAPS). Both are derivatives of adenosine diphosphate (ADP), with the second phosphate of ADP being replaced by sulfate. (b) Schemes of assimilative and dissimilative sulfate reduction.

Biochemistry and Energetics of Sulfate Reduction

The reduction of SO_4^{2-} to H_2S, an eight-electron reduction (Table 17.3), proceeds through a number of intermediate stages. The sulfate ion is stable and cannot be reduced without first being activated. Sulfate is activated by means of ATP. The enzyme *ATP sulfurylase* catalyzes the attachment of the sulfate ion to a phosphate of ATP, leading to the formation of **adenosine phosphosulfate (APS)** as shown in Figure 17.38●. In dissimilative sulfate reduction, the sulfate in APS is reduced directly to sulfite (SO_3^{2-}) by the enzyme *APS reductase* with the release of AMP. In assimilative reduction, another P is added to APS to form *phosphoadenosine phosphosulfate (PAPS)* (Figure 17.38*a*), and only then is the sulfate reduced. In both cases, the first product of sulfate reduction is *sulfite*, SO_3^{2-}. Once SO_3^{2-} is formed, sulfide is formed by the enzyme *sulfite reductase* (Figure 17.38*b*).

In the process of dissimilative sulfate reduction, electron transport reactions occur leading to proton motive force formation, and this drives ATP synthesis by an ATPase. A major electron carrier is cytochrome c_3, a periplasmic low potential cytochrome (Figure 17.39●). Cytochrome c_3 accepts electrons from a periplasmically located hydrogenase (see below) and transfers these electrons to a membrane-associated protein complex.

● Figure 17.39 Electron transport and energy conservation in sulfate-reducing bacteria. In addition to external hydrogen (H_2), H_2 originating from the catabolism of organic compounds such as lactate and pyruvate can fuel hydrogenase. The enzymes hydrogenase (H_2ase), cytochrome c_3, and a cytochrome complex (Hmc) are periplasmic proteins. A separate protein functions to shuttle electrons across the cytoplasmic membrane from Hmc to a cytoplasmic iron-sulfur protein that supplies electrons to APS reductase (forming SO_3^{2-}) and sulfite reductase (forming H_2S see Figure 17.38*b*). LDH, lactate dehydrogenase.

This complex, called *Hmc*, carries the electrons across the cytoplasmic membrane, thus making them available to APS reductase and sulfite reductase, which are cytoplasmic enzymes (Figure 17.39).

The enzyme hydrogenase plays a central role in sulfate reduction whether an organism like *Desulfovibrio* is growing on H_2, *per se*, or on an organic compound, like lactate. This is because lactate is converted through pyruvate to acetate (the latter is mainly excreted because *Desulfovibrio* is a nonacetate-oxidizing sulfate reducer, ⚭ Section 12.18), with the production of H_2. The H_2 produced crosses the cytoplasmic membrane and is oxidized by the periplasmic hydrogenase to initiate a proton motive force (Figure 17.39). Growth yields of sulfate-reducing bacteria suggest that one ATP is produced for each SO_4^{2-} reduced to HS^-. With H_2 the reaction is:

$$4 H_2 + SO_4^{2-} + H^+ \longrightarrow HS^- + 4 H_2O$$
$$\Delta G^{0\prime} = -152 \text{ kJ}$$

When lactate or pyruvate is the electron donor, not only is ATP produced from the proton motive force, but additional ATP is produced during the oxidation of pyruvate to acetate plus CO_2 via acetyl-CoA and acetyl-phosphate (see Section 17.19 for further discussion of this).

Acetate Use and Autotrophy

Many sulfate reducers can completely oxidize acetate to CO_2 as an electron donor for sulfate reduction (⚭ Section 12.18):

$$CH_3COO^- + SO_4^{2-} + 3 H^+ \longrightarrow$$
$$2 CO_2 + H_2S + 2 H_2O \quad \Delta G^{0\prime} = -57.5 \text{ kJ}$$

Although the energetics of this process are not as well understood as H_2 or lactate metabolism, the mechanism for acetate oxidation is known. With few exceptions acetate is oxidized to CO_2 by the *acetyl-CoA pathway*, a series of reversible reactions used by a wide variety of anaerobes for acetate synthesis or acetate oxidation (see Section 17.16). This pathway employs the key enzyme *carbon monoxide dehydrogenase* and was first discovered in homoacetogenic bacteria that make acetate from $H_2 + CO_2$ as a mechanism of energy conservation (see Section 17.16). A few sulfate-reducing bacteria can also grow autotrophically in an anoxic mineral salts medium containing H_2 (as electron donor), SO_4^{2-} (as electron acceptor), and CO_2 (as carbon source). When growing under these conditions, autotrophic sulfate reducers use the acetyl-CoA pathway as a means of producing cell material. The acetate-oxidizing sulfate-re-ducing bacterium *Desulfobacter* lacks acetyl-CoA pathway enzymes and oxidizes acetate through the citric acid cycle (⚭Figure 5.22), but this seems to be the exception rather than the rule.

Sulfur Disproportionation

Certain sulfate-reducing bacteria are capable of a form of energy metabolism called **disproportionation**, using sulfur compounds of intermediate oxidation state. The term *disproportionation* refers to the splitting of a compound into two new compounds, one of which is *more oxidized* and one of which is *more reduced* than the original substrate. For example, *Desulfovibrio sulfodismutans* can disproportionate thiosulfate as follows:

$$S_2O_3^{2-} + H_2O \longrightarrow SO_4^{2-} + H_2S$$
$$\Delta G^{0\prime} = -21.9 \text{ kJ/reaction}$$

Note that in this reaction one sulfur atom of $S_2O_3^{2-}$ becomes more oxidized (forming SO_4^{2-}) while the other becomes more reduced (forming H_2S). The oxidation of thiosulfate by *D. sulfodismutans* drives formation of a proton motive force that is used by the organism to make ATP. Other reduced sulfur compounds such as sulfite (SO_3^{2-}) and sulfur (S^0) can also be disproportionated by one or another sulfate reducer, as well. These forms of metabolism allow sulfate-reducing bacteria to recover energy from sulfur intermediates produced from the oxidation of H_2S by sulfur chemolithotrophs that coexist with them in nature (⚭ Section 19.13).

Phosphite Oxidation

At least one sulfate-reducing bacterium is capable of coupling phosphite (HPO_3^-) oxidation to sulfate reduction. The reaction is chemolithotrophic, and the products are phosphate and sulfide:

$$4 HPO_3^- + SO_4^{2-} + H^+ \longrightarrow 4 HPO_4^{2-} + HS^-$$
$$\Delta G^{0\prime} = -364 \text{ kJ}$$

The organism involved, *Desulfotignum phosphitoxidans*, requires only CO_2 for its carbon needs (it is thus an *autotroph*). It is a strict anaerobe, which it must be since phosphite spontaneously oxidizes in air. The source of phosphite in nature is not clear. However, since *D. phosphitoxidans* can use phosphite as sole energy source, it is likely produced in anoxic environments, probably from the degradation of organophosphates. Along with sulfur disproportionation (also a chemolithotrophic process) and H_2 utilization, phosphite oxidation emphasizes the chemolithotrophic diversity of sulfate-reducing bacteria.

 17.15 Concept Check

The sulfate-reducing bacteria reduce sulfate to hydrogen sulfide. The reduction of sulfate first requires activation by a reaction with ATP to form the compound adenosine phosphosulfate (APS). Electron donors for sulfate reduction include H_2, organic compounds, and even phosphite. Disproportionation of sulfur compounds is an additional energy-yielding strategy for certain members of this group.

- ◆ Identify the following: S^0, SO_4^{2-}, SO_3^{2-}, $S_2O_3^{2-}$, H_2S.
- ◆ How is sulfate converted to sulfite during dissimilative sulfate reduction?
- ◆ Why is H_2 of importance to sulfate-reducing bacteria?
- ◆ Give an example of disproportionation.

17.16 Acetogenesis

Carbon dioxide, CO_2, is common in nature and usually abundant in anoxic habitats because it is a major product of energy metabolism of chemoorganotrophs. Two major groups of strictly anaerobic prokaryotes can use CO_2 as an electron acceptor in energy metabolism, **homoacetogens** and **methanogens**. Hydrogen is a major electron donor for both of these organisms, and an overview of the processes of methanogenesis and acetogenesis is shown in Figure 17.40●. Both processes result in the generation of ion gradients either of H^+ or Na^+, which fuel ATPases in the membrane. Acetogenesis also involves energy conservation via substrate-level phosphorylation.

Organisms and Pathway

Homoacetogens can carry out the reaction:

$$4\,H_2 + H^+ + 2\,HCO_3^- \longrightarrow CH_3COO^- + 4\,H_2O$$

In addition to H_2, electron donors for acetogenesis include a variety of C_1 compounds, sugars, organic and amino acids, alcohols, and certain nitrogen bases, depending on the organism. Many homoacetogens can also reduce NO_3^- and $S_2O_3^{2-}$. However, CO_2 reduction is probably the major reaction of ecological significance.

The major unifying thread among homoacetogens is the pathway of CO_2 reduction. Homoacetogens convert CO_2 to acetate by the **acetyl-CoA pathway**, and in many homoacetogens autotrophic growth via this pathway also occurs. Table 17.4 lists the major organisms that produce acetate or oxidize acetate via the acetyl-CoA pathway. Organisms such as *Acetobacterium woodii* and *Clostridium aceticum* can grow either chemoorganotroph-

Table 17.4	Organisms employing the acetyl-CoA pathway of CO₂ fixation
I.	**Acetate synthesis, the result of energy metabolism**
	Acetoanaerobium noterae
	Acetobacterium woodii
	Acetobacterium wieringae
	Acetogenium kivui
	Acetitomaculum ruminis
	Clostridium aceticum
	Clostridium thermaceticum
	Clostridium formicaceticum
	Desulfotomaculum orientis
	Sporomusa paucivorans
	Eubacterium limosum (also produces butyrate)
	Treponema primitia (from termite hindguts)
II.	**Acetate synthesis in autotrophic metabolism**
	Autotrophic homoacetogenic bacteria
	Autotrophic methanogens
	Autotrophic sulfate-reducing bacteria
III.	**Acetate oxidation in energy metabolism**
	Reaction: Acetate + 2 H₂O → 2 CO₂ + 8 H
	Group II sulfate reducers (other than *Desulfobacter*)
	Reaction: Acetate → CO₂ + CH₄
	Acetotrophic methanogens (*Methanosarcina, Methanosaeta*)

ically by fermentation of sugars (reaction 1) or chemolithotrophically through the reduction of CO_2 to acetate with H_2 (reaction 2) as electron donor. In either case, the major product is acetate:

$$(1)\ C_6H_{12}O_6 \longrightarrow 3\,CH_3COO^- + 3\,H^+$$

$$(2)\ 2\,HCO_3^- + 4\,H_2 + H^+ \longrightarrow CH_3COO^- + 4\,H_2O$$

Homoacetogens ferment glucose via the glycolytic pathway converting glucose to two molecules of pyruvate and two molecules of NADH (the equivalent of 4 H). From this point, two molecules of acetate are produced:

$$(3)\ 2\,\text{pyruvate}^- \longrightarrow 2\,\text{acetate}^- + 2\,CO_2 + 4\,H$$

The third acetate of the homoacetate fermentation comes from the reduction of the two molecules of CO_2 generated in reaction (3), using the four electrons generated from glycolysis *plus* the four electrons produced from the oxidation of two pyruvates to two acetates [reaction (3)]. Starting from pyruvate, then, the overall production of acetate can be written as

$$2\,\text{pyruvate}^- + 4\,H \longrightarrow 3\,\text{acetate}^- + H^+$$

Most homoacetogenic bacteria that produce and excrete acetate in energy metabolism are gram-positive, and many are classified in the genus *Clostridium*. A few other gram-positive and many different gram-negative bacteria use the acetyl-CoA pathway for autotrophic purposes, reducing CO_2 to acetate for cell carbon.

● **Figure 17.40 The contrasting processes of methanogenesis and acetogenesis.** Note the difference in free energy released in the reactions.

The acetyl-CoA pathway functions in autotrophic growth for certain sulfate-reducing bacteria (∞ Sections 12.18 and 19.13). It is also used by the methanogens, most of which can grow autotrophically on $H_2 + CO_2$ (∞ Sections 13.4, 17.17, and 19.10). By contrast, certain bacteria employ the reactions of the acetyl-CoA pathway primarily in the *reverse* direction as a means of oxidizing acetate to CO_2. These include acetotrophic methanogens (∞ Section 13.4) and sulfate-reducing bacteria (∞ Section 12.18).

Reactions of the Acetyl-CoA Pathway

Unlike other autotrophic pathways such as the Calvin cycle (see Section 17.6) or the reverse citric acid or hydroxypropionate cycles (see Section 17.7), the acetyl-CoA pathway of CO_2 fixation is *not* a cycle. Instead it involves the reduction of CO_2 via two linear pathways— one molecule of CO_2 is reduced to the methyl group of acetate, and the other molecule of CO_2 is reduced to the carbonyl group. They are assembled at the end to form acetyl-CoA (Figure 17.41●).

A key enzyme of the acetyl-CoA pathway is *carbon monoxide (CO) dehydrogenase*. CO dehydrogenase is a complex enzyme that contains the metals Ni, Zn, and Fe as cofactors. CO dehydrogenase catalyzes the following reaction:

$$CO_2 + H_2 \longrightarrow CO + H_2O$$

Net: $4 H_2 + 2 CO_2 \longrightarrow$ Acetate$^-$ + $2 H_2O$ + H^+

● **Figure 17.41 Reactions of the acetyl-CoA pathway.** THF, Tetrahydrofolate; B_{12}, vitamin B_{12} in an enzyme–bound intermediate. CO is bound to an Fe atom in CO dehydrogenase, and the CH_3 group to a nickel atom in an organic nickel compound in CO dehydrogenase. Note how formation of acetyl-CoA powers a Na^+ pump that is used to drive ATP synthesis and that ATP synthesis also occurs in the conversion of acetyl-CoA to acetate.

and the CO produced ends up in the *carbonyl* position of acetate (Figure 17.41). The methyl group of acetate originates from the reduction of CO_2 by a series of reactions involving the coenzyme *tetrahydrofolate* (Figure 17.41). The methyl group that is formed is then transferred from tetrahydrofolate to an enzyme containing vitamin B_{12} as cofactor (Figure 17.41). In the final step of the pathway, the CH_3 group is combined with CO in CO dehydrogenase to form acetate. Interestingly, the reaction mechanism here involves the CH_3 group, which is attached to an atom of nickel in the enzyme, combining with CO, which is bound to an atom of Fe in the enzyme, along with coenzyme A to form the final product, acetyl-CoA. This mechanism was the first alkyl nickel reaction to be discovered in biochemistry.

Because homoacetogens can grow at the expense of reactions of the acetyl-CoA pathway, this reaction sequence must be overall an energy-conserving one (Figure 17.41). One site of ATP synthesis is during the conversion of acetyl-CoA to acetate plus ATP (via acetyl-P) (see Section 17.19). However, additional energy-conserving steps occur because a sodium motive force (analogous to a proton motive force but involving Na^+ instead of H^+) is established across the cytoplasmic membrane during acetogenesis. This energized state of the membrane allows for energy conservation via the activity of a Na^+-powered ATPase. A similar situation occurs in the succinate fermenter *Propionigenium*, whose energy conservation from Na^+ gradients will be discussed in Section 17.20.

17.16 Concept Check

Homoacetogens are anaerobes that reduce CO_2 to acetate, usually with H_2 as electron donor. The mechanism of acetate formation is the acetyl-CoA pathway, a series of reactions widely distributed in obligate anaerobes as either a mechanism of autotrophy or for acetate catabolism.

◆ Draw the structure of acetate and identify the carbonyl group and the methyl group. What key enzyme of the acetyl-CoA pathway produces the *carbonyl* group of acetate?

◆ How do homoacetogens make ATP from the synthesis of acetate?

◆ If catabolism of fructose via glycolysis yields only *two* molecules of acetate, how can *Clostridium aceticum* ferment fructose by this pathway and produce *three* molecules of acetate?

17.17 Methanogenesis

The biological production of methane—**methanogenesis**— is carried out by a group of strictly anaerobic *Archaea* called the *methanogens*. We considered the basic properties, phylogeny, and taxonomy of the methanogens in Section 13.4; here we focus on their biochemistry and bioenergetics. The biological production of methane occurs through a series of reactions involving novel coenzymes and amazing complexity. We begin our discussion with a consideration of the coenzymes, which are central to methanogenic reactions, and on the production of methane starting from the substrates $H_2 + CO_2$.

C₁ Carriers in Methanogenesis

The key coenzymes in methanogenesis can be divided into two classes: (1) those involved in carrying the C_1 unit from the initial substrate, CO_2, to the final product; and (2) those that function in redox reactions to supply the electrons necessary for the reduction of CO_2 to CH_4 (Figure 17.42●, and see Figure 17.44).

The coenzyme **methanofuran** is involved in the first step of methanogenesis. Methanofuran contains the five-membered furan ring and an amino nitrogen atom that binds CO_2 (Figure 17.42a). **Methanopterin** (Figure 17.42b) is a methanogenic coenzyme that resembles the vitamin folic acid (⬤ Figure 20.17c) and is the C_1

● **Figure 17.42 Coenzymes of methanogenic Archaea.** The atoms shaded in brown or yellow are the sites of oxidation–reduction reactions (F_{420}—brown) or the position to which the C_1 moiety is attached during the reduction of CO_2 to CH_4 (methanofuran, methanopterin, and coenzyme M—yellow). The colors used to highlight a particular coenzyme itself (CoB is orange, for example) are used throughout in Figures 17.44–17.46 and can be used to follow the reactions in each figure.

carrier in the intermediate steps of CO_2 reduction to CH_4. **Coenzyme M (CoM)** (Figure 17.42c) is a small molecule that is involved in the terminal step of methanogenesis, the conversion of a methyl group (CH_3) to CH_4. Although not a C_1 carrier, the nickel-containing tetrapyrrole **coenzyme F_{430}** (Figure 17.42d) is also involved in the terminal step of methanogenesis as part of the methyl reductase enzyme complex (see later).

Redox Coenzymes

The coenzymes F_{420} and **7-mercaptoheptanoylthreonine phosphate**, or **coenzyme B (CoB)**, are electron *donors* in methanogenesis. Coenzyme F_{420} (Figure 17.42e) is a flavin derivative, structurally resembling the common flavin coenzyme FMN (∞ Figure 5.15). F_{420} also plays a role in methanogenesis as the electron donor in several steps of CO_2 reduction (see Figure 17.44). The oxidized form of F_{420} absorbs light at 420 nm and fluoresces blue-green. Such fluorescence is useful for the microscopic identification of an organism as a methanogen (Figure 17.43●). CoB is involved in the terminal step of methanogenesis catalyzed by the **methyl reductase enzyme complex**. As shown in Figure 17.42f, the structure of CoB resembles the vitamin pantothenic acid (which is part of acetyl-CoA) (∞ Figure 5.12).

Biochemistry of CO_2 Reduction to CH_4

The reduction of CO_2 to CH_4 generally occurs from H_2, but formate, carbon monoxide, and even certain organic compounds such as alcohols can also supply the electrons for CO_2 reduction. Figure 17.44● shows the steps in CO_2 reduction by H_2:

1. CO_2 is activated by a methanofuran-containing enzyme and subsequently reduced to the formyl level.

2. The formyl group is transferred from methanofuran to an enzyme containing methanopterin (MP in Figure 17.44). It is subsequently dehydrated and reduced in two separate steps to the methylene and methyl levels.

3. The methyl group is transferred from methanopterin to an enzyme containing CoM.

4. Methyl-CoM is reduced to methane by the methyl reductase system in which F_{430} and CoB are intimately involved. Coenzyme F_{430} removes the CH_3 group from CH_3-CoM, forming a Ni^{2+}-CH_3 complex. This is reduced by electrons from CoB, generating CH_4 and a disulfide complex of CoM and CoB (CoM-S—S-CoB).

5. Free CoM and CoB are regenerated by the reduction of CoM-S—S-CoB with H_2. As we will see, it is this reaction that allows for energy conservation in methanogenesis.

(a)　(b)

● **Figure 17.43 Green fluorescence due to the methanogenic coenzyme F_{420}.** (a) Autofluorescence in cells of the methanogen *Methanosarcina barkeri* due to the presence of the unique electron carrier F_{420}. A single cell is about 1.7 μm in diameter. The organisms were made visible with blue light in a fluorescence microscope (∞ Section 4.1 and Figure 4.6). (b) F_{420} fluorescence in cells of the methanogen *Methanobacterium formicicum*. A single cell is about 0.6 μm in diameter.

● **Figure 17.44 Pathway of methanogenesis from CO_2 using H_2 as electron donor.** MF, Methanofuran; MP, methanopterin; CoM, coenzyme M; F_{420red}, reduced coenzyme F_{420}; F_{430}, coenzyme F_{430}; CoB, coenzyme B. The carbon atom reduced is shown in yellow, and the source of electrons is highlighted in brown. See Figure 17.42 for the structures of the coenzymes and the text for discussion of the reversible Na^+ pump. The immediate electron donor in the first step of methanogenesis is unknown. Electrons for CO_2 reduction generally come from H_2 but in certain methanogens, a few organic compounds can be oxidized to yield electrons for CO_2 reduction (∞ Section 13.4).

(a) Methanol to CH$_4$

(b) Acetate to CH$_4$

Methanogenesis from Methyl Compounds and Acetate

We learned in Section 13.4 that in addition to H$_2$ + CO$_2$, methane can be formed from a variety of methylated compounds. Methyl compounds such as methanol are catabolized by donating methyl groups to a corrinoid protein to form CH$_3$-corrinoid (Figure 17.45●). Corrinoids are the parent structures of such compounds as vitamin B$_{12}$ and contain a porphyrin-like corrin ring with a central cobalt atom (∞ Figure 30.12). The CH$_3$-corrinoid complex donates the methyl group to CoM, yielding CH$_3$-CoM from which methane is formed in the same way as in the terminal step of CO$_2$ reduction just described (compare Figures 17.44 and 17.45a). If reducing power (such as H$_2$) is not available to drive the terminal step, some of the methanol must be oxidized to CO$_2$ to yield electrons. This occurs by reversal of steps in methanogenesis (Figure 17.45a).

When *acetate* is the substrate for methanogenesis, it is first activated to acetyl-CoA. The latter interacts with carbon monoxide dehydrogenase of the acetyl-CoA pathway (see Section 17.16). Then, the methyl group of acetate is transferred to the corrinoid enzyme to yield CH$_3$-corrinoid, and from there it goes through the CoM-mediated terminal step of methanogenesis (Figure 17.45b). Simultaneously, the CO group is oxidized to CO$_2$.

Autotrophy

Autotrophy in methanogens occurs via the acetyl-CoA pathway discussed in Section 17.16. As we have seen, parts of this pathway are already integrated into the catabolism of methanol and acetate (Figure 17.45). However, methanogens lack the tetrahydrofolate-driven series of reactions of the acetyl-CoA pathway that lead to the production of a methyl group (see Figure 17.41). But this is not a problem since methanogens either derive methyl groups directly from their electron donors (Figure 17.45) or make methyl groups during methanogenesis from H$_2$ + CO$_2$ (Figure 17.44). Thus methyl groups are abundant in the cell to begin with. The carbonyl group of the acetate produced during autotrophic growth of methanogens is derived from the enzyme CO dehydrogenase, and the terminal step in acetate synthesis occurs as described for homoacetogens (see Section 17.16 and Figure 17.41).

Energy Conservation in Methanogenesis

Under standard conditions, the free energy change in the reduction of CO$_2$ to CH$_4$ with H$_2$ is −131 kJ/mol. This is sufficient for the synthesis of at least one ATP. We mentioned that energy conservation in methanogenesis is linked to the terminal step, the *methyl reductase* step (Figure 17.44). The interaction of CoB with CH$_3$—CoM in this terminal step forms CH$_4$ and a heterodisulfide, CoM-S—S-CoB. The latter complex is then reduced with electrons from F$_{420}$ to regenerate

● **Figure 17.45** **The conversion of methanol and acetate to methane.** Parts of the acetyl-CoA pathway (see Figure 17.41) are involved in both reaction series. For growth on methanol, most methanol carbon is converted to CH$_4$, while a smaller amount is converted to either CO$_2$ or, via formation of acetyl-CoA, is assimilated into cell material. Abbreviations and color coding are as in Figures 17.42 and 17.44; Corr, corrinoid-containing protein. CODH, carbon monoxide dehydrogenase.

CoM-SH and CoB-SH (Figure 17.44). This reduction, carried out by the enzyme *heterodisulfide reductase*, is exergonic and is associated with the extrusion of protons across the membrane, creating a proton motive force (Figure 17.46●).

Dissipation of the proton gradient by a proton-translocating ATPase (∞ Section 5.12 discusses ATPases) drives ATP synthesis during methanogenesis in the same way that this process occurs in other forms of respiration. Electron flow to the heterodisulfide reductase involves a unique membrane-integrated electron carrier, a phenazine compound called *methanophenazine* (Figure 17.46). In the electron transport process in the terminal step, methanophenazine is alternately reduced (by F_{420}) and then oxidized (by a *b*-type cytochrome), which is the ultimate donor of electrons to the heterodisulfide reductase (Figure 17.46).

Methanogenesis from methyl compounds is also linked to the heterodisulfide reductase proton pump, but an additional factor is involved. As previously mentioned, in the absence of H_2, methanogenesis from compounds like CH_3OH requires that some of the CH_3OH be oxidized to CO_2 to generate the electrons needed for methyl reduction to methane (Figure 17.45). This re-

quires an energy input and occurs at the expense of a Na^+ motive force (a Na^+ energized membrane potential, see Section 17.20). The energy inherent in this potential is derived from the conversion of CH_3-MP to CH_3-CoM during methanogenesis. The reverse reaction consumes energy and is driven by the Na^+ pump. Further oxidative steps in the conversion of methyl groups to CO_2 proceed by reversal of the enzymatic steps leading to CH_4 formation from CO_2 (see Figure 17.44).

In methanogens we thus see *two* types of ATPases: a typical proton pump used to drive adenosine triphosphate (ATP) synthesis and a Na^+ pump that functions to drive methyl group oxidation.

17.17 Concept Check

Methanogenesis is the biological production of CH_4 from either CO_2 plus H_2 or from methylated compounds. A variety of unique coenzymes are involved in methanogenesis, and the process is strictly anaerobic. Energy conservation in methanogenesis involves both proton and sodium ion gradients.

◆ What coenzymes function as C_1 carriers in methanogenesis? As electron donors?

◆ Why are the steps in CH_3OH catabolism during methanogenesis different if H_2 is present than when it is not present?

◆ How is a proton motive force produced in methanogenesis?

17.18 Ferric Iron, Manganese, Chlorate, and Organic Electron Acceptors

In addition to the electron acceptors for anaerobic respiration discussed thus far, ferric iron (Fe^{3+}), manganic ion (Mn^{4+}), chlorate (ClO_3^-), and various organic compounds are important electron acceptors for bacteria in nature (Figure 17.47●). A wide diversity of bacteria are able to reduce these acceptors, especially Fe^{3+}, and many are able to reduce other acceptors as well, such as NO_3^- and S^0 (see Sections 17.14 and 17.15).

Ferric Iron Reduction

Ferric iron is an electron acceptor for energy metabolism in a wide variety of both chemoorganotrophic and chemolithotrophic bacteria, and because Fe^{3+} is abundant in nature, its reduction is a major form of anaerobic respiration. The reduction potential of the Fe^{3+}/Fe^{2+} couple is somewhat electropositive ($E_0' = +0.2$ V at pH 7), and because of this, Fe^{3+} reduction can be coupled to the oxidation of several organic and inorganic electron donors. Various organic compounds, including aromatic compounds, can be oxidized anaerobically by ferric iron reducers; in all cases electrons travel through electron transport chains that terminate in a ferric iron reductase system. The electron flow establishes a proton motive force that can be used to generate ATP. Much research on

● **Figure 17.46 Energy conservation in methanogenesis.** (a) Structure of methanophenazine (MPH in part b), an electron carrier in the electron transport chain leading to ATP synthesis. The central ring of the molecule can be alternately reduced and oxidized. (b) Steps in electron transport. Electrons originating from H_2 reduce F_{420} and then methanophenazine. The latter, through a cytochrome of the *b* type, reduces heterodisulfide reductase with the extrusion of protons to the outside of the membrane. In the final step, heterodisulfide reductase reduces Co-M-S—S-CoB to HS-CoM and HS-CoB. Refer to Figure 17.42 for the structures of CoM and CoB.

Acceptor/Product (E_0')	Reaction
Chlorate/ Chloride (+1.03)	$ClO_3^- \xrightarrow[6\,H^+]{6\,e^-} Cl^- + 3\,H_2O$
Manganic ion/ Manganous ion (+0.798)	$Mn^{4+} \xrightarrow{2\,e^-} Mn^{2+}$
Selenate/ Selenite (+0.475)	$^-O{-}\underset{\underset{O}{\|}}{\overset{\overset{O}{\|}}{Se}}{-}O^- \xrightarrow[2\,H^+]{2\,e^-} \underset{O^-}{Se}{=}O + H_2O$
Ferric ion/ Ferrous ion (+0.2)	$Fe^{3+} \xrightarrow{e^-} Fe^{2+}$
Dimethyl sulfoxide (DMSO)/ Dimethyl sulfide (DMS) (+0.16)	$H_3C{-}\underset{\underset{O}{\downarrow}}{S}{-}CH_3 \xrightarrow[2\,H^+]{2\,e^-} (CH_3)_2S + H_2O$
Arsenate/ Arsenite (+0.139)	$^-O{-}\underset{O^-}{As}{=}O \xrightarrow[2\,H^+]{2\,e^-} \underset{O^-}{As}{-}O^- + H_2O$
Trimethylamine-N-oxide (TMAO)/ Trimethylamine (TMA) (+0.13)	$H_3C{-}\underset{\underset{O}{\downarrow}}{\overset{\overset{CH_3}{\|}}{N}}{-}CH_3 \xrightarrow[2\,H^+]{2\,e^-} (CH_3)_3N + H_2O$
Fumarate/ Succinate (+0.03)	$^-O{-}\overset{O}{\overset{\|}{C}}{-}CH{=}CH{-}\overset{O}{\overset{\|}{C}}{-}O^- \xrightarrow[2\,H^+]{2\,e^-} {}^-O{-}\overset{O}{\overset{\|}{C}}{-}CH_2{-}CH_2{-}\overset{O}{\overset{\|}{C}}{-}O^-$

● **Figure 17.47 Some alternative electron acceptors for anaerobic respirations.** Note the difference in E_0' in the different acceptors.

the energetics of ferric iron reduction has been done with the gram-negative bacterium *Shewanella putrefaciens*, in which Fe^{3+}-dependent anaerobic growth occurs with various organic electron donors. Other important Fe^{3+} reducers include *Geobacter*, *Geospirillum*, and *Geovibrio*.

Geobacter metallireducens has been a model for study of the physiology of Fe^{3+} reduction. This organism can oxidize acetate with Fe^{3+} as an acceptor as follows:

$$Acetate^- + 8\,Fe^{3+} + 4\,H_2O \longrightarrow$$
$$2\,HCO_3^- + 8\,Fe^{2+} + 9\,H^+ \qquad \Delta G^{0'} = -233\,kJ$$

Geobacter can also use H_2 or organic electron donors, including the aromatic hydrocarbon toluene (see Figure 17.56c for the structure of toluene). This may be of environmental significance because toluene from accidental spills or leakage from hydrocarbon storage tanks often contaminates ferric-rich aquifers, and organisms such as *Geobacter* may be natural cleanup agents in such environments.

Reduction of Manganese (Mn^{4+}) and Other Inorganic Substances

The metal manganese has a number of oxidation states, of which Mn^{4+} and Mn^{2+} are the most stable and biologically relevant. Anoxic reduction of Mn^{4+} to Mn^{2+} is carried out

by a variety of microorganisms, mostly chemoorganotrophs. In *Shewanella putrefaciens* and a few other bacteria, anoxic growth on acetate and several other nonfermentable carbon sources occurs with Mn^{4+} as electron acceptor. The reduction potential of the Mn^{4+}/Mn^{2+} couple is extremely high (Figure 17.47); thus, several compounds should be able to donate electrons to Mn^{4+} reduction. This is also the case for chlorate, because its reduction potential is very positive (Figure 17.47). Several chlorate-reducing bacteria have been isolated, and most of them are facultative and thus also capable of aerobic growth.

Other inorganic substances can function as electron acceptors for anaerobic respiration. These include selenium and arsenic compounds (Figure 17.47). Although usually not present in large amounts in natural systems, arsenic and selenium compounds are occasional pollutants and can support anoxic growth of various bacteria. The reduction of SeO_4^{2-} to SeO_3^{2-} and eventually to Se^0 (metallic selenium) is an important method of selenium removal from water and has been used as a means of cleaning up (bioremediation) (∞ Section 19.19) of selenium-contaminated soils. Conversely, the reduction of arsenate to arsenite can actually *create* a toxicity problem. Some groundwaters flow through rocks containing relatively insoluble arsenate minerals. However, if the arsenate is reduced to arsenite by bacteria, the arsenite becomes more mobile and can contaminate groundwater. This has been a serious problem in arsenic contamination of well water in Bangladesh in recent years.

Other forms of arsenate reduction are beneficial. For example, the sulfate-reducing bacterium *Desulfotomaculum* can reduce arsenate (AsO_4^{3-}) to arsenite (AsO_3^{3-}), along with SO_4^{2-} (to HS^-). During this process a mineral complex of arsenic and sulfide—As_2S_3—precipitates spontaneously (Figure 17.48●). The mineral is formed both

● **Figure 17.48 Production of the mineral arsenic trisulfide (As_2S_3) during arsenate reduction by the sulfate-reducing bacterium *Desulfotomaculum auripigmentum*.** Left, appearance of culture bottle after inoculation. Right, following growth for two weeks. Center, synthetic sample of As_2S_3. Mineral production by microorganisms is called *biomineralization*.

Dianne K. Newman and Stephen Tay

intracellularly as well as extracellularly, and its formation is an example of *biomineralization*, the formation of a mineral by bacterial action. In this case As_2S_3 formation (Figure 17.48) also functions as a means of detoxifying what would otherwise be a toxic compound (arsenic), and such activities may have practical applications for the microbial cleanup of arsenic-containing toxic wastes.

Organic Electron Acceptors

Several organic compounds can participate as external electron acceptors in anaerobic respirations. Of those listed in Figure 17.47, the compound that has been most extensively studied is **fumarate**, which is reduced to **succinate**. An examination of the *citric acid cycle* (∞ Figure 5.22) indicates that fumarate and succinate are important intermediates. The role of fumarate as an electron acceptor for anaerobic respiration derives from the fact that the fumarate-succinate couple has a reduction potential near 0 V (see Figure 17.35), which allows coupling of fumarate reduction to NADH or H_2 oxidation. The energy yield is sufficient for the synthesis of one ATP. Bacteria able to use fumarate as an electron acceptor include *Wolinella succinogenes* (which can grow on H_2 as electron donor using fumarate as electron acceptor), *Desulfovibrio gigas* (a sulfate-reducing bacterium that can also grow under non-sulfate-reducing conditions), some clostridia, *Escherichia coli*, and many other bacteria.

The compound **trimethylamine oxide** (Figure 17.47) is an important organic electron acceptor. Trimethylamine oxide (TMAO) is a product of marine fish, where it functions as a means of excreting excess nitrogen. A variety of bacteria are able to reduce TMAO to trimethylamine (TMA), which has a strong odor and flavor. Some of the odor that occurs in spoiled marine fish is due to TMA produced by bacterial action. A variety of facultatively aerobic bacteria are able to use TMAO as an alternate electron acceptor. In addition, several phototrophic purple bacteria (∞ Section 12.2) are able to use TMAO as an electron acceptor for anaerobic metabolism in darkness.

A compound analogous to TMAO is **dimethyl sulfoxide** (DMSO), which is reduced by a variety of bacteria to dimethyl sulfide (DMS). DMSO is a common natural product and is found in both marine and freshwater environments. DMS has a strong, pungent odor, and bacterial reduction of DMSO to DMS is signaled by the presence of the characteristic odor of DMS. A variety of bacteria, including *Campylobacter*, *Escherichia*, and many purple bacteria, are able to use DMSO as an electron acceptor in energy generation (∞ Section 19.13 for further discussion of DMSO metabolism).

The reduction potentials of the TMAO/TMA and DMSO/DMS couples are similar, near $+0.15$ V, which means that any electron transport chain that ends with TMAO or DMSO reduction must be rather short. As in fumarate reduction, in most instances of TMAO and DMSO reduction, cytochromes of the *b* type (with reduction potentials near 0 V) have been identified as terminal oxidases.

Halogenated Compounds as Electron Acceptors: Reductive Dechlorination

Several chlorinated compounds can function as electron acceptors for anaerobic respiration in the process called **reductive dechlorination** or **dehalorespiration**. For example, the sulfate-reducing bacterium *Desulfomonile* grows anaerobically on H_2 or organic compounds as electron donors with chlorobenzoate as an electron acceptor:

$$\underset{\text{3-Chlorobenzoate}}{C_7H_4O_2Cl^-} + 2\,H \longrightarrow \underset{\text{Benzoate}}{C_7H_5O_2^-} + HCl$$

Besides *Desulfomonile*, which is also a sulfate-reducing bacterium (Table 17.5), a wide variety of other bacteria can re-

Table 17.5	**Characteristics of major genera of bacteria capable of reductive dechlorination**			
	Genus			
Property	*Dehalobacter*	*Desulfomonile*	*Desulfitobacterium*	*Dehalococcoides*
Electron donors	H_2	H_2, formate, pyruvate, lactate, benzoate	H_2, formate, pyruvate, lactate	H_2, lactate
Electron acceptors	Trichloroethylene, tetrachloroethylene	Metachlorobenzoates, tetrachloroethylene, SO_4^{2-}, SO_3^{2-}, $S_2O_3^{2-}$	Ortho-, meta- or para-chlorophenols, NO_3^-, fumarate, SO_3^{2-}, $S_2O_3^{2-}$, S^0	Trichloroethylene, tetrachloroethylene
Product of reduction of tetrachloroethylene	Dichloroethylene	Dichloroethylene	Trichloroethylene	Ethene
Other properties[a]	Contains cytochrome *b*	Contains cytochrome c_3; requires organic carbon source; can grow by fermentation of pyruvate	Can also grow by fermentation	Lacks peptidoglycan
Phylogeny[b]	Related to low GC gram-positive *Bacteria*	Related to delta Proteobacteria	Related to low GC gram-positive *Bacteria*	Related to green non-sulfur *Bacteria* (*Chloroflexus* group)

[a] All organisms are obligate anaerobes.
[b] See Chapter 12 and Figure 12.1 for discussion of prokaryotic phylogeny.

ductively dechlorinate; some of these are restricted to chlorinated compounds as electron acceptors (Table 17.5). Many of the chlorinated compounds reduced are toxic to fish and other animal life while many of the products of dechlorination are less toxic or even completely nontoxic. For example, the bacterium *Dehalococcoides* reduces tri- and tetrachloroethylene to the harmless gas ethene (Table 17.5). The dehalogenating organism *Dehalobacterium* converts the toxic compound dichloromethane (CH_2Cl_2) into the fatty acids acetate and formate. Thus, reductive dechlorination is not only a form of energy metabolism, but an environmentally significant bioremedial process (∞ Section 19.18). Many reductive dechlorinators are also capable of reducing nitrate or various reduced sulfur compounds (Table 17.5), and thus the group consists of both specialist and opportunist species.

 17.18 **Concept Check**

Besides inorganic nitrogen and sulfur compounds or CO_2, a variety of other substances, both organic and inorganic, can function as electron acceptors for anaerobic respiration. These include in particular Fe^{3+}, Mn^{4+}, fumarate, and certain chlorinated compounds.

◆ With H_2 as electron donor why is reduction of Fe^{3+} a more favorable reaction than reduction of fumarate?

◆ Give an example of *biomineralization*.

◆ Where in nature might TMAO-degrading bacteria be abundant? Why?

◆ What is *reductive dechlorination*?

IV THE ANAEROBIC WAY OF LIFE: FERMENTATIONS AND SYNTROPHY

17.19 Fermentations: Energetic and Redox Considerations

Because oxygen is not highly soluble in water (9.6 mg/l distilled water in equilibrium with air at 25° C), many environments easily become anoxic. In such environments, decomposition of organic material must occur anaerobically. If adequate supplies of electron acceptors like SO_4^{2-}, NO_3^-, Fe^{3+}, and the others considered in previous sections are not available in such anoxic environments, much of the carbon will be catabolized by **fermentation**. (CO_2 as an electron acceptor is an exception. It is rarely limiting in anoxic habitats but its conversion to methane requires H_2, and the latter is itself a product of fermentative reactions.) We discussed the overall process of fermentation in Sections 5.9 and 5.10. There we saw how fermentation is an internally balanced oxidation-reduction process in which the fermentable substrate becomes both oxidized and reduced (Figure 17.49●).

● **Figure 17.49 Overall process of fermentation.** In a typical fermentation, most of the carbon is excreted as a partially reduced end product of energy metabolism and only a small amount is used in biosynthesis.

An organism faces two problems if it is to catabolize organic compounds in energy-yielding metabolism: (1) conserving some of the energy released as ATP, and (2) disposing of electrons removed from the electron donor. In fermentation, ATP synthesis typically occurs by way of *substrate-level phosphorylation*, a mechanism by which energy-rich phosphate bonds from organic intermediates of the fermentation are transferred to ADP (∞ Section 5.9). The second problem, that of redox balance, is solved by the production and excretion of *fermentation products* generated from the original substrate (Figures 17.49 and 17.50●).

Energy-Rich Compounds and Substrate-Level Phosphorylation

Energy conservation from substrate-level phosphorylation can occur in many different ways. However, central to the mechanism of ATP synthesis is the production of energy-rich compounds. These are organic compounds containing a phosphate group or a coenzyme-A molecule, the hydrolysis of which is highly exergonic. Table 17.6 lists some energy-rich intermediates formed during biochemical processes. Most of the compounds listed in Table 17.6 can couple directly to ATP synthesis (−31.8 kJ/mol). If an organism can form one of these compounds during fermentative metabolism, it can thus make ATP. Substrate-level phosphorylation is a more direct way of making ATP than by oxidative phosphorylation (∞ Figure 5.13), but requires that the energy source couple directly to an energy-rich intermediate. Several pathways for the anaerobic breakdown of various fermentable substances to energy-rich intermediates are summarized in Figure 17.50.

Energy Yields of Fermentative Organisms

How much ATP can be produced by a fermentative organism? As we have seen, glucose fermenters produce 2–3 ATPs per glucose fermented in glycolysis (∞ Figure 5.14). This is about the maximum amount of ATP produced by fermentation; many other substrates provide less energy. The potential energy released from a particular fermentation can be calculated from the balanced reaction and from the free energy values given in

Table 17.6	**Energy-rich compounds involved in substrate-level phosphorylation**[a]
Compound	**Free energy of hydrolysis, $\Delta G^{0\prime}$ (kJ/mol)**[b]
Acetyl-CoA	−35.7
Propionyl-CoA	−35.6
Butyryl-CoA	−35.6
Succinyl-CoA	−35.1
Acetylphosphate	−44.8
Butyrylphosphate	−44.8
1,3-Bisphosphoglycerate	−51.9
Carbamyl phosphate	−39.3
Phosphoenolpyruvate	−51.6
Adenosine-phosphosulfate (APS)	−88
N^{10}-formyltetrahydrofolate	−23.4
Energy of hydrolysis of ATP ($ATP \rightarrow ADP + P_i$)	−31.8

[a] Data from Thauer, R. K., K. Jungermann, and K. Decker. 1977. Energy conservation in chemotrophic anaerobic bacteria. *Bacteriol. Rev.* 41: 100–180.

[b] The $\Delta G^{0\prime}$ values shown here are for "standard conditions," which are not necessarily those of cells. Including heat loss, the energy costs of making an ATP are more like 60 kJ than 32 kJ, and the energy of hydrolysis of the energy-rich compounds shown here is thus likely higher. But for simplicity and comparative purposes, the values in this table will be taken as the actual energy released per reaction.

Appendix 1. For instance, the fermentation of glucose to ethanol and CO_2 has a theoretical energy yield of −235 kJ/mol, enough to produce about 7 ATPs. However, only 2 ATPs are actually produced, which implies that the organism operates at considerably less than 100% efficiency, some energy being lost as heat.

Oxidation-Reduction Balance and H_2 Production

In any fermentation reaction, there must be a *balance* between oxidation and reduction. The total number of electrons in the products on the right side of the equation must balance the number in the substrates on the left side of the equation. When fermentations are studied experimentally in the laboratory, it is conventional to calculate a **fermentation balance** to make certain that no products are missed. The fermentation balance can also be calculated theoretically from the oxidation states of the substrates and products (see Appendix 1 for the procedure for calculating oxidation states).

In a number of fermentations, electron balance is maintained by the production of molecular hydrogen, H_2. In H_2 production, protons derived from water serve as electron acceptors. Production of H_2 is generally associated with the

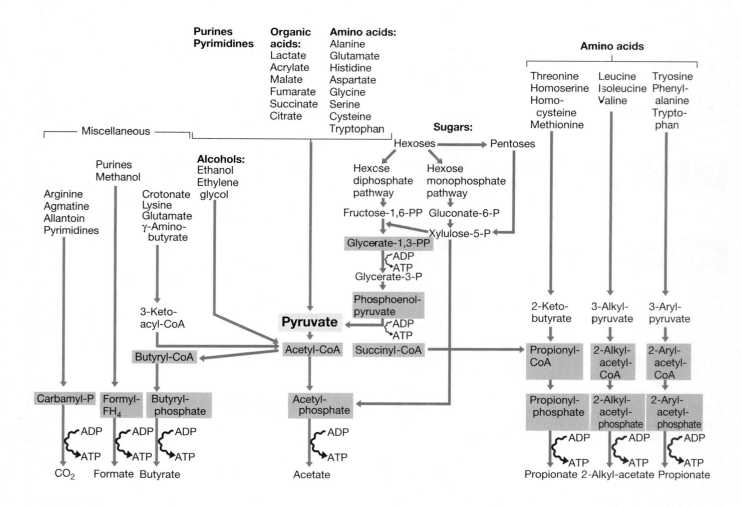

● **Figure 17.50   Major routes of the anaerobic breakdown of various fermentable substances.** The sites of substrate–level phosphorylation are shown. Energy-rich CoA derivatives and other key high energy compounds are highlighted in blue. Refer back to Table 17.6.

presence in the organism of an iron-sulfur protein called *ferredoxin*, a very low potential electron carrier. The transfer of electrons from ferredoxin to H^+ is catalyzed by the enzyme **hydrogenase**, as illustrated in Figure 17.51●. The energetics of hydrogen production are actually somewhat unfavorable, and so most fermentative organisms produce only a relatively small amount of hydrogen along with other fermentation products. Hydrogen production thus functions primarily to maintain redox balance.

Many anaerobic bacteria produce *acetate* as one of the products of fermentation. The production of acetate or certain other fatty acids (see Table 17.6) is energetically advantageous because it allows the organism to make ATP by substrate-level phosphorylation. The key intermediate generated in acetate production is acetyl-CoA (see Table 17.6 and Figure 17.50), an energy-rich intermediate. Acetyl-CoA can be converted to acetyl phosphate (also listed in Table 17.6), and the phosphate group of acetyl phosphate subsequently transferred to adenosine diphosphate (ADP) by the enzyme acetate kinase, yielding ATP.

One of the main substrates that is converted to acetyl-CoA is *pyruvate*, a major product of glycolysis. Its conversion to acetyl-CoA is an oxidation reaction (Figure 17.51), and the excess electrons generated must be used either to make a more reduced end product or be released as H_2.

 17.19 Concept Check

In the absence of an external electron acceptor, organic compounds can be catabolized only by fermentation. Only certain compounds are fermentable, and a requirement for most fermentations is that an energy-rich organic intermediate be formed that can yield ATP by substrate-level phosphorylation. Redox balance must also be achieved in fermentations, and H_2 production is one means of disposing of excess electrons.

◆ What is *substrate-level phosphorylation*?

◆ Why is acetate formation in fermentation energetically beneficial?

● **Figure 17.51 Production of molecular hydrogen and acetate from pyruvate.** Note how production of acetate leads to ATP synthesis from hydrolysis of the energy-rich intermediate, acetyl phosphate (see Table 17.6).

17.20 Fermentative Diversity

Fermentations are classified in terms of either the substrate fermented or the fermentation products formed. Some specific fermentation reactions were discussed when the individual groups of prokaryotes were examined in Chapters 12 and 13. Here we present an overview of common fermentations.

Diversity of Fermentations

Table 17.7 summarizes some of the main types of fermentations as classified on the basis of *products formed*. Note some of the broad categories, such as alcohol, lactic acid, propionic acid, mixed acid, butyric acid, and homoacetic acid.

A number of fermentations are classified on the basis of the *substrate fermented* rather than the fermentation product. For instance, many of the endospore-forming anaerobic bacteria (genus *Clostridium*) ferment *amino acids* with the production of acetate, ammonia, and H_2 (∞ Figure 12.59). Other *Clostridium* species, such as *C. acidiurici* and *C. purinolyticum*, ferment the *purines* xanthine or adenine with the formation of acetate, formate, CO_2, and ammonia. Still other anaerobes ferment *aromatic* compounds. For example, the bacterium *Pelobacter acidigallici* ferments the aromatic compound *phloroglucinol* (1,3,5-benzenetriol, $C_6H_6O_3$) to acetate, conserving energy in the conversion of acetyl-CoA to acetate (Table 17.7).

Many unusual fermentations are carried out by a very restricted group of anaerobes, and in some cases, by only a single known bacterium. Some examples are listed in Table 17.8. Many of these bacteria can be considered metabolic specialists, having evolved biochemical capabilities to catabolize a substrate or substrates not catabolized by other bacteria. However, as for the substances listed in Table 17.7, fermentation of these substrates requires that the organism be able to produce an energy-rich intermediate of the type listed in Table 17.6 in order to conserve some of the energy released as ATP.

Fermentations Lacking Substrate-Level Phosphorylation

With certain substrates there is insufficient energy released to couple to the synthesis of ATP directly by substrate-level phosphorylation, yet these compounds still support growth of an organism. In these cases, catabolism of the substrate is linked to ion pumps that establish a proton or sodium gradient across the membrane. Examples of this include the fermentations by *Propionigenium modestum* and *Oxalobacter formigenes*; both of these organisms couple the fermentation of dicarboxylic acids to membrane-bound energy-linked ion pumps.

Propionigenium modestum carries out the following reaction:

$$\text{Succinate}^{2-} + H_2O \longrightarrow \text{propionate}^- + HCO_3^-$$
$$\Delta G^{0\prime} = -20.5 \text{ kJ}$$

Table 17.7	Examples of common bacterial fermentations and some of the organisms carrying them out	
Type	**Overall reaction[a]**	**Organisms**
Alcoholic	Hexose → 2 Ethanol + 2 CO_2	Yeast *Zymomonas*
Homolactic	Hexose → 2 Lactate$^-$ + 2 H$^+$	*Streptococcus* Some *Lactobacillus*
Heterolactic	Hexose → Lactate$^-$ + Ethanol + CO_2 + H$^+$	*Leuconostoc* Some *Lactobacillus*
Propionic acid	Lactate$^-$ → Propionate$^-$ + Acetate$^-$ + CO_2	*Propionibacterium* *Clostridium propionicum*
Mixed acid	Hexose → Ethanol + 2, 3-Butanediol + Succinate^{2-} + Lactate$^-$ + Acetate$^-$ + Formate$^-$ + H$_2$ + CO_2	Enteric bacteria[b] *Escherichia* *Salmonella* *Shigella* *Klebsiella* *Enterobacter*
Butyric acid	Hexose → Butyrate$^-$ + Acetate$^-$ + H$_2$ + CO_2	*Clostridium butyricum*
Butanol	Hexose → Butanol + Acetate$^-$ + Acetone + Ethanol + H$_2$ + CO_2	*Clostridium acetobutylicum*
Caproate	Ethanol + Acetate$^-$ + CO_2 → Caproate$^-$ + Butyrate$^-$ + H$_2$	*Clostridium kluyveri*
Homoacetogenic	Fructose → 3 Acetate$^-$ + 3 H$^+$ 2 H$_2$O 4 H$_2$ + 2 CO_2 + H$^+$ → Acetate$^-$ +	*Clostridium aceticum* *Acetobacterium*
Methanogenic	Acetate$^-$ + H$_2$O → CH$_4$ + HCO$_3^-$	*Methanosaeta* *Methanosarcina*

[a] Reactions are intended as an overview of the process and are not necessarily balanced.
[b] Not all organisms produce all products. In particular, butanediol production is limited to only certain enteric bacteria.

This overall reaction yields insufficient free energy to couple to ATP synthesis directly by substrate-level phosphorylation, but nevertheless it serves as the sole energy-yielding reaction for growth of the organism. This is possible because the decarboxylation of succinate (via methylmalonyl-CoA and its membrane-bound decarboxylase) by *Propionigenium modestum* is coupled to the export of sodium (Na$^+$) across the cytoplasmic membrane (Figure 17.52a●). A Na$^+$-translocating ATPase in the membrane of *P. modestum* dissipates this Na$^+$ gradient to drive ATP synthesis (Figure 17.52a).

Oxalobacter formigenes carries out the fermentation of oxalate:

$$Oxalate^{2-} + H_2O \longrightarrow formate^- + HCO_3^-$$

$$\Delta G^{0\prime} = -26.7 \text{ kJ}$$

At neutral pH, oxalate exists in the ionized form as oxalate^{2-}. Its decarboxylation to formate$^-$ consumes one proton. The subsequent export of formate from the cell then builds a proton motive force that can be coupled to ATP synthesis by a proton-translocating ATPase in the membrane (Figure 17.52b).

Table 17.8	Some unusual bacterial fermentations	
Type	**Overall balanced reaction**	**Organisms**
Acetylene	2 C$_2$H$_2$ + 3 H$_2$O → Ethanol + Acetate$^-$ + H$^+$	*Pelobacter acetylenicus*
Glycerol	4 Glycerol + 2 HCO$_3^-$ → 7 Acetate$^-$ + 5 H$^+$ + 4 H$_2$O	*Acetobacterium* sp.
Resorcinol (an aromatic compound)	2 C$_6$H$_4$(OH)$_2$ + 6 H$_2$O → 4 Acetate$^-$ + Butyrate$^-$ + 5 H$^+$	*Clostridium* sp.
Phloroglucinol (an aromatic compound)	C$_6$H$_6$O$_3$ + 3 H$_2$O → 3 Acetate$^-$ + 3 H$^+$	*Pelobacter massiliensis* *Pelobacter acidigallici*
Putrescine	10 C$_4$H$_{12}$N$_2$ + 26 H$_2$O → 6 Acetate$^-$ + 7 Butyrate$^-$ + 20 NH$_4^+$ + 16 H$_2$ + 13 H$^+$	Unclassified gram-positive nonsporing anaerobes
Citrate	Citrate^{3-} + 2 H$_2$O → Formate$^-$ + 2 Acetate$^-$ + HCO$_3^-$ + H$^+$	*Bacteroides* sp.
Aconitate	Aconitate^{3-} + H$^+$ + 2 H$_2$O → 2 CO_2 + 2 Acetate$^-$ + H$_2$	*Acidaminococcus fermentans*
Glyoxylate	4 Glyoxylate$^-$ + 3 H$^+$ + 3 H$_2$O → 6 CO_2 + 5 H$_2$ + Glycolate$^-$	Unclassified gram-negative bacterium
Succinate	Succinate^{2-} + H$_2$O → Propionate$^-$ + HCO$_3^-$	*Propionigenium modestum*
Oxalate	Oxalate^{2-} + H$_2$O → Formate$^-$ + HCO$_3^-$	*Oxalobacter formigenes*
Malonate	Malonate^{2-} + H$_2$O → Acetate$^-$ + HCO$_3^-$	*Malonomonas rubra* *Sporomusa malonica*
Benzoate	2 Benzoate$^-$ → Cyclohexane carboxylate$^-$ + 3 Acetate$^-$ + HCO$_3^-$ + 3H$^+$	*Syntrophus aciditrophicus*

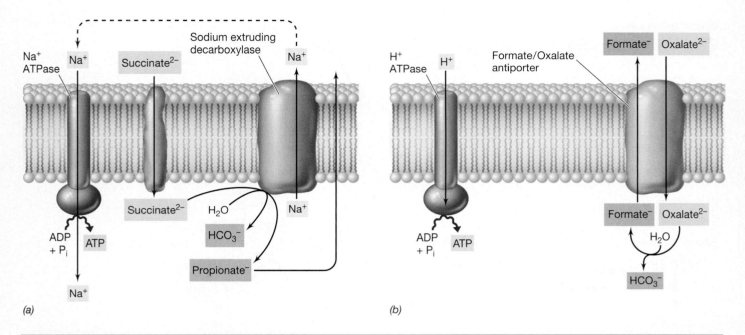

● **Figure 17.52   The unique fermentations of succinate and oxalate.** (a) Succinate fermentation by *Propionigenium modestum*. A sodium-translocating ATPase produces ATP; sodium export is linked to the energy released by succinate decarboxylation. (b) Oxalate fermentation by *Oxalobacter formigenes*. Oxalate import and formate export by a formate–oxalate antiporter consume protons. ATP synthesis is linked to a proton-driven ATPase. All substrates and products of a given reaction are shown in contrasting colors.

The interesting and unique aspect of the fermentations of both *Propionigenium* and *Oxalobacter* is that ATP synthesis occurs without substrate-level phosphorylation *or* electron transport occurring. However, chemiosmotic ATP formation still occurs in both cases as a result of a Na⁺ or H⁺ pump linked to decarboxylation of organic acids. These organisms have thus taught us an important lesson in bioenergetics: any chemical reaction that yields less than the 31.8 kJ required to make one ATP (see Table 17.6) cannot be ruled out as a growth-supporting reaction for a bacterium. If the reaction can be coupled to an ion gradient, ATP production remains a possibility. However, because the influx of approximately $3 H^+$ (or $3 Na^+$) is required to drive ATP formation by a membrane-associated ATPase, a reaction must yield at least the energy required to pump a single H^+ or Na^+ to be theoretically capable of supporting growth.

 17.20   *Concept Check*

A wide variety of fermentations are known, and in many cases the product of one organism's fermentation is fermented by a second organism. Some fermentations employ ion gradients (H^+ or Na^+) as the basis of their energetics.

◆ What is unusual about the fermentation of succinate and oxalate?

◆ A common product of the fermentations shown in Table 17.7 and 17.8 is a fatty acid such as acetate. Why is this important energetically?

17.21   **Syntrophy**

There are many examples in microbiology of **syntrophy**, a situation where two different organisms degrade a substance—and conserve energy doing it—that neither could degrade separately. We will see in Section 19.10 how syntrophy is extremely important in anoxic catabolism leading to the production of methane. Here we consider the microbiology and energetic aspects of syntrophy.

Hydrogen Consumption in Syntrophic Reactions

In most cases the heart of a syntrophic reaction involves H_2 production by one partner in the syntrophic relationship linked to H_2 consumption by the other. Thus, syntrophy has also been called **interspecies H_2 transfer**. The H_2 consumer can be any of a number of organisms we have already considered: denitrifying bacteria, sulfate-reducing bacteria, homoacetogens, or methanogens.

Consider the case of syntrophy involving ethanol fermentation to acetate with the eventual production of methane (Figure 17.53●). As seen, the ethanol fermenter carries out a reaction that has an unfavorable (that is, positive) standard free-energy change. However, the H_2 produced by the ethanol fermenter can be used as an electron donor for methanogenesis by a methanogen (Figure 17.53). When the two reactions are summed, the overall reaction is exergonic (Figure 17.53) and supports growth of both partners in the syntrophic mixture.

Ethanol fermentation:

2 CH_3CH_2OH + 2 H_2O → 4 H_2 + 2 CH_3COO^- + 2 H^+

$$\Delta G^{0\prime} = +19.4 \text{ kJ/reaction}$$

Methanogenesis:

4 H_2 + CO_2 → CH_4 + 2 H_2O

$$\Delta G^{0\prime} = -130.7 \text{ kJ/reaction}$$

Coupled reaction:

2 CH_3CH_2OH + CO_2 → CH_4 + 2 CH_3COO^- + 2 H^+

$$\Delta G^{0\prime} = -111.3 \text{ kJ/reaction}$$

(a) Reactions

Ethanol fermenter   **Methanogen**

2 Ethanol → Interspecies hydrogen transfer → 4 H_2

CO_2

2 Acetate

CH_4

(b) Syntrophic transfer of H_2

● **Figure 17.53 Interspecies H_2 transfer.** Fermentation of ethanol to methane and acetate by syntrophic association of an ethanol-oxidizing bacterium and a H_2-consuming partner bacterium, in this case, a methanogen. (a) Reactions involved. The two organisms thus share the energy released in the coupled reaction. (b) Nature of the syntrophic transfer of H_2.

Another good example of syntrophy is the oxidation of butyrate to acetate plus H_2 by the fatty acid-oxidizing syntroph *Syntrophomonas* (Figure 17.54●):

$$\text{Butyrate}^- + 2\,H_2O \longrightarrow 2\text{ acetate}^- + H^+ + 2\,H_2$$

$$\Delta G^{0\prime} = +48.2 \text{ kJ}$$

The free energy change of this reaction is highly unfavorable, and in pure culture *Syntrophomonas* will not grow on butyrate. But if the H_2 produced in the reaction is consumed by a partner organism (like a methanogen), *Syntrophomonas* grows luxuriantly in coculture with the H_2 consumer.

How can chemical reactions whose free energy changes are positive support growth of an organism? We examine this conundrum now.

Energetics of Syntrophy

If growth of syntrophic organisms occurs only when H_2 is removed by a partner organism, the removal itself must obviously affect the energetics of the reaction. How does this happen? A brief review of the principles of free energy given in Appendix 1 indicates that the *actual* concentration of reactants and products in a reaction can drastically change its energetics. This is because $\Delta G^{0\prime}$ is calculated on the basis of *standard* conditions—one molar concentration of products and reactants—whereas the re-

(a) **Syntrophic culture**

1. Crotonate oxidation

$CH_3HC=CH-\overset{\displaystyle O}{\underset{\displaystyle O^-}{C}}$ (Crotonate⁻) + H_2O → 2 acetate⁻ + H_2 + H^+

2. Crotonate reduction

$CH_3HC=CH-\overset{\displaystyle O}{\underset{\displaystyle O^-}{C}}$ + H_2 → butyrate⁻

Proton motive force

Sum: 2 crotonate⁻ + H_2O → 2 acetate⁻ + butyrate⁻ + H^+ $\Delta G^{0\prime} = -340$ kJ

(b) **Pure culture**

● **Figure 17.54 Energetics of growth of *Syntrophomonas wolfei* in syntrophic culture and in pure culture.** (a) In syntrophic culture, growth is dependent on the presence of a H_2-consuming organism such as a methanogen. H_2 production probably involves proton motive force-driven reverse electron flow (see Section 17.4) because the $E_0\prime$ of FAD/FADH₂ or NAD⁺/NADH is more electropositive than that of 2 H^+/H_2 (∞ Figure 5.9). (b) In pure culture, energy production is linked to crotonate reduction to butyrate.

lated term ΔG is calculated on the basis of *actual* concentrations of products and reactants that are present.

In anoxic habitats, H_2, because it is such a powerful electron donor for anaerobic respirations, is immediately consumed. This keeps H_2 concentrations extremely low, typically below 10^{-4} atm (∞ Table 19.4). If the concentration of H_2 is very low and ΔG is used to calculate the energetics of a reaction (see Appendix 1 for how to calculate ΔG), the free-energy change associated with the oxidation of substances such as ethanol or fatty acids to acetate plus H_2 *becomes exergonic*. For example, if the concentration of H_2 is kept extremely low by the activities of a H_2-consuming partner organism, the oxidation of butyrate by *Syntrophomonas* yields about -18 kJ.

The mechanism of ATP production by syntrophs involves both substrate-level and oxidative phosphorylations, depending on how the organisms are grown. Substrate-level phosphorylation can occur during the conversion of acetyl-CoA (generated by β-oxidation of ethanol or the fatty acid) to acetate (Figure 17.54a); this may be the *only* way ATP is produced under syntrophic conditions. However, *Syntrophomonas* can also carry out an anaerobic respiration, because it can be grown in pure culture by disproportionation of unsaturated fatty acids. Crotonate, for example, supports growth of *Syntrophomonas*, with some of the crotonate being oxidized to acetate and some being reduced to butyrate (Figure 17.54b). Crotonate reduction by *Syntrophomonas* is coupled to formation of a proton motive force and ATP synthesis as in other anaerobic respirations employing organic electron acceptors, such as fumarate reduction to succinate (see Section 17.18).

Regardless of how ATP is made in the syntrophic situation, there are additional energetic burdens for *Syntrophomonas* in this growth mode because the organism must somehow produce H_2 from more electropositive electron donors such as $FADH_2$ and $NADH_2$ (Figure 17.54a). This suggests that part of the ATP produced may be needed to drive reverse electron flow reactions (see Section 17.5) to yield H_2. It is clear that syntrophic alcohol and fatty acid-oxidizing bacteria are, at least from an energetic standpoint, living on the edge of existence. Indeed, the energy available from syntrophic reactions, even under coculture conditions, may be only a few kilojoules.

Ecologically, syntrophic bacteria are key links in the carbon cycle. These organisms have evolved mechanisms for using the fermentation products of primary fermenters and for cooperating with other anaerobes to supply them with a necessary substrate, H_2. By contrast, for *aerobic* organisms syntrophy is not an issue, because with O_2 as electron acceptor the energetics of the reaction is much more favorable than for fermentative catabolism of the same substrate. Thus, syntrophic relationships are characteristic of anoxic transformations in which the energy available is only very small and the organisms highly specialized for exploiting energetically marginal reactions.

17.21 Concept Check

Syntrophy involves two organisms working together to degrade some compound that neither can degrade alone. This process usually involves H_2 production by one organism that is consumed by the partner. H_2 consumption can affect the energetics of the reaction carried out by the H_2 producer, allowing it to make ATP where it otherwise could not.

◆ Give an example of syntrophy. Why can it be said that both organisms benefit in this example?

◆ Predict how ATP is made during the syntrophic degradation of ethanol shown in Figure 17.53.

◆ Can a fatty acid-oxidizing syntroph be grown in pure culture? Give an example of this and describe how the organism makes ATP under these conditions.

V HYDROCARBON OXIDATION AND THE ROLE OF O₂ IN THE CATABOLISM OF ORGANIC COMPOUNDS

Hydrocarbons are both abundant on Earth and excellent substrates for various chemoorganotrophs. We will see that their aerobic degradation, however, poses special biochemical problems. In addition, we briefly consider the aerobic degradation of polysaccharides, fats, and a few other substances as further examples of metabolic diversity. We begin with a consideration of the O_2 requirements for some of these reactions.

17.22 Molecular Oxygen (O₂) as a Reactant in Biochemical Processes

We have discussed the role of O_2 as an *electron acceptor* in energy-generating reactions (∞ Section 5.11). Although this is by far the most important role of O_2 in cellular metabolism, O_2 plays an interesting and important role as a *direct reactant* in certain types of anabolic and catabolic processes.

Oxygenases

Oxygenases are enzymes that catalyze the incorporation of oxygen from O_2 into organic compounds. There are two classes of oxygenases: *dioxygenases*, which catalyze the incorporation of *both* atoms of O_2 into the molecule, and *monooxygenases*, which catalyze the transfer of *only one* of the two oxygen atoms in O_2 to an organic compound; the second atom of O_2 is reduced to water, H_2O. For most monooxygenases, the electron donor is NADH or NADPH, although the direct coupling to O_2 is through a flavin that is reduced by the NADH or NADPH donor (Figure 17.55●). In the case of ammonia monooxygenase discussed previously (see Section 17.12), the electron donor was cytochrome c.

Several types of reactions in living organisms require O_2 as a reactant. One of the best examples is the involvement of O_2 in sterol biosynthesis. The formation of the fused sterol ring system (∞ Figure 4.18) requires the participation of molecular oxygen. Such a reaction obviously cannot take place under anoxic conditions, so organisms that grow anaerobically must either dispense with this reaction or obtain the required substance (sterol) preformed from their environment. The requirement of O_2 in biosynthesis is of considerable evolutionary significance, as O_2 was originally absent from the atmosphere of

$$C_7H_{15}-CH_3 \quad + \quad \boxed{NADH} \quad + \quad \boxed{O:O}$$
$$\text{n-Octane} \qquad\qquad\qquad \text{(O}_2\text{)}$$

↓ Monooxygenase

$$C_7H_{15}CH_2OH \quad + \quad NAD^+ \quad + \quad H_2O$$
$$\text{n-Octanol}$$

NAD⁺ ↘
↘ NADH

$$\begin{array}{c} H \\ | \\ C_7H_{15}C=O \end{array}$$
n-Octanal

H₂O ↘ NAD⁺
↘ NADH

$$\begin{array}{c} OH \\ | \\ C_7H_{15}C=O \end{array}$$
n-Octanoic acid

ATP ↘ CoA
AMP + P$_i$P ↙

β-Oxidation
to acetyl-CoA (see Figure 17.69)

● **Figure 17.55 Monooxygenase activity.** Steps in oxidation of an aliphatic hydrocarbon, the first of which is catalyzed by a monooxygenase. Some sulfate-reducing and denitrifying bacteria can degrade aliphatic hydrocarbons under anoxic conditions.

Earth when life first evolved and became available only after the evolution of cyanobacteria (⚬⚬ Chapter 11).

 17.22 Concept Check

In addition to its role as an electron acceptor, oxygen is also a chemical reactant in certain biochemical processes. Enzymes called oxygenases introduce O_2 into a biochemical compound.

◆ How do *monooxygenases* differ in function from *dioxygenases*?

17.23 Hydrocarbon Oxidation

Hydrocarbons are organic compounds containing only carbon and hydrogen and are highly insoluble in water. Low-molecular-weight hydrocarbons are gases, whereas those of higher molecular weight are liquids or solids at room temperature. In *aliphatic* hydrocarbons, the carbon atoms are joined in open chains. There is a tremendous variation among aliphatic hydrocarbons in chain length, degree of branching, and number of double bonds. Aromatic hydrocarbons contain the aromatic ring and can be viewed as derivatives of benzene (see Figure 17.56).

Aliphatic Hydrocarbon Metabolism

A number of bacteria and several molds and yeasts can use hydrocarbons as electron donors to support growth under aerobic conditions. The initial oxidation step of saturated aliphatic hydrocarbons involves molecular oxygen (O_2) as a reactant, and one of the atoms of the oxygen molecule is incorporated into the oxidized hydrocarbon, typically at the terminal carbon atom. This reaction is carried out by a monooxygenase (see Section 17.22), and a typical reaction sequence is that shown in Figure 17.55. The end product of the reaction sequence is acetyl-CoA, and this is further catabolized in the citric acid cycle.

Aliphatic hydrocarbons, both saturated and unsaturated, can be oxidized to CO_2 under anoxic conditions as well. This is done by a restricted group of denitrifying and sulfate-reducing bacteria. Obviously monooxygenases cannot be involved here. The mechanism of degradation involves *carboxylation* (addition of CO_2) of the hydrocarbon to form a fatty acid. Following this, further degradation occurs by β-oxidation, a process that does not require O_2 (see Figure 17.69).

Aromatic Hydrocarbons

Many aromatic hydrocarbons can be used as electron donors aerobically by microorganisms, of which bacteria of the genus *Pseudomonas* have been the best studied. The metabolism of these compounds, some of which are quite large molecules, typically has as its initial stage the formation of *catechol* or a structurally related compound, as shown in Figure 17.56a●.

These single-ring compounds are referred to as *starting substrates* because oxidative catabolism proceeds only after the large aromatic molecules have been converted to these more simple forms. Protocatechuate and catechol may then be further degraded to compounds that can enter the citric acid cycle: succinate, acetyl-CoA, and pyruvate (⚬⚬ Figure 5.22). Several steps in the catabolism of aromatic hydrocarbons usually require oxygenases. Figures 17.56a, b, and c show three different oxygenase-catalyzed reactions, one using a monooxygenase and two using a dioxygenase.

Anoxic Degradation of Aromatic Compounds

Aromatic compounds can also be degraded anaerobically by facultatively aerobic bacteria if the aromatic compound already contains an atom of oxygen. Such phenolic compounds are degraded by certain denitrifying, phototrophic, ferric iron-reducing, and sulfate-reducing bacteria. From a biochemical standpoint, the anoxic catabolism of aromatic compounds proceeds by *reductive* rather than oxidative ring cleavage (Figure 17.57●). This involves *ring reduction* followed by *ring cleavage* to yield a straight-chain fatty acid or dicarboxylic acid. These intermediates can be converted to acetyl-CoA and used for both biosynthetic and energy-yielding purposes. Benzoate and benzoate derivatives are common natural products and are readily degraded anaerobically.

(a)

Benzene → Benzene epoxide → Benzenediol → Catechol + NADH

O:O, NADH, Benzene monooxygenase; $H_2O + NAD^+$; H_2O; NAD^+

Monooxygenase

(b)

Catechol → Catechol dioxetane (hypothetical) → Cis, cis-muconate

O:O, Catechol 1,2-dioxygenase

Dioxygenase

(c)

Toluene → → Methyl catechol →

O:O, NADH, Toluene dicxygenase; NAD^+; O:O, Methyl catechol 2,3-dioxygenase

Sequential dioxygenases

● **Figure 17.56 Roles of oxygenases in catabolism of aromatic compounds.** (a) Hydroxylation of benzene to catechol by a monooxygenase in which NADH is an electron donor. (b) Cleavage of catechol to *cis,cis*-muconate by a dioxygenase. Reactant oxygen atoms are shown in color in both reactions to demonstrate the different mechanisms. (c) The activity of toluene dioxygenase and methyl catechol 2,3-dioxygenase in the degradation of toluene. The oxygen atoms that each enzyme introduces are distinguished by different colors. Catechol and proto-catechuate are common intermediates in aerobic aromatic catabolism. Proto-catechuate contains a carboxyl group (COO^-) two carbon atoms removed from one of the OH groups of catechol.

Anoxic degradation of benzene and toluene, aromatic compounds lacking an oxygen atom (see Figure 17.56 for structures), also occurs. Catabolism of benzene occurs by certain denitrifying bacteria and toluene by certain ferric iron-reducing, phototrophic purple, and denitrifying bacteria. Biochemically, toluene is eventually converted to the benzoate derivative benzoyl-CoA and then presumably further catabolized via ring reduction as shown in Figure 17.57.

 17.23 Concept Check

Many microorganisms can degrade aliphatic and aromatic hydrocarbons. Catabolism aerobically involves the activity of oxygenase enzymes. Anoxic aromatic degradation proceeds by reductive rather than oxidative pathways.

◆ Draw the chemical structure for benzene. Do the same for benzoate. Are these compounds aliphatic or aromatic?

◆ What fundamental difference exists in the *anaerobic* degradation of an aromatic compound compared with its *aerobic* metabolism? Give an example of this.

17.24 Methanotrophy and Methylotrophy

Section 12.6 considered the unique situation, relative to carbon metabolism, of *methanotrophs* and *methylotrophs*. Although not autotrophs, **methylotrophs** use C_1 compounds (or other organic compounds lacking C—C bonds) for energy metabolism and biosynthesis. We focus here on the physiology of this process.

Biochemistry of Methane Oxidation

The individual steps in methane oxidation to CO_2 can be summarized as:

$$CH_4 \rightarrow CH_3OH \rightarrow CH_2O \rightarrow HCOO^- \rightarrow CO_2$$

Methanotrophs are those methylotrophs that can use CH_4 (◯◯ Section 12.6). Methanotrophs assimilate either all or one-half of their carbon (depending on the pathway they use) at the level of *formaldehyde* (CH_2O). We will see that this affects a major energy savings compared with autotrophs,

Benzoate → Benzoyl CoA → → → → → → Pimelyl-CoA → 3 Acetate + CO_2

CoA, ATP; 6 H; 2 H; H_2O; 2 H; H_2O; β-Oxidation

● **Figure 17.57 Anoxic degradation of benzoate by reductive ring cleavage.** Note that all intermediates of the pathway are bound to coenzyme A. The acetate produced is further catabolized in the citric acid cycle (◯◯ Section 5.13 and Figure 5.22).

where carbon is assimilated from the more oxidized CO_2. But here our focus is on energy conservation.

The initial step in the oxidation of methane under oxic conditions involves an enzyme called **methane monooxygenase**. As we discussed in Section 17.22, oxygenase enzymes catalyze the incorporation of oxygen from O_2 into carbon compounds (and some nitrogen compounds, see Section 17.12) and seem to be widely involved in the metabolism of hydrocarbons. Monooxygenases incorporate one atom of O_2 into the substrate while the second atom is reduced to H_2O. In the methanotroph *Methylosinus*, where the methane oxidation process has been thoroughly studied, the electrons needed for the oxidation of CH_4 to CH_3OH come from cytochrome *c* (Figure 17.58●.) This need for reducing power precludes synthesis of ATP during the oxidation of methane to methanol. In agreement with this, growth yields (grams of cells produced per mole of substrate consumed) of methanotrophs are the same whether methane or methanol is used as the growth substrate.

The other oxidation steps from CH_3OH to CO_2 use typical redox cofactors, such as quinones or NADH, which enter the electron transport chain. From these, ATP is made from the resulting proton motive force generated (Figure 17.58).

In addition to the *aerobic* catabolism of CH_4, anoxic oxidation of this hydrocarbon can occur. This is a syntrophic process involving the cooperation of sulfate-reducing bacteria and methanogens, both strict anaerobes, as partner organisms. We explore this process in Section 19.10.

C_1 Assimilation into Cell Material

As was noted in Section 12.6, two classes of methanotrophs are known, *type I* and *type II*. Members of each class share a number of phenotypic properties in common, and each also shows a distinct mechanism for C_1 assimilation.

The **serine pathway**, utilized by type II methanotrophs, is outlined in Figure 17.59●. In this pathway, a two-carbon unit, *acetyl-CoA*, is synthesized from one molecule of formaldehyde (produced from the oxidation of CH_4, see Figure 17.58) and one molecule of CO_2. The serine pathway requires reducing power and energy in the form of two molecules each of NADH and ATP, respectively, for each acetyl-CoA synthesized. The serine pathway employs a number of enzymes of the citric acid cycle and one enzyme, *serine transhydroxymethylase*, unique to the pathway (Figure 17.59).

The **ribulose monophosphate pathway**, used by type I methanotrophs, is outlined in Figure 17.60●. It is more efficient than the serine pathway because *all* of the carbon atoms for cell material are derived from formaldehyde. And, since formaldehyde is at the same

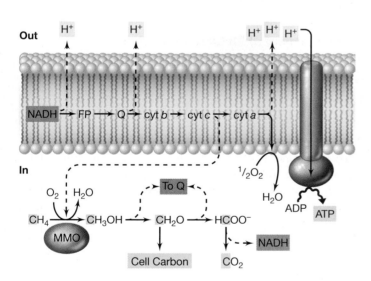

● **Figure 17.58 Oxidation of methane by methanotrophic bacteria.** Methane (CH_4) is converted to methanol (CH_3OH) by the enzyme *methane monooxygenase*. The electrons needed to drive this first step come from cytochrome *c*, and no energy is conserved in this reaction. A proton motive force is established from electron flow in the membrane, and this fuels ATPase. Note how carbon for biosynthesis comes primarily from formaldehyde (CH_2O). MMO, methane monooxygenase; FP, flavoprotein; cyt, cytochrome; Q, quinone. Although not depicted as such, MMO is actually a membrane-associated enzyme.

Overall: Formaldehyde + CO_2 + CoA + 2 NADH + 2 H⁺ + 2 ATP⟶ Acetyl∼CoA + 2 NAD⁺ + 2 ADP + 2 P_i + 2 H_2O

● **Figure 17.59 The serine pathway for the assimilation of C_1 units into cell material by type II methylotrophic bacteria.** The product of the pathway, acetyl–CoA, is used as the starting point for making new cell material. The key enzyme of the pathway is *serine transhydroxymethylase*.

Overall: 3 Formaldehyde + ATP ⟶ glyceraldehyde-3-P + ADP

● **Figure 17.60 The ribulose monophosphate pathway for assimilation of one-carbon compounds, as found in type I methylotrophic bacteria.** The complete name of the hexulose sugar is D-erythro-L-glycero-3-hexulose 6-phosphate. Three formaldehydes are needed to carry the cycle to completion, with the net result being one molecule of glyceraldehyde-3-P. The key enzyme of this pathway is *hexulose-P-synthase*. The "sugar rearrangements" involve enzymes of the pentose phosphate pathway (see Figure 17.65).

oxidation level as cell material, *no reducing power is needed*. The ribulose monophosphate pathway requires one molecule of ATP for each molecule of glyceraldehyde-3-phosphate synthesized (Figure 17.60). Consistent with the lower energy requirements of the ribulose monophosphate pathway, the cell yield (grams of cells produced per mole of CH_4 oxidized) of type I methanotrophs is higher than for type II methanotrophs.

The enzymes *hexulosephosphate synthase*, which condenses one molecule of formaldehyde with one molecule of ribulose-5-phosphate, and *hexulose-6-P isomerase* (Figure 17.60) are unique to the ribulose monophosphate pathway. The remaining enzymes of this pathway are involved in sugar rearrangements in many different organisms. It should also be noted that the substrate for the initial reaction in this pathway, *ribulose-5-P*, is almost identical to the C_1 acceptor in the Calvin cycle, *ribulose 1,5-bisphosphate* (⟳ Section 17.6), an indication that these two cycles probably share evolutionary roots.

 17.24 Concept Check

Methanotrophy is the use of CH_4 as a carbon and energy source. The enzyme *methane monooxygenase* is a key enzyme in the catabolism of methane. C_1 units get assimilated into cell material at the level of formaldehyde in methanotrophs by either the ribulose monophosphate pathway or the serine pathway.

◆ What are the energy and reducing power requirements for the ribulose monophosphate pathway? For the serine pathway? Why do they differ?

◆ Why does the oxidation of CH_4 to CH_3OH require reducing power?

◆ Which pathway, the Calvin cycle or the ribulose monophosphate pathway, requires the greater energy input? Why?

17.25 Hexose, Pentose, and Polysaccharide Metabolism

We complete our discussion of chemoorganotrophic metabolism with consideration of a few special aspects of the catabolism of organic compounds, especially the use of *polymeric* substances that must first be hydrolyzed to monomeric units before energy-generating mechanisms can be employed. We also examine a special pathway employed by cells to produce and catabolize pentose sugars. We begin with the microbial degradation of polysaccharides.

Hexose and Polysaccharide Utilization

Sugars with six carbon atoms, called **hexoses**, are the most important electron donors for many chemoorganotrophs and are also important structural components of microbial cell walls, capsules, slime layers, and storage products. The most common hexose sources in nature are listed in Table 17.9, from which it can be seen that most are polysaccharides, although a few are disaccharides. *Cellulose* and *starch* are two of the most important natural polysaccharides.

Although both starch and cellulose are composed of glucose units, they are bonded differently (Table 17.9, ⟳ Figure 3.6), and this profoundly affects their properties. Cellulose is much more insoluble than starch and is usually less rapidly digested. Cellulose forms long fibrils, and organisms that digest cellulose are often found closely associated with them (Figure 17.61●). Many fungi are able to digest cellulose, and these are mainly responsible for decomposition of plant materials on the forest floor. Among bacteria, however, cellulose digestion is restricted to relatively few groups, of which the gliding bacteria such as *Sporocytophaga* and *Cytophaga* (Figures 17.61 and 17.62●; ⟳ Figure 12.90), clostridia, and actinomycetes are among the most common.

Anoxic digestion of cellulose is carried out by a few *Clostridium* species, which are common in lake sediments, animal intestinal tracts, and systems for anoxic

Table 17.9	Naturally occurring polysaccharides yielding hexose and pentose sugars[a]		
Substance	**Composition**	**Sources**	**Catabolic enzymes**
Cellulose	Glucose polymer (β-1,4-)	Plants (leaves, stems)	Cellulases (β, 1-4-glucanases)
Starch	Glucose polymer (α-1,4-)	Plants (leaves, seeds)	Amylase
Glycogen	Glucose polymer (α-1,4- and α-1,6-)	Animals (muscle)	Amylase, phosphorylase
Laminarin	Glucose polymer (β-1,3-)	Marine algae (Phaeophyta)	β-1,3-Glucanase (laminarinase)
Paramylon	Glucose polymer (β-1,3-)	Algae (Euglenophyta and Xanthophyta)	β-1,3-Glucanase
Agar	Galactose and galacturonic acid polymer	Marine algae (Rhodophyta)	Agarase
Chitin	N-Acetylglucosamine polymer (β-1,4-)	Fungi (cell walls) Insects (exoskeletons)	Chitinase
Pectin	Galacturonic acid polymer (from galactose)	Plants (leaves, seeds)	Pectinase (polygalacturonase)
Dextran	Glucose polymer	Capsules or slime layers of bacteria	Dextranase
Xylan	Heteropolymer of xylose and other sugars (β-1,4- and α-1,2 or α-1,3 side groups)	Plants	Xylanases
Sucrose	Glucose-fructose disaccharide	Plants (fruits, vegetables)	Invertase
Lactose	Glucose-galactose disaccharide	Milk	β-Galactosidase

[a] Each of these is subject to degradation by microorganisms.

sewage digestion. Cellulose digestion is also a major process in the rumen of ruminant animals where *Fibrobacter* and *Ruminococcus* species actively degrade cellulose (Section 19.11).

Starch is digestible by many fungi and bacteria; this is illustrated for a laboratory culture in Figure 17.63●. Starch-digesting enzymes, called *amylases*, are of considerable practical utility in many industrial situations where starch must be digested, such as the textile, laundry, paper, and food industries, and fungi and bacteria are the commercial sources of these enzymes (Section 30.9).

All the polysaccharides occurring extracellularly and used as substrates are first broken down to monomeric units by hydrolysis. In contrast, the polysaccharides formed within cells as storage products are broken down not by hydrolysis, but by **phosphorolysis**. This process, involving the addition of *inorganic* phosphate, results in the formation of hexose phosphate rather than the free hexose and may be summarized as follows for the degradation of starch, an α-1,4 polymer of glucose:

$$(C_6H_{12}O_6)_n + P_i \longrightarrow$$
$$(C_6H_{12}O_6)_{n-1} + \text{glucose 1-phosphate}$$

Because glucose 1-phosphate can be easily converted to glucose 6-phosphate—a key intermediate in glycolysis

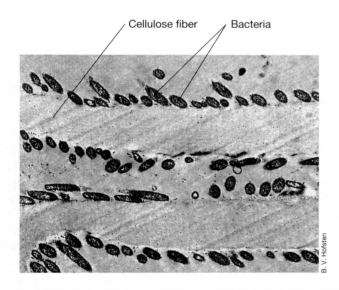

● **Figure 17.61 Cellulose digestion.** Transmission electron micrograph showing attachment of cellulose-digesting bacteria, *Sporocytophaga myxococcoides*, to cellulose fibers. Cells are about 0.5 μm in diameter (Figure 12.90).

● **Figure 17.62** *Cytophaga hutchinsonii* **colonies on a cellulose-agar plate.** Clear areas are where cellulose has been hydrolyzed.

● **Figure 17.64** **Slime formation.** Slimy colony formed by the dextran-producing bacterium *Leuconostoc mesenteroides*, growing on a sucrose-containing medium. When the same organism is grown on glucose, the colonies are small and not slimy because dextran synthesis requires sucrose (∞ Section 21.3 for further discussion).

● **Figure 17.63** **Demonstration of hydrolysis of starch by colonies of *Bacillus subtilis*.** After incubation, the plate was flooded with Lugol's iodine solution. Where starch hydrolysis occurred, the characteristic purple color of the starch–iodine complex is absent. Hydrolysis of starch occurs at some distance from the bacterial colonies because of the production of extracellular amylase, which diffuses into the surrounding medium.

(∞ Figure 5.14)—with no energy expenditures, phosphorolysis represents a net energy savings to the cell.

Disaccharides

Many microorganisms can use *disaccharides* for growth (Table 17.9). *Lactose* utilization by microorganisms is of considerable economic importance because milk-souring organisms produce lactic acid from lactose. *Sucrose*, the common disaccharide of higher plants, is usually first hydrolyzed to its component monosaccharides (glucose and fructose) by the enzyme *invertase*, and the monomers are then metabolized by normal pathways. *Cellobiose*, β-1,4-diglucose and a major product of cellulose digestion, is also readily degraded by a variety of bacteria that cannot degrade the cellulose polymer itself.

The microbial polysaccharide *dextran* is synthesized by some bacteria using the enzyme *dextransucrase* and sucrose as starting material:

$$n \text{ sucrose} \longrightarrow \underset{\text{dextran}}{(\text{glucose})_n} + n \text{ fructose}$$

Dextran is formed in this way by the bacterium *Leuconostoc mesenteroides* and a few others, and the polymer formed accumulates around the cells as a massive slime layer or capsule (Figure 17.64●). Because sucrose is required for dextran formation, no dextran is formed when the bacterium is cultured on a medium with glucose or fructose. In nature, when cells containing dextran or other polysaccharide capsules die, these materials once again become available for attack by fermentative or other chemoorganotrophic microorganisms.

The Pentose Phosphate Pathway

Pentose sugars are often available in nature. But if they are not available, they must be made, since they form the backbone of the nucleic acids. Pentoses are made from hexose sugars, and the major pathway for this process is the **pentose phosphate pathway**.

Figure 17.65● summarizes the pentose phosphate pathway. Several important features should be noted in this pathway. First, *glucose can be converted into a pentose* by loss of one carbon atom by oxidation followed by decarboxylation. This generates NADPH and the key intermediate of the pathway, *ribulose-5-phosphate* (Figure 17.65). From the latter, ribose (and from it deoxyribose, ∞ Section 5.16) are formed to supply the needs of nucleic acid synthesis. Pentose sugars as electron donors can also feed into the pentose phosphate pathway, typically becoming phosphorylated to form ribose phosphate or a related compound (Figure 17.65) before being further catabolized.

A second important feature of the pentose phosphate pathway is the *generation of sugar diversity*. A variety of sugar derivatives—C_4, C_5, C_6, and C_7—are formed in ancillary reactions of the pathway (Figure 17.65). This allows for pentose sugars to eventually yield hexoses for either catabolic purposes or for biosynthesis (gluconeogenesis, ∞ Section 5.15).

A final important aspect of the pentose phosphate pathway is that it *generates NADPH*. This form of the reduced coenzyme is used for reductive biosyntheses in the cell. Although most cells have a mechanism for converting NADH into NADPH, the pentose phosphate pathway is the major means for direct synthesis of this important coenzyme.

 17.25 Concept Check

Polysaccharides are abundant in nature and can be broken down, usually by phosphorolysis, into hexose or pentose monomers and used as energy sources. Starch and cellulose are common polysaccharides. The pentose phosphate pathway is the major means for generating pentose sugars for biosynthesis.

● **Figure 17.65   The pentose phosphate pathway.** The pathway can be used to: (1) generate pentoses from hexoses; (2) generate hexoses from pentoses (gluconeogenesis); (3) catabolize pentoses as an energy source; and (4) generate NADPH. Some key enzymes of the pathway are indicated.

◆ What is *phosphorolysis*?

◆ What disaccharides are common in nature?

◆ What functions does the pentose phosphate pathway play in the cell?

17.26   Organic Acid Metabolism

A variety of organic acids can be used by microorganisms as carbon sources and electron donors. The acids of the citric acid cycle, such as *citrate, malate, fumarate,* and *succinate,* are common natural products formed by plants and are also fermentation products of microorganisms. Because the citric acid cycle has major *biosynthetic* (Section 5.16) as well as *energetic* (Section 5.13) functions, the complete cycle or major portions of it, are nearly universal in microorganisms. Thus, it is not surprising that many microorganisms are able to use these acids as electron donors and carbon sources. Aerobic utilization of four-, five-, and six-carbon acids can be accomplished by means of enzymes of the citric acid cycle, with ATP formation by oxidative phosphorylation.

Anaerobic utilization of organic acids typically involves conversion to pyruvate followed by formation of acetate via acetyl phosphate with ATP production by substrate-level phosphorylation (see Section 17.19).

Glyoxylate Cycle

Utilization of two- or three-carbon acids as carbon sources cannot occur by means of the citric acid cycle alone. The cycle can continue to operate only if the acceptor molecule, the four-carbon acid *oxalacetate,* is regenerated at each turn; any removal of carbon compounds for biosynthetic reactions would prevent completion of the cycle (Figure 5.22).

When acetate is used, the oxalacetate needed to continue the cycle is produced through the **glyoxylate cycle** (Figure 17.66●), so named because glyoxylate is a key intermediate. This cycle is composed of most of the citric acid cycle reactions plus two additional enzymes: *isocitrate lyase,* which splits isocitrate into succinate and glyoxylate, and *malate synthase,* which converts glyoxylate and acetyl-CoA to malate (Figure 17.66).

Biosynthesis through the glyoxylate cycle occurs as follows. The splitting of isocitrate into succinate and glyoxylate allows the succinate molecule (or another citric acid cycle intermediate derived from it) to be drawn off for biosynthesis because glyoxylate combines with acetyl-CoA

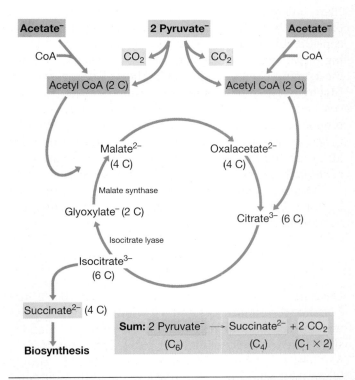

● **Figure 17.66   The glyoxylate cycle, leading to the synthesis of oxalacetate from acetate.** Two unique enzymes, *isocitrate lyase* and *malate synthase,* operate with a majority of the citric acid cycle reactions. In addition to growth on pyruvate, the glyoxylate cycle can also operate during growth on acetate.

to yield malate. Malate can be converted to oxalacetate to maintain the citric acid cycle after a C_4 intermediate (succinate) has been drawn off. The succinate molecule can be used directly in the production of porphyrins (needed for cytochromes, chlorophyll, and other tetrapyrroles). It can also be oxidized to oxalacetate as a carbon skeleton for C_4 amino acids, or it can be converted (via oxalacetate and phosphoenolpyruvate) to glucose.

Pyruvate and C_3 Utilization

Three-carbon compounds such as pyruvate or compounds that can be converted to pyruvate (for example, lactate or carbohydrates) also cannot be used as energy sources through the citric acid cycle alone. Because some of the citric acid cycle intermediates are used for biosynthesis, the oxalacetate needed to keep the cycle going is synthesized from pyruvate or phosphoenolpyruvate by the addition of a carbon atom from CO_2. In some organisms this step is catalyzed by the enzyme *pyruvate carboxylase*:

$$\text{Pyruvate} + \text{ATP} + CO_2 \longrightarrow \text{oxalacetate} + \text{ADP} + P_i$$

whereas in others it is catalyzed by *phosphoenolpyruvate carboxylase*:

$$\text{Phosphoenolpyruvate} + CO_2 \longrightarrow \text{oxalacetate} + P_i$$

These reactions replace oxalacetate that is lost when intermediates of the citric acid cycle are removed for use in biosynthesis, and the cycle can continue to function.

 17.26 Concept Check

Organic acids are frequently metabolized through the citric acid cycle or through the glyoxylate cycle. Isocitrate lyase and malate synthase are the key enzymes of the glyoxylate cycle.

◆ Why is the glyoxylate cycle necessary for growth on acetate but not on succinate?

17.27 Lipids as Microbial Nutrients

Lipids are abundant in nature. The cytoplasmic membranes of all cells contain lipids and many microorganisms, as well as macroorganisms, produce lipid storage materials. These substances are all biodegradable and are excellent substrates for microbial energy-yielding metabolism. When cells die, their lipids are thus catabolized, with CO_2 being the final product.

Fat and Phospholipid Hydrolysis

Fats are esters of glycerol and fatty acids (∞ Section 3.4) and are readily available from the release of lipids from dead organisms. Microorganisms use fats only after hydrolysis of the ester bond, and extracellular enzymes called **lipases** are responsible for the reaction (Figure 17.67●). The end result is formation of glycerol and free fatty acids

● **Figure 17.67 Phospholipase activity.** Action of phospholipase around a streak of *Clostridium perfringens* growing on an agar medium containing egg yolk. On half of the plate, an inhibitor of phospholipase was added, preventing action of the enzyme.

(Figures 17.67 and 17.68●). Lipases will attack fats containing fatty acids of various chain lengths. Phospholipids are hydrolyzed by enzymes called *phospholipases*, given different letter designations depending on which ester bond they cleave (Figure 17.68). Phospholipases A and B cleave fatty acid esters while phospholipases of the C and D types cleave phosphate ester bonds and hence are different types of enzymes. The result of lipase activity is the release of free fatty acids and glycerol. All of these substances can then be attacked by various chemoorganotrophic microorganisms.

Fatty Acid Oxidation

Fatty acids are oxidized by a process called *beta oxidation*, in which two carbons of the fatty acid are split off at a time (Figure 17.69●). In eukaryotes the enzymes are in the mitochondria, whereas in prokaryotes they are cytoplasmic. The fatty acid is first activated with coenzyme A; oxidation results in the release of *acetyl-CoA* and the formation of a fatty acid shorter by two carbons (Figure 17.69). The process of beta oxidation is then repeated, and another acetyl-CoA molecule is released.

Two separate dehydrogenation reactions occur in beta oxidation. In the first, electrons are transferred to flavin-adenine dinucleotide (FAD), whereas in the second they are transferred to NAD^+. Most fatty acids have an even number of carbon atoms, and complete oxidation yields acetyl-CoA. The acetyl-CoA formed is then oxidized by way of

● **Figure 17.68** **Lipases.** (a) Activity of lipases on a fat. (b) Phospholipase activity on phospholipid. The cut sites of the four distinct phospholipases A, B, C, and D are shown. X refers to a number of small organic molecules that may be at this position in different phospholipids. Compare this diagram to the more complete figure of a phospholipid in Figure 3.7.

the citric acid cycle or is converted to hexose and other cell constituents via the glyoxylate cycle.

Fatty acids are good electron donors. For example, the oxidation of the 16-carbon fatty acid palmitic acid results in the net synthesis of 129 ATP molecules. These include electron transport phosphorylation from electrons generated during the formation of acetyl-CoA from beta oxidations and from oxidation of the acetyl-CoA units themselves through the citric acid cycle (Section 5.13).

 17.27 **Concept Check**

Fats are metabolized by hydrolysis by lipases or phospholipases to free fatty acids. The latter are oxidized by beta oxidation to acetyl-CoA units, which are subsequently oxidized to CO_2 by the citric acid cycle.

◆ What are the functions of phospholipases?

◆ What is meant by the term *beta oxidation*?

VI NITROGEN FIXATION

The utilization of N_2 as a source of cell nitrogen is called **nitrogen fixation**. The ability to fix N_2 frees an organism from dependence on fixed forms of nitrogen. Since the latter are in high demand in microbial ecosystems, the capacity to fix N_2 confers a significant ecological advantage on those cells capable of the process. Moreover, certain forms of nitrogen fixation are of enormous agricultural importance, driving the cultivation of key row crops, such as soybeans.

● **Figure 17.69** **Beta oxidation.** Mechanism of beta oxidation of a fatty acid, which leads to successive formation of two-carbon fragments of acetyl-CoA.

17.28 Nitrogenase and the Process of Nitrogen Fixation

Only certain prokaryotes—some aerobes and some anaerobes—can fix N_2. An abbreviated list of nitrogen-fixing organisms is given in Table 17.10. Some nitrogen-fixing bacteria are free-living and require no host in order to carry out the process. By contrast, others are *symbiotic* and fix nitrogen only in association with certain plants (Section 19.22). But it is the *bacterium*, not the *plant*, that fixes the N_2. As far as is known, no eukaryotic organisms fix nitrogen.

Nitrogenase

In nitrogen fixation N_2 is *reduced* to ammonia and the ammonia converted to organic form. The reduction process is

Table 17.10 Some nitrogen-fixing organisms[a]

Free-living aerobes

Chemo-organotrophs	Phototrophs	Chemo-lithotrophs
Bacteria:	Cyanobacteria	Alcaligenes
Azotobacter		Thiobacillus
Azomonas		Acidithiobacillus
Agrobacterium		Streptomyces
Klebsiella[b]		thermoau-
Beijerinckia		totrophicus
Bacillus polymyxa		
Mycobacterium flavum		
Azospirillum lipoferum		
Citrobacter freundii		
Acetobacter diazotrophicus		
Methylomonas		
Methylococcus		
Methylosinus		
Pseudomonas		

Free-living anaerobes

Chemo-organotrophs	Phototrophs	Chemo-lithotrophs
Bacteria:	Bacteria:	Archaea:
Clostridium	Chromatium	Methanosarcina
Desulfovibrio	Ectothiorhodospira	Methanococcus
Desulfobacter	Thiocapsa	Methanobacterium
Desulfotomaculum	Chlorobium	Methanospirillum
	Chlorobaculum	Methanolobus
	Rhodospirillum	
	Rhodopseudomonas	
	Rhodomicrobium	
	Rhodopila	
	Rhodobacter	
	Heliobacterium	
	Heliobacillus	
	Heliophilum	

Symbiotic

Leguminous plants	Nonleguminous plants
Soybeans, peas, clover, locust, and so on, in association with a bacterium of the genus Rhizobium, Bradyrhizobium, Sinorhizobium, or Azorhizobium	Alnus, Myrica, Ceanothus, Comptonia, Casuarina; in association with actinomycetes of the genus Frankia

[a] For some genera listed, N_2 fixation occurs in only one or a few species.
[b] N_2 fixation occurs only under anoxic conditions.

catalyzed by the enzyme complex **nitrogenase**, which consists of two separate proteins called *dinitrogenase* and *dinitrogenase reductase*. Both components contain iron, and dinitrogenase contains molybdenum as well. The iron and molybdenum in dinitrogenase are contained within a cofactor known as *FeMo-co* (Figure 17.70●), and the actual reduction of N_2 occurs on this iron-molybdenum center. The composition of FeMo-co is $MoFe_7S_8 \cdot$ homocitrate (Figure 17.70), and FeMo-co is present in two copies per molecule of nitrogenase.

● **Figure 17.70 Structure of FeMo-co, the iron-molybdenum cofactor from nitrogenase.** On the top is the Fe_7S_8 cube that binds to molybdenum along with oxygen atoms from homocitrate (bottom, all oxygen atoms shown in green) and N and S atoms from dinitrogenase. Two molecules of FeMo-co are present per molecule of dinitrogenase. The interactions of the components of nitrogenase—dinitrogenase and dinitrogenase reductase—are shown in Figure 17.71.

Owing to the stability of the dinitrogen triple bond, N_2 is extremely inert; its activation is therefore a very energy-demanding process. Six electrons must be transferred to reduce N_2 to NH_3. The three successive reduction steps occur directly on nitrogenase with no free intermediates accumulating (Figure 17.71●).

Nitrogen fixation is inhibited by oxygen because dinitrogenase reductase is rapidly and irreversibly inactivated by O_2. This is true even when this enzyme is isolated from *aerobic* nitrogen fixers. In aerobic nitrogen-fixing bacteria, N_2 fixation occurs in the presence of O_2 in cells but not in purified enzyme preparations, and nitrogenase in such organisms is protected from O_2 inactivation by one of several different mechanisms, including the rapid removal of O_2 by respiration, the production of O_2-retarding slime layers (Figure 17.72●), or, in certain cyanobacteria, by compartmentalization of nitrogenase in a special type of cell (the heterocyst) (Section 12.25). In addition, although N_2 fixation does not occur in cell extracts exposed to oxygen, in aerobic nitrogen fixers like *Azotobacter*, nitrogenase is protected from oxygen inactivation by complexing with a specific protein; this is called *conformational protection*.

(a)

(a)   *(b)*

● **Figure 17.72** **Induction of slime formation by O_2 in nitrogen-fixing cells of *Azotobacter vinelandii*.** (a) Transmission electron micrograph of cells grown under micro-oxic conditions on 2.5% O_2; very little slime is evident. (b) Cells grown in air (21% O_2). Note the extensive darkly staining slime layer (arrow). The slime retards diffusion of O_2 into the cell, thus preventing nitrogenase inactivation by oxygen.

Overall reaction
$$8\,H^+ + 8\,e^- + N_2 \longrightarrow 2\,NH_3 + H_2$$
(16–24 ATP〰16–24 ADP + 16–24 P_i)

(b)

● **Figure 17.71** **Nitrogenase function.** (a) Shown are the steps in N_2 fixation starting from pyruvate. Electrons are supplied from dinitrogenase reductase to dinitrogenase one at a time, and each electron supplied is associated with the hydrolysis of 2–3 ATPs. (b) Hypothetical steps in N_2 reduction showing the H_2 evolution step, and a summary of nitrogenase activity.

Electron Flow in Nitrogen Fixation

The sequence of electron transfer in nitrogenase is as follows: electron donor → dinitrogenase reductase → dinitrogenase → N_2 (Figure 17.71). The electrons for nitrogen reduction are transferred to dinitrogenase reductase from ferredoxin or flavodoxin, both of which are low potential iron-sulfur proteins (⬤⬤ Section 5.11). In *Clostridium pasteurianum*, ferredoxin is the electron donor and is reduced by the splitting of pyruvate to acetyl-CoA + CO_2. In addition to reduced ferredoxin or flavodoxin, ATP is required for N_2 fixation. In each cycle of electron transfer, dinitrogenase reductase is reduced by ferredoxin/flavodoxin and binds two molecules of ATP. ATP binding alters the conformation of dinitrogenase reductase and lowers its reduction potential, allowing it to interact with dinitrogenase. Upon electron transfer to dinitrogenase, the ATP is hydrolyzed and dinitrogenase reductase dissociates from dinitrogenase and begins another cycle of reduction and

ATP binding (Figure 17.71). When appropriately reduced, dinitrogenase then reduces N_2 to NH_3, with the actual reduction occurring at the FeMo-co center (Figure 17.70).

Although only *six* electrons are necessary to reduce N_2 to two NH_3, *eight* electrons are actually consumed in the process, *two* electrons being lost as H_2 for each mole of N_2 reduced (Figure 17.71). The reason for this apparent waste is not known, but it is clear that H_2 evolution is an intimate part of the reaction mechanism of nitrogenase.

Alternative Nitrogenases

Some nitrogen-fixing bacteria produce non-molybdenum nitrogenases under certain growth conditions. These so-called *alternative nitrogenases* contain either vanadium and iron or iron only in place of molybdenum. Cofactors similar to FeMo-co are present in these alternative nitrogenases. FeVa-co occurs in the vanadium nitrogenase, and an iron-sulfur cluster resembling FeMo-co and FeVa-co but lacking both Mo and Va is present in the iron nitrogenase. Alternative nitrogenases are not synthesized when sufficient molybdenum is present, as the molybdenum nitrogenase is the main nitrogenase in the cell. Alternative nitrogenases function as a backup to ensure that N_2 fixation can still occur when molybdenum is limiting in the habitat (⬤⬤ Section 12.9). Nitrogen-fixing *Archaea* lack the Mo- and Va-based systems and contain iron-only nitrogenases.

The Unique Nitrogenase of *Streptomyces thermoautotrophicus*

A structurally and functionally novel molybdenum nitrogenase is present in the thermophilic streptomycete, *Streptomyces thermoautotrophicus*. This organism is a thermophilic (optimum temperature 65°C), filamentous member of the Actinobacteria (∞ Section 12.24). *S. thermoautotrophicus* is a hydrogen bacterium that can also use carbon monoxide (CO) as an electron donor. *S. thermoautotrophicus* fixes nitrogen and its nitrogenase contains Mo. But unlike classic Mo nitrogenase, the *S. thermoautotrophicus* nitrogenase is completely *insensitive* to O_2. The dinitrogenase component of the *S. thermoautotrophicus* nitrogenase, called *Str1*, contains three different polypeptides that show some structural similarity to dinitrogenase polypeptides from other nitrogen-fixers. In contrast, the dinitrogenase reductase component, called *Str2*, shows no similarity to other dinitrogenase reductases. Str2 is, however, highly homologous to manganese-containing enzymes called *superoxide dismutases* (∞ Section 6.16), and this is its role in this unusual nitrogenase (Figure 17.73●).

There is a pattern to electron flow in the *S. thermoautotrophicus* nitrogenase system that mimics that of classical nitrogenase. Str2 supplies electrons to Str1 in the *S. thermoautotrophicus* nitrogenase. The source of the electrons is superoxide (O_2^-), and the O_2^- is formed from the reduction of O_2 by a CO dehydrogenase (Figure 17.73). Thus, in analogy to the pyruvate → flavodoxin → dinitrogenase reductase → dinitrogenase reaction sequence in classical nitrogen fixation (Figure 17.71), the *S. thermoautotrophicus* sequence is $CO \rightarrow O_2^- \rightarrow Str2 \rightarrow Str1$. And remarkably, instead of oxygen *inhibiting* nitrogenase (as it does in every classical nitrogenase that has ever been examined), in *S. thermoautotrophicus* nitrogenase, oxygen is actually *required* in the reaction mechanism of the enzyme.

Other unique properties of the *S. thermoautotrophicus* nitrogenase include the fact that the enzyme consumes less than half of the ATP of classical nitrogenases and does not reduce other triply bonded compounds, such as acetylene (see Figure 17.74). Collectively, these properties, especially that of oxygen insensitivity, lend hope to plant biotechnologists trying to engineer nitrogen fixation into crop plants such as corn (maize) that do not harbor known nitrogen fixing symbionts.

Assaying Nitrogenase: Acetylene Reduction

Nitrogenase is not entirely specific for N_2 because it also reduces other triply bonded compounds such as cyanide (CN^-) and acetylene ($HC\equiv CH$). The reduction of acetylene by nitrogenase is only a two-electron process, and *ethylene* ($H_2C=CH_2$) is produced. The reduction of acetylene to ethylene provides a simple and rapid way of measuring the activity of nitrogen-fixing systems by gas chromatography (Figure 17.74●). This technique, known as the **acetylene reduction technique**, is widely used to detect nitrogen fixation in unknown systems.

Definitive proof for N_2 fixation is obtained using an isotope of nitrogen, ^{15}N, as a tracer. (^{15}N is not a radioisotope but a stable isotope. It is detected with a mass spectrometer.) The gas phase of a culture is enriched with $^{15}N_2$, and after incubation, the cells and medium are digested, the ammonia produced is distilled off, and it is assayed for its ^{15}N content. If there has been a significant production of ^{15}N-labeled NH_3, it is proof of nitrogen fixation.

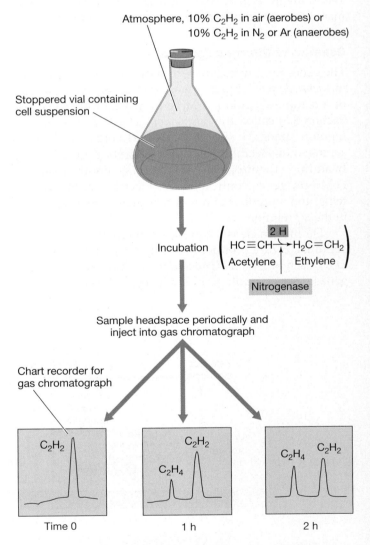

● **Figure 17.74** **The acetylene reduction assay for nitrogenase activity.** The results show no C_2H_4 when the experiment begins (time 0), but increasing production of C_2H_4 as the assay proceeds. Note how as C_2H_4 is produced, C_2H_2 is consumed. If the vial contained an enzyme extract, conditions would be anoxic, even if the nitrogenase was from an aerobic bacterium.

● **Figure 17.73** **Reactions of nitrogen fixation in *Streptomyces thermoautotrophicus*.** Although quite different proteins than dinitrogenase reductase and dinitrogenase, Str2 and Str1 are functionally equivalent, respectively, to these proteins.

Although $^{15}N_2$ assimilation is considered the "gold standard" for N_2 fixation, the acetylene reduction method is a more rapid and sensitive, albeit indirect, way of measuring this process. The sample, which may be soil, water, a culture, or a cell extract, is incubated with acetylene, and the gas phase of the reaction mixture is later analyzed by gas chromatography for production of ethylene (Figure 17.73). This method is far simpler and faster than other methods and can easily be adapted for field use in ecological studies of N_2-fixing bacteria directly in their habitats.

17.29 Genetics and Regulation of N₂ Fixation

Because the process of N_2 fixation is highly energy demanding, the synthesis and activity of nitrogenase is highly regulated. These regulatory events as well as the enzymes involved in N_2 fixation itself (Figure 17.71) require many genes to encode them, which we discuss here.

Genetics of Nitrogen Fixation

The genes for dinitrogenase and dinitrogenase reductase in *Klebsiella pneumoniae*, a well-studied N_2 fixer, are part of a complex regulon (a large network of operons, ∞ Section 8.6) called the *nif* regulon. The *K. pneumoniae nif* regulon spans 24 kbp of DNA and contains 20 genes arranged in several transcriptional units (Figure 17.75●). In addition to nitrogenase structural genes, the genes for FeMo-co, genes controlling the electron transport proteins, and a number of regulatory genes are also present in the *nif* regulon.

Dinitrogenase is a complex protein made up of two subunits, α (the product of *nifD*) and β (the product of *nifK*), each of which is present in two copies in the nitrogenase enzyme complex. Dinitrogenase reductase is a protein dimer consisting of two identical subunits, the product of *nifH*. FeMo-co is synthesized through the participation of several genes, including *nifN, V, Z, W, E,* and *B*, as well as *Q*, which encodes a product involved in molybdenum processing. The *nifA* gene encodes a positive regulatory protein that activates transcription of other *nif* genes (Figure 17.75).

Nitrogenase is a highly conserved protein, and the *nifHDK* genes that encode its components have been used as molecular probes to screen DNA from various prokaryotes for homologous genes, signaling nitrogen fixation capacity. In all nitrogen fixers examined (except for the highly unusual *Streptomyces thermoautotrophicus*, see Section 19.28), *nifHDK*-like genes are present, suggesting that the genetic requirements for nitrogenase are rather specific. Alternative nitrogenases (∞Section 17.28) are encoded by their own genes, *vnfHDK* in the vanadium system and *anfHDK* in the iron-only system, that show significant sequence homology to *nifHDK*.

Regulation of Nitrogen Fixation

Nitrogenase is subject to strict regulatory controls. Nitrogen fixation is repressed by O_2 and by fixed nitrogen, including NH_3, NO_3^-, and certain amino acids. A major part of this regulation is at the level of transcription, as shown in Figure 17.75. While transcription of the *nif* structural genes is *activated* by the NifA protein (positive regulation), NifL is a *negative* regulator of *nif* gene expression and contains a molecule of FAD (recall that FAD is a redox coenzyme for flavoproteins, ∞ Section 5.11) that is involved in O_2-sensing by the protein. In the presence of sufficient O_2, NifL functions to shut down transcription of *nif* genes in order to prevent synthesis of the oxygen labile nitrogenase.

Ammonia represses N_2 fixation through a second protein, called *NtrC*, whose activity is regulated by the

● **Figure 17.75** **The *nif* regulon in *Klebsiella pneumoniae*, the best-studied nitrogen-fixing bacterium.** The functions of some of the genes are uncertain. The mRNA transcripts (transcriptional units) are shown below the genes; arrows indicate the direction of transcription. Proteins involved in FeMo-co synthesis are shown in yellow. Other colors match those of Figure 17.71. Although capable of growth aerobically with ammonia as nitrogen source, nitrogen fixation in *Klebsiella* only occurs under anoxic conditions.

nitrogen status of the cell. When ammonia is limiting, NtrC is active and promotes transcription of *nifA*. This produces NifA, the nitrogen fixation activator protein, and *nif* transcription begins.

The ammonia produced by nitrogenase does not repress enzyme synthesis because as soon as it is made, it is incorporated into organic form and used in biosynthesis. But when ammonia is in excess (as in natural environments or culture media high in ammonia), nitrogenase synthesis is quickly repressed. In this way, ATP is not wasted in making a product already available in ample amounts.

In certain nitrogen-fixing bacteria, nitrogenase *activity* is also regulated by ammonia, a phenomenon called the *ammonia switch-off effect*. In this case, excess ammonia causes a covalent modification (∞ Section 8.3) of dinitrogenase reductase, which results in a loss of enzyme activity. When ammonia again becomes limiting, this modified protein is converted back to the active form and N_2 fixation resumes. Ammonia switch-off is thus a rapid and reversible method of controlling ATP consumption by nitrogenase. Not all nitrogen-fixing bacteria have the switch-off system, but switch-off is widespread in certain groups such as the phototrophic bacteria, where it was first discovered.

 17.28–17.29 Concept Checks

Nitrogen fixation—the reduction of N_2 to NH_3—requires an enzyme complex called nitrogenase, which consists of dinitrogenase and dinitrogenase reductase. Most nitrogenases contain molybdenum or vanadium plus iron as metal cofactors, and the process of nitrogen fixation is highly energy-demanding. Nitrogenase and most associated regulatory proteins are encoded by the *nif* regulon. Certain substances that are structurally similar to N_2, such as acetylene and cyanide, are also reduced by nitrogenase. Nitrogenase and nitrogen fixation are both highly regulated processes, with O_2 and NH_3 being the two main regulatory effectors.

◆ Write a balanced equation for the reaction carried out by the enzyme *nitrogenase*.

◆ What is *FeMo-co* and what metals does it contain?

◆ How is C_2H_2 useful for studies of nitrogen fixation?

◆ What chemical and physical factors affect the function of nitrogenase? How does the *Streptomyces thermoautotrophicus* nitrogenase system differ in this regard?

REVIEW QUESTIONS

1. What are the major differences between oxygenic and anoxygenic phototrophs (∞ Sections 17.1–17.5)?

2. What are the functions of light-harvesting and reaction center chlorophylls? Why would a mutant incapable of making light-harvesting chlorophylls (such mutants can be readily isolated in the laboratory) probably not be a successful competitor in nature (∞ Section 17.2)?

3. Where are the photosynthetic pigments located in a purple bacterium? A cyanobacterium? A green alga? Considering the function of chlorophyll pigments, why can't they be located elsewhere in the cell, for example, in the cytoplasm or in the cell wall (∞ Section 17.3)?

4. What accessory pigments are present in phototrophs and what are their functions (∞ Section 17.3)?

5. How does light result in ATP production in an *anoxygenic phototroph*? In what ways are photosynthetic and respiratory electron flow similar? In what ways do they differ (∞ Section 17.4)?

6. How is reducing power made for autotrophic growth in a purple bacterium? In a cyanobacterium (∞ Section 17.4)?

7. How does the reduction potential of chlorophyll *a* in photosystem I and photosystem II differ? Why must the reduction potential of photosystem II chlorophyll be so highly electropositive (∞ Section 17.5)?

8. What two enzymes are unique to organisms that carry out the Calvin cycle? What reactions do these enzymes carry out? What would be the consequences if a mutant arose that lacked either of these enzymes (∞ Section 17.6)?

9. What organisms employ the hydroxypropionate or reverse citric acid cycles as autotrophic pathways (∞ Section 17.7)?

10. Compare and contrast the utilization of H_2S by a purple phototrophic bacterium and by a colorless sulfur bacterium like *Beggiatoa*. What role does H_2S play in the metabolism of each organism (∞ Section 17.8)?

11. Which inorganic electron donors are used by the organisms *Ralstonia* and *Thiobacillus* (∞ Sections 17.9 and 17.10)?

12. What is unusual about the environment of *Acidithiobacillus ferrooxidans* that affects the energetics of its metabolism (∞ Section 17.11)?

13. Contrast classical nitrification with anammox in terms of oxygen requirements, organisms involved, and the need for monooxygenases (∞ Section 17.12).

14. In *Escherichia coli* synthesis of the enzyme *nitrate reductase* is repressed by oxygen. On the basis of bioenergetic arguments, why do you think this repression phenomenon might have evolved (∞ Sections 17.13 and 17.14)?

15. Why is hydrogenase a constitutive enzyme in *Desulfovibrio* (∞ Section 17.15)?

16. Compare and contrast *homoacetogens* with *methanogens* in terms of (1) substrates and products of their energy metabolism, (2) ability to use organic compounds as electron donors in energy metabolism, (3) mechanism of autotrophy, and (4) ability to grow by aerobic respiration (∞ Sections 17.16 and 17.17; you may also want to review material in Section 13.4 before answering this question).

17. Compare and contrast ferric iron reduction with reductive dechlorination in terms of (1) product of the reduction, and (2) environmental significance (◯◯◯ Section 17.18).

18. Define the term *substrate-level phosphorylation*. How does it differ from *oxidative phosphorylation*? Assuming an organism is facultative, what basic nutritional conditions dictate whether the organism obtains energy from substrate-level rather than oxidative phosphorylation (◯◯◯ Section 17.19)?

19. Although many different compounds are theoretically fermentable, in order to support a fermentative process, most organic compounds must eventually be converted to one of a relatively small group of molecules. What are these molecules and why must they be produced (◯◯◯ Sections 17.19 and 17.20)?

20. How can fermentations occur in the absence of substrate-level phosphorylation (◯◯◯ Section 17.20)?

21. Why is syntrophy sometimes called "interspecies H_2 transfer" (◯◯◯ Section 17.21)?

22. How do *monooxygenases* differ from *dioxygenases* in the reactions they catalyze (◯◯◯ Sections 17.22 and 17.23)?

23. How does a *methanotroph* differ from a *methanogen*? How do type I and type II methanotrophs differ in their carbon assimilation patterns (◯◯◯ Section 17.24)?

24. Compare and contrast the conversion of cellulose and intracellular starch to glucose units. What enzymes are involved and which process is the more energy efficient (◯◯◯ Section 17.25)?

25. What is the major function of the glyoxalate cycle (◯◯◯ Section 17.26)?

26. Why are oxygenases needed for the aerobic catabolism of hexadecane (C_{16}) but not for the corresponding C_{16} fatty acid palmitate (◯◯◯ Sections 17.22 and 17.27)?

27. Write out the reaction catalyzed by the enzyme *nitrogenase*. How many electrons are required in this reaction? How many are actually used? Explain (◯◯◯ Section 17.28).

28. How does the *Streptomyces thermoautotrophicus* nitrogenase differ from that of *Azotobacter* (◯◯◯ Section 17.28)?

29. Which genes in the *nif* regulon actually encode nitrogenase proteins (◯◯◯ Section 17.29)?

APPLICATION QUESTIONS

1. Compare and contrast the absorption spectrum of chlorophyll *a* and bacteriochlorophyll *a*. What wavelengths are preferentially absorbed by each pigment and how do the absorption properties of these molecules compare with the regions of the spectrum visible to our eye? Why are most plants green?

2. The growth rate of the phototrophic purple bacterium *Rhodobacter* is about twice as fast when the organism is grown *phototrophically* in a medium containing malate as carbon source as when it is grown with CO_2 as carbon source (with H_2 as electron donor). Discuss the reasons why this is true and list the nutritional class we would place *Rhodobacter* in when growing under each of the two different conditions.

3. Although physiologically distinct, chemolithotrophs and chemoorganotrophs share a number of features with respect to the production of ATP. Discuss these common features along with reasons why the growth yield (grams of cells per mole of substrate) of a chemoorganotroph respiring glucose is so much higher than for a chemolithotroph respiring sulfur.

4. When methane is made from CO_2 (plus H_2) or from methanol (in the absence of H_2), various steps in the pathway shown in Figures 17.44 and 17.45 are used. Compare and contrast methanogenesis from these two substrates and discuss why they must be metabolized in *opposite* directions.

5. Although dextran is a glucose polymer, glucose cannot be used to make dextran. Explain. How is dextran synthesis important in oral hygiene (◯◯◯ Section 21.3)?

6. *Pseudomonas fluorescens* can grow on benzoate aerobically, whereas the phototrophic bacterium *Rhodopseudomonas palustris* can grow on benzoate anaerobically. Compare and contrast the metabolism of benzoate by these two species, focusing on the following considerations: requirement for oxygenases, initial reactions leading to the opening of the ring, and product(s) formed that can feed into central metabolic pathways.

18

METHODS IN MICROBIAL ECOLOGY

Fluorescent stains can be used in both qualitative and quantitative ways in microbial ecology. In some cases, very specific stains are required to positively identify specific organisms, while in others, nonspecific stains are sufficient for enumerating cells in the environment. Shown here are cells stained with a nonspecific stain called DAPI, which makes the DNA in the cells stain blue and the cells easier to see for counting purposes.

WORKING GLOSSARY

Denaturing gradient gel electrophoresis (DGGE) an electrophoretic technique capable of separating nucleic acid fragments of the same size but that differ in sequence

DAPI a nonspecific fluorescent dye used to stain bacteria in a natural sample to obtain total cell counts

Enrichment bias a problem with liquid enrichment cultures in which certain species tend to dominate in the enrichment, often excluding the most abundant or ecologically significant organisms in the inoculum

Enrichment culture a means of obtaining laboratory cultures of microorganisms from a natural sample by using highly selective culture methods

Environmental genomics the use of genomic methods (sequencing and analyzing genomes) to characterize natural microbial communities

Fluorescent *in situ* hybridization (FISH) a method employing a fluorescent dye covalently bonded to a specific nucleic acid probe for identifying or tracking organisms in the enrichment

Green fluorescent protein a fluorescing protein that renders a cell green and can be used to track an organism in nature

Isotope fractionation discrimination by enzymes against the *heavier* isotope of the various isotopes of C or S, leading to enrichment in the lighter isotopes of each during the metabolism of CO_2 or SO_4^{2-}

Laser tweezers a device used to obtain pure cultures in which a single cell is optically trapped with a laser beam and moved away from surrounding cells into sterile growth medium

Metagenome the entire complement of genes present in a given environment

Microautoradiography (MAR) measurement of the uptake of radioactive substrates by visually observing the cells in an exposed photograph emulsion

Microbial ecology study of the interaction of microorganisms with each other and their environment

Microelectrode a small glass electrode for measuring pH or specific compounds such as O_2 or H_2S that can be immersed into a microbial habitat at microscale intervals

Most probable number (MPN) serial dilution of a natural sample to determine the highest dilution yielding growth

Nucleic acid probe an oligonucleotide, usually 10–20 bases in length, complementary in base sequence to a sequence in a target gene

Winogradsky column a column packed with mud and overlaid with water to mimic a lake, in which various bacteria develop over a period of months

---

Microbial ecology deals with how microorganisms interact with each other and their environment. The focus of microbial ecology is *biodiversity* and *microbial activity*. To study **biodiversity**, microbial ecologists must be able to isolate, identify, and quantify microorganisms in various habitats. To study **microbial activity**, techniques must be available to measure what microorganisms are *doing* in their habitats. This chapter explores the tools microbial ecologists use to assess both microbial diversity and activity.

The methods described in this chapter include *enrichment* and *isolation*, both nonspecific and specific *cell-staining* methods, *gene isolation and characterization*, and methods for measuring the *activities* of microorganisms *in situ* (in the natural environment). Chapter 19 will then discuss the ecology of both large-scale microbial communities (for example, soil or water) and more specialized microbial habitats (for example, the rumen of ruminant animals or the root nodules of leguminous plants). With the tools to be described in this chapter, microbial ecologists can ask sophisticated questions about the structure and function of microbial communities.

CULTURE-DEPENDENT ANALYSES OF MICROBIAL COMMUNITIES

Rarely does a natural environment contain only a single type of microorganism. Instead, microorganisms exist in nature in *microbial communities* (∞ Section 19.1 and

Figure 19.1). Microbial ecologists have developed cultural methods to dissect microbial communities through isolation and characterization of their components.

18.1 Enrichment and Isolation

One way to study a microbial ecosystem is to isolate microorganisms from it and study their properties in laboratory cultures. This process involves teasing the organisms of interest out from the community, a procedure called **enrichment**, and then isolating the organism of interest in pure culture. Isolation is important because it yields cultures for detailed and controlled laboratory studies and for applications in the fields of biotechnology and industrial and environmental microbiology.

A common method for isolating microorganisms from nature is the *enrichment culture technique*. In an **enrichment culture**, a medium and a set of incubation conditions are established that are *selective* for the desired organism and *counterselective* for undesired organisms. Effective enrichment cultures duplicate as closely as possible the *resources* and *conditions* of a particular niche. Literally hundreds of different enrichment strategies have been devised, and Table 18.1 provides an overview of some successful ones.

Inocula and Enrichment Culture Outcomes

Successful enrichment cultures require that an appropriate **inoculum** (sample) containing the organism be

Table 18.1 Some enrichment culture methods for prokaryotes[a] (continues on next page)

Light-phototrophic bacteria: main C source, CO_2

Incubation in air	Organisms enriched	Inoculum
N_2 as nitrogen source	Cyanobacteria	Pond or lake water; sulfide-rich muds; stagnant water; raw sewage; moist, decomposing leaf litter; moist soil exposed to light
NO_3^- as nitrogen source, 55° C	Thermophilic cyanobacteria	Hot spring microbial mat
Anoxic incubation		
H_2 or organic acids; N_2 as sole nitrogen source	Purple nonsulfur bacteria, heliobacteria	Same as above plus hypolimnetic lake water; pasteurized soil (heliobacteria)
H_2S as electron donor	Purple and green sulfur bacteria	

Dark-chemolithotrophic bacteria: main C source, CO_2 (medium must lack organic C)

Incubation in air: aerobic respiration

Electron donor	Electron acceptor	Organisms enriched	Inoculum
NH_4^+	O_2	Nitrosifying bacteria (*Nitrosomonas*)	Soil, mud; sewage effluent
NO_2^-	O_2	Nitrifying bacteria (*Nitrobacter*)	
H_2	O_2	Hydrogen bacteria (various genera)	
$H_2S, S^0, S_2O_3^{2-}$	O_2	*Thiobacillus* spp.	
Fe^{2+}, low pH	O_2	*Acidithiobacillus ferrooxidans*	

Anoxic incubation

Electron donor	Electron acceptor	Organisms enriched	Inoculum
$S^0, S_2O_3^{2-}$	NO_3^-	*Thiobacillus denitrificans*	Mud, lake sediments, soil
H_2	NO_3^- + yeast extract	*Paracoccus denitrificans*	

Dark-chemoorganotrophic bacteria and methanogens: main C source, organic compounds

Incubation in air: aerobic respiration

Electron donor and nitrogen source	Electron acceptor	Typical organisms enriched	Inoculum
Lactate + NH_4^+	O_2	*Pseudomonas fluorescens*	Soil, mud; lake sediments; decaying vegetation; pasteurize inoculum (80° C for 15 min) for all *Bacillus* enrichments
Benzoate + NH_4^+	O_2	*Pseudomonas fluorescens*	
Starch + NH_4^+	O_2	*Bacillus polymyxa*, other *Bacillus* spp.	
Ethanol (4%) + 1% yeast extract, pH 6.0	O_2	*Acetobacter, Gluconobacter*	
Urea (5%) + 1% yeast extract	O_2	*Sporosarcina ureae*	
Hydrocarbons (e.g., mineral oil, gasoline, toluene) + NH_4^+	O_2	*Mycobacterium, Nocardia, Pseudomonas* (Figure 19.42)	
Cellulose + NH_4^+	O_2	*Cytophaga, Sporocytophaga* (Figure 12.90)	
Mannitol or benzoate, N_2 as N source	O_2	*Azotobacter*	

available. Thus, enrichment cultures begin by sampling the appropriate habitat (Table 18.1). Enrichment cultures are established by placing the inoculum directly into a highly selective medium and incubating under specific conditions. In this way, many common prokaryotes can be isolated. For example, the great Dutch microbiologist Martinus Beijerinck, who pioneered the enrichment culture technique (Section 1.6), first isolated the N_2-fixing bacterium *Azotobacter* using a classic enrichment strategy (Figure 18.1●). Because *Azotobacter* is capable of N_2 fixation in air, enrichment using media devoid of ammonia or other forms of fixed nitrogen and incubated in air selects strongly for this organism. Non–nitrogen fixers or

Table 18.1 Some enrichment culture methods for prokaryotes[a] (continued)

Anoxic incubation: anaerobic respiration

Electron donor	Electron acceptor	Organisms enriched	Inoculum
Organic acids	KNO_3 (0.2%)	*Pseudomonas* (denitrifying species)	Soil, mud; lake sediments
Yeast extract	KNO_3 (.1%)	*Bacillus* (denitrifying species)	
Organic acids	Na_2SO_4	*Desulfovibrio, Desulfotomaculum*	
Acetate, propionate, butyrate	Na_2SO_4	Fatty-acid oxidizing sulfate reducers	As above; or sewage digestor sludge; rumen contents; marine sediments
Acetate, ethanol	S^0	*Desulfuromonas*	
Acetate	Fe^{3+}	*Geobacter, Geospirillum*	
Acetate	ClO_3^-	Various chlorate-reducing bacteria	
H_2	Na_2CO_3	Methanogens (chemolithotrophic species only), homoacetogens	Mud, sediments, sewage sludge
CH_3OH	Na_2CO_3	*Methanosarcina barkeri*	
CH_3NH_2 or CH_3OH	KNO_3	*Hyphomicrobium*	

Anoxic incubation: fermentation

Electron donor and nitrogen source	Electron acceptor	Organisms enriched	Inoculum
Glutamate or histidine	No exogenous electron acceptors added	*Clostridium tetanomorphum*	Mud, lake sediments; rotting plant material; dairy products (lactic and propionic acid bacteria); rumen or intestinal contents (enteric bacteria); sewage sludge; soil
Starch + NH_4^+	None	*Clostridium* spp.	
Starch + N_2 as N source	None	*Clostridium pasteurianum*	
Lactate + yeast extract	None	*Veillonella* spp.	
Glucose or lactose + NH_4^+	None	*Escherichia, Enterobacter,* other fermentative organisms	
Glucose + yeast extract (pH 5)	None	Lactic acid bacteria (*Lactobacillus*)	
Lactate + yeast extract	None	Propionic acid bacteria	
Succinate + NaCl	None	*Propionigenium*	
Oxalate	None	*Oxalobacter*	

[a] All media must contain an assortment of mineral salts including N, P, S, Mg^{2+}, Mn^{2+}, Fe^{2+}, Ca^{2+}, and other trace elements (∞ Sections 5.1–5.3). Certain organisms may have requirements for vitamins or other growth factors. This table is meant as an overview of enrichment methods and does not speak to the effect incubation temperature might have in isolating thermophilic (high temperature), hyperthermophilic (very high temperature), and psychrophilic (low temperature) species, or the effect that extremes of pH or salinity might have, assuming an appropriate inoculum was available.

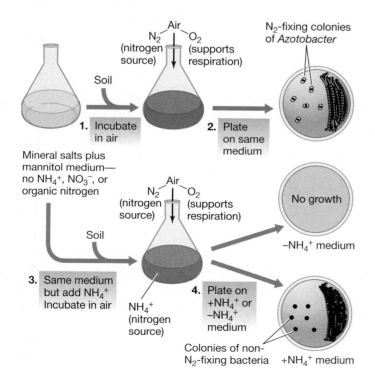

anaerobic nitrogen-fixing bacteria are counterselected in this instance (Figure 18.1).

It should be emphasized that both the culture medium *and* the incubation conditions are critical to a successful enrichment culture. That is, both *resources* and *conditions* must be optimized to have the best chance of enriching the organism of interest. Some enrichment cultures yield nothing. This may mean that an organism showing the desired set of properties is absent from the habitat. Alternatively, it could mean that such an organism exists but conditions for its culture were insufficient to enrich it. Thus enrichment culture can prove a positive but never a negative. Moreover, the isolation of the desired organism from an enrich-

● **Figure 18.1** **The isolation of *Azotobacter*.** Selection for aerobic N_2-fixing bacteria usually results in the isolation of *Azotobacter* or its relatives. By contrast, enrichment with fixed forms of nitrogen such as NH_4^+ rarely results in N_2-fixing bacteria because the selective pressure for nitrogen fixation is lost.

ment culture says nothing about the *ecological importance* or *abundance* of the organism in the inoculum; a positive enrichment only proves that the organism was present in the sample and, in theory, requires that only a single viable cell be present.

The Winogradsky Column

The **Winogradsky column** is a miniature anoxic ecosystem that can be used as a long-term source of bacteria for enrichment culture purposes. It has been used to isolate purple and green phototrophic bacteria, sulfate-reducing bacteria, and many other anaerobes (Figure 18.2●). Named for the famous Russian microbiologist Sergei Winogradsky (∞ the Microbial Sidebar, Winogradsky's Legacy, Chapter 17, and Section 1.7), the column was first used by him in the late nineteenth century to study soil microorganisms.

A Winogradsky column is prepared by filling a glass cylinder about one-half full with organic-rich, preferably sulfide-containing, mud (Figure 18.2). Carbon substrates are first mixed into the mud. The substrates will determine which organisms are enriched, but fermentative substrates (that can lead to excessive gas formation) are avoided. The mud is supplemented with $CaCO_3$ and $CaSO_4$ as a buffer and as a source of sulfate, respectively. The mud is packed tightly in the container, taking care to avoid trapping air (Figure 18.2). The mud is then covered with lake, pond, or ditch water (or seawater if a ma-

rine column is established), and the top of the cylinder is covered to prevent evaporation. The cylinder is placed near a window receiving diffuse sunlight and left to develop for a period of weeks.

In a typical Winogradsky column a mixture of organisms develops. Algae and cyanobacteria appear quickly in the upper portions of the water column; by producing O_2 these organisms help to keep this zone oxic. Decomposition processes in the mud lead to the production of organic acids, alcohols, and H_2, suitable substrates for sulfate-reducing bacteria (∞ Sections 12.18, 13.7, 17.15, and 19.13). Sulfide from the sulfate reducers triggers development of purple and green sulfur bacteria (anoxygenic phototrophs, ∞ Sections 12.2 and 12.32) that use sulfide as a photosynthetic electron donor. These organisms typically appear in patches in the mud on the sides of the column, but may bloom in the water itself if oxygenic phototrophs are scarce (Figure 18.2). Sampling of the column for phototrophic bacteria is performed by inserting a long, thin pipette and removing some colored mud or water. Such samples can be used for microscopy and to inoculate enrichment media (Table 18.1) for isolation and characterization.

Winogradsky columns have been used to enrich a variety of prokaryotes, both aerobes and anaerobes. Besides supplying a ready source of inocula for enrichment cultures, columns can also be supplemented with a particular compound to test the hypothesis that an

(a)

(b)

● **Figure 18.2 The Winogradsky column.** (a) Schematic view of a typical column. The column is placed to receive subdued sunlight. Chemoorganotrophic bacteria grow throughout the column, aerobes and microaerophiles in the upper regions, anaerobes in the zones containing H_2S. Anoxic decomposition leading to sulfate reduction creates the gradient of H_2S. Green and purple sulfur bacteria stratify according to their tolerance for H_2S. (b) Photo of Winogradsky columns that have remained anoxic up to the top where blooms of three different phototrophic bacteria have occurred in the mud and up into the water column. Left to right: *Thiospirillum jenense, Chromatium okenii,* (both purple sulfur bacteria, Section 12.2), and *Chlorobium limicola* (green sulfur bacteria, ∞ Section 12.32).

organism or organisms exist in the inoculum that can degrade it. Once a crude enrichment has been established, pure cultures can be pursued, as will be discussed in Section 18.2.

Enrichment Bias

Although the enrichment culture is a powerful tool, in most enrichments there exists a bias, and sometimes a very severe bias, in the outcome. In typical liquid enrichment cultures the most fit organism for the chosen set of conditions becomes dominant. Unfortunately, however, the most fit organism for laboratory culture may be only a minor component of the microbial ecosystem rather than the dominant or most ecologically relevant organism.

Enrichment bias can be demonstrated by comparing the results obtained in dilution cultures (see Section 18.2) with classical liquid enrichment. Dilution of the inoculum followed by enrichment or direct plating often yields different organisms than liquid enrichment using undiluted inoculum. It is thought that dilution eliminates the quantitatively insignificant but rapidly growing "weed" species from the inoculum to allow other members of the microbial community a chance to develop. Dilution of the inoculum is thus a common practice in enrichment culture microbiology today.

 18.1 Concept Check

The enrichment culture technique is a means of obtaining microorganisms from natural samples. Specific reactions can be investigated by enrichment methods by choosing media and incubation conditions selective for particular organisms.

◆ Describe the enrichment strategy behind Beijerinck's isolation of *Azotobacter*.

◆ Why is sulfate (SO_4^{2-}) added to a Winogradsky column?

◆ What is *enrichment bias*?

18.2 Isolation in Pure Culture

Pure cultures—cultures containing a single kind of microorganism—can be obtained from enrichments in many ways. Common isolation procedures include the streak plate, the agar shake, and liquid dilution. For organisms that grow well on agar plates, the streak plate is quick, easy, and the method of choice (Figure 18.3a●). By repeated picking and restreaking of a well-isolated colony, a pure culture can be obtained that can then be transferred to a liquid medium. With proper incubation facilities (for example, anoxic jars for anaerobes ∞ Section 6.13), it is possible to purify both aerobes and anaerobes on agar plates by the streak plate method.

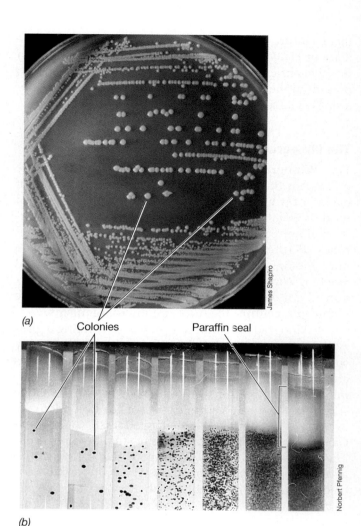

(a) Colonies   Paraffin seal

(b)

● **Figure 18.3 Pure culture methods.** (a) Organisms that form distinct colonies on plates such as those shown here are usually very easy to purify. (b) Colonies of oxygen-sensitive anaerobes using the agar shake technique. A dilution series was established from right to left, eventually yielding well-isolated colonies. The tubes are sealed with a sterile mixture of paraffin and mineral oil to maintain anaerobiosis.

Agar Shakes and Most Probable Numbers (MPN)

The agar shake-tube method involves the dilution of a mixed culture in tubes of molten agar, resulting in colonies *embedded* in the agar rather than growing on the surface of an agar plate (Figure 18.3b). Shake tubes are useful for purifying anaerobic prokaryotes such as phototrophic sulfur bacteria and sulfate-reducing bacteria (∞ Figure 12.50g) from Winogradsky columns (see Figure 18.2). Purification is obtained by successive dilutions of cell suspensions in tubes of liquid medium (Figure 18.3b). By repeating this procedure using a colony from a tube at the highest dilution as inoculum for a new set of dilutions, pure cultures are eventually obtained.

A similar procedure is serial dilution of an inoculum in a *liquid* medium until the final tube or tubes in the series show no growth (Figure 18.4●). Using 10-fold serial dilutions, for example, the last tube showing growth should have originated from 10 or fewer cells. By repeating this process several times, pure cultures can be obtained. This method, known as the **most probable number** (or MPN) technique (Figure 18.4), has been used for estimating the numbers of microorganisms in foods, wastewater, and other samples, where cell numbers need to be assessed routinely. The MPN can be done using highly selective media and incubation conditions to target one or a small group of organisms, such as a particular pathogen, or using complex media to get an estimate of total cell numbers (but see the caveat that applies here in Section 6.7)

Regardless of the methods used to purify a culture, once a putative pure culture has been obtained, it is essential to check its purity. This should be done through a combination of (1) *microscopy*, (2) *the observation of colony characteristics* on plates or in shake tubes, and (3) *testing the culture for growth* in other media. In the latter regard, it is important to test the culture for growth in media in which it is suspected

that the culture will grow poorly or not at all but contaminants will grow vigorously. In the final analysis, the microscopic observation of a single morphological type of cell coupled with uniform colony characteristics and the absence of contamination in growth tests with various culture media, is good evidence that a culture is *axenic* (pure).

Probing for Pure Cultures—The Laser Tweezers

In addition to the more classical methods just described, technological advances in microscopy have yielded new tools for obtaining pure cultures, such as the **laser tweezers**.

Laser tweezers consist of an inverted light microscope equipped with a strongly focused infrared laser and a micromanipulation device (Figure 18.5●). A single cell in a viewing field can be optically trapped by the laser beam and moved away from contaminating organisms. Trapping occurs because the laser creates a

(a)

(b)

● **Figure 18.5   The laser tweezers for the isolation of bacteria.** (a) Principle of the laser tweezers. A strongly focused laser beam on an object as small as a bacterial cell creates downward radiation forces (F_a, F_b) that allows the cell to be dragged in any direction. (b) Isolation. The laser beam can lock onto a single cell present in a mixture in a capillary tube and drag the optically trapped cell (in the example here, the desired cell is dragged from right to left) away from the other cells. Once the desired cell is far enough away from the other cells, the capillary is severed and the cell flushed into a tube of sterile medium.

● **Figure 18.4   Procedure for a most-probable number (MPN) analysis.** An existing enrichment culture or a natural sample of water, soil, etc., is added to a selective culture medium, and serial dilutions are made. The last tube showing growth (in the example shown it is the 10^{-5} dilution) should have developed from 10 or fewer cells. As a means of obtaining pure cultures, the highest dilution showing growth is used as inoculum, and the procedure is repeated. As a means of estimating the numbers of a particular type of bacteria in a sample, the MPN is defined as the highest dilution showing growth. If the medium used in this case were highly selective for sulfate-reducing bacteria, for example, one could conclude that $10^5 - 10^6$ culturable cells of such organisms were present per gram of sample. Use of several replicate tubes at each dilution improves accuracy of the final MPN obtained.

force that pushes down on small objects (like microbial cells) and holds them in place. When the laser beam is moved, the trapped bacterial cell moves along with it. If the sample is in a capillary tube, a single cell can be trapped and then isolated by breaking the tube at a point between the cell and the rest of the population. The trapped cell can then be flushed into a small tube of sterile medium to initiate a pure culture (Figure 18.5).

Laser tweezers technology is especially useful for isolating slow-growing bacteria that may be overgrown in enrichment culture or for organisms present in such low numbers that they would be missed using dilution-based enrichment methods. Coupled with the use of specific dyes that can identify particular organisms in a microscope field (see Section 18.4), the laser tweezers can be used to select the organisms of interest from a mixture for purification and further laboratory study.

18.2 Concept Check

Once a successful enrichment culture has been established, a pure culture can be obtained by conventional microbiological procedures, including streak plates, agar shakes, and dilution methods. Laser tweezers allow one to "pick" a cell from a microscope field and literally move it away from contaminants.

◆ How does the agar shake method differ from streaking to obtain isolated colonies?

◆ In an MPN (see Figure 18.4), why is the actual number of cultivable cells per sample likely to be between the reciprocal of the highest dilution showing growth and the next highest dilution?

◆ How might you isolate a morphologically unique bacterium present in an enrichment culture in relatively low numbers?

 # II MOLECULAR (CULTURE-INDEPENDENT) ANALYSES OF MICROBIAL COMMUNITIES

Microbial ecologists often need to quantify cells in a microbial habitat to know whether particular organisms are present and in what numbers. Such information yields insight on the dynamics of an ecosystem and helps pinpoint the most ecologically relevant organisms. Cell stains help microbiologists obtain these numbers, and we detail these methods here.

Organisms in natural environments can also be detected by assaying their *genes*. Genes encoding either ribosomal RNA (co Sections 11.5–11.7) or a specific metabolic process are the usual targets in these studies, although the field of environmental genomics (see Sec-

tion 18.6) is opening up the collective genomes of microbial communities in new and exciting ways.

18.3 Viability and Quantification Using Staining Techniques

A number of staining methods have been devised in microbiology suitable for quantifying microorganisms in natural samples. Although these methods do not reveal the physiology or phylogeny of the cells, they are widely used and are fairly reliable for obtaining quantitative information about the *total numbers* of cells in a particular habitat.

Fluorescent Staining Using DAPI

Fluorescent dyes have been widely used to stain microorganisms in opaque habitats. **DAPI** (4′,6-diamido-2-phenylindole) is a popular stain for this purpose. Cells stained with DAPI fluoresce bright blue and are easy to see and enumerate (Figure 18.6●). DAPI staining is widely used for the enumeration of microorganisms in environmental, food, and clinical samples.

Depending on the sample, nonspecific background staining can occasionally be a problem with fluorescent stains, but DAPI, which stains nucleic acids, is for the most part unreactive with inert matter. Thus, for many samples, soil as well as aquatic, DAPI staining can give a reasonable estimate of the cell numbers present. For dilute aquatic samples cells can be stained on the surface of a filter after filtering a given volume of liquid. Alternatively, cells can be concentrated by centrifugation. On the one hand, staining techniques such as DAPI have the advantage of being nonspecific: They stain *all* microorganisms in a sam-

Nancy J. Trun

● **Figure 18.6 Cells of *Escherichia coli* made fluorescent by staining with DAPI.** This staining technique is typically used to make total microbial counts. DAPI works by penetrating cells and binding to DNA; this typically makes prokaryotic cells fluoresce uniformly.

ple. However, on the other hand, they fail to differentiate between living and dead cells or between different species and thus cannot track *specific* organisms in an environment.

Viability Staining

A fluorescent staining method has been developed for differentiating live from dead bacterial cells. This gives information not only on the *number* of organisms present in a sample, but also on their *viability*. The basis of the differentiation concerns whether the cytoplasmic membrane of the cells is intact. Two dyes, colored green and red, are added to a sample; the green fluorescing dye penetrates all cells, viable or not, while the red dye, which contains propidium iodide, penetrates only those cells whose cell membrane is no longer intact and are therefore dead. Thus, *green* cells are *alive* and *red* cells are *dead*, giving an instant assessment of viability (Figure 18.7●). Although useful for research situations using laboratory cultures, the live/dead staining method is not suitable for use in the microscopic examination of most natural habitats because of problems with nonspecific staining of background materials.

Fluorescent Antibodies

Staining techniques can be made more specific by using *fluorescent antibodies*. Antibodies are specific biological reagents, and we discuss the theory of fluorescent antibodies in Section 24.9. The great specificity of antibodies against surface constituents of a particular organism can be exploited as a means of identifying or tracking an organism in a complex habitat containing a mixture of many organisms, such as soil (Figure 18.8●) or a clinical sample. The method requires the preparation of specific antibodies against the organism of interest, often a time-consuming

● **Figure 18.8 Fluorescent antibodies as a cell tag.** Visualization of cells in bacterial microcolonies (bacterial cells appear as greenish-yellow dots) of the hyperthermophilic archaeon, *Sulfolobus acidocaldarius* on the surface of solfatara soil particles by use of the fluorescent antibody technique. Cells are about 1 μm in diameter.

and laborious procedure. For clinically relevant microorganisms, however, highly specific antibodies can be purchased commercially and are therefore widely used in clinical microbiology laboratories (◠◠ Section 24.9 and Figure 24.9).

Green Fluorescent Protein as a Cell Tag

Bacterial cells can be altered by genetic engineering to make them fluoresce. As will be discussed in Chapter 31, a gene encoding a protein called the **green fluorescent protein (GFP)** can be inserted into the genome of virtually any bacterium. When expressed, GFP-containing cells appear green when observed with ultraviolet microscopy (Figure 18.9●).

Although not useful for the study of natural populations (because these cells lack the GFP gene), GFP-tagged cells can be *introduced* into an environment, such

● **Figure 18.7 Viability staining.** Live (green) and dead (red) cells of *Micrococcus luteus* (cocci) and *Bacillus cereus* (rods) stained by the LIVE/DEAD Bac Light™ Bacterial Viability Stain.

● **Figure 18.9 The green fluorescent protein (GFP).** Cells of *Pseudomonas fluorescens* genetically engineered to express the green fluorescent protein and observed by confocal laser microscopy (◠◠ Section 4.2). The cells are attached to barley roots and fluoresce green. The blue-staining cells are part of the normal flora of barley roots and have been stained with DAPI (see Section 18.3 and Figure 18.6). Using GFP one can track cells introduced into the environment.

as plant roots (Figure 18.9), and the tagged strain tracked by microscopy. Using this method, microbial ecologists can study microbial competition between the native microflora and the GFP-tagged introduced strain *in situ*, or assess the effect of perturbations in the environment on introduced bacteria.

The gene for the GFP has also been used extensively in laboratory cultures of various bacteria as a *reporter gene*. When fused with an operon under the control of a specific repressor, transcriptional control of the genes can be studied using fluorescence as an assay of transcription. That is, when the genes containing the fused GFP gene are transcribed, the GFP gene is also transcribed. Following translation to produce the green fluorescent protein, cells fluoresce green (◌◌ Section 31.4 and Figure 31.7).

Limitations of Microscopy

The microscope is a valuable tool for enumerating and identifying microorganisms in natural samples, but can the microscope suffice for the study of microorganisms in their natural habitats? The answer is no, for several reasons.

Small cells can be a major problem. Prokaryotes vary greatly in size (◌◌ Section 4.4 and Table 4.1), and some are near the limits of resolution of the light microscope. When observing natural samples, such organisms can easily be overlooked, especially if the sample contains high levels of particulate matter or high numbers of large cells. Also, in natural samples it is often difficult to differentiate live cells from dead cells or cells from nonliving matter. Even if the goal is simple quantification, all of these factors can affect the results, sometimes in dramatic ways.

However, perhaps the biggest limitation with the stains we have discussed thus far is that they are unable to measure the *diversity* of microorganisms in a natural habitat. In other words, *which species of microorganisms* are we observing as we count cells under the microscope? It is easy to be fooled by simple microscopic observations. For example, two morphologically similar cells may in fact be genetically distinct species that play quite different roles in the ecosystem (Figure 18.10●). Even simple staining methods cannot resolve this. Thus, when viewing unstained or nonspecifically stained natural populations of microorganisms under the microscope, it should be with the realization that the sample almost certainly contains a genetically diverse community, even if many cells "look" the same (Figure 18.10). As we will see soon, specific staining methods can reveal this (Figure 18.10*b*).

Although the microscope is an important tool of the microbial ecologist, it must be supplemented with cultural and molecular tools if the microbial ecosystem is to be truly understood. Molecular methods target

(a)   *(b)*

● **Figure 18.10 Morphology and genetic diversity.** The tandem photos shown here, (a) phase contrast, and (b) phylogenetic staining (see Section 18.4), are of the same field of cells. Although the large oval-shaped cells are of a rather unusual morphology and size for prokaryotic cells and look similar by phase-contrast microscopy, the phylogenetic stains reveal that these are two genetically distinct cell types (one stains yellow and one stains blue). Simple light microscopic observations of natural samples must therefore be interpreted with caution, because it is easy to assume that cells of an unusual morphology in a single microscopic field are all of the same organism. The oval cells stained in blue or yellow are about 2.25 μm in diameter. The green cells that appear in pairs or clusters are about 1 μm in diameter

get specific genes that are linked to specific microorganisms; the presence of the gene implies the presence of the organism. We consider these molecular methods now, some of which also involve microscopy (see Figure 18.10).

18.3 Concept Check

DAPI is a general stain for identifying microorganisms in natural samples. Some stains can differentiate live versus dead cells, and fluorescent antibodies that are specific for one or a small group of related cells can be prepared. The green fluorescent protein makes cells autofluorescent and is a means for tracking cells introduced into the environment. Unlike pure cultures, in natural samples, morphologically similar cells may actually be quite different genetically.

◆ Why is it incorrect to say that the GFP is a "staining" method?

◆ How do stains like DAPI differ from fluorescent antibodies in terms of specificity?

18.4 Genetic Stains

Nucleic acid probes are powerful methods for identifying and quantifying microorganisms in nature. Recall that a **nucleic acid probe** is a DNA or RNA oligonucleotide complementary to a sequence in a target gene and that can *hybridize* to the target gene (∞ Sections 7.7 and 11.7). As discussed in Section 11.6, oligonucleotide probes can be made fluorescent with certain dyes. These now-fluorescent probes can be used to identify organisms containing a nucleic acid sequence complementary to the probe. This technique is called **fluorescent *in situ* hybridization (FISH,** ∞ Section 11.6), and three different applications of this method are described here.

Phylogenetic Staining Using FISH

Phylogenetic stains are fluorescing oligonucleotides complementary in base sequence to signature sequences in 16S (prokaryotes) or 18S (eukaryotes) ribosomal RNA (∞ Section 11.7). Phylogenetic stains work by penetrating the cell and hybridizing with ribosomal RNA directly in the ribosomes (∞ Figure 11.14). Because phylogenetic probes can be designed for individual microbial species as well as for entire domains of organisms, the degree of specificity of a phylogenetic stain can be controlled by the sequence of the probe. Using FISH, one can then identify or track a particular organism (or group of related organisms, depending on the specificity of the phylogenetic probe) in a natural sample.

Natural samples can also be treated with *multiple* phylogenetic probes. Using a suite of probes, each designed to react with a particular organism or group and each containing its own fluorescent dye, FISH can be used to phylogenetically characterize an entire habitat (Figure 18.11●). If FISH is combined with confocal microscopy (∞ Section 3.2), it is possible to explore microbial populations in a sample *with depth*, as for example, in a biofilm (∞ Section 19.3 and Figures 18.11 and 19.4). Besides microbial ecology, FISH technology has become popular as a tool to identify specific pathogens in the food industry and in clinical diagnostics.

Chromosome Painting and *In Situ* Reverse Transcription

Two other FISH technologies are *chromosome painting* and *in situ reverse transcription*. The former technique is used to identify specific genes in the microflora of natural samples, whereas the latter targets specific mRNA (that is, *expressed* genes).

Chromosome painting begins with fluorescent labeling of an oligonucleotide probe or an entire gene, set of genes, or even an entire genome digested in such a way as to yield small probes. The probes are then hybridized to DNA from organisms in a natural sample (Figure 18.12a●). Chromosome painting is typically

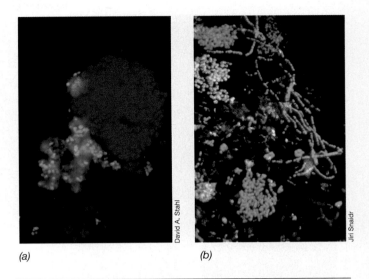

(a)   (b)

● **Figure 18.11 FISH analysis of sewage.** (a) Nitrifying bacteria in sewage sludge. Red, ammonia-oxidizing ($NH_3 \rightarrow NO_2^-$) bacteria; green, nitrite-oxidizing ($NO_2^- \rightarrow NO_3^-$) bacteria. Note the close proximity of the two metabolic types, presumably to allow for cross-feeding of NO_2^- from producers (ammonia-oxidizers) to consumers (nitrite-oxidizers). (b) Confocal laser scanning micrograph of a sewage sludge sample. The sample was treated with three phylogenetic probes, each containing a different fluorescent dye (green, red, or purple) and each targeting a different group of Proteobacteria (one of several major lineages in the domain *Bacteria*, ∞ Chapter 12). Green-, red-, or purple-stained cells reacted with only a single probe while blue- and yellow-stained cells reacted with two probes to give the different colors.

used to identify bacteria that contain specific, highly conserved genes, such as those that encode the enzyme nitrogenase (responsible for nitrogen fixation, ∞ Section 17.28), genes encoding the photosynthetic reaction center (anoxygenic photosynthesis, ∞ Section 17.4), or genes encoding specific autotrophic pathways (CO_2 fixation, ∞ Sections 17.6 and 17.7). The numbers of fluorescing cells in a sample would be an estimate of the numbers of nitrogen-fixing bacteria, phototrophic bacteria, or autotrophic bacteria, respectively, in that natural sample.

In contrast to chromosome painting, *in situ* **reverse transcription (ISRT) FISH** is used to measure *gene expression* in cells in a natural sample (Figure 18.12*b,c*). The method involves the use of a probe that hybridizes with a specific mRNA previously transcribed by cells in a natural sample. Once the probe reacts, reverse transcription (∞ Section 9.13) is performed to produce a complementary DNA, and the latter is amplified by the polymerase chain reaction (PCR, ∞ Section 7.9). The amplified DNA is then reacted with a fluorescent probe (Figure 18.12*b, c*). ISRT FISH is typically used to explore the factors that control gene expression in natural populations and to observe how experimental perturbations in a sample affect the transcription of specific genes.

(a)

Brian D. Lanoil

(b) *(c)*

Feng Chen   Feng Chen

● **Figure 18.12 Fluorescent techniques employing bacterial chromosome painting and *in situ* reverse transcription (ISRT).** (a) Chromosome painting. Mixture of cells of *Escherichia coli* (red) and *Oceanospirillum linum* (blue). Each fluorescent dye was linked to DNA oligonucleotides formed from genomic DNA from the respective organism. Because of the significant sequence differences in the genomes of each organism, the labeled genes only hybridize with genomic DNA from their respective organisms. (b, c) ISRT. The two photos show a mixed culture of the bacteria *Microbulbifer hydrolyticus* (short, fat rods) and *Sagittula stellata* (long, thin rods). Reverse transcription (⚬⚬ Section 9.13) of a specific mRNA produced only by *M. hydrolyticus* led to production of a cDNA that was subsequently amplified by PCR and hybridized by a fluorescently labeled probe. (b) All cells stained with DAPI, a nonspecific stain (see Figure 18.6). (c) Cells stained with the ISRT probe. ISRT data are useful for understanding which organisms in a microbial community are carrying out a specific metabolic activity.

 18.4 Concept Check

A variety of fluorescent-staining methods employ the power of nucleic acid probes and thus are highly specific in their staining properties. These include phylogenetic staining, chromosome painting, and reverse transcription fluorescent *in situ* hybridization (FISH).

◆ What part of the cell is the target for fluorescent probes in phylogenetic FISH?

◆ Chromosome painting and ISRT FISH reveal different things about cells in nature. Explain.

◆ Which fluorescent technique would you use to enumerate nitrogen-fixing bacteria in a soil sample?

18.5 Linking Specific Genes to Specific Organisms Using PCR

For many ecological studies it is unnecessary to isolate organisms or even to quantitate or identify them microscopically using the stains described in Sections 18.3 and 18.4. Indeed, it is possible to survey the biodiversity of a habitat *without* culturing or otherwise observing the cells. For these purposes, specific *genes* are characterized as a measure of biodiversity in the microbial community. The rationale here is as follows. Specific genes are often linked to specific organisms. Therefore, detection of the *gene* implies that the *organism* is present. The major techniques employed in this type of microbial community analysis are the *polymerase chain reaction* (PCR), *denaturing gradient gel electrophoresis* (DGGE) or *molecular cloning*, and *DNA sequencing and analysis*. In addition, as we will see, genomic techniques can also be used to characterize the composition of microbial communities.

Polymerase Chain Reaction and Microbial Community Analysis

We discussed the principle of PCR in Chapter 7 (⚬⚬ Section 7.9). Recall the major steps in PCR: (1) two nucleic acid primers are hybridized to a complementary sequence in a target gene, (2) DNA polymerase copies the target gene, and (3) multiple copies of the target gene are made by repeated melting of complementary strands, binding of primers, and new synthesis (⚬⚬ Figure 7.28). From a single copy of a gene, several million copies can be made (Figure 18.13●).

Which genes are used as target genes? A commonly amplified gene is that encoding *16S ribosomal RNA*. As we have seen, sequence analyses of 16S rRNA genes can be used to derive phylogenetic relationships in pure cultures of microorganisms (⚬⚬ Section 11.5 and Figure 11.13). Likewise, sequence analyses of 16S genes amplified from a microbial community can paint a phylogenetic picture of that community (Figure 18.13). Alternatively, metabolic genes that encode proteins unique to the metabolism of a specific organism (or group of related organisms) can be the gene targets. However, regardless of the gene (or genes) chosen for PCR amplification, *specificity of the primers* is essential. With natural samples as sources of DNA, primer design is critical for obtaining PCR results that are unambiguous and readily interpretable.

Microbial community

Extract DNA

Total community DNA

PCR

Amplify 16S rRNA genes using general or specific primers

Sample
1 2 3 4

All 16S rRNA genes

Excise bands and clone 16S rRNA genes

Sample
1 2 3 4

Different 16S rRNA genes

DGGE

Excise bands

Sequence

Generate tree

Bacillus subtilis
Unknown *Bacillus* species
Bacillus cereus
Bacillus megaterium
Unknown "low GC gram-positive"
Clostridium histolyticum
Unknown *Clostridium* species

Sequence

Generate tree

● **Figure 18.13 Steps in PCR-driven biodiversity analysis of a microbial community.** From total community DNA, 16S rRNA genes are amplified using, in this experiment, primers that target only low GC gram-positive *Bacteria* (∞ Section 12.20). The PCR bands are excised and the different 16S genes separated by either cloning or by DGGE (see Figure 18.14b). Note how in the DGGE gel, samples 1, 2, and 4 share a common band (gene), whereas samples 2 and 3 each contain one unique band. Following sequencing, a phylogenetic tree is generated. Note that many of the sequences do not match the 16S rRNA sequence of any recognized species and therefore are "new" species that await isolation. Such findings are very common in molecular community analyses, emphasizing the fact that many microorganisms present in nature have yet to be cultured.

In a typical community analysis experiment, total community DNA is isolated from a microbial habitat (Figure 18.13). Commercially available kits that yield highly purified DNA from soil and other complex habitats are available for this purpose. The DNA obtained is a mixture of genomic DNA from all of the microorganisms that were originally present in the habitat (Figure 18.13). It is the job of PCR to "fish out" the targeted gene(s) of interest from this mixture and make copies sufficient for sequence analysis.

Figure 18.13 shows the series of events in molecular microbial community analysis. The end result is bands

of the target gene on a gel (Figures 18.13 and 18.14●). Ideally only a *single* band of a given size will appear (the size of the DNA fragment amplified is determined by the distance between the primers). If sequencing is the ultimate goal, the PCR products first need to be sorted out. A major method for separating PCR products is called *denaturing gradient gel electrophoresis*, to which we turn now.

Denaturing Gradient Gel Electrophoresis (DGGE): Separating Similar Genes

PCR amplification from natural microbial communities using a single set of 16S rRNA primers typically generates a single gel band containing amplified DNA fragments of a single size (Figures 18.13 and 18.14a). However, despite the appearance of purity, the band typically contains many highly related but not identical genes. This is because although priming sites in the target gene are highly conserved, the nucleotide sequence *between* the priming sites can vary as a result of evolutionary divergence of the target gene in the different species of organism that contain it. Thus, if the goal is to sequence amplified genes, as in phylogenetic analyses, an additional step is needed to resolve

(a)

(b)

Jennifer A. Fagg and Michael J. Ferris

● **Figure 18.14 PCR and DGGE analyses.** Bulk DNA was isolated from a microbial community and PCR amplified using primers for 16S rRNA genes of *Bacteria* (a; lanes 1 and 8). The PCR products yielded only a single band, but these products actually consisted of *six* distinct 16S rRNA gene sequences as determined by DGGE (b; lanes 1 and 8). Each of these six bands was then purified, reamplified by PCR (a; lanes 2–7), and then run on DGGE (b; lanes 2–7). Note how all bands run to the same location in the PCR gel in (a), since although they differ in *sequence*, they are all of the same *size*. Each band in lanes 2–7 of part (b) could also be sequenced and a phylogenetic tree generated as shown in Figure 18.13.

the different forms of the gene before sequencing can proceed.

One way to do this is by molecular cloning (Figure 18.13, ⚭ Section 10.15). Alternatively, one can use **denaturing gradient gel electrophoresis (DGGE)**, a gel electrophoresis method that separates genes of the same *size* that differ in their melting (denaturing) profile because of differences in *base sequence* (Figures 18.13 and 18.14*b*). DGGE employs a gradient of a DNA denaturant, such as a mixture of urea and formamide. When a double-stranded DNA fragment moving through the gel reaches a region containing sufficient denaturant, the strands begin to "melt," at which point their migration stops (Figures 18.13 and 18.14*b*). Differences in melting properties are to a large degree controlled by differences in base sequence. Thus, the different bands observed in a DGGE gel are different forms of a given gene that vary—and sometimes only slightly—in their sequences.

Once DGGE has been performed, individual bands can be excised and sequenced (Figures 18.13 and 18.14). Using 16S rRNA genes, for example, DGGE analyses provide a detailed description of the number of **phylotypes** (distinct 16S rRNA genes) present in a habitat (Figure 18.14*b*). By sequencing these bands, the actual species present in the community can be determined by comparison of the sequences with those of known species available from appropriate data bases (Figure 18.13*b*). Using primers specific for genes other than 16S rRNA, such as metabolic genes, provides information about the phylotypes present in the community that contain the specific gene. The number of bands obtained on a DGGE gel is a clue to the biocomplexity of the habitat with respect to a given gene (Figures 18.13 and 18.14*b*).

Results of PCR Phylogenetic Analyses

Phylogenetic analyses of microbial communities have been refined to allow for both qualitative (that is, "who's out there?") as well as quantitative (that is, "how many of each type are out there?") analyses. They have also yielded some surprising results. Using 16S rRNA as the target gene, most natural microbial communities have been shown to contain several phylogenetically distinct organisms whose ribosomal RNA sequences do not match any of those from laboratory cultures (Figures 18.13 and 18.14). In fact, in many cases it turns out that the most abundant members of the natural community are species that have so far defied laboratory culture! These results make it clear that our knowledge of microbial diversity from enrichment culture studies is very incomplete and that enrichment bias (see Section 18.1) is likely a serious problem in culture-dependent biodiversity studies.

18.5 Concept Check

The polymerase chain reaction (PCR) can be used to amplify specific target genes such as small subunit ribosomal RNA genes or key metabolic genes. Denaturing gradient gel electrophoresis (DGGE) can be used to resolve slightly to greatly different versions of these genes present in the different species inhabiting a natural sample.

◆ Why can it be said that primer specificity is the key to successful microbial community analysis by PCR?

◆ What could you conclude from PCR/DGGE analysis of a sample that yielded one band by PCR and one band by DGGE? One band by PCR and four bands by DGGE?

◆ What surprising finding has come out of many molecular studies of natural habitats using 16S ribosomal RNA as the target gene?

18.6 Environmental Genomics (Metagenomics)

A more encompassing genetic approach to the molecular study of microbial communities is **environmental genomics**, also called **metagenomics**. In this approach, random shotgun sequencing (⚭ Section 15.2) of total DNA cloned from a microbial community is used to reveal the entire gene complement of that community (Figure 18.15●).

The goal of environmental genomics is not to generate complete and finished genome sequences, as has been done for many pure cultures of microorganisms (⚭ Chapter 15). Instead, the idea is to detect as many genes as possible that encode recognizable proteins and then determine to which phylogenetic "scaffold" they belong. This is done by sequencing overlaps to the genes that include phylogenetic markers, such as 16S rRNA genes (Figure 18.15). It is also important to distinguish environmental genomics from molecular community analysis, discussed in the previous section. The latter is focused on the biodiversity of a *single gene*. By contrast, in environmental genomics, *all* genes in the microbial community—the **metagenome**—are sampled, leading to a much more robust genetic picture of the community.

Despite the much larger scale and cost of environmental genomics compared with single gene analyses, the approach is inherently more in-depth. Moreover, metagenomics can reveal features of a microbial community missed by single gene approaches. For example, in a metagenomics study of prokaryotes in the Sargasso Sea, an enormous diversity was discovered that was previously missed by rRNA-based community analyses. This is because not all 16S rRNA genes are amplified with the universal or domain-specific primers used in PCR analyses. Those genes that do not amplify, of course, remain undetected. Environmental

Microbial
community

Extract DNA

Total community DNA

**Community
sampling approach**

Amplify single gene,
for example, gene
encoding 16S rRNA

Sequence and
generate tree

**Environmental
genomics approach**

Restriction digest total
DNA and then shotgun
sequence

Assembly and
annotation

Outcomes

Genomes

Phylogenetic tree

1. Phylogenetic snapshot
of most members of
the community
2. Identification of novel
phylotypes

Total gene pool of the community

1. Identification of all gene categories
2. Discovery of new genes
3. Linking genes to particular
phylotypes

● **Figure 18.15** **Comparison of single gene and environmental
genomics (metagenomics) approaches to microbial community
analysis.** Note that in the metagenomics approach, all community
DNA is sequenced; however, assembly and annotation are not
performed to the complete "finished genome" state, as is usually done
with pure cultures (Section 15.4). Thus the chromosomes
assembled will have small gaps.

genomics sidesteps this problem by shotgun sequencing community DNA directly without first amplifying by PCR (Figure 18.15). Therefore, *all* genes are sequenced whether they are amplifiable or not.

The Potential of Environmental Genomics

Among many other things, environmental genomics can detect new genes in known organisms and known genes in new organisms. In the Sargasso Sea study, for example, genes encoding specific metabolisms were sometimes found within the genomes of phylogenetic groups not previously known to carry them out. For instance, genes encoding the enzyme *ammonia monooxygenase*, the key enzyme of ammonia-oxidizing *Bacteria* (Sections 12.3 and 17.12), were discovered within the genomic scaffold of *Archaea*. Thus, although ammonia-oxidizing *Archaea* have never been described, environmental genomics has shown us that they almost certainly exist.

In a second example, genes encoding *proteorhodopsin*, the light-mediated proton pump found in certain Proteobacteria (Section 13.3), were found to be linked to several new phylogenetic groups of prokaryotes inhabiting the Sargasso Sea. Since cultures of these organisms have not yet been obtained, connections between this form of metabolism and these specific phylogenies would not otherwise have been suspected. These examples show how environmental genomics offers microbial ecology a new approach to exploring microbial diversity not possible by other means.

Environmental genomics is a powerful tool for assessing both the phylogenetic diversity and metabolic diversity of microbial communities, both of which are major goals of microbial ecology. As sequencing costs drop, more microbial community analyses will likely turn to genomics approaches. However, in addition to whole genome shotgun sequencing, other, more targeted environmental genomics approaches are possible. These include in particular the use of *microarrays* for assessing gene expression in microbial communities (Section 15.11). Using microarrays, one can measure the effect of environmental parameters on transcription of individual components of the community, defined groups, or even the entire microbial community, depending on how the microarray is constructed.

18.6 Concept Check

Environmental genomics involves shotgun sequencing and analysis of the collective genomes of the organisms present in a microbial community.

◆ What is a metagenome?

◆ How do environmental genomic approaches differ from microbial community analyses, such as that based on 16S rRNA gene analysis?

 III MEASURING MICROBIAL ACTIVITIES IN NATURE

So far in this chapter our discussion has focused on biodiversity. We now turn to measurements of *microbial activities*, using (1) radioisotopes, (2) microelectrodes, and (3) stable isotopes.

Activity measurements in a natural sample are *collective* estimates for the entire microbial community, although one technique, FISH-MAR, allows for a more targeted assessment of activity. However, well-designed activity measurements can reveal both the *types* and *rates* of major metabolic reactions in a habitat. Along with biodiversity estimates, these parameters define the microbial ecology of that habitat. They also provide valuable information for the design of enrichment cultures.

18.7 Radioisotopes and Microelectrodes

In many situations, direct chemical measurements of microbial transformations are sufficient for assessing microbial activities in an environment. For example, the fate of lactate oxidation by sulfate-reducing bacteria in a sediment sample can easily be followed. Lactate is consumed and SO_4^{2-} is reduced to H_2S (Figure 18.16a●). All of these substances are easily measured chemically. However, when very high sensitivity is required, turnover rates need to be determined, or the fate of portions of a molecule needs to be followed, *radioisotopes* are very useful. For instance, if photoautotrophy is to be measured, the light-dependent uptake of $^{14}CO_2$ into microbial cells can be measured (Figure 18.16b). If sulfate reduction is of interest, the rate of conversion of $^{35}SO_4^{2-}$ to $H_2^{35}S$ can be assessed (Figure 18.16c). Chemoorganotrophic activity can easily be assessed by tracking the release of $^{14}CO_2$ from ^{14}C-labeled organic compounds (Figure 18.16d), and so on.

Isotope methods are widely used in microbial ecology. These must, however, employ proper controls, because some transformation of labeled compound might be due to a strictly chemical (rather than microbial) process. The *killed cell control* is key here. It is absolutely essential to know that the transformation being measured is prevented by chemical agents or by heat treatments that either block microbial

action or actually kill the organisms. Formalin at a final concentration of 4% is commonly used as a chemical sterilant in microbial ecology studies. This level of formalin kills all cells, and transformations of radiolabeled materials that occur in its presence can be ascribed to nonliving processes (Figure 18.16d).

Radioisotopes in Combination with FISH: FISH-MAR

Radioisotopes can be used as measures of microbial activity in a microscopic technique called microautoradiography (MAR). In this method, cells, either of a pure culture or from a microbial community, are exposed to a radioisotope, such as an organic compound for chemoorganotrophs or CO_2 for an autotroph. Following exposure, cells are then affixed to a slide and the slide dipped in photographic emulsion. When left in darkness for a period, radioactive decay from the incorporated substrate exposes silver grains in the emulsion. These appear as black dots within and around the cells (Figure 18.17a●).

MAR can be combined with FISH (see Section 18.4) in **FISH-MAR**, a powerful technique for combining identification and activity measurements. FISH-MAR allows one to both identify which organisms in a natural sample are metabolizing a particular radiolabeled substance (by MAR) and to identify the organism(s) doing so (by FISH) (Figure 18.17b). FISH-MAR thus goes a major step beyond simple identification. The technique reveals valuable physiological information on the components as well. Such information is useful not only in understanding the activity of the microbial ecosystem, but also helps in guiding enrichment cultures. For example, knowledge of which organisms are metabolizing which substrates can be used for the design of specific enrichment protocols.

Microelectrodes

Microbial ecologists have used small glass electrodes, called **microelectrodes**, to study the activity of microorganisms in nature. The most useful microelectrodes have been those that measure pH, oxygen, N_2O, CO_2, H_2, or H_2S. As the name implies, microelectrodes are very small, the tips of the electrode ranging in diameter from 2 to 100 μm (Figure 18.18a●). The electrodes must be carefully inserted into the habitat in increments of a millimeter or less in order to follow microbial activities at the microenvironmental level (Figure 18.18b).

Microelectrodes have been used extensively in the study of microbial activities, including photosynthesis in microbial mats. **Microbial mats** are layered microbial communities typically containing cyanobacteria in the uppermost layers, anoxygenic phototrophic bacteria in subsequent layers (until the mat becomes light-limited)

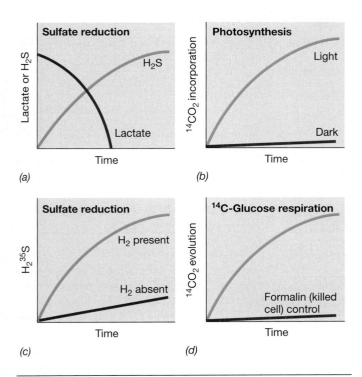

● **Figure 18.16 Microbial activity measurements.** (a) Chemical assay for lactate and H_2S during sulfate reduction. (b–d) Use of radioisotopes. (b) Photosynthesis measured in natural seawater with $^{14}CO_2$. (c) Sulfate reduction in mud measured with $^{35}SO_4^{2-}$. (d) Production of $^{14}CO_2$ from ^{14}C-glucose.

(a)

(b)   *(c)*

● **Figure 18.17 FISH-MAR:** Fluorescence *in situ* hybridization (FISH) combined with microautoradiography (MAR). (a) An uncultured filamentous cell belonging to the gamma Proteobacteria (by FISH analysis) and also an autotroph (by MAR-measured uptake of $^{14}CO_2$). Radioactive decay exposes the photographic emulsion, generating the black dots. (b) Uptake of ^{14}C-glucose by a mixed culture of *Escherichia coli* (yellow cells) and *Herpetosiphon aurantiacus* (filamentous, green cells). (c) MAR of the same field of cells shown in (b). Incorporated radioactivity exposes the film and shows that glucose was assimilated by the cells of *E. coli* but not by the filaments of *H. aurantiacus*.

and chemoorganotrophic, especially sulfate-reducing, bacteria in the lower layers (Figures 18.18*b* and 18.19●). Microbial mats develop in a variety of environments, especially in hot springs (Figure 18.19) and marine intertidal zones.

With the use of O_2 microelectrodes, oxygen concentrations in microbial mats (Figure 18.19*b*) or soil particles (∞ Figure 19.2) can be very accurately measured over extremely fine intervals. A *micromanipulator* is used to immerse electrodes gradually through a sample such that readings can be taken every 0.05–0.1 mm (Figures 18.18*b* and 18.19*b*). Using a bank of microelectrodes, each sensitive to a different chemical, simultaneous measurements of several microbial transformations can be made at the same time. For example, oxygen and sulfide measurements are often taken together because gradients of both form in many microbial environments as a result of photosynthesis and sulfate re-

(a) ⊢ 5 μm ⊣   *(b)*

● **Figure 18.18 Microelectrodes.** (a) Schematic drawing of an oxygen microelectrode. The platinum rod functions as a cathode, and when voltage is applied, O_2 is reduced to H_2O, generating a current. The current resulting from the reduction of the O_2 at the gold surface of the cathode is proportional to the oxygen concentration in the sample. Note the scale of the electrode. (b) Photo of microelectrodes being used in a microbial mat (see Figure 18.19*a*).

duction, respectively (Figure 18.18*b*). Near the zone where O_2 and H_2S begin to mix, intense activity by phototrophic and chemolithotrophic sulfur bacteria (∞ Sections 12.2 and 12.4) may consume both of these substrates, at least to the limits of detection of the microelectrodes (Figure 18.19*b*).

Using microelectrodes as assay tools one can perturb the habitat while the electrodes are immersed, for instance by adding an organic compound or shading the mat surface, and measure what effects these perturbations have on the production or consumption of specific compounds in real time. From studies such as these, it is often possible to deduce the kinds of microbial interactions taking place in an environment. It is also possible to ask specific questions approachable by experiment. In this way, it may be possible to model the *in situ* metabolism of a microbial environment.

⬢ **18.7 Concept Check**

The activity of microorganisms in natural samples can be assessed very sensitively using radioisotopes and/or microelectrodes. In most cases, measurements are of the net activity of a microbial community rather than of a population of a single species.

◆ Why are radioisotopes so useful in measuring microbial activities?

◆ What does the acronym *MAR* stand for, and how are MAR reactions detected?

◆ Explain how a microelectrode works.

(a)

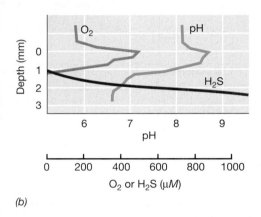

(b)

● **Figure 18.19 Microbial mats and the use of microelectrodes to study them.** (a) Photograph of a core taken through the kind of hot spring microbial mat used in the experiment shown in part (b). Upper layer (dark green) contains cyanobacteria (⟨∞ Section 12.25), beneath which are several layers of anoxygenic phototrophic bacteria (orange), primarily *Chloroflexus* (⟨∞ Section 12.35). The whole thickness of the mat is about 2 cm. (b) Oxygen, sulfide, and pH microprofiles in a hot spring microbial mat. Note the millimeter scale on the ordinate.

18.8 Stable Isotopes

For many elements several *isotopes* exist, differing in the number of neutrons present. Certain isotopes are unstable and break down as a result of radioactive decay. Others, called **stable isotopes**, are not radioactive but are metabolized differentially by microorganisms. Stable isotopes can be used to study various microbial transformations in nature (Figure 18.20●).

The two elements that have proven most useful for stable isotope studies in microbial ecology are *carbon*, especially CO_2, and *sulfur*, especially SO_4^{2-} and H_2S. Carbon exists in nature primarily as ^{12}C, but a small amount (about 5%) exists as ^{13}C. Likewise, sulfur exists primarily as ^{32}S, although some sulfur is found as ^{34}S. The natural abundance of these isotopes in a given compound changes when C or S is metabolized by organisms because biochemical reactions

● **Figure 18.20 Mechanism of isotopic fractionation using carbon as an example.** Although the ratio of natural abundance of $^{12}CO_2$ to $^{13}CO_2$ is about 19:1, enzymes that fix CC_2 preferentially fix the lighter isotope (^{12}C). This results in fixed carbon being enriched in ^{12}C and depleted in ^{13}C relative to the starting substrate. The degree of ^{13}C depletion is calculated as an isotopic fractionation (see legend to Figure 18.21 for calculation). The size of the arrows indicates the relative abundance of each isotope of carbon.

favor the *lighter* isotope. That is, the heavier isotope is *discriminated against* relative to the lighter isotope when the compounds are metabolized by enzymes (Figure 18.20). For example, when CO_2 is fed to a phototrophic organism, the cellular carbon becomes *enriched* in ^{12}C and *depleted* in ^{13}C, relative to a carbon standard. Likewise, sulfide produced from the bacterial reduction of sulfate is "lighter" than sulfide that has formed geochemically. This process, known as **isotopic fractionation** (Figure 18.20), is a uniquely biological phenomenon and can be used as an assay for whether a particular transformation is microbiological or not.

Use of Isotopic Fractionation in Microbial Ecology

The isotopic composition of a sample contains a record of its past biological activity. In the case of carbon (Figure 18.21●), plant material and petroleum (which is derived from plant material) have similar isotopic compositions. Carbon from both sources is isotopically lighter than a nonbiological standard because it was fixed as CO_2 by a pathway that discriminated against $^{13}CO_2$ (Figure 18.21). Moreover, methane of microbiological origin is extremely light, indicating that methanogenic *Archaea* (⟨∞ Section 13.4) discriminate against ^{13}C in a major way. By contrast, marine carbonates are clearly of geological origin (Figure 18.21). Because of the differences in the proportion of ^{12}C and ^{13}C in carbon of biological and geological origin, the $^{13}C/^{12}C$ isotopic ratio of geological strata has been used to detect the onset of living processes in ancient rocks. Interestingly, organic carbon in rocks as old as 3.5 billion years shows some fractionation (Figure 18.21), supporting the idea that life existed at this time.

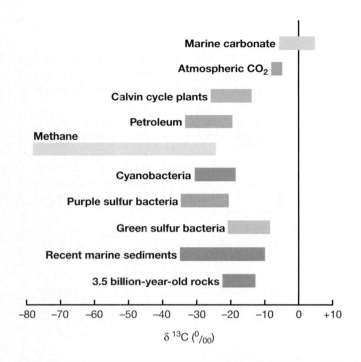

● **Figure 18.21 Carbon isotopic compositions of various substances.** The values are given in parts per thousand (‰) and were calculated using the formula:

$$\frac{(^{13}C/^{12}C\ \text{sample}) - (^{13}C/^{12}C\ \text{standard})}{(^{13}C/^{12}C\ \text{standard})} \times 1000$$

The standard is a belemnite sample from the PeeDee rock formation. Note that carbon fixed by autotrophic organisms is *enriched* in ^{12}C and *depleted* in ^{13}C.

The activities of sulfate-reducing bacteria in nature are easy to recognize from the fractionation of $^{34}S/^{32}S$ (Figure 18.22●). As compared with a sulfide standard, sedimentary sulfide is highly enriched in ^{32}S, while nonbiogenic sulfide (as, for instance, in igneous rocks from volcanic deposits) is significantly heavier (Figure 18.22). Fractionation allows one to monitor these important links in the sulfur cycle, and it has been widely used to trace the activities of sulfur-cycling prokaryotes through geological strata. Sulfur isotope analyses have also been used as evidence for the lack of life on the Moon. For example, the data in Figure 18.22 show that the isotopic composition of sulfides in lunar rocks closely approximates that of the sulfide standard and not that of biogenic sulfide.

Oxygen isotopic analyses ($^{18}O/^{16}O$) have been used to trace Earth's transition from an anoxic to an oxic environment, since Earth's molecular oxygen originated from oxygenic photosynthesis by cyanobacteria (∞ Section 11.1). Using this tool it has been possible to show that Earth was oxygenated only very gradually and that current oxygen levels have been

● **Figure 18.22 Summary of the isotope geochemistry of sulfur, indicating the range of values for ^{34}S and ^{32}S in various sulfur-containing substances.** The values are given in parts per thousand ‰ and were calculated using the formula:

$$\frac{(^{34}S/^{32}S\ \text{sample}) - (^{34}S/^{32}S\ \text{standard})}{(^{34}S/^{32}S\ \text{standard})} \times 1000$$

The standard is an iron sulfide mineral from the Canyon Diablo meteorite. Note that sulfide and sulfur of *biogenic* origin tend to be *depleted* in ^{34}S (enriched in ^{32}S).

present for only the past 750 million years or so (∞ Section 11.3 and Figure 11.8).

Epilogue

With an understanding of some of the techniques in the microbial ecologist's toolbox, we move on to Chapter 19 to explore the discipline of microbial ecology itself. In that chapter we will consider the composition of microbial communities and their major activities in nature. In many cases we will use one or more of the tools described in this chapter to help illustrate important principles and to identify the most active microbial species in highly diverse microbial communities.

 18.8 Concept Check

Isotopic fractionation can reveal the biological origin of various substances. Fractionation is a result of the activity of enzymes that discriminate against the heavier form of an element when binding their substrates.

◆ How can the $^{12}C/^{13}C$ composition of a substance reveal its possible biological origin?

◆ What is the simplest explanation for why lunar sulfides are isotopically heavy?

REVIEW QUESTIONS

1. What is the basis of the *enrichment culture technique*? Why is an enrichment medium usually suitable for the enrichment of only a certain group or groups of bacteria (∞ Section 18.1)?

2. What is the principle of the Winogradsky column and what types of organisms does it serve to enrich? How might a Winogradsky column be used to enrich organisms present in an extreme environment, like a hot spring microbial mat (∞ Section 18.1)?

3. Describe the principle of MPN for enumerating bacteria from a natural sample (∞ Section 18.2).

4. Why would the laser tweezers be a superior method to dilution and liquid enrichment for obtaining an organism present in a sample in *low* numbers (∞ Section 18.2)?

5. Compare and contrast the use of fluorescent dyes and fluorescent antibodies for use in enumerating bacteria in natural environments. What advantages and limitations do each of these methods have (∞ Section 18.3)?

6. Can nucleic acid probes in microbial ecology be as sensitive as culturing methods? What advantages do nucleic acid methods have over culture methods? What disadvantages (∞ Section 18.4)?

7. How does *chromosomal painting* work? Why might it be a good method for enumerating microorganisms in a habitat capable of carrying out some specific metabolic process (∞ Section 18.4)?

8. What is the *green fluorescent protein*? In what ways does a green fluorescing cell differ from a cell fluorescing from, for example, staining with a phylogenetic strain (∞ Section 18.4)?

9. How can a phylogenetic picture of a microbial community be obtained without culturing the inhabitants (∞ Section 18.5)?

10. After PCR amplification of total community DNA using a specific primer set, why is it usually necessary to either clone or run DGGE on the products before sequencing them (∞ Section 18.5)?

11. Give an example of how environmental genomics has discovered a known metabolism in a new organism (∞ Section 18.6).

12. What are the major advantages of *radioisotopic methods* in the study of microbial ecology? What type of controls (discuss at least two) would you include in a radioisotopic experiment to show $^{14}CO_2$ incorporation by phototrophic bacteria or to show $^{35}SO_4^{2-}$ reduction by sulfate-reducing bacteria (∞ Section 18.7)?

13. What can FISH-MAR tell you that FISH alone cannot (∞ Section 18.7)?

14. Will autotrophic organisms contain *more* or *less* ^{12}C than the CO_2 that feeds them (∞ Section 18.8)?

APPLICATION QUESTIONS

1. Why is the enrichment for nitrifying bacteria as described in Table 18.1? Why are each of the resources and conditions necessary? (You might want to review material in Sections 12.3 and 17.12 before answering.)

2. Design an experiment for measuring the activity of sulfur-oxidizing bacteria in soil. If only certain species of the sulfur-oxidizers present were metabolically active, how could you tell this? How would you prove that your activity measurement was due to biological activity?

3. You wish to know whether *Archaea* exist in a lake water sample but are unsuccessful in culturing any. Using techniques described in this chapter, how could you determine whether *Archaea* existed in the sample and if they did, what proportion of the cells in the lake water were *Archaea*?

4. You wish to identify whether organisms capable of autotrophic growth using the Calvin cycle exist in various soil samples (∞ Section 17.6). This pathway requires a unique enzyme, *ribulose bisphosphate carboxylase* (RubisCO). Describe two ways you could do this, one using a microscopic method and another using a method that does not involve either culturing or microscopic methods.

5. Design an experiment to solve the following problem. Determine the rate of methanogenesis ($CO_2 + 4 H_2 \rightarrow CH_4 + H_2O$) in anoxic lake sediments and whether or not it is H_2-limited. Also, determine the morphology of the dominant methanogen (recall that these are *Archaea*, ∞ Section 13.4). Finally, calculate what percentage the dominant methanogen is of the *total archaeal* and *total prokaryotic* populations in the sediments.

19

MICROBIAL ECOLOGY

Microorganisms interact in nature with other microorganisms and with plants and animals. A beneficial association is seen here between the water fern *Azolla* and nitrogen-fixing symbiotic cyanobacteria of the genus *Anabaena*, cells of which live in cavities within the leaves of the plant.

WORKING GLOSSARY

Acid mine drainage acidic water containing H_2SO_4 derived from the microbial oxidation of iron sulfide minerals

Anoxic an oxygen-free environment, usually also highly reducing (low E_0')

Bacteroid morphologically misshapen *Rhizobium* cells inside a leguminous plant root nodule; can fix N_2

Barophilic an organism that grows best when placed under a pressure greater than 1 atm

Barotolerant an organism that can grow under elevated pressures but that grows best at atmospheric pressure

Biochemical oxygen demand (BOD) the microbial oxygen-consuming properties of a water sample

Biofilm colonies of microbial cells encased in a porous slime and attached to a surface

Biogeochemistry study of biologically mediated chemical transformations

Bioremediation the cleanup of oil, toxic chemicals, and other pollutants by microorganisms

Black smoker an extremely hot (250–350° C) deep-sea hot spring emitting both hot water and various minerals

Cometabolism metabolism of a compound in the presence of a second organic compound, which is used as the primary energy source

Denitrification the microbial process by which NO_3^- is reduced to gaseous nitrogenous compounds

Ecosystem a community of organisms and their natural environment

Guild a population of metabolically related microorganisms

Hydrothermal vent a deep-sea warm or hot spring

Infection thread in the formation of root nodules, a cellulosic tube through which *Rhizobium* cells can travel to reach and infect root cells

Leghemoglobin an O_2-binding protein found in root nodules

Lichen a fungus and an alga (or cyanobacterium) living in symbiotic association

Microbial leaching the removal of valuable metals such as copper from sulfide ores by microbial activities

Microenvironment the immediate environmental surroundings of a microbial cell or group of cells

Mycorrhizae a symbiotic association between a fungus and the roots of a plant

Nitrification the process by which NH_3 is oxidized to NO_3^-

Nitrogen fixation the microbiological reduction of N_2 to NH_3

Oxic an oxygen-containing environment, typically with a high E_0'

Primary producer an organism that uses light to synthesize new organic material from CO_2

Proteorhodopsin a light-sensitive protein present in some open ocean *Bacteria* that catalyzes ATP formation

Pyrite a common iron-containing ore, FeS_2

Reductive dechlorination removal of Cl as Cl^- from an organic compound by reducing the carbon atom from C—Cl to C—H

Rhizosphere the region immediately adjacent to plant roots

Root nodule a tumorlike growth on plant roots that contains symbiotic nitrogen-fixing bacteria

Rumen the first vessel in the multichambered stomach of ruminant animals in which cellulose digestion occurs

Syntrophy a metabolic process in which two or more microorganisms cooperate to carry out a process that could not be carried out by a single organism

Ti plasmid a conjugative plasmid present in the bacterium *Agrobacterium tumefaciens* that can transfer genes into plants

Volatile fatty acids (VFAs) the major fatty acids (acetate, propionate, and butyrate) produced during the rumen fermentation

Xenobiotic a totally synthetic product not naturally occurring in nature

icroorganisms are not alone in nature. Each microorganism in an ecosystem interacts with both its surroundings and with other organisms. These interactions can result in significant chemical and physical changes in the environment that may be beneficial or detrimental to other organisms. Whether harmful, beneficial, or neutral, the collective activities of microorganisms to a great extent control the workings of the biosphere.

In this chapter we will examine these key microbial activities. We will look first at microbial ecosystems in a general way and then explore some different environments for microorganisms—soil, freshwater, and the oceans. We will then examine some key reactions in microbial ecology, focusing on the nutrient cycles that help

sustain plant life on Earth. We finish up with coverage of some of the special relationships microorganisms have with plants and animals.

MICROBIAL ECOSYSTEMS

19.1 Populations, Guilds, and Communities

In nature, individual microbial cells grow to form **populations**. Metabolically related populations are called **guilds**, and sets of guilds interact in **microbial communities** (Figure 19.1●). Microbial communities in turn interact with communities of macroorganisms and the environment to define the entire **ecosystem**.

● **Figure 19.1 Populations, guilds, and communities—an example of microbial community structure in a lake ecosystem.** Microbial communities consist of populations of cells of various species. In a lake ecosystem various communities would exist as shown here. For the sediment community, major guilds are indicated. The reduction of CO_2, SO_4^{2-}, S^0, NO_3^-, and Fe^{3+} are examples of anaerobic respirations, discussed in Sections 17.13–17.18.

Energy enters ecosystems in the form of sunlight, organic carbon, and reduced inorganic substances. Light is used by phototrophic organisms (∞ Sections 17.1–17.5) to synthesize new organic matter (Figure 19.1●) containing not only carbon but also nitrogen, sulfur, phosphorus, iron, and other elements. This newly synthesized organic material, along with organic matter that enters the ecosystem from the outside (called *allochthonous* organic matter) drives the metabolic activities of chemoorganotrophic organisms. Chemolithotrophs, by contrast, obtain their energy from inorganic electron donors, such as H_2, Fe^{2+}, S^0, or NH_3 (∞ Chapter 17).

Biogeochemical Cycles

Microorganisms play an important role in the recycling of several elements in the environment, in particular carbon, sulfur, nitrogen, and iron. The study of these chemical transformations is called **biogeochemistry**. A **biogeochemical cycle** describes the transformations catalyzed by both biological and chemical agents during the cycling of these key elements of living systems. Often these cycles involve oxidation–reduction reactions (∞ Section 5.6) as the element moves through the ecosystem (Figure 19.1). Sulfur, for example, in the

form of hydrogen sulfide (H_2S), can be oxidized by a variety of microorganisms, both phototrophic and chemotrophic, to sulfur (S^0) and sulfate (SO_4^{2-}). Sulfate is a key nutrient for plants. Sulfate, in turn, can be reduced to sulfide by activities of the sulfate-reducing bacteria. This closes the biogeochemical cycle for sulfur by regenerating sulfide (see Figure 19.29).

Microorganisms of various types are intimately involved in biogeochemical cycling and in many instances are the only biological agents capable of regenerating forms of the elements needed by other organisms, particularly plants. We will discuss many biogeochemical cycles in this chapter, including those of carbon, nitrogen, sulfur, and iron.

 19.1 Concept Check

Microbial communities consist of guilds of metabolically related organisms. Microorganisms play major roles in energy transformations and biogeochemical processes that result in the recycling of elements essential to living systems.

◆ How does a microbial *guild* differ from a microbial *community*?

◆ What is a *biogeochemical cycle*?

19.2 Environments and Microenvironments

Microorganisms inhabit every conceivable habitat that will sustain life. Microbial habitats are exceedingly diverse and often rapidly changing. Besides the common habitats of soil and water, microorganisms inhabit the surfaces of higher organisms and in many cases actually live *within* them. Microorganisms frequently reach large numbers in such habitats and may benefit the plant or animal in a nutritionally significant way.

The Microorganism and the Microenvironment

The growth of microorganisms in nature, as in laboratory culture, depends on the *resources* (nutrients) available and on the *growth conditions* (Table 19.1). Differences in the type and quantity of resources and the physiochemical conditions of a habitat define the **niche** for each particular microorganism. Ecological theory states that for every organism there exists at least one niche, the *prime niche*, in which that organism is most successful. The organism dominates the prime niche but may also inhabit other niches; in these it is less ecologically successful than in its prime niche. Countless microbial niches exist on Earth and are in part responsible for evolution of the great metabolic diversity (∞ Chapter 17) and biodiversity (∞ Chapters 12–14) of microorganisms we see today.

Because microorganisms are so small, their habitats are also small. Microbial ecologists use the term **microenvironment** to describe where a microorganism actually lives

Table 19.1	**Major resources and conditions that govern microbial growth in nature**
Resources	**Conditions**
Carbon (organic, CO_2)	Temperature: cold → warm → hot
Nitrogen (organic, inorganic)	Water potential: dry → moist → wet
Other macronutrients (S, P, K, Mg)	pH: 0 → 7 → 14
Micronutrients (Fe, Mn, Co, Cu, Zn, Mn, Ni)	O_2: oxic → microoxic → anoxic
O_2 and other electron acceptors (NO_3^-, SO_4^{2-}, Fe^{3+}, etc.)	Light: bright light → dim light → dark
Inorganic electron donors (H_2, H_2S, Fe^{2+}, NH_4^+, NO_2^-, etc.)	Osmotic conditions: freshwater → marine → hypersaline

and metabolizes within its habitat. Microbiologists must thus learn to "think small" when considering the microorganism in its environment. For example, for a typical 3-μm rod-shaped bacterium, a distance of 3 mm is huge. It is equivalent to that which a human experiences over a distance of 2 km! Across the 3-mm distance, chemical and physical gradients may exist that can greatly affect the activities of the organism.

In a 3-mm particle of soil, for example, several different microenvironments could exist, differing chemically and physically in many ways. Visualize, for example, the distribution of an important microbial nutrient like oxygen in a soil particle. Microelectrodes (∞ Section 18.7) can measure oxygen concentrations throughout small soil particles. As shown in data from an actual experiment (Figure 19.2●), soil particles are not homogeneous in terms of their oxygen content. The outer zones of a small soil particle may be fully **oxic**, meaning that O_2 is present, whereas the center, only a very short distance away, can remain completely **anoxic** (O_2-free) (Figure 19.2).

We thus see that different niches can exist across a very small spatial dimension, and this is why various physiological types of microorganisms can coexist in a single "soil particle," such as that shown in Figure 19.2. Anaerobic organisms can be active near the center of the particle, microaerophiles (aerobes that require very low oxygen levels) active further out, and obligately aerobic organisms can metabolize in the outermost and fully oxic region of the particle. Facultatively aerobic bacteria can be distributed throughout the particle (∞ Section 6.15).

Physiochemical conditions in a microenvironment can change rapidly, both temporally and spatially. For example, the oxygen concentrations shown in Figure 19.2 represent "instantaneous" measurements. Measurements taken following a period of microbial respiration or after an increase in soil water content could show a drastically different gradient of oxygen across the microenvironment. It can thus be concluded that microenvironments are *heterogeneous* and that conditions in a given microenvironment can change rapidly. In this way microenvironments promote high microbial diversity in a relatively small physical space.

Nutrient Levels and Growth Rates

Nutrients (or in ecological terms, *resources*, see Table 19.1) typically enter an ecosystem in intermittent fashion. A

large pulse of nutrients—for example, an input of leaf litter or the carcass of a dead fish or animal—may be followed by a period of nutrient deprivation. Microorganisms in nature often face "feast-or-famine," and thus many microorganisms produce *storage polymers* as reserve materials when resources permit. Reserve polymers store excess nutrients present under favorable growth conditions for use during periods of nutrient deprivation. Examples of storage materials are poly-β-hydroxybutyrate and other alkanoates, polysaccharides, polyphosphate, and so on (∞ Section 4.11).

Extended periods of exponential growth of microorganisms in nature are rare. Growth typically occurs in spurts, linked closely to the availability of resources. Because physiochemical conditions in nature are rarely optimal all at the same time, growth rates of microorganisms in nature are thus well below the maximum growth rates recorded in the laboratory most of the time. For instance,

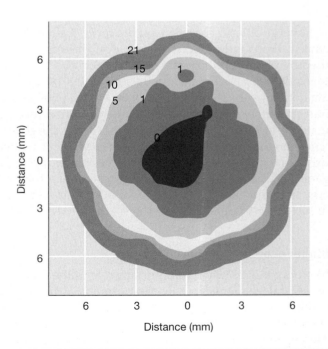

● **Figure 19.2 Contour map of O_2 concentrations in a soil particle.** The axes show the dimensions of the particle. The numbers on the contours are O_2 concentrations (in percentage, air is 21% O_2). In terms of oxygen relationships for microorganisms, each zone can be considered a different microenvironment.

the generation time of *Escherichia coli* in the intestinal tract is about 12 hours (two doublings per day), whereas in pure culture it grows much faster, with a minimum generation time of 20 minutes under the best of conditions.

Moreover, estimates have shown that typical soil bacteria grow in nature at less than 1% of the maximal growth rate measured in the laboratory. On average, these slow growth rates reflect the fact that (1) nutrients (resources, Table 19.1) are frequently in low supply and/or growth conditions suboptimal; (2) the distribution of nutrients throughout the microbial habitat is not uniform; and (3) except for rare instances, microorganisms grow in mixed populations rather than pure culture in natural environments. Thus an organism that grows rapidly in pure culture may do poorly in a natural environment where it must compete with other organisms better suited to the resources and growth conditions available.

Microbial Competition and Cooperation

Competition among microorganisms for available resources may be intense, with the outcome dependent on several factors, including rates of nutrient uptake, inherent metabolic rates and, ultimately, growth rates. A typical habitat will contain a mixture of different microorganisms (Figure 19.1), with the density of each population dependent on how closely the habitat resembles its prime niche.

Instead of directly competing for the same resources, some microorganisms work together to carry out transformations that neither could accomplish alone. These types of microbial partnerships, called **syntrophy** (∞ Section 17.21), are crucial to the success of certain anaerobic bacteria, as will be described in Section 19.10. Metabolic cooperation can also be seen in the activities of organisms that carry out *complementary* metabolisms. For example, in Chapter 17 we discussed metabolic transformations that involved two distinct groups of organisms, such as those of the *nitrosifying* and the *nitrifying* bacteria. Together, these organisms can oxidize NH_3 to NO_3^-, although neither group is capable of doing this alone (∞ Section 17.12). Because the product of the nitrosifying bacteria (NO_2^-) is the substrate for the nitrifying bacteria, such organisms typically live in nature in tight association (∞ Figure 18.12a).

19.2 Concept Check

Microorganisms are very small and their habitats are likewise small. The microenvironment is the place in which the microorganism actually lives. Microorganisms in nature often live a feast-or-famine existence such that only the best-adapted species thrive in a given niche. Cooperation among microorganisms is also important in many microbial interrelationships.

♦ What aspects define the niche of a particular microorganism?

♦ Why can many different physiological groups of organisms live in a single habitat?

19.3	**Microbial Growth on Surfaces and Biofilms**

Surfaces are important microbial habitats because nutrients can adsorb to them. In the microenvironment of a surface, nutrient levels may be much higher than they are in the bulk solution. As a consequence, microbial numbers and activities are usually much greater on surfaces than in water.

Microscope slides can be used as experimental surfaces to which organisms can attach and grow. Typically, a slide is immersed in a microbial habitat, left for a period of time, and then retrieved and examined by microscopy (Figure 19.3a●). Microcolonies readily develop on such surfaces, much as they do on natural surfaces in nature. In fact, periodic microscopic examination of immersed microscope slides has been used to measure growth rates of attached organisms in nature (∞ Figure 6.20b).

A surface may itself also be a nutrient, such as a particle of organic matter, where attached microorganisms catabolize nutrients directly from the surface of the particle. Plant materials, for example, are rapidly colonized by microorganisms in soil, and simple staining techniques can detect microbial populations attached to the solid surface (Figure 19.3b).

Biofilms

Oftentimes microorganisms grow on surfaces enclosed in **biofilms**—assemblages of bacterial cells attached to a surface and enclosed in adhesive polysaccharides excreted by the cells (Figure 19.4●). Biofilms trap nutrients for

(a) (b)

● **Figure 19.3 Microorganisms on surfaces.** (a) Bacterial microcolonies developing on a microscope slide immersed in a small river. The bright particles are mineral matter. The short, rod-shaped cells are about 3 μm long. (b) Fluorescence photomicrograph of a natural microbial community colonizing plant roots in soil. Note microcolony development. The preparation has been stained with acridine orange.

(a)

(b)

C.-T. Huang, Karen Xu, Gordon McFeters, and Philip S. Stewart

Cindy E. Morris

(c)

J.M. Sánchez, J.J. deLope, and Ricardo Amils

● **Figure 19.4 Microbial biofilms.** (a) A cross-sectional view of an experimental biofilm made up of cells of *Pseudomonas aeruginosa*. The yellow-green layer (about 15 μm in depth) contains cells and is stained by an enzyme activity stain for the enzyme alkaline phosphatase. The red area is the surface. (b) Confocal laser scanning microscopy (∞ Section 4.2) of a natural biofilm (top view) that developed on a leaf surface. The color of the cells indicates their depth in the biofilm: red, cells on the surface; green, 9 μm depth; blue, 18 μm depth. (c) A biofilm of iron-oxidizing prokaryotes on the surface of rocks in the iron-rich Rio Tinto, Spain. As Fe^{2+} rich water passes over and through the biofilm, iron-oxidizing microorganisms convert Fe^{2+} to Fe^{3+} to obtain energy (∞ Section 17.11).

growth of the microbial population and help prevent detachment of cells on surfaces present in flowing systems (Figure 19.5●). Biofilms typically contain many porous layers, and microscopic examination of the microorganisms in each layer can be done using scanning laser confocal microscopy (∞ Section 4.2) (Figure 19.4).

Cell-to-cell communication is critical in the development and maintenance of a biofilm. Attachment of a cell to a surface is a signal for the expression of *biofilm-specific genes*. These genes encode proteins that synthesize cell-to-cell signaling molecules and begin polysaccharide formation (Figure 19.5a). In *Pseudomonas aeruginosa*, a notorious biofilm former (Figure 19.4a), the major signaling molecules are compounds called *homoserine lactones*. As these molecules accumulate, they function as chemotactic agents to recruit nearby *P. aeruginosa* cells (a mechanism called *quorum sensing*, ∞ Section 8.10), and the biofilm

develops (Figure 19.4a). *P. aeruginosa* has been implicated in the disease *cystic fibrosis*, where a tenacious biofilm forms in the lungs, leading to symptoms of pneumonia (∞ Sections 8.10 and 31.7).

Why Do Bacteria Form Biofilms?

Bacteria form biofilms because the biofilm mode of growth undoubtedly improves survival and growth of the organisms. At least four reasons underlie the formation of biofilms. First, biofilms are a type of *defense*. Biofilms resist physical forces that could sweep unattached cells away, phagocytosis by cells of the immune system, or the penetration of toxic molecules, such as antibiotics. Second, biofilm formation *allows cells to remain in a favorable niche*. Biofilms attached to nutrient-rich surfaces, such as animal tissues, or to surfaces in flowing systems, such as a rock in a stream (Figure 19.4c), fix bacterial cells in a location where nutrients are often more

Attachment
(adhesion of a few cells to a suitable solid surface)

Colonization
(intercellular communication, growth and polysaccharide formation)

Development
(more growth and polysaccharide)

FLOW
Cell
Water channels
Polysaccharide
Surface

(a)

(b)

Rodney M. Donlan and Emerging Infectious Diseases

● **Figure 19.5 Biofilms.** (a) Formation of a bacterial biofilm. The biofilm begins with attachment of a few cells, after which growth and intercellular communication occurs. Polysaccharide formation occurs and becomes more extensive as the biofilm grows. (b) Photomicrograph of a DAPI-stained (∞ Section 18.3) biofilm that developed on a stainless steel pipe. Note the water channels.

abundant or are constantly replenished. Third, biofilms form because they *allow bacterial cells to live in close association* with each other. As we have already seen in the case of *Pseudomonas aeruginosa*, this facilitates intercellular communication via sensory molecules, which may ultimately benefit the cells. Moreover, when cells are in close proximity, opportunities for genetic exchange improve. And fourth, biofilms may be the *typical way bacterial cells grow in nature*, where nutrient concentrations are not nearly what they are in culture media. In other words, the biofilm may be the "default" mode of growth for prokaryotes in nature versus the more artificial rich medium liquid culture approach taken in the laboratory.

Biofilm Control

Biofilms have significant implications in human medicine and commerce. In the body, bacterial cells within a biofilm are protected from attack by the immune system, and antibiotics and other antimicrobial agents often fail to pierce the biofilm. Biofilms have been implicated in a variety of medical and dental conditions besides cystic fibrosis, including periodontal disease, kidney stones, tuberculosis, Legionnaire's disease, and *Staphylococcus* infections. Medical implants are excellent surfaces for biofilm development. These include both short-term devices, such as a urinary catheter, as well as long-term implants, such as artificial joints. Estimates are that as many as 10 million people a year in the United States alone experience biofilm infections from implants or intrusive medical procedures. Biofilms explain why routine oral hygiene is so important. Dental plaque is a typical biofilm and contains acid-producing bacteria responsible for dental caries (∞ Section 21.3).

In industrial situations biofilms can slow the flow of water, oil, or other liquids through pipelines and can accelerate corrosion of the pipes themselves. Biofilms also initiate degradation of submerged objects, such as structural components of offshore oil rigs, boats, and shoreline installations. The safety of drinking water may be compromised by biofilms that develop in water distribution pipes, many of which in the United States are nearly 100 years old. Although water pipe biofilms mostly contain harmless bacteria, if pathogens successfully colonize the biofilm, standard chlorination practices may be insufficient to kill them. Periodic releases of cells can then lead to outbreaks of disease. There is some concern that *Vibrio cholerae*, the causative agent of cholera (∞ Section 28.5), may be propagated in this manner.

Biofilm control is a big business and thus far only a limited repertoire of tools is available. Collectively, industries spend billions of dollars each year treating pipes and other surfaces to keep them free of biofilms. New antibiotics that can penetrate biofilms and other drugs that prevent biofilm formation by interfering with intercellular communication are being developed. A class of chemicals called *furanones*, for example, has

shown promise as biofilm preventatives in tests on abiotic surfaces. Because furanones are stable and nontoxic in humans, they may also have applications as antibiofilm agents in human medicine.

 19.3 Concept Check

Biofilms are bacterial assemblages encased in slime that form on surfaces. Biofilms can lead to the destruction of inert as well as living surfaces from the excretory products of the bacterial cells. Biofilm formation is a complex process involving cell-to-cell communication.

◆ What is the chemical nature of the biofilm matrix?

◆ Why might a biofilm be a good habitat for bacterial cells living in a flowing system?

◆ Give an example of a medically relevant biofilm that most humans likely harbor.

 SOIL AND FRESHWATER MICROBIAL HABITATS

Among the major habitats of microorganisms are soils and freshwater, the latter including lakes, ponds, and streams. Soils and waters vary in their physical structure, nutrient composition, temperature, and water potential. All of these factors combine to influence the diversity and numbers of microorganisms present.

19.4 Terrestrial Environments

The term *soil* refers to the loose outer material of Earth's surface, a layer distinct from bedrock that lies underneath (Figure 19.6●). Soil development involves complex interactions among the parent material (rock, sand, glacial drift, and so on), the topography, climate, and living organisms. Soils can be divided into two broad groups—**mineral soils** and **organic soils**—depending on whether they derive initially from the weathering of rock and other inorganic material or from sedimentation in bogs and marshes, respectively. Our discussion will concentrate on *mineral soils*, the predominant soil in most terrestrial environments.

Soil Formation

Soil forms as a result of combined physical, chemical, and biological processes. An examination of almost any exposed rock reveals the presence of algae, lichens (see Section 19.20), or mosses. These organisms are dormant or nearly so on dry rock and then grow when moisture is present. They are phototrophic and produce organic matter, which supports the growth of chemoorganotrophic bacteria and fungi.

O horizon
-Layer of undecomposed plant materials

A horizon
Surface soil (high in organic matter, dark in color, is tilled for agriculture; plants and large numbers of microorganisms grow here; microbial activity high)

B horizon
Subsoil (minerals, humus, and so on, leached from soil surface accumulate here; little organic matter; microbial activity detectable but lower than at A horizon)

C horizon
Soil base (develops directly from underlying bedrock; microbial activity generally very low)

(a)

(b)

● **Figure 19.6** **Soil.** (a) Profile of a mature soil. The soil horizons are soil zones as defined by soil scientists. (b) Photo of a soil profile showing O, A, and B horizons. This soil from Carbondale, Illinois, is rich in clay, and thus contains very small particles and is very compact. Such soils are not as well drained as those containing sand as a major component.

The numbers of chemoorganotrophs increase directly with the degree of plant cover. Carbon dioxide produced during respiration by chemoorganotrophs is converted to carbonic acid ($CO_2 + H_2O \rightleftharpoons H_2CO_3$), which is important in the dissolution of rocks, especially limestone. Many chemoorganotrophs also excrete organic acids, which further promote the dissolution of rock into smaller particles.

Freezing, thawing, and other physical processes lead to cracks in the rocks. A raw soil forms in these crevices, and pioneering plants can develop there. The plant roots penetrate farther into crevices and increase the fragmentation of the rock, and their excretions promote the development of a microflora in the **rhizosphere** (the soil that surrounds plant roots, see Figure 19.3*b*). When the plants die, their remains are added to the soil and become nutrients for an even more extensive microbial development. Minerals are further rendered soluble, and as water percolates, it carries some of these chemical substances deeper.

As weathering proceeds, the soil increases in depth, thus permitting the development of larger plants and trees. Soil animals become established and play an important role in keeping the upper layers of the soil mixed and aerated. Eventually, the movement of materials downward results in the formation of layers, called a *soil profile* (Figure 19.6). The rate of development of a typical soil profile depends on climatic and other factors, but it typically takes hundreds of years.

Soil as a Microbial Habitat

The most extensive microbial growth takes place on the *surfaces* of soil particles, usually within the rhizosphere

(see Figures 19.2, 19.3*b*, and 19.6*b*). As pointed out in Section 19.2, even a small soil aggregate can have many different microenvironments (compare Figures 19.2 and 19.7●), and thus, several different types of microorganisms may be present. To examine soil particles directly for microorganisms, fluorescence microscopes are often used, the organisms in the soil being stained with a dye that fluoresces (see Figure 19.3*b*). To observe a *specific* microorganism in a soil particle, fluorescent-antibody staining (∞ Figure 18.8) or phylogenetic stains (∞ Section 18.4) can be used. Microorganisms can also be seen on soil surfaces by means of the scanning electron microscope (Figure 19.8●). The scanning electron microscope gives excellent information on the morphology of

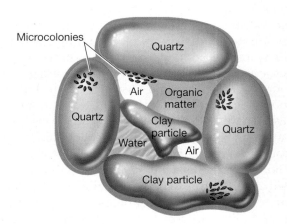

● **Figure 19.7** **A soil aggregate composed of mineral and organic components, showing the localization of soil microorganisms.** Very few microorganisms are found free in the soil solution; most of them occur in microcolonies attached to the soil particles.

(a)   (b)   (c)

T. R. G. Gray

● **Figure 19.8 Visualization of microorganisms on the surface of soil particles by use of the scanning electron microscope.** (a) A microcolony of short, rod-shaped bacteria. (b) Actinomycete spores (∞ Section 12.24). The cells in (a) and the spores in (b) are about $1-2$ μm wide. (c) Fungal hyphae. The fungal hyphae are about 4 μm wide and are coated with mineral matter.

soil bacteria and can also be used to enumerate cells on soil particle surfaces.

One of the major factors affecting microbial activity in soil is the availability of *water*, and we have previously discussed the importance of water to microbial growth (∞ Section 6.14). Water is a highly variable component of soil, depending on soil composition, rainfall, drainage, and plant cover. Water is held in the soil in two ways—by *adsorption* onto surfaces or as *free water* in thin sheets or films between soil particles.

The water present in soils has a variety of materials dissolved in it, and the mixture is called the *soil solution*. In well-drained soils, air penetrates readily, and the oxygen concentration of the soil solution can be high. In waterlogged soils, however, the only oxygen present is that dissolved in the water, and this is soon consumed by microorganisms. Such soils quickly become anoxic, showing profound changes in their biological properties. We discussed oxygen relationships in soil microenvironments in Section 19.2 and Figure 19.2.

The *nutrient status* (resources, Table 19.1) of a soil is the other major factor affecting microbial activity. The greatest microbial activity is in the organic-rich surface layers, especially in and around the rhizosphere (see Figure 19.6b). The numbers and activity of soil microorganisms depend to a great extent on the balance of nutrients present. In some soils carbon is not the limiting nutrient, but instead the availability of *inorganic* nutrients such as phosphorus and nitrogen limit microbial productivity.

Deep Subsurface Microbiology

Interest in the chemistry of groundwater and the potential leaching of pollutants and their transfer in groundwater aquifers has led to studies of the role microorganisms play in the *deep subsurface* terrestrial environment. The deep soil subsurface, which can extend for several hundred meters below the soil surface, is not a biological wasteland. Although cell numbers are much lower than in the A soil horizon (Figure 19.6), a variety of microorganisms, primarily bacteria, are present in deep subsurface soils. For example, in samples collected asep-

tically from bore holes drilled down to 300 m, a diverse array of prokaryotes has been found, including anaerobes such as sulfate-reducing bacteria, methanogens, and homoacetogens (∞ Sections 17.15–17.17), as well as various aerobes and facultative aerobes.

Microorganisms in the deep subsurface presumably have access to nutrients because groundwater flows through their habitats, but activity measurements suggest that metabolic rates of these bacteria are rather low in their natural habitats (see the Microbial Sidebar, Microbial Life Deep Underground). Compared to microorganisms in the upper layers of soil, the biogeochemical significance of deep subsurface microorganisms may thus be minimal. However, over very long periods, the metabolic activities of these buried microorganisms may be responsible for the slow mineralization of organic compounds and the release of products into groundwater.

The potential for *in situ* bioremediation (see Section 19.18) of toxic substances (for example, benzenes and agricultural chemicals) leached from soil into groundwater by deep subsurface microorganisms is of particular current interest. The resulting addition of inorganic nutrients may stimulate biodegradation of these chemicals, and the introduction of specific microorganisms into contaminated aquifers may promote biodegradation.

⬣ *19.4 Concept Check*

The soil is a complex habitat with numerous microenvironments and niches. Microorganisms are present in the soil primarily attached to soil particles. The most important factor influencing microbial activity in surface soil is the availability of water, whereas in deep soil (the subsurface environment) nutrient availability plays a major role.

◆ What is the difference between *mineral* soils and *organic* soils?

◆ What factors govern the extent and type of microbial activity in soils?

◆ Which region of soil is the most microbially active?

Microbial Sidebar ◆ Microbial Life Deep Underground

Microbiologists studying the deep terrestrial subsurface have found viable prokaryotes at depths of several *thousand* meters below the surface, existing in different physical and chemical environments. How are these microorganisms making a living? Initial findings indicated that these buried microorganisms were very slow growing *chemoorganotrophic* bacteria surviving by the slow catabolism of organic carbon deposited within the sediments. However, studies on the microbial ecology of deep basalt aquifers[1] have shown that sluggish chemoorganotrophs are not the only prokaryotes that live deep underground. *Chemolithotrophic* prokaryotes are present there, too, and may actually dominate microbial populations underground.

Basalts are iron-rich volcanic rocks essentially devoid of organic matter. In certain basalts up to 1500 m deep from the Columbia River Basin (Washington, USA), large numbers of anaerobic, chemolithotrophic bacteria have been found (Fig. 1a), including sulfate-reducing bacteria, methanogens, and homoacetogens.[1] Thus, both species of *Bacteria* and *Archaea* are found in the deep subsurface ecosystem. The methanogens were shown to be metabolically active by carbon-stable isotope analyses of methane (CH_4) present in the rocks and surrounding groundwaters. Measurements showed a strong enrichment in the lighter isotope of carbon (^{12}C), typical of biological methanogenesis (◌ Section 18.8). If

the methanogens are active, it is likely that other groups are active as well. However, their products, H_2S from the sulfate-reducers and acetate from the homoacetogens (Fig. 1b), do not accumulate because they are consumed by other microorganisms. These include sulfur-oxidizing chemolithotrophs and various chemoorganotrophic prokaryotes that can use acetate.

A common metabolic link among these anaerobes is the utilization of H_2 as an electron donor for their energy-yielding metabolisms (◌ Sections 13.3, 17.9, 17.14–17.18, and Figure 1b below). Hydrogen is a common product of the fermentative decomposition of organic matter (◌ Sections 17.19–17.21). But if basalts contain very little organic material, where does the H_2 come from to support metabolism of the H_2 consumers? Hydrogen in the Columbia River basalts apparently originates from the *strictly chemical* interaction of water with iron minerals in the rocks (see proposed reaction in Fig. 1b). Such reactions are known from inorganic chemistry, and in laboratory studies in which crushed Columbia River basalt was mixed with sterile water under anoxic conditions, rapid H_2 evolution occurred.[2] H_2 was also detected *in situ* in groundwater percolating through the basalts.

It is therefore possible that H_2 formed in deep underground basalts is the electron donor that supports growth of the substantial populations of anaerobic prokaryotes found there. If this is true, these organisms would be living a

strictly *geochemical* existence because their electron acceptor (CO_2 in the case of methanogens and homoacetogens and SO_4^{2-} in the case of sulfate-reducing bacteria) and electron donor, H_2, are all derived from inorganic materials. These buried chemolithotrophs would also be novel in the sense that they are totally divorced from photosynthetic primary production. The latter is the source of the molecular oxygen and/or organic matter required by virtually all surface-dwelling organisms.

As sometimes happens in science, additional research has questioned the Columbia River basalt findings. Although there is not much organic matter in these basalts, there is some, and thus fermentation could be the source of some of the H_2 used by the various subsurface H_2 chemolithotrophs.[2]

How widespread and ecologically significant H_2-based microbial ecosystems deep underground are, regardless of the source of H_2, awaits further research. But it is clear that metabolically active H_2-consuming microorganisms inhabit the deep subsurface, and this in turn suggests that they may be important biogeochemical agents in these environments. Indeed, we now suspect that total prokaryotic numbers in Earth's deep subsurface are very large (◌ Section 1.3), larger in fact than prokaryotes in virtually all other environments. The activities of these deep subsurface organisms are still very much a mystery, but they may have important implications for nutrient cycling on Earth. ■

Figure 1 Microbial life in the deep subsurface. *(a) Laser confocal photomicrograph of a microbial biofilm attached to the surface of basalt chips from a depth of 1500 meters. Green is reflected light from the basalt surface while the red color is from Nile red-stained bacterial cells. The cells in the biofilm were grown on H_2 from basalt as depicted in the bottom reaction of Fig. 1 b. (b) Key metabolic reactions of anaerobic prokaryotes growing in anoxic deep basalt aquifers.*

(a)

Todd O. Stevens

Methanogenesis: $4 H_2 + CO_2 \rightarrow CH_4 + 2 H_2O$

Acetogenesis: $4 H_2 + 2 HCO_3^- + H^+ \rightarrow CH_3COO^- + 4 H_2O$

Sulfate reduction: $4 H_2 + SO_4^{2-} + H^+ \rightarrow HS^- + 4 H_2O$

Proposed inorganic H_2 production: $FeO + H_2O \rightarrow H_2 + FeO_2$

(b)

[1]Stevens, T. O., and J. P. McKinley. 1995. Lithoautotrophic microbial ecosystems in deep basalt aquifers. *Science* 270:450–454.

[2]Anderson, R. T., F. H. Chapelle, and D. R. Lovely. 1998. Evidence against hydrogen-based microbial ecosystems in basalt aquifers. *Science* 281:976–977.

19.5 Freshwater Environments

Typical freshwater environments are lakes, ponds, rivers, and springs. Freshwater environments differ considerably from one another in their available resources and other chemical and physical properties. Thus it is not surprising that their microbial diversity varies. The predominant phototrophic organisms in most aquatic environments are microorganisms. In **oxic** areas cyanobacteria and algae prevail, and in **anoxic** areas anoxygenic phototrophic bacteria dominate.

Algae floating or suspended freely in the water are called **phytoplankton**; those attached to the bottom or sides are called **benthic algae**. Because these phototrophic organisms use energy from light in the initial production of organic matter, they are called **primary producers**. In the final analysis, the microbiological activity of an aquatic ecosystem depends on the rate of primary production by the phototrophic organisms. Oxygenic phototrophs produce both new organic material as well as oxygen. If photosynthetic activity is very high, the excessive organic matter leads to O_2 depletion and anoxic conditions. This in turn triggers alternative forms of metabolism among prokaryotes, in particular, anaerobic respiration and fermentation (☞ Sections 17.13–17.21).

Oxygen Relationships in Lakes and Rivers

We discussed oxygen requirements and anaerobiosis in Section 6.15, the production of oxygen via photosynthesis in Section 17.5, and oxygen in microenvironments in Section 19.2. Although oxygen is one of the most plentiful gases in the atmosphere (~21% of air), it has limited solubility in water, and in a large water mass its exchange with the atmosphere is slow. Significant photosynthetic production of oxygen occurs only in the *surface layers* of a lake or ocean, where light is available (see Figure 19.1). Organic matter that is not consumed in these surface layers sinks to the depths and is decomposed by facultative microorganisms, using oxygen dissolved in the water.

In freshwater lakes, once the oxygen is consumed, the deep layers become anoxic. Here strictly aerobic organisms such as higher plants and animals cannot exist, and the bottom layers contain anaerobic prokaryotes and a few kinds of anaerobic eukaryotes, in particular, certain protozoa. In addition, there is a conversion from respiratory to fermentative and methanogenic metabolisms, with important consequences for the carbon cycle and other nutrient cycles (see Sections 19.9–19.11).

Whether a body of water becomes depleted of oxygen depends on several factors, including the amount of organic matter present and mixing. If organic matter is sparse, as it is in pristine lakes or in the open ocean, there may be insufficient substrate available for chemoorganotrophs to consume all the oxygen. The microorganisms present in such environments are typically *oligotrophs*,

organisms adapted to growth under very dilute conditions. Where strong currents or turbulence occurs because of wind mixing, the water mass may be well mixed and consequently oxygen may be transferred to the deeper layers.

In contrast, in many lakes in temperate climates, the water mass becomes *stratified* during the summer, with the warmer and less dense surface layers, called the **epilimnion**, separated from the colder and denser bottom layers (the **hypolimnion**) (Figure 19.9●). The **thermocline** is the zone of transition from the epilimnion to the hypolimnion. After stratification sets in, usually in early summer, the bottom layers become anoxic (Figure 19.9). Such lakes often contain much higher levels of dissolved organic matter from nutrient run-off from land surrounding the lakes. This can trigger blooms of algae and cyanobacteria, that increase the organic content of the lake.

In temperate lakes in the late fall and early winter, the surface waters become colder and thus more dense than the bottom layers. This causes the water to "turn over," leading to reaeration of the bottom waters. This annual cycle allows the bottom layers of water to pass from oxic to anoxic and back to oxic. Microbial activity is altered with these changes in oxygen content, but other factors, especially temperature, strongly govern microbial activity as well.

Rivers

The oxygen relations in a river are of particular interest, especially in regions where rivers receive much organic

● **Figure 19.9 Development of anoxic conditions in the depths of a temperate lake as a result of summer stratification.** The colder bottom waters are more dense and contain H_2S from bacterial sulfate reduction. The zone of rapid temperature change is referred to as the thermocline. Typically, as surface waters cool in the fall and early winter they reach the temperature and density of hypolimnetic waters and sink, displacing bottom waters and effecting "lake turnover." Data from a small freshwater lake in northern Wisconsin (USA).

matter in the form of sewage, agricultural, and industrial pollution. Even though a river may be well mixed because of rapid water flow and turbulence, large amounts of added organic matter can lead to a marked oxygen deficit from bacterial respiration (Figure 19.10a●). As the water moves away from a sewage outfall, organic matter is gradually consumed, and the oxygen content returns to normal.

Oxygen depletion in a body of water is undesirable because many aquatic animals die under even very temporary anoxic conditions. Furthermore, conversion to anoxia results in the production by anaerobic bacteria of odorifer-

ous compounds (for example, amines, H_2S, mercaptans, fatty acids), some of which are also toxic to higher organisms. As in lakes, nutrient inputs from sewage or other pollutants can also trigger mass blooms of cyanobacteria and algae (Figure 19.10b). This increased load of organic matter eventually leads to oxygen depletion.

Biochemical Oxygen Demand

The microbial oxygen-consuming property of a body of water is its **biochemical oxygen demand (BOD)**. The BOD is determined by taking a sample of water, aerating it well, placing it in a sealed bottle, incubating for a standard period of time (usually 5 days at 20° C), and determining the residual oxygen in the water at the end of incubation. A BOD determination thus gives a measure of the amount of organic material in the water that could be oxidized by microorganisms. As a river recovers from contamination with an organic pollutant, the drop in BOD is accompanied by a corresponding increase in dissolved oxygen (Figure 19.10a).

We thus see that in a water body, the oxygen and carbon cycles are related, with the levels of organic carbon and oxygen being inversely related. This is particularly evident in *anoxic* environments, which are frequently rich in organic carbon (see Sections 19.9–19.11).

(a)

(b)

● **Figure 19.10 Effect of the input of sewage or other organic-rich wastewaters into aquatic systems.** (a) In a river, an increase in bacterial numbers and a decrease in O_2 levels occur immediately upon a spike of organic matter. If NH_4^+ is present in the input, for example from sewage, NH_4^+ is oxidized to NO_3^- by nitrifying bacteria (◯◯ Sections 12.3, 17.12, and 19.12). Note how the rise in NH_4^+ is followed shortly by the rise in NO_3^-, as the two-stage process of nitrification proceeds. The rise in numbers of algae and cyanobacteria is primarily a response to inorganic nutrients, especially PO_4^{3-}. (b) A eutrophic (nutrient-rich) lake, Lake Mendota, Madison, Wisconsin, showing algae, cyanobacteria, and macrophytes (aquatic weeds) that develop in response to pollution by inorganic nutrients from agricultural runoff. Dead plant and algal material contains a very large BOD, such that although these phototrophs contribute oxygen by photosynthesis, their degradation results in large–scale O_2 depletion.

19.5 *Concept Check*

In aquatic ecosystems phototrophic microorganisms are usually the main primary producers. Most of the organic matter produced is consumed by bacteria, which can lead to depletion of oxygen in the environment. BOD is a measure of the oxygen-consuming properties of a water sample.

◆ What is a *primary producer*?

◆ In a lake, where is the *epilimnion* and where is the *hypolimnion*?

◆ Will addition of organic matter to a water sample *increase* or *decrease* its BOD?

 III **MARINE MICROBIOLOGY**

The oceans differ from freshwater environments in many ways, including salinity, average temperature, depth, and nutrient status. Through use of the molecular tools of microbial ecology, especially genetic stains and gene sequencing (◯◯ Sections 18.4–18.6), much new information is emerging about marine microorganisms. We focus here on three key topics: (1) prokaryotes in the open oceans, (2) deep-sea microbiology, and (3) the extensive microbial ecosystems supporting animal life near deep-sea hydrothermal vents.

19.6 Marine Habitats and Microbial Distribution

Compared with typical freshwater environments, nutrient levels in the open ocean are often very low. This is especially true of such key inorganic nutrients as nitrogen, phosphorus, and iron. The activity of primary producers (mainly, photosynthetic CO_2 fixation) is limited by these nutrient deficiencies, and thus cell numbers are typically lower in the ocean than in freshwater ecosystems. However, because the oceans are so large, the collective photosynthesis and oxygen production that takes place there are major factors in Earth's carbon balance. Marine microbial communities influence everything from the marine food chain to the Earth's climate.

Primary Productivity

Much of the primary productivity in the open oceans, even at significant depths, is due to the photosynthetic activities of **prochlorophytes**, tiny oxygenic prokaryotes that contain chlorophylls *a* and *b*. The organism *Prochlorococcus* (a relative of the cyanobacteria) is a particularly important primary producer (Figure 19.11*a*● and ∞ Section 12.26). In tropical and subtropical oceans the planktonic filamentous marine cyanobacterium *Trichodesmium* (Figure 19.11*b*) is also widespread. Cells of *Trichodesmium* form tufts of filaments that constitute a significant fraction of the biomass suspended in these waters. *Trichodesmium* is a *nitrogen-fixing* cyanobacterium, and the production of fixed nitrogen by this organism is thought to be a major link in the nitrogen cycle (see Section 19.12) in marine environments.

Along with prochlorophytes in coastal waters, very small phototrophic *eukaryotes* have been found, and some of these are among the smallest eukaryotic cells known. *Ostreococcus*, for example, is an extremely small alga. It measures only about 0.7 μm in diameter (Figure 19.11*c*), which is smaller than a cell of *Escherichia coli*!

Inshore ocean areas are typically more nutritionally rich than open ocean waters and therefore support more dense populations of phototrophic microorganisms (Figure 19.12●). This in turn supports higher densities of chemotrophic bacteria and aquatic animals, such as fish and shellfish. Marine bays and inlets receiving sewage or industrial waste runoff can have very high phytoplankton and bacterial populations. If the pollution is severe enough, shallow marine waters can actually become anoxic and toxic to marine life from respiratory removal of O_2 by aerobic and facultative bacteria and the production of H_2S by sulfate-reducing bacteria (see Section 19.13).

Open Ocean Microbiology

Despite low inorganic nutrient and organic carbon levels, significant numbers—between 10^5 and 10^6/ml—of prokaryotic cells are suspended in open ocean waters. In addition, about 10^4 very small eukaryotic cells are

(a)   (b)

(c)

● **Figure 19.11 Marine oxygenic phototrophs.** (a) FISH phylogenetic stain (∞ Section 18.4) of cells of the prochlorophyte *Prochlorococcus*, the dominant phototroph in subtropical open oceanic waters. (b) The nitrogen-fixing cyanobacterium *Trichodesmium*. Cells form tufts that actively fix N_2 in tropical waters worldwide. A filament of *Trichodesmium* is about 6 μm in diameter. For more coverage of prochlorophytes, algae, and cyanobacteria, see Sections 12.26, 14.13, and 12.25, respectively. (c) Transmission electron micrograph of a cell of *Ostreococcus*, a very small alga (eukaryote), found in substantial numbers in marine coastal waters. The arrow points to the chloroplast. Cells of both *Prochlorococcus* and *Ostreococcus* are about 0.7 μm in diameter.

present per milliliter. How do these organisms make a living? Accumulating evidence points to light-driven energy metabolism as the key to supporting much of this open ocean microflora. Some of these organisms, such as *Prochlorococcus*, *Trichodesmium*, and the tiny phototrophic eukaryotes have already been discussed (see Figure 19.11).

Other phototrophs are present that carry out *anoxygenic* photosynthesis. However, these are not the

Great Lakes

Chesapeake Bay

Otis Brown and Robert Evans, NASA

● **Figure 19.12   Distribution of chlorophyll in the western North Atlantic Ocean as recorded by satellite.** The east coast of the United States from mid-Florida to northern Maine is shown in dotted outline. Near the center of the photograph is Chesapeake Bay; the Great Lakes are at the upper left. Areas rich in phytoplankton are shown in red (>1 mg chlorophyll/m^3); blue and purple areas have lower chlorophyll concentrations (<0.01 mg/m^3). Note the high primary productivity of coastal areas and the Great Lakes.

basis of energy metabolism for these marine bacteria, releasing them from a dependence on scarce organic carbon for an energy source.

Environmental genomics (∞ 18.6) has shown that genes encoding a diverse array of rhodopsin-like proteins exist in many phyla of *Bacteria* other than Proteobacteria in the open oceans. Thus, this form of metabolism in particular seems to be widely exploited among *Bacteria* and may have evolved (or have been obtained by lateral gene transfer) because of the very low concentrations of organic or inorganic electron donors available in marine waters.

Archaea/Bacteria Distribution

Numbers of prokaryotes in the open oceans *decrease* with depth. As previously mentioned, in surface waters cell numbers average between 10^5–10^6 cells/ml. Below 1000 m, however, total numbers fall to between 10^3–10^5/ml. Using phylogenetic stains to differentiate cells of each phylogenetic domain (∞ Section 18.4), an interesting distribution of *Bacteria* and *Archaea* with depth has been found in the open oceans. In general, species of *Bacteria* predominate in upper waters (<1000), while numbers are about equal or show a slight predominance of

typical purple and green bacteria (∞ Sections 12.2 and 12.33) whose anoxygenic photosynthesis occurs under *anoxic* conditions. Instead, these are the so-called **aerobic anoxygenic phototrophs**—organisms that carry out anoxygenic photosynthesis but only under *aerobic* conditions. These include organisms like *Erythrobacter* and *Roseobacter*, genera of Proteobacteria (∞ Chapter 12), that can carry out photosynthesis and presumably use the ATP produced to support their metabolism and growth. There appears to be a great diversity of such organisms in marine waters, especially in in-shore regions. Oligotrophic freshwater lakes also appear to be populated with aerobic phototrophs.

In addition to these phototrophs, many prokaryotes in the ocean's photic zone (top 300 meters) contain a form of the visual pigment rhodopsin. The cells use this pigment to convert light energy into ATP. In Section 13.3 we discussed the well-studied case of *bacteriorhodopsin*, present in the extreme halophile *Halobacterium* (a species of *Archaea*), and how this molecule was involved in ATP synthesis. The form of rhodopsin found in open ocean prokaryotes is very similar to bacteriorhodopsin but is present in cells that are phylogenetically *Bacteria*, not *Archaea*. Called **proteorhodopsin**, because the organisms that contain it are *Proteobacteria* (∞ Section 12.1), this molecule is presumably the

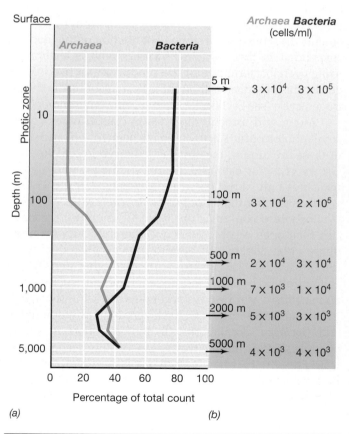

● **Figure 19.13   Percentage of total prokaryotes belonging to either the *Archaea* or the *Bacteria* in North Pacific Ocean water.** (a) Distribution of *Archaea* and *Bacteria* with depth. (b) Absolute numbers of *Archaea* and *Bacteria* with depth (per milliliter). Adapted from *Nature* 409:507–510 (2000).

Archaea in lower waters (Figure 19.13●). The *Archaea* present in lower waters are almost exclusively species of the *Crenarchaecta*, a phylum of *Archaea* that includes the hyperthermophiles (∞ Sections 13.8–13.11). Extrapolating from the data of Figure 19.13, it has been estimated that 1.3×10^{28} and 3.1×10^{28} cells of *Archaea* and *Bacteria*, respectively, exist in the world's oceans. This indicates that the oceans contain the largest microbial biomass on the surface of the Earth (∞ Section 1.3).

 19.6 Concept Check

Marine waters are more nutrient deficient than many freshwaters, yet substantial numbers of microorganisms exist there. Many of these use light to drive ATP synthesis. In terms of prokaryotes, species of the domain *Bacteria* tend to predominate in oceanic surface waters whereas *Archaea* are more prevalent in deeper waters.

◆ How does the organism *Prochlorococcus* contribute to both the carbon and oxygen cycles in the oceans?

◆ What is proteorhodopsin and why was it so named? How did the discovery of proteorhodopsin help solve the mystery of how large numbers of bacterial cells could exist in open ocean waters essentially devoid of dissolved organic matter?

◆ How are *Bacteria* and *Archaea* differentially distributed in ocean waters?

19.7 Deep-Sea Microbiology

Visible light penetrates no further than about 300 m in open ocean waters; this illuminated region is called the **photic zone** (Figure 19.13). Beneath the photic zone down to a depth of about 1000 m, considerable biological activity still occurs as a result of the action of animals and chemoorganotrophic microorganisms. Water at depths greater than 1000 m is, by comparison, relatively biologically inactive and has come to be known as the **deep sea**. Greater than 75% of all ocean water is deep-sea water, primarily at depths between 1000 and 6000 m.

Conditions in the Deep Sea

Organisms that inhabit the deep sea are faced with three major environmental extremes: *low temperature, high pressure*, and *low nutrient levels*. Below depths of about 100 m ocean water stays a constant 2–3° C. We discussed the responses of microorganisms to changes in temperature in Section 6.10. As would be expected, bacteria isolated from depths below 100 m are *psychrophilic* (cold-loving). Some are *extreme* psychrophiles, growing only in a narrow range near the *in situ* temperature. Deep-sea microorganisms must also be able to withstand the enormous hydrostatic pressures associated with great

depths. Pressure increases by *1 atm* for every *10 m* depth. Thus, an organism growing at a depth of 5000 m must be able to withstand pressures of 500 atm.

Barotolerant and Barophilic Bacteria

Different responses to pressure are evident in different deep-sea prokaryotes. Some organisms simply *tolerate* high pressure; they are called **barotolerant**. By contrast, others actually *grow best* under pressure; these are called **barophilic**.

Organisms isolated from depths down to about 3000 m and tested for growth or metabolic activity as a function of pressure are typically *barotolerant*. In barotolerant organisms, higher metabolic rates are observed at 1 atm than at 300 atm, although growth rates at the two pressures are about the same (Figure 19.14●). However, barotolerant isolates do not grow at pressures above 500 atm. By contrast, cultures derived from samples taken at greater depths, 4000–6000 m, are typically *barophilic*, growing optimally at pressures of about 400 atm (Figure 19.14). However, although barophiles grow best under pressure, they retain the ability to grow at 1 atm (Figure 19.14).

In even deeper water (10,000 m) **extreme (obligate) barophiles** can be found. These organisms have major pressure requirements for growth. For example, the bacterium *Moritella*, isolated from the Mariana Trench (Pacific Ocean, >10,000 m depth), grows fastest at a pressure of 700–800 atm and nearly as well at 1035 atm, the pressure it experiences in its natural habitat (Figures 19.14 and 19.15●).

● **Figure 19.14 Growth of barotolerant, barophilic, and extremely barophilic bacteria.** The extreme barophile (*Moritella*) was isolated from the Mariana Trench (see Figure 19.15). Note the much slower growth rate (at any pressure) of the extreme barophile (right ordinate) as compared to the barotolerant and barophilic bacteria (left ordinate). Note also the inability of the extreme barophile to grow at low pressures.

But unlike barotolerant or barophilic prokaryotes, *Moritella* will not grow at pressures of less than about 400 atm (Figure 19.14). Interestingly, however, *Moritella* is not killed by decompression because it can tolerate moderate periods of decompression. However, viability is lost when the culture is left for several hours in a decompressed state. *Moritella* is also sensitive to temperature. Its optimal growth temperature is its environmental temperature (2°C), and temperatures above 10° C significantly reduce viability.

Molecular Effects of High Pressure

Pressure affects cellular physiology and biochemistry in part because it decreases the binding capacity of enzymes for their substrates. Thus, the enzymes of extreme barophiles must fold so as to minimize these pressure-related effects. Other potential pressure-sensitive events include protein synthesis and membrane activities, such as transport. An organism grown under high pressure has a higher proportion of unsaturated fatty acids in its cytoplasmic membrane. This change is presumably of adaptive significance because it makes the membrane less likely to gel at high pressures. The rather slow growth rates of extreme barophiles such as *Moritella* (see Figure 19.14) are probably due to a combination of pressure effects on cellular biochemistry and low temperatures, where reaction rates decrease considerably to begin with (∞ Figure 6.16).

In barophiles capable of growth up to 500–600 atm (Figure 19.14), it has been shown that growth at high pressure is accompanied by changes in the protein composition of the cell wall. In one barophile studied in detail, a specific outer membrane protein (∞ Section 4.9) called *OmpH* (*o*uter *m*embrane *p*rotein H) is synthesized in cells grown under pressure but not in cells grown at 1

atm. The OmpH protein is a type of *porin*. Porins are structural proteins that form channels for the diffusion of organic molecules through the outer membrane and into the periplasm (∞ Section 4.9 and Figure 4.35*b*). Presumably, the porin present in cells of the barophile grown at low pressures cannot function properly at high pressure, and thus a new type of porin molecule must be synthesized. Curiously, relatively few proteins seem to be controlled by pressure in barophiles, because many proteins are the same in cells grown at both high and low pressure. Cell wall and related structural proteins, and transport proteins seem to vary the most in response to pressure, suggesting that they are likely the keys to a barophilic lifestyle.

 19.7 **Concept Check**

The deep sea is a cold, dark habitat where high hydrostatic pressure and low nutrient availability prevails. Barophiles grow best under pressure, and extreme barophiles, obtained from the greatest depths, require high pressure for growth.

◆ How does pressure change with depth?

◆ What molecular adaptations occur in barophiles that allow them to grow optimally under pressure?

19.8 Hydrothermal Vents

Although we think of the deep sea as a remote, low-temperature, high-pressure environment capable of supporting the slow growth of barotolerant and barophilic bacteria, there are some amazing exceptions. Dense, thriving *animal* communities, supported by the activities of microorganisms, cluster about thermal springs in deep-sea waters throughout the world. Geologically, these springs are associated with *sea floor spreading centers* (*rifts*), regions where hot basalt and magma near the sea floor cause the floor to slowly drift apart. Seawater seeping into these cracked regions mixes with hot minerals and is emitted from the springs (Figure 19.16●). These underwater hot springs are known as **hydrothermal vents**.

Two major types of vents have been found. *Warm vents* emit hydrothermal fluid at temperatures of 6–23° C (into seawater at 2° C). *Hot vents*, referred to as **black smokers** because the mineral-rich hot water forms a dark cloud of precipitated material upon mixing with seawater, emit hydrothermal fluid at 270–380° C (Figure 19.16, and see Figure 19.19). Warm vents emit fluid at 0.5–2 cm/s, whereas hot vents have higher flow rates, about 1–2 m/s.

● **Figure 19.15** **Sampling the deep sea.** The sampling arm of the unmanned submersible *Kaiko* inserting a sampling tube into sediment on the sea floor of the Mariana Trench (off the Philippines, Pacific Ocean) at a depth of 10,897 m and collecting a sample. The tubes of sediment are then retrieved and used for enrichment and isolation of barophilic bacteria.

Animals Living at Hydrothermal Vents

Using small pressurized submarines, it is possible to visit hydrothermal vents and study the organisms associated with them. Thriving invertebrate communities are present near hydrothermal vents, including *tube worms* over

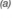

● **Figure 19.16 Deep sea hot springs: Hydrothermal vents.**
Schematic diagram showing the geological formations and major chemical species occurring at warm vents and black smokers. At warm vents (see Figure 19.17), the hot hydrothermal fluid is cooled by cold 2–3° C seawater permeating the sediments. In black smokers (see Figure 19.19), hot hydrothermal fluid near 350° C reaches the sea floor directly. Warm vents and black smokers are typically found at about 2000 in depth. Such depths can be explored by humans in small submersibles such as *Alvin*, used by researchers at the Woods Hole Oceanographic Institution, Woods Hole, MA.

2 m in length and large numbers of giant clams and mussels (Figure 19.17●). Considering that other locations in the deep sea are so biologically unproductive, how do dense animal communities exist in the absence of phototrophic primary producers? What are their energy source(s)?

Chemical analyses of hydrothermal fluid show large amounts of reduced inorganic materials, including H_2S, Mn^{2+}, H_2, and CO. Some vents contain little H_2S but have high levels of NH_4^+. Organic matter is not present in the fluid emitted from the hydrothermal vents. It is clear that the animals are dependent on the activities of *chemolithotrophic* bacteria (🔗 Sections 12.3–12.5 and 17.8–17.12), which grow at the expense of the inorganic energy sources emitted from the vents. Carbon dioxide, abundant in seawater as CO_3^{2-} and HCO_3^-, is fixed into organic carbon by the chemolithotrophs, and some of it is used to feed the vent animals.

Microorganisms in Hydrothermal Vents

Large numbers of sulfur-oxidizing chemolithotrophs are present in and around sulfide-emitting vents. Samples collected near such vents have yielded cultures of *Thiobacillus*,

(a)

(b)

(c)

● **Figure 19.17 Invertebrates from habitats near deep-sea thermal vents.** (a) Tube worms (family *Pogonophora*), showing the sheath (white) and plume (red) of the worm bodies. (b) Close-up photograph showing worm plume. The plume collects nutrients from hydrothermal vents, including in particular, H_2S. The latter is transported to symbiotic sulfide-oxidizing bacteria that live within the worm. See text for details. (c) Mussel bed in vicinity of a warm vent. Note yellow deposition of elemental sulfur (from the oxidation of H_2S by chemolithotrophic symbionts) in and around mussels. See Table 19.2 for a list of chemolithotrophic prokaryotes found near hydrothermal vents. Some vent animals may rely on nitrifying bacteria (🔗Sections 12.3 and 17.12) or methylotrophic bacteria (🔗 Sections 12.6 and 17.24) for their nutrients.

Thiomicrospira, Thiothrix, and *Beggiatoa* (∞ Sections 12.4 and 17.10), and field experiments have shown fixation of CO_2 and oxidation of H_2S and $S_2O_3^{2-}$ by natural populations of these bacteria. Some vents contain nitrifying, hydrogen-oxidizing, iron- and manganese-oxidizing bacteria, or methylotrophic bacteria, the latter presumably growing on the methane and carbon monoxide (CO) emitted from the vents (∞ Chapter 12 discusses these physiological groups). Table 19.2 summarizes the electron donors and electron acceptors for chemolithotrophs that likely play a role in hydrothermal vent animal ecology.

Nutrition of Animals Living Near Hydrothermal Vents

Various chemolithotrophs have been found to live in symbiotic association with animals of the thermal vents. For example, the 2-m-long tube worms (see Figure 19.17) lack a mouth, gut, or anus, but contain a modified gastrointestinal tract consisting primarily of spongy tissue called the **trophosome**. This structure, which makes up about half the worm's weight, is loaded with sulfur granules and large numbers of prokaryotic cells (Figure 19.18●), an average of 3.7×10^9 cells/g of trophosome tissue. The large spherical cells are structurally similar to the marine sulfur-oxidizing bacterium *Thiovulum*. The bacterial cells also show activity of the enzyme RubisCO and other enzymes of the *Calvin cycle,* the pathway by which most autotrophic organisms fix CO_2 into cellular material (∞ Section 17.6).

The chemolithotrophic bacteria thus supply the worm with its nourishment, the animal probably living off the excretory products and dead cells of its chemolithotrophic symbionts. The bright red plume of the tube worm (Figure 19.17*b*) is rich in blood vessels and serves as a trap for O_2 and H_2S (see the next paragraph) for transport to chemolithotrophs in the trophosome. Giant clams and mussels (see Figure 19.17*c*) are also present around the vents, and sulfur-oxidizing bacterial communities are found here as well, localized in the gill tissues of these animals. Phylogenetic analyses (∞ Chapter 11) have shown that each vent animal har-

(a)   (b)

● **Figure 19.18 Chemolithotrophic sulfur-oxidizing bacteria associated with the trophosome tissue of tube worms from hydrothermal vents.** (a) Scanning electron microscopy of trophosome tissue showing spherical chemolithotrophic sulfur-oxidizing bacteria. Cells are 3–5 μm in diameter. (b) Transmission electron micrograph of bacteria in sectioned trophosome tissue. The cells are frequently enclosed in pairs by an outer membrane of unknown origin. Chemolithotrophic sulfur bacteria are discussed in Sections 12.4, 17.8, and 17.10. Reprinted with permission from *Science 213*:340–342 (1981), © AAAS.

bors one major species of bacterial symbiont and that the bacterial species differ among the different animal types.

Tube worms contain unusual hemoglobins that bind H_2S as well as O_2; they transport both substrates to the trophosome where they are released to the bacterial symbiont. Trapping and transporting sulfide are necessary to prevent the H_2S from poisoning the animal. The CO_2 content of tube worm blood is also high, some 20–30 mM, and presumably this is released in the trophosome as a carbon source for the symbionts. In addition, stable isotope analyses (∞ Section 18.8) of the elemental sulfur found within the bacterial symbionts have shown the $^{34}S/^{32}S$ isotope composition to be the same as the sulfide emitted from the vent. This ratio is distinct from that of seawater sulfate and is additional proof that geothermal sulfide is actually entering the worm.

Other marine animals may rely on chemolithotrophs for their nutrition, as well. For example, methanotrophic

Table 19.2	**Chemolithotrophic prokaryotes of potential significance to hydrothermal vent primary productivity**[a]		
Chemolithotroph	**Electron donor**	**Electron acceptor**	**Product from donor**
Sulfur-oxidizing	$HS^-, S^0, S_2O_3^{2-}$	O_2, NO_3^-	S^0, SO_4^-
Nitrifying	NH_4^+, NO_2^-	O_2	NO_2^-, NO_3^-
Sulfate-reducing	H_2	S^0, SO_4^{2-}	H_2S
Methanogenic	H_2	CO_2	CH_4
Hydrogen-oxidizing	H_2	O_2, NO_3^-	H_2O
Iron and manganese-oxidizing	Fe^{2+}, Mn^{2+}	O_2	Fe^{3+}, Mn^{4+}
Methylotrophic[b]	CH_4, CO	O_2	CO_2

[a] ∞ Sections 12.3–12.5 and 17.8–17.12 for a discussion of chemolithotrophic metabolism.

[b] ∞ Sections 12.6 and 17.24 for a discussion of methylotrophy.

symbionts have been shown to play a nutritional role for animals living in symbiotic association with giant clams near natural gas seeps at relatively shallow depths in the Gulf of Mexico (⊂⊃ Figure 12.16). Although not truly autotrophs (CH_4 is an organic compound), these symbionts support growth of the animal, in this case by the oxidation of CH_4 as an energy source.

Superheated Water: Black Smokers

The great depths of the deep sea create huge hydrostatic pressures. At a depth of 2600 m, for example, water does not boil until it reaches a temperature of 450° C. At certain vent sites superheated hydrothermal fluid is emitted at temperatures up to 350°C (see Figure 19.16). This superheated but not boiling water could theoretically be a habitat for hyperthermophilic bacteria (⊂⊃ Sections 6.10 and 6.12, and Chapter 13). The hydrothermal fluid emitted from black smokers contains abundant metal sulfides, especially iron sulfides, and cools quickly as it mixes with cold seawater. The precipitated metal sulfides form a tower called a "chimney" about the source (Figure 19.19●).

Although it is clear that prokaryotes do not live in the superheated hydrothermal fluid itself, thermophilic and hyperthermophilic bacteria live in the seawater or hydrothermal fluid *gradient* that forms as the hot water blends with cold ocean water. For example, the walls of smoker chimneys are teeming with hyperthermophilic prokaryotes such as *Methanopyrus*, an organism that oxidizes H_2 and makes methane (CH_4) (⊂⊃ Sections 13.4

and 13.6). FISH technology (⊂⊃ Section 18.4) has detected cells of both *Bacteria* and *Archaea* in smoker chimney walls (Figure 19.20●). The most thermophilic of all known prokaryotes, species of *Pyrolobus* and *Pyrodictium* (⊂⊃ Sections 13.10–13.13), also reside in smoker chimney walls.

When smokers plug up from mineral debris, hyperthermophiles presumably drift away to colonize newly formed smokers and become integrated into the growing chimney wall. Surprisingly, although requiring very high temperatures for growth, hyperthermophiles are remarkably tolerant of cold temperatures and oxygen. Thus, transport of cells from one thermal feature to another in cold oxic seawater is apparently not a problem.

◉ 19.8 Concept Check

Hydrothermal vents are deep-sea hot springs where volcanic activity generates fluids containing large amounts of inorganic energy sources that can be used by chemolithotrophic bacteria. The latter fix CO_2 into organic carbon, some of which is then used by the deep-sea animals. Deep-sea hydrothermal vents are habitats where the primary producers are chemolithotrophic rather than phototrophic.

◆ How does a *warm hydrothermal vent* differ from a *black smoker*, both chemically and physically?

◆ How do giant tube worms receive their nutrition?

◆ Why is 350°C water emitting from a black smoker not boiling?

● **Figure 19.19 A hydrothermal vent black smoker emitting sulfide- and mineral-rich water at temperatures of 350° C.** The walls of the black smoker chimneys display a steep temperature gradient and contain several types of prokaryotes (see Figure 19.20).

● **Figure 19.20 Phylogenetic staining of black smoker chimney material from a smoker at the Snake Pit vent field of the Mid-Atlantic Ridge (3500 m deep).** A green fluorescing dye was conjugated to a probe that reacts with the 16S rRNA of all *Bacteria* and a red dye to a 16S rRNA probe for *Archaea* (⊂⊃ Sections 11.7 and 18.4). Cell numbers tend to be highest in the outer regions of the chimney wall (closest to 2°C seawater) and lowest in the inner region. The hydrothermal fluid going through the center of the chimney was 300°C. This water is sterile. The current consensus is that microbial life can exist up to about 150°C but that above this temperature, the key biomolecules of life are unstable (⊂⊃ Section 13.12).

IV THE CARBON AND OXYGEN CYCLES

Global carbon cycling involves the activities of both microorganisms and macroorganisms and is intimately tied to the oxygen cycle. Due to the acceleration of CO_2 inputs into the atmosphere by human activities, the carbon cycle has long been a subject of scientific interest. Scientists study the carbon cycle to better understand the magnitude of carbon reservoirs, the major sinks for CO_2, and the rates of cycling within and between compartments, in order to model carbon transformations and prevent planetary catastrophes, such as severe global warming.

Table 19.3	Major carbon reservoirs on Earth	
Reservoir	**Carbon (gigatons)[a]**	**Percent of total carbon on Earth**
Oceans	38×10^3 (>95% is inorganic C)	0.05
Rocks and sediments	75×10^6 (>80% is inorganic C)	>99.5[b]
Terrestrial biosphere	2×10^3	0.003
Aquatic biosphere	1–2	0.000002
Fossil fuels	4.2×10^3	0.006
Methane hydrates	10^4	0.014

[a] One gigaton is 10^9 tons. Data adapted from *Science* 290:291–295 (2000).
[b] Much of the organic carbon is in prokaryotic cells.

19.9 The Carbon Cycle

On a global basis, carbon is cycled through all Earth's major carbon reservoirs: the atmosphere, the land, the oceans and other aquatic environments, sediments and rocks, and biomass (Figure 19.21●). As we have already seen in the case of aquatic environments, the carbon and oxygen cycles are intimately intertwined. This is true of soil and all other environments on Earth as well, as we see now.

Carbon Reservoirs

The largest carbon reservoir is in the sediments and rocks of Earth's crust (Table 19.3), but the turnover time is so long here that flux out of this compartment is relatively insignificant on a human time scale. From the viewpoint of living organisms, a large amount of organic carbon is found in land plants. This represents the carbon of forests and grasslands and constitutes the major site of photosynthetic CO_2 fixation. However, more carbon is present in dead organic material, called **humus**, than in living organisms.

Humus is a complex mixture of organic materials. It is derived partly from the protoplasmic constituents of soil microorganisms that have resisted decomposition and partly from resistant plant materials. Some humic substances are fairly stable, with a global turnover time of about 40 years, although certain other humic components decompose much more rapidly than this.

The most rapid means of global transfer of carbon is via the CO_2 of the atmosphere. Carbon dioxide is removed from the atmosphere primarily by photosynthesis of land plants and is returned to the atmosphere by respiration of animals and chemoorganotrophic microorganisms. An analysis of the various processes suggests that the single most important contribution of CO_2 to the atmosphere is microbial decomposition of dead organic material, including humus. In recent times, however, human activities have added tremendously to the atmospheric CO_2 reservoir. For example, in the past 40 years alone, CO_2 levels in the atmosphere have risen nearly 14%.

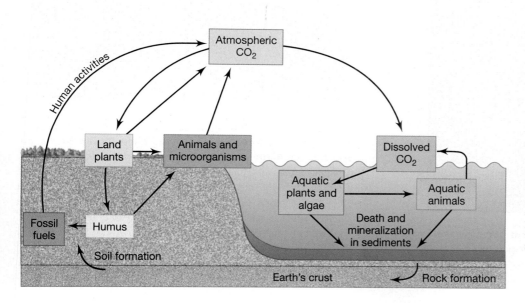

● **Figure 19.21 The carbon cycle.** The carbon and oxygen cycles are closely connected, as oxygenic photosynthesis both removes CO_2 and produces O_2 while respiratory processes both produce CO_2 and remove O_2 (see also Section 19.5). Carbon cycling in the deep subsurface may be as significant as carbon cycling on Earth's surface (see the Microbial Sidebar, Microbial Life Deep Underground), but no reliable estimates of this are available. Most prokaryotic cells on Earth are thought to lie in the terrestrial and oceanic subsurfaces (➤ Sections 1.3 and 19.4).

Importance of Photosynthesis in the Carbon Cycle

The only major ways in which new organic carbon is synthesized on Earth are via photosynthesis and chemosynthesis (CO_2 fixation by chemolithotrophs); most organic carbon comes from photosynthesis. Phototrophic organisms are therefore the foundation of the carbon cycle (Figure 19.21). Phototrophic organisms are found in nature almost exclusively in habitats where light is available. Thus, the deep sea and other permanently dark habitats are devoid of indigenous phototrophs.

Oxygenic phototrophic organisms can be divided into two major groups: *higher plants* and *microorganisms*. Higher plants are the dominant phototrophic organisms of terrestrial environments, whereas phototrophic microorganisms are the most abundant photosynthesizers of aquatic environments.

The redox cycle for carbon (Figure 19.22●) begins with photosynthesis:

$$CO_2 + H_2O \longrightarrow (CH_2O) + O_2$$

light

Here (CH_2O) represents organic matter at the oxidation state of cell material, such as polysaccharides (the main form in which photosynthesized organic matter is stored in the cell). Phototrophic organisms also carry out respiration, both in the light and the dark. The overall equation for respiration is the reverse of oxygenic photosynthesis:

$$(CH_2O) + O_2 \longrightarrow CO_2 + H_2O$$

light or dark

where (CH_2O) again represents storage polysaccharides. If a phototrophic organism is to increase in cell number or mass, then the rate of photosynthesis must exceed the rate of respiration. If growth occurs, then some of the carbon fixed from CO_2 into polysaccharide can become the starting material for biosynthesis. The whole carbon cycle is built on a net positive balance of the rate of photosynthesis over the rate of respiration.

Decomposition

Photosynthetically fixed carbon is eventually degraded by microorganisms and two major forms of carbon result: methane (CH_4) and carbon dioxide (CO_2) (Figure 19.22). These two gaseous products are formed from the activities of methanogens (CH_4) or from chemoorganotrophs via fermentation, anaerobic respiration, or aerobic respiration (CO_2). In anoxic habitats CH_4 is produced from both the reduction of CO_2 with H_2 and from certain organic compounds like acetate. However, virtually *any* organic compound can eventually be converted to CH_4 from the combined activities of *syntrophic* bacteria and methanogens: H_2 generated from the fermentative degradation of organic compounds gets consumed by methanogens and converted to CH_4 (👓 Section 17.17 and see next section). Methane produced in anoxic habitats is insoluble and migrates to oxic environments where it is oxidized to CO_2 by methanotrophs (Figure 19.22). Hence, all organic carbon eventually returns to CO_2, and the carbon cycle is complete.

The balance between the oxidative and reductive portions of the carbon cycle is critical: the products of metabolism of some organisms are the substrates for others. Thus, the cycle needs to keep in balance if it is to continue as it has for billions of years. Any significant changes in levels of gaseous forms of carbon may have serious global consequences (as we are already experiencing in global warming from the increasing CO_2 levels in the atmosphere caused by deforestation and the burning of fossil fuels). In terms of decomposition, CO_2 release by microbial activities far exceeds that of eukaryotes, and this is especially true of anoxic environments, which we consider next.

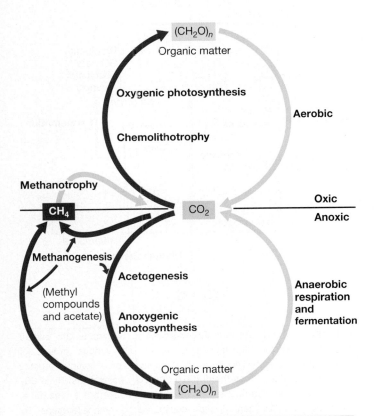

● **Figure 19.22 Redox cycle for carbon.** The figure contrasts autotrophic ($CO_2 \rightarrow$ organic compounds) and heterotrophic (organic compounds $\rightarrow CO_2$) processes. Yellow arrows indicate oxidations; red arrows indicate reductions.

 19.9 Concept Check

The oxygen and carbon cycles are interconnected through the complementary activities of autotrophic and heterotrophic organisms. Microbial decomposition is the single largest source of CO_2 released to the atmosphere.

◆ How is new organic matter made in nature?

◆ In what ways are oxygenic photosynthesis and respiration related?

19.10 Syntrophy and Methanogenesis

Methane from biological methanogenesis is of great importance to carbon flow in many anoxic habitats. **Methanogenesis** is carried out by a group of *Archaea*, the methanogens, which are strict anaerobes. We discussed the biochemistry of methanogenesis in Section 17.17 and methanogens themselves in Section 13.4. Most methanogens use CO_2 as a terminal electron acceptor in anaerobic respiration, reducing it to methane with H_2 (Figure 19.22). Only a very few other substrates, acetate being chief among them, can be directly converted to methane by methanogens. Thus, for the conversion of most organic compounds to CH_4, methanogens must team up with partner organisms that can supply them with their needed substrates. This is the job of the *syntrophs*.

Anoxic Decomposition and Syntrophy

In Section 17.21 we discussed the concept of **syntrophy**, a process in which two or more organisms cooperate in the degradation of some compound, and focused on the energetics behind the process. Here we consider the *ecological interactions* of syntrophic bacteria with other organisms and their significance for the anoxic carbon cycle. Our focus is on anoxic freshwater sediments, where much anaerobic metabolism takes place.

Polymeric substances such as polysaccharides, proteins, and fats are the products of dead organisms, and they are converted to CH_4 and CO_2 by the cooperative interaction of several physiological groups of prokaryotes. For the breakdown of a typical polysaccharide such as cellulose (Figure 19.23● and Table 19.4), the process begins with *cellulolytic bacteria*, which cleave the cellulose molecule into cellobiose (glucose—glucose) and from there into free glucose. Glucose is then fermented by *primary* fermenters into a variety of fermentation products, including acetate, propionate, butyrate, succinate, alcohols, H_2, and CO_2. The H_2 produced in primary fermentation is removed by H_2 *consumers*, such as methanogens, homoacetogens, or sulfate-reducing bacteria (the latter in environments containing sufficient levels of sulfate). In addition, acetate can be converted to methane by certain methanogens. But this leaves a large amount of carbon in the form of fatty acids and alcohols. The catabolism of these compounds occurs by way of the syntrophs (Figure 19.23).

Role of the Syntrophs

Key organisms in the conversion of complex organic materials to methane are the syntrophs. These organisms are examples of *secondary* fermenters. Syntrophs ferment the fermentation products of primary fermenters, producing H_2 and other products. For example, *Syntrophomonas wolfei* oxidizes C_4 to C_8 fatty acids yielding acetate, CO_2 (if the fatty acid contained an odd number of carbon atoms), and H_2 (Table 19.4). Other species of *Syntrophomonas* use fatty acids up to C_{18} in length, including some unsaturated fatty acids. *Syntrophobacter wolinii* specializes in propionate oxidation and generates acetate, CO_2, and H_2, while *Syntrophus gentianae* degrades the aromatic compound benzoate to acetate, H_2, plus CO_2 (Table 19.4). Indeed, with the right combination of organisms, virtually any organic compound can be converted into methane. However, recall that the H_2 producers—the syntrophs—are unable to carry out their syntrophic reaction in pure culture; their

● **Figure 19.23 Anoxic decomposition.** Shown is the overall process of anoxic decomposition, in which various groups of fermentative anaerobes cooperate in the conversion of complex organic materials ultimately to methane (CH_4) and CO_2. Acetate and $H_2 + CO_2$ from primary fermentations can be directly converted to methane, although $H_2 + CO_2$ can also be consumed by homoacetogens. By activities of the syntrophs, fatty acids and alcohols are converted to the substrates for methanogenesis and acetogenesis. This picture holds for environments in which sulfate-reducing bacteria play only a minor role, for example, in freshwater lake sediments, sewage sludge bioreactors, or the rumen.

Table 19.4 Major reactions occurring in the anoxic conversion of organic compounds to methane[a]

Reaction type	Reaction	Free-energy change (kJ/reaction) $\Delta G^{0\prime b}$	ΔG^c
Fermentation of glucose to acetate, H_2, and CO_2	Glucose $+ 4 H_2O \rightarrow 2$ Acetate$^- + 2 HCO_3^- + 4 H^+ + 4 H_2$	−207	−319
Fermentation of glucose to butyrate, CO_2, and H_2	Glucose $+ 2 H_2O \rightarrow$ Butyrate$^- + 2 HCO_3^- + 2 H_2 + 3 H^+$	−135	−284
Fermentation of butyrate to acetate and H_2	Butyrate$^- + 2 H_2O \rightarrow 2$ Acetate$^- + H^+ + 2 H_2$	+48.2	−17.6
Fermentation of propionate to acetate, CO_2, and H_2	Propionate$^- + 3 H_2O \rightarrow$ Acetate$^- + HCO_3^- + H^+ + H_2$	+76.2	−5.5
Fermentation of ethanol to acetate and H_2	2 Ethanol $+ 2 H_2O \rightarrow 2$ Acetate$^- + 4 H_2 + 2 H^+$	+19.4	−37
Fermentation of benzoate to acetate, CO_2, and H_2	Benzoate$^- + 7 H_2O \rightarrow 3$ Acetate$^- + 3 H^+ + HCO_3 + 3 H_2$	+70.1	−18
Methanogenesis from $H_2 + CO_2$	$4 H_2 + HCO_3^- + H^+ \rightarrow CH_4 + 3 H_2O$	−136	−3.2
Methanogenesis from acetate	Acetate$^- + H_2O \rightarrow CH_4 + HCO_3^-$	−31	−24.7
Acetogenesis from $H_2 + CO_2$	$4 H_2 + 2 HCO_3^- + H^+ \rightarrow$ Acetate$^- + 4 H_2O$	−105	−7.1

[a] Data adapted from Zinder, S. 1984. Microbiology of anaerobic conversion of organic wastes to methane: Recent developments. *Am. Soc. Microbiol. News* 50:294–298.
[b] Standard conditions: solutes, 1 M; gases, 1 atm.
[c] Concentrations of reactants in typical anoxic freshwater ecosystem: fatty acids, 1 mM; HCO_3^-, 20 mM; glucose, 10 μM; CH_4, 0.6 atm; H_2, 10^{-4} atm. For calculating ΔG from $\Delta G^{0\prime}$, refer to Appendix 1.

growth requires a H_2-consuming partner organism. This requirement has to do with the energetics of syntrophic processes.

As was explained in Section 17.21, *H_2 consumption by the partner organism* is absolutely essential for growth of the syntrophs. When the reactions in Table 19.4 are written with all reactants at *standard conditions* (solutes, 1 M; gases, 1 atm), the reactions yield free-energy changes that are positive in arithmetic sign. That is, the $\Delta G^{0\prime}$ (⚭⚭ Section 5.4) of these reactions is not exergonic (Table 19.4). But H_2 consumption by partner bacteria dramatically changes the energetic picture, allowing for sufficient energy conservation to support growth of the syntroph. This can be seen in Table 19.4, where the ΔG values (free-energy change measured under *actual conditions* in the habitat) are favorable for energy conservation if H_2 concentrations are kept very low by the activities of the partner organism. The final products of this metabolic cooperation are CO_2 and CH_4. The list of syntrophically catabolized substances is nearly endless and even includes saturated hydrocarbons, as we see now.

Anaerobic Metabolism of Methane

In freshwater ecosystems, methane, produced in the anoxic sediments, is oxidized to CO_2 by methanotrophs when it reaches oxic zones (Figure 19.22). These methane-oxidizing prokaryotes require O_2 for the catabolism of methane because the first step in methane oxidation employs a monooxygenase enzyme (⚭⚭ Sections 17.22 and 17.24 and Figures 17.55 and 17.58). However, methane can be oxidized under *anoxic* conditions in marine sediments by cell aggregates that contain certain sulfate-reducing bacteria and methanogens (Figure 19.24a●).

How methane is oxidized by the aggregates is not yet clear, but two alternatives have been suggested. In one, the methanogens oxidize methane to CO_2 by reversing the

steps of methanogenesis (⚭⚭ Section 17.17; Figure 19.24b). This reaction is energetically unfavorable unless the H_2 produced is removed by a second organism, a syntrophic type of metabolism. This would be the job of the sulfate reducer, consuming H_2 in the production of H_2S

(a)

Antje Boetius and Armin Gieseke

Reaction	Organism	$\Delta G^{0\prime}$(kJ)
$CH_4 + 2 H_2O \longrightarrow CO_2 + 4 H_2$	Methanogen	+131
$SO_4^{2-} + 4 H_2 + H^+ \longrightarrow HS^- + 4 H_2O$	Sulfate-reducer	−156
Sum: $SO_4^{2-} + CH_4 \longrightarrow HCO_3^- + HS^- + H_2O$	Syntrophic reaction	−25

(b)

● **Figure 19.24 Anoxic methane oxidation.** (a) Methane-oxidizing cell aggregates from marine sediments. The aggregates contain methanogenic bacteria (red) surrounded by sulfate-reducing bacteria (green). Each cell type is stained by a different phylogenetic FISH stain (⚭⚭ Section 18.4). The diameter of the aggregate is about 30 μm. (b) Possible mechanism for syntrophic anoxic methane oxidation in the cell aggregates.

(Figure 19.24). This yields a very small amount of energy (Figure 19.24b), and how this energy could be split between the components in the aggregate is unclear.

Alternatively, the methanogenic component may convert CH_4 into acetate with the sulfate-reducing bacterium oxidizing the acetate to CO_2 during sulfate reduction. Regardless of how this partnership works, anoxic methane oxidation is clearly a syntrophic process whereby two organisms cooperate to carry out a reaction that neither could do alone. Recent environmental genomic (⊙⊙ Section 18.6) analyses of the anoxic methane-oxidizing cell aggregates favor the reverse methanogenesis hypothesis.

Methanogenic Habitats

Despite the obligate anaerobiosis and specialized metabolism of methanogens, they are widespread on Earth. Although high levels of methanogenesis occur only in anoxic environments, such as swamps and marshes, or in the rumen (see Section 19.11), the process also occurs in habitats that normally might be considered oxic, such as forest and grassland soils. In such habitats methanogenesis occurs in anoxic microenvironments, for example in the midst of soil crumbs (see Figure 19.2). Table 19.5 shows the rates of methanogenesis in different habitats. Note that biogenic production of methane exceeds the production rate from gas wells and other abiogenic sources. Eructation by ruminants

(see Section 19.11) and CH_4 released from termites, paddy fields, and natural wetlands are the largest sources of biogenic methane.

Methanogens have also been found as endosymbionts of certain protozoa. Several types of protozoa, including free-living aquatic amebas and flagellates found in the insect gut, have been shown to harbor methanogens. In termites, for example, methanogens are present primarily within cells of trichomonal protozoa inhabiting the termite hindgut (Figure 19.25●). Methanogenic symbionts of protozoa resemble rod-shaped species of the genus *Methanobacterium* or *Methanobrevibacter* (⊙⊙ Sec-tion 13.4), but their exact relationship to free-living methanogens is unclear. In the termite hindgut, endosymbiotic methanogens are thought to benefit their protozoan hosts by consuming H_2 gen-

(a)

(b)   *(c)*

● **Figure 19.25 Termites and their carbon metabolism.** (a) A common eastern (USA) subterranean termite worker larva shown beneath a hindgut extracted from a separate worker. The animal is about 0.5 cm long. Acetogenesis is the major form of anoxic carbon metabolism in these termites although methanogenesis also occurs. (b, c) Microorganisms from the hindgut of the termite *Zootermopsis angusticolis*. A single microscope field was photographed by two different methods. (b) Phase contrast. (c) Epifluorescence, showing color typical of methanogens due to the high content of the fluorescent coenzyme F_{420} (⊙⊙ Section 17.17 and Figures 17.42 and 17.43). The methanogens are inside cells of the protozoan *Tricercomitis* sp. Plant particles fluoresce yellow. The average diameter of a protozoan cell is $15-20$ μm. See Section 4.1 for an explanation of florescence microscopy.

Table 19.5	Estimates of CH_4 released into the atmosphere[a]	
Source	**CH_4 emission (10^{12} g/year)**	
Biogenic		
Ruminants	80–100	
Termites	25–150[b]	
Paddy fields	70–120	
Natural wetlands	120–200	
Landfills	5–70	
Oceans and lakes	1–20	
Tundra	1–5	
Abiogenic		
Coal mining	10–35	
Natural gas flaring and venting	10–30	
Industrial and pipeline losses	15–45	
Biomass burning	10–40	
Methane hydrates	2–4	
Volcanoes	0.5	
Automobiles	0.5	
Total	350–820	
Total biogenic	302–665	81–86% of total
Total abiogenic	48–155	13–19% of total

[a] Data adapted from estimates in Tyler, S. C. 1991. The global methane budget, pp. 7–58, in E. J. Rogers and W. B. Whitman (eds.), *Microbial Production and Consumption of Greenhouse Gases: Methane, Nitrogen Oxides, and Halomethanes*, American Society for Microbiology, Washington, DC.

[b] More recent estimates indicate that the lower value is probably the more accurate.

erated from glucose fermentation by cellulolytic protozoans. As was shown in Table 19.5, termites can be a major source of biogenic CH_4.

Acetogenic Habitats

A competing process to methanogenesis is **acetogenesis** (⚭ Section 17.16). In some habitats, for example the rumen (see Section 19.11), homoacetogens appear to be poor competitors with methanogens and thus methanogenesis is the dominant H_2-consuming process. But in other habitats, such as the termite hindgut, where some methanogenesis occurs (Table 19.5 and Figure 19.25), acetogenesis is a much more important process.

What governs the dominant form of anaerobic metabolism in the termite hindgut? Based on simple energetic considerations, methanogenesis from H_2 is more favorable than acetogenesis (-131 kJ versus -105 kJ, respectively) and thus methanogens should have a competitive advantage. However, in termites this is not the case. There are three reasons why this is thought to occur, each supported by experimental evidence. First, homoacetogens are able to position themselves in the termite gut nearer to the source of H_2 than methanogens and thus consume the majority of H_2 produced from cellulose fermentation. Second, unlike methanogens, homoacetogens can ferment glucose (obtained from cellulose). And third, termites eat wood, which contains a high content of lignin mixed in with cellulose, and this type of material may be degraded in such a way that favors acetogenesis. Thus, the combination of factors in the special habitat of the termite hindgut likely contributes to the dominance of acetogenesis over methanogenesis, an atypical situation for anoxic environments.

Methanogenesis versus Sulfidogenesis

Methanogenesis and acetogenesis are more extensive in anoxic freshwater and terrestrial environments than in marine sediments. This is because sulfate-reducing bacteria are abundant in marine sediments and out compete methanogens and homoacetogens for H_2. The biochemical basis for this lies in the fact that the energetics of sulfate reduction with H_2 is better than the reduction of CO_2 with H_2 to either CH_4 or acetate (⚭ Sections 17.15–17.17); for certain biochemical reasons, this favors sulfate reduction. In freshwater, however, where sulfidogenesis is typically very low because sulfate is limiting, methanogenesis and acetogenesis dominate.

 19.10 Concept Check

Under anoxic conditions, organic matter is degraded principally to CH_4 and CO_2. Much CH_4 is formed from the reduction of CO_2 by H_2 supplied by H_2-producing syntrophic bacteria; these organisms depend on H_2 consumption to balance their energetics. On a global basis, biogenic CH_4 is a much larger source than abiogenic CH_4.

◆ What kinds of organisms can grow in coculture with *Syntrophomonas*?

◆ What is the final product of acetogenesis? What anoxic habitat shows greater acetogenesis than methanogenesis?

◆ Why is methanogenesis from H_2 not a major process in marine sediments?

◆ How is CH_4 oxidized in the absence of O_2?

19.11 Carbon Cycling in Ruminant Animals

Ruminants are herbivorous mammals that possess a special digestive organ, the **rumen**. Within the rumen, the digestion of cellulose and other plant polysaccharides occurs through the activity of microorganisms. Some of the most important domestic animals—cows, sheep, and goats—are ruminants. Because the human food economy depends to a great extent on these animals, rumen microbiology is of considerable economic significance.

Rumen Anatomy and Action

The bulk of the organic matter in terrestrial plants is present in insoluble polysaccharides, of which *cellulose* is the most important. Mammals—and indeed almost all animals—lack the enzymes necessary to digest cellulose. But ruminants can metabolize cellulose by using microorganisms as digestive agents. Unique features of the rumen as a site of cellulose digestion are its relatively large size (100–150 liters in a cow, 6 liters in a sheep) and its position in the alimentary tract *before* the acidic stomach. The warm and constant temperature (39° C), constant pH (6.5), and anoxic nature of the rumen are also important factors in overall rumen function.

Figure 19.26a● shows the relationship of the rumen to other parts of the ruminant digestive system. The digestive processes and microbiology of the rumen are well understood, in part because it is possible to create a sampling port, called a *fistula*, into the rumen of a cow (Figure 19.26b) or a sheep and remove samples for analysis. Food first enters the reticulum and is quickly pumped into the rumen, where it is mixed with saliva containing bicarbonate and churned in a rotary motion. This peristaltic action grinds the cellulose into a fine suspension that assists in microbial attachment and fermentation. The food mass then migrates into the reticulum, where it is formed into small clumps called *cuds*; these are regurgitated into the mouth and chewed again. The now finely divided solids, well mixed with saliva, are swallowed again, but this time the material passes mainly to the *omasum*, and from there to the *abomasum*, an organ more like a true (acidic) stomach. In the abomasum, chemical digestive processes begin that continue in the small and large intestine.

(a)

(b)

Sharisa D. Beek, Dept. Animal Science, Southern Illinois Univ.

● **Figure 19.26** **The rumen.** (a) Schematic diagram of the rumen and gastrointestinal system of a cow. Food travels from the esophagus to the rumen and is then regurgitated and travels to the reticulum, omasum, abomasum, and intestines, in that order. The abomasum is an acidic vessel, analogous to the stomach of monogastric animals like pigs and humans. (b) Photo of a fistulated Holstein cow. The fistula, shown unplugged, is a sampling port that allows access to the rumen. Fistulated cows and sheep have been very useful for the study of both rumen microbiology and ruminant nutrition.

● **Figure 19.27** **Biochemical reactions in the rumen.** The major starting substrate, glucose (from cellulose), and end products are highlighted; dashed lines indicate minor pathways. Approximate steady-state rumen levels of volatile fatty acids (VFAs) are acetate, 60 mM; propionate, 20 mM; butyrate, 10 mM. VFAs are consumed by the ruminant and converted into animal proteins. For this reason, syntrophic fatty acid-oxidizing bacteria (see Section 19.10) are not needed in the rumen microbial ecosystem.

Microbial Fermentation in the Rumen

Food remains in the rumen about 9–12 hours. During this period cellulolytic microorganisms hydrolyze cellulose to the disaccharide cellobiose and to free glucose. The released glucose then undergoes a bacterial fermentation with the production of **volatile fatty acids** (VFAs), primarily *acetic, propionic,* and *butyric,* and the gases *carbon dioxide* and *methane* (Figure 19.27●). The fatty acids pass through the rumen wall into the bloodstream and are oxidized by the animal as its main source of energy.

The rumen contains enormous numbers of prokaryotes (10^{10}–10^{11} bacteria/g rumen contents). In addition to their digestive functions, rumen microorganisms synthesize amino acids and vitamins that are the main source of these essential nutrients for the animal. Most of the bacteria are adhered tightly to plant materials and feed particles. These materials proceed through the gastrointestinal tract of the animal where they undergo further digestive processes similar to those of nonruminants. Many micro-

bial cells from the rumen are digested in the abomasum and thus are a major source of protein and vitamins for the animal. Because this microbial protein can be recovered and used by the animal, a ruminant is thus nutritionally superior to a nonruminant when subsisting on foods that are deficient in protein, such as grasses.

Rumen Bacteria

The biochemical reactions occurring in the rumen are complex and involve the combined activities of a variety of microorganisms. Because the rumen is anoxic, anaerobic bacteria naturally dominate. Furthermore, because the conversion of cellulose to CO_2 and CH_4 involves a multistep microbial food chain, a variety of anaerobes can be expected (Table 19.6).

Several different rumen bacteria hydrolyze polymers such as cellulose to sugars and ferment the sugars

Table 19.6 Characteristics of some rumen prokaryotes

Organism	Gram stain	Phylogenetic domain[a]	Morphology	Motility	Fermentation products	DNA (mol % GC)
Cellulose decomposers						
Fibrobacter succinogenes[b]	Negative	B	Rod	−	Succinate, acetate, formate	45–51
Butyrivibrio fibrisolvens[c]	Negative	B	Curved rod	+	Acetate, formate, lactate, butyrate, H_2, CO_2	41
Ruminococcus albus[b]	Positive	B	Coccus	−	Acetate, formate, H_2, CO_2	43–46
Clostridium lochheadii	Positive	B	Rod (endospores)	+	Acetate, formate, butyrate, H_2, CO_2	—
Starch decomposers						
Prevotella ruminicola	Negative	B	Rod	−	Formate, acetate, succinate	40–42
Ruminobacter amylophilus	Negative	B	Rod	−	Formate, acetate, succinate	49
Selenomonas ruminantium	Negative	B	Curved rod	+	Acetate, propionate, lactate	49
Succinomonas amylolytica	Negative	B	Oval	+	Acetate, propionate, succinate	—
Streptococcus bovis	Positive	B	Coccus	−	Lactate	37–39
Lactate decomposers						
Selenomonas lactilytica	Negative	B	Curved rod	+	Acetate, succinate	50
Megasphaera elsdenii	Positive	B	Coccus	−	Acetate, propionate, butyrate, valerate, caproate, H_2, CO_2	54
Succinate decomposer						
Schwartzia succinovorans	Negative	B	Rod	+	Propionate, CO_2	46
Pectin decomposer						
Lachnospira multiparus	Positive	B	Curved rod	+	Acetate, formate, lactate, H_2, CO_2	—
Methanogens						
Methanobrevibacter ruminantium	Positive	A	Rod	−	CH_4 (from H_2 + CO_2 or formate)	31
Methanomicrobium mobile	Negative	A	Rod	+	CH_4 (from H_2 + CO_2 or formate)	49

[a] B, *Bacteria*; A, *Archaea*

[b] These species also degrade xylan, a major plant cell wall polysaccharide (Section 17.25).

[c] Also degrades starch

to volatile fatty acids. *Fibrobacter succinogenes* and *Ruminococcus albus* are the two most abundant cellulolytic rumen anaerobes. Although both organisms produce cellulases, *Fibrobacter*, a gram-negative bacterium, contains a periplasmic cellulase (Section 4.9) to break down cellulose. Because of this, cells of *Fibrobacter* must remain attached to the cellulose fibril while digesting it. *Ruminococcus*, on the other hand, produces a cellulase that is excreted into the rumen contents (thus, it is an *exoenzyme*) where it degrades cellulose outside the bacterial cell proper. However, the end result is the same in both cases: Free glucose is made available for fermentative anaerobes.

If a ruminant is gradually switched from cellulose to a diet high in starch (grain, for instance), then starch-digesting bacteria such as *Ruminobacter amylophilus* and *Succinomonas amylolytica* develop to high numbers in the rumen. On a low-starch diet these organisms are minor components. If an animal is fed legume hay, which is high in the polysaccharide pectin, then the pectin-digesting bacterium *Lachnospira multiparus* (Table 19.6) is a common member of the rumen flora.

Occasionally, changes in the microbial composition of the rumen cause illness or even death of the an-

imal. For example, if a cow is changed abruptly from forage to a completely grain diet, an explosive growth of *Streptococcus bovis* is observed in the rumen. The normal level of *S. bovis*, about 10^7 cells/g (insignificant in terms of total rumen bacterial numbers), quickly expands to over 10^{10} cells/g. This occurs because *S. bovis* grows rapidly on starch, and grain contains high levels of starch, whereas grasses contain mainly cellulose. Being a lactic acid bacterium (Section 12.19), *S. bovis* produces large amounts of lactate from the fermentation of starch. This acidifies the rumen (a condition called *acidosis*), killing off the normal rumen microflora. Severe acidosis can cause death of the animal. To avoid acidosis, animals are switched from forage rations to grain *gradually* over a period of a few days. A slow introduction of starch selects for volatile fatty acid-producing starch degraders (Table 19.6) instead of *S. bovis*, and thus normal rumen biochemical processes are not disrupted.

Some of the fermentation products of the saccharolytic rumen microflora are used as energy sources by secondary fermenters in the rumen. Thus, *succinate* is fermented to *propionate* and CO_2 (Figure 19.27) by *Schwartzia*, and *lactate* is fermented to *acetic* and other acids by *Selenomonas* and

Megasphaera (Table 19.6). *Hydrogen* produced in the rumen by fermentative processes never accumulates because it is quickly consumed by methanogens. Despite the high VFA content of the rumen, syntrophs (∞ Section 19.10) do not play a major role there because the animal itself is the major sink for fatty acids (Figure 19.27). That is, with propionate and butyrate being consumed by the animal (Figure 19.27), syntrophic conversion processes (Table 19.4 and Figure 19.23) are unnecessary in the rumen.

Rumen Protozoa and Fungi

In addition to prokaryotes, the rumen has a characteristic protozoal fauna (about 10^6/ml), composed almost exclusively of ciliates (∞ Section 14.10). Many of these protozoa are obligate anaerobes, a property that is rare among eukaryotes. Although protozoa are not essential for the rumen fermentation, they contribute to the overall process. In fact, some protozoa are able to hydrolyze cellulose and starch and ferment glucose with the production of the same organic acids formed by the bacteria (Table 19.6). Rumen protozoa also ingest rumen bacteria as food sources and are thought to play a role in controlling bacterial densities in the rumen.

Anaerobic fungi also inhabit the rumen and are known to play a role in ruminal digestive processes. Rumen fungi are generally species that alternate between a flagellated and a thallus form, and studies with pure cultures show that they can ferment cellulose to VFAs. *Neocalimastix*, for example, is an obligately anaerobic fungus that ferments glucose to formate, acetate, lactate, ethanol, CO_2 and H_2. Although a eukaryote, this fungus lacks mitochondria and cytochromes and thus lives an obligately fermentative existence. However, *Neocallimastix* cells contain a redox organelle called the *hydrogenosome* that functions to evolve H_2 and has thus far only been found in certain phylogenetically "early branching" *Eukarya* (∞ Sections 14.2, 14.4, and 14.9). Rumen fungi play an important role in the degradation of polysaccharides other than cellulose as well, including a partial degradation of lignin (the strengthening agent in the cell walls of woody plants), hemicellulose, and pectins.

Ruminant Feeding Practices and Foodborne Illness

In recent years, safety concerns have arisen over feeding practices with ruminant animals, especially cattle, and their impact on the animal's microflora. Humans have drastically altered the diet of domesticated ruminant animals from one of forage, the diet the rumen evolved to handle, to more starch-based grain diets. Grain promotes rapid weight gain and tends to make meat from the animal more tender. Thus, grain-fed animals get to market sooner, and their meat has more value. However, a high-starch diet not only affects the *rumen* microflora, as we have seen, but it also affects the microbial composition of the intestinal tract *downstream* of the rumen (Figure 19.26*a*). In particular, the pH of the intestinal tract is more

acidic in grain-fed animals. This condition allows acid-tolerant enteropathogenic strains of *Escherichia coli*, like *E. coli* strain O157:H7 (∞ Section 29.8), to predominate over less acid-tolerant nonpathogenic strains. When grain-fed animals are slaughtered, these *E. coli* cells can adhere to the carcass and be introduced into meat products, particularly ground meat products like ground beef and beef sausage, thus increasing the risk of foodborne illness (∞ Section 29.8).

Other Herbivorous Animals: Cecal Animals

The familiar ruminants are cows and sheep. Goats, camels, buffalo, deer, reindeer, caribou, and elk are also ruminants. However, although horses and rabbits are herbivorous mammals, they are not ruminants. Instead, these animals have only one stomach but use an organ called the **cecum**, a digestive organ located posterior to the small intestine and anterior to the large intestine, as their cellulolytic fermentation vessel. The cecum contains cellulolytic microorganisms and digestion of cellulose occurs there. Nutritionally, ruminants have an advantage over cecal animals in that the cellulolytic microflora of the ruminant eventually passes through a true (acidic) stomach and as such is killed and is a protein source for the animal. By contrast, in horses and rabbits the cellulolytic microflora is passed out of the animal in the feces.

 19.11 Concept Check

Ruminants are animals that have a special digestive organ, the rumen, that is a unique ecosystem in which anaerobic microorganisms digest insoluble feed components such as cellulose and starch. Bacteria, protozoa, and fungi of the rumen produce volatile fatty acids that are used by the ruminant. In addition to their role in the digestive process, rumen microorganisms synthesize vitamins and amino acids and are also a major source of protein for the ruminant.

◆ What physical and chemical conditions prevail in the rumen?

◆ What are *VFAs* and of what value are they to the ruminant?

◆ Why is the metabolism of *Streptococcus bovis* of special concern to ruminant nutrition? How can foodborne illness in humans be due to ruminant feeding practices?

◆ How do cecal animals differ from ruminants in their digestive tract anatomy?

V OTHER KEY NUTRIENT CYCLES

Although on a quantitative basis carbon is the major cycled element in nature, many other elements important in the metabolism of plants and animals are cycled by microorganisms. These include *nitrogen*, *sulfur*, and *iron*, and we examine the cycling of these key nutrients now.

19.12 The Nitrogen Cycle

The element nitrogen, N, a key constituent of protoplasm, exists in a number of oxidation states (∽ Table 17.2). We have discussed three major processes of microbial nitrogen transformation: *nitrification* (∽ Section 17.12), *denitrification* (∽ Section 17.14), and *nitrogen fixation* (∽ Section 17.28). These and several other nitrogen transformations are summarized in the redox cycle shown in Figure 19.28●.

Nitrogen Fixation

Several of the key redox reactions of nitrogen are carried out in nature almost exclusively by prokaryotes (Figure 19.28). *Nitrogen gas*, N_2, is the most stable form of nitrogen and a major reservoir for nitrogen on Earth. However, only a relatively small number of organisms are able to use N_2 as a nitrogen source in the process of **nitrogen fixation** $(N_2 + 8 H \rightarrow 2 NH_3 + H_2)$ (∽ Section 17.28). The recycling of nitrogen on Earth involves to a great extent the more easily available fixed forms of nitrogen, such as ammonia and nitrate. In many environments, however, the short supply of such compounds puts a premium on biological nitrogen fixation. We revisit this important process later in this chapter when we describe symbiotic N_2 fixation in leguminous plants (see Section 19.22).

Denitrification

We discussed the role of *nitrate* as an alternative electron acceptor in anaerobic respiration in Section 17.14. Under most conditions, the end product of nitrate reduction is N_2 or N_2O. The conversion of nitrate to gaseous nitrogen compounds, called **denitrification** (Figure 19.28), is the main means by which gaseous N_2 is formed biologically. On the other hand, denitrification is a detrimental process. For example, fields fertilized with nitrate fertilizer typically become waterlogged following heavy spring rains. Anoxic conditions then rapidly develop, and denitrification can be extensive. This removes fixed nitrogen from the soil. On the other hand, denitrification can aid in wastewater treatment (∽ Section 28.2). By removing nitrate, demitrification minimizes algal growth when the water is discharged into lakes and streams.

Ammonia Fluxes, Nitrification, and Anammox

Ammonia is produced during the decomposition of organic nitrogen compounds (**ammonification**, Figure 19.28) such as amino acids and nucleotides. At neutral pH, ammonia exists as the ammonium ion (NH_4^+). Under anoxic conditions, ammonia is relatively stable (but see Section 17.12), and it is in this form that nitrogen predominates in most anoxic sediments. In soils, much of the ammonia released by aerobic decomposition is rapidly recycled and converted to amino acids in plants and microorganisms. Because ammonia is volatile, some loss can occur from soils (especially highly alkaline soils) by vaporization, and major losses of ammonia to the atmosphere occur in areas of dense animal populations (for example, cattle feedlots). On a global basis, however, ammonia constitutes only about 15% of the nitrogen released to

Key Processes and Prokaryotes in the Nitrogen Cycle	
Processes	**Example organisms**
Nitrification ($NH_4^+ \rightarrow NO_3^-$)	
$NH_4^+ \rightarrow NO_2^-$	*Nitrosomonas*
$NO_2^- \rightarrow NO_3^-$	*Nitrobacter*
Denitrification ($NO_3^- \rightarrow N_2$)	*Bacillus, Paracoccus, Pseudomonas*
N_2 Fixation ($N_2 + 8H \rightarrow NH_3 + H_2$)	
Free-living	
Aerobic	*Azotobacter* Cyanobacteria
Anaerobic	*Clostridium*, purple and green bacteria
Symbiotic	*Rhizobium Bradyrhizobium Frankia*
Ammonification (organic-N $\rightarrow NH_4^+$)	
	Many organisms can do this
Anammox ($NO_2^- + NH_3 \rightarrow 2N_2$)	*Brocadia*

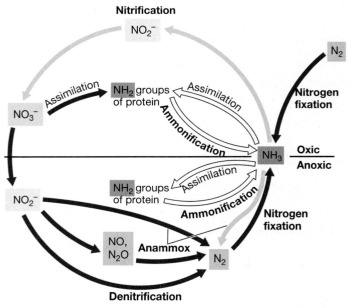

● **Figure 19.28 Redox cycle for nitrogen.** Oxidation reactions are represented by yellow arrows and reductions in red. Reactions in which no redox changes occur are in white. For a more complete list of nitrogen-fixing prokaryotes, ∽Table 17.10.

the atmosphere, the rest being primarily N_2 or N_2O (from denitrification).

Nitrification, the oxidation of NH_3 to NO_3^-, occurs readily in well-drained soils at neutral pH by the activities of the nitrifying bacteria (⚬ Sections 12.3 and 17.12) (Figure 19.28). While denitrification *consumes* nitrate, nitrification *produces* nitrate. If materials high in protein, such as manure or sewage, are added to soils, the rate of nitrification increases. Although nitrate is readily assimilated by plants, it is very water-soluble and is rapidly leached from soils receiving high rainfall. Consequently, nitrification is not beneficial to plant agriculture. Ammonia, on the other hand, is positively charged and consequently is strongly adsorbed to negatively charged clay-rich soils.

Anhydrous ammonia is used extensively as a nitrogen fertilizer, and chemicals are commonly added to the fertilizer to inhibit the nitrification process. One of the most common inhibitors of nitrification is a substituted pyridine compound called *nitrapyrin* (2-chloro-6-trichloromethylpyridine, also known as N-SERVE). Nitrapyrin specifically inhibits the first step in nitrification, the oxidation of NH_3 to NO_2^- (⚬ Section 17.12). This effectively inhibits both steps in the nitrification process since the second step, $NO_2^- \rightarrow NO_3^-$, depends on the first. The addition of nitrification inhibitors has greatly increased the efficiency of fertilization and has helped prevent pollution of waterways from nitrate leached from fertilized soils.

Ammonia can be catabolized anaerobically by *Brocadia* and related organisms in the process called **anammox**. In this reaction, ammonia is oxidized with ni-trite (NO_2^-) as electron acceptor, forming N_2 as the final product (Figure 19.28). The microbiology and biochemistry of anammox was discussed in Section 17.12.

 19.12 Concept Check

The principal form of nitrogen on Earth is nitrogen gas (N_2), which can be used as a nitrogen source only by the nitrogen-fixing bacteria. Ammonia produced by nitrogen fixation or by ammonification from organic nitrogen compounds can be assimilated into organic matter or it can be oxidized to nitrate by the nitrifying bacteria. Losses of nitrogen from the biosphere occur as a result of denitrification, in which nitrate is converted back to N_2.

◆ What is *nitrogen fixation* and why is it important?

◆ What is the process called that results in $NO_3^- \rightarrow N_2$?

◆ How does the compound *nitrapyrin* benefit both agriculture and the environment?

◆ How do the processes of *nitrification* and *denitrification* differ? How do *nitrification* and *anammox* differ?

19.13 The Sulfur Cycle

Sulfur transformations are even more complex than those of nitrogen because of the variety of oxidation states of sulfur (Figure 19.29●; ⚬ Table 17.3). In addition, unlike nitrogen, some transformations of sulfur occur *chemically* as well as biologically. Sulfate reduction and chemolithotrophic sulfur oxidation were covered in Sections 17.15

Key Processes and Prokaryotes in the Sulfur Cycle	
Process	**Organisms**
Sulfide/sulfur oxidation ($H_2S \rightarrow S^0 \rightarrow SO_4^{2-}$)	
Aerobic	Sulfur chemolithotrophs (*Thiobacillus, Beggiatoa*, many others)
Anaerobic	Purple and green phototrophic bacteria, some chemolithotrophs
Sulfate reduction (anaerobic) ($SO_4^{2-} \rightarrow H_2S$)	*Desulfovibrio, Desulfobacter*
Sulfur reduction (anaerobic) ($S^0 \rightarrow H_2S$)	*Desulfuromonas*, many hyperthermophilic *Archaea*
Sulfur disproportionation ($S_2O_3^{2-} \rightarrow H_2S + SO_4^{2-}$)	*Desulfovibrio*, and others
Organic sulfur compound oxidation or reduction ($CH_3SH \rightarrow CO_2 + H_2S$) ($DMSO \rightarrow DMS$)	Many organisms can do this
Desulfurylation (organic–S $\rightarrow H_2S$)	Many organisms can do this

● **Figure 19.29 Redox cycle for sulfur.** Oxidations are shown in yellow arrows and reductions in red. Reactions in which no redox changes occurs, are in white. DMSO, dimethylsulfoxide; DMS, dimethylsulfide.

and 17.10, respectively. The redox cycle for sulfur and the involvement of microorganisms in sulfur transformations are given in Figure 19.29. Although a number of oxidation states of sulfur are possible, only three are significant in nature, -2 (sulfhydryl, R—SH, and sulfide, HS$^-$), 0 (elemental sulfur, S^0), and $+6$ (sulfate, SO$_4^{2-}$). The bulk of the sulfur on Earth is found in sediments and rocks in the form of sulfate minerals (primarily gypsum, CaSO$_4$) and sulfide minerals (primarily pyrite, FeS$_2$), although the oceans constitute the most significant reservoir of sulfur (as sulfate) for the biosphere.

Hydrogen Sulfide and Sulfate Reduction

A major volatile sulfur gas is hydrogen sulfide (H$_2$S). Sulfide is produced from bacterial sulfate reduction (SO$_4^{2-}$ + 8 H$^+$ → H$_2$S + 2 H$_2$O + 2 OH$^-$) (Figure 19.29) or is emitted from geochemical sources in sulfide springs and volcanoes. Although H$_2$S is volatile, the form of sulfide present in an environment is pH dependent: H$_2$S predominates below pH 7 while HS$^-$ and S^{2-} predominate above pH 7.

Sulfate-reducing bacteria are a highly diverse group (∞ Section 12.18) and are widespread in nature. However, in many anoxic habitats, such as freshwaters and many soils, their activities are limited by the low levels of sulfate present. Moreover, because of the necessity for organic electron donors (or molecular hydrogen, which is a product of the fermentation of organic compounds) to drive sulfate reduction, the latter only occurs where significant amounts of organic material are present. In most marine sediments, the rate of sulfate reduction is carbon-limited and can be greatly increased by the addition of organic matter. This is important because disposal of sewage, sewage sludge, and garbage in the sea can lead to marked increases in organic matter in the sediments, triggering marine sulfide pollution.

Since sulfide (HS$^-$) is a toxic substance to many organisms, formation of HS$^-$ by sulfate reduction is potentially detrimental. Sulfide is toxic because it combines with the iron of cytochromes and other essential iron-containing compounds in the cell. Sulfide is commonly detoxified in the environment by combination with iron, forming the insoluble FeS and FeS$_2$ (pyrite), and accounts for the black color of many sediments.

Sulfide and Elemental Sulfur Oxidation/Reduction

Under oxic conditions, sulfide (HS$^-$) rapidly oxidizes spontaneously at neutral pH (∞ Section 17.10). Sulfur-oxidizing bacteria, most of which are aerobes, can catalyze the oxidation of sulfide. Because of the rapid spontaneous reaction, however, significant bacterial oxidation of sulfide occurs only in areas in which H$_2$S emerging from anoxic areas meets O$_2$ from oxic areas. In addition, if light is available, anoxic oxidation of HS$^-$ can also occur, catalyzed by the phototrophic sulfur bacteria (∞ Sections 12.2, 12.32, and 17.4).

Elemental sulfur, S^0, is chemically stable but is readily oxidized by sulfur-oxidizing bacteria, such as *Thiobacillus* and *Acidithiobacillus* (∞ Sections 12.4 and 17.10). Elemental sulfur is insoluble, and the bacteria that oxidize it attach firmly to the sulfur crystals (∞ Figure 17.26b). Oxidation of elemental sulfur results in the formation of sulfate (SO$_4^{2-}$) and protons, and thus sulfur oxidation characteristically results in a *lowering* of the pH. Elemental sulfur is sometimes added to alkaline soils to effect a lowering of the pH, reliance being placed on the ubiquitous thiobacilli to carry out the acidification process.

S^0 can be reduced as well as oxidized. Sulfur reduction to sulfide (a form of anaerobic respiration, ∞ Section 17.15) is a major ecological process, especially among hyperthermophilic *Archaea* (∞ Chapter 13). Although sulfate-reducing bacteria can also carry out this reaction, the bulk of S^0 reduction in nature probably occurs by the phylogenetically distinct S^0 reducers that are unable to reduce SO$_4^{2-}$ to H$_2$S. However, the habitats of the S^0 reducers are generally those of the sulfate reducers, so from an ecological standpoint, the two groups form a guild (see Section 19.1) and coexist.

Organic Sulfur Compounds

In addition to the *inorganic* forms of sulfur whose biogeochemistry was just discussed, a vast array of *organic* sulfur compounds are also synthesized by living organisms, and these enter into biogeochemical sulfur cycling as well. Many of these foul-smelling compounds are highly volatile and can thus enter the atmosphere. The most abundant organic sulfur compound in nature is dimethyl sulfide (H$_3$C—S—CH$_3$). It is produced primarily in marine environments as a degradation product of dimethylsulfoniopropionate, a major osmoregulatory solute in marine algae (∞ Section 6.14). Dimethylsulfoniopropionate can be used as a carbon and energy source by microorganisms and is catabolized to dimethyl sulfide and acrylate. The latter, a derivative of the fatty acid propionate, is used to support growth.

Dimethyl sulfide released to the atmosphere undergoes photochemical oxidation to methane sulfonate (CH$_3$SO$_3^-$), SO$_2$, and SO$_4^{2-}$. By contrast, dimethyl sulfide produced in anoxic habitats can be transformed microbially in at least three ways: (1) in methanogenesis (yielding CH$_4$ and H$_2$S), (2) as an electron donor for photosynthetic CO$_2$ fixation in phototrophic purple bacteria [yielding dimethyl sulfoxide (DMSO)], and (3) as an electron donor in energy metabolism in certain chemoorganotrophs and chemolithotrophs (also yielding DMSO). The DMSO produced can be an electron acceptor for anaerobic respiration (∞ Section 17.18), once again yielding dimethyl sulfide. Many other organic sulfur compounds affect the global sulfur cycle, including methanethiol (CH$_3$SH), dimethyl disulfide (H$_3$C—S—S—CH$_3$), and carbon disulfide (CS$_2$), but on a global basis, *dimethyl sulfide* is the most significant.

19.13 Concept Check

Bacteria play major roles in both the oxidative and reductive sides of the sulfur cycle. Sulfur- and sulfide-oxidizing bacteria *produce* sulfate, while sulfate-reducing bacteria *consume* sulfate as an electron acceptor in anaerobic respiration, producing hydrogen sulfide. Because sulfide is toxic and also reacts with various metals, sulfate reduction is an important biogeochemical process. Dimethyl sulfide is the major organic sulfur compound of ecological significance in nature.

◆ Is H_2S a *substrate* or a *product* of the sulfate-reducing bacteria? Chemolithotrophic bacteria?

◆ Why does the bacterial oxidation of sulfur result in a pH drop?

◆ What organic sulfur compound is most abundant in nature?

19.14 The Iron Cycle

Iron is one of the most abundant elements in Earth's crust. On the surface of the Earth, iron exists naturally in two oxidation states, ferrous (Fe^{2+}) and ferric (Fe^{3+}). Fe^0 is a major product only of human activities in the smelting of iron ores to form cast iron. In nature then, the element iron cycles primarily between the ferrous and ferric forms. The reduction of Fe^{3+} occurs both chemically and as a form of anaerobic respiration, and the oxidation of Fe^{2+} occurs both chemically and as a form of chemolithotrophic metabolism (Figure 19.30●).

Bacterial Iron Reduction

A number of organisms can use ferric iron as an electron acceptor (⬡⬡ Section 17.18). Ferric iron reduction is common in waterlogged soils, bogs, and anoxic lake sediments. Movement of iron-rich groundwater from anoxic bogs or waterlogged soils can result in the transport of large amounts of ferrous iron. When this iron-laden water reaches oxic regions, the ferrous iron is oxidized chemically or by iron bacteria (⬡⬡ Section 17.11). Ferric compounds then precipitate, leading to the formation of brown iron deposits (⬡⬡ Figure 17.28c):

$$Fe^{2+} + \tfrac{1}{4}O_2 + 2\tfrac{1}{2}H_2O \longrightarrow Fe(OH)_3 + 2\,H^+$$

The ferric hydroxide precipitate can interact with other nonbiological substances, such as humics (see Section 19.9), to reduce Fe^{3+} back to Fe^{2+} (Figure 19.30). Ferric iron can also form complexes with various organic constituents. In this way, it becomes solubilized and once again available to ferric iron-reducing bacteria as an electron acceptor.

Ferrous Iron and Pyrite Oxidation at Acid pH

The only electron acceptor able to spontaneously oxidize Fe^{2+} is O_2. In neutral habitats Fe^{2+} can be oxidized by iron bacteria such as *Gallionella* and *Leptothrix* (⬡⬡ Sections 12.15 and 12.16). This occurs primarily at interfaces between ferrous-rich anoxic ground waters and air. However, the most extensive bacterial iron oxidation occurs at acidic pH, where Fe^{2+} is stable to spontaneous oxidation. In extremely acidic habitats, the acidophilic chemolithotroph *Acidithiobacillus ferrooxidans* and related acidophilic iron oxidizers oxidize Fe^{2+} to Fe^{3+} (Figure 19.31●). Very little energy is generated in the oxidation of ferrous to ferric iron (⬡⬡ Section 17.11), and so these bacteria must oxidize large amounts of iron in order to grow; consequently, even a low number of cells can be responsible for precipitating a large amount of iron. *A. ferrooxidans* is a strict acidophile and is very common in acid mine drainages and in acid springs; it is probably responsible for most of the ferric iron precipitated under moderately acidic (pH 2–4) conditions.

Acidithiobacillus ferrooxidans and *Leptospirillum ferrooxidans* live in environments in which sulfuric acid is the dominant acid and large amounts of sulfate are present. At 20–30° C and moderately acidic pH, *A. ferrooxidans* seems to dominate, while at 30–50° C and more acidic pH (1–2), *L. ferrooxidans* is the dominant organism. Under these conditions, ferric iron does not precipitate as the hydroxide but as a complex sulfate

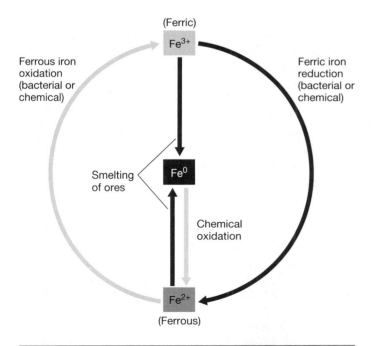

● **Figure 19.30   The redox cycle of iron.** The major forms of iron in nature are Fe^{2+} and Fe^{3+}; Fe^0 is primarily a product of human activities in the smelting of iron ores. Ferrous iron oxidation occurs aerobically by the iron chemolithotrophs (or chemically at neutral pH) and anaerobically by certain anoxygenic phototrophic bacteria and denitrifying bacteria. Oxidations are shown with yellow arrows and reductions with red.

(a)

(b)

Ricardo Amils

● **Figure 19.31 Oxidation of Fe²⁺.** (a) Oxidation of ferrous iron as a function of pH and the presence of the bacterium *Acidithiobacillus ferrooxidans*. Note how Fe^{2+} is stable under acidic conditions in the absence of the bacterial cells. (b) A microbial mat containing acidophilic green algae and various iron-oxidizing prokaryotes in the Rio Tinto, Spain. The river is highly acidic and contains high levels of dissolved metals, in particular Fe^{2+}. The red-brown precipitates contain Fe^{3+}.

(a)

Ravin Donald

(b)

T. D. Brock

● **Figure 19.32 Pyrite and coal.** (a) Pyrite in coal that can be oxidized by sulfur- and iron-oxidizing bacteria. Section through a piece of coal from the Black Mesa formation in northern Arizona (USA). The gold-colored spherical discs (about 1 mm in diameter) are sectioned particles of pyrite, FeS_2. (b) A coal seam in a surface coal mining operation. Exposing the coal to oxygen and moisture stimulates the activities of iron-oxidizing bacteria growing on the pyrite in the coal.

mineral called *jarosite* $[HFe_3(SO_4)_2(OH)_6]$. Jarosite is a yellowish or brownish precipitate and is one of the major pollutants in acid mine drainage, an unsightly yellow stain called "yellow boy" by U.S. miners (◐◐ Figure 17.28*a* and see also Figure 19.34). *A. ferrooxidans* and *L. ferrooxidans* are phylogenetically and morphologically distinct, but are also distinct in their metabolism. While *L. ferrooxidans* can grow only on Fe^{2+}, *A. ferrooxidans* grows chemolithotrophically on either Fe^{2+} or S^0.

One of the most common forms of iron in nature is **pyrite** (FeS_2). Pyrite is formed from the reaction of sulfur with ferrous sulfide (FeS) to form an insoluble crystalline structure and is very common in bituminous coals and in many ore bodies (Figure 19.32*a*●). The bacterial oxidation of pyrite is of great significance in the develop-

ment of acidic conditions in coal mining operations (Figure 19.32*b*). Additionally, oxidation of pyrite by bacteria is of considerable importance in the *microbial leaching of ores* (see Section 19.15). The oxidation of pyrite is a combination of chemically and bacterially catalyzed reactions. Two electron acceptors are involved in this process: molecular oxygen (O_2) and ferric ions (Fe^{3+}).

When pyrite is first exposed, as in a mining operation (Figure 19.32*b*), a slow chemical reaction with molecular oxygen occurs, as shown in Figure 19.33●. This reaction, called the *initiator reaction*, leads to the oxidation of sulfide to sulfate and the development of acidic conditions under which the ferrous iron released is

$$FeS_2 \text{ (pyrite)} + 3\tfrac{1}{2} O_2 + H_2O \rightarrow Fe^{2+} + 2 SO_4^{2-} + 2 H^+$$

(a) **Initiator reaction**

Spontaneous (bacteria may also catalyze)

(b) **Propagation cycle**

Slow spontaneous, bacteria catalyze — Fe^{2+} — O_2 — Fe^{3+} — FeS_2 — Fast spontaneous (bacteria may also catalyze)

● **Figure 19.33 Role of iron-oxidizing bacteria in oxidation of the mineral pyrite.** (a) The primarily nonbiological initiator reaction. (b) The propagation cycle, which includes biotic and abiotic components.

relatively stable in the presence of oxygen. *Acidithiobacillus ferrooxidans* and *L. ferrooxidans* then catalyze the oxidation of ferrous to ferric ions. The ferric ions formed under these acidic conditions, being soluble, can react spontaneously with more pyrite and oxidize it to ferrous ions plus sulfate ions:

$$FeS_2 + 14\,Fe^{3+} + 8\,H_2O \longrightarrow 15\,Fe^{2+} + 2\,SO_4^{2-} + 16\,H^+$$

The ferrous ions formed are again oxidized to ferric ions by the bacteria, and these ferric ions again react with more pyrite. Thus, there is a progressive, rapidly increasing rate at which pyrite is oxidized, called the *propagation cycle*, as illustrated in Figure 19.33. Under natural conditions some of the ferrous iron generated by the bacteria leaches away, being carried by groundwater into surrounding streams. However, because oxygen is present in the aerated drainage, bacterial oxidation of the ferrous iron takes place in these outflows and an insoluble ferric precipitate is formed.

Acid Mine Drainage

Bacterial oxidation of sulfide minerals is the major factor in **acid mine drainage**, an environmental problem in coal-mining regions (Figure 19.34●). Mixing of acidic mine waters with natural waters in rivers and lakes causes serious degradation in the quality of the natural water because both the acid and the dissolved metals are toxic to aquatic organisms (∞ Figure 17.28a). In addition, such polluted waters are unsuitable for human consumption and industrial use. The breakdown of pyrite leads ultimately to the formation of sulfuric acid and ferrous iron, and pH values can be lower than pH 1. The acid formed attacks other minerals associated with the

coal and pyrite, causing breakdown of the whole rock fabric. A major rock-forming element, aluminum, is soluble only at low pH, and often high levels of Al^{3+}, which can be highly toxic to aquatic organisms, are present in acid mine waters.

The requirement for O_2 in the oxidation of ferrous to ferric iron helps to explain how acid mine drainage develops. As long as the coal is unmined, oxidation of pyrite cannot occur because neither air, water, nor the bacteria can reach it. When the coal seam is exposed (Figure 19.32b), it quickly becomes contaminated with *Acidithiobacillus ferrooxidans*, and O_2 and water are introduced, making oxidation of pyrite possible. The acid formed can then leach into the surrounding streams (Figure 19.34).

Where acid mine drainage is extensive, a strongly acidophilic species of *Archaea*, *Ferroplasma*, is typically present. This aerobic iron-oxidizing prokaryote is capable of growth at pH 0 and at temperatures to 50° C. *F. acidarmanus* forms slimy cell masses attached to pyrite surfaces in iron ore deposits (Figure 19.35a●). At Iron Mountain, California (Figure 19.35a), a particularly well-studied acid mine drainage site, the thick biofilm of

(a)

(b)

● **Figure 19.35** *Ferroplasma acidarmanus,* an extremely acidophilic iron-oxidizing archaeon responsible for severe acid mine drainage. (a) Streamers of *Ferroplasma* cells growing in an acidic (pH near 0) mine drainage stream, Iron Mountain, CA. (b) Scanning electron micrograph of a cell of *F. acidarmanus* among mineral matter.

● **Figure 19.34 Acid mine drainage from a bituminous coal region.** Note the yellowish-red color due to precipitated iron oxides (∞ Figure 17.28a).

F. acidarmanus maintains the pH near 0. With iron concentrations of nearly 30 g/l at this site, the *F. acidarmanus* mat is constantly bathed in fresh substrate (Fe^{2+}) from which more acid is generated by the reactions described previously. *Ferroplasma* is a cell wall-less prokaryote morphologically and is phylogenetically related to *Thermoplasma* (Figure 19.35*b* and Section 13.5).

19.14 Concept Check

Iron exists in nature primarily in two oxidation states, ferrous (Fe^{2+}) and ferric (Fe^{3+}), and bacterial and chemical transformation of these metals is of geological and ecological importance. Bacterial ferric iron reduction occurs in anoxic environments and results in the mobilization of iron from swamps, bogs, and other iron-rich aquatic habitats. Bacterial oxidation of ferrous iron occurs on a large scale at low pH and is very common in coal-mining regions, where it results in a type of pollution called acid mine drainage.

◆ What oxidation state is iron in the mineral $Fe(OH)_3$? FeS? How is $Fe(OH)_3$ formed?

◆ Why does biological Fe^{2+} oxidation under *oxic* conditions occur mainly at *acidic* pH?

VI MICROBIAL BIOREMEDIATION

The collective biogeochemical potential of microorganisms is enormous. Indeed, microorganisms are Earth's greatest chemists. As such, they have been used to help extract valuable metals from low-grade ores (microbial leaching) and to clean up the environment. The term **bioremediation** refers to the cleanup of oil, toxic chemicals, or other pollutants by microorganisms. Bioremediation is a cost effective method for cleaning up pollutants, and in some cases, is the only practical way to get the job done.

19.15 Microbial Leaching of Ores

Acid production and mineral dissolution by acidophilic bacteria (see Section 19.14) can play a beneficial role in the mining of metals. Sulfide forms highly insoluble minerals with many metals, and many ores used as sources of these metals are sulfides. If the concentration of metal in the ore is low, it may not be economically feasible to concentrate the mineral by conventional chemical means. Under these conditions, **microbial leaching** is practiced. Leaching is especially useful for *copper* ores because copper sulfate, formed during oxidation of copper sulfide ores, is very water-soluble. Indeed, approximately one-fourth of all copper mined worldwide is obtained by leaching.

We noted that sulfide oxidizes spontaneously in air. Most metal sulfides also oxidize spontaneously, but much more slowly than H_2S or HS^-. However, *Acidithiobacillus ferrooxidans* and other metal-oxidizing prokaryotes can catalyze a much faster rate of oxidation of the sulfide minerals, thus aiding in solubilization of the metal (Figure 19.33). The relative rates of oxidation of a copper mineral in the presence and absence of bacteria is illustrated in Figure 19.36●. The susceptibility to oxidation also varies among minerals, and those minerals that are most readily oxidized are most amenable to microbial leaching. Thus, iron and copper sulfide ores such as pyrrhotite (FeS) and covellite (CuS) are readily leached, whereas lead and molybdenum ores are much less so.

The Leaching Process

In microbial leaching, low-grade ore or waste rock is dumped in a large pile (the leach dump) and a dilute sulfuric acid solution (pH about 2) is percolated down through the pile (Figure 19.37*a*●). The liquid coming out of the bottom of the pile (Figure 19.37*b*), rich in dissolved metals (called the pregnant liquor solution), is collected and transported to a precipitation plant (Figure 19.37*c*) where the desired metal is reprecipitated and purified. The pregnant liquor is then pumped back to the top of the pile and the cycle is repeated. As needed, more acid is added to maintain the low pH.

There are various mechanisms by which bacteria can catalyze oxidation of the sulfide minerals. To illustrate, we use the example of *covellite*, CuS, in which copper has a valence of +2. As illustrated in Figure 19.38●, *Acidithiobacillus ferrooxidans* can oxidize the sulfide in CuS to SO_4^{2-}. This reaction can also occur spontaneously.

However, the most important reaction in copper leaching operations involves *chemical* oxidation of the copper ore with *ferric* ions formed by the bacterial oxidation

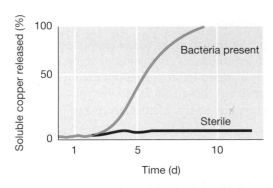

● **Figure 19.36 Effect of the bacterium *Acidithiobacillus ferrooxidans* on the leaching of copper from the mineral covellite (CuS).** The leaching was done in a laboratory column, and the acid leach solution contained inorganic nutrients necessary for growth of the bacterium. The leaching activity was monitored by assaying for soluble copper in the leach solution at the bottom of the column. The leach solution was continuously recirculated, maintaining an essentially closed system.

(a)

(b)

(c)

(d)

● **Figure 19.37   The leaching of low-grade copper ores using bacteria.** (a) A typical leaching dump. The low-grade ore has been crushed and dumped in a large pile in such a way that the surface area exposed is as high as possible. Pipes distribute the acidic leach water over the surface of the pile. The acidic water slowly percolates through the pile and exits at the bottom. (b) Effluent from a copper leaching dump. The acidic water is very rich in dissolved copper. (c) Recovery of dissolved copper by passage of the copper-rich water over metallic iron in a long flume. (d) A small pile of recovered copper metal removed from the flume, ready for further purification.

of ferrous ions (Figure 19.38). In almost any ore, pyrite (FeS_2) is present, and its oxidation leads to the formation of ferric iron (see Figure 19.33). Ferric iron is an excellent electron acceptor for sulfide minerals, and the spontaneous reaction of CuS with ferric iron results in solubilization of the copper and the formation of ferrous iron. Importantly, this can happen deep in the leach dump where conditions are anoxic.

Metal Recovery

The precipitation plant is where the soluble copper from the leaching solution is recovered (Figure 19.37*c, d*). Scrap iron, Fe^0, is added to recover copper from the leach liquid by the reaction shown in the lower part of Figure 19.38, and this results in the formation of Fe^{2+}. This Fe^{2+}-rich

● **Figure 19.38   Arrangement of a leaching pile and reactions involved in the microbial leaching of copper sulfide minerals to yield Cu⁰** (copper metal) (see Figure 19.37). Reaction 1 occurs both biologically and chemically. Reaction 2 is strictly chemical, but is the most important reaction in copper-leaching processes. Note how it is essential for Reaction 2 to proceed that the Fe^{2+} produced (from the oxidation of sulfide in CuS to sulfate) be oxidized back to Fe^{3+} by *Acidithiobacillus ferrooxidans* and *Leptospirillum ferrooxidans*.

Sprinkling of acidic leach liquor on copper ore

Low grade copper ore: copper sulfide (CuS)

Copper ore can be oxidized by oxygen-dependent (1) and oxygen-independent (2) reactions, solubilizing the copper:

1. $CuS + 2\ O_2 \rightarrow Cu^{2+} + SO_4^{2-}$

2. $CuS + 8\ Fe^{3+} + 4\ H_2O$
   $Cu^{2+} + 8\ Fe^{2+} + SO_4^{2-} + 8\ H^+$

Copper in solution — Cu^{2+}

Acidic leach liquor pumped back to top of leach dump

H_2SO_4 addition

Recovery of copper metal (Cu^0)
$Fe^0 + Cu^{2+} \rightarrow Cu^0 + Fe^{2+}$
(Fe^0 from steel cans)

Acidic leach liquor

$Fe^{2+} + \frac{1}{4}O_2 + H^+ \rightarrow Fe^{3+} + \frac{1}{2}H_2O$
Leptospirillum ferrooxidans
Acidithiobacillus ferrooxidans
Oxidation pond

Copper metal (Cu^0)

liquid is then transferred to an oxidation pond where *Acidithiobacillus ferrooxidans* grows and oxidizes it to Fe^{3+}. Acid is added to the pond to keep the pH low— keeping the Fe^{3+} in solution—and this ferric-rich liquid is then pumped to the top of the pile and the Fe^{3+} used to oxidize more sulfide mineral (Figure 19.38). Thus, the entire leaching operation is maintained by the oxidation of Fe^{2+} to Fe^{3+} by bacteria.

High temperature can be a problem with leaching operations. *A. ferrooxidans* is a mesophile, but temperatures inside a leach dump often rise spontaneously as a result of heat generated by microbial activities. Thus, thermophilic iron-oxidizing chemolithotrophs such as thermophilic *Thiobacillus* species, *Leptospirillum ferrooxidans*, and *Sulfobacillus* and, at higher temperatures (60–80° C), the organism *Sulfolobus* (∞ Section 13.9), are also important in the leaching of ores.

Other Leaching Processes: Uranium and Gold

Microorganisms are also used in the leaching of uranium- and gold-containing ores. *Acidithiobacillus ferrooxidans* can oxidize U^{4+} to U^{6+} with O_2 as an electron acceptor. However, uranium leaching probably depends more on the chemical oxidation of uranium by Fe^{3+}, with *A. ferrooxidans* contributing mainly through the re-oxidation of Fe^{2+} to Fe^{3+} as in copper leaching (see Figure 19.38). The reaction observed is

$$UO_2 + Fe_2(SO_4)_3 \longrightarrow UO_2SO_4 + 2\ FeSO_4$$
$$(U^{4+}) \quad\ (Fe^{3+}) \qquad\qquad (U^{6+}) \qquad (Fe^{2+})$$

Unlike UO_2, the oxidized uranium mineral is soluble and can be recovered by other processes.

Gold is typically found in nature associated with minerals containing arsenic and pyrite. *A. ferrooxidans* and its relatives can attack and solubilize the arsenopyrite minerals, releasing the trapped gold (Au):

$$2\ FeAsS[Au] + 7\ O_2 + 2\ H_2O + H_2SO_4 \longrightarrow$$
$$Fe_2(SO_4)_3 + 2\ H_3AsO_4 + [Au]$$

The gold is then complexed with cyanide by traditional gold-mining methods. Unlike copper, where leaching occurs in a huge leach dump (see Figure 19.37a), gold leaching takes place in relatively small *bioreactor tanks* (Figure 19.39●). Such bioleaching releases more than 95% of the trapped gold. Moreover, although arsenic and cyanide are toxic residues from the mining process, both are removed in the gold-leaching bioreactor. Arsenic is removed as a ferric precipitate and cyanide (CN^-) by its microbial oxidation to CO_2 plus urea in later stages of the gold recovery process. Small-scale microbial gold leaching has thus become popular as an alternative to the more costly and environmentally damaging conventional gold-mining techniques. Pilot processes are also being developed for the bioleaching of zinc, lead, and nickel.

● **Figure 19.39 Gold bioleaching.** Photo of the gold leaching tanks at the Ashanti Gold-fields, Ghana, Africa. Within the tanks, a mixture of *Acidithiobacillus ferrooxidans, Acidithiobacillus thiooxidans,* and *Leptospirillum ferrooxidans* solubilizes the pyrite/arsenic mineral containing trapped gold and releases the gold. See text for the reactions involved.

 19.15 Concept Check

Oxidation of copper ores by bacteria can lead to the solubilization of copper, a process called *microbial leaching*. Leaching is important in the recovery of copper, uranium, and gold from low-grade ores. Bacterial oxidation of iron in the iron sulfide mineral pyrite is also an important part of the microbial-leaching process because the ferric iron produced is itself an oxidant of ores.

◆ How can CuS be oxidized under *anoxic* conditions?

◆ Why is it important to keep the leach liquor acidic in the copper ore leaching process?

◆ From the standpoint of metal oxidation, how does the copper leach process differ from the gold leach process?

19.16 Mercury and Heavy Metal Transformations

Metals are present in low concentrations in rocks, soils, waters, and the atmosphere. Some of these are *trace elements* needed by cells (for example, cobalt, copper, zinc, nickel, molybdenum, ∞ Section 5.1). In high concentrations, however, these same metals can be toxic to organisms. Of these, several are sufficiently volatile to contaminate the air as well. These include mercury, lead, arsenic, cadmium, and selenium.

Many heavy metals undergo redox reactions catalyzed by microorganisms. Several are also converted to organic form via microbial action. Because of environmental concern and significant microbial involvement, we focus our discussion on the biogeochemistry of the element *mercury*.

Global Cycling of Mercury and Methylmercury

Although mercury is present in extremely low concentrations in most natural environments, averaging about 1 nanogram per liter, it is a widely used industrial product and is the active component of many pesticides. Because of its unusual propensity to concentrate in living tissues and its high toxicity, mercury is of considerable environmental importance. For example, the mining of mercury ores and the burning of fossil fuels release about 40,000 *tons* of mercury into the environment each year. An even greater amount is released by natural geochemical processes. Mercury is also a by-product of the electronics industry, especially battery and wiring production, the chemical industry, and the burning of municipal wastes.

The major form of mercury in the atmosphere is elemental mercury (Hg^0), which is volatile and is oxidized to mercuric ion (Hg^{2+}) photochemically. Most mercury entering aquatic environments is thus as Hg^{2+} (Figure 19.40●). Mercuric ion readily adsorbs to particulate matter and can be metabolized from there by microorganisms. Microbial activity results in the *methylation* of mercury, yielding **methylmercury**, CH_3Hg^+ (Figure 19.40). Metabolically, methylation of mercury occurs by donation of methyl groups from methyl-B_{12}. Methylmercury is particularly toxic because it can be absorbed through the skin. But in addition, it is soluble and can be concentrated in the aquatic food chain, primarily in fish. Methymercury can be further methylated by microorganisms to yield the volatile compound dimethylmercury, $CH_3-Hg-CH_3$.

Both methylmercury and dimethylmercury accumulate in animal tissues, especially muscle. Methylmercury is about 100 times more toxic than Hg^0 or Hg^{2+} and concentrates in fish, where it is a potent neurotoxin. Methylmercury is thus a major environmental toxin, and its accumulation seems to be a particular problem in freshwater lakes where enhanced levels of methylmercury have been observed in fish caught for human consumption.

Mercury can also cause liver and kidney damage in humans and other animals.

Several other mercury transformations occur on a global scale, including reactions involving sulfate-reducing bacteria ($H_2S + Hg^{2+} \rightarrow HgS$) and methanogens ($CH_3Hg^+ \rightarrow CH_4 + Hg^0$) (see Figure 19.40). The solubility of HgS is very low, so in anoxic sulfate-reducing sediments, most mercury is found as HgS. But on aeration, oxidation of HgS can occur, primarily by thiobacilli, leading to the formation of Hg^{2+} and SO_4^{2-}, and eventually, methylmercury.

Mercuric Resistance

At sufficiently high concentrations, Hg^{2+} and CH_3Hg^+ can be toxic not only to higher organisms but also to microorganisms. However, several bacteria can carry out the biotransformation of toxic forms of mercury to nontoxic forms. In mercury-resistant gram-negative bacteria an NADPH-linked enzyme called *mercuric reductase* reduces Hg^{2+} to Hg^0. The Hg^0 produced in this reaction is volatile but is nontoxic to humans and microorganisms. Bacterial conversion of Hg^{2+} to Hg^0 then allows for more CH_3Hg^+ to be converted to Hg^{2+} (Figure 19.41●).

In the gram-negative bacterium *Pseudomonas aeruginosa* genes for mercury resistance reside on a plasmid.

● **Figure 19.40 Biogeochemical cycling of mercury.** The major reservoirs of mercury are in water and in sediments where it can be concentrated in animal tissues or precipitated out as HgS. The various forms of mercury commonly found in aquatic environments are each shown in a different color.

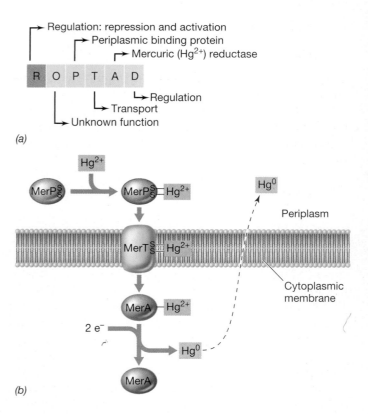

● **Figure 19.41 Mechanism of Hg^{2+} reduction to Hg^0 in *Pseudomonas aeruginosa*.** (a) The *mer* operon. MerR can function as either a repressor (absence of Hg^{2+}) or transcriptional activator (presence of Hg^{2+}). (b) Transport and reduction of Hg^{2+}. Hg^{2+} is bound by cysteine residues in both proteins MerP and MerT.

These genes, called *mer* genes, are arranged in an operon and are under control of the regulatory protein MerR (Figure 19.41*a*). Interestingly, MerR functions as both a repressor and an activator (∞ Chapter 8). In the absence of Hg^{2+}, MerR binds to the operator region of the *mer* operon and prevents transcription of *merTPCAD* genes (Figure 19.41*a*). However, when Hg^{2+} is present, it forms a complex with MerR, which then functions as an *activator* of transcription of the *mer* operon.

The protein MerP is a periplasmic Hg^{2+} binding protein (Figure 19.41*b*). MerP binds Hg^{2+} and transfers it to a membrane protein MerT, which transports Hg^{2+} into the cell for reduction by mercuric reductase (Figure 19.41*b*). The final result is reduction of Hg^{2+} to Hg^0, which is volatile and is released from the cell (Figure 19.41*b*).

Resistance to Other Heavy Metals

A variety of plasmids (∞ Section 10.8) isolated from both gram-positive and gram-negative *Bacteria* encode resistance to heavy metals. Certain antibiotic resistance plasmids also have genes for resistance to mercury and arsenic. Other plasmids encode only heavy-metal resistances, typically multiple resistances. The mechanism of resistance to any specific metal varies. For example, arsenate and cadmium resistances are due to the activity of enzymes that pump out any arsenate or cadmium incorporated, thus preventing the metals from denaturing proteins. Other resistances may involve redox changes (Figure 19.41). Bacteria naturally resistant to high levels of several metals often contain multiple plasmids encoding the resistances. Such multiply-resistant organisms are common in wastewaters of the metal processing industry or in mining operations, where heavy metals are leached out along with iron or copper (see Section 19.15).

 19.16 Concept Check

A major toxic form of mercury is methylmercury. The latter can yield Hg^{2+}, which is reduced by bacteria to Hg^0. The ability of bacteria to resist the toxicity of heavy metals is often due to the presence of specific plasmids that encode enzymes capable of detoxifying or pumping out the metals.

◆ What forms of mercury are most toxic to organisms?

◆ How is mercury detoxified by bacteria?

19.17 Petroleum Biodegradation

Petroleum is a rich source of organic matter, and because of this, microorganisms readily attack hydrocarbons aerobically when it is brought into contact with air and moisture. Under some circumstances, such as in bulk storage tanks, microbial growth is not desirable. However, in spills, bioremediation of petroleum is desirable and

can be promoted by the addition of inorganic nutrients. Indeed, the importance of prokaryotes to petroleum bioremediation has been demonstrated in several major crude oil spills in recent years (see Figure 19.43).

The biochemistry of petroleum biodegradation was covered in Section 17.23. There we emphasized the important role *oxygenase enzymes* play in introducing oxygen atoms into the hydrocarbon (∞ Figure 17.55). Our discussion here will thus focus on aerobic processes where the activity of such enzymes are relevant and where hydrocarbon oxidation in nature is quantitatively most significant.

Hydrocarbon Decomposition

A wide variety of bacteria, several molds and yeasts, and certain cyanobacteria and green algae can oxidize petroleum products aerobically. Small-scale oil pollution of aquatic and terrestrial ecosystems from human as well as natural activities is common. Thus a diverse microbial community exists in these environments that uses oil as electron donors. Methane, the simplest hydrocarbon, is only catabolized by a specialized group of bacteria, the *methanotrophic* bacteria (∞ Sections 12.6 and 17.24), and these organisms do not grow on higher hydrocarbons.

Oil-oxidizing microorganisms develop rapidly on oil films and slicks. Oil-oxidizing activity is then most extensive if temperature and inorganic nutrients (primarily nitrogen and phosphorous), are adequate (see Figure 19.43). Because oil is insoluble in water and is less dense, it floats to the surface and forms slicks. Oil-oxidizing bacteria often attach to insoluble oil droplets in large numbers (Figure 19.42●). These bacteria eventually decompose the oil and disperse the slick.

Microorganisms bioremediate oil (Figure 19.43●) by oxidizing it to CO_2. In large spills, volatile hydrocarbon fractions evaporate quickly, leaving medium- to longer-chain aliphatic and aromatic components for cleanup crews or microorganisms to tackle. By promoting bioremediation activities, oil-oxidizing bacteria can increase in number by a factor of 10^3–10^6 shortly after an oil spill.

Experiments using radioisotopic hydrocarbons as tracers or O_2 uptake as a measure of respiratory activity have shown that under ideal conditions, up to 80% of the

● **Figure 19.42 Hydrocarbon-oxidizing bacteria in association with oil droplets.** The bacteria are concentrated in large numbers at the oil–water interface but are not within the droplet itself.

(a)

(b)

● **Figure 19.43** **Environmental consequences of large oil spills in the marine environment and the effect of bioremediation.** (a) A contaminated beach along the coast of Alaska containing oil from the *Exxon Valdez* spill of 1989. (b) The center rectangular plot (arrow) was treated with inorganic nutrients to stimulate bioremediation of spilled oil by microorganisms, whereas areas to the left and right were untreated.

nonvolatile components in oil are oxidized by bacteria within a year of the spill. Certain oil fractions, such as branched-chain and polycyclic hydrocarbons, however, remain in the environment much longer. Spilled oil that travels to the sediments is only slowly degraded, and it can have a significant long-term impact on fisheries and related activities that depend on unpolluted waters for productive yields.

Interfaces where oil and water meet often occur on a large scale. For example, it is virtually impossible to keep moisture from accumulating in bulk-fuel storage tanks, because water forms in a layer beneath the petroleum. Gasoline and crude oil storage tanks (Figure 19.44●) are thus potential habitats for hydrocarbon-oxidizing microorganisms. If sufficient sulfate is present in the water, sulfate-reducing bacteria can grow in the tanks, consuming hydrocarbons under anoxic conditions (∞ Section 17.23). The sulfide (H_2S) produced is highly corrosive and causes pitting and subsequent leakage of the tanks.

Petroleum Production

Although microbial degradation of petroleum can be quite extensive, microbial *production* of hydrocarbons

● **Figure 19.44** **Oil biodegradation.** Bulk fuel-storage tanks, where massive microbial growth may occur at oil-water interfaces.

also occurs, particularly in certain green algae. For example, in the colonial alga *Botryococcus braunii*, growth is accompanied by the excretion of long-chain hydrocarbons C_{30}–C_{36} that have the consistency of oil (Figure 19.45●). In *B. braunii* about 30% of the cell dry weight is petroleum, and there has been interest in using this and other oil-producing algae as renewable sources of petroleum. There is even evidence that oil in certain types of oil shale formations originated from green algae like *B. braunii* that grew in lake beds in ancient times.

⬢ *19.17* *Concept Check*

Hydrocarbons are subject to microbial attack. Hydrocarbon-oxidizing microorganisms are used for bioremediation of spilled oil and their activities are assisted by addition of inorganic nutrients to balance the influx of carbon from the oil. Some algae can produce hydrocarbons.

◆ What is *bioremediation*?

● **Figure 19.45** **Phototrophic oil production.** Photomicrograph by Nomarski interference contrast of cells of the green alga *Botryococcus braunii*. Note oil droplets, produced and excreted by this alga, at the margin of the cells.

◆ Why might the addition of inorganic nutrients stimulate oil degradation while the addition of glucose might not?

19.18 Biodegradation of Xenobiotics

Xenobiotics are synthetic chemicals that are not naturally occurring substances. Xenobiotics include pesticides, polychlorinated biphenyls (PCBs, used in the electric-generating and related industries), munitions, dyes, and chlorinated solvents, among other things. Many xenobiotics are structurally related to natural compounds, and can thus be slowly degraded by enzymes that already exist to degrade these natural compounds. However, some xenobiotics differ chemically in such a major way from anything organisms have naturally experienced that they degrade extremely slowly, if at all. We focus here on pesticides as an example of the potential of microbial degradation.

Pesticide Catabolism

Some of the most widely distributed xenobiotics are the *pesticides*, which are common components of toxic wastes. Over 1000 pesticides have been marketed for chemical pest control purposes. Pesticides include *herbicides, insecticides,* and *fungicides*. Pesticides are of a wide variety of chemical types, including chlorinated compounds, aromatic rings, nitrogen- and phosphorus-containing compounds, and others (Figure 19.46●). Some of these substances are suitable as carbon sources and electron donors for certain soil microorganisms, whereas others are not.

If a substance can be attacked by microorganisms, it will eventually disappear from the soil. Such degradation in the soil is desirable because toxic accumulations of the compound are avoided. However, even closely related compounds may differ remarkably in their degradability, as is shown in Table 19.7. Thus some compounds persist relatively unaltered for years, while others are significantly degraded in only weeks or months.

The relative persistence rates of xenobiotics are only approximate because a variety of environmental factors—temperature, pH, aeration, and organic matter content of the soil—also influence decomposition. Some of the chlorinated insecticides (Figure 19.46) are so recalcitrant that they can persist for over 10 years. However, even complete disappearance of a pesticide from an ecosystem does not necessarily mean that it was degraded by microorganisms; pesticide loss can also occur by volatilization, leaching, or spontaneous chemical breakdown. *Bioavailability* to some extent also governs microbial attack on xenobiotic compounds. Many xenobiotics are quite hydrophobic and thus not very soluble in water, and adsorption of these compounds to organic matter and clay in soils and sediments prevents access to the organism. Addition of surfactants or emulsifiers often increases bioavailability, and ultimately biodegradation, of the xenobiotic compound.

● **Figure 19.46** **Examples of xenobiotic compounds.** Although none of these compounds exist naturally, various microorganisms exist that will break them down (see persistence data in Table 19.7 and the pathway for biodegradation of 2,4,5-T in Figure 19.47).

Table 19.7	Persistence of herbicides and insecticides in soils	
Substance	**Time for 75–100% disappearance**
Chlorinated insecticides	
DDT [1,1,1-trichloro-2,2-bis-(ρ-chlorophenyl)ethane][a]	4 years
Aldrin	3 years
Chlordane	5 years
Heptachlor	2 years
Lindane (hexachlorocyclohexane)	3 years
Organophosphate insecticides	
Diazinon	12 weeks
Malathion[a]	1 week
Parathion	1 week
Herbicides	
2,4-D (2,4-dichlorophenoxyacetic acid)	4 weeks
2,4,5-T (2,4,5-trichlorophenoxyacetic acid)	20 weeks
Dalapon	8 weeks
Atrazine[a]	40 weeks
Simazine	48 weeks
Propazine	1.5 years

[a] Structure shown in Figure 19.46.

Many organisms metabolize pesticides and herbicides, including genera of both bacteria and fungi. Some pesticides can be both carbon and energy sources and are oxidized completely to CO_2. However, other compounds are attacked only slightly or not at all. Some compounds may be degraded either partially or totally provided that some other organic material is present as primary energy source. This is a phenomenon called **cometabolism** or **cooxidation**. However, if the breakdown is only partial, the microbial degradation product of a pesticide may sometimes be even more toxic than the original compound.

Reductive Dechlorination

In Chapter 17 we discussed a form of anaerobic respiration called **reductive dechlorination**. In this process, chlorinated organic compounds are used as terminal electron acceptors under anoxic conditions (∞ Section 17.18). A laboratory model for studying reductive dechlorination is the reduction of chlorobenzoate to benzoate by the bacterium *Desulfomonile*:

$$C_7H_4O_2Cl^- + 2\,H \longrightarrow C_7H_5O_2^- + HCl$$

From the standpoint of bioremediation, other chlorinated compounds are ecologically much more important than chlorobenzoate. For example, reductive dechlorination can occur with the compounds dichloro-, trichloro-, and tetrachloro- (perchloro-) ethylene, chloroform, dichloromethane, certain polychlorinated biphenyls (PCBs; Figure 19.46), and various brominated and fluorinated compounds. These toxic compounds, some of which (particularly trichloroethylene, Figure 19.46) have been linked to cancer, are widely used as industrial solvents, degreasing agents, and insulators in electrical transformers. They enter anoxic environments through accidental spills or from slow leakage of storage containers or abandoned electrical transformers. Eventually they migrate into groundwater, where they are the most frequently detected groundwater contaminants in the United States.

Several genera of reductive dechlorinators (∞ Section 17.18) can transform a wide variety of chlorinated compounds into harmless metabolites. There is currently great interest in stimulating the bioremediation activities of reductive dechlorinating prokaryotes as a strategy for the removal of these highly toxic compounds in anoxic environments.

Aerobic Dechlorination

Aerobic dechlorination of chlorinated compounds is also possible (Figure 19.47●), but here, biochemical mechanisms involving oxygen can come into play. Under these conditions, the degradation of chlorinated aromatic

(a)

(b)

● **Figure 19.47 Biodegradation of the herbicide 2,4,5-T.** (a) Growth of *Burkholderia cepacia* on 2,4,5-T as sole source of carbon and energy. The strain was enriched from nature using a chemostat to keep the concentration of herbicide low. Growth here is aerobic on 1.5 g/l of 2,4,5,-T. The release of chloride from the molecule is indicative of biodegradation. (b) Pathway of aerobic 2,4,5-T biodegradation. The mechanism of action of a dioxygenase is discussed in Chapter 17 (∞ Figure 17.56c). Several bacteria can catabolize aromatic compounds under *anoxic* conditions, but the mechanism employed is quite different from that shown here (∞ Figure 17.57).

compounds occurs by way of *oxygenase* enzymes (∞ Section 17.22). For example, in the aerobic degradation of the pesticide 2,4,5-T, following dechlorination, a dioxygenase enzyme breaks down the aromatic ring. This generates compounds that can be metabolized by the citric acid cycle (Figure 19.47b).

Although the aerobic breakdown of chlorinated xenobiotics is undoubtedly of ecological importance, reductive dechlorination is of particular environmental interest because of the rapidity with which anoxic conditions can develop in polluted microbial habitats in nature and the biochemical constraints (Figure 19.47b) this puts on aerobic organisms that could otherwise degrade the compound.

Biodegradable Plastics

A major area of environmental concern besides the biodegradation of toxic wastes like pesticides and other chlorinated hydrocarbons is the disposal of *solid wastes*—in particular, plastics. Typical landfills contain large amounts of paper, food, construction and demolition debris, and plastics. Degradation rates of these materials are frequently very low because the conditions, especially the lack of moisture and oxygen, are not suitable to rapid microbial activity. In addition, products such as plastics are inherently recalcitrant.

The plastics industry currently produces nearly 40 *billion* kilograms of plastic per year, almost half of which is discarded in landfills. Plastics are xenobiotic polymers of various types; *polyethylene, polypropylene,* and *polystyrene* are typical examples (Figure 19.48●). Many of these synthetic polymers remain essentially unaltered for decades in landfills and refuse dumps. This problem has fueled the search for *biodegradable* alternatives—called **biopolymers**—to the synthetic polymers now in use. Some success stories include photodegradable and starch-linked polymers, and microbial plastics.

Photobiodegradable plastics are polymers whose structure is altered by exposure to ultraviolet radiation (from sunlight), generating modified polymers amenable to microbial attack. Starch-based plastics incorporate starch (👓 Figure 3.6*a*) as a link to short fragments of a second biodegradable polymer. This design accelerates biodegradation, because starch-digesting bacteria in soil attack the starch, releasing polymer fragments that are then degraded by other microorganisms. Microbially synthesized plastics are based on the storage polymer poly-β-hydroxyalkanoate (PHAs, 👓 Section 4.11) (Figure 19.49●). PHAs have many of the general properties of synthetic plastics and can be biosynthesized in various chemical forms. A *copolymer* containing approximately equal amounts of PHB (poly-

● Figure 19.49 Bacterial plastics. (a) PHV/PHB copolymer. (b) A brand of shampoo marketed in Europe and packaged in a bottle made of a copolymer of poly-β-hydroxybutyrate (PHB) and poly-β-hydroxyvalerate (PHV, see *a*). Because this material is a natural product, the bottle readily degrades both aerobically and anaerobically.

(b)

-β-hydroxybutyrate) and PHV (poly-β-hydroxyvalerate) (Figure 19.49*a*), has had the greatest market success thus far. However, because they are more cost-effective, petroleum-based plastics make up virtually the entire plastics market today.

19.18 Concept Check

Many chemically synthesized compounds such as insecticides, herbicides, and plastics (all called xenobiotics) are completely foreign to microorganisms but can often be degraded by one or another prokaryote nonetheless. Both aerobic and anaerobic mechanisms are known.

◆ What chemical features are shared by the compounds shown in Figure 19.46?

◆ What is *reductive dechlorination* and how does it differ from the reactions shown in Figure 19.47?

◆ What advantages do *biopolymers* have over synthetic polymers?

● Figure 19.48 Chemistry of synthetic polymers. Shown is the monomeric structure of a number of common synthetic polymers.

VII MICROBIAL INTERACTIONS WITH PLANTS

We close this chapter with three examples of microbial interactions with plants. Many beneficial symbiotic associations exist, such as with lichens, mycorrhizae, and the root nodules of leguminous plants. There are also destructive microorganisms that can cause plant diseases. Of the latter, we consider here only crown gall because of the unique features of the disease and the microbial mode of transmission.

19.19 The Plant Environment

As microbial habitats, plants are clearly vastly different from animals. Compared with warm-blooded animals, plants vary greatly in temperature, both diurnally and throughout the year. Compared with the complex circulatory system of animals, the internal communication system of the plant is only poorly developed, and so transfer of microorganisms within the plant is relatively inefficient.

The above-ground parts of the plant, especially the leaves and stems, are subjected to intermittent drying. For this reason many plants contain waxy coatings on their leaves that retain moisture and keep out microorganisms. The roots, on the other hand, are a main area of microbial activity. Here moisture is less variable and nutrient concentrations are higher.

Rhizosphere and Phyllosphere

The **rhizosphere** is the region immediately outside the root (see Figure 19.6*b*); it is a zone where microbial activity is usually high. The **rhizoplane** is the actual root surface. The bacterial count is almost always higher in the rhizosphere/rhizoplane than it is in regions of the soil devoid of roots, often many times higher (Figure 19.50●). This is because roots excrete significant amounts of sugars, amino acids, hormones, and vitamins. Bacteria and fungi thus often form microcolonies on the root surface.

The **phyllosphere** is the surface of the plant leaf. Under conditions of high humidity, as in wet forests in tropical and temperate zones, the microflora of leaves may be quite high and include fungi (Figure 19.50*a*) as well as bacteria. Many bacteria on leaves fix nitrogen (∞ Section 17.28, and see Section 19.22), while the leaves in turn provide carbohydrates and other nutrients to the bacteria.

 19.19 Concept Check

Key microbial habitats on plants include the rhizoplane/rhizosphere and the phyllosphere.

◆ Why might the roots of plants be a more attractive environment for bacteria than in free soil away from plants?

19.20 Lichens and Mycorrhizae

Lichens are leafy or encrusting growths often found growing on bare rocks, tree trunks, house roofs, and the surfaces of bare soils (Figure 19.51●). Lichens consist of a symbiosis of two organisms, a fungus and an alga or cyanobacterium. However, a given fungus can establish the lichen symbiosis with several different phototrophs, and vice versa.

The alga is phototrophic and able to produce organic matter, which is then used for nutrition of the fungus. The fungus, unable to carry out photosynthesis, provides a firm anchor within which the alga can grow pro-

(a)

(b)

(c)

● **Figure 19.50 Examples of phyllosphere and rhizosphere microbial life.** (a) Fluorescent micrograph of the fungus *Aureobasidium pullulans* on the phyloplane (leaf surface) of apple. Cells were stained by FISH (∞ Section 18.4), and both fungal filaments and spores show the green stain. A fungal filament is about 7 μm in diameter. (b) Laser-scanning confocal (∞ Section 4.2) micrograph of bacterial cells on the rhizosphere/rhizoplane of clover. The association of nitrogen-fixing bacteria with the roots of leguminous plants is discussed in Section 19.22. (c) Scanning electron micrograph of cells of *Rhizobium leguminosarum* biovar *trifolii* attached to the root hair of clover. The bacterial cells in (b) and (c) are about 1 μm in diameter.

tected from erosion by rain or wind. In addition, the fungus facilitates the uptake of water. From the rock or other substrate on which the lichen is living, the fungus also absorbs inorganic nutrients essential for the growth of the alga. Lichens are usually found on surfaces where other organisms do not grow, and their success in colonizing such environments is due to the mutual interrelationships between the alga and fungus partners.

Lichen Structure and Ecology

Lichens consist of a tight association of many fungal cells within which the phototroph cells are embedded (Figure 19.52●). The shape of the lichen is determined primarily by the fungal partner, and a wide variety of fungi are able to form lichen associations. The diversity of algal types is much smaller, and many different kinds

● **Figure 19.51 Lichens.** (a) A lichen growing on a branch of a dead tree. (b) Lichens coating the surface of a large rock.

● **Figure 19.52 Lichen structure.** Photomicrograph of a cross section through a lichen. The algal layer is positioned within the lichen structure so as to receive the most sunlight.

of lichens may have the same algal component. Some lichens contain cyanobacteria, frequently N_2-fixing species, instead of algae as the phototrophic component. The algae or cyanobacteria are usually present in defined layers or clumps within the lichen structure (Figure 19.52).

The fungus clearly benefits from associating with the alga, but how does the alga benefit? *Lichen acids*, complex organic compounds excreted by the fungus, promote the dissolution and chelation of inorganic nutrients needed by the alga. Another role of the fungus is to protect the phototroph from drying; most of the habitats in which lichens live are dry (rock, bare soil, roof tops) (Figure 19.52), and fungi are, in general, much better able to tolerate dry conditions than are algae.

Most lichens grow extremely slowly. For example, a 2-cm lichen observed on the surface of a rock may actually be several years old. Measurements of lichen growth vary from 1 mm or less per year to over 3 cm/year, depending on the organisms composing the symbiosis, and the amount of rainfall and sunlight received.

Mycorrhizae

Mycorrhizae (literally, "root fungus") refers to the symbiotic association between plant roots and fungi. The roots of the majority of terrestrial plants are mycorrhizal. There are two classes of mycorrhizae. In **ectomycorrhizae** (Figure 19.53●), fungal cells form an extensive sheath around the outside of the root with only little penetration into the root tissue itself. In **endomycorrhizae**, the fungal mycelium becomes deeply embedded within the root tissue.

Ectomycorrhizae are found mainly in forest trees, especially conifers, beeches, and oaks, and are most highly developed in boreal and temperate forests. In such forests, almost every root of every tree is mycorrhizal. The root system of a mycorrhizal tree such as pine (genus *Pinus*) is composed of both long and short roots. The short roots, which are characteristically dichotomously branched in *Pinus* (Figure 19.53*a*), show typical fungal colonization, and long roots are also frequently colonized. Endomycorrhizae are even more common than ectomycorrhizae. *Arbuscular mycorrhizae*, a type of endomycorrhizae, are found in the roots of over 80% of all terrestrial plant species so far examined.

Most mycorrhizal fungi do not catabolize cellulose and leaf litter, but instead use simple carbohydrates for growth and usually have one or more vitamin requirements. They obtain their carbon from root secretions but get inorganic minerals from the soil. Mycorrhizal fungi are rarely found in nature except in association with roots and hence most can be considered obligate symbionts. Mycorrhizal fungi produce plant growth substances that induce morphological alterations in the roots, stimulating formation of the mycorrhizal state. However, despite the close relationship between fungus and root, a single species of pine can form a mycorrhizal association with over 40 species of fungi.

The beneficial effect of the mycorrhizal fungus on the plant is best observed in poor soils, where trees that are mycorrhizal thrive but nonmycorrhizal ones do not. For example, when trees are planted in prairie soils, which ordinarily lack a suitable fungal inoculum, trees that are artificially inoculated at the time of planting grow much more rapidly than uninoculated trees (Figure 19.54●). The mycorrhizal plant can absorb nutrients

(a)

● **Figure 19.53 Mycorrhizae.** (a) Typical ectomycorrhizal root of the pine, *Pinus rigida*, with rhizomorphs of the fungus *Thelophora terrestis*. (b) Seedling of *Pinus contorta* (lodgepole pine), showing extensive development of the absorptive mycelium of its fungal associate *Suillus bovinus*. This grows in a fanlike formation from the ectomycorrhizal roots to capture nutrients from the soil.

(b)

● **Figure 19.54 Effect of mycorrhizal fungi on plant growth.** Six-month-old seedlings of Monterey pine (*Pinus radiata*) growing in prairie soil: left, nonmycorrhizal; right, mycorrhizal.

from its environment more efficiently, and thus has a competitive advantage. This improved nutrient absorption is due to the greater surface area provided by the fungal mycelium. For example, in the pine seedling shown in Figure 19.53b, the ectomycorrhizal fungal mycelium makes up the overwhelming part of the absorptive area of the plant root system.

In addition to simply helping plants absorb nutrients, mycorrhizae also appear to play a significant role in controlling plant diversity. Indeed, field experiments have shown that there is a positive correlation between (1) the abundance and diversity of mycorrhizae in a soil, and (2) the extent of the plant diversity that develops in it. Thus, mycorrhizae are a prime example of a plant-microorganism symbiosis that benefits both partners: the mycorrhizal plant is better able to function physiologically and compete successfully in a species-rich plant community, while the fungus benefits from a steady supply of organic nutrients.

19.20 Concept Check

Lichens are symbiotic associations between a fungus and an alga or cyanobacterium. Mycorrhizae are formed from fungi that associate with plant roots and improve their ability to absorb nutrients. Mycorrhizae have a great beneficial effect on plant health and competitiveness.

◆ How do *endomycorrhizae* differ from *ectomycorrhizae*?

◆ Why are mycorrhizal associations with plants considered a type of symbiosis?

19.21 *Agrobacterium* and Crown Gall Disease

Some microorganisms are plant pathogens; that is, they cause plant disease. The genus *Agrobacterium*, a close relative of the root nodule bacterium *Rhizobium* (see next section), comprises organisms that cause the formation of tumorous growths on a wide variety of plants. The two species most widely studied are *A. tumefaciens*, which causes **crown gall disease**, and *A. rhizogenes*, which causes **hairy root disease**.

Although plants often form a benign accumulation of tissue, called a *callus*, when wounded, the growth induced by *A. tumefaciens* (Figure 19.55●) is different in that the callus shows uncontrolled growth. It thus resembles tumor growth in animals. Once induced, these tumors continue to grow in the absence of *Agrobacterium* cells. That is, once *Agrobacterium* has brought about the induction of the tumorous condition, its presence is no longer necessary.

A large plasmid called the **Ti** (*tumor induction*) **plasmid** (Figure 19.56●) must be present in the *Agrobacterium* cells if they are to induce tumor formation (Figure 19.55). In *Agrobacterium rhizogenes*, a similar plasmid called the *Ri plasmid* is necessary for induction of hairy root. Following infection, a part of the Ti plasmid called the *transfer DNA* (T-DNA), is integrated into the plant's genome. T-DNA carries the genes for tumor formation and also for the production of a number of modified amino acids called **opines**. *Octopine* [N^2-(1,3-dicarboxyethyl)-*L*-arginine] and *nopaline* [N^2-(1,3-dicarboxypropyl)-*L*-arginine] are two common opines. Opines are produced by plant cells transformed by T-DNA and are a source of carbon and nitrogen for *Agrobacterium* cells.

Recognition and T-DNA Transfer

To initiate the tumorous state, cells of *Agrobacterium* must first attach to a wound site on the plant. The recognition of *Agrobacterium* by plant tissue involves complementary receptor molecules on the surfaces of the bacterial and plant cells. It is thought that the plant receptor molecule is a type of *pectin* (a complex polysaccharide) and that the bacterial receptor is a type of polysaccharide containing β-glucans, embedded in the cell wall lipopolysaccharide.

Studies with nontumorigenic mutants of *Agrobacterium tumefaciens* have shown that most functions necessary for attachment of the bacterium to the plant are borne on the bacterial chromosome. Following attachment, rapid synthesis of cellulose microfibrils by the bacterium anchors the cells to the wound site and literally entraps the bacterial cells, forming large bacterial aggregates on the plant cell surface. This sets the stage for plasmid transfer from bacterium to plant.

The structure of the Ti plasmid is given in Figure 19.56. Although a number of genes are needed for infectivity, only a small region of the Ti plasmid, the *T-DNA* (Figure 19.56), is actually transferred to the plant. The T-DNA con-

● **Figure 19.55 Crown gall.** Photograph of a crown gall tumor on a tobacco plant caused by crown gall bacteria of the genus *Agrobacterium*. *A. tumefaciens* causes these brown, woody-like growths on a wide variety of plants. Crown gall is a particular problem on fruit trees and various ornamental flowers, such as roses and daisies. The disease usually does not kill the plant but may weaken it and make it more susceptible to drought and other diseases. Detached galls serve as a reservoir of *A. tumefaciens* cells to maintain the infection.

tains genes that induce tumorigenesis. The *vir* genes that reside on the Ti plasmid encode proteins that are essential for T-DNA transfer (Figure 19.56). Expression of *vir* genes is induced by plant signal molecules synthesized by wounded plant tissues. Examples of inducers include the

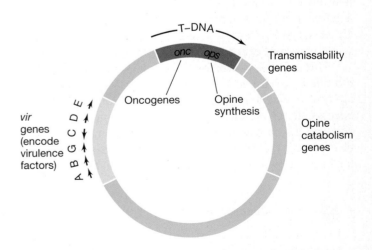

● **Figure 19.56 Structure of the Ti plasmid of *Agrobacterium tumefaciens*.** T-DNA is the region actually transferred to the plant. Arrows indicate the direction of transcription of each gene. The entire Ti plasmid is about 200 kbp of DNA, and the T-DNA is about 20 kbp (∞ Section 31.10 for further discussion of Ti).

phenolic compounds acetosyringone, *p*-hydroxybenzoic acid, and vanillin.

The *vir* genes are the key to T-DNA transfer. The *virA* gene encodes a protein kinase (VirA) that interacts with inducer molecules and then phosphorylates the product of the *virG* gene (Figure 19.57●). The latter is activated by phosphorylation and functions to activate other *vir* genes. The product of the *virD* gene (VirD) has endonuclease activity and nicks DNA in the Ti plasmid in a region adjacent to the T-DNA (Figure 19.57). The product of the *virE* gene is a DNA-binding protein that binds the *single strand* of T-DNA generated from endonuclease activity and transports this small fragment of DNA into the plant cell. VirB, located in the bacterial membrane, mediates transfer of the single strand of DNA between bacterium and plant.

T-DNA transfer (Figure 19.57) thus resembles bacterial conjugation (👓 Figure 10.11). The T-DNA becomes inserted into the nuclear genome of the plant and integration occurs wherever specific inverted or direct tandem repeats are present. The tumorigenesis (*onc*) genes of the Ti plasmid (Figure 19.56) encode enzymes involved in plant hormone production, as well as at least one key enzyme of opine biosynthesis. Expression of these genes leads to tumor formation. The Ri plasmid involved in hairy root disease also contains *onc* genes. However, in this case the genes confer increased auxin responsiveness to the plant; this leads to overproduction of root tissue, resulting in the symptoms of the disease. The Ri plasmid also encodes several opine biosynthetic enzymes.

Genetic Engineering with the Ti Plasmid

From the standpoint of microbiology and plant pathology, both crown gall and hairy root disease involve a unique type of interaction in which bacterial DNA is physically transferred to plant cells. However, once the mechanism of DNA transmission was understood, it quickly became clear that the Ti system could be used as a vector to introduce genetically engineered DNA into plants. In other words, Ti is a *natural plant transformation system*. Thus, the focus of the Ti/crown gall system has shifted away from the disease itself to new applications in biotechnology.

Through the power of genetic engineering a host of modified ("disarmed") Ti plasmids that lack disease genes are now available for the production of transgenic plants. Success stories are already recorded in the areas of herbicide and insect resistance, and many other areas remain to be explored. We discuss the use of the Ti plasmid as a vector in plant biotechnology in more detail in Section 31.10.

19.21 *Concept Check*

The crown gall bacterium *Agrobacterium* enters into a unique relationship with higher plants. A plasmid in the bacterium (the Ti plasmid) is able to transfer part of itself into the genome of the plant, in this way bringing about the production of crown gall disease. The crown gall plasmid has also found extensive use in the genetic engineering of crop plants.

● **Figure 19.57 Mechanism of transfer of T-DNA to the plant cell by *Agrobacterium tumefaciens*.** (a) VirA activates VirG by phosphorylation, and VirG activates transcription of other *vir* genes. (b) VirD is an endonuclease. (c) VirE is a single-strand DNA-binding protein. (d) VirB functions as a conjugation bridge between *Agrobacterium* and the plant cell. Plant DNA polymerase produces the complementary strand to the single strand of T-DNA that is transferred before the DNA gets integrated into the plant genome.

◆ What are *opines* and why are they produced?

◆ How do the *vir* genes differ from *T-DNA* in the Ti plasmid?

◆ How has an understanding of crown gall disease benefited the area of plant molecular biology?

Web Tutorial 19.1 Root Nodule Bacteria and Symbiosis with Legumes

19.22 Root Nodule Bacteria and Symbiosis with Legumes

One of the most important plant bacterial interactions is that between leguminous plants and certain gram-negative *nitrogen-fixing* bacteria. Legumes are defined as plants that bear seeds in pods and are the third largest family of flowering plants. This large group includes such agriculturally important plants as *soybeans, clover, alfalfa, beans,* and *peas.* These plants are key commodities for the soy-processing industry and the feeding of domesticated animals, as well as major vegetables for human nutrition. Several agricultural industries revolve around leguminous crops, and the ability of legumes to grow without nitrogen fertilizer saves farmers millions of dollars in fertilizer costs yearly.

Rhizobium, Bradyrhizobium, Sinorhizobium, Mesorhizobium, Azorhizobium, and *Photorhizobium* are gram-negative motile rod-shaped Proteobacteria that can grow free-living in soil or can infect leguminous plants and establish a symbiotic existence. Infection of the roots of a legume with the appropriate species of one of these genera leads to the formation of **root nodules** (Figure 19.58●) that fix nitrogen (⌀⌀ Section 17.28). Nitrogen fixation by these symbioses are of considerable agricultural importance, as it leads to significant increases in combined nitrogen in the soil. Because nitrogen deficiencies often occur in unfertilized bare soils, nodulated legumes can grow well in areas where other plants cannot (Figure 19.59●).

Leghemoglobin and Cross-Inoculation Groups

Under normal conditions, neither the legume plant nor its bacterial symbiont can fix nitrogen. How then does interaction between the two lead to nitrogen fixation? In pure culture, rhizobia are able to fix N_2 alone when grown under *microaerophilic* conditions. Apparently the rhizobia need some O_2 to generate energy for N_2 fixation, but their nitrogenases (like those of other nitrogen-fixing organisms, ⌀⌀ Section 17.28) are inactivated by O_2.

In the nodule, precise O_2 levels are controlled by the O_2-binding protein **leghemoglobin.** This red, iron-containing protein is present in healthy N_2-fixing nodules (Figure 19.60●) and is induced through the interaction of the plant host and the bacterial symbiont. Leghemoglobin functions as an "oxygen buffer," cycling between the oxidized (Fe^{3+}) and reduced (Fe^{2+}) forms to keep free O_2 levels within the nodule low. The ratio of leghemoglobin-bound O_2 to free O_2 in the root nodule is on the order of 10,000:1.

About 90% of all leguminous plant species are capable of nodulation. However, there is a marked specificity between species of legume and rhizobial species. A single rhizobial species is generally able to infect certain species of legumes and not others. A species or group of rhizobia able to infect a group of related

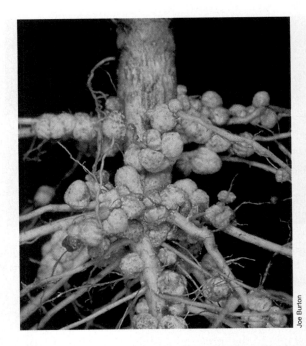

● **Figure 19.58 Soybean root nodules.** The nodules develop by infection with *Bradyrhizobium japonicum.* The main stem of this soybean plant is about 0.5 cm in diameter.

Joe Burton

legumes is called a **cross-inoculation group** (Table 19.8). However, even if a rhizobial strain can infect a certain legume, it is not always able to bring about the production of nitrogen-fixing nodules. If the strain is genetically *ineffective,* the nodules formed will be small, greenish-white, and incapable of fixing nitrogen. By contrast, if the strain is *effective,* the nodule will be large, reddish (Figure 19.60), and nitrogen-fixing. Effectiveness is determined by genes in the bacterium (see the discussion of *nod* genes later in this section) and can be measured by acetylene reduction (⌀⌀ Section 17.28).

● **Figure 19.59 Effect of nodulation on plant growth.** A field of unnodulated (left) and nodulated (right) soybean plants growing in nitrogen-poor soil.

Ben B. Bohlool

● **Figure 19.60** **Root nodule structure.** Sections of root nodules from the legume *Coronilla varia*, showing the reddish pigment leghemoglobin.

Stages in Root Nodule Formation

The stages in the infection and development of root nodules are now fairly well understood (Figure 19.61●). The stages are:

1. **Recognition** of the correct partner on the part of both plant and bacterium and **attachment** of the bacterium to root hairs;

2. **Excretion** of nod factors by the bacterium;

3. **Bacterial invasion** of the root hair;

4. **Travel** to the main root via the infection thread;

5. **Formation of modified bacterial cells, bacteroids**, within the plant cells and development of the nitrogen-fixing state; and

6. **Continued plant and bacterial division** and formation of the mature **root nodule**.

We now explore nodulation in more detail.

1. Recognition and attachment (rhicadhesin-mediated)

Root hair

Rhizobial cell

2. Excretion of nod factors by bacterium causing root hair curling

3. Invasion. Rhizobia penetrate root hair and multiply within an "infection thread"

4. Bacteria in infection thread grow toward root cell

Infection thread

Invaded plant cells and those nearby are stimulated to divide

5. Formation of bacteroid state within plant cell

Nodules

6. Continued plant and bacterial cell division

● **Figure 19.61** **Steps in the formation of a root nodule in a legume infected by *Rhizobium*.** For a photomicrograph of an infection thread, see Figure 19.62*a*. For photos of root nodules, see Figures 19.58 and 19.60. Formation of the bacteroid state is a prerequisite for nitrogen fixation.

Table 19.8	Major cross-inoculation groups of leguminous plants
Host plant	**Nodulated by**
Pea	*Rhizobium leguminosarum* biovar *viciae*[a]
Bean	*Rhizobium leguminosarum* biovar *phaseoli*[a]
Bean	*Rhizobium tropici*
Lotus	*Mesorhizobium loti*
Clover	*Rhizobium leguminosarum* biovar *trifolii*[a]
Alfalfa	*Sinorhizobium meliloti*
Soybean	*Bradyrhizobium japonicum*
Soybean	*Bradyrhizobium elkanii*
Soybean	*Rhizobium fredii*
Sesbania rostrata (a tropical legume)	*Azorhizobium caulinodans*

[a] Several varieties (biovars) of *Rhizobium leguminosarum* exist, each capable of nodulating a different legume.

Attachment and Infection

The roots of leguminous plants secrete a variety of organic compounds that stimulate the growth of a rhizosphere microflora. This stimulation is not restricted to the rhizobia but occurs with a variety of rhizosphere bacteria. If there are rhizobia in the soil, they grow in the rhizosphere and build up to high population densities. Attachment of bacterium to plant in the legume-*Rhizobium* symbiosis is the first step in the formation of nodules (Figure 19.61). A specific adhesion protein called *rhicadhesin* is present on the surfaces of species of *Rhizobium* and *Bradyrhizobium*. Rhicadhesin is a calcium-binding protein and may function by binding calcium complexes on the root hair surface. Other substances, such as carbohydrate-containing proteins called *lectins* and specific receptors in the plant cell membrane, also play roles in plant-bacterium attachment.

Initial penetration of rhizobial cells into the root hair is via the root hair tip. Following binding, the root hair curls as a result of substances excreted by the bacterium called *nod factors* (see later). Following curling, the bacteria enter the root hair and induce formation by the plant of a cellulosic tube, called the **infection thread** (∞ Figure 19.62*a*), which spreads down the root hair. Root cells adjacent to the root hairs subsequently become infected by rhizobia, and nod factors stimulate plant cell division. Continued plant cell division leads to formation of the nodule (Figures 19.62● and 19.63●, see also Figures 19.58 and 19.60).

Bacteroids

The rhizobia multiply rapidly within the plant cells and are transformed into swollen, misshapen, and branched forms called **bacteroids**. Bacteroids become surrounded singly or in small groups by portions of the plant cell membrane to form structures called the **symbiosome** (see Figure 19.67). Only after the formation of the symbiosome does nitrogen fixation begin. (Nitrogen-fixing nodules can be detected by acetylene reduction to ethylene, ∞ Section 17.28 and Figure 17.71.) When the plant dies, the nodule deteriorates, releasing bacteria into the soil. Bacteroids are incapable of division, but a small number of dormant rod-shaped rhizobial cells are always present in the nodule. These now proliferate, using some of the products of the deteriorating nodule as nutrients. The bacteria can then initiate the infection in other roots or maintain a free-living existence in the soil.

Nodule Formation: Nod Genes, Nod Proteins, and Nod Factors

Genes that direct the steps in nodulation of a legume by a specific rhizobial strain are called *nod genes* (Figure 19.64●). Many *nod* genes from different rhizobia are highly conserved and are contained on large plasmids called *Sym* (for *symbiosis*) *plasmids*. In addition to *nod* genes, which direct specific nodulation events, Sym plasmids contain *specificity genes*; these restrict a particular species of rhizobia to a particular host plant. Indeed, cross-inoculation group specificity (Table 19.8) can be transferred across species of rhizobia by simply transferring the respective Sym plasmid.

In the Sym plasmid of *Rhizobium leguminosarum* biovar *viciae*, *nod* genes are located between two clusters of genes for nitrogen fixation, the *nif* genes (Figure 19.64). Ten *nod* genes have been identified in this species. The *nodABC* genes are involved in the production of oligosaccharides, called *nod factors*, which induce root hair curling and trigger plant cell division, eventually leading to formation of the nodule (Figure 19.62). Nod factors consist of a backbone of *N*-acetylglucosamine to which various substituents are linked (Figure 19.65●). Host specificity is determined by the precise structure of the nod factor

(a)

Ben B. Bohlool

(c)

Jacques Vasse, Jean Dénarié, and Georges Truchet

(b)

(d)

Jacques Vasse, Jean Dénarié, and Georges Truchet

●**Figure 19.62 The infection thread and formation of root nodules.** (a) An infection thread formed by cells of *Rhizobium leguminosarum* biovar *trifolii* formed on a root hair of white clover (*Trifolium repens*). The infection thread consists of a cellulosic tube through which bacteria move to root cells. (b–d) Nodules from alfalfa roots infected with cells of *Sinorhizobium meliloti* shown at different stages of development. Cells of both *R. leguminosarum* biovar *trifolii* and *Sinorhizobium meliloti* are about 2 μm long. The time course of nodulation events from infection to effective nodule is about one month in both soybean and alfalfa. Photos b–d reprinted with permission from *Nature* 351:670–673 (1991), © Macmillan Magazines Ltd.

P. J. Dart and F. V. Mercer

● **Figure 19.63** **Cross-section through a legume root nodule.** Electron micrograph of a thin section through a single bacteria-filled cell of a subterranean clover nodule. Cells of *Rhizobium leguminosarum* biovar *trifolii* are about 2 μm long.

(a)

Species	R_1	R_2
Sinorhizobium meliloti	C16:2 or C16:3	SO_4^{2-}
Rhizobium leguminosarum biovar *viciae*	C18:1 or C18:4	H or Ac

(b)

● **Figure 19.65** **Nod factors.** (a) The general structure of the nod factor produced by *Sinorhizobium meliloti* and *Rhizobium leguminosarum* biovar *viciae* and (b) a table of the structural differences (R_1, R_2) that define the precise nod factor of each species. The central hexose unit can repeat up to three times. C16:2, palmitic acid with two double bonds; C16:3, palmitic acid with three double bonds; C18:1, oleic acid with one double bond; C18:4, oleic acid with four double bonds; Ac, acetyl.

produced by a given species of *Rhizobium*. Thus, besides the *nodABC* genes, which are universal and whose products synthesize the nod backbone, each species of *Rhizobium* contains additional *nod* genes that encode proteins that introduce chemical variations on the basic nod factor backbone to form the species-specific nod factors. (Figure 19.65).

In *Rhizobium leguminosarum* biovar *viciae*, the gene *nodD* encodes a regulatory protein, NodD, which controls transcription of other *nod* genes (Figure 19.64). NodF encodes a specific acyl carrier protein used by NodA to acylate the nod factor (Figure 19.65b), NodE and NodL are involved in host range (cross-inoculation groups), NodM is a glucosamine synthase involved in nod factor synthesis, and NodI and NodJ are membrane proteins that function to export nod factors from the bacterial cells.

NodD binds to DNA upstream of *nod* structural gene operons, and after interacting with inducer molecules, promotes transcription; thus NodD is a type of positive regulatory protein (∞ Section 8.6). Key inducers are plant *flavonoids*, organic molecules that are widespread

plant products (Figure 19.66a●). Interestingly, some flavonoids that are structurally very closely related to *nodD* inducers in *R. leguminosarum* biovar *viciae* (Figure 19.66a) function to *inhibit* induction of *nod* genes in other rhizobial species (Figure 19.66b). This indicates that part of the specificity observed between plant and bacterium in the *Rhizobium*-legume symbiosis lies in the chemical nature of the flavonoids excreted by a particular plant.

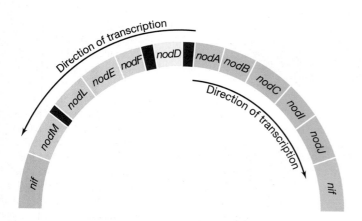

● **Figure 19.64** **Nod genes.** Organization of the *nod* gene cluster on the Sym plasmid of *Rhizobium leguminosarum* biovar *viciae*, the species that nodulates peas. The product of *nodD* controls transcription of other *nod* genes. The *nod* boxes are highlighted in red, and the arrows indicate the direction of transcription of *nod* genes. The genes labeled *nif* encode proteins for nitrogen fixation (∞ Section 17.28).

5, 7, 3′, 4′-Tetrahydroxyflavone
(a)

5, 7, 4′-Trihydroxyisoflavone
(b)

Inducer
Inhibitor

● **Figure 19.66** **Plant flavonoids and nodulation.** Structure of a flavonoid molecule that is (a) an inducer of *nod* gene expression and (b) an inhibitor of *nod* gene expression in *Rhizobium leguminosarum* biovar *viciae*, the species that nodulates peas. Note the similarities in the structures of both molecules. The common name of the structure shown in (a) is *luteolin*, and it is a flavone derivative. The structure in (b) is called *genistein*, and it is an isoflavone derivative.

Biochemistry of Nitrogen Fixation in Nodules

As discussed in Section 17.28, nitrogen fixation requires the enzyme **nitrogenase**, a large two-component protein containing iron and molybdenum. Nitrogenase in root nodules has characteristics similar to the enzyme from free-living N_2-fixing bacteria, including O_2 sensitivity and the ability to reduce acetylene as well as N_2 (Figure 17.71). Nitrogenase is localized within the bacteroids themselves and is not released into the plant cytoplasm.

Bacteroids are totally dependent on the plant for the energy sources for N_2 fixation. The major organic compounds transported across the symbiosome membrane and into the bacteroid proper are citric acid cycle intermediates—in particular, the C_4 organic acids *succinate, malate,* and *fumarate* (Figure 19.67●). These are used as electron donors for ATP production and, following conversion to pyruvate, as the ultimate source of electrons for the reduction of N_2 (Figure 17.71).

The first stable product of N_2 fixation is *ammonia,* and assimilation of ammonia into organic nitrogen compounds in the root nodule is carried out primarily by the plant. Although bacteroids can assimilate some ammonia into organic form, the levels of ammonia-assimilatory enzymes in bacteroids are quite low. By contrast, the ammonia-assimilating enzyme *glutamine synthetase* (Figure 5.27*b*) is present in high levels in the plant cell cytoplasm. Hence, ammonia transported from the bacteroid to the plant cell can be assimilated by the plant as the amino acids *glutamine* and *asparagine* (see Figure 19.67).

Stem-Nodulating Rhizobia

Although most leguminous plants form nitrogen-fixing nodules on their *roots,* a few legume species bear nodules on their *stems.* Stem-nodulated leguminous plants are quite widespread in tropical regions where soils are often nitrogen-deficient because of leaching and intense biological activity. The major experimental system here has been the tropical aquatic legume *Sesbania,* which is nodulated by the bacterium *Azorhizobium caulinodans* (Figure 19.68●).

Stem nodules typically form in the submerged portion of the stems or just above the water level (Figure 19.68). The general sequence of events in formation of stem nodules in *Sesbania* strongly resembles that of root nodules. An infection thread is formed, and bacteroid formation precedes the N_2-fixing state.

Curiously, some stem-nodulating rhizobia produce bacteriochlorophyll *a* and thus have the potential to carry out anoxygenic photosynthesis (Section 17.4). Bacteriochlorophyll-containing rhizobia, grouped in the genus *Photorhizobium,* are widespread in nature, particularly in association with tropical legumes. In these

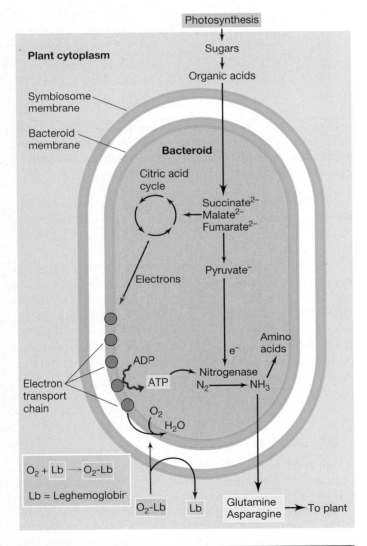

● **Figure 19.67 The root nodule bacteroid.** Schematic diagram of major metabolic reactions and nutrient exchanges occurring in the bacteroid. The symbiosome is a collection of bacteroids surrounded by a single membrane originating from the plant.

● **Figure 19.68 Stem nodules caused by stem-nodulating *Azorhizobium.*** The photograph shows the stem of the tropical legume *Sesbania rostrata.* On the left side of the stem are uninoculated sites, on the right identical sites inoculated with stem–nodulating rhizobia.

species, light could drive at least part of the energy-demanding process of N₂ fixation.

Nonlegume Nitrogen-Fixing Symbioses: *Azolla-Anabaena* and *Frankia*

Nitrogen-fixing symbioses involving microorganisms other than rhizobia occur in a variety of nonleguminous plants. For example, nitrogen-fixing cyanobacteria form symbioses with a variety of plants. The water fern *Azolla* contains a species of heterocystous N₂-fixing cyanobacteria called *Anabaena azollae* within small pores of its fronds (Figure 19.69●). *Azolla* has been used for centuries to enrich rice paddies with fixed nitrogen. Before planting rice, the farmer allows the surface of the rice paddy to become densely covered with *Azolla*. As the rice plants grow, they eventually crowd out the *Azolla*, leading to death of the fern and release of its nitrogen, which is assimilated by the rice plants. By repeating this process each growing season, the farmer can obtain high yields of rice without the need for nitrogenous fertilizers.

The alder tree (genus *Alnus*) has nitrogen-fixing root nodules (Figure 19.70*a*●) that harbor a filamentous, streptomycete-like, nitrogen-fixing organism called *Frankia*. Although when assayed in cell extracts the nitrogenase of *Frankia* is sensitive to molecular oxygen, like intact cells of *Azotobacter* (⬤◯◯ Section 12.9), intact cells of *Frankia* fix N₂ at full oxygen tensions. This is because *Frankia* protects its nitrogenase by localizing the enzyme in terminal swellings on the cells called *vesicles* (Figure 19.70b). The vesicles contain thick walls that retard O₂ diffusion, thus maintaining the O₂ tension within vesicles at levels compatible with nitrogenase activity. In this regard,

(a)

(b)

● **Figure 19.70** *Frankia* **nodules and** *Frankia* **cells.** (a) Root nodules of the common alder *Alnus glutinosa*. (b) *Frankia* culture purified from nodules of *Comptonia peregrina*. Note vesicles (spherical structures) on the tips of hyphal filaments.

Frankia vesicles resemble the heterocysts produced by some filamentous cyanobacteria as localized sites of N₂ fixation (see Figure 19.69b; ⬤◯◯ Section 12.25 and Figure 12.80).

Alder is a characteristic pioneer tree able to colonize bare soils at nutrient-poor sites, probably because of its ability to enter into a symbiotic nitrogen-fixing relationship with *Frankia*. A number of other small woody plants are nodulated by *Frankia*. Unlike the *Rhizobium*–legume relationship, a single strain of *Frankia* can form nodules on several different species of plants, suggesting that the *Frankia*/root nodule sym-biosis is less specific in this regard.

(a)

(b)

● **Figure 19.69** *Azolla-Anabaena* **symbiosis.** (a) Intact association showing a single plant of *Azolla pinnata*. The diameter of the plant is approximately 1 cm. (b) Cyanobacterial symbiont *Anabaena azollae* as observed in crushed leaves of *A. pinnata*. Single cells of *Anabaena azollae* are about 5 μm wide. Note the oval-shaped *heterocysts* (lighter color), the site of nitrogen fixation in the cyanobacterium. For discussion of the heterocyst see Section 12.25 and Figure 12.80.

⬟ **19.22** *Concept Check*

One of the most widespread and important plant-microbial symbioses is that between legumes and certain nitrogen-fixing bacteria. The bacteria induce the formation of root nodules within which the nitrogen-fixing process occurs. The plant provides the energy source needed by the root nodule bacteria, and the bacteria provide fixed nitrogen for the growth of the plant. Legume root nodule bacteria play an important agricultural role because many important crop plants are legumes. Other nitrogen-fixing symbioses include the water fern *Azolla* and the nodule-forming *Frankia*.

◆ What effects do nod factors have on the legume plant?

◆ What is *leghemoglobin* and what is its function?

◆ What is a *bacteroid* and what occurs within it?

◆ What are the major similarities and differences between *Rhizobium* and *Frankia*?

REVIEW QUESTIONS

1. List some of the key *resources* and *conditions* that microorganisms need to thrive in their habitats (⚭ Sections 19.1 and 19.2).

2. Explain why both obligately *anaerobic* and obligately *aerobic* bacteria can be isolated from the same soil sample (⚭ Section 19.2).

3. The surface of a rock in a flowing stream will often contain a biofilm. What advantages could be conferred on bacteria growing in a biofilm compared with growth in a flowing stream (⚭ Section 19.3)?

4. How can biofilms complicate treatment of infectious diseases (⚭ Section 19.3)?

5. In what soil horizon are microbial numbers and activities the highest, and why (⚭ Section 19.4)?

6. How and in what way does an input of organic matter, such as sewage, effect the O_2 content of a river or stream (⚭ Section 19.5)?

7. Assuming proteorhodopsin to be a functional analog of bacteriorhodopsin, describe how this molecule might benefit a cell living in a very organically deficient environment such as the open ocean (⚭ Section 19.6).

8. What is the difference between *barotolerant* and *barophilic* bacteria? Between these two groups and *extreme barophiles*? What properties do barotolerant, barophilic, and extremely barophilic bacteria have in common (⚭ Section 19.7)?

9. What evidence from hydrothermal vents exists to support the idea that prokaryotes are growing at extremely high temperatures (⚭ Section 19.8)?

10. Why can it be said that the oxygen and carbon cycles are interconnected (⚭ Section 19.9)?

11. How can organisms such as *Syntrophobacter* and *Syntrophomonas* grow when their metabolism is based on thermodynamically unfavorable reactions? How are these organisms grown in laboratory culture (⚭ Section 19.10)?

12. What is a *rumen* and how do the digestive processes operate in the ruminant digestive tract? What are the major benefits and the disadvantages of a rumen system? How does a cecal animal compare with a ruminant (⚭ Section 19.11)?

13. Compare and contrast the processes of *nitrification* and *denitrification* in terms of the organisms involved, the environmental conditions that favor each process, and the changes in nutrient availability that accompany each process (⚭ Section 19.12).

14. Why is sulfate reduction the main form of anaerobic respiration in marine environments, whereas methanogenesis dominates in freshwater? Does any methanogenesis occur in the marine environment? If so, how (⚭ Section 19.13)?

15. What organisms are involved in cycling sulfur compounds anoxically? If sulfur chemolithotrophs had never evolved, would there be a problem in the microbial cycling of sulfur compounds? What *organic* sulfur compounds are of interest in nature (⚭ Section 19.13)?

16. Why are most iron-oxidizing chemolithotrophs obligate aerobes, and why are most iron oxidizers acidophilic (⚭ Section 19.14)?

17. Explain how spontaneous chemical reactions can acidify a coal seam and how both chemical reactions and *Acidithiobacillus ferrooxidans* continue the production of acid thereafter (⚭ Section 19.14).

18. How is *Acidithiobacillus ferrooxidans* useful in the mining of copper ores? What crucial step in the indirect oxidation of copper ores is carried out by *A. ferrooxidans*? How is copper recovered from copper solutions produced by leaching (⚭ Section 19.15)?

19. How is mercury detoxified by the *mer* system (⚭ Section 19.16)?

20. What physical and chemical conditions are necessary for the rapid microbial degradation of oil in aquatic environments? Design an experiment that would allow you to test which conditions optimized the oil oxidation process (⚭ Section 19.17).

21. What are *xenobiotic compounds* and why might microorganisms have difficulty catabolizing them (⚭ Section 19.18)?

22. How are reductive dechlorinators of benefit in solving environmental pollution problems (⚭ Section 19.18)?

23. How do mycorrhizae improve the growth of trees (⚭ Section 19.20)?

24. What genetic information resides on the Ti plasmid? Which gene(s) could be deleted from the Ti plasmid without affecting tumorigenesis (⚭ Section 19.21)?

25. Compare and contrast the production of a plant tumor by *Agrobacterium tumefaciens* and a root nodule by a *Rhizobium* species. In what ways are these structures similar? In what ways are they different? Of what importance are plasmids to the development of both structures (⚭ Sections 19.21 and 19.22)?

26. Describe the steps in the development of root nodules on a leguminous plant. What is the nature of the recognition between plant and bacterium and how do nod factors help control this? How does this compare with recognition in the *Agrobacterium*–plant system (⚭ Section 19.22)?

APPLICATION QUESTIONS

1. Compare and contrast a lake ecosystem with a hydrothermal vent ecosystem. How does energy enter each ecosystem? What is the basis of primary production in each ecosystem? What nutritional classes of organisms exist in each ecosystem, and how do they feed themselves?

2. Imagine a sewage plant that is releasing sewage containing high levels of ammonia and phosphate, but very little organic carbon. Growth of what types of organisms might be triggered by this sewage? How would the oxygen profile near and beyond the dump point differ from that shown in Figure 19.10a?

3. Look at the data of Figure 19.13. Keeping in mind that the open ocean waters are highly oxic, predict the possible metabolic lifestyles of open ocean *Archaea* and *Bacteria*. Why might proteorhodopsin be present only in one group of organisms but not the other?

4. ^{14}C-Labeled cellulose is added to a vial containing a small amount of sewage sludge and sealed under anoxic conditions. A few hours later $^{14}CH_4$ appears in the vial. Discuss what has happened to yield such a result.

5. Imagine that you have discovered a new animal that consumes only grass in its diet. You suspect it to be a ruminant and have available a specimen for anatomical inspection. If this animal is a ruminant, describe the position and basic components of the digestive tract you would expect to find and any key microorganisms and substances you might look for.

6. Acid mine drainage is in part a chemical process and in part a biological process. Discuss the chemistry and microbiology that lead up to acid mine drainage and point out the key reaction(s) that are biological.

20

MICROBIAL GROWTH CONTROL

A variety of physical, chemical, and biological methods are used to control microbial growth. Filters, shown here trapping a population of microorganisms, are often used to isolate microorganisms from liquid media or to sterilize liquids.

WORKING GLOSSARY

Aminoglycoside an antibiotic such as streptomycin, containing amino sugars linked by glycosidic bonds

Antibiotic chemical substance produced by a microorganism that kills or inhibits the growth of another microorganism

Antimicrobial drug resistance the acquired ability of a microorganism to grow in the presence of an antimicrobial drug to which the microorganism is usually susceptible

Antimicrobial agent a chemical compound that kills or inhibits the growth of microorganisms

Antiseptic (germicide) chemical agent that kills or inhibits growth of microorganisms and is sufficiently nontoxic to be applied to living tissues

Autoclave a sterilizer that destroys microorganisms with temperature and steam under pressure

β-lactam antibiotic an antibiotic, including penicillin, that contains the four-membered heterocyclic β-lactam ring

Bacteriocidal agent an agent that kills bacteria

Bacteriostatic agent an agent that inhibits bacterial growth

Broad-spectrum antibiotic an antibiotic that acts on both gram-positive and gram-negative *Bacteria*

Chemotherapeutic agent an antimicrobial agent that can be used internally

Decontamination treatment that renders an object or inanimate surface safe to handle

Disinfectant an antimicrobial agent used only on inanimate objects

Disinfection the elimination of microorganisms from inanimate objects or surfaces

Fungicidal agent an agent that kills fungi

Fungistatic agent an agent that inhibits fungal growth

Fusion inhibitor a peptide that blocks the fusion of viral and target cell membranes

Germicide (antiseptic) chemical agent that kills or inhibits growth of microorganisms and is sufficiently nontoxic to be applied to living tissues

Growth factor analog a chemical agent that is related to and blocks the uptake of a growth factor

HEPA filter a *high efficiency particulate air* filter used in laboratories and industry to remove particles, including microbes, from intake or exhaust air flow

Inhibition the reduction of microbial growth because of a decrease in the number of organisms present or alterations in the microbial environment

Interferon a cytokine protein produced by virus-infected cells that induces signal transduction in nearby cells, resulting in transcription of antiviral genes and expression of antiviral proteins

Lysis loss of cellular integrity with release of cytoplasmic contents

MIC minimum inhibitory concentration—the minimum concentration of a substance necessary to prevent microbial growth

NNRTI non-nucleoside reverse transcriptase inhibitor

NRTI nucleoside reverse transcriptase inhibitor

Pasteurization reduction of the microbial load in heat-sensitive liquids to kill disease-producing microorganisms and reduce the number of spoilage microorganisms

Penicillin a class of antibiotics that inhibit bacterial cell wall synthesis, characterized by a β-lactam ring

PI protease inhibitor

Quinolone a synthetic antibacterial compound that interacts with DNA gyrase and prevents supercoiling of bacterial DNA

Sanitizer agent that reduces, but may not eliminate, microbial numbers to a safe level

Semisynthetic penicillin a natural penicillin that has been chemically altered

Sterilant (sterilizer) (sporicide) a chemical agent that destroys all forms of microbial life

Sterilization the killing or removal of all living organisms and their viruses from a growth medium

Tetracycline an antibiotic characterized by the four-ring naphthacene structure

Viricidal agent an agent that stops viral replication and activity

Viristatic agent an agent that inhibits viral replication

W ith this chapter we begin to study the relationships between microorganisms and humans. We start with the agents and methods used for control of microbial growth. In some cases, we can eliminate microbial growth entirely, with **sterilization**—the killing or removal of all viable organisms within a growth medium. Agents that destroy or kill bacteria, for example, are called **bacteriocidal**. Often, however, sterility is not attainable, but we can still control microorganisms effectively by limiting their growth, the process of **inhibition**. Agents that inhibit bacterial growth are said to be **bacteriostatic**.

Methods for inhibiting rapid microbial growth include **decontamination** and **disinfection**. *Decontamination* is the treatment of an object or surface to make it safe to handle. For example, simply wiping a table after a meal removes contaminating microorganisms and their potential nutrients. *Disinfection*, in contrast, directly targets pathogens, although it may not eliminate all microorganisms. Specialized chemical or physical agents called **disinfectants** can kill microorganisms or inhibit microbial growth. Dilute bleach (sodium hypochlorite) solutions, for example, are routinely used to clean and disinfect food preparation areas.

At times, it may still be necessary to destroy all microorganisms. Sterilization, though difficult to achieve, completely prevents contamination and growth of microorganisms. Such measures are necessary, for instance, when making microbiological media or preparing surgical instruments. The goal for all decontamination and sterilization procedures is to reduce or eliminate the *microbial load*, or the number of viable microorganisms.

Microbial control *in vivo* is much more difficult: Clinically useful bacteriocidal or bacteriostatic agents must reduce or prevent microbial growth, while causing no harm to the host cell. This goal is achieved by a wide variety of selective natural and synthetic chemotherapeutic agents.

In this chapter we will first examine methods of microbial control that are used *in vitro*. We will then discuss antimicrobial drugs used *in vivo* in humans.

PHYSICAL ANTIMICROBIAL CONTROL

Physical methods are often used to achieve microbial decontamination, disinfection, and sterilization. Heat, radiation, and filtration can destroy or remove undesirable microorganisms. These methods prevent microbial growth or decontaminate areas or materials harboring microbes. Here we discuss how these methods work and discuss some practical examples.

20.1 Heat Sterilization

Perhaps the most widespread method used for controlling microbial growth is the application of heat, and we consider the process of heat sterilization here.

Measuring Heat Sterilization

For all microorganisms, there is a maximum temperature for growth, beyond which viability decreases. At very high temperatures, virtually all macromolecules lose their structure and their ability to function, a process known as *denaturation*. As shown in Figure 20.1●, death from heating is an exponential (first-order) function, occurring more rapidly as the temperature rises. This means that the rate of death is proportional at any time to the concentration of organisms at that time; the time necessary to kill a definite fraction (for example, 90%) of viable cells is independent of the initial cell concentration. Thus, if we wish to *sterilize* a microbial population, it will take longer at lower temperatures than at higher temperatures. It is therefore necessary to adjust the time and temperature to achieve sterilization for each specific

set of conditions. The *nature* of the heat is also important: Moist heat has better penetrating power than dry heat and produces a faster reduction in the number of living organisms at a given temperature.

The time required for a 10-fold reduction in the population density at a given temperature, called the *decimal reduction time* or *D*, is the most precise way to characterize heat sterilization. Over the range of temperatures usually used in food preparation (*i.e.*, cooking and canning), the relationship between *D* and temperature is essentially exponential. Thus, when the logarithm of *D* is plotted against temperature, a straight line is obtained (Figure 20.2●). The slope of the line indicates the sensitivity of the organism to heat under the conditions employed, and the graph can be used to calculate process times to achieve sterilization, for instance in canning operations (∞ Section 29.2).

Determination of a decimal reduction time requires a large number of viable count measurements (∞ Section 6.5). An easier way of characterizing the heat sensitivity of an organism is to measure the *thermal death time*, the time it takes to kill all cells at a given temperature. To determine the thermal death time, samples of a cell suspension are heated for different times, mixed with culture medium, and incubated. If all the cells have been killed, no growth is observed in the incubated samples. The thermal death time depends on the size of the population tested because a longer time is required to kill all cells in a large population than in a small one. When the number of cells is standardized, it is possible to compare the heat sensitivities of different organisms by comparing their thermal death times at a given temperature.

● **Figure 20.1** **The effect of temperature on the viability of a mesophilic bacterium.** The decimal reduction time, *D*, was obtained for the same mesophilic organism at three different temperatures. *D* is the time at which only 10% of the original population of organisms remains viable at a given temperature. For 70° C, *D* = 3 min; for 60° C, *D* = 12 min; for 50° C, *D* = 42 min.

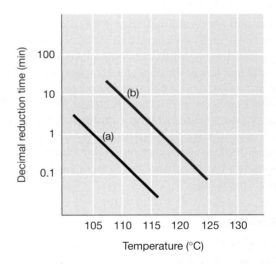

● **Figure 20.2** **The relationship between temperature and the rate of killing as indicated by the decimal reduction time for two different microorganisms.** Data were obtained for decimal reduction times, *D*, at several different temperatures, as in Figure 20.1. For organism (a), a typical mesophile, exposure to 110° C for less than 20 sec resulted in a decimal reduction, while for organism (b), a thermophile, 10 min were required to achieve a decimal reduction.

Endospores and Heat Sterilization

The heat resistance of vegetative cells and bacterial endospores from the same organism varies considerably. For instance, in the autoclave (see below) a temperature of 121° C is normally reached. Under these conditions, endospores may require 4–5 minutes for a decimal reduction, whereas vegetative cells may require only 0.1–0.5 min at 65° C. Because of this variation, effective heat sterilization procedures must be designed to destroy endospores.

Bacterial endospores are able to survive heat that would rapidly kill vegetative cells of the same species. A major factor in heat resistance is the amount and state of *water* within the endospore. During endospore formation, the protoplasm is reduced to a minimum volume as a result of the accumulation of Ca^{2+}-dipicolinic acid complexes and *small acid-soluble spore proteins (SASP)* (∞ Section 4.13). This mixture forms a gel in the cytoplasm, and a thick cortex then forms around the protoplast core. Contraction of the cortex results in a shrunken, dehydrated protoplast with a water content of only 10–30% of a vegetative cell.

The water content of the protoplast coupled with the concentration of SASPs determines the heat resistance of the endospore. If endospores have a low concentration of SASPs and high water content, they have low heat resistance. If they have a high concentration of SASPs and low water content, they have high heat resistance. Water moves freely in and out of endospores, so it is not the impermeability of the endospore coat that excludes water, but the gel-like material in the endospore protoplast.

The nature of the medium in which heating takes place also influences the killing of both vegetative cells and endospores. Microbial death is more rapid at acidic pH, and acid foods such as tomatoes, fruits, and pickles are much easier to sterilize than neutral pH foods such as corn and beans. High concentrations of sugars, proteins, and fats decrease heat penetration and usually increase the resistance of organisms to heat, whereas high salt concentrations may either increase or decrease heat resistance, depending on the organism. Dry cells (and endospores) are more heat-resistant than moist ones; consequently, heat sterilization of dry objects always requires higher temperatures and longer times than sterilization of moist objects.

The Autoclave

The **autoclave** is a sealed heating device that allows the entrance of steam under pressure (Figure 20.3●). Killing of the heat-resistant endospores requires heating at temperatures above boiling with the use of steam under pressure (Figure 20.3a). The usual procedure is to heat at 1.1 kilograms/square centimeter (kg/cm^2) [15 pounds/square inch (lb/in^2)] steam pressure, which yields a temperature of 121° C. At 121° C, the time to achieve sterilization is generally 10–15 minutes (Figure 20.3b).

If bulky objects are being sterilized, heat transfer to the interior will be retarded, and the total heating time must be extended to ensure that the *entire* object is at 121° C for 10–15 minutes. Extended times are also required when large volumes of liquids are being autoclaved because large volumes take longer to reach sterilization temperatures. Note that it is not the *pressure* inside the autoclave that kills the microorganisms but the *high temperature* that can be achieved when steam is placed under pressure.

Pasteurization

Pasteurization uses precisely controlled heat to reduce the microbial populations (*microbial load*) in milk and other heat-sensitive liquids. The process is named for Louis Pasteur, who first used heat for controlling the spoilage of wine. Pasteurization does not kill all organisms and is therefore not synonymous with sterilization. Originally, pasteurization of milk was used to kill pathogenic bacteria, especially the organisms causing tuberculosis, brucellosis, Q fever, and typhoid fever. While these pathogens are no longer common in food in developed countries, pasteurization also prevents the spread of pathogens such as *Listeria monocytogenes*, *Campylobacter* species, *Salmonella* species, and *Escherichia coli* O157:H7 (∞ Sections 29.7–29.10) from common sources such as dairy products and juices. In addition, pasteurization retards the growth of spoilage organisms, dramatically increasing the shelf life of perishable liquids (∞ Sections 29.1 and 29.2).

Pasteurization of milk is usually achieved by passing the milk through a heat exchanger. The milk is pumped through tubing that is in contact with a heat source. Careful control of the milk flow rate and the size and temperature of the heat source raises the temperature of the milk to 71° C for 15 seconds. The milk is then rapidly cooled. The whole process is aptly called *flash pasteurization*.

Milk can also be heated in large vats to 63–66° C for 30 minutes. However, this *bulk pasteurization* method is less satisfactory because the milk heats and cools slowly and must be held at high temperatures for longer times. Flash pasteurization, sometimes done at even higher temperatures and shorter times, alters the flavor less, kills heat-resistant organisms more effectively, and can be done on a continuous-flow basis, making it more adaptable to large modern dairy operations.

 20.1 Concept Check

Sterilization is the killing of all organisms including viruses. Heat is the most widely used method of sterilization. The temperature must eliminate the most heat-resistant organisms, usually bacterial endospores. An autoclave permits applications of steam heat under pressure at temperatures above the boiling point of water, killing endospores. Pasteurization does not sterilize liquids, but reduces microbial load, killing most pathogens and inhibiting the growth of spoilage microorganisms.

♦ Why is heat an effective sterilizing agent?

● **Figure 20.3 Use of the autoclave for sterilization.** (a) The flow of steam through an autoclave. (b) A typical autoclave cycle. Shown is the sterilization of a fairly bulky object. The temperature of the object rises more slowly than the temperature of the autoclave. (c) A modern research autoclave. Note the pressure–lock door and the automatic cycle controls on the right panel. The steam inlet and exhaust fittings are on the right side of the autoclave.

◆ What steps are necessary to ensure the sterility of material that may be contaminated with bacterial endospores?

◆ Distinguish between the need to *sterilize* microbiological media and the need to *pasteurize* apple juice.

20.2 Radiation Sterilization

Heat is just one type of electromagnetic radiation that can sterilize or reduce microbial load. Microwaves, ultraviolet (UV) radiation, X-rays, gamma rays (γ-rays), and electrons are all forms of electromagnetic radiation (∞ Section 10.4) and all can affect microbial growth. However, each type of radiation has a different mode of action. For example, the antimicrobial effects of microwaves are due, at least in part, to thermal effects. UV radiation between 220 and 300 nm in wavelength has sufficient energy to cause modifications or breaks in DNA, sometimes leading to disruption of genetic material and death of the exposed organism (∞ Section 10.4). This "near-visible" UV light is useful for disinfecting surfaces, air, and materials such as water that do not absorb the UV waves. For example, laboratory biological containment cabinets are equipped with a "germicidal" UV light to decontaminate the surface after use (Figure 20.4●). UV radiation cannot penetrate solid, opaque, or light-absorbing surfaces, limiting its use to disinfection of exposed surfaces.

● **Figure 20.4   A biological safety cabinet.** The cabinet is shown with an ultraviolet (UV) radiation source (mercury vapor lamp), which is used for decontamination of the inside surfaces.

Table 20.1	Radiation sensitivity of microorganisms and biological functions	
Species or function	**Type of microorganism**	**$D10^a$(Gy)**
Clostridium botulinum	Gram-positive anaerobic sporulating Bacteria	3300
Clostridium tetani	Gram-positive anaerobic sporulating Bacteria	2400
Bacillus subtilis	Gram-positive aerobic sporulating Bacteria	600
Salmonella typhimurium	Gram-negative Bacteria	200
Lactobacillus brevis	Gram-positive Bacteria	1200
Deinococcus radiodurans	Gram-negative radiation-resistant Bacteria	2200
Aspergillus niger	Mold	500
Saccharomyces cerevisiae	Yeast	500
Foot-and-mouth	Virus	13,000
Coxsackie	Virus	4500
Enzyme inactivation		20,000–50,000
Insect deinfestation	—	1000–5000

^{a} D10 is the amount of radiation necessary to reduce the initial population or activity level 10-fold (one logarithm). Gy = grays. 1 gray = 100 rads.

Ionizing Radiation

Ionizing radiation is electromagnetic radiation of sufficient energy to produce ions and other reactive molecular species from molecules with which the radiation particles collide. Ionizing radiation generates electrons, e^-, hydroxyl radicals, $OH\bullet$; and hydride radicals, $H\bullet$ (∞ Section 6.16). Each of these highly reactive molecules is capable of altering and disrupting biopolymers such as deoxyribonucleic acid (DNA) and protein. The ionization and subsequent degradation of biologically important molecules such as DNA and proteins leads to the death of irradiated cells. Several radiation sources are potentially useful for sterilization.

The unit of radiation is the *roentgen*, which is a measure of the radiation energy output from a source. The standard for biological applications such as sterilization is the *absorbed radiation dose*, measured in *rads* (100 erg/g), or *grays* (1 Gy = 100 rad). Certain microorganisms are much more resistant to radiation than others. Table 20.1 shows the dose of radiation necessary for a 10-fold (one log) reduction in the numbers of selected microorganisms or biological functions.

For example, the amount of energy necessary to achieve a *D10* or *decimal* (10-fold) *reduction* of radiation-sensitive gram-negative *Bacteria* such as *Salmonella typhimurium* is 200 Gy. The data in Table 20.1 provide information similar to the decimal reduction time for heat sterilization (see Section 20.1): The relationship of the survival fraction plotted on a semilogarithmic scale versus the radiation dose in grays is essentially linear (Figure 20.5●).

A standard *killing* dose for radiation sterilization is 12 *D10* requirements. For destruction of the radioresistant endospores of *Clostridium botulinum*, for example, 39,600 Gy are required (Table 20.1). For *Salmonella typhimurium*, the killing dose is 2400 Gy. In general, microorganisms are much more resistant to ionizing radiation than multicellular organisms. For example, the lethal radiation dose for humans is only 10 Gy!

Radiation Practice

Common sources of ionizing radiation include X-ray machines, cathode ray tubes, and radioactive nuclides. These sources produce either X-rays or γ-rays, both of which have sufficient energy and penetrating power to efficiently inhibit microbial growth in solid and liquid media. The principal sources of commercially useful ion-

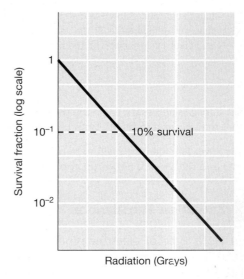

● **Figure 20.5   Relationship between the survival fraction and the radiation dose.** The *D*10, or decimal reduction dose, can be interpolated from the data as shown. The dose in Grays differs significantly for each organism.

izing radiation are radioactive nuclides that emit γ rays. The two most commonly used radioisotopes are ^{60}Co and ^{137}Cs, both relatively inexpensive by-products of nuclear fission.

Radiation is currently used for sterilization and decontamination in the medical supplies and food industries. In the United States, the Food and Drug Administration has approved the use of radiation for sterilization of such diverse items as surgical supplies, disposable labware, drugs, and even tissue grafts (Table 20.2). However, because of the costs and hazards of radiation equipment, this type of sterilization is limited to large industrial applications or very specialized facilities. Food and food products are also routinely irradiated (∞ Section 29.2). In addition to sterilization, pasteurization and insect deinfestation can be accomplished by adjusting the dose of radiation applied.

Radiation is approved by the World Health Organization for decontamination of foods particularly susceptible to microbial contamination, especially spices and fresh meat products such as hamburger and chicken, and is approved in the United States for decontamination of ground meat (∞ Section 29.2). The use of radiation for all these purposes is an established and accepted technology in many countries. However, the practice has not been readily accepted in other countries because of fears of possible radioactive contamination, alteration in nutritional value, production of toxic or carcinogenic products, and perceived production of "off" tastes in irradiated food.

⬡ 20.2 *Concept Check*

Controlled doses of electromagnetic radiation effectively inhibit microbial growth. Ultraviolet radiation is used for decontaminating surfaces and materials that do not absorb light, such as air and water. Ionizing radiation, necessary to penetrate solid or light-absorbing materials, is widely used for sterilization and decontamination in the medical and food industries.

◆ Define the decimal reduction dose and the lethal dose for radiation treatment of microorganisms.

◆ Why is ionizing radiation more effective than UV radiation for sterilization of food products?

Table 20.2	Medical and laboratory products sterilized by radiation	
Tissue grafts	**Drugs**	**Medical and laboratory supplies**
Cartilage	Chloramphenicol	Disposable labware
Tendon	Ampicillin	Culture media
Skin	Tetracycline	Syringes
Heart valve	Atropine	Surgical equipment
	Vaccines	Sutures
	Ointments	

20.3 Filter Sterilization

As we have seen, heat is a very effective way to sterilize most liquids. However, heat-sensitive liquids may be sterilized by *filtration*. A *filter* is a device with pores too small for the passage of microorganisms but large enough to allow the passage of the liquid or gas. The selection of filters for sterilization must account for the size range of the contaminants to be excluded. Some of the largest microbial cells are greater than 10 μm in diameter, but the smallest bacteria are less than 0.3 μm in diameter. Historically, selective filtration methods were used to define and isolate viruses, most of which range from 25 nm to 200 nm in diameter (∞ Section 9.2). Figure 20.6● illustrates these three major types of filters.

Depth Filters

One of the simplest types of filters is the *depth filter*. A depth filter is a fibrous sheet or mat made from a random

(a)

(b)

(c)

● **Figure 20.6 Microbiological filters.** The structure of (a) a depth filter, (b) a conventional membrane filter, and (c) a Nucleopore filter. Depth filters are used as prefilters and for the filtration of liquids with a high amount of suspended particles. Membrane filters are used in many applications in the laboratory and in industry because they are readily available in a wide variety of sizes and porosities, economical, and applicable to nearly all filtration needs. Nucleopore, or nucleation track, filters are useful for isolating specimens for microscopy because filtered material is captured in a single plane on the filter surface.

array of overlapping paper, asbestos, or borosilicate (glass) fibers (Figure 20.6*a*). The depth filter traps particles in the network of fibers that forms throughout the depth of the structure. Because they are rather porous, depth filters are often used as *prefilters* to remove larger particles from liquid suspensions so that the final filter in the sterilization process is not clogged. Depth filters are also used for the filter sterilization of air in industrial processes (∞ Section 30.4). In the home, most forced air heating and cooling systems use a simple depth filter to trap particulate matter including dust, spores, and allergens.

Depth filters are important for a variety of biosafety applications. For example, manipulations of cell cultures, microbial cultures, and growth media require that contamination of both the operator and the experimental materials are minimized. These operations can be efficiently performed in a biological safety cabinet with air flow, both in and out of the cabinet, directed through a **high-efficiency particulate air**, or **HEPA** filter (Figure 20.4).

On a larger scale, control of airborne particulate materials with the HEPA filters allows the construction of "clean rooms" and isolation rooms for quarantine, as well as specialized biological safety laboratories (∞ Section 24.4). HEPA filters generally remove 0.3-μm test particles with an efficiency of at least 99.97%; they effectively remove both smaller and larger particles from the airstream. A typical HEPA filter is a single sheet of borosilicate (glass) fibers that has been treated with a water-repellant binder. The filter is pleated to increase the overall surface area and mounted inside a rigid frame to prevent the pleats from collapsing. HEPA filters come in a variety of shapes and sizes, from several square centimeters for vacuum cleaners, to several square meters for biological containment hoods (Figure 20.4) and room air systems.

Membrane Filters

The most common type of filter for liquid sterilization in the microbiology laboratory is the *membrane filter* (Figure 20.6*b*). Membrane filters are composed of polymers with high tensile strength such as cellulose acetate, cellulose nitrate, or polysulfone, manufactured in such a way as to contain a large number of tiny holes. By adjusting the polymerization conditions during manufacture, the size of the holes in the membrane (and thus the size of the molecules that can pass through) can be precisely controlled. The membrane filter differs from the depth filter because it functions more like a sieve, trapping particles on the filter surface. About 80–85% of the membrane surface area consists of the open pores. The porosity provides for a relatively high fluid flow rate.

Membrane filters for the sterilization of a liquid are illustrated in Figure 20.7●. The filter apparatus is generally sterilized separately from the filter, and the apparatus is assembled aseptically at the time of filtration. The arrangement shown in Figure 20.7*a* is suitable for small volumes of liquid. For large-volume sterile filtration, the membrane filter is assembled in a cartridge and placed

(a)

(b)

● **Figure 20.7 Membrane filters.** (a) Assembly of a reusable membrane filter apparatus. (b) Disposable, presterilized, and assembled membrane filter units. Left: a filter system designed for small volumes. Right: a filter system designed for larger volumes.

in a stainless steel housing. Large-volumes of heat-sensitive fluids are commonly sterilized by filtration in the pharmaceutical industry.

Presterilized membrane filter assemblies for sterilization of small to medium volumes are routinely used in research and clinical laboratories (Figure 20.7*b*). Filtration is accomplished by using a syringe, pump, or vacuum to force the liquid through the filtration apparatus into a sterile collection vessel.

Another type of membrane filter in common use is the **nucleation track (Nucleopore) filter**. These filters are created by treating very thin polycarbonate films (10 μm) with nuclear radiation and then etching the film with a chemical. The radiation causes localized damage to the film, and the etching chemical enlarges these damaged locations into holes. The sizes of the holes can be precisely controlled by the strength of the etching solution and the etching time. A typical nucleation track filter has very uniform holes arranged almost vertically through the thin film (Figure 20.6*c*). Nucleopore filters are commonly used in scanning electron microscopy. An organism can be removed from liquid and concentrated

(a)

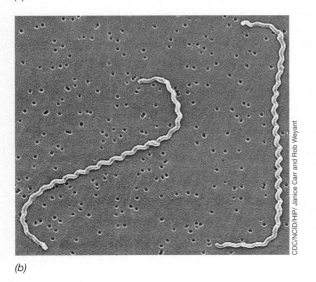

(b)

Carlos Pedrós-Alió and T. D. Brock

CDC/NCID/HIP/ Janice Carr and Rob Weyant

● **Figure 20.8 Scanning electron micrographs of bacteria trapped on Nucleopore membrane filters.** (a) Aquatic bacteria and algae. The pore size is 5 μm. (b) *Leptospira interrogans*. The bacterium is about 0.1 μm in diameter and up to 20 μm in length. The pore size of the filter is 0.2 μm.

in a single plane on the top of the filter, where it can be observed with the microscope (Figure 20.8●).

 20.3 Concept Check

Filters remove microorganisms from air or liquids. Depth filters, including HEPA filters, are used to remove microorganisms and other contaminants from liquids or air. Membrane filters are used for sterilization of heat-sensitive liquids, and nucleation filters are used to isolate specimens for electron microscopy.

◆ Describe the potential use of HEPA filters for maintaining clean air in hospitals.

◆ Why are membrane filters used for sterilization of heat-sensitive liquids?

II CHEMICAL ANTIMICROBIAL CONTROL

In the home, workplace, and laboratory, chemical agents are routinely used to control microbial growth. An **antimicrobial agent** is a natural or synthetic chemical that kills or inhibits the growth of microorganisms. Agents that kill organisms are called *cidal agents*, with a prefix indicating the kind of organism killed. Thus, these agents are termed **bacteriocidal**, **fungicidal**, and **viricidal agents**, killing bacteria, fungi, and viruses, respectively. Agents that do not kill but only inhibit growth are called *static agents*, and these include **bacteriostatic**, **fungistatic**, and **viristatic agents**.

20.4 Chemical Growth Control

Antimicrobial agents vary with regard to *selective toxicity*. Some are nonselective and have similar effects on all cells. Others are far more selective and are more toxic to microorganisms than to animal tissues. Antimicrobial agents with selective toxicity are especially useful for treating infectious diseases because they kill selected microorganisms *in vivo* without harming the host. They will be described later in this chapter. Here we discuss chemical agents that have relatively broad toxicity, but are useful for limiting microbial growth *in vitro*.

Effect of Antimicrobial Agents on Growth

Three distinct effects can be observed when an antimicrobial agent is added to an exponentially growing bacterial culture: *bacteriostatic, bacteriocidal,* and *bacteriolytic* (Figure 20.9●). *Bacteriostatic* agents are frequently inhibitors of protein synthesis and act by binding to ribosomes. If the concentration of the agent is lowered, the agent is released from the ribosome and growth is resumed (Figure 20.9a). Many antibiotics work by this mechanism and will be discussed in Sections 20.6–20.9. *Bacteriocidal* agents bind tightly to their cellular targets and are not removed by dilution, but **lysis**, the loss of cell integrity and release of contents, does not occur (Figure 20.9b). *Bacteriolytic* agents induce killing by cell lysis, which can be observed as a decrease in cell number or in turbidity after the agent is added (Figure 20.9c). Bacteriolytic agents include antibiotics that inhibit cell wall synthesis, such as penicillin (∞ Section 4.8 and see Section 20.8), as well as chemicals like detergents that rupture the cytoplasmic membrane.

Measuring Antimicrobial Activity

Antimicrobial activity is measured by determining the smallest amount of agent needed to inhibit the growth of a test organism, a value called the **minimum inhibitory concentration (MIC)**. To determine the MIC for a given

(a)

(b)

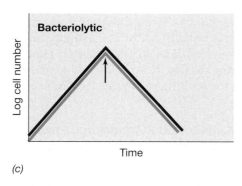

(c)

● **Figure 20.9   Three types of action of antimicrobial agents.** At the time indicated by the arrow, a growth-inhibitory concentration was added to the exponentially growing culture. Note the relationships between viable and total cell counts.

agent, a series of culture tubes is prepared and inoculated, with each tube containing medium with a different concentration of the agent. After incubation, the tubes are checked for visible growth (turbidity). The MIC is the *lowest concentration of agent that completely inhibits the growth of the test organism* (Figure 20.10●). This simple and effective procedure is often called the *tube dilution technique*.

The MIC is not a constant for a given agent, because it is affected by the nature of the test organism used, the inoculum size, the composition of the culture medium, the incubation time, and the conditions of incubation such as temperature, pH, and aeration. However, when all conditions are rigorously standardized, different antimicrobial agents can be compared to determine which is most effective against a given organism. A standard method used to compare antimicrobial agents is the *phenol coefficient method*, a ratio of the MIC of the test

● **Figure 20.10   Antibiotic susceptibility assay by dilution methods.** The assay permits detection of the *minimum inhibitory concentration* (MIC). A series of increasing concentrations of antibiotic is prepared in the culture medium. Each tube is inoculated, and incubation is allowed to proceed. Growth (turbidity) occurs in those tubes with antibiotic concentrations below the MIC.

agent to the MIC of phenol, using the same organism, media, and growth conditions.

Another common assay for antimicrobial activity is *agar diffusion* (Figure 20.11●). A Petri plate containing an agar medium overlayed with a culture of the test organ-

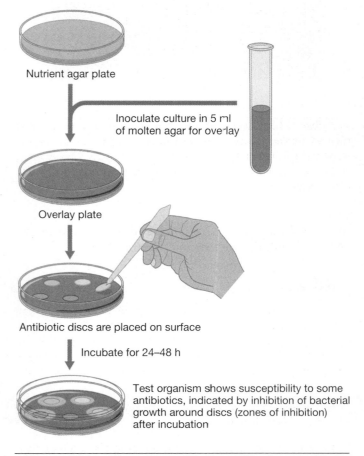

● **Figure 20.11   Measuring antibiotic activity.** An agar diffusion method for assaying antibiotic susceptibility.

ism is prepared. Known amounts of the antimicrobial agent are added to filter paper disks, which are then placed on the surface of the agar. During incubation, the agent diffuses from the filter paper into the agar; the further from the filter paper, the lower the concentration of the agent. At some distance from the disk, the effective MIC is reached. Distal to this point growth occurs, but closer to the disk, growth is absent. A *zone of inhibition* is created with a diameter proportional to the amount of antimicrobial agent added to the disk, the solubility of the agent, the diffusion coefficient, and the overall effectiveness of the agent. This method is routinely used to test for antibiotic susceptibility in pathogens (∞ Section 24.3).

 20.4 Concept Check

Chemicals are often used to control microbial growth. Chemicals that kill organisms are called *cidal* agents; those that inhibit growth are called *static* agents. The effectiveness of an antibacterial chemical agent is assessed by determining the minimum concentration necessary to completely inhibit bacterial growth.

◆ With regard to antibacterial agents, distinguish between *static*, *cidal*, and *lytic* agents.

◆ Describe how the *minimum inhibitory concentration* of an antibacterial agent is determined.

20.5 Chemical Antimicrobial Agents for External Use

Chemical antimicrobial agents are divided into two categories. The first category contains antimicrobial products that are used to control microorganisms not pathogenic in humans. These agents may be used to control microorganisms that affect animal health, but also include industrial products used to control microbial contamination of foods, air-conditioning cooling towers, textile and paper

products, and even fuel tanks! Table 20.3 provides some examples of industrial applications for chemicals used to control microbial growth.

The second category of chemical antimicrobial agents contains products intended to prevent growth of human pathogens in inanimate environments. This category is subdivided into *sterilants, disinfectants, sanitizers,* and *antiseptics*.

Sterilants

Chemical **sterilants**, also called **sterilizers** or **sporicides** destroy all forms of microbial life, including endospores. Chemical sterilants have wide use in situations where it is impractical to use heat (see Section 20.1) or radiation (see Section 20.2) for decontamination or sterilization. Hospitals and laboratories must be able to sterilize heat-sensitive materials, such as thermometers, lensed instruments, polyethylene tubing, catheters, and reusable medical equipment such as respirometers. Usually, some form of *cold sterilization* is used for these purposes. Cold sterilization is performed in enclosed devices that resemble autoclaves, but employ a gaseous chemical agent such as ethylene oxide, formaldehyde, peroxyacetic acid, or hydrogen peroxide. Liquid sterilants such as a sodium chlorite solution or amylphenol are used for delicate instruments that cannot withstand high temperatures or gas (Table 20.4).

Disinfectants, Sanitizers, and Antiseptics

Disinfectants are chemicals that kill microorganisms, but not necessarily endospores, and are also used on inanimate objects. For example, hospital-type disinfectants such as ethanol and cationic detergents, critical for infection control, are used to decontaminate floors, tables, bench tops, walls, and so on in medical settings. General disinfectants are used in households, swimming pools, and water purifiers (Table 20.4).

Table 20.3	Industrial uses of antimicrobial chemicals	
Industry	**Chemicals**	**Use**
Paper	Organic mercurials, phenols[a], methylisothiazolinone	To prevent microbial growth during manufacture
Leather	Heavy metals, phenols[a]	Antimicrobial agents are present in the final product
Plastic	Cationic detergents	To prevent growth of bacteria on aqueous dispersions of plastics
Textile	Heavy metals, phenols[a]	To prevent microbial deterioration of fabrics exposed in the environment such as awnings and tents
Wood	Heavy metals, phenols[a]	To prevent deterioration of wooden structures
Metal working	Cationic detergents	To prevent growth of bacteria in aqueous cutting emulsions
Petroleum	Mercurics, phenols, cationic detergents, methylisothiazolinone	To prevent growth of bacteria during recovery and storage of petroleum and petroleum products
Air conditioning	Chlorine, phenols[a], methylisothiazolinone	To prevent growth of bacteria (for example, *Legionella*) in cooling towers
Electrical power	Chlorine	To prevent growth of bacteria in condensers and cooling towers
Nuclear	Chlorine	To prevent growth of radiation-resistant bacteria in nuclear reactors

[a]Heavy metal (mercury and arsenic) compounds and phenolic compounds produce environmentally hazardous waste products and may create health hazards.

Table 20.4	Antiseptics, sterilants, disinfectants, and sanitizers	
Agent	**Use**	**Mode of action**
Antiseptics		
Alcohol (60–85% ethanol or isopropanol in water)[a]	Topical antiseptic	Lipid solvent and protein denaturant
Phenol-containing compounds (hexachlorophene, triclosan, chloroxylenol, chlorhexidine)[b]	Soaps, lotions, cosmetics, body deodorants, topical disinfectants	Disrupts cell membrane
Cationic detergents, especially quaternary ammonium compounds (benzalkonium chloride)	Soaps, lotion, topical disinfectants	Interact with phospholipids of cell membrane
Hydrogen peroxide[a] (3% solution)	Topical antiseptic	Oxidizing agent
Iodine-containing iodophor compounds in solution[a] (Betadine®)	Topical antiseptic	Iodinates tyrosine residues of proteins; oxidizing agent
Octenidine	Topical antiseptic	Disrupts cell membrane
Silver nitrate	Eyes of newborn to prevent blindness due to infection by *Neisseria gonorrhoeae*	Protein precipitant
Sterilants, Disinfectants and Sanitizers[c]		
Alcohol (60–85% ethanol or isopropanol in water)[a]	Disinfectant for medical instruments and laboratory surfaces	Lipid solvent and protein denaturant
Cationic detergents (quaternary ammonium compounds)	Disinfectant and sanitizer for medical instruments, food and dairy equipment	Interact with phospholipids
Chlorine gas	Disinfectant for purification of water supplies	Oxidizing agent
Chlorine compounds (chloramines, sodium hypochlorite, sodium chlorite chlorine dioxide)	Disinfectant and sanitizer for dairy and food industry equipment, and water supplies	Oxidizing agent
Copper sulfate	Algicide disinfectant in swimming pools and water supplies	Protein precipitant
Ethylene oxide (gas)	Sterilant for temperature-sensitive materials such as plastics and lensed instruments	Alkylating agent
Formaldehyde	3%–8% solution used as surface disinfectant, 37% (formalin) or vapor used as sterilant	Alkylating agent
Glutaraldehyde	2% solution used as high-level disinfectant or sterilant	Alkylating agent
Hydrogen peroxide[a]	Vapor used as sterilant	Oxidizing agent
Iodine-containing iodophor compounds in solution[a] (Wescodyne®)	Disinfectant for medical instruments and laboratory surfaces	Iodinates tyrosine residues
Mercuric dichloride[b]	Disinfectant for laboratory surfaces	Combines with -SH groups
OPA (orthophalaldehyde)	High-level disinfectant for medical instruments	Alkylating agent
Ozone	Disinfectant for drinking water	Strong oxidizing agent
Peroxyacetic acid	Solution used as high-level disinfectant or sterilant	Strong oxidizing agent
Phenolic compounds[b]	Disinfectant for laboratory surfaces	Protein denaturant

[a] Alcohols, hydrogen peroxide, and iodine-containing iodophor compounds can act as antiseptics, disinfectants, sanitizers, or sterilants depending on concentration, length of exposure and form of delivery.

[b] Use of heavy metal (mercury) compounds and phenolic compounds may produce environmentally hazardous waste products and may create health hazards.

[c] Many water-soluble antimicrobial compounds, with the exception of those containing heavy metals, can be used as sanitizers for food and dairy equipment and preparation areas, provided their use is followed by adequate draining before food contact.

Sanitizers are agents that reduce, but may not eliminate, microbial numbers to a level considered to be safe. *Food contact sanitizers* are widely used in the food industry to treat surfaces such as mixing and cooking equipment, dishes, and utensils. A separate category of *non-food contact sanitizers* are used to treat surfaces such as counters, floors, walls, carpets, air, and laundry (Table 20.4).

Antiseptics and **germicides** are chemical agents that kill or inhibit growth of microorganisms and are sufficiently nontoxic to be applied to living tissues. Most of the compounds that fall into this category are used for handwashing or for treating surface wounds (Table 20.4). Under some circumstances, certain antiseptics are also effective disinfectants. These products are considered drugs and fall under regulation by the Food and Drug Administration in the United States.

Antimicrobial Efficacy

Several factors affect the efficacy of these chemical antimicrobial agents. For example, many disinfectants are neutralized by organic materials, reducing disinfectant concentrations and microbial killing capacity. Furthermore, pathogens are often encased in particles or grow in large numbers as *biofilms*, covering the surfaces of tissue with several layers of microbial cells (∞ Section 19.3). As a result, penetration of a chemical agent to the viable cells may be slowed or even completely prevented.

Only sterilants are effective against bacterial endospores. Endospores are much more resistant to other agents than are vegetative cells because of their low water availability and reduced metabolism (see Section 20.1 and ∞ Section 4.13). Certain vegetative cells such as those of *Mycobacterium tuberculosis*, the causal agent of tuberculosis, are resistant to the action of common disinfectants because of the waxy nature of their cell wall (∞ Sections 12.23 and 26.5). Thus, the efficacy of antimicrobial treatment is an empirical process that must be determined under the actual conditions of use.

 20.5   Concept Check

Sterilants, disinfectants, and sanitizers are compounds used to decontaminate nonliving material. Antiseptics and germicides are used to reduce microbial growth on living tissues. Antimicrobial compounds have commercial, health care, and industrial applications.

♦ Distinguish between sterilizer, a disinfectant, a sanitizer, and an antiseptic.

♦ What disinfectants are routinely used for sterilization of water? Why are these disinfectants not harmful to humans?

III ANTIMICROBIAL AGENTS USED *IN VIVO*

A number of antimicrobial agents, both synthetic and naturally occurring, are used to treat microbial infections *in vivo*. These compounds have revolutionized the treatment of infectious diseases.

20.6   Synthetic Antimicrobial Drugs

The previous section dealt with chemical agents used to inhibit microbial growth *outside* the human body. Most of the chemicals mentioned are too toxic to be used inside the body; even the relatively mild antiseptics can be used only on the skin. For control of infectious disease, chemical compounds that can be used internally are essential. Such compounds are called antimicrobial **chemotherapeutic agents**, and they play major roles in clinical and veterinary medicine, as well as in agriculture.

Work on antimicrobial chemotherapeutic agents started with the German scientist Paul Ehrlich. In the early 1900s, Ehrlich developed the concept of **selective toxicity**, the ability to inhibit or kill pathogenic microorganisms without adversely affecting the host. In his

search for a "magic bullet" that would kill only pathogens, Ehrlich tested large numbers of chemical dyes for selective toxicity and discovered the first effective antimicrobial drugs, of which *Salvarsan*, an arsenic-containing compound used for the cure of syphilis, was the most successful (Figure 20.12●).

Chemotherapeutic agents are classified based on their structure (Figure 20.13●), mechanism of action (Figure 20.14●), and spectrum of antimicrobial activity (Figure 20.15●). Worldwide, more than 500 metric tons of various chemotherapeutic agents are manufactured annually (Figure 20.16●). Antimicrobial agents fall into two broad categories, *synthetic agents* and *antibiotics*. Here we will concentrate on the synthetic agents. We discuss antibiotics in the next three sections.

Growth Factor Analogs

In Section 5.1 we defined growth factors as specific chemical substances *required* in the medium because the organisms cannot synthesize them. **Growth factor analogs** are synthetic compounds that are structurally similar to a growth factor, but subtle structural differences between the analogs and the authentic growth factors prevent the analogs from functioning in the cell. Analogs are known for vitamins, amino acids, purines, pyrimidines, and other compounds. We first discuss *bacterial* growth factor analogs. Growth factor analogs effective for the treatment of *viral* and *fungal* infections will be discussed in Sections 20.10 and 20.11.

Sulfa Drugs

Discovered by Gerhard Domagk in the 1930s, the *sulfa drugs* were the first widely used growth factor analogs shown to specifically inhibit the growth of bacteria. The discovery of the first sulfa drug resulted from the large-scale screening of chemicals for activity in curing streptococcal diseases in experimental animals.

● **Figure 20.12   Salvarsan.** This arsenic-containing compound was one of the first useful antimicrobial chemotherapeutic drugs. Salvarsan was used in the early 1900s to treat syphilis.

Antibiotic classification	Subclassification	Example	Representative structure
I. **Carbohydrate-containing compounds**	Pure sugars Aminoglycosides Orthosomycins N-Glycosides C-Glycosides Glycolipids	Nojirimycin Streptomycin Everninomicin Streptothricin Vancomycin Moenomycin	
II. **Macrocyclic lactones**	Macrolide antibiotics Polyene antibiotics Ansamycins Macrotetrolides	Erythromycin Candicidin Rifampin Tetranactin	
III. **Quinones and related compounds**	Tetracyclines Anthracyclines Naphthoquinones Benzoquinones	Tetracycline Adriamycin Actinorhodin Mitomycin	
IV. **Amino acid and peptide analogs**	Amino acid derivatives β-Lactam antibiotics Peptide antibiotics Chromopeptides Depsipeptides Chelate-forming peptides	Cycloserine Penicillin, ceftriaxone Bacitracin Actinomycin Valinomycin Bleomycin	
V. **Heterocyclic compounds containing nitrogen**	Nucleoside antibiotics	Polyoxins	
VI. **Heterocyclic compounds containing oxygen**	Polyether antibiotics	Monensin	
VII. **Alicyclic derivatives**	Cycloalkane derivatives Steroid antibiotics	Cycloheximide Fusidic acid	
VIII. **Aromatic compounds**	Benzene derivatives Condensed aromatics Aromatic ether	Chloramphenicol Griseofulvin Novobiocin	
IX. **Aliphatic compounds**	Compounds containing phosphorus	Fosfomycin	
X. **Quinolone compounds**	4-Quinolone Fluoro-4-quinolones	Nalidixic acid Ciprofloxacin	
XI. **Oxazolidinone**	Cyclic lactone	2-Oxazolidinone	

● **Figure 20.13 Antibacterial chemotherapeutic agents.** The agents are classified according to chemical structure. A representative example is shown for each group.

● **Figure 20.14 Mode of action of major antimicrobial chemotherapeutic agents.** THF, Tetrahydrofolate; DHF, dihydrofolate; mRNA, messenger RNA; tRNA, transfer RNA.

Sulfanilamide, the simplest sulfa drug, is an analog of *p-aminobenzoic acid*, which is itself a part of the vitamin folic acid, a nucleic acid precursor (Figure 20.17●). Sulfanilamide blocks the synthesis of folic acid, inhibiting nucleic acid synthesis. Sulfanilamide is active only in *Bacteria* because *Bacteria* synthesize their own folic acid, whereas most animals obtain folic acid from their diet. Widely used for treatment of streptococcal infections (∞ Section 26.2), drug resistance to the clinically useful sulfonamides is now quite common, usually because the resistant microorganisms have developed the ability to use exogenous sources of preformed folic acid (see Section 20.12). However, treatment with sulfamethaxazole (a clinically useful sulfa drug) plus trimethoprim, a related folic acid synthesis competitor, remains useful because the drug combination produces sequential blocking of the folic acid synthesis pathway; resistance to the combination requires two mutations in the same organism.

Isoniazid

Isoniazid (∞ Figure 26.9) is an important growth factor analog with a narrow spectrum of activity (Figure 20.15). Effective only against *Mycobacterium tuberculosis*, isoniazid interferes with the synthesis of mycolic acid, a mycobacterial cell wall component. Isoniazid is a nicotinamide (vitamin) analog and is the most effective single drug used for control and treatment of tuberculosis (Figure 20.15, ∞ Section 26.5).

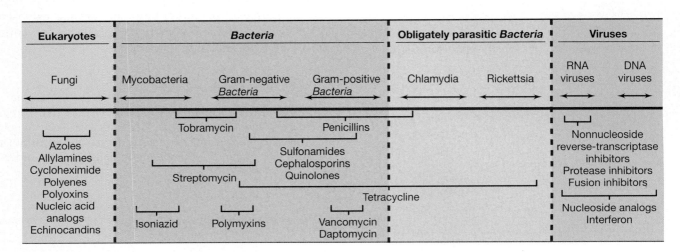

● **Figure 20.15 Antimicrobial spectrum of action.** Antimicrobial chemotherapeutic agents each affect a limited group of microorganisms.

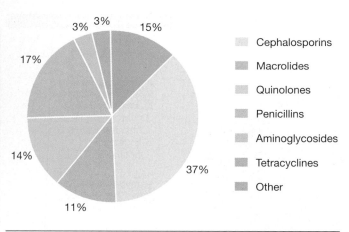

● **Figure 20.16 Annual worldwide production and use of antibiotics.** Each year more than 500 metric tons of antimicrobial chemotherapeutic agents are manufactured.

Nucleic Acid Analogs

In the examples shown in Figure 20.18●, analogs of nucleic acid bases have been formed by addition of a flourine or a bromine atom. Fluorine is a relatively small atom and does not alter the overall shape of the bases, but changes the chemical properties such that the compound does not function in cell metabolism. Fluorouracil resembles the nucleic acid base, uracil; bromouracil resembles another base, thymine. Growth factor analogs that resemble nucleic acids are used in the treatment of viral and fungal infections and are also used as mutagens (see Sections 20.10 and 20.11, ⟳ Section 10.4).

Quinolones

The **quinolones** are a separate class of synthetic antibacterial compounds that interact with bacterial DNA gyrase and prevent the gyrase from supercoiling bacterial DNA, a required step for packaging DNA in the bacteri-

● **Figure 20.18 Growth factors.** Structurally similar analogs are shown for comparison. Growth factors are discussed in Section 5.1 and Table 5.3.

al cell (⟳ Section 7.3). Nalidixic acid is the prototype quinolone (Figure 20.13 and Figure 20.14). Fluoroquinolone derivatives of nalidixic acid such as ciprofloxacin (Figure 20.19●) are routinely used to treat urinary tract infections in humans. Ciprofloxacin is also the drug of choice for treating infections from penicillin-resistant strains of *Bacillus anthracis*, the cause of anthrax (⟳ Section 25.13). Because DNA gyrase is found in all *Bacteria*, the fluoroquinolones are effective for treating both gram-positive and gram-negative bacterial infections (Figure 20.15). Fluoroquinolones are also extensively used in the beef and poultry industries for prevention and treatment of respiratory diseases.

20.6 Concept Check

Synthetic antimicrobial agents are selective for *Bacteria*, viruses, and fungi. Growth factor analogs such as sulfa drugs, isoniazid, and nucleic acid analogs are synthetic metabolic inhibitors. Quinolones inhibit the action of DNA gyrase in *Bacteria*.

◆ What is *selective toxicity*?

◆ Distinguish synthetic chemotherapeutic agents from antiseptics and disinfectants.

◆ Describe the action of any one of the growth factor analogs.

(a) Sulfanilamide

(b) p-Aminobenzoic acid

(c) Folic acid

● **Figure 20.17 Sulfa drugs.** (a) The simplest sulfa drug, sulfanilamide. (b) Sulfanilamide is an analog of p-aminobenzoic acid, a precursor of (c) folic acid, a growth factor (⟳ Section 5.1).

● **Figure 20.19 Ciprofloxacin, a quinolone antibiotic.** Ciprofloxacin is a fluorinated derivative of nalidixic acid (Figure 20.13). The derivatives are more soluble than nalidixic acid and reach clinically therapeutic levels in blood and tissues. Ciprofloxacin is used to treat urinary tract infections and anthrax caused by penicillin-resistant *Bacillus anthracis*.

20.7 Naturally Occurring Antimicrobial Drugs: Antibiotics

Antibiotics are natural rather than synthetic antimicrobial compounds. They are produced by a wide range of fungi and bacteria and inhibit or kill other microorganisms. A large number of antibiotics have been identified in nature, but less than 1% are clinically useful. However, the useful antibiotics have had a dramatic impact on the treatment of infectious diseases. Many natural antibiotics have been structurally modified in the laboratory to enhance their efficacy. These are said to be *semisynthetic antibiotics*. The large-scale industrial production of antibiotics will be discussed in Chapter 30.

Antibiotics and Selective Antimicrobial Toxicity

The susceptibility of microorganisms to individual antibiotics and other chemotherapeutic agents varies significantly (Figure 20.15). For example, gram-positive *Bacteria* and gram-negative *Bacteria* may differ in their susceptibility to an individual antibiotic, although certain **broad-spectrum antibiotics** are effective on both groups. In general, a broad-spectrum antibiotic finds wider medical use than a *narrow-spectrum antibiotic*. An antibiotic with a limited spectrum of activity may, however, be quite valuable for the control of microorganisms that fail to respond to other antibiotics. An example is *vancomycin*, a narrow-spectrum glycopeptide that is a highly effective bacteriocidal agent for gram-positive penicillin-resistant *Bacteria* from the genera *Staphylococcus*, *Bacillus*, and *Clostridium* (Figures 20.13, 20.14, and 20.15).

Important targets of antibiotics in *Bacteria* are ribosomes (translation), the cell wall, the cytoplasmic membrane, and DNA replication and transcription (Figure 20.14). Here we discuss antibiotic mechanisms that inhibit protein synthesis and transcription.

Antibiotics Affecting Protein Synthesis

Many antibiotics inhibit protein synthesis by interacting with the ribosome (Figure 20.14). These interactions are quite specific and many involve rRNA. Several of these antibiotics are medically useful and several are also effective research tools because they block individual steps in protein synthesis (⌘ Section 7.16). For instance, streptomycin inhibits protein chain *initiation*, whereas puromycin, chloramphenicol, cycloheximide, and tetracycline inhibit protein *elongation*.

Even when two antibiotics inhibit the same step in protein synthesis, the mechanisms of inhibition can be quite different. For example, puromycin binds to the A site on the ribosome, and the growing polypeptide chain is transferred to puromycin instead of the amino acyl–transfer RNA complex (⌘Figures 7.37 and 7.38). The puromycin–peptide complex is then released from the ribosome, prematurely halting elongation. By contrast, chloramphenicol inhibits elongation by blocking formation of the peptide bond.

Many antibiotics specifically inhibit ribosomes of organisms from only one or two phylogenetic domains. For example, chloramphenicol and streptomycin are specific for ribosomes of *Bacteria*, while cycloheximide only affects ribosomes from *Eukarya*. Since mitochondria and chloroplasts have ribosomes of the prokaryotic type, antibiotics that inhibit protein synthesis in *Bacteria* also inhibit protein synthesis in these organelles (⌘ Section 14.4). Several of these antibiotics, such as tetracycline, are medically useful because the eukaryotic mitochondria are not affected at the concentrations used for antimicrobial therapy.

Antibiotics Affecting Transcription

A number of antibiotics specifically inhibit transcription by inhibiting RNA synthesis (⌘ Section 7.10 and Figure 20.14). For example, rifampin and the streptovaricins inhibit RNA synthesis by binding to the β subunit of RNA polymerase. These antibiotics have specificity for *Bacteria*, chloroplasts, and mitochondria. Actinomycin inhibits RNA synthesis by combining with DNA and blocking RNA elongation. Actinomycin binds most strongly to DNA at guanine–cytosine base pairs, fitting into the major groove in the double strand where RNA is synthesized.

Some of the most useful antibiotics are directed against unique structural features of *Bacteria*, such as their cell walls. We discuss a number of these antibiotics and their targets in the next section.

⬢ **20.7 Concept Check**

Antibiotics are a chemically diverse group of antimicrobial compounds that are produced by a variety of microorganisms. Although many antibiotics are known, most are not useful in humans or animals because of poor uptake or toxicity. Many antibiotics function by inhibiting transcription or translation in the target microorganisms.

♦ Distinguish *antibiotics* from *growth factor analogs*.

♦ What is a *broad-spectrum antibiotic*?

♦ Identify the potential targeted sites for antibiotics that inhibit protein synthesis and transcription.

20.8 β-Lactam Antibiotics: Penicillins and Cephalosporins

One of the most important groups of antibiotics, both historically and medically, is the β-lactam group. **β-lactam antibiotics** include the medically important *penicillins, cephalosporins,* and *cephamycins.* These antibiotics share a characteristic structural component, the β-lactam ring (Figure 20.20●). Together, the penicillins and cephalosporins account for over one-half of all of the antibiotics produced and used worldwide (Figure 20.16).

Penicillins

In 1929, the British scientist Alexander Fleming characterized an antibacterial product of the fungus *Penicillium chrysogenum* as the β-lactam antibiotic *penicillin G* (Figure 20.20). Penicillin G was the first antibiotic discovered and became the first clinically effective antibiotic. Even with the use of sulfa drugs, most infectious diseases were not under control in the 1930s. However, in 1939, Howard Florey and his colleagues, spurred on by the beginning of World War II, developed a process for large-scale production of **penicillin**. The new antibiotic was dramatically effective in controlling staphylococcal and pneumococcal infections among military personnel and was more effective for treating streptococcal infections than the sulfa drugs. By 1945 and the end of World War II, penicillin became available for general use, and pharmaceutical companies began to look for other antibiotics, leading to the discovery of drugs that revolutionized treatment of infectious diseases.

Penicillin G is active primarily against gram-positive *Bacteria* because gram-negative *Bacteria* are impermeable to the antibiotic. However, many **semisynthetic penicillins** are quite effective against gram-negative *Bacteria.* Figure 20.20 shows the structures of some of the penicillins. Modifications of the penicillin G structure significantly changes the properties of the resulting antibiotics. For example, the semisynthetic penicillins ampicillin and carbenicillin are effective against some gram-negative *Bacteria.* The structural differences in the *N*-acyl groups of these semisynthetic penicillins allow them to be transported inside the gram-negative outer membrane (∞ Section 4.9) where they inhibit cell wall synthesis. Note also that penicillin G is sensitive to β-lactamase, an enzyme produced by a number of penicillin-resistant *Bacteria* (see Section 20.12). The semisynthetic penicillins oxacillin and methicillin are β-lactamase sensitive.

Mechanism of Action

The β-lactam antibiotics are potent inhibitors of cell wall synthesis. An important feature of cell wall synthesis is the transpeptidation reaction, which results in the cross-linking of two glycan-linked peptide chains (∞ Figure 6.5). The transpeptidase enzymes are capable of binding to penicillin or other β-lactam antibiotics. Thus, these transpeptidases are known as *penicillin binding proteins* (PBPs). When the PBPs bind penicillin, they cannot cat-

Designation	N-Acyl group
NATURAL PENICILLIN Benzylpenicillin (penicillin G) Gram-positive activity β-lactamase-sensitive	—CH₂—CO—
SEMISYNTHETIC PENICILLINS Methicillin acid-stable, β-lactamase-resistant	OCH₃ ... —CO— ... OCH₃
Oxacillin acid-stable, β-lactamase-resistant	—CO— ... N—O ... CH₃
Ampicillin broadened spectrum of activity (especially against gram-negative *Bacteria*), acid-stable, β-lactamase-resistant	—CH—CO— D(–) ... NH₂
Carbenicillin broadened spectrum of activity (especially against *Pseudomonas aeruginosa*), acid-stable but ineffective orally, β-lactamase-sensitive	—CH—CO— ... COOH

● **Figure 20.20 Penicillins.** The red arrow (top panel) is the site of action for most β-lactamases.

alyze the transpeptidase reaction, but the cell wall continues to be formed. The newly synthesized wall is no longer cross-linked and cannot maintain strength as the defective peptidoglycan backbone is laid down. In addition, the antibiotic-PBP complex stimulates the release of autolysins that digest the existing cell wall. The result is a weakened, self-degrading cell wall (∞ Sections 4.8 and 6.3). Under normal circumstances, the osmotic pressure differences inside the cell, as compared with outside, lyse the cell. By contrast, *vancomycin,* a glycopeptide (Figure 20.13), does not bind to PBPs but binds directly to the terminal D-alanyl-D-alanine peptide on the peptidoglycan precursors (∞ Section 6.3 and Figure 6.5), blocking the transpeptidase reaction.

Because the cell wall and its synthesis mechanisms are unique to *Bacteria,* the β-lactam antibiotics have very high selectivity and are not toxic to host cells. However, because of the complex structure of these antibiotics, some individuals develop allergies to individual β-lactam compounds after repeated courses of antibiotic therapy (∞ Section 22.15).

Cephalosporins

The cephalosporins are another group of clinically important antibiotics that contain the β-lactam ring. Cephalosporin is produced by the fungus *Cephalosporium* sp. (∞ Section 30.6). Cephalosporins differ structurally from the penicillins. They retain the β-lactam ring but have a six-member dihydrothiazine ring instead of the five-member thiazolidine ring. The cephalosporins have the same mode of action as the penicillins; they bind irreversibly to the PBPs and prevent the cross-linking of peptidoglycan. Clinically important cephalosporins are semisynthetic antibiotics with a broader spectrum of antibiotic activity than the penicillins. In addition, they are generally more resistant to the enzymes that destroy β-lactam rings, the β-lactamases. For example, ceftriaxone (Figure 20.13) is highly resistant to β-lactamases and has replaced penicillin for treatment of infections due to *Neisseria gonorrhoeae* (see Section 20.12, ∞ Section 26.12) because many *N. gonorrhoeae* strains are now penicillin resistant.

 20.8 Concept Check

The β-lactam compounds, including the penicillins and the cephalosporins, are the most important clinical antibiotics. These antibiotics target cell wall synthesis in *Bacteria*. They have low host toxicity and a broad spectrum of activity.

◆ Draw the structure of the β-lactam ring and indicate the site of β-lactamase activity.

◆ How do the β-lactam antibiotics function?

◆ What advantages do the cephalosporins have as compared to the penicillins for treating infections due to gram-negative *Bacteria*?

20.9 Antibiotics from Prokaryotes

Many antibiotics active against *Bacteria* are also produced by *Bacteria*. These include the *aminoglycosides*, the *macrolides*, and the *tetracyclines*. Many of these antibiotics have major clinical applications, and we discuss their general properties here.

Aminoglycoside Antibiotics

Aminoglycoside antibiotics contain amino sugars bonded by glycosidic linkages (∞ Section 3.3) to other amino sugars. Clinically useful aminoglycosides include *streptomycin* (produced by *Streptomyces griseus*) (Figure 20.13) and its relatives, *kanamycin* (Figure 20.21●), *neomycin, gentamicin, tobramycin, netilmicin, spectinomycin*, and *amikacin*. The aminoglycosides inhibit protein synthesis at the 30S

● **Figure 20.21 Kanamycin, an aminoglycoside antibiotic.** The amino sugars are in yellow. The site of modification by an *N*-acetyltransferase, encoded by a resistance plasmid, is indicated (see Section 20.12). Following acetylation, the antibiotic is inactive.

subunit of the ribosome (Figure 20.14) and are useful clinically against gram-negative *Bacteria* (Figure 20.15).

Streptomycin was the first effective antibiotic used for the treatment of tuberculosis. However, none of the aminoglycoside antibiotics are widely used today, and together the aminoglycosides account for only about 3% of the total of all antibiotics produced and used (Figure 20.16). Streptomycin has been supplanted by several synthetic antimicrobials for tuberculosis treatment because of serious side effects such as neurotoxicity and nephrotoxicity (kidney toxicity). Bacterial resistance also develops readily. The use of aminoglycosides for treatment of gram-negative infections has decreased since the development of the semisynthetic penicillins (see Section 20.8) and the tetracyclines (see later in this section). Aminoglycoside antibiotics are now considered reserve antibiotics used primarily when other antibiotics fail.

Macrolide Antibiotics

Macrolide antibiotics such as erythromycin contain large lactone rings connected to sugar molecules (Figure 20.22●). Variations in both the macrolide ring

● **Figure 20.22 Erythromycin, a macrolide antibiotic.** Erythromycin is a widely used broad-spectrum antibiotic.

and the sugars exist, creating a large variety of macro-lides. The best-known macrolide antibiotic is *erythro-mycin* (produced by *Streptomyces erythreus*). Other clinically useful macrolides include *dirithromycin, clarithromycin,* and *azithromycin.* The macrolides account for 11% of the total world production and use of antibiotics (Figure 20.16). Erythromycin is a protein synthesis inhibitor at the level of the 50S subunit of the ribosome (Figure 20.14). Erythromycin is commonly used clinically in place of penicillin in patients allergic to penicillin or other β-lactam antibiotics and has been particularly useful for treating legionellosis (⚬⚬ Section 28.7).

Tetracyclines

The **tetracyclines**, produced by several members of the genus *Streptomyces* (⚬⚬Section 30.6), are an important group of antibiotics that find widespread medical use in humans. They were some of the first broad-spectrum antibiotics, inhibiting almost all gram-positive and gram-negative *Bacteria*. The basic structure of the tetracyclines consists of a naphthacene ring system (Figure 20.23●). The naphthacene ring structure can be substituted at several positions to form new tetracycline analogs. *Chlortetracycline*, for instance, has a chlorine atom, whereas *oxytetracycline* has an additional hydroxyl (OH) group and no chlorine.

These three tetracyclines are produced naturally, but semisynthetic tetracyclines having substitutions in the naphthacene ring system have also been developed. Like erythromycin and the aminoglycoside antibiotics, tetracycline is a protein synthesis inhibitor, interfering with 30S ribosomal subunit function (see Figure 20.14).

The tetracyclines, with the β-lactam antibiotics, comprise the two most important groups of antibiotics in the medical field. The tetracyclines are also widely used in veterinary medicine, and in some countries are used as nutritional supplements for poultry and swine. Because

extensive nonmedical uses of medically important antibiotics have contributed to widespread antibiotic resistance, this use is now discouraged (see Section 20.12).

Daptomycin

Daptomycin is another antibiotic produced by a member of the *Streptomyces* genus. This novel antibiotic is a cyclic lipopeptide with a unique mode of action (Figure 20.14). Used mainly to treat infections by gram-positive *Bacteria* such as the pathogenic staphylococci and streptococci, daptomycin binds to bacterial cell membranes and induces rapid depolarization of the membrane potential. The depolarized cell loses its ability to synthesize macromolecules such as nucleic acids and proteins, resulting in cell death. Because of this unique membrane-binding mechanism, there is no identifiable, transferable mechanism of resistance.

⬣ 20.9 Concept Check

The aminoglycosides, macrolides, and tetracycline antibiotics are structurally complex molecules produced by *Bacteria* and are active against other *Bacteria*. All of these work by interfering with protein synthesis. Daptomycin, a novel antibiotic, depolarizes the cell membrane

◆ What are the biological sources of aminoglycosides, tetracyclines, macrolides, and daptomycin?

◆ How does the activity of each class of antibiotics lead to death of the affected cell?

IV CONTROL OF VIRUSES AND EUKARYOTIC PATHOGENS

Drugs that control growth of viruses and eukaryotic pathogens such as fungi often affect eukaryotic host cells as well. As a result, selective toxicity for eukaryotic pathogens is very difficult to attain; only chemotherapeutic agents that preferentially affect pathogen-specific metabolic pathways or structural components are useful. There are a limited number of these drugs, and here we discuss the mode of action of some of them.

Tetracycline analog	R₁	R₂	R₃	R₄
Tetracycline	H	OH	CH₃	H
7-Chlortetracycline (aureomycin)	H	OH	CH₃	Cl
5-Oxytetracycline (terramycin)	OH	OH	CH₃	H

● **Figure 20.23 Tetracycline.** Semisynthetic analogs of tetracycline are important antibiotics.

20.10 Antiviral Drugs

The therapeutic successes achieved against bacteria using selective antibacterial agents have not been matched by similar achievements with antiviral agents because viruses use their eukaryotic hosts to reproduce and perform metabolic functions (⚬⚬ Chapter 9). Most drugs that control viruses target host structures and are thus toxic for the host. However, several agents are *more* toxic for viruses than for the host, and a few agents specifically target viruses. Largely because of efforts to

Table 20.5 Antiviral chemotherapeutic compounds

Category/drug	Mechanism of action	Virus affected
Fusion inhibitor Enfuvirtide	Blocks HIV-T lymphocyte membrane fusion	HIV[a]
Interferons Interferon α Interferon β Interferon γ	Induces proteins that inhibit viral replication	Broad spectrum (host specific)
Neuraminidase inhibitors Oseltamivir Zanamivir	Block active site of influenza neuraminidase	Influenza A and B Influenza A and B
Nonnucleoside reverse transcriptase inhibitor (NNRTI) Nevirapine	Reverse transcriptase inhibitor	HIV
Nucleoside analogs Acyclovir Ganciclovir Trifluridine Valacyclovir Vidarabine	Viral polymerase inhibitors	Herpes viruses, *Varicella zoster* Cytomegalovirus Herpesvirus Herpesvirus Herpesvirus, vaccinia, hepatitis B virus
Abacavir (ABC) Didanosine (dideoxyinosine or ddI) Emtricitabine (FTC) Lamivudine (3TC) Stavudine (d4T) Zalcitabine (ddC) Zidovudine (AZT) (Figure 26.38)	Reverse transcriptase inhibitors	HIV HIV HIV HIV, hepatitis B virus HIV HIV HIV
Ribavirin	Blocks capping of viral RNA	Respiratory syncytial virus, influenza A and B, Lassa fever
Nucleotide analogs Cidofovir Tenofovir (TDF)	Viral polymerase inhibitor Reverse transcriptase inhibitor	Cytomegalovirus, herpesviruses HIV
Protease inhibitors Amprenavir Indinavir (Figure 20.28) Lopinavir Nelfinavir Saquinavir (Figure 20.28)	Viral protease inhibitor	HIV HIV HIV HIV HIV
Pyrophosphate analog Phosphonoformic acid (Foscarnet)	Viral polymerase inhibitor	Herpesviruses, HIV, hepatitis B virus
RNA polymerase inhibitor Rifamycin	RNA polymerase inhibitor	Vaccinia, pox viruses
Synthetic amines Amantadine Rimantadine	Viral uncoating blocker	Influenza A Influenza A

[a] Human immunodeficiency virus

find effective measures to control the human immunodeficiency virus (HIV), the causative agent of AIDS (Section 26.14), significant achievements have been made in controlling viruses with chemical agents.

Antiviral Chemotherapeutic Agents

The most successful and commonly used agents for antiviral chemotherapy are the *nucleoside analogs* (Table 20.5). The first compound to gain universal acceptance in this category was zidovudine, or azidothymidine (AZT). AZT inhibits retroviruses such as HIV (Figure 26.38). Azidothymidine is chemically related to thymidine but is a dideoxy derivative, lacking the 3'-hydroxyl group. AZT inhibits multiplication of retroviruses by blocking reverse transcription and production of the virally encoded DNA intermediate. This process inhibits multiplication of HIV. A number of other nucleoside analogs having similar mechanisms have been developed for the treatment of HIV and other viruses, as shown in Table 20.5.

Nearly all nucleoside analogs, or **nucleoside reverse transcriptase inhibitors (NRTI)**, work by the same mechanism, inhibiting elongation of the viral nucleic acid chain by a nucleic acid polymerase. The *nucleotide* analog cidofovir works in the same way (Table 20.5). Because the normal cell function of nucleic acid replication is targeted, these drugs usually induce a level of host toxicity. Many NRTIs also lose their antiviral potency with time due to the emergence of drug-resistant viruses (◯ Section 26.14).

Several other antiviral agents target reverse transcriptase. *Nevirapine*, a **non-nucleoside reverse transcriptase inhibitor (NNRTI)**, binds directly to reverse transcriptase and inhibits reverse transcription. *Phosphonoformic acid* is an analog of inorganic pyrophosphate that inhibits normal internucleotide linkages, preventing synthesis of viral nucleic acids. As with the NRTIs, the NNRTIs generally induce some level of host toxicity because their action also affects normal host cell nucleic acid synthesis.

A newer class of antiviral drugs is the **protease inhibitors (PI)** (Table 20.5 and see Figure 20.28). These drugs are particularly effective for treatment of HIV. They prevent viral replication by binding the active site of HIV protease, inhibiting processing of viral polypeptides and virus maturation (◯ Section 26.15 and Section 20.13).

A final category of anti-HIV drugs is represented by a single drug, *enfuvirtide*, a **fusion inhibitor** composed of a 36-amino acid synthetic peptide that binds to the gp41 membrane protein of HIV (Table 20.5 and ◯ Section 26.14). Binding of the protein stops the conformational changes necessary for the fusion of viral and target T lymphocyte cell membranes.

Influenza Antiviral Agents

Two categories of drugs effectively limit influenza infections. The adamantane derivatives *amantadine* and *rimantadine* are synthetic amines that interfere with an influenza A ion transport protein, inhibiting virus uncoating and subsequent replication. The neuraminidase inhibitors *oseltamivir* and *zanamivir* block the active site of neuraminidase in influenza A and B viruses, inhibiting virus release from infected cells. Both categories of drugs are used for treatment and chemoprophylaxis of influenza infection (◯ Section 26.8).

Interferon

Interferons are low-molecular-weight proteins (17,000 MW) that prevent viral multiplication by stimulating the production of antiviral proteins in normal cells. Interferons from virus-infected cells interact with receptors on noninfected cells, promoting the synthesis of antiviral proteins that function to prevent further virus infection. Interferons are *cytokines* and are produced in three molecular forms: *IFN-α* is produced by leukocytes, *IFN-β* is produced by fibroblasts, and *IFN-γ* is produced by immune lymphocytes (◯ Section 23.11). All three forms are effective viral inhibitors.

Studies of virus interference, a phenomenon in which infection with one virus interferes with subsequent infection by another virus, identified several small proteins as the cause of the interference. These proteins, later named *interferons*, are formed in response to live virus, viral nucleic acids, and virus inactivated by radiation. Interferon is produced in larger amounts by cells infected with viruses of low virulence, but little is produced against highly virulent viruses. Highly virulent viruses apparently inhibit cell protein synthesis before interferon can be produced. Interferon is also induced by a variety of double-stranded RNA (dsRNA) molecules, either natural or synthetic. In nature, dsRNA exists only as the replicative form in cells infected with certain RNA viruses such as rhinoviruses (cold viruses) (◯ Sections 16.10 and 26.8). Double-stranded RNA apparently signals virus infection in the animal cell and stimulates interferon production.

Interferon activity is host-specific rather than virus-specific. Interferon produced by a member of one species recognizes specific receptors only on cells of the same species. As a result, interferon produced by cells of an animal in response to, for example, a rhinovirus, could also inhibit multiplication of influenza viruses in cells within the same species, but has no effect on the multiplication of any virus in cells from other animal species.

Interferons have potential as possible antiviral and anticancer agents. There are now several approved recombinant interferon preparations available. However, the use of interferons as general chemotherapeutic agents is not widespread because interferon must be delivered locally in high concentrations to stimulate the production of antiviral proteins in uninfected host cells. Thus, the clinical utility of these antiviral agents depends on our ability to deliver interferon to local areas in the host through injections or aerosols. Alternatively, appropriate interferon-stimulating signals (*i.e.*, stimulation with viral nucleotides, nonvirulent viruses, or even synthetic nucleotides) given to host cells prior to viral infection might achieve the same goal.

 20.10 Concept Check

Effective antiviral agents must target virus-specific enzymes and processes. Clinically effective antiviral agents include nucleoside analogs and other drugs that inhibit nucleic acid polymerases and viral genome replication. Agents such as the protease inhibitors interfere with viral maturation steps. Host cells also produce the antiviral *interferon* proteins that stop viral replication.

◆ Why are there relatively few effective antiviral chemotherapeutic agents? Why aren't such agents used to treat common viral illnesses such as colds?

◆ What steps in the viral maturation process are inhibited by nucleoside analogs? By protease inhibitors? By interferons?

20.11 Antifungal Drugs

Like viruses, fungi pose special problems for successful chemotherapy. Since fungi are *Eukarya*, much of their cellular machinery is the same as in animals and humans. Thus, chemotherapeutic agents that affect metabolic pathways in fungi often affect corresponding pathways in host cells, resulting in drug toxicity. As a result, many antifungal drugs can be used only for topical (surface) applications. However, a few drugs are selectively toxic for fungi because they target unique fungal structures or metabolic processes. Drugs for fungal treatment are becoming increasingly important as fungal infections in immunocompromised individuals become more prevalent (⌾ Sections 26.14 and 27.8). We will look in detail at the selective toxicity of several classes of chemicals that are effective against fungi.

Ergosterol Inhibitors

Ergosterol in fungal cell membranes replaces cholesterol in higher eukaryotic cytoplasmic membranes (⌾ Section 4.5). Two groups of antifungal compounds work by interacting with ergosterol or inhibiting its synthesis (Table 20.6). The first group includes the *polyenes*, another group of antibiotics produced by members of the genus *Streptomyces*. Polyenes bind to ergosterol, disrupting membrane function, causing membrane permeability and cell death (Figure 20.24●). A second major group of antifungal compounds includes the *azoles* and the *allylamines*, synthetic agents that selectively inhibit ergosterol biosynthesis and therefore have broad antifungal activity. Treatment with azoles results in the inability of fungi to produce a normal membrane, leading to membrane damage and alteration of critical membrane transport activities. Allylamines also inhibit ergosterol

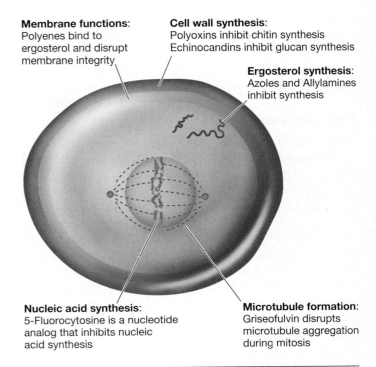

Membrane functions:
Polyenes bind to ergosterol and disrupt membrane integrity

Cell wall synthesis:
Polyoxins inhibit chitin synthesis
Echinocandins inhibit glucan synthesis

Ergosterol synthesis:
Azoles and Allylamines inhibit synthesis

Nucleic acid synthesis:
5-Fluorocytosine is a nucleotide analog that inhibits nucleic acid synthesis

Microtubule formation:
Griseofulvin disrupts microtubule aggregation during mitosis

● **Figure 20.24 Sites of action of some antifungal chemotherapeutic agents.** Because fungi are eukaryotic cells, antibacterial antibiotics are generally ineffective.

biosynthesis but are restricted to topical use because they are not readily taken up by animal cells and tissues.

Echinocandins

Echinocandins are a new class of antifungal drugs (Table 20.6). They act by inhibiting 1,3 β-D glucan synthase, the enzyme that forms glucan polymers in the fungal cell wall (Figure 20.24). Since mammalian cells

Table 20.6	Antifungal agents		
Category	**Target**	**Examples**	**Use**
Allylamines	Ergosterol synthesis	Terbenafine	Oral, topical
Aromatic antibiotic	Mitosis inhibitor	Griseofulvin	Oral
Azoles	Ergosterol synthesis	Clortrimazole	Topical
		Fluconazole	Oral
		Itraconazole	Oral
		Ketoconazole	Oral
		Miconazole	Topical
		Posoconazole	Experimental
		Ravuconazole	Experimental
		Voriconazole	Oral
Chitin synthesis inhibitor	Chitin synthesis	Nikkomycin Z	Experimental
Echinocandins	Cell wall synthesis	Caspofungin	Intravenous
Nucleic acid analogs	DNA synthesis	5-Fluorocytosine	Oral
Polyenes	Ergosterol synthesis	Amphotericin B	Oral, intravenous
		Nystatin	Oral, topical
Polyoxins	Chitin synthesis	Polyoxin A	Agricultural
		Polyoxin B	Agricultural

do not have 1,3 β-D glucan synthase, the action of these agents is specific, resulting in rapid cell death. These agents are used to treat infections with fungi as *Candida* and are active against some fungi that are resistant to other agents (∞ Sections 26.14 and 27.8.)

Other Antifungal Agents

Other antifungal drugs interfere with fungus-specific structures and functions (Table 20.6). For example, fungal cell walls contain *chitin*, a polymer of *N*-acetylglucosamine found only in fungi and insects (∞ Table 17.9). Several *polyoxins* inhibit cell wall synthesis by interfering with chitin biosynthesis. Polyoxins are widely used as agricultural fungicides, but are not used clinically. Other drugs inhibit folate biosynthesis, interfere with DNA topology during replication, or, like *griseofulvin*, disrupt microtubule aggregation during mitosis (∞ Section 14.7). The nucleic acid analog *5-fluorocytosine* is an effective nucleic acid synthesis inhibitor in fungi. Some very effective antifungal drugs also have other biological applications. For example, *vincristine, vinblastin*, and *toxol* are effective antifungal agents and have known anticancer properties.

Unfortunately, the use of antifungal drugs has resulted in the emergence of populations of resistant fungi and the emergence of "new" fungal pathogens. For example, *Candida* species, which are normally not pathogenic, now produce disease in immunocompromised individuals who have been treated with antifungal drugs; several pathogenic *Candida* strains are now resistant to all of the currently used antifungal agents.

 20.11 Concept Check

Antifungal agents fall into a wide variety of chemical categories. Because fungi are *Eukarya*, selective toxicity is hard to achieve, but some effective chemotherapeutic agents are available. Treatment of fungal infections is an emerging human health issue.

◆ Why are there very few clinically effective antifungal antibiotics?

◆ What factors are contributing to the apparent rise in fungal infections?

V ANTIMICROBIAL DRUG RESISTANCE AND DRUG DISCOVERY

Antimicrobial drug resistance is a major problem when dealing with many pathogenic microorganisms, especially in health-care settings. Here we explore some of the reasons for drug resistance in microorganisms and present strategies for developing new antimicrobial agents. Practical methods for preserving and enhancing the efficacy of currently used antimicrobial drugs are discussed in the Microbial Sidebar "Preventing Antibiotic Resistance."

20.12 Antimicrobial Drug Resistance

Antimicrobial drug resistance is the acquired ability of a microorganism to resist the effects of a chemotherapeutic agent to which it is normally susceptible. Antimicrobial drug resistance does not involve the host, but is a function of the microorganisms that reside in the host. As we have discussed, antibiotic producers are microorganisms, and in order to survive, the antibiotic-producing microorganisms developed resistance mechanisms to neutralize or destroy their own antibiotics. In addition, genes encoding these resistance mechanisms can be transferred to other, usually related, organisms. As a result, most antimicrobial drug resistance involves *resistance genes* that are transferred between and among microorganisms by genetic exchange.

Resistance Mechanisms

No antibiotic inhibits all microorganisms and some microorganisms are *naturally resistant* to certain antibiotics. There are several reasons why microorganisms may have an inherent natural resistance to an antibiotic. (1) The organism may lack the structure an antibiotic inhibits. For instance, some bacteria, such as mycoplasmas, lack a typical bacterial cell wall and are resistant to penicillins. (2) The organism may be impermeable to the antibiotic. For example, most gram-negative *Bacteria* are impermeable to penicillin G. (3) The organism may be able to alter the antibiotic to an inactive form. Many staphylococci contain β-lactamases that cleave the β-lactam ring of most penicillins (Figure 20.25●). (4) The organism may modify the *target* of the antibiotic. (5) The organism may develop a resistant biochemical pathway. For example, many pathogens develop resistance to sulfonamide drugs (see Section 20.6 and Figure 20.17) that inhibit the production of folic acid in *Bacteria*. Resistant *Bacteria* modify their metabolism to take up preformed folic acid from the environment, avoiding the need for the pathway blocked by sulfonamides. (6) The organism may be able to pump out an antibiotic entering the cell (efflux). Some specific examples of bacterial resistance to antibiotics are shown in Table 20.7.

As discussed in Sections 10.9 and 10.10, antibiotic resistance can be genetically encoded by the microorganism at either the chromosomal or the plasmid level on so-called *resistance plasmids* (*R factors*). Specific types of resistance typically have a genetic basis in one location or the other (Table 20.7). Because of the development of antibiotic resistance, bacteria isolated from clinical specimens must be tested for antibiotic susceptibility using the MIC method or the agar diffusion method (see Section 20.4 and Figures 20.10 and 20.11). Details of the

Table 20.7 Mechanisms of bacterial resistance to antibiotics

Resistance mechanism	Antibiotic example	Genetic basis of resistance	Mechanism present in:
Reduced permeability	Penicillins	Chromosomal	*Pseudomonas aeruginosa* Enteric *Bacteria*
Inactivation of antibiotic (for example, penicillinase; modifying enzymes methylases, acetylases, such as phosphorylases, and others)	Penicillins	Plasmid and chromosomal	*Staphylococcus aureus* Enteric *Bacteria* *Neisseria gonorrhoeae*
	Chloramphenicol	Plasmid and chromosomal	*Staphylococcus aureus* Enteric *Bacteria*
Alteration of target (for example, RNA polymerase, rifamycin; ribosome, erythromycin, and streptomycin; DNA gyrase, quinolones)	Aminoglycosides	Plasmid	*Staphylococcus aureus*
	Erythromycin	Chromosomal	*Staphylococcus aureus*
	Rifamycin		Enteric *Bacteria*
	Streptomycin		Enteric *Bacteria*
	Norfloxacin		Enteric *Bacteria* *Staphylococcus aureus*
Development of resistant biochemical pathway	Sulfonamides	Chromosomal	Enteric *Bacteria* *Staphylococcus aureus*
Efflux (pumping out of cell)	Tetracyclines	Plasmid	Enteric *Bacteria*
	Chloramphenicol	Chromosomal	*Staphylococcus aureus* *Bacillus subtilis*
	Erythromycin	Chromosomal	*Staphylococcus* spp.

Streptomycin

Phosphorylation
Adenylation

β-Lactamase

Penicillin

Chloramphenicol

Acetylation

antibiotic susceptibility testing of clinical isolates are described in Section 24.3.

In the *laboratory*, antibiotic-resistant cells are often isolated from cultures that were uniformly susceptible to the antibiotic. The resistance of these isolates is usually due to mutations in *chromosomal* genes. In most cases, antibiotic resistance mediated by chromosomal genes arises because of a modification of the *target* of antibiotic activity (for example, a ribosome).

Mechanism of Resistance Mediated by R Plasmids

Most drug-resistant bacteria isolated from *patients* contain drug-resistance genes located on R (resistance) plasmids, rather than on the chromosome (∞ Section 10.10). R plasmid resistance is usually due to the presence in the R plasmid of genes encoding new enzymes that *inactivate* the drug (Figure 20.25) or genes that encode enzymes that either prevent drug uptake or actively pump it out. For instance, the aminoglycoside antibiotics streptomycin, neomycin, kanamycin, and spectinomycin have similar chemical structures. Strains carrying R plasmids for these drugs can make enzymes that chemically modify the antibiotics either by phosphorylation, acetylation, or adenylylation. The modified drug then lacks antibiotic activity.

In the case of the penicillins, R plasmids encode a penicillinase enzyme (β-lactamase) that splits the β-lactam ring, inactivating the antibiotic. Chloramphenicol resistance is due to an R plasmid-encoded enzyme

● **Figure 20.25 Sites at which antibiotics are attacked by enzymes encoded by R plasmid genes.** Antibiotics may be selectively inactivated by chemical modification or cleavage (see also Figure 20.20).

Microbial Sidebar ◆ Preventing Antimicrobial Drug Resistance

Antimicrobial drug resistance is acquired resistance to the compounds used for prevention and treatment of infectious diseases. Antimicrobial drugs act by disrupting a single metabolic pathway, often in a small number of microorganisms. Microorganisms develop or acquire drug resistance by modification or acquisition of a single gene, often very quickly. For example, for antimicrobial agents used in the 1930s (sulfanilamide) and 1940s (penicillin), many microorganisms exhibited resistance by the 1950s.

The Centers for Disease Control and Prevention (CDC) in Atlanta, Georgia (USA) suggest that resistance is widespread and will soon affect everyone, either as a health care provider or a patient. Each year, nearly 2 million patients in the United States develop a hospital-acquired (*nosocomial*) infection. Nosocomial infections are difficult to treat because up to 70% of the infecting microorganisms are resistant to antimicrobial drugs. For *Staphylococcus aureus*, the species that causes over 10% of these infections in intensive care units, the figure shows a two-fold increase in drug resistance to methicillin and oxacillin, the newest penicillin derivatives, over a 13-year period. Since *S. aureus* isolates may be resistant to a number of other drugs as well, the appearance of the methicillin/oxacillin resistant *S. aureus* (MRSA) strains limits the choice of effective therapeutic drugs. At least one strain each of *Mycobacterium tuberculosis* and *Enterococcus faecium* is resistant to even the most advanced antibiotics in use; some pathogens may no longer be controllable by antibiotic therapy.

A 12-step CDC instruction program to prevent antimicrobial resistance is summarized below (*http://www.cdc.gov/drugresistance/healthcare/ha/12steps_HA.htm*).

12 Steps to Prevent Antimicrobial Resistance

Prevent Infection

1. **Immunize to prevent common diseases.** Keep immunizations up to date, especially for likely disease exposures. In addition to *required* vaccinations (∞ Section 22.12), this should include yearly influenza vaccination for nearly everyone, meningitis vaccines, and pneumococcal immunizations for health care providers or those exposed to large numbers of people, as in schools, colleges, and the military.

2. **Avoid unnecessary introduction of parenteral devices such as IV tubes and catheters.** All present a risk of introducing infectious agents into the body. If such devices are necessary, remove them as soon as possible.

Diagnose and Treat Infection Effectively

3. **Target the pathogen.** Culture the infection to recover the pathogen whenever possible. Provide antimicrobial drug treatment only for the most likely pathogens. After positive culture results, adjust the definitive therapy to the known pathogen. Monitor the response to therapy and adjust treatment when needed. Complete the full course of therapy.

4. **Access the experts.** For serious infections, follow up with a health-care provider, and get a second opinion if conditions do not rapidly improve after treatment has begun.

Use Antimicrobial Agents Wisely

5. **Practice antimicrobial control.** Be aware and current in knowledge of appropriate antimicrobial drugs and their use. Be sure the treatment offered is current and recommended for the pathogen.

6. **Use local data.** Obtain and understand data for the antibiogram (antibiotic susceptibility) profile for the infectious agent from local health care sources.

7. **Treat infection, not contamination.** Antiseptic techniques must be followed to obtain appropriate samples from infected tissues. Contaminating organisms may be present in catheters or IV lines. Obtain cultures only from the site of infection. When in doubt, reculture the infection site.

8. **Treat infection, not colonization.** Treat the pathogen and not other colonizing

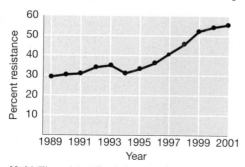

microorganisms that are not causing disease. For example, cultures from normal skin and throat are often colonized with potential pathogens such as *Staphylococcus* species that may have nothing to do with the current infection.

9. **Treat with the least exotic antimicrobial agent that will eliminate the pathogen.** Treatment with the latest broad-spectrum antibiotic, while efficacious, may not be warranted if other drugs are still effective. The more a drug is used, the greater the chance that resistant microorganisms will develop. For example, some *Enterococcus* isolates are already resistant to vancomycin, a relatively new broad-spectrum antibiotic, largely because the drug was over-prescribed to treat MRSA infections when it was first licensed for use.

Stop Antimicrobial Treatment When Unnecessary.

10. Antimicrobial treatment for a diagnosed infection should be discontinued as soon as the prescribed course of treatment is completed. If an infection cannot be diagnosed, treatment should be discontinued. For example, in the case of pharyngitis (sore throat), antibiotic treatment for *Streptococcus pyogenes* infection (*strep throat*) is often started before throat culture results are confirmed. If throat cultures are negative for *S. pyogenes*, treatment with antibiotics should be stopped. Antibiotics are ineffective for the most probable cause of pharyngitis, a virus.

Prevent Pathogen Transmission

11. **Isolate the pathogen.** Keep areas around infected persons free from contamination. Clean up and contain body fluids appropriately. Decontaminate linens, clothes, and other potential sources of contamination. In a hospital setting, infection control experts should be consulted.

12. **Break the chain of contagion.** Avoid contact with others, even at work or school, when sick. Maintain cleanliness, especially by handwashing, if you are sick, or are caring for sick persons.

These steps are guidelines for the intelligent use of antimicrobial drugs. Such measures can slow development of resistance and prolong the useful lifespan of many drugs. ■

Methicillin-resistant Staphylococcus aureus *nosocomial infections among intensive care unit patients from 1989–2001.* Data are from the National Nosocomial Infections Surveillance (NNIS) System of the Centers for Disease Control and Prevention(CDC), Atlanta, Georgia, USA.

that acetylates the antibiotic. Many R plasmids confer multiple antibiotic resistance because a single R plasmid may contain several different genes, each encoding a different antibiotic-inactivating enzyme (∞ Section 10.10).

Origin of Resistance Plasmids

Although specific data for the origin of multiple drug-resistance R plasmids is not obtainable, evidence suggests that R plasmids existed before the antibiotic era. For example, a strain of *Escherichia coli* that was freeze-dried in 1946 contained a plasmid with genes conferring resistance to both tetracycline and streptomycin, even though neither of these antibiotics was used clinically until several years later. Also, strains carrying R plasmid genes for resistance to *semisynthetic penicillins* were shown to exist before the semisynthetic penicillins had been synthesized.

Of perhaps even more ecological significance, R plasmids with antibiotic resistance genes have been detected in some nonpathogenic gram-negative soil *Bacteria*. In the soil, R plasmids may confer selective advantages because major antibiotic-producing organisms (*Streptomyces* and *Penicillium*) are also normal soil organisms. Thus, it seems that R plasmids existed in the natural microbial population before antibiotics were discovered and used in medicine or agriculture.

Spread of Antimicrobial Drug Resistance

The widespread use of medical, veterinary, and agricultural antibiotics continue to provide selective conditions for the spread of R plasmids with antibiotic resistance genes, conferring a selective advantage to bacteria containing R plasmids. Antibiotic resistance due to R plasmids is thus a predictable outcome of natural selection. The R plasmids and other sources of resistance genes pose significant limits for the long-term use of any single antibiotic as an effective chemotherapeutic agent.

Inappropriate, extensive use of antimicrobial drugs is the leading cause of rapid development of drug-specific resistance in disease-causing microorganisms. The discovery and clinical use of the many known antibiotics has been paralleled by the emergence of bacteria that resist them. Figure 20.26a● shows a correlation between the number of tons of antibiotics used and the percentage of bacteria resistant to each antibiotic.

There are numerous examples of the overuse of antibiotics, resulting in development of resistance. Increasingly, the antimicrobial agent prescribed for treatment of a particular infection must be changed because of increased resistance of the microorganism causing the disease. A classic example is the development of resistance to penicillin in *Neisseria gonorrhoeae*, the bacterium that causes gonorrhea (Figure 20.26b). Prior to 1980, penicillin was the accepted treatment for gonorrhea, and penicillin had been in continuous use for this application since it became available in the 1940s. However, penicillin is no longer a useful antibiotic for treatment of gon-

(a)

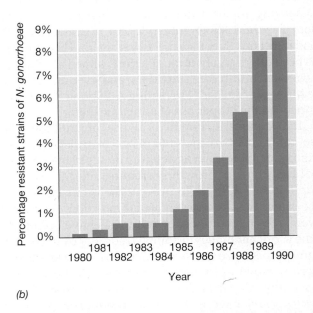

(b)

● **Figure 20.26 The emergence of antimicrobial drug-resistant bacteria.** (a) Relationship between antibiotic use and the percentage of antibiotic-resistant bacteria isolated from diarrheal patients. Those agents that have been used in the largest amounts, as indicated by the amount produced commercially, are those for which drug-resistant strains are most frequent. (b) Percentage of reported cases of gonorrhea caused by drug-resistant strains. The actual number of reported drug-resistant cases in 1985 was 9000. This number rose to 59,000 in 1990. Greater than 95% of the reported drug-resistant cases are due to penicillinase-producing strains of *Neisseria gonorrhoeae*. Since 1990, penicillin has not been recommended for treatment of gonorrhea because of emerging drug resistance. (Source: Centers for Disease Control and Prevention, Atlanta, GA).

orrhea because a large percentage of the clinical *N. gonorrhoeae* isolates now produce β-lactamase, conferring penicillin resistance. Virtually all resistant strains have developed since 1980. The current drug of choice is ceftriaxone, but treatment guidelines are updated nearly every year, simply to control the emerging drug-resistant strains of *N. gonorrhoeae* (∞ Section 26.12).

A number of surveys worldwide suggest that antibiotics are used in clinical practice far more often than is necessary. Data indicate that antibiotic treatment is warranted in 20% of individuals who are seen for clinical infectious disease. Yet, antibiotics are prescribed up to 80% of the time. Furthermore, in up to 50% of cases, recommended doses or duration of treatments are not correct. This is compounded by patient noncompliance: Many patients stop taking medications, particularly antibiotics, as soon as they "feel better." For example, the emergence of isoniazid-resistant tuberculosis correlates with a patient's failure to take the oral medication daily for the full six to nine months prescribed (👓 Section 26.5). Thus, virulent pathogens are often subjected to sublethal doses of antibiotics for short periods of time, selecting for resistant strains.

Other recent studies, however, indicate that this trend is changing in the United States. Physicians prescribe about one-third fewer antibiotics for treatment of childhood infections than they did 10 years ago. This reduction has been done largely through efforts aimed at educating physicians, health care providers, and even patients (see the Microbial Sidebar "Preventing Antibiotic Resistance") concerning the proper use of antibiotic therapy.

Indiscriminant, nonmedical use of antibiotics has also contributed to the emergence of resistant strains. Antibiotics are used in agriculture as supplements to animal feeds both as growth-promoting substances and as prophylactic additives to prevent the occurrence of disease rather than to treat an actual infection. As a result, several recent food infection outbreaks have been traced to the use of antibiotics in animal feeds. For example, fluoroquinolones have been extensively used for less than 20 years as growth-promoting and prophylactic agents in agriculture. However, fluoroquinolone-resistant *Campylobacter jejuni* have emerged as foodborne pathogens (👓 Section 29.9), presumably because of the routine treatment of poultry flocks with fluoroquinolones to prevent respiratory diseases. Voluntary guidelines are being used by both poultry and drug producers to monitor the use of fluoroquinolones in agriculture and prevent rapid emergence of resistance to new fluoroquinolones.

Antibiotic-Resistant Pathogens

Largely as a result of failures to properly use antibiotics and monitor resistance, almost all pathogenic microorganisms have developed resistance to some chemotherapeutic agents since widespread use of antimicrobial chemotherapy began in the 1950s (Figure 20.27●). Penicillin and sulfa drugs, the first widely used chemotherapeutic agents, are not as widely used today because many pathogens have acquired some resistance. Even the organisms that are still uniformly sensitive to penicillin, such as *Streptococcus pyogenes* (the bacterium that causes strep throat, scarlet fever, and rheumatic fever; 👓 Section 26.2), typically require

● Figure 20.27 The appearance of antimicrobial drug resistance in some human pathogens. The asterisk * indicates that some multi-drug resistant strains of these organisms are now untreatable with known antimicrobial drugs.

larger doses of penicillin for successful treatment now as compared to a decade ago.

A few pathogens have developed resistance to all known antimicrobial agents (Figure 20.27). Among these are several isolates of **methicillin-resistant *Staphylococcus aureus* (MRSA)** (👓 Section 26.9) and **vancomycin-resistant *Enterococcus faecium* (VRE)**, as well as isolates of *Mycobacterium tuberculosis* (👓 Section 26.5). An increasing number of independently derived MRSA strains have developed reduced susceptibility to vancomycin and are termed "vancomycin intermediate *Staphylococcus aureus*" (VISA) strains.

Antibiotic resistance can be minimized if drugs are used only for treatment of susceptible diseases and are given in sufficiently high doses and for sufficient lengths of time such that the microbial population is reduced before resistant mutants have a chance to appear. Combining two unrelated chemotherapeutic agents may also reduce resistance; it is less likely that a mutant strain resistant to one antibiotic will also be resistant to the second antibiotic. However, the prevalence of R plasmids conferring multiple drug resistance makes multiple antibiotic therapy less useful as a clinical treatment strategy.

A few reports suggest that if the use of a particular antibiotic is stopped, the resistance to that antibiotic will be reversed over the course of several years. On the other hand, antibiotic-resistant organisms may persist in the gut for some time. This information implies that the efficacy of some antibiotics may be reestablished by withdrawing the antibiotic from use, but only by following a carefully monitored plan of prudent use upon reintroduction. Finally, as we discuss below, new chemotherapeutic agents are constantly being produced using several methods for discovering and designing new drugs.

 20.12 Concept Check

The use of antimicrobial drugs has fostered the development of resistance in the targeted microorganisms. Resistance results from the selection of resistance genes and can be accelerated by the indiscriminate use of antimicrobial drugs. A few organisms have developed resistance to all known antimicrobial drugs.

♦ Identify the six basic mechanisms of antibiotic resistance among bacteria.

♦ Identify the primary sources of antibiotic resistance genes.

♦ What practices encourage the development of antibiotic-resistant pathogens?

20.13 The Search for New Antimicrobial Drugs

As we have just discussed, given sufficient drug exposure and time, resistance will develop to all known antimicrobial drugs. As a result, conservative, appropriate use of antibiotics is absolutely necessary to prolong the useful clinical life of these drugs. However, the long-term solution to microbial drug resistance is to develop new antimicrobial drugs. Several strategies are used to identify and produce useful analogs of existing compounds or to design or discover novel antimicrobial compounds.

New Analogs of Existing Antimicrobial Compounds

The production of new analogs of existing antimicrobial compounds is generally straightforward and often productive, largely because new compounds that are structural mimics of older ones have a predictable mechanism of action. In many cases, parameters such as solubility and affinity can be optimized by introducing minor modifications to the chemical structure of a drug without altering structures critical to drug action. The new compound may actually be more potent than the parent compound, and, because resistance is based on structural recognition, the new compound may not be recognized by resistance factors. For example, Figure 20.23 shows the basic structure of tetracycline and two bioactive derivatives. Using tetracycline as a *lead compound* (a biologically active compound on which modifications will be made), systematic chemical substitutions can be made at the four R group sites, generating an almost endless series of tetracycline analogs. Using this basic strategy, new β-lactam antibiotics (see Section 20.8), new tetracycline-related compounds (see Section 20.9), and new analogs of vancomycin (Figure 20.13), some up to 100 times as potent as the parent compound, are routinely synthesized and tested.

The application of automated robotic chemistry methods to drug discovery has dramatically increased our ability to rapidly generate potential new antimicrobial compounds. The automated robotic methods, referred to as *combinatorial chemistry*, apply systematic modifications of a known antimicrobial product to generate large numbers of new analogs. For instance, using automated combinatorial chemistry methods, and starting with the tetracycline lead compound (Figure 20.23), five different reagents might be used to introduce substitutions at the four different tetracycline R groups. The substituted sites would yield $5 \times 5 \times 5 \times 5$ (five derivatives at each of four sites), or 625 different tetracycline derivatives from only six different reagents, all in a few hours! These compounds are then assayed for *in vitro* biological activity on different microorganisms using automated testing methods for antibiotic susceptibility (see Section 20.4). The automated synthesis and screening processes dramatically shorten drug discovery time and increase the number of new candidate drugs by 10 times or more each year.

According to pharmaceutical industry estimates, about 7 *million* candidate compounds must be screened to yield a single useful clinical drug. Drugs effective in the laboratory must then be tested for efficacy and toxicity in animals and finally in clinical trials in humans. Animal testing usually involves multiple trials over several years to ensure that the candidate drug is effective and safe. Clinical trials in humans to check efficacy and safety take several years for each drug. Each year, the pharmaceutical industry spends up to $4 billion on new antimicrobial drug development. The total time for discovery and development for each drug typically takes 10 to 25 years before approval for clinical use. The cost of discovery and development, from the laboratory through clinical trials, is estimated to be at least $500 million for each new antibiotic approved for human use.

Computerized Drug Design

Novel antimicrobial compounds are much more difficult to identify than analogs of existing drugs because new antimicrobial compounds must work at unique sites in metabolism and biosynthesis, or be structurally dissimilar to existing compounds to avoid inducing known resistance mechanisms. In the past, candidate drugs were isolated from natural sources such as *Streptomyces* or *Penicillium* cultures and systematically screened for antimicrobial activity to find new antimicrobial compounds. However, recent advances in computer and structural graphics technology now make it possible to design a drug to interact with specific microbial structures. Drug discovery can now begin at the computer, where new drugs can be rapidly "created" and "tested" for binding and efficacy in the computer environment at relatively low cost (∞ Sections 30.5 and 26.8).

One of the most dramatic recent successes in computer-directed drug design is the development of *saquinavir*, a protease inhibitor that is used to slow the growth of the human immunodeficiency virus (HIV) in infected individuals (Figure 20.28●). Saquinavir was designed by computer to bind the active site of the enzyme *HIV protease*, based on the known three-dimensional structure of the protease-substrate complex. The HIV protease normally cleaves a virus-encoded precursor protein to produce the mature viral core and activate the reverse transcriptase enzyme necessary for replication (∞ Section 16.15). Saquinavir is a peptide analog that displaces the HIV precursor protein, inhibiting virus maturation and slowing its growth in the human host. A number of other computer-designed protease inhibitors like saquinavir are in use as chemotherapeutic drugs for the treatment of AIDS (Table 20.5, Figure 20.28, and ∞ Figure 26.38). Computer design and testing based on structural and biochemical modeling is a practical, rapid, and cost-effective method for designing antimicrobial drugs. Of course, these drugs must undergo animal testing and clinical trials before they can be used therapeutically.

Novel Antimicrobial Agents

Bacteriophages are viruses that target individual bacterial species (∞ Sections 16.1–16.6). *Bacteriophage therapy* has been used on a limited basis for over 80 years to treat infections in animals and in a few instances in humans. Phages bind bacteria by interacting with specific bacterial cell surface components. The binding is reasonably specific, generally targeting individual bacterial species. The attached phage enters the bacterial cell, replicates, and kills the bacterial host in the process. The efficacy and efficiency of these agents for human treatment is largely untested and somewhat controversial, although clinical trials of several products are ongoing. Because bacteria can become resistant to a phage through mutations that alter receptors or reduce the susceptibility of the cell wall to phage enzymes, this therapy will likely not be immune to the problem of resistance.

 20.13   Concept Check

New antimicrobial compounds are constantly being discovered and developed to deal with drug-resistant pathogens, and enhance our ability to treat infectious diseases. Analogs of existing drugs are often developed to be used as next generation antimicrobial compounds. Computer drug design is an important new tool for drug discovery.

◆   Explain the advantages and disadvantages of developing new drugs based on existing drug analogs.

◆   How can computer drug design save time and money in the search for new drugs?

(a)

Saquinavir

Indinavir

(b)

● **Figure 20.28   Computer-generated antiviral drugs.** (a) The HIV protease homodimer. Individual polypeptide chains are shown in green and blue. A peptide (yellow) is bound by the catalytic site. This protease cleaves an HIV precursor protein, a necessary step in virus maturation (∞ Section 16.15 and Figure 16.26). Blocking of the protease site by the peptide shown inhibits precursor processing and HIV maturation. This structure is derived from information in the Protein Data Bank. (b) These anti-HIV drugs are peptide analogs that were designed by computer to block the active site of HIV protease. The areas highlighted in orange show the regions analogous to peptide bonds. Binding of these compounds by the HIV protease prevents HIV precursor processing and virus maturation. These compounds represent a class of therapeutic drugs called *protease inhibitors*. The concentration of these compounds in HIV-infected cells, coupled with their strong affinity for HIV protease, makes them very potent competitive inhibitors for the active sites of the protease and prevents virus maturation. These protease inhibitors are widely used for treatment of HIV infection (see Table 20.5 and ∞ Section 26.14).

REVIEW QUESTIONS

1. Why is the decimal reduction time (*D*) important in heat sterilization? How would the presence of bacterial endospores affect *D* (⌦ Section 20.1)?

2. Describe the effects of a lethal dose of ionizing radiation at the molecular level (⌦ Section 20.2).

3. What are the principal advantages of using membrane filters instead of depth filters (⌦ Section 20.3)?

4. Why are Nucleopore filters particularly useful for isolating specimens for microscopy (⌦ Section 20.4)?

5. Describe the procedure for obtaining the minimum inhibitory concentration (MIC) for a chemical that is bacteriocidal for *Escherichia coli* (⌦ Section 20.5).

6. Contrast the action of disinfectants and antiseptics. Why can't disinfectants normally be used on living tissue (⌦ Section 20.6)?

7. Growth factor analogs are generally distinguished from antibiotics by a single important criterion. Explain (⌦ Section 20.7).

8. Describe the mode of action that characterizes a β-lactam antibiotic (⌦ Section 20.8).

9. Distinguish between the mode of action of at least three of the protein synthesis-inhibiting antibiotics (⌦ Section 20.9).

10. Why do antiviral drugs generally exhibit host toxicity (⌦ Section 20.10)?

11. Identify the targets that allow selective toxicity of chemotherapeutic agents in fungi (⌦ Section 20.11).

12. Identify six mechanisms responsible for antibiotic resistance (⌦ Section 20.12).

13. Starting with a lead compound, describe *combinatorial chemistry* methods used for the production of new drug analogs (⌦ Section 20.13).

APPLICATION QUESTIONS

1. What are some potential drawbacks to the use of radiation in food preservation? Do you think these drawbacks could be manifested as health hazards? Why or why not? How would you distinguish between radiation-damaged and radiation-contaminated food?

2. Filtration is an acceptable means of pasteurization for some liquids. Design a filtration system for pasteurization of a heat-sensitive liquid. Why might a filtration system be desirable over a heat pasteurization system?

3. Although growth factor analogs may inhibit microbial metabolism, only a few of these agents are practically useful. Many potential agents, and some that are in wide use, such as azidothymidine (see Table 20.5), exhibit significant host cell toxicity. Describe a growth factor analog that is effective and has low toxicity for host cells. Why is the toxicity low for the agent you chose? Also describe a growth factor analog that is effective against an infectious disease but exhibits toxicity for host cells. Why might a toxic agent such as AZT still be used in certain situations to treat infectious diseases? What precautions would you take to limit the toxic effects of such a drug, while maximizing the therapeutic activity? Explain your answer.

4. Design an experiment to examine microorganisms for production of novel antibiotics. Which group or groups of microorganisms would you choose to screen for antibi-otic production? Where could you find and isolate these organisms in a natural environment? What advantage would the production of an antibiotic provide for these organisms in nature? What *in vitro* methods would you use to test the efficacy of your potential new antibiotics?

5. Although the β-lactam antibiotics demonstrate clear selective toxicity for *Bacteria*, many groups of *Bacteria* are innately resistant to their effects. Indicate why gram-negative *Bacteria* are resistant to the effects of most, but not all, β-lactam antibiotics. Further explain why some β-lactam antibiotics are useful against these organisms.

6. List the features of an ideal antiviral drug, especially with regard to selective toxicity. Do such drugs exist? What factors might limit the use of such a drug?

7. Like viruses, fungi present special chemotherapeutic problems. Explain the problems inherent in chemotherapy of both groups and explain whether or not you agree with the preceding statement. Give specific examples and suggest at least one group of chemotherapeutic agents that might target both types of infectious agents.

8. Explain the genetic basis of acquired resistance to β-lactam antibiotics in *Staphylococcus aureus*. Design experiments to reverse resistance to the β-lactam antibiotics. Do you think this can be done in the laboratory? Can your experiment be applied "in the field" to promote de-selection of antibiotic-resistant organisms?

MICROBIAL INTERACTIONS WITH HUMANS

21

Bacillus anthracis, the cause of anthrax, produces a thick polysaccharide capsule, stained green in the photo. This capsule prevents killing of the pathogen by interfering with host defense systems.

WORKING GLOSSARY

Attenuation decrease or loss of virulence
Bacteremia the presence of microorganisms in the blood
Capsule dense, well-defined polysaccharide or protein layer closely surrounding a cell
Colonization multiplication of a pathogen after it has gained access to host tissues
Dental caries tooth decay resulting from bacterial infection
Dental plaque bacterial cells encased in a matrix of extracellular polymers and salivary products, found on the teeth
Disease injury to the host that impairs host function
Endotoxin the lipopolysaccharide portion of the cell envelope of certain gram-negative *Bacteria*, which is a toxin when solubilized
Enterotoxin protein released extracellularly by a microorganism as it grows that produces immediate damage to the small intestine of the host

Exotoxin protein released extracellularly by a microorganism as it grows that produces immediate host cell damage
Hemolysins bacterial toxins capable of hemolysis, the lysis of red blood cells
Host an organism that harbors a parasite
Infection growth of organisms in the host
Invasiveness pathogenicity caused by the ability of a pathogen to enter the body and spread
Lower respiratory tract trachea, bronchi, and lungs
Mucous membrane layers of epithelial cells that interact with the external environment
Mucus soluble glycoproteins secreted by epithelial cells that coat the mucous membranes
Normal microbial flora microorganisms that are usually found associated with healthy body tissue

Nosocomial infection infection contracted in a hospital or health-care setting
Opportunistic pathogen an organism that causes disease in the absence of normal host resistance
Parasite an organism that grows in or on a host and causes disease
Pathogen an organism, usually a microorganism, that causes disease
Pathogenicity the ability of a pathogen to cause disease
Slime layer a diffuse layer of polymer fibers, typically polysaccharides, that forms an outer surface layer on the cell
Toxicity pathogenicity caused by toxins produced by ▨▨▨▨▨n
Upper respi▨▨▨▨t the nasopharynx, oral cavity, ▨▨▨
Virulence th▨▨▨▨▨ pathogenicity displayed by a▨▨▨

The human body has an extensive population of microorganisms on the skin and the mucous membranes lining the mouth, gut, excretory, and reproductive systems. These microorganisms are often beneficial and sometimes necessary to maintain good health. However, another group of microorganisms uses direct and indirect means to colonize, invade, and damage the human body, a process called infectious disease. These microorganisms, called pathogens, use several strategies to gain access to nutrients in a host. These strategies include the production of specialized attachment structures, unique growth factors, invasive enzymes, and potent biological toxins. These factors often lead to damage and occasionally death of the host.

In Chapter 20, we saw how microbiologists have developed countermeasures to destroy or inhibit growth of microorganisms. In addition, our bodies have developed effective countermeasures to suppress or destroy most microbial invaders. Nonspecific physical, anatomical, and biochemical processes make microbial infectious disease a relatively infrequent event.

⦿ BENEFICIAL MICROBIAL INTERACTIONS WITH HUMANS

We begin with microorganisms that inhabit the healthy human body and contribute to overall good health under normal circumstances.

21.1 Overview of Human–Microbial Interactions

The human body is constantly exposed to microorganisms. Through normal everyday activities, we are exposed to countless microorganisms in our environment. In addition, hundreds of species and billions of individual microorganisms, collectively referred to as the **normal microbial flora**, grow on or in the human body. Most, but not all, microorganisms are benign, and a few contribute directly to our health.

Pathogens

Organisms that live on or in a host organism, causing damage to the **host**, are called **parasites**. Microbial parasites are often called **pathogens**. The outcome of a host–parasite relationship depends on the **pathogenicity** of the parasite. The ability of a parasite to inflict damage on the host varies considerably, as does the *resistance* or *susceptibility* of the host to the parasite. An **opportunistic pathogen** causes disease only in the absence of normal host resistance.

Pathogenicity varies markedly for individual pathogens. The quantitative measure of pathogenicity is termed **virulence**. Virulence can be expressed as the cell number that will elicit a pathogenic response in a host within a given time period. Neither the virulence of the pathogen nor the relative resistance of the host is a constant factor. The host–parasite interaction is a dynamic relationship between the two organisms, influenced by changing conditions in either the pathogen or the host.

Infection and Disease

Infection refers to any situation in which a microorganism is established and growing in a host, whether or not the host is harmed. **Disease** is damage or injury to the host that impairs host function. *Infection is not synonymous with disease* because growth of a microorganism on a host does not always cause host damage. Thus, members of the normal microbial flora may produce infections, but seldom cause disease. However, the normal flora microorganisms sometimes cause disease if host resistance is compromised, as happens in diseases such as cancer and AIDS (∞Section 26.14).

Host–Parasite Interactions

Animal bodies act as hosts and provide favorable environments for the growth of many microorganisms. They are rich in the organic nutrients and growth factors required by chemoorganotrophs and provide locally controlled pH, osmotic pressure, and temperature. However, the animal body is not a uniform microbial environment. Each region or organ differs chemically and physically from others and thus provides a selective environment where the growth of certain microorganisms is favored.

For example, the skin, respiratory tract, and gastrointestinal tract provide a wide variety of chemical and physical environments in which different microorganisms can grow selectively. The relatively dry environment of the skin favors the growth of organisms with substantial barriers to dehydration such as the gram-positive *Staphylococcus aureus* (∞Section 26.9); the highly oxygenated environment of the lungs favors the growth of the obligately aerobic *Mycobacterium tuberculosis* (∞Section 26.5); and the anaerobic environment of the large intestine supports the growth of members of the obligately anaerobic *Clostridium* genus (∞Section 12.20). Animals also possess a variety of defense mechanisms that collectively prevent or inhibit microbial invasion and growth. The microorganisms that ultimately colonize the host successfully are those that have developed ways of circumventing these defense mechanisms.

Infections frequently begin at sites in the animal's **mucous membranes**. Mucous membranes consist of single or multiple layers of *epithelial cells*, tightly packed cells that interface with the external environment. They are found throughout the body and include the mouth, pharynx, esophagus, and urogenital, respiratory, and gastrointestinal tracts. Mucous membranes are frequently coated with a protective layer of viscous soluble glycoproteins called **mucus**. When bacteria contact host tissues at mucous membranes, they may associate either loosely or firmly. If they associate loosely with the mucosal surface, they are usually swept away by physical processes, but they may also adhere to the epithelial surface as a result of specific cell–cell recognition between pathogen and host. Tissue infection may follow, breaching the mucosal barrier and allowing the pathogen to invade deeper (submucosal) tissues (Figure 21.1●).

Microorganisms are almost always found on surfaces of the body exposed to the environment, such as the skin, oral cavity, respiratory tract, intestinal tract, and urogenital tract. They are not normally found in the internal organs, or in the blood, lymph, or nervous systems of the body. The growth of microorganisms in these usually sterile environments indicates serious infectious disease.

Table 21.1 shows some of the major types of microorganisms normally found in association with body surfaces. The mucosal surfaces have a large variety of associated microorganisms because they offer a sheltered, moist environment and a huge overall surface area of about 400 m². For example, the specialized function of a mucosal organ such as the small intestine requires a large surface area for nutrient transport, and this surface is also a site for microbial growth. Here we examine some normal microbial–host interactions at mucosal surfaces.

Table 21.1	**Representative genera of microorganisms in the normal flora of humans**
Anatomical site	**Genera[a]**
Skin	*Acinetobacter, Corynebacterium, Enterobacter, Klebsiella, Malassezia (f), Micrococcus, Pityrosporum (f), Propionibacterium, Proteus, Pseudomonas, Staphylococcus*
Mouth	*Streptococcus, Lactobacillus, Fusobacterium, Veillonella, Corynebacterium, Neisseria, Actinomyces, Geotrichum (f), Candida (f), Capnocytophaga, Eikenella, Prevotella,* spirochetes (several genera)
Respiratory tract	*Streptococcus, Staphylococcus, Corynebacterium, Neisseria, Haemophilus*
Gastrointestinal tract	*Lactobacillus, Streptococcus, Bacteroides, Bifidobacterium, Eubacterium, Peptococcus, Peptostreptococcus, Ruminococcus, Clostridium, Escherichia, Klebsiella, Proteus, Enterococcus. Staphylococcus*
Urogenital tract	*Escherichia, Klebsiella, Proteus, Neisseria, Lactobacillus, Corynebacterium, Staphylococcus, Candida (f), Prevotella, Clostridium, Peptostreptococcus, Ureaplasma, Mycoplasma. Mycobacterium, Streptococcus, Torulopsis (f)*

[a] This list is not meant to be exhaustive, and not all of these organisms are found in every individual. Some organisms are more prevalent at certain ages (adults vs. children). Distribution may also vary between sexes. Most of these organisms can be opportunistic pathogens under certain conditions. Several genera can be found in more than one body area. (f)–fungi.

● **Figure 21.1** **Bacterial interactions with mucous membranes.** (a) Loose association. (b) Adhesion. (c) Invasion into submucosal epithelial cells.

⬡ *21.1* **Concept Check**

Animal bodies are favorable environments for the growth of microorganisms, most of which do no harm. Microorganisms that cause harm are called *pathogens*. Pathogen growth on the surface of a host, often on the mucous membranes, may result in infection and disease. The ability of a microorganism to cause or prevent disease is influenced by complex host–parasite interactions.

◆ Distinguish between *infection* and *disease*.

◆ Why might one area of the body be more suitable for microbial growth than another?

21.2 Normal Microbial Flora of the Skin

An average adult human has about 2 m² of skin that varies greatly in chemical composition and moisture content. Figure 21.2● shows the anatomy of the skin and regions in which microorganisms may live. The skin surface (epidermis) is not a favorable place for abundant microbial growth, as it is subject to periodic drying.

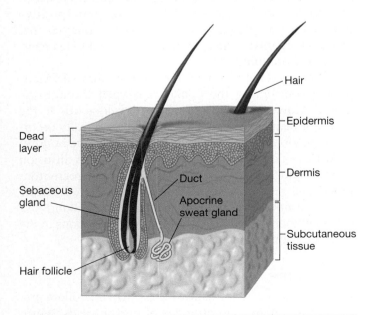

● **Figure 21.2** **The human skin.** Microorganisms are associated primarily with the sweat ducts and the hair follicles.

Most skin microorganisms are associated directly or indirectly with the *apocrine glands* (sweat glands). These are secretory glands found mainly in the underarm and genital regions, the nipples, and the umbilicus. They are inactive in childhood and become fully functional only at puberty. Bacterial populations on the surface of the skin in these warm, humid places are relatively high in contrast to the situation on the smooth, dry skin surface. Underarm odor develops as a result of bacterial activity in the apocrine secretions; aseptically collected apocrine secretion is odorless but develops odor on inoculation with bacteria. Similarly, each hair follicle is associated with a *sebaceous gland*, which secretes a lubricant fluid. Hair follicles provide an attractive habitat for microorganisms in the area just below the surface of the skin. The secretions of the skin glands are rich in microbial nutrients such as urea, amino acids, salts, lactic acid, and lipids. The pH of human secretions is almost always acidic, the usual range being between pH 4 and 6.

The normal flora of the skin consists of either *transient* or *resident* populations of microorganisms. The skin is continually being inoculated with *transient* microorganisms, virtually all of which are unable to multiply and usually die. *Resident* microorganisms are able to multiply, not merely survive, on the skin. The normal flora of the skin consists primarily of gram-positive *Bacteria* restricted to a few groups (Table 21.1). These include several species of *Staphylococcus* and a variety of both aerobic and anaerobic corynebacteria. Of the latter, *Propionibacterium acnes* can contribute to the condition known as *acne*.

Gram-negative *Bacteria* are occasional constituents of the normal skin flora because such intestinal organisms as *Escherichia coli* are being continually inoculated onto the surface of the skin by fecal contamination. Few of these gram-negative *Bacteria* actually *grow* on skin, due to their inability to compete with gram-positive organisms that are better adapted to the dry conditions. However, there are exceptions, and the gram-negative rod *Acinetobacter* is commonly found on skin. Yeasts are uncommon in large numbers on the skin surface or mucous membranes, although the lipophilic yeast *Pityrosporum ovalis* is occasionally found on the scalp. In the absence of host resistance, as in patients with acquired immunodeficiency syndrome (AIDS) (∞Section 26.14), or in the absence of normal bacterial flora, yeasts such as *Candida* and other fungi may grow and cause serious infections on the skin surface.

Although the resident microflora remains somewhat constant, various factors influence its normal composition. (1) The *weather* may cause an increase in skin temperature and moisture, which increases the density of the skin microflora. (2) The *age* of the host has an effect, and young children have a more varied microflora and carry more potentially pathogenic gram-negative *Bacteria* than adults. (3) *Personal hygiene* influences the resident microflora, and unclean individuals usually have higher microbial population densities on their skin. Organisms that cannot survive on the skin generally succumb from either the skin's low moisture content or low pH (due to organic acid content).

21.2 *Concept Check*

The skin is a generally dry, acidic environment that does not support the growth of most microorganisms. However, moist areas, especially around sweat glands, are colonized by gram-positive *Bacteria* and other members of the skin normal flora. Environmental and host factors influence the quantity and quality of the normal skin microflora.

◆ Compare the surface area of the skin to that of the mucosal tissues.

◆ Describe the properties of microorganisms that grow well on the skin.

21.3 Normal Microbial Flora of the Oral Cavity

The oral cavity is a very complex and heterogeneous microbial habitat. On the one hand, saliva contains a variety of microbial nutrients, but it is not an especially good growth medium because the nutrients are present in low concentration and saliva also contains a number of antibacterial substances. For example, saliva contains *lysozyme*, an enzyme that cleaves glycosidic linkages in the peptidoglycan of the bacterial cell wall, leading to weakening of the wall and cell lysis (⌬ Section 4.8). *Lactoperoxidase*, an enzyme in both milk and saliva, kills bacteria by a reaction in which singlet oxygen is generated (⌬ Sections 6.15 and 22.2). Despite the activity of these antibacterial substances, the presence of food particles and epithelial debris provides high concentrations of nutrients on and around oral surfaces such as teeth and gums, creating favorable conditions for extensive local microbial growth, tissue damage, and disease.

The Teeth and Dental Plaque

The tooth consists of a mineral matrix of calcium phosphate crystals (enamel) surrounding the living tooth tissue (dentin and pulp) (Figure 21.3●). Bacteria found in the mouth during the first year of life (when teeth are absent) are predominantly aerotolerant anaerobes such as streptococci and lactobacilli. However, a variety of other bacteria, including some aerobes, occur in small numbers. When the teeth appear, there is a pronounced shift in the balance of the microflora toward anaerobes that are specifically adapted for growth on surfaces of the teeth and in the gingival crevices.

Bacterial colonization of tooth surfaces begins with the attachment of single bacterial cells. Even on a freshly cleaned tooth surface, a thin organic film several micrometers thick forms quickly as a result of the attachment of acidic glycoproteins from the saliva. This film provides a firm attachment site for bacterial micro-

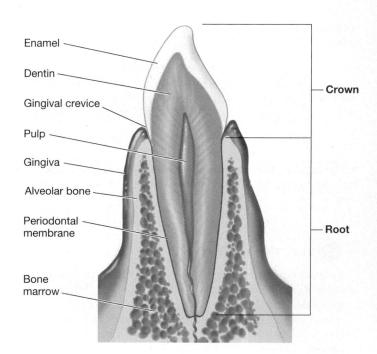

● **Figure 21.3   Section through a tooth.** The diagram shows the tooth architecture and the surrounding tissues that anchor the tooth in the gum.

colonies (Figure 21.4●). Colonization of the glycoprotein film involves only a few species of *Streptococcus* (primarily *S. sanguis*, *S. sobrinus*, *S. mutans*, and *S. mitis*). Extensive growth of these organisms results in a thick bacterial layer called **dental plaque** (Figures 21.5● and 21.6●).

If plaque continues to form, filamentous *Fusobacterium* species begin to grow. The filamentous bacteria embed in the matrix formed by the streptococci and extend perpendicular to the tooth surface, making an ever-thicker bacterial layer. Associated with the filamentous bacteria, in addition to the streptococci, are spirochetes such as *Borrelia* species (⌬ Section 12.33), gram-positive rods, and gram-negative cocci. In heavy plaque, filamentous organisms such as *Actinomyces*, obligate anaerobes, may predominate.

The anaerobic nature of the oral flora may seem surprising considering the intake of oxygen through the mouth. However, anoxia probably develops due to the action of facultative bacteria growing aerobically on organic materials at the tooth surface. The plaque buildup produces a dense matrix that decreases oxygen diffusion to the tooth surface, forming an anoxic microenvironment. The microbial populations within dental plaque exist in a microenvironment of their own making and maintain themselves in the face of wide variations in the macroenvironment of the oral cavity.

Dental Caries

As dental plaque accumulates, the resident microflora produce locally high concentrations of organic acids, which cause decalcification of the tooth enamel (Figure 21.3), and

(a)

(b)

T. Lie

● **Figure 21.4 Microcolonies of bacteria.** (a) The colonies are growing on a model tooth surface inserted into the mouth for 6 h. (b) Higher magnification of the preparation in (a). Note the diverse morphology of the organisms present and the slime layer (arrows) holding the organisms together.

dental caries (tooth decay) results. Thus, tooth decay is an infectious disease caused by microorganisms. The smooth surfaces of the teeth are relatively resistant to decay as they are exposed to frequent cleaning by the tongue, cheek, saliva, or toothbrush or to the abrasive action of food mastication. The tooth surfaces in and near the gingival crevice, where food particles can be retained, are the sites where tooth decay usually occurs.

Diets high in sucrose (table sugar) are especially *cariogenic* because lactic acid bacteria ferment sugars to lactic acid. Once breakdown of the hard tissue (tooth enamel) has begun, proteolysis of the tooth enamel matrix occurs through the action of bacterial proteolytic

Day 1 1436 mm²

Day 10 22,522 mm²

● **Figure 21.5 Distribution of dental plaque.** Plaque is revealed by use of a disclosing agent, on brushed (top) and unbrushed (bottom) teeth. The numbers give the total area of dental plaque. The purple areas indicate plaque. Note that the plaque builds preferentially near the gum line, first occurring directly adjacent to the mucous membranes of the gingiva.

(a)　　　　　　　(b)

C. Lai, M.A. Listgarten, and B. Rosan

● **Figure 21.6 Electron micrographs of thin sections of dental plaque.** Bottom is the base of the plaque; top is the portion exposed to the oral cavity. (a) Low-power electron micrograph. Organisms are predominantly streptococci. The species *Streptococcus sobrinus* has been labeled by an antibody-microchemical technique, and these cells appear darker than the rest. They are seen as two distinct chains (arrows). The total thickness of the plaque layer shown is about 50 μm. (b) Higher-power electron micrograph showing the region with *S. sobrinus* cells (dark, arrow). Note the extensive slime layer (see Section 21.6) surrounding the *S. sobrinus* cells. Individual cells are about 1 μm in diameter.

enzymes. Microorganisms penetrate further into the decomposing matrix, but the later stages of the process may be exceedingly slow and complex. The structure of the calcified tissue also plays a role in the extent of dental caries. For example, incorporation of fluoride into the calcium phosphate crystal matrix increases resistance to acid decalcification. Thus, fluorides in drinking water and dentifrices inhibit tooth decay.

Two organisms implicated in dental caries are *Streptococcus sobrinus* and *Streptococcus mutans*, both lactic acid-producing bacteria. *S. sobrinus* is probably the primary organism involved in decay of smooth surfaces because of its specific affinity for salivary glycoproteins found on smooth tooth surfaces (Figure 21.6). *S. mutans* is found predominantly in crevices and small fissures, and its ability to attach to tooth surfaces results from production of a dextran polysaccharide that is strongly adhesive (Figure 21.7●). *S. mutans* produces dextran only when sucrose (table sugar) is present, by means of the enzyme *dextransucrase*:

$$n \text{ Sucrose} \xrightarrow{\text{Dextransucrase}} \text{Dextran } (n \text{ Glucose}) + n \text{ Fructose}$$

Susceptibility to tooth decay varies and is affected by inherent traits in the individual as well as by diet and other extraneous factors. For example, sucrose is found in the diet of most individuals in developed countries and is highly cariogenic because it is a substrate for dextransucrase. Studies of the distribution of the cariogenic oral streptococci show a direct correlation between the presence of *S. mutans* and *S. sobrinus* and the extent of dental caries. In the United States and Western Europe, for example, 80–90% of all people have their teeth colonized by *S. mutans*, and dental caries is nearly universal. By contrast, *S. mutans* is absent from the plaque of Tanzanian children, and dental caries does not occur, presumably because sucrose is almost completely absent in their diets.

Microorganisms in the mouth can also cause other infections. The areas along the periodontal membrane at or below the gingival crevice (*periodontal pockets*) (Figure 21.3)

● **Figure 21.7 Scanning electron micrograph of the cariogenic bacterium *Streptococcus mutans*.** The sticky dextran material holds the cells together as filaments. Individual cells are about 1 μm in diameter.

can be infected with a variety of microorganisms, causing inflammation of the gum tissues (*gingivitis*) leading to tissue and bone-destroying *periodontal disease*. Some of the genera involved include *fusiform* bacteria (long, thin, gram-negative rods with tapering ends) such as the facultative aerobic *Capnocytophaga*. The aerobic *Rothia* and an anaerobic methanogen and member of the *Archaea, Methanobrevibacter*, may also be present (∞Section 13.4).

 21.3 Concept Check

Bacteria can grow on tooth surfaces in thick layers called plaque. Plaque microorganisms produce adherent substances. Acid produced by microorganisms in plaque damages tooth surfaces, and dental caries result. A variety of microorganisms contribute to caries and periodontal disease.

◆ How do anaerobic microorganisms become established in the mouth?

◆ Is dental caries an infectious disease? Give at least one reason for your answer.

◆ Identify the contribution of the lactic acid bacteria to tooth decay.

21.4 Normal Microbial Flora of the Gastrointestinal Tract

The human gastrointestinal tract (Figure 21.8●), the site of food digestion, consists of the stomach, small intestine, and large intestine. Because stomach fluids are highly acidic (about pH 2), the stomach is a chemical barrier to entry of microorganisms into the intestinal tract. As a result, the bacterial count of the stomach contents is generally very low. However, organisms such as *Helicobacter pylori* can colonize the stomach wall in normal individuals, but can cause ulcers in susceptible hosts (∞Section 26.10). Distal to the stomach, the intestinal tract consists of the *small intestine* and the *large intestine*, each of which is divided into different anatomical regions.

The composition of the gut flora in humans varies considerably and is somewhat dependent on diet. For example, persons who consume a considerable amount of meat show higher numbers of the highly proteolytic *Bacteroides* and lower numbers of coliforms and lactic acid bacteria than individuals with a vegetarian diet. Representative genera of microorganisms found in the gut are shown in Figure 21.8.

The Small Intestine

The *small intestine* is separated into two parts, the *duodenum* and the *ileum*, with the *jejunum* connecting them. The duodenum, adjacent to the stomach, is fairly acidic and resembles the stomach in its lack of a microbial flora. From the duodenum to the ileum, the pH gradually becomes less acidic and bacterial numbers increase.

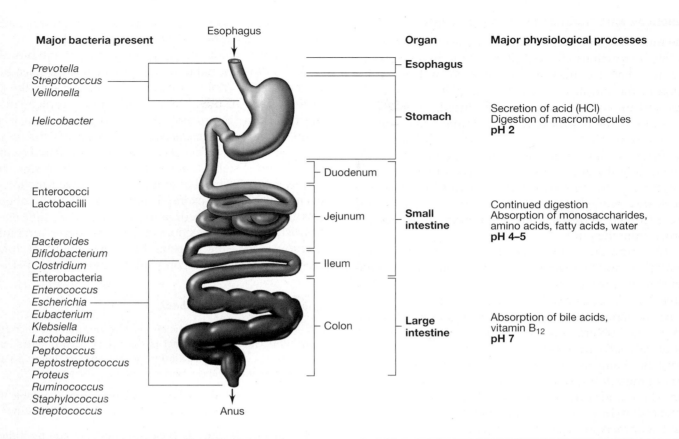

Major bacteria present

Esophagus

Organ

Major physiological processes

Prevotella
Streptococcus
Veillonella

Helicobacter

Esophagus

Stomach

Secretion of acid (HCl)
Digestion of macromolecules
pH 2

Enterococci
Lactobacilli

Duodenum

Jejunum

Small
intestine

Continued digestion
Absorption of monosaccharides,
amino acids, fatty acids, water
pH 4–5

Bacteroides
Bifidobacterium
Clostridium
Enterobacteria
Enterococcus
Escherichia
Eubacterium
Klebsiella
Lactobacillus
Peptococcus
Peptostreptococcus
Proteus
Ruminococcus
Staphylococcus
Streptococcus

Ileum

Colon

Large
intestine

Absorption of bile acids,
vitamin B_{12}
pH 7

Anus

● **Figure 21.8 The human gastrointestinal tract.** The functions and the distribution of nonpathogenic microorganisms usually found in normal (healthy) individuals are shown.

In the lower ileum, bacteria are found in the intestinal lumen, mixed with digestive material. Cell numbers of $10^5–10^7$/gram of intestinal contents are common.

The Large Intestine

The ileum empties into the *cecum*, the connecting portion of the *large intestine*. The *colon* makes up the rest of the large intestine. In the colon, bacteria are present in enormous numbers. The colon acts as a specialized fermentation vessel. Many bacteria live within the lumen, probably using as nutrients some products of the digestion of food (Table 21.1). Facultative aerobes such as *Escherichia coli* are present but in smaller numbers than other bacteria; total counts of facultative aerobes are generally less than 10^7/gram of intestinal contents. The facultative aerobes consume all remaining oxygen, making the environment of the large intestine strictly anoxic. This condition promotes abundant growth of obligate anaerobes. Many of these are fusiform anaerobes and are attached end-on to the intestinal wall (Figure 21.9●). Other obligate anaerobes include species of *Clostridium* and *Bacteroides*. The total number of obligate anaerobes in the colon is enormous. Counts of $10^{10}–10^{11}$ cells/gram of intestinal contents are normal, with *Bacteroides* species accounting for the majority. *Enterococcus faecalis* is also usually present in significant numbers.

(a)   (b)

● **Figure 21.9 Scanning electron micrographs of the microbial community on the surface of the columnar epithelium in the mouse ileum.** (a) An overview at low magnification. Note the long, filamentous *fusiform* bacteria lying on the surface. (b) Higher magnification, showing several filaments attached at a single depression. Note that the attachment is at the end of the filaments only. Individual cells are 10–15 μm long.

Functions and Products of Intestinal Flora

The intestinal flora carry out a wide variety of essential metabolic reactions (Table 21.2) resulting in a variety of products. For all of these metabolic products, the composition of the intestinal flora and the diet influence the type and amount of compounds produced.

Among these products are vitamins B_{12} and K. These essential vitamins are not made by humans, but are made by the indigenous microbial flora and absorbed from the gut. Steroids, produced in the liver and released into the intestine from the gall bladder as bile acids, are modified in the intestine by the microbial flora. The modified active steroid compounds are also absorbed from the gut.

Other products generated by the activities of fermentative and methanogenic microorganisms include gas (*flatus*) and the odor-producing substances listed in Table 21.2. The gas results from the metabolism of fermentative and methanogenic microorganisms. Normal adults expel several hundred milliliters of gas, of which about half is N_2 from swallowed air, from the intestines every day. Some foods metabolized by fermentative bacteria in the intestines result in the production of hydrogen (H_2) and carbon dioxide (CO_2). Methanogens (∞Section 13.4), found in the intestines of over one-third of normal adults, convert H_2 and CO_2 produced by the other intestinal microorganisms to methane (CH_4). The methanogens in the rumen of cattle produce significant amounts of methane, up to a quarter of the total global production (∞Section 19.10).

During the passage of food through the gastrointestinal tract, water is absorbed from the digested material, which gradually becomes more concentrated and is converted to feces. *Bacteria* make up about one-third of the weight of fecal matter. Organisms living in the lumen of the large intestine are continuously displaced downward by the flow of material, and bacteria that are lost are continuously replaced by new growth. Thus, the large intestine resembles a chemostat (∞Section 6.9). The time needed for passage of material through the complete gastrointestinal

tract is about 24 hours in humans; the growth rate of bacteria in the lumen is one to two doublings per day.

When an antibiotic is given orally, it inhibits the growth of the normal flora as well as pathogens, leading to the loss of antibiotic-susceptible bacteria in the intestinal tract. In the absence of the full complement of normal flora, opportunistic microorganisms such as antibiotic-resistant *Staphylococcus*, *Proteus*, *Clostridium difficile*, or the yeast *Candida albicans* occasionally become established. Establishment of opportunistic pathogens can lead to a harmful alteration in digestive function or even to disease. For example, antibiotic treatment allows less susceptible microorganisms such as *C. difficile* to flourish, causing infection and colitis. After antibiotic therapy stops, the normal flora eventually becomes reestablished, but often only after a considerable period.

21.4 Concept Check

The stomach is very acidic and is a barrier to most microbial growth. The intestinal tract is slightly acidic to neutral and supports a diverse population of microorganisms in a variety of nutritional and environmental conditions.

◆ Why might the small intestine be more suitable for growth of facultative aerobes than the large intestine?

◆ Identify several essential compounds made by indigenous intestinal microorganisms. What would happen if the microorganisms were completely eliminated from the body by the use of antibiotics?

21.5 Normal Microbial Flora of Other Body Regions

Each individual mucous membrane supports the growth of a specialized group of microorganisms. These organisms are part of the normal local environment and are characteristic of healthy tissue. In many cases, pathogenic microorganisms cannot colonize mucous membranes because of the presence of the normal resident population of microorganisms. Here we discuss two mucosal environments and their resident microorganisms.

Respiratory Tract

The anatomy of the respiratory tract is shown in Figure 21.10● (Figure 26.2). In the **upper respiratory tract** (nasopharynx, oral cavity, and throat) microorganisms live in areas bathed with the secretions of the mucous membranes. Bacteria enter the upper respiratory tract from the air during breathing, but most of them are trapped in the nasal passages and expelled again with the nasal secretions. The resident organisms most commonly found are staphylococci, streptococci, diphtheroid bacilli, and gram-negative cocci. Potentially harmful bacteria such as *Staphylococcus aureus* and *Streptococcus pneumoniae* are often part of the normal flora of the nasopharynx of healthy individuals (Table 21.1). These individuals are *carriers* of the

Table 21.2	Biochemical/metabolic contributions of intestinal microorganisms
Vitamin synthesis	Product: thiamine, riboflavin, pyridoxine, B_{12}, K
Gas production	Product: CO_2, CH_4, H_2
Odor production	Product: H_2S, NH_3, amines, indole, skatole, butyric acid
Organic acid production	Product: acetic, propionic, butyric acids
Glycosidase reactions	Enzyme: β-glucuronidase, β-galactosidase, β-glucosidase, α-glucosidase, α-galactosidase
Steroid metabolism (bile acids)	Process: esterification, dehydroxylation, oxidation, reduction, inversion

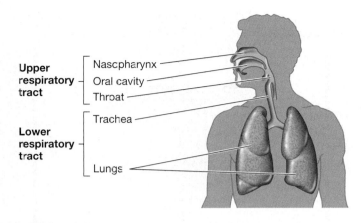

Upper respiratory tract
- Nascpharynx
- Oral cavity
- Throat

Lower respiratory tract
- Trachea
- Lungs

● **Figure 21.10 The respiratory tract.** The upper respiratory tract is populated with a large variety and number of microorganisms, but the lower respiratory tract has relatively few microbial inhabitants, unless there is an ongoing active infection (∞Figure 26.2).

and components of the adaptive immune system such as IgA (∞Section 22.9) are particularly active at mucosal surfaces and may also inhibit the growth of pathogens.

The normal **lower respiratory tract** (trachea, bronchi, and lungs) has no resident microflora, despite the large numbers of organisms potentially able to reach this region during breathing. Dust particles, which are fairly large, settle out in the upper respiratory tract. As the air passes into the lower respiratory tract, the flow rate decreases markedly, and organisms settle onto the walls of the passages. The walls of the entire respiratory tract are lined with ciliated epithelium, and the cilia, beating upward, push bacteria and other particulate matter toward the upper respiratory tract where they are then expelled in the saliva and nasal secretions. Only particles smaller than about 10 μm in diameter reach the lungs.

pathogens but do not normally acquire disease, presumably because the other resident microorganisms compete successfully for resources and limit pathogen growth. The innate immune system (∞Section 22.2 and Section 23.1)

Urogenital Tract

In the male and female urogenital tracts (Figure 21.11a●), the bladder itself is typically sterile, but the epithelial cells lining the urethra are colonized by facultatively aerobic gram-negative rods and cocci (Table 21.1). Microorganisms

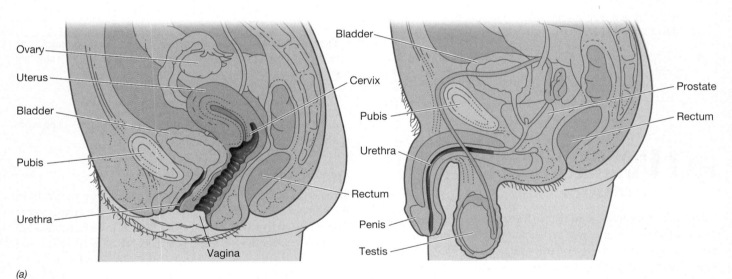

Ovary
Uterus
Bladder
Pubis
Urethra
Vagina
Bladder
Cervix
Pubis
Urethra
Rectum

Bladder
Prostate
Rectum
Penis
Testis

(a)

John Durham

(b)

● **Figure 21.11 Microbial growth in the genitourinary tract.** (a) The genitourinary tracts of the human female and male, showing regions (red) where microorganisms often grow. Note that the upper regions of the genitourinary tracts of both males and females are sterile in normal individuals. (b) Gram stain of *Lactobacillus acidophilus*, the predominant organism in the vagina of women between the onset of puberty and the end of menopause. Individual rods are 3–4 μm long. The genitourinary tract of older and younger women is less acidic and is populated by a much more heterogeneous group of microorganisms that grow at neutral to slightly alkaline pH.

such as *Escherichia coli*, *Proteus mirabilis*, and others, normally present in the body or in the local environment, influenced by local factors such as pH changes, can multiply and become pathogenic. Such organisms are a frequent cause of urinary tract infections, especially in women.

The vagina of the adult female is weakly acidic and contains significant amounts of the polysaccharide glycogen. *Lactobacillus acidophilus*, a resident organism in the vagina, ferments glycogen to produce lactic acid and maintain the acidic conditions (Figure 21.11*b*) (∞ Section 12.19). Other organisms such as yeasts (*Torulopsis* and *Candida* species), streptococci, and *E. coli* may also be present. Before puberty, the female vagina is alkaline and does not produce glycogen, *L. acidophilus* is absent, and the flora consists predominantly of staphylococci, streptococci, diphtheroids, and *E. coli*. After menopause, glycogen production ceases, the pH rises, and the flora again resembles that found before puberty.

21.5 Concept Check

The presence of a population of normal nonpathogenic microorganisms in the respiratory and urogenital tracts is essential for normal organ function and often prevents the colonization of pathogens.

◆ Potential pathogens are often found in the normal flora of the upper respiratory tract. Why do they not cause disease in most cases?

◆ Why is *Lactobacillus* found in the urogenital tract of normal adult women? Why is it not found in postmenopausal women?

II HARMFUL MICROBIAL INTERACTIONS WITH HUMANS

Microbial interactions may be harmful to the host and cause disease. Here we will examine mechanisms of pathogenesis, the ability of microorganisms to cause disease. Microbial pathogenesis begins with exposure and adherence of the microorganisms to host cells, followed by invasion, colonization, and growth. Unchecked growth of the pathogen then results in host damage. Disease-producing microorganisms use several different strategies to establish **virulence**, the relative ability of a pathogen to cause disease (Figure 21.12●). Here we sequentially consider the factors responsible for establishing virulence.

21.6 Entry of the Pathogen into the Host

A pathogen must usually gain access to host tissues and multiply before damage can be done. In most cases, this requires that the organisms penetrate the skin, mucous

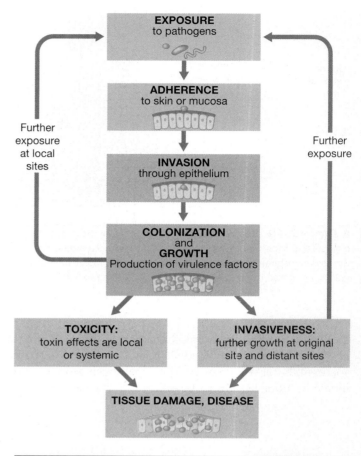

● **Figure 21.12 Microorganisms and pathogenesis.** The presence and even the growth of microorganisms on the host does not necessarily lead to disease.

membranes, or intestinal epithelium, surfaces that normally act as microbial barriers.

Specific Adherence

Most microbial infections begin at breaks or wounds in the skin or on the mucous membranes of the respiratory, digestive, or genitourinary tract. Bacteria or viruses able to initiate infection often adhere *specifically* to epithelial cells (Figure 21.13●) through macromolecular interactions on the surfaces of the pathogen and the host cell.

An infecting microorganism does not adhere to all epithelial cells equally but selectively adheres to cells in a particular region of the body. For example, *Neisseria gonorrhoeae*, the causative agent of the sexually transmitted disease gonorrhea (∞ Section 26.12), adheres much more strongly to urogenital epithelia than to other tissues through a surface protein called *Opa* (*o*pacity *a*ssociated *p*roteins). Host cells bind specifically to Opa with a protein called CD66, a protein found only on the surface of human epithelial cells. Thus, *N. gonorrhoeae* interacts exclusively with target cells through a cell surface receptor–ligand pair.

The species of the host also influences specificity. In many cases, a bacterial strain that normally infects hu-

(a)   (b)

● **Figure 21.13 Adherence of pathogens to animal tissues.** (a) Transmission electron micrograph of a thin section of *Vibrio cholerae* adhering to the brush border of rabbit villi. Note the absence of a capsule. (b) Enteropathogenic *Escherichia coli* in a fatal model of infection in the newborn calf. The bacterial cells are attached to the brush border of calf villi via a well-defined capsule. The rods are about 0.5 μm in diameter.

mans adheres more strongly to the appropriate human cells than to similar cells in another animal (for example, the rat), and vice versa.

Some macromolecules responsible for bacterial adherence are not covalently attached to the bacteria. These are usually polysaccharides, proteins, or protein-carbohydrate mixtures synthesized and secreted by the bacteria (∞Section 4.10). A loose network of polymer fibers extending outward from a cell is known as a **slime layer** (Figure 21.4b). A polymer coat consisting of a dense, well-defined layer surrounding the cell is known as a **capsule** (Figures 21.13 and 21.14●). These structures may be important for adherence not only to host tissues, but also between other bacteria. In addition, these structures can protect bacteria from host defense mechanisms such as phagocytosis (∞Section 22.2).

Fimbriae and *pili* (∞Section 4.10) are bacterial cell surface protein structures that may also function in the attachment process. For instance, the pili of *Neisseria gonorrhoeae* play a key role in the attachment of this organism to urogenital epithelium, and fimbriated strains of *Escherichia coli* (Figure 21.15●) are much more frequent causes of urinary tract infections than strains lacking fimbriae. Among the best-characterized fimbriae are the so-called *type I fimbriae* of enteric bacteria (*Escherichia, Klebsiella, Salmonella, Shigella*). Type I fimbriae are uniformly distributed on the surface of cells. Pili are generally longer than fimbriae, with fewer pili found on the cell surface. Both pili and fimbriae function by binding host cell glycoproteins, initiating the attachment event. Flagella (∞Section 4.14) also increase adherence to host cells and can lead to enhanced virulence (see Figure 21.17).

(a)

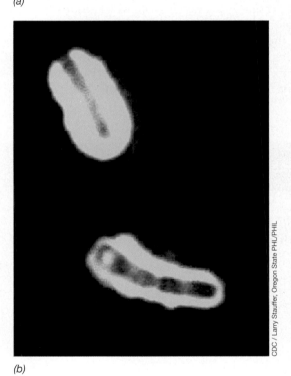

(b)

● **Figure 21.14** *Bacillus anthracis* **capsules.** (a) Capsules of *B. anthracis* on bicarbonate agar media. Encapsulated colonies are typically very large and mucoid in appearance. The individual encapsulated colonies are 0.5 cm in diameter. (b) Direct immunofluorescent stain of *B. anthracis* capsules. Antibodies coupled to fluorescein isothiocyanate (FITC) (∞Section 24.9) stain the capsule bright green, indicating that the capsule extends about 1 μm from the cell, which is about 0.5 μm in diameter. Further coverage of *B. anthracis*, the disease anthrax, and the use of *B. anthracis* as a biological weapon can be found in Sections 25.12 and 25.13.

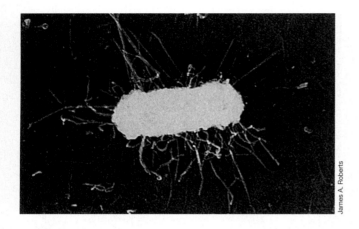

● **Figure 21.15 Fimbriae.** Shadow-cast electron micrograph of the bacterium *Escherichia coli* showing type P fimbriae. Type P fimbriae resemble type I fimbriae but are somewhat longer. The cell shown is about 0.5 μm wide.

Evidence for specific interactions between the mucosal epithelium and pathogens comes from studies of diarrhea caused by *Escherichia coli*. Most strains of *E. coli* are nonpathogenic and are part of the normal flora, inhabiting the cecum and the colon (see Figure 21.8). Several strains of *E. coli* are usually present in the body at the same time, and large numbers of these nonpathogens routinely pass through the body and are eliminated in feces. However, *enterotoxic* strains of *E. coli* express fimbrial proteins called *CFA (colonization factor antigens)* that adhere specifically to cells in the *small intestine*. From here, they can colonize and produce enterotoxins (see Section 21.11) that cause diarrhea as well as other illnesses (∞Section 29.8). The nonpathogenic strains of *E. coli* seldom have the CFA proteins. Some major factors important in microbial adherence are shown in Table 21.3.

Invasion

A few microorganisms are pathogenic solely because of the toxins they produce. These organisms do not need to gain access to host tissues, and we will discuss them separately (see Sections 21.10 and 21.11). However, most pathogens penetrate the epithelium to initiate pathogenicity, a process called *invasion*. At the point of entry, usually at small breaks or lesions in the skin or in mucosal surfaces, growth is established. Growth may also begin on intact mucosal surfaces, especially if the normal flora is altered or eliminated, for example, by antimicrobial chemotherapy. Pathogens may then more readily colonize the tissue and begin the invasion process. Pathogen growth may also be established at sites distant from the original point of entry. Access to distant, usually interior, sites is through the blood or lymphatic circulatory system (∞Section 22.1).

Table 21.3	Major adherence factors used to facilitate attachment of microbial pathogens to host tissues[a]	
Factor	**Example**	
Capsule/ slime layer (∞Section 4.10, Figure 21.4, Figure 21.13, Figure 21.14)	Pathogenic *Escherichia coli*—capsule promotes adherence to the brush border of intestinal villi	
	Streptococcus mutans—dextran slime layer promotes binding to tooth surfaces	
Adherence proteins	*Streptococcus pyogenes*—M protein on the cell binds to receptors on respiratory mucosa	
	Neisseria gonorrhoeae—Opa protein on the cell binds to receptors on urogenital epithelium	
Lipoteichoic acid (∞Section 4.8 and Figure 4.31)	*Streptococcus pyogenes*—facilitates binding to respiratory mucosal receptor (along with M protein)	
Fimbriae (pili) (∞Section 4.10 and Figure 21.15)	*Neisseria gonorrhoeae*—pili facilitate binding to urogenital epithelium	
	Salmonella species—type I fimbriae facilitate binding to epithelium of small intestine	
	Pathogenic *Escherichia coli*— colonization factor antigens (CFAs) facilitate binding to epithelium of small intestine	

[a] Most receptor sites on host tissues are glycoproteins or complex lipids such as gangliosides or globosides.

 21.6 Concept Check

Pathogens gain access to host tissues by adherence at mucosal surfaces through interactions between pathogen and host macromolecules. Pathogen invasion starts at the site of adherence and may spread throughout the host via the circulatory systems.

◆ How do CFA molecules on *Escherichia coli* and Opa proteins on *Neisseria gonorrhoeae* influence adherence to mucosal tissues?

◆ How does adherence initiate invasion?

21.7 Colonization and Growth

If a pathogen gains access to tissues, it may multiply, a process called **colonization**. Because the initial inoculum of a pathogen is too small to cause damage, a pathogen must find appropriate nutrients and environmental conditions to grow in the host. Temperature, pH, and presence or absence of oxygen affect pathogen growth, but the availability of microbial nutrients is most important.

Although a vertebrate host might seem to be a nutritional paradise for microorganisms, not all nutrients are plentiful. Soluble nutrients such as sugars, amino acids, and organic acids are limited, and so organisms able to use complex nutrients such as glycogen may be favored. Not all vitamins and growth factors are in adequate supply in all tissues at all times. *Brucella abortus*, for example, grows very slowly in most tissues of infected cattle but grows very rapidly in the placenta, where it causes abortion. This is due to the elevated concentration of erythritol, a nutrient that enhances growth of *B. abortus* when it infects the placenta (see Table 21.6).

Trace elements may also be in short supply and can influence establishment of the pathogen. For example, the concentration of *iron* greatly influences microbial growth (∞Section 5.1). Specific proteins called *transferrin* and *lactoferrin*, present in animals, bind iron tightly and transfer it through the body. These proteins have such high affinity for iron that microbial iron deficiency may be common; an iron salt solution given to an infected animal greatly increases the virulence of some pathogens.

As we noted in Section 5.1, many bacteria produce iron-chelating compounds (*siderophores*) that help them obtain iron, a growth-limiting micronutrient, from the environment. Some iron chelators isolated from pathogenic bacteria are so efficient that they can remove iron from animal iron-binding proteins. For example, a siderophore called *aerobactin*, produced by certain strains of *Escherichia coli* and encoded by the Col V plasmid (∞Section 10.9), readily removes iron bound to transferrin. Likewise, *Neisseria* species produce a transferrin-specific receptor and can remove iron from the bound transferrin.

Localization in the Body

After initial entry, the organism often remains localized and multiplies, producing a small focus of infection such as the boil that may arise from *Staphylococcus* skin infections (∞Section 26.9). Alternatively, the organisms may enter the lymphatic vessels and be deposited in lymph nodes. If an organism reaches the blood, it will be distributed to distant parts of the body, usually concentrating in the liver or spleen. Spread of the pathogen through the blood and lymph systems can result in a generalized (systemic) infection of the body, with the organism growing in a variety of tissues. If extensive bacterial growth in tissues occurs, some of the organisms are usually shed into the bloodstream in large numbers, a condition called **bacteremia**. Widespread infections of this type almost always start as a localized infection in a specific organ such as the kidney, intestine, or lung.

 21.7 *Concept Check*

A pathogen must gain access to nutrients and appropriate growth conditions before it can colonize and grow in substantial numbers in host tissue. Organisms may grow locally at the site of invasion, or may spread through the body.

◆ Why are colonization and growth necessary for the success of most pathogens?

◆ What host factors limit or accelerate colonization and growth of a microorganism at a local site?

21.8 Virulence

Virulence is the relative ability of a parasite to cause disease, and here we discuss some basic methods used to measure it. We will then provide specific examples of particularly virulent organisms, highlighting the strategies used to enhance virulence.

Measuring Virulence

The virulence of a pathogen can be estimated from experimental studies of the LD_{50} (lethal dose$_{50}$), the dose of an agent that kills 50% of the animals in a test group. Highly virulent pathogens frequently show little difference in the number of cells required to kill 100% of the population as compared with the number required to kill 50% of the population. This is illustrated in Figure 21.16● for experimental *Streptococcus* and *Salmonella* infections in mice. Only a few cells of *Streptococcus pneumoniae* are required to establish a fatal infection and kill all members of a test population of mice once the virulence of a particular strain has been established. In fact, the LD_{50} for *S. pneumoniae* is hard to determine because so few organisms are needed to produce a lethal infection. By contrast, the LD_{50} for *Salmonella typhimurium*, a much less virulent pathogen, is much higher than for *S. pneumoniae*. The number of cells of *Salmonella typhimurium* cells required to kill 100% of the population is more than 100 times greater than the number of cells needed to achieve the LD_{50}.

When pathogens are kept in laboratory culture and not passed through animals, their virulence is often decreased or even lost completely. Such organisms are said to be *attenuated*. **Attenuation** probably occurs because nonvirulent mutants may grow faster and, after successive transfers to fresh media, such mutants are selectively

● **Figure 21.16 Microbial virulence.** Comparison of differences in microbial virulence based on the number of cells of *Streptococcus pneumoniae* or *Salmonella typhimurium* required to kill mice.

favored. Attenuation often occurs more readily when culture conditions are not optimal for the species. If an attenuated culture is reinoculated into an animal, organisms sometimes regain virulence, but in many cases loss of virulence is permanent. Attenuated strains are often used for production of vaccines, especially viral vaccines (∞Section 22.13). Measles, mumps, and rubella vaccines, for example, contain attenuated live viruses.

Toxicity and Invasiveness

Virulence is due to the ability of a pathogen to cause host damage through toxicity and invasiveness. Each pathogen uses these properties to cause disease.

Toxicity is the ability of an organism to cause disease by means of a preformed toxin that inhibits host cell function or kills host cells. For example, the disease *tetanus* is caused by a potent exotoxin produced by *Clostridium tetani* (see Section 21.10). *C. tetani* cells rarely leave the wound where they were first introduced, growing relatively slowly at the wound site. Yet *C. tetani* is able to bring about disease because it produces tetanus toxin that moves to distant parts of the body and initiates irreversible muscle contraction and often death of the host.

Invasiveness is the ability of an organism to grow in host tissue in such large numbers that the pathogen inhibits host function. A microorganism may still be able to produce disease through invasiveness even if it produces no toxin. For example, the major virulence factor for *Streptococcus pneumoniae* is the polysaccharide capsule that prevents the phagocytosis of pathogenic strains (see Section 21.6, ∞Section 26.2, and Figure 26.3), defeating a major defense mechanism used by the host to prevent invasion. Encapsulated strains of *S. pneumoniae* are able to cause extensive host damage because they are highly invasive. They grow in lung tissues in enormous numbers where they initiate host responses that lead to pneumonia (∞Section 26.2). Nonencapsulated strains are quickly and efficiently taken up and destroyed by phagocytes.

Clostridium tetani and *Streptococcus pneumoniae* exemplify the extremes of *toxicity* and *invasiveness*, respectively. Most pathogens fall between these extremes and use a combination of toxins and invasiveness to cause disease.

Virulence in *Salmonella*

Salmonella species employ a mixture of toxins, invasiveness, and other virulence factors to enhance pathogenicity (Figure 21.17●). First, several toxins contribute to the virulence of *Salmonella*, and at least three toxins are produced: *enterotoxin* (Table 21.4), *endotoxin* (∞Section 4.9 and Figure 4.35), and *cytotoxin*. Cytotoxin acts by inhibiting host cell protein synthesis and allowing Ca²⁺ to escape from host cells. A number of virulence factors are involved in invasion by *Salmonella*. The cell surface polysaccharide O antigen (∞Figure 4.35), the flagellar H antigen, and fimbriae enhance adherence. The capsular Vi polysaccharide inhibits complement binding and antibody-mediated killing (∞Section 22.11). The *inv* (invasion) genes of *Salmonella* encode at least 10 different proteins involved in invasion. For example, *invH* encodes a surface adhesin protein, while *invC*, *invG*, *invI*, and *invJ* encode proteins involved in assembly of the special *surface appendages* used for host-cell binding.

Salmonella readily establish infections through intracellular parasitism and grow inside the cells that line the intestine as well as in macrophages, white blood cells that normally ingest and kill bacteria (∞Section 22.2). The toxic oxygen products of macrophages are neutralized by proteins induced by the *Salmonella oxyR* gene.

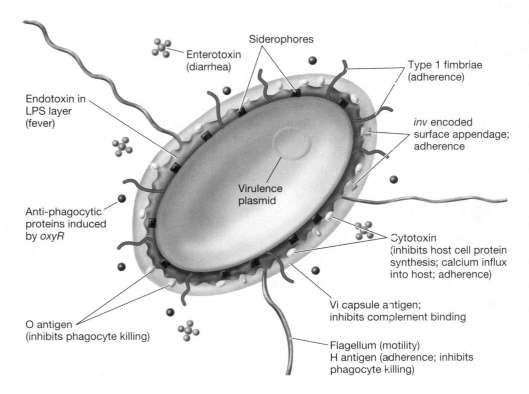

● **Figure 21.17 Virulence factors in *Salmonella* pathogenesis.** Structural elements known to be important in pathogenesis are shown. Protein products of the *pho* gene system have not been identified, but are known to neutralize the effects of the detergent-like, macrophage-produced defensins.

Siderophores

Enterotoxin (diarrhea)

Type 1 fimbriae (adherence)

Endotoxin in LPS layer (fever)

inv encoded surface appendage; adherence

Virulence plasmid

Anti-phagocytic proteins induced by *oxyR*

Cytotoxin (inhibits host cell protein synthesis; calcium influx into host; adherence)

Vi capsule antigen; inhibits complement binding

O antigen (inhibits phagocyte killing)

Flagellum (motility) H antigen (adherence; inhibits phagocyte killing)

Table 21.4	Exotoxins and extracellular virulence factors produced by human pathogens		
Organism	**Disease**	**Toxin or factor**[a]	**Action**
Bacillus anthracis	Anthrax	Lethal factor (LF) Edema factor (EF) Protective antigen (PA) (AB)	PA is the cell-binding B component, EF causes edema, LF causes cell death
Bacillus cereus	Food poisoning	Enterotoxin complex	Induces fluid loss from intestinal cells
Bordetella pertussis	Whooping cough	Pertussis toxin (AB)	Blocks G protein signal transduction, kills cells
Clostridium botulinum	Botulism	Neurotoxin (AB)	Flaccid paralysis (see Figure 21.20)
Clostridium tetani	Tetanus	Neurotoxin (AB)	Spastic paralysis (see Figure 21.21)
Clostridium perfringens	Gas gangrene, food poisoning	α-Toxin (CT)	Hemolysis (lecithinase, see Figure 21.18b)
		β-Toxin (CT)	Hemolysis
		γ-Toxin (CT)	Hemolysis
		δ-Toxin (CT)	Hemolysis (cardiotoxin)
		κ-Toxin (E)	Collagenase
		λ-Toxin (E)	Protease
		Enterotoxin (CT)	Alters permeability of intestinal epithelium
Corynebacterium diphtheriae	Diphtheria	Diphtheria toxin (AB)	Inhibits protein synthesis in eukaryotes (Figure 21.19)
Escherichia coli (enterotoxigenic strains only)	Gastroenteritis	Enterotoxin (AB)	Induces fluid loss from intestinal cells
Haemophilus ducreyi	Chancroid	Cytolethal distending toxin[b] (CDT) (AB)	Genotoxin (DNA lesions cause apoptosis in host cells)
Pseudomonas aeruginosa	*P. aeruginosa* infections	Exotoxin A (AB)	Inhibits protein synthesis
Salmonella sp.	Salmonellosis, typhoid fever, paratyphoid fever	Enterotoxin (AB)	Inhibits protein synthesis and lyses host cells
		Cytotoxin (CT)	Induces fluid loss from intestinal cells
Shigella dysenteriae	Bacterial dysentery	Shiga toxin (AB)	Inhibits protein synthesis
Staphylococcus aureus	Pyogenic (pus-forming) infections (boils, and so on), respiratory infections, food poisoning, toxic shock syndrome, scalded skin syndrome	α-Toxin (CT)	Hemolysis
		Toxic shock syndrome toxin (SA)	Systemic shock
		Exfoliating toxin A and B (SA)	Peeling of skin, shock
		Leukocidin (CT)	Destroys leukocytes
		β-Toxin (CT)	Hemolysis
		γ-Toxin (CT)	Kills cells
		δ-Toxin (CT)	Hemolysis, leukolysis
		Enterotoxin A, B, C, D, and E (SA)	Induce vomiting, diarrhea, shock
		Coagulase (E)	Induces fibrin clotting
Streptococcus pyogenes	Pyogenic infections, tonsillitis, scarlet fever	Streptolysin O (CT)	Hemolysin
		Streptolysin S (CT)	Hemolysin (see Figure 21.18a)
		Erythrogenic toxin (SA)	Causes scarlet fever rash
		Streptokinase (E)	Dissolves fibrin clots
		Hyaluronidase (E)	Dissolves hyaluronic acid in connective tissue
Vibrio cholerae	Cholera	Enterotoxin (AB)	Induces fluid loss from intestinal cells (Figure 21.22)

[a] AB, A-B toxin; CT, cytolytic toxin; E, enzymatic virulence factor; SA, superantigen toxin.

[b] Cytolethal distending toxin is found in other gram-negative pathogens including *Actinobacillus actinomycetemcomitans, Campylobacter* sp., *Escherichia coli, Helicobacter* sp., *Salmonella typhi,* and *Shigella dysenteriae.*

Macrophage-produced antibacterial molecules called *defensins* are neutralized by the products of the *phoP* and *phoQ* genes. Thus, the *oxy* and *pho* gene products of *Salmonella* enhance pathogenicity by fostering methods of intracellular invasion, neutralizing host defenses that normally inhibit intracellular bacterial growth.

Several plasmid-borne virulence factors are also involved in persistence and spread in most *Salmonella* species. For example, antibiotic resistance is encoded on a plasmid. Finally, *Salmonella* also produce *siderophores*, bacterial iron-chelating proteins (∞ Section 5.1 and see Section 21.7). Thus, *Salmonella*, and probably most other pathogens, use several factors to establish virulence and initiate disease. These virulence factors are shared between and among various members of the enteric bacteria (∞Section 12.11).

21.8 Concept Check

Virulence is determined by invasiveness, toxigenicity, and other factors produced by a pathogen. In most pathogens, a number of factors contribute to virulence. Attenuation is loss of virulence. *Salmonella* displays a wide variety of traits that enhance virulence.

◆ Distinguish between *toxicity* and *invasiveness*. Give examples of organisms that rely almost exclusively on *toxicity* or *invasiveness* to promote virulence.

◆ Explain how an organism may become attenuated. Discuss the role of attenuated organisms in vaccine production.

III VIRULENCE FACTORS AND TOXINS

Extracellular virulence factors and toxins produced by microorganisms promote pathogenesis. A variety of these proteins are produced by different pathogens, but many share molecular characteristics and modes of action. Here we examine representative examples of several virulence factors and toxins with an eye toward common mechanisms.

21.9 Virulence Factors

Pathogen-produced extracellular proteins that aid in the establishment and maintenance of disease are called *virulence factors*. Most virulence factors are *enzymes* that enhance pathogen colonization and growth. For example, streptococci, staphylococci, and certain clostridia produce **hyaluronidase** (Table 21.4), an enzyme that promotes spreading of organisms in tissues by breaking down hyaluronic acid, a polysaccharide that functions as intercellular cement. Hyaluronidase digestion of the intercellular matrix enables these organisms to spread from an initial site. Streptococci and staphylococci also produce an array of proteases, nucleases, and lipases that depolymerize host proteins, nucleic acids, and lipids. Similarly, clostridia that cause gas gangrene produce collagenase, or κ-toxin (Table 21.4), which breaks down the tissue-supporting collagen network, enabling these organisms to spread through the body.

Fibrin, Clots, and Virulence

Fibrin clots are often formed at a site of microbial invasion by the host. The clotting mechanism, triggered by tissue injury, isolates the pathogens, limiting infection to a local region. Some organisms produce fibrinolytic enzymes that dissolve the clots and make further invasion possible. One fibrinolytic substance produced by *Streptococcus pyogenes* is known as **streptokinase** (Table 21.4).

By contrast, other organisms produce enzymes that *promote* the formation of fibrin clots, causing localization and protection of the organism rather than spread of the organism. The best-studied fibrin-clotting enzyme is **coagulase** (Table 21.4), produced by pathogenic *Staphylococcus aureus*. Coagulase causes fibrin to be deposited on *S. aureus* cells and protects the coated bacteria from attack by host cells. The fibrin matrix produced as a result of coagulase activity probably accounts for the extremely localized nature of many staphylococcal infections, as in boils and pimples (∞Figure 26.21). Coagulase-positive *Staphylococcus aureus* is considered to be more virulent than coagulase-negative strains.

21.9 Concept Check

Pathogens produce a variety of enzymes that enhance virulence by breaking down or altering host tissue to provide access and nutrients. Still other pathogen-produced virulence factors provide protection to the pathogen by interfering with normal host defense mechanisms. These factors enhance colonization and growth of the pathogen.

◆ What advantage does the pathogen gain by producing enzymes that digest structural components of host tissues?

◆ How can the activity of coagulase work to enhance the growth of *Staphylococcus aureus*?

21.10 Exotoxins

Exotoxins are toxic proteins released extracellularly as the organism grows. These toxins may travel from a limited focus of infection and cause damage at distant sites. Table 21.4 provides a summary of the properties and actions of some of the best-known exotoxins as well as other extracellular virulence factors.

Most exotoxins fall into one of three categories; the *cytolytic toxins*, the *A-B toxins*, or the *superantigen toxins*. The cytolytic toxins work by enzymatically attacking cell constituents, causing lysis. The A-B toxins consist of two covalently bonded subunits, A and B. The B component generally binds to a cell surface receptor, allowing the transfer of the A subunit across the targeted cell membrane, where it functions to damage the cell. The superantigens work by stimulating large numbers of immune response cells, resulting in extensive inflammatory reactions (∞Section 22.16).

Cytolytic Toxins

Various pathogens produce proteins that damage the host cytoplasmic membrane, causing cell lysis and death. Because the activity of these toxins is most easily

● **Figure 21.18 Hemolysis.** (a) Zones of hemolysis around colonies of *Streptococcus pyogenes* growing on a blood agar plate. (b) Action of lecithinase, a phospholipase, around colonies of *Clostridium perfringens* growing on an agar medium containing egg yolk, a source of lecithin. Lecithinase dissolves the membranes of red blood cells, leading to the clearing zones shown.

(a)

(b)

detected with red blood cells (erythrocytes), they are called **hemolysins** (Table 21.4). However, they also work on cells other than erythrocytes. The production of hemolysins is demonstrated in the laboratory by streaking the organism on a *blood agar plate*. During growth of the colonies, hemolysin is released and lyses the surrounding red blood cells, typically creating a zone of hemolysis (Figure 21.18a●).

Some hemolysins attack the phospholipid of the host cytoplasmic membrane. Because the phospholipid lecithin (phosphatidylcholine) is often used as a substrate, these enzymes are called **lecithinases** or **phospholipases**. An example is the α-toxin of *Clostridium perfringens*, a lecithinase that dissolves membrane lipids, resulting in cell lysis (Figure 21.18b) (Table 21.4). Since the cytoplasmic membranes of all organisms contain phospholipids, phospholipases sometimes destroy bacterial as well as animal cytoplasmic membranes.

Some hemolysins, however, are not phospholipases. Streptolysin O, a hemolysin produced by streptococci, affects the sterols of the host cytoplasmic membrane. *Leukocidins* (Table 21.4) are lytic agents capable of lysing white blood cells and may decrease host resistance (∞Section 22.2).

Diphtheria Toxin

Diphtheria toxin, produced by *Corynebacterium diphtheriae*, is an important virulence factor in the pathogenesis of diphtheria (∞Section 26.3). Rats and mice are relatively resistant to diphtheria toxin, but human, rabbit, guinea pig, and bird cells are very susceptible, with only a single toxin molecule required to kill each cell. Diphtheria toxin is an A-B toxin secreted by cells of *C. diphtheriae* as a single polypeptide. Fragment B promotes specific binding of the toxin to a host cell receptor (Figure 21.19●). After binding, proteolytic cleavage between Fragment A and B

Key
A — B Diptheria toxin
Amino acid

EF—2

t-RNA

Ribosome

(a) **Normal protein synthesis**

EF—2 → EF—2*

EF—2*

t-RNA

(b) **Protein synthesis stops**

● **Figure 21.19 The action of diphtheria toxin from *Corynebacterium diphtheriae*.** (a) Elongation factor 2 (EF-2) normally binds to the ribosome and brings an amino acid–charged t-RNA to the ribosome, causing protein elongation. (b) Diphtheria toxin binds to the cell membrane, where it is cleaved and the A peptide is internalized. The A peptide catalyzes the ADP–ribosylation of elongation factor 2 (EF-2*). The modified elongation factor no longer aids transfer of amino acids to the growing polypeptide chain, resulting in shutdown of protein synthesis and death of the cell.

Web Tutorial 21.1 Diphtheria and Cholera Toxins

allows entry of Fragment A into the host cytoplasm. Fragment A then disrupts protein synthesis by blocking transfer of an amino acid from a tRNA (transfer ribonucleic acid) to the growing polypeptide chain (∞Section 7.15). The toxin specifically inactivates elongation factor 2, a protein in eukaryotic cells involved in growth of the polypeptide chain, by catalyzing the attachment of adenosine diphosphate (ADP) ribose from NAD^+. Following ADP-ribosylation, the activity of the modified elongation factor 2 decreases dramatically and protein synthesis stops.

Diphtheria toxin is formed only by strains of *C. diphtheriae* that are lysogenized by a bacteriophage called phage β, which carries the toxin-encoding *tox* gene. Nontoxigenic, nonpathogenic strains of *C. diphtheriae* can be converted to pathogenic strains by infection with phage β (the process of *phage conversion*) (∞Section 10.8).

Exotoxin A of *Pseudomonas aeruginosa* (Table 21.4) functions similarly to diphtheria toxin, also modifying elongation factor 2 by ADP-ribosylation.

Tetanus and Botulinum Toxins

These toxins are produced by two species of obligately anaerobic bacteria, *Clostridium tetani* and *Clostridium botulinum*, normal soil organisms that occasionally cause disease in animals. Neither species is invasive, and virtually all pathogenic effects are due to toxicity. *C. botulinum* sometimes grows directly in the body, causing infant or wound botulism, and also grows and produces toxin in improperly preserved foods (∞Section 29.6). Death

from **botulism** is usually due to respiratory failure from flaccid muscle paralysis. *C. tetani* grows in the body in deep wound punctures that become anoxic, and although *C. tetani* does not invade the body from the initial site of infection, the toxin can spread via the neural cells and cause spastic paralysis, the hallmark of **tetanus**, often leading to death (∞Section 27.9).

Botulinum toxin consists of seven related A-B toxins that are the most potent biological toxins known. One milligram of botulinum toxin is enough to kill more than 1 million guinea pigs. Of the seven distinct botulinum toxins known, at least two are encoded on lysogenic bacteriophages specific for *Clostridium botulinum*. The major toxin is a protein that forms complexes with nontoxic botulinum proteins to give a heteromeric bioactive macromolecule. The bioactive toxin complex binds to presynaptic membranes on the termini of the stimulatory motor neurons at the neuromuscular junction, blocking the release of *acetylcholine*. Because transmission of the nerve impulse to the muscle is through acetylcholine interaction with a muscle receptor, the botulinum-poisoned muscle cannot receive an excitatory signal and contraction is prevented, causing flaccid paralysis and death by suffocation characteristic of *botulism* (Figure 21.20●).

Tetanus toxin is also an A-B protein toxin. On contact with the central nervous system, this toxin is transported through the motor neurons back to the spinal cord, where it binds specifically to ganglioside lipids at the termini of the inhibitory interneurons. The inhibitory interneurons normally work by releasing an inhibitory

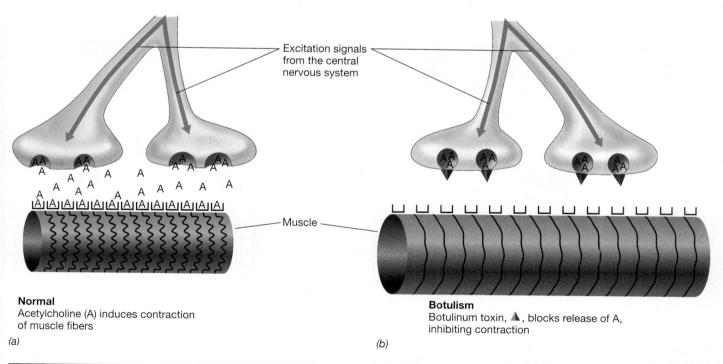

Normal
Acetylcholine (A) induces contraction of muscle fibers

(a)

Botulism
Botulinum toxin, ▲, blocks release of A, inhibiting contraction

(b)

● **Figure 21.20** **The action of botulinum toxin from *Clostridium botulinum*.** (a) Upon stimulation of peripheral and cranial nerves, acetylcholine (A) is normally released from vesicles at the neural side of the motor end plate. Acetylcholine then binds to specific receptors on the muscle, inducing contraction. (b) Botulinum toxin acts at the motor end plate to prevent release of acetylcholine (A) from vesicles, resulting in a lack of stimulus to the muscle fibers, irreversible relaxation of the muscles, and flaccid paralysis (∞Section 29.6).

neurotransmitter, usually glycine, that binds to receptors on the motor neurons. Normally, glycine from the inhibitory interneurons stops the release of acetylcholine by the motor neurons and inhibits muscle contraction, allowing relaxation of the muscle fibers. However, if tetanus toxin blocks glycine release, the motor neurons cannot be inhibited, resulting in *tetanus*, continual release of acetylcholine and uncontrolled contraction of the poisoned muscles (Figure 21.21●). The outcome is a spastic, twitching paralysis, and affected muscles are constantly contracted. If the muscles of the mouth are involved, the prolonged contractions restrict the mouth's movement, resulting in the condition known as *trismus* or *lockjaw*. If respiratory muscles are involved, prolonged contraction may result in death due to asphyxiation.

Tetanus toxin and botulinum toxin both block release of neurotransmitters involved in muscle control, but symptoms are quite different, depending on the particular neurotransmitters involved.

 21.10 Concept Check

The most potent biological toxins are the exotoxins produced by microorganisms. Each exotoxin affects specific host cells, causing specific impairment of a major host cell function.

◆ What key features are shared by all exotoxins? What features are unique for the A-B exotoxins?

◆ Are bacterial growth and infection in the host necessary for the production of toxins? Explain your answer.

21.11 Enterotoxins

Enterotoxins are exotoxins whose activity affects the *small* intestine, generally causing massive secretion of fluid into the intestinal lumen, leading to vomiting and diarrhea. Enterotoxins are produced by a variety of bacteria, including the food-poisoning organisms *Staphylococcus aureus*, *Clostridium perfringens*, and *Bacillus cereus*, and the intestinal pathogens *Vibrio cholerae*, *Escherichia coli*, and *Salmonella enteritidis*.

Cholera Toxin

The enterotoxin produced by *Vibrio cholerae*, the causal agent of cholera (◥Section 28.5) is the best understood enterotoxin. Cholera toxin is an A-B toxin (see Section 21.10) and consists of an A subunit and five B subunits. The B subunit contains the site by which the cholera toxin binds specifically with the ganglioside GM1 (a complex glycolipid) in the cytoplasmic membrane of epithelial cells (Figure 21.22●), but the B subunit itself does not cause an alteration in membrane permeability. Rather, the toxic action is a function of

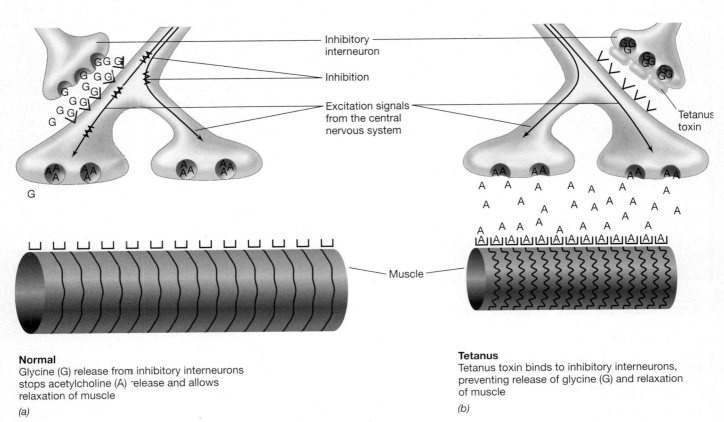

Normal
Glycine (G) release from inhibitory interneurons stops acetylcholine (A) release and allows relaxation of muscle

(a)

Tetanus
Tetanus toxin binds to inhibitory interneurons, preventing release of glycine (G) and relaxation of muscle

(b)

● **Figure 21.21 The action of tetanus toxin from *Clostridium tetani.*** (a) Muscle relaxation is normally induced by glycine (G) release from inhibitory interneurons. Glycine acts on the motor neurons to block excitation and release of acetylcholine (A) at the motor end plate. (b) Tetanus toxin binds to the interneuron to prevent release of glycine from vesicles, resulting in a lack of inhibitory signals to the motor neurons, constant release of acetylcholine to the muscle fibers, irreversible contraction of the muscles, and spastic paralysis (◥Section 27.9 and Figure 27.19).

1. Normal ion movement, Na⁺ from lumen to blood, no net Cl⁻ movement

2. Colonization and toxin production

3. Activation of epithelial adenyl cyclase by cholera toxin

4. Na⁺ movement blocked, net Cl⁻ movement to lumen

5. Massive water movement to the lumen

● **Figure 21.22   The action of cholera enterotoxin.** (1) The normal process of ion movement in the intestine. (2) Adherence and colonization of the intestinal microvilli by *Vibrio cholerae*, followed by production and binding of the A-B enterotoxin by specific interaction with the GM1 ganglioside on host cells. The B subunits (green) of A-B toxin bind GM1, allowing the release of the A subunit (red) across the cell membrane. The internalized A subunit activates adenyl cyclase, (3) leading to disruption of normal sodium (Na⁺) influx, (4) loss of H₂O into the lumen and diarrhea. Treatment for cholera is by ion replacement and rehydration therapy. Antibiotic treatment may shorten the course of the disease by limiting *V. cholerae* growth, but has no effect on toxin that has already been produced.

the A chain, which crosses the cell membrane and activates the enzyme *adenyl cyclase*, converting adenosine triphosphate (ATP) to cyclic adenosine monophosphate (cAMP).

As discussed in Section 8.7, cyclic AMP is a specific mediator of a variety of regulatory systems in cells, including ion balance. The increased cyclic AMP levels induced by the cholera enterotoxin bring about the secretion of chloride and bicarbonate ions from the mucosal cells into the intestinal lumen. This change in ion concentrations leads to the secretion of large amounts of water into the lumen (Figure 21.22). In the acute phase of cholera, the rate of water loss into the small intestine is greater than the possible reabsorption of water by the large intestine, and so massive net fluid loss occurs. Cholera victims generally die from dehydration, and the best treatment for the disease is oral fluid replacement with solutions containing electrolytes and other solutes.

Because cholera enterotoxin activates adenyl cyclase in a variety of cells and tissues, pathological manifestations of cholera toxin are related more to the specific site at which it binds, the epithelial cells of the small intestine, than to toxin activation of adenyl cyclase. Indeed, purified B subunits devoid of adenyl cyclase activity can actually *prevent* the activity of cholera enterotoxin if the B subunits are administered first. The B subunits bind to the specific cholera receptors on the mucosal cells and block the binding of the complete toxin containing the cytotoxic A subunit.

Expression of the cholera enterotoxin genes *ctxA* and *ctxB* is controlled by a positive regulatory protein encoded by the *toxR* gene. The *toxR* gene product is a transmembrane protein that controls cholera toxin production and other important virulence factors, such as the outer membrane proteins and pili required for successful attachment and colonization of *Vibrio cholerae* in the small intestine.

Other Enterotoxins

Enterotoxins produced by enterotoxigenic *Escherichia* and *Salmonella* are functionally, structurally, and evolutionarily related to cholera toxin. These toxins are also produced in the gut by colonizing bacteria. However, enterotoxins produced by some of the food-poisoning bacteria (*Staphylococcus aureus, Clostridium perfringens, Bacillus cereus*) have quite different modes of action. For example, *Clostridium perfringens* enterotoxin is a cytotoxin and *Staphylococcus aureus* enterotoxin is a superantigen (Table 21.4). Superantigens have a completely different mode of action, stimulating large numbers of immune lymphocytes and causing systemic as well as intestinal inflammatory responses (∞Section 22.16). All of the food-poisoning toxins can be acquired by the ingestion of preformed toxin and do not require the growth of bacteria in the body.

21.12 Endotoxins

Many gram-negative *Bacteria* produce potentially toxic lipopolysaccharides as part of the outer layer of their cell envelope (∞Section 4.9). These lipopolysaccharides are called **endotoxins**; they are cell-bound and released in large amounts only when cells lyse. In contrast, all *exotoxins* are secreted products of *living* cells. Endotoxins have been studied primarily in *Escherichia*, *Shigella*, and especially *Salmonella*. The properties of exotoxins and endotoxins are compared in Table 21.5.

Endotoxin Structure and Function

The structure of lipopolysaccharide (LPS) was diagrammed in Figures 4.34 and 4.35. Lipopolysaccharide consists of three covalently linked subunits; *lipid A, a core polysaccharide*, and the *O-polysaccharide*.

Endotoxins cause a variety of physiological effects. *Fever* is an almost universal symptom because endotoxin stimulates host cells to release proteins called *endogenous pyrogens*, which affect the temperature-controlling center of the brain. In addition, endotoxins can cause diarrhea, rapid decrease in lymphocyte, leukocyte, and platelet numbers, release of cytokines, and generalized inflammation (∞ Section 22.3). Large doses of endotoxin can cause death from hemorrhagic shock and tissue necrosis. The toxicity of endotoxins is, however, much lower than that of exotoxins. For instance, in mice the LD_{50} for endotoxin is $200–400\ \mu g$ per animal, whereas the LD_{50} for botulinum toxin is about 25 picograms (pg), about 10 million times less!

Studies indicate that the lipid A portion of LPS is responsible for toxicity (∞Figure 4.35), while the polysaccharide fraction makes the complex water soluble and immunogenic (∞Section 22.5). Animal studies indicate that both the lipid and polysaccharide fractions are necessary for an *in vivo* toxic effect.

Limulus Amebocyte Lysate Assay for Endotoxin

Because endotoxins are pyrogens, pharmaceuticals such as antibiotics and intravenous solutions must be endotoxin-free. An endotoxin assay of very high sensitivity has been developed using lysates of amebocytes from the horseshoe crab, *Limulus polyphemus*. Endotoxin specifically causes lysis of amebocytes (Figure 21.23●). In the standard *Limulus* amebocyte lysate (LAL) assay, *Limulus* amebocyte extracts are mixed with the solution to be tested. If endotoxin is present, the amebocyte extract gels and precipitates, causing a change in turbidity. This reaction can be measured quantitatively with a spectrophotometer with as little as 10 pg/ml of lipopolysaccharide. The *Limulus* assay is used to detect endotoxin in serum, cerebrospinal fluid, drinking water, and fluids used for injection and drug formulation.

The *Limulus* test is so sensitive that considerable care must be taken to avoid contamination of the equipment, solutions, and reagents with gram-negative *Bacteria* in the laboratory and clinical environment, for example, as contaminants in distilled water. Detection of endotoxin by the *Limulus* assay in serum or cerebrospinal fluid is presumptive evidence of gram-negative infection.

Table 21.5	**Properties of exotoxins and endotoxins**	
Property	**Exotoxins**	**Endotoxins**
Chemical properties	Proteins, excreted by certain gram-positive or gram-negative *Bacteria*; generally heat-labile	Lipopolysaccharide–lipoprotein complexes (∞Figures 4.34 and 4.35); released on cell lysis as part of the outer membrane of gram-negative *Bacteria*; extremely heat-stable
Mode of action; symptoms	Specific; usually binds to specific cell receptors or structures; either cytotoxin, enterotoxin, or neurotoxin with defined specific action on cells or tissues	General; fever, diarrhea, vomiting
Toxicity	Often highly toxic, sometimes fatal	Weakly toxic, rarely fatal
Immunogenicity response	Highly immunogenic; stimulate the production of neutralizing antibody (antitoxin)	Relatively poor immunogen; immune response not sufficient to neutralize toxin
Toxoid potential	Treatment of toxin with formaldehyde will destroy toxicity, but treated toxin (toxoid) remains immunogenic	None
Fever potential	Does not produce fever in host	Pyrogenic, often induces fever in host

(a) (b)

● **Figure 21.23** *Limulus* **amebocytes.** (a) Normal amebocytes from the horseshoe crab, *Limulus polyphemus.* (b) Amebocytes following exposure to bacterial lipopolysaccharide. Treatment with lipopolysaccharide causes degranulation of the cells, and this response can be used as an assay for lipopolysaccharide content.

21.12 Concept Check

Endotoxins are lipopolysaccharides derived from the outer membrane of gram-negative *Bacteria*. Released upon lysis of the *Bacteria*, endotoxins cause fever and other systemic toxic effects in the host. Endotoxins are generally less toxic than exotoxins. The presence of endotoxin detected by the *Limulus* amebocyte lysate assay indicates contamination of a substance by gram-negative *Bacteria*.

◆ Why do gram-positive *Bacteria* not produce endotoxins?

◆ Why is it necessary to test drug preparations for endotoxin?

 IV HOST FACTORS IN INFECTION

Host factors can either limit or contribute to the pathogenicity of a microorganism. Here we first present some important risk factors that contribute to infection. Risk factors such as age, genetic makeup, or tissue specificity, cannot be controlled by the host. However, factors such as diet and stress are potentially controllable. We conclude this chapter with a discussion of the physical and chemical barriers built into human anatomy that work to limit infection and colonization. In the following chapter we will deal with host responses to pathogen contact, the immune response.

21.13 Host Risk Factors for Infection

A number of factors contribute to the susceptibility of the host to infection and disease. Here we will discuss some of these factors and relate them to the infectious disease process.

Age, Stress, and Diet

Age is an important factor for determining susceptibility to infectious disease. Infectious diseases are more common in the very young and in the very old. In the infant, for example, development of an intestinal microflora occurs quite quickly, but the normal flora of an infant is not the same as that of the adult. Before the development of an adult flora, and especially in the days immediately following birth, pathogens have a greater opportunity to become established and produce disease. Thus, diarrhea caused by pathogenic strains of *Escherichia coli* (◌◌ Section 29.8) or *Pseudomonas aeruginosa* is frequently encountered in infants under the age of 1 year.

Infant botulism results from an intestinal infection with *Clostridium botulinum* (◌◌Section 29.6). As the pathogen colonizes and grows, it secretes botulinum toxin, leading to flaccid paralysis (see Section 21.10).This disease, contracted after ingestion of *C. botulinum* from soil, air, or raw honey, is found almost exclusively in infants under one year of age, presumably because establishment of the intestinal normal flora in older children and adults prevents colonization by *C. botulinum*.

In individuals over 65 years of age, infectious diseases are much more common than in younger adults. For example, the elderly are much more susceptible to respiratory infections, particularly influenza (◌◌Section 26.8), probably because of a declining ability to make an effective immune response to respiratory pathogens.

Anatomical changes associated with age may also encourage infection. Enlargement of the prostate gland, a common condition in men over the age of 50, frequently leads to a decreased urinary flow rate, allowing pathogens to colonize the male urinary tract (see Figure 21.11) more readily, leading to an increase in urinary tract infections.

Stress can predispose a normally healthy individual to disease. In studies with rats and mice, physiological stressors such as fatigue, exertion, poor diet, dehydration, or drastic climate changes increase the incidence and severity of infectious diseases. For example, rats subjected to intense physical activity for long periods of time show a higher mortality rate from experimental *Salmonella* infections compared with well-rested control animals. Hormones that are produced under stress influence the immune system and may play a role in stress-mediated disease. The hormone *cortisone*, for example, produced at much higher levels in times of stress than during normal periods, is an effective anti-inflammatory agent. However, suppression of inflammation inhibits the activation of the immune response, removing one of the normal defenses against disease (◌◌Section 22.2).

Diet plays a role in host susceptibility to infection. The association between famine and infectious disease has been known for centuries. Diets low in protein and calories alter the composition of the normal flora, allowing opportunistic pathogens a better chance to multiply and increasing susceptibility of the host to known pathogens. For example, the number of *Vibrio cholerae* necessary to produce cholera in an exposed individual is drastically reduced if the individual is malnourished. The number of organisms necessary to induce disease is also reduced when the *V. cholerae* is ingested in food, presumably because the food neutralizes the stomach acids that would normally destroy the pathogen (∞Section 28.5).

In some cases, absence of a particular dietary substance may prevent disease by depriving a pathogen of critical nutrients. The best example here is the effect sucrose has on the development of dental caries. As explained in Section 21.3, dietary restriction of sucrose, along with good oral hygiene, virtually eliminates tooth decay. Without dietary sucrose, the highly cariogenic bacteria *Streptococcus mutans* and *S. sobrinus* are unable to synthesize the polysaccharide slime layer needed to keep the bacterial cells attached to the teeth.

The Compromised Host

The term *compromised host* refers to hosts in which one or more resistance mechanisms are inactive and in which the probability of infection is therefore increased.

Many hospital patients with noninfectious diseases (for example, cancer and heart disease) acquire microbial infections because they are compromised hosts (∞ Section 25.7). Such hospital-acquired infections are called **nosocomial infections** and affect up to 2 million individuals each year in the United States, causing up to 100,000 deaths. Hospital procedures such as catheterization, hypodermic injection, spinal puncture, biopsy, and surgery may unintentionally introduce microorganisms into the patient. The stress of surgery may also lower patient resistance. Anti-inflammatory drugs given to reduce pain and swelling may also reduce host resistance (∞ Sections 22.2 and 22.3). Organ transplant patients are treated with immunosuppressive drugs. These drugs suppress rejection of the transplant but also reduce the ability of the patient to resist infection.

Compromised hosts exist even outside the hospital. Smoking, excess consumption of alcohol, intravenous drug use, lack of sleep, poor nutrition, and acute or chronic infection with another agent are conditions that compromise host resistance. For example, infection with HIV (human immunodeficiency virus) predisposes a patient to a variety of infections from microorganisms that are not pathogens in uninfected individuals. HIV causes acquired immunodeficiency syndrome (AIDS) by destroying one type of immune cell, the CD4 T lymphocytes (∞Section 22.7), involved in the immune response. Because of preexisting host damage from HIV

infection, AIDS patients are unable to mount effective resistance to infection. Death is generally due to an infectious disease caused by an *opportunistic pathogen*, a microorganism that does not ordinarily cause disease in an uncompromised host (∞Sections 25.6 and 26.14).

Finally, certain genetic conditions compromise the host. For example, genetic diseases that eliminate important parts of the immune system predispose individuals to infections. Individuals with such conditions frequently die at an early age, not from the genetic condition itself but from microbial infection.

21.13 Concept Check

Conditions of age, stress, diet, general health, lifestyle, prior or concurrent disease, and genetic makeup may compromise the host's ability to resist infection.

◆ List important factors that control susceptibility to infection and cannot be controlled by the host.

◆ List important factors that control susceptibility to infection and can be controlled by the host.

21.14 Innate Resistance to Infection

Most animals have an innate first line of defense against pathogens. These physical and chemical barriers including tissue specificity act nonspecifically to inhibit invasion by pathogens and prevent disease (Figure 21.24● and Table 21.6).

Natural Host Resistance

Under certain circumstances, closely related species, or even members of the same species, may have different susceptibilities to a particular pathogen. The ability of a particular pathogen to cause disease in an individual animal species is highly variable. In *rabies*, for instance, death usually occurs in all species of mammals once symptoms of the disease develop. Nevertheless, certain animal species are much more susceptible to rabies than others. Raccoons and skunks, for example, are extremely susceptible to rabies infection as compared with opossums, which rarely acquire the disease. *Anthrax* infects a variety of animals and causes disease symptoms varying from mild pustules in humans to a fatal blood poisoning in cattle. However, *pulmonary*, or airborne anthrax, such as that induced by weaponized strains used for bioterrorism, is almost universally fatal in humans (∞Section 25.13). In contrast, birds are totally resistant to anthrax. As another example, diseases of warm-blooded animals are rarely transmitted to cold-blooded species, and vice versa.

As a final example, hosts and pathogens have sometimes evolved to interact specifically and exclusively with one another. Infections by HIV, for instance, occur only in higher-order primates, including the great apes

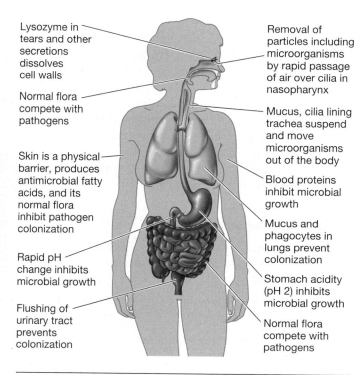

Lysozyme in tears and other secretions dissolves cell walls

Normal flora compete with pathogens

Skin is a physical barrier, produces antimicrobial fatty acids, and its normal flora inhibit pathogen colonization

Rapid pH change inhibits microbial growth

Flushing of urinary tract prevents colonization

Removal of particles including microorganisms by rapid passage of air over cilia in nasopharynx

Mucus, cilia lining trachea suspend and move microorganisms out of the body

Blood proteins inhibit microbial growth

Mucus and phagocytes in lungs prevent colonization

Stomach acidity (pH 2) inhibits microbial growth

Normal flora compete with pathogens

● **Figure 21.24 Physical, chemical, and anatomical barriers to infection.** These barriers provide inborn resistance to microbial colonization and infection.

and humans. This is because the CXCR4 protein on human T cells and the CCR5 protein on human macrophages are uniquely expressed in primates. These proteins interact with and bind the gp120 protein of HIV and are the only known cell surface receptors for HIV. Other animals lack these receptors, cannot bind HIV, and are thus protected from HIV infection (⌒⌒Section 26.14).

Physical and Chemical Defenses

The structural integrity of tissue surfaces poses a barrier to penetration by microorganisms. In the skin and mucosal tissues, potential pathogens must not only adhere

to tissue surfaces but must also grow at these sites before traveling elsewhere in the body. Intact surfaces usually prevent colonization, but damaged surfaces (for example, a puncture wound) are readily colonized, promoting invasion. Resistance to colonization and invasion is due to the production of host defense substances and to various anatomical mechanisms. A summary of the major anatomical defenses is shown in Figure 21.24.

The *skin* is an effective barrier to the penetration of microorganisms. *Sebaceous glands* in the skin (see Figure 21.2) secrete fatty acids and lactic acid, lowering the pH of the skin to about pH 5 and inhibiting colonization of many pathogenic bacteria (blood and internal organs are about pH 7.4). Microorganisms inhaled through the *nose* or *mouth* are removed by *ciliated epithelial cells* on the mucosal surfaces of the nasopharynx and trachea. Cilia push bacterial cells upward until they are caught in oral secretions and are either expectorated or swallowed and killed in the stomach.

Potential pathogens entering the stomach must survive the *acidity* (approximately pH 2) and then successfully compete with the increasingly abundant resident microflora present in the small intestine (approximately pH 5) and finally in the large intestine (pH 6–7). The large intestine contains bacterial numbers of 10^{10} or more per gram of intestinal contents in a normal adult (see Section 21.4), making establishment of new microorganisms very difficult.

The lumen of the kidney and the surface of the eye are constantly bathed with secretions containing *lysozyme*, an enzyme that markedly reduces microbial populations (⌒⌒Section 4.8). Extracellular fluids such as blood plasma also contain bactericidal substances. For example, blood proteins called *β-lysins* function by binding and disrupting the bacterial cytoplasmic membrane, leading to leakage of cytoplasmic constituents and cell death.

Tissue Specificity

Most pathogens must first adhere and colonize at the site of exposure. Even if pathogens adhere to an exposure

Table 21.6	Tissue specificity in infectious disease	
Disease	**Tissue infected**	**Organism**
Acquired immunodeficiency syndrome (AIDS)	T helper lymphocytes	Human immunodeficiency virus (HIV)
Botulism	Motor end plate	*Clostridium botulinum*
Cholera	Small intestine epithelium	*Vibrio cholerae*
Dental caries	Oral epithelium	*Streptococcus mutans, S. sobrinus, S. sanguis, S. mitis*
Diphtheria	Throat epithelium	*Corynebacterium diphtheriae*
Gonorrhea	Urogenital epithelium	*Neisseria gonorrhoeae*
Malaria	Blood (erythrocytes)	*Plasmodium* sp.
Pyelonephritis	Kidney medulla	*Proteus* sp.
Spontaneous abortion (cattle)	Placenta	*Brucella abortus*
Tetanus	Inhibitory interneuron	*Clostridium tetani*

site, if the site is not compatible with their nutritional and metabolic needs, the organisms cannot colonize. Thus, if *Clostridium tetani* were ingested, tetanus would not result because the pathogen is killed by the acidity of the stomach. If, on the other hand, *C. tetani* cells or endospores were introduced into a deep wound, the organism would grow and produce tetanus toxin in the anoxic zones created by local tissue death (see Sections 21.10 and 27.9). By contrast, enteric bacteria such as *Salmonella* and *Shigella* do not cause wound infections but successfully colonize the intestinal tract. Table 21.6 presents examples of tissue specificity.

Here we have identified passive host resistance mechanisms. In the next chapter, we will introduce the concepts of immunity and discuss the nonspecific and pathogen-specific methods used by the host to recognize and target the pathogen.

 21.14 Concept Check

Nonspecific physical, anatomical, and chemical barriers prevent colonization of the host by most pathogens. Lack of these defenses results in susceptibility to infection and colonization by a pathogen.

◆ Identify physical and chemical barriers to pathogens. How might these barriers be compromised?

◆ How might preexisting infection compromise an otherwise healthy host?

REVIEW QUESTIONS

1. Distinguish between a *parasite* and a *pathogen*. Distinguish between *infection* and *disease* (∞Section 21.1).

2. Which organs of the human body are normally colonized by microorganisms? What do these organs have in common? Which organs are normally devoid of microorganisms? What do these organs have in common (∞Section 21.2)?

3. Distinguish between *resident* and *transient* microorganisms at a body site. How could you distinguish between resident and transient microorganisms experimentally (∞Section 21.3)?

4. Why are members of the genus *Streptococcus* instrumental in forming dental caries? Why are they more capable of causing caries than other organisms (∞Section 21.4)?

5. How does pH affect the types of microorganisms that grow in each different region of the gastrointestinal tract? How does oxygen concentration affect the types of microorganisms that grow in each definable region of the gastrointestinal tract (∞Section 21.5)?

6. Describe the relationship between *Lactobacillus acidophilus* and glycogen in the vaginal tract. Why do adult females have a different normal flora in the vagina than juvenile females (∞Section 21.6)?

7. Distinguish between glycocalyx, the capsule, and the slime layer. How do these unique structures contribute to microbial adherence (∞Section 21.7)?

8. Give an example of a microorganism that is pathogenic almost solely because of its *toxin-producing* ability. Define the toxin and its mode of action. Give an example of a microorganism that is pathogenic almost solely because of its invasive characteristics. What factors confer invasive qualities on this microorganism (∞Section 21.8)?

9. Distinguish between A-B toxins, cytotoxins, and superantigens. Give an example of each category of toxin. How does each toxin category promote disease (∞Section 21.9)?

10. Define the mechanisms of action for tetanus toxin and botulinum toxin. With regard to their mechanisms of action, why are these toxins so dangerous (∞Section 21.10)?

11. Review the mode of action of cholera enterotoxin (Figure 21.22). What is the appropriate therapy for this disease, and why is antibiotic treatment generally not effective? Review the mode of action of staphylococcal enterotoxin. What is the appropriate therapy for this disease, and why is antibiotic treatment generally not effective (∞Section 21.11)?

12. Describe the structure of a typical *endotoxin*. How does endotoxin induce fever? What microorganisms produce endotoxin (∞Section 21.12)?

13. Identify common factors that lead to host compromise. Indicate which factors are controllable by the host and which factors are not controllable (∞Section 21.13).

14. How does pH work to limit bacterial infections? Which locations in or on the body might pH values differ from standard body conditions? Which organisms might benefit from differences in pH (∞Section 21.14)?

APPLICATION QUESTIONS

1. Mucous membranes are effective barriers against colonization and growth of microorganisms. However, mucous membranes, for example in the throat, are colonized with a variety of different microorganisms that occasionally cause disease. Explain how normally nonpathogenic microorganisms can become pathogenic under certain circumstances. Be sure to describe at least one set of circumstances that might encourage pathogenicity.

2. Antibiotic therapy can significantly reduce the number of microorganisms residing in the gastrointestinal tract. What physiological symptoms might the reduction of normal flora produce in the host? Long-term antimicrobial therapy is often followed by infections due to opportunistic pathogens, many of which are caused by organisms causing opportunistic infections in individuals with AIDS (⌘Section 26.14). What pathogens might be involved? Why are individuals who have undergone antibiotic therapy susceptible to these pathogens?

3. Identify potential selection pressures that might result in an increase in the virulence and pathogenicity of *Streptococcus pneumoniae*. Identify potential selection pressures that might result in an increase in the virulence and pathogenicity of *Clostridium tetani*. Would an increase in virulence by either of these organisms result in selective advantages for the organism? For each organism, be sure to consider the natural habitat.

4. Coagulase is a virulence factor for *Staphylococcus aureus* that acts by causing clot formation at the site of *S. aureus* growth. Streptokinase is a virulence factor for *Streptococcus pyogenes* that acts by dissolving clots at the site of *S. pyogenes* growth. Reconcile these opposing strategies for enhancing pathogenicity.

5. Although mutants incapable of producing exotoxins are relatively easy to isolate, mutants incapable of producing endotoxins are much harder to isolate. From what you know of the structure and function of these types of toxins, explain the differences in mutant recovery.

22

ESSENTIALS OF IMMUNOLOGY

Immunity starts with innate responses of *phagocytes*, shown here in human tissue as they take up cells of *Neisseria gonorrhoeae*. Whether innate or adaptive, immune mechanisms initiate *inflammatory responses* that limit pathogen spread.

WORKING GLOSSARY

Adaptive immunity or antigen-specific immunity is the acquired ability to recognize and destroy an individual pathogen or its products and relies on previous exposure to the pathogen or its products

Antibody a soluble protein, produced by B cells, that interacts with antigen; also called immunoglobulin

Antibody-mediated immunity (humoral immunity) immunity resulting from direct interaction with antibodies

Antigen a molecule capable of interacting with specific components of the immune system

Antigenic determinant (epitope) that portion of an antigen that is reactive with a specific antibody or T-cell receptor

Antigen-presenting cell (APC) a macrophage, dendritic cell, or B cell that presents processed antigen peptides to a T cell

Autoantibody an antibody that reacts to self antigens

B cell a lymphocyte that has immunoglobulin surface receptors, produces immunoglobulin, and may present antigens to T cells

Cell-mediated immunity (CMI) immunity resulting from direct interaction with antigen-specific T cells

Class I MHC protein antigen-presenting molecule found on all nucleated vertebrate cells

Class II MHC protein antigen-presenting molecule found primarily on macrophages, B cells, and dendritic cells

Complement a series of proteins that react in a sequential manner with antibody–antigen complexes to amplify or potentiate antibody activity

Cytokine a soluble immune response modulator produced by leukocytes

Dendritic cell a type of leukocyte having phagocytic and antigen-presenting properties, found in lymph nodes and spleen

Domain a region of a protein having a defined structure and function

Hapten a low-molecular-weight molecule that combines with specific antibodies but that is incapable of eliciting an immune response by itself

Hypersensitivity an immune response leading to damage to host tissues

Immunity the ability of an organism to resist infection

Immunization (vaccination) inoculation of a host with inactive or weakened pathogens or pathogen products to stimulate protective immunity

Immunogen a molecule capable of eliciting an immune response

Immunoglobulin (Ig) a soluble protein produced by B cells and plasma cells that interacts with antigens; also called antibody

Inflammation a nonspecific reaction to noxious stimuli such as toxins and pathogens, characterized by redness (erythema), swelling (edema), pain, and heat (fever), usually localized at the site of infection

Innate immunity or nonspecific immunity is the noninducible ability to recognize and destroy an individual pathogen or its products and does not rely on previous exposure to a pathogen or its products

Interleukin (IL) a soluble cytokine or chemokine secreted by leukocytes

Leukocyte a nucleated cell found in the blood (white blood cells)

Lymph a fluid similar to blood which lacks red blood cells and travels through a separate circulatory system (the lymphatic system) containing lymph nodes

Lymphocyte a subset of nucleated cells found in the blood that are involved in the immune response

Macrophage a type of large leukocyte found in tissues that has phagocytic and antigen-presenting capabilities

Major histocompatibility complex (MHC) a genetic complex responsible for encoding several cell surface proteins important in antigen presentation

Memory (immunologic memory) ability to rapidly produce large quantities of specific immune cells or antibodies after subsequent exposure to a previously encountered antigen

Memory B cells long-lived cells responsive to an individual antigen

Natural killer (NK) cell a specialized lymphocyte that recognizes and destroys foreign cells or infected host cells in a nonspecific manner

Neutrophil (polymorphonuclear leukocyte) (PMN) a type of leukocyte exhibiting phagocytic properties, a granular cytoplasm (granulocyte), and a multilobed nucleus

Pathogen-associated molecular pattern (PAMP) unique structural components of a microbe or virus recognized by a pattern recognition molecule

Pattern recognition molecule (PRM) a membrane-bound protein that recognizes a pathogen-associated molecular pattern, such as a unique component of a microbial cell surface structure

Phagocyte one of a group of cells that recognizes, ingests, and degrades pathogens and pathogen products

Plasma the liquid portion of the blood with cells removed and clotting proteins deactivated

Plasma cell a differentiated B cell that produces large amounts of antibodies

Primary antibody response antibodies made on first exposure to antigen; mostly of the IgM class

Secondary antibody response antibody made on subsequent exposure to antigen; mostly of the IgG class

Serology the study of antigen–antibody reactions *in vitro*

Serum the liquid portion of the blood with clotting proteins and cells removed

Specificity the ability of the immune response to interact with individual antigens

Stem cell a cell that can develop into a number of final cell types

Superantigen a pathogen product capable of eliciting an inappropriately strong immune response by stimulating greater than normal numbers of T cells

T cell a lymphocyte responsible for antigen-specific cellular interactions; T cells are divided into functional subsets including T_C cytotoxic T cells and T_H helper T cells. T_H cells are further subdivided into T_H1 inflammatory cells and T_H2 helper cells, which aid B cells in antibody formation

T-cell receptor (TCR) antigen-specific receptor protein on the surface of T cells

Tolerance inability to produce an immune response to specific antigens

Toll-like receptor (TLR) one of a family of pattern recognition molecules found on phagocytes that recognize a pathogen-associated molecular pattern (PAMP)

Toxoid an attenuated form of a toxin that retains immunogenicity while losing toxicity

Vaccination (immunization) inoculation of a host with inactive or weakened pathogens or pathogen products to stimulate protective immunity

Vaccine an inactivated or weakened pathogen, or an innocuous pathogen product used to stimulate protective immunity.

All multicellular organisms depend on the interactions of a variety of cells and their products to defend against invasion and infection by pathogens. This process is known as **immunity**. In Chapter 21, we discussed *passive* physical and chemical protection against pathogen invasion and disease. Here we will examine *active* resistance to pathogens. We start with **innate immunity**, or *nonspecific immunity*, the body's built in ability to recognize and destroy pathogens or their products. Innate immunity does not rely on previous exposure to a pathogen or its products and is largely a function of **phagocytes**, cells that recognize, engulf, kill, and digest most pathogens. Unfortunately, innate immunity is not always effective, and dangerous infections sometimes still occur. However, phagocytes can also activate another defense mechanism called **adaptive immunity**, or *specific immunity*, to deal with these infections.

Adaptive immunity is the acquired ability to recognize and destroy a pathogen or its products, and requires exposure of the immune system to the pathogen. Phagocytes use partially digested molecules from pathogens to activate *lymphocytes*, specialized cells that are responsible for specific immunity. Each lymphocyte is genetically programmed to recognize a single molecule, called an **antigen**, on the pathogen. An antigen is any molecule that interacts with specific receptors on lymphocytes (see Section 22.4). When the lymphocyte interacts with the antigen, it grows and divides rapidly, forming copies of itself. The copies, or *clones*, of antigen-reactive lymphocytes either destroy the individual antigen directly or produce soluble antigen-specific proteins called *antibodies* that target the antigens. This ability to recognize a single antigen is called immune *specificity*.

Some antigen-reactive lymphocytes live for years. If we are exposed to the same pathogen at a later time, the specific lymphocyte numbers expand rapidly, producing an immediate and vigorous immune response. The ability to react specifically and more strongly to subsequent antigen exposures is called immune *memory*.

The immune system reacts very strongly with antigens from pathogens, but it also has a safety mechanism that prevents destruction of our own cells. The ability of the immune system to destroy dangerous nonself pathogen antigens while ignoring benign self antigens is called *immune tolerance*. Unfortunately, tolerance occasionally fails, and the immune system responds inappropriately, damaging host tissues and producing *autoimmune disease*.

In this chapter we will first see how innate immunity triggered by nonspecific phagocytes deals with most pathogens. We will then look at how these same phagocytes present antigens and activate lymphocytes, inducing an adaptive immune response. The activated immune lymphocytes recognize individual antigens (**specificity**), respond vigorously when re-exposed to antigen (**memory**), and do not harm host cells (**tolerance**). Innate and adaptive immunity have evolved to protect animals from dangerous nonself pathogens and are critical components of the pathogen defense system.

OVERVIEW OF THE IMMUNE SYSTEM

The immune response has evolved to recognize any foreign molecule, especially antigen molecules derived from pathogens. The antigen-specific response targets reactive molecules for destruction. Here we introduce some of the important features of innate and adaptive host responses to pathogens and other foreign substances. *Innate* responses recognize all foreign molecules and pathogens, while *adaptive* responses are directed at individual molecules on pathogens. We begin with the cells and organs of the immune system and then consider the cells and mechanisms involved in innate immunity. We finish with an overview of specific, or adaptive, immunity, the focus of the rest of the chapter.

22.1 Cells and Organs of the Immune System

Innate and adaptive immunity result from the actions of cells that circulate in the *blood* and *lymph*, body fluids that directly or indirectly interact with every major organ system. All of the cells involved in immunity develop from common precursors called **stem cells**, found in the bone marrow.

Blood and Lymph Components

Blood consists of cellular and noncellular components and contains many cells and molecules involved in the immune response. Because blood can be easily and safely obtained from patients, it is a valuable source of material for immune response assays. The most numerous cells in human blood are *erythrocytes* (red blood cells), nonnucleated cells that function to carry oxygen from the lungs to the tissues (Table 22.1). However, about 0.1% of the cells in blood are nucleated white blood cells, or *leukocytes*. Leukocytes include a variety of phagocytic cells such as *monocytes*, as well as cells called *lymphocytes*, which are involved in antibody production and cell-mediated immunity. **Lymph** is a fluid similar to blood, but which lacks red blood cells.

As shown in Figure 22.1●, common stem cells in the bone marrow are the progenitors of blood cells and leukocytes. Stem cells differentiate to produce mature cells under the influence of a group of soluble cell proteins called **cytokines** (⌘Section 23.11).

Whole blood is composed of **plasma**, a liquid containing proteins and a variety of other solutes and suspended cells. Outside the body, whole blood or plasma quickly forms an insoluble clot, caused by the conversion of a soluble plasma protein, *fibrinogen*, to an insoluble protein, *fibrin*. Plasma or whole blood remains liquid

Table 22.1	Major cells found in normal human blood
Cell type	**Cells per milliliter**
Erythrocytes	$4.2–6.2 \times 10^9$
Leukocytes[a]	$4.5–11 \times 10^6$
Lymphocytes	$1.0–4.8 \times 10^6$
Myeloid cells	Up to 7.0×10^6

[a] Leukocytes include all nucleated blood cells and are subdivided into the lymphocytes and myeloid cells (monocytes and granulocytes).
Source: Henry, J. B. 1996. *Clinical Diagnosis and Management by Laboratory Methods*, 19th edition. W. B. Saunders, Philadelphia.

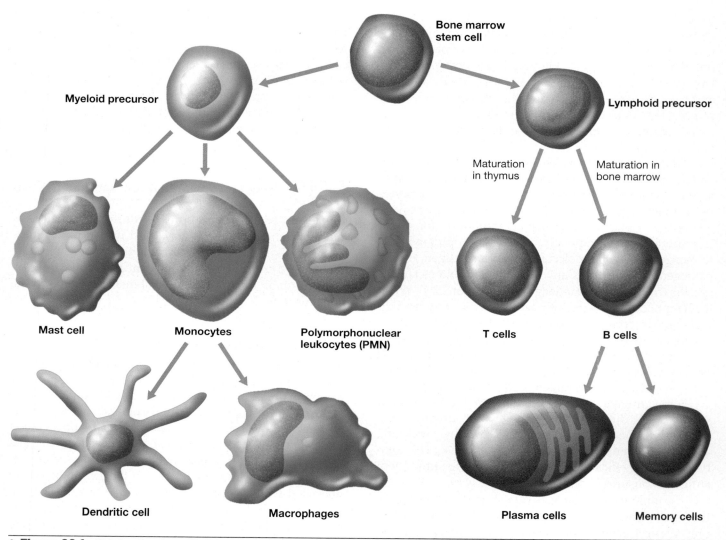

● **Figure 22.1 Origin of major cells involved in the immune response.** Cells involved in immunity develop from two major precursors. The myeloid precursor generates *phagocytes*, cells involved in innate immunity and antigen presentation. The lymphoid precursor generates *T* and *B lymphocytes*, the cells that participate directly in the antigen-specific immune response.

only when an anticoagulant is added. The addition of anticoagulant chemicals such as potassium oxalate, potassium citrate, or heparin prevent the conversion of fibrinogen to fibrin, stopping the clotting process. If clotting occurs, the insoluble proteins trap the blood cells in a large, insoluble mass. After clotting, the remaining fluid, called **serum**, contains no cells or clotting proteins. Serum does, however, contain a high concentration of other proteins, including soluble antibody proteins, and is widely used in immunological investigations (∞ Section 24.7). The use of serum antibodies to detect antigens *in vitro* is called **serology**.

Blood and Lymph Circulation

Blood is pumped by the heart through arteries and capillaries throughout the body and is returned through the veins (Figure 22.2*a*, *b*●). Figure 22.2*b* and *c* show the capillary beds where leukocytes may pass to and from the blood into the *lymphatic system*, a separate circulatory system through which lymph circulates.

Lymph fluid drains from extravascular tissues into lymphatic capillaries and then into *lymph nodes* (Figure 22.2*d*) found throughout the lymph system. Lymph nodes contain high concentrations of leukocytes and phagocytes, arranged in such a way that they encounter microorganisms and antigens as they enter the nodes via the lymph ducts. The spleen serves an analogous function in the blood circulatory system. The lymph nodes and the spleen are the sites of most adaptive immune responses. Lymph, carrying antibodies and immune cells, eventually flows back into the circulatory system via the thoracic lymph duct.

Leukocytes

Leukocytes are nucleated white blood cells found in the blood and the lymph. There are several distinct kinds of leukocytes (Table 22.1 and Figure 22.1), and all participate in innate or adaptive immune functions. **Lymphocytes** are specialized leukocytes involved in the specific immune response. Mature lymphocytes are concentrated in the

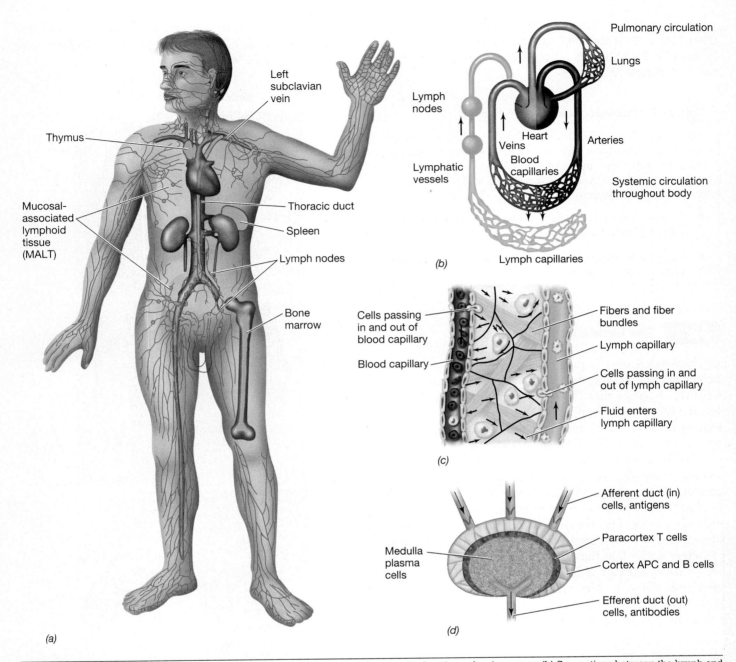

● **Figure 22.2 The blood and lymph systems.** (a) The lymph system, showing the locations of major organs. (b) Connections between the lymph and blood systems. Blood flows from the veins to the heart, then to the lungs where it becomes oxygenated, and then through the arteries to the tissues. Lymph drains from the thoracic duct into the left subclavian vein of the blood circulatory system. (c) Connection between the blood and lymph systems is shown microscopically. Both blood and lymph capillaries are closed vessels, but cells and fluids can pass from one vessel to another by a process known as extravasation. (d) A lymph node. The diagram depicts major anatomical areas and identifies immune cells present in those areas.

lymph nodes and spleen where they interact with antigens. The lymphocytes are further divided into **B cells** and **T cells** (Figure 22.1). *B cells* originate and mature in the bone marrow and are the precursors of the antibody-producing plasma cells. *T cells* originate in the bone marrow, but travel to the thymus to mature. Myeloid cells are derived from a myeloid precursor (Figure 22.1) and are divided into two large categories, based on functional differences. Monocytes are precursors of specialized cells called **macrophages** and **dendritic cells**, found in abundance in tissues, spleen, and lymph nodes (Figure 22.1).

These are called **antigen-presenting cells**, or **APCs**, because of their specialized ability to engulf, process, and present antigens to lymphocytes (see Section 22.5). Granulocytes are cells containing cytoplasmic inclusions, or granules, that can be visualized using staining techniques. *Polymorphonuclear leukocytes*, also called *neutrophils* or *PMNs*, are phagocytic granulocytes (see Section 22.2). *Basophils* are granulocytes with granules containing toxic chemicals. Release of these granules, called *degranulation*, can cause allergy-like symptoms (see Section 22.15).

Leukocytes can travel throughout the body and pass from blood, to interstitial spaces, to lymph, and back to the blood circulatory system, a process called *extravasation* (Figure 22.2c).

⬡ *22.1* **Concept Check**

All the cells involved in immunity originate from common stem cells in the bone marrow. The blood and lymph systems circulate cells and proteins that are important for a functional immune system. A variety of leukocytes participate in immune responses.

◆ Describe the circulation of a leukocyte from the blood to the lymph and back to the blood.

◆ Trace the development of B cells, T cells, and macrophages from the common stem cell.

22.2 The Innate Immune Response

On rare occasions, pathogens break through the host physical and chemical defense mechanisms described in Chapter 21 (∞Section 21.14). The pathogen can then infect the host (∞Section 21.7). At this point, the immune system must be mobilized, and the *innate* or *nonspecific* immune response is the first line of defense.

The starting point for innate immunity is contact of a foreign substance such as a pathogen or a pathogen product with a **phagocyte**. A phagocyte is one of several leukocytes with the capacity to engulf and destroy foreign substances. The usual result of such an interaction is a complex host-mediated process called **inflammation**.

Here we look at the cells and molecules involved in recognizing pathogens and initiating an inflammatory response. We will also examine mechanisms used by pathogens to defeat phagocytes. In the next section, we will examine the inflammatory response in detail.

Phagocytes

The cell type involved initially in the inflammatory response is usually a **phagocyte** (literally, a cell that eats). The primary function of a phagocyte is to engulf and destroy pathogens. In this process, some phagocytes also act as **antigen-presenting cells (APCs)** and generate the peptide antigens that activate the adaptive immune response.

Phagocytes include macrophages, monocytes, neutrophils, and dendritic cells. These cells are found in tissues and fluids throughout the body. Phagocytes are usually motile, moving by ameboid action. Most have granular inclusions called *lysosomes*, which contain bactericidal substances such as hydrogen peroxide, lysozyme, proteases, phosphatases, nucleases, and lipases. Phagocytes trap a pathogen on a surface such as a blood vessel wall or a fibrin clot and engulf the pathogen. The membrane surrounding the pathogen pinches off, forms a *phagosome*, and moves inside the cell, where it fuses with a lysosome, forming a

phagolysosome. The toxic substances and enzymes inside the phagolysosome usually kill and digest the engulfed microbial cell (Figure 22.3●).

● **Figure 22.3** **Phagocytosis.** Time-lapse photomicrographs of the engulfment and digestion of a chain of *Bacillus megaterium* cells by a human macrophage, observed by phase contrast microscopy. The bacterial chain is about 18–20 μm long. The macrophage is one of a group of cells that ingests and degrades pathogens and pathogen products.

One group of phagocytes, the **neutrophils**, or **polymorphonuclear leukocytes (PMNs)**, are actively motile granulocytes containing large numbers of lysosomes (Figure 22.4a●). Derived from the myeloid stem cell (Figure 22.1), neutrophils are found predominantly in the bloodstream and bone marrow, from where they migrate to sites of active infection in tissues. Large numbers of neutrophils in the blood or at a site of inflammation usually indicate an active infection.

Monocytes are circulating precursors of macrophages and dendritic cells, major phagocytic cell types (Figures 22.4a and 22.3). Macrophages are large cells found in tissues such as lymph nodes and spleen and are capable of ingesting and destroying most pathogens and antigens, as well as interacting with lymphocytes to initiate adaptive immune responses. Macrophages are important APCs that present peptide antigens to T cells, the first step in activating an adaptive immune response. Dendritic cells (dendrocytes) (Figure 22.1) are also phagocytes involved in the immune response. Derived from the same monocyte progenitor as macrophages, dendritic cells are also found in lymph nodes and spleen, where they are efficient APCs. The specialized antigen-presenting properties of macrophages and dendritic cells make them very important for initiation of adaptive immunity, and we will examine their role as APCs in more detail in Section 22.5.

Phagocytes and Pathogen Recognition

Phagocytes have a general pathogen-recognition system designed to trigger a timely and appropriate inflammatory response. This system employs several evolutionarily conserved **pattern-recognition molecules (PRMs)**. PRMs are membrane-bound phagocyte proteins that recognize a **pathogen-associated molecular pattern (PAMP)**, a unique structural component on a microbial cell or virus (Figure 22.5●). PRMs were first observed in *Drosophila*, the fruit fly, where they are called *Toll receptors*. Each **Toll-like receptor (TLR)** on a human phagocyte recognizes a specific PAMP. For example, TLR-4 recognizes the lipopolysaccharide (LPS) present in cell walls of all gram-negative bacteria (∞Section 4.9 and Section 23.1), inducing a level of immunity to all gram-negative pathogens. Other TLRs recognize PAMPS such as cell-surface components of yeast (zymosan), peptidoglycan, the unmethylated CpG oligonucleotides in bacteria, and bacterial peptidoglycan (∞Section 23.1).

Interaction of a PAMP with the phagocyte PRM triggers a transmembrane signal, initiating signal transduction events (∞Section 8.12) that result in activation of the phagocyte. These pathogen-induced signals enhance the phagocytic and destructive abilities of the phagocyte, and here we will concentrate on the production of toxic oxygen products that can lead to pathogen death.

Oxygen-Dependent Phagocyte Activation

Ingestion of a pathogen activates phagocytes, enhancing their phagocytic and destructive abilities. As discussed in Section 6.15, various reactions can lead to the formation of toxic oxygen-containing compounds including hydrogen peroxide (H_2O_2), superoxide anions (O_2^-), hydroxyl radicals ($OH\cdot$), singlet oxygen ($^1O_2$), hypochlorous acid (HOCl), and nitric oxide (NO). The acidic conditions found in the phagolysosome aid in the generation and reactivity of the toxic oxygen compounds. Phagocytic cells make use of toxic forms of oxygen to kill ingested bacterial cells. The combined activity of these oxygen-dependent phagocyte enzymes forms sufficient levels of toxic oxygen compounds to kill ingested bacterial cells by oxidizing key cellular constituents. These reactions occur within the phagocytic cell itself, which is not damaged by the toxic oxygen products. The activity of phagocytic cells in oxygen-mediated killing is summarized in Figure 22.6●. Activated phagocytes must take up larger than normal quantities of O_2 over a short time period to produce these toxic oxygen

(a)

(b)

J. Martinko and M. T. Madigan

● **Figure 22.4 Major immune cell types.** (a) Phagocytic cells. The nucleated cell in the lower left center is a neutrophil (PMN), characterized by a segmented nucleus (violet stain) and granular cytoplasm. The nucleated cell to the right and slightly above the PMN is a monocyte. Phagocytes are 12–15 μm in diameter. The nonnucleated red blood cells are about 6 μm in diameter. (b) The nucleated cell is a circulating lymphocyte. The lymphocyte has almost no visible cytoplasm and is smaller than the phagocytes, about 10 μm in diameter.

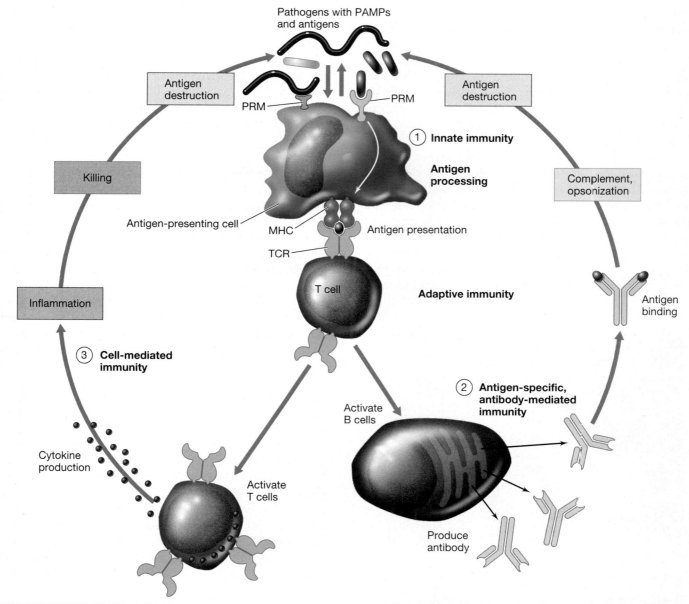

● **Figure 22.5** **Overview of the immune response.** Pathogens are destroyed by three mechanisms. ① *Innate immunity* results from interactions between pathogen–associated molecular patterns (PAMPs) found as cell-surface components of pathogens, and pattern recognition molecules (PRMs) found on phagocytes. *Adaptive immunity*, orchestrated by antigen-specific T cells, results in two distinct effector pathways: ② *Antibody-mediated immunity* results from soluble antigen-specific antibody proteins, products of antigen-stimulated B lymphocytes; ③ *Cell-mediated immunity* is mediated by antigen-specific T effector cells.

compounds. This increased rate of O_2 uptake by activated phagocytes is known as the *respiratory burst*.

Inhibiting Phagocytes

Some pathogens have developed mechanisms for neutralizing toxic phagocyte products, for killing the phagocyte, or for avoiding phagocytosis. For example, *Staphylococcus aureus* (∞Section 26.9) produces pigmented compounds called *carotenoids*, which neutralize singlet oxygen and prevent killing (∞Section 6.15). Intracellular pathogens such as *Mycobacterium tuberculosis* (the cause of tuberculosis) grow and persist within phagocytic cells (∞Section 26.5). They apparently use cell wall glycolipids (∞Section 12.23) to scavenge toxic oxygen com-

pounds. These glycolipids remove hydroxyl radicals and superoxide anions, the most lethal toxic oxygen species produced by phagocytic cells.

Some intracellular pathogens produce phagocyte-killing proteins called *leukocidins*. In such cases, the pathogen is ingested as usual, but kills the phagocyte and is then released. *Streptococcus pyogenes* and *Staphylococcus aureus* are major leukocidin producers. Dead phagocytes make up much of the material of *pus*; organisms that produce leukocidins are therefore usually *pyogenic* (pus-forming). Localized infections by pyogenic bacteria often result in boils or abscesses (∞Section 26.9).

Another important microbial defense against phagocytosis is the bacterial capsule (∞Section 4.10). Encapsulated bacteria are often highly resistant to phagocytosis,

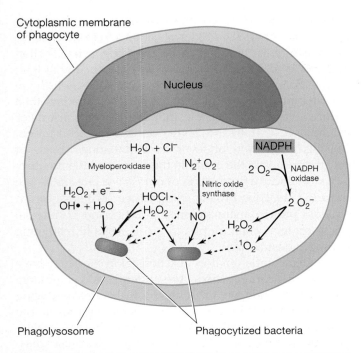

● **Figure 22.6 Action of phagocyte enzymes in generating toxic oxygen species.** These include hydrogen peroxide (H_2O_2), the hydroxyl radical ($OH\bullet$), hypochlorous acid ($HOCl$), the superoxide anion (O_2^-), singlet oxygen ($^1O_2$), and nitric oxide (NO). Formation of these toxic compounds requires a substantial increase in the uptake and utilization of molecular oxygen, O_2. This increase in oxygen uptake and consumption by activated phagocytes is known as the *respiratory burst*. For more discussion of toxic oxygen species, see Section 6.15.

apparently because the capsule prevents adherence of the phagocyte to the bacterial cell. The clearest case of the importance of a capsule that prevents phagocytosis is that of *Streptococcus pneumoniae*. Fewer than 10 cells of an encapsulated strain of *S. pneumoniae*, when injected, can kill a mouse in a few days. On the other hand, nonencapsulated mutant strains are completely avirulent (∞ Figure 21.16). Surface components other than capsules can also inhibit phagocytosis. For instance, pathogenic *Streptococcus pyogenes* produces a specific substance, M-protein (∞ Section 26.2), that alters the surface of the bacterial cell and inhibits phagocytosis.

Antibodies to capsules or other cell surface molecules often reverse the protective effect of these bacterial defense mechanisms and enhance phagocytosis, a process known as *opsonization* (see Section 22.11).

 22.2 Concept Check

The innate immune response is mediated by phagocytes. Phagocytes recognize pathogen-associated molecular patterns (PAMPs) via a family of membrane-bound pattern-recognition molecules, or PRMs. Interaction of the PAMPs with PRMs activates phagocytes to produce metabolic products that kill the pathogen or limit its effects. Many pathogens have developed mechanisms to inhibit phagocytes.

◆ Describe the cellular location and molecular specificity of PAMPs and PRMs.

◆ Identify toxic oxygen products and explain the respiratory burst that takes place in activated phagocytes.

◆ Identify at least one mechanism used by pathogens to inhibit phagocytosis.

22.3 Inflammation, Fever, and Septic Shock

Inflammation is a nonspecific reaction to noxious stimuli such as toxins and pathogens. Inflammation results in *redness (erythema), swelling (edema), pain,* and *heat (fever),* usually localized at the site of infection. The molecular mediators of inflammation include a group of proteins called *cytokines* and *chemokines* (∞ Section 23.11), which are produced by a variety of cells, but especially by phagocytes (Section 22.1).

Inflammation is the usual outcome of either an innate or an adaptive immune response; both branches of immunity induce the same inflammatory mediators that recruit and activate the same effector cells and molecules. An effective inflammatory response isolates and limits tissue damage, destroying pathogen invaders as well as damaged cells at local sites. In some cases, however, inflammation can result in considerable systemic host cell damage.

Inflammatory Cells and Local Inflammation

The first inflammatory cell to arrive at the scene of an infection or tissue injury is the *neutrophil* (Figure 22.7a●). These phagocytic cells are attracted to the site of an active infection or tissue injury by soluble chemoattractants called chemokines (∞ Section 23.11). For example, IL-8 (**interleukin**-8), a small protein in the chemokine family, is produced by damaged host cells. Neutrophils migrate toward the cells secreting IL-8 and are activated by the interaction with IL-8. Activated neutrophils are important phagocytes, and they also recruit macrophages (Figure 22.7b) by secreting chemokines such as MIP-α and MIP-β (macrophage inflammatory proteins). These chemokines are chemoattractants for macro-phages, which migrate up the chemokine gradient to the neutrophils, already at the site of infection or injury. The macrophages in turn become activated, enhancing their phagocytic abilities (see Section 22.2) and secreting a number of *cytokines* (∞ Section 23.11), soluble mediators that activate other cell types. The major macrophage cytokines involved in inflammation are IL-1, IL-6, and TNF-α (tumor necrosis factor-alpha). These and other soluble mediators released by injured cells, and phagocytes contribute to inflammation. For example, IL-1, IL-6, and TNF-α all act to increase vascular permeability, increasing volume (edema), and causing the reddening (erythema) and local heating associated with inflammation. The edema stimulates local neurons, causing pain.

The usual outcome of the inflammatory response is a rapid localization and destruction of the pathogen by the

(a)

(b)

● **Figure 22.7 Activated phagocytes.** (a) Neutrophils actively ingest and kill *Neisseria gonorrhoeae*. The neutrophils are about 12–15 μm in diameter. Note the multilobed nucleus in each cell. Not all cells have ingested the bacteria. (b) A skin macrophage that has taken up numerous *Leishmania* (arrows). Activated macrophages are capable of killing these intracellular protozoan parasites.

neutrophils and macrophages. As the pathogen is destroyed, the inflammatory cells are no longer stimulated, their numbers at the site are decreased, cytokine production is reduced, and inflammation subsides.

In addition to their role in innate immunity, the macrophages may also initiate an adaptive immune response. They have the capacity to act as antigen-presenting cells, secreting cytokines to attract T cells and then presenting antigens to the T cells (see Section 22.5).

Systemic Inflammation and Septic Shock

In some cases, the inflammatory response fails to localize the pathogen and the reaction becomes widespread, with inflammatory cells and mediators activated on a larger scale. A systemic inflammatory response, spreading inflammatory cells and mediators through the circulatory and lymphatic systems (Section 22.1), can lead to *septic shock*, a life-threatening condition. The most common cause of septic shock is systemic infection by gram-negative enteric *Bacteria* such as *Salmonella* or *Escherichia coli*, often caused by a ruptured or leaking bowel that releases the gram-negative organisms into the intraperitoneal cavity or the bloodstream. The primary infection is often cleared by the phagocytes or is treated successfully

with antibiotics. However, the endotoxic outer membrane lipopolysaccharides (LPS) (∞Section 4.9) from these organisms interact with TLR-4 on the phagocytes (Section 22.2 and ∞Section 23.1) and induce production of IL-1, IL-6, IL-8, and TNF-α inflammatory cytokines, which are released into the systemic circulation. The cytokines then induce a variety of systemic antipathogen responses. For example, IL-1, IL-6, and TNF-α are *endogenous pyrogens*, causing production of fever by stimulating release of prostaglandins in the brain. In the liver, the inflammatory cytokines induce production of *acute phase proteins*. These include C-reactive protein, complement proteins, serum amyloid protein, mannan-binding lectin, and fibrinogen. These proteins enhance phagocytosis, usually mediating pathogen destruction.

Systemic inflammatory reactions, however, may have very serious consequences. The systemic release of large quantities of endogenous pyrogens, instead of producing localized heating, induces uncontrollable high fever. The same mechanism that caused local edema due to vasodilation and increased vascular permeability now causes massive efflux of fluids from the central vascular tissue, resulting in a precipitous loss of blood pressure and edema of the surrounding tissues. The resulting septic shock, characterized by the loss of blood volume as well as the high fever, is catastrophic and causes death in up to 30% of affected individuals. Uncontrolled systemic inflammation can be more dangerous than the original infection.

 22.3 Concept Check

Inflammation is characterized by *pain, swelling (edema), redness (erythema),* and *heat*. The inflammatory response is a normal and generally desirable outcome of an immune response. Uncontrolled systemic inflammation, called *septic shock*, can lead to serious illness and death.

◆ Identify the cells involved in the production of inflammation.

◆ Identify the molecular mediators and their individual roles in inflammation.

22.4 The Adaptive Immune Response

The overall *adaptive*, or *specific*, immune response is outlined in Figure 22.5. Phagocytes such as macrophages and dendritic cells, as well as B lymphocytes, take up and digest pathogens. These cells are antigen-presenting cells (APCs), and they function to present the digested peptide antigens to lymphocytes called T cells. T cells recognize the peptide antigen through antigen-specific **T-cell receptors (TCRs)** located on their surface. The TCRs on each T cell interact specifically with a *single peptide antigen*. Some T cells, the T_C (T-cytotoxic) cells, *directly* attack and destroy antigen-bearing cells. Other antigen-activated T cells, the T_H1 or T-helper 1 cells, act *indirectly* by secreting proteins called *cytokines* that activate other cells such as macrophages to destroy the antigen-bearing cells. This **cell-mediated immunity** leads to killing of pathogen-infected cells through recognition of pathogen antigens found on infected host cells.

Another subset of T cells, the T_H2 cells, interact with antigen-specific *B lymphocytes* or *B cells* and stimulate the B cells to make *antibodies* (also called *immunoglobulins*). Each B cell and its antibody product is specific for a single antigen. Antibodies are soluble proteins that interact specifically with antigens in the circulatory system or body fluids, where the antibodies neutralize antigens or target them for destruction. **Antibody-mediated immunity** is particularly effective against pathogens such as viruses and bacteria in the blood or lymph and also against soluble pathogen products such as toxins.

While innate immunity is directed against common pathogen features, specific immunity is directed to interactions with unique macromolecules. Specific immunity is defined by the properties of *specificity, memory*, and *tolerance*.

Specificity

The **specificity** of the antigen–antibody or antigen–T-cell interaction is unlike the innate host resistance mechanisms we have discussed. The innate host response challenges virtually any invading microorganism, even those pathogens the host has never before encountered. In the adaptive immune response, immunity cannot be detected for several days after the first contact with the pathogen. However, once the adaptive immune response is triggered, it is exclusively and specifically directed to the eliciting pathogen through recognition of unique molecular antigen features (Figure 22.8a●).

Memory

Once the immune system produces an antigen-specific antibody or T cell, further exposure to the same microorganism stimulates very rapid production of large quantities of antigen-reactive T cells or immunoglobulins. This capacity to respond more quickly and vigorously after further exposure or challenge with the eliciting antigen is known as **immunologic memory** (Figure 22.8b). Memory allows the host to specifically resist disease caused by previously encountered pathogens. We take advantage of immunologic memory by immunizing (inoculating, vaccinating) susceptible individuals with dead or weakened pathogens (or their products) to artificially stimulate and enhance immunity for a number of dangerous pathogens (see Section 22.11).

Tolerance

Tolerance is the acquired inability to make an adaptive immune response to an individual's own antigens. Tolerance is necessary because all macromolecules in the host

Specificity: Immune cells recognize and react with individual molecules (antigens) via direct molecular interactions.

Memory: The immune response to a specific antigen is *faster* and *stronger* upon subsequent exposure because the initial antigen exposure induced growth and division of antigen-reactive cells, resulting in multiple copies of antigen-reactive cells.

Tolerance: Immune cells are not able to react with self antigen. Self-reactive cells are destroyed during development of the immune response.

● **Figure 22.8 The adaptive immune response.** Key features of antibody and cell-mediated immunity are shown.

are potential antigens. Therefore, the immune system must "learn" to *not* recognize host macromolecules because host cells would be damaged if they were recognized by antibodies or T cells (Figure 22.8c). The adaptive immune response has the ability to discriminate between foreign (nonself and dangerous) antigens and host (self and not dangerous) antigens and react appropriately.

22.4 Concept Check

Nonspecific phagocytes present antigen to specific T cells, triggering the production of effector T cells and antibodies. Immune T cells and antibodies react directly or indirectly to neutralize or destroy the antigen. The adaptive immune response is characterized by *specificity* for the antigen, the ability to respond more vigorously when reexposed to the same antigen (*memory*), and the ability to discriminate self antigens from nonself antigens (*tolerance*).

♦ Identify the antigen-specific cells involved in the cell-mediated and antibody-mediated immune response.

◆ What effects would a breakdown of specificity, memory, or tolerance have on the ability to respond to a pathogen?

II ANTIGENS, T CELLS, AND CELLULAR IMMUNITY

The adaptive immune response has evolved to recognize a broad range of pathogen-derived molecules and structures. *Antigens*, the molecules recognized by the adaptive immune response, are first recognized by antigen-specific T cells. Here we first discuss antigens and then focus on T cells. These include antigen-specific reactions such as cell-mediated killing, inflammatory responses, and providing "help" for antibody-producing B cells. In the absence of T cells, there is no effective antigen-specific adaptive immunity.

22.5 Immunogens and Antigens

Antigens are substances that react with antibodies or T-cell receptors (TCRs). Most, but not all, antigens are **immunogens**, substances that induce an immune response. Here we examine the features of effective immunogens and then define the features of antigens that interact with antibodies and TCRs.

Intrinsic Properties of Immunogens

All immunogens must meet several molecular criteria to be effectively recognized by antibodies or TCRs. Immunogens must meet minimum values for the intrinsic properties of *size, complexity*, and *form*.

Molecular size is an important component of immunogenicity. For example, low-molecular-weight compounds called **haptens** cannot induce an immune response but can bind to antibodies. Because haptens are bound by antibodies, they are antigens even though they are not immunogenic. Haptens can include sugars, amino acids, and a wide variety of low-molecular-weight organic compounds. When coupled to a larger protein *carrier*, haptens become effective immunogens. Most immunogens have a molecular weight of 10,000 or greater. Thus, sufficient *molecular size* is an indication of potential immunogenicity.

Complex, nonrepeating polymers such as proteins are usually effective immunogens. Complex carbohydrates can also be very effective immunogens. In contrast, nucleic acids, simple polysaccharides, and lipids, because they are composed of repeating monomers, tend to be poor immunogens. Thus, the *molecular complexity* of a substance is another predictor of immunogenicity.

Complex macromolecules in insoluble or aggregated form (for example, proteins precipitated by heating) are usually excellent immunogens. The insoluble material is readily taken up by phagocytes, leading to an immune response. By contrast, the soluble form of the same molecule is often a very poor immunogen because the soluble molecule is not taken up by phagocytes. Thus, appropriate *physical form* is another condition of immunogenicity.

Extrinsic Properties of Immunogens

Although many substances are intrinsically immunogenic, several *extrinsic* factors also influence immunogenicity. These include the *dose* of the immunogen, the *route* of administration, and the *foreign* nature of the immunogen with respect to the host.

The *dose* of an immunogen administered to a host can be important for an effective immune response, but a broad range of doses ordinarily provides satisfactory immunity. In general, doses of 10 μg to 1 g are effective in most mammals. Doses of immunogen higher than 1 g or lower than 10 μg may not stimulate an immune response; extremely high or low doses may actually suppress a specific immune response and cause tolerance.

The *route* of administration of an immunogen is also important. Immunizations given by parenteral (outside of the gastrointestinal tract) routes, usually by injection, are normally more effective than those given topically or orally. Significant antigen degradation may occur before phagocyte contact when using oral or topical routes.

The final and most important extrinsic feature of an immunogen is that an effective immunogen must be *foreign* with respect to the host. The adaptive immune system recognizes and eliminates only foreign (nonself) antigens. Self antigens are not recognized and thus individuals are *tolerant* to their own self molecules, even though these same molecules have the capacity to be immunogens in other individuals of the same species.

Antigen Binding by Antibodies and T-Cell Receptors

The antibody or TCR does not interact with the antigenic macromolecule as a whole but only against a distinct portion of the molecule called an **antigenic determinant** or *epitope* (Figure 22.9●). Antigenic determinants may include sugars, amino acids, and other organic molecules.

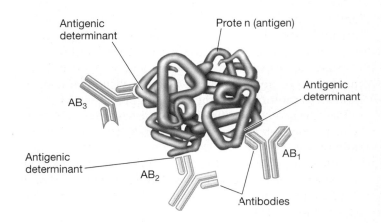

● **Figure 22.9 Antigens and antigenic determinants for antibodies.** Antigens may contain several different antigenic determinants, each capable of reacting with a specific antibody. The antigenic determinant recognized by AB₁ is a *conformational determinant* consisting of two different parts of the same polypeptide chain. The polypeptide chain is folded to bring two distant parts of the protein together to make a single determinant.

Antibodies interact with accessible surface epitopes on antigens. A sequence of four to six amino acids is the optimal size of an antigenic determinant on a protein. Thus, proteins, usually consisting of hundreds or even thousands of amino acids, can be thought of as arrays of overlapping antigenic epitopes. In some cases, antibodies may even recognize epitopes that are composed of amino acids from two portions of the molecule that are distant in terms of their primary structure, but are brought together by the secondary, tertiary, or quaternary structure of a macromolecule (Figure 22.9, ∞Sections 3.7 and 3.8). These *conformational determinants* add to the antigenic complexity of macromolecules. The surface of a bacterial cell or virus consists of a mosaic of proteins, polysaccharides, and other macromolecules, all with individual epitopes. Thus, the antigenic makeup of a typical microorganism, or even of a single macromolecule, is extremely complex.

While antibodies generally recognize epitopes expressed on macromolecular surfaces, TCRs recognize epitopes only after the immunogens have been partially degraded. This degraded or *processed* antigen is then presented to T cells on the surface of specialized APCs or target cells (see Section 22.5 and see Figure 22.12). Antigen processing normally destroys the conformational structure of an antigen, since the antigens are normally processed into peptides of less than 20 amino acids. As a result, T-cell epitopes consist of sequential *linear determinants* rather than the conformational determinants recognized by antibodies.

Antibody or TCR specificity is sensitive enough to distinguish between closely related epitopes. For example, antibodies can distinguish between the sugars glucose and galactose, which differ only in the orientation of a single hydroxyl group. However, *specificity is not absolute*, and an individual antibody or TCR may react to some extent with several different but structurally similar epitopes. The antigen that induced the antibody or TCR is called the *homologous antigen*, and the nonhomologous antigens that react with the antibody are called *heterologous antigens*. The interaction between an antibody or TCR and a heterologous antigen is called a *cross-reaction*.

22.5 Concept Check

Immunogens are foreign macromolecules that induce an immune response. Molecular size, complexity, and physical form are intrinsic properties of immunogens. When foreign immunogens are introduced into a host in an appropriate dose and route, they initiate an immune response. Antigens are molecules recognized by antibodies or TCRs. Antibodies recognize conformational determinants; TCRs recognize linear peptide determinants.

◆ Distinguish between *immunogens* and *antigens*.

◆ Identify the *intrinsic* and *extrinsic* features of an immunogen.

◆ Describe an antibody epitope and compare it to a TCR epitope.

22.6 Presentation of Antigen to T Lymphocytes

T lymphocytes interact specifically with antigens through their cell-surface T-cell receptor. At the molecular level, TCRs interact with peptide antigens bound by *major histocompatibility complex (MHC) proteins* on antigen presenting phagocyte cells or on infected target cells.

The T-Cell Receptor

The TCR is a membrane-spanning protein that extends from the T-cell surface into the extracellular environment. Each T cell has thousands of copies of the same TCR on its surface. A functional TCR consists of two proteins, an α chain and a β chain. Each of these chains has a variable (V) domain and a constant (C) domain (Figure 22.10●). The Vα and Vβ domains interact cooperatively to form a complete antigen-binding site. As we shall see (∞ Section 23.7), the adaptive immune response can generate TCRs that will bind nearly every known peptide antigen. Other antigens, such as complex polysaccharides, are not recognized by TCRs, but may be bound by the immunoglobulin receptors on B cells (see Section 22.10). TCRs can only recognize and bind a peptide antigen if the peptide is first bound to a *self* protein known as a *major histocompatibility complex (MHC) protein*.

Major Histocompatibility Complex (MHC) Proteins

MHC proteins are encoded by a genetic region called the **major histocompatibility complex (MHC)** found in all vertebrates. MHC proteins are produced by a number of genes in this complex and are collectively called *human leukocyte antigens* or HLAs. The MHC proteins were first discovered as the major target molecules for transplantation rejection. If tissues from a donor animal are immunologically rejected when transplanted to a recipient animal, then their MHC proteins are different. We now know that

● **Figure 22.10   Structure of the T-cell receptor (TCR).** The V domains of the α chain and β chain combine to form the peptide antigen-binding site.

MHC proteins function primarily as antigen-presenting molecules that interact with both the antigen and the TCR.

MHC genes encode both *class I* and *class II* MHC proteins. **Class I MHC proteins** are found on the surfaces of *all* nucleated cells. **Class II MHC proteins** are found only on the surface of B lymphocytes, macrophages, and dendritic cells, all of which are antigen-presenting cells (APCs). The differential cellular distribution of these molecules is linked to their individual functions.

Class I MHC proteins consist of two polypeptides (Figure 22.11*a*●), a membrane-embedded alpha (α) chain encoded in the MHC gene region, and a smaller protein called *beta-2 microglobulin* ($\beta_2 m$) encoded by a non-MHC gene. Class II MHC proteins consist of two noncovalently linked polypeptides called α and β. Like class I α chains, these polypeptides are embedded in the cytoplasmic membrane and project outward from the cell surface (Figure 22.10*b*).

MHC proteins are not structurally identical, even within a given species. Different individuals often show subtle differences in the amino acid sequence of their MHC proteins. These limited sequence variations are called *polymorphisms*. There are several hundred different MHC genes in humans. These MHC polymorphisms are the major antigenic barrier for tissue transplantation from one individual to another. As a result, tissue transplants not matched for MHC identity are recognized as nonself and are therefore rejected. The detailed molecular structure and genetic organization of the MHC genes and proteins are presented in Chapter 23.

Antigen Presentation

MHC proteins function as molecular reference points that permit T cells to identify foreign antigens. T cells continually sample the molecular landscape on the surface of other cells to identify cells carrying nonself antigens. The TCR on a given T cell binds only to MHC molecules having foreign antigens embedded in the MHC structure; a T cell does not recognize a foreign antigen unless it is presented in the context of an MHC protein.

How does this happen? A variety of host cells can acquire nonself antigens through infection or phagocytosis. The host cells then degrade (process) the antigens and load the processed antigen peptides into the MHC protein. The MHC–peptide complex is passed through the cytoplasmic membrane and presented on the surface of the cell to be recognized by T cells. Two distinct antigen-processing schemes are known, one for MHC I antigen presentation and one for MHC II antigen presentation (Figure 22.12●).

MHC I proteins present peptide antigens derived from pathogen proteins found in the cytoplasm of non-phagocytic cells (Figure 22.12*a*). The foreign pathogen proteins result from intracellular infections by viruses and other intracellular pathogens. Proteins derived, for example, from infecting viruses, are taken up and digested in the cytoplasm in a structure called the *proteosome*. Peptide antigens of about 10 amino acids in length are transported via an energy-dependent reaction into the endoplasmic reticulum (ER) through a pore formed by two proteins, called the *transporters associated with antigen processing (TAP)*. Once the peptides have entered the ER, they are bound by the MHC I protein, held in place near the TAP site by a group of *chaperone* proteins. These chaperone proteins direct the MHC I molecule to the TAP site, holding it in place until a peptide is bound. The MHC I–peptide complex is then released from the chaperones and moves to the cell surface where it integrates into the membrane and can be recognized by T_C cells.

Thus, the MHC proteins act as a *platform* on which the foreign viral antigen is bound. Next, the TCR on the surface of a T cell interacts with both antigen (nonself) and MHC protein (self) on the surface of the target cell. This T cell–target cell interaction induces specialized T-cytotoxic (T_C) cells to produce cytotoxic proteins called *perforins* (see Section 22.6) that kill the virus-infected target cell.

A second antigen presentation scheme involves the MHC class II proteins (Figure 22.12*b*). MHC II proteins are expressed exclusively in the phagocytic APCs, where they function to present peptides from engulfed extracellular pathogens such as bacteria. MHC class II proteins are initially assembled in the ER, much like MHC I proteins. However, unlike MHC I proteins, a chaperone protein called *Ii*, or *invariant chain*, binds to the MHC II protein, blocking the MHC II peptide-binding site and preventing the MHC II proteins from binding peptides inside the ER.

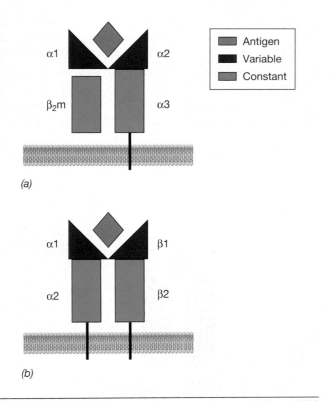

(a)

(b)

● **Figure 22.11 Structures of the MHC proteins.** (a) Class I MHC protein. The α1 and α2 domains interact to form the peptide antigen-binding site. (b) Class II MHC protein. The α1 and β1 domains combine to form the peptide antigen-binding site

(a)   *(b)*

● **Figure 22.12** **Antigen presentation by MHC I and MHC II proteins.** (a) In the MHC I antigen presentation pathway, the membrane-bound MHC I proteins are made and assembled in the endoplasmic reticulum. Chaperone proteins stabilize MHC I until antigen is bound. ① Protein antigens manufactured within the cell, for instance from viruses, are degraded by the proteosome in the cytoplasm and transported across the endoplasmic reticulum membrane through a pore formed by the TAP proteins. ② The peptides then bind to MHC I, are transported to the cell surface, and ③ interact with T-cell receptors (TCRs) on the surface of T_C cells. ④ The CD8 coreceptor on the T_C cell engages the class I MHC, resulting in a stronger complex. The T_C cells then release cytokines and cytotoxins, killing the target cell. Any nucleated cell can act as a target cell for T cells recognizing peptide–MHC I complexes. (b) In the MHC II antigen presentation pathway, ①️ MHC II proteins are produced in the endoplasmic reticulum and are assembled with a blocking protein, Ii (invariant chain), preventing MHC II from complexing with peptides found in the endoplasmic reticulum. ② Lysosomes containing MHC II then fuse with phagosomes, forming phagolysosomes where the Ii and foreign proteins, imported from outside the cell by endocytosis, are digested. ③ The MHC II protein then binds to the digested foreign peptides, and the complex is transported to the cell surface ④ where it interacts with TCRs and ⑤ the CD4 coreceptor on T_H cells. The T_H cells then release cytokines that act on other cells to activate an immune response. Only APCs can be targets for T cells recognizing the peptide–MHC II complex. The APCs are macrophages, dendritic cells, and B cells.

These MHC II-Ii complexes are transported from the ER to the cell vacuoles (lysosomes) (see Section 22.2). The phagosome containing the foreign antigen fuses with the lysosome to form a phagolysosome, and the foreign antigens, as well as the Ii peptide, are digested by lysosomal enzymes. The foreign peptides, generally about 11 to 15 amino acids in length (slightly larger than MHC I-binding peptides), are then bound in the newly opened MHC II antigen-binding site. The complex is then transported to the cytoplasmic membrane, where it is presented to specialized T-helper (T_H) cells. The T_H cells, through the TCR, recognize the MHC II-peptide complex. This interaction activates the T_H cells to secrete cytokines that either stimulate antibody production by specific B-cell clones or cause inflammation (see Sections 22.8 and 22.10).

CD4 and CD8 Coreceptors

In addition to the TCR, each T cell expresses a unique cell surface protein that functions as a *coreceptor*. T_H cells express a CD4 protein coreceptor, while T_C cells express a CD8 protein coreceptor (Figure 22.12). When the TCR binds to the peptide–MHC complex, the coreceptor on the T cell also binds to the MHC protein, strengthening the molecular interactions and enhancing activation of the T cell. CD4 binds only to the class II protein, strengthening T_H cell interaction with APCs that express MHC II protein. Likewise, CD8 binds only to the MHC I protein, enhancing the binding of T_C cells to MHC I-bearing target cells. The CD4 and CD8 proteins are used *in vitro* as markers to differentiate T_H cells from T_C cells (∞Section 24.9).

22.6 Concept Check

T cells recognize antigens presented by APCs or by pathogen-infected cells. At the molecular level, TCRs bind peptide antigens presented by MHC proteins. These molecular interactions stimulate T cells to kill antigen-bearing cells or to produce cell-stimulating proteins known as cytokines.

◆ Identify the cells that display MHC I and MHC II proteins as surface components.

◆ Define the process of antigen presentation for both intracellular and extracellular pathogens.

22.7 T-Cytotoxic Cells and Natural Killer Cells

In the previous section, we introduced two sets of T lymphocytes, the *T-cytotoxic cells* and the *T-helper cells*. Here we examine the cell-killing function of the T-cytotoxic cells in detail. We also introduce the *natural killer cell*, a cell that also recognizes and kills cells infected with intracellular pathogens.

T-Cytotoxic Cells

T-cytotoxic cells (T_C), also known as cytotoxic T lymphocytes (CTLs), directly mediate the destruction of cells that display surface foreign antigens. As we discussed, T_C cells recognize foreign antigens embedded in MHC I proteins. A cell displaying a foreign antigen embedded in MHC I, such as viral peptides displayed on virus-infected cells, can be killed by T_C cells (see Section 22.6).

Contact between a T_C cell and the target cell is required for cell death. The contact is initiated by the TCR: peptide–MHC I complex (Figure 22.12a). On contact with the target cell, granules in the T_C cell migrate to the contact site, where the contents of the granules are released (*degranulation*). The granules contain *perforin*, which enters the membrane of the target cell and forms a pore. In addition to the perforins, T_C granules contain *granzymes*, proteins that cause *apoptosis*, or programmed cell death. When granzymes enter the target cell through the pores created by perforins, the target cell undergoes apoptosis, characterized by death and degradation of the cell from within (Figure 22.13a●). The T_C cells, however, remain unaffected; their membranes are not damaged by perforin. The T_C cells kill only those cells displaying the foreign antigen because degranulation occurs only at the contact surface between the T_C and the antigen-bearing target cell. Cells lacking the antigen recognized by the T_C cells do not make contact and are not killed.

Natural Killer Cells

Natural killer cells (NK cells) are a class of lymphocytes that are distinct from cytotoxic T cells. However, NK cells also play a role in destroying cells infected with intracellular pathogens. NK cells are neither T cells nor

● **Figure 22.13 Effector T cells.** (a) T-cytotoxic cells, or T_C cells, are activated by antigens presented on any cell in the context of MHC I protein. The T_C cells respond by releasing granules that contain perforin and granzymes, cytotoxins which perforate the target cell and cause apoptosis, respectively. (b) T-inflammatory cells, or T_H1 cells, are activated by antigens presented on macrophages in the context of MHC II protein. The T_H1 cells respond by producing cytokines that stimulate the macrophages to increase phagocyte activity and promote inflammation.

B cells. Their numbers are not enhanced, nor do they exhibit memory after stimulation. Nevertheless, NK cells resemble T_C cells in their ability to destroy infected cells. For example, NK cells also use perforin and granzymes to kill their targets. However, NK cells differ from T_C cells in that they kill targets in the *absence* of a specific antigen. NK cells are capable of destroying malignant and virus-infected cells *in vitro* without previous exposure or contact with the foreign antigen.

The molecular target for NK cells seems to be the *lack* of appropriate MHC class I proteins. NK cells recognize normal cells and their MHC I protein through a set of special MHC I receptors. Binding of the receptors to MHC I *deactivates* the killing mechanism; in the absence of binding, the NK cell kills the unrecognized target. The main targets of NK cells, tumor cells, and virus-infected cells often have reduced or altered MHC I protein expression.

 22.7 Concept Check

T-cytotoxic (T$_C$) cells recognize antigens on virus-infected host cells and tumor cells through antigen-specific TCRs. Antigen-specific recognition triggers killing via perforin and granzymes. Natural killer (NK) cells use the same effectors to kill virus-infected cells and tumors. However, NK cells do not require stimulation, nor do they exhibit memory. NK cells respond in the *absence* of MHC proteins.

◆ Identify and compare the target recognition mechanisms used by T$_C$ and NK cells.

◆ Describe the common effector system (the cell-killing system) used by T$_C$ and NK cells.

22.8 T-Helper Cells: Activating the Immune Response

Here we will define the functions of the T-helper cells and their pivotal role in producing cytokines to foster an effective immune response. T-helper cells are divided into two subsets, T$_H$1 cells and T$_H$2 cells. T$_H$1 cells play a role in macrophage activation and inflammation and T$_H$2 cells interact with B lymphocytes and stimulate antibody production.

T$_H$1 Cells and Macrophage Activation

Macrophages play a central role as APCs in both antibody-mediated and cell-mediated immunity. As was illustrated in Figure 22.12*b*, macrophages bind, process, and present antigen to T$_H$ cells. As phagocytic cells, however, macrophages also take up and kill certain foreign cells by themselves, an ability stimulated by T$_H$1 cells. T$_H$1-activated macrophages can kill intracellular bacteria that normally multiply in unactivated macrophages or other cell types. As we have noted (see Section 22.2), some bacteria survive and multiply within macrophages, whereas most bacteria taken into macrophages are killed and digested. Bacteria multiplying within macrophages include *Mycobacterium tuberculosis, M. leprae* (causal agents of tuberculosis and leprosy, respectively), and *Listeria monocytogenes* (causal agent of listeriosis, ⬯Section 29.10). Animals given a moderate dose of *M. tuberculosis* are able to overcome the infection and develop resistance because of the T cell-mediated immune response. The cells involved are the T-inflammatory cells, the T$_H$1 subset. They activate macrophages and other nonspecific phagocytes by secreting a

number of cytokines, including IFN-γ (interferon gamma), GM-CSF (granulocyte-monocyte colony-stimulating factor), and TNF-α (Figure 22.13*b*). Surprisingly, such immunized animals also phagocytize and kill unrelated organisms such as *Listeria*, because macrophages in the immunized animal have been activated so that they more readily kill any secondary invader.

T$_H$1-activated macrophages not only kill foreign pathogens but are also involved in the destruction of foreign mammalian cells. For example, tumor cells often have specific antigens not found on normal cells. Most tumor cells are recognized as foreign and are destroyed primarily by macrophages activated by cytokines produced by tumor-specific T$_H$1 cells. Transplantation rejection, also mediated by macrophages activated by T$_H$1 cells, is a major problem encountered when transplanting organs and tissues from one person to another. In this case, T$_H$1 cells recognize the MHC proteins of the transplant, triggering macrophage activation and transplant destruction.

T$_H$2 Cells and B Cell Activation

T$_H$2 cells play a pivotal role in B cell activation and antibody production. As discussed in Section 22.3, B cells make antibodies. How are T$_H$2 cells involved? Mature B cells are coated with antibodies that act as antigen receptors. However, when antigen binds to the B cell antigen receptors, the B cell does not immediately produce soluble antibodies. Instead the B cell functions as an APC and interacts with a T$_H$2 cell. More specifically, the bound antigen is first endocytosed and degraded in the B cell. Peptides from the degraded antigen are then presented on a MHC II protein to a T$_H$2 cell (Figures 22.12*b* and 22.14●). The T$_H$2 cell then responds by producing IL-4 (*interleukin-4*) and IL-5, cytokines that stimulate production and secretion of soluble antibodies by the B cell, as we will discuss in Section 22.10. Thus, the T$_H$2 cell is a "helper cell" that activates antibody production by B cells.

 22.8 Concept Check

T$_H$1 and T$_H$2 cells play pivotal roles in cell-mediated and antibody-mediated immune responses. T$_H$1 inflammatory and T$_H$2 helper cells each stimulate effector cells through the action of cytokines.

◆ Describe the role of T$_H$1 cells in activation of macrophages.

◆ Describe the role of T$_H$2 cells in activation of B cells.

III ANTIBODIES AND IMMUNITY

Here we concentrate on the role of B cells and antibodies in immunity. Antibodies are important components of an effective immune response and provide antigen-specific immunity to extracellular pathogens and dangerous soluble proteins such as toxins. After considering

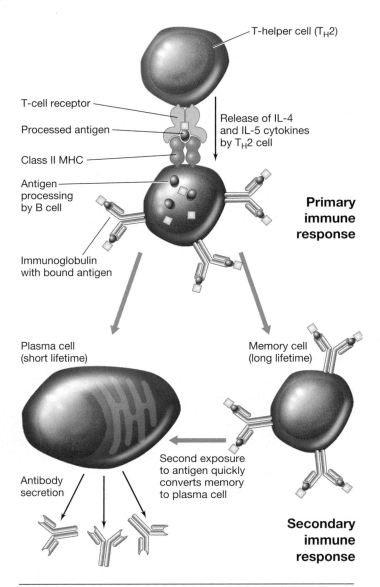

● **Figure 22.14   T cell–B cell interaction and antibody production.** The B cell functions as an antigen-presenting cell (APC), first recognizing antigen using antigen-specific immunoglobulin receptors. After processing, antigen is presented to the T_H2 cell by a class II MHC molecule. The T_H2 cell then signals the same B cell to proliferate and form plasma cells (antibody producers) or memory cells. After subsequent antigen exposure, memory cells quickly convert to plasma cells.

the molecular nature of antibodies, we will look at the production of antibodies by B cells. We will conclude by discussing the ability of antibodies to neutralize or destroy antigens *in vivo*.

22.9   Antibodies (Immunoglobulins)

Antibodies, or **immunoglobulins** (Ig), are protein molecules that are able to combine with antigenic determinants. They are found in the serum and other body fluids such as gastric secretions and milk. Serum containing antigen-specific antibody is called antiserum. Immunoglobulins (Igs) can be separated into five major classes on the basis of

their physical, chemical, and immunological properties: *IgG, IgA, IgM, IgD*, and *IgE* (Table 22.2). In most individuals, about 80% of the serum immunoglobulins are IgG proteins, and these have been studied extensively.

Immunoglobulin G Structure

IgG is the most common circulating antibody. IgG has a molecular weight of about 150,000 and is composed of four polypeptide chains (Figure 22.15●). Interchain disulfide bridges (S–S bonds) connect the individual chains. In an individual IgG protein, two identical light chains of 25,000 molecular weight are paired with two identical heavy chains of 50,000 molecular weight. Each light chain has about 220 amino acids, and each heavy chain has about 440 amino acids. A functional IgG antibody consists of two antigen-binding units, each unit consisting of one heavy chain and one light chain. Therefore, antibodies are *bivalent* and can bind two identical epitopes.

Light Chains

Each IgG light chain consists of two protein domains of equal size (Figure 22.15). The amino-terminal region is a *variable* **domain**, meaning that the amino acid sequence in

● **Figure 22.15   Immunoglobulin G structure.** IgG consists of two heavy chains (50,000 molecular weight) and two light chains (25,000 molecular weight), with a total molecular weight of 150,000. One heavy and one light chain interact to form an antigen-binding unit. The variable domains of the heavy and light chain (V_H and V_L) bind antigen and show sequence differences in each different immunoglobulin. The constant domains (C_H1, C_H2, C_H3) are identical in all IgG proteins. The chains are covalently joined by disulfide bonds.

Table 22.2 **Properties of human immunoglobulins**

Class/ H chain[a]	Molecular weight/ formula[b]	Serum (mg/ml)	Antigen-binding sites	Properties	Distribution
IgG γ	150,000 2(H + L)	13.5	2	Major circulating antibody; four subclasses: IgG$_1$, IgG$_2$, IgG$_3$, IgG$_4$; IgG$_1$ and IgG$_3$ activate complement	Extracellular fluid; blood and lymph; crosses placenta
IgM μ	970,000 (pentamer) 5[2(H + L)] + J	1.5	10	First antibody to appear after immunization; strong complement activator	Blood and lymph; monomer is B cell-surface receptor
	175,000 (monomer) 2(H + L)	0	2		
IgA α	150,000 2(H + L)	3.5	2	Important circulating antibody	Secretions (saliva, colostrum, cellular and blood fluids); monomer in blood and dimer in secretions
	385,000 (secreted dimer) 2[2(H + L)] + J + SC	0.05	4	Major secretory antibody	
IgD δ	180,000 2(H + L)	0.03	2	Minor circulating antibody	Blood and lymph; B lymphocyte surfaces
IgE ε	190,000 2(H + L)	0.00005	2	Involved in allergic reactions; C$_H$4 contains mast cell binding fragment	Blood and lymph; binds to mast cell surfaces

[a] All immunoglobulins may have either λ or κ light chain types, but not both.

[b] Based on the number and arrangement of heavy (H) and light (L) chains in each functional molecule. J is a joining protein present in serum IgM and secretory IgA. SC is the secretory component found in secreted IgA. See Figures 22.15, 22.17, 22.18, and 22.19 for a diagram of each immunoglobulin.

this structural region differs in each different antibody. The variable domain is involved in antigen binding. By contrast, the carboxy-terminal domain is a *constant domain*. The amino acid sequence in this domain does not differ among light chains of the same type.

Heavy Chains

Each IgG heavy chain has an amino-terminal variable domain and three constant domains of about 110 amino acids each. As is the case for light chains, the three constant domains of each heavy chain are identical in antibodies of the same class, while the variable region differs in each antibody.

The Antigen–Binding Site

The antigen-binding site of IgG and all other Igs is formed by cooperative interaction between the variable regions of both heavy and light chains (Figure 22.16●). The variable regions of both chains interact to form a molecular site that binds strongly but noncovalently with the antigen. The strength of this binding is termed *antigen-binding affinity*. An antibody is said to bind with

high affinity if the antigen interacts strongly and is held tightly by the antibody.

Each individual's immune system has the capacity to recognize, or bind, countless antigens, and each antigen is recognized by a unique antigen-binding site. To accommodate all possible antigens, each individual can produce billions of different antigen-binding sites in the form of antibodies. How is this diversity in the antigen-binding site generated? As we discuss in Section 22.10 and the next chapter, new antibodies are constantly created by randomly recombining and mutating the 300 or more genes that encode the light and heavy chain variable domains. Each Ig gene construct encodes a unique antigen-binding Ig expressed on a B cell. Antigen interaction with these unique B-cell Ig stimulates the B cell to produce copies of the Ig, secreted as soluble antibodies (∞Section 23.5).

Other Classes of Immunoglobulins

Immunoglobulins of the other classes differ from IgG. The heavy-chain *constant* domains of a given Ig molecule define its class and can have one of five amino acid sequences: gamma (γ), alpha (α), mu (μ), delta (δ), or ep-

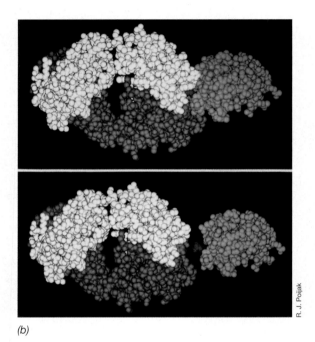

(a)

Richard J. Feldman

(b)

R. J. Poljak

● **Figure 22.16 Immunoglobulin structure and the antigen-binding site.** (a) Three–dimensional view of an IgG molecule. The heavy chains are shown in red and dark blue. The light chains are in green and light blue. (b) Structure of the combining site of an antigen and Ig. The antigen (lysozyme) is in green. The variable region of the Ig heavy chain is shown in blue, and the light chain in yellow. The amino acid shown in red is a glutamine residue of lysozyme. The glutamine residue fits into a pocket on the Ig molecule, but the overall antigen-antibody recognition involves contacts made between several other amino acids on both the Ig and the antigen. Reprinted with permission from *Science* 233:747 (1986) ©AAAS.

silon (ε). The constant domain sequences constitute the carboxy-terminal three-fourths of the heavy chains of Igs of IgG, IgA, and IgD, respectively, and four-fifths of the heavy chains of IgM and IgE (Figure 22.17●). Each antibody of the IgM class, for example, contains amino acids in its heavy chain constant domains that constitute the *mu* sequence.

In a typical immune response, two Igs of different classes, often IgM and IgG, may bind the same epitope. In this case, the *variable* (antigen-binding) domains of the heavy and light chains from both IgM and IgG may be identical, but the class-determining *constant* domains of the

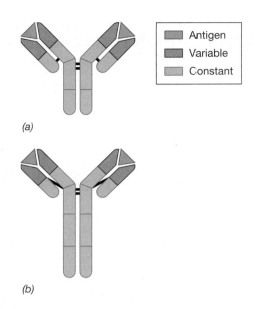

(a)

(b)

● **Figure 22.17 Immunoglobulin classes.** All classes of Igs have V_H and V_L domains (red) that bind antigen (brown). (a) IgG, IgA, and IgD have three constant domains (blue). (b) IgM and IgE each have a fourth constant domain.

heavy chains are different, with IgM expressing a mu (μ) heavy chain and IgG expressing a gamma (γ) heavy chain.

The structure of IgM is shown in Figure 22.18●. IgM is usually found as an aggregate of five immunoglobulin molecules attached by at least one *J chain*. Each heavy chain of IgM contains a fourth constant domain

J chain

● **Figure 22.18 Immunoglobulin M.** IgM is found in serum as a pentameric protein consisting of five IgM proteins covalently linked to one another via disulfide bonds and a J chain protein. Because it is a pentamer, IgM can bind up to 10 antigens, as shown.

(C_H4). IgM is the first class of Ig made in a typical immune response to a bacterial infection, but IgMs are generally of low affinity. Antigen-binding strength is enhanced to some degree, however, by the high *valence* of the pentameric IgM molecule; 10 binding sites are available for interaction with antigen (Table 22.2 and Figure 22.18). The term *avidity* is used to describe the *strength of binding* by multivalent antigen-binding molecules. Thus, IgM has *low* affinity but *high* avidity for antigen. About 10% of serum antibodies are IgM. IgM monomers are also found on the surface of B cells, where they bind antigen.

IgA, in the dimeric form (Figure 22.19●), is present in body fluids such as saliva, tears, breast milk colostrum, and mucosal secretions from the gastrointestinal, respiratory, and genitourinary tracts. All of these mucosal surfaces are associated with mucosal-associated lymphoid tissue (MALT) (Figure 22.2) that produces IgA. These mucosal surfaces total about 400 m^2, and large amounts of secretory IgA are produced—about 10 g per day. By contrast, the serum IgG produced in an individual is about 5 g per day. Thus the total amount of *secretory* IgA produced by the body is higher than the amount of *serum* IgG. The secretory form consists of two IgA molecules covalently linked by a J (joining) chain peptide and a protein *secretory component* that aids in transport of IgA across membranes (Figure 22.19). IgA is also present in serum as a monomer (Table 22.2).

IgE is found in serum in extremely small amounts (about 1 of every 50,000 serum Ig molecules is IgE). Despite its low concentration, IgE is important because immediate-type hypersensitivities (allergies) are mediated by IgE (see Section 22.15). The molecular weight of an IgE molecule is significantly higher than most other Igs (Table 22.2) because, like IgM, IgE has a fourth constant domain (Figure 22.17). This additional constant region functions to bind IgE to mast cell surfaces (see Figure 22.25), an important prerequisite for allergic reactions.

IgD, present in serum in low concentrations, has no known function. However, IgD, with IgM, is abundant on the surfaces of B cells, and IgD is especially abundant on memory B cells. Thus, IgD surface Igs may be important for the secondary antibody response.

● **Figure 22.19   Immunoglobulin A.** Secretory IgA (sIgA) is often found in body secretions as a dimer consisting of two IgA proteins covalently linked to one another via a J chain protein. A secretory protein component, not shown, aids in transport of IgA across membranes.

 22.9   Concept Check

Ig (antibody) proteins consist of four chains, two heavy and two light. The antigen-binding site is formed by the interaction of variable regions of heavy and light chains. Each class of Ig has different structural and functional characteristics.

◆   Identify the Ig domains involved in antigen binding.

◆   Differentiate between Ig classes using functional and structural characteristics.

22.10   Antibody Production

In this section, we will examine a typical antibody response, first reviewing the complex cell interactions necessary to produce effective antigen-specific immunity. We will then look briefly at the unique genetic mechanism used to generate antibody diversity. Finally, we will review the salient features leading up to and including antibody production in response to antigen exposure.

T Cell–B Cell Interactions

The production of immunoglobulins in response to antigen involves interactions between T cells and B cells through their respective antigen-specific cell surface molecules, the TCR on the T cell, and the surface Ig on the B cell (Figure 22.14). Each step in the antibody production mechanism is *highly* specific. Upon initial antigen exposure, the antigen is bound by a B cell with an antigen-specific surface Ig. The antigen-Ig complex is then internalized, and the antigen is processed and loaded onto MHC II proteins. Thus, the B cell first functions as an antigen-presenting cell, using its surface Ig to capture antigen. The MHC–antigen complex then moves to the cell surface, displaying the antigen for interaction with a T_H2 cell having an antigen-specific TCR on its surface (see Figure 22.12*b*). Binding of the B-cell MHC–antigen complex activates the antigen-specific T_H2 cell, stimulating cytokine production. The released cytokines stimulate the nearby antigen-specific B cells, activating them to produce and secrete the antigen-specific antibody.

The Genetic Mechanism of Antibody and TCR Diversity

Each individual is capable of producing billions of different antibodies. How does the immune system choose the correct antibody for each antigen? Early theories postulated that individual genes encoded each antibody, but we now know that only a small number of genes are used to encode this immense antibody diversity. Antibody production starts with stepwise rearrangements of the genes necessary to encode an Ig. During development of lymphocytes in the bone marrow, both heavy and light chain gene rearrangements occur in B cells. The genes are reassorted—individual gene pieces are mixed and matched in a variety of combinations—by gene splicing and rearrangements in the maturing lymphocytes, a process known as *somatic recombination*. Figure 22.20● shows a typical rearrangement and expression pattern for the human κ light chain. The heavy chain genes rearrange in an analogous, but more complex fashion. In each mature B cell, the final result is a single functional heavy chain gene and a single functional light chain gene. Each of these rearranged genes is transcribed and expressed as an Ig antigen receptor protein on the surface of the B cell. Upon antigen exposure, as we have seen, soluble Ig is produced by the B cell. In addition, antigen induces genetic hypermutation in the Ig genes, further modifying and diversifying the expressed antibodies. The large number of possible gene rearrangements, coupled with the somatic mutations after antigen exposure, allow almost unlimited diversity in Igs (∞Section 23.6).

Similar rearrangements also occur during T-cell development and result in the generation of considerable diversity in T-cell receptors. However, T cells do not use the hypermutation mechanism to expand diversity (∞Section 23.8).

Antibody Production and Immune Memory

Starting with a B cell that expresses a surface Ig, antibody production begins with antigen exposure and culminates with the production and secretion of an antigen-specific antibody according to the following sequence:

1. Antigens are spread via the lymphatic and blood circulatory systems to neighboring secondary lymphoid organs such as lymph nodes, spleen, or mucosal-associated lymphoid tissue (MALT) (Section 22.1 and Figure 22.2). The route of antigen exposure influences the nature of the antibodies produced. Intravenously injected antigen travels via the blood to the spleen, where mostly IgM and IgG and serum IgA antibodies are formed. If antigen is introduced subcutaneously, intradermally, topically, or intraperitoneally, the lymphatic system carries antigen to the nearest lymph nodes, again stimulating production of mostly IgM, IgG, and serum IgA. Antigen introduced to mucosal surfaces is delivered to the nearest MALT. For example, antigen delivered by mouth is delivered to the MALT in the intestinal tract, preferentially stimulating specific antibody production in the gut, with a primary IgM response and a secondary antigen-specific secretory IgA response.

2. Following the initial antigen exposure, each antigen-stimulated B cell multiplies and differentiates to form both antibody-secreting **plasma cells** and **memory cells** (Figure 22.14). *Plasma cells* are relatively short-lived (less than 1 week), but produce and secrete large amounts of IgM antibody in this **primary antibody response** (Figure 22.21●). There is a latent period before specific antibody appears in the blood, followed by a gradual increase in *antibody titer* (antibody quantity), and then a slow fall in the primary antibody response.

3. The **memory B cells** generated by the initial exposure to antigen may live for years. If reexposure to the immunizing antigen occurs at a later time, memory cells need no T-cell activation; they quickly transform to plasma cells and begin producing IgG. Upon reexposure, the antibody titer rises rapidly to a level 10–100 times greater than the titer achieved following the first exposure. This rise in antibody titer is referred to as the **secondary antibody response** (Figure 22.21). The secondary response is the result of *immunologic memory* and produces a more rapid, more abundant antibody response when compared with the primary response. The secondary response is also characterized by a switch from IgM to IgG production, a phenomenon called *class switching* (Figure 22.21).

4. Over time the titer slowly decreases, but subsequent exposure to the same antigen can cause another secondary response. The secondary response is the basis for the immunization procedure known as a "booster shot" (for example, the yearly rabies shot given to do-

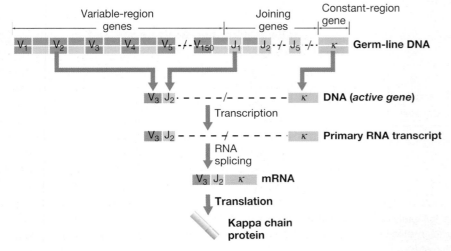

● **Figure 22.20 Immunoglobulin kappa chain gene rearrangement in human B cells.** The gene segments are arranged in tandem in the kappa (κ) light-chain complex on chromosome 2. DNA rearrangements are completed in the maturing B cell. Any one of the 150 V (variable) sequences may combine with any one of the 5 J sequences. Thus, 750 (150 V × 5) recombinations are possible, encoding 750 distinct kappa chains, but only one rearrangement actually occurs in each cell. An analogous process occurs in the heavy chain genes in each B cell. The heavy chain gene complex has even more gene segments, allowing greater numbers of recombinations and potential heavy chains, but again only one heavy chain rearrangement is found in each mature B cell. The mature Ig protein is made by combining two light chain proteins with two heavy chain proteins after translation (∞Section 23.6).

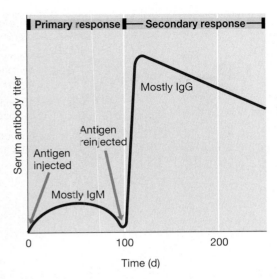

● Figure 22.21 Primary and secondary antibody responses in serum. The antigen injected at day 0 and day 100 must be identical to induce the secondary response. The secondary response, also called a booster response, may be more than tenfold greater than the primary response. Note the class switch from IgM production in the primary response to IgG production in the secondary response.

mestic animals). Periodic reimmunization maintains high levels of circulating antibody specific for a certain antigen and provides long-term active protection against individual infectious diseases (see Section 22.12).

 22.10 Concept Check

Antibody production is initiated by antigen contact with an antigen-specific B cell that processes the antigen and presents it to an antigen-specific T$_H$2 cell. The activated T$_H$2 cell then signals the antigen-specific B cell to produce antibody. Activated B cells live for years as memory cells and can rapidly produce large quantities (high titers) of antibodies upon re-exposure to antigen.

◆ How do B cells act as APCs?

◆ How do T$_H$2 cells activate antigen-specific B cells?

◆ Explain the rationale for a "booster" immunization.

22.11 Complement, Antibodies, and Pathogen Destruction

A related group of sequentially interacting proteins called **complement** play an important role as an effector in both innate and adaptive immunity. Complement proteins are often activated by interactions with antigen–antibody complexes, but can also be activated by innate mechanisms in the absence of specific antibodies. Complement proteins can react sequentially to cause *lysis* of bacterial cells or can label antigen-expressing cells for

enhanced recognition by phagocytes, accelerating destruction of the antigens.

Classical Complement Activation and Cell Damage

Complement is made up of a number of proteins, many with enzymatic activity. These proteins are activated in a sequential, ordered fashion by antigen–antibody complexes on bacterial cells. The result is cell membrane damage, leakage of contents, and possibly lysis of the bacterial cell. Complement proteins are found at comparable levels in the serum of all individuals. A major function of antibody is to recognize invading cells and activate the complement system for attack. Thus, many antibodies, each specific for a single antigen, can recruit the ubiquitous complement proteins; the immune system does not need individual enzymes to attack each invading agent.

The individual proteins of complement are designated C1, C2, C3, and so on. Classical activation of complement occurs only with antibodies of the IgG and IgM classes (see Table 22.2). When IgG or IgM antibodies bind antigens, especially on cell surfaces, the antibodies fix (bind) the ever-present complement proteins (Figure 22.22●). The complement proteins react in a sequential cascade, with activation of one complement component leading to activation of the next, and so on. The key steps start with (1) binding of antibody to antigen (initiation); (2) binding of C1 components (C1q, C1r, and C1s) to the antibody–antigen complex, leading to C4–C2 binding at an adjacent membrane site, and activation and binding of C3; (3) membrane-bound C3 catalyzes formation of a C5-C6-C7 complex at a second membrane site, and C8 and C9 are then deposited with the C5–C6–C7 complex, resulting in membrane damage and cell lysis (Figure 22.23●). The C5-9 components are known as the *membrane attack complex (MAC)* because they interact at a single membrane site to form a pore.

By-products of complement activation include chemoattractants called *anaphylatoxins*. They cause inflammatory reactions at the site of complement deposition (see Section 22.15). Reactions involving C3 result in chemotactic attraction and activation of phagocytes, resulting in increased phagocytosis. Reactions involving C5 lead to T-cell attraction and cytokine release.

Complement is a bactericidal and lytic agent for many gram-negative *Bacteria* when used in conjunction with specific antibodies. However, gram-positive *Bacteria* are not killed by complement, even in the presence of specific antibodies. Gram-positive *Bacteria* can, however, be destroyed through opsonization.

Opsonization

When antibody binds antigen on the surface of a cell, the cell is much more likely to be phagocytized. When complement binds an antibody–antigen complex on a cell surface, the cell is even more likely to be phagocytized. This is because most phagocytes, including macrophages and B cells, have antibody receptors (*FcR*) as well as C3 receptors

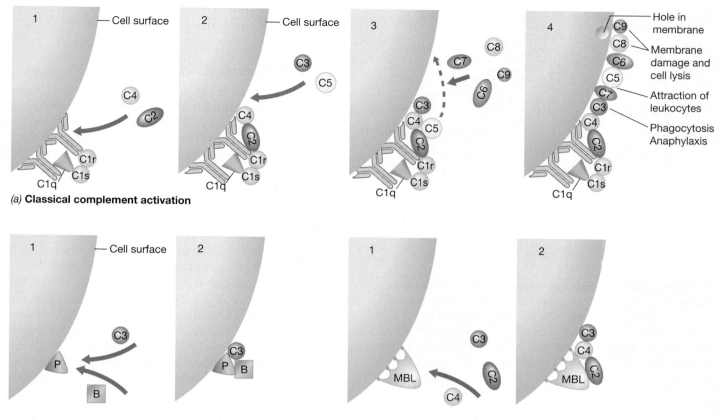

(a) **Classical complement activation**

(b) **Alternate pathway activation by Properdin (P)**

(c) **Mannan-binding lectin (MBL) pathway activation**

(d)

● **Figure 22.22 Complement activation.** (a) The sequence, orientation, and activity of the components of the classical complement pathway as they interact to lyse a cell. Panel 1: Binding of the antibody and the C1 protein complex (Clq, Clr, and Cls). Panel 2: the C42 complex interacts with C3. Panel 3: the C423 complex activates C5 which binds an adjacent membrane site. Panel 4: Sequential binding of C6, C7, C8, and C9 to C5, producing a pore in the membrane. C5-9 is the membrane attack complex (MAC). (b) The alternate pathway. Properdin (P) binds to the membrane, followed by the C3B complex. The complex activates C5, as in panel 3 above, and initiates formation of the MAC (panel 4 above) (c) The mannan-binding lectin (MBL) pathway. MBL binds to the mannose on the bacterial membrane, and anchors formation of a C423 complex. The complex activates C5, as in panel 3 above, and initiates formation of the MAC (panel 4 above) (d) A schematic view of the membrane pore formed by complement components C5 through C9.

(*C3R*). These receptors bind the antibody constant domain and C3 complement protein, respectively. Normal phagocytic processes are enhanced about 10-fold by antibody binding and amplified another 10-fold by C3 fixation (binding). This enhancement of phagocytosis by antibody or complement binding is called *opsonization*.

Gram-positive *Bacteria* can be opsonized in this way, leading to enhanced phagocytosis and pathogen destruction. But, as was the case with gram-negative *Bacteria*, antibodies must first be bound to surface antigens on the gram-positive cell to activate the classical complement pathway and promote opsonization.

Non-Antibody Dependent Complement Activation

The *alternate pathway* is a nonspecific method of complement activation using many of the classical complement pathway components to induce opsonization and activate the C5-9 MAC. The alternate pathway activates complement through the actions of several serum proteins not associated with the classical complement pathway (Figure 22.22b). In place of antibodies, the alternate pathway uses unique alternate pathway proteins to target molecules on the surface of both gram-positive and gram-negative *Bacteria*. The first step in alternate pathway activation is the binding of a serum protein, properdin (P), to the bacterial cell surface. The membrane-bound P binds C3B (a soluble complex consisting of the complement protein C3 and the alternate pathway serum protein factor B). The C3BP complex, now fixed on the cell, acts like the membrane-bound C423 complex of the classical complement pathway and catalyzes formation of the C5-9 MAC, resulting in cell destruction. The C3 fixed to the surface of the

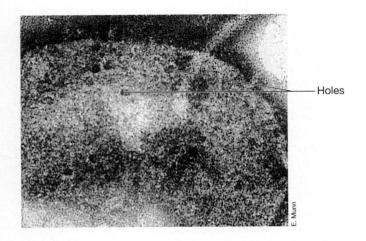

— Holes

E. Munn

● **Figure 22.23 Complement activity on bacterial cells.** Shown is an electron micrograph of *Salmonella paratyphi*. The negatively stained preparation shows holes created in the bacterial cell envelope as a result of a reaction involving cell envelope antigens, specific antibody, and complement.

cell by the alternate pathway can enhance phagocytosis through opsonization via phagocyte C3 receptors.

Another method of activating the complement system is through the *mannan-binding lectin (MBL) pathway*. This pathway depends on the activity of a serum MBL protein that binds to the unique mannose-containing polysaccharides found only on bacterial cell surfaces (Figure 22.22c). The MBL–polysaccharide complex resembles the C1 complexes of the classical complement system and fixes the C4, C2, and C3 components to the cell surface, again catalyzing formation of the C5-9 MAC, leading to lysis or opsonization of the bacterial cell.

Both the alternate pathway and the MBL pathway non-specifically target bacterial invaders, and lead to activation of the membrane attack complex and enhanced opsonization. As with Toll-like receptors, properdin and MBL target pathogen-associated molecular patterns, as both are pattern recognition molecules (PRMs) (see Section 22.2 and ⟳Section 23.1). Moreover, both are part of the innate immune system; neither the properdin pathway nor the MBL pathway requires prior antigen exposure or the presence of antibodies. As in the classical complement pathway, the alternate and MBL pathways activate C3, triggering formation of the C5-9 MAC and depositing C3-membrane complexes, enhancing opsonization.

 22.11 Concept Check

The complement system catalyzes bacterial cell destruction and opsonization. Complement is triggered by antibody interactions or by interactions with nonspecific activators. Complement is a critical component of both innate and adaptive host defense.

◆ Identify the immunoglobulin classes that fix complement.

◆ Identify the complement components that promote opsonization and those that induce cell lysis.

◆ Identify two mechanisms that induce complement activation by nonspecific means.

IV IMMUNITY AND PREVENTION OF INFECTIOUS DISEASE

Now that we have described the basic features of natural immunity, we switch to immunity acquired by artificial means. We discuss both standard and experimental techniques used in the preparation of immunogens. Application of these methods has revolutionized our ability to safely induce adaptive immune responses to a variety of pathogens and their products. Artificial immunization remains our best weapon against many infectious diseases.

22.12 Natural Immunity

The role of both the innate and adaptive immune response in the body is to protect the host from the consequences of infection. Both innate and adaptive immunity are essential for survival. For example, individuals with innate immune genetic defects that prevent neutrophil or macrophage development produce nonfunctional phagocytes that cannot ingest and destroy pathogens. As a result, these individuals cannot live without medical intervention such as isolation from all environmental exposure. In a normal environment, they develop recurrent infections from a variety of bacteria, viruses, and fungi and usually die at a young age. The role of adaptive immunity is no less essential.

Adaptive Immunity and Prevention of Infectious Disease

Animals normally develop *natural active immunity* to a variety of diseases by acquiring a natural infection that initiates an adaptive immune response. Natural active immunity is the outcome of exposure to antigens through infection and usually results in protective immunity conferred by antibodies as well as T cells.

The importance of active immunity in disease resistance is shown dramatically in individuals with certain genetic disorders. For example, *agammaglobulinemia* is a disease in which patients cannot produce antibodies because of genetic defects in their B cells. Such individuals suffer from recurrent, life-threatening *bacterial* infections, but develop normal immune responses to viruses. Thus, antibody-mediated immunity is essential for protection from extracellular pathogens, especially bacteria.

Individuals with *DiGeorge's syndrome*, a developmental defect that prevents maturation of the thymus and, therefore, production of mature T cells, suffer from serious recurrent infections with viruses and other intracellular pathogens. Thus, the T-cell immune deficiency problems seen in DiGeorge's syndrome define the essential protective role for T-cell immunity: protection from intracellular pathogens.

The general lack of an adaptive immune response is also observed in individuals with *acquired immunodeficiency syndrome (AIDS)*. In AIDS patients, infection with the *human immunodeficiency virus (HIV)*, if not controlled, can cause nearly total depletion of CD4 (T_H) cells, resulting in a lack of effective antibody-mediated and cellular (T cell-mediated) immunity. Such individuals suffer from recurrent, severe infections due to viral, bacterial, and fungal pathogens. Death from AIDS is characteristically due to one or more of these infections (∞Section 26.14).

 22.12 Concept Check

Innate and adaptive immune responses are necessary for survival. Lack of innate immunity results in death due to recurrent, uncontrollable infections. Lack of adaptive immunity also results in uncontrollable infectious disease.

◆ Describe the effect of a lack of innate immunity.

◆ Describe the effects of a lack of T or B cell immunity.

22.13 Artificial Immunity

The purposeful artificial induction of specific immunity to infectious diseases is a major weapon for the treatment and prevention of infectious diseases. This can occur in many ways. (1) An individual may be exposed to an antigen to induce formation of antibodies, a type of immunity known as *artificial active immunity* because the individual produced the antibodies. This process of **immunization** is commonly known as **vaccination**. (2) Alternatively, an individual may receive injections of an antiserum or purified immunoglobulin (gamma globulin) derived from another individual who previously formed antibodies against the antigen in question. This is called *artificial passive immunity* because the individual receiving the antibodies played no active part in the antibody-producing process. (3) In some cases, *natural passive immunity* can occur. For several months after birth, newborns have maternal IgG antibodies in their blood. These antibodies, transferred across the placenta before birth, provide disease protection while the immune system of the newborn matures. Active and passive immunity are contrasted in Table 22.3.

In *active* immunity, introduction of antigen produces fundamental changes in the host: The immune cells produce large quantities of antigen-specific immune effector molecules (Igs) and cells. A second ("booster") dose of the same antigen results in a faster, higher titer secondary response (see Figure 22.21). Active immunity often remains throughout life. A *passively* immunized individual never has more antibodies than received in the initial injection, and these antibodies gradually disappear from the body. Moreover, a later exposure to the antigen does not elicit a secondary response.

Passive artificial immunity is usually *therapeutic*, designed to cure a person who is presently suffering from the disease. For example, tetanus toxoid (see the following section) actively immunizes an individual against future encounters with *Clostridium tetani* exotoxin, whereas tetanus antiserum is administered to passively immunize an individual suspected of being exposed to *C. tetani*. Active artificial immunity, in contrast, is usually used as a *prophylactic* measure, to protect a person against future attack by a pathogen.

Immunization

The material used to induce artificial active immunity, the antigen or mixture of antigens, is known as a **vaccine** or an immunogen, and the process of generating such an immune response is immunization. Immunization with a vaccine designed to produce active immunity can introduce risks for infection and for other adverse reactions. To reduce risks, pathogens or their products are often inactivated. For example, many immunogens consist of pathogenic bacteria killed by chemical agents such as phenol or formaldehyde, or physical agents such as heat. Formaldehyde is also used to inactivate viruses for vaccines, such as in the *Salk polio vaccine*. Likewise, for exotoxin-caused diseases, the active form of the exotoxin is not used as an immunogen. Many exotoxins can be modified chemically so they retain their antigenicity but are no longer toxic. Such a modified exotoxin is called a **toxoid**. Toxoids are usually not as efficient as the original exotoxin for producing immunity, but they can be given safely and in high doses.

Immunization with live cells or virus is usually more effective than with dead or inactivated material. It is often possible to isolate a mutant strain of a pathogen

Table 22.3	Active and passive immunity	
Active immunity	**Passive immunity**
Exposure to antigen; immunity achieved by injecting antigen	No exposure to antigen; immunity achieved by injecting antibodies or antigen-reactive T cells
Specific response made by individual achieving immunity	Specific immune response made in a secondary host
Immune system activated to antigen; immunological memory in effect	No immune system activation; no immunological memory
Immune response can be maintained via stimulation of memory cells (*i.e.*, booster immunization)	Immunity cannot be maintained and decays rapidly
Immune state develops over a period of weeks	Immunity develops immediately

that has lost its virulence but still retains the immunizing antigens; strains of this type are called *attenuated strains* (∞Section 21.8). However, because attenuated strains of pathogens are still viable, some individuals, especially those who are immunocompromised, occasionally acquire active disease caused by the immunizing pathogen. Serious cases of vaccine-acquired infectious diseases have occurred, for example, from attenuated poliovirus vaccines and smallpox virus vaccines (∞Section 25.11).

A summary of vaccines available for use in humans is given in Table 22.4. Most effective *viral* vaccines are in an attenuated form. Attenuated vaccines tend to provide long-lasting T cell-mediated immunity, as well as a vigorous antibody response and a strong secondary response upon reimmunization. However, the attenuated strains are diffi-

cult to select, standardize, and maintain. Live vaccines also have a limited shelf life and usually require refrigeration for adequate storage. Killed virus vaccines, on the other hand, tend to provide short-lived immune responses, without the development of a long-term memory response.

Bacterial vaccines are nearly always provided in an inactivated form. Inactivated bacterial vaccines induce long-term antibody-mediated protection without exposing recipients to the risk of infection.

Immunization Practices

Infants possess natural passive immunity in the form of maternal antibodies transferred across the placenta or in breast milk. As a result, infants are relatively immune to infectious disease during the first 6 months of life. However, infants should be immunized to prevent key infectious diseases as soon as possible so that their own *active* immunity can replace the maternal *passive* immunity. As discussed in Section 22.10, a single exposure to antigen does not lead to a high antibody titer. After an initial immunization, a series of "booster" immunizations are given to produce a secondary response and a high antibody titer. The recommended vaccine schedule for children and adults in the United States is shown in Figure 22.24●. With most recommended

Table 22.4	Vaccines for infectious diseases in humans
Disease	**Type of vaccine used**
Bacterial diseases	
Anthrax	Toxoid
Diphtheria	Toxoid
Tetanus	Toxoid
Pertussis	Killed bacteria (*Bordetella pertussis*) or acellular proteins
Typhoid fever	Killed bacteria (*Salmonella typhi*)
Paratyphoid fever	Killed bacteria (*Salmonella paratyphi*)
Cholera	Killed cells or cell extract (*Vibrio cholerae*)
Plague	Killed cells or cell extract (*Yersinia pestis*)
Tuberculosis	Attenuated strain of *Mycobacterium tuberculosis* (BCG)
Meningitis	Purified polysaccharide from *Neisseria meningitidis*
Bacterial pneumonia	Purified polysaccharide from *Streptococcus pneumoniae*
Typhus fever	Killed bacteria (*Rickettsia prowazekii*)
Haemophilus influenzae meningitis	Conjugated vaccine (polysaccharide of *Haemophilus influenzae* conjugated to protein)
Viral diseases	
Yellow fever	Attenuated virus
Measles	Attenuated virus
Mumps	Attenuated virus
Rubella	Attenuated virus
Polio	Attenuated virus (Sabin) or inactivated virus (Salk)
Influenza	Inactivated virus
Rabies	Inactivated virus (human) or attenuated virus (dogs and other animals)
Smallpox	Cross-reacting virus (vaccinia) (∞Section 16.13)
Hepatitis A	Recombinant DNA vaccine
Hepatitis B	Recombinant DNA vaccine or inactivated virus
Varicella (chickenpox)	Attenuated virus

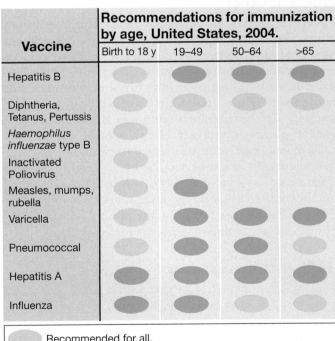

Vaccine	Recommendations for immunization by age, United States, 2004.			
	Birth to 18 y	19–49	50–64	>65
Hepatitis B				
Diphtheria, Tetanus, Pertussis				
Haemophilus influenzae type B				
Inactivated Poliovirus				
Measles, mumps, rubella				
Varicella				
Pneumococcal				
Hepatitis A				
Influenza				

Recommended for all.

Recommended for individuals with predictable risk (medical, behavioral, occupational, or other indicators of enhanced risk).

● **Figure 22.24 Recommended immunizations for children and adults in the United States.** This course of immunizations is specified by the Centers for Disease Control and Prevention, Atlanta, GA, in 2004. For more specific recommendations for any age group, including recommendations for travelers, consult the CDC Website at http://www.cdc.gov.

vaccines, periodic reimmunization is necessary to maintain effective immunity.

The importance of immunization in controlling infectious diseases is well established. Introduction of an effective vaccine into a population often dramatically reduces the incidence of the disease (∞Figures 26.13 and 26.23). The degree of immunity obtained by vaccination, however, varies greatly with the individual as well as with the quality and quantity of the vaccine. Lifelong immunity is rarely achieved by means of a single injection, or even a series of injections, and the immune cells induced by immunization gradually disappear from the body. On the other hand, antigenic stimulation often occurs even in the absence of artificial immunization through natural infections. A natural infection induces a secondary response, leading to an increase in production of antibody. In the complete absence of antigenic stimulation, the length of effective immunity varies considerably with different antigens. For example, effective immunity to tetanus toxoid may last many years, but immunity induced by a particular influenza virus vaccine may disappear within a year or two in the absence of a secondary stimulation.

Immunizations not only benefit the individual but are effective public health measures because disease spreads poorly through a population with a large proportion of immune individuals (∞Section 25.5).

Passive Immunity

Passive immunity is introduced by injecting preformed antibodies. The antibody-containing preparation is known as a *serum*, an *antiserum*, or an *antitoxin* (antibodies directed against a toxin). Antisera are obtained from immunized animals, such as horses, or from humans with high antibody titers. These individuals are said to be hyperimmune. The antiserum or antitoxin is standardized to contain a known antibody titer; a sufficient number of units of antiserum must be injected to neutralize any antigen that might be present in the body. The immunoglobulin fraction separated from serum pooled from a number of individuals is also used for passive immunization. Pooled sera contain a variety of antibodies induced by artificial or natural exposure to various antigens.

 22.13 Concept Check

Immunity to infectious disease can be either passive or active, natural or artificial. Immunization, a form of artificial active immunity, is widely employed to prevent infectious diseases. Most agents used for immunization are either attenuated or inactivated pathogens or inactivated forms of natural microbial products.

◆ Provide an example of natural passive immunity. How does natural passive immunity benefit the immunized individual?

◆ Provide an example of artificial active immunity. How does artificial active immunity benefit the immunized individual?

22.14 New Immunization Strategies

Most immunization preparations are produced from whole organisms or toxoids, as described in the previous section. However, there are several other methods for producing antigens suitable for immunization.

Synthetic and Genetically Engineered Immunizing Agents

The simplest alternate approach to vaccine development is the use of *synthetic peptides*. To make a vaccine, a peptide can be synthesized that corresponds to a known epitope on an infectious agent. For example, the structure of the protein antigen responsible for immunity to foot-and-mouth virus, an important animal pathogen, is known. A synthetic peptide of 20 amino acids constituting an important antigenic determinant of the protein has been made and attached to suitable carrier molecules. This synthetic vaccine evokes an excellent neutralizing antibody response to foot-and-mouth virus. However, as a general method, this approach has one major problem: The entire sequence, representing the complete antigenic profile of the protein, must be known to make an effective vaccine. This condition has been met for foot-and-mouth virus, and a variety of other pathogens have now been sequenced, providing the potential for identifying the specific antigenic profile necessary to design an effective vaccine (∞Section 15.3).

Using information derived from genomics of pathogens, molecular biology techniques can be used to make vaccines. For example, genes that encode antigens from virtually any virus can be cloned into the vaccinia virus genome and expressed. Inoculation with the *genetically engineered* vaccinia virus can then be used to induce immunity to the product of the cloned gene. Such a preparation is a *recombinant-vector vaccine*. This method depends on the identification and cloning of the gene that encodes the antigen and also on the ability of the vaccinia virus to express the cloned gene as an antigenic protein. An effective recombinant vaccinia-rabies vaccine has been developed for use in animals. Recombinant DNA methods to develop vaccines will be discussed in Section 31.6.

Another immunization strategy involves the production of recombinant DNA proteins as immunogens. First, a pathogen gene must be cloned in a suitable microbial host. The host, chosen for its ability to express protein encoded by the cloned gene, will then express the pathogen protein. The pathogen protein can then be harvested and used as a vaccine. This is known as a *recombinant antigen vaccine*. For example, the current hepatitis B virus vaccine is a major hepatitis surface protein antigen (HbsAg) expressed by yeast cells.

DNA Vaccines

A novel method for immunization is based on expression of cloned genes in host cells. Bacterial plasmids containing cloned DNA are injected intramuscularly into a host animal. After several weeks, the host responds with T_C cells, T_H1 cells, and antibodies directed to the protein encoded by the cloned DNA. Taken up by host cells, the DNA is transcribed and translated to produce immunogenic proteins, triggering a conventional immune response. These plasmids are called DNA vaccines.

DNA vaccines provide considerable advantages over many conventional immunization strategies. For instance, since only a single foreign pathogen gene is normally cloned and injected, there is no chance of an infection as there might be with an attenuated vaccine. Second, genes for individual antigens such as a tumor-specific antigen, or even a single antigenic determinant can be cloned, targeting the immune response to a particular cell component. The response can also be targeted directly to APCs by including a MHC class II promoter in the gene construct. The promoter assures selective expression in dendritic cells, B cells, and macrophages, the only cells that express the MHC II protein. The expressed and processed antigen can then be presented on both MHC I and MHC II proteins. Thus, a single bioengineered plasmid can encode an antigen and elicit a complete immune response, inducing immune T cells and antibodies.

22.14 Concept Check

Alternate immunization strategies using bioengineered molecules eliminate exposure to microorganisms and, in some cases, even to protein antigen. Application of these strategies may provide safer and more targeted vaccines.

◆ Provide two examples of alternate immunization strategies.

◆ What is the advantage of each alternative immunization strategy over current immunization procedures?

V IMMUNE RESPONSE DISEASES

Immune reactions can cause host cell damage and disease. **Hypersensitivity** responses are inappropriate immune responses that result in host damage. Hypersensitivity diseases are categorized according to the antigens and effector mechanisms that produce disease (Table 22.5). **Superantigens** are proteins produced by certain bacteria and viruses that cause widespread stimulation of immune cells, resulting in host damage by activating a massive inflammatory response.

22.15 Allergy, Hypersensitivity, and Autoimmunity

Antibody-mediated *immediate hypersensitivity* is commonly called *allergy*. Cell-mediated reactions also cause disease in the form of *delayed-type hypersensitivity*. *Autoimmune diseases* are directed against self antigens.

Immediate Hypersensitivity (Type I Hypersensitivity)

Immediate hypersensitivity or Type I hypersensitivity is mediated by release of vasoactive products from IgE antibody-coated mast cells (Table 22.5). Immediate hypersensitivity reactions, commonly called *allergies*, occur within minutes after exposure to antigen. Depending on the individual and the antigen, immediate hypersensitivity reactions can be very mild or can cause extremely severe life-threatening reactions, a condition know as *anaphylaxis*. Antigens that cause Type I hypersensitivities are called *allergens*.

About 20% of the population suffers from immediate hypersensitivities involving allergic (*anaphylactic*) reactions to specific allergens such as pollens, animal dander, certain foods, and a variety of other agents (Table 22.6). Almost all allergens are delivered at the surface of mucous membranes such as the lungs or the gut. There they stimulate T_H2 cells to produce cytokines that preferentially induce B cells to make IgE antibodies. Rather than circulating like IgG or IgM, the allergen-specific IgE antibodies bind via a cell-membrane-binding constant domain (see Section 22.8) to IgE receptors on mast cells (Figure 22.25●). Mast cells are nonmotile granulocytes (see Section 22.2) associated with

Table 22.5	**Hypersensitivity**				
Classification	**Description**	**Immune mechanism**	**Time of latency**	**Examples**
Type I	Immediate	IgE sensitization of mast cells	Minutes	Reaction to bee venom (sting) Hay fever
Type II	Cytotoxic[a]	IgG interaction with cell surface antigen	Hours	Drug reactions (penicillin)
Type III	Immune complex	IgG interaction with soluble or circulating antigen	Hours	Systemic lupus erythematosis (SLE)
Type IV	Delayed type	T_H1 inflammatory cells	Days	Poison ivy Tuberculin test

[a] Autoimmune diseases may be caused by Type II, Type III, or Type IV reactions.

Table 22.6	Immediate hypersensitivity allergens

Pollen and fungal spores (hay fever)
Insect venoms (bee sting)
Certain foods
Animal dander
Mites in house dust

the connective tissue adjacent to capillaries throughout the body. With any subsequent exposure to the immunizing allergen, the cell-bound IgE molecules bind the antigen. Cross-linking of two or more IgEs by an antigen triggers the release of several soluble allergic mediators from the mast cells, a process called *degranulation*. These mediators cause allergic symptoms within minutes of antigen exposure. In general, these symptoms are relatively short-lived, but once initially sensitized by an allergen, an individual can respond to each subsequent exposure to the antigen.

The primary chemical mediators released from mast cells are histamine and serotonin, modified amino acids that cause rapid dilation of blood vessels and contraction of smooth muscle, initiating the symptoms of systemic anaphylaxis. These symptoms include vasodilation (causing a sharp drop in blood pressure), severe respiratory distress caused by bronchial edema, flushed skin, mucus production, sneezing, and itchy, watery eyes. If severe cases of anaphylaxis are not treated immediately with adrenalin to counter smooth muscle contraction, increase blood pressure, and promote breathing, death can

occur due to anaphylactic shock. Fortunately, the magnitude of most allergic reactions is limited to mild local anaphylaxis involving symptoms such as itchy, watery eyes. Less serious allergic symptoms are treated with drugs called *antihistamines* that neutralize the histamine mediators. Treatment for more serious symptoms may also include general antiinflammatory drugs such as steroids. Finally, immunization with escalating doses of the allergen may be done to shift antibody production from IgE to IgG. The IgG interacts with antigen, preventing interactions with the relatively scarce IgE, stopping allergic symptoms and inhibiting production of more IgE. This procedure is called *desensitization*.

Delayed-Type Hypersensitivity (Type IV Hypersensitivity)

Type IV hypersensitivity or delayed-type hypersensitivity (DTH) is cell-mediated hypersensitivity characterized by tissue damage due to inflammatory responses produced by T_H1 inflammatory cells (Table 22.5). Symptoms begin to appear several hours after secondary exposure to the eliciting antigen, with a *maximal* response usually occurring in 24–48 hours. Typical antigens include certain microorganisms, a few self-antigens (Table 22.7), and several chemicals that covalently bind to the skin, creating new antigens. Hypersensitivity to these newly created antigens is known as *contact dermatitis* and results in skin reactions to poison ivy (Figure 22.26●), jewelry, cosmetics, latex, and other chemicals. Within several hours after exposure to the agent, the skin feels itchy at the site of contact, and reddening and

● Figure 22.25 **Immediate hypersensitivity.** IgE binds to mast cells by means of a high-affinity surface receptor for the C_H4 domain. Binding arms the mast cell. Antigen contact and cross-linking of the IgE proteins initiates release of vasoactive mediators, resulting in systemic symptoms ranging from mild allergies to life-threatening anaphylaxis.

Table 22.7 Autoimmune diseases of humans

Disease	Organ or area affected	Mechanism (hypersensitivity type)
Juvenile diabetes (insulin-dependent diabetes mellitus)	Pancreas	Cell-mediated immunity and autoantibodies against surface and cytoplasmic antigens of islets of Langerhans (II and IV)
Myasthenia gravis	Skeletal muscle	Autoantibodies against acetylcholine receptors on skeletal muscle (II)
Goodpasture's syndrome	Kidney	Autoantibodies against basement membrane of kidney glomeruli (II)
Rheumatoid arthritis	Cartilage	Autoantibodies against self IgG antibodies, which form complexes deposited in joint tissue, causing inflammation and cartilage destruction (III)
Hashimoto's disease (hypothyroidism)	Thyroid	Autoantibodies to thyroid surface antigens (II)
Male infertility (some cases)	Sperm cells	Autoantibodies agglutinate host sperm cells (II)
Pernicious anemia	Intrinsic factor	Autoantibodies prevent absorption of vitamin B_{12} (III)
Systemic lupus erythematosis	DNA, cardiolipin, nucleoprotein, blood clotting proteins	Massive autoantibody response to various cellular constituents results in immune complex formation (III)
Addison's disease	Adrenal glands	Autoantibodies to adrenal cell antigens (II)
Allergic encephalitis	Brain	Cell-mediated response against brain tissue (IV)
Multiple sclerosis	Brain	Cell-mediated and autoantibody response against central nervous system (II and IV)

swelling appear, indicative of a general inflammatory response. Localized tissue destruction, often in the form of blistering, occurs as a result of the activities of immune cells, again with a maximal reaction in about 24–48 hours.

An example of delayed-type hypersensitivity is the development of immunity to the causal agent of tuberculosis, *Mycobacterium tuberculosis* (Figure 22.24). This protective cellular immune response was discovered by Robert Koch in his classical studies on tuberculosis (∞ Section 1.6) and has been widely studied. Antigens derived from the bacterium, when injected subcutaneously into an animal previously infected with *M. tuberculosis*, elicit a characteristic skin reaction that develops fully only after a period of 24–48 hours. By contrast, skin reactions to IgE-mediated immediate hypersensitivity responses develop within minutes after antigen injection. In the region of the introduced antigen, T_H1 cells become stimulated by the antigen and release cytokines that attract and activate large numbers of macrophages, which in turn produce a skin reaction at the site of injection. The reaction is characterized by induration (hardening), edema (swelling), erythema (reddening), pain, and localized heating, common symptoms of an inflammatory response (see Section 22.3). The activated macrophages then ingest and destroy the invading antigen. This skin response is the basis for the tuberculin test used to determine prior exposure to *M. tuberculosis* (Figure 22.26).

A number of microbial infections elicit delayed-type hypersensitivity reactions. In addition to tuberculosis, these include leprosy, brucellosis, psittacosis (all caused by species of *Bacteria*), mumps (caused by a virus), and coccidioidomycosis, histoplasmosis, and blastomycosis (caused by fungi). In all of these cellular immune reactions, visible antigen-specific skin responses resembling the tuberculin reaction (Figure 22.26) occur after injection of antigens derived from the pathogens, indicating prior exposure to the pathogen.

In addition, T_H1-mediated delayed hypersensitivity can also be involved in autoimmune responses directed against self-antigens. For example, T_H1 cells are involved in the pathogenesis of allergic encephalitis and Type I (juvenile) diabetes mellitus (Table 22.6). However, many autoimmune diseases are mediated by antibodies, as we will now discuss.

Autoimmune Diseases (Type II and Type III Hypersensitivities)

T and B cells destined to react with self-antigens are normally eliminated during the process of lymphocyte maturation. However, in some individuals, T and B lymphocytes can be activated to produce immune reactions against self-proteins, leading to autoimmune disease (Table 22.7).

Depending on the specific disorder, autoimmune disease may involve **autoantibodies**, antibodies that interact with self-antigens, or a cellular immune response to self constituents. Certain autoimmune diseases are highly organ-specific. For example, in *Hashimoto's disease*, autoantibodies are made against thyroid gland products such as thyroglobulin. The disease affects thyroid function and is classified as a Type II hypersensitivity: Antibodies interact with antigens found on the surface of host cells and cause destruction of the tissue (Table 22.5). In this case, antibodies to thyroglobulin fix complement and cause destruction of the thyroid gland cells. In *juvenile diabetes* (insulin-dependent diabetes mellitus), autoantibodies against the insulin-producing cells in the pancreas are observed, but tissue destruction occurs primarily through inflammatory reactions mediated by T_H1 cells.

(a)

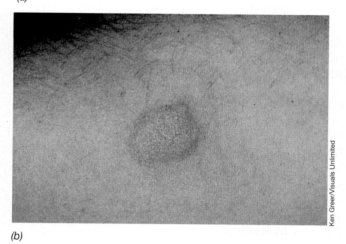

(b)

● **Figure 22.26 Cell-mediated immunity.** (a) Poison ivy blisters on an arm. The raised rash appears maximally in sensitized individuals about 24–48 hours after exposure to plants of the genus *Rhus*. (b) A positive tuberculin test, typical for delayed hypersensitivity. The raised area of inflammation on the forearm is 1.5 cm in diameter. Both reactions result from the action of antigen-specific T_H1 cells.

Systemic lupus erythematosis (SLE) is a clear example of a Type III hypersensitivity: Antibodies bind to soluble proteins, producing insoluble immune complexes, leading to complement deposition and inflammation. Thus, Type III hypersensitivity is an *immune complex disorder* (Table 22.5). This disease and others like it involve a large-scale production of autoantibodies against soluble, circulating self-antigens. In the case of SLE, the antigens include nucleoproteins and DNA. The disease symptoms are induced by circulating antigen–antibody complexes that deposit in different body tissues such as the kidney,

lungs, and spleen. Complement fixation and the resulting lytic and inflammatory responses cause local, often severe, cell damage at the site of the complex deposition.

Organ-specific autoimmune diseases are sometimes more easily controlled clinically because the product of organ function, such as thyroxin in autoimmune hypothyroidism or insulin in juvenile diabetes, can often be supplied in pure form from another source. SLE and other diseases that affect multiple organs can be controlled only by general immunosuppressive therapy, such as the use of steroid drugs. However, steroid therapy carries its own risks. General immunosuppression significantly increases chances of opportunistic infections.

Heredity has an important influence on the incidence, type, and severity of autoimmune diseases. Many autoimmune diseases correlate strongly with the presence of certain major histocompatibility complex (MHC) antigens (see Section 22.6; ∞Section 23.3). Studies of model autoimmune diseases in mice support such a genetic link, but the precise conditions necessary for developing autoimmunity may also depend on other factors, such as gender. Women, for example, are up to 20 times more likely to develop SLE than are men.

 22.15 Concept Check

Hypersensitivity results when foreign antigens induce cellular or antibody immune responses, leading to host tissue damage. Autoimmunity occurs when the immune response is directed against self-antigens, resulting in host tissue damage.

◆ Discriminate between *immediate hypersensitivity* and *delayed hypersensitivity* with respect to antigens and effectors.

◆ Identify the two main categories of autoimmune disease with respect to antigens and immune effectors.

22.16 Superantigens

We discussed the mechanisms of action for several different categories of bacterial toxins in Chapter 21. Most toxins interact directly with host cells to cause tissue damage. Endotoxins, for example, interact directly with a large variety of cell types, causing release of endogenous pyrogens and other soluble mediators, and producing fever and general inflammation (see Section 22.2). Most exotoxins also interact directly with cells to cause cell damage (∞Sections 21.10 and 21.11). However, certain exotoxins, the superantigens, act *indirectly* on host cells, using a novel immune mechanism to cause extensive host tissue damage.

Superantigen Activation of T Cells

Superantigens are proteins capable of eliciting a very strong response because they activate more T cells than a normal immune response. The superantigens differ from conventional antigens in their mode of binding to the TCR. Most superantigens are produced by viruses and bacteria. Streptococci and staphylococci, in particular, produce a wide variety of different and very potent superantigens (Table 21.4).

Conventional foreign antigens bind to the TCR at the variable (V) domain antigen-binding site. In a typical immune response, less than 0.01% of all T cells normally interact with a conventional foreign antigen (see Section 22.5). However, superantigens bind to a site on the Vβ domain of the TCR that is *outside* the antigen-specific TCR binding site. Superantigen binds to *all* TCRs with a shared structure, and many different TCRs share the same structure *outside* the antigen-binding site. In some cases, superantigen exposure can result in the binding and activation of 5% to 25% of all T cells. The superantigens also bind to class II MHC molecules on APCs, again at a specific site outside the normal peptide binding site (Figure 22.27●). These cell surface interactions, mimicking conventional antigen presentation (see Section 22.7), activate large numbers of T cells. The activated T cells, in turn, produce cytokines. The cytokines then stimulate other cells, in particular, macrophages and other phagocytes. Because of the extensive cytokine production, this cell-mediated response is characterized by *systemic* inflammatory reactions, often resulting in fever, diarrhea, vomiting, mucus production, and systemic shock. In extreme cases, exposure to superantigens can be fatal. Superantigen shock is virtually indistinguishable from septic shock as discussed in Section 22.3.

Several diseases can be attributed to superantigens and were mentioned previously (Section 21.10 and Table 21.4). Notable superantigen diseases include *Staphylococcus aureus* food poisoning, characterized by fever, vomiting, and diarrhea, caused by one of several superantigens that act as staphylococcal enterotoxins. *Staphylococcus aureus* also produces the superantigen responsible for *toxic shock syndrome*. *Streptococcus pyogenes*

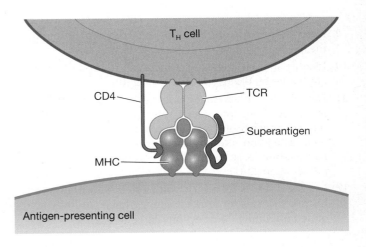

● **Figure 22.27 Bacterial superantigens.** Superantigens act by binding to both the MHC protein and the TCR at positions outside the normal binding site. Because the superantigen binds to conserved regions of MHC and TCR proteins, the superantigen can interact with large numbers of cells, stimulating massive T-cell activation, cytokine release, and even systemic inflammation.

produces erythrogenic toxin, the superantigen responsible for scarlet fever (Figure 26.5).

 22.16 Concept Check

Superantigens bind and activate large numbers of T cells in a novel fashion. Superantigen-activated T cells are capable of producing systemic diseases characterized by massive inflammatory reactions.

◆ Discriminate between normal and superantigen activation of T cells.

◆ Identify the binding site for superantigens on both T cells and APCs.

REVIEW QUESTIONS

1. What is the origin of the phagocytic and antigen-specific cells involved in the immune response? Track the maturation of B cells and T cells (Section 22.1).

2. What are some pathogen-associated molecular patterns (PAMPs) that are recognized by pattern recognition molecules (PRMs)? What is the significance of the interactions between these molecules (Section 22.2)?

3. Explain how phagocytes engulf and kill microorganisms, with particular attention to oxygen-dependent mechanisms (Section 22.3).

4. Identify the most important features of the adaptive immune response (Section 22.4).

5. What molecules induce immune responses? What properties are necessary for a molecule to induce an immune response (Section 22.5)?

6. Describe the basic structure of class I and class II major histocompatibility complex (MHC) proteins. In what functional ways do they differ (Section 22.6)?

7. Differentiate between T_C cells and NK cells. What is the activation signal for each cell type (Section 22.7)?

8. How do T_H cells differ from T_C cells? Differentiate between the functional roles of T_H1 and T_H2 cells (Section 22.8).

9. Describe the structural and functional differences among the five major antibody classes (∞Section 22.9).

10. Identify the cell interactions involved in production of antibodies by B cells (∞Section 22.10).

11. Describe the complement cascade. Is the order of protein interactions important? Why or why not? Identify the components of the alternate pathway for complement activation. Identify the components of the mannan-binding lectin pathway for complement activation (∞Section 22.11).

12. List the diseases for which you have been immunized. List the diseases for which you may have naturally acquired immunity (∞Sections 22.12 and 22.13).

13. Describe at least one of the biotechnology-based immunization strategies that we described. How does this strategy differ from a conventional immunization strategy (∞Section 22.14)?

14. Define the differences between immediate and delayed-type hypersensitivity in terms of immune effectors, target tissues, antigens, and clinical outcome (∞Section 22.15).

15. Describe the general mechanism used by superantigens to activate T cells. How does superantigen activation differ from T-cell activation by conventional antigens (∞Section 22.16)?

APPLICATION QUESTIONS

1. Describe the potential problems that would arise if an individual had an acquired *inability* to phagocytize pathogens. Could the individual survive in a normal environment such as a college campus? What defects in the phagocyte might cause lack of phagocytosis? Explain.

2. Specificity and tolerance are necessary qualities for an adaptive immune response. However, memory seems to be less critical, at least at first glance. Define the role of immune memory and explain how the production and maintenance of memory cells might benefit the host in the long term.

3. Describe the potential problems that would arise if an individual had a hereditary deficiency that resulted in inability to present antigens to T_C cells. To T_H cells? To both T_C and T_H cells? What molecules might be deficient in each individual? Could any of these individuals survive in a normal environment? Explain.

4. Antibodies of the IgA class are probably more prevalent than those of the IgG class. Explain why this is true and what advantage this provides for the host.

5. Complement is regarded as a critical humoral defense mechanism. Do you agree with this statement? Explain your answer. What might happen to individuals who lack complement component C3? C5? Properdin (alternate pathway)? MBL?

6. Many infectious diseases have no effective vaccines. Pick several of these diseases (for example, AIDS, malaria, the common cold) and explain why current vaccine strategies have not been effective. Prepare some alternate strategies for immunization against the diseases you have chosen.

23

MOLECULAR IMMUNOLOGY

Immune recognition involves molecular interactions at the surface of cells. Major histocompatibility proteins, found on all cells, bind peptides as shown here. This complex is, in turn, bound by T-cell receptors, triggering an antigen-specific immune response.

WORKING GLOSSARY

Agretope the portion of a processed antigen that is recognized by MHC protein

Anergy the inability to produce an immune response to specific antigens due to neutralization of effector cells

Antibody a soluble protein, produced by B cells, that interacts with antigen; also called immunoglobulin

Apoptosis programmed cell death

Chemokine a small soluble protein produced by a variety of cells that modulates inflammatory reactions and immunity in target cells

Class I MHC protein antigen-presenting protein found on all nucleated vertebrate cells

Class II MHC protein antigen-presenting protein found on macrophages, B cells, and dendritic cells (antigen-presenting cells)

Clonal deletion for T-cell selection in the thymus, the killing of useless or self-reactive clones

Clonal selection each B or T cell produces copies of itself when stimulated with antigen

Complementarity-determining region (CDR) a variation in amino acid sequence within the variable domains of immunoglobulins or TCRs, providing most of the molecular contacts with antigen (also known as a hypervariable region)

Cytokine a small, soluble protein produced by a leukocyte that modulates inflammatory reactions and immunity in target cells

Domain a region of a protein having a defined structure and function

Epitope the portion of an antigen that is recognized by an immunoglobulin or a T-cell receptor

HLA human leukocyte antigen, the MHC of humans

Hypervariable region variation in amino acid sequence within the variable domains of immunoglobulins or TCRs, providing most of the molecular contacts with antigen (also known as a complementarity-determining region)

Immunoglobulin (Ig) a soluble protein, produced by B cells, that interacts with antigen; also called antibody

Immunoglobulin gene superfamily a family of genes that are evolutionarily, structurally, and functionally related to immunoglobulins

Major histocompatibility complex (MHC) a genetic region responsible for encoding several cell-surface proteins important for antigen processing and presentation

Motif a conserved amino acid sequence found in all peptide antigens that bind to a given MHC protein

Negative selection in T-cell selection, T cells that interact with self antigens in the thymus are deleted (see clonal deletion)

Pathogen-associated molecular pattern (PAMP) a structural component of a pathogen or pathogen product that is recognized by a pattern recognition molecule

Pattern recognition molecule (PRM) a protein that recognizes a pathogen-associated molecular pattern, such as a component of a microbial cell surface structure

Polymorphism the occurrence of multiple alleles at a locus in frequencies that cannot be explained by the occurrence of recent random mutations

Positive selection in T-cell selection, T cells that interact with self MHC protein in the thymus are stimulated to grow and develop

Somatic hypermutation the mutation of immunoglobulin genes at rates higher than these observed in other genes

T-cell receptor (TCR) antigen-specific receptor protein on the surface of T cells

Tolerance inability to produce an immune response to a specific antigen

Toll-like receptor (TLR) a pattern-recognition molecule found on phagocytes, structurally and functionally related to Toll receptors in *Drosophila*

We have seen that the immune response employs a number of proteins to target pathogens. This chapter discusses these proteins and their complex molecular interactions. We will first examine the pattern recognition molecules (PRMs), proteins that interact with antigens in the innate immune response. We will then shift our attention to the adaptive immune response and the immunoglobulin superfamily, concentrating on the major histocompatibility complex (MHC) proteins, T-cell receptors (TCRs), and immunoglobulins (Igs), all proteins that interact directly with antigen. In the final section, we will concentrate on the molecular interactions that control cell signaling and activate the immune response.

RECEPTORS AND IMMUNITY

23.1 Innate Immunity and Pattern Recognition

Phagocytes such as macrophages and neutrophils are the cells that initiate the innate immunity, generally leading to *phagocytosis*, the engulfment and destruction of the pathogen. Phagocytes express a number of proteins that interact with molecules shared by broad groups of pathogens.

Pathogen–Associated Molecular Patterns (PAMPs), Pattern Recognition Molecules (PRMs), and Toll-Like Receptors (TLRs)

Phagocytes have acquired a number of evolutionarily conserved membrane-bound proteins called **pattern-recognition molecules (PRMs)** to identify pathogens. A *PRM* is a protein that recognizes a pathogen-associated molecular pattern, such as a component of a microbial cell surface structure. We have already discussed two PRMs, the soluble proteins *mannan-binding lectin (MBL)* and *properdin* (factor P), in the context of the complement system (∞Section 22.11). These PRMs activate the complement system, inducing target cell lysis or opsonization. MBL and properdin act in much the same fashion, binding to cell surface components on microbes, and activating complement through the MBL or alternate pathway (∞Section 22.11) (Table 23.1). The targets of these proteins are **pathogen-associated molecular patterns (PAMPs)**, unique structural components found in particular groups of organisms or infectious agents. The PAMP recognized by MBL is the sugar mannose, found almost exclusively as a repeating subunit in bacterial

Table 23.1 Receptors and targets in the innate immune response

Pattern recognition molecules (PRM)	Pathogen-associated molecular patterns (PAMP)	Result of interaction
MBL (mannan-binding lectin) (soluble)	Mannose-containing cell-surface microbial components	Complement activation via the MBL pathway (∞ Section 22.11)
Properdin (soluble)	Zymosan and other microbial cell-surface molecular complexes	Complement activation via the alternate pathway (∞ Section 22.11)
TLR-2[a] (Toll-like receptor 2) (membrane bound)	Peptidoglycan on gram-positive bacteria Zymosan on yeast cell walls	For all TLRs, interaction with pathogen-restricted macromolecules can result in signal transduction, leading to phagocyte activation, inflammation, and pathogen destruction
TLR-3 (membrane bound)	ds RNA (viral replication)	
TLR-4 (membrane bound)	LPS (lipopolysaccharide)	
TLR-5 (membrane bound)	Flagellin on bacteria	
TLR-9 (membrane bound)	Unmethylated CpG oligonucleotides in bacteria	

[a] The Toll-like receptors are found almost exclusively in phagocytes.

polysaccharides. Properdin recognizes zymosan, a protein-polysaccharide complex present as an integral part of the cell wall of a number of yeast (fungal) pathogens.

Other PRMs are membrane-integrated proteins found almost exclusively on phagocytes. First discovered as the Toll receptors in *Drosophila* (the fruit fly), the evolutionarily and functionally related transmembrane proteins are called **Toll-like receptors (TLRs)** in mammals (see the Microbial Sidebar "Toll Receptors and *Drosophila*—An Ancient Response to Infections"). There are now 10 known TLRs in humans, recognizing PAMPs from viruses, bacteria, and fungi, including pathogen cell-surface molecules and several soluble products of pathogen metabolism. Table 23.1 summarizes some of the known interactions of TLRs and other PRMs with a variety of PAMPs.

Several TLRs are involved in responses to more than one PAMP. For example, TLR-4 is involved in the innate immune response to gram-negative LPS and also responds to the presence of a host-response molecule called *heat shock protein (HSP)*. Neither LPS nor HSP interact directly with the TLR-4, but rather interact via a receptor protein that, in turn, interacts with TLR-4. In other cases, the TLR binds directly to the PAMP, as is the case for flagellin and TLR-5.

Interaction of a PAMP with the phagocyte TLR either directly or indirectly triggers a transmembrane signal transduction event, initiating translation of a variety of proteins and resulting in activation of the phagocyte. Activation can result in enhanced phagocytosis and killing of pathogens, or contribute to production of inflammation (∞ Sections 22.2 and 22.3).

⬣ **23.1 Concept Check**

Interactions between PAMPs and host PRMs are integral components of the innate immune response. PRMs interact with PAMPS shared by a variety of pathogens, activating complement and phagocyte effector mechanisms to target and destroy pathogens.

◆ Identify the target of the MBL pathway and the outcome for the targeted pathogen.

◆ Identify the cell carrying TLR-4 and the bacterial PAMP recognized by the TLR-4-bearing cell.

23.2 ## Adaptive Immunity and the Immunoglobulin Superfamily

The **immunoglobulin gene superfamily** consists of genes and their protein products that share structural, evolutionary, and/or functional features with immunoglobulin genes and proteins. The antigen-binding proteins involved in the adaptive immune response are members of this extended gene family.

As we have discussed in Chapter 22, three different cell-surface proteins interact specifically with antigens during the adaptive immune response. These are the **immunoglobulins** (Igs or antibodies), the **T-cell receptors** (TCRs), and the MHC proteins. Each of these antigen-binding proteins has a different location, structure, and function. MHC proteins, found on the surface of antigen-presenting cells (APCs) and target cells, present antigens to TCRs, found exclusively on T cells (∞ Section 22.6). Igs, found on B lymphocytes, interact directly with extracellular antigens (∞ Section 22.9).

Structure and Evolution of Antigen-Binding Proteins

Ig, TCR, and MHC proteins share structural features and have evolved by duplication and selection of primordial antigen-receptor genes. Selected Ig superfamily proteins are shown in Figure 23.1●. The proteins consist of a number of discrete **domains,** regions of the protein having a defined structure and function. Each protein has at least one region of highly conserved amino acid sequence, a *constant (C) domain* of about 100 amino acids in length with an intrachain disulfide bond spanning 50–70 amino acids.

Microbial Sidebar ◆ Toll Receptors and *Drosophila*—An Ancient Response to Infections

The fruit fly, *Drosophila melanogaster*, has been an experimental model for biologists for nearly a century. Now, details of the immune system of *Drosophila*, an invertebrate, have provided insight into innate immune mechanisms in vertebrates, other invertebrates, and even in plants. Several proteins that are involved in fruit fly development also serve as important receptors for invading bacteria, acting as *pattern-recognition molecules (PRMs)*. The best example is the *Drosophila* Toll, a cell-surface receptor involved in development of the dorso-ventral axis formation as well as in the innate immune response of the fly.

Toll immune signaling is initiated by the interaction of a pathogen or its components with the Toll protein. The Toll protein is found as a cell-surface receptor on phagocytes. *Drosophila* Toll itself, however, does not interact directly with the pathogen. Signal transduction events start with the binding of a pathogen-derived molecule, like the lipopolysaccharide (LPS) of gram-negative bacteria (◯◯ Section 4.9), by one or more accessory proteins (see the figure for the analogous TLR-4 system in humans). The LPS-accessory protein complex then binds to Toll. The membrane-integrated Toll protein transduces a signal to the cytoplasm, activating a nuclear transcription factor and inducing transcription of several antimicrobial peptide genes. In fruit flies, Toll-associated transcription factors induce expression of antimicrobial peptides including *drosomycin*, an antifungal peptide, *diptericin*, active against gram-negative bacteria, and *defensin*, active against gram-positive bacteria. The peptides, produced in the liver-like fat body, are released into the fly's circulation where they penetrate the target organism and cause lysis.

Structurally, the Toll proteins are related to lectins, a group of proteins found in virtually all multicellular organisms, including vertebrates and plants. Lectins interact very specifically with certain oligosaccharide monomers. In humans, *Toll-like receptors (TLRs)* react with a wide variety of *pathogen-associated molecular patterns (PAMPs)*, as we describe in Table 23.1. For example, human TLR-4, provides innate immunity against the gram-negative bacteria through interactions with LPS. As with *Drosophila* Toll, the figure shows that TLR-4 does not interact directly with the pathogen or with LPS. Rather, a soluble host-defense protein, *LPS-binding protein (LBP)*, interacts with LPS in circulation. This complex binds to CD14, a protein on the surface of phagocytes. The LPS-CD14 complex then binds to TLR-4. The transmembrane TLR-4 protein initiates a signal cascade, activating the nuclear transcription factor NFκB. NFκB activates transcription of cytokines and other phagocyte proteins involved in the host responses. The activated phagocyte can then destroy pathogens, as well as release inflammatory proteins.

Drosophila Toll is a functional, evolutionary, and structural ancestor to the analogous Toll-like receptor in higher vertebrates, including humans. The Toll proteins and their counterparts are evolutionarily ancient, highly conserved components of the innate immune system in a wide variety of animals and may even be found in plants. ■

Gram-negative bacteria
▲ Lipopolysaccharide (LPS)
LPS-binding protein (LBP)

CD14 — — TLR-4

NF-κB — Signal transduction

Transcription

Translation

Inflammation
IL-1, IL-6, IL-8, TNFα
activation

Figure 1 TLR-4 activation of macrophages in humans. *Lipopolysaccharide (LPS) released from gram-negative bacteria binds LPS-binding protein (LBP) in body fluids. The LPS-LBP complex then binds to CD14. LBP is released and the LPS-CD14 complexes with TLR-4. The new TLR-4–LPS-CD14 complex induces signal transduction through TLR-4, activating NFκB (a transcription factor), leading to the transcription and translation of genes encoding inflammatory cytokines and chemokines such as IL-1, IL-6, IL-8, and TNF-α. These mediators in turn activate other cells and produce inflammation, pathogen uptake, and destruction (◯◯ Sections 22.2 and 22.3).*

Beta-2 microglobulin (β_2m), part of the class I MHC protein (see Section 23.2), consists of a single C domain.

The *variable (V) domains* of TCR, Ig, and MHC proteins are about the same size as the constant domains, but V domain structures can be considerably different from one another and from the C domains. C domains provide structural integrity for the antigen-binding molecules, attach the V domains to the membrane, and give each protein its characteristic shape. C domains also act as targets for accessory molecules. For example, constant domains

● **Figure 23.1 Immunoglobulin gene superfamily proteins.** Constant domains (C) have homologous amino acid sequence and higher-order structure. The Ig-like C domains identify the proteins as members of the Ig gene superfamily, indicating evolutionary relationships. The variable domains (V) of Igs and TCRs and the antigen-binding domains of MHC class I and class II proteins cannot be identified as members of the Ig superfamily because of their nonconserved structure. Presumably, these antigen-binding domains evolved from Ig-like domains, but evolutionary selection to bind certain forms of antigen has diluted structural homology. A number of other proteins involved in immunity including some of the cytokines and complement proteins are also members of the Ig gene superfamily, but only the proteins shown here interact directly with antigens.

of IgG and IgM bind complement (∞Section 22.11), MHC class I constant domains bind to the accessory CD8 protein on T-cytotoxic (T_C) cells, and homologous MHC class II constant domains bind CD4 on T_H cells (∞Section 22.6). The V domains, however, have evolved to interact with a wide variety of antigens.

TCR, Ig, and MHC proteins each consist of two nonidentical polypeptides. These *heterodimers* associate to form a functional protein that is expressed on a cell surface as an antigen receptor. However, the specific functions of these molecules are quite different. *Igs* or *antibodies* can be anchored on B-cell surfaces where they bind to pathogens and their products such as toxins. Igs are also produced in large quantities as soluble serum proteins. *TCRs*, found exclusively on T lymphocytes, interact with antigenic peptides presented by target cells or APCs in the context of *MHC proteins*. The antigen-reactive T cells then kill the interacting cell or produce proteins that activate the immune response (∞Sections 22.7 and 22.8).

 23.2 Concept Check

The Ig gene superfamily encodes proteins that are evolutionarily, structurally, and functionally related to Igs. The antigen-binding Igs, TCRs, and MHC proteins are members of this family, as are other proteins involved in the immune response.

◆ Describe the structural features of the Ig constant domain.

◆ Why do the V domains have structures that are dissimilar to the C domains?

◆ Identify the antigens bound by Igs, TCRs, and MHC proteins.

II THE MAJOR HISTOCOMPATIBILITY COMPLEX (MHC)

The **major histocompatibility complex (MHC)** is a group of genes found in all vertebrates. The MHC spans about 4 million base pairs on human chromosome 6 and is known as the **HLA (human leukocyte antigen)** complex (Figure 23.2●). MHC proteins are involved in antigen processing and presentation.

23.3 MHC Protein Structure

MHC proteins were discovered because of their role as the major antigen barriers for tissue transplantation, hence their name. However, the real biological function of MHC proteins is to bind and present peptide antigens to T cells (∞Section 22.6).

MHC proteins are divided into two structural classes. *Class I MHC proteins* are found on the surfaces of *all nucleated cells*. As a rule, the class I proteins present peptide antigens to *T-cytotoxic cells* (T_C). If the class I embedded peptides are recognized as foreign, the antigen-containing cell is targeted and directly destroyed by the T_C cell (∞Section 22.7). *Class II MHC proteins* are found only on the surface of B lymphocytes, macrophages, and dendritic cells, the *dedicated APCs* (∞Section 22.6). Through the class II proteins, the APCs are involved in presentation of antigens to the *T-helper cells*, stimulating either inflammatory reactions or antibody responses (∞Section 22.8).

DPA DPB   DRB   C4   B
     DQB   DQA   Bf   TNF   C   A
DPA   DQA   DRA   C2

Gene
complex

Region | **Class II** | **Class III** | **Class I**
Size (Mbp)   1   2   3   4

● **Figure 23.2   The human major histocompatibility complex (MHC) gene map.** The human MHC is called the HLA (human leukocyte antigen) complex. The HLA complex is located on chromosome 6 and is more than 4 million bases in length. Class II genes DPA and DPB encode class II proteins DPα and DPβ; DQA and DQB encode DQα and DQβ; DRA and DRB encode DRα and DRβ. Class III genes encode several proteins associated with immune recognition functions, as well as other unrelated proteins. C4 and C2 encode complement proteins C4 and C2 (∞ Section 22.11). The TNF gene encodes a cytokine, tumor necrosis factor (see Section 23.10). The class I MHC proteins HLA-B, HLA-C, and HLA-A are encoded by genes B, C, and A. Over 200 genes, many of which are involved in antigen recognition, processing, and presentation are located in the HLA complex.

Class I MHC proteins

Class I MHC proteins consist of two polypeptides (Figures 23.1 and 23.3●). The membrane-integrated alpha (α) chain is encoded in the MHC gene region on chromosome 6. The other class I polypeptide is the non-covalently associated beta-2 microglobulin (β₂m). The three-dimensional structure of class I MHC protein reveals a distinctive shape that suggests how this protein interacts with the antigen peptide and the TCR simultaneously (Figure 23.3). The class I α chain folds to form a large groove between the α1 and α2 domains, and it is within this groove that the MHC molecule binds to peptide antigens. The groove is closed on both ends, restricting the size of bound peptides to 8–10 amino acids. The folded MHC I protein consists of the two α-helices straddling a β-sheet. This structure binds the peptide and

forms a peptide-MHC complex that interacts with the TCR, as we will discuss (see Section 23.7).

Class II MHC proteins

Class II MHC proteins consist of two noncovalently linked, membrane-integrated polypeptides, called α and β (Figures 23.1 and 23.3). Class II proteins are often arranged in *pairs*, enhancing their ability to bind to TCRs. The α1 and β1 domains of the class II protein interact to form a peptide-binding site similar to the class I peptide-binding site. However, the ends of the groove are *open*, permitting the class II protein to bind and display peptides that may be significantly longer than 10 amino acids (Figure 23.3). The class II-binding peptides are generally proteolytic fragments derived from exogenous pathogen proteins processed in the APCs and

● **Figure 23.3   Three-dimensional structures of MHC proteins.** (a) The class I protein. (b) The class II protein in its dimeric form. The yellow arrows indicate the peptide and their position in the binding site positions. [Reprinted with permission from *Nature* 364:33 (1993), © Macmillan Magazines.] (c) A space-filling model of a class I protein with a bound peptide, as seen from above. A 9-amino acid peptide is shown as a stick structure, embedded in the mouse class I protein. The class I proteins have closed ends and bind peptides of about 8–10 amino acids in length. The open-ended class II proteins can bind peptides of up to about 20 amino acids in length.

ranging in size from 10 to 20 or more amino acids in length (Section 22.6).

23.3 Concept Check

Class I MHC proteins are expressed on all cells and function to present cytosol-derived antigenic peptides to TCRs on T$_C$ cells. Class II MHC proteins are expressed only on APCs. They function to present exogenously derived peptide antigens to TCRs on T$_H$ cells.

◆ Compare the class I and class II MHC protein structures. How do they differ? How are they similar?

◆ Compare the peptide-binding sites of class I and class II MHC proteins. How do they differ? How are they similar?

23.4 MHC Genes and Polymorphism

There are at least three distinct MHC class I genes, HLA-A, -B, and -C, and all show a high degree of *polymorphism*. **Polymorphism** is the occurrence of multiple alleles at a locus (the location of the gene on the chromosome) in frequencies that cannot be explained by the occurrence of recent random mutations. For example, there are 95, 207, and 50 different alleles at the HLA-A, HLA-B, and HLA-C loci, respectively. Thus, within the human species, there are multiple polymorphisms at each locus. However, each individual has only two of these alleles at each locus (one allele is of paternal origin and one is of maternal origin). The two allelic variant proteins are expressed *codominantly* (equally).

Likewise, highly polymorphic loci encode class II proteins HLA-DR, -DP, and -DQ. Again, the class II gene products are expressed codominantly. Thus, an individual usually displays *six* genetically and structurally distinct class I proteins and *six* distinct class II proteins. These polymorphic variations in MHC proteins are the major barriers to successful tissue transplants because the MHC proteins on the donor tissue (graft) are recognized as foreign antigens by the recipient's immune system. An immune response directed against the graft MHC proteins causes cell death and rejection of the graft.

Each MHC gene encodes a single protein. However, the polymorphic nature of the MHC genes ensures that, *within the human population*, there are a large number of MHC proteins available for antigen presentation, and each MHC allele encodes a protein with a different amino acid sequence. Because the amino acid sequence variations in MHC proteins are concentrated in the peptide-binding groove (Figure 23.3), each polymorphic variation of the MHC protein binds a different set of peptide antigens.

The peptides bound by a single MHC protein share a common structural patterns, or **motif,** and each different MHC protein binds a different motif. For example, all of the 8-amino acid peptides bound by a certain class I protein may have a phenylalanine at position 5 and a leucine at position 8. Thus, all peptides with the sequence X-X-X-

X-phenylalanine-X-X-leucine (where X is *any* amino acid) would be bound and presented by that MHC protein. Another MHC class I protein encoded by a polymorphic allele might preferentially bind a different motif, with a valine at position 4 and a proline at position 8 (X-X-X-valine-X-X-X-proline). The invariant amino acids in each motif are known as *anchor residues* because they bind directly and specifically within an individual MHC peptide-binding groove. Thus, an individual MHC protein can bind and present a large number of different peptides, as long as the peptides contain the appropriate anchor residues. Each MHC protein binds a different motif and, hence, different anchor residues. In this manner, the six possible MHC I proteins in an individual bind and present a large number of different peptide antigens. An analogous situation occurs with MHC II proteins and antigen presentation. However, the other antigen-binding proteins, notably the Igs and TCRs, interact very *specifically* with a nearly infinite number of antigens and therefore employ a different genetic mechanism to generate even greater receptor diversity (see Sections 23.5 and 23.7).

23.4 Concept Check

The MHC is a group of genes encoding proteins involved in antigen processing and presentation. Class I and class II MHC genes are the most polymorphic genes known. MHC class I and class II alleles encode proteins that bind and present peptides with conserved structural motifs.

◆ Define and explain *polymorphism* in MHC genes.

◆ How do polymorphic MHC proteins facilitate the presentation of a large number of peptides to T-cell receptors?

 III ANTIBODIES

Antibodies or immunoglobulins (Ig) are found as cell-surface antigen receptor proteins on B cells, or as soluble proteins in serum and other body fluids, where Ig functions to neutralize and opsonize foreign antigens (Sections 22.11 and 24.6). In this section, we look at the structure, antigen-binding function, and genetic organization of the infinitely variable Igs.

23.5 Antibody Proteins and Antigen Binding

As we discussed above (see Section 23.1) and in Chapter 22, functional Igs consist of four polypeptides, two heavy chains and two light chains. The functional antigen-binding unit consists of a heavy-light chain heterodimer. The heavy and light chains are further divided into C (constant) and V (variable) domains, with the C domains responsible for common functions such as complement binding, while the V domains of H and L chains interact to form the antigen-binding site (Figure 23.4●). Here we examine the structural features of the V domains and the antigen-binding site.

Variable Domains

Variations in amino acid sequence occur in the *variable domains* (V domains) of different Igs (Figure 23.4). Amino acid variability is especially apparent in several

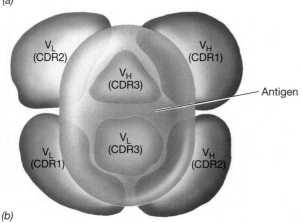

● **Figure 23.4 The variable regions of immunoglobulin light and heavy chains.** (a) One-half of a typical Ig is shown schematically. C_H and C_L are constant domains of the heavy and light chain, respectively. (b) The complementarity-determining regions (CDRs) from both heavy and light chains in (a) are conformed to make a single antigen-binding site, or pocket, on the Ig. The site is shown from above. Red binding areas are from the heavy chain, and blue binding areas are from the light chain. Binding areas are numbered to indicate the CDR and chain. For example, L1 is the CDR1 from the light chain. The highly variable CDR3s from both heavy and light chains cooperate at the center of the site. An antigen is shown in gray, overlaying the site and contacting all CDRs. The actual shape of the site may be a shallow groove or a deep pocket, depending on the antibody-antigen pair chosen.

so-called **hypervariable** or **complementary-determining regions (CDR)**. The CDRs provide most of the molecular contacts with antigen. Each V domain in the light and heavy chains has three CDRs. CDR1 and CDR2 domains differ somewhat between different immunoglobulins, but the CDR3s differ very dramatically from one another. The CDR3 of the heavy chain has a particularly complex structure. Protein sequence comparisons indicate that CDR3 consists of the carboxy-terminal portion of the V domain, followed by a short "diversity" (D) segment of about 3 amino acids, and a longer "joining" (J) region of about 13–15 amino acids in length. The light chain has a similar arrangement for its CDR3, but lacks the D region. All of the CDRs are involved in antigen binding.

Antigen Binding

An Ig three-dimensional structure was shown in Figure 22.16. The principle of all antibody reactions lies in the *specific* combination of determinants on the antigen with the *variable* region. The antigen-binding site of an antibody molecule is formed by the association of heavy and light chains and measures about 2×3 nm. This site is capable of binding an epitope that approximates a 10 to 15-amino acid peptide. Antigen binding is ultimately a function of the Ig folding pattern of the heavy and light polypeptide chains. The Ig folds of the V region bring all six CDRs (CDR1, 2, and 3 from both heavy and light chains) together at the end of the Ig protein. The result is a *unique* and *specific* antigen-binding site (∞Figure 22.16 and Figure 23.4*b*). In the following section, we will examine the genetic mechanisms that produce the tremendous diversity found in the Ig proteins.

 23.5 *Concept Check*

The antigen-binding site of an Ig is composed of the V (variable) domains of one heavy chain and one light chain. Each heavy and light chain contains three complementarity-determining regions, or CDRs, that are folded together to form the antigen-binding site.

◆ Describe the contributions of the V, D, and J regions of the heavy chain and the V and J regions of the light chain to the CDRs.

◆ Draw a complete Ig molecule and identify the antibody antigen-binding sites.

23.6 Antibody Genes and Diversity

For most proteins, one gene encodes one protein. It may therefore seem that if one B lymphocyte produces one Ig, one gene should encode the light chain and one gene should encode the heavy chain. However, this is *not* the case. Because antibodies must specifically recognize and bind a nearly infinite variety of molecular structures, the immune system has evolved to produce almost unlimited variations of antibodies using a limited number of genes. *Somatic recombination, random heavy and light chain reassortment,* and *hypermutation* all contribute to the nearly infinite diversity generated from a relatively small, fixed number of Ig genes.

Immunoglobulin Genes in Pieces

A single light or heavy chain is encoded by *several gene segments* that undergo a series of *somatic rearrangements* (recombination followed by deletion of intervening sequences) as B cells develop. For Ig genes, somatic rearrangements allow the formation of a large number of antibodies generated by recombinations among about 400 different gene pieces. Molecular

studies verified the "genes in pieces" hypothesis by demonstrating that genes for V, D, J, and C regions are separated from one another in the genome, but are brought together to form a single mature Ig gene in each developed B cell (Figure 23.5●). For the heavy chain, the V gene encodes CDR1 and CDR2, while CDR3 is encoded by a mosaic of the 3' end of the V gene, followed by the entire D and J genes. Finally, the class-defining constant domain of the Ig molecule is encoded by the C gene. Thus, four different genes, V, D, J, and C, recombine to form one functional heavy chain gene. Similarly, light chains are encoded by recombination of light chain V, J, and C genes.

The genes required for all Igs exist in each developing B lymphocyte. As shown in Figure 23.5, each B cell

● **Figure 23.5   Immunoglobulin gene rearrangement in human B cells.** Ig genes are arranged in tandem on three different chromosomes. (a) The heavy-chain (H) gene complex on chromosome 14. The filled boxes represent Ig coding genes. The broken lines indicate intervening sequences and are not shown to scale. (b) The kappa (κ) light-chain complex on chromosome 2. The lambda (λ) light-chain genes are in a similar complex on chromosome 22. (c) Assembly of one-half of an antibody molecule.

contains about 150 tandemly arranged light-chain V genes and 5 distinct J genes. About 200 tandem V genes, 50 D genes, and 4 J genes exist for the heavy chains. In addition, the heavy-chain constant domain (C_H) genes and the light-chain constant domain genes (C_L) are present. The V, D, J, and C genes are separated by noncoding sequences (introns) typical of gene arrangements in eukaryotes (∞Section 7.1 and 14.8). During maturation of B lymphocytes, genetic recombination occurs in each B cell. Randomly selected V, D, and J segments are fused, resulting in an active heavy-chain gene and an active light-chain gene. The active gene (containing an intervening sequence between the VDJ gene segment and the C gene segment) is transcribed, and the resulting primary RNA transcript is spliced to yield the final messenger mRNA. The mRNA is then translated to make the heavy and light chains of the Ig molecule.

Reassortment and VDJ (VJ) Joining

Up to this point, all Ig diversity has been generated from recombination of existing genes. The final light chain and heavy chain genes expressed by a given B cell is due to *chance reassortment* of these rearranged heavy chains and light chains. For example, based on the numbers of genes at the kappa (κ) light-chain loci, there are 150 V × 5 J possible rearrangements or 750 possible light chains (∞ Section 22.10 and Figure 22.20). At the heavy-chain (H-chain) loci, there are approximately 200 V × 50 D × 4 J, or 4000 possible heavy chains. Assuming that each heavy chain and light chain has an equal chance to be expressed in each cell, there are 750 × 4000, or 3,000,000 possible antibodies that can be expressed (Figure 23.5).

Another diversity-generating mechanism involves the DNA-joining mechanism. DNA joining of the V-D-J or V-J genes is imprecise and frequently varies the site of VDJ (or VJ) fusions by a few nucleotides. This genetic imprecision is sufficient to change an amino acid or two and leads to even greater antibody diversity. In each B cell, only a single protein-producing rearrangement occurs in the heavy- and light-chain genes. Called *allelic exclusion*, this mechanism ensures that *each B cell produces only one Ig.*

Hypermutation

Additional antibody diversity is generated in B cells by **somatic hypermutation,** the mutation of Ig genes at higher rates than the mutation rates observed in other genes. Somatic hypermutation of Ig genes is usually evident after a second exposure to the immunizing antigen. As we saw, a second exposure to antigen results in a change in the predominant antibody class produced, with a switch from IgM to IgG production (∞Section 22.10). In addition, class switching is accompanied by an increase in antigen-binding strength (affinity). This *affinity maturation* is one of the factors responsible for the dramatically stronger secondary response in immunity (∞Section 22.10 and Figure 22.21). Affinity maturation adds virtually unlimited possibilities to the generation of Ig diversity, producing at least 10^{12} possible Igs.

23.6 Concept Check

Recombination allows shuffling of various pieces of the final Ig genes. Random reassortment of the heavy- and light-chain genes maximizes genetically encoded diversity. Imprecise joining of VDJ and VJ segments as well as hypermutation and affinity maturation also contribute to virtually unlimited immunoglobulin diversity.

◆ Describe the recombination events that produce a mature heavy-chain gene.

◆ Describe the somatic mutation events that further enhance antibody diversity.

◆ How many possible Ig-binding sites can be produced?

 IV **T-CELL RECEPTORS**

TCRs are cell-surface antigen receptors on T lymphocytes that recognize peptide antigens embedded in MHC proteins (∞Section 22.6). In this section, we look at the structure, antigen-binding function, and genetic organization of the TCRs.

23.7 TCR Proteins and Antigen Binding

The TCR is a membrane-integrated protein found only on the surface of T cells. TCRs consist of two polypeptides, the alpha (α) chain and the beta (β) chain. The α:β TCR specifically binds foreign peptides that are embedded in MHC molecules on the surface of APCs or target cells (∞Section 22.6). Thus, *the TCR binds both a self-MHC protein and a foreign peptide.* The TCRs accomplish this dual binding function through a binding site composed of the V domains of the α chain and β chain. As with Igs, the α-chain and β-chain V domains of TCRs contain CDR1, CDR2, and CDR3 segments that bind to antigen.

The three-dimensional structure of the TCR-peptide-MHC protein complex is shown in Figure 23.6●. Both TCR and MHC proteins bind directly to peptide antigen. The MHC protein binds one face of the peptide, the **agretope,** whereas the TCR interacts with the other peptide face, the **epitope.** The CDR regions are directly involved in the binding of the MHC-peptide complex, and each CDR has a specific binding function. The CDR3 regions of the TCR α chain and β chain make contact with the peptide epitope. The CDR1 and CDR2 regions of the TCR α and β chains contact the peptide-binding α-helices of the MHC proteins.

23.7 Concept Check

The TCR is a protein that binds to peptide antigens presented by MHC proteins. The CDR3 regions of both the α chain and the β chain bind to the peptide epitope, whereas the CDR1 and CDR2 regions bind to the MHC protein.

TCR Cβ

TCR Vβ

TCR Vα

Peptide antigen

MHC α1

MHC α2

β₂m

MHC α3

(a)

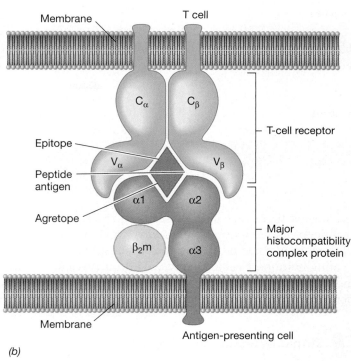

Membrane

T cell

Cα Cβ

T-cell receptor

Epitope

Peptide antigen

Vα Vβ

Agretope

α1 α2

β₂m α3

Major histocompatibility complex protein

Membrane

Antigen-presenting cell

(b)

● **Figure 23.6 The TCR-peptide-MHC I protein complex.** (a) The three-dimensional structure, showing the orientation of TCR, peptide (brown), and MHC. This structure was derived from data deposited in the Protein Data Bank. (b) A diagram-matic interpretation of the MHC–peptide–TCR structure.

◆ Distinguish among the functions of the TCR CDR1, CDR2, and CDR3 segments.

◆ Distinguish between an *agretope* and an *epitope*.

23.8 TCR Genes and Diversity

The TCR α and β chains are encoded by distinct gene segments for constant and variable domains. Much like Ig genes, TCR V region genes are arranged as a number of tandemly arranged genes. The α chain has about 80 variable (V) genes and 61 joining (J) segments, whereas the β chain has 50 variable (V) genes, 2 diversity (D) genes, and 13 joining (J) segments (Figure 23.7●). The β chain V, D, and J genes and the α chain V and J genes undergo recombination to form functional V-region genes. Additional diversity called *N-region diversity* is generated by random additions of one to six nucleotides between V and D gene segments and between D and J gene segments. Finally, the D region of the β chain can be transcribed in all three reading frames, leading to production of three separate transcripts from each D-region gene and creating greater diversity than would be expected from the D gene segments alone. As we discussed for reassortment of Ig H and L chains, individual α and β chains are produced by each T cell at random and joined to form a complete α:β heterodimer. Somatic hypermutation mechanisms are not involved in generating

TCR diversity. However, the potential diversity is still enormous, and theoretically on the order of 10^{15} different TCRs can be generated.

23.8 Concept Check

The V domain of the β chain of the TCR is encoded by V, D, and J gene segments. The V domain of the α chain of the TCR is encoded by V and J gene segments. Diversity generated by recombination, reassortment of gene products, transcription of D regions in three reading frames, and random N nucleotide addition ensure practically unlimited antigen-binding TCRs.

◆ Explain the contributions to diversity for reading of the D region in all three reading frames.

◆ Explain the contributions of the N region additions to TCR diversity.

V MOLECULAR SIGNALS IN IMMUNITY

Antigen recognition through Igs and TCRs is the *first signal* and the most important molecular interaction for activating antigen-specific T and B lymphocytes. Other molecular signals, however, need to be transmitted to lymphocytes before they can respond to antigens. Here we examine some of the other molecular signals necessary to activate the immune response. First, we explore

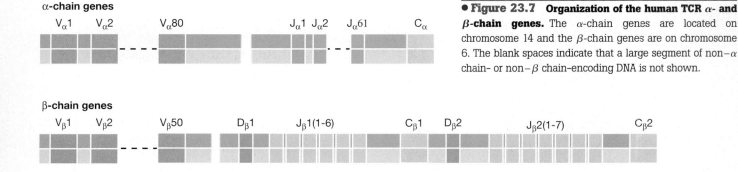

● **Figure 23.7 Organization of the human TCR** α**- and** β**-chain genes.** The α-chain genes are located on chromosome 14 and the β-chain genes are on chromosome 6. The blank spaces indicate that a large segment of non–α chain- or non–β chain-encoding DNA is not shown.

the mechanisms that *select* immune cells to react with foreign antigens and to ignore self proteins. Next, we examine the molecular *second signals* that are responsible for activating immune T cells. Finally, we introduce *cytokines* and *chemokines*, soluble proteins capable of activating other cells in the immune response.

23.9 Clonal Selection and Tolerance

For an effective immune response, T cells must discriminate between the *dangerous non-self antigens* and the *nondangerous self antigens* that compose our body tissue. Thus, T cells must achieve **tolerance,** or specific unresponsiveness to self antigens. As we achieve tolerance, we select immune cells that interact only with dangerous non-self antigens.

Clonal Selection

Clonal selection is a hypothesis stating that each antigen-reactive B cell or T cell has a cell-surface receptor for a single antigen. When stimulated by interaction with that antigen, each cell can replicate, and antigen-stimulated B and T cells grow and differentiate. As a result, antigen-stimulated cells divide and produce a pool of cells that express the same antigen-specific receptors. These copies of the original antigen-reactive cell are known as *clones* (Figure 23.8●). Cells that have not interacted with antigen do not grow.

To respond to the seemingly infinite variety of antigens, a large number of antigen-reactive cells are needed in the body. However, antigen-reactive cells must not provoke immune reactions against self antigens in the host. To meet these special requirements, the immune system selects clones that may be useful against non-self antigens and eliminates or suppresses self-reactive clones.

T-Cell Selection and Tolerance

T cells undergo *selection for* antigen-reactive T cells and *selection against* clones that react with self antigens. Selection against self-reactive clones results in the development of *tolerance,* or specific immune unresponsiveness to self antigens. Previously, we discussed the failure of tolerance and the subsequent development of autoimmunity (⌒Section 22.15).

Lymphocytes that will become T cells leave the bone marrow and enter the thymus, a primary lymphoid organ, via the lymphatic ducts (Figure 23.9●, ⌒Figure 22.2). In the first T-cell maturation stage in the thymus, called **positive selection**, immature T cells interact with the thymic epithelium, using their newly developed TCRs to bind to MHC proteins on the thymus. The T cells that do not bind MHC proteins are programmed to die, a process called **apoptosis**; the T cells that bind thymic MHC proteins continue to grow. Thus, positive selection retains T cells that recognize self-MHC proteins and deletes T cells that do not recognize self-MHC proteins.

In the second stage of T-cell maturation, **negative selection**, the positively selected T-cells continue to interact with thymic MHC proteins, which are complexed with antigens. The antigens in the thymus are mostly of self origin. T cells that react with the thymic self antigens are potentially dangerous because they can react with self tissue antigens (autoimmunity). Therefore, these self-reactive T cells must be eliminated. The self-reactive T cells bind tightly to thymus cells and eventually die. However, the T cells that are destined to interact with non-self antigens do not bind as tightly, presumably because the non-self antigens are not found in the thymus. These T cells do not die, but leave the thymus and migrate to the spleen and lymph nodes where they can contact foreign antigens presented by B lymphocytes and other APCs.

This two-stage mechanism for selecting for antigen-reactive T cells while inducing tolerance is called **clonal deletion**. Precursors of T-cell clones that are either useless or harmful die, and more than 99% of all T cells that enter the thymus do not survive the selection process. T cells that survive positive and negative selection then leave the thymus and travel to the secondary lymphoid organs, where they can participate in the immune response to foreign antigens.

B-Cell Tolerance

The acquisition of immune tolerance in B cells is also necessary because antibodies produced by self-reactive B cells (autoantibodies) may damage host tissue (⌒Section 22.15). B cells also undergo clonal deletion. Many self-reactive B cells are eliminated during development in the bone marrow, the primary lymphoid organ responsible for B-cell development in humans (⌒Section 22.1).

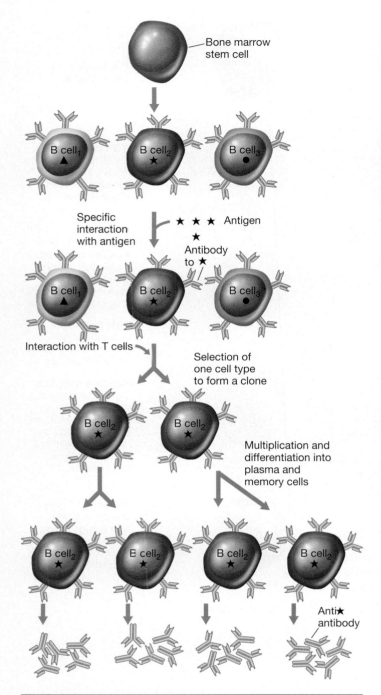

● **Figure 23.8 Clonal selection.** Individual B cells, specific for a single antigen, proliferate and expand to form a clone after interaction with the specific antigen. The antigen acts as the selection agent, driving selection and then proliferation of the individual antigen-specific B cell. Clonal copies of the antigen-reactive cell all have the same antigen-specific surface antibody. Continued exposure to antigen results in continued expansion of the clone. An analogous situation exists for T cells.

In addition to the clonal deletion events, **clonal anergy** (unresponsiveness) also plays a role in final selection. Immature B cells that are self-reactive do not develop, even when exposed to high concentrations of self antigens in the bone marrow. Although self antigens may occupy some B-cell Ig receptors, the B cell is dependent on T cells for a second activation signal, as we

shall see (see Section 23.10). If no antigen-reactive T cell is available to provide help because of elimination during T-cell selection, the B cell remains in a state of anergy.

 23.9 Concept Check

The thymus is a primary lymphoid organ that provides an environment for the maturation of antigen-reactive T cells. Immature T cells that do not interact with MHC protein (positive selection) or react strongly with self antigens (negative selection) are eliminated by *clonal deletion* in the thymus. T cells that survive positive and negative selection leave the thymus and can participate in an effective immune response. B cell reactivity to self antigens is controlled through clonal deletion, selection, and anergy.

◆ Provide an example of a self protein to which you are tolerant and another human protein to which you would *not* be tolerant.

◆ Distinguish between *positive* and *negative* T-cell selection.

◆ For B cells, distinguish between clonal deletion and clonal anergy.

23.10 Second Signals

As we have seen, T cells selected for reactivity against dangerous non-self antigens in the thymus are also selected for *tolerance*, or unresponsiveness to self antigens, and T cells that react with self antigens are deleted in the thymus. However, many self antigens are not expressed in the thymus. T-cell clones responsive to these non-thymus antigens can avoid clonal deletion, but may become unresponsive to self antigen through interactions in the secondary lymphoid organs, developing *clonal anergy*. Analogous to the situation discussed above for B cells, the key to inducing clonal anergy in T cells lies in the two-signal mechanism used to activate T cells.

T-Cell Activation and the Second Signal

When selected mature T cells leave the thymus, they travel to the secondary lymphoid organs (lymph nodes, spleen, and mucosal-associated lymphoid tissue, or MALT), where they take up residence. These T cells have not yet been exposed to antigen and are therefore called naive or uncommitted T cells. Uncommitted T cells must be activated by APC (Section 22.6) before they are fully competent as effector cells.

The first step in activation of uncommitted T cells is binding of the peptide: MHC protein complex on the APC by the TCR (Figure 23.10●). This is *signal 1* and is absolutely required for activation. Without signal 1, a T cell cannot be activated.

The next step in activation of the T cell involves the interaction of two additional proteins, one found on the APC called *B7* and one found only on T cells, called *CD28*. The interaction and binding of B7 to CD28 is

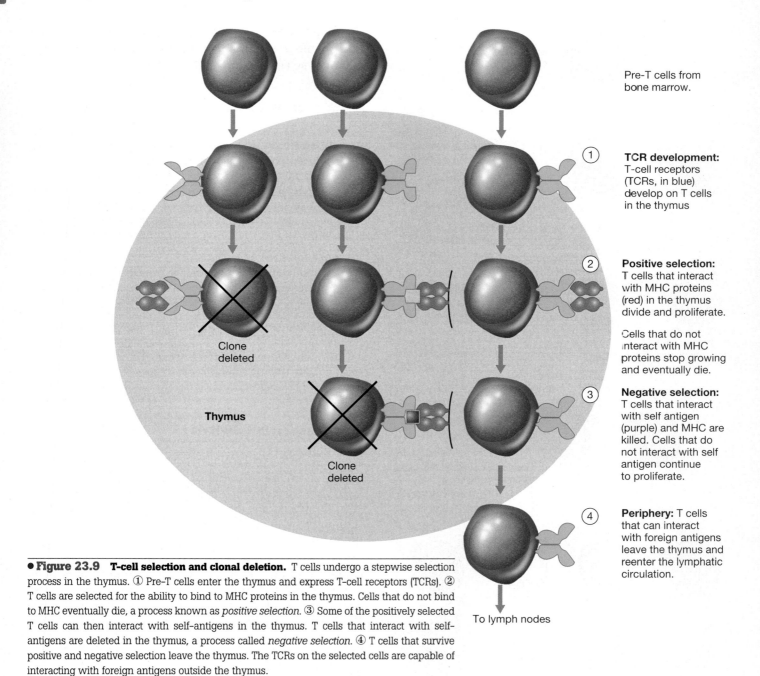

Pre-T cells from bone marrow.

①
TCR development:
T-cell receptors (TCRs, in blue) develop on T cells in the thymus

②
Positive selection:
T cells that interact with MHC proteins (red) in the thymus divide and proliferate.

Cells that do not interact with MHC proteins stop growing and eventually die.

③
Negative selection:
T cells that interact with self antigen (purple) and MHC are killed. Cells that do not interact with self antigen continue to proliferate.

④
Periphery: T cells that can interact with foreign antigens leave the thymus and reenter the lymphatic circulation.

Clone deleted

Thymus

Clone deleted

To lymph nodes

● **Figure 23.9 T-cell selection and clonal deletion.** T cells undergo a stepwise selection process in the thymus. ① Pre-T cells enter the thymus and express T-cell receptors (TCRs). ② T cells are selected for the ability to bind to MHC proteins in the thymus. Cells that do not bind to MHC eventually die, a process known as *positive selection*. ③ Some of the positively selected T cells can then interact with self-antigens in the thymus. T cells that interact with self-antigens are deleted in the thymus, a process called *negative selection*. ④ T cells that survive positive and negative selection leave the thymus. The TCRs on the selected cells are capable of interacting with foreign antigens outside the thymus.

signal 2 and activates the T cell. In the absence of signal 2, the T cell cannot be activated (Figure 23.11●). The activated T cell then becomes an effector cell. A T_C cell that is activated will kill any target cell that displays antigen, even those target cells that do not display CD28. Activated effector cells need only signal 1 (peptide-MHC) to induce effector activity (Figure 23.10).

This requirement for a second activation signal has major implications for establishing and maintaining clonal anergy. For example, an uncommitted T_C cell that interacts with a self antigen on a non-APC will receive only signal 1, since non-APCs do not display the B7 protein necessary to complete signal 2. In the absence of signal 2, this T_C is anergized and cannot be activated (Figure 23.10). Thus, the B7:CD28 second signal is absolutely required for activation. Absence of signal 2 in the presence of signal 1 induces per-

manent anergy. Uncommitted T_H1 and T_H2 lymphocytes are activated in the same way, also using the B7-CD28 co-receptor second signal.

A different second signal is used to activate B cells and other immune response effectors, as we will see in the next section.

 23.10 Concept Check

Many self-reactive T cells are deleted during development and maturation in the thymus. Uncommitted T cells are activated in the secondary lymphoid organs by first binding peptide:MHC with their TCRs (signal 1), followed by binding of the B7 APC protein to the CD28 T-cell protein (signal 2). Uncommitted self-reactive T cells are anergized in the secondary lymphoid organs if they interact with signal 1 in the absence of signal 2.

● **Figure 23.10   Signal 1 and signal 2 are required to activate naive T cells.** (a) A naive T_c cell interacts via TCR with the peptide-MHC complex on an APC. This is signal 1. The T_c cell also has a CD28 protein that interacts with a B7 protein on the APC. This is signal 2. The simultaneous interactions of the T_c cell and APC via signal 1 and signal 2 activate the naive T cell. (b) This permanently activated T_c cell is then capable of killing any target cell as long as signal 1 interactions take place. (c) A naive T_c cell interacts via the TCR with the peptide-MHC complex on any cell. Although the conditions for signal 1 (interactions via TCR with the peptide-MHC complex) are met, signal 2 cannot be generated because only APCs display the B7 protein. (d) In the absence of signal 2, the T_c cell becomes permanently unresponsive, or anergized.

◆   Define signal 1 and signal 2 for an *uncommitted* T cell.

◆   Identify the signal(s) necessary to induce effector function in an *activated* T cell.

23.11   Cytokines and Chemokines

Intercellular communication, necessary for activation in the immune system, is accomplished through a family of soluble proteins known as **cytokines.** Cytokines, produced by leukocytes, regulate a variety of cellular functions in immune cells. Some cytokines have very broad effects, activating a variety of cell types. However, the

● **Figure 23.11   Cytokines and antibody production.** Some major effects and interactions of IL–1, IL–2, and IL–4 cytokines in the antibody response. Cytokines provide signals for macrophages, T cells, and B cells. Cytokines stimulate enhanced antigen uptake by macrophages, as well as activation, differentiation, and expansion of both T and B lymphocytes.

unifying feature of these heterogeneous proteins is that they are all produced by leukocytes. The cytokines produced by lymphocytes are often called *lymphokines*. The cytokines called *interleukins (IL)* mediate interactions between leukocytes.

In general, cytokines are secreted from one cell and bind specific receptors on a target cell. Some cytokines bind to receptors on the cell that produced them. Thus, these cytokines have *autocrine* (self-stimulatory) abilities. Binding the cytokine receptors generally activates a signal transduction pathway (∞Section 8.12), relaying information across the cell membrane to control activities such as protein synthesis and cell division. These signals can ultimately result in cell growth, differentiation, and clonal proliferation.

Table 23.2 lists some important cytokines, their major producer cells, their most common target cells, and their biological effects. In all, there are nearly 40 known cytokines, most of which are produced by either T_H cells or monocytes and macrophages. We now examine the activity of three cytokines involved in the induction of an antigen-specific antibody-mediated immune response.

IL-1, IL-2, and IL-4

As we have seen, macrophages are responsible for antigen uptake, processing, and presentation (◌◌◌ Section 22.16). In addition, they secrete a cytokine known as IL-1 that affects several different cell types (Table 23.2 and Figure 23.11). IL-1 is a key component of the immune response because T_H cells are activated by IL-1. Because of the proximity of T_H cells and macrophages in the lymph nodes, nearby T_H2 cells are very likely to be activated. IL-1 binding by the IL-1 receptors (IL-1R) on the T_H2 cell acts as an activation signal. The activated T_H2 cell, in turn, responds by producing IL-2, which is secreted and bound by the IL-2R on the surface of the T_H2 cells. Thus, IL-2 can activate the same cell that secreted it. Under the influence of IL-2, the cell divides, making clonal copies. In the process, the T_H2 cell also makes other cytokines, in this case IL-4, which then binds to the IL-4R on B cells. The IL-4:IL-4R complex stimulates the B cells to differentiate into plasma cells, which ultimately produce antibodies (◌◌◌Section 22.10). Thus, IL-1, IL-2, and IL-4 cytokines are soluble mediators and activators for macrophages, T lymphocytes, and B cells, the cells that interact to produce the antibody-mediated immune response.

Other Cytokines

Table 23.2 shows the activity of several other cytokines. Many of these proteins affect cells involved in specific immunity. For example, IL-4 primarily affects B cells. However, several cytokines do not affect T or B lymphocytes, but act on other cells; the cytokine-activated cells in turn serve as important modulators of innate host responses. For example, interferons (IFN-α and IFN-γ) are produced by leukocytes and inhibit viral replication in virtually any cell in the body. Tumor necrosis factors TNF-α and TNF-β can kill a variety of tumors if the TNF-producing cells have access to the tumor. TNF-α is also a critical activator of inflammation (◌◌◌Section 22.3). Interferons and TNFs appear to have no target cell specificity, but are produced by T cells as well as phagocytes and can amplify the effects of immune cells.

Chemokines

Chemokines are a group of small proteins that function as chemoattractants for phagocytic cells and T cells. They are produced by lymphocytes and a wide variety of other cells in response to bacterial products, viruses, and other agents that cause damage to host cells. Chemokines attract phagocytes and T cells to the site of injury, stimulating an inflammatory response as well as potentiating a specific immune response.

About 40 chemokines are known. Perhaps the best-studied chemokines are IL-8 and *macrophage chemoattractant protein-1 (MCP-1)* (Table 23.2). IL-8 is produced by a wide variety of cell types including monocytes, macrophages, fibroblasts (connective tissue cells), and keratinocytes (skin cells) in response to tissue injury or contact

Table 23.2	Properties of some major cytokines and chemokines		
Cytokine	**Producers**	**Major targets**	**Effect**
IL-1[a]	Monocytes	T_H	Activation[a]
IL-2	Activated T cells	T cells	Growth, differentiation
IL-3	T_H1	Hematopoietic stem cells	Growth factor
IL-4	T_H2	B cells	IgG1 and IgE synthesis
IL-5	T_H2	B cells	IgA synthesis
IL-10	T_H2	T_H1	Inhibits T_H1
IL-12	Macrophages, dendritic cells	T_H1, NK cells	Differentiation, activation
IFN-α[b]	Leukocytes	Normal cells	Anti-viral
IFN-γ	T_H1	Macrophages	Activation
GM-CSF[c]	T_H1	Myeloid stem cells	Differentiation to granulocytes, monocytes
TGF-β[d]	T_H1 and T_H2	Macrophages	Inhibits activation
TNF-α[e]	T_H1, macrophages, NK cells	Macrophages	Activation
TNF-β	T_H1	Macrophages	Activation
Chemokines			
IL-8	Macrophages, fibroblasts, keratinocytes	Neutrophils, T cells	Attractant and activator
MCP-1[f]	Macrophages, fibroblasts, keratinocytes	Macrophages, T cells	Attractant and activator

[a]IL, interleukin; [b]IFN, interferon; [c]GM-CSF, granulocyte, monocyte-colony stimulating factor; [d]TGF, T-cell growth factor; [e]TNF, tumor necrosis factor, [f]MCP, macrophage chemoattractant protein.

with pathogens. IL-8 is secreted by the affected cells and binds to the surrounding tissue, where it is a chemoattractant for T cells and neutrophils (⚭ Section 22.1). This results in a neutrophil-mediated inflammatory response followed by a specific immune response mediated by the attracted T cells. As is the case for the cytokine receptors, engaged chemokine receptors on the target cells act through signal transduction pathways (⚭ Section 8.12) to induce activation of the target phagocytes or T cells.

MCP-1 is also produced by a variety of cells and attracts macrophages and T cells, again stimulating production of inflammatory mediators and potentially organizing an antigen-specific immune response. Thus, chemokines are potent initiators of nonspecific inflammatory reactions that lead to recruitment of T cells and antigen-specific immune reactions.

⬣ 23.11 Concept Check

Cytokines are soluble mediators produced by leukocytes that regulate interactions between cells. Several cytokines such as IL-1, IL-2, and IL-4 affect leukocytes and are critical components in the generation of specific immune responses. Other cytokines such as IFN and TNF affect a wide variety of cell types. *Chemokines* are produced by a variety of cell types in response to injury and are potent attractants for nonspecific inflammatory cells and T cells.

◆ Compare the target cells for IL-1, IL-4, IL-12, IFN-γ, and TNF-β.

◆ How do cytokines and chemokines differ with respect to their cell sources? Their cell targets?

◆ What events stimulate cytokine production? What events stimulate chemokine production?

REVIEW QUESTIONS

1. Identify the pattern recognition molecules (PRMs) and their interacting pathogen-associated molecular patterns (PAMPs) (⚭ Section 23.1).

2. Define the criteria used to assign a gene and its encoded protein to the Ig gene superfamily (⚭ Section 23.2).

3. Identify the major structural features of class I and class II MHC proteins (⚭ Section 23.3).

4. Polymorphism implies that each different MHC protein binds a different peptide motif. For the HLA class I polymorphisms, how many different HLA proteins are expressed in an individual? By the entire human population (⚭ Section 23.4)?

5. Which Ig chains are used to construct a complete antigen-binding site? Which domains? Which CDRs (⚭ Section 23.5)?

6. Calculate the total number of V_H and V_L domains that can be constructed from the available Ig genes. How many complete Ig proteins can be produced from the reassortment of all possible heavy chains and light chains (⚭ Section 23.6)?

7. Describe the interaction of the TCR with peptide antigen and MHC protein. Be sure to identify the roles of the CDRs in the TCR (⚭ Section 23.7).

8. In TCRs, diversity can be generated by recombination and reassortment events such as in Igs. However, additional diversity is accomplished by the use of N-region nucleotide additions and reading of the D (diversity) segment in all three reading frames. Explain how these diversity-generating mechanisms work (⚭ Section 23.8).

9. Explain *positive* and *negative* selection of T cells (⚭ Section 23.9).

10. What molecular interactions are necessary for activation of naive T cells? For activation of effector cells (⚭ Section 23.10)?

11. What are the chief effects of cytokines and chemokines? What are the chief differences between cytokines and chemokines (⚭ Section 23.11)?

APPLICATION QUESTIONS

1. Identify the consequences of a genetic mutation that eliminates a PRM by predicting the outcome for the host. Do this for each PRM.

2. Construct a table that lists the common features of proteins encoded by members of the Ig gene superfamily. For Igs, TCRs, and MHC proteins, identify the structural components that fit these common features.

3. Polymorphism implies that each different MHC protein binds a different peptide motif. However, for the class I proteins, only 6 peptide motifs can be recognized in an individual, while over 350 motifs can be recognized by the entire human population. What advantage does this have for the population? For the individual?

4. While genetic recombination events are important for generating significant diversity in the antigen-binding site of Igs, postrecombination somatic events may be even more important in achieving overall Ig diversity. Do you agree or disagree with this statement? Explain.

5. What would happen to the T-cell repertoire in the absence of positive selection? In the absence of negative selection?

6. What would be the result of activation of all peripheral T cells that contact antigen? How does the second signal scheme prevent this from happening?

DIAGNOSTIC MICROBIOLOGY AND IMMUNOLOGY

24

Precautions must be taken to protect laboratory personnel from laboratory infections. The figure shows a worker in a maximum-containment biosafety level-4 (BSL-4) facility working with a lethal human pathogen.

WORKING GLOSSARY

Agglutination reaction between antibody and particle-bound antigen, resulting in visible clumping of the particles

Antibiogram a report indicating the sensitivity of clinically isolated microorganisms to the antibiotics in current use

Bacteremia the presence of bacteria in the blood

Differential media growth media that allows identification of microorganisms based on phenotypic properties

ELISA *enzyme-linked immunosorbent assay*

Enriched media media that allow metabolically fastidious microorganisms to grow because of the addition of specific growth factors

Enrichment culture the use of selected culture media and incubation conditions to isolate microorganisms from natural samples

Fluorescent antibody covalent modification of an antibody molecule with a fluorescent dye; the dye makes the antibody visible under fluorescent light

General purpose media growth media that support the growth of most aerobic and facultatively anaerobic organisms

Immunoblot (Western blot) electrophoresis of proteins followed by transfer to a membrane and detection by addition of specific antibodies

Monoclonal antibody antibody made by a single B cell clone

Neutralization interaction of antibody with antigen that reduces or blocks the biological activity of the antigen

Nucleic acid probe an oligonucleotide of unique sequence used as a hybridization probe for identifying specific genes

Polyclonal antibodies antibodies made by many different B cell clones

Precipitation reaction between antibody and a soluble antigen resulting in a visible, insoluble complex

RIA *radioimmunoassay*

Selective media media that enhance the growth of certain organisms while retarding the growth of others due to an added media component

Sensitivity the lowest amount of antigen that can be detected

Septicemia blood infection

Serology the study of antigen–antibody reactions *in vitro*

Specificity the ability of an antibody to recognize a single antigen

Titer in an immunological context, the quantity of antibody present in a solution

The microbiologist plays a critical role in identifying the agents that cause infectious disease. This area of microbiology is called *diagnostic* or *clinical microbiology*. Clinical laboratories can grow, isolate, and identify most routinely encountered pathogenic bacteria within 48 hours of sampling. Immunologic and molecular methods can also be used to identify many pathogens without culturing the organism. These methods are particularly important for the diagnosis of viral and protozoal infections, diseases that are a challenge to identify because of the difficulty of culturing the causal agent.

 GROWTH-DEPENDENT DIAGNOSTIC METHODS

We begin with a discussion of the principles involved in isolation and growth of pathogens from host tissue. Growth-dependent methods are important for identifying many pathogens.

24.1 Isolation of Pathogens from Clinical Specimens

The physician, following examination of a patient, may suspect that an infectious disease is present. Samples of infected tissues or fluids are then collected for microbiological, immunological, and molecular biological analyses (Figure 24.1●). Materials collected may include blood, urine, feces, sputum, cerebrospinal fluid, or pus from a wound. A sterile swab may be used to sample a suspected infected area (Figure 24.2●). The swab is then used to inoculate the surface of an agar plate or a tube of liquid culture medium. In some cases, small pieces of living tissue may be sampled for culture. Table 24.1 summarizes recommendations for initial culture of organisms isolated from typical clinical specimens.

If clinically relevant organisms are to be isolated and identified, the specimen must be properly obtained. The clinician must ensure that the specimen is removed from the *actual site of the infection*. In addition, recovery of pathogens may not be possible if insufficient inoculum is taken. The sample must also be taken under aseptic conditions so that contamination is avoided. Care must also be taken to ensure that metabolic requirements for certain organisms, such as anoxic conditions, are maintained. Once obtained, the sample must be analyzed as soon as possible.

Growth Media and Culture

Enrichment culture, the use of selected culture media and incubation conditions to isolate microorganisms from samples (∞Section 18.1), is an important part of clinical microbiology. Through the use of various specialized growth media, most microorganisms of clinical importance can be grown, isolated, and identified. Most clinical samples are first grown on **general purpose media**, media such as blood agar (∞Figure 21.18 and Table 24.1) that support the growth of most aerobic and facultatively anaerobic organisms. **Enriched media** that allow metabolically fastidious organisms to grow because of the addition of specific growth factors are often necessary to enhance the growth of certain pathogens, such as *Neisseria gonorrhoeae*, the organism that causes gonorrhea (Table 24.1).

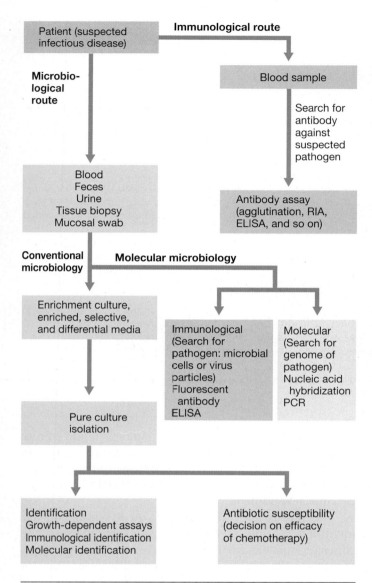

● **Figure 24.1   Isolation and identification of pathogens.** Clinical and diagnostic methods used for isolation and identification of infectious pathogens.

● **Figure 24.2   Methods for obtaining specimens from the upper respiratory tract.** (a) Throat swab. (b) Nasopharyngeal swab passed through the nose. (c) Swabbing the inside of the nose.

Certain media are **selective media**, media that enhance the growth of certain organisms while retarding the growth of others due to an added media component (Table 24.1). Finally, **differential media** are specialized media that allow identification of organisms based on their growth and appearance on the media (see Figure 24.7).

Blood Cultures

Bacteremia is the presence of bacteria in the blood (⚬⚬ Section 21.7). Bacteremia is extremely uncommon in healthy individuals, normally occurring only transiently in response to invasive procedures like dental work or trauma. The prolonged presence of bacteria in the blood is generally indicative of systemic infection. The most common pathogens found in blood include *Pseudomonas aeruginosa*, enteric bacteria, especially *Escherichia coli* and

Klebsiella pneumoniae, and the gram-positive cocci *Staphylococcus aureus* and *Streptococcus pyogenes*.

Septicemia is a blood infection resulting from the growth of a virulent organism entering the blood from a focus of infection, multiplying, and traveling to various body tissues to initiate new infections. Septicemia results in severe systemic symptoms, including fever and chills, followed by prostration. Severe cases of septicemia may result in *septic shock*, a life-threatening systemic condition characterized by severe reduction in blood pressure and multiple organ failures, including heart, kidneys, and lungs. Blood cultures provide the only immediate way of isolating and identifying the causal agent, and diagnosis therefore depends on careful and proper blood culture.

The standard blood culture procedure is to draw 20 ml of blood aseptically from a vein and inject it into two blood culture bottles containing an anticoagulant and an all-purpose culture medium. One bottle is incubated aerobically and one anaerobically. Blood culture bottles are incubated at 35°C and examined several times each hour for up to five days in most automated systems. Most clinically significant bacteria are recovered within two days, while more fastidious organisms may be recovered in three to five days. Some blood culture systems employ a chemical that lyses red and white blood cells, releasing intracellular pathogens. Microorganisms from blood cultures are commonly detected by indicators of microbial growth in automated systems, microscopic examination, and subculture. Automated blood culture systems detect growth by monitoring carbon dioxide production and turbidity as often as every 10 minutes.

A certain amount of contamination from the normal flora of the skin is unavoidable when blood is drawn, and a contamination rate of 2–3% can be expected. Certain organisms commonly found on the skin, such as *Staphylococcus epidermidis*, coryneform bacteria, or propionibacteria, are common contaminants. However, these organisms can occasionally cause infection of the heart (subacute bacterial endocarditis) or colonization of intravascular devices such as artificial heart valves. Thus, considerable microbiological and clinical experience is necessary when interpreting blood culture results.

Table 24.1   Recommended enriched and selective media for primary isolation of pathogens

Specimen	Media[a]				
	Blood agar	Enteric agar	CA	MTM	ANA
Fluids from chest, abdomen, pericardium, joint	+	+	+	−	+
Feces: rectal or enteric transport swabs[b]	+	+	+	−	−
Surgical tissue biopsies	+	+	−	−	+
Throat, sputum, tonsil, nasopharynx, lung, lymph nodes	+	+	+	−	−
Urethra, vagina, cervix	+	+	+	+	−
Urine	+	+	−	−	−
Blood[c]	+	+	+	−	+
Wounds, abscesses, exudates	+	+	+	−	+

SOURCE: Adapted from Murray, P. R., E. J. Baron, J. H. Jorgenson, M. A. Pfaller, and R. H. Yolken. 2003. *Manual of Clinical Microbiology*, 8th edition. American Society for Microbiology, Washington, DC..

[a] Blood agar, 5% whole sheep blood added to trypticase soy agar; enteric agar, either eosin-methylene blue (EMB) agar or MacConkey agar; CA, chocolate (heated blood) agar; MTM, modified Thayer-Martin agar; ANA, anaerobic agar, thioglycolate-containing blood agar or supplemented thioglycolate agar incubated anaerobically.

[b] Special enteric pathogen media, SMAC (MacConkey agar with sorbitol), is also used to culture fecal and enteric samples. SMAC is a selective and differential medium used for the isolation and identification of sorbitol-negative enteric pathogens such as enteropathogenic *Escherichia coli*.

[c] Blood is cultured initially in broth. Depending on the Gram stain characteristics of isolates, subculturing is done on MacConkey agar (gram-negative) or chocolate agar (gram-positive)

Urine Cultures

Urinary tract infections are very common, especially in females. Because the causal agents are often members of the normal flora (for example, *Escherichia coli*), considerable care must be taken in the bacteriological analysis of urine. In most cases, urinary tract infection occurs as a result of an organism ascending the urethra from the outside, infecting the bladder. Urinary tract infections are also the most common form of *nosocomial* (hospital-acquired) infection (∞Section 25.7).

Significant urinary infection generally results in bacterial counts of 10^5 or more organisms per milliliter of a clean-voided midstream specimen. In the absence of infection, contamination of the urine from the external genitalia (almost unavoidable to some extent) results in less than 10^3 organisms per milliliter. The most common urinary tract pathogens are members of the enteric bacteria, with *E. coli* accounting for about 90% of the cases. Other urinary tract pathogens include *Klebsiella, Enterobacter, Proteus, Pseudomonas, Staphylococcus saprophyticus*, and *Enterococcus faecalis. Neisseria gonorrhoeae*, the causal agent of gonorrhea, does not grow in the urine itself, but on the urethral epithelium, and is diagnosed by different methods (see later).

Direct microscopic examination of urine may be used to indicate *bacteriuria*, the presence of abnormal numbers of bacteria in the urine. However, because nearly all urine contains some level of bacterial growth, bacteriuria is most commonly monitored by using commercially available dipstick tests. For example, one dipstick test monitors the reduction of nitrate by detecting the reduction product, nitrite. A positive test is indicated by a color change on the dipstick (Figure 24.3●). Since significant nitrite production occurs in urine only when large numbers ($>10^5$ per milliliter) of enteric organisms

are present, the method is a very rapid check for urinary tract infections. Other dipstick tests for urinary tract infections, often used in conjunction with nitrate reduction, detect esterase (produced by leukocytes, ∞ Section 22.1) and peroxidase (produced by a variety of bacteria, ∞Sections 6.15 and 17.22). Positive dipstick tests are followed by a urine culture.

A Gram stain (∞Section 4.1) may also be done directly on urine samples exhibiting bacteriuria to identify characteristic morphotypes of potential urinary tract pathogens. These include gram-negative rods (enteric *Bacteria*, ∞Section 12.11), gram-negative cocci (*Neisseria*, ∞ Section 12.10), and gram-positive cocci (*Enterococcus*, ∞ Section 12.19). The Gram stain and other direct staining methods are also useful for direct detection of bacteria in other body fluids such as sputum (∞Section 26.5).

John Martinko and Cheryl Broadie

● **Figure 24.3   Urinalysis dipstick test.** A control strip is shown underneath the test strip. From left to right, the strip measures abnormal levels of glucose, bilirubin, ketones, specific gravity, blood, pH, protein, urobilinogen, nitrite, and leukocytes (esterase) in a urine sample. Abnormal readings for esterase (trace positive, far right) and nitrite (strong positive, second from right) indicate bacteriuria. Subsequent culture of this sample indicated the presence of *Escherichia coli*.

To culture potential urinary tract pathogens, two media are normally used: (1) blood agar as a nonselective general medium, and (2) a medium selective for enteric bacteria, such as MacConkey or eosin-methylene blue agar (EMB) (see Section 24.2 and Figure 24.4●). These specialized enteric media permit the initial differentiation of lactose fermenters from nonfermenters, while the growth of gram-positive organisms such as *Staphylococcus* spp. (common skin contaminants) is inhibited. Experienced clinical microbiologists may make a tentative identification of an isolate by observing the color and morphology of colonies of the suspected pathogen growth on various media as described in Table 24.2. Such an identification must be followed with more detailed tests, but clinical microbiologists use this information in conjunction with further test results, discussed throughout the remainder of this chapter, to make a positive identification.

Urine cultures can be done quantitatively by counting colonies on blood agar or a selective agar medium, using a calibrated amount of urine, usually 1 μl, as the inoculum for a plate.

Finally, if no bacterial growth is obtained despite persistent urinary tract infection symptoms, a clinician may request direct cultures for fastidious organisms such as *Neisseria gonorrhoeae, Chlamydia trachomatis, Branhamella* species, mycoplasma, or several anaerobic organisms.

Fecal Cultures

Proper collection and preservation of feces is important in the isolation of intestinal pathogens. During storage, fecal acidity increases so extended delay between sampling and sample processing must be avoided. This is especially critical for the isolation of *Shigella* and *Salmonella* species, both of which are sensitive to acid pH.

Freshly collected fecal samples are placed in a vial containing phosphate buffer for transport to the lab. Bloody or pus-containing stools as well as stools from patients with suspected foodborne or waterborne infections are inoculated into a variety of selective media (see Section 24.2) for isolation of individual bacteria. Intestinal parasites are identified by observing cysts microscopically in the stool sample or through antigen-detection assays (see Section 24.5 through 24.11) rather than by culture methods. Many laboratories also use a variety of selective and differential media and incubation conditions to identify *Escherichia coli* O157:H7 and *Campylobacter*, two important intestinal pathogens generally acquired from contaminated food or water (∞Sections 29.8 and 29.9).

Wounds and Abscesses

Infections associated with traumatic injuries such as animal or human bites, burns, cuts, or the penetration of foreign objects must be carefully sampled to recover the relevant pathogen, and results must be interpreted carefully. Wound infections and abscesses are frequently contaminated with normal flora, and swab samples from such lesions are frequently misleading. For abscesses and other purulent lesions, the best sampling method is to aspirate pus with a sterile syringe and needle following disinfection of the skin surface. Internal purulent lesions are sampled by biopsy or from tissues removed in surgery.

A variety of pathogens are associated with wound infections. Since some of these pathogens are anaerobes, proper evaluation requires that samples be obtained, transported, and cultured under anaerobic as well as aerobic conditions. For example, potential pathogens commonly associated with purulent discharges from wound infections are *Staphylococcus aureus*, enteric bacteria, *Pseudomonas aeruginosa*, and anaerobes from the genera *Bacteroides* and *Clostridium*. The major isolation media are blood agar, several selective media for enteric bacteria (Tables 24.1 and 24.2), and blood agar containing additional supplements and reducing agents for obligate anaerobes. Gram stains from such specimens are examined directly by microscopy.

Genital Specimens and the Laboratory Diagnosis of Gonorrhea

In males, a purulent urethral discharge is the classic symptom of the sexually transmitted disease *gonorrhea* (∞Section 26.12). If no discharge is present, a sample can be obtained using a sterile narrow-diameter cotton swab that is inserted into the anterior urethra, left in place a few seconds to absorb any exudate, and then removed for identification of *Neisseria gonorrhoeae*, the causative agent of gonorrhea. Alternatively, a sample of the first early morning urine of an infected individual

John Martinko and Cheryl Broadie

● **Figure 24.4   An eosin-methylene blue (EMB) agar plate.** The plate shows a lactose fermenter, *Escherichia coli* (left), and a nonlactose fermenter, *Pseudomonas aeruginosa* (right). Note the green metallic sheen of the *E. coli* colonies.

Table 24.2 | **Colony characteristics of frequently isolated gram-negative rods cultured on various clinically useful media**

| Organism | Agar media[a] | | | | |
	EMB	MC	SS	BS	HE
Escherichia coli	Dark center with greenish metallic sheen (see Figure 24.4)	Red or pink	Red to pink	Mostly inhibited	Yellow-pink
Enterobacter	Similar to *E. coli*, but colonies are larger	Red or pink	White or beige	Mucoid colonies with silver sheen	Yellow-pink
Klebsiella	Large, mucoid, brownish	Pink	Red to pink	Mostly inhibited	Yellow-pink
Proteus	Translucent, colorless	Transparent, colorless	Black center, clear periphery	Green	Clear
Pseudomonas	Translucent, colorless to gold (see Figure 24.4)	Transparent, colorless	Mostly inhibited	No growth	Clear
Salmonella	Translucent, colorless to gold	Translucent, colorless	Opaque	Black to dark green	Green or transparent with black centers
Shigella	Translucent, colorless to gold	Transparent, colorless	Opaque	Brown or inhibited	Green or transparent

SOURCE: Adapted from Murray, P. R., E. J. Baron, J. H. Jorgenson, M. A. Pfaller, and R. H. Yolken. 2003. *Manual of Clinical Microbiology*, 8th edition. American Society for Microbiology, Washington, DC.

[a] BS, Bismuth sulfite agar; EMB, eosin-methylene blue agar; MC, MacConkey agar; SS, *Salmonella-Shigella* agar; HE, Hektoen enteric agar.

usually contains viable cells of *N. gonorrhoeae*. In females suspected of having gonorrhea or other genital infections, samples are usually obtained by swab from the cervix and the urethra.

Clinical microbiology procedures are central to the diagnosis of gonorrhea. *N. gonorrhoeae* (referred to clinically as *gonococcus*) colonizes mucosal surfaces of the urethra, uterine cervix, anal canal, throat, and conjunctiva. The organism is sensitive to drying and therefore is transmitted almost exclusively by direct person-to-person contact, usually by sexual intercourse. Public health measures to control gonorrhea involve identification of asymptomatic carriers, and this requires microbiological analysis.

Neisseria gonorrhoeae is usually found as a gram-negative diplococcus, but can be quite pleomorphic. No similar microorganisms are observed among the normal flora of the urogenital tract. Thus, direct microscopy of a gram-stained vaginal or cervical smear showing gram-negative diplococci is a probable indication of gonorrhea. In acute gonorrhea, microscopic examination of purulent discharges usually reveals phagocytized gram-negative diplococci in the neutrophils (∞Section 22.2) (Figure 24.5a●).

Most laboratory testing of urogenital samples for *N. gonorrhoeae* (and the often-associated *Chlamydia trachomatis*, ∞Section 26.13) is done using nucleic acid probe or

Cells of
N. gonorrhoeae

Theodor Rosebury

(a)

Leon J. LeBeau

(b)

● **Figure 24.5** **Identification of *Neisseria gonorrhoeae*.** (a) Photomicrograph of *Neisseria gonorrhoeae* within human polymorphonuclear leukocytes from a urethral exudate. Note the paired diplococci (leader). (b) *N. gonorrhoeae* growing on Thayer-Martin agar. The plate has been stained in the middle with a reagent that turns colonies blue if cells contain cytochrome *c* (the oxidase test). *N. gonorrhoeae* cells are oxidase-positive.

amplification procedures (see Section 24.12). However, specimens obtained from nonurogenital sources, such as the eyes and rectum, should also be cultured.

Nonselective *enrichment* media for the isolation of *N. gonorrhoeae* contain heat-lysed blood and are called *chocolate agar* because of the deep brown appearance. The heated blood interacts with the media components, absorbing compounds that are normally toxic for *N. gonorrhoeae*. One of several *selective* media used for primary isolation is modified Thayer-Martin (MTM) agar (Figure 24.5). This medium incorporates the antibiotics vancomycin, nystatin, trimethoprim, and colistin to suppress the growth of normal flora, but these antibiotics do not affect either *N. gonorrhoeae* or *N. meningitidis*, the cause of bacterial meningitis (◘Section 26.6).

Inoculated plates are incubated in a humid environment in an atmosphere containing 3–7% CO_2, required for growth of gonococci. The plates are examined after 24 and 48 hours and tested by the oxidase test because all *Neisseria* are oxidase-positive (Figure 24.5*b*; see Section 24.2). Oxidase-positive gram-negative diplococci growing on chocolate agar or selective media are presumed to be gonococci if the inoculum was derived from genitourinary sources, but definitive identification requires determination of carbohydrate utilization patterns and immunological or nucleic acid probe tests (see Sections 24.5–24.12).

Culture of Anaerobes

Obligately anaerobic bacteria are common causes of infection. Isolation and identification of anaerobic pathogens requires special isolation and culture methods. In general, media for anaerobes do not differ greatly from those used for aerobes, except that they are (1) usually richer in organic constituents, (2) contain reducing agents (usually cysteine or thioglycolate), and (3) contain a redox indicator (◘Section 6.15). Specimen collection, handling, and processing are designed to exclude possible oxygen contamination because oxygen is toxic to obligately anaerobic organisms.

There are several habitats in the body (for example, portions of the oral cavity and the lower intestinal tract) (Sections 21.3 and 21.4) that are generally anoxic and in which obligately anaerobic bacteria are part of the normal flora. However, other parts of the body can also become anoxic as a result of tissue injury or trauma, reducing blood supply and oxygen perfusion to the injured site. These anaerobic sites can then be colonized by obligate anaerobes. In general, pathogenic anaerobic bacteria are part of the normal flora and are opportunistic pathogens. Two important exceptions are the pathogenic anaerobes *Clostridium tetani* (causal agent of tetanus) (◘Section 27.9) and *Clostridium perfringens* (causal agent of gas gangrene and a type of food poisoning) (◘Section 29.6), both endospore-forming *Bacteria* that are predominantly soil organisms.

Anaerobic culture methods involve the usual problems of specimen contamination as well as the addition-al challenges of maintaining an anoxic growth environment. Samples collected by syringe aspiration or biopsy must be immediately placed in a tube containing oxygen-free gas, usually with a dilute salt solution containing a reducing agent such as thioglycolate and the redox indicator dye *resazurin*. This dye is colorless when reduced and becomes pink when oxidized, indicating oxygen contamination of the specimen If an anaerobic transport tube is not available, the syringe itself can be used to transport the specimen; the needle is discarded and the syringe is plugged with a rubber stopper so that no air enters the syringe.

For anaerobic incubation, agar plates are placed in a sealed jar, which is made anoxic by either replacing the atmosphere in the jar with an oxygen-free gas mixture (usually a mixture of N_2 and CO_2) or by adding some compound to the enclosed vessel that removes O_2 from the atmosphere. For example, as shown in Figure 24.6, H_2 is generated and, in the presence of a suitable catalyst, usually palladium, the H_2 is combined with free O_2 to form H_2O, thus removing the contaminating oxygen. Alternate means for providing anaerobic conditions include the use of culture media containing reducing agents or the use of anoxic "glove boxes" filled with an oxygen-free gas such as nitrogen or hydrogen (◘Figure 6.26*b*).

⬡ *24.1* **Concept Check**

Proper sampling and culture of the suspected pathogen is the most reliable way to identify an organism that causes a disease. The selection of appropriate sampling and culture conditions requires knowledge of bacterial ecology, physiology, and nutrition.

● **Figure 24.6 Sealed jar for incubating cultures under anoxic conditions.** The catalyst and hydrogen generator packet produce and maintain a reducing (anoxic) environment.

◆ Why are urine cultures almost always positive for bacterial growth?

◆ Describe precautions required for successful isolation of anaerobic pathogens.

24.2 Growth-Dependent Identification Methods

If the inoculation of a primary medium results in bacterial growth, the clinical microbiologist must identify the organism or organisms present. Identification of a clinical isolate can frequently be made using a variety of growth-dependent assays. We discuss some of these methods here.

Growth on Selective and Differential Media

Based on growth characteristics on primary isolation media, an unknown pathogen is usually subcultured onto media that are designed to measure one of many different biochemical reactions. Some of the most important biochemical tests are listed in Table 24.3. Many of the specialized media are available in miniaturized kits containing a number of media, all in separate wells, all of which can be inoculated at one time (Figure 24.7●).

The media employed are selective, differential, or both. A *selective medium* contains compounds that inhibit the growth of certain microorganisms. A *differential medium* contains an indicator, usually a dye, that allows differentiation between chemical reactions carried out during growth. Eosin-methylene blue (EMB) agar, for example, is a widely used selective *and* differential medium. EMB agar is used for the isolation of gram-negative enteric *Bacteria*. Methylene blue dye inhibits the growth of most gram-positive *Bacteria*. Eosin is a dye that responds to changes in pH, going from colorless to black under acidic conditions. EMB agar contains lactose and sucrose, but not glucose, as energy sources. Lactose-fermenting bacteria such as *Escherichia coli*, *Klebsiella*, and *Enterobacter* acidify the medium and the colonies appear black with a greenish sheen. Colonies of lactose nonfermenters, such as *Salmonella*, *Shigella*, and *Pseudomonas*, are translucent or pink (Figure 24.4). Thus, EMB preferentially *selects* for the growth of gram-negative *Bacteria*, and *differentiates* among several genera of the selected gram-negative *Bacteria*.

These individual biochemical tests measure the presence or absence of *enzymes* involved in catabolism of the substrate or substrates in the differential medium. Fermentation of sugars is measured by incorporating pH indicator dyes that change color on acidification

Table 24.3	Important clinical diagnostic tests for bacteria (continued on next page)		
Test	**Principle**	**Procedure**	**Most common use**
Carbohydrate fermentation	Acid and/or gas produced during fermentative growth with sugars or sugar alcohols	Broth medium with carbohydrate and phenol red as pH indicator; inverted tube for gas	Enteric bacteria differentiation
Catalase	Enzyme decomposes hydrogen peroxide, H_2O_2	Add drop of H_2O_2 to dense culture and look for bubbles (O_2) (Figure 6.29)	*Bacillus* (+) from *Clostridium* (−); *Streptococcus* (−) from *Micrococcus-Staphylococcus* (+)
Citrate utilization	Utilization of citrate as sole carbon source, results in alkalinization of medium	Citrate medium with bromthymol blue as pH indicator. Look for intense blue color (alkaline pH)	*Klebsiella-Enterobacter* (+) from *Escherichia* (−), *Edwardsiella* (−) from *Salmonella* (+) (Figure 24.7)
Coagulase	Enzyme causes clotting of blood plasma	Mix dense liquid suspension of bacteria with plasma, incubate, and look for fibrin clot	*Staphylococcus aureus* (+) from *S. epidermidis* (−)
Decarboxylases (lysine, ornithine, arginine)	Decarboxylation of amino acid releases CO_2 and amine	Medium enriched with amino acids. Bromcresol purple pH indicator becomes purple (alkaline pH) if there is enzyme action	Aid in determining bacterial group among the enteric bacteria
β-Galactosidase (ONPG) test	Orthonitrophenyl-β-galactoside (ONPG) is an artificial substrate for the enzyme. When hydrolyzed, nitrophenol (yellow) is formed.	Incubate heavy suspension of lysed culture with ONPG Look for yellow color	*Citrobacter* (+) from *Salmonella* (−). Identifying some *Shigella* and *Pseudomonas* species
Gelatin liquefaction	Many proteases hydrolyze gelatin and destroy the gel	Incubate in broth with 12% gelatin. Cool to check for gel formation. If gelatin is hydrolyzed, tube remains liquid on cooling	To aid in identification of *Serratia*, *Pseudomonas*, *Flavobacterium*, *Clostridium*

Table 24.3 Important clinical diagnostic tests for bacteria (continued)

Test	Principle	Procedure	Most common use
Hydrogen sulfide (H_2S) production	H_2S produced by breakdown of sulfur amino acids or reduction of thiosulfate	H_2S detected in iron-rich medium from formation of black ferrous sulfide (many variants: Kliger's iron agar and triple sugar iron agar also detect carbohydrate fermentation)	In enteric bacteria, to aid in identifying *Salmonella*, *Edwardsiella*, and *Proteus* (Figure 24.7)
Indole test	Tryptophan from proteins converted to indole	Detect indole in culture medium with dimethyl-aminobenzaldehyde (red color) or in colony smeared on paper containing dimethylaminocinnamaldehyde (spot test; blue color)	To distinguish *Escherichia* (+) from most *Klebsiella* (−) and *Enterobacter* (−); *Edwardsiella* (+) from *Salmonella* (−) *Proteus vulgaris* (+) from *Proteus mirabilis* (−)
Methyl red test	Mixed-acid fermenters produce sufficient acid to lower pH below 4.3	Glucose-broth medium. Add methyl red indicator to a sample after incubation	To differentiate *Escherichia* (+, culture red) from *Enterobacter* and *Klebsiella* (usually −, culture yellow)
Nitrate reduction	Nitrate (NO_3^-) as alternate electron acceptor, reduced to NO_2^- or N_2	Broth with nitrate. After incubation, detect nitrite with α-naphthylamine-sulfanilic acid (red color). If negative, confirm that NO_3^- is still present by adding zinc dust to reduce NO_3^- to NO_2^-. If no color after zinc, then $NO_3^- \rightarrow N_2$	To aid in identification of enteric bacteria (usually +)
Oxidase test	Cytochrome *c* oxidizes artificial electron acceptor: tetramethyl *p*-phenylenediamine (Kovac's reagent), or dimethyl *p*-phenylenediamine (Gordon and McLeod's reagent).	Colonies are smeared on paper impregnated with reagent. Oxidase-positive colonies produce dark purple-black color in 10–15 s with Kovac's reagent and blue color in 10–30 m with Gordon and McLeod's reagent.	To differentiate *Neisseria* and *Moraxella* (+) from *Acinetobacter* (−), pseudomonads (+) and *Vibrionaceae* (+) from *Enterobacteriaceae* (−) To aid in identification of *Aeromonas* (+)
Oxidation-fermentation (O/F) test	Some organisms produce acid only when growing aerobically	Acid production in top part of sugar-containing culture tube; soft agar used to restrict mixing during incubation	To differentiate *Micrococcus* (acid produced aerobically only) from *Staphylococcus* (acid produced anaerobically). To characterize *Pseudomonas* (aerobic acid production) from enteric bacteria (acid produced anaerobically)
Phenylalanine deaminase test	Deamination produces phenylpyruvic acid, which is detected in a colorimetric test	Medium enriched in phenylalanine. After growth, add ferric chloride reagent and look for green color	To characterize the genus *Proteus* and the *Providencia*
Starch hydrolysis	Iodine-iodide gives blue color with starch	Grow organism on plate containing starch. Flood plate with Gram's iodine and look for clear zones around colonies	To identify typical starch hydrolyzers such as *Bacillus* spp.
Urease test	Urea, $H_2N-CO-NH_2$, split to $2 NH_3 + CO_2$	Medium with 2% urea and phenol red indicator. Ammonia release raises pH, intense pink-red color	To distinguish *Klebsiella* (+) from *Escherichia* (−), and *Proteus* (+) from *Providencia* (−). To identify *Helicobacter pylori* (+).
Voges-Proskauer test	Acetoin produced from sugar fermentation	Chemical test for acetoin using α-naphthol	To separate *Klebsiella* and *Enterobacter* (+) from *Escherichia* (−). To characterize members of the genus *Bacillus*

(a)

(b)

(c)

(d)

(e)

● **Figure 24.7 Growth-dependent diagnostic methods used for the identification of clinical isolates by color changes in various diagnostic media.** (a) Use of a differential medium to assess sugar fermentation. Acid production is indicated by color change of the pH-indicating dye added to the liquid medium. If gas production occurs, a bubble appears in the inverted vial in each tube. From left to right: acid, acid and gas, negative, uninoculated. (b) A conventional diagnostic test for enteric bacteria in a medium called *triple sugar iron (TSI) agar*. The medium is inoculated both on the surface of the slant and by stabbing into the solid agar butt. The medium contains a small amount of glucose and a large amount of lactose and sucrose. Organisms able to ferment only the glucose cause acid formation only in the butt, whereas lactose or sucrose-fermenting organisms also cause acid formation throughout the slant. Gas formation is indicated by the breaking up of the agar in the butt. Hydrogen sulfide formation (either from protein degradation or from reduction of thiosulfate in the medium) is indicated by a blackening due to reaction of the H_2S with ferrous iron in the medium. From left to right: fermentation of glucose only; no reaction; hydrogen sulfide formation; fermentation of glucose and another sugar. (c) Measurement of citrate utilization by *Salmonella* on Simmons citrate agar. The change in pH causes a change in the color of the indicating dye. From left to right: positive, negative, uninoculated. (d) Media kits used for the rapid identification of clinical isolates. The principle is the same as in (a), but the whole arrangement has been miniaturized so that a number of tests can be run at the same time. Four separate strips, each with a separate culture, are shown. (e) Another arrangement of a miniaturized test kit. This one defines sugar utilization in nonfermentative organisms.

(Figure 24.7a). Production of hydrogen and/or carbon dioxide during sugar fermentation is assayed by observing gas production either in gas collection vials or in agar (Figure 24.7a,b). Hydrogen sulfide (H_2S) production is indicated following growth in a medium containing ferric iron. If sulfide is produced, ferric iron reacts with H_2S to form a black precipitate of iron sulfide (Figure 24.7b). Utilization of citric acid, an organic acid with three carboxylic acid groups, is accompanied by a pH increase, and a specific dye in the citric acid test medium changes color as conditions become alkaline (Figure 24.7c). Hundreds of differential tests have been developed for clinical use, but only about 20 are used routinely (Figure 24.7d).

The biochemical patterns for various pathogens are stored in a computer databank. The results of the differential tests on an unknown pathogen are entered, and the computer makes the best identification by comparing the characteristics of the unknown to metabolic patterns in the pathogen database. For many organisms, as few as three or four key tests are sufficient to make an unambiguous identification. In cases of a dubious match, however, more sophisticated identification procedures may be required.

Clinical Diagnosis

Growth-dependent rapid identification systems are often used to identify enteric bacteria because these organisms are frequently implicated in routine urinary tract and intestinal infections (Figure 24.7) (see Section 24.1). Other growth-dependent rapid identification kits are available for other bacterial groups or even for single bacterial species. For example, kits containing a battery of tests have been developed for *Staphylococcus aureus*, *Streptococcus pyogenes*, *Neisseria gonorrhoeae*, *Haemophilus influenzae*, and *Mycobacterium tuberculosis*. Other kits are available for identification of the pathogenic fungi *Candida albicans* and *Cryptococcus neoformans* (∞Section 27.8).

The decision to use a specific diagnostic test is usually made by the clinical microbiologist who must consider the nature of the clinical specimen, the basic characteristics (morphology and Gram stain) of pure cultures obtained, and previous experience with similar cases.

 24.2 Concept Check

Traditional methods for identifying pathogens depend on observing metabolic changes induced as a result of growth. These growth-dependent methods provide rapid and accurate pathogen identification.

◆ Distinguish between *selective* and *differential* media. Give an example of a medium used for each purpose.

◆ Suggest appropriate general, selective, and differential media for preliminary isolation of pathogens from (1) a urine culture, and (2) a blood culture.

24.3 Antimicrobial Drug Susceptibility Testing

Identification of microbial pathogens isolated from clinical specimens is used to confirm medical diagnoses and to guide antimicrobial therapy. Determination of the susceptibility of microbial isolates to antimicrobial agents is a critical health care issue.

We discussed some principles for the measurement of antimicrobial activity in Chapter 20. The susceptibility of a culture can be most easily determined by an agar diffusion method or by using a tube dilution technique to determine the minimum inhibitory concentration (MIC) of an agent that is necessary to inhibit growth (∞ Section 20.4). Food and Drug Administration (FDA) regulations control the automated instruments used for susceptibility testing in the United States. Procedures and standards, including experimental endpoints for each organism and antibiotic, are constantly updated by the National Committee for Clinical Laboratory Standards (NCCLS), a nonprofit, standards-developing organization that establishes voluntary consensus standards for antibiotic testing as well as other healthcare technologies (http://www.nccls.org).

The MIC procedure for antibiotic susceptibility testing involves an *antibiotic dilution assay*, either in culture tubes (∞Figure 20.10) or in the wells of a microtiter plate (Figure 24.8e●). Wells containing serial dilutions of antibiotics are inoculated with a standard amount of the test organism. Growth in the presence of the various antibiotics is observed by measuring turbidity. Antibiotic susceptibility is usually expressed as the *highest dilution* (lowest concentration) of antibiotic that completely inhibits growth. This defines the value of the MIC.

The standard procedure that assesses antimicrobial activity is called the *Kirby–Bauer method* (Figure 24.8). Agar media are inoculated by evenly spreading a defined density of a suspension of the pure culture on the agar surface. Filter paper disks containing a defined quantity (μg/disk) of the antimicrobial agents are then placed on the inoculated agar. After a specified period of incubation, the diameter of the inhibition zone around each disk is measured. Table 24.4 presents zone sizes for several antibiotics. Inhibition zone diameters are then interpreted into susceptibility categories based on zone size. Interpretive criteria for the various agents for different bacteria are provided either by the FDA or the NCCLS.

Etest® (AB BIODISK, Solna, Sweden) is a non-diffusion based technique that employs a preformed and predefined gradient of an antimicrobial agent immobilized on a plastic strip. The concentration gradient covers a MIC range across 15 two-fold dilutions of the conventional method. When applied to the surface of an inoculated agar plate, the gradient is transferred from the strip to the agar and remains stable for a period that covers the wide variation of critical times associated with the growth characteristics of different microorganisms. After overnight incubation or longer, an elliptical zone of inhi-

Table 24.4 Standards for dilution for some antimicrobial disk diffusion susceptibility tests[a]

Antibiotic	Amount on disk	Inhibition zone diameter (mm)[b]		
		Resistant	Intermediate	Susceptible
Ampicillin[c]	10 μg	13 or less	14–16	17 or more
Ampicillin[d]	10 μg	28 or less	—	29 or more
Ceftriaxone	30 μg	13 or less	14–20	21 or more
Chloramphenicol	30 μg	12 or less	13–17	18 or more
Clindamycin	2 μg	14 or less	15–20	21 or more
Erythromycin	15 μg	13 or less	14–22	23 or more
Gentamicin	10 μg	12 or less	13–14	15 or more
Methicillin[c]	5 μg	9 or less	10–13	14 or more
Nitrofurantoin	300 μg	14 or less	15–16	17 or more
Penicillin G[e]	10 units	28 or less	—	29 or more
Penicillin G[f]	10 units	14 or less	—	15 or more
Streptomycin	10 μg	6 or less	7–9	10 or more
Sulfonamide	10 μg	12 or less	13–16	17 or more
Tetracycline	30 μg	14 or less	15–18	19 or more
Trimethoprim-sulfamethoxazole	1.25/23.75 μg	10 or less	11–15	16 or more
Tobramycin	10 μg	12 or less	13–14	15 or more
Vancomycin[g]	10 μg	14 or less	15–16	17 or more
Vancomycin[h]	10 μg	14 or less (test for MIC)	—	15 or more

[a] Standards are defined and updated by NCCLS, an international nonprofit organization that develops voluntary consensus standards for antibiotic testing and other healthcare technologies (http://www.nccls.org)

[b] See Figure 24.8 for illustrations of standard susceptibility testing methods.

[c] For Enterobacteriaceae.

[d] For staphylococci and highly penicillin-sensitive organisms.

[e] For staphylococci.

[f] For organisms such as enterococci that may cause some systemic infections treatable with high doses of penicillin G.

[g] For enterococci.

[h] For staphylococci.

bition centered along the axis of the strip develops. The MIC value in μg/ml can be read at the point where the ellipse edge intersects the precalibrated Etest strip, providing a precise MIC. This value can then be interpreted using FDA or NCCLS MIC data (Figure 24.8f).

Because of the widespread occurrence of antimicrobial drug resistance (∞Section 20.12), an antibiotic susceptibility test is recommended for most pathogens isolated from clinical specimens. Data such as those in Table 24.4 are useful for choosing the best antibiotic for a specific bacterial infection. Although most community-acquired pathogens are susceptible to a number of different antibiotics, some pathogens, for example, *Pseudomonas aeruginosa*, are susceptible to very few antimicrobial drugs.

Nosocomial pathogens are those acquired in hospitals. Many nosocomial pathogens have developed antibiotic resistance, and a few are resistant to all known antibiotics (∞Section 20.12). Thus, antibiotic susceptibility testing for hospital-acquired pathogens is essential for effective chemotherapy. Using the drug susceptibility information gathered in this fashion, the hospital infection control microbiologist generates periodic reports called **antibiograms** that indicate the susceptibility of clinically isolated organisms to the antibiotics in current local use. Antibiograms are valuable for monitoring control of known pathogens and for tracking the emergence of new antibiotic-resistant strains of pathogens, especially at the local level.

 24.3 Concept Check

Antimicrobial drugs are widely used for the treatment of infectious diseases. Pathogens should be tested for susceptibility to individual antibiotics to ensure appropriate chemotherapy. This rigorous approach to antimicrobial drug treatment is usually applied only in health care settings.

◆ Describe the Kirby–Bauer technique. For an individual organism and antimicrobial agent, what do the results indicate?

◆ What is the value of antimicrobial drug susceptibility testing for the microbiologist, the physician, and the patient, both for community-based and nosocomial infections?

24.4 Safety in the Microbiology Laboratory

Microbiology laboratories, and especially clinical laboratories, pose significant potential for infectious biological hazards to workers and those who come into contact with workers. Hence, a defined protocol of standard laboratory practices for handling clinical samples has been established to avoid laboratory accidents. In the United States, every clinical and research institution that deals with human or primate tissue is required by law to have an occupational exposure control plan for the handling of all

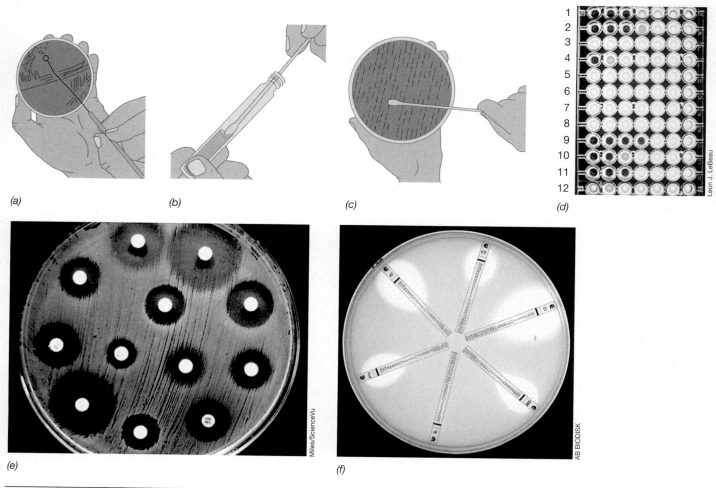

(a) (b) (c) (d)

(e) (f)

● **Figure 24.8 Antibiotic susceptibility testing.** (a–e) The Kirby–Bauer procedure for determining the susceptibility of an organism to antibiotics. (a) Isolated pure colonies are homogenized in a tube with an appropriate liquid medium to achieve a specified density as judged by comparison to a turbidity standard. (b) A sterile cotton swab is dipped into the bacterial suspension and excess fluid removed by pressing the swab against the side of the tube. (c) The swab is streaked evenly over the surface of an appropriate agar medium. (d) Antibiotic susceptibility determined by the broth dilution method. The organism is *Pseudomonas aeruginosa*. Each row has a different antibiotic. The microtiter plate enables automation of these tests. The end point is the first well with the lowest concentration of antibiotic that shows no visible bacterial growth. The highest concentration of antibiotic is in the well at the left; serial two-fold dilutions are made in the wells to the right. For example, in rows 1 and 2, the end point is the third well. In row 3, the antibiotic is ineffective at the concentrations tested, since there is bacterial growth in all the wells. In row 4, the end point is in the first well. (e) Disks containing known amounts of different antibiotics are placed on the agar surface. After incubation, inhibition zones are seen. The susceptibility category of the organism is determined by reference to an interpretive chart of zone sizes (Table 24.4). (f) Antibiotic susceptibility determined by the Etest® (AB BIODISK, Solna, Sweden) for different antibiotics (from 8 o'clock, PTc-piperacillin/tazobactam; AT-aztreonam; CT-cefotaxime; CI-ciprofloxacin; GM-gentamicin; IP-imipenem). Each strip is calibrated in terms of the minimum inhibitory concentration (MIC) in μg/ml starting with the lowest concentration from the center of the plate. The lowest concentration of antibiotic that inhibits bacterial growth is the MIC value for that particular agent (⊂⊃Section 20.4). For example, the MIC for cefotaxime (CT) is 16 μg/ml. This organism is resistant to imipenem (IP); MIC > 32 μg/ml.

bloodborne pathogens. This law was specifically designed to protect workers from infection by hepatitis B virus (HBV) (⊂⊃Section 26.11) and human immunodeficiency virus (HIV) (⊂⊃Section 26.14), but effectively limits infection by all pathogens because of the implementation of stringent infection controls in the laboratory.

The two most common causes of laboratory accidents are ignorance and carelessness. Appropriate training and enforcement of established safety procedures, however, can prevent most accidents. Unfortunately, most laboratory-acquired infections do not result from identifiable exposures or accidents, but rather from routine handling of patient specimens. Infectious aerosols,

generated during processing of the specimen, are the most common cause of laboratory infections. Clinical laboratories follow the safety rules outlined here (required by law in the United States) to minimize the exposure of health care workers to infectious agents and thereby reduce the numbers of nonaccident-associated laboratory infections.

1. Laboratories handling hazardous materials *must* restrict access to include only laboratory and support personnel. These individuals *must* have knowledge of the biological risks involved in the laboratory and act accordingly.

2. Effective procedures for decontaminating infectious materials or wastes, including specimens, syringes and needles, inoculated media, bacterial cultures, tissue cultures, experimental animals, glassware, instruments, and surfaces *must* be in place and be practiced without compromise. A 5.25% (full strength) chlorine bleach solution or other approved disinfectant is *recommended* for decontaminating spilled infectious material. All potentially infectious waste must be burned in a certified incinerator or handled by a licensed waste handler.

3. Personnel working with hazardous infectious agents or vaccines (for example, rabies, polio, or diphtheria-pertussis-tetanus vaccines) *must* be properly vaccinated against the agent. Persons working with human or primate tissue *must* be vaccinated against HBV.

4. All clinical specimens *should* be considered infectious and handled appropriately. This is especially important for preventing laboratory-acquired hepatitis because of the relative frequency with which hepatitis viruses are present in blood specimens (⌘Section 26.11).

5. All pipetting *must* be done with mechanical pipetting devices (not by mouth).

6. Animals *should* be handled only by trained laboratory personnel, and anesthetics and/or tranquilizers *should* be used to avoid injury to both personnel and animals.

7. Laboratory personnel *must* wear laboratory coats or gowns, sealed shoes, rubber gloves, masks, eye protection, respiratory devices when needed, and other barrier protection as deemed appropriate by the level of exposure and the severity of the potential infection. These barrier devices *must* also be properly stored and decontaminated after use. Laboratory personnel *must* also practice good personal hygiene with respect to hand washing. *Eating and drinking, applying cosmetics or lip balm, or wearing contact lenses is never permitted in the clinical laboratory.*

8. Because of the special risks associated with AIDS, all clinical (human) specimens *should* be treated as if they contain HIV. Protective gloves *should* be worn whenever handling specimens of *any* kind. Masks and/or full-face shields *must* be worn any time there is a possibility of generating an aerosol during specimen preparation. Needles *must* not be resheathed, bent, or broken; they should be placed in a labeled container designated expressly for this purpose that can be sealed and decontaminated before disposal.

These safety rules should be in effect in all laboratories that handle potential infectious agents. Specialized clinical laboratories may have additional rules to ensure a safe work environment, as we will discuss below. In the final analysis, however, it is the responsibility of the personnel to assure safety in the laboratory. Any clinical laboratory is a potentially hazardous place for untrained personnel or those unwilling to take the necessary steps to prevent laboratory-acquired infection.

Biological Containment and Laboratory Biosafety Levels

The level of containment used to prevent accidental infections or accidental environmental contamination (escape) in clinical, research, and teaching laboratories must be adjusted to counter the biohazard potential of the organisms handled in the laboratory. Laboratories are classified according to their containment potential, or *biosafety level (BSL)*, and are designated as *BSL-1, BSL-2, BSL-3,* or *BSL-4*. Laboratories working at all biosafety levels must follow good laboratory practices that ensure basic cleanliness and limit contamination. For example, no food or drink is to be consumed in the laboratory, and laboratory surfaces must be decontaminated after each work shift or whenever spills occur. Personnel must wash hands when leaving the laboratory, and access to the laboratory is restricted.

BSL-1 laboratories are the lowest level of containment. Work can be done on the open bench with organisms that present a *low risk of infection*; these organisms are not pathogens in normal individuals and include organisms such as *Bacillus subtilis*. An example of a BSL-1 facility would be a basic microbiology student teaching laboratory that does not use pathogens.

BSL-2 laboratories are designed to contain organisms that present a *moderate risk of infection* due to accidental ingestion, percutaneous injection, or exposure to mucous membranes via aerosols. Work with pathogens such as *Escherichia coli* or *Streptococcus pyogenes* is done in a BSL-2 laboratory or at higher containment levels. Normal procedures can be performed on bench tops, but barrier protection such as face and eye protection, gloves, and lab coats or gowns (barrier protection devices) are used when appropriate. Most microbiology research, clinical, and instructional laboratories maintain BSL-2 containment standards.

BSL-3 laboratories are designed to contain pathogens that have a *very high potential for causing infections, especially from aerosols*. For example, if laboratory personnel handle extremely infectious airborne pathogens such as *Mycobacterium tuberculosis*, the causative agent of tuberculosis, the laboratory should be fitted with special features such as negatively pressurized rooms and air filters to prevent accidental release of the pathogen from the laboratory. Biological safety cabinets (⌘Figure 20.4) are required for manipulations; work must not be done on the open bench. In some special cases, organisms that can normally be handled at BSL-2 must be handled at BSL-3. For example, *Staphylococcus aureus* can be handled on culture plates at BSL-2. However, when large quantities are grown, and especially when such quantities are centrifuged, work must be done in a BSL-3 facility to contain potential infectious aerosols. Specialized research and teaching facilities often have a BSL-3 laboratory.

BSL-4 laboratories are designed for *maximum containment of life-threatening pathogens that have a high probability of transmission by aerosols and for which there is no effective immunization or cure.* Physical containment in BSL-4 facilities must include some form of total isolation of the organism, such as manipulation through gloves in a sealed biological safety cabinet or by personnel wearing full-body, positive pressure suits with air supplies (Figure 24.9●). Some examples of pathogens that must be manipulated in a BSL-4 facility include hemorrhagic fever viruses (Lassa, Marburg, Ebola, ∞Section 25.11), variola virus (smallpox, ∞Section 25.12), and drug-resistant *Mycobacterium tuberculosis* (∞Section 26.5). BSL-4 laboratories are found only in research and development facilities that are specifically designed and equipped to handle the most dangerous pathogens.

24.4 *Concept Check*

Safety in the clinical laboratory requires effective training, planning, and care to prevent the infection of laboratory workers with pathogens. Materials such as live cultures, inoculated culture media, used hypodermic needles, and patient specimens require specific precautions for safe handling.

♦ What are the major precautions necessary to prevent spread of a bloodborne pathogen to laboratory personnel?

♦ What are the major causes of laboratory infections?

♦ Identify the basic features of BSL containment laboratories, from BSL-1 to BSL-4.

II IMMUNOLOGY AND CLINICAL DIAGNOSTIC METHODS

Immunoassays are widely used in clinical, reference, and research laboratories for the detection of specific pathogens or pathogen products. They are used to confirm infections when the agent is not or cannot be cultured. When culture methods for pathogens are not routinely available or are prohibitively difficult to perform, as is the case with most viral infections, including infection with the human immunodeficiency virus (HIV), immunoassays often provide an effective and relatively simple means of identifying a specific pathogen.

24.5 Immunoassays for Infectious Disease

Specific immune antibodies to pathogens are utilized *in vitro* to detect infectious diseases. In some cases, the antibody or cellular immune response of a patient is used to indicate infection by a pathogen. In other cases, antibodies are used in tests to identify pathogens *in vitro*.

Innate Immunity

The immune response was discussed in Chapter 22. The major aspects of immunity are summarized in Figure

● **Figure 24.9 A worker in a BSL-4 (biological safety level-4) laboratory.** BSL-4 is the highest level of biological control, affording maximum worker protection and pathogen containment. The worker has a whole-body sealed suit with an outside air supply and ventilation system. Air locks control all access to the laboratory. All material leaving the laboratory is autoclaved or chemically decontaminated.

24.10●. For a pathogen that the body has never before encountered, the pathogen must first be recognized, usually by cells called phagocytes (∞Section 22.1). A group of specialized receptors on the phagocytes recognize one or more repeating macromolecular patterns, such as the repetitive chains of saccharide groups on the cell wall (∞Section 22.2). Recognition of macromolecular patterns shared by a number of pathogen types or species activates the phagocyte to ingest and destroy the targeted pathogens, a process called *phagocytosis.* Pattern recognition followed by destruction of pathogens is called *innate immunity* or *nonspecific immunity* and is the cornerstone for the ability to resist invasive pathogens.

Antigen-Specific Immunity and Antibodies

In the second phase of immunity, the phagocytes present pathogen-derived *antigens* (proteins obtained from the destroyed pathogen) to antigen-specific lymphocytes known as T cells (Figure 24.10 and ∞Section 22.1). T cells known as *T helper* (T_H) cells do not act directly on the pathogen but recruit and stimulate (help) other cells. T_H2 cells, a particular T_H subset, activate other antigen-specific lymphocytes, the *B cells*. The B cells then respond by producing soluble, antigen-binding proteins known as *antibodies* (Figure 24.10 and ∞Sections 22.9 and 22.10). A *primary antibody response* generally occurs within five days, but antibodies do not reach peak quantities for several weeks. The antibody proteins are pathogen-specific, exclusively recognizing individual antigens from particular pathogens.

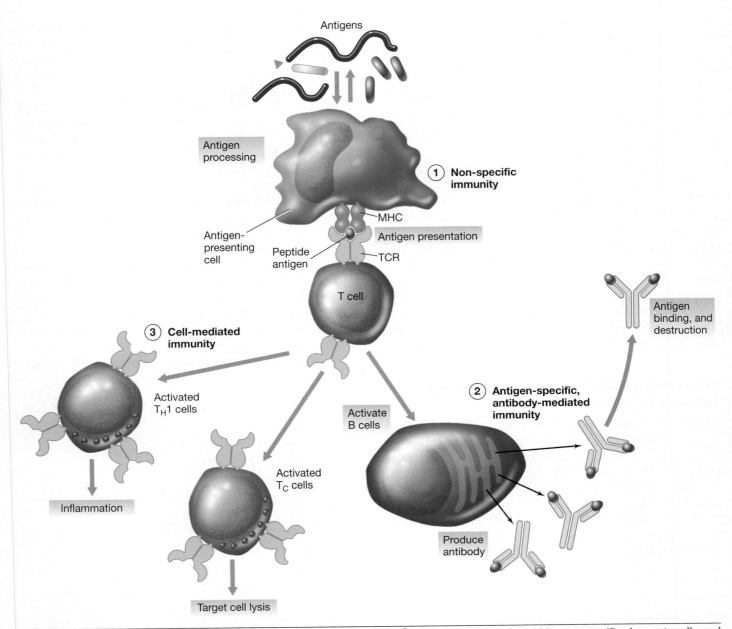

● **Figure 24.10** **The immune response.** Exposure to pathogens stimulates ① innate immunity, mediated by non–specific phagocyte cells, and antigen-specific immunity. Specific immunity is geared to recognize individual pathogen antigens and is mediated by ② antigen-reactive antibodies or ③ by antigen-reactive T cells, including T inflammatory cells (T_H1) and T cytotoxic cells (T_C).

The specific antibodies interact with the antigen on target cells, but cannot kill the cells. A group of proteins, known collectively as complement (⌐ Section 22.11), may attach to antibodies bound to the pathogen and lyse all cells with attached antibody. The complement–antibody interaction affects only those cells that have been targeted by antibodies. For example, antibodies specific for cell surface proteins of *Salmonella* sp. interact only with *Salmonella*: Complement may cause lysis of the antibody-sensitized *Salmonella* cell, but not of a neighboring *Escherichia coli* cell that is not antibody-sensitized. Thus, the immune response is *specific* for individual antigens by virtue of specific antibodies, but the final effect may occur by means of nonspecific mechanisms such as complement.

Cell-Mediated Immunity and Immune Memory

In many cases, antibody-mediated immunity is not an effective mechanism for controlling the spread of infection. Some infectious agents parasitize the body from *within* cells. For example, animal viruses reproduce using host cell systems and, therefore, spend a large portion of their life cycle within the host cells (⌐ Section 9.12). Likewise, bacteria such as *Mycobacterium tuberculosis*, the causative agent of tuberculosis, live in phagocytes (⌐ Sections 22.2 and 26.5). Because antibodies are geared to recognize pathogens in the blood or at mucosal cell surfaces, the infected host cells must be identified and destroyed by other means, usually involving the cell-to-cell interactions of the *cell-mediated immunity*. One type of T_H cell, the antigen-specific T_H1 cell, attracts and activates phagocytes such as macrophages and neutrophils, causing inflammatory reactions and limiting infection (∞Section 22.8). In addition, intracellular pathogens produce antigens that are in turn presented on the surface of the infected cells. *T cytotoxic cells* (T_C) recognize the antigen and act directly on the infected *target cell* by secreting cytolytic proteins called *perforins*, which lyse the infected cell (Figure 24.10 and ∞Section 22.7).

No antigen-specific immunity exists before exposure to antigen, but after the first antigen exposure, antigen-reactive immune T and B cells are produced and may persist for years, conferring long-term specific immunity. More importantly, a second antigen stimulation of these antigen-reactive cells generates a very rapid and very strong immune response that peaks within several days (∞Section 22.10). The T cell and antibody products of this *secondary response* quickly target and destroy the pathogen. Thus, the immune response has *memory*. Memory is often characterized by a rapid increase in immunity, and this increase is used to track infections.

Antibody Titers, Skin Tests, and the Diagnosis of Infectious Disease

Isolation of a pathogen is not always possible or practical to confirm diagnosis of an infectious disease. An alternative is to measure antibody **titer** (quantity) for a suspected pathogen. As we discussed, if an individual is infected with a suspected pathogen, the immune response—in this case, the antibody titer—to that pathogen should be elevated. Antibody titer can be measured by precipitation, agglutination, or any of the methods discussed in Sections 24.7–24.12. The general procedure is to prepare serial dilutions of patient serum and then determine the *highest* dilution at which the antigen–antibody reaction occurs (Figure 24.11●). These methods are all termed *serological tests* because they make use of antibody-containing patient *serum*.

A single measure of antibody titer cannot be used to indicate active infection because many antibodies remain at high titer for long periods after a previous infection has been resolved. To establish that an acute illness is due to a particular pathogen, it is essential to show a *rise* in antibody titer in serum samples taken from a patient during the acute disease and later during the convalescent phase of the disease. Frequently, the antibody titer is low during the acute stage of the infection and rises during convalescence (Figure 24.11). A rise in antibody titer is the best indication that the illness is due to the suspected agent. In some cases, however, the mere presence of antibody may be sufficient to indicate infection. This is true for pathogens rarely found in a population. The presence of antibody to such a pathogen indicates a recent or ongoing infection, as is the case for acquired immunodeficiency syndrome (AIDS) (∞ Section 26.14). We will discuss methods for determining HIV antibody levels in Sections 24.11 and 24.12.

Unfortunately, not all infections result in formation of systemic immunity. If a pathogen is extremely localized, there may be little induction of an immunological response and no rise in antibody titer even if the pathogen is proliferating profusely at the site of infection. A good example is the disease gonorrhea. Infection with *Neisseria gonorrhoeae*, the causative agent of gonorrhea, does not elicit a systemic or protective immune response, and thus reinfection of a cured individual is common (see Section 24.1) (∞Section 26.12).

In some cases, the presence of antibody in the serum may be due to a recent immunization. In fact, measurement of the rise in antibody titer following immunization is one of the best ways of determining that the immunization was effective.

Skin testing is another method for determining exposure to a pathogen. The most commonly used skin test is the *tuberculin test*, which consists of an intradermal injection of a soluble extract from cells of *Mycobacterium tuberculosis*. A positive inflammatory reaction at the site of injection within 48 hours indicates current infection or previous exposure to *M. tuberculosis*. This test identifies delayed-type

● **Figure 24.11 The course of infection in a typical untreated typhoid fever patient.** Measurement of body temperature provides a measure of the course of clinical symptoms. The antibody titer was measured by determining the highest serum dilution (twofold series) causing agglutination of a test strain of *Salmonella typhi*. Titer is shown as the *reciprocal* of the highest dilution showing an agglutination reaction. Presence of viable bacteria in blood, feces, and urine was determined from periodic cultures. Note that the pathogen clears from the blood as the antibody titer rises, and clearance from feces and urine requires a longer time. Body temperature gradually drops to normal as the antibody titer rises. The data given do not represent a single patient but are a composite of the pattern seen in large numbers of patients.

hypersensitivity responses caused by pathogen-specific T_H1 cells (Section 22.15). Skin tests are routinely used for diagnosing tuberculosis, leprosy (Section 26.5), a variety of fungal diseases (Section 27.8), and other diseases in which the antibody response is weak or nonexistent. Common immunodiagnostic tests for pathogens are shown in Table 24.5.

24.5 Concept Check

An immune response is a natural outcome of infection. Specific immune responses, particularly antibody titers and skin tests, can be monitored to provide information concerning past infections, current infections, and convalescence.

◆ Describe the change in antibody titer to an infectious agent from the acute phase through the convalescent phase of the infection.

◆ Describe the method, time frame, and rationale for the TB skin test. What component of the immune response does this test detect?

24.6 Polyclonal and Monoclonal Antibodies

The immune response to an antigen usually results in the production of immunoglobulin (Ig) molecules (Sections 22.9 and 22.10) directed at the numerous determinants present on the antigen (Section 22.5). Only a few of the many Igs are directed toward each antigen determinant. The resulting antiserum, a mixture of different antibodies, is known as a polyclonal antiserum. Polyclonal antisera consist of antibodies that provide adequate immune protection to the host, but are usually specific for a variety of determinants. These antisera are not precisely reproducible because they are the sum of the antibody response produced by an individual animal at a single time.

However, each Ig is produced by a single B cell (Section 22.10), and B cells cloned *in vitro* can produce limitless supplies of a single monospecific immunoglobulin. Antibodies made by a single cloned B-cell are called **monoclonal antibodies**. Long-lived B-cell clones can be stored as frozen cells and later reconstituted, providing a reproducible source of specific antibodies. As a

Table 24.5 Immunological procedures for identification of infectious agents

Pathogen/disease	Antigen	Procedure[a]
HIV (AIDS)	Human immunodeficiency virus (HIV)	ELISA
Borrelia burgdorferi (Lyme disease)	Flagellin	ELISA
	Surface proteins	Immunoblot
		Bactericidal test (Section 27.4)
Brucella (brucellosis)	Cell wall antigen	Agglutination
Candida albicans (yeast infections)	Soluble extract of fungal proteins	Skin test
Corynebacterium diphtheriae (diphtheria)	Toxin	Skin test (Schick test)
Influenza virus (influenza)	Influenza virus suspensions	Complement-based assay
	Nasopharynx cells containing influenza virus	Immunofluorescence
Mycobacterium leprae (leprosy)	Lepromin (soluble extract of bacterial proteins)	Skin test
Mycobacterium tuberculosis (tuberculosis)	Tuberculin (purified protein derivative, PPD)	Skin test
Neisseria meningitidis (meningitis)	Capsular polysaccharide	Passive hemagglutination (*N. meningitidis* polysaccharide adsorbed to red blood cells)
Pneumocystis carinii (lung infection)	*P. carinii* cells	Immunofluorescence
Rickettsial diseases (Q fever, typhus, Rocky Mountain spotted fever)	Killed rickettsial cells	Complement-based assay or cell agglutination tests
		ELISA
Salmonella (gastroenteritis)	O and H antigen	Agglutination (Widal test)
		ELISA
Streptococcus (group A) (strep throat, scarlet fever)	Streptolysin O (exotoxin), DNase (extracellular protein)	Neutralization of hemolysis / Neutralization of enzyme
Treponema pallidum (syphilis)	Cardiolipin-lecithin-cholesterol	Flocculation [Venereal Disease Research Laboratory (VDRL) test]
Vibrio cholerae (cholera)	O antigen	Agglutination
		Bactericidal test (in presence of complement)
		ELISA

[a] Immunofluorescence tests use preformed antibody to detect the presence of the indicated pathogen in a patient specimen. Skin tests for *C. albicans, M. tuberculosis,* and *M. leprae* indicate T_H1-mediated delayed-type hypersensitivity. The *C. diphtheriae* Schick test detects serum antibodies with a toxin-neutralization skin test. All other tests measure serum antibody levels.

result, monoclonal antibody techno-logy has supplanted standard polyclonal techniques for many immunodiagnostic applications. Table 24.6 compares the properties of **polyclonal antibodies** with the properties of monoclonal antibodies.

Monoclonal Antibodies and Hybridomas

Antibody-producing B lymphocytes normally die after several weeks in cell culture (*in vitro*). Therefore, antibody-producing B lymphocytes are fused with B-cell tumors called *myelomas*. These myelomas are capable of dividing indefinitely and are therefore often called immortal cell lines. The immortal cell lines that result from the B cell–myeloma fusion are hybrid cell lines called *hybridomas*. The hybridoma cell lines share the properties of both fusion partners. They grow indefinitely *in vitro* and produce antibodies (Figure 24.12●).

To produce a monoclonal antibody, a mouse is immunized with the antigen of interest. During the next several weeks, antigen-specific B cells proliferate and begin producing antibodies in the mouse (∞ Section 22.10). Spleen tissue, rich in B-lymphocytes, is then removed from the mouse, and the B cells are fused with myeloma cells (Figure 24.12). Even though many cells fuse in culture and begin to grow, only a small fraction are viable antibody-producing hybridomas. Hybridomas are first selected from other cells by addition of *h*ypoxanthine, *a*minopterin, and *t*hymidine to the *in vitro* cell culture medium (HAT medium). The HAT medium stops the growth of unfused myeloma cells because the myeloma cells, though able to grow indefinitely in cell culture, are unable to use the metabolites hypoxanthine and thymidine to bypass a metabolic block caused by aminopterin, a cell poison. By contrast, fused hybridoma cells can use hypoxanthine and thymidine to bypass the aminopterin block and grow normally in HAT medium; they receive the genetic information for use of hypoxanthine and thymidine from the B-cell fusion partner. Any unfused B cells die in a few days because they cannot grow in culture. Following

Table 24.6	Characteristics of monoclonal and polyclonal antibody production
Polyclonal	**Monoclonal**
Contains many antibodies recognizing many determinants on an antigen	Contains a single antibody recognizing only a single determinant
Various classes of antibodies are present (IgG, IgM, and so on)	Single class of antibody produced
Can make a specific antibody using only a highly purified antigen	Can make a specific antibody using an impure antigen
Reproducibility and standardization difficult	Highly reproducible

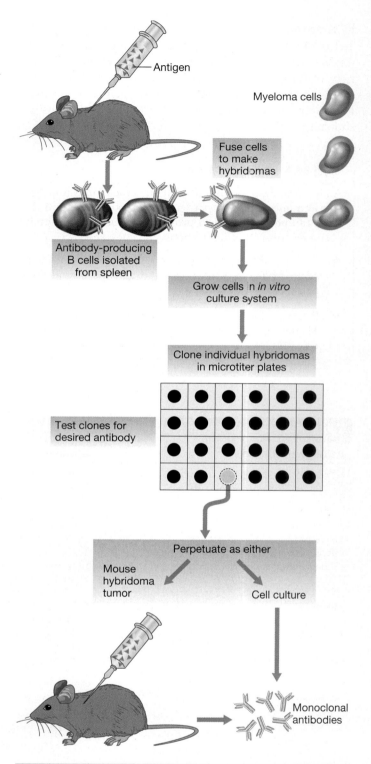

● **Figure 24.12 The hybridoma technique and production of monoclonal antibodies.** The hybridoma can be indefinitely cultured or passed through animals as a tumor. The hybridoma cells can also be stored as frozen tumor cells and reconstituted when needed in tissue culture or in a suitable animal host.

fusion, the antibody-producing hybridoma clones must be identified.

An ELISA (enzyme-linked immunoassay) test (see Section 24.10) can be used to identify hybridomas that

Table 24.7 | **Types of antigen–antibody reactions**

Location of antigen	Accessory factors required	Reaction observed
Soluble	None	Precipitation (Section 24.7)
On cell or inert particle	None	Agglutination (Section 24.8)
Flagellum	None	Immobilization or agglutination (Section 24.8)
On bacterial cell	Complement	Lysis (∞ Section 22.11)
On bacterial cell	Complement	Killing (∞ Section 22.11)
On erythrocyte	Complement	Hemolysis (∞ Section 22.11)
Toxin	None	Neutralization (Section 24.7)
Virus	None	Neutralization (Section 24.7)
On bacterial cell	Phagocyte, complement	Phagocytosis (opsonization; ∞ Section 22.11)

produce monoclonal antibodies. From a typical fusion, several distinct clones are isolated, each making a monoclonal antibody. Once the clones of interest are identified, they can be grown *in vivo* as an antibody-producing tumor or in cell culture. Antibody can be harvested from the tumor or from the culture supernatant. Hybridomas can grow indefinitely or can be stored as frozen cells.

Diagnostic and Therapeutic Uses

Monoclonal antibodies provide a limitless supply of highly specific biological reagents with unlimited applications. Monoclonal antibodies are widely used for clinical diagnostic tests, immunological typing of bacteria, and for the identification of cells containing foreign surface antigens (for example, a virus-infected cell). Monoclonal antibodies have also been used in genetic engineering for identifying and measuring levels of gene products not detectable by other methods and also show great promise for increasing the specificity of existing clinical tests including blood and tissue typing.

Because of their specificity, monoclonal antibodies are used to detect and treat human cancers. Malignant cells contain a variety of surface antigens not expressed by normal cells. These *tumor antigens* are unique, tumor-specific cell proteins. Monoclonal antibodies prepared against the tumor antigens specifically target the malignant cells and have been used to deliver toxins directly and specifically to malignant cells. Tumor-specific monoclonal antibodies covalently linked to toxins are now undergoing clinical trials. The specificity of monoclonal antibody treatments may greatly improve cancer chemotherapy by offering an alternative to chemical and radiation treatments that damage normal host cells as well as cancer cells.

24.6 Concept Check

Polyclonal and monoclonal antibodies are used for research and clinical applications. Hybridoma technology provides reproducible, monospecific antibodies for a wide range of clinical, diagnostic, and research purposes.

◆ How can a polyclonal antibody preparation recognize a variety of determinants?

◆ What advantages do monoclonal antibodies have as compared with polyclonal antibodies? What advantages do polyclonal antibodies offer?

24.7 *In Vitro* Antigen–Antibody Reactions: Serology

The study of antigen–antibody reactions *in vitro* is called **serology**. Serological reactions are the basis for all diagnostic immunology tests. The principles of antigen–antibody reactions lie in the specific interaction of determinants on the antigen with the *variable* region of the antibody molecule.

A variety of serological tests are used to identify antigens, depending on the properties of the antigen and on the conditions chosen for reaction (Table 24.7).

Specificity and Sensitivity

The usefulness of a serological test for diagnostic purposes is dependent on the test's specificity and sensitivity. **Specificity** is the ability of an antibody preparation to recognize a single antigen. Optimal specificity implies that the antibody is specific for a single antigen, will not cross-react with any other antigen, and therefore, will not provide *false positive* results. Specificity *must* be defined in terms of reactions with positive and negative control antigens. Specificity for each test must be determined experimentally and verified every time the test is used.

Sensitivity defines the lowest amount of an antigen that can be detected. The highest level of sensitivity requires that the antibody in a test be capable of identifying a single antigen molecule. High sensitivity prevents *false negative* reactions. The sensitivity of some common tests is shown in Table 24.8 in terms of the amount of antibody necessary to detect antigen. The amount of antigen detected by each test system is proportional to the amount of antibody used. For example, immune precipitation reactions require a large amount of antibody and generally detect 0.1 to 1.0 mg quantities of antigen. Thus, precipitation tests are the least sensitive serological tests. By contrast, enzyme-linked immunosorbent assays, or ELISA

Table 24.8 Sensitivity of immunodiagnostic assays

Assay	Sensitivity (μg antibody/ml)[a]
Precipitin reaction	
In fluids	24–160
In gels (double immunodiffusion)	24–160
Agglutination reactions	
Direct	0.4
Passive	0.08
Radioimmunoassay (RIA)	0.0008–0.008
Enzyme-linked immunosorbent assay (ELISA)	0.0008–0.008
Immunofluorescence	8.0

[a] The smallest amount of antibody necessary to give a positive reaction in the presence of antigen.

tests (see Section 24.10) require 100,000 times less antibody and detect 1 million times less antigen (0.1 to 1.0 ng quantities) than precipitation antigens. Thus, ELISA tests are among the most sensitive serological tests.

Neutralization

Neutralization is the interaction of antibody with antigen to block or distort the antigen sufficiently to reduce or eliminate its biological activity. Neutralization reactions can occur *in vitro* or *in vivo*.

For example, neutralization of microbial toxins by specific antibody occurs when toxin and specific antibody combine in such a way that the active portion of the toxin is blocked (Figure 24.13●). Neutralization reactions of this type occur for many bacterial exotoxins, including many of those listed in Table 21.4. An antiserum containing an antibody that neutralizes a toxin is referred to as an *antitoxin*. Antitoxin therapy is used to treat botulism (∞ Section 29.6), tetanus (∞ Section 27.9), and diphtheria (∞ Section 26.3), all diseases that result from bacterial production of exotoxins.

Neutralization reactions may also occur when viruses are bound by specific antibodies. For example, antibodies directed against the hemagglutinin and neuraminidase proteins of influenza viruses prevent the adsorption of the viruses to specific receptors on host cells (∞ Section 26.8). Neutralization reactions can be used for *in vitro* testing using patient serum, but are not used in routine diagnostic laboratories because neutralization tests require biologically active systems.

Precipitation

Precipitation results from the interaction of a *soluble* antibody with a *soluble* antigen to form an *insoluble* complex. Antibody molecules generally have two antigen-binding sites (that is, they are bivalent) (∞ Section 22.9). Therefore, it is possible for each site to combine with a separate antigen molecule. If the antigen also has more

than one available determinant, a precipitate may develop from aggregates of antibody and antigen molecules (Figure 24.14a●). Because they are easily observed *in vitro*, precipitation reactions are very informative serological tests, especially for the quantitative measurement of antibody concentrations. Precipitation occurs maximally only when there are optimal proportions of the two reacting substances. The presence of either antigen or antibody in excess quantities results in the formation of small, soluble immune complexes.

Precipitation reactions carried out in agar gels, referred to as *immunodiffusion* tests, are used to study the specificity of antigen–antibody reactions. Both antigen and antibody diffuse outward from separate wells cut in an agar gel, and precipitation bands form in the region where antibody and antigen interact in optimal proportions (Figure 24.14 b). The precipitation bands are characteristic for the reacting substances; two antigens reacting with an antiserum can be tested for molecular relationships by observing the bands formed when the two antigens are placed in adjacent wells near the antiserum well. For example, if two antigens in adjacent wells are identical, they will form a single, fused, precipitin band. This is referred to as a line of *identity*. If, on the other hand, adjacent wells contain one antigen in common, but one well contains a second antigen, a line of *partial identity* will form (Figure 24.14b). The extension of the precipitin line (representing a reaction between the antiserum and the second antigen) is referred to as a *spur*. Immunodiffusion is used as a tool in biochemical research to assess the relatedness of proteins obtained from different sources.

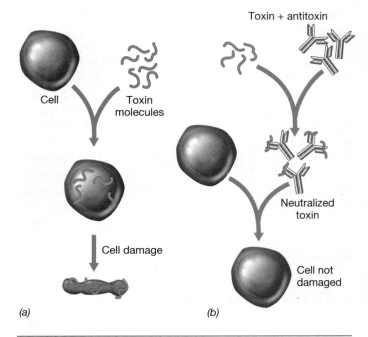

● **Figure 24.13 Neutralization of an exotoxin by antibody.** Antitoxins are used to treat acute cases of diphtheria (∞ Section 26.3), tetanus (∞ Section 27.9), and botulism (∞ Section 29.6). (a) Untreated toxin results in cell destruction. (b) Antitoxin neutralizes toxin and prevents cell destruction.

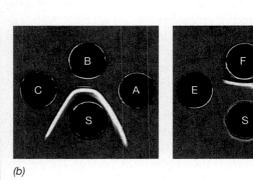

C. Weibull, W. D. Bickel, W. T. Hashius, K. C. Milner, and E. Ribi

(b)

● **Figure 24.14 Precipitation reactions between soluble antigen and antibody.** The graph (a) shows the extent of precipitation as a function of antigen and antibody concentration. (b) Precipitation in agar gel, a process called immunodiffusion. Wells labeled S contain antibodies to cells of *Proteus mirabilis*. Wells labeled A, B, and C contain soluble extracts of *Proteus mirabilis*. A line of identity is observed in the wells on the left. On the right, antigen E does not react and antigen A shows partial identity with antigen F (see leader to spur).

Unfortunately, the readily visible precipitation reactions are not very sensitive. Microgram quantities of specific antibody are necessary to visualize a precipitate (Table 24.8) and most useful diagnostic tests require sensitivity at nanogram levels. Consequently, precipitation reactions are normally used only in research and reference laboratories.

 24.7 Concept Check

Antigen–antibody reactions require that antibody bind to antigen. Specificity and sensitivity define the accuracy of individual serological tests. Neutralization and precipitation reactions are examples of antigen-binding tests that produce visible results involving antigen–antibody interactions.

◆ In serological reactions, high specificity prevents false-positive reactions. High sensitivity prevents false-negative reactions. Explain.

◆ Explain the principles of a neutralization reaction.

◆ What are the minimum antigen and antibody requirements for a precipitation reaction?

24.8 Agglutination

Agglutination is the visible clumping of a particulate antigen when mixed with antibodies specific for the particulate antigens. Although not as sensitive as some other serological tests, agglutination tests are about 100 times more sensitive than precipitation tests (Table 24.8). Agglutination tests are widely used in clinical and diagnostic laboratories because they are simple to perform, highly specific, inexpensive, rapid, and reasonably sensitive. Standardized tests are available for the identification of blood group antigens as well as many pathogens and pathogen products.

Direct Agglutination

Direct agglutination results when soluble antibody causes clumping due to interaction with an antigen that is an integral part of the surface of a cell or other insoluble particle. Direct agglutination procedures are used for the classification of antigens found on the surface of red blood cells (erythrocytes). Agglutination of red blood cells is referred to as *hemagglutination*, and is the basis for *blood typing*.

Red blood cells exhibit a variety of cell-surface antigens, and individuals vary considerably with respect to the antigens present on their red blood cells. The major human red cell antigens are called A, B, and D (also known as Rh).

A and B antigens and antibodies are the basis for the ABO blood-typing assay. Antibodies specific for particular erythrocyte surface antigens cause red blood cells to visibly clump when a small drop of blood is mixed on a microscope slide with antiserum reactive against either the A or B antigens of human erythrocytes (Figure 24.15●). The antisera are obtained from human donors who have been immunized to A or B antigens by natural or artificial means.

For the A, B, and O blood types, individuals express codominant A and B alleles as one of the following antigen phenotypes: A, B, AB (one allele expressing the A antigen and one expressing the B antigen), or O (the absence of either A or B antigens). In addition, individuals make antibodies to most nonself blood group antigens. Type A individuals make antibodies to group B antigens, while type B individuals make antibodies to group A antigens. Type AB individuals have neither A nor B antibodies, whereas type O individuals have antibodies to both A and B antigens (Figure 24.15). Antibodies against A and B antigens are called *natural antibodies* because they appear to be produced by most individuals in response to ubiquitous antigen sources such as enteric *Bacteria*.

Blood typing must be performed before blood transfusion to prevent red blood cell destruction that would occur if antibodies in the recipient's blood reacted with the red

(a)

Blood type	Percentage of U.S. population	Serum	
		Anti A	Anti B
Type O	47	No aggl.	No aggl.
Type A	42	Aggl.	No aggl.
Type B	8	No aggl.	Aggl.
Type AB	3	Aggl.	Aggl.

(b)

● **Figure 24.15 Direct agglutination of human red blood cells for ABO blood typing.** (a) The reaction on the left shows no agglutination. The reaction in the center shows the diffuse agglutination pattern that indicates a positive reaction for the B blood group. The reaction on the right shows the strong agglutination pattern with large, clumped agglutinates typical for the A blood group. (b) Table of expected blood grouping results for the U.S. population.

blood cells in the transfused blood, or vice versa. Antibody-coated red blood cells would likely undergo *hemolysis* (lysis of red blood cells) through the action of complement (oo Section 22.11), resulting in severe anemia.

Passive Agglutination

Passive agglutination is the agglutination of soluble antigens or antibodies that have been adsorbed or chemically coupled to cells or insoluble particles such as latex beads or charcoal particles. The insolubilized antigen or antibody can then be detected by agglutination reactions. The cell or particle serves as an inert carrier. Passive agglutination reactions can be up to five times more sensitive than direct agglutination tests (Table 24.8), significantly increasing sensitivity.

The agglutination of antigen-coated or antibody-coated latex beads by complementary antibody or antigen from a patient is a typical method of rapid diagnosis. Small (0.8 μm) latex beads coated with a specific antigen are mixed with patient serum on a microscope slide and incubated for a short period. If patient antibody binds the antigen on the bead surface, the milky white latex suspension will become visibly clumped, indicating a positive agglutination reaction. Latex agglutination is also used to detect bacterial surface antigens by mixing a small amount of a bacterial colony with antibody-coated latex beads. For example, a commercially available sus-

pension of latex beads coated with antibodies to protein A and clumping factor, two molecules found exclusively on the surface of *Staphylococcus aureus*, is virtually 100% specific in identifying clinical isolates of *S. aureus*. Unlike traditional growth-dependent tests for *S. aureus*, identification of *S. aureus* by the latex bead assay takes only 30 seconds (Figure 24.16●). Other latex bead agglutination assays have been developed to identify various pathogenic beta-hemolytic streptococci including Lancefield groups A (*Streptococcus pyogenes*), B, C, D, F, and G (oo Section 12.19), *Neisseria gonorrhoeae* and *Neisseria meningitidis*, *Haemophilus influenza*, *Escherichia coli* O157: H7, and the fungi *Cryptococcus neoformans* and *Candida albicans*.

A widely employed latex agglutination test is used for detecting specific serum antibodies for *rheumatoid factor*, an antibody directed against the body's own Ig and associated with the autoimmune disease *rheumatoid arthritis* (oo Section 22.15). Latex beads coated with human Ig are mixed with whole blood or serum, and agglutination is compared with positive and negative controls.

Passive agglutination assays require no expensive equipment or particular expertise and can be highly specific and very sensitive. In addition, the inexpensive nature of the assays makes them suitable for large-scale screening programs. These tests are widely used in clinical and research laboratories.

⬟ 24.8 Concept Check

Direct agglutination tests are widely used for determination of blood types. A number of passive agglutination tests are available for identification of a variety of pathogens and pathogen-related products. Agglutination tests are rapid, relatively sensitive, highly specific, simple to perform, and inexpensive.

◆ Distinguish between *direct* and *passive* agglutination. Which tests are more sensitive?

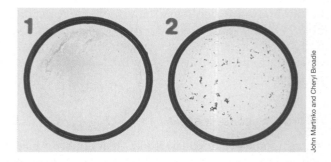

● **Figure 24.16 Latex bead agglutination test for *Staphylococcus aureus*.** Panel 1 shows a negative control. Note the uniform pink color of the suspended latex beads coated with antibodies to protein A and clumping factor, two antigens found exclusively on the surface of *S. aureus* cells. Panel 2 shows the same suspension after a loopful of material from a bacterial colony was mixed into the suspension. The bright red clumps indicate a positive agglutination reaction took place and indicates that the colony is *S. aureus*.

◆ What advantages do agglutination tests have over other immunoassays? What disadvantages?

24.9 Fluorescent Antibodies

Antibodies can be chemically modified with fluorescent dyes. These modified antibodies are used to detect antigens on intact cells. **Fluorescent antibodies** are widely used for diagnostic and research applications.

Fluorescent Methods

Antibodies can be covalently modified by fluorescent dyes such as rhodamine B, which fluoresces red, or fluorescein isothiocyanate, which fluoresces yellow-green. This does not alter the specificity of the antibody but makes it possible to detect the antibody bound to cell or tissue surface antigens by use of a fluorescence microscope (Figure 24.17●). Cell-bound fluorescent antibodies emit a bright fluorescent color when excited with light of particular wavelengths. The emitted fluorescent light is usually red-orange or yellow-green, depending on the dye used. Fluorescent antibodies are used in diagnostic microbiology because they permit the identification of a microorganism directly in a patient specimen (*in situ*), by-passing the need for the isolation and culturing of the organism (see below). The fluorescent antibody technique is also very useful in microbial ecology as a method for directly viewing and identifying microbial cells without the need to isolate and culture them (∞Section 18.3).

Distinct *direct* and *indirect* fluorescent staining methods are used. In the direct method, the antibody directed to the surface antigen is covalently linked with the fluorescent dye. In the indirect method, the presence of a nonfluorescent antibody on the surface of a cell is detected by the use of a fluorescent antibody directed against the nonfluorescent antibody (Figure 24.18●).

Applications

In a typical test using fluorescent antibodies, a specimen containing a suspected pathogen is allowed to react with a specific fluorescent antibody and observed with a fluorescent microscope. If the pathogen contains surface antigens reactive with the antibody, the cells will fluoresce (Figure 24.19●).

Fluorescent antibodies can be applied directly to infected host tissues, permitting diagnosis long before

(a)

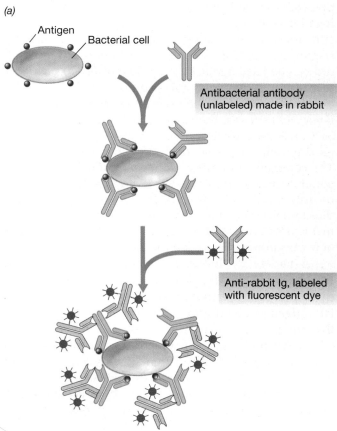

(b)

● **Figure 24.18 Fluorescent antibody methods for detection of microbial surface antigens.** (a) Direct staining method. (b) Indirect staining method.

● **Figure 24.17 Fluorescent antibody reactions.** Cells of *Clostridium septicum* were stained with antibody conjugated with fluorescein isothiocyanate, which fluoresces yellow-green. Cells of *Clostridium chauvoei* were stained with antibody conjugated with rhodamine B, which fluoresces red-orange.

(a)   (b)

William B. Cherry

Dharam Ablashi and Robert C. Gallo

● **Figure 24.19 Examples of the use of fluorescent antibodies in clinical microbiology.** (a) Immunofluorescent stained cells of *Legionella pneumophila*, the cause of legionellosis. The specimen was taken from biopsied lung tissue. The individual organisms are 2–5 μm in length. (b) Detection of virus-infected cells by immunofluorescence. Human B lymphotrophic virus (HBLV)-infected spleen cells were incubated with serum containing antibodies to HBLV from a patient with a lymphoproliferative disorder. Cells were then treated with fluorescein isothiocyanate-conjugated anti-human IgG antibodies. HBLV-infected cells fluoresce bright yellow. Cells in the background did not react with the patient's serum. Individual cells are about 10 μm in diameter.

(a)

(b)

G. Bradley

● **Figure 24.20 Use of fluorescent antibodies in noninfectious disease diagnostics.** (a) Human leukemic cells, some of which are sensitive to a toxic anticancer drug and some of which are not, appear indistinguishable. (b) When the cells in (a) are treated with a fluorescent monoclonal antibody that binds specifically to a protein found only on the surface of drug-resistant cells, the latter fluoresce whereas drug-sensitive cells do not. Individual cells are about 10–12 μm in diameter.

primary isolation techniques yield a suspected pathogen. For example, in diagnosing legionellosis (◐Section 28.7), a positive identification can be made by staining biopsied lung tissue with fluorescent antibodies specific for cell wall antigens of *Legionella pneumophila* (Figure 24.19*a*). Likewise, a direct fluorescent antibody test can be used against the capsule of *Bacillus anthracis* to confirm a diagnosis for anthrax (◐Figure 21.14). This methodology is also used to determine infection by *Bordetella pertussis*, the cause of whooping cough (◐ Section 26.4). Direct fluorescent antibody tests are also used to help diagnose viral infections (Figure 24.19*b*). The common respiratory pathogens influenza A and B, parainfluenza, respiratory syncytial virus (RSV), and adenovirus are identified from respiratory tract specimens by direct fluorescent antibody methods. Fluorescent antibody methods are also used to identify viruses grown in tissue or organ culture (see Section 24.12).

A variety of fluorescent antibody assays are also used to aid in the diagnosis of some noninfectious diseases. Fluorescent antibodies that interact with a particular antigen can be used to identify cell types expressing that antigen. For example, fluorescent antibodies directed against tumor-specific antigens found on malignant cells may be used to identify malignant cells and monitor the course of the disease (Figure 24.20●).

Fluorescent antibodies can also be used to separate mixtures of cells into relatively pure populations or to define the numbers of certain cell types in complex mixtures such as blood. Fluorescent-labeled monoclonal antibodies directed against the CD4 and CD8 surface antigens of T lymphocytes (◐Section 22.6) are routinely used to identify and enumerate these cells in the blood leukocyte population (Figure 24.21●). For example, the

definition of acquired immunodeficiency syndrome (AIDS) includes a reduction in CD4 cell numbers. In addition, the CD4 T cell number changes during the progression of AIDS and is diagnostic for the disease. Thus, by defining the CD4 numbers, the clinician can identify the reduction in CD4 cells compared with normal values and, with successive assays over time, follow the progress of the disease (◐Section 26.14).

Fluorescent antibody–labeled cells can be visualized, counted, and separated with an instrument called a *fluorescence cytometer*, often referred to as a fluorescence-activated cell sorter (FACS). The FACS uses a laser beam to activate fluorescent antibody bound to cells, placing a charge on the labeled cells. An electric field is then applied to the cell mixture. Fluorescing and nonfluorescing cells are then deflected to opposite poles of the electric field, where each cell population is counted and deposited in a tube. The use of several antibodies, each labeled with a different fluorescent dye, can result in the simultaneous identification of several cell markers. A typical application used for identifying CD3 and CD4 surface proteins on T cells in normal and AIDS patients is shown in Figure 24.22●.

● **Figure 24.21 T lymphocytes stained with fluorescent-tagged monoclonal antibodies to specific surface markers.** Yellow-green cells are T_C (CD8) cells; red-orange cells are T_H (CD4) cells. The different-colored cells can be separated from one another by a fluorescence-activated cell sorter to yield enriched populations of different cell types. Individual cells are about 10–12 μm in diameter. Reprinted with permission from Science 239: Cover (Feb. 12, 1988), © AAAS.

FACS analysis is also useful for research applications. For example, immunologists routinely use FACS methods to separate complex mixtures of immune cells.

They can then study the properties of the highly enriched cell populations.

Under appropriate conditions, fluorescent antibodies yield rapid, highly specific information about a variety of clinical conditions. However, fluorescent antibody techniques are not without their pitfalls. Nonspecific staining can be a problem because of surface antigens that may *cross-react* between various bacterial species, some of which may be members of the normal flora. This is a major problem among enteric bacteria, where lipopolysaccharide antigens (∞Section 4.9) are sufficiently similar among species to cause binding of the fluorescent probe by several different organisms. The clinical microbiologist must therefore perform controls using nonspecific sera and confirm all positive immunofluorescent findings by other immunological or microbiological tests.

⬢ **24.9 Concept Check**

Fluorescent antibodies are used for quick, accurate identification of pathogens and other antigenic substances in tissue samples and other complex environments. Fluorescent antibody-based methods can be used for identification, quantitative enumeration, and sorting of a variety of cell types.

◆ Explain and compare the principles, advantages, and disadvantages of direct and indirect fluorescent antibody assays.

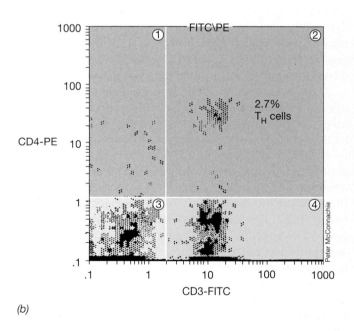

● **Figure 24.22 CD3 and CD4 cell enumeration.** Samples were obtained from a healthy human (a) and from a human with acquired immunodeficiency syndrome (AIDS) (b) using a fluorescence-activated cell sorter (FACS). Each dot represents a single cell. Peripheral blood cells were simultaneously labeled with monoclonal antibody to CD4 conjugated to phycoerythrin (PE) and with monoclonal antibody to CD3 conjugated to fluorescein isothiocyanate (FITC). CD3 is found on all T cells. CD4 is found on T-helper (T_H) cells only. Quadrant 3 shows cells that were stained with neither antibody. Quadrant 1 shows cells stained with only anti-CD4. Quadrant 4 shows cells stained with only anti-CD3. Quadrant 2 shows cells stained with both anti-CD3 and anti-CD4. (a) Results from a healthy human. In this case, 56.3% of the T cells were T_H cells. Thus, quadrant 2 shows a dense staining pattern. (b) Results from a patient with clinical AIDS. In this case, only 2.7% of the total T cells are T_H cells. This is indicated by the very light staining pattern in quadrant 2. Original data from Peter McConnachie, used with permission.

◆ How are fluorescent antibodies used to identify specific cells in complex mixtures such as blood?

24.10 Enzyme–Linked Immunosorbent Assay and Radioimmunoassay

It is desirable to use immunoassays that have high *sensitivity*, the ability to detect very low quantities of antigen–antibody complexes (see Section 24.5 and Table 24.8). Because of their high sensitivity, **enzyme-linked immunosorbent assay (ELISA)** and **radioimmunoassay (RIA)** methods are widely used. ELISA and RIA employ enzymes and radioisotopes, respectively, to label antibody molecules used for antigen detection. The covalent attachment of radioactive molecules or enzymes to antibody molecules decreases the amount of antigen–antibody complex required to detect a reaction. This increased sensitivity has been used in clinical diagnostics and research and has allowed the development of a variety of new immunological tests.

ELISA

The covalent attachment of enzymes to antibody molecules creates an immunological tool possessing both high specificity and high sensitivity. The technique, called *ELISA*, makes use of antibodies to which enzymes have been covalently bound such that the enzyme's catalytic properties and the antibody's specificity are unaltered. Typical bound enzymes include peroxidase, alkaline phosphatase, and β-galactosidase, all of which catalyze reactions that develop colored products that can be detected in very low amounts with a spectrophotometer.

Two basic ELISA methodologies have been developed, one for detecting antigen (*direct* ELISA) and the other for detecting antibodies (*indirect* ELISA). For detecting antigens such as virus particles from a blood or fecal sample, the direct ELISA method is used. In this procedure the antigen is "trapped" between two layers of antibodies (Figure 24.23●). Thus, this method is sometimes called the *sandwich ELISA*. The specimen is added to the wells of a microtiter plate previously coated with antibodies specific for the antigen to be detected. If an antigen (virus particle) is present in a sample, it will be trapped by the antigen-binding sites on the antibodies. After washing unbound material away, a second antibody containing a conjugated enzyme is added. The second antibody is also specific for the antigen: It binds to other exposed determinants. Following a wash, the enzyme activity of the bound material in each microtiter well is determined by adding the substrate for the enzyme. The color produced is proportional to the amount of antigen present (Figure 24.23).

To detect *antibodies* in human serum, an indirect ELISA is employed. An indirect ELISA is widely used to detect antibodies to human immunodeficiency virus (HIV) in human body fluids, and we will discuss this test in detail because the principles involved apply to all indirect ELISA tests (Figure 24.24●).

Modified rapid ELISA procedures use reagents adsorbed to a fixed support material such as paper strips, nitrocellulose or plastic membranes, or plastic "dipsticks." These tests cause a color change on the strip or stick in a very short time. These rapid "point of care" tests are used for diagnosis of infectious diseases such as HIV-AIDS (⌀Section 26.14) or "strep throat" (pharyngeal infection with *Streptococcus pyogenes*, ⌀Section 26.2).

Nondisease applications for rapid tests include pregnancy testing and drug testing (Figure 24.25●). In most of these tests, a body fluid, generally urine or blood, is applied to the reagent-support matrix. After reacting with the body fluids, the support is washed and developed with a second reagent that identifies antigen (direct test) or antibody (indirect test such as HIV-AIDS) bound to the matrix. These tests are especially valuable where a small number of samples are to be analyzed at the site where the urine or blood is collected (for instance, away from a clinical laboratory) and can be performed by individuals lacking certified laboratory skills. Results can be reported at the site, avoiding the need for delays in patient care or for follow-up visits to obtain test results. The drawback to these tests, however, is that they tend to be less specific than more elaborate tests. As a result, these point-of-care tests often need to be confirmed by standard laboratory tests.

The HIV ELISA

The virus that causes HIV-AIDS, the *human immunodeficiency virus (HIV)* (⌀Section 26.14), is transmitted by body fluids including blood. Sensitive, specific, rapid, and cost-effective screening tools are needed to test blood samples from individuals exposed to HIV and to ensure that HIV is not being inadvertently transmitted during blood transfusions or through the transfer of blood products. An ELISA test is used for the routine screening of blood to test for infection with HIV.

The HIV ELISA test is an *indirect* ELISA designed to measure *antibodies* to HIV present in serum. Initial infection with HIV leads to the production of antibodies to several HIV antigens, in particular, those of the HIV envelope. These antibodies can be detected by the HIV ELISA test (Figure 24.24).

To carry out an HIV ELISA test, microtiter plates are first coated with a preparation of disrupted HIV particles; about 200 ng of disrupted HIV is required in each well. A diluted patient serum sample is then added, and the mixture is incubated to allow HIV-specific antibodies to bind to HIV antigens. To detect the presence of antigen–antibody complexes, a second antibody is then added. This second antibody is an enzyme-conjugated anti-human IgG preparation. The anti-human IgG antibodies bind to any HIV-specific IgG antibodies now bound to the HIV antigen preparation. Following addi-

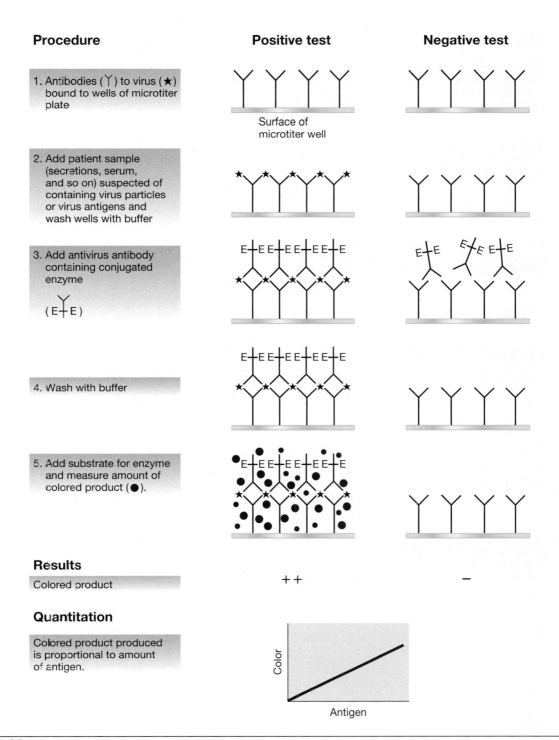

● **Figure 24.23 The direct ELISA test.** A direct ELISA test can be used to detect antigenic pathogen components or antigenic metabolites in blood, urine, and other body fluids.

tion of the second antibody, the enzyme substrate is added and the enzyme activity is assayed. The color obtained in the enzyme assay is proportional to the amount of anti-human IgG antibody bound (Figure 24.24). The binding of the second antibody is an indication that antibodies from the patient's serum recognized the HIV antigens and that the patient has antibodies to HIV and therefore has been infected with HIV. Control sera (known to be HIV-positive or HIV-negative) are assayed in parallel with patient samples to establish specificity

(positive control) and measure the extent of background absorbance in the assay (negative control).

The HIV ELISA test is a rapid, highly sensitive, specific method for detecting exposure to HIV. Since ELISAs in general are highly adaptable to mass screening and automation, the HIV ELISA test is used as a standard screening method for blood. However, this test method can give erroneous results under certain circumstances. For example, the test occasionally gives *false-positive* results. Because a number of factors can contribute to these results,

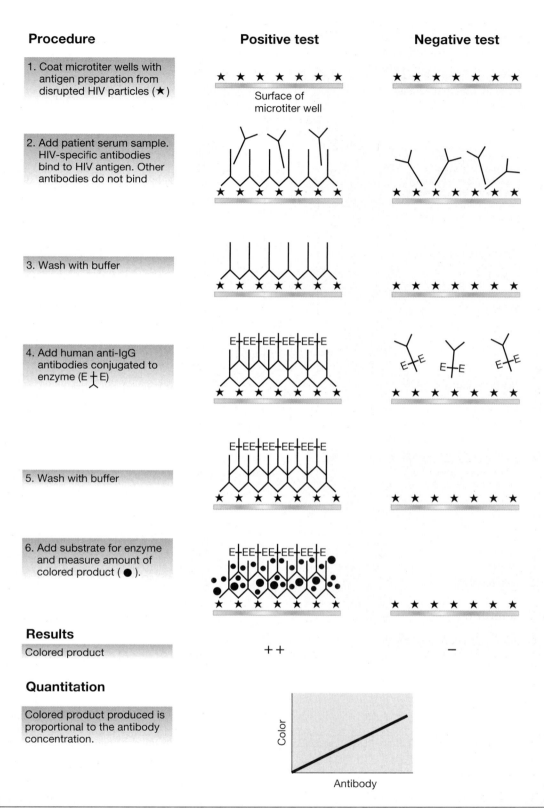

● **Figure 24.24 Indirect ELISA test.** An indirect test is used for detecting antibodies to human immunodeficiency virus (HIV), the causal agent of acquired immunodeficiency syndrome (AIDS).

none of which are related to exposure to HIV, all positive HIV ELISA tests *must* be confirmed by another independent test, usually the Western blot (immunoblot) test (see Section 24.11). A positive HIV Western blot test after a positive HIV ELISA test is considered proof of HIV infection.

The other drawback to the HIV ELISA test is the possibility of obtaining *false-negative* results. The immune system develops a detectable antibody response titer only several weeks after antigen exposure (see Section 24.5). In the case of HIV infection, this lag time is esti-

(a)

John M. Martinko

(b)

Keystone Diagnostics, Inc.

● **Figure 24.25 Rapid ELISA-based assay kits.** Pregnancy test kits (a) and the drug-testing kit (b) are available for point-of-care testing for these applications. Other kits are used for diagnosis of infectious diseases including streptococcal sore throat and HIV-AIDS.

mated to be six weeks to a year. Therefore, individuals who have been recently infected with HIV may not yet be producing detectable amounts of antibody when they are tested. Another reason for a false-negative result in the HIV ELISA test is the total destruction of the immune system seen in advanced cases of AIDS; if no immune cells are left in the body, no antibodies can be made and the ELISA test is not useful. However, at this stage of disease, an AIDS diagnosis can be based on clinical information (∞Section 26.14), and the ELISA test is useful only as a confirmatory indicator.

Other ELISA Tests of Clinical Importance

Besides the ELISA test for HIV, literally hundreds of clinically useful ELISAs have been developed. Some of these are direct ELISAs for detecting antigens, including bacterial toxins such as cholera toxin, enteropathogenic *Escherichia coli* toxin, and *Staphylococcus aureus* enterotoxin. Viruses currently detected using direct ELISA techniques include rotavirus, hepatitis viruses, rubella virus, bunyavirus, measles virus, mumps virus, and parainfluenza virus.

Indirect ELISAs have been developed for detecting antibodies to a variety of clinically important bacteria.

Although not meant to be a complete list, ELISAs for detecting serum antibodies to *Salmonella* (gastrointestinal diseases), *Yersinia* (plague), *Brucella* (brucellosis), a variety of rickettsias (Rocky Mountain spotted fever, typhus, Q fever), *Vibrio cholerae* (cholera), *Mycobacterium tuberculosis* (tuberculosis), *Mycobacterium leprae* (leprosy), *Legionella pneumophila* (legionellosis), *Borrelia burgdorferi* (Lyme disease), and *Treponema pallidum* (syphilis) have been developed. ELISAs have also been developed for detecting antibodies to *Candida* (yeast) and a variety of parasites, including those causing amebiasis, Chagas' disease, schistosomiasis, toxoplasmosis, and malaria.

The speed, low cost, lack of radioactive waste, and long shelf life make ELISA tests particularly attractive for many laboratories. But it is the high *sensitivity* that makes the ELISA a very important immunodiagnostic tool.

Radioimmunoassay

Radioimmunoassay (RIA) employs radioisotopes as antibody or antigen conjugates, in contrast to the enzymes used in ELISA. The isotope iodine-125 (^{125}I) is commonly used as the conjugate because antibodies or antigens can be readily iodinated without disrupting their immune specificity. RIA is used clinically to measure rare serum proteins such as human growth hormone, glucagon, vasopressin, testosterone, and insulin present in humans in extremely small amounts (Figure 24.26●). RIAs are also used in some tests for illegal drugs.

In most cases, a *direct* RIA is employed. The direct assay is a two-step procedure. First, a series of microtiter wells are prepared containing known concentrations of pure antigen (such as a hormone). Radiolabeled antigen-specific antibodies are added to the wells, and the radioactivity bound in each of these standard wells is measured. These data establish a standard curve for antigen concentrations. Next, the antigen sample from a patient is allowed to bind to another well, and radioactive antibodies are added and measured, as before. The amount of radioactivity bound by the patient sample is then compared to a standard plot generated from the binding data obtained using the pure antigen, and the concentration of antigen in the patient serum is interpolated from the standard plot (Figure 24.26).

RIA has the same sensitivity range as ELISA (Table 24.8) and can also be performed very rapidly. However, the instruments used to detect radioactivity are quite specialized and expensive. RIA generates a considerable amount of radioactive waste, and the radioactive decay time (half life) of the radioisotopes used for detection may limit the useful life of a test kit. As a result, RIA is often used only when ELISA is not sufficiently accurate or sensitive. For example, RIA is often more useful than ELISA for detecting serum protein levels (as described earlier) because serum components inhibit some ELISA enzyme–substrate

1. Bind insulin to wells of microtiter plate

Insulin

Microtiter wells

1   2   4   Serum sample

Increasing concentration ⟶

2. Add excess anti-insulin antibodies that are labeled with ^{125}I; wash to remove unbound antibody

Radioactive antibody

1   2   4   Serum sample

3. Count radioactivity in gamma radiation counter. Wells labeled 1, 2, and 4 establish a standard curve with known amounts of antigen (insulin). The radioactivity in the last well indicates, by comparison to the standard curve, how much insulin is present in a known amount of serum.

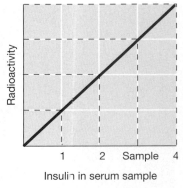

Radioactivity

1   2   Sample   4

Insulin in serum sample

● **Figure 24.26** **Radioimmunoassay (RIA).** Using RIA to detect insulin levels in human serum. Following establishment of a standard curve, the insulin concentration in a serum sample can be estimated.

or antigen–antibody interactions. Thus, for certain applications, each test system has clear advantages.

 24.10 Concept Check

ELISA and RIA methods are the most sensitive immunoassay techniques. Both involve linking a detection system, either an enzyme or a radioactive molecule, to an antibody or antigen, significantly enhancing sensitivity. ELISA and RIA are used for clinical and research work; tests have been designed to detect either antibody or antigen in many applications.

◆ Why are ELISA and RIA techniques more sensitive than immunoassays such as precipitation and agglutination?

◆ Compare ELISA and RIA with respect to their relative uses, advantages, and disadvantages.

24.11 Immunoblot Procedures

Antibodies can also be used in the **immunoblot** method to identify individual specific *proteins* associated with specific pathogens, even in complex mixtures such as cell lysates or blood. The immunoblot method employs three techniques discussed previously: (1) the separation of proteins on polyacrylamide gels, (2) the transfer (blotting) of proteins from gels to a nitrocellulose or nylon membrane, and (3) identification of the proteins by specific antibodies. Protein blotting and the subsequent identification of the proteins by specific antibodies is also called the *Western blot* technique to distinguish it from the (DNA) *Southern blot* technique (∞Section 7.7).

In the first step of an immunoblot, a protein mixture is subjected to electrophoresis on a polyacrylamide gel. This separates the proteins into several distinct bands, each of which represents a single protein of specific molecular weight (Figure 24.27●). The proteins are then transferred to the membrane by an electrophoretic transfer process that elutes the proteins from the gel onto the membrane. At this point, antibodies raised against the pathogen components are added to the blot. Following a short incubation period to allow the antibodies to bind, a radioactive marker that binds antigen–antibody complexes is added. A common radioactive marker is *Staphylococcus* protein A iodinated with radioactive iodine, ^{125}I. Protein A has a strong affinity for antibody and binds firmly. Once the radioactive marker has bound, its location on the blot can be detected by exposing the membrane to X-ray film; the gamma rays emitted by the ^{125}I expose the film only in the region where the radioactive antibody has bound to antigen–antibody complexes (Figure 24.27).

For many clinical applications, immunoblots employ enzyme-linked immunosorbent assay (ELISA) technology (see Section 24.10) for detection of bound antigen–antibody complexes. Following treatment of the blotted proteins with specific antibody, the membrane is washed and then treated with a second antibody, which binds to the first. Covalently attached to this second antibody is an enzyme. The original antigen–antibody complexes are visualized when the enzyme is exposed to substrate: The product of the enzyme reaction leaves a colored product on the membrane at any spot where the enzyme-labeled secondary antibodies are bound to the antigen-reactive antibodies. By comparing the location of the color bands on the blot with the position of colored bands from control samples, a protein associated with a given pathogen can be positively identified.

The immunoblot procedure can be used to detect either antigen (*direct* evidence for pathogen presence) or antibody (*indirect* evidence for pathogen exposure).

1. Denature proteins by boiling in detergent and subject to electrophoresis; proteins separate by molecular weight

2. Blot the separated proteins from the gel to membrane

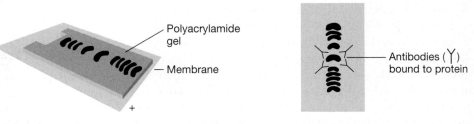

Polyacrylamide gel

Membrane

3. Treat membrane containing blotted proteins with antibodies; each antibody recognizes and binds to a specific protein

Antibodies (Y) bound to protein

4. Add marker to bind to antigen–antibody complexes, either (left) radioactive *Staphylococcus* protein A–¹²⁵I, or (right) antibody containing conjugated enzyme

^{125}I ^{125}I

E E
E E

5. Expose to film (¹²⁵I) or enzyme substrate and develop to reveal antibody-labeled protein

X-ray film

Membrane with enzyme-produced colored spot

(a)

GP41-45   P24

(b)

● **Figure 24.27 The Western blot (immunoblot) and its use in the diagnosis of human immunodeficiency virus (HIV) infection.** (a) Protocol for an immunoblot. (b) Developed HIV immunoblot. The proteins P24 and GP41–45 are coat proteins of the virus and are diagnostic for HIV. Lane 1, Positive control serum (from known AIDS patients); lane 2, negative control serum (from healthy volunteer); lane 3, strong positive from patient sample; lane 4, weak positive from patient sample; lane 5, reagent blank to check for background binding.

The HIV Immunoblot

Immunoblots have had a significant clinical impact on the diagnosis and confirmation of HIV exposure. Because an immunoblot is more laborious, more time-consuming, less sensitive, and more costly than the ELISA test, HIV-ELISA tests have been widely used for screening purposes. However, the HIV-ELISA test occasionally yields false-positive results. Thus, the highly specific immunoblot is used to confirm positive ELISA results.

Like the HIV ELISA, the HIV immunoblot is designed to detect the presence of *antibodies* to HIV in a serum sample. To perform the immunoblot, a purified preparation of HIV is treated with sodium dodecyl sulfate (SDS), a detergent that solubilizes HIV proteins and also inactivates the virus. The HIV proteins are then resolved by polyacrylamide gel electrophoresis and blotted from the gel onto membranes (Figure 24.26). At least seven major HIV proteins are resolved by electrophoresis, and two of them, designated P24

and GP41-45, are used as specific diagnostic proteins in the HIV immunoblot. Protein P24 is the HIV core protein, and proteins GP41-45 are HIV coat proteins (∞Section 16.15).

Following blotting of the proteins, the membrane strips are incubated with the test serum sample. If the sample is HIV-positive, antibodies against HIV proteins will be present and will bind to the HIV proteins on the membrane (Figure 24.27). To detect whether antibodies from the serum sample have bound to HIV antigens, a *detecting antibody*, anti-human IgG conjugated to the enzyme peroxidase, is added to the strips. If the detecting antibody binds, the activity of the conjugated enzyme, after addition of substrate, will form a brown band on the strip at the site of antibody binding. The patient is confirmed as HIV-positive if the position of the bands in the patient serum and a positive control serum are identical; negative control serum is also analyzed in parallel and must show no bands (Figure 24.27).

Although the intensity of the bands obtained in the HIV immunoblot varies somewhat from sample to sample (Figure 24.27 b), the interpretation of an immunoblot is generally unequivocal, and thus the test is valuable for confirming HIV-ELISA positives and eliminating false positives. To make the HIV immunoblot clinically accessible, membrane strips containing inactivated HIV antigens (previously separated by electrophoresis) are available commercially. Separate strips can be incubated directly with the patient and control serum samples and subsequently treated with the detecting antibody.

The immunoblot technique is also used to confirm the specificity of antibody tests for Lyme disease (see Table 24.5). However, because of the expense, technical requirements, lower *sensitivity* (but higher *specificity*), and time involved, immunoblot tests are less useful than the rapid, low-cost ELISA methods for general screening purposes.

 24.11 Concept Check

Immunoblot procedures are used to detect antibodies to specific antigens or to detect the presence of the antigens themselves. The antigens are separated by electrophoresis, transferred (blotted) to a membrane, and exposed to antibody. Immune complexes are visualized with enzyme-labeled or radioactive secondary antibodies. Immunoblots are extremely specific, but procedures are complex and time-consuming.

◆ What advantage does the immunoblot have over immunoassays such as ELISA and RIA?

◆ Why is the immunoblot not used for general screening for HIV exposure?

III MOLECULAR AND VISUAL DIAGNOSTIC METHODS

Extremely sensitive methods using molecular or electron microscopy methods are available for detecting pathogens. These methods do not depend on pathogen isolation or growth, or on the detection of an immune response to the pathogen. Since it is often difficult and sometimes impossible to grow and identify pathogens in culture, growth-independent identification methods based on molecular or physical methods are widely used.

24.12 Nucleic Acid Methods

Molecular methods are widely used in diagnostic microbiology. These methods use *genotypic* rather than *phenotypic* characteristics to identify specific pathogens. The success of genetic or DNA-based diagnostic procedures is based on several principles: (1) Nucleic acids can be readily isolated from infected tissues; (2) nucleic acids can be readily visualized and measured; (3) the nucleic acid sequence of an individual pathogen genome is unique, and so nucleic acid analysis can be used for unequivocal identification; and (4) nucleic acid sequences can be amplified to increase the amount of material available for analysis.

Nucleic Acid Probes in Diagnostics

One of the most powerful analytical tools available to clinical microbiologists is *nucleic acid hybridization*. Instead of detecting a whole organism or its products, hybridization detects the presence of *specific DNA sequences* associated with a specific organism. To identify a microorganism through DNA analysis, the clinical microbiologist must have available a **nucleic acid probe** for that microorganism. Nucleic acid probes normally consist of a *single strand* of DNA with a sequence unique to the gene of interest. A typical probe may be several kilobases in length. If a microorganism from a clinical specimen contains DNA or RNA sequences complementary to the probe, the probe sequence can hybridize (following appropriate sample preparation to yield single-stranded DNA from the microorganism), forming a *double-stranded* molecule (Figure 24.28●). To detect a reaction, the probe is labeled with a *reporter molecule*, a radioisotope, an enzyme, or a fluorescent compound that can be detected following hybridization. Depending on the reporter (radioisotopes are the most sensitive), as little as 0.25 μg of DNA per sample can be used for identification.

Nucleic acid probes offer many advantages over immunological assays. Nucleic acids are much more stable than proteins at high temperatures and at high pH and are more resistant to organic solvents and other chemicals. Because of the relative chemical stability of the target nucleic acids, nucleic acid probe technology can even be used to positively identify organisms that are no longer alive. Additionally, some nucleic acid probes may be more specific than antibodies and can detect single-nucleotide differences between DNA sequences.

Clinical Laboratory Probes

In most clinical probe assays, colonies from plates or pieces of infected tissue are treated with strong alkali, usually NaOH, to lyse the cells and partially denature the DNA, forming single-stranded molecules (Figure 24.28a). This mixture is then affixed to a matrix (filter or dipstick) or left in solution and the labeled probe added. Hybridization is allowed to occur at a temperature at which sequence homology between target DNA and probe DNA is necessary to form a stable duplex (the actual temperature used in a given probe assay is governed by the length and nucleic acid composition of the probe and target DNA).

Following a wash to remove unhybridized probe DNA, the extent of hybridization is measured using the reporter molecule attached to the probe. Depending on how the probe was labeled, this involves measurement of radioactivity, enzyme activity, or fluorescence. Nucleic acid probes have been marketed for the identification of several major microbial pathogens and are in widespread use for the detection of *Neisseria gonorrhoeae* and *Chlamydia trachomatis* (Table 24.9 and ∞Sections 26.12 and 26.13).

In addition to their use in clinical diagnostics, molecular probes are finding widespread application in food industries and in food regulatory agencies. Probe detection systems can be used to monitor foods for their content of important pathogens such as *Salmonella* and *Staphylococcus*. In probe assays of food, an enrichment period is usually employed to allow low numbers of cells in the food to multiply to a sufficient number to be detectable by the probe. Probes designed for use in the food industry employ dipsticks coated with pathogen-specific DNA to hybridize with pathogen DNA from solution. Two-component probes are used here, one serving as a *reporter probe* and the other as a *capture probe* (Figure 24.28 b). Following hybridization of the reporter or capture probe to target DNA, the dipstick, which contains a sequence complementary to the capture probe (usually poly dT to capture poly dA on the probe) is inserted into the hybridization solution, and it traps hybridized DNA for removal and measurement.

Advances in bacterial phylogenetics based on 16S ribosomal RNA (rRNA) sequences (∞Section 11.6) have allowed the construction of species-specific and even strain-specific nucleic acid probes. Typing based on differences in rRNA sequences is called *ribotyping*. Ribotyping visualizes the unique DNA restriction patterns of the rRNA genes when DNA from a particular organism is digested by restriction endonucleases (∞Section 7.7). The digested DNA is separated on an agarose gel, and a labeled rRNA probe is used to visualize the unique restriction patterns of the genes encoding rRNA (∞Section 11.11 and Figure 11.23). Within a *species*, and especially within a *strain*, the restriction pattern is highly conserved and is a molecular fingerprint for the organism. Because all organisms have ribosomal RNA genes, ribotyping has been applied in a wide variety of clinical as well as phylogenetic studies.

Nucleic acid probes detecting less than 1 μg of nucleic acid per sample can identify DNA extracted from about 10^6 bacterial cells or about 10^8 virus particles. Although molecular probes are not as sensitive as direct

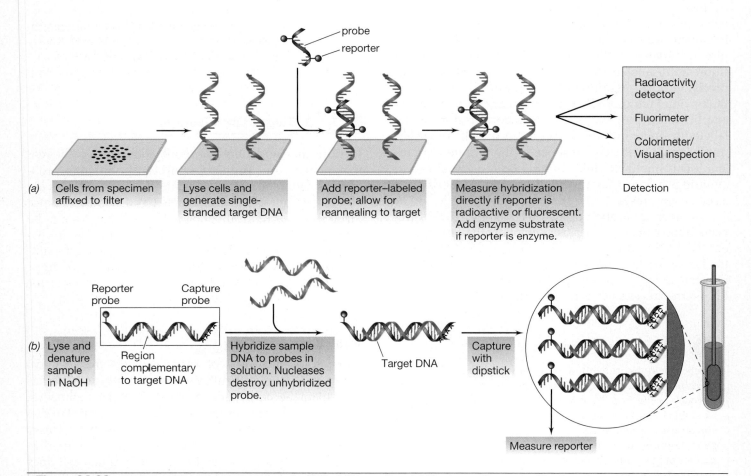

● **Figure 24.28 Nucleic acid probe methodology in clinical diagnostics.** (a) Membrane filter assay. The detecting system (reporter) can be a radioisotope, a fluorescent dye, or an enzyme. (b) Dipstick assay. In the dipstick assay a dual reporter or capture probe is used. The capture probe contains a poly-dA tail that hybridizes to a poly-dT oligonucleotide affixed to the dipstick, binding the oligonucleotide–target-reporter complex. The complex can be detected as in (a) above.

culture (where as few as 1–10 cells per sample can be detected), probe methods are useful in situations where culture of organisms is difficult or even impossible. However, other applications of DNA technology rival the sensitivity of the culture method.

The Polymerase Chain Reaction in Diagnostics

Extremely short sequence-specific nucleic acid oligonucleotides (usually 15–24 nucleotides in length) can act as primers for the polymerase chain reaction (PCR) amplification of DNA from specific pathogens. PCR is widely used in biology research and is also a major clinical diagnostic tool (∞Section 7.9). A typical PCR procedure is designed to amplify DNA about 1 billion-fold. Theoretically, this allows the visualization of a single DNA molecule obtained from a single bacterial cell. For example, sequence-specific primers for a pathogen-specific gene might be used to examine DNA derived from suspected infected tissue, even in the absence of an observable, culturable pathogen. These methods are particularly useful for identifying viral and intracellular infections. The presence of an appropriate amplified gene segment (Figure 24.29●) confirms the presence of the pathogen.

Several specific organisms for which either hybridization or PCR diagnostic methods are in use are listed in Table 24.9. Various PCR applications are used in clinical laboratories. Currently, many PCR tests employ *real-time PCR*. Real-time PCR involves the use of fluorescent-labeled PCR primers. PCR machines with precision optics monitor incorporation of the labeled PCR primers into the PCR products at every cycle. An amplification plot monitors the incorporation of the labeled primers into the amplified product. An amplified product indicates presence of the targeted organism. Real-time PCR yields an immediate visual result, avoiding the need for further detection methods such as electrophoresis.

Another application of PCR technology, *RT-PCR* (reverse-transcriptase PCR) is used to follow the progression of HIV-AIDS and is discussed in Section 26.14. This method incorporates production of cDNA directly from samples containing pathogen-specific RNA, as in the case of RNA retroviruses such as HIV (∞Section 9.13).

24.12 *Concept Check*

Nucleic acid hybridization is a powerful laboratory tool used for identification of microorganisms. A nucleic acid sequence specific for the microorganism of interest must be available in order to design a probe. Perhaps the most widespread use of probe-based technology is in the application of gene amplification (PCR) methods. Various DNA-based methodologies are currently used in clinical, food, and research laboratories.

◆ What advantage does nucleic acid hybridization have over standard culture methods for identification of microorganisms? What disadvantages?

● **Figure 24.29 Polymerase chain reaction (PCR) analysis of patient sputum for *Mycobacterium tuberculosis* in the diagnosis of tuberculosis.** Sputum samples from patients were used as a source of DNA. Amplification was initiated with a primer pair, which produced the indicated 439-base pair product when a pure culture of *M. tuberculosis* was used as the DNA source (lane 15). Lanes 2–9, 11 and 12 are from sputums positive for *M. tuberculosis* (lane 12 is a weak positive). Lanes 13 and 14 are from *M. tuberculosis*-negative sputum samples. Lanes 1 and 10 are molecular weight reference markers. For a description of PCR technology, see Section 10.17.

◆ Cite an example where information about a microorganism can be obtained with a nucleic acid probe or PCR assay in the absence of standard growth-dependent assays.

24.13 Diagnostic Virology

Identification and diagnosis of infections with pathogenic animal viruses presents significantly different problems as compared with bacterial pathogens. Viruses are all obligate intracellular parasites (∞Chapter 9). Since viruses cannot replicate outside the cell, they cannot be directly cultured on artificial media, but must be grown in mammalian cells. However, viruses or their cytopathic effects can be observed directly, and a variety of direct and indirect *in vitro* test procedures are available for identification of viral pathogens.

Virus Growth *In Vitro*

Laboratory cultivation of animal viruses from clinical materials is more difficult, time-consuming, and specialized than the cultivation of most bacterial pathogens. This is because viruses grow only in living cells. We discussed the use of cell cultures for the growth of viruses in Section 9.3, and such cultures are commonly used in diagnostic virology. Several long-lived human cell lines are available for propagation of viruses. These cell lines, often termed *immortal cell lines*, grow rapidly and divide indefinitely in cell culture media and some are readily infected by mammalian viruses.

Table 24.9	Pathogens identified with nucleic acid probe and PCR methods
Pathogen	**Diseases**
Bacteria	
Campylobacter sp.	Food infections
Chlamydia trachomatis	Venereal syndromes; trachoma
Enterococcus sp.	Nosocomial infections
Escherichia coli (enteropathogenic strains)	Gastrointestinal disease
Haemophilus influenzae	Infectious meningitis
Legionella pneumophila	Pneumonia
Listeria monocytogenes	Listeriosis
Mycobacterium avium	Tuberculosis
Mycobacterium tuberculosis	Tuberculosis
Mycoplasma hominis	Urinary tract infection; pelvic inflammatory disease
Mycoplasma pneumoniae	Pneumonia
Neisseria gonorrhoeae	Gonorrhea
Neisseria meningitidis	Meningitis
Rickettsia sp.	Typhus, hemorrhagic fever, etc.
Salmonella sp.	Gastrointestinal disease
Shigella sp.	Gastrointestinal disease
Staphylococcus aureus	Purulent discharges (boils, blisters, pus-forming skin infections)
Streptococcus pyogenes	Scarlet fever; rheumatic fever; strep throat
Streptococcus pneumoniae	Pneumonia
Treponema pallidum	Syphilis
Fungi	
Blastomyces dermatitidis	Blastomycosis
Candida sp.	Candidiasis, thrush
Coccidioides immitis	Coccidioidomycosis
Histoplasma capsulatum	Histoplasmosis
Viruses	
Cytomegalovirus	Congenital viral infections
Epstein-Barr virus	Burkitt's lymphoma; mononucleosis
Hepatitis viruses A, B. C, D, E	Hepatitis
Herpes virus (types I and II)	Cold sores; genital herpes
Human immunodeficiency virus (HIV)	Acquired immunodeficiency syndrome (AIDS)
Human papilloma virus	Genital warts; cervical cancer
Influenza	Respiratory disease
Polyoma virus	Neurological disease
Rotavirus	Gastrointestinal disease
Protozoa	
Leishmania donovani	Leishmaniasis
Plasmodium sp.	Malaria
Pneumocystis carinii	Pneumonia
Trichomonas vaginalis	Trichomoniasis
Trypanosoma sp.	Trypanosomiasis

In addition to these immortal cell lines, viruses can also be grown in Rhesus monkey kidney cell lines. Monkey kidney cell lines are called *primary* cell lines because they die after a limited number of cell divisions and cannot be maintained indefinitely in the laboratory. Primary monkey kidney cells support growth of a number of pathogenic viruses and are valuable for the initial isolation of unknown viruses.

Electron Microscopy

Diagnostic virology can also be done by electron microscopy. Because many viruses have distinctive morphologies (∞Chapter 9), they can often be detected in clinical samples by observing the sample with an elec-

tron microscope (Figure 24.30●). In most specimens, the virus particles must first be concentrated and separated from human tissues, and a variety of techniques, generally employing centrifugation and filtration methods, are used to obtain a virus-enriched sample. Although not as reliable as immunologic or nucleic acid probe methods, the observation of virus particles of a specific morphology in a particular type of human tissue is presumptive evidence for infection.

Antibodies directed against particular viruses can be used to increase the sensitivity and specificity of this method. Agglutinating antibodies may cause the viral particles to aggregate, making them easier to distinguish from cellular debris under the electron microscope. Viruses can also be visualized by treatment of

(a)

S. E. Miller

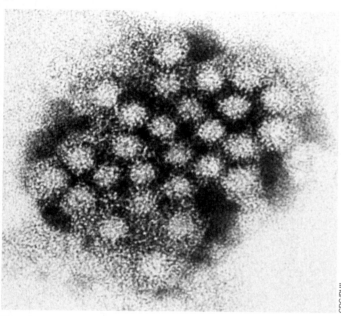

(b)

CDC/PHIL

● **Figure 24.30 Electron microscopic observation of clinical specimens to detect viruses.** (a) Human rotavirus from a fecal sample. The distinct spherical nature of the virus when found in fecal matter is a highly diagnostic criterion. Each rotavirus particle is approximately 75 nm in diameter. (b) Human Norwalk-like virus isolated from a patient with diarrhea. The individual viral particles are characterized as members of the Norwalk-like group due to their indistinct rough outer edge and small size of about 27 nm in diameter.

specimens with antiviral antibodies conjugated to heavy metals (see Section 24.9). With the use of negative staining techniques (Figure 24.30), results from electron microscopic analysis can be available 20 minutes after collection of the specimen. However, this method re-

quires an expensive and complex tool—the electron microscope—for analysis.

Other Virus Diagnostic Tests

A variety of laboratory procedures used in diagnostic virology are shown in Table 24.10. In many cases, immunodiagnostic methods (see Sections 24.5–24.12) and nucleic acid hybridization methods (see Section 24.13) are used for viral identification.

Many immunologic tests are either direct enzyme-linked immunosorbent assays (ELISAs) that detect viral particles (see Section 24.11 and Figure 24.24) or fluorescent antibody methods in which antibodies made against viral antigens are used to detect cells containing viruses (see Section 24.10 and Figure 24.20 *b*). In indirect tests, the virus itself is not detected, but antibodies to the virus are regarded as presumptive evidence for viral infection. For example, routine testing for human immunodeficiency virus (HIV) infection involves the ELISA test and the immunoblot test (see Section 24.10 and Section 24.11). Both the ELISA and the immunoblot detect *antibodies* to HIV, not HIV itself. However, the *viral load* test discussed in Section 26.15 is used to detect the presence of HIV. Agglutination tests may also detect virus directly. Antiviral antibodies conjugated to latex beads or activated charcoal (see Section 24.8) can be used for agglutination of viral particles released from lysed tissue samples.

Nucleic acid probes and PCR assays for detection and identification of clinically significant viruses are also a major diagnostic tool in clinical virology (see Section 24.11 and Table 24.9).

Many viral diagnostic procedures are used only under special circumstances. For routine virus infections, diagnoses are made by assessing clinical symptoms. Because of the technical expertise and expense involved in many viral diagnostic procedures (for example, laboratories that do primary culture must maintain or purchase Rhesus monkeys and have expertise in cell culture methods), most diagnostic virology laboratories are located at specialized government reference laboratories or major clinical research institutions.

 24.13 Concept Check

Virus propagation *in vitro* can be accomplished only in tissue culture. Therefore, most diagnostic techniques for viral identification are not growth-dependent, but routinely rely on immunoassays and nucleic acid–based techniques. Electron microscopy techniques are useful for direct observation of viruses in host samples.

◆ Why must viruses be grown in tissue culture?

◆ How can individual pathogenic viruses be identified in the laboratory?

Table 24.10 Laboratory procedures used in diagnostic virology[a]

Condition	Possible viral cause	Sample source	Cell lines/observation techniques
Upper respiratory infection	Rhinovirus	Nasopharyngeal or tracheal fluid (aspirate)	Human fibroblast culture
	Coronavirus		
	Adenovirus		
Pneumonia	Influenza	Nasopharyngeal fluid or swab	Human fibroblast cultures or embryonated eggs
Measles	Measles virus	Nasopharyngeal fluid or swab	Monkey kidney cells
Vesicular rash	Herpes simplex	Vesicular fluid by aspiration	Human fibroblast culture
Diarrhea	Rotavirus (infants)	Feces or rectal swab	Observe characteristic virus particles using electron microscope (Figure 24.30)
	Norwalk-like viruses		
Nonbacterial meningitis	Enterovirus	Spinal fluid	Human fibroblast or monkey kidney cultures
	Mumps		
	Herpes simplex		

[a] Immunological methods and nucleic acid probe methods are widely used in the diagnosis of viral infections (see Sections 24.5–24.12). The methods listed here are used for initial isolation and characterization of infectious viruses in research and reference laboratories.

REVIEW QUESTIONS

1. Describe the standard procedure for obtaining and culturing a throat culture and a blood sample. What special precautions must be taken while obtaining the blood culture (∞Section 24.1)?

2. Why are bacteria nearly always cultured from a urine specimen? Why is the *number* of bacterial cells in urine of significance? What organism is responsible for most urinary tract infections? Why (∞Section 24.1)?

3. Describe the procedures used for culturing anaerobic microorganisms. Why is it important to process all clinical specimens quickly? What special procedures and precautions are necessary for the isolation and culture of anaerobes (∞Section 24.1)?

4. Differentiate between *selective* and *differential* media. Is eosin-methylene blue (EMB) agar a selective medium or a differential medium? How and why is it used in a clinical laboratory (∞Section 24.2)?

5. Describe the Kirby–Bauer test for antibiotic sensitivity. Why should potential pathogens from patient isolates be tested by this method (∞Section 24.3)?

6. How are most laboratory-associated infections contracted? What action can be taken to prevent laboratory infections (∞Section 24.4)?

7. Why does the antibody titer rise after infection? Is a high antibody titer indicative of an ongoing infection? Explain. Why is it necessary to obtain an acute and a convalescent blood sample to monitor infections (∞Section 24.5)?

8. What advantages do *monoclonal* antibodies have over *polyclonal* antibody preparations, especially with regard to standardization of antibody preparations (∞Section 24.6)?

9. Describe a neutralization reaction with reference to microbial toxins and antisera (∞Section 24.7).

10. Agglutination tests are more widely used for clinical diagnostic purposes than are precipitation tests. Why is this the case (∞Section 24.8)?

11. How are fluorescent antibodies used for the diagnosis of viral diseases? What advantages do fluorescent antibodies have over unlabeled antibodies (∞Section 24.9)?

12. Radioimmunoassay (RIA) and enzyme-linked immunosorbent assay (ELISA) tests are extremely sensitive, as compared with agglutination. Why is this the case (∞Section 24.10)?

13. Why is the immunoblot (Western blot) procedure used to confirm positive human immunodeficiency virus (HIV)-ELISA results (∞Section 24.11)?

14. What information is essential for the design of a pathogen-specific nucleotide probe? Where can one obtain such information? Is this information available for all pathogens (∞Section 24.12)?

15. What is a *primary cell line*? Why do animal viruses grow only in tissue culture (∞Section 24.13)?

APPLICATION QUESTIONS

1. A blood culture is positive for *Staphylococcus epidermidis*. Interpret and explain the finding. Is it likely that the patient has *S. epidermidis* bacteremia? Prepare a list of possibilities and questions for a discussion with the physician in charge. What additional information will be needed to confirm or rule out a bacteremia?

2. With respect to both short-term and the long-term consequences, why is it a common medical practice to treat an infectious disease with antibiotics before isolating the suspected pathogen? After a pathogen has been isolated and identified, what further steps should be taken to confirm appropriate antibiotic sensitivity? Why are these measures rarely employed away from a hospital environment?

3. Explain the rationale for collecting serum specimens from patients during an acute infectious disease and about two weeks later (∞ Section 24.5). What information would you expect to obtain from the serum of a recovering patient?

4. What are the advantages of rapid identification systems such as agglutination tests and immune-based detection systems such as ELISA tests as compared with growth-dependent clinical diagnostic procedures? What are the potential disadvantages of the non–culture-based tests?

5. What are the major advantages of using DNA probes in diagnostic microbiology? What information is needed to design sequence-specific polymerase chain reaction (PCR) assay probes for a microorganism? Where can you find this information?

6. Define the procedures you would use to isolate and identify a new viral pathogen. Be sure to include both growth-dependent assays, molecular assays, and visual methods such as electron microscopy. Where would you report your findings? Which of your assays could be adapted to be used as a routine, high-throughput test for rapid clinical diagnosis?

25

EPIDEMIOLOGY

Severe acute respiratory syndrome, or SARS, is an emerging respiratory disease. The SARS coronavirus, shown in the photo, causes serious respiratory infections.

WORKING GLOSSARY

Acute short-term infection usually characterized by dramatic onset

Biological warfare the use of biological agents to incapacitate or kill humans

Carrier subclinically infected individual who may spread a disease

Centers for Disease Control and Prevention (CDC) an agency of the United States Public Health Service that tracks disease trends, provides disease information to the public and to health-care professionals, and forms public policy regarding disease prevention and intervention

Chronic long-term infection

Common-source epidemic an epidemic resulting from infection of a large number of people from a single contaminated source

Emerging infection infectious disease whose incidence has increased in the past 20 years or whose incidence threatens to increase in the near future

Endemic disease constantly present, usually in low numbers

Epidemic the occurrence of a disease in unusually high numbers in a localized population

Epidemiology the study of the occurrence, distribution, and control of diseases

Fomite inanimate object that, when contaminated with a viable pathogen, can transfer the pathogen to a host

Herd immunity resistance of a population to a pathogen as a result of the immunity of a large portion of the population

Host-to-host epidemic an epidemic resulting from person-to-person contact, characterized by a gradual rise and fall in number of cases

Incidence the number of cases of disease in a population

Morbidity incidence of illness in a population

Mortality incidence of death in a population

Nosocomial infection hospital-acquired infection

Outbreak the occurrence of a large number of cases of a disease in a short period of time

Pandemic a worldwide epidemic

Prevalence the proportion or percentage of individuals in the population having a disease

Public health the health of the population as a whole

Quarantine the practice of restricting the movement of individuals with highly contagious serious infections to prevent spread of the disease

Reemerging infection infectious disease, thought to be under control, that produces a new epidemic

Reservoir site in which viable infectious agents remain and from which infection of individuals may occur

Surveillance observation, recognition, and reporting of diseases as they occur

Vector a living agent that transfers a pathogen (note alternative usage in Chapter 31)

Vehicle nonliving source of pathogens that infect large numbers of individuals; common vehicles are food and water

Zoonosis a disease that occurs primarily in animals but can be transmitted to humans

---

PRINCIPLES OF EPIDEMIOLOGY

Infectious disease affects individuals, but infections are not usually acquired unless the individual is a part of a *population*. Here we consider how pathogens spread. **Epidemiology** is the study of the occurrence, distribution, and control of disease in populations. It thus deals with *public health*. In the next four chapters we will survey the diseases themselves. Here we look first at the principles of epidemiology and its importance in the control of infectious diseases.

Infectious disease control has been very successful in developed countries. Figure 1.7 compared the current causes of death in the United States with those at the beginning of the twentieth century. In most developed countries, microbial diseases cause few deaths as compared to noninfectious agents. Worldwide, however, infectious diseases account for nearly 30% of the 56 million annual deaths and remain serious public health problems in developing countries. Even in developed countries, new infectious diseases such as severe acute respiratory syndrome (SARS) are continuously emerging, and previously controlled diseases such as tuberculosis are reemerging. In the United States, deaths due to infectious diseases are increasing (Figure 25.1●). Effective control of infectious diseases remains a worldwide challenge that requires scientific, medical, economic, political, and educational solutions.

25.1 The Science of Epidemiology

In order to cause disease, a pathogen must grow and reproduce in the host. For this reason, epidemiologists seek to track the natural history of the pathogen. In

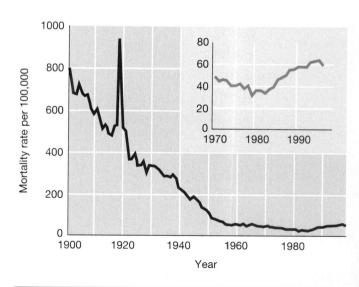

●**Figure 25.1** **Deaths due to infectious disease in the United States.** Although infectious disease death rates steadily declined throughout most of the twentieth century (except for the large numbers of deaths in 1918–1919 due to the influenza pandemic), the death rate has increased significantly since 1980. Adapted from "Hughes, J.M. 2001. Emerging Infectious Diseases: A CDC Perspective. *Emerg. Infect. Dis. 17: 494–496.*

many cases the pathogen cannot grow outside the host; if the host dies, the pathogen will also die. Pathogens that kill the host before they move to a new host will become extinct. How, then, do pathogens continue to exist if they kill the host?

A well-adapted pathogen actually lives in balance with its host, taking what it needs for existence and causing only a minimum of harm. Such pathogens may cause a **chronic**, or long-term infection in the host. When equilibrium between host and pathogen exists, both host and pathogen survive. On the other hand, serious host damage often occurs when new natural pathogens arise for which the host has not developed resistance, or when the natural resistance of the host decreases because of factors such as poor diet, age, and other stressors (∞Section 21.13). Thus, pathogens can be selective forces in the evolution of the host, just as hosts are selective forces in the evolution of pathogens. Such pathogens often cause **acute** infections, characterized by rapid and dramatic onset and, often, rapid recovery.

In cases where the pathogen is not dependent on the host for survival, the pathogen can cause devastating disease. Organisms in the genus *Clostridium*, for example, ubiquitous inhabitants of the soil (∞Section 12.20), are occasional accidental human pathogens and cause such diseases such as tetanus and botulism (∞Sections 21.10, 27.9, and 29.6).

The epidemiologist traces the spread of a disease to identify its origin and mode of transmission. Epidemiologic data are obtained from clinical studies, disease reporting surveys, insurance questionnaires, and interviews with patients to define common factors that constitute a disease. This is in contrast to the clinical or laboratory study of disease, where the focus is on treating the *individual* patient. Knowledge of both the clinical aspects and ecological aspects of a given disease are important if public health measures to control diseases are to be effective.

25.1 Concept Check

Epidemiology is the study of disease in populations. To understand infectious disease, the epidemiologist studies the interactions of the pathogen with the host population.

◆ How does an *epidemiologist* differ from a *microbiologist*?

◆ Why do epidemiologists acquire data for infectious diseases within a population?

25.2 The Vocabulary of Epidemiology

A number of terms have specific meaning to the epidemiologist. The **prevalence** of a disease in a population is defined as the *proportion* of diseased individuals in a population in a given time period. The **incidence** of a disease is the *number* of cases of an individual disease in a population in a given time period. A disease is said to be **epidemic** when it occurs in an unusually high number of individuals in a population at the same time; a **pandemic** is a widespread, usually worldwide, epidemic (Figure 25.2●). By contrast, an **endemic** disease is one that is constantly present, usually at low incidence, in a population. In an endemic disease, the pathogen may not be highly virulent, or the majority of the individuals may be immune, resulting in low disease incidence. However, as long as an endemic situation lasts, a few infected individuals remain who serve as *reservoirs* of infection.

Sporadic cases of disease may occur when individual cases are recorded in geographically separated areas, implying that the incidents are not related. A disease **outbreak**, on the other hand, occurs when a number of cases are observed, usually in a relatively short period of time, in an area previously experiencing only sporadic cases of the disease. Finally, diseased individuals who show no symptoms or only mild symptoms have *subclinical infections*. Subclinically infected individuals are frequently identified as **carriers** of a particular disease, because

● **Figure 25.2 Classification of disease by incidence.** Each dot represents several cases of a particular disease. Endemic diseases are always present in a population in a given geographical area. Epidemic diseases show high incidence in a wider area, usually developing from an endemic focus. Pandemic diseases are distributed worldwide. Diseases such as influenza are endemic in certain areas and often develop into epidemics under appropriate circumstances, such as crowding. With influenza, annual epidemics may also develop into ocassional pandemics.

(a) Endemic disease   (b) Epidemic disease   (c) Pandemic disease even though they themselves show few (or perhaps no) symptoms, they may be actively carrying and shedding the pathogen (see Section 25.3).

Mortality and Morbidity

The incidence and prevalence of disease is determined from statistical analysis of illnesses and deaths. From these data, a picture of the public health in a population can be obtained. The population might be the total global population of humans or the population of a localized region, such as a city, state, or country. Public health concerns vary with location and time. The assessment of public health at a given moment provides only a snapshot of a dynamic situation. By continuing to examine health statistics over many years, it is possible to assess the value of various public health policies that may influence the incidence of disease.

Mortality is the incidence of *death* in the population. Infectious diseases were the major causes of death in 1900 in developed countries (∞Figure 1.7), but they are now much less significant. Noninfectious "lifestyle" diseases such as heart disease and cancer are of greater importance. However, the current situation could rapidly change if a breakdown in public health measures was to occur. In developing countries, infectious diseases are still major causes of mortality (Table 25.1 and Section 25.10).

Morbidity refers to the incidence of *disease* in populations and includes both fatal and nonfatal diseases. Morbidity statistics define the health of the population more precisely than mortality statistics because many diseases

that affect health have relatively low mortality. The major causes of illness are quite different from the major causes of death. Major illnesses include acute respiratory diseases (the common cold, for instance) and acute digestive system conditions, which are often due to infectious agents and seldom cause death in developed countries.

Disease Progression

In terms of clinical symptoms, the course of a typical acute infectious disease can be divided into stages:

1. **Infection:** The organism begins to grow in the host.

2. **Incubation period:** The time between infection and the appearance of disease symptoms. Some diseases, like influenza (∞Section 26.8), have short incubation periods, measured in days; others, like AIDS, have longer ones, measured in years (∞Section 26.14). The incubation period for a given disease is determined by inoculum size, virulence and life cycle of the pathogen, resistance of the host, and distance of the site of entrance from the focus of infection (∞Sections 21.6. 21.7, and 21.8). At the end of incubation, the first symptoms, such as headache and a feeling of illness, appear.

3. **Acute period:** The disease is at its height, with overt symptoms such as fever and chills.

4. **Decline period:** Disease symptoms are subsiding, the temperature falls, usually following a period of intense sweating, and a feeling of well-being develops. The decline may be rapid (within 1 day), in which

Table 25.1	Worldwide deaths due to infectious diseases, 2002	
Disease	**Deaths**	**Causative agent(s)**
Acute respiratory infections[a, b]	3,963,000	Bacteria, viruses, fungi
Acquired immunodeficiency syndrome (AIDS)	2,777,000	Virus
Diarrheal diseases	1,798,000	Bacteria, viruses
Tuberculosis[a]	1,566,000	Bacterium
Malaria	1,272,000	Protozoa
Measles[a]	611,000	Virus
Pertussis (whooping cough)[a]	294,000	Bacterium
Tetanus[a]	214,000	Bacterium
Meningitis, bacterial[a]	173,000	Bacterium
Hepatitis (all types)[a, c]	157,000	Viruses
Syphilis	153,000	Bacterium
Leishmaniasis	51,000	Protozoan
Trypanosomiasis (sleeping sickness)	48,000	Protozoan
Chlamydia	16,000	Bacterium
Schistosomiasis	15,000	Parasitic worm
Chagas disease	14,000	Parasitic worms
Japanese encephalitis	14,000	Virus
Dengue	13,000	Virus
Intestinal nematode infections	12,000	Parasitic worms
Other communicable diseases	1,700,000	Various agents

Globally, there were about 57 million deaths from all causes in 2002. About 14.9 million deaths were from communicable infectious diseases, nearly all in developing countries. Data show the 20 leading causes of death due to infectious diseases. The world population in 2002 was estimated at 6.2 billion.

Data are from the World Health Organization (WHO), Geneva, Switzerland.

[a]Diseases for which effective vaccines are available.

[b]For some acute respiratory agents such as influenza and *Streptococcus pneumoniae* there are effective vaccines; for others, such as colds, there are no vaccines.

[c]Vaccines are available for hepatitis A virus and hepatitis B virus. There are no vaccines for other hepatitis agents.

case it is said to occur by *crisis*, or it may be slower, extending over several days, in which case it is said to be by *lysis*.

5. **Convalescent period:** The patient regains strength and returns to normal.

During the later stages of the infection cycle, the immune mechanisms of the host become increasingly important, and in most cases complete recovery from the disease requires (and results in) active immunity.

⬣ 25.2 Concept Check

An endemic disease is constantly present at low incidence in a specific population. In epidemics, an unusually high incidence of disease occurs in a specific population. Infectious diseases cause morbidity (illness) and may cause mortality (death). An infectious disease follows a predictable clinical pattern in the host.

◆ Distinguish between an *endemic* disease, an *epidemic* disease, and a *pandemic* disease.

◆ Distinguish between *morbidity* and *mortality*. How might high host *morbidity* be advantageous for the pathogen? Is host *mortality* advantageous for the pathogen?

25.3 Disease Reservoirs and Epidemics

Reservoirs are sites in which infectious agents remain viable and from which infection of individuals may occur. Reservoirs may be either *animate* or *inanimate*. Table 25.2 lists some common human infectious diseases and their reservoirs. Some pathogens are primarily *saprophytic* (living on dead matter) and only incidentally infect humans and cause disease. For example, *Clostridium tetani* (the causal agent of tetanus) normally inhabits the soil. Infection of animals by this organism is an accidental event: Infection of a host is not essential for its continued existence and in the absence of susceptible hosts, *C. tetani* would survive in nature.

For many other pathogens, however, living organisms are the only reservoirs. In these cases, the reservoir host is essential for the life cycle of the infectious agent. A number of pathogens live only in humans, and maintenance of the pathogen involves person-to-person transmission (∞Chapter 26). This is common for viral and bacterial respiratory pathogens, sexually transmitted pathogens, staphylococci and streptococci, diphtheria, typhoid fever, and mumps. As we shall see, pathogens that live their entire life cycle dependent on a single host can be eradicated (see Section 25.8).

Zoonosis

A number of infectious diseases occur both in humans and in animals. A disease that occurs primarily in animals but is occasionally transmitted to humans is called a **zoonosis**. Because public health measures for animal populations are much less developed than for humans, the infection rate for many diseases is much higher in animals, and animal-to-animal transmission is the rule. Occasionally, transmission is from animal to human; person-to-person transfer of these pathogens is rare (but see Section 25.8). As a result, control of a zoonosis in the human population does not eliminate it as a potential public health problem.

Eradication of the human form of a zoonotic disease can generally be achieved only through elimination of the disease in the animal reservoir. This is because maintenance of the pathogen in nature depends on animal-to-animal transfer and humans are incidental, nonessential hosts. For example plague (∞Section 27.7) is primarily a disease of rodents, and control of the infected rodent population and the insect vector controls epidemic plague far better than interventions such as vaccines in the incidental human host. Considerable success has also been achieved in the control of zoonotic bovine tuberculosis, a disease indistinguishable from human tuberculosis (∞Section 26.5), that was often spread from infected cattle to humans. Control was achieved primarily by identifying and destroying infected animals. Pasteurization of milk was also of considerable importance because milk was the main vehicle of bovine tuberculosis transmission to humans.

Certain infectious diseases, particularly those caused by organisms such as protozoa, have more complex cycles, involving an obligate transfer from animal to human to animal (for example, malaria, ∞Section 27.5). In such cases, the disease may potentially be controlled in either humans or the alternate animal host.

Carriers

A **carrier** is a pathogen-infected individual showing no signs of clinical disease. Carriers are potential sources of infection for others and are critically important for the spread of disease. Carriers may be individuals in the incubation period of the disease, in which case the carrier state precedes the development of actual symptoms. These individuals are prime sources of respiratory infections because they are not yet aware of their infection and so are not taking any precautions against infecting others. For such *acute carriers*, the carrier state lasts for only a short time. On the other hand, *chronic carriers* may remain infected and carry disease for extended periods of time. Chronic carriers usually appear perfectly healthy. They may be individuals who have recovered from a clinical disease, but still harbor viable pathogens, or they may be individuals with inapparent infections.

Carriers can be identified in populations by using a variety of diagnostic techniques, such as culture surveys (∞Section 24.1) or serological (antibody) surveys (∞Section 24.5). For example, tests using *Mycobacterium tuberculosis* antigens to test for delayed hypersensitivity (and thus exposure and previous or current infection) are widely used to define the carrier state for tuberculosis

Table 25.2 **Epidemic diseases: Agents, sources, reservoirs, and control (continued on next page)**

Disease	Causative agent[a]	Infection sources	Reservoirs	Control measures
Common-source epidemics[b]				
Anthrax	*Bacillus anthracis* (B)	Milk or meat from infected animals	Cattle, swine, goats, sheep, horses	Destruction of infected animals
Bacillary dysentery	*Shigella dysenteriae* (B)	Fecal contamination of food and water	Humans	Detection and control of carriers; oversight of food handlers; decontamination of water supplies
Botulism	*Clostridium botulinum* (B)	Soil-contaminated food	Soil	Proper preservation of food
Brucellosis	*Brucella melitensis* (B)	Milk or meat from infected animals	Cattle, swine, goats, sheep, horses	Pasteurization of milk; control of infection in animals
Cholera	*Vibrio cholerae* (B)	Fecal contamination of food and water	Humans	Decontamination of public water sources; immunization
E. coli O157:H7 food infection	*Escherichia coli* O157:H7 (B)	Fecal contamination of food and water	Humans, cattle	Decontamination of public water sources; oversight of food handlers; pasteurization of beverages
Giardiasis	*Giardia* spp. (P)	Fecal contamination of water	Wild mammals	Decontamination of public water sources
Hepatitis	Hepatitis A, B, C, D, E (V)	Infected humans	Humans	Decontamination of contaminated fluids and fomites; immunization if available (A and B)
Legionnaire's disease	*Legionella pneumophila* (B)	Contaminated water	High-moisture environments	Decontamination of air conditioning cooling towers, etc.
Paratyphoid	*Salmonella paratyphi* (B)	Fecal contamination of food and water	Humans	Decontamination of public water sources; oversight of food handlers; immunization
Typhoid fever	*Salmonella typhi* (B)	Fecal contamination of food and water	Humans	Decontamination of public water sources; oversight of food handlers; pasteurization of milk; immunization
Host-to-host epidemics				
Respiratory diseases				
Diphtheria	*Corynebacterium diphtheriae* (B)	Human cases and carriers; infected food and fomites	Humans	Immunization; quarantine of infected individuals
Hantavirus pulmonary syndrome	Hantavirus (V)	Inhalation of contaminated fecal material; contact	Rodents	Control of rodent population and exposure
Hemorrhagic fever	Ebola virus (V)	Infected body fluids	Unknown	Quarantine of active cases
Meningococcal meningitis	*Neisseria meningitidis* (B)	Human cases and carriers	Humans	Exposure treated with sulfadiazine for susceptible strains; immunization
Pneumococcal pneumonia	*Streptococcus pneumoniae* (B)	Human carriers	Humans	Antibiotic treatment; isolation of cases for period of communicability
Tuberculosis	*Mycobacterium tuberculosis* (B)	Sputum from human cases; infected milk	Humans, cattle	Treatment with isoniazid; pasteurization of milk
Whooping cough	*Bordetella pertussis* (B)	Human cases	Humans	Immunization; case isolation
German measles	Rubella virus (V)	Human cases	Humans	Immunization; avoid contact between infected individuals and pregnant women
Influenza	Influenza virus (V)	Human cases	Humans, animals	Immunization
Measles	Measles virus (V)	Human cases	Humans	Immunization
Sexually transmitted diseases[c]				
Acquired immunodeficiency syndrome (AIDS)	Human immunodeficiency virus (HIV)	Infected body fluids, especially blood and semen	Humans	Treatment with reverse transcriptase inhibitors, protease inhibitors (not curative) (➥Section 26.14)

Table 25.2	Epidemic diseases: Agents, sources, reservoirs, and control (continued)			
Disease	**Causative agent[a]**	**Infection sources**	**Reservoirs**	**Control measures**
Chlamydia	*Chlamydia trachomatis* (B)	Urethral, vaginal, and anal secretions	Humans	Testing for organism during routine pelvic examinations; chemotherapy of carriers and potential contacts; case tracing and treatment
Gonorrhea	*Neisseria gonorrhoeae* (B)	Urethral and vaginal secretions	Humans	Chemotherapy of carriers and potential contacts; case tracing and treatment
Syphilis	*Treponema pallidum* (B)	Infected exudate or blood	Humans	Identification by serological tests; antibiotic treatment of seropositive individuals
Trichomoniasis	*Trichomonas vaginalis* (P)	Urethral, vaginal, and prostate secretions	Humans	Chemotherapy of infected individuals and contacts
Vectorborne diseases				
Epidemic typhus	*Rickettsia prowazekii* (B)	Bite from infected louse	Humans, lice	Control louse population
Lyme disease	*Borrelia burgdorferi* (B)	Bite from infected tick	Rodents, deer, ticks	Avoid tick exposure; treat infected individuals with antibiotics
Malaria	*Plasmodium* spp. (P)	Bite from *Anopheles* mosquito	Humans, mosquito	Control mosquito population; treat infected humans with antimalarial drugs
Plague	*Yersinia pestis* (B)	Bite from flea	Wild rodents	Control rodent populations; immunization
Rocky Mountain spotted fever	*Rickettsia rickettsii* (B)	Bite from infected tick	Ticks, rabbits, mice	Avoid tick exposure; treat infected individuals with antibiotics
Direct-contact diseases				
Psittacosis	*Chlamydia psittaci* (B)	Contact with birds or bird excrement	Wild and domestic birds	Avoid contact with birds; treat infected individuals with antibiotics
Rabies	Rabies virus (V)	Bite by carnivores, contact with infected neural tissue	Wild and domestic carnivores	Avoid animal bites; immunization of animal handlers and exposed individuals
Tularemia	*Francisella tularensis* (B)	Contact with rabbits	Rabbits	Avoid contact with rabbits; treat infected individuals with antibiotics

[a] B, Bacteria; V, virus; P, protozoan.

[b] Some common-source diseases can also be spread from host-to-host.

[c] Sexually transmitted diseases can also be controlled by effective use of condoms and by sexual abstinence.

(◯Sections 22.15 and 26.5). Other diseases in which carriers are important for the spread of infection include hepatitis (◯Section 26.11) and typhoid fever (◯ Section 29.7). Culture or serology surveys of food handlers and health-care workers are sometimes used to identify individuals who are carriers and pose a risk as common sources of infection.

The classic example of a chronic carrier was the woman known as "Typhoid Mary," a cook in New York City in the early part of the twentieth century. Typhoid Mary (her real name was Mary Mallon) was employed as a cook during a typhoid fever epidemic in 1906. Investigations of the typhoid outbreaks revealed that Mary was associated with all of them. She was the likely source of infection because her feces contained large numbers of the typhoid bacterium, *Salmonella typhi*. She remained a carrier throughout her life, probably because her gallbladder

was infected and continuously secreted organisms into her intestine. She refused to have her gallbladder removed and was imprisoned. Released on the pledge that she would not cook or handle food for others, Mary disappeared, changed her name, and continued to cook in restaurants and public institutions, leaving behind epidemic outbreaks of typhoid fever. After several years, she was again arrested and imprisoned and remained in custody until her death in 1938.

 25.3 **Concept Check**

To understand how diseases spread, the pathogen reservoir must be known. Some pathogens exist in soil, water, or animals. Other pathogens exist only in humans and are maintained solely by person-to-person contact. An understanding of disease carriers and pathogen life cycles is critical for controlling disease.

◆ What is a *disease reservoir*?

◆ Distinguish between *acute* and *chronic* carriers. Provide an example of each.

25.4 Infectious Disease Transmission

Epidemiologists follow the transmission of a disease by correlating geographical, seasonal, and age group incidence of a disease with possible modes of transmission. A disease limited to a restricted geographical location may suggest a particular vector. Malaria, for example, is transmitted by a mosquito found mainly in tropical regions. A marked seasonality to a disease is often indicative of certain modes of transmission, such as in the case of influenza, where the number of cases jumps sharply when children enter school and come in close contact.

Different pathogens have different modes of transmission, which are usually related to the habitats of the organisms in the body. For instance, respiratory pathogens are generally airborne, whereas intestinal pathogens are spread through contaminated food or water. If the pathogen is to survive, it must be transmitted from one host to another. Even environmental factors may play a role in survival of the pathogen, and such variables as weather patterns may influence exposure to a pathogen. For example, California encephalitis, caused by single-stranded RNA Bunyaviruses (∞Section 16.9), occurs primarily during the summer and fall months, and disappears every winter, in a predictable cyclical pattern (Figure 25.3●). The virus is transmitted from mosquitoes that, of course, are prevalent only during warmer months. Disappearance of the insect vector causes the disease to disappear until the vector reappears and retransmits the virus in the summer months. Virtually all mosquito-transmitted encephalitis viruses follow the same seasonal pattern.

Pathogens can be classified by their mechanism of transmission, but all mechanisms have several stages in common: (1) *escape from the host*, (2) *travel*, and (3) *entry into a new host*. We give here a brief overview of common pathogen transmission mechanisms.

Direct Host-to-Host Transmission

Host-to-host transmission occurs whenever an infected host transmits the disease directly to a susceptible host (∞Chapter 26). Transmission by the respiratory route and by direct contact is very common. Transmission by infectious droplets is the most frequent means by which upper respiratory infections such as the common cold and influenza are propagated. However, some pathogens are very sensitive to environmental influences, are unable to survive for significant periods of time away from the host, and must be transmitted from host to host by direct contact. These pathogens include those responsible for sexually transmitted diseases such as *Treponema pallidum* (syphilis) and *Neisseria gonorrhoeae* (gonorrhea).

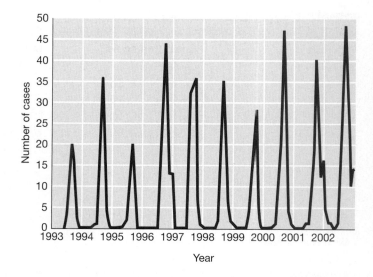

● **Figure 25.3 The incidence of California encephalitis in the United States, by month of onset.** Note the sharp rise in late summer, followed by a complete decline in every winter. The disease cycle follows the yearly cycle of the mosquito vector prevalence. In 2002 there were 167 cases in 16 states. A number of mosquito-borne viral diseases have similar seasonal infection patterns in humans in the United States. These include Western equine encephalitis, Eastern equine encephalitis, St. Louis encephalitis, and West Nile encephalitis. Data are from the Centers for Disease Control and Prevention, Atlanta, GA, USA.

These agents are extremely sensitive to drying and do not survive away from the body. The viable pathogen is transmitted only if there is intimate person-to-person contact and exchange of body fluids, such as in sexual intercourse.

Direct contact is also involved in the transmittal of skin pathogens such as staphylococci (boils and pimples) and fungi (ringworm). However, these pathogens are relatively resistant to environmental influences such as drying, and intimate person-to-person contact is not the only means of transmission. Many respiratory pathogens are also transmitted by direct means because they are spread by droplets resulting from sneezing or coughing. However, many of these droplets do not remain airborne for long. Transmission, therefore, requires close, although not necessarily intimate, person-to-person contact.

Indirect Host-to-Host Transmission

Indirect transmission can be facilitated by either living or inanimate agents. Living agents transmitting pathogens are called **vectors**. Commonly, arthropods (insects, mites, ticks, or fleas) or vertebrates (dogs, cats, or rodents) (∞Chapter 27) act as vectors. Arthropod vectors may not be hosts for the disease but simply carry the agent from one host to another. Large numbers of arthropods obtain nourishment by biting and sucking blood, and if the pathogen is present in the blood, the arthropod vector may ingest the pathogen and transmit it when biting another individual. In some cases, the pathogen replicates in the arthropod, which is then considered an alternate host. Such replication leads to an increase in pathogen numbers, increasing the probability that a subsequent bite will lead to infection.

Inanimate agents such as bedding, toys, books, and surgical instruments can also transmit disease. These inanimate objects are collectively referred to as **fomites.** Food and water are referred to as disease **vehicles.** Fomites can also be disease vehicles, but major epidemics originating from a single source are usually traced to food or water because these are actively consumed in large amounts by a number of individuals in a population (⊙Chapters 28 and 29).

Epidemics

Major epidemics are usually classified as *common-source* or *host-to-host* epidemics. These two types of epidemic propagation are contrasted in Figure 25.4●. Table 25.2 summarizes some of the key epidemiological features of some major epidemic diseases.

A **common-source epidemic** arises as the result of infection (or intoxication) of a large number of people from a contaminated common source, such as food or water. Usually such contamination occurs because of a malfunction in the sanitation of a central food or water distribution system. Foodborne and waterborne common-source

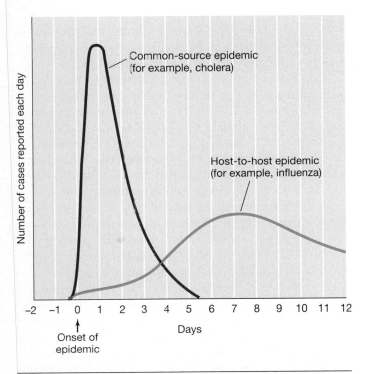

● **Figure 25.4** **Origins of epidemics.** The shape of the epidemic curve helps to distinguish the likely origin. In a common-source epidemic, such as from contaminated food or water, the curve is characterized by a sharp rise to a peak, with a rapid decline, which is less abrupt than the rise. Cases continue to be reported for a period approximately equal to the duration of one incubation period of the disease. In a host-to-host epidemic, the curve is characterized by a relatively slow, progressive rise, and cases continue to be reported over a period equivalent to several incubation periods of the disease.

epidemics are primarily *intestinal* diseases; the pathogen leaves the body in fecal material, contaminates food or water supplies due to improper sanitary procedures, and then enters the intestinal tract of the recipient during ingestion. Foodborne and waterborne diseases are generally amenable to control by public health measures, which we discuss further in Chapters 28 and 29. A classic example of a common-source epidemic is that of cholera. The spread of cholera through drinking water was first shown in 1855 by British physician John Snow. His studies demonstrated that cholera spread is related to fecal contamination of a water supply. In this case, the contaminated *water* was a common source of infection. The infectious agent, *Vibrio cholerae*, was transmitted through consumption of the contaminated common-source vehicle (Figure 25.4).

The disease incidence for a common-source outbreak is characterized by a rapid rise to a peak because a large number of individuals ingest contaminated food or water and become ill within a relatively brief period of time (Figure 25.4). The common-source illness also declines rapidly, although the decline is less rapid than the rise. Cases continue to be reported for a period of time approximately equal to the duration of one incubation period of the disease.

In a **host-to-host epidemic**, the disease incidence shows a relatively slow, progressive rise (Figure 25.4) and a gradual decline. Cases continue to be reported over a period of time equivalent to several incubation periods of the disease. A host-to-host epidemic can be initiated by the introduction of a single infected individual into a susceptible population, with this individual infecting one or more people in the population. The pathogen then replicates in susceptible individuals, reaches a communicable stage, and is transferred to other susceptible individuals, where it again replicates and becomes communicable. Chapter 26 discusses a number of diseases propagated by host-to-host transmission.

 25.4 Concept Check

A pathogen can be transmitted directly from one host to another, or indirectly by living vectors or inanimate objects (fomites) and common vehicles such as food and water. Epidemics may be of common-source or host-to-host origin.

◆ Distinguish between a *common-source* epidemic and a *host-to-host* epidemic. Cite at least one example of each.

◆ Suggest one method for halting the spread of a common-source epidemic and a host-to-host epidemic.

25.5 The Host Community

The colonization of a susceptible, unimmunized host by a parasite may first lead to explosive infections, transmission to uninfected hosts, and an epidemic. As the host population develops resistance, however, the spread of the parasite is checked, and eventually a *balance* is reached in which host and parasite are in equilibrium. In an extreme case, failure to reach equilibrium could result in death and eventual extinction of the host species. If the pathogen has no other host, then the extinction of the host could also result in extinction of the pathogen. Thus, the evolutionary success of a pathogen is best measured by its ability to establish a balanced equilibrium with the host, rather than by a pathogen's ability to destroy the host. In effect, the host and parasite are affecting each other's evolution; that is, the host and parasite are *coevolving*.

Coevolution of a Host and a Parasite

An excellent example of coevolution of host and parasite occurred when a virus was intentionally introduced for purposes of population control for feral rabbits in Australia. Rabbits introduced into Australia from Europe in 1859 spread until they were over-running large parts of the continent.

Myxoma virus was introduced into Australia in 1950 to control the rabbit population. This virus, spread very rapidly among rabbits by mosquitoes and other biting insects, is extremely virulent and usually causes a fatal infection. Within several months, the virus spread over a large area, rising to a peak in the summer when the mosquito vectors were present and declining in the winter. However, virus isolates from the feral rabbits, characterized for virulence in newborn feral rabbits and laboratory rabbits, indicated that the virus isolated was of decreased virulence. During the first year of the epidemic, over 95% of the infected rabbits died. However, within six years, rabbit mortality dropped to about 84%. The virus isolated was of decreased virulence and the resistance of the feral rabbits had increased dramatically (Figure 25.5●). In time, all of the surviving feral rabbits acquired the resistance factors. By the 1980s the rabbit population in Australia was nearing the premyxomatosis levels, with widespread environmental destruction and pressure on native plants and animals.

In 1995, Australian authorities began controlled releases of another highly virulent rabbit pathogen, the rabbit hemorrhagic disease virus (RHDV), a single-stranded, positive-sense RNA virus (Section 16.8). Because RHDV is spread by direct contact and kills animals within days of initial infection, authorities believed the infections would kill *all* rabbits in a local population, preventing the emergence of resistance. Thus, they reasoned, the virus-host relationship could be maintained in favor of the pathogen more reliably than the arthropod-borne myxomatosis virus. Initial reports indicated that RHDV was very effective at reducing local rabbit populations. However, infection of some rabbits with an indigenous hemorrhagic fever virus conferred immune cross-resistance to the introduced RHDV. This unpredictable immune response limited the effectiveness of the control program in certain areas of Australia. The establishment

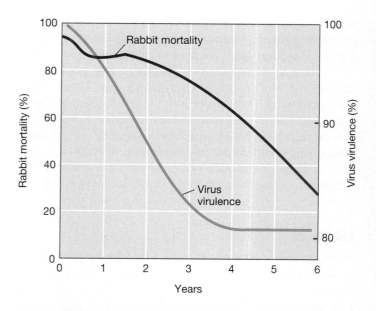

● **Figure 25.5 Changes in virulence of myxoma virus and in susceptibility of the Australian rabbit.** Data is from the years after the virus was introduced into Australia in 1950. Virus virulence is given as the average mortality in standard laboratory rabbits for virus recovered from the field each year. Rabbit susceptibility was determined by removing young rabbits from their dens and infecting them with a virus strain of moderately high virulence, which killed 90–95% of normal laboratory rabbits.

of immune rabbit populations indicates that the host is developing resistance to the control agent, again moving the host-pathogen balance toward equilibrium.

While coevolution of host and pathogen may be the norm for diseases that rely on host-to-host transmission, for pathogens that do not rely on host-to-host transmission, as we have already mentioned for *Clostridium* spp. (see Section 25.1), there is no selection for decreased virulence to support mutual coexistence. Vectorborne pathogens normally transmitted by the bite of arthropods or ticks are also under no evolutionary pressure to spare the human host. As long as the vector can obtain its blood meal before the host dies, the pathogen can maintain a high level of virulence, decimating the human host in the process of infection. For example, the malaria parasites *Plasmodium* spp. show antigenic variations in their coat proteins that aid in avoiding the immune response of the host. This genetic ability to avoid the host responses *increases* pathogen virulence without regard to the susceptibility of the host. However, as we shall see for malaria, the host may develop disease-specific resistance under the constant evolutionary pressure exerted by a highly virulent pathogen (Section 27.5).

Other evidence for the phenomenon of continually increasing pathogen virulence comes from studies of super-virulent diarrheal diseases in newborns. In hospital situations, *Escherichia coli* can cause severe diarrheal illness and even death, and virulence seems to increase with each passage of the pathogen through a hospital patient. The *E. coli* organisms replicate in one host and are then transferred through carriers such as hospital attendants, or on fomites such as soiled bedding and furniture, to another patient. Even if the host dies or cannot contact others to transfer the disease, the virulent *E. coli* strain infects others through transmission by other than the person-to-person route. Extraordinary efforts such as completely washing the nursery and furniture with disinfectant and transferring staff are sometimes necessary to interrupt the cycle of these super-virulent infections.

Herd Immunity

Herd immunity is the resistance of a group to infection due to immunity of a high proportion of the members of the group. Assessment of the immune state of a group is thus of great importance in understanding the development of epidemics. If a high proportion of individuals is immune, then the whole population will be protected. The proportion of immune individuals necessary to prevent an epidemic is higher for a highly virulent agent or one with a long period of infectivity and lower for a mildly virulent agent or one with a brief period of infectivity.

The proportion of the population that must be immune to prevent infection in the rest of the population can be estimated from data derived from immunization programs. For example, for poliovirus immunization in the United States, epidemiological studies of the incidence of polio in large populations indicate that if a population is 70% immunized, polio will be essentially absent in the population. The immunized individuals protect the rest of the population because they cannot acquire and pass on the pathogen, thus breaking the cycle of infection (Figure 25.6●). For highly infectious diseases such as influenza and measles, the proportion of immune individuals necessary to confer herd immunity is higher, about 90–95%.

A value of about 70% immunized has also been estimated for diphtheria, but studies of several small diphtheria outbreaks indicate that in densely populated areas a much higher proportion of susceptible individuals must be immunized to prevent an epidemic. Person-to-person transmission occurs even if the agent is not highly infectious if susceptible hosts come into repeated or constant contact with an infected individual. In the case of diphtheria, an additional complication arises because immunized persons can still harbor the pathogen (inapparent infection) and act as chronic carriers. This is because immunization protects against the effects of the diphtheria toxin, but not against diphtheria infection (Section 26.3).

Cycles of Disease

The principles of epidemics and herd immunity explain why certain diseases occur in *cycles*. A good example of a cyclical disease is *influenza*, which occurs in a high proportion of school children. Because the influenza virus is transmitted by the respiratory route (Section 26.8), its infectivity is high in crowded situations such as schools.

(a)   (b)

● **Figure 25.6 Herd immunity and transmission of infection.** Immunity in some individuals protects nonimmune individuals from infection. (a) In an unprotected population, an infected individual (red) can successfully infect (arrows) all of the susceptible individuals (blue). Newly infected individuals will in turn transfer the disease directly to other susceptible individuals. (b) For the case of a moderately transmissible pathogen such as *Corynebacterium diphtheriae* (diphtheria) in a population of moderate density, the infected individual (red) cannot transfer the disease to all susceptible individuals because resistant individuals (yellow), immune by virtue of previous exposure or immunization, break the cycle of pathogen transmission. Even if susceptible individual A (blue) acquires the disease, the other susceptible individuals, B and C, are protected. For a moderately transmissible pathogen such as *C. diphtheriae*, 70% immunity confers resistance to the entire population. For highly transmissible pathogens such as chickenpox (varicella) (Section 26.7), higher levels of immunity, on the order of 90%, are needed to stop transmission.

Since the epidemic strains of influenza virus change virtually every year, most children are susceptible. On the introduction of virus into the school, an explosive propagated epidemic results. Virtually every individual becomes infected, but then develops immunity, and as the immune population builds up, the epidemic subsides.

 25.5 Concept Check

Hosts and pathogens coevolve with time and arrive at a steady state that favors the continued survival of both. With herd immunity, a large fraction of a population is immune to a given disease, and it is difficult for the disease to spread. Disease cycles occur when a large, recurring, nonimmune population such as children entering school is exposed to a pathogen.

◆ Explain coevolution of host and pathogen. Cite a specific example.

◆ How does herd immunity prevent a nonimmune individual from acquiring a disease?

 CURRENT EPIDEMICS

In this section we will examine data collected by several disease-tracking programs based on the epidemiological principles just described. These data document emerging disease patterns for AIDS, nosocomial infections, and SARS, and provide information to health-care workers so that prevention and treatment strategies may be implemented.

25.6 The AIDS Pandemic

Acquired immunodeficiency syndrome (AIDS) is a viral disease that attacks the immune system (∞Section 26.14). The first reported cases were diagnosed in the United States in 1981. Through 2003, 928,407 cases have been reported in the United States, with 531,669 deaths through 2002. About 40,000 new AIDS cases continue to be diagnosed in the United States annually (Figure 25.7●). Worldwide, from 1981 to 2003, at least 70 million individuals have been infected with human immunodeficiency virus (HIV), the virus that causes AIDS. Twenty-five million people have already died from AIDS, and 40 million are currently living with the disease. Globally, another 5 million individuals are infected each year. North America has about 1 million infected individuals. Sub-Saharan Africa has 26.6 million infected people (Table 25.3). In the African countries of Botswana and Swaziland, 39% of the adult population is infected with HIV. AIDS caused about 3 million deaths in 2003, with 2.3 million of those deaths occurring in sub-Saharan Africa.

Tracking the Epidemic

Initial case studies in the United States suggested an unusually high AIDS prevalence among homosexual men and intravenous drug abusers. This indicated a transmis-

Table 25.3	HIV/AIDS infections, worldwide, 2003
Location	**HIV/AIDS Infections**
North America	1 million
Caribbean	460,000
Latin America	1.6 million
Western Europe	600,000
Eastern Europe and Central Asia	1.5 million
North Africa and Middle East	600,000
Sub-Saharan Africa	26.6 million
East Asia and Pacific	1 million
South and Southeast Asia	6.4 million
Australia and New Zealand	15,000

The total number of individuals infected with HIV/AIDS is estimated to be 40 million. Data are from the World Health Organization and the Joint United Nations Programme on HIV/AIDS.

sible agent, presumably transferred during sexual activity or by contaminated needles. Individuals receiving blood or blood products were also at high risk: Hemophiliacs who required infusions of blood products and a small number of individuals who received blood transfusions or tissue transplants before 1982 acquired AIDS (today less than 1% of the total current AIDS cases are attributable to these modes of transmission). The connection between transfer of AIDS with blood or tissue further reinforced the case for an infectious transmissible agent.

Soon after the discovery of HIV, laboratory tests were developed to detect antibodies to the virus in serum (∞Sections 24.10 and 24.11). This allowed extensive surveys of HIV incidence in different populations and also served as a screening method to ensure that new cases of AIDS were not transmitted by blood transfusions. The pattern illustrated in Figure 25.8 is typical of an agent transmissible by sexual activity or by blood. The identification of well-defined high-risk groups implied that AIDS was *not* transmitted from person to person by casual contact, such as the respiratory route, or by contaminated food or water. Instead, body fluids, primarily blood and semen, were identified as the major vehicles for transmission of HIV.

In the United States, the number of AIDS cases has been disproportionately high in homosexual men (Figure 25.8●) but the patterns in women and in certain racial and ethnic minorities indicate that homosexuality is not the predisposing factor for acquiring AIDS. For example, among women the heterosexual contact group is now the largest risk group (Figure 25.8), while in African-American and Hispanic men, intravenous drug use is nearly as often implicated in transmission of HIV as homosexual activity. If we consider all risk groups, heterosexual activity is the fastest growing risk category for new AIDS cases among adults. This diverse array of individuals who are at high risk for acquiring AIDS prompts us to look for a set of factors common to all, and

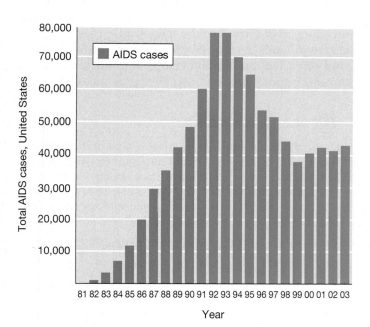

● **Figure 25.7 Annual diagnosed cases of acquired immunodeficiency syndrome (AIDS), since 1981 in the United States.** Cumulatively there have been 928,407 cases of AIDS and an estimated 531,669 deaths due to AIDS through 2003. Data are from the Centers for Disease Control and Prevention, Division of HIV/AIDS Prevention.

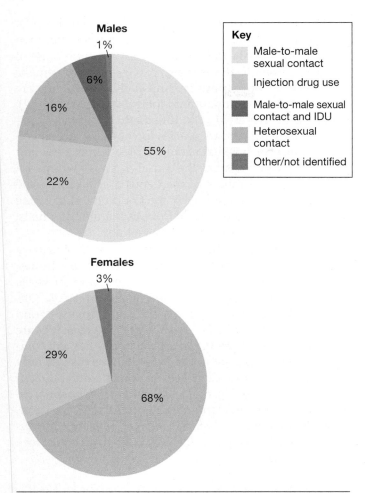

● **Figure 25.8 Distribution of AIDS cases by risk group and sex in adolescents and adults in the United States for the year 2002.** The total number of cases reported was 42,745. AIDS in males, N = 31,712. AIDS in women, N = 11,059. Data are from the Centers for Disease Control and Prevention, Division of HIV/AIDS Prevention. IDU, injection drug use.

virtually all individuals who are at risk for acquiring HIV share two specific behavior patterns. First, they engage in activities (sex or drug use) that involve *transfer of body fluids*, usually semen or blood. Second, individuals who acquire HIV often exchange body fluids with *multiple partners*, either through sexual activity or through needle-sharing drug activity (or both). Thus, they increase their probability of exchanging body fluids with an HIV-infected individual and therefore their chance of acquiring HIV infection and AIDS.

The incidence of AIDS in hemophiliacs and blood transfusion recipients has been nearly eliminated. This is due to rigorous screening of the blood supply and also because many blood clotting factors needed by hemophiliacs can withstand a heat treatment sufficient to inactivate HIV or are available as genetically engineered products (∞Section 31.7). In 2002, there were 158 new cases of pediatric AIDS. HIV can be transmitted to the fetus by infected mothers and probably also in mother's milk. Infants born to HIV-infected mothers have maternally de-

rived antibodies to HIV in their blood, but a positive diagnosis of HIV infection in infants must wait a year or more after birth because about 70% of infants showing maternal HIV antibodies at birth are not infected with HIV.

Epidemiological studies of AIDS in Africa have confirmed that transmission of AIDS is not linked to particular sexual practices, such as homosexuality, because heterosexual transmission of AIDS is the norm. In some regions, fewer men than women are infected with HIV. The identification of high-risk groups such as individuals involved in the sex trade led to the development of health education campaigns. These campaigns inform the public of AIDS transmission methods and define high-risk behaviors. Because no cure or effective immunization for AIDS is available, public health education offers the most effective approach to the control of AIDS and is the major weapon for preventing the spread of infection. We discuss the pathology and therapy of AIDS in Section 26.15.

 25.6 Concept Check

AIDS is one of the most thoroughly studied disease pandemics. AIDS will continue to be a major public health problem, especially in developing countries. There is no effective cure or immunization to prevent AIDS, although we now know a great deal about its pathology and spread.

◆ Describe the major risk factors for acquiring AIDS. Tailor your answer to your country of origin.

◆ Estimate the total number of individuals in the United States who now have AIDS and predict how many will be living with AIDS in the next two years.

25.7 Hospital-Acquired (Nosocomial) Infections

A hospital is usually a place where sick people get well, but may also be a place where sick people get sicker. Cross-infection from patient to patient or from hospital personnel to patients presents a constant hazard. Hospital infections are called *nosocomial infections (nosocomium* is the Latin word for "hospital") and occur in about 5% of all patients admitted. In certain clinical services, such as intensive care units, up to 10% of the patients acquire a nosocomial infection. In all, there are about 2 million nosocomial infections each year in the United States, leading directly or indirectly to 80,000 deaths.

Hospital infections are partly due to the prevalence of diseased patients but are often due to the presence of pathogenic microorganisms that are selected and maintained within the hospital environment. For example, multiple-drug-resistant organisms are often spread from host to host as part of the normal flora in hospital situations. Nosocomial pathogens are often found as normal flora in either patients or hospital staff.

The Hospital Environment

Infectious diseases are spread easily and rapidly in hospital environments for several reasons. (1) Many patients have weakened resistance to infectious disease because of their illness (compromised hosts) (∞Section 21.13). (2) Hospitals treat patients suffering from infectious disease, and these patients may be reservoirs of highly virulent pathogens. (3) The housing of multiple patients in rooms and wards increases the chance of cross-infection. (4) Hospital personnel move from patient to patient, increasing the probability of transfer of pathogens. (5) Many hospital procedures, such as hypodermic injection, spinal puncture, and removal of tissue samples (biopsy) or fluids (drawing blood), breach the skin barrier and carry with them the risk of introducing pathogens to the patient. (6) In maternity wards of hospitals, newborn infants are unusually susceptible to certain infections because they lack well-developed defense mechanisms. (7) Surgical procedures expose internal organs to sources of contamination and the stress of surgery often diminishes the resistance of the patient to infection. (8) Certain therapeutic drugs, such as steroid drugs used for controlling inflammation, increase susceptibility to infection. (9) Use of antibiotics to control infections selects for antibiotic-resistant organisms (∞ Section 20.12). Table 25.4 summarizes information concerning prevalent hospital-acquired infections.

Hospital Pathogens

Hospital pathogens preferentially infect several sites, notably the urinary tract, blood, and the respiratory tract. A relatively small number of pathogens cause the majority of nosocomial infections at these sites (Table 25.4).

Table 25.4	Number of intensive care unit nosocomial infections in the U.S., by site and organism		
	Bloodstream	**Pneumonia**	**Urinary Tract**
Pathogen	**Number**	**Number**	**Number**
Enterobacter spp.	1,083	4,444	1,560
Escherichia coli	514	1,725	5,393
Klebsiella pneumoniae	735	2,865	1,891
Haemophilus influenzae		1,738	
Pseudomonas aeruginosa	841	6,752	3,365
Staphylococcus aureus	2,758	7,205	497
Staphylococcus spp.	8,181		838
Enterococcus spp.	2,967	682	4,226
Candida albicans	1,090	1,862	4,856
Other pathogens	3,774	12,537	8,075
Total number[a]	21,943	39,810	30,701
Total %	23.7	43.1	33.2

[a]The total number of nosocomial infections in intensive care units during a recent eight-year time period was 92,454.

Source: National Nosocomial Infections Surveillance System Report, Centers for Disease Control and Prevention, Atlanta, Georgia, USA.

One of the most important and widespread hospital pathogens is *Staphylococcus aureus*. It is the most common cause of pneumonia and the third most common cause of blood infections. *S. aureus* is also particularly problematic in nurseries. Many strains are unusually virulent and are also resistant to common antibiotics (∞Section 20.12), making their treatment very difficult. In addition to *S. aureus*, other *Staphylococcus* species are now the largest collective cause of hospital-acquired blood infections and are also very prevalent as the causal agents of wound infections. Most members of the genus *Staphylococcus* are found in the upper respiratory tract or on the skin where they are a part of the normal flora in many individuals, including hospital patients and personnel.

Escherichia coli is the most common cause of urinary tract infections in hospitals, but *Enterococcus* species, *Pseudomonas aeruginosa*, *Candida albicans*, and *Klebsiella pneumoniae* infections are also very common. While *Enterococcus*, *E. coli*, and *K. pneumoniae* are normally found only in the human body, *Candida* and *Pseudomonas* are good examples of *opportunistic pathogens*: They are commonly found in the environment and cause disease only in individuals whose defenses are somehow debilitated (∞Section 21.13). *P. aeruginosa* isolates from hospital infections are often resistant to multiple antibiotics, complicating treatment. *E. coli*, *Staphylococcus*, and *Enterococcus* also have the potential for multiple drug resistance (∞Section 20.12).

 25.7 Concept Check

Many common microorganisms have the potential to be pathogens in a hospital environment. Hospital patients are unusually susceptible to infectious disease and are exposed to a variety of infectious agents, including opportunistic pathogens, in the hospital environment. Treatment of these infections is complicated by antibiotic resistance.

◆ Why are hospital patients more susceptible than normal individuals to pathogens?

◆ Why is antibiotic resistance a major problem in hospital environments?

◆ What is the source of opportunistic pathogens?

25.8 Severe Acute Respiratory Syndrome

Severe acute respiratory syndrome (SARS) is an emerging infectious zoonotic disease that first appeared in November 2002 in the Guandong Province of China. By February, 2003, the virus had spread to 32 countries. Global travel has provided the vehicle for SARS dissemination. Its origins have been tentatively traced back to a coronavirus derived from an animal source. Evidence indicates that exotic animals such as civets, kept in China for exotic food, may be the animal source, and civets are clearly one animal reservoir.

SARS-Coronavirus

The severe acute respiratory syndrome coronavirus (SARS-CoV) is the agent that causes SARS. Coronaviruses are single-stranded RNA viruses that usually cause mild upper respiratory tract, coldlike infections in humans, and respiratory or intestinal diseases in animals (∞Section 9.12). SARS-CoV causes an unusually severe respiratory illness, often resulting in pneumonia. Figure 25.9● shows SARS-CoV, a spherical virus of about 120–160 nm in diameter with characteristic protein spikes, the corona, arranged around the sphere. The small panel shows the typical morphology of the isolated virus. The larger panel shows the localization of the virus in the host-cell cytoplasm, where it replicates.

The SARS-CoV genome sequence, completed within two months of the virus isolation, indicates that SARS-CoV is not a mutant of a known coronavirus, nor is it a recombinant between known human and animal coronaviruses. Because SARS-CoV exhibits so little homology to any known single strain of virus, it is likely not the product of a genetic engineering or bioterrorism experiment. SARS-CoV most probably evolved over an extended period of time in an animal host and developed, quite by accident, the ability to infect humans.

SARS Epidemiology

In humans, SARS-CoV spreads by droplet secretions, fomites, person-to-person contact, or through fecal contamination. Much like common cold viruses, SARS-CoV is relatively hardy, easily spread, and difficult to contain.

● **Figure 25.9 Severe acute respiratory virus syndrome corona virus (SARS-CoV).** The upper left panel shows isolated SARS-CoV. An individual virion is about 125 nm in diameter. The large panel shows coronaviruses within the cytoplasmic membrane-bound vacuoles and in the rough endoplasmic reticulum of host cells. The virus replicates in the cytoplasm and exits the cell through the cytoplasmic vacuoles.

Ordinarily, a new coldlike virus would be of little concern, but SARS-CoV causes infections with significant morbidity and mortality. There have been about 8459 known SARS-CoV respiratory infections and over 800 deaths, for an overall mortality rate of nearly 10%. About 20% of all infections have been in health-care workers, demonstrating the high infectivity of the virus. Standard containment and infection control methods practiced by health-care personnel have not been effective in controlling spread of the disease. In persons over 65 years of age, the mortality rate approaches 50%, attesting to its virulence as a human pathogen. Because of the high mortality and ease of spread, SARS-CoV control is a critical issue. Patients with SARS are confined for the course of the disease in strict isolation in negative-pressure rooms. To prevent infection, health-care workers wear respirators when working with SARS patients or SARS-CoV–contaminated materials.

SARS-CoV infection is diagnosed by already standard methods: detection of antibody, a rise in antibody titer over time (∞Section 24.7), isolation and identification of the virus in cell culture from a clinical specimen (∞Section 24.13), or a positive RT-PCR test (∞Section 24.12).

SARS is an example of a serious infection that has emerged very rapidly from a unique source. Its spread has been accelerated due to dissemination by international travel. On the other hand, rapid identification of the pathogen, nearly instant development of worldwide notification procedures, rapid development of diagnostic tests, and a concerted effort to understand the biology and genetics of this novel pathogen provide hope for rapid control of the disease.

 25.8 Concept Check

Infection with a novel zoonotic virus, SARS-CoV, causes severe acute respiratory syndrome (SARS). Person-to-person spread is by respiratory means. Control of this high mortality virus is through rapid diagnosis and isolation of victims.

◆ Identify the suspected source of the SARS-CoV.

◆ Define methods used to contain SARS-CoV spread.

III EPIDEMIOLOGY AND PUBLIC HEALTH

Here we identify some of the methods used to identify, contain, and eradicate infectious diseases within populations. We will also identify some important current and future public health threats from infectious diseases.

25.9 **Public Health Measures for the Control of Disease**

Public health refers to the health of the general population and to the activities of public health authorities in the control of disease. The incidence of many infectious diseases has dropped dramatically over the past 100 years, especially in developed countries, largely because of universal improvements in basic living conditions. Better nutrition, access to clean potable water, improved public sewage treatment, less crowded living quarters, and lighter workloads, have contributed immeasurably to control diseases, primarily by reducing the risk factors related to disease (∞Section 21.13). However, individual diseases such as smallpox, typhoid fever, diphtheria, brucellosis, and poliomyelitis have been controlled largely by active, disease-specific public health measures such as the administration of disease-specific vaccines.

Controls Directed Against the Reservoir

If the disease reservoir is primarily in *domestic animals*, then infection of humans can be prevented if the disease is eliminated from the infected animal population. Immunization or destruction of infected animals may eliminate the disease in animals and, consequently, in humans. These procedures have nearly eliminated *brucellosis* and *bovine tuberculosis* in humans. Recently, these procedures have been used to eliminate bovine spongiform encephalitis (mad cow disease) in cattle in the United Kingdom, Canada, and the United States. Not incidentally, the health of the domestic animal population is also enhanced, with likely long-term economic benefits to the farmer and the consumer.

When the reservoir is a *wild animal*, eradication is much more difficult. *Rabies* is a disease that occurs in both wild and domestic animals but is transmitted to domestic animals primarily by wild animals. Thus *control* of rabies in domestic animals and in humans can be achieved by immunization of domestic animals. However, since the majority of rabies cases are in wild rather than domestic animals, at least in the United States (∞Section 27.1), *eradication* of rabies would require the immunization or destruction of all wild animal reservoirs, including such diverse species as raccoons, bats, skunks, and foxes. Although oral rabies immunization is practical and recommended for rabies control in restricted wild animal populations, its efficacy is untested in large, diverse animal populations such as the wild animal reservoir in the United States.

If the reservoir is an *insect* (such as mosquitoes in malaria and West Nile fever) (∞Sections 27.5 and 27.6) effective control of the disease can be accomplished by eliminating the reservoir with chemical insecticides or other lethal agents. The use of toxic or carcinogenic chemicals, however, must be balanced with environmental concerns. In some cases the elimination of one public health problem only creates another. For example, the insecticide dichlorodiphenyltrichloroethane (DDT) (∞ Section 19.18) is very effective against mosquitoes and is credited with eradicating yellow fever and malaria in North America. However, its use is currently banned in the United States because of environmental concerns. DDT is still used in many developing countries to con-trol mosquito-borne diseases but its use is declining worldwide.

When *humans* are the reservoir (for example, AIDS), control and eradication can be difficult, especially if there are asymptomatic carriers. On the other hand, diseases limited to humans that have no asymptomatic phase and can be prevented through immunization or treated with chemotherapy, can be eradicated if each case and all possible contacts are strictly quarantined, immunized, and treated. Such a strategy was successfully employed by the World Health Organization to eradicate smallpox and is currently being used to eradicate polio (see below).

Controls Directed Against Transmission of the Pathogen

Pathogens transmitted via *food or water* can be eliminated by preventing contamination of these common-source vehicles by destroying the pathogen in the vehicle. Water purification methods (∞Section 28.3) have dramatically reduced the incidence of typhoid fever, and the pasteurization of milk has helped control bovine tuberculosis in humans. Food protection laws have greatly decreased the probability of transmission of a number of enteric pathogens to humans (∞Chapter 29).

Transmission of respiratory pathogens is much more difficult to prevent. Attempts at chemical disinfection of air have been unsuccessful. Air filtration is a viable method, but is limited to small enclosed areas (∞Section 20.3). In Japan, many individuals wear face masks when they have upper respiratory infections to prevent transmission to others, but such methods, although effective, are voluntary and would be difficult to institute as public health measures.

Immunization

Smallpox, diphtheria, tetanus, pertussis (whooping cough), measles, mumps, rubella, and poliomyelitis have been controlled primarily by means of immunization. Effective vaccines are available for a number of other infectious diseases (∞Section 22.13). As we discussed in Section 25.5, 100% immunization is not necessary for disease control in a population, although the percentage needed to ensure disease control varies with the infectivity and virulence of the pathogen and with the living conditions of the population (for example, crowding).

Measles epidemics offer an example of the effects of herd immunity. The occasional resurgence of the highly contagious measles virus emphasizes the importance of maintaining appropriate immunization levels for a given pathogen. Until 1963, the year an effective measles vaccine was licensed, nearly every child in the United States acquired measles through natural infections, resulting in over 400,000 annual cases. After introduction of the vaccine, the number of annual measles infections dropped precipitously (∞Section 26.7). Case numbers reached a low of 1497 by 1983. However, by 1990, the percentage of children immunized against measles fell to 70%, and the number of new cases rose to 27,786. Within three years, a concerted effort to increase measles immunization levels to above 90% virtually eliminated indigenous measles transmission in the United States, and a total of only 312 measles cases were reported in 1993. Currently, about 100 cases of measles are reported each year in the United States, over half due to infections imported by visitors from other countries.

In the United States, virtually all children are now adequately immunized, but up to 80% of adults lack effective immunity to important infectious diseases because immunity from childhood vaccinations declines with time. When childhood diseases occur in adults, they can have devastating effects. For example, if a woman contracts *rubella* (a viral disease) (∞ Section 26.7) during pregnancy, the unborn child can be affected by serious developmental and neurological disorders. Measles, mumps, and chickenpox are also more serious diseases in adults than in children.

All adults are advised to review their immunization status and check their medical records (if available) to ascertain dates of immunizations. *Tetanus* immunizations, for example, must be renewed at least every 10 years to provide effective immunity. Surveys of adult populations have shown that more than 10% of adults under the age of 40 and over 50% of those over 60 are not adequately immunized. *Measles* immunity in adults should also be reviewed. People born before 1957 probably had measles as children and are immune. Those born after 1956 may have been immunized, but the effectiveness of early vaccines was variable and effective immunity may not be present, especially if the immunization was given before one year of age. Reimmunization for polio is not recommended for adults unless they are traveling to countries in western Africa and Asia where polio may still be endemic.

Recommendations for immunization were discussed in Section 22.12, and those for particular infections will be discussed in Chapters 26 through 29.

Quarantine

Quarantine involves restricting the movement of an individual with active infection to prevent spread of disease to other members of the population. The *time limit for quarantine is the longest period of communicability of the given disease.* Quarantine measures must prevent the infected individual from contacting unexposed individuals. Quarantine is not as severe a measure as strict isolation, which is used for unusually infectious diseases in hospital situations.

By international agreement, six diseases require quarantine: smallpox, cholera, plague, yellow fever, typhoid fever, and relapsing fever. Each is considered a highly serious, particularly communicable disease. Spread of certain other highly contagious diseases, such as Ebola

hemorrhagic fever and meningitis, may also be controlled by quarantine as outbreaks occur (see Table 25.5 and Section 25.10).

Surveillance

Surveillance is the observation, recognition, and reporting of diseases as they occur. Table 25.5 lists the diseases currently under surveillance in the United States. Several epidemic diseases listed in Table 25.2 and Table 25.8 are not on the surveillance list. However, many other diseases are surveyed through regional laboratories that identify *index cases*—those cases that exhibit new syndromes, characteristics, or pathogens indicating high potential for new epidemics.

Table 25.5	Reportable infectious agents and diseases in the United States
Diseases caused by *Bacteria*	**Diseases caused by fungi (molds, yeast)**
Anthrax	Coccidiomycosis
Botulism	Cryptosporidiosis
Brucellosis	**Diseases caused by viruses**
Chancroid	Acquired immunodeficiency
Chlamydia trachomatis	syndrome
Cholera	(AIDS) and pediatric
Diphtheria	HIV infection
Ehrlichiosis	Encephalitis/meningitis
Enterohemorrhagic *Escherichia coli*	(mosquito-borne)
Escherichia coli O157:H7	California serogroup
Gonorrhea	Eastern equine
Haemophilus influenzae, invasive disease	Powassan
Hansen's disease (leprosy)	St. Louis
Hemolytic uremic syndrome	Western equine
Legionellosis	West Nile
Listeriosis	Hantavirus pulmonary
Lyme disease	syndrome
Meningococcal disease	Hepatitis A, B, C
Pertussis	HIV infection
Plague	Adult
Psittacosis	Pediatric (<13 yrs)
Q fever	Measles
Rocky Mountain spotted fever	Mumps
Salmonellosis	Poliomyelitis, paralytic
Shigellosis	Rabies, animal, human
Streptococcal diseases, invasive, Group A	Rubella, acute and congenital syndrome
Streptococcal toxic shock syndrome	Severe acute respiratory
Streptococcus pneumoniae, drug-resistant and invasive disease	syndrome (SARS)
Syphilis	Smallpox
Tetanus	Varicella
Toxic shock syndrome	Yellow fever
Tuberculosis	**Diseases caused by protozoans**
Tularemia	Cyclosporiasis
Typhoid fever	Malaria
Vancomycin Intermediate *Staphylococcus aureus* (VISA)	Giardiasis
Vancomycin Resistant *Staphylococcus aureus* (VRSA)	**Disease caused by a helminth**
	Trichinosis

The **Centers for Disease Control and Prevention (CDC)** in the United States, through the National Center for Infectious Diseases (NCID), operates a number of surveillance programs, as shown in Table 25.6. In many cases, diseases are reportable to more than one surveillance network. While redundant reporting may at first seem unnecessary, a number of diseases fall into several categories that may affect health-care plans and policies. Cross-referencing and reporting of data from infections, for example, of vancomycin-resistant staphylococci with the National Nosocomial Infections Surveillance System (NNIS), and with CDC as a notifiable disease (Table 25.5) provides the hospital infection control team with a national data base. Using this information, health-care providers can formulate and implement rational plans for isolation, diagnosis, and drug-sensitivity testing of staphylococcal infections to identify antibiotic resistant strains and to begin appropriate treatment.

Pathogen Eradication

Disease eradication can be accomplished in specific cases and has been successful in the eradication of naturally occurring smallpox. As we mentioned earlier in this section, smallpox was a disease with a reservoir consisting solely of the individuals with acute smallpox infections, and transmission of smallpox was exclusively person-to-person. Infected individuals transmitted the disease through

Table 25.6	National Center for Infectious Diseases (NCID) surveillance systems for infectious disease notification and tracking, United States, 2004[a]
Surveillance System (acronym)	**Disease Surveillance Responsibility**
121 Cities Mortality Reporting System	Influenza, pneumonia, all deaths
Active Bacterial Core Surveillance	Invasive bacterial diseases
BaCon Study	Bacterial contamination associated with blood transfusion
Border Infectious Disease Surveillance Project (BIDS)	Infectious disease along the U.S.-Mexican border
Dialysis Survey Network (DSN)	Vascular access infections and bacterial resistance in hemodialysis patients
Electronic Foodborne Outbreak Investigation and reporting System (EFORS)	Foodborne outbreaks
EMERGEncy ID NET	Emerging infectious diseases
Foodborne Diseases Active Surveillance Network (FOODNET)	Foodborne disease
Global Emerging Infections Sentinel Network (GeoSentinel)	Global emerging diseases
Gonococcal Isolate Surveillance Project (GISP)	Antimicrobial resistance in *Neisseria gonorrhceae*
Integrated Disease Surveillance and Response (IDSR)	World Health Organization (WHO/AFRO) initiative for infectious diseases in Africa
Intensive Care Antimicrobial Resistance Epidemiology (ICARE)	Antimicrobial resistance and antimicrobial use in health-care settings
International Network for the Study and Prevention of Emerging Antimicrobial Resistance (INSPEAR)	Global emerging of drug-resistant organisms
Measles Laboratory Network	Measles in the Americas and the Caribbean
National Antimicrobial Resistance Monitoring System: Enteric Bacteria (NARMS)	Antimicrobial resistance in human nontyphoid *Salmonella*, *Escherichia coli* O157:H7, and *Campylobacter* isolates from agricultural and food sources
National Malaria Surveillance	Malaria in the United States
National Molecular Subtyping Network for Foodborne Disease Surveillance (PulseNet)	Molecular fingerprinting of foodborne bacteria
National Nosocomial Infections Surveillance System (NNIS)	Hospital-acquired infections
National Notifiable Diseases Surveillance System (NNDSS)	Notifiable diseases (see Table 25.5)
National Respiratory and Enteric Virus Surveillance System (NREVSS)	Respiratory syncytial virus (RSV), human parainfluenza viruses, respiratory and enteric adenoviruses, and rotavirus
National Surveillance System for Health Care Workers (NaSH)	Health-care worker occupational infections
National Tuberculosis Genotyping and Surveillance Network	Tuberculosis genotyping repository
National West Nile Virus Surveillance System	West Nile virus
Public Health Laboratory Information System (PHLIS)	Notifiable diseases
Surveillance for Emerging Antimicrobial Resistance Connected to Healthcare (SEARCH)	Emerging antimicrobial resistance in health-care settings
Unexplained Deaths and Critical Illnesses Surveillance System	Emerging infectious diseases worldwide
United States Influenza Sentinel Physicians Surveillance Network	260 clinical sites that report incidence and prevalence of influenza infections
Viral Hepatitis Surveillance Program (VHSP)	Viral hepatitis
Waterborne-Disease Outbreak Surveillance System	Waterborne diseases

[a]Contact information for these surveillance systems is available at: *http://www.cdc.gov/ncidod/osr/site/surv_resources/surv_sys.htm.*

direct contact with previously unexposed members of the population. Although smallpox, a viral disease, cannot be treated, immunization practices were very effective: Vaccination with a related viral strain conferred virtually complete immunity. In 1967, the World Health Organization (WHO) developed a plan to eradicate smallpox. Because of the success of vaccination programs worldwide, endemic smallpox was then confined to Africa, the Middle East, and the Indian subcontinent. After a preliminary program to vaccinate everyone in remaining endemic areas, every suspected smallpox outbreak was targeted by a team of WHO personnel who traveled to the outbreak site, quarantined individuals with active disease, and vaccinated all contacts. To break the chain of possible infection, they then immunized everyone who had contact with the contacts. This aggressive policy resulted in the elimination of the active natural disease within a decade, and the WHO announced the eradication of smallpox in 1980.

Polio, another viral disease with a very effective immunization program, is also targeted for eradication (endemic polio has been eradicated from the Western Hemisphere). Using much the same strategy to target polio as was used for smallpox, in the 1990s the WHO undertook a massive immunization program directed toward remaining endemic areas. By 2004, known *endemic* polio was restricted to Nigeria (with spread to nearby western African nations in 2003), India, and Pakistan. Recent sporadic outbreaks have also occurred in Afghanistan, and a single case was reported in Egypt in 2003.

Leprosy, another disease restricted to humans, is also targeted for eradication. Active cases of leprosy can now be effectively treated with a multidrug therapy that cures the patient and also prevents spread of *Mycobacterium leprae*, the causal agent (∞ Section 26.5).

Other diseases that are being targeted for eradication are Chagas' disease (treat active cases and destroy the insect vector of the *Trypanosoma cruzi* parasite in the American tropics) and dracunculiasis (treat drinking water to prevent transmission of *Dracunculus medinensis*, the Guinea helminth parasite in Africa, Arabia, Pakistan, and other places in Asia). Candidates for eradication also include syphilis (∞ Section 26.12) and rabies (∞ Section 27.1).

 25.9 Concept Check

Food and water purity regulations, vector control, immunization, quarantine, disease surveillance, and pathogen eradication are public health measures that play a major role in reduction of disease incidence.

◆ Compare public measures for controlling infectious disease caused by insect reservoirs and by human carriers.

◆ Identify public health methods used to halt the spread of an epidemic disease.

25.10 Global Health Considerations

The World Health Organization has divided the world into six geographic regions for the purpose of collecting and reporting health information, such as reports of morbidity and mortality. These geographic regions are Africa, the Americas (North America, the Caribbean, Central America, and South America), the eastern Mediterranean, Europe, Southeast Asia, and the Western Pacific. Here we compare mortality data from a developed region, the Americas, to that from a developing region, Africa.

Infectious Disease in the Americas and Africa: A Comparison

About 853 million people live in the Americas. Each year there are about 6 million deaths, or about 7 deaths per 1000 inhabitants per year. In Africa, there are about 672 million people and about 10.7 million annual deaths, or about 15.9 deaths per 1000 inhabitants per year. Although these statistics alone are cause for concern, examination of the causes of mortality in these regions is even more disturbing. Figure 25.10● indicates that most African

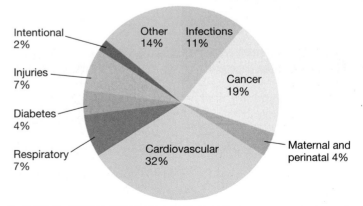

● **Figure 25.10 Causes of death in the Americas and Africa, 2002, by percentage of cause.** There were 10.7 million deaths in Africa, 6.7 million due to infectious diseases. There were 6 million deaths in the Americas, 623,000 due to infectious diseases. Intentional deaths include murder, suicide, and war.

deaths are due to infectious diseases, while in the Americas, cancer and cardiovascular diseases are the leading causes of mortality. In Africa, there are over 10 times as many deaths due to infectious diseases as compared to the Americas. Based on experiences in developed countries over the last century (∞Figure 1.7), these differences in death rates from infection are due to differences in public health services. Lack of resources in developing regions limits access to health care, medicines, safe food and water, and immunization.

Travel to Endemic Areas

The high incidence of disease in many parts of the world is a concern for people traveling to such areas. However, travelers can be immunized against many of the diseases that are endemic in foreign countries. Some typical recommendations for immunization for those traveling abroad are shown in Table 25.7. Many foreign countries currently require immunization certificates for yellow fever, but most other nonstandard immunizations are recommended only for people who are expected to be at high risk. In many parts of the world, travelers may be exposed to diseases for which there are no effective immunizations (for example, AIDS, Ebola hemorrhagic fever, dengue fever, amebiasis, encephalitis, malaria, and typhus). Travelers should take reasonable precautions such as avoiding unprotected sex, avoiding insect and animal bites, drinking only water that has been properly treated, eating properly stored and prepared food, and undergoing antibiotic and chemotherapeutic prophylactic programs or when exposure is suspected.

25.10 Concept Check

Infectious diseases account for nearly 30% of all worldwide mortality. Most infectious diseases occur in developing countries. Travelers to endemic disease areas should be immunized when possible and should take appropriate precautions to prevent infection.

◆ Contrast mortality due to infectious diseases in Africa and the Americas.

◆ List a series of infectious diseases for which you have *not* been immunized and with which you could come into contact next year.

25.11 Emerging and Reemerging Infectious Diseases

Infectious diseases are *global*, dynamic health problems. Here we examine some recent patterns of infectious disease, some reasons for the changing patterns, and the methods used by epidemiologists to identify and deal with new threats to public health.

The worldwide distribution of diseases can change dramatically and rapidly. Alterations in the pathogen, the environment, or the host population contribute to the spread of new diseases, with potential for high morbidity and mortality. Diseases that suddenly become prevalent are **emerging** diseases. Emerging infections are not limited to "new" diseases but also include **reemergence** of diseases thought to be controlled, especially when antibiotics become less effective and public health systems fail. Recent, dramatic examples of global emerging and reemerging disease are shown in Figure 25.11●. Diseases with potential for emergence or reemergence are described in Table 25.8. In addition, the epidemic diseases listed in Table 25.2 have the potential to emerge or reemerge as widespread epidemics and pandemics.

The phenomenon of suddenly emerging epidemic diseases is not new. Some of the diseases that suddenly emerged into prominence in the past were syphilis (caused by *Treponema pallidum*) (∞Section 26.12) and plague (caused by *Yersinia pestis*) (∞Section 27.7). In the Middle Ages, up to one-third of all living humans were killed by the plague epidemics that swept Europe, Asia, and Africa. Influenza caused a devastating worldwide epidemic in 1918–1919 (Figure 25.1 and ∞Section 26.8). In the 1980s, legionellosis (caused by *Legionella pneumophila*) (∞Section 28.7), acquired immunodeficiency syndrome (AIDS) (∞Section 26.14), and Lyme disease (∞Section 27.4) emerged as major new diseases. Current

Table 25.7	Immunizations required or recommended for international travel[a]	
Disease	**Destination**	**Recommendation[b]**
Yellow fever	Tropical and subtropical countries, especially in sub-Saharan Africa and South America	*Immunization required* for entry
Rabies	Rural, mountainous, and upland areas	*Immunization recommended* if direct contact with wild carnivores is anticipated
Typhoid fever	Many African, Asian, Central, and South American countries	*Immunization recommended* in areas endemic for typhoid fever

[a]*National Center for Infectious Diseases Travelers' Health*, U.S. Department of Health and Human Services, *http://www.cdc.gov/travel*.

[b]Vaccinations are generally recommended for diphtheria, pertussis, hepatitis A, hepatitis B, tetanus, polio, measles, mumps, rubella, and influenza as appropriate for the age of the traveler (∞Section 22.13). Most U.S. citizens are immunized against these diseases through normal immunization practices. Requirements and recommendations for specific vaccinations for each country are found at the website. Recommendations are also made for other appropriate infectious disease prevention measures, such as prophylactic drug therapy for malaria and plague prevention when visiting endemic areas. There are no current requirements for reentry to the United States.

● **Figure 25.11   Recent outbreaks of emerging and reemerging infectious diseases.** Emerging and reemerging diseases are first recognized as local epidemics. Recent epidemic outbreaks of significant but rare diseases are recorded here. All of these diseases are capable of producing widespread epidemics and even pandemics. The distribution of established pandemic diseases such HIV/AIDS and annual predictable epidemic diseases such as human influenza is not shown here.

emerging pathogens in the United States include West Nile virus (∞Section 27.6). On a worldwide scale, the rapid emergence of severe acute respiratory syndrome (SARS) has the potential to create a disastrous pandemic, especially considering the ease of spread of this highly virulent disease (Section 25.8).

Emergence Factors

Some factors responsible for emergence of new pathogens are (1) human demographics and behavior; (2) technology and industry; (3) economic development and land use; (4) international travel and commerce; (5) microbial adaptation and change; (6) breakdown of public health measures; and (7) abnormal natural occurrences that upset the usual host-pathogen balance.

The demographics of human populations have changed dramatically in the last two centuries. In 1800, less than 2% of the world's population lived in urban areas. By contrast, today nearly one-half of the world's population lives in cities. The numbers, sizes, and population densities of modern urban centers makes disease transmission much easier. For example, dengue fever (Table 25.8) is now recognized as a serious hemorrhagic disease in tropical cities, largely due to the spread of dengue virus in the mosquito *Aedes aegypti*. The disease now spreads as an epidemic in tropical urban areas. Prior to 1950, dengue fever was rare, presumably because the virus was not easily spread among a more dispersed, smaller population.

Human behavior, especially in large population centers, also contributes to disease spread. For example, sex-

ually promiscuous practices in population centers have been a major contributing factor to the spread of hepatitis and AIDS (Table 25.6; ∞Sections 26.11 and 26.14).

Technological advances and industrial development have had a generally positive impact on living standards worldwide, but in some cases these advances have contributed to the spread of diseases. For example, while tremendous technological advances have been made in health care during the twentieth century, there has been a dramatic increase in nosocomial infections (see Section 25.7). Antibiotic resistance in microorganisms is another negative outcome of modern health care practices. For example, vancomycin-resistant enterococci and multiple drug-resistant *Streptococcus pneumoniae* are important emerging diseases in developed countries.

Transportation, bulk processing, and central distribution methods have become increasingly important for quality assurance and economy in the food industry. However, these same factors can increase the potential for common-source epidemics when sanitation measures fail. For example, a single meat-processing plant spread *Escherichia coli* O157:H7 (Table 25.8) to at least 500 individuals in four states in the United States. The contaminated food source, ground beef, was recalled and the epidemic was eventually stopped, but not before several people died (∞Section 29.8). SARS spread rapidly to 32 countries through international travel from the point of origin of the disease, near the southern Chinese city of Guangzhou (see Section 25.8).

Economic development and changes in land use also can potentially promote disease spread. For example,

Table 25.8 Emerging and reemerging epidemic infectious diseases (continued on pages 839 and 840)

Agent	Disease and symptoms	Mode of transmission	Cause(s) of emergence
Bacteria, Rickettsias, and Chlamydias			
Bacillus anthracis	Anthrax: respiratory distress, hemorrhage	Inhalation or contact with endospores	Bioterrorism
Borrelia burgdorferi	Lyme disease: rash, fever, neurological and cardiac abnormalities, arthritis	Bite of infective *Ixodes* tick	Increase in deer and human populations in wooded areas
Campylobacter jejuni	Campylobacter enteritis: abdominal pain, diarrhea, fever	Ingestion of contaminated food, water, or milk; fecal-oral spread from infected person or animal	Increased recognition; consumption of undercooked poultry
Chlamydia trachomatis	Trachoma, genital infections, conjunctivitis, infant pneumonia	Sexual intercourse	Increased sexual activity; changes in sanitation
Escherichia coli O157:H7	Hemorrhagic colitis; thrombocytopenia; hemolytic uremic syndrome	Ingestion of contaminated food, especially undercooked beef and raw milk	Development of a new pathogen
Haemophilus influenzae biogroup *aegyptus*	Brazilian purpuric fever; purulent conjunctivitis, fever, vomiting	Discharges of infected persons; flies are suspected vectors	Possible increase in virulence due to mutation
Helicobacter pylori	Gastritis, peptic ulcers, possibly stomach cancer	Contaminated food or water, especially unpasteurized milk; contact with infected pets	Increased recognition
Legionella pneumophila	Legionnaires' disease: malaise, myalgia, fever, headache, respiratory illness	Air-cooling systems, water supplies	Recognition in an epidemic situation
Mycobacterium tuberculosis	Tuberculosis: cough, weight loss, lung lesions; infection can spread to other organ systems	Sputum droplets (exhaled through a cough or sneeze) of a person with active disease	Immunosuppression, immunodeficiency
Neisseria meningitidis	Bacterial meningitis	Person-to-person contact	Urbanization, breakdown or lack of local public health surveillance
Staphylococcus aureus	Abscesses, pneumonia, endocarditis, toxic shock	Contact with the organism in a purulent lesion or on the hands	Recognition in an epidemic situation; possibly mutation
Streptococcus pyogenes	Scarlet fever, rheumatic fever, toxic shock	Direct contact with infected persons or carriers; ingestion of contaminated foods	Change in virulence of the bacteria; possibly mutation
Vibrio cholerae	Cholera: severe diarrhea, rapid dehydration	Water contaminated with the feces of infected persons; food exposed to contaminated water	Poor sanitation and hygiene; possibly introduced via bilge water from cargo ships
Viruses			
Dengue	Hemorrhagic fever	Bite of an infected mosquito (primarily *Aedes aegypti*)	Poor mosquito control; increased urbanization in tropics; increased air travel
Filoviruses (Marburg, Ebola)	Fulminant, high mortality, hemorrhagic fever	Direct contact with infected blood, organs, secretions, and semen	Unknown; in Europe and the United States, virus-infected monkeys shipped from developing countries via air
Hendravirus	Respiratory and neurological disease in horses and humans	Contact with infected bats, horses	Human intrusion into natural environment
Hantaviruses	Abdominal pain, vomiting, hemorrhagic fever	Inhalation of aerosolized rodent urine and feces	Human intrusion into virus or rodent ecological niche
Hepatitis B	Nausea, vomiting, jaundice; chronic infection leads to hepatocellular carcinoma and cirrhosis	Contact with saliva, semen, blood, or vaginal fluids of an infected person; mode of transmission to children not known	Probably increased sexual activity and intravenous drug abuse; transfusion (before 1978)
Hepatitis C	Nausea, vomiting, jaundice; chronic infection leads to hepatocellular carcinoma and cirrhosis	Exposure (percutaneous) to contaminated blood or plasma; sexual transmission	Recognition through molecular virology applications; blood transfusion practices, especially in Japan
Hepatitis E	Fever, abdominal pain, jaundice	Contaminated water	Newly recognized

Table 25.8	**Emerging and reemerging epidemic infectious diseases (continued)**		
Agent	**Disease and symptoms**	**Mode of transmission**	**Cause(s) of emergence**
Viruses			
Human immuno-deficiency viruses: HIV-1 and HIV-2	HIV disease, including AIDS: severe immune system dysfunction, opportunistic infections	Sexual contact with or exposure to blood or tissues of an infected person; vertical transmission	Urbanization; changes in lifestyle or mores; increased intravenous drug use; international travel; medical technology (transfusions and transplants)
Human papillomavirus	Skin and mucous membrane lesions (often, warts); strongly linked to cancer of the cervix and penis	Direct contact (sexual contact or contact with contaminated surfaces)	Newly recognized; perhaps changes in sexual lifestyle
Human T-cell lymphotrophic viruses (HTLV-I and HTLV-II)	Leukemias and lymphomas	Vertical transmission through blood or breast milk; exposure to contaminated blood products; sexual transmission	Increased intravenous drug abuse; medical technology (transfusion and transplantation)
Influenza	Fever, headache, cough, pneumonia	Airborne; especially in crowded, enclosed spaces	Animal-human virus reassortment; antigenic shift
Lassa	Fever, headache, sore throat, nausea	Contact with urine or feces of infected rodents	Urbanization and conditions favoring infestation by rodents
Measles	Fever, conjunctivitis, cough, red blotchy rash	Airborne; direct contact with respiratory secretions of infected persons	Deterioration of public health infrastructure supporting immunization
Monkey pox	Rash, lymphadenopathy, pulmonary distress	Direct contact with infected primates and other hosts	Travel to endemic areas, consumption and handling of infected primates and other hosts
Nipah virus	Hemorrahagic fever	Close contact with bats and pigs in Malaysia	Exposure to infected animals
Norwalk and Norwalk-like agents	Gastroenteritis; epidemic diarrhea	Most likely fecal-oral; vehicles may include drinking and swimming water, and uncooked foods	Increased recognition
Rabies	Acute viral encephalomyelitis	Bite of a rabid animal; contact with infected neural tissue	Introduction of infected host reservoir to new areas
Rift Valley	Febrile illness	Bite of an infective mosquito	Importation of infected mosquitoes and/or animals; development (dams, irrigation)
Rotavirus	Enteritis: diarrhea, vomiting, dehydration, and low grade fever	Primarily fecal-oral; fecal-respiratory transmission can also occur	Increased recognition
Severe acute respiratory syndrome coronavirus (SARS-CoV)	Respiratory infection and pneumonia	Original zoonotic infection now spread person to person via infected droplets	Zoonotic spread from captured exotic animals (civet)
Venezuelan equine encephalitis	Encephalitis	Bite of an infective mosquito	Movement of mosquitoes and hosts (horses)
West Nile virus	Meningitis, encephalitis	*Culex pipiens* mosquito and avian hosts	Agricultural development, increase in mosquito breeding areas, rapid spread to nonimmune populations
Yellow fever	Fever, headache, muscle pain, nausea, vomiting	Bite of an infective mosquito (*Aedes aegypti*)	Lack of effective mosquito control and widespread vaccination; urbanization in tropics; increased air travel
Protozoa and Fungi			
Candida	Candidiasis: fungal infections of the gastrointestinal tract, vagina, and oral cavity	Endogenous flora; contact with secretions or excretions from infected persons	Immunosuppression; medical management (catheters); antibiotic use
Cryptococcus	Meningitis; sometimes infections of the lungs, kidneys, prostate, liver	Inhalation	Immunosuppression
Cryptosporidium	Cryptosporidiosis: infection of epithelial cells in the gastrointestinal and respiratory tracts	Fecal-oral, person to person, waterborne	Development near watershed areas; immunosuppression

Table 25.8	Emerging and reemerging epidemic infectious diseases (continued)		
Agent	**Disease and symptoms**	**Mode of transmission**	**Cause(s) of emergence**
Protozoa and Fungi (cont.)			
Giardia lamblia	Giardiasis; infection of the upper small intestine, diarrhea, bloating	Ingestion of fecally contaminated food or water	Inadequate control in some water supply systems; immunosuppression; international travel
Microsporidia	Gastrointestinal illness, diarrhea; wasting in immunosuppressed persons	Unknown; probably ingestion of fecally contaminated food or water	Immunosuppression; recognition
Plasmodium	Malaria	Bite of an infective *Anopheles* mosquito	Urbanization; changing parasite biology; environmental changes; drug resistance; air travel
Pneumocystis carinii	Acute pneumonia	Unknown; possibly reactivation of latent infection	Immunosuppression
Toxoplasma gondii	Toxoplasmosis; fever, lymphadenopathy, lymphocytosis	Exposure to feces of cats carrying the protozoan; sometimes foodborne	Immunosuppression; increase in cats as pets
Other Agents			
Bovine prions	Bovine spongiform encephalitis (BSE, animal) and variant Creutzfeld-Jacob disease (vCJD, human)	Foodborne	Consumption of contaminated beef

Rift Valley fever, a mosquito-borne viral infection, has been on the increase since the completion of the Aswan High Dam in Egypt in 1970. The dam flooded 2 million acres, and the enlarged shoreline increased breeding grounds for mosquitoes at the edge of the new reservoir. The first major epidemic of Rift Valley fever occurred in Egypt in 1977, when an estimated 200,000 people became ill and 598 died. Several epidemic outbreaks have occurred in the area since then, and the disease has become endemic near the reservoir.

Lyme disease, the most common vectorborne disease in the United States, is on the rise largely due to changes in land use patterns. Reforestation and the resulting increase in the numbers of deer and mice (the natural reservoirs for the disease-producing *Borrelia burgdorferi*) have resulted in greater numbers of infected ticks, the arthropod vector (∞Section 27.4). In addition, larger numbers of homes and recreational areas in and near forests increase contact between the infected ticks and humans, consequently increasing disease incidence.

International travel and commerce also affect the spread of pathogens. For example, filoviruses (*Filoviridae*), a group of RNA viruses, cause fevers culminating in hemorrhagic disease in infected hosts. These untreatable viral diseases generally have a mortality rate of greater than 20%. Most outbreaks have been restricted to equatorial central Africa, where the primate natural hosts and other vectors live. Travel of potential hosts to or from endemic areas is usually implicated in disease transmission. For example, one of the filoviruses was imported into Marburg, Germany, with a shipment of African green monkeys used for laboratory work. The virus quickly spread from the primate vector to some of the human handlers. Twenty-five people were initially infected, and six more developed disease as a result of contact with the human cases. Seven people died in this outbreak of what became known as the *Marburg virus*. Another shipment of laboratory monkeys brought a different filovirus to Reston, Virginia, in the United States. Fortunately, the virus was not pathogenic for humans, but due to its respiratory transmission mode, the Reston virus infected and killed most of the monkeys at the Reston facility within days. These two filoviruses are closely related to the Ebola virus (Table 25.8).

Sporadic Ebola outbreaks in central Africa, often characterized by mortality rates greater than 50%, highlight a group of pathogens for which there is no immunity or therapy. These pathogens could potentially be spread via air travel throughout the world in a matter of days. A highly contagious respiratory agent like the Reston virus that also possesses the high mortality potential of the Ebola virus could devastate population centers worldwide in a matter of weeks.

Microbial adaptation and change can contribute to pathogen emergence. For example, nearly all RNA viruses, including influenza and HIV, undergo rapid, unpredictable genetic mutations. RNA viruses lack correction mechanisms for replication steps, and so they incorporate mutations in their genome at an extremely high rate compared with most of the DNA viruses. The RNA viruses are considered to be major epidemiological problems because of their constantly changing genomes.

Bacterial genetic mechanisms are capable of enhancing virulence and promoting emergence of new epidemics. One group of virulence-enhancing mechanisms are the mobile genetic elements, bacteriophages, plasmids and transposons (⌘Sections 16.1–16.5, 10.9, and 10.11). Table 25.9 lists some virulence factors carried on these mobile genetic elements that contribute to pathogen emergence.

Antibiotic resistance is another factor in bacterial pathogen resurgence (⌘Section 20.12) and in virus emergence. Although several drugs are effective against certain viral diseases (⌘Section 20.10), resistance to these drugs is very common, especially among the RNA viruses. For example, many strains of HIV develop resistance to azidothymidine (AZT) unless it is used in combination with other drugs (⌘Section 26.14).

A breakdown of public health measures is sometimes responsible for the emergence or resurgence of diseases. For instance, cholera (caused by *Vibrio cholerae*, ⌘Section 28.5) can be adequately controlled, even in endemic areas, by providing proper sewage disposal and water treatment. However, in 1991 an outbreak of cholera due to contaminated municipal water supplies in Peru was one of the first indications that the current cholera pandemic had reached the Americas (⌘Section 28.5). In 1993, the municipal water supply of Milwaukee, Wisconsin, was contaminated with the chlorine-resistant protozoan *Cryptosporidium*, resulting in over 400,000 cases of intestinal disease, 4000 of which required hospitalization. Enhanced filtration systems were required to rid the water supply of the pathogen (⌘Section 28.6).

Inadequate public vaccination programs can lead to the resurgence of previously controlled diseases. For example, recent outbreaks of diphtheria (caused by *Corynebacterium diphtheriae*) (⌘Section 26.3) in the former Soviet Union result from inadequate immunization of susceptible children due to the breakdown in public health infrastructures. Pertussis, another vaccine-preventable childhood respiratory disease (caused by *Bordetella pertussis*) (⌘Section 26.4), has increased re-

cently in eastern Europe and in the United States due to inadequate immunization.

Finally, abnormal natural occurrences sometimes upset the usual host-pathogen balance. For example, hantavirus is a well-known human pathogen that occurs in many rodent populations, even in laboratory animals (⌘Section 27.2). A number of lethal cases of hantavirus infection and disease were reported in 1993 in the American Southwest and were linked to exposure to wild animal droppings. The likelihood of exposure to mice and droppings was increased due to a larger than normal wild mouse population resulting from near-record rainfall, a long growing season, and a mild winter.

Addressing Emerging Diseases

Many of the emerging diseases we have discussed are absent from the official notifiable disease list for the United States (Table 25.5). How then do public health officials define emerging diseases and prevent major epidemics? The keys for addressing emerging diseases are *recognition* of the disease and *intervention* to prevent disease transmission.

The first step in disease recognition is *surveillance*. Epidemic diseases that exhibit particular *clinical syndromes* warrant intensive public health surveillance. These syndromes are (1) acute respiratory diseases, (2) encephalitis and aseptic meningitis, (3) hemorrhagic fever, (4) acute diarrhea, (5) clusterings of high fever cases, (6) unusual clusterings of any disease or deaths, and (7) resistance to common drugs or treatment. Thus, new diseases are recognized because of their epidemic incidence, clusterings, and syndromes. As the prevalence and pathology of an emerging disease are recognized, the disease is added to the notifiable disease list.

Table 25.9	Virulence factors encoded by bacteriophages, plasmids, and transposons[a]	
Genetic element	**Organism**	**Virulence factors**
Bacteriophage	*Streptococcus pyogenes*	Erythrogenic toxin
	Escherichia coli	Shiga-like toxin
	Staphylococcus aureus	Enterotoxins A, D, E, staphylokinase, toxic shock syndrome toxin-1 (TSST-1)
	Clostridium botulinum	Neurotoxins C, D, E
	Corynebacterium diphtheriae	Diphtheria toxin
Plasmid	*Escherichia coli*	Enterotoxins, pili colonization factor, hemolysin, urease, serum resistance factor, adherence factors, cell invasion factors
	Bacillus anthracis	Edema factor, lethal factor, protective antigen, poly-D-glutamic acid capsule
	Yersinia pestis	Coagulase, fibrinolysin, murine toxin
Transposon	*Escherichia coli*	Heat-stable enterotoxins, aerobactin siderophores, hemolysin and pili operons
	Shigella dysenteriae	Shiga toxin
	Vibrio cholerae	Cholera toxin

[a] For discussion of bacteriophages, plasmids, and transposons, see Sections 9.8–9.11 and 16.1–16.5, 10.9, and 10.14, respectively.

For example, AIDS was recognized as a disease in 1981 and was added to the notifiable disease list (see Table 25.5) in 1984. Lyme disease was first recognized as a separate clinical disease in the 1980s and added to the notifiable disease list in 1991. Likewise, outbreaks of gastrointestinal disease due to enteropathogenic *Escherichia coli* O157:H7 have been increasing in recent years, and the strain was added to the notifiable disease list in 1995.

Intervention to prevent spread of emerging infections must be a public health response involving a variety of methods. Disease-specific intervention is the key to controlling individual outbreaks. Methods such as quarantine, immunization, and drug treatment must be applied to contain and isolate outbreaks of specific diseases. Finally, for vectorborne and zoonotic diseases, the nonhuman host or vector must be identified to allow intervention in the life cycle of the pathogen and interrupt transfer to humans.

 25.11 Concept Check

Changes in host, vector, or pathogen conditions, whether natural or artificial, can result in conditions that encourage the explosive emergence or reemergence of certain infectious diseases. Global surveillance and intervention programs must be developed to prevent new epidemics and pandemics.

◆ What factors are important in the emergence or reemergence of potential pathogens?

◆ Indicate general and specific methods that would be useful for dealing with emerging infectious diseases.

25.12 Biological Warfare and Biological Weapons

Biological warfare is the use of biological agents to incapacitate or kill a military or civilian population in an act of war or terrorism. Biological weapons have been used against targets in the United States, and weapons-making facilities are suspected to be in the hands of several governments as well as extremist groups.

Characteristics of Biological Weapons

Biological weapons must be organisms or toxins that are (1) easy to produce and deliver, (2) safe for use by the offensive soldiers, and (3) able to incapacitate or kill individuals under attack in a reproducible and consistent manner. Many organisms or biological toxins fit these rather general criteria, and we will discuss several of these below.

Although bioweapons are potentially useful in the hands of conventional military forces, the greatest likelihood of bioweapons use is probably by terrorist groups. This is in part due to the availibility and low cost of producing and propagating many of the organisms useful for biological warfare. Biological weapons are accessible to nearly every government and even to well-financed private organizations.

Candidate Biological Weapons

Virtually all pathogenic bacteria or viruses are potentially useful for biological warfare, and several of the most likely candidate organisms are relatively simple to grow and disseminate. Commonly considered biological weapons agents are listed in Table 25.10. The most commonly mentioned candidate bioweapon is *Bacillus anthracis*, the causal agent of anthrax. We will discuss anthrax separately.

Other important bacterial bioweapons candidates include *Yersinia pestis*, the organism responsible for plague (∞Section 27.7), *Brucella abortus* (fever and bacteremia; brucellosis), *Francisella tularensis* ("rabbit fever"), and *Salmonella* (foodborne and waterborne illnesses) (∞ Section 29.7). Viral pathogens with bioweapons potential include hemorrhagic fever viruses and encephalitis viruses. These agents cause diseases associated with significant morbidity, and some have very high mortality rates.

Table 25.10 **Potential bioterrorism agents and diseases**
Bacteria and rickettsias
Bacillus anthracis (anthrax)
Brucella sp. (brucellosis)
Burkholderia mallei (glanders)
Burkholderia pseudomallei (melioidosis)
Chlamydia psittaci (psittacosis)
Vibrio cholerae (cholera)
Clostridium botulinum toxin (botulism[a])
Clostridium perfringens (Epsilon toxin[a])
Coxiella burnetii (Q fever)
Escherichia coli O157:H7 (gastrointestinal disease)
Francisella tularensis (tularemia)
Yersinia pestis (plague)
Staphylococcus aureus enterotoxin B[a]
Salmonella Typhi (typhoid fever)
Salmonella sp. (salmonellosis)
Shigella (shigellosis)
Rickettsia prowazekii (typhus)
Viral agents
Variola major (smallpox)
Alphaviruses (viral encephalitis)
Venezuelan equine encephalitis virus
Eastern equine encephalitis virus
Western equine encephalitis virus
Nipah virus
Viral hemorrhagic fevers viruses
filoviruses; Ebola, Marburg
arenaviruses; Lassa, Machupo
hantaviruses
Protozoa
Cryptosporidium parvum (waterborne gastroenteritis)
Plants
Ricinus communis (ricin toxin from castor bean[a])

[a]Preformed toxin; all other agents require infection.

Source: Information is from the Centers for Disease Control and Prevention, Atlanta, GA, USA.

Bacterial toxins such as the botulinum toxin of *Clostridium botulinum* are also possible bioweapons (Table 25.8) (∞Sections 21.10 and 29.5). Large amounts of the preformed toxin delivered to a population through a common vehicle such as drinking water could have devastating consequences: The lethal dose of botulinum toxin for a human is 2 μg or less.

Smallpox

Smallpox virus (∞Section 16.13) is an intimidating potential biological warfare agent because it can be easily spread by contact or aerosol and it has a mortality rate of 30% or more. Its potential for use as a biowarfare agent is considered low, partly because the only *known* stocks of smallpox virus are in guarded repositories in the United States and Russia. However, a finite possibility remains for terrorist groups or military forces to gain access to smallpox virus. Because of this possibility, the U.S. government has made provisions to immunize front-line health-care and public safety personnel for smallpox. Although an extremely effective smallpox vaccine exists, this live *vaccinia virus* vaccine has not been in general use for almost 30 years because wild smallpox was eradicated worldwide by 1977 and because the vaccine can have serious side effects. As a result, over 90% of the current worldwide population is now inadequately vaccinated and susceptible to the disease. Preparations for a potential smallpox attack in the United States have included controversial recommendations for immunization of selected individuals: persons having close contact with smallpox patients; workers involved in the direct evaluation, care, or transportation of smallpox patients; laboratory personnel handling clinical specimens from smallpox patients; and other persons such as housekeeping personnel who would have increased contact with infectious materials from smallpox patients.

Although vaccinia immunization is very effective, it carries significant risk. Normal vaccine reactions include formation of a pustule, with a scab that falls off in two to three weeks, leaving a small scar. Mild adverse reactions occur in a large number of individuals and commonly include fevers and rashes. Vaccination is not currently recommended for persons with eczema or other chronic or acute skin conditions, heart disease, pregnant women, and those with reduced immune competence, such as individuals using anti-inflammatory steroid medications and those with HIV/AIDS.

About 1 in 1000 vaccinated individuals develop serious complications from the vaccine. Serious reactions include myocarditis and *erythema multiforme*, a toxic or allergic response to the vaccine. *Generalized vaccinia* (systemic vaccinia infection) occasionally occurs in individuals with skin conditions such as eczema. Very serious, life-threatening *progressive vaccinia* sometimes occurs, especially in individuals who are immunosuppressed due to therapy or disease. About one to two individuals per million will die from this vaccination.

Delivery of Biological Weapons

Most organisms suitable for bioweapons use can be spread in an aerosolized form, providing simple, rapid, widespread dissemination and infection. Examples of several exposures involving aerosols are instructive.

In 1962, one of the last outbreaks of smallpox in a developed country occurred in Germany. A German worker developed smallpox after returning from Pakistan, a country with endemic smallpox at the time. The individual was immediately hospitalized and quarantined, but the patient had a cough, and the aerosolized virus caused illness in 19 *vaccinated* individuals; at least one individual died from the resulting infection.

Planned bioterrorist attacks occurred even before the anthrax attacks of 2001 (see Section 25.12). In 1984 in The Dalles, Oregon (United States), cultists inoculated a salad bar with a *Salmonella typhimurium* culture in aerosol form at 10 local restaurants, causing 751 cases of foodborne salmonellosis in a region that usually has less than 10 cases per year (∞Section 29.7). In 1995, a radical political group released Saran nerve gas into a Tokyo subway, killing several people and injuring scores of others. Although this was a chemical weapon, this group also possessed anthrax cultures, bacteriological media, drone airplanes, and spray tanks.

In terms of strategic warfare, delivery of preformed bacterial toxins such as botulinum toxin or staphylococcal enterotoxin (Table 25.5) to large populations is somewhat impractical because most potent exotoxins are proteins that would lose effectiveness as they are diluted or are destroyed in common sources such as drinking water. However, delivery of toxins could be aimed at selected individuals, small groups, or even at random to instigate panic in a population.

Prevention and Response to Biological Weapons

Proactive measures against bioweapons have already begun with periodic planned efforts to update the international agreements of the 1972 Biological and Toxic Weapons Convention. The fifth and most recent update was in 2002. At the practical level, governments are now supporting the large-scale production and distribution of vaccines and the development of strategic and tactical plans to prevent and contain the effects of bioweapons.

The U.S. government, through the CDC, has devised and enhanced the Select Agent Program surveillance systems to monitor possession and use of potential bioterrorism agents. The CDC Laboratory Response Network and the Health Alert Network have been upgraded to enhance diagnostic capabilities and increase the reporting abilities of local and regional health-care centers to more quickly identify bioterrorism events as well as emerging diseases.

25.12 Concept Check

Bioterrorism is a threat in a world of rapid international travel and easily accessible technical information. Biological agents can be used as weapons by contemporary military forces or by terrorist groups. Aerosols or common sources such as food and water are the most likely mode of inoculation. Prevention and containment measures rely on a well-prepared public health infrastructure.

◆ What characteristics make a pathogen or its products particularly useful as a bioweapon?

◆ Identify two infectious agents that could be effective bioweapons. How could the agents be disseminated?

(a)

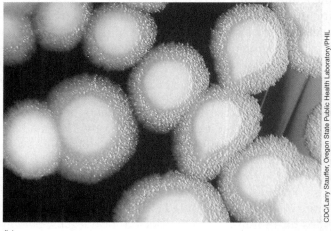

(b)

● **Figure 25.12   Bacillus anthracis.** (a) *B. anthracis* is a gram-positive endospore-forming rod (∞Section 12.20) approximately 1 μm in diameter and 3–4 μm in length. Note the formation of endospores (arrows). Endospore formation (∞Section 4.13) enhances the ability to disseminate *B. anthracis* in aerosols. (b) *B. anthracis* colonies on blood agar. The nonhemolytic colonies take on a characteristic "ground glass" appearance.

25.13 Anthrax as a Biological Weapon

Bacillus anthracis is a preferred agent for biowarfare and bioterrorism. Here we discuss its unique properties, the diseases it causes, and methods for prevention, diagnosis, and treatment.

Biology and Growth

Bacillus anthracis is a ubiquitous saprophytic soil inhabitant. It grows as an aerobic gram-positive rod, 1 μm in diameter by 3–4 μm in length. As with other members of the genus *Bacillus*, it produces endospores resistant to heat and drying (∞Sections 4.13 and 12.20) (Figure 25.12*a*●). Viable endospores are sometimes recovered from contaminated animal products such as hides and fur. Growth on blood agar results in large colonies with a characteristic "ground glass" appearance (Figure 25.12*b*). Strains having a poly-D-glutamic acid capsule are resistant to phagocytosis (∞Section 21.6).

Infection and Pathogenesis

Bacillus anthracis endospores are the standard means of acquiring anthrax. The disease usually affects domestic animals, especially ungulates—cows, sheep, and goats. The number of infections in animals, while considerable, is not known. The animals acquire the disease from plants or soil in pastures. In humans and animals, three forms of the disease can occur. *Cutaneous anthrax* results when abraded skin is contaminated by *B. anthracis* endospores (Figure 25.13*a*●). *Gastrointestinal anthrax* occurs due to consumption of endospore-contaminated plants and, potentially, from anthrax-infected carcasses. Human gastrointestinal anthrax is rarely seen. *Pulmonary anthrax* occurs when the endospores are inhaled. Inhalation of the endospores or the live bacteria results in pulmonary infections characterized by pulmonary and cerebral hemorrhage (Figure 25.13*b*). Untreated pulmonary anthrax infections have a mortality rate of nearly 100%. Cutaneous anthrax cases are rare in the United States, and pulmonary anthrax cases, even in agricultural workers, are extremely rare. The last natural pulmonary anthrax case occurred in 1976. However, several cases of pulmonary anthrax were identified in the United States in 2001 due to bioterrorism events.

Pathogenesis results from inhalation of 8000–50,000 spores of an encapsulated toxigenic strain. Pathogenic *B. anthracis* produces three proteins—*protective antigen (PA)*, *lethal factor (LF)*, and *edema factor (EF)*. PA and LF form *lethal toxin*, while PA and EF form *edema toxin*. PA is the cell-binding B component of these A-B toxins (∞ Section 21.10 and Table 21.4). EF cause edema and LF causes cell death. Growth of *B. anthracis* in the lymph nodes and lymphatic tissues draining the lungs leads to edema and cell death, culminating in tissue destruction, shock, and death.

(a)

(b)

CDC/James H. Steele/Public Health Image Library

CDC/Public Health Image Library

● **Figure 25.13 Anthrax.** (a) Cutaneous anthrax. The blackened lesion on the forearm of a patient, about 2 cm in diameter, results from tissue necrosis. Cutaneous anthrax, even when untreated, usually is a localized, nonlethal infection. (b) Inhalation anthrax. Fixed and sectioned brain showing hemorrhagic meningitis (dark coloration) due to a fatal case of inhalation anthrax.

Clinical symptoms can start with sore throat, fever, and muscle aches. After several days, symptoms progress to include difficulty in breathing, followed by systemic shock. Fatality rates can approach 90% even when exposure is suspected, and can be nearly 100% for cases where treatment is not started at or before the onset of symptoms.

Weaponized Anthrax

The term *weaponized* is applied to strains and preparations of *B. anthracis*, usually in endospore form, that exhibit properties that enhance dissemination and use as biological weapons. Such strains and preparations were developed by several governments in the post-World War II era, but overt development of new biological weapons was halted by international treaty in 1972. The physical characteristics of the weaponized anthrax preparations generally include a small particle size, usually interspersed with a very fine particulate agent such as talc. This small-particle, powdery form ensures that the endospores will spread easily by air currents. Thus, opening an envelope containing endospores or releasing the powder–spore mixture into a ventilation system or other air current has the potential to contaminate surrounding areas and personnel.

A weaponized form of anthrax was used as a vehicle for a series of bioterrorism attacks in the United States in 2001. These incidents were attributed to envelopes or packages containing weaponized anthrax endospores. The attacks were apparently directed at the news media (Florida), and the government (Washington, D.C. area). A third focus of attack, the Pennsylvania-New Jersey-New York area, had no defined single target, but disrupted mail service in the Northeast; some anthrax-contaminated mail facilities were still not in use in 2003. In all, there were 22 anthrax infections. Eleven were cutaneous anthrax. Of eleven cases of inhalation anthrax, five cases resulted in death. The bioterrorists were never identified.

The incidents in the United States were not the first or the most serious anthrax bioweapons infections. In a previous incident, *B. anthracis* spores were inadvertently released into the atmosphere from a bioweapons facility in Sverdlovsk, Russia in 1979. Less than 1 g of spores was released, and everyone in the area surrounding the facility was immunized and given prophylactic antibiotic therapy as soon as the first anthrax case was diagnosed. However, 77 individuals outside the facility contracted pulmonary anthrax and 66 died.

Vaccination, Prophylaxis, Treatment, and Diagnosis

Vaccination for anthrax has been restricted to individuals who are considered at risk. This includes agricultural animal workers and military personnel. The current vaccine, called *AVA (anthrax vaccine adsorbed)*, is prepared from a cell-free *B. anthracis* culture filtrate.

Treatment of infection, which seems to have a minimum incubation time of about eight days, is usually done with antibiotics. Ciprofloxacin, a broad-spectrum quinolone antibiotic (Figure 20.19) is used for strains that are penicillin-resistant, as are many known laboratory and bioweapons strains. Ciprofloxacin is also used as a prophylactic measure to treat potentially exposed individuals.

A number of rapid diagnostic tests are available to detect microbial endospores. However, positive identification of *B. anthracis* relies on culture techniques and direct observation of either infected tissues or cultured organisms. The characteristic ground-glass appearance on blood agar, coupled with isolation of gram-positive rods growing in extended chains is presumptive evidence for *B. anthracis* (Figure 25.12).

25.13 *Concept Check*

Bacillus anthracis has emerged as an important pathogen because of its use as a bioweapon. Highly infective weaponized endospore preparations have been used as bioterror agents. Inhalation anthrax has a fatality rate of about 90% in untreated individuals. Effective treatment relies on timely observation and diagnosis of symptoms. Treatment does not guarantee survival for inhalation anthrax.

◆ What factors contribute to the preferred use of *B. anthracis* as a bioweapon?

◆ Indicate the steps you would use to define and treat a bioterror attack using *B. anthracis*.

REVIEW QUESTIONS

1. List the five most common causes of mortality due to infectious diseases throughout the world. Are any of these diseases preventable by immunization (∞Section 25.1)?

2. Distinguish between *mortality* and *morbidity, prevalence* and *incidence*, and *epidemic* and *pandemic*, as these terms relate to infectious disease (∞Section 25.2).

3. Explain the difference between a *chronic* carrier and an *acute* carrier of an infectious disease (∞Section 25.3).

4. Give examples of host-to-host transmission of disease via direct contact. Also give examples of indirect host-to-host transmission of disease via vector agents and fomites (∞Section 25.4).

5. How can immunity to a pathogen by a large proportion of the population protect the nonimmune members of the population from acquiring a disease? Will this herd immunity work for diseases that have a common source, such as water? Why or why not (∞Section 25.5)?

6. Identify the major risk factors for acquiring human immunodeficiency virus (HIV) infection in the United States. Does this pattern hold for all geographic regions (∞Section 25.6)?

7. Hospital environments are conducive to the spread of infectious diseases. Review the reasons for the enhanced spread of infection in hospitals. What are the sources of most nosocomial infections (∞Section 25.7)?

8. Describe the source, the pathogen, and the treatment for severe acute respiratory syndrome (SARS). Does SARS have the potential to become a major epidemic (∞Section 25.8)?

9. Describe the major medical and public health measures developed in the twentieth century that were instrumental for controlling the spread of infectious diseases in developed countries (∞Section 25.9).

10. Compare the role of infectious diseases on mortality in developed and developing countries (∞Section 25.10).

11. Review the major reasons for the emergence of new infectious diseases. What methods are available for identifying and controlling the emergence of new infectious diseases (∞Section 25.11)?

12. Describe the general properties of an effective biological warfare agent. How does *smallpox* meet these criteria? Identify other organisms that meet the basic requirements for a bioweapon (∞Section 25.12).

13. Describe the use of *Bacillus anthracis* as a bioweapon. Devise a plan to protect yourself against a *B. anthracis* attack (∞Section 25.13).

APPLICATION QUESTIONS

1. Smallpox, a disease that was limited to humans, was eradicated. Plague, a disease with a zoonotic reservoir in rodents (Table 25.2) can never be eradicated. Explain this statement and why you agree or disagree with the possibility of eradicating plague on a global scale. Devise a plan to eradicate plague in a limited environment such as a town or city. Be sure to use methods that involve the reservoir, the pathogen, and the host.

2. Acquired immunodeficiency syndrome (AIDS) is a candidate for a disease that can be eliminated because it is propagated by person-to-person contact and there are no known animal reservoirs. Design a program for eliminating AIDS in a developed country and in a developing country. How would these programs differ from one another? What factors would work against the success of your program, both in terms of human behavior and in terms of the AIDS disease itself? Why are the numbers of HIV-infected and AIDS patients continuing to grow, especially in developing countries?

3. Travel to developing countries involves some exposure to infectious diseases. What general precautions should you take before, during, and after visits to developing countries? Where can you obtain information on the infectious disease status in a specific foreign country? When you return from a foreign country, are you a disease risk to your family or your associates? Explain.

4. Identify a specific pathogen that would be a suitable agent for effective biological warfare. Describe the properties of the pathogen in the context of its use as a bioweapon. Describe conditions for growing large amounts of the pathogen. Identify a suitable delivery method. Since you will propagate and deliver the pathogen, describe the precautions you will take to protect yourself. Now reverse your role. As a public health official in a large city, describe how you would recognize and diagnose the disease caused by the agent. Indicate the measures you would take to treat the illnesses caused by the agent. How could you best limit the damage? Would quarantine and isolation methods be useful? What about immunization and antibiotics?

26

PERSON-TO-PERSON MICROBIAL DISEASES

Mycobacterial infections are characterized and identified by the red (acid-fast) staining properties of the pathogen in tissue and sputum samples.

WORKING GLOSSARY

Antigenic drift minor changes in antigens due to gene mutation in influenza virus

Antigenic shift major changes in antigens due to gene reassortment in influenza virus

Cirrhosis breakdown of the normal liver architecture resulting in fibrosis

Congenital syphilis syphilis contracted by an infant from its mother during birth

Fusion inhibitor a synthetic polypeptide that binds to viral glycoproteins, inhibiting fusion of viral and host cell membranes

Hepatitis a liver inflammation commonly caused by an infectious agent

Jaundice production and release of excess bilirubin in the liver due to destruction of liver cells, resulting in yellowing of the skin and whites of the eye

Meningitis inflammation of the meninges (brain tissue), sometimes caused by *Neisseria meningitidis* and characterized by sudden onset of headache, vomiting, and stiff neck, often progressing to coma within hours

Meningococcemia fulminant disease caused by *Neisseria meningitidis* and characterized by septicemia, intravascular coagulation, and shock

Non-nucleoside reverse transcriptase inhibitor (NNRTI) a non-nucleoside compound that inhibits the action of viral reverse transcriptase by binding directly to the catalytic site

Nucleoside reverse transcriptase inhibitor (NRTI) a nucleoside analog compound that inhibits the action of viral reverse transcriptase by competing with nucleosides

Opportunistic infection an infection usually observed only in an individual with a dysfunctional immune system

Protease inhibitor a compound that inhibits the action of viral protease by binding directly to the catalytic site, preventing viral protein processing

Rheumatic fever an inflammatory autoimmune disease triggered by an immune response to infection by *Streptococcus pyogenes*

Scarlet fever characteristic reddish rash resulting from an exotoxin produced by *Streptococcus pyogenes*

Sexually transmitted infection (STI) an infection that is usually transmitted by sexual contact

Toxic shock syndrome (TSS) acute systemic shock resulting from a host response to an exotoxin produced by *Staphylococcus aureus*

Tuberculin test a skin test for previous infection with *Mycobacterium tuberculosis*

Viral load a quantitative assessment of the amount of virus in a host organism, usually in the blood

More than 500,000 microbial species exist in nature (Chapter 11), but only a few hundred species cause disease. Most microorganisms carry out essential activities independent of interactions with other organisms, and many microorganisms are closely associated with plants or animals in stable, beneficial relationships (◠◠ Chapter 19). However, pathogenic species have profoundly negative effects on host organisms. In the next four chapters, representative human pathogens and their biology are examined, as well as the pathology, diagnosis, treatment, and prevention of the diseases they cause. This discussion is organized based on the pathogen's *mode of transmission* and presents infectious disease in relation to the ecology of the pathogen. In this chapter disease transmission via direct *person-to-person* interactions is considered. In Chapters 27 through 29, diseases whose modes of transmission involve animal or arthropod vectors or common sources such as soil, water, or food are examined.

Connections between and among seemingly unrelated organisms can be made by examining pathogens according to their modes of transmission and the diseases they cause. For example, influenza virus and streptococci produce diseases with overlapping symptoms, although the causal agents, one viral and one bacterial, are markedly different. Here these pathogens are discussed together because they are spread person-to-person via a respiratory route. Using this approach, connections can be established between biologically diverse but ecologically and pathogenically related agents.

AIRBORNE TRANSMISSION OF DISEASES

Aerosols, such as those generated by the sneeze shown in Figure 26.1●, are important for person-to-person transmission of many infectious diseases. Most respira-

● **Figure 26.1 High speed photograph of an unstifled sneeze.**

tory diseases are spread almost exclusively in this fashion. For example, *Mycobacterium tuberculosis* has successfully used this strategy to infect at least one-third of the world's population (⟨∞⟩ Section 25.1). Influenza and cold viruses are passed by respiratory routes so commonly that virtually everyone acquires more than one cold or case of influenza every year.

26.1   Airborne Pathogens

Air is not a growth medium for microorganisms. Thus, microorganisms found in air are derived from soil, water, plants, animals, people, or other sources. In outdoor air, soil organisms predominate. Indoors, the concentration of microorganisms is considerably higher than outdoors, especially for those originating in the human respiratory tract.

Most microorganisms survive poorly in air, and so effective transmittal to another human occurs only over short distances. However, certain human pathogens (*Staphylococcus, Streptococcus*) survive under dry conditions fairly well and remain alive in dust for long periods of time. Gram-positive *Bacteria* are in general more resistant to drying than gram-negative *Bacteria* because of their thick, rigid cell wall. Likewise, the waxy layer of mycobacterial cell walls (see Section 26.5) resists drying. The endospores of endospore-forming *Bacteria* are extremely resistant to drying but are not generally passed from human to human in the endospore form.

An enormous number of moisture droplets are expelled during sneezing (Figure 26.1), and a considerable number are expelled during coughing or talking. Each infectious droplet has a diameter of about 10 μm and contains one or two microbial cells or virions. The initial speed of the droplet movement is about 100 m/sec (more than 200 mi/h) in a sneeze and ranges from 16 to 48 m/sec during coughing or shouting. The number of bacteria in a single sneeze varies from 10,000 to 100,000. Because of their small size, the moisture droplets evaporate quickly in the air, leaving behind a nucleus of organic matter and mucus to which bacterial cells are attached.

Respiratory Infections

Humans breathe about 500 million liters of air in a lifetime, much of it containing microorganism-laden dust, a potential source of inoculum for respiratory infections. The speed at which air moves through the respiratory tract varies, and in the lower respiratory tract the rate is quite slow. As air slows down, particles in it stop moving and settle. Large particles settle first and the smaller ones later, and only particles smaller than 3 μm travel as far as the bronchioles in the lower respiratory tract (Figure 26.2●). Of course, most pathogens are much smaller than this and different organisms characteristically colonize the respiratory tract at different levels. This is due to the unique metabolic requirements and virulence factors associated with each pathogen. The upper and lower respiratory tracts offer decidedly different environments, favoring certain microorganisms.

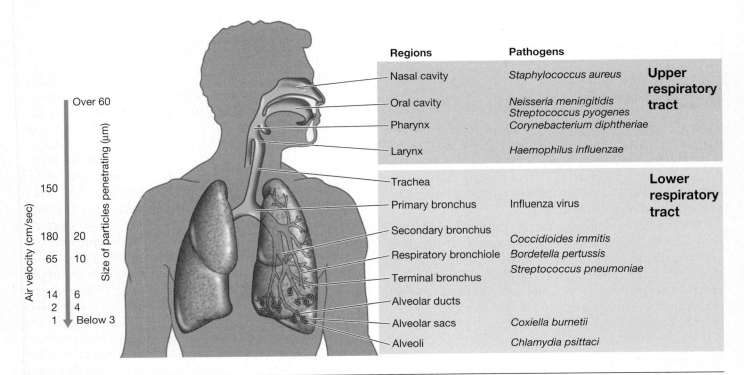

● **Figure 26.2   The respiratory system of humans.** Pathogenic microorganisms generally initiate infections at the indicated locations.

Bacterial Respiratory Pathogens

We begin here with a consideration of some common bacterial respiratory pathogens. We then examine viral respiratory pathogens and the much less treatable and preventable diseases they cause.

Most bacterial pathogens affecting the respiratory tract inhabit only humans and are normally transmitted from person to person. Since humans are the only reservoir for these pathogens, their survival is dependent on person-to-person transmission. A few respiratory pathogens such as *Legionella pneumophila*, transmitted primarily from water or soil do not require person-to-person propagation, and we discuss these later (see Chapter 28). As we explained above, many person-to-person respiratory pathogens are gram-positive *Bacteria*. Bacterial respiratory infections, while serious by themselves, often initiate secondary problems that can be life-threatening. Thus, accurate and rapid diagnosis and treatment of bacterial respiratory infections is necessary to limit host damage. Fortunately, most respiratory bacterial pathogens respond readily to antibiotic therapy, and many can also be controlled by immunization.

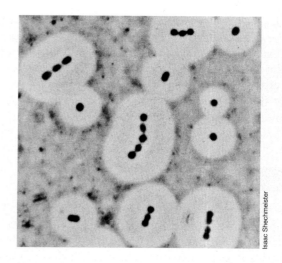

● **Figure 26.3   India ink negatively stained preparation of cells of** *Streptococcus pneumoniae.* Note the extensive capsule surrounding the cells. The cells are about 0.5 μm in diameter.

 26.1   Concept Check

Many respiratory pathogens are gram-positive *Bacteria*. Because gram-positive *Bacteria* are resistant to drying, they are easily transmitted in air. Most respiratory pathogens are transferred from person to person via respiratory aerosols generated by coughing, sneezing, talking, or breathing.

◆ What physical features of gram-positive *Bacteria* allow them to survive for long periods in air and dust?

◆ Identify pathogens more commonly found in the upper respiratory tract. Identify pathogens more commonly found in the lower respiratory tract.

26.2   Streptococcal Diseases

Streptococcus pyogenes and *Streptococcus pneumoniae* are important human respiratory pathogens. *S. pyogenes* is transmitted by the respiratory route. *S. pneumoniae* is found in the respiratory flora of up to 40% of healthy individuals, and endogenous strains can cause severe respiratory disease in compromised individuals.

Streptococci are nonsporulating, homofermentative, aerotolerant, anaerobic gram-positive cocci (∞ Section 12.19). Cells of *Streptococcus pyogenes* typically grow in elongated chains (∞ Figure 12.54). Pathogenic strains of *Streptococcus pneumoniae* typically grow in pairs or short chains and virulent strains typically produce an extensive polysaccharide capsule (Figure 26.3●).

Streptococcus pyogenes: Epidemiology and Pathogenesis

Streptococcus pyogenes is frequently isolated from the upper respiratory tract of healthy adults. Although num-

bers of endogenous *S. pyogenes* are usually low, if host defenses are weakened or a new, highly virulent strain is introduced, acute streptococcal infections are possible. *S. pyogenes* is the cause of streptococcal pharyngitis, also called *strep throat* (Figure 26.2). Most isolates from clinical cases of strep throat produce a toxin that lyses red blood cells, a condition called β-*hemolysis* (∞ Figure 21.18). Streptococcal pharyngitis is characterized by a severe sore throat, enlarged tonsils, tonsillar exudate, tender cervical lymph nodes, a mild fever, and general malaise. *S. pyogenes* can also cause related infections of the inner ear (*otitis media*), the mammary glands (*mastitis*), and infections of the superficial layers of the skin, a condition referred to as *impetigo* (impetigo can also be caused by *Staphylococcus aureus*) (Figure 26.4●).

About half of the clinical cases of severe sore throat are due to *Streptococcus pyogenes*, with the remainder of viral origin. An accurate, prompt diagnosis is important because if the sore throat is due to a virus, treatment with antibacterial drugs (antibiotics) will be useless, whereas if the sore throat is due to *S. pyogenes*, immediate antibacterial therapy is indicated. Rapid, complete treatment of streptococcal sore throat is important because it can occasionally lead to more serious streptococcal syndromes such as scarlet fever, rheumatic fever, acute glomerulonephritis, and streptococcal toxic shock syndrome.

Certain strains of *Streptococcus pyogenes* carry a lysogenic bacteriophage that encodes production of streptococcal pyrogenic exotoxin A, an exotoxin responsible for most of the symptoms of streptococcal toxic shock syndrome and **scarlet fever**. Exotoxin A is a *superantigen* and acts by recruiting massive numbers of T cells to the infected tissues (∞ Section 22.16). The T cells then secrete *cytokines*, which activate large numbers of effector cells, resulting in systemic inflammation and tissue destruction. Streptococcal toxic shock syndrome is the term given to this systemic, often life-threatening condition. Exotoxin A

analyzed in minutes. These rapid diagnostic procedures allow the physician to immediately initiate appropriate antibiotic therapy in order to avoid complications such as rheumatic fever.

A more accurate confirmation of infection by pathogenic streptococci is a positive *S. pyogenes* culture from the throat done on sheep blood agar (⚬⚬ Figure 21.18). While the RAD tests are nearly as *specific* as throat cultures, they can be up to 40% less *sensitive*, leading to false negative reports (⚬⚬ Section 24.7). Throat cultures take up to two days to process, hence the popularity of the RAD tests. The most sensitive methods for identifying recent streptococcal infections are serology tests, where patients are examined for the presence or increase (rise in titer) of antibodies to various streptococcal antigens (⚬⚬ Section 24.7). The presence of new antibodies or an increase in the quantity of an existing antibody confirms a very recent streptococcal infection.

Streptococcus pneumoniae

The other major pathogenic streptococcal species, *Streptococcus pneumoniae*, causes lung infections that often develop as secondary infections to other respiratory disorders. Capsulated strains of *S. pneumoniae* are particularly pathogenic because they are potentially very invasive. Cells invade alveolar tissues (lower respiratory tract) of the lung where the capsule enables the cells to resist phagocytosis and elicit a strong host inflammatory response. Reduced lung function (pneumonia) can result from accumulation of recruited phagocytic cells and fluid. The *S. pneumoniae* cells can then spread from the focus of infection as a bacteremia, sometimes resulting in bone infections, inner ear infections, and endocarditis. Pneumococcal pneumonia is a serious infection and untreated cases have a mortality rate of about 30%. Even with aggressive antimicrobial treatment, individuals hospitalized with pneumococcal pneumonia have up to 10% mortality.

Laboratory diagnosis of *S. pneumoniae* involves the culture of gram-positive diplococci from either patient sputum or blood. There are over 90 different serotypes (antigenic capsule variants), and, as for *S. pyogenes*, infection induces immunity to only the infecting serotype of *S. pneumoniae*.

Prevention and Treatment

An effective multivalent vaccine is available for prevention of infection by at least two-thirds of the 90 known strains of *Streptococcus pneumoniae*, including all common pathogenic strains. The vaccine consists of a mixture of the capsular polysaccharides from the most prevalent pathogenic strains. The vaccine is recommended for the elderly, health-care providers, individuals with compromised immunity, and others at high risk for respiratory infections (⚬⚬ Section 22.13).

Penicillin and its semisynthetic derivatives (⚬⚬ Sections 20.8 and 30.6) are the agents of choice for treating *S. pyogenes* infections. Erythromycin and other antibacterial drugs are used in individuals who have acquired penicillin allergies (⚬⚬ Section 22.15).

Most strains of *S. pneumoniae* respond to penicillin therapy. However, there are penicillin-resistant strains, especially among strains causing hospital-acquired infections (⚬⚬ Section 25.7). Thus individual isolates must be tested for penicillin susceptibility (⚬⚬ Section 24.3). Erythromycin is the drug of choice for penicillin-resistant organisms, but cephalosporin, fluoroquinolone, ceftriaxone, cefotaxime, or vancomycin (⚬⚬ Sections 20.6–20.9) may also be used. However, some strains have acquired resistance to each of these drugs, and some strains have acquired multiple drug resistance, underscoring the need to test each isolate individually.

26.2 Concept Check

Diseases caused by streptococci include streptococcal sore throat and pneumococcal pneumonia. Occasionally, *Streptococcus pyogenes* infections develop from pharyngitis into serious conditions such as scarlet fever and rheumatic fever. Pneumonia caused by *Streptococcus pneumoniae* is a serious disease with high mortality. Definitive diagnosis for both pathogens is by culture. Infections with both pathogens are treatable with antimicrobial drugs, but drug-resistant strains are known, especially for *Streptococcus pneumoniae*.

◆ How does *Streptococcus pyogenes* infection cause rheumatic fever?

◆ What is the primary virulence factor for *Streptococcus pneumoniae*?

26.3 Corynebacterium **and Diphtheria**

Corynebacterium diphtheriae causes diphtheria, a severe respiratory disease that usually infects children. Diphtheria is preventable and treatable. *C. diphtheriae* is a gram-positive, nonmotile, aerobic bacterium that forms irregular rods that may appear as club-shaped cells during growth (Figure 26.6*a*●) (⚬⚬ Section 12.22).

Epidemiology and Pathology

Corynebacterium diphtheriae enters the body via the respiratory route with cells lodging in the throat and tonsils. Infection is usually spread from healthy carriers or infected individuals to susceptible individuals by airborne droplets. Previous infection or immunization (see below) provides resistance to the effects of the potent exotoxin. Although limited information is available concerning the mechanism of adherence of *C. diphtheriae* to these tissues, the organism produces a neuraminidase capable of splitting *N*-acetylneuraminic acid (a component of glycoproteins found on animal cell surfaces), and this may enhance the invasion process. The inflammatory response of throat tissues to *C. diphtheriae* infection results in formation of a characteristic lesion called a *pseudomembrane* (Figure 26.6*b*), which consists of damaged host cells and cells of *C. diphtheriae*. As described in Section 10.9, certain strains of *C. diphtheriae* are lysogenized by bacteriophage β. These lysogenized toxin-producing strains are pathogenic because they produce a powerful exotoxin, the *diphtheria*

(a)

(b)

● **Figure 26.6 Diphtheria.** (a) Cells of *Corynebacterium diphtheriae* showing typical club-shaped appearance. The cells are 0.5 to 1.0 μm in diameter and may be several micrometers in length. (b) Pseudomembrane (arrows) in an active case of diphtheria caused by the bacterium *C. diphtheriae*.

toxin. Diphtheria toxin inhibits eukaryotic protein synthesis and thus kills cells (∞ Section 21.10).

Tissue death due to absorption of the toxin causes the appearance of a pseudomembrane in the patient's throat. The pseudomembrane may block the passage of air, and death from diphtheria is usually due to a combination of the effects of partial suffocation and tissue destruction by exotoxin. In untreated infections, the toxin can cause systemic damage to the heart (about 25% of diphtheria patients develop myocarditis), kidneys, liver, and adrenal glands.

Although diphtheria was once a major childhood disease, it is now rarely encountered because an effective vaccine is available. In developed countries like the United States, the disease is virtually unknown. Worldwide, there are still more than 50,000 cases of diphtheria per year, largely because of a lack of effective immunization programs. For example, diphtheria outbreaks have recently occurred in refugee camps in Afghanistan in individuals lacking immunization.

Diagnosis, Prevention, and Treatment

Corynebacterium diphtheriae isolated from the throat is diagnostic for diphtheria. Nasal or throat swabs are used to inoculate blood agar, tellurite medium, and the selective Loeffler's medium that inhibits the growth of most other respiratory pathogens.

Prevention of diphtheria is accomplished with a highly effective vaccine. The vaccine is made by treating the diphtheria exotoxin with formalin to yield an immunogenic toxoid preparation. Diphtheria toxoid is part of the *DTaP* (*d*iphtheria, *t*etanus, *a*cellular *p*ertussis) vaccine (∞ Section 22.13).

A patient diagnosed with diphtheria is treated simultaneously with antibiotics and diphtheria antitoxin (an antitoxin contains neutralizing antibodies formed in another animal) (∞ Sections 24.7 and 22.13). Penicillin, erythromycin, and gentamicin are generally effective for stopping *C. diphtheriae* growth and further toxin production, but do not alter the effects of preformed toxin. Early administration of both antibiotics and antitoxin is necessary for effective treatment of the disease.

⬣ 26.3 Concept Check

Diphtheria is an acute respiratory disease caused by the gram-positive bacterium *Corynebacterium diphtheriae*. Early childhood immunization (DTP) is very effective for preventing this very serious respiratory disease.

◆ Is the pathogenesis of diphtheria due to infection?

◆ How can the spread of diphtheria be prevented?

26.4 *Bordetella* and Whooping Cough

Whooping cough is a potentially serious childhood respiratory disease caused by infection with *Bordetella pertussis*. *B. pertussis* is a small, gram-negative, aerobic coccobacillus and was linked to whooping cough by Bordet and Gengou in 1906.

Epidemiology and Pathology

Whooping cough or **pertussis** is an acute, highly infectious respiratory disease observed most often in children under 5 years of age. *Bordetella pertussis* attaches to cells of the upper respiratory tract by producing a specific adherence factor called *filamentous hemagglutinin antigen*, which recognizes a complementary molecule on the surface of host cells. Once attached, *B. pertussis* grows and produces pertussis exotoxin that induces synthesis of cyclic adenosine monophosphate (cyclic AMP) (∞ Section 8.7), which is at least partially responsible for the events that lead to host tissue damage. *B. pertussis* also produces an endotoxin, which also may induce some of the symptoms of whooping cough. Clinically, whooping cough is characterized by a recurrent, violent cough that can last up to 6 weeks. The spasmodic coughing gives the disease its name, for a whooping sound results from the patient inhaling in deep breaths to obtain sufficient air. Worldwide, there are up to 50 million cases and 350,000 deaths due to pertussis each year.

In recent years, there has been an alarming upward trend in the cases of pertussis in the United States. Starting in the 1980s, there has been a consistent upward trend of *Bordetella pertussis* infections and disease, reversing a general downward trend that started with the introduction of an effective pertussis vaccine. In 1976, the year of lowest prevalence and incidence, there were only 1010 reported cases of pertussis. In 2003, there were 8489 cases. Up to 60%, or over 4000 cases per year, are in individuals over 5 years of age, including many adolescents and adults who lack appropriate immunity. About 24% of cases were in children less than six months of age who had not yet received all of the recommended doses of pertussis vaccine. The threat of this very communicable disease remains high, as illustrated by recent sporadic outbreaks in cities in the United States, and epidemic outbreaks in refugee camps in Afghanistan. Up to 32% of coughs lasting one to two weeks or longer may be caused by *B. pertussis*. Worldwide, research indicates that immunization programs should still be targeted to children, but immunization of adolescents and adults should also be a priority to build herd immunity (∞ Section 25.4) and prevent serious infections by this endemic respiratory pathogen.

Diagnosis, Prevention, and Treatment

Diagnosis of whooping cough can be made by fluorescent antibody staining of a nasopharyngeal swab specimen or by culture of the organism. For best recovery of *Bordetella pertussis*, a nasopharyngeal aspirate is inoculated directly onto a blood–glycerol–potato extract agar plate (although not selective, this medium supports good recovery of *B. pertussis*). β-hemolytic colonies containing small gram-negative coccobacilli are tested for *B. pertussis* by a latex bead agglutination test or are stained with an anti-*B. pertussis* fluorescent antibody for positive identification (∞ Sections 24.8 and 24.9). If available, a polymerase chain reaction test (PCR) is considered the most sensitive and preferred diagnostic test (∞ Section 24.12). Improved diagnostic and reporting techniques may be one reason for the recent observed increase in pertussis cases in the United States.

A vaccine consisting of proteins derived from *B. pertussis* is part of the routinely administered DTaP (*d*iphtheria, *t*etanus, *a*cellular *p*ertussis) vaccine. This vaccine is effective only when given to children at appropriate intervals beginning soon after birth (∞ Section 22.13). In the United States, up to 50% of children who acquire whooping cough have not been properly immunized, and current immunization preparations are only 60–90% effective.

Undesirable side effects of pertussis vaccination include local swelling, redness, and fever, and occur in up to one in four vaccinated individuals. Occasional more serious problems such as seizures (1 in 14,000), uncontrollable crying (1 in 1000), or very high fever (1 in 16,000) also occur. Overall, the risk of infection and serious consequences of acquiring whooping cough far outweigh the risks of immunization. Currently, about eight

deaths occur per year. No deaths have been attributed to the acellular pertussis vaccine.

Cultures of *B. pertussis* are killed by ampicillin, tetracycline, and erythromycin, although antibiotics alone do not seem to be sufficient to kill the pathogen *in vivo*. Because a patient with whooping cough remains infectious for up to 2 weeks following commencement of antibiotic therapy, the immune response may be as important, if not more so, than antibiotics in the elimination of *B. pertussis* from the body.

 ### 26.4 Concept Check

In the United States, there has been an increase in the number of annual cases of whooping cough. From an average of less than 2000 pertussis cases per year in the 1970s, the number of cases has now risen to over 8,000 per year. Inadequately immunized children are at high risk for acquiring pertussis.

◆ What measures can be taken to decrease the current incidence of pertussis in a population?

◆ Indicate potential problems with the use of pertussis vaccines.

26.5 Mycobacterium, Tuberculosis, and Leprosy

Tuberculosis is caused by the gram-positive, acid-fast bacillus *Mycobacterium tuberculosis* (∞ Section 12.23). The famous German microbiologist Robert Koch isolated and described the causative agent of tuberculosis, *M. tuberculosis*, in 1882 (∞ Section 1.6). The acid-fast properties of all mycobacteria result from the waxy mycolic acid constituent of the cell wall. Shared by all members of the genus, mycolic acid allows these organisms to retain carbolfuchsin, a red dye, even after washing in 3% hydrochloric acid in alcohol (Figure 26.7●).

Epidemiology

Mycobacterium tuberculosis is transmitted by the respiratory route, and even the simple act of talking can spread the organism from person to person. At one time, tuberculosis was the most important infectious disease of humans and accounted for one-seventh of all deaths worldwide. Presently, over 11,000 new cases of tuberculosis and about 750 deaths occur each year in the United States. Worldwide, tuberculosis still accounts for almost *1.6 million deaths per year*, about 11% of all deaths due to infectious disease. Up to one-third of the world's population have been infected with *M. tuberculosis* (∞ Table 25.1). In recent years, many of the new tuberculosis cases in the United States have occurred in acquired immunodeficiency syndrome (AIDS) patients.

Pathology

The interaction of the human host and *Mycobacterium tuberculosis* is extremely complex, determined both by the virulence of the strain and the resistance of the host. Cell-

CDC/Dr. Edwin P. Ewing, Jr./PHIL

● **Figure 26.7** *Mycobacterium avium* **in an acid-fast stained lymph node biopsy from a patient with AIDS.** Multiple bacilli are evident inside each cell. The individual rods, stained red with carbol fuchsin (arrows), are about 0.4 μm in diameter and up to 4 μm in length.

mediated immunity plays a critical role in the prevention of active disease after infection. Tuberculosis can be classified as a *primary* infection (initial infection) or *postprimary* infection (reinfection). Primary infection usually results from inhalation of droplets or dust particles containing viable *M. tuberculosis* bacteria from an individual with an active pulmonary infection. The inhaled bacteria settle in the lungs and grow. The host responds with an immune response to *M. tuberculosis*, resulting in a delayed-type hypersensitivity reaction (∞ Section 22.15) and the formation of aggregates of activated macrophages, called *tubercles* (∞ Figure 1.13). The bacteria often survive and grow within the macrophages, even with an ongoing immune response. In individuals with low resistance, the bacteria are not controlled, and an acute pulmonary infection occurs, leading to the extensive destruction of lung tissue, the spread of the bacteria to other parts of the body, and death. In these cases, *M. tuberculosis* is able to survive both the low pH and the effects of the oxidative antibacterial products found in the lysosomes of phagocytes such as macrophages (∞ Section 22.2).

In most cases of tuberculosis, however, acute infection does not occur. The infection remains localized, is usually inapparent, and appears to end. But this initial infection hypersensitizes the individual to the bacteria or their products and consequently alters the response of the individual to

subsequent *M. tuberculosis* exposures. A diagnostic test, called the **tuberculin test**, can be used to measure this hypersensitivity. When *tuberculin*, a protein fraction extracted from *M. tuberculosis*, is injected intradermally into a hypersensitive individual, it elicits a localized immune reaction within 1–3 days at the site of injection. The reaction is characterized by *induration* (hardening) and *edema* (swelling) (∞ Figure 22.26). An individual exhibiting this reaction is said to be *tuberculin-positive*, and many healthy adults show positive reactions as a result of previous inapparent infections. A positive tuberculin test does not indicate active disease but only that the individual has been exposed to the organism in the past and has generated a cell-mediated immune response.

For most individuals, this immunity is protective and lifelong. However, some tuberculin-positive patients develop postprimary tuberculosis through reinfection from outside sources or as a result of reactivation of bacteria that have remained alive but dormant in lung macrophages, often for years. Because of the latent nature of tuberculosis infection, individuals who have a positive tuberculin test are generally treated with antibiotics (see below). Factors such as aging, malnutrition, overcrowding, stress, and hormonal changes may reduce effective immunity in untreated individuals and allow reactivation of dormant infections.

Secondary pulmonary infections often progress to chronic infections that result in destruction of lung tissue, followed by partial healing and calcification at the infection site. Thus, chronic postprimary tuberculosis often results in a gradual spread of tubercular lesions in the lungs. Bacteria are found in the sputum in individuals with active disease and areas of destroyed tissue can be seen in X-rays (Figure 26.8●).

(a)   (b)

Aaron Friedman   Aaron Friedman

● **Figure 26.8** **Tuberculosis and diagnostic X-ray.** (a) Normal chest X-ray. The faint white lines are arteries and other blood vessels. The heart is visible as a white bulge in the lower right quadrant. (b) An advanced case of pulmonary tuberculosis; white patches (arrows) indicate areas of disease. These patches, or tubercles, may contain live *Mycobacterium tuberculosis*. Lung tissue and function is permanently destroyed by these lesions.

Prevention and Treatment

Individuals who have active cases of tuberculosis may spread the disease simply by coughing or speaking near uninfected individuals. Because tuberculosis is highly contagious, the United States Occupational Safety and Health Administration has stringent requirements for the protection of health-care workers who are responsible for tuberculosis patient care. For example, patients with infectious tuberculosis must be hospitalized in negative-pressure rooms. In addition, health-care workers who have patient contact must be provided with personally fitted face masks with high-efficiency particulate air (HEPA) filters. These special filters prevent the passage of *Mycobacterium tuberculosis* in sputum or on dust particles.

Chemotherapy of tuberculosis has been a major factor in control of the disease. Initial success in chemotherapy occurred with the introduction of streptomycin, but the real revolution in tuberculosis treatment came with the discovery of *isonicotinic acid hydrazide (isoniazid or INH)* (Figure 26.9●). This drug, virtually specific for mycobacteria, is effective, inexpensive, and is readily absorbed when given orally. Although the mode of action of isoniazid is not completely understood, it affects the synthesis of mycolic acid by *Mycobacterium* (mycolic acid is a lipid that complexes with peptidoglycan in the mycobacterial cell wall) (⚬⚬ Section 12.23). Isoniazid may mimic the activity of a structurally related molecule, nicotinamide (Figure 26.9), becoming incorporated in place of nicotinamide and thus inactivating enzymes requiring this compound for activity.

Treatment of mycobacteria with very small amounts of isoniazid (as little as 5 picomoles [pmol] per 10^9 cells) results in complete inhibition of mycolic acid synthesis, and continued incubation results in loss of outer membrane areas of the cell, a loss of cellular integrity, and death. Following treatment with isoniazid, mycobacteria lose their acid-alcohol fastness, in keeping with the role of mycolic acid in this staining property (⚬⚬ Section 12.23). However, mycobacterial resistance to isoniazid and other drugs is increasing at an alarming rate, especially in AIDS patients (see Section 26.14).

Treatment typically involves daily doses of isoniazid and rifampin for 2 months, followed by biweekly doses for a total of 9 treatment months, to eradicate the tubercle bacilli and prevent emergence of antibiotic-resistant organisms.

Failure to complete the entire prescribed treatment plan may allow the infection to be reactivated, and the reactivated organisms often have acquired resistance to the original treatment drugs. Inadequate treatment encourages antibiotic resistance because a high number of mutations spontaneously occur in *M. tuberculosis*, rapidly conferring resistance to single antibiotics. To ensure treatment and thus discourage development of antibiotic resistant organisms, direct observation of treatment (DOT) may be necessary for noncompliant individuals.

In populations such as hospitals and nursing homes where resistant strains are most likely to be present, patients are routinely treated with up to 4 antimycobacterial drugs for 2 months, followed by rifampin-isoniazid treatment for a total of 6 months. Multiple drug therapy reduces the possibility that strains will emerge having resistance to more than one drug.

In many countries, immunization with an attenuated strain of *M. bovis*, the *bacillus Calmette-Guerin (BCG)* strain, is routine for prevention of tuberculosis. However, in the United States and other countries where the prevalence of tuberculosis is low, immunization with BCG is usually discouraged. The live BCG vaccine induces a delayed-type hypersensitivity response (⚬⚬ Section 22.15), and all individuals who receive it develop a positive tuberculin test, neutralizing the value of the tuberculin test as a diagnostic and epidemiologic indicator for the spread of *M. tuberculosis* infection.

Mycobacterium leprae and Hansen's Disease (Leprosy)

Mycobacterium leprae, discovered by G. A. Hansen in 1873, is the causative agent of *Hansen's disease*, or *leprosy. M. leprae* is the only *Mycobacterium* species that has not been grown on artificial media. The only experimental animal that has been successfully used to grow *M. leprae* and reproduce a similar disease is the armadillo.

The most serious form of leprosy is characterized by folded, bulblike lesions on the body, especially on the face and extremities (Figure 26.10●), due to growth of *M. leprae* cells in the skin. The lesions contain up to 10^9 bacterial cells per gram of tissue. Like other mycobacteria, *M. leprae* from the lesions stain deep red with carbol fuchsin in the acid-fast staining procedure, providing a rapid, definitive demonstration of active infection (⚬⚬ Section 12.23). This *lepromatous* form of leprosy has a very poor prognosis. In severe cases the disfiguring lesions lead to destruction of peripheral nerves and loss of motor function. Many patients exhibit less pronounced lesions from which no bacterial cells can be recovered. These individuals have the *tuberculoid* form of the disease. *Tuberculoid leprosy* is characterized by a vigorous delayed-type hypersensitivity response (⚬⚬ Section 22.15) and a good prognosis for spontaneous recovery. Hansen's disease of either form, and the continuum of intermediate forms, is treated using a multiple drug therapy (MDT) protocol, which includes some combina-

Isoniazid   Nicotinamide

● **Figure 26.9 Structure of isoniazid (isonicotinic acid hydrazide).** Isoniazid is an effective chemotherapeutic agent for tuberculosis. Note the structural similarity to nicotinamide.

● **Figure 26.10 Lepromatous leprosy lesions on the skin.** Lepromatous leprosy is due to infection with *Mycobacterium leprae*. The lesions can contain up to 10^9 bacterial cells per gram of tissue, indicating an active uncontrolled infection with a poor prognosis.

tion of dapsone (4, 4'-sulfonylbisbenzeneamine), rifampin, and clofazimine. As in the case of tuberculosis, drug-resistant organisms have appeared, especially after treatment with single drugs or inadequate treatment. Extended drug therapy of up to 1 year with a MDT protocol is required for eradication of the organism.

The pathogenicity of *M. leprae* is due to a combination of delayed hypersensitivity (∞ Section 22.15) and the invasiveness of the organism. Transmission involves both direct contact and respiratory routes, and incubation time varies from several weeks to years or even decades. *M. leprae* grows within macrophages, causing an intracellular infection that can result in the large numbers of bacteria within the skin.

In many areas of the world, the incidence of Hansen's disease is very low. Worldwide, there were 763,917 new cases of leprosy reported in 2002, but only 96 cases in the United States. Ninety percent of all cases occur in Madagascar, Mozambique, Tanzania, and Nepal. Up to 2 million people are permanently disabled as a result of leprosy, but, because of the chronic nature and long latent period of the disease, leprosy may go unrecognized and unreported in as many as 12 million people.

Other Pathogenic *Mycobacterium* Species

A common pathogen of dairy cattle, *Mycobacterium bovis*, is pathogenic for humans as well as other animals. *M. bovis* enters humans via the intestinal tract, typically from the ingestion of raw milk. After a localized intestinal infection, the organism eventually spreads to the respiratory tract and initiates the classic symptoms of tuberculosis. *M. bovis* is a different organism from *M. tuberculosis*, but the two organisms are highly homologous at the DNA level. Al-

though *M. bovis* has several gene deletions, there is no observed difference in their infectivity and pathogenesis in humans. Pasteurization of milk and elimination of diseased cattle have eradicated bovine-to-human transmission of tuberculosis in developed countries.

A number of other *Mycobacterium* species are also occasional human pathogens. For example, *M. kansasii*, *M. scrofulaceum*, *M. chelonae*, and other members of the genus (∞ Section 12.23) cause disease. Respiratory disease due to any one of the *M. avium* complex (MAC) group of organisms is particularly pathogenic in AIDS patients as compared with members of the normal population (Figure 26.7) (see Section 26.14).

⬢ 26.5 Concept Check

Tuberculosis is one of the most prevalent and dangerous single diseases in the world. Its incidence is on the increase in developed countries, in part because of the emergence of drug-resistant strains. The pathology of tuberculosis and leprosy is influenced by the cellular immune response.

◆ Why is *Mycobacterium tuberculosis* such a widespread respiratory pathogen?

◆ Describe factors that contribute to the incidence of drug resistance in mycobacterial infections.

◆ Identify other species of the genus *Mycobacterium* that can cause human disease.

26.6 *Neisseria meningitidis*, Meningitis, and Meningococcemia

Meningitis is an inflammation of the *meninges*, or the membranes that line the central nervous system, especially the spinal cord and brain. Meningitis can be caused by viral, bacterial, fungal, or protozoan infections. Here we will deal with infectious bacterial meningitis caused by *Neisseria meningitidis* and a related infection, **meningococcemia**.

Neisseria meningitidis, often called meningococcus, is a gram-negative, nonsporulating, obligately aerobic, oxidase positive, encapsulated diplococcus (Figure 26.11●) of about 0.6–1.0 μm in diameter. At least 13 different pathogenic strains of *N. meningitidis* are recognized, based on antigenic differences in the capsular polysaccharides.

Epidemiology and Pathology

Meningococcal meningitis often occurs in epidemics, usually in closed populations such as military installations and college campuses. It normally affects older school-age children and young adults. Up to 30% of individuals normally carry *Neisseria meningitidis* in the nasopharynx with no apparent harmful effects. In epidemic situations, the prevalence of carriers may rise to 80%. The trigger for conversion from the asymptomatic carrier state to pathogenic acute infection is not known.

CDC/Dr. M.S. Mitchell/PHIL

● **Figure 26.11 Fluorescent antibody stain of *Neisseria menin-gitidis.*** The organism causes adult meningitis and meningococcemia, as observed in cerebrospinal fluid of an infected patient. The individual cocci are about 0.6–1.0 μm in diameter.

In an acute meningococcus infection, the bacterium is transmitted to the host, usually via the airborne route. There it attaches to the cells of the nasopharynx and gains access to the bloodstream, causing bacteremia and simple upper respiratory tract symptoms. The bacteremia sometimes leads to fulminant meningococcemia, characterized by septicemia, intravascular coagulation, shock, and death in over 10% of cases. Meningitis is another possible serious outcome of infection. Meningitis is characterized by sudden onset of headache, vomiting, and stiff neck, and can progress to coma and death in a matter of hours. Death occurs in up to 3% of acute meningococcal meningitis victims.

In the United States, there were 1814 cases of serious meningococcal disease in 2002, the lowest number of cases since 1977. The pattern of decreased incidence indicates the success of widespread vaccination in susceptible populations. However, the mortality rate was almost 12%.

Diagnosis, Prevention, and Treatment

Specimens isolated from nasopharyngeal swabs, blood, or cerebrospinal fluid are inoculated onto modified Thayer-Martin medium (MTM) (∞ Section 24.1), a selective medium that suppresses the growth of most normal flora, but allows the growth of *Neisseria meningitidis* and *Neisseria gonorrhoeae*. Colonies showing a gram-negative diplococcus coupled with a positive oxidase test (∞ Table 24.3) are presumptively identified as *Neisseria*. However, due to the rapid onset of life-threatening symptoms, preliminary diagnosis is often based on clinical symptoms and treatment is started before culture tests confirm infection with *N. meningitidis*.

Penicillin G is the drug of choice for treatment of *Neisseria meningitidis* infections. However, resistant strains

have been reported. Chloramphenicol is the accepted alternative agent for treatment of infections in penicillin-sensitive individuals. A number of broad-spectrum cephalosporins are also effective (∞ Section 20.8).

Naturally occurring strain-specific antibodies acquired by subclinical infections are effective for preventing infections in most adults. Unfortunately, there are no long-term vaccines available for the prevention of meningitis or meningococcemia. However, vaccines consisting of purified polysaccharides from some of the most prevalent pathogenic strains are available and are used to immunize close contacts when epidemics occur. The vaccine is also used to prevent infection in certain susceptible populations such as the military and students living in dormitories. In addition, rifampin is often used as a chemoprophylactic antibiotic to prevent disease in close contacts and family members of infected individuals.

Other Causes of Meninigitis

A number of other organisms can also cause meningitis. Acute meningitis is usually caused by one of the pyogenic bacteria such as *Staphylococcus, Streptococcus,* or *Haemophilus influenzae*. *H. influenzae* primarily infects young children. An effective vaccine for preventing *H. influenzae* meningitis is available and is required in the United States for school-age children (∞ Section 22.13).

Several viruses also cause meningitis. Among these are herpes simplex virus (HSV), lymphocytic choriomeningitis virus (LCM), mumps virus, and the enteroviruses. In general, viral meningitis is less severe than bacterial meningitis.

 26.6 *Concept Check*

Neisseria meningitidis is a common cause of meningococcemia and meningitis in young adults and occasionally occurs in epidemics in closed populations. Bacterial meningitis and meningococcemia are serious diseases with very high mortality rates. Treatment and prevention strategies are in place to deal with epidemic outbreaks, but an effective universal vaccine is not yet available.

◆ Describe the infection by *Neisseria meningitidis* and the resulting development of meningococcemia.

◆ Can meningococcemia be prevented? Explain.

26.7 Viruses and Respiratory Infections

As we discussed (∞ Section 20.10), viruses are less easily controlled by chemotherapeutic means than bacteria or other microorganisms because the growth of viruses is intimately tied to host cell functions. Most chemotherapeutic agents that specifically attack viruses cause at least some harm to host cells as well. Not surprisingly, therefore, the most prevalent infectious diseases, especially in developed countries, are caused by viruses. Most viral diseases are acute, self-limiting infections, but some can be problematic in normal healthy adults. In addition, serious viral diseases such as smallpox and rabies have been effectively controlled by immunization. We begin here by describing measles, mumps, rubella, and chickenpox, all viral diseases transmitted in infectious droplets by an airborne route.

Measles

Measles (*rubeola* or *7-day measles*) is an acute highly infectious childhood disease characterized by nasal discharges, redness of the eyes, cough, and fever. The measles virus is a *paramyxovirus* (◌ Section 16.9) that enters the nose and throat by airborne transmission, quickly leading to systemic viremia. As the disease progresses, fever and cough appear and rapidly intensify, and a rash appears (Figure 26.12●); in most cases measles symptoms last for 7–10 days. Circulating antibodies to measles virus are measurable about 5 days after initiation of infection, and both serum antibodies and cytotoxic T lymphocytes (◌ Sections 22.9 and 22.7) combine to eliminate the virus from the system. A variety of complications may occur due to measles infection, including inner ear infection, pneumonia, and, in rare cases, measles encephalomyelitis. Encephalomyelitis can cause neurological disorders and a form of epilepsy, and has a mortality rate of nearly 20%.

Although once a common childhood illness, measles generally occurs nowadays in rather isolated outbreaks because of widespread immunization programs begun in the mid-1960s (Figure 26.13a●). Because of the highly infectious nature of the disease, all public school systems in the United States require proof of immunization before children can enroll. Active immunization is done with the MMR (measles, mumps, and rubella) vaccine (◌ Section 22.13). A childhood case of measles generally confers lifelong immunity to reinfection.

Mumps

Mumps, like measles, is caused by a paramyxovirus and is also highly infectious. Mumps is spread by airborne droplets, and the disease is characterized by inflammation of the salivary glands leading to swelling of the jaws and neck (Figure 26.14●). The virus spreads through the bloodstream and may infect other organs including the brain, testes, and pancreas. Severe complications may include encephalitis and, very rarely, sterility. The host immune response produces antibodies to mumps virus surface proteins, and this generally leads to a quick recovery. An attenuated vaccine is highly effective for preventing mumps (Figure 26.13b). Hence, the prevalence of mumps in developed countries has been greatly reduced in the last three decades, with disease usually restricted to individuals who did not receive the MMR vaccine.

(a)

(b)

● **Figure 26.12 Measles in children.** (a) The light pink rash starts on the head and neck, and (b) spreads to the chest, trunk, and limbs. Discrete papules coalesce into blotches as the rash progresses for several days.

Rubella

Rubella (*German measles* or *3-day measles*) is caused by a single-stranded positive sense RNA virus of the togavirus group (◌ Section 16.8). The symptoms of the disease resemble those of measles but are generally milder. Rubella is less contagious than true measles, and thus a good proportion of the population has never been infected. However, during the first 3 months of pregnancy, rubella virus can infect the fetus by placental transmission and cause serious fetal abnormalities.

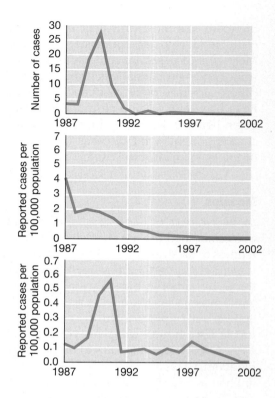

● **Figure 26.13 Viral diseases and vaccines.** Major childhood viral diseases are now controlled by the MMR (measles, mumps, rubella) vaccine in the United States. (a) Measles. (b) Mumps. (c) Rubella. The insets show a more recent picture of these diseases. Data were obtained from the Centers for Disease Control and Prevention, Atlanta, GA, USA.

Rubella can cause stillbirth, deafness, heart and eye defects, and brain damage in live births. Thus, pregnant women should not be immunized with the rubella vaccine or contract rubella during this period. For this reason, routine childhood immunization against rubella should be practiced. An attenuated virus is adminis-

tered as part of the MMR vaccine (see Figure 26.13c and Section 22.13).

Chickenpox and Shingles

Chickenpox (varicella) is a common childhood disease caused by the Varicella-zoster virus (VZV), a herpesvirus (∞ Section 16.12). VZV is highly contagious and is transmitted by infectious droplets, especially when susceptible individuals are in close contact. In school children, for example, close confinement during the winter months leads to the spread of chickenpox through airborne droplets from infected classmates and through contact with contaminated fomites. The virus enters the respiratory tract, multiplies, and is quickly disseminated via the bloodstream, resulting in a systemic papular rash that quickly heals, rarely leaving disfiguring marks (Figure 26.15●). An attenuated virus vaccine is now recommended for use in the United States (∞ Section 22.13). The current reported annual incidence of chickenpox is now about one-third of the number of cases reported prior to 1994, the year immunization became widespread.

The VZV virus can remain dormant in nerve cells for years with no apparent symptoms. The virus occasionally migrates from this reservoir to the skin surface, causing a painful skin eruption referred to as *shingles* (zoster). Shingles most commonly strikes immunosuppressed individuals or the elderly. Studies with human volunteers suggest that T cells are important in destroying the virus.

Centers for Disease Control

● **Figure 26.14 Mumps.** Glandular swelling characterizes infection with the mumps virus.

● **Figure 26.15   Chickenpox.** Mild papular rash associated with the infection by Varicella–zoster virus (VZV), the herpesvirus that causes chickenpox.

The prophylactic use of human hyperimmune globulin prepared against the virus is useful for preventing the onset of symptoms of shingles. Such therapy is advised only for patients where secondary infections occasionally associated with shingles, such as pneumonia or encephalitis, may be life-threatening.

26.7   Concept Check

Viral respiratory diseases are highly infectious and may cause serious health problems. However, the common childhood viral diseases measles, mumps, rubella, and chickenpox are all controllable with appropriate immunization procedures.

◆   Describe the potential serious outcomes of infection by each of these viruses.

◆   Identify the date immunization was made available for measles, mumps, rubella, and chickenpox.

26.8   Colds and Influenza

Colds and influenza are the most common infectious diseases. As shown in Figure 26.16●, there are about two cases of influenza and about 15 colds for every other infectious disease. These viral infections are transmitted

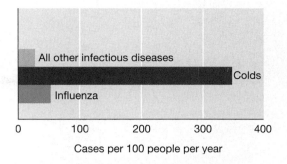

● **Figure 26.16   Colds and influenza.** These viral diseases are the leading causes of acute infectious disease in the United States. These data are typical for recent years.

via droplets spread from person to person in coughs, sneezes, and respiratory secretions.

Symptoms of a common cold and symptoms of "the flu" (influenza) often seem similar, but the two diseases are symptomatically distinct and caused by quite different viruses. A typical common cold, caused by a rhinovirus, is associated with nasal discharges, cough, chills, and perhaps a sore throat. Influenza, caused by an orthomyxovirus, is generally associated with a different set of symptoms. Although either condition may cause illness, colds are usually of shorter duration and the symptoms are milder. Table 26.1 compares the symptoms of colds and influenza.

The Common Cold

Each person probably averages more than three colds per year throughout his or her lifetime (Figure 26.16). Cold symptoms include rhinitis (inflammation of the nasal region, especially the mucous membranes), nasal obstruction, watery nasal discharges, and a general feeling of malaise, usually without an accompanying fever. *Rhinoviruses*, single-stranded RNA viruses of the *picornavirus* group (see Figure 26.17a● and Section 16.8), are the most common causes of colds. At least 115 different serotypes of rhino-viruses have been identified. Another group of single-stranded RNA viruses, the *coronaviruses* (Figure 26.17b), are responsible for about 15% of all colds in adults. A variety of other viruses including adenoviruses, coxsackie viruses, respiratory syncytial viruses (RSV), and orthomyxoviruses, are responsible for about 10% of common colds. Colds generally induce a specific, local, neutralizing IgA

Table 26.1	Colds and influenza	
Symptoms	**Common cold**	**Influenza**
Fever	Rare	Common (39–40°C) sudden onset
Headache	Rare	Common
General malaise	Slight	Common; often quite severe; can last several weeks
Nasal discharge	Common and abundant	Less common; usually not abundant
Sore throat	Common	Less common
Vomiting and/or diarrhea	Rare	Common in children

(a)

B. Dowsett and D. Tyrell

(b)

Heather Davies and D. Tyrell

● **Figure 26.17 Electron micrographs of some common cold viruses.** (a) Human rhinovirus. (b) Human coronavirus. Each rhinovirus virion is about 30 nm in diameter. Each coronavirus virion is about 60 nm in diameter.

response (∞ Section 22.9). However, the number of potential infectious agents makes immunity via immunization or previous exposure to a given virus very unlikely.

Aerosol transmission of the virus is probably the major means of spreading colds, although experiments with volunteers suggest that direct contact and/or fomite contact is also an important method of transmission. Most antiviral drugs are ineffective, but a pyrazidine derivative (Figure 26.18*a*●) has shown promise for preventing colds after virus exposure. In addition, new experimental antiviral drugs are being designed based on information derived from three-dimensional structures. For example, the antirhinovirus drug WIN 52084 (Figure 26.18*b*) binds to the virus, changing its three-dimensional surface configuration and disrupting rhinovirus binding to the host cell receptor, *ICAM-1 (intercellular adhesion molecule-1)*, thus preventing infection. Interferon-α, a cytokine (∞ Section 23.11), is also effective in preventing the onset of colds. Thus, there are several experimental possibilities for cold prevention and treatment, although none are widely accepted as effective and safe. Because colds are generally brief and self-limiting, the usual treatment is aimed at symptoms, especially nasal discharges, with a variety of antihistamine and decongestant drugs.

(a)

(b)

● **Figure 26.18 Experimental antirhinovirus drugs.** (a) The structure of 3-methoxy-6-[4-(3-methylphenyl)]-1-piper-azinyl. (b) The structure of WIN 52084, a receptor-blocking drug.

Influenza

Influenza is caused by an RNA virus of the orthomyxovirus group (∞ Section 16.9). Influenza virus is a single-stranded, negative-sense, helical RNA genome surrounded by an envelope made up of protein, a lipid bilayer, and external glycoproteins (Figure 26.19● and ∞ Figure 16.17). Three different types of influenza viruses exist: influenza A, influenza B, and influenza C. We limit our discussion here to influenza A since it is the most important human pathogen.

The genome of influenza A virus is single-stranded RNA that is arranged in a highly unusual manner. As discussed in Section 16.9, the influenza virus genome is *segmented*, with genes found on each of eight distinct fragments of its single-stranded RNA (∞ Figure 16.17). This arrangement allows the reassortment of gene fragments between different strains of influenza virus *if* more than one strain of influenza infects a cell at one time. Thus, in a cell infected with two strains of virus, packaging of the RNA segments will occur in a random

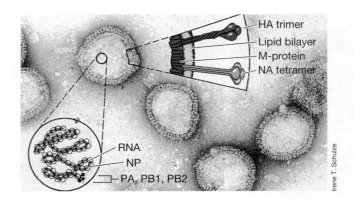

Irene T. Schulze

● **Figure 26.19 Electron micrograph of the influenza virus.** The photo shows the location of the major viral coat proteins and the nucleic acid. Each virion is about 100 nm in diameter. HA, hemagglutinin (three copies make up the HA coat spike); NA, neuraminidase (four copies make up the NA coat spike); M, coat protein; NP, nucleoprotein; PA, PB1, PB2, other internal proteins, some of which may have enzymatic functions.

pattern, with each complete virus containing a random assortment of gene segments from each virus. *The unique reassorted viruses are new viral strains.* Reassortment results in the phenomenon called **antigenic shift**. Antigenic shift refers to *major* modifications in the protein coat of the virions, especially to the two proteins important in the attachment and eventual release of virus from host cells, hemagglutinin (HA or H) and neuraminidase (NA or N) (⚭ Figures 16.17 and 26.19). The H and N protein antigens can also acquire minor antigenic changes due to genetic variations such as point mutations in a gene segment, altering one or more amino acids in the H or N protein. This phenomenon is known as **antigenic drift**.

Influenza Epidemiology

Human influenza virus is transmitted from person to person through the air, primarily in droplets expelled during coughing and sneezing. The virus infects the mucous membranes of the upper respiratory tract and occasionally invades the lungs. Symptoms include a low-grade fever lasting for 3–7 days, chills, fatigue, headache, and general aching (Table 26.1). Recovery is usually spontaneous and rapid. Most of the serious consequences of influenza infection occur from bacterial secondary infections in persons whose resistance has been lowered by the influenza infection. Especially in infants and elderly people, influenza is often followed by bacterial pneumonia; death, if it occurs, is usually due to the bacterial infection. Annually, influenza causes 3–5 million cases of severe illness and 250,000–500,000 deaths worldwide.

Most infected individuals develop immunity to the infecting virus, and it is impossible for a strain of similar antigenic type to cause an epidemic for about 2 to 3 years. Immunity is largely dependent on the production of secretory antibody (IgA) (⚭ Section 22.9), especially to antigenic determinants of the hemagglutinin (H) and neuraminidase (N) proteins.

Influenza exists in human populations as an *endemic viral disease* and widespread outbreaks occur every year from late autumn through the winter months. Antigenic *drift* results in a reduction of immunity in the population and is responsible for the recurrence of *epidemics*, severe localized influenza outbreaks, occurring in a 2- to 3-year cycle. *Pandemics*, worldwide epidemics, occur much less frequently, being from 10 to 40 years apart, and are the result of antigenic *shift* (⚭ Section 25.3).

The 1957 outbreak of the so-called Asian flu provided an opportunity to study the development of a pandemic (Figure 26.20●). The pandemic probably arose when a virulent mutant virus strain, differing antigenically from all previous strains, appeared in the population. Since immunity to this strain was not present, the virus spread rapidly throughout the world. It first appeared in the interior of China in February 1957 and by April had spread to Hong Kong. From Hong Kong, the virus was spread by naval ships to San Diego, California. In May, an outbreak occurred in Newport, Rhode Island, on a naval vessel. From that time, outbreaks continuously occurred in various parts of the United States. Peak incidence occurred in October, when 22 million new cases developed. In the pandemic of 1918, the influenza A "Spanish Flu" killed 230 million people

Key
- ▨ Country of origin
- ▨ Countrywide epidemic
- • Localized outbreaks
- → Routes of spread

● **Figure 26.20 An influenza pandemic.** The spread of the Asian influenza pandemic of 1957 is shown. The original epidemic focus of this pandemic was probably in China. Evidence suggests that agricultural practices involving poultry and swine, and human interactions with these animals, allowed the reassortment of influenza viral genomes from the three host species. New strains for which there is no immune memory in humans are sometimes formed, leading to rapid and uncontrollable spread and pandemic disease.

worldwide (Figure 25.1). Although pandemics occur periodically, none have been as catastrophic as the 1918 flu. The basis for the virulence of the 1918 pandemic is not understood.

Evidence links the pandemics of 1918 and 1957 to the reassortment of avian influenza viruses and human viruses in swine. Swine cells have receptors for both avian and human influenza orthomyxoviruses and can bind and propagate both avian and human influenza strains. If swine are infected with both human and avian strains at the same time, the two unrelated viruses can reassort (antigenic shift), and a unique influenza can be produced that will infect humans and for which there has been no previous antigen exposure. This reassortment from animal strains and infection into humans occurs periodically, but unpredictably, constantly presenting the possibility of a rapidly emerging, highly virulent influenza strain for which there is no preexisting immunity in the human population.

A new avian influenza strain appeared in Hong Kong in 1997, apparently jumping directly from the avian host to humans. The resulting strain, the Influenza A H5N1 strain, also called *avian influenza*, caused numerous infections and several deaths and reemerged in Viet Nam in 2004. This strain is still spread directly from an avian host to humans and can only be spread among humans after prolonged close contact. However, there is some indication that H5N1 has infected pigs, setting the stage for reassortment with human influenza strains that also infect pigs. This reassortment could create a new virus, initiating a new influenza pandemic. Vaccines are being developed to provide appropriate immunity to this potential emergent strain.

Influenza Prevention and Treatment

Influenza epidemics can be controlled by immunization. However, the choice of appropriate vaccines is complicated by the large number of existing strains and the continuing ability of these strains to undergo antigenic drift or antigenic shift. When new strains evolve, vaccines are not immediately available, but through careful worldwide surveillance (Section 25.9), samples of the major emerging strains of influenza virus are usually obtained before epidemic outbreaks occur. In the United States, candidate strains are chosen at the end of each influenza season, are grown in eggs, and inactivated. The inactivated viral strains are mixed to prepare a polyvalent vaccine used for immunization prior to the next influenza season. Influenza immunization is recommended for those individuals most likely to succumb or be exposed to serious secondary illnesses. Influenza immunization is recommended for everyone over 50 years of age, for those suffering from chronic debilitating diseases (e.g., AIDS patients, see Section 26.14), and for health-care workers. Effective artificial immunity from the inactivated influenza vaccine lasts only a few years, and is strain-specific. Therefore, immunization preparations are updated annually.

Influenza A may also be controlled by use of *amantadine* and *rimantadine*, synthetic amines that inhibit viral re-

plication. Both influenza A and B virus release are blocked by the neuraminidase inhibitors *oseltamivir* and *zanavir* (Section 20.10 and Table 20.5). These drugs are all used to treat ongoing influenza and shorten the course and severity of infection. They are most effective when given very early in the course of the infection. Amantadine, rimantadine, and oseltamivir are also effective chemoprophylactic agents that prevent the onset and spread of influenza infection.

Treatment of influenza symptoms with aspirin, especially in children, is not recommended. There is evidence of a link between aspirin treatment of influenza and *Reye's syndrome*, a rare but occasionally fatal complication involving the central nervous system.

🛑 *26.8 Concept Check*

Colds and influenza, or flu, are the most common infectious diseases. While they are not usually life-threatening diseases by themselves, they can lower resistance and allow serious secondary bacterial infections. Influenza outbreaks occur annually and more serious epidemics and pandemics occur periodically.

◆ Define and compare the cause and symptoms of influenza and of common colds, and discuss the possibilities for effective immunization programs for each.

◆ Distinguish between *antigenic drift* and *antigenic shift*.

⬛ III DIRECT CONTACT TRANSMISSION OF DISEASES

A number of diseases are spread primarily by direct contact with an infected person or by contact with blood or excreta from the infected person. Many of the respiratory diseases we have discussed can also be spread by direct contact. Here we discuss staphylococcal infections, ulcers, and hepatitis, diseases spread primarily person to person through direct contact with infected individuals.

26.9 *Staphylococcus*

The genus *Staphylococcus* contains common pathogens of humans and animals, occasionally causing life-threatening disease. Staphylococci commonly infect the skin and wounds. Most staphylococcal infections result from the transfer of staphylococci in normal flora from an infected but asymptomatic individual to a susceptible individual.

Staphylococci are gram-positive cocci of about 0.8–1.0 μm in diameter that divide in several planes to form irregular clumps (Section 12.19 and Figure 12.51). Staphylococci are nonsporulating but are resistant to drying and are readily dispersed in dust particles through the air and on surfaces. In humans, two species are important: *Staphylococcus epidermidis*, a nonpigmented form usually found on the skin or mucous membranes, and *Staphylo-*

coccus aureus, a yellow-pigmented form. While both species are potential pathogens, *S. aureus* is more commonly associated with human disease. Both species frequently occur in the normal microbial flora of the upper respiratory tract or on the skin (Figure 26.2).

Epidemiology and Pathogenesis

Staphylococci cause a variety of diseases including acne, boils (Figure 26.21●), pimples, impetigo (Figure 26.4), pneumonia, osteomyelitis, carditis, meningitis, and arthritis. Many of these diseases cause the production of pus, so they are said to be *pyogenic* (pus-forming). The most common habitats of *Staphylococcus aureus* are the upper respiratory tract, especially the nose and throat, and the surface of the skin. Healthy individuals are often carriers, and the resident staphylococci do not cause disease. Infants often become infected during the first week of life from the mother or from another close human contact. Serious staphylococcal infections often occur when the resistance of the host is low because of hormonal changes, debilitating illness, wounds, or treatment with steroids or other drugs that compromise the immune system.

Those strains of *S. aureus* most frequently causing human disease produce a number of extracellular enzymes or toxins (∞ Section 21.9). At least four different *hemolysins* have been recognized, and a single strain often produces several. The production of hemolysins is responsible for the hemolysis seen around colonies on blood agar plates. *S. aureus* is also capable of producing an *enterotoxin* associated with foodborne illness (∞ Sections 21.11, 22.16, and 29.5). All staphylococci also produce *catalase*, an enzyme that converts H_2O_2 to H_2O and O_2. Catalase is not considered a virulence factor, but the catalase test is used to distinguish staphylococci from streptococci, which do not produce catalase.

Another substance produced by *S. aureus* is *coagulase*, an enzyme that causes fibrin to coagulate and form a clot (∞ Section 21.9). The production of coagulase is generally associated with pathogenicity. Clotting induced by coagulase results in the accumulation of fibrin around the bacterial cells making it difficult for host defense agents to come into contact with the bacteria and preventing phagocytosis (Figure 26.21). Most *S. aureus* strains also produce *leukocidin*, which causes the destruction of leukocytes. Production of leukocidin in skin lesions such as boils and pimples results in considerable host cell destruction and is one of the factors responsible for pus formation (Figure 26.21). Some strains of *S. aureus* also produce other extracellular virulence factors, including proteolytic enzymes, hyaluronidase, fibrinolysin, lipase, ribonuclease, and deoxyribonuclease (∞ Table 21.4).

Certain strains of *S. aureus* have been implicated as the agents responsible for **toxic shock syndrome (TSS)**, a serious outcome of staphylococcal infection characterized by high fever, rash, vomiting, diarrhea, and occasionally death. Toxic shock was first recognized in menstruating women and was associated with use of highly absorbent tampons. In menstruating females, blood and mucus in the vagina can become colonized by *S. aureus* from the skin, and the presence of a tampon concentrates this material, creating ideal microbial growth conditions. Largely through education and alterations in materials used in tampons, toxic shock due to tampon use is now relatively rare. However, toxic shock syndrome is still seen in both men and women as a result of staphylococcal infections following surgery.

The symptoms of TSS result indirectly from an exotoxin called *toxic shock syndrome toxin (TSST)*. TSST is a superantigen (∞ Section 22.16). TSST is released by the growing staphylococci, causing a massive T-cell reaction

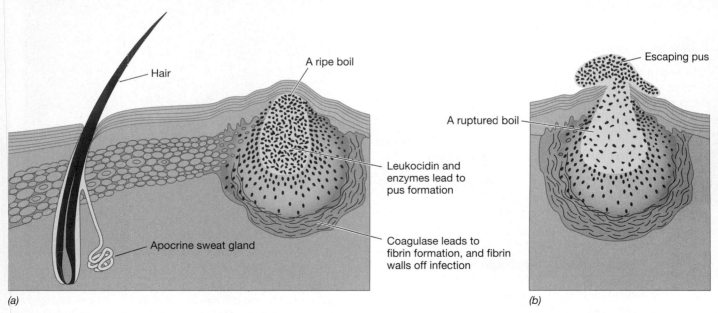

(a)
Hair
A ripe boil
Leukocidin and enzymes lead to pus formation
Coagulase leads to fibrin formation, and fibrin walls off infection
Apocrine sweat gland

(b)
Escaping pus
A ruptured boil

● **Figure 26.21** **The structure of a boil.** (a) Staphylococci initiate a localized infection of the skin and become walled off by coagulated blood and fibrin through the action of coagulase. (b) The rupture of the boil releases pus and bacteria.

resulting in the inflammatory response characteristic of superantigen reactions. A TSS-like disease may also be caused by different superantigens from other pathogenic *Bacteria*, including *Streptococcus pyogenes* (see Section 26.2).

Staphylococcal enterotoxin A, another superantigen, causes a form of food poisoning. After ingestion of toxin-contaminated food, the toxin stimulates T cells localized along the intestine, resulting in a massive T-cell response, release of mediators, and increased permeability of the intestine. The final outcome is the severe but short-lived diarrhea and vomiting associated with staphylococcal food poisoning (⌒⌒ Section 29.5).

Treatment and Prevention

Extensive use of antibiotics has resulted in the selection of resistant strains of *Staphylococcus aureus* and *Staphylococcus epidermidis*. Hospital (nosocomial) infections with antibiotic-resistant staphylococci often occur in patients whose resistance is lowered due to other diseases, surgical procedures, or drug therapy (⌒⌒ Section 25.7). Patients often acquire staphylococci from hospital personnel who are asymptomatic carriers of drug-resistant strains. As a result, appropriate antimicrobial drug therapy for *S. aureus* infections is a major problem in hospital environments. Although some community-acquired staphylococcal infections are treatable with penicillin, disease-producing isolates of *S. aureus* must be individually checked for antibiotic susceptibility (⌒⌒ Section 24.3).

Prevention of staphylococcal infections is problematic because most individuals are asymptomatic carriers, and diseases such as acne and impetigo can be transmitted by simple contact with contaminated fingers. In hospital environments such as surgical wards and nurseries, carriers of known pathogenic strains must either be excluded or be treated with topical or systemic antimicrobial drugs to eradicate the carrier state.

 26.9 Concept Check

Although staphylococci are usually harmless inhabitants of the upper respiratory tract and skin, several serious diseases can result from pyogenic infection, including some caused by staphylococcal superantigens.

◆ What is the normal habitat of *Staphylococcus aureus*? How is *S. aureus* spread from person to person?

◆ Identify the mechanisms of staphylococcal food poisoning and toxic shock syndrome.

26.10 *Helicobacter pylori* and Gastric Ulcers

Helicobacter pylori was first identified in human intestinal biopsies in 1983. This organism is a pathogen associated with gastritis, ulcers, and gastric cancers. *H. pylori* is a gram-negative, highly motile, spiral-shaped bacterium that is related to *Campylobacter* (Figure 26.22●) (⌒⌒ Section 29.9). It is 2.5–3.5 μm long and 0.5–1.0 μm in diameter and has one to six polar flagella at one end. *H. pylori* colonizes the non-acid-secreting mucosa of the stomach and the upper intestinal tract, including the duodenum (⌒⌒ Section 21.4).

Epidemiology

Up to 80% of gastric ulcer patients have concomitant *Helicobacter pylori* infections, and up to 50% of asymptomatic adults in developing countries are chronically infected. Person-to-person contact and ingestion of contaminated food or water are the probable transmission methods for *H. pylori*. Although there is no known nonhuman reservoir of *H. pylori*, the organism has occasionally been recovered from cats kept as household pets, indicating that it can be spread to or from animals in close contact with humans. Infection occurs in high incidence in certain families, and the overall prevalence in the population increases with age. These factors suggest a host-to-host type of transmission (⌒⌒ Sections 25.3, 25.4, and 25.5). However, infections with *H. pylori* sometimes also occur in epidemic clusters, suggesting that a common source such as food or water may sometimes be involved.

Pathology, Diagnosis, and Treatment

Although the causal relationship between *Helicobacter pylori* infection and ulcers has not been unequivocally

● **Figure 26.22   Scanning electron micrograph of *Helicobacter*.** Members of the genus range in size from 1.5–10 μm in length and 0.3–1 μm in diameter. *Helicobacter* spp. are found in the intestinal and hepatic tracts of mammals and birds. Note the sheathed flagella. The organism in the lower left is 3.2 μm in length.

established, current evidence indicates that *H. pylori* is a major preventable and treatable cause of many gastric ulcers. The bacterium is slightly invasive and colonizes the surfaces of the gastric mucosa, where it is protected from the effects of stomach acids by the gastric mucus layer. After mucosal colonization, a combination of pathogen products and host responses causes inflammation, tissue destruction, and ulceration. Pathogen products such as *vacA* (a cytotoxin; ⚭ Table 21.4), *urease*, and *lipopolysaccharide* may contribute to localized tissue destruction and ulceration. Antibodies to *H. pylori* are usually present in infected individuals, but are not protective and do not prevent colonization. Individuals who acquire *H. pylori* tend to have chronic, long-term infections unless they are treated with antibiotics. Chronic gastritis due to untreated *H. pylori* infection may lead to the development of gastric cancers.

Clinical signs of *H. pylori* infection include belching and stomach (epigastric) pain. Definitive diagnosis involves the recovery and culture or observation of *Helicobacter pylori* from a gastric ulcer biopsy. Serum antibodies indicate *H. pylori* infection, but since infections seem to be chronic and antibodies may persist for months after a given infection (⚭ Section 24.7), *H. pylori* antibodies are not reliable indicators of acute, active disease. A simple *in vivo* test, the *urease test*, is used as a metabolic test for *H. pylori*. In this test, the patient ingests ^{13}C or ^{14}C-labeled urea. The presence of labeled CO_2 in the patient's exhaled breath indicates the presence of urease, produced almost exclusively by *H. pylori*. Recovery of *H. pylori* organisms or antigens from the stool is also indicative of infection.

Evidence for a causal association between *Helicobacter pylori* and gastric ulcers comes from antibiotic treatments for the disease. Long-term treatment of ulcers with antacid preparations has seldom been successful. Most patients relapse within 1 year. However, by treating ulcers as an infectious disease, permanent cures are often obtained. Treatment of *H. pylori* infection usually consists of a combination of drugs including metranidazole, a second antibiotic such as tetracycline or amoxycillin, and a bismuth-containing antacid preparation. The combination treatment, administered for 14 days, abolishes the *H. pylori* infection and provides a long-term cure.

26.10 Concept Check

Helicobacter pylori infection appears to be the most common cause of gastric ulcers. Treatment of gastric ulcers now involves antibiotics, which seem to promote a permanent cure.

◆ Describe the infection by *H. pylori* and the resulting development of an ulcer.

◆ Describe evidence indicating that *H. pylori* infections are spread from person to person and also from common sources.

26.11 Hepatitis Viruses

Hepatitis is a liver inflammation commonly caused by an infectious agent. Hepatitis sometimes results in acute illness followed by destruction of functional liver anatomy and cells, a condition known as **cirrhosis**. Hepatitis due to an infection can cause chronic or acute disease and some forms lead to liver cancer. Although many viruses and a few bacteria can cause hepatitis, a restricted group of viruses is often associated with liver disease. Hepatitis viruses are diverse, and none of these viruses are genetically related, but all infect cells in the liver, causing hepatitis. Table 26.2 defines the five known hepatitis viruses.

Epidemiology

Hepatitis A virus (HAV) is transmitted from person to person or by ingestion of fecally contaminated food or water. Often called *infectious hepatitis*, the virus often causes mild, even subclinical infections, but rare cases of severe liver disease can occur. The most significant food vehicles for hepatitis A are shellfish, usually oysters and clams harvested from water polluted by human fecal material. In recent years however, HAV has also been transmitted in fresh produce. In 2003, a significant outbreak in the eastern United States was traced to eating uncooked or undercooked green onions. The general trend for numbers of HAV infections has moved downward (Figure 26.23●), partly due to the availability of an effective vaccine. HAV causes more cases of viral hepatitis than any other virus, and over 30% of individuals in the United States have antibodies to HAV, indicating they have been infected at some time in their lives.

Table 26.2	Hepatitis viruses				
Disease	**Virus and genome**	**Vaccine**	**Disease**		**Transmission route**
Hepatitis A	*Hepatovirus* (HAV) ss RNA	Yes	Acute		Enteric
Hepatitis B	*Orthohepadnavirus* (HBV) ds DNA	Yes	Acute, chronic, oncogenic		Parenteral, sexual
Hepatitis C	*Hepacivirus* (HCV) ss RNA	No	Chronic, oncogenic		Parenteral
Hepatitis D	*Deltavirus* (HDV) ss RNA	No	Fulminant, only with HBV		Parenteral
Hepatitis E	Calciviridae family (HEV) ss RNA	No	Fulminant disease in pregnant women		Enteric
Hepatitis G	Flaviviridae family (HGV) ss RNA	No	Asymptomatic		Parenteral

● **Figure 26.23 Hepatitis in the United States.** The prevalence of hepatitis is shown by viral agent. In 2002 there were 8795 cases of hepatitis A, 7996 cases of hepatitis B, and 1835 cases of other hepatitis, mostly caused by HCV. Data were obtained from the Centers for Disease Control and Prevention, Atlanta, GA, USA.

Infection due to hepatitis B virus (HBV) is often called *serum hepatitis*. HBV is a hepadnavirus (∞ Section 16.15), a partially double-stranded DNA virus. The mature virus particle containing the viral genome is called a Dane particle (Figure 26.24●). HBV causes acute, often severe disease that can lead to liver failure and death. Chronic HBV infection can lead to cirrhosis and liver cancer. HBV is usually transmitted by a *parenteral* (outside the gut) route, such as blood transfusion or through shared hypodermic needles contaminated with infected blood. HBV may also be transmitted through exchanges of body fluids, as in sexual intercourse. The numbers of new HBV infections is decreasing, again due to an effective vaccine. However, approximately 5,000

people die each year due to complications such as cancers generated by chronic HBV infection.

Hepatitis D virus (HDV) is a *defective virus* that lacks genes for its own protein coat (∞ Section 16.15). HDV is also transmitted by parenteral routes, but since it is a defective virus, it cannot replicate and express a complete virus unless the cell is also infected with HBV: The HDV genome replicates independently but uses the protein coat of HBV for expression. Thus, HDV infections are always seen as coinfections with HBV.

Hepatitis C virus (HCV) is also transmitted parenterally. HCV generally produces a mild or even asymptomatic disease at first, but up to 85% of individuals develop chronic hepatitis, with up to 20% leading to chronic liver disease and cirrhosis. Chronic infection leads to *hepatocarcinoma* (liver cancer) in 3–5% of infected individuals each year. The latency period for development of cancer can be several decades after the primary infection. The *reported* new cases of HCV in the United States (Figure 26.23) are only a fraction of the approximately 25,000 new infections annually. Large numbers of HCV-related deaths occur annually due to chronic HCV infections that develop into liver cancer. HCV-induced liver disease is the most common liver disease currently seen in clinical settings in the United States and accounts for up to 10,000 of the 25,000 annual deaths due to liver cancer, other chronic liver disease, and cirrhosis.

Hepatitis E virus (HEV) transmits hepatitis via an enteric route. HEV causes an acute, self-limiting hepatitis that varies in severity from case to case, but is often the cause of rapid fulminant disease in pregnant women. HEV is endemic in Mexico as well as in tropical and subtropical regions of Africa and Asia.

Hepatitis G virus (HGV) is commonly found in the blood of patients with other forms of acute hepatitis, but HGV alone seems to cause very mild disease or is completely asymptomatic. Several different samples of volunteer blood donors tested as high as 8.1% positive for HGV, but since HGV is not associated with demonstrable clinical disease, the significance of these findings is not clear.

Pathology and Diagnosis

Hepatitis is an acute disease of the liver. Symptoms include fever, **jaundice** (production and release of excess bilirubin by the liver due to destruction of liver cells, resulting in yellowing of the skin), *icterus* (yellowing of the whites of the eyes), *hepatomegaly* (liver enlargement), and *cirrhosis* (breakdown of the normal liver tissue architecture, including fibrosis). Mild hepatitis may be characterized by relatively minor elevation of liver enzymes such as alanine aminotransferase (ALT). Fulminant disease is characterized by rapid onset of severe symptoms such as jaundice and cirrhosis and is often a life-threatening condition. The various hepatitis viruses cause similar acute clinical disease and cannot be readily distinguished based on the clinical findings alone. Chronic hepatitis infections, usually caused by HBV or HCV, are often asymptomatic or produce very mild

● **Figure 26.24 Hepatitis B virus (HBV).** The arrow indicates a complete HBV particle, which is about 42 nm in diameter and is called a Dane particle (∞ Section 16.15 and Figure 16.27).

symptoms but can cause serious liver disease, even in the absence of hepatocarcinoma.

Diagnosis of hepatitis is based primarily on clinical findings and laboratory tests that determine liver function problems. Cirrhosis is diagnosed by visual examination of biopsied liver tissue. A number of virus-specific assays are also used to confirm diagnosis, identify the infectious agent, and determine a course of treatment. Direct culture of hepatitis viruses is usually not used for identification purposes, and HCV and HGV have not been successfully cultured.

Some of the most widely used methods for determining hepatitis identity are enzyme-linked immunoassay (ELISA) tests (∞ Section 24.10). Most ELISA tests are designed to identify viral proteins in blood specimens. However, tests are available that indicate the presence of IgM or IgG *antibodies* to HBV. IgM is associated with the primary immune response to HBV and IgG is associated with the secondary response to HBV. Therefore, identification of the antibody class can determine whether or not the HBV infection is a new infection (IgM) or whether the antibody is due to a secondary response to a chronic or latent HBV infection (IgG) (∞ Section 22.9). Other immune-based tests used for the detection of hepatitis viruses include immunoblots (∞ Section 24.11), immunoelectron microscopy (IEM) (∞ Section 24.13), and immunofluorescence (∞ Section 24.9).

PCR-based tests and dot-blot DNA hybridization tests (∞ Section 24.12) are also used for the detection of the viral genome in blood or in liver tissue obtained by biopsy.

Prevention and Treatment

Infection with HAV or HBV can be prevented with effective vaccines. HBV vaccination is recommended and in most cases is required for school-age children in the United States (∞ Section 22.13). No effective vaccines are available for the other hepatitis viruses.

"Universal precautions" for prevention of disease spread by bloodborne pathogens are mandated by law primarily to protect against HBV infection and the human immunodeficiency virus (HIV) (∞ Section 24.4). These standards mandate precautions for personnel when handling infectious waste and body fluids and are designed to prevent infection by all parenterally transmitted hepatitis viruses (HBV, HCV, HDV, and HGV). The precautions prescribe a high level of vigilance and aseptic handling and containment procedures to deal with patients, body fluids, and infected waste materials.

Hepatitis A can be spread through contamination of common sources such as food and water, and epidemic outbreaks of hepatitis A can be prevented by maintaining pathogen-free food and water supplies (∞ Chapter 28 and Chapter 29).

Postexposure treatment of hepatitis is sometimes successful. Pooled human immune gamma globulin can be used to prevent HAV infection if given soon after exposure.

For postexposure prevention of HBV infection, specific hepatitis B immune globulin, coupled with administration of the HBV vaccine, has been effective (∞ Section 22.13).

Most treatment of hepatitis is supportive, providing rest and time to allow liver damage to resolve and be repaired. In some cases, some antiviral drugs are effective for treatment. Interferon α is effective against HCV when combined with ribavirin in some patients. HBV can be treated with the antiviral drugs foscarnet, ribavirin, lamivudine, or ganciclovir (∞ Section 20.10).

⬣ 26.11 *Concept Check*

Hepatitis caused by viruses can cause cirrhosis, an acute liver disease. HBV and HCV can cause chronic infections leading to liver cancer. Vaccines are available for HAV and HBV. The overall prevalence of hepatitis has decreased significantly in the last 20 years in the United States, but viral hepatitis is still a major public health problem because of the high infectivity of the viruses.

◆ Describe the mode of transmission for hepatitis A virus, hepatitis B virus, and hepatitis C virus.

◆ Describe potential prevention and treatment methods for hepatitis A virus and hepatitis B virus.

III SEXUALLY TRANSMITTED INFECTIONS

Several important human pathogens are transmitted almost exclusively by sexual contact. These pathogens cause **sexually transmitted infections**, or **STIs**.

STIs, also called *sexually transmitted diseases (STDs)* or *venereal diseases*, are caused by a wide variety of bacteria, viruses, protozoa, and even fungi (Table 26.3). Unlike respiratory pathogens that are shed constantly in large numbers by an infected individual, sexually transmitted pathogens are generally found only in body fluids from the genitourinary tract that are exchanged during sexual activity. This is because sexually transmitted pathogens are very sensitive to drying and other environmental stresses such as heat and light. Their habitat, the human genitourinary tract, is a protected, moist environment. Thus, these organisms preferentially and sometimes exclusively colonize the genitourinary tract.

Effectively diagnosing and treating STIs is very difficult for a number of social and biological reasons. First, up to one-third of all STIs involve teenagers with multiple sex partners, making it difficult to identify the infection source and stop its spread. Second, many STIs have minor symptoms and infected individuals do not seek treatment. Third, social stigmas still attached to STIs prevent many individuals from seeking prompt treatment. However, prompt effective treatment of STIs is desirable for a number of reasons. First, most STIs are curable and virtually all are controllable with appropriate medical intervention.

Table 26.3	Sexually transmitted diseases and treatment guidelines	
Disease	**Causative organism(s)[a]**	**Recommended treatment[b]**
Gonorrhea	*Neisseria gonorrhoeae* (B)	Cefixime or ceftriaxone, *and* azithromycin or doxycycline
Syphilis	*Treponema pallidum* (B)	Benzathine penicillin G
Chlamydia trachomatis infections	*Chlamydia trachomatis* (B)	Doxycycline or azithromycin
Nongonococcal urethritis	*C. trachomatis* (B) or *Ureaplasma urealyticum* (B) or *Mycoplasma genitalium* (B) or *Trichomonas vaginalis* (P)	Azithromycin
Lymphogranuloma venereum	*C. trachomatis* (B)	Doxycycline
Chancroid	*Haemophilus ducreyi* (B)	Azithromycin
Genital herpes	Herpes simplex type 2 (V)	No known cure; symptoms can be controlled with topical application of acyclovir (Figure 26.31).
Genital warts	Papilloma virus (certain strains)	No known cure; symptomatic warts can be removed surgically, chemically, or by cryotherapy.
Trichomoniasis	*Trichomonas vaginalis* (P)	Metronidazole
Acquired immunodeficiency syndrome (AIDS)	Human immunodeficiency virus (HIV)	No known cure; nucleotide base analogs, protease inhibitors, fusion inhibitors, and nonnucleoside reverse transcriptase inhibitors are clinically useful in some treatments (see Table 26.4).
Pelvic inflammatory disease	*N. gonorrhoeae* (B) or *C. trachomatis* (B)	Cefotetan
Vulvovaginal candidiasis	*Candida albicans* (F)	Butoconazole

[a] B, bacterium; V, virus; P, protozoan; F, fungus.
[b] Recommendations of the U.S. Department of Health and Human Services, Public Health Service. For many drugs, there are a number of possible alternatives.

Second, delay or lack of treatment can lead to long-term problems such as infertility, cancer, heart disease, degenerative nerve disease, birth defects, or stillbirth.

Because transmission of STIs is limited to intimate physical contact, generally during sexual intercourse, the spread of venereal diseases can be controlled by sexual abstinence (no exchange of body fluids) or by the use of barriers such as condoms that stop the exchange of body fluids during sexual activity.

STIs are very common and continue to pose social as well as medical problems. We deal here with a discussion of a number of prevalent STIs.

26.12 Gonorrhea and Syphilis

Gonorrhea and syphilis are preventable, treatable bacterial STIs. Because of differences in their symptoms, the overall pattern of disease prevalence is quite different. Gonorrhea is very prevalent, and often asymptomatic, especially in women. As a result, the disease is often unrecognized and remains untreated. Syphilis, on the other hand, now has a low prevalence. This is partly because the primary stage of syphilis exhibits very obvious symptoms and infected individuals usually seek immediate treatment (Figure 26.25●).

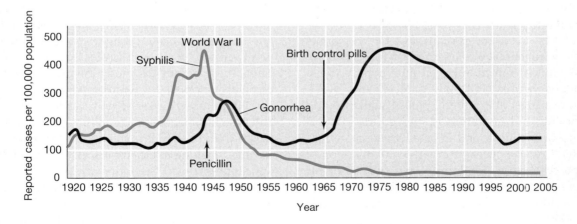

● **Figure 26.25 Reported cases of gonorrhea and syphilis in the United States.** Note the downward trend in disease incidence after the introduction of antibiotics and the upward trend in the incidence of gonorrhea after the introduction of birth control pills. In 2002 there were 351,852 cases of gonorrhea and only 6862 cases of syphilis in the United States.

Gonorrhea

Neisseria gonorrhoeae, often called *gonococcus*, causes gonorrhea. *N. gonorrhoeae* is a gram-negative, nonsporulating, obligately aerobic, oxidase-positive diplococcus (∞ Section 12.10) related biochemically and phylogenetically to *Neisseria meningitidis* (see Section 26.6). *N. gonorrhoeae* is very sensitive to drying and normally does not survive away from the mucus membranes of the genitourinary tract (Figure 26.26●). Gonococci are killed rapidly by drying, sunlight, and ultraviolet light. Because of its extreme sensitivity to environmental conditions, *N. gonorrhoeae* can only be transmitted by intimate person-to-person contact. The pathogen enters the body by way of the mucus membranes of the genitourinary tract.

The symptoms of gonorrhea are quite different in the male and female. In females, gonorrhea is characterized by a mild vaginitis that is difficult to distinguish from vaginal infections caused by other organisms, and thus, the infection may easily go unnoticed. Complications from untreated gonorrhea in females can lead, however, to a condition known as *pelvic inflammatory disease (PID)*. PID is a chronic inflammatory disease that can lead to long-term complications such as sterility. In the male, the organism causes a painful infection of the urethral canal (∞ Figure 21.11). Complications from untreated gonorrhea affecting both males and females include damage to heart valves and joint tissues due to immune complex deposition (∞ Section 22.15).

In addition to gonorrhea, the organism also causes eye infections in newborns. Infants born of infected mothers may acquire eye infections during birth. There-fore, prophylactic treatment of the eyes of all newborns with an ointment containing erythromycin is generally mandatory to prevent gonococcal infection in infants. We discussed the clinical microbiology and diagnosis of gonorrhea in Section 24.1.

Treatment of gonorrhea with penicillin has been successful in the past. However, strains of *N. gonorrhoeae* resistant to penicillin arose in the 1980s and are now widespread. Fortunately, the penicillinase-producing strains respond to alternative antibiotic therapy, with a single dose of the β-lactam antibiotics cefixime or ceftriaxone. The quinolones ciprofloxacin, oflaxacin, or levofloxacin can also be used. By 2001, however, 14.3% of *N. gonorrhoeae* strains isolated from Hawaii were resistant to quinolones, underscoring the need to test isolates for antibiotic sensitivity. An antichlamydial agent, normally azithromycin or doxycycline, is often given at the same time because nearly 50% of gonorrhea patients are also infected with the hard-to-diagnose *Chlamydia trachomatis* (Table 26.3).

The incidence of gonococcus infection remains relatively high for the following reasons: (1) Acquired immunity does not exist; hence repeated reinfection is possible. Although antibodies are produced, they are either not protective, or they are strain-specific and provide no cross immunization. Within a single *Neisseria gonorrhoeae* strain, antigenic switches can occur, changing *opacity protein antigens (Opa)* and *surface pilin antigens*. These changes create new serotypes and apparently prevent effective immunity. (2) The use of oral contraceptives alters the local mucosal environment in favor of the pathogen. Oral contraceptives induce the body to mimic pregnancy, which results, among other things, in a lack of glycogen production in the vagina and a rise in the vaginal pH. Lactic acid bacteria normally found in the adult vagina fail to develop under such circumstances, allowing colonization by *N. gonorrhoeae* transmitted from an infected partner (∞ Section 21.5). (3) Symptoms in the female are so mild that the disease may be unrecognized, and a promiscuous infected female can infect many males. The disease can be controlled if the sexual contacts of infected persons are quickly identified and treated, but it is often difficult to obtain this information and even more difficult to arrange treatment.

Syphilis

Syphilis is caused by a spirochete, *Treponema pallidum*. *T. pallidum* is about 10–15 μm in length and about 0.15 μm in diameter (Figure 26.27●). The spirochete is extremely sensitive to environmental stress and, therefore, syphilis is normally transmitted from person to person by intimate sexual contact.

Syphilis, often transmitted with *N. gonorrhoeae*, is potentially much more serious than gonorrhea. Largely because of differences in the symptoms and pathobiology of the two diseases, the incidence of syphilis in the United States is much lower than the incidence of gonorrhea (Figure 26.25).

Morris D. Cooper

● **Figure 26.26 The causative agent of gonorrhea, *Neisseria gonorrhoeae*.** The scanning electron micrograph of the microvilli of human fallopian tube mucosa shows how cells of *N. gonorrhoeae* attach to the surface of epithelial cells. Note the distinct diplococcus morphology. Cells of *N. gonorrhoeae* are about 0.8 μm in diameter.

(a)

(b)

● **Figure 26.27** **The syphilis spirochete, *Treponema pallidum*.** (a) Dark-field microscopy of an exudate. *Treponema pallidum* cells measure 0.15 μm wide and 10–15 μm long. (b) Shadow-cast electron micrograph of a cell of *T. pallidum*. Note the endoflagella, typical of spirochetes (∞ Section 12.33).

Syphilis does not pass through unbroken skin, and initial infection most probably takes place through tiny breaks in the epidermal layer. In the male, initial infection is usually on the penis; in the female it is most often in the vagina, cervix, or perineal region. In about 10% of cases, infection is extragenital, usually in the oral region. During pregnancy, the organism can be transmitted from an infected woman to the fetus; the disease acquired by the infant is called **congenital syphilis**.

Syphilis is an extremely complex disease and, in an individual patient, may progress into any of three stages, but the disease always begins with a localized infection called *primary syphilis*. In primary syphilis, *T. pallidum* multiplies at the initial site of entry, and a characteristic *primary* lesion known as a *chancre* (Figure 26.28●) is formed within 2 weeks to 2 months. Darkfield microscopy of the exudate from syphilitic chancres often reveals the actively motile spirochetes (Figure 26.27a). In most cases the chancre heals spontaneously and the organisms disappear from the site. Some cells, however, spread from the initial site to various parts of the body, such as the mucous membranes, the eyes, joints, bones, or central nervous system, and extensive multiplication occurs. A hypersensitivity reaction to the treponeme often takes place, revealed by the development of a generalized skin rash; this rash is the key symptom of *secondary syphilis*. At first, the patient may be highly infectious, but eventually *T. pallidum* is cleared from the secondary lesions and infectivity is reduced.

The subsequent course of the disease in the absence of treatment is highly variable. About one-fourth of infected individuals appear to undergo a spontaneous cure as demonstrated by a decrease in antibody titer, and another one-fourth do not exhibit any further symptoms, although static or elevated antibody titers indicate a persistent, active infection. About half of these patients develop *tertiary syphilis*, with symptoms ranging from relatively mild infections of the skin and bone to serious and even fatal infections of the cardiovascular system or central nervous system. Involvement of the nervous system can cause generalized paralysis or other severe neurological damage. Relatively low numbers of organisms are present in individuals with tertiary syphilis, and most of the symptoms probably result from inflammation due to delayed hypersensitivity reactions to the spirochetes (∞ Sections 22.3 and 22.15).

The clinical immunology and microbiology as well as the laboratory diagnosis of syphilis were discussed in Section 24.1. The single most important physical sign of a primary syphilis infection, the chancre, is also diag-

(a)

(b)

● **Figure 26.28** **Primary syphilis lesions.** (a) Chancre on lip. (b) Several chancres on penis. The chancre is the characteristic lesion of primary syphilis at the site of infection by *Treponema pallidum*. Patients who acquire such lesions generally seek medical intervention, and the obvious chancre hastens diagnosis and treatment. Syphilis is treatable and curable with a single injection of penicillin G (∞ Table 26.3). The biology of the genus *Treponema* is discussed in Section 12.33.

nostic for the disease. Infected individuals generally seek treatment for syphilis because of the highly visible chancre.

Penicillin is highly effective in syphilis therapy, and the primary and secondary stages of the disease can usually be controlled by a single injection of benzathine penicillin G. In tertiary syphilis, penicillin treatment must be extended for longer periods of time. The incidence of primary and secondary syphilis in the United States has decreased significantly over the last two decades and is now at the lowest level since record keeping began.

⬡ 26.12 Concept Check

Gonorrhea and syphilis, caused by *Neisseria gonorrhoeae* and *Treponema pallidum*, respectively, are STIs with potential serious consequences if not treated. Although the incidence of these diseases has generally declined in recent years, there are still over 350,000 cases of gonorrhea and 6000 cases of syphilis annually in the United States.

◆ Explain at least one potential reason for the high incidence of gonorrhea as compared with syphilis.

◆ Describe the progression of untreated gonorrhea and untreated syphilis. Do treatments produce a cure for each disease?

26.13 Chlamydia, Herpes, and Trichomoniasis

Chlamydia, herpes, and trichomoniasis are important STIs transmitted by a bacterium, a virus, and a protozoan, respectively. These diseases are very prevalent in the population and are much more difficult to diagnose and treat than are syphilis and gonorrhea.

Chlamydia

A number of sexually transmitted diseases can be ascribed to infection by the obligate intracellular bacterium *Chlamydia trachomatis* (Figure 26.29● and ⟨∞⟩ Section 12.27). The total incidence of sexually transmitted *C. trachomatis* infections (a reportable disease, ⟨∞⟩ Table 25.3) probably greatly outnumbers the incidence of gonorrhea. There are over 600,000 *reported* cases and there may be up to 3 million new *C. trachomatis* sexually transmitted infections every year, making this organism the most prevalent cause of venereal disease. *C. trachomatis* also causes a serious eye infection called *trachoma* (⟨∞⟩ Section 12.27), but the strains of *C. trachomatis* responsible for venereal infections are distinct from those causing trachoma. Chlamydial infections may also be transmitted congenitally to the newborn in the birth canal, causing newborn conjunctivitis and pneumonia.

Nongonococcal urethritis (NGU) due to *C. trachomatis* is one of the most frequently observed sexually transmitted diseases in males and females, but the infections are often inapparent. In a small percentage of cases, chlamydial NGU leads to serious acute complications, including testicular swelling and prostate inflammation in men, and cervicitis, pelvic inflammatory disease, and fallopian tube damage in women; cells of *C. trachomatis* attach to microvilli of fallopian tube cells, enter, multiply, and eventually lyse the cells (Figure 26.29). Untreated NGU can cause infertility.

Chlamydia NGU is relatively difficult to diagnose by traditional isolation and identification methods. To expedite

(a)

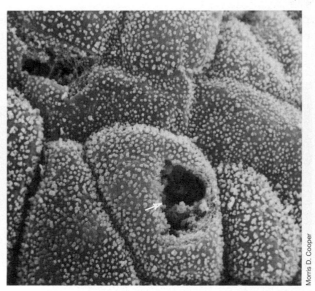

(b)

● **Figure 26.29 Cells of *Chlamydia trachomatis* (arrows) attached to human fallopian tube tissues.** (a) Cells attached to the microvilli of a fallopian tube. (b) A damaged fallopian tube containing a cell of *C. trachomatis* (arrow) in the lesion.

diagnoses, a variety of tests have been developed for identifying *C. trachomatis* from a vaginal or pelvic swab or from discharges. These clinical tests include nucleic acid probe tests as well as fluorescent monoclonal antibodies and various enzyme-linked immunosorbent assay (ELISA) tests for detecting specific *C. trachomatis* antigens. If a chlamydial infection is suspected, treatment is initiated with the antibiotics azithromycin or doxycycline (Table 26.3).

Chlamydial NGU is frequently observed as a secondary infection following gonorrhea. If both *Neisseria gonorrhoeae* and *Chlamydia trachomatis* are transmitted to a new host in a single event, treatment of gonorrhea with cefixime or ceftriaxone is usually successful but does not eliminate the chlamydia. Although cured of gonorrhea, such patients, still infected with chlamydia, eventually experience an apparent recurrence of gonorrhea that is instead a case of chlamydial NGU. Thus, patients treated for gonorrhea are also given azithromycin or doxycycline to treat the potential coinfection with the usually undiagnosed *C. trachomatis*.

Lymphogranuloma venereum is a sexually transmitted disease caused by distinct strains of *C. trachomatis* (LGV 1, 2, or 3). The disease occurs most frequently in males and consists of a swelling of the lymph nodes in and about the groin. From the infected lymph nodes, chlamydial cells may travel to the rectum and cause a painful inflammation of rectal tissues called *proctitis*. Because of the potential for regional lymph node damage and the complications of proctitis, lymphogranuloma venereum is also a serious sexually transmitted chlamydial infection. It is the only chlamydial infection that invades beyond the epithelial cell layer. Other *Chlamydia* infect and complete their development in the epithelial cells.

Herpes

Herpesviruses are a large group of complex double stranded DNA viruses (∞ Section 16.12), many of which are human pathogens. A subgroup of herpesviruses, the *herpes simplex viruses*, are responsible for cold sores and genital infections.

Herpes simplex virus type 1 (HSV-1) infects the epithelial cells around the mouth and lips, causing cold sores (fever blisters) (Figure 26.30*a*●). HSV-1 may, however, occasionally infect other body sites including the anogenital regions. HSV-1 is spread via direct contact or through saliva. The incubation period of HSV-1 infections is short (3–5 days), and the lesions heal without treatment in 2–3 weeks. Relapses of HSV-1 infections are relatively common, and it is thought that the virus is spread primarily by contact with infectious lesions. Latent herpes infections are apparently quite common, with the virus persisting in low numbers in nerve tissue. Recurrent acute herpes infections are due to a periodic triggering of virus activity by unknown or indeterminant causes such as coinfections and stress. Oral herpes caused by HSV-1 is quite common and apparently has no harmful effects on the host beyond the discomfort of the oral blisters.

Herpes simplex virus 2 (HSV-2) infections are associated primarily with the anogenital region, where the virus causes painful blisters on the penis of males or on the cervix, vulva, or vagina of females (Figure 26.30*b*). HSV-2 infections are generally transmitted by direct sexual contact, and the disease is most easily transmitted during the active blister stage rather than during periods of inapparent (presumably latent) infection. HSV-2 occasionally infects other body sites such as the mucous membranes of the mouth. HSV-2 can also be transmitted to a newborn at birth by contact with herpetic lesions in the birth canal. The disease in the newborn varies from latent infections with no apparent damage to systemic disease resulting in brain damage or death. To avoid herpes infections in newborns, delivery by caesarean section is advised for pregnant women with genital herpes infections. The long-term effects of genital herpes infections are not yet fully understood. However, studies have indicated a significant correla-

(a)   *(b)*   *(c)*

● **Figure 26.30 Herpesvirus and *Trichomonas*.** (a) A severe case of herpes blisters on the face due to infection with herpes simplex virus type 1. (b) Genital herpes due to infection on the penis with herpes simplex virus type 2. (c) Cells of the flagellated protozoan *Trichomonas vaginalis*.

tion between *genital* herpes infections and cervical cancer in females.

Genital herpes infections are presently incurable, although a limited number of drugs have been successful in controlling the infectious blister stages. The guanine analog acyclovir (Figure 26.31●), given orally and also applied topically, is particularly effective in limiting the shed of active virus from blisters and promoting the healing of blistering lesions. Acyclovir specifically interferes with herpesvirus DNA polymerase, inhibiting viral DNA replication.

Trichomoniasis

Nongonococcal urethritis may also be caused by infections with the protozoan *Trichomonas vaginalis* (Figure 26.30c). The biology of *Trichomonas* is discussed in Sections 14.2, 14.9, and 14.10. *T. vaginalis* does not produce the resting cells or *cysts* important to the life cycle of many protozoans. As a result, transmission is usually from person to person, generally by sexual intercourse. However, cells of *T. vaginalis* can survive for 1–2 hours on moist surfaces, 30–40 minutes in water, and up to 24 hours in urine or semen. Thus, transmission of *T. vaginalis* by contaminated toilet seats, sauna benches, and paper towels occasionally occurs. *T. vaginalis* infects the vagina in women, the prostate and seminal vesicles of men, and the urethra of both males and females.

Many cases of trichomoniasis are totally asymptomatic in males. In women trichomoniasis is characterized by a vaginal discharge, vaginitis, and painful urination. The infection is more common in females; surveys indicate that 25–50% of sexually active women are infected, while only about 5% of men are infected because of the killing action of prostatic fluids. The male partner of an infected female should be examined for *T. vaginalis* and treated if necessary because promiscuous asymptomatic males can transmit the infection to several females. Trichomoniasis is diagnosed by observation of the motile protozoan in a wet mount of fluid discharged from the patient (Figure 26.30c). The antiprotozoal drug *metronidazole* is particularly effective in treating trichomoniasis (Table 26.3).

 26.13 *Concept Check*

Chlamydia, the most prevalent STI, is caused by infection with the bacterium *Chlamydia trachomatis*. Untreated chlamydial nongonococcal urethritis causes serious complications in males and females. Herpes lesions can also be transmitted sexually and are caused by herpes simplex virus type 1 and herpes simplex virus type 2. HSV-2 is generally associated with sexual transmission and infection of the anogenital regions. *Trichomonas vaginalis* is a protozoan responsible for trichomoniasis, another STI. In general, these STIs are widespread and are more difficult to diagnose and treat than gonorrhea or syphilis. There is no cure for herpes.

● **Figure 26.31 Guanine and the guanine analog acyclovir.** Acyclovir has been used therapeutically to control genital herpes (HSV-2) blisters.

◆ Describe pertinent clinical features and treatment protocols for chlamydia, herpes, and trichomoniasis.

◆ Why are these diseases more difficult to diagnose than gonorrhea or syphilis?

26.14 Acquired Immunodeficiency Syndrome: AIDS and HIV

Acquired immunodeficiency syndrome (AIDS) was recognized as a distinct disease in 1981. More than 900,000 cases of AIDS have been reported since then in the United States alone, and more than 500,000 people have died (∞ Section 25.6). Over 1.4 million in the United States have been infected with HIV and about 1 million are living with the infection. Worldwide, the outlook is even more serious, with more than 50 million people already infected with the human immunodeficiency virus (HIV), the causative agent of AIDS. Over 16 million people have already died from AIDS.

HIV is divided into two major types, HIV-1 and HIV-2. HIV-1 is genetically similar, but distinct from HIV-2. HIV-2, discovered in West Africa in 1985, has reduced virulence as compared with HIV-1, and causes a milder AIDS-like disease. Currently, more than 99% of global AIDS cases are due to HIV-1, and our discussion of AIDS will center on infections with HIV-1.

AIDS is caused by the human immunodeficiency virus (HIV-1), a retrovirus (∞ Section 16.15). The genome contains 9749 nucleotides in each of its two identical single-stranded RNA genomes. The *reverse transcriptase* enzyme present in the intact virion catalyzes formation of a complementary single-stranded DNA molecule using RNA as a template. The enzyme then converts the complementary DNA (cDNA) into double-stranded DNA, which can integrate into the host cell genome.

The numbers of HIV-infected individuals will continue to rise unless effective treatment or prevention methods are discovered. In the United States, the numbers of newly diagnosed HIV infections are increasing, and declines in HIV morbidity and mortality attributable to combination antiretroviral therapy have ended, reversing the trends of decreasing AIDS prevalence and severity seen in the 1990s. We have already discussed the epidemiology of AIDS (∞ Section 25.6) and the clinical diagnostic methods for identifying and tracking

HIV infection (∞ Sections 24.10–24.11). In this section, we will focus on the pathogenesis of AIDS, concentrating on the natural course of HIV infection and the effects of HIV on the immune system.

A Definition of AIDS

AIDS was first suspected of being a disease affecting the immune system because a large number of opportunistic infections were observed in certain populations (∞ Section 25.6). **Opportunistic infections** are infections usually observed only in individuals with a dysfunctional immune system. This definition was adopted in 1993 by the Centers for Disease Control and Prevention (Atlanta, Georgia, USA) and is used to define AIDS in the United States.

The current case definition for acquired immunodeficiency syndrome (AIDS) includes those who (a) test positive for human immunodeficiency virus (HIV) and (b) meet one of the two following criteria.

1. They have a CD4 T-cell number of less than $200/mm^3$ of whole blood (the normal count is $600–1000/mm^3$) or a CD4 T-cell/total lymphocytes percentage of less than 14%; or

2. They may have a CD4 T-cell number of more than $200/mm^3$ *and* any of the following conditions: fungal diseases including candidiasis (Figure 26.32a●), coccidiomycosis, cryptococcosis (Figure 26.32b), histoplasmosis (Figure 26.32c), isosporiasis, *Pneumocystis carinii* pneumonia (Figure 26.32d), cryptosporidiosis (Figure 26.32e), or toxoplasmosis (Figure 26.32f) of the brain; bacterial diseases including pulmonary tuberculosis or other *Mycobacterium* spp. infections (Figure 26.32g), or recurrent *Salmonella* septicemia; viral diseases including cytomegalovirus infection, HIV-related encephalopathy, HIV wasting syndrome, chronic ulcers, or bronchitis due to *herpes simplex* (∞ Section 16.12); or progressive multifocal leukoencephalopathy, malignant diseases such as invasive cervical cancer, Kaposi's sarcoma (Figure 26.33●), Burkitt's lymphoma, primary lymphoma of the brain, or immunoblastic lymphoma; recurrent pneumonia due to any agent.

The most common opportunistic infection in AIDS patients is by the protozoan *Pneumocystis carinii* pneumonia. A frequent non-microbial disease in AIDS patients is Kaposi's sarcoma, an atypical cancer observed in high frequency in HIV-infected homosexual males. Kaposi's sarcoma is a cancer of the cells lining the blood vessels and is

(a) (b) (c)

(d) (e) (f)

(g)

● **Figure 26.32 Opportunistic pathogens associated with cases of acquired immunodeficiency syndrome (AIDS).** (a) *Candida albicans*, from heart tissue of patient with systemic *Candida* infection. (b) *Cryptococcus neoformans*, from liver tissue of a patient with cryptococcosis. (c) *Histoplasma capsulatum*, from liver tissue of patient with histoplasmosis. (d) *Pneumocystis carinii*, from patient with pulmonary pneumocytosis. (e) *Cryptosporidium* sp. from small intestine of a patient with cryptosporidiosis. (f) *Toxoplasma gondii*, from brain tissue of patient with toxoplasmosis. (g) *Mycobacterium* spp. infection of small bowel, acid-fast stain.

(a)

(b)

Centers for Disease Control

● **Figure 26.33 Kaposi's sarcoma.** Lesions are shown as they appear on (a) the heel and lateral foot, and (b) the distal leg and ankle.

Mature form

Budding particles

Centers for Disease Control

● **Figure 26.34 Transmission electron micrograph of a thin section of a lymphocyte releasing human immunodeficiency virus (HIV).** Cells were from a hemophiliac patient who developed AIDS. HIV particles are 90–120 nm in diameter. Compare with Figure 9.25.

characterized by purple patches on the surface of the skin, especially in the extremities (Figure 26.33). Kaposi's sarcoma is caused by coinfection of HIV and human herpesvirus 8 (HHV-8) and is 20,000-fold more prevalent in AIDS patients than in the general population.

In summary, an individual has AIDS if he or she (1) tests positive for HIV or HIV antibodies *and* (2) has a drastically reduced T-helper lymphocyte count, *or* (3) has at least one of a number of opportunistic infections or atypical cancers.

HIV Pathogenesis

HIV has the ability to infect cells displaying the CD4 cell-surface protein. The two cell types most commonly infected are macrophages and T-helper (T_H) cells, both of which are important components of the immune system (∞ Sections 22.1 and 22.8). Infected macrophages and T cells produce and release large numbers of HIV particles, which in turn infect other cells that display CD4 (Figure 26.34●).

In addition to CD4, HIV must interact with coreceptors on target cells. HIV infection normally occurs first in macrophages, a type of antigen-presenting cell (APC) that has a very low level of CD4 on its surface (Figure 26.35●). At the cell surface, the macrophage CD4 molecule binds to the gp120 protein of HIV. The viral gp120 protein then interacts with another macrophage protein, the membrane-spanning chemokine receptor CCR5 (∞ Section 23.11). CCR5 is a coreceptor for HIV and, together with CD4, forms the docking site

where the HIV envelope fuses with the host cell membrane; this allows the insertion of the viral nucleocapsid into the cell. The CCR5 coreceptor is required for HIV binding to macrophages. Individuals who express a variant CCR5 protein do not bind HIV and do not acquire HIV infections. After HIV has infected the macrophage APCs, a different form of gp120 is made, which in turn binds to a different coreceptor, the CXCR4 chemokine receptor on T cells.

HIV then enters and destroys the CD4 T-helper lymphocytes, the T_H1 and T_H2 cells that are responsible for cell-mediated inflammatory responses and B-cell help, respectively (∞ Section 22.8). Thus, HIV starts as a macrophage infection and progresses to a T-cell infection. The result of HIV infection is the systematic destruction of macrophages and T cells, leading to a catastrophic breakdown of immunity. Specific knowledge of the coreceptors involved in HIV infection in macrophages and T cells may be used to design HIV-specific chemokine-receptor blocking agents to prevent the attachment of HIV to either the macrophage or the T cell, thus preventing infection.

In cases of AIDS, CD4 lymphocytes are greatly reduced in number. However, HIV infection does not immediately kill the host cell. HIV can exist as a provirus and not an infectious virion after reverse transcription to produce DNA from the RNA genome. The reverse-transcribed HIV genome is integrated as viral cDNA into host chromosomal DNA, The cell may show no outward sign of infection, and HIV DNA can remain in this latent state for long periods, replicating as the host DNA replicates. Eventually, virus synthesis occurs and new HIV particles are produced and released from the cell. T cells producing HIV no longer divide and eventually die.

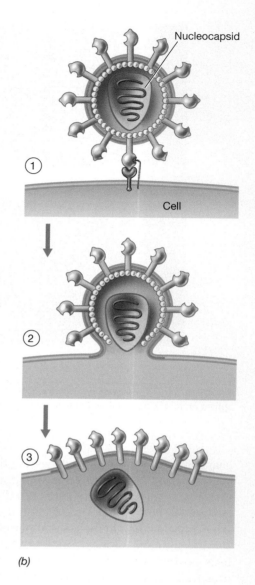

● **Figure 26.35 Infection of a CD4 target cell with the human immunodeficiency virus (HIV).** (a) Interaction of HIV with a target cell through specific binding of HIV gp120 to the CD4 receptor and a coreceptor on target cells such as macrophages or T cells. The coreceptor shown is the membrane-spanning chemokine receptor CCR5 found on macrophages. A similar coreceptor, CXCR4, is found on CD4 T cells. (b) Fusion of the HIV envelope with the host cell begins with ① the interaction of the virus with a receptor/coreceptor pair on the host cell. Cells lacking either the receptor or the coreceptor cannot bind HIV and are not infected. Following binding at the cell surface, ② the viral envelope and host cell membrane coalesce. Finally, ③ the nucleocapsid is inserted into the host cell, beginning the active viral infection. For coverage of the biology and replication of retroviruses such as HIV, see Sections 9.13 and 16.15 and Figures 9.25, 9.26, 16.25, and 16.26.

Accelerated destruction of CD4 cells occurs following the processing of HIV antigens by infected T cells. Such cells insert molecules of gp120 from HIV particles into their cell surfaces. The embedded gp120 protein on the infected cells then sticks to uninfected T cells by binding to the CD4 molecule. The infected and uninfected cells can then fuse to produce multinucleate giant cells called *syncytia*. One HIV-infected T cell may eventually bind and fuse with up to 50 uninfected T cells. Shortly after syncytia formation occurs, the fused cells lose immune function and die.

The HIV infection results in a progressive decline in CD4 cell number. In a normal human, CD4 cells constitute about 70% of the total T-cell pool; in AIDS patients, the number of CD4 cells steadily decreases, and by the time opportunistic infections become established, CD4 cells may be almost absent (∞ Figure 24.22 and Figure 26.36●).

The dramatic reduction of CD4 T cells has serious health consequences in AIDS patients. As CD4 cells decline in number, cytokine production falls, leading to the gradual reduction of uninfected T cells as well as all other lymphocytes. In the process, both the humoral and cellular immune system in AIDS patients are effectively destroyed. Systemic infections by fungi and mycobacteria (Figure 26.30) point to a loss in T_H1 cellular immunity (∞ Section 22.8). Opportunistic viral and bacterial infections associated with AIDS indicate a decline in antibody production due to the loss of the T_H2 cells necessary to stimulate antibody production by B cells (∞ Section 22.8).

The progression of untreated HIV infection to AIDS follows a predictable pattern. During the clinical latency period, a very active infection is in progress. First, there is an intense immune response to HIV: About 1 billion virions are destroyed each day and HIV numbers drop. However, this means that HIV is replicating at a very high rate, and this replication results in the corresponding destruction of about 100 million CD4 T cells each day. Eventually, the immune response is simply overwhelmed, HIV levels increase, and the T cells are completely destroyed, crippling the immune response and

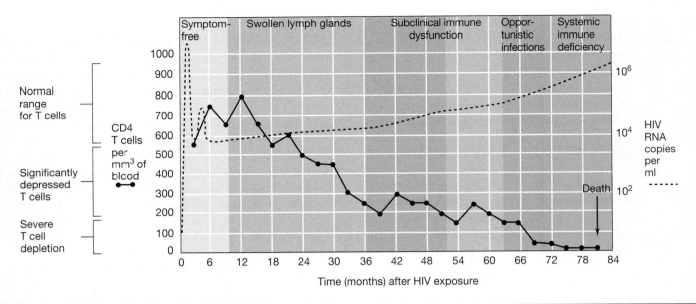

● **Figure 26.36   Decline of CD4 T lymphocytes and progress of HIV infection.** During the typical progression of untreated AIDS, there is a gradual loss in the number and functional ability of the CD4 T cells, while the *viral load*, measured as HIV-specific RNA copies per milliliter of blood, gradually increases after an initial decline.

allowing the emergence of opportunistic infections. The example in Figure 26.36 documents T-cell destruction and the increase in HIV over a typical time course for untreated HIV infection culminating in AIDS.

Diagnosis

HIV infection can be diagnosed by immunological means. Both radioimmunoassay (RIA) and enzyme-linked immunosorbent assay (ELISA) tests (☜ Section 24.10) have been developed for screening blood samples to detect anti-HIV antibodies. The ELISA test has proven particularly valuable for large-scale screening of donated blood to prevent transfusion-associated HIV. Statistics have shown that about 0.25% (2–3 per 1000) of all blood donated by volunteer donors in the United States tests HIV-positive in the ELISA assay. A positive HIV-ELISA test must be confirmed by a second procedure called an immunoblot (Western blot), a technique that combines the analytical tools of protein purification and immunology (☜ Section 24.11).

A number of other rapid tests are being developed and marketed to identify individuals infected with HIV. One test uses a single drop of patient blood and a single reagent. The reagent is a bioengineered antibody in which one binding site is directed to a red blood cell antigen, while the other site is directed to the gp41 HIV surface antigen. In a positive test, the bifunctional antibody cross-links the red blood cells to the HIV, resulting in a visible agglutination (☜ Section 24.8). In another test, saliva is used as a source of secretory antibody to HIV. The saliva is expelled onto a cartridge containing immobilized HIV antigens. A second antibody, reactive

with the bound antibody and conjugated to an enzyme, is then added. After addition of the enzyme substrate, a positive reaction shows a colored product, as in the ELISA methods (☜ Section 24.10). The rapid tests are designed to provide maximum convenience, speed (*minutes* instead of the hours or days required for ELISA or immunoblot), extended shelf life, portability, and ease of use and interpretation. However, in general, the rapid tests are not as sensitive or accurate as the standard HIV-ELISA and HIV-immunoblot tests (☜ Section 24.10 and 24.11).

All tests, no matter how sensitive or accurate, fail to detect HIV-positive individuals who have recently acquired the virus but have not yet made a detectable antibody response. This period may be more than 6 weeks after exposure to HIV. Despite this drawback, these tests ensure the general safety of the blood supply, and statistics indicate that the risk of contracting HIV through contaminated blood or blood products is now very low. Sexual promiscuity and group intravenous drug use are the major routes of HIV infection today (☜ Figure 25.8).

The presence of HIV antibodies in the blood is the usual indicator for HIV exposure and infection (☜ Sections 24.10 and 24.11). In addition, several laboratory tests have been developed, based on a virus-specific *reverse transcriptase-polymerase chain reaction (RT-PCR)* (☜ Section 18.5). These tests identify HIV RNA directly and quantitatively from blood samples (☜ Section 24.12). The RT-PCR estimates the number of viruses present in the blood, or the **viral load**. The RT-PCR test indicates the magnitude of HIV replication and correlates with the rate of CD4 T cell destruction. The CD4 test indicates the extent of HIV-induced immune damage already suffered, a direct indicator of the magnitude of destruction of the immune system. The RT-PCR test is not routinely used to screen for HIV because it is costly and technically demanding. After initial discovery of infection, however, the RT-PCR test is used to monitor progression of AIDS and the effectiveness of chemotherapy (Figure 26.37●).

The prognosis of an *untreated* HIV-infected individual is poor. Opportunistic pathogens or malignancies (Figures 26.32 and 26.33) eventually kill most AIDS patients. Long-term studies of AIDS patients indicate that the average person infected with HIV progresses through several stages of decreasing immune function, with CD4 cells dropping from a normal range of $600-1000/mm^3$ of blood to near zero over a period of 5–7 years (Figure 26.36). Although the *rate* of decline in immune function varies from one HIV-infected individual to another, it is rare for an HIV-positive individual to live for more than 10 years without some form of chemotherapeutic intervention.

Treatment

Several drugs have been identified that delay symptoms of AIDS and can significantly prolong the life of those infected with HIV (Table 26.4). Therapy is aimed at reducing the viral load of HIV-infected individuals to below detectable levels. The strategy used to accomplish this aim is called *highly active antiretroviral therapy (HAART)* and is accomplished by giving two or more antiretroviral drugs at once to inhibit the development of drug resistance. A *cure* for HIV infection has not been accomplished through drug therapy. Even in individuals who have no detectable viral load after drug treatment, a significant viral load returns if HAART therapy is interrupted or discontinued, or if multiple drug resistance develops.

Effective anti-HIV drugs fall into four categories. The first of these are a group of *nucleoside analogs* that act as *reverse transcriptase inhibitors. Reverse transcriptase* is the enzyme that converts the single-stranded RNA genetic information into complementary DNA (∞ Section 16.15). The oldest effective anti-HIV drug, *azido-*

(a)

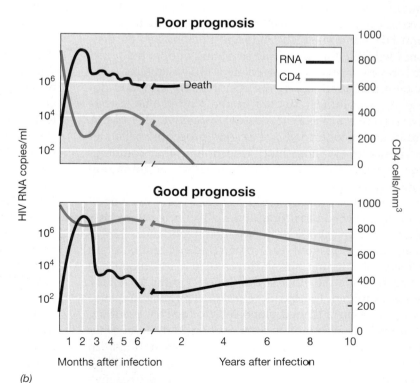

(b)

● **Figure 26.37 Monitoring of HIV load.** (a) Detection of HIV by the RT-PCR test (reverse transcription–polymerase chain reaction). The HIV copies obtained are compared quantitatively with DNA copies from a control template that is amplified in the same PCR amplification. HIV load is expressed as the number of HIV copies per milliliter of patient plasma. For a discussion of reverse transcription and the PCR technique, see Sections 16.15 and 7.9, respectively. (b) Time course for HIV infection as monitored by HIV RT-PCR. Progression of infection is estimated based on viral load at successive times after infection. CD4 T cell counts are measured in cells per cubic millimeter (see Figure 24.22). In the upper panel, a viral load of greater than 10^4 copies per milliliter correlates with below normal CD4 cell numbers (normal = $600-1500/mm^3$), indicating a poor prognosis and early death of the patient. In the lower panel, a viral load of less than 10^4 copies per milliliter correlates with normal CD4 cell numbers, indicating a good prognosis and extended survival of the patient. This test is used to monitor the course of HIV infection and is particularly useful for tracking the efficacy of drug treatment protocols. Data are adapted from the Centers for Disease Control and Prevention, Atlanta, GA, USA.

Table 26.4 Chemotherapeutic agents approved for HIV/AIDS treatment

Drug	Mechanism of activity
Azidothymidine (AZT, ZDV, or Zidovudine) (Figure 26.38a)	*Nucleoside analog*; reverse transcriptase inhibitor; nucleotide chain synthesis terminator; increases survival time and reduces incidence of opportunistic infection in AIDS patients; toxic to bone marrow cells; may be used in combination with other drugs in multiple drug treatment protocols.
Dideoxycytidine (ddC or zalcitabine) Dideoxyinosine (ddI or didanosine) Stavudine (d4T) Lamivudine (3TC)	*Nucleoside analogs*; reverse transcriptase inhibitors; mechanism of action and effects are the same as AZT; may have less toxicity than AZT in some patients; may be used in combination with other drugs in multiple drug treatment protocols.
Efavirenz Nevirapine (Figure 26.38b) Delavirdine	*Nonnucleoside reverse transcriptase inhibitors* (NNRTIs); bind directly to reverse transcriptase and disrupt the catalytic site; do not compete with nucleosides; may be used in combination with other drugs in multiple-drug treatment protocols.
Indinavir Nelfinavir Saquinavir (Figure 26.38c)	*Protease inhibitors*; computer-designed peptide analogs designed to bind to the active site of HIV protease, inhibiting processing of viral polypeptides and virus maturation; may be used in combination with other drugs in multiple drug treatment protocols.
Enfuvirtide	*Fusion inhibitor*; synthetic polypeptide that binds to the gp41 protein and inhibits the fusion of HIV membranes with host cell membranes.

thymidine (AZT), is an effective inhibitor of HIV replication because it closely resembles the nucleoside thymidine but lacks the correct attachment site for the next base in a replicating nucleotide chain, resulting in termination of the growing DNA chain. Thus, AZT and the other nucleoside analogs are **nucleoside reverse transcriptase inhibitors (NRTIs)**, and they stop HIV replication (Figure 26.38a●, Table 26.4). When these drugs are given to HIV-infected individuals, the result is a rapid decrease in viral load. However, within several weeks, drug resistant strains of HIV arise in each patient as a result of mutation and selection. Although this process seems very rapid, HIV replicates very quickly (see above) and as little as four single-nucleotide mutations result in resistance to a given nucleoside analog.

The second category of anti-HIV drugs are the **nonnucleoside reverse transcriptase inhibitors (NNRTIs)** (Figure 26.38b; Table 26.4). These compounds directly inhibit the action of reverse transcriptase by interacting with the protein and altering the conformation of the catalytic site. Unfortunately, a single mutation in the reverse transcriptase gene is often sufficient to reduce the effectiveness of these drugs.

Another category of anti-HIV drugs are the **protease inhibitors** (Figure 26.38c; Table 26.4). The protease inhibitors are computer-designed peptide analogs that inhibit processing of viral polypeptides by binding to the active site of the processing enzyme, HIV protease (∞ Section 20.13). This action inhibits virus maturation. However, as with the other enzyme-targeted chemotherapy strategy (reverse transcriptase; see above), a single mutation in the HIV protease gene is capable of rendering these drugs nonfunctional.

● **Figure 26.38 Examples of HIV/AIDS chemotherapeutic drugs.** (a) Azidothymidine (AZT), a nucleoside analog. The missing OH group on the 3′ carbon causes nucleotide chain elongation to terminate when this analog is incorporated, inhibiting virus replication. (b) Nevirapine, a non-nucleoside reverse transcriptase inhibitor, binds directly to the catalytic site, also inhibiting elongation of the nucleotide chain. (c) Saquinavir, a protease inhibitor, was designed by computer modeling to fit the active site of the HIV protease (∞ Section 20.13). Saquinavir is a peptide analog: The tan highlighted area shows the region analogous to peptide bonds. Blocking the activity of HIV protease prevents the processing of HIV proteins and maturation of the virus (∞ Figure 20.28).

A final category of anti-HIV drugs is represented by a single drug, *enfuvirtide*, a **fusion inhibitor** composed of a 36–amino acid synthetic peptide that acts by binding to the gp41 membrane protein of HIV (Table 26.4). Binding of the protein stops the conformational changes necessary for the fusion of viral and CD4 cell membranes. Enfuvirtide is prescribed to HIV-positive individuals who have developed antiretroviral drug resistance or increased viral load after conventional HAART therapy.

Because of the problems with drug resistance, a typical recommended protocol for treatment of an individual with established HIV infection includes at least one protease inhibitor *plus* a combination of two nucleoside analogs (Table 26.4). Multiple drug therapy reduces the possibility that a drug-resistant virus could emerge because the virus would need to develop resistance to three drugs simultaneously. This combination therapy is then monitored by RT-PCR to track changes in viral load. An effective protocol reduces viral load to nondetectable levels (less than 500 copies of HIV per milliliter of blood) within several days (Figure 26.37). The therapy is continued and monitored for viral load indefinitely. If the viral load again reaches detectable limits, the drug cocktail is changed because an increase in viral load indicates the emergence of a drug-resistant strain of HIV.

In addition to drug resistance, some of the antiviral drugs are toxic to the host. In many cases, nucleoside analogs are not well tolerated by patients, presumably because they interfere with host functions such as cell division (Table 26.4). In general, the NNRTIs and the protease inhibitors are better tolerated because they interfere with only virus-specific functions. However, drug resistance coupled with host toxicity are major problems in HIV therapy, and new chemotherapeutic agents and drug protocols are constantly being developed and tailored to the needs of individual patients.

AIDS Immunization

The genetic variability of HIV has thus far hampered the development of an AIDS vaccine. One vaccine strategy is to induce antibodies to the envelope protein, gp120. These antibodies would then block CD4-gp120 interactions (Figure 26.35) and thus block infection. A gp120 subunit vaccine is now in phase III human clinical trials, the step before general public use. The vaccine consists of two gp120 variants known to be predominant in the local population in Thailand, where it is being tested. Unfortunately, this approach probably will not be successful for a universal vaccine because the gene encoding gp120 mutates frequently, forming antigenic variants of the protein that are not recognized by antibodies made to a different form.

Clinical immunization trials are proceeding with *subunit vaccines* (◌ Section 31.8). In these vaccines, genes for several HIV envelope proteins have been engineered into vaccinia virus or adenovirus particles. Using these harmless viruses as expression vectors and vehicles for delivery of HIV antigens, several vaccines elicit a potent humoral and cellular immune response to HIV.

Other potential immunization candidates include *killed intact HIV*. These inactivated vaccines are restricted to use in HIV-infected individuals because inactivation procedures may not deactivate 100% of the HIV; it would be unethical to expose uninfected individuals to even a small risk of HIV infection. Next, some laboratories are exploring the possibilities of producing *live attenuated virus* for use as an immunizing agent. This strategy is bolstered by the finding that individuals infected with HIV-2, a related virus that causes a milder form of AIDS with a very long latent period, prevents infection with HIV-1, the strain responsible for severe AIDS. However, there are serious potential risks. For example, integrated virus could cause cancer, mutations might reactivate virulence, and so on.

In all, there are about 20 different AIDS immunization protocols undergoing development, but no immunization has yet been found to be protective and safe. Additionally, AIDS immunization would most probably not be useful for *treating* most patients that already have the disease because these individuals lack significant immune function and could not respond to a vaccine. Thus, despite considerable knowledge of the structural and functional mechanisms of HIV infection and a clear understanding of the AIDS disease process, there is no proven medical intervention strategy for the prevention or cure of HIV/AIDS. Public education and avoidance of high risk behavior remain the major tools used to prevent HIV/AIDS (see the Microbial Sidebar, Sexual Activity and AIDS).

 26.14 *Concept Check*

AIDS is now one of the most prevalent infectious diseases in the human population. HIV destroys the immune system, and opportunistic pathogens then kill the host. There is still no effective vaccine for HIV. However, several antiviral drugs slow the progress of AIDS. The only prevention for the spread of HIV infection is through avoidance of behavior such as intravenous drug use (needle sharing) and unsafe sexual practices.

◆ Review the definition of AIDS. What diagnostic features are shared by all AIDS patients?

◆ What is the current treatment for HIV infection? Is it effective? What are the side effects?

Microbial Sidebar ◆ Sexual Activity and AIDS

Sexual promiscuity has always been associated with sexually transmitted diseases, but the acquired immunodeficiency syndrome (AIDS) epidemic discussed in this chapter and elsewhere in this book has focused attention on the dangers of multiple sex partners and on the high risk associated with certain sex practices. HIV/AIDS is only one type of sexually transmitted disease. Others include gonorrhea, syphilis, herpes simplex, nonspecific urethritis (caused by *Chlamydia*), protozoal vaginitis (caused by *Trichomonas vaginalis*), fungal vaginitis (caused by *Candida albicans*), and venereal warts (caused by the human papilloma virus). Some of these sexually transmitted diseases have been associated with human society for all of recorded history. However, AIDS is unique. There are no drugs or immunizations to cure or prevent AIDS. Drugs used for treatment are costly and available only in developed countries, so 90% of the 50 million or more HIV-infected individuals do not have access to therapy. AIDS kills more than 2 million people every year, and this number will grow as people who have harbored and spread HIV for up to 10 years develop full-blown AIDS.

Because AIDS is linked to certain sex practices, prevention means avoidance of these sex practices. The United States Surgeon General has issued a report that makes specific recommendations that individuals can follow if they wish to reduce the likelihood of AIDS infection. Among the recommendations are

1. Avoid mouth contact with penis, vagina, or rectum.
2. Avoid all sexual activities that could cause cuts or tears in the linings of the rectum, vagina, or penis.
3. Avoid sexual activities with individuals from high-risk groups. These include prostitutes (both male and female), promiscuous homosexual men, bisexual individuals, and intravenous drug users.
4. If a person has had sex with a member of one of the high-risk groups, a blood test should be done to determine if infection with HIV has occurred. Keep in mind that the blood test should be repeated at intervals for a year or more, because of the lag time in the immune response. If the test is positive, the sexual partners of the HIV-positive individual must be protected by use of a condom during sexual intercourse.

AIDS is not just a disease of male homosexuals. In certain cultures, AIDS is as common in women as in men, and in the United States the fastest growing method of transmission is between heterosexual partners. The disease is linked to promiscuous sexual activities and other activities that involve exchange of body fluids, which include not only male homosexuality but also female prostitution and intravenous drug use.

Is it possible to have sex without incurring the risk of AIDS? Certain sex practices are inherently much safer than others. Safe sex practices include dry kissing (mouths closed), mutual masturbation (in the absence of breaks in the skin), and intercourse protected by a condom. Dangerous sex practices include wet kissing (mouths open), masturbation where breaks in the skin occur, oral sex (either male or female), and unprotected sexual intercourse (either vaginal or anal). The U.S. Surgeon General has recommended that if the health status of the partner is unknown, a condom be used for all sex practices in which exchange of body fluids occurs.

The AIDS epidemic has focused new attention on the condom (see photo). Condoms have always played two roles in sexual activity: disease protection and prevention of pregnancy. Although the best way to avoid AIDS is to avoid dangerous sex practices, a latex condom should be used when having sexual intercourse with any individual whose infection status is unknown. The U.S. Surgeon General strongly recommends the use of condoms for all extramarital sexual activity. In certain countries, widespread advertising campaigns very explicitly promote the use of condoms.

Statements and programs advocating monogamy, abstinence, avoidance of sexual activity outside of matrimony, and even warnings about the potential fatal consequences of HIV infection will not by themselves control the AIDS epidemic. Studies of other sexually transmitted diseases indicate that fear of disease is not sufficient to prevent risky sexual activities. The sex drive in some individuals is so strong that it will suppress the fear of even a disease like AIDS. Every individual must therefore take the responsibility for protecting himself or herself from infection.

For more information about AIDS, the Public Health Service has a toll-free telephone resource, the CDC National STD and AIDS Hotline at 800-342-2437. The Website is at *http://www.ashastd.org/nah/*. ■

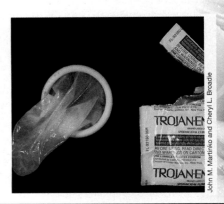

John M. Martinko and Cheryl L. Broadie

REVIEW QUESTIONS

1. Why do gram-*positive* bacteria cause respiratory diseases more frequently than gram-*negative* bacteria (∞ Section 26.1)?

2. What are the typical symptoms of a streptococcal respiratory infection? Why should streptococcal infections be treated promptly (∞ Section 26.2)?

3. Describe the causal agent and the symptoms of diphtheria. Are the disease symptoms due to infection or to toxicity? Explain. How is diphtheria prevented (∞ Section 26.3)?

4. Describe the causal agent and the symptoms of whooping cough. Describe the changes in vaccine technology that have led to safer pertussis vaccines (∞ Section 26.4).

5. Describe the process of infection by *Mycobacterium tuberculosis*. Does infection always lead to active tuberculosis? Why or why not? Why are individuals in the United States not vaccinated with the BCG vaccine (co Section 26.5)?

6. Describe the symptoms of meningococcemia and meningitis. How are these diseases treated? What is the prognosis for each (co Section 26.6)?

7. Compare and contrast measles, mumps, and rubella. Include in your discussion a description of the pathogen, major symptoms encountered, and any potential consequences of these infections. Why is it important that women be vaccinated against rubella *before* puberty (co Section 26.7)?

8. Why are colds and influenza such common respiratory diseases? Give at least two reasons for the high incidence of each of these diseases (co Section 26.8).

9. Distinguish between pathogenic staphylococci and those that are part of the normal flora (co Section 26.9).

10. Describe the evidence linking *Helicobacter pylori* to gastric ulcers. How would you treat an ulcer patient (co Section 26.10)?

11. Describe the major pathogenic hepatitis viruses. How are they related to one another? How are they spread (co Section 26.11)?

12. Why did the incidence of gonorrhea rise dramatically in the mid-1960s, while the incidence of syphilis actually decreased at the same time (co Section 26.12)?

13. For the sexually transmitted diseases chlamydia, herpes, and trichomoniasis, describe the methods of treating each infection. In each case, is the treatment an effective cure? Why or why not (co Section 26.13)?

14. Describe how human immunodeficiency virus (HIV) effectively shuts down both humoral immunity and cell-mediated immunity. Why are vaccines to HIV so difficult to develop (co Section 26.14)?

APPLICATION QUESTIONS

1. How can an epidemic of whooping cough be controlled? How can it be prevented? Since the incidence of this disease is no longer decreasing, apply your prevention methods to the current "mini-epidemic" (Section 26.4). Would your methods be cost-effective?

2. Why does tuberculosis often lead to a permanent reduction in lung capacity, whereas most other respiratory diseases cause only temporary respiratory problems?

3. Your college roommate goes home for the weekend, becomes extremely ill, and is diagnosed with bacterial meningitis at a local hospital. Because he was away, university officials are not aware of his illness. What should you do to protect yourself against meningitis? Should you notify university health officials?

4. Measles, mumps, and rubella were once very common childhood diseases. However, outbreaks of these diseases are now regarded as serious incidents requiring immediate attention from public health officials. Explain this shift in attitudes in the context of disease prevalence, availability of vaccines, and the potential health consequences of each disease.

5. Discuss the molecular biology of *antigenic shift* in influenza viruses and comment on the immunologic consequences for the host. Why does antigenic shift prevent the production of a single universally effective vaccine for influenza control? Next, compare antigenic *shift* to antigenic *drift*. Which mechanism is more important for the evolution of the influenza virus? Which causes the greatest antigenic change? Which creates the biggest problems for vaccine developers? Why?

6. Arrange the hepatitis viruses in order of disease severity, both in the short term and in the long term.

7. As the director of your dormitory's public health advisory group, you are charged to present information on chlamydia, herpes, and trichomoniasis, all STIs. Besides this textbook, where can you get reliable information about STIs? Present information on prevention, symptoms, and treatment for each STI. Will your program for each disease overlap? For each of the diseases, discuss the social, legal, and public health issues that must be considered for identifying the sexual partners of infected individuals.

27

ANIMAL-TRANSMITTED, ARTHROPOD-TRANSMITTED, AND SOILBORNE MICROBIAL DISEASES

The mosquito transmits malaria, the biggest infectious disease killer of all time, and West Nile virus, an emerging pathogen.

WORKING GLOSSARY

Ehrlichiosis one of a group of emerging tick-transmitted diseases caused by rickettsias of the *Ehrlichia* genus

Hantavirus pulmonary syndrome (HPS) an emerging acute viral disease characterized by respiratory pneumonia, obtained by transmission of hantavirus from rodents

Lyme disease an emerging tick-transmitted disease caused by the spirochete *Borrelia burgdorferi*

Malaria an insect-transmitted disease characterized by recurrent episodes of fever and anemia caused by the protozoan *Plasmodium* spp., usually transmitted between mammals through the bite of the *Anopheles* mosquito

Mycosis infection caused by a fungus

Plague an endemic disease in rodents caused by *Yersinia pestis* that is occasionally transferred to humans through the bite of a flea

Rabies a usually fatal neurological disease caused by the rabies virus that is usually transmitted by the bite or saliva of an infected animal

Rickettsia obligate intracellular parasite genus responsible for diseases including typhus, Rocky Mountain spotted fever, and ehrlichiosis

Rocky Mountain spotted fever a tick-transmitted disease caused by *Rickettsia rickettsii*, causing fever, headache, rash, and gastrointestinal symptoms

Sickle cell anemia a genetic trait that confers resistance to malaria but causes a reduction in the efficiency of red blood cells by reducing the oxygen-binding affinity of hemoglobin

Thalasemia a genetic trait that confers resistance to malaria but causes a reduction in the efficiency of red blood cells by altering a red blood cell enzyme

Tetanus a disease involving rigid paralysis of the voluntary muscles, caused by an exotoxin produced by *Clostridium tetani*

Typhus a louse-transmitted disease caused by *Rickettsia prowazekii*, causing fever, headache, weakness, rash, and damage to the central nervous system and internal organs

West Nile fever a neurological disease caused by West Nile virus and transmitted by mosquito from birds to humans

Zoonosis an animal disease transmitted to humans

Through the ages, insect-transmitted diseases such as plague and malaria have killed millions, changing the course of human history and evolution. Today, vectorborne diseases such as Lyme disease, hantavirus pulmonary syndrome, and West Nile fever are key emerging diseases, even in highly developed regions of the world. Soilborne diseases also present major problems because organisms in soil cannot be effectively controlled. Tetanus, for example, while totally preventable, remains a serious, often life-threatening disease.

All of these are diseases with *nonhuman* reservoirs. In this chapter, we look at animal-transmitted, vectorborne, and soilborne pathogens and the diseases they cause. The natural host for the *animal-transmitted pathogens* is a nonhuman vertebrate. When infected animal populations come in contact with humans, the result is often human infection. Vectorborne pathogens are spread to new hosts via the bite of an arthropod vector that last fed on an infected host. Humans are often accidental hosts in the life cycle of these pathogens, but they may also act as a disease reservoir, as in the case of malaria. Soilborne pathogens include a variety of fungi, as well as the members of the genus *Clostridium*.

ANIMAL-TRANSMITTED DISEASES

A **zoonosis** is an animal disease transmissible to humans, generally by direct contact, aerosols, or bites. Immunization and veterinary care prevent most infectious diseases in domesticated animals, preventing much zoonotic disease transfer to humans. However, *feral* (wild) animals cannot be immunized, nor do they receive veterinary care. Diseases in these populations often occur on a periodic, cyclic basis. In the next sections, we will look at two important examples of animal-transmitted diseases—rabies and hantavirus pulmonary syndrome.

27.1 Rabies

Rabies is one of a handful of zoonotic diseases that occurs primarily in animals but is spread to humans under certain conditions. The major reservoir of rabies in the United States is in wild animals, primarily raccoons, skunks, coyotes, foxes, and bats. However, a small but significant number of cases are also seen in domestic animals (Figure 27.1●).

Epidemiology

Rabies is a major preventable infectious disease in humans worldwide. Nevertheless, about 52,000 people die every year from rabies, primarily in developing countries where it is endemic in domestic animals such as dogs. Annually, about 1 million people worldwide receive rabies treatment for animal bites.

Rabies is caused by a member of the Rhabdovirus family (a negative-stranded RNA virus) (∞ Section 16.9) that infects cells in the central nervous system of most warm-blooded animals, almost invariably leading to death if not treated. The virus, present in the saliva of rabid animals, enters the body through a wound from a bite or through contamination of mucous membranes by the infected saliva.

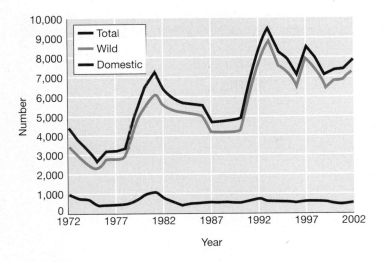

● **Figure 27.1 Rabies cases in wild and domestic animals in the United States.** Since 1990, there have been 30 human rabies cases and eight deaths. Rabies is endemic in wild animal populations, especially in raccoons in the eastern United States. The periodic rise in reported cases following several years of decline is the result of fluctuations in the host population due to deaths from rabies. High case numbers indicate a large infected population. The population is decimated by the infection, and for several years the lower host numbers prevent rabies spread, and case numbers are reduced. As the population recovers, rabies reemerges, case numbers increase, and the host population is again decimated, continuing the cycle. Over 500 cases of rabies occur annually in domestic animals, nearly all acquired from contact with wild animals. Data are from the Centers for Disease Control and Prevention, Atlanta, GA, USA.

Rabies virus multiplies at the site of inoculation and then travels to the central nervous system. The incubation period for the onset of symptoms is highly variable, depending on the animal, the size, location, and depth of the wound, and the number of viral particles transmitted in the bite. In dogs, the incubation period averages 10–14 days. In humans, nine months or more may pass before rabies symptoms become apparent.

The virus proliferates in the brain (especially in the thalamus and hypothalamus), leading to fever, excitation, dilation of the pupils, excessive salivation, and anxiety. A fear of swallowing (hydrophobia) develops from uncontrollable spasms of the throat muscles. Death eventually results from respiratory paralysis. In humans, an *untreated* rabies infection progressing to the symptomatic stage is nearly always fatal.

Diagnosis, Treatment, and Prevention

Rabies is diagnosed in the laboratory by examining tissue samples. Fluorescent antibody tests (∞ Section 24.9) or immunoperoxidase tests using monoclonal antibodies that recognize rabies virus-infected brain or corneal tissue are used for confirming a clinical diagnosis of rabies, either in a potentially rabid animal or in postmortem examination of a human case. In addition, characteristic virus inclusion bodies in the cytoplasm of nerve cells,

called *Negri bodies* (Figure 27.2●), obtained by biopsy or postmortem sampling, are taken as confirmation of rabies. Reverse transcriptase–PCR testing and sequencing can also be done to confirm the presence a of particular rabies strain in a clinical specimen.

Because of the lethal nature of rabies, any contact with potentially rabid animals must be taken seriously. A wild animal suspected of being rabid should be captured, sacrificed, and immediately examined for evidence of rabies. If a domestic animal, generally a dog, cat, or ferret, bites a human, especially if the bite is unprovoked, the animal is typically held 10 days to check for clinical signs of rabies. If the animal is wild, exhibits rabies symptoms, or a determination cannot be made after 10 days, the patient will be passively immunized with rabies immune globulin (purified anti-rabies virus antibodies obtained from a hyperimmune individual) (∞ Section 22.13), injected at both the site of the bite and intramuscularly. The patient will also be immunized with an inactivated rabies virus preparation. A summary of guidelines for treating possible human exposure to rabies is shown in Table 27.1. Because of the very slow progression of rabies in humans, this combination immune therapy is nearly 100% effective, stopping the onset of the active disease.

Rabies is prevented largely through immunization. Inactivated rabies vaccines are used in the United States for both human and domestic animal immunizations, and a variety of inactivated and attenuated virus preparations are also used worldwide. Because of the long incubation period, passive and active immunization of potentially exposed individuals is sufficient to prevent disease. Therefore, prophylactic rabies immunization is

CDC/Mekonnen Fekadu/PHIL

● **Figure 27.2 Tissue section from the brain of a human rabies victim, stained with hematoxylin and eosin.** The rabies virus causes characteristic cytoplasmic inclusions known as Negri bodies. Here they are seen as dark-stained, sharply differentiated, roughly spherical masses of about 2–10 μm in diameter, distributed throughout the cytoplasm. The Negri bodies contain rabies virus antigens.

Table 27.1	Guidelines for treating possible human exposure to rabies virus

Unprovoked bite by a domestic animal

Animal suspected of rabies
1. Sacrifice animal and test for rabies.
2. Begin treatment of human immediately.[a]

Animal not suspected of rabies
1. Hold for 10 days. If no symptoms, do not treat human.
2. If symptoms develop, treat human immediately.[a]

Bite by wild carnivore (for example, skunk, bat, fox, raccoon, coyote)
Regard animal as rabid
1. Sacrifice animal and test for rabies.
2. Begin treatment of human immediately.[a]

Bite by wild rodent, squirrel, livestock, rabbit
Consult local or state public health officials about possible recent cases of rabies transmitted by these animals (these animals rarely transmit rabies). If no reports, do not treat human.

[a] All bites should be thoroughly cleansed with soap and water. Treatment is generally a combination of rabies immunoglobulin and human diploid cell rabies vaccine (five injections intramuscularly).

generally not recommended for humans, except for individuals at high risk, such as veterinarians and animal control personnel.

This rabies prevention strategy has been extremely successful, and fewer than three cases of human rabies are reported in the United States each year, nearly always the result of bites by wild animals. Since domestic animals often have exposure to wild animals, all dogs and cats should be vaccinated beginning at 3 months of age, and booster inoculations should be given yearly or triannually. Other domestic animals, including large farm animals, are also often immunized with rabies vaccines. However, the key to effective rabies prevention and possible eradication, at least in the United States, lies in control of the disease in the large rabies virus reservoir in wild animals (Figure 27.1). If all or even most members of the disease reservoir are immune, the disease can be stopped and possibly even eradicated. Currently, subunit vaccines consisting of rabies virus genes that encode rabies coat proteins expressed in vaccinia virus (➤ Section 31.8) are available. Because effective doses can be given orally, subunit vaccines can be included in food "baits" and used to immunize local populations of susceptible wild animals, reducing the incidence and spread of rabies. Such vaccines are a completely safe means of controlling rabies in the wild animal reservoir and could lead to eventual eradication of the disease.

 27.1 Concept Check

Rabies occurs primarily in wild animals and is an important endemic zoonotic disease in developing countries. In the U.S. rabies can be transmitted from the wild animal reservoir to domestic animals or, very rarely, to humans. Vaccination of dogs and cats is important for the control of rabies.

◆ What is the procedure for treating a human bitten by an animal *if* the animal cannot be found?

◆ What advantages might an oral vaccine have over a parenteral vaccine for rabies control in wild animals?

27.2	Hantavirus Pulmonary Syndrome

Hantavirus is the cause of an acute respiratory syndrome, **hantavirus pulmonary syndrome** (HPS). Hantavirus is named for Hantaan, Korea, the site of a hemorrhagic fever outbreak where the virus was first recognized as a human pathogen. The 1993 hantavirus outbreak in the United States occurred in the Four Corners region of Arizona, Colorado, New Mexico, and Utah, and was eventually traced to the deer mouse (*Peromyscus maniculatis*) population in the area. The outbreak caused 32 deaths in 53 infected adults, and it underscored the potential danger of outbreaks due to diseases that are directly transmitted from a variety of different animal reservoirs, sometimes under new or unusual circumstances.

Hantaviruses are related to hemorrhagic fever viruses such as Lassa fever virus and Ebola virus (➤ Section 25.11), and all are occasionally transmitted to humans from animal reservoirs. The genus *Hantavirus* is a member of the Bunyaviridae, a family of enveloped single-stranded RNA viruses (➤ Section 16.8) (Figure 27.3●). The family includes a number of viruses that cause either *hantavirus pulmonary syndrome (HPS)* or *hemorrhagic fever with renal syndrome (HFRS)*.

Epidemiology

Hantaviruses are usually found in rodents, including mice and rats of several species, lemmings, and voles, and are occasionally found in other animals. The HFRS strains are more common in the Eastern Hemisphere, and have been implicated in a number of HFRS outbreaks in recent years. Up to 200,000 cases per year are recognized, chiefly in China, Korea, and Russia. The HPS strains are more prevalent in the Western Hemisphere, and continued investigation of the ecology of these viruses is likely to uncover a number of other pathogenic strains.

Hantaviruses are most commonly transmitted by inhalation of virus-contaminated rodent excreta. Hu-

(a)

(b)

● **Figure 27.3   Hantavirus.** (a) An electron micrograph of the Sin Nombre hantavirus. The arrow indicates one of several virions. The virus is approximately 0.1 μm in diameter. (b) Immunostaining of Andes hantavirus antigens in alveolar macrophages. Each granular dark blue stained area indicates cellular infection of an individual macrophage (approximately 15 μm in diameter). Hantaviruses belong to the Bunyaviridae and, like many phylogenetically similar viruses (for example, Rift Valley virus and Ebola virus), cause hemorrhagic fevers with very high human mortality rates.

mans seem to be an accidental host and are infected only when they come into contact with rodents and their feces. For example, the individuals who acquired HPS in the Four Corners outbreak had been exposed to mice or their droppings, the result of a warm winter and an unusual increase in numbers of rodents in 1993 (∞ Section 25.11). The *aerosol route* of infection is most common, with most aerosols being in the form of dust generated from mouse droppings or dried urine. However, there are rare reports of person-to-person transmission, as well as a few incidents where HPS or HFRS was spread by a rodent bite.

Pathology, Diagnosis, Treatment, and Prevention

Hantavirus pulmonary syndrome is characterized by a sudden onset of fever, myalgia (muscle pain), thrombocytopenia (reduction in the number of blood platelets), leukocytosis (an increase in the number of circulating leukocytes), and pulmonary capillary leakage. Death occurs in a matter of several days in about 50% of cases, usually due to shock and cardiac complications precipitated by pulmonary edema (leakage of fluid into the lungs, causing suffocation). These symptoms are typical of the Sin Nombre hantavirus, which caused the Four Corners outbreak, but a variety of other symptoms may be evident, depending on the strain of virus causing the disease. For example, the Bayou strain common in rodents in the southeastern United States also causes kidney failure.

If hantavirus from candidate infections can be grown in tissue culture (∞ Section 9.3), the strain can be identified by serological techniques including a virus plaque-reduction neutralization assay. In this assay, patient serum is tested for antibodies that inhibit the formation of viral plaques in tissue culture (∞

Section 9.3). More commonly, ELISAs (enzyme-linked immunoassays) are performed on patient blood to identify antibodies, indicating exposure and an immune response (∞ Section 24.10). The presence of the viral genome, indicating infection, is detected based on a polymerase chain reaction (PCR) assay using patient tissue or blood specimens (∞ Section 24.12).

There is no virus-specific treatment or vaccine for any of the hantaviruses. However, the disease can be prevented by avoiding contact with rodents and rodent habitat, since the mode of transmission is usually through exposure to rodent excreta. Reduction in exposure can be accomplished by destroying mouse habitat, restricting food supplies (for example, keeping food in sealed containers), and aggressive rodent extermination measures. The long-term prognosis for disease eradication is poor because a considerable percentage of the rodent population in a given geographical area is generally infected with the local hantavirus strain. For example, retrospective serological testing of the deer mice in the Four Corners area in 1993 indicated that 30% of the local mouse population carried the Sin Nombre hantavirus.

⬤ 27.2 Concept Check

Hantaviruses occur worldwide in rodent populations and cause serious diseases such as hantavirus pulmonary syndrome (HPS) when accidentally spread to humans. In the United States, hantavirus infections were not recognized until 1993.

◆ Why were hantaviruses not recognized as a major public health problem until 1993 in the United States?

◆ Describe the spread of hantaviruses to humans. What are some effective measures for preventing infection by hantaviruses?

ARTHROPOD-TRANSMITTED DISEASES

Pathogens can be spread to hosts from the bite of a pathogen-infected arthropod vector. In many cases, such as in the rickettsial illnesses, humans are accidental hosts for the pathogen. However, infected humans can also play a critical role in the life cycle of the pathogen, as is the case for malaria.

27.3 Rickettsial Diseases

The **rickettsias** are small bacteria that have a strictly intracellular existence in vertebrates, usually in mammals, and are also associated at some point in their natural cycle with blood-sucking arthropods such as fleas, lice, or ticks. We discussed the biology of rickettsias in Section 12.13. Rickettsias cause a variety of diseases in humans and animals, of which the most important are typhus fever, Rocky Mountain spotted fever, and ehrlichiosis. Rickettsias take their name from Howard Ricketts, a scientist at the University of Chicago who first provided evidence for their existence and who died from infection with the rickettsia that causes typhus fever, *Rickettsia prowazekii*. Rickettsias have not been cultured in artificial media but can be cultured in laboratory animals, lice, mammalian tissue culture cells, and the yolk sac of chick embryos. In animals, growth takes place primarily in phagocytes (Section 22.1).

Although the rickettsias have not been grown in pure culture, the 1.1 Mb genome of *Rickettsia prowazekii* has been sequenced (Section 15.5). Based on homology with other genomic sequences, these intracellular parasites are closely related to human mitochondria. Like the mitochondria, the rickettsial genome has been reduced in size to contain a set of genes tailored for intracellular dependency. The rickettsias do not have many of the genes necessary for independent energy metabolism and structural biosynthesis. The rickettsial genome also contains virulence genes closely related to the *virB* operon of the plant pathogen *Agrobacterium tumefaciens* (Section 19.21). This operon encodes components of virulence factors involved in DNA transfer and protein export systems. Thus, the genomic sequence of *R. prowazekii* provides evidence for the intracellular dependence and the virulence of these pathogens.

Rickettsias are divided into three groups, based loosely on the clinical diseases they produce. The groups are (1) the *typhus group*, typified by *Rickettsia prowazekii*; (2) the *spotted fever group*, typified by *Rickettsia rickettsii*; and (3) the *ehrlichiosis group*, characterized by *Ehrlichia chaffeensis*. Here, we examine one example of a pathogen from each group.

The Typhus Group: *Rickettsia prowazekii*

Typhus is caused by *Rickettsia prowazekii*. Epidemic typhus is transmitted from human to human by the common body or head louse. Humans are the only known mammalian host for typhus. During World War I, an epidemic of typhus spread throughout eastern Europe and caused almost 3 million deaths. Typhus has frequently been a problem among military troops during wartime. Because of the unsanitary, cramped conditions characteristic of wartime military infantry operations, lice are spread easily among soldiers and typhus is spread in epidemic proportions. Up until World War II, typhus caused more military deaths than combat.

Cells of *R. prowazekii* are introduced through the skin when the puncture caused by the louse bite becomes contaminated with louse feces, the major source of rickettsial cells. During an incubation period of 1–3 weeks, the organism multiplies inside cells lining the small blood vessels. Symptoms of typhus (fever, headache, and general body weakness) then begin to appear. Five to nine days later a characteristic *rash* is observed in the armpits and generally spreads over the body *except* for the face, palms of the hands, and soles of the feet. Complications from untreated typhus involve damage to the central nervous system, lungs, kidneys, and heart. Epidemic typhus has a mortality rate of 6–30%. Tetracycline and chloramphenicol are most commonly used to control *R. prowazekii*.

Rickettsia typhi, the organism that causes murine typhus, is another important pathogen in the typhus group.

The Spotted Fever Group: *Rickettsia rickettsii*

Rocky Mountain spotted fever was first recognized in the western United States in about 1900 but is more prevalent today in the southeastern United States. Rocky Mountain spotted fever is caused by *Rickettsia rickettsii* and is transmitted to humans by various ticks, most commonly the dog and wood ticks. The incidence of Rocky Mountain spotted fever is relatively low but between 500 and 1000 people acquire the disease every year. Humans acquire the pathogen from tick fecal matter, which is injected into the body during a bite, or by rubbing infectious material into the skin by scratching. Cells of *R. rickettsii*, unlike other rickettsias, grow within the nucleus of the host cell as well as in host cell cytoplasm (Figures 27.4a● and 27.4b). Following an incubation period of 3–12 days, an abrupt onset of symptoms occurs, including fever and a severe headache. Within 3 to 5 days, a rash occurs on the whole body (Figure 27.4c). Gastrointestinal problems such as diarrhea and vomiting are usually observed as well, and the clinical symptoms of Rocky Mountain spotted fever may persist for over 2 weeks if the disease is untreated. Tetracycline or chloramphenicol generally promotes a prompt recovery from Rocky Mountain spotted fever if administered early in the course of the infection, but even treated patients suffer about 5% mortality. There is 30% mortality in untreated cases.

(a)

(b)

(c)

● **Figure 27.4** *Rickettsia rickettsii*, the causative agent of **Rocky Mountain spotted fever.** (a) Cells of *R. rickettsii*, growing in the cytoplasm and nucleus of tick hemocytes. Individual cells are about 0.4 μm in diameter. (b) Cells of *R. rickettsii* in a granular hemocyte of an infected wood tick, *Dermacentor andersoni*. Transmission electron micrograph. (c) Rash of the disease on the feet. The appearance of a rash covering the whole body is indicative of Rocky Mountain spotted fever and helps distinguish this disease clinically from typhus, in which the rash does not cover the whole body. Rickettsias are small *Bacteria* that are dependent on intracellular growth in host mammalian cells. Their metabolism relies on host functions and is discussed in Section 12.13. The 1.1 Mbp genome of *Rickettsia rickettsii* has been sequenced (∞Section 15.5).

The Ehrlichiosis Group: *Ehrlichia*

The genus *Ehrlichia* (∞ Section 12.13) is responsible for two emerging tickborne diseases in the United States, human monocytic **ehrlichiosis** (HME) and **human granulocytic ehrlichiosis (HGE)**. The rickettsias that cause the diseases are *Ehrlichia chaffeensis* and an organism similar or identical to *Ehrlichia equi*, respectively.

The onset of these clinically indistinguishable ehrlichioses is characterized by flulike symptoms that may include fever, headache, malaise, and frequently leukopenia (decreased number of leukocytes) or thrombocytopenia. Laboratory findings frequently document changes in liver function, characterized by an increase in the enzyme hepatic transaminase. In addition, peripheral blood leukocytes have visible inclusions of *Ehrlichia* cells (Figure 27.5●). The symptoms, except for the inclusions, are similar to other rickettsioses, but the *Ehrlichia* genus is antigenically distinct from members of the other major rickettsial groups. Infections range from subclinical to fatal. Long-term complications for progressive untreated cases may include respiratory and renal insufficiency and serious neurological involvement.

Diagnosis is based on an indirect fluorescence antibody assay (∞ Section 24.9) of patient serum and also

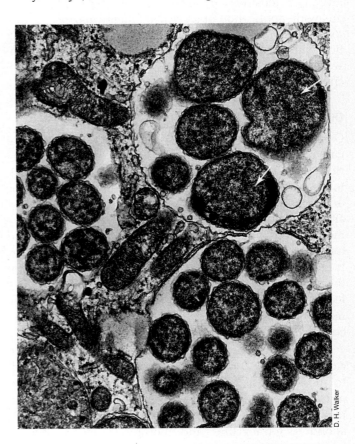

● **Figure 27.5** *Ehrlichia chaffeensis*, the causative agent of **human monocytic ehrlichiosis (HME).** The electron micrograph shows inclusions in a human monocyte which contains large numbers of *E. chaffeensis*. The arrows indicate two of the many bacteria in each inclusion. *E. chaffeensis* cells range from about 300–900 nm in diameter.

on polymerase chain reaction (PCR) tests of whole blood or serum to detect the presence of *Ehrlichia* DNA (∞ Section 24.12).

Ehrlichiosis is spread by the bites of infected ticks. The mammalian reservoirs include deer and possibly rodents, in addition to the human hosts. Retrospective serological analyses in areas with relatively high incidence of tickborne disease indicate that HGE may be a more prevalent disease than Rocky Mountain spotted fever. Many infections are not properly identified because of the variable nature of the symptoms. However, since 1998, ehrlichiosis has been a reportable disease in the United States (∞ Section 25.9). On average, about 500 cases are reported each year. HGE comprises about 60% of total cases while HME is responsible for 40%. Ehrlichiosis will be reported more frequently as physicians become more familiar with this emerging tickborne disease.

As with other tickborne illnesses, outdoor activities in tick-infested habitat are the major predisposing factors in acquiring ehrlichiosis. Golfers and hikers are particularly prone to infection. Prevention of ehrlichiosis involves reducing exposure to ticks and tick bites by avoiding tick habitat, wearing tick-proof clothing, and applying appropriate insect repellents such as those containing diethyl-*m*-toluamide (DEET). At the community level, tick densities can be successfully reduced through areawide application of *acaricides* (chemicals specifically toxic for ticks and related arthropods) and removal of tick habitat such as leaves and brush. Doxycycline, a semisynthetic tetracycline derivative, is the antibiotic of choice for the treatment of ehrlichiosis.

Other Rickettsial Diseases

Q fever is a pneumonia-like infection caused by an obligate intracellular parasite, *Coxiella burnetii*, related to the rickettsias (∞ Section 12.13). Although not transmitted to *humans* directly by an insect bite, the agent of Q fever is transmitted to *animals* by insect bites, and various arthropod species serve as a reservoir of infection. Domestic animals generally have inapparent infections, but may shed large quantities of *C. burnetii* cells in their urine, feces, milk, and other body fluids. Contact with the animals or animal products serves as a source of infection for humans. The resulting influenza-like illness may progress to include prolonged fever, headache, chills, chest pains, pneumonia, and endocarditis.

Laboratory diagnosis of infection with *C. burnetti* can be made by a variety of immunologic tests designed to measure host antibodies to the pathogen. An immunofluorescence antibody test (IFA) is the serological test of choice (∞ Section 24.9). *C. burnetii* infections respond to tetracycline, and therapy is usually begun quickly in any suspected human case of Q fever in order to prevent heart damage. Finally, Q fever is one of the infectious diseases that has been studied as a possible agent for biological warfare (∞ Section 25.12).

Scrub typhus, or *tsutsugamushi disease*, is restricted to Asia, the Indian subcontinent, and Australia, and is caused by *Orientia tsutsugamushi*. Although the disease is similar to typhus, it is transmitted by *mites* to its normal rodent hosts.

Diagnosis and Control

In the past, rickettsial infections have been difficult to diagnose because the characteristic rash associated with many rickettsial diseases may be mistaken for measles, scarlet fever, or adverse drug reactions. Clinical confirmation of rickettsial diseases has now been greatly aided by the introduction of specific immunological and molecular biology reagents. These include antibody-based tests that detect rickettsial surface antigens by latex bead agglutination assays, immunofluorescence assays, ELISA analyses (∞ Sections 24.8, 24.9, and 24.10), and PCR assays (∞ Section 24.12). Control of most rickettsial diseases requires control of the vectors: lice, fleas, and ticks. For humans traveling in wooded or grassy areas, the use of insect repellants containing diethyl-*m*-toluamide (DEET) usually prevents tick attachment. Firmly attached ticks should be removed gently with forceps, care being taken to remove all the mouth parts. A solvent such as ethanol applied to a tick with a saturated swab usually expedites removal. Although a vaccine is available for the prevention of typhus, the few cases reported do not warrant its general administration in the United States. No vaccines are currently available for the prevention of Rocky Mountain spotted fever or ehrlichiosis.

 27.3 Concept Check

Rickettsias are obligate intracellular parasitic *Bacteria* that are transmitted by arthropods. Most rickettsial infections are readily controlled by antibiotic therapy, but prompt recognition and diagnosis of these diseases is still difficult.

◆ What are the arthropod vectors and animal hosts for typhus, Rocky Mountain spotted fever, and ehrlichiosis?

◆ What precautions can be taken to prevent rickettsial infections?

27.4 Lyme Disease

Lyme disease is an emerging tickborne disease that affects humans and other animals. Lyme disease was named for Old Lyme, Connecticut, where cases were first recognized, and is the most prevalent tickborne disease in the United States. Lyme disease is caused by a spirochete, *Borrelia burgdorferi* (Figure 27.6●; ∞ Section 12.33), which is spread primarily by the deer tick, *Ixodes scapularis* (Figure 27.7●). The ticks that carry *B. burgdorferi* feed on the blood of birds, domesticated animals, various wild animals, and occasionally humans.

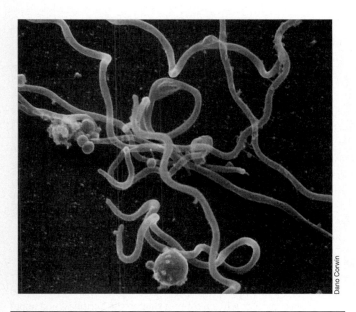

● **Figure 27.6 Electron micrograph of the Lyme spirochete,**
Borrelia burgdorferi. The diameter of a single cell is approximately
0.4 μm.

Epidemiology

Deer and the white-footed field mouse are prime mammalian reservoirs of *B. burgdorferi* in the northeastern United States. However, in other parts of the country, different species of rodents and ticks are involved in the transmission of Lyme disease. In the western United States, *Ixodes pacificus* and the wood rat are common vectors and hosts.

Lyme disease has also been identified in Europe and Asia. In Europe, the tick vector is *Ixodes ricinus*, which may also harbor *Borrelia garinii*, another organism that causes Lyme diseaselike symptoms. In Asian countries, *Borrelia afzelii* is transmitted by *Ixodes persulcatus* and causes Lyme disease. In all cases, different local rodent reservoirs have been identified. Thus, Lyme disease seems to

● **Figure 27.7 Deer ticks (Ixodes scapularis), the major vectors**
of Lyme disease. Left to right, male and female adult ticks, nymph,
and larval forms. The length of an adult female is about 3 mm. All forms
feed on humans and are capable of transmitting *Borrelia burgdorferi*.

have a broad geographic distribution and is transmitted to humans by closely related *Borrelia* and tick species that have a variety of rodent and other mammalian reservoirs.

The deer tick and other members of the genus are smaller than many other species of ticks, making them easy to overlook (Figure 27.7). Unlike the case with other tickborne diseases, a very high percentage of deer ticks (up to 50% in certain regions of the Northeast) carry *B. burgdorferi* cells. Extended contact with the infected tick vector increases the probability of disease transmission.

In the United States most cases of Lyme disease have been reported from the Northeast and upper Midwest, but cases have been observed in nearly every state, and Lyme disease is spreading west and south. Figure 27.8● shows the spread of Lyme disease across the continental United States and the rapid rise in the total number of annual cases.

Pathology

Cells of *Borrelia burgdorferi* are transmitted to humans while the tick is obtaining a blood meal (Figure 27.9*a*●). A systemic infection develops, leading to the main symptoms of Lyme disease, which include headache, backache, chills, and fatigue. In about 75% of all cases, a large rash, known as *erythema migrans (EM)*, is observed at the site of the tick bite (Figure 27.9*b*). At this time, Lyme disease is treatable with tetracycline or penicillin. Untreated Lyme disease may progress to a chronic stage beginning weeks to months after the initial tick bite. Chronic Lyme disease is characterized by arthritis in 40–60% of patients. Neurological involvement such as palsy, weakness in the limbs, and facial ticks occurs in 15–20% of patients. Cardiac damage occurs in about 8% of all cases. In untreated cases, cells of *B. burgdorferi* infecting the central nervous system may lie dormant for long periods before eliciting a variety of additional chronic symptoms, including visual disturbances, facial paralysis, and seizures.

No toxins or other virulence factors have yet been identified in Lyme disease pathogenesis. In many respects the latent symptoms of Lyme disease resemble those of syphilis, caused by a different spirochete, *Treponema pallidum*. Indeed, some of the neurological symptoms of Lyme disease resemble those of chronic syphilis (∞ Section 26.12). However, unlike syphilis, Lyme disease is not spread by sexual intercourse or other types of human contact. Small numbers of *Borrelia burgdorferi* cells are shed in the urine of infected individuals, and there is some indication that Lyme disease can spread from domestic animal populations, particularly cattle, by infected urine.

Diagnosis

Serological tests have been developed for detection of antibodies to *Borrelia burgdorferi*. Antibodies appear about 4–6 weeks after infection and can be detected by an indirect enzyme-linked immunosorbent assay (ELISA) or a

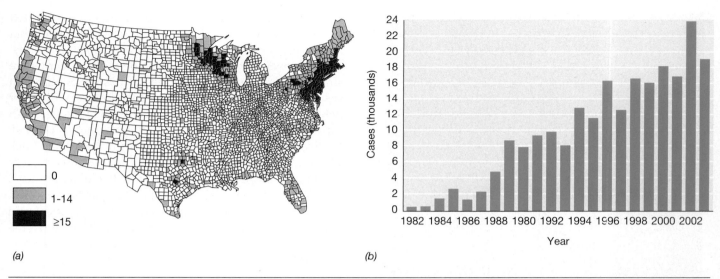

(a)

(b)

● **Figure 27.8   Lyme disease in the United States.** (a) Incidence of Lyme disease in the United States in 2002. Each county that reported Lyme disease in 2002 is shaded, with counties reporting more than 15 cases shown in red. Lyme disease, while most prevalent in the Northeast and upper Midwest, is found throughout the United States. (b) Number of reported cases of Lyme disease by year in the United States. In 2002, there were 23,763 cases, more than in any previous year. Lyme disease is reported through the National Notifiable Diseases Surveillance System of the Centers for Disease Control and Prevention.

(a)

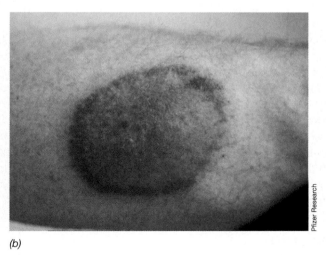

(b)

● **Figure 27.9   Lyme disease infection.** (a) Deer tick obtaining a blood meal from a human. (b) Characteristic circular rash associated with Lyme disease. The rash, known as *erythema migrans* (EM), typically starts at the site of the bite and grows in a circular fashion over a period of several days. This typical EM example is about 5 cm in diameter.

fluorescent antibody assay (⟳ Sections 24.10 and 24.9). However, the most definitive serological test for Lyme disease is a Western blot (⟳ Section 24.11). Because antibodies to the Lyme spirochete antigens persist for years after infection, the presence of antibodies does not necessarily confer immunity to the disease or indicate recent infection.

A polymerase chain reaction (PCR) assay (⟳ Section 24.12) has also been developed for the detection of *Borrelia burgdorferi* from many body fluids and tissues. While they are rapid and sensitive, the PCR methods cannot differentiate between live *B. burgdorferi* in active disease and dead *B. burgdorferi* found in treated or inactive disease. *B. burgdorferi* can also be cultured from nearly 80% of the original erythema migrans lesions (Figure 27.9*b*), but culture is usually not done because of the long latent period before the organism grows on a highly specialized medium.

In the end, Lyme disease is usually diagnosed clinically. If a patient has Lyme disease symptoms and has had a recent tick exposure, especially if followed by erythema migrans, then a presumptive diagnosis of Lyme disease is made and antibiotic treatment is initiated.

Prevention and Treatment

Prevention of Lyme disease requires proper precautions to prevent tick attachment. In tick-infested areas such as woods, tall grass, and brush, it is advisable to wear protective clothing such as shoes, long pants, and a long-sleeved shirt with a snug collar and cuffs. Tucking the pants into tight-fitting socks worn with boots forms an effective barrier to tick attachment. After spending time in a tick-infested environment, individuals should check themselves carefully for ticks and gently remove any attached ticks (including the head). Insect repellants containing diethyl-*m*-toluamide (DEET) are very effective if applied to both skin and clothing. An effective human Lyme disease vaccine was withdrawn from the market in 2001 due to poor sales and possible adverse reactions. Lyme disease vaccines are available for veterinary use.

Treatment of acute Lyme disease can be with doxycycline (a tetracycline derivative, ∞ Section 20.9) or with amoxicillin (a β-lactam antibiotic, ∞ Section 20.8) for 20–30 days. Established Lyme arthritis or other symptoms indicating chronic *B. burgdorferi* infection are treated with high doses of penicillin or ceftriaxone, a β-lactam antibiotic that crosses the blood-brain barrier and attacks spirochetes residing in the central nervous system. Long-term Lyme arthritis is treated with doxycycline or amoxicillin plus probenicid, an agent that helps retain high serum levels of the antibiotic, for 30 days or more.

 27.4 Concept Check

Lyme disease is now the most prevalent arthropod-borne disease in the United States. It is transmitted from several mammalian host vectors to humans by ticks. Prevention and treatment of Lyme disease are straightforward, but accurate and timely diagnosis is a major problem.

◆ What are the primary symptoms of Lyme disease?

◆ What antibiotics can be used to treat Lyme disease?

27.5 Malaria

Malaria is a disease caused by a protozoan, a member of the Sporozoa group (∞ Section 14.10). The malaria parasite is one of the most important human pathogens and has played an important role in the development and spread of human culture. As we will see, malaria has even affected human genetics and evolution. Malaria is still a significant human disease even though there are several effective treatments available. Over *100 million* people worldwide have malaria, and each year over 1 million of these will die (∞ Section 25.1). The mammalian reservoir for malaria is humans. Four species of sporozoa infect humans. The most widespread is *Plasmodium vivax* and the most serious is *Plasmodium falciparum*. This parasite carries out part of its life cycle in human reservoir and part in the mosquito vector, which spreads the parasite from person to person. Only female mosquitoes of the genus *Anopheles* transmit malaria (see Figure 27.12*a*).

Epidemiology

Anopheles mosquitoes inhabit warmer parts of the world; therefore, malaria occurs predominantly in the tropics and subtropics. Malaria did not exist in the northern regions of North America prior to settlement by Europeans but was a major problem in areas such as the southern United States, where appropriate mosquito habitat existed. The disease is associated with wet low-lying areas, and the name *malaria* is derived from the Italian words for "bad air."

The life cycle of the malaria parasite is complex (Figure 27.10●). First, the human host is infected by plasmodial *sporozoites*, small, elongated cells produced in the mosquito, which localize in the salivary gland of the insect. The mosquito injects saliva (containing an anticoagulant) along with the sporozoites. The sporozoites travel

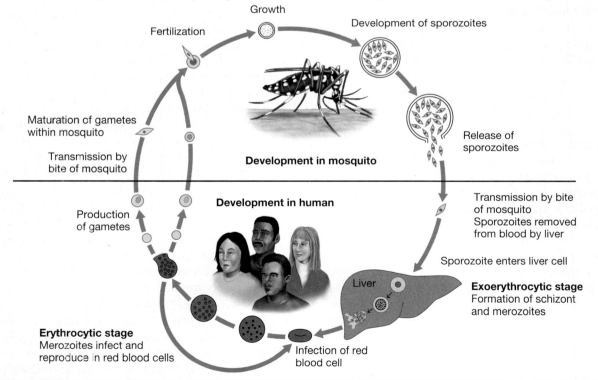

● **Figure 27.10   Life cycle of the malaria parasite, *Plasmodium vivax*.** The *Plasmodium* genera of protozoans (∞ Section 14.10) comprises the malarial pathogens, all of which have a life cycle dependent on growth in both a warm-blooded host and the mosquito vector. Transmission of the protozoan to and from the warm-blooded host is done by the bite of a mosquito.

through the bloodstream to the liver, where they may remain quiescent, or they may replicate and become enlarged in a stage known as a *schizont*. The schizonts then segment into a number of small cells called *merozoites*, and these cells are liberated from the liver into the bloodstream. Some of the merozoites then infect red blood cells (erythrocytes). The cycle in erythrocytes proceeds as in the liver and, in the case of *P. vivax*, usually repeats at synchronized intervals of 48 hours. During this 48-hour period, the defining clinical symptoms of malaria occur, characterized by *chills* followed by *fever* of up to 40°C (104°F). The chill–fever pattern occurs with the release of *P. vivax* cells from the erythrocytes during the synchronized asexual reproduction cycle. Vomiting and severe headache may accompany the chill–fever cycles, and asymptomatic periods generally alternate with periods in which the characteristic symptoms are present. Because of the loss of red blood cells, malaria generally causes anemia and some enlargement of the spleen.

Not all protozoal cells liberated from red blood cells are able to infect other erythrocytes. The protozoal cells that cannot infect erythrocytes are called *gametocytes* and are infective only for the mosquito. These gametocytes are ingested when another *Anopheles* mosquito bites the infected person; they mature within the mosquito into *gametes*. Two gametes fuse, and a zygote is formed. The zygote then migrates by ameboid motility to the outer wall of the insect's intestine where it enlarges and forms a number of sporozoites. These are released and some reach the salivary gland of the mosquito, from where they can be inoculated into another human, and the cycle begins again.

Diagnosis and Treatment

Conclusive diagnosis of malaria in humans requires the identification of *Plasmodium*-infected erythrocytes in blood smears (Figure 27.11●). Fluorescent nucleic acid stains, nucleic acid probes, polymerase chain reaction (PCR) methods, and antigen-detection methods may all be used to verify *Plasmodium* infections, or to differentiate between infections with various *Plasmodium* species.

Prophylaxis (when traveling to endemic areas) and treatment of malaria are usually accomplished with *chloroquine*. Chloroquine is the drug of choice for treating parasites within red cells, but does not kill malarial parasites outside the cells. The closely related drug *primaquine* eliminates sporozoites, merozoites, and gametes outside the cells. Treatment with both chloroquine and primaquine produces a cure. However, even in individuals who have undergone drug treatment, malaria may recur years after the primary infection. Apparently, small numbers of sporozoites survive in the liver and can reinitiate malaria months or years later by releasing merozoites.

In many parts of the world *Plasmodium* strains have developed resistance to chloroquine or primaquine or both, and some strains have developed resistance to other drugs as well. For use in areas with known drug-resistant strains, mefloquin or doxycycline is prescribed for prophylaxis and malarone, a combination of ato-

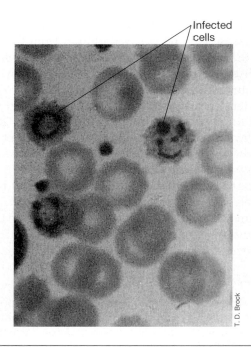

Infected cells

T. D. Brock

● **Figure 27.11** *Plasmodium vivax,* **the causative agent of malaria.** The parasite is shown growing inside human red blood cells.

vaquone and proguanil, is recommended for both treatment and prophylaxis. A new category of antimalarial drugs are synthetic derivatives of *artemisinin*, a natural compound containing reactive peroxide groups that form free radicals. These compounds are anti-parasitic agents *in vivo* and are now undergoing clinical trials.

Prevention

Antimalarial drug treatment is an expensive and short-term solution to malaria prevention and control, and drug-resistant strains of the parasite complicate matters even further. The most effective control measure is to interrupt the life cycle of the parasite by eliminating one of the obligate hosts, the *Anopheles* mosquito.

Two approaches to mosquito control are possible: (1) *elimination of habitat* by drainage of swamps and similar breeding areas, or (2) *elimination of the mosquito* by insecticides, followed by treatment of patients with primaquine, thereby breaking the *Plasmodium* life cycle. During the 1930s, about 33,000 miles of ditches were constructed in 16 southern states in the United States, removing 544,000 acres of mosquito breeding area. Millions of gallons of oil were also spread on swamps to reduce the oxygen supply to mosquito larvae. With the discovery of the insecticide dichlorodiphenyltrichloroethane (DDT) (∞ Figure 19.49), chemical control of both larvae and adult mosquitoes was possible. During World War II, the Public Health Service organized an Office of Mosquito Control in War Areas, and because many U.S. military bases were in the southern states, this organization carried out an extensive eradication program in the United States as well as overseas. In 1946, a year in which there were 48,610 cases of malaria in the United States, Congress established a five-year malaria eradication program. In endemic areas, the program involved drug prophylaxis and treatment for individuals, along with DDT treatment of mosquito infestations. By 1953 there were only 1310 malaria cases. In 1935 there were about 4000 deaths from malaria; in 1952 there were only 25 deaths.

Although the overall public health threat from malaria in the United States is now minimal, very low numbers of *endemic* malaria cases have resurfaced in recent years as far north as New York City. Significant periodic increases in malaria incidence also occur due to cases imported by soldiers or immigrants from malaria-endemic areas. On average, about 1,500 cases of malaria and five deaths occur in the United States each year.

In other parts of the world, eradication has been much slower, but the same control measures are used and are still effective. Reduction of mosquito habitat, control of mosquitoes by insecticides, and treatment of infected individuals with drugs both for cure and prophylaxis are still the major strategies for controlling malaria. Several malaria vaccines are in development, including synthetic peptide vaccines, recombinant particle vaccines, and DNA vaccines (⌀ Section 22.13 and 31.8).

Malaria and Human Evolution

Malaria has undoubtedly been endemic in Africa for thousands of years. In West Africans, resistance to malaria caused by *Plasmodium falciparum* is associated with an altered red blood cell protein, hemoglobin S, which differs from normal hemoglobin A at only a single amino acid. In the beta chain of hemoglobin S, the neutral amino acid *valine* is substituted for the *glutamic acid* of hemoglobin A. As a result, hemoglobin S binds oxygen less efficiently than hemoglobin A. Under conditions of low oxygen concentration, hemoglobin S forms long, thin aggregates that cause the red cell to change from a biconcave round cell (⌀ Figure 22.4) to an elongated C-shaped cell. Because of the shape of the cell, this condition is known as a *sickle cell*. Individuals who are *homozygous* for the sickle cell trait are particularly susceptible to changes in oxygen concentrations and suffer from **sickle cell anemia**.

Individuals who are heterozygous for hemoglobin S have the *sickle cell trait*, but have increased resistance to malaria. With the sickle cell trait, hemoglobin S can still produce sickled cells, but not as readily as in the case of the homozygotes. However, the growth of *P. falciparum* inside the red cell causes the heterozygous cells to sickle more easily than uninfected cells. The aggregated hemoglobin S in sickled cells apparently disrupts the membrane of red cells, allowing potassium to diffuse from the cell. *P. falciparum* cannot grow in the low-potassium environment of the disrupted cell. Thus, persons with the sickle cell trait are resistant to malaria.

In certain Mediterranean regions where malaria is endemic, resistance to *P. falciparum* is associated with a deficiency in the red blood cells of the enzyme *glucose-6-phosphate dehydrogenase (G6PD)*, an enzyme that acts as an intracellular antioxidant (reducing) compound. The faulty G6PD

leads to higher levels of intracellular oxidants such as the H_2O_2 produced inside the red cell by the growing *P. falciparum*. The increased levels of oxidants, normally removed by the activity of functional G6PD, damage parasite membranes and limit parasite growth.

In many Mediterranean populations, a diverse group of genetic abnormalities affect hemoglobin production and efficiency. These are known collectively as the **thalassemias**. The thalassemias are also statistically and geographically associated with increased resistance to malaria, and, like the G6PD deficiency, are associated with decreased levels of antioxidants in red cells.

Hemoglobin S, G6PD deficiency, and thalassemias are the result of genetic mutations that confer resistance to malaria infections and thus are selected in the population, although the mutations also confer red blood cell abnormalities and oxygen-processing deficiencies.

Another case in which the malaria parasite influences evolution involves the major histocompatibility complex (MHC) and the immune system (⌀ Sections 23.3 and 23.4). As discussed previously, the MHC class I and class II proteins present antigens to T cells for initiation of an immune response. In malaria-prone equatorial West Africa, individuals are very likely to have one particular MHC class I gene and one particular set of class II genes. These selected MHC genes are more common in the West African population and are virtually unknown in other human population groups. Individuals who express these genes have as much resistance to severe fatal malaria infections as those with the hemoglobin S trait. The MHC proteins encoded by these selected genes are exceptionally good antigen-presenting molecules for certain malarial antigens and initiate a strong protective immune response to *Plasmodium* spp. infection. As is the case for selection of the hemoglobin variants, the parasite is a selection factor for MHC genes that enhance host survival. Individuals with selected MHC genes that confer malaria resistance have a measurable survival advantage and are more likely to live and pass the resistance-conferring genes to their descendents.

Thus, in several ways malaria has been a selective agent in human evolution. Other pathogens, such as *Mycobacterium tuberculosis* (tuberculosis, ⌀ Section 26.5) and *Yersinia pestis* (plague, see also Section 27.7), may also have promoted selective changes in humans, but in no case is the evidence as clear as it is for malaria.

 ### 27.5 *Concept Check*

Malaria is a widespread, mosquito-borne infectious disease occurring mainly in tropical and subtropical regions of the world. It is a major cause of morbidity and mortality in developing countries and has been a selection factor for several resistance genes in humans. The disease is preventable with a combination of public health and chemotherapy measures.

◆ What is the natural reservoir for *Plasmodium* species? How can malaria be prevented or eradicated?

◆ Review genetic mechanisms responsible for malaria resistance. Why are antimalarial genes not found in all humans?

| **27.6** | **West Nile Virus** |

West Nile virus **(WNV)** causes **West Nile fever**, a rapidly emerging human viral disease transmitted through the bite of a mosquito. WNV is a member of the flavivirus group and has a symmetrical, enveloped icosahedral capsid with a positive-sense, single-stranded RNA genome of about 11,000 nucleotides (⟳ Section 16.8). The icosahedron is about 40–60 nm in diameter, and the virus can invade the nervous system of its warm-blooded host (Figure 27.12●).

Epidemiology

WNV infection in humans was first identified in Uganda in 1937. By the 1950s, the virus had spread to Egypt and Israel. In the 1990s, WNV outbreaks occurred in horses, birds, and humans in a number of African and European countries. In 1999, the first cases were reported in the United States, and early cases were centered in the Northeast, around New York. From 1999 through 2001, there were 149 confirmed cases of human WNV disease, including 18 deaths. Moving with the seasonal appearance and disappearance of the mosquito vectors, by 2002 this emerging disease had shifted from the East Coast to the Midwest, with a peak reported incidence of 884 in Illinois and total nationwide case totals of 4156. By 2003, there were 9186 confirmed cases, now centered in the upper Midwest and West. Colorado had the highest incidence of 2477 cases (Figure 27.13●). Illinois had only 53 cases. Deaths due to WNV infections *decreased* from 284 in 2002 to 231 in 2003, even with over twice as many diagnosed cases! This was probably due to increased recognition and testing for WNV infections in individuals displaying even mild symptoms (see below).

WNV normally causes active disease in birds, and is transferred to susceptible hosts by the bite of an infected mosquito. A number of mosquito species are known vectors, and at least 130 species of birds are known host reservoirs. The infected birds develop a viremia lasting 1–4 days, and survivors develop life-long immunity. Mosquitoes feeding on viremic birds are infected and can then infect susceptible birds, renewing the cycle. The incidence of disease in the avian population in a given area decreases as susceptible avian hosts die or recover and develop immunity. However, the mosquito vectors transmit the WNV to new susceptible hosts in new areas, moving the epidemic in a wavelike fashion across the continent. Incomplete data from 2004 indicate that the epidemic has already shifted to the West Coast and Southwest. The highest incidence of human disease in 2004 occurred in California and Arizona, with significantly lower numbers

CDC/Cynthia Goldsmith/PHIL

CDC/W. Brogdon, J. Gethany/PHIL

(a)

(b)

● **Figure 27.12   West Nile virus.** (a) The mosquito *Culex quinque-fasciatus*, shown here engorged with human blood, is a West Nile virus vector. (b) An electron micrograph of the West Nile virus. The icosohedral virion is about 40–60 nm in diameter.

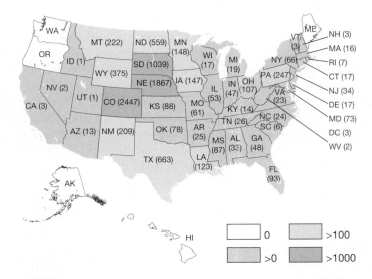

● **Figure 27.13   West Nile virus in the United States in 2003.** The virus caused 9186 cases of human disease, including 231 deaths. From 1999 through 2003, the peak incidence of the disease in the United States shifted from the East Coast, through the Midwest, and to the upper Midwest and West. Each year, the focus of cases moved as the virus infected new, susceptible host populations. Data are from the Centers for Disease Control and Prevention, Atlanta, GA, USA of cases in the upper Midwest and Colorado. Within the next several years, the disease will cease to be an epidemic in the contiguous United States, and WNV will be an endemic pathogen in North American bird populations.

Humans and other animals are terminal hosts, since they do not develop the viremia necessary to infect mosquitoes. The human mortality rate for *diagnosed* infections is very significant at 2–3%, and horses have mortality rates of up to 40%. Most human infections are asymptomatic or very mild. About 20% of infected individuals develop a mild illness called *West Nile fever*, with an incubation period of 3–14 days and a duration of 3–6 days. Fever may be accompanied by headache, nausea, myalgia, rash, lymphadenopathy (swelling of lymph nodes), and malaise. Less than 1% of infected individuals develop serious neurological diseases such as West Nile encephalitis or meningitis. Adults over age 50 appear to be more susceptible to neurological complications than do others. Diagnosis of WNV disease includes assessment of clinical symptoms followed by confirmation with a positive ELISA test for WNV antibodies in serum or cerebrospinal fluid (∞ Section 24.10).

Prevention and Control

Like St. Louis encephalitis virus and other mosquito-borne viruses that cause encephalitis (∞ Figure 25.3), transmission of WNV is seasonal in the United States, and is dependent on exposure to the mosquito population. The primary means of control for WNV spread is by limiting exposure to the disease vector. This can be accomplished individually by avoiding mosquito habitat, remaining indoors between dusk and dawn (the prime hours for mosquito activity), wearing appropriate mosquito-resistant clothing, and application of repellents containing DEET, as for Lyme disease (see Section 27.4). At the community level, controlling the mosquito vector population by destroying mosquito habitat and applying appropriate insecticides are important public health measures to control WNV. There is no effective human vaccine, although several candidates are in development. Veterinary vaccines are in use, but their efficacy is unknown. Treatment, as for most viral illnesses, is centered around supportive therapy such as rest, fluids, and symptomatic relief of fever and pain. There are no antiviral drugs known to be effective *in vivo*.

 27.6 Concept Check

West Nile fever is an emerging mosquito-borne viral disease. The natural cycle of the disease involves infections of birds by the bite of an infected mosquito. Humans and other vertebrates are occasional terminal hosts. While most human infections are asymptomatic and undiagnosed, complications in diagnosed infections can cause up to 3% mortality due to encephalitis and meningitis.

◆ Identify the vector and reservoir for West Nile virus.

◆ Trace the progress of West Nile virus in the United States since 1999.

27.7 Plague

Pandemic occurrences of **plague** have been directly responsible for more human deaths than any other infectious disease except malaria and tuberculosis. Plague killed between 25% and 33% of Europe's population in individual epidemics in the Middle Ages.

Plague is caused by *Yersinia pestis*, a gram-negative facultatively aerobic rod that is a member of the enteric bacteria group (∞ Section 12.11) (Figure 27.14●). Plague is a natural disease of domestic and wild rodents; rats are the primary disease reservoir. Humans are only accidental hosts and are not critical for the maintenance of the disease. Fleas are intermediate hosts and act as vectors, spreading plague between the mammalian hosts (Figure 27.15●). Most infected rats die soon after symptoms appear, but a low proportion survive and develop a chronic infection, providing a persistent reservoir of virulent *Y. pestis*.

Epidemiology

The majority of cases of human plague in the United States occur in the southwestern states, where the disease, called *sylvatic plague*, is endemic among wild rodents. Plague is transmitted by the rat flea (*Xenopsylla cheopis*), which ingests *Y. pestis* cells by sucking blood from an infected animal. Cells multiply in the flea's intestine and can be transmitted to a healthy animal in the next bite. As the disease spreads, rat mortality becomes so great that infected fleas seek new hosts, including humans. Once in humans, cells of *Y. pestis* usually travel to the lymph nodes, where they cause the formation of swollen areas referred to as *buboes*. For this reason the disease is frequently referred to as *bubonic plague* (Figure 27.14*b*). The buboes become filled with *Y. pestis* and the capsule on cells of *Y. pestis* prevents phagocytosis by cells of the immune system (∞ Section 22.2). Secondary buboes form in peripheral lymph nodes, and cells eventually enter the bloodstream, causing a generalized septicemia. Multiple hemorrhages produce dark splotches on the skin giving plague its historical name, the "Black Death" (Figure 27.14*c*). If not treated prior to the septicemic stage, the symptoms of plague (lymph node swelling and pain, prostration, shock, and delirium) progress and usually cause death within 3–5 days.

Pathology

The pathogenesis of plague is not clearly understood, but cells of *Yersinia pestis* produce a number of virulence factors, including toxins, that contribute to the disease process. The V and W antigens of *Y. pestis* cell walls are protein-lipoprotein complexes that inhibit phagocytosis. Other envelope proteins are also present. An exotoxin called *murine toxin*, because of its extreme toxicity for mice, is produced by virulent strains of *Y. pestis*. Murine

(a)

(c)

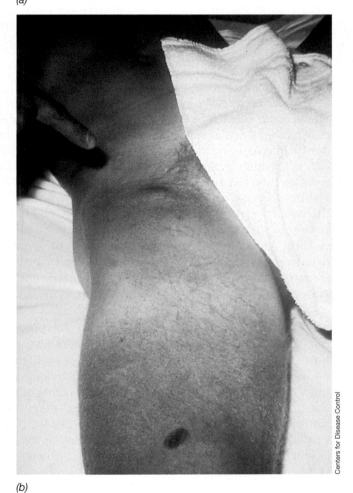

(b)

● **Figure 27.14   Plague in humans.** (a) *Yersinia pestis*, the causative agent of plague is a gram–negative rod, about 2 μm in length and up to 1 μm in diameter. The organisms in this blood smear (arrows) show the characteristic bipolar staining pattern. (b) A bubo formed in the groin. (c) Gangrene and sloughing of skin in the hand of a plague victim. Bubonic plague can be controlled by antibiotic therapy if diagnosed early in the course of the disease. Pneumonic plague and septicemic plague cannot be controlled by antibiotics, and so mortality is very high.

Pneumonic plague occurs when cells of *Yersinia pestis* are either inhaled directly or reach the lungs during bubonic plague. Symptoms are usually absent until the last day or two of the disease when large amounts of bloody sputum are produced. Untreated individuals rarely survive more than 2 days. Pneumonic plague is highly contagious and can spread rapidly via the person-to-person respiratory route if infected individuals are not immediately quarantined. *Septicemic plague* involves the rapid spread of *Y. pestis* throughout the body via the bloodstream without the formation of buboes and usually causes death before a diagnosis can be made.

Treatment and Control

Bubonic plague can be successfully treated if rapidly diagnosed. *Yersinia pestis* infection is treated with streptomycin or gentamycin, given parenterally. Alternatively, doxycycline, ciprofloxacin, or chloramphenicol may be given intravenously. If treatment is started promptly, mortality from bubonic plague can be reduced to 1–5% of those infected. Pneumonic and septicemic plague can also be treated, but these forms progress so rapidly that antibiotic therapy, even if begun when symptoms first appear, is usually too late. Although potentially a devastating disease, there have been only 97 cases of plague in the United States since 1990. Unfortunately, eight of these patients died. Worldwide, there are usually fewer than 1500 confirmed cases and 300 deaths per year. *Y. pestis* is an organism that could be used for a bioterrorism attack (∞ Section 25.12), and oral doxycycline and ciprofloxacin are recommended as prophylactic antibiotics in this setting.

toxin is a respiratory inhibitor that blocks mitochondrial electron transport reactions at the point of coenzyme Q (∞ Section 5.11). Although it is not clear that murine toxin is involved in the pathogenesis of human plague (murine toxin is highly toxic for certain animal species but not for others), it produces systemic shock, liver damage, and respiratory distress in mice. These symptoms are all seen in human plague as well. *Y. pestis* also produces a highly immunogenic *endotoxin* that may also play a role in the disease process.

● **Figure 27.15 The epidemiology of plague due to *Yersinia pestis.*** Plague in most wild rodents is generally a mild, self–limiting infection. Plague in rats and humans is frequently fatal. Infected fleas desert the dead host and look for other hosts, such as humans, an accidental host.

Plague control is accomplished through surveillance of infected animals, vectors, and human contacts. Plague-infected animal populations must be destroyed when identified. Undoubtedly, improved public health practices and the overall control of rodent populations have limited human exposure to plague.

◾ **27.7   Concept Check**

Plague is largely confined to individuals who come into contact with rodent populations that are endemic reservoirs for *Yersinia pestis.* A disseminated systemic infection or a pneumonic infection leads to rapid death, but the bubonic form is treatable with antibiotics if rapidly diagnosed.

◆ Distinguish among *sylvatic, bubonic, septicemic,* and *pneumonic* plague.

◆ What is the insect reservoir, the natural host, and the treatment for plague?

III SOILBORNE DISEASES

We now investigate several diseases that are normally spread through soil. Fungi are common and ubiquitous soil microorganisms found worldwide, and a few can be pathogens. Some bacteria are also important soilborne pathogens. In contrast to many person-to-person or vector-borne pathogens, soilborne pathogens are accidental agents of infection, with no life-cycle dependency on the accidental host. Soil is an unlimited reservoir of these pathogens and, thus, these pathogens cannot be eliminated.

27.8 The Pathogenic Fungi

Fungi in some form grow in nearly every ecological niche, but are most commonly found in nature as free-living saprophytes (⊂⊃ Section 14.12). However, a few fungi are occasionally found as accidental, often opportunistic, pathogens.

The fungi include the eukaryotic organisms commonly known as *yeasts*, which normally grow as single cells (Figure 27.16a●), and *molds*, which grow in branching

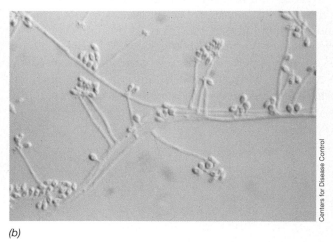

(a)   (b)

● **Figure 27.16   Typical forms of pathogenic fungi.** (a) Yeast form of *Cryptococcus neoformans*, stained with India ink to show the capsule. The cells are from 4 to 20 μm in diameter (b) *Sporothrix schenckii*, showing the branching, or hyphae, characteristic of the mold form of fungi. The round basidiospores are about 2 μm in diameter. Fungi are discussed in Section 14.12.

chains called *hyphae* (Figure 27.16*b*). The taxonomy and biological diversity of these organisms were discussed in Section 14.12. Fortunately, most fungi are harmless to humans. Only about 50 species cause human disease, and the overall incidence of serious fungal infections is rather low, although certain superficial fungal infections are quite common.

Epidemiology

Fungi cause disease through three major mechanisms. First, some fungi trigger immune responses that result in allergic (hypersensitivity) reactions following exposure to specific fungal antigens (⌦ Section 22.15). Reexposure to the same fungi, whether growing on the host or in the environment, may cause allergic symptoms. For example, *Aspergillus* spp. (⌦ Figure 14.31), a common saprophyte often found in nature as a leaf mold, produces potent allergens, often causing asthma and other hypersensitivity reactions. *Aspergillus* also has other mechanisms for producing disease.

A second fungal disease-producing mechanism involves the production and activity of *mycotoxins*, a large, diverse group of fungal exotoxins. The best-known examples of mycotoxins are produced by *Aspergillus flavus*, a species that commonly grows on improperly stored food such as grain. The toxins produced by *A. flavus* are known as *aflatoxins* (Figure 27.17●). Aflatoxins are highly toxic and induce tumors in some animals, especially in birds that feed on contaminated grain. Their direct role in human disease is not well defined.

The third fungal disease-producing mechanism is through infections called *mycoses*.

Mycosis

The growth of a fungus on or in the body is called a **mycosis** (plural, **mycoses**). Mycoses are fungal infections that range in severity from relatively innocuous, superficial lesions to serious, life-threatening diseases.

Mycoses fall into three categories. The first of these are the *superficial mycoses*. These diseases involve fungal colonization of the skin, hair, or nails, and infect only the surface layers (Figure 27.18*a*●). Table 27.2 lists some of the common superficial fungi. In general, these diseases are relatively benign and self-limiting. Some, such as *Trichophyton* infections of the feet (*athlete's foot*), are quite

common. Spread is by personal contact with an infected person or by contact with contaminated surfaces such as bathtubs, shower stalls, or floors, or by contact with contaminated shared articles such as towels or bed linens. Treatment for severe cases is with topical application of miconazole nitrate or griseofulvin. Griseofulvin can also be administered orally. After entering the bloodstream, it passes to the skin where it can inhibit fungal growth.

The *subcutaneous mycoses* are a second category of fungal infections. They involve deeper layers of skin (Figure 27.18*b*) and a different group of organisms (Table 27.2). One disease in this category is *sporotrichosis*, an occupational hazard of agricultural workers, miners, and others who come into contact with the soil. The causative organism is found as a ubiquitous saprophyte on wood and in soil. The lesions are usually initiated by fungal infection of a small wound or abrasion. *Sporothrix schenckii* can readily be isolated from the lesion and cultured *in vitro*. Treatment is with oral potassium iodide or oral ketoconazole.

The *systemic mycoses* are the third and most serious category of fungal infections. They involve fungal growth in internal organs of the body and are subclassified as *primary* or *secondary* infections. A primary infection is one resulting directly from the fungal pathogen in otherwise normal, healthy individuals. A secondary infection involves infection in hosts with a predisposing condition such as antibiotic therapy or immunosuppression. In the United States, the most widespread primary fungal infections are *histoplasmosis*, caused by *Histoplasma capsulatum*, and *coccidioidomycosis* (San Joaquin Valley fever), caused by *Coccidioides immitis*. Both of these organisms normally live in soil. Both are respiratory diseases in which the host becomes infected by inhaling airborne spores, which germinate and grow in the lungs. Histoplasmosis is primarily a disease of rural areas in the midwestern United States, especially in the Ohio and Mississippi River valleys. Most cases are mild and are often mistaken for more common respiratory infections. San Joaquin Valley fever is generally restricted to the desert regions of the southwestern United States. The fungus lives in desert soils, and the spores are disseminated on dry, windblown particles that are inhaled. In some areas in the southwestern United States, as many as 80% of the inhabitants may be infected, although most individuals suffer no apparent ill effects.

A number of systemic fungal infections, including histoplasmosis and coccidioidomycosis, are especially serious and common in individuals whose immune systems have been impaired, for example, by acquired immunodeficiency syndrome (AIDS) (⌦ Section 26.14) or by immunosuppressive drugs. These are *secondary* fungal diseases because normal individuals either do not get the disease or generally have a less severe form. Examples of other fungal organisms involved as secondary pathogens are given in Table 27.2. These fungi are known as *opportunistic* pathogens because of their particular ability to cause serious infections only in individu-

● **Figure 27.17   Structure of aflatoxin B1.** This toxin is one of a group of related compounds produced by *Aspergillus flavus*.

(a) *(b)*

● **Figure 27.18 Fungal infections.** (a) Superficial mycosis of the foot (athlete's foot) due to infection with *Trichophyton rubrum*. (b) Sporotrichosis, a subcutaneous infection due to *Sporothrix schenckii*.

als with impaired defense mechanisms (in particular AIDS patients) (Section 26.14).

Treatment and Control

Effective chemotherapy against systemic fungal infections is very difficult (∞ Section 20.11). Most antibiotics that inhibit fungi also harm other eukaryotic organisms, including the human host. One of the most effective antibiotics, amphotericin B, is widely used to treat systemic fungal infections of humans, but serious side effects such as kidney toxicity may occur.

Control of infections by elimination of fungal pathogens from the environment is impractical. As with many common-source pathogens, control of fungal growth is very difficult because there is a limitless reservoir. Exposure to fungi cannot be eliminated, but risks can be reduced by decontamination and air filtration in restricted local environments.

■ **27.8 Concept Check**

A variety of soilborne fungi produce disease in humans. Superficial, subcutaneous, and systemic mycoses are infections that are difficult to control because of a lack of specific antifungal drugs and the ubiquitous nature of the pathogens. Fungal infections may cause serious systemic disease, often in individuals with impaired immunity, such as in AIDS patients.

◆ Describe *superficial, subcutaneous,* and *systemic* mycoses.

◆ Distinguish between a *primary* and a *secondary* fungal disease.

27.9 Tetanus

Tetanus is a serious, often life-threatening disease. Although tetanus is preventable through immunization (∞ Section 22.13), 473 individuals have acquired tetanus in the United States within the last decade and 68 have died.

Table 27.2	Some pathogenic fungi and the diseases they cause	
Disease	**Causal organism**	**Main disease foci**
Superficial mycoses (dermatomycoses)		
Ringworm	*Microsporum*	Scalp of children
Favus	*Trichophyton*	Scalp
Athlete's foot	*Epidermophyton, Trichophyton*	Between toes, skin
Jock itch	*Trichophyton, Epidermophyton*	Genital region
Subcutaneous mycoses		
Sporotrichosis	*Sporothrix schenckii*	Arms, hands
Chromoblastomycosis	Several fungal genera	Legs, feet
Systemic mycoses		
Cryptococcosis[a]	*Cryptococcus neoformans*	Lungs, meninges
Coccidioidomycosis[a]	*Coccidioides immitis*	Lungs
Histoplasmosis[a]	*Histoplasma capsulatum*	Lungs
Blastomycosis	*Blastomyces dermatitidis*	Lungs, skin
Candidiasis[a]	*Candida albicans*	Oral cavity, intestinal tract

[a] Considered opportunistic pathogens and have been implicated in the pathogenesis of AIDS (∞ Section 26.14).

Biology and Epidemiology

Tetanus is caused by an exotoxin produced by *Clostridium tetani*, a motile, anaerobic, endospore-forming rod (∞ Section 12.20). The natural reservoir of *C. tetani* is soil, where it is a ubiquitous resident, although it is occasionally found in the gut of mammals, as are other *Clostridium* species. *C. tetani* normally gains access to the body through a soil-contaminated wound, typically a deep puncture. In the wound, anoxic conditions allow germination of endospores and growth of the organism. *C. tetani* produces a potent exotoxin, the *tetanus toxin*. The organism is noninvasive, and so its sole method of causing disease is through the direct action of tetanus toxin on host cells. The incubation time is variable and may take from 4 days to several weeks, depending on the number of endospores inoculated at the time of injury. Tetanus is not transmitted from person to person.

Pathogenesis

We have already examined the activity of tetanus toxin at the cellular and molecular level (∞ Section 21.10). The toxin directly affects the release of inhibitory signaling molecules in the nervous system. These inhibitory signals control the "relaxation" phase of muscle contraction. The overall result is rigid paralysis of the voluntary muscles, often called *lockjaw* because it is observed first in the muscles of the jaw and face (Figure 27.19●). Death is usually due to respiratory failure, and the mortality rate is very high (15% over the last decade in the United States).

Diagnosis, Control, Prevention, and Treatment

Diagnosis of tetanus is based on exposure, clinical symptoms, and, rarely, identification of the toxin in the blood or tissues of the patient. The organism may also be cultured from the wound, but success is highly variable.

The natural reservoir of *Clostridium tetani* is the soil. Since *C. tetani* is an accidental pathogen in humans and is not dependent on humans or other animals for its propagation, there is no possibility for eradication. Therefore, control measures must focus on methods of prevention.

Tetanus is a preventable disease. The existing toxoid vaccine (∞ Section 22.13) is completely effective for disease prevention. Virtually all tetanus cases occur in individuals who were inadequately immunized. The fastest growing age group for contracting tetanus is, surprisingly, individuals from 25–59 years of age, presumably because public health immunization programs target infants, school-age individuals, and seniors 60 years of age and older.

Appropriate treatment of serious cuts, lacerations, and punctures includes administration of a "booster" tetanus toxoid immunization. If the wound is severe and is contaminated by soil, treatment should also include administration of an antitoxin preparation, especially if the patient's immunization status is unknown or is out of date. The antitoxin is a preparation of antibodies to tetanus, usually made in horses, that neutralizes the tetanus toxin as it is released (∞ Section 22.13). These measures prevent active tetanus from occurring.

Treatment of acute symptomatic tetanus involves administration of antibiotics, usually penicillin, to stop growth and toxin production by *C. tetani*; administration of antitoxin to prevent binding of newly released toxin to cells; and supportive therapy such as sedation, muscle relaxants, and mechanical respiration to control the effects of paralysis. Treatment at this level cannot provide a reversal of symptoms, since toxin that is already bound to tissues cannot be neutralized. Even with antitoxin, antibiotics, and supportive therapy, tetanus patients have significant morbidity and mortality.

Several species of *Clostridium* are pathogens and virtually all exist as normal members of the microbial community in soil, making their eradication impossible. All cause disease because of their production of potent exotoxins. *Clostridium tetani* is found almost exclusively in soil, but *C. botulinum* and *C. difficile* are also occasionally found in the gut of humans and other animals as part of the normal microbial flora (∞ Section 21.4). *C. perfringens* and *C. botulinum*, as we shall see in Chapter 29, are important potential pathogens in common-source foodborne diseases. In each case, the source of the *Clostridium* is food contaminated with soil containing clostridial endospores. The endospores then germinate because of inadequate decontamination and food preservation methods, and produce potent exotoxins that cause disease symptoms.

 27.9 Concept Check

Clostridium tetani is a ubiquitous soilborne microorganism that can cause *tetanus*, a disease characterized by toxin production and rigid paralysis. Tetanus is preventable with appropriate immunization. Treatment for acute tetanus is generally unsatisfactory, with significant morbidity and mortality.

◆ Describe infection by *C. tetani* and the elaboration of tetanus toxin.

◆ Describe the overall effect of tetanus toxin on the host and outline the steps taken to prevent tetanus in an individual who has sustained a puncture wound.

● **Figure 27.19 A soldier dying from tetanus.** Note the rigid paralysis. The painting by Charles Bell is in the Royal College of Surgeons, Edinburgh, Scotland.

REVIEW QUESTIONS

1. Identify the animals most likely to carry rabies in the United States. Which immunization programs are in place for the treatment of rabies? For the prevention of rabies (⊙ Section 27.1)?

2. Why has hantavirus pulmonary syndrome (HPS) emerged as a human pathogen in the United States? How can HPS be prevented (⊙ Section 27.2)?

3. What are the three major categories of organisms that cause rickettsial diseases? For typhus, Rocky Mountain spotted fever, and ehrlichiosis, identify the most common reservoir and vector (⊙ Section 27.3).

4. Identify the most common reservoir and vector for Lyme disease in the United States. How can the spread of Lyme disease be controlled? How can Lyme disease be treated (⊙ Section 27.4)?

5. Malaria involves severe long-term fever followed by chills. These symptoms are related to activities of the pathogen *Plasmodium vivax* or *Plasmodium falciparum*. Describe the growth stages of this pathogen in the human host and relate them to the fever–chill pattern.

Why might a person of western European descent be more susceptible to malaria than a person of African or Mediterranean descent (⊙ Section 27.5)?

6. Explain the life cycle and the role of the mosquito in spreading West Nile virus. What animals are the primary hosts? Are humans productive alternative hosts? Explain (⊙ Section 27.6).

7. For a potentially serious disease like bubonic plague, why aren't vaccines provided for the general population? What public health measures are used to control this disease (⊙ Section 27.7)?

8. What is the reservoir for most fungal pathogens? How can fungal exposure be controlled? What particular problems, especially in terms of therapy, do fungi pose for the clinician (⊙ Section 27.8)?

9. Describe the invasiveness and toxicity of *Clostridium tetani*. Discuss the major mechanism of pathogenesis for tetanus and define measures for prevention and treatment of tetanus (⊙ Section 27.9).

APPLICATION QUESTIONS

1. Describe the sequence of events you would take if a child received a bite (provoked or unprovoked) from a stray dog with no record of rabies immunization. Present one scenario where you were able to capture and detain the dog and another for a dog that escaped. How would these procedures differ from a situation in which the child was bitten by a dog that had documented, up-to-date rabies immunizations?

2. Discuss at least three common properties of the disease agents and review the disease process for Rocky Mountain spotted fever, typhus, and ehrlichiosis. Why is ehrlichiosis emerging as an important rickettsial disease? Compare its emergence to that of Lyme disease.

3. Malaria eradication has been a goal of public health programs for at least 100 years. What factors preclude our ability to eradicate malaria? If an effective vaccine was developed, could malaria be eradicated? Compare this possibility to the possibility of eradicating plague.

4. Devise a plan to prevent the spread of West Nile virus to humans in your community. Identify the costs involved in such a plan, both at the individual level and at the community level. Find out if a mosquito abatement program is active in your community. What methods are used for the reduction of mosquito populations?

28

WASTEWATER TREATMENT, WATER PURIFICATION, AND WATERBORNE MICROBIAL DISEASES

Water is the most important common source of infectious disease. Water treatment standards are designed to prevent infections from bacteria, viruses, and protozoans such as the cell of the protozoan *Giardia* shown in the photo.

WORKING GLOSSARY

Aerobic secondary wastewater treatment digestive reactions carried out by microorganisms under aerobic conditions to treat wastewater containing low levels of organic materials

Anoxic secondary wastewater treatment digestive and fermentative reactions carried out by microorganisms under anoxic conditions to treat wastewater containing high levels of insoluble organic materials

Biochemical oxygen demand (BOD) the relative amount of dissolved oxygen consumed by microorganisms for complete oxidation of organic and inorganic material in a water sample

Chloramine a chemical manufactured on site by combining chlorine and ammonia at precise ratios

Chlorine a chemical fed in its gaseous state to disinfect water. A residual level is maintained throughout the distribution system

Clarifier (coagulation basin) a reservoir in which the suspended solids of raw water are coagulated and removed

Coagulation the formation of large insoluble particles from much smaller, colloidal particles by the addition of aluminum sulfate and anionic polymers

Coliform facultatively aerobic, gram-negative, nonspore-forming lactose-fermenting bacterium

Cyst an infectious form of a protozoan parasite that is encased in a thick-walled, chemically and physically resistant coating

Distribution system water pipes, storage reservoirs, tanks, and other means used to deliver drinking water to consumers or store it before delivery

Effluent water treated wastewater discharged from a wastewater treatment facility

Filtration the removal of suspended particles from water by passing it through one or more permeable membranes or media (e.g., sand, anthracite, or diatomaceous earth)

Finished water water delivered to the distribution system after treatment

Flocculation the water-treatment process after coagulation that uses gentle stirring to cause suspended particles to form larger, aggregated masses (floc)

Meningoencephalitis invasion, inflammation, and destruction of brain tissue, by the ameba *Naegleria fowleri* or a variety of other pathogens

Polymer in water purification, a chemical in liquid form used as a coagulant in the clarification process to produce flocculation

Potable drinkable; safe for human consumption

Primary wastewater treatment physical separation of wastewater contaminants, usually by separation and settling

Raw water surface water or groundwater that has not been treated in any way (also called untreated water)

Sediment soil, sand, minerals, and other large particles found in raw water

Sewage liquid effluents contaminated with human or animal fecal material

Suspended solid small particle of solid pollutant that resists separation by ordinary physical means

Tertiary wastewater treatment physicochemical processing of wastewater to reduce levels of inorganic nutrients

Turbidity a measurement of suspended solids in water

Untreated water surface water or groundwater that has not been treated in any way (also called raw water)

Wastewater liquid derived from domestic sewage or industrial sources, which cannot be discarded in untreated form into lakes or streams

---

Wastewater is liquid derived from domestic sewage or industrial sources that cannot be discarded in untreated form into lakes or streams due to public health, economic, environmental, and aesthetic considerations. Clean, pure water is absolutely essential to public health, and procedures to monitor, assess, and remediate water quality are thus necessary. However, water quality is sometimes compromised, even in publicly operated wastewater and drinking water systems. Lapses in water quality can promote dramatic and even life-threatening spread of infectious disease. This chapter examines standard methods of water treatment and purification and a variety of waterborne diseases.

WASTEWATER MICROBIOLOGY AND WATER PURIFICATION

Water is the most important potential *common source* of infectious diseases, and water purification is the most important single measure available for ensuring public health. The methods commonly used to assess water quality depend on standardized microbiology techniques. In addition to physical and chemical purification procedures, water treatment and purification schemes use a variety of standardized methods that use microorganisms to identify, remove, or degrade pollutants.

28.1 Public Health and Water Quality

How can we routinely ensure that drinking water is safe? Even water that looks clear and clean may be contaminated with pathogenic microorganisms and may pose a serious health hazard. Unfortunately, it is impractical to screen water for every pathogenic organism that may be present, and a few nonpathogenic microorganisms are generally tolerable, and even unavoidable, in a water supply. It is, however, possible to sample water supplies for the presence of certain groups of microorganisms that may include pathogens. We discuss here the general methods used to identify potentially harmful microorganisms in water.

Coliforms

The presence of specific *indicator organisms* signals that a given water source might be contaminated with pathogens. The most widely used indicator for microbial water contamination is the **coliform** group of microorganisms. Coliforms are used as indicators of water contamination because many of them inhabit the intestinal tract of humans and other animals in large numbers. Thus, their presence in water indicates fecal contamination. Coliforms are defined as facultatively aerobic, gram-negative, nonspore-forming, rod-shaped *Bacteria* that ferment lactose with gas formation within 48 hours at 35° C. However, this is an operational rather than a taxonomic definition, and the coliform group actually includes a variety of organisms. Most coliforms are members of the enteric bacterial group (∞ Section 12.11). For example, the coliform group includes the organism *Escherichia coli*, a common intestinal organism, and the organism *Klebsiella pneumoniae*, a less common pathogenic intestinal inhabitant. However, *Enterobacter aerogenes*, an organism not found in the enteric group or in the intestine, is also classified as a coliform because of its fermentative properties.

In general, we assume that the presence of coliform organisms in a water sample indicates fecal contamination and makes the water unsafe for human consumption. When excreted into water, the coliforms eventually die, but they do not die as quickly as some pathogens. The coliforms and the pathogens behave similarly during water purification.

The Coliform Test

Two procedures are commonly used to test for coliforms in water samples. These are the *most-probable-number (MPN)* procedure and the *membrane filter (MF)* procedure. The MPN procedure employs liquid culture medium in test tubes in which samples of drinking water are added to the media. Growth in the culture vessels indicates microbial contamination of the water supply. For the more common MF procedure, at least 100 ml of the water sample is passed through a sterile membrane filter, which removes the bacteria (∞ Section 20.3). The filter is placed on a plate of eosin-methylene blue (EMB) culture medium, which is highly selective for coliform organisms (Figure 28.1●; ∞ Section 24.2). Following incubation, coliform colonies are counted, and from this value the number of coliform *Bacteria* in the original water sample can be calculated. In well-regulated water supply systems, coliform tests should be negative. If coliform tests are positive, a breakdown has occurred in the purification or distribution systems.

Drinking water standards in the United States are specified under the Safe Drinking Water Act, which provides a framework for the development of drinking water standards. For the membrane filter (MF) technique, 100 ml samples are filtered. To be considered safe, the number of coliform bacteria in drinking water samples cannot exceed any of the following levels: (1) 1 per 100 ml as the arithmetic mean of all samples examined per month; (2) 4 per 100 ml in more than one sample when fewer than 20 are examined per month; or (3) 4 per 100 ml in more than 5% of the samples when 20 or more samples are examined per month. Water utilities report their results to the United States Environmental Protection Agency, and if they do not meet the prescribed stan-

● **Figure 28.1   Coliform colonies growing on a membrane filter.** A drinking water sample was passed through the filter. The filter was then placed on eosin-methylene blue (EMB) medium that is both selective and differential for lactose-fermenting bacteria (coliforms) (∞ Section 24.2). The dark color of the colonies is characteristic of coliforms. Each colony developed from one viable coliform cell present in the original sample.

dards, the utilities must notify the public and take steps to correct the problem. Many smaller communities and even large cities sometimes fail to meet these standards.

Public Health and Drinking Water Purification

Today the incidence of waterborne disease in developed countries is so low that it is difficult to directly measure the effectiveness of treatment practices using drinking water standards. Most intestinal infections in developed countries are no longer transmitted by water but by food (∞ Chapter 29). Highly effective water treatment practices, however, were not in place until the twentieth century.

Microbial culture methods for evaluating the health significance of polluted drinking water were not practiced until coliform-counting procedures were developed and adapted in about 1905. Until then, water purification was limited to *filtration* to reduce turbidity. Although filtration significantly decreased the microbial load of water, many microorganisms still passed through the filters. In about 1910, **chlorine** was discovered to be an extremely efficient disinfectant for large water supplies. Chlorine gas was so effective and so inexpensive that its use as a general disinfectant for drinking water spread widely and was of major significance in reducing the incidence of waterborne disease. Figure 28.2● illustrates the dramatic drop in incidence of typhoid fever (infection with *Salmonella typhi*) in a major American city after purification procedures using filtration and chlorination were introduced. Similar results were obtained in other cities.

The first dramatic improvements in public health in the United States, starting near the beginning of the twentieth century, were largely due to the establishment

Figure 28.2   The effect of water purification on the incidence of waterborne disease. The graph shows the incidence of typhoid fever in Philadelphia (Pennsylvania, USA) before and after the introduction of filtration and chlorination processes for drinking water supplies. Note the dramatic reduction in the incidence of typhoid fever after the introduction of both filtration and chlorination. This result parallels the results obtained for reduction of other waterborne diseases.

of water treatment procedures employed in large-scale, publicly operated, wastewater and drinking water treatment facilities. The effectiveness of filtration and chlorination could not have been assessed if standard methods for determining the coliform content of drinking water had not been developed. Thus, public works engineering, microbiology, and public health moved forward together.

28.1   Concept Check

Drinking water quality is determined by counting coliform bacteria. Strict adherence to uniform microbiologic standards makes this method a reliable and reproducible indicator of fecal contamination in all public water supplies in the United States. Filtration and chlorination of water supplies significantly decreases microbial load. Application of water purification methods to drinking water is the most important public health measure ever devised.

◆ Why do the bacterial colonies recovered from drinking water and grown on EMB media indicate fecal contamination of the water supply?

◆ What general procedures are used to reduce microbial load in water supplies?

28.2   Wastewater and Sewage Treatment

Wastewater and sewage treatment involves a large-scale use of microorganisms for bioconversion on an industrial scale. Wastewater enters a treatment plant and, following treatment, the **effluent water**—treated wastewater discharged from the wastewater treatment facility—is

suitable for release into surface waters such as lakes and streams or to drinking water purification facilities.

Wastewater and Sewage

Wastewater from domestic sewage or industrial sources cannot be discarded in untreated form into lakes or streams. **Sewage** is liquid effluent contaminated with human or animal fecal materials. Wastewater commonly contains potentially harmful inorganic and organic compounds as well as pathogenic microorganisms. Complete wastewater treatment involves physical, chemical, and biological (microbiological) treatments to remove or neutralize contaminants.

On average, individuals in the United States use 100–200 gallons of water per day for washing, cooking, drinking, and sanitary needs. The wastewater must then be treated to remove contaminants before it can be released into surface waters. About 16,000 publicly owned treatment works (POTW) operate in the United States. Most POTWs are fairly small, treating 1 million gallons (3.8 million liters) or less of wastewater per day. Collectively, however, these plants treat about 32 *billion* gallons of wastewater daily. Wastewater plants are usually constructed to handle both domestic and industrial wastes. *Domestic wastewater* is made up of sewage, "gray water" (the water resulting from washing, bathing, and cooking), and wastewater from food processing.

Industrial wastewater includes liquid discharged from the petrochemical, pesticide, food and dairy, plastics, pharmaceutical, and metallurgical industries. Industrial wastewater may contain toxic substances and some industrial facilities are required by the United States Environmental Protection Agency (EPA) to pretreat toxic or heavily contaminated discharges before they enter POTWs. Pretreatment may simply involve mechanical processes in which debris that could clog equipment in the wastewater treatment plant is removed. However, certain wastewaters are pretreated biologically or chemically to remove highly toxic substances such as cyanide, heavy metals such as arsenic, lead and mercury, or organic materials such as acrylamide, atrazine (a herbicide), and benzene. These substances are converted to less toxic forms by treatment with chemicals or microorganisms capable of neutralizing, oxidizing, precipitating, or volatilizing these wastes. The pretreated wastewater can then be released to the POTW.

Wastewater Treatment and Biochemical Oxygen Demand

The goal of a wastewater treatment facility is to reduce organic and inorganic materials in wastewater to a level that no longer supports microbial growth and to eliminate other potentially toxic materials. The efficiency of treatment is expressed in terms of a reduction in the **biochemical oxygen demand (BOD)**, the relative amount of dissolved oxygen consumed by microorganisms to completely oxidize all organic and inorganic matter in a water sample (◌◦ Section 19.5). Higher levels of oxidizable organic and inorganic materials in the wastewater result in a higher BOD.

Typical values for domestic wastewater, including sewage, are approximately 200 BOD units. For industrial wastewater, for example, from sources such as dairy plants, the values can be as high as 1500 BOD units. An efficient wastewater treatment facility reduces levels to less than 5 BOD units in the water released from the treatment plant. Wastewater facilities must be designed to treat both low-BOD sewage and high-BOD industrial wastes.

Treatment is a multistep operation employing a number of independent physical and biological processes (Figure 28.3●). Primary, secondary, and sometimes tertiary treatments are employed to reduce fecal and chemical contamination in the incoming water. Each level of treatment employs more complex and more expensive technologies.

Primary Wastewater Treatment

Primary wastewater treatment uses only *physical separation* methods to separate solid and particulate organic and inorganic materials from wastewater. Wastewater

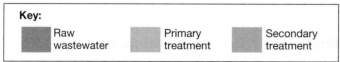

● Figure 28.3 Wastewater treatment processes. Effective water treatment plants use the primary and secondary treatment methods shown here. Wastewater can contain from 200 BOD units for sewage and domestic waste to 1500 or more BOD units for industrial waste. Effluent from the treatment plant usually contains 5 BOD units or less. Tertiary treatment of the effluent water to further reduce BOD is rarely done because it is not cost effective.

entering the treatment plant is passed through a series of grates and screens that remove large objects. The effluent is left to settle for several hours to allow solids to settle to the bottom of the separation reservoir (Figure 28.4●).

Municipalities that provide only primary treatment suffer from extremely polluted water when the effluent is discharged into adjacent waterways because high levels of soluble and suspended organic matter and other nutrients remain in water following primary treatment. These nutrients can propagate undesirable microbial growth (◌◦ Figure 19.10), further reducing water quality. Therefore, most treatment plants employ secondary and even tertiary treatments to reduce the organic content of the wastewater before release to natural waterways. Secondary treatment processes use microbial digestion both in the presence and absence of oxygen to further reduce organic nutrients in wastewater.

Anoxic Secondary Wastewater Treatment

Anoxic secondary wastewater treatment involves a series of digestive and fermentative reactions carried out by a number of bacterial and archaeal species under anoxic conditions. Anoxic treatment is typically used to treat wastewater containing large quantities of insoluble organic matter (and, hence, very high BOD), such as fiber and cellulose waste from food- and dairy-processing plants. The anoxic degradation process itself is carried out in large enclosed tanks called *sludge digesters* or *bioreactors* (Figure 28.5a and b●). The process requires the collective activities of many different types of microorganisms. The major reactions are summarized in Figure 28.5c.

Through the action of the resident anoxic microorganisms, the macromolecular waste components are first digested by polysaccharases, proteases, and lipases into soluble components. These soluble components are then fermented to yield a mixture of fatty acids, H_2,

● Figure 28.4 Primary treatment of wastewater. Wastewater is pumped into the reservoir (left) where settling of solids occurs. As the water level rises, the water spills through the grates to successively lower levels. Water at the lowest level, now free of solids, enters the spillway (arrow) and is pumped to a secondary treatment facility.

(a)

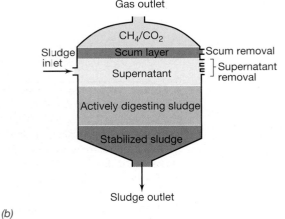

(b)

Complex polymers
(polysaccharides, lipids, proteins)

Hydrolysis by
microbial enzymes

Monomers
(sugars, fatty acids, amino acids)

Fermentation   Fermentation

Acetate $H_2 + CO_2$

Methanogenesis

$CH_4 + CO_2$ CH_4

(c)

● **Figure 28.5 Anoxic secondary wastewater treatment.** (a) Anoxic sludge digester. Only the top of the tank is shown; the remainder is underground. (b) Inner workings of a sludge digestor. (c) Major microbial processes occurring during anoxic sludge digestion. Methane (CH_4) and carbon dioxide (CO_2) are the major products of anaerobic biodegradation (⚮ Section 17.17). Figure 19.23 shows the major microbial groups involved in anoxic decomposition as well as the interactions among them that lead to methanogenesis.

and CO_2; the fatty acids are further fermented to acetate, CO_2, and H_2. These products are then used as substrates by methanogenic *Archaea* (⚮ Sections 12.6 and 17.17), capable of carrying out the reactions $CH_3COOH \rightarrow CH_4 + CO_2$ and $4 H_2 + CO_2 \rightarrow CH_4 + 2 H_2O$ (Figure 28.5c). Thus, the major products of anoxic sewage treatment are CH_4 (methane) and CO_2. The methane can be collected and either burned off or used as fuel to heat and power the treatment plant.

Aerobic Secondary Wastewater Treatment

Aerobic secondary wastewater treatment uses digestive reactions carried out by microorganisms under aerobic conditions to treat wastewater containing low levels of organic materials. In general, nonindustrial wastewater can be treated efficiently using only aerobic secondary treatment. Several kinds of aerobic decomposition processes are used for wastewater treatment, but the trickling filter and activated sludge methods are the most common. A trickling filter (Figure 28.6a●) is a bed of crushed rocks, about 2 m thick, on top of which the wastewater is sprayed. The liquid slowly passes through the bed, the organic matter adsorbs to the rocks, and microbial growth occurs on the large exposed surface of the rocks. The complete mineralization of organic matter to

carbon dioxide, ammonia, nitrate, sulfate, and phosphate takes place in the microbial biofilm on the rocks.

The most common aerobic treatment system is the activated sludge process. Here, the wastewater to be treated is mixed and aerated in a large tank (Figure 28.6b). Slime-forming bacteria, including *Zoogloea ramigera*, among others, grow and form *flocs* (larger, aggregated masses) (Figure 28.7●). These flocs form a substratum to which protozoa, small animals, and sometimes filamentous bacteria and fungi attach. The basic process of oxidation is similar to the trickling filter. The aerated effluent containing the flocs is pumped into a holding tank or *clarifier* where the flocs settle. Some of the floc material (called *activated sludge*) is then returned to the aerator to serve as inoculum, and the rest is pumped to the anoxic sludge digestor (Figure 28.5) or is removed, dried, and burned or used for fertilizer.

Wastewater normally stays in an activated sludge tank for 5 to 10 hours, a time too short for complete oxidation of all organic matter. However, during this time much of the soluble organic matter is adsorbed to the floc and incorporated by the microbial cells. The BOD of the liquid effluent is considerably reduced (up to 95%) by this process, with most of the BOD now contained in the settled flocs. Nearly complete BOD reduction can occur if the flocs are then transferred to the anoxic sludge digestor for conversion to CO_2 and CH_4.

(a)

(b)

Wastewater from primary treatment

Aeration tank

Settling tank

Clear effluent

Activated sludge

Air

Anoxic sludge digestor

Activated sludge return

Excess sludge

(c)

● **Figure 28.6** **Aerobic secondary wastewater treatment processes.** A treatment facility for a small city, Carbondale, Illinois. (a) Trickling filter. The booms rotate, distributing wastewater slowly and evenly on the rock bed. The rocks are 10–15 cm in diameter, and the bed is 2 m deep. (b–c) The activated sludge process. (b) Aeration tank of an activated sludge installation in a metropolitan wastewater treatment plant. The tank is 30 m long, 10 m wide, and 5 m deep. (c) Wastewater flow through an activated sludge installation. Recirculation of activated sludge to the aeration tank introduces microorganisms responsible for oxic digestion of the organic components of the wastewater. The anoxic sludge digestor is detailed in Figure 28.5 and the major microbial interactions are described in Figure 19.23.

Tertiary Wastewater Treatment

Tertiary wastewater treatment is any physicochemical or biological process employing bioreactors, precipitation, filtration, or chlorination procedures similar to those employed for drinking water purification (see Section 28.3). The goal is to sharply reduce the levels of inorganic nutrients, especially phosphate, nitrite, and nitrate, from the final effluent.

Most treatment plants now chlorinate the effluent at the end of the secondary treatment procedures (to further reduce the possibility of biological contamination) and discharge the treated water into streams or lakes. In the eastern United States, many wastewater treatment facilities are using ultraviolet (UV) radiation to disinfect effluent water. Ozone (O_3), a strong oxidizing agent that is an effective bactericide and viricide, is also used at over 40 plants in the United States for wastewater disinfection.

Wastewater receiving proper tertiary treatment contains so few nutrients that it cannot support extensive microbial growth. Tertiary treatment is the most complete method of treating sewage but has not been widely adopted due to expense.

🛑 28.2 Concept Check

Wastewater treatment is primarily concerned with treating sewage and industrial wastes to reduce the BOD (biochemical oxygen demand) to acceptable levels. Primary, secondary, and tertiary treatment of water involve physical, biological, and physicochemical processes. After secondary or optional tertiary treatment, water may be suitable for release directly to a water purification plant.

◆ What is biochemical oxygen demand (BOD)? Why is BOD reduction necessary in wastewater treatment?

◆ Identify the processes in primary, secondary (anoxic and oxic), and tertiary sewage treatment.

◆ Identify the final products of wastewater treatment. How might these end products be used?

● **Figure 28.7** **A floc formed by the bacterium *Zoogloea ramigera*.** The floc found in the activated sludge process consists of a large number of small, rod-shaped cells of *Z. ramigera* surrounded by a polysaccharide slime layer, arranged in characteristic fingerlike projections. Negative stain using India ink. The biology of *Zoogloea* is discussed in Section 12.7.

28.3 Drinking Water Purification

Wastewaters treated by secondary methods are generally of such quality that they can be directly discharged into rivers and streams. However, such water is not **potable**, or safe for human consumption. The production of potable water requires further treatment to remove potentially pathogenic microorganisms, eliminate taste and odor, reduce nuisance chemicals such as iron and manganese, and decrease **turbidity** (a measure of *suspended solids*). **Suspended solids** are small particles of solid pollutants that resist separation by ordinary physical means.

Physical and Chemical Purification

A typical drinking water treatment installation for a medium-sized city is shown in Figure 28.8*a*●. Figure 28.8*b* traces the flow of **raw water (untreated water)** through a typical treatment scheme. Raw water is first pumped from the source, in this case a river, to a *sedi-*

mentation basin where anionic **polymers**, alum (aluminum sulfate), and chlorine are added. **Sediment**, soil, sand, mineral particles, and other large particles, settle out. The sediment-free water is then pumped to a **clarifier** or **coagulation basin**, a large holding tank where **coagulation** takes place. The alum and anionic polymers form large particles from the much smaller *suspended solids*. After mixing, the particles continue to interact, forming large, aggregated masses, a process known as **flocculation**. The large, aggregated particles, called floc, settle out by gravity, trapping microorganisms and absorbing suspended organic matter and sediment.

After coagulation and flocculation, the clarified water undergoes **filtration**. The water is passed through a series of filters designed to remove any remaining organic or inorganic solutes, as well as any suspended particles and microorganisms. The filters usually consist of thick layers of sand, activated charcoal, and ion exchange filtration media. When combined with previous purification steps, the filtered water is free of all particulate matter, most organic and inorganic chemicals, and nearly all microorganisms.

Disinfection

Clarified, filtered water must then be disinfected before it is released to the supply system as pure, potable **finished water**. Chlorination is the most common method of disinfection. In sufficient doses, chlorine kills microorganisms within 30 minutes (certain pathogenic protozoa such as *Cryptosporidium* are not easily killed by chlorine treatment and thus can be important waterborne pathogens; see Section 28.6). In addition to killing microorganisms, chlorine reacts with organic compounds, oxidizing and effectively neutralizing them. Therefore, since most taste- and odor-producing compounds are organic in nature, chlorine treatment also improves water taste and smell. Chlorine is added to water either from a concentrated solution of sodium hypochlorite or calcium hypochlorite or as a gas from pressurized tanks. The latter method is used most commonly in large water treatment plants because it is most amenable to automatic control.

Chlorine is consumed when it reacts with organic materials. Therefore, sufficient quantities of chlorine must be added to water containing organic materials so that a small amount, called the *chlorine residual*, remains to react with the microorganisms after all reactions with organic materials have occurred. The water plant operator performs chlorine analyses on the treated water to determine the residual level of chlorine. A residual chlorine level of $0.2-0.6$ $\mu g/ml$ is suitable for most water supplies. After chlorine treatment, the now potable water is pumped to storage tanks from which it flows by gravity or pumps through a **distribution system** of storage tanks and supply lines to the consumer. Residual chlorine levels ensure that the finished water will reach the

● **Figure 28.8** **Water purification plant.** (a) Aerial view of a water treatment plant in Louisville, Kentucky, USA. The arrows indicate direction of flow of water through the plant. (b) Schematic overview of a typical community water purification system.

consumer without becoming contaminated (assuming that there is no catastrophic failure, such as a broken pipe, in the distribution system). Chlorine gas, even when dissolved in water, is extremely volatile and can dissipate within hours from treated water. To further ensure that residual chlorine levels are maintained throughout the distribution system, most municipal water treatment plants also introduce ammonia gas with the chlorine to form the stable, nonvolatile chlorine-containing compound **chloramine**, $HOCl + NH_3 \rightarrow NH_2Cl + H_2O$.

Ultraviolet (UV) radiation is also used as an effective means of disinfection. As we discussed in Section 22.2, UV radiation is used to treat secondarily treated effluent from water treatment plants. In Europe, UV irradiation is commonly used for drinking water applications and is being considered for use in the United States. For disinfection purpose, UV light is generated from mercury vapor lamps. Their major energy output is at 253.7 nm, a wavelength that is bactericidal and viricidal (Section 10.4). However, standard application of UV radiation does not kill cysts and oocysts of protozoans like *Giardia* and *Cryptosporidium* (see Section 28.6).

The advantages of using UV radiation over chemical disinfection procedures like chlorination are several. First, UV irradiation is a physical process that introduces no chemicals to the water. Second, short contact times allow it to be used in existing flow systems, keeping capital costs very low. Third, numerous studies indicate that no disinfection by-products are formed. Especially in smaller systems where finished water is not pumped long distances or held for long periods (reducing the need for residual chlorine), UV disinfection may be preferable to chlorination.

28.3 Concept Check

Drinking water plants employ industrial-scale physical and chemical systems that remove or neutralize biological, inorganic, and organic contaminants from a variety of natural, community, and industrial sources. Water purification plants employ clarification, filtration, and disinfection processes to produce potable water. Finished potable water is free of chemical and biological contamination.

◆ Trace the treatment of water through a drinking water treatment plant, from the inlet to the final distribution point (faucet).

◆ What specific purposes do *sedimentation, coagulation, filtration,* and *disinfection* accomplish in the drinking water treatment process?

 II WATERBORNE MICROBIAL DISEASES

Common-source infectious diseases are caused by microbial contamination of materials shared by a large number of individuals. The most important common source of infectious disease is contaminated water. This is because the failure of a single step in drinking water treatment may result in exposure of thousands or even millions of individuals to an infectious agent.

Common-source waterborne diseases are a very significant source of morbidity and mortality, especially in developing countries. A variety of bacteria, viruses, and protozoa cause waterborne infectious diseases. Waterborne diseases begin as infections. Water may cause infection even if only a small number of microorganisms are present. The exact numbers of pathogens necessary to cause disease are functions of the virulence of the pathogen and the general ability of the host to resist infection (∞ Sections 21.8 and 21.14).

28.4 Sources of Waterborne Infection

Human pathogens can be transmitted through improperly treated water used for drinking and cooking. Another very common source of disease transmission is through pathogen-contaminated water used for swimming and bathing.

Potable Water

Because everyone consumes water through drinking and cooking, water is a common source of pathogen dissemination and has a very high potential for the catastrophic spread of epidemic disease. As discussed previously, water supplies in developed countries usually meet rigid quality standards, effectively limiting the sp-read of waterborne diseases. There are, however, occasional waterborne disease outbreaks in developing countries due to lapses in water quality. Isolated outbreaks affecting low numbers of individuals also occur from consumption of contaminated water from nonregulated sources such as private wells or from untreated water from streams or lakes. These sources may be contaminated by fecal material from humans or animals.

Microorganisms transmitted in water generally grow in the intestines and leave the body in feces. Fecal pollution of water supplies may then occur. In addition, if the contamination is not identified (for example, by the coliform test; see Section 28.1) and eliminated by disinfection (see Section 28.3), then a new host may consume the water and the pathogen may colonize the intestine and cause disease. In the United States, a number of different bacterial and protozoan pathogens are occasionally transmitted in drinking water (Table 28.1). We will discuss giardiasis and cryptosporidiosis in Section 28.7, but we will save our discussions of salmonellosis and *Escherichia coli* O157:H7 infections for Chapter 29, where we introduce these pathogens as agents of foodborne infections, their most common mode of transmission.

Recreational Water

Recreational water includes freshwater recreational areas such as ponds, streams, and lakes, as well as public swimming and wading pools. The operation, disinfection, and filtration of public swimming and wading pools are regulated by state and local health departments. In the United States, the Environmental Protection Agency has established guidelines for recreational fresh water (monthly geometric mean of ≤33/100 ml for enterococci or ≤126/100 ml for *Escherichia coli*), although local and state authorities can set standards above or below the guidelines. Private and therefore unregulated swimming pools, spas, and

Table 28.1	Infectious disease outbreaks associated with drinking water in the United States[a]		
Disease	**Agent**	**Outbreaks**	**Cases**
Salmonellosis	*Salmonella* species	2	208
Giardiasis	*Giardia intestinalis*	6	52
Cryptosporidiosis	*Cryptosporidium parvum*	1	5
Acute gastro-intestinal illness	*Escherichia coli* O157:H7	4	60
	Campylobacter jejuni	2	117
	E. coli O157:H7 and *C. jejuni*	1	781
	Small round virus	1	70
	Norwalk-like viruses	3	356
	Unknown	17	416

[a] Compiled from data provided by the Centers for Disease Control and Prevention for 1999–2000. There were a total of 37 outbreaks and 2065 cases of infectious disease due to drinking water contamination by infectious agents. Regulated community-owned water systems were responsible for 237 cases (11.5%). Noncommunity water systems such as those in some factories, schools, and on cruise ships accounted for 1425 cases (69%). Individual water supply systems such as wells, springs, and streams accounted for 403 cases (19.5%).

hot tubs are also occasional sources of outbreaks of waterborne diseases.

Over the last 12 years for which data are complete, an average of 13 waterborne disease outbreaks have occurred annually from recreational waters in the United States. Table 28.2 categorizes these outbreaks according to the diseases produced.

Waterborne Infections in Developing Countries

Worldwide, waterborne infections are a much larger problem than in the United States and other developed countries. Developing countries often have inadequate water and sewage treatment facilities, and access to safe, potable water is limited. As a result, diseases such as cholera (Section 28.5), typhoid fever, and amebiasis (Section 28.8) are important public health problems worldwide.

 28.4 Concept Check

Drinking water and recreational water may both be sources of waterborne pathogens. In the United States, the number of disease outbreaks due to either of these sources is relatively small in relation to the large number of exposures to water. Worldwide, lack of adequate water treatment facilities and access to clean water contribute significantly to the spread of infectious diseases.

◆ Identify the microorganism most commonly responsible for disease outbreaks due to drinking water contamination.

◆ Identify the microorganism most commonly responsible for disease incidence (numbers of infections) due to drinking water contamination.

◆ Identify the organisms most likely to cause an outbreak of disease due to contaminated recreational water.

Table 28.2	Infectious disease outbreaks associated with recreational water in the United States[a]	
Disease	**Number of outbreaks**	**Percent**
Gastroenteritis[b]	74	46.8
Dermatitis/keratitis[c]	50	31.6
Meningoencephalitis[d]	22	13.9
Other[e]	12	7.6

[a] Compiled from data provided by the Centers for Disease Control and Prevention for 1989–2000.
There were 158 outbreaks of recreational waterborne disease, or about 13 outbreaks per year.

[b] Most cases of gastroenteritis were due to *Cryptosporidium parvum* (Section 28.6), *Escherichia coli* O157:H7 (∞ Section 29.8), or a Norwalk-like virus (Section 28.8).

[c] Most cases of dermatitis were caused by *Pseudomonas aeruginosa*.

[d] Meningoencephalitis was caused by the ameba *Naegleria fowleri* (Section 28.8).

[e] Other diseases include leptospirosis caused by *Leptospira interrogans*, Pontiac fever due to infection by *Legionella* (Section 28.7), and acute respiratory infections of unknown cause.

28.5 Cholera

Cholera is a severe diarrheal disease that is now largely restricted to the developing parts of the world. Cholera is an example of a major waterborne disease that can be controlled by application of appropriate water treatment measures.

Biology and Epidemiology

Cholera is caused by *Vibrio cholerae*, a gram-negative, curved rod (∞ Section 12.12) transmitted through ingestion of contaminated water. However, as with many waterborne diseases, cholera is also associated with food consumption. For example, in the Americas, consumption of raw shellfish and raw vegetables has been associated with cholera. Presumably, the vegetables were washed in contaminated water and the shellfish beds were contaminated by untreated sewage.

Since 1817, cholera has swept the world in seven major pandemics. Two distinct strains of *V. cholerae* are recognized, known as the *classic* and the *El Tor* biotypes. The *V. cholerae* O1 El Tor biotype started the seventh pandemic in Indonesia in 1961, and its spread continues to the present. In 1992, a genetic variant, the *V. cholerae* O139 Bengal serotype, arose in Bangladesh and caused an extensive epidemic. In 2001, there were 184,311 reported cases and 2728 deaths worldwide, with over 94% of disease occurring in Africa. The 1961 pandemic has caused over 5 million cases of cholera and more than 250,000 deaths.

Cholera is endemic in Africa, Southeast Asia, the Indian subcontinent, and Central and South America. Epidemic cholera occurs frequently in areas where sewage treatment is either inadequate or absent. Even in devel-

oped countries, the disease is a threat. Sporadic endemic outbreaks of cholera (fewer than 250 total cases) have been reported in the United States, mostly along the Gulf Coast, over the last decade. Currently less than five cases are reported each year in the United States, and the cases are rarely caused by drinking water. Raw shellfish seems to be the most common vehicle, presumably because *V. cholerae* appears to be endemic and free-living in coastal waters, adhering to normal flora (Figure 28.9●).

Pathogenesis

Following ingestion of a substantial inoculum, the *Vibrio cholerae* cells take up residence in the small intestine. Studies in human volunteers have shown that stomach acidity is responsible for the large inoculum needed to initiate cholera. The ingestion of $10^8 - 10^9$ cholera vibrios is generally required to cause disease, but human volunteers given bicarbonate to neutralize gastric acidity developed cholera when given as few as 10^4 cells. Even lower cell numbers can initiate infection if *V. cholerae* is ingested with food, presumably because the food protects the vibrios from stomach acidity.

In the small intestine *V. cholerae* attaches to epithelial cells, where it grows and releases enterotoxin (∞ Section 21.11). Cholera enterotoxin causes severe diarrhea that can result in dehydration and death unless the patient is given fluid and electrolyte therapy. The enterotoxin causes fluid losses of up to 20 liters (20 kg or 44 lb) per day. The mortality rate from *untreated* cholera is generally 25 to 50%, but can be much greater under conditions of severe crowding and malnutrition.

Diagnosis, Prevention, and Treatment

Cholera is diagnosed by the presence of the gram-negative comma-shaped *Vibrio cholerae* bacilli in the "rice water" stools of patients with severe diarrhea (Figure 28.10●).

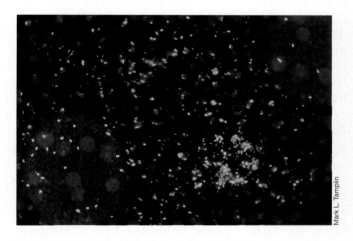

● **Figure 28.9** Cells of *Vibrio cholerae* attached to the surface of *Volvox*, a freshwater alga. The isolate was from a cholera-endemic area in Bangladesh. The *V. cholerae* cells are stained green by a monoclonal antibody to bacterial cell surface proteins. The red color is due to the fluorescence of chlorophyll *a* in the algae.

CDC/PHIL

● **Figure 28.10 A stool sample from a cholera patient.** The "rice-water" stool is nearly liquid. The solid material that has settled in a bottom layer is mucus. The stool from cholera patients is essentially isotonic with blood, containing high amounts of Na^+, K^+, and HCO_3^- (bicarbonate) ions, as well as large numbers of *Vibrio cholerae* cells. The biology and metabolism of *Vibrio cholerae* is discussed in Section 12.12, and the action of the cholera toxin is shown in Figure 21.22.

Immunization is not normally recommended for cholera prevention. Natural infection and the current vaccines offer only limited, short-duration immunity. No vaccine protects against the *V. cholerae* O139 serotype. Public health measures such as adequate sewage treatment and a reliable source of safe drinking water are the most important measures for preventing cholera. *V. cholerae* is completely eliminated from wastewater during proper sewage treatment and drinking water purification procedures. For individuals traveling in cholera-endemic areas, untreated water, raw foods, fish, and shellfish should be avoided.

Intravenous or oral liquid and electrolyte replacement therapy (20 g glucose, 4.2 g NaCl, 4.0 g NaHCO₃, 1.8 g KCl, dissolved in one liter of water) is the most effective means of cholera treatment. Oral treatment is preferred since no special equipment or sterile precautions are necessary. Effective treatment reduces the mortality rate to about 1%. Streptomycin or tetracycline may shorten the course of cholera, but antibiotics are of little benefit without simultaneous fluid and electrolyte replacement.

28.6 Giardiasis and Cryptosporidiosis

Giardiasis and cryptosporidiosis are diseases caused by the protozoans *Giardia intestinalis* and *Cryptosporidium parvum*, respectively. These organisms continue to be problematic even in well-regulated water supplies because they are found in nearly all surface waters and are highly resistant to chlorine.

Giardiasis

Giardia intestinalis is a flagellated protozoan (∞ Section 14.10) that is usually transmitted to humans in fecally contaminated water, although foodborne and sexual transmission of giardiasis has also been documented. Giardiasis is an acute gastroenteritis caused by the protozoan parasite (Figure 28.11●). The protozoal cells, called trophozoites (Figure 28.11*a*) produce a resting stage called a **cyst** (Figure 28.11*b*). The cyst has a thick protective wall that allows the pathogen to resist drying and chemical disinfection. After ingestion in contaminated water, the cysts germinate, attach to the intestinal wall, and cause the symptoms of giardiasis: an explosive, foul-smelling, watery diarrhea, intestinal cramps, flatulence, nausea, weight loss, and malaise. Symptoms may be acute or chronic. The foul-smelling diarrhea and the absence of blood or mucus in the stool are diagnostically helpful in distinguishing giardiasis from diarrhea of bacterial or viral origin. Many individuals who are infected exhibit no symptoms, simply acting as infected carriers.

Giardia intestinalis was implicated in 6 of the 37 recent drinking water infectious disease outbreaks in the United States (Table 28.1). Giardiasis can also occur after accidental ingestion of water from infected swimming pools or lakes. *Giardia* cysts have been found in 97% of surface water sources (lakes, ponds, and streams) in the United States. The thick-walled cysts are resistant to chlorine and ultraviolet radiation, and many outbreaks have been associated with water systems that use only chlorination as a means of water purification. Water

(a) *(b)*

● **Figure 28.11** **Scanning electron micrographs of the parasite *Giardia*.** (a) Motile trophozoite. The trophozoite is about 15 μm in length. (b) Cyst. The cyst is about 11 μm in length. The interesting phylogeny of this protozoan is discussed in Section 14.9. The biology of protozoa in general is discussed in Section 14.10.

subjected to proper clarification and filtration followed by chlorination or other disinfection (see Section 28.3) is generally free of *Giardia* cysts.

Isolated cases of giardiasis have also been associated with untreated drinking water in wilderness areas. Beavers and muskrats are frequent carriers of *Giardia* and may transmit cells or cysts to water supplies, causing human infection when ingested. As a safety precaution, all water consumed from rivers and streams, for example, during a camping or hiking trip, should be filtered *and* treated with iodine or chlorine, or filtered *and* boiled. Boiling is the preferred method to assure that water is free of pathogens.

Laboratory diagnostic methods include the demonstration of *Giardia* cysts in the stool or the demonstration of *Giardia* antigens in the stool using a direct ELISA (enzyme-linked immunosorbent assay; ⚬⚬ Section 24.10). The drugs quinacrine, furazolidone, and metronidazole are useful in treating acute disease.

Cryptosporidiosis

The protozoan *Cryptosporidium parvum* occurs in nature as a parasite in a variety of warm-blooded animals. The parasites are small 2–5 μm round *coccidia* that invade and grow intracellularly in mucosal epithelial cells of the stomach and intestine. The protozoan produces thick-walled, chlorine-resistant, infective oocysts, which are shed into water in high numbers in the feces of infected warm-blooded animals (Figure 28.12●). The infection is passed on when other animals consume the fecally contaminated water. *Cryptosporidium* cysts are highly resistant to chlorine (up to 14 times as resistant as the chlorine-resistant *Giardia*) and ultraviolet radiation disinfection, and therefore sedimentation and filtration methods must be used to remove *Cryptosporidium* from water supplies.

Cryptosporidium parvum was responsible for the largest single common-source outbreak of a waterborne disease ever recorded in the United States. In Milwau-

kee, Wisconsin, in the spring of 1993, over 403,000 people in the population of 1.6 million developed a diarrheal illness that was traced to the municipal water supply. Spring rains and runoff from surrounding farmland had drained into Lake Michigan and overburdened the water purification system, leading to contamination by *C. parvum*. The protozoan is a significant intestinal parasite in dairy cattle, the likely source of the outbreak.

Cryptosporidiosis is usually a self-limiting mild diarrhea that subsides in 2 weeks or less in normal individuals. However, individuals with impaired immunity such as acquired immunodeficiency syndrome (AIDS) patients or the very young or old can develop serious complications. In the Milwaukee outbreak, about 4400 people

● **Figure 28.12** **An acid-fast stained fecal smear showing *Cryptosporidium* oocysts.** The thick-walled oocysts, stained red, are about 3–5 μm in diameter. *Cryptospordium*, an obligate intracellular parrasite, is a member of the Sprorozoa, a group of protozoans discussed in Section 14.10.

required hospital care, and several died of complications from the disease, including severe dehydration.

The Milwaukee outbreak highlights the vulnerability of water purification systems, the need for constant water monitoring and surveillance, and the catastrophic consequences of the failure of a large water supply system. From an economic point of view, the epidemic cost the inhabitants of Milwaukee over $96 million in medical costs and lost productivity.

Laboratory diagnostic methods for cryptosporidiosis include the demonstration of *Cryptosporidium* oocysts in the stool (Figure 28.12). Treatment is unnecessary for those with normal immunity. For individuals undergoing immunosuppressive therapy (e.g., prednisone), discontinuation of immunosuppressive drugs is recommended. Immunocompromised individuals should be given supportive therapy (for example, intravenous fluids and electrolytes).

 28.6 Concept Check

Giardiasis and cryptosporidiosis are spread by the chlorine-resistant cysts of *Giardia intestinalis* and *Cryptosporidium parvum*, respectively, in drinking water and recreational water contaminated by the feces of infected humans or animals. Infection with either parasite causes diarrhea and may lead to more serious disease in compromised individuals.

◆ Explain the importance of cysts in the survival and infectivity of both *Giardia intestinalis* and *Cryptosporidium parvum*.

◆ Why are protozoans often associated with waterborne diseases, even in developed countries? Outline steps to reduce their impact.

28.7 Legionellosis (Legionnaires' Disease)

Legionella pneumophila, the bacterium that causes legionellosis, is an important waterborne pathogen normally transmitted in aerosols rather than through drinking water or recreational water (⌘ Section 25.11).

Biology and Epidemiology

Legionella pneumophila was first discovered as the pathogen that caused an outbreak of pneumonia during an American Legion convention in Philadelphia (USA) in the summer of 1976. *L. pneumophila* is a thin, gram-negative obligately aerobic rod (Figure 28.13●) with complex nutritional requirements, including an unusually high iron requirement. *L. pneumophila* can be detected by immunofluorescence techniques (⌘ Section 24.9) and can be isolated from terrestrial and aquatic habitats as well as from patients suffering from legionellosis.

L. pneumophila is present in small numbers in lakes, streams, and soil. It is relatively resistant to heating and chlorination, so it can spread through water distribution

● **Figure 28.13** *Legionella pneumophila* **and Legionnaires' disease.** Transmission electron micrograph of a thin section of *Legionella pneumophila* cells. Cells are 0.3–0.6 μm in diameter and up to 2 μm in length. The clear areas in the cells are polymers of β-hydroxybutyrate, a carbon storage polymer (⌘ Section 4.11).

systems. It is commonly found in large numbers in cooling towers and evaporative condensers of large air conditioning systems. The pathogen grows in the water and is disseminated in humidified aerosols. Human infection is via airborne droplets, but the infection is not spread person to person. Multiple outbreaks of legionellosis tend to peak in mid-to late summer months when air conditioners are extensively used.

L. pneumophila has also been found in hot water tanks and whirlpool spas where it grows to high numbers in warm (35–45°C), stagnant water. Epidemiological studies indicate that *L. pneumophila* infections occur at all times of the year, primarily as a result of aerosols generated by heating/cooling systems and common practices such as showering or bathing. Overall, the incidence of *reported* cases of legionellosis has been about 4–6 cases per million in the United States, but up to 90% of actual cases are probably not diagnosed or properly reported. Prevention of legionellosis can be accomplished by improving the maintenance and design of water-dependent cooling and heating systems and water delivery systems. The pathogen can be eliminated from water supplies by hyperchlorination or by heating to >63°C.

Pathogenesis

L. pneumophila is an intracellular parasite that invades and grows in alveolar macrophages and monocytes (⌘ Section 22.2). Infections are often asymptomatic or produce a mild cough, sore throat, mild headache, and fever. These mild, self-limiting cases, called *Pontiac fever*, need no treatment, and resolve in 2–5 days. However, more serious infections resulting in cases of pneumonia often occur in elderly individuals whose resistance has been previously compromised. Certain serotypes of *L. pneumophila* (more than 10 are known) are strongly associated with the pneumonic form of the infection. Prior to the onset of pneumonia, intestinal disorders are common, followed by high fever, chills, and muscle aches. These symptoms precede the dry cough and chest and abdominal pains typical of legionellosis. Death occurs in up to 10% of cases and is usually due to respiratory failure.

Diagnosis and Treatment

Clinical detection of *L. pneumophila* is usually done by culture from bronchial washings, pleural fluid, or other body fluids. Serological (antibody) tests are used as retrospective evidence for *Legionella* infection (∞ Section 24.7). As an aid in diagnosis, *L. pneumophila* antigens can sometimes be detected in patient urine. *Legionella pneumophila* is sensitive to the antibiotics rifampin and erythromycin. Intravenous administration of erythromycin is the treatment of choice in most cases.

 28.7 Concept Check

Legionella pneumophila is a respiratory pathogen that causes legionellosis and Pontiac fever. *L. pneumophila* grows to high numbers in warm water and is spread via aerosols. The prevalence of legionellosis is not decreasing and infections are underreported.

◆ Indicate the source of *Legionella pneumophila*.

◆ Identify specific measures for control of *Legionella pneumophila*.

28.8 Typhoid Fever and Other Waterborne Diseases

A variety of bacteria, viruses, and protozoa can transmit common-source waterborne diseases. These diseases are a significant source of morbidity, especially in developing countries. Here we briefly discuss several other important waterborne diseases.

Typhoid Fever

On a global scale, probably the most important pathogenic *Bacteria* transmitted by the water route are *Salmonella typhi*, the organism causing typhoid fever, and *Vibrio cholerae*, the organism causing cholera, as discussed above. Although *Salmonella typhi* may also be transmitted by contaminated food (∞ Section 29.7) and by direct contact from infected individuals (∞ Section 25.4), the most common and serious means of transmission worldwide is through water. Fortunately, typhoid fever has been virtually eliminated in developed countries, primarily due to effective water treat-

ment procedures (see Figure 28.2). In the United States, there are fewer than 400 cases in most years, but, as was discussed in Section 28.1, typhoid fever was a major public health threat before drinking water was routinely filtered and chlorinated. However, breakdown of water treatment methods, contamination of water during floods, earthquakes, and other disasters, or cross-contamination of water supply pipes from leaking sewer lines, can propagate epidemics of typhoid fever, even in developed countries.

Viruses

Viruses can also be transmitted in water and cause human disease. Quite commonly, enteroviruses such as poliovirus (∞ Section 16.8), Norwalk-like virus (∞ Table 25.8), and hepatitis A virus (∞ Section 26.11) are shed into the water in fecal material. The most serious of these is poliovirus, but wild poliovirus has been eliminated from the Western Hemisphere (∞ Section 25.9). Although viruses can survive in water for relatively long periods, they are inactivated by water disinfection agents such as chlorine. The maintenance of 0.6 parts per million (ppm) of chlorine in water (Section 28.3) ensures the neutralization of viruses in the water supply.

Amebiasis

A number of amoebae inhabit the tissues of humans and other vertebrates, usually in the oral cavity or intestinal tract, and some of these are pathogenic. We discussed the general properties of ameboid protozoa in Section 14.10.

Worldwide, *Entamoeba histolytica* is a common pathogenic protozoan transmitted to humans, primarily by contaminated water and occasionally through contaminated food (Figure 28.14●). *E. histolytica* is an *anaerobic* ameba; the trophozoites lack mitochondria. Like *Giardia*, the trophozoites of *E. histolytica* produce cysts. Ingested cysts germinate in the intestine, where amebic cells grow both on and in intestinal mucosal cells. Many infections are asymptomatic, but continued growth may lead to invasion and ulceration of the intestinal mucosa, causing diarrhea and severe intestinal cramps. Diarrhea can progress to invasion of the intestinal wall, a condition known as *dysentery*, characterized by intestinal inflammation, fever, and the passage of intestinal exudates, including blood and mucus. If not treated, invasive trophozoites of *E. histolytica* can invade the liver, and occasionally the lung and brain. Growth in these tissues can cause severe abscesses and even death. Worldwide, up to 100,000 individuals die each year from invasive amebic dysentery. The disease is extremely common in tropical and subtropical countries worldwide, with at least 50 million people developing symptomatic diarrhea annual-

● **Figure 28.14 The trophozoite of *Entamoeba histolytica*, the ameba that causes amebiasis.** Note the discrete, darkly stained nucleus. The small red structures are red blood cells. The trophozoites range from 12–60 μm in length. The biology of the amoebae is discussed in Section 14.10.

ly and up to tenfold more having asymptomatic disease. In the United States, there are several hundred cases per year, with most occurring near international borders in the Southwest.

E. histolytica amebiasis can be treated with the drug *dehydroemetine* for invasive disease and diloxanide furoate for certain asymptomatic cases, such as in immune-compromised individuals, but amebicidal drugs are not universally effective. Spontaneous cures do occur, suggesting that the host immune system plays some role in ending the infection. However, protective immunity is not an outcome of primary infection, and reinfection is common. The disease occurs at very low incidence in regions that practice adequate sewage treatment. Ineffective sewage treatment and use of untreated surface waters for drinking purposes are the usual scenarios for cases of amebiasis.

Laboratory diagnosis involves the demonstration of *E. histolytica* cysts in the stool, trophozoites in tissue, or the demonstration of antibodies to *E. histolytica* in the blood using an ELISA (enzyme-linked immunosorbent assay; ∞ Section 24.10).

Naegleria fowleri can also cause amebiasis, but in a very different form. *N. fowleri* is a free-living ameba found in soil and in water runoff. *N. fowleri* infections usually result from swimming or bathing in warm, soil-contaminated water sources such as hot springs or lakes and streams in the summer. This free-living ameba enters the body through the nose and burrows directly into the brain. Here, the organism propagates, causing extensive hemorrhage and brain damage (Figure 28.15●). This condition is called **meningoencephalitis**. Death usually results within a week. In the past 12 years, there have been 22 outbreaks of meningoencephalitis from recreational waters in the United States (Table 28.2). From 1999 through 2000, there were

● **Figure 28.15 Trophozoites of *Naegleria fowleri*, an ameba that causes meningoencephalitis, in brain tissue.** A large number of oval to round and ameboid (irregularly shaped) trophozoites are present as dark-stained structures with densely stained nuclei. The photograph shows extensive destruction of the surrounding brain tissue. Individual trophozoites are 10–35 μm long. The biology of the amoebae is discussed in Section 14.10.

4 outbreaks, each involving a single individual who was infected by swimming or wading in a lake, pond, or stream in summer.

Diagnosis of *N. fowleri* infection requires observation of the amoebae in the cerebrospinal fluid. If a definitive diagnosis can be done quickly, the drug amphotericin B can be successfully used to treat infections.

⬡ **28.8 Concept Check**

Typhoid fever, viral infections, and amebiasis are important waterborne diseases. Waterborne typhoid fever and viral illnesses, while still common diseases in developing countries, have been controlled by effective water treatment in developed countries. Amebic dysentery caused by *Entamoeba histolytica* is a worldwide problem that affects millions of people. Meningoencephalitis is a rare but serious condition caused by *Naegleria* amebiasis.

◆ Explain the impact of effective water hygiene on the spread of human diseases such as typhoid fever and polio spread by fecal contamination of water supplies.

◆ Describe public health measures that could be used to eliminate or reduce the number of cases of amebiasis due to *Entamoeba histolytica* or meningoencephalitis due to *Naegleria fowleri*.

REVIEW QUESTIONS

1. Define the term *coliform* and explain the coliform test. Why is the coliform test used to assess the purity of drinking water (◯◯◯ Section 28.1)?

2. Trace the purification of wastewater in a typical treatment plant. What is the overall reduction in the BOD (◯◯◯ Section 28.2)?

3. Identify (stepwise) the methods used to process drinking water. What important contaminants are targeted by each step in the process (◯◯◯ Section 28.3)?

4. Identify the major sources of waterborne infection. Why are common sources of infection so dangerous to public health (◯◯◯ Section 28.4)?

5. Why are antibiotics ineffective for the treatment of cholera? What methods are commonly used to treat cholera victims (◯◯◯ Section 28.5)?

6. Giardiasis and cyptosporidiosis remain significant public health problems even in areas with stringent water-quality standards. Explain (◯◯◯ Section 28.6).

7. Describe the main features of legionellosis. What distinguishes this disease from other waterborne diseases (◯◯◯ Section 28.7)?

8. Indicate the methods that have been used to control salmonellosis, Norwalk virus, and amebiasis in water systems in developed countries (◯◯◯ Section 28.8).

APPLICATION QUESTIONS

1. Why is reduction in the BOD of wastewater the primary goal of wastewater treatment? What occurs if the BOD of wastewater is not significantly reduced before it is distributed to local water sources such as lakes or streams?

2. In the United States, the federal government has defined a strict set of drinking water standards. These standards are enforced by law. However, recreational water standards are not regulated as tightly. Explain why recreational water standards are more flexible and devise recreational water standards for the area in which you live.

3. Cholera has a long and complex history of interaction with humans, and remains today as a very serious waterborne disease. Worldwide, we are in the midst of an ongoing pandemic. Using sources such as the World Health Organization and the Centers for Disease Control and Prevention, define the status of the current pandemic with regard to its geographic distribution and most recent outbreaks. Comment on methods that could be used to decrease the spread of cholera and control the annual epidemic outbreaks that occur in endemic areas. Can cholera be eradicated?

4. As a visitor to a country in which cholera is an endemic disease, what specific steps would you take to reduce your risk of cholera exposure? Will these precautions also prevent you from contracting other waterborne diseases? Which ones? Identify waterborne diseases for which your precautions will not prevent infection. Present further recommendations for preventing those diseases.

5. Why are surface waters contaminated with the cysts of various protozoans? What steps might public health officials take to remedy this problem?

6. Discuss the wastewater and drinking water treatment schemes that must be in place to control such diseases as typhoid fever. Is it possible to eliminate *Salmonella typhi* as has been effectively done for poliovirus? Would it be possible, or even prudent, to eliminate *Naegleria fowleri* or *Entamoeba histolytica*? Explain.

29

Food infections and food poisoning are important causes of morbidity and mortality worldwide. Food infections with *Campylobacter jejuni*, shown in this electron micrograph, are due largely to contaminated raw and ready-to-eat foods.

FOOD PRESERVATION AND FOODBORNE MICROBIAL DISEASES

WORKING GLOSSARY

Botulism food poisoning due to ingestion of food containing botulinum toxin produced by *Clostridium botulinum*

Canning the process of sealing food in a closed container and heating to destroy living organisms

Fermentation the anaerobic catabolism of organic compounds, generally carbohydrates, in the absence of an external electron acceptor

Food infection disease caused by active infection resulting from ingestion of pathogen-contaminated food

Food poisoning (food intoxication) disease caused by the ingestion of food that contains preformed microbial toxins

Food spoilage a change in the appearance, smell, or taste of a food that makes it unacceptable to the consumer

Irradiation the exposure of food to ionizing radiation for the purpose of inhibiting growth of microorganisms and insect pests or to retard ripening

Listeriosis gastrointestinal food infection caused by *Listeria monocytogenes* that may lead to bacteremia and meningitis

Lyophilization (freeze-drying) the process of removing all water from frozen food under vacuum

Nonperishable (stable) food food of low water activity that has an extended shelf life and is resistant to spoilage by microorganisms

Perishable food fresh food generally of high water activity that has a very short shelf life due to potential for spoilage by growth of microorganisms

Pickling the process of acidifying food to prevent microbial growth and spoilage

Salmonellosis enterocolitis caused by any of over 2000 variants of *Salmonella* spp.

Semiperishable food food of intermediate water activity that has a limited shelf life due to potential for spoilage by growth of microorganisms

Water activity (a_w) the availability of water for use in metabolic processes

---

Foodborne disease is an emerging area of concern in microbiology. Microorganisms are important factors in our food supply. Microorganisms are ubiquitous in our environment and are found in water, air, and even in food. Fresh foods, most prepared foods, and even some preserved foods are contaminated with microorganisms. However, microbial activity is highly desirable in certain foods. A variety of foods are produced or enhanced by microbial action. For example, dairy products such as cheese, buttermilk, sour cream, and yogurt are all produced by microbial fermentation. Sauerkraut is a fermented vegetable food. Meat products including certain sausages, pates, and liver spreads are produced using microbial fermentation techniques. Cider vinegar is produced by lactic acid bacteria (∞ Section 30.11), and alcoholic beverages are produced by fermentation processes using yeast (∞ Section 30.13). We discuss the use of microorganisms to produce edible foods on an industrial scale in Chapter 30.

In this chapter, however, we will concentrate on the negative aspects of microbial growth in food. Uncontrolled and unwanted microbial growth destroys vast quantities of food, causing significant economic loss as well as a tremendous loss of nutrients. Consumption of food contaminated with particular microorganisms or microbial products can also cause food infections or food poisoning.

▌FOOD PRESERVATION AND MICROBIAL GROWTH

First we discuss microorganisms that are important spoilage agents in foods. We then present a variety of methods used for preserving foods by controlling unwanted microbial growth.

29.1 Microbial Growth and Food Spoilage

A wide variety of microorganisms, including a few human pathogens, colonize and grow on common foods. Most foods provide a suitable medium for the growth of some microorganisms, and microbial growth usually reduces food quality and availability.

Food Spoilage

Food spoilage is any change in the appearance, smell, or taste of a food product that makes it unacceptable to the consumer. Spoiled food is not necessarily unsafe to eat, but spoiled food is generally regarded as unpalatable and will not be purchased or consumed. Food spoilage causes economic loss to producers, distributors, and consumers in the form of reduced quality and quantity as well as higher prices.

Since foods are organic material, they provide nutrients for the growth of a wide variety of chemoorganotrophic bacteria. The physical and chemical characteristics of the food determine its degree of susceptibility to microbial activity. With respect to spoilage, foods can be classified into three major categories: (1) **perishable food**, including many fresh food items; (2) **semiperishable food**, such as potatoes and nuts; and (3) **stable** or **nonperishable food,** such as flour and sugar (Table 29.1).

These food categories differ largely with regard to *moisture content*, which is related to **water activity**, a_w (∞ Section 6.14). Water activity is the availability of water for use in metabolic processes. *Nonperishable foods* have *low water activity* and can generally be stored for considerable lengths of time without spoilage. *Perishable* and *semiperishable foods* are those with *higher water activity*. These foods must be stored under conditions that slow or stop microbial growth.

Table 29.1	Food classification by storage potential
Food classification	**Examples**
Perishable	Meats, fish, poultry, eggs, milk, most fruits and vegetables
Semiperishable	Potatoes, some apples, and nuts
Nonperishable	Sugar, flour, rice, and dry beans

Fresh foods are spoiled by a variety of bacteria and fungi, and each type of fresh food is typically colonized by particular microorganisms (Table 29.2). Because the chemical properties of foods vary widely, different foods are colonized by the indigenous spoilage organisms that are best able to use the available nutrients.

For example, enteric bacteria such as *Salmonella, Shigella,* and *Escherichia,* all potential pathogens that live in the gut of animals, are rarely implicated in fruit or vegetable spoilage, but often contaminate and spoil meat. At slaughter, intestinal contents, including live bacteria, can leak and contaminate the meat. Likewise, lactic acid bacteria, the most common microorganisms in dairy products, are the major spoilers of milk and milk products. *Pseudomonas* species are found in both soil and animals and are thus widely involved in the spoilage of fresh foods.

Microbial growth in foods follows the normal pattern for bacterial growth (Sections 6.4–6.6). The *lag phase* may be of variable duration in a food, depending on the contaminating organism and its previous growth history. The time required for the population density to reach a significant level in a given food product depends on both the size of the initial inoculum and the rate of growth during the *exponential phase*. The rate of growth during the exponential phase depends on the *temperature*, the *nutrient value of the food*, and other conditions of growth. Only when the microbial population density reaches a substantial level are spoilage effects usually observed. Throughout much of the exponential growth phase, population densities may be so low that no effect can be observed, and only the last few doublings lead to observable spoil-

lage (Section 6.6). Thus, for much of the period of microbial growth in a food, there is no visible or easily detectable change in food quality.

 29.1 Concept Check

Foods often spoil due to growth of contaminating microorganisms. Foods vary considerably in their sensitivity to microbial growth, depending on their nutrient value and water content. Perishable and semiperishable foods have limited shelf life due to spoilage. A variety of microorganisms induce spoilage, and some food spoilage microorganisms are also potential pathogens.

◆ List the major categories of food with respect to water availability. Suggest storage conditions for each food type.

◆ Identify at least three bacterial *genera* that cause both food spoilage and human disease.

29.2 Food Preservation

We now examine food storage and preservation processes. In many cases these methods slow the growth of microorganisms that produce food spoilage or are foodborne human pathogens. In other cases, methods of preservation are used to kill all microorganisms.

Cold

Besides moisture, one of the most crucial factors affecting microbial growth is temperature (Section 6.10). In general, a *lower* storage temperature results in less microbial growth and slower spoilage. A number of *psychrotolerant* (cold tolerant) microorganisms (Section 6.11), however, can survive and grow at refrigerator temperatures. Therefore, storage of perishable food products for long periods of time (greater than several days) is possible only at temperatures below freezing. Freezing and thawing alter the physical structure of many foods. Therefore, freezing is not an acceptable preservation method for many fresh foods, but is widely used for the

Table 29.2	Microbial spoilage of fresh food[a]	
Food product	**Type of microorganism**	**Common spoilage organisms, by genus**
Fruits and vegetables	Bacteria	*Erwinia,* **Pseudomonas, Corynebacterium** (mainly vegetable pathogens; rarely spoil fruit)
	Fungi	*Aspergillus, Botrytis, Geotrichum, Rhizopus, Penicillium, Cladosporium, Alternaria, Phytophthora,* various yeasts
Fresh meat, poultry, and seafood	Bacteria	*Acinetobacter, Aeromonas,* **Pseudomonas,** *Micrococcus, Achromobacter, Flavobacterium,* **Proteus, Salmonella, Escherichia, Campylobacter, Listeria**
	Fungi	*Cladosporium, Mucor, Rhizopus, Penicillium, Geotrichium,* **Sporotrichium, Candida,** *Torula, Rhodotorula*
Milk	Bacteria	**Streptococcus,** *Leuconostoc, Lactococcus, Lactobacillus,* **Pseudomonas, Proteus**
High-sugar foods	Bacteria	**Clostridium, Bacillus,** *Flavobacterium*
	Fungi	*Saccharomyces, Torula, Penicillium*

[a] The organisms listed are the most commonly observed spoilage agents of fresh, perishable foods. Genera in bold face include possible human pathogens.

preservation of meats and many fruits and vegetables. Freezers providing a temperature of −20°C are most commonly used. At −20°C, storage for weeks or months is possible, but microbial growth may still occur in pockets of liquid water trapped within the frozen mass. For long-term storage, temperatures such as −80°C (dry ice temperature) are necessary. Maintenance of such low temperatures is expensive and consequently is not used for routine food storage.

Pickling and Acidity

Another major factor affecting microbial growth in food is pH or acidity. Foods vary somewhat in pH, but most are neutral or acidic. Microorganisms differ in their ability to grow under acidic conditions, but conditions of pH 5 or less inhibit the growth of most spoilage organisms (∞ Section 6.13). Therefore, acid is often used in food preservation, a process called **pickling**. Vinegar, which is dilute acetic acid, is usually added in the pickling process (vinegar is a fermentation product of the acetic acid bacteria; its industrial production will be discussed in Section 30.11). In addition to vinegar, pickling methods usually include the addition of large amounts of salt or sugar to decrease water availability (see below) and further inhibit microbial growth. Common pickled foods include cucumbers (sweet, sour, and dill pickles), peppers, meats, fish, and fruits.

Drying and Dehydration

Water activity, or a_w, is a measure of the availability of water for use by microorganisms in metabolic processes. In pure water, the a_w is 1.00 and the molecules are loosely ordered and rearrange freely. When solute is added, the a_w is decreased. As water molecules reorder around the solute, the free rearrangement of the solute-bound water molecules becomes energetically unfavorable. The microbial cells must then compete with solute for the reduced amount of free water. In general, bacteria are poor competitors for the remaining free water, but fungi are good competitors. In practice, this means that the addition of high concentrations of solutes such as sugars or salts reduces a_w and bacterial growth is inhibited. For example, nearly all bacterial growth is inhibited at a concentration of 7.5% NaCl (a_w of 0.957) with the exception of some of the gram-positive cocci such as *Staphylococcus* species. On the other hand, molds compete well for free water under conditions of low a_w and often grow well in foods such as sugary syrups.

A number of commercially important foods are preserved by addition of salt or sugar. Foods preserved by addition of sugar are mainly fruits (jams, jellies, and preserves). Salted products are primarily meats and fish. Sausage and ham are preserved by various curing salts, including NaCl, and some also undergo a smoking process. Individual products vary widely in a_w, depending on how much salt is added and how much the meat

has been dried. Some cured meat products, such as "country ham," can be kept at room temperature for extended periods of time. Others with higher a_w still require refrigeration for long-term storage

Microbial growth can also be controlled by lowering the available water content of the food by *drying*. Milk, meat, fish, vegetables, fruit, eggs, and other economically important foods are commonly preserved by some form of drying. The least damaging physical method used to dry foods is the process of **lyophilization (freeze-drying)**, where foods are frozen and water is removed under vacuum. This method is very expensive, however, and is used mainly for specialized applications like preparation of military rations, where long-term storage under adverse conditions may be necessary.

Heating

The application of heat to reduce the bacterial load or to kill all bacteria is also employed for preservation of liquids and wet foods. *Pasteurization*, a process in which liquids are heated to a specified temperature for a precise time, was discussed in Section 20.1. Pasteurization does not sterilize liquids, but reduces the bacterial load of both spoilage organisms and pathogens and significantly extends the shelf life of the liquid.

Spray drying is the process of spraying, or atomizing, liquids such as milk in a heated atmosphere. The atomization reduces surface area on the droplets, allowing rapid drying under controlled heating conditions without destroying the food. This technology is widely used in the production of powdered milk, certain concentrated liquid dairy products, and food ingredients such as flavorings.

Canning is a process in which food is sealed in a container such as a can or glass jar and heated. The process kills all living organisms or at least ensures that there will be no growth of residual organisms. When the can is properly sealed and heated, the food should remain stable and unspoiled indefinitely, even when stored at ambient temperatures.

The temperature–time relationships for canning depend on the type of food, its pH, the size of the container, and the consistency or density of the food. Because heat must penetrate completely to the center of the food within the can, heating times must be longer for large cans or very dense foods. Acid foods can often be canned effectively by heating just to boiling, 100°C, whereas nonacid foods must be heated to autoclave temperatures (121°C). Unfortunately, heating times long enough to guarantee absolute *sterility* of every can (for example, with the use of an autoclave, ∞ Section 20.1) would change the food so greatly that it would likely be unpalatable and lose nutritional value. Therefore, even properly canned foods may not be sterile.

Microbial growth in a sealed can of food is frequently the result of organisms that produce extensive amounts of gas. This can build pressure inside the can,

(a)   (b)   (c)   (d)

● **Figure 29.1   Changes in sealed tin cans as a result of microbial spoilage.** (a) Normal can; the top of the can is slightly indented due to negative pressure (vacuum) inside. (b) Swelling resulting from minimal gas production. Note that the top is slightly distended. (c) Severe swelling due to extensive gas production. (d) The can shown in (c) was dropped and the gas pressure resulted in a violent explosion, tearing the lid apart.

resulting in bulges or, in severe cases, explosion of the can (Figure 29.1●). The environment inside a can is anoxic, and some of the anoxic bacteria that grow in canned foods are toxin producers of the genus *Clostridium* (∞ Sections 12.20, 21.10 and 29.6). Therefore, food from a can that is visibly altered should never be eaten. On the other hand, the lack of obvious gas production is not an absolute guarantee that the food is safe to consume.

Chemical Preservation

Over 3000 different compounds are used as food additives. These chemical additives are classified by the U.S. Food and Drug Administration as *"generally recognized as safe" (GRAS)* and find wide application in the food industry for enhancing or preserving texture, color, freshness, or flavor. A small number of these compounds are used to control microbial growth in food (Table 29.3). Many of these microbial growth inhibitors, like *sodium*

Table 29.3	Chemical food preservatives
Chemical	**Foods**
Sodium or calcium propionate	Bread
Sodium benzoate	Carbonated beverages, fruit, fruit juices, pickles, margarine, preserves
Sorbic acid	Citrus products, cheese, pickles, salads
Sulfur dioxide, sulfites, bisulfites	Dried fruits and vegetables, wine
Formaldehyde (from food-smoking process)	Meat, fish
Ethylene and propylene oxides	Spices, dried fruits, nuts
Sodium nitrite	Smoked ham, bacon

propionate, have been used for many years with no evidence of human toxicity. Others, like *nitrites* (carcinogen precursor), *ethylene oxide* and *propylene oxide* (mutagens; ∞ Sections 10.4 and 10.5), are more controversial food additives because of statistical evidence that these compounds may negatively affect human health. However, the use of spoilage-retarding additives significantly extends the useful shelf life of finished foods. Thus, chemical food additives contribute significantly to an increase in quantity and in the perceived quality of available food items.

Because of lengthy and costly testing programs for any new chemical proposed as a food preservative, very few new compounds will likely be added to the list of safe and approved chemical food preservatives in Table 29.3 in the near future.

Irradiation

Irradiation of food with ionizing irradiation is now a standard method for reducing contamination by bacteria, fungi, and even insects (∞ Section 20.2). Table 29.4 lists foods for which radiation treatment has been approved. Foods such as spices are routinely irradiated. In the United States, fresh meat products such as hamburger and poultry can now be irradiated to limit contamination by *Escherichia coli* O157:H7 and other enteric pathogens (hamburger) and *Campylobacter jejuni* (poultry). For food irradiation, gamma rays are generated from ^{60}Co or ^{137}Cs sources, or from high-energy electrons produced by linear accelerators. The food products receive a controlled radiation dose. This dose varies considerably by each food category and purpose. For example, a dose of 44 kilograys (kGys) is used to sterilize meat products used on United States NASA space flights and is nearly 10

Table 29.4 Irradiated foods by category and purpose[a]	
Food category	**Purpose for irradiation**
Fresh pork	Control of *Trichinella spiralis* parasite
Fresh fruits and vegetables	Inhibition of growth and maturation (ripening)
Dried spices, herbs, and flavoring mixtures	Microbial disinfection
Refrigerated or frozen uncooked meat products, including ground meat	Control of foodborne pathogens
	Extension of shelf life
Packaged frozen meats used in the National Aeronautics and Space Administration (NASA) flight program	Sterilization
Dry or dehydrated enzyme preparations (e.g., meat tenderizer)	Microbial disinfection
Frozen, uncooked poultry and poultry products	Control of foodborne pathogens

[a] Consumer labeling laws in the United States require that all irradiated foods must be marked with the radura (Figure 29.2) and be conspicuously labeled "Treated with radiation" or "Treated by irradiation" in addition to information required by other regulations.

times higher than the dose of 4.5 kGys used for control of pathogens in hamburger (Table 29.4). In the United States, a consumer product information label must be affixed to foods that are irradiated (Figure 29.2●).

 29.2 Concept Check

Food microbiology deals with methods for limiting spoilage and the growth of disease-causing microorganisms in food during processing and storage. Foods vary considerably in their sensitivity to microbial growth, depending on their nutrient content, water availability, and pH. The growth of microorganisms in perishable foods can be controlled by refrigeration, freezing, canning, pickling, dehydration, chemical preservation, or irradiation.

◆ Outline at least four methods of food preservation. How does each method limit growth of microorganisms?

◆ Identify food spoilage microorganisms that are also pathogens.

● **Figure 29.2 The radura, the international symbol for radiation.** Packaging of foods treated with radiation must be labeled with the radura as well as the statement "treated by irradiation" or "treated with radiation."

29.3 Fermented Foods

Here we consider foods that are preserved, produced, or enhanced through the actions of microorganisms. Microorganisms are instrumental in the preparation of a wide variety of common foods and beverages. Microbial processes can produce significant alterations in raw products, and the product is called a *fermented food*. **Fermentation** is the anaerobic catabolism of organic compounds, generally carbohydrates, in the absence of an external electron acceptor (∞ Sections 5.9 and 5.10). The microorganisms involved in food fermentations include the lactic acid bacteria (∞ Section 12.19), the acetic acid bacteria (∞ Section 12.8), and the propionic acid bacteria (∞ Section 12.22) (Table 29.5). These bacteria do not grow below about pH 4, so food fermentation is a self-limiting process.

One of the most common fermented foods is yeast bread, in which the fermentation of simple sugars and grain carbohydrates by the yeast *Saccharomyces cerevisiae* (Table 29.5) produces carbon dioxide, raising the bread and producing the holes in the finished loaf (Figure 29.3●). Fermented beverage products such as wine, beer, and whiskey will be discussed in Chapter 30, as large-scale industrial applications of microbial processes (∞ Microbial Sidebar, Chapter 5).

Dairy Products

Fermented dairy products were originally developed as methods to preserve milk, an economically important fresh food that normally undergoes rapid spoilage (Section 29.2). Dairy products include cheese and other fermented milk products such as yogurt, buttermilk, and sour cream (Figure 29.3). Milk contains the disaccharide *lactose*. Lactose can be hydrolyzed by the enzyme *lactase* into glucose and galactose. These monosaccharides are fermented to the final product of lactic acid (∞ Figure 5.14). The result of this fermentation is a significant decrease in pH from neutral or slightly basic in raw milk to less than 5.3 in cheeses and less than 4.6 in other fermented milk products. The organisms that accomplish this fermentation are members of the lactic acid bacteria, and all produce lactic acid as a major fermentation product (∞ Section 12.19) (Table 29.5).

Starter cultures of various lactic acid bacteria are introduced into the raw milk and fermentation proceeds for various times, depending on the desired product (Table 29.5). For some cheeses, a second inoculum may be introduced to produce another fermentation. For example, Swiss-type cheeses are inoculated with *Propionibacterium*. The secondary fermenter catabolizes lactic acid to propionic acid, acetic acid, and CO_2 (∞ Section 12.22 and Figure 12.68). The carbon dioxide produces the large holes or eyes that characterize Swiss-type cheeses. The secondary addition of *Lactobacillus* and the mold *Penicillium roqueforti* produce the blue veins and distinctive taste and aroma of blue cheese. Each cheese type is produced under carefully controlled conditions. Time of fermentation, temperature, the extent of aging, and the types of fermentative microorganisms must be rigidly controlled to ensure a distinctive and reproducible product.

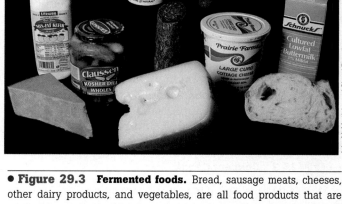

● **Figure 29.3 Fermented foods.** Bread, sausage meats, cheeses, other dairy products, and vegetables, are all food products that are produced or enhanced by fermentation reactions catalyzed by microorganisms.

Table 29.5	Fermented foods and primary fermentation microorganisms[a]
Food category	**Primary fermenting microorganism**
Dairy foods	
Cheeses	*Lactococcus*
	Lactobacillus
	Streptococcus thermophilus
Fermented milk products	
Buttermilk	*Lactococcus*
Sour cream	*Lactococcus*
Yogurt	*Lactobacillus*
	Streptococcus thermophilus
Alcoholic beverages	*Zymomonas*
	Saccharomyces[b]
Yeast breads	*Saccharomyces cerevisiae*[c]
Meat products	
Dry sausages (pepperoni, salami)	*Pediococcus*
and semidry sausages	*Lactobacillus*
(summer sausage, bologna)	*Micrococcus*
	Staphylococcus
Vegetables	
Cabbage (sauerkraut)	*Leuconostoc*
	Lactobacillus
Cucumbers (pickles)	Lactic acid bacteria
Soy sauce	*Aspergillus*
	Tetragenococcus halophilus
	Yeasts

[a] Unless otherwise noted, these are all members of the lactic acid bacteria (∞ Section 12.19). *Zymomonas* is related to *Pseudomonas* (∞ Section 12.7).

[b] Yeast (∞ Section 30.13). A variety of *Saccharomyces* species are used in alcohol fermentations.

[c] Baker's yeast.

Meat Products

Meat products fall into several categories. *Sausages* are generally made from pork or beef and, occasionally, from poultry meats. The most common are the dry sausages, such as salami and pepperoni, and the semidry sausages such as bolognas and summer sausages (Figure 29.3).

Sausages are made using a uniformly blended mixture of meat, salt, and seasonings. A starter culture of lactic acid bacteria is added, and fermentation proceeds to reduce the pH in the sausages to below 5. After fermentation, sausages are often smoked and then dried. The moisture content of dry sausages is about 30%. After final processing, dry sausages can be held at room temperature for extended periods of time. The semidry sausages have a final moisture content of about 50% and are less resistant to spoilage if not refrigerated.

Fish, often mixed with rice, shrimp, and spices, are also fermented to make a wide variety of fish pastes and fish-flavoring products.

Vegetables and Vegetable Products

A number of vegetable food products are made using fermentation processes, and the variety of specialty fermented vegetable foods is practically endless. The most economically important fermented vegetable foods are sauerkraut (fermented cabbage) and some pickles (fermented cucumbers). Olives, onions, tomatoes, and peppers, as well as a variety of fruits, are also fermented.

Vegetables are often fermented in salt brine to aid in preservation and enhance flavor. Fermentation may also improve digestibility by breaking down plant tissues. For example, fermented legume products (peas, beans, lentils) have a marked reduction in the flatulence-producing oligosaccharides that characterize fresh legumes.

Soy sauce is a complex fermentation product made by fermentation of soy beans and wheat. Cooked soybeans are mixed with an enzyme preparation derived from a culture of *Aspergillus* (Table 29.5) on rice or other cereals. This starter culture is spread on a wheat-soybean mixture and is then grown for 2–3 days. This preparation, called *koji*, is mixed with a brine (17–19% NaCl) and fermentation is allowed to proceed for 2–4 months or more. Various microbial species, including members of the *Lactobacillus* and *Pediococcus* genera, as well as several yeasts and molds, produce fermentation products that contribute to the desirable characteristics of the final product. After fermentation, the liquid sauce is decanted, filtered, pasteurized (∞ Section 20.1), and bottled as soy sauce.

⬡ **29.3** *Concept Check*

Microbial fermentation is an important process used for preserving and enhancing a number of foods, including breads, dairy products, meats, and vegetables.

◆ Identify important dairy, meat, and vegetable products that are produced or enhanced by microbial fermentation processes.

◆ Identify the microbial group or groups that are most important for food fermentations.

 II MICROBIAL SAMPLING AND FOOD POISONING

Failure to adequately decontaminate and preserve food may allow the growth of pathogens, resulting in diseases with significant morbidity and mortality. Like waterborne diseases, foodborne illnesses are usually common-source diseases. A single contaminated food source at a food-processing plant or a restaurant, for instance, may affect a large number of individuals. Each year in the United States, there are nearly 25,000 *reported* foodborne disease outbreaks, causing an estimated 13 million foodborne illnesses. Most of these reported outbreaks affect very small numbers of individuals, occurring in isolated instances in the home. These outbreaks are usually the result of improper food handling and preparation at the consumer level. Most foodborne outbreaks and diseases remain unreported because the connection between food and illness is not made.

29.4 Foodborne Diseases and Microbial Sampling

Table 29.6 summarizes the most prevalent foodborne diseases in the United States and the microorganisms that cause them. These common diseases can be separated into two categories, *food poisoning* and *food infection*. Specialized microbial sampling techniques are necessary to isolate the pathogens responsible for foodborne diseases.

Foodborne Diseases

Food poisoning or **food intoxication** is disease that results from ingestion of foods containing preformed microbial toxin. The microorganisms that produced the toxins do not have to grow in the host and are often not alive at the time the contaminated food is consumed. The illness is due to ingestion and subsequent action of the preformed bioactive toxin. We previously discussed some of these toxins, notably the exotoxin of *Clostridium botulinum* (∞ Section 21.10) and the superantigen toxins of *Staphylococcus aureus* (∞ Section 22.16). **Food infection** is an infection resulting from ingestion of pathogen-contaminated food. We will specifically discuss major foodborne infections in Sections 29.7–29.11.

Microbial Sampling for Foodborne Disease

As discussed in Section 29.1, microorganisms are always present in fresh foods. Because pathogens may be present along with many harmless organisms, rapid non–culture-based methods have been developed to detect important pathogens such as *Escherichia coli* O157:H7, *Salmonella, Staphylococcus*, and *Clostridium botulinum*. A number of molecular and immunology-based methods are in use to test for both toxins and pathogen contamination of foods and other commodities such as drugs and cosmetics. We discussed in Section 24.12 the use of nucleic acid probes and PCR for the detection of specific pathogens, including foodborne pathogens.

The presence of a potential foodborne pathogen or toxin is, however, often not enough to link particular foods or pathogens to a specific foodborne disease outbreak. To complete investigations of foodborne illness outbreaks, it is desirable to isolate and positively identify the causal organism. Isolation and growth of pathogens from nonliquid foods usually requires preliminary treatment to suspend microorganisms embedded or entrapped within the food. The most suitable method is with a specialized blender called a *stomacher*. The stomacher blends and homogenizes using paddles to crush and extrude the food sample. The food should be examined as soon after sampling as possible; if examination cannot begin within 1 hour of sampling, the food should be refrigerated. A frozen food should be thawed in its original container in a refrigerator and examined or cultured as soon as thawing is complete. In addition to recovering agents from the food, it is also necessary to recover foodborne pathogens from the patients to establish a cause and effect relationship between the pathogen and the illness. In many cases, fecal samples can be cultured to recover organisms for testing (∞ Section 24.1).

Food samples or patient samples can be inoculated onto enriched media, followed by transfer to differential or selective media for isolation and identification, as we described for human pathogens (∞ Section 24.2). Final identification of foodborne pathogens by growth charac-

Table 29.6	Annual foodborne disease estimates for the United States[a]		
Organism	**Disease**[b]	**Number per year**	**Foods**
Bacteria			
Bacillus cereus	FP and FI	27,000	Rice and starchy foods, high-sugar foods, meats, gravies, pudding, dry milk
Campylobacter jejuni	FI	1,963,000	Poultry, dairy
Clostridium perfringens	FP and FI	248,000	Cooked and reheated meats and meat products
Escherichia coli O157:H7	FI	63,000	Meat, especially ground meat
Other enteropathogenic Escherichia coli	FI	110,000	Meat, especially ground meat
Listeria monocytogenes	FI	2,500	Meat and dairy
Salmonella spp.	FI	1,340,000	Poultry, meat, dairy, eggs
Staphylococcus aureus	FP	185,000	Meat, desserts
Streptococcus spp.	FI	50,000	Dairy, meat
Yersinia enterocolitica	FI	87,000	Pork, milk
All other bacteria	FP and FI	102,000	
Total bacteria		**4,177,500**	
Protozoa			
Cryptosporidium parvum	FI	30,000	Raw and undercooked meat
Cyclospora cayetanensis	FI	16,000	Fresh produce
Giardia lamblia	FI	200,000	Contaminated or infected meat
Toxoplasma gondii	FI	113,000	Raw and undercooked meat
Total protozoa		**359,000**	
Viruses			
Norwalk-like viruses	FI	9,200,000	Shellfish, many other foods
All other viruses	FI	82,000	
Total viruses		**9,282,000**	
Total Annual Foodborne Diseases		**13,818,500**	

[a] Estimates are based on data provided by the Centers for Disease Control and Prevention, Atlanta, GA, USA and are typical of recent years.

[b] FP, food poisoning: FI, food infection.

teristics as well as the use of genetic fingerprinting techniques such as PCR, nucleic acid probes, nucleic acid sequencing, and other molecular diagnostic methods (oo Section 24.12) may be necessary to link specific organisms to foodborne diseases and individual illnesses.

 29.4 Concept Check

Foodborne diseases include food poisoning and food infection. Food poisoning results from the action of microbial toxins, and food infections are due to the growth of microorganisms in the body. Specialized techniques are used to sample microorganisms in food.

◆ Distinguish between food infection and food poisoning.

◆ Describe the sampling of a solid food such as meat for the presence of microorganisms.

◆ What methods are used for the identification of foodborne pathogens?

29.5 Staphylococcal Food Poisoning

A very common food *poisoning* is caused by *Staphylococcus aureus*. This organism and other members of the genus are small, gram-positive cocci (Figure 29.4●) (oo

Section 12.19). As we discussed in Section 26.9, staphylococci are found on the skin and in the respiratory tract of nearly all humans and are often opportunistic pathogens. *S. aureus* is frequently associated with food poisoning because it can grow in many common foods and some strains produce several heat-stable protein superantigen enterotoxins (oo Section 22.16). If food that contains toxin is ingested, gastroenteritis characterized by nausea, vomiting, and diarrhea, occurs within 1–6 hours.

Epidemiology

Each year, an estimated 185,000 cases of staphylococcal food poisoning occur in the United States (Table 29.6). The foods most commonly involved are custard- and cream-filled baked goods, poultry, meat and meat products, gravies, egg and meat salads, puddings, and creamy salad dressings. If such foods are kept refrigerated after preparation, they usually remain safe because *S. aureus* growth is significantly reduced at low temperatures. Foods of this type, however, are often kept in kitchens at room temperature or outdoors at picnics. The food, if inoculated with *S. aureus* from an infected food handler, supports rapid bacterial growth and enterotoxin production. Even if the toxin-containing foods are reheated again before eating, the toxin is relatively

CDC/PHIL

● **Figure 29.4** *Staphylococcus aureus.* In this gram stain, the individual gram-positive cocci are about 0.8 μm in diameter. Staphylococci can divide on several different axes, giving rise to the typical "grape-like" clusters of cells.

heat-stable and may remain active. Live *S. aureus* need not be present in foods causing illness: The illness is solely due to the preformed toxin.

Staphylococcus aureus strains produce at least seven different but related enterotoxins. Most strains of *S. aureus* produce only one or two of these toxins, and some produce none. However, any one of these toxins can cause staphylococcal food poisoning. These enterotoxins are further classified as *superantigens*. Superantigens work by stimulating large numbers of T cells, which in turn release intercellular mediators called cytokines, activating a general inflammatory response in the intestine that results in gastroenteritis, including massive loss of fluids from the intestine (∞ Section 22.16).

The most common enterotoxin produced by *S. aureus* is enterotoxin A, a peptide encoded by a chromosomal gene. Cloning and sequencing of this gene, the *entA* gene, and of several other *S. aureus* enterotoxin genes, show that the toxins are genetically related. Although the *entA* gene is on the bacterial chromosome, type B and C *S. aureus* enterotoxins may be encoded on plasmids, transposons, or lysogenic bacteriophages. We discussed the importance of accessory genetic elements such as plasmids and bacteriophages as vectors for toxin production in Sections 21.10 and 21.11.

Diagnosis, Treatment, and Prevention

Assays based on the detection of either enterotoxin (ELISA detection of enterotoxin; ∞ Section 24.10) or *S. aureus* exonuclease (an enzyme that degrades DNA) are available to detect *S. aureus* in food. However, these rapid tests are qualitative, only confirming the presence or absence of *S. aureus* products above the detection limits of the assay. In addition, the exonuclease test does not indicate the presence of exotoxin, and thus is not predictive for food poisoning. To obtain quantitative data and determine the extent of bacterial contamination, bacterial plate counts are required. For staphylococcal counts, a high-salt medium (either sodium chloride or lithium

chloride at a final concentration of 7.5%) is used. Compared to most bacteria present in foods, staphylococci can thrive in habitats with high-salt and low-water activity (Section 29.1).

S. aureus food poisoning can be quite severe, but is self-limiting, usually resolving within 48 hours after onset. Severe cases may require treatment for dehydration. Treatment with antibiotics is not useful because the disease is caused by a preformed toxin, not an active infection.

Staphylococcal food poisoning can be prevented by careful sanitation and hygiene measures both in food production and food preparation steps and by storing foods at low temperatures to inhibit bacterial growth. Foods susceptible to colonization by *S. aureus* and kept for several hours above 4°C (refrigerator temperature) should be discarded rather than eaten.

⬡ 29.5 Concept Check

Staphylococcal food poisoning results from the ingestion of preformed enterotoxin, a superantigen produced by *Staphylococcus aureus* when growing in foods. In many cases, *S. aureus* cannot be cultured from the contaminated food.

◆ Identify the symptoms of staphylococcal food poisoning and explain the activity of staphylococcal enterotoxins.

◆ Will antibiotics affect the outcome or the severity of staphylococcal food poisoning? Explain.

29.6 Clostridial Food Poisoning

Both *Clostridium perfringens* and *Clostridium botulinum* cause serious food poisoning. Members of the *Clostridium* genus are anaerobic endospore formers (∞ Section 12.20). Canning and cooking procedures kill living organisms but do not kill endospores. Under appropriate anaerobic conditions, the endospores germinate and toxin is produced.

Clostridium perfringens Food Poisoning

Clostridium perfringens is an anaerobic, gram-positive spore-forming rod commonly found in soil (Figure 29.5●). It also lives in small numbers in the intestinal tract of many animals and humans and is therefore found in sewage (∞ Section 12.20). *C. perfringens* is the most prevalent reported cause of *food poisoning* in the United States, with an estimated 248,000 annual cases (Table 29.6).

The disease results from the ingestion of a large dose of *Clostridium perfringens* ($>10^8$ cells) in contaminated cooked and uncooked foods, especially meat, poultry, and fish. Large numbers of *C. perfringens* can grow in meat dishes cooked in bulk (heat penetration in these situations is often slow and insufficient) and then left at 20–40°C for short time periods. Endospores of *C. perfringens* germinate under anoxic conditions, such as in a sealed container, and grow quickly in the meat. However, the toxin is not yet present.

● **Figure 29.5 Gram stain of *Clostridium perfringens*.** The individual gram-positive rods are about 1 μm in diameter.

After consumption of the contaminated food, the living *C. perfringens* begins to sporulate in the intestine, triggering production of the perfringens enterotoxin (∞ Table 21.4). The enterotoxin alters the permeability of the intestinal epithelium, leading to diarrhea and intestinal cramps, usually with no fever or vomiting. The onset of perfringens food poisoning begins about 7–15 hours after consumption of the contaminated food, but usually resolves within 24 hours, and fatalities are rare.

Diagnosis, Treatment, and Prevention

Diagnosis of perfringens food poisoning is made by isolation of *C. perfringens* from the gut or, more reliably, by a direct enzyme-linked immunosorbent assay (ELISA) to detect *C. perfringens* enterotoxin in feces (∞ Section 24.10). Because *C. perfringens* food poisoning is self-limiting, treatment is usually not necessary, although antitoxins are available (∞ Section 22.13). Prevention of perfringens food poisoning requires measures to prevent contamination of raw and cooked foods and control of cooking and canning procedures to ensure proper heat treatment of all foods. Cooked foods should be refrigerated as soon as possible to rapidly lower temperatures and inhibit *C. perfringens* growth.

Botulism

Botulism is a severe food poisoning; it is often fatal and occurs following the consumption of food containing the exotoxin produced by the anaerobic, gram-positive rod *Clostridium botulinum*. This bacterium normally inhabits soil or water, but its endospores may contaminate raw foods before harvest or slaughter. If the foods are properly processed so that the *C. botulinum* endospores are removed or killed, no problem arises; but if viable endospores are present, they may initiate growth and toxin production. Even a small amount of the resultant neurotoxin can be poisonous.

We discussed the nature and activity of botulinum toxin in Section 21.10 (∞ Figure 21.20). Botulinum toxin

is a neurotoxin that causes flaccid paralysis, usually affecting the autonomic nerves that control body functions such as respiration and heartbeat. At least seven distinct types of botulinum toxin are known. The toxins are destroyed by high heat (80°C for 10 minutes), and so thoroughly cooked food, even if contaminated with toxin, *may* be harmless.

Most cases of botulism occur as a result of eating foods that are not cooked after processing (Figure 29.6a●). For example, nonacid, home-canned vegetables (e.g., home-canned corn and beans) are often used without cooking when making cold salads. Smoked and fresh fish, vacuum packed in plastic, are also often eaten without cooking. Under such conditions, *C. botulinum* endospores germinate, and the resulting cells produce toxin. If these foods are consumed, then ingestion of even

(a) Foodborne botulism

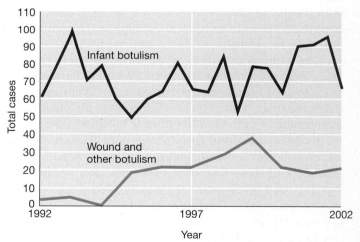

(b) Infant, wound, and other botulism

● **Figure 29.6 The incidence of botulism in the United States.** (a) Foodborne botulism. In years with high numbers of cases, major outbreaks that account for the increase are indicated. (b) Other botulism, including infant botulism. More than half of the cases of infant botulism in the United States occur in California. Wound botulism and other botulism includes all other sources. Data are from the Centers for Disease Control and Prevention, Atlanta, GA, USA.

a small amount will result in this severe and highly dangerous type of food poisoning.

Infant botulism occurs when endospores of *Clostridium botulinum* are ingested, sometimes from raw honey, but more often from no identifiable source (Figure 29.6b). If the infant's normal flora is not well developed or if the infant is undergoing antibiotic therapy, endospores may germinate in the infant's intestine and *C. botulinum* cells may grow and release toxin. Most cases of infant botulism occur between the first week of life and 2 months of age; infant botulism is rare in children older than 6 months when the normal intestinal flora is more developed (∞ Section 21.4). About 70% of all botulism cases in the United States are infant botulism. Wound botulism also occasionally occurs, presumably due to endospores introduced by introduction of a foreign substance by a parenteral route. Wound botulism is commonly associated with illicit injectable drug use (Figure 29.6b).

All forms of botulism are quite rare, with at most six cases occurring per 10 million individuals per year in the United States in recent years. Over the last decade there have been fewer than 155 cases or less each year, but up to 25% of all cases are fatal. Death occurs from respiratory paralysis or cardiac arrest due to the paralyzing action of the botulinum neurotoxin (∞ Section 21.10).

Diagnosis, Treatment, and Prevention

Diagnosis of botulism is by demonstrating botulinum toxin in patient serum or by finding toxin or live *Clostridium botulinum* in suspected food products. Laboratory findings are coupled with clinical observations including neurological signs of localized paralysis (impaired vision and speech) beginning 18–24 hours after ingestion of contaminated food. Treatment involves administration of antitoxin if the diagnosis is early (∞ Section 22.13) and mechanical ventilation for symptoms of flaccid respiratory paralysis (∞ Section 21.10). Prevention requires maintaining careful controls over canning and preservation methods. Heating susceptible foods to destroy endospores, or boiling for 20 minutes destroys the toxin. Home-prepared foods are the most common source of individual foodborne botulism outbreaks.

In infant botulism, *C. botulinum* and toxin are often found in bowel contents. Infant botulism is usually self-limiting, and most infants recover with only supportive therapy, such as assisted ventilation. Occasional deaths may occur due to respiratory failure. Honey is an occasional source of *C. botulinum* endospores. Therefore, feeding honey to children under 2 years of age is not recommended.

 ### 29.6 *Concept Check*

Clostridium food poisoning results from ingestion of toxins produced by microbial growth in foods or due to microbial growth and toxin production in the body. Perfringens food poisoning is quite common and is usually a self-limiting gastrointestinal disease. Botulism is a rare but very serious disease, with significant mortality.

◆ Describe the events that lead to *C. perfringens* food poisoning. What is the likely outcome of the poisoning?

◆ Describe the events that lead to botulism. What is the likely outcome of botulism?

◆ Why is infant botulism more prevalent than other types of botulism?

III FOOD INFECTION

Food infection is active infection resulting from ingestion of pathogen-contaminated food. Food may contain sufficient numbers of viable pathogens to cause infection and disease in the host. Food infection is a very common type of foodborne illness (Table 29.6), and we begin with *Salmonella* food infection, a typical and widespread example. Many food infection agents also cause waterborne diseases (∞ Chapter 28).

29.7 Salmonellosis

Although sometimes called food poisoning, **salmonellosis** is a gastrointestinal disease due to foodborne *Salmonella* infection. Symptoms begin only after the pathogen colonizes the intestinal epithelium.

Salmonella are gram-negative facultatively aerobic rods related to *Escherichia coli*, *Shigella*, and other enteric *Bacteria* (∞ Section 12.11). *Salmonella* normally inhabit the gut of animals and are thus found in sewage. Virtually all *Salmonella* are pathogenic for humans. One, *S. typhi*, causes the serious human disease typhoid fever, but it is very rare in the United States, with most of the 500 foodborne cases imported from other countries. However, a number of *Salmonella* species cause foodborne gastroenteritis. In all, over 2000 *serovars*, or individual variants, of various *Salmonella* species are known to be pathogenic for humans. *S. typhimurium* is the most common agent of salmonellosis.

Epidemiology

The incidence and prevalence of *reported* salmonellosis has been steady over the last decade, with about 40,000–45,000 documented cases each year (Figure 29.7●). However, less than 4% of the total cases of salmonellosis are probably reported, and the estimated number of actual cases of salmonellosis is over 1.3 million every year (Table 29.6).

The ultimate sources of the foodborne salmonellas are the intestinal tracts of humans and warm-blooded animals, and several mechanisms may introduce these organisms into the food supply. The organism may reach food by fecal contamination from food handlers. Food production animals such as chickens and cattle may also harbor *Salmonella* strains that are pathogenic to humans and may pass the bacteria to finished fresh foods such as

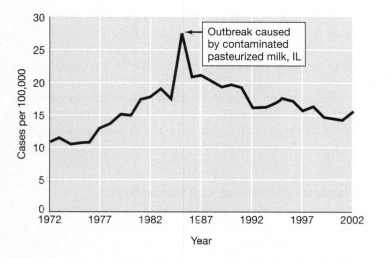

● **Figure 29.7 Prevalence of salmonellosis in the United States, 1972–2002.** The data are presented as the number of cases per 100,000 individuals. The total number of *reported* cases in 2002 was 44,264. Epidemiologic investigations suggest that only about 3% of all cases of salmonellosis are properly identified and reported, indicating that there are about 1.3 million cases of salmonellosis each year. The high incidence in 1984 resulted from several thousand cases caused by contamination of pasteurized milk that was mixed with raw (unprocessed) milk in a major dairy. About 95% of all cases are foodborne. Data are from the Centers for Disease Control and Prevention, Atlanta, GA, USA.

eggs, meat, and dairy products. *Salmonella* food infections are often traced to products such as custards, cream cakes, meringues, pies, and eggnog made with uncooked eggs. Other foods commonly implicated in salmonellosis outbreaks are meats and meat products such as meat pies, cured but uncooked sausages and meats, poultry, milk, and milk products.

The most common salmonellosis is a *Salmonella*-induced *enterocolitis*. Ingestion of food containing viable *Salmonella* results in colonization of the small and large intestine. Onset of the disease occurs 8–48 hours after ingestion. Symptoms include the sudden onset of headache, chills, vomiting, and diarrhea, followed by a fever that lasts a few days. The disease normally resolves without intervention in 2 to 3 days. However, even after recovery, patients shed *Salmonella* in feces for several weeks. Some patients recover and remain asymptomatic, but shed organisms for months or even years, resulting in a chronic carrier condition (Section 25.3).

Salmonellosis may also cause septicemia (a blood infection) and enteric or *typhoid fever*, a disease characterized by systemic infection and high fever lasting several weeks. Mortality can approach 15% in untreated typhoid fever.

Diagnosis, Treatment, and Prevention

Diagnosis of foodborne salmonellosis is made by observation of clinical symptoms, history of recent food consumption, and by culture of the organism from feces. Several selective media (Section 24.2) and tests for the presence of *Salmonella* are commonly done on animal

food products, such as raw meat, poultry, eggs, and powdered milk, because *Salmonella* from food production animals is the usual source of food contamination.

For enterocolitis, treatment is usually unnecessary, and antibiotic treatment does not shorten the course of the disease or eliminate the carrier state. Antibiotic treatment, however, significantly reduces the length and severity of septicemia and typhoid fever. Mortality due to typhoid fever can be reduced to less than 1% with appropriate antibiotic therapy. Several strains of *Salmonella* are resistant to multiple antimicrobial drugs.

Cooked or canned foods that become contaminated by an infected food handler can support the growth of *Salmonella* if the foods are held for long periods of time without heating or refrigeration. Cooked foods heated to 70°C for at least 10 minutes are considered safe if consumed immediately, or if held at 50°C or stored at 10°C or lower. *Salmonella* infections are more common in summer than in winter, probably because warm environmental conditions favor the growth of microorganisms in foods (Figure 29.4).

Although local laws and enforcement vary, because of the lengthy carrier state, infected individuals are often banned from work as food handlers until their feces are negative for *Salmonella* in three successive cultures.

29.7 Concept Check

There are more than 1.3 million cases of salmonellosis every year in the United States. The disease results from infection with ingested *Salmonella* introduced into the food chain from food production animals or food handlers.

◆ Describe salmonellosis food infection. How does it differ from food poisoning?

◆ How might *Salmonella* contamination of food production animals be contained?

29.8 Pathogenic *Escherichia coli*

Most strains of *Escherichia coli* are not pathogenic and are common commensals found in the intestines of humans. A few strains, however, are potential foodborne pathogens. All pathogenic strains act on the intestine and several are characterized by their ability to produce potent enterotoxins (Section 21.11). The short, gram-negative rods are classified as enteric *Bacteria* (Section 12.11). There are about 200 known pathogenic *E. coli* that can cause life-threatening diarrheal disease and urinary tract infections. The pathogenic strains are divided into several categories based primarily on the toxins they produce and the diseases they cause.

Enterohemorrhagic *Escherichia coli* (EHEC)

Enterohemorrhagic *Escherichia coli* (EHEC) produce *verotoxin*, an enterotoxin similar to one produced by *Shigella dysenteriae*, the Shiga toxin (Table 21.4). After ingestion of food or water containing one well-known EHEC strain, *E. coli* O157:H7, the organism grows in the small intestine and produces the verotoxin. Verotoxin causes both hemorrhagic (bloody) diarrhea and kidney failure. *E. coli* O157:H7 causes at least 60,000 infections and 50 deaths from foodborne disease in the United States each year (Table 29.5). This pathogen is a leading cause of kidney failure in children. The most common cause of this infection is the consumption of contaminated uncooked or undercooked meat, particularly mass-processed ground meat.

In several major outbreaks in the United States involving *E. coli* O157:H7–infected ground beef, regional distribution centers were the source of the contaminated meat. Infected meat products caused disease in several states. Another outbreak involved processed, cured, but uncooked beef in ready-to-eat sausages (Section 29.3). The source of contamination was the beef, and the *E. coli* O157:H7 probably originated from slaughtered beef carcasses. In 2001, there were 16 documented food infection outbreaks in the United States due to *E. coli* O157:H7. Five of these were conclusively linked to contaminated beef. In 2003, the Food Safety and Inspection Service of the U.S. Department of Agriculture reported that there were 20 positive results of 6584 samples (0.03%) of ground beef analyzed for *E. coli* O157:H7. *E. coli* O.157:H7 has also been implicated in food infection outbreaks involving dairy products, fresh fruit, and raw vegetables.

Since *E. coli* O.157:H7 grows in the intestines and is found in fecal material, it is also a potential source of waterborne disease. There have been several cases of serious *E. coli* O157:H7 infections from fecally contaminated public swimming areas (∞ Table 28.2). Several outbreaks have also been reported in day care settings, where the presumed route of exposure is by oral-fecal contamination.

Other Pathogenic *Escherichia coli*

Diarrheal diseases often occur in children in developing countries. It also occurs as "traveler's diarrhea," an extremely common enteric infection causing watery diarrhea in travelers to developing countries. The primary causal agents are the enterotoxigenic *Escherichia coli* (ETEC). The ETEC strains usually produce one of two heat-labile diarrhea-producing enterotoxins. In studies done with U.S. citizens traveling in Mexico, the infection rate with ETEC is often greater than 50%. The prime vehicles are foods such as fresh vegetables (for example, lettuce in salads) and water. The very high infection rate in travelers is due to contamination of local public water supplies. The local population is usually resistant to the infecting strains, undoubtedly because they have lived with the agent for a long period of time. Secretory antibodies (∞ Section 22.9) present in the bowel may prevent successful colonization of the pathogen in local residents, but the organism readily colonizes the intestine of a nonimmune person, causing disease.

Enteropathogenic *E. coli* (EPEC) cause diarrheal diseases in infants and small children, but do not cause invasive disease or produce toxins. Enteroinvasive *E. coli* (EIEC) strains cause invasive disease in the colon, producing watery to bloody diarrhea. The cells are taken up by phagocytes, where they escape lysis in the phagolysosomes (∞ Section 22.2), grow in the cytoplasm, and move into other cells. This invasive disease causes diarrhea and is common in developing countries.

Diagnosis, Treatment, and Prevention

Escherichia coli O157:H7 illness is a reportable infectious disease in the United States (∞ Table 25.5). The general pattern established for diagnosis, treatment, and prevention of infection by *Escherichia coli* O157:H7 reflects current procedures used for all of the pathogenic *E. coli* strains. Diagnosis of infection by *E. coli* O157:H7 involves culture from the feces and identification of the O and H antigens and toxins by serology (∞ Sections 4.9 and 24.7). Subtyping of strains is also done using molecular methods such as restriction fragment-length polymorphism (RFLP) and pulse-field gel electrophoresis (PFGE) (∞ Section 24.12).

Treatment of all pathogenic *E. coli* infections involves supportive therapy and, in severe cases, antimicrobial drugs to shorten and eliminate infection.

The most effective way to prevent infection with foodborne enteropathogenic *E. coli* O157:H7 is to make sure that meat is cooked thoroughly, which means that it should appear gray or brown and juices should be clear. As we discussed above (Section 29.2), the United States has approved the irradiation of ground meat as an acceptable means of eliminating or reducing food infection bacteria, largely because *E. coli* O157:H7 has been implicated in several foodborne epidemics. Thus, contamination from the meat of one animal can potentially contaminate the meat of several animals when the meat is mixed in the grinding process. Penetrating radiation is considered the only effective means to ensure decontamination after the grinding process because grinding may distribute the pathogens throughout meat, not simply on the surface.

In general, proper food handling, water purification, and appropriate hygiene habits will prevent the spread of pathogenic *E. coli*. Traveler's diarrhea can be prevented by avoiding local water sources and uncooked foods.

 29.8 Concept Check

Enteropathogenic *Escherichia coli* can cause serious food infections. Specific measures, such as radiation of ground beef, have been implemented to curb spread of these pathogens. Large-scale processing methods for meats and meat products allow contaminants from a small number of individual carcasses to contaminate/infect large numbers of products.

◆ Describe the pathology of *Escherichia coli* food infections due to EHEC, ETEC, EPEC, and EIEC strains.

◆ How might *Escherichia coli* contamination of food-production animals be prevented?

◆ Why is *Escherichia coli* O157:H7 considered a dangerous and reportable pathogen?

29.9 *Campylobacter*

Campylobacter spp. cause the *most prevalent bacterial foodborne infections* in the United States. *Campylobacter* species are gram-negative, motile, curved rod-to-spirillar-shaped organisms that grow at reduced oxygen tension as microaerophiles (Section 12.14). Several pathogenic species, *Campylobacter jejuni* (Figure 29.8●), *C. coli*, and *C. fetus*, are recognized. *C. jejuni* and *C. coli* account for about 2 million annual cases of bacterial diarrhea (Table 29.6). *Campylobacter fetus* is economically important because it is a major cause of sterility and spontaneous abortion in cattle and sheep.

Epidemiology

Campylobacter is transmitted to humans via contaminated food, most frequently in poultry, pork, raw clams, and other shellfish, or in surface waters not subjected to chlorination. *C. jejuni* is a normal resident in the intestinal tract of poultry, and virtually all chickens and turkeys normally have this organism. According to the U.S. Department of Agriculture, up to 90% of turkey carcasses, 88% of chicken carcasses, and 32% of hog carcasses may be contaminated with *Campylobacter*. Beef, on the other hand, is rarely a vehicle. *Campylobacter* species also infect domestic animals such as dogs, causing a milder form of diarrhea than that observed in humans. Infant cases of *Campylobacter* infection are frequently traced to infected domestic animals, especially dogs.

After ingesting cells of *Campylobacter*, the organism multiplies in the small intestine, invades the epithelium, and causes inflammation, resulting in disease. Since *C. jejuni* is sensitive to gastric acid, numbers as high as 10^4 may be required to initiate infection. However, ingestion of the pathogen directly in food, or ingestion by individuals taking medication to reduce stomach acid produc-

● **Figure 29.8 Scanning electron micrograph of a cell of *Campylobacter jejuni*.** The individual gram-positive curved rods are about 1 μm in diameter.

tion, may reduce this number to less than 500 bacteria. The symptoms of *Campylobacter* infection include a high fever (usually greater than 104°F or 40°C), headache, malaise, nausea, abdominal cramps, and profuse diarrhea with watery, frequently bloody, stools. The disease subsides in about 7–10 days. Spontaneous recovery from *Campylobacter* infections is often complete, but relapses occur in up to 25% of cases.

Diagnosis, Treatment, and Prevention

Diagnosis requires isolation of the organism from stool samples and identification by growth-dependent tests or immunological assays. Because of the frequency with which *C. jejuni* infections are observed in infants, a variety of selective media and highly specific immunological methods have been developed for positive identification of this organism. Treatment of infections with erythromycin does not shorten the acute diarrhea, but may shorten the time during which patients shed *Campylobacter* in their feces. Personal hygiene, proper washing of uncooked poultry (and any kitchenware coming in contact with uncooked poultry), and thorough cooking of meat eliminate the possibility of *Campylobacter* infection.

■ 29.9 Concept Check

Campylobacter infection is by far the most prevalent foodborne bacterial infection. Though usually self-limiting, this disease affects nearly 2 million people per year.

◆ Describe the pathology of *Campylobacter* food infection. What is the likely outcome?

◆ How might *Campylobacter* contamination of food production animals be controlled?

29.10 Listeriosis

Listeria monocytogenes causes **listeriosis**, a gastrointestinal food infection that may lead to bacteremia and meningitis. It is a short, gram-positive, nonspore-forming rod that is acid-tolerant, *psychrotolerant* (cold-tolerant), facultatively aerobic, and salt-tolerant (Section 12.19) (Figure 29.9●).

Epidemiology

Listeria monocytogenes is found widely in soil and water, and virtually no fresh food source is safe from possible *L. monocytogenes* contamination. Fresh food can become contaminated at any stage during food growth or processing. Methods such as refrigeration, which ordinarily slow microbial growth, are ineffective in limiting growth of this psychrotolerant organism. Thus, meat, dairy products, and fresh produce can be contaminated with this pathogen. Common sources of listeriosis outbreaks

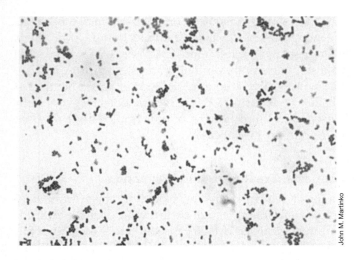

John M. Martinko

● **Figure 29.9 Gram stain of *Listeria monocytogenes.*** The short, gram-positive rods are about 0.5 µm in diameter.

are ready-to-eat processed foods such as meat products and unpasteurized dairy products that are stored for long periods, even at refrigerator temperature (4°C).

Listeria monocytogenes is an intracellular pathogen. It enters the body through the gastrointestinal tract after ingestion of contaminated food. Uptake of the pathogen by phagocytes results in growth and proliferation of the bacterium, lysis of the phagocyte, and spread to surrounding cells. Immunity to *L. monocytogenes* is mainly cell-mediated via T_H1 cells (∞ Section 22.8). Individuals having weakened cellular immunity, including the elderly, neonates, patients undergoing immunosuppressive drug treatment (e.g., steroid treatment), or those who have immunosuppressive diseases such as AIDS, have increased susceptibility to listeriosis (∞ Section 26.14).

Although exposure to *L. monocytogenes* is undoubtedly very common, acute *listeriosis* is quite rare. The acute disease is characterized by septicemia, often leading to meningitis. Acute listeriosis has a mortality rate of about 20%. Although there are only about 2500 cases of acute listeriosis each year, about 500 cases end in death. Nearly all diagnosed cases require hospitalization.

Diagnosis, Treatment, and Prevention

The diagnosis of listeriosis is accomplished by culturing *Listeria monocytogenes* from the blood or spinal fluid. *L. monocytogenes* can be identified in food by direct culture or by a variety of molecular methods such as ribotyping (∞ Section 11.11) and the polymerase chain reaction (PCR) (∞ Section 7.9, Section 24.12, and Table 24.9). Antibiotic treatment with penicillin, ampicillin, or a trimethoprim-sulfamethoxazole combination may be effective.

Prevention measures include recalling contaminated food and taking steps to limit *L. monocytogenes* contamination at the food-processing site. Since *L. monocytogenes* is susceptible to heat and radiation, raw food and food-

handling equipment can be readily decontaminated. However, without sterilizing the finished food product, the risk of food contamination cannot be completely eliminated because of the widespread distribution of the pathogen.

Individuals who are immunocompromised are usually advised to avoid nonpasteurized dairy products and ready-to-eat processed meat. Spontaneous abortion is also a frequent outcome of listeriosis. Therefore, to protect the fetus, pregnant women may also be advised to avoid foods that may transmit *L. monocytogenes*.

 29.10 Concept Check

Listeria monocytogenes is an environmentally ubiquitous microorganism. In normal individuals, *Listeria* seldom causes infection. However, in immunocompromised individuals, *Listeria* can cause serious disease and even death.

◆ What is the likely outcome of *Listeria* exposure in normal individuals?

◆ What populations are most susceptible to *Listeria* infection? Why?

◆ Describe the pathology of listeriosis.

29.11 Other Foodborne Infectious Diseases

A number of other microorganisms and infectious agents—bacteria, viruses, parasites, and otherwise—contribute to foodborne diseases, and we consider a few of them here.

Bacteria

Table 29.6 lists several other bacteria that cause human foodborne disease. *Yersinia enterocolitica* is commonly found in the intestines of domestic animals and causes foodborne infections due to contaminated meat and dairy products. *Y. enterocolitica* causes enteric fever, a severe life-threatening infection. *Bacillus cereus* produces two enterotoxins that cause diarrhea and vomiting. *B. cereus* grows in high-carbohydrate foods such as rice. Endospores of this gram-positive rod germinate and, as the organism grows in food that is left at room temperature, pathogenic amounts of toxin are produced. Reheating may kill the *B. cereus*, but the toxin may remain active. *B. cereus* may also cause a food infection similar to that caused by *Clostridium perfringens* (Section 29.6). *Shigella* spp. cause nearly 100,000 cases of severe foodborne invasive gastroenteritis called *shigellosis* each year. Several members of the *Vibrio* genus cause food poisoning after consumption of contaminated shellfish.

Viruses

The largest number of annual foodborne infections are thought to be caused by viruses. In general, viral foodborne illness consists of gastroenteritis characterized by diarrhea, often accompanied by nausea and vomiting. Recovery is spontaneous and rapid, usually within 24–48 hours ("24-hour bug"). *Norwalk-like viruses* (∞ Table 25.8 and Section 28.8) are responsible for most of these mild foodborne infections in the United States (Table 29.6), accounting for over 9 million of the estimated 13 million cases of food infection per year. Rotavirus, astrovirus, and hepatitis A (∞ Section 26.11) collectively cause 100,000 cases of foodborne disease each year. These viruses inhabit the gut and are often transmitted to food or water with fecal matter. As with many foodborne infections, proper food handling, handwashing, and a source of clean water to prepare fresh foods are essential to prevent infection.

Protozoa

Important foodborne protozoan diseases are listed in Table 29.6. Protozoan parasites including *Giardia lamblia*, *Cryptosporidium parvum*, and *Cyclospora cayetanensis* can be spread via food, presumably contaminated by fecal matter in untreated water used to wash, irrigate, or spray crops. Fresh foods such as fruits are often implicated as the source of these protozoans. We discussed giardiasis and cryptosporidiosis in the previous chapter (∞ Section 28.6 and Figures 28.11 and 28.12). Cyclosporiasis is an acute gastroenteritis and is an important emerging disease. In the United States, most cases appear to be transmitted by eating fresh produce, often imported from other countries.

Toxoplasma gondii is a protozoan spread through cat feces, but also found in raw or undercooked meat. In most individuals, the *toxoplasmosis* infection causes a mild, self-limiting gastroenteritis. However, prenatal infection can lead to a variety of complications, including blindness and stillbirth. Immunocompromised patients also exhibit signs of acute toxoplasmosis.

Prions, BSE, and nvCJD Disease

Prions are proteins, presumably of host origin, that adopt novel conformations, inhibiting normal protein function and causing disruption in neural tissue (∞ Section 9.14). Human prion diseases are characterized by a number of neurological symptoms including depression, loss of motor coordination, and dementia.

A foodborne variety of prion disease in humans is known as "new variant Creutzfeldt-Jakob Disease" (nvCJD) and has been linked to consumption of meat products from cattle afflicted with *bovine spongiform encephalopathy (BSE)*, a prion disease commonly called "mad cow disease." The nvCJD is a slow-acting degenerative nervous system disorder with a latent period that extends for years after exposure to the BSE prion (∞ Section 9.14). Nearly 200 people in the United Kingdom and other European countries have acquired nvCJD. However, nvCJD linked to domestic meat consumption has not been observed in the United States. Although the mechanism of prion infection is not entirely clear, BSE prions consumed in the meat products from affected cattle trigger structurally and functionally related human proteins to assume an altered conformation, resulting in protein dysfunction and disease (∞ Figure 9.29). The terminal stages of both BSE and nvCJD are characterized by large vacuoles in brain tissue, giving the brain a "spongy" appearance, from which BSE derives its name (Figure 29.10●).

In the United Kingdom and Europe, about 180,000 cattle have been diagnosed with BSE. Several cattle with BSE have been found in Canada. Small numbers of affected cattle have also been found in the United States during routine testing of brains from slaughtered cattle. In Europe and North America, all cattle known or suspected to have BSE have been destroyed. Bans on feeding cattle with meat and bone meal appear to have stopped the development of new cases of BSE in Europe and have kept the incidence of this disease very low in North America. The infecting prions were probably transferred to food production animals through meat and bone meal feed derived from cattle or other animals not approved for human consumption.

Diagnosis of BSE is done by testing using a prion-susceptible mouse strain or by immunohistochemical or micrographic analysis of biopsied neural tissue (Figure 29.10).

 29.11 Concept Check

Over 200 different infectious agents cause foodborne disease. Viruses cause the vast majority of foodborne illnesses. A number of bacteria, protozoans, and prions also cause foodborne illnesses.

◆ Identify the viruses most likely to be involved in foodborne illnesses.

◆ How might *prion* contamination of food production animals be prevented in the United States?

USDA/A. Jenny/PHIL

● **Figure 29.10 A brain section from a cow with bovine spongiform encephalopathy (BSE).** Note the vacuoles appearing as holes in the brain, giving it a characteristic spongelike appearance.

REVIEW QUESTIONS

1. Identify and define the three major categories of food with respect to their perishability (Section 29.1).

2. Identify the major methods used to preserve food. Provide an example of a food preserved by each method (Section 29.2).

3. Identify the major categories of fermented foods (Section 29.3).

4. Distinguish between foodborne *infection* and foodborne *poisoning* (Section 29.4).

5. Outline the pathogenesis of staphylococcal food poisoning. Suggest methods for prevention of this disease (Section 29.5).

6. Identify the two major types of clostridial food poisoning. Which is most prevalent? Which is most dangerous? Why (Section 29.6)?

7. What are the possible sources of *Salmonella* spp. that cause food infections (Section 29.7)?

8. What measures are used to control the growth of *Escherichia coli* O157:H7 in ground meat in production plants? By consumers (Section 29.8)?

9. *Campylobacter* causes more foodborne infections than any other bacterium. Identify at least one reason why this is true (Section 29.9).

10. Identify the food sources of *Listeria monocytogenes* infections. Identify the individuals who are at high risk for listeriosis (Section 29.10).

11. Why are viral agents so commonly associated with foodborne disease (Section 29.11)?

APPLICATION QUESTIONS

1. Identify optimum storage conditions for perishable, semiperishable and nonperishable food products. Consider economic factors such as the cost of storage and the value of the food item.

2. For a food of your choice, devise a way to preserve the food by lowering the water activity without drying.

3. Perfringens food poisoning involves ingestion of *Clostridium perfringens* followed by growth and sporulation in the intestine of the host. Sporulation triggers toxin production. Is this disease truly a food poisoning, or might it be classified as a food infection?

4. Improperly handled potato salads are often the source of staphylococcal food poisoning or salmonellosis. Explain several means by which a potato salad could become inoculated with either *Staphylococcus aureus* or *Salmonella* spp.

5. *Clostridium botulinum* requires an anoxic environment for production of botulinum toxin. Identify methods of food preservation that create the anoxic environment necessary for growth of *C. botulinum*. Conversely, identify methods of food preservation that create an oxic environment and discourage the growth of *C. botulinum*. What other factors influence the growth of *C. botulinum*?

6. Why aren't antibiotics generally used to treat salmonellosis? Explain your answer based on the habitat of the organism and access to antibiotics. Also consider the issue of potential antibiotic resistance.

7. Indicate the precautions necessary to prevent infection with pathogenic *Escherichia coli*. Concentrate on *E. coli* O157:H7 and safe food handling, cooking, and consumption.

8. Devise a plan to eliminate *Campylobacter* contamination from a poultry flock or from the finished poultry product. Explain the benefits of *Campylobacter*-free poultry and explain the problems that your plans might encounter.

9. Listeriosis normally occurs only when there is a breakdown in T_H1 cell-mediated immunity. Indicate why this is so. Devise a vaccine to protect against listeriosis. Would your vaccine be of use in the listeriosis-prone population?

10. Indicate potential reasons for the high incidence of viral foodborne disease, especially with Norwalk-like viruses. Devise a plan to eliminate Norwalk-like viruses from the food supply.

11. Indicate the problems inherent in tracking a latent infectious agent like the BSE prion. Can prion diseases be eliminated, and, if so, how?

30

INDUSTRIAL MICROBIOLOGY

In industrial microbiology, economical production is only possible if it is done on a very large scale. In order to make a competitively priced product by microbial fermentation, the growth medium used must be inexpensive and the growth vessels extremely large. Shown here is the inside of a large industrial fermentor complete with fittings for cooling and mixing.

WORKING GLOSSARY

Aspartame a non-nutritive sweetener composed of the amino acids aspartate and phenylalanine, the latter as a methyl ester

β-Lactam antibiotic a member of a group of antibiotics including penicillin, containing the four-membered heterocyclic β-lactam ring

Biocatalysis the use of microorganisms to carry out a specific chemical synthesis

Biosynthetic penicillin production of a particular form of penicillin by supplying the organism with specific side chain precursors

Biotransformation the use of microorganisms to carry out a chemical reaction that is more costly or not feasible nonbiologically

Brewing the manufacture of alcoholic beverages such as beer from the fermentation of malted grains

Broad-spectrum antibiotic an antimicrobial drug useful in treating a wide variety of bacterial diseases

Commodity chemical a chemical such as ethanol that has low monetary value and is thus sold primarily in bulk

Distilled beverage a beverage containing alcohol concentrated by distillation

Exoenzyme an enzyme produced by a microorganism and then excreted into the environment

Extremozyme an enzyme able to function in the presence of one or more chemical or physical extremes, for example, high temperature or low pH

Fermentation in an industrial context, any large-scale microbial process whether carried out aerobically or anaerobically

Fermentor the tank in which an industrial fermentation is carried out

Immobilized enzyme an enzyme attached to a solid support over which substrate is passed and converted to product

Industrial microbiology the large-scale use of microorganisms to produce products of commercial value

Natural penicillin the parent penicillin structure, produced by cultures of *Penicillium* not supplemented with side-chain precursors

Primary metabolite a metabolite excreted during the growth phase

Protease an enzyme that can degrade proteins by hydrolysis

Scale-up conversion of an industrial process from a small laboratory setup to a large commercial fermentation

Secondary metabolite a metabolite excreted at the end of the primary growth phase and into the stationary phase

Semisynthetic penicillin a penicillin produced using components derived from both microbial fermentation and chemical syntheses

Tetracycline a member of a class of antibiotics containing the four-membered naphthacene ring

INDUSTRIAL MICROORGANISMS AND PRODUCT FORMATION

Industrial microbiology uses microorganisms, typically grown on a large scale, to produce valuable commercial products or to carry out important chemical transformations. Industrial microbiological processes are enhancements of *metabolic reactions* that microorganisms are already capable of carrying out, with the goal in most cases of simply overproducing the product of interest. Industrial microbiology thus differs from *microbial biotechnology*. In biotechnology, methods for gene manipulation are used to yield new microbial products, most of which are not naturally produced by microorganisms (Chapter 31).

Biocatalysis is the word used to describe the actual reactions carried out by microorganisms in industrial microbiology. In this chapter we discuss several industrial biocatalyses along with some of the problems large-scale microbial culture entails. We begin with an overview of industrial organisms and products. Industrial microbiology originated with alcoholic fermentation processes, such as those for making beer and wine. Subsequently, microbial processes were developed for the production of pharmaceuticals (such as antibiotics), food additives (such as amino acids), enzymes, and chemicals such as butanol and citric acid. We shall see how industrial microbiologists have made microbial processes economically possible on an industrial scale.

30.1 Industrial Microorganisms and Their Products

The major organisms used in industrial microbiology are fungi (yeasts and molds) (Chapter 14) and certain prokaryotes, in particular members of the genus *Streptomyces* (Section 12.24). Industrial microorganisms are metabolic specialists, capable of synthesizing one or more products in high yield. To achieve this high metabolic specialization, scientists often genetically alter strains of industrial microorganisms by mutation or recombination. These changes work to increase the *yield* of the particular product that a given organism can produce.

The ultimate source of all strains of microorganisms used in biocatalytic processes is nature. However, actual industrial strains are usually far removed from the wild-type strain first isolated. Once valuable industrial microorganisms have been developed, high-yielding strains are conserved by microbiology laboratories in industry. In addition, large national microbial culture collections, such as the American Type Culture Collection (ATCC) in the United States or the Deutsche Sammlung von Mikroorganismen und Zellkulturen (DSMZ) in Germany, supply cultures for educational, research, and industrial purposes.

When a new biocatalytic process is patented, a strain capable of carrying out the process must be deposited into a national collection. However, for several reasons, proprietary rights chief among them, the strains deposited are *not* the actual high-yielding production strain(s) but instead a strain or strains that carry out the process at lower yield.

Properties of a Useful Industrial Microorganism

A microorganism suitable for an industrial process must have other features besides just being able to produce the substance of interest. First and foremost, the organism must be capable of growth and product formation in large-scale culture. Moreover, it should preferably produce spores or some other reproductive cell form so that it can be easily inoculated into large fermentors. It must also grow rapidly and produce the desired product in a relatively short period of time.

An industrially useful organism must also be able to grow in a relatively inexpensive liquid culture medium obtainable in bulk quantities. Many industrial microbiological processes use waste carbon from other industries as major or supplemental ingredients for large-scale culture media. These include *corn steep liquor* (a product of the corn wet milling industry that is rich in nitrogen and growth factors), and *whey* (a waste liquid of the dairy industry containing lactose and minerals).

An industrial microorganism should not be pathogenic, especially to humans or economically important animals or plants. Because of the high cell densities in the industrial fermentor and the virtual impossibility of avoiding contamination of the environment outside the fermentor, a pathogen would present potentially disastrous problems.

Finally, an industrial microorganism should be amenable to genetic manipulation. In industrial microbiology, increased yields have often been obtained genetically by means of mutation and selection. A genetically stable and easily manipulable microorganism is thus a clear advantage.

Examples of Industrial Products

Microbial products of industrial interest (Figure 30.1●) include the microbial cells themselves—for example, yeast cultivated for food, baking, or brewing (see Figure 30.24)—and substances produced by cells. Examples of the latter include enzymes such as glucose isomerase, important in the production of high-fructose syrups, pharmacologically active agents such as antibiotics, steroids and alkaloids, specialty chemicals and food additives such as aspartame (a food and drink sweetener), and **commodity chemicals**—inexpensive chemicals produced in bulk—such as ethanol, citric acid, and many others.

30.1 Concept Check

An industrial microorganism must produce the product of interest in high yield, grow rapidly on inexpensive culture media available in bulk quantities, be amenable to genetic manipulation, and, if possible, be nonpathogenic. Industrial products are many and include both cells and substances made by cells.

◆ Why should industrial microorganisms be genetically manipulable?

◆ List three important products of industrial biocatalysis.

30.2 Primary and Secondary Metabolites

When in the growth cycle is the industrially useful metabolite produced? In Section 6.1 we discussed microbial growth and described the various stages: *lag, exponential,* and *stationary*. Here we describe microbial growth and product formation in an industrial context, and will see that some products are formed during exponential growth while others are produced only after exponential growth has ceased.

There are two basic types of microbial metabolites: *primary* and *secondary*. A **primary metabolite** is one that is formed during the growth phase of the microorganism. In contrast, a **secondary metabolite** is one that is formed near the end of the growth phase, frequently at, near, or in the stationary phase of growth (Figure 30.2●). A typical primary metabolite would be *alcohol*. Ethanol is a product of anaerobic metabolism of yeast and certain bacteria (∞ Section 5.10) and is formed as part of energy metabolism. Because growth can occur only if energy production can occur, ethanol formation takes place in parallel with growth (Figure 30.2a).

In contrast to ethanol production by yeast, in some biocatalyses, the desired product is not made during the active growth phase but only in the *stationary* phase. These *secondary metabolites* are some of the most common and important metabolites of industrial interest (Figure 30.2b). All share the following characteristics:

1. Secondary metabolites are *not essential for growth* and reproduction.

Cells

Supplements (for example, yeast extract) **Yeast cells**

Biotransformation

Cells

Substrate

Product (for example, steroid biotransformations)

Products from cells

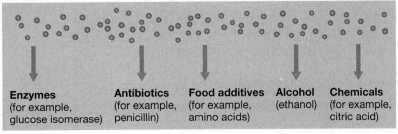

Enzymes (for example, glucose isomerase) **Antibiotics** (for example, penicillin) **Food additives** (for example, amino acids) **Alcohol** (ethanol) **Chemicals** (for example, citric acid)

● **Figure 30.1 Products of industrial microbiology/biocatalysis.** The products may be the cells themselves or products made from cells. In the case of biotransformation, cells are used to chemically convert a specific substance from one form to another.

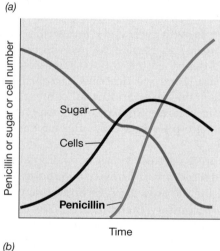

● **Figure 30.2 Contrast between production of primary and secondary metabolites.** (a) Alcohol formation by yeast—an example of a *primary* metabolite. (b) Penicillin production by the mold *Penicillium chrysogenum*—an example of a *secondary* metabolite. Note how penicillin is not made until after mid-log phase (⌘ Figure 6.8).

2. The formation of secondary metabolites is extremely *dependent on growth conditions*, especially on the composition of the medium. Repression of secondary metabolite formation frequently occurs.

3. Secondary metabolites are often *produced as a group of closely related compounds*. For instance, a single strain of a species of *Streptomyces* has been found to produce over 30 related but different anthracycline antibiotics.

4. It is often possible to get dramatic *overproduction* of secondary metabolites, whereas primary metabolites, linked as they are to primary metabolism, usually cannot be significantly overproduced (Figure 30.2).

Primary and Secondary Metabolism Pathways

Most secondary metabolites are complex organic molecules that require a large number of specific enzymatic reactions for synthesis. For instance, it is known that at least 72 separate enzymatic steps are involved in synthesis of the antibiotic *tetracycline* (see Section 30.6) and over

25 steps in the synthesis of *erythromycin*. None of these reactions occur during primary metabolism. However, the metabolic pathways of these secondary metabolites arise out of primary metabolism because their starting materials originate from the major biosynthetic pathways. Figure 30.3● shows the interrelationship of the main primary metabolic pathway for aromatic amino acid synthesis with the secondary metabolic pathways for a variety of antibiotics. As can be seen, many of these antibiotics originate from primary metabolites (Figure 30.3).

⬡ **30.2 Concept Check**

Primary and secondary metabolites are produced during active cell growth or near the onset of stationary phase, respectively. Many economically valuable microbial products are secondary metabolites.

◆ Is penicillin a *primary* or a *secondary* metabolite? Why?

◆ What type of metabolite, primary or secondary, can be more easily overproduced? Why?

● **Figure 30.3 Aromatic antibiotics.** Relationship of the primary metabolic pathway for the synthesis of aromatic amino acids (⌘ Section 5.16) and formation of a variety of secondary metabolite antibiotics containing aromatic rings. This is a composite scheme of processes occurring in a variety of microorganisms. No one organism produces all these secondary metabolites, and many individual steps exist between amino acid and antibiotic in all cases.

30.3 Characteristics of Large-Scale Fermentations

The vessel in which the industrial process is carried out is called a **fermentor**. In industrial microbiology, the term **fermentation** refers to *any* large-scale microbial process, whether or not it is biochemically a fermentation. In fact, most industrial fermentations are aerobic processes. Thus, while the *tank* in which the industrial fermentation is carried out is called a fermentor, the *microorganism* involved is the fermenter.

Fermentors can vary in size from the small 5- to 10-l laboratory scale (Figure 30.4*a*●) to the enormous 500,000-l industrial scale. The size of the fermentor used depends on the process and how it is operated. A summary of fermentor sizes for some common microbial fermentations is given in Table 30.1.

Industrial fermentors can be divided into two major classes, those for *anaerobic* processes and those for *aerobic* processes. *Anaerobic fermentors* require little special equipment except for removal of heat generated during the fermentation. Aerobic fermentors, however, require much more elaborate equipment to ensure that mixing and adequate aeration are achieved. Because most industrial fermentations are aerobic, we focus on aerobic fermentors here.

Construction of an Aerobic Fermentor

Large-scale industrial fermentors are almost always constructed of stainless steel. Such a fermentor is essentially a large cylinder, closed at the top and bottom, into which various pipes and valves have been fitted (Figure 30.4*b*). Because sterilization of the culture medium and removal

● **Figure 30.4 Fermentors.** (a) A small research fermentor. The volume is 5 l. (b) Diagram of a fermentor, illustrating construction and facilities for aeration and process control. (c) Inside of an industrial fermentor, showing the impeller and internal heating and cooling coils. In a typical industrial fermentation, aeration and cooling are the key components for online monitoring and adjustment. Nutrient levels and pH are also closely monitored and adjustments made automatically when needed.

Table 30.1	Fermentor sizes for various industrial processes
Size of fermentor (liters)	**Product**
1–20,000	Diagnostic enzymes, substances for molecular biology
40–80,000	Some enzymes, antibiotics
100–150,000	Penicillin, aminoglycoside antibiotics, proteases, amylases, steroid transformations, amino acids, wine, beer
200,000–500,000	Amino acids (glutamic acid), wine, beer

of heat are vital for successful operation, the fermentor is fitted with an external *cooling jacket* through which steam (for sterilization) or cooling water (for cooling) can be run. For very large fermentors, insufficient heat transfer occurs through the jacket, and so *internal coils* must be provided through which either steam or cooling water can be piped.

A critical part of the fermentor is the *aeration system*. With large-scale equipment, transfer of oxygen throughout the growth medium is critical, and elaborate precautions must be taken to ensure proper aeration. Oxygen is poorly soluble in water, and in a fermentor with a high microbial population density, there is a tremendous oxygen demand by the culture.

Two separate installations are used to ensure adequate aeration: an aeration device, called a *sparger*, and a stirring device, called an *impeller* (Figure 30.4*b*). The sparger is typically just a series of holes in a metal ring or a nozzle through which filter-sterilized air (or oxygen-enriched air) can be passed into the fermentor under high pressure. The air enters the fermentor as a series of tiny bubbles from which the oxygen passes by diffusion into the liquid. In small fermentors use of a sparger alone may be sufficient to ensure adequate aeration. But in industrial-size fermentors, *stirring* of the fermentor with an impeller is essential (Figure 30.4*c*). Stirring accomplishes two things: It *mixes the gas bubbles* through the liquid and it *mixes the organism* through the liquid, thus ensuring uniform access of microbial cells to the nutrients.

Fermentation Control and Monitoring

Any microbial fermentation must be monitored to ensure that it is proceeding properly, but it is especially important that industrial fermentors be monitored carefully because there is such a major expense involved. In most cases, it is necessary not only to measure growth and product formation but also to *control* the process by altering environmental parameters as the process proceeds. Environmental factors that are frequently controlled include temperature, oxygen concentration, pH, cell mass, levels of key nutrients, and product concentration.

During growth and product formation in a large-scale fermentation, it is essential to obtain data in real time (see Figure 30.5*b*). For instance, it may be desirable to

alter one or more of the environmental parameters as the fermentation progresses or to feed a nutrient at a rate that exactly balances growth. Computers process these types of data on-line and then respond accordingly by adding nutrients at the right time to maintain high product yield.

Computers are also used to *model* fermentation processes. Mathematical models can be used to test the effect of various parameters on growth and product yield quickly and interactively. With the models, industrial microbiologists can modify the parameters to see how each affects the process. In this way, many variations in the fermentation can be studied inexpensively on screen, rather than expensively at the pilot plant stage or in the industrial plant.

 30.3 Concept Check

Large-scale industrial fermentations present several engineering problems. Aerobic processes require mechanisms for stirring and aeration. The microbial process must be continuously monitored to ensure satisfactory yields of the desired product.

◆ What types of devices are used to ensure proper aeration in a large-scale fermentation?

◆ What parameters in an industrial fermentation need to be monitored and what adjustments need to be made by computerized control?

30.4 Fermentation Scale-Up

An important aspect of industrial microbiology is the transfer of a process from small-scale laboratory equipment to large-scale commercial equipment, a procedure called **scale-up**. An understanding of the problems in scale-up is extremely important because biocatalytic processes rarely behave the same way in large-scale fermentors (Figure 30.5●) as in small-scale laboratory equipment (Figure 30.6●). Scale-up of an industrial process is the task of the *biochemical engineer*, one who is familiar with gas transfer, fluid dynamics, mixing, and thermodynamics. Scale-up requires knowledge not only of the biology of the producing organism, but also of the physics of fermentor design and operation.

Much of the problem in scale-up involves proper aeration and mixing. These essential ingredients in the industrial process are much easier to accomplish in the small laboratory flask than in the large industrial fermentor. Oxygen transfer especially is much more difficult to obtain in a large fermentor, and because most industrial fermentations are aerobic, effective oxygen transfer is essential. With the rich culture media used in industrial processes, a high biomass is obtained, leading to a high oxygen demand. If aeration is reduced, even for a short period, the culture may experience temporary anoxic conditions, with serious metabolic consequences and reduction in product yield.

(a)

(b)

● **Figure 30.5 Industrial scale fermentations.** (a) A large industrial fermentation plant. Only the tops of the fermentors, which can be several stories high, are visible. (b) Computer control room for a large fermentation plant.

The Scale-Up Process

Transferring an industrial process from the laboratory to the commercial fermentor involves several stages. Things begin in the *laboratory flask*, a very small-scale operation but typically the first indication that a process of commercial interest is possible. From here things are transferred to the *laboratory fermentor,* a small-scale fermentor, generally of glass and of 1 to 10 l in size, in which the first efforts at scale-up are made (Figures 30.4a and see 30.6a). In the laboratory fermentor, it is possible to test variations in medium, temperature, pH, and so on, inexpensively because little cost is involved for either equipment or culture medium.

When tests in the laboratory fermentor are successful, the process moves into the *pilot plant stage*, usually carried out in equipment 300–3000 l in size. Here, the conditions more closely approach the commercial scale; however, cost is not yet a major factor. Finally, the process is moved to the *commercial fermentor* itself, typically 10,000–500,000 l in volume (Figures 30.5a and 30.6b). In all stages, aeration is very closely monitored. As scale-up proceeds from flask to production fermentor, oxygen dynamics are carefully measured at each

step to determine how volume increases affect oxygen demand in the fermentation.

 30.4 Concept Check

Scale-up is the process of gradually converting a useful industrial fermentation from laboratory scale to production scale. Aeration is a particularly critical aspect to monitor during scale-up studies.

◆ What are the differences in size among a typical laboratory fermentor, a pilot plant fermentor, and a commercial fermentor?

II MAJOR INDUSTRIAL PRODUCTS FOR THE HEALTH INDUSTRY

We now consider the products of industrial microbiology, beginning with the antibiotics. Of the microbial products manufactured commercially, the most important for the health industry are the antibiotics. Antibiotic production is a huge industry worldwide and one where many important principles for growing large-scale microbial cultures were first developed.

30.5 Antibiotics: Isolation and Characterization

As discussed in Chapter 20, antibiotics are chemicals produced by microorganisms that kill or inhibit the growth of other microorganisms. The development of antibiotics as agents for treatment of infectious disease has probably had more impact on the practice of medicine than any other single development. Antibiotics are typical secondary metabolites (see Section 30.2). Commercially useful antibiotics are produced primarily by filamentous fungi and by *Bacteria* of the actinomycete group (∞ Section 12.24). Table 30.2 lists the most important antibiotics produced by large-scale industrial fermentation.

Search for New Antibiotics

Although pharmaceutical companies currently do much of their drug discovery using computer modeling (∞ Section 20.13), traditionally, antibiotics were discovered by *screening*. In this approach, a large number of isolates of possible antibiotic-producing microorganisms are obtained from nature in pure culture (Figure 30.7a●), and these isolates are then tested for antibiotic production by seeing whether they produce any diffusible materials that are inhibitory to the growth of test bacteria. The test bacteria used are selected from a variety of bacterial types but are chosen to be representative of, or related to, bacterial pathogens.

The classical procedure for testing new microbial isolates for antibiotic production is the *cross-streak* method. Those isolates that show evidence of antibiotic

(a)

(b)

● **Figure 30.6** **Research and production fermentors.** (a) A bank of small research fermentors used in process development. The fermentors are the glass vessels with the stainless steel tops. The small plastic bottles are to collect overflow. (b) A large bank of outdoor industrial-scale fermentors (240 m³) used in commercial production of alcohol in Japan. Because of the great difference in their size, the same microbial fermentation would probably operate quite differently in the two different sizes of fermentors.

production are then studied further to determine if the antibiotics they produce are new. Most of the isolates obtained produce known antibiotics, so the industrial microbiologist must quickly identify such organisms so that time and resources are not wasted in studying them. Once an organism producing a new antibiotic is discovered, the antibiotic is produced in sufficient amounts for structural analyses and then tested for toxicity and therapeutic activity in infected animals. Unfortunately, most new antibiotics fail these animal tests, but a few prove to be medically useful and are produced commercially. However, with estimates of the number of different antibiotics produced by species of the genus *Streptomyces* alone at over 100,000, research to discover new antibiotics occurs on a continuous basis.

Purification

An antibiotic that is to be produced commercially must first be produced successfully in large-scale industrial fermentors (scale-up, see Section 30.4). The next challenge is to *purify* the product efficiently. Because of the relatively small amounts of antibiotic present in the fermentation liquid, elaborate methods for extraction and purification of the antibiotic are necessary (Figure 30.8●). If the antibiotic is soluble in an organic solvent, it may be relatively simple to purify it by extraction into a small volume of the solvent. If the antibiotic is not solvent soluble, then it must be removed from the fermentation liquid by adsorption, ion exchange, or chemical precipitation (Figure 30.8). In all cases, the goal is to eventually obtain a crystalline product of high purity. Depending on the process, further purification steps may be necessary to remove traces of microbial cell material if it copurifies with the antibiotic.

Increasing the Yield

Rarely do antibiotic-producing strains just isolated from nature produce the desired antibiotic at sufficiently high concentration that commercial production can begin

Table 30.2	**Some antibiotics produced commercially**[a]
Antibiotic	**Producing microorganism**[b]
Bacitracin	*Bacillus licheniformis* (EFB)
Cephalosporin	*Cephalosporium* spp. (F)
Chloramphenicol	Chemical synthesis (formerly produced microbially by *Streptomyces venezuelae*) (A)
Cycloheximide	*Streptomyces griseus* (A)
Cycloserine	*Streptomyces orchidaceus* (A)
Erythromycin	*Streptomyces erythreus* (A)
Griseofulvin	*Penicillium griseofulvin* (F)
Kanamycin	*Streptomyces kanamyceticus* (A)
Lincomycin	*Streptomyces lincolnensis* (A)
Neomycin	*Streptomyces fradiae* (A)
Nystatin	*Streptomyces noursei* (A)
Penicillin	*Penicillium chrysogenum* (F)
Polymyxin B	*Bacillus polymyxa* (EFB)
Streptomycin	*Streptomyces griseus* (A)
Tetracycline	*Streptomyces rimosus* (A)

[a] See Chapter 12 and Figure 20.12 for information on the structures of these antibiotics.

[b] EFB, endospore-forming bacterium; F, fungus; A, actinomycete.

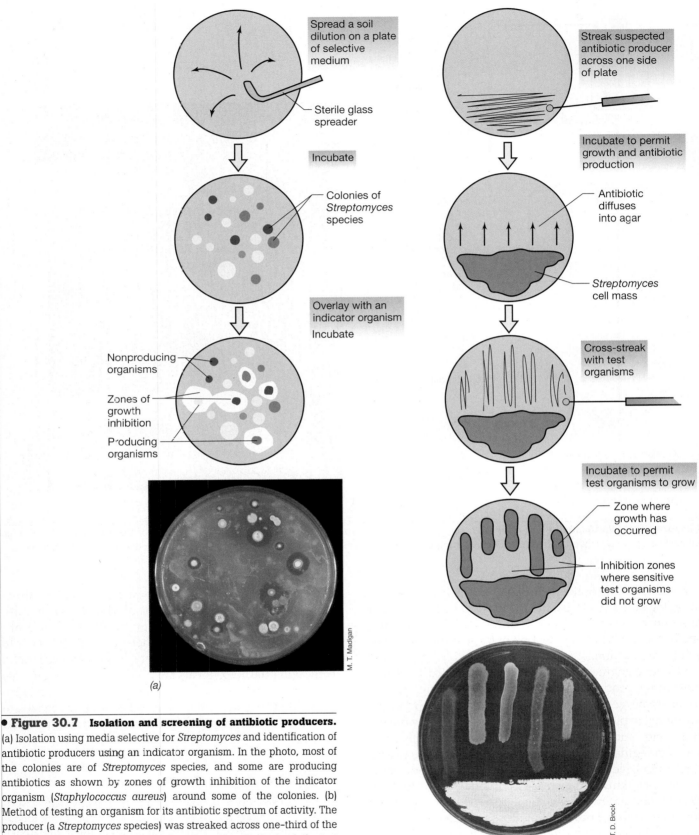

● **Figure 30.7 Isolation and screening of antibiotic producers.**
(a) Isolation using media selective for *Streptomyces* and identification of antibiotic producers using an indicator organism. In the photo, most of the colonies are of *Streptomyces* species, and some are producing antibiotics as shown by zones of growth inhibition of the indicator organism (*Staphylococcus aureus*) around some of the colonies. (b) Method of testing an organism for its antibiotic spectrum of activity. The producer (a *Streptomyces* species) was streaked across one-third of the plate and the plate incubated. After good growth was obtained, the test bacteria were streaked perpendicular to the *Streptomyces*, and the plate was further incubated. The failure of several organisms to grow near the mass growth of *Streptomyces* indicates that the *Streptomyces* produced an antibiotic active against these bacteria. Test organisms (left to right): *Escherichia coli, Bacillus subtilis, Staphylococcus aureus, Klebsiella pneumoniae, Mycobacterium smegmatis.*

● **Figure 30.8 Purification of an antibiotic.** (a) Overall process of extraction and purification. (b) Installation for the solvent extraction of an antibiotic from fermentation broth. Effective engineering is as important as microbiological factors in the successful production of an antibiotic.

immediately. One of the major tasks of the industrial microbiologist is thus to isolate *high-yielding strains*. Strain selection involves mutagenesis of the initial culture, plating of mutant types, and testing of these mutants for antibiotic production.

Genetic engineering has greatly improved the time-consuming process of obtaining high-yielding strains. For example, *gene amplification* makes it possible to insert additional copies of genes of interest into a cell by means of a vector such as a plasmid (∞ Chapter 31). Alterations in regulatory processes also may permit increased yields. However, one difficulty with using genetic engineering procedures to increase antibiotic yield is that the biosynthetic pathways for the synthesis of most antibiotics involve large numbers of steps with many genes. It is often unclear which genes should be altered or increased in number to increase yields. Thus, it is important first to identify the *rate-limiting* or other key regulatory steps in a given biochemical pathway through basic research. Meanwhile, however, since this can be a lengthy process, empirical studies to obtain a high-yielding strain continue.

Final yield is a critical issue with virtually all pharmaceuticals. Even after commercial production of an antibiotic or other product has begun, research often continues to identify or produce higher-yielding strains or to modify the process in some way so as to increase yield. Although pharmaceuticals are produced on a small scale compared with commodity chemicals or agricultural products, there are nevertheless good economic reasons to seek the highest possible yields in the shortest period of time.

🛑 30.5 Concept Check

The industrial production of antibiotics begins with screening for antibiotic producers. Once new producers are identified, purification and chemical analyses of the antimicrobial agent are performed. If the new antibiotic is biologically active *in vivo*, the industrial microbiologist may genetically modify the producing strain to increase yields to levels acceptable for commercial development.

◆ What is the natural habitat of most antibiotic-producing microorganisms?

◆ What is meant by the word *screening* in the context of finding new antibiotics?

Industrial Production of Penicillins and Tetracyclines

Once a new antibiotic has been structurally characterized and has proven medically effective and nontoxic in tests on experimental animals, it is ready for clinical trials on humans. In practice, all of the events leading up to clinical trials can take many years. If the new drug proves clinically effective, it is ready to be produced commercially and marketed. For antibiotics such as penicillin and tetracycline, these hurdles were cleared long ago. Today, literally *tons* of these two antibiotics are produced for medical and veterinary use, and we thus focus on their production here.

β-Lactam Antibiotics: Penicillin and Its Relatives

As discussed in Section 20.8, the penicillins are a class of **β-lactam antibiotics** characterized by the *β-lactam ring* and are produced by a variety of molds (eukaryotes) of the genera *Penicillium* and *Aspergillus* and by certain prokaryotes (Figure 30.9●). Clinically useful penicillins are of several different types, and many are the product of biocatalytic reactions and later chemical modification by the organic chemist. This allows for the synthesis of a suite of penicillins, each with its own special clinical properties (Figure 30.9).

The parent structure of all penicillins is *6-aminopenicillanic acid* (6-APA), which consists of a thiazolidine ring with a condensed β-lactam ring (Figure 30.9). The 6-APA carries a variable side chain in position 6. If the penicillin fermentation is carried out without addition of side-chain precursors, the **natural penicillins** are produced (Figure 30.9). The fermentation can be more directed by adding to the broth a *side-chain precursor* so that only one desired penicillin is produced. The product

formed under these conditions is referred to as a **biosynthetic penicillin** (Figure 30.9).

To produce the most useful penicillins, those with activity against *gram-negative Bacteria*, a combined fermentation and chemical approach is used that leads to the production of **semisynthetic penicillins** (Figure 30.9). In this case, a microbially produced natural penicillin is split either chemically or enzymatically to yield 6-APA; the latter is then chemically modified by the addition of a side chain (Figure 30.9). Semisynthetic penicillins have many significant clinical advantages. These include, among other features, a typically broad spectrum of activity and the fact that many of them can be taken orally and thus do not require injection. The widely prescribed drug *ampicillin* is a good example here. For these reasons, semisynthetic penicillins make up the bulk of the penicillin market today.

Production of β-Lactam Antibiotics

Penicillin G is produced by the mold *Penicillium chrysogenum* in fermentors of 40,000–200,000 l in size. Penicillin production is a highly aerobic process, and efficient aeration is thus necessary. Penicillin is a typical secondary metabolite. During the growth phase, very little penicillin is produced, but once the carbon source has been nearly exhausted, the penicillin production phase begins (Figure 30.10●). By supplying additional carbon and nitrogen, the production phase can be extended for several days (Figure 30.10).

A major ingredient of most penicillin production media is *corn steep liquor*, which contains nitrogen as well as several growth factors. The carbon source is generally *lactose* (Figure 30.10), obtained from whey. High levels of glucose tend to repress penicillin production, but high levels of lactose do not. Thus lactose is the initial carbon source present in large amounts. As the lactose becomes limiting and biomass levels in the fermentor become very high, "feedings" with glucose later in the fermentation maximize penicillin yield. Penicillin is excreted into the medium, and after the cells are removed by filtration, the pH of the medium is lowered and the antibiotic extracted with an organic

Penicillin fermentation

6-Aminopenicillanic acid

● **Figure 30.9 Industrial production of penicillins.** The β-lactam ring is shown in dark brown. The normal fermentation leads to the *natural* penicillins. If specific precursors are added during the fermentation, various *biosynthetic* penicillins are formed. *Semisynthetic* penicillins are produced by chemically adding a specific side chain to the 6-aminopenicillanic acid nucleus on the "R" group shown in purple. Semisynthetic penicillins have the greatest clinical usefulness because they are typically active against gram-negative *Bacteria* and can be administered orally. Commercial penicillin production in the United States occurs using high-yielding strains of the mold *Penicillium chrysogenum.*

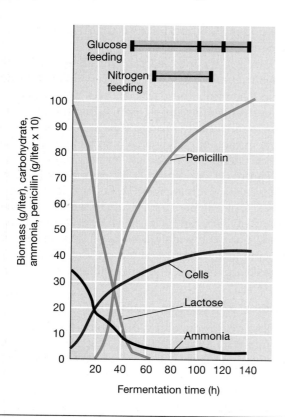

● **Figure 30.10 Kinetics of the penicillin fermentation with** *Penicillium chrysogenum.* Note how the production of penicillin occurs as cells are entering stationary phase and most of the carbon and nitrogen is exhausted. Nutrient "feedings" keep penicillin production high.

● **Figure 30.11 Production scheme for chlortetracycline with** *Streptomyces aureofaciens.* The structure of chlortetracycline is shown on the bottom right. Growth temperature, 28° C throughout.

solvent. After concentration into the solvent, the antibiotic is back-extracted into an alkaline aqueous medium, concentrated further, and crystallized. Highly purified penicillin can be readily obtained in this way.

Other β-Lactam Antibiotics

Other β-lactam antibiotics include the *cephalosporins*, drugs which contain a dihydrothiazine instead of a thiazolidine ring system (∞ Section 20.8 and Figure 20.12). Cephalosporins were first discovered as products of the fungus *Cephalosporium acremonium*, but a number of other fungi as well as some prokaryotes also produce antibiotics with this ring system. In addition, a number of semisynthetic cephalosporins, analogous to the semisynthetic penicillins (see Figure 30.9), are produced. Cephalosporins are valued clinically not only because of their low toxicity but also because they are **broad-spectrum antibiotics** (∞ Section 20.8), useful against a wide variety of bacterial pathogens.

Production of Tetracyclines

The biosynthesis of **tetracyclines** involves a large number of enzymatic steps. In the case of chlortetracycline (Figure 30.11●), for example, as many as 72 intermediate products may be involved, most of which are known in only a general way. Studies on the genetics of *Streptomyces aureofa-*

ciens, the producer of chlortetracycline, have shown that a total of more than 300 genes are involved!

With such a large number of genes, regulation of biosynthesis of this antibiotic is obviously quite complex. However, a few regulatory signals are known and production schemes are well worked out. For example, repression of chlortetracycline synthesis by both glucose and phosphate occurs. Phosphate repression is especially significant, and so the medium used in commercial production contains low phosphate concentrations. Figure 30.11 shows one such production scheme. As in penicillin production, corn steep liquor is used in the large-scale production of chlortetracycline, but sucrose (not lactose) is the typical carbon source. Glucose is avoided because in the tetracycline fermentation it causes catabolite repression (∞ Section 8.7) of antibiotic production.

30.6 Concept Check

Major antibiotics of clinical significance include the β-lactam antibiotics penicillin and cephalosporin and the tetracyclines. All of these antibiotics are typical secondary metabolites, and their industrial production is well worked out despite the fact that the biochemistry and genetics of their biosynthesis are only partially understood.

◆ What chemical structure is common to both penicillin and cephalosporin?

◆ In terms of penicillin production, what is meant by the term *semisynthetic*? *Biosynthetic*?

30.7 Vitamins and Amino Acids

Vitamins and amino acids are growth factors that are often used pharmaceutically or are added to foods. Several important vitamins and amino acids are produced commercially by biocatalytic processes.

Vitamins

Vitamins are used as supplements for human food and animal feeds, and production of vitamins is second only to that of antibiotics in terms of total sales of pharmaceuticals. Most vitamins are made commercially by chemical synthesis. However, a few are too complicated to be synthesized inexpensively but can be made by biocatalysis. Vitamin B_{12} and riboflavin are the most important of this class of vitamins.

Vitamin B_{12} (Figure 30.12*a*●) is synthesized in nature exclusively by microorganisms. As a coenzyme, vitamin B_{12} plays an important role in animal biochemistry in various intramolecular rearrangements in which a hydrogen atom on one carbon atom and a substituent on the adjacent carbon atom exchange places. In humans, a major deficiency of vitamin B_{12} leads to a severe condition called *pernicious anemia*, characterized by low production of red blood cells and nervous system disorders. The requirements of animals for vitamin B_{12} are satisfied by food intake or by absorption of the vitamin produced in the gut of the animal by intestinal microorganisms. Plants do not produce or use vitamin B_{12}.

For industrial production of vitamin B_{12}, microbial strains are employed that have been specifically selected for their high yields of the vitamin. Members of the bacterial genera *Propionibacterium* and *Pseudomonas* are the main commercial producers. Cobalt is a metal found in vitamin B_{12} (Figure 30.12*a*) and yields of the vitamin are greatly increased by addition of small amounts of cobalt to the culture medium.

Riboflavin (Figure 30.12*b*) is the parent compound of the flavins, FAD and FMN, coenzymes that play important roles in enzymes involved in oxidation-reduction reactions in virtually all organisms (⚮ Section 5.11). Riboflavin is synthesized by many microorganisms, including bacteria, yeasts, and fungi. The fungus *Ashbya gossypii* naturally produces huge amounts of this vitamin (up to 7 g/l) and is therefore used for most of the microbial production processes. Despite this good yield, there is great economic competition between the microbiological process and strictly chemical synthesis.

(a) **B₁₂**

(b) **Riboflavin**

● **Figure 30.12 Vitamins produced by microorganisms on an industrial scale.** (a) Vitamin B_{12}. Shown is the structure of cobalamin; note the central cobalt atom. The coenzyme form of vitamin B_{12} contains a deoxyadenosyl group attached to Co above the plane of the ring. (b) Riboflavin (vitamin B_2; ⚮ Section 5.11 and Figure 5.15).

Amino Acids

Amino acids have extensive uses in the food industry, as feed additives, in medicine, and as starting materials in the chemical industry (Table 30.3). The most important commercial amino acid is **glutamic acid**, which is used as a flavor enhancer [monosodium glutamate (MSG)]. Two other important amino acids, *aspartic acid* and *phenylalanine*, are the ingredients of the artificial sweetener **aspartame**, a non-nutritive sweetener of diet soft drinks and other foods sold as low-calorie or sugar-free products. Aspartame is a dipeptide of aspartate and the methyl ester of phenylalanine. *Lysine*, an essential amino acid for humans and certain farm animals, is commercially produced by the bacterium *Brevibacterium flavum* for use as a food additive, and we focus on its production here.

Because amino acids are used by microorganisms as building blocks of enzymes and other proteins, strict

Table 30.3	Amino acids used in the food industry[a]		
Amino acid[b]	**Annual production worldwide (metric tons)**	**Uses**	**Purpose**
L-Glutamate (monosodium glutamate, MSG)	370,000	Various foods	Flavor enhancer; meat tenderizer
L-Aspartate and alanine	5,000	Fruit juices	"Round off" taste
Glycine	6,000	Sweetened foods	Improves flavor; starting point for organic syntheses
L-Cysteine	700	Bread	Improves quality
		Fruit juices	Antioxidant
L-Tryptophan + L-Histidine	400	Various foods, dried milk	Antioxidant, prevent rancidity; nutritive additives
Aspartame (made from L- phenylalanine + L-aspartic acid)	7,000	Soft drinks, chewing gum, many other "sugar-free" products	Low-calorie sweetener
L-Lysine	70,000	Bread (Japan), feed additives	Nutritive additive
DL-Methionine	70,000	Soy products, feed additives	Nutritive additive

[a] Data from Glazer, A. N., and H. Mikaido. 1995. *Microbial Biotechnology*, W. H. Freeman, New York.
[b] The structures of these amino acids are shown in Figure 3.12.

cellular regulation of their production generally occurs (👓 Chapter 8). However, for the industrial production of an amino acid, ways to circumvent these regulatory mechanisms are necessary in order to obtain an *overproducing strain* capable of producing the amino acid economically.

The production of lysine in *Brevibacterium flavum* is biochemically controlled at the level of the enzyme aspartokinase; excess lysine feedback inhibits activity of this enzyme (Figure 30.13a●) (the general phenomenon of feedback inhibition was described in Section 8.2). However, overproduction of lysine can be obtained by isolating mutants of *B. flavum* in which aspartokinase is no longer subject to feedback inhibition. This is done by isolating mutants resistant to the lysine analog S-aminoethylcysteine (AEC), which binds to the allosteric site of aspartokinase and shuts down activity of the enzyme (Figure 30.13b). AEC-resistant mutants, which are easily obtained by positive selection, produce a modified form of aspartokinase with an allosteric site that no longer recognizes AEC *or* lysine, and thus feedback inhibition by lysine is greatly reduced. Such mutants of *B. flavum* can produce over 60 g of lysine per liter in industrial fermentors, a concentration sufficiently high to make the process commercially viable.

 30.7 Concept Check

Vitamins produced microbially include vitamin B_{12} and riboflavin, whereas the most important amino acids produced commercially are glutamic acid, aspartic acid, phenylalanine, and lysine. High yields of amino acids are obtained by modifying regulatory signals that control synthesis of the particular amino acid such that overproduction occurs.

◆ Which amino acid is commercially produced in the greatest amounts?

◆ How can overcoming feedback inhibition improve the yield of an amino acid?

● **Figure 30.13 Industrial production of lysine using *Brevibacterium flavum*.** (a) Biochemical pathway leading from aspartate to lysine; note that lysine can feedback inhibit (👓 Section 8.2) activity of the enzyme aspartokinase, leading to cessation of lysine production. (b) Structure of lysine and the lysine analog S-aminoethylcysteine (AEC). The molecules are identical except for the regions shown in color. AEC normally inhibits growth, but AEC-resistant mutants of *B. flavum* have an altered allosteric site on their aspartokinase and grow and overproduce lysine because feedback inhibition no longer occurs.

30.8 Steroids and the Biotransformation Process

We discussed the role of sterols in eukaryotic membranes in Section 4.5. **Steroids** are derivatives of sterols, and are important animal hormones that regulate various metabolic processes. Some steroids are also used as drugs in human medicine. Members of one group, the *corticosteroids*, reduce inflammation. Hence, these compounds are effective therapeutic agents in controlling the symptoms of arthritis and allergies. Members of another group, the *estrogens* and *androgenic steroids*, play a role in human fertility, and some of them can be used therapeutically in the control of fertility or as stimulants for building muscle mass.

Steroids can be obtained by complete chemical synthesis, but this is a complicated and expensive process. Certain key steps in chemical synthesis can be carried out more efficiently by microorganisms, and thus commercial production of steroids typically involves at least one microbial step. Most steroids are produced industrially by the process of **biotransformation**. This involves growth of the organism in large fermentors, followed by the addition at the appropriate time of a sterol precursor. The latter is obtained from inexpensive starting materials, typically stigmasterol, a by-product of the soybean industry. Following an incubation period during which the sterol precursor is biotransformed, the fermentation broth is extracted and the sterol purified (Figure 30.14●).

Cortisone and Hydrocortisone

In the production of hydrocortisone and cortisone, corticosteroids used to reduce swelling and itching from minor skin irritations, the fungus *Rhizopus nigricans* carries out a key biotransformation, the stereospecific hydroxylation of a cortisone precursor (Figure 30.14). Most steroid biotransformations involve hydroxylations of this type, and a variety of different fungi are used industrially to carry out the stereospecific hydroxylation of the steroid at a particular site on the steroid molecule.

Although most biotransformations are not beyond the capabilities of organic chemistry, microorganisms carry out the reactions more economically. Steroid production is currently a big business, as worldwide sales of the four major corticosteroids, *hydrocortisone, cortisone, prednisone,* and *prednisolone*, amount to over 800 tons/year.

30.8 Concept Check

Microbial biotransformation employs microorganisms to biocatalyze a specific step or steps in an otherwise strictly chemical synthesis.

◆ Give an example of a microbial biotransformation. Why is it necessary?

◆ Describe two ways in which a biotransformation differs from a typical fermentation, such as the production of antibiotics.

30.9 Enzymes as Industrial Products

Every organism produces a large variety of enzymes, most of which are made in only small amounts and are involved in cellular processes (∞ Section 5.5). However, certain enzymes are produced in much larger amounts by some organisms, and instead of being held within the cell, they are excreted into the medium. These extracellular enzymes, called **exoenzymes**, are capable of digesting insoluble polymers such as cellulose, protein, and starch. The products of digestion are then transported into the cell where they are used as nutrients for growth.

Enzymes are especially useful biocatalysts because they typically target single chemical functional groups and can distinguish between similar functional groups on a single molecule. Moreover, in many cases (for example with the sterols, just discussed), they catalyze reactions in a *stereospecific manner*, producing only one of two possible enantiomers (∞ Section 3.6). Various enzymes are used in the food and health industries, primarily as dietary supplements, while many others are produced for laundry applications or in the textile industry (Table 30.4). We briefly discuss their production and usage now.

Progesterone   11α-Hydroxyprogesterone   Hydrocortisone   **Cortisone**

● **Figure 30.14 Cortisone production using a microorganism: an example of a biotransformation.** The first reaction in cortisone production is a typical microbial biotransformation, the formation of 11 α-hydroxyprogesterone from progesterone. This highly specific (and stereospecific) oxidation, carried out by the fungus *Rhizopus nigricans*, bypasses a difficult chemical synthesis. All the other steps from 11α-hydroxyprogesterone to the steroid hormone cortisone are performed chemically.

Proteases, Amylases, and High-Fructose Syrup

Enzymes are produced commercially from both fungi and bacteria. The microbial enzymes produced in the largest amounts on an industrial basis are the bacterial **proteases**, used as additives in laundry detergents. Most laundry detergents today contain enzymes, chiefly proteases, but also *amylases, lipases, reductases*, and others. Many of these enzymes are isolated from alkaliphilic bacteria (Section 6.13), mainly species of *Bacillus* such as *Bacillus licheniformis* (Table 30.4). These enzymes, which have pH optima between 9 and 10, remain active at the alkaline pH of laundry detergent solutions.

Other important enzymes manufactured commercially are amylases and glucoamylases, which are used in the production of glucose from starch. The glucose so produced can then be converted by the enzyme *glucose isomerase* to produce fructose, which is about twice as sweet as glucose. The final result is the production of a *high-fructose syrup* from corn, wheat, or potato starch. High-fructose syrups are a big business, since they have a major market in the production of soft drinks. Worldwide production of high-fructose syrups is nearly 10 billion kilograms per year.

Extremozymes: Enzymes from Prokaryotes That Inhabit Extreme Environments

In Chapters 2 and 6 we considered aspects of microbial growth at high temperature and discovered that some prokaryotes, called *hyperthermophiles*, grow optimally at very high temperatures. Hyperthermophiles can grow at such high temperatures because they produce heat-stable macromolecules (Sections 6.10 and 6.12) including enzymes (see Figure 30.15b). The term **extremozyme** has been coined to describe enzymes that function at some environmental extreme, such as high temperatures or low pH (Figure 30.15●). The organisms that produce extremozymes are called *extremophiles* (Table 2.1).

Because many industrial processes operate best at high temperatures, extremozymes from hyperthermophiles are widely used in both industry and research. Besides the *Taq* and *Pfu* DNA polymerases used in the polymerase chain reaction (PCR) described in Section 7.9, extremely thermostable proteases, amylases, cellulases, pullulanases (Figure 30.15b), and xylanases have been isolated and characterized from various hyperthermophiles. However, it is not only thermal stable enzymes that have found a market. Extremozymes that are cold-active (from psychrophiles), active in the presence of high salt (from halophiles), or active at high or low pH (from alkaliphiles and acidophiles, respectively) (Figure 30.15a) are known, and several have been commercialized. The great specificity of enzymes and their ability to distinguish between chiral isomers make those that also function at environmental extremes particularly useful in various industrial applications.

Table 30.4	**Microbial enzymes and their applications**		
Enzyme	**Source**	**Application**	**Industry**
Amylase (starch-digesting)	Fungi	Bread	Baking
	Bacteria	Starch coatings	Paper
	Fungi	Syrup and glucose manufacture	Food
	Bacteria	Cold-swelling laundry starch	Starch
	Fungi	Digestive aid	Pharmaceutical
	Bacteria	Removal of coatings (desizing)	Textile
	Bacteria	Removal of stains; detergents	Laundry
Protease (protein-digesting)	Fungi	Bread	Baking
	Bacteria	Spot removal	Dry cleaning
	Bacteria	Meat tenderizing	Meat
	Bacteria	Wound cleansing	Medicine
	Bacteria	Desizing	Textile
	Bacteria	Household detergent	Laundry
Invertase (sucrose-digesting)	Yeast	Soft-center candies	Candy
Glucose oxidase	Fungi	Glucose removal, oxygen removal	Food
		Test paper for diabetes	Pharmaceutical
Glucose isomerase	*Bacteria*	High-fructose corn syrup	Soft drink
Pectinase	Fungi	Pressing, clarification	Wine, fruit juice
Rennin	Fungi	Coagulation of milk	Cheese
Cellulase	*Bacteria*	Fabric softening, brightening; detergent	Laundry
Lipase	Fungi	Breaks down fat	Dairy, laundry
Lactase	Fungi	Breaks down lactose to glucose and galactose	Dairy, health foods
DNA polymerase	*Bacteria*	DNA replication in polymerase chain	Biological research;
	Archaea	reaction (PCR) technique (Section 7.9)	forensics

(a)

(b)

● **Figure 30.15 Examples of extremozymes, enzymes that function under environmentally extreme conditions.** (a) Acid-tolerant enzymes. An enzyme mixture used as a feed supplement for poultry. The enzymes function in the stomach of the bird to digest fibrous materials in the feed, thereby improving the nutritional value of the feed and promoting more rapid growth of the bird. (b) Thermostable enzymes. Thermostability of the enzyme pullulanase from *Pyrococcus woesei*, a hyperthermophile whose growth temperature optimum is 100°C (∞∞ Section 13.6). Calcium improves the heat stability of this enzyme; however, at 110°C, the enzyme denatures (∞∞ Section 3.8).

Immobilized Enzymes

For some biocatalytic processes it is desirable to fix soluble enzymes onto a solid surface. These are called **immobilized enzymes**. Immobilization not only makes it easier to carry out the enzymatic reaction under large-scale continuous conditions, but also helps stabilize the en-

zyme to denaturation. A number of different immobilized enzymes have been employed in industrial processes, and a good example is in the starch-processing industry. As previously described, starch can be converted to high-fructose corn syrup by sequential treatment with amylase and glucose isomerase. The practical conversion of glucose to fructose is about 50%, but this can be increased somewhat by separating out the fructose and recycling the remaining glucose.

There are three basic approaches to enzyme immobilization (Figure 30.16●):

1. *Bonding of the enzyme to a carrier.* The bonding can be through adsorption, ionic bonding, or covalent bonding. Carriers used include modified celluloses, activated carbon, clay minerals, aluminum oxide, and glass beads (Figure 30.16).

2. *Cross-linking (polymerization) of enzyme molecules.* Linking of enzyme molecules with each other is usually done by chemical reaction with a cross-linking agent such as glutaraldehyde. Cross-linking of enzymes involves the chemical reaction of amino groups of the enzyme protein with glutaraldehyde. Aldehydes are also denaturing agents, but if the reaction is carried out properly, the enzyme molecules can be linked in such a way that most enzymatic activity is maintained.

3. *Enzyme inclusion, the incorporation of the enzyme into a semipermeable membrane.* Enzymes can be enclosed in microcapsules, gels, semipermeable polymer membranes, or fibrous polymers such as cellulose acetate (Figure 30.16).

Carrier-bound enzyme

Cross-linked enzyme

Enzyme inclusion in fibrous polymers

Enzyme inclusion in microcapsules

● **Figure 30.16 Procedures for the immobilization of enzymes.** In all cases, enzyme molecules are shown in red. The actual procedure used varies with the enzyme, the reaction, and the scale.

Each of these methods for the use of immobilized enzymes has advantages and disadvantages, and the procedure used depends on the enzyme, the particular industrial application, and the scale of the operation.

30.9 *Concept Check*

Microorganisms are ideal for the large-scale production of enzymes. Many enzymes are used in the laundry industry to remove stains from clothing, and thermostable and alkali-stable enzymes have many advantages in these markets. Enzymes from extremophiles are desirable for biocatalyses under extreme conditions. When an enzyme is used in a large-scale process, it may be desirable to immobilize it by bonding it to an inert substrate.

◆ How are enzymes of use in the laundry industry?

◆ What key enzyme is needed to produce high-fructose syrups?

◆ What is an *extremozyme*?

III MAJOR INDUSTRIAL PRODUCTS FOR THE FOOD AND BEVERAGE INDUSTRIES

Although industrial microbiology as a discipline can trace its roots to the antibiotic era, antibiotics are only one class of products made on an industrial scale. Many food products and supplements are made by industrial processes, as we have already seen. We consider some others of these now, beginning with the production of alcohol for the beverage industry and for use as an industrial solvent.

30.10 Alcohol and Alcoholic Beverages

Most fruit juices and grains will undergo a natural fermentation catalyzed by wild yeasts present on these materials. From these natural fermentations, strains of yeasts have been selected for more controlled production, and today alcoholic beverage production is a large industry worldwide. The most important alcoholic beverages are *wine*, produced by the fermentation of fruit juice; *beer*, or *ale*, produced by the fermentation of malted grains; and *distilled beverages*, produced by concentrating alcohol from a fermentation by distillation. The biochemistry of alcohol fermentation by yeast was discussed in Section 5.10.

Wine Varieties and Production

Most wine is made from grapes, and thus wine manufacture occurs in parts of the world where high-quality grapes can be grown most economically. These include many countries of the European Union, the United States (Figure 30.17●), New Zealand and Australia, and South America.

There are a great number of different wines, and their quality and character vary considerably. *Dry wines* are wines in which the sugars of the juice are practically all fermented, whereas in *sweet wines* some of the sugar is left or additional sugar is added after the fermentation. A *fortified wine* is one to which brandy or some other alcoholic spirit is added after the fermentation; sherry and port are the best-known fortified wines. A *sparkling wine*, such as champagne, is one in which considerable carbon dioxide is present, arising from a final fermentation by the yeast directly in the sealed bottle.

The production of wine begins in the early fall with the harvesting of grapes. The grapes are crushed and the juice, called *must*, is squeezed out. Depending on the grapes used and on how the must is prepared, either white or red wine may be produced (Figure 30.18●). A white wine is made either from white grapes or from the juice of red grapes from which the skins, containing the red coloring matter, have been removed. In the making of red wine, the *pomace* (skins, seeds, and pieces of stem) is left in during the fermentation. In addition to the color difference, red wine has a stronger flavor than white because of larger amounts of *tannins*, chemicals that are extracted into the juice from the grape skins during the fermentation.

The yeasts involved in wine fermentation are of two types: *wild yeasts*, which are present on the grapes as they are taken from the field and are transferred to the juice, and the *cultivated wine yeast, Saccharomyces ellipsoideus*, which is added to the juice to begin the fermentation. Wild yeasts are less alcohol-tolerant than commercial wine yeasts and can also produce undesirable compounds affecting quality of the final product. Thus, it is the practice in most wineries to kill the wild yeasts present in the must by adding sulfur dioxide (listed on the bottle as "sulfites") at a level of about 100 parts per million (ppm). *Saccharomyces ellipsoideus* is resistant to this concentration of sulfur dioxide and is added as a starter culture from a pure culture grown on sterilized grape juice. The fermentation is carried out in vats of various sizes, from 200-l casks to 200,000-l tanks made of oak, cement, stone, or glass-lined metal (see Figure 30.17b). The fermentor must be constructed so that the large amount of carbon dioxide produced during the fermentation can escape but air cannot enter, and this is accomplished by fitting the vessel with a special one-way valve.

With a red wine, after 3–5 days of fermentation, sufficient tannin and color have been extracted from the pomace and the wine is drawn off for further fermentation in a new tank, usually for another week or two. The next step is called *racking*; the wine is separated from the sediment, which contains yeast cells and precipitate, and then stored at lower temperature for aging, flavor development, and further clarification. The final clarification may be hastened by the addition of materials called *fining agents*, such as casein, tannin, or bentonite clay. Alternatively, the wine may be filtered through diatomaceous earth, asbestos, or membrane filters. The wine is then bottled and either stored for further aging or sold.

● **Figure 30.17 Commercial wine making in California (USA).** (a) Equipment for transporting grapes to the winery for crushing. (b) Large tanks where the main wine fermentation takes place. (c) Barrels where the aging process takes place. Wooden cask storage can last for several years in the case of certain red wines.

Red wine is usually aged for several years or more (Figure 30.17c), but white wine is typically sold without much aging. During aging, complex chemical changes occur, including reduction of bitter components, resulting in improvement in flavor and odor, or *bouquet*. The final alcohol content of wine varies from 6 to 15% depending on the sugar content of the grapes, length of the fermentation, and strain of wine yeast used.

Brewing: Making the Wort

The manufacture of alcoholic beverages made from malted grains is called **brewing** (Figure 30.19●). Typical malt beverages include beer, ale, porter, and stout. *Malt* is prepared from germinated barley seeds, and it contains natural enzymes that digest the starch of grains and convert it to sugar. Since brewing yeasts are unable to digest starch (∞ Figure 3.6b), the malting process is essential for the generation of fermentable substrates from cereal grains.

The fermentable liquid from which beer and ale are made is prepared by a process called *mashing*. The grain of the mash may consist only of malt, or other grains such as corn, rice, or wheat may be added. The mixture of ingredients in the mash is cooked and allowed to steep in a large mash tub at warm temperatures. During the heating period, enzymes from the malt cause digestion of the starches and liberate sugars, which will be fermented by the yeast. Proteins and amino acids are also liberated into the liquid, as are other nutrient ingredients necessary for the growth of yeast.

After cooking, the aqueous extract, called *wort*, is separated by filtration from the husks and other grain residues of the mash. *Hops*, an herb derived from the female flowers of the hops plant, are added to the wort at this stage. Hops is a flavoring ingredient, but it also has antimicrobial properties, which probably help to prevent contamination in the subsequent fermentation. The wort is then boiled for several hours, usually in large copper kettles (Figure 30.19a,b), during which time desired ingredients are extracted from the hops, proteins present in the wort that are undesirable from the point of view of beer stability are coagulated and removed, and the wort is sterilized. The wort is filtered again, cooled, and transferred to the fermentation vessel.

Brewing: The Fermentation Process

Brewery yeast strains are of two major types: *top fermenting* and *bottom fermenting*. The main distinction between the two is that top-fermenting yeasts remain uniformly distributed in the fermenting wort and are carried to the top by the CO_2 gas generated during the fermentation, whereas bottom-fermenting yeasts settle to the bottom. Top yeasts are used in the brewing of *ales*, and bottom yeasts are used to make *lager beers*. Bottom yeasts are usually given the species designation *Saccharomyces carlsbergensis*, and top yeasts are called *Saccharomyces cerevisiae*. Fermentation by top yeasts usually occurs at higher temperatures (14–23°C) than that by bottom yeasts (6–12°C) and is accomplished in a shorter period of time (5–7 days for top fermentation versus 8–14 days for bottom fermentation).

After completion of lager beer fermentation by bottom yeasts, the beer is pumped off into large tanks where it is stored at a cold temperature (about −1°C) for several weeks (Figure 30.19c). Following this, the beer is filtered and placed in storage tanks (Figure 30.19d) from which packaging occurs. Top-fermented ale is stored for only short periods at a higher temperature (4–8°C), which assists in development of the characteristic ale flavor. Some *(a)* **White wine** *(b)* **Red wine**

● **Figure 30.18 Wine production.** (a) White wine. White wines can vary from nearly colorless to straw-colored depending on the grapes used. (b) Red wine. Red wines can vary in color from a faint red to a deep, rich burgundy. The photograph shows a glass of chenin blanc, a typical white wine (left), and a glass of a light red wine (rosé, right). Other typical varieties of white wine include chablis, Rhine wine, sauterne, and chardonnay, while other typical red wines include burgundy, chianti, claret, zinfandel, cabernet, and merlot.

additional details on the brewing process are discussed in the Microbial Sidebar, Home Brew.

Distilled Alcoholic Beverages

Distilled alcoholic beverages are made by heating a fermented liquid at a high temperature that volatilizes most of the alcohol. The alcohol is then condensed and collected, a process called *distilling*. A product much higher in alcohol content can be obtained by this process than is possible by direct fermentation. Virtually any alcoholic liquid can be distilled, and each yields a characteristic distilled beverage. The distillation of malt brews yields *whiskey*, distilled wine yields *brandy*, distillation of fermented molasses yields *rum*, distillation of fermented grain or potatoes yields *vodka*, and distillation of grain and juniper berries yields *gin* (Figure 30.20●).

The distillate contains not only alcohol but also other volatile products arising either from the yeast fermentation or from the ingredients themselves. Some of these other products are desirable flavor ingredients, whereas others are undesirable. To eliminate the latter, the distilled product is almost always aged, usually in wooden barrels. During the aging process, undesirable products are removed and desirable new flavor ingredients develop. The fresh distillate is usually colorless, whereas the aged product is often brown or yellow (Figure 30.20). The character of the final product is partly determined by the manner and length of aging (aging times of 10 years or more are not uncommon for some distilled spirits), and the whole process of manufacturing distilled alcoholic beverages is highly complex. To a great extent, the process is carried out by traditional methods that have been found

(a)   (b)

(c)   (d)

● **Figure 30.19 Brewing beer in a commercial brewery.** (a, b) The copper brew kettle is where the wort is mixed with hops and then boiled. From the brew kettle the liquid is passed to large fermentation tanks where yeast ferments glucose to ethanol plus CO_2. (c) If a lager, the beer is then stored for several weeks at low temperature in tanks where settling of particulate matter including yeast cells occurs. (d) The beer is then filtered and placed in storage tanks from which it is packaged into kegs, bottles, or cans.

to yield a particular product, rather than by scientifically proven methods.

Commodity Ethanol

Production of alcohol (ethanol) as a *commodity chemical* is a major biocatalytic process, and today over 30 billion liters of alcohol are produced yearly worldwide from the fermentation of a variety of materials. In the United States most ethanol is obtained by yeast fermentation from corn starch. The starch is treated with enzymes (see Section 30.9) to yield glucose, which is the fermentable substrate. Ethanol is stripped from the fermentation broth by distillation (Figure 30.21●). In other countries, for example in Brazil, a major ethanol producer, corn and a variety of other fermentable materials are used. These include sugar cane, whey, sugar beets, and even wood chips and waste paper. The cellulose in wood is treated to release glucose, which is then fermented to alcohol.

Ethanol is used as an industrial solvent and also for the production of a gasoline supplement. In the United States *gasohol* is produced by adding 10% ethanol to lead-free gasoline. The combustion of gasohol produces lower amounts of carbon monoxide and nitrogen oxides than pure gasoline. Hence gasohol is marketed as a cleaner burning fuel and gasoline extender and its use is encouraged in major cities where automobile pollution is extensive. Various yeasts have been used in commodity ethanol production, including species of *Saccharomyces*, *Kluyveromyces*, and *Candida*, but most ethanol in the United States is produced by *Saccharomyces*.

● **Figure 30.20 Color of distilled spirits.** Gin or vodka (not shown) are colorless. However, aging in wooden casks yields a distinctive amber or yellow color to certain distilled spirits. Left to right, dark rum, brandy, whiskey.

 30.10 Concept Check

Alcoholic beverages are produced by yeast from the fermentation of sugar to ethyl alcohol and CO_2. Wine is produced from grape juice, beer from malted grain, and distilled spirits from the distillation of fermented solutions. Commodity alcohol is used as a gasoline additive and industrial solvent.

◆ How do wines differ from beer in terms of the amount of alcohol they contain?

◆ What are the major differences between a *beer* and an *ale*?

Microbial Sidebar ◆ Home Brew[a]

The amateur brewer can make many kinds of beer, from English bitters and India pale ale to German bock and Russian Imperial stout. The necessary equipment and supplies, including yeast, can be purchased from a local beer and winemakers shop. The Home Wine and Beer Trade Association, P.O. Box 1373, Valrico, FL 33595, (*http://www.hwbta.org*) can supply the address of a nearby shop.

The brewing process can be divided into three basic stages: (1) *making the wort*, (2) *carrying out the fermentation*, and (3) *bottling and aging*. The character of a brew depends on many factors: the proportion of malt, sugar, hops, and grain; the kind of yeast; the temperature and duration of the fermentation; and how the aging process is carried out. The fermentor itself consists of a 20-l (5-gal) glass jar or carboy that can be fitted with a tightly fitting closure. In order to have a good quality beer, it is essential that *everything* be sterilized that comes into contact with the wort. This includes the fermentor, tubing, stirring spoon, and bottles. An easy procedure is to use a sterilizing rinse consisting of 50–60 ml of liquid bleach in 20 l of water. Soak the items for 15 minutes and then rinse lightly with hot water or air dry.

1. **Making the wort** (Figure 1). In commercial brewing, the wort is made by extracting fermentable sugars and yeast nutrients from malt, sugar, and hops. Many home brewers make their own wort from malt, but a reasonably satisfactory beer can be made with hop-flavored malt extract purchased ready-made. Malt extracts come in a variety of flavors and colors, and the kind of beer depends on the type of malt extract used. A simple recipe for making the wort uses 5–6 lb (2.25–2.75 kg) of hop-flavored malt extract and 20 l of water. The malt extract and 5–6 l of water are brought to a boil for 15 minutes in an enamel or stainless steel container (aluminum heating kettles must be avoided because of the inhibiting action of metals leached from aluminum containers) (Figure 1). The hot wort is then poured into 14–15 l of clean, cold water that has already been added to the fermentor. After the temperature has dropped below 30°C, the yeast is added to initiate the fermentation.

2. **Carrying out the fermentation** (Figure 2). Add two packs of fresh beer yeast to the cooled wort and cover the fermentor with a rubber stopper into which a plastic hose has been inserted. The hose is directed into a bucket containing water. During the initial 2–3 days of the fermentation, large amounts of CO_2 will be given off, which will exit through the hose. The water trap is to prevent wild yeasts or bacteria from the air from getting back into the fermentor. After about 3 days, the activity will diminish as the fermentable sugars are used up. At this time, the rubber stopper and hose are replaced with an inexpensive fermentation lock. The fermentation lock, which can be purchased at the home brew store, prevents contamination while permitting the small amount of gas still being produced to escape. Allow the beer to ferment for 7–10 days at 10–15°C or higher.

3. **Bottling and aging** (Figure 3). The fermentation should be allowed to proceed for the full 7–10 days, even if the vigorous fermentation action ceases earlier. By this time, most of the yeast should have settled to the bottom of the fermentor. Carefully siphon the beer off the yeast layer, allowing the beer but not the sediment to run into sanitized glass beer bottles. The bottles used should accept standard crown caps, and new, clean caps should be used. Before capping, add three-quarters of a teaspoon of corn syrup to each 12-to 16-ounce (350-to 465-ml) bottle. Once the bottles are capped, turn each one upside down once to mix the syrup and then allow the beer to age upright at room temperature for at least 7–10 days. After this aging period, the beer may be stored at a cooler temperature.

All homemade beer will improve if it is allowed to age for several weeks. Aging tends to make beer smoother and remove bitter components. Using the same basic production equipment, several different types of beer can be made, each with its own distinctive taste and character (the reference listed in the footnote contains a number of beer recipes). Dark beers, which typically contain more alcohol than lighter beers, require more malt for their production, and are usually brewed from a combination of different malts such as those obtained from darker varieties of grain or those that have been roasted to caramelize the sugars and yield a darker color. A typical American style light lager (Figure 4, left) contains about 3.5% alcohol (by volume), whereas a Munich style dark (Figure 4, right) contains 4.25% alcohol, and bock beers contain about 5% alcohol.

The trend toward "individuality" in beer is evident not only by the growing number of home brewers, but also by the fact that major brewers in the United States are feeling more and more competition from new, usually very small, breweries called *microbreweries*. Although total production by a microbrewery may pale by comparison to that of a major brewer, the products themselves often have their own distinctive character and local appeal. Part of these differences probably has to do with the smaller scale on which the brewing takes place but also undoubtedly has to do with the use of different sources of ingredients, water, yeast strains, and brewing procedures. ■

Figure 1

Figure 2

Figure 3

Figure 4

[a]Source: Burch, B. 1992. *Brewing Quality Beers—The Home Brewer's Essential Guidebook*, 2nd edition. Joby Books, Fulton, CA

● **Figure 30.22 The chemistry of vinegar production.** Oxidation of ethanol to acetic acid, the key process in the production of vinegar. UQ, ubiquinone.

● **Figure 30.21 Ethanol production plant, Nebraska, USA.** Here, glucose obtained from corn is fermented by the yeast, *Saccharomyces cerevisiae*, to ethanol plus CO_2. The large tank in the left foreground is the ethanol storage tank and the tanks and pipes in the background are for distilling ethanol from the fermentation liquor.

30.11 Vinegar Production

Vinegar is the product resulting from the conversion of ethyl alcohol to the key ingredient, *acetic acid*, by the **acetic acid bacteria**. Key genera of acetic acid bacteria include *Acetobacter* and *Gluconobacter*. Vinegar can be produced from any substance that contains ethanol, although the usual starting material is wine, beer, or alcoholic apple juice (hard cider). Vinegar can also be produced from a mixture of pure alcohol in water, in which case it is called *distilled vinegar*, the term *distilled* referring to the alcohol from which the product is made rather than the vinegar itself. Vinegar is used as a flavoring ingredient in salads and other foods, and because of its acidity, it is also used in pickling. Meats and vegetables properly pickled in vinegar can be stored unrefrigerated for years.

The acetic acid bacteria are an interesting group of prokaryotes (⟨∞⟩ Section 12.8). These are strictly aerobic bacteria that differ from most other aerobes in that some of them, such as species of *Gluconobacter*, do not oxidize their organic electron donors completely to CO_2 and water (Figure 30.22●). Thus, when provided with ethyl alcohol as electron donor, they oxidize it via quinones only to acetic acid, which accumulates in the medium. Acetic acid bacteria are quite acid tolerant and are not killed by the acidity that they produce. There is a high oxygen demand during growth, and the main problem in the production of vinegar is to ensure sufficient aeration of the medium.

Vinegar Production

There are three different processes for the production of vinegar. The *open-vat (Orleans) method*, which is the original process, is still used in France where it was first devel-

oped. Wine is placed in shallow vats with considerable exposure to the air, and the acetic acid bacteria develop as a slimy layer on the top of the liquid. This process is not very efficient because the bacteria come in contact with both the air and the substrate only at the surface.

The second process is the *trickle (Quick vinegar) method*, in which the contact between the bacteria, air, and substrate is increased by trickling the alcoholic liquid over beechwood twigs or wood shavings packed loosely in a vat or column while a stream of air enters at the bottom and passes upward. The bacteria grow on the surface of the wood shavings and thus are maximally exposed both to air and liquid. The vat is called a *vinegar generator* (Figure 30.23●), and the whole process is operated in a

● **Figure 30.23 Diagram of a vinegar generator.** The alcoholic juice is allowed to trickle through the wood shavings, and air is passed up through the shavings from the bottom. Acetic acid bacteria develop on the wood shavings and convert alcohol to acetic acid. The acetic acid solution accumulates in the collecting chamber and is recycled through the generator until the acetic acid content reaches at least 4%, the minimum for a product to be labeled as "vinegar."

continuous fashion. The life of the wood shavings in a vinegar generator is long, from 5 to 30 years, depending on the kind of alcoholic liquid used in the process.

Finally, the *bubble method* is basically a submerged fermentation such as was already described for antibiotic production. With proper aeration, the efficiency of the bubble method is high, and 90–98% of the alcohol is converted to acetic acid.

Although acetic acid can be easily made chemically from alcohol, the microbial product, *vinegar*, is a distinctive material, the flavor being due in part to other substances present in the starting material and produced in the fermentation. For this reason, the microbial process, especially using the vinegar generator, has not been supplanted by a chemical process.

30.11 Concept Check

The active ingredient in vinegar is acetic acid, which is produced by acetic acid bacteria oxidizing an alcohol-containing fruit juice. Adequate aeration is the most important consideration in ensuring a successful vinegar process.

◆ Why is O_2 necessary in vinegar production?

◆ Why does vinegar produced by the trickle method have a more distinctive taste than vinegar produced by the bubble method?

30.12 Citric Acid and Other Organic Compounds

Many organic chemicals are produced by microorganisms in sufficient yields that they can be manufactured commercially by fermentation. For example, *citric acid*, is widely used in the food industry as a supplement in beverages, confectionaries, and other foods, and in the leavening of bread, and is also used industrially in the treatment of certain metals, as a detergent additive in place of phosphates, and in various pharmaceutical applications. Other acids include *itaconic acid*, used in the manufacture of acrylic resins, and *gluconic acid*, used in the form of calcium gluconate to treat calcium deficiencies in humans and industrially as a washing and water-softening agent. All of these acids are produced by fungi.

Important bacterially produced organic chemicals include *sorbose*, which is produced when *Acetobacter* oxidizes sorbitol and is used in the manufacture of *ascorbic acid* (vitamin C), and *lactic acid*, used in the food industry to acidify foods and beverages. Lactic acid is produced by various species of lactic acid bacteria (∞ Section 12.19).

We focus here on the citric acid fermentation. Historically, the citric acid fermentation was of great importance because it was the first submerged-culture *aerobic* industrial fermentation. As the citric acid fermentation was perfected, the lessons learned were applied to the

penicillin and other important antibiotic fermentations. Thus to some degree, we owe our current success with the large-scale production of antibiotics to the pioneering work done on the citric acid fermentation. Overall, citric acid production is a big business, as over 550,000 tons of citric acid are produced per year worldwide with a total value approaching $1 billion U.S.

Citric Acid Production: The Link to Iron

Citric acid is produced microbiologically by a fermentation using the mold *Aspergillus niger*. Although citric acid is normally considered in connection with the citric acid cycle (∞ Section 5.13), in certain organisms such as *A. niger*, excretion of large amounts of citric acid can be obtained. The fermentation is carried out aerobically in large fermentors, and a key requirement for high citric acid yield is that the medium be *iron deficient*. The iron deficiency makes cells of *A. niger* overproduce citric acid as a chelator to scavenge iron (Figure 30.24a●; see also Figure 5.1). Therefore, the medium used for citric acid production is treated to remove most of the iron, and the fermentors themselves are made of stainless steel or are glass lined to prevent leaching of iron from the fermentor walls at the low pH values generated by citric acid accumulation.

● **Figure 30.24 Citric acid fermentation.** (a) Structure of citric acid. Note how the ionized form, citrate, contains three carboxylic acid groups, which can chelate ferric iron (Fe^{3+}). (b) Kinetics of the citric acid fermentation. Sucrose is degraded by the enzyme *sucrase* to yield glucose plus fructose. See text for further details.

Citric Acid Production: Growth Media, Conditions, and Purification

Growth media used for citric acid production contain any of a variety of starting materials including starch from potatoes, starch hydrolysates, glucose syrup from saccharified starch, sucrose (Figure 30.24b), sugarcane syrup, sugarcane molasses, and sugar beet molasses. If starch is used, amylase (Table 30.4), formed by the producing fungus or added to the fermentation broth, hydrolyzes the starch to sugars. The sugars are catabolized through the glycolytic pathway (∞ Section 5.10) and enter the citric acid cycle where citrate production occurs.

Most citric acid today is produced by submerged processes in large fermentors. However, because *A. niger* is a strict aerobe, it is crucial to this fermentation to make sure that the culture stays properly aerated. Citric acid is produced in this way as a typical secondary metabolite (see Figure 30.2). During the growth phase, sucrose is broken down into glucose and fructose, and by the time the stationary phase is reached, large amounts of these hexoses remain and are converted to citric acid to counter iron starvation (Figure 30.24).

Citric acid is purified following removal of fungal biomass by adding lime (CaO); this precipitates citric acid as calcium citrate. The latter is concentrated by filtration and treated with sulfuric acid to form a solution of citric acid and calcium sulfate ($CaSO_4$, a solid). Following a second filtration to remove crystals of $CaSO_4$, crystals of citric acid are formed by evaporation.

⬤ 30.12 Concept Check

A number of organic chemicals are produced commercially by use of microorganisms, of which the most important economically is citric acid, produced by *Aspergillus niger*.

◆ Why is citric acid considered a *secondary* metabolite (see Figure 30.2b)?

◆ What is the relationship between iron and citric acid production by *A. niger*?

30.13 Yeast as a Food and Food Supplement

Yeasts are among the most extensively used microorganisms in industry. Besides their importance to the baking and beverage industries, yeasts are grown for the cells themselves or for compounds extracted from the cells (Table 30.5). Unlike in the brewing and baking industries, where *fermentation* is important (∞ Microbial Sidebar: The Products of Yeast Fermentation, Chapter 5), the production of yeast cells requires *respiration* for maximum production of cell material. For food and food supplement yeast, derivatives of *Saccharomyces cerevisiae* are used almost exclusively, and we consider their large scale production here.

Table 30.5	**Industrial uses of yeast and yeast products[a]**

Production of yeast cells
Baker's yeast, for bread making
Dried food yeast, for food supplements
Dried feed yeast, for animal feeds

Yeast products
Yeast extract, for microbial culture media
B vitamins, vitamin D
Enzymes for food industry: invertase (sucrase), galactosidase
Biochemicals for research: ATP, NAD^+, RNA

Fermentation products from yeast
Ethanol, for industrial alcohol and as a gasoline extender
Glycerol

Beverage alcohol
Beer, wine

Distilled beverages
Whiskey, brandy, vodka, rum

[a] ∞ Microbial Sidebar, The Products of Yeast Fermentation, Chapter 5 and Sections 5.10 and 5.13.

Yeast Cell Production

Yeast for baking or nutritional purposes is cultured in large aerated fermentors in a medium containing molasses as a major ingredient. Molasses contains large amounts of sugar as the source of carbon and energy, and also contains minerals, vitamins, and amino acids used by the yeast. To make a complete medium for yeast growth, phosphoric acid (a phosphorus source) and ammonium sulfate (a source of nitrogen and sulfur) are added.

Fermentation vessels for yeast production range from 40,000 to 200,000 liters. Beginning with the pure stock culture, several intermediate stages are needed to scale up the inoculum to a size sufficient to inoculate the final stage (Figure 30.25a●). It is undesirable to add all the molasses to the fermentor at once because this results in a sugar excess. The yeast then ferments much of the sugar to alcohol plus CO_2 rather than turning it into yeast cells. Therefore, only a small amount of the molasses is added initially, and then as the yeast culture grows and consumes this sugar, more molasses is added in controlled "feedings."

At the end of the growth period, the yeast cells are recovered from the broth by centrifugation. The cells are then washed by dilution with water and recentrifuged until they are light in color. Baker's yeast is marketed in two ways, either as compressed cakes or as a dry powder. *Compressed yeast cakes* (Figure 30.25b) are made by mixing the centrifuged yeast with emulsifying agents, starch, and other additives that give it a suitable consistency and reasonable shelf life, and the product is then formed into cubes or blocks of various sizes for domestic or commercial use. A compressed yeast cake contains about 70% moisture and thus must be stored in the refrigerator so its activity is maintained.

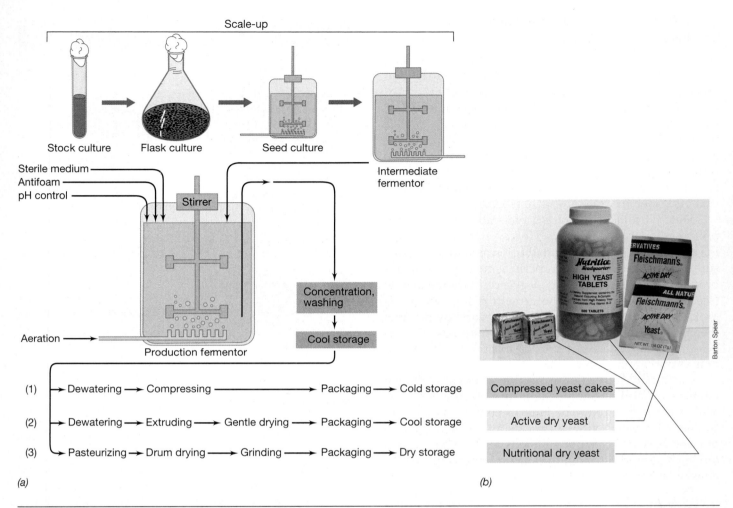

● **Figure 30.25   Industrial production of yeast cells.** (a) Stages in production. Antifoaming agents are added to the fermentor to prevent aeration and stirring from creating excessive foam on the surface of the medium. (b) Photograph of common yeast products for consumers: yeast cakes, packages of active dry yeast, bottle of nutritional yeast.

Yeast marketed in the dry state for baking is usually called *active dry yeast* (Figure 30.25*b*). The washed yeast cells are mixed with additives and dried under vacuum until their moisture is reduced to about 8%. It is then packed in airtight containers such as fiber drums, cartons, or multiwall bags, sometimes under a nitrogen atmosphere to promote long shelf life. Active dry yeast does not exhibit as great a leavening action as compressed fresh yeast, but has a much longer shelf life.

Nutritional yeast, marketed as a food supplement (Figure 30.25*b*), is heat-killed and usually dried. Yeast cells are rich in B vitamins and in protein, except for sulfur-containing amino acids. Yeast is added to wheat or corn flour to increase the nutritional value of these foods and is also sold in pelleted form as a health food (Figure 30.25*b*).

30.13   *Concept Check*

Yeast cells are grown for use in the baking and food industries. Commercial yeast is produced in large-scale aerated fermentors using molasses as the main carbon and energy source.

◆ Why is it important when growing yeast *for cells* to maintain oxic conditions in the fermentor?

◆ What different forms of yeast are commercially available?

30.14   Mushrooms as a Food Source

Several kinds of *fungi* are sources of human food, of which the most important are the mushrooms. Mushrooms are a group of filamentous fungi that form large, edible structures called fruiting bodies (Figure 30.26●). The fruiting body is commonly called the **mushroom** and is formed through the association of a large number of individual hyphae to form a mycelium. We discussed the basic biology of mushrooms as well as yeasts in Section 14.12.

Commercial Growth of Mushrooms

The mushroom commercially grown in most parts of the world is *Agaricus bisporus*, and it is generally cultivated in

(a)

(b)

● **Figure 30.26   Commercial mushroom production.** (a) Close-up of a flush of the mushroom *Agaricus bisporus*. (b) Close-up of the shiitake mushroom, *Lentinus edulus*.

so-called "mushroom farms." The organism is grown in special beds, usually in buildings where temperature and humidity are carefully controlled (Figure 30.26a). Beds are prepared by mixing soil with a material very rich in organic matter, such as horse manure, and the beds are then inoculated with mushroom *spawn*. The spawn is actually a pure culture of the mushroom fungus that has been grown in large bottles on an organic-rich medium.

In the bed, the mycelium grows and spreads through the substrate, and after several weeks it is ready for the next step, the induction of mushroom formation. This is accomplished by adding to the surface of the bed a layer of soil called *casing soil*. The appearance of mushrooms on the surface of the bed is called a *flush* (Figure 30.26a), and when flushing occurs, the mushrooms must be collected immediately while still fresh. After collection they are packaged and kept cool until brought to market.

Another widely cultured mushroom is the shiitake, *Lentinus edulus*. The most widely cultivated mushroom in Asia, shiitake is now finding expanding demand in

North America. Shiitake is a cellulose-digesting fungus that grows well on hardwood trees and is cultivated on small logs (Figure 30.26b). The logs are soaked in water to hydrate them and then inoculated by inserting plugs of spawn into small holes drilled in them. The fungus grows through the log, and after about a year forms a flush of fruiting bodies (Figure 30.26b). The shiitake mushroom is generally considered to have a superior taste to *Agaricus bisporus*, and because of this, it commands a substantially greater price.

 30.14   Concept Check

The most important food produced from a microorganism is the mushroom, which is produced not for its protein but for its flavor.

◆ Why are mushrooms considered microorganisms?

◆ What is a *mushroom flush?*

REVIEW QUESTIONS

1. In what ways do industrial microorganisms differ from conventional microorganisms? In what ways are they similar (∞ Section 30.1)?

2. List three major types of industrial products that can be obtained with microorganisms and give two examples of each (∞ Section 30.1).

3. Compare and contrast *primary* and *secondary* metabolites and give an example of each. List at least two molecular explanations for why some metabolites are secondary rather than primary (∞ Section 30.2).

4. How does an industrial fermentor differ from a laboratory culture vessel? How does a fermentor differ from a fermenter (∞ Section 30.3)?

5. Discuss the problems of scale-up from the viewpoints of *aeration, sterilization,* and *process control*. Why is sterility so important in an industrial fermentor (∞ Section 30.4)?

6. List three examples of *antibiotics* that are important industrially. For each of these antibiotics, list the producing organisms, the general chemical structure, and the mode of action (∞ Section 30.5).

7. Compare and contrast the production of *natural, biosynthetic,* and *semisynthetic* β-lactam antibiotics (∞ Section 30.6).

8. Addition of what metal to the fermentation medium can markedly improve production of vitamin B_{12} (∞ Section 30.7)?

9. What unusual characteristics must an organism have if it is to overproduce and excrete an amino acid such as *lysine* (∞ Section 30.7)?

10. Define *microbial biotransformation* and give an example. Explain why the chemical reactions involved in microbial

biotransformations are preferably carried out microbially rather than chemically (⚬⚬ Section 30.8).

11. List three different kinds of enzymes that are produced commercially. For each enzyme, list the organism used in commercial production, the activity of the enzyme, and how the enzyme is used commercially (⚬⚬ Section 30.9).

12. What is *high-fructose syrup*, how is it produced, and what is it used for in the food industry (⚬⚬ Section 30.9)?

13. What are extremozymes? What industrial uses do they have (⚬⚬ Section 30.9)?

14. In what way is the manufacture of *beer* similar to the manufacture of *wine*? In what ways do these two processes differ? How does the production of *distilled alcoholic beverages* differ from that of beer and wine (⚬⚬ Section 30.10)?

15. What is the active ingredient in vinegar? From what substance is it made? Why is *Gluconobacter* suitable for the production of vinegar, whereas *Escherichia coli* is not (*Hint*: think of metabolic differences here) (⚬⚬ Section 30.11)?

16. Give two reasons why stainless steel fermentors are used in the industrial production of citric acid (⚬⚬ Section 30.12).

17. Why are yeasts of such great industrial importance (⚬⚬ Section 30.13)?

18. What part of the mushroom is actually consumed as food? What is contained within this structure (⚬⚬ Sections 30.14 and 14.12)?

APPLICATION QUESTIONS

1. As a researcher in a pharmaceutical company you are assigned the task of finding and developing an antibiotic effective against a new bacterial pathogen. Outline a plan for this process, starting from isolation of the low-yield producing organism to high-yield industrial production of the new antibiotic.

2. A partially consumed bottle of an "organic" (containing no preservatives) red wine is recapped and stored under refrigeration for 2 months. On tasting the wine again, you notice a distinct bitter taste, making the wine undrinkable. Using information presented in this chapter and in Section 12.8, describe (a) what microbially-mediated process occurred in the wine, and (b) a very easy way in which this process could have been prevented.

3. You wish to produce high yields of the amino acid phenylalanine for use in production of the sweetener *aspartame*. The overproducing organism you wish to use is not subject to feedback inhibition by phenylalanine but is subject to typical repression of phenylalanine biosynthesis enzymes by excess phenylalanine. Applying the principles of enzyme regulation studied in Chapter 8 and microbial genetics in Chapter 10, describe *two classes* of mutants you could isolate that would overcome this problem and detail the genetic lesions each would have.

31

GENETIC ENGINEERING AND BIOTECHNOLOGY

From a scientific standpoint, the ability to manipulate DNA in a test tube and then place the altered DNA back into an organism and have it expressed, is an extremely useful and powerful technique. Here, through genetic engineering techniques, soybean plants have been made resistant to the commonly used herbicide, glyphosate.

WORKING GLOSSARY

Biotechnology use of genetically engineered organisms to carry out defined chemical processes for industrial application

Cloning vector a genetic element, typically a plasmid or a bacteriophage, into which genes can be recombined and replicated

DNA fingerprinting use of the techniques of genetic engineering to determine the origin of DNA in a sample of tissue

DNA vaccine using the genetic material of a pathogen to elicit an immune response

Expression vector a cloning vector that contains the necessary regulatory sequences to allow transcription and translation of cloned genes

Gene therapy treatment of a disease caused by a dysfunctional gene by introduction of a normally functioning copy of the gene

Genetically modified organism (GMO) an organism whose genome has been altered using genetic engineering. The abbreviation *GM* is also used in constructions such as *GM crops* and *GM foods*

Genetic engineering the use of *in vitro* techniques in the isolation, manipulation, recombination, and expression of DNA, and in the development of *genetically modified organisms*

Integrating vector a cloning vector that becomes integrated into a host chromosome

Molecular cloning isolation and incorporation of a fragment of DNA into a vector where it can be replicated

Nucleic acid probe a strand of nucleic acid that can be labeled and used to hybridize to a complementary molecule from a mixture of other nucleic acids

Reporter gene a gene incorporated into a vector because the product it encodes is easy to detect

Reverse translation the mental process of using a codon table and the amino acid sequence of a protein to obtain a possible sequence of the mRNA or the gene that encoded the protein

Shuttle vector a cloning vector that can replicate in two or more dissimilar hosts

T-DNA the segment of the *Agrobacterium* Ti plasmid that is transferred to plant cells

Ti plasmid a plasmid in *Agrobacterium* species capable of transferring genes from bacteria to plants

Transgenic organism a plant or animal that stably passes on cloned DNA that has been inserted into it

B iotechnology is the use of living organisms to carry out chemical processes for industrial or commercial application. In this sense, biotechnology refers to all the industrial microbial processes discussed in Chapter 30—and much more. However, the word *biotechnology* implies that the producing organism has been manipulated using *in vitro* genetic techniques.

As was discussed in Chapter 10, these techniques make it possible to isolate and manipulate DNA, as well as to control its expression. In Chapter 15 we saw how these techniques can be used to examine entire genomes. These *in vitro* techniques, particularly molecular cloning (⌀Section 10.15), result in **genetically modified organisms** (**GMOs**). The use of these techniques is called **genetic engineering**.

In Chapters 10 and 15 our coverage of these techniques was limited to basic research. However, genetic engineering strives toward *practical* applications. In this chapter we will look first at the range of techniques to clone genes and to express them efficiently for biotechnological applications. We will then see how genetically engineered microorganisms can produce valuable products such as human insulin, human growth hormone, interferon, vaccines, and industrial enzymes.

THE TECHNIQUES OF GENETIC ENGINEERING

The basic techniques of genetic engineering are also the basic techniques of modern microbial genetics, and we introduced most of them in Chapters 7–10. However, ge-

nomics and bioinformatics (⌀Chapter 15) also are beginning to play increasingly important roles in biotechnology. We cannot review all of these techniques here, but we can summarize some of the principles underlying them. In addition, we will expand our discussion of certain techniques because they are more directly applied to biotechnology.

31.1 Review of Principles Underlying Genetic Engineering

Much of genetic engineering is based on **molecular cloning** (⌀Section 10.15). In molecular cloning, a double-stranded DNA fragment from any source is recombined with a vector and introduced into a suitable host. Commonly employed **cloning vectors** include plasmids and bacteriophages (⌀Sections 10.16 and 10.17).

In many cases the specific biotechnological application depends not just on the ability to identify and clone a gene, but also on *manipulating the expression* of the gene to produce, identify, and purify its protein product. This requires knowledge of all aspects of molecular biology, and Figure 31.1● summarizes the important concepts:

1. **DNA chemistry**: isolation, sequencing, and synthesis of DNA (⌀Section 7.8);

2. **DNA enzymology**: restriction endonucleases, DNA ligases, and DNA polymerases (⌀Sections 7.5 and 7.7);

3. **DNA replication**: DNA replication, genetic elements, and DNA cloning vectors capable of inde-

pendent replication, and the *polymerase chain reaction* (*PCR*) (∞Sections 7.4–7.9);

4. **Plasmids and conjugation**: plasmids, plasmid replication, and conjugation (∞Sections 10.9 and 10.10);

5. **Temperate bacteriophage**: phage replication and integration, temperate bacteriophages, lysogeny, and specialized transducing phages (∞Sections 9.10 and 10.8, and 10.17);

6. **Transformation**: methods for getting free DNA into cells (∞Section 10.7);

7. **RNA chemistry and enzymology**: messenger RNA, RNA polymerase, differences between prokaryotic and eukaryotic mRNA, and RNA processing (∞Sections 14.7 and 14.8);

8. **Reverse transcription**: activity of the enzyme *reverse transcriptase* for transcribing information from mRNA back into DNA (∞Sections 9.13 and 16.15);

9. **Regulation**: regulation of transcription, including promoter sites and operon control (∞Sections 7.10–7.13 and Chapter 8);

10. **Translation**: translational processes, the importance of ribosome binding sites on mRNA, the role of the initiation codon, and importance of the proper reading frame (∞Sections 7.14–7.16);

11. **Protein chemistry**: isolation, purification, assay, and sequencing of proteins;

12. **Protein excretion and posttranslational modification**: signal sequences (∞Section 7.17) and other kinds of posttranslational modification of proteins; and

13. **The genetic code**: the nearly universal genetic code and differences in codon usage (∞Section 7.14).

⬡ 31.1 Concept Check

The techniques of genetic engineering are based on fundamental concepts in molecular genetics and biochemistry. Successful genetic engineering depends not only on being able to carry out molecular cloning, but also on knowledge of replication, transcription, and translation, and the regulatory aspects that control all of these processes.

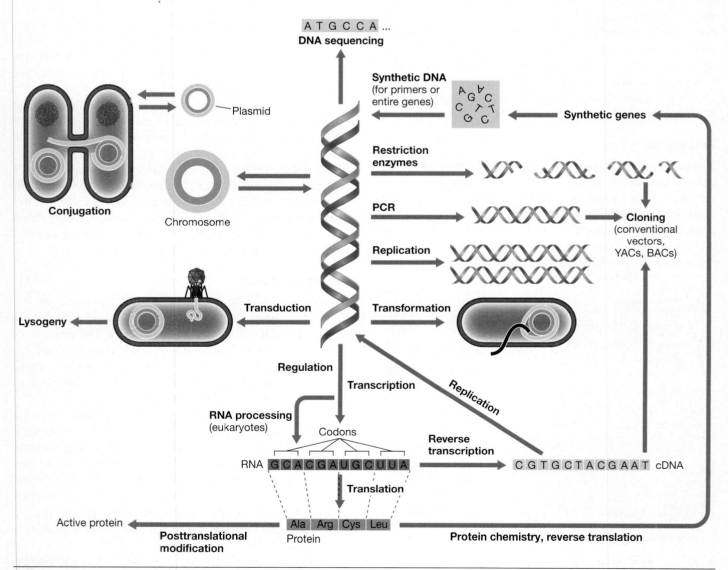

● **Figure 31.1 Summary of the fundamental processes underlying genetic engineering.** All techniques and processes are labeled in bold.

◆ What is a *cloning vector*?

◆ How can DNA sequencing tell us about protein structure?

31.2 Hosts for Cloning Vectors

When *in vitro* genetic techniques were introduced in Chapter 10, it was in the context of the genetic analysis of *Bacteria*. The primary goal there was to understand the organism. By contrast, in biotechnology genetic engineering is used for commercial ends. For example, we may wish to produce large amounts of a valuable protein, produce a new vaccine, or introduce a specific gene into a plant or an animal. In these cases it is essential to consider the *host* that we will use, which will also guide our choice of a vector.

To obtain large amounts of cloned DNA, an ideal host should grow rapidly in an inexpensive culture medium. The host should also be nonpathogenic, be capable of taking up DNA, be genetically stable in culture, and have the appropriate enzymes to allow replication of the vector. It is also helpful if there is considerable genetic information on the host already available and a wealth of tools for its genetic manipulation. The most useful hosts for cloning are typically microorganisms that grow well and for which we have much information. These include the bacteria *Escherichia coli* and *Bacillus subtilis,* and the yeast *Saccharomyces cerevisiae* (Figure 31.2●). Complete genome sequences are available for all of these organisms (◯◯Chapter 15).

Prokaryotic Hosts

Although most molecular cloning has been done in *Escherichia coli*, this host has potential disadvantages (Figure 31.2). *E. coli* is an excellent choice for initial cloning work, but this bacterium is problematic as a producing organism because it is found in the human intestinal tract,

and wild-type strains are potentially pathogenic. Also, even nonpathogenic strains produce endotoxins (◯◯ Sections 4.9 and 21.12) that can contaminate products, an especially bad situation with pharmaceutical injectables. In addition, *E. coli* retains extracellular proteins in its periplasmic space, making isolation and purification of recombinant proteins potentially difficult. However, modified *E. coli* strains have been developed for which most of these problems have been eliminated. Thus, because of the extensive knowledge of its genetics and biochemistry (◯◯Section 10.19), *E. coli* remains the organism of choice for most cloning studies.

The gram-positive organism *Bacillus subtilis* is also used as a cloning host (Figure 31.2). *B. subtilis* is not a potential pathogen, does not produce endotoxin, does produce spores (which is helpful in industrial microbial processes ◯◯Section 30.1), lacks a periplasm, and naturally secretes proteins into the medium. Although the technology for cloning in *B. subtilis* is not nearly as well developed as that for *E. coli*, several plasmids and phages suitable for cloning have been developed, and transformation (◯◯Section 10.7) is a well-developed procedure in *B. subtilis*. Disadvantages of using *B. subtilis* as a cloning host, however, include plasmid instability: It can be difficult to maintain plasmid replication over many subcultures of the organism. Also, foreign DNA is not well maintained in *B. subtilis* cells and so the cloned DNA is often unexpectedly lost.

Often organisms used as hosts for cloning must have specific genotypes to be effective. For instance, if the vector carries the gene for β-galactosidase as a selectable marker (◯◯Sections 10.16 and 15.1), then the host must have a mutation disabling this gene. Because the bacteriophage M13 infects only bacteria with F pili (◯◯Sections 10.8 and 16.3), hosts used with M13-derived vectors must contain and express genes on the F plasmid. These types of considerations and others, such as the ability to select for transformants, must be taken into account when selecting a host.

Eukaryotic Hosts

Cloning in eukaryotic microorganisms has some important applications, especially in understanding the details of gene regulation in eukaryotic systems. The yeast *Saccharomyces cerevisiae* has been well studied and is used extensively as a cloning host (Figure 31.2). Plasmid vectors, as well as yeast artificial chromosomes (YACs, ◯◯Section 15.1), have been developed for yeast.

One important advantage of eukaryotic cells as hosts for cloning vectors is that they already possess the complex RNA and posttranslational processing systems required for the production of gene products in higher organisms. Thus, unlike when eukaryotic DNA is cloned into prokaryotes, these systems do not have to be engineered into the vector. Posttranslational processing, in particular, can create some molecular cloning problems (see Section 31.5).

For many applications, gene cloning in *mammalian cells* is desirable. Mammalian cell culture systems can be

Cloning Hosts

Prokaryotes		Eukaryote
Escherichia coli	*Bacillus subtilis*	*Saccharomyces cerevisiae*
Well-developed genetics / Many strains available / Best-known prokaryote	Easily transformed / Nonpathogenic / Naturally secretes proteins / Endospore formation simplifies culture	Well-developed genetics / Nonpathogenic / Can process mRNA and proteins / Easy to grow
Potentially pathogenic / Periplasm traps proteins	Genetically unstable / Genetics less developed than in *E. coli*	Plasmids unstable / Will not replicate most prokaryotic plasmids

☐ Advantages   ▨ Disadvantages

● **Figure 31.2 Hosts for molecular cloning.** A summary of advantages and disadvantages of common cloning hosts.

handled in some ways like microbial cultures and find wide use in research on human genetics, cancer, infectious disease, and physiology. A disadvantage of mammalian cells as hosts is that they are expensive and difficult to produce under large-scale conditions, and expression levels of cloned genes are often low. Insect cell lines are simpler to grow, and vectors have been developed from an insect DNA virus, the baculovirus (see Section 31.4).

For some applications it is important that the eukaryotic host be a plant cell tissue culture, or even an entire plant. Indeed, genetic engineering has a large number of applications in plant agriculture (see Section 31.10). Regardless of host type, it is necessary that the host be able to take up DNA. We discussed this problem in prokaryotic hosts earlier (∞Section 10.6), and here we turn our attention to eukaryotic hosts.

Transfection of Eukaryotic Cells

Eukaryotic microorganisms and animal and plant cells can take up DNA in a process that resembles bacterial transformation (∞Section 10.7). Because the word *transformation* in mammalian cells is used to describe the conversion of cells to the malignant (tumorous, cancerous) state (∞Section 9.12), the introduction of DNA into mammalian cells has been called *transfection*.

Transfection of cultured animal cells was originally accomplished by precipitating DNA in such a way that the cells would take it up by phagocytosis (∞Sections 14.10 and 22.2) because they do not have cell walls. In yeast, which is an important organism for genetic engineering, transfection at low efficiencies can be mediated by various methods. However, a technique called *electroporation* is widely applied to all types of eukaryotic cells and can be used whether or not the cell wall of the organism is removed. Electroporation involves exposing host cells to pulsed electrical fields in the presence of cloned DNA. The electrical treatment opens small pores in their membranes through which the DNA can enter. The pores generated are insufficient to cause cell damage or lysis, but sufficient to allow cloned DNA to enter.

In addition to electroporation, a high velocity microprojectile "gun" can be used to incorporate DNA into cells. The *particle gun* operates somewhat like a shotgun. A small steel cylinder containing a gunpowder charge is used to fire nucleic acid–coated particles at the target cells (Figure 31.3●). The particles bombard the cell, piercing cell walls and membranes without actually killing the cells. The nucleic acid entering the cells can then recombine with host DNA. The particle gun has been used successfully to transfect yeast, algae, a variety of plant cells, and even mitochondria and chloroplasts. Moreover, unlike electroporation, the particle gun can be used to get DNA into intact tissue, such as plant seeds. Finally, in animal cells, DNA can be injected into the nucleus using micropipets, a technique called *microinjection*.

● **Figure 31.3 Nucleic acid gun for transfection of certain eukaryotic cells.** The inner workings of the gun show how nucleic acids coating metal pellets are projected at target cells.

 31.2 Concept Check

The choice of a cloning host depends on the final application. In many cases the host can be a prokaryote, but in others it is essential that the host be a eukaryote. Any host must be able to take up DNA, and there are a variety of techniques by which this can be accomplished, both natural and artificial.

♦ Why does molecular cloning require a host?

♦ Describe three mechanisms by which cells can take up DNA.

31.3 Finding the Right Clone

Genetic engineering begins with the isolation of a clone containing a gene of interest. A variety of manipulations can then be performed on this gene, depending on the application, before the final result is achieved. But first it is necessary to identify the host containing the original clone.

Methods to clone DNA include making *gene libraries* from total genomic DNA (∞Section 10.15), or cloning a DNA fragment made by PCR (∞Section 7.9). One usually clones from PCR products when the gene of interest has already been identified and the goal is simply to obtain a single clone. However, in a gene library there are thousands or tens of thousands of clones, and typically, only one or a few may be the genes of interest.

One can select for hosts containing a plasmid vector by selecting for a vector marker, such as antibiotic resistance, so that only these cells form colonies (∞Section 10.15). For host cells containing a viral vector, one simply looks for plaques (∞Section 10.16). These colonies or plaques can also be screened for vectors that contain foreign DNA inserts by looking for the inactivation of a vector gene (∞Figures 10.37 and 15.1). In cloning a single DNA fragment generated by PCR or purified by some other means, these simple selections or screenings are usually sufficient. However, with a heterogeneous collection of DNA fragments, as in a gene library, identifying cells carrying cloned DNA is only the first step. The biggest challenge remains *finding the clone* that has the gene of interest. One must examine colonies of bacteria or plaques from viral-infected cells growing on agar plates and detect those few that contain the gene of interest. Sometimes the cloned gene is *expressed* (that is, the protein is synthesized) in the cloning host. Often however, the cloned gene is *not* expressed. Then the only option is to look for the DNA itself.

Foreign Genes Expressed in the Cloning Host

If the foreign gene is expressed in the cloning host, we can look for the presence of this protein in colonies of the host organism. The cloning host itself must *not* produce the protein being studied. If the protein is one that the cloning host normally produces, then the host used must be defective—that is, carry a mutation in the gene of interest. When the foreign gene is incorporated, the expression of this foreign gene can be detected by complementation (∞Section 10.13). If the protein is not normally produced in host bacteria, then the host can be considered naturally defective. This makes selection of cells containing cloned genes particularly easy. A striking example is the cloning of luciferase genes into other organisms. This allows the organism to glow in the dark (∞Figure 8.22).

Antibody as a Method of Detecting the Protein

Antibodies can be used as detection reagents for the protein of interest. Recall from Chapter 22 that an antibody is a serum protein produced by an animal that combines in a highly specific way with another protein, the *antigen* (∞Section 22.4). In the present case, the protein product encoded by the cloned gene is the antigen, and this protein is used to produce an antibody in an experimental animal. Since the antibody combines specifically with the antigen, when the antigen is present in one or more colonies on the plate, then the locations of these colonies can be determined by observing the binding of the antibody.

Because only a small amount of the protein (antigen) is present in the colonies, only a small amount of antibody is bound, and so a highly sensitive procedure for detecting bound antibody must be available. In practice, detection relies on radioisotopes, chemiluminescent chemicals, or enzymes. These and other extremely sensitive techniques for detecting antigens were previously discussed (∞Sections 24.7–24.12).

This method of detection involves *screening*, not selection, and so thousands of clones may need to be examined. These can be colonies containing plasmids or plaques containing viruses that produce the cloned product. The procedure using a radioactive detection system is outlined in Figure 31.4*a*●. As seen, the replica-plating procedure (∞Figure 10.2) is used to make a duplicate of the master plate, but the duplication is done onto a membrane filter and all further manipulations are done with this filter. The duplicate colonies are lysed to release the protein (antigen) of interest. (Screening for expression in phage vectors eliminates this step because the bacteria are already lysed.) The antibody is then added, and the antibody–antigen reaction allowed to

proceed. Unbound antibody is then washed off, and a radioactive agent is added that is specific for the antibody. A sheet of X-ray film is placed over the filter and exposed. If a radioactive colony is present, a spot on the X-ray film will be observed after it is developed (Figure 31.4*a*). The location of this spot on the film corresponds to a location on the master plate where a colony is present that produces the protein. This colony can then be picked from the master plate and subcultured.

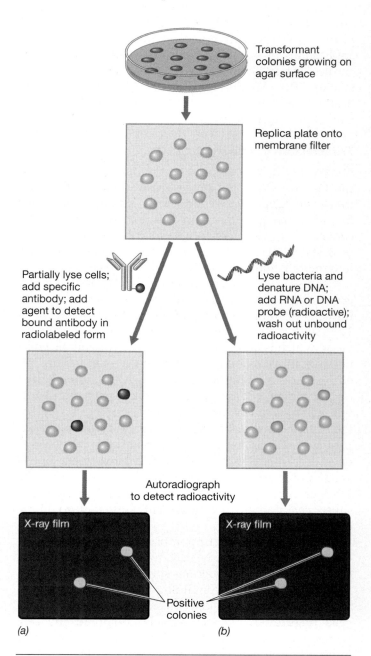

● **Figure 31.4 Techniques for finding the right clone.** (a) Method for detecting production of protein by use of specific antibody. (b) Method for detection of recombinant clones by colony hybridization with a radioactive nucleic acid probe. Formation of the DNA duplex affixes the DNA probe to the particular spot on the membrane. Although both parts of the figure show detection involving radioactivity, many other types of nonradioactive detection systems are now being employed. These include chemiluminescent chemicals that will also expose X-ray film and enzymes that release a colored product.

One limitation of this procedure is that an antibody must be available that is *specific* for the protein in question. As we saw in Chapter 22, antibodies can be readily produced by injecting the protein (antigen) into an animal, but the protein injected must be pure; otherwise antibodies of several specificities will be formed. Thus, one must have previously purified the protein of interest or false-positive clones will make selection difficult.

Nucleic Acid Probes: Searching for the Gene Itself

Suppose that the gene is not expressed in the cloning host or that no assay or antibody is available for the gene product. How does one detect its presence in colonies? This is done by using a **nucleic acid probe**, a DNA molecule containing a key part of the base sequence of the gene of interest. As we have discussed (oooSection 7.7), nucleic acid hybridization can be used as a specific means of detecting polynucleotides with specific sequences. Either DNA or RNA can be employed as a probe. The general procedure is to label the nucleic acid probe, typically with radioactive phosphate ($^{32}PO_4^{2-}$), but nonisotopic techniques are increasingly being used. Following this, one allows the single-stranded probe to hybridize with single-stranded nucleic acid derived from the cloned DNA. Because of complementary base pairing, two single-stranded polynucleotides will hybridize only if their sequences are complementary. By using appropriate hybridization conditions, it is possible to adjust the "stringency" of the hybridization, such that complementarity must be nearly exact; this helps avoid false positives.

In *colony hybridization*, a nucleic acid probe can be used to detect the presence of recombinant DNA in colonies, as shown in Figure 31.4b. This procedure again makes use of replica plating to produce a duplicate of the master plate on a membrane filter. (The same procedure can be carried out with virus vectors by blotting the plaques onto a membrane.) The cells on the filter are lysed in place to release their nucleic acid and to convert the DNA into a single-stranded form and fix it to the filter. This filter is then treated with a radioactive nucleic acid probe to allow hybridization, and after removal of unbound radioactive nucleic acid, the filter is overlaid with X-ray film. After development, the X-ray film is examined for spots (Figure 31.4b). These correspond to locations on the membrane where the radioactive probe hybridized the DNA from a particular colony. Colonies corresponding to these spots are then picked and studied further.

● **31.3** *Concept Check*

Special procedures are needed for detecting the foreign gene in the cloning host. If the gene is expressed, the presence of the foreign protein itself, as detected either by its activity or by reaction with specific antibodies, is evidence that the gene is present. However, if the gene is not expressed, then its presence can be detected with a nucleic acid probe.

♦ Does use of nucleic acid probes depend on gene expression? Explain.

♦ Why is it necessary to lyse cells containing plasmids in order to detect the product of the cloned gene?

31.4 Specialized Vectors

If the purpose of cloning a gene is to achieve a high level of its expression in a suitable host, then it is insufficient simply to locate a cloned copy of the gene. Just as specialized vectors have been developed in genomics for working with large fragments of DNA, such as bacterial artificial chromosomes and yeast artificial chromosomes (BACs and YACs, oooSection 15.1), so specialized vectors have been developed for use as specific tools in biotechnology. We consider two such classes of vectors here, *shuttle vectors* and *expression vectors*.

Shuttle and Expression Vectors

Shuttle vectors allow cloned DNA to be moved between unrelated organisms. Thus, a shuttle vector is a cloning vector that can stably replicate in two different organisms. Shuttle vectors have been developed that replicate in both *Escherichia coli* and *Bacillus subtilis*, *E. coli* and yeast, and *E. coli* and mammalian cells, as well as in many other pairs of organisms. The importance of a shuttle vector is that DNA cloned in one organism can be replicated in a second host without modifying the vector in any way to do so.

Even more important than shuttle vectors for many purposes are cloning vectors that facilitate the *expression* of cloned DNA. Organisms have complex regulatory systems (oooChapter 8), and one would expect that many cloned genes will not be expressed in a foreign host. This obstacle can be overcome by the use of **expression vectors**. An expression vector is a vector that can be used not only to clone the desired gene, but can also contain the necessary *regulatory sequences* so that expression of the gene may be manipulated.

Regulation of Transcription from Expression Vectors

A key element in an expression vector is a mechanism for transcriptional control. For high levels of expression, it is essential to produce high levels of mRNA. The promoter region is the site at which binding of RNA polymerase first occurs (oooSection 7.10). For *Bacteria*, the DNA region around 10 and 35 nucleotides before the start of transcription (called the −10 and −35 regions) (oooFigure 7.30) is especially important in the promoter. The native promoter of a cloned gene may work poorly in the new host. For example, promoters from eukaryotes or even from other prokaryotes function poorly or not at all in *Escherichia coli*. Even some *E. coli* promoters function at low levels in *E. coli* because their sequences are not as close a match to the consensus sequence as that of "strong" promoters (∞Section 7.10). For this reason, the expression vector must contain a promoter that will function efficiently in the host and one that is correctly positioned so that it can permit the transcription of the cloned gene. Promoters from *E. coli* that have been used in the construction of expression vectors include *lac* (the *lac* operon promoter), *trp* (the *trp* operon promoter), *tac* and *trc* (synthetic hybrids of the *trp* and *lac* promoters), and lambda P_L (the leftward lambda promoter; ∞Section 9.11). Each of these promoters is a "strong" promoter and can be specifically regulated (∞ Sections 8.4–8.9, 9.11).

In almost all cases it is important to be able to regulate the expression of the cloned gene. That is, although one typically wants to produce very high levels of mRNA (and have it translated), it is usually undesirable to design a vector that permits the gene to be transcribed to high levels at all times. Indeed, some proteins that are of commercial value are toxic to the host, and so in the early stages of growth of the culture it may be important that the gene not be transcribed at all. Ideally, the culture containing the expression vector should grow until a large population of cells is obtained, each containing a large copy number of the vector. Then it should be possible to express the genes in all copies simultaneously by manipulation of a regulatory switch.

We discussed regulatory controls of gene expression in Chapter 8. Recall the major importance of the *repressor-operator system* in regulating gene transcription (∞ Section 8.5). A strong repressor can completely block the synthesis of the proteins under its control by binding to the operator region. However, repressor function can be turned off by adding an *inducer*, allowing transcription of the genes controlled by the operator. For the repressor-operator system to work as a regulatory switch for the production of a foreign protein, the expression vector is designed such that the cloned gene is *fused* to a promoter and operator region. This permits proper arrangement of the sequence of genetic elements: promoter-operator-ribosome binding site-structural gene, such that efficient transcription and translation occur. In most cases, the operator and promoter correspond to each other (for instance, the *lac* operator is used with the *lac* promoter), but this is not always the case. For example, a vector could easily be constructed to contain a *trp* promoter under the control of a *lac* operator.

Figure 31.5● shows an expression vector that uses the synthetic hybrid promoter *trc* under the control of the *lac* operator (*lacO*). This plasmid also contains a copy of the *lacI* gene that encodes the *lac* repressor. The level of repressor in a cell containing this plasmid is sufficient to prevent transcription from the *trc* promoter until inducer is added. Addition of lactose or related *lac* inducers triggers transcription of cloned DNA (Figure 31.5). In addition to a strong and easily regulated promoter, most expression vectors contain an effective transcription terminator (∞Section 7.12). This prevents transcription of the entire vector, which could interfere with vector sta-

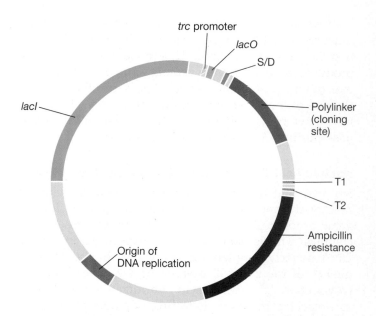

● **Figure 31.5 A partial genetic map of the expression vector pSE420.** This vector is sold by Invitrogen Corp., a biotechnology company. The polylinker is a site containing many different restriction enzyme recognition sequences to facilitate cloning (∞Section 10.15). This region (and the cloned gene) would be transcribed by the *trc* promoter, which is immediately upstream of the *lac* operator (*lacO*). Immediately upstream of the polylinker is a sequence that encodes a Shine–Dalgarno (S/D) site on the resulting mRNA (∞Section 7.16). Down-stream of the polylinker are two transcription terminators (T1 and T2). The plasmid also contains the *lacI* gene, which encodes the *lac* re-pressor, and a gene conferring resistance to the antibiotic ampicillin. These two genes are under the control of their own promoters, which are not shown.

bility. The expression vector shown in Figure 31.5 has strong transcription terminators to terminate transcription downstream from the cloned gene.

Vector Control Using Bacteriophage T7 Control Elements

In some cases the transcriptional control system may not be a normal part of the host at all. An example of this is the use of the bacteriophage T7 promoter and RNA polymerase as a regulatory system in an expression vector. When T7 infects *Escherichia coli*, it encodes its own RNA polymerase; the latter recognizes only T7 promoters, thus effectively shutting down host transcription (∞Section 16.4). In expression vectors it is possible to place expression of cloned genes under control of a T7 promoter, thereby limiting transcription only to the cloned genes. However, in order to do this, it is necessary to engineer into the plasmid the gene for T7 RNA polymerase as well. The latter is placed under control of an easily regulated system, such as *lac*. Expression of the cloned genes occurs shortly after T7 RNA polymerase transcription has been switched on by a *lac* inducer. However, because it recognizes only T7 promoters, T7 RNA polymerase transcribes *only the cloned genes*; all other host genes remain silent (untranscribed), and thus the cells stop growing. Protein synthesis in such cells then targets only the protein of interest. Large amounts of the protein of interest can be obtained using the T7 control system.

Translation of the Cloned Gene

Expression vectors must also be designed to ensure that the mRNA produced can be efficiently *translated*. To synthesize protein from an mRNA, it is essential that the ribosomes bind at the correct site and begin reading in the correct frame. In prokaryotes this is accomplished by having a ribosome binding site (Shine–Dalgarno sequence) (⊂◯⊃ Section 7.16) and a nearby start codon on the mRNA. Bacterial ribosome binding sites are not found in eukaryotic genes, and it is thus essential that such be engineered into the vector if high levels of eukaryotic gene expression are to be obtained. Once again, the vector shown in Figure 31.5 has such a site.

Often, adjustments have to be made to ensure high efficiency translation *after* the gene has been cloned. For example, *codon usage* can be an obstacle. More than one codon exists for most of the 20 amino acids (⊂◯⊃ Table 7.3), and some codons are used more frequently than others. Codon usage differs between organisms. Codon usage is partly a function of the concentration of the appropriate tRNA in the cell. Therefore, if a cloned gene has a codon usage pattern considerably different from that of its expression host, it may be translated in that host inefficiently. Insertion of the appropriate codons would be difficult because it would have to be changed at all locations in the gene. However, if necessary, synthetic DNA and site-directed mutagenesis (⊂◯⊃ Sections 7.8 and 10.18) can be used to change select codons in the gene, making it more amenable to the codon usage pattern of the host.

Finally, if the cloned gene contains introns, as eukaryotic genes typically do (⊂◯⊃ Sections 14.7 and 14.8), the correct protein product will not be made if the host is a prokaryote. This problem can also be corrected by site-specific mutagenesis or synthetic DNA, but there are other methods for creating an intron-free gene (see Section 31.5).

Eukaryotic Vectors

The previous discussion dealt with vectors that replicate in *Bacteria*. However, it is often desirable to clone and express genes in eukaryotes. Several vectors are available and their important features are shown in Figure 31.6. Major features include *origins of replication* in both the eukaryote and the prokaryote (for shuttle vector purposes), *selectable markers* for both hosts, and a *multiple cloning site* (Figure 31.6●).

We discussed the use of *yeast artificial chromosomes* (YACs) as cloning vectors in *Saccharomyces cerevisiae* to clone very large fragments of DNA (⊂◯⊃ Section 15.1). Many other vectors are also available for cloning into yeast. Yeast is one of the few eukaryotes that has a plasmid, called the *two-micron circle* (2-μ circle) because of its

● **Figure 31.6 Features of an expression vector for use in a eukaryotic host.** The vector contains components that allow for it to shuttle between *Escherichia coli* and yeast and be selectable in each organism: *ori*^Ec, origin of replication in *Escherichia coli*; *ori*^y, origin of replication in yeast; Amp^r, ampicillin resistance; ESM, eukaryotic selectable marker. The polylinker is the multiple cloning site. Other features: p, promoter; t/pa, transcription termination/polyadenylation signals (⊂◯⊃ Section 14.8). Arrows indicate the direction of transcription.

size, and many yeast vectors are based on this plasmid. Although yeast is an extremely useful organism both for genetic studies and commercial applications, it is often important to use other eukaryotes as hosts for cloned DNA. Cloning vectors have been developed for many different eukaryotes, including plants (see Section 31.10).

Virus vectors are commonly used in higher eukaryotes. For example, the DNA virus SV40 (⊂◯⊃ Section 16.11), a virus causing tumors in primates, has been engineered as a cloning vector for human tissue culture lines. Derivatives of SV40 that do not induce tumors have been developed for cloning mammalian genes and also for the expression of these genes. There are also mammalian vectors that use *adenovirus* (⊂◯⊃ Section 16.14) or *vaccinia virus* (⊂◯⊃ Section 16.13). Vaccinia virus vectors in particular have been used for the development of recombinant vaccines (see Section 31.8). Vectors derived from *baculovirus*, a DNA virus that replicates in insect cells, can be used to make large quantities of the products of cloned genes.

Other expression vectors have been developed specifically to get cloned genes stably maintained and expressed in an organism or tissue. These **integrating vectors**, as they are called, are designed to be maintained in very low copy number (typically one copy per genome) and have been developed in eukaryotes ranging from yeast to mammalian cells, as well as in certain prokaryotes. Integrating vectors have uses in both basic science and in applications, such as gene therapy (see Section 31.9). The *retroviruses* can also be used to introduce genes into mammalian cells because these viruses replicate through a DNA form that becomes integrated into the host chromosome (⚭Sections 9.13 and 16.15).

Reporter Genes

Just as there are specialized vectors for certain applications, there are also specialized *genes* or portions of genes that are useful in genetic engineering. We have already seen how certain promoters and operators can be used in the construction of expression vectors (Figures 31.5 and 31.6). We have also discussed the usefulness of having a cloning site within a gene whose activity was easy to assay. The system we described in an earlier chapter was one in which the gene encoding β-galactosidase was used (⚭Figure 15.1). The activity of the enzyme β-galactosidase can be detected easily on indicator plates or by simple liquid assays. Therefore, *lacZ*, the gene that encodes β-galactosidase, can be used as a *reporter gene*.

Reporter genes are incorporated into vectors because they encode proteins that are readily detected. These genes can be used to signal the presence or absence of a particular genetic element or its location. They can also be fused to other genes or to the promoter of other genes so that expression can be studied. There are many choices for reporter genes besides *lacZ*. The luciferase enzyme, which we discussed earlier, makes cells expressing it luminescent (⚭Figures 1.1*d* and 8.22). One can then detect colonies containing this reporter system on agar plates by their luminescence against a large background of other colonies. Production and detection of luciferase usually involves more than one gene and accessory factors. However, a gene encoding a fluorescent protein, called the *green fluorescent protein (GFP)*, has been cloned from the jellyfish *Aequorea victoria*. GFP needs no accesso-

ry factors and has been used as a reporter or tag in a wide variety of organisms (Figure 31.7●). If expression of the cloned gene of interest is linked to that of GFP, expression of the latter signals that expression of the cloned gene has occurred (Figure 31.7).

31.4 Concept Check

Many cloned genes are not expressed efficiently in a new host. Expression vectors have been developed for both prokaryotic and eukaryotic hosts that contain genes that will increase the level of transcription of the cloned gene and make its transcription subject to specific regulation. Signals to improve the efficiency of translation may also be present in the expression vector.

◆ Describe some of the elements on an expression vector that will improve transcription of the cloned gene.

◆ Reporter genes are often included as a part of a cloning vector. What is a *reporter* gene?

31.5 Expression of Mammalian Genes in Bacteria

In the past, one of the greatest challenges facing the genetic engineer wishing to clone and express a particular mammalian gene was simply to find the correct clone. We have discussed some of the methods that can be used to do this (see Section 31.3). However, there are still obstacles to be faced, even if one already has a mammalian gene cloned in an expression vector. One of the more formidable obstacles is the presence of *introns* (⚭Section 14.8). An intron in a protein-encoding mammalian gene cannot be removed by a prokaryotic host. Since almost all mammalian genes contain introns, the genes will therefore be nonfunctional in a prokaryotic host unless

(a)

(b)

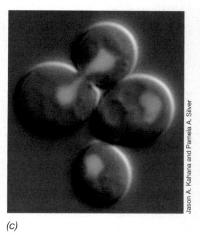

(c)

●**Figure 31.7 The green fluorescent protein (GFP).** GFP can be used as a tag for protein localization *in vivo*. In this example, the gene encoding Pho2, a DNA-binding protein from the yeast *Saccharomyces cerevisiae*, was fused to the gene encoding GFP. The recombinant gene was transformed into budding yeast cells, which could then express the fluorescent fusion protein that has localized to the nucleus. (a) Cells expressing Pho2-GFP were observed by differential interference contrast microscopy and (b) by epifluorescence microscopy (⚭Sections 4.1 and 4.2). (c) This panel is an overlay of the images in a and b.

they are removed. We discuss two such methods here. In both cases the introns are not removed from the cloned genes, but instead are removed *before cloning*.

Cloning the Gene via Messenger RNA (mRNA)

One approach to isolating a functional eukaryotic gene is to clone it through its mature mRNA. A major advantage of using mature mRNA is that any introns originally present have been removed. The isolated mRNA is used to make complementary DNA (cDNA) by means of reverse transcription (∞Sections 9.13 and 16.15). Most often a tissue expressing the gene of interest contains large amounts of the desired mRNA, although other mRNAs are present as well. In certain situations, however, a single mRNA dominates in a tissue type, and extraction of bulk mRNA from that tissue provides a useful starting point for gene cloning.

In a typical mammalian cell, about 80–85% of the RNA is ribosomal RNA, 10–15% is transfer RNA, and 1–5% is messenger RNA. Although low in abundance, the mRNA in a eukaryote is identifiable because of the poly-A tails found at the 3′ end (∞Section 14.8). If a cell extract is passed over a chromatographic column containing poly-T fragments (linked to a cellulose support), most of the mRNA of the cell can be separated from the other cellular RNA by the specific pairing of A and T bases. Elution of the RNA from the column then gives a preparation greatly enriched in mRNA.

Once mRNA has been isolated, it is necessary to convert the information to DNA. This is accomplished by use of the enzyme *reverse transcriptase*. This enzyme, an essential component of retrovirus replication (∞Sections 9.13 and 16.15), copies information from RNA into DNA, a process called *reverse transcription* (Figures 31.1 and 31.8●). As we noted, reverse transcriptase requires a primer in order for it to begin working (in retrovirus infection the primer is a tRNA). In cloning from messenger RNA, an oligo-dT primer is used that is complementary to the poly-A tail of the mRNA. The oligo-dT primer is hybridized with the mRNA, and to this mixture, reverse transcriptase is added.

As seen in Figure 31.8, the newly synthesized complementary DNA (cDNA) has a hairpin loop at its end that is synthesized because after the enzyme completes copying the mRNA, it starts to copy the newly synthesized DNA. This hairpin loop provides a convenient primer for synthesis of the complementary strand of DNA. The resultant double-stranded DNA, with the hairpin loop intact, is then cleaved by a single strand–specific nuclease to produce the desired double-stranded DNA, one strand of which is complementary to the mRNA (Figure 31.8). This double-stranded DNA encodes the gene of interest and can be inserted into a plasmid or other vector for cloning. The detection of specific clones was discussed in Section 31.3.

The cDNA obtained in this way encodes the protein of interest and contains no introns. Although there is a

● **Figure 31.8 Complementary DNA (cDNA).** Steps in the synthesis of cDNA from an isolated mRNA using the retroviral enzyme reverse transcriptase. The poly-A tail is typical of eukaryotic messenger RNAs (∞Section 14.8).

start codon, there are no promoters because these are not transcribed, and therefore their sequence will not be in the mRNA (∞Section 7.8). The requirements for achieving high levels of expression with genes made in this manner were previously discussed (see Section 31.4).

Reaching the Gene via the Protein

Knowledge of a gene's sequence facilitates its detection in a gene library or its cloning using PCR. This is because the gene sequence can be used to construct a synthetic and complementary DNA molecule to use as a *probe*

(∞Section 7.8). However, knowledge of the amino acid sequence of a protein can be useful, too, as it can be used to literally *construct* a gene *de novo*.

One can use the amino acid sequence of a protein to design and synthesize an oligonucleotide probe that encodes it. This process is in effect **reverse translation** and is illustrated in Figure 31.9●. Unfortunately, degeneracy of the genetic code (∞Section 7.14) complicates this picture somewhat. Most amino acids are encoded by more than one codon (∞Table 7.5), and codon usage varies from organism to organism. Thus, the best region of a gene to synthesize as a probe is one that encodes part of the protein rich in amino acids specified by only a *single codon* (for example, methionine, AUG; tryptophan, UGG) or at most two codons (for example, phenylalanine, UUU, UUC; tyrosine, UAU, UAC; histidine, CAU, CAC). This strategy increases the chances that the probe will be exactly complementary or nearly complementary to the mRNA or gene of interest. If the complete amino acid sequence of the protein is not known, then one typically uses the sequence at the amino terminus of the protein, where sequencing of the protein begins.

For certain small proteins there may be good reason to synthesize the entire gene. Many mammalian proteins (including high-value therapeutic peptide hormones) are the products of posttranslational processing (∞Section 8.3) and might, therefore, be quite small. Thus, to produce a peptide hormone, it would be more efficient to construct a gene that encoded just the final hormone and not the larger precursor protein from which it was derived. In addition,

chemical synthesis not only permits the acquisition of genes that cannot be obtained otherwise but also allows synthesis of *modified genes* that may make useful new proteins. Techniques for the synthesis of DNA molecules are now well developed, and it is possible to synthesize genes encoding proteins over 200 amino acid residues in length (600 nucleotides). The synthetic approach was first used in a major way for production of the human hormone insulin in bacteria, as will be discussed in Section 31.6. It should also be noted that a constructed gene is free of introns, and thus the mRNA made from it does not need to be processed. Also, promoters and other regulatory sequences can easily be built into the gene upstream of the coding sequences and codon bias (∞Section 7.14) accounted for.

With the use of all these techniques, a large number of different human and viral proteins have been expressed at high yield under the control of bacterial regulatory systems. These include, among others, *insulin, virus antigens, interferon*, and *somatostatin* (see Sections 31.6 and 31.7).

Protein Folding and Stability

The synthesis of a protein in a new host is sometimes accompanied by problems other than those whose solutions we have already discussed. For example, some proteins are susceptible to degradation by intracellular proteases and may be destroyed before they can be isolated. Moreover, some eukaryotic proteins are toxic to the prokaryotic host, and therefore the host for the cloning vector may be killed before a sufficient amount of the product is synthesized. Further engineering of either the host or the vector may be necessary to eliminate these problems.

Sometimes when foreign proteins are massively overproduced they form *inclusion bodies* inside the host. Although inclusion bodies are relatively easy to purify because of their size, the protein found in these bodies can be very difficult to solubilize. In many cases these bodies form because the protein is incorrectly folded. One potential solution to this problem is to use a host that overproduces molecular chaperones that aid in folding (∞Section 7.17). However, folding problems can often be solved if the protein from the cloned gene is made as a **fusion protein** along with a protein encoded by the vector. This involves fusing the two genes to yield, following transcription and translation, a single, but cleavable protein. Fusion proteins often simplify purification of the protein of interest if the other part of the fusion protein is one for which rapid and inexpensive purification techniques are available.

Several special *fusion vectors* are now available to encode fusion proteins. The "cloned protein" is then released from the fusion protein after purification by special proteases. Figure 31.10● shows an example of a fusion vector that is also an expression vector. In some cases the cloned protein can also be removed from the fusion protein by specific chemical treatment. Fusion systems can also be used for purposes other than achieving increased protein stability. One advantage of making a fusion pro-

Protein

H₂N – ·· Met – Trp – Tyr – Glu – His – Lys – Glu – COOH

Possible mRNA codons

5′ – · AUG UGG UAU GAG CAU AAG GAG – – 3′
            C   C   A   C   A   A

DNA oligonucleotides (possible probes)

T A C A C C A T A C T C G T A T T C C T C 5′

T A C A C C A T G C T C G T A T T C C T C 5′

T A C A C C A T A C T T G T A T T C C T C 5′

T A C A C C A T A C T C G T G T T C C T C 5′

and so on

Preferred DNA sequence (based on the organism's codon bias)

T A C A C C A T G C T C G T A T T C C T C 5′

● **Figure 31.9 Deducing the best sequence of an oligonucleotide probe from the amino acid sequence of the protein: Reverse translation.** Because of degeneracy of the genetic code (∞Section 7.14), many probes are possible. If codon usage by the same organism is known, then a preferred sequence can be selected. It is not essential that complete accuracy be achieved because a small amount of mismatch can be tolerated, especially with long probes.

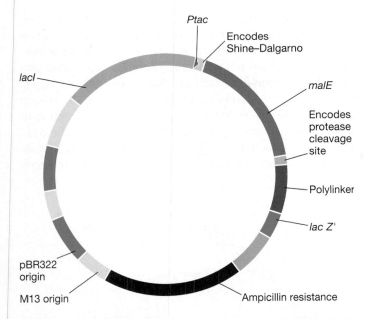

Ptac
Encodes
Shine–Dalgarno

lacI

malE

Encodes
protease
cleavage
site

Polylinker

lac Z'

pBR322
origin

M13 origin

Ampicillin resistance

● **Figure 31.10   An expression vector for fusions.** This vector was developed by the New England Biolabs Company. The gene to be cloned is inserted at the polylinker site (∞Figure 15.1), so it is in frame with the *malE* gene, which encodes the maltose-binding protein. This insertion inactivates the *lacZ'* gene (∞Figure 15.1). The fused gene is under control of the hybrid *tac* promoter (*Ptac*). The plasmid also contains the *lacI* gene, which encodes the *lac* repressor. Therefore, an inducer must be added to the cells in order to turn on the *tac* promoter. The fusion protein is easily purified by methods involving the affinity of the protein for maltose. Once purified, the two portions of the fusion protein can be separated by a very specific protease (factor Xa). The plasmid contains a gene conferring ampicillin resistance on its host. In addition to the plasmid origin of replication, there is a bacteriophage M13 origin. Therefore, this is a *phagemid* and can be propagated either as a plasmid or as a phage.

tein is that the host portion can be engineered to contain the bacterial sequence encoding the *signal peptide* that enables transport of the protein across the cytoplasmic membrane (∞Section 7.17). This makes possible a bacterial expression system that not only synthesizes the mammalian protein, but also actually *excretes* it. By employing the right strains and vectors, the desired protein can exceed 200,000 molecules per cell and make up as much as 40% of the protein molecules in a cell.

⬣ 31.5 Concept Check

It is possible to achieve very high levels of expression of mammalian genes in prokaryotes. However, the expressed gene must be free of introns. This can be accomplished by using reverse transcriptase to synthesize cDNA from the mature mRNA encoding the protein of interest. It can also be accomplished by synthesizing an entirely synthetic gene from knowledge of the amino acid sequence of the protein of interest. Fusion proteins are often used to stabilize or solubilize the cloned protein.

◆ What major advantage does cloning mammalian genes from mRNA or synthetic genes have over PCR amplification and cloning of the native gene?

◆ How is a fusion protein made?

II PRACTICAL APPLICATIONS OF GENETIC ENGINEERING

We present here only a few of the many applications of genetic engineering in biotechnology. These will focus on existing applications that are of importance to medicine and agriculture. In addition to these, many other exciting applications are emerging or in development. These include, in particular, the enhancement of industrial fermentations (∞Chapter 30) and the use of genetically engineered prokaryotes in bioremediation (∞Section 19.18) and other areas of environmental biotechnology.

31.6 Production of Insulin: The Beginnings of Commercial Biotechnology

The most economically robust area of biotechnology is the production of *human proteins*. Many proteins from mammalian cells have high pharmaceutical value. However, these proteins are typically present in very small amounts in normal tissue, and it is therefore extremely costly to purify them. Moreover, even if the protein can be produced in cell cultures (∞Section 9.3), this is a much more expensive and difficult method of obtaining it than growing microbial cultures that produce it in high yield. Therefore, the biotechnology industry has genetically engineered microorganisms to produce these proteins. A classic example is *insulin* and is discussed in detail here.

The Biology of Insulin

An early and dramatic commercial success in biotechnology was the production of the hormone **insulin**. Many hormones are small proteins. These molecules are critical for the proper control of mammalian metabolism and have important therapeutic uses. Insulin is a protein produced in the pancreas that is vital for the regulation of glucose metabolism in the body. *Juvenile diabetes*, a disease characterized by insulin deficiency, afflicts millions of people (∞Section 22.15). The standard treatment for juvenile diabetes is periodic injections or oral administration of insulin.

Because insulins of most mammals are similar in structure, it has been possible in the past to treat human diabetes by use of insulin isolated commercially from beef or pork pancreas. However, nonhuman insulin is not as effective as human insulin, and the isolation process is expensive and complex. Cloning of a human insulin gene in bacteria solved this problem. Cloning and production began using *Escherichia coli*, but today, most genetically engineered insulin is produced in yeast. The final product, biosynthetic human insulin, is identical in every respect to insulin purified from the human pancreas.

Despite being a small protein, insulin was a particular challenge to the biotechnology industry. Insulin in its active form consists of two polypeptides (A and B) connected by disulfide bridges (Figure 31.11a●). These two polypeptides are encoded by separate parts of a single insulin gene. The insulin gene encodes *preproinsulin*, a polypeptide containing a signal sequence (involved in excretion of the protein) (◌◌Section 7.17), the A and B polypeptides of the active insulin molecule, and a connecting polypeptide that is absent from active insulin. *Proinsulin* is formed by the processing of preproinsulin, and the conversion of proinsulin to insulin involves enzymatic cleavage of the connecting polypeptide from the A and B chains.

Genetically Engineered Insulin

The production of human insulin in bacteria illustrates some of the previous principles we have discussed. Because the insulin protein is fairly small, the entire gene could be synthesized. So this approach was taken rather than cloning the gene from human DNA and having to deal with the preproinsulin-to-proinsulin processing step. There are only 63 bases (for 21 amino acids) encoding the A chain and 90 bases (for 30 amino acids) encoding the B chain. In proinsulin there are an additional 105 bases for the peptide connecting the A and B chains (Figure 31.11b). When the polynucleotides were synthesized, suitable restriction enzyme sites were engineered at each end so the polynucleotides could be ligated into a plasmid vector.

To obtain effective expression, the synthesized insulin genes were inserted downstream from a suitable *Escherichia coli* promoter but in a manner such that the insulin fragment was synthesized as part of a fusion protein (see Section 31.5). An important advantage of making the fusion protein is that the fusion product was much more stable in *E. coli* than insulin itself. Finally, a codon for methionine was added at the junction joining the insulin gene to the upstream part of the fusion gene (Figure 31.11b). This was done because a chemical, *cyanogen bromide*, specifically cleaves polypeptide chains at methionine residues, permitting recovery of insulin from the fused protein. Insulin itself lacks methionine and hence, is unaffected by cyanogen bromide treatment.

Proinsulin (Figure 31.11a), isolated from producing cells following cyanogen bromide treatment of the fusion protein, is converted to insulin by disulfide bond formation, followed by enzymatic removal of the connecting peptide of proinsulin. Starting from proinsulin was necessary because proinsulin naturally folds so that the critical cysteine residues involved in disulfide bond formation in insulin are opposite each other (Figure 31.11a). A chemical treatment is used to catalyze the formation of disulfide cross-links. Once this has been accomplished, the connect-

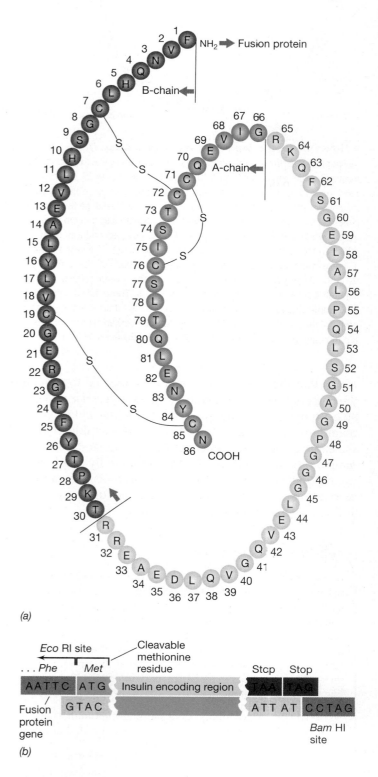

(a)

(b)

● **Figure 31.11   Genetic engineering for the production of human insulin in bacteria.** (a) Structure of human proinsulin. Amino acids in the protein are labeled by their one letter abbreviations (◌◌Figure 3.12). The peptide shown in yellow must be removed from between the A and B chains in order to make insulin. (b) Chemical synthesis of the insulin gene and suitable linkers, permitting cloning and expression. The synthesized fragments were linked via restriction sites *Eco*RI and *Bam*HI in a plasmid vector in such a way that the insulin chains are formed as a fusion protein (see Section 31.5) with a portion of a gene found on the vector (note that the *Eco*RI site is part of this coding region). The methionine coding sequence was inserted to permit chemical cleavage of the A and B chains from the fused protein made in the bacteria because the reagent cyanogen bromide specifically cleaves at methionine residues and insulin does not contain methionine. Two stop codons were incorporated at the downstream end of the coding sequence to ensure that translation would cease.

ing peptide is removed by treatment with the proteolytic enzymes trypsin and carboxypeptidase B, which have no effect on insulin itself, and the final product, human biosynthetic insulin, is ready for human use.

Many of the tricks learned in the genetic engineering of insulin were applied to other products. Insulin production today stands as one of the crowning achievements of this technology, made even more impressive considering that it was accomplished in the very early days of commercial genetic engineering.

⬡ 31.6 Concept Check

The first human protein made commercially using engineered bacteria was human insulin, but numerous other hormones and other human proteins are now being produced. In addition, many recombinant vaccines have been produced.

◆ Why did the genetic engineering of insulin dispense with the preproinsulin step but require that proinsulin formation occur before insulin formation?

◆ What is the importance of cyanogen bromide in the production of biosynthetic insulin?

31.7 Other Mammalian Proteins and Products

Besides insulin, many other mammalian proteins and products are produced by genetic engineering (Table 31.1). These include, in particular, an assortment of hormones and proteins involved in blood clotting and other blood processes. For example, *tissue plasminogen activator* (*TPA*) is a blood protein that scavenges and dissolves blood clots that can occur in the final stages of the healing process. The clinical usefulness of TPA is primarily in heart patients or others suffering from poor circulation because of excessive clotting. TPA is administered following heart attacks and cardiac bypass, transplant, or other heart surgeries to prevent the development of pulmonary embolisms, which can be life-threatening. Heart disease is a leading cause of death in many developed countries, especially in the United States (⚫⚫Figure 1.7), so microbially produced TPA is in high demand.

Blood clotting factors VII, VIII, and IX are important genetically engineered products. Unlike TPA, these proteins are critically important for the *formation* of blood clots. Hemophiliacs suffer from a deficiency of one or more clotting factors and can be treated with microbially produced clotting factors. Recombinant clotting factors take on added significance when one considers that hemophiliacs have in the past been treated with concentrated clotting factor extracts from pooled human blood, some of which was contaminated with the AIDS virus, putting hemophiliacs at high risk for contracting AIDS. Recombinant clotting factors have eliminated this health risk.

A number of proteins have roles as anticancer agents or immune modulators. Chief among these are the *inter-*

Table 31.1	A few therapeutic products made by genetic engineering
Product	**Function**
Blood proteins	
Erythropoietin	Treats certain types of anemia
Factors VII, VIII, IX	Promotes clotting
Tissue plasminogen activator	Dissolves clots
Urokinase	Blood clotting
Human hormones	
Epidermal growth factor	Wound healing
Follicle-stimulating hormone	Treatment of reproductive disorders
Insulin	Treatment of diabetes
Nerve growth factor	Possible treatment of degenerative neurological disorders and stroke
Relaxin	Facilitates childbirth
Somatotropin (growth hormone)	Treatment of some types of growth failure and short stature
Immune modulators	
α-Interferon	Antiviral, antitumor agent
β-Interferon	Treatment of multiple sclerosis
Colony-stimulating factor	Treatment of infections and cancer
Interleukin-2	Treatment of certain cancers
Lysozyme	Anti-inflammatory
Tumor necrosis factor	Antitumor agent, potential treatment of arthritis
Replacement enzymes	
β-glucocerebrosidase	Treatment of Gaucher disease, an inherited neurological disease
Therapeutic enzymes	
Human DNase I	Treatment of cystic fibrosis
Alginate lyase	Treatment of cystic fibrosis

ferons, a series of proteins made by animal cells in response to viral infection (⚫⚫Section 23.10) or immune activation in the case of one type of interferon. Although in most instances interferon treatments have met with mixed clinical success, it is still hopeful that certain interferons may have specific therapeutic applications.

Nonblood Proteins

Some mammalian proteins made by genetic engineering are enzymes rather than hormones (Table 31.1). For instance, *human DNase I* is being produced and used to treat the build-up of DNA-containing mucus in patients with **cystic fibrosis**. The mucus forms because cystic fibrosis is accompanied by life-threatening lung infections by the bacterium *Pseudomonas aeruginosa*. The bacterial cells form biofilms (⚫⚫Section 19.3) within the lungs that make drug treatment difficult, and DNase digests the DNA released when these bacterial cells lyse. A second enzyme, called *alginate lyase* and also produced by genetic engineering, shows promise in cystic fibrosis patients for treatment of polysaccharide produced by *P. aeruginosa* cells. Like DNA from lysed cells, this polymer contributes to lung mucus. Thus, its hydrolysis can relieve respiratory symptoms. Over 30,000 active cases of cystic fibrosis are known in the United States alone. Treatment of cystic fibrosis with DNase was approved in 1994, and sales today of this life-saving enzyme exceed $100 million.

Not all the enzymes produced by genetic engineering have therapeutic uses. Many commercial enzymes (∞Section 30.9) are produced in this way as well. Non-human hormones are also produced. For example, *bovine somatotropin* produced by recombinant techniques is used to increase milk production in cattle in the United States (transgenic cattle, see Section 31.9). Often the benefits of genetic engineering can be quite unexpected. *Rennin*, which is used to make cheese, is an animal product, and thus not consumed by strict vegetarians (vegans). However, in Great Britain a "vegetarian cheese" containing a recombinant protein produced in a microorganism is being marketed and has found wide acceptance. Added applications come from being able to use site-directed mutagenesis (∞Section 10.18) on existing cloned genes such that new products with new properties can be generated. Also, since molecules such as the antibiotics are synthesized in cells in biochemical pathways using a series of enzymes (∞Chapter 30), these enzymes can be modified by genetic engineering to produce a variety of unnatural, yet clinically effective, antibiotics.

 31.7 *Concept Check*

Many proteins found in humans that were formerly extremely expensive to produce because they were found in human tissues in only small amounts can now be made in large amounts from the cloned gene in a suitable expression system.

◆ Contrast the activity of TPA and blood factors VII, VIII, and IX.

◆ Explain how a DNA-degrading enzyme can be useful in treating a bacterial infection, such as occurs in cystic fibrosis.

31.8 Genetically Engineered Vaccines

Vaccines are suspensions of killed or modified pathogenic microorganisms (or specific fractions isolated from the microorganisms) that, when injected into an animal, produce immunity to a particular disease (∞Sections 22.13 and 22.14). Often the substance that elicits the immune response is a surface protein, for instance, a virus coat protein. Genetic engineering can be applied in many different ways to the production of vaccines (Table 31.2).

Recombinant Vaccines

Recombinant DNA techniques can be used to modify a pathogen. For instance, in some cases one can simply *delete* genes involved in virulence but leave those whose products elicit an immune response. This yields a **recombinant live attenuated vaccine**. Using recombinant technology one can also *add* genes to a virus that will specifically confer immunity to a viral disease. For example, a recombinant virus vaccine is used in poultry against both fowlpox (a disease that reduces weight gain and egg production) and Newcastle disease (a viral disease that is often fatal). To do this, the fowlpox virus, a typical pox virus (∞Section 16.13), was first modified to delete genes that cause disease (but not those that elicit immunity). Then immunity-inducing genes from the Newcastle virus were added. This resulted in a *polyvalent vaccine*, a single virus that can confer immunity to two different diseases.

A vector widely used to prepare live recombinant vaccines for potential human use is *vaccinia virus* (∞Section 16.12). Such vaccines have been called *vector vaccines*. Vaccinia virus itself is generally not pathogenic for humans (vaccinia virus was originally used as a vaccine against the related virus smallpox, ∞Section 16.13). Cloning in vaccinia virus is done using an *Escherichia coli* plasmid containing a fragment of the vaccinia virus *thymidine kinase* gene (Figure 31.12*a*). An appropriate foreign DNA is inserted into this plasmid, for example, genes encoding

Table 31.2	Some genetically engineered vaccines	
Disease	**Causative organism**	**Vaccine function**
Viral diseases		
Hepatitis A	Hepatitis A virus	Prevention of infectious hepatitis (∞Section 26.11)
Hepatitis B	Hepatitis B virus	Prevention of serum hepatitis (∞Sections 16.15 and 26.11)
Measles	Measles virus	Prevention of measles (∞Section 26.7)
Rabies	Rabies virus	Prevention of rabies (∞Section 27.1)
Influenza	Influenza virus	Prevention of influenza (∞Section 26.8)
Polio	Polio virus	Prevention of polio (∞Sections 16.8, 25.9, and 28.8)
Bacterial Diseases		
Tetanus	*Clostridium tetani*	Prevention of tetanus (∞Section 27.9)
Diphtheria	*Corynebacterium diphtheriae*	Prevention of diphtheria (∞Section 26.3)
Chlamydial syndromes	*Chlamydia trachomatis*	Prevention of pelvic inflammatory disease (∞Section 26.13)
Protozoal Diseases		
Malaria	*Plasmodium vivax*	Prevention of malaria (∞Section 27.5)

the coat protein(s) of a viral pathogen, and the recombinant plasmid is transformed into a host cell whose own thymidine kinase gene is inactive but that has previously been infected with wild-type vaccinia virus (Figure 31.12*b●*). If homologous recombination occurs between plasmid DNA and vaccinia genomic DNA (Figure 31.12*c*), recombinant virions can be obtained containing an *inactivated* thymidine kinase gene. An *active* thymidine kinase leads to growth inhibition by the compound 5-bromodeoxyuridine. Therefore, recombinant vaccinia virions can be selected because

of their replication in the presence of this inhibitor (Figure 31.12*d*).

Although such recombinant viruses no longer express thymidine kinase, they can still infect human cells and express the foreign genes that have been cloned into them. Indeed, some recombinant vaccinia viruses can be constructed to carry genes from multiple viruses (that is, they are *polyvalent* vaccines). Currently, several vaccinia vector vaccines have been developed and licensed for veterinary use, including a vaccine for rabies. Many human subunit vaccines are at the clinical trial stage. Thus far, vaccinia vaccines seem sufficiently benign yet highly immunogenic in humans, and their use will likely be even more widespread in coming years.

Subunit Vaccines

Recombinant vaccines do not have to include the entire pathogenic organism. **Subunit vaccines** may contain only a specific protein from a pathogenic organism. For viruses this protein is often the coat protein, since these are typically highly immunogenic (∞Section 22.13). The coat proteins are purified and used in high dosage to elicit a rapid and high level of immunity. Subunit vaccines are currently very popular because recombinant DNA techniques can be used to produce large amounts of the specific proteins and there is no possibility that the purified products will contain the entire pathogenic organism in even minute amounts.

The steps in preparing a subunit vaccine are as follows: fragmentation of viral DNA by restriction enzymes; cloning viral coat protein genes into a suitable vector; providing for proper promoters, reading frame, and ribosome binding sites; and reinsertion and expression of the viral genes in a microorganism. Sometimes only certain *domains* of the protein are expressed, rather than the entire protein, since immune cells and antibodies typically react with only small portions of the protein (∞Section 22.5).

When *Escherichia coli* is used as the cloning host, subunit vaccines are often poorly immunogenic and fail to protect in experimental tests of infection with the virus. The problem is that many viral coat proteins are *posttranslationally modified*, typically by the addition of sugar residues (glycosylation), when the virus replicates in the animal. The recombinant proteins produced by bacteria are unglycosylated, and glycosylation is necessary for the proteins to be immunologically active. To solve the glycosylation problem, a eukaryotic host is used. For example, the first recombinant subunit vaccine approved for use in humans was made in yeast. The gene encoding a surface protein from hepatitis B virus was cloned and expressed in yeast. The protein was produced and formed aggregates very similar to those found in patients infected with the virus. These aggregates were purified and used to effectively vaccinate humans against hepatitis B virus infection.

Employing genetic engineering, subunit vaccines against a large variety of viruses and pathogenic organisms

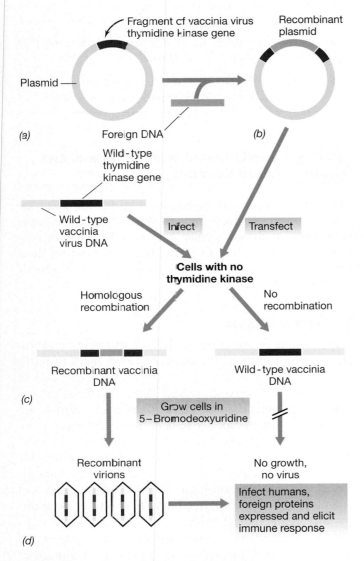

● **Figure 31.12 Production of recombinant vaccinia virus and its use as a recombinant vaccine.** (a) Foreign DNA is cloned into a plasmid containing a small piece of the vaccinia virus thymidine kinase gene. (b) Recombinant plasmid formed. (c) The recombinant plasmid is then used to transfect host cells already infected with wild-type vaccinia. If recombination occurs, recombinant vaccinia DNA can be produced. (d) The cells are then placed in the presence of 5-bromodeoxyuridine, a compound that is toxic to cells having an active thymidine kinase. Only recombinant virions develop under these conditions. If the recombinant vaccinia virions contain genes for other viral coat proteins, these may be expressed.

are being developed. Insect cells and cultured mammalian cells are often used as hosts to prepare these recombinant vaccines. To obtain the correct pattern of glycosylation or other modifications of the protein, it is often important to use a host that is closely related to humans. However, vaccines can also be produced in plants (see Section 31.10); because they are eukaryotes, proper glycosylation is usually obtained.

The Future of Recombinant Vaccines

Genetically engineered recombinant vaccines will likely become increasingly common for several reasons. First, they are safer than normal attenuated or killed vaccines; it is impossible to transmit the disease in the vaccine. Second, they are more reproducible because their genetic makeup can be carefully monitored. And third, recombinant vaccines can be administered in high doses without fear of side effects.

In addition, recombinant vaccines can usually be prepared much more quickly than those produced by more traditional methods. For some diseases, such as influenza, timing is critical. Recombinant vaccines using cloned influenza virus hemagglutinin genes (co Section 16.9) can be made in just 2 or 3 months. This is in contrast to the 6–9 months needed to make an attenuated intact virus influenza vaccine. Preparation time can be an important factor in responding to an epidemic caused by a new strain of virus, a common cause of outbreaks of influenza (co Section 26.8). And finally, recombinant vaccines are typically much less expensive than those produced in traditional ways, and this is also an important factor in health care today.

DNA Vaccines

Although recombinant and traditional vaccines have been extremely successful in the fight against a variety of infectious diseases, in some cases vaccines may be difficult to produce. For such important diseases as malaria (co Section 27.5) and AIDS (co Section 26.14), "standard" genetic engineering techniques might yet yield effective vaccines. But while we await breakthroughs in this area, an exciting new direction has opened up— *DNA vaccines,* also known as *genetic vaccines.*

DNA vaccines use the genetic material of the pathogen itself to immunize the individual. This may be in the form of defined fragments of the pathogen's genome or only certain genes that encode immunogenic proteins. The key genes are cloned into a plasmid or viral vector and delivered by injection. When DNA is taken up by animal cells it may be degraded, or it may be transcribed and translated. If the latter occurs and the protein produced is immunogenic, the animal will be effectively immunized against the pathogen in the process. Thus, a DNA vaccine is really an immune response made against the *protein encoded by* the vaccine DNA. The DNA itself is not immunogenic.

A number of DNA vaccines are undergoing clinical trials and many more are being developed. DNA vaccines have the advantage that they are both safe and inexpensive. Unlike viral vaccines, a DNA vaccine would escape surveillance by the host immune system, since nucleic acids themselves are poorly immunogenic. The latter prevents the animal from suffering autoimmune effects, whereby antibodies and immune cells attack self components (co Section 22.15).

31.8 Concept Check

Many recombinant vaccines have been produced. These include live recombinant, vector, subunit, and DNA vaccines.

◆ Explain why recombinant vaccines might be safer than some vaccines produced by traditional methods.

◆ Distinguish between a *recombinant live attenuated vaccine,* a *vector vaccine,* a *subunit vaccine,* and a *DNA vaccine.*

31.9 Genetic Engineering in Animal and Human Genetics

This section covers only a few highlights of some of the uses of genetic engineering in animal and human genetics. Some of these applications have to do with products, but more have to do with understanding gene function in mammals and in curing or treating genetic diseases.

Transgenic Animals

With the use of recombinant DNA technology and microinjection techniques to deliver cloned genes to fertilized eggs, many foreign genes have been expressed in both laboratory research animals and in species important in commercial animal industries. Such animals are **transgenic organisms** because of the permanent presence of foreign genes. Transgenic animals have become increasingly important in basic biomedical research for studying gene regulation and developmental biology. The production of transgenic animals for research and commercial purposes seems likely to continue as an important area in biotechnology.

One application is to improve the productivity or disease resistance of the animal, as in the case of transgenic plants used in agriculture. However, transgenic animals can also be used to produce proteins of pharmaceutical value—a process that has been called *pharming.* Transgenic animals may be useful for producing human proteins that require specific posttranslational modifications for activity, such as certain blood clotting enzymes; many proteins of this type are not produced in an active form by microorganisms or by plants. Also, some proteins have been engineered to be secreted in high yield into the animals' milk, a substance that can be readily collected in large volumes and processed. Thus far these proteins include *α-1-antitrypsin* (used to treat lung disease) produced in sheep and *tissue plasminogen activator* (used to dissolve blood clots, see Section 31.7) produced in goats.

Human Genetics

Conventional genetics, involving genetic crosses or mutagenesis, obviously cannot be done with humans. Therefore, our understanding of human genetics has lagged behind our understanding of the genetics of many other organisms, until the era of genetic engineering. However, recombinant DNA technology has proved valuable in human genetics, especially in identification techniques and in medicine.

As mentioned in Chapter 15, a draft sequence of the human genome was released in 2000. From this draft, the techniques of genetic engineering have been widely used to identify individuals in a process called **DNA fingerprinting** (see the Microbial Sidebar, DNA Fingerprinting). However, a major goal of sequencing the human genome was to better understand human genetic diseases, with a goal of better treating human genetic and other diseases and curing them if possible.

A vast number of human genetic diseases are known. By use of recombinant DNA technology, coupled with conventional genetic studies (following family inheritance, and so on), it is possible to localize particular defects to particular chromosomes and to particular locations on chromosomes. With the use of recombinant DNA technology, it is possible to clone the region containing the genetic defect and compare the base sequence in the normal gene and the defective gene. From such studies, even in the absence of knowledge of the enzyme defect, it has been possible to obtain information about the genetic change. Many genes, including those for *Huntington's disease, hemophilia types A and B, cystic fibrosis, Duchenne's muscular dystrophy, multiple sclerosis,* and *breast cancer* have been localized with these techniques and the mutation(s) in the defective genes identified. With this in mind, how can genetic engineering be used to treat or cure these diseases?

Gene Therapy

Genetic engineering has been employed to provide treatment for some human genetic diseases as well as cancers using **gene therapy**. In gene therapy, a nonfunctional or dysfunctional gene in an individual is augmented or replaced by a functional gene. Major obstacles to this approach exist in trying to target the correct cells for gene therapy and in successfully transfecting cell lines that will perpetuate the genetic alteration.

The first genetic disease for which an approved gene therapy technique was used was one form of *severe combined immune deficiency* (SCID). This disease is caused by the absence of adenosine deaminase (ADA), an enzyme involved in purine metabolism, in bone marrow cells and leads to an essentially undeveloped immune system. The gene therapy procedure involved using a retrovirus as a vector to insert an unmutated copy of the ADA gene into T lymphocytes (cells that are part of the immune system; ⬿Section 22.1) removed from the patient, and then placing these "corrected" cells back in the body (the retrovirus also carries a marker gene, resistance to neomycin, so cells carrying the inserted retrovirus can be selected and identified). However, since T lymphocytes have a limited life span, it is necessary to repeat the therapy every month or two. Thus, attempts have been made to insert the gene into the stem cells of the bone marrow (which continue to divide, ⬿Section 22.1) and effect a true cure for the disease. In at least one case of SCID, this has been successful.

Several other gene therapy treatments, some using other virus vectors, are currently being tested with various levels of success. Since the first gene therapy experiment with ADA in 1990, there have been no striking practical breakthroughs until 2000. Another form of SCID, involving a different gene, was successfully treated in three different patients. It seems likely that this very rare form of the disease can now be successfully treated using gene therapy.

Though gene therapy has tremendous practical potential, most applications remain a distant prospect. Some of the current difficulties are related to the vectors being used. Although transduction using retroviral vectors gives stable integration of the gene, the site of insertion is unpredictable, the amount of cloned DNA is limited, and expression of the cloned gene is often transient. The vectors also have limited infectivity and are rapidly inactivated in the host. Many nonretroviral vectors, such as the adenovirus vector (⬿Section 16.14), have similar problems, and adverse reactions to the vector can also be a severe problem. However, promising new vectors include human artificial chromosomes and highly modified versions of the HIV virus (⬿Sections 9.13 and 16.15).

It is important to note that in the gene therapy protocols being tested, the defective gene is not replaced. Rather, the retrovirus (and the good copy of the gene) simply integrates somewhere in the human genome of these cells. Actual gene replacements in germ line cells (cells that give rise to gametes) can be accomplished in experimental animals, although the techniques of isolating individual animals with these changes cannot readily be applied to humans. Moreover, attempts to change the germ cells of humans would raise many ethical questions beyond the simple questions of experimental protocols.

 31.9 Concept Check

Genetic engineering can be used to develop transgenic animals capable of producing proteins of pharmaceutical value. The techniques of genetic engineering are also applied to identifying individuals using DNA fingerprinting. One of the great hopes of genetic engineering is in gene therapy, where functional copies of a gene can be supplied to an individual to treat human genetic disease.

Microbial Sidebar ◆ DNA Fingerprinting

The techniques used in molecular genetics and genetic engineering can be used to differentiate between very closely related organisms—even between humans—in a process called **DNA fingerprinting**.

DNA fingerprinting is made possible by the fact that higher organisms contain *repetitive DNA sequences* that exist in different numbers and patterns. The genomes of higher eukaryotes contain a very large amount of repetitive DNA. Some of these repeats exist in families of related sequences scattered around the genome, and members of these families have been cloned and sequenced. One class of these repetitive sequences was found to not only have unique sequences, but also to vary as to *how many repeats* of a given sequence occurred at a single site on a particular chromosome. These sequences are called *variable number of tandem repeats* (VNTR), and several have been identified.

The use of VNTRs in DNA fingerprinting is illustrated in Figure 1*a*. It shows two different alleles (alternative forms of the same gene) of a eukaryotic chromosome that differ only in how many copies of the repeated sequence are present. Since the VNTR DNA has been sequenced, it is known which restriction enzyme sites are *not* present in a particular VNTR. Di-

gestion with such an enzyme then releases the complete VNTR intact. When the DNA from two chromosomes with a different number of repeats at this particular locus is digested by a restriction enzyme, the restriction fragments containing this DNA differ in size. Such a difference is called a *restriction fragment length polymorphism*, or RFLP. The RFLP can be separated by gel electrophoresis and the VNTR-containing fragments detected by Southern blotting (⬮ Section 7.7) using a probe made from the cloned VNTR sequence. Even if multiple alleles are present in a population (rather than just two as in the example), differences at a single site are insufficient to unequivocally identify an individual. Many individuals in a population would be expected to have the same pattern at this site. However, it is possible to probe the digested DNA simultaneously for several different VNTR markers. With the use of these methods, the probability of identifying a particular individual by comparing two different DNA samples is extremely high.

The polymerase chain reaction (PCR) can simplify the identification process. Although PCR does not need to be used in DNA fingerprinting, *per se*, it is an essential tool when the amount of DNA in the sample is very small—such as that found in the cells of a single hair.

But in addition to getting enough DNA to work with, the use of PCR directly in DNA fingerprinting is also possible (Figure 1*b*). To use PCR, it is necessary to know the sequences surrounding a particular site that contains a VNTR so that specific primers can be synthesized. However, since PCR amplifies *only the DNA between the primers*, one does not have to cut the DNA with restriction enzymes before running it on a gel. With enough cycles of amplification, it is possible to detect the PCR-generated bands by simply staining the gels rather than using a hybridization probe; the bands run to different regions of the gel based on their size (number of repeats). By testing enough different VNTRs, a unique pattern can be found for any individual.

VNTRs are not restricted to higher organisms. The genome of *Bacillus anthracis*, the species of *Bacteria* that causes the disease anthrax, contains several different VNTRs. This allowed strains of *B. anthracis* used as biological weapons (⬮ Sections 25.11 and 25.13) to be identified and tracked. Such tracking can assist those in law enforcement identify the source of the *B. anthracis* strains or can be used as evidence that a particular strain was used in a bioterrorism incident. ∎

Figure 1 DNA fingerprinting. *(a) Two different alleles of a region of a single chromosome. The alleles differ only in the number of repeats in the VNTR. DNA from cells containing these chromosomes can be cut with the restriction enzyme EcoRI (which does not cut within the VNTR) and the fragments separated on an agarose gel (smaller fragments run further on a gel than do larger fragments). The fragments containing the VNTR are then identified after Southern blotting by hybridization with a probe specific to the VNTR. The figure shows only the result from individuals whose two chromosomes have different alleles at this site. If an individual had the same allele on both chromosomes, only a single band would be observed. (b) The same alleles, but using specific primers to amplify the VNTR segments by PCR. The products of the PCR reaction can be loaded directly onto the gel without restriction digestion.*

◆ What is *pharming*?

◆ What is a VNTR?

◆ A person treated successfully by gene therapy will still have a defective copy of the gene. Explain.

31.10 Genetic Engineering in Plant Agriculture: Transgenic Plants

Genetic improvement of plants by traditional methods has a long history but recombinant DNA technology has led to revolutionary changes. It is possible to use genetic engineering to modify plant DNA and then transform plant cells with the DNA by either electroporation or particle gun methods (see Section 31.2). Alternatively, one can use vectors from the bacterium *Agrobacterium tumefaciens*, which can transfer DNA directly into certain plants (∞Section 19.21). It is possible to use plant tissue culture techniques to select clones of plant cells that have been genetically altered using *in vitro* techniques and then, with proper treatments, induce these cell cultures to make whole plants that can be propagated vegetatively or by seeds.

In contrast to plants whose properties have been improved by traditional plant genetics, genetically engineered plants are *transgenic*. The public knows these plants as *genetically modified* (GM) *plants*. Although the techniques to generate transgenic plants or transgenic animals (see Section 31.9) are virtually identical to those used to generate microorganisms expressing foreign genes, the use of the term *transgenic* is confined to multicellular organisms. In this section we will see how the foreign gene (called a *transgene*) is inserted into a plant genome and how transgenic plants can be used.

The Ti Plasmid and Transgenic Plants

The gram-negative plant pathogen *Agrobacterium tumefaciens* contains a large plasmid, called the **Ti plasmid,** which is responsible for its virulence. The plasmid contains genes that mobilize DNA for transfer to the plant (for details of the disease process and genetic events, ∞Section 19.21). The segment of the Ti plasmid DNA that is actually transferred to the plant is called **T-DNA**. The sequences at the ends of the T-DNA are essential for transfer, and the DNA to be transferred must be between these ends.

One common Ti-vector system that has been constructed and used for the transfer of genes to plants is a two plasmid system called a *binary vector*. It contains the two ends of the T-DNA on either side of the site used for cloning and an antibiotic resistance marker that can be used in plants. It also contains an origin of replication so that it can replicate in both *A. tumefaciens* and *Escherichia coli* (the latter serves as a host for cloning work), and another antibiotic resistance marker that is expressed in bacteria (Figure 31.13●). The DNA to be cloned is inserted into the vector, which is then transformed into *E. coli*. It is then transferred to *A. tumefaciens* by conjugation.

This cloning vector lacks the genes necessary for transfer of T-DNA to a plant, so the *Agrobacterium* into which it is transferred must contain the other member of the binary vector system. This other plasmid contains the virulence (*vir*) region of a Ti plasmid but lacks disease-causing genes. Although it can direct the transfer of DNA into a plant, it no longer has genes that cause disease. This so-called "disarmed" plasmid, called *D-Ti*, supplies all the genes needed to transfer the T-DNA from the cloning vector. The cloned DNA and the kanamycin resistance marker of the vector can be mobilized by D-Ti and transferred into a plant cell (Figure 31.13). Following recombination

● **Figure 31.13 Production of transgenic plants using a binary vector system in *Agrobacterium tumefaciens*.** (a) Generalized plant cloning vector containing ends of T-DNA (in red), foreign DNA (in yellow), origin of replication elements for both *E. coli* and *A. tumefaciens* (in purple), and spectinomycin and kanamycin resistance markers. The kanamycin resistance marker can be selected for in plants. (b) The vector can be put into cells of *E. coli* for cloning purposes and then transferred to *A. tumefaciens* by conjugation. (c) The resident Ti plasmid used for transferring the vector to the plant (D-Ti) is itself genetically engineered to remove key pathogenesis genes. (d) However, D-Ti can mobilize the T-DNA region of the vector for transfer to plant cells grown in tissue culture. From the recombinant cell, whole plants can be regenerated. Details of the transfer of DNA from *Agrobacterium tumefaciens* to plants can be found in Section 19.21 and Figures 19.55–19.57.

with a host chromosome, the foreign DNA can be expressed and can confer new properties on the plant.

With the use of *Agrobacterium tumefaciens*, a number of transgenic plants have been produced. Most successes have come with broadleaf crop plants (dicots) such as tomato, potato, tobacco, soybean, alfalfa, and cotton. *A. tumefaciens* has also been used to produce transgenic trees, such as walnut and apple. Transgenic crop plants from the grass family (monocots) have been more difficult to generate using *A. tumefaciens*, but other methods of introducing DNA, such as the particle gun (see Section 31.2), have been used successfully here.

Herbicide and Insect Resistance

Major areas targeted for genetic improvement in plants include herbicide, insect, and microbial disease resistance, as well as improved product quality. More than a thousand different field trials on more than 30 different plant species have been carried out in recent years. The first genetically modified (GM) crop to be grown commercially was tobacco grown in China in 1992. By the year 2000, over 100 million acres (44 million hectares) of GM crops were planted worldwide. Of these, 58% were in soybeans, 23% were in corn, 12% were in cotton, and 6% were in canola. Almost all the GM soybeans and canola planted were herbicide resistant, whereas corn and cotton were herbicide resistant or insect resistant, or both.

Herbicide resistance is genetically engineering into a crop plant so that it will not be killed by the toxic chemical. Many herbicides inhibit a key plant enzyme or protein necessary for growth. For example, the herbicide *glyphosate* kills plants by inhibiting the activity of an enzyme necessary for making aromatic amino acids. Such a herbicide kills both weeds and crop plants and thus must be used as a "preemergence herbicide," that is, before the crop plants emerge from the ground. However, some bacteria contain a form of the enzyme that is naturally resistant to glyphosate. A gene encoding a resistant enzyme from *Agrobacterium* has been cloned, modified for expression in plants, and transferred into important crop plants, such as soybeans. When sprayed with glyphosate, plants containing the bacterial gene grow as well as unsprayed control plants (Figure 31.14●). Herbicide resistant soybeans are now widely planted in the United States.

Genetic engineering has also been used to protect plants from virus infection. For example, it has been discovered that transgenic plants that express the coat protein gene of a virus become resistant to infection by that virus. Although the mechanism of resistance is unknown, the presence of viral coat protein in plant cells apparently interferes with the uncoating of viral particles containing that coat protein, and this interrupts the virus replication cycle.

● **Figure 31.14   Transgenic plants: Herbicide resistance.** The photograph shows a portion of a field of soybeans that has been treated with *Roundup*, a glyphosate–based herbicide manufactured by Monsanto. The plants on the right are normal soybeans; those on the left have been genetically engineered to express glyphosate resistance.

Insect Resistance: Bt Toxin

Insect resistance has also been genetically introduced into plants. One example already in wide use involves the genes encoding the toxic protein of *Bacillus thuringiensis*. This bacterium produces a crystalline protein called *Bt-toxin* (Section 12.20) that is toxic to moth and butterfly larvae. Certain strains of *B. thuringiensis* produce additional proteins toxic to beetle and fly larvae and mosquitoes.

Biotechnologists are using several different approaches to enhance the use of Bt-toxin for pest control in plants. One approach is to develop a single Bt-toxin that is effective against many different insects. This can be done because the protein has separate domains for specificity and toxicity. The toxic domain is highly conserved in all the various Bt-toxins. Genetic engineers can thus make a gene that encodes the toxic domain and one of several different specificity domains, to yield a series of toxins, each best suited for a particular plant or pest situation.

An effective approach to achieve transgene expression and stability is to transfer the gene directly into the plant genome. For example, a natural Bt-toxin gene has been cloned into a plasmid vector under control of a chloroplast rRNA promoter and transferred into tobacco plant chloroplasts by particle bombardment (see Section 31.2). With this methodology, transgenic plants were obtained that expressed this protein at levels that were extremely toxic to larvae from a number of insect species (Figure 31.15●).

Although transgenic Bt-toxin looked at first to be a great success, there have been some reports of insect resistance. Resistance to insecticides and herbicides is a

(a)

(b)

Kevin McBride, Calgene, Inc.

● **Figure 31.15 Transgenic plants: Insect resistance** (a) The results of two different assays to determine the effect of beet armyworm larvae on tobacco leaves from normal plants. (b) The results of similar assays, but using tobacco leaves taken from transgenic plants that express Bt-toxin in their chloroplasts.

common problem in agriculture, and the fact that a product has been produced by genetic engineering does not protect it from this problem. This emphasizes that many approaches must be used for pest control in agriculture, and Bt-toxin is just one of many tools being developed by agricultural biotechnologists. Nevertheless, Bt-toxin-resistant crops are still widely planted in the United States.

Other Uses of Plant Biotechnology

Not all genetic engineering is directed toward making plants disease resistant. Genetic engineering can be used in a variety of ways for developing GM plants with desirable consumer-oriented characteristics. For instance, the first GM food grown for sale in the U.S. market was a tomato that underwent delayed spoilage. This benefit increased the shelf life of the vegetable. In addition, transgenic plants can be genetically engineered to produce commercial or pharmaceutical products, as has been done with microorganisms (see Section 31.6) and animals (see Section 31.9). For example, crop plants such as tobacco and tomatoes have been engineered to produce a number of different products, such as the human protein interferon. Transgenic crop plants can also be used

to produce human antibodies efficiently and inexpensively. These antibodies, coined "plantibodies," have potential as anticancer or antivirus drugs, and some are undergoing clinical trials. Plants are useful in producing these types of products because they typically modify proteins correctly and because crop plants can be efficiently grown and harvested in large amounts.

Crop plants are also being developed for the production of vaccines. For instance, a recombinant tobacco mosaic virus (∞Section 16.7) has been engineered whose coat contains antigens of *Plasmodium vivax*, the organism that causes malaria (∞Section 27.5). This recombinant virus could be used to produce a malaria vaccine in large amounts and at low cost by simply harvesting infected tobacco or tomato plants grown in fields. Another interesting approach is to produce a vaccine in an edible plant product. Such *edible vaccines* now under development could immunize humans against diseases caused by enteric bacteria, including cholera and diarrhea (∞Section 28.5).

Although public acceptance of GM crops remains high in the United States, there have been some concerns over the contamination of human food with GM corn, so far approved only for animal food. In Europe there has been considerable public concern over GM organisms, and hence several EU countries have imposed bans on the growth or importation of GM crops. Some countries that still accept GM foods have introduced legislation to require all GM foods to be labeled as such. What these practices show is that even good science and potential commercial success does not always ensure public acceptance of an agricultural product.

Because of increased public concern, the acreage planted in many GM crops in the United States has leveled off in recent years. Most concerns center around either the perception of adverse effects of foreign genes on humans or domesticated animals, or the potential "escape" of transgenes from transgenic plants into native plants. At present, supporting evidence for either of these scenarios is scant. However, concerns remain and have served to control the speed at which transgenic plants enter the marketplace.

 31.10 Concept Check

Genetic engineering is being used to make plants resistant to disease, to improve product quality, and to use crop plants as a source of recombinant proteins and even vaccines. One commonly used cloning vector for plants is the Ti plasmid of the bacterium *Agrobacterium tumefaciens*. This plasmid can transfer DNA into plant cells. Commercial plants whose genomes have been modified using *in vitro* genetic techniques are called *genetically modified (GM)* organisms.

◆ Transfer of DNA to plants by the Ti plasmid most resembles what form of bacterial gene transfer (∞Chapter 10)?

◆ What is a *transgenic plant*?

REVIEW QUESTIONS

1. Genetic engineering depends on many different *in vitro* genetic techniques. Describe the techniques used in molecular cloning. Include in your description the terms *cloning vector*, *restriction endonuclease*, and *transformation* (∞Section 31.1).

2. Describe two prokaryotic cloning hosts and the beneficial and detrimental features of each (∞Section 31.2).

3. How could you detect a colony containing a cloned gene if you already knew the sequence of the gene? If you didn't know the gene sequence but had available purified protein encoded by the gene (∞Section 31.3)?

4. Describe the similarities and differences between *expression* vectors, *shuttle* vectors, and *integrating* vectors (∞ Section 31.4).

5. How has bacteriophage T7 been used in expressing foreign genes in *Escherichia coli* and what desirable features does this regulatory system possess (∞Section 31.4)?

6. Explain why the use of a *regulatory switch* is desirable for the large-scale production of a protein (∞Section 31.5).

7. What product was the first commercial "success story" in commercial biotechnology? How was the gene for this product obtained (∞Section 31.6)?

8. What is a *subunit vaccine* and why are subunit vaccines considered a safer way of conferring immunity to viral pathogens than attenuated virus vaccines (∞Section 31.7)?

9. How has genetic engineering benefited the treatment of *cystic fibrosis* (∞Section 31.8)?

10. How do transgenic plants and animals differ from plants and animals modified by conventional breeding techniques (∞Section 31.9)?

11. What is the Ti plasmid and how has it been of use in genetic engineering (∞Section 31.10)?

APPLICATION QUESTIONS

1. Suppose you are given the task of constructing a plasmid expression vector suitable for molecular cloning in an organism of industrial interest. List the characteristics such a plasmid should have. List the steps you would use to develop such a plasmid.

2. Suppose you have just determined the DNA base sequence for an especially strong promoter in *Escherichia coli* and you are interested in incorporating this sequence into an expression vector. Describe the steps you would use. What precautions would be necessary to be sure that this promoter actually works as expected in its new location?

3. You have just discovered a protein in mice that may be an effective cure for cancer, but it is present only in extremely small amounts. Describe the steps you would use to obtain production of this protein. Which host would you want to *clone* the gene into and why? Which host would you want to use to *express* the protein in and why?

4. Gene therapy is used to treat people who have a genetic disease and, if successful, will cure them. However, such people will still be able to pass on the genetic disease to their offspring. Explain. Why do you believe this might be an area of research that is not attracting as much attention as treatment of the individual?

The information in Appendix 1 is intended to help calculate changes in free energy accompanying chemical reactions carried out by microorganisms. It begins with definitions of the terms required to make such calculations and proceeds to show how knowledge of redox state, atomic and charge balance, and other factors are necessary to calculate free-energy problems successfully.

I. Definitions

1. ΔG^0 = standard free-energy change of the reaction at 1 atm pressure and 1 M concentrations; ΔG = free-energy change under the conditions specified; $\Delta G^{0\prime}$ = free-energy change under standard conditions at pH 7.

2. Calculation of ΔG^0 for a chemical reaction from the free energy of formation, $G_f{}^0$, of products and reactants:

   $$\Delta G^0 = \sum \Delta G_f{}^0(\text{products}) - \sum \Delta G_f{}^0(\text{reactants})$$

   That is, sum the $\Delta G_f{}^0$ of products, sum the $\Delta G_f{}^0$ of reactants, and subtract the latter from the former.

3. For energy-yielding reactions involving H^+, converting from standard conditions (pH 0) to cellular conditions (pH 7):

   $$\Delta G^{0\prime} = \Delta G^0 + m\Delta G_f{}^0(H^+)$$

   where m is the net number of protons in the reaction (m is negative when more protons are consumed than formed) and $\Delta G_f{}^0(H^+)$ is the free energy of formation of a proton at pH 7 (-39.83 kJ) at 25°C.

4. Effect of concentrations on ΔG: With soluble substrates, the concentration ratios of products formed to exogenous substrates used are generally equal to or greater than 10^{-2} at the beginning of growth and equal to or less than 10^{-2} at the end of growth. From the relation between ΔG and the equilibrium constant (see item 8), it can be calculated that ΔG for the free-energy yield in practical situations differs from the free-energy yield under standard conditions by at most 11.7 kJ, a rather small amount, and so for a first approximation, standard free-energy yields can be used in most situations. However, with H_2 as a product, H_2-consuming bacteria present may keep the concentration of H_2 so low that the free-energy yield is significantly affected. Thus, in the fermentation of ethanol to acetate and H_2 by syntrophic bacteria $(C_2H_5OH + H_2O \longrightarrow C_2H_3O_2{}^- + 2 H_2 + H^+)$, the $\Delta G^{0\prime}$ at 1 atm H_2 is $+9.68$ kJ, but at 10^{-4} atm H_2 it is -36.03 kJ. With H_2-consuming bacteria present, therefore, the ethanol fermentation becomes exergonic. (See also item 9.)

5. Reduction potentials: by convention, electrode equations are written in the direction, oxidant + $ne^- \longrightarrow$ reductant (that is, as reductions), where n is the number of electrons transferred. The standard potential (E_0) of the hydrogen electrode, $2 H^+ + 2 e^- \longrightarrow H_2$ is set by definition at 0.0 V at 1 atm pressure of H_2 gas and 1.0 M H^+, at 25°C. $E_0{}'$ is the standard reduction potential at pH 7. See also Table A1.2.

6. Relation of free energy to reduction potential:

   $$\Delta G^{0\prime} = -nF\Delta E_0{}'$$

where n is the number of electrons transferred, F is the Faraday constant (96.48 kJ/V), and $\Delta E_0{}'$ is the $E_0{}'$ of the electron-*accepting* couple minus the $E_0{}'$ of the electron-*donating* couple.

7. Equilibrium constant, K. For the generalized reaction $aA + bB \rightleftharpoons cC + dD$,

   $$K = \frac{[C]^c[D]^d}{[A]^a[B]^b}$$

   where A, B, C, and D represent reactants and products; a, b, c, and d represent number of molecules of each; and brackets indicate concentrations. This is true only when the chemical system is in equilibrium.

8. Relation of equilibrium constant, K, to free-energy change. At constant temperature, pressure, and pH,

   $$\Delta G = \Delta G^{0\prime} + RT \ln K$$

   where R is a constant (8.29 J/mol/°K) and T is the absolute temperature (in °K).

9. Two substances can react in a redox reaction even if the standard potentials are unfavorable, provided that the concentrations are appropriate.

   Assume that normally the reduced form of A would donate electrons to the oxidized form of B. However, if the concentration of the reduced form of A was low and the concentration of the reduced form of B was high, it would be possible for the reduced form of B to donate electrons to the oxidized form of A. Thus, the reaction would proceed in the direction opposite that predicted from standard potentials. A practical example of this is the utilization of H^+ as an electron acceptor to produce H_2. Normally, H_2 production in fermentative bacteria is not extensive because H^+ is a poor electron acceptor; the $E_0{}'$ of the $2 H^+/H_2$ pair is -0.41 V. However, if the concentration of H_2 is kept low by continually removing it (a process done by methanogenic prokaryotes, which use $H_2 + CO_2$ to produce methane, CH_4, or by many other anaerobes capable of consuming H_2 anaerobically), the potential will be more positive and then H^+ will be a suitable electron acceptor.

II. Oxidation State or Number

1. The oxidation state of an element in an elementary substance (for example, H_2, O_2) is zero.
2. The oxidation state of the ion of an element is equal to its charge (for example, $Na^+ = +1$, $Fe^{3+} = +3$, $O^{2-} = -2$).
3. The sum of oxidation numbers of all atoms in a neutral molecule is zero. Thus, H_2O is neutral because it has two H at $+1$ each and one O at -2.
4. In an ion, the sum of oxidation numbers of all atoms is equal to the charge on that ion. Thus, in the OH^- ion, $O(-2) + H(+1) = -1$.
5. In compounds, the oxidation state of O is virtually always -2 and that of H is $+1$.
6. In simple carbon compounds, the oxidation state of C can be calculated by adding up the H and O atoms present and using the oxidation states of these elements as given in item 5,

because in a neutral compound the sum of all oxidation numbers must be zero. Thus, the oxidation state of carbon in methane, CH_4, is -4 (4 H at $+1$ each $= +4$); in carbon dioxide, CO_2, the oxidation state of carbon is $+4$ (2 O at -2 each $= -4$).

7. In organic compounds with more than one C atom, it may not be possible to assign a specific oxidation number to each C atom, but it is still useful to calculate the oxidation state of the compound as a whole. The same conventions are used. Thus, the oxidation state of carbon in glucose, $C_6H_{12}O_6$, is zero (12 H at $+1 = 12$; 6 O at $-2 = -12$) and the oxidation state of carbon in ethanol, C_2H_6O, is -2 each (6 H at $+1 = +6$; one O at -2).

8. In all oxidation-reduction reactions there is a balance between the oxidized and reduced products. To calculate an oxidation-reduction balance, the number of molecules of each product is multiplied by its oxidation state. For instance, in calculating the oxidation-reduction balance for the alcoholic fermentation, there are two molecules of ethanol at $-4 = -8$ and two molecules of CO_2 at $+4 = +8$ so the net balance is zero. When constructing model reactions, it is useful to first calculate redox balances to be certain that the reaction is possible.

III. Calculating Free-Energy Yields for Hypothetical Reactions

Energy yields can be calculated either from free energies of formation of the reactants and products or from differences in reduction potentials of electron-donating and electron-accepting partial reactions.

Calculations from Free Energy

Free energies of formation are given in Table A1.1. The procedure to use for calculating energy yields of reactions follows.

1. *Balancing reactions.* In all cases, it is essential to ascertain that the coupled oxidation-reduction reaction is balanced. Balancing involves three things: (*a*) the *total number of each kind of atom* must be identical on both sides of the equation; (*b*) there must be an *ionic balance* so that when positive and negative ions are added up on the right side of the equation, the total ionic charge (whether positive, negative, or neutral) exactly balances the ionic charge on the left side of the equation; and (*c*) there must be an *oxidation-reduction balance* so that all the electrons removed from one substance are transferred to another substance. In general, when constructing balanced reactions, one proceeds in the reverse of the three steps just listed. Usually, if steps (*c*) and (*b*) have been properly handled, step (*a*) becomes correct automatically.

2. *Examples:* (*a*) What is the balanced reaction for the oxidation of H_2S to SO_4^{2-} with O_2? First, decide how many electrons are involved in the oxidation of H_2S to SO_4^{2-}. This can be most easily calculated from the oxidation states of the compounds, using the rules given previously. Because H has an oxidation state of $+1$, the oxidation state of S in H_2S is -2. Because O has an oxidation state of -2, the oxidation state of S in SO_4^{2-} is $+6$ (because it is an ion, using the rules given in items 4 and 5 of the previous section). Thus, the oxidation of H_2S to SO_4^{2-} involves an *eight-electron transfer* (from -2 to

$+6$). Because each O atom can accept two electrons (the oxidation state of O in O_2 is zero, but in H_2O is -2), this means that two molecules of molecular oxygen, O_2, are required to provide sufficient electron-accepting capacity. Thus, at this point, we know that the reaction requires 1 H_2S and 2 O_2 on the left side of the equation, and 1 SO_4^{2-} on the right side. To achieve an ionic balance, we must have two positive charges on the right side of the equation to balance the two negative charges of SO_4^{2-}. Thus, 2 H^+ must be added to the right side of the equation, making the overall reaction.

$$H_2S + 2 O_2 \longrightarrow SO_4^{2-} + 2 H^+$$

By inspection, it can be seen that this equation is also balanced in terms of the total number of atoms of each kind on each side of the equation.

(*b*) What is the balanced reaction for the oxidation of H_2S to SO_4^{2-} with Fe^{3+} as electron acceptor? We have just ascertained that the oxidation of H_2S to SO_4^{2-} is an eight-electron transfer. Because the reduction of Fe^{3+} to Fe^{2+} is only a one-electron transfer, 8 Fe^{3+} will be required. At this point, the reaction looks like

$$H_2S + 8 Fe^{3+} \longrightarrow 8 Fe^{2+} + SO_4^{2-} \text{ (not balanced)}$$

We note that the ionic balance is incorrect. We have 24 positive charges on the left and 14 positive charges on the right (16+ from Fe, 2− from sulfate). To equalize the charges, we add 10 H^+ on the right. Now our equation looks like

$$H_2S + 8 Fe^{3+} \longrightarrow 8 Fe^{2+} + 10 H^- + SO_4^{2-}$$
$$\text{(not balanced)}$$

To provide the necessary hydrogen for the H^+ and oxygen for the sulfate, we add 4 H_2O to the left and find that the equation is now balanced:

$$H_2S + 4 H_2O + 8 Fe^{3+} \longrightarrow 8 Fe^{2+} + 10 H^+ + SO_4^{2-}$$
$$\text{(balanced)}$$

In general, in microbiological reactions, ionic balance can be achieved by adding H^+ or OH^- to the left or right side of the equation, and because all reactions take place in an aqueous medium, H_2O molecules can be added where needed. Whether H^+ or OH^- is added generally depends on whether the reaction is taking place under acidic or alkaline conditions.

3. *Calculation of energy yield for balanced equations from free energies of formation.* Once an equation has been balanced, the free-energy yield can be calculated by inserting the values for the free energy of formation of each reactant and product from Table A1.1 and using the formula in item 2 of the first section of this appendix.

For instance, for the equation

$$H_2S + 2 O_2 \longrightarrow SO_4^{2-} + 2 H^+$$
$$G_f^0 \text{ values} \longrightarrow (-27.87) + (0)(-744.6) + 2(-39.83)$$
$$\text{(assuming pH 7)}$$

$$\Delta G^{0\prime} = -796.39 \text{ kJ}$$

The G_f^0 values for the products (right side of equation) are summed and subtracted from the G_f^0 values for the reactants (left side of equation), taking care to ensure that the arithmetic signs are correct. From the data in Table A1.1, a wide variety of free-energy yields for reactions of microbiological interest can be calculated.

Calculation of Free-Energy Yield from Reduction Potential

Reduction potentials of some important redox pairs are given in Table A1.2. The amount of energy that can be released from two half reactions can be calculated from the *differences* in reduction potentials of the two reactions and from the number of electrons transferred. The further apart the two half reactions are, and the greater the number of electrons transferred, the more energy released.

The conversion of potential difference to free energy is given by the formula $\Delta G^{0\prime} = -nF\Delta E_0{}'$, where n is the number of electrons, F is the Faraday constant (96.48 kJ/V), and $\Delta E_0{}'$ is the difference in potentials. Thus, the $2\,H^+/H_2$ couple has a potential of -0.41 V and the $\frac{1}{2}O_2/H_2O$ pair has a potential of $+0.82$ V, and so the potential difference is 1.23 V, which (be-cause two electrons are involved) is equivalent to a free-energy yield (ΔG^0) of -237.34 kJ. On the other hand, the potential difference between the $2\,H^+/H_2$ and the $NO_3{}^-/NO_2{}^-$ reactions is less, 0.84 V, which is equivalent to a free-energy yield of -162.08 kJ.

Because many biochemical reactions are two-electron transfers, it is often useful to give energy yields for two-electron reactions, even if more electrons are involved. Thus, the $SO_4{}^{2-}/H_2$ redox pair involves eight electrons, and complete reduction of $SO_4{}^{2-}$ with H_2 requires $4\,H_2$ (equivalent to eight electrons). From the reduction potential difference between $2\,H^+/H_2$ and $SO_4{}^{2-}/H_2S$ (0.19 V), a free-energy yield of -146.64 kJ is calculated, or -36.66 kJ per two electrons. By convention, reduction potentials are given for conditions in which equal concentrations of oxidized and reduced forms are present. In actual practice, the concentrations of these two forms may be quite different. As discussed earlier in this appendix (item 9, Section I), it is possible to couple half reactions even if the potential difference is unfavorable, providing the concentrations of the reacting species are appropriate.

Table A1.1	Free energies of formation ($G_f{}^0$) for some substances (kJ/mol)[a]			
Carbon compound	**Carbon compound**	**Metal**	**Nonmetal**	**Nitrogen compound**
CO, -137.34	Glutamine, -529.7	Cu^+, $+50.28$	H_2, 0	N_2, 0
CO_2, -394.4	Glyceraldehyde, -437.65	Cu^{2+}, $+64.94$	H^+, 0 at pH 0;	NO, $+86.57$
CH_4, -50.75	Glycerate, -658.1	CuS, -49.02	-39.83 at pH 7	NO_2, $+51.95$
H_2CO_3, -623.16	Glycerol, -488.52	Fe^{2+}, -78.87	(-5.69 per pH unit)	$NO_2{}^-$, -37.2
$HCO_3{}^-$, -586.85	Glycine, -314.96	Fe^{3+}, -4.6	O_2, 0	$NO_3{}^-$, -111.34
$CO_3{}^{2-}$, -527.90	Glycolate, -530.95	$FeCO_3$, -673.23	OH^-, -157.3 at pH 14;	NH_3, -26.57
Acetaldehyde -139.9	Glyoxalate, -468.6	FeS_2, -150.84	-198.76 at pH 7;	$NH_4{}^+$, -79.37
Acetate, -369.41	Guanine, $+46.99$	$FeSO_4$, -829.62	-237.57 at pH 0	N_2O, $+104.18$
Acetone, -161.17	α-Ketoglutarate, -797.55	PbS, -92.59	H_2O, -237.17	
Alanine, -371.54	Lactate, -517.81	Mn^{2+}, -227.93	H_2O_2, -134.1	
Arginine, -240.2	Lactose, -1515.24	Mn^{3+}, -82.12	$PO_4{}^{3-}$, -1026.55	
Aspartate, -700.4	Malate, -845.08	$MnO_4{}^-$, -506.57	Se^0, 0	
Benzene, $+124.5$	Mannitol, -942.61	MnO_2, -456.71	H_2Se, -77.09	
Benzoic acid, -245.6	Methanol, -175.39	$MnSO_4$, -955.32	$SeO_4{}^{2-}$, -439.95	
n-Butanol, -171.84	Methionine, -502.92	HgS, -49.02	S^0, 0	
Butyrate, -352.63	Methylamine, -40.0	MoS_2, -225.42	$SO_3{}^{2-}$, -486.6	
Caproate, -335.96	Oxalate, -674.04	ZnS, -198.60	$SO_4{}^{2-}$, -744.6	
Citrate, -1168.34	Phenol, -47.6		$S_2O_3{}^{2-}$, -513.4	
o-Cresol -37.1	*n*-Propanol, -175.81		H_2S, -27.87	
Crotonate, -277.4	Propionate, -361.08		HS^-, $+12.05$	
Cysteine, -339.8	Pyruvate, -474.63		S^{2-}, $+85.8$	
Dimethylamine, -3.3	Ribose, -757.3			
Ethanol, -181.75	Succinate, -690.23			
Formaldehyde, -130.54	Sucrose, -370.90			
Formate, -351.04	Toluene, $+114.22$			
Fructose, -915.38	Trimethylamine, -37.2			
Fumarate, -604.21	Tryptophan, -112.6			
Gluconate, -1128.3	Urea, -203.76			
Glucose, -917.22	Valerate, -344.34			
Glutamate, -699.6				

[a] Values for free energy of formation of various compounds can be found in Dean, J. A. 1973. *Lange's Handbook of Chemistry*, 11th edition. McGraw-Hill, New York; Garrels, R. M., and C. L. Christ. 1965. *Solutions, Minerals, and Equilibria*. Harper and Row, New York; Burton, K. 1957. In Krebs, H. A., and H. L. Komberg. Energy transformations in living matter, *Ergebnisse der Physiologie* (appendix): Springer-Verlag, Berlin; and Thauer, R. K., K. Jungermann, and H. Decker. 1977. Energy conservation in anaerobic chemotrophic bacteria. *Bacteriol Rev* 41:100–180.

Table A1.2	Microbiologically important reduction potentials[a]

Redox pair	$E_0{'}$ (V)
SO_4^{2-}/HSO_3^-	−0.52
CO_2/formate	−0.43
$2\,H^+/H_2$	−0.41
$S_2O_3^{2-}/HS^- + HSO_3^-$	−0.40
Ferredoxin ox/red	−0.39
Flavodoxin ox/red[b]	−0.37
$NAD^+/NADH$	−0.32
Cytochrome c_3 ox/red	−0.29
CO_2/acetate$^-$	−0.29
S^0/HS^-	−0.27
CO_2/CH_4	−0.24
FAD/FADH	−0.22
SO_4^{2-}/HS^-	−0.217
Acetaldehyde/ethanol	−0.197
Pyruvate$^-$/lactate$^-$	−0.19
FMN/FMNH	−0.19
Dihydroxyacetone phosphate/glycerolphosphate	−0.19
$HSO_3^-/S_3O_6^{2-}$	−0.17
Flavodoxin ox/red[b]	−0.12
HSO_3^-/HS^-	−0.116
Menaquinone ox/red	−0.075
APS/AMP + HSO_3^-	−0.060
Rubredoxin ox/red	−0.057
Acrylyl-CoA/propionyl-CoA	−0.015
Glycine/acetate$^-$ + NH_4^+	−0.010
$S_4O_6^{2-}/S_2O_3^{2-}$	+0.024
Fumarate^{2-}/succinate^{2-}	+0.033
Cytochrome b ox/red	+0.035
Ubiquinone ox/red	+0.113
AsO_4^{3-}/AsO_3^{3-}	+0.139
Dimethyl sulfoxide (DMSO)/dimethylsulfide (DMS)	+0.16
$Fe(OH)_3 + HCO_3^-/FeCO_3$	+0.20
$S_3O_6^{2-}/S_2O_3^{2-} + HSO_3^-$	+0.225
Cytochrome c_1 ox/red	+0.23
NO_2^-/NO	+0.36
Cytochrome a_3 ox/red	+0.385
Chlorobenzoate$^-$/benzoate$^-$ + HCl	+0.297
NO_3^-/NO_2^-	+0.43
SeO_4^{2-}/SeO_3^{2-}	+0.475
Fe^{3+}/Fe^{2+}	+0.77
Mn^{4+}/Mn^{2+}	+0.798
O_2/H_2O	+0.82
ClO_3^-/Cl^-	+1.03
NO/N_2O	+1.18
N_2O/N_2	+1.36

[a] Data from Thauer, R. K., K. Jungermann, and K. Decker, 1977. Energy conservation in anaerobic chemotrophic bacteria. *Bacteriol. Rev.* 41:100–180.
[b] Separate potentials are given for each electron transfer in this potentially two-electron transfer.

LIST OF GENERA AND HIGHER ORDER TAXA[a]

Domain *Archaea*
Phylum AI. *Crenarchaeota*
  Class I. *Thermoprotei*
    Order I. *Thermoproteales*
      Family I. *Thermoproteaceae*
        Genus I. *Thermoproteus*
        Genus II. *Caldivirga*
        Genus III. *Pyrobaculum*
        Genus IV. *Thermocladium*
        Genus V. *Vulcanisaeta*
      Family II. *Thermofilaceae*
        Genus I. *Thermofilum*
    Order II. *Caldisphaerales*
      Family I. *Caldisphaeraceae*
        Genus I. *Caldisphaera*
    Order III. *Desulfurococcales*
      Family I. *Desulfurococcaceae*
        Genus I. *Desulfurococcus*
        Genus II. *Acidilobus*
        Genus III. *Aeropyrum*
        Genus IV. *Ignicoccus*
        Genus V. *Staphylothermus*
        Genus VI. *Stetteria*
        Genus VII. *Sulfophobococcus*
        Genus VIII. *Thermodiscus*
        Genus IX. *Thermosphaera*
      Family II. *Pyrodictiaceae*
        Genus I. *Pyrodictium*
        Genus II. *Hyperthermus*
        Genus III. *Pyrolobus*
    Order IV. *Sulfolobales*
      Family I. *Sulfolobaceae*
        Genus I. *Sulfolobus*
        Genus II. *Acidianus*
        Genus III. *Metallosphaera*
        Genus IV. *Stygiolobus*
        Genus V. *Sulfurisphaera*
        Genus VI. *Sulfurococcus*
Phylum AII. *Euryarchaeota*
  Class I. *Methanobacteria*
    Order I. *Methanobacteriales*
      Family I. *Methanobacteriaceae*
        Genus I. *Methanobacterium*
        Genus II. *Methanobrevibacter*
        Genus III. *Methanosphaera*
        Genus IV. *Methanothermobacter*
      Family II. *Methanothermaceae*
        Genus I. *Methanothermus*
  Class II. *Methanococci*
    Order I. *Methanococcales*
      Family I. *Methanococcaceae*
        Genus I. *Methanococcus*
        Genus II. *Methanothermococcus*
      Family II. *Methanocaldococcaceae*
        Genus I. *Methanocaldococcus*
        Genus II. *Methanotorris*
  Class III. *Methanomicrobia*
    Order I. *Methanomicrobiales*
      Family I. *Methanomicrobiaceae*
        Genus I. *Methanomicrobium*
        Genus II. *Methanoculleus*
        Genus III. *Methanofollis*
        Genus IV. *Methanogenium*
        Genus V. *Methanolacinia*
        Genus VI. *Methanoplanus*
      Family II. *Methanocorpusculaceae*
        Genus I. *Methanocorpusculum*

      Family III. *Methanospirillaceae*
        Genus I. *Methanospirillum*
      Genera incertae sedis[b]
        Genus I. *Methanocalculus*
    Order II. *Methanosarcinales*
      Family I. *Methanosarcinaceae*
        Genus I. *Methanosarcina*
        Genus II. *Methanococcoides*
        Genus III. *Methanohalobium*
        Genus IV. *Methanohalophilus*
        Genus V. *Methanolobus*
        Genus VI. *Methanomethylovorans*
        Genus VII. *Methanimicrococcus*
        Genus VIII. *Methanosalsum*
      Family II. *Methanosaetaceae*
        Genus I. *Methanosaeta*
  Class IV. *Halobacteria*
    Order I. *Halobacteriales*
      Family I. *Halobacteriaceae*
        Genus I. *Halobacterium*
        Genus II. *Haloarcula*
        Genus III. *Halobaculum*
        Genus IV. *Halobiforma*
        Genus V. *Halococcus*
        Genus VI. *Haloferax*
        Genus VII. *Halogeometricum*
        Genus VIII. *Halomicrobium*
        Genus IX. *Halorhabdus*
        Genus X. *Halorubrum*
        Genus XI. *Halosimplex*
        Genus XII. *Haloterrigena*
        Genus XIII. *Natrialba*
        Genus XIV. *Natrinema*
        Genus XV. *Natronobacterium*
        Genus XVI. *Natronococcus*
        Genus XVII. *Natronomonas*
        Genus XVIII. *Natronorubrum*
  Class V. *Thermoplasmata*
    Order I. *Thermoplasmatales*
      Family I. *Thermoplasmataceae*
        Genus I. *Thermoplasma*
      Family II. *Picrophilaceae*
        Genus I. *Picrophilus*
      Family III. *Ferroplasmaceae*
        Genus I. *Ferroplasma*
  Class VI. *Thermococci*
    Order I. *Thermococcales*
      Family I. *Thermococcaceae*
        Genus I. *Thermococcus*
        Genus II. *Palaeococcus*
        Genus III. *Pyrococcus*
  Class VII. *Archaeoglobi*
    Order I. *Archaeoglobales*
      Family I. *Archaeoglobaceae*
        Genus I. *Archaeoglobus*
        Genus II. *Ferroglobus*
        Genus III. *Geoglobus*
  Class VIII. *Methanopyri*
    Order I. *Methanopyrales*
      Family I. *Methanopyraceae*
        Genus I. *Methanopyrus*

Domain *Bacteria*
Phylum BI. *Aquificae*
  Class I. *Aquificae*
    Order I. *Aquificales*
      Family I. *Aquificaceae*
        Genus I. *Aquifex*
        Genus II. *Calderobacterium*
        Genus III. *Hydrogenobaculum*

[a]The list of genera and higher order taxa shown here are the organisms recognized as of 2005. Genera or higher-order taxa in quotation marks are recognized taxa whose names have not yet been validated. Because bacterial taxonomy is a work in progress, updates to the list shown here occur as new genera and species are described and as new data support new taxonomic arrangements. For further discussion on bacterial taxonomy, see Sections 11.10–11.13.
[b]Taxa of uncertain affiliation.

Genus IV. *Hydrogenobacter*
Genus V. *Hydrogenothermus*
Genus VI. *Persephonella*
Genus VII. *Sulfurihydrogenibium*
Genus VIII. *Thermocrinis*
Genera incertae sedis[b]
Genus I. *Balnearium*
Genus II. *Desulfurobacterium*
Genus III. *Thermovibrio*
Phylum BII. *Thermotogae*
Class I. *Thermotogae*
Order I. *Thermotogales*
Family I. *Thermotogaceae*
Genus I. *Thermotoga*
Genus II. *Fervidobacterium*
Genus III. *Geotoga*
Genus IV. *Marinitoga*
Genus V. *Petrotoga*
Genus VI. *Thermosipho*
Phylum BIII. *Thermodesulfobacteria*
Class I. *Thermodesulfobacteria*
Order I. *Thermodesulfobacteriales*
Family I. *Thermodesulfobacteriaceae*
Genus I. *Thermodesulfobacterium*
Genus II. *Thermodesulfatator*
Phylum BIV. *Deinococcus-Thermus*
Class I. *Deinococci*
Order I. *Deinococcales*
Family I. *Deinococcaceae*
Genus I. *Deinococcus*
Order II. *Thermales*
Family I. *Thermaceae*
Genus I. *Thermus*
Genus II. *Marinithermus*
Genus III. *Meiothermus*
Genus IV. *Oceanithermus*
Genus V. *Vulcanithermus*
Phylum BV. *Chrysiogenetes*
Class I. *Chrysiogenetes*
Order I. *Chrysiogenales*
Family I. *Chrysiogenaceae*
Genus I. *Chrysiogenes*
Phylum BVI. *Chloroflexi*
Class I. *Chloroflexi*
Order I. *Chloroflexales*
Family I. *Chloroflexaceae*
Genus I. *Chloroflexus*
Genus II. *Chloronema*
Genus III. *Heliothrix*
Genus IV. *Roseiflexus*
Family II. *Oscillochloridaceae*
Genus I. *Oscillochloris*
Order II. *Herpetosiphonales*
Family I. *Herpetosiphonaceae*
Genus I. *Herpetosiphon*
Class II. *Anaerolieae*
Order I. *Anaerolinaeles*
Family I. *Anaerolinaceae*
Genus I. *Anaerolinea*
Genus II. *Caldilinea*
Phylum BVII. *Thermomicrobia*
Class I. *Thermomicrobia*
Order I. *Thermomicrobiales*
Family I. *Thermomicrobiaceae*
Genus I. *Thermomicrobium*
Phylum BVIII. *Nitrospirae*
Class I. *Nitrospira*
Order I. *Nitrospirales*
Family I. *Nitrospiraceae*
Genus I. *Nitrospira*
Genus II. *Leptospirillum*
Genus III. *Magnetobacterium*
Genus IV. *Thermodesulfovibrio*
Phylum BIX. *Deferribacteres*

Class I. *Deferribacteres*
Order I. *Deferribacterales*
Family I. *Deferribacteraceae*
Genus I. *Deferribacter*
Genus II. *Denitrovibrio*
Genus III. *Flexistipes*
Genus IV. *Geovibrio*
Genera incertae sedis[b]
Genus I. *Synergistes*
Genus II. *Caldithrix*
Phylum BX. *Cyanobacteria*
Class I. *Cyanobacteria*
Subsection I. *Subsection 1*
Family I. Family 1.1
Form genus I. *Chamaesiphon*[c]
Form genus II. *Chroococcus*
Form genus III. *Cyanobacterium*
Form genus IV. *Cyanobium*
Form genus V. *Cyanothece*
Form genus VI. *Dactylococcopsis*
Form genus VII. *Gloeobacter*
Form genus VIII. *Gloeocapsa*
Form genus IX. *Gloeothece*
Form genus X. *Microcystis*
Form genus XI. *Prochlorococcus*
Form genus XII. *Prochloron*
Form genus XIII. *Synechococcus*
Form genus XIV. *Synechocystis*
Subsection II. *Subsection 2*
Family I. Family 2.1
Form genus I. *Cyanocystis*
Form genus II. *Dermocarpella*
Form genus III. *Stanieria*
Form genus IV. *Xenococcus*
Family II. Family 2.2
Form genus I. *Chroococcidiopsis*
Form genus II. *Myxosarcina*
Form genus III. *Pleurocapsa*
Subsection III. *Subsection 3*
Family I. Family 3.1
Form genus I. *Arthrospira*
Form genus II. *Borzia*
Form genus III. *Crinalium*
Form genus IV. *Geitlerinema*
Genus V. *Halospirulina*
Form genus VI. *Leptolyngbya*
Form genus VII. *Limnothrix*
Form genus VIII. *Lyngbya*
Form genus IX. *Microcoleus*
Form genus X. *Oscillatoria*
Form genus XI. *Planktothrix*
Form genus XII. *Prochlorothrix*
Form genus XIII. *Pseudanabaena*
Form genus XIV. *Spirulina*
Form genus XV. *Starria*
Form genus XVI. *Symploca*
Genus XVII. *Trichodesmium*
Form genus XVIII. *Tychonema*
Subsection IV. *Subsection 4*
Family I. Family 4.1
Form genus I. *Anabaena*
Form genus II. *Anabaenopsis*
Form genus III. *Aphanizomenon*
Form genus IV. *Cyanospira*
Form genus V. *Cylindrospermopsis*
Form genus VI. *Cylindrospermum*
Form genus VII. *Nodularia*
Form genus VIII. *Nostoc*
Form genus IX. *Scytonema*
Family II. Family 4.2
Form genus I. *Calothrix*
Form genus II. *Rivularia*
Form genus III. *Tolypothrix*
Subsection V. *Subsection 5*

---

[c]The taxonomic position of the cyanobacteria in *Bergey's Manual* is left open. The term "form genus" refers to a group of cyanobacteria with very characteristic morphology found worldwide. However, not all isolates of such a type may actually fit into the same genus. In some cases, pure cultures of form genera are not available.

Family I. Subsection 5.1
    Form genus I. *Chlorogloeopsis*
    Form genus II. *Fischerella*
    Form genus III. *Geitleria*
    Form genus IV. *Iyengariella*
    Form genus V. *Nostochopsis*
    Form genus VI. *Stigonema*

Phylum BXI. *Chlorobi*
Class I. *Chlorobia*
    Order I. *Chlorobiales*
      Family I. *Chlorobiaceae*
        Genus I. *Chlorobium*
        Genus II. *Ancalochloris*
        Genus III. *Chlorobaculum*
        Genus IV. *Chloroherpeton*
        Genus V. *Pelodictyon*
        Genus VI. *Prosthecochloris*

Phylum BXII. *Proteobacteria*
Class I. *Alphaproteobacteria*
    Order I. *Rhodospirillales*
      Family I. *Rhodospirillaceae*
        Genus I. *Rhodospirillum*
        Genus II. *Azospirillum*
        Genus III. *Inquilinus*
        Genus IV. *Magnetospirillum*
        Genus V. *Phaeospirillum*
        Genus VI. *Rhodocista*
        Genus VII. *Rhodospira*
        Genus VIII. *Rhodovibrio*
        Genus IX. *Roseospira*
        Genus X. *Skermanella*
        Genus XI. *Thalassospira*
        Genus XII. *Tistrella*
      Family II. *Acetobacteraceae*
        Genus I. *Acetobacter*
        Genus II. *Acidiphilium*
        Genus III. *Acidisphaera*
        Genus IV. *Acidocella*
        Genus V. *Acidomonas*
        Genus VI. *Asaia*
        Genus VII. *Craurococcus*
        Genus VIII. *Gluconacetobacter*
        Genus IX. *Gluconobacter*
        Genus X. *Kozakia*
        Genus XI. *Muricoccus*
        Genus XII. *Paracraurococcus*
        Genus XIII. *Rhodopila*
        Genus XIV. *Roseococcus*
        Genus XV. *Rubritepids*
        Genus XVI. *Stella*
        Genus XVII. *Teichococcus*
        Genus XVIII. *Zavarzinia*
    Order II. *Rickettsiales*
      Family I. *Rickettsiaceae*
        Genus I. *Rickettsia*
        Genus II. *Orientia*
      Family II. *Anaplasmataceae*
        Genus I. *Anaplasma*
        Genus II. *Aegyptianella*
        Genus III. *Cowdria*
        Genus IV. *Ehrlichia*
        Genus V. *Neorickettsia*
        Genus VI. *Wolbachia*
        Genus VII. *Xenohaliotis*
      Family III. *Holosporaceae*
        Genus I. *Holospora*
      Genera incertae sedis[b]
        Genus I. *Caedibacter*
        Genus II. *Lyticum*
        Genus III. *Odyssella*
        Genus IV. *Pseudocaedibacter*
        Genus V. *Symbiotes*
        Genus VI. *Tectibacter*
    Order III. *Rhodobacterales*
      Family I. *Rhodobacteraceae*
        Genus I. *Rhodobacter*

        Genus II. *Ahrensia*
        Genus III. *Albidovulum*
        Genus IV. *Amaricoccus*
        Genus V. *Antarctobacter*
        Genus VI. *Gemmobacter*
        Genus VII. *Hirschia*
        Genus VIII. *Hyphomonas*
        Genus IX. *Jannaschia*
        Genus X. *Ketogulonicigenium*
        Genus XI. *Leisingera*
        Genus XII. *Maricaulis*
        Genus XIII. *Methylarcula*
        Genus XIV. *Oceanicaulis*
        Genus XV. *Octadecabacter*
        Genus XVI. *Pannonibacter*
        Genus XVII. *Paracoccus*
        Genus XVIII. *Pseudorhodobacter*
        Genus XIX. *Rhodobaca*
        Genus XX. *Rhodothalassium*
        Genus XXI. *Rhodovulum*
        Genus XXII. *Roseibium*
        Genus XXIII. *Roseinatronobacter*
        Genus XXIV. *Roseivivax*
        Genus XXV. *Roseobacter*
        Genus XXVI. *Roseovarius*
        Genus XXVII. *Rubrimonas*
        Genus XXVIII. *Ruegeria*
        Genus XXIX. *Sagittula*
        Genus XXX. *Silicibacter*
        Genus XXXI. *Staleya*
        Genus XXXII. *Stappia*
        Genus XXXIII. *Sulfitobacter*
    Order IV. *Sphingomonadales*
      Family I. *Sphingomonadaceae*
        Genus I. *Sphingomonas*
        Genus II. *Blastomonas*
        Genus III. *Erythrobacter*
        Genus IV. *Erythromicrobium*
        Genus V. *Erythromonas*
        Genus VI. *Novosphingobium*
        Genus VII. *Porphyrobacter*
        Genus VIII. *Rhizomonas*
        Genus IX. *Sandaracinobacter*
        Genus X. *Sphingobium*
        Genus XI. *Sphingopyxis*
        Genus XII. *Zymomonas*
    Order V. *Caulobacterales*
      Family I. *Caulobacteraceae*
        Genus I. *Caulobacter*
        Genus II. *Asticcacaulis*
        Genus III. *Brevundimonas*
        Genus IV. *Phenylobacterium*
    Order VI. *Rhizobiales*
      Family I. *Rhizobiaceae*
        Genus I. *Rhizobium*
        Genus II. *Agrobacterium*
        Genus III. *Allorhizobium*
        Genus IV. *Carbophilus*
        Genus V. *Chelatobacter*
        Genus VI. *Ensifer*
        Genus VII. *Sinorhizobium*
      Family II. *Aurantimonadaceae*
        Genus I. *Aurantimonas*
        Genus II. *Fulvimarina*
      Family III. *Bartonellaceae*
        Genus I. *Bartonella*
      Family IV. *Brucellaceae*
        Genus I. *Brucella*
        Genus II. *Mycoplana*
        Genus III. *Ochrobactrum*
      Family V. *Phyllobacteriaceae*
        Genus I. *Phyllobacterium*
        Genus II. *Aminobacter*
        Genus III. *Aquamicrobium*
        Genus IV. *Fluvibacter*
        Genus V. *Candidatus Liberibacter*[d]

[d]In bacterial taxonomy, candidatus status is for organisms known to exist by 16S rRNA gene sequencing and other key properties, but which are not yet in pure culture.

Genus VI. *Mesorhizobium*
Genus VII. *Nitratireductor*
Genus VIII. *Pseudaminobacter*
Family VI. *Methylocystaceae*
Genus I. *Methylocystis*
Genus II. *Albibacter*
Genus III. *Methylopila*
Genus IV. *Methylosinus*
Genus V. *Terasakiella*
Family VII. *Beijerinckiaceae*
Genus I. *Beijerinckia*
Genus II. *Chelatococcus*
Genus III. *Methylocapsa*
Genus IV. *Methylocella*
Family VIII. *Bradyrhizobiaceae*
Genus I. *Bradyrhizobium*
Genus II. *Afipia*
Genus III. *Agromonas*
Genus IV. *Blastobacter*
Genus V. *Bosea*
Genus VI. *Nitrobacter*
Genus VII. *Oligotropha*
Genus VIII. *Rhodoblastus*
Genus IX. *Rhodopseudomonas*
Family IX. *Hyphomicrobiaceae*
Genus I. *Hyphomicrobium*
Genus II. *Ancalomicrobium*
Genus III. *Ancylobacter*
Genus IV. *Angulomicrobium*
Genus V. *Aquabacter*
Genus VI. *Azorhizobium*
Genus VII. *Blastochloris*
Genus VIII. *Devosia*
Genus IX. *Dichotomicrobium*
Genus X. *Filomicrobium*
Genus XI. *Gemmiger*
Genus XII. *Labrys*
Genus XIII. *Methylorhabdus*
Genus XIV. *Pedomicrobium*
Genus XV. *Prosthecomicrobium*
Genus XVI. *Rhodomicrobium*
Genus XVII. *Rhodoplanes*
Genus XVIII. *Seliberia*
Genus XIX. *Starkeya*
Genus XX. *Xanthobacter*
Family X. *Methylobacteriaceae*
Genus I. *Methylobacterium*
Genus II. *Microvirga*
Genus III. *Protomonas*
Genus IV. *Roseomonas*
Family XI. *Rhodobiaceae*
Genus I. *Rhodobium*
Genus II. *Roseospirillum*
Order VII. *Parvularculales*
Family I. *Parvularculaceae*
Genus I. *Parvularcula*
Class II. *Betaproteobacteria*
Order I. *Burkholderiales*
Family I. *Burkholderiaceae*
Genus I. *Burkholderia*
Genus II. *Cupriavidus*
Genus III. *Lautropia*
Genus IV. *Limnobacter*
Genus V. *Pandoraea*
Genus VI. *Paucimonas*
Genus VII. *Polynucleobacter*
Genus VIII. *Ralstonia*
Genus IX. *Thermothrix*
Genus X. *Wautersia*
Family II. *Oxalobacteraceae*
Genus I. *Oxalobacter*
Genus II. *Duganella*
Genus III. *Herbaspirillum*
Genus IV. *Janthinobacterium*
Genus V. *Massilia*
Genus VI. *Oxalicibacterium*
Genus VII. *Telluria*

Family III. *Alcaligenaceae*
Genus I. *Alcaligenes*
Genus II. *Achromobacter*
Genus III. *Bordetella*
Genus IV. *Brackiella*
Genus V. *Derxia*
Genus VI. *Kerstersia*
Genus VII. *Oligella*
Genus VIII. *Pelistega*
Genus IX. *Pigmentiphaga*
Genus X. *Sutterella*
Genus XI. *Taylorella*
Family IV. *Comamonadaceae*
Genus I. *Comamonas*
Genus II. *Acidovorax*
Genus III. *Alicycliphilus*
Genus IV. *Brachymonas*
Genus V. *Caldimonas*
Genus VI. *Delftia*
Genus VII. *Diaphorobacter*
Genus VIII. *Hydrogenophaga*
Genus IX. *Hylemonella*
Genus X. *Lampropedia*
Genus XI. *Macromonas*
Genus XII. *Ottowia*
Genus XIII. *Polaromonas*
Genus XIV. *Ramlibacter*
Genus XV. *Rhodoferax*
Genus XVI. *Variovorax*
Genus XVII. *Xenophilus*
Genera incertae sedis[b]
Genus I. *Aquabacterium*
Genus II. *Ideonella*
Genus III. *Leptothrix*
Genus IV. *Roseateles*
Genus V. *Rubrivivax*
Genus VI. *Schlegelella*
Genus VII. *Sphaerotilus*
Genus VIII. *Tepidimonas*
Genus IX. *Thiomonas*
Genus X. *Xylophilus*
Order II. *Hydrogenophilales*
Family I. *"Hydrogenophilaceae"*
Genus I. *Hydrogenophilus*
Genus II. *Thiobacillus*
Order III. *Methylophilales*
Family I. *Methylophilaceae*
Genus I. *Methylophilus*
Genus II. *Methylobacillus*
Genus III. *Methylovorus*
Order IV. *Neisseriales*
Family I. *Neisseriaceae*
Genus I. *Neisseria*
Genus II. *Alysiella*
Genus III. *Aquaspirillum*
Genus IV. *Chromobacterium*
Genus V. *Eikenella*
Genus VI. *Formivibrio*
Genus VII. *Iodobacter*
Genus VIII. *Kingella*
Genus IX. *Laribacter*
Genus X. *Microvirgula*
Genus XI. *Morococcus*
Genus XII. *Prolinoborus*
Genus XIII. *Simonsiella*
Genus XIV. *Vitreoscilla*
Genus XV. *Vogesella*
Order V. *Nitrosomonadales*
Family I. *Nitrosomonadaceae*
Genus I. *Nitrosomonas*
Genus II. *Nitrosolobus*
Genus III. *Nitrosospira*
Family II. *Spirillaceae*
Genus I. *Spirillum*
Family III. *Gallionellaceae*
Genus I. *Gallionella*
Order VI. *Rhodocyclales*

Family I. *Rhodocyclaceae*
    Genus I. *Rhodocyclus*
    Genus II. *Azoarcus*
    Genus III. *Azonexus*
    Genus IV. *Azospira*
    Genus V. *Azovibrio*
    Genus VI. *Dechloromonas*
    Genus VII. *Dechlorosoma*
    Genus VIII. *Ferribacterium*
    Genus IX. *Propionibacter*
    Genus X. *Propionivibrio*
    Genus XI. *Quadricoccus*
    Genus XII. *Sterolibacterium*
    Genus XIII. *Thauera*
    Genus XIV. *Zoogloea*
Order VII. *Procabacteriales*
  Family I. *Procabacteriaceae*
    Genus I. *Procabacter*

Class III. *Gammaproteobacteria*
Order I. *Chromatiales*
  Family I. *Chromatiaceae*
    Genus I. *Chromatium*
    Genus II. *Allochromatium*
    Genus III. *Amoebobacter*
    Genus IV. *Halochromatium*
    Genus V. *Isochromatium*
    Genus VI. *Lamprobacter*
    Genus VII. *Lamprocystis*
    Genus VIII. *Marichromatium*
    Genus IX. *Nitrosococcus*
    Genus X. *Pfennigia*
    Genus XI. *Rhabdochromatium*
    Genus XII. *Rheinheimera*
    Genus XIII. *Thermochromatium*
    Genus XIV. *Thioalkalicoccus*
    Genus XV. *Thiobaca*
    Genus XVI. *Thiocapsa*
    Genus XVII. *Thiococcus*
    Genus XVIII. *Thiocystis*
    Genus XIX. *Thiodictyon*
    Genus XX. *Thioflavicoccus*
    Genus XXI. *Thiohalocapsa*
    Genus XXII. *Thiolamprovum*
    Genus XXIII. *Thiopedia*
    Genus XXIV. *Thiorhodococcus*
    Genus XXV. *Thiorhodovibrio*
    Genus XXVI. *Thiospirillum*
  Family II. *Ectothiorhodospiraceae*
    Genus I. *Ectothiorhodospira*
    Genus II. *Alcalilimnicola*
    Genus III. *Alkalispirillum*
    Genus IV. *Arhodomonas*
    Genus V. *Halorhodospira*
    Genus VI. *Nitrococcus*
    Genus VII. *Thioalkalispira*
    Genus VIII. *Thialkalivibrio*
    Genus IX. *Thiorhodospira*
  Family III. *Halothiobacillaceae*
    Genus I. *Halothiobacillus*
Order II. *Acidithiobacillales*
  Family I. *Acidithiobacillaceae*
    Genus I. *Acidithiobacillus*
  Family II. *Thermithiobacillaceae*
    Genus I. *Thermithiobacillus*
Order III. *Xanthomonadales*
  Family I. *Xanthomonadaceae*
    Genus I. *Xanthomonas*
    Genus II. *Frateuria*
    Genus III. *Fulvimonas*
    Genus IV. *Luteimonas*
    Genus V. *Lysobacter*
    Genus VI. *Nevskia*
    Genus VII. *Pseudoxanthomonas*
    Genus VIII. *Rhodanobacter*
    Genus IX. *Schineria*
    Genus X. *Stenotrophomonas*
    Genus XI. *Thermomonas*
    Genus XII. *Xylella*
Order IV. *Cardiobacteriales*
  Family I. *Cardiobacteriaceae*
    Genus I. *Cardiobacterium*
    Genus II. *Dichelobacter*
    Genus III. *Suttonella*
Order V. *Thiotrichales*
  Family I. *Thiotrichaceae*
    Genus I. *Thiothrix*
    Genus II. *Achromatium*
    Genus III. *Beggiatoa*
    Genus IV. *Leucothrix*
    Genus V. *Thiobacterium*
    Genus VI. *Thiomargarita*
    Genus VII. *Thioploca*
    Genus VIII. *Thiospira*
  Family II. *Francisellaceae*
    Genus I. *Francisella*
  Family III. *Piscirickettsiaceae*
    Genus I. *Piscirickettsia*
    Genus II. *Cycloclasticus*
    Genus III. *Hydrogenovibrio*
    Genus IV. *Methylophaga*
    Genus V. *Thioalkalimicrobium*
    Genus VI. *Thiomicrospira*
Order VI. *Legionellales*
  Family I. *Legionellaceae*
    Genus I. *Legionella*
  Family II. *Coxiellaceae*
    Genus I. *Coxiella*
    Genus II. *Aquicella*
    Genus III. *Rickettsiella*
Order VII. *Methylococcales*
  Family I. *Methylococcaceae*
    Genus I. *Methylococcus*
    Genus II. *Methylobacter*
    Genus III. *Methylocaldum*
    Genus IV. *Methylomicrobium*
    Genus V. *Methylomonas*
    Genus VI. *Methylosarcina*
    Genus VII. *Methylosphaera*
Order VIII. *Oceanospirillales*
  Family I. *Oceanospirillaceae*
    Genus I. *Oceanospirillum*
    Genus II. *Balneatrix*
    Genus III. *Marinomonas*
    Genus IV. *Marinospirillum*
    Genus V. *Neptunomonas*
    Genus VI. *Oceanobacter*
    Genus VII. *Oleispira*
    Genus VIII. *Pseudospirillum*
    Genus IX. *Thalassolituus*
  Family II. *Alcanivoraceae*
    Genus I. *Alcanivorax*
    Genus II. *Fundibacter*
  Family III. *Hahellaceae*
    Genus I. *Hahella*
    Genus II. *Zooshikella*
  Family IV. *Halomonadaceae*
    Genus I. *Halomonas*
    Genus II. *Carnimonas*
    Genus III. *Chromohalobacter*
    Genus IV. *Cobetia*
    Genus V. *Deleya*
    Genus VI. *Zymobacter*
  Family V. *Oleiphilaceae*
    Genus I. *Oleiphilus*
  Family VI. *Saccharospirillaceae*
    Genus I. *Saccharospirillum*
Order IX. *Pseudomonadales*
  Family I. *Pseudomonadaceae*
    Genus I. *Pseudomonas*
    Genus II. *Azomonas*
    Genus III. *Azotobacter*
    Genus IV. *Cellvibrio*
    Genus V. *Chryseomonas*
    Genus VI. *Flavimonas*

Genus VII. *Mesophilobacter*
Genus VIII. *Rhizobacter*
Genus IX. *Rugamonas*
Genus X. *Serpens*
Family II. *Moraxellaceae*
Genus I. *Moraxella*
Genus II. *Acinetobacter*
Genus III. *Psychrobacter*
Family III. Incertae sedis[b]
Genus I. *Enhydrobacter*
Order X. *Alteromonadales*
Family I. *Alteromonadaceae*
Genus I. *Alteromonas*
Genus II. *Aestuariibacter*
Genus III. *Alishewanella*
Genus IV. *Colwellia*
Genus V. *Ferrimonas*
Genus VI. *Glaciecola*
Genus VII. *Idiomarina*
Genus VIII. *Marinobacter*
Genus IX. *Marinobacterium*
Genus X. *Microbulbifer*
Genus XI. *Moritella*
Genus XII. *Pseudoalteromonas*
Genus XIII. *Psychromonas*
Genus XIV. *Shewanella*
Genus XV. *Thalassomonas*
Family II. Incertae sedis[b]
Genus I. *Teredinibacter*
Order XI. *Vibrionales*
Family I. *Vibrionaceae*
Genus I. *Vibrio*
Genus II. *Allomonas*
Genus III. *Catenococcus*
Genus IV. *Enterovibrio*
Genus V. *Grimontia*
Genus VI. *Listonella*
Genus VII. *Photobacterium*
Genus VIII. *Salinivibrio*
Order XII. *Aeromonadales*
Family I. *Aeromonadaceae*
Genus I. *Aeromonas*
Genus II. *Oceanimonas*
Genus III. *Oceanisphaera*
Genus IV. *Tolumonas*
Family II. Incertae sedis: *Succinivibrionaceae*[b]
Genus I. *Succinivibrio*
Genus II. *Anaerobiospirillum*
Genus III. *Ruminobacter*
Genus IV. *Succinimonas*
Order XIII. *Enterobacteriales*
Family I. *Enterobacteriaceae*
Genus I. *Escherichia*
Genus II. *Alterococcus*
Genus III. *Arsenophonus*
Genus IV. *Brenneria*
Genus V. *Buchnera*
Genus VI. *Budvicia*
Genus VII. *Buttiauxella*
Genus VIII. *Calymmatobacterium*
Genus IX. *Cedecea*
Genus X. *Citrobacter*
Genus XI. *Edwardsiella*
Genus XII. *Enterobacter*
Genus XIII. *Erwinia*
Genus XIV. *Ewingella*
Genus XV. *Hafnia*
Genus XVI. *Klebsiella*
Genus XVII. *Kluyvera*
Genus XVIII. *Leclercia*
Genus XIX. *Leminorella*
Genus XX. *Moellerella*
Genus XXI. *Morganella*
Genus XXII. *Obesumbacterium*
Genus XXIII. *Pantoea*
Genus XXIV. *Pectobacterium*
Genus XXV. *Phlomobacter*

Genus XXVI. *Photorhabdus*
Genus XXVII. *Plesiomonas*
Genus XXVIII. *Pragia*
Genus XXIX. *Proteus*
Genus XXX. *Providencia*
Genus XXXI. *Rahnella*
Genus XXXII. *Raoultella*
Genus XXXIII. *Saccharobacter*
Genus XXXIV. *Salmonella*
Genus XXXV. *Samsonia*
Genus XXXVI. *Serratia*
Genus XXXVII. *Shigella*
Genus XXXVIII. *Sodalis*
Genus XXXIX. *Tatumella*
Genus XL. *Trabulsiella*
Genus XLI. *Wigglesworthia*
Genus XLII. *Xenorhabdus*
Genus XLIII. *Yersinia*
Genus XLIV. *Yokenella*
Order XIV. *Pasteurellales*
Family I. *Pasteurellaceae*
Genus I. *Pasteurella*
Genus II. *Actinobacillus*
Genus III. *Gallibacterium*
Genus IV. *Haemophilus*
Genus V. *Lonepinella*
Genus VI. *Mannheimia*
Genus VII. *Phocoenobacter*
Class IV. *Deltaproteobacteria*
Order I. *Desulfurellales*
Family I. *Desulfurellaceae*
Genus I. *Desulfurella*
Genus II. *Hippea*
Order II. *Desulfovibrionales*
Family I. *Desulfovibrionaceae*
Genus I. *Desulfovibrio*
Genus II. *Bilophila*
Genus III. *Lawsonia*
Family II. *Desulfomicrobiaceae*
Genus I. *Desulfomicrobium*
Family III. *Desulfohalobiaceae*
Genus I. *Desulfohalobium*
Genus II. *Desulfomonas*
Genus III. *Desulfonatronovibrio*
Genus IV. *Desulfothermus*
Family IV. *Desulfonatronumaceae*
Genus I. *Desulfonatronum*
Order III. *Desulfobacterales*
Family I. *Desulfobacteraceae*
Genus I. *Desulfobacter*
Genus II. *Desulfatibacillum*
Genus III. *Desulfobacterium*
Genus IV. *Desulfobacula*
Genus V. *Desulfobotulus*
Genus VI. *Desulfocella*
Genus VII. *Desulfococcus*
Genus VIII. *Desulfofaba*
Genus IX. *Desulfofrigus*
Genus X. *Desulfomusa*
Genus XI. *Desulfonema*
Genus XII. *Desulforegula*
Genus XIII. *Desulfosarcina*
Genus XIV. *Desulfospira*
Genus XV. *Desulfotignum*
Family II. *Desulfobulbaceae*
Genus I. *Desulfobulbus*
Genus II. *Desulfocapsa*
Genus III. *Desulfofustis*
Genus IV. *Desulforhopalus*
Genus V. *Desulfotalea*
Family III. *Nitrospinaceae*
Genus I. *Nitrospina*
Order IV. *Desulfarcales*
Family I. *Desulfarculaceae*
Genus I. *Desulfarculus*
Order V. *Desulfuromonales*
Family I. *Desulfuromonaceae*

Genus I. *Desulfuromonas*
Genus II. *Desulfuromusa*
Genus III. *Malonomonas*
Genus IV. *Pelobacter*
Family II. *Geobacteraceae*
Genus I. *Geobacter*
Genus II. *Trichlorobacter*
Order VI. *Syntrophobacterales*
Family I. *Syntrophobacteraceae*
Genus I. *Syntrophobacter*
Genus II. *Desulfacinum*
Genus III. *Desulforhabdus*
Genus IV. *Desulfovirga*
Genus V. *Thermodesulforhabdus*
Family II. *Syntrophaceae*
Genus I. *Syntrophus*
Genus II. *Desulfobacca*
Genus III. *Desulfomonile*
Genus IV. *Smithella*
Order VII. *Bdellovibrionales*
Family I. *Bdellovibrionaceae*
Genus I. *Bdellovibrio*
Genus II. *Bacteriovorax*
Genus III. *Micavibrio*
Genus IV. *Vampirovibrio*
Order VIII. *Myxococcales*
Suborder I. *Cystobacterineae*
Family I. *Cystobacteraceae*
Genus I. *Cystobacter*
Genus II. *Anaeromyxobacter*
Genus III. *Archangium*
Genus IV. *Hyalangium*
Genus V. *Melittangium*
Genus VI. *Stigmatella*
Family II. *Myxococcaceae*
Genus I. *Myxococcus*
Genus II. *Corallococcus*
Genus III. *Pyxicoccus*
Suborder II. *Sorangineae*
Family I. *Polyangiaceae*
Genus I. *Polyangium*
Genus II. *Byssophaga*
Genus III. *Chondromyces*
Genus IV. *Haploangium*
Genus V. *Jahnia*
Genus VI. *Sorangium*
Suborder III. *Nannocystineae*
Family I. *Nannocystaceae*
Genus I. *Nannocystis*
Genus II. *Plesiocystis*
Family II. *Haliangiaceae*
Genus I. *Haliangium*
Family III. *Kofleriaceae*
Genus I. *Kofleria*
Class V. *Epsilonproteobacteria*
Order I. *Campylobacterales*
Family I. *Campylobacteraceae*
Genus I. *Campylobacter*
Genus II. *Arcobacter*
Genus III. *Dehalospirillum*
Genus IV. *Sulfurospirillum*
Family II. *Helicobacteraceae*
Genus I. *Helicobacter*
Genus II. *Sulfurimonas*
Genus III. *Thiovulum*
Genus IV. *Wolinella*
Family III. *Nautiliaceae*
Genus I. *Nautilia*
Genus II. *Caminibacter*
Family IV. *Hydrogenimonaceae*
Genus I. *Hydrogenimonas*
Phylum BXIII. *Firmicutes*
Class I. *Clostridia*
Order I. *Clostridiales*
Family I. *Clostridiaceae*
Genus I. *Clostridium*
Genus II. *Acetivibrio*

Genus III. *Acidaminobacter*
Genus IV. *Alkaliphilus*
Genus V. *Anaerobacter*
Genus VI. *Anaerotruncus*
Genus VII. *Bryantella*
Genus VIII. *Caminicella*
Genus IX. *Caloramator*
Genus X. *Caloranaerobacter*
Genus XI. *Coprobacillus*
Genus XII. *Dorea*
Genus XIII. *Faecalibacterium*
Genus XIV. *Hespellia*
Genus XV. *Natronincola*
Genus XVI. *Oxobacter*
Genus XVII. *Parasporobacterium*
Genus XVIII. *Sarcina*
Genus XIX. *Soehngenia*
Genus XX. *Tepidibacter*
Genus XXI. *Thermobrachium*
Genus XXII. *Thermohalobacter*
Genus XXIII. *Tindallia*
Family II. *Lachnospiraceae*
Genus I. *Lachnospira*
Genus II. *Acetitomaculum*
Genus III. *Anaerofilum*
Genus IV. *Anaerostipes*
Genus V. *Butyrivibrio*
Genus VI. *Catenibacterium*
Genus VII. *Catonella*
Genus VIII. *Coprococcus*
Genus IX. *Johnsonella*
Genus X. *Lachnobacterium*
Genus XI. *Pseudobutyrivibrio*
Genus XII. *Roseburia*
Genus XIII. *Ruminococcus*
Genus XIV. *Shuttleworthia*
Genus XV. *Sporobacterium*
Family III. *Peptostreptococcaceae*
Genus I. *Peptostreptococcus*
Genus II. *Anaerococcus*
Genus III. *Filifactor*
Genus IV. *Finegoldia*
Genus V. *Fusibacter*
Genus VI. *Gallicola*
Genus VII. *Helcococcus*
Genus VIII. *Micromonas*
Genus IX. *Peptoniphilus*
Genus X. *Sedimentibacter*
Genus XI. *Sporanaerobacter*
Genus XII. *Tissierella*
Family IV. *Eubacteriaceae*
Genus I. *Eubacterium*
Genus II. *Acetobacterium*
Genus III. *Anaerovorax*
Genus IV. *Mogibacterium*
Genus V. *Pseudoramibacter*
Family V. *Peptococcaceae*
Genus I. *Peptococcus*
Genus II. *Carboxydothermus*
Genus III. *Dehalobacter*
Genus IV. *Desulfotobacterium*
Genus V. *Desulfonispora*
Genus VI. *Desulfosporosinus*
Genus VII. *Desulfotomaculum*
Genus VIII. *Pelotomaculum*
Genus IX. *Syntrophobotulus*
Genus X. *Thermoterrabacterium*
Family VI. *Heliobacteriaceae*
Genus I. *Heliobacterium*
Genus II. *Heliobacillus*
Genus III. *Heliophilum*
Genus IV. *Heliorestis*
Family VII. *Acidaminococcaceae*
Genus I. *Acidaminococcus*
Genus II. *Acetonema*
Genus III. *Allisonella*
Genus IV. *Anaeroarcus*

Genus V. *Anaeroglobus*
Genus VI. *Anaeromusa*
Genus VII. *Anaerosinus*
Genus VIII. *Anaerovibrio*
Genus IX. *Centipeda*
Genus X. *Dendrosporobacter*
Genus XI. *Dialister*
Genus XII. *Megasphaera*
Genus XIII. *Mitsuokella*
Genus XIV. *Papillibacter*
Genus XV. *Pectinatus*
Genus XVI. *Phascolarctobacterium*
Genus XVII. *Propionispira*
Genus XVIII. *Propionispora*
Genus XIX. *Quinella*
Genus XX. *Schwartzia*
Genus XXI. *Selenomonas*
Genus XXII. *Sporomusa*
Genus XXIII. *Succiniclasticum*
Genus XXIV. *Succinispira*
Genus XXV. *Veillonella*
Genus XXVI. *Zymophilus*
Family VIII. *Syntrophomonadaceae*
Genus I. *Syntrophomonas*
Genus II. *Acetogenium*
Genus III. *Aminobacterium*
Genus IV. *Aminomonas*
Genus V. *Anaerobaculum*
Genus VI. *Anaerobranca*
Genus VII. *Caldicellulosiruptor*
Genus VIII. *Carboxydocella*
Genus IX. *Dethiosulfovibrio*
Genus X. *Pelospora*
Genus XI. *Syntrophospora*
Genus XII. *Syntrophothermus*
Genus XIII. *Thermaerobacter*
Genus XIV. *Thermanaerovibrio*
Genus XV. *Thermohydrogenium*
Genus XVI. *Thermosyntropha*
Order II. *Thermoanaerobacteriales*
Family I. *Thermoanaerobacteriaceae*
Genus I. *Themoanaerobacterium*
Genus II. *Ammonifex*
Genus III. *Caldanaerobacter*
Genus IV. *Carboxydibrachium*
Genus V. *Coprothermobacter*
Genus VI. *Gelria*
Genus VII. *Moorella*
Genus VIII. *Sporotomaculum*
Genus IX. *Thermacetogenium*
Genus X. *Thermanaeromonas*
Genus XI. *Thermoanaerobacter*
Genus XII. *Thermoanaerobium*
Genus XIII. *Thermovenabulum*
Family II. *Thermodesulfobiaceae*
Genus I. *Thermodesulfobium*
Order III. *Halanaerobiales*
Family I. *Halanaerobiaceae*
Genus I. *Halanaerobium*
Genus II. *Halocella*
Genus III. *Halothermothrix*
Family II. *Halobacteroidaceae*
Genus I. *Halobacteroides*
Genus II. *Acetohalobium*
Genus III. *Halanaerobacter*
Genus IV. *Halonatronum*
Genus V. *Natroniella*
Genus VI. *Orenia*
Genus VII. *Selenihalanaerobacter*
Genus VIII. *Sporohalobacter*
Class II. *Mollicutes*
Order I. *Mycoplasmatales*
Family I. *Mycoplasmataceae*
Genus I. *Mycoplasma*
Genus II. *Eperythrozoon*
Genus III. *Haemobartonella*
Genus IV. *Ureaplasma*

Order II. *Entomoplasmatales*
Family I. *Entomoplasmataceae*
Genus I. *Entomoplasma*
Genus II. *Mesoplasma*
Family II. *Spiroplasmataceae*
Genus I. *Spiroplasma*
Order III. *Acholeplasmatales*
Family I. *Acholeplasmataceae*
Genus I. *Acholeplasma*
Genus II. *Phytoplasma*
Order IV. *Anaeroplasmatales*
Family I. *Anaeroplasmataceae*
Genus I. *Anaeroplasma*
Genus II. *Asteroleplasma*
Order V. Incertae sedis[b]
Family I. *Erysipelotrichaceae*
Genus I. *Erysipelothrix*
Genus II. *Bulleidia*
Genus III. *Holdemania*
Genus IV. *Solobacterium*

Class III. *Bacilli*
Order I. *Bacillales*
Family I. *Bacillaceae*
Genus I. *Bacillus*
Genus II. *Amphibacillus*
Genus III. *Anoxybacillus*
Genus IV. *Exiguobacterium*
Genus V. *Filobacillus*
Genus VI. *Geobacillus*
Genus VII. *Gracilibacillus*
Genus VIII. *Halobacillus*
Genus IX. *Jeotgalibacillus*
Genus X. *Lentibacillus*
Genus XI. *Marinibacillus*
Genus XII. *Oceanobacillus*
Genus XIII. *Paraliobacillus*
Genus XIV. *Saccharococcus*
Genus XV. *Salibacillus*
Genus XVI. *Ureibacillus*
Genus XVII. *Virgibacillus*
Family II. *Alicyclobacillaceae*
Genus I. *Alicyclobacillus*
Genus II. *Pasteuria*
Genus III. *Sulfobacillus*
Family III. *Caryophanaceae*
Genus I. *Caryophanon*
Family IV. *Listeriaceae*
Genus I. *Listeria*
Genus II. *Brochothrix*
Family V. *Paenibacillaceae*
Genus I. *Paenibacillus*
Genus II. *Ammoniphilus*
Genus III. *Aneurinibacillus*
Genus IV. *Brevibacillus*
Genus V. *Oxalophagus*
Genus VI. *Thermicanus*
Genus VII. *Thermobacillus*
Family VI. *Planococcaceae*
Genus I. *Planococcus*
Genus II. *Filibacter*
Genus III. *Kurthia*
Genus IV. *Planomicrobium*
Genus V. *Sporosarcina*
Family VII. *Sporolactobacillaceae*
Genus I. *Sporolactobacillus*
Genus II. *Marinococcus*
Family VIII. *Staphylococcaceae*
Genus I. *Staphylococcus*
Genus II. *Gemella*
Genus III. *Jeotgalicoccus*
Genus IV. *Macrococcus*
Genus V. *Salinicoccus*
Family IX. *Thermoactinomycetaceae*
Genus I. *Thermoactinomyces*
Family X. *Turicibacteraceae*
Genus I. *Turicibacter*
Order II. *Lactobacillales*

Family I. *Lactobacillaceae*
  Genus I. *Lactobacillus*
  Genus II. *Paralactobacillus*
  Genus III. *Pediococcus*
Family II. *Aerococcaceae*
  Genus I. *Aerococcus*
  Genus II. *Abiotrophia*
  Genus III. *Dolosicoccus*
  Genus IV. *Eremococcus*
  Genus V. *Facklamia*
  Genus VI. *Globicatella*
  Genus VII. *Ignavigranum*
Family III. *Carnobacteriaceae*
  Genus I. *Carnobacterium*
  Genus II. *Agitococcus*
  Genus III. *Alkalibacterium*
  Genus IV. *Allofustis*
  Genus V. *Alloiococcus*
  Genus VI. *Desemzia*
  Genus VII. *Dolosigranulum*
  Genus VIII. *Granulicatella*
  Genus IX. *Isobaculum*
  Genus X. *Lactosphaera*
  Genus XI. *Marinilactibacillus*
  Genus XII. *Trichococcus*
Family IV. *Enterococcaceae*
  Genus I. *Enterococcus*
  Genus II. *Atopobacter*
  Genus III. *Melissococcus*
  Genus IV. *Tetragenococcus*
  Genus V. *Vagococcus*
Family V. *Leuconostocaceae*
  Genus I. *Leuconostoc*
  Genus II. *Oenococcus*
  Genus III. *Weissella*
Family VI. *Streptococcaceae*
  Genus I. *Streptococcus*
  Genus II. *Lactococcus*
Family VII. Incertae sedis[b]
  Genus I. *Acetoanaerobium*
  Genus II. *Oscillospira*
  Genus III. *Syntrophococcus*
Phylum BXIV. *Actinobacteria*
  Class I. *Actinobacteria*
    Subclass I. *Acidimicrobidae*
      Order I. *Acidimicrobiales*
        Suborder IV. *Acidimicrobineae*
          Family I. *Acidimicrobiaceae*
            Genus I. *Acidimicrobium*
    Subclass II. *Rubrobacteridae*
      Order I. *Rubrobacterales*
        Suborder V. *Rubrobacterineae*
          Family I. *Rubrobacteraceae*
            Genus I. *Rubrobacter*
            Genus II. *Conexibacter*
            Genus III. *Solirubrobacter*
            Genus IV. *Thermoleophilum*
    Subclass III. *Coriobacteridae*
      Order I. *Coriobacteriales*
        Suborder VI. *Coriobacterineae*
          Family I. *Coriobacteriaceae*
            Genus I. *Coriobacterium*
            Genus II. *Atopobium*
            Genus III. *Collinsella*
            Genus IV. *Cryptobacterium*
            Genus V. *Denitrobacterium*
            Genus VI. *Eggerthella*
            Genus VII. *Olsenella*
            Genus VIII. *Slackia*
    Subclass IV. *Sphaerobacteridae*
      Order I. *Sphaerobacterales*
        Suborder VII. *Sphaerobacterineae*
          Family I. *Sphaerobacteraceae*
            Genus I. *Sphaerobacter*
    Subclass V. *Actinobacteridae*
      Order I. *Actinomycetales*
        Suborder VIII. *Actinomycineae*

Family I. *Actinomycetaceae*
  Genus I. *Actinomyces*
  Genus II. *Actinobaculum*
  Genus III. *Arcanobacterium*
  Genus IV. *Mobiluncus*
  Genus V. *Varibaculum*
Suborder IX. *Micrococcineae*
  Family I. *Micrococcaceae*
    Genus I. *Micrococcus*
    Genus II. *Arthrobacter*
    Genus III. *Citricoccus*
    Genus IV. *Kocuria*
    Genus V. *Nesterenkonia*
    Genus VI. *Renibacterium*
    Genus VII. *Rothia*
    Genus VIII. *Stomatococcus*
    Genus IX. *Yania*
  Family II. *Bogoriellaceae*
    Genus I. *Bogoriella*
  Family III. *Rarobacteraceae*
    Genus I. *Rarobacter*
  Family IV. *Sanguibacteraceae*
    Genus I. *Sanguibacter*
  Family V. *Brevibacteriaceae*
    Genus I. *Brevibacterium*
  Family VI. *Cellulomonadaceae*
    Genus I. *Cellulomonas*
    Genus II. *Oerskovia*
    Genus III. *Tropheryma*
  Family VII. *Dermabacteraceae*
    Genus I. *Dermabacter*
    Genus II. *Brachybacterium*
  Family VIII. *Dermatophilaceae*
    Genus I. *Dermatophilus*
    Genus II. *Kineosphaera*
  Family IX. *Dermacoccaceae*
    Genus I. *Dermacoccus*
    Genus II. *Demetria*
    Genus III. *Kytococcus*
  Family X. *Intrasporangiaceae*
    Genus I. *Intrasporangium*
    Genus II. *Arsenicicoccus*
    Genus III. *Janibacter*
    Genus IV. *Knoellia*
    Genus V. *Ornithinicoccus*
    Genus VI. *Ornithinimicrobium*
    Genus VII. *Nostocoidia*
    Genus VIII. *Terrabacter*
    Genus IX. *Terracoccus*
    Genus X. *Tetrasphaera*
  Family XI. *Jonesiaceae*
    Genus I. *Jonesia*
  Family XII. *Microbacteriaceae*
    Genus I. *Microbacterium*
    Genus II. *Agreia*
    Genus III. *Agrococcus*
    Genus IV. *Agromyces*
    Genus V. *Aureobacterium*
    Genus VI. *Clavibacter*
    Genus VII. *Cryobacterium*
    Genus VIII. *Curtobacterium*
    Genus IX. *Frigoribacterium*
    Genus X. *Leifsonia*
    Genus XI. *Leucobacter*
    Genus XII. *Mycetocola*
    Genus XIII. *Okibacterium*
    Genus XIV. *Plantibacter*
    Genus XV. *Rathayibacter*
    Genus XVI. *Rhodoglobus*
    Genus XVII. *Salinibacterium*
    Genus XVIII. *Subtercola*
  Family XIII. *Beutenbergiaceae*
    Genus I. *Beutenbergia*
    Genus II. *Georgenia*
    Genus III. *Salana*
  Family XIV. *Promicromonosporaceae*
    Genus I. *Promicromonospora*

Genus II. *Cellulosimicrobium*
Genus III. *Xylanibacterium*
Genus IV. *Xylanimonas*
Suborder X. *Corynebacterineae*
Family I. *Corynebacteriaceae*
Genus I. *Corynebacterium*
Family II. *Dietziaceae*
Genus I. *Dietzia*
Family III. *Gordoniaceae*
Genus I. *Gordonia*
Genus II. *Skermania*
Family IV. *Mycobacteriaceae*
Genus I. *Mycobacterium*
Family V. *Nocardiaceae*
Genus I. *Nocardia*
Genus II. *Rhodococcus*
Family VI. *Tsukamurellaceae*
Genus I. *Tsukamurella*
Family VII. *Williamsiaceae*
Genus I. *Williamsia*
Suborder XI. *Micromonosporineae*
Family I. *Micromonosporaceae*
Genus I. *Micromonospora*
Genus II. *Actinoplanes*
Genus III. *Asanoa*
Genus IV. *Catellatospora*
Genus V. *Catenuloplanes*
Genus VI. *Couchioplanes*
Genus VII. *Dactylosporangium*
Genus VIII. *Pilimelia*
Genus IX. *Spirilliplanes*
Genus X. *Verrucosispora*
Genus XI. *Virgisporangium*
Suborder XII. *Propionibacterineae*
Family I. *Propionibacteriaceae*
Genus I. *Propionibacterium*
Genus II. *Luteococcus*
Genus III. *Microlunatus*
Genus IV. *Propioniferax*
Genus V. *Propionimicrobium*
Genus VI. *Tessaracoccus*
Family II. *Nocardioidaceae*
Genus I. *Nocardioides*
Genus II. *Aeromicrobium*
Genus III. *Actinopolymorpha*
Genus IV. *Friedmanniella*
Genus V. *Hongia*
Genus VI. *Kribbella*
Genus VII. *Micropruina*
Genus VIII. *Marmoricola*
Genus IX. *Propionicimonas*
Suborder XIII. *Pseudonocardineae*
Family I. *Pseudonocardiaceae*
Genus I. *Pseudonocardia*
Genus II. *Actinoalloteichus*
Genus III. *Actinopolyspora*
Genus IV. *Amycolatopsis*
Genus V. *Crossiella*
Genus VI. *Kibdelosporangium*
Genus VII. *Kutzneria*
Genus VIII. *Prauserella*
Genus IX. *Saccharomonospora*
Genus X. *Saccharopolyspora*
Genus XI. *Streptoalloteichus*
Genus XII. *Thermobispora*
Genus XIII. *Thermocrispum*
Family II. *Actinosynnemataceae*
Genus I. *Actinosynnema*
Genus II. *Actinokineospora*
Genus III. *Lechevalieria*
Genus IV. *Lentzea*
Genus V. *Saccharothrix*
Suborder XIV. *Streptomycineae*
Family I. *Streptomycetaceae*
Genus I. *Streptomyces*
Genus II. *Kitasatospora*
Genus III. *Streptoverticillium*

Suborder XV. *Streptosporangineae*
Family I. *Streptosporangiaceae*
Genus I. *Streptosporangium*
Genus II. *Acrocarpospora*
Genus III. *Herbidospora*
Genus IV. *Microbispora*
Genus V. *Microtetraspora*
Genus VI. *Nonomuraea*
Genus VII. *Planobispora*
Genus VIII. *Planomonospora*
Genus IX. *Planopolyspora*
Genus X. *Planotetraspora*
Family II. *Nocardiopsaceae*
Genus I. *Nocardiopsis*
Genus II. *Streptomonospora*
Genus III. *Thermobifida*
Family III. *Thermomonosporaceae*
Genus I. *Thermomonospora*
Genus II. *Actinomadura*
Genus III. *Spirillospora*
Suborder XVI. *Frankineae*
Family I. *Frankiaceae*
Genus I. *Frankia*
Family II. *Geodermatophilaceae*
Genus I. *Geodermatophilus*
Genus II. *Blastococcus*
Genus III. *Modestobacter*
Family III. *Microsphaeraceae*
Genus I. *Microsphaera*
Family IV. *Sporichthyaceae*
Genus I. *Sporichthya*
Family V. *Acidothermaceae*
Genus I. *Acidothermus*
Family VI. *Kineosporiaceae*
Genus I. *Kineosporia*
Genus II. *Cryptosporangium*
Genus III. *Kineococcus*
Suborder XVII. *Glycomycineae*
Family I. *Glycomycetaceae*
Genus I. *Glycomyces*
Order II. *Bifidobacteriales*
Family I. *Bifidobacteriaceae*
Genus I. *Bifidobacterium*
Genus II. *Aeriscardovia*
Genus III. *Falcivibrio*
Genus IV. *Gardnerella*
Genus V. *Parascardovia*
Genus VI. *Scardovia*
Family II. Incertae sedis[b]
Genus I. *Actinobispora*
Genus II. *Actinocorallia*
Genus III. *Excellospora*
Genus IV. *Pelczaria*
Genus V. *Turicella*

Phylum BXV. *Planctomycetes*
Class I. *Planctomycetacia*
Order I. *Planctomycetales*
Family I. *Planctomycetaceae*
Genus I. *Planctomyces*
Genus II. *Gemmata*
Genus III. *Isosphaera*
Genus IV. *Pirellula*
Phylum BXVI. *Chlamydiae*
Class I. *Chlamydiae*
Order I. *Chlamydiales*
Family I. *Chlamydiaceae*
Genus I. *Chlamydia*
Genus II. *Chlamydophila*
Family II. *Parachlamydiaceae*
Genus I. *Parachlamydia*
Genus II. *Neochlamydia*
Family III. *Simkaniaceae*
Genus I. *Simkania*
Genus II. *Rhabdochlamydia*
Family IV. *Waddliaceae*
Genus I. *Waddlia*
Phylum BXVII. *Spirochaetes*

Class I. *Spirochaetes*
  Order I. *Spirochaetales*
    Family I. *Spirochaetaceae*
      Genus I. *Spirochaeta*
      Genus II. *Borrelia*
      Genus III. *Brevinema*
      Genus IV. *Clevelandina*
      Genus V. *Cristispira*
      Genus VI. *Diplocalyx*
      Genus VII. *Hollandina*
      Genus VIII. *Pillotina*
      Genus IX. *Treponema*
    Family II. *Serpulinaceae*
      Genus I. *Serpulina*
      Genus II. *Brachyspira*
    Family III. *Leptospiraceae*
      Genus I. *Leptospira*
      Genus II. *Leptonema*
Phylum BXVIII. *Fibrobacteres*
  Class I. *Fibrobacteres*
    Order I. *Fibrobacterales*
      Family I. *Fibrobacteraceae*
        Genus I. *Fibrobacter*
Phylum BXIX. *Acidobacteria*
  Class I. *Acidobacteria*
    Order I. *Acidobacteriales*
      Family I. *Acidobacteriaceae*
        Genus I. *Acidobacterium*
        Genus II. *Geothrix*
        Genus III. *Holophaga*
Phylum BXX. *Bacteroidetes*
  Class I. *Bacteroidetes*
    Order I. *Bacteroidales*
      Family I. *Bacteroidaceae*
        Genus I. *Bacteroides*
        Genus II. *Acetofilamentum*

        Genus III. *Acetomicrobium*
        Genus IV. *Acetothermus*
        Genus V. *Anaerophaga*
        Genus VI. *Anaerorhabdus*
        Genus VII. *Megamonas*
      Family II. *Rikenellaceae*
        Genus I. *Rikenella*
        Genus II. *Alistipes*
        Genus III. *Marinilabilia*
      Family III. *Porphyromonadaceae*
        Genus I. *Porphyromonas*
        Genus II. *Dysgonomonas*
        Genus III. *Tannerella*
      Family IV. *Prevotellaceae*
        Genus I. *Prevotella*
  Class II. *Flavobacteria*
    Order I. *Flavobacteriales*
      Family I. *Flavobacteriaceae*
        Genus I. *Flavobacterium*
        Genus II. *Aequorivita*
        Genus III. *Arenibacter*
        Genus IV. *Bergeyella*
        Genus V. *Capnocytophaga*
        Genus VI. *Cellulophaga*
        Genus VII. *Chryseobacterium*
        Genus VIII. *Coenonia*
        Genus IX. *Croceibacter*
        Genus X. *Empedobacter*
        Genus XI. *Gelidibacter*
        Genus XII. *Gillisia*
        Genus XIII. *Mesonia*
        Genus XIV. *Muricauda*
        Genus XV. *Myroides*
        Genus XVI. *Ornithobacterium*
        Genus XVII. *Polaribacter*

GLOSSARY

Only the major terms and concepts are included. If a term is not here, consult the index.

ABC transporter A membrane transport system consisting of three proteins, one of which hydrolyzes ATP, one of which binds the substrate, and one of which functions as the transport channel through the membrane.

Abscess A localized infection characterized by production of pus.

Acetotrophic Splitting of acetate into CH_4 plus CO_2 by certain methanogens.

Acetyl-CoA pathway A pathway of autotrophic CO_2 fixation widespread in obligate anaerobes including methanogens, homoacetogens, and sulfate-reducing bacteria.

Acetylene reduction assay Method of measuring activity of nitrogenase by substituting acetylene for the natural substrate of the enzyme, N_2. Acetylene is reduced to ethylene or ethane, depending on the nitrogenase system involved.

Acid fastness A staining property of *Mycobacterium* species where cells stained with hot carbol fuchsin do not decolorize with acid-alcohol.

Acid mine drainage Acidic water containing H_2SO_4 derived from the microbial oxidation of iron sulfide minerals.

Acidophile An organism that grows best at acidic pH values.

Activation energy Energy needed to make substrate molecules more reactive; enzymes function by lowering activation energy.

Activator protein A regulatory protein that binds to specific sites on DNA and stimulates transcription; involved in positive control.

Active immunity An immune state achieved by self-production of antibodies. Compare with *Passive immunity*.

Active site The portion of an enzyme that is directly involved in binding substrate(s).

Active transport The energy-dependent process of transporting substances into or out of the cell in which the transported substances are chemically unchanged.

Acute In reference to infections, short-term, usually characterized by dramatic onset and rapid recovery.

Adaptive immunity (antigen-specific immunity) The acquired ability to recognize and destroy an individual pathogen or its products relying on previous exposure to the pathogen or its products.

Adherence A property of bacteria that allows them to stick to host surfaces.

Aerobic secondary wastewater treatment Digestive reactions carried out by microorganisms under aerobic conditions to treat wastewater containing low levels of organic materials.

Aerobe An organism that grows in the presence of O_2; may be facultative, obligate, or microaerophilic.

Aerosol Suspension of particles in airborne water droplets.

Aerotolerant Of an anaerobe, not being inhibited by O_2.

Agglutination Reaction between antibody and particle-bound antigen resulting in visible clumping of the particles.

Agretope The portion of a processed antigen that is recognized by MHC protein.

Algae Phototrophic eukaryotic microorganisms.

Alkaliphile An organism that grows best at high pH.

Allergy A harmful immune reaction, usually caused by a foreign antigen in food, pollen, or chemicals, which results in immediate-type or delayed-type hypersensitivity.

Allosteric enzyme An enzyme that contains two combining sites, the active site (where the substrate binds) and the allosteric site (where an effector molecule binds).

Ameboid movement A type of motility in which cytoplasmic streaming moves the organism forward.

Aminoacyl-tRNA synthetase An enzyme that catalyzes the attachment of the correct amino acid to the correct tRNA.

Aminoglycoside An antibiotic such as streptomycin, containing amino sugars linked by glycosidic bonds.

Anabolism The biochemical processes involved in the synthesis of cell constituents from simpler molecules, usually requiring energy.

Anaerobe An organism that grows in the absence of O_2; some may even be killed by O_2 (obligate or strict anaerobes).

Anaerobic respiration Use of an electron acceptor other than O_2 in an electron transport-based oxidation leading to a proton motive force.

Anammox The term that describes anoxic ammonia oxidation.

Anaphylatoxins The C3a and C5a fractions of complement that act to mimic some of the reactions of anaphylaxis.

Anaphylaxis (anaphylactic shock) A violent allergic reaction caused by an antigen–antibody reaction.

Anergy The inability to produce an immune response to specific antigens due to neutralization of effector cells.

Anoxic Absence of oxygen. Usually used in reference to a microbial habitat.

Anoxic secondary wastewater treatment Digestive and fermentative reactions carried out by microorganisms under anoxic conditions to treat wastewater containing high levels of insoluble organic materials.

Anoxygenic photosynthesis Use of light energy to synthesize ATP by cyclic photophosphorylation without O_2 production.

Antibiogram A report indicating the sensitivity of clinically isolated microorganisms to the antibiotics in current use.

Antibiotic A chemical substance produced by a microorganism that kills or inhibits the growth of another microorganism.

Antibiotic resistance The acquired ability of a microorganism to grow in the presence of an antibiotic to which the microorganism is usually sensitive.

Antibody A protein, produced by B lymphocytes, present in serum or other body fluid that combines specifically with antigen. An immunoglobulin.

Antibody-mediated immunity Immunity resulting from direct interaction with antibodies; also called humoral immunity.

Anticodon A sequence of three bases in transfer RNA that base-pairs with a codon in messenger RNA during protein synthesis.

Antigen A molecule capable of interacting with specific components of the immune system.

Antigen-presenting cell (APC) A macrophage, dendritic cell, or B cell that presents processed antigen peptides to a T cell.

Antigenic determinant The portion of an antigen that interacts with an immunoglobulin or T cell receptor. Also called an *epitope*.

Antigenic drift In influenza virus, minor changes in viral proteins (antigens) due to gene mutation.

Antigenic shift In influenza virus, major changes in viral proteins (antigens) due to gene reassortment.

Antimicrobial Harmful to microorganisms by either killing or inhibiting growth.

Antimicrobial agent A chemical that kills or inhibits the growth of microorganisms.

Antimicrobial drug resistance The acquired ability of a microorganism to grow in the presence of an antimicrobial drug to which the microorganism is usually susceptible.

Antiparallel In reference to nucleic acids; one strand runs $5' \rightarrow 3'$, the other $3' \rightarrow 5'$.

Antiseptic (germicide) A chemical agent that kills or inhibits growth of microorganisms and is sufficiently nontoxic to be applied to living tissues.

Antiserum A serum containing antibodies.

Antitoxin An antibody that specifically interacts with and neutralizes a toxin.

Apoptosis Programmed cell death.

Archaea A phylogenetic domain of prokaryotes consisting of the methanogens, most extreme halophiles and hyperthermophiles, and extreme acidophiles such as *Thermoplasma*.

Artificial chromosomes Cloning vectors which can carry very large inserts of foreign DNA and exist in the cell very much like a cellular chromosome. The most widely used are bacterial artificial chromosomes (BACs) and yeast artificial chromosomes (YACs).

Aseptic technique Manipulation of sterile instruments or culture media in such a way as to maintain sterility.

ATP Adenosine triphosphate, the principal energy carrier of the cell.

Attenuation In a pathogen, a decrease or loss of virulence. Also, a mechanism for controlling gene expression. Typically transcription is terminated after initiation but before a full-length mRNA is produced.

Autoantibody An antibody that reacts to self antigens.

Autoclave A sterilizer that destroys microorganisms with temperature and steam under pressure.

Autoimmunity Immune reactions of a host against its own self antigens.

Autolysis The lysis of a cell brought about by the activity of the cell itself.

Autoradiography Detection of radioactivity in a sample, for example, a cell or gel, by placing it in contact with a photographic film.

Autotroph An organism able to use CO_2 as a sole source of carbon.

Auxotroph An organism that has developed a nutritional requirement through mutation. Contrast with a *Prototroph*.

B lymphocyte (B cell) A lymphocyte that has immunoglobulin surface receptors, produces immunoglobulin, and may present antigens to T cells.

Bacteremia The transient appearance of bacteria in the blood.

Bacteria All prokaryotes that are not members of the domain *Archaea*.

Bacteriocidal Capable of killing bacteria.

Bacteriocins Agents produced by certain bacteria that inhibit or kill closely related species.

Bacteroid A swollen, deformed *Rhizobium* cell found in the root nodule; capable of nitrogen fixation.

Bacteriophage A virus that infects prokaryotic cells.

Bacteriorhodopsin A protein containing retinal that is found in the membranes of certain extremely halophilic *Archaea* and that is involved in light-mediated ATP synthesis.

Bacteriostatic Capable of inhibiting bacterial growth without killing.

Barophile An organism that lives optimally at high hydrostatic pressure.

Barotolerant An organism able to tolerate high hydrostatic pressure, although growing better at 1 atm.

Base composition In reference to nucleic acids, the proportion of the total bases consisting of guanine plus cytosine or thymine plus adenine base pairs. Usually expressed as a guanine + cytosine (GC) value, for example, 60% GC.

Batch culture A closed-system microbial culture of fixed volume.

Beta-lactam antibiotic An antibiotic such as penicillin that contains the four-membered heterocyclic β-lactam ring.

Binomial nomenclature system The system for naming organisms in which an organism is given a genus name and a species epithet.

Biocatalysis The use of microorganisms to synthesize a product or carry out a specific chemical transformation.

Biochemical oxygen demand (BOD) The amount of dissolved oxygen consumed by microorganisms for complete oxidation of organic and inorganic material in a water sample.

Biofilm Microbial colonies encased in an adhesive, usually polysaccharide material and attached to a surface.

Biogeochemistry Study of microbially mediated chemical transformations of geochemical interest, for example, nitrogen or sulfur cycling.

Bioinformatics The use of computer programs to analyze, store, and access DNA and protein sequences.

Biological warfare The use of biological agents to kill or incapacitate a population.

Bioremediation Use of microorganisms to remove or detoxify toxic or unwanted chemicals in an environment.

Biosynthesis The production of needed cellular constituents from other (usually simpler) molecules.

Biosynthetic penicillin Production of a particular form of penicillin by supplying the producing organism with specific side chain precursors.

Biotechnology The use of living organisms to carry out defined chemical processes for industrial application.

Biotransformation In industrial microbiology, use of microorganisms to convert a substance to a chemically modified form.

Black smoker A deep-sea hydrothermal vent emitting superheated 250–400°C water and minerals.

Botulism Food poisoning due to ingestion of food containing botulism toxin produced by *Clostridium botulinum*.

Brewing The manufacture of alcoholic beverages such as beer from the fermentation of malted grains.

Broad-spectrum antibiotic An antibiotic that acts on both gram-positive and gram-negative *Bacteria*.

Calvin cycle The biochemical route of CO_2 fixation in many autotrophic organisms.

Canning The process of sealing food in a closed container and heating to destroy living organisms.

Capsid The protein coat of a virus.

Capsomere An individual protein subunit of the virus capsid.

Capsule A dense, well-defined polysaccharide or protein layer closely surrounding a cell.

Carboxysomes Polyhedral cellular inclusions of crystalline ribulose bisphosphate carboxylase (RubisCO), the key enzyme of the Calvin cycle.

Carcinogen A substance that causes the initiation of tumor formation. Frequently a mutagen.

Carrier An individual that harbors infectious organisms but does not show symptoms of disease.

Catabolism The biochemical processes involved in the breakdown of organic or inorganic compounds, usually leading to the production of energy.

Catabolite repression Repression of a variety of unrelated enzymes when cells are grown in a medium containing glucose.

Catalysis Increase in rate of a chemical reaction.

Catalyst A substance that promotes a chemical reaction without itself being changed in the end.

CD4 cells T helper cells. They are targets for HIV infection.

Cell The fundamental unit of life.

Cell-mediated immunity An immune response generated by interactions with antigen-specific T cells. Compare with *Humoral immunity*.

Centers for Disease Control and Prevention (CDC) An agency of the United States Public Health Service that tracks disease trends, provides disease information to the public and to health-care professionals, and forms public policy regarding disease prevention and intervention.

Chemiosmosis The use of ion gradients, especially proton gradients, across membranes to generate ATP. See *Proton motive force*.

Chemokine A small soluble protein produced by a variety of cells that modulates inflammatory reactions and immunity in target cells.

Chemolithotroph An organism obtaining its energy from the oxidation of inorganic compounds.

Chemoorganotroph An organism obtaining its energy from the oxidation of organic compounds.

Chemostat A continuous culture device controlled by the concentration of limiting nutrient and dilution rate.

Chemotaxis Movement toward or away from a chemical.

Chemotherapeutic agent An antimicrobial agent that can be used internally.

Chemotherapy Treatment of infectious disease with chemicals or antibiotics.

Chloramine A water purification chemical made by combining chlorine and ammonia at precise ratios.

Chlorination A highly effective disinfectant procedure for drinking water using chlorine gas or other chlorine-containing compounds as disinfectant.

Chlorine A chemical fed in its gaseous state to disinfect water. A residual level is maintained throughout the distribution system.

Chlorophyll and bacteriochlorophyll Pigments of phototrophic organisms consisting of light-sensitive magnesium tetrapyrroles.

Chloroplast The chlorophyll-containing organelle of phototrophic eukaryotes.

Chlorosomes Cigar-shaped structures enclosed by a nonunit membrane and containing the light-harvesting bacteriochlorophyll (c, c_s, d, or e) in green sulfur bacteria and in *Chloroflexus*.

Chromogenic Producing color; a chromogenic colony is a pigmented colony.

Chromosome A genetic element carrying genes essential to cellular function. Prokaryotes typically have a single chromosome consisting of a circular DNA molecule. Eukaryotes typically have several chromosomes, each containing a linear DNA molecule.

Chronic In reference to infections, long-term, usually characterized by gradual onset, mild or subclinical disease, and lack of permanent recovery.

Cidal Lethal or killing.

Cilium Short, filamentous structure that beats with many others to make a cell move.

Cirrhosis Breakdown of the normal liver architecture resulting in fibrosis.

Citric acid cycle A cyclical series of reactions resulting in the conversion of acetate to CO_2 and NADH. Also called the *Tricarboxylic acid cycle* or the *Kreb's cycle*.

Clarifier (coagulation basin) A reservoir in which the suspended solids of raw water are coagulated and removed.

Class I MHC protein Antigen-presenting molecule found on all nucleated vertebrate cells.

Class II MHC protein Antigen-presenting molecule found on macrophages, B lymphocytes, and dendritic cells in vertebrates.

Clonal deletion For T-cell selection in the thymus, the killing of useless or self-reactive clones.

Clonal selection A theory that each B or T lymphocyte, when stimulated by antigen, divides to form a clone of itself.

Clone A population of cells all descended from a single cell. Also, a number of copies of a DNA fragment obtained by allowing an inserted DNA fragment to be replicated by a phage or plasmid.

Cloning vectors Genetic elements into which genes can be recombined and replicated.

Coagulation The formation of large insoluble particles from much smaller, colloidal particles by the addition of aluminum sulfate and anionic polymers.

Coccoid Sphere-shaped.

Coccus A spherical bacterium.

Codon A sequence of three bases in messenger RNA that encodes a specific amino acid.

Coenzyme A low-molecular-weight molecule that participates in an enzymatic reaction by accepting and donating electrons or functional groups. Examples: NAD^+, FAD.

Coliform Gram-negative, nonsporing, facultative rod that ferments lactose with gas formation within 48 hours at 35°C.

Colonization Multiplication of a microorganism after it has attached to host tissues or other surfaces.

Colony A macroscopically visible population of cells growing on solid medium, arising from a single cell.

Cometabolism The metabolic transformation of a substance while a second substance serves as primary energy or carbon source.

Commodity chemicals Chemicals such as ethanol that have low monetary value and thus are sold primarily in bulk.

Common-source epidemic An epidemic resulting from infection of a large number of people from a single contaminated source.

Compatible solutes Organic compounds that serve as cytoplasmic solutes to balance water relations for cells growing in environments of high salt or sugar.

Competence Ability to take up DNA and become genetically transformed.

Complement A series of proteins that react in a sequential manner with antibody–antigen complexes to amplify or potentiate antibody activity.

Complement fixation The consumption of complement by an antibody–antigen reaction.

Complementary Nucleic acid sequences that can base-pair with each other.

Complementarity-determining regions (CDRs) Variations in amino acid sequence that occur within the variable domains of Igs or TCRs that provide most of the molecular contacts with antigen (also known as *hypervariable regions*).

Complex media Culture media whose precise chemical composition is unknown. Also called *undefined media*.

Concatemer A DNA molecule consisting of two or more separate molecules linked end to end to form a long, linear structure.

Congenital syphilis Syphilis contracted by an infant from its mother during birth.

Conjugation Transfer of genes from one prokaryotic cell to another by a mechanism involving cell-to-cell contact.

Consensus sequence A nucleic acid sequence in which the base present in a given position is that base most commonly found when many experimentally determined sequences are compared.

Consortium A two-membered (or more) bacterial culture (or natural assemblage) in which each organism benefits from the others.

Contagious Transmissible.

Cortex The region inside the spore coat of an endospore, around the core.

Covalent bond A nonionic chemical bond formed by a sharing of electrons between two atoms.

Crista Inner membrane in a mitochondrion; site of respiration.

Culture A particular strain or kind of organism growing in a laboratory medium.

Culture medium An aqueous solution of various nutrients suitable for the growth of microorganisms.

Cutaneous Relating to the skin.

Cyanobacteria Prokaryotic oxygenic phototrophs containing chlorophyll *a* and phycobilins.

Cyst A resting stage formed by some bacteria and protozoa in which the whole cell is surrounded by a thick-walled chemically and physically resistant coating; not the same as a spore or endospore.

Cytochrome Iron-containing porphyrin complexed with proteins, which functions as an electron carrier in the electron transport system.

Cytokine A small, soluble protein produced by a leukocyte that modulates inflammatory reactions and immunity in target cells.

Cytoplasm Cellular contents inside the cytoplasmic membrane, excluding the nucleus.

Cytoplasmic membrane The permeability barrier of the cell, separating the cytoplasm from the environment.

Cytoskeleton Cellular scaffolding in which microfilaments define the shape of eukaryotic cells.

Decontamination Treatment that renders an object or inanimate surface safe to handle.

Defined media Culture media whose exact chemical composition is known. Compare with *Complex media*.

Degeneracy In relation to the genetic code, the fact that more than one codon can code for the same amino acid.

Deletion Removal of a portion of a gene.

Denaturation Irreversible destruction of a macromolecule, as for example, the destruction of a protein by heat.

Denaturing gradient gel electrophoresis (DGGE) An electrophoretic technique that allows resolving nucleic acid fragments of the same size but that differ in sequence.

Dendritic cell A type of leukocyte having phagocytic and antigen-presenting properties, found in lymph nodes and spleen.

Denitrification Microbial conversion of nitrate into nitrogen gases under anoxic conditions.

Dental caries Tooth decay resulting from bacterial infection.

Dental plaque Bacterial cells encased in a matrix of extracellular polymers and salivary products, found on the teeth.

Deoxyribonucleic acid (DNA) A polymer of nucleotides connected via a phosphate–deoxyribose sugar backbone; the genetic material of cells and some viruses.

Desiccation Drying.

Dideoxynucleotide A nucleotide lacking the 3′-hydroxyl group on the deoxyribose sugar. Used in the Sanger method of DNA sequencing.

Differential media A growth medium that allows identification of microorganisms based on phenotypic properties.

Differentiation The modification of a cell in terms of structure and/or function occurring during the course of development.

Diploid In eukaryotes, an organism or cell with two chromosome complements, one derived from each haploid gamete.

Disease Injury to the host that impairs host function.

Disinfectant An antimicrobial agent used only on inanimate objects.

Disinfection The elimination of microorganisms from inanimate objects or surfaces.

Disproportionation The splitting of a chemical compound into two new compounds, one more oxidized and one more reduced than the original compound.

Distribution system Water pipes, storage reservoirs, tanks, and other means used to deliver drinking water to consumers or store it before delivery.

DNA fingerprinting Use of genetic engineering to determine the origin of DNA in a sample of tissue.

DNA gyrase An enzyme found in most prokaryotes that introduces negative supercoils in DNA.

DNA library See *gene library*.

DNA polymerase An enzyme that synthesizes a new strand of DNA in the 5′ → 3′ direction using an antiparallel DNA strand as a template.

Domain The highest level of biological classification. The three domains of biological organisms are the *Bacteria*, the *Archaea*, and the *Eukarya*. Also used to describe a region of a protein having a defined structure and function.

Doubling time The time needed for a population to double. See also *Generation time*.

Downstream position Refers to nucleic acid sequences on the 3′ side of a given site on the DNA or RNA molecule. Compare with *Upstream position*.

Early protein Proteins synthesized soon after virus infection.

Ecology Study of the interrelationships between organisms and their environments.

Ecosystem A community of organisms and their natural environment.

Ecotype A population of genetically identical cells sharing a particular resource within an ecological niche.

Effluent water Treated wastewater discharged from a wastewater treatment facility.

Ehrlichiosis One of a group of emerging tick-transmitted disease caused by rickettsias of the *Ehrlichia* genus.

Electron acceptor A substance that accepts electrons during an oxidation–reduction reaction.

Electron donor A compound that donates electrons in an oxidation–reduction reaction.

Electron transport phosphorylation Synthesis of ATP involving a membrane-associated electron transport chain and the creation of a proton motive force. Also called *Oxidative phosphorylation*. See also *Chemiosmosis*.

Electrophoresis Separation of charged molecules in an electric field.

Electroporation The use of an electric pulse to enable cells to take up DNA.

ELISA Enzyme-linked immunosorbent assay. An immunoassay that uses specific antibodies to detect antigens or antibodies in body fluids. The antibody-containing complexes are visualized through enzyme coupled to the antibody. Addition of substrate to the enzyme–antibody–antigen complex results in a colored product.

Emerging infection An infectious disease that has increased in incidence in the last 20 years or threatens to increase in incidence in the future.

Enantiomer One form of a molecule that is the mirror image of another form of the same molecule.

Endemic A disease that is constantly present in low numbers in a population, usually in low numbers. Compare with *Epidemic*.

Endergonic reaction A chemical reaction requiring an input of energy to proceed.

Endocytosis A process in which a particle such as a virus is taken intact into an animal cell. Phagocytosis and pinocytosis are two kinds of endocytosis.

Endoplasmic reticulum An extensive array of internal membranes in eukaryotes.

Endospore A differentiated cell formed within the cells of certain gram-positive bacteria that is extremely resistant to heat as well as to other harmful agents.

Endosymbiosis The process by which mitochondria, chloroplasts, and hydrogenosomes originated from the descendants of ancient prokaryotic organisms from the domain *Bacteria*.

Endotoxin The lipopolysaccharide portion of the cell envelope of certain gram-negative *Bacteria*, which is a toxin when solubilized. Compare with *Exotoxin*.

Enriched media Media that allow metabolically fastidious organisms to grow because of the addition of specific growth factors.

Enrichment bias The skewing of enrichment culture results toward "weed" species.

Enrichment culture Use of selective culture media and incubation conditions to isolate specific microorganisms from natural samples.

Enteric Intestinal.

Enterotoxin A protein released extracellularly by a microorganism as it grows that produces immediate damage to the small intestine of the host.

Entropy A measure of the degree of disorder in a system; entropy always increases in a closed system.

Enzyme A catalyst, usually composed of protein, that promotes specific reactions or groups of reactions.

Epidemic A disease occurring in an unusually high number of individuals in a population at the same time. Compare with *Endemic*.

Epidemiology The study of the occurrence, distribution, and control of diseases.

Epitope The portion of an antigen that is recognized by an immunoglobulin or a T-cell receptor.

***Escherichia coli* O157:H7** An enterotoxigenic strain of *E. coli* spread by fecal contamination of animal or human origin to food and water.

Eukarya The phylogenetic domain containing all eukaryotic organisms.

Eukaryote A cell or organism having a unit membrane-enclosed (true) nucleus and usually other organelles.

Evolution Change in a line of descent over time leading to the production of new species or varieties within a species.

Evolutionary distance In phylogenetic trees, the sum of the physical distance on a tree separating organisms; this distance is inversely proportional to evolutionary relatedness.

Exergonic reaction A chemical reaction that proceeds with the liberation of energy.

Exons The coding sequences in a split gene. Contrast with *Introns*, the intervening noncoding regions.

Exotoxin A protein released extracellularly by a microorganism as it grows that produces immediate host cell damage. Compare with *Endotoxin*.

Exponential growth Growth of a microorganism where the cell number doubles within a fixed time period.

Exponential phase A period during the growth cycle of a population in which growth increases at an exponential rate.

Expression The ability of a gene to function within a cell in such a way that the gene product is formed.

Expression vector A cloning vector that contains the necessary regulatory sequences allowing transcription and translation of a cloned gene or genes.

Extreme halophile An organism whose growth is dependent on large amounts (generally >10%) of NaCl.

Extremophile An organism that grows optimally under one or more chemical or physical extremes, such as high or low temperature or pH.

Facultative A qualifying adjective indicating that an organism is able to grow in either the presence or absence of an environmental factor (for example, "facultative aerobe").

FAME Fatty acid methyl ester.

Feedback inhibition A decrease in the activity of the first enzyme of a biochemical pathway caused by buildup of the final product of the pathway.

Fermentation Catabolic reactions producing ATP in which organic compounds serve as both primary electron donor and ultimate electron acceptor and ATP is produced by substrate-level phosphorylation.

Fermentation (industrial) A large-scale microbial process.

Fermenter An organism that carries out the process of fermentation.

Fermentor A growth vessel, usually quite large, used to culture microorganisms for the production of some commercially valuable product.

Ferredoxin An electron carrier of low reduction potential; small protein containing iron-sulfur clusters.

Fever An abnormal increase in body temperature.

Filamentous In the form of very long rods, many times longer than wide.

Filtration The removal of suspended particles from water by passing it through one or more permeable membranes or media (e.g., sand, anthracite, or diatomaceous earth).

Fimbria (plural fimbriae) Short, filamentous structures on a bacterial cell; although flagella-like in structure, generally present in many copies and not involved in motility. Plays a role in adherence to surfaces and in the formation of pellicles. See also *Pilus*.

Finished water Water delivered to the distribution system after treatment.

FISH Fluorescent *in-situ* hybridization; a process in which a cell is made fluorescent by labeling it with a specific nucleic acid probe that contains an attached fluorescent dye.

Flagellum (plural flagella) A thin, filamentous organ of motility in prokaryotes that functions by rotating. In motile eukaryotes, the flagellum, if present, moves by a whip-like motion.

Flavoprotein A protein containing a derivative of riboflavin, which functions as electron carrier in the electron transport system.

Flocculation The water-treatment process after coagulation that uses gentle stirring to cause suspended particles to form larger, aggregated masses (floc).

Fluorescent Having the ability to emit light of a certain wavelength when activated by light of another wavelength.

Fluorescent antibody Covalent modification of an antibody molecule with a fluorescent dye; the dye makes the antibody visible under fluorescent light.

Fomite Inanimate object that, when contaminated with a viable pathogen, can transfer the pathogen to a host.

Food infection Microbial infection resulting from ingestion of pathogen-contaminated food.

Food poisoning (food intoxication) Disease caused by the ingestion of food that contains preformed microbial toxins.

Food spoilage Any change in a food product that makes it unacceptable to the consumer.

Frameshift A type of mutation. Because the genetic code is read three bases at a time, if reading begins at either the second or third base of a codon, a faulty product usually results.

Free energy Energy available to do useful work.

Fruiting body A macroscopic reproductive structure produced by some fungi (for example, mushrooms) and some *Bacteria* (for example, myxobacteria), each distinct in size, shape, and coloration.

Fungi Nonphototrophic eukaryotic microorganisms that contain rigid cell walls.

Fungicidal agent An agent that kills fungi.

Fungistatic agent An agent that inhibits fungal growth.

Fusion inhibitor A short synthetic peptide like enfuvirtide that acts by binding to the gp41 membrane protein of HIV, inhibiting fusion of viral and CD4 cell membranes.

Fusion protein The result of translation of two or more genes joined such that they retain their correct reading frames but make a single protein.

GC base ratio In DNA (or RNA) from any organism, the percentage of the total nucleic acid that consists of guanine plus cytosine bases (expressed as mol% GC).

Gametes In eukaryotes, the haploid germ cells that result from meiosis.

Gas vesicle A gas-filled structure made of protein that when present in large numbers, confers buoyancy on a cell.

Gel An inert polymer, usually made of agarose or polyacrylamide, used for separating macromolecules such as nucleic acids and proteins by electrophoresis.

Gene A unit of heredity; a segment of DNA specifying a particular protein or polypeptide chain, a tRNA or an rRNA.

Gene cloning See *Molecular cloning*.

Gene disruption Use of genetic techniques to inactivate a gene by inserting within it a DNA fragment containing an easily selectable marker. The inserted fragment is called a *cassette*, and the process of insertion, *cassette mutagenesis*.

Gene family Genes that are related by sequence to other genes within the organism.

Gene library A collection of cloned DNA fragments that contains all the genetic information for a particular organism.

Gene therapy Treatment of a disease caused by a dysfunctional gene by introduction of a normally functioning copy of the gene.

General purpose medium A growth medium that supports the growth of most aerobic and facultatively aerobic organisms.

Generation time Time needed for a population to double. See also *Doubling time*.

Genetically modified organism (GMO) An organism whose genome has been altered using genetic engineering. The abbreviation is also used in constructions such as *GM crops* and *GM foods*.

Genetic engineering The use of *in vitro* techniques in the isolation, manipulation, recombination, and expression of DNA, and in the development of *genetically modified organisms* (GMO).

Genetic map The arrangement of genes on a chromosome.

Genetics Heredity and variation of organisms.

Genome The complete set of genes present in an organism.

Genomics The discipline involving mapping, sequencing, and analyzing genomes.

Genotype The precise genetic constitution of an organism. Compare with *Phenotype*.

Genus A taxonomic group of related species.

Germicide (antiseptic) A chemical agent that kills or inhibits growth of microorganisms and is sufficiently nontoxic to be applied to living tissues.

Glycocalyx A term that describes polysaccharide components outside of the bacterial cell wall; usually a loose network of polymer fibers extending outward from the cell.

Glycolysis Reactions of the Embden–Meyerhof pathway in which glucose is converted to pyruvate.

Glycosidic bond A type of covalent bond that links sugar units together in a polysaccharide.

Gonococcus *Neisseria gonorrhoeae*, the gram-negative diplococcus that causes the disease gonorrhea.

Gram-negative cell A prokaryotic cell whose cell wall contains relatively little peptidoglycan but has an outer membrane composed of lipopolysaccharide, lipoprotein, and other complex macromolecules.

Gram-positive cell A prokaryotic cell whose cell wall consists chiefly of peptidoglycan and lacks the outer membrane of gram-negative cells.

Growth In microbiology, an increase in cell number.

Growth-factor analog A chemical agent that is related to and blocks the uptake or utilization of a growth factor.

Growth rate The rate at which growth occurs, usually expressed as the generation time.

Guild A group of metabolically related organisms.

HAART; highly active antiretroviral therapy Treatment of HIV infection with two or more antiretroviral drugs at once to inhibit the development of drug resistance.

Habitat The location in nature where an organism resides.

Halophile An organism requiring salt (NaCl) for growth.

Halotolerant Capable of growing in the presence of NaCl, but not requiring it.

Hantavirus pulmonary syndrome (HPS) An emerging acute viral disease characterized by respiratory pneumonia, obtained by transmission of hantavirus from rodents.

Haploid An organism or cell containing only one set of chromosomes.

Hapten A low-molecular-weight substance not inducing antibody formation itself but still able to combine with a specific antibody.

Helix A spiral structure in a macromolecule that contains a repeating pattern.

Hemagglutination Agglutination of red blood cells.

Hemolysins Bacterial toxins capable of lysing red blood cells.

Hemolysis Lysis of red blood cells.

HEPA filter A *high efficiency particulate air* filter used in laboratories and industry to remove particles, including microorganisms, from intake or exhaust air flow.

Hepatitis Liver inflammation commonly caused by an infectious agent.

Herd immunity Resistance of a group to a pathogen as a result of the immunity of a large proportion of the group to that pathogen.

Heterocyst A differentiated cyanobacterial cell that carries out nitrogen fixation.

Heteroduplex A double-stranded DNA in which one strand is from one source and the other strand is from another, usually related, source.

Heterofermentation Fermentation of glucose or another sugar to a mixture of reduced products.

Heterotroph Chemoorganotroph.

HLA Human leukocyte antigen, the MHC of humans.

Homoacetogens *Bacteria* that produce acetate as the sole product of sugar fermentation or from $H_2 + CO_2$.

Homofermentation Fermentation of glucose or other sugar leading to a single product, lactic acid.

Homologous antigen An antigen that reacts with the antibody it has induced.

Horizontal (lateral) gene transfer A gene in an organism that originated by transfer from another organism.

Host An organism capable of supporting the growth of a virus or other parasite.

Host-to-host epidemic An epidemic resulting from host-to-host contact, characterized by a gradual rise and fall in disease incidence.

Humoral immunity An immune response involving antibodies.

Hybridization The natural formation or artificial construction of a duplex nucleic acid molecule by complementary base pairing between two nucleic acid strands derived from different sources.

Hybridoma The fusion of an immortal (tumor) cell with a lymphocyte to produce an immortal lymphocyte.

Hydrogen bond A weak chemical bond between a hydrogen atom and a second, more electronegative element, usually an oxygen or nitrogen atom.

Hydrogenosome An organelle of endosymbiotic origin in the cytoplasm of certain anaerobic eukaryotes that functions to oxidize pyruvate to $H_2 + CO_2 +$ acetate.

Hydrolysis Breakdown of a polymer into smaller units, usually monomers, by addition of water; digestion.

Hydrophobic interactions Attractive forces between molecules due to the close positioning of nonhydrophilic portions of the two molecules.

Hydrothermal vents Warm or hot water-emitting springs associated with crustal spreading centers on the sea floor.

Hypersensitivity An immune reaction causing damage to the host, caused either by antigen-antibody reactions or cellular immune processes. See *Allergy*.

Hyperthermophile A prokaryote having a growth temperature optimum of 80°C or higher.

Hypervariable region Variation in amino acid sequence within the variable domains of Igs or TCRs that provide most of the molecular contacts with antigen (also known as *complementarity-determining regions*).

Icosahedron A geometrical shape occurring in many virus particles, with 20 triangular faces and 12 corners.

Immobilized enzyme An enzyme attached to a solid support over which substrate is passed and converted to product.

Immune Able to resist infectious disease.

Immunity The ability of an organism to resist infection.

Immunization (vaccination) Inoculation of a host with inactive or weakened pathogens or pathogen products to stimulate protective immunity.

Immunoblot (Western blot) Detection of proteins immobilized on a filter by complementary reaction with specific antibody. Compare with *Southern blot* and *Northern blot*.

Immunodeficiency Having dysfunctional or completely nonfunctional immune system.

Immunogen A molecule capable of eliciting an immune response.

Immunoglobulin (Ig) Antibody.

Immunoglobulin gene superfamily A family of genes that are evolutionarily, structurally, and functionally related to immunoglobulins.

Immunologic memory The ability to rapidly produce large quantities of specific immune cells or antibodies following reexposure to a previously encountered antigen.

In silico The use of computers to perform sophisticated analyses.

In vitro In glass, away from the living organism.

In vivo In the body, in a living organism.

Incidence In reference to disease transmission, the number of cases of the disease in a specific subset of the population.

Induced enzyme An enzyme subject to induction.

Induction The process by which an enzyme is synthesized in response to the presence of an external substance, the inducer.

Infection Growth of an organism within a host.

Infection thread In the formation of root nodules, a cellulosic tube through which *Rhizobium* cells travel to reach and infect root cells.

Inflammation Host response to injury or infection, characterized by redness, swelling, heat, and pain.

Inhibition In reference to growth, the reduction of microbial growth because of a decrease in the number of organisms present or alterations in the microbial environment.

Innate immunity (nonspecific immunity) The noninducible ability to recognize and destroy an individual pathogen or its products that does not rely on previous exposure to a pathogen or its products.

Inoculum Cell material used to initiate a microbial culture.

Insertion A genetic phenomenon in which a piece of DNA is inserted into the middle of a gene.

Insertion sequence (IS elements) The simplest type of transposable element. Has only genes involved in transposition.

Integrating vector A cloning vector that becomes integrated into a host chromosome.

Integration The process by which a DNA molecule becomes incorporated into another genome.

Interferons Cytokine proteins produced by virus-infected cells that induce signal transduction in nearby cells, resulting in transcription of antiviral genes and expression of antiviral proteins.

Interleukin (IL) Soluble cytokine or chemokine mediator secreted by leukocytes.

Interspecies hydrogen transfer The process by which organic matter is degraded by the interaction of several groups of microorganisms in which H_2 production and H_2 consumption are closely coupled.

Introns The intervening noncoding sequences in a split gene. Contrasted with *Exons*, the coding sequences.

Invasiveness Pathogenicity caused by the ability of a pathogen to enter the body and spread.

Ionophore A compound that can cause the leakage of ions across membranes.

Irradiation In food microbiology, the exposure of food to ionizing radiation to inhibit microorganisms and insect pests, or to retard growth or ripening.

Isomers Two molecules with the same molecular formula but which differ structurally.

Isotopes Different forms of the same element containing the same number of protons and electrons but differing in the number of neutrons.

Jaundice Production and release of excess bilirubin in the liver due to destruction of liver cells, resulting in yellowing of the skin and whites of the eye.

Joule (J) A unit of energy equal to 10^7 ergs; 1000 Joules equal 1 kilojoule (kJ).

Kilobase (kb) A 1000-base fragment of nucleic acid. A *kilobase pair* (kbp) is a fragment containing 1000 base pairs.

Kinase An enzyme that adds a phosphoryl group, usually from ATP, to a compound.

Koch's Postulates A set of criteria for proving that a given microorganism causes a given disease.

Lag phase The period after inoculation of a culture before growth begins.

Laser tweezers A device used to obtain pure cultures in which a single cell is optically trapped with a laser and moved away from contaminating organisms into sterile growth medium.

Late protein Proteins synthesized toward the end of virus infection.

Latent virus A virus present in a cell, yet not causing any detectable effect.

Leaching Removal of valuable metals from ores by microbial action.

Leukocidin A substance able to destroy phagocytes.

Leukocyte A nucleated cell found in the blood. A white blood cell.

Lichen A fungus and an alga (or a cyanobacterium) living in symbiotic association.

Lipid Water-insoluble organic molecules important in structure of the cytoplasmic membrane and (in some organisms) the cell wall. See also *Phospholipid*.

Lipopolysaccharide (LPS) Complex lipid structure containing unusual sugars and fatty acids found in most gram-negative *Bacteria* and constituting the chemical structure of the outer membrane.

Listeriosis Gastrointestinal food infection caused by *Listeria monocytogenes* that may lead to bacteremia and meningitis.

Lophotrichous Having a tuft of polar flagella.

Lower respiratory tract Trachea, bronchi, and lungs.

Luminescence Production of light.

Lyme disease An emerging tick-transmitted disease caused by the spirochete *Borrelia burgdorferi*.

Lymph A fluid similar to blood that lacks red blood cells and travels through a separate circulatory system (the lymphatic system) containing lymph nodes.

Lymphocyte A white blood cell involved in antibody formation or cellular immune responses.

Lyophilization (freeze-drying) The process of removing all water from frozen food under vacuum.

Lysin An antibody that induces lysis.

Lysis Loss of cellular integrity with release of cytoplasmic contents.

Lysogen A prokaryote containing a prophage. See also *Temperate virus*.

Lysogenic pathway A series of steps that after virus infection, leads to a genetic state (lysogeny) where the viral genome is replicated as a prophage along with that of the host.

Lysosome A cell organelle containing digestive enzymes.

Lytic pathway A series of steps after virus infection that lead to virus replication and the destruction (lysis) of the host cell.

Macromolecule A large molecule (polymer) formed by the connection of a number of small molecules (monomers).

Macrophage Large, noncirculating phagocytic cells involved in both phagocytosis and the antibody production process.

Magnetosomes Small particles of Fe_3O_4 present in cells that exhibit magnetotaxis (magnetic bacteria).

Magnetotaxis Directed movement of bacterial cells by a magnetic field.

Major histocompatibility complex (MHC) A gene cluster, or complex, encoding cell surface proteins important for antigen presentation to T cells in the immune response.

Malaria An insect-transmitted disease characterized by recurrent episodes of fever and anemia caused by the protozoan *Plasmodium* spp., usually transmitted between mammals through the bite of the *Anopheles* mosquito.

Malignant In reference to a tumor, an infiltrating metastasizing growth no longer under normal growth control.

Mast cells Tissue cells adjoining blood vessels throughout the body that contain granules with inflammatory mediators.

Medium (plural media) In microbiology, the nutrient solution(s) used to grow microorganisms.

Megabase (Mb) One million nucleotide bases (or base pairs, abbreviated Mbp).

Meiosis In eukaryotes, reduction division, the process by which the change from diploid to haploid occurs.

Membrane Any thin sheet or layer. See especially *Cytoplasmic membrane*.

Memory cell A differentiated B lymphocyte capable of rapid conversion to an antibody-producing plasma cell on subsequent stimulation with antigen.

Meningitis Inflammation of the meninges (brain tissue), sometimes caused by *Neisseria meningitidis* and characterized by sudden onset of headache, vomiting, and stiff neck, often progressing to coma within hours.

Meningococcemia Fulminant disease caused by *Neisseria meningitidis* and characterized by septicemia, intravascular coagulation, and shock.

Meningoencephalitis Invasion, inflammation, and destruction of brain tissue by the ameba *Naegleria fowlerii* or a variety of other pathogens.

Mesophile Organism living in the temperature range near that of warm-blooded animals and usually showing a growth temperature optimum between 25 and 40°C.

Messenger RNA (mRNA) An RNA molecule transcribed from DNA that contains the genetic information necessary to encode a particular protein.

Metabolism All biochemical reactions in a cell, both anabolic and catabolic.

Metazoa Multicellular organisms.

Methanogen A methane-producing prokaryote; member of the *Archaea*.

Methanogenesis The biological production of methane (CH_4).

Methanotroph An organism capable of oxidizing methane.

Methylotroph An organism capable of oxidizing organic compounds that do not contain carbon–carbon bonds; if able to oxidize CH_4, also a methanotroph.

MIC Minimum inhibitory concentration—the minimum concentration of a substance necessary to prevent microbial growth.

Microaerophilic An organism requiring O_2 but at a level lower than that in air.

Microenvironment The immediate physical and chemical surroundings of a microorganism.

Micrometer One-millionth of a meter, or 10^{-6} m (abbreviated μm), the unit used for measuring microorganisms.

Microorganism A microscopic organism consisting of a single cell or cell cluster, also including the viruses, which are not cellular.

Microtubules Tubes that are the structural entity for eukaryotic flagella, play a role in maintaining cell shape and function as mitotic spindle fibers.

Minus (negative)-strand nucleic acid An RNA or DNA strand that has the opposite sense of (would be complementary to) the mRNA of a virus.

Mitochondrion Eukaryotic organelle responsible for the processes of respiration and electron transport phosphorylation.

Mitosis A highly ordered process by which the nucleus divides in eukaryotes.

Mixotroph An organism able to assimilate organic compounds as carbon sources while using inorganic compounds as electron donors for energy metabolism.

Molds Filamentous fungi.

Molecular chaperone A protein that helps other proteins fold or refold properly.

Molecular cloning Isolation and incorporation of a fragment of DNA into a vector where it can be replicated.

Molecule Two or more atoms chemically bonded to one another.

Monoclonal antibody An antibody produced from a single clone of B cells. This antibody has uniform structure and specificity.

Monocytes Circulating white blood cells that contain many lysosomes and can differentiate into macrophages.

Monomer A building block of a polymer.

Monotrichous Having a single polar flagellum.

Morbidity Incidence and prevalence of disease in a population, including both fatal and nonfatal cases.

Morphology The shape of an organism.

Mortality Incidence and prevalence of death in a population.

Most probable number (MPN) Serial dilution of a natural sample to determine the highest dilution yielding growth.

Motif A conserved amino acid sequence found in all peptide antigens that bind to a given MHC protein.

Motility The property of movement of a cell under its own power.

Mucous membrane Layers of epithelial cells that interact with the external environment.

Mucus Soluble glycoproteins secreted by epithelial cells that coat the mucous membrane.

Mushrooms Filamentous fungi that produce large, often edible structures called fruiting bodies.

Mutagen An agent that induces mutation, such as radiation or certain chemicals.

Mutant A strain differing from its parent because of mutation.

Mutation An inheritable change in the base sequence of the genome of an organism.

Mycorrhiza A symbiotic association between a fungus and the roots of a plant.

Mycosis Infection caused by fungi.

Myeloma A malignant tumor of a plasma cell (antibody-producing cell).

Natural killer (NK) cell A specialized lymphocyte that recognizes and destroys foreign cells or infected host cells in a nonspecific manner.

Negative selection In T-cell selection, T cells that interact with self antigens in the thymus are deleted. See *Clonal deletion*.

Neutralization Interaction of antibody with antigen that reduces or blocks the biological activity of the antigen.

Neutrophil (polymorphonuclear leukocyte) (PMN) A type of leukocyte exhibiting phagocytic properties, a granular cytoplasm (granulocyte), and a multilobed nucleus.

Neutrophile An organism that grows best around pH 7.

Nitrification The microbiological conversion of ammonia to nitrate.

Nitrogen fixation Reduction of nitrogen gas to ammonia ($N_2 + 8\,H \longrightarrow 2\,NH_3 + H_2$) by the enzyme nitrogenase.

Nodule A tumorlike structure produced by the roots of symbiotic nitrogen-fixing plants. Contains the nitrogen-fixing microbial component of the symbiosis.

Nonnucleoside reverse transcriptase inhibitor (NNRTI) A nonnucleoside compound that inhibits the action of retroviral reverse transcriptase by binding directly to the catalytic site.

Nonperishable (stable) foods Foods of low water activity that have an extended shelf life and are resistant to spoilage by microorganisms.

Nonpolar Possessing hydrophobic (water-repelling) characteristics and not easily dissolved in water.

Nonsense mutation A mutation that changes a sense codon into one that does not code for an amino acid.

Normal microbial flora Microorganisms that are usually found associated with healthy body tissue.

Northern blot Hybridization of a single strand of nucleic acid (DNA or RNA) to RNA fragments immobilized on a filter. Compare with *Southern blot* and *Western blot*.

Nosocomial infection Hospital-acquired infection.

Nucleic acid   A polymer of nucleotides. See *Deoxyribonucleic acid* and *Ribonucleic acid*.

Nucleic acid probe   A strand of nucleic acid that can be labeled and used to hybridize to a complementary molecule from a mixture of other nucleic acids. In clinical microbiology or microbial ecology, a short oligonucleotide of unique sequence used as a hybridization probe for identifying specific genes.

Nucleocapsid   The complex of nucleic acid and proteins of a virus, not including, if present, an envelope.

Nucleoid   The aggregated mass of DNA that makes up the chromosome of prokaryotic cells.

Nucleoside   A nucleotide minus phosphate.

Nucleoside reverse transcriptase inhibitor (NRTI)   A nucleoside analog compound that inhibits the action of viral reverse transcriptase by competing with nucleosides.

Nucleotide   A monomeric unit of nucleic acid, consisting of a sugar, a phosphate, and a nitrogenous base.

Nucleus   A membrane-enclosed structure in eukaryotes containing the genetic material (DNA) organized in chromosomes.

Nutrient   A substance taken by a cell from its environment and used in catabolic or anabolic reactions.

Obligate   A qualifying adjective referring to an environmental condition always required for growth (for example, "obligate anaerobe").

Oligonucleotide   A short nucleic acid molecule, either obtained from an organism or synthesized chemically.

Oligotrophic   Describing a habitat in which nutrients are in low supply.

Oncogene   A gene whose expression causes formation of a tumor.

Open reading frame (ORF)   A sequence of DNA which, if transcribed, could be translated to yield a protein of known length and composition. A *functional* ORF is one that actually encodes a protein in the cell.

Operator   A specific region of the DNA at the initial end of a gene, where the repressor protein binds and blocks mRNA synthesis.

Operon   A cluster of genes whose expression is controlled by a single operator. Typical of prokaryotic cells.

Opportunistic infection   An infection usually observed only in an individual with a dysfunctional immune system.

Opportunistic pathogen   An organism that causes disease in the absence of normal host resistance.

Opsonization   Promotion of phagocytosis by a specific antibody in combination with complement.

Organelle   A membrane-enclosed structure found in eukaryotic cells.

Orthologs   Genes found in one organism that are similar to those in another organism but differ because of speciation. See also *Paralogs*.

Osmosis   Diffusion of water through a membrane from a region of low solute concentration to one of higher concentration.

Outbreak   The occurrence of a large number of cases of a disease in a short period of time.

Oxic   Containing oxygen; aerobic. Usually used in reference to a microbial habitat.

Oxidation   A process by which a compound gives up electrons (or H atoms) and becomes oxidized.

Oxidation-reduction (redox) reaction   A pair of reactions in which one compound becomes oxidized while another becomes reduced and takes up the electrons released in the oxidation reaction.

Oxidative (electron transport) phosphorylation   The nonphototrophic production of ATP at the expense of a proton motive force formed by electron transport.

Oxygenic photosynthesis   Use of light energy to synthesize ATP and NADPH by noncyclic photophosphorylation with the production of O_2 from water.

Palindrome   A nucleotide sequence on a DNA molecule in which the same sequence is found on each strand but in the opposite direction.

Pandemic   A worldwide epidemic.

Paralogs   Genes within an organism whose similarity is the result of gene duplication at some time during the evolution of the organism. See also *Orthologs*.

Parasite   An organism able to live in or on a host and cause disease.

Passive immunity   Immunity resulting from transfer of antibodies or immune cells from an immune to a nonimmune individual.

Pasteurization   Reduction of the microbial load in heat-sensitive liquids to kill disease-producing microorganisms and reduce the number of spoilage microorganisms.

Pathogen   An organism, usually a microorganism, that causes disease.

Pathogen-associated molecular pattern (PAMP)   A unique structural components of a microbe or virus recognized by a pattern recognition molecule.

Pathogenicity   The ability of a pathogen to cause disease.

Pattern recognition molecule (PRM)   A membrane-bound protein that recognizes a pathogen-associated molecular pattern, such as a unique component of a microbial cell surface structure.

Penicillin   A class of antibiotics that inhibit bacterial cell wall synthesis; characterized by a β-lactam ring.

Peptide bond   A type of covalent bond joining amino acids in a polypeptide.

Peptidoglycan   The rigid layer of the cell walls of *Bacteria*, a thin sheet composed of *N*-acetylglucosamine, *N*-acetylmuramic acid, and a few amino acids.

Perishable food   Fresh food generally of high water activity that has a very short shelf life due to potential for spoilage by growth of microorganisms.

Periplasmic space   The area between the cytoplasmic membrane and the outer membrane in gram-negative *Bacteria*.

Peritrichous flagellation   In flagellar arrangements, having flagella attached to many places on the cell surface.

Phage   See *Bacteriophage*.

Phagemid   A cloning vector that can replicate either as a plasmid or as a bacteriophage.

Phagocyte   One of a group of cells that recognizes, ingests, and degrades pathogens and pathogen products.

Phagocytosis   Ingestion of particulate material such as bacteria by protozoa and phagocytic cells of higher organisms.

Phenotype   The observable characteristics of an organism. Compare with *Genotype*.

Phosphodiester bond   A type of covalent bond linking nucleotides together in a polynucleotide.

Phospholipid   Lipids containing a substituted phosphate group and two fatty acid chains on a glycerol backbone.

Photoautotroph   An organism able to use light as its sole source of energy and CO_2 as its sole carbon source.

Photoheterotroph An organism using light as a source of energy and organic materials as carbon source.

Photophosphorylation Synthesis of energy-rich phosphate bonds in ATP using light energy.

Photosynthesis The use of light energy to drive the incorporation of CO_2 into cell material. See also *Anoxygenic photosynthesis* and *Oxygenic photosynthesis*.

Phototaxis Movement of a cell toward light.

Phototroph An organism that obtains energy from light.

Phylogeny The ordering of species into higher taxa and the construction of evolutionary trees based on evolutionary (natural) relationships.

Phytanyl A branched-chain hydrocarbon containing 20 carbon atoms, commonly found in the lipids of *Archaea*.

Pickling The process of acidifying food, typically with acetic acid, to prevent microbial growth and spoilage.

Pilus A fimbria-like structure that is present on fertile cells, both Hfr and F$^+$, and is involved in DNA transfer during conjugation. Sometimes called a *sex pilus*. See also *Fimbria*.

Pinocytosis In eukaryotes, phagocytosis of soluble molecules.

Plague An endemic disease in rodents caused by *Yersinia pestis* that is occasionally transferred to humans through the bite of a flea.

Plaque A zone of lysis or cell inhibition caused by virus infection on a lawn of cells.

Plasma The liquid portion of the blood with cells removed and clotting proteins deactivated.

Plasma cell A large, differentiated, short-lived B lymphocyte specializing in abundant (but short-term) antibody production.

Plasmid An extrachromosomal genetic element that is not essential for growth and has no extracellular form.

Platelet A noncellular disc-shaped structure containing protoplasm found in large numbers in blood and functioning in the blood-clotting process.

Plus-strand nucleic acid An RNA or DNA strand that has the same sense as the mRNA of a virus.

Point mutation A mutation that involves one or only a very few base pairs.

Polar Possessing hydrophilic characteristics and generally water-soluble.

Polar flagellation In flagellar arrangements, having flagella attached at one end or both ends of the cell.

Poly-β-hydroxybutyrate (PHB) A common storage material of prokaryotic cells consisting of a polymer of β-hydroxybutyrate (PHB) or other β-alkanoic acids (PHA).

Polyclonal antibodies A mixture of antibodies made by many different B cell clones.

Polyclonal antiserum A mixture of antibodies to a variety of antigens or to a variety of determinants on a single antigen.

Polymer A large molecule formed by polymerization of monomeric units. In water purification, a chemical in liquid form used as a coagulant to produce flocculation in the clarification process.

Polymerase chain reaction (PCR) A method used to amplify a specific DNA sequence *in vitro* by repeated cycles of synthesis using specific primers and DNA polymerase.

Polymorphism The occurrence of multiple alleles at a locus in frequencies that cannot be explained by the occurrence of recent random mutations.

Polymorphonuclear leukocyte (PMN) Motile white blood cells containing many lysosomes and specializing in phagocytosis. Characterized by a distinct segmented nucleus. Also, a neutrophil.

Polynucleotide A polymer of nucleotides bonded to one another by phosphodiester bonds.

Polypeptide Several amino acids linked together by peptide bonds.

Polysaccharide A long chain of monosaccharides (sugars) linked by glycosidic bonds.

Porins Protein channels in the outer membrane of gram-negative *Bacteria* through which small to medium-sized molecules can flow.

Porters Membrane proteins that function to transport substances into and out of the cell.

Positive selection In T-cell selection, T cells that interact with self MHC protein in the thymus are stimulated to grow and develop.

Potable In water purification, drinkable; safe for human consumption.

Precipitation A reaction between antibody and soluble antigen resulting in visible antibody–antigen complexes.

Prevalence The *proportion* of individuals in a population having a disease.

Pribnow box The consensus sequence TATAAT located approximately 10 base pairs upstream from the transcriptional start site. A binding site for RNA polymerase.

Primary antibody response Antibodies made on first exposure to antigen; mostly of the class IgM.

Primary metabolite A metabolite excreted during the growth phase.

Primary producer An organism that uses light or an inorganic compound to synthesize new organic material from CO_2.

Primary structure In an informational macromolecule, such as a polypeptide or a nucleic acid, the *precise sequence* of monomeric units.

Primary transcript An unprocessed RNA molecule that is the direct product of transcription.

Primary wastewater treatment Physical separation of wastewater contaminants, usually by separation and settling.

Primer A molecule (usually a polynucleotide) to which DNA polymerase can attach the first deoxyribonucleotide during DNA replication.

Prion An infectious agent whose extracellular form may contain no nucleic acid.

Probe See *Nucleic acid probe*.

Prochlorophyte A prokaryotic oxygenic phototroph that contains chlorophylls *a* and *b* but lacks phycobilins.

Prokaryote A cell or organism lacking a nucleus and other membrane-enclosed organelles and usually having its DNA in a single circular molecule. Members of the *Bacteria* and the *Archaea*.

Promoter The site on DNA where the RNA polymerase binds and begins transcription.

Prophage The state of the genome of a temperate virus when it is replicating in synchrony with that of the host, typically integrated into the host genome.

Prophylactic Treatment, usually immunological or chemotherapeutic, designed to protect an individual from a future attack by a pathogen.

Prostheca A cytoplasmic extrusion from a cell such as a bud, hypha, or stalk.

Prosthetic group The tightly bound, nonprotein portion of an enzyme; not the same as a *Coenzyme*.

Protease inhibitor A compound that inhibits the action of viral protease by binding directly to the catalytic site, preventing viral protein processing.

Protein A polymeric molecule consisting of one or more polypeptides.

Proteobacteria A large phylum of *Bacteria* that includes many of the common gram-negative bacteria, including *Escherichia coli*.

Proteome The total complement of proteins present in a cell, tissue, or organism at any one time.

Proteomics The large scale or genomewide study of the structure, function, and regulation of the proteins of an organism.

Proteorhodopsin A light-sensitive retinal-containing protein found in some marine *Bacteria* that catalyzes ATP formation.

Proton motive force An energized state of a membrane created by a protein gradient and usually formed through action of an electron transport chain. See also *Chemiosmosis*.

Protoplasm The complete cellular contents, cytoplasmic membrane, cytoplasm, and nucleus/nucleoid.

Protoplast A cell from which the wall has been removed.

Prototroph The parent from which an auxotrophic mutant has been derived. Contrast with *Auxotroph*.

Protozoa Unicellular eukaryotic microorganisms that lack cell walls.

Provirus See *Prophage*.

Psychrophile An organism able to grow at low temperatures and showing a growth temperature optimum of <15°C.

Psychrotolerant Able to grow at low temperature but having a growth temperature optimum of >15°C.

Public health The health of the population as a whole.

Pure culture A culture containing a single kind of microorganism.

Pyogenic Pus-forming; causing abscesses.

Pyrite A common iron ore, FeS_2.

Pyrogenic Fever-inducing.

Quarantine The practice of restricting the movement of individuals with highly contagious serious infections to prevent spread of the disease.

Quaternary structure In proteins, the number and arrangement of individual polypeptides in the final protein molecule.

Quinolones Synthetic antibacterial compounds that interact with DNA gyrase and prevent supercoiling of bacterial DNA.

Quorum sensing Regulatory networks in prokaryotes that respond to population density.

Rabies A usually fatal neurological disease caused by the rabies virus that is usually transmitted by the bite or saliva of an infected carnivore.

Raw water Surface water or groundwater that has not been treated in any way (also called untreated water).

Radioimmunoassay (RIA) An immunological assay employing radioactive antibody or antigen for the detection of antigen or antibody binding.

Radioisotope An isotope of an element that undergoes spontaneous decay with the release of radioactive particles.

Reaction center A photosynthetic complex containing chlorophyll (or bacteriochlorophyll) and other components, within which occurs the initial electron transfer reactions of photophosphorylation.

Reading-frame shift See *Frameshift*.

Recalcitrant Resistant to microbial attack.

Recombinant DNA A DNA molecule containing DNA originating from two or more sources.

Recombination Process by which genetic elements in two separate genomes are brought together in one unit.

Redox See *Oxidation–reduction reaction*.

Reduction A process by which a compound accepts electrons to become reduced.

Reduction potential (E_0') The inherent tendency, measured in volts, of the oxidized compound of a redox pair to become reduced.

Reductive dechlorination Removal of Cl as Cl^- from an organic compound by reducing the carbon atom from C—Cl to C—H.

Reemerging infection Infectious disease, thought to be under control, that produces a new epidemic.

Regulation Processes that control the rates of synthesis of proteins, such as induction and repression.

Regulon A set of operons that are all controlled by the same regulatory protein (repressor or activator).

Replacement vector A cloning vector, such as a bacteriophage, in which some of the DNA of the vector can be replaced with foreign DNA.

Replication Synthesis of DNA using DNA as a template.

Replicative form A double-stranded DNA molecule that is an intermediate in the replication of single-stranded DNA viruses.

Reporter gene A gene incorporated into a vector because the product it encodes is easy to detect.

Repression The process by which the synthesis of an enzyme is inhibited by the presence of an external substance, the repressor.

Repressor protein A regulatory protein that binds to specific sites on DNA and blocks transcription; involved in negative control.

Reservoir In epidemiology, sites in which viable infectious agents remain and from which infection of individuals may occur.

Respiration Catabolic reactions producing ATP in which either organic or inorganic compounds are primary electron donors and organic or inorganic compounds are ultimate electron acceptors.

Response regulator protein One of the members of a two-component system; a regulatory protein that is phosphorylated by a sensor protein (see *Sensor protein*).

Restriction endonucleases (restriction enzymes) Enzymes that recognize and cleave specific DNA sequences, generating either blunt or single-stranded (sticky) ends.

Retrovirus A virus containing single-stranded RNA as its genetic material, which produces a complementary DNA by activity of the enzyme reverse transcriptase.

Reverse electron transport The energy-dependent movement of electrons against the thermodynamic gradient to form a strong electron donor from a weaker electron donor.

Reverse gyrase A topoisomerase present in hyperthermophilic prokaryotes that introduces positive supercoils in DNA.

Reverse transcription The process of copying information found in RNA into DNA.

Reverse translation The mental process of using a codon table and the amino acid sequence of a protein to obtain a possible sequence of the mRNA or the gene that encoded the protein.

Rheumatic fever An inflammatory autoimmune disease triggered by an immune response to infection by *Streptococcus pyogenes*.

Rhizosphere The region immediately adjacent to plant roots.

Ribonucleic acid (RNA) A polymer of nucleotides connected via a phosphate–ribose backbone; involved in protein synthesis or as genetic material of some viruses.

Ribosomal RNA (rRNA) Type of RNA found in the ribosome; some rRNAs participate actively in the process of protein synthesis.

Ribosome A cytoplasmic particle composed of ribosomal RNA and protein, which is part of the protein-synthesizing machinery of the cell.

Riboswitch A messenger RNA that can bind a specific small molecule near its 5' end that alters its secondary structure and makes it unavailable for translation.

Ribozyme An RNA molecule that can catalyze a chemical reaction.

Rickettsia Obligate intracellular parasites that cause disease, including typhus, Rocky Mountain spotted fever, and ehrlichiosis.

RNA editing Modification of the RNA transcript of a protein-encoding gene by a process other than splicing, yielding a molecule with the required coding properties.

RNA life A hypothetical ancient life form lacking DNA and protein, in which RNA had both a genetic coding and a catalytic function.

RNA polymerase An enzyme that synthesizes RNA in the $5' \rightarrow 3'$ direction using an antiparallel $3' \rightarrow 5'$ DNA strand as a template.

RNA processing The conversion of a precursor RNA to its mature form.

Rocky Mountain spotted fever A tick-transmitted disease caused by *Rickettsia rickettsii*, causing fever, headache, rash, and gastrointestinal symptoms.

Root nodule A tumor-like growth on certain plant roots that contains symbiotic nitrogen-fixing bacteria.

Rumen The forestomach of ruminant animals in which cellulose digestion occurs.

S-layer A paracrystalline outer wall layer composed of protein or glycoprotein and found in many prokaryotes.

Salmonellosis Enterocolitis caused by any of more than 2000 variants of *Salmonella* species.

Sanitizers Agents that reduce, but may not eliminate, microbial numbers to a safe level.

Scale-up Conversion of an industrial process from a small laboratory setup to a large commercial fermentation.

Scarlet fever Disease characterized by high fever and a reddish skin rash resulting from an exotoxin produced by cells of *Streptococcus pyogenes*.

Screening Any of a number of procedures that permits the sorting of organisms by phenotype or genotype by allowing growth of some types but not others.

Secondary antibody response Antibody made on second (subsequent) exposure to antigen; mostly of the class IgG.

Secondary metabolite A product excreted by a microorganism near the end of the growth phase or during the stationary phase.

Secondary structure The initial pattern of folding of a polypeptide or a polynucleotide, usually the result of hydrogen bonding.

Secondary treatment In sewage treatment, either the aerobic or anaerobic decomposition of sewage following the removal of nondegradable objects by primary treatment.

Secretion vector A DNA vector in which the protein product is both expressed in and secreted (excreted) from the cell.

Sediment In water purification, soil, sand, minerals, and other large particles found in raw water. In large bodies of water (lakes, the oceans), the materials (mud, rock, and the like) that form the bottom surface of the water body.

Selection Placing organisms under conditions where the growth of those with a particular genotype will be favored.

Selective medium A growth medium that enhances the growth of certain organisms while inhibiting the growth of others due to an added media component.

Semiconservative replication DNA synthesis yielding new double helices, each consisting of one parental and one progeny strand.

Semiperishable food Food of intermediate water activity that has a limited shelf life due to potential for spoilage by growth of microorganisms.

Semisynthetic penicillin A natural penicillin that has been chemically altered.

Sensitivity In immunodiagnostics, the lowest amount of antigen that can be detected in an immunological assay.

Sensor protein One of the members of a two-component system; a kinase found in the cell membrane that phosphorylates itself in response to an external signal and then passes the phosphoryl group to a response regulator protein (see *Response regulator protein*).

Septicemia Infection of the bloodstream by microorganisms.

Serology The study of antigen–antibody reactions *in vitro*.

Serum Fluid portion of blood remaining after the blood cells and materials responsible for clotting are removed.

Sewage Liquid effluents contaminated with human or animal fecal material.

Sexually transmitted infection (STI) An infection that is usually transmitted by sexual contact.

Shine–Dalgarno sequence A short stretch of nucleotides on a prokaryotic mRNA molecule upstream of the translational start site that binds to ribosomal RNA and thereby brings the ribosome to the initiation codon on the mRNA.

Shotgun sequencing Sequencing of DNA from previously cloned small fragments of a genome in a random fashion followed by computational methods to reconstruct the entire genome sequence.

Shuttle vector A cloning vector that can replicate in two different organisms; used for moving DNA between unrelated organisms.

Sickle-cell anemia A genetic trait that confers resistance to malaria but causes a reduction in the number and efficiency of red blood cells by reducing the oxygen-binding affinity of hemoglobin.

Siderophore An iron chelator that can bind iron present at very low concentrations.

Signal sequence A short stretch of amino acids found at the beginning of proteins destined to be excreted from the cell. The signal sequence is usually rich in hydrophobic amino acids, which helps transport the entire polypeptide through the membrane.

Signature sequence Short oligonucleotides of unique sequence found in 16S ribosomal RNA of a particular group of prokaryotes.

Single-cell protein Protein derived from microbial cells for use as food or a food supplement.

Site-directed mutagenesis A technique whereby a gene with a specific mutation can be constructed *in vitro*.

16S rRNA A large polynucleotide (~1500 bases) that functions as a part of the small subunit of the ribosome of prokaryotes (*Bacteria* and *Archaea*) and from whose sequence evolutionary relationships can be obtained; eukaryotic counterpart, 18S rRNA.

Slime layer A diffuse layer of polymer fibers, typically polysaccharides, that forms an outer surface layer on the cell.

Slime molds Nonphototrophic eukaryotic microorganisms lacking cell walls, which aggregate to form fruiting structures (cellular slime molds) or simply masses of protoplasm (acellular slime molds).

Solfatara A hot, sulfur-rich, generally acidic environment commonly inhabited by hyperthermophilic *Archaea*.

Southern blot Hybridization of a single strand of nucleic acid (DNA or RNA) to DNA fragments immobilized on a filter. Compare with *Northern blot* and *Western blot*.

Species Of prokaryotes, a collection of closely related (>97% 16S rRNA sequence homology and >70% genomic hybridization) strains sufficiently different from all other strains to be recognized as a distinct unit.

Specificity The ability of the immune response to interact with individual antigens.

Spheroplast A spherical, osmotically sensitive cell derived from a bacterium by loss of some but not all of the rigid wall layer. If all the rigid wall layer has been completely lost, the structure is called a *Protoplast*.

Splicing The RNA-processing step by which introns are removed and exons joined.

Spontaneous generation The hypothesis that living organisms can originate from nonliving matter.

Spore A general term for resistant resting structures formed by many prokaryotes and fungi.

Sporozoa Nonmotile parasitic protozoa.

Stalk An elongate structure, either cellular or excreted, that anchors a cell to a surface.

Static Inhibitory.

Stationary phase The period during the growth cycle of a microbial population in which growth ceases.

Stereoisomers Mirror image forms of two molecules having the same molecular and structural formulas.

Sterilant (sterilizer) (sporicide) A chemical agent that destroys all foms of microbial life.

Sterile Free of living organisms and viruses.

Sterilization The killing or removal of all living organisms and their viruses from a growth medium.

Sterols Hydrophobic multiringed structures that strengthen the cytoplasmic membrane of eukaryotic cells and a few prokaryotes.

Strain A population of cells of a single species all descended from a single cell; a clone.

Stromatolites Laminated microbial mats, typically built from layers of filamentous and other microorganisms that can become fossilized.

Substrate The molecule that undergoes a specific reaction with an enzyme.

Substrate-level phosphorylation Synthesis of high-energy phosphate bonds through reaction of inorganic phosphate with an activated organic substrate.

Supercoil Highly twisted form of circular DNA.

Superoxide anion (O_2^-) A derivative of O_2 capable of oxidative destruction of cell components.

Suppressor A mutation that restores a wild-type phenotype without altering the original mutation, usually arising by mutation in another gene.

Surveillance Observation, recognition, and reporting of diseases as they occur.

Suspended solid A small particle of solid pollutant that resists separation by ordinary physical means.

Symbiosis A relationship between two organisms.

Synthetic DNA A DNA molecule that has been made by a chemical process in a laboratory.

Syntrophy A nutritional situation in which two or more organisms combine their metabolic capabilities to catabolize a substance not capable of being catabolized by either one alone.

Systemic Not localized in the body; an infection disseminated widely through the body.

T cell A lymphocyte responsible for antigen-specific cellular interactions. T cells are divided into functional subsets including T_C (cytotoxic) T cells and T_H (helper) T cells. T_H cells are further subdivided into T_H1 (inflammatory) and T_H2 (helper cells) that aid B cells in antibody formation.

T-cell receptor The antigen-specific receptor protein on the surface of T lymphocytes.

T-DNA The segment of the *Agrobacterium* Ti plasmid that is transferred to plant cells.

Taxis Movement toward or away from a stimulus.

Taxonomy The study of scientific classification and nomenclature.

Temperate virus A virus that on infection of a host does not necessarily cause lysis but whose genome may replicate in synchrony with that of the host. See *Lysogen*.

Tertiary structure The final folded structure of a polypeptide that has previously attained secondary structure.

Tertiary wastewater treatment Physicochemical processing of wastewater to reduce levels of inorganic nutrients.

Tetanus A disease involving rigid paralysis of the voluntary muscles caused by an exotoxin produced by *Clostridium tetani*.

Tetracycline A member of a class of antibiotics characterized the four-membered naphthacene ring.

Thalassemia A genetic trait that confers resistance to malaria, but causes a reduction in the efficiency of red blood cells by altering a red blood cell enzyme.

Thermocline Zone of water in a stratified lake in which temperature and oxygen concentration drop precipitously with depth. If stratification is based on chemical differences, called the *chemocline*.

Thermophile An organism with a growth temperature optimum between 45 and 80° C.

Thylakoids Layers of membranes containing the photosynthetic pigments in chloroplasts and in cyanobacteria.

Ti plasmid A conjugative plasmid present in the bacterium *Agrobacterium tumefaciens* that can transfer genes into plants.

Titer In immunology, the quantity of antibody present in a solution.

Tolerance Inability to produce an immune response to a specific antigen.

Toll-like receptor (TLR) One of a family of pattern recognition molecules (PRMs) found on phagocytes, structurally and functionally related to Toll receptors in *Drosophila*, that recognize a pathogen-associated molecular pattern (PAMP).

Toxic shock syndrome (TSS) Acute systemic shock resulting from host response to an exotoxin produced by *Staphylococcus aureus*.

Toxicity Pathogenicity caused by toxins produced by a pathogen.

Toxigenicity   The degree to which an organism is able to elicit toxic symptoms.

Toxin   A microbial substance able to induce host damage.

Toxoid   A toxin modified so that it is no longer toxic but is still able to induce antibody formation.

Transcription   Synthesis of an RNA molecule complementary to one of the two strands of a double-stranded DNA molecule.

Transduction   Transfer of host genes from one cell to another by a virus.

Transfection   The transformation of a prokaryotic cell by DNA or RNA from a virus. Used also to describe the process of genetic transformation in eukaryotic cells.

Transfer RNA (tRNA)   A type of RNA that carries amino acids to the ribosome during translation; contains the anticodon.

Transformation   Transfer of genetic information via free DNA. Also, a process, sometimes initiated by infection with certain viruses, whereby a normal animal cell becomes a cancer cell.

Transgenic organisms   Plants or animals that stably pass on cloned DNA that has been inserted into them.

Translation   The synthesis of protein using the genetic information in a messenger RNA as a template.

Transpeptidation   The formation of peptide bonds between the short peptides present in the cell wall polymer, peptidoglycan.

Transposable element   A genetic element that has the ability to move (transpose) from one site on a chromosome to another.

Transposon   A type of transposable element that, in addition to genes involved in transposition, carries other genes; often genes conferring selectable phenotypes such as antibiotic resistance.

Transposon mutagenesis   Insertion of a transposon into a gene; this inactivates the host gene, leading to a mutant phenotype, and also confers the phenotype associated with the transposon gene.

Tuberculin test   A test for previous infection with *Mycobacterium tuberculosis* characterized by an inflammatory cell-mediated immune response.

Turbidity   A measurement of suspended solids in water.

Two-component regulatory system   A regulatory system containing a sensor protein and a response regulator protein (see *Sensor protein* and *Response regulator protein*).

Typhus   A louse-transmitted disease caused by *Rickettsia prowazekii*, causing fever, headache, weakness, rash, and damage to the central nervous system and internal organs.

Untreated water   Surface water or groundwater that has not been treated in any way (also called raw water).

Upper respiratory tract   The nasopharynx, oral cavity, and throat.

Upstream position   Refers to nucleic acid sequences on the 5′ side of a given site on a DNA or RNA molecule. Compare with *Downstream position*.

Vaccination   Inoculation of a host with inactive, killed, or weakened pathogens or pathogen products to stimulate protective immunity.

Vaccine   Material used to induce specific protective immunity to a pathogen.

Vacuole   A small space in a cell that contains fluid and is surrounded by a membrane. In contrast to a vesicle, a vacuole is not rigid.

Vector   An agent, usually an insect or other animal, able to carry pathogens from one host to another. Also, a genetic element able to incorporate DNA and cause it to be replicated in another cell.

Vehicle   Nonliving source of pathogens that infect large numbers of individuals; common vehicles are food and water.

Viable   Alive; able to reproduce.

Viable count   Measurement of the concentration of live cells in a microbial population.

Viral load   The number of viral genome copies in the tissue of an infected host, providing a quantitative assessment of the amount of virus in the host.

Viricidal agent   An agent that stops viral replication and activity.

Virion   A virus particle; the virus nucleic acid surrounded by a protein coat and in some cases other material.

Viristatic agent   An agent that inhibits viral replication.

Viroid   A small RNA molecule with viruslike properties.

Virulence   Degree of pathogenicity produced by a pathogen.

Virus   A genetic element containing either DNA or RNA that replicates in cells but is characterized by having an extracellular state.

Wastewater   Liquid derived from domestic sewage or industrial sources, which cannot be discarded in untreated form into lakes or streams

Water activity (a_w)   An expression of the relative availability of water in a substance. Pure water has an a_w of 1.000.

Western blot   See *Immunoblot*.

Wild type   A strain of microorganism isolated from nature. The usual or native form of a gene or organism.

Wobble   In reference to reading the genetic code, the concept that nonstandard base pairing is allowed between the anticodon and the third position of the codon.

Xenobiotic   A completely synthetic chemical compound not naturally occurring on Earth.

Xerophile   An organism adapted to growth at very low water potentials.

Yeasts   Unicellular fungi.

Zoonosis   A disease primarily of animals that is occasionally transmitted to humans.

Zygote   In eukaryotes, the single diploid cell resulting from the union of two haploid gametes.

INDEX

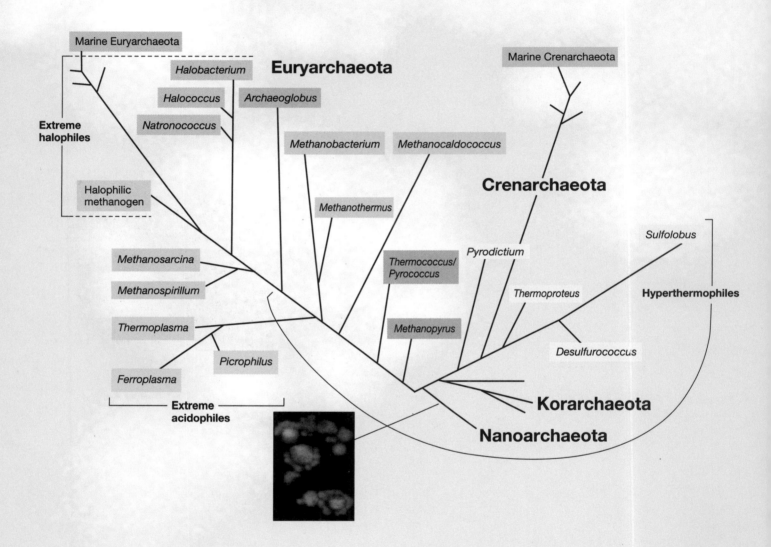

PHYLOGENETIC TREE OF *ARCHAEA*. This tree is derived from 16S ribosomal RNA sequences. Four major phyla of *Archaea* can be defined: the Crenarchaeota, which consists of both hyperthermophiles and cold-dwelling species; the Euryarchaeota, which contains methanogenic and extremely halophilic prokaryotes; the Korarchaeota, which are, as far as is known, hyperthermophiles; and the Nanoarchaeota, tiny, parasitic hyperthermophiles. See Sections 11.5–11.9 for further information on ribosomal RNA-based phylogenies. *Data for the tree obtained from the Ribosomal Database project* **http://rdp.cme.msu.edu**

PHYLOGENY OF THE LIVING WORLD–*EUKARYA*

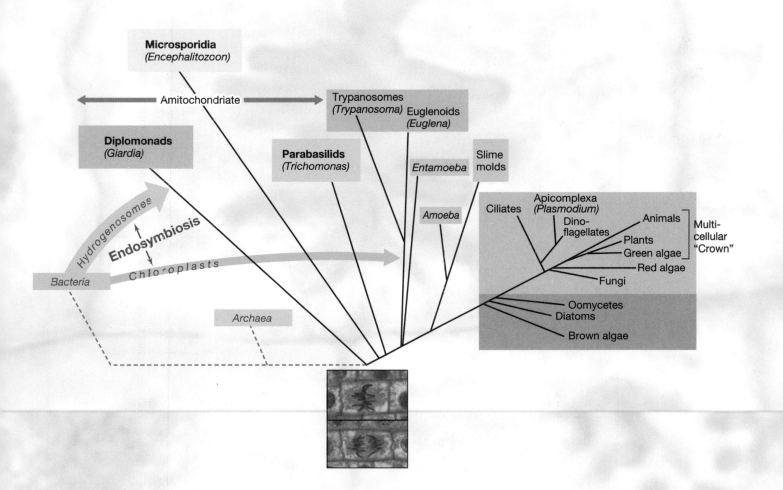

PHYLOGENETIC TREE OF *EUKARYA*, THE EUKARYOTES. This tree is based on 18S rRNA sequences obtained from the small subunit of cytoplasmic ribosomes. Note the close phylogenetic relationship between animals and plants but the distant relationship between these groups and organisms such as *Giardia*. See Sections 11.5-11.9 for further information on ribosomal RNA-based phylogenies. *Data for the tree obtained from the Ribosomal Database project* **http://rdp.cme.msu.edu**